T0333239

Your Guide to Success

- Implement the study skills that appear in the *Reach for Success* feature at the beginning and end of each chapter, and complete the activities that accompany them.

Reach for Success Organizing Your Time

Life during your college years should be fun but it also requires time-management skills if you want to be academically successful. Although you previously examined a typical 24-hour day and a week, looking at a longer period will help you identify potential time conflicts.

© iStockphoto.com/Ryan Balderas

Populate a two-week calendar with all the graded assignments from your syllabus in each of your classes.

Begin by filling in the dates in the chart. Then

1. Record all exams in the boxes.
2. Include due dates for any papers.
3. Add any labs/homework assignments.
4. Record employment hours including any scheduled overtime.

- Review each **EXAMPLE** and complete the corresponding SELF CHECK to be sure you understand the concepts.

EXAMPLE 4 Solve: $x + 4 = 9$

Solution To isolate x on one side of the $=$ sign, we will use the subtraction property of equality to undo the addition of 4 by subtracting 4 from both sides of the equation.

COMMENT Note that Example 4 can be solved by using the addition property of equality. We would add -4 to both sides to undo the addition of 4.

$$x + 4 = 9$$
$$x + 4 - 4 = 9 - 4 \quad \text{Subtract 4 from both sides.}$$
$$x + 0 = 5 \quad \text{Apply the additive inverse property.}$$
$$x = 5 \quad \text{Apply the additive identity property.}$$

We can check by substituting 5 for x in the original equation and simplifying.

$$x + 4 = 9$$
$$5 + 4 \stackrel{?}{=} 9 \quad \text{Substitute 5 for } x.$$
$$9 = 9 \quad \text{True.}$$

Since the solution 5 checks, the solution set is $\{5\}$.

SELF CHECK 4 Solve: $a + 175 = 122$

- Solve the **NOW TRY THIS** problems at the end of the section to strengthen and deepen what you have practiced and to help you transition to skills needed in future sections and chapters.

NOW TRY THIS

Simplify each expression. Write your answer with positive exponents only. Assume no variables are 0.

1. $-2(x^2y^5)^0$
2. $-3x^{-2}$
3. $9^2 - 9^0$
4. Explain why the instructions above include the statement "Assume no variables are 0."

- Work the Warm-Up, Review, and Vocabulary Exercises to ensure you are prepared for the remainder of the exercise set. Guided Practice Exercises literally "guide" you through the section's content with references to the Examples and Objectives, should you encounter difficulty. Additional Exercises provide you with the opportunity to select the correct procedure to work an exercise from the many you have learned throughout the section. Finally, Applications give you the opportunity to apply these skills in real-world situations.

- Repeat the above process to continually *Reach for Success* throughout the course!

Important Concepts Guide

CHAPTER 1 REAL NUMBERS AND THEIR BASIC PROPERTIES

Natural numbers: $\{1, 2, 3, 4, 5, . . .\}$

Whole numbers: $\{0, 1, 2, 3, 4, 5, . . .\}$

Integers: $\{. . . , -3, -2, -1, 0, 1, 2, 3, . . .\}$

Rational numbers: All numbers that can be written as a fraction with an integer numerator and a nonzero integer denominator

Real numbers: All numbers that are either a rational number or an irrational number

Prime numbers: $\{2, 3, 5, 7, 11, 13, 17, . . .\}$

Composite numbers: $\{4, 6, 8, 9, 10, 12, 14, 15, . . .\}$

Even integers: $\{. . . , -6, -4, -2, 0, 2, 4, 6, . . .\}$

Odd integers: $\{. . . , -5, -3, -1, 1, 3, 5, . . .\}$

Fractions:
If there are no divisions by 0, then

$$\frac{ax}{bx} = \frac{a}{b} \qquad \frac{a}{b} \cdot \frac{c}{d} = \frac{ac}{bd}$$

$$\frac{a}{b} \div \frac{c}{d} = \frac{ad}{bc} \qquad \frac{a}{d} + \frac{b}{d} = \frac{a+b}{d}$$

$$\frac{a}{d} - \frac{b}{d} = \frac{a-b}{d}$$

Exponents and order of operations:
If n is a natural number, then

$$x^n = \overbrace{x \cdot x \cdot x \cdot \cdots \cdot x}^{n \text{ factors of } x}$$

To simplify expressions, do all calculations within each pair of grouping symbols, working from the innermost pair to the outermost pair.

1. Find the values of any exponential expressions.
2. Do all multiplications and divisions from left to right.
3. Do all additions and subtractions from left to right.

In a fraction, simplify the numerator and denominator separately and then simplify the fraction, if possible.

Figure	Perimeter	Area
Square	$P = 4s$	$A = s^2$
Rectangle	$P = 2l + 2w$	$A = lw$
Triangle	$P = a + b + c$	$A = \frac{1}{2}bh$
Trapezoid	$P = a + b + c + d$	$A = \frac{1}{2}h(b + d)$
Circle	$C = \pi D = 2\pi r$	$A = \pi r^2$

Figure	Volume
Rectangular solid	$V = lwh$
Cylinder	$V = Bh$*
Pyramid	$V = \frac{1}{3}Bh$*
Cone	$V = \frac{1}{3}Bh$*
Sphere	$V = \frac{4}{3}\pi r^3$

*B is the area of the base.

If a, b, and c are real numbers, then

Closure properties:

$a + b$ is a real number.

$a - b$ is a real number.

ab is a real number.

$\frac{a}{b}$ is a real number $(b \neq 0)$.

Commutative properties:

$a + b = b + a$ for addition

$ab = ba$ for multiplication

Associative properties:

$(a + b) + c = a + (b + c)$ for addition

$(ab)c = a(bc)$ for multiplication

Distributive property:

$a(b + c) = ab + ac$

CHAPTER 2 EQUATIONS AND INEQUALITIES

Let a, b, and c be real numbers.

If $a = b$, then $a + c = b + c$.

If $a = b$, then $a - c = b - c$.

If $a = b$, then $ca = cb$.

If $a = b$, then $\frac{a}{c} = \frac{b}{c}$ $(c \neq 0)$.

Sale price = regular price − markdown

Retail price = wholesale cost + markup

Amount = rate · base

Solving inequalities:
Let a, b, and c be real numbers.

If $a < b$, then $a + c < b + c$.

If $a < b$, then $a - c < b - c$.

If $a < b$, and $c > 0$, then $ac < bc$.

(*continues on the back endsheet*)

7th Edition

Beginning and Intermediate Algebra

A Guided Approach

Rosemary M. Karr
Collin College

Marilyn B. Massey
Collin College

R. David Gustafson
Rock Valley College

CENGAGE

Australia • Brazil • Canada • Mexico • Singapore • United Kingdom • United States

CENGAGE

***Beginning and Intermediate Algebra: A Guided Approach*, Seventh Edition**
Rosemary M. Karr, Marilyn B. Massey and R. David Gustafson

Product Manager: Marc Bove
Content Developer: Stefanie Beeck, Cynthia Ashton
Content Coordinator: Lauren Crosby
Product Assistant: Kathryn Clark
Media Developer: Bryon Spencer
Brand Manager: Gordon Lee
Market Development Manager: Mark Linton
Content Project Manager: Jennifer Risden
Art Director: Vernon Boes
Manufacturing Planner: Becky Cross
Rights Acquisitions Specialist: Tom McDonough
Production Service: Lachina Publishing Services
Photo Researcher: Q2A/Bill Smith Group
Text Researcher: Pablo D'Stair
Copy Editor: Lachina Publishing Services
Illustrator: Lori Heckelman; Lachina Publishing Services
Text Designer: Terri Wright
Cover Designer: Terri Wright
Cover Image: Kevin Twomey
Compositor: Lachina Publishing Services
Design images: Fig leaf (chapter opener): © Michael Breuer/Masterfile; leaf (front matter): © JoLin/Shutterstock.com; Green background: © John Fox/Getty Images

© 2015, 2011 Cengage Learning, Inc.

ALL RIGHTS RESERVED. No part of this work covered by the copyright herein may be reproduced, transmitted, stored, or used in any form or by any means graphic, electronic, or mechanical, including but not limited to photocopying, recording, scanning, digitizing, taping, Web distribution, information networks, or information storage and retrieval systems, except as permitted under Section 107 or 108 of the 1976 United States Copyright Act, without the prior written permission of the publisher.

For product information and technology assistance, contact us at
Cengage Customer & Sales Support, 1-800-354-9706.
For permission to use material from this text or product, submit all requests online at **www.cengage.com/permissions.**
Further permissions questions can be e-mailed to
permissionrequest@cengage.com.

Library of Congress Control Number: 2013932723

Student Edition:
ISBN-13: 978-1-4354-6253-3
ISBN-10: 1-4354-6253-X

Loose-leaf Edition:
ISBN-13: 978-1-305-08118-5
ISBN-10: 1-305-08118-8

Cengage
200 Pier 4 Boulevard
Boston, MA 02210
USA

Cengage is a leading provider of customized learning solutions with office locations around the globe, including Singapore, the United Kingdom, Australia, Mexico, Brazil, and Japan. Locate your local office at **www.cengage.com/global.**

To learn more about Cengage platforms and services, register or access your online learning solution, or purchase materials for your course, visit **www.cengage.com.**

Printed in the United States of America
25 24 23 22 21   4 5 6 7 8 9 10 11 12 13

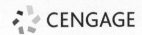

About the Authors

ROSEMARY M. KARR graduated from Eastern Kentucky University (EKU) with a Bachelor's degree in Mathematics, attained her Master of Arts degree at EKU in Mathematics Education, and earned her Ph.D. from the University of North Texas. After two years of teaching high school mathematics, she joined the faculty at Eastern Kentucky University, where she earned tenure as Assistant Professor of Mathematics. A professor at Collin College in Plano, Texas, since 1990, Professor Karr has written more than 10 solutions manuals, presented numerous papers, and been an active member in several educational associations including President of the National Association for Developmental Education. She has been honored several times by Collin College, and has received such national recognitions as U.S. Professor of the Year (2007) and induction as Fellow by the Council of Learning Assistance and Developmental Education Associations (2012).

MARILYN B. MASSEY teaches mathematics at Collin College in McKinney, Texas, where she joined the faculty in 1991. She has been President of the Texas Association for Developmental Education, served as academic chair of the Developmental Mathematics Department, and received an Excellence in Teaching Award from the National Conference for College Teaching and Learning. Professor Massey has presented at numerous state and national conferences. She earned her Bachelor's degree from the University of North Texas and Master of Arts degree in Mathematics Education from the University of Texas at Dallas.

R. DAVID GUSTAFSON is Professor Emeritus of Mathematics at Rock Valley College in Illinois and also has taught extensively at Rockford College and Beloit College. He is coauthor of several best-selling mathematics textbooks, including Gustafson/Frisk/Hughes, *College Algebra*; Gustafson/Karr/Massey, *Beginning Algebra, Intermediate Algebra, Beginning and Intermediate Algebra: A Combined Approach*; and the Tussy/Gustafson and Tussy/Gustafson/Koenig developmental mathematics series. His numerous professional honors include Rock Valley Teacher of the Year and Rockford's Outstanding Educator of the Year. He has been very active in AMATYC as a Midwest Vice-president and has been President of IMACC, AMATYC's Illinois affiliate. He earned a Master of Arts degree in Mathematics from Rockford College in Illinois, as well as a Master of Science degree from Northern Illinois University.

To my husband and family, for their
unwavering support of my work
—R.M.K.

To Ron for his unconditional love
and support of my projects
—M.B.M.

To Craig, Jeremy, Paula, Gary, Bob, Jennifer,
John-Paul, Gary, and Charlie
—R.D.G.

Contents

1 Real Numbers and Their Basic Properties 1

REACH FOR SUCCESS 2
1.1 Real Numbers and Their Graphs 3
1.2 Fractions 13
1.3 Exponents and Order of Operations 29
1.4 Adding and Subtracting Real Numbers 39
1.5 Multiplying and Dividing Real Numbers 47
1.6 Algebraic Expressions 54
1.7 Properties of Real Numbers 62
Projects 69
REACH FOR SUCCESS EXTENSION 70
CHAPTER REVIEW 71
CHAPTER TEST 77

2 Equations and Inequalities 79

REACH FOR SUCCESS 80
2.1 Solving Basic Linear Equations in One Variable 81
2.2 Solving More Linear Equations in One Variable 93
2.3 Simplifying Expressions to Solve Linear Equations in One Variable 102
2.4 Formulas 109
2.5 Introduction to Problem Solving 116
2.6 Motion and Mixture Applications 124
2.7 Solving Linear Inequalities in One Variable 132
Projects 139
REACH FOR SUCCESS EXTENSION 140
CHAPTER REVIEW 141
CHAPTER TEST 146
CUMULATIVE REVIEW FOR CHAPTERS 1–2 147

3 Graphing; Writing Equations of Lines; Functions; Linear Inequalities in Two Variables 149

REACH FOR SUCCESS 150
3.1 The Rectangular Coordinate System 151
3.2 Graphing Linear Equations 161
3.3 Slope of a Line 175
3.4 Point-Slope Form 187
3.5 Slope-Intercept Form 194
3.6 Functions 204
3.7 Solving Linear Inequalities in Two Variables 211
Projects 219
REACH FOR SUCCESS EXTENSION 221
CHAPTER REVIEW 222
CHAPTER TEST 227

4 Polynomials 229

REACH FOR SUCCESS 230
4.1 Natural-Number Exponents 231
4.2 Zero and Negative-Integer Exponents 239
4.3 Scientific Notation 244
4.4 Polynomials 251
4.5 Adding and Subtracting Polynomials 261
4.6 Multiplying Polynomials 268
4.7 Dividing Polynomials by Monomials 277
4.8 Dividing Polynomials by Polynomials 283
4.9 Synthetic Division 289
Projects 295
REACH FOR SUCCESS EXTENSION 296
CHAPTER REVIEW 297
CHAPTER TEST 302
CUMULATIVE REVIEW FOR CHAPTERS 1–4 303

5 Factoring Polynomials and Solving Equations by Factoring 305

REACH FOR SUCCESS 306
5.1 Factoring Out the Greatest Common Factor; Factoring by Grouping 307
5.2 Factoring the Difference of Two Squares 315
5.3 Factoring Trinomials with a Leading Coefficient of 1 320
5.4 Factoring Trinomials with a Leading Coefficient Other Than 1 329

5.5 Factoring the Sum and Difference of Two Cubes 338
5.6 Summary of Factoring Techniques 342
5.7 Solving Equations by Factoring 346
5.8 Solving Applications 352
■ *Projects* 359
REACH FOR SUCCESS EXTENSION 360
CHAPTER REVIEW 361
CHAPTER TEST 366

6 Rational Expressions and Equations; Proportion and Variation 367

REACH FOR SUCCESS 368
6.1 Simplifying Rational Expressions 369
6.2 Multiplying and Dividing Rational Expressions 378
6.3 Adding and Subtracting Rational Expressions 386
6.4 Simplifying Complex Fractions 398
6.5 Solving Equations That Contain Rational Expressions 405
6.6 Solving Applications Whose Models Contain Rational Expressions 411
6.7 Proportion and Variation 418
■ *Projects* 431
REACH FOR SUCCESS EXTENSION 432
CHAPTER REVIEW 433
CHAPTER TEST 439
CUMULATIVE REVIEW FOR CHAPTERS 1–6 440

7 Transitioning to Intermediate Algebra 441

REACH FOR SUCCESS 442
7.1 Review of Solving Linear Equations and Inequalities in One Variable 443
7.2 Review of Graphing Linear Equations, Finding the Slopes of Lines, and Writing Equations of Lines 455
7.3 Review of Functions 475
7.4 Review of Factoring and Solving Quadratic Equations 484
7.5 Review of Rational Expressions and Solving Rational Equations 497
7.6 Solving Equations in One Variable Containing an Absolute Value Expression 510
7.7 Solving Inequalities in One Variable Containing an Absolute Value Expression 516
■ *Projects* 523
REACH FOR SUCCESS EXTENSION 525
CHAPTER REVIEW 526
CHAPTER TEST 538

8 Solving Systems of Linear Equations and Inequalities 541

REACH FOR SUCCESS 542
8.1 Solving Systems of Linear Equations by Graphing 543
8.2 Solving Systems of Linear Equations by Substitution and Elimination 553
8.3 Solving Applications of Systems of Linear Equations in Two Variables 567
8.4 Solving Systems of Three Linear Equations in Three Variables 579
8.5 Solving Systems of Linear Equations Using Matrices 589
8.6 Solving Systems of Linear Equations Using Determinants 597
8.7 Solving Systems of Linear Inequalities in Two Variables 608
8.8 Solving Systems Using Linear Programming 621
■ *Projects* 631
REACH FOR SUCCESS EXTENSION 633
CHAPTER REVIEW 634
CHAPTER TEST 642

9 Radicals and Rational Exponents 645

REACH FOR SUCCESS 646
9.1 Radical Expressions 647
9.2 Applications of the Pythagorean Theorem and the Distance Formula 659
9.3 Rational Exponents 666
9.4 Simplifying and Combining Radical Expressions 674
9.5 Multiplying Radical Expressions and Rationalizing 684
9.6 Radical Equations 693
9.7 Complex Numbers 702
■ *Projects* 712
REACH FOR SUCCESS EXTENSION 714
CHAPTER REVIEW 715
CHAPTER TEST 722

10 Quadratic and Other Nonlinear Functions and Inequalities 725

REACH FOR SUCCESS 726
10.1 Solving Quadratic Equations Using the Square-Root Property and by Completing the Square 727
10.2 Solving Quadratic Equations Using the Quadratic Formula 738
10.3 The Discriminant and Equations That Can Be Written in Quadratic Form 745

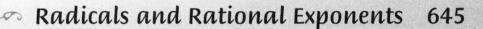

10.4 Graphs of Quadratic Functions 754
10.5 Graphs of Other Nonlinear Functions 770
10.6 Solving Quadratic and Other Nonlinear Inequalities 782
Projects 792
REACH FOR SUCCESS EXTENSION 794
CHAPTER REVIEW 795
CHAPTER TEST 801
CUMULATIVE REVIEW FOR CHAPTERS 1–10 802

11 Algebra, Composition, and Inverses of Functions; Exponential and Logarithmic Functions 805

REACH FOR SUCCESS 806
11.1 Algebra and Composition of Functions 807
11.2 Inverses of Functions 814
11.3 Exponential Functions 823
11.4 Base-*e* Exponential Functions 834
11.5 Logarithmic Functions 842
11.6 Natural Logarithms 852
11.7 Properties of Logarithms 858
11.8 Exponential and Logarithmic Equations 868
Projects 879
REACH FOR SUCCESS EXTENSION 881
CHAPTER REVIEW 882
CHAPTER TEST 890

12 Conic Sections, Systems of Equations and Inequalities, and More Graphing 893

REACH FOR SUCCESS 894
12.1 The Circle and the Parabola 895
12.2 The Ellipse 908
12.3 The Hyperbola 918
12.4 Solving Systems of Equations and Inequalities Containing One or More Second-Degree Terms 929
12.5 Piecewise-Defined Functions and the Greatest Integer Function 936
Project 942
REACH FOR SUCCESS EXTENSION 943
CHAPTER REVIEW 944
CHAPTER TEST 952
CUMULATIVE REVIEW FOR CHAPTERS 1–12 953

13 Miscellaneous Topics 955

REACH FOR SUCCESS 956
13.1 The Binomial Theorem 957
13.2 Arithmetic Sequences and Series 965
13.3 Geometric Sequences and Series 974
13.4 Infinite Geometric Sequences and Series 982
13.5 Permutations and Combinations 988
13.6 Probability 997
■ *Projects* 1003
REACH FOR SUCCESS EXTENSION 1005
CHAPTER REVIEW 1006
CHAPTER TEST 1011
CUMULATIVE REVIEW FOR CHAPTERS 1–13 1011

GLOSSARY G-1

APPENDIX 1 **Measurement Conversions** A-1

APPENDIX 2 **Symmetries of Graphs** A-13

APPENDIX 3 **Tables** A-19

APPENDIX 4 **Answers to Selected Exercises** A-23

INDEX I-1

Preface

To the Instructor

This seventh edition of *Beginning and Intermediate Algebra: A Guided Approach* is an exciting and innovative revision. The new edition reflects a thorough update, has new pedagogical features that make the text easier to read, and has an entirely new and fresh interior design. This series is known for its integrated approach, for the clarity of its writing, for making algebra relevant and engaging, and for developing student skills. The revisions to this already successful text will further promote student achievement. Co-authors Rosemary Karr and Marilyn Massey joined David Gustafson and have now assumed primary responsibility, bringing more experience in developmental education.

This new edition has expanded on the learning plan that helps students transition to the next level, teaching them the problem-solving strategies that will serve them well in their everyday lives. Most textbooks share the goals of clear writing, well-developed examples, and ample exercises, whereas the Karr/Massey/Gustafson series develops student success beyond the demands of traditional required coursework. The seventh edition's learning tools have been developed with your students in mind.

Through their collective teaching experience, the authors have developed an acute awareness of students' approaches to homework and have determined that exercise sets should serve as more than just a group of problems to work. The authors' philosophy is to guide the student through new material in a gentle progression of thought development that slowly reduces the student's dependence on external factors and relates new concepts to previously learned material. They have written the textbook to guide students through the material while providing a decreasing level of support throughout each section. Initially, the authors provide a map to the content through learning objectives that serve as advanced organizers for what students can expect to learn that day. The vocabulary encourages students to speak the language of mathematics. *Getting Ready* exercises at the beginning of each section prepares students for the upcoming concepts by reviewing relevant previous skills. The instructor may guide the students through the examples, but the students will independently attempt the Self Checks. Students will begin to use their "mathematical voice" to explain problems to one another or work collectively to find a solution to a problem not previously encountered, a primary goal of the *Now Try This* feature. The guidance shifts from instructor to fellow students to individual through carefully designed exercise sets.

New to This Edition

The major changes to this edition are the inclusion of a true "transition" chapter, a reorganization of many topics to more closely meet the needs of most beginning and intermediate algebra curricula, the inclusion of *Self Checks* for ALL examples, study skills strategies for each chapter, modification of the *Warm-Up* exercises, and the revision of many exercises.

- The organizational changes are as follows:
    - The section "Solving Linear Inequalities in Two Variables" has been moved to Chapter 3 ("Graphing; Writing Equations of Lines; Functions; Linear Inequalities in Two Variables").
    - The section "Synthetic Division" has been moved to Chapter 4 ("Polynomials").

    - A new section "Review of Graphing Linear Equations, Finding the Slopes of Lines, and Writing Equations of Lines" has been added to Chapter 7 ("Transitioning to Intermediate Algebra).
    - A new section "Solving Systems Using Linear Programming" has been added to Chapter 8 ("Solving Systems of Linear Equations and Inequalities").
    - The section "Graphs of Other Nonlinear Functions" has been moved to Chapter 10 ("Quadratic and Other Nonlinear Functions and Inequalities").
    - Sections 10.6 & 10.7 ("Algebra and Composition of Functions" and "Inverses of Functions") have been moved to the beginning of Chapter 11 ("Algebra, Composition, and Inverses of Functions; Exponential and Logarithmic Functions").
    - The section "Solving Systems of Equations and Inequalities Containing One or More Second-Degree Terms" has been moved to Chapter 12 ("Conic Sections, Systems of Equations and Inequalities, and More Graphing").
- The *Warm-Ups* have been significantly revised to reflect the original intention of this group of exercises. We wanted the *Warm-Ups* to be just that, warm-ups. Thus, most have been revised to include skills needed for the development of section concepts.
- A set of interactive study skills strategies have been incorporated throughout the textbook called *Reach for Success* for a more holistic learning experience. The authors feel that most students already know effective study strategies and could list many of the successful techniques. It is the authors' intent to encourage the student's thoughtful consideration and implementation of these skills. The interactive approach to these skills engages the student, encouraging an active participation to develop and reinforce the skills. The incorporation of student advising and the development of study skills is critical for academic success. That is why this feature is titled *Reach for Success*. The authors hope to help the students do just that. These worksheets (and others) are available online for flexibility in the order of assignment.
- *Everyday Connections* have been revised to more accurately reflect a topic of the selected chapter.
- *Self Checks* have been included for *every* Example in the textbook, especially significant for applications where students need to develop problem-solving strategies.
- Additional *Teaching Tips* have been included at the request of our reviewers.

Calculators

The use of calculators is assumed throughout the text although the calculator icon makes the exercises easily identifiable for possible omission if the instructor chooses to do so. We believe that students should learn calculator skills in the mathematics classroom. They will then be prepared to use calculators in science and business classes and for nonacademic purposes.

Since most intermediate algebra students now have graphing calculators, keystrokes are given for both scientific and graphing calculators. Removable cards for the ***Basic Calculator Keystroke Guide for the TI-83/84 Family of Calculator*** and ***Basic Calculator Keystroke Guide for the fx-9750GII CASIO*** are bound into the book as resources for those students learning how to use a graphing calculator.

A Guided Tour

Chapter Openers showcase the variety of career paths available in the world of mathematics. We include a brief overview of each career, as well as job outlook statistics from the U.S. Department of Labor, including potential job growth and annual earning potential.

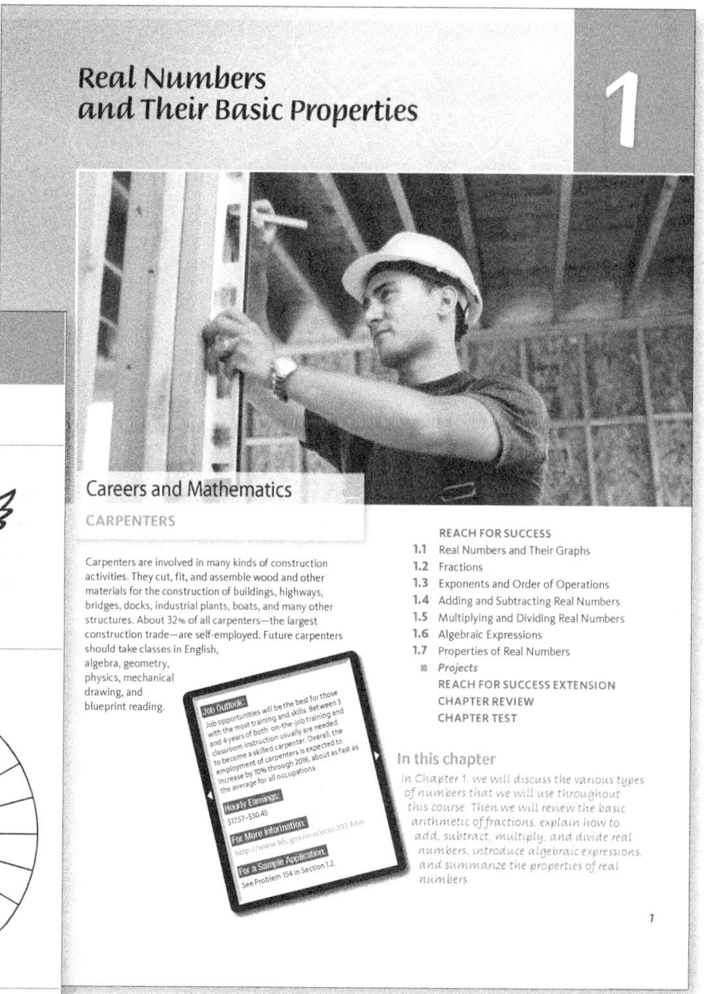

Real Numbers and Their Basic Properties

1

Careers and Mathematics

CARPENTERS

Carpenters are involved in many kinds of construction activities. They cut, fit, and assemble wood and other materials for the construction of buildings, highways, bridges, docks, industrial plants, boats, and many other structures. About 32% of all carpenters—the largest construction trade—are self-employed. Future carpenters should take classes in English, algebra, geometry, physics, mechanical drawing, and blueprint reading.

Job Outlook:
Job opportunities will be the best for those with the most training and skills. Between 3 and 4 years of both on-the-job training and classroom instruction usually are needed to become a skilled carpenter. Overall, the employment of carpenters is expected to increase by 10% through 2016, about as fast as the average for all occupations.

Hourly Earnings:
\$17.57–\$30.45

For More Information:
http://www.bls.gov/oco/ocos202.htm

For a Sample Application:
See Problem 154 in Section 1.2.

REACH FOR SUCCESS
1.1 Real Numbers and Their Graphs
1.2 Fractions
1.3 Exponents and Order of Operations
1.4 Adding and Subtracting Real Numbers
1.5 Multiplying and Dividing Real Numbers
1.6 Algebraic Expressions
1.7 Properties of Real Numbers
Projects
REACH FOR SUCCESS EXTENSION
CHAPTER REVIEW
CHAPTER TEST

In this chapter

In Chapter 1, we will discuss the various types of numbers that we will use throughout this course. Then we will review the basic arithmetic of fractions, explain how to add, subtract, multiply, and divide real numbers, introduce algebraic expressions, and summarize the properties of real numbers.

1

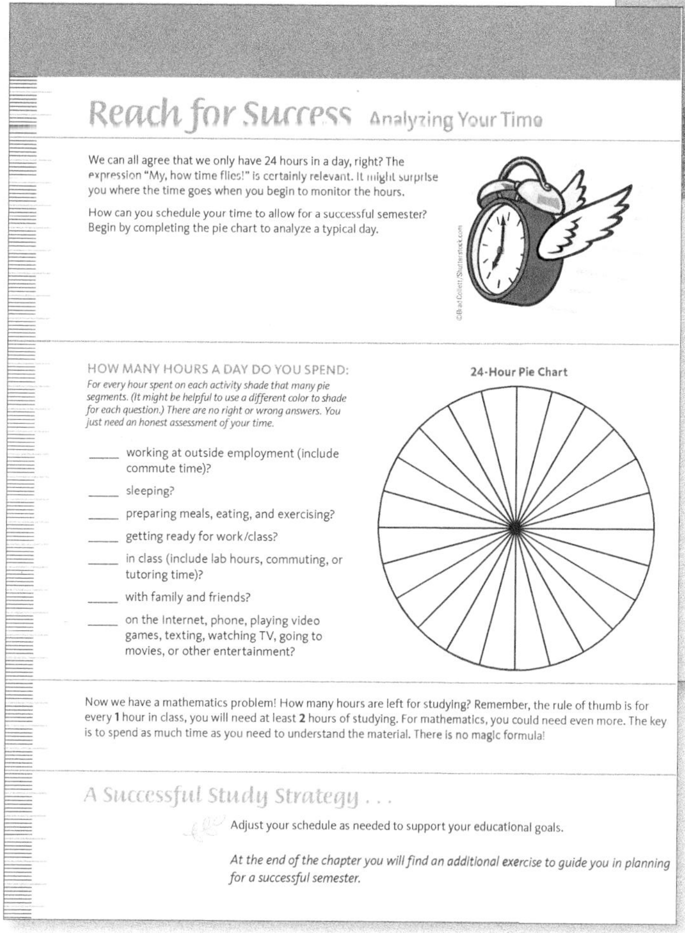

Reach for Success Analyzing Your Time

We can all agree that we only have 24 hours in a day, right? The expression "My, how time flies!" is certainly relevant. It might surprise you where the time goes when you begin to monitor the hours.

How can you schedule your time to allow for a successful semester? Begin by completing the pie chart to analyze a typical day.

©Brad Collett/Shutterstock.com

HOW MANY HOURS A DAY DO YOU SPEND:

For every hour spent on each activity shade that many pie segments. (It might be helpful to use a different color to shade for each question.) There are no right or wrong answers. You just need an honest assessment of your time.

_____ working at outside employment (include commute time)?
_____ sleeping?
_____ preparing meals, eating, and exercising?
_____ getting ready for work/class?
_____ in class (include lab hours, commuting, or tutoring time)?
_____ with family and friends?
_____ on the Internet, phone, playing video games, texting, watching TV, going to movies, or other entertainment?

24-Hour Pie Chart

Now we have a mathematics problem! How many hours are left for studying? Remember, the rule of thumb is for every **1** hour in class, you will need at least **2** hours of studying. For mathematics, you could need even more. The key is to spend as much time as you need to understand the material. There is no magic formula!

A Successful Study Strategy . . .

Adjust your schedule as needed to support your educational goals.

At the end of the chapter you will find an additional exercise to guide you in planning for a successful semester.

Reach for Success is a new feature where study skills have been written for inclusion in every chapter. This feature is designed as an opening activity to help the student prepare for a successful semester.

Appearing at the beginning of each section, ***Learning Objectives*** are mapped to the appropriate content, as well as to relevant exercises in the ***Guided Practice*** section. Measurable objectives allow students to identify specific mathematical processes that may need additional reinforcement. For the instructor, homework assignments can be developed more easily with problems keyed to objectives, thus facilitating identification of appropriate exercises.

In order to work mathematics, one must be able to speak the language. Not only are ***Vocabulary*** words identified at the beginning of each section, these words are also bolded within the section. Exercises include questions on the vocabulary words, and a glossary has been included to facilitate the students' reference to these words. An optional Spanish glossary is available upon request.

Getting Ready questions appear at the beginning of each section, linking past concepts to the upcoming material.

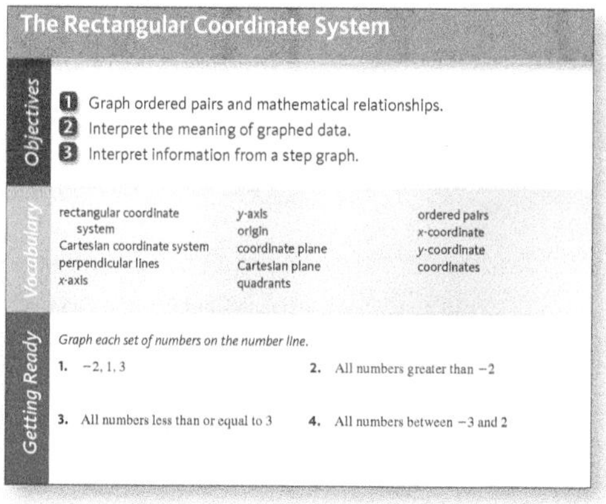

The Rectangular Coordinate System

Objectives
1 Graph ordered pairs and mathematical relationships.
2 Interpret the meaning of graphed data.
3 Interpret information from a step graph.

Vocabulary

rectangular coordinate system	y-axis	ordered pairs
Cartesian coordinate system	origin	x-coordinate
perpendicular lines	coordinate plane	y-coordinate
x-axis	Cartesian plane	coordinates
	quadrants	

Getting Ready

Graph each set of numbers on the number line.

1. $-2, 1, 3$
2. All numbers greater than -2
3. All numbers less than or equal to 3
4. All numbers between -3 and 2

EXAMPLE 7 **ISOSCELES TRIANGLES** The vertex angle of an isosceles triangle is 56°. Find the measure of each base angle.

Analyze the problem: An **isosceles triangle** has two sides of equal length, which meet to form the **vertex angle**. See Figure 2-10. The angles opposite those sides, called **base angles**, have equal measures. If we let x represent the measure of one base angle, the measure of the other base angle is also x.

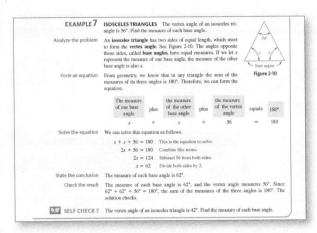

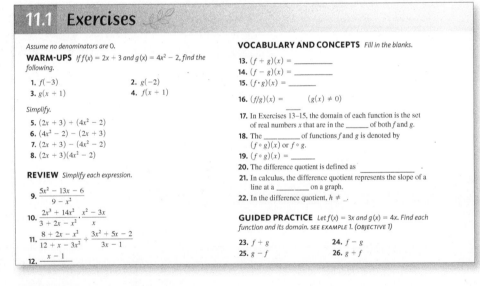

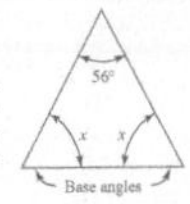

Figure 2-10

Form an equation: From geometry, we know that in any triangle the sum of the measures of its three angles is 180°. Therefore, we can form the equation.

The measure of one base angle	plus	the measure of the other base angle	plus	the measure of the vertex angle	equals	180°.
x	+	x	+	56	=	180

Solve the equation: We can solve this equation as follows.

$x + x + 56 = 180$ This is the equation to solve.
$2x + 56 = 180$ Combine like terms.
$2x = 124$ Subtract 56 from both sides.
$x = 62$ Divide both sides by 2.

State the conclusion: The measure of each base angle is 62°.

Check the result: The measure of each base angle is 62°, and the vertex angle measures 56°. Since 62° + 62° + 56° = 180°, the sum of the measures of the three angles is 180°. The solution checks.

SELF CHECK 7 The vertex angle of an isosceles triangle is 42°. Find the measure of each base angle.

Examples are worked out in each chapter, highlighting the concept being discussed. We include Author notes in many of the text's examples, giving students insight into the thought process one goes through when approaching a problem and working toward a solution.

All examples end with a ***Self Check***, so that students can measure their reading comprehension. Answers to each section's Self Checks are found at the end of that section for ease of reference.

Each section ends with ***Now Try This,*** exercises intended to increase conceptual understanding through active classroom participation and involvement. These problems transition to the Exercise Sets, as well as to material in future sections and can be worked independently or in small groups.

NOW TRY THIS

A cell phone is shaped like a rectangle and its longer edge is one inch shorter than twice its shorter edge. If the area of the phone is 10 square inches, find the dimensions of the phone.

The exercises reinforce topics, digging a little deeper than the examples. To discourage a student from simply looking up the answer and trying to find the process that will produce the answer, answers to these problems will be provided only in the *Annotated Instructor's Edition* of the text.

11.1 Exercises

Assume no denominators are 0.

WARM-UPS *If $f(x) = 2x + 3$ and $g(x) = 4x^2 - 2$, find the following.*

1. $f(-3)$
2. $g(-2)$
3. $g(x + 1)$
4. $f(x + 1)$

Simplify.

5. $(2x + 3) + (4x^2 - 2)$
6. $(4x^2 - 2) - (2x + 3)$
7. $(2x + 3) - (4x^2 - 2)$
8. $(2x + 3)(4x^2 - 2)$

REVIEW *Simplify each expression.*

9. $\dfrac{5x^2 - 13x - 6}{9 - x^2}$
10. $\dfrac{2x^3 + 14x^2}{3 + 2x - x^2} \cdot \dfrac{x^2 - 3x}{x}$
11. $\dfrac{8 + 2x - x^2}{12 + x - 3x^2} \div \dfrac{3x^2 + 5x - 2}{3x - 1}$
12. $\dfrac{x - 1}{}$

VOCABULARY AND CONCEPTS *Fill in the blanks.*

13. $(f + g)(x) =$ ______
14. $(f - g)(x) =$ ______
15. $(f \cdot g)(x) =$ ______
16. $(f/g)(x) =$ ______ $(g(x) \neq 0)$
17. In Exercises 13–15, the domain of each function is the set of real numbers x that are in the ______ of both f and g.
18. The ______ of functions f and g is denoted by $(f \circ g)(x)$ or $f \circ g$.
19. $(f \circ g)(x) =$ ______
20. The difference quotient is defined as ______.
21. In calculus, the difference quotient represents the slope of a line at a ______ on a graph.
22. In the difference quotient, $h \neq$ _.

GUIDED PRACTICE *Let $f(x) = 3x$ and $g(x) = 4x$. Find each function and its domain.* SEE EXAMPLE 1. (OBJECTIVE 1)

23. $f + g$
24. $f - g$
25. $g - f$
26. $g + f$

The ***Exercise Sets*** transition students through progressively more difficult homework problems. Students are initially asked to work quick, basic problems on their own, then proceed to work exercises keyed to examples and/or objectives, and finally to complete applications and critical-thinking questions on their own.

Warm-Ups get students into the homework mindset, asking quick memory-testing questions.

Review exercises are included to remind students of previous skills.

Vocabulary and Concepts exercises emphasize the main concepts taught in this section.

Guided Practice exercises are keyed to the objectives to increase student success by directing students to the concept covered in that group of exercises. Should a student encounter difficulties working a problem, a specific example within the objective is also cross-referenced.

Additional Practice problems are mixed and not linked to objectives or examples, providing the student with the opportunity to distinguish among problem types and to select an appropriate problem-solving strategy. This will help students prepare for the format generally seen on exams.

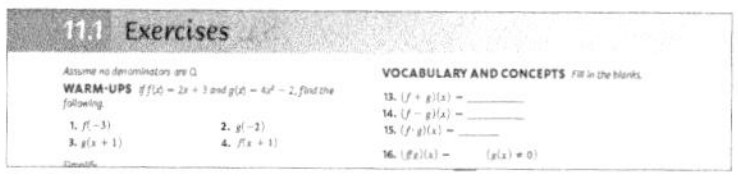
11.1 Exercises

Applications ask students to apply their new skills in real-life situations, *Writing About Math* problems build students' mathematical communication skills, and *Something to Think About* encourages students to consider what they have learned in a section, and apply those concepts to a new situation.

Many exercises are available online through Enhanced WebAssign®. These homework problems are algorithmic, ensuring that your students will learn mathematical processes, not just how to work with specific numbers.

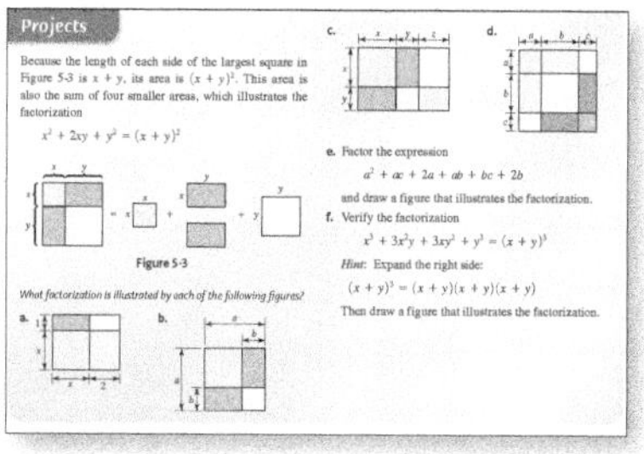
Projects

Because the length of each side of the largest square in Figure 5-3 is $x + y$, its area is $(x + y)^2$. This area is also the sum of four smaller areas, which illustrates the factorization

$x^2 + 2xy + y^2 = (x + y)^2$

Figure 5-3

What factorization is illustrated by each of the following figures?

e. Factor the expression $a^2 + ac + 2a + ab + bc + 2b$ and draw a figure that illustrates the factorization.

f. Verify the factorization $x^3 + 3x^2y + 3xy^2 + y^3 = (x + y)^3$

Hint: Expand the right side: $(x + y)^3 = (x + y)(x + y)(x + y)$

Then draw a figure that illustrates the factorization.

Chapter-ending *Projects* encourage in-depth exploration of key concepts.

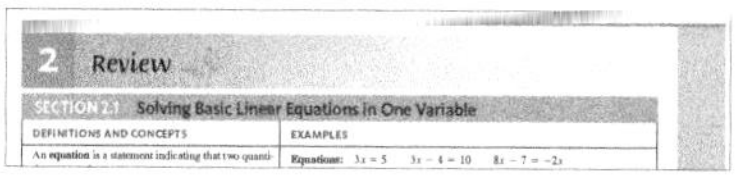
2 Review

SECTION 2.1 Solving Basic Linear Equations in One Variable

DEFINITIONS AND CONCEPTS | EXAMPLES

A *Chapter Review* grid presents the material cleanly and simply, giving students an efficient means of reviewing the chapter.

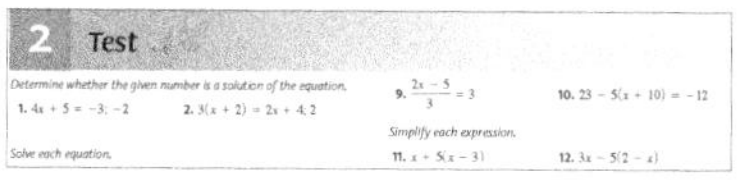
2 Test

Determine whether the given number is a solution of the equation.

Simplify each expression.

Solve each equation.

Chapter Tests allow students to pinpoint their strengths and challenges with the material. Answers are included at the back of the book.

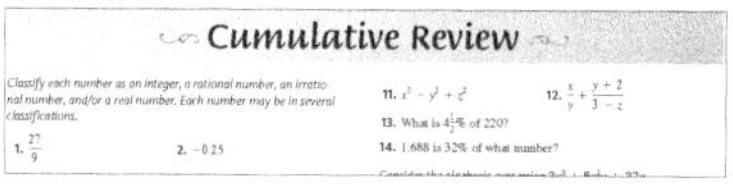
Cumulative Review

Cumulative Review Exercises follow the end of chapter material and keep students' skills current before moving on to the next topic.

A *Glossary* has been included to function as a student reference to the vocabulary words and definitions.

Additional Features

Everyday Connections boxes reveal the real-world power of mathematics. Each Everyday Connections box invites students to see how the material covered in the chapter is relevant to their lives.

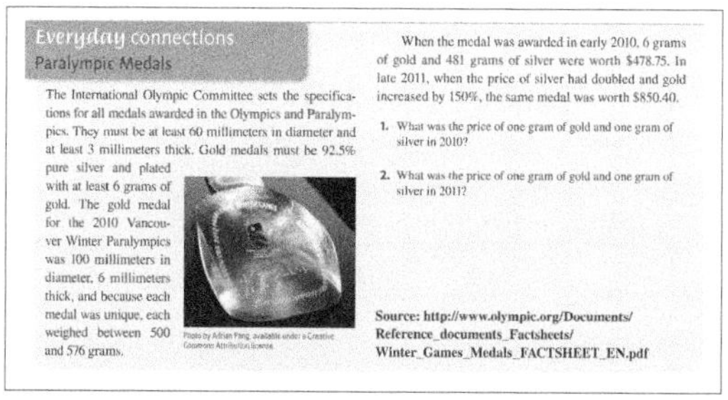
Everyday connections
Paralympic Medals

The International Olympic Committee sets the specifications for all medals awarded in the Olympics and Paralympics. They must be at least 60 millimeters in diameter and at least 3 millimeters thick. Gold medals must be 92.5% pure silver and plated with at least 6 grams of gold. The gold medal for the 2010 Vancouver Winter Paralympics was 100 millimeters in diameter, 6 millimeters thick, and because each medal was unique, each weighed between 500 and 576 grams.

When the medal was awarded in early 2010, 6 grams of gold and 481 grams of silver were worth \$478.75. In late 2011, when the price of silver had doubled and gold increased by 150%, the same medal was worth \$850.40.

1. What was the price of one gram of gold and one gram of silver in 2010?

2. What was the price of one gram of gold and one gram of silver in 2011?

Source: http://www.olympic.org/Documents/Reference_documents_Factsheets/Winter_Games_Medals_FACTSHEET_EN.pdf

Perspective boxes highlight interesting facts from mathematics history or important mathematicians, past and present. These brief but interesting biographies connect students to discoveries of the past and their importance to the present.

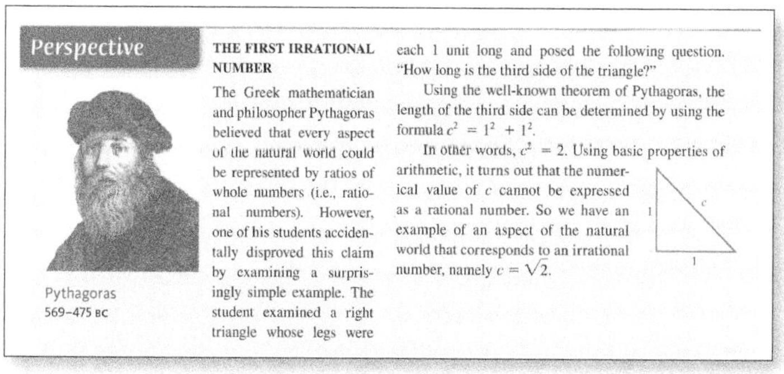
Perspective

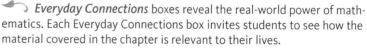

Pythagoras
569–475 BC

THE FIRST IRRATIONAL NUMBER

The Greek mathematician and philosopher Pythagoras believed that every aspect of the natural world could be represented by ratios of whole numbers (i.e., rational numbers). However, one of his students accidentally disproved this claim by examining a surprisingly simple example. The student examined a right triangle whose legs were each 1 unit long and posed the following question. "How long is the third side of the triangle?"

Using the well-known theorem of Pythagoras, the length of the third side can be determined by using the formula $c^2 = 1^2 + 1^2$.

In other words, $c^2 = 2$. Using basic properties of arithmetic, it turns out that the numerical value of c cannot be expressed as a rational number. So we have an example of an aspect of the natural world that corresponds to an irrational number, namely $c = \sqrt{2}$.

Accent on technology

▸ Approximating Zeros of Polynomials

COMMENT The x-coordinates of the x-intercepts of the graph represent what we call the real zeros of a polynomial.

We can use a graphing calculator to approximate the real zeros of a polynomial function $f(x)$. For example, to find the real zeros of $f(x) = 2x^3 - 6x^2 + 7x - 21$, we graph the function as in Figure 4-7 on the next page.

It appears from the figure that the function f has a zero at $x = 3$. To be certain, find $f(3)$.

$$f(3) = 2(3)^3 - 6(3)^2 + 7(3) - 21 = 0 \quad \text{Substitute 3 for } x.$$

From the factor theorem, we know that if $P(3) = 0$, then $(x - 3)$ is a factor of the polynomial. To find the other factors, we can synthetically divide by 3.

$$\begin{array}{r|rrrr} 3 & 2 & -6 & 7 & -21 \\ & & 6 & 0 & 21 \\ \hline & 2 & 0 & 7 & 0 \end{array}$$

Thus, $f(x) = (x - 3)(2x^2 + 7)$. Since $(2x^2 + 7)$ cannot be factored over the real numbers, we can conclude that 3 is the only real zero of the polynomial function.

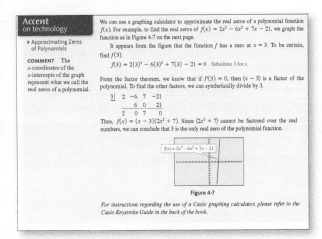

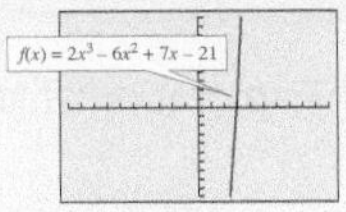

Figure 4-7

For instructions regarding the use of a Casio graphing calculator, please refer to the Casio Keystroke Guide in the back of the book.

Accent on Technology boxes teach students the calculator skills to prepare them for using these tools in science and business classes, as well as for nonacademic purposes. Calculator examples are given in these boxes, and keystrokes are given for both scientific and graphing calculators. For instructors who do not use calculators in the classroom, the material on calculators is easily omitted without interrupting the flow of ideas.

Comment notations alert students to common errors as well as provide helpful and pertinent information about the concepts they are learning.

COMMENT Remember that when you multiply one side of an equation by a nonzero number, you must multiply the other side by the same number to maintain the equality.

For the instructor, ***Teaching Tips*** are provided in the margins of the *Annotated Instructor's Edition* as interesting historical information, alternative approaches for teaching the material, and class activities.

A ***Spanish Glossary*** is available for inclusion upon request.

Teaching Tip

If your class is full and the chairs are arranged in rows and columns, introduce graphing this way:

1. Number the rows and columns.
2. Ask row 3, column 1 to stand. This is the point (3, 1).
3. Ask (1, 3) to stand. That this is a different point emphasizes the ordered pair idea.
4. Repeat with more examples.

Supplements

FOR THE STUDENT	FOR THE INSTRUCTOR
	Annotated Instructor Edition (ISBN: 978-1-4354-6254-0) The **Annotated Instructor Edition** features answers to all problems in the book.
Student Solutions Manual (ISBN: 978-1-285-84642-2) *Author: Michael Welden, Mt. San Jacinto College* The **Student Solutions Manual** provides worked-out solutions to the odd-numbered problems in the book.	***Complete Solutions Manual*** (ISBN: 978-1-285-84639-2) *Author: Michael Welden, Mt. San Jacinto College* The **Complete Solutions Manual** provides worked-out solutions to all of the problems in the text.
Student Workbook (ISBN: 978-1-285-84645-3) *Author: Maria H. Andersen, former math faculty at Muskegon Community College and now working in the learning software industry* The **Student Workbook** contains all of the assessments, activities, and worksheets from the **Instructor's Resource Binder** for classroom discussions, in-class activities, and group work.	***Instructor's Resource Binder*** (ISBN: 978-0-538-73675-6) *Author: Maria H. Andersen, former math faculty at Muskegon Community College and now working in the learning software industry* The **Instructor's Resource Binder** contains uniquely designed Teaching Guides, which include instruction tips, examples, activities worksheets, overheads, and assessments, with answers provided.
	Instructor Companion Website Everything you need for your course in one place! This collection of book-specific lecture and class tools is available online via **www.cengage.com/login**. Formerly found on the PowerLecture, access and download PowerPoint® presentations, images, and more.
	Solution Builder This online instructor database offers complete worked solutions to all exercises in the text, allowing you to create customized, secure solutions printouts (in PDF format) matched exactly to the problems you assign in class. For more information, visit **www.cengage.com/solutionbuilder**.
ENHANCED WebAssign ***Enhanced WebAssign®*** (Printed Access Card ISBN: 978-1-285-85770-1 Online Access Code ISBN: 978-1-285-85773-2) **Enhanced WebAssign** (assigned by the instructor) provides instant feedback on homework assignments to students. This online homework system is easy to use and includes a multimedia eBook, video examples, and problem-specific tutorials.	ENHANCED WebAssign ***Enhanced WebAssign®*** (Printed Access Card ISBN: 978-1-285-85770-1 Online Access Code ISBN: 978-1-285-85773-2) Instant feedback and ease of use are just two reasons why **WebAssign** is the most widely used homework system in higher education. **WebAssign's** homework delivery system allows you to assign, collect, grade, and record homework assignments via the web. And now this proven system has been enhanced to include a multimedia eBook, video examples, and problem-specific tutorials. **Enhanced WebAssign** is more than a homework system—it is a complete learning system for math students.
	Text-Specific Videos *Author: Rena Petrello* These videos are available at no charge to qualified adopters of the text and feature 10–20 minute problem-solving lessons that cover each section of every chapter.

To the Student

Congratulations! You now own a state-of-the-art textbook that has been written especially for you. We have tried to write a book that you can read and understand. The text includes carefully written narrative and an extensive number of worked examples with corresponding **Self Checks. Now Try This** exercises can be worked with your classmates, and **Guided Practice** exercises tell you exactly which example to use as a resource for each question. These are just a few of the many features included in this text with your success in mind. Perhaps the biggest change is the inclusion of Study Skill strategies at the start and end of each chapter. We urge you to take the time to complete these worksheets to increase your understanding of what it takes to be a successful student.

To get the most out of this course, you must read and study the textbook carefully. We recommend that you work the examples on paper first, and then work the Self Checks. Only after you thoroughly understand the concepts taught in the examples should you attempt to work the exercises. A **Student Solutions Manual** is available, which contains the worked-out solutions to the odd-numbered exercises.

Since the material presented in **Beginning and Intermediate Algebra, Seventh Edition,** may be of value to you in later years, we suggest that you keep this text. It will be a good reference in the future and will keep at your fingertips the material that you have learned here.

Hints on Studying Algebra

The phrase "practice makes perfect" is not quite true. It is "*perfect* practice that makes perfect." For this reason, it is important that you learn how to study algebra to get the most out of this course.

Although we all learn differently, here are some hints on studying algebra that most students find useful.

Planning a Strategy for Success To get where you want to be, you need a goal and a plan. Your goal should be to pass this course with a grade of A or B. To earn one of these grades, you must have a plan to achieve it. A good plan involves several points:

Getting ready for class,
Attending class,
Doing homework,
Making use of the extensive extra help available, including WebAssign if your instructor has set up a course, and
Having a strategy for taking tests.

Getting Ready for Class To get the most out of every class period, you will need to prepare for class. One of the best things you can do is to preview the material in the text that your instructor will be discussing in class. Perhaps you will not understand all of what you read, but you will be better able to understand your instructor when he or she discusses the material in class.

Do your work every day. If you fall behind, you will become frustrated and discouraged. Make a promise that you will always prepare for class, and then keep that promise.

Attending Class The classroom experience is your opportunity to learn from your instructor and interact with your classmates. Make the most of it by attending every class. Sit near the front of the room where you can see and hear easily. Remember that it is your responsibility to follow the discussion, even though it takes concentration and hard work.

Pay attention to your instructor, and jot down the important things that he or she says. However, do not spend so much time taking notes that you fail to concentrate on what your instructor is explaining. Listening and understanding the big picture is much more important than just copying solutions to problems.

Do not be afraid to ask questions when your instructor asks for them. Asking questions will make you an active participant in the class. This will help you pay attention and keep you alert and involved.

Doing Homework It requires practice to excel at tennis, master a musical instrument, or learn a foreign language. In the same way, it requires practice to learn mathematics. Since practice in mathematics is homework, homework is your opportunity to practice your skills and experiment with ideas. Consider creating note cards for important concepts.

It is important for you to pick a definite time to study and do homework. Set a formal schedule and stick to it. Try to study in a place that is comfortable and quiet. If you can, do some homework shortly after class, or at least before you forget what was discussed in class. This quick follow-up will help you remember the skills and concepts your instructor taught that day.

Each formal study session should include three parts:

1. Begin every study session with a review period. Look over previous chapters and see if you can do a few problems from previous sections. Keeping old skills alive will greatly reduce the amount of time you will need to prepare for tests.
2. After reviewing, read the assigned material. Resist the temptation to dive into the problems without reading and understanding the examples. Instead, work the examples and Self Checks with pencil and paper. Only after you completely understand the underlying principles behind them should you try to work the exercises.

   Once you begin to work the exercises, check your answers with the printed answers in the back of your text. If one of your answers differs from the printed answer, see if the two can be reconciled. Sometimes answers have more than one form. If you decide that your answer is incorrect, compare your work to the example in the text that most closely resembles the exercise, and try to find your mistake. If you cannot find an error, consult the *Student Solutions Manual.* If nothing works, mark the problem and ask about it in your next class meeting.
3. After completing the written assignment, preview the next section. This preview will be helpful when you hear the material discussed during the next class period.

You probably already know the general rule of thumb for college homework: two to three hours of practice for every hour you spend in class. If mathematics is difficult for you, plan on spending even more time on homework.

To make doing homework more enjoyable, study with one or more friends. The interaction will clarify ideas and help you remember them. If you choose to study alone, a good study technique is to explain the material to yourself out loud or use a white board where you can stand back and look at the big picture.

Accessing Additional Help Access any help that is available from your instructor. Often, your instructor can clear up difficulties in a short period of time. Find out whether your college has a free tutoring program. Peer tutors often can be a great help or consider setting up your own study group.

Taking Tests Students often become nervous before taking a test because they are afraid that they will do poorly.

To build confidence in your ability to take tests, rework many of the problems in the exercise sets, work the exercises in the Chapter Reviews, and take the Chapter Tests. Check all answers with the answers printed at the back of your text.

Guess what the instructor will ask, build your own tests, and work them. Once you know your instructor, you will be surprised at how good you can get at selecting test questions. With this preparation, you will have some idea of what will be on the test, and you will have more confidence in your ability to do well.

When you take a test, work slowly and deliberately. Write down any formulas at the top of the page. Scan the test and work the problems you find easiest first. Tackle the hardest problems last.

We wish you well.

Acknowledgements

We are grateful to the following people who reviewed the new editions of this series of texts. They all had valuable suggestions that have been incorporated into the texts.

Joan M. Haig, *University of Alaska Anchorage*
Harvey Hanna II, *Ferris State University*
Michael McComas, *Mountwest Community & Technical College*
Daisy McCoy, *Lyndon State College*

Additional Acknowledgments

We also thank the following people who reviewed previous editions.
Kent Aeschliman, *Oakland Community College*
Carol Anderson, *Rock Valley College*
Cynthia Broughtou, *Arizona Western College*
David Byrd, *Enterprise State Junior College*
Pablo Chalmeta, *New River Community College*
Jerry Chen, *Suffolk County Community College*
Michael F. Cullinan, *Glendale Community College*
Amy Cuneo, *Santiago Canyon College*
Lou D'Alotto, *York College-CUNY*
Thomas DeAgostino, *Jackson Community College*
Kristin Dillard, *San Bernardino Valley College*
Kirsten Dooley, *Midlands Technical College*
Karen Driskell, *Calhoun Community College*
Joan Evans, *Texas Southern University*
Hamidullah Farhat, *Hampton University*
Harold Farmer, *Wallace Community College-Hanceville*
Mark Fitch, *University of Alaska, Anchorage*
Mark Foster, *Santa Monica College*
Tom Fox, *Cleveland State Community College*
Jeremiah Gilbert, *San Bernardino Valley College*
Harvey Hanna, *Ferris State University*
Jonathan P. Hexter, *Piedmont Virginia Community College*
Kathy Holster, *South Plains College*
Dorothy K. Holtgrefe, *Seminole Community College*
Mark Hopkins, *Oakland Community College*
Mike Judy, *Fullerton College*
Lynette King, *Gadsden State Community College*
Janet Mazzarella, *Southwestern College*
Donald J. McCarthy, *Glendale Community College*
Robert McCoy, *University of Alaska Anchorage*
Andrew P. McKintosh, *Glendale Community College*
Christian R. Miller, *Glendale Community College*
Feridoon Moinian, *Cameron University*
Brent Monte, *Irvine Valley College*
Daniel F. Mussa, *Southern Illinois University*
Joanne Peeples, *El Paso Community College*

Mary Ann Petruska, *Pensacola Junior College*
Linda Pulsinelli, *Western Kentucky University*
Kimberly Ricketts, *Northwest-Shoals Community College*
Janet Ritchie, *SUNY-Old Westbury*
Joanne Roth, *Oakland Community College*
Richard Rupp, *Del Mar College*
Rebecca Sellers, *Jefferson State Community College*
Kathy Spradlin, *Liberty University*
John Squires, *Cleveland State Community College*
April D. Strom, *Glendale Community College*
Robert Vilardi, *Troy University*
Victoria Wacek, *Missouri Western State College*
Judy Wells, *University of Southern Indiana*
Hattie White, *St. Phillip's College*
George J. Witt, *Glendale Community College*
Margaret Yoder, *Eastern Kentucky University*

We are grateful to the staff at Cengage Learning, especially our publisher Charlie Van Wagner and our editor Marc Bove. We also thank Vernon Boes, Jennifer Risden, Stefanie Beeck, and Bryon Spencer.

We are indebted to Lachina Publishing Services, our production service; to Rhoda Oden, who read the entire manuscript and worked every problem; and to Michael Welden, who prepared the *Student Solutions Manual.*

—Rosemary M. Karr
Marilyn B. Massey
R. David Gustafson

Index of Applications

Examples that are applications are shown with boldface numbers.
Exercises that are applications are shown with regular face numbers.

Architecture
Building construction, 148, 430
Designing a patio, 973–974
Drafting, 428
Enclosing swimming pools, 454
Triangular bracing, 123
Trusses, 123
Window designs, 122

Business
Advertising, **97**
Auto sales, 148
Bus fares, 744
Calculating SEP contributions, 115
Car repairs, 203
Cell phone growth, **828–829**
Cell phone usage, 833
Choosing salary plans, 567
Clearance sales, 101
Closing real estate transactions, 47
Comparing bids, 28
Cost of a car, 148
Cost of carpet, **183**
Cost of rain gutters, 203
Cost of telephone service, 211
Costs of a trucking company, 430
Customer satisfaction, 92
Depreciating word processors, 203
Earning money, **214–215**, 611–612
Electric bills, 193
Employee discounts, 304
Excess inventory, 101
Figuring taxes, 218
Food service, 552
Furnace repairs, 417
Furniture pricing, 148
Home prices, 92
Installing gutters, 148
Installing solar heating, 148
Inventories, 53, 607, 630
Inventory, 620
Inventory costs, 28
Investing, 620
Lawn care, **569–570**
Loss of revenue, 53
Making bicycles, 565
Making clothes, 27
Making sporting goods, 218, 620
Making statues, 587
Making tires, 578
Making water pumps, 566
Managing a computer store, 566
Managing a makeup studio, 566
Manufacturing, 139, **574–575**
Manufacturing footballs, 587
Manufacturing hammers, **584–585**
Manufacturing profits, 28
Merchandising, 565
Monthly sales, 101
Mowing lawns, 53
Newspaper declines, **480**
Operating costs, 769
Planning for growth, 28
Printer charges, 203
Printing books, 566
Production, 630–631
Production planning, 619
Quality control, 28
Rate of decrease, 187
Rate of growth, 187
Real estate, 61, 194, 203
Real estate listings, 474
Retail sales, 142, 143, **559**
Retailing, 552
Running a record company, 566
Running a small business, 566
Sales, 417
Sales growth, 530
Salvage value, 834
Salvage values, 203
Selling DVD players, 483
Selling ice cream, 578
Selling microwave ovens, 92
Selling radios, 578
Selling real estate, 92
Selling shirts, 428
Selling tires, 483
Selling trees, 218
Shopper dissatisfaction, 93
Shopper satisfaction, 93
Small business, 47
Telephone costs, 193–194
Toy inventories, 53
Wages, 160
Wages and commissions, **446**

Education
Absenteeism, 73
Calculating grades, 138
College tuition, 46, 483
Course loads, 61
Educational costs, 173
Getting an A, 101
Grades, **136**, 455
Grading papers, 416
Homework, 139
Off-campus housing, 92
Saving for college, **830**
Saving for school, 54
School enrollment, 769
Staffing, 428
Student enrollment, 483
Study times, 73
Taking a test, 997
Teacher pensions, 203

Electronics
Broadcast ranges, 907
Circuit boards, 123
Computer repairs, 194

Computers, 996
Electronics, 430, 565, 658, 712
Information access, 942
Ohm's law, 114
Radio frequencies, 567
Resistance, 38
Satellite antennas, 907
Signum function, 941
Sorting records, 61
TV translators, **899–900**

Entertainment
Buying tickets, 577
Call letters, 997
Concert tickets, 219, 588, 744
Concert tours, 122–123
Downloading music, 455
DVDs, **735**
At the movies, 577
Production schedules, **627–628**
Royalties, 942
Theater seating, 596
Ticket sales, 101
TV programming, 578, **991**

Farming
Buying fencing, 38
Dairy production, 28
Farming, 73, 429, 565, **568–569**
Feeding dairy cows, 28
Fencing a field, 769
Fencing a garden, 27
Fencing land, **21**
Fencing pastures, 454
Fencing pens, 454
Planting corn, 1008
Raising livestock, 578
Spring plowing, 27
Wheat harvest, 483

Finance
Account balances, **44**
Amount of an annuity, **979–980**
Annuities, 982
Appreciation, 93, 202, 203
Auto loans, **25**
Balancing the books, 47
Banking, 46, 92
Boat depreciation, 1012
Break-point analysis, **561–562**, 566
Buying stock, 46
Calculating compound interest, **829**
Car depreciation, 1008
Comparing assets, 61
Comparing interest rates, 416, 833
Comparing investments, 61, **414–415**, 416
Comparing savings plans, 833
Compound interest, 833, 879, 885
Compounding methods, 841
Computing revenue, 260
Computing taxes, **939**
Continuous compound interest, 840, 879, 886
Cost and revenue, 552
Declining savings, 982
Demand equations, 472
Depreciation, 93, 160, 202, 203, 225, 530, 701, 841, 851
Depreciation equations, 474
Depreciation rates, **698–699**
Determining a previous balance, 840
Determining the initial deposit, 840
Doubling money, 857
Doubling time, **856**
Earnings per share, 54
Equilibrium price, 579
Figuring inheritances, 577
Financial planning, 631
Finding profit, 455
Frequency of compounding, 833
Growth of money, 114
House appreciation, 472, 982
Income, **625–626**
Installment loans, 973
Interest rates, 744
Investing, **90**, 145, 147, 219, 238, 455, **571–572**, 607
Investing money, 577, 578, 935
Investment income, 565
Investments, **121**, 123–124
Linear depreciation, **199–200**
Maximizing revenue, 769
Motorboat depreciation, 982
Percent of decrease, 143
Percent of increase, 143
Piggy banks, 596
Present value, 243, 244
Profit, 744
Property values, **265**
Retirement income, 28
Rule of seventy, 879
Saving money, **735**, 737, 973
Savings, 885
Savings accounts, 701
Savings growth, 982
Stock appreciation, 1008
Stock market, 46, 53
Stock reports, **51**
Stock splits, 93
Stock valuations, **56**
Supply equations, 473
T-bills, 61
Time for money to grow, 851
Tripling money, 857
Value of a boat, 429
Value of a house, 267
Value of two houses, 267

Geometry
Angles, 123
Area of a circle, 429
Area of a track, 918
Area of a triangle, 357
Area of an ellipse, 918
Area of many cubes, 665
Base of a triangle, 366, 744
Circles, **32–33**
Circumference of a circle, 114
Complementary angles, **119**
Concentric circles, 304
Curve fitting, **585–586**, 587, 588, 596
Dimensions of a parallelogram, 357
Dimensions of a rectangle, 365, **742**, 744, 796
Dimensions of a triangle, 357
Equilateral triangles, 123, **136–137**
Geometry, 114–115, **118**, 138, 139, 429, 430, 538, 565, 566, 577, 587, 596, 597, 664, 665, **680–681**, 935
Height of a triangle, 744
Inscribed squares, 982
Integers, 101, **352–353**, 356, 578, 587, 744, 935
Interior angles, 974
Isosceles triangles, **120**
Numbers, 365, 416, 417, 935
Palindromes, 996
Parallelograms, **563**

Perimeter of a rectangle, 744
Radius of a circle, 658
Rectangles, **119–120**, 145, **354–355**
Side of a square, 744
Supplementary angles, **119**
Surface area, 304
Surface area of a cube, 665
Triangles, **355–356**
Volume of a classroom, 38
Volume of a cone, **112–113**, 114, **425–426**
Volume of a pyramid, 358
Volume of a solid, 358
Volume of a tank, 38

Home Management and Shopping
Avoiding service charges, 139
Buying baseball equipment, 578
Buying boats, 620
Buying cameras, **99**
Buying carpets, 93
Buying cars, **88**
Buying CDs, 620
Buying clothes, 578
Buying contact lens cleaner, 578
Buying furniture, **87**, 620
Buying paint, 93
Buying painting supplies, 579
Buying real estate, 93
Choosing a furnace, 28, 578
Choosing housekeepers, 218
Electric bills, 148
Furniture pricing, 101, 148
Furniture sales, 101
Grocery shopping, **420**
Ordering furnace equipment, 621
Value of coupons, 101
Wallpapering, 38

Medicine
Alcohol absorption, 841
Body mass, 428, 665
Causes of death, 577
Diet, 630
Epidemics, 841
Forensic medicine, 356, 858
Hospital costs, 187
Hospitals, 92
Medical technology, **573–574**
Medicine, 131, 658, 701, 737, 878, 879, 1002
Mixing pharmaceuticals, 578
Nutritional planning, 587
Physical fitness, 473
Physical therapy, 566, 596
Pulse rates, 693
Recommended dosage, 428
Red blood cells, 250
Supplements, **626–627**

Miscellaneous
Accidents, 737
Antique cars, 61, **98**
Antiseptic solutions, 131
Apartment rentals, 101
Arranging an evening, 996
Arranging appointments, 996
Arranging books, 996
Auto repairs, 101
Birthday parties, 168–169
Blending gourmet tea, 131
Boarding dogs, 101
Bookbinding, 276
Bouncing balls, 238, 982
Building a dog run, **447**
Building construction, 148
Buying airplanes, 93
Calculating clearance, 918
Cannon fire, 366
Car repair(s), 358, 474, 552
Cards, 1012
Carpentry, 122, 145, 357, 664, 701
Cats and dogs, 596
Chainsaw sculpting, 588
Charities, 93
Choosing books, 997
Choosing clothes, 997
Choosing people, 1010
Coffee blends, 131
Combination locks, 996
Communications, 717
Comparing weights, 139
Controlling exotic plants, 203
Controlling moths, 982
Conveyor belts, 417
Cooking, 428
Cost of a car, 148
Crafts, 630
Cutting beams, 455
Cutting boards, 455
Cutting lumber, 577
Cutting pipe, 538, 577
Cutting ropes, **57**, 61
Designing an underpass, 918
Designing tents, 357
Dieting, 46, 53
Dimensions of a painting, 274
Dimensions of a window, 744
Discharging a battery, 833
Dolphins, 483
DVD rentals, 159
Electric service, 664
Falling balloons, 260
Filling a pool, 416, 417
Filling an oil tank, **413**
Finding dimensions, 454
Finding the variance, 769
Fitness equipment, 917
Fleet averages, 139
Focal length, 38
Forming a committee(s), 997, 1012
Framing a picture(s), 123, 744
Freeze-drying, 28
Furniture, 630
Gardening, 365, **570–571**
Gateway Arch, **902–903**
Generating power, 701
Getting exercise, 139
Global warming, 473
Grade of a road, 187, 473
Graphing points, **153**
Grass seed mixture, 578
Guy wires, 123
Hardware, 683
Having babies, 238
Heaviside unit step function, 941
Height of a bridge, **695**
Height of a flagpole, 429
Height of a rocket, 260
Height of a tree, **421–422**, 429
Heights of trees, 61
Hiring babysitters, 619
Hobbies, 428
Horizon distance, 700
House construction, 358
Housing, 483
Installing carpet, 38
Installing gutters, 148

Installing solar heating, 148
Insulation, 357
Invisible tape, 61
Ironing boards, 684
Jogging, 437
Labor force, 745
Land areas, 123
Land elevations, 139
Landscape design, **914–915**
Landscaping, 552, **615**
Lawn seed blends, 131
Lining up, 996, 1010, 1012
Making cottage cheese, 131
Making Jell-O, 858
Making license plates, 996
Manufacturing concrete, 148
Marketing, 701
Maximizing area, **764–765**, 769
Melting iron, 139
Meshing gears, 907
Metal fabrication, 744
Milling brass plates, 566
Millstones, 276
Mixing acid, **127–128**
Mixing candy, 131, 146, 565
Mixing chemicals, 578
Mixing coffee, 131
Mixing fuel, 428
Mixing fuels, 131
Mixing milk, 146
Mixing nuts, **128–129**, 131, 147, 578, 588
Mixing paint, 131
Mixing peanuts and candy, 578
Mixing photographic chemicals, 131
Mixing solutions, 147, **559–561**, 565
Model railroading, 428
Oil reserves, 250
Organ pipes, 430
Packaging, 73
Painting houses, 437
Pendulums, 723, 737
Period of a pendulum, **654**, 658
Petroleum storage, 74
Phone numbers, 996
Phonograph records, 139
Photography, 146, **690–691**, 693
Pitch of a roof, 187
Planning a picnic, 997
Plumbing, 61, **116–117**, 122
Pool tables, 918
Postage rates, 159
Printing schedules, 416
Printing stationery, **938–939**
Publishing, 122
Publishing directory of members, 782
Pulley designs, 276
Pumping a basement, 437
Pythons, 139
Quality control, **90**, 1002
Rational expressions, 693
Reach of a ladder, 664
Reading graphs, **155–156**
Renting a car, **94**
Renting a ski jet, 941
Riding bicycles, 145
Riding in a taxi, 941
Roofing a house, 416
Ropes courses, 357
Sealing asphalt, 28
Selecting people, 1011
Sharing costs, 417
Shipping crates, 722
Shipping packages, 665
Shipping pallets, 357
Signaling, 607
Slope of a ladder, 187
Slope of a ramp, 187
Slope of a roof, 473
Splitting the lottery, 577
Statue of Liberty, 122
Step graphs, **157**
Storing oil, 430
Storing solvents, 38
Supporting a weight, 664
Swimming pool borders, 357
Swimming pools, 123, 146
Targets, 693
Telephone charges, 101
Telephone service, 664
Tornado damage, 358
Tornadoes, 145–146
Trail mix, 131
Union membership, 93
Value of a car, 160
Value of antiques, 194
Variance, **765**
View from a submarine, 717
Water pressure, 160
Wheelchair ramps, 473
Width of a river, 429
Width of a walkway, 907
Winning a lottery, 47, 997

Politics, Government, and the Military

2010 gubernatorial elections, **24**
Artillery, 935
Artillery fire, 483
Battling ships, 159
Building highways, **426**
City planning, **837**
Cleaning highways, 439
Congress, **993**
Crime prevention, 473
Disaster relief, 27
Doubling time, **855**
Fighting fires, **660**
Flags, 737
Forming committees, 1010
Government, 577
Growth of a town, **978–979**
Highway design, 700, 907
Income taxes, **55**
Law enforcement, 658, 737
Louisiana Purchase, 834
Making a ballot, 996
Malthusian population growth, **874**, 886
Marine Corps, 566–567
Military history, 46
Minority population, 27
National debt, 250
Paving highways, 130
Paying taxes, 28
Police investigations, 769
Pony Express, 417
Population decline, 841, 981
Population growth, **838**, 840–841, 857, **875**, 878, 981
Predicting burglaries, 473
Predicting fires, 194
Rodent control, 878
Sales taxes, 92
Selecting committees, **993**, 1011
Sewage treatment, 358, 417
Space program, 745
Tax rates, 143
Taxes, 93, 439
Town population, 833

U.S. population, 886, 888
Violent crime, 483
Water billing, 101, **191**
World population growth, 840

Science
Alloys, 813
Alpha particles, 928
Angstroms, 250
Aquariums, 867
Astronomy, 53, 304
Atomic structure, **925–926**
Bacteria cultures, 833, 878
Ballistics, 356, 483, **763–764**, 768, 769, 796
Biology, 244
Carbon-14 dating, **874**, 878, 890
Change in intensity, 868
Change in loudness, 868, 889
Chemistry, 131, 745
Comparing temperatures, 139
Conversion of degrees, 483
dB gain, 888
dB gain of an amplifier, 851
decibel voltage gain, **847**
Distance between Mercury and the Sun, 250
Distance to Alpha Centauri, 249
Distance to Mars, 249
Distance to Venus, 249
Earthquakes, 851, 888
Earth's atmosphere, 588
Electrostatic repulsion, 928
Falling objects, 429, 658, 737, 974, 1008
Finding pH of a solution, **864–865**
Finding the constant of variation, 430
Finding the gain of an amplifier, 851
Finding the hydrogen ion concentration, **865**
Finding the speed of a current, 578
Force of gravity, 115
Free-falling objects, 841
Gas pressure, 429, 430
Genealogy, 982
Generation time, **875–876**
Growth of money, 114
Half-life, 878
Half-life of radon-22, **873**
History, 46
Hydrogen ion concentration, 867
Lead decay, 879
Length of one meter, 249
Light intensity, **424–425**
Light year, 250
LORAN, 928
Measuring earthquakes, **848**
Oceanography, 879
Optics, 411
Orbits, **153–154**
Path of a comet, 907
pH of a solution, 867
pH of grapefruit, 889
pH of pickles, 868
Physiology, 174
Projectiles, 907
Radioactive decay, 833, **839**, 841, 878, 886
Research, 174
Robotics, 122
Sonic boom, 928
Speed of sound, 249, 250
Statistics, 658
Temperature change, **811–812**
Temperature changes, **44**
Temperatures, 46, 53
Tension, 430
Thermodynamics, 115
Thorium decay, 879
Tritium decay, 878
Wavelengths, 250
Weather forecasting, 813
Weber–Fechner law, **865**
Wind power, **178–179**

Sports
Baseball, 276, 664
Bowling, 665
Buying tickets, 577, 620
Diagonal of a baseball diamond, 658
Exhibition diving, 356
Football, 46
Golf swings, 160
NBA Salaries, **258**
NFL records, 588
Packing a tennis racket, 665
Road Rallies, 417
Sailing, 664, 717
Ski runs, 429
Skydiving, 841
Speed skating, 28
Track and field, 27, **413–414**

Travel
Average speeds, 131
Aviation, 46, 130, 578
Band trips, 146
Biking, 130
Boating, 130, 417, 439, **572–573**, 577
Chasing a bus, 130
Comparing travel, 416
Cost of gasoline, 28
Direct variation, **423**
Driving rates, 935
Finding distance, 429
Flight path, 439
Flight paths, 429
Flying objects, **353–354**
Flying speeds, 416
Gas consumption, 428
Gas mileage, 160
Group rates, 173
Hiking, 130
Hot pursuit, 130
Motion, 578
Mountain climbing, 46
Navigation, 552
Plane altitudes, 139
Planning a trip, 1010
Rate of descent, **183–184**
Rates, 744
Riding bicycles, 145
River tours, 417
Road maps, 159
Shipping, **125–126**
Speed of airplanes, 130
Speed of trains, 130
Stopping distance, 260–261
Stowing baggage, 565
Time of flight, 356
Touring, 416
Travel, **16–17**, 565
Travel choices, 996
Travel time(s), 130, 417
Traveling, **125**, **126–127**, 147
Vacation driving, 131
Wind speed, 416, 437
Winter driving, **35**

Real Numbers and Their Basic Properties

1

Dusan Petkovic/Shutterstock.com

Careers and Mathematics

CARPENTERS

Carpenters are involved in many kinds of construction activities. They cut, fit, and assemble wood and other materials for the construction of buildings, highways, bridges, docks, industrial plants, boats, and many other structures. About 32% of all carpenters—the largest construction trade—are self-employed. Future carpenters should take classes in English, algebra, geometry, physics, mechanical drawing, and blueprint reading.

REACH FOR SUCCESS
1.1 Real Numbers and Their Graphs
1.2 Fractions
1.3 Exponents and Order of Operations
1.4 Adding and Subtracting Real Numbers
1.5 Multiplying and Dividing Real Numbers
1.6 Algebraic Expressions
1.7 Properties of Real Numbers
■ *Projects*
REACH FOR SUCCESS EXTENSION
CHAPTER REVIEW
CHAPTER TEST

In this chapter

In Chapter 1, we will discuss the various types of numbers that we will use throughout this course. Then we will review the basic arithmetic of fractions, explain how to add, subtract, multiply, and divide real numbers, introduce algebraic expressions, and summarize the properties of real numbers.

Reach for Success Analyzing Your Time

We can all agree that we only have 24 hours in a day, right? The expression "My, how time flies!" is certainly relevant. It might surprise you where the time goes when you begin to monitor the hours.

How can you schedule your time to allow for a successful semester? Begin by completing the pie chart to analyze a typical day.

©Brad Collett/Shutterstock.com

HOW MANY HOURS A DAY DO YOU SPEND:

For every hour spent on each activity shade that many pie segments. (It might be helpful to use a different color to shade for each question.) There are no right or wrong answers. You just need an honest assessment of your time.

_____ working at outside employment (include commute time)?

_____ sleeping?

_____ preparing meals, eating, and exercising?

_____ getting ready for work/class?

_____ in class (include lab hours, commuting, or tutoring time)?

_____ with family and friends?

_____ on the Internet, phone, playing video games, texting, watching TV, going to movies, or other entertainment?

24-Hour Pie Chart

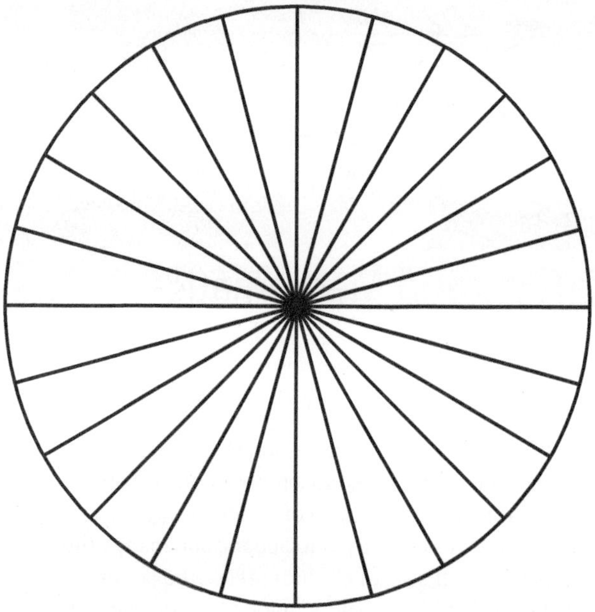

Now we have a mathematics problem! How many hours are left for studying? Remember, the rule of thumb is for every **1** hour in class, you will need at least **2** hours of studying. For mathematics, you could need even more. The key is to spend as much time as you need to understand the material. There is no magic formula!

A Successful Study Strategy . . .

Adjust your schedule as needed to support your educational goals.

At the end of the chapter you will find an additional exercise to guide you in planning for a successful semester.

Unless otherwise noted, all content on this page is © Cengage Learning.

Section 1.1 Real Numbers and Their Graphs

Objectives

1. List the numbers in a set of real numbers that are natural, whole, integers, rational, irrational, composite, prime, even, or odd.
2. Insert a symbol $<$, $>$, or $=$ to define the relationship between two rational numbers.
3. Graph a real number or a subset of real numbers on the number line.
4. Find the absolute value of a real number.

Vocabulary

set	irrational numbers	inequality symbols
natural numbers	real numbers	variables
positive integers	prime numbers	number line
whole numbers	composite numbers	origin
ellipses	even integers	coordinate
negative numbers	odd integers	negatives
integers	sum	opposites
subsets	difference	intervals
set-builder notation	product	absolute value
rational numbers	quotient	

Getting Ready

1. Give an example of a number that is used for counting.
2. Give an example of a number that is used when dividing a pizza.
3. Give an example of a number that is used for measuring temperatures that are below zero.
4. What other types of numbers can you think of?

We will begin by discussing various sets of numbers.

1 List the numbers in a set of real numbers that are natural, whole, integers, rational, irrational, composite, prime, even, or odd.

A **set** is a collection of objects. For example, the set

$\{1, 2, 3, 4, 5\}$ Read as "the set with elements 1, 2, 3, 4, and 5."

contains the numbers 1, 2, 3, 4, and 5. The *members*, or *elements*, of a set are listed within braces { }.

Two basic sets of numbers are the **natural numbers** (often called the **positive integers**) and the **whole numbers**.

THE SET OF NATURAL NUMBERS (POSITIVE INTEGERS)

$\{1, 2, 3, 4, 5, 6, 7, 8, 9\ 10, \ldots\}$

THE SET OF WHOLE NUMBERS

$$\{0, 1, 2, 3, 4, 5, 6, 7, 8, 9, 10, \ldots\}$$

The three dots in the previous definitions, called **ellipses**, indicate that each list of numbers continues on forever.

We can use whole numbers to describe many real-life situations. For example, some cars might get 30 miles per gallon (mpg) of gas, and some students might pay \$1,750 in tuition.

Numbers that show a loss or a downward direction are called **negative numbers**, and they are denoted with a $-$ sign. For example, a debt of \$1,500 can be denoted as $-\$1{,}500$, and a temperature of 20° below zero can be denoted as $-20°$.

The negatives of the natural numbers and the whole numbers together form the set of **integers**.

THE SET OF INTEGERS

$$\{\ldots, -5, -4, -3, -2, -1, 0, 1, 2, 3, 4, 5, \ldots\}$$

Because the set of natural numbers and the set of whole numbers are included within the set of integers, these sets are called **subsets** of the set of integers.

Integers cannot describe every real-life situation. For example, a student might study $3\frac{1}{2}$ hours, or a TV set might cost \$217.37. To describe these situations, we need fractions, more formally called *rational numbers*.

We cannot list the set of rational numbers as we have listed the previous sets in this section. Instead, we will use **set-builder notation**. This notation uses a variable (or variables) to represent the elements in a set and a rule to determine the possible values of the variable.

THE SET OF RATIONAL NUMBERS

Rational numbers are fractions that have an integer numerator and a nonzero integer denominator. Using set-builder notation, the rational numbers are

$$\left\{\frac{a}{b}\ \middle|\ a \text{ is an integer and } b \text{ is a nonzero integer.}\right\}$$

COMMENT Because division by 0 is undefined, expressions such as $\frac{6}{0}$ and $\frac{8}{0}$ do not represent any number.

The previous notation is read as "the set of all numbers $\frac{a}{b}$ such that a is an integer and b is a nonzero integer."

Some examples of rational numbers are

$$\frac{3}{2}, \frac{17}{12}, 5, -\frac{43}{8}, 0.25, \text{ and } -0.66666\ldots$$

The decimals 0.25 and $-0.66666\ldots$ are rational numbers, because 0.25 can be written as the fraction $\frac{1}{4}$, and $-0.66666\ldots$ can be written as the fraction $-\frac{2}{3}$.

Since every integer can be written as a fraction with a denominator of 1, every integer is also a rational number. Since every integer is a rational number, the set of integers is a subset of the rational numbers.

Since π and $\sqrt{2}$ cannot be written as fractions with an integer numerator and a nonzero integer denominator, they are not rational numbers. They are called **irrational numbers**. We can find their decimal approximations with a calculator. For example,

$\pi \approx 3.141592654$ Using a scientific calculator, press π. Using a graphing calculator, press 2nd ^ (π) ENTER. Read $\approx$ as "is approximately equal to."

$\sqrt{2} \approx 1.414213562$ Using a scientific calculator, press 2 $\sqrt{\ }$. Using a graphing calculator, press 2nd x^2 ($\sqrt{\ }$) 2 ENTER.

If we combine the rational and the irrational numbers, we have the set of **real numbers**.

THE SET OF REAL NUMBERS

$\{x|x$ is either a rational number or an irrational number.$\}$

COMMENT The symbol ℝ is often used to represent the set of real numbers.

The previous notation is read as "the set of all numbers x such that x is either a rational number or an irrational number."

Figure 1-1 illustrates how the various sets of numbers are interrelated.

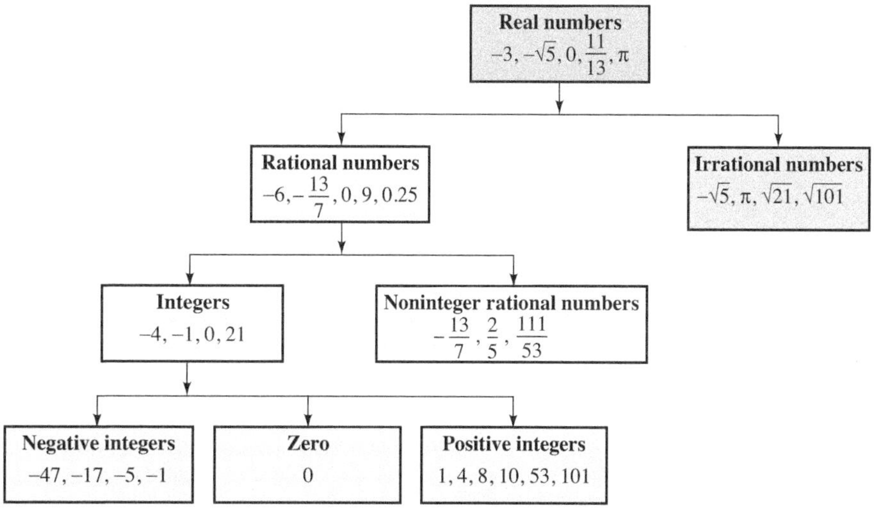

Figure 1-1

EXAMPLE 1 Which numbers in the set $\{-3, 0, \frac{1}{2}, 1.25, \sqrt{3}, 5\}$ are **a.** natural numbers **b.** whole numbers **c.** negative integers **d.** rational numbers **e.** irrational numbers **f.** real numbers?

Solution

a. The only natural number is 5.

b. The whole numbers are 0 and 5.

c. The only negative integer is -3.

d. The rational numbers are $-3, 0, \frac{1}{2}$, 1.25, and 5. (1.25 is rational, because 1.25 can be written in the form $\frac{125}{100}$.)

e. The only irrational number is $\sqrt{3}$.

f. All of the numbers are real numbers.

SELF CHECK 1 Which numbers in the set $\{-2, 0, 1.5, \sqrt{5}, 7\}$ are **a.** natural numbers **b.** rational numbers?

Unless otherwise noted, all content on this page is © Cengage Learning.

Perspective

Pythagoras
569–475 BC

THE FIRST IRRATIONAL NUMBER

The Greek mathematician and philosopher Pythagoras believed that every aspect of the natural world could be represented by ratios of whole numbers (i.e., rational numbers). However, one of his students accidentally disproved this claim by examining a surprisingly simple example. The student examined a right triangle whose legs were each 1 unit long and posed the following question. "How long is the third side of the triangle?"

Using the well-known theorem of Pythagoras, the length of the third side can be determined by using the formula $c^2 = 1^2 + 1^2$.

In other words, $c^2 = 2$. Using basic properties of arithmetic, it turns out that the numerical value of c cannot be expressed as a rational number. So we have an example of an aspect of the natural world that corresponds to an irrational number, namely $c = \sqrt{2}$.

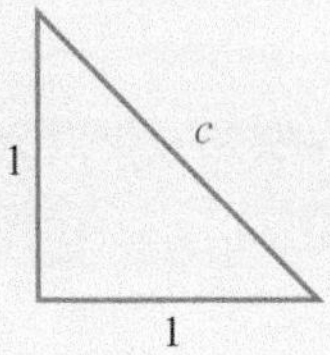

A natural number greater than 1 that can be divided evenly only by 1 and itself is called a **prime number**.

The set of prime numbers:

$\{2, 3, 5, 7, 11, 13, 17, 19, 23, 29, \ldots\}$

A nonprime natural number greater than 1 is called a **composite number**.

The set of composite numbers:

$\{4, 6, 8, 9, 10, 12, 14, 15, 16, 18, 20, 21, 22, \ldots\}$

An integer that can be divided evenly by 2 is called an **even integer**. An integer that cannot be divided evenly by 2 is called an **odd integer**.

The set of even integers:

$\{\ldots, -10, -8, -6, -4, -2, 0, 2, 4, 6, 8, 10, \ldots\}$

The set of odd integers:

$\{\ldots, -9, -7, -5, -3, -1, 1, 3, 5, 7, 9, \ldots\}$

EXAMPLE 2 Which numbers in the set $\{-3, -2, 0, 1, 2, 3, 4, 5, 9\}$ are
a. prime numbers **b.** composite numbers **c.** even integers **d.** odd integers?

Solution
a. The prime numbers are 2, 3, and 5.
b. The composite numbers are 4 and 9.
c. The even integers are $-2, 0, 2$, and 4.
d. The odd integers are $-3, 1, 3, 5$, and 9.

SELF CHECK 2 Which numbers in the set $\{-5, 0, 1, 2, 4, 5\}$ are
a. prime numbers **b.** even integers?

Unless otherwise noted, all content on this page is © Cengage Learning.

2 Insert a symbol $<$, $>$, or $=$ to define the relationship between two rational numbers.

To show that two expressions represent the same number, we use an $=$ sign. Since $4 + 5$ and 9 represent the same number, we can write

$4 + 5 = 9$ Read as "the sum of 4 and 5 is equal to 9." The answer to any addition problem is called a ***sum***.

Likewise, we can write

$5 - 3 = 2$ Read as "the difference between 5 and 3 equals 2," or "5 minus 3 equals 2." The answer to any subtraction problem is called a ***difference***.

$4 \cdot 5 = 20$ Read as "the product of 4 and 5 equals 20," or "4 times 5 equals 20." The answer to any multiplication problem is called a ***product***.

and

$30 \div 6 = 5$ Read as "the quotient obtained when 30 is divided by 6 is 5," or "30 divided by 6 equals 5." The answer to any division problem is called a ***quotient***.

We can use **inequality symbols** to show that expressions are not equal.

Symbol	Read as	Symbol	Read as
$\approx$	"is approximately equal to"	$\neq$	"is not equal to"
$<$	"is less than"	$>$	"is greater than"
$\leq$	"is less than or equal to"	$\geq$	"is greater than or equal to"

EXAMPLE 3 **Inequality symbols**

a. $\pi \approx 3.14$ Read as "pi is approximately equal to 3.14."

b. $6 \neq 9$ Read as "6 is not equal to 9."

c. $8 < 10$ Read as "8 is less than 10."

d. $12 > 1$ Read as "12 is greater than 1."

e. $5 \leq 5$ Read as "5 is less than or equal to 5." (Since $5 = 5$, this is a true statement.)

f. $9 \geq 7$ Read as "9 is greater than or equal to 7." (Since $9 > 7$, this is a true statement.)

SELF CHECK 3 Determine whether each statement is true or false.

a. $12 \neq 12$ **b.** $7 \geq 7$ **c.** $125 < 137$

Inequality statements can be written so that the inequality symbol points in the opposite direction. For example,

$5 < 7$ and $7 > 5$

both indicate that 5 is less than 7. Likewise,

$12 \geq 3$ and $3 \leq 12$

both indicate that 12 is greater than or equal to 3.

In algebra, we use letters, called **variables**, to represent real numbers. For example,

- If x represents 4, then $x = 4$.
- If y represents any number greater than 3, then $y > 3$.
- If z represents any number less than or equal to -4, then $z \leq -4$.

COMMENT In algebra, we usually do not use the times sign ($\times$) to indicate multiplication. It might be mistaken for the variable x.

3 Graph a real number or a subset of real numbers on the number line.

COMMENT The number 0 is neither positive nor negative.

We can use the **number line** shown in Figure 1-2 to represent sets of numbers. The number line continues forever to the left and to the right. Numbers to the left of 0 (the **origin**) are negative, and numbers to the right of 0 are positive.

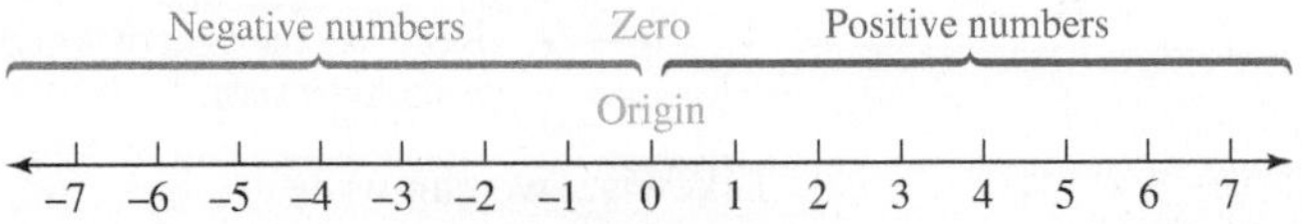

Figure 1-2

The number that corresponds to a point on the number line is called the **coordinate** of that point. For example, the coordinate of the origin is 0.

Many points on the number line do not have integer coordinates. For example, the point midway between 0 and 1 has the coordinate $\frac{1}{2}$, and the point midway between -3 and -2 has the coordinate $-\frac{5}{2}$ (see Figure 1-3).

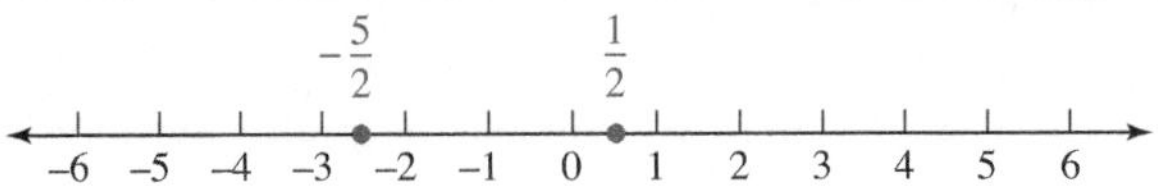

Figure 1-3

Numbers represented by points that lie on opposite sides of the origin and at equal distances from the origin are called **negatives** (or **opposites**) of each other. For example, 5 and -5 are negatives (or opposites). We need parentheses to express the opposite of a negative number. For example, $-(-5)$ represents the opposite of -5, which we know to be 5. Thus,

$$-(-5) = 5$$

This suggests the following rule.

DOUBLE NEGATIVE RULE If x represents a real number, then

$$-(-x) = x$$

If one point lies to the *right* of a second point on a number line, its coordinate is the *greater*. Since the point with coordinate 1 lies to the right of the point with coordinate -2 (see Figure 1-4(a)), it follows that $1 > -2$.

If one point lies to the *left* of another, its coordinate is the *smaller* (see Figure 1-4(b)). The point with coordinate -6 lies to the left of the point with coordinate -3 so it follows that $-6 < -3$.

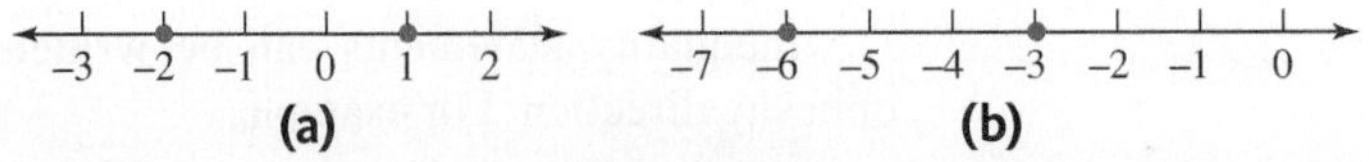

Figure 1-4

Figure 1-5 shows the graph of the natural numbers from 2 to 8. The points on the line are called graphs of their corresponding coordinates.

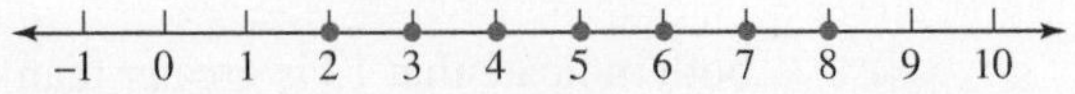

Figure 1-5

EXAMPLE 4 Graph the set of integers between -3 and 3.

Solution The integers between -3 and 3 are $-2, -1, 0, 1$, and 2. The graph is shown in Figure 1-6.

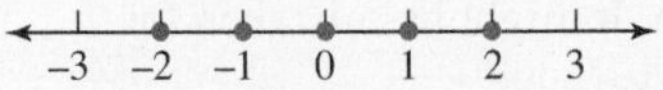

Figure 1-6

SELF CHECK 4 Graph the set of integers between -4 and 0.

Unless otherwise noted, all content on this page is © Cengage Learning.

Leonardo Fibonacci
(late 12th and early 13th centuries)

Fibonacci, an Italian mathematician, is also known as Leonardo da Pisa. In his work *Liber abaci,* he advocated the adoption of Arabic numerals, the numerals that we use today. He is best known for a sequence of numbers that bears his name. Can you find the pattern in this sequence?

1, 1, 2, 3, 5, 8, 13, . . .

Graphs of many sets of real numbers are **intervals** on the number line. For example, two graphs of all real numbers x such that $x > -2$ are shown in Figure 1-7. The parenthesis and the open circle at -2 show that this point is not included in the graph. The arrow pointing to the right shows that all numbers to the right of -2 are included.

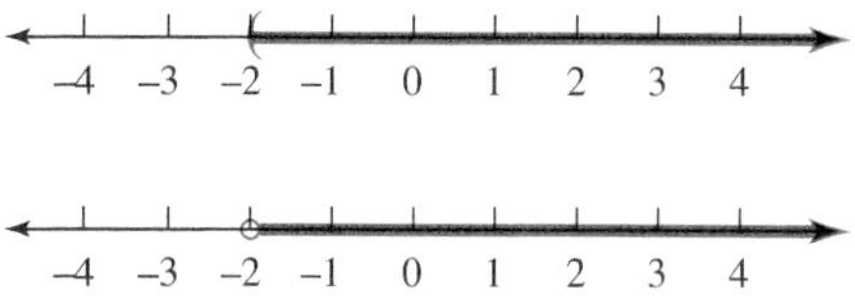

Figure 1-7

Figure 1-8 shows two graphs of the set of real numbers x between -2 and 4. This is the graph of all real numbers x such that $x > -2$ and $x < 4$. The parentheses or open circles at -2 and 4 show that these points are not included in the graph. However, all the numbers between -2 and 4 are included.

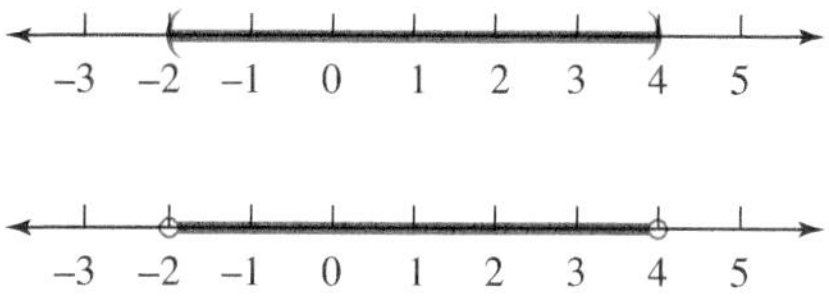

Figure 1-8

EXAMPLE 5 Graph all real numbers x such that $x < -3$ or $x > 1$.

Solution The graph of all real numbers less than -3 includes all points on the number line that are to the left of -3. The graph of all real numbers greater than 1 includes all points that are to the right of 1. The two graphs are shown in Figure 1-9.

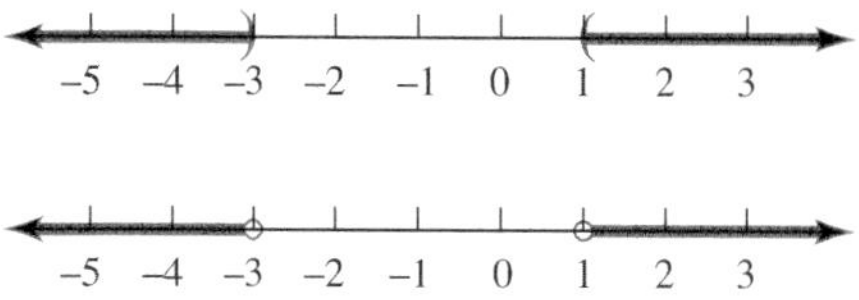

Figure 1-9

SELF CHECK 5 Graph all real numbers x such that $x < -1$ or $x > 0$. Use parentheses.

Perspective

Algebra is an extension of arithmetic. In algebra, the operations of addition, subtraction, multiplication, and division are performed on numbers and letters, with the understanding that the letters represent numbers.

The origins of algebra are found in a papyrus written before 1600 BC by an Egyptian priest named Ahmes. This papyrus contains 84 algebra problems and their solutions.

Further development of algebra occurred in the ninth century in the Middle East. In AD 830, an Arabian mathematician named al-Khowarazmi wrote a book called *Ihm al-jabr wa'l muqabalah.* This title was shortened to *al-Jabr.* We now know the subject as *algebra.* The French mathematician François Vieta (1540–1603) later simplified algebra by developing the symbolic notation that we use today.

The Ahmes Papyrus

© The Trustees of the British Museum/Art Resource, NY

Unless otherwise noted, all content on this page is © Cengage Learning.

EXAMPLE 6 Graph the set of all real numbers from -5 to -1.

Solution The set of all real numbers from -5 to -1 includes -5 and -1 and all the numbers in between. In the graphs shown in Figure 1-10, the brackets or the solid circles at -5 and -1 show that these points are included.

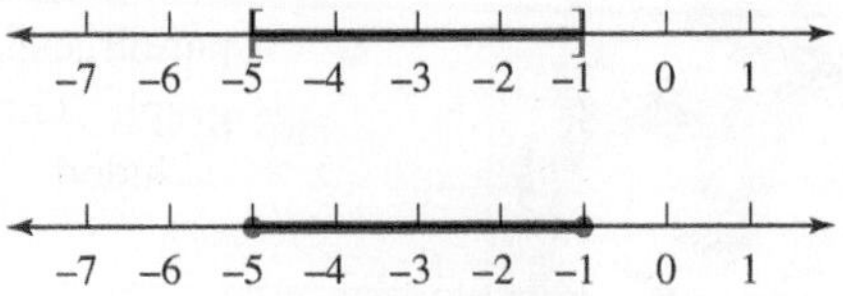

Figure 1-10

SELF CHECK 6 Graph the set of real numbers from -2 to 1.

4 Find the absolute value of a real number.

On a number line, the distance between a number x and 0 is called the **absolute value** of x. For example, the distance between 5 and 0 is 5 units (see Figure 1-11). Thus, the absolute value of 5 is 5:

$|5| = 5$ Read as "The absolute value of 5 is 5."

Since the distance between -6 and 0 is 6,

$|-6| = 6$ Read as "The absolute value of -6 is 6."

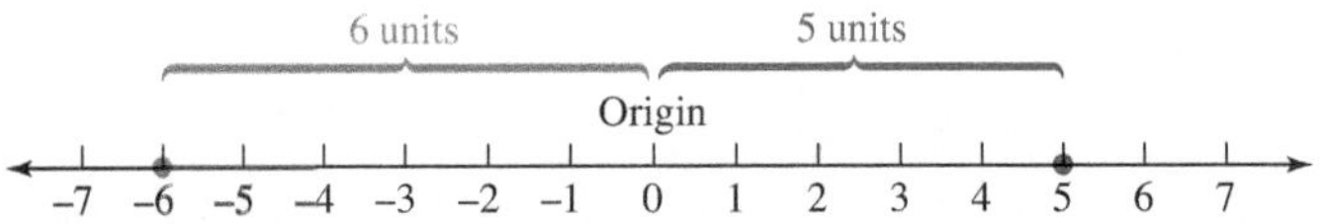

Figure 1-11

Because the absolute value of a real number represents that number's distance from 0 on the number line, the absolute value of every real number x is either positive or 0. In symbols, we say

$|x| \geq 0$ for every real number x

EXAMPLE 7 Evaluate: **a.** $|6|$ **b.** $|-3|$ **c.** $|0|$ **d.** $-|2 + 3|$

Solution
a. $|6| = 6$, because 6 is six units from 0.
b. $|-3| = 3$, because -3 is three units from 0.
c. $|0| = 0$, because 0 is zero units from 0.
d. $-|2 + 3| = -|5| = -5$, because the opposite of the absolute value of 5 is -5.

SELF CHECK 7 Evaluate: **a.** $|8|$ **b.** $|-8|$ **c.** $-|-8|$

SELF CHECK ANSWERS

1. a. 7 **b.** $-2, 0, 1.5, 7$ **2. a.** 2, 5 **b.** 0, 2, 4 **3. a.** false **b.** true **c.** true
4. 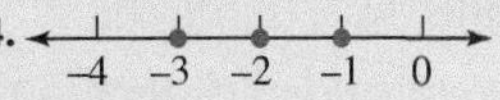 **5.** 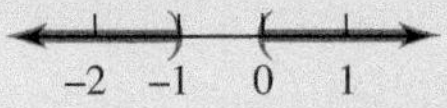 **6.** 
7. a. 8 **b.** 8 **c.** -8

NOW TRY THIS

Given the set $\{\sqrt{10}, 4.2, \sqrt{16}, 0, |-1|, 9\}$, *list*

1. the integer(s)
2. the irrational number(s)

Unless otherwise noted, all content on this page is © Cengage Learning.

3. the rational number(s)
4. the prime number(s)
5. the composite number(s)

1.1 Exercises

WARM-UPS *Describe each set of numbers.*

1. Natural numbers
2. Whole numbers
3. Integers
4. Rational numbers
5. Real numbers
6. Prime numbers
7. Composite numbers
8. Even integers
9. Odd integers
10. Irrational numbers

Find each value.

11. $-|-7|$
12. $|-12|$

VOCABULARY AND CONCEPTS *Fill in the blanks.*

13. A ___ is a collection of objects.
14. The numbers 1, 2, 3, 4, 5, . . . form the set of _______ numbers. This set is also called the set of _______________.
15. The set of _____ numbers is the set {0, 1, 2, 3, 4, 5, . . .}.
16. The dots following the sets in Exercises 14 and 15 are called ______.
17. The set of _______ is the set {. . . , −3, −2, −1, 0, 1, 2, 3, . . .}.
18. Numbers that show a loss or a downward direction are called ______________.
19. Since every whole number is also an integer, the set of whole numbers is called a _____ of the set of integers.
20. The set $\{x \mid x$ is a whole number.$\}$ is read as "_______________________________."
21. Fractions that have an integer numerator and a nonzero integer denominator are called _______ numbers.
22. $\sqrt{2}$ is an example of an ________ number.
23. The set that includes the rational and irrational numbers is called the set of ___ numbers.
24. If a natural number is greater than 1 and can be divided exactly only by 1 and itself, it is called a _____ number.
25. A composite number is a ______ number that is greater than 1 and is not _____.
26. An integer that can be evenly divided by 2 is called an ____ integer.
27. An integer that cannot be evenly divided by 2 is called an ____ integer.
28. The symbol $\neq$ means __________.
29. The symbol _ means "is less than."
30. The symbol $\geq$ means _________________.
31. In algebra, we use letters, called _______, to represent real numbers.
32. The figure 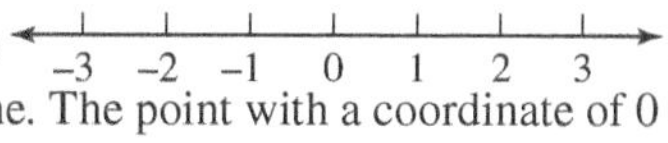

is called a ______ line. The point with a coordinate of 0 is called the _____.
33. The negative, or opposite, of −7 is _.
34. The graphs of inequalities are _______ on the number line.
35. A _________ or ____ circle shows that a point is not included in a graph.
36. A ______ or _____ circle shows that a point is included in a graph.
37. The ______ between 0 and 6 on a number line is called the absolute value of _.
38. The result of an addition is called the ____. The result of a subtraction is called a ________. The result of a multiplication is called a ______. The result of a division is called a _______.

GUIDED PRACTICE *Which numbers in the set* $\{-3, -\frac{1}{2}, -1, 0, 1, 2, \frac{5}{3}, \sqrt{7}, 3.25, 6, 9\}$ *are in each category?* SEE EXAMPLES 1–2. (OBJECTIVE 1)

39. natural numbers
40. whole numbers
41. positive integers
42. negative integers
43. integers
44. rational numbers
45. real numbers
46. irrational numbers
47. odd integers
48. even integers
49. composite numbers
50. prime numbers

Place one of the symbols =, <, or > in each box to make a true statement. SEE EXAMPLE 3. (OBJECTIVE 2)

51. 7 ☐ 10
52. 3 ☐ 2 + 1
53. 9 ☐ 2 + 5
54. −5 ☐ −4
55. −6 ☐ −8
56. 2 + 3 ☐ 17
57. 5 + 7 ☐ 10
58. 3 + 3 ☐ 9 − 3

Graph each pair of numbers on a number line. In each pair, indicate which number is the greater and which number lies farther to the right. (OBJECTIVE 3)

59. 2, 4
60. 5, 9
61. 11, 6
62. 15, 10

Unless otherwise noted, all content on this page is © Cengage Learning.

63. $-5, -2$

64. $4, 10$

65. $8, 0$

66. $-7, -1$

Graph each set of numbers on a number line. Use brackets or parentheses where applicable. SEE EXAMPLES 4–6. (OBJECTIVE 3)

67. The natural numbers between 2 and 8

68. The prime numbers between 5 and 15

69. The real numbers between 3 and 8

70. The odd integers between -5 and 5 that are exactly divisible by 3

71. The real numbers greater than or equal to 8

72. The real numbers greater than or equal to 3 or less than or equal to -3

73. The odd numbers from 10 to 20

74. The even integers greater than or equal to 10 and less than or equal to 20.

Find each absolute value. SEE EXAMPLE 7. (OBJECTIVE 4)

75. $|36|$

76. $|-17|$

77. $|0|$

78. $|120|$

79. $-|-23|$

80. $|18 - 12|$

81. $|12 - 4|$

82. $|100 - 100|$

ADDITIONAL PRACTICE *Simplify each expression. Then classify the result as a natural number, an even integer, an odd integer, a prime number, a composite number, and/or a whole number.*

83. $6 + 3$

84. $7 - 2$

85. $15 - 15$

86. $13 - 6$

87. $3 \cdot 8$

88. $6 \cdot 12$

89. $24 \div 8$

90. $7 \div 7$

Place one of the symbols =, <, or > in each box to make a true statement.

91. $5 + 6$ ☐ $13 - 1$

92. $19 - 3$ ☐ $8 + 6$

93. $4 \cdot 3$ ☐ $3 \cdot 4$

94. $7 \cdot 9$ ☐ $9 \cdot 6$

95. $0 \div 6$ ☐ 1

96. $2 + 7$ ☐ $7 + 2$

97. $45 \div 9$ ☐ $36 \div 12$

98. $5 \cdot 12$ ☐ $300 \div 5$

99. $3 + 2 + 5$ ☐ $5 + 2 + 3$

100. $8 + 5 + 2$ ☐ $5 + 2 + 8$

Write each sentence as a mathematical expression.

101. Nine is greater than four.

102. Five is less than thirty-two.

103. Eight is less than or equal to eight.

104. Twenty-five is not equal to twenty-three.

105. The sum of adding three and four is equal to seven.

106. Thirty-seven is greater than the product of multiplying three and four.

107. $\sqrt{2}$ is approximately equal to 1.41.

108. x is greater than or equal to 5.

Write each inequality as an equivalent inequality in which the inequality symbol points in the opposite direction.

109. $3 \le 7$

110. $5 > 2$

111. $6 > 0$

112. $34 \le 40$

113. $3 + 8 > 8$

114. $8 - 3 < 8$

115. $6 - 2 < 10 - 4$

116. $8 \cdot 2 \ge 8 \cdot 1$

117. $2 \cdot 3 < 3 \cdot 4$

118. $8 \div 2 \ge 9 \div 3$

119. $\frac{12}{4} < \frac{24}{6}$

120. $\frac{2}{3} \le \frac{3}{4}$

Graph each set of numbers on a number line. Use brackets or parentheses where applicable.

121. The even integers that are also prime numbers

122. The numbers that are whole numbers but not natural numbers

123. The natural numbers between 15 and 25 that are multiples of 6

124. The real numbers greater than -2 and less than 3

125. The real numbers greater than or equal to -5 and less than 4

126. The real numbers between -7 and 7, including -7 and 7

Find each absolute value.

127. $|21 - 19|$

128. $|25 - 21|$

WRITING ABOUT MATH

129. Explain why there is no greatest natural number.

130. Explain why 2 is the only even prime number.

131. Explain how to determine the absolute value of a number.

132. Explain why zero is an even integer.

SOMETHING TO THINK ABOUT *Consider the following sets: the integers, natural numbers, even and odd integers, positive and negative numbers, prime and composite numbers, and rational numbers.*

133. Find a number that fits in as many of these categories as possible.

134. Find a number that fits in as few of these categories as possible.

Section 1.2 Fractions

Objectives

1. Simplify a fraction.
2. Multiply and divide two fractions.
3. Add and subtract two or more fractions.
4. Add and subtract two or more mixed numbers.
5. Add, subtract, multiply, and divide two or more decimals.
6. Round a decimal to a specified number of places.
7. Use the appropriate operation for an application.

Vocabulary

numerator	proper fraction	mixed number
denominator	improper fraction	terminating decimal
lowest terms	reciprocal	repeating decimal
simplest form	equivalent fractions	divisor
factors of a product	least (or lowest) common denominator	dividend
prime-factored form		percent

Getting Ready

1. Add: $\begin{array}{r}132\\45\\\underline{73}\end{array}$

2. Subtract: $\begin{array}{r}321\\\underline{173}\end{array}$

3. Multiply: $\begin{array}{r}437\\\underline{38}\end{array}$

4. Divide: $37\overline{)3{,}885}$

In this section, we will review arithmetic fractions. This will help us prepare for algebraic fractions, which we will encounter later in the book.

1 Simplify a fraction.

In the fractions

$$\frac{1}{2}, \frac{3}{5}, \frac{2}{17}, \quad \text{and} \quad \frac{37}{7}$$

the number above the bar is called the **numerator**, and the number below the bar is called the **denominator**.

We often use fractions to indicate parts of a whole. In Figure 1-12(a) on the next page, a rectangle has been divided into 5 equal parts, and 3 of the parts are shaded. The fraction $\frac{3}{5}$ indicates how much of the figure is shaded. In Figure 1-12(b), $\frac{5}{7}$ of the rectangle is shaded. In either example, the denominator of the fraction shows the total number of equal parts into which the whole is divided, and the numerator shows how many of these equal parts are being considered.

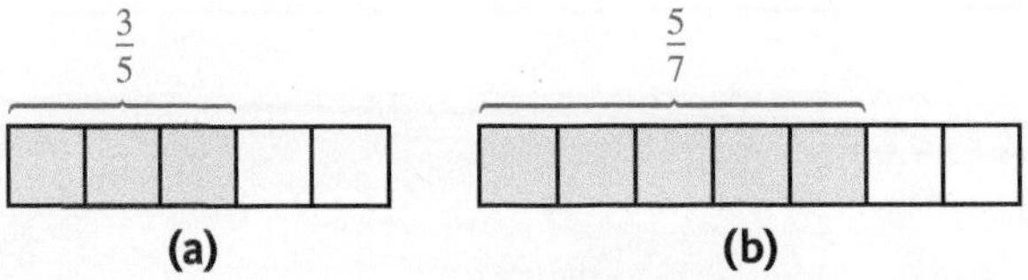

Figure 1-12

We can also use fractions to indicate division. For example, the fraction $\frac{8}{2}$ indicates that 8 is to be divided by 2:

$$\frac{8}{2} = 8 \div 2 = 4$$

COMMENT Note that $\frac{8}{2} = 4$, because $4 \cdot 2 = 8$, and that $\frac{0}{7} = 0$, because $0 \cdot 7 = 0$. However, $\frac{6}{0}$ is undefined, because no number multiplied by 0 gives 6. Remember that the denominator of a fraction cannot be 0.

A fraction is said to be in **lowest terms** (or **simplest form**) when no integer other than 1 will divide both its numerator and its denominator exactly. The fraction $\frac{6}{11}$ is in lowest terms because only 1 divides both 6 and 11 exactly. The fraction $\frac{6}{8}$ is not in lowest terms, because 2 divides both 6 and 8 exactly.

We can simplify a fraction that is not in lowest terms by dividing its numerator and its denominator by the same number. For example, to simplify $\frac{6}{8}$, we divide the numerator and the denominator by 2.

$$\frac{6}{8} = \frac{6 \div 2}{8 \div 2} = \frac{3}{4}$$

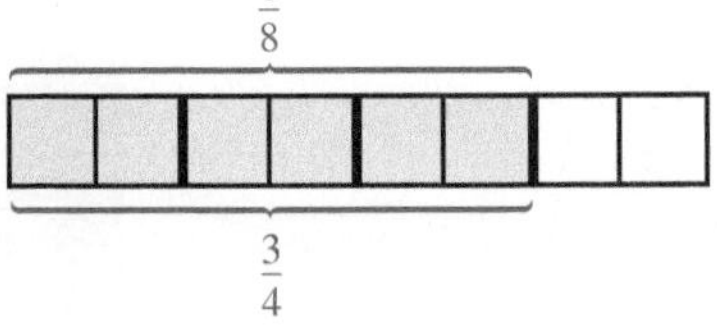

Figure 1-13

From Figure 1-13, we see that $\frac{6}{8}$ and $\frac{3}{4}$ are equal fractions, because each one represents the same part of the rectangle.

When a composite number has been written as the product of other natural numbers, we say that it has been factored. For example, 15 can be written as the product of 5 and 3.

$$15 = 5 \cdot 3$$

The numbers 5 and 3 are called **factors** of 15. When a composite number is written as the product of prime numbers, we say that it is written in **prime-factored form**.

EXAMPLE 1 Write 210 in prime-factored form.

Solution We can write 210 as the product of 21 and 10 and proceed as follows:

$$210 = \mathbf{21 \cdot 10}$$

$$210 = \mathbf{3 \cdot 7 \cdot 2 \cdot 5} \quad \text{Factor 21 as } 3 \cdot 7 \text{ and factor 10 as } 2 \cdot 5.$$

Since 210 is now written as the product of prime numbers, its prime-factored form is $210 = 2 \cdot 3 \cdot 5 \cdot 7$.

SELF CHECK 1 Write 70 in prime-factored form.

To simplify a fraction, we factor its numerator and denominator and divide out all factors that are common to the numerator and denominator. For example,

$$\frac{6}{8} = \frac{3 \cdot 2}{4 \cdot 2} = \frac{3 \cdot \overset{1}{\cancel{2}}}{4 \cdot \underset{1}{\cancel{2}}} = \frac{3}{4} \quad \text{and} \quad \frac{15}{18} = \frac{5 \cdot 3}{6 \cdot 3} = \frac{5 \cdot \overset{1}{\cancel{3}}}{6 \cdot \underset{1}{\cancel{3}}} = \frac{5}{6}$$

COMMENT Remember that a fraction is in lowest terms only when its numerator and denominator have no common factors.

Unless otherwise noted, all content on this page is © Cengage Learning.

EXAMPLE 2 Simplify, if possible: **a.** $\frac{6}{30}$ **b.** $\frac{33}{40}$

Solution **a.** To simplify $\frac{6}{30}$, we factor the numerator and denominator and divide out the common factor of 6.

$$\frac{6}{30} = \frac{6 \cdot 1}{6 \cdot 5} = \frac{\overset{1}{\cancel{6}} \cdot 1}{\underset{1}{\cancel{6}} \cdot 5} = \frac{1}{5}$$

b. To simplify $\frac{33}{40}$, we factor the numerator and denominator and divide out any common factors.

$$\frac{33}{40} = \frac{3 \cdot 11}{2 \cdot 2 \cdot 2 \cdot 5}$$

Since the numerator and denominator have no common factors, $\frac{33}{40}$ is in lowest terms.

SELF CHECK 2 Simplify, if possible: $\frac{14}{35}$

The preceding examples illustrate the *fundamental property of fractions.*

THE FUNDAMENTAL PROPERTY OF FRACTIONS

If a, b, and x are real numbers,

$$\frac{a \cdot x}{b \cdot x} = \frac{a}{b} \quad (b \neq 0 \text{ and } x \neq 0)$$

2 Multiply and divide two fractions.

To multiply fractions, we use the following rule.

MULTIPLYING FRACTIONS

To multiply fractions, we multiply their numerators and multiply their denominators. In symbols, if a, b, c, and d are real numbers,

$$\frac{a}{b} \cdot \frac{c}{d} = \frac{a \cdot c}{b \cdot d} \quad (b \neq 0 \text{ and } d \neq 0)$$

For example,

$$\frac{4}{7} \cdot \frac{2}{3} = \frac{4 \cdot 2}{7 \cdot 3} = \frac{8}{21} \qquad \frac{4}{5} \cdot \frac{13}{9} = \frac{4 \cdot 13}{5 \cdot 9} = \frac{52}{45}$$

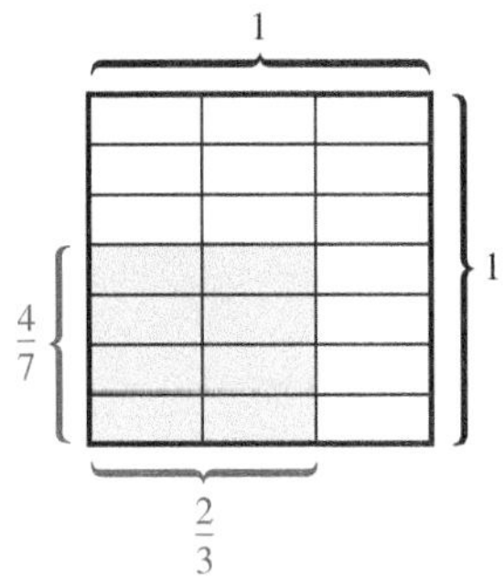

Figure 1-14

To justify the rule for multiplying fractions, we consider the square in Figure 1-14. Because the length of each side of the square is 1 unit and the area is the product of the lengths of two sides, the area is 1 square unit.

If this square is divided into 3 equal parts vertically and 7 equal parts horizontally, it is divided into 21 equal parts, and each represents $\frac{1}{21}$ of the total area. The area

Unless otherwise noted, all content on this page is © Cengage Learning.

of the shaded rectangle in the square is $\frac{8}{21}$, because it contains 8 of the 21 parts. The width, w, of the shaded rectangle is $\frac{4}{7}$; its length, l, is $\frac{2}{3}$; and its area, A, is the product of l and w:

$$A = l \cdot w$$
$$\frac{8}{21} = \frac{2}{3} \cdot \frac{4}{7}$$

This suggests that we can find the product of

$$\frac{4}{7} \text{ and } \frac{2}{3}$$

by multiplying their numerators and multiplying their denominators.

Fractions whose numerators are less than their denominators, such as $\frac{8}{21}$, are called **proper fractions**. Fractions whose numerators are greater than or equal to their denominators, such as $\frac{52}{45}$, are called **improper fractions**.

EXAMPLE 3 Perform each multiplication.

a. $\frac{3}{7} \cdot \frac{13}{5} = \frac{3 \cdot 13}{7 \cdot 5}$ Multiply the numerators and multiply the denominators. There are no common factors.

$= \frac{39}{35}$ Multiply in the numerator and multiply in the denominator.

b. $5 \cdot \frac{3}{15} = \frac{5}{1} \cdot \frac{3}{15}$ Write 5 as the improper fraction $\frac{5}{1}$.

$= \frac{5 \cdot 3}{1 \cdot 15}$ Multiply the numerators and multiply the denominators.

$= \frac{5 \cdot 3}{1 \cdot 5 \cdot 3}$ To simplify the fraction, factor the denominator.

$= \frac{\overset{1}{\cancel{5}} \cdot \overset{1}{\cancel{3}}}{1 \cdot \underset{1}{\cancel{5}} \cdot \underset{1}{\cancel{3}}}$ Divide out the common factors of 3 and 5.

$= 1$ $\frac{1 \cdot 1}{1 \cdot 1 \cdot 1} = 1$

SELF CHECK 3 Multiply: $\frac{5}{9} \cdot \frac{7}{10}$

EXAMPLE 4 **TRAVEL** Out of 36 students in a history class, three-fourths have signed up for a trip to Europe. If there are 28 places available on the flight, will there be room for one more student?

Solution We first find three-fourths of 36.

$\frac{3}{4} \cdot 36 = \frac{3}{4} \cdot \frac{36}{1}$ Write 36 as $\frac{36}{1}$.

$= \frac{3 \cdot 36}{4 \cdot 1}$ Multiply the numerators and multiply the denominators.

$= \frac{3 \cdot 4 \cdot 9}{4 \cdot 1}$ To simplify, factor the numerator.

$= \frac{3 \cdot \overset{1}{\cancel{4}} \cdot 9}{\underset{1}{\cancel{4}} \cdot 1}$ Divide out the common factor of 4.

$$= \frac{27}{1}$$

$$= 27$$

Twenty-seven students plan to go on the trip. Since there is room for 28 passengers, there is room for one more.

SELF CHECK 4 If seven-ninths of the 36 students had signed up, would there be room for one more?

One number is called the **reciprocal** of another if their product is 1. For example, $\frac{3}{5}$ is the reciprocal of $\frac{5}{3}$, because

$$\frac{3}{5} \cdot \frac{5}{3} = \frac{15}{15} = 1$$

DIVIDING FRACTIONS

To divide two fractions, we multiply the first fraction by the reciprocal of the second fraction. In symbols, if a, b, c, and d are real numbers,

$$\frac{a}{b} \div \frac{c}{d} = \frac{a}{b} \cdot \frac{d}{c} = \frac{a \cdot d}{b \cdot c} \quad (b \neq 0, c \neq 0, \text{and } d \neq 0)$$

EXAMPLE 5 Perform each division.

a. $\frac{3}{5} \div \frac{6}{5} = \frac{3}{5} \cdot \frac{5}{6}$ Multiply $\frac{3}{5}$ by the reciprocal of $\frac{6}{5}$.

$= \frac{3 \cdot 5}{5 \cdot 6}$ Multiply the numerators and multiply the denominators.

$= \frac{3 \cdot 5}{5 \cdot 2 \cdot 3}$ Factor the denominator.

$= \frac{\overset{1}{\cancel{3}} \cdot \overset{1}{\cancel{5}}}{\underset{1}{\cancel{5}} \cdot 2 \cdot \underset{1}{\cancel{3}}}$ Divide out the common factors of 3 and 5.

$= \frac{1}{2}$

b. $\frac{15}{7} \div 10 = \frac{15}{7} \div \frac{10}{1}$ Write 10 as the improper fraction $\frac{10}{1}$.

$= \frac{15}{7} \cdot \frac{1}{10}$ Multiply $\frac{15}{7}$ by the reciprocal of $\frac{10}{1}$.

$= \frac{15 \cdot 1}{7 \cdot 10}$ Multiply the numerators and multiply the denominators.

$= \frac{3 \cdot \overset{1}{\cancel{5}}}{7 \cdot 2 \cdot \underset{1}{\cancel{5}}}$ Factor the numerator and the denominator, and divide out the common factor of 5.

$= \frac{3}{14}$

SELF CHECK 5 Perform the division: $\frac{13}{6} \div \frac{26}{8}$

3 Add and subtract two or more fractions.

To add fractions with like denominators, we use the following rule.

ADDING FRACTIONS WITH THE SAME DENOMINATOR

To add fractions with the same denominator, we add the numerators and keep the common denominator. In symbols, if a, b, and d are real numbers,

$$\frac{a}{d} + \frac{b}{d} = \frac{a+b}{d} \quad (d \neq 0)$$

For example,

$$\frac{3}{7} + \frac{2}{7} = \frac{3+2}{7} \quad \text{Add the numerators and keep the common denominator.}$$
$$= \frac{5}{7}$$

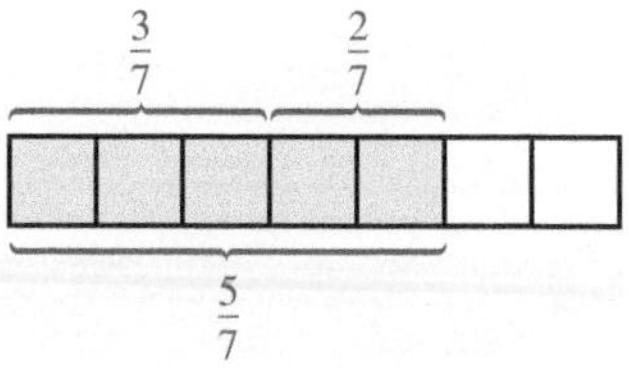

Figure 1-15

Figure 1-15 illustrates why $\frac{3}{7} + \frac{2}{7} = \frac{5}{7}$.

To add fractions with unlike denominators, we write the fractions so that they have the same denominator. For example, we can multiply both the numerator and denominator of $\frac{1}{3}$ by 5 to obtain an **equivalent fraction** with a denominator of 15:

$$\frac{1}{3} = \frac{1 \cdot \mathbf{5}}{3 \cdot \mathbf{5}} = \frac{5}{15}$$

To write $\frac{1}{5}$ as an equivalent fraction with a denominator of 15, we multiply the numerator and the denominator by 3:

$$\frac{1}{5} = \frac{1 \cdot \mathbf{3}}{5 \cdot \mathbf{3}} = \frac{3}{15}$$

Since 15 is the smallest number that can be used as a common denominator for $\frac{1}{3}$ and $\frac{1}{5}$, it is called the **least** (or **lowest**) **common denominator** (the **LCD**).

To add the fractions $\frac{1}{3}$ and $\frac{1}{5}$, we write each fraction as an equivalent fraction having a denominator of 15, and then we add the results:

$$\frac{1}{3} + \frac{1}{5} = \frac{1 \cdot \mathbf{5}}{3 \cdot \mathbf{5}} + \frac{1 \cdot \mathbf{3}}{5 \cdot \mathbf{3}}$$
$$= \frac{5}{15} + \frac{3}{15}$$
$$= \frac{5+3}{15}$$
$$= \frac{8}{15}$$

In the next example, we will add the fractions $\frac{3}{10}$ and $\frac{5}{28}$.

EXAMPLE 6 Add: $\frac{3}{10} + \frac{5}{28}$

Solution To find the LCD, we find the prime factorization of each denominator and use each prime factor the greatest number of times it appears in either factorization.

$$\left.\begin{aligned} 10 &= 2 \cdot 5 \\ 28 &= 2 \cdot 2 \cdot 7 \end{aligned}\right\} \quad \text{LCD} = 2 \cdot 2 \cdot 5 \cdot 7 = 140$$

Unless otherwise noted, all content on this page is © Cengage Learning.

Since 140 is the smallest number that 10 and 28 divide exactly, we write both fractions as fractions with denominators of 140.

$$\frac{3}{10} + \frac{5}{28} = \frac{3 \cdot 14}{10 \cdot 14} + \frac{5 \cdot 5}{28 \cdot 5}$$ Write each fraction as a fraction with a denominator of 140.

$$= \frac{42}{140} + \frac{25}{140}$$ Do the multiplications.

$$= \frac{42 + 25}{140}$$ Add the numerators and keep the denominator.

$$= \frac{67}{140}$$

Since 67 is a prime number, it has no common factor with 140. Thus, $\frac{67}{140}$ is in lowest terms.

SELF CHECK 6 Add: $\frac{3}{8} + \frac{5}{12}$

To subtract fractions with like denominators, we use the following rule.

SUBTRACTING FRACTIONS WITH THE SAME DENOMINATOR

To subtract fractions with the same denominator, we subtract their numerators and keep their common denominator. In symbols, if a, b, and d are real numbers,

$$\frac{a}{d} - \frac{b}{d} = \frac{a - b}{d} \quad (d \neq 0)$$

For example,

$$\frac{7}{9} - \frac{2}{9} = \frac{7 - 2}{9} = \frac{5}{9}$$

To subtract fractions with unlike denominators, we write them as equivalent fractions with a common denominator. For example, to subtract $\frac{2}{5}$ from $\frac{3}{4}$, we write $\frac{3}{4} - \frac{2}{5}$, find the LCD of 4 and 5, which is 20, and proceed as follows:

$$\frac{3}{4} - \frac{2}{5} = \frac{3 \cdot 5}{4 \cdot 5} - \frac{2 \cdot 4}{5 \cdot 4}$$ Write each fraction as a fraction with a denominator of 20.

$$= \frac{15}{20} - \frac{8}{20}$$ Do the multiplications.

$$= \frac{15 - 8}{20}$$ Add the numerators and keep the denominator.

$$= \frac{7}{20}$$

EXAMPLE 7 Subtract 5 from $\frac{23}{3}$.

Solution

$$\frac{23}{3} - 5 = \frac{23}{3} - \frac{5}{1}$$ Write 5 as the improper fraction $\frac{5}{1}$.

$$= \frac{23}{3} - \frac{5 \cdot 3}{1 \cdot 3}$$ Write $\frac{5}{1}$ as a fraction with a denominator of 3.

$$= \frac{23}{3} - \frac{15}{3}$$ Do the multiplications.

$$= \frac{23 - 15}{3}$$ Subtract the numerators and keep the denominator.

$$= \frac{8}{3}$$

SELF CHECK 7 Subtract: $\frac{5}{6} - \frac{3}{4}$

4 Add and subtract two or more mixed numbers.

The **mixed number** $3\frac{1}{2}$ represents the sum of 3 and $\frac{1}{2}$. We can write $3\frac{1}{2}$ as an improper fraction as follows:

$$3\frac{1}{2} = 3 + \frac{1}{2}$$

$$= \frac{6}{2} + \frac{1}{2} \qquad 3 = \frac{6}{2}$$

$$= \frac{6 + 1}{2} \qquad \text{Add the numerators and keep the denominator.}$$

$$= \frac{7}{2}$$

To write the fraction $\frac{19}{5}$ as a mixed number, we divide 19 by 5 to get 3, with a remainder of 4.

$$\frac{19}{5} = 3 + \frac{4}{5} = 3\frac{4}{5}$$

EXAMPLE 8 Add: $2\frac{1}{4} + 1\frac{1}{3}$

Solution We first change each mixed number to an improper fraction.

$$2\frac{1}{4} = 2 + \frac{1}{4} \qquad 1\frac{1}{3} = 1 + \frac{1}{3}$$

$$= \frac{8}{4} + \frac{1}{4} \qquad = \frac{3}{3} + \frac{1}{3}$$

$$= \frac{9}{4} \qquad = \frac{4}{3}$$

Then we add the fractions.

$$2\frac{1}{4} + 1\frac{1}{3} = \frac{9}{4} + \frac{4}{3}$$

$$= \frac{9 \cdot 3}{4 \cdot 3} + \frac{4 \cdot 4}{3 \cdot 4} \qquad \text{Write each fraction with the LCD of 12.}$$

$$= \frac{27}{12} + \frac{16}{12}$$

$$= \frac{43}{12}$$

Finally, we change $\frac{43}{12}$ to a mixed number.

$$\frac{43}{12} = 3 + \frac{7}{12} = 3\frac{7}{12}$$

SELF CHECK 8 Add: $5\frac{1}{7} + 4\frac{2}{3}$

EXAMPLE 9 **FENCING LAND** How much fencing will be needed to enclose the area within the triangular lot shown in Figure 1-16?

Solution We can find the sum of the lengths by adding the whole-number parts and the fractional parts of the dimensions separately.

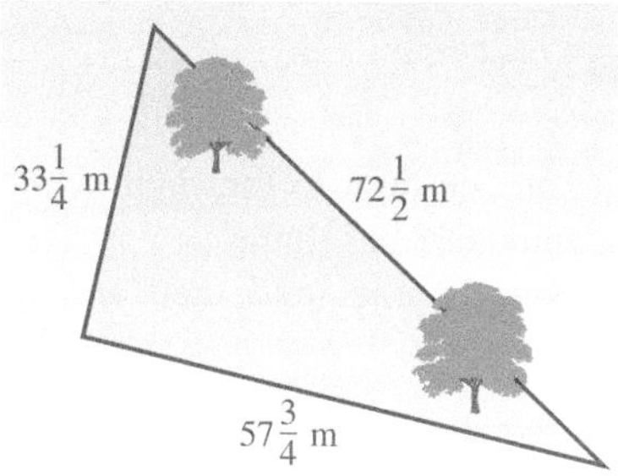

Figure 1-16

$$33\frac{1}{4} + 57\frac{3}{4} + 72\frac{1}{2} = 33 + 57 + 72 + \frac{1}{4} + \frac{3}{4} + \frac{1}{2}$$

$$= 162 + \frac{1}{4} + \frac{3}{4} + \frac{2}{4}$$ Write $\frac{1}{2}$ as $\frac{2}{4}$ to obtain a common denominator.

$$= 162 + \frac{6}{4}$$ Add the fractions by adding the numerators and keeping the common denominator.

$$= 162 + \frac{3}{2}$$ $\frac{6}{4} = \frac{2 \cdot 3}{2 \cdot 2} = \frac{\overset{1}{\cancel{2}} \cdot 3}{\underset{1}{\cancel{2}} \cdot 2} = \frac{3}{2}$

$$= 162 + 1\frac{1}{2}$$ Write $\frac{3}{2}$ as a mixed number.

$$= 163\frac{1}{2}$$

To enclose the area, $163\frac{1}{2}$ meters of fencing will be needed.

COMMENT Remember to include the proper units in your answer. The Mars Climate Orbiter crashed due to lack of unit communication between the Jet Propulsion Lab and Lockheed/Martin engineers, resulting in a loss of $125 million.

SELF CHECK 9 Find the length of fencing needed to enclose a rectangular plot that is $85\frac{1}{2}$ feet wide and $140\frac{2}{3}$ feet deep.

5 Add, subtract, multiply, and divide two or more decimals.

Rational numbers can always be changed to decimal form. For example, to write $\frac{1}{4}$ and $\frac{5}{22}$ as decimals, we use long division.

```
   0.25            0.22727 ...
4)1.00          22)5.00000
   8                44
   20                60
   20                44
                     160
                     154
                       60
                       44
                      160
```

The decimal 0.25 is called a **terminating decimal**. The decimal 0.2272727. . . (often written as $0.2\overline{27}$) is called a **repeating decimal**, because it repeats the block of digits 27. Every rational number can be changed into either a terminating or a repeating decimal.

Unless otherwise noted, all content on this page is © Cengage Learning.

Terminating decimals	*Repeating decimals*
$\frac{1}{2} = 0.5$	$\frac{1}{3} = 0.33333\ldots$ or $0.\overline{3}$
$\frac{3}{4} = 0.75$	$\frac{1}{6} = 0.16666\ldots$ or $0.1\overline{6}$
$\frac{5}{8} = 0.625$	$\frac{5}{22} = 0.2272727\ldots$ or $0.2\overline{27}$

The decimal 0.5 has one *decimal place*, because it has one digit to the right of the decimal point. The decimal 0.75 has two decimal places, and 0.625 has three.

To *add* or *subtract* decimals, we align their decimal points and then add or subtract.

EXAMPLE 10 Add 25.568 and 2.74 using a vertical format.

Solution We align the decimal points and add the numbers, column by column,

$$\begin{array}{r} 25.568 \\ +\ \ 2.74 \\ \hline 28.308 \end{array}$$

SELF CHECK 10 Subtract 2.74 from 25.568 using a vertical format.

To perform the previous operations with a calculator, we enter these numbers and press these keys:

25.568 + 2.74 = and 25.568 − 2.74 = Using a scientific calculator

25.568 + 2.74 ENTER and 25.568 − 2.74 ENTER Using a graphing calculator

To *multiply* decimals, we multiply the numbers and place the decimal point so that the number of decimal places in the answer is equal to the sum of the decimal places in the factors.

EXAMPLE 11 Multiply: 9.25 by 3.453

Solution We multiply the numbers and place the decimal point so that the number of decimal places in the answer is equal to the sum of the decimal places in the factors.

$$\begin{array}{r} 3.453 \\ \times\ \ 9.25 \\ \hline 17265 \\ 6906 \\ 31\,077 \\ \hline 31.94025 \end{array}$$

3.453: Here there are three decimal places.
9.25: Here there are two decimal places.
31.94025: The product has $3 + 2 = 5$ decimal places.

SELF CHECK 11 Multiply: 2.45 by 9.25

To perform the multiplication of Example 11 with a calculator, we enter these numbers and press these keys:

3.453 × 9.25 = Using a scientific calculator

3.453 × 9.25 ENTER Using a graphing calculator

To *divide* decimals, we move the decimal point in the **divisor** to the right to make the divisor a whole number. We then move the decimal point in the **dividend** the same number of places to the right.

EXAMPLE 12 Divide: 30.258 by 1.23

Solution We will write the division using a long division format in which the divisor is 1.23 and the dividend is 30.258.

$1.23\overline{)30.258}$ Move the decimal point in both the divisor and the dividend two places to the right.

We align the decimal point in the quotient with the repositioned decimal point in the dividend and use long division.

$$\begin{array}{r} 24.6 \\ 123\overline{)3025.8} \\ \underline{246} \\ 565 \\ \underline{492} \\ 73\ 8 \\ \underline{73\ 8} \end{array}$$

SELF CHECK 12 Divide 579.36 by 1.2.

To perform the previous division with a calculator, we enter these numbers and press these keys:

30.258 ÷ 1.23 = Using a scientific calculator

30.258 ÷ 1.23 **ENTER** Using a graphing calculator

6 Round a decimal to a specified number of places.

We often round long decimals to a specific number of decimal places. For example, the decimal 25.36124 rounded to one place (or to the nearest tenth) is 25.4. Rounded to two places (or to the nearest one-hundredth), the decimal is 25.36.

Throughout this text, we use the following procedures to round decimals.

ROUNDING DECIMALS

1. Determine to how many decimal places you want to round.
2. Look at the first digit to the right of that decimal place.
3. If that digit is 4 or less, drop it and all digits that follow. If it is 5 or greater, add 1 to the digit in the position to which you want to round, and drop all digits that follow.

EXAMPLE 13 Round 2.4863 to two decimal places.

Solution Since we are to round to two digits, we look at the digit to the right of the 8, which is 6. Since 6 is greater than 5, we add 1 to the 8 and drop all of the digits that follow. The rounded number is 2.49.

SELF CHECK 13 Round 6.5731 to three decimal places.

Everyday connections
2010 Gubernatorial Elections

In the 2010 gubernatorial elections, six of the most closely contested races were in Connecticut, Florida, Illinois, Minnesota, Ohio, and Oregon.*

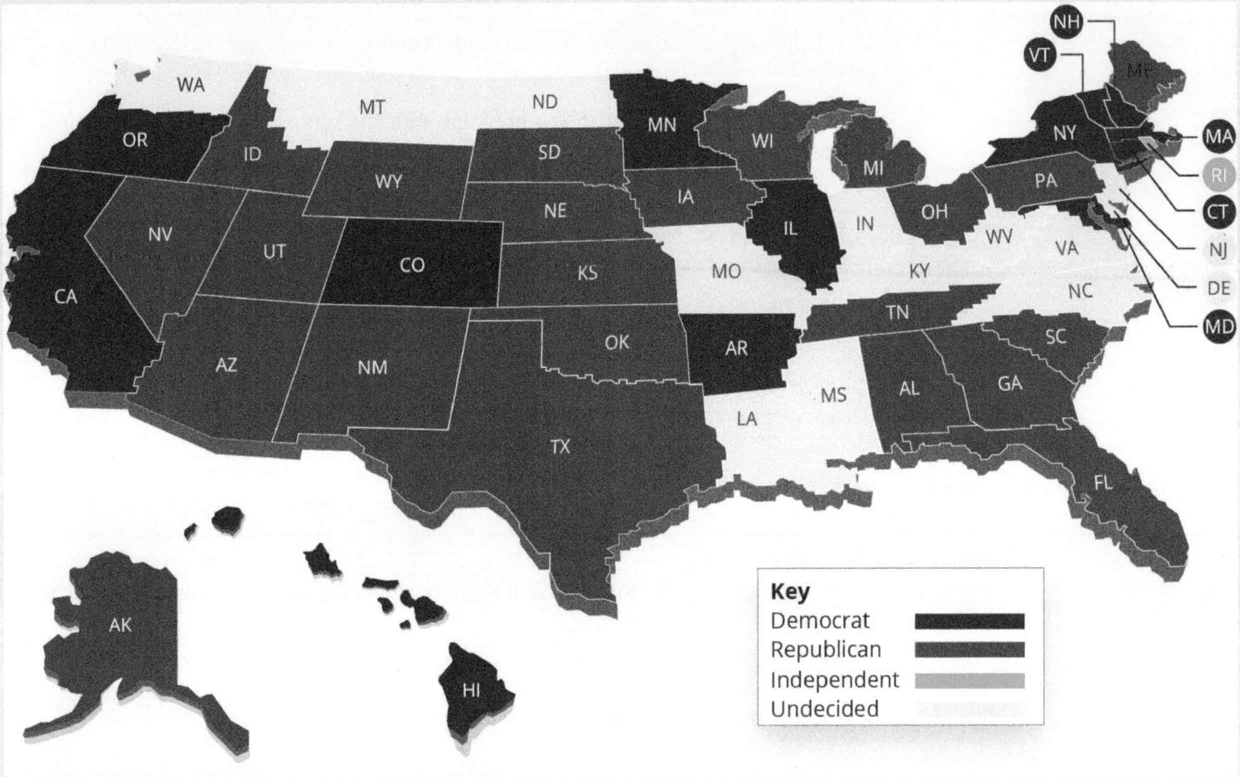

From The Washington Post (11/10/2010). Copyright © 2010 the Washington Post. Used by permission and protected by the Copyright Laws of the United States. www.washingtonpost.com

State	Democrats	Republicans
Connecticut	564,885	557,123
Florida	2,522,857	2,589,915
Illinois	1,721,812	1,702,399
Minnesota	919,231	910,480
Ohio	1,752,507	1,849,609
Oregon	680,840	665,930

1. Find the number of votes cast for Democrats and the number of votes cast for Republicans in the six states.

2. Find the average number of votes cast for Democrats and the average number of votes cast for Republicans, rounded to the nearest vote.

3. Find the difference in the number of Democratic and Republican votes cast in Connecticut, Illinois, and Minnesota.

4. Use the differences found in Question 3 to determine the percent of the total number of votes cast in each of the three states. This percentage is referred to as the margin of victory.

5. Which of the three states in Question 3 had the most closely contested race?

*All of these states had third-, and some fourth-, party candidates whose votes are not considered here.

7 Use the appropriate operation for an application.

A **percent** is the numerator of a fraction with a denominator of 100. For example, $6\frac{1}{4}$ percent, written $6\frac{1}{4}\%$, is the fraction $\frac{6.25}{100}$, or the decimal 0.0625. In problems involving percent, the word *of* usually indicates multiplication. For example, $6\frac{1}{4}\%$ of 8,500 is the product 0.0625(8,500).

EXAMPLE 14 **AUTO LOANS** Juan signs a one-year note to borrow \$8,500 to buy a car. If the rate of interest is $6\frac{1}{4}\%$, how much interest will he pay?

Solution For the privilege of using the bank's money for one year, Juan must pay $6\frac{1}{4}\%$ of \$8,500. We calculate the interest, i, as follows:

$$\begin{aligned} i &= 6\tfrac{1}{4}\% \text{ of } 8{,}500 \\ &= 0.0625 \cdot 8{,}500 && \text{In this case, the word } \textit{of} \text{ means } \textit{times.} \\ &= 531.25 \end{aligned}$$

Juan will pay \$531.25 interest.

SELF CHECK 14 If the rate is 9%, how much interest will he pay?

SELF CHECK ANSWERS

1. $2 \cdot 5 \cdot 7$ **2.** $\frac{2}{5}$ **3.** $\frac{7}{18}$ **4.** no **5.** $\frac{2}{3}$ **6.** $\frac{19}{24}$ **7.** $\frac{1}{12}$ **8.** $9\frac{17}{21}$ **9.** $452\frac{1}{3}$ ft **10.** 22.828 **11.** 22.6625 **12.** 482.8 **13.** 6.573 **14.** \$765

NOW TRY THIS

Perform each operation.

1. $\frac{7}{3} + \frac{7}{9} - \frac{5}{6} - \frac{41}{18}$

2. $25 - 13.583$

3. Robert's answer to a problem asking to find the length of a piece of lumber is $\frac{5}{2}$ feet. Is this the best form for the answer given the context of the problem? If not, write the answer in the most appropriate form.

4. $\frac{5}{x-3} - \frac{1}{x-3}$ $(x \neq 3)$

1.2 Exercises

WARM-UPS *Find the largest common factor of each pair of numbers.*

1. 3, 6

2. 5, 10

3. 12, 18

4. 15, 27

Perform each operation.

5. $\frac{3}{4} \cdot \frac{1}{2}$

6. $\frac{5}{6} \cdot \frac{5}{7}$

7. $\frac{3}{4} \div \frac{4}{3}$

8. $\frac{3}{5} \div \frac{5}{2}$

9. $\frac{4}{9} + \frac{7}{9}$
10. $\frac{10}{11} - \frac{2}{11}$
11. $\frac{2}{3} - \frac{1}{2}$
12. $\frac{3}{4} + \frac{1}{2}$
13. $5.1 + 0.62$
14. $3.45 - 2.21$
15. $0.2 \cdot 2.5$
16. $0.4 \cdot 16$

Round each decimal to two decimal places.

17. 5.165329
18. 5.164493

REVIEW *Determine whether the following statements are true or false.*

19. 6 is an integer.
20. $\frac{1}{2}$ is a natural number.
21. 21 is a prime number.
22. No prime number is an even number.
23. $-5 > -2$
24. $-3 < -2$
25. $9 \leq |-9|$
26. $|-11| \geq 10$

Place an appropriate symbol in each box to make the statement true.

27. $3 + 7 \;\square\; 10$
28. $\frac{3}{7} \;\square\; \frac{2}{7} = \frac{1}{7}$
29. $|-2| \;\square\; 2$
30. $4 + 8 \;\square\; 11$

VOCABULARY AND CONCEPTS *Fill in the blanks.*

31. The number above the bar in a fraction is called the ________.
32. The number below the bar in a fraction is called the ________.
33. The fraction $\frac{17}{0}$ is said to be ________.
34. To ______ a fraction, we divide its numerator and denominator by the same number.
35. To write a number in prime-factored form, we write it as the product of _____ numbers.
36. If the numerator of a fraction is less than the denominator, the fraction is called a _____ fraction.
37. If the numerator of a fraction is greater than the denominator, the fraction is called an _______ fraction.
38. A fraction is written in _________ or simplest form when its numerator and denominator have no common factors.
39. If the product of two numbers is _, the numbers are called reciprocals.
40. $\frac{ax}{bx} = \_$
41. To multiply two fractions, _______ the numerators and multiply the denominators.
42. To divide two fractions, multiply the first fraction by the ________ of the second fraction.
43. To add fractions with the same denominator, add the ________ and keep the common _________.
44. To subtract fractions with the same __________, subtract the numerators and keep the common denominator.
45. To add fractions with unlike denominators, first find the ____________________ and write each fraction as an ________ fraction.
46. $75\frac{2}{3}$ means 75 ____ $\frac{2}{3}$. The number $75\frac{2}{3}$ is called a _____ number.
47. 0.75 is an example of a _________ decimal and it has _ decimal places.
48. $5.3\overline{27}$ is an example of a _______ decimal.
49. In the figure $2\overline{)6}$ with 3 above, 2 represents the ______, 6 represents the _______, and 3 represents the _______.
50. A ______ is the numerator of a fraction whose denominator is 100.

GUIDED PRACTICE *Write each number in prime-factored form.* SEE EXAMPLE 1. (OBJECTIVE 1)

51. 24
52. 105
53. 48
54. 315

Write each fraction in lowest terms. If the fraction is already in lowest terms, so indicate. SEE EXAMPLE 2. (OBJECTIVE 1)

55. $\frac{6}{12}$
56. $\frac{3}{9}$
57. $\frac{15}{20}$
58. $\frac{33}{55}$
59. $\frac{27}{18}$
60. $\frac{35}{14}$
61. $\frac{72}{64}$
62. $\frac{26}{21}$

Perform each multiplication. Simplify each result when possible. SEE EXAMPLE 3. (OBJECTIVE 2)

63. $\frac{1}{3} \cdot \frac{2}{5}$
64. $\frac{3}{4} \cdot \frac{5}{7}$
65. $\frac{4}{3} \cdot \frac{6}{5}$
66. $\frac{7}{8} \cdot \frac{6}{15}$
67. $12 \cdot \frac{5}{6}$
68. $10 \cdot \frac{5}{12}$
69. $\frac{10}{21} \cdot 14$
70. $\frac{5}{24} \cdot 16$

Perform each division. Simplify each result when possible. SEE EXAMPLE 5. (OBJECTIVE 2)

71. $\frac{2}{5} \div \frac{3}{2}$
72. $\frac{4}{5} \div \frac{3}{7}$
73. $\frac{3}{4} \div \frac{6}{5}$
74. $\frac{3}{8} \div \frac{15}{28}$
75. $9 \div \frac{3}{8}$
76. $23 \div \frac{46}{5}$
77. $\frac{54}{20} \div 3$
78. $\frac{39}{27} \div 13$

Perform each operation. Simplify each result when possible. SEE EXAMPLES 6–7. (OBJECTIVE 3)

79. $\frac{3}{5}+\frac{3}{5}$

80. $\frac{4}{7}-\frac{2}{7}$

81. $\frac{5}{17}-\frac{3}{17}$

82. $\frac{2}{11}+\frac{9}{11}$

83. $\frac{1}{42}+\frac{1}{6}$

84. $\frac{17}{25}-\frac{2}{5}$

85. $\frac{7}{10}-\frac{1}{14}$

86. $\frac{8}{25}+\frac{1}{10}$

Perform each operation. Simplify each result when possible. SEE EXAMPLE 8. (OBJECTIVE 4)

87. $4\frac{3}{5}+\frac{3}{5}$

88. $2\frac{1}{8}+\frac{3}{8}$

89. $3\frac{1}{3}-1\frac{2}{3}$

90. $6\frac{1}{5}-4\frac{2}{5}$

91. $3\frac{3}{4}-2\frac{1}{2}$

92. $15\frac{5}{6}+11\frac{5}{8}$

93. $8\frac{2}{9}-7\frac{2}{3}$

94. $3\frac{4}{5}-3\frac{1}{10}$

Change each fraction to decimal form and determine whether the decimal is a terminating or repeating decimal. (OBJECTIVE 5)

95. $\frac{3}{5}$

96. $\frac{5}{9}$

97. $\frac{9}{22}$

98. $\frac{8}{5}$

Perform each operation. SEE EXAMPLES 10–12. (OBJECTIVE 5)

99. $43.54+315.7$

100. $345.213-27.35$

101. $67.235-22.45$

102. $21.36+4.573$

103. $7.2\cdot 15.6$

104. $4.21\cdot 2.73$

105. $0.23\overline{)1.0465}$

106. $4.7\overline{)10.857}$

Round each of the following to two decimal places and then to three decimal places. SEE EXAMPLE 13. (OBJECTIVE 6)

107. 496.2583

108. 13.0547

109. 6,025.3982

110. 1.6048

ADDITIONAL PRACTICE *Perform each operation.*

111. $\frac{5}{12}\cdot\frac{18}{5}$

112. $\frac{5}{4}\cdot\frac{12}{10}$

113. $\frac{17}{34}\cdot\frac{3}{6}$

114. $\frac{21}{14}\cdot\frac{3}{6}$

115. $\frac{2}{13}\div\frac{8}{13}$

116. $\frac{4}{7}\div\frac{20}{21}$

117. $\frac{21}{35}\div\frac{3}{14}$

118. $\frac{23}{25}\div\frac{46}{5}$

119. $\frac{3}{5}+\frac{2}{3}$

120. $\frac{4}{3}+\frac{7}{2}$

121. $\frac{9}{4}-\frac{5}{6}$

122. $\frac{2}{15}+\frac{7}{9}$

123. $3-\frac{3}{4}$

124. $5+\frac{21}{5}$

125. $\frac{17}{3}+4$

126. $\frac{13}{9}-1$

Use a calculator to perform each operation and round each answer to two decimal places.

127. $474.81+23.4532$

128. $843.45213-712.765$

129. $25.25\cdot 132.179$

130. $234.874\cdot 242.46473$

131. $0.456\overline{)4.5694323}$

132. $43.225\overline{)32.465748}$

133. $55.77443-0.568245$

134. $0.62317+1.3316$

APPLICATIONS SEE EXAMPLES 4, 9, AND 14. (OBJECTIVE 7)

135. Spring plowing A farmer has plowed $12\frac{1}{3}$ acres of a $43\frac{1}{2}$-acre field. How much more needs to be plowed?

136. Fencing a garden The four sides of a garden measure $7\frac{2}{3}$ feet, $15\frac{1}{4}$ feet, $19\frac{1}{2}$ feet, and $10\frac{3}{4}$ feet. Find the length of the fence needed to enclose the garden.

137. Making clothes A designer needs $4\frac{1}{3}$ yards of material for each dress he makes. How much material will he need to make 15 dresses?

138. Track and field Each lap around a stadium track is $\frac{1}{4}$ mile. How many laps would a runner have to complete to run 26 miles?

139. Disaster relief After hurricane damage estimated at \$187.75 million, a county sought relief from three agencies. Local agencies gave \$46.8 million and state agencies gave \$72.5 million. How much must the federal government contribute to make up the difference?

140. Minority population 26.5% of the 12,419,000 citizens of Illinois are nonwhite. How many are nonwhite?

The following circle graph shows the various sources of retirement income for a typical retired person. Use this information in Exercises 141–142.

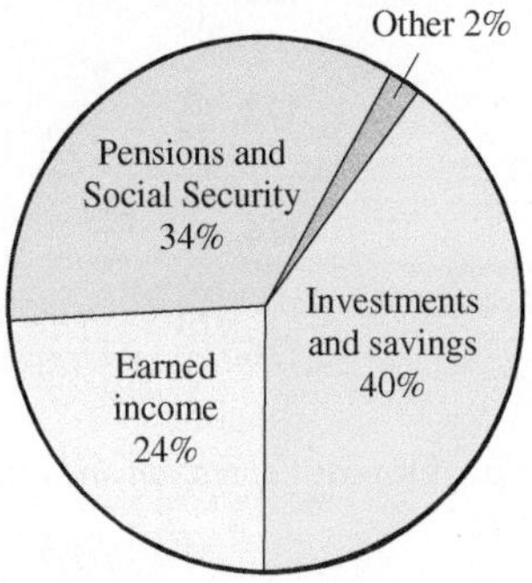

141. Retirement income If a retiree has \$36,000 of income, how much is expected to come from pensions and Social Security?

142. Retirement income If a retiree has \$52,000 of income, how much is expected to come from earned income?

143. Quality control In the manufacture of active-matrix color LCD computer displays, many units must be rejected as defective. If 23% of a production run of 17,500 units is defective, how many units are acceptable?

144. Freeze-drying Almost all of the water must be removed when food is preserved by freeze-drying. Find the weight of the water removed from 750 pounds of a food that is 36% water.

145. Planning for growth This year, sales at Positronics Corporation totaled \$18.7 million. If the projection of 12% annual growth is true, what will be next year's sales?

146. Speed skating In tryouts for the Olympics, a speed skater had times of 44.47, 43.24, 42.77, and 42.05 seconds. Find the average time. Give the result to the nearest hundredth. (*Hint:* Add the numbers and divide by 4.)

147. Cost of gasoline Samuel drove his car 16,275.3 miles last year, averaging 25.5 miles per gallon of gasoline. If the average cost of gasoline was \$3.45 per gallon, find the fuel cost to drive the car.

148. Paying taxes A woman earns \$48,712.32 in taxable income. She must pay 15% tax on the first \$23,000 and 28% on the rest. In addition, she must pay a Social Security tax of 15.4% on the total amount. How much tax will she need to pay?

149. Sealing asphalt A rectangular parking lot is 253.5 feet long and 178.5 feet wide. A 55-gallon drum of asphalt sealer covers 4,000 square feet and costs \$97.50. Find the cost to seal the parking lot. (Sealer can be purchased only in full drums.)

150. Inventory costs Each TV a retailer buys costs \$3.25 per day for warehouse storage. What does it cost to store 37 TVs for three weeks?

151. Manufacturing profits A manufacturer of computer memory boards has a profit of \$37.50 on each standard-capacity memory board, and \$57.35 on each high-capacity board. The sales department has orders for 2,530 standard boards and 1,670 high-capacity boards. Which order will produce the greater profit?

152. Dairy production A Holstein cow will produce 7,600 pounds of milk each year, with a $3\frac{1}{2}$% butterfat content. Each year, a Guernsey cow will produce about 6,500 pounds of milk that is 5% butterfat. Which cow produces more butterfat?

153. Feeding dairy cows Each year, a typical dairy cow will eat 12,000 pounds of food that is 57% silage. To feed 30 cows, how much silage will a farmer use in a year?

154. Comparing bids Two carpenters bid on a home remodeling project. The first bids \$9,350 for the entire job. The second will work for \$27.50 per hour, plus \$4,500 for materials. He estimates that the job will take 150 hours. Which carpenter has the lower bid?

155. Choosing a furnace A high-efficiency home heating system can be installed for \$4,170, with an average monthly heating bill of \$57.50. A regular furnace can be installed for \$1,730, but monthly heating bills average \$107.75. After three years, which system has cost more altogether?

156. Choosing a furnace Refer to Exercise 155. Decide which furnace system will have cost more after five years.

WRITING ABOUT MATH

157. Describe how you would find the common denominator of two fractions.

158. Explain how to convert an improper fraction into a mixed number.

159. Explain how to convert a mixed number into an improper fraction.

160. Explain how you would decide which of two decimal fractions is the larger.

SOMETHING TO THINK ABOUT

161. In what situations would it be better to leave an answer in the form of an improper fraction?

162. When would it be better to change an improper-fraction answer into a mixed number?

163. Can the product of two proper fractions be larger than either of the fractions?

164. How does the product of one proper and one improper fraction compare with the two factors?

Unless otherwise noted, all content on this page is © Cengage Learning.

Section 1.3 Exponents and Order of Operations

Objectives

1. Identify the base and the exponent to simplify an exponential expression.
2. Evaluate a numeric expression following the order of operations.
3. Use the correct geometric formula for an application.

Vocabulary

base	perimeter	radius
exponent	area	volume
exponential expression	circumference	linear units
power of *x*	diameter	square units
grouping symbol	center	cubic units

Getting Ready

Perform each operation.

1. $2 \cdot 2$
2. $3 \cdot 3$
3. $3 \cdot 3 \cdot 3$
4. $2 \cdot 2 \cdot 2$
5. $\frac{1}{2} \cdot \frac{1}{2}$
6. $\frac{1}{3} \cdot \frac{1}{3} \cdot \frac{1}{3}$
7. $\frac{2}{5} \cdot \frac{2}{5} \cdot \frac{2}{5}$
8. $\frac{3}{10} \cdot \frac{3}{10} \cdot \frac{3}{10}$

In algebra we encounter many expressions that contain exponents, a shortcut method of showing repeated multiplication. In this section, we will introduce exponential notation and discuss the rules for the order of operations.

1 Identify the base and the exponent to simplify an exponential expression.

To show how many times a number is to be used as a factor in a product, we use exponents. In the expression 2^3, 2 is called the **base** and 3 is called the **exponent**.

$$\text{Base} \rightarrow 2^3 \leftarrow \text{Exponent}$$

The exponent of 3 indicates that the base of 2 is to be used as a factor three times:

$$2^3 = \overbrace{2 \cdot 2 \cdot 2}^{\text{3 factors of 2}} = 8$$

COMMENT Note that $2^3 = 8$. This is not the same as $2 \cdot 3 = 6$.

In the expression x^5 (called an **exponential expression** or a **power of *x***), *x* is the base and 5 is the exponent. The exponent of 5 indicates that a base of *x* is to be used as a factor five times.

$$x^5 = \overbrace{x \cdot x \cdot x \cdot x \cdot x}^{\text{5 factors of } x}$$

In expressions such as 7, *x*, or *y*, the exponent is understood to be 1:

$$7 = 7^1 \qquad x = x^1 \qquad y = y^1$$

In general, we have the following definition.

NATURAL-NUMBER EXPONENTS

If n is a natural number, then

$$x^n = \underbrace{x \cdot x \cdot x \cdot \cdots \cdot x}_{n \text{ factors of } x}$$

EXAMPLE 1 Write each expression without exponents.

a. $4^2 = 4 \cdot 4 = 16$ — Read 4^2 as "4 squared" or as "4 to the second power."

b. $5^3 = 5 \cdot 5 \cdot 5 = 125$ — Read 5^3 as "5 cubed" or as "5 to the third power."

c. $6^4 = 6 \cdot 6 \cdot 6 \cdot 6 = 1{,}296$ — Read 6^4 as "6 to the fourth power."

d. $\left(\frac{2}{3}\right)^5 = \frac{2}{3} \cdot \frac{2}{3} \cdot \frac{2}{3} \cdot \frac{2}{3} \cdot \frac{2}{3} = \frac{32}{243}$ — Read $\left(\frac{2}{3}\right)^5$ as "$\frac{2}{3}$ to the fifth power."

SELF CHECK 1 Write each expression without exponents: **a.** 7^2 **b.** $\left(\frac{3}{4}\right)^3$

We can find powers using a calculator. For example, to find 2.35^4, we enter these numbers and press these keys:

2.35 y^x 4 = — Using a scientific calculator

2.35 ∧ 4 **ENTER** — Using a graphing calculator

Either way, the display will read 30.49800625 . Some scientific calculators have an x^y key rather than a y^x key.

In the next example, the base of an exponential expression is a variable.

EXAMPLE 2 Write each expression without exponents.

a. $y^6 = y \cdot y \cdot y \cdot y \cdot y \cdot y$ — Read y^6 as "y to the sixth power."

b. $x^3 = x \cdot x \cdot x$ — Read x^3 as "x cubed" or as "x to the third power."

c. $z^2 = z \cdot z$ — Read z^2 as "z squared" or as "z to the second power."

d. $a^1 = a$ — Read a^1 as "a to the first power."

e. $2(3x)^2 = 2(3x)(3x)$ — Read $2(3x)^2$ as "2 times $(3x)$ to the second power."

SELF CHECK 2 Write each expression without exponents. **a.** a^3 **b.** b^4

2 Evaluate a numeric expression following the order of operations.

Suppose you are asked to contact a friend if you see a Rolex watch for sale while traveling in Switzerland. After locating the watch, you send the following message to your friend.

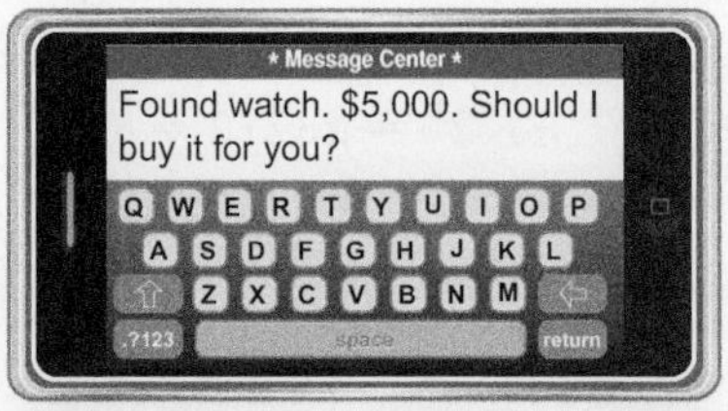

You receive this response.

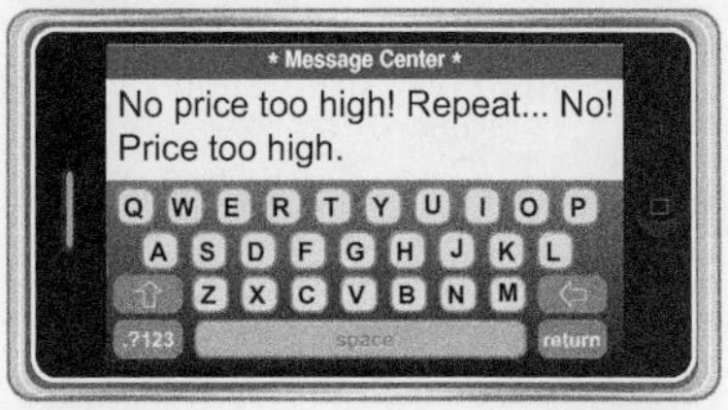

Unless otherwise noted, all content on this page is © Cengage Learning.

The first statement says to buy the watch at any price. The second says not to buy it, because it is too expensive. The placement of the exclamation point makes these statements read differently, resulting in different interpretations.

When reading a mathematical statement, the same kind of confusion is possible. To illustrate, we consider the expression $2 + 3 \cdot 4$, which contains the operations of addition and multiplication. We can calculate this expression in two different ways. We can perform the multiplication first and then perform the addition. Or we can perform the addition first and then perform the multiplication. However, we will get different results.

Multiply first		***Add first***	
$2 + \mathbf{3 \cdot 4} = 2 + \mathbf{12}$	Multiply 3 and 4.	$\mathbf{2 + 3} \cdot 4 = \mathbf{5} \cdot 4$	Add 2 and 3.
$= 14$	Add 2 and 12.	$= 20$	Multiply 5 and 4.

Different results

To eliminate the possibility of getting different answers, we will agree to perform multiplications before additions. The correct calculation of $2 + 3 \cdot 4$ is

$$2 + \mathbf{3 \cdot 4} = 2 + \mathbf{12} \quad \text{Do the multiplication first.}$$
$$= 14$$

To indicate that additions are to be done before multiplications, we use **grouping symbols** such as parentheses ( ), brackets [ ], or braces {}. The operational symbols $\sqrt{\ }$, | |, and fraction bars are also grouping symbols. In the expression $(2 + 3)4$, the parentheses indicate that the addition is to be done first:

$$(\mathbf{2 + 3})4 = \mathbf{5} \cdot 4 \quad \text{Do the addition within the parentheses first.}$$
$$= 20$$

To guarantee that calculations will have one correct result, we will always perform calculations in the following order.

RULES FOR THE ORDER OF OPERATIONS

Use the following steps to perform all calculations within each pair of grouping symbols, working from the innermost pair to the outermost pair.

1. Find the values of any exponential expressions.
2. Perform all multiplications and divisions, working from left to right.
3. Perform all additions and subtractions, working from left to right.
4. Because a fraction bar is a grouping symbol, simplify the numerator and the denominator in a fraction separately. Then simplify the fraction, whenever possible.

COMMENT Note that $4(2)^3 \neq (4 \cdot 2)^3$:

$$4(2)^3 = 4 \cdot 2 \cdot 2 \cdot 2 = 4(8) = 32 \quad \text{and} \quad (4 \cdot 2)^3 = 8^3 = 8 \cdot 8 \cdot 8 = 512$$

Likewise, $4x^3 \neq (4x)^3$ because

$$4x^3 = 4xxx \quad \text{and} \quad (4x)^3 = (4x)(4x)(4x) = 64xxx$$

EXAMPLE 3 Evaluate: $5^3 + 2(8 - 3 \cdot 2)$

Solution We perform the work within the parentheses first and then simplify.

$$5^3 + 2(8 - \mathbf{3 \cdot 2}) = 5^3 + 2(8 - \mathbf{6}) \quad \text{Do the multiplication within the parentheses.}$$
$$= 5^3 + 2(2) \quad \text{Do the subtraction within the parentheses.}$$
$$= 125 + 2(2) \quad \text{Find the value of the exponential expression.}$$

$= 125 + 4$ Do the multiplication.

$= 129$ Do the addition.

SELF CHECK 3 Evaluate: $5 + 4 \cdot 3^2$

EXAMPLE 4 Evaluate: $\dfrac{3(3 + 2) + 5}{17 - 3(4)}$

Solution We simplify the numerator and denominator separately and then simplify the fraction.

$$\frac{3(3 + 2) + 5}{17 - 3(4)} = \frac{3(5) + 5}{17 - 3(4)} \quad \text{Do the addition within the parentheses.}$$

$$= \frac{15 + 5}{17 - 12} \quad \text{Do the multiplications.}$$

$$= \frac{20}{5} \quad \text{Do the addition and the subtraction.}$$

$$= 4 \quad \text{Do the division.}$$

SELF CHECK 4 Evaluate: $\dfrac{4 + 2(5 - 3)}{2 + 3(2)}$

EXAMPLE 5 Evaluate: $\dfrac{3(4^2) - 2(3)}{2(4 + 3)}$

Solution

$$\frac{3(4^2) - 2(3)}{2(4 + 3)} = \frac{3(16) - 2(3)}{2(7)} \quad \text{Find the value of } 4^2 \text{ in the numerator and do the addition in the denominator.}$$

$$= \frac{48 - 6}{14} \quad \text{Do the multiplications.}$$

$$= \frac{42}{14} \quad \text{Do the subtraction.}$$

$$= 3 \quad \text{Do the division.}$$

SELF CHECK 5 Evaluate: $\dfrac{2^2 + 6(5)}{2(2 + 5) + 3}$

3 Use the correct geometric formula for an application.

To find perimeters and areas of geometric figures, we often must substitute numbers for variables in a formula. The **perimeter** of a geometric figure is the distance around it, and the **area** of a geometric figure is the amount of surface that it encloses. The perimeter of a circle is called its **circumference**.

EXAMPLE 6 **CIRCLES** Use the information in Figure 1-17 to find:

a. the circumference **b.** the area of the circle.

Solution **a.** The formula for the circumference of a circle is

$$C = \pi D$$

where C is the circumference, π can be approximated by $\frac{22}{7}$, and D is the **diameter**—a line segment that passes through the center of the circle and joins two points on the

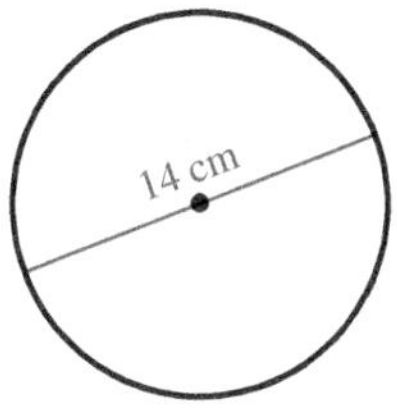

Figure 1-17

circle. We can approximate the circumference by substituting $\frac{22}{7}$ for π and 14 for D in the formula and simplifying.

$$C = \pi D$$

$$C \approx \frac{22}{7} \cdot 14 \quad \text{Read } \approx \text{ as "is approximately equal to."}$$

$$C \approx \frac{22 \cdot \overset{2}{\cancel{14}}}{\underset{1}{\cancel{7}} \cdot 1} \quad \text{Multiply the fractions and simplify.}$$

$$C \approx 44$$

The circumference is approximately 44 centimeters. To use a calculator, we enter these numbers and press these keys:

π × 14 = Using a scientific calculator

2nd ∧ (π) × 14 **ENTER** Using a graphing calculator

Either way, the display will read 43.98229715. The result is not 44, because a calculator uses a better approximation for π than $\frac{22}{7}$.

COMMENT A segment drawn from the **center** of a circle to a point on the circle is called a **radius**. Since the diameter D of a circle is twice as long as its radius r, we have $D = 2r$. If we substitute $2r$ for D in the formula $C = \pi D$, we obtain an alternate formula for the circumference of a circle: $C = 2\pi r$.

b. The formula for the area of a circle is

$$A = \pi r^2$$

where A is the area, $\pi \approx \frac{22}{7}$, and r is the radius of the circle. We can approximate the area by substituting $\frac{22}{7}$ for π and 7 for r in the formula and simplifying.

$$A = \pi r^2$$

$$A \approx \frac{22}{7} \cdot 7^2$$

$$A \approx \frac{22}{7} \cdot \frac{49}{1} \quad \text{Evaluate the exponential expression.}$$

$$A \approx \frac{22 \cdot \overset{7}{\cancel{49}}}{\underset{1}{\cancel{7}} \cdot 1} \quad \text{Multiply the fractions and simplify.}$$

$$A \approx 154$$

The area is approximately 154 square centimeters.

To use a calculator, we enter these numbers and press these keys:

π × 7 x^2 = Using a scientific calculator

2nd ∧ (π) × 7 x^2 **ENTER** Using a graphing calculator

The display will read 153.93804.

SELF CHECK 6 Given a circle with a diameter of 28 meters, find an estimate to the nearest whole number of

a. the circumference **b.** the area.

(Use $\frac{22}{7}$ to estimate π.) Check your results with a calculator.

Table 1-1 shows the formulas for the perimeter and area of several geometric figures.

Unless otherwise noted, all content on this page is © Cengage Learning.

Euclid
325–265 BC
Although Euclid is best known for his study of geometry, many of his writings deal with number theory. In about 300 BC, the Greek mathematician Euclid proved that the number of prime numbers is unlimited—that there are infinitely many prime numbers. This is an important branch of mathematics called number theory.

TABLE 1-1

Figure	Name	Perimeter	Area
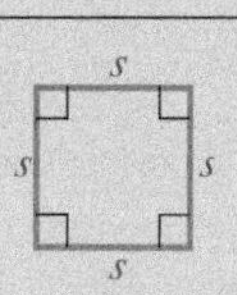	Square	$P = 4s$	$A = s^2$
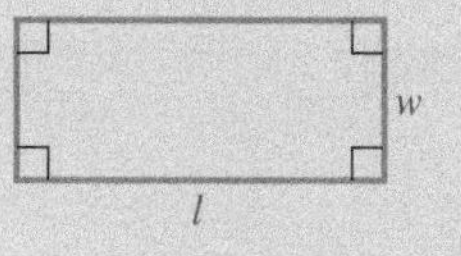	Rectangle	$P = 2l + 2w$	$A = lw$
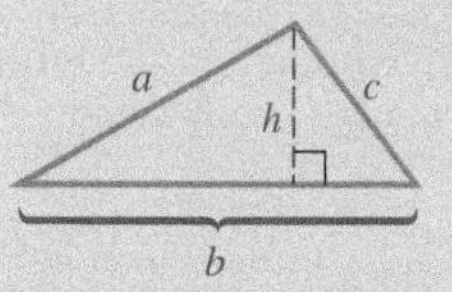	Triangle	$P = a + b + c$	$A = \frac{1}{2}bh$
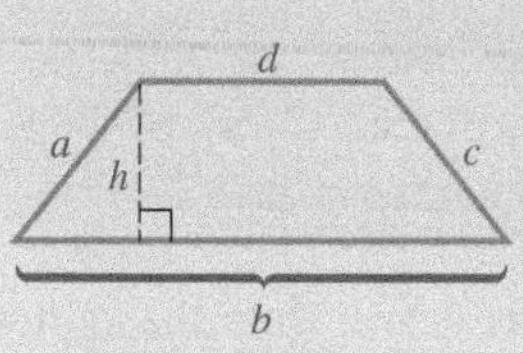	Trapezoid	$P = a + b + c + d$	$A = \frac{1}{2}h(b + d)$
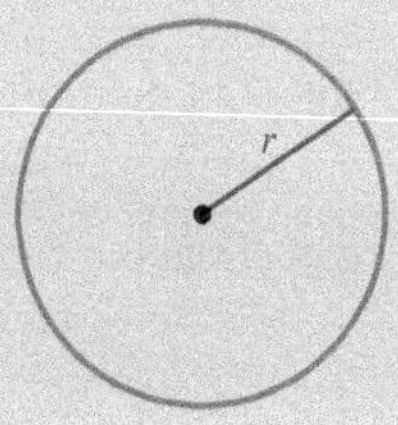	Circle	$C = \pi D = 2\pi r$ $(D = 2r)$	$A = \pi r^2$

The **volume** of a three-dimensional geometric solid is the amount of space it encloses. Table 1-2 shows the formulas for the volume of several solids.

TABLE 1-2

Figure	Name	Volume
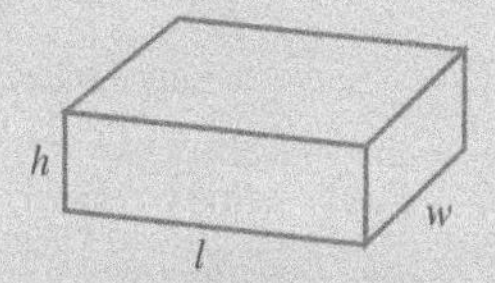	Rectangular solid	$V = lwh$
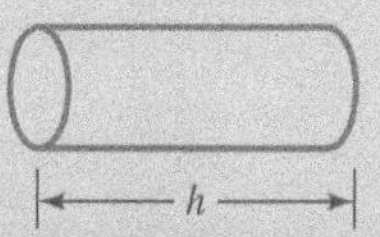	Cylinder	$V = Bh$, where B is the area of the base
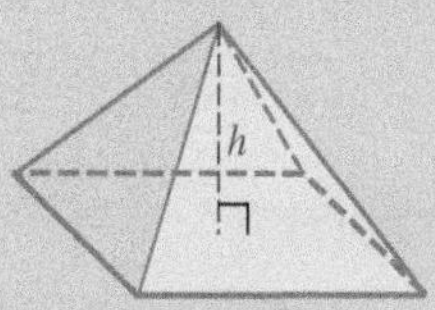	Pyramid	$V = \frac{1}{3}Bh$, where B is the area of the base
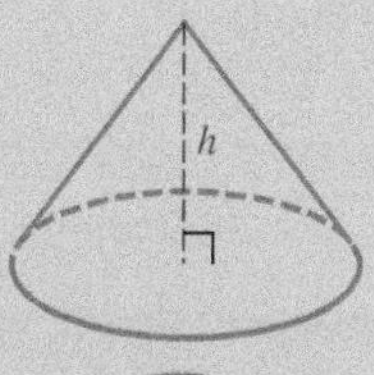	Cone	$V = \frac{1}{3}Bh$, where B is the area of the base
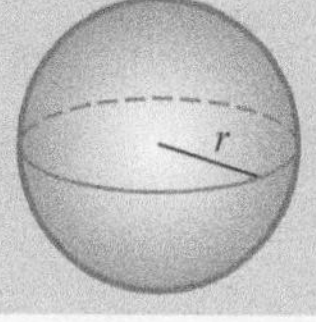	Sphere	$V = \frac{4}{3}\pi r^3$

Unless otherwise noted, all content on this page is © Cengage Learning.

When working with geometric figures, measurements are often given in **linear units** such as feet (ft), centimeters (cm), or meters (m). If the dimensions of a two-dimensional geometric figure are given in feet, we can calculate its perimeter by finding the sum of the lengths of its sides. This sum will be in feet.

If we calculate the area of a two-dimensional figure, the result will be in **square units**. For example, if we calculate the area of the figure whose sides are measured in centimeters, the result will be in square centimeters (cm^2).

If we calculate the volume of a three-dimensional figure, the result will be in **cubic units**. For example, the volume of a three-dimensional geometric figure whose sides are measured in meters will be in cubic meters (m^3).

EXAMPLE 7 **WINTER DRIVING** Find the number of cubic feet of road salt in the conical pile shown in Figure 1-18. Round the answer to two decimal places.

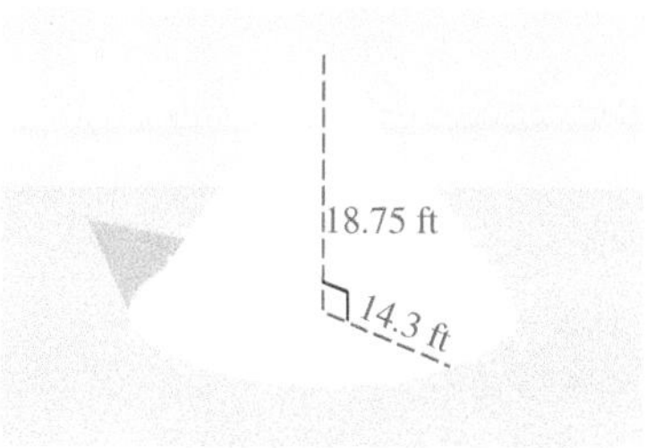

Figure 1-18

Solution We can find the area of the circular base by substituting $\frac{22}{7}$ for π and 14.3 for the radius.

$$A = \pi r^2$$
$$\approx \frac{22}{7}(14.3)^2$$
$$\approx 642.6828571 \quad \text{Use a calculator.}$$

We then substitute 642.6828571 for B and 18.75 for h in the formula for the volume of a cone.

$$V = \frac{1}{3}Bh$$
$$\approx \frac{1}{3}(642.6828571)(18.75)$$
$$\approx 4{,}016.767857 \quad \text{Use a calculator.}$$

To two decimal places, there are 4,016.77 cubic feet of salt in the pile.

SELF CHECK 7 To the nearest hundredth, find the number of cubic feet of water that can be contained in a spherical tank that has a radius of 9 feet. (Use $\pi \approx \frac{22}{7}$.)

SELF CHECK ANSWERS

1. a. 49 **b.** $\frac{27}{64}$ **2. a.** $a \cdot a \cdot a$ **b.** $b \cdot b \cdot b \cdot b$ **3.** 41 **4.** 1 **5.** 2 **6. a.** 88 m **b.** 616 m^2
7. 3,054.86 ft^2

Unless otherwise noted, all content on this page is © Cengage Learning.

NOW TRY THIS

Simplify each expression.

1. $28 - 7(4 - 1)$
2. $\dfrac{5 - |4 - 1|}{2}$
3. Insert the appropriate operations and one set of parentheses (if necessary) so that the expression yields the given value.
   a. 16 3 5 = 2
   b. 4 2 6 = 12

1.3 Exercises

WARM-UPS *Perform each operation.*

1. $2 \cdot 2 \cdot 2 \cdot 2 \cdot 2$
2. $3 \cdot 3 \cdot 3 \cdot 3$
3. $4 \cdot 4 \cdot 4$
4. $5 \cdot 5 \cdot 5$
5. $\frac{2}{3} \cdot \frac{2}{3} \cdot \frac{2}{3}$
6. $\frac{4}{5} \cdot \frac{4}{5}$

Identify the base in each expression.

7. y^3
8. $(2x)^4$
9. $7(4x)^2$
10. $3y^2$

REVIEW

11. On the number line, graph the prime numbers between 10 and 20.
12. Write the inequality $7 \le 12$ as an inequality using the symbol $\ge$.
13. Classify the number 17 as a prime number or a composite number.
14. Evaluate: $\frac{3}{5} - \frac{1}{2}$

VOCABULARY AND CONCEPTS *Fill in the blanks.*

15. An _______ indicates how many times a base is to be used as a factor in a product.
16. In the exponential expression (power of x) x^7, x is called the ____ and 7 is called an _______.
17. Parentheses, brackets, and braces are called _______ symbols.
18. A line segment that passes through the center of a circle and joins two points on the circle is called a _______. A line segment drawn from the center of a circle to a point on the circle is called a _____.
19. The distance around a rectangle is called the _______, and the distance around a circle is called the __________.
20. The region enclosed by a two-dimensional geometric figure is called the ____ and is designated by _____ units, and the region enclosed by a three-dimensional geometric figure is called the ______ and is designated by ____ units.

Write the appropriate formula to find each quantity and state the correct units.

21. The perimeter of a square ______; ____
22. The area of a square _____; __________
23. The perimeter of a rectangle __________; ____
24. The area of a rectangle ______; __________
25. The perimeter of a triangle ____________; ____
26. The area of a triangle ________; __________
27. The perimeter of a trapezoid _________________; ____
28. The area of a trapezoid ____________; __________
29. The circumference of a circle _________________; ____
30. The area of a circle ________; __________
31. The volume of a rectangular solid ________; __________
32. The volume of a cylinder _______; _________
33. The volume of a pyramid ________; _________
34. The volume of a cone ________; _________
35. The volume of a sphere ________; _________
36. In Exercises 32–34, B is the ____ of the base.

GUIDED PRACTICE *Write each expression without using exponents and find the value of each expression.* SEE EXAMPLE 1. *(OBJECTIVE 1)*

37. 6^2
38. 9^2
39. $\left(-\frac{1}{5}\right)^4$
40. $\left(\frac{1}{2}\right)^6$

Write each expression without using exponents. SEE EXAMPLE 2. *(OBJECTIVE 1)*

41. x^3
42. y^4
43. $8z^4$
44. $5t^2$
45. $(4x)^3$
46. $(3z)^4$
47. $3(6y)^2$
48. $2(4t)^3$

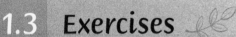

Find the value of each expression. SEE EXAMPLES 3–5. (OBJECTIVE 2)

49. $4(3^2)$

50. $4(2^3)$

51. $(2 \cdot 5)^4$

52. $(2 \cdot 2)^3$

53. $5(4)^2$

54. $4(5)^2$

55. $(3 \cdot 2)^3$

56. $(2 \cdot 3)^2$

57. $3 \cdot 5 - 4$

58. $3 + 6 \cdot 4$

59. $3(5 - 4)$

60. $3(5 + 8)$

61. $2 + 3 \cdot 5 - 4$

62. $10 + 2 \cdot 4 + 3$

63. $48 \div (4 + 2)$

64. $16 \div (5 + 3)$

65. $3^2 + 2(1 + 4) - 2$

66. $4 \cdot 3 + 2(5 - 2) - 2^3$

67. $\frac{3}{5} \cdot \frac{10}{3} + \frac{1}{2} \cdot 12$

68. $\frac{15}{4}\left(1 + \frac{3}{5}\right)$

69. $\left[\frac{1}{3} - \left(\frac{1}{2}\right)^2\right]^2$

70. $\left[\left(\frac{2}{3}\right)^2 - \frac{1}{3}\right]^2$

71. $\frac{(3 + 5)^2 + 2}{2(8 - 5)}$

72. $\frac{25 - (2 \cdot 3 - 1)}{2 \cdot 9 - 8}$

73. $\frac{(5 - 3)^2 + 2}{4^2 - (8 + 2)}$

74. $\frac{(4^2 - 2) + 7}{5(2 + 4) - 3^2}$

75. $\frac{3 \cdot 7 - 5(3 \cdot 4 - 11)}{4(3 + 2) - 3^2 + 5}$

76. $\frac{2 \cdot 5^2 - 2^2 + 3}{2(5 - 2)^2 - 11}$

Find the perimeter of each figure. (OBJECTIVE 3)

77.

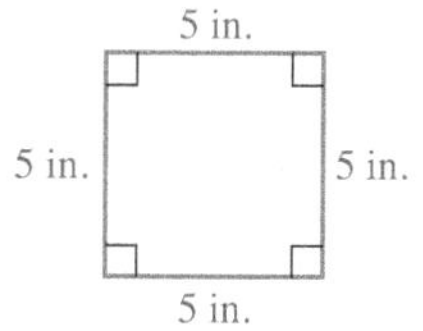

78.

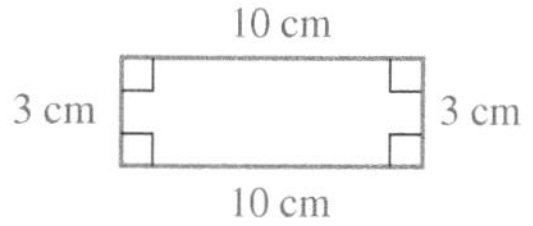

79.

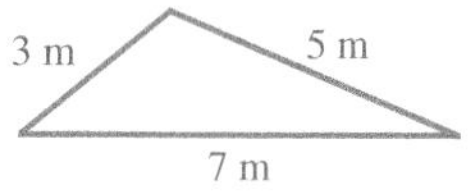

80.

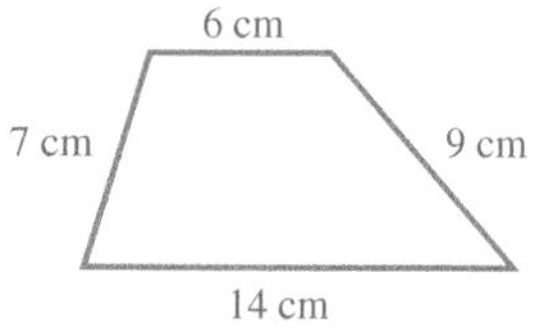

Find the area of each figure. (OBJECTIVE 3)

81.

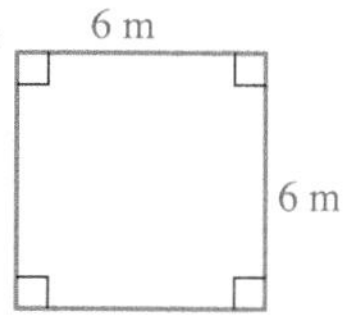

82.

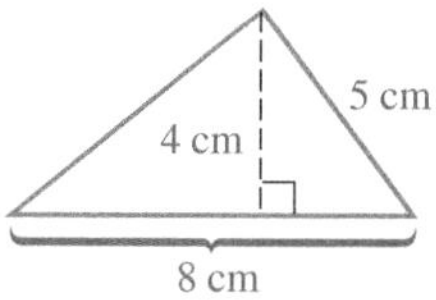

83.

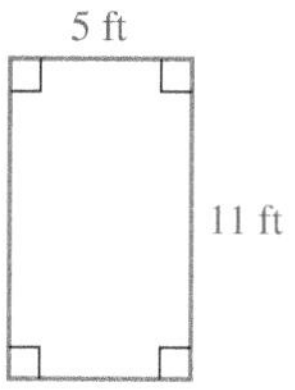

84.

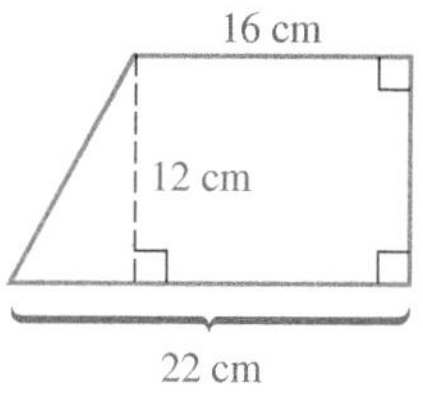

Find the circumference of each circle. Use $\pi \approx \frac{22}{7}$. SEE EXAMPLE 6. (OBJECTIVE 3)

85.

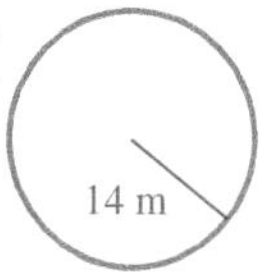

86.

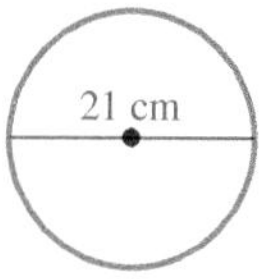

Find the area of each circle. Use $\pi \approx \frac{22}{7}$. SEE EXAMPLE 6. (OBJECTIVE 3)

87.

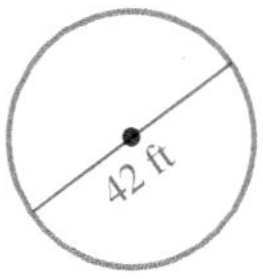

88.

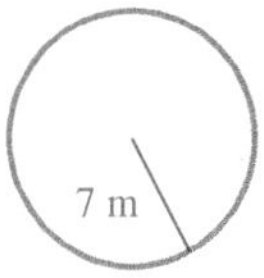

Find the volume of each solid. Use $\pi \approx \frac{22}{7}$ *where applicable.* SEE EXAMPLE 7. (OBJECTIVE 3)

89.

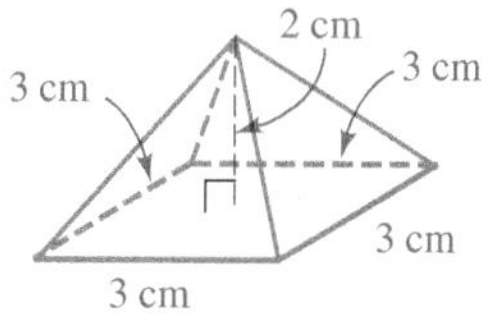

90.

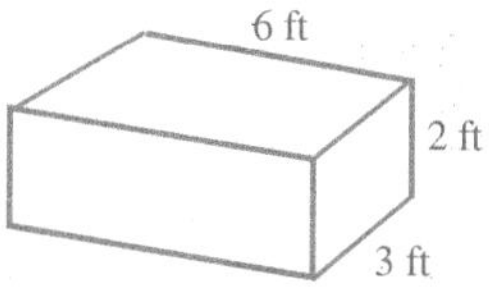

91.

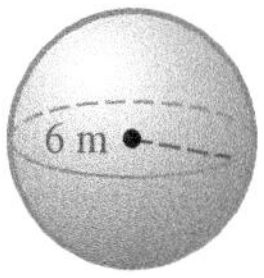

92.

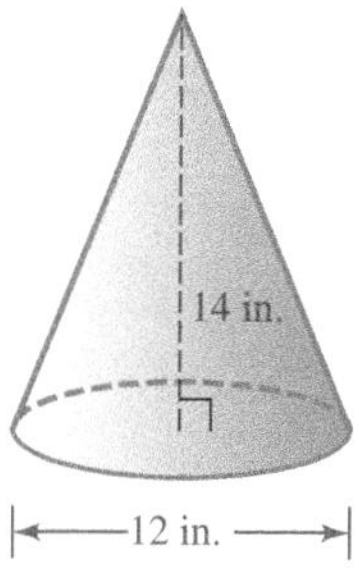

93.

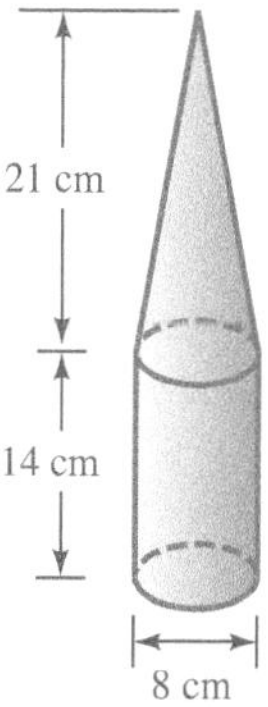

94.

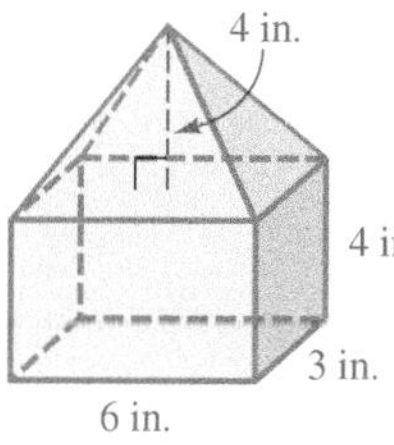

Unless otherwise noted, all content on this page is © Cengage Learning.

ADDITIONAL PRACTICE *Simplify each expression.*

95. 6^2

96. 7^3

97. $2 + 4^2$

98. $4^2 - 2^2$

99. $(2 + 4)^2$

100. $(7 - 3)^3$

101. $(7 + 9) \div (2 \cdot 4)$

102. $(7 + 9) \div 2 \cdot 4$

103. $(5 + 7) \div 3 \cdot 4$

104. $(5 + 7) \div (3 \cdot 4)$

105. $24 \div 4 \cdot 3 + 3$

106. $36 \div 9 \cdot 4 - 2$

107. $6^2 - (8 - 3)^2$

108. $3^3 + (3 - 1)^3$

109. $(2 \cdot 3 - 4)^3$

110. $(3 \cdot 5 - 2 \cdot 6)^2$

111. $\dfrac{2[4 + 2(3 - 1)]}{3[3(2 \cdot 3 - 4)]}$

112. $\dfrac{3[9 - 2(7 - 3)]}{(8 - 5)(9 - 7)}$

Use a calculator to find each power.

113. 7.9^3

114. 0.45^4

115. 25.3^2

116. 7.567^3

Insert parentheses in the expression $3 \cdot 8 + 5 \cdot 3$ *to make its value equal to the given number.*

117. 39

118. 117

119. 87

120. 69

APPLICATIONS *Use a calculator. For π, use the π key. Round to two decimal places.* SEE EXAMPLE 7. (OBJECTIVE 3)

121. Volume of a tank Find the number of cubic feet of water in the spherical tank at the top of the water tower.

122. Storing solvents A hazardous solvent fills a rectangular tank with dimensions of 12 inches by 9.5 inches by 7.3 inches. For disposal, it must be transferred to a cylindrical canister 7.5 inches in diameter and 18 inches high. How much solvent will be left over?

123. Buying fencing How many meters of fencing are needed to enclose the square pasture shown in the illustration?

124. Installing carpet What will it cost to carpet the area shown in the illustration with carpet that costs \$29.79 per square yard? (One square yard is 9 square feet.)

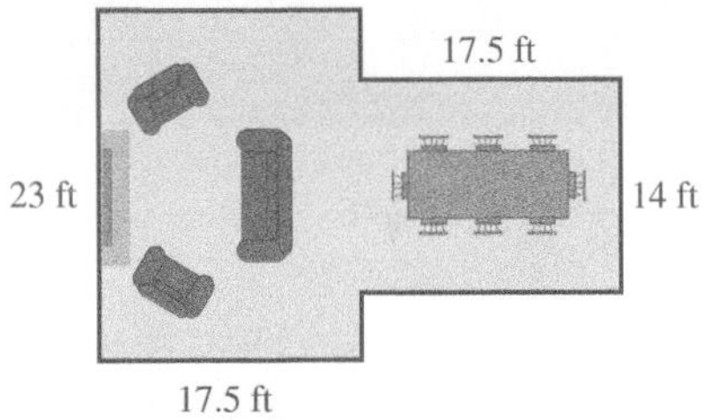

125. Volume of a classroom Thirty students are in a classroom with dimensions of 40 feet by 40 feet by 9 feet. How many cubic feet of air are there for each student?

126. Wallpapering One roll of wallpaper covers about 33 square feet. At \$27.50 per roll, how much would it cost to paper two walls 8.5 feet high and 17.3 feet long? (*Hint:* Wallpaper can be purchased only in full rolls.)

127. Focal length The focal length f of a double-convex thin lens is given by the formula

$$f = \frac{rs}{(r + s)(n - 1)}$$

If $r = 8$, $s = 12$, and $n = 1.6$, find f.

128. Resistance The total resistance R of two resistors in parallel is given by the formula

$$R = \frac{rs}{r + s}$$

If $r = 170$ and $s = 255$, find R.

WRITING ABOUT MATH

129. Explain why the symbols $3x$ and x^3 have different meanings.

130. Students often say that x^n means "x multiplied by itself n times." Explain why this is not correct.

SOMETHING TO THINK ABOUT

131. If x were greater than 1, would raising x to higher and higher powers produce bigger numbers or smaller numbers?

132. What would happen in Exercise 131 if x were a positive number that was less than 1?

Unless otherwise noted, all content on this page is © Cengage Learning.

Section 1.4 Adding and Subtracting Real Numbers

Objectives

1. Add two or more real numbers with like signs.
2. Add two or more real numbers with unlike signs.
3. Subtract two real numbers.
4. Use signed numbers and one or more operations to model an application.
5. Use a calculator to add or subtract two real numbers.

Vocabulary

like signs   unlike signs

Getting Ready

Perform each operation.

1. $14.32 + 3.2$
2. $5.54 - 2.6$
3. $4.2 - (3 - 0.8)$
4. $(5.42 - 4.22) - 0.2$
5. $(437 - 198) - 143$
6. $437 - (198 - 143)$

In this section, we will discuss how to add and subtract real numbers. Recall that the result of an addition is called a *sum* and the result of a subtraction is called a *difference.* To develop the rules for adding real numbers, we will use the number line.

1 Add two or more real numbers with like signs.

Since the positive direction on the number line is to the right, positive numbers can be represented by arrows pointing to the right. Negative numbers can be represented by arrows pointing to the left.

To add +2 and +3, we can represent +2 with an arrow the length of 2, pointing to the right. We can represent +3 with an arrow of length 3, also pointing to the right. To add the numbers, we place the arrows end to end, as in Figure 1-19. Since the endpoint of the second arrow is the point with coordinate +5, we have

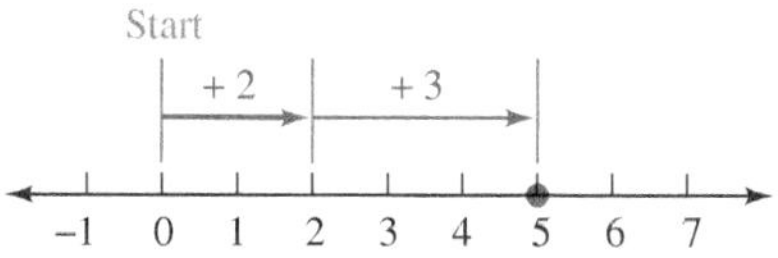

Figure 1-19

$$(+2) + (+3) = +5$$

COMMENT
$0 + a = a + 0 = a$

As a check, we can think of this problem in terms of money. If you had \$2 and earned \$3 more, you would have a total of \$5.

The addition

$$(-2) + (-3)$$

Unless otherwise noted, all content on this page is © Cengage Learning.

can be represented by the arrows shown in Figure 1-20. Since the endpoint of the final arrow is the point with coordinate −5, we have

$$(-2) + (-3) = -5$$

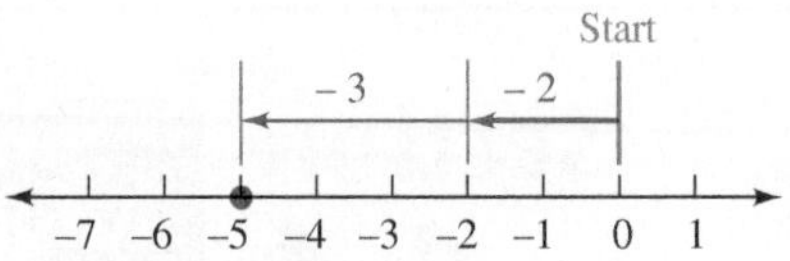

Figure 1-20

As a check, we can think of this problem in terms of money. If you lost \$2 and then lost \$3 more, you would have lost a total of \$5.

Because two real numbers with **like signs** can be represented by arrows pointing in the same direction, we have the following rule.

ADDING REAL NUMBERS WITH LIKE SIGNS

1. *To add two positive numbers*, add their absolute values and the answer is positive.
2. *To add two negative numbers*, add their absolute values and the answer is negative.

EXAMPLE 1 Add:

a. $(+4) + (+6) = +(4 + 6)$
$= 10$

b. $(-4) + (-6) = -(4 + 6)$
$= -10$

c. $+5 + (+10) = +(5 + 10)$
$= 15$

d. $-\frac{1}{2} + \left(-\frac{3}{2}\right) = -\left(\frac{1}{2} + \frac{3}{2}\right)$
$= -\frac{4}{2}$
$= -2$

SELF CHECK 1 Add: **a.** $(+0.5) + (+1.2)$ **b.** $(-3.7) + (-2.3)$

2 Add two or more real numbers with unlike signs.

Real numbers with **unlike signs** can be represented by arrows on a number line pointing in opposite directions. For example, the addition

$$(-6) + (+2)$$

can be represented by the arrows shown in Figure 1-21. Since the endpoint of the final arrow is the point with coordinate −4, we have

$$(-6) + (+2) = -4$$

COMMENT We do not need to write a + sign in front of a positive number.

$+4 = 4$ and $+5 = 5$

However, we must always write a − sign in front of a negative number.

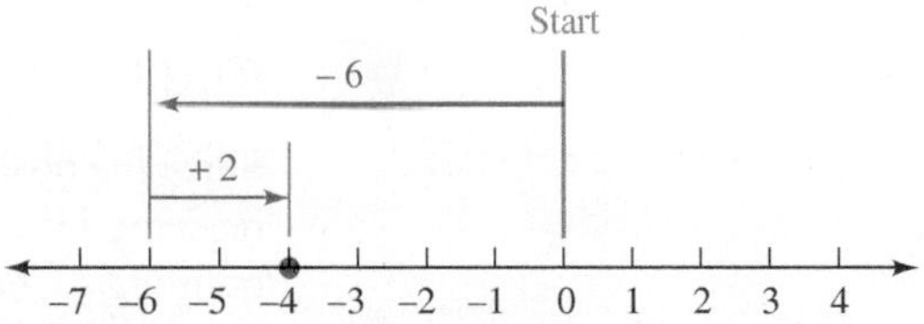

Figure 1-21

As a check, we can think of this problem in terms of money. If you lost \$6 and then earned \$2, you would still have a loss of \$4.

The addition

$$(+7) + (-4)$$

can be represented by the arrows shown in Figure 1-22. Since the endpoint of the final arrow is the point with coordinate +3, we have

$$(+7) + (-4) = +3$$

COMMENT On a number line, a negative number moves to the left, and a positive number moves to the right.

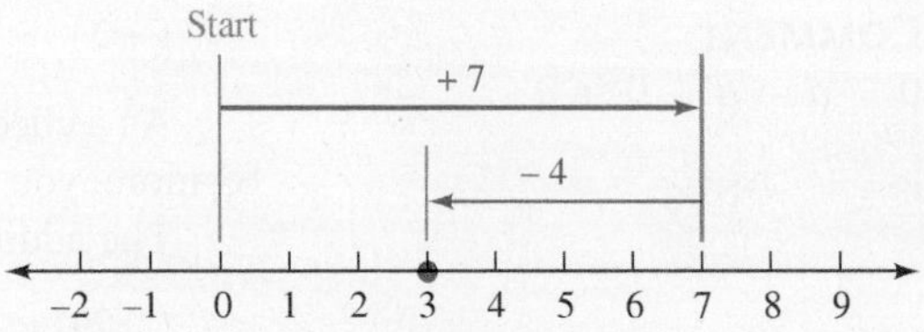

Figure 1-22

Unless otherwise noted, all content on this page is © Cengage Learning.

As a check, you can think of this problem in terms of money. If you had $7 and then lost $4, you would still have a gain of $3.

Because two real numbers with unlike signs can be represented by arrows pointing in opposite directions, we have the following rule.

ADDING REAL NUMBERS WITH UNLIKE SIGNS

To add a positive and a negative number, subtract the smaller absolute value from the larger.

1. If the positive number has the larger absolute value, the answer is positive.
2. If the negative number has the larger absolute value, the answer is negative.
3. If a number and its opposite are added, the answer is zero.

EXAMPLE 2 Add:

a. $(+6) + (-5) = +(6 - 5)$
$= 1$

b. $(-2) + (+3) = +(3 - 2)$
$= 1$

c. $+6 + (-9) = -(9 - 6)$
$= -3$

d. $-\frac{2}{3} + \left(+\frac{1}{2}\right) = -\left(\frac{2}{3} - \frac{1}{2}\right)$
$= -\left(\frac{4}{6} - \frac{3}{6}\right)$
$= -\frac{1}{6}$

SELF CHECK 2 Add: **a.** $(+3.5) + (-2.6)$ **b.** $(-7.2) + (+4.7)$

When adding three or more real numbers, we use the rules for the order of operations.

EXAMPLE 3 Add:

a. $[(+3) + (-7)] + (-4) = [-4] + (-4)$ Do the work within the brackets first.
$= -8$

b. $-3 + [(-2) + (-8)] = -3 + [-10]$ Do the work within the brackets first.
$= -13$

c. $2.75 + [8.57 + (-4.8)] = 2.75 + 3.77$ Do the work within the brackets first.
$= 6.52$

SELF CHECK 3 Add: $-2 + [(+5.2) + (-12.7)]$

Sometimes numbers are added vertically, as shown in the next example.

EXAMPLE 4 Add:

a. $\begin{array}{r} +5 \\ \underline{+2} \\ +7 \end{array}$ **b.** $\begin{array}{r} +5 \\ \underline{-2} \\ +3 \end{array}$ **c.** $\begin{array}{r} -5 \\ \underline{+2} \\ -3 \end{array}$ **d.** $\begin{array}{r} -5 \\ \underline{-2} \\ -7 \end{array}$

SELF CHECK 4 Add: **a.** $\begin{array}{r} +3.2 \\ \underline{-5.4} \end{array}$ **b.** $\begin{array}{r} -13.5 \\ \underline{-4.3} \end{array}$

3 Subtract two real numbers.

In arithmetic, subtraction is a take-away process. For example,

$$7 - 4 = 3$$

can be thought of as taking 4 objects away from 7 objects, leaving 3 objects.

For algebra, a better approach treats the subtraction problem

$$7 - 4$$

as the equivalent addition problem:

$$7 + (-4)$$

In either case, the answer is 3.

$$7 - 4 = 3 \text{ and } 7 + (-4) = 3$$

Thus, to subtract 4 from 7, we can add the negative (or opposite) of 4 to 7. In general, to subtract one real number from another, we add the negative (or opposite) of the number that is being subtracted.

SUBTRACTING REAL NUMBERS

If a and b are two real numbers, then

$$a - b = a + (-b)$$

EXAMPLE 5 Evaluate: **a.** $12 - 4$ **b.** $-13 - 5$ **c.** $-14 - (-6)$

Solution

a. $12 - 4 = 12 + (-4)$ To subtract 4, add the opposite of 4.
$= 8$

b. $-13 - 5 = -13 + (-5)$ To subtract 5, add the opposite of 5.
$= -18$

c. $-14 - (-6) = -14 + [-(-6)]$ To subtract -6, add the opposite of -6. The opposite of -6 is 6.
$= -14 + 6$
$= -8$

SELF CHECK 5 Evaluate: **a.** $-12.7 - 8.9$ **b.** $15.7 - (-11.3)$

To use a vertical format for subtracting real numbers, we add the opposite of the number that is to be subtracted by changing the sign of the lower number and proceeding as in addition.

EXAMPLE 6 Perform each subtraction by doing an equivalent addition.

a. The subtraction $\begin{array}{r} 5 \\ -\underline{-4} \end{array}$ becomes the addition $\begin{array}{r} 5 \\ +\underline{+4} \\ 9 \end{array}$

b. The subtraction $\begin{array}{r} -8 \\ -\underline{+3} \end{array}$ becomes the addition $\begin{array}{r} -8 \\ +\underline{-3} \\ -11 \end{array}$

SELF CHECK 6 Perform the subtraction: $\begin{array}{r} 5.8 \\ -\underline{-4.6} \end{array}$

When adding or subtracting three or more real numbers, we use the order of operations.

EXAMPLE 7 Simplify: **a.** $3 - [4 + (-6)]$ **b.** $[-5 + (-3)] - [-2 - (+5)]$

Solution

a. $3 - [4 + (-6)] = 3 - (-2)$ Do the addition within the brackets first.

$= 3 + [-(-2)]$ To subtract -2, add the opposite of -2.

$= 3 + 2$ $\quad -(-2) = 2$

$= 5$

b. $[-5 + (-3)] - [-2 - (+5)]$

$= [-5 + (-3)] - [-2 + (-5)]$ To subtract -5, add the opposite of 5.

$= -8 - (-7)$ Do the work within the brackets.

$= -8 + [-(-7)]$ To subtract -7, add the opposite of -7.

$= -8 + 7$ $\quad -(-7) = 7$

$= -1$

SELF CHECK 7 Simplify: $[7.2 - (-3)] - [3.2 + (-1.7)]$

EXAMPLE 8 Evaluate: **a.** $\dfrac{-3 - (-5)}{7 + (-5)}$ **b.** $\dfrac{6 + (-5)}{-3 - (-5)} - \dfrac{-3 - 4}{7 + (-5)}$

Solution

a. $\dfrac{-3 - (-5)}{7 + (-5)} = \dfrac{-3 + [-(-5)]}{7 + (-5)}$ To subtract -5, add the opposite of -5.

$= \dfrac{-3 + 5}{2}$ $\quad -(-5) = 5; 7 + (-5) = 2$

$= \dfrac{2}{2}$

$= 1$

b. $\dfrac{6 + (-5)}{-3 - (-5)} - \dfrac{-3 - 4}{7 + (-5)} = \dfrac{1}{-3 + 5} - \dfrac{-3 + (-4)}{2}$ $\quad 6 + (-5) = 1; -(-5) = +5; -3 - 4 = -3 + (-4); 7 + (-5) = 2$

$= \dfrac{1}{2} - \dfrac{-7}{2}$ $\quad -3 + (-4) = -7; -3 + 5 = 2$

$= \dfrac{1 - (-7)}{2}$ Subtract the numerators and keep the denominator.

$= \dfrac{1 + [-(-7)]}{2}$ To subtract -7, add the opposite of -7.

$= \dfrac{1 + 7}{2}$ $\quad -(-7) = 7$

$= \dfrac{8}{2}$

$= 4$

SELF CHECK 8 Evaluate: $\dfrac{7 - (-3)}{-5 - (-3) + 3}$

4 Use signed numbers and one or more operations to model an application.

Words such as *found, gain, credit, up, increase, forward, rises, in the future,* and *to the right* indicate a positive direction. Words such as *lost, loss, debit, down, decrease, backward, falls, in the past,* and *to the left* indicate a negative direction.

EXAMPLE 9 **ACCOUNT BALANCES** The treasurer of a math club opens a checking account by depositing \$350 in the bank. The bank debits the account \$9 for check printing, and the treasurer writes a check for \$22. Find the balance after these transactions.

Solution The deposit can be represented by $+350$. The debit of \$9 can be represented by -9, and the check written for \$22 can be represented by -22. The balance in the account after these transactions is the sum of 350, -9, and -22.

$$\begin{aligned} \mathbf{350} + (\mathbf{-9}) + (-22) &= \mathbf{341} + (-22) && \text{Work from left to right.} \\ &= 319 \end{aligned}$$

The balance is \$319.

SELF CHECK 9 Find the balance if another deposit of \$17 is made.

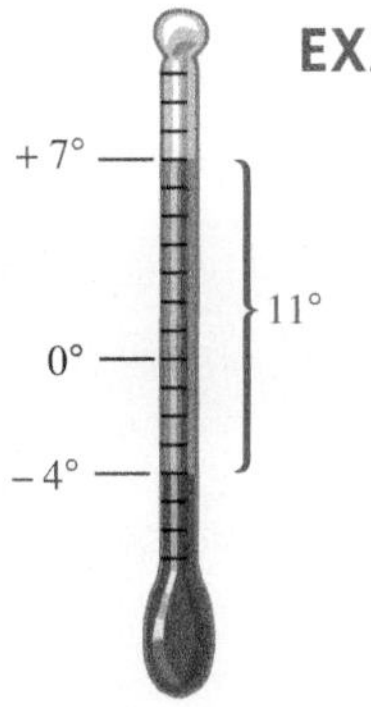

Figure 1-23

EXAMPLE 10 **TEMPERATURE CHANGES** At noon, the temperature was 7° above zero. At midnight, the temperature was 4° below zero. Find the difference between these two temperatures.

Solution A temperature of 7° above zero can be represented as $+7$. A temperature of 4° below zero can be represented as -4. To find the difference between these temperatures, we can set up a subtraction problem and simplify.

$$\begin{aligned} 7 - (-4) &= 7 + [-(-4)] && \text{To subtract } -4\text{, add the opposite of } -4. \\ &= 7 + 4 && -(-4) = 4 \\ &= 11 \end{aligned}$$

The difference between the temperatures is 11°. Figure 1-23 shows this difference.

SELF CHECK 10 Find the difference between temperatures of 32° and $-10°$.

5 Use a calculator to add or subtract two real numbers.

COMMENT A common error is to use the subtraction key $-$ on a calculator rather than the negative key $(-)$.

A calculator can add positive and negative numbers.

- You do not have to do anything special to enter positive numbers. When you press 5, for example, a positive 5 is entered.
- To enter -5 into a calculator with a $+/-$ key, called the *plus-minus* or *change-of-sign* key, you must enter 5 and then press the $+/-$ key. To enter -5 into a calculator with a $(-)$ key, you must press the $(-)$ key and then press 5.

EXAMPLE 11 To evaluate $-345.678 + (-527.339)$, we enter these numbers and press these keys:

345.678 +/− + 527.339 +/− = Using a calculator with a +/− key

(−) 345.678 + (−) 527.339 ENTER Using a graphing calculator

The display will read -873.017.

SELF CHECK 11 Evaluate: $-783.291 - (-28.3264)$

Unless otherwise noted, all content on this page is © Cengage Learning.

SELF CHECK ANSWERS

1. a. 1.7 **b.** −6 **2. a.** 0.9 **b.** −2.5 **3.** −9.5 **4. a.** −2.2 **b.** −17.8 **5. a.** −21.6 **b.** 27 **6.** 10.4 **7.** 8.7 **8.** 10 **9.** $336 **10.** 42° **11.** −754.9646

NOW TRY THIS

1. Evaluate each expression.

a. $-2 - |5 - 8|$ **b.** $\dfrac{|6 - (-4)|}{|-1 - 9|}$

2. Determine the signs necessary to obtain the given value.

a. $\_\ 3 + (\_)5 = -2$
b. $\_\ 6 + (\_)8 = -14$
c. $\_\ 56 + (\_)24 = -32$

1.4 Exercises

WARM-UPS *Find each value.*

1. $2 + 3$
2. $2 + (-5)$
3. $-4 + 7$
4. $-5 + (-6)$
5. $6 - 2$
6. $-8 - 4$
7. $-5 - (-7)$
8. $12 - (-4)$

REVIEW *Simplify each expression.*

9. $5 + 3(7 - 2)$
10. $(5 + 3)(7 - 2)$
11. $5 + 3(7) - 2$
12. $(5 + 3)7 - 2$

VOCABULARY AND CONCEPTS *Fill in the blanks.*

13. Positive and negative numbers can be represented by ______ on the number line.

14. The numbers +5 and +8 and the numbers −5 and −8 are said to have ____ signs.

15. The numbers +7 and −9 are said to have ______ signs.

16. To find the sum of two real numbers with like signs, ____ their absolute values and ____ their common sign.

17. To find the sum of two real numbers with unlike signs, _______ their absolute values and use the sign of the number with the ______ absolute value.

18. $a - b =$ ________

19. To subtract a number, we ____ its ________.

20. $a +$ ____ $= (-a) +$ _ $= 0$.

GUIDED PRACTICE *Find each sum.* SEE EXAMPLE 1. *(OBJECTIVE 1)*

21. $5 + 9$
22. $(-6) + (-4)$
23. $(-7) + (-2)$
24. $(+4) + 11$
25. $\frac{1}{5} + \left(+\frac{1}{7}\right)$
26. $\left(-\frac{3}{4}\right) + \left(-\frac{1}{4}\right)$
27. $44.902 + 33.098$
28. $-421.377 + (-122.043)$

Find each sum. SEE EXAMPLE 2. *(OBJECTIVE 2)*

29. $7 + (-3)$
30. $8 + (-5)$
31. $(-0.4) + 0.9$
32. $(-1.2) + (-5.3)$
33. $\frac{2}{3} + \left(-\frac{1}{4}\right)$
34. $-\frac{1}{2} + \frac{1}{3}$
35. $73.82 + (-108.4)$
36. $-721.964 + (38.291)$

Evaluate each expression. SEE EXAMPLE 3. *(OBJECTIVES 1 AND 2)*

37. $5 + [4 + (-2)]$
38. $-2 + [(-5) + 3]$
39. $-2 + (-4 + 5)$
40. $5 + [-4 + (-6)]$
41. $(-7 + 5) + 2$
42. $-12 + (-2 + 10)$
43. $-9 + [-6 + (-4)]$
44. $-27 + [-12 + (-13)]$

Add vertically. SEE EXAMPLE 4. *(OBJECTIVES 1 AND 2)*

45. $\begin{array}{r} 5 \\ +\underline{-4} \end{array}$
46. $\begin{array}{r} -18 \\ +\underline{-11} \end{array}$
47. $\begin{array}{r} -1.3 \\ +\underline{\ 3.5} \end{array}$
48. $\begin{array}{r} 1.3 \\ +\underline{-2.5} \end{array}$

Find each difference. SEE EXAMPLE 5. *(OBJECTIVE 3)*

49. $8 - 4$
50. $-8 - 4$
51. $8 - (-4)$
52. $-8 - (-4)$
53. $0 - (-5)$
54. $0 - 75$
55. $\frac{5}{3} - \frac{7}{6}$
56. $-\frac{5}{9} - \frac{5}{3}$

Subtract vertically. SEE EXAMPLE 6. *(OBJECTIVE 3)*

57. $\begin{array}{r} 8 \\ \underline{-4} \end{array}$
58. $\begin{array}{r} 8 \\ -\underline{-3} \end{array}$
59. $\begin{array}{r} -10 \\ -\underline{-3} \end{array}$
60. $\begin{array}{r} -13 \\ -\underline{\ 5} \end{array}$

Simplify each expression. SEE EXAMPLE 7. (OBJECTIVE 3)

61. $5 - [(-2) - 4]$
62. $-3 - [5 - (-4)]$
63. $4 - [(-3) - 5]$
64. $(3 - 5) - [5 - (-3)]$

Simplify each expression. SEE EXAMPLE 8. (OBJECTIVE 3)

65. $\dfrac{5 - (-4)}{3 - (-6)}$
66. $\dfrac{2 + (-3)}{-3 - (-4)}$
67. $\dfrac{-6 - (-3)}{5 + (-8)}$
68. $\dfrac{2 + (-3)}{-3 - (-5)} + \dfrac{-4 + 1}{8 + (-6)}$

Use a calculator to evaluate each quantity. Round the answers to two decimal places. SEE EXAMPLE 11. (OBJECTIVE 5)

69. $4.26 - 6.34 + 0.56$
70. $6.34 - 0.56 - 4.26$
71. $(2.34)^2 - (3.47)^2 - (0.72)^2$
72. $(0.72)^2 - (2.34)^2 + (3.47)^3$

ADDITIONAL PRACTICE *Simplify each expression.*

73. $\left(\frac{5}{2} - 3\right) - \left(\frac{3}{2} - 5\right)$
74. $\left(\frac{7}{3} - \frac{5}{6}\right) - \left[\frac{5}{6} - \left(-\frac{7}{3}\right)\right]$
75. $(5.2 - 2.5) - (5.25 - 5)$
76. $(3.7 - 8.25) - (3.75 + 2.5)$
77. $4 + (-12)$
78. $11 + (-15)$
79. $[-4 + (-3)] + [2 + (-2)]$
80. $[3 + (-1)] + [-2 + (-3)]$
81. $-4 + (-3 + 2) + (-3)$
82. $5 + [2 + (-5)] + (-2)$
83. $-|8 + (-4)| + 7$
84. $\left|\frac{3}{5} + \left(-\frac{4}{5}\right)\right|$
85. $-5.2 + |-2.5 + (-4)|$
86. $6.8 + |8.6 + (-1.1)|$
87. $-3\frac{1}{2} - 5\frac{1}{4}$
88. $2\frac{1}{2} - \left(-3\frac{1}{2}\right)$
89. $-6.7 - (-2.5)$
90. $25.3 - 17.5$
91. $\dfrac{-4 - 2}{-[2 + (-3)]}$
92. $\dfrac{-3 + (-2)}{2 - (-1)} - \dfrac{1 - 7}{-4 - (-7)}$
93. $\left(\frac{3}{4} - \frac{4}{5}\right) - \left(\frac{2}{3} + \frac{1}{4}\right)$
94. $\left(3\frac{1}{2} - 2\frac{1}{2}\right) - \left[5\frac{1}{3} - \left(-5\frac{2}{3}\right)\right]$

APPLICATIONS *Use the appropriate signed numbers and operations for each.* SEE EXAMPLES 9–10. (OBJECTIVE 4)

95. College tuition A student owes $735 in tuition. If she is awarded a scholarship that will pay $500 of the bill, what will she still owe?

96. Dieting Scott weighed 212 pounds but lost 24 pounds during a three-month diet. What does Scott weigh now?

97. Temperatures The temperature rose 13 degrees in 1 hour and then dropped 4 degrees in the next hour. What signed number represents the net change in temperature?

98. Mountain climbing A team of mountaineers climbed 2,347 feet one day and then came down 597 feet to make camp. What signed number represents their net change in altitude?

99. Temperatures The temperature fell from zero to 14° below one night. By 5:00 P.M. the next day, the temperature had risen 10 degrees. What was the temperature at 5:00 P.M.?

100. History In 1897, Joseph Thompson discovered the electron. Fifty-four years later, the first fission reactor was built. Nineteen years before the reactor was erected, James Chadwick discovered the neutron. In what year was the neutron discovered?

101. History The Greek mathematician Euclid was alive in 300 BC. The English mathematician Sir Isaac Newton was alive in AD 1700. How many years apart did they live?

102. Banking A student deposited $415 in a new checking account, wrote a check for $176, and deposited another $212. Find the balance in his account.

103. Military history An army retreated 2,300 meters. After regrouping, it moved forward 1,750 meters. The next day it gained another 1,875 meters. What was the army's net gain?

104. Football A football player gained and lost the yardage shown in the illustration on six consecutive plays. How many total yards were gained or lost on the six plays?

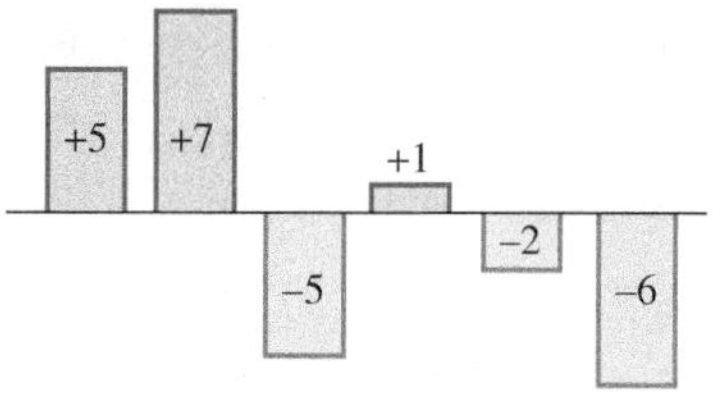

Gains and Losses

105. Aviation A pilot flying at 32,000 feet is instructed to descend to 28,000 feet. How many feet must he descend?

106. Stock market Tuesday's high and low prices for Transitronics stock were 37.125 and 31.625. Find the range of prices for this stock.

107. Temperatures Find the difference between a temperature of 32° above zero and a temperature of 27° above zero.

108. Temperatures Find the difference between a temperature of 3° below zero and a temperature of 21° below zero.

109. Stock market At the opening bell on Monday, the Dow Jones Industrial Average was 12,153. At the close, the Dow was down 23 points, but news of a half-point drop in interest rates on Tuesday sent the market up 57 points. What was the Dow average after the market closed on Tuesday?

110. Stock market On a Monday morning, the Dow Jones Industrial Average opened at 11,917. For the week, the Dow rose 29 points on Monday and 12 points on Wednesday. However, it fell 53 points on Tuesday and 27 points on both Thursday and Friday. Where did the Dow close on Friday?

111. Buying stock A woman owned 500 shares of Microsoft stock, bought another 500 shares on a price dip, and then sold 300 shares when the price rose. How many shares does she now own?

Unless otherwise noted, all content on this page is © Cengage Learning.

112. Small business Maria earned \$2,532 in a part-time business. However, \$633 of the earnings went for taxes. Find Maria's net earnings.

Use a calculator to help answer each question.

113. Balancing the books On January 1, Sally had \$437.45 in the bank. During the month, she had deposits of \$25.17, \$37.93, and \$45.26, and she had withdrawals of \$17.13, \$83.44, and \$22.58. How much was in her account at the end of the month?

114. Small business The owner of a small business has a gross income of \$97,345.32. However, he paid \$37,675.66 in expenses plus \$7,537.45 in taxes, \$3,723.41 in health-care premiums, and \$5,767.99 in pension payments. What was his profit?

115. Closing real estate transactions A woman sold her house for \$115,000. Her fees at closing were \$78 for preparing a deed, \$446 for title work, \$216 for revenue stamps, and a sales commission of \$7,612.32. In addition, there was a deduction of \$23,445.11 to pay off her old mortgage. As part of the deal, the buyer agreed to pay half of the title work. How much money did the woman receive after closing?

116. Winning the lottery Mike won \$500,000 in a state lottery. He will get $\frac{1}{20}$ of the sum each year for the next 20 years. After he receives his first installment, he plans to pay off a car loan of \$7,645.12 and give his son \$10,000 for college. By paying off the car loan, he will receive a rebate of 2% of the loan. If he must pay income tax of 28% on his first installment, how much will he have left to spend?

WRITING ABOUT MATH

117. Explain why the sum of two negative numbers is always negative, and the sum of two positive numbers is always positive.

118. Explain why the sum of a negative number and a positive number could be either negative or positive.

SOMETHING TO THINK ABOUT

119. Think of two numbers. First, add the absolute values of the two numbers, and write your answer. Second, add the two numbers, take the absolute value of that sum, and write that answer. Do the two answers agree? Can you find two numbers that produce different answers? When do you get answers that agree, and when don't you?

120. "Think of a very small number," requests the teacher. "One one-millionth," answers Charles. "Negative one million," responds Mia. Explain why either answer might be considered correct.

Section 1.5 Multiplying and Dividing Real Numbers

Objectives

1. Multiply two or more real numbers.
2. Divide two real numbers.
3. Use signed numbers and an operation to model an application.
4. Use a calculator to multiply or divide two real numbers.

Getting Ready

Find each product or quotient.

1. $8 \cdot 7$	**2.** $9 \cdot 6$	**3.** $8 \cdot 9$	**4.** $7 \cdot 9$
5. $\frac{81}{9}$	**6.** $\frac{48}{8}$	**7.** $\frac{64}{8}$	**8.** $\frac{56}{7}$

In this section, we will develop the rules for multiplying and dividing real numbers. We will see that the rules for multiplication and division are very similar.

1 Multiply two or more real numbers.

Because the times sign, $\times$, looks like the letter x, it is seldom used in algebra. Instead, we will use a dot, parentheses, or no symbol at all to denote multiplication. Each of the following expressions indicates the *product* obtained when two real numbers x and y are multiplied.

$$x \cdot y \quad (x)(y) \quad x(y) \quad (x)y \quad xy$$

To develop rules for multiplying real numbers, we rely on the definition of multiplication. The expression $5 \cdot 4$ indicates that 4 is to be used as a term in a sum five times.

$$5(4) = 4 + 4 + 4 + 4 + 4 = 20 \quad \text{Read 5(4) as "5 times 4."}$$

Likewise, the expression $5(-4)$ indicates that -4 is to be used as a term in a sum five times.

$$5(-4) = (-4) + (-4) + (-4) + (-4) + (-4) = -20 \quad \text{Read } 5(-4) \text{ as "5 times negative 4."}$$

If multiplying by a positive number indicates repeated addition, it is reasonable that multiplication by a negative number indicates repeated subtraction. The expression $(-5)4$, for example, means that 4 is to be used as a term in a repeated subtraction five times.

$$\begin{aligned}(-5)4 &= -(4) - (4) - (4) - (4) - (4)\\ &= (-4) + (-4) + (-4) + (-4) + (-4)\\ &= -20\end{aligned}$$

Likewise, the expression $(-5)(-4)$ indicates that -4 is to be used as a term in a repeated subtraction five times.

$$\begin{aligned}(-5)(-4) &= -(-4) - (-4) - (-4) - (-4) - (-4)\\ &= -(-4) + [-(-4)] + [-(-4)] + [-(-4)] + [-(-4)]\\ &= 4 + 4 + 4 + 4 + 4\\ &= 20\end{aligned}$$

The expression $0(-2)$ indicates that -2 is to be used zero times as a term in a repeated addition. Thus,

$$0(-2) = 0$$

Finally, the expression $(-3)(1) = -3$ suggests that the product of any number and 1 is the number itself.

The previous results suggest the following rules.

RULES FOR MULTIPLYING SIGNED NUMBERS

To multiply two real numbers, multiply their absolute values.

1. *If the numbers are positive*, the product is positive.
2. *If the numbers are negative*, the product is positive.
3. *If one number is positive and the other is negative*, the product is negative.
4. Any number multiplied by 0 is 0: $a \cdot 0 = 0 \cdot a = 0$
5. Any number multiplied by 1 is the number itself: $a \cdot 1 = 1 \cdot a = a$

EXAMPLE 1 Find each product: **a.** $4(-7)$ **b.** $(-5)(-4)$ **c.** $(-7)(6)$ **d.** $8(6)$ **e.** $(-3)^2$ **f.** $(-3)^3$ **g.** $(-3)(5)(-4)$ **h.** $(-4)(-2)(-3)$

Solution

a. $4(-7) = (-4 \cdot 7)$
$= -28$

b. $(-5)(-4) = +(5 \cdot 4)$
$= +20$

c. $(-7)(6) = -(7 \cdot 6)$
$= -42$

d. $8(6) = +(8 \cdot 6)$
$= +48$

e. $(-3)^2 = (-3)(-3)$
$= +9$

f. $(-3)^3 = (-3)(-3)(-3)$
$= 9(-3)$
$= -27$

g. $(-3)(5)(-4) = (-15)(-4)$
$= +60$

h. $(-4)(-2)(-3) = 8(-3)$
$= -24$

SELF CHECK 1 Find each product: **a.** $-7(5)$ **b.** $-12(-7)$ **c.** $(-5)^2$ **d.** $-2(-4)(-9)$

EXAMPLE 2 Evaluate: **a.** $2 + (-3)(4)$ **b.** $-3(2 - 4)$ **c.** $(-2)^2 - 3^2$ **d.** $-(-2)^2 + 4$

Solution

a. $2 + (-3)(4) = 2 + (-12)$
$= -10$

b. $-3(2 - 4) = -3[2 + (-4)]$
$= -3(-2)$
$= 6$

c. $(-2)^2 - 3^2 = 4 - 9$
$= -5$

d. $-(-2)^2 + 4 = -4 + 4$
$= 0$

SELF CHECK 2 Evaluate: **a.** $-4 - (-3)(5)$ **b.** $(-3.2)^2 - 2(-5)^3$

EXAMPLE 3 Find each product: **a.** $\left(-\frac{2}{3}\right)\left(-\frac{6}{5}\right)$ **b.** $\left(\frac{3}{10}\right)\left(-\frac{5}{9}\right)$

Solution

a. $\left(-\frac{2}{3}\right)\left(-\frac{6}{5}\right) = +\left(\frac{2}{3}\cdot\frac{6}{5}\right)$
$= \frac{2\cdot \overset{2}{\cancel{6}}}{\underset{1}{\cancel{3}}\cdot 5}$
$= \frac{4}{5}$

b. $\left(\frac{3}{10}\right)\left(-\frac{5}{9}\right) = -\left(\frac{3}{10}\cdot\frac{5}{9}\right)$
$= -\frac{\overset{1}{\cancel{3}}\cdot\overset{1}{\cancel{5}}}{\underset{2}{\cancel{10}}\cdot\underset{3}{\cancel{9}}}$
$= -\frac{1}{6}$

SELF CHECK 3 Evaluate: **a.** $\frac{3}{5}\left(-\frac{10}{9}\right)$ **b.** $-\left(\frac{15}{8}\right)\left(-\frac{16}{5}\right)$

2 Divide two real numbers.

Recall that the result in a division is called a *quotient.* We know that 8 divided by 4 has a quotient of 2, and 18 divided by 6 has a quotient of 3.

$\frac{8}{4} = 2$, because $2\cdot 4 = 8$ $\qquad$ $\frac{18}{6} = 3$, because $3\cdot 6 = 18$

These examples suggest that the following rule

$$\frac{a}{b} = c \quad \text{if and only if} \quad c\cdot b = a$$

is true for the division of any real number a by any nonzero real number b as illustrated in the following.

$\frac{+10}{+2} = +5$, because $(+5)(+2) = +10$.

$\frac{-10}{-2} = +5$, because $(+5)(-2) = -10$.

$$\frac{+10}{-2} = -5, \text{ because } (-5)(-2) = +10.$$

$$\frac{-10}{+2} = -5, \text{ because } (-5)(+2) = -10.$$

Furthermore,

$\frac{-10}{0}$ is undefined, because no number multiplied by 0 gives -10.

However,

$$\frac{0}{-10} = 0, \text{ because } 0(-10) = 0.$$

These examples suggest the rules for dividing real numbers.

RULES FOR DIVIDING SIGNED NUMBERS

To divide two real numbers, find the quotient of their absolute values.

1. *If the numbers are positive*, the quotient is positive.
2. *If the numbers are negative*, the quotient is positive.
3. *If one number is positive and the other is negative*, the quotient is negative.
4. $\frac{a}{0}$ is undefined.
5. $\frac{a}{1} = a$.
6. If $a \neq 0$, then $\frac{0}{a} = 0$.
7. If $a \neq 0$, then $\frac{a}{a} = 1$.

EXAMPLE 4 Find each quotient: **a.** $\frac{36}{18}$ **b.** $\frac{-44}{11}$ **c.** $\frac{27}{-9}$ **d.** $\frac{-64}{-8}$

Solution

a. $\frac{36}{18} = +\frac{36}{18} = 2$ The quotient of two numbers with like signs is positive.

b. $\frac{-44}{11} = -\frac{44}{11} = -4$ The quotient of two numbers with unlike signs is negative.

c. $\frac{27}{-9} = -\frac{27}{9} = -3$ The quotient of two numbers with unlike signs is negative.

d. $\frac{-64}{-8} = +\frac{64}{8} = 8$ The quotient of two numbers with like signs is positive.

COMMENT
$$\frac{-a}{b} = \frac{a}{-b} = -\frac{a}{b}$$

SELF CHECK 4 Find each quotient: **a.** $\frac{-72.6}{12.1}$ **b.** $\frac{-24.51}{-4.3}$

EXAMPLE 5 Evaluate: **a.** $\frac{16(-4)}{-(-64)}$ **b.** $\frac{(-4)^3(16)}{-64}$

Solution

a. $\frac{16(-4)}{-(-64)} = \frac{-64}{+64}$
$= -1$

b. $\frac{(-4)^3(16)}{-64} = \frac{(-64)(16)}{(-64)}$
$= 16$

SELF CHECK 5 Evaluate: $\frac{-64 + 16}{-(-4)^2}$

When multiplying or dividing three or more real numbers, we use the order of operations.

EXAMPLE 6 Evaluate: **a.** $\dfrac{(-50)(10)(-5)}{-50-5(-5)}$ **b.** $\dfrac{3(-50)(10)+2(10)(-5)}{2(-50+10)}$

Solution **a.** $\dfrac{(-50)(10)(-5)}{-50-5(-5)} = \dfrac{(-500)(-5)}{-50+25}$ Multiply.

$= \dfrac{2{,}500}{-25}$ Multiply and add.

$= -100$ Divide.

b. $\dfrac{3(-50)(10)+2(10)(-5)}{2(-50+10)} = \dfrac{-150(10)+(20)(-5)}{2(-40)}$ Multiply and add.

$= \dfrac{-1{,}500-100}{-80}$ Multiply

$= \dfrac{-1{,}600}{-80}$ Subtract.

$= 20$ Divide.

SELF CHECK 6 Evaluate: $\dfrac{2(-50)(10)-3(-5)-5}{3[10-(-5)]}$

3 Use signed numbers and an operation to model an application.

EXAMPLE 7 **STOCK REPORTS** In its annual report, a corporation reports its performance on a per-share basis. When a company with 35 million shares loses $2.3 million, find the per-share loss.

Solution A loss of $2.3 million can be represented by −2,300,000. Because there are 35 million shares, the per-share loss can be represented by the quotient $\frac{-2{,}300{,}000}{35{,}000{,}000}$.

$\dfrac{-2{,}300{,}000}{35{,}000{,}000} \approx -0.065714286$ Use a calculator.

The company lost about 6.6¢ per share.

SELF CHECK 7 If the company earns $1.5 million in the following year, find its per-share gain for that year.

4 Use a calculator to multiply or divide two real numbers.

A calculator can be used to multiply and divide positive and negative numbers. To evaluate $(-345.678)(-527.339)$, we enter these numbers and press these keys:

345.678 +/− × 527.339 +/− = Using a calculator with a +/− key

(−) 345.678 × (−) 527.339 **ENTER** Using a graphing calculator

The display will read 182289.4908.

To evaluate $\frac{-345.678}{-527.339}$, we enter these numbers and press these keys:

345.678 +/− ÷ 527.339 +/− = Using a calculator with a +/− key

(−) 345.678 ÷ (−) 527.339 **ENTER** Using a graphing calculator

The display will read 0.655513816.

SELF CHECK ANSWERS

1. a. −35 **b.** 84 **c.** 25 **d.** −72 **2. a.** 11 **b.** 260.24 **3. a.** $-\frac{2}{3}$ **b.** 6 **4. a.** −6 **b.** 5.7 **5.** 3 **6.** −22 **7.** about 4.3¢

NOW TRY THIS

Perform each operation. If the result is undefined, so indicate.

1. $-2 - 3(1 - 6)$
2. $-3^2 - 4(3)(-1)$
3. $\dfrac{5^2 - 2(6)(-1)}{45 - 5 \cdot 9}$

Determine the value of x that will make each fraction undefined.

4. $\dfrac{12}{x}$ 5. $\dfrac{7}{x + 1}$

1.5 Exercises

WARM-UPS *Find each product or quotient.*

1. 1(3) **2.** 2(5)
3. 2(3)(4) **4.** 5(3)(2)
5. $\dfrac{12}{6}$ **6.** $\dfrac{10}{2}$
7. $\dfrac{3(6)}{2}$ **8.** $\dfrac{2 \cdot 3}{6}$
9. $12 \div 4(3)$ **10.** $16 \div 2(4)$

REVIEW

11. A concrete block weighs $37\frac{1}{2}$ pounds. How much will 30 of these blocks weigh?
12. If one brick weighs 1.3 pounds, how much will 500 bricks weigh?
13. Evaluate: $3^3 - 8(3)^2$
14. Place <, =, or > in the box to make a true statement:
$-2(-3 + 4) \;\square\; -3[3 - (-4)]$

VOCABULARY AND CONCEPTS *Fill in the blanks.*

15. The product of two positive numbers is ______.
16. The product of a ______ number and a negative number is negative.
17. The product of two negative numbers is ______.
18. The quotient of a ______ number and a positive number is negative.
19. The quotient of two negative numbers is ______.
20. Any number multiplied by _ is 0.
21. $a \cdot 1 =$ _
22. The quotient $\frac{a}{0}$ is ______.
23. If $a \neq 0$, $\frac{0}{a} =$ _. **24.** If $a \neq 0$, $\frac{a}{a} =$ _.

GUIDED PRACTICE *Perform each operation.* SEE EXAMPLE 1. (OBJECTIVE 1)

25. (4)(9) **26.** (−5)(−6)
27. (−8)(−7) **28.** (9)(−6)
29. (−10)(+9) **30.** (−3)(11)
31. (−32)(−14) **32.** (−27)(14)
33. (−2)(3)(4) **34.** (5)(0)(−3)
35. $(-5)^2$ **36.** $(-2)^3$
37. $(-4)^3$ **38.** $(-6)^2$
39. (−3)(5)(−6) **40.** (−1)(−3)(−6)

Perform each operation. SEE EXAMPLE 2. (OBJECTIVE 1)

41. $2 + (-1)(-3)$ **42.** $-3 - (-1)(2)$
43. $(-1 + 2)(-3)$ **44.** $3[-2 - (-4)]$
45. $[-1 - (-3)][-1 + (-3)]$
46. $[2 + (-3)][-1 - (-3)]$
47. $2(-1)^2 - 3(-2)^2$
48. $(-1)^2(3) + (-3)(2)$

Perform each operation. SEE EXAMPLE 3. (OBJECTIVE 1)

49. $\left(\frac{2}{3}\right)(-36)$ **50.** $\left(-\frac{3}{4}\right)(12)$
51. $\left(-\frac{20}{3}\right)\left(-\frac{3}{5}\right)$ **52.** $\left(-\frac{2}{5}\right)\left(\frac{15}{2}\right)$

Perform each operation. SEE EXAMPLE 4. (OBJECTIVE 2)

53. $\dfrac{80}{-20}$ **54.** $\dfrac{-66}{33}$
55. $\dfrac{-110}{-55}$ **56.** $\dfrac{200}{40}$
57. $\dfrac{-120}{30}$ **58.** $\dfrac{-250}{-25}$
59. $\dfrac{320}{-16}$ **60.** $\dfrac{180}{-36}$

Perform each operation. SEE EXAMPLE 5. (OBJECTIVE 2)

61. $\dfrac{-3(6)}{-(-2)}$

62. $\dfrac{4(-3)^2}{-2}$

63. $\dfrac{(-2)^3(10)}{-(-5)}$

64. $\dfrac{-18}{-2(3)}$

Perform each operation. If the result is undefined, so indicate. SEE EXAMPLE 6. (OBJECTIVE 2)

65. $\dfrac{18-20}{-2}$

66. $\dfrac{16-2}{2-9}$

67. $\dfrac{-3(-2)(-4)}{-4-2(-5)}$

68. $\dfrac{2(15)^2-2}{-2^3+1}$

69. $\dfrac{6-3(2)^2}{-1(7-4)}$

70. $\dfrac{2(-25)(10)+4(5)(-5)}{5(125-25)}$

71. $\dfrac{-4(5)(2)+2(-10)(3)}{-2(-4)-8}$

72. $\dfrac{-5(-2)+4}{-4(2)+8}$

Use a calculator to evaluate each expression. Refer to the calculator tear out card for entering fractions. (OBJECTIVE 4)

73. $\dfrac{(-6)+4(-3)}{4-6}$

74. $\dfrac{4-2(4)(-3)+(-3)}{4-(-6)-3}$

75. $\dfrac{4(-6)^2(-3)+4^2(-6)}{2(-6)-2(-3)}$

76. $\dfrac{[4^2-2(-6)](-3)^2}{-4(-3)}$

ADDITIONAL PRACTICE *Simplify each expression.*

77. $-4\left(\dfrac{-3}{4}\right)$

78. $(5)\left(-\dfrac{2}{5}\right)$

79. $(-1)(2^3)$

80. $[2(-3)]^2$

81. $(-2)(-2)(-2)(-3)(-4)$

82. $(-5)(4)(3)(-2)(-1)$

83. $(2)(-5)(-6)(-7)$

84. $(-3)(-5)(-5)(-2)$

85. $(-7)^2$

86. $(-2)^3$

87. $-(-3)^2$

88. $-(-1)(-3)^2$

89. $(-1)^2[2-(-3)]$

90. $2^2[-1-(-3)]$

91. $-3(-1)-(-3)(2)$

92. $-1(2)(-3)+6$

93. $(-1)^3(-2)^2+(-3)^2$

94. $(-2)^3[3-(-5)]$

95. $\dfrac{4+(-12)}{(-2)^2-4}$

96. $\dfrac{-2(3)(4)}{3-1}$

97. $\dfrac{-2(5)(4)}{-3+1}$

98. $\dfrac{-3+2-(-10)}{4(-3)+2(6)}$

99. $\dfrac{1}{2}-\dfrac{2}{3}-\dfrac{3}{4}$

100. $-\dfrac{2}{3}+\dfrac{1}{2}+\dfrac{3}{4}$

101. $\dfrac{1}{2}-\dfrac{2}{3}$

102. $-\dfrac{2}{3}-\dfrac{3}{4}$

103. $\left(\dfrac{1}{2}-\dfrac{2}{3}\right)\left(\dfrac{1}{2}+\dfrac{2}{3}\right)$

104. $\left(\dfrac{1}{2}+\dfrac{3}{4}\right)\left(\dfrac{1}{2}-\dfrac{3}{4}\right)$

105. $\left(\dfrac{1}{4}-\dfrac{2}{3}\right)\left(\dfrac{3}{4}-\dfrac{1}{3}\right)$

106. $\left(\dfrac{2}{5}-\dfrac{1}{4}\right)\left(\dfrac{1}{5}-\dfrac{3}{4}\right)$

APPLICATIONS *Use signed numbers and one or more operations to answer each question.* SEE EXAMPLE 7. (OBJECTIVE 3)

107. **Loss of revenue** A manufacturer's website normally produces sales of \$425 per hour, but was offline for 12 hours due to a systems virus. How much revenue was lost?

108. **Mowing lawns** Justin worked all day mowing lawns and was paid \$8 per hour. If he had \$94 at the end of an 8-hour day, how much did he have before he started working?

109. **Temperatures** Suppose that the temperature is dropping at the rate of 3 degrees each hour. If the temperature has dropped 18 degrees, what signed number expresses how many hours the temperature has been falling?

110. **Dieting** A man lost 37.5 pounds. If he lost 2.5 pounds each week, how long has he been dieting?

111. **Inventories** A spreadsheet is used to record inventory losses at a warehouse. The items, their cost, and the number missing are listed in the table.

a. Find the value of the lost MP3 players.
b. Find the value of the lost cell phones.
c. Find the value of the lost GPS systems.
d. Find the total losses.

	A	B	C	D
	Item	**Cost**	**Number of units**	**\$ Losses**
1	MP3 player	75	−32	
2	Cell phone	57	−17	
3	GPS system	87	−12	

112. **Toy inventories** A spreadsheet is used to record inventory losses at a warehouse. The item, the number of units, and the dollar losses are listed in the table.

a. Find the cost of a truck.
b. Find the cost of a drum.
c. Find the cost of a ball.

	A	B	C	D
	Item	**Cost**	**Number of units**	**\$ Losses**
1	Truck		−12	−\$60
2	Drum		−7	−\$49
3	Ball		−13	−\$39

Use a calculator to help answer each question.

113. **Stock market** Over a 7-day period, the Dow Jones Industrial Average had gains of 26, 35, and 17 points. In that period, there were also losses of 25, 31, 12, and 24 points. What is the average daily performance over the 7-day period?

114. **Astronomy** Light travels at the rate of 186,000 miles per second. How long will it take light to travel from the Sun to Venus? (*Hint*: The distance from the Sun to Venus is

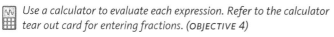

115. **Saving for school** A student has saved \$15,000 to attend graduate school. If she estimates that her expenses will be \$613.50 a month while in school, does she have enough to complete an 18-month master's degree program?

116. **Earnings per share** Over a five-year period, a corporation reported profits of \$19 million, \$15 million, and \$12 million. It also reported losses of \$11 million and \$39 million. What is the average gain (or loss) each year?

WRITING ABOUT MATH

117. Explain how you would decide whether the product of several numbers is positive or negative.

118. Describe two situations in which negative numbers are useful.

SOMETHING TO THINK ABOUT

119. If the quotient of two numbers is undefined, what would their product be?

120. If the product of five numbers is negative, how many of the factors could be negative?

121. If x^5 is a negative number, can you determine whether x is also negative?

122. If x^6 is a positive number, can you determine whether x is also positive?

Section 1.6 Algebraic Expressions

Objectives

1. Translate an English phrase into an algebraic expression.
2. Evaluate an algebraic expression when given values for its variables.
3. Identify the number of terms in an algebraic expression and identify the numerical coefficient of each term.

Vocabulary

algebraic expression | constant | term | numerical coefficient

Getting Ready

Identify each of the following as a sum, difference, product, or quotient.

1. $x + 3$
2. $57x$
3. $\frac{x}{9}$
4. $19 - y$
5. $\frac{x - 7}{3}$
6. $x - \frac{7}{3}$
7. $5(x + 2)$
8. $5x + 10$

Algebraic expressions are a fundamental concept in the study of algebra. They convey mathematical operations and are the building blocks of many equations, the main topic of the next chapter.

1 Translate an English phrase into an algebraic expression.

Variables and numbers can be combined with the operations of arithmetic to produce algebraic expressions. For example, if x and y are variables, the **algebraic expression** $x + y$ represents the sum of x and y, and the algebraic expression $x - y$ represents their difference.

There are many other ways to express addition or subtraction with algebraic expressions, as shown in Tables 1-3 and 1-4.

TABLE 1-3

The phrase	translates into the algebraic expression
the *sum* of t and 12	$t + 12$
5 *plus s*	$5 + s$
7 *added to a*	$a + 7$
10 *more than q*	$q + 10$
12 *greater than m*	$m + 12$
l increased by m	$l + m$
exceeds p by 50	$p + 50$

TABLE 1-4

The phrase	translates into the algebraic expression
the *difference* of 50 and r	$50 - r$
1,000 *minus q*	$1{,}000 - q$
15 *less than w*	$w - 15$
t decreased by q	$t - q$
12 *reduced by m*	$12 - m$
l subtracted from 250	$250 - l$
2,000 *less p*	$2{,}000 - p$

EXAMPLE 1 Let x represent a certain number. Write an expression that represents

a. the number that is 5 more than x **b.** the number 12 decreased by x.

Solution **a.** The number "5 more than x" is the number found by adding 5 to x. It is represented by $x + 5$.

b. The number "12 decreased by x" is the number found by subtracting x from 12. It is represented by $12 - x$.

SELF CHECK 1 Let y represent a certain number. Write an expression that represents y increased by 25.

EXAMPLE 2 **INCOME TAXES** Bob worked x hours preparing his income tax return. He worked 3 hours less than that on his son's return. Write an expression that represents

a. the number of hours he spent preparing his son's return

b. the total number of hours he worked.

Solution **a.** Because he worked x hours on his own return and 3 hours less on his son's return, he worked $(x - 3)$ hours on his son's return.

b. Because he worked x hours on his own return and $(x - 3)$ hours on his son's return, the total time he spent on taxes was $[x + (x - 3)]$ hours.

SELF CHECK 2 Javier deposited \$$d$ in a bank account. Later, he withdrew \$500. Write an expression that represents the number of dollars in his account.

There are several ways to indicate the product of two numbers with algebraic expressions, as shown in Table 1-5.

TABLE 1-5

The phrase	translates into the algebraic expression
the *product* of a and b	ab
25 *times B*	$25B$
twice x	$2x$
$\frac{1}{2}$ *of z*	$\frac{1}{2}z$
12 *multiplied by m*	$12m$

EXAMPLE 3 Let x represent a certain number. Denote a number that is

a. twice as large as x **b.** 5 more than 3 times x **c.** 4 less than $\frac{1}{2}$ of x.

Solution **a.** The number "twice as large as x" is found by multiplying x by 2. It is represented by $2x$.

b. The number "5 more than 3 times x" is found by adding 5 to the product of 3 and x. It is represented by $3x + 5$.

c. The number "4 less than $\frac{1}{2}$ of x" is found by subtracting 4 from the product of $\frac{1}{2}$ and x. It is represented by $\frac{1}{2}x - 4$.

SELF CHECK 3 Find the product of 40 and t.

EXAMPLE 4 **STOCK VALUATIONS** Jim owns x shares of Transitronics stock, valued at \$29 a share; y shares of Positone stock, valued at \$32 a share; and 300 shares of Baby Bell, valued at \$42 a share.

a. How many shares of stock does he own?

b. What is the value of his stock?

Solution **a.** Because there are x shares of Transitronics, y shares of Positone, and 300 shares of Baby Bell, his total number of shares is $x + y + 300$.

b. The value of x shares of Transitronics is \$29x, the value of y shares of Positone is \$32y, and the value of 300 shares of Baby Bell is \$42(300). The total value of the stock is $\$(29x + 32y + 12{,}600)$.

SELF CHECK 4 If water softener salt costs \$$p$ per bag, find the cost of 25 bags.

There are also several ways to indicate the quotient of two numbers with algebraic expressions, as shown in Table 1-6.

TABLE 1-6

The phrase	translates into the algebraic expression
the *quotient* of 470 and A	$\frac{470}{A}$
B *divided by* 9	$\frac{B}{9}$
the *ratio* of h *to* 5	$\frac{h}{5}$
x *split into* 5 equal parts	$\frac{x}{5}$

EXAMPLE 5 Let x and y represent two numbers. Write an algebraic expression that represents the sum obtained when 3 times the first number is added to the quotient obtained when the second number is divided by 6.

Solution Three times the first number x is denoted as $3x$. The quotient obtained when the second number y is divided by 6 is the fraction $\frac{y}{6}$. Their sum is expressed as $3x + \frac{y}{6}$.

SELF CHECK 5 If the cost c of a meal is split equally among 4 people, what is each person's share?

EXAMPLE 6 **CUTTING ROPES** A 5-foot section is cut from the end of a rope that is l feet long. If the remaining rope is divided into three equal pieces, find an expression for the length of each of the equal pieces.

Solution After a 5-foot section is cut from one end of l feet of rope, the rope that remains is $(l - 5)$ feet long. When that remaining rope is cut into 3 equal pieces, each piece will be $\frac{l-5}{3}$ feet long. See Figure 1-24.

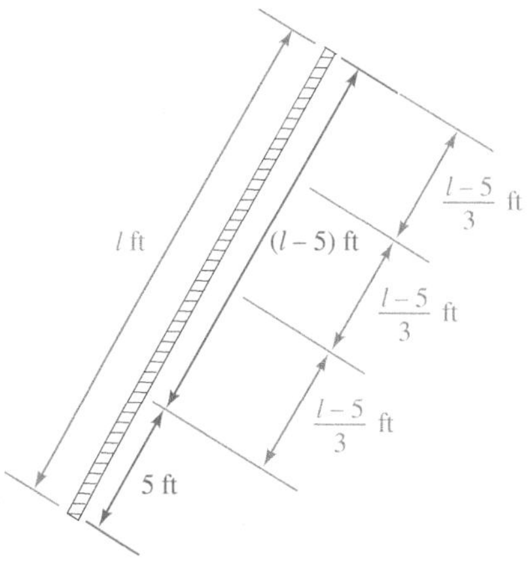

Figure 1-24

SELF CHECK 6 If a 7-foot section is cut from a rope that is l feet long and the remaining rope is divided into two equal pieces, find an expression for the length of each piece.

2 Evaluate an algebraic expression when given values for its variables.

Since variables represent numbers, algebraic expressions also represent numbers. We can evaluate algebraic expressions when we know the values of the variables.

EXAMPLE 7 If $x = 8$ and $y = 10$, evaluate: **a.** $x + y$ **b.** $y - x$ **c.** $3xy$ **d.** $\frac{5x}{y - 5}$

Solution We substitute 8 for x and 10 for y in each expression and simplify.

a. $x + y = \mathbf{8} + \mathbf{10}$
$= 18$

b. $y - x = \mathbf{10} - \mathbf{8}$
$= 2$

c. $3xy = (3)(\mathbf{8})(\mathbf{10})$
$= (24)(10)$ Do the multiplications from left to right.
$= 240$

d. $\frac{5x}{y - 5} = \frac{5(\mathbf{8})}{\mathbf{10} - 5}$
$= \frac{40}{5}$ Simplify the numerator and the denominator separately.
$= 8$ Simplify the fraction.

COMMENT When substituting a number for a variable in an expression, it is a good idea to write the number within parentheses. This will avoid mistaking 5(8) for 58.

SELF CHECK 7 If $a = 2$ and $b = 5$, evaluate: $\frac{6b + 18}{a + 2b}$

Unless otherwise noted, all content on this page is © Cengage Learning.

EXAMPLE 8 If $x = -4$, $y = 8$, and $z = -6$, evaluate: **a.** $\dfrac{7x^2y}{2(y - z)}$ **b.** $\dfrac{3xz^2}{y(x + z)}$

Solution We substitute -4 for x, 8 for y, and -6 for z in each expression and simplify.

a. $\dfrac{7x^2y}{2(y - z)} = \dfrac{7(-4)^2(8)}{2[8 - (-6)]}$

$= \dfrac{7(16)(8)}{2(14)}$ $\quad (-4)^2 = 16; 8 - (-6) = 14$

$= \dfrac{\overset{1}{\cancel{7}}(\overset{1}{\cancel{2}})(\overset{1}{\cancel{2}})(4)(8)}{\underset{1}{\cancel{2}}(\underset{1}{\cancel{2}})(\underset{1}{\cancel{7}})}$ $\quad$ Factor the numerator and denominator and divide out all common factors.

$= 32$ $\quad 4 \cdot 8 = 32$

b. $\dfrac{3xz^2}{y(x + z)} = \dfrac{3(-4)(-6)^2}{8[-4 + (-6)]}$

$= \dfrac{3(-4)(36)}{8(-10)}$ $\quad (-6)^2 = 36; -4 + (-6) = -10$

$= \dfrac{3(\overset{1}{\cancel{-2}})(\overset{1}{\cancel{2}})(\overset{1}{\cancel{4}})(9)}{\underset{1}{\cancel{2}}(\underset{1}{\cancel{4}})(\underset{1}{\cancel{-2}})(5)}$ $\quad$ Factor the numerator and denominator and divide out all common factors.

$= \dfrac{27}{5}$ $\quad 3(9) = 27; 1(5) = 5$

SELF CHECK 8 If $a = -3$, $b = -2$, and $c = -5$, evaluate: $\dfrac{b(a + c^2)}{abc}$

3 Identify the number of terms in an algebraic expression and identify the numerical coefficient of each term.

Numbers without variables, such as 7, 21, and 23, are called **constants**. Expressions such as 37, xyz, and $32t$, which are constants, variables, or products of constants and variables, are called algebraic **terms**.

- The expression $3x + 5y$ contains two terms. The first term is $3x$, and the second term is $5y$.
- The expression $xy + (-7)$ contains two terms. The first term is xy, and the second term is -7.
- The expression $3 + x + 2y$ contains three terms. The first term is 3, the second term is x, and the third term is $2y$.

Numbers and variables that are part of a product are called factors. For example,

- The product $7x$ has two factors, which are 7 and x.
- The product $-3xy$ has three factors, which are -3, x, and y.
- The product $\frac{1}{2}abc$ has four factors, which are $\frac{1}{2}$, a, b, and c.

The number factor of a product is called its **numerical coefficient**. The numerical coefficient (or just the *coefficient*) of $7x$ is 7. The coefficient of $-3xy$ is -3, and the coefficient of $\frac{1}{2}abc$ is $\frac{1}{2}$. The coefficient of terms such as x, ab, and rst is understood to be 1.

$$x = 1x, \qquad ab = 1ab, \qquad \text{and} \qquad rst = 1rst$$

EXAMPLE 9

a. The expression $5x + y$ has two terms. The coefficient of its first term is 5. The coefficient of its second term is 1.

b. The expression $-17wxyz$ has one term, which contains the five factors -17, w, x, y, and z. Its coefficient is -17.

c. The expression 37 has one term, the constant 37.

d. The expression $3x^2 - 2x$ has two terms. The coefficient of the first term is 3. Since $3x^2 - 2x$ can be written as $3x^2 + (-2x)$, the coefficient of the second term is -2.

SELF CHECK 9 How many terms does the expression $3x^2 - 2x + 7$ have? Find the sum of the coefficients.

SELF CHECK ANSWERS

1. $y + 25$ **2.** $d - 500$ **3.** $40t$ **4.** $\$25p$ **5.** $\frac{c}{4}$ **6.** $\frac{l-7}{2}$ ft **7.** 4 **8.** $\frac{22}{15}$ **9.** 3; 8

NOW TRY THIS

If $a = -2$, $b = -1$, and $c = 8$, evaluate each expression.

1. $3a^2$
2. $\dfrac{a-b}{c-a}$
3. $b^2 - 4ac$
4. Write $2(a - b)$ as an English phrase.

1.6 Exercises

WARM-UPS *Identify each as a sum, difference, product, or quotient.*

1. 5 more than p
2. twice y
3. $\frac{1}{2}$ of q
4. 6 less than y
5. the ratio of 7 to z
6. x divided by 12
7. 8 decreased by x
8. r greater than 9

REVIEW *Evaluate each expression.*

9. $0.14 \cdot 3{,}800$
10. $\dfrac{3}{5} \cdot 4{,}765$
11. $\dfrac{-4 + (7 - 9)}{(-9 - 7) + 4}$
12. $\dfrac{5}{4}\left(1 - \dfrac{3}{5}\right)$

VOCABULARY AND CONCEPTS *Fill in the blanks.*

13. The answer to an addition problem is called a ____.
14. The answer to a __________ problem is called a difference.
15. The answer to a ____________ problem is called a product.
16. The answer to a division problem is called a ________.
17. An ________ expression is a combination of variables, numbers, and the operation symbols for addition, subtraction, multiplication, or division.
18. To _______ an algebraic expression, we substitute values for the variables and simplify.
19. A ____ is the product of constants and/or variables and the numerical part is called the _________.
20. Terms that have no variables are called ________.

GUIDED PRACTICE *Let x and y represent two real numbers. Write an algebraic expression to denote each quantity.* SEE EXAMPLE 1. *(OBJECTIVE 1)*

21. The sum of x and y
22. The sum of twice x and twice y
23. The number that is 3 less than x
24. The difference obtained when twice x is subtracted from y

Let x, y, and z represent three real numbers. Write an algebraic expression to denote each quantity. SEE EXAMPLE 3. *(OBJECTIVE 1)*

25. The product of twice x and y
26. The product of x and twice y
27. The product of 3, x, and y
28. The product of 3 and $2z$

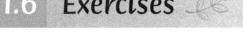

Let x, y, and z represent three real numbers. Write an algebraic expression to denote each quantity. Assume that no denominators are 0. SEE EXAMPLE 5. (OBJECTIVE 1)

29. The quotient obtained when y is divided by x

30. The quotient obtained when the sum of x and y is divided by y

31. The quotient obtained when the product of 3 and z is divided by the product of 4 and x

32. The quotient obtained when the sum of x and y is divided by the sum of y and z

Evaluate each expression if $x = -2$, $y = 5$, and $z = -3$. SEE EXAMPLES 7–8. (OBJECTIVE 2)

33. $x + y$

34. $x - z$

35. $4xyz$

36. $2x^2z$

37. $\frac{x^2y}{z - 1}$

38. $\frac{xy - 2}{z}$

39. $\frac{4z^2y}{3(x - z)}$

40. $\frac{x + y + z}{4y^2x}$

41. $\frac{x(y + z) - 25}{(x + z)^2 - y^2}$

42. $\frac{(x + y)(y + z)}{x + z + y}$

43. $\frac{3(x + z^2) + 4}{y(x - z)}$

44. $\frac{x(y^2 - 2z) - 1}{z(y - x^2)}$

Give the number of terms in each algebraic expression and also give the numerical coefficient of the first term. SEE EXAMPLE 9. (OBJECTIVE 3)

45. $-7c$

46. $4c - 9d$

47. $-xy - 5z + 8$

48. cd

49. $-3xy + yz - zw + 5$

50. $-2xyz + cde - 14$

51. $9abc - 5ab - c$

52. $5uvw - 4uv + 8uw$

53. $5x - 4y + 3z + 2$

54. $7abc - 9ab + 2bc + a - 1$

ADDITIONAL PRACTICE *Let x, y, and z represent three real numbers. Write an algebraic expression to denote each quantity. Assume that no denominators are 0.*

55. The sum obtained when the quotient of x divided by y is added to z

56. z decreased by 3

57. z less the product of x and y

58. z less than the product of x and y

59. The quotient obtained when the product of x and y is divided by the sum of x and z

60. The sum of the product xy and the quotient obtained when y is divided by z

61. The number obtained when x decreased by 4 is divided by the product of 3 and y

62. The number obtained when $2z$ minus $5y$ is divided by the sum of x and $3y$

Let x, y, and z represent three real numbers. Write each algebraic expression as an English phrase. Assume that no denominators are 0.

63. $y + 4$

64. $x - 5$

65. $xy(x + y)$

66. $(x + y + z)(xyz)$

67. $\frac{x + 2}{z}$

68. $5 + \frac{y}{z}$

69. $\frac{y}{z}$

70. xy

71. $2xy$

72. $\frac{x + y}{2}$

73. $\frac{5}{x + y}$

74. $\frac{3x}{y + z}$

Let $x = 8$, $y = 4$, and $z = 2$. Write each phrase as an algebraic expression, and evaluate it. Assume that no denominators are 0.

75. The sum of x and z

76. The product of x, y, and z

77. z less than y

78. The quotient obtained when y is divided by z

79. 3 less than the product of y and z

80. 7 less than the sum of x and y

81. The quotient obtained when the product of x and y is divided by z

82. The quotient obtained when 10 greater than x is divided by z

Consider the algebraic expression $29xyz + 23xy + 19x$.

83. What are the factors of the third term?

84. What are the factors of the second term?

85. What factor is common to the first and third terms?

86. What factor is common to all three terms?

Consider the algebraic expression $3xyz + 5xy + 17xz$.

87. What are the factors of the first term?

88. What are the factors of the second term?

89. What are the factors of the third term?

90. What factor is common to all three terms?

Consider the algebraic expression $5xy + yt + 8xyt$.

91. Find the numerical coefficients of each term.

92. What factor is common to all three terms?

93. What factors are common to the first and third terms?

94. What factors are common to the second and third terms?

Consider the algebraic expression $3xy + y + 25xyz$.

95. Use the numerical coefficient of each term to find their product.

96. Use the numerical coefficient of each term to find their sum.

97. What factors are common to the first and third terms?

98. What factor is common to all three terms?

APPLICATIONS

Write an algebraic expression to denote each quantity. Assume that no denominators are 0. SEE EXAMPLES 2, 4, AND 6. (OBJECTIVE 1)

99. Course loads A man enrolls in college for c hours of credit, and his sister enrolls for 6 more hours than her brother. Write an expression that represents the number of hours the sister is taking.

100. Antique cars An antique Ford has 25,000 more miles on its odometer than a newer car. If the newer car has traveled m miles, find an expression that represents the mileage on the Ford.

101. Heights of trees

a. If h represents the height (in feet) of the oak tree, write an expression that represents the height of the crab apple tree.

b. If c represents the height (in feet) of the crab apple tree, write an expression that represents the height of the oak.

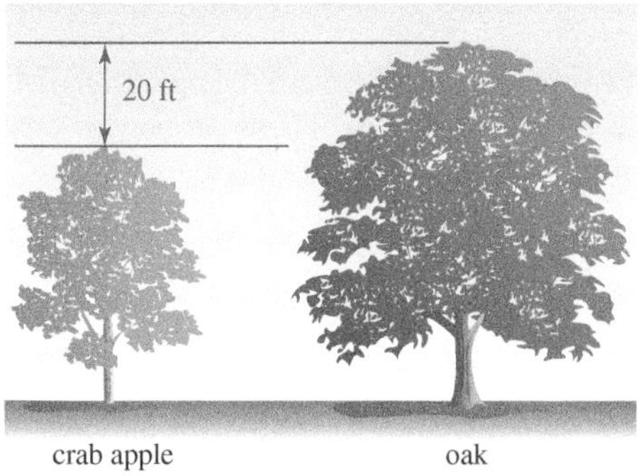

102. T-bills Write an expression that represents the value of t T-bills, each worth \$9,987.

103. Real estate Write an expression that represents the value of n vacant lots if each lot is worth \$35,000.

104. Cutting ropes A rope x feet long is cut into 5 equal pieces. Find an expression for the length of each piece.

105. Invisible tape If x inches of tape have been used off the roll shown below, how many inches of tape are left on the roll?

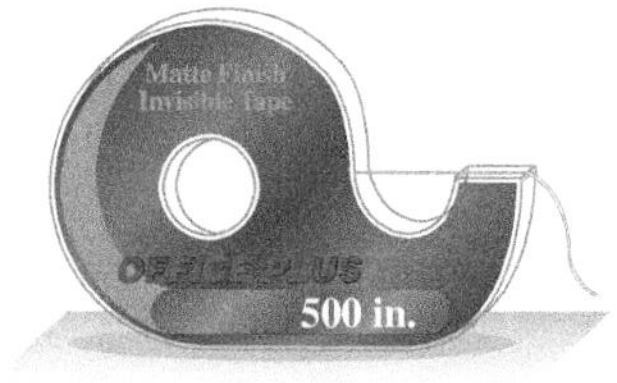

106. Plumbing A plumber cuts a pipe that is 12 feet long into x equal pieces. Find an expression for the length of each piece.

107. Comparing assets A girl had d dollars, and her brother had \$5 more than three times that amount. How much did the brother have?

108. Comparing investments Wendy has x shares of stock. Her sister has 2 fewer shares than twice Wendy's shares. How many shares does her sister have?

109. Sorting records In electronic data processing, the process of sorting records into sequential order is a common task. One sorting technique, called a **selection sort**, requires C comparisons to sort N records, where C and N are related by the formula

$$C = \frac{N(N-1)}{2}$$

How many comparisons are necessary to sort 10,000 records?

110. Sorting records How many comparisons are necessary to sort 50,000 records? See Exercise 109.

WRITING ABOUT MATH

111. Distinguish between the meanings of these two phrases: "3 less than x" and "3 is less than x."

112. Distinguish between *factor* and *term*.

113. What is the purpose of using variables? Why aren't ordinary numbers enough?

114. In words, xy is "the product of x and y." However, $\frac{x}{y}$ is "the quotient obtained when x is divided by y." Explain why the extra words are needed.

SOMETHING TO THINK ABOUT

115. If the value of x were doubled, what would happen to the value of $37x$?

116. If the values of both x and y were doubled, what would happen to the value of $5xy^2$?

Unless otherwise noted, all content on this page is © Cengage Learning.

Section 1.7 Properties of Real Numbers

Objectives

1. Apply the closure properties by evaluating an expression for given values for variables.
2. Apply the commutative and associative properties.
3. Apply the distributive property of multiplication over addition to rewrite an expression.
4. Recognize the identity elements and find the additive and multiplicative inverse of a nonzero real number.
5. Identify the property that justifies a given statement.

Vocabulary

closure properties
commutative properties
associative properties
distributive property
identity elements
additive inverse
reciprocal
multiplicative inverse

Getting Ready

Perform each operation.

1. $3 + (5 + 9)$
2. $(3 + 5) + 9$
3. $23.7 + 14.9$
4. $14.9 + 23.7$
5. $7(5 + 3)$
6. $7 \cdot 5 + 7 \cdot 3$
7. $125.3 + (-125.3)$
8. $125.3\left(\frac{1}{125.3}\right)$
9. $777 + 0$
10. $777 \cdot 1$

To understand algebra, we must know the properties that govern the operations of addition, subtraction, multiplication, and division of real numbers. These properties enable us to write expressions in equivalent forms, often making our work easier.

1 Apply the closure properties by evaluating an expression for given values for variables.

The **closure properties** guarantee that the sum, difference, product, or quotient (except for division by 0) of any two real numbers is also a real number.

CLOSURE PROPERTIES

If a and b are real numbers, then

$a + b$ is a real number. $a - b$ is a real number.

ab is a real number. $\frac{a}{b}$ is a real number $(b \neq 0)$.

EXAMPLE 1 Let $x = 8$ and $y = -4$. Find the real-number answers to show that each expression represents a real number.

a. $x + y$ **b.** $x - y$ **c.** xy **d.** $\frac{x}{y}$

Solution We substitute 8 for x and -4 for y in each expression and simplify.

a. $x + y = 8 + (-4)$
$= 4$

b. $x - y = 8 - (-4)$
$= 8 + 4$
$= 12$

c. $xy = 8(-4)$
$= -32$

d. $\frac{x}{y} = \frac{8}{-4}$
$= -2$

SELF CHECK 1 Let $a = -6$ and $b = 3$. Find the real-number answers to show that each expression represents a real number.
a. $a - b$ **b.** $\frac{a}{b}$

2 Apply the commutative and associative properties.

The **commutative properties** (from the word *commute*, which means to go back and forth) guarantee that addition or multiplication of two real numbers can be done in either order.

COMMUTATIVE PROPERTIES

If a and b are real numbers, then

$a + b = b + a$	commutative property of addition
$ab = ba$	commutative property of multiplication

EXAMPLE 2 Let $x = -3$ and $y = 7$. Show that **a.** $x + y = y + x$ **b.** $xy = yx$

Solution
a. We can show that the sum $x + y$ is the same as the sum $y + x$ by substituting -3 for x and 7 for y in each expression and simplifying.

$$x + y = -3 + 7 = 4 \quad \text{and} \quad y + x = 7 + (-3) = 4$$

b. We can show that the product xy is the same as the product yx by substituting -3 for x and 7 for y in each expression and simplifying.

$$xy = -3(7) = -21 \quad \text{and} \quad yx = 7(-3) = -21$$

COMMENT Since $5 - 3 \neq 3 - 5$ and $5 \div 3 \neq 3 \div 5$, the commutative property cannot be applied to a subtraction or a division.

SELF CHECK 2 Let $a = 6$ and $b = -5$. Show that
a. $a + b = b + a$
b. $ab = ba$

The **associative properties** guarantee that three real numbers can be regrouped in an addition or multiplication.

ASSOCIATIVE PROPERTIES

If a, b, and c are real numbers, then

$(a + b) + c = a + (b + c)$	associative property of addition
$(ab)c = a(bc)$	associative property of multiplication

Because of the associative property of addition, we can group (or associate) the numbers in a sum in any way that we wish. For example,

$$(3+4)+5 = 7+5 \qquad\qquad 3+(4+5) = 3+9$$
$$= 12 \qquad\qquad\qquad\qquad = 12$$

The answer is 12 regardless of how we group the three numbers.

COMMENT Since $(2-5)-3 \neq 2-(5-3)$ and $(2 \div 5) \div 3 \neq 2 \div (5 \div 3)$, the associative property cannot be applied to subtraction or division.

The associative property of multiplication permits us to group (or associate) the numbers in a product in any way that we wish. For example,

$$(3 \cdot 4) \cdot 7 = 12 \cdot 7 \qquad\qquad 3 \cdot (4 \cdot 7) = 3 \cdot 28$$
$$= 84 \qquad\qquad\qquad\qquad = 84$$

The answer is 84 regardless of how we group the three numbers.

3 Apply the distributive property of multiplication over addition to rewrite an expression.

The **distributive property** shows how to multiply the sum of two numbers by a third number. Because of this property, we can often add first and then multiply, or multiply first and then add.

For example, $2(3+7)$ can be calculated in two different ways. We will add and then multiply, or we can multiply each number within the parentheses by 2 and then add.

$$2(3+7) = 2(10) \qquad\qquad 2(3+7) = 2 \cdot 3 + 2 \cdot 7$$
$$= 20 \qquad\qquad\qquad\qquad = 6 + 14$$
$$= 20$$

Either way, the result is 20.

In general, we have the following property.

DISTRIBUTIVE PROPERTY OF MULTIPLICATION OVER ADDITION

If a, b, and c are real numbers, then

$$a(b+c) = ab + ac$$

Because multiplication is commutative, the distributive property also can be written in the form

$$(b+c)a = ba + ca$$

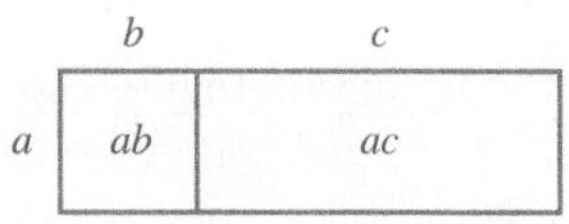

Figure 1-25

We can interpret the distributive property geometrically. Since the area of the largest rectangle in Figure 1-25 is the product of its width a and its length $b+c$, its area is $a(b+c)$. The areas of the two smaller rectangles are ab and ac. Since the area of the largest rectangle is equal to the sum of the areas of the smaller rectangles, we have $a(b+c) = ab + ac$.

The previous discussion shows that multiplication distributes over addition. Multiplication also distributes over subtraction. For example, $2(3-7)$ can be calculated in two different ways. We will subtract and then multiply, or we can multiply each number within the parentheses by 2 and then subtract.

$$2(3-7) = 2(-4) \qquad\qquad 2(3-7) = 2 \cdot 3 - 2 \cdot 7$$
$$= -8 \qquad\qquad\qquad\qquad = 6 - 14$$
$$= -8$$

Either way, the result is -8. In general, we have

$$a(b-c) = ab - ac$$

Unless otherwise noted, all content on this page is © Cengage Learning.

EXAMPLE 3 Evaluate each expression in two different ways:

a. $3(5 + 9)$ **b.** $4(6 - 11)$ **c.** $-2(-7 + 3)$

Solution

a. $3(5 + 9) = 3(14) = 42$

$3(5 + 9) = 3 \cdot 5 + 3 \cdot 9 = 15 + 27 = 42$

b. $4(6 - 11) = 4(-5) = -20$

$4(6 - 11) = 4 \cdot 6 - 4 \cdot 11 = 24 - 44 = -20$

c. $-2(-7 + 3) = -2(-4) = 8$

$-2(-7 + 3) = -2(-7) + (-2)(3) = 14 + (-6) = 8$

SELF CHECK 3 Evaluate $-5(-7 + 20)$ in two different ways.

The distributive property can be extended to three or more terms. For example, if a, b, c, and d are real numbers, then

$$a(b + c + d) = ab + ac + ad$$

EXAMPLE 4 Write $3.2(x + y + 2.7)$ without using parentheses.

Solution

$$3.2(x + y + 2.7) = 3.2x + 3.2y + (3.2)(2.7) \quad \text{Distribute the multiplication by 3.2.}$$
$$= 3.2x + 3.2y + 8.64$$

SELF CHECK 4 Write $-6.3(a + 2b + 3.7)$ without using parentheses.

4 Recognize the identity elements and find the additive and multiplicative inverse of a nonzero real number.

The numbers 0 and 1 play special roles in mathematics. The number 0 is the only number that can be added to another number (say, a) and give an answer that is the same number a:

$$0 + a = a + 0 = a$$

The number 1 is the only number that can be multiplied by another number (say, a) and give an answer that is the same number a:

$$1 \cdot a = a \cdot 1 = a$$

Because adding 0 to a number or multiplying a number by 1 leaves that number the same (identical), the numbers 0 and 1 are called **identity elements**.

IDENTITY ELEMENTS

0 is the **identity element for addition (additive identity)**.

1 is the **identity element for multiplication (multiplicative identity)**.

If the sum of two numbers is 0, the numbers are called **negatives** (or **opposites** or **additive inverses**) of each other. Since $3 + (-3) = 0$, the numbers 3 and -3 are negatives (or opposites or additive inverses) of each other. In general, because

$$a + (-a) = 0$$

the numbers represented by a and $-a$ are negatives (or opposites or additive inverses) of each other.

If the product of two numbers is 1, the numbers are called **reciprocals**, or **multiplicative inverses**, of each other. Since $7(\frac{1}{7}) = 1$, the numbers 7 and $\frac{1}{7}$ are reciprocals. Since $(-0.25)(-4) = 1$, the numbers -0.25 and -4 are reciprocals. In general, because

$$a\left(\frac{1}{a}\right) = 1 \quad \text{provided } a \neq 0$$

the numbers represented by a and $\frac{1}{a}$ are reciprocals (or multiplicative inverses) of each other.

ADDITIVE AND MULTIPLICATIVE INVERSES

Because $a + (-a) = 0$, the numbers a and $-a$ are called **negatives, opposites,** or **additive inverses.**

Because $a\left(\frac{1}{a}\right) = 1 \quad (a \neq 0)$, the numbers a and $\frac{1}{a}$ are called **reciprocals** or **multiplicative inverses.**

EXAMPLE 5 Find the additive and multiplicative inverses of $\frac{2}{3}$.

Solution The additive inverse of $\frac{2}{3}$ is $-\frac{2}{3}$ because $\frac{2}{3} + \left(-\frac{2}{3}\right)$.

The multiplicative inverse of $\frac{2}{3}$ is $\frac{3}{2}$ because $\frac{2}{3}\left(\frac{3}{2}\right) = 1$.

SELF CHECK 5 Find the additive and multiplicative inverses of $-\frac{1}{5}$.

5 Identify the property that justifies a given statement.

EXAMPLE 6 The property in the right column justifies the statement in the left column.

a. $3 + 4$ is a real number.	closure property of addition
b. $\frac{8}{3}$ is a real number.	closure property of division
c. $3 + 4 = 4 + 3$	commutative property of addition
d. $-3 + (2 + 7) = (-3 + 2) + 7$	associative property of addition
e. $(5)(-4) = (-4)(5)$	commutative property of multiplication
f. $(ab)c = a(bc)$	associative property of multiplication
g. $3(a + 2) = 3a + 3 \cdot 2$	distributive property
h. $3 + 0 = 3$	additive identity property
i. $3(1) = 3$	multiplicative identity property
j. $2 + (-2) = 0$	additive inverse property
k. $\left(\frac{2}{3}\right)\left(\frac{3}{2}\right) = 1$	multiplicative inverse property

SELF CHECK 6 Which property justifies each statement?

a. $a + 7 = 7 + a$

b. $3(y + 2) = 3y + 3 \cdot 2$

c. $3 \cdot (2 \cdot p) = (3 \cdot 2) \cdot p$

The properties of the real numbers are summarized as follows.

PROPERTIES OF REAL NUMBERS

For all real numbers a, b, and c,

	Addition	Multiplication
Closure properties	$a + b$ is a real number. $a - b$ is a real number.	$a \cdot b$ is a real number. $a \div b$ is a real number $(b \neq 0)$.
Commutative properties	$a + b = b + a$	$a \cdot b = b \cdot a$
Associative properties	$(a + b) + c = a + (b + c)$	$(ab)c = a(bc)$
Identity properties	$a + 0 = a$	$a \cdot 1 = a$
Inverse properties	$a + (-a) = 0$	$a\left(\frac{1}{a}\right) = 1 \quad (a \neq 0)$
Distributive property	$a(b + c) = ab + ac$	

SELF CHECK ANSWERS

1. a. -9 **b.** -2 **2. a.** $a + b = 1$ and $b + a = 1$ **b.** $ab = -30$ and $ba = -30$ **3.** -65 **4.** $-6.3a - 12.6b - 23.31$ **5.** $\frac{1}{5}, -5$ **6. a.** commutative property of addition **b.** distributive property **c.** associative property of multiplication

NOW TRY THIS

1. Give the additive and multiplicative inverses of 1.2.
2. Use the commutative property of multiplication to write the expression $x(y + w)$.
3. Simplify: $-4(2a - 3b + 5)$
4. Use the distributive property to complete the following multiplication:
   $9x - 12y - 3 = 3(\ \square - \square - \square\ )$
5. Find the additive inverse of $x - y$. Try to find a second way to write it (there are three).

1.7 Exercises

WARM-UPS *Give an example of each property.*

1. The associative property of multiplication
2. The additive identity property
3. The distributive property
4. The inverse property for multiplication

Provide an example to illustrate each statement.

5. Subtraction is not commutative.
6. Division is not associative.

REVIEW

7. Write as a mathematical inequality: The sum of x and the square of y is greater than or equal to z.
8. Write as an English phrase: $3(x + z)$

Fill each box with an appropriate symbol.

9. For any number x, $|x| \ \square\ 0$.
10. $x - y = x + (\ \square\ )$

Fill in the blanks.

11. The product of two negative numbers is a ______ number.
12. The sum of two negative numbers is a ______ number.

VOCABULARY AND CONCEPTS *Fill in the blanks.*

13. Closure property: If a and b are real numbers, $a + b$ is a ___ number.
14. Closure property: If a and b are real numbers, $\frac{a}{b}$ is a real number, provided that _____.
15. Commutative property of addition: $a + b = b +$ _
16. Commutative property of multiplication: $a \cdot b =$ _ $\cdot a$
17. Associative property of addition: $(a + b) + c = a +$ ______
18. Associative property of multiplication: $(ab)c =$ _ $\cdot (bc)$
19. Distributive property: $a(b + c) = ab +$ __
20. $0 + a =$ _
21. $a \cdot 1 =$ _
22. 0 is the ______ element for ______.
23. 1 is the identity ______ for ___________.
24. If $a + (-a) = 0$, then a and $-a$ are called ______ inverses.
25. If $a\left(\frac{1}{a}\right) = 1$, then _ and _ are called *reciprocals* or __________ inverses.
26. $a(b + c + d) = ab +$ ______

GUIDED PRACTICE *Let $x = 12$ and $y = -2$. Show that each expression represents a real number by finding the real-number answer.* SEE EXAMPLE 1. (OBJECTIVE 1)

27. $x + y$
28. $y - x$
29. xy
30. $\frac{x}{y}$
31. x^2
32. y^2
33. $\frac{x}{y^2}$
34. $\frac{2x}{3y}$

Let $x = 5$, $y = 7$, and $z = -1$. Show that the two expressions have the same value. SEE EXAMPLE 2. (OBJECTIVE 2)

35. $x + y; y + x$
36. $xy; yx$
37. $3x + 2y; 2y + 3x$
38. $3xy; 3yx$
39. $x(x + y); (x + y)x$
40. $xy + y^2; y^2 + xy$
41. $x^2(yz^2); (x^2y)z^2$
42. $x(y^2z^3); (xy^2)z^3$

Use the distributive property to write each expression without parentheses. Simplify each result, if possible. SEE EXAMPLES 3–4. (OBJECTIVE 3)

43. $3(x + 5)$
44. $7(y + 2)$
45. $5(z - 4)$
46. $4(a - 3)$
47. $-2(3x + y)$
48. $-3(4a + b)$
49. $x(x + 3)$
50. $y(y + z)$
51. $-x(a + b)$
52. $-a(x + y)$
53. $-4(x^2 + x + 2)$
54. $-2(a^2 - a + 3)$

Give the additive and the multiplicative inverses of each number, if possible. SEE EXAMPLE 5. (OBJECTIVE 4)

55. 5
56. 3
57. $\frac{1}{3}$
58. $-\frac{1}{3}$
59. 0
60. -4
61. $-\frac{2}{3}$
62. 0.5
63. -0.2
64. 0.75
65. $\frac{5}{4}$
66. -1.25

Use the given property to rewrite the expression in a different form. SEE EXAMPLE 6. (OBJECTIVE 5)

67. $8(x + 2)$; distributive property
68. $a + b$; commutative property of addition
69. xy^3; commutative property of multiplication
70. $2 + (5 + 3)$; associative property of addition
71. $(x + y)z$; commutative property of addition
72. $7(x + 2)$; distributive property
73. $(xy)z$; associative property of multiplication
74. $1x$; multiplicative identity property

ADDITIONAL PRACTICE

Let $x = 2$, $y = -3$, and $z = 1$. Show that the two expressions have the same value.

75. $(x + y) + z; x + (y + z)$
76. $(xy)z; x(yz)$
77. $(xz)y; x(yz)$
78. $(x + y) + z; y + (x + z)$

Use the distributive property to write each expression without parentheses.

79. $-6(a + 4)$
80. $2x(a - x)$
81. $-3x(x - a)$
82. $-a(a + b)$

Which property of real numbers justifies each statement?

83. $3 + x = x + 3$
84. $(3 + x) + y = 3 + (x + y)$
85. $xy = yx$
86. $(4)(-7) = (-7)(4)$
87. $-5(x + 4) = -5x + (-5)(4)$
88. $x(y + z) = (y + z)x$
89. $(x + y) + z = z + (x + y)$
90. $3(x + y) = 3x + 3y$
91. $5 \cdot 1 = 5$
92. $x + 0 = x$

93. $3 + (-3) = 0$

94. $9 \cdot \frac{1}{9} = 1$

95. $0 + x = x$

96. $5 \cdot \frac{1}{5} = 1$

WRITING ABOUT MATH

97. Explain why division is not commutative.

98. Describe two ways of calculating the value of $3(12 + 7)$.

SOMETHING TO THINK ABOUT

99. Suppose there were no numbers other than the odd integers.
- Would the closure property for addition still be true?
- Would the closure property for multiplication still be true?
- Would there still be an identity for addition?
- Would there still be an identity for multiplication?

100. Suppose there were no numbers other than the even integers. Answer the four parts of Exercise 99 again.

Projects

PROJECT 1

The circumference of any circle and its diameter are related. When you divide the circumference by the diameter, the quotient is always the same number, **pi**, denoted by the Greek letter π.

- Carefully measure the circumference of several circles—a quarter, a dinner plate, a bicycle tire—whatever you can find that is round. Then calculate approximations of π by dividing each circle's circumference by its diameter.
- Use the π key on the calculator to obtain a more accurate value of π. How close were your approximations?

PROJECT 2

a. The fraction $\frac{22}{7}$ is often used as an approximation of π. To how many decimal places is this approximation accurate?

b. Experiment with your calculator and try to do better. Find another fraction (with no more than three digits in either its numerator or its denominator) that is closer to π. Who in your class has done best?

PROJECT 3

Write an essay answering this question.

When three professors attending a convention in Las Vegas registered at the hotel, they were told that the room rate was \$120. Each professor paid his \$40 share.

Later the desk clerk realized that the cost of the room should have been \$115. To fix the mistake, she sent a bellhop to the room to refund the \$5 overcharge. Realizing that \$5 could not be evenly divided among the three professors, the bellhop refunded only \$3 and kept the other \$2.

Since each professor received a \$1 refund, each paid \$39 for the room, and the bellhop kept \$2. This gives \$39 + \$39 + \$39 + \$2, or \$119. What happened to the other \$1?

Reach for Success

EXTENSION OF ANALYZING YOUR TIME

Now that you've analyzed a day, let's move on to take a look at a typical week. Fill in the chart below to account for every hour of every day. To simplify the process, you can use the following abbreviations:

W (work time including commute), **S** (sleeping), **P** (preparing for work; preparing meals, eating, exercising), **C** (class time including commute), **F** (time with family and friends), **ST** (study time), and **E** (entertainment)

Remember, there are no right or wrong answers. This information is to give you a complete picture of how you are spending your time in a typical week.

	SUNDAY	MONDAY	TUESDAY	WEDNESDAY	THURSDAY	FRIDAY	SATURDAY
6:00–7:00							
7:00–8:00							
8:00–9:00							
9:00–10:00							
10:00–11:00							
11:00–Noon							
Noon–1:00							
1:00–2:00							
2:00–3:00							
3:00–4:00							
4:00–5:00							
5:00–6:00							
6:00–7:00							
7:00–8:00							
8:00–9:00							
9:00–10:00							
10:00–11:00							
11:00–12:00							
12:00–1:00							
1:00–2:00							
2:00–3:00							
3:00–4:00							
4:00–5:00							
5:00–6:00							

After reviewing your weekly schedule, do you have enough time to meet the "rule of thumb" of studying at least 2 hours per week for every 1 hour you are in class? If yes, congratulations!

If not, is it possible for you to find the additional hours to help you be successful in this and all your other classes?

Are you able/willing to adjust your schedule to find this time? ______ Explain your answer.

1 Review

SECTION 1.1 Real Numbers and Their Graphs

DEFINITIONS AND CONCEPTS	EXAMPLES
Natural numbers: $\{1, 2, 3, 4, 5, \ldots\}$ **Whole numbers:** $\{0, 1, 2, 3, 4, 5, \ldots\}$ **Integers:** $\{\ldots, -3, -2, -1, 0, 1, 2, 3, \ldots\}$ **Rational numbers:** $\left\{\frac{a}{b} \middle\vert a \text{ is an integer and } b \text{ is a nonzero integer.}\right\}$ **Irrational numbers:** $\{x \vert x$ is a number such as π or $\sqrt{2}$ that cannot be written as a fraction with an integer numerator and a nonzero integer denominator.$\}$ **Real numbers:** {Rational numbers or irrational numbers} **Prime numbers:** $\{2, 3, 5, 7, 11, 13, 17, \ldots\}$ **Composite numbers:** $\{4, 6, 8, 9, 10, 12, 14, 15, \ldots\}$ **Even integers:** $\{\ldots, -6, -4, -2, 0, 2, 4, 6, \ldots\}$ **Odd integers:** $\{\ldots, -5, -3, -1, 1, 3, 5, \ldots\}$	Which numbers in the set $\{-5, 0, \frac{2}{3}, 1.5, \sqrt{9}, \pi, 6\}$ are **a.** natural numbers **b.** whole numbers **c.** integers **d.** rational numbers **e.** irrational numbers **f.** real numbers **g.** prime numbers **h.** composite numbers **i.** even integers **j.** odd integers? **a.** $\sqrt{9}, 6,$ ($\sqrt{9}$ is a natural number since $\sqrt{9} = 3$.) **b.** $0, \sqrt{9}, 6$ **c.** $-5, 0, \sqrt{9}, 6$ **d.** $-5, 0, \frac{2}{3}, 1.5, \sqrt{9}, 6$ **e.** π **f.** $-5, 0, \frac{2}{3}, 1.5, \sqrt{9}, \pi, 6$ **g.** $\sqrt{9}$ **h.** 6 **i.** $0, 6$ **j.** $-5, \sqrt{9}$
Double negative rule: $-(-x) = x$	$-(-3) = 3$
Sets of numbers can be graphed on the number line.	**1.** Graph the set of integers between -2 and 4. 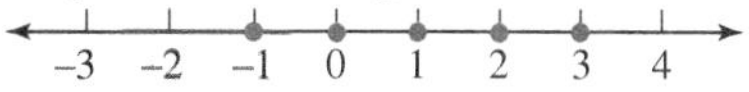 **2.** Graph all real numbers x such that $x < -2$ or $x > 1$. 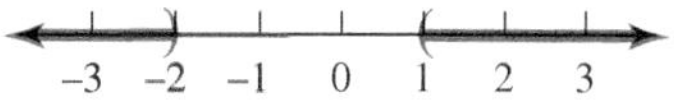
The **absolute value** of x, denoted as $\lvert x \rvert$, is the distance between x and 0 on the number line. $\lvert x \rvert \geq 0$	Evaluate: $-\lvert -8 \rvert$ $-\lvert -8 \rvert = -(8) = -8$

REVIEW EXERCISES

Consider the set $\{0, 1, 2, 3, 4, 5\}$.

1. Which numbers are natural numbers?
2. Which numbers are prime numbers?
3. Which numbers are odd natural numbers?
4. Which numbers are composite numbers?

Consider the set $\{-6, -\frac{2}{3}, 0, \sqrt{2}, 2.6, \pi, 5\}$.

5. Which numbers are integers?
6. Which numbers are rational numbers?
7. Which numbers are prime numbers?
8. Which numbers are real numbers?
9. Which numbers are even integers?
10. Which numbers are odd integers?
11. Which numbers are irrational?
12. Which numbers are negative numbers?

Place one of the symbols $=$, $<$, *or* $>$ *in each box to make a true statement.*

13. $-3 \ \square \ 5 - 5$
14. $\frac{12}{4} \ \square \ 7$
15. $\frac{36}{4} \ \square \ -2$
16. $2 - 2 \ \square \ 8 - \frac{24}{3}$

Unless otherwise noted, all content on this page is © Cengage Learning.

Simplify each expression.

17. $-(-9)$ **18.** $-(12 - 4)$

Draw a number line and graph each set of numbers.

19. The composite numbers from 14 to 20

20. The whole numbers between 19 and 25

21. The real numbers less than or equal to -3 or greater than 2

22. The real numbers greater than -4 and less than 3

Find each absolute value.

23. $|29 - 24|$ **24.** $|-25|$

SECTION 1.2 Fractions

DEFINITIONS AND CONCEPTS	EXAMPLES
To simplify a fraction, factor the numerator and the denominator. Then divide out all common factors.	Simplify: $\frac{12}{32}$ $\frac{12}{32} = \frac{4 \cdot 3}{4 \cdot 8} = \frac{\overset{1}{\cancel{4}} \cdot 3}{\underset{1}{\cancel{4}} \cdot 8} = \frac{3}{8}$
To multiply two fractions, multiply their numerators and multiply their denominators.	$4 \cdot \frac{5}{6} = \frac{4}{1} \cdot \frac{5}{6}$ $= \frac{4 \cdot 5}{1 \cdot 6}$ $= \frac{\overset{1}{\cancel{2}} \cdot 2 \cdot 5}{1 \cdot \underset{1}{\cancel{2}} \cdot 3}$ $= \frac{10}{3}$
To divide two fractions, multiply the first by the reciprocal of the second.	$\frac{2}{3} \div \frac{5}{6} = \frac{2}{3} \cdot \frac{6}{5}$ $= \frac{2 \cdot 6}{3 \cdot 5}$ $= \frac{2 \cdot 2 \cdot \overset{1}{\cancel{3}}}{\underset{1}{\cancel{3}} \cdot 5}$ $= \frac{4}{5}$
To add (or subtract) two fractions with like denominators, add (or subtract) their numerators and keep their common denominator.	$\frac{9}{11} + \frac{2}{11} = \frac{9 + 2}{11}$ $= \frac{11}{11}$ $= 1$
To add (or subtract) two fractions with unlike denominators, find equivalent fractions with the same denominator (LCD), add (or subtract) their numerators, and keep the common denominator.	Subtract: $\frac{11}{12} - \frac{3}{4}$ Begin by finding the LCD. $\left.\begin{array}{l} 12 = 2 \cdot 2 \cdot 3 \\ 4 = 2 \cdot 2 \end{array}\right\} \text{LCD} = 2 \cdot 2 \cdot 3 = 12$ Write $\frac{3}{4}$ as a fraction with a denominator of 12 and then do the subtraction. $\frac{11}{12} - \frac{3}{4} = \frac{11}{12} - \frac{3 \cdot 3}{4 \cdot 3}$ $= \frac{11}{12} - \frac{9}{12}$ $= \frac{11 - 9}{12}$

	$= \frac{2}{12}$ $= \frac{\overset{1}{\cancel{2}}}{\underset{1}{\cancel{2}} \cdot 6}$ $= \frac{1}{6}$
Before working with mixed numbers, convert them to improper fractions.	Write $5\frac{7}{9}$ as an improper fraction. $5\frac{7}{9} = 5 + \frac{7}{9} = \frac{45}{9} + \frac{7}{9} = \frac{52}{9}$
A **percent** is the numerator of a fraction with a denominator of 100.	$5\frac{1}{2}\%$ can be written as $\frac{5.5}{100}$, or as the decimal 0.055.

REVIEW EXERCISES

Simplify each fraction.

25. $\frac{45}{27}$

26. $\frac{48}{18}$

Perform each operation and simplify the answer, if possible.

27. $\frac{31}{15} \cdot \frac{10}{62}$

28. $\frac{25}{36} \cdot \frac{12}{15} \cdot \frac{3}{5}$

29. $\frac{18}{21} \div \frac{6}{7}$

30. $\frac{14}{24} \div \frac{7}{12} \div \frac{2}{5}$

31. $\frac{7}{12} + \frac{9}{12}$

32. $\frac{13}{24} - \frac{5}{24}$

33. $\frac{1}{5} + \frac{1}{4}$

34. $\frac{5}{7} + \frac{4}{9}$

35. $\frac{2}{3} - \frac{1}{7}$

36. $\frac{4}{5} - \frac{2}{3}$

37. $3\frac{2}{3} + 5\frac{1}{4}$

38. $7\frac{5}{12} - 4\frac{1}{2}$

Perform each operation.

39. $48.29 + 31.90$

40. $36.85 - 15.86$

41. $4.32 \cdot 1.5$

42. $21.83 \div 5.9$

Perform each operation and round to two decimal places.

43. $2.7(4.92 - 3.18)$

44. $\frac{3.3 + 2.5}{0.22}$

45. $\frac{12.5}{14.7 - 11.2}$

46. $(3 - 0.7)(3.63 - 2)$

47. Farming One day, a farmer plowed $17\frac{1}{2}$ acres and on the second day, $15\frac{3}{4}$ acres. How much is left to plow if the fields total 100 acres?

48. Study times Four students recorded the time they spent working on a take-home exam: 5.2, 4.7, 9.5, and 8 hours. Find the average time spent.

49. Absenteeism During the height of the flu season, 20% of the 425 university faculty members were sick. How many were ill?

50. Packaging Four steel bands surround the shipping crate in the illustration. Find the total length of strapping needed.

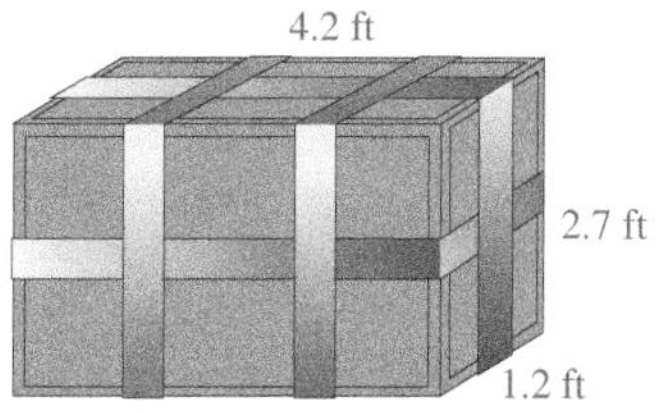

SECTION 1.3 Exponents and Order of Operations

DEFINITIONS AND CONCEPTS	EXAMPLES
If n is a natural number, then $x^n = \underbrace{x \cdot x \cdot x \cdot x \cdot \cdots \cdot x}_{n \text{ factors of } x}$	$x^5 = x \cdot x \cdot x \cdot x \cdot x$ $b^7 = b \cdot b \cdot b \cdot b \cdot b \cdot b \cdot b$
Order of operations Within each pair of grouping symbols (working from the innermost pair to the outermost pair), perform the following operations: **1.** Evaluate all exponential expressions. **2.** Perform multiplications and divisions, working from left to right. **3.** Perform additions and subtractions, working from left to right.	Evaluate: $6^2 - 5(12 - 2 \cdot 5)$ $6^2 - 5(12 - 2 \cdot 5) = 6^2 - 5(12 - 10)$ Do the multiplication within the parentheses. $= 6^2 - 5(2)$ Do the subtraction within the parentheses. $= 36 - 5(2)$ Find the value of the exponential expression. $= 36 - 10$ Do the multiplication. $= 26$ Do the subtraction.

Unless otherwise noted, all content on this page is © Cengage Learning.

4. Because the bar in a fraction is a grouping symbol, simplify the numerator and the denominator of a fraction separately. Then simplify the fraction, whenever possible.	To simplify $\frac{2^3 + 4\cdot 2}{2 + 6}$, we first simplify the numerator and the denominator. $\frac{2^3 + 4\cdot 2}{2+6} = \frac{8+8}{8}$ Find the power. Then find the product. Find the sum in the denominator. $= \frac{16}{8}$ Find the sum in the numerator. $= 2$ Find the quotient.
To find perimeters, areas, and volumes of geometric figures, substitute numbers for variables in the formulas. Be sure to include the proper units in the answer.	Find the perimeter of a rectangle whose length is 4 feet and whose width is 1 foot. $P = 2l + 2w$ This is the formula for the perimeter of a rectangle. $= 2(4) + 2(1)$ Substitute 4 for l and 1 for w. $= 8 + 2$ $= 10$ The perimeter is 10 feet.

REVIEW EXERCISES

Find the value of each expression.

51. 3^4

52. $\left(\frac{2}{3}\right)^2$

53. $(0.5)^2$

54. $5^2 + 2^3$

55. $3^2 + 4^2$

56. $(3 + 4)^2$

57. Geometry Find the area of a triangle with a base of $6\frac{1}{2}$ feet and a height of 7 feet.

58. **Petroleum storage** Find the volume of the cylindrical storage tank in the illustration. Round to the nearest tenth.

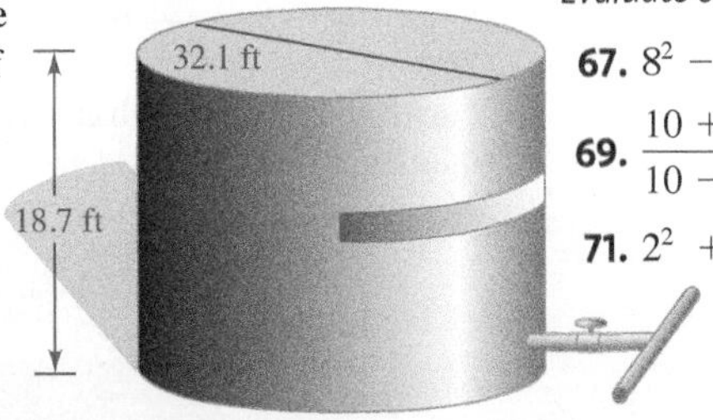

Simplify each expression.

59. $7 + 3^3$

60. $6 + 2\cdot 4$

61. $5 + 6 \div 2$

62. $(8 + 6) \div 2$

63. $5^3 - \frac{81}{3}$

64. $(5 - 2)^2 + 5^2 + 2^2$

65. $\frac{4\cdot 3 + 3^4}{31}$

66. $\frac{4}{3}\cdot\frac{9}{2} + \frac{1}{2}\cdot 18$

Evaluate each expression.

67. $8^2 - 6$

68. $(8 - 6)^2$

69. $\frac{10 + 2}{10 - 6}$

70. $\frac{6(8) - 12}{4 + 8}$

71. $2^2 + 2(3^2)$

72. $\frac{2^2 + 3}{2^3 - 1}$

SECTION 1.4 Adding and Subtracting Real Numbers

DEFINITIONS AND CONCEPTS	EXAMPLES
To add two positive numbers, add their absolute values and make the answer positive. To add two negative numbers, add their absolute values and make the answer negative.	$(+1) + (+6) = +7$ $(-1) + (-6) = -7$
To add a positive and a negative number, subtract the smaller absolute value from the larger. **1.** If the positive number has the larger absolute value, the answer is positive. **2.** If the negative number has the larger absolute value, the answer is negative.	$(-1) + (+6) = +5$ $(+1) + (-6) = -5$
If a and b are two real numbers, then $a - b = a + (-b)$	$-8 - 2 = -8 + (-2)$ To subtract 2, add the opposite of 2. $= -10$

Unless otherwise noted, all content on this page is © Cengage Learning.

REVIEW EXERCISES

Simplify each expression.

73. $(+15) + (+9)$

74. $(-17) + (-16)$

75. $(-2.7) + (-3.8)$

76. $\frac{1}{2} + \left(-\frac{1}{6}\right)$

77. $(+12) + (-24)$

78. $(-44) + (+60)$

79. $3.7 + (-2.5)$

80. $-5.6 + (+2.06)$

81. $15 - (-4)$

82. $-8 - (-15)$

83. $[-5 + (-5)] - (-5)$

84. $1 - [5 - (-3)]$

85. $-\frac{7}{10} - \left(-\frac{2}{5}\right)$

86. $\frac{2}{3} - \left(\frac{1}{3} - \frac{2}{3}\right)$

87. $\left|\frac{3}{7} - \left(-\frac{4}{7}\right)\right|$

88. $\frac{3}{7} - \left|-\frac{4}{7}\right|$

SECTION 1.5 Multiplying and Dividing Real Numbers

DEFINITIONS AND CONCEPTS	EXAMPLES
To multiply two real numbers, multiply their absolute values.	
1. If the numbers are positive, the product is positive.	$3(7) = 21$
2. If the numbers are negative, the product is positive.	$-3(-7) = 21$
3. If one number is positive and the other is negative, the product is negative.	$-3(7) = -21$ $3(-7) = -21$
4. $a \cdot 0 = 0 \cdot a = 0$	$5(0) = 0$
5. $a \cdot 1 = 1 \cdot a = a$	$-7(1) = -7$
To divide two real numbers, find the quotient of their absolute values.	
1. If the numbers are positive, the quotient is positive.	$\frac{+6}{+2} = +3$ because $(+2)(+3) = +6$
2. If the numbers are negative, the quotient is positive.	$\frac{-6}{-2} = +3$ because $(-2)(+3) = -6$
3. If one number is positive and the other is negative, the quotient is negative.	$\frac{+6}{-2} = -3$ because $(-2)(-3) = +6$
4. $\frac{a}{0}$ is undefined.	$\frac{-6}{+2} = -3$ because $(+2)(-3) = -6$
5. If $a \neq 0$, then $\frac{0}{a} = 0$.	$\frac{2}{0}$ is undefined because no number multiplied by 0 gives 2
	$\frac{0}{2} = 0$ because $(2)(0) = 0$

REVIEW EXERCISES

Simplify each expression.

89. $(+5)(+8)$

90. $(-5)(-12)$

91. $\left(-\frac{3}{14}\right)\left(-\frac{7}{6}\right)$

92. $(3.75)(0.37)$

93. $5(-7)$

94. $(-15)(7)$

95. $\left(-\frac{1}{2}\right)\left(\frac{4}{3}\right)$

96. $(2.1)(-8.2)$

97. $\frac{+36}{+12}$

98. $\frac{-14}{-2}$

99. $\frac{(-2)(-7)}{4}$

100. $\frac{-22.5}{-3.75}$

101. $\frac{(-2)(-9)}{-3}$

102. $\frac{(-6)(12)}{-4}$

103. $\left(\frac{-10}{2}\right)^2 - (-1)^3$

104. $\frac{[-3 + (-4)]^2}{10 + (-3)}$

105. $\left(\frac{-3 + (-3)}{3}\right)\left(\frac{-15}{5}\right)$

106. $\frac{-2 - (-8)}{5 + (-1)}$

SECTION 1.6 Algebraic Expressions

DEFINITIONS AND CONCEPTS	EXAMPLES
Variables and numbers can be combined with operations of arithmetic to produce **algebraic expressions**.	$5x \quad 3x^2 + 7x \quad 5(3x - 8)$
We can **evaluate** algebraic expressions when we know the values of the variables.	Evaluate: $5x - 2$ when $x = 3$ $5x + 2 = 5(\mathbf{3}) - 2$ Substitute 3 for x. $= 15 - 2$ $= 13$
Numbers written without variables are called **constants**.	Identify the constant in $6x^2 - 4x + 2$ The constant is 2.
Expressions that are constants, variables, or products of constants and variables are called **algebraic terms**.	Identify the algebraic terms: $6x^2 - 4x + 2$ The terms are $6x^2$, $-4x$, and 2.
Numbers and variables that are part of a product are called **factors**.	Identify the factors in $7x$. The factors are 7 and x.
The number factor of a product is called its **numerical coefficient**.	Identify the numerical coefficient of $7x$. The numerical coefficient is 7.

REVIEW EXERCISES

Let x, y, and z represent three real numbers. Write an algebraic expression that represents each quantity.

107. The product of x and z

108. The sum of x and twice y

109. Twice the sum of x and y

110. x decreased by the product of y and z

Write each algebraic expression as an English phrase.

111. $5xz$

112. $5 - yz$

113. $xy - 4$

114. $\dfrac{x + y + z}{2xyz}$

Let x = 2, y = −3, and z = −1 and evaluate each expression.

115. $x + z$

116. $x + y + z$

117. $5x + (y - z)$

118. $z^2 - y$

119. $x - (y - z)$

120. $(x - y) - z$

Let x = 2, y = −3, and z = −1 and evaluate each expression.

121. yz

122. xyz

123. $(x + y)(y + z)$

124. $\dfrac{3(x - y)}{x + (y - z)}$

125. $y^2z + x$

126. $yz^3 + (xy)^2$

127. $\dfrac{2y^2}{3x - 6}$

128. $\dfrac{|xy|}{3z}$

129. How many terms does the expression $3x + 4y + 9$ have?

130. What is the numerical coefficient of the term $7xy$?

131. What is the numerical coefficient of the term xy?

132. Find the sum of the numerical coefficients in $2x^3 + 4x^2 + 3x$.

SECTION 1.7 Properties of Real Numbers

DEFINITIONS AND CONCEPTS	EXAMPLES
The closure properties:	
$a + b$ is a real number.	$5 + (-2) = 3$ is a real number.
$a - b$ is a real number.	$5 - 2 = 3$ is a real number.
ab is a real number.	$5(-2) = -10$ is a real number.
$\dfrac{a}{b}$ is a real number $(b \neq 0)$.	$\dfrac{10}{-5} = -2$ is a real number.

The commutative properties: $a + b = b + a$ of addition $ab = ba$ of multiplication	 The commutative property of addition justifies the statement $x + 3 = 3 + x$. The commutative property of multiplication justifies the statement $x \cdot 3 = 3 \cdot x$.
The associative properties: $(a + b) + c = a + (b + c)$ of addition $(ab)c = a(bc)$ of multiplication	 The associative property of addition justifies the statement $(x + 3) + 4 = x + (3 + 4)$. The associative property of multiplication justifies the statement $(x \cdot 3) \cdot 4 = x \cdot (3 \cdot 4)$.
The distributive property of multiplication over addition: $a(b + c) = ab + ac$ $a(b - c) = ab - ac$	Use the distributive property to write the expression $7(2x - 8)$ without parentheses. $7(2x - 8) = 7(2x) - 7(8) = 14x - 56$
The identity elements: 0 is the identity for addition (additive identity). 1 is the identity for multiplication (multiplicative identity).	 $0 + 5 = 5$ $1 \cdot 5 = 5$
The additive and multiplicative inverse properties: $a + (-a) = 0$ $a\left(\frac{1}{a}\right) = 1 \quad (a \neq 0)$	 The additive inverse of -2 is 2 because $-2 + 2 = 0$. The multiplicative inverse of -2 is $-\frac{1}{2}$ because $-2\left(-\frac{1}{2}\right) = 1$.

REVIEW EXERCISES

Determine which property of real numbers justifies each statement. Assume that all variables represent real numbers.

133. $b + c$ is a real number

134. $3 \cdot (4 \cdot 5) = (4 \cdot 5) \cdot 3$

135. $3 + (4 + 5) = (3 + 4) + 5$

136. $5(x + 2) = 5 \cdot x + 5 \cdot 2$

137. $b + c = c + b$

138. $3 \cdot (4 \cdot 5) = (3 \cdot 4) \cdot 5$

139. $3 + (x + 1) = (x + 1) + 3$

140. $x \cdot 1 = x$

141. $8 + (-8) = 0$

142. $x + 0 = x$

1 Test

1. List the prime numbers between 30 and 50.

2. What is the only even prime number?

3. Graph the composite numbers less than 10 on a number line.

4. Graph the real numbers from 5 to 15 on a number line.

5. Evaluate: $-|-17|$

6. Evaluate: $-|9| + |-9|$

Place one of the symbols $=$, $<$, or $>$ in each box to make a true statement.

7. $3(4 - 2) \ \square \ -2(2 - 5)$

8. $1 + 4 \cdot 3 \ \square \ -2(-7)$

9. 25% of 136 $\square$ $\frac{1}{2}$ of 66

10. $-8.5 \ \square \ -|-8.5|$

Simplify each expression.

11. $\frac{26}{40}$

12. $\frac{9}{11} \cdot \frac{44}{45}$

13. $\frac{14}{21} \div \frac{28}{9}$

14. $\frac{24}{16} + 3$

15. $\frac{17 - 5}{36} - \frac{2(13 - 5)}{12}$

16. $\frac{|-7 - (-6)|}{-7 - |-6|}$

17. Find 13% of 256 and round the answer to one decimal place.

18. Find the area of a rectangle 18.9 feet wide and 21.25 feet long. Round the answer to two decimal places.

19. Find the area of the triangle in the illustration.

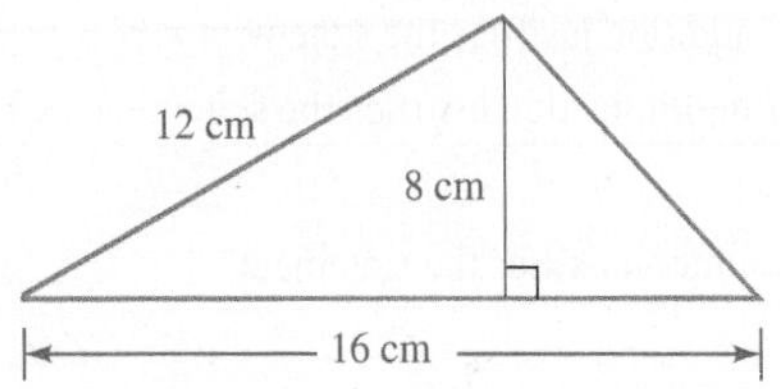

20. To the nearest cubic inch, find the volume of the solid in the illustration.

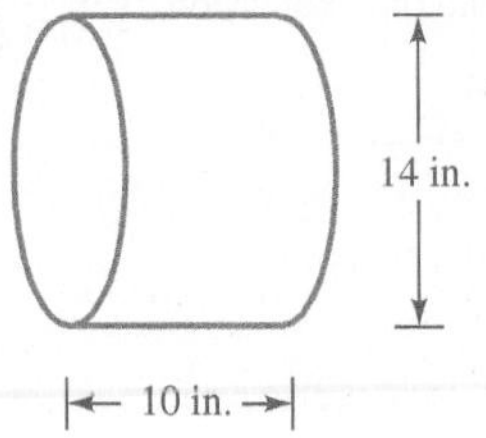

Let $x = -2$, $y = 3$, and $z = 4$. Evaluate each expression.

21. $xy + z$

22. $x(y + z)$

23. $\frac{z + 4y}{2x}$

24. $|x^3 - z|$

25. $x^3 + y^2 + z$

26. $|x| - 3|y| - 4|z|$

27. Let x and y represent two real numbers. Write an algebraic expression to denote the quotient obtained when the product of the two numbers is divided by their sum.

28. Let x and y represent two real numbers. Write an algebraic expression to denote the difference obtained when the sum of x and y is subtracted from the product of 5 and y.

29. A man lives 12 miles from work and 7 miles from the grocery store. If he made x round trips to work and y round trips to the store, write an expression to represent how many miles he drove.

30. A baseball costs \$$a$ and a glove costs \$$b$. Write an expression to represent how much it will cost a community center to buy 12 baseballs and 8 gloves.

31. What is the numerical coefficient of the term $-5x^2y^3$?

32. How many terms are in the expression $3x^2y + 5xy^2 + x + 7$?

Write each expression without using parentheses.

33. $3(x + 2)$

34. $-p(r - t)$

35. What is the identity element for addition?

36. What is the multiplicative inverse of $\frac{1}{5}$?

Determine which property of the real numbers justifies each statement.

37. $(xy)z = z(xy)$

38. $3(x + y) = 3x + 3y$

39. $2 + x = x + 2$

40. $7 \cdot \frac{1}{7} = 1$

Unless otherwise noted, all content on this page is © Cengage Learning.

Equations and Inequalities

2

© iStockphoto.com/John Cowie

Careers and Mathematics

SECURITIES AND FINANCIAL SERVICES SALES AGENTS

Many investors use securities and financial sales agents when buying or selling stocks, bonds, shares in mutual funds, annuities, or other financial products. The overwhelming majority of people in this occupation are college graduates, with courses in business administration, economics, mathematics, and finance. After working for a few years, many agents get a Master's degree in Business Administration (MBA).

Job Outlook:
Employment of people in this field is expected to grow rapidly over the next decade, especially in banking. However, there will be keen competition for these jobs.
Annual Earnings:
$42,630–$126,290
For More Information:
http://www.bls.gov/oco/ocos122.htm
For a Sample Application:
See Problem 59 in Section 2.5.

REACH FOR SUCCESS

2.1 Solving Basic Linear Equations in One Variable

2.2 Solving More Linear Equations in One Variable

2.3 Simplifying Expressions to Solve Linear Equations in One Variable

2.4 Formulas

2.5 Introduction to Problem Solving

2.6 Motion and Mixture Applications

2.7 Solving Linear Inequalities in One Variable

■ ***Projects***

REACH FOR SUCCESS EXTENSION

CHAPTER REVIEW

CHAPTER TEST

CUMULATIVE REVIEW

In this chapter

In this chapter, we will learn how to solve basic linear equations and apply that knowledge to solving many types of problems. We also will consider special equations called formulas and conclude by solving linear inequalities.

Reach for Success Reading This Mathematics Textbook

You might be asking yourself, "Who reads their math book anyway?" The better question might be, "Why should I read it?" or "How do I read it?" Why is easy to answer . . . to improve understanding and, thus, your grade! How is not as easy to answer. It takes practice.

Get acquainted with some of the features of this textbook.

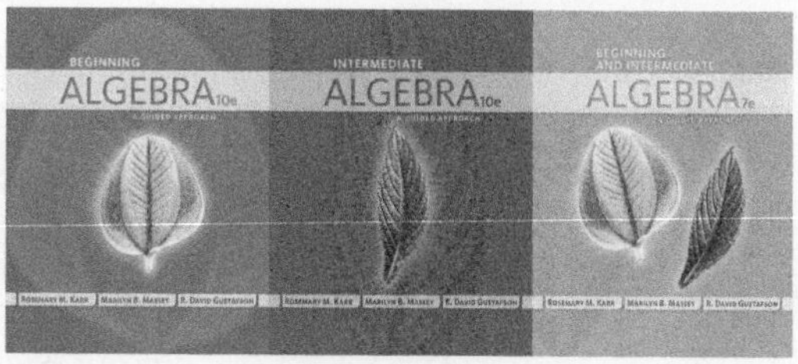

Turn to the first page of Section 2.2 in this chapter. On which page did you find this? ______

Note that the objectives are provided at the beginning of each section. Why do you think these are listed?

______

Note the vocabulary list at the beginning of each section. Where might you find the definitions?

1. ______ 2. ______

What is the purpose of the self-check exercises that directly follow each section's worked-out examples?

______

A mathematics textbook is structured differently than texts in other subjects. Usually, each chapter is divided into 5 to 7 sections, consisting of 6 to 10 pages each. Thus, you only need to study a few pages at a time. Can you see an advantage to this structure? ______

______

______

______

One advantage to fewer pages is that you can study a smaller "chunk" of mathematics at a time. Use your syllabus (or ask your instructor) to identify which objectives are covered in each section for your course.

Which objectives are you responsible for learning in the second section of this chapter? ______

______

Did you know that the answers to the odd-numbered exercises are in the back of the book? ______
Why could this be important to you? ______

______

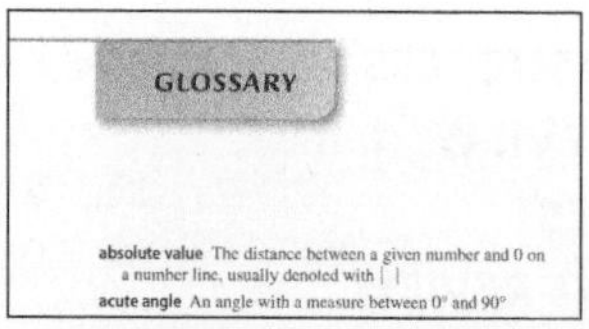

What is the purpose of the glossary? ______

When would you use it?

______

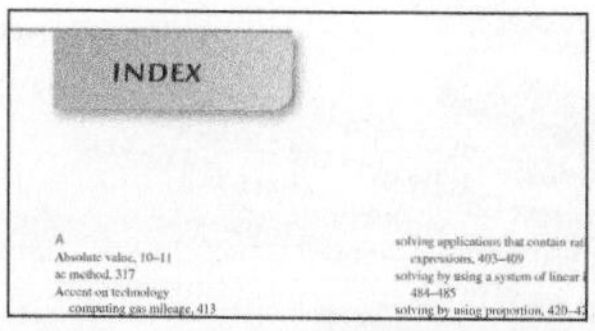

What is the purpose of the index? ______

When would you use it?

______

A Successful Study Strategy . . .

Read each textbook section prior to the classroom discussion. Even having an idea of the topic for the day will better prepare you for learning.

At the end of the chapter you will find an additional exercise to help guide you to planning for a successful semester.

Unless otherwise noted, all content on this page is © Cengage Learning.

Section 2.1

Solving Basic Linear Equations in One Variable

Objectives

1. Determine whether a statement is an expression or an equation.
2. Determine whether a number is a solution of an equation.
3. Solve a linear equation in one variable by applying the addition or subtraction property of equality.
4. Solve a linear equation in one variable by applying the multiplication or division property of equality.
5. Solve a linear equation in one variable involving markdown and markup.
6. Solve a percent problem involving a linear equation in one variable using the formula $rb = a$.
7. Solve an application involving percents.

Vocabulary

equation	linear equation	discount
expression	addition property of equality	markup
variable	equivalent equations	rate
solution	multiplication property of equality	base
root	markdown	amount
solution set		

Getting Ready

Fill in the blanks.

1. $3 + \underline{\quad} = 0$
2. $(-7) + \underline{\quad} = 0$
3. $(-x) + \underline{\quad} = 0$
4. $\frac{1}{3} \cdot 3 = \underline{\quad}$
5. $x \cdot \underline{\quad} = 1 \quad x \neq 0$
6. $\frac{-6}{-6} = \underline{\quad}$
7. $\frac{4(2)}{\underline{\quad}} = 2$
8. $5 \cdot \frac{4}{5} = \underline{\quad}$
9. $\frac{-5(3)}{-5} = \underline{\quad}$

To answer questions such as "How many?," "How far?," "How fast?," and "How heavy?," we will often use mathematical statements called *equations*. In this chapter, we will discuss this important concept.

1 Determine whether a statement is an expression or an equation.

An **equation** is a statement indicating that two quantities are equal. Some examples of equations are

$$x + 5 = 21 \qquad 2x - 5 = 11 \qquad \text{and} \qquad 3x^2 - 4x + 5 = 0$$

The **expression** $3x + 2$ is not an equation, because we do not have two quantities that are being compared to one another. Some examples of expressions are

$$6x - 1 \qquad 3x^2 - x - 2 \qquad \text{and} \qquad -8(x + 1)$$

EXAMPLE 1 Determine whether the following are expressions or equations.

a. $9x^2 - 5x = 4$ **b.** $3x + 2$ **c.** $6(2x - 1) + 5$

Solution **a.** $9x^2 - 5x = 4$ is an equation because it indicates that $9x^2 - 5x$ is equal to 4.

b. $3x + 2$ is an expression because we do not have $3x + 2$ set equal to another quantity.

c. $6(2x - 1) + 5$ is an expression because we do not have $6(2x - 1) + 5$ set equal to another quantity.

SELF CHECK 1 Is $8(x + 1) = 4$ an expression or an equation?

2 Determine whether a number is a solution of an equation.

In the equation $x + 5 = 21$, the expression $x + 5$ is called the *left side* and 21 is called the *right side*. The letter x is called the **variable** (or the **unknown**).

An equation can be true or false. The equation $16 + 5 = 21$ is true, but the equation $10 + 5 = 21$ is false. The equation $2x - 5 = 11$ might be true or false, depending on the value of x. For example, when $x = 8$, the equation is true, because when we substitute 8 for x we obtain 11.

$$\begin{aligned} 2(8) - 5 &= 16 - 5 \\ &= 11 \end{aligned}$$

Any number that makes an equation true when substituted for its variable is said to *satisfy* the equation. A number that makes an equation true is called a **solution** or a **root** of the equation. Since 8 is the only number that satisfies the equation $2x - 5 = 11$, it is the only solution.

The **solution set** of an equation is the set of numbers that make the equation true. In the previous equation, the solution set is $\{8\}$.

EXAMPLE 2 Determine whether 6 is a solution of $3x - 5 = 2x$.

Solution To see whether 6 is a solution, we can substitute 6 for x and simplify.

$$\begin{aligned} 3x - 5 &= 2x \\ 3 \cdot \mathbf{6} - 5 &\stackrel{?}{=} 2 \cdot \mathbf{6} && \text{Substitute 6 for } x. \\ 18 - 5 &\stackrel{?}{=} 12 && \text{Do the multiplication.} \\ 13 &= 12 && \text{False.} \end{aligned}$$

Since $13 = 12$ is a false statement, 6 is not a solution.

SELF CHECK 2 Determine whether 1 is a solution of $2x + 3 = 5$.

3 Solve a linear equation in one variable by applying the addition or subtraction property of equality.

To solve an equation means to find its solutions. To develop an understanding of how to solve basic equations of the form $ax + b = c$, $a \neq 0$, called **linear equations**, we will refer to the scales shown in Figure 2-1. We can think of the scale shown in Figure 2-1(a) as representing the equation $x - 5 = 2$. The weight on the left side of the scale is $(x - 5)$ grams, and the weight on the right side is 2 grams. Because these weights are equal, the scale is in balance. To find the value of x, we need to isolate it by adding 5 grams to the left side of the scale. To keep the scale in balance, we must also add 5 grams to the right side. After adding 5 grams to both sides of the scale, we can see from Figure 2-1(b) that x grams will be balanced by 7 grams. We say that we have solved the equation and that the solution is 7, or we can say that the solution set is $\{7\}$.

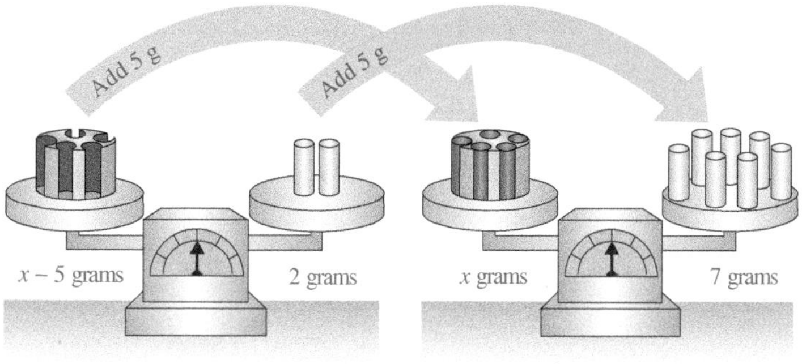

(a) (b)

Figure 2-1

Figure 2-1 suggests the **addition property of equality**: *If the same quantity is added to equal quantities, the results will be equal quantities.*

We can think of the scale shown in Figure 2-2(a) as representing the equation $x + 4 = 9$. The weight on the left side of the scale is $(x + 4)$ grams, and the weight on the right side is 9 grams. Because these weights are equal, the scale is in balance. To find the value of x, we need to isolate it by removing 4 grams from the left side. To keep the scale in balance, we must also remove 4 grams from the right side. In Figure 2-2(b), we can see that x grams will be balanced by 5 grams. We have found that the solution is 5, or that the solution set is $\{5\}$.

François Vieta (Viete)
1540–1603
By using letters in place of unknown numbers, Vieta simplified algebra and brought its notation closer to the notation that we use today. The one symbol he didn't use was the equal sign.

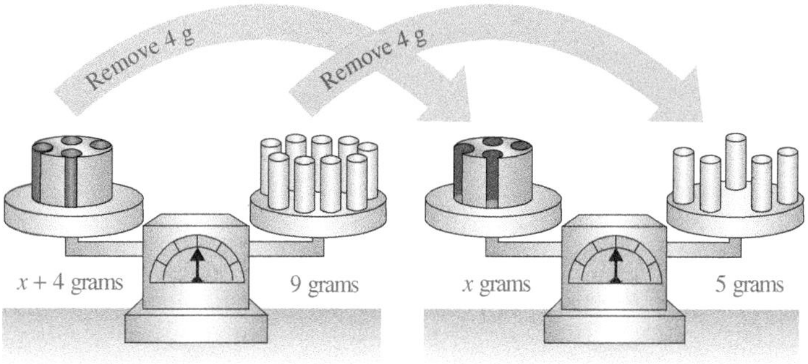

(a) (b)

Figure 2-2

Figure 2-2 suggests the *subtraction property of equality: If the same quantity is subtracted from equal quantities, the results will be equal quantities.*

The previous discussion justifies the following properties.

ADDITION PROPERTY OF EQUALITY

Suppose that a, b, and c are real numbers.

If $a = b$, then $a + c = b + c$.

SUBTRACTION PROPERTY OF EQUALITY

Suppose that a, b, and c are real numbers.

If $a = b$, then $a - c = b - c$.

COMMENT The subtraction property of equality is a special case of the addition property. Instead of subtracting a number from both sides of an equation, we could add the opposite of the number to both sides.

When we use the properties described above, the resulting equation will have the same solution set as the original one. We say that the equations are *equivalent*.

Unless otherwise noted, all content on this page is © Cengage Learning.

EQUIVALENT EQUATIONS Two equations are called **equivalent equations** when they have the same solution set.

Using the scales shown in Figures 2-1 and 2-2, we found that $x - 5 = 2$ is equivalent to $x = 7$ and $x + 4 = 9$ is equivalent to $x = 5$. In the next two examples, we use properties of equality to solve these equations algebraically.

EXAMPLE 3 Solve: $x - 5 = 2$

Solution To isolate x on one side of the = sign, we will use the addition property of equality to undo the subtraction of 5 by adding 5 to both sides of the equation.

$$x - 5 = 2$$
$$x - 5 + \mathbf{5} = 2 + \mathbf{5} \quad \text{Add 5 to both sides of the equation.}$$
$$x + 0 = 7 \quad \text{Apply the additive inverse property.}$$
$$x = 7 \quad \text{Apply the additive identity property.}$$

We check by substituting 7 for x in the original equation and simplifying.

$$x - 5 = 2$$
$$7 - 5 \stackrel{?}{=} 2 \quad \text{Substitute 7 for } x.$$
$$2 = 2 \quad \text{True.}$$

Since the previous statement is true, 7 is a solution. The solution set of this equation is $\{7\}$.

SELF CHECK 3 Solve: $b - 14 = 6$

EXAMPLE 4 Solve: $x + 4 = 9$

Solution To isolate x on one side of the = sign, we will use the subtraction property of equality to undo the addition of 4 by subtracting 4 from both sides of the equation.

COMMENT Note that Example 4 can be solved by using the addition property of equality. We would add -4 to both sides to undo the addition of 4.

$$x + 4 = 9$$
$$x + 4 - \mathbf{4} = 9 - \mathbf{4} \quad \text{Subtract 4 from both sides.}$$
$$x + 0 = 5 \quad \text{Apply the additive inverse property.}$$
$$x = 5 \quad \text{Apply the additive identity property.}$$

We can check by substituting 5 for x in the original equation and simplifying.

$$x + 4 = 9$$
$$5 + 4 \stackrel{?}{=} 9 \quad \text{Substitute 5 for } x.$$
$$9 = 9 \quad \text{True.}$$

Since the solution 5 checks, the solution set is $\{5\}$.

SELF CHECK 4 Solve: $a + 175 = 122$

4 Solve a linear equation in one variable by applying the multiplication or division property of equality.

We can think of the scale shown in Figure 2-3(a) as representing the equation $\frac{x}{3} = 12$. The weight on the left side of the scale is $\frac{x}{3}$ grams, and the weight on the right side is 12 grams. Because these weights are equal, the scale is in balance. To find the value of x, we can triple (or multiply by 3) the weight on each side. When we do this, the scale will remain in balance. From the scale shown in Figure 2-3(b), we can see that x grams will be balanced by 36 grams. Thus, $x = 36$. Since 36 is the solution of the equation, the solution set is $\{36\}$.

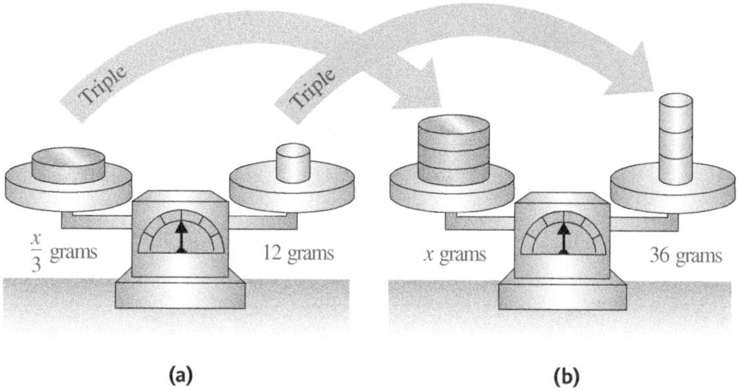

Figure 2-3

Perspective

To answer questions such as "How many?," "How far?," "How fast?," and "How heavy?," we often make use of equations. This concept has a long history, and the techniques that we will study in this chapter have been developed over many centuries.

The mathematical notation that we use today to solve equations is the result of thousands of years of development. The ancient Egyptians used a word for variables, best translated as *heap*. Others used the word *res*, which is Latin for *thing*. In the fifteenth century, the letters *p*: and *m*: were used for *plus* and *minus*. What we would now write as $2x + 3 = 5$ might have been written by those early mathematicians as

2 *res p*:3 *aequalis* 5

Figure 2-3 suggests the **multiplication property of equality**: *If equal quantities are multiplied by the same quantity, the results will be equal quantities.*

We will now consider how to solve the equation $2x = 6$. Since $2x$ means $2 \cdot x$, the equation can be written as $2 \cdot x = 6$. We can think of the scale shown in Figure 2-4(a) as representing this equation. The weight on the left side of the scale is $2 \cdot x$ grams, and the weight on the right side is 6 grams. Because these weights are equal, the scale is in balance. To find the value of x, we remove half of the weight from each side. This is equivalent to dividing the weight on both sides by 2. When we do this, the scale will remain in balance. From the scale shown in Figure 2-4(b), we can see that x grams will be balanced by 3 grams. Thus, $x = 3$. Since 3 is a solution of the equation, the solution set is $\{3\}$.

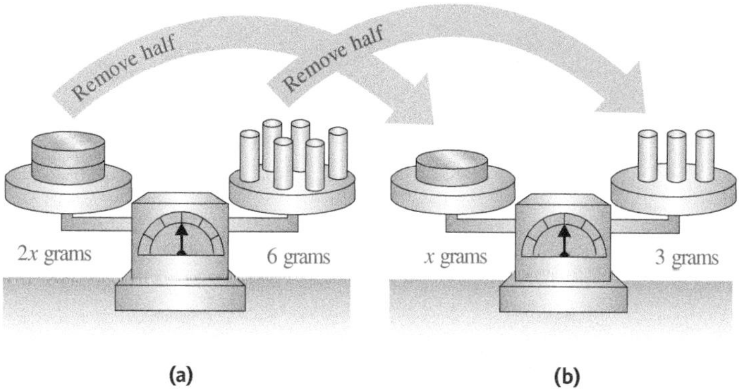

Figure 2-4

Figure 2-4 suggests the *division property of equality: If equal quantities are divided by the same quantity, the results will be equal quantities.*

Unless otherwise noted, all content on this page is © Cengage Learning.

The previous discussion justifies the following properties.

MULTIPLICATION PROPERTY OF EQUALITY

Suppose that a, b, and c are real numbers.

If $a = b$, then $ca = cb$.

DIVISION PROPERTY OF EQUALITY

Suppose that a, b, and c are real numbers and $c \neq 0$.

If $a = b$, then $\frac{a}{c} = \frac{b}{c}$.

COMMENT Since dividing by a number is the same as multiplying by its reciprocal, the division property is a special case of the multiplication property. However, because the reciprocal of 0 is undefined, we must exclude the possibility of division by 0.

When we use the multiplication and division properties, the resulting equations will be equivalent to the original ones.

To solve the previous equations algebraically, we proceed as in the next examples.

EXAMPLE 5 Solve: $\frac{x}{3} = 12$

Solution To isolate x on one side of the = sign, we use the multiplication property of equality to undo the division by 3 by multiplying both sides of the equation by 3.

$$\frac{x}{3} = 12$$

$$\mathbf{3} \cdot \frac{x}{3} = \mathbf{3} \cdot 12 \quad \text{Multiply both sides by 3.}$$

$$x = 36 \quad 3 \cdot \tfrac{x}{3} = x \text{ and } 3 \cdot 12 = 36.$$

Since 36 is a solution, the solution set is $\{36\}$. Verify that the solution checks.

SELF CHECK 5 Solve: $\frac{x}{5} = -7$

EXAMPLE 6 Solve: $2x = 6$

Solution To isolate x on one side of the = sign, we use the division property of equality to undo the multiplication by 2 by dividing both sides by 2.

$$2x = 6$$

$$\frac{2x}{\mathbf{2}} = \frac{6}{\mathbf{2}} \quad \text{Divide both sides by 2.}$$

$$x = 3 \quad \tfrac{2}{2} = 1 \text{ and } \tfrac{6}{2} = 3.$$

Since 3 is a solution, the solution set is $\{3\}$. Verify that the solution checks.

SELF CHECK 6 Solve: $-5x = 15$

COMMENT Note that we could have solved the equation in Example 6 by using the multiplication property of equality. To isolate x, we could have multiplied both sides by $\frac{1}{2}$.

EXAMPLE 7 Solve: $3x = \frac{1}{5}$

Solution To isolate x on the left side of the equation, we could undo the multiplication by 3 by dividing both sides by 3. However, it is easier to isolate x by multiplying both sides by the reciprocal of 3, which is $\frac{1}{3}$.

$$3x = \frac{1}{5}$$

$$\frac{1}{3}(3x) = \frac{1}{3}\left(\frac{1}{5}\right) \quad \text{Multiply both sides by } \tfrac{1}{3}.$$

$$\left(\frac{1}{3} \cdot 3\right)x = \frac{1}{15} \quad \text{Apply the associative property of multiplication.}$$

$$1x = \frac{1}{15} \quad \text{Apply the multiplicative inverse property.}$$

$$x = \frac{1}{15} \quad \text{Apply the multiplicative identity property.}$$

Since the solution is $\frac{1}{15}$, the solution set is $\left\{\frac{1}{15}\right\}$. Verify that the solution checks.

SELF CHECK 7 Solve: $-5x = \frac{1}{3}$

5 Solve a linear equation in one variable involving markdown and markup.

When the price of merchandise is reduced, the amount of reduction is called the **markdown** or the **discount**. To find the sale price of an item, we subtract the markdown from the regular price.

EXAMPLE 8 **BUYING FURNITURE** A sofa is on sale for \$650. If it has been marked down \$325, find its regular price.

Solution We can let r represent the regular price and substitute 650 for the sale price and 325 for the markdown in the following formula.

Sale price	equals	regular price	minus	markdown.
650	=	r	−	325

We can use the addition property of equality to solve the equation.

$$650 = r - 325$$

$$650 + \mathbf{325} = r - 325 + \mathbf{325} \quad \text{Add 325 to both sides.}$$

$$975 = r \quad 650 + 325 = 975 \text{ and } -325 + 325 = 0.$$

The regular price is \$975.

SELF CHECK 8 Find the regular price of the sofa if the discount is \$275.

To make a profit, a merchant must sell an item for more than he paid for it. The retail price of the item is the sum of its wholesale cost and the **markup**.

EXAMPLE 9 **BUYING CARS** A car with a sticker price of \$17,500 has a markup of \$3,500. Find the invoice price (the wholesale price) to the dealer.

Solution We can let w represent the wholesale price and substitute 17,500 for the retail price and 3,500 for the markup in the following formula.

Retail price	equals	wholesale cost	plus	markup.
17,500	$=$	w	$+$	3,500

We can use the subtraction property of equality to solve the equation.

$$17{,}500 = w + 3{,}500$$

$$17{,}500 - \mathbf{3{,}500} = w + 3{,}500 - \mathbf{3{,}500} \quad \text{Subtract 3,500 from both sides.}$$

$$14{,}000 = w \quad 17{,}500 - 3{,}500 = 14{,}000 \text{ and } 3{,}500 - 3{,}500 = 0.$$

The invoice price is \$14,000.

SELF CHECK 9 Find the invoice price of the car if the markup is \$6,700.

6 Solve a percent problem involving a linear equation in one variable using the formula $rb = a$.

A percent is the numerator of a fraction with a denominator of 100. For example, $6\frac{1}{4}$ percent (written as $6\frac{1}{4}\%$) is the fraction $\frac{6.25}{100}$, or the decimal 0.0625. In problems involving percent, the word *of* usually means multiplication. For example, $6\frac{1}{4}\%$ of 8,500 is the product of 0.0625 and 8,500.

$$6\tfrac{1}{4}\% \text{ of } 8{,}500 = 0.0625 \cdot 8{,}500$$
$$= 531.25$$

In the statement $6\frac{1}{4}\%$ of 8,500 = 531.25, the percent $6\frac{1}{4}\%$ is called a **rate**, 8,500 is called the **base**, and their product, 531.25, is called the **amount**. Every percent problem is based on the equation rate · base = amount.

PERCENT FORMULA If r is the rate, b is the base, and a is the amount, then

$$rb = a.$$

COMMENT Note that the percent formula can be written in the equivalent form $a = rb$.

Percent problems involve questions such as the following.

- What is 30% of 1,000? We must find the amount.
- 45% of what number is 405? We must find the base.
- What percent of 400 is 60? We must find the rate.

When we substitute the values of the rate, base, and amount into the percent formula, we will obtain an equation that we can solve.

EXAMPLE 10 What is 30% of 1,000?

Solution In this example, the rate r is 30% and the base is 1,000. We must find the amount.

Rate	·	base	=	amount.
30%	of	1,000	is	the amount.

We can substitute these values into the percent formula and solve for a.

$$rb = a$$
$$\mathbf{30\% \cdot 1{,}000} = a \quad \text{Substitute 30\% for } r \text{ and 1,000 for } b.$$
$$0.30 \cdot 1{,}000 = a \quad \text{Change 30\% to the decimal 0.30.}$$
$$300 = a \quad \text{Multiply.}$$

Thus, 30% of 1,000 is 300.

SELF CHECK 10 Find 45% of 800.

EXAMPLE 11 45% of what number is 405?

Solution In this example, the rate r is 45% and the amount a is 405. We must find the base.

Rate	·	base	=	amount.
45%	of	what number	is	405?

We can substitute these values into the percent formula and solve for b.

$$rb = a$$
$$\mathbf{45\%} \cdot b = \mathbf{405} \quad \text{Substitute 45\% for } r \text{ and 405 for } a.$$
$$0.45 \cdot b = 405 \quad \text{Change 45\% to a decimal.}$$
$$\frac{0.45b}{\mathbf{0.45}} = \frac{405}{\mathbf{0.45}} \quad \text{To undo the multiplication by 0.45, divide both sides by 0.45.}$$
$$b = 900 \quad \frac{0.45}{0.45} = 1 \text{ and } \frac{405}{0.45} = 900.$$

Thus, 45% of 900 is 405.

SELF CHECK 11 35% of what number is 280?

EXAMPLE 12 What percent of 400 is 60?

Solution In this example, the base b is 400 and the amount a is 60. We must find the rate.

Rate	·	base	=	amount.
What percent	of	400	is	60?

We can substitute these values in the percent formula and solve for r.

$$rb = a$$
$$r \cdot \mathbf{400} = \mathbf{60} \quad \text{Substitute 400 for } b \text{ and 60 for } a.$$
$$\frac{400r}{\mathbf{400}} = \frac{60}{\mathbf{400}} \quad \text{To undo the multiplication by 400, divide both sides by 400.}$$
$$r = 0.15 \quad \frac{400}{400} = 1 \text{ and } \frac{60}{400} = 0.15.$$
$$r = 15\% \quad \text{To change the decimal into a percent, we multiply by 100 and insert a \% sign.}$$

Thus, 15% of 400 is 60.

SELF CHECK 12 What percent of 600 is 150?

7 Solve an application involving percents.

The ability to solve linear equations enables us to solve many applications. This is what makes the algebra relevant to our lives.

EXAMPLE 13 **INVESTING** At a stockholders meeting, members representing 4.5 million shares voted in favor of a proposal for a mandatory retirement age for the members of the board of directors. If these shares represented 75% of the number of shares outstanding, how many shares were outstanding?

Solution Let b represent the number of outstanding shares. Then 75% of b is 4.5 million. We can substitute 75% for r and 4.5 million for a in the percent formula and solve for b.

$$rb = a$$

$$\mathbf{75\%} \cdot b = \mathbf{4{,}500{,}000} \quad \text{4.5 million} = 4{,}500{,}000.$$

$$0.75b = 4{,}500{,}000 \quad \text{Change 75\% to a decimal.}$$

$$\frac{0.75b}{\mathbf{0.75}} = \frac{4{,}500{,}000}{\mathbf{0.75}} \quad \text{To undo the multiplication of 0.75, divide both sides by 0.75.}$$

$$b = 6{,}000{,}000 \quad \text{Divide.}$$

There were 6 million shares outstanding.

SELF CHECK 13 If the 4.5 million shares represented 60% of the number of shares outstanding, how many shares were outstanding?

EXAMPLE 14 **QUALITY CONTROL** After examining 240 sweaters, a quality-control inspector found 5 with defective stitching, 8 with mismatched designs, and 2 with incorrect labels. What percent were defective?

Solution Let r represent the percent that are defective. Then the base b is 240 and the amount a is the number of defective sweaters, which is $5 + 8 + 2 = 15$. We can find r by using the percent formula.

$$rb = a$$

$$r \cdot \mathbf{240} = \mathbf{15} \quad \text{Substitute 240 for } b \text{ and 15 for } a.$$

$$\frac{240r}{\mathbf{240}} = \frac{15}{\mathbf{240}} \quad \text{To undo the multiplication of 240, divide both sides by 240.}$$

$$r = 0.0625 \quad \text{Divide.}$$

$$r = 6.25\% \quad \text{To change 0.0625 to a percent, multiply by 100 and insert a \% sign.}$$

The defect rate is 6.25%.

SELF CHECK 14 If the inspector examined 400 sweaters and found the same number of defects as in Example 14, what percent were defective?

SELF CHECK ANSWERS

1. equation **2.** yes **3.** 20 **4.** −53 **5.** −35 **6.** −3 **7.** $-\frac{1}{15}$ **8.** \$925 **9.** \$10,800 **10.** 360 **11.** 800 **12.** 25% **13.** 7,500,000 or 7.5 million **14.** 3.75%

NOW TRY THIS

Solve each equation.

1. $\frac{x}{12} = 0$
2. $\frac{2}{3}x = 24$
3. $-25 + x = 25$

2.1 Exercises

WARM-UPS *State the property of equality you would use to solve each equation. Do not solve.*

1. $x - 5 = 15$
2. $x - 3 = 13$
3. $w + 5 = 7$
4. $x + 32 = 36$
5. $-8x = -24$
6. $-7x = 14$
7. $\frac{x}{5} = 2$
8. $\frac{x}{2} = -10$

REVIEW *Perform the operations. Simplify the result when possible.*

9. $\frac{4}{5} + \frac{2}{3}$
10. $\frac{5}{6} \cdot \frac{12}{25}$
11. $\frac{5}{9} \div \frac{3}{5}$
12. $\frac{15}{7} - \frac{10}{3}$
13. $3 + 5 \cdot 6$
14. $3 \cdot 4^2$
15. $3 + 4^3(-5)$
16. $\frac{5(-4) - 3(-2)}{10 - (-4)}$

VOCABULARY AND CONCEPTS *Fill in the blanks.*

17. An _______ is a statement that two quantities are equal. An _________ is a mathematical statement without an = sign.
18. A _______ or ____ of an equation is a number that satisfies the equation.
19. If two equations have the same solutions, they are called _________ equations.
20. To solve a linear equation, we isolate the _______, or unknown, on one side of the equation.
21. If the same quantity is added to _____ quantities, the results will be equal quantities.
22. If the same quantity is subtracted from equal quantities, the results will be _____ quantities.
23. If equal quantities are multiplied or divided by the same nonzero quantity, the results are _____ quantities.
24. An equation in the form $x + b = c$ is called a _____ equation.
25. Sale price = ___________ − markdown
26. Retail price = wholesale cost + _______
27. A *percent* is the numerator of a fraction whose denominator is ___.
28. Rate · ____ = amount

GUIDED PRACTICE *Determine whether each statement is an expression or an equation.* SEE EXAMPLE 1. (OBJECTIVE 1)

29. $x = -4$
30. $y = 3$
31. $6x + 7$
32. $2(x - 3) + 1$
33. $x^2 + 2x = 3$
34. $3(x + 1) = 9$
35. $3(x - 4)$
36. $5(2 + x)$

Determine whether the given number is a solution of the equation. SEE EXAMPLE 2. (OBJECTIVE 2)

37. $x + 3 = 6; 3$
38. $x + 5 = 8; 3$
39. $2y - 5 = y; 4$
40. $x - 7 = 2; 9$
41. $\frac{y}{7} = 4; 28$
42. $\frac{c}{-5} = -2; -10$
43. $\frac{x}{5} = x; 0$
44. $\frac{x}{7} = 7x; 0$
45. $3k + 5 = 5k - 1; 3$
46. $2s - 1 = s + 7; 6$
47. $\frac{5 + x}{10} - x = \frac{1}{2}; 0$
48. $\frac{x - 5}{6} = 12 - x; 11$

Use the addition property of equality to solve each equation. Check all solutions. SEE EXAMPLE 3. (OBJECTIVE 3)

49. $a - 6 = 9$
50. $y - 9 = 20$
51. $b - 5 = -19$
52. $m - 5 = -12$
53. $4 = c - 9$
54. $1 = y - 5$
55. $r - \frac{1}{5} = \frac{3}{10}$
56. $\frac{4}{3} = -\frac{2}{3} + x$

Use the subtraction property of equality to solve each equation. Check all solutions. SEE EXAMPLE 4. (OBJECTIVE 3)

57. $y + 8 = 11$
58. $x + 4 = 12$
59. $a + 9 = -12$
60. $c + 6 = -9$
61. $41 = 45 + q$
62. $0 = r + 10$
63. $k + \frac{2}{3} = \frac{1}{5}$
64. $b + \frac{4}{7} = \frac{15}{14}$

Use the multiplication property of equality to solve each equation. Check all solutions. SEE EXAMPLE 5. (OBJECTIVE 4)

65. $\frac{x}{6} = 3$
66. $\frac{y}{11} = 4$
67. $\frac{b}{3} = 5$
68. $\frac{a}{5} = -3$
69. $\frac{a}{3} = \frac{1}{9}$
70. $\frac{a}{13} = \frac{1}{26}$

71. $\frac{u}{5} = -\frac{3}{10}$

72. $\frac{t}{-7} = \frac{1}{2}$

Use the division property of equality to solve each equation. Check all solutions. SEE EXAMPLE 6. (OBJECTIVE 4)

73. $7x = 28$

74. $25x = 625$

75. $11x = -121$

76. $-8a = -32$

77. $-4x = 36$

78. $-16y = 64$

79. $4w = 108$

80. $-66 = -6w$

Use the multiplication or division property of equality to solve each equation. Check all solutions. SEE EXAMPLE 7. (OBJECTIVE 4)

81. $5x = \frac{5}{8}$

82. $6x = \frac{2}{3}$

83. $\frac{1}{7}w = 14$

84. $-19x = -57$

85. $-1.2w = -102$

86. $1.5a = -15$

87. $0.25x = 1{,}228$

88. $-0.2y = 51$

Solve each application involving markdown or markup. SEE EXAMPLES 8–9. (OBJECTIVE 5)

89. Buying boats A boat is on sale for $7,995. Find its regular price if it has been marked down $1,350.

90. Buying houses A house that was priced at $105,000 has been discounted $7,500. Find the new asking price.

91. Buying clothes A sport jacket that sells for $175 has a markup of $85. Find the wholesale price.

92. Buying vacuum cleaners A vacuum that sells for $97 has a markup of $37. Find the wholesale price.

Use the formula $rb = a$ or $a = rb$ to find each value. SEE EXAMPLES 10–12. (OBJECTIVE 6)

93. What number is 40% of 200?

94. What number is 45% of 340?

95. What number is 50% of 38?

96. What number is 25% of 300?

97. 35% of what number is 182?

98. 26% of what number is 78?

99. 48 is 15% of what number?

100. 13.3 is 3.5% of what number?

101. 28% of what number is 42?

102. 44% of what number is 143?

103. What percent of 357.5 is 71.5?

104. What percent of 254 is 13.208?

ADDITIONAL PRACTICE

Solve each equation. Be sure to check each answer.

105. $p + 0.27 = 3.57$

106. $m - 5.36 = 1.39$

107. $\frac{x}{15} = -4$

108. $\frac{y}{16} = -5$

109. $-57 = b - 29$

110. $-93 = 67 + y$

111. $y - 2.63 = -8.21$

112. $s + 8.56 = 5.65$

113. $\frac{y}{-3} = -\frac{5}{6}$

114. $\frac{y}{-8} = -\frac{3}{16}$

115. $-18 + y = 18$

116. $-43 + a = -43$

117. $-3 = \frac{x}{11}$

118. $\frac{w}{-12} = 4$

119. $b + 7 = \frac{20}{3}$

120. $x + \frac{5}{7} = -\frac{2}{7}$

121. $3x = -\frac{1}{4}$

122. $-8x = -8$

123. $-\frac{3}{5} = x - \frac{2}{5}$

124. $d + \frac{2}{3} = \frac{3}{2}$

125. $\frac{1}{7}x = \frac{5}{7}$

126. $-17x = -51$

127. $-27w = 81$

128. $15 = \frac{r}{-5}$

129. $18x = -9$

130. $-12x = 3$

Find each value.

131. 0.48 is what percent of 8?

132. 3.6 is what percent of 28.8?

133. 34 is what percent of 17?

134. 39 is what percent of 13?

APPLICATIONS

Solve each application involving percents. SEE EXAMPLES 13–14. (OBJECTIVE 7)

135. Selling microwave ovens The 5% sales tax on a microwave oven amounts to $13.50. What is the microwave's selling price?

136. Hospitals 18% of hospital patients stay for less than 1 day. If 1,008 patients in January stayed for less than 1 day, what total number of patients did the hospital treat in January?

137. Sales taxes Sales tax on a $12 compact disc is $0.72. At what rate is sales tax computed?

138. Home prices The average price of homes in one neighborhood decreased 8% since last year, a drop of $7,800. What was the average price of a home last year?

Solve.

139. Banking The amount A in an account is given by the formula

$$A = p + i$$

where p is the principal and i is the interest. How much interest was earned if an original deposit (the principal) of $4,750 has grown to be $5,010?

140. Selling real estate The money m received from selling a house is given by the formula

$$m = s - c$$

where s is the selling price and c is the agent's commission. Find the selling price of a house if the seller received $217,000 and the agent received $13,020.

141. Customer satisfaction One-third of the movie audience left the theater in disgust. If 78 angry patrons walked out, how many were there originally?

142. Off-campus housing One-seventh of the senior class is living in off-campus housing. If 217 students live off campus, how large is the senior class?

143. Shopper dissatisfaction Refer to the survey results shown in the table. What percent of those surveyed were not pleased?

Shopper survey results	
First-time shoppers	1,731
Major purchase today	539
Shopped within previous month	1,823
Satisfied with service	4,140
Seniors	2,387
Total surveyed	9,200

144. Shopper satisfaction Refer to the survey results shown in the table above. What percent of those surveyed were satisfied with their service?

145. Union membership If 2,484 union members represent 90% of a factory's work force, how many workers are employed?

146. Charities Out of $237,000 donated to a certain charity, $5,925 is used to pay for fund-raising expenses. What percent of the donations is overhead?

147. Stock splits After a 3-for-2 stock split, each shareholder will own 1.5 times as many shares as before. If 555 shares are owned after the split, how many were owned before?

148. Stock splits After a 2-for-1 stock split, each shareholder owned twice as many shares as before. If 2,570 shares are owned after the split, how many were owned before?

149. Depreciation Find the original cost of a car that is worth $10,250 after depreciating $7,500.

150. Appreciation Find the original purchase price of a house that is worth $150,000 and has appreciated $57,000.

151. Taxes Find the tax paid on an item that was priced at $37.10 and cost $39.32.

152. Buying carpets How much did it cost to install $317 worth of carpet that cost $512?

153. Buying paint After reading this ad, a decorator bought 1 gallon of primer, 1 gallon of paint, and a brush. If the total cost was $30.44, find the cost of the brush.

154. Painting a room After reading the ad above, a woman bought 2 gallons of paint, 1 gallon of primer, and a brush. If the total cost was $46.94, find the cost of the brush.

155. Buying real estate The cost of a condominium is $57,595 less than the cost of a house. If the house costs $202,744, find the cost of the condominium.

156. Buying airplanes The cost of a twin-engine plane is $175,260 less than the cost of a two-seater jet. If the jet cost $321,435, find the cost of the twin-engine plane.

WRITING ABOUT MATH

157. Explain what it means for a number to satisfy an equation.

158. How can you tell whether a number is the solution to an equation?

SOMETHING TO THINK ABOUT

159. The Ahmes papyrus mentioned on page 9 contains this statement: *A circle nine units in diameter has the same area as a square eight units on a side.* From this statement, determine the ancient Egyptians' approximation of π.

160. Calculate the Egyptians' *percent of error*: What percent of the actual value of π is the difference of the estimate obtained in Exercise 159 and the actual value of π?

Section 2.2 Solving More Linear Equations in One Variable

Objectives

1. Solve a linear equation in one variable requiring more than one property of equality.
2. Solve an application requiring more than one property of equality.
3. Solve an application involving percent of increase or decrease.

Unless otherwise noted, all content on this page is © Cengage Learning.

Vocabulary

clearing fractions | percent of increase | percent of decrease

Getting Ready

Perform the operations.

1. $7 + 3 \cdot 5$
2. $3(5 + 7)$
3. $\dfrac{3 + 7}{2}$
4. $3 + \dfrac{7}{2}$
5. $\dfrac{3(5 - 8)}{9}$
6. $3 \cdot \dfrac{5 - 8}{9}$
7. $\dfrac{3 \cdot 5 - 8}{9}$
8. $3 \cdot \dfrac{5}{9} - 8$

We have solved linear equations in one variable by using one of the addition, subtraction, multiplication, and division properties of equality. To solve some equations, we need to use several of these properties in succession.

1 Solve a linear equation in one variable requiring more than one property of equality.

In the following examples, we will combine the addition or subtraction property with the multiplication or division property to solve more complicated equations.

Everyday connections
Renting a Car

Rental car rates for various cars are given for three different companies. In each formula, the constant represents the base charge.

Economy car from Dan's Rentals

$C = 19.95x + 39.99$

Luxury car from Spencer's Cars

$C = 55x + 124.15$

SUV from Tyler's Auto Rentals

$C = 35.88x + 89.95$

1. Interpret the meaning of the variables x and C.

2. What does the coefficient of x represent in each equation?

We can solve an equation to compare the companies to one another.

3. Suppose you have \$900 available to spend on a rental car. Find the number of days you can afford to rent from each company. (*Hint*: Substitute 900 for C.)

Dan's Rentals

Spencer's Cars

Tyler's Auto Rentals

EXAMPLE 1 Solve: $-12x + 5 = 17$

Solution The left side of the equation indicates that x is to be multiplied by -12 and then 5 is to be added to that product. To isolate x, we must undo these operations in the reverse order.

- To undo the addition of 5, we subtract 5 from both sides.
- To undo the multiplication by -12, we divide both sides by -12.

Unless otherwise noted, all content on this page is © Cengage Learning.

$$-12x + 5 = 17$$

$$-12x + 5 - 5 = 17 - 5 \quad \text{To undo the addition of 5, subtract 5 from both sides.}$$

$$-12x = 12 \quad 5 - 5 = 0 \text{ and } 17 - 5 = 12.$$

$$\frac{-12x}{-12} = \frac{12}{-12} \quad \text{To undo the multiplication by } -12\text{, divide both sides by } -12.$$

$$x = -1 \quad \frac{-12}{-12} = 1 \text{ and } \frac{12}{-12} = -1.$$

Check:

$$-12x + 5 = 17$$

$$-12(-1) + 5 \stackrel{?}{=} 17 \quad \text{Substitute } -1 \text{ for } x.$$

$$12 + 5 \stackrel{?}{=} 17 \quad \text{Multiply.}$$

$$17 = 17 \quad \text{True.}$$

Since $17 = 17$, the solution -1 checks and the solution set is $\{-1\}$.

SELF CHECK 1 Solve: $2x + 3 = 15$

EXAMPLE 2 Solve: $\frac{x}{3} - 7 = -3$

Solution The left side of the equation indicates that x is to be divided by 3 and then 7 is to be subtracted from that quotient. To isolate x, we must undo these operations in the reverse order.

- To undo the subtraction of 7, we add 7 to both sides.
- To undo the division by 3, we multiply both sides by 3.

$$\frac{x}{3} - 7 = -3$$

$$\frac{x}{3} - 7 + 7 = -3 + 7 \quad \text{To undo the subtraction of 7, add 7 to both sides.}$$

$$\frac{x}{3} = 4 \quad -7 + 7 = 0 \text{ and } -3 + 7 = 4.$$

$$3 \cdot \frac{x}{3} = 3 \cdot 4 \quad \text{To undo the division by 3, multiply both sides by 3.}$$

$$x = 12 \quad 3 \cdot \frac{x}{3} = x \text{ and } 3 \cdot 4 = 12.$$

Check:

$$\frac{x}{3} - 7 = -3$$

$$\frac{12}{3} - 7 \stackrel{?}{=} -3 \quad \text{Substitute 12 for } x.$$

$$4 - 7 \stackrel{?}{=} -3 \quad \text{Simplify.}$$

$$-3 = -3$$

Since $-3 = -3$, the solution 12 checks and the solution set is $\{12\}$.

SELF CHECK 2 Solve: $\frac{x}{4} - 3 = 5$

EXAMPLE 3 Solve: $\dfrac{x-7}{3}=9$

Solution The left side of the equation indicates that 7 is to be subtracted from x and that the difference is to be divided by 3. To isolate x, we must undo these operations in the reverse order.

- To undo the division by 3, we multiply both sides by 3.
- To undo the subtraction of 7, we add 7 to both sides.

$$\frac{x-7}{3}=9$$

$$3\left(\frac{x-7}{3}\right)=3(9) \quad \text{To undo the division by 3, multiply both sides by 3.}$$

$$x-7=27 \quad 3\cdot\tfrac{1}{3}=1 \text{ and } 3(9)=27.$$

$$x-7+7=27+7 \quad \text{To undo the subtraction of 7, add 7 to both sides.}$$

$$x=34 \quad -7+7=0 \text{ and } 27+7=34.$$

Since the solution is 34, the solution set is $\{34\}$. Verify that the solution checks.

SELF CHECK 3 Solve: $\dfrac{a-3}{5}=-2$

EXAMPLE 4 Solve: $\dfrac{3x}{4}+\dfrac{2}{3}=-7$

Solution The left side of the equation indicates that x is to be multiplied by 3, then $3x$ is to be divided by 4, and then $\frac{2}{3}$ is to be added to that result. To isolate x, we must undo these operations in the reverse order.

COMMENT We undo the multiplication by 3 by multiplying both sides by $\frac{1}{3}$ to simplify the arithmetic. Dividing both sides by 3 would introduce a complex fraction.

- To undo the addition of $\frac{2}{3}$, we subtract $\frac{2}{3}$ from both sides.
- To undo the division by 4, we multiply both sides by 4.
- To undo the multiplication by 3, we multiply both sides by $\frac{1}{3}$.

$$\frac{3x}{4}+\frac{2}{3}=-7$$

$$\frac{3x}{4}+\frac{2}{3}-\frac{2}{3}=-7-\frac{2}{3} \quad \text{To undo the addition of } \tfrac{2}{3}\text{, subtract } \tfrac{2}{3} \text{ from both sides.}$$

$$\frac{3x}{4}=-\frac{23}{3} \quad \tfrac{2}{3}-\tfrac{2}{3}=0 \text{ and } -7-\tfrac{2}{3}=-\tfrac{23}{3}.$$

$$4\left(\frac{3x}{4}\right)=4\left(-\frac{23}{3}\right) \quad \text{To undo the division by 4, multiply both sides by 4.}$$

$$3x=-\frac{92}{3} \quad 4\cdot\tfrac{3x}{4}=3x \text{ and } 4\left(-\tfrac{23}{3}\right)=-\tfrac{92}{3}.$$

$$\frac{1}{3}(3x)=\frac{1}{3}\left(-\frac{92}{3}\right) \quad \text{To undo the multiplication by 3, multiply both sides by } \tfrac{1}{3}.$$

$$x=-\frac{92}{9} \quad \tfrac{1}{3}\cdot 3x=x \text{ and } \tfrac{1}{3}\left(-\tfrac{92}{3}\right)=-\tfrac{92}{9}.$$

Since the solution is $-\frac{92}{9}$, the solution set is $\left\{-\frac{92}{9}\right\}$. Verify that the solution checks.

SELF CHECK 4 Solve: $\dfrac{2x}{3}-\dfrac{4}{5}=3$

An alternate method for solving Example 4 is to **clear fractions** first. In this method, we multiply both sides of the equation by the least common multiple of the denominators.

Solve: $\frac{3x}{4} + \frac{2}{3} = -7$

The least common multiple for 4 and 3 is 12. We will multiply both sides of the equation by 12.

$$12\left(\frac{3x}{4} + \frac{2}{3}\right) = \mathbf{12}(-7) \quad \text{Multiply both sides by 12.}$$

$$12\left(\frac{3x}{4}\right) + 12\left(\frac{2}{3}\right) = \mathbf{12}(-7) \quad \text{Use the distributive property to remove the parentheses.}$$

$$\overset{3}{\cancel{12}}\left(\frac{3x}{\cancel{4}_1}\right) + \overset{4}{\cancel{12}}\left(\frac{2}{\cancel{3}_1}\right) = \mathbf{-84} \quad \text{Simplify each fraction.}$$

$$9x + 8 = -84 \quad \text{Simplify.}$$

$$9x + 8 \mathbf{- 8} = -84 \mathbf{- 8} \quad \text{Subtract 8 from both sides.}$$

$$9x = \mathbf{-92} \quad \text{Simplify.}$$

$$\frac{9x}{\mathbf{9}} = \frac{-92}{\mathbf{9}} \quad \text{Divide both sides by 9.}$$

$$x = -\frac{92}{9}$$

This method is one you will use when solving rational equations later in this textbook.

2 Solve an application requiring more than one property of equality.

EXAMPLE 5 **ADVERTISING** A store manager hires a student to distribute advertising circulars door to door. The student will be paid \$24 a day plus 12¢ for every ad she distributes. How many ads must she distribute to earn \$42 in one day?

Solution We can let a represent the number of ads that the student must distribute. Her earnings can be expressed in two ways: as \$24 plus the 12¢-apiece pay for distributing the ads, and as \$42.

\$24	plus	*a* ads at \$0.12 each	is	\$42	12¢ = \$0.12
24	+	0.12*a*	=	42	

We can solve this equation as follows:

$$24 + 0.12a = 42$$

$$24 \mathbf{- 24} + 0.12a = 42 \mathbf{- 24} \quad \text{To undo the addition of 24, subtract 24 from both sides.}$$

$$0.12a = 18 \quad 24 - 24 = 0 \text{ and } 42 - 24 = 18.$$

$$\frac{0.12a}{\mathbf{0.12}} = \frac{18}{\mathbf{0.12}} \quad \text{To undo the multiplication by 0.12, divide both sides by 0.12.}$$

$$a = 150 \quad \frac{0.12}{0.12} = 1 \text{ and } \frac{18}{0.12} = 150.$$

The student must distribute 150 ads. Check the result.

SELF CHECK 5 How many ads must the student deliver in one day to earn \$48?

3 Solve an application involving percent of increase or decrease.

We have seen that the retail price of an item is the sum of the cost and the markup.

Retail price equals cost plus markup.

Often, the markup is expressed as a percent of the cost.

Markup equals Percent of markup times cost.

Suppose a store manager buys toasters for \$21 and sells them at a 17% markup. To find the retail price, the manager begins with his cost and adds 17% of that cost.

$$\begin{aligned}\text{Retail price} &= \text{cost} + \text{markup.}\\ &= \text{cost} + \text{Percent of markup} \cdot \text{cost}\\ &= 21 + 0.17 \cdot 21\\ &= 21 + 3.57\\ &= 24.57\end{aligned}$$

The retail price of a toaster is \$24.57.

EXAMPLE 6 **ANTIQUE CARS** In 1956, a Chevrolet BelAir automobile sold for \$4,000. Today, it is worth about \$28,600. Find the percent that its value has increased, called the **percent of increase**.

Solution We let p represent the percent of increase, expressed as a decimal.

Current price equals original price plus p(original price).

$$28{,}600 = 4{,}000 + p(4{,}000)$$

$$28{,}600 - \mathbf{4{,}000} = 4{,}000 - \mathbf{4{,}000} + 4{,}000p$$ To undo the addition of 4,000, subtract 4,000 from both sides.

$$24{,}600 = 4{,}000p$$ $28{,}600 - 4{,}000 = 24{,}600$ and $4{,}000 - 4{,}000 = 0$.

$$\frac{24{,}600}{\mathbf{4{,}000}} = \frac{4{,}000p}{\mathbf{4{,}000}}$$ To undo the multiplication by 4,000, divide both sides by 4,000.

$$6.15 = p$$ Simplify.

To convert 6.15 to a percent, we multiply by 100 and insert a % sign. Since the percent of increase is 615%, the car has appreciated 615%.

SELF CHECK 6 Find the percent of increase if the car sells for \$30,000.

We have seen that when the price of merchandise is reduced, the amount of reduction is the markdown (also called the *discount*).

Sale price equals regular price minus markdown.

Usually, the markdown is expressed as a percent of the regular price.

Markdown equals percent of markdown times regular price.

Suppose that a TV set that regularly sells for \$570 has been marked down 25%. That means the customer will pay 25% less than the regular price. To find the sale price, we use the formula

Sale price	=	regular price	−	markdown		
	=	regular price	−	percent of markdown	·	regular price
	=	\$570	−	25%	of	\$570

$= \$570 - (0.25)(\$570)$ Write 25% as a decimal.

$= \$570 - \$142.50$ Multiply.

$= \$427.50$ Subtract.

The TV set is selling for \$427.50.

EXAMPLE 7 **BUYING CAMERAS** A camera that was originally priced at \$452 is on sale for \$384.20. Find the percent of markdown.

Solution We let p represent the percent of markdown, expressed as a decimal, and substitute \$384.20 for the sale price and \$452 for the regular price.

Sale price	equals	regular price	minus	percent of markdown	times	regular price.
384.20	=	452	−	p	·	452

$384.20 - \mathbf{452} = 452 - \mathbf{452} - p(452)$ To undo the addition of 452, subtract 452 from both sides.

$-67.80 = -p(452)$ $384.20 - 452 = -67.80$; $452 - 452 = 0$

$\frac{-67.80}{\mathbf{-452}} = \frac{-p(452)}{\mathbf{-452}}$ To undo the multiplication by −452, divide both sides by −452.

$0.15 = p$ $\frac{-67.80}{-452} = 0.15$ and $\frac{-452}{-452} = 1$.

The camera is on sale at a 15% markdown.

SELF CHECK 7 If the camera is reduced another \$22.60, find the percent of markdown.

COMMENT When a price increases from \$100 to \$125, the percent of increase is 25%. When the price *decreases* from \$125 to \$100, the **percent of decrease** is 20%. These different results occur because the percent of increase is a percent of the original (smaller) price, \$100. The percent of decrease is a percent of the original (larger) price, \$125.

SELF CHECK ANSWERS **1.** 6 **2.** 32 **3.** −7 **4.** $\frac{57}{10}$ **5.** 200 **6.** 650% **7.** 20%

NOW TRY THIS

Solve each equation.

1. $\frac{2}{7}x + 3 = 3$

2. $10 - \frac{2}{3}x = -6$

3. $-0.2x - 4.3 = -10.7$

2.2 Exercises

WARM-UPS *What would you do first when solving each equation?*

1. $6x - 9 = -15$
2. $15 = \frac{x}{5} + 3$
3. $\frac{x}{7} - 3 = 0$
4. $\frac{x - 3}{7} = -7$
5. $\frac{x - 7}{3} = 5$
6. $\frac{3x - 5}{2} + 2 = 0$

Solve each equation.

7. $7z - 7 = 14$
8. $\frac{p - 1}{2} = 6$

REVIEW *Refer to the formulas given in Section 1.3.*

9. Find the perimeter of a rectangle with sides measuring 8.5 and 16.5 cm.
10. Find the area of a rectangle with sides measuring 2.3 in. and 3.7 in.
11. Find the area of a trapezoid with a height of 8.5 in. and bases measuring 6.7 in. and 12.2 in.
12. Find the volume of a rectangular solid with dimensions of 8.2 cm by 7.6 cm by 10.2 cm.

VOCABULARY AND CONCEPTS *Fill in the blanks.*

13. Retail price = ____ + markup
14. Markup = percent of markup · ____
15. Markdown = ______ of markdown · regular price
16. Another word for markdown is _______.
17. The percent that an object has increased in value is called the ______________.
18. The percent that an object has deceased in value is called the ______________.

GUIDED PRACTICE *Solve each equation. Check all solutions. SEE EXAMPLE 1. (OBJECTIVE 1)*

19. $5x - 1 = 4$
20. $5x + 3 = 8$
21. $-6x + 2 = 14$
22. $4x - 4 = 8$
23. $6x + 2 = -4$
24. $4x - 4 = 4$
25. $3x - 8 = 1$
26. $7x - 19 = 2$
27. $4x - 7 = 5$
28. $-8x + 5 = 21$
29. $-3x - 6 = 12$
30. $5x + 9 = -16$
31. $-2x - 8 = -2$
32. $-3x + 17 = -4$

Solve each equation. Check all solutions. SEE EXAMPLE 2. (OBJECTIVE 1)

33. $\frac{z}{9} + 5 = -1$
34. $\frac{y}{5} - 3 = 3$
35. $\frac{x}{4} + 7 = 3$
36. $\frac{a}{5} - 3 = -4$
37. $\frac{x}{3} - 10 = -1$
38. $\frac{x}{7} + 3 = 5$
39. $\frac{p}{11} + 9 = 6$
40. $\frac{r}{12} + 2 = 4$

Solve each equation. Check all solutions. SEE EXAMPLE 3. (OBJECTIVE 1)

41. $\frac{b + 5}{3} = 11$
42. $\frac{a + 2}{13} = 3$
43. $\frac{x + 5}{2} = 4$
44. $\frac{r - 3}{8} = -2$
45. $\frac{3x - 12}{2} = 9$
46. $\frac{5x + 10}{7} = 0$
47. $\frac{4k - 1}{5} = 3$
48. $\frac{2k - 1}{3} = -5$

Solve each equation. Check all solutions. SEE EXAMPLE 4. (OBJECTIVE 1)

49. $\frac{k}{5} - \frac{1}{2} = \frac{3}{2}$
50. $\frac{y}{3} - \frac{6}{5} = -\frac{1}{5}$
51. $\frac{w}{16} + \frac{5}{4} = 1$
52. $\frac{m}{7} - \frac{1}{14} = \frac{1}{14}$
53. $\frac{3x}{2} - 6 = 9$
54. $\frac{5x}{7} + 3 = 8$
55. $\frac{9y}{2} + 3 = -15$
56. $\frac{5z}{3} + 3 = -2$

ADDITIONAL PRACTICE *Solve each equation. Check all solutions.*

57. $43p + 72 = 158$
58. $96q + 23 = -265$
59. $-47 - 21n = 58$
60. $-151 + 13m = -229$
61. $2y - \frac{5}{3} = \frac{4}{3}$
62. $9y + \frac{1}{2} = \frac{3}{2}$
63. $-0.4y - 12 = -20$
64. $-0.8y + 64 = -32$
65. $\frac{2x}{3} + \frac{1}{2} = 3$
66. $\frac{4x}{5} - \frac{1}{3} = 1$
67. $\frac{3x}{4} - \frac{2}{5} = 2$
68. $\frac{5x}{6} + \frac{3}{5} = 3$
69. $\frac{u - 4}{7} = 1$
70. $\frac{v - 7}{3} = -1$
71. $\frac{x - 5}{3} = -4$
72. $\frac{3 + y}{5} = -3$
73. $\frac{3z + 2}{17} = 0$
74. $\frac{8n - 7}{4} = -1$
75. $\frac{17k - 28}{21} + \frac{4}{3} = 0$
76. $\frac{5a - 2}{3} = \frac{1}{6}$

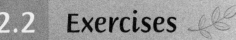

77. $-\frac{x}{3} - \frac{1}{2} = -\frac{5}{2}$

78. $\frac{15 - 5a}{3} = 3$

79. $\frac{10 - 3w}{9} = \frac{2}{3}$

80. $\frac{3p - 5}{5} + \frac{1}{2} = -\frac{19}{2}$

APPLICATIONS *Solve.* SEE EXAMPLE 5. (OBJECTIVE 2)

81. Apartment rentals A student moves into a bigger apartment that rents for \$450 per month. That rent is \$200 less than twice what she had been paying. Find her former rent.

82. Auto repairs A mechanic charged \$20 an hour to repair the water pump on a car, plus \$95 for parts. If the total bill was \$155, how many hours did the repair take?

83. Boarding dogs A sportsman boarded his dog at a kennel for a \$20 registration fee plus \$14 a day. If the stay cost \$104, how many days was the owner gone?

84. Water billing The city's water department charges \$7 per month, plus 42¢ for every 100 gallons of water used. Last month, one homeowner used 1,900 gallons and received a bill for \$17.98. Was the billing correct?

Solve. SEE EXAMPLES 6–7. (OBJECTIVE 3)

85. Clearance sales Sweaters already on sale for 20% off the regular price cost \$36 when purchased with a promotional coupon that allows an additional 10% discount. Find the original price. (*Hint:* When you save 20%, you are paying 80%.)

86. Furniture sales A \$1,250 sofa is marked down to \$900. Find the percent of markdown.

87. Value of coupons The percent discount offered by this coupon depends on the amount purchased. Find the range of the percent discount.

Value coupon

on purchases of \$100 to \$250.

88. Furniture pricing A bedroom set selling for \$1,900 cost \$1,000 wholesale. Find the percent markup.

Solve.

89. Integers Six less than 3 times a number is 9. Find the number.

90. Integers Seven less than 5 times a number is 23. Find the number.

91. Integers If a number is increased by 7 and that result is divided by 2, the number 5 is obtained. Find the original number.

92. Integers If twice a number is decreased by 5 and that result is multiplied by 4, the result is 36. Find the number.

93. Telephone charges A call to Tucson from a pay phone in Chicago costs 85¢ for the first minute and 27¢ for each additional minute or portion of a minute. If a student has an \$8.68 balance on a phone card, how long can she talk?

94. Monthly sales A clerk's sales in February were \$2,000 less than 3 times her sales in January. If her February sales were \$7,000, by what amount did her sales increase?

95. Ticket sales A music group charges \$1,500 for each performance, plus 20% of the total ticket sales. After a concert, the group received \$2,980. How much money did the ticket sales raise?

96. Getting an A To receive a grade of A, the average of four 100-point exams must be 90 or better. If a student received scores of 88, 83, and 92 on the first three exams, what minimum score does he need on the fourth exam to earn an A?

97. Getting an A The grade in history class is based on the average of five 100-point exams. One student received scores of 85, 80, 95, and 78 on the first four exams. With an average of 90 needed, what chance does he have for an A?

98. Excess inventory From the portion of the following ad, determine the sale price of a shirt.

Clearance Sale
Save 40%
Regularly
Sale
30% savings
Sweaters
\$45.95
\$27.57
Shirts
\$37.50
\$

WRITING ABOUT MATH

99. In solving the equation $5x - 3 = 12$, explain why you would add 3 to both sides first, rather than dividing by 5 first.

100. To solve the equation $\frac{3x - 4}{7} = 2$, what operations would you perform, and in what order?

SOMETHING TO THINK ABOUT

101. Suppose you must solve the following equation but you can't read one number. If the solution of the equation is 1, what is the equation?

$$\frac{7x + \square}{22} = \frac{1}{2}$$

102. A store manager first increases his prices by 30% to get a new retail price and then advertises as shown at the right. What is the real percent discount to customers?

Unless otherwise noted, all content on this page is © Cengage Learning.

Section 2.3 Simplifying Expressions to Solve Linear Equations in One Variable

Objectives

1. Simplify an expression using the order of operations and combining like terms.
2. Solve a linear equation in one variable requiring simplifying one or both sides.
3. Solve a linear equation in one variable that is an identity or a contradiction.

Vocabulary

coefficient
like terms
unlike terms
conditional equation
identity
contradiction
empty set

Getting Ready

Use the distributive property to remove parentheses.

1. $(3 + 4)x$ **2.** $(7 + 2)x$

3. $(8 - 3)w$ **4.** $(10 - 4)y$

Simplify each expression by performing the operations within the parentheses.

5. $(3 + 4)x$ **6.** $(7 + 2)x$

7. $(8 - 3)w$ **8.** $(10 - 4)y$

When algebraic expressions with the same variables occur, we can combine them.

1 Simplify an expression using the order of operations and combining like terms.

Recall that a *term* is either a number or the product of numbers and variables. Some examples of terms are $7x$, $-3xy$, y^2, and 8. The number part of each term is called its **coefficient**.

- The coefficient of $7x$ is 7.
- The coefficient of $-3xy$ is -3.
- The coefficient of y^2 is the understood factor of 1.
- The coefficient of 8 is 8.

LIKE TERMS

Like terms, or *similar terms*, are terms with the same variables having the same exponents.

The terms $3x$ and $5x$ are like terms, as are $9x^2$ and $-3x^2$. The terms $4xy$ and $3x^2$ are **unlike terms**, because they have different variables. The terms $4x$ and $5x^2$ are unlike terms, because the variables have different exponents.

COMMENT Terms are separated by + and − signs.

The distributive property can be used to combine terms of algebraic expressions that contain sums or differences of like terms. For example, the terms in $3x + 5x$ and $9xy^2 - 11xy^2$ can be combined as follows:

$$\overbrace{3x + 5x}^{\text{expressions with like terms}} = (3 + 5)x = 8x \qquad \overbrace{9xy^2 - 11xy^2}^{\text{expressions with like terms}} = (9 - 11)xy^2 = -2xy^2$$

These examples suggest the following rule.

COMBINING LIKE TERMS To combine like terms, add their coefficients and keep the same variables and exponents.

COMMENT If the terms of an expression are unlike terms, they cannot be combined. For example, since the terms in $9xy^2 - 11x^2y$ have variables with different exponents, they are unlike terms and cannot be combined.

EXAMPLE 1 Simplify: $3(x + 2) + 2(x - 8)$

Solution To simplify the expression, we will use the distributive property to remove parentheses and then combine like terms.

$3(x + 2) + 2(x - 8)$

$= 3x + 3 \cdot 2 + 2x + 2 \cdot (-8)$ Use the distributive property to remove parentheses.

$= 3x + 6 + 2x - 16$ $3 \cdot 2 = 6$ and $2 \cdot 8 = 16$.

$= 3x + 2x + 6 - 16$ Use the commutative property of addition: $6 + 2x = 2x + 6$

$= 5x - 10$ Combine like terms.

SELF CHECK 1 Simplify: $-5(a + 3) + 2(a - 5)$

EXAMPLE 2 Simplify: $3(x - 3) - 5(x + 4)$

Solution To simplify the expression, we will use the distributive property to remove parentheses and then combine like terms.

$3(x - 3) - 5(x + 4)$

$= 3(x - 3) + (-5)(x + 4)$ $a - b = a + (-b)$.

$= 3x - 3 \cdot 3 + (-5)x + (-5)4$ Use the distributive property to remove parentheses.

$= 3x - 9 + (-5x) + (-20)$ $3 \cdot 3 = 9$ and $(-5)(4) = -20$.

$= -2x - 29$ Combine like terms.

SELF CHECK 2 Simplify: $-3(b - 2) - 4(b - 4)$

COMMENT In algebra, you will simplify expressions and solve equations. Recognizing which one to do is a skill that we will apply throughout this course.

An expression can be simplified only by combining its like terms. Since an equation contains two expressions set equal to each other, it can be solved. Remember that

Expressions are to be simplified. Equations are to be solved.

2 Solve a linear equation in one variable requiring simplifying one or both sides.

To solve a linear equation in one variable, we must isolate the variable on one side. This is often a multistep process that may require combining like terms. As we solve equations, we will follow these steps, if necessary.

SOLVING EQUATIONS

1. Clear the equation of any fractions or decimals.
2. Use the distributive property to remove any grouping symbols.
3. Combine like terms on each side of the equation.
4. Undo the operations of addition and subtraction to collect the variables on one side and the constants on the other.
5. Combine like terms and undo the operations of multiplication and division to isolate the variable.
6. Check the solution in the original equation.

EXAMPLE 3 Solve: $3(x + 2) - 5x = 0$

Solution To solve the equation, we will remove parentheses, combine like terms, and solve for x.

$$3(x + 2) - 5x = 0$$

$$3x + 3 \cdot 2 - 5x = 0 \quad \text{Use the distributive property to remove parentheses.}$$

$$3x - 5x + 6 = 0 \quad \text{Use the commutative property of addition and simplify.}$$

$$-2x + 6 = 0 \quad \text{Combine like terms.}$$

$$-2x + 6 - 6 = 0 - 6 \quad \text{Subtract 6 from both sides.}$$

$$-2x = -6 \quad \text{Combine like terms.}$$

$$\frac{-2x}{-2} = \frac{-6}{-2} \quad \text{Divide both sides by } -2.$$

$$x = 3 \quad \text{Simplify.}$$

Check:

$$3(x + 2) - 5x = 0$$

$$3(3 + 2) - 5 \cdot 3 \stackrel{?}{=} 0 \quad \text{Substitute 3 for } x.$$

$$3 \cdot 5 - 5 \cdot 3 \stackrel{?}{=} 0 \quad \text{Perform the operation inside the parentheses.}$$

$$15 - 15 \stackrel{?}{=} 0 \quad \text{Multiply.}$$

$$0 = 0 \quad \text{True.}$$

Since the solution 3 checks, the solution set is $\{3\}$.

SELF CHECK 3 Solve: $-2(y - 3) - 4y = 0$

In the next example, we will isolate the variable on the right side of the equation to keep the coefficient of x positive.

EXAMPLE 4 Solve: $3(x - 5) = 4(x + 9)$

Solution To solve the equation, we will remove parentheses, collect all like terms involving x on one side, combine like terms, and solve for x.

$$3(x - 5) = 4(x + 9)$$

$$3x - 15 = 4x + 36 \quad \text{Use the distributive property to remove parentheses.}$$

$$3x - 15 - 3x = 4x + 36 - 3x \quad \text{Subtract } 3x \text{ from both sides.}$$

$$-15 = x + 36 \quad \text{Combine like terms.}$$

$$-15 - \mathbf{36} = x + 36 - \mathbf{36} \quad \text{Subtract 36 from both sides.}$$
$$-51 = x \quad \text{Combine like terms.}$$

Check:
$$3(x - 5) = 4(x + 9)$$
$$3(\mathbf{-51} - 5) \stackrel{?}{=} 4(\mathbf{-51} + 9) \quad \text{Substitute } -51 \text{ for } x.$$
$$3(-56) \stackrel{?}{=} 4(-42)$$
$$-168 = -168 \quad \text{True.}$$

Since the solution -51 checks, the solution set is $\{-51\}$.

SELF CHECK 4 Solve: $4(z + 3) = -3(z - 4)$

Note: The solution to an equation is the same whether the variable is isolated on the left or the right side of the equation.

EXAMPLE 5 Solve: $\dfrac{3x + 11}{5} = x + 3$

Solution We first multiply both sides by 5 to clear the equation of fractions. When we multiply the right side by 5, we must multiply the *entire* right side by 5.

COMMENT Remember that when you multiply one side of an equation by a nonzero number, you must multiply the other side by the same number to maintain the equality.

$$\frac{3x + 11}{5} = x + 3$$
$$\mathbf{5}\left(\frac{3x + 11}{5}\right) = \mathbf{5}(x + 3) \quad \text{Multiply both sides by 5.}$$
$$3x + 11 = 5x + 15 \quad \text{Use the distributive property to remove parentheses.}$$
$$3x + 11 - \mathbf{11} = 5x + 15 - \mathbf{11} \quad \text{Subtract 11 from both sides.}$$
$$3x = 5x + 4 \quad \text{Combine like terms.}$$
$$3x - \mathbf{5x} = 5x + 4 - \mathbf{5x} \quad \text{Subtract } 5x \text{ from both sides.}$$
$$-2x = 4 \quad \text{Combine like terms.}$$
$$\frac{-2x}{\mathbf{-2}} = \frac{4}{\mathbf{-2}} \quad \text{Divide both sides by } -2.$$
$$x = -2 \quad \text{Simplify.}$$

Check:
$$\frac{3x + 11}{5} = x + 3$$
$$\frac{3(\mathbf{-2}) + 11}{5} \stackrel{?}{=} (\mathbf{-2}) + 3 \quad \text{Substitute } -2 \text{ for } x.$$
$$\frac{-6 + 11}{5} \stackrel{?}{=} 1 \quad \text{Simplify.}$$
$$\frac{5}{5} \stackrel{?}{=} 1$$
$$1 = 1 \quad \text{True.}$$

Since the solution -2 checks, the solution set is $\{-2\}$.

SELF CHECK 5 Solve: $\dfrac{2x - 5}{4} = x - 2$

EXAMPLE 6 Solve: $0.2x + 0.4(50 - x) = 19$

Solution Since $0.2 = \frac{2}{10}$ and $0.4 = \frac{4}{10}$, this equation contains fractions. To clear the fractions, we will multiply both sides by 10.

$$\begin{aligned}
0.2x + 0.4(50 - x) &= 19 \\
\mathbf{10}[0.2x + 0.4(50 - x)] &= \mathbf{10}(19) && \text{Multiply both sides by 10.} \\
\mathbf{10}[0.2x] + \mathbf{10}[0.4(50 - x)] &= \mathbf{10}(19) && \text{Use the distributive property on the left side.} \\
2x + 4(50 - x) &= 190 && \text{Multiply.} \\
2x + 200 - 4x &= 190 && \text{Use the distributive property to remove parentheses.} \\
-2x + 200 &= 190 && \text{Combine like terms.} \\
-2x &= -10 && \text{Subtract 200 from both sides.} \\
x &= 5 && \text{Divide both sides by } -2.
\end{aligned}$$

Since the solution is 5, the solution set is $\{5\}$. Verify that the solution checks.

SELF CHECK 6 Solve: $0.3(20 - x) + 0.5x = 15$

3 Solve a linear equation in one variable that is an identity or a contradiction.

The equations solved in Examples 3–6 are called **conditional equations**. For these equations, each has exactly one solution.

An equation that is true for all values of its variable is called an **identity**. For example, the equation $x + x = 2x$ is an identity because it is true for all values of x. The solution of an identity is the set of *all real numbers* and is denoted by the symbol $\mathbb{R}$.

An equation that is not true for any value of its variable is called a **contradiction**. For example, the equation $x = x + 1$ is a contradiction because there is no value of x that will make the statement true. Since there are no solutions to a contradiction, its set of solutions is empty. This is denoted by the symbol $\varnothing$ or $\{\ \}$ and is called the **empty set**. Table 2.1 summarizes the three possibilities.

TABLE 2-1

Type of equation	Examples		Solution sets
Conditional	$2x + 4 = 8$	$\frac{x}{2} - 4 = 12$	$\{2\}$ and $\{32\}$
Identity	$x + x = 2x$	$2(x + 3) = 2x + 6$	$\mathbb{R}$ and $\mathbb{R}$
Contradiction	$x - 1 = x$	$2(x + 3) = 2x + 5$	$\varnothing$ and $\varnothing$

EXAMPLE 7 Solve: $3(x + 8) + 5x = 2(12 + 4x)$

Solution To solve this equation, we will remove parentheses, combine terms, and solve for x.

$$\begin{aligned}
3(x + 8) + 5x &= 2(12 + 4x) \\
3x + 24 + 5x &= 24 + 8x && \text{Use the distributive property to remove parentheses.} \\
8x + 24 &= 24 + 8x && \text{Combine like terms.} \\
8x + 24 - \mathbf{8x} &= 24 + 8x - \mathbf{8x} && \text{Subtract } 8x \text{ from both sides.} \\
24 &= 24 && \text{Combine like terms.}
\end{aligned}$$

Since the result $24 = 24$ is true for every number x, every number is a solution of the original equation. This equation is an identity. The solution set is the set of real numbers, $\mathbb{R}$.

SELF CHECK 7 Solve: $-2(x - 3) - 18x = 2(3 - 10x)$

EXAMPLE 8 Solve: $3(x + 7) - x = 2(x + 10)$

Solution To solve this equation, we will remove parentheses, combine terms, and solve for x.

$$3(x + 7) - x = 2(x + 10)$$

$$3x + 21 - x = 2x + 20$$ Use the distributive property to remove parentheses.

$$2x + 21 = 2x + 20$$ Combine like terms.

$$2x + 21 - 2x = 2x + 20 - 2x$$ Subtract $2x$ from both sides.

$$21 = 20$$ Combine like terms.

Since the result $21 = 20$ is false, the original equation is a contradiction. Since the original equation has no solution, the solution set is $\varnothing$.

SELF CHECK 8 Solve: $5(x - 2) - 2x = 3(x + 7)$

SELF CHECK ANSWERS

1. $-3a - 25$ 2. $-7b + 22$ 3. 1 4. 0 5. $\frac{3}{2}$ 6. 45 7. identity, $\mathbb{R}$ 8. contradiction, $\varnothing$

NOW TRY THIS

Identify each of the following as an expression or an equation. Simplify or solve as appropriate.

1. $4\left(x - \frac{7}{4}\right) + 3(x + 2)$
2. $4\left(x - \frac{7}{4}\right) = 3(x + 2)$
3. $6x - 2(3x - 9)$

2.3 Exercises

WARM-UPS *Identify each statement as an expression or an equation.*

1. $5x - 2$
2. $2x + 3 = 5$
3. $3x + 7 = -1$
4. $9 + x$
5. $6 - 4x = 7$
6. $7x + 8$

Identify each equation as an identity or a contradiction.

7. $2x + 5 = 2x + 5$
8. $-3x - 9 = -3x + 9$
9. $6x - 2 = 6x + 4$
10. $4x + 9 = 4x + 9$

REVIEW *Evaluate each expression when $x = -3$, $y = -5$, and $z = 0$.*

11. $x^2z(y^3 - z)$
12. $y - x^3$
13. $\dfrac{x - y^2}{2y - 1 + x}$
14. $\dfrac{3y + x^2}{x} + z$

Perform the operations.

15. $\frac{6}{7} - \frac{5}{8}$
16. $\frac{6}{7} \cdot \frac{5}{8}$
17. $\frac{6}{7} \div \frac{5}{8}$
18. $\frac{6}{7} + \frac{5}{8}$

VOCABULARY AND CONCEPTS *Fill in the blanks.*

19. If terms have the same ________ with the same exponents, they are called ___ terms. Terms that have different variables or have a variable with different exponents are called _____ terms. The number part of a term is called its _________.
20. To combine like terms, ___ their coefficients and ____ the same variables and exponents.
21. If an equation is true for all values of its variable, it is called an ______. If an equation is true for no values of its variable, it is called a __________.
22. If an equation is true for some values of its variable, but not all, it is called a _________ equation.

GUIDED PRACTICE *Simplify each expression, when possible.* SEE EXAMPLE 1. (OBJECTIVE 1)

23. $8x + 12x$
24. $12y - 15y$
25. $8x^2 - 5x^2$
26. $17x^2 + 3x^2$
27. $9x + 3y$
28. $5x + 5y$
29. $4(x + 3) - 2x$
30. $9(y - 3) + 2y$

Simplify each expression. SEE EXAMPLE 2. (OBJECTIVE 1)

31. $5(z - 3) + 2z$
32. $7(y + 4) - 10y$
33. $12(x + 11) - 11$
34. $-3(3 + z) + 2z$
35. $6(y - 2) - 3(y + 1)$
36. $9(z + 2) + 5(3 - z)$
37. $5x - 2(y - x) + 4y$
38. $3y - 6(y + z) + y$

Solve each equation. Check all solutions. SEE EXAMPLE 3. (OBJECTIVE 2)

39. $8(x + 5) + 6(7 - x) = 0$
40. $3(x + 15) + 4(11 - x) = 0$
41. $12x - 4(5 + x) = 4$
42. $5(x - 6) - 8x = 15$

Solve each equation. Check all solutions. SEE EXAMPLE 4. (OBJECTIVE 2)

43. $3x + 2 = 2x$
44. $7x + 5 = 6x$
45. $5x - 3 = 4x$
46. $4x + 3 = 5x$
47. $9y - 3 = 6y$
48. $8y + 4 = 4y$
49. $10y - 10 = 5y$
50. $9y - 8 = y$
51. $3(a + 2) = 4a$
52. $4(a - 5) = 3a$
53. $6(b + 4) = 8b$
54. $7(b - 3) = 10b$
55. $2 + 3(x - 5) = 4(x - 1)$
56. $2 - (4x + 7) = 3 + 2(x + 2)$
57. $3(a + 2) = 2(a - 7)$
58. $9(n - 1) = 6(n + 2) - n$

Solve each equation. Check all solutions. SEE EXAMPLE 5. (OBJECTIVE 2)

59. $\frac{3(t - 7)}{2} = t - 6$
60. $\frac{4(p + 8)}{3} = p - 4$
61. $\frac{2(t - 1)}{6} - 2 = \frac{t + 2}{6}$
62. $\frac{2(2r - 1)}{6} + 5 = \frac{3(r + 7)}{6}$

Solve each equation. Check all solutions. SEE EXAMPLE 6. (OBJECTIVE 2)

63. $3.1(x - 2) = 1.3x + 2.8$
64. $0.6x - 0.8 = 0.8(2x - 1) - 0.7$
65. $2.7(y + 1) = 0.3(3y + 33)$
66. $1.5(5 - y) = 3y + 12$

Solve each equation. If it is an identity or a contradiction, so indicate. SEE EXAMPLES 7–8. (OBJECTIVE 3)

67. $7x + 5(3 - x) = 2(x + 5) + 5$
68. $21(b - 1) + 3 = 3(7b - 6)$
69. $2(s + 2) = 2(s + 1) + 3$
70. $4(2z + 3) = 2(4z - 6) + 11$
71. $\frac{5(x + 3)}{3} - x = \frac{2(x + 8)}{3}$
72. $5(x + 2) = 5x - 2$
73. $x + 7 = \frac{2x + 6}{2} + 4$
74. $2(y - 3) - \frac{y}{2} = \frac{3}{2}(y - 4)$

ADDITIONAL PRACTICE *Identify each statement as an expression or an equation, and then either simplify or solve as appropriate.*

75. $2(x - y) - (x + y) + y$
76. $3z + 2(y - z) + y$
77. $\frac{4(2x - 10)}{3} = 2(x - 4)$
78. $\frac{11(x - 12)}{2} = 9 - 2x$
79. $2\left(4x + \frac{9}{2}\right) - 3\left(x + \frac{2}{3}\right)$
80. $\frac{5(2 - m)}{3} = m + 6$
81. $\frac{8(5 - q)}{5} = -2q$
82. $\frac{20 - a}{2} = \frac{3}{2}(a + 4)$
83. $\frac{3x + 14}{2} = x - 2 + \frac{x + 18}{2}$
84. $7\left(3x - \frac{2}{7}\right) - 5\left(2x - \frac{3}{5}\right) + x$
85. $6 - 5r = 7r$
86. $y + 4 = -7y$
87. $22 - 3r = 8r$
88. $14 + 7s = s$
89. $8(x + 3) - 3x$
90. $2x + 2(x + 3)$
91. $19.1x - 4(x + 0.3) = -46.5$
92. $18.6x + 7.2 = 1.5(48 - 2x)$
93. $3.2(m + 1.3) - 2.5(m - 7.2)$
94. $6.7(t - 2.1) + 5.5(t + 1)$
95. $14.3(x + 2) + 13.7(x - 3) = 15.5$
96. $1.25(x - 1) = 0.5(3x - 1) - 1$
97. $10x + 3(2 - x) = 5(x + 2) - 4$
98. $19.1x - 4(x + 0.3)$

Solve each equation and round the result to the nearest tenth.

99. $\frac{3.7(2.3x - 2.7)}{1.5} = 5.2(x - 1.2)$
100. $\frac{-2.1(1.7x + 0.9)}{3.1} = -7.1(x - 1.3)$

WRITING ABOUT MATH

101. Explain why $3x^2y$ and $5x^2y$ are like terms.
102. Explain why $3x^2y$ and $3xy^2$ are unlike terms.
103. Discuss whether $7xxy^3$ and $5x^2yyy$ are like terms.
104. Discuss whether $\frac{3}{2}x$ and $\frac{3x}{2}$ are like terms.

SOMETHING TO THINK ABOUT

105. What number is equal to its own double?
106. What number is equal to one-half of itself?

Section 2.4 Formulas

Objectives

1. Solve a formula for an indicated variable using the properties of equality.
2. Evaluate a formula for specified values for the variables.
3. Solve an application using a given formula and specified values for the variables.

Vocabulary

literal equations   formulas

Getting Ready

Fill in the blanks.

1. $\frac{3x}{\square} = x$
2. $\frac{-5y}{\square} = y$
3. $\frac{rx}{\square} = x$
4. $\frac{-ay}{\square} = y$
5. $\square \cdot \frac{x}{7} = x$
6. $\square \cdot \frac{y}{12} = y$
7. $\square \cdot \frac{x}{d} = x$
8. $\square \cdot \frac{y}{s} = y$

Equations with several variables are called **literal equations**. Often these equations are **formulas** such as $A = lw$, the formula for finding the area of a rectangle.

Suppose that we want to find the lengths of several rectangles whose areas and widths are known. It would be tedious to substitute values for A and w into the formula and then repeatedly solve the formula for l. It would be much easier to solve the formula $A = lw$ for l first, then substitute values for A and w, and compute l directly.

1 Solve a formula for an indicated variable using the properties of equality.

To *solve a formula for a variable* means to isolate that variable on one side of the equation, with all other numbers and variables on the other side. We can isolate the variable by using the equation-solving techniques we have learned in the previous three sections.

EXAMPLE 1 Solve $A = lw$ for l.

Solution To isolate l, we undo the multiplication by w by dividing both sides of the equation by w.

$$A = lw$$

$$\frac{A}{w} = \frac{lw}{w} \quad \text{To undo the multiplication by } w \text{, divide both sides by } w.$$

$$\frac{A}{w} = l \quad \frac{w}{w} = 1.$$

SELF CHECK 1 Solve $A = lw$ for w.

EXAMPLE 2 Recall that the formula $A = \frac{1}{2}bh$ gives the area of a triangle with base b and height h. Solve the formula for b.

Solution To isolate b, we will clear the fraction by multiplying both sides by 2. Then we will undo the multiplication by h by dividing both sides by h.

$$A = \frac{1}{2}bh$$

$$2 \cdot A = 2 \cdot \left(\frac{1}{2}bh\right) \quad \text{To clear the fraction, multiply both sides by 2.}$$

$$2A = bh \quad 2 \cdot \tfrac{1}{2} = 1.$$

$$\frac{2A}{h} = \frac{bh}{h} \quad \text{To undo the multiplication by } h\text{, divide both sides by } h.$$

$$\frac{2A}{h} = b \quad \tfrac{h}{h} = 1.$$

If the area A and the height h of a triangle are known, the base b is given by the formula $b = \frac{2A}{h}$.

SELF CHECK 2 Solve $A = \frac{1}{2}bh$ for h.

EXAMPLE 3 The formula $C = \frac{5}{9}(F - 32)$ is used to convert Fahrenheit temperature readings into their Celsius equivalents. Solve the formula for F.

Solution To isolate F, we will undo the multiplication by $\frac{5}{9}$ by multiplying both sides by the reciprocal of $\frac{5}{9}$, which is $\frac{9}{5}$. Then we will use the distributive property to remove parentheses and finally undo the subtraction of 32 by adding 32 to both sides.

$$C = \frac{5}{9}(F - 32)$$

$$\frac{9}{5} \cdot C = \frac{9}{5} \cdot \left[\frac{5}{9}(F - 32)\right] \quad \text{To eliminate } \tfrac{5}{9}\text{, multiply both sides by } \tfrac{9}{5}.$$

$$\frac{9}{5}C = \left(\frac{9}{5} \cdot \frac{5}{9}\right)(F - 32) \quad \text{Apply the associative property of multiplication.}$$

$$\frac{9}{5}C = 1(F - 32) \quad \tfrac{9}{5} \cdot \tfrac{5}{9} = \tfrac{9 \cdot 5}{5 \cdot 9} = 1.$$

$$\frac{9}{5}C = F - 32 \quad \text{Use the distributive property to remove parentheses.}$$

$$\frac{9}{5}C + 32 = F - 32 + 32 \quad \text{To undo the subtraction of 32, add 32 to both sides.}$$

$$\frac{9}{5}C + 32 = F \quad \text{Combine like terms.}$$

The formula $F = \frac{9}{5}C + 32$ is used to convert degrees Celsius to degrees Fahrenheit.

SELF CHECK 3 Solve $x = \frac{2}{3}(y + 5)$ for y.

EXAMPLE 4 Recall that the area A of the trapezoid shown in Figure 2-5 is given by the formula

$$A = \frac{1}{2}h(B + b)$$

where B and b are its bases and h is its height. Solve the formula for b.

Solution There are two different ways to solve this formula.

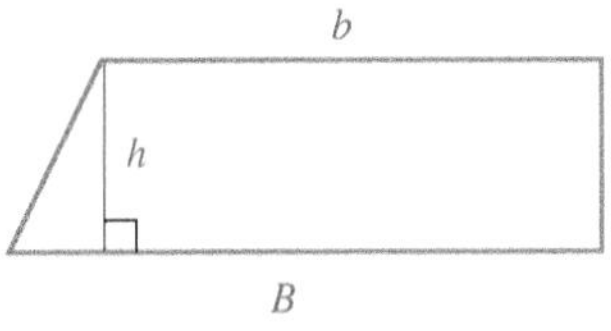

Figure 2-5

Method 1:

$$A = \frac{1}{2}(B + b)h$$

$$\mathbf{2} \cdot A = \mathbf{2} \cdot \left[\frac{1}{2}(B + b)h\right] \quad \text{Multiply both sides by 2.}$$

$$2A = \left(2 \cdot \frac{1}{2}\right)(B + b)h \quad \text{Apply the associative property of multiplication.}$$

$$2A = Bh + bh \quad \text{Simplify and use the distributive property to remove parentheses.}$$

$$2A - \mathbf{Bh} = Bh + bh - \mathbf{Bh} \quad \text{Subtract } Bh \text{ from both sides.}$$

$$2A - Bh = bh \quad \text{Combine like terms.}$$

$$\frac{2A - Bh}{\mathbf{h}} = \frac{bh}{\mathbf{h}} \quad \text{Divide both sides by } h.$$

$$\frac{2A - Bh}{h} = b \quad \frac{h}{h} = 1.$$

Method 2:

$$A = \frac{1}{2}(B + b)h$$

$$\mathbf{2} \cdot A = \mathbf{2} \cdot \left[\frac{1}{2}(B + b)h\right] \quad \text{Multiply both sides by 2.}$$

$$2A = (B + b)h \quad \text{Simplify.}$$

$$\frac{2A}{\mathbf{h}} = \frac{(B + b)h}{\mathbf{h}} \quad \text{Divide both sides by } h.$$

$$\frac{2A}{h} = B + b \quad \frac{h}{h} = 1.$$

$$\frac{2A}{h} - \mathbf{B} = B + b - \mathbf{B} \quad \text{Subtract } B \text{ from both sides.}$$

$$\frac{2A}{h} - B = b \quad \text{Combine like terms.}$$

Although they look different, the results of Methods 1 and 2 are equivalent.

SELF CHECK 4 Solve $A = \frac{1}{2}h(B + b)$ for B.

2 Evaluate a formula for specified values for the variables.

EXAMPLE 5 Solve the formula $P = 2l + 2w$ for l. Evaluate that formula for l when $P = 56$ and $w = 11$.

Solution We first solve the formula $P = 2l + 2w$ for l.

$$P = 2l + 2w$$

$$P - \mathbf{2w} = 2l + 2w - \mathbf{2w} \quad \text{Subtract } 2w \text{ from both sides.}$$

Unless otherwise noted, all content on this page is © Cengage Learning.

Albert Einstein
1879–1955
Einstein was a theoretical physicist best known for his theory of relativity. Although Einstein was born in Germany, he became a Swiss citizen and earned his doctorate at the University of Zurich in 1905. In 1910, he returned to Germany to teach. He fled Germany because of the Nazi government and became a United States citizen in 1940. He is famous for his formula $E = mc^2$.

$$P - 2w = 2l \quad \text{Combine like terms.}$$

$$\frac{P - 2w}{2} = \frac{2l}{2} \quad \text{Divide both sides by 2.}$$

$$\frac{P - 2w}{2} = l \quad \tfrac{2}{2} = 1.$$

We will then substitute 56 for P and 11 for w and simplify.

$$l = \frac{P - 2w}{2}$$

$$l = \frac{56 - 2(11)}{2}$$

$$= \frac{56 - 22}{2}$$

$$= \frac{34}{2}$$

$$= 17$$

Thus, $l = 17$.

SELF CHECK 5 Solve $P = 2l + 2w$ for w. Evaluate that fomula for w when $P = 46$ and $l = 16$.

3 Solve an application using a given formula and specified values for the variables.

EXAMPLE 6 Recall that the volume V of the right-circular cone shown in Figure 2-6 is given by the formula

$$V = \frac{1}{3}Bh$$

where B is the area of its circular base and h is its height. Solve the formula for h and find the height of a right-circular cone with a volume of 64 cubic centimeters and a base area of 16 square centimeters.

Solution We first solve the formula for h.

$$V = \frac{1}{3}Bh$$

$$\mathbf{3} \cdot V = \mathbf{3} \cdot \left(\frac{1}{3}Bh\right) \quad \text{Multiply both sides by 3.}$$

$$3V = Bh \quad 3 \cdot \tfrac{1}{3} = 1.$$

$$\frac{3V}{\mathbf{B}} = \frac{Bh}{\mathbf{B}} \quad \text{Divide both sides by } B.$$

$$\frac{3V}{B} = h \quad \tfrac{B}{B} = 1.$$

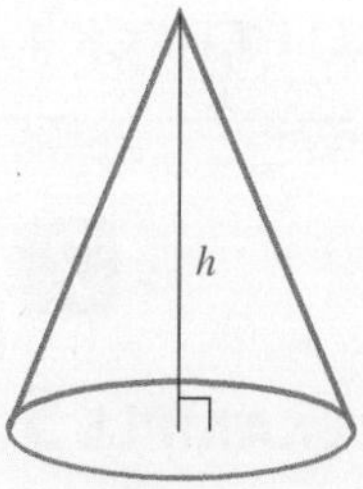

Figure 2-6

Unless otherwise noted, all content on this page is © Cengage Learning.

We then substitute 64 for V and 16 for B and simplify.

$$h = \frac{3V}{B}$$

$$h = \frac{3(64)}{16}$$

$$= 3(4)$$

$$= 12$$

The height of the cone is 12 centimeters.

SELF CHECK 6 Solve $V = \frac{1}{3}Bh$ for B, and find the area of the base when the volume is 42 cubic feet and the height is 6 feet.

SELF CHECK ANSWERS

1. $w = \frac{A}{l}$ **2.** $h = \frac{2A}{b}$ **3.** $y = \frac{3}{2}x - 5$ **4.** $B = \frac{2A - hb}{h}$ or $B = \frac{2A}{h} - b$ **5.** $w = \frac{P - 2l}{2}$, 7 **6.** $B = \frac{3V}{h}$, 21 ft^2

NOW TRY THIS

A student's test average for four tests can be modeled by the equation

$$A = \frac{T_1 + T_2 + T_3 + T_4}{4}$$

where T_1 is the grade for Test 1, T_2 is the grade for Test 2, and so on.

1. Solve the equation for T_4.

2. Julio has test grades of 82, 88, and 71. What grade would he need on Test 4 to have a test average of 80?

3. Leslie has test grades of 75, 80, and 89. What grade would Leslie need on Test 4 to have a test average of 90? Interpret your answer.

2.4 Exercises

WARM-UPS *Solve the equation $ab + c = 0$.*

1. for a

2. for c

Solve the equation $a = \frac{b}{c}$.

3. for b

4. for c

REVIEW *Simplify each expression, if possible.*

5. $7x - 4y - 4x$

6. $3ab^2 + 7a^2b$

7. $\frac{2}{3}(a + 3) - \frac{5}{3}(6 + a)$

8. $\frac{2}{11}(22x - y) + \frac{9}{11}y$

VOCABULARY AND CONCEPTS *Fill in the blanks.*

9. Equations that contain several variables are called _____ equations.

10. The equation $A = lw$ is an example of a _______.

11. To solve a formula for a variable means to ______ the variable on one side of the equation.

12. To solve the formula $d = rt$ for t, divide both sides of the formula by _.

13. To solve $A = p + i$ for p, _______ i from both sides.

14. To solve $t = \frac{d}{r}$ for d, _______ both sides by r.

GUIDED PRACTICE *Solve for the indicated variable.* SEE EXAMPLE 1. (OBJECTIVE 1)

15. $E = IR$ for I
16. $i = prt$ for t
17. $V = lwh$ for w
18. $C = 2\pi r$ for r
19. $x = y + 12$ for y
20. $P = a + b + c$ for c

Solve for the indicated variable. SEE EXAMPLE 2. (OBJECTIVE 1)

21. $V = \frac{1}{3}Bh$ for h
22. $V = \frac{1}{3}Bh$ for B
23. $V = \frac{1}{3}\pi r^2 h$ for h
24. $I = \frac{E}{R}$ for R

Solve for the indicated variable. SEE EXAMPLE 3. (OBJECTIVE 1)

25. $y = \frac{1}{2}(x + 2)$ for x
26. $x = \frac{1}{5}(y - 7)$ for y
27. $A = \frac{5}{2}(B + 3)$ for B
28. $y = \frac{5}{2}(x - 10)$ for x

Solve for the indicated variable. SEE EXAMPLE 4. (OBJECTIVE 1)

29. $p = \frac{h}{2}(q + r)$ for q
30. $p = \frac{h}{2}(q + r)$ for r
31. $G = 2b(r - 1)$ for r
32. $F = f(1 - M)$ for M

Solve each formula for the indicated variable. Then evaluate the new formula for the values given. SEE EXAMPLE 5. (OBJECTIVE 2)

33. $d = rt$ Find t if $d = 455$ and $r = 65$.
34. $d = rt$ Find r if $d = 275$ and $t = 5$.
35. $P = a + b + c$ Find b if $P = 37$, $a = 15$, and $c = 6$.
36. $y = mx + b$ Find x if $y = 30$, $m = 3$, and $b = 0$.

ADDITIONAL PRACTICE *Solve each formula for the indicated variable.*

37. $3x + 2y = 5$ for y
38. $y = mx + b$ for b
39. $C = \pi d$ for d
40. $P = I^2R$ for R
41. $P = 2l + 2w$ for w
42. $V = lwh$ for l
43. $A = P + Prt$ for t
44. $A = \frac{1}{2}(B + b)h$ for h
45. $K = \frac{wv^2}{2g}$ for w
46. $V = \pi r^2 h$ for h
47. $K = \frac{wv^2}{2g}$ for g
48. $P = \frac{RT}{mV}$ for V
49. $F = \frac{GMm}{d^2}$ for M
50. $C = 1 - \frac{A}{a}$ for A

51. Given that $i = prt$, find p if $i = 90$, $t = 4$, and $r = 0.03$.

52. Given that $i = prt$, find r if $i = 120$, $p = 500$, and $t = 6$.

53. Given that $K = \frac{1}{2}h(a + b)$, find h if $K = 48$, $a = 7$, and $b = 5$.

54. Given that $\frac{x}{2} + y = z^2$, find x if $y = 3$ and $z = 3$.

APPLICATIONS *Solve.* SEE EXAMPLE 6 (OBJECTIVE 3)

55. Volume of a cone The volume V of a cone is given by the formula $V = \frac{1}{3}\pi r^2 h$. Solve the formula for h, and then calculate the height h if V is 36π cubic inches and the radius r is 6 inches.

56. Circumference of a circle The circumference C of a circle is given by $C = 2\pi r$, where r is the radius of the circle. Solve the formula for r, and then calculate the radius of a circle with a circumference of 14.32 feet. Round to the nearest hundredth of a foot.

57. Ohm's law The formula $E = IR$, called **Ohm's law,** is used in electronics. Solve for I, and then calculate the current I if the voltage E is 48 volts and the resistance R is 12 ohms. Current has units of *amperes.*

58. Growth of money At a simple interest rate r, an amount of money P grows to an amount A in t years according to the formula $A = P(1 + rt)$. Solve the formula for P. After $t = 3$ years, a girl has an amount $A = \$4,357$ on deposit. What amount P did she start with? Assume an interest rate of 6%.

59. Power loss The power P lost when an electric current I passes through a resistance R is given by the formula $P = I^2R$. Solve for R. If P is 2,700 watts and I is 14 amperes, calculate R to the nearest hundredth of an ohm.

60. Geometry The perimeter P of a rectangle with length l and width w is given by the formula $P = 2l + 2w$. Solve this formula for w. If the perimeter of a certain rectangle is 58.37 meters and its length is 17.23 meters, find its width. Round to two decimal places.

61. Force of gravity The masses of the two objects in the illustration are m and M. The force of gravitation F between the masses is given by

$$F = \frac{GmM}{d^2}$$

where G is a constant and d is the distance between them. Solve for m.

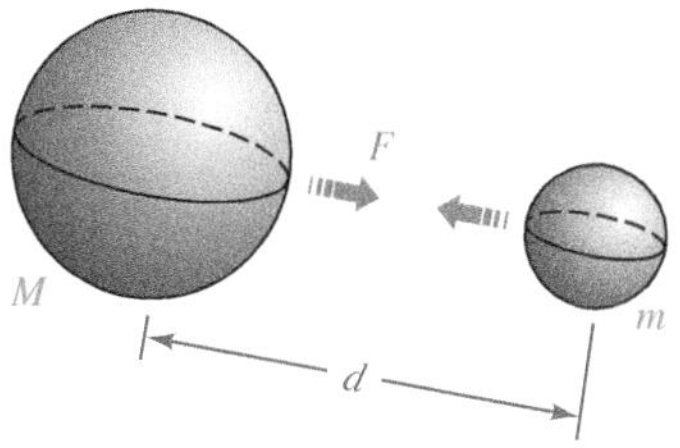

62. Thermodynamics In thermodynamics, the Gibbs free-energy equation is given by

$$G = U - TS + pV$$

Solve this equation for the pressure, p.

63. Pulleys The approximate length L of a belt joining two pulleys of radii r and R feet with centers D feet apart is given by the formula

$$L = 2D + 3.25(r + R)$$

Solve the formula for D. If a 25-foot belt joins pulleys with radii of 1 foot and 3 feet, how far apart are their centers?

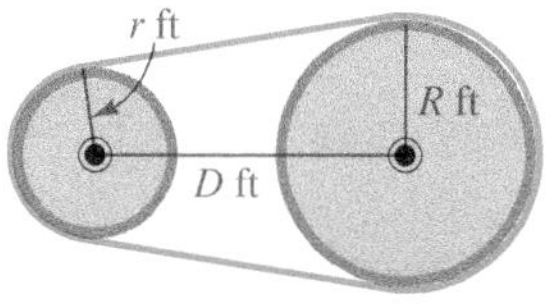

64. Geometry The measure a of an interior angle of a regular polygon with n sides is given by $a = 180°(1 - \frac{2}{n})$. Solve the formula for n. How many sides does a regular polygon have if an interior angle is 108°? (*Hint:* Distribute first.)

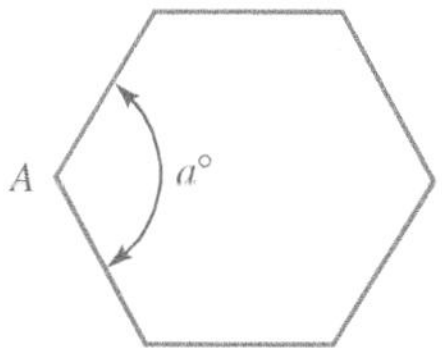

One common retirement plan for self-employed people is called a Simplified Employee Pension Plan. It allows for a maximum annual contribution of 15% of taxable income (earned income minus deductible expenses). However, since the Internal Revenue Service considers the SEP contribution to be a deductible expense, the taxable income must be reduced by the amount of the contribution. Therefore, to calculate the maximum contribution C, we take 15% of what's left after we subtract the contribution C from the taxable income T.

$$C = 0.15(T - C)$$

65. Calculating SEP contributions Find the maximum allowable contribution to a SEP plan by solving the equation $C = 0.15(T - C)$ for C.

66. Calculating SEP contributions Find the maximum allowable contribution to a SEP plan for a person who earns $75,000 and has deductible expenses of $27,540. See Exercise 65.

WRITING ABOUT MATH

67. The formula $P = 2l + 2w$ is also an equation, but an equation such as $2x + 3 = 5$ is not a formula. What equations do you think should be called formulas?

68. To solve the equation $s - A(s - 5) = r$ for the variable s, one student simply added $A(s - 5)$ to both sides to get $s = r + A(s - 5)$. Explain why this is not correct.

SOMETHING TO THINK ABOUT

69. The energy of an atomic bomb comes from the conversion of matter into energy, according to Einstein's formula $E = mc^2$. The constant c is the speed of light, about 300,000 meters per second. Find the energy in a mass m of 1 kilogram. Energy has units of **joules**.

70. When a car of mass m collides with a wall, the energy of the collision is given by the formula $E = \frac{1}{2}mv^2$. Compare the energy of two collisions: a car striking a wall at 30 mph, and at 60 mph.

Unless otherwise noted, all content on this page is © Cengage Learning.

Section 2.5 Introduction to Problem Solving

Objectives

1. Solve a number application using a linear equation in one variable.
2. Solve a geometry application using a linear equation in one variable.
3. Solve an investment application using a linear equation in one variable.

Vocabulary

angle	adjacent angles	vertex angle of an isosceles triangle
degree	complementary angles	base angles of an isosceles triangle
right angle	supplementary angles	
straight angle	isosceles triangle	

Getting Ready

1. If one part of a pipe is x feet long and the other part is $(x + 2)$ feet long, find an expression that represents the total length of the pipe.
2. If one part of a board is x feet long and the other part is three times as long, find an expression that represents the length of the board.
3. What is the formula for the perimeter of a rectangle?
4. Define a triangle.

In this section, we will use the equation-solving skills we have learned in the previous four sections to solve many types of applications. The key to successful problem solving is to understand the situation thoroughly and then devise a plan to solve it. To do so, we will use the following problem-solving strategy.

PROBLEM SOLVING

1. **Analyze the problem** and **identify a variable** by asking yourself "What am I asked to find?" Choose a variable to represent the quantity to be found and then express all other unknown quantities in the problem as expressions involving that variable.
2. **Form an equation** by expressing a quantity in two different ways. This may require reading the problem several times to understand the given facts. What information is given? Is there a formula that applies to this situation? Often a sketch, chart, or diagram will help you visualize the facts of the problem.
3. **Solve the equation** found in Step 2.
4. **State the conclusion.**
5. **Check the result** to be certain it satisfies the given conditions.

In this section, we will use this five-step strategy to solve many types of applications.

1 Solve a number application using a linear equation in one variable.

EXAMPLE 1 **PLUMBING** A plumber wants to cut a 17-foot pipe into three parts. (See Figure 2-7.) If the longest part is to be 3 times as long as the shortest part, and the middle-sized part is to be 2 feet longer than the shortest part, how long should each part be?

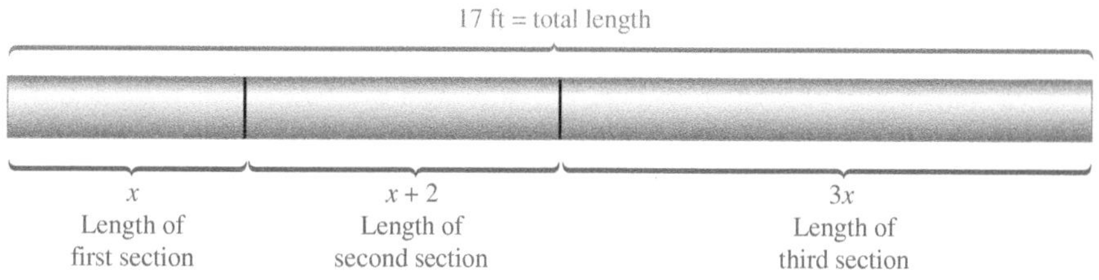

Figure 2-7

Analyze the problem We are asked to find the length of three pieces of pipe. The information is given in terms of the length of the shortest part. Therefore, we let x represent the length, in feet, of the shortest part and express the other lengths in terms of x. Then $3x$ represents the length of the longest part, and $x + 2$ represents the length of the middle-sized part.

Form an equation The sum of the lengths of these three parts is equal to the total length of the pipe.

The length of part 1	plus	the length of part 2	plus	the length of part 3	equals	the total length.
x	$+$	$x + 2$	$+$	$3x$	$=$	17

Solve the equation We can solve this equation as follows.

$$x + x + 2 + 3x = 17 \quad \text{This is the equation to solve.}$$
$$5x + 2 = 17 \quad \text{Combine like terms.}$$
$$5x = 15 \quad \text{Subtract 2 from both sides.}$$
$$x = 3 \quad \text{Divide both sides by 5.}$$

State the conclusion The shortest part is 3 feet long. Because the middle-sized part is 2 feet longer than the shortest, it is 5 feet long. Because the longest part is 3 times longer than the shortest, it is 9 feet long.

Check the result Because the sum of 3 feet, 5 feet, and 9 feet is 17 feet, the solution checks.

SELF CHECK 1 A plumber wants to cut a 17-foot pipe into three parts. If the longest part is to be 2 times as long as the shortest part, and the middle-sized part is to be 1 foot longer than the shortest part, how long should each part be?

COMMENT Remember to include any units (feet, inches, pounds, etc.) when stating the conclusion to an application.

2 Solve a geometry application using a linear equation in one variable.

The geometric figure shown in Figure 2-8(a) is an **angle**. Angles are measured in **degrees**. The angle shown in Figure 2-8(b) measures 45 degrees (denoted as 45°). If an angle measures 90°, as in Figure 2-8(c), it is a **right angle**. If an angle measures 180°, as in Figure 2-8(d), it is a **straight angle**. **Adjacent angles** are two angles that share a common side.

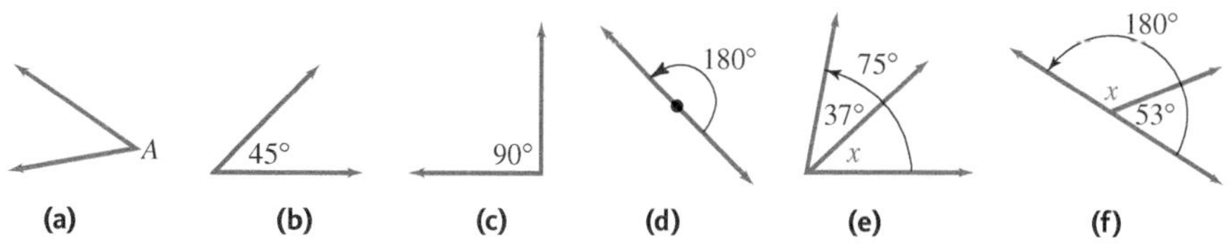

Figure 2-8

Unless otherwise noted, all content on this page is © Cengage Learning.

EXAMPLE 2 **GEOMETRY** Refer to Figure 2-8(e) on the previous page and find the value of x.

Analyze the problem — In Figure 2-8(e), we have two adjacent angles. The unknown angle measure is designated as x degrees.

Form an equation — From the figure, we can see that the sum of their measures is 75°. Since the sum of x and 37° is equal to 75°, we can form the equation.

The angle that measures $x°$	plus	the angle that measures 37°	equals	the angle that measures 75°.
x	$+$	37	$=$	75

Solve the equation — We can solve this equation as follows.

$$x + 37 = 75 \quad \text{This is the equation to solve.}$$
$$x + 37 - \mathbf{37} = 75 - \mathbf{37} \quad \text{Subtract 37 from both sides.}$$
$$x = 38 \quad 37 - 37 = 0 \text{ and } 75 - 37 = 38.$$

State the conclusion — The value of x is 38°.

Check the result — Since the sum of 38° and 37° is 75°, the solution checks.

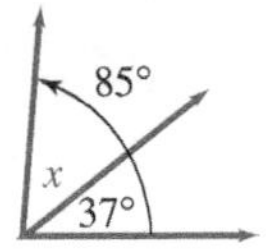

SELF CHECK 2 Find the value of x in the figure to the right.

EXAMPLE 3 **GEOMETRY** Refer to Figure 2-8(f) and find the value of x.

Analyze the problem — In Figure 2-8(f), we have two adjacent angles. The unknown angle measure is designated as x degrees.

Form an equation — From the figure, we can see that the sum of their measures is 180°. Since the sum of x and 53° is equal to 180°, we can form the equation.

The angle that measures $x°$	plus	the angle that measures 53°	equals	the angle that measures 180°.
x	$+$	53	$=$	180

Solve the equation — We can solve this equation as follows.

$$x + 53 = 180 \quad \text{This is the equation to solve.}$$
$$x + 53 - \mathbf{53} = 180 - \mathbf{53} \quad \text{Subtract 53 from both sides.}$$
$$x = 127 \quad 53 - 53 = 0 \text{ and } 180 - 53 = 127.$$

State the conclusion — The value of x is 127°.

Check the result — Since the sum of 127° and 53° is 180°, the solution checks.

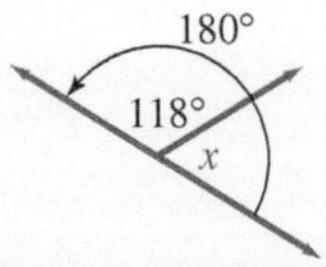

SELF CHECK 3 Find the value of x in the figure to the right.

If the sum of two angles is 90°, the angles are **complementary angles** and either angle is the *complement* of the other. If the sum of two angles is 180°, the angles are **supplementary angles** and either angle is the *supplement* of the other.

Unless otherwise noted, all content on this page is © Cengage Learning.

EXAMPLE 4 **COMPLEMENTARY ANGLES** Find the complement of an angle measuring 30°.

Analyze the problem: To find the complement of a 30° angle, we must find an angle whose measure plus 30° equals 90°. We can let x represent the complement of 30°.

Form an equation: Since the sum of two complementary angles is 90°, we can form the equation.

The angle that measures $x°$	plus	the angle that measures 30°	equals	90°.
x	$+$	30	$=$	90

Solve the equation: We can solve this equation as follows.

$$x + 30 = 90 \quad \text{This is the equation to solve.}$$
$$x + 30 - \mathbf{30} = 90 - \mathbf{30} \quad \text{Subtract 30 from both sides.}$$
$$x = 60 \quad 30 - 30 = 0 \text{ and } 90 - 30 = 60.$$

State the conclusion: The complement of a 30° angle is a 60° angle.

Check the result: Since the sum of 60° and 30° is 90°, the solution checks.

SELF CHECK 4 Find the complement of an angle measuring 40°.

EXAMPLE 5 **SUPPLEMENTARY ANGLES** Find the supplement of an angle measuring 50°.

Analyze the problem: To find the supplement of a 50° angle, we must find an angle whose measure plus 50° equals 180°. We can let x represent the supplement of 50°.

Form an equation: Since the sum of two supplementary angles is 180°, we can form the equation.

The angle that measures $x°$	plus	the angle that measures 50°	equals	180°.
x	$+$	50	$=$	180

Solve the equation: We can solve this equation as follows.

$$x + 50 = 180 \quad \text{This is the equation to solve.}$$
$$x + 50 - \mathbf{50} = 180 - \mathbf{50} \quad \text{Subtract 50 from both sides.}$$
$$x = 130 \quad 50 - 50 = 0 \text{ and } 180 - 50 = 130.$$

State the conclusion: The supplement of a 50° angle is a 130° angle.

Check the result: Since the sum of 50° and 130° is 180°, the solution checks.

SELF CHECK 5 Find the supplement of an angle measuring 80°.

EXAMPLE 6 **RECTANGLES** The length of a rectangle is 4 meters longer than twice its width. If the perimeter of the rectangle is 26 meters, find its dimensions.

Analyze the problem: Because we are asked to find the dimensions of the rectangle, we will need to find both the width and the length. If we let w represent the width of the rectangle in meters, then $(4 + 2w)$ will represent its length.

Form an equation: To visualize the problem, we sketch the rectangle as shown in Figure 2-9 on the next page. Recall that the formula for finding the perimeter of a rectangle is $P = 2l + 2w$. Therefore, the perimeter of the rectangle in the figure is $2(4 + 2w) + 2w$. We also are told that the perimeter is 26.

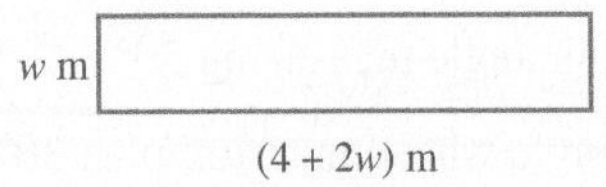

Figure 2-9

We can form the equation as follows.

2	times	the length	plus	2	times	the width	equals	the perimeter.
2	$\cdot$	$(4 + 2w)$	$+$	2	$\cdot$	w	$=$	26

Solve the equation — We can solve this equation as follows.

$$2(4 + 2w) + 2w = 26 \quad \text{This is the equation to solve.}$$
$$8 + 4w + 2w = 26 \quad \text{Use the distributive property to remove parentheses.}$$
$$6w + 8 = 26 \quad \text{Combine like terms.}$$
$$6w = 18 \quad \text{Subtract 8 from both sides.}$$
$$w = 3 \quad \text{Divide both sides by 6.}$$

State the conclusion — The width of the rectangle is 3 meters, and the length, $4 + 2w$, is 10 meters.

Check the result — If the rectangle has a width of 3 meters and a length of 10 meters, the length is 4 meters longer than twice the width $(4 + 2 \cdot 3 = 10)$, and the perimeter is 26 meters. The solution checks.

SELF CHECK 6 The length of a rectangle is 1 meter longer than twice its width. If the perimeter of the rectangle is 32 meters, find its dimensions.

EXAMPLE 7

ISOSCELES TRIANGLES The vertex angle of an isosceles triangle is 56°. Find the measure of each base angle.

Analyze the problem — An **isosceles triangle** has two sides of equal length, which meet to form the **vertex angle**. See Figure 2-10. The angles opposite those sides, called **base angles**, have equal measures. If we let x represent the measure of one base angle, the measure of the other base angle is also x.

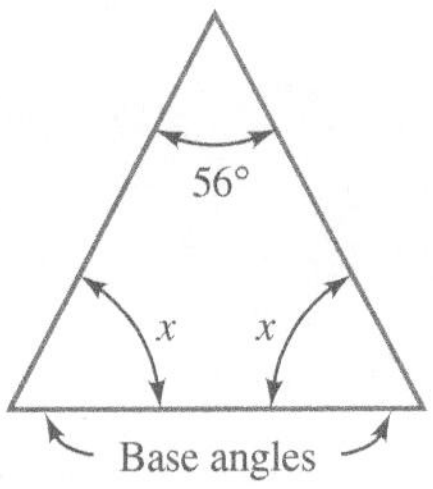

Figure 2-10

Form an equation — From geometry, we know that in any triangle the sum of the measures of its three angles is 180°. Therefore, we can form the equation.

The measure of one base angle	plus	the measure of the other base angle	plus	the measure of the vertex angle	equals	180°.
x	$+$	x	$+$	56	$=$	180

Solve the equation — We can solve this equation as follows.

$$x + x + 56 = 180 \quad \text{This is the equation to solve.}$$
$$2x + 56 = 180 \quad \text{Combine like terms.}$$
$$2x = 124 \quad \text{Subtract 56 from both sides.}$$
$$x = 62 \quad \text{Divide both sides by 2.}$$

State the conclusion — The measure of each base angle is 62°.

Check the result — The measure of each base angle is 62°, and the vertex angle measures 56°. Since $62° + 62° + 56° = 180°$, the sum of the measures of the three angles is 180°. The solution checks.

SELF CHECK 7 The vertex angle of an isosceles triangle is 42°. Find the measure of each base angle.

Unless otherwise noted, all content on this page is © Cengage Learning.

3 Solve an investment application using a linear equation in one variable.

EXAMPLE 8 **INVESTMENTS** A teacher invests part of \$12,000 at 6% annual simple interest, and the rest at 9%. If the annual income from these investments was \$945, how much did the teacher invest at each rate?

Analyze the problem We are asked to find the amount of money the teacher has invested in two different accounts. If we let x represent the amount of money invested at 6% annual interest, the remainder, $(12{,}000 - x)$, represents the amount invested at 9% annual interest.

Form an equation The interest i earned by an amount p invested at an annual rate r for t years is given by the formula $i = prt$. In this example, $t = 1$ year. Hence, if x dollars were invested at 6%, the interest earned would be $0.06x$ dollars. If x dollars were invested at 6%, the rest of the money, $(12{,}000 - x)$ dollars, would be invested at 9%. The interest earned on that money would be $0.09(12{,}000 - x)$ dollars. The total interest earned in dollars can be expressed in two ways: as 945 and as the sum $0.06x + 0.09(12{,}000 - x)$.

We can form an equation as follows.

The interest earned at 6%	plus	the interest earned at 9%	equals	the total interest.
$0.06x$	$+$	$0.09(12{,}000 - x)$	$=$	945

Solve the equation We can solve this equation as follows.

$0.06x + 0.09(12{,}000 - x) = 945$	This is the equation to solve.
$6x + 9(12{,}000 - x) = 94{,}500$	Multiply both sides by 100 to clear the equation of decimals.
$6x + 108{,}000 - 9x = 94{,}500$	Use the distributive property to remove parentheses.
$-3x + 108{,}000 = 94{,}500$	Combine like terms.
$-3x = -13{,}500$	Subtract 108,000 from both sides.
$x = 4{,}500$	Divide both sides by -3.

State the conclusion The teacher invested \$4,500 at 6% and \$12,000 − \$4,500 or \$7,500 at 9%.

Check the result The first investment earned 6% of \$4,500, or \$270. The second investment earned 9% of \$7,500, or \$675. Because the total return was \$270 + \$675, or \$945, the solutions check.

SELF CHECK 8 A teacher invests part of \$15,000 at 6% annual simple interest, and the rest at 9%. If the annual income from these investments was \$945, how much did the teacher invest at each rate?

SELF CHECK ANSWERS

1. 4 ft, 8 ft, 5 ft **2.** 48° **3.** 62° **4.** 50° **5.** 100° **6.** 5 meters by 11 meters **7.** 69° **8.** \$13,500 at 6%, \$1,500 at 9%

NOW TRY THIS

1. Mark invested \$100,000 in two accounts. Part was in bonds that paid 7% annual interest and the rest in stocks that lost 5% of their value. How much did he originally invest in each account if his total earned interest for the year was \$2,200? How much money does he have in each account now?

2. Sharlette invested \$80,000 in two accounts. Part was in an account paying 6.25% and the rest in an account paying 4.48%. If her total interest was \$4,115, how much was invested in each account?

2.5 Exercises

WARM-UPS

If the length of a board is 12 feet and one part of it is x feet,

1. Write an expression for the other part.

2. Write an expression for a length that is 5 feet shorter than twice the first part.

If an instructor invests \$18,000, \$$x$ in an account paying 4%, and the rest in an account paying 7%,

3. Write an expression for the interest earned on the money in the 7% account.

4. Write an expression for the total interest earned.

REVIEW *Refer to the formulas in Section 1.3.*

5. Find the volume of a pyramid that has a height of 6 centimeters and a square base, 10 centimeters on each side.

6. Find the volume of a cone with a height of 6 centimeters and a circular base with radius 6 centimeters. Use $\pi \approx \frac{22}{7}$.

Simplify each expression.

7. $5(x - 2) + 3(x + 4)$ **8.** $6(y + 2) - 4(y - 3)$

9. $\frac{1}{2}(x + 1) - \frac{1}{2}(x + 4)$ **10.** $\frac{3}{2}\left(x + \frac{2}{3}\right) + \frac{1}{2}(x + 8)$

11. The amount A on deposit in a bank account bearing simple interest is given by the formula

$$A = P + Prt$$

Find A when $P = \$1{,}200$, $r = 0.08$, and $t = 3$.

12. The distance s that a certain object falls from a height of 350 ft in t seconds is given by the formula

$$s = 350 - 16t^2 + vt$$

Find s when $t = 3$ and $v = -4$.

VOCABULARY AND CONCEPTS *Fill in the blanks.*

13. The perimeter of a rectangle is given by the formula $P =$ _______.

14. An _______ triangle is a triangle with two sides of equal length.

15. The sides of equal length of an isosceles triangle meet to form the _____ angle.

16. The angles opposite the sides of equal length of an isosceles triangle are called ____ angles.

17. Angles are measured in _______.

18. If an angle measures 90°, it is called a _____ angle.

19. If an angle measures 180°, it is called a _______ angle.

20. If the sum of the measures of two angles is 90°, the angles are called _____________ angles.

21. If the sum of the measures of two angles is 180°, the angles are called ____________ angles.

22. The sum of the measures of the angles of any triangle is ____.

APPLICATIONS *SEE EXAMPLE 1. (OBJECTIVE 1)*

23. Carpentry The 12-foot board in the illustration has been cut into two parts, one twice as long as the other. How long is each part?

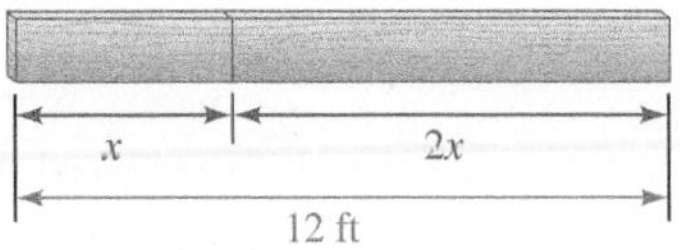

24. Plumbing A 20-foot pipe has been cut into two parts, one 3 times as long as the other. How long is each part?

25. Robotics If the robotic arm shown in the illustration will extend a total distance of 30 feet, how long is each section?

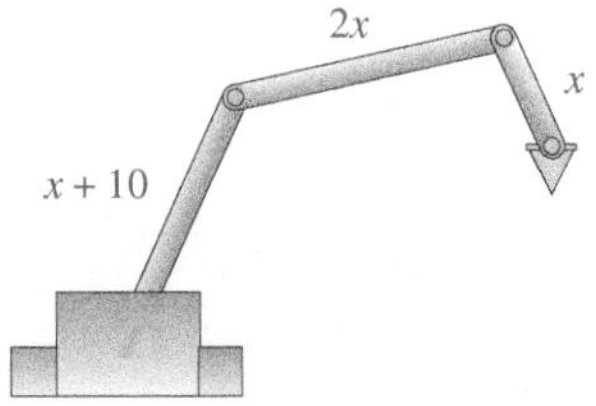

26. Statue of Liberty If the figure part of the Statue of Liberty is 3 feet shorter than the height of its pedestal base, find the height of the figure.

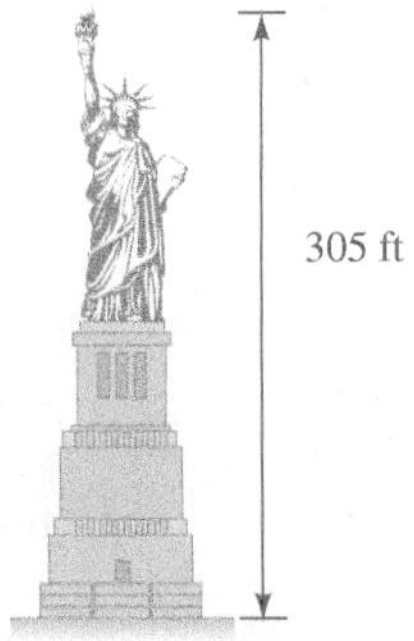

27. Window designs The perimeter of the triangular window shown in the illustration is 24 feet. How long is each section?

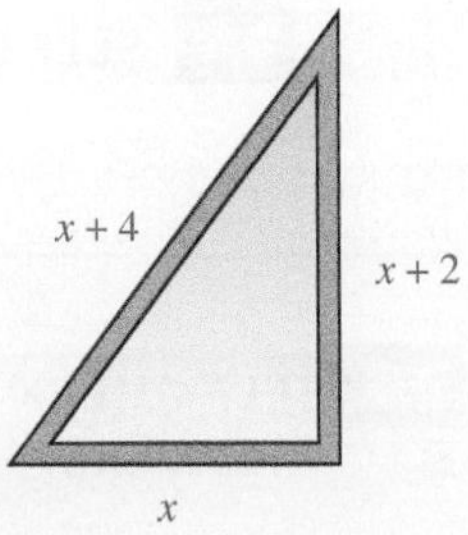

28. Football In 1967, Green Bay beat Kansas City by 25 points in the first Super Bowl. If a total of 45 points were scored, what was the final score of the game?

29. Publishing A book can be purchased in hardcover for \$16.95 or in paperback for \$5.95. How many of each type were printed if 14 times as many paperbacks were printed as hardcovers and a total of 210,000 books were printed?

30. Concert tours A rock group plans three concert tours over a period of 26 weeks. The tour in Britain will be twice as long

Unless otherwise noted, all content on this page is © Cengage Learning.

as the tour in France and the tour in Germany will be 2 weeks shorter than the tour in France. How many weeks will they be in France?

Find the value of x. SEE EXAMPLE 2. (OBJECTIVE 2)

31.

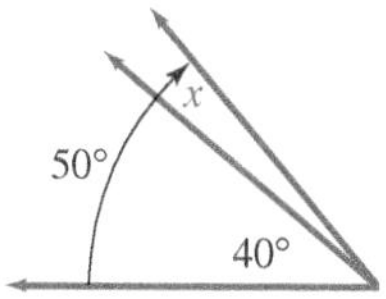

32.

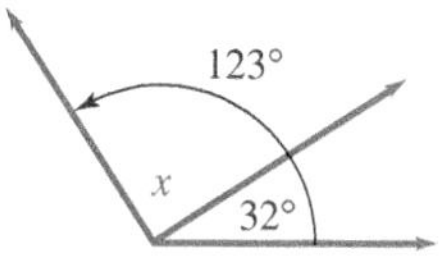

33.

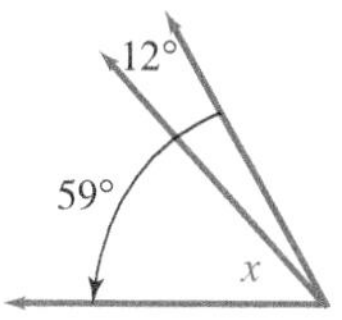

34.

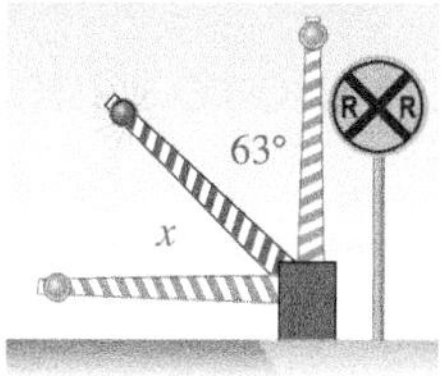

Find the value of x. SEE EXAMPLE 3. (OBJECTIVE 2)

35.

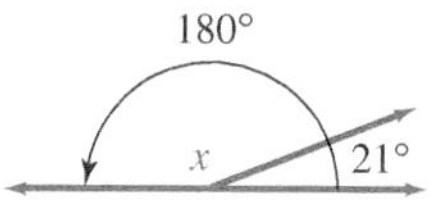

36.

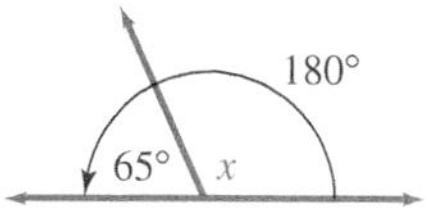

37.

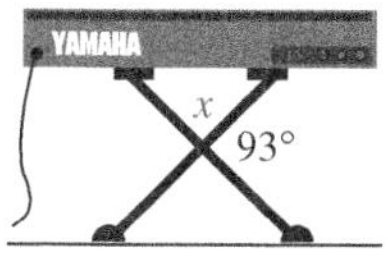

38.

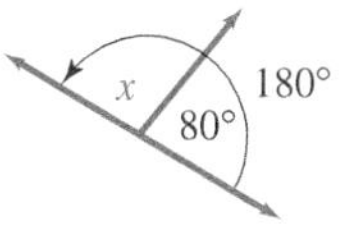

Find each value. SEE EXAMPLES 4–5. (OBJECTIVE 2)

39. Find the complement of an angle measuring 46°.

40. Find the supplement of an angle measuring 46°.

41. Find the supplement of the complement of an angle measuring 40°.

42. Find the complement of the supplement of an angle measuring 125°.

Solve. SEE EXAMPLE 6. (OBJECTIVE 2)

43. Circuit boards The perimeter of the circuit board in the illustration is 90 centimeters. Find the dimensions of the board.

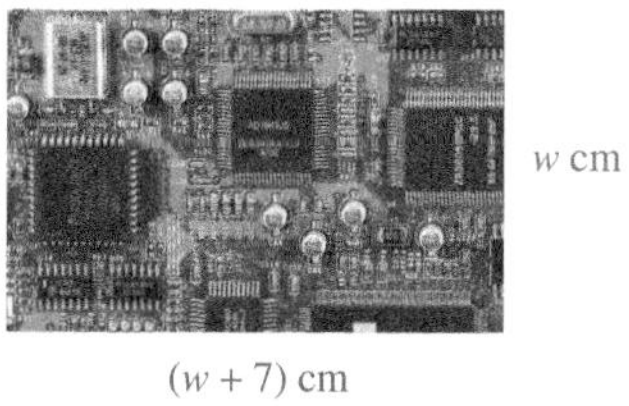

© DariushM/ Shutterstock.com

44. Swimming pools The width of a rectangular swimming pool is 11 meters less than the length, and the perimeter is 94 meters. Find its dimensions.

45. Framing pictures The length of a rectangular picture is 6 inches less than twice the width. If the perimeter is 60 inches, find the dimensions of the frame.

46. Land areas The perimeter of a square piece of land is twice the perimeter of an equilateral (equal-sided) triangular lot. If one side of the square is 60 meters, find the length of a side of the triangle.

Solve. SEE EXAMPLE 7. (OBJECTIVE 2)

47. Triangular bracing The outside perimeter of the triangular brace shown in the illustration is 57 feet. If all three sides are of equal length, find the length of each side.

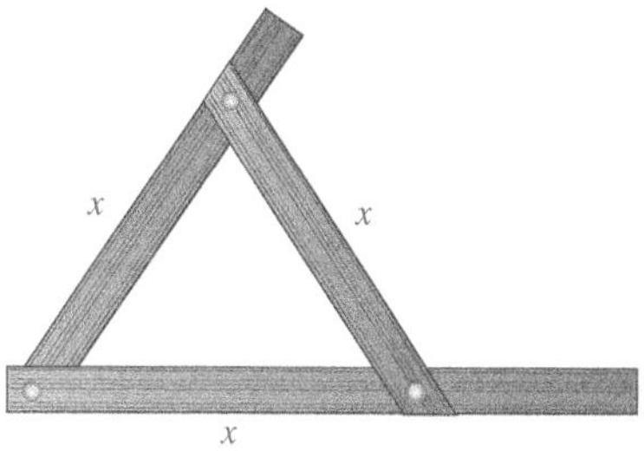

48. Trusses The truss in the illustration is in the form of an isosceles triangle. Each of the two equal sides is 5 feet less than the third side. If the perimeter is 29 feet, find the length of each side.

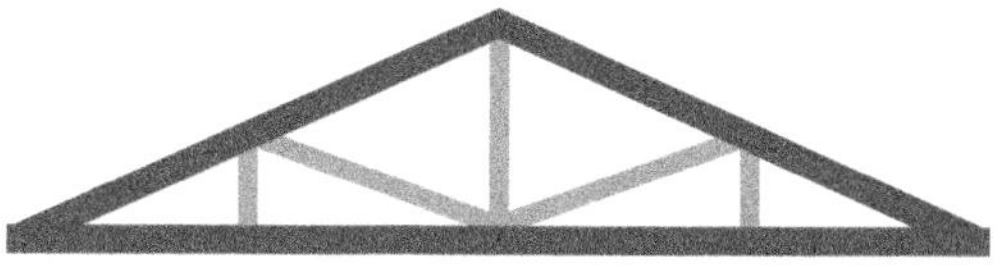

49. Guy wires The two guy wires in the illustration form an isosceles triangle. One of the two equal angles of the triangle is 4 times the third angle (the vertex angle). Find the measure of the vertex angle.

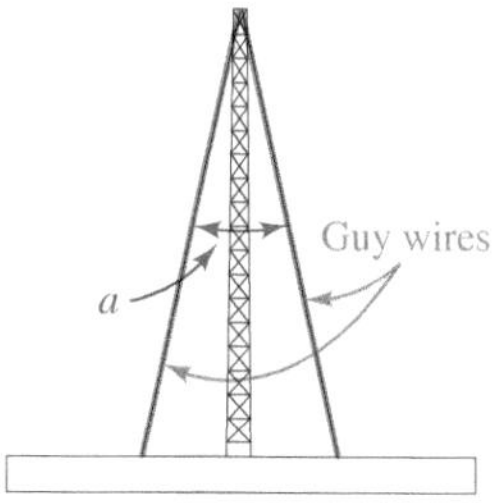

50. Equilateral triangles Find the measure of each angle of an equilateral triangle. (*Hint*: The three angles of an equilateral triangle are equal.)

Solve. SEE EXAMPLE 8. (OBJECTIVE 3)

51. Investments A student invested some money at an annual rate of 5%. If the annual income from the investment is $300, how much did he invest?

52. Investments A student invested 90% of her savings in the stock market. If she invested $4,050, what are her total savings?

53. Investments A broker invested $24,000 in two mutual funds, one earning 9% annual interest and the other earning

Unless otherwise noted, all content on this page is © Cengage Learning.

14%. After 1 year, his combined interest is \$3,135. How much was invested at each rate?

54. Investments A rollover IRA of \$18,750 was invested in two mutual funds, one earning 12% interest and the other earning 10%. After 1 year, the combined interest income is \$2,117. How much was invested at each rate?

55. Investments One investment pays 8% and another pays 11%. If equal amounts are invested in each, the combined interest income for 1 year is \$712.50. How much is invested at each rate?

56. Investments When equal amounts are invested in each of three accounts paying 7%, 8%, and 10.5%, one year's combined interest income is \$1,249.50. How much is invested in each account?

57. Investments A college professor wants to supplement her retirement income with investment interest. If she invests \$15,000 at 6% annual interest, how much more would she have to invest at 7% to achieve a goal of \$1,250 in supplemental income?

58. Investments A teacher has a choice of two investment plans: an insured fund that has paid an average of 11% interest per year, or a riskier investment that has averaged a 13% return. If the same amount invested at the higher rate would generate an extra \$150 per year, how much does the teacher have to invest?

59. Investments A financial counselor recommends investing twice as much in CDs (certificates of deposit) as in a bond fund. A client follows his advice and invests \$21,000 in CDs paying 1% more interest than the fund. The CDs would generate \$840 more interest than the fund. Find the two rates. (*Hint*: 1% = 0.01.)

60. Investments The amount of annual interest earned by \$8,000 invested at a certain rate is \$200 less than \$12,000 would earn at a 1% lower rate. At what rate is the \$8,000 invested?

WRITING ABOUT MATH

61. Write a paragraph describing the problem-solving process.

62. List as many types of angles as you can think of. Then define each type.

SOMETHING TO THINK ABOUT

63. If two lines intersect as in the illustration, angle 1 (denoted as $\angle 1$) and $\angle 2$, and $\angle 3$ and $\angle 4$, are called **vertical angles**. Let the measure of $\angle 1$ be various numbers and compute the values of the other three angles. What do you discover?

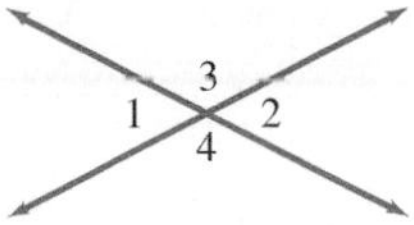

64. If two lines meet and form a right angle, the lines are said to be **perpendicular**. See the illustration. Find the measures of $\angle 1$, $\angle 2$, and $\angle 3$. What do you discover?

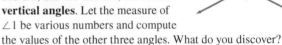

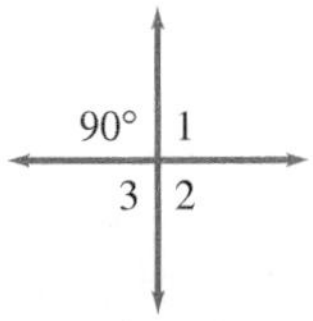

Section 2.6 Motion and Mixture Applications

Objectives

1. Solve a motion application using a linear equation in one variable.
2. Solve a liquid mixture application using a linear equation in one variable.
3. Solve a dry mixture application using a linear equation in one variable.

Getting Ready

1. At 30 mph, how far would a bus go in 2 hours?
2. At 55 mph, how far would a car travel in 7 hours?
3. If 8 gallons of a mixture of water and alcohol is 70% alcohol, how much alcohol does the mixture contain?
4. At \$7 per pound, how many pounds of chocolate would be worth \$63?

In this section, we consider uniform motion and mixture applications. In these problems, we will use the following three formulas:

$r \cdot t = d$	The rate multiplied by the time equals the distance.
$r \cdot b = a$	The rate multiplied by the base equals the amount.
$v = p \cdot n$	The value equals the price multiplied by the number.

Unless otherwise noted, all content on this page is © Cengage Learning.

1 Solve a motion application using a linear equation in one variable.

EXAMPLE 1 **TRAVELING** Chicago and Green Bay are about 200 miles apart. If a car leaves Chicago traveling toward Green Bay at 55 mph at the same time as a truck leaves Green Bay bound for Chicago at 45 mph following the same route, how long will it take them to meet?

Analyze the problem: We are asked to find the amount of time it takes for the two vehicles to meet, so we will let t represent the time in hours.

Form an equation: Motion applications are based on the relationship $d = rt$, where d is the distance traveled, r is the rate, and t is the time. We can organize the information of this problem in a chart or a diagram, as shown in Figure 2-11.

	r	$\cdot\ t$	$= d$
Car	55	t	$55t$
Truck	45	t	$45t$

(a)

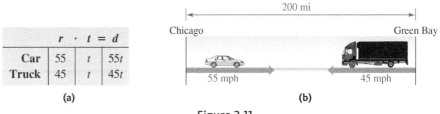

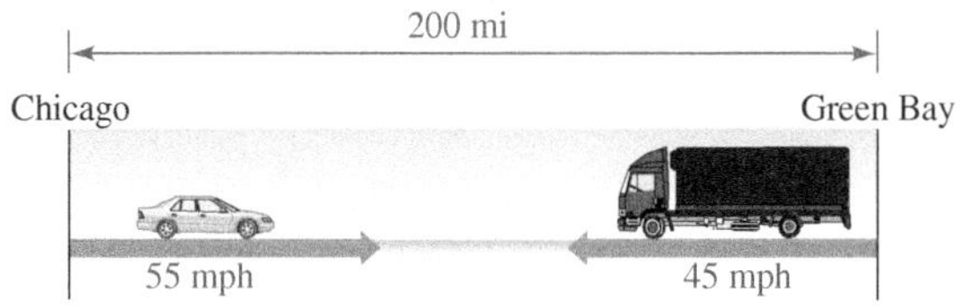

(b)

Figure 2-11

We know that the two vehicles travel for the same amount of time, t hours. The faster car will travel $55t$ miles, and the slower truck will travel $45t$ miles. At the time they meet, the total distance traveled can be expressed in two ways: as the sum $55t + 45t$, and as 200 miles.

After referring to Figure 2-11, we can form the equation.

The distance the car travels	plus	the distance the truck travels	equals	the total distance traveled.
$55t$	$+$	$45t$	$=$	200

Solve the equation: We can solve this equation as follows.

$$55t + 45t = 200 \quad \text{This is the equation to solve.}$$
$$100t = 200 \quad \text{Combine like terms.}$$
$$t = 2 \quad \text{Divide both sides by 100.}$$

State the conclusion: The vehicles will meet in 2 hours.

Check the result: In 2 hours, the car will travel $55 \cdot 2 = 110$ miles, while the truck will travel $45 \cdot 2 = 90$ miles. The total distance traveled will be $110 + 90 = 200$ miles. Since this is the total distance between Chicago and Green Bay, the solution checks.

SELF CHECK 1 Dallas and Austin are about 200 miles apart. If a car leaves Dallas traveling toward Austin at 65 mph at the same time as a truck leaves Austin bound for Dallas at 60 mph following the same route, how long will it take them to meet?

EXAMPLE 2 **SHIPPING** Two ships leave port, one heading east at 12 mph and one heading west at 10 mph. How long will it take before they are 33 miles apart?

Analyze the problem: We are asked to find the amount of time in hours, so we will let t represent the time.

Form an equation: The ships leave port at the same time and travel in opposite directions. We know that both travel for the same amount of time, t hours. The faster ship will travel $12t$ miles, and the slower ship will travel $10t$ miles. We can organize the information of this example in a chart or a diagram, as shown in Figure 2-12 on the next page. When the ships are 33 miles apart, the total distance traveled can be expressed in two ways: as the sum $(12t + 10t)$, and as 33 miles.

Unless otherwise noted, all content on this page is © Cengage Learning.

	r	$\cdot\ t$	$= d$
Faster ship	12	t	$12t$
Slower ship	10	t	$10t$

(a)

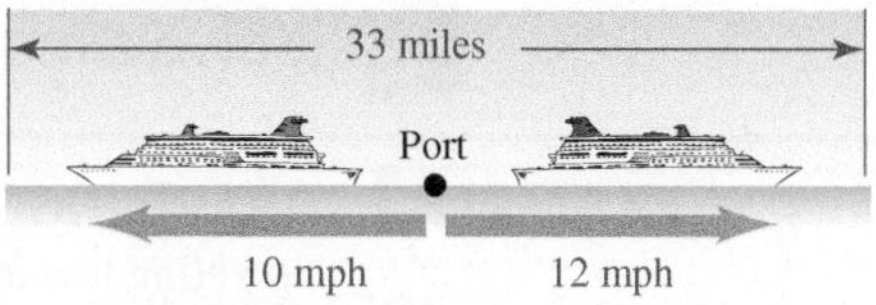

(b)

Figure 2-12

After referring to Figure 2-12, we can form the equation.

The distance the faster ship travels	plus	the distance the slower ship travels	equals	the total distance traveled.
$12t$	$+$	$10t$	$=$	33

Solve the equation We can solve this equation as follows.

$$12t + 10t = 33 \quad \text{This is the equation to solve.}$$

$$22t = 33 \quad \text{Combine like terms.}$$

$$t = \frac{33}{22} \quad \text{Divide both sides by 22.}$$

$$t = \frac{3}{2} \quad \text{Simplify the fractions: } \frac{33}{22} = \frac{3 \cdot \overset{1}{\cancel{11}}}{2 \cdot \underset{1}{\cancel{11}}} = \frac{3}{2}$$

State the conclusion The ships will be 33 miles apart in $\frac{3}{2}$ hours (or $1\frac{1}{2}$ hours).

Check the result In 1.5 hours, the faster ship travels $12 \cdot 1.5 = 18$ miles, while the slower ship travels $10 \cdot 1.5 = 15$ miles. Since the total distance traveled is $18 + 15 = 33$ miles, the solution checks.

SELF CHECK 2 Two ships leave port, one heading east at 15 mph and one heading west at 12 mph. How long will it take before they are 54 miles apart?

EXAMPLE 3

TRAVELING A car leaves Beloit, heading east at 50 mph. One hour later, a second car leaves Beloit, heading east along the same route at 65 mph. How long will it take for the second car to overtake the first car?

Analyze the problem The cars travel different amounts of time. In fact, the first car travels for one extra hour because it had a 1-hour head start. It is convenient to let the variable t represent the time traveled by the second car. Then $(t + 1)$ represents the number of hours the first car travels.

Form an equation We know that car 1 travels at 50 mph and car 2 travels at 65 mph. Using the formula $r \cdot t = d$, when car 2 overtakes car 1, car 2 will have traveled $65t$ miles and car 1 will have traveled $50(t + 1)$ hours.

We can organize the information in a chart or a diagram, as shown in Figure 2-13.

	r	$\cdot\ t$	$= d$
Car 1	50	$(t + 1)$	$50(t + 1)$
Car 2	65	t	$65t$

(a)

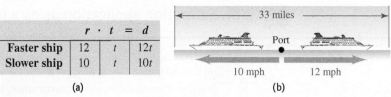

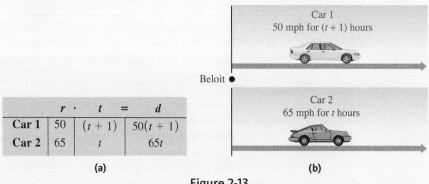

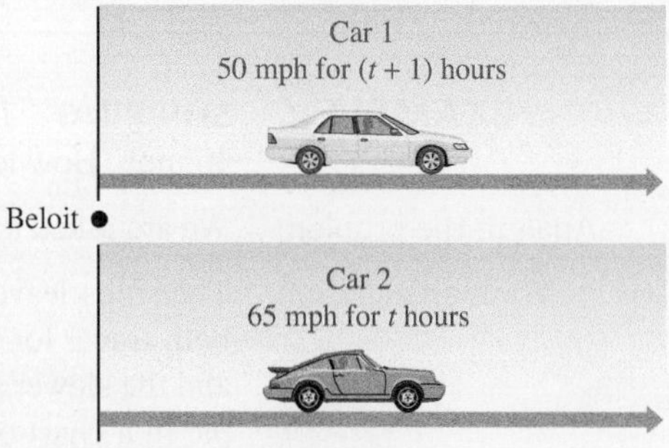

(b)

Figure 2-13

Unless otherwise noted, all content on this page is © Cengage Learning.

The distance the cars travel can be expressed in two ways: as $50(t + 1)$ miles, and as $65t$ miles.

Since these distances are equal when car 2 overtakes car 1, we can form the equation:

The distance that car 1 goes	equals	the distance that car 2 goes.
$50(t + 1)$	$=$	$65t$

Solve the equation

We can solve this equation as follows.

$$50(t + 1) = 65t \quad \text{This is the equation to solve.}$$
$$50t + 50 = 65t \quad \text{Use the distributive property to remove parentheses.}$$
$$50 = 15t \quad \text{Subtract } 50t \text{ from both sides.}$$
$$\frac{50}{15} = t \quad \text{Divide both sides by 15.}$$
$$t = \frac{10}{3} \quad \text{Simplify the fraction: } \frac{50}{15} = \frac{10 \cdot \overset{1}{\cancel{5}}}{3 \cdot \underset{1}{\cancel{5}}} = \frac{10}{3}$$

State the conclusion

Recall that t represents the time car 2 travels. Car 2 will overtake car 1 in $\frac{10}{3}$, or $3\frac{1}{3}$ hours.

Check the result

In $3\frac{1}{3}$ hours, car 2 will have traveled $65\left(\frac{10}{3}\right)$, or $\frac{650}{3}$, miles. With a 1-hour head start, car 1 will have traveled $50\left(\frac{10}{3} + 1\right) = 50\left(\frac{13}{3}\right)$, or $\frac{650}{3}$, miles. Since these distances are equal, the solution checks.

SELF CHECK 3 A car leaves Plano, heading east at 55 mph. One hour later, a second car leaves Plano, heading east along the same route at 65 mph. How long will it take for the second car to overtake the first car?

COMMENT In the previous example, we could let t represent the time traveled by the first car. Then $(t - 1)$ would represent the time traveled by the second car.

2 Solve a liquid mixture application using a linear equation in one variable.

EXAMPLE 4 **MIXING ACID** A chemist has one solution that is 50% sulfuric acid and another that is 20% sulfuric acid. How much of each should she use to make 12 liters of a solution that is 30% sulfuric acid?

Analyze the problem

We will let x represent the number of liters of the 50% sulfuric acid solution. Since there must be 12 liters of the final mixture, $(12 - x)$ represents the number of liters of 20% sulfuric acid solution to use.

Form an equation

Liquid mixture applications are based on the relationship $rb = a$, where b is the base, r is the rate, and a is the amount.

If x represents the number of liters of 50% solution to use, the amount of sulfuric acid in the solution will be $0.50x$ liters. The amount of sulfuric acid in the 20% solution will be $0.20(12 - x)$ liters. The amount of sulfuric acid in the final mixture will be $0.30(12)$ liters. We can organize this information in a chart or a diagram, as shown in Figure 2-14 on the next page.

	r	$\cdot$	b	$=$	a
50% solution	0.50		x		$0.50x$
20% solution	0.20		$12 - x$		$0.20(12 - x)$
30% solution	0.30		12		$0.30(12)$

(a)

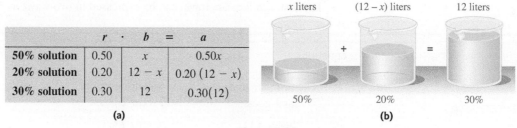

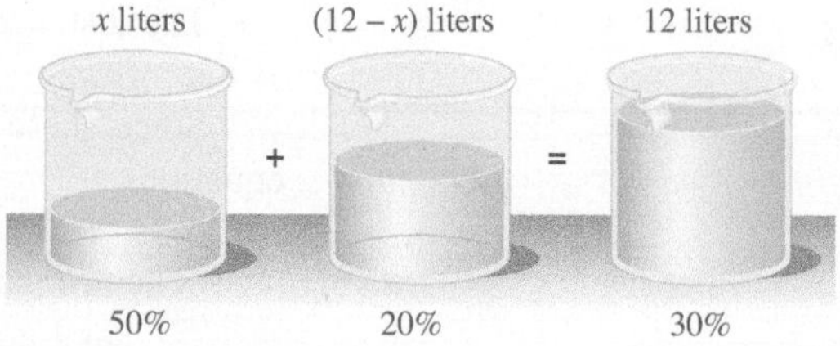

(b)

Figure 2-14

Since the number of liters of sulfuric acid in the 50% solution plus the number of liters of sulfuric acid in the 20% solution will equal the number of liters of sulfuric acid in the mixture, we can form the equation.

The amount of sulfuric acid in the 50% solution	plus	the amount of sulfuric acid in the 20% solution	equals	the amout of sulfuric acid in the final mixture.
50% of x	$+$	20% of $(12 - x)$	$=$	30% of 12

Solve the equation We can solve this equation as follows.

$0.5x + 0.2(12 - x) = 0.3(12)$	This is the equation to solve written in decimal form.
$5x + 2(12 - x) = 3(12)$	Multiply both sides by 10 to clear the equation of decimals.
$5x + 24 - 2x = 36$	Use the distributive property to remove parentheses.
$3x + 24 = 36$	Combine like terms.
$3x = 12$	Subtract 24 from both sides.
$x = 4$	Divide both sides by 3.

State the conclusion Recall that we let x represent the number of liters of the 50% sulfuric acid solution. The chemist must mix 4 liters of the 50% solution and $12 - 4 = 8$ liters of the 20% solution.

Check the result The amount of acid in 4 liters of 50% solution is $4(0.50) = 2$ liters.

The amount of acid in 8 liters of 20% solution is $8(0.20) = 1.6$ liters.

The amount of acid in 12 liters of 30% solution is $12(0.30) = 3.6$ liters.

Since $2 + 1.6 = 3.6$, the results check.

SELF CHECK 4 A chemist has one solution that is 40% sulfuric acid and another that is 10% sulfuric acid. How much of each should she use to make 20 liters of a solution that is 28% sulfuric acid?

3 Solve a dry mixture application using a linear equation in one variable.

EXAMPLE 5 **MIXING NUTS** Fancy cashews are not selling at \$9 per pound. However, filberts are selling well at \$6 per pound. How many pounds of filberts should be combined with 50 pounds of cashews to obtain a mixture that can be sold at \$7 per pound?

Analyze the problem We will let x represent the number of pounds of filberts in the mixture. Since we will be adding the filberts to 50 pounds of cashews, the total number of pounds of the mixture will be $(50 + x)$.

Form an equation Dry mixture applications are based on the relationship $v = pn$, where v is the value of the mixture, p is the price per pound, and n is the number of pounds. At \$6 per pound, x pounds of the filberts are worth \$$6x$. At \$9 per pound, the 50 pounds of cashews are

Unless otherwise noted, all content on this page is © Cengage Learning.

worth $9 · 50, or $450. The mixture will weigh $(50 + x)$ pounds, and at $7 per pound, it will be worth 7(50 + x)$. The *value* of the filberts (in dollars), $6x$, plus the *value* of the cashews (in dollars), 450, is equal to the *value* of the mixture (in dollars), $7(50 + x)$. We can organize this information in a table or a diagram, as shown in Figure 2-15.

	p ·	n =	v
Filberts	6	x	$6x$
Cashews	9	50	$9(50)$
Mixture	7	$50 + x$	$7(50 + x)$

(a)

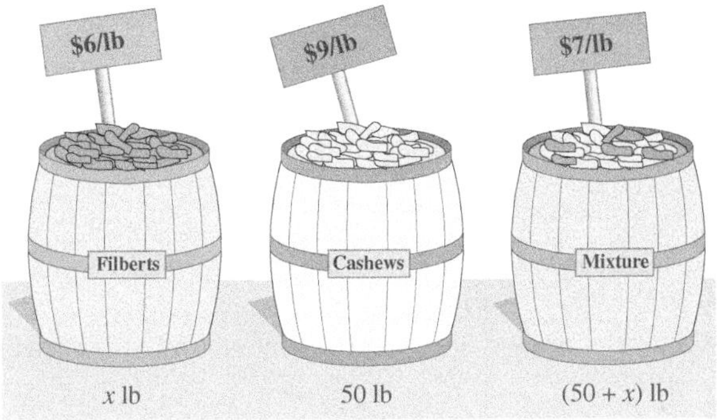

(b)

Figure 2-15

We can form the equation:

The value of the filberts	plus	the value of the cashews	equals	the value of the mixture.
$6x$	$+$	$9(50)$	$=$	$7(50 + x)$

Solve the equation We can solve this equation as follows.

$$6x + 9(50) = 7(50 + x)$$ This is the equation to solve.

$$6x + 450 = 350 + 7x$$ Use the distributive property to remove parentheses and simplify.

$$100 = x$$ Subtract $6x$ and 350 from both sides.

State the conclusion Recall that we let x represent the number of pounds of filberts to use. There should be 100 pounds of filberts in the mixture.

Check the result The value of 100 pounds of filberts at $6 per pound is $ 600

The value of 50 pounds of cashews at $9 per pound is $ 450

The value of the mixture is $ 1,050

The value of 150 pounds of mixture at $7 per pound is also $1,050.

SELF CHECK 5 Cashews are selling at $8 per pound, and filberts sell at $5 per pound. How many pounds of filberts should be combined with 40 pounds of cashews to obtain a mixture that can be sold at $6 per pound?

SELF CHECK ANSWERS

1. 1.6 hours **2.** 2 hours **3.** $5\frac{1}{2}$ hours **4.** 12 liters of 40% sulfuric acid solution, 8 liters of 10% sulfuric acid solution **5.** 80 pounds of filberts

NOW TRY THIS

1. A nurse has 5 ml of a 10% solution of benzalkonium chloride. If a doctor orders a 40% solution, how much pure benzalkonium chloride must he add to the solution to obtain the desired strength?
2. A paramedic has 5 ml of a 5% saline solution. If she needs a 4% saline solution, how much distilled water must she add to obtain the desired strength?

Unless otherwise noted, all content on this page is © Cengage Learning.

2.6 Exercises

WARM-UPS

1. How far will a car travel in h hours at a speed of 60 mph?
2. Two cars leave McKinney at the same time, one at 60 mph and the other at 70 mph. If they travel in the same direction, how far apart will they be in h hours?
3. How many ounces of alcohol are there in 20 ounces of a solution that is 25% alcohol?
4. Find the value of 10 pounds of coffee worth $\$d$ per pound.

REVIEW *Simplify each expression.*

5. $5 + 2(-7)$
6. $\dfrac{-5(3) - 2(-2)}{6 - (-5)}$
7. $3^3 - 5^2$
8. $4^2 + 2(5) - (-3)$

Solve each equation.

9. $-6x + 4 = -8$
10. $\dfrac{1}{4}y - 3 = 6$
11. $\dfrac{2}{3}p + 1 = 5$
12. $3(z - 4) = 5(z + 4)$

VOCABULARY AND CONCEPTS *Fill in the blanks.*

13. Motion problems are based on the formula _____.
14. Liquid mixture problems are based on the formula ______.
15. Dry mixture problems are based on the formula ______.
16. The information in motion and mixture problems can be organized in the form of a ____ or a _______.

APPLICATIONS *Solve. SEE EXAMPLES 1–2. (OBJECTIVE 1)*

17. **Travel times** Ashford and Bartlett are 315 miles apart. A car leaves Ashford bound for Bartlett at 50 mph. At the same time, another car leaves Bartlett bound for Ashford at 55 mph. How long will it take them to meet?

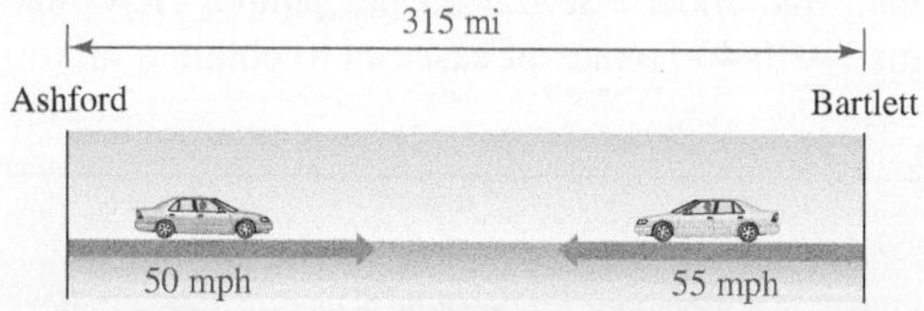

18. **Travel times** Granville and Preston are 535 miles apart. A car leaves Preston bound for Granville at 47 mph. At the same time, another car leaves Granville bound for Preston at 60 mph. How long will it take them to meet?
19. **Paving highways** Two crews working toward each other are 9.45 miles apart. One crew paves 1.5 miles of highway per day, and the other paves 1.2 miles per day. How long will it take them to meet?
20. **Biking** Two friends who live 20 miles apart ride bikes toward each other. One averages 11 mph, and the other averages 9 mph. How long will it take for them to meet?
21. **Travel times** Two cars leave Peoria at the same time, one heading east at 60 mph and the other west at 50 mph. How long will it take them to be 715 miles apart?

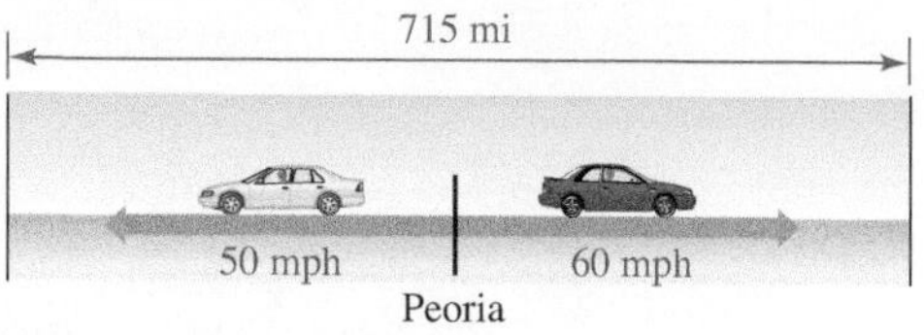

22. **Boating** Two boats leave port at the same time, one heading north at 35 knots (nautical miles per hour) and the other south at 47 knots. How long will it take them to be 738 nautical miles apart?
23. **Hiking** Two boys with two-way radios that have a range of 2 miles leave camp and walk in opposite directions. If one boy walks 3 mph and the other walks 4 mph, how long will it take before they lose radio contact?
24. **Biking** Two cyclists leave a park and ride in opposite directions, one averaging 9 mph and the other 6 mph. If they have two-way radios with a 5-mile range, for how many minutes will they remain in radio contact?

Solve. SEE EXAMPLE 3. (OBJECTIVE 1)

25. **Chasing a bus** Complete the table and compute how long it will take the car to overtake the bus if the bus had a 2-hour head start.

	r ·	t =	d
Car	60 mph	t	
Bus	50 mph	$t + 2$	

26. **Hot pursuit** Two crooks rob a bank and flee to the east at 66 mph. In 30 minutes, the police follow them in a helicopter, flying at 132 mph. How long will it take for the police to overtake the robbers?
27. **Travel times** Two cars start together and head east, one averaging 42 mph and the other averaging 53 mph. See the illustration. In how many hours will the cars be 82.5 miles apart?

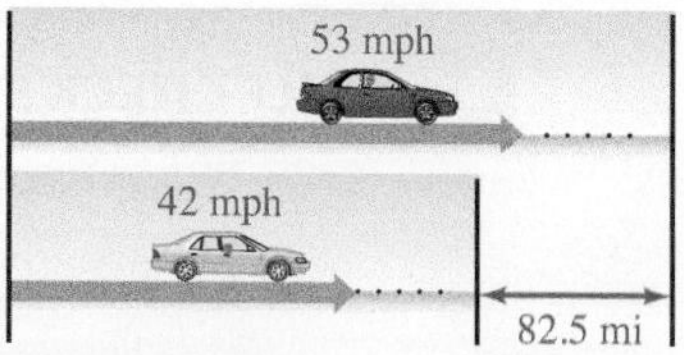

28. **Aviation** A plane leaves an airport and flies south at 180 mph. Later, a second plane leaves the same airport and flies south at 450 mph. If the second plane overtakes the first one in $1\frac{1}{2}$ hours, how much of a head start did the first plane have?
29. **Speed of trains** Two trains are 330 miles apart, and their speeds differ by 20 mph. They travel toward each other and meet in 3 hours. Find the speed of each train.
30. **Speed of airplanes** Two planes are 6,000 miles apart, and their speeds differ by 200 mph. They travel toward each other and meet in 5 hours. Find the speed of the slower plane.

Unless otherwise noted, all content on this page is © Cengage Learning.

31. **Average speeds** An automobile averaged 40 mph for part of a trip and 50 mph for the remainder. If the 5-hour trip covered 210 miles, for how long did the car average 40 mph?

32. **Vacation driving** A family drove to the Grand Canyon, averaging 45 mph. They returned using the same route, averaging 60 mph. If they spent a total of 7 hours of driving time, how far is their home from the Grand Canyon?

Solve. SEE EXAMPLE 4. (OBJECTIVE 2)

33. **Chemistry** A solution contains 0.3 liters of sulfuric acid. If this represents 12% of the total amount, find the total amount.

34. **Medicine** A laboratory has a solution that contains 3 ounces of benzalkonium chloride. If this is 15% of the total solution, how many ounces of solution does the lab have?

35. **Mixing fuels** How many gallons of fuel costing $3.35 per gallon must be mixed with 20 gallons of a fuel costing $3.85 per gallon to obtain a mixture costing $3.55 per gallon?

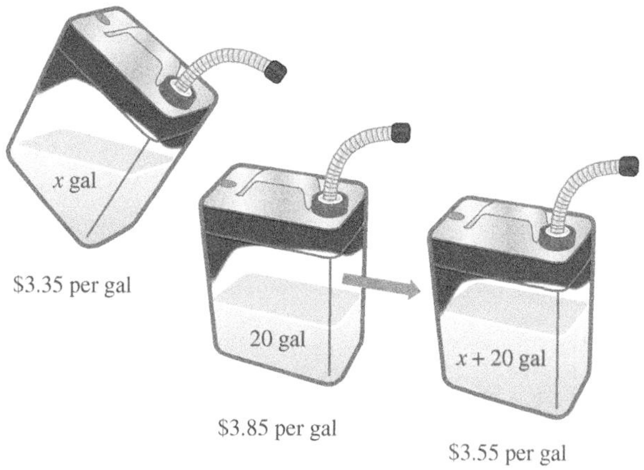

36. **Mixing paint** Paint costing $19 per gallon is to be mixed with 5 gallons of paint thinner costing $3 per gallon to make a paint that can be sold for $14 per gallon. Refer to the table and compute how much paint will be produced.

	p ·	n =	r
Paint	$19	x gal	$19x$
Thinner	$3	5 gal	$3(5)
Mixture	$14	$(x + 5)$ gal	$14(x + 5)$

37. **Brine solutions** How many gallons of a 3% salt solution must be mixed with 50 gallons of a 7% solution to obtain a 5% solution?

38. **Making cottage cheese** To make low-fat cottage cheese, milk containing 4% butterfat is mixed with 10 gallons of milk containing 1% butterfat to obtain a mixture containing 2% butterfat. How many gallons of the fattier milk must be used?

39. **Antiseptic solutions** A nurse wants to add water to 30 ounces of a 10% solution of benzalkonium chloride to dilute it to an 8% solution. How much water must she add?

40. **Mixing photographic chemicals** A photographer wants to mix 2 liters of a 5% acetic acid solution with a 10% solution to get a 7% solution. How many liters of 10% solution must be added?

Solve. SEE EXAMPLE 5. (OBJECTIVE 3)

41. **Mixing candy** Lemon drops are to be mixed with jelly beans to make 100 pounds of mixture. Refer to the illustration and compute how many pounds of each candy should be used.

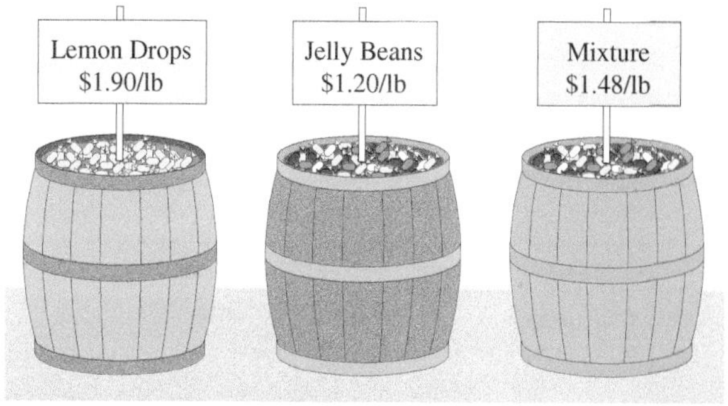

42. **Blending gourmet tea** One grade of tea, worth $3.20 per pound, is to be mixed with another grade worth $2 per pound to make 20 pounds that will sell for $2.72 per pound. How much of each grade of tea must be used?

43. **Mixing nuts** A bag of peanuts is worth 30¢ less than a bag of cashews. Equal amounts of peanuts and cashews are used to make 40 bags of a mixture that sells for $1.05 per bag. How much would a bag of cashews be worth?

44. **Mixing candy** Twenty pounds of lemon drops are to be mixed with cherry chews to make a mixture that will sell for $1.80 per pound. How much of the more expensive candy should be used? See the table.

Candy	Price per pound
Peppermint patties	$1.35
Lemon drops	$1.70
Licorice lumps	$1.95
Cherry chews	$2.00

45. **Coffee blends** A store sells regular coffee for $7 a pound and a gourmet coffee for $12 a pound. To reduce the inventory by 30 pounds of the gourmet coffee, the shopkeeper plans to make a gourmet blend to sell for $9 a pound. How many pounds of regular coffee should be used?

46. **Lawn seed blends** A garden store sells Kentucky bluegrass seed for $6 per pound and ryegrass seed for $3 per pound. How much rye must be mixed with 100 pounds of bluegrass to obtain a blend that will sell for $5 per pound?

47. **Mixing coffee** A shopkeeper sells chocolate coffee beans for $7 per pound. A customer asks the shopkeeper to mix 2 pounds of chocolate beans with 5 pounds of hazelnut coffee beans. If the customer paid $6 per pound for the mixture, what is the price per pound of the hazelnut beans?

48. **Trail mix** Fifteen pounds of trail mix are made by mixing 2 pounds of raisins worth $3 per pound with peanuts worth $4 per pound and M&Ms worth $5 per pound. How many pounds of peanuts must be used if the mixture is to be worth $4.20 per pound?

Unless otherwise noted, all content on this page is © Cengage Learning.

WRITING ABOUT MATH

49. Describe the steps you would use to analyze and solve a problem.

50. Create a mixture problem that could be solved by using the equation $4x + 6(12 - x) = 5(12)$.

51. Create a mixture problem of your own, and solve it.

52. In mixture problems, explain why it is important to distinguish between the quantity and the value (or strength) of the materials being combined.

SOMETHING TO THINK ABOUT

53. Is it possible for the equation of a problem to have a solution, but for the problem to have no solution? For example, is it possible to find two consecutive even integers whose sum is 16?

54. Invent a motion problem that leads to an equation that has a solution, although the problem does not.

55. Consider the problem: How many gallons of a 10% and a 20% solution should be mixed to obtain a 30% solution? Without solving it, how do you know that the problem has no solution?

56. What happens if you try to solve Exercise 55?

Section 2.7 Solving Linear Inequalities in One Variable

Objectives

1. Solve a linear inequality in one variable using the properties of inequality and graph the solution on a number line.
2. Solve a compound linear inequality in one variable.
3. Solve an application involving a linear inequality in one variable.

Vocabulary

inequality
solution of an inequality
addition property of inequality
multiplication property of inequality
double inequality
compound inequality
interval

Getting Ready

Graph each set on the number line.

1. All real numbers greater than -1

2. All real numbers less than or equal to 5

3. All real numbers between -2 and 4

4. All real numbers less than -2 or greater than or equal to 4

Many times, we will encounter mathematical statements indicating that two quantities are not necessarily equal. These statements are called *inequalities.*

1 Solve a linear inequality in one variable using the properties of inequality and graph the solution on a number line.

Recall the meaning of the following symbols.

INEQUALITY SYMBOLS

$<$ means "is less than"
$>$ means "is greater than"
$\leq$ means "is less than or equal to"
$\geq$ means "is greater than or equal to"

An **inequality** is a statement that indicates that two quantities are not necessarily equal. A **solution of an inequality** is any number that makes the inequality true. The number 2 is a solution of the inequality

$$x \leq 3$$

because $2 \leq 3$.

This inequality has many more solutions, because any real number that is less than or equal to 3 will satisfy it. We can use a graph on the number line to represent the solutions of the inequality. The red arrow in Figure 2-16 indicates all those points with coordinates that satisfy the inequality $x \leq 3$.

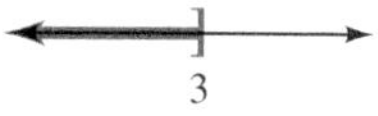
Figure 2-16

The bracket at the point with coordinate 3 indicates that the number 3 is a solution of the inequality $x \leq 3$.

The graph of the inequality $x > 1$ appears in Figure 2-17. The red arrow indicates all those points whose coordinates satisfy the inequality. The parenthesis at the point with coordinate 1 indicates that 1 is not a solution of the inequality $x > 1$.

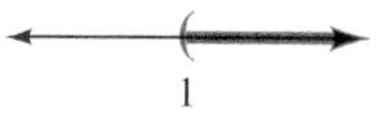
Figure 2-17

To solve more complicated inequalities, we need to use the addition, subtraction, multiplication, and division properties of inequalities. When we use any of these properties, the resulting inequality will have the same solutions as the original one.

ADDITION PROPERTY OF INEQUALITY

Suppose a, b, and c are real numbers.

If $a < b$, then $a + c < b + c$.

SUBTRACTION PROPERTY OF INEQUALITY

Suppose a, b, and c are real numbers.

If $a < b$, then $a - c < b - c$.

Similar statements can be made for the symbols $>$, $\leq$, and $\geq$.

The **addition property of inequality** can be stated this way: *If any quantity is added to both sides of an inequality, the resulting inequality has the same direction as the original inequality.*

The subtraction property of inequality can be stated this way: *If any quantity is subtracted from both sides of an inequality, the resulting inequality has the same direction as the original inequality.*

COMMENT The subtraction property of inequality is included in the addition property: To *subtract* a number a from both sides of an inequality, we *add* the *negative* of a to both sides.

EXAMPLE 1 Solve $2x + 5 > x - 4$ and graph the solution on a number line.

Solution To isolate the x on the left side of the $>$ sign, we proceed as if we were solving an equation.

$$2x + 5 > x - 4$$

$$2x + 5 - \mathbf{5} > x - 4 - \mathbf{5} \quad \text{Subtract 5 from both sides.}$$

$$2x > x - 9 \quad \text{Combine like terms.}$$

$$2x - \mathbf{x} > x - 9 - \mathbf{x} \quad \text{Subtract } x \text{ from both sides.}$$

$$x > -9 \quad \text{Combine like terms.}$$

Figure 2-18

The graph of the solution (see Figure 2-18) includes all points to the right of -9 but does not include -9 itself. For this reason, we use a parenthesis at -9.

SELF CHECK 1 Graph the solution of $2x - 2 < x + 1$.

Unless otherwise noted, all content on this page is © Cengage Learning.

If both sides of the true inequality $6 < 9$ are multiplied or divided by a *positive* number, such as 3, another true inequality results.

$$6 < 9$$
$$3 \cdot 6 < 3 \cdot 9 \quad \text{Multiply both sides by 3.}$$
$$18 < 27 \quad \text{True.}$$

$$6 < 9$$
$$\frac{6}{3} < \frac{9}{3} \quad \text{Divide both sides by 3.}$$
$$2 < 3 \quad \text{True.}$$

The inequalities $18 < 27$ and $2 < 3$ are true.

However, if both sides of $6 < 9$ are multiplied or divided by a negative number, such as -3, the direction of the inequality symbol must be reversed to produce another true inequality.

$$6 < 9$$
$$-3 \cdot 6 > -3 \cdot 9 \quad \text{Multiply both sides by } -3 \text{ and reverse the direction of the inequality.}$$
$$-18 > -27 \quad \text{True.}$$

$$6 < 9$$
$$\frac{6}{-3} > \frac{9}{-3} \quad \text{Divide both sides by } -3 \text{ and reverse the direction of the inequality.}$$
$$-2 > -3 \quad \text{True.}$$

The inequality $-18 > -27$ is true, because -18 lies to the right of -27 on the number line. The inequality $-2 > -3$ is true, because -2 lies to the right of -3 on the number line. This example suggests the multiplication and division properties of inequality.

MULTIPLICATION PROPERTY OF INEQUALITY

Suppose a, b, and c are real numbers.

If $a < b$ and $c > 0$, then $ac < bc$.

If $a < b$ and $c < 0$, then $ac > bc$.

DIVISION PROPERTY OF INEQUALITY

Suppose a, b, and c are real numbers.

If $a < b$ and $c > 0$, then $\frac{a}{c} < \frac{b}{c}$.

If $a < b$ and $c < 0$, then $\frac{a}{c} > \frac{b}{c}$.

Similar statements can be made for the symbols $>$, $\leq$, and $\geq$.

COMMENT In the previous definitions, we did not consider the case of $c = 0$. If $a < b$ and $c = 0$, then $ac = bc$, but $\frac{a}{c}$ and $\frac{b}{c}$ are not defined.

The **multiplication property of inequality** can be stated this way:

If unequal quantities are multiplied by the same positive quantity, the results will be unequal and in the same direction as the original inequality.

If unequal quantities are multiplied by the same negative quantity, the results will be unequal but in the opposite direction of the original inequality.

The division property of inequality can be stated this way:

If unequal quantities are divided by the same positive quantity, the results will be unequal and in the same direction as the original inequality.

If unequal quantities are divided by the same negative quantity, the results will be unequal but in the opposite direction of the original inequality.

To *divide* both sides of an inequality by a nonzero number c, we could instead *multiply* both sides by $\frac{1}{c}$.

COMMENT Remember that if both sides of an inequality are multiplied by a *positive* number, the direction of the resulting inequality remains the same. However, if both sides of an inequality are multiplied by a *negative* number, the direction of the resulting inequality must be reversed.

Note that the procedures for solving inequalities are the same as for solving equations, except that we must reverse the inequality symbol whenever we multiply or divide by a negative number.

EXAMPLE 2 Solve $3x + 7 \le -5$ and graph the solution on the number line.

Solution To isolate x on the left side, we proceed as if we were solving an equation.

$$3x + 7 \le -5$$
$$3x + 7 - 7 \le -5 - 7 \quad \text{Subtract 7 from both sides.}$$
$$3x \le -12 \quad \text{Combine like terms.}$$
$$\frac{3x}{3} \le \frac{-12}{3} \quad \text{Divide both sides by 3.}$$
$$x \le -4$$

Figure 2-19

The solution consists of all real numbers that are less than or equal to -4. The bracket at -4 in the graph of Figure 2-19 indicates that -4 is one of the solutions.

SELF CHECK 2 Graph the solution of $2x - 5 \ge -3$ on the number line.

EXAMPLE 3 Solve $5 - 3x \le 14$ and graph the solution on the number line.

Solution To isolate x on the left side, we proceed as if we were solving an equation. This time, we will have to reverse the inequality symbol.

$$5 - 3x \le 14$$
$$5 - 3x - 5 \le 14 - 5 \quad \text{Subtract 5 from both sides.}$$
$$-3x \le 9 \quad \text{Combine like terms.}$$
$$\frac{-3x}{-3} \ge \frac{9}{-3} \quad \text{Divide both sides by } -3 \text{ and reverse the direction of the} \le \text{symbol.}$$
$$x \ge -3$$

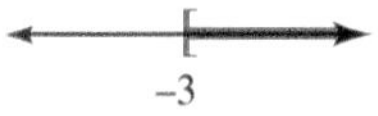

Figure 2-20

Since both sides of the inequality were divided by -3, the direction of the inequality was *reversed*. The graph of the solution appears in Figure 2-20. The bracket at -3 indicates that -3 is one of the solutions.

SELF CHECK 3 Graph the solution of $6 - 7x \ge -15$ on the number line.

2 Solve a compound linear inequality in one variable.

Two inequalities often can be combined into a **double inequality** or **compound inequality** to indicate that numbers lie *between* two fixed values. For example, the inequality $2 < x < 5$ indicates that x is greater than 2 and that x is also less than 5. The solution of $2 < x < 5$ consists of all numbers that lie *between* 2 and 5. The graph of this set (called an **interval**) appears in Figure 2-21.

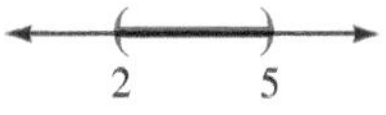

Figure 2-21

EXAMPLE 4 Solve $-4 < 2(x - 1) \le 4$ and graph the solution on the number line.

Solution To isolate x in the center, we proceed as if we were solving an equation with three parts: a left side, a center, and a right side.

$$-4 < 2(x - 1) \le 4$$
$$-4 < 2x - 2 \le 4 \quad \text{Use the distributive property to remove parentheses.}$$
$$-2 < 2x \le 6 \quad \text{Add 2 to all three parts.}$$
$$-1 < x \le 3 \quad \text{Divide all three parts by 2.}$$

Unless otherwise noted, all content on this page is © Cengage Learning.

The graph of the solution appears in Figure 2-22.

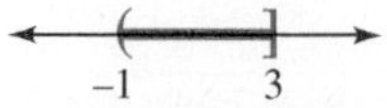

Figure 2-22

SELF CHECK 4 Graph the solution of $0 \le 4(x + 5) < 26$ on the number line.

3 Solve an application involving a linear inequality in one variable.

When solving applications, there are certain words that help us translate a sentence into a mathematical inequality.

Words	Sentence	Inequality
at least	To earn a grade of A, you must score at least 90%.	$S \ge 90\%$
is less than	The perimeter is less than 30 feet.	$P < 30$ ft
is no less than	The perimeter is no less than 100 centimeters.	$P \ge 100$ cm
is more than	The area is more than 30 square inches.	$A > 30$ sq in.
exceeds	The car's speed exceeded the limit of 45 mph.	$S > 45$ mph
cannot exceed	The salary cannot exceed $50,000.	$S \le \$50{,}000$
at most	The perimeter is at most 75 feet.	$P \le 75$ ft
is between	The altitude is between 10,000 and 15,000 feet.	$10{,}000 < A < 15{,}000$

EXAMPLE 5 **GRADES** A student has scores of 72, 74, and 78 points on three mathematics examinations. How many points does he need on his last exam to earn a B or better, an average of at least 80 points?

Solution We can let x represent the score on the fourth (last) exam. To find the average grade, we add the four scores and divide by 4. To earn a B, this average must be greater than or equal to 80 points.

The average of the four grades	is greater than or equal to	80.
$\frac{72 + 74 + 78 + x}{4}$	$\ge$	80

We can solve this inequality for x.

$$\frac{224 + x}{4} \ge 80 \quad \text{Add.}$$

$$224 + x \ge 320 \quad \text{Multiply both sides by 4.}$$

$$x \ge 96 \quad \text{Subtract 224 from both sides.}$$

To earn a B, the student must score at least 96 points.

SELF CHECK 5 A student has scores of 70, 75, and 77 points on three mathematics examinations. How many points does he need on his last exam to earn a B or better, an average of at least 80 points?

EXAMPLE 6 **EQUILATERAL TRIANGLES** If the perimeter of an equilateral triangle is less than 15 feet, how long could each side be?

Solution Recall that each side of an equilateral triangle is the same length and that the perimeter of a triangle is the sum of the lengths of its three sides. If we let x represent the length of one of the sides in feet, then $(x + x + x)$ represents the perimeter. Since the perimeter is to be less than 15 feet, we have the following inequality:

Unless otherwise noted, all content on this page is © Cengage Learning.

$$x + x + x < 15$$
$$3x < 15 \quad \text{Combine like terms.}$$
$$x < 5 \quad \text{Divide both sides by 3.}$$

Each side of the triangle must be less than 5 feet long.

SELF CHECK 6 If the perimeter of an equilateral triangle is less than 21 feet, how long could each side be?

SELF CHECK ANSWERS

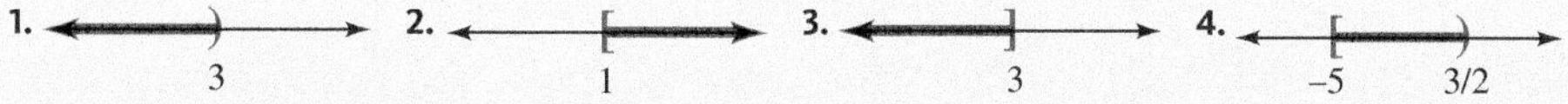

5. at least 98 points **6.** less than 7 feet long

NOW TRY THIS

Solve each inequality and graph the solution.

1. $2(x - 3) \le 2x - 1$
2. $-5x - 7 > 5(3 - x)$
3. A person's body-mass index (BMI) determines the amount of body fat. BMI is represented by the formula $B = 703\frac{w}{h^2}$, where w is weight (in pounds) and h is height (in inches). A 5-foot 8-inch gymnast must maintain a normal body-mass index. If the normal range for men is represented by $18.5 < 703\frac{w}{h^2} < 25$, within what range should the gymnast maintain his weight? Give the answer to the nearest tenth of a pound.

2.7 Exercises

WARM-UPS *Determine whether the sign would stay the same or need to be reversed if the variable remains on the left side of each inequality when solving.*

1. $2x < 4$

2. $x + 5 \ge 6$

3. $-3x \le -6$

4. $-x > 2$

5. $2x - 5 < 7$

6. $5 - 2x < 7$

REVIEW *Simplify each expression.*

7. $3x^2 - 2(y^2 - x^2)$

8. $5(xy + 2) - 3xy - 8$

9. $\frac{1}{3}(x + 6) - \frac{4}{3}(x - 9)$

10. $\frac{4}{5}x(y + 1) - \frac{9}{5}y(x - 1)$

VOCABULARY AND CONCEPTS *Fill in the blanks.*

11. The symbol $<$ means ________. The symbol $>$ means ________.

12. The symbol __ means "is greater than or equal to." The symbol __ means "is less than or equal to."

13. Two inequalities often can be combined into a ______ ________ or *compound inequality.*

14. The graph of the solution of $2 < x < 5$ on the number line is called an ______.

15. An ________ is a statement indicating that two quantities are not necessarily equal.

16. A ______ of an inequality is any number that makes the inequality true.

GUIDED PRACTICE *Solve each inequality and graph the solution on the number line.* SEE EXAMPLE 1. (OBJECTIVE 1)

17. $x + 5 > 8$

18. $x + 5 \ge 2$

19. $3x - 6 \le 2x - 7$

20. $3 + x < 2$

Solve each inequality and graph the solution on the number line. SEE EXAMPLE 2. (OBJECTIVE 1)

21. $3x - 10 \le 2$

22. $9x + 13 \ge 8x$

Unless otherwise noted, all content on this page is © Cengage Learning.

23. $6x - 2 > 3x - 11$

24. $7x + 6 \geq 4x$

25. $9x + 4 > 4x - 1$

26. $5x + 7 < 2x + 1$

27. $\frac{5}{2}(7x - 15) + x \geq \frac{13}{2}x - \frac{3}{2}$

28. $\frac{5}{3}(x + 1) \leq -x + \frac{2}{3}$

Solve each inequality and graph the solution on the number line. SEE EXAMPLE 3. (OBJECTIVE 1)

29. $-x - 3 \leq 7$

30. $-x - 9 > 3$

31. $-3x - 5 < 4$

32. $3x + 7 \leq 4x - 2$

33. $-5x + 17 > 37$

34. $7x - 9 > 5$

35. $-3x - 7 > -1$

36. $-2x - 5 \leq 4x + 1$

37. $9 - 2x > 24 - 7x$

38. $13 - 17x < 34 - 10x$

39. $2(x + 7) \leq 4x - 6$

40. $9(x - 11) > 13 + 7x$

Solve each inequality and graph the solution on the number line. SEE EXAMPLE 4. (OBJECTIVE 2)

41. $2 < x - 5 < 5$

42. $3 < x - 2 < 7$

43. $-4 < 2(x + 1) \leq 12$

44. $-9 \leq 3(x - 2) < 6$

45. $0 \leq x + 10 \leq 10$

46. $-8 < x - 8 < 8$

47. $-6 < 3(x + 2) < 9$

48. $-18 \leq 9(x - 5) < 27$

ADDITIONAL PRACTICE *Solve each inequality and graph the solution on the number line.*

49. $8 + x > 7$

50. $7x - 16 < 6x$

51. $7 - x \leq 3x - 1$

52. $2 - 3x \geq 6 + x$

53. $8(5 - x) \leq 10(8 - x)$

54. $17(3 - x) \geq 3 - 13x$

55. $\frac{3x - 3}{2} < 2x + 2$

56. $\frac{x + 7}{3} \geq x - 3$

57. $\frac{2(x + 5)}{3} \leq 3x - 6$

58. $\frac{3(x - 1)}{4} > x + 1$

59. $9 < -3x < 15$

60. $-4 \leq -4x < 12$

61. $-3 \leq \frac{x}{2} \leq 5$

62. $-12 \leq \frac{x}{3} < 0$

63. $3 \leq 2x - 1 < 5$

64. $4 < 3x - 5 \leq 7$

65. $0 < 10 - 5x \leq 15$

66. $1 \leq -7x + 8 \leq 15$

67. $-4 < \frac{x - 2}{2} < 6$

68. $-1 \leq \frac{x + 1}{3} \leq 3$

APPLICATIONS *Express each solution as an inequality.* SEE EXAMPLES 5–6. (OBJECTIVE 3)

69. Calculating grades A student has test scores of 68, 75, and 79 points. What must she score on the fourth exam to have an average score of at least 80 points?

70. Calculating grades A student has test scores of 84, 89, and 93 points. What must he score on the fourth exam to have an average score of at least 90 points?

71. Geometry The perimeter of a square is no less than 68 centimeters. How long can a side be?

72. **Geometry** The perimeter of an equilateral triangle is at most 57 feet. What could be the length of a side? (*Hint*: All three sides of an equilateral triangle are equal.)

Express each solution as an inequality.

73. **Fleet averages** An automobile manufacturer produces three light trucks in equal quantities. One model has an economy rating of 17 miles per gallon, and the second model is rated for 19 mpg. If the manufacturer is required to have a fleet average of at least 21 mpg, what economy rating is required for the third model?

74. **Avoiding service charges** When the average daily balance of a customer's checking account is less than $500 in any business week, the bank assesses a $5 service charge. Bill's account balances for the week were as shown in the table. What must Friday's balance be to avoid the service charge?

Monday	$540.00
Tuesday	$435.50
Wednesday	$345.30
Thursday	$310.00

75. **Land elevations** The land elevations in Nevada fall from the 13,143-foot height of Boundary Peak to the Colorado River at 470 feet. To the nearest tenth, what is the range of these elevations in miles? (*Hint*: 1 mile is 5,280 feet.)

76. **Homework** A teacher requires that students do homework at least 2 hours a day. How many minutes should a student work each week?

77. **Plane altitudes** A pilot plans to fly at an altitude of between 17,500 and 21,700 feet. To the nearest tenth, what will be the range of altitudes in miles? (*Hint*: There are 5,280 feet in 1 mile.)

78. **Getting exercise** A certain exercise program recommends that your daily exercise period should exceed 15 minutes but should not exceed 30 minutes per day. In hours, find the range of exercise time for one week.

79. **Comparing temperatures** To hold the temperature of a room between 23° and 26° Celsius, what Fahrenheit temperatures must be maintained? (*Hint:* Fahrenheit temperature (F) and Celsius temperature (C) are related by the formula $C = \frac{5}{9}(F - 32)$.)

80. **Melting iron** To melt iron, the temperature of a furnace must be at least 1,540°C but at most 1,650°C. What range of Fahrenheit temperatures must be maintained?

81. **Phonograph records** The radii of old phonograph records lie between 5.9 and 6.1 inches. What variation in circumference can occur? (*Hint*: The circumference of a circle is given by the formula $C = 2\pi r$, where r is the radius. Use 3.14 to approximate π.)

82. **Pythons** A large snake, the African Rock Python, can grow to a length of 25 feet. To the nearest hundredth, find the snake's range of lengths in meters. (*Hint*: There are about 3.281 feet in 1 meter.)

83. **Comparing weights** The normal weight of a 6 foot 2 inch man is between 150 and 190 pounds. To the nearest hundredth, what would such a person weigh in kilograms? (*Hint*: There are approximately 2.2 pounds in 1 kilogram.)

84. **Manufacturing** The time required to assemble a television set at the factory is 2 hours. A stereo receiver requires only 1 hour. The labor force at the factory can supply at least 644 and at most 805 hours of assembly time per week. When the factory is producing 3 times as many television sets as stereos, how many stereos could be manufactured in 1 week?

85. **Geometry** A rectangle's length is 3 feet less than twice its width, and its perimeter is between 24 and 48 feet. What might be its width?

86. **Geometry** A rectangle's width is 8 feet less than 3 times its length, and its perimeter is between 8 and 16 feet. What might be its length?

WRITING ABOUT MATH

87. Explain why multiplying both sides of an inequality by a negative constant reverses the direction of the inequality.

88. Explain the use of parentheses and brackets in the graphing of the solution of an inequality.

SOMETHING TO THINK ABOUT

89. To solve the inequality $1 < \frac{1}{x}$, one student multiplies both sides by x to get $x < 1$. Why is this not correct?

90. Find the solution of $1 < \frac{1}{x}$. (*Hint*: Will any negative values of x work?)

Projects

PROJECT 1

Build a scale similar to the one shown in Figure 2-1. Demonstrate to your class how you would use the scale to solve the following equations.

a. $x - 4 = 6$ **b.** $x + 3 = 2$ **c.** $2x = 6$

d. $\frac{x}{2} = 3$ **e.** $3x - 2 = 5$ **f.** $\frac{x}{3} + 1 = 2$

PROJECT 2

Use a calculator to determine whether the following statements are true or false.

a. $7^5 = 5^7$ **b.** $2^3 + 7^3 = (2 + 7)^3$

c. $(-4)^4 = -4^4$ **d.** $\frac{10^3}{5^3} = 2^3$

e. $8^4 \cdot 9^4 = (8 \cdot 9)^4$ **f.** $2^3 \cdot 3^3 = 6^3$

g. $\frac{3^{10}}{3^2} = 3^5$ **h.** $[(1.2)^3]^2 = [(1.2)^2]^3$

i. $(7.2)^2 - (5.1)^2 = (7.2 - 5.1)^2$

Unless otherwise noted, all content on this page is © Cengage Learning.

Reach for Success

EXTENSION OF READING THIS MATHEMATICS TEXTBOOK

READING THIS MATHEMATICS TEXTBOOK

Now that we have discussed the overall structure of this textbook, let's move on to some special features designed to help you work through each section.

Prior to the homework exercise set, you will find the following features:	
• Getting Ready	What does this mean to you? ______________
• Examples All examples are cross-referenced with the objectives. Thus, if your instructor does not cover a particular objective, you may omit that example.	How many objectives are in this section? ______ How many of these are you required to know? ______ Which examples may you omit? ______
• Self Checks After each numbered example, you will find a Self Check practice exercise so that you may immediately determine your reading comprehension.	Explain how the Self Checks might help you understand and retain the text material. ______________
• Now Try This Created to provide collaborative exercises to be discussed in small groups, these problems are slightly more difficult than the examples and encourage you to deepen your understanding or provide transition to a future topic.	Explain why the Now Try This might be considered transitional group-work exercises. ______________

Match each exercise set feature with its description.	
_____ 1. Additional Practice	A. Gets you ready for homework
_____ 2. Applications	B. Keeps previously learned skills current
_____ 3. Guided Practice	C. Helps you speak the language
_____ 4. Review	D. Provides references for assistance
_____ 5. Something to Think About	E. Begins to duplicate the exam format
_____ 6. Vocabulary	F. Shows relevance of the topics to the world
_____ 7. Warm-up	G. Increases communication skills
_____ 8. Writing About Math	H. Transitions students' concepts
When beginining the exercises, be sure to *read the instructions* carefully.	What is the importance of understanding the instructions in the following two exercises? 1. Find the perimeter of the figure. 2. Find the area of the figure. 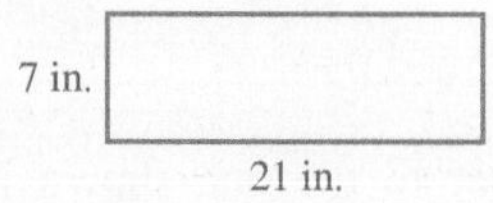 ______________

Don't forget the computer software, Enhanced WebAssign. The exercises there can be done on the computer with additional help options, an excellent resource particularly during late evening when your other resources are not available!

Unless otherwise noted, all content on this page is © Cengage Learning.

2 Review

SECTION 2.1 Solving Basic Linear Equations in One Variable

DEFINITIONS AND CONCEPTS	EXAMPLES
An **equation** is a statement indicating that two quantities are equal. An **expression** is a mathematical statement that does not compare two quantities.	**Equations:** $3x = 5$ $\quad 3x - 4 = 10$ $\quad 8x - 7 = -2x$ **Expressions:** $5x + 1$ $\quad 5x^2 + 3x - 2$ $\quad -8(2x - 4)$
A number is said to *satisfy* an equation if it makes the equation true when substituted for the variable.	To determine whether 3 is a solution of the equation $2x + 5 = 11$, substitute 3 for x and determine whether the result is a true statement. $2x + 5 = 11$ $2(\mathbf{3}) + 5 \stackrel{?}{=} 11$ $6 + 5 \stackrel{?}{=} 11$ $11 = 11$ Since the result is a true statement, 3 satisfies the equation.
Addition and subtraction properties of equality: Any real number can be added to (or subtracted from) both sides of an equation to form another equation with the same solutions as the original equation.	To solve $x - 3 = 8$, add 3 to both sides. To solve $x + 3 = 8$, subtract 3 from both sides. $x - 3 = 8$ $\qquad x + 3 = 8$ $x - 3 + \mathbf{3} = 8 + \mathbf{3}$ $\qquad x - 3 - \mathbf{3} = 8 - \mathbf{3}$ $x = 11$ $\qquad x = 5$ Verify that each result satisfies its corresponding equation.
Two equations are **equivalent equations** when they have the same solutions.	$3x + 4 = 10$ and $3x = 6$ are equivalent equations because 2 is the only solution of each equation.
Multiplication and division properties of equality: Both sides of an equation can be multiplied (or divided) by any *nonzero* real number to form another equation with the same solutions as the original equation.	To solve $\frac{x}{3} = 4$, multiply both sides by 3. To solve $3x = 12$, divide both sides by 3. $\frac{x}{3} = 4$ $\qquad 3x = 12$ $\mathbf{3}\left(\frac{x}{3}\right) = \mathbf{3}(4)$ $\qquad \frac{3x}{\mathbf{3}} = \frac{12}{\mathbf{3}}$ $x = 12$ $\qquad x = 4$ Verify that each result satisfies its corresponding equation.
Sale price = regular price − markdown	If a coat regularly costs \$150 and is marked down \$25, its selling price is \$150 − \$25 = \$125.
Retail price = wholesale cost + markup	If the wholesale cost of a TV is \$500 and it is marked up \$200, its retail price is \$500 + \$200 = \$700.
A **percent** is the numerator of a fraction with a denominator of 100.	$6\% = \frac{6}{100} = 0.06$ $\qquad 8\% = \frac{8}{100} = 0.08$
Amount = rate · base	An amount of \$150 will be earned when a base of \$3,000 is invested at a rate of 5%. $a = rb$ $\mathbf{150} = \mathbf{0.05} \cdot \mathbf{3{,}000}$ $150 = 150$

REVIEW EXERCISES

Determine whether each statement is an expression or an equation.

1. $3(x + 4)$

2. $x + 4 = 3$

Determine whether the given number is a solution of the equation.

3. $6 - 4x = 2; -1$

4. $2(x + 3) = 3(x - 4); 18$

Solve each equation and check all solutions.

5. $x - 6 = -7$

6. $-y - 3 = 10$

7. $p + 5 = 9$

8. $c + 9 = -6$

9. $p + \frac{1}{2} = -\frac{1}{2}$

10. $x + \frac{5}{7} = \frac{5}{7}$

11. $z + \frac{2}{3} = \frac{1}{3}$

12. $b - \frac{1}{4} = -\frac{3}{4}$

13. Retail sales A necklace is on sale for \$69.95. If it has been marked down \$35.45, what is its regular price?

14. Retail sales A suit that has been marked up \$115.25 sells for \$212.95. Find its wholesale price.

Solve each equation and check all solutions.

15. $3x = 15$

16. $8r = -16$

17. $-10z = 5$

18. $14q = 21$

19. $\frac{y}{3} = 6$

20. $\frac{w}{7} = -5$

21. $\frac{a}{-7} = \frac{1}{14}$

22. $\frac{p}{12} = \frac{1}{2}$

Solve.

23. What number is 35% of 700?

24. 72% of what number is 936?

25. What percent of 2,300 is 851?

26. 72 is what percent of 576?

SECTION 2.2 Solving More Linear Equations in One Variable

DEFINITIONS AND CONCEPTS	EXAMPLES
Solving a linear equation may require the use of several properties of equality. If the equation contains any fractions, consider clearing the fractions first.	To solve $\frac{x}{3} - 4 = -8$, proceed as follows: $\frac{x}{3} - 4 + 4 = -8 + 4$ To undo the subtraction of 4, add 4 to both sides. $\frac{x}{3} = -4$ Simplify. $3\left(\frac{x}{3}\right) = 3(-4)$ To undo the division of 3, multiply both sides by 3. $x = -12$
Retail price = cost + $\text{percent of markup} \cdot \text{cost}$ Markup = percent of markup · cost Sale price = regular price − percent of markdown · regular price Markdown (discount) = percent of markdown · regular price	A wholesale cost of a necklace is \$125. If its retail price is \$150, find the percent of markup. $150 = 125 + p \cdot 125$ $25 = 125p$ Subtract 125 from both sides. $0.20 = p$ Divide both sides by 125. The percent of markup is 20%. A used textbook that was originally priced at \$95 is now priced at \$57. Find the percent of markdown. $57 = 95 - p \cdot 95$ $-38 = -95p$ Subtract 95 from both sides. $0.40 = p$ Divide both sides by −95. The used textbook has a markdown of 40%.

REVIEW EXERCISES

Solve each equation and check all solutions.

27. $5y + 6 = 21$

28. $5y - 9 = 1$

29. $-12z + 4 = -8$

30. $17z + 3 = 20$

31. $13 - 13p = 0$

32. $10 + 7p = -4$

33. $23a - 43 = 3$

34. $84 - 21a = -63$

35. $3x + 7 = 1$

36. $7 - 9x = 16$

37. $\frac{b + 3}{4} = 2$

38. $\frac{b - 7}{2} = -2$

39. $\frac{3y-2}{4} = -5$

40. $\frac{3x+10}{2} = -1$

41. $\frac{x}{2} + 7 = 11$

42. $\frac{r}{3} - 3 = 7$

43. $\frac{2x}{3} - 4 = 6$

44. $\frac{y}{4} - \frac{6}{5} = -\frac{1}{5}$

45. $\frac{a}{2} + \frac{3}{4} = 6$

46. $\frac{x}{8} - 2.3 = 3.2$

47. **Retail sales** An iPhone is on sale for \$240, a 25% savings from the regular price. Find the regular price.

48. **Tax rates** A \$38 dictionary costs \$40.47 with sales tax. Find the tax rate.

49. **Percent of increase** A Turkish rug was purchased for \$560. If it is now worth \$1,100, find the percent of increase to the nearest 10th.

50. **Percent of decrease** A clock on sale for \$221.84 was regularly priced at \$470. Find the percent of decrease.

SECTION 2.3 Simplifying Expressions to Solve Linear Equations in One Variable

DEFINITIONS AND CONCEPTS	EXAMPLES
Like terms are terms with the same variables having the same exponents. They can be combined by adding their numerical coefficients and using the same variables and exponents.	Combine like terms. $4(x+3) + 6(x-5)$ $= 4x + 12 + 6x - 30$ Use the distributive property to remove parentheses. $= 10x - 18$ Combine like terms: $4x + 6x = 10x$, $12 - 30 = -18$.
An **identity** is an equation that is true for all values of its variable. A **contradiction** is an equation that is true for no values of its variable.	Show that the following equation is an identity. $2(x-5) + 6x = 8(x-1) - 2$ $2x - 10 + 6x = 8x - 8 - 2$ Use the distributive property to remove parentheses. $8x - 10 = 8x - 10$ Combine like terms. $-10 = -10$ Subtract $8x$ from both sides. Since the final result is always true, the equation is an identity and its solution set is $\mathbb{R}$. Show that the following equation is a contradiction. $6x - 2(x+5) = 4x - 1$ $6x - 2x - 10 = 4x - 1$ Use the distributive property to remove parentheses. $4x - 10 = 4x - 1$ Combine like terms. $-10 = -1$ Subtract $4x$ from both sides. Since the final result is false, the equation is a contradiction and its solution set is $\varnothing$.

REVIEW EXERCISES

Simplify each expression, if possible.

51. $5x + 9x$

52. $7a + 12a$

53. $18b - 13b$

54. $21x - 23x$

55. $5y - 7y$

56. $19x - 19$

57. $7(x+2) + 2(x-7)$

58. $2(3-x) + x - 6x$

Solve each equation and check all solutions.

59. $10y - 14 = 3y$

60. $6(a+4) = 3a$

61. $2x - 19 = 2 - x$

62. $5b - 19 = 2b + 20$

63. $3x + 20 = 5 - 2x$

64. $0.9x + 10 = 0.7x + 1.8$

65. $10(p-3) = 3(p+11)$

66. $2(5x-7) = 2(x-35)$

67. $\frac{3u-6}{5} = 3$

68. $\frac{5v-35}{3} = -5$

69. $\frac{2(b+4)}{3} = b - 2$

70. $\frac{3(x-1)}{6} - 5 = \frac{2(x+3)}{6}$

Classify each equation as an identity or a contradiction and give the solution.

71. $4x - 2 = x + 3(x+1) - 5$

72. $-3(a+1) - a = -4a + 3$

73. $2(x-1) + 4 = 4(1+x) - (2x+2)$

74. $3(2x+1) + 3 = 9(x+2) + 9 - 3x$

SECTION 2.4 Formulas

DEFINITIONS AND CONCEPTS	EXAMPLES
A **literal equation** or **formula** often can be solved for any of its variables.	Solve $2x + 3y = 6$ for y. $2x + 3y = 6$ $3y = -2x + 6$ Subtract $2x$ from both sides. $y = -\frac{2}{3}x + 2$ Divide both sides by 3.

REVIEW EXERCISES

Solve each equation for the indicated variable.

75. $E = IR$ for R

76. $i = prt$ for t

77. $P = I^2R$ for R

78. $V = \frac{1}{3}Bh$ for B

79. $p = a + b + c$ for c

80. $y = mx + b$ for m

81. $V = \pi r^2h$ for h

82. $A = \frac{3}{2}(B + 4)$ for B

83. $F = \frac{GMm}{d^2}$ for G

84. $P = \frac{RT}{mV}$ for m

SECTION 2.5 Introduction to Problem Solving

DEFINITIONS AND CONCEPTS	EXAMPLES
To solve applications, follow these steps: **1.** Analyze the situation and choose a variable. **2.** Form an equation. **3.** Solve the equation. **4.** State the conclusion. **5.** Check the result to be certain it satisfies the given conditions.	The length of a rectangular frame is 4 in. longer than twice the width. If the perimeter is 38 in., find the width of the frame. **1.** Let w represent the width of the frame in inches. **2.** The width of the frame is w and because the length is 4 in. longer than twice the width, the length is $(2w + 4)$. Since the frame is a rectangle, its perimeter is the sum of two widths and two lengths. This perimeter is 38. So $2w + 2(2w + 4) = 38$ **3.** To solve the equation, proceed as follows: $2w + 2(2w + 4) = 38$ $2w + 4w + 8 = 38$ $6w + 8 = 38$ $6w = 30$ $w = 5$ **4.** The frame is 5 in. wide. **5.** If the width is 5 in., the length is $2 \cdot 5 + 4 = 14$ in. The perimeter is $2 \cdot 5 + 2 \cdot 14 = 10 + 28 = 38$. The result checks.
If the sum of the measures of two angles is 90°, the angles are called **complementary angles**.	Find the complement of an angle measuring 42°. $x + 42 = 90$ $x + 42 - 42 = 90 - 42$ $x = 58°$
If the sum of the measures of two angles is 180°, the angles are called **supplementary angles**.	Find the supplement of an angle measuring 42°. $x + 42 = 180$ $x + 42 - 42 = 180 - 42$ $x = 138°$

REVIEW EXERCISES

85. Carpentry A carpenter wants to cut an 8-foot board into two pieces so that one piece is 7 feet shorter than twice the longer piece. Where should he make the cut?

86. Find the value of x.

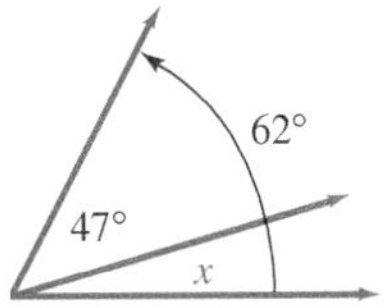

87. Find the value of x.

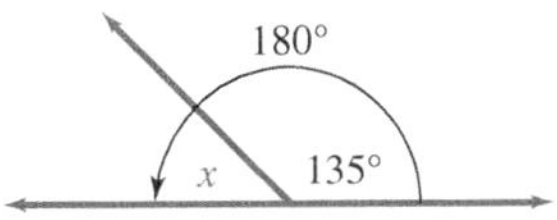

88. Find the complement of an angle that measures 12°.

89. Find the supplement of an angle that measures 75°.

90. Rectangles If the length of the rectangular painting in the illustration is 3 inches more than twice the width, how wide is the rectangle?

© Laurin Rinder/Shutterstock.com

91. Investing A woman has \$27,000. Part is invested for 1 year in a certificate of deposit paying 7% interest, and the remaining amount in a cash management fund paying 9%. The total interest on the two investments is \$2,110. How much does she invest at each rate?

SECTION 2.6 Motion and Mixture Applications

DEFINITIONS AND CONCEPTS

Distance = rate · time

$d = rt$

EXAMPLES

Two cars leave McKinney at the same time traveling in opposite directions. If the speed of one car is 62 mph and the cars are 268 miles apart after two hours, what is the speed of the second car?

Let x represent the speed, in mph, of the second car.

$$\begin{aligned} 2(62) + 2x &= 268 \\ 124 + 2x &= 268 \\ 2x &= 268 - 124 \\ 2x &= 144 \\ x &= 72 \text{ mph} \end{aligned}$$

The second car is traveling at 72 mph.

DEFINITIONS AND CONCEPTS

Value = price · number

$v = pn$

EXAMPLES

How many pounds of candy priced at \$2.50 per pound must be mixed with 30 pounds of another candy priced at \$3.75 per pound to make a mixture that would sell for \$3.25 per pound?

Let x represent the number of pounds of candy priced at \$2.50 per pound.

$$\begin{aligned} 2.50x + 3.75(30) &= 3.25(x + 30) \\ 2.50x + 112.50 &= 3.25x + 97.50 \\ 2.50x - 2.50x + 112.50 &= 3.25x - 2.50x + 97.50 \\ 112.50 &= 0.75x + 97.50 \\ 112.50 - 97.50 &= 0.75x + 97.50 - 97.50 \\ 15 &= 0.75x \\ 20 &= x \end{aligned}$$

There should be 20 pounds of the \$2.50 per pound candy.

REVIEW EXERCISES

92. Riding bicycles A bicycle path is 7 miles long. A man walks from one end at the rate of 4 mph. At the same time, a friend bicycles from the other end, traveling at 10 mph. In how many minutes will they meet?

93. Tornadoes During a storm, two teams of scientists leave a university at the same time in specially designed vans to search for tornadoes. The first team travels east at 20 mph

Unless otherwise noted, all content on this page is © Cengage Learning.

and the second travels west at 25 mph. If their radios have a range of up to 90 miles, how long will it be before they lose radio contact?

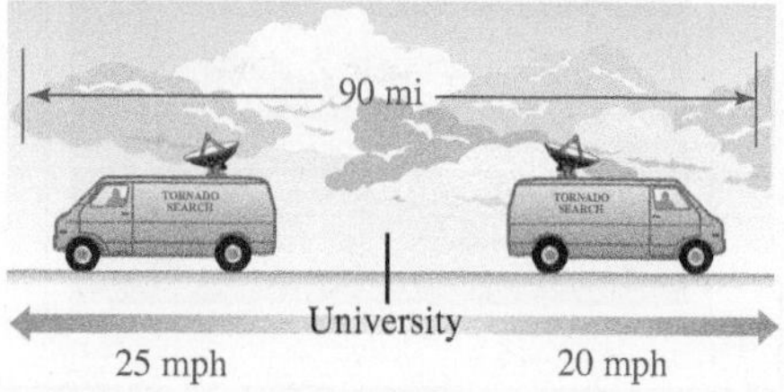

94. Band trips A bus carrying the members of a marching band and a truck carrying their instruments leave a high school at the same time and travel in the same direction. The bus travels at 60 mph and the truck at 50 mph. In how many hours will they be 90 miles apart?

95. Mixing milk A container is partly filled with 12 liters of whole milk containing 4% butterfat. How much 1% milk must be added to get a mixture that is 2% butterfat?

96. Photography A photographer wants to mix 2 liters of a 6% acetic acid solution with a 12% solution to get an 8% solution. How many liters of 12% solution must be added?

97. Mixing candy A store manager mixes candy worth 90¢ per pound with gumdrops worth \$1.50 per pound to make 20 pounds of a mixture worth \$1.20 per pound. How many pounds of each kind of candy must he use?

SECTION 2.7 Solving Linear Inequalities in One Variable

DEFINITIONS AND CONCEPTS	EXAMPLES
Inequalities are solved by techniques similar to those used to solve equations, with this exception: *If both sides of an inequality are multiplied or divided by a negative number, the direction of the inequality must be reversed.*	To solve the inequality $-3x - 8 < 7$, proceed as follows: $-3x - 8 < 7$ $-3x < 15$ $x > -5$ Divide both sides by -3 and reverse the inequality symbol.
The solution of an inequality can be graphed on the number line.	The graph of $x > -5$ is 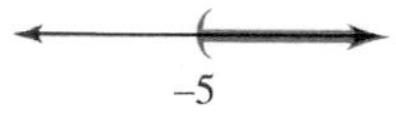

REVIEW EXERCISES

Graph the solution to each inequality on a number line.

98. $3x + 2 < 5$

99. $-5x - 8 > 7$

100. $5x - 3 \geq 2x + 9$

101. $7x + 1 \leq 8x - 5$

102. $5(3 - x) \leq 3(x - 3)$

103. $8(2 - x) > 4 - 2x$

104. $8 < x + 2 < 13$

105. $4 \leq 2 - 2x < 8$

106. Swimming pools By city ordinance, the perimeter of a rectangular swimming pool cannot exceed 68 feet. The width is 6 feet shorter than the length. What possible lengths will meet these conditions?

2 Test

Determine whether the given number is a solution of the equation.

1. $4x + 5 = -3; -2$

2. $3(x + 2) = 2x + 4; 2$

Solve each equation.

3. $x + 17 = -19$

4. $a - 15 = 32$

5. $12x = -144$

6. $\frac{x}{7} = -1$

7. $8x + 2 = -14$

8. $3 = 5 - 2x$

9. $\frac{2x - 5}{3} = 3$

10. $23 - 5(x + 10) = -12$

Simplify each expression.

11. $x + 5(x - 3)$

12. $3x - 5(2 - x)$

13. $-3(x + 3) + 3(x - 3)$

14. $-4(2x - 5) - 7(4x + 1)$

Unless otherwise noted, all content on this page is © Cengage Learning.

Solve each equation.

15. $8x + 6 = 2(4x + 3)$

16. $2(x + 6) = 2(x - 2)$

17. $\frac{3x - 18}{2} = 6x$

18. $\frac{7}{8}(x - 4) = 5x - \frac{7}{2}$

Solve each equation for the variable indicated.

19. $d = rt$ for t

20. $S = 2\pi rh + 2\pi r^2$ for h

21. $A = 2\pi rh$ for h

22. $x + 2y = 5$ for y

23. Find the value of x.

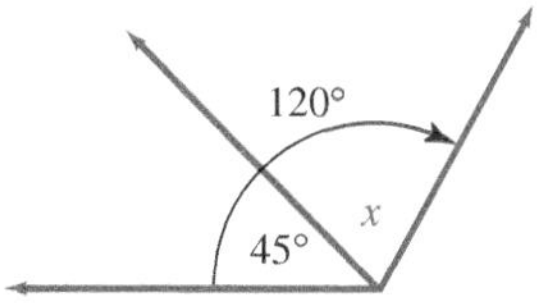

24. Find the supplement of a 79° angle.

25. Investing A student invests part of \$10,000 at 6% annual interest and the rest at 5%. If the annual income from these investments is \$560, how much was invested at each rate?

26. Traveling A car leaves Rockford at the rate of 65 mph, bound for Madison. At the same time, a truck leaves Madison at the rate of 55 mph, bound for Rockford. If the cities are 72 miles apart, how long will it take for the car and the truck to meet?

27. Mixing solutions How many liters of water must be added to 30 liters of a 10% brine solution to dilute it to an 8% solution?

28. Mixing nuts Twenty pounds of cashews are to be mixed with peanuts to make a mixture that will sell for \$4 per pound. How many pounds of peanuts should be used?

Nut	Price per pound
Cashews	\$6
Peanuts	\$3

Graph the solution of each inequality.

29. $8x - 20 \geq 4$

30. $x - 2(x + 7) > 14$

31. $-4 \leq 2(x + 1) < 10$

32. $-10 < 2(4 - x) \leq 12$

Cumulative Review

Classify each number as an integer, a rational number, an irrational number, and/or a real number. Each number may be in several classifications.

1. $\frac{27}{9}$

2. -0.25

Graph each set of numbers on the number line.

3. The natural numbers between 2 and 7

4. The real numbers between -2 and 5

Simplify each expression.

5. $\frac{|-3| - |3|}{|-3 - 3|}$

6. $\frac{14}{15} \cdot \frac{3}{4}$

7. $2\frac{3}{5} + 5\frac{1}{2}$

8. $35.7 - 0.05$

Let $x = -5$, $y = 3$, and $z = 0$, and evaluate each expression.

9. $(2z - 3x)y$

10. $\frac{x - 3y + |z|}{2 - x}$

11. $x^2 - y^2 + z^2$

12. $\frac{x}{y} + \frac{y + 2}{3 - z}$

13. What is $4\frac{1}{2}\%$ of 220?

14. 1,688 is 32% of what number?

Consider the algebraic expression $3x^3 + 5x^2y + 37y$.

15. Identify the coefficient of the second term.

16. List the factors of the third term.

Simplify each expression.

17. $4x + 7y - 9x$

18. $3(x - 7) + 2(8 - x)$

19. $2x^2y^3 - 4x^2y^3$

20. $3(5 - x) - 6(x + 2)$

Solve each equation and check the result.

21. $2(x - 7) + 5 = 3x$

22. $\frac{x - 5}{3} - 5 = 7$

23. $\frac{2x - 1}{5} = \frac{1}{2}$

24. $2(a - 3) - 3(a - 2) = -a$

Unless otherwise noted, all content on this page is © Cengage Learning.

Solve each formula for the variable indicated.

25. $A = \frac{1}{2}h(b + B)$ for h

26. $y = mx + b$ for x

27. Auto sales An auto dealer's promotional ad appears in the illustration. One car is selling for \$23,499. What was the dealer's invoice?

700 cars to choose from!
Buy at
3%
over dealer invoice!

28. Furniture pricing A sofa and a \$300 chair are discounted 35%, and are priced at \$780 for both. Find the original price of the sofa.

29. Cost of a car The total cost of a new car, including an 8.5% sales tax, is \$24,618.65. Find the cost before tax.

30. Manufacturing concrete Concrete contains 3 times as much gravel as cement. How many pounds of cement are in 500 pounds of dry concrete mix?

31. Building construction A 35-foot beam, 1 foot wide and 2 inches thick, is cut into three sections. One section is 14 feet long. Of the remaining two sections, one is twice as long as the other. Will the shortest section span an 8-foot-wide doorway?

32. Installing solar heating One solar panel in the illustration is 3.4 feet wider than the other. Find the width of each panel.

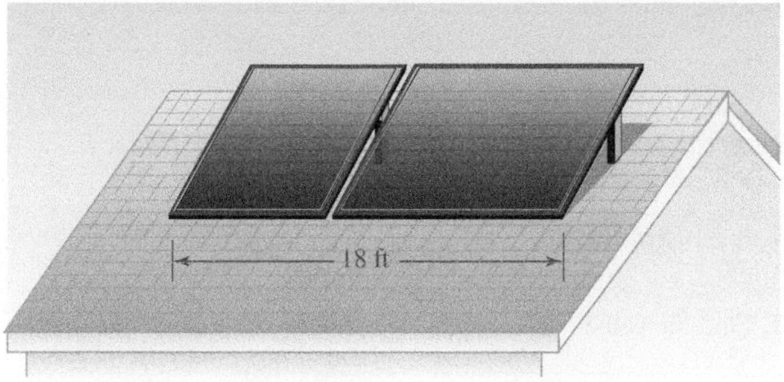

33. Electric bills An electric company charges \$52.50 per month, plus 28¢ for every kWh of energy used. One resident's bill was \$203.70. How many kWh were used that month?

34. Installing gutters A contractor charges \$47.75 for the installation of rain gutters, plus \$2.05 per foot. If one installation cost \$427, how many feet of gutter were required?

Evaluate each expression.

35. $4^2 - 5^2$

36. $(4 - 5)^2$

37. $5(4^3 - 2^3)$

38. $-2(5^4 - 7^3)$

Graph the solutions of each inequality.

39. $8(4 + x) > 10(6 + x)$

40. $-9 < 3(x + 2) \leq 3$

Unless otherwise noted, all content on this page is © Cengage Learning.

Graphing; Writing Equations of Lines; Functions; Linear Inequalities in Two Variables

3

© Jess Alford/Jupiterimages

Careers and Mathematics

PSYCHOLOGIST

Psychologists study the human mind and human behavior. Research psychologists investigate the physical, cognitive, emotional, or social aspects of human behavior. Psychologists in health service fields provide mental health care in hospitals, clinics, schools, or private settings. Psychologists employed in applied settings, such as business, industry, or government, provide training, conduct research, and design organizational systems.

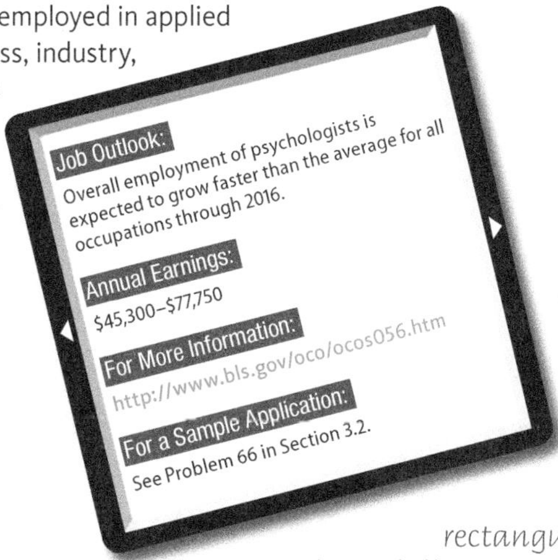
Job Outlook:
Overall employment of psychologists is expected to grow faster than the average for all occupations through 2016.
Annual Earnings:
$45,300–$77,750
For More Information:
http://www.bls.gov/oco/ocos056.htm
For a Sample Application:
See Problem 66 in Section 3.2.

REACH FOR SUCCESS
3.1 The Rectangular Coordinate System
3.2 Graphing Linear Equations
3.3 Slope of a Line
3.4 Point-Slope Form
3.5 Slope-Intercept Form
3.6 Functions
3.7 Solving Linear Inequalities in Two Variables
■ *Projects*
REACH FOR SUCCESS EXTENSION
CHAPTER REVIEW
CHAPTER TEST

In this chapter

In this chapter, we will discuss equations with two variables and we will see that the relationships set up by these equations also can be expressed in tables or graphs.

We begin the chapter by using the rectangular coordinate system to plot ordered pairs and graph linear equations. After explaining how to find the slope of a line, we will show how to write the equation of a line that passes through two known points. We then discuss functions, one of the most important ideas in mathematics, and conclude the chapter with solving linear inequalities in two variables.

Reach for Success Understanding Your Syllabus

When you go on a trip, you have a plan, an itinerary. You might even have a travel representative along on the trip as a guide to handle the details and any unexpected situations that might occur. Your course syllabus is similar to your itinerary and your instructor is similar to the travel guide.

A syllabus is mandatory for all college courses and serves as a contract between the student and the instructor. Examine the syllabus you received for this course to identify important information.

Course Syllabus

Course Title: Beginning Algebra
Instructor's Office: B232 University Building
Office Hours: MTWR: 8:30am - 10:00am plus other times by appointment
Course Resources: The college provides a Math Lab at no charge to support student success.
Supplies: Textbook: Beginning and Intermediate Algebra: A Guided Approach, 7th edition by Karr/Massey/Gustafson
A graphing calculator is recommended. The TI 84 and Casio fx-9760GII are each supported by a tear-out card in the back of the textbook which provide keystrokes for calculator concepts covered in the text.
Access to homework will be through WebAssign. The course code required to register for this site will be provided the first day of the semester.

Method of Evaluation:

4 Tests (10% each)	40%
20 Homework Assignments	10%
10 Online Labs	10%
Midterm	20%
Comprehensive Final Exam	20%

90 - 100	80 - 89	70 - 79	60 - 69	Below 60
A	B	C	D	F

Tests

Tests count 40% of your final grade.
Tests are given in class on the date indicated in your schedule. There are no make-ups. However, if you miss a test, your Final Exam grade will be used to replace it. If you take all tests, the Final Exam grade, if higher, can be used to replace a lower test grade.

Homework

Homework counts 10% of your final grade.
You have a 3 calendar day "extension" after the deadline in which to submit homework. There is, however, a 10% penalty each day but ONLY for the problems that were not completed by the deadline.

LABS

Labs count 10% of your final grade.
Late labs are not accepted, i.e., there is no 3-day "extension."

MIDTERM

The Midterm counts 20% of your final grade. If you miss the midterm, your Final Exam grade will be used to replace it.

FINAL EXAM

The Final Exam counts 20% of your final grade.
The final exam is cumulative, meaning that it includes all topics covered during the semester.

Your syllabus will include your instructor's contact information, office location, and e-mail address. It typically contains the instructor's office hours, which are used to meet with and help students.

Write the information regarding your instructor:

Office location: ____________________

E-mail address: ____________________

Office hours: ____________________

Write down at least one listed office hour during which you could meet with your instructor if necessary. ____________________

Does your instructor include a statement such as "and other times by appointment" with the office hours?

If your schedule absolutely prohibits you from meeting with your instructor during office hours, write down at least two times during the week that you are available and could request an appointment.

____

Most instructors prefer that students be proactive. For example, it is important to contact your instructor prior to any known life event that may interfere with meeting due dates.

Write down any issues that might interfere with your work this semester and that you might wish to discuss with your instructor.

____

Your syllabus will include requirements of the course and how each will be used to determine your grade. Knowing exactly how your grade will be computed, you can estimate your average at any time.

Write down how the following will be used to determine your grade.

Tests (how many?) ______ ______ points or ______ percentage
Homework/daily work ______ points or ______ percentage
Attendance ______ points or ______ percentage
Quizzes ______ points or ______ percentage
Other ______ points or ______ percentage

How would you use this information to calculate your grade?

____

A Successful Study Strategy . . .

Know your instructor's office hours so you can have a plan of action if it becomes necessary to meet. In addition, knowing how your grade will be determined early in the semester can eliminate surprises at the end of the semester.

At the end of the chapter you will find additional exercises to guide you to planning for a successful semester.

Unless otherwise noted, all content on this page is © Cengage Learning.

Section 3.1 The Rectangular Coordinate System

Objectives

1. Graph ordered pairs and mathematical relationships.
2. Interpret the meaning of graphed data.
3. Interpret information from a step graph.

Vocabulary

rectangular coordinate system	*y*-axis	ordered pairs
Cartesian coordinate system	origin	*x*-coordinate
perpendicular lines	coordinate plane	*y*-coordinate
x axis	Cartesian plane	coordinates
	quadrants	

Getting Ready

Graph each set of numbers on the number line.

1. $-2, 1, 3$

2. All numbers greater than -2

3. All numbers less than or equal to 3

4. All numbers between -3 and 2

It is often said, "A picture is worth a thousand words." In this section, we will show how numerical relationships can be described by using mathematical pictures called *graphs.* We also will show how we can obtain important information by reading graphs.

1 Graph ordered pairs and mathematical relationships.

When designing the Gateway Arch in St. Louis, shown in Figure 3-1(a) on the next page, architects created a mathematical model of the arch called a *graph.* This graph, shown in Figure 3-1(b), is drawn on a grid called the **rectangular coordinate system**. This coordinate system is sometimes called a **Cartesian coordinate system** after the 17th-century French mathematician René Descartes.

A rectangular coordinate system (see Figure 3-2 on the next page) is formed by two perpendicular number lines. Recall that **perpendicular lines** are lines that meet at a 90° angle.

- The horizontal number line is called the ***x*-axis**.
- The vertical number line is called the ***y*-axis**.

The positive direction on the *x*-axis is to the right, and the positive direction on the *y*-axis is upward. The scale on each axis should fit the data. For example, the axes of the graph of the arch shown in Figure 3-1(b) are scaled in units of 100 feet. If no scale is indicated on the axes, we assume that the axes are scaled in units of 1.

René Descartes
1596–1650
Descartes is famous for his work in philosophy as well as for his work in mathematics. His philosophy is expressed in the words "I think, therefore I am." He is best known in mathematics for his invention of a coordinate system and his work with conic sections.

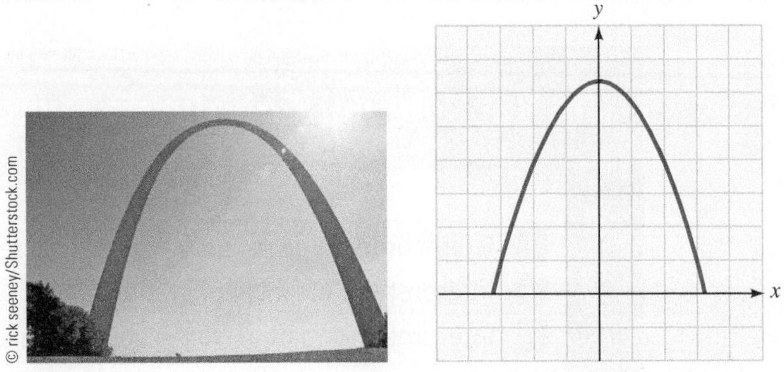

Scale: 1 unit = 100 ft

(a) (b)

Figure 3-1

The point where the axes cross is called the **origin**. This is the 0 point on each axis. The two axes form a **coordinate plane** (often referred to as the **Cartesian plane**) and divide it into four regions called **quadrants**, which are numbered as shown in Figure 3-2.

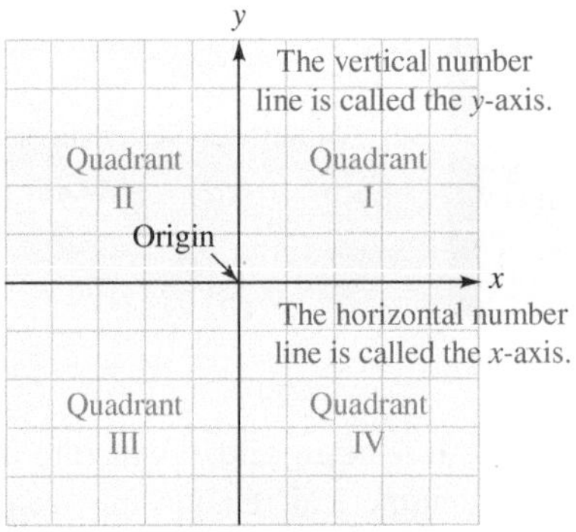

Figure 3-2

Each point in a coordinate plane can be identified by a pair of real numbers x and y, written as (x, y) and called **ordered pairs**. The first number in the pair is the ***x*-coordinate**, and the second number is the ***y*-coordinate**. The numbers are called the **coordinates** of the point. Some examples of ordered pairs are $(3, -4)$, $\left(-1, -\frac{3}{2}\right)$, $(0, 2.5)$, and the origin $(0, 0)$.

$$(3, -4)$$
$$\uparrow \quad \uparrow$$

In an ordered pair, the x-coordinate is listed first. The y-coordinate is listed second.

The process of locating a point in the coordinate plane is called *graphing* or *plotting* the point. In Figure 3-3(a), we show how to graph the point A with coordinates of $(3, -4)$. Since the x-coordinate is positive, we start at the origin and move 3 units to the right along the x-axis. Since the y-coordinate is negative, we then move down 4 units to locate point A. Point A is the *graph* of $(3, -4)$ and lies in quadrant IV.

To plot the point $B(-4, 3)$, we start at the origin, move 4 units to the left along the x-axis, and then move up 3 units to locate point B. Point B lies in quadrant II.

Unless otherwise noted, all content on this page is © Cengage Learning.

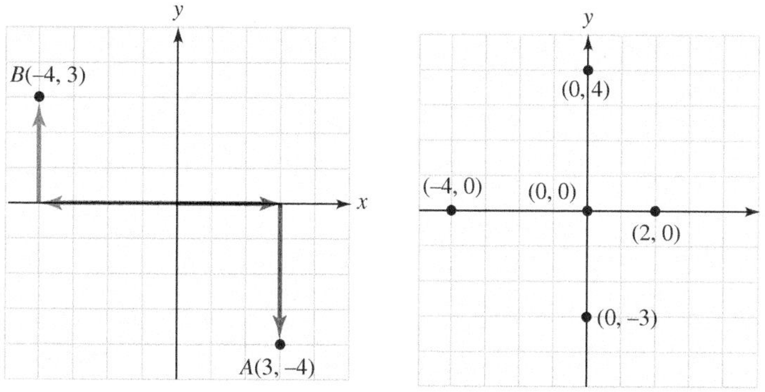

(a) (b)

Figure 3-3

COMMENT Note that point A with coordinates of $(3, -4)$ is not the same as point B with coordinates $(-4, 3)$. Since the order of the coordinates of a point is important, we call them ordered pairs.

In Figure 3-3(b), we see that the points $(-4, 0)$, $(0, 0)$, and $(2, 0)$ lie on the x-axis. In fact, all points with a y-coordinate of 0 will lie on the x-axis.

From Figure 3-3(b), we also see that the points $(0, -3)$, $(0, 0)$, and $(0, 4)$ lie on the y-axis. All points with an x-coordinate of 0 lie on the y-axis. From the figure, we also can see that the coordinates of the origin are $(0, 0)$ and will lie on both the x- and y-axes.

EXAMPLE 1 GRAPHING POINTS

Plot the points.

a. $A(-2, 3)$ **b.** $B\left(-1, -\frac{3}{2}\right)$ **c.** $C(0, 2.5)$ **d.** $D(4, 2)$

Solution

a. To plot point A with coordinates $(-2, 3)$, we start at the origin, move 2 units to the *left* on the x-axis, and move 3 units *up*. Point A lies in quadrant II. (See Figure 3-4.)

b. To plot point B with coordinates of $\left(-1, -\frac{3}{2}\right)$, we start at the origin and move 1 unit to the *left* and $\frac{3}{2}$ $\left(\text{or } 1\frac{1}{2}\right)$ units *down*. Point B lies in quadrant III, as shown in Figure 3-4.

c. To graph point C with coordinates of $(0, 2.5)$, we start at the origin and move 0 units on the x-axis and 2.5 units *up*. Point C lies on the y-axis, as shown in Figure 3-4.

d. To graph point D with coordinates of $(4, 2)$, we start at the origin and move 4 units to the *right* and 2 units *up*. Point D lies in quadrant I, as shown in Figure 3-4.

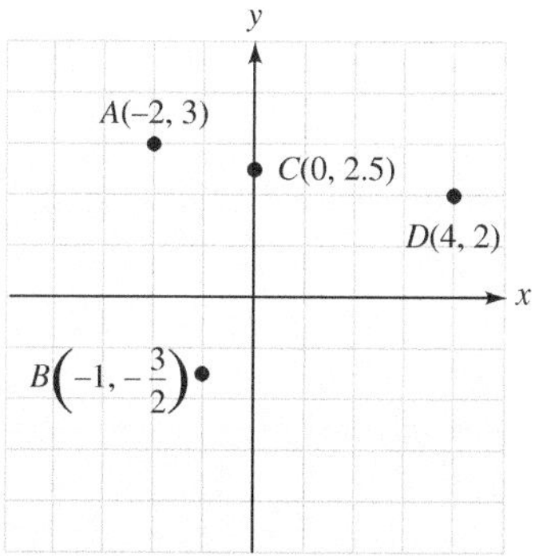

Figure 3-4

SELF CHECK 1 Plot the points. **a.** $E(2, -2)$ **b.** $F(-4, 0)$ **c.** $G\left(1.5, \frac{5}{2}\right)$ **d.** $H(0, 5)$

EXAMPLE 2 ORBITS

The circle shown in Figure 3-5 is an approximate graph of the orbit of Earth. The graph is made up of infinitely many points, each with its own x and y-coordinates. Use the graph to find the approximate coordinates of Earth's position during the months of February, May, and August.

Unless otherwise noted, all content on this page is © Cengage Learning.

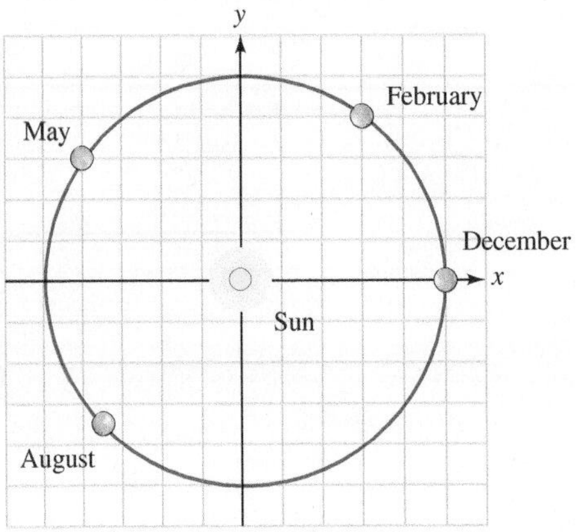

Figure 3-5

Solution To find the coordinates of each position, we start at the origin and move left or right along the x-axis to find the x-coordinate and then up or down to find the y-coordinate. See Table 3-1.

TABLE 3-1

Month	Position of Earth on the graph	Coordinates
February	3 units to the *right*, then 4 units *up*	$(3, 4)$
May	4 units to the *left*, then 3 units *up*	$(-4, 3)$
August	3.5 units to the *left*, then 3.5 units *down*	$(-3.5, -3.5)$

SELF CHECK 2 From Figure 3-5, find the coordinates of Earth's position in December.

Perspective

As a child, René Descartes was frail and often sick. To improve his health, eight-year-old René was sent to a Jesuit school. The headmaster encouraged him to sleep in the morning as long as he wished. As a young man, Descartes spent several years as a soldier and world traveler, but his interests included mathematics and philosophy, as well as science, literature, writing, and taking it easy. The habit of sleeping late continued throughout his life. He claimed that his most productive thinking occurred when he was lying in bed. According to one story, Descartes first thought of analytic geometry as he watched a fly walking on his bedroom ceiling.

Descartes might have lived longer if he had stayed in bed. In 1649, Queen Christina of Sweden decided that she needed a tutor in philosophy, and she requested the services of Descartes. Tutoring would not have been difficult, except that the queen scheduled her lessons before dawn in her library with her windows open. The cold Stockholm mornings were too much for a man who was used to sleeping past noon. Within a few months, Descartes developed a fever and died, probably of pneumonia.

Every day, we deal with quantities that are related.

- The distance that we travel depends on how fast we are going.
- Our weight depends on how much we eat.
- The amount of water in a tank depends on how long the water has been running.

We often can use graphs to visualize relationships between two quantities. For example, suppose that we know the number of gallons of water that are in a tank at several time intervals after the water has been turned on. We can list that information in a *table of values.* (See Figure 3-6.)

Unless otherwise noted, all content on this page is © Cengage Learning.

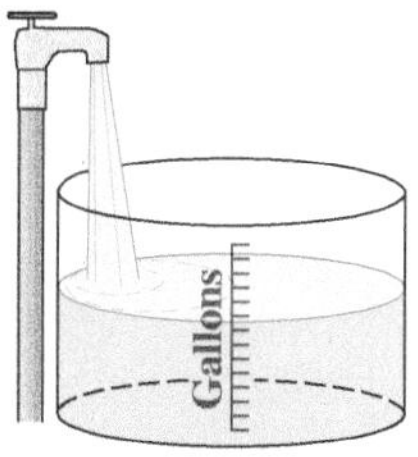

At various times, the amount of water in the tank was measured and recorded in the table of values.

Time (minutes)	Water in tank (gallons)	
0	0	→ (0, 0)
1	3	→ (1, 3)
3	9	→ (3, 9)
4	12	→ (4, 12)
↑ x-coordinate	↑ y-coordinate	↑ The data in the table can be expressed as ordered pairs (x, y).

Figure 3-6

The information in the table can be used to construct a graph that shows the relationship between the amount of water in the tank and the time the water has been running. Since the amount of water in the tank *depends* on the time, we will associate *time* with the x-axis and the *amount of water* with the y-axis.

To construct the graph in Figure 3-7, we plot the four ordered pairs and draw a line through the resulting data points.

COMMENT Note that the scale for the gallons of water (y-axis) is $1\frac{1}{2}$ units while the scale for minutes (x-axis) is 1 unit. The scales on both axes do not have to be the same, but remember to label them!

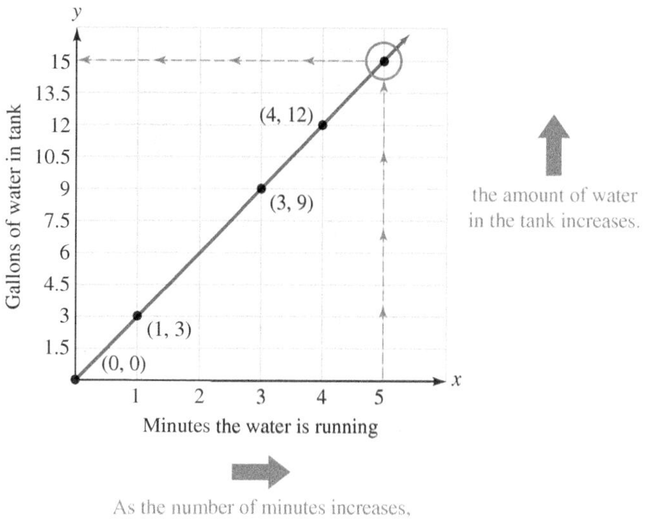

Figure 3-7

From the graph, we can see that the amount of water in the tank increases as the water is allowed to run. We also can use the graph to make observations about the amount of water in the tank at other times. For example, the dashed line on the graph shows that in 5 minutes, the tank will contain 15 gallons of water.

2 Interpret the meaning of graphed data.

In the next example, we show that valuable information can be obtained from reading a graph.

EXAMPLE 3 **READING GRAPHS** The graph in Figure 3-8 on the next page shows the number of people in an audience before, during, and after the taping of a television show. On the x-axis, 0 represents the time when taping began. Use the graph to answer the following questions, and record each result in a table of values.

a. How many people were in the audience when taping began?

b. What was the size of the audience 10 minutes before taping began?

c. At what times were there exactly 100 people in the audience?

Unless otherwise noted, all content on this page is © Cengage Learning.

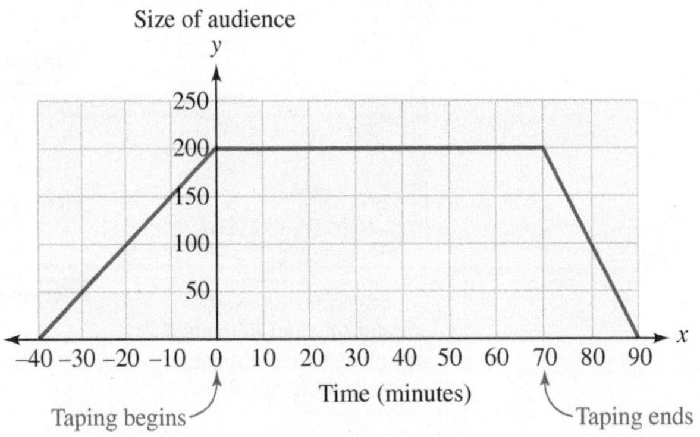

Figure 3-8

Solution

Time	Audience
0	200

a. The time when taping began is represented by 0 on the x-axis. Since the point on the graph directly above 0 has a y-coordinate of 200, the point $(0, 200)$ is on the graph. The y-coordinate of this point indicates that 200 people were in the audience when the taping began.

Time	Audience
−10	150

b. Ten minutes before taping began is represented by -10 on the x-axis. Since the point on the graph directly above -10 has a y-coordinate of 150, the point $(-10, 150)$ is on the graph. The y-coordinate of this point indicates that 150 people were in the audience 10 minutes before the taping began.

Time	Audience
−20	100
80	100

c. We can draw a horizontal line passing through 100 on the y-axis. Since this line intersects the graph twice, there were two times when 100 people were in the audience. The points $(-20, 100)$ and $(80, 100)$ are on the graph. The y-coordinates of these points indicate that there were 100 people in the audience 20 minutes before and 80 minutes after taping began.

SELF CHECK 3 Use the graph in Figure 3-8 to answer the following questions.

a. What was the size of the audience that watched the taping?

b. How long did it take for the audience to leave the studio after taping ended?

c. At what times were there exactly 50 people in the audience?

3 Interpret information from a step graph.

The graph in Figure 3-9 shows the cost of renting a trailer for different periods of time. For example, the cost of renting the trailer for 4 days is $60, which is the y-coordinate of the point with coordinates of $(4, 60)$. For renting the trailer for a period lasting over 4 and up to 5 days, the cost jumps to $70. Since the jumps in cost form steps in the graph, we call the graph a *step graph*.

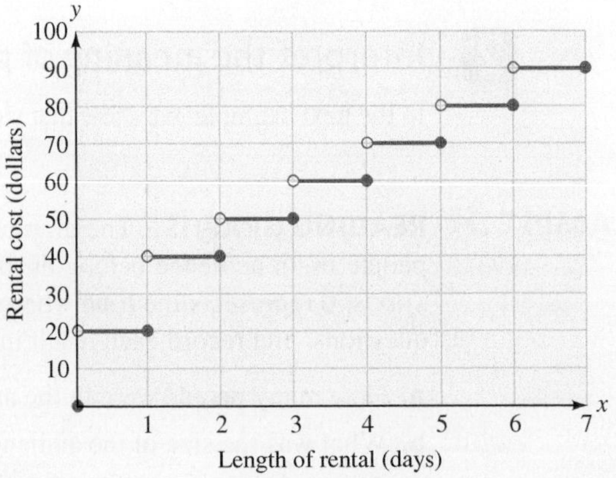

Figure 3-9

Unless otherwise noted, all content on this page is © Cengage Learning.

EXAMPLE 4 **STEP GRAPHS** Use the information in Figure 3-9 to answer the following questions. Write the results in a table of values.

a. Find the cost of renting the trailer for 2 days.

b. Find the cost of renting the trailer for $5\frac{1}{2}$ days.

c. How long can you rent the trailer if you have \$50?

d. Is the rental cost per day the same?

Solution **a.** We locate 2 days on the x-axis and move up to locate the point on the graph directly above the 2. Since the point has coordinates $(2, 40)$, a two-day rental would cost \$40. We enter this ordered pair in Table 3-2.

b. We locate $5\frac{1}{2}$ days on the x-axis and move straight up to locate the point on the graph with coordinates $\left(5\frac{1}{2}, 80\right)$, which indicates that a $5\frac{1}{2}$-day rental would cost \$80. We enter this ordered pair in Table 3-2.

c. We draw a horizontal line through the point labeled 50 on the y-axis. Since this line intersects one step of the graph, we can look down to the x-axis to find the x-values that correspond to a y-value of 50. From the graph, we see that the trailer can be rented for more than 2 and up to 3 days for \$50. We write $(3, 50)$ in Table 3-2.

d. No, the cost per day is not the same. If we look at the y-coordinates, we see that for the first day, the rental fee is \$20. For the second day, the cost jumps another \$20. For the third day, and all subsequent days, the cost jumps only \$10.

TABLE 3-2

Length of rental (days)	Cost (dollars)
2	40
$5\frac{1}{2}$	80
3	50

SELF CHECK 4 Use the information in Figure 3-9 to answer the following questions. Write the results in a table of values.

a. Find the cost of renting the trailer for 4 days.

b. Find the cost of renting the trailer for $2\frac{1}{2}$ days.

c. How long can you rent the trailer if you have \$80?

SELF CHECK ANSWERS

1.

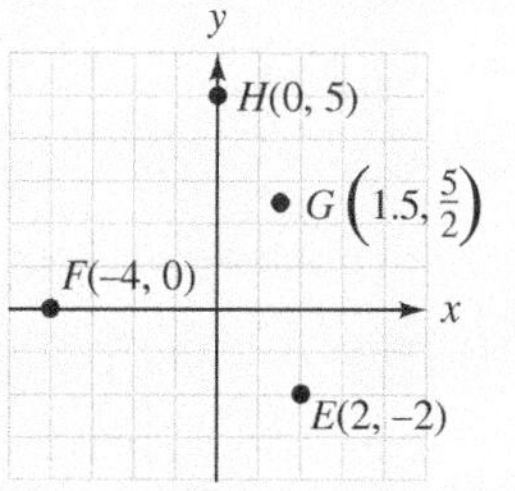

2. $(5, 0)$ **3. a.** 200 people **b.** 20 min **c.** 30 min before and about 85 min after taping began

4.

Length of rental (days)	Cost (dollars)
4	60
$2\frac{1}{2}$	50
6	80

NOW TRY THIS

1. Find three ordered pairs that represent the information stated below.

Damon paid \$1,150 for 2 airline tickets. Javier paid \$1,250 for 3 tickets, and Caroline paid \$1,400 for 4 tickets.

2. Because the size of some data is large, we sometimes insert a // (break) symbol on the x- and/or y-axis of the rectangular coordinate system near the origin to indicate that the designated scale does not begin until the first value is listed.

Plot the points from Problem 1 on a single set of coordinate axes with an appropriate scale.

Unless otherwise noted, all content on this page is © Cengage Learning.

3.1 Exercises

WARM-UPS

1. In which quadrant does the graph of $(-2, -6)$ lie?
2. At what point do the coordinate axes intersect?
3. In which quadrant does the graph of $(3, -5)$ lie?
4. On which axis does the point $(2, 0)$ lie?

REVIEW

5. Evaluate: $-5 - 5(-4)$
6. Evaluate: $(-3)^2 + (-10)$
7. What is the opposite of -8?
8. Simplify: $-|-2 - 10|$
9. Solve: $-4x + 7 = -21$
10. Solve $P = 2l + 2w$ for w.
11. Evaluate $(x + 1)(x + y)^2$ for $x = -2$ and $y = -5$.
12. Simplify: $-5(x + 2) - 3(4 - x)$

VOCABULARY AND CONCEPTS *Fill in the blanks.*

13. The pair of numbers $(-1, -5)$ is called an __________.
14. In the __________ $\left(-\frac{3}{2}, -5\right)$, $-\frac{3}{2}$ is called the _coordinate and -5 is called the _coordinate.
15. The point with coordinates $(0, 0)$ is the _____.
16. The x-and y-axes divide the _____________ into four regions called ________.
17. The point with coordinates $(4, 2)$ can be graphed on a _________ or ________ coordinate system.
18. The rectangular coordinate system is formed by two ___________ number lines called the _ and _axes.
19. The values x and y in the ordered pair (x, y) are called the _________ of its corresponding point.
20. The process of locating the position of a point on a coordinate plane is called _______________ the point.

Answer the question or fill in the blanks.

21. Do $(3, 2)$ and $(2, 3)$ represent the same point?
22. In the ordered pair $(2, 3)$, is 3 associated with the horizontal or the vertical axis?
23. To plot the point with coordinates $(-5, 4.5)$, we start at the _____, move 5 units to the ___, and then move 4.5 units __.
24. To plot the point with coordinates $\left(-\frac{3}{2}, -4\right)$, we start at the _____, move $\frac{3}{2}$ units to the ___, and then move 4 units _____.
25. In which quadrant do points with a negative x-coordinate and a positive y-coordinate lie?
26. In which quadrant do points with a positive x-coordinate and a negative y-coordinate lie?

GUIDED PRACTICE *Plot each point on a coordinate grid.* SEE EXAMPLE 1. (OBJECTIVE 1)

27. $A(-3, 4)$, $B(4, 3.5)$, $C\left(-2, -\frac{5}{2}\right)$, $D(0, -4)$, $E\left(\frac{3}{2}, 0\right)$, $F(3, -4)$

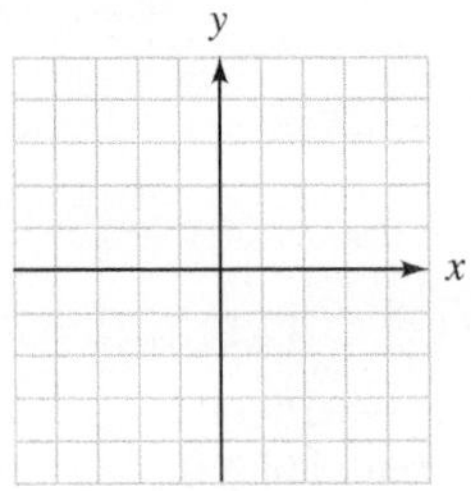

28. $G(4, 4)$, $H(0.5, -3)$, $I(-4, -4)$, $J(0, -1)$, $K(0, 0)$, $L(0, 3)$, $M(-2, 0)$

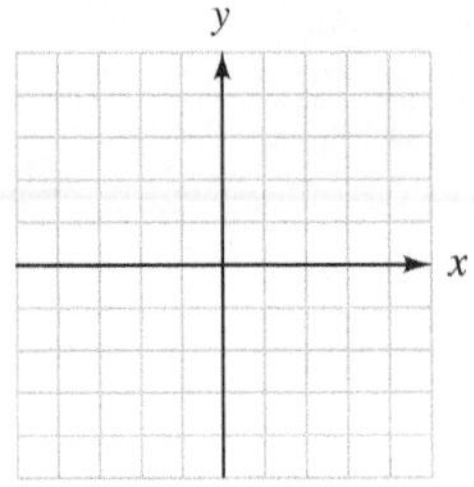

Use each graph to complete the table. SEE EXAMPLE 2. (OBJECTIVE 1)

29.

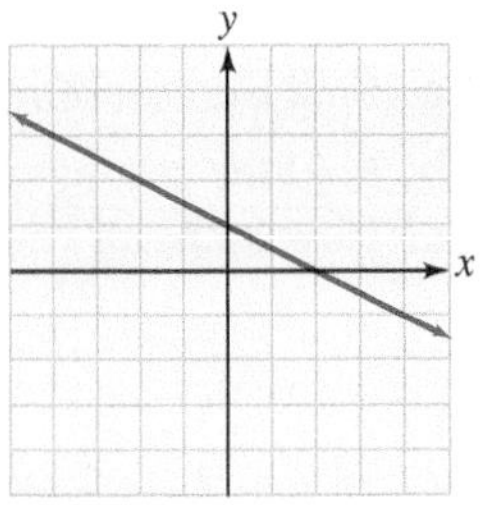

x	y
	0
	2
	-1
-4	
	1

30.

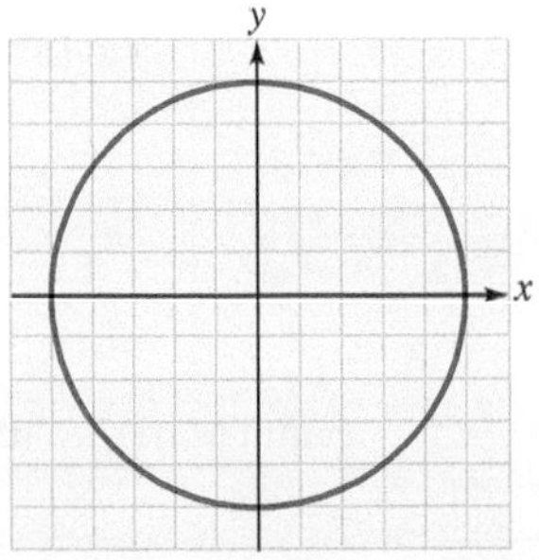

x	y
4	
0	
-3	
	0
-4	
3	

The graph gives the heart rate of a woman before, during, and after an aerobic workout. Use the graph to answer the following questions. SEE EXAMPLE 3. (OBJECTIVE 2)

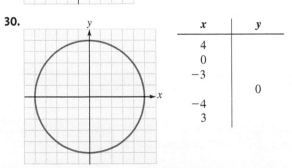

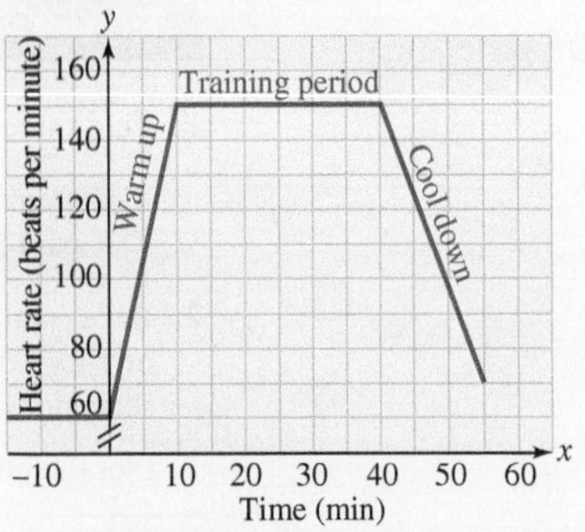

Unless otherwise noted, all content on this page is © Cengage Learning.

31. What information does the point $(-10, 60)$ give us?

32. After beginning the workout, how long did it take the woman to reach her training-zone heart rate?

33. What was her heart rate one-half hour after beginning the workout?

34. For how long did she work out at the training-zone level?

35. At what times was her heart rate 100 beats per minute?

36. How long was her cooldown period?

37. What was the difference in her heart rate before the workout and after the cooldown period?

38. What was her approximate heart rate 5 minutes after beginning her cooldown?

Use the graph to answer the questions. SEE EXAMPLE 4. (OBJECTIVE 3)

DVD rentals *The charges for renting a movie are shown in the graph.*

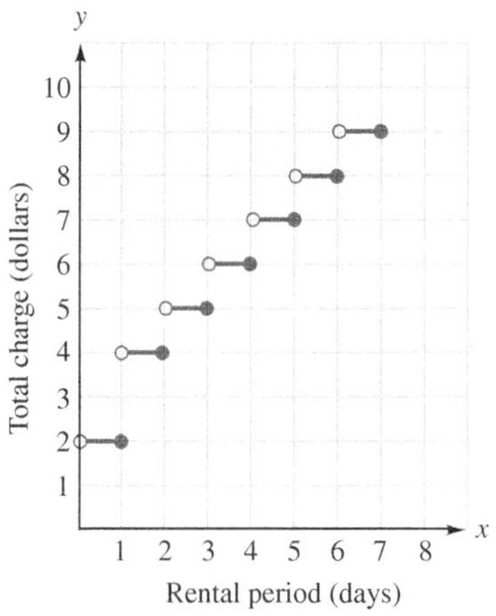

39. Find the charge for a 1-day rental.

40. Find the charge for a 3-day rental.

41. Find the charge if the DVD is kept for 5 days.

42. Find the charge if the DVD is kept for a week.

ADDITIONAL PRACTICE *Write the coordinates of each point.*

43.

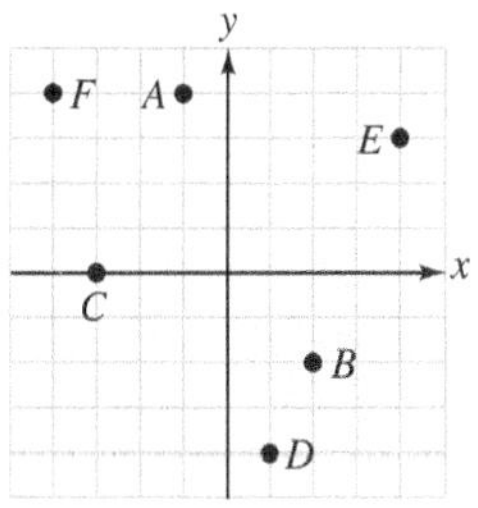

44.

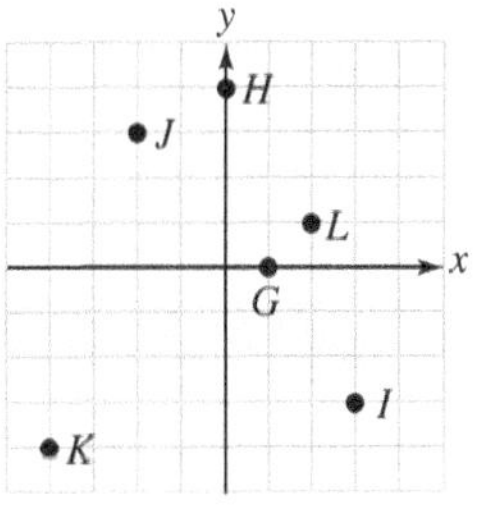

Postage rates *The graph gives the first-class postage rates for mailing letters weighing up to 3.5 ounces.*

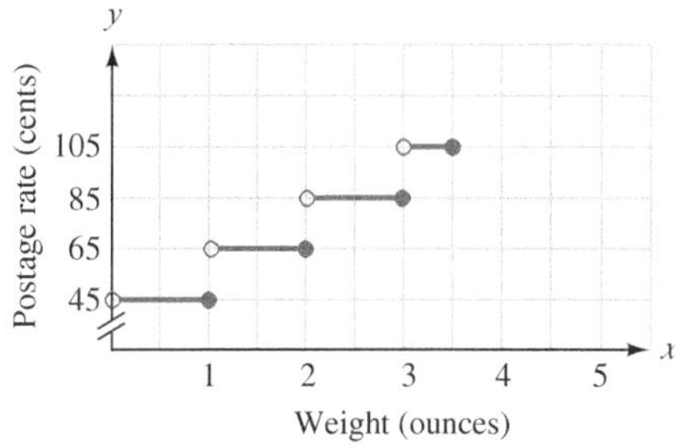

45. Find the cost of postage to mail each of the following letters first class: 1-ounce; $2\frac{1}{2}$-ounce.

46. Find the cost of postage to mail each of the following letters first class: 1.5-ounce; 3.25-ounce.

47. Find the difference in postage for a 0.75-ounce letter and a 2.75-ounce letter.

48. What is the heaviest letter that can be mailed for $1.05?

APPLICATIONS *Use the graphs to answer the questions.* SEE EXAMPLES 3 AND 4. (OBJECTIVES 2–3)

49. Road maps Road maps usually have a coordinate system to help locate cities. Use the map to locate Carbondale, Champaign, Chicago, Peoria, and Rockford. Express each answer in the form (number, letter).

50. Battling ships In a computer game of battling ships, players use coordinates to drop depth charges from a battleship to hit a hidden submarine. What coordinates should be used to make three hits on the exposed submarine shown in the illustration? Express each answer in the form (letter, number).

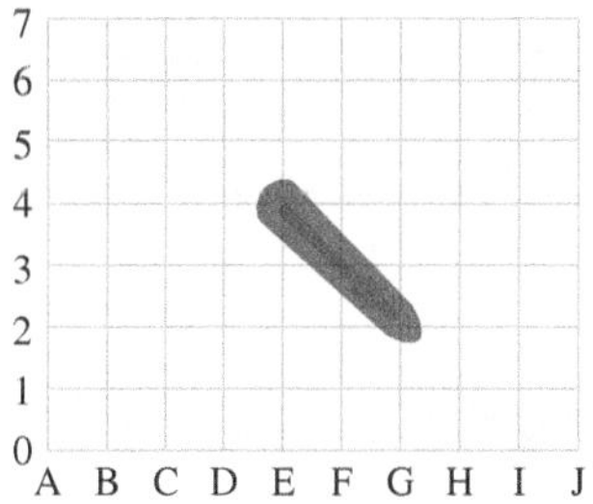

Unless otherwise noted, all content on this page is © Cengage Learning.

51. Water pressure The graphs show the paths of two streams of water from the same hose held at two different angles.

a. At which angle does the stream of water shoot higher? How much higher?

b. At which angle does the stream of water shoot out farther? How much farther?

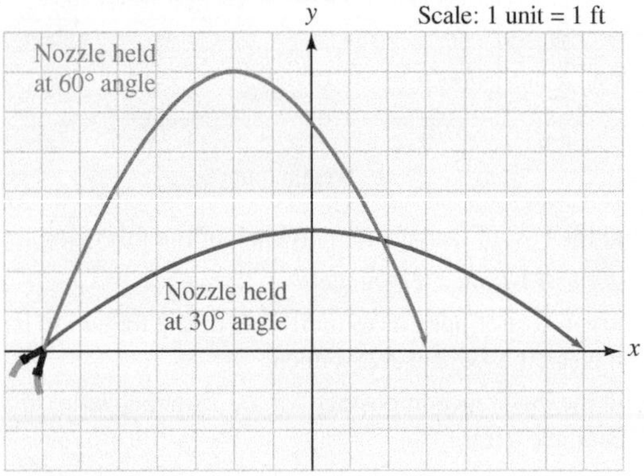

52. Golf swings To correct her swing, a golfer was videotaped and then had her image displayed on a computer monitor so that it could be analyzed by a golf pro. See the illustration. Give the coordinates of the points that are highlighted on the arc of her swing.

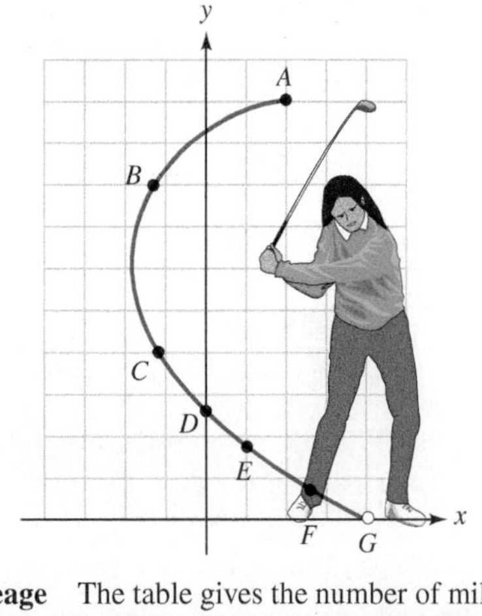

53. Gas mileage The table gives the number of miles (y) that a truck can be driven on x gallons of gasoline. Plot the ordered pairs and draw a line connecting the points.

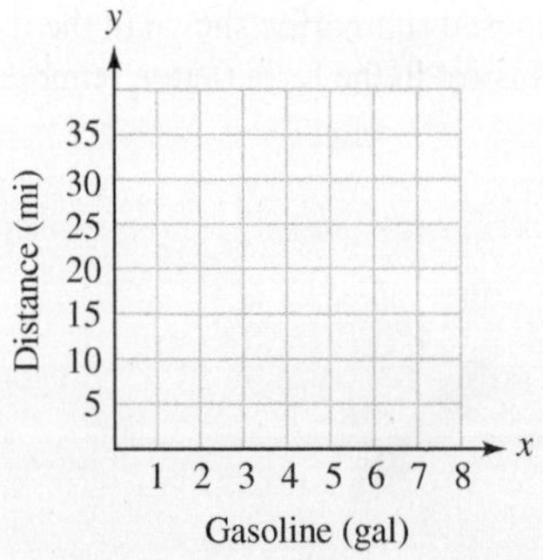

Gallons	Miles
2	10
3	15
5	25

a. Estimate how far the truck can go on 6 gallons of gasoline.

b. How many gallons of gas are needed to travel a distance of 40 miles?

c. Estimate how far the truck can go on 6.5 gallons of gasoline.

54. Wages The table gives the amount y (in dollars) that a student can earn by working x hours. Plot the ordered pairs and draw a line connecting the points.

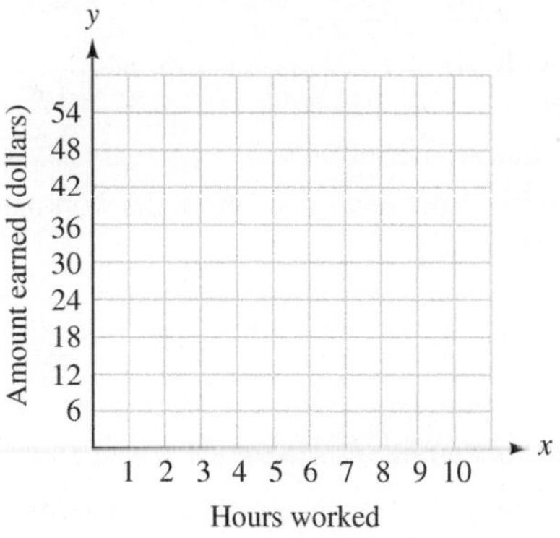

Hours	Dollars
3	18
6	36
7	42

a. How much will the student earn in 5 hours?

b. How long would the student have to work to earn $12?

c. Estimate how much the student will earn in 3.5 hours.

55. Value of a car The table shows the value y (in thousands of dollars) of a car that is x years old. Plot the ordered pairs and draw a line connecting the points.

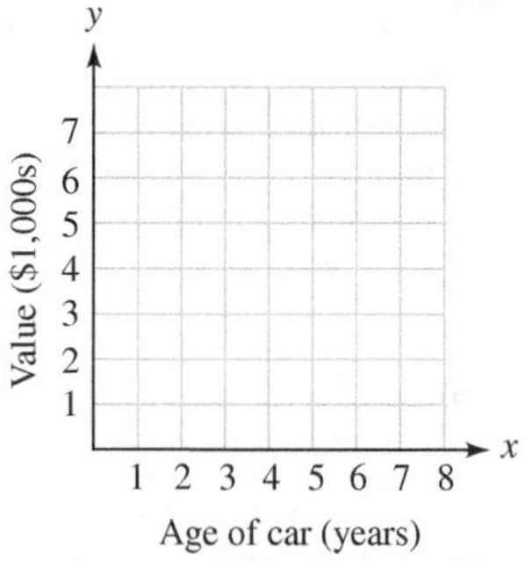

Years	Value (in thousands)
3	7
4	5.5
5	4

a. What does the point $(3, 7)$ on the graph tell you?

b. Estimate the value of the car when it is 7 years old.

c. After how many years will the car be worth $2,500?

56. Depreciation As a piece of farm machinery gets older, it loses value. The table shows the value y of a tractor that is x years old. Plot the ordered pairs and draw a line connecting them.

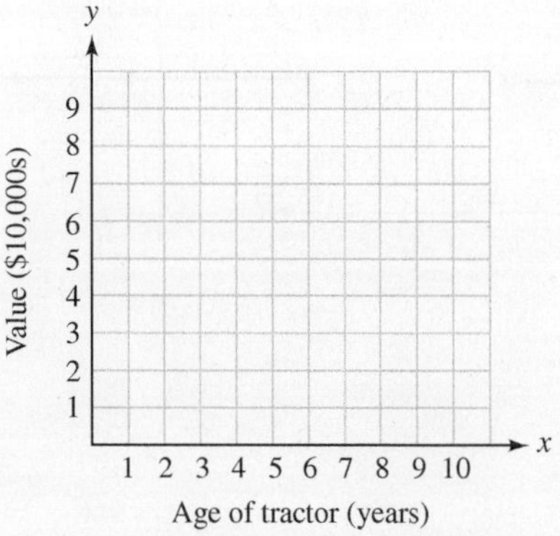

Years	Value
0	9
6	5
9	3

Unless otherwise noted, all content on this page is © Cengage Learning.

a. What does the point $(0, 9)$ on the graph tell you?

b. Estimate the value of the tractor in 3 years.

c. When will the tractor's value fall below \$30,000?

WRITING ABOUT MATH

57. Explain why the point with coordinates $(-3, 3)$ is not the same as the point with coordinates $(3, -3)$.

58. Explain what is meant when we say that the rectangular coordinate graph of the St. Louis Arch is made up of *infinitely many* points.

59. Explain how to plot the point with coordinates $(-2, 5)$.

60. Explain why the coordinates of the origin are $(0, 0)$.

SOMETHING TO THINK ABOUT

61. Could you have a coordinate system in which the coordinate axes were not perpendicular? How would it be different?

62. René Descartes is famous for saying, "I think, therefore I am." What do you think he meant by that?

Section 3.2 Graphing Linear Equations

Objectives

1. Determine whether an ordered pair satisfies an equation in two variables.
2. Construct a table of values given an equation.
3. Graph a linear equation in two variables by constructing a table of values.
4. Graph a linear equation in two variables using the intercept method.
5. Graph a horizontal line and a vertical line given an equation.
6. Write a linear equation in two variables from given information, graph the equation, and interpret the graphed data.

Vocabulary

input value	dependent variable	*y*-intercept
output value	independent variable	intercept method
linear equation in two variables	*x*-intercept	general form

Getting Ready

In Problems 1–4, let $y = 2x + 1$.

1. Find the value of y when $x = 0$.

2. Find the value of y when $x = 2$.

3. Find the value of y when $x = -2$.

4. Find the value of y when $x = \frac{1}{2}$.

5. Find five pairs of numbers with a sum of 8.

6. Find five pairs of numbers with a difference of 5.

In this section, we will discuss how to graph linear equations in two variables. We will then show how to create tables and graphs using a graphing calculator.

1 Determine whether an ordered pair satisfies an equation in two variables.

The equation $x + 2y = 5$ contains the two variables x and y. The solutions of such equations are ordered pairs of numbers. For example, the ordered pair $(1, 2)$ is a solution, because the equation is satisfied when $x = 1$ and $y = 2$.

$x + 2y = 5$

$1 + 2(2) = 5$ Substitute 1 for x and 2 for y.

$1 + 4 = 5$ Multiply

$5 = 5$

EXAMPLE 1 Is the pair $(-2, 4)$ a solution of $y = 3x + 9$?

Solution We substitute -2 for x and 4 for y and determine whether the resulting equation is true.

$y = 3x + 9$ This is the original equation.

$4 \stackrel{?}{=} 3(-2) + 9$ Substitute -2 for x and 4 for y.

$4 \stackrel{?}{=} -6 + 9$ Do the multiplication: $3(-2) = -6$.

$4 = 3$ Do the addition: $-6 + 9 = 3$.

Since the equation $4 = 3$ is false, the pair $(-2, 4)$ is not a solution.

SELF CHECK 1 Is $(-1, -5)$ a solution of $y = 5x$?

2 Construct a table of values given an equation.

To find solutions of equations in x and y, we can pick numbers at random, substitute them for x, and find the corresponding values of y. For example, to find some ordered pairs that satisfy $y = 5 - x$, we can let $x = 1$ (called the **input value**), substitute 1 for x, and solve for y (called the **output value**).

(1) $y = 5 - x$

x	y	(x, y)
1	4	$(1, 4)$

$y = 5 - x$ This is the original equation.

$y = 5 - 1$ Substitute the input value of 1 for x.

$y = 4$ The output is 4.

The ordered pair $(1, 4)$ is a solution. As we find solutions, we will list them in a *table of values* like Table (1) at the left.

If $x = 2$, we have

(2) $y = 5 - x$

x	y	(x, y)
1	4	$(1, 4)$
2	3	$(2, 3)$

$y = 5 - x$ This is the original equation.

$y = 5 - 2$ Substitute the input value of 2 for x.

$y = 3$ The output is 3.

A second solution is $(2, 3)$. We list it in Table (2) at the left.

If $x = 5$, we have

(3) $y = 5 - x$

x	y	(x, y)
1	4	$(1, 4)$
2	3	$(2, 3)$
5	0	$(5, 0)$

$y = 5 - x$ This is the original equation.

$y = 5 - 5$ Substitute the input value of 5 for x.

$y = 0$ The output is 0.

A third solution is $(5, 0)$. We list it in Table (3) at the left.

If $x = -1$, we have

$y = 5 - x$ This is the original equation.

$y = 5 - (-1)$ Substitute the input value of -1 for x.

$y = 6$ The output is 6.

A fourth solution is $(-1, 6)$. We list it in Table (4) at the left.

(4) $y = 5 - x$

x	y	(x, y)
1	4	(1, 4)
2	3	(2, 3)
5	0	(5, 0)
−1	**6**	**(−1, 6)**

If $x = 6$, we have

$y = 5 - x$ This is the original equation.

$y = 5 - 6$ Substitute the input value of 6 for x.

$y = -1$ The output is -1.

A fifth solution is $(6, -1)$. We list it in Table (5) at the left.

(5) $y = 5 - x$

x	y	(x, y)
1	4	(1, 4)
2	3	(2, 3)
5	0	(5, 0)
−1	6	(−1, 6)
6	**−1**	**(6, −1)**

Since we can choose any real number for x, and since any choice of x will give a corresponding value of y, we can see that the equation $y = 5 - x$ has *infinitely many solutions*.

3 Graph a linear equation in two variables by constructing a table of values.

A *linear equation* is any equation that can be written in the form $Ax + By = C$, where A, B, and C are real numbers and A and B are not both 0. To graph the equation $y = 5 - x$, we plot the ordered pairs listed in the table on a rectangular coordinate system, as in Figure 3-10. From the figure, we can see that the five points lie on a line.

$y = 5 - x$

x	y	(x, y)
1	4	(1, 4)
2	3	(2, 3)
5	0	(5, 0)
−1	6	(−1, 6)
6	−1	(6, −1)

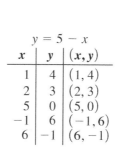

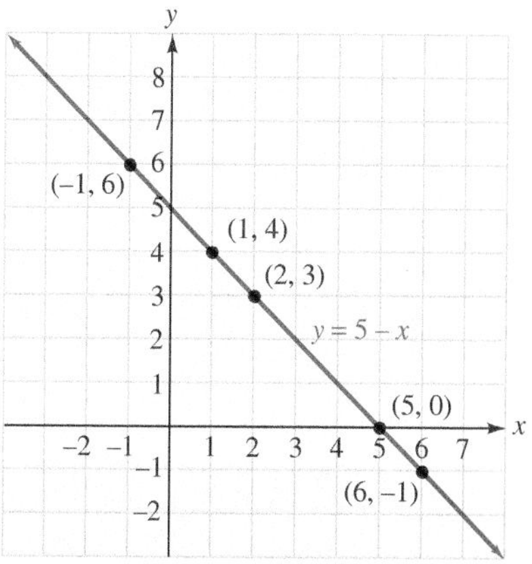

Figure 3-10

We draw a line through the points. The arrowheads on the line show that the graph continues forever in both directions. Since the graph of any solution of $y = 5 - x$ will lie on this line, the line is a picture of all of the solutions of the equation. The line is said to be the *graph* of the equation.

Any equation, such as $y = 5 - x$, whose graph is a line is called a **linear equation in two variables**. Any point on the line has coordinates that satisfy the equation, and the graph of any pair (x, y) that satisfies the equation is a point on the line.

Since we usually will choose a number for x first and then find the corresponding value of y, the value of y depends on x. For this reason, we call y the **dependent variable** and x the **independent variable**. The value of the independent variable is the input value, and the value of the dependent variable is the output value.

Although only two points are needed to graph a linear equation, we often plot a third point as a check. If the three points do not lie on a line, at least one of them is in error.

Unless otherwise noted, all content on this page is © Cengage Learning.

COMMENT The equation $y = 5 - x$ can be written as $x + y = 5$, which is in the form $Ax + By = C$.

GRAPHING LINEAR EQUATIONS IN TWO VARIABLES

1. Find two ordered pairs (x, y) that satisfy the equation by choosing arbitrary input values for x and solving for the corresponding output values of y. A third point may be used to provide a check.
2. Plot each resulting pair (x, y) on a rectangular coordinate system. If they do not lie on a line, check your calculations.
3. Draw the line passing through the points.

EXAMPLE 2 Graph by constructing a table of values and plotting points: $y = 3x - 4$

Solution We find three ordered pairs that satisfy the equation.

If $x = 1$	*If* $x = 2$	*If* $x = 3$
$y = 3x - 4$	$y = 3x - 4$	$y = 3x - 4$
$y = 3(1) - 4$	$y = 3(2) - 4$	$y = 3(3) - 4$
$y = -1$	$y = 2$	$y = 5$

We enter the results in a table of values, plot the points, and draw a line through the points. The graph appears in Figure 3-11.

$y = 3x - 4$

x	y	(x, y)
1	−1	(1, −1)
2	2	(2, 2)
3	5	(3, 5)

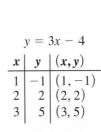

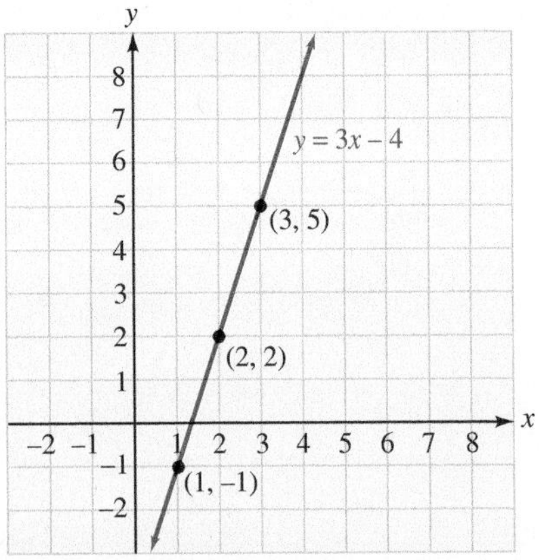

Figure 3-11

SELF CHECK 2 Graph: $y = 3x$

EXAMPLE 3 Graph by constructing a table of values and plotting points: $y = -0.4x + 2$

Solution We find three ordered pairs that satisfy the equation.

If $x = -5$	*If* $x = 0$	*If* $x = 5$
$y = -0.4x + 2$	$y = -0.4x + 2$	$y = -0.4x + 2$
$y = -0.4(-5) + 2$	$y = -0.4(0) + 2$	$y = -0.4(5) + 2$
$y = 2 + 2$	$y = 2$	$y = -2 + 2$
$y = 4$		$y = 0$

We enter the results in a table of values, plot the points, and draw a line through the points. The graph appears in Figure 3-12.

Unless otherwise noted, all content on this page is © Cengage Learning.

$y = -0.4x + 2$

x	y	(x, y)
-5	4	$(-5, 4)$
0	2	$(0, 2)$
5	0	$(5, 0)$

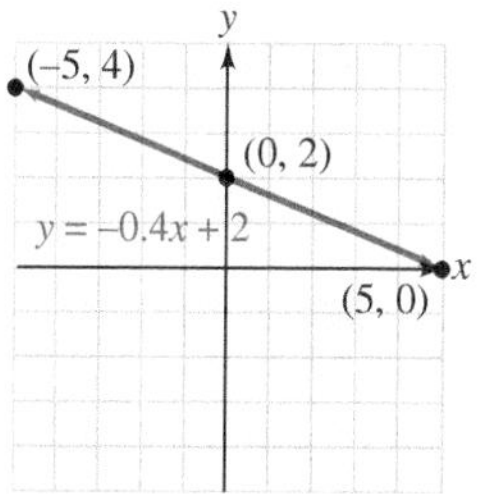

Figure 3-12

SELF CHECK 3 Graph: $y = 1.5x - 2$

EXAMPLE 4 Graph by constructing a table of values and plotting points: $y - 4 = \frac{1}{2}(x - 8)$

Solution We first solve for y and simplify.

$$y - 4 = \frac{1}{2}(x - 8)$$

$$y - 4 = \frac{1}{2}x - 4 \quad \text{Use the distributive property to remove parentheses.}$$

$$y = \frac{1}{2}x \quad \text{Add 4 to both sides.}$$

We now find three ordered pairs that satisfy the equation.

***If* $x = 0$**	***If* $x = 2$**	***If* $x = -4$**
$y = \frac{1}{2}x$	$y = \frac{1}{2}x$	$y = \frac{1}{2}x$
$y = \frac{1}{2}(\mathbf{0})$	$y = \frac{1}{2}(\mathbf{2})$	$y = \frac{1}{2}(\mathbf{-4})$
$y = 0$	$y = 1$	$y = -2$

We enter the results in a table of values, plot the points, and draw a line through the points. The graph appears in Figure 3-13.

$y - 4 = \frac{1}{2}(x - 8)$

x	y	(x, y)
0	0	$(0, 0)$
2	1	$(2, 1)$
-4	-2	$(-4, -2)$

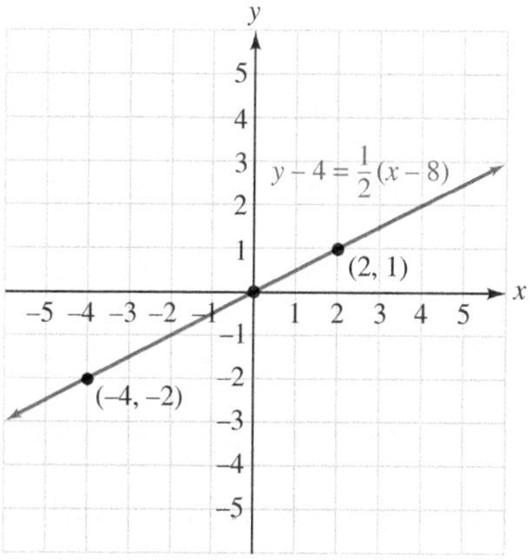

Figure 3-13

SELF CHECK 4 Graph: $y + 3 = \frac{1}{3}(x - 6)$

Unless otherwise noted, all content on this page is © Cengage Learning.

4 Graph a linear equation in two variables using the intercept method.

The points where a line intersects the x- and y-axes are called **intercepts** of the line.

x- AND y-INTERCEPTS

The ***x*-intercept** of a line is the point $(a, 0)$ where the line intersects the x-axis. (See Figure 3-14.) To find a, substitute 0 for y in the equation of the line and solve for x.

A ***y*-intercept** of a line is the point $(0, b)$ where the line intersects the y-axis. To find b, substitute 0 for x in the equation of the line and solve for y.

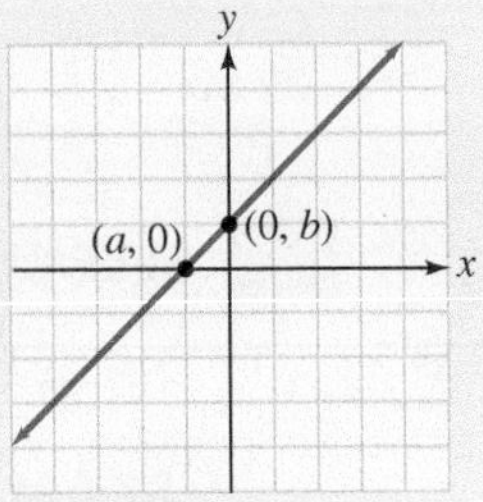

Figure 3-14

Plotting the x- and y-intercepts and drawing a line through them is called the **intercept method** of graphing a line. This method is useful for graphing equations written in *general form*.

GENERAL FORM OF THE EQUATION OF A LINE

If A, B, and C are real numbers and A and B are not both 0, then the equation

$$Ax + By = C$$

is called the **general form** of the equation of a line.

COMMENT Whenever possible, we will write the general form $Ax + By = C$ so that A, B, and C are integers and $A \geq 0$. We also will make A, B, and C as small as possible. For example, the equation $6x + 12y = 24$ can be written as $x + 2y = 4$ by dividing both sides by 6.

EXAMPLE 5 Graph using the intercept method: $3x + 2y = 6$

Solution To find the y-intercept, we let $x = 0$ and solve for y.

$$3x + 2y = 6$$
$$3(\mathbf{0}) + 2y = 6 \quad \text{Substitute 0 for } x.$$
$$2y = 6 \quad \text{Simplify.}$$
$$y = 3 \quad \text{Divide both sides by 2.}$$

The y-intercept is the point with coordinates $(0, 3)$.

To find the x-intercept, we let $y = 0$ and solve for x.

$$3x + 2y = 6$$
$$3x + 2(\mathbf{0}) = 6 \quad \text{Substitute 0 for } y.$$
$$3x = 6 \quad \text{Simplify.}$$
$$x = 2 \quad \text{Divide both sides by 3.}$$

The x-intercept is the point with coordinates $(2, 0)$.

As a check, we plot one more point. If $x = 4$, then

$$3x + 2y = 6$$
$$3(4) + 2y = 6 \quad \text{Substitute 4 for } x.$$
$$12 + 2y = 6 \quad \text{Simplify.}$$
$$2y = -6 \quad \text{Subtract 12 from both sides.}$$
$$y = -3 \quad \text{Divide both sides by 2.}$$

Unless otherwise noted, all content on this page is © Cengage Learning.

The point $(4, -3)$ is on the graph. We plot these three points and join them with a line. The graph of $3x + 2y = 6$ is shown in Figure 3-15.

$3x + 2y = 6$

x	y	(x, y)
0	3	(0, 3)
2	0	(2, 0)
4	−3	(4, −3)

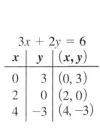

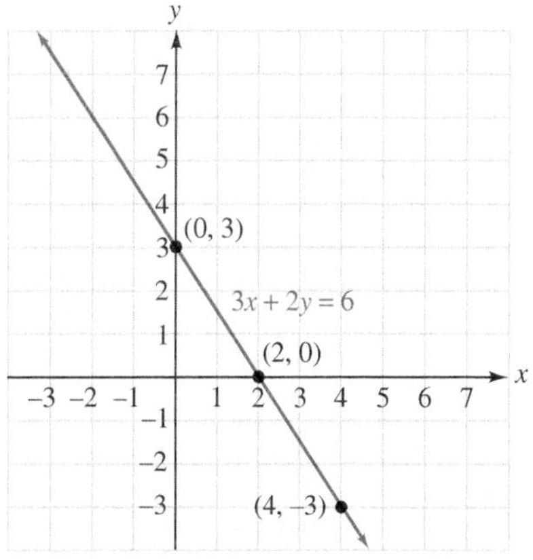

Figure 3-15

SELF CHECK 5 Graph: $4x + 3y = 6$

5 Graph a horizontal line and a vertical line given an equation.

Equations such as $y = 3$ and $x = -2$ are linear equations, because they can be written in the general form $Ax + By = C$.

$y = 3$	is equivalent to	$0x + 1y = 3$
$x = -2$	is equivalent to	$1x + 0y = -2$

Next, we discuss how to graph these types of linear equations.

EXAMPLE 6 Graph: **a.** $y = 3$ **b.** $x = -2$

Solution **a.** We can write the equation $y = 3$ in general form as $\mathbf{0}x + y = 3$. Since the coefficient of x is 0, the numbers chosen for x have no effect on y. The value of y is always 3. For example, if we substitute -3 for x, we get

$$0x + y = 3$$
$$0(-3) + y = 3$$
$$0 + y = 3$$
$$y = 3$$

The table in Figure 3-16(a) on the next page gives several pairs that satisfy the equation $y = 3$. After plotting these pairs and joining them with a line, we see that the graph of $y = 3$ is a horizontal line that intersects the y-axis at 3. The y-intercept is $(0, 3)$. There is no x-intercept.

b. We can write $x = -2$ in general form as $x + \mathbf{0}y = -2$. Since the coefficient of y is 0, the values of y have no effect on x. The value of x is always -2. A table of values and the graph are shown in Figure 3-16(b). The graph of $x = -2$ is a vertical line that intersects the x-axis at -2. The x-intercept is $(-2, 0)$. There is no y-intercept.

Unless otherwise noted, all content on this page is © Cengage Learning.

$y = 3$

x	y	(x, y)
−3	3	(−3, 3)
0	3	(0, 3)
2	3	(2, 3)
4	3	(4, 3)

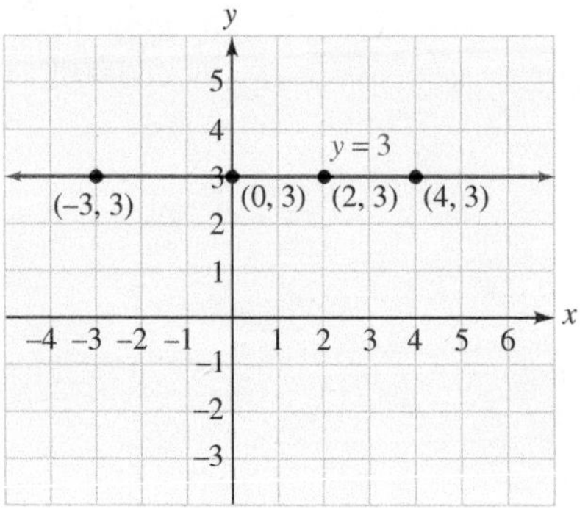

(a)

$x = -2$

x	y	(x, y)
−2	−2	(−2, −2)
−2	0	(−2, 0)
−2	2	(−2, 2)
−2	3	(−2, 3)

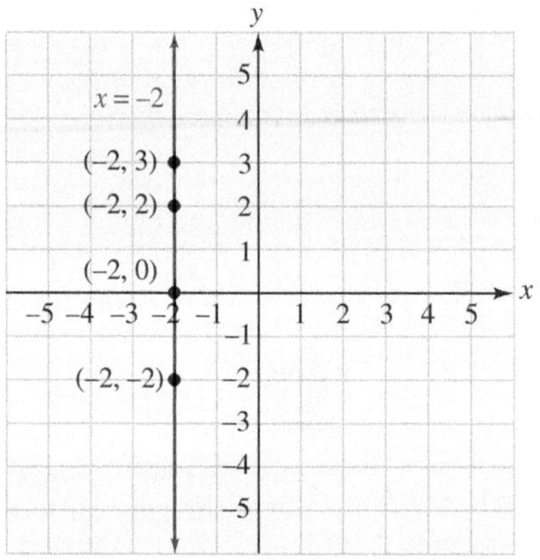

(b)

Figure 3-16

SELF CHECK 6 Identify the graph of each equation as a horizontal or a vertical line.
a. $x = 5$ **b.** $y = -3$ **c.** $x = 0$

From the results of Example 6, we have the following facts.

EQUATIONS OF HORIZONTAL AND VERTICAL LINES

Suppose a and b are real numbers.

The equation $y = b$ represents a horizontal line that intersects the y-axis at $(0, b)$.

If $b = 0$, the line is the x-axis.

The equation $x = a$ represents a vertical line that intersects the x-axis at $(a, 0)$.

If $a = 0$, the line is the y-axis.

6 Write a linear equation in two variables from given information, graph the equation, and interpret the graphed data.

In Chapter 2, we solved applications using one variable. In the next example, we will write an equation containing two variables to describe an application and then graph the equation.

EXAMPLE 7 **BIRTHDAY PARTIES** A restaurant offers a party package that includes food, drinks, cake, and party favors for a cost of \$25 plus \$3 per child. Write a linear equation that will give the cost for a party of any size. Graph the equation and determine the meaning of the y-intercept in the context of this problem.

Solution We can let c represent the cost of the party and n represent the number of children attending. Then c will be the sum of the basic charge of \$25 and the cost per child times the number of children attending.

Unless otherwise noted, all content on this page is © Cengage Learning.

The cost	equals	the basic $25 charge	plus	$3	times	the number of children.
c	$=$	25	$+$	3	$\cdot$	n

For the equation $c = 25 + 3n$, the independent variable (input) is n, the number of children. The dependent variable (output) is c, the cost of the party. We will find three points on the graph of the equation by choosing n-values of 0, 5, and 10 and finding the corresponding c-values. The results are recorded in the table.

$c = 25 + 3n$

n	c
0	25
5	40
10	55

If $\boldsymbol{n = 0}$
$c = 25 + 3(\mathbf{0})$
$c = 25$

If $\boldsymbol{n = 5}$
$c = 25 + 3(\mathbf{5})$
$c = 25 + 15$
$c = 40$

If $\boldsymbol{n = 10}$
$c = 25 + 3(\mathbf{10})$
$c = 25 + 30$
$c = 55$

Next, we graph the points in Figure 3-17 and draw a line through them. We don't draw an arrowhead on the left, because it doesn't make sense to have a *negative* number of children attend a party.

From the graph, we can determine the y-intercept is $(0, 25)$. The $25 represents the setup cost for a party with no attendees.

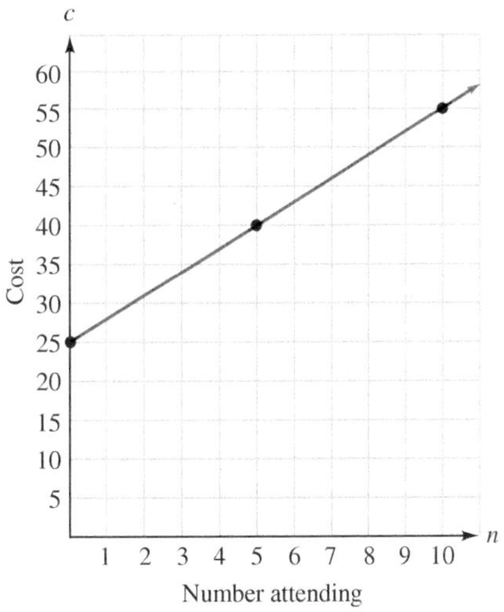

Figure 3-17

COMMENT The scale for the cost (y-axis) is 5 units and the scale for the number attending (x-axis) is 1. Since the scales on the x- and y-axes are not the same, you must label them!

SELF CHECK 7 Write a linear equation that will give the cost for a party of any size if the party package costs $30 plus $4 per child. Then graph the equation.

Accent on technology

▸ Creating Tables and Graphs

So far, we have graphed equations by creating tables and plotting points. This method is often tedious and time-consuming. Fortunately, creating tables and graphing equations is easier when we use a graphing calculator.

Although we will use calculators to generate tables and graph equations, we will show complete keystrokes only for the TI 83-84 family of calculators. For keystrokes of other calculators, please consult your owner's manual.

All graphing calculators have a *viewing window* that is used to display graphs. We will first discuss how to generate tables and then discuss how to draw graphs.

Unless otherwise noted, all content on this page is © Cengage Learning.

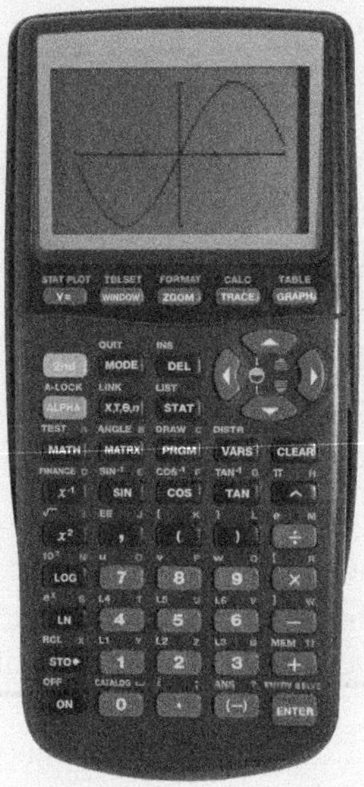
James Hoenstine/Shutterstock.com

TABLES To construct a table of values for the equation $y = x^2$, press the **Y =** key, enter the expression x^2 by pressing **X, T, θ, n** **x^2** keys, and press the **2nd** and **GRAPH** keys to obtain a screen similar to Figure 3-18(a). You can use the up and down keys to scroll through the table to obtain a screen like Figure 3-18(b).

X	Y1	
0	0	
1	1	
2	4	
3	9	
4	16	
5	25	
6	36	
X = 0		

(a)

X	Y1	
−5	25	
−4	16	
−3	9	
−2	4	
−1	1	
0	0	
1	1	
X = −5		

(b)

Figure 3-18

GRAPHS To see an accurate graph, we must often set the minimum and maximum values for the x- and y-coordinates. The standard window settings of

Xmin = −10 Xmax = 10 Ymin = −10 Ymax = −10

indicate that −10 is the minimum x- and y-coordinate to be used in the graph, and that 10 is the maximum x- and y-coordinate to be used. We will usually express window values in interval notation. In this notation, the standard settings are

$$X = [-10, 10] \quad Y = [-10, 10]$$

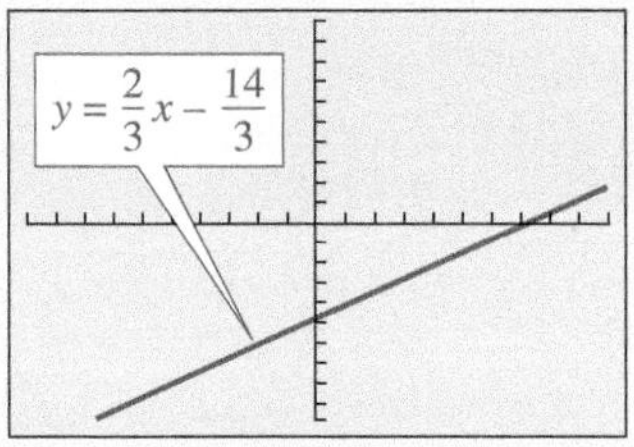

Figure 3-19

To graph the equation $2x - 3y = 14$ with a calculator, we must first solve the equation for y.

$$2x - 3y = 14$$

$$-3y = -2x + 14 \quad \text{Subtract } 2x \text{ from both sides.}$$

$$y = \frac{2}{3}x - \frac{14}{3} \quad \text{Divide both sides by } -3.$$

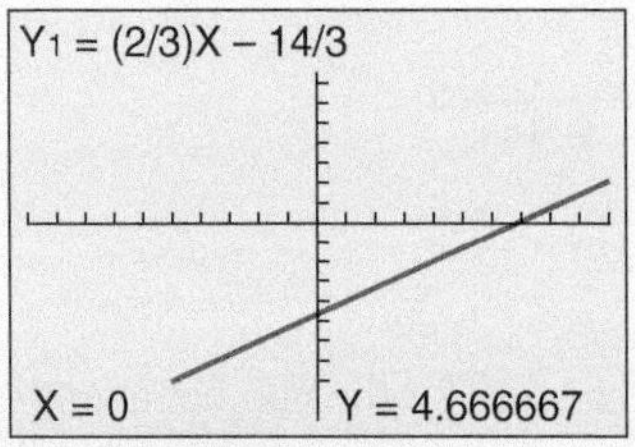

Figure 3-20

Press the **Y =** key and enter the equation as $(2/3)x - 14/3$ by pressing **(2 ÷ 3) X, T, θ, n − 1 4 ÷ 3**, and press **GRAPH** to obtain the line shown in Figure 3-19.

COMMENT To graph an equation with a graphing calculator, the equation must be solved for y.

We can find the y-intercept by using the **TRACE** feature of the calculator. From the graph, press **TRACE**. The y-intercept will be highlighted and its coordinates will be displayed at the bottom of the screen. See Figure 3-20.

For instructions regarding the use of a Casio graphing calculator, please refer to the Casio Keystroke Guide in the back of the book.

SELF CHECK ANSWERS

1. yes 2.

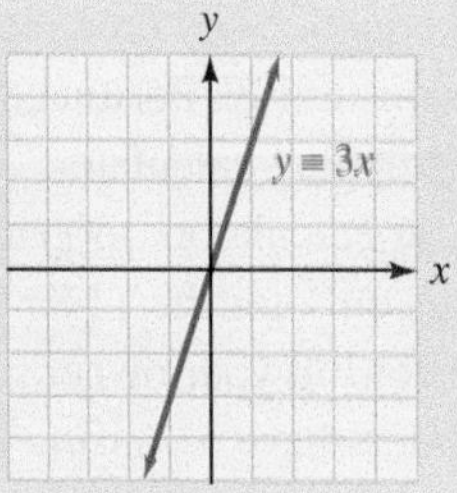

3.

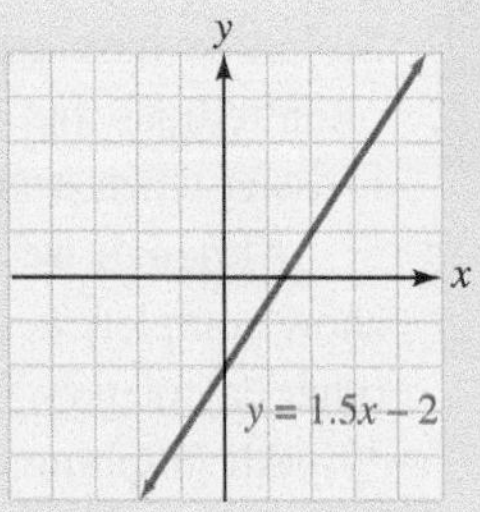

4.

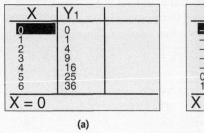

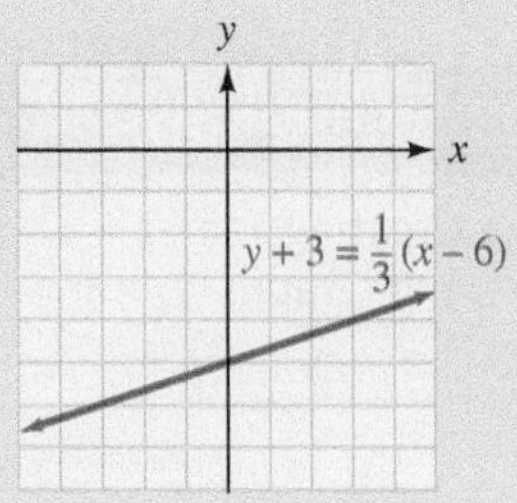

Unless otherwise noted, all content on this page is © Cengage Learning.

5.

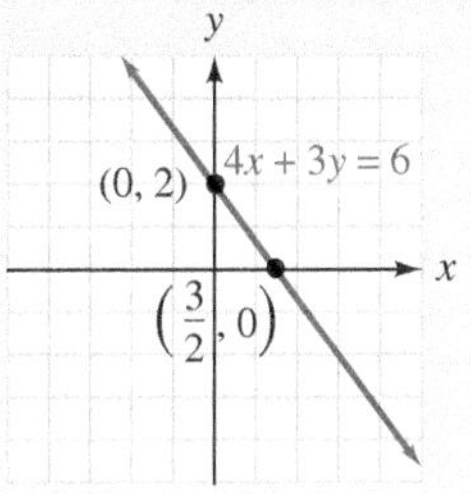

6. a. vertical
b. horizontal
c. vertical

7. $c = 30 + 4n$

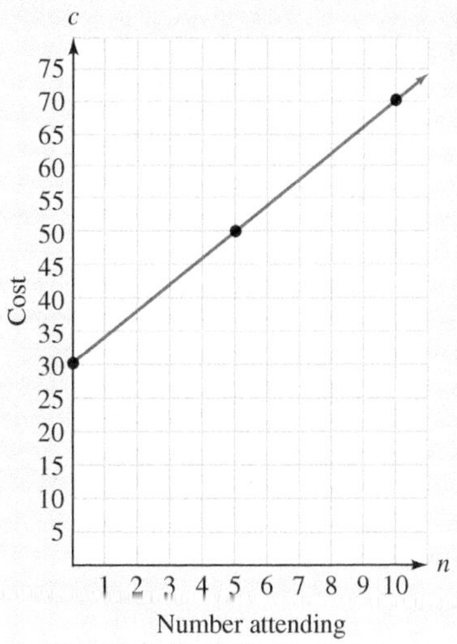

NOW TRY THIS

1. Given $8x - 7y = 12$, complete the ordered pair $(-2, \quad)$ that satisfies the equation.
2. Graph: $y - 5 = 0$
3. Graph: $y = x$
4. Identify the x-intercept and the y-intercept of $y = \frac{2}{3}x + 8$.

3.2 Exercises

WARM-UPS

1. How many points determine a line?
2. Define the intercepts of a line.
3. If $y = 3x + 2$, find the value of y when $x = 1$.
4. Find three pairs (x, y) with a sum of 7.
5. Which lines have no y-intercepts?
6. Which lines have no x-intercepts?

REVIEW

7. Solve: $\frac{2}{3}x = -12$
8. Combine like terms: $3t - 4T + 5T - 6t$
9. Is $\frac{x + 5}{6}$ an expression or an equation?
10. Write the formula used to find the perimeter of a rectangle.
11. What number is 0.5% of 250?
12. Solve: $6 - 4x > 22$
13. Subtract: $-2.5 - (-2.6)$
14. Evaluate: $(-5)^3$

VOCABULARY AND CONCEPTS *Fill in the blanks.*

15. The equation $y = x + 1$ is a _____ equation in ___ variables.
16. An ordered pair is a _______ of an equation if the numbers in the ordered pair satisfy the equation.
17. In equations containing the variables x and y, x is called the __________ variable and y is called the _________ variable.
18. When constructing a table of values, the values of _ are the input values and the values of _ are the output values.
19. An equation whose graph is a line and whose variables are to the first power is called a _____ equation.
20. The equation $Ax + By = C$ is the ________________ form of the equation of a line.
21. The _________ of a line is the point $(0, b)$ where the line intersects the y-axis.
22. The _________ of a line is the point $(a, 0)$ where the line intersects the x-axis.

Unless otherwise noted, all content on this page is © Cengage Learning.

GUIDED PRACTICE *Determine whether the ordered pair satisfies the equation.* SEE EXAMPLE 1. (OBJECTIVE 1)

23. $x - 2y = -4$; $(4, 4)$

24. $y = -7x - 3$; $(4, -30)$

25. $y = \frac{2}{3}x + 5$; $(6, 12)$

26. $y = -\frac{2}{5}x - 1$; $(5, -3)$

Complete each table of values. Check your work with a graphing calculator. (OBJECTIVE 2)

27. $y = x - 3$

x	y	(x, y)
0		
1		
−2		
−4		

28. $y = x - 2$

x	y	(x, y)
0		
−1		
−2		
3		

29. $y = -2x$

x	y	(x, y)
0		
1		
3		
−2		

30. $y = -1.7x + 2$

x	y	(x, y)
−3		
−1		
0		
3		

Graph each equation by constructing a table of values and then plotting the points. SEE EXAMPLE 2. (OBJECTIVE 3)

31. $y = 2x$

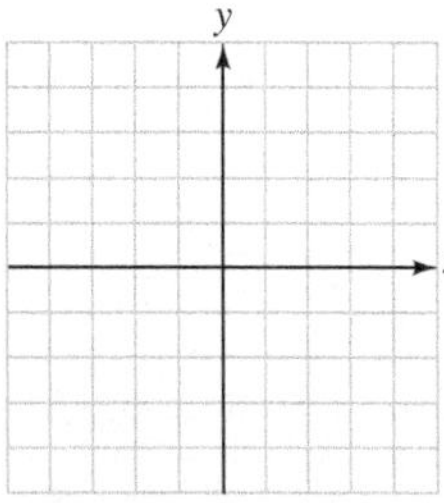

32. $y = -\frac{1}{2}x$

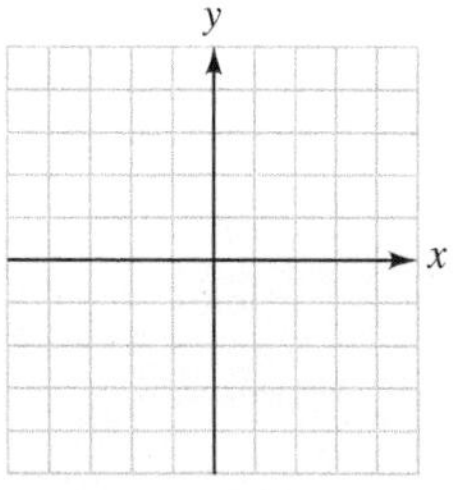

33. $y = 2x - 1$

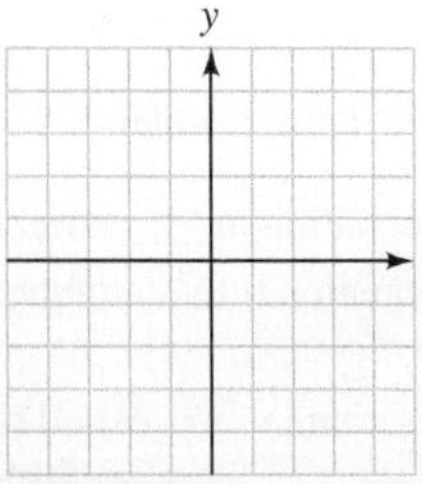

34. $y = 3x + 1$

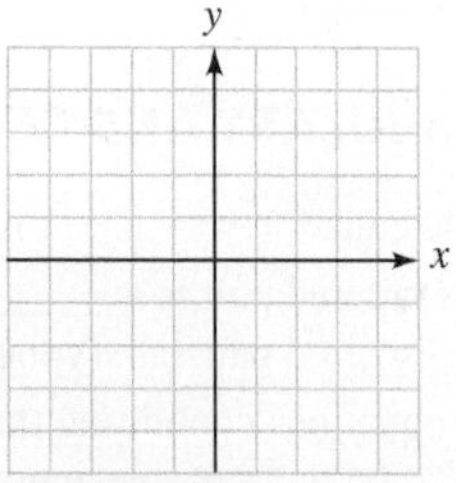

Graph each equation by constructing a table of values and then plotting the points. SEE EXAMPLE 3. (OBJECTIVE 3)

35. $y = 1.2x - 2$

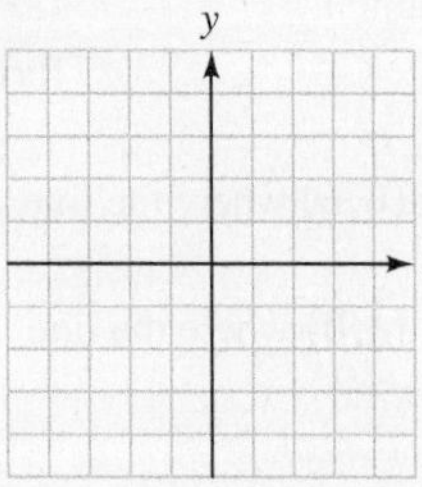

36. $y = -2.4x + 1$

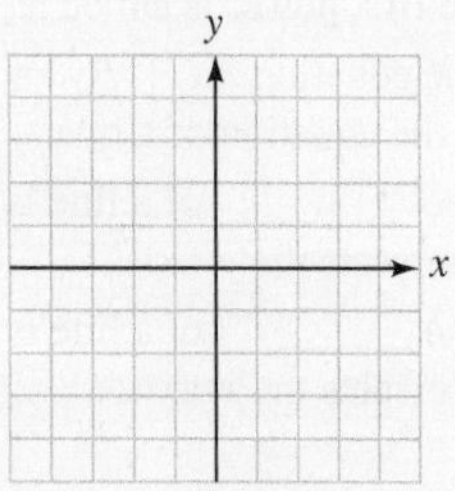

37. $y = 2.5x - 5$

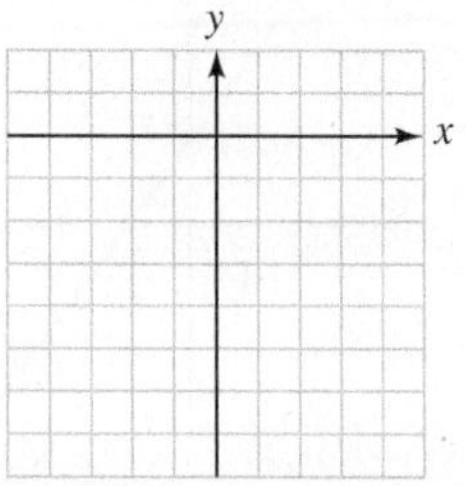

38. $y = x$

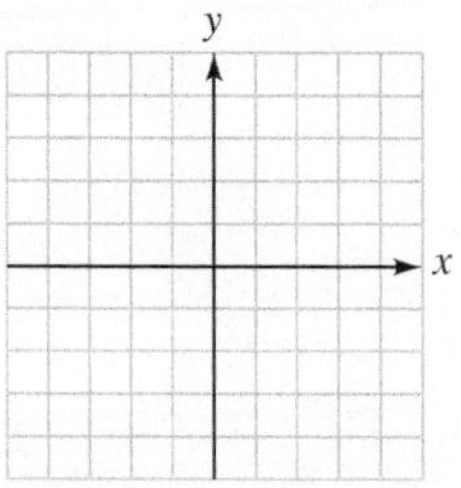

Graph each equation by constructing a table of values and then plotting the points. SEE EXAMPLE 4. (OBJECTIVE 3)

39. $y = \frac{x}{2} - 2$

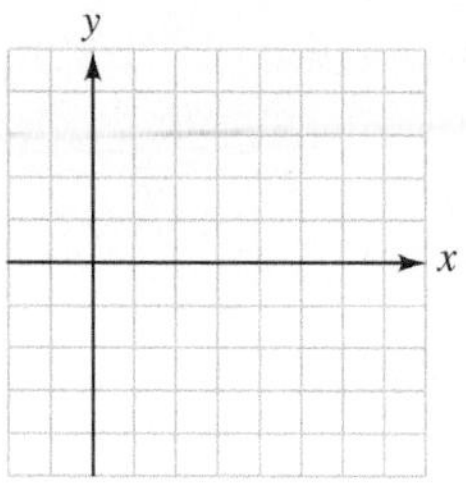

40. $y = \frac{x}{3} - 3$

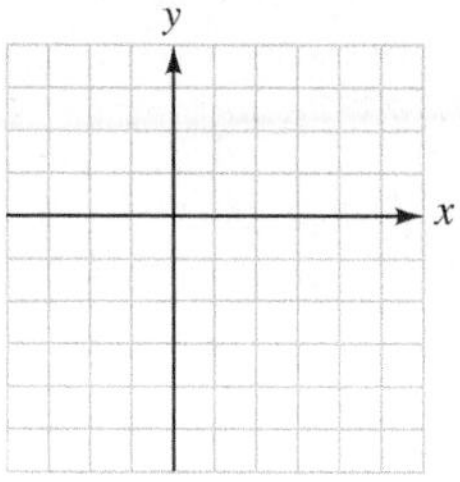

41. $y - 3 = -\frac{1}{2}(2x + 4)$

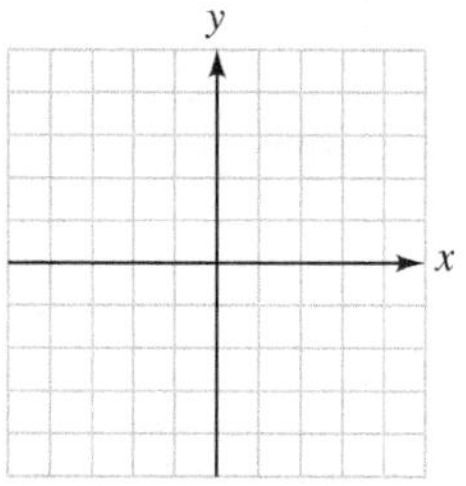

42. $y + 1 = 3(x - 1)$

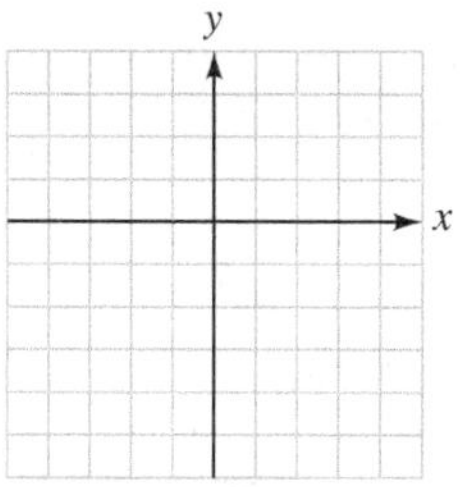

Graph each equation using the intercept method. Write the equation in general form, if necessary. SEE EXAMPLE 5. (OBJECTIVE 4)

43. $x + y = 7$

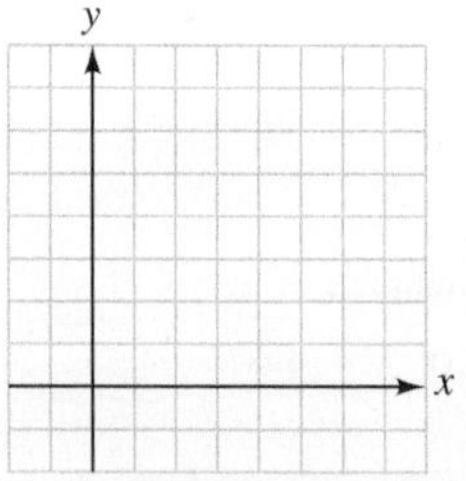

44. $x + y = -2$

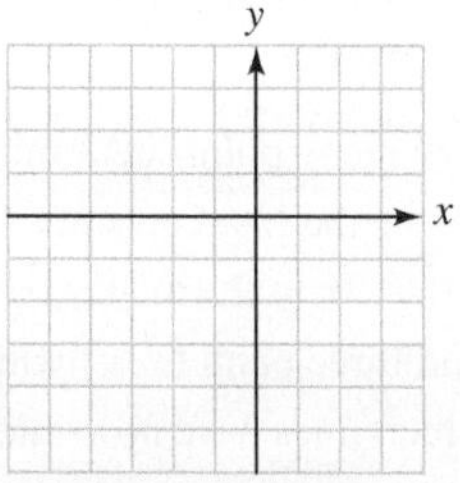

45. $2x + 3y = 12$

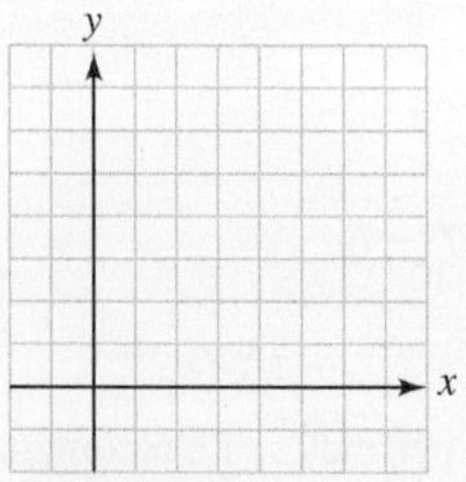

46. $3x - 2y = 6$

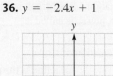

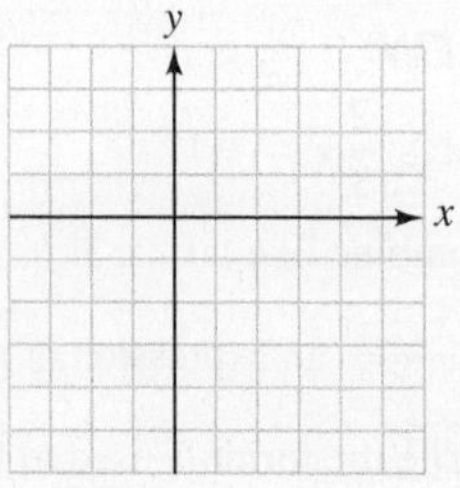

Unless otherwise noted, all content on this page is © Cengage Learning.

Graph each equation. SEE EXAMPLE 6. (OBJECTIVE 5)

47. $y = -5$

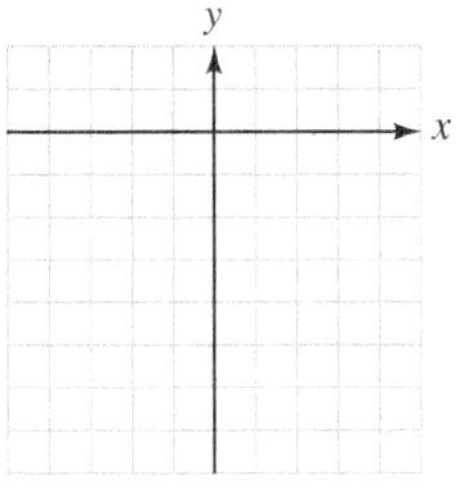

48. $x = 4$

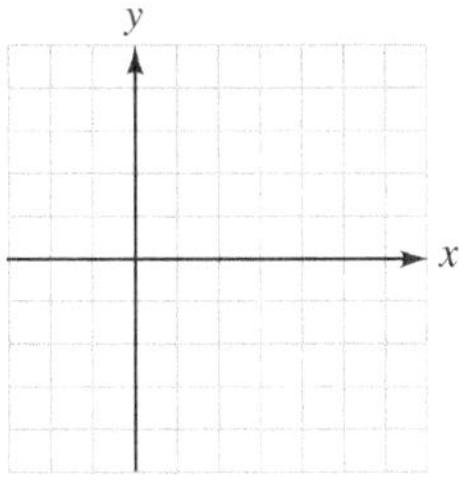

49. $x = 5$

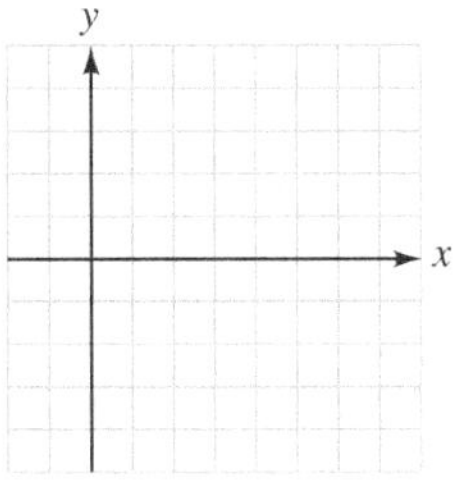

50. $y = 4$

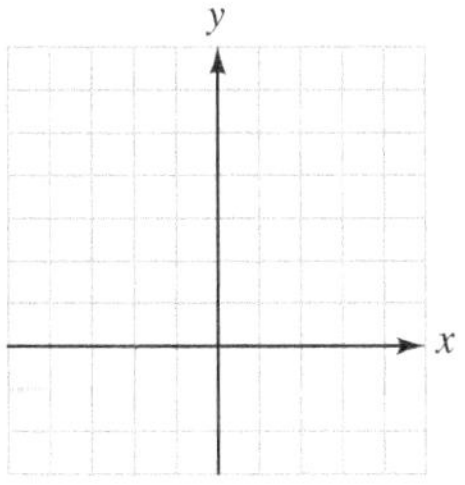

59. $y = -3x$

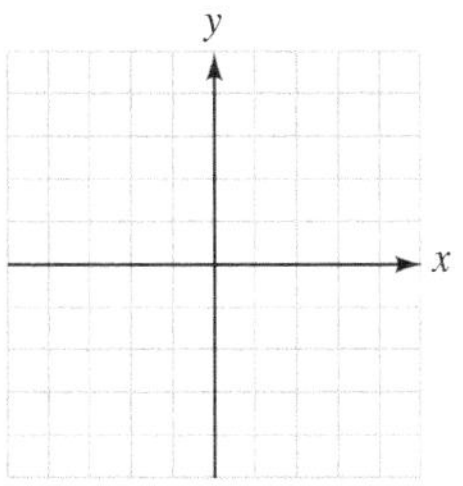

60. $x + y = -2$

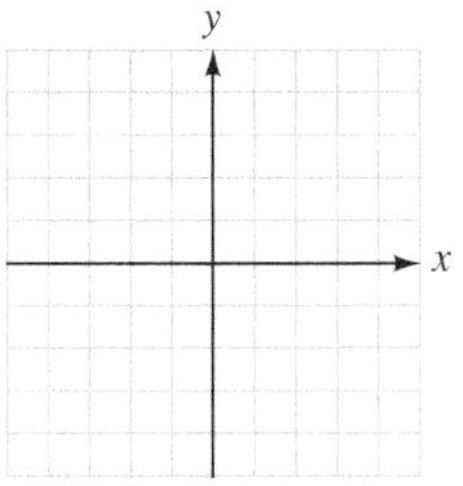

61. $y + 2 = \frac{3}{4}(4x + 8)$

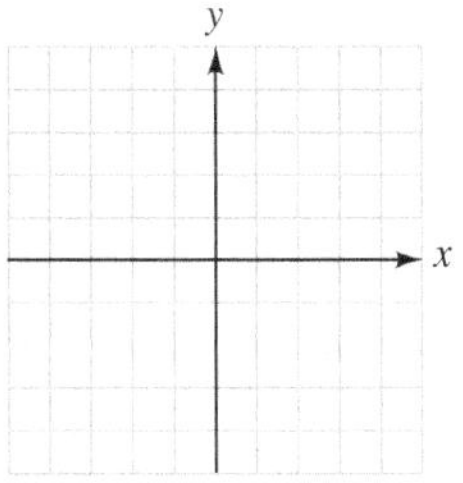

62. $y = 4.5x + 2$

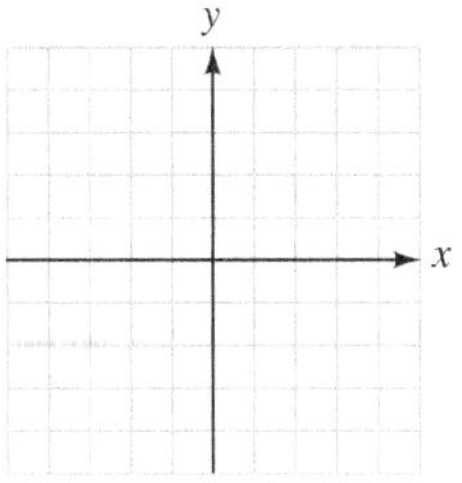

ADDITIONAL PRACTICE *Graph each equation using any method.*

51. $y = -3x - 1$

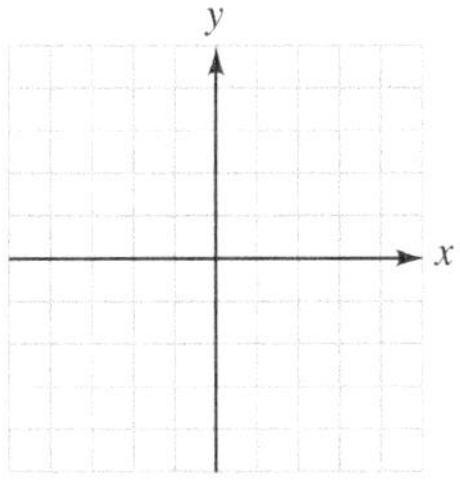

52. $2x = 5$

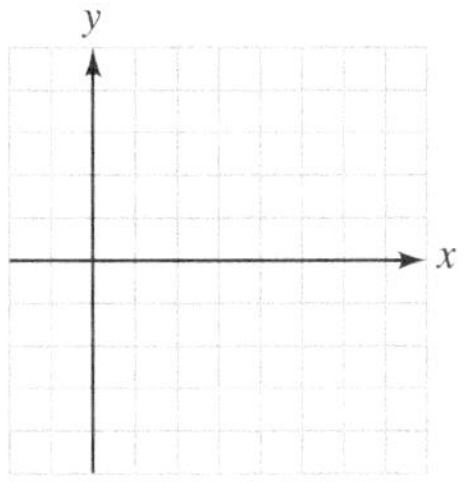

53. $x - y = -2$

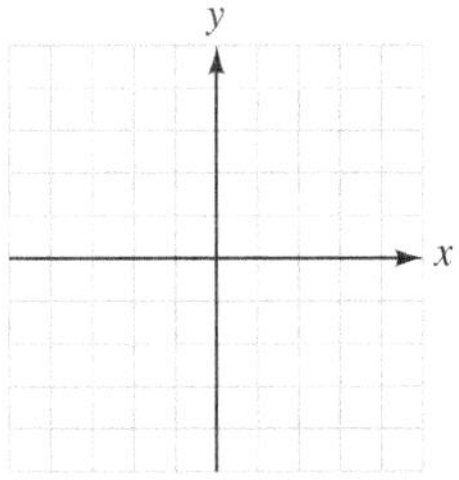

54. $y = 0$

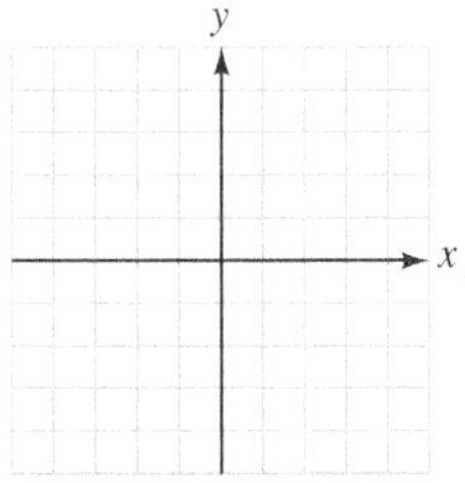

55. $3y = 7$

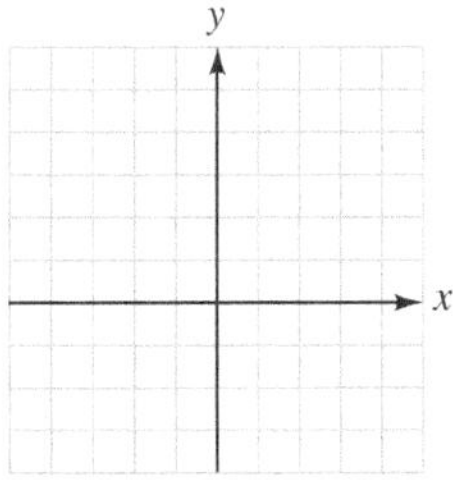

56. $x = 0$

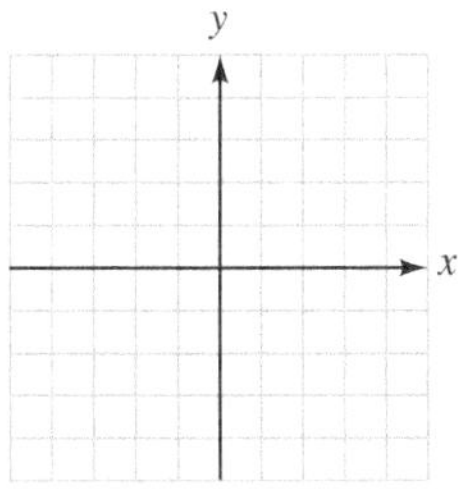

57. $x - y = 7$

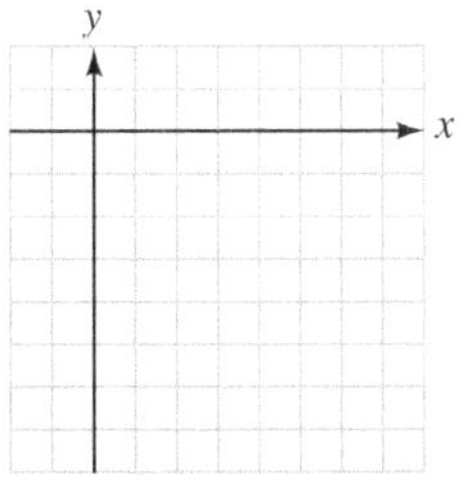

58. $y = -2x + 5$

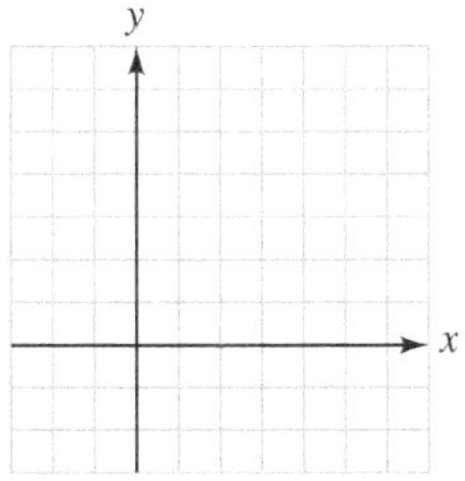

APPLICATIONS SEE EXAMPLE 7. (OBJECTIVE 6)

63. Educational costs Each semester, a college charges a service fee of \$50 plus \$25 for each unit taken by a student.

a. Write a linear equation that gives the total enrollment cost c for a student taking u units.

b. Complete the table of values and graph the equation. See the illustration.

c. What does the y-intercept of the line tell you?

d. Use the graph to find the total cost for a student taking 18 units the first semester and 12 units the second semester.

u	c	(u, c)
4		
8		
14		

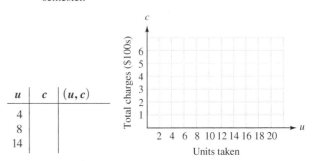

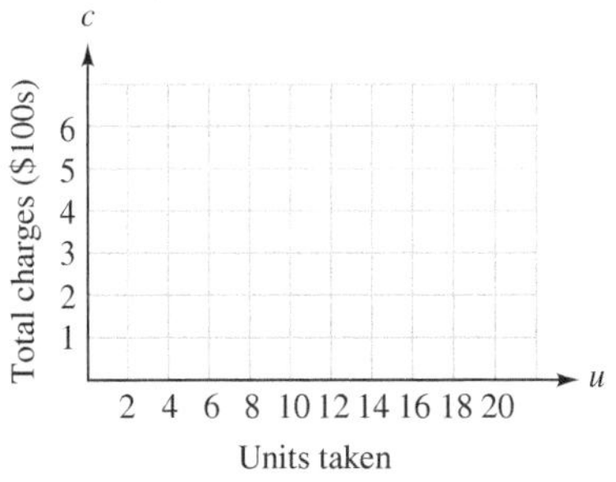

64. Group rates To promote the sale of tickets for a cruise to Alaska, a travel agency reduces the regular ticket price of \$3,000 by \$5 for each individual traveling in the group.

a. Write a linear equation that would find the ticket price T for the cruise if a group of p people travel together.

b. Complete the table of values and then graph the equation. See the illustration on the following page.

c. As the size of the group increases, what happens to the ticket price?

d. Use the graph to determine the cost of an individual ticket if a group of 40 will be traveling together.

Unless otherwise noted, all content on this page is © Cengage Learning.

p	T	(p, T)
10		
30		
60		

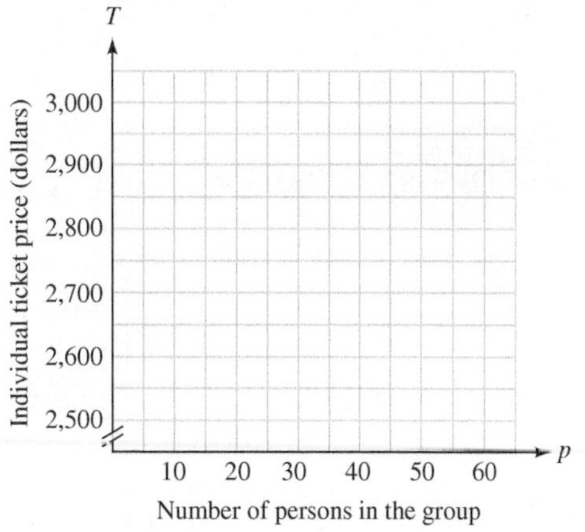

65. Physiology Physiologists have found that a woman's height h in inches can be approximated using the linear equation $h = 3.9r + 28.9$, where r represents the length of her radius bone in inches.

a. Complete the table of values (round to the nearest tenth), and then graph the equation.
b. Complete this sentence: From the graph, we see that the longer the radius bone, the . . .
c. From the graph, estimate the height of a girl whose radius bone is 10 inches long.

r	h	(r, h)
7		(7,)
8.5		(8.5,)
9		(9,)

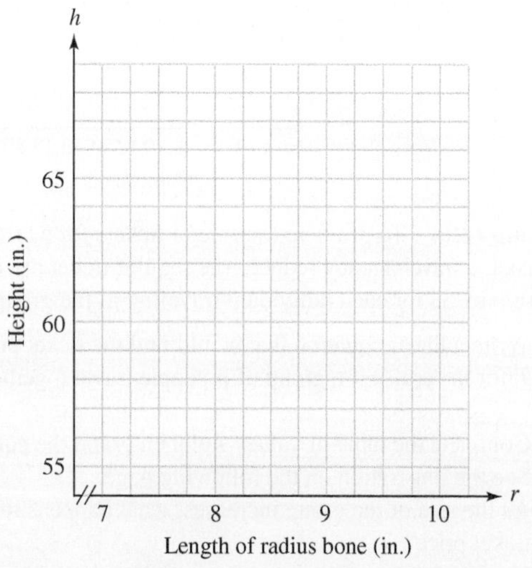

66. Research A psychology major found that the time t in seconds that it took a white rat to complete a maze was related to the number of trials n the rat had been given by the equation $t = 25 - 0.25n$.

a. Complete the table of values and then graph the equation.
b. Complete this sentence: From the graph, we see that the more trials the rat had, the . . .
c. From the graph, estimate the time it will take the rat to complete the maze on its 32nd trial.
d. Interpret the meaning of the y-intercept.

n	t	(n, t)
4		
12		
16		

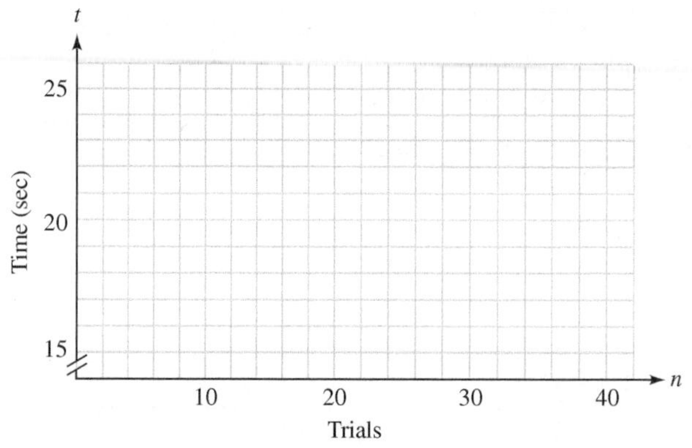

WRITING ABOUT MATH

67. From geometry, we know that two points determine a line. Explain why it is good practice when graphing linear equations to find and plot three points instead of just two.

68. Explain the process used to find the x- and y-intercepts of the graph of a line.

69. What is a table of values? Why is it often called a table of solutions?

70. When graphing an equation in two variables, how many solutions of the equation must be found?

71. Give examples of an equation in one variable and an equation in two variables. How do their solutions differ?

72. What does it mean when we say that an equation in two variables has infinitely many solutions?

SOMETHING TO THINK ABOUT

If points $P(a, b)$ and $Q(c, d)$ are two points on a rectangular coordinate system and point M is midway between them, then point M is called the midpoint of the line segment joining P and Q. (See the illustration on the following page.) To find the coordinates of the midpoint $M(x_M, y_M)$ of the segment PQ, we find the average of the x-coordinates and the average of the y-coordinates of P and Q.

$$x_M = \frac{a + c}{2} \quad \text{and} \quad y_M = \frac{b + d}{2}$$

Unless otherwise noted, all content on this page is © Cengage Learning.

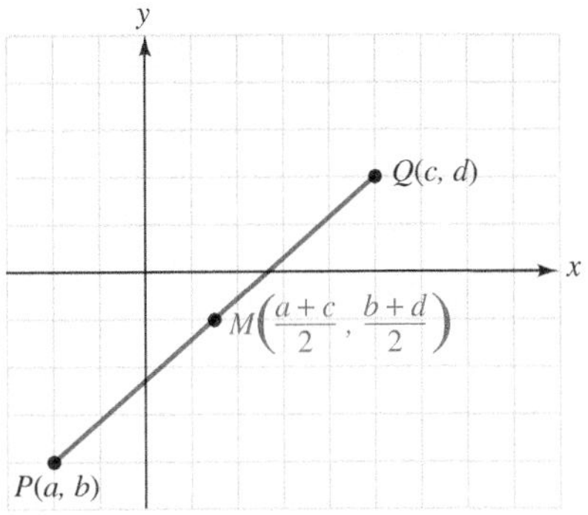

Find the coordinates of the midpoint of the line segment with the given endpoints.

73. $P(5, 3)$ and $Q(7, 9)$

74. $P(3, 8)$ and $Q(9, -2)$

75. $P(2, -7)$ and $Q(-3, 12)$

76. $P(-8, 12)$ and $Q(3, -9)$

77. $A(4, 6)$ and $B(10, 6)$

78. $A(8, -6)$ and the origin

79. $A(x, 3)$ and $B(x - 1, -4)$

80. $A(-2, y + 1)$ and $B(6, y - 1)$

Section 3.3 Slope of a Line

Objectives

1. Find the slope of a line given a graph.
2. Find the slope of a line passing through two specified points.
3. Find the slope of a line given an equation.
4. Identify the slope of a horizontal and vertical line.
5. Determine whether two lines are parallel, perpendicular, or neither parallel nor perpendicular.
6. Interpret slope in an application.

Vocabulary

slope
subscript notation
rise
run
hypotenuse
parallel lines
negative reciprocals

Getting Ready

Simplify each expression, if possible.

1. $\dfrac{6 - 2}{12 - 8}$ **2.** $\dfrac{-12 - 3}{11 - 8}$ **3.** $\dfrac{16 - 16}{6 - 2}$ **4.** $\dfrac{2 - 9}{7 - 7}$

We have seen that two points can be used to graph a line. We can also graph a line if we know the coordinates of only one point and the slant (or steepness) of the line. A measure of this slant is called the **slope** of the line. Slope will be discussed in this section.

Unless otherwise noted, all content on this page is © Cengage Learning.

1 Find the slope of a line given a graph.

A research service offered by an Internet company costs \$2 per month plus \$3 for each hour of connect time. The table shown in Figure 3-21(a) gives the cost y for different hours x of connect time. If we construct a graph from this data, we obtain the line shown in Figure 3-21(b).

x (hours)	0	1	2	3	4	5
y (\$)	2	5	8	11	14	17

(a)

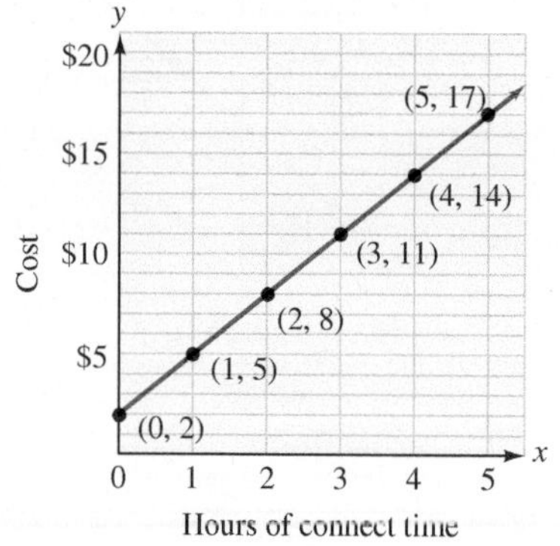

(b)

Figure 3-21

From the graph, we can see that if x changes from 0 to 1, y changes from 2 to 5. As x changes from 1 to 2, y changes from 5 to 8, and so on. The ratio of the change in y divided by the change in x is the constant 3.

$$\frac{\text{Change in } y}{\text{Change in } x} = \frac{5-2}{1-0} = \frac{8-5}{2-1} = \frac{11-8}{3-2} = \frac{14-11}{4-3} = \frac{17-14}{5-4} = \frac{3}{1} = 3$$

The ratio of the change in y divided by the change in x between any two points on any line is always a constant. This constant *rate of change* is called the slope of the line and usually is denoted by the letter m.

To distinguish between the coordinates of points P and Q in Figure 3-22, we use **subscript notation**. Point P is denoted as $P(x_1, y_1)$ and is read as "point P with coordinates of x sub 1 and y sub 1." Point Q is denoted as $Q(x_2, y_2)$ and is read as "point Q with coordinates of x sub 2 and y sub 2."

As a point on the line in Figure 3-22 moves from P to Q, its y-coordinate changes by the amount $y_2 - y_1$, and its x-coordinate changes by $x_2 - x_1$. The change in y is often called the **rise** of the line between points P and Q, and the change in x is often called the **run**.

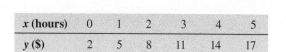

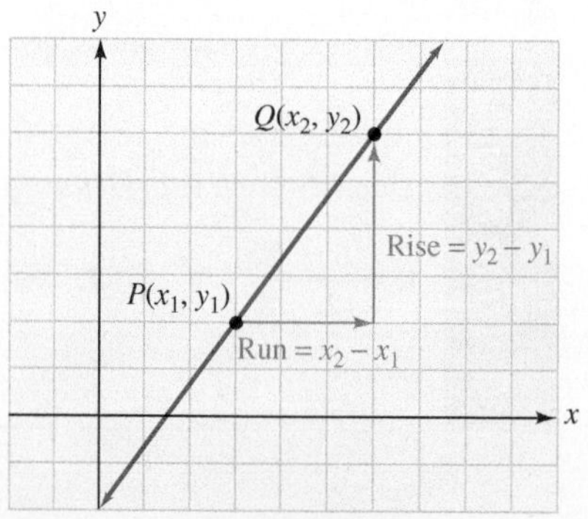

Figure 3-22

Unless otherwise noted, all content on this page is © Cengage Learning.

EXAMPLE 1 Find the slope of the line shown in Figure 3-23(a).

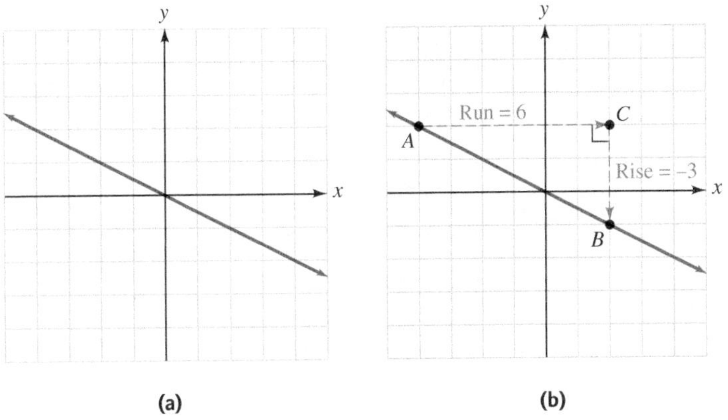

Figure 3-23

Solution In Figure 3-23(b), we choose two points on the line and call them A and B. Then we draw right triangle ABC, having a horizontal leg and a vertical leg. The longest side AB of the right triangle is called the **hypotenuse**. As we move from A to B, we move to the right, a run of 6, and then down, a rise of -3. To find the slope of the line, we write a ratio.

$m = \frac{\text{rise}}{\text{run}}$ The slope of a line is the ratio of the rise to the run.

$m = \frac{-3}{6}$ From Figure 3-23(b), the rise is -3 and the run is 6.

$m = -\frac{1}{2}$ Simplify the fraction.

The slope of the line is $-\frac{1}{2}$.

SELF CHECK 1 Find the slope of the line shown in Figure 3-23(a) using two points different from those used in Example 1.

2 Find the slope of a line passing through two specified points.

Once we know the coordinates of two points on a line, we can substitute those coordinates into the *slope formula*.

SLOPE OF A NONVERTICAL LINE

The slope of the nonvertical line passing through points $P(x_1, y_1)$ and $Q(x_2, y_2)$ is

$$m = \frac{\text{change in } y}{\text{change in } x} = \frac{rise}{run} = \frac{y_2 - y_1}{x_2 - x_1} \quad (x_2 \neq x_1)$$

COMMENT You can use the coordinates of any two points on a line to compute the slope of the line and obtain the same result.

EXAMPLE 2 Use the two points shown in Figure 3-24 on the next page to find the slope of the line passing through the points $P(-3, 2)$ and $Q(2, -5)$.

Solution We can let $P(x_1, y_1) = P(-3, 2)$ and $Q(x_2, y_2) = Q(2, -5)$. Then $x_1 = -3$, $y_1 = 2$, $x_2 = 2$, and $y_2 = -5$. To find the slope, we substitute these values into the slope formula and simplify.

Unless otherwise noted, all content on this page is © Cengage Learning.

$$m = \frac{\text{change in } y}{\text{change in } x}$$

$$= \frac{y_2 - y_1}{x_2 - x_1}$$

$$= \frac{-5 - 2}{2 - (-3)} \quad \text{Substitute } -5 \text{ for } y_2,\ 2 \text{ for } y_1,\ 2 \text{ for } x_2, \text{ and } -3 \text{ for } x_1.$$

$$= \frac{-7}{5} \quad \text{Simplify.}$$

$$= -\frac{7}{5}$$

The slope of the line is $-\frac{7}{5}$. We would obtain the same result if we had let $P(x_1, y_1) = P(2, -5)$ and $Q(x_2, y_2) = Q(-3, 2)$.

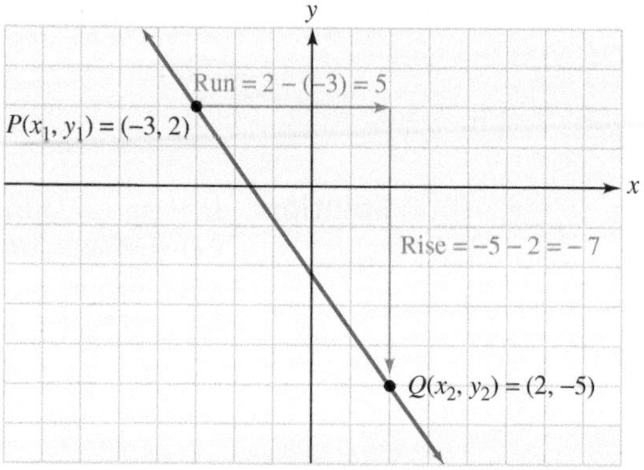

Figure 3-24

SELF CHECK 2 Find the slope of the line passing through points $P(-2, 4)$ and $Q(3, -4)$.

COMMENT When calculating slope, always subtract the y-values and the x-values *in the same order.*

$$m = \frac{y_2 - y_1}{x_2 - x_1} \qquad m = \frac{y_1 - y_2}{x_1 - x_2} \qquad \text{True.}$$

However, the following are not true:

$$m = \frac{y_2 - y_1}{x_1 - x_2} \qquad m = \frac{y_1 - y_2}{x_2 - x_1}$$

Everyday connections
Wind Power

The chart on the right reflects the capacity of California and Texas to produce megawatts of electricity from wind turbines. Although the states have the capacity to produce large amounts of electricity from wind power, the actual output is much less.

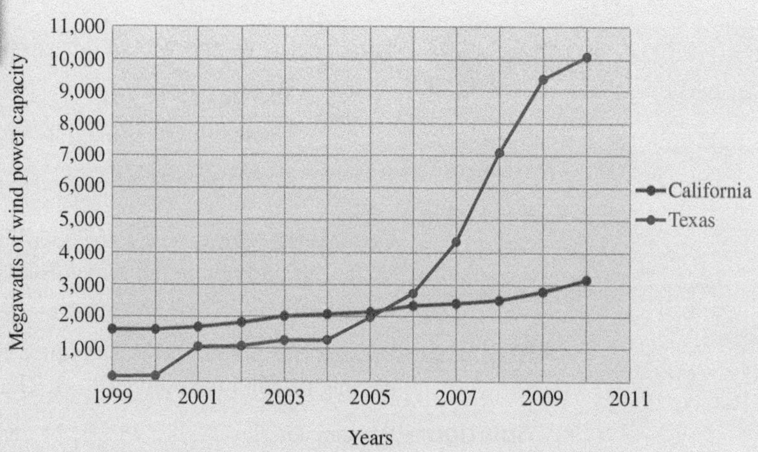

Source: http://www.windpoweringamerica.gov/wind_installed_capacity.asp

Unless otherwise noted, all content on this page is © Cengage Learning.

Use the data from the graph to answer the questions.

1. About how many megawatts of wind power could California have produced in 2003? Texas?
2. Compute the rate of change in the capacity of megawatts power in Texas between 2000 and 2001.
3. Between what two years was there the greatest rate of change in Texas? Estimate this rate.
4. What was the overall rate of change for Texas between 1999 and 2010?
5. If the actual production of megawatts is roughly 30% of the capacity, how many megawatts of power were produced by California and Texas in 2010?

3 Find the slope of a line given an equation.

If we need to find the slope of a line from a given equation, we could graph the line and count squares to determine the rise and the run. Instead, we could find the x- and y-intercepts and then use the slope formula.

In Example 3, we will find the slope of the line determined by the equation $3x - 4y = 12$. To do so, we will find the x- and y-intercepts of the line and use the slope formula.

EXAMPLE 3 Find the slope of the line determined by $3x - 4y = 12$.

Solution We first find the coordinates of the intercepts of the line.

- If $x = 0$, then $y = -3$, and the point $(0, -3)$ is on the line.
- If $y = 0$, then $x = 4$, and the point $(4, 0)$ is on the line.

We then refer to Figure 3-25 and use the slope formula to find the slope of the line passing through $(0, -3)$ and $(4, 0)$.

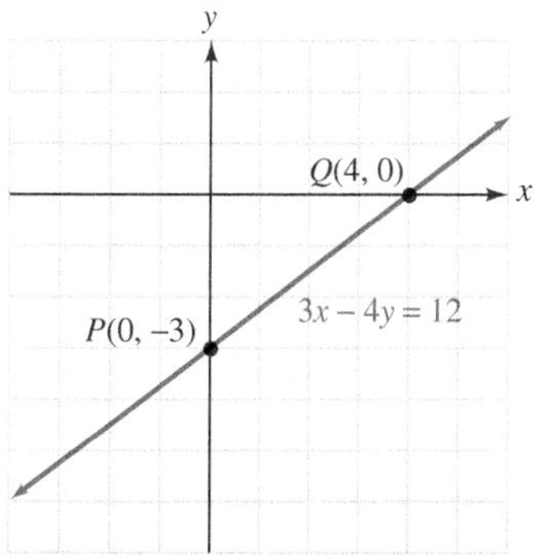

Figure 3-25

$$m = \frac{\text{change in } y}{\text{change in } x}$$

$$= \frac{y_2 - y_1}{x_2 - x_1}$$

$$= \frac{0 - (-3)}{4 - 0} \quad \text{Substitute 0 for } y_2, -3 \text{ for } y_1, 4 \text{ for } x_2, \text{ and 0 for } x_1.$$

$$= \frac{3}{4} \quad \text{Simplify.}$$

The slope of the line is $\frac{3}{4}$.

SELF CHECK 3 Find the slope of the line determined by $4x + 3y = 12$.

4 Identify the slope of a horizontal and vertical line.

If $P(x_1, y_1)$ and $Q(x_2, y_2)$ are points on the horizontal line shown in Figure 3-26(a) on the next page, then $y_1 = y_2$, and the numerator of the slope formula

$$\frac{y_2 - y_1}{x_2 - x_1} \quad \text{On a horizontal line, } x_2 \neq x_1.$$

is 0. Thus, the value of the fraction is 0, and the slope of the horizontal line is 0.

Unless otherwise noted, all content on this page is © Cengage Learning.

If $P(x_1, y_1)$ and $Q(x_2, y_2)$ are two points on the vertical line shown in Figure 3-26(b), then $x_1 = x_2$, and the denominator of the slope formula

$$\frac{y_2 - y_1}{x_2 - x_1} \quad \text{On a vertical line, } y_2 \neq y_1.$$

is 0. Since the denominator cannot be 0, a vertical line has an undefined slope.

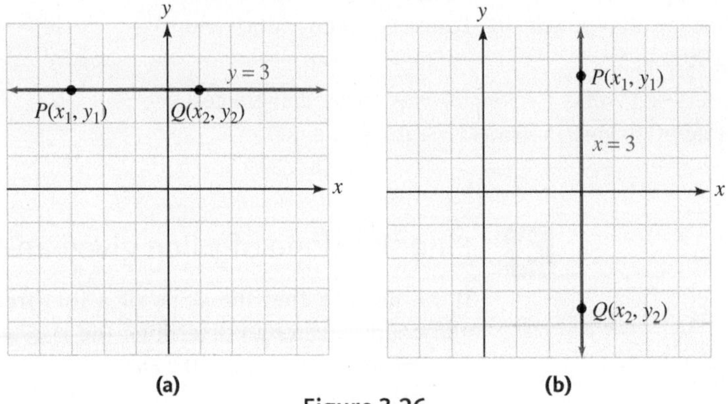

Figure 3-26

SLOPES OF HORIZONTAL AND VERTICAL LINES

All horizontal lines (lines with equations of the form $y = b$) have a slope of 0.

Vertical lines (lines with equations of the form $x = a$) have an undefined slope.

COMMENT It is also true that if a line has a slope of 0, it is a horizontal line with an equation of the form $y = b$. Also, if a line has an undefined slope, it is a vertical line with an equation of the form $x = a$.

If a line rises as we follow it from left to right as in Figure 3-27(a), the line is said to be *increasing* and its slope is positive. If a line drops as we follow it from left to right, as in Figure 3-27(b), it is said to be *decreasing* and its slope is negative. If a line is horizontal, as in Figure 3-27(c), it is said to be *constant* and its slope is 0. If a line is vertical, as in Figure 3-27(d), it has an undefined slope.

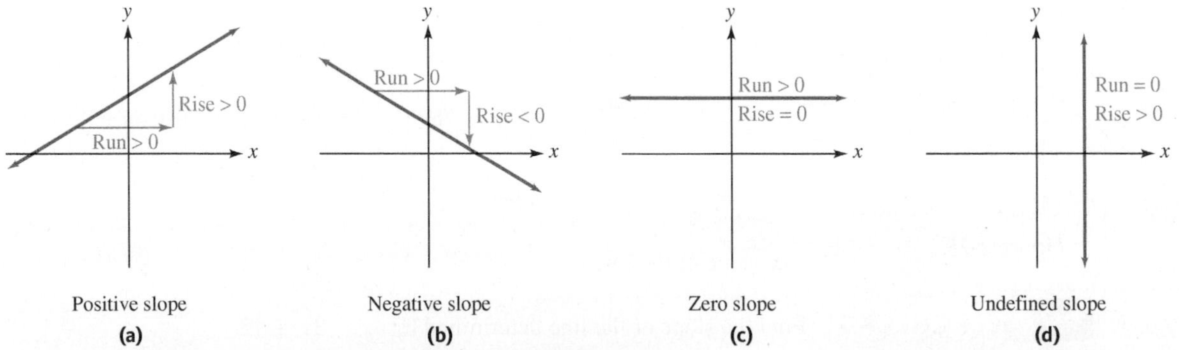

Figure 3-27

5 Determine whether two lines are parallel, perpendicular, or neither parallel nor perpendicular.

To see a relationship between parallel lines and their slopes, we refer to the parallel lines l_1 and l_2 shown in Figure 3-28, with slopes of m_1 and m_2, respectively. Because right triangles ABC and DEF are similar, it follows that

COMMENT The triangle is an uppercase delta in the Greek alphabet.

$$m_1 = \frac{\Delta y \text{ of } l_1}{\Delta x \text{ of } l_1} \quad \text{Read } \Delta y \text{ as "the change in } y\text{." Read } \Delta x \text{ as "the change in } x\text{."}$$

$$= \frac{\Delta y \text{ of } l_2}{\Delta x \text{ of } l_2}$$

$$= m_2$$

Unless otherwise noted, all content on this page is © Cengage Learning.

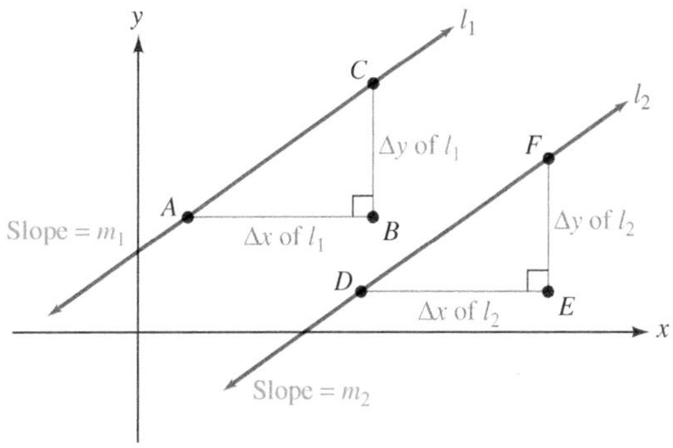

Figure 3-28

This illustrates that if two nonvertical lines are parallel, they have the same slope. It is also true that when two distinct nonvertical lines have the same slope, they are **parallel lines**.

SLOPES OF PARALLEL LINES

Nonvertical parallel lines have the same slope.

Two distinct nonvertical lines having the same slope are parallel.

Since vertical lines are parallel, two distinct lines each with an undefined slope are parallel.

EXAMPLE 4 The lines in Figure 3-29 are parallel. Find the slope of line l_2.

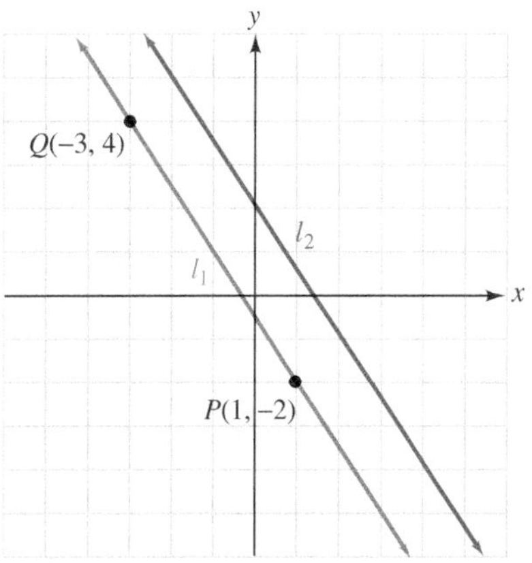

Figure 3-29

Solution From the information in the figure, we can find the slope of line l_1. Since the lines are parallel, they will have equal slopes. Therefore, the slope of line l_2 will be equal to the slope of line l_1.

$$m = \frac{y_2 - y_1}{x_2 - x_1}$$

$$m = \frac{-2 - 4}{1 - (-3)}$$

$$= \frac{-6}{4}$$

$$= -\frac{3}{2}$$

The slope of line l_1 is $-\frac{3}{2}$. Because the lines are parallel, the slope of line l_2 is also equal to $-\frac{3}{2}$.

SELF CHECK 4 Find the slope of any line parallel to a line with a slope of 6.

Unless otherwise noted, all content on this page is © Cengage Learning.

Two real numbers a and b are called **negative reciprocals** or opposite reciprocals if $a \cdot b = -1$. For example,

$$-\frac{4}{3} \quad \text{and} \quad \frac{3}{4}$$

are negative reciprocals, because $-\frac{4}{3}\left(\frac{3}{4}\right) = -1$.

The following relates perpendicular lines and their slopes.

SLOPES OF PERPENDICULAR LINES

If two nonvertical lines are perpendicular, their slopes are negative reciprocals.

If the slopes of two lines are negative reciprocals, the lines are perpendicular.

Because a horizontal line is perpendicular to a vertical line, a line with a slope of 0 is perpendicular to a line with an undefined slope.

EXAMPLE 5 Are the lines shown in Figure 3-30 perpendicular?

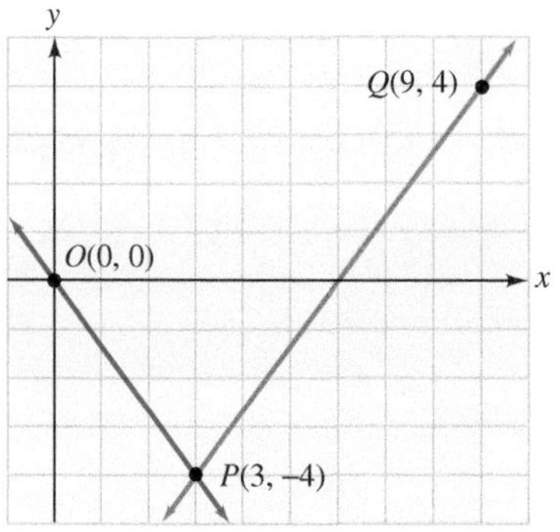

Figure 3-30

Solution We find the slopes of the lines to see whether they are negative reciprocals.

$$\text{Slope of } OP = \frac{\Delta y}{\Delta x} = \frac{y_2 - y_1}{x_2 - x_1} = \frac{\mathbf{-4 - 0}}{\mathbf{3 - 0}} = -\frac{4}{3}$$

$$\text{Slope of } PQ = \frac{\Delta y}{\Delta x} = \frac{y_2 - y_1}{x_2 - x_1} = \frac{\mathbf{4 - (-4)}}{\mathbf{9 - 3}} = \frac{8}{6} = \frac{4}{3}$$

Since their slopes are not negative reciprocals, the lines are not perpendicular.

SELF CHECK 5 In Figure 3-30, is the line PQ perpendicular to a line passing through the points $(3, -4)$ and $(0, -1)$?

6 Interpret slope in an application.

Many applications involve equations of lines and their slopes.

Unless otherwise noted, all content on this page is © Cengage Learning.

EXAMPLE 6 **COST OF CARPET** If carpet costs \$25 per square yard, the total cost c of n square yards is the price per square yard times the number of square yards purchased.

c	equals	the cost per square yard	times	the number of square yards.
c	$=$	25	$\cdot$	n

Graph the equation $c = 25n$ and interpret the slope of the line.

Solution We can graph the equation on a coordinate system with a vertical c-axis and a horizontal n-axis. Figure 3-31 shows a table of ordered pairs and the graph.

$c = 25n$

n	c	(n, c)
10	250	(10, 250)
20	500	(20, 500)
30	750	(30, 750)
40	1,000	(40, 1,000)
50	1,250	(50, 1,250)

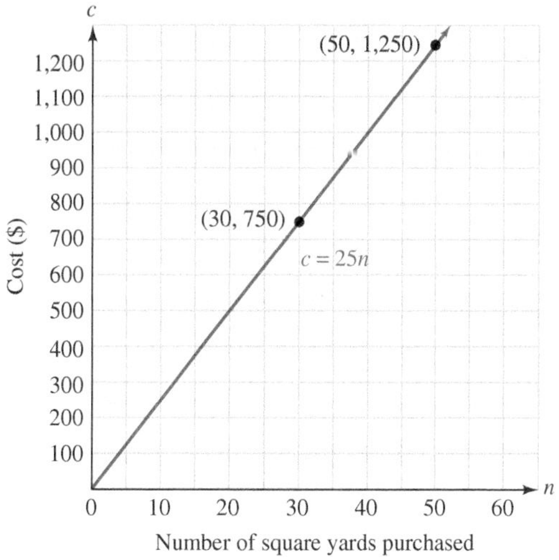

Figure 3-31

COMMENT Recall that any two points on a line can be used to find the slope. In this example, we chose (30, 750) and (50, 1,250).

If we choose the points (30, 750) and (50, 1,250) and substitute into the slope formula, we have

$$m = \frac{\text{change in } c}{\text{change in } n}$$

$$= \frac{c_2 - c_1}{n_2 - n_1}$$

$$= \frac{1{,}250 - 750}{50 - 30} \quad \text{Substitute 1,250 for } c_2\text{, 750 for } c_1\text{, 50 for } n_2\text{, and 30 for } n_1.$$

$$= \frac{500}{20} \quad \text{Subtract.}$$

$$= 25$$

The slope of 25 is the ratio of the change in the cost to the change in the number of square yards purchased. As a rate of change, the cost of carpet is 25 dollars per square yard.

SELF CHECK 6 If carpet costs \$30 per square yard, the total cost c of n square yards is $c = 30n$. Graph and interpret the slope of the line.

EXAMPLE 7 **RATE OF DESCENT** It takes a skier 25 minutes to complete the course shown in Figure 3-32 on the next page. Find his average rate of descent in feet per minute.

Solution To find the average rate of descent, we must find the ratio of the change in altitude to the change in time. To find this ratio, we calculate the slope of the line passing through the points (0, 12,000) and (25, 8,500).

Unless otherwise noted, all content on this page is © Cengage Learning.

$$\text{Slope} = \frac{12{,}000 - 8{,}500}{0 - 25} = \frac{3{,}500}{-25} = -140$$

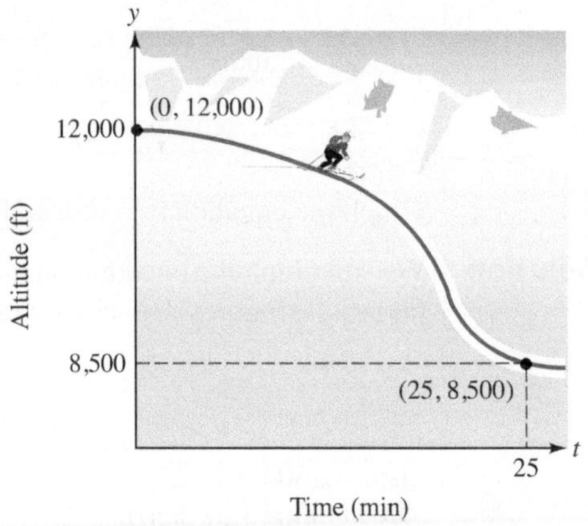

Figure 3-32

Although the slope is −140, the rate of change of descent is 140 ft/min. The term *descent* automatically determines the direction (negative).

SELF CHECK 7 In Figure 3-32, interpret the ordered pair (0, 12,000).

SELF CHECK ANSWERS

1. $-\frac{1}{2}$ **2.** $-\frac{8}{5}$ **3.** $-\frac{4}{3}$ **4.** 6 **5.** no **6.** The slope of 30 is the ratio of the change in the cost to the change in the number of square yards purchased. As a rate of change, the cost of carpet is 30 dollars per square yard.

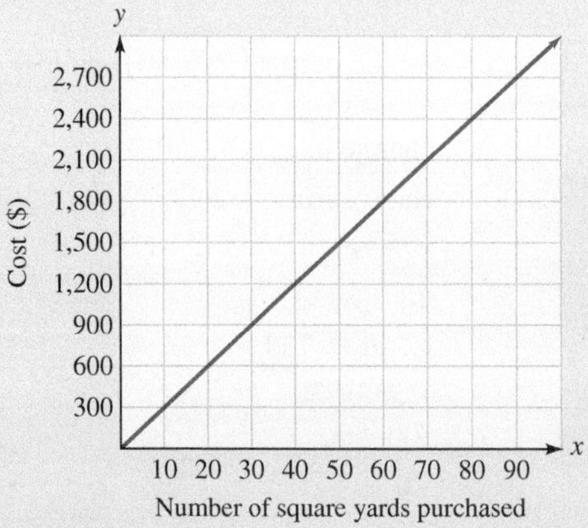

7. The skier begins at an elevation of 12,000 feet.

NOW TRY THIS

1. Interpret the sign of the slope of the three sections of the graph shown at right.

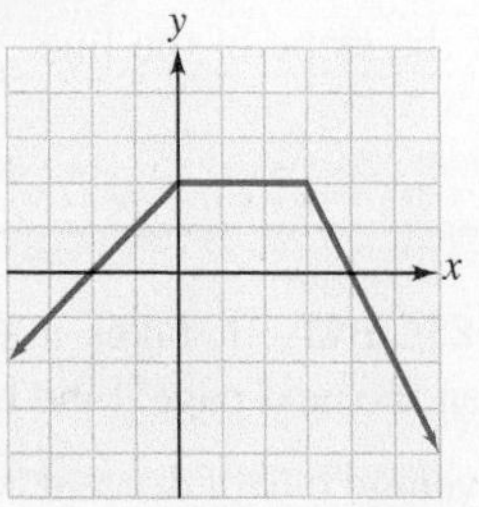

2. Find the slope of any line parallel to the line shown.

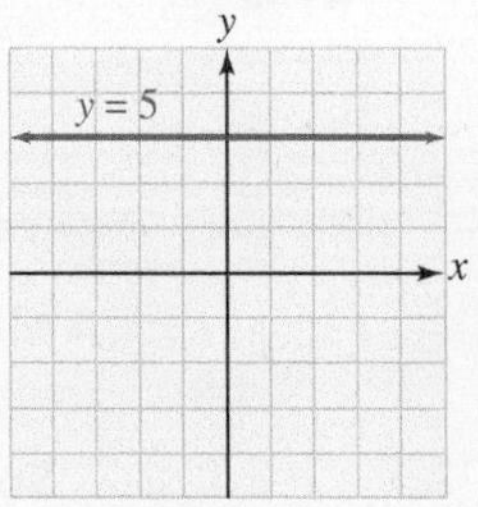

3. Find the slope, if any, of any line perpendicular to the graph shown in Problem 2.

Unless otherwise noted, all content on this page is © Cengage Learning.

4. The lines in the graph are parallel. Find the value of y.

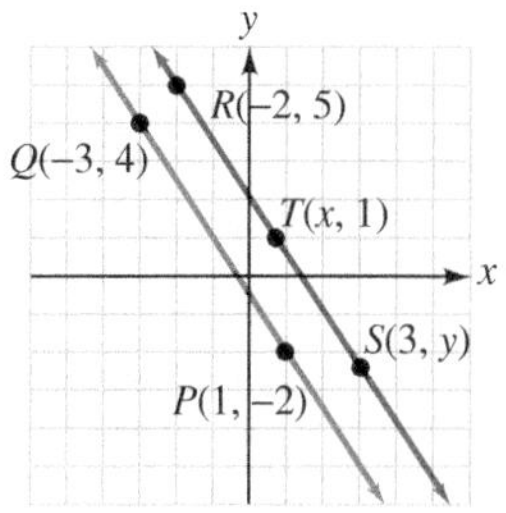

5. Graph the line passing through the points $(2.1, -3)$ and $(4.7, 1.6)$. Find the slope of this line.

3.3 Exercises

WARM-UPS *Simplify each expression.*

1. $\dfrac{3-7}{-6-4}$

2. $\dfrac{8-(-1)}{-5-(-2)}$

3. $\dfrac{9-(-8)}{-2-(-2)}$

4. $\dfrac{-7+7}{12-8}$

REVIEW *Simplify each expression.*

5. $4(a-3)+2a$

6. $5(y-8)-2y$

7. $4z-6(z+w)+2w$

8. $7(b-3)-(b+2)$

9. $3(a-b)-2(a+b)$

10. $2m-4(m-n)+2n$

VOCABULARY AND CONCEPTS *Fill in the blanks.*

11. The slope of a line is the change in _ divided by the change in _.
12. The point (x_1, y_1) is read as "the ordered pair x ___ 1, y ___ 1."
13. The vertical change between two points is called the ___. The horizontal change between two points is called the ___. Slope is sometimes defined as ___ over ___.
14. The slope of a ________ line is 0. The slope of a vertical line is ________.
15. The longest side of a right triangle is called the _________.
16. ______ lines are lines that have the same slope.
17. Slopes of nonvertical ___________ lines are negative reciprocals.
18. If a line rises as x gets larger, the line has a ______ slope.
19. If a line has a positive slope, the line is said to be ________. If a line has a negative slope, the line is said to be ________.
20. If the product of two numbers is -1, the numbers are said to be _______________.

GUIDED PRACTICE

Find the slope of each line. SEE EXAMPLE 1. (OBJECTIVE 1)

21.

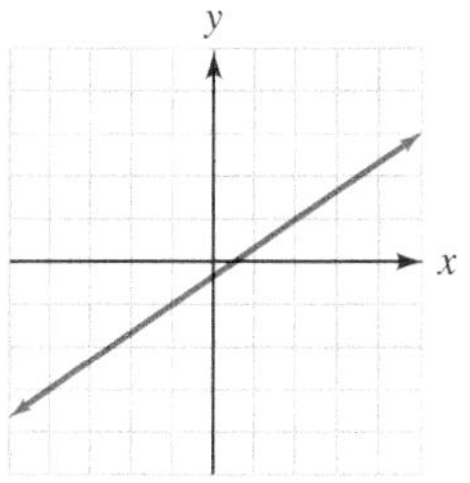

22.

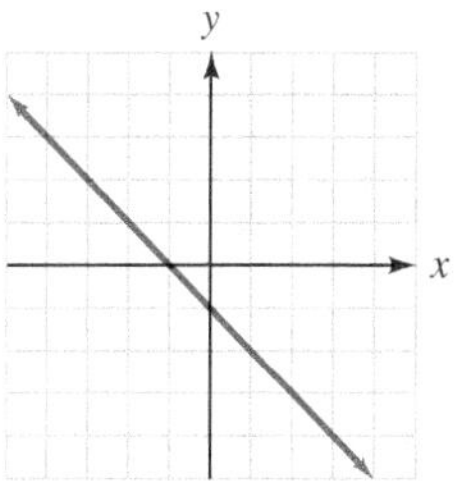

23.

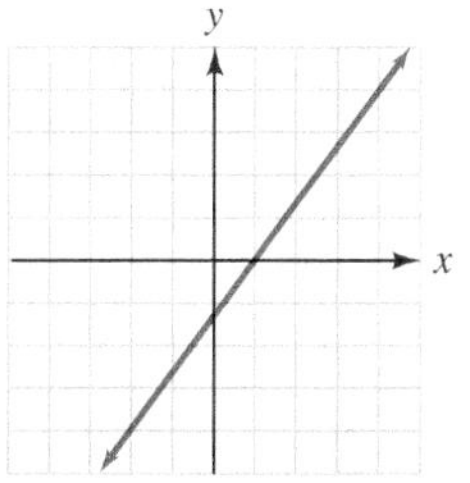

24.

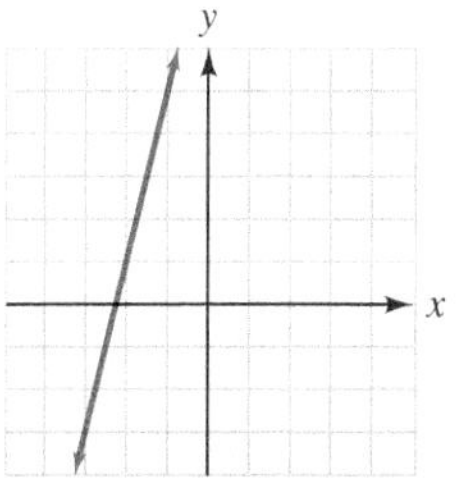

25.

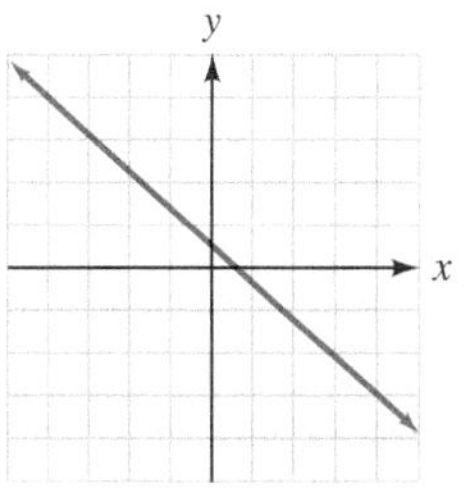

26.

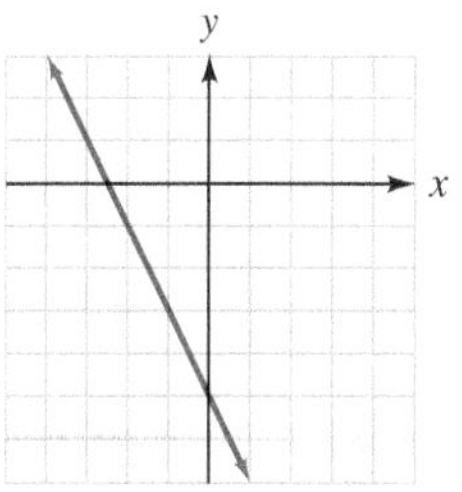

Unless otherwise noted, all content on this page is © Cengage Learning.

27.

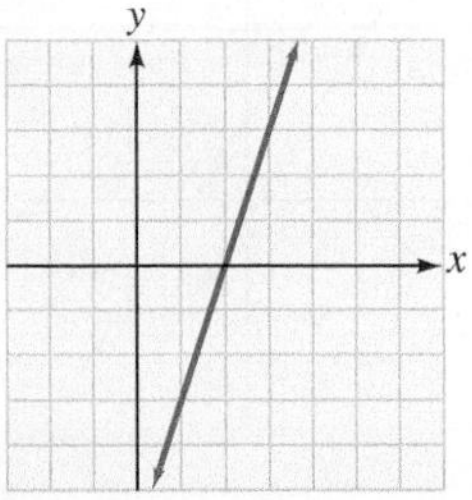

28.

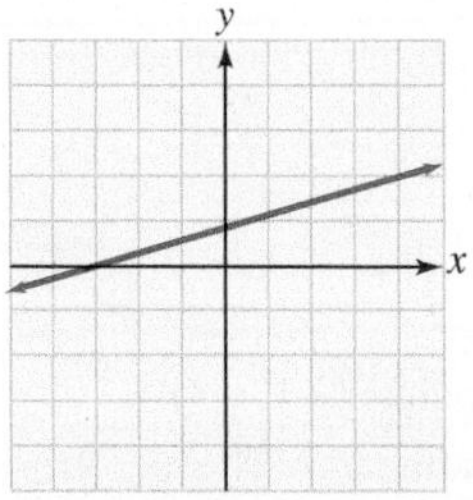

Find the slope of the line that passes through the given points. SEE EXAMPLE 2. (OBJECTIVE 2)

29. $(0, 0), (3, 9)$ **30.** $(9, 6), (0, 0)$
31. $(2, -5), (4, -3)$ **32.** $(-2, -8), (3, 2)$
33. $(3, -1), (-6, 2)$ **34.** $(0, -8), (-5, 0)$
35. $(4, 15), (-6, -11)$ **36.** $(-6, -2), (-1, -9)$

Find the slope of the line determined by each equation. SEE EXAMPLE 3. (OBJECTIVE 3)

37. $3x + 2y = 12$ **38.** $2x - y = 6$
39. $2x = 5y - 10$ **40.** $x = y$
41. $y = \frac{x - 4}{2}$ **42.** $x = \frac{3 - y}{4}$
43. $y = 5x - 4$ **44.** $y = -\frac{2}{3}x + 6$

Find the slope of each vertical or horizontal line. (OBJECTIVE 4)

45.

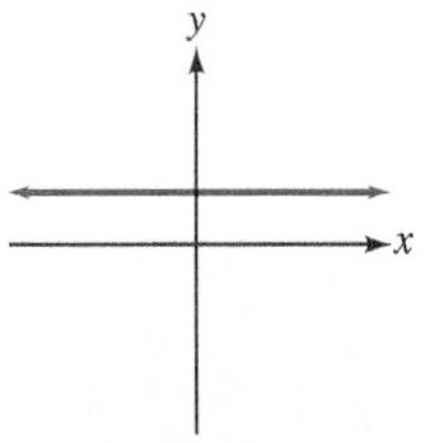

46.

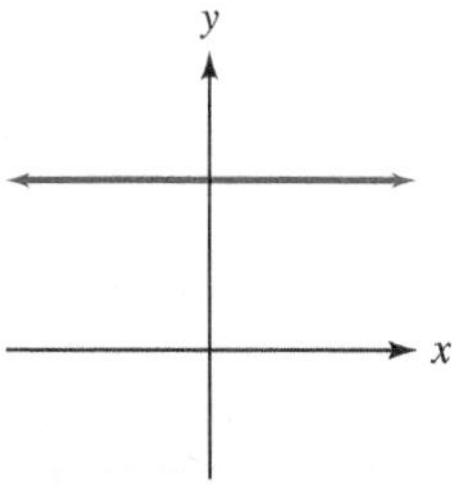

47. passing through $(-7, -5)$ and $(-7, -2)$
48. passing through $(3, -5)$ and $(3, 14)$
49. passing through $(6, -2)$ and $(-4, -2)$
50. passing through $(-3, 5)$ and $(9, 5)$
51. $4x - 2 = 12$ **52.** $-8y + 7 = 15$
53. $2(y + 1) = 3y$ **54.** $5x + 2 = 6(x + 1)$
55. $x - y = \frac{7 - 5y}{5}$ **56.** $x + y = \frac{2 + 3x}{3}$

Determine whether the lines with the given slopes are parallel, perpendicular, or neither parallel nor perpendicular. SEE EXAMPLES 4–5. (OBJECTIVE 5)

57. $m_1 = 3, m_2 = -\frac{1}{3}$ **58.** $m_1 = \frac{1}{4}, m_2 = 4$
59. $m_1 = 4, m_2 = 0.25$ **60.** $m_1 = -5, m_2 = \frac{1}{-0.2}$
61. $m_1 = \frac{2}{3}, m_2 = \frac{4}{6}$ **62.** $m_1 = \frac{5}{6}, m_2 = \frac{6}{5}$
63. $m_1 = 0$, m_2 is undefined **64.** $m_1 = 0, m_2 = 0$

ADDITIONAL PRACTICE *Find the slope of the line that passes through the given points.*

65. $(2, -1), (-3, 2)$ **66.** $(2, -8), (3, -8)$
67. $(2, -3), (-3, 2)$ **68.** $(-4, 1), (-1, 4)$

Determine whether the slope of the line in each graph is positive, negative, 0, or undefined.

69.

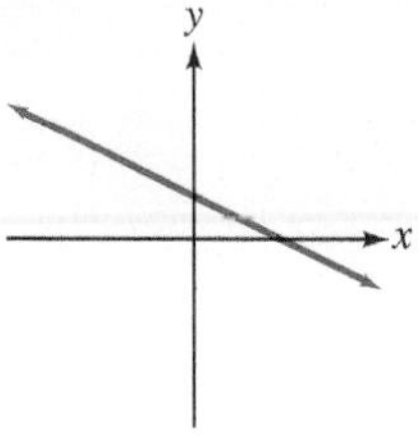

70.

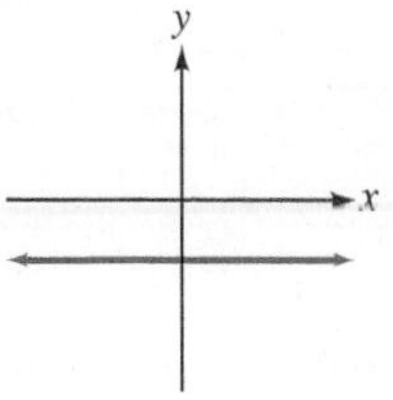

71.

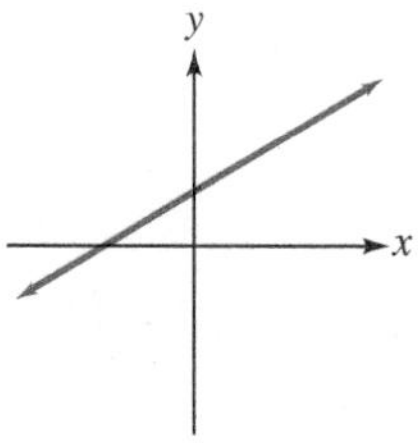

72.

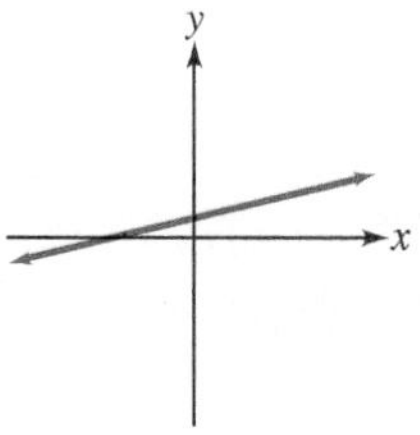

73.

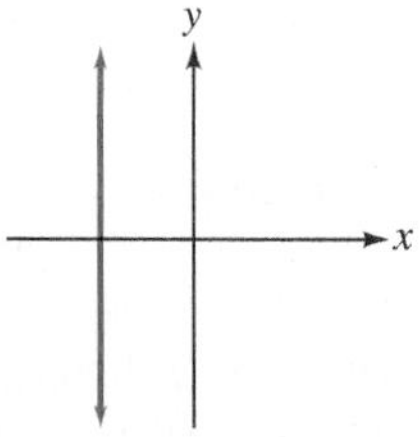

74.

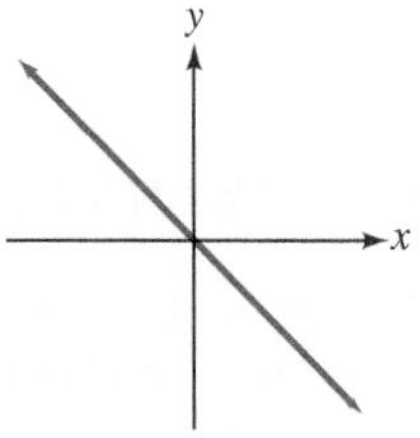

Find the slopes of lines PQ and PR and determine whether the points P, Q, and R lie on the same line. (Hint: Two lines with the same slope and a point in common must be the same line.)

75. $P(-2, 4), Q(4, 8), R(8, 12)$
76. $P(6, 10), Q(0, 6), R(3, 8)$
77. $P(-4, 10), Q(-6, 0), R(-1, 5)$
78. $P(-10, -13), Q(-8, -10), R(-12, -16)$
79. $P(-2, 4), Q(0, 8), R(2, 12)$
80. $P(8, -4), Q(0, -12), R(8, -20)$

Determine whether the line PQ is parallel, perpendicular, or neither to a line with a slope of −2.

81. $P(-2, -3), Q(-4, 1)$ **82.** $P(-5, 3), Q(-7, 2)$
83. $P(-2, 1), Q(6, 5)$ **84.** $P(3, 4), Q(-3, -5)$
85. $P(4, -3), Q(-2, 0)$ **86.** $P(-2, 3), Q(4, -9)$

Unless otherwise noted, all content on this page is © Cengage Learning.

87. Find the equation of the x-axis and its slope.

88. Find the equation of the y-axis and its slope, if any.

APPLICATIONS *SEE EXAMPLES 6–7. (OBJECTIVE 6)*

89. Grade of a road If the vertical rise of the road shown in the illustration is 24 feet for a horizontal run of 1 mile, find the slope of the road. (*Hint*: 1 mile = 5,280 feet.)

90. Pitch of a roof If the rise of the roof shown in the illustration is 5 feet for a run of 12 feet, find the pitch of the roof.

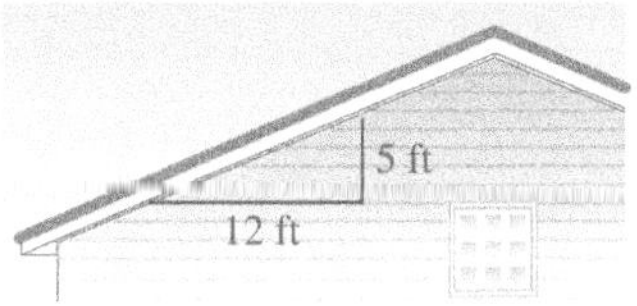

91. Slope of a ramp If a ramp rises 3 feet over a run of 15 feet, find its slope.

92. Slope of a ladder A ladder leans against a building and reaches a height of 24 feet. If its base is 10 feet from the building, find the slope of the ladder.

93. Rate of growth When a college started an aviation program, the administration agreed to predict enrollments using a straight-line method. If the enrollment during the first year was 12, and the enrollment during the fifth year was 26, find the rate of growth per year (the slope of the line). See the illustration.

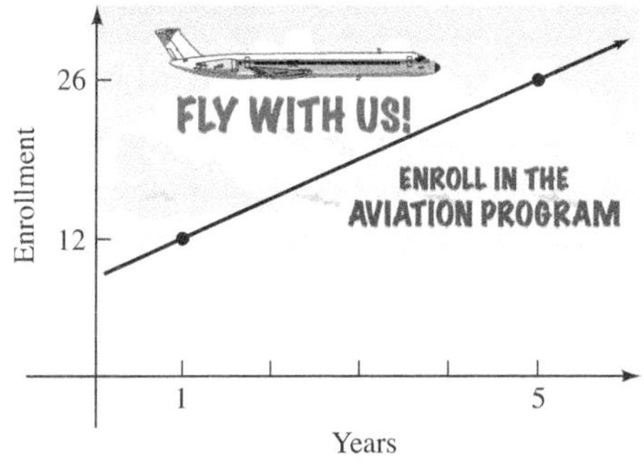

94. Rate of growth A small business predicts sales according to a straight-line method. If sales were \$50,000 in the first year and \$110,000 in the third year, find the rate of growth in sales per year (the slope of the line).

95. Rate of decrease The price of computer equipment has been dropping steadily for the past ten years. If a desktop PC cost \$6,700 ten years ago, and the same computing power cost \$2,200 three years ago, find the rate of decrease per year. (Assume a straight-line model).

96. Hospital costs The table shows the changing mean daily cost for a hospital room. Find the rate of change per year of the portion of the room cost that was absorbed by the hospital between 2000 and 2010.

	Cost passed on to the patient	Total cost to the hospital
2000	\$130	\$245
2005	214	459
2010	295	670

WRITING ABOUT MATH

97. Explain why a vertical line has no defined slope.

98. Explain how to determine from their slopes whether two lines are parallel, perpendicular, or neither.

SOMETHING TO THINK ABOUT

99. The points $(3, a)$, $(5, 7)$, and $(7, 10)$ lie on a line. Find the value of a.

100. The line passing through points $A(1, 3)$ and $B(-2, 7)$ is perpendicular to the line passing through points $C(4, b)$ and $D(8, -1)$. Find the value of b.

Section 3.4 Point-Slope Form

Objectives

1. Write the point-slope form of the equation of a line with a given slope that passes through a specified point.
2. Determine the slope of a line from a graph and use it with a specified point to write an equation for the line.
3. Graph a line given the point-slope form of an equation.
4. Find an equation of the line representing real-world data.

Unless otherwise noted, all content on this page is © Cengage Learning.

Vocabulary

point-slope form

Getting Ready

Solve each equation for y.

1. $2x + y = 12$
2. $3x - y = 7$
3. $4x + 2y = 9$
4. $5x - 4y = 12$

Earlier in the chapter, we were given a linear equation and saw how to construct its graph. In this section, we will start with a graph and see how to write its equation.

1 Write the point-slope form of the equation of a line with a given slope that passes through a specified point.

Consider the line shown in Figure 3-33(a). From the graph, we can determine that the slope is 2 and passes through the point $(3, 1)$. If the point (x, y) is to be a second point on the line, it must satisfy the equation

$$\frac{y - 1}{x - 3} = 2$$ By formula, the slope of the line is $\frac{y-1}{x-3}$, which is given to be 2.

After multiplying both sides of this equation by $(x - 3)$, we have

(1) $$y - 1 = 2(x - 3)$$

We now consider the line with slope m shown in Figure 3-33(b). From the graph, we can see that it passes through the point (x_1, y_1). If the point (x, y) is to be a second point on the line, it must satisfy the equation

$$\frac{y - y_1}{x - x_1} = m$$ This is the slope formula.

After multiplying both sides by $(x - x_1)$, we have

(2) $$y - y_1 = m(x - x_1)$$

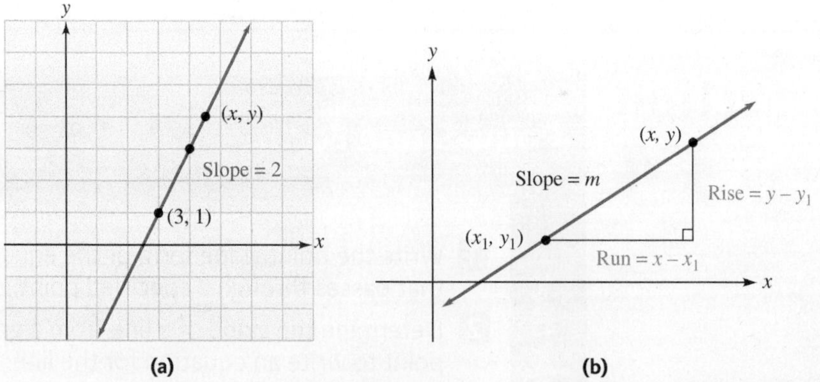

Figure 3-33

Because Equations 1 and 2 display the coordinates of a point on a line and the slope of the line, they are written in a form called **point-slope form**.

Unless otherwise noted, all content on this page is © Cengage Learning.

POINT-SLOPE FORM OF THE EQUATION OF A LINE

The equation of the line that has a slope of m and passes through the point (x_1, y_1) is

$$y - y_1 = m(x - x_1)$$

EXAMPLE 1 Find the point-slope equation of the line that has a slope of $-\frac{3}{5}$ and passes through the point $(3, -6)$.

Solution We substitute $-\frac{3}{5}$ for m, 3 for x_1, and -6 for y_1 into the point-slope equation.

$y - y_1 = m(x - x_1)$ This is the formula for point-slope form.

$y - (-6) = -\frac{3}{5}(x - 3)$ Substitute -6 for y_1, $-\frac{3}{5}$ for m, and 3 for x_1.

$y + 6 = -\frac{3}{5}(x - 3)$ $-(-6) = 6$

SELF CHECK 1 Find the point-slope equation of the line that has a slope of 3 and passes through the point $(-2, 8)$.

EXAMPLE 2 Find the point-slope equation of the line that has a slope of $\frac{2}{3}$ and passes through the point $(-4, 2)$. Then solve it for y.

Solution We substitute $\frac{2}{3}$ for m, -4 for x_1, and 2 for y_1 into the point-slope form.

$y - y_1 = m(x - x_1)$ This is the point-slope equation.

$y - 2 = \frac{2}{3}[x - (-4)]$ Substitute -4 for x_1, 2 for y_1, and $\frac{2}{3}$ for m.

$y - 2 = \frac{2}{3}(x + 4)$ $-(-4) = 4$

To solve the equation for y, we proceed as follows.

$y - 2 = \frac{2}{3}x + \frac{8}{3}$ Use the distributive property to remove parentheses.

$y = \frac{2}{3}x + \frac{14}{3}$ Add 2 to both sides and simplify.

The equation is $y = \frac{2}{3}x + \frac{14}{3}$.

SELF CHECK 2 Find the point-slope equation of the line that has a slope of $\frac{3}{4}$ and passes through the point $(-6, -2)$. Then solve it for y.

2 Determine the slope of a line from a graph and use it with a specified point to write an equation for the line.

In the next example, we are given a graph and asked to write its equation.

EXAMPLE 3 Find the point-slope equation of the line shown in Figure 3-34 on the next page. Then solve it for y.

Solution To find the equation of the line that passes through two known points, we must find the slope by substituting -6 for y_2, 4 for y_1, 8 for x_2, and -5 for x_1.

$$m = \frac{y_2 - y_1}{x_2 - x_1}$$ This is the slope formula.

$$= \frac{-6 - 4}{8 - (-5)}$$ Substitute -6 for y_2, 4 for y_1, 8 for x_2, and -5 for x_1.

$$= -\frac{10}{13}$$

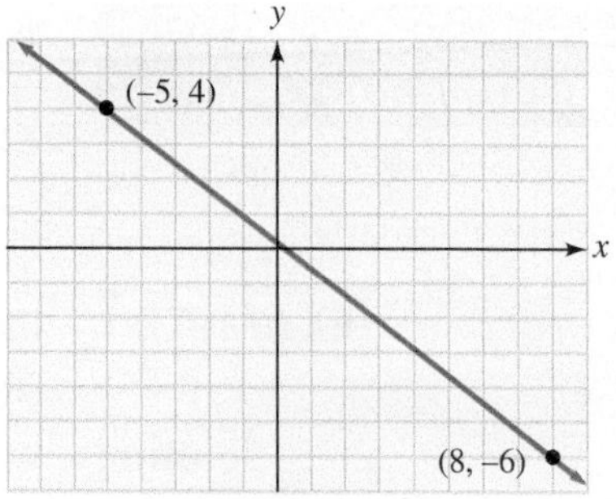

Figure 3-34

Since the line passes through both points, we can choose either one and substitute its coordinates into the point-slope equation. If we choose $(-5, 4)$, we substitute -5 for x_1, 4 for y_1, and $-\frac{10}{13}$ for m.

$$y - y_1 = m(x - x_1)$$ This is point-slope form.

$$y - 4 = -\frac{10}{13}[x - (-5)]$$ Substitute.

$$y - 4 = -\frac{10}{13}(x + 5)$$ $-(-5) = 5$

To solve for y, we proceed as follows:

$$y - 4 = -\frac{10}{13}x - \frac{50}{13}$$ Use the distributive property to remove parentheses.

$$y = -\frac{10}{13}x + \frac{2}{13}$$ Add 4 to both sides and simplify.

The equation is $y = -\frac{10}{13}x + \frac{2}{13}$.

SELF CHECK 3 Find the point-slope equation of the line that passes through $(-2, 5)$ and $(5, -2)$. Then solve it for y.

3 Graph a line given the point-slope form of an equation.

Graphing a linear equation written in point-slope form does not require solving for one of the variables. For example, to graph

$$y - 2 = \frac{3}{4}(x - 1)$$

we compare the equation to point-slope form

$$y - y_1 = m(x - x_1)$$

and note that the slope of the line is $m = \frac{3}{4}$ and that it passes through the point $(x_1, y_1) = (1, 2)$. Because the slope is $\frac{3}{4}$, we can start at the point $(1, 2)$ and locate another point on the line by counting 4 units to the right and 3 units up as shown in Figure 3-35. The change in x from point P to point Q is 4, and the corresponding change in y is 3. The line joining these points is the graph of the equation.

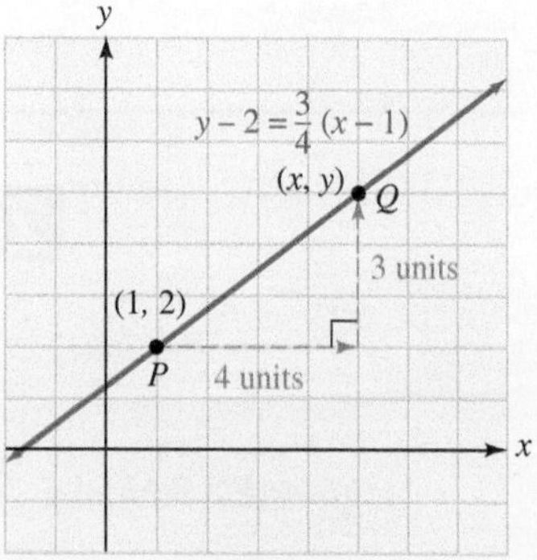

Figure 3-35

Unless otherwise noted, all content on this page is © Cengage Learning.

4 Find an equation of the line representing real-world data.

EXAMPLE 4 **WATER BILLING** In a city, the monthly cost for water is related to the number of gallons used by a linear equation. If a customer is charged \$12 for using 1,000 gallons and \$16 for using 1,800 gallons, find the cost of 2,000 gallons.

Solution When a customer uses 1,000 gallons, the charge is \$12. When he uses 1,800 gallons, the charge is \$16. Since y is related to x by a linear equation, the points $(1{,}000, 12)$ and $(1{,}800, 16)$ will lie on a line as shown in Figure 3-36. To write the equation of the line passing through these points, we first find the slope of the line.

$$m = \frac{y_2 - y_1}{x_2 - x_1} \quad \text{This is the slope formula.}$$

$$= \frac{16 - 12}{1{,}800 - 1{,}000} \quad \text{Substitute.}$$

$$= \frac{4}{800}$$

$$= \frac{1}{200}$$

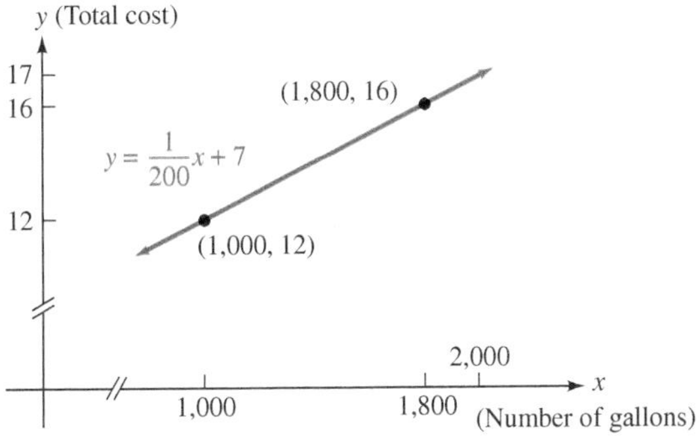

Figure 3-36

We then substitute $\frac{1}{200}$ for m and the coordinates of one of the known points, say (1,000, 12) into the point-slope equation of a line and solve for y.

$$y - y_1 = m(x - x_1) \quad \text{This is point-slope form.}$$

$$y - 12 = \frac{1}{200}(x - 1{,}000) \quad \text{Substitute.}$$

$$y - 12 = \frac{1}{200}x - \frac{1{,}000}{200} \quad \text{Use the distributive property to remove parentheses.}$$

$$y - 12 = \frac{1}{200}x - 5 \quad \frac{1{,}000}{200} = 5$$

(3) $$y = \frac{1}{200}x + 7 \quad \text{Add 12 to both sides and simplify.}$$

To find the charge for 2,000 gallons, we substitute 2,000 for x in Equation 3 and find the value of y.

$$y = \frac{1}{200}(2{,}000) + 7 \quad \text{Substitute 2,000 for } x.$$

$$y = 10 + 7 \quad \text{Multiply.}$$

$$y = 17 \quad \text{Simplify.}$$

The charge for 2,000 gallons of water is \$17.

SELF CHECK 4 Find the cost for 5,000 gallons of water.

Unless otherwise noted, all content on this page is © Cengage Learning.

SELF CHECK ANSWERS 1. $y - 8 = 3(x + 2)$ 2. $y = \frac{3}{4}x + \frac{5}{2}$ 3. $y = -x + 3$ 4. $32

NOW TRY THIS

1. Find the point-slope form of the equation of the line passing through the point $(-3, 2)$ that is parallel to the line with slope of 2. Solve this equation for y.
2. Find the point-slope form of the equation of the line perpendicular to a line with a slope of $\frac{1}{4}$ passing through the point $(-1, -4)$.
3. Write the point-slope form of the equation of the line shown.

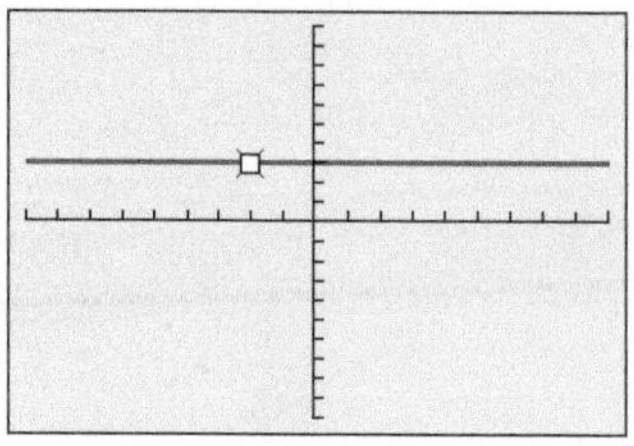

3.4 Exercises

WARM-UPS *Find the slope and one point on each line.*

1. $y - 5 = 2(x - 3)$
2. $y - (-2) = -4[x - (-1)]$
3. $y - (-5) = \frac{7}{8}[x - (-4)]$
4. $y - (-8) = -\frac{4}{5}(x - 3)$

REVIEW *Solve each equation.*

5. $3(x + 2) + x = 5x$
6. $12b + 6(3 - b) = b + 3$
7. $\frac{5(2 - x)}{3} - 1 = x + 5$
8. $r - 1 = \frac{r + 2}{2} + 6$
9. **Junk mail** According to the U.S. Postal Service, people open 53% of the junk mail they receive. If a carpet cleaning business sends out 15,000 promotional ads, how many are likely to be read?
10. **Mixing coffee** To make a mixture of 80 pounds of coffee worth $272, a grocer mixes coffee worth $3.25 a pound with coffee worth $3.85 a pound. How many pounds of cheaper coffee should the grocer use?

VOCABULARY AND CONCEPTS *Fill in the blanks.*

11. Write the point-slope form of the equation of a line.

    ____
12. The graph of the equation $y - 3 = 7[x - (-2)]$ has a slope of _ and passes through _______.
13. To graph the equation $y - 2 = \frac{3}{2}(x - 1)$, we start at the point ______ and count _ units to the right and _ units up to locate a second point on the line. The graph is the line joining the two points.
14. To graph the equation $y - 4 = -\frac{2}{3}(x - 2)$, we start at the point ______ and count _ units to the right and _ units down to locate a second point on the line. The graph is the line joining the two points.

GUIDED PRACTICE *Write the point-slope equation of the line with the given slope that passes through the given point.* SEE EXAMPLE 1. (OBJECTIVE 1)

15. $m = 3, (0, 0)$
16. $m = -5, (2, 1)$
17. $m = -7, (-1, -2)$
18. $m = -4, (0, 0)$
19. $m = 2, (-5, 3)$
20. $m = -\frac{5}{6}, (12, 7)$
21. $m = -\frac{6}{7}, (6, 5)$
22. $m = -3, (0, 7)$

Write the point-slope equation of the line with the given slope that passes through the given point. Solve each equation for y. SEE EXAMPLE 2. (OBJECTIVE 1)

23. $m = -5, (2, -3)$
24. $m = 7, (1, 4)$

Unless otherwise noted, all content on this page is © Cengage Learning.

25. $m = 5, (0, 7)$

26. $m = -8, (0, -2)$

27. $m = -3, (2, 0)$

28. $m = 4, (-5, 0)$

29. $m = \frac{1}{3}, (6, -2)$

30. $m = -\frac{3}{5}, (5, 7)$

Use point-slope form to write an equation of the given line. Solve each equation for y. SEE EXAMPLE 3. (OBJECTIVE 2)

31.

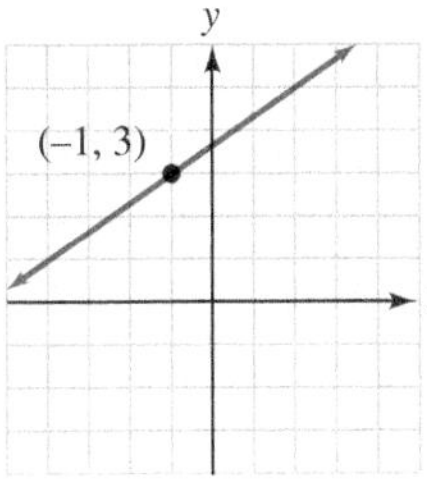

32.

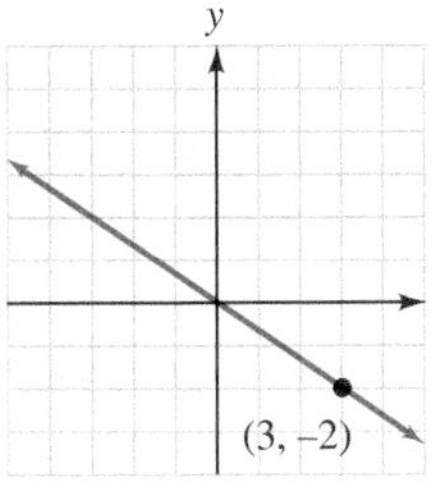

33.

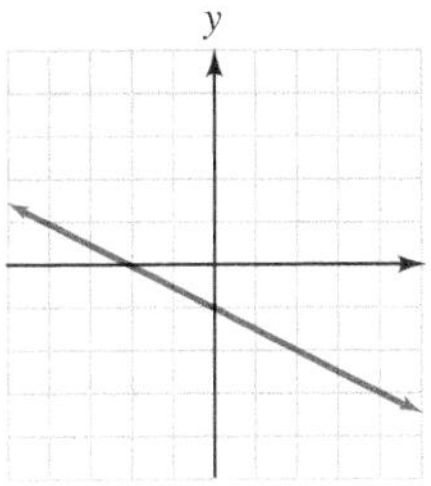

34.

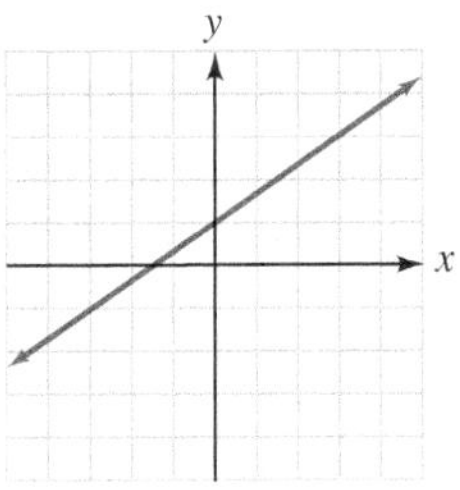

Graph each equation. On the graph, label the ordered pair and the slope identified in the given point-slope equation. (OBJECTIVE 3)

35. $y - 3 = 2(x - 1)$

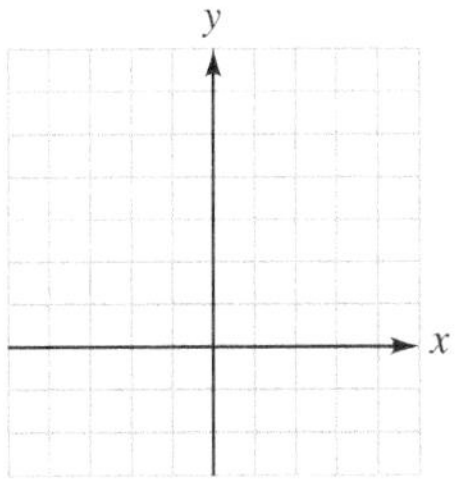

36. $y - 1 = -2(x - 3)$

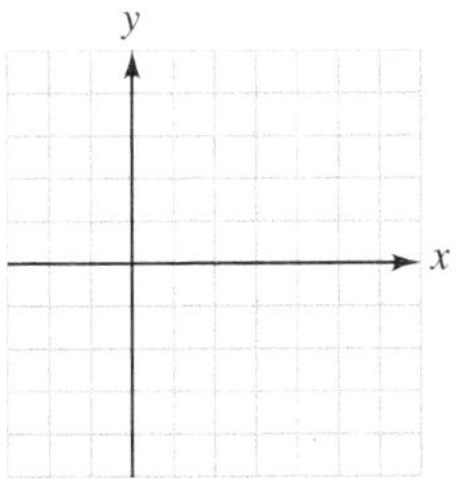

37. $y - 4 = -\frac{2}{3}(x - 2)$

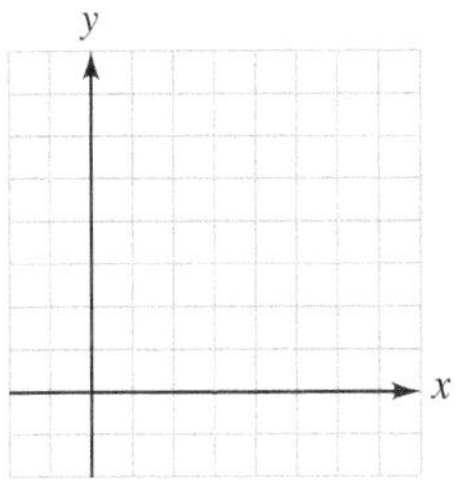

38. $y - (-2) = \frac{3}{2}(x - 1)$

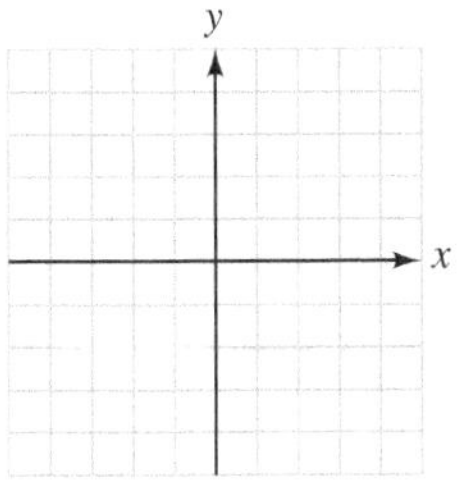

ADDITIONAL PRACTICE *Write the point-slope equation of the line with the given slope that passes through the given point.*

39. $m = 0.5, (-1, -8)$

40. $m = \frac{8}{7}, (5, -3)$

41. $m = -4, (-3, 2)$

42. $m = 1.5, (-3, -9)$

Use point-slope form to write an equation of the given line. Solve each equation for y.

43. $(0, 0), (4, 4)$

44. $(-2, -5), (4, -2)$

45. $(3, 4), (0, -3)$

46. $(4, 0), (6, -8)$

Graph each equation. On the graph, label the ordered pair and the slope identified in the given point-slope equation.

47. $y - 1 = -\frac{1}{2}(x + 2)$

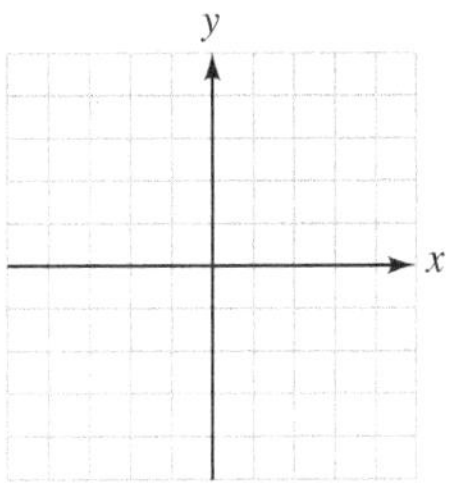

48. $y - (-2) = \frac{3}{2}(x + 3)$

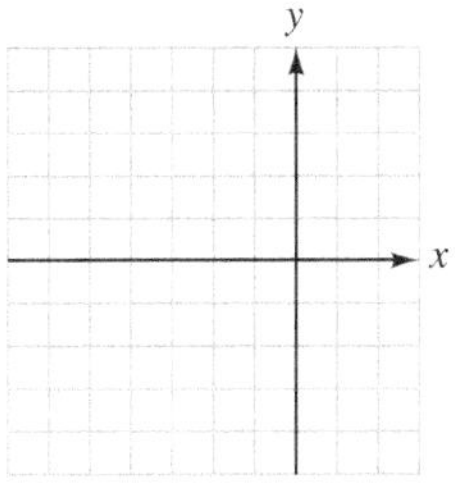

49. $y + 2 = 3(x - 2)$

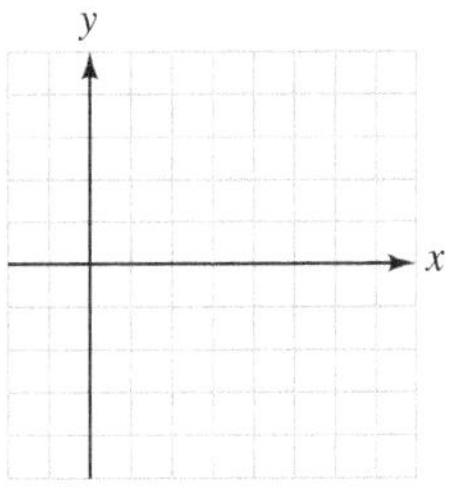

50. $y + 3 = -2(x - 1)$

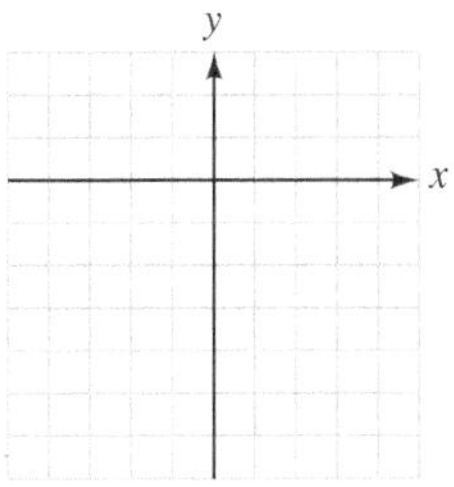

Write the point-slope equation of the line with the given properties. Solve each equation for y.

51. $m = -\frac{2}{3}, (3, -1)$

52. $m = -\frac{5}{3}, (6, -7)$

53. $m = 0.5, (10, 8)$

54. $m = -0.25, (8, -7)$

55. $(1, 2), (-3, -4)$

56. $(3, -4), (5, -1)$

APPLICATIONS *Assume that x and y are related by a linear equation.* SEE EXAMPLE 4. (OBJECTIVE 4)

57. Electric bills One month, a customer received an electric bill for \$200 when she used 1,450 kilowatt hours (kWh) of electricity. The next month, she received a bill for \$250 when she used 1,850 kWh. Compute her bill for the month when she used 1,500 kWh.

58. Telephone costs One month, a customer received a phone bill for \$30.37 when he made 67 calls. The next month, he

Unless otherwise noted, all content on this page is © Cengage Learning.

received a bill for $31.69 when he made 79 calls. Compute his bill for a month when he made 47 calls.

59. Value of antiques An antique table was purchased for $370 and is expected to be worth $450 in 2 years. What will it be worth in 13 years?

60. Value of antiques An antique clock is expected to be worth $350 after 2 years and $530 after 5 years. What will the clock be worth after 7 years?

61. Computer repairs A repair service charges a fixed amount to repair a computer, plus an hourly rate. Use the information in the table to find the hourly rate.

62. Predicting fires A local fire department recognizes that city growth and the number of reported fires are related by a linear equation. City records show that 300 fires were reported in a year when the local population was 57,000 people, and 325 fires were reported in a year when the population was 59,000 people. How many fires can be expected when the population reaches 100,000 people?

63. Real estate Three years after a cottage was purchased it was appraised at $147,700. The property is now 10 years old and is worth $172,200. Find its original purchase price.

64. Rate of depreciation A truck that cost $27,600 when new is expected to be worthless in 12 years. How much will it be worth in 9 years?

WRITING ABOUT MATH

65. Explain how to find the equation of a line passing through two given points.

66. Explain how to graph an equation written in point-slope form.

SOMETHING TO THINK ABOUT

67. Can the equation of a vertical line be written in point-slope form? Explain.

68. Can the equation of a horizontal line be written in point-slope form? Explain.

Section 3.5 Slope-Intercept Form

Objectives

1. Find the slope and y-intercept given a linear equation.
2. Write an equation of the line that has a given slope and passes through a specified point.
3. Use the slope-intercept form to write an equation of the line that passes through two given points.
4. Graph a linear equation using the slope and y-intercept.
5. Determine whether two linear equations define lines that are parallel, perpendicular, or neither parallel nor perpendicular.
6. Write an equation of the line passing through a specified point and parallel or perpendicular to a given line.
7. Write an equation of a line representing real-world data.

Vocabulary

slope-intercept form linear depreciation annual depreciation rate

Unless otherwise noted, all content on this page is © Cengage Learning.

Getting Ready

Solve each equation for y.

1. $4x + 8y = 12$ **2.** $6x - 5y = 7$

In this section, we will discuss another form of an equation of a line called *slope-intercept form*.

1 Find the slope and y-intercept given a linear equation.

Since the y-intercept of the line shown in Figure 3-37 is the point $(0, b)$, we can write its equation by substituting 0 for x_1 and b for y_1 in the point-slope equation and simplifying.

$$y - y_1 = m(x - x_1)$$
$$y - \mathbf{b} = m(x - \mathbf{0})$$
$$y - b = mx$$
$$(1) \qquad y = mx + b$$

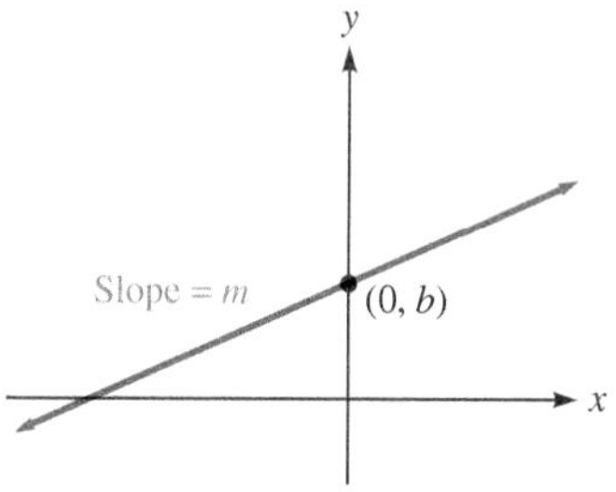

Figure 3-37

Because Equation 1 displays the slope m and the y-coordinate b of the y-intercept, it is called the **slope-intercept form** of the equation of a line.

SLOPE-INTERCEPT FORM OF THE EQUATION OF A LINE

The equation of the line with slope m and y-intercept $(0, b)$ is

$$y = mx + b$$

EXAMPLE 1 Find the slope and y-intercept of the line with the given equation.

a. $y = \frac{2}{3}x + 5$ **b.** $3x - 4y = 9$

Solution **a.** Since the equation is solved for y, we can identify the slope and the y-intercept from the equation $y = \frac{2}{3}x + 5$.

$$y = \frac{\mathbf{2}}{\mathbf{3}}x + \mathbf{5}$$

The slope is $\frac{2}{3}$, and the y-intercept is $(0, 5)$.

b. To write the equation $3x - 4y = 9$ in slope-intercept form, we solve it for y.

$$3x - 4y - 9$$
$$-4y = -3x + 9 \quad \text{Subtract } 3x \text{ from both sides.}$$
$$y = \frac{3}{4}x - \frac{9}{4} \quad \text{Divide both sides by } -4.$$

We can read the slope and the y-intercept from the equation because it is now in $y = mx + b$ form.

Unless otherwise noted, all content on this page is © Cengage Learning.

$$y = \frac{3}{4}x + \left(-\frac{9}{4}\right)$$

The slope is $\frac{3}{4}$. The y-intercept is $\left(0, -\frac{9}{4}\right)$.

SELF CHECK 1 Find the slope and y-intercept of $2x + 5y = 12$.

2 Write an equation of the line that has a given slope and passes through a specified point.

If we know the slope of a line and the coordinates of one point on the line, we can write the equation of the line.

EXAMPLE 2 Write the slope-intercept equation of the line that has a slope of 5 and passes through the point $(-2, 9)$.

Solution We are given that $m = 5$ and that the ordered pair $(-2, 9)$ satisfies the equation. To find b, we can substitute -2 for x, 9 for y, and 5 for m in the equation $y = mx + b$.

$y = mx + b$

$9 = 5(-2) + b$ Substitute 9 for y, 5 for m, and -2 for x.

$9 = -10 + b$ Multiply.

$19 = b$ Add 10 to both sides.

Since $m = 5$ and $b = 19$, the equation in slope-intercept form is $y = 5x + 19$.

SELF CHECK 2 Write an equation of the line that has a slope of -2 and passes through the point $(2, 3)$.

COMMENT Note that Example 2 could be written using the point-slope form, since the given information contained a point and the slope.

3 Use the slope-intercept form to write an equation of the line that passes through two given points.

EXAMPLE 3 Use the slope-intercept form to write an equation of the line that passes through the points $(-3, 4)$ and $(5, -8)$.

Solution We first find the slope of the line.

$m = \dfrac{y_2 - y_1}{x_2 - x_1}$ This is the formula for slope.

$= \dfrac{-8 - 4}{5 - (-3)}$ Substitute -8 for y_2, 4 for y_1, 5 for x_2, and -3 for x_1.

$= \dfrac{-12}{8}$ $-8 - 4 = -12, 5 - (-3) = 8$

$= -\dfrac{3}{2}$ Simplify.

Then we write an equation of the line that has a slope of $-\frac{3}{2}$ and passes through either of the given points, say $(-3, 4)$. To do so, we substitute $-\frac{3}{2}$ for m, -3 for x, 4 for y, and solve the equation $y = mx + b$ for b.

$$y = mx + b$$

$$4 = -\frac{3}{2}(-3) + b \quad \text{Substitute.}$$

$$4 = \frac{9}{2} + b \quad -\frac{3}{2}(-3) = \frac{9}{2}$$

$$-\frac{1}{2} = b \quad \text{Subtract } \frac{9}{2} \text{ from both sides and simplify.}$$

Since $m = -\frac{3}{2}$ and $b = -\frac{1}{2}$, an equation is $y = -\frac{3}{2}x + \left(-\frac{1}{2}\right)$, or more simply, $y = -\frac{3}{2}x - \frac{1}{2}$.

SELF CHECK 3 Use the slope-intercept form to write an equation of the line that passes through the points $(2, -1)$ and $(-3, 7)$.

4 Graph a linear equation using the slope and y-intercept.

We can graph linear equations when they are written in the slope-intercept form. For example, to graph $y = \frac{4}{3}x - 2$, we note that $b = -2$ and that the y-intercept is $(0, b) = (0, -2)$. (See Figure 3-38.)

Since the slope of the line is $\frac{4}{3}$, we can locate another point on the line by starting at the point $(0, -2)$ and counting 3 units to the right and 4 units up. The change in x is 3, and the corresponding change in y is 4. The line joining the two points is the graph of the equation.

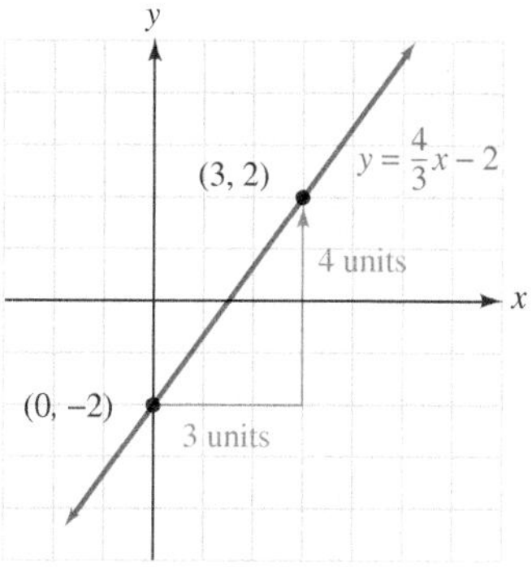

Figure 3-38

EXAMPLE 4 Find the slope and the y-intercept of the line with the equation $2(x - 3) = -3(y + 5)$. Then graph the line.

Solution We write the equation in the form $y = mx + b$ to find the slope m and the y-intercept $(0, b)$.

$$2(x - 3) = -3(y + 5)$$

$$2x - 6 = -3y - 15 \quad \text{Use the distributive property to remove parentheses.}$$

$$2x + 3y - 6 = -15 \quad \text{Add } 3y \text{ to both sides.}$$

$$3y - 6 = -2x - 15 \quad \text{Subtract } 2x \text{ from both sides.}$$

$$3y = -2x - 9 \quad \text{Add 6 to both sides.}$$

$$y = -\frac{2}{3}x - 3 \quad \text{Divide both sides by 3.}$$

The slope is $-\frac{2}{3}$ and the y-intercept is $(0, -3)$. To draw the graph, we plot the y-intercept $(0, -3)$ and then locate a second point on the line by moving 3 units to the right and 2 units down. We draw a line through the two points to obtain the graph shown in Figure 3-39.

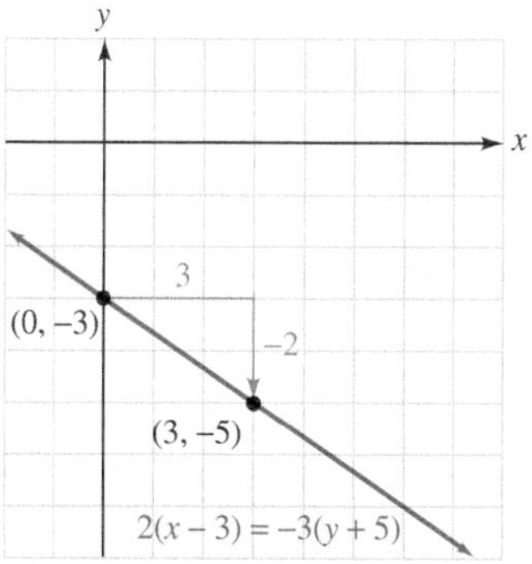

Figure 3-39

SELF CHECK 4 Find the slope and the y-intercept of the line with the equation $-3(x + 2) = 2(y - 3)$.

Unless otherwise noted, all content on this page is © Cengage Learning.

5 Determine whether two linear equations define lines that are parallel, perpendicular, or neither parallel nor perpendicular.

Recall from an earlier section that different lines having the same slope are parallel. Also recall that if the slopes of two lines are negative reciprocals, the lines are perpendicular. We will use these facts in the next examples.

EXAMPLE 5 Show that the lines represented by $4x + 8y = 16$ and $x = 6 - 2y$ are parallel.

Solution We solve each equation for y to see whether their slopes are equal and then determine if the lines are different.

$$\begin{aligned} 4x + 8y &= 16 \\ 8y &= -4x + 16 \\ y &= -\frac{1}{2}x + 2 \end{aligned} \qquad \begin{aligned} x &= 6 - 2y \\ 2y &= -x + 6 \\ y &= -\frac{1}{2}x + 3 \end{aligned}$$

Since the slope of each line is $-\frac{1}{2}$, the lines are possibly parallel, but we need to know whether the lines are distinct. Since the values of b in these equations are different, the lines are distinct and, therefore, parallel.

SELF CHECK 5 Are the lines represented by $3x + y = 2$ and $2y = 6x - 3$ parallel?

EXAMPLE 6 Show that the lines represented by $4x + 8y = 16$ and $4x - 2y = 22$ are perpendicular.

Solution We solve each equation for y to see whether the slopes of their graphs are negative reciprocals.

$$\begin{aligned} 4x + 8y &= 16 \\ 8y &= -4x + 16 \\ y &= -\frac{1}{2}x + 2 \end{aligned} \qquad \begin{aligned} 4x - 2y &= 22 \\ -2y &= -4x + 22 \\ y &= 2x - 11 \end{aligned}$$

Since the slopes of $-\frac{1}{2}$ and 2 are negative reciprocals $\left(-\frac{1}{2}\cdot 2 = -1\right)$, the lines are perpendicular.

SELF CHECK 6 Are the lines represented by $y = 2x + 3$ and $2x + y = 7$ perpendicular?

6 Write an equation of the line passing through a specified point and parallel or perpendicular to a given line.

We will now use the slope properties of parallel and perpendicular lines to write more equations of lines.

EXAMPLE 7 Write an equation of the line passing through $(-3, 2)$ and parallel to the line $y = 8x - 5$.

Solution Since $y = 8x - 5$ is written in slope-intercept form, the slope of its graph is the coefficient of x, which is 8. The desired equation is to have a graph that is parallel to the line, thus its slope must be 8 as well.

To find the equation of the line, we substitute -3 for x, 2 for y, and 8 for m in the slope-intercept equation and solve for b.

$$y = mx + b$$
$$2 = 8(-3) + b \quad \text{Substitute.}$$
$$2 = -24 + b \quad \text{Multiply.}$$
$$26 = b \quad \text{Add 24 to both sides.}$$

Since the slope of the desired line is 8 and the y-intercept is $(0, 26)$, the equation is $y = 8x + 26$.

SELF CHECK 7 Write an equation of the line passing through $(0, 0)$ and parallel to the line $y = 8x - 3$.

EXAMPLE 8 Write an equation of the line passing through $(-2, 5)$ and perpendicular to the line $y = 8x - 3$.

Solution Since the slope of the given line is 8, the slope of the desired line must be $-\frac{1}{8}$, which is the negative reciprocal of 8.

To find the equation of the desired line, we substitute -2 for x, 5 for y, and $-\frac{1}{8}$ for m in the slope-intercept form and solve for b.

$$y = mx + b$$
$$5 = -\frac{1}{8}(-2) + b \quad \text{Substitute.}$$
$$5 = \frac{1}{4} + b \quad \text{Multiply.}$$
$$\frac{19}{4} = b \quad \text{Subtract } \tfrac{1}{4} \text{ from both sides.}$$

Since the slope of the desired line is $-\frac{1}{8}$ and the y-intercept is $\left(0, \frac{19}{4}\right)$, the equation is $y = -\frac{1}{8}x + \frac{19}{4}$.

SELF CHECK 8 Write the equation of the line passing through $(0, 0)$ and perpendicular to the line $y = 8x - 3$.

7 Write an equation of a line representing real-world data.

As machinery wears out, it becomes worth less. Accountants often estimate the decreasing value of aging equipment with **linear depreciation**, a method based on linear equations.

EXAMPLE 9 **LINEAR DEPRECIATION** A company buys a \$12,500 computer with an estimated life of 6 years. The computer can then be sold as scrap for an estimated *salvage value* of \$500. If y represents the value of the computer after x years of use and y and x are related by the equation of a line,

a. find an equation of the line and graph it.

b. find the value of the computer after 2 years.

c. find the economic meaning of the y-intercept of the line.

d. find the economic meaning of the slope of the line.

Solution **a.** To find an equation of the line, we calculate its slope and then use the slope-intercept form to find its equation.

When the computer is new, its age x is 0 and its value y is the purchase price of \$12,500. We can represent this information as the point with coordinates $(0, 12{,}500)$. When it is six years old, $x = 6$ and $y = 500$, its *salvage value*. We can represent this information as the point with coordinates $(6, 500)$. Since the line passes through these two points as shown in Figure 3-40, the slope of the line is

$$m = \frac{y_2 - y_1}{x_2 - x_1}$$
$$= \frac{500 - 12{,}500}{6 - 0}$$
$$= \frac{-12{,}000}{6}$$
$$= -2{,}000$$

The slope is $-2{,}000$.

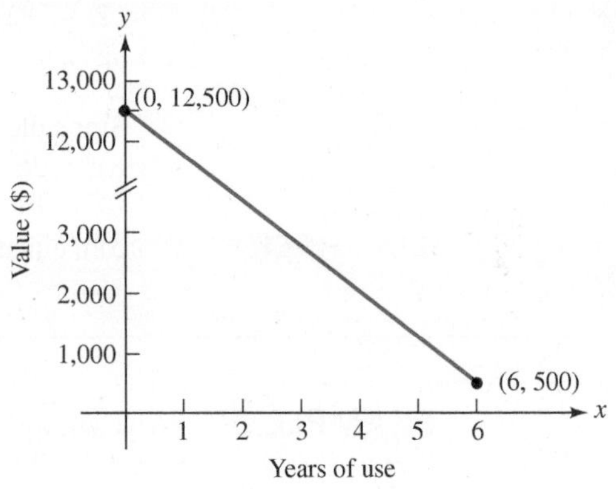

Figure 3-40

To find an equation of the line, we substitute $-2{,}000$ for m, 0 for x, and 12,500 for y in the slope-intercept form and solve for b.

$$y = mx + b$$
$$12{,}500 = -2{,}000(0) + b$$
$$12{,}500 = b$$

The current value of the computer is related to its age by the equation $y = -2{,}000x + 12{,}500$.

b. To find the value of the computer after 2 years, we substitute 2 for x in the equation $y = -2{,}000x + 12{,}500$.

$$y = -2{,}000x + 12{,}500$$
$$y = -2{,}000(2) + 12{,}500$$
$$= -4{,}000 + 12{,}500$$
$$= 8{,}500$$

In 2 years, the computer will be worth \$8,500.

c. Since the y-intercept of the graph is $(0, 12{,}500)$, the y-coordinate of the y-intercept is the computer's original purchase price.

d. Each year, the value decreases by \$2,000, shown by the slope of the line, $-2{,}000$. The slope of the depreciation line is called the **annual depreciation rate**.

SELF CHECK 9 Find the equation of the line if the computer was sold as scrap for a value of \$350. Find the economic meaning of the slope of the line.

SELF CHECK ANSWERS

1. $-\frac{2}{5}, \left(0, \frac{12}{5}\right)$ **2.** $y = -2x + 7$ **3.** $y = -\frac{8}{5}x + \frac{11}{5}$ **4.** $-\frac{3}{2}, (0, 0)$ **5.** no **6.** no **7.** $y = 8x$ **8.** $y = -\frac{1}{8}x$ **9.** $y = -2{,}025x + 12{,}500$; each year, the value decreases by \$2,025, shown by the slope of the line, $-2{,}025$

Unless otherwise noted, all content on this page is © Cengage Learning.

NOW TRY THIS

1. Find the slope and y-intercept of the line $y = x$.
2. Write an equation of the line passing through $(2, -6)$ and perpendicular to $2x + 3y = 12$.
3. In one state, approximately 82,000 fires were reported in a year when the population was 20,000,000 and 94,000 were reported in a year when the population was 23,000,000. Interpret the slope in the context of this problem. Predict the number of fires that will occur when the population of the state is 27,000,000.

3.5 Exercises

WARM-UPS *Solve each equation for y.*

1. $2x + 4y = 12$ **2.** $4x - 2y = 5$

Solve each equation for b.

3. $2 = \frac{3}{5}(10) + b$ **4.** $-4 = \frac{3}{2}(-5) + b$

REVIEW *Solve each equation.*

5. $2x + 3 = 7$ **6.** $3x - 2 = 7$

7. $3(y - 2) = y + 1$ **8.** $-4z - 6 = 2(z + 3)$

VOCABULARY AND CONCEPTS *Fill in the blanks.*

9. The formula for the slope-intercept form of the equation of a line is __________. The slope of the graph of $y = -3x + 7$ is ___ and the y-intercept is _____.

10. In the straight-line method, the declining value of aging equipment is called ________________. The slope of the depreciation line is called the annual depreciation ___.

11. The numbers 3 and $-\frac{1}{3}$ are called negative __________.

12. If two distinct lines have the same slope, they are _______.

GUIDED PRACTICE *Find the slope and the y-intercept of the line with the given equation.* SEE EXAMPLE 1. (OBJECTIVE 1)

13. $y = 7x - 5$ **14.** $y = -2x + 5$

15. $y = -\frac{2}{5}x + 6$ **16.** $y = \frac{3}{7}x - 8$

17. $3x - 2y = 8$ **18.** $x - 2y = -6$

19. $-2x - 6y = 5$ **20.** $10x - 15y = 4$

Write the slope-intercept equation of the line that has the given slope and passes through the given point. SEE EXAMPLE 2. (OBJECTIVE 2)

21. $m = 12, (0, 0)$ **22.** $m = -6, (0, 0)$

23. $m = -5, (0, -4)$ **24.** $m = -2, (0, 11)$

25. $m = -7, (7, 5)$ **26.** $m = 3, (-2, -5)$

27. $m = 0, (3, -5)$ **28.** $m = -7, (1, -10)$

Write the slope-intercept equation of the line that passes through the given points. SEE EXAMPLE 3. (OBJECTIVE 3)

29. $(1, -5), (3, -11)$ **30.** $(0, 7), (-2, 3)$

31. $(-8, 10), (8, 2)$ **32.** $(-4, 5), (2, -6)$

Write each equation in slope-intercept form to find the slope and the y-intercept. Then use the slope and y-intercept to graph the line. SEE EXAMPLE 4. (OBJECTIVE 4)

33. $x - y = 1$

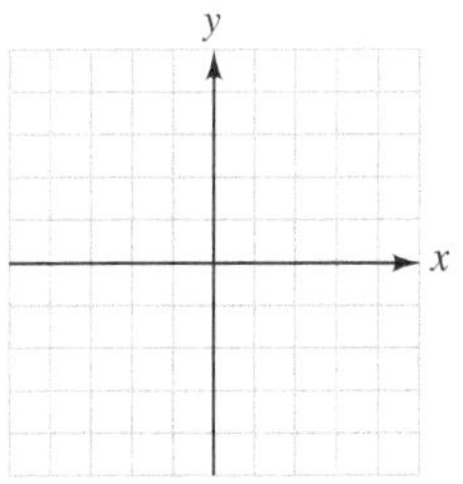

34. $x + y = 2$

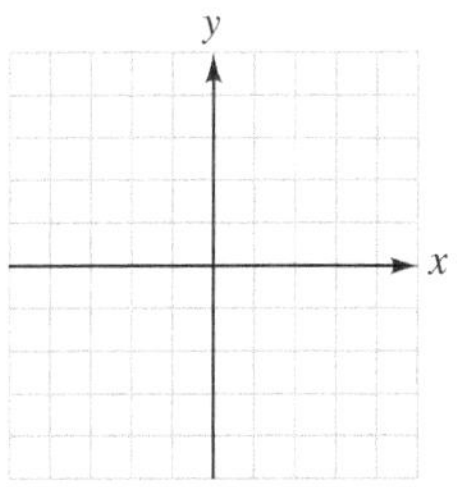

35. $2x = 3y - 6$

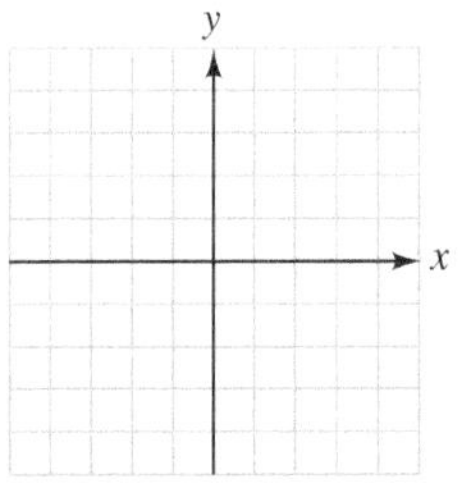

36. $5x = -4y + 10$

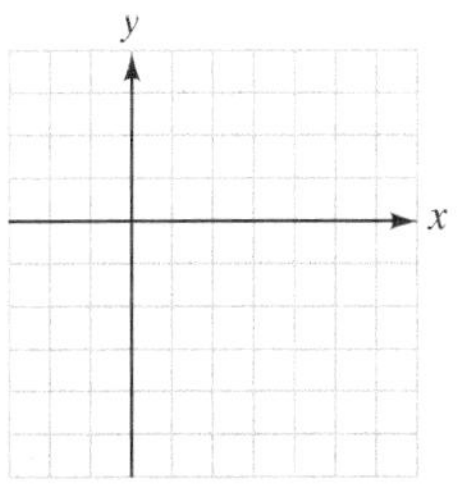

Unless otherwise noted, all content on this page is © Cengage Learning.

37. $3y = -2x + 18$

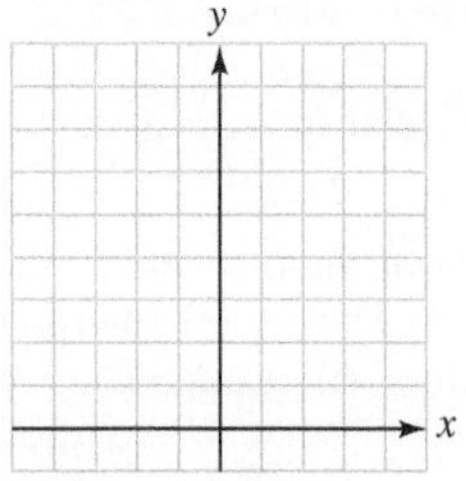

38. $-8x = 9y + 36$

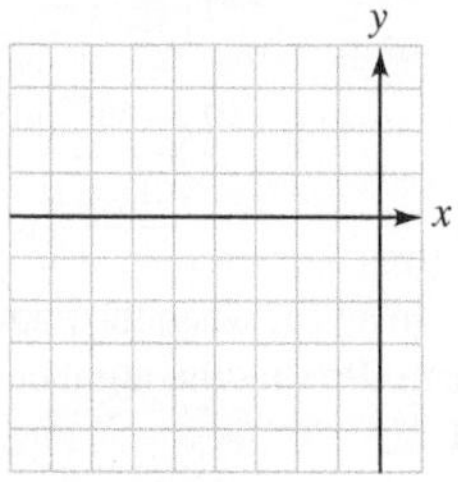

Determine whether the graphs represented by each pair of equations are parallel, perpendicular, or neither. SEE EXAMPLES 5–6. (OBJECTIVE 5)

39. $y = -6x + 3, y = -6x - 4$

40. $y = 5x - 13, y = \frac{1}{5}x + 28$

41. $x + y = 7, y = x - 3$

42. $x = y + 5, y = x + 8$

43. $y = 3x + 9, 2y = 6x - 10$

44. $2x + 3y = 11, 3x - 2y = 15$

45. $x = 3y + 9, y = -3x + 2$

46. $3x + 6y = 7, y = \frac{1}{2}x$

Write the slope-intercept equation of the line that passes through the given point and is parallel to the given line. SEE EXAMPLE 7. (OBJECTIVE 6)

47. $(0,0), y = 4x - 9$

48. $(-3,2), x = -3y + 8$

49. $(2,5), 4x - y = 7$

50. $(-6,3), y + 3x = -12$

Write the slope-intercept equation of the line that passes through the given point and is perpendicular to the given line. SEE EXAMPLE 8. (OBJECTIVE 6)

51. $(0,0), y = 4x - 9$

52. $(0,-4), x = -3y - 2$

53. $(2,5), 4x - y = 7$

54. $(-6,3), y + 3x = -12$

ADDITIONAL PRACTICE

Find the slope and the y-intercept of the line determined by the given equation.

55. $7x = 2y - 4$

56. $15x = -2y - 14$

Write an equation of the line with the following properties. Write the equation in slope-intercept form.

57. $m = \frac{2}{3}$, passing through $(-3,4)$

58. $m = 0$, passing through $(5,3)$

59. $m = -\frac{4}{3}$, passing through $(6,-2)$

60. $m = \frac{3}{4}$, passing through $(8,5)$

61. passing through $(3,-1), (-3,-9)$

62. passing through $(9,8), (-6,-2)$

63. passing through $(4,-2)$ and parallel to $x = \frac{5}{4}y - 2$

64. passing through $(4,-2)$ and perpendicular to $x = \frac{5}{4}y - 2$

65. passing through $(1,-5)$ and perpendicular to $x = -\frac{3}{4}y + 5$

66. passing through $(1,-5)$ and parallel to $x = -\frac{3}{4}y + 5$

67. perpendicular to the line $y = 5$ and passing through $(-2,7)$.

68. parallel to the line $y = 7$ and passing through $(-2.5,3)$.

69. parallel to the line $x = 8$ and passing through $(5,2)$.

70. perpendicular to the line $x = 5$ and passing through $(1,-3)$.

For Exercises 71–78, determine whether the graphs represented by each pair of equations are parallel, perpendicular, or neither.

71. $y = 8, x = 4$

72. $x = -3, x = -7$

73. $3x = y - 2, 3(y - 3) + x = 0$

74. $2y = 8, x = y$

75. $4x + 5y = 20, 5x - 4y = 20$

76. $6x - 15y = 11, 2x - 5y = 12$

77. $2x + 3y = 12, 6x + 9y = 32$

78. $5x + 6y = 30, 6x + 5y = 24$

APPLICATIONS

For Exercises 79–85, assume straight-line depreciation or straight-line appreciation. SEE EXAMPLE 9. (OBJECTIVE 7)

79. Depreciation A taxicab was purchased for \$24,300. Its salvage value at the end of its 7-year useful life is expected to be \$1,900. Find the depreciation equation.

80. Depreciation A small business purchases the computer shown in the illustration. It will be depreciated over a 4-year period, when its salvage value will be \$300. Find the depreciation equation.

81. Appreciation An apartment building was purchased for \$450,000. The owners expect the property to double in value in 12 years. Find the appreciation equation.

Unless otherwise noted, all content on this page is © Cengage Learning.

82. Appreciation A house purchased for \$112,000 is expected to double in value in 12 years. Find its appreciation equation.

83. Depreciation Find the depreciation equation for the TV in the want ad.

> **For Sale:** 3-year-old 54-inch TV, \$1,900 new. Asking \$1,190. Call 875-5555. Ask for Mike.

84. Depreciating word processors A word processor cost \$555 when new and is expected to be worth \$80 after 5 years. What will it be worth after 3 years?

85. Salvage values A copier cost \$1,050 when new and will have a salvage value of \$90 when it is replaced in 8 years. Find its annual depreciation rate.

86. Car repairs An auto shop charges an hourly rate for repairs. If it costs \$69 for a $1\frac{1}{2}$-hour radiator repair and \$230 for a 5-hour transmission overhaul, what is the hourly rate?

87. Printer charges To print advertising brochures, a printer charges a fixed setup cost, plus \$50 for every 100 brochures. If 700 brochures cost \$375, and 1, 000 brochures cost \$525, what is the setup cost?

88. Cost of rain gutters An installer of rain gutters charges a service charge, plus a dollar amount per foot. If one neighbor installed 250 feet and paid \$435, and another neighbor installed 300 feet and paid \$510, what is the service charge?

89. Teacher pensions Average pensions for Illinois teachers nearly doubled from 1993 to 2002, as shown in the illustration. If pensions grew in a linear fashion over the next 10 years, what was the average pension in 2012? (*Hint*: Label 1993 as year 0, 1994 as year 1, etc.)

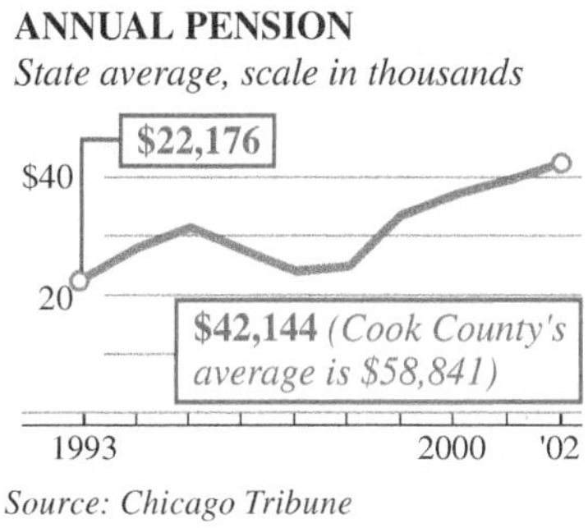

90. Controlling exotic plants Eurasian water milfoil, an undesirable aquatic plant, has infested the waters of northern Wisconsin. To control the weed, lakes are seeded with a chemical that kills it. Four lakes on the Eagle River chain that were treated for milfoil are shown in the following table. If there is a cost per acre to treat a lake, and a state permit costs \$800, complete the table.

Lake	Acres treated	Cost
Catfish	36	\$22,400
Cranberry	42	\$26,000
Eagle	53	
Yellow Birch	17	

Source: Eagle River Chain of Lakes Association

An investor bought an office building for \$465,000 excluding the value of the land. At that time, a real-estate appraiser estimated that the building would retain 80% of its value after 40 years. For tax purposes, the investor used linear depreciation to depreciate the building over a period of 40 years.

91. Real estate Find the estimated value of the property after 40 years. Then find the slope of the straight-line depreciation graph of the building by finding the slope of the line passing through $(0, 465{,}000)$ and $(40, ?)$. Write the equation of the depreciation graph.

92. Real estate See Exercise 91. Find the estimated value of the property after 35 years. If the investor sells the building for \$400,000 at that time, find the taxable capital gain.

WRITING ABOUT MATH

93. Explain how to use the slope-intercept form to graph the equation $y = -\frac{3}{5}x + 2$.

94. In straight-line depreciation, explain why the slope of the line is called the *rate of depreciation*.

SOMETHING TO THINK ABOUT

95. Solve $Ax + By = C$ for y and thereby show that the slope of its graph is $-\frac{A}{B}$ and its y-intercept is $\left(0, \frac{C}{B}\right)$.

96. Show that the x-intercept of the graph of $Ax + By = C$ is $\left(\frac{C}{A}, 0\right)$.

97. Can the equation of a vertical line be written in slope-intercept form? Explain.

98. Can the equation of a horizontal line be written in slope-intercept form? Explain.

99. If the graph of $y = ax + b$ passes through quadrants I, II, and IV, what can be known about the constants a and b?

100. The graph of $Ax + By = C$ passes through the quadrants I and IV, only. What is known about the constants A, B, and C?

Investigate the properties of slope and y-intercept by experimenting with the following problems.

101. Graph $y = mx + 2$ for several positive values of m. What do you notice?

102. Graph $y = mx + 2$ for several negative values of m. What do you notice?

103. Graph $y = 2x + b$ for several increasing positive values of b. What do you notice?

104. Graph $y = 2x + b$ for several decreasing negative values of b. What do you notice?

105. How will the graph of $y = \frac{1}{2}x + 5$ compare to the graph of $y = \frac{1}{2}x - 5$?

106. How will the graph of $y = \frac{1}{2}x - 5$ compare to the graph of $y = \frac{1}{2}x$?

Unless otherwise noted, all content on this page is © Cengage Learning.

Section 3.6 Functions

Objectives

1. Find the domain and range of a set of ordered pairs.
2. Determine whether a given equation defines y to be a function of x.
3. Evaluate a function written in function notation.
4. Graph a function and determine its domain and range.
5. Determine whether a graph represents a function.

Vocabulary

relation
domain
range
function
linear function
vertical line test

Getting Ready

Let $y = 2x - 1$. *Find the value of* y *when*

1. $x = 0$
2. $x = 2$
3. $x = -1$
4. $x = -2$

Let $y = -3x + 2$. *Find the value of* y *when*

5. $x = 0$
6. $x = 2$
7. $x = -1$
8. $x = -2$

In this section, we will discuss *relations* and *functions.* We include these concepts in this chapter because they involve ordered pairs.

1 Find the domain and range of a set of ordered pairs.

Table 3-3 shows the number of medals won by United States athletes during five Winter Olympics.

TABLE 3-3
USA Winter Olympic Medal Count

Year	1992	1994	1998	2002	2006	2010
Medals	11	13	13	34	25	37

We can display the data in the table as a set of ordered pairs, where the *first component* represents the year and the *second component* represents the number of medals won by U.S. athletes.

$$\{(1992, 11), (1994, 13), (1998, 13), (2002, 34), (2006, 25), (2010, 37)\}$$

A set of ordered pairs, such as this, is called a **relation**. The set of all first components is called the **domain** of the relation, and the set of all second components is called the **range** of the relation.

EXAMPLE 1 Find the domain and range of the relation $\{(-2, -5), (4, 7), (8, 9)\}$.

Solution The domain is the set of first components of the ordered pairs:

$$\{-2, 4, 8\}$$

The range is the set of second components of the ordered pairs:

$$\{-5, 7, 9\}$$

SELF CHECK 1 Find the domain and range of the relation $\{(-3, -2), (-1, 3), (4, 5)\}$.

When to each first component in a relation there corresponds exactly one second component, the relation is called a **function**.

FUNCTION

A function is a set of ordered pairs (a relation) in which to each first component, there corresponds exactly one second component.

2 Determine whether a given equation defines *y* to be a function of *x*.

Earlier in the chapter, we constructed the following table of ordered pairs for the equation $y = 3x - 4$ by substituting specific values for x and computing the corresponding values of y. Since the equation determines a set of ordered pairs, the equation determines a relation.

$y = 3x - 4$

x	y	(x, y)	
-2	-10	$(-2, -10)$	If $x = -2$, then $y = 3(-2) - 4 = -10$.
-1	-7	$(-1, -7)$	If $x = -1$, then $y = 3(-1) - 4 = -7$.
0	-4	$(0, -4)$	If $x = 0$, then $y = 3(0) - 4 = -4$.
1	-1	$(1, -1)$	If $x = 1$, then $y = 3(1) - 4 = -1$.
2	2	$(2, 2)$	If $x = 2$, then $y = 3(2) - 4 = 2$.
↑ Input values	↑ Output values		

From the table or equation, we can see that each *input value x* determines exactly one *output value y*. Because this is true, the equation also defines *y* to be a *function* of *x*. This leads to the following definition.

y* IS A FUNCTION OF *x

Any equation in *x* and *y* where each value of *x* (the *input*) determines exactly one value of *y* (the *output*) is called a function. In this case, we say that ***y*** is a function of ***x***.

The set of all input values *x* is called the domain of the function, and the set of all output values *y* is called the range of the function.

COMMENT A function is always a relation, but a relation is not necessarily a function. For example, the relation $\{(2, 1), (2, 3)\}$ is not a function because the input 2 determines two different values in the range: 1 and 3.

Since each value of y in a function depends on a specific value of x, y is the *dependent variable* and x is the *independent variable.* The graph of a function is the graph of the equation that defines the function. In Example 2, we graph a linear equation. Any function that is defined by a linear equation is called a **linear function**.

EXAMPLE 2 Determine whether the equations define y to be a function of x.

a. $y = x^2$ **b.** $x = y^2$

Solution **a.** We construct a table of ordered pairs for the equation $y = x^2$ by substituting values for x and computing the corresponding values.

$y = x^2$

x	y	(x, y)	
−2	4	(−2, 4)	If $x = -2$, then $y = (-2)^2 = 4$.
−1	1	(−1, 1)	If $x = -1$, then $y = (-1)^2 = 1$.
0	0	(0, 0)	If $x = 0$, then $y = 0$.
1	1	(1, 1)	If $x = 1$, then $y = (1)^2 = 1$.
2	4	(2, 4)	If $x = 2$, then $y = (2)^2 = 4$.
3	9	(3, 9)	If $x = 3$, then $y = (3)^2 = 9$.
↑ Input values	↑ Output values		

From the table we can see that each input value x determines exactly one output value y. The relation is a function.

b. We construct a table of ordered pairs for the equation $x = y^2$. Because y is squared, it will be more convenient to substitute values for y and compute the corresponding values for x.

$x = y^2$

x	y	(x, y)	
4	−2	(4, −2)	If $y = -2$, then $x = (-2)^2 = 4$.
1	−1	(1, −1)	If $y = -1$, then $x = (-1)^2 = 1$.
0	0	(0, 0)	If $y = 0$, then $x = 0$.
1	1	(1, 1)	If $y = 1$, then $x = (1)^2 = 1$.
4	2	(4, 2)	If $y = 2$, then $x = (2)^2 = 4$.
9	3	(9, 3)	If $y = 3$, then $x = (3)^2 = 9$.
↑ Input values	↑ Output values		

From the table we can see that each input value x does not determine exactly one output value y. The relation is not a function.

SELF CHECK 2 Determine whether the equations define y to be a function of x.

a. $y = -x^2$ **b.** $x = -y^2$

3 Evaluate a function written in function notation.

There is a special notation for functions that uses the symbol $f(x)$, read as "f of x."

FUNCTION NOTATION The notation $y = f(x)$ denotes that the variable y is a function of x.

COMMENT The notation $f(x)$ does not mean "f times x."

The notation $y = f(x)$ provides a way to denote the values of y in a function that correspond to individual values of x. For example, if $y = f(x)$, the value of y that is determined by $x = 3$ is denoted as $f(3)$. Similarly, $f(-1)$ represents the value of y that corresponds to $x = -1$.

EXAMPLE 3 Let $f(x) = 2x - 3$ and find **a.** $f(3)$ **b.** $f(-1)$ **c.** $f(0)$ **d.** the value of x for which $f(x) = 5$.

Solution **a.** We replace x with 3.

$$f(x) = 2x - 3$$
$$f(3) = 2(3) - 3$$
$$= 6 - 3$$
$$= 3$$

b. We replace x with -1.

$$f(x) = 2x - 3$$
$$f(-1) = 2(-1) - 3$$
$$= -2 - 3$$
$$= -5$$

c. We replace x with 0.

$$f(x) = 2x - 3$$
$$f(0) = 2(0) - 3$$
$$= 0 - 3$$
$$= -3$$

d. We replace $f(x)$ with 5 and solve for x.

$$f(x) = 2x - 3$$
$$5 = 2x - 3$$
$$8 = 2x \quad \text{Add 3 to both sides.}$$
$$4 = x \quad \text{Divide both sides by 2.}$$

SELF CHECK 3 Use the function of Example 3 and find **a.** $f(-2)$ **b.** $f\left(\frac{3}{2}\right)$ **c.** $f(-5)$ **d.** the value for which $f(x) = 11$.

4 Graph a function and determine its domain and range.

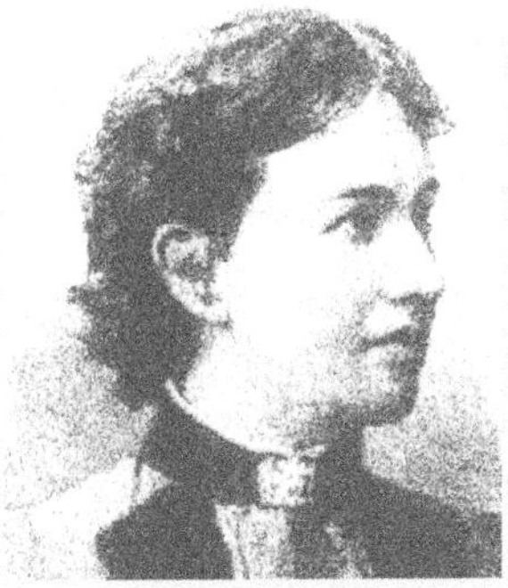

Sonya Kovalevskaya
1850–1891
This talented young Russian woman hoped to study mathematics at the University of Berlin, but strict rules prohibited women from attending lectures. Undaunted, she studied privately with the great mathematician Karl Weierstrauss and published several important papers.

For the graph of the function shown in Figure 3-41, the domain (the set of input values x) is shown on the x-axis, and the range (the set of output values y) is shown on the y-axis. The domain of this function is the set of real numbers $\mathbb{R}$. The range is also the set of real numbers $\mathbb{R}$.

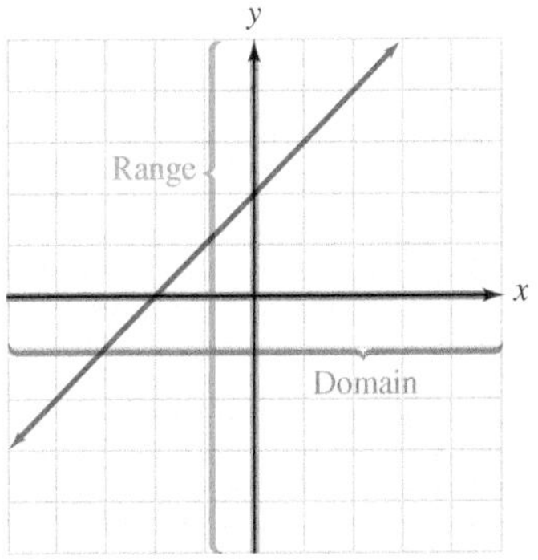

Figure 3-41

Not all functions are linear functions. For example, the graph of the *absolute value function* defined by the equation $y = |x|$ is not a line, as the next example will show.

EXAMPLE 4 Graph $f(x) = |x|$ and determine the domain and range.

Solution We begin by setting up a table of values.

Unless otherwise noted, all content on this page is © Cengage Learning.

$y = f(x) = |x|$

x	y	(x, y)
−2	2	(−2, 2)
−1	1	(−1, 1)
0	0	(0, 0)
1	1	(1, 1)
2	2	(2, 2)

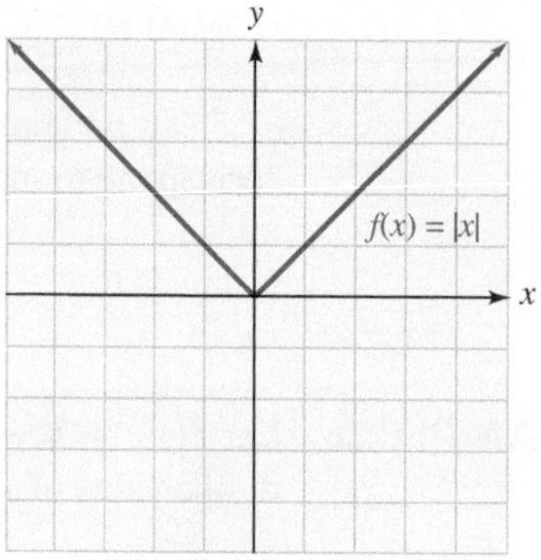

Figure 3-42

From Figure 3-42, we can determine that the domain is the set of all real numbers $\mathbb{R}$. This makes sense because we can find the absolute value of any real number.

From the graph, we can see that the range is the set of values where $y \geq 0$. We write this as $\{y \mid y \text{ is a real number and } y \geq 0\}$. The absolute value of any real number is always positive or 0.

SELF CHECK 4 Graph $f(x) = -|x|$ and determine the domain and range.

5 Determine whether a graph represents a function.

A **vertical line test** can be used to determine whether the graph of an equation represents a function. If any vertical line intersects a graph more than once, the graph cannot represent a function, because to one number x, there would correspond more than one value of y.

The graph in Figure 3-43(a) represents a function, because every vertical line that intersects the graph does so exactly once. The graph in Figure 3-43(b) does not represent a function, because some vertical lines intersect the graph more than once.

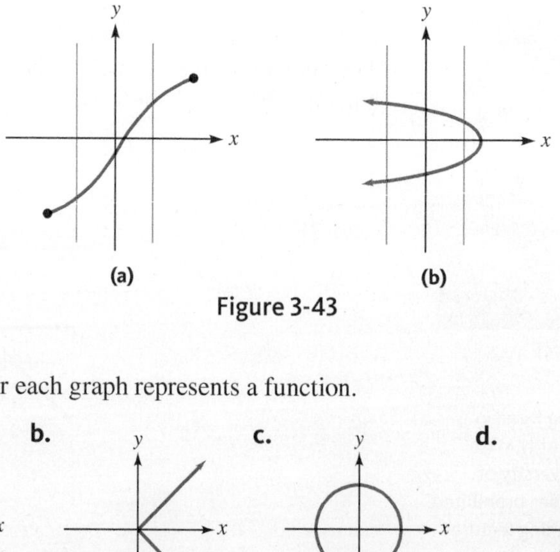

Figure 3-43

EXAMPLE 5 Determine whether each graph represents a function.

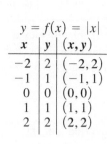

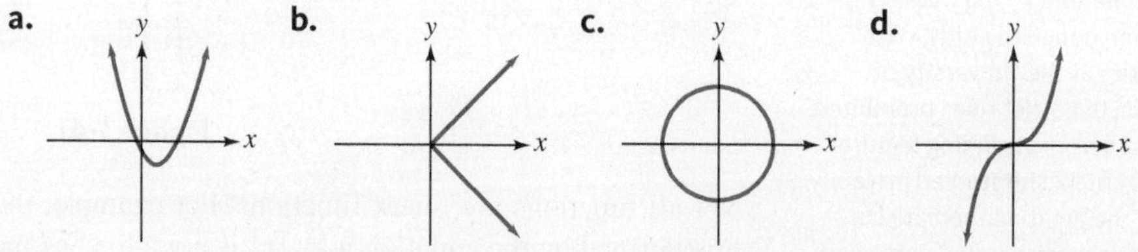

Solution

a. Since any vertical line that intersects the graph does so only once, the graph represents a function.

b. Since some vertical lines that intersect the graph do so twice, the graph does not represent a function.

Unless otherwise noted, all content on this page is © Cengage Learning.

c. Since some vertical lines that intersect the graph do so twice, the graph does not represent a function.

d. Since any vertical line that intersects the graph does so only once, the graph represents a function.

SELF CHECK 5 Determine whether each graph represents a function.

a.

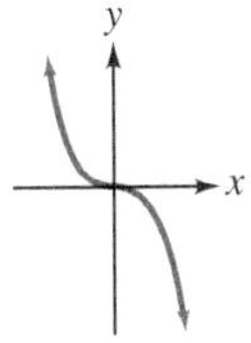

b.

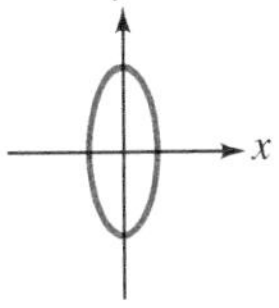

SELF CHECK ANSWERS

1. domain: $\{-3, -1, 4\}$; range: $\{-2, 3, 5\}$ **2. a.** yes **b.** no **3. a.** -7 **b.** 0 **c.** -13 **d.** $x = 7$
4. domain: $\mathbb{R}$; range: $\{y \mid y \text{ is a real number and } y \leq 0\}$

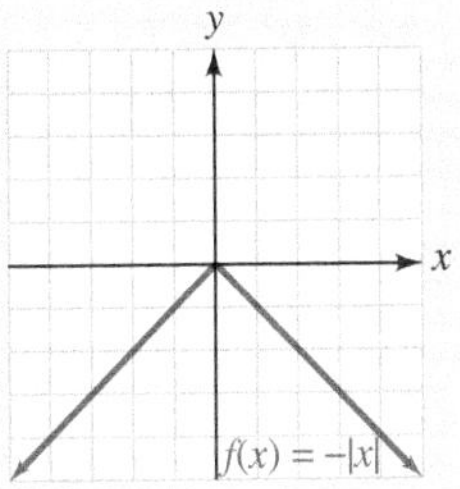

5. a. yes **b.** no

NOW TRY THIS

Refer to the graph and find:

1. $f(1)$

2. x when $f(x) = -3$

3. the domain

4. the range.

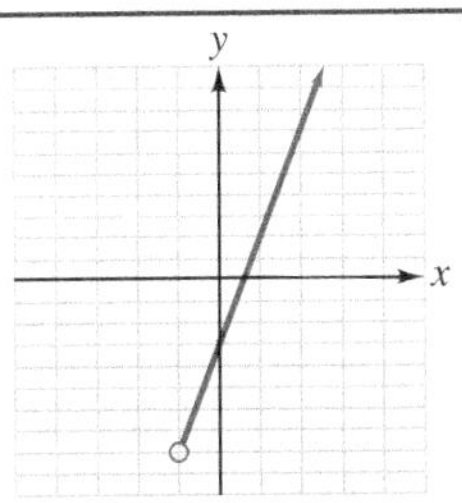

3.6 Exercises

WARM-UPS *Let $y = 2x + 1$. Find the value of y when*

1. $x = 0$
2. $x = 1$
3. $x = -1$
4. $x = -2$

REVIEW *Solve each equation.*

5. $4x = 3(x + 2)$
6. $6y + 3 = 7(y - 1)$
7. $5(2 - a) = 3(a + 6)$
8. $3x = 1.5(5 - x) - 12$

VOCABULARY AND CONCEPTS *Fill in the blanks.*

9. A ______ is a set of ordered pairs.

10. In a relation, the set of first components is called the ______ of the relation. The set of second components is called the ______.

11. Any equation in x and y where each ____ x determines exactly one output y is called a ______.

12. In a function, the set of all inputs is called the ______ of the function.

Unless otherwise noted, all content on this page is © Cengage Learning.

13. In a function, the set of all outputs is called the _____ of the function.

14. In the function $y = f(x)$, y is called the _________ variable.

15. In the function $y = f(x)$, x is called the _________ variable.

16. The function $y = f(x) = mx + b$ is a _____ function and $y = f(x) = |x|$ is called the _______ value function.

17. If a vertical line intersects a graph more than once, the graph ______ represent a function.

18. If $y = f(x)$, the value of y that is determined by $x = 4$ is denoted as ____.

GUIDED PRACTICE *Find the domain and range of each relation.* SEE EXAMPLE 1. (OBJECTIVE 1)

19. $\{(-3, -1), (1, 2), (3, 7)\}$

20. $\{(-5, -2), (0, 3), (3, 9), (5, 12)\}$

21. $\{(0, 5), (2, 7), (-3, -8), (4, 0)\}$

22. $\{(3, 7), (-2, -5), (3, -1), (2, -5)\}$

Determine whether the equation defines y to be a function of x. SEE EXAMPLE 2. (OBJECTIVE 2)

23. $y = -x + 1$

24. $y = \frac{1}{2}x - 3$

25. $x = -y^2$

26. $x = |y - 1|$

Find $f(3)$, $f(0)$, $f(-1)$, and the value of x for which $f(x) = 3$. SEE EXAMPLE 3. (OBJECTIVE 3)

27. $f(x) = -3x$

28. $f(x) = -4x$

29. $f(x) = 2x - 3$

30. $f(x) = 3x - 5$

31. $f(x) = 7 + 5x$

32. $f(x) = 3 - 3x$

33. $f(x) = 9 - 2x$

34. $f(x) = 12 + 3x$

35. $f(x) = \frac{1}{2}x + \frac{3}{2}$

36. $f(x) = -\frac{1}{3}x + \frac{1}{3}$

Graph each function and state its domain and range. SEE EXAMPLE 4. (OBJECTIVE 4)

37. $y = x + 2$

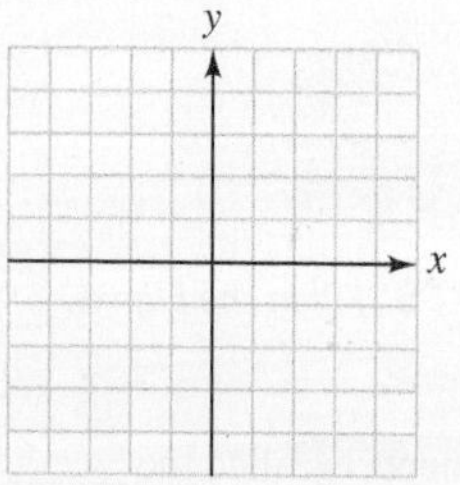

38. $y = -x + 1$

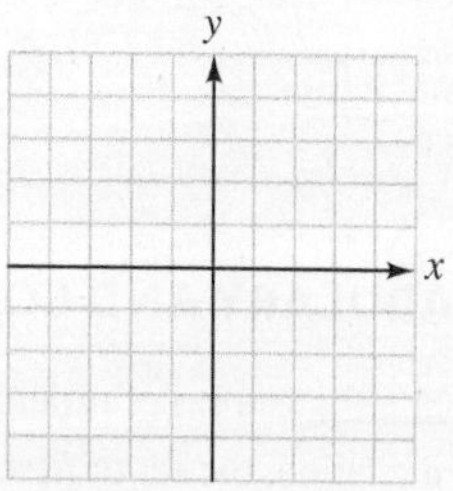

39. $f(x) = \frac{1}{2}|x|$

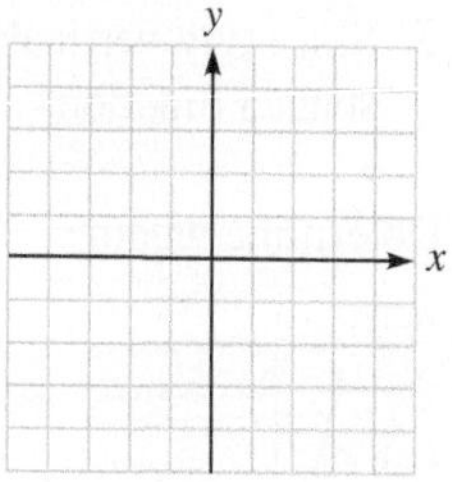

40. $f(x) = -2|x|$

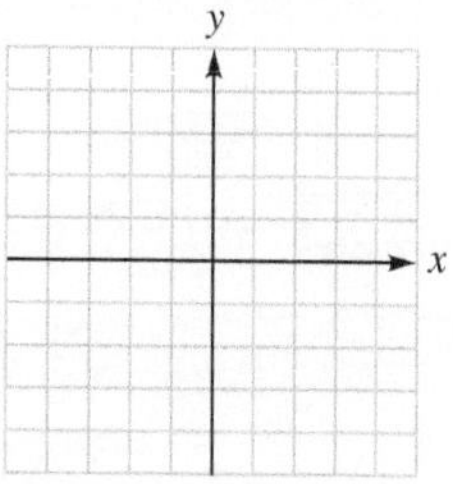

Determine whether each graph represents a function. SEE EXAMPLE 5. (OBJECTIVE 5)

41.

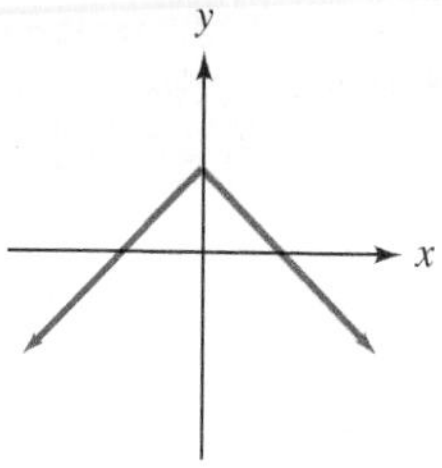

42.

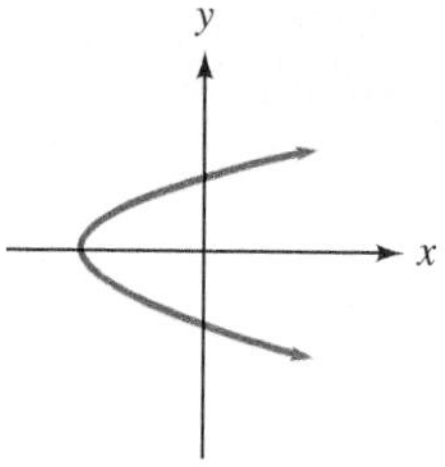

43.

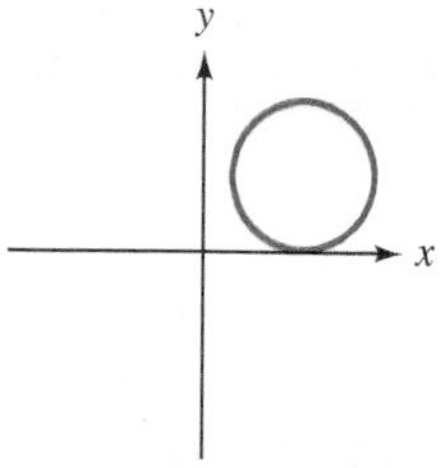

44.

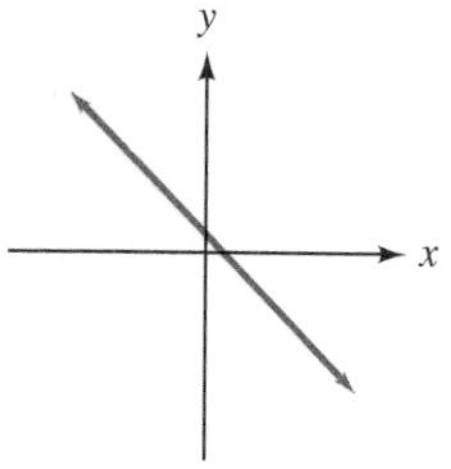

ADDITIONAL PRACTICE *Find $f(1)$, $f(-2)$, and $f(3)$.*

45. $f(x) = x^2 + 1$

46. $f(x) = x^2 - 2$

47. $f(x) = x^3 - 1$

48. $f(x) = x^3$

49. $f(x) = (x - 1)^2$

50. $f(x) = (x + 3)^2$

51. $f(x) = 3x^2 - 2x$

52. $f(x) = 5x^2 + 2x - 1$

53. $f(x) = \frac{1}{2}x^2 - 2x + 3$

54. $f(x) = \frac{3}{4}x^2 + 4x - 5$

Graph each function and state the domain and range.

55. $y = -\frac{1}{2}x + 2$

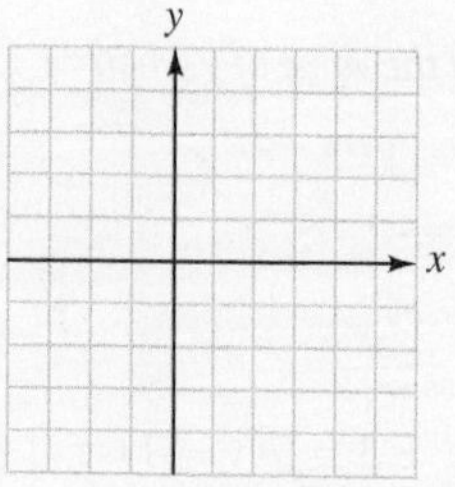

56. $y = \frac{1}{2}x - 3$

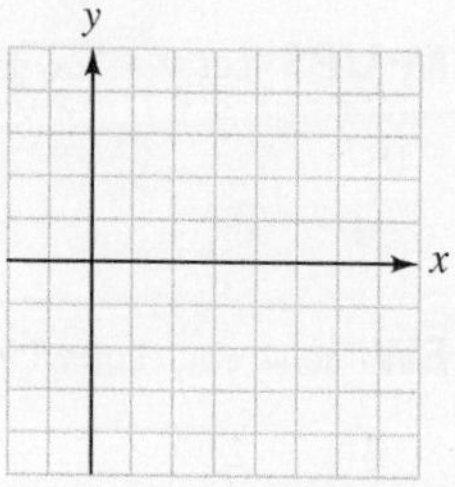

Unless otherwise noted, all content on this page is © Cengage Learning.

57. $f(x) = |x| - 1$

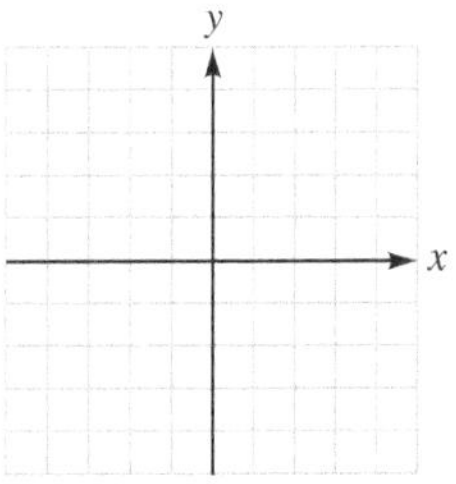

58. $f(x) = -|x| + 2$

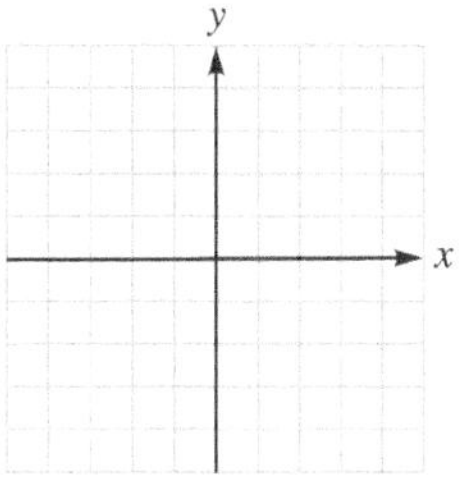

APPLICATIONS *In a certain city, the monthly cost of telephone service is given by the function $C = 0.10n + 12$, where n is the number of calls, 10¢ is the cost per call, and \$12 is a fixed charge.*

59. Find the cost of making 20 calls in a month.

60. Find the cost of making 60 calls in a month.

61. Find the cost of making 100 calls in a month.

62. Find the cost of making 400 calls in a month.

In a certain city, the monthly cost of residential electric service is given by the function $C = 0.08n + 17$, where n is the number of kilowatt hours (kWh) used, 8¢ is the cost per kWh, and \$17 is a fixed charge.

63. Find the monthly bill if 500 kWh were used.

64. Find the monthly bill if 600 kWh were used.

65. Find the monthly bill if 800 kWh were used.

66. Find the monthly bill if 1,000 kWh were used.

WRITING ABOUT MATH

67. Define the domain and range of a function.

68. Explain why the vertical line test works.

SOMETHING TO THINK ABOUT *Let $f(x) = 2x + 1$ and $g(x) = x$. Assume $f(x) \neq 0$ and $g(x) \neq 0$.*

69. Is $f(x) + g(x) = g(x) + f(x)$?

70. Is $f(x) - g(x) = g(x) - f(x)$?

71. Is $f(x) \cdot g(x) = g(x) \cdot f(x)$?

72. Is $f(x) + g(x) = f(x) \cdot g(x)$?

Section 3.7 Solving Linear Inequalities in Two Variables

Objectives

1. Solve a linear inequality in two variables.
2. Solve a compound linear inequality in two variables.
3. Solve an application requiring the use of a linear inequality in two variables.

Vocabulary

linear inequalities in two variables
boundary line
half-plane
test point
edge

Getting Ready

Do the coordinates of the given points satisfy the inequality $y < 5x + 3$?

1. $(0, 2)$ **2.** $\left(-\frac{2}{5}, 0\right)$ **3.** $(3, 18)$ **4.** $(-3, -13)$

Unless otherwise noted, all content on this page is © Cengage Learning.

In this section, we will show the solutions of linear inequalities in two variables. We begin by discussing their graphs.

1 Solve a linear inequality in two variables.

Inequalities such as

$$2x + 3y < 6 \quad \text{or} \quad 3x - 4y \geq 9$$

are called **linear inequalities in two variables**. In general, we have the following definition:

LINEAR INEQUALITIES IN TWO VARIABLES

A linear inequality in x and y is any inequality that can be written in the form

$$Ax + By < C, \quad Ax + By > C, \quad Ax + By \leq C, \quad \text{or} \quad Ax + By \geq C$$

where A, B, and C are real numbers and A and B are not both 0.

The *graph of a linear inequality* in x and y is the graph of all ordered pairs (x, y) that satisfy the inequality.

The inequality $y > 3x + 2$ is an example of a linear inequality in two variables because it can be written in the form $-3x + y > 2$. To graph it, we first graph the related equation $y = 3x + 2$ as shown in Figure 3-44(a). This **boundary line**, sometimes called an **edge**, divides the coordinate plane into two **half-planes**, one on either side of the line.

To find which half-plane is the graph of $y > 3x + 2$, we select a **test point** not on the line and substitute the coordinates into the inequality to determine if it results in a true statement. In this case, a convenient test point is the origin, $(0, 0)$.

$$y > 3x + 2$$
$$\mathbf{0} \stackrel{?}{>} 3(\mathbf{0}) + 2 \quad \text{Substitute 0 for } x \text{ and 0 for } y.$$
$$0 \not> 2$$

Since the coordinates do not satisfy $y > 3x + 2$, the origin is not part of the graph. Thus, the half-plane on the other side of the dashed line is the graph, which is shown in Figure 3-44(b). We use a *dashed* line to indicate that the boundary line is not included in the graph.

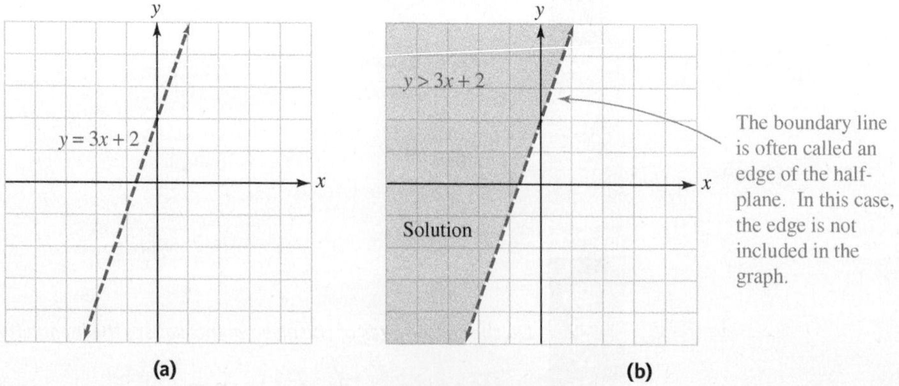
y
x
y = 3x + 2
(a)
y
x
y > 3x + 2
Solution
(b)
The boundary line is often called an edge of the half-plane. In this case, the edge is not included in the graph.

Figure 3-44

EXAMPLE 1 Solve: $2x - 3y \leq 6$

Solution We start by graphing the related equation $2x - 3y = 6$ to find the boundary line. This time, we draw the *solid* line shown in Figure 3-45(a) on the next page, because equality is included in the original inequality. To determine which half-plane represents $2x - 3y < 6$, we select a test point, $(0, 0)$, to determine whether the coordinates satisfy the inequality.

Unless otherwise noted, all content on this page is © Cengage Learning.

$$2x - 3y < 6$$
$$2(\mathbf{0}) - 3(\mathbf{0}) < 6 \quad \text{Substitute 0 for } x \text{ and 0 for } y.$$
$$0 < 6$$

Since the coordinates satisfy the inequality, the origin is in the half-plane that is the graph of $2x - 3y < 6$. The graph is shown in Figure 3-45(b).

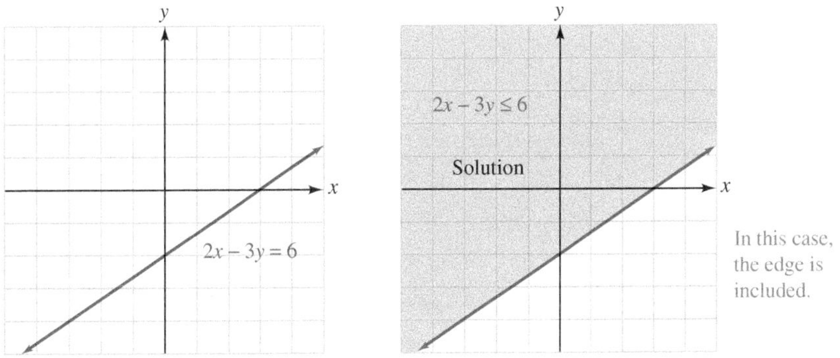

(a) (b)

Figure 3-45

SELF CHECK 1 Solve: $3x - 2y \geq 6$

EXAMPLE 2 Solve: $y < 2x$

Solution We start by graphing the related equation $y = 2x$, as shown in Figure 3-46(a). Because it is not part of the inequality, we draw the edge as a dashed line.

To determine which half-plane is the graph of $y < 2x$, we select a test point. We cannot use the origin because the edge passes through it. We select a different test point—say, $(3, 1)$.

$$y < 2x$$
$$\mathbf{1} \stackrel{?}{<} 2(\mathbf{3}) \quad \text{Substitute 1 for } y \text{ and 3 for } x.$$
$$1 < 6$$

Since $1 < 6$ is a true inequality, the point $(3, 1)$ satisfies the inequality and is in the graph, which is shown in Figure 3-46(b).

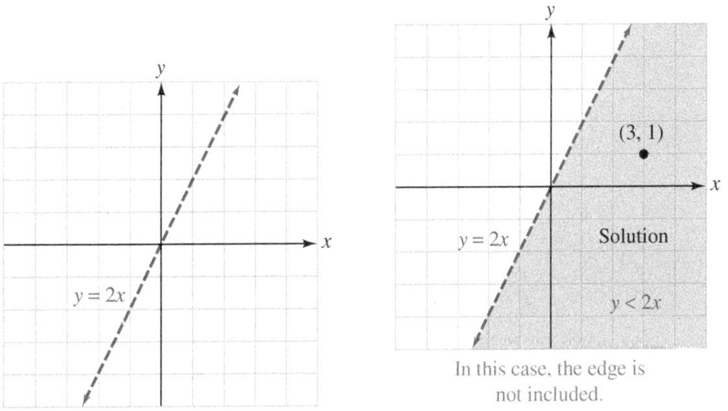

(a) (b)

Figure 3-46

SELF CHECK 2 Solve: $y > 2x$

Unless otherwise noted, all content on this page is © Cengage Learning.

COMMENT As the first two examples suggest, we draw a boundary line as a solid line when the inequality symbol is $\leq$ or $\geq$. We draw a dashed line when the inequality symbol is $<$ or $>$.

2 Solve a compound linear inequality in two variables.

EXAMPLE 3 Solve: $2 < x \leq 5$

Solution The inequality $2 < x \leq 5$ is equivalent to the following two inequalities:

$$2 < x + 0y \quad \text{and} \quad x \leq 5 + 0y$$

Its graph will contain all points in the plane that satisfy the inequalities $2 < x + 0y$ and $x \leq 5 + 0y$ simultaneously. These points are in the shaded region of Figure 3-47.

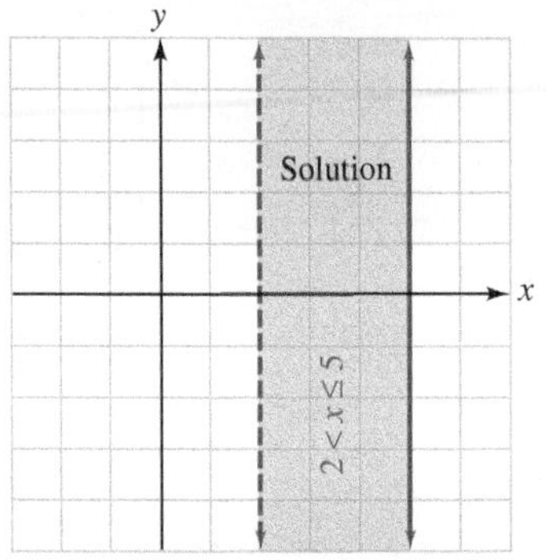

Figure 3-47

SELF CHECK 3 Solve: $-2 \leq x < 3$

3 Solve an application requiring the use of a linear inequality in two variables.

EXAMPLE 4 **EARNING MONEY** Rick has two part-time jobs, one paying \$7 per hour and the other paying \$5 per hour. He must earn at least \$140 per week to pay his expenses while attending college. Write and graph an inequality that shows the various ways he can schedule his time to achieve his goal.

Analyze the problem: If we let x represent the number of hours per week he works on the first job, he will earn $\$7x$ per week on the first job. If we let y represent the number of hours per week he works on the second job, he will earn $\$5y$ per week on the second job.

Form an inequality: To achieve his goal, the sum of these two incomes must be at least \$140.

The hourly rate on the first job	times	the hours worked on the first job	plus	the hourly rate on the second job	times	the hours worked on the second job	is greater than or equal to	\$140.
\$7	$\cdot$	x	$+$	\$5	$\cdot$	y	$\geq$	\$140

Solve the inequality: The graph of $7x + 5y \geq 140$ is shown in Figure 3-48 on the next page. Any point in the shaded region indicates a way that he can schedule his time and earn \$140 or more per week. For example, if he works 10 hours on the first job and 15 hours on the second job, he will earn

$$\begin{aligned} \$7(10) + \$5(15) &= \$70 + \$75 \\ &= \$145 \end{aligned}$$

Unless otherwise noted, all content on this page is © Cengage Learning.

COMMENT We could let x represent the number of hours he works on the second job and y represent the number of hours he works on the first job. The resulting graph would be different, but the combination of hours would remain the same.

If he works 5 hours on the first job and 25 hours on the second job, he will earn

$$\$7(5) + \$5(25) = \$35 + \$125$$
$$= \$160$$

Since Rick cannot work a negative number of hours, the graph has no meaning when x or y is negative, so only the first quadrant of the graph is shown.

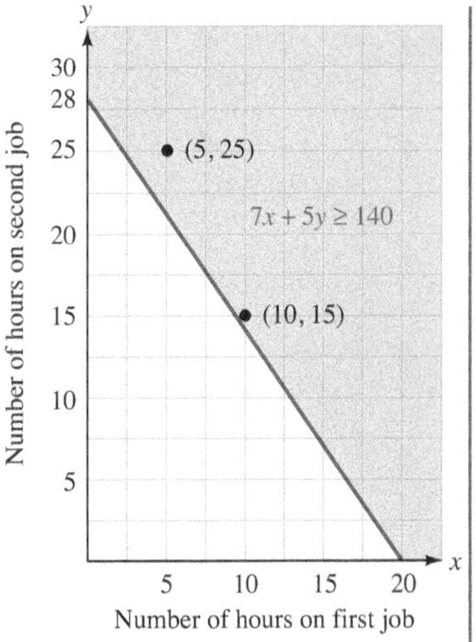

Figure 3-48

SELF CHECK 4 If Rick needs at least \$175 per week, write and graph an inequality that shows the various ways he can schedule his time.

Accent on technology

▸ Graphing Inequalities

The TI-84 graphing calculator has a graphing-style icon in the Y = editor. See Figure 3-49(a). Some of the different graphing styles are as follows.

\	line	A straight line or curved graph is shown.
◥	above	Shading covers the area above a graph.
◣	below	Shading covers the area below a graph.

We can change the icon by placing the cursor on it and pressing **ENTER** until the desired style is visible.

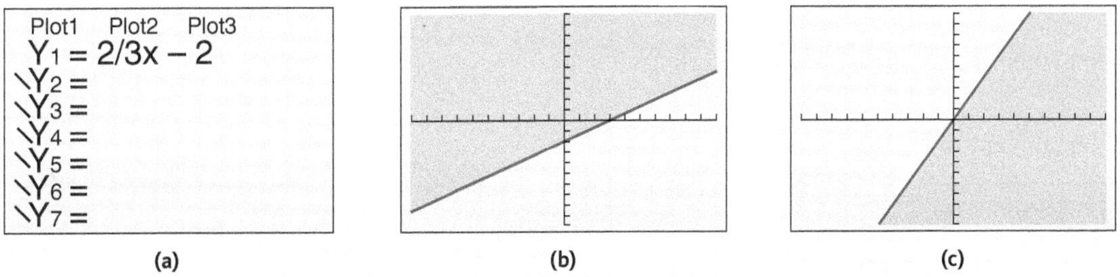

Figure 3-49

To graph the inequality of Example 1, $2x - 3y \leq 6$, using window settings of $x = [-10, 10]$ and $y = [-10, 10]$, we solve for y in the inequality $2x - 3y \leq 6$, $-3y \leq -2x + 6$, $y \geq \frac{2}{3}x - 2$. We enter the equation $2x - 3y = 6$ as $y = \frac{2}{3}x - 2$, change the graphing-style icon to "above" (◥), and press **GRAPH** to obtain Figure 3-49(b).

To graph the inequality of Example 2, $y < 2x$, we change the graphing-style icon to "below" (◣), enter the equation $y = 2x$, and press **GRAPH** to get Figure 3-49(c).

Note that graphing calculators do not distinguish between solid and dashed lines to show whether or not the edge of a region is included within the graph.

For instructions regarding the use of a Casio graphing calculator, please refer to the Casio Keystroke Guide in the back of the book.

Unless otherwise noted, all content on this page is © Cengage Learning.

SELF CHECK ANSWERS

1.
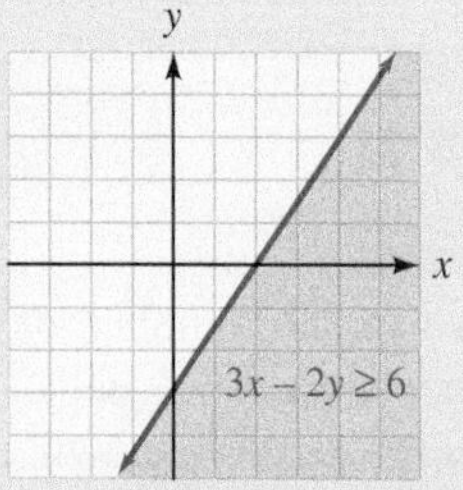

2.
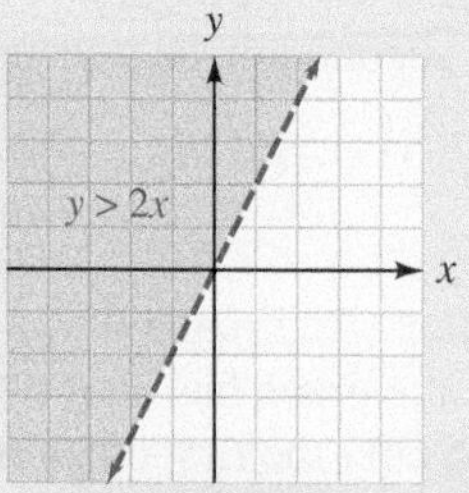

3.
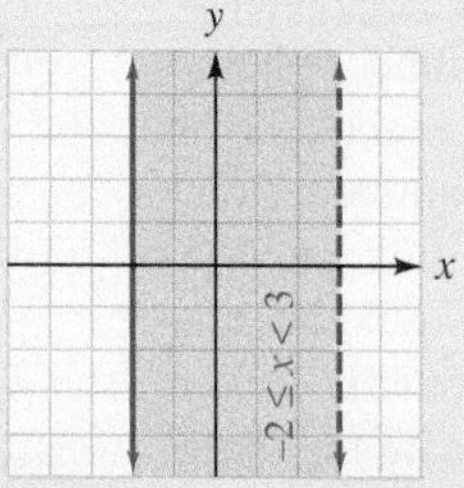

4. $7x + 5y \geq 175$
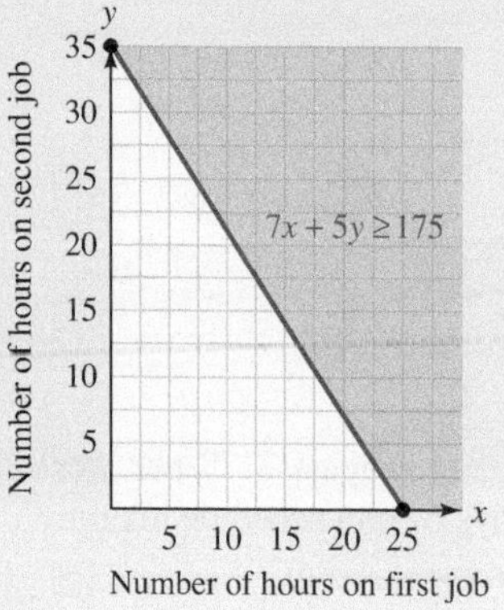

NOW TRY THIS

Solve each of the following by graphing.

1. $x \geq 2$ and $y < -1$
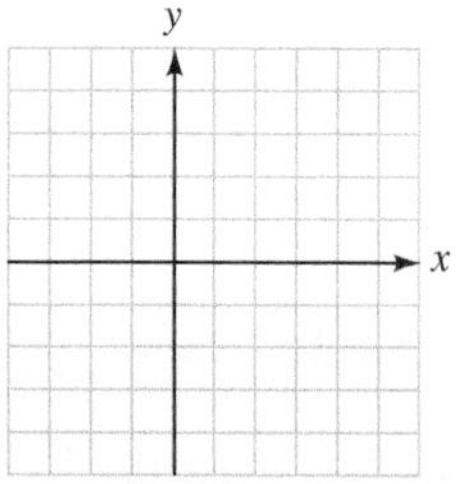

2. $x = 2$ and $y = -1$
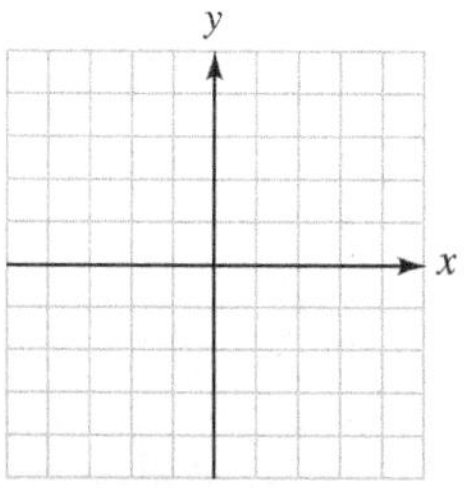

3.7 Exercises

WARM-UPS *Graph $2x + 5y = 10$. State whether the points with the given coordinates lie above or below the graph of the line.*

1. $(0, 0)$
2. $(2, 3)$
3. $(6, 1)$
4. $(-1, 1)$

Graph $x = 3$. State whether the points with the given coordinates lie to the right or left of the graph of the line.

5. $(0, 0)$
6. $(4, 1)$

Graph $y = -2$. State whether the points with the given coordinates lie above or below the graph of the line.

7. $(0, 0)$
8. $(1, -3)$

REVIEW *Solve each system.*

9. $\begin{cases} x + y = 3 \\ x - y = 5 \end{cases}$
10. $\begin{cases} 5x + 3y = -2 \\ x = -3y + 2 \end{cases}$
11. $\begin{cases} 4x + 3y = 12 \\ 5x - 2y = 15 \end{cases}$
12. $\begin{cases} 3x + 5y = 21 \\ 4x + 2y = 14 \end{cases}$

VOCABULARY AND CONCEPTS *Fill in the blanks.*

13. $3x + 2y < 12$ is an example of a _____ inequality in two variables.
14. Graphs of linear inequalities in two variables are __________.
15. The boundary line of a half-plane is called an ____.
16. If an inequality involves $\leq$ or $\geq$, the boundary line of its graph will be ____.

Unless otherwise noted, all content on this page is © Cengage Learning.

17. If an inequality involves $<$ or $>$, the boundary line of its graph will be ______.

18. If $y < \frac{1}{2}x - 2$ and $y > \frac{1}{2}x - 2$ are false, then __________.

GUIDED PRACTICE *Solve each inequality.* SEE EXAMPLE 1. (OBJECTIVE 1)

19. $y \geq x$

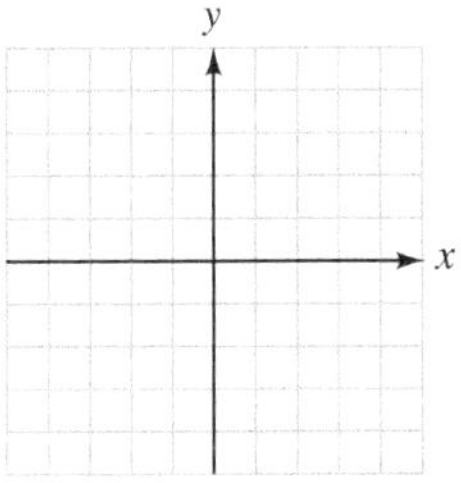

20. $y \leq 2x$

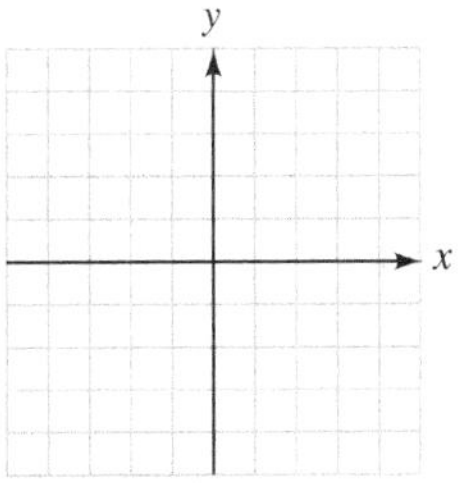

21. $y \geq -2$

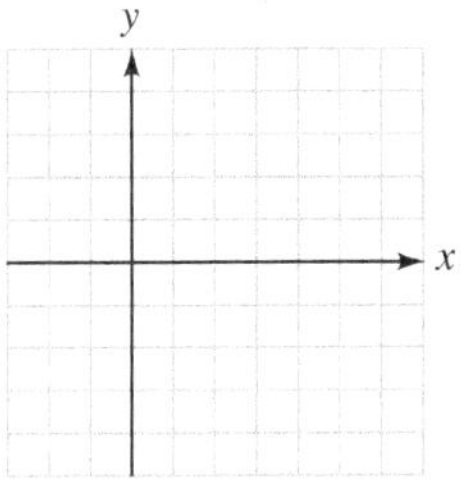

22. $x \geq -3$

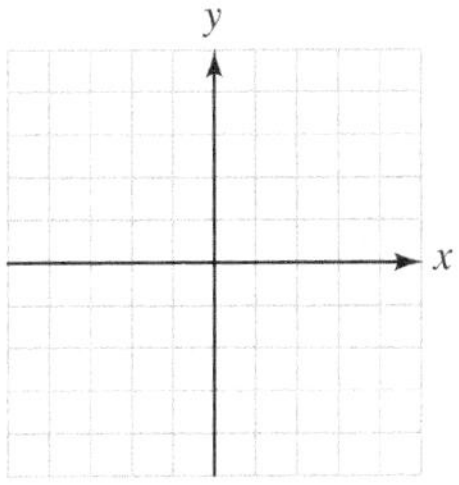

Solve each inequality. SEE EXAMPLE 2. (OBJECTIVE 1)

23. $y > x + 1$

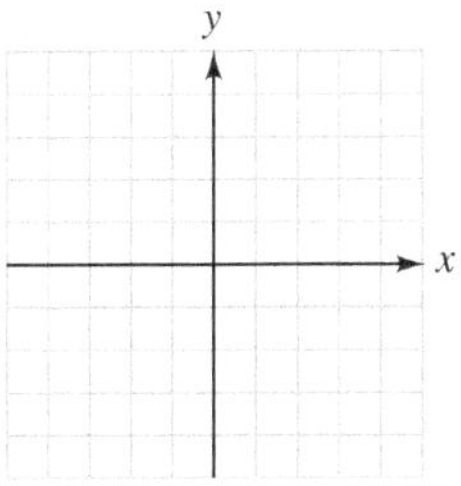

24. $y < 2x - 1$

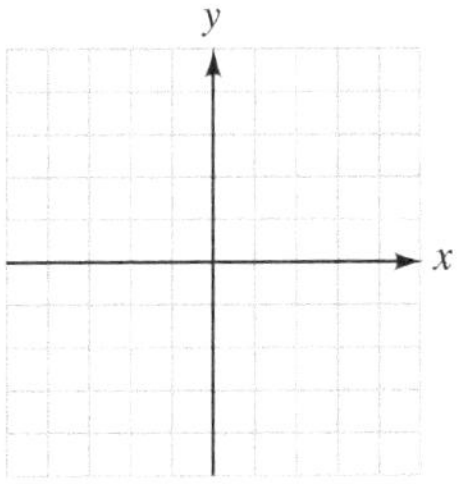

25. $x < 4$

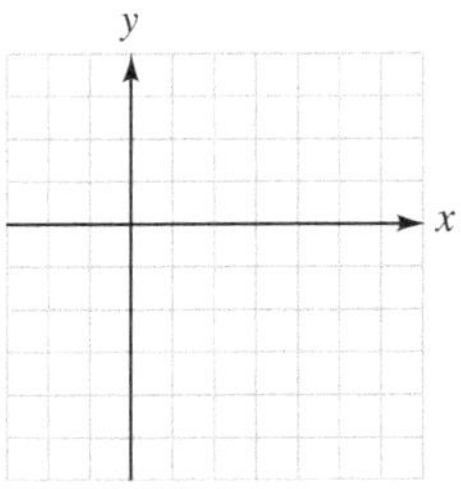

26. $y < 3$

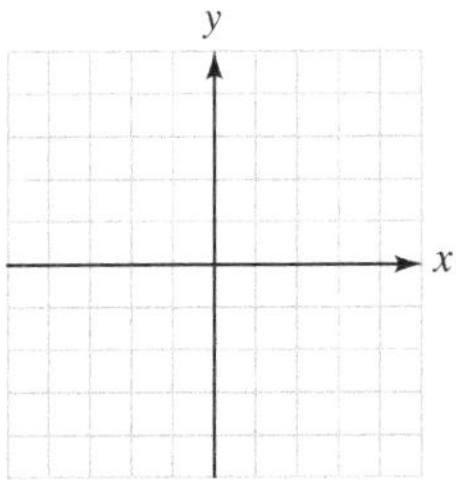

Solve each inequality. SEE EXAMPLE 3. (OBJECTIVE 2)

27. $-2 \leq x < 0$

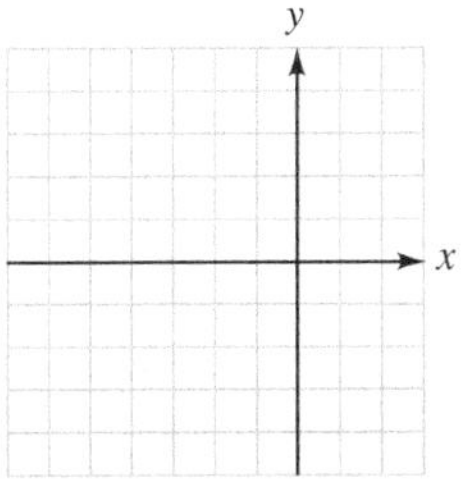

28. $-3 < y \leq -1$

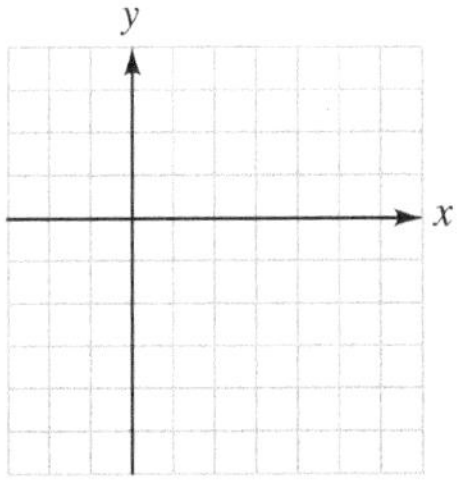

29. $y < -2$ or $y > 3$

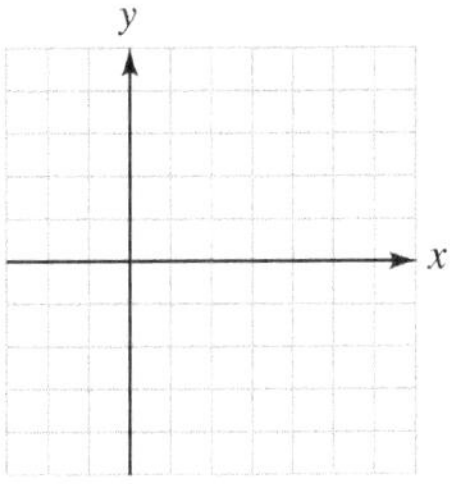

30. $-x \leq 1$ or $x \geq 2$

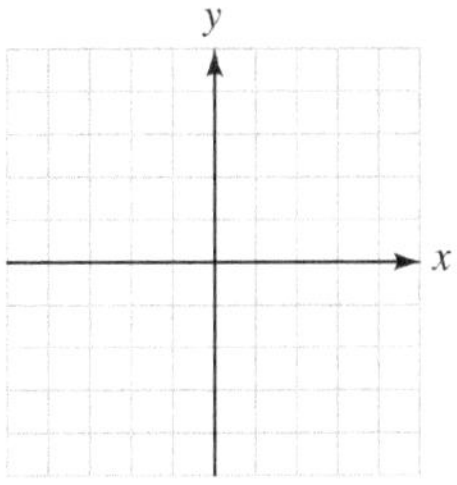

ADDITIONAL PRACTICE *Solve each inequality.*

31. $y \geq 1 - \dfrac{3}{2}x$

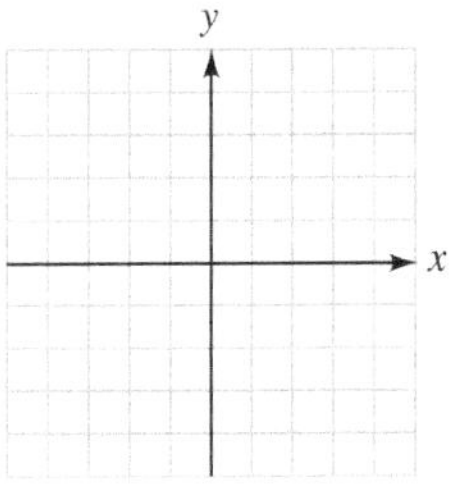

32. $x - 2y \geq 4$

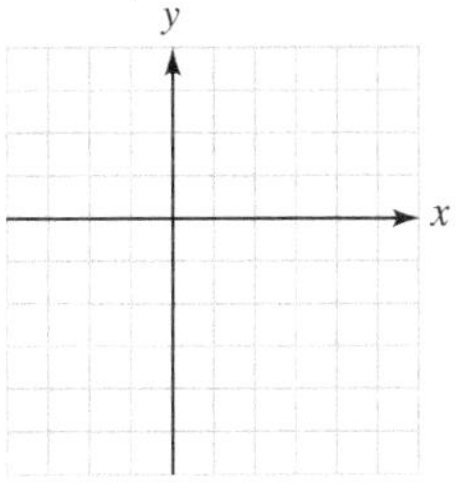

33. $0.5x + y > 1.5 + x$

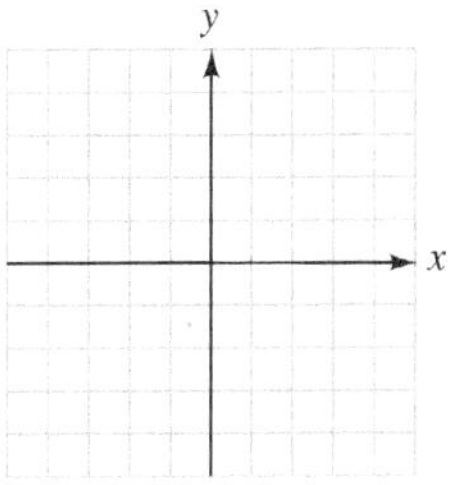

34. $0.5x + 0.5y \leq 2$

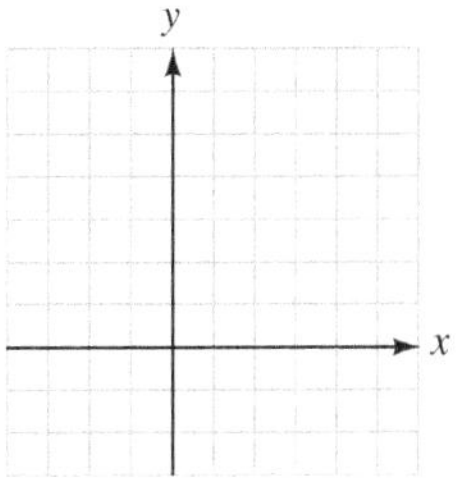

35. $y < \dfrac{1}{3}x - 1$

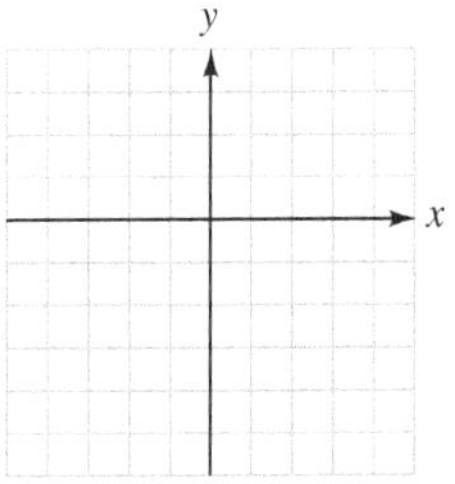

36. $2x + y \leq 6$

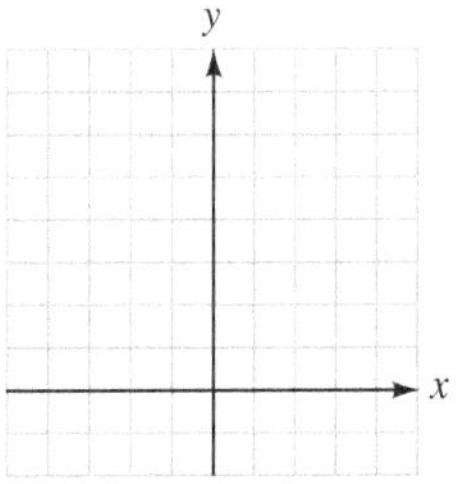

37. $3x \geq -y + 3$

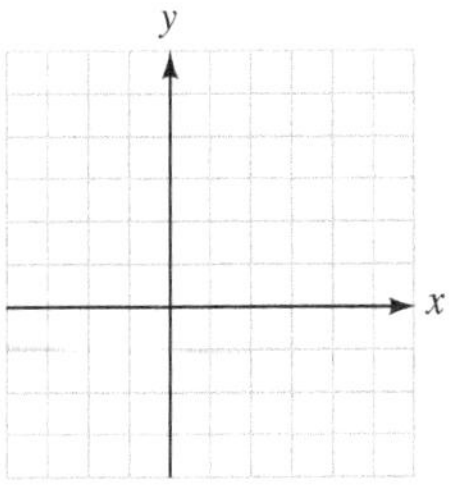

38. $2x \leq -3y - 12$

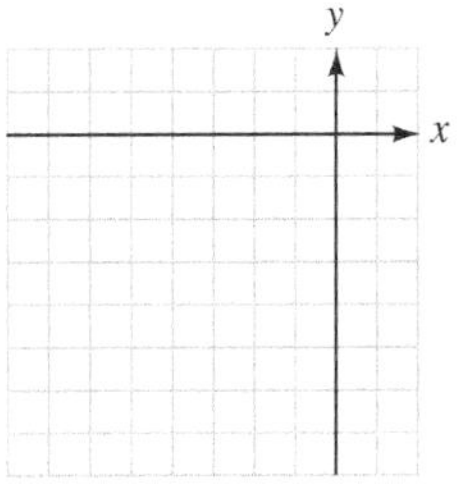

Unless otherwise noted, all content on this page is © Cengage Learning.

Use a graphing calculator to solve each inequality.

39. $y < 0.27x - 1$

40. $y > -3.5x + 2.7$

41. $y \geq -2.37x + 1.5$

42. $y \leq 3.37x - 1.7$

Find the equation of the boundary line or lines. Then write the inequality whose graph is shown.

43.

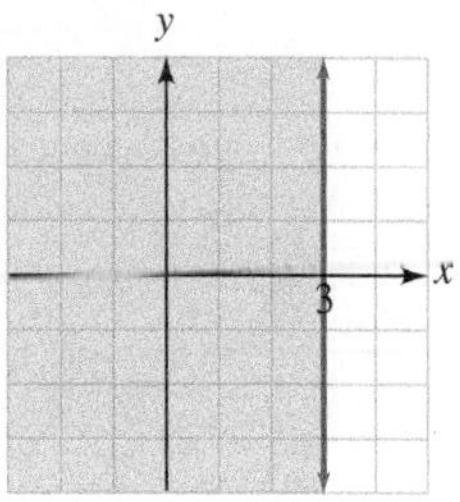

44.

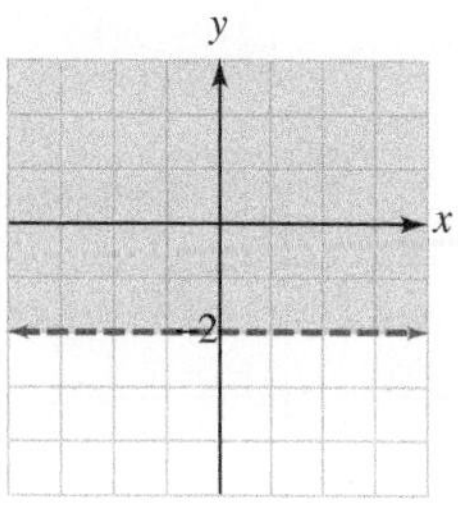

45.

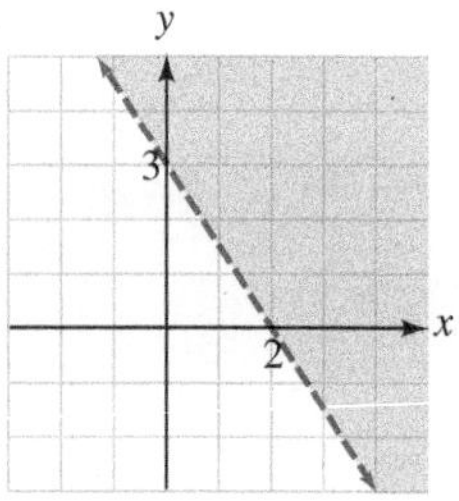

46.

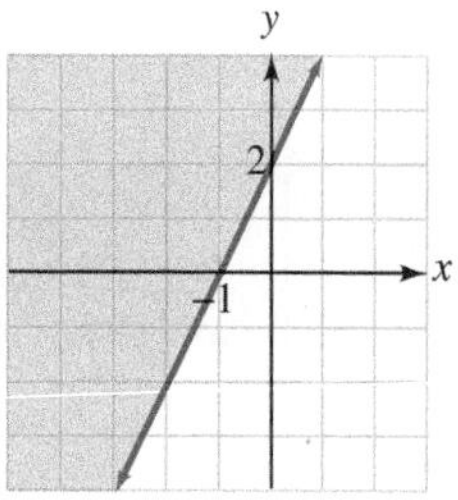

47.

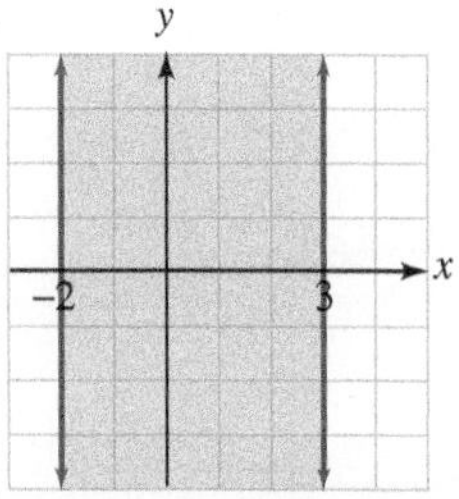

48.

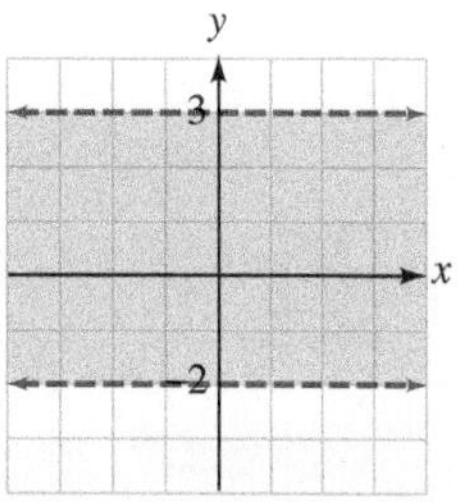

49.

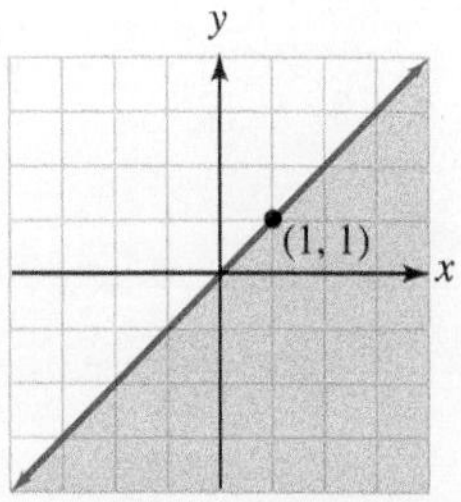

50.

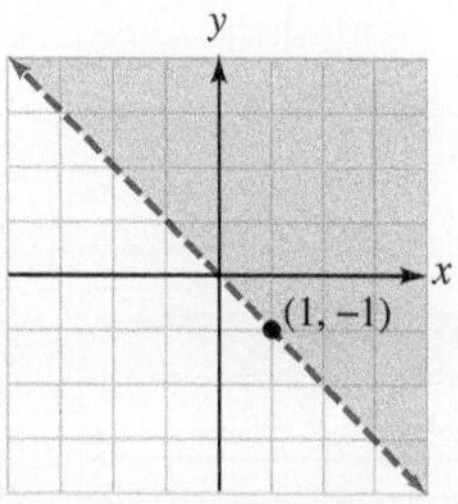

51.

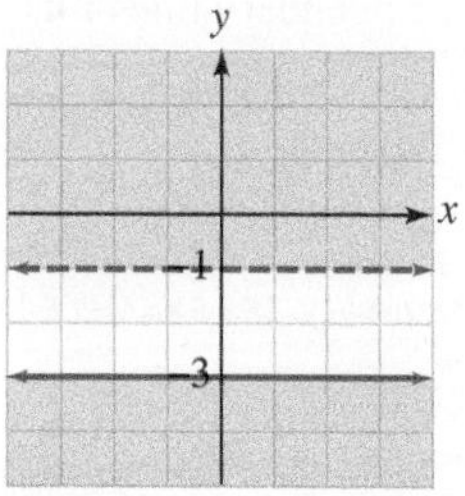

52.

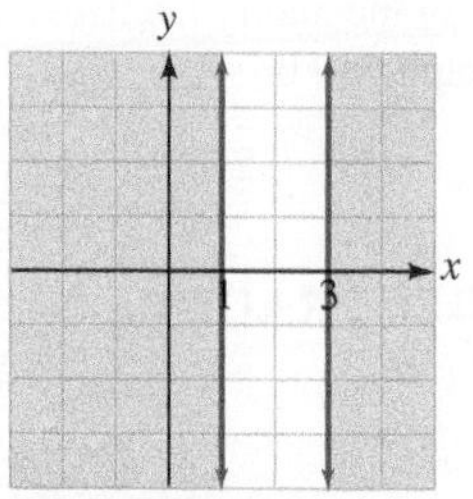

APPLICATIONS

Graph each inequality for nonnegative values of x and y. Then state three ordered pairs that satisfy the inequality. SEE EXAMPLE 4. (OBJECTIVE 3)

53. Figuring taxes On average, it takes an accountant 1 hour to complete a simple tax return and 3 hours to complete a complicated return. If the accountant wants to work no more than 9 hours per day, graph an inequality that shows the possible ways that the number of simple returns (x) and complicated returns (y) can be completed each day.

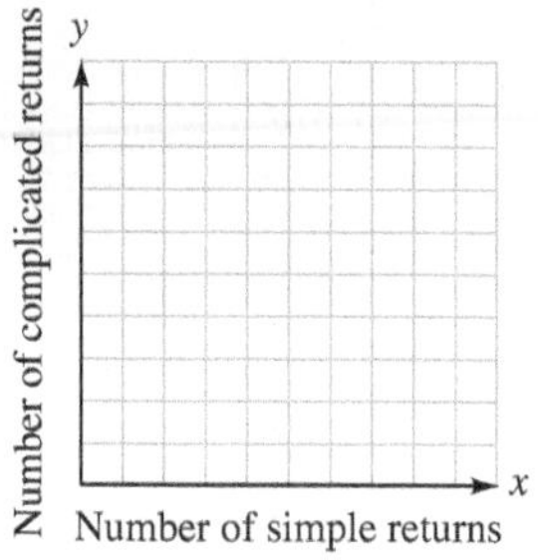

54. Selling trees During a sale, a garden store sold more than \$2,000 worth of trees. If a 6-foot maple costs \$100 and a 5-foot pine costs \$125, graph an inequality that shows the possible ways that the number of maple trees (x) and pine trees (y) were sold.

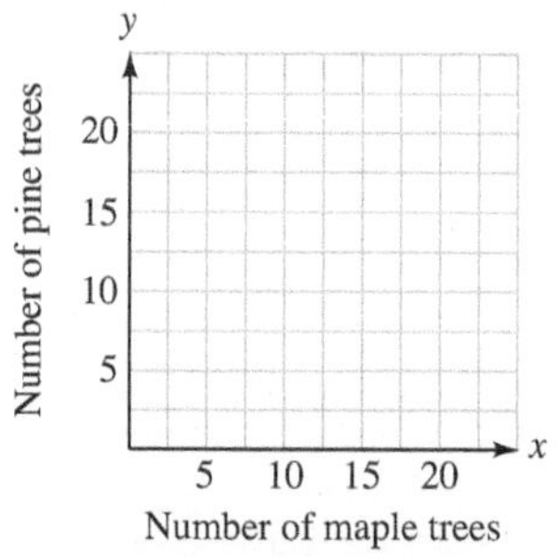

55. Choosing housekeepers One housekeeper charges \$6 per hour, and another charges \$7 per hour. If Sarah can afford no more than \$42 per week to clean her house, graph an inequality that shows the possible ways that she can hire the first housekeeper for x hours and the second housekeeper for y hours.

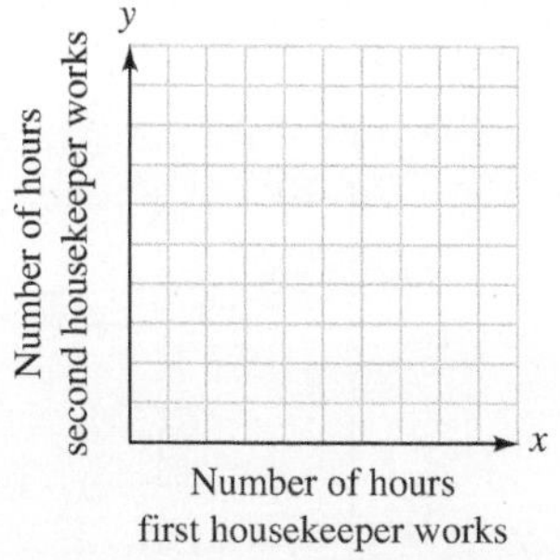

56. Making sporting goods A sporting goods manufacturer allocates at least 1,200 units of time per day to make fishing rods and reels. If it takes 10 units of time to make a rod and 15 units of time to make a reel, graph an inequality that shows the possible ways to schedule the time to make x rods and y reels.

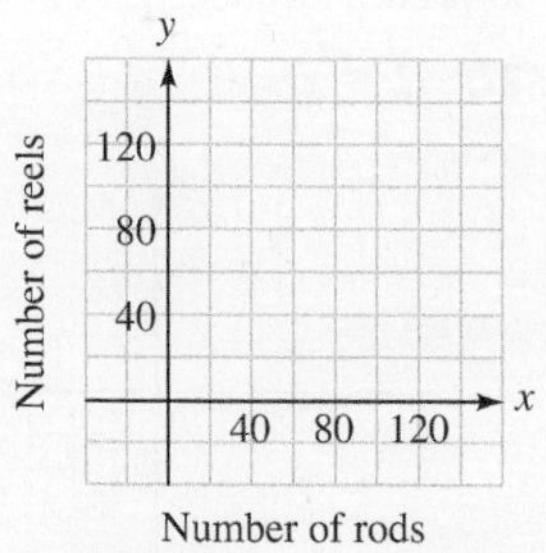

Unless otherwise noted, all content on this page is © Cengage Learning.

57. **Investing** A woman has up to $6,000 to invest. If stock in Traffico sells for $50 per share and stock in Cleanco sells for $60 per share, graph an inequality that shows the possible ways that she can buy x shares of Traffico and y shares of Cleanco.

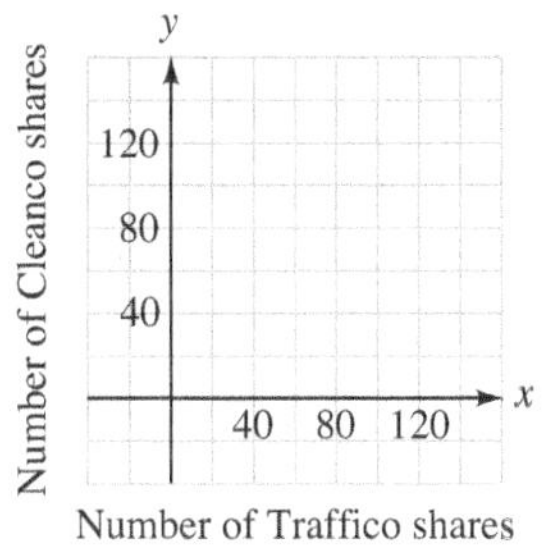

58. **Concert tickets** Tickets to a children's matinee concert cost $6 for reserved seats and $4 for general admission. If receipts must be at least $10,200 to meet expenses, graph an inequality that shows the possible ways that the box office can sell x reserved seats and y general admission tickets to meet expenses.

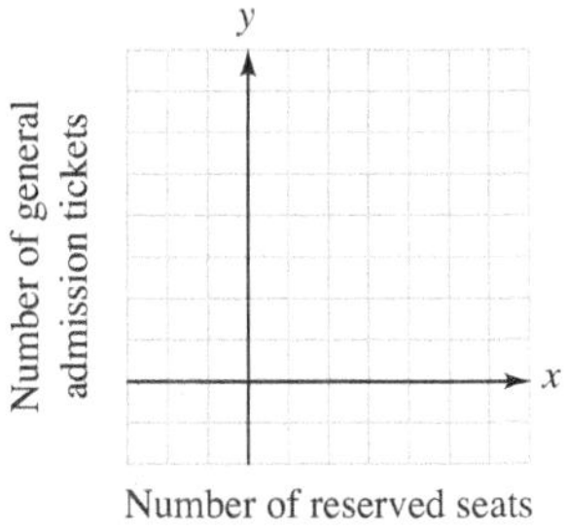

WRITING ABOUT MATH

59. Explain how to decide where to draw the boundary of the graph of a linear inequality, and whether to draw it as a solid or a broken line.
60. Explain how to decide which side of the boundary of the graph of a linear inequality should be shaded.

SOMETHING TO THINK ABOUT

61. Can an inequality be an identity, one that is satisfied by all (x, y) pairs? Illustrate.
62. Can an inequality have no solutions? Illustrate.

Projects

PROJECT 1

In Section 3.6, we defined functions in terms of an equation in x and y, where each input number x determines a single output number y. In this case, all functions were functions whose domains and ranges were subsets of the real numbers. The concept of function can be defined in a more general way.

> *A **function** is a correspondence between a set of input values x (called the **domain**) and a set of output values y (called the **range**), where to each x-value in the domain there corresponds exactly one y-value in the range.*

Using this definition, the domain and range of a function do not have to be sets of real numbers. For example, the correspondence formed by the set of states in the U.S. and the set of governors determines a function because:

To every state, there corresponds exactly one governor.

In this function, the domain is the set of states and the range is the set of governors.

However, many correspondences in the real world do not determine functions. For example, the correspondence formed by mothers and their children does not determine a function because:

To every mother, there corresponds one or more children.

a. Think of five correspondences in the real world that determine functions.

b. Think of five correspondences in the real world that do not determine functions.

PROJECT 2

Graphs are often used in newspapers and magazines to convey complex information at a glance. Unfortunately, it is easy to use graphs to convey misleading information. For example, the profit percents of a company for several years are given in the table and two graphs in Figure 3-50 on the next page.

The first graph in the figure accurately indicates the company's steady performance over five years. But because the vertical axis of the second graph does not start at zero, the performance appears deceptively erratic.

Year	Profit
2006	6.2%
2007	6.0%
2008	6.2%
2009	6.1%
2010	6.3%
2011	6.6%

(continued)

Unless otherwise noted, all content on this page is © Cengage Learning.

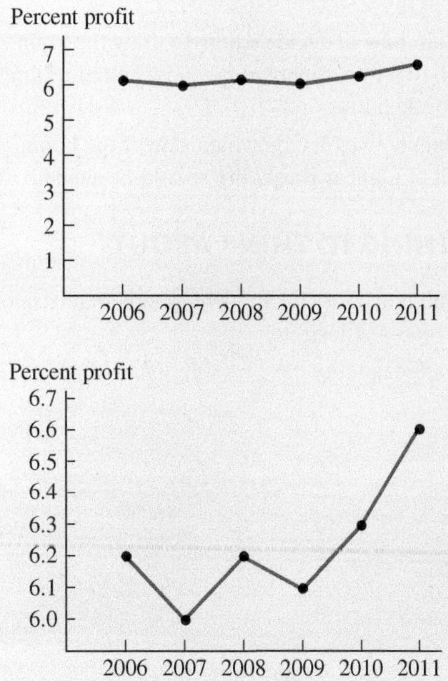

Figure 3-50

As your college's head librarian, you spend much of your time writing reports, either trying to make the school library look good (for college promotional literature) or bad (to encourage greater funding). In 2006, the library held a collection of 17,000 volumes. Over the years, the library has acquired many new books and has retired several old books. The details appear in the table.

Using the data in the table,

- Draw a misleading graph that makes the library look good.
- Draw a misleading graph that makes the library look bad.
- Draw a graph that accurately reflects the library's condition.

Year	Volumes acquired	Volumes removed
2006	215	137
2007	217	145
2008	235	185
2009	257	210
2010	270	200
2011	275	180

Unless otherwise noted, all content on this page is © Cengage Learning.

Reach for Success

EXTENSION OF UNDERSTANDING YOUR SYLLABUS

Now that you know where your instructor's office is, how to get in touch with him or her, and how your grade will be determined, take a look at other information in the syllabus that may be important to know before you get too far into the semester.

On your syllabus, locate your instructor's make-up policy. If you know of any conflicts in meeting established deadlines, contact your instructor *prior* to the due date.	Does your instructor accept late work? ______________ If so, under what conditions? State the work that is accepted late and any penalty assessed. Test make-up policy ______________ Penalty assessed ______________ Other requirements make-up policy ______________ Penalty assessed ______________ If not, try to reschedule any personal obligations.
On your syllabus, locate your instructor's attendance policy.	What is your instructor's attendance policy? ______________ What does the syllabus indicate about arriving late or leaving class early? ______________

Understanding any penalties for late work can make a difference in the successful completion of the course. Do not hesitate to contact your instructor if you are unclear regarding any of the policies stated on the syllabus.

3 Review

SECTION 3.1 The Rectangular Coordinate System

DEFINITIONS AND CONCEPTS

Any **ordered pair of real numbers** represents a point on the rectangular coordinate system.

The point where the axes cross, $(0, 0)$, is called the **origin**.
The four regions of a coordinate plane are called **quadrants**.

EXAMPLES

Plot $(2, 6), (-2, 6), (-2, -6), (2, -6)$, and $(0, 0)$.

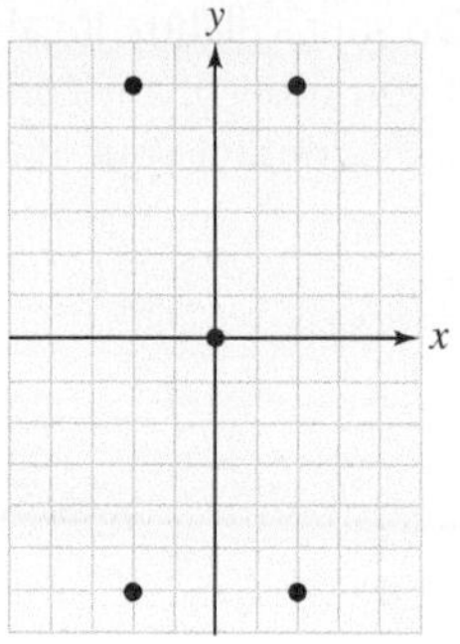

The origin is represented by the ordered pair $(0, 0)$.

The ordered pair $(2, 6)$ is found in quadrant I.

The ordered pair $(-2, 6)$ is found in quadrant II.

The ordered pair $(-2, -6)$ is found in quadrant III.

The ordered pair $(2, -6)$ is found in quadrant IV.

REVIEW EXERCISES

Plot each point on the rectangular coordinate system in the illustration.

1. $A(1, 3)$
2. $B(1, -3)$
3. $C(-3, 1)$
4. $D(-3, -1)$
5. $E(0, 5)$
6. $F(-5, 0)$

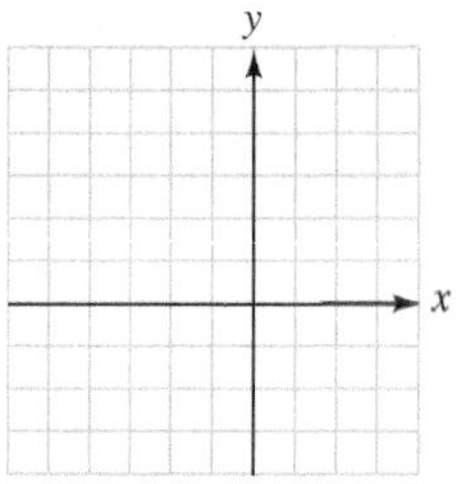

Find the coordinates of each point in the illustration.

7. A
8. B
9. C
10. D
11. E
12. F
13. G
14. H

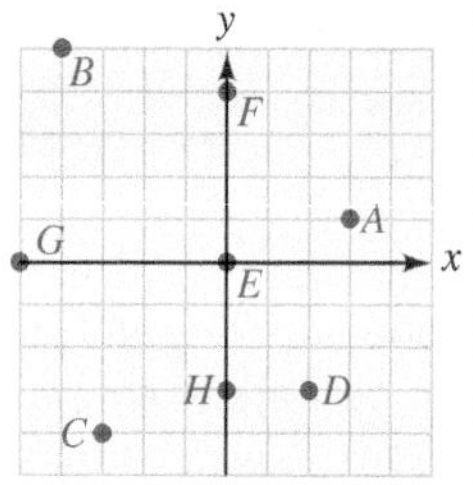

SECTION 3.2 Graphing Linear Equations

DEFINITIONS AND CONCEPTS

An ordered pair of real numbers is a **solution** to an equation in two variables if it satisfies the equation.

EXAMPLES

The ordered pair $(-1, 5)$ satisfies the equation $x - 2y = -11$.

$$(-1) - 2(5) \stackrel{?}{=} -11 \quad \text{Substitute } -1 \text{ for } x \text{ and 5 for } y.$$
$$-1 - 10 \stackrel{?}{=} -11$$
$$-11 = -11 \quad \text{True.}$$

Since the results are equal, $(-1, 5)$ is a solution.

DEFINITIONS AND CONCEPTS

To graph a linear equation,

1. Find two ordered pairs (x, y) that satisfy the equation.
2. Plot each pair on the rectangular coordinate system.
3. Draw a line passing through the two points.

General form of an equation of a line:

$Ax + By = C$
(A and B are not both 0.)

EXAMPLES

Graph: $x - y = -2$.

Create a table of values by either using the original equation or by solving for y first.

$$x - y = -2 \quad \text{This is the original equation.}$$
$$-y = -x - 2 \quad \text{Subtract } x \text{ from both sides.}$$
$$y = x + 2 \quad \text{Divide both sides by } -1.$$

Unless otherwise noted, all content on this page is © Cengage Learning.

Then find two ordered pairs that satisfy the equation. Use a third as a check.

$x - y = -2$

x	y	(x, y)
1	3	(1, 3)
2	4	(2, 4)
−3	−1	(−3, −1)

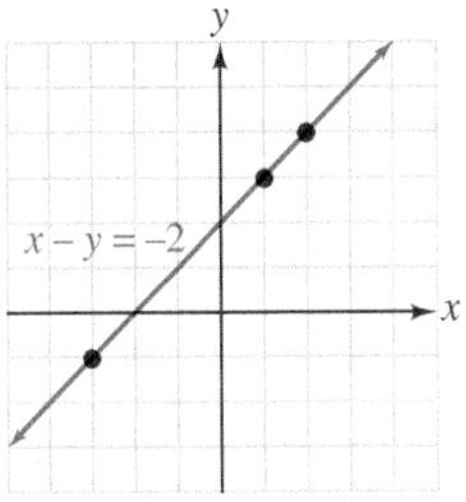

We then plot the points and draw a line passing through them.

The equation $y = b$ represents a horizontal line that intersects the y-axis at $(0, b)$.

The graph of $y = 5$ is a horizontal line passing through $(0, 5)$.

The equation $x = a$ represents a vertical line that intersects the x-axis at $(a, 0)$.

The graph of $x = 3$ is a vertical line passing through $(3, 0)$.

REVIEW EXERCISES

Determine whether each pair satisfies the equation $3x - 4y = 12$.

15. $(0, -3)$

16. $\left(\frac{4}{3}, 2\right)$

Graph each equation on a rectangular coordinate system.

17. $y = x - 5$

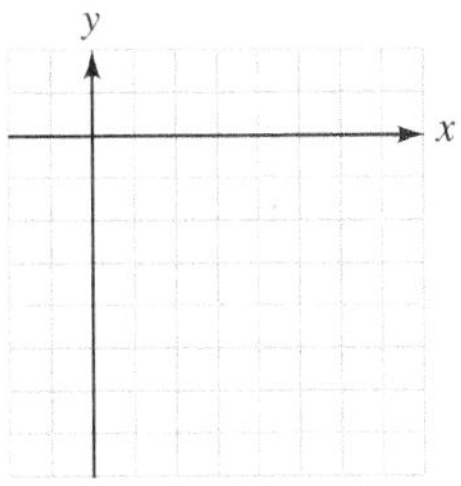

18. $y = 2x + 1$

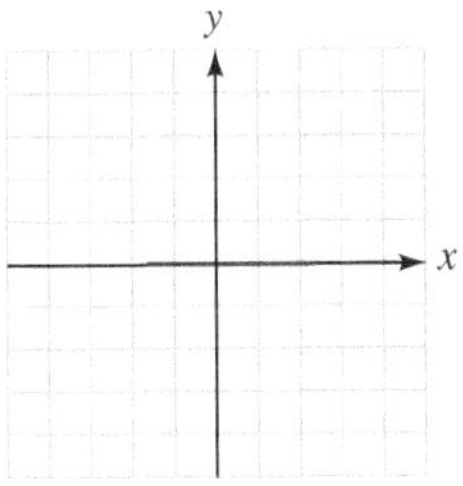

19. $y = \frac{x}{2} + 2$

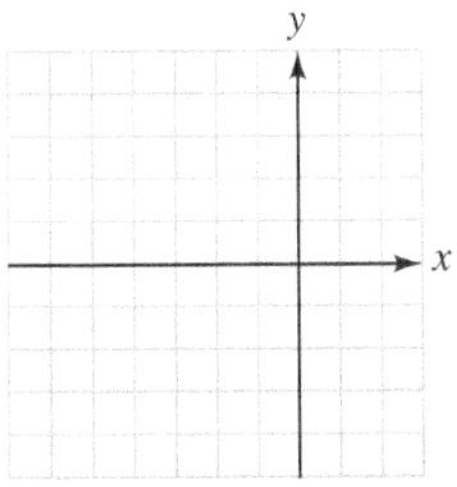

20. $y = 3$

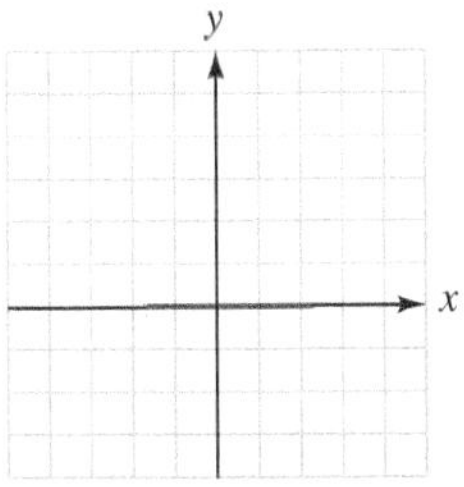

21. $x + y = 4$

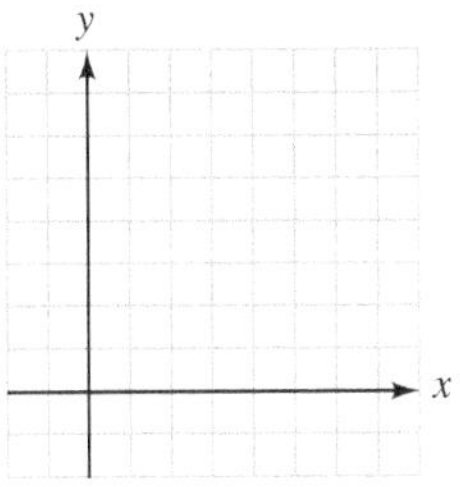

22. $x - y = -3$

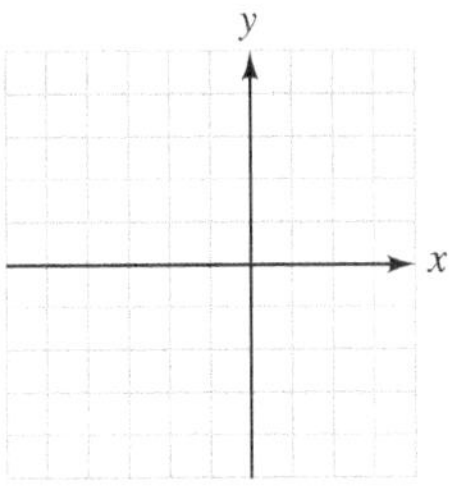

23. $3x + 5y = 15$

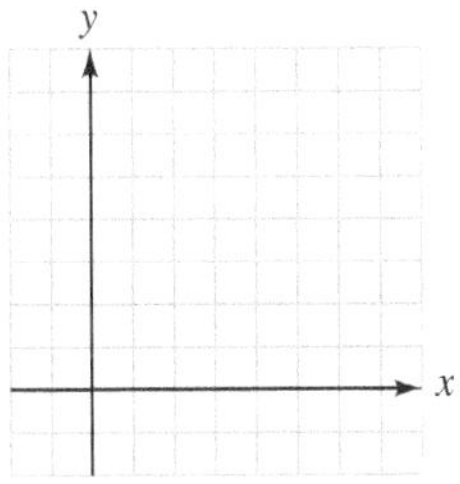

24. $7x - 4y = 28$

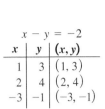

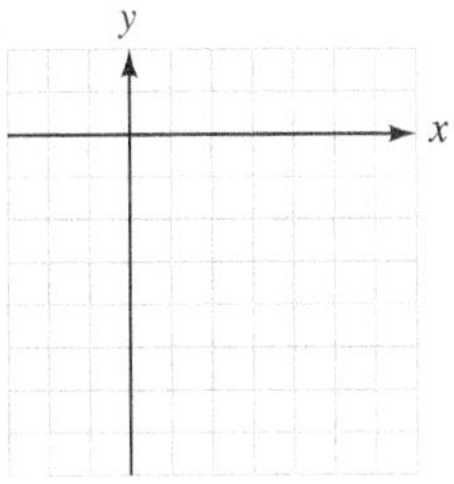

SECTION 3.3 Slope of a Line

DEFINITIONS AND CONCEPTS

The **slope** of a line passing through (x_1, y_1) and (x_2, y_2) is given by the formula

$$m = \frac{y_2 - y_1}{x_2 - x_1} \quad (x_2 \neq x_1)$$

EXAMPLES

The slope of a line passing through $(-1, 4)$ and $(5, -3)$ is given by

$$m = \frac{y_2 - y_1}{x_2 - x_1} = \frac{-3 - 4}{5 - (-1)} = \frac{-7}{6} = -\frac{7}{6}$$

Unless otherwise noted, all content on this page is © Cengage Learning.

Horizontal lines have a slope of 0.	$y = 5$ is a horizontal line with slope 0.
Vertical lines have an undefined slope.	$x = 5$ is a vertical line with an undefined slope.
Parallel lines have the same slope.	Two lines with slopes $\frac{2}{3}$ and $\frac{2}{3}$ are parallel.
The product of the slopes of nonvertical perpendicular lines is -1.	Two lines with slopes 3 and $-\frac{1}{3}$ are perpendicular.

REVIEW EXERCISES

Find the slope of the line passing through the given points.

25. $(3, 8), (7, 2)$
26. $(-1, 3), (3, -2)$
27. $(-2, -5), (-4, 9)$
28. $(-8, 2), (3, 2)$

Find the slope of the line determined by each equation.

29. $5x - 2y = 10$
30. $x = -4$

Determine whether the slope of each line is positive, negative, 0, or undefined.

31.

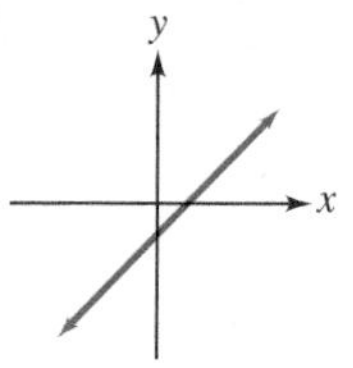

32.

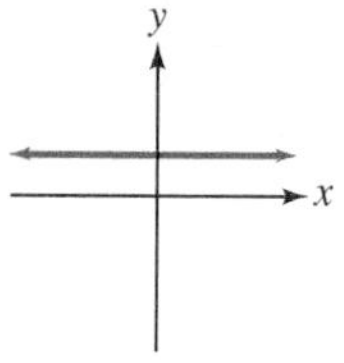

33.

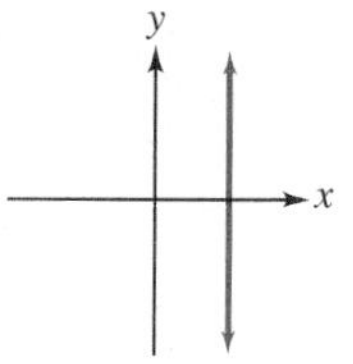

34.

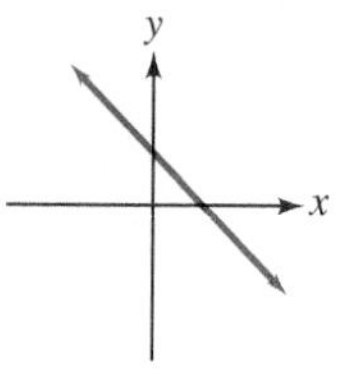

Determine whether lines with the given slopes are parallel, perpendicular, or neither parallel nor perpendicular.

35. 4 and $-\frac{1}{4}$
36. 0.2 and $\frac{1}{5}$
37. -5 and $-\frac{1}{5}$
38. $\frac{1}{2}$ and 2
39. -7 and $\frac{1}{7}$
40. 0.25 and $\frac{1}{4}$

41. Find the slope of a roof if it rises 4 feet for every run of 12 feet.

42. Find the average rate of growth of a business if sales were $25,000 the first year and $66,000 the third year.

SECTION 3.4 Point-Slope Form

DEFINITIONS AND CONCEPTS	EXAMPLES
Point-slope form of a line: $y - y_1 = m(x - x_1)$ where m is the slope and (x_1, y_1) is a point on the line.	To write the point-slope form of a line passing through $(6, 5)$ with slope $\frac{1}{2}$, proceed as follows: $y - y_1 = m(x - x_1)$ This is point-slope form. $y - \mathbf{5} = \frac{\mathbf{1}}{\mathbf{2}}(x - \mathbf{6})$ Substitute. $y - 5 = \frac{1}{2}(x - 6)$

REVIEW EXERCISES

Use the point-slope form of a linear equation to find the equation of each line. Then solve the equation for y.

43. $m = 5$ and passing through $(-2, 3)$

44. $m = -\frac{1}{3}$ and passing through $\left(1, \frac{2}{3}\right)$

45. $m = \frac{1}{9}$ and passing through $(-27, -2)$

46. $m = -\frac{3}{5}$ and passing through $\left(1, -\frac{1}{5}\right)$

Unless otherwise noted, all content on this page is © Cengage Learning.

SECTION 3.5 Slope-Intercept Form

DEFINITIONS AND CONCEPTS

Slope-intercept form of a line:

$$y = mx + b$$

where m is the slope and $(0, b)$ is the y-intercept.

EXAMPLES

To find the slope and y-intercept of the graph of $5x - 2y = 8$, we solve the equation for y.

$$5x - 2y = 8$$

$$-2y = -5x + 8 \quad \text{Subtract } 5x \text{ from both sides.}$$

$$y = \frac{5}{2}x - 4 \quad \text{Divide both sides by } -2.$$

The slope is $\frac{5}{2}$ and the y-intercept is $(0, -4)$.

To find the slope-intercept form of a line with slope $\frac{1}{2}$ and y-intercept $(0, 5)$, proceed as follows:

$$y = \mathbf{m}x + \mathbf{b} \quad \text{This is slope-intercept form.}$$

$$y = \mathbf{\tfrac{1}{2}}x + \mathbf{5} \quad \text{Substitute.}$$

$$y = \frac{1}{2}x + 5$$

REVIEW EXERCISES

Find the slope and the y-intercept of the line defined by each equation.

47. $y = -\frac{x}{3} + 6$ **48.** $3x - 6y = 9$

49. $2x + 5y = 1$ **50.** $x + 3y = 1$

Use the slope-intercept form to find the equation of each line.

51. $m = -3$, y-intercept $(0, 2)$

52. $m = 0$, y-intercept $(0, -7)$

53. $m = 7$, y-intercept $(0, 0)$

54. $m = \frac{1}{2}$, y-intercept $\left(0, -\frac{3}{2}\right)$

Graph the line that passes through the given point and has the given slope.

55. $(-1, 3), m = -2$

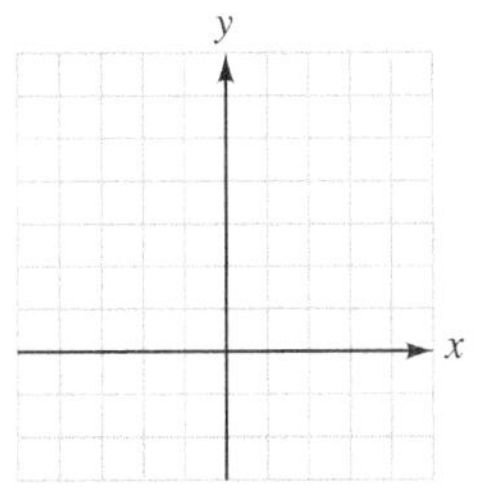

56. $(1, -2), m = 2$

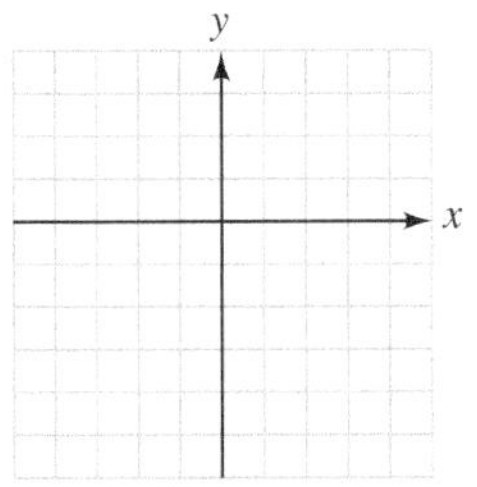

57. $\left(0, \frac{1}{2}\right), m = \frac{3}{2}$

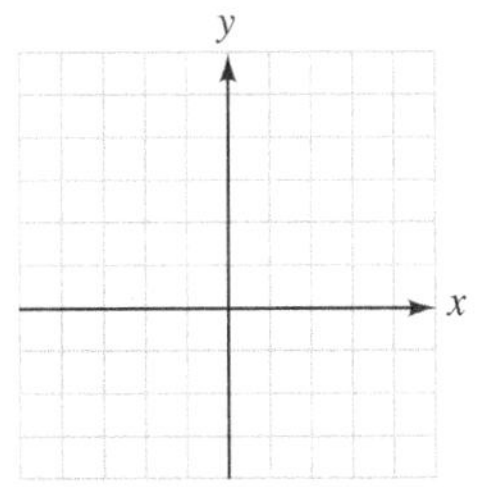

58. $(-3, 0), m = -\frac{5}{2}$

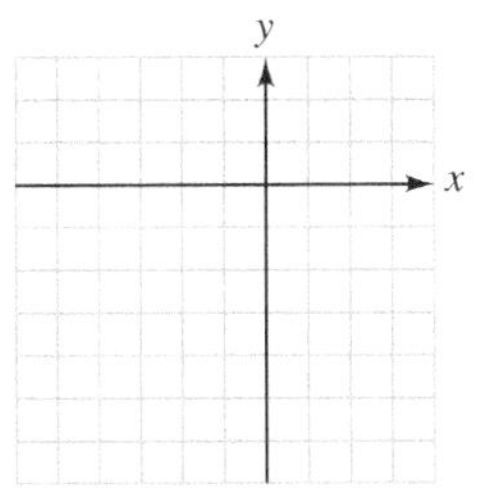

Determine whether the graphs of the equations are parallel, perpendicular, or neither parallel nor perpendicular.

59. $y = 3x, x = 3y$

60. $3x = y, x = -3y$

61. $x + 2y = y - x, 2x + y = 3$

62. $3x + 2y = 7, 2x - 3y = 8$

Write the equation of each line in slope-intercept form.

63. parallel to $y = 7x - 18$, passing through $(2, 5)$

64. parallel to $3x + 2y = 7$, passing through $(-3, 5)$

65. perpendicular to $2x - 5y = 12$, passing through $(0, 0)$

66. perpendicular to $y = \dfrac{x}{3} + 17$, y-intercept at $(0, -4)$

Find the slope-intercept equation of a line that passes through the given points.

67. $(-2, 5), (1, -1)$

68. $(10, 8), (2, 4)$

69. Depreciation A company buys a fax machine for \$2,700 and will sell it for \$200 at the end of its useful life. If the company depreciates the machine linearly over a 5-year period, what will the machine be worth after 3 years?

Unless otherwise noted, all content on this page is © Cengage Learning.

SECTION 3.6 Functions

DEFINITIONS AND CONCEPTS	EXAMPLES
A **relation** is any set of ordered pairs.	The equation $y = \lvert x - 1 \rvert$ represents a relation because it determines a set of ordered pairs (x, y).
Any equation in x and y where each value of x determines one value of y is called a **function**.	The equation $y = \lvert x - 1 \rvert$ also represents a function because each value of x determines one value of y.
The set of all inputs into a function is called the **domain**. The set of all outputs is called the **range**.	The graph of $y = 4x - 1$ is a line with slope 4 and y-intercept $(0, -1)$. The domain is the set of real numbers $\mathbb{R}$, and the range is the set of real numbers $\mathbb{R}$.
The notation $y = f(x)$ denotes that the variable y (**dependent variable**) is a function of x (**independent variable**).	If $y = f(x) = -7x + 4$, find $f(2)$. $f(2) = -7(2) + 4$ Substitute 2 for x. $= -14 + 4$ $= -10$ In ordered pair form, we can write $(2, -10)$.
If any vertical line intersects a graph more than once, the graph does not represent a function. This is called the vertical line test.	Determine whether each graph represents a function. **a.** 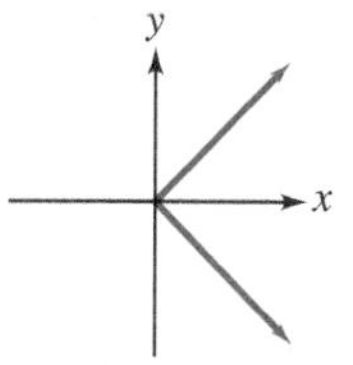 **b.** 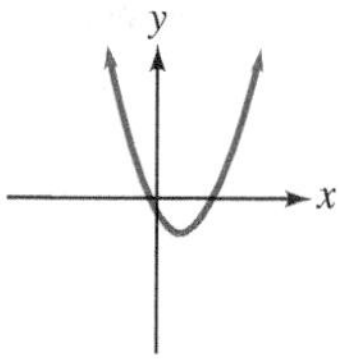 The graph in part **a** does not represent a function because it does not pass the vertical line test. The graph in part **b** represents a function because it passes the vertical line test.

REVIEW EXERCISES

70. Find the domain and range of the relation $\{(-3, -2), (0, -1), (-3, 5)\}$.

71. Determine whether the equation $y = \frac{x}{2} - 1$ defines y to be a function of x.

Let $f(x) = -2x + 5$ and find each value.

72. $f(0)$ **73.** $f(-5)$

74. $f\left(-\frac{1}{2}\right)$ **75.** $f(6)$

76. Graph $f(x) = \lvert x \rvert - 3$ and state its domain and range.

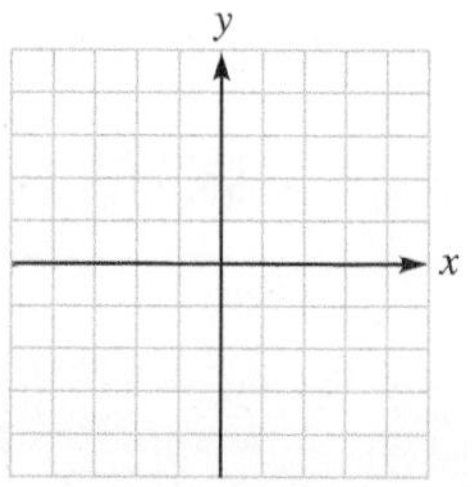

Determine whether each graph represents a function.

77.

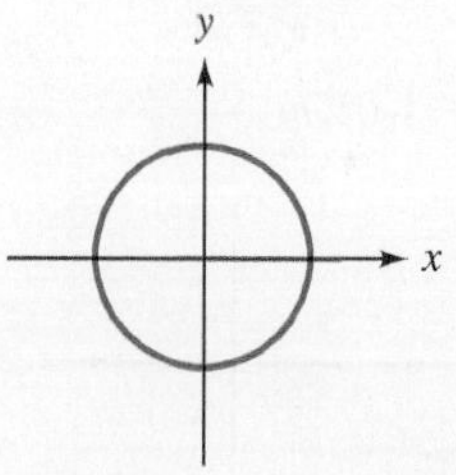

78.

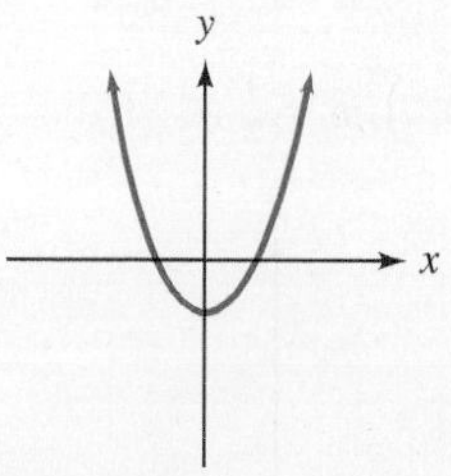

Unless otherwise noted, all content on this page is © Cengage Learning.

SECTION 3.7 Solving Linear Inequalities in Two Variables

DEFINITIONS AND CONCEPTS	EXAMPLES
To graph a linear inequality in x and y, graph the boundary line, and then use a test point to determine which side of the boundary should be shaded.	To graph the inequality $3x - 2y > 6$, we start by graphing $3x - 2y = 6$ to find the boundary line. We draw a dashed line because equality is not included. To determine which half-plane represents $3x - 2y > 6$, we select a test point not on the line and determine whether the coordinates satisfy the inequality. A convenient test point is the origin, $(0, 0)$.
If the original inequality is $<$ or $>$, the boundary line will be dashed because the points that lie on the line are not included in the solution. If the original inequality is $\leq$ or $\geq$, the boundary line will be solid.	$3x - 2y > 6$ $3(\mathbf{0}) - 2(\mathbf{0}) \overset{?}{>} 6$ Substitute 0 for x and 0 for y. $0 > 6$ Since the coordinates do not satisfy the inequality, the origin is not in the half-plane that is the graph of $3x - 2y > 6$. The graph is shown below.

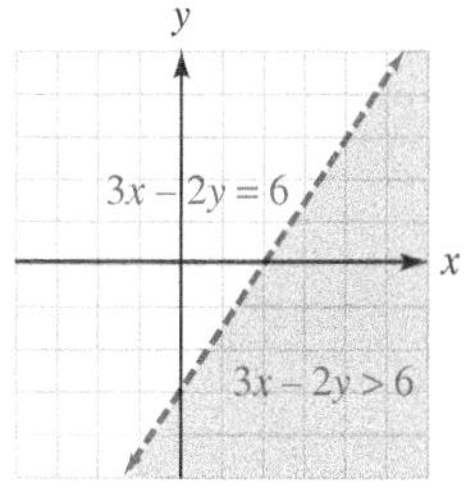

REVIEW EXERCISES

Graph each inequality on the coordinate plane.

79. $2x + 3y > 6$

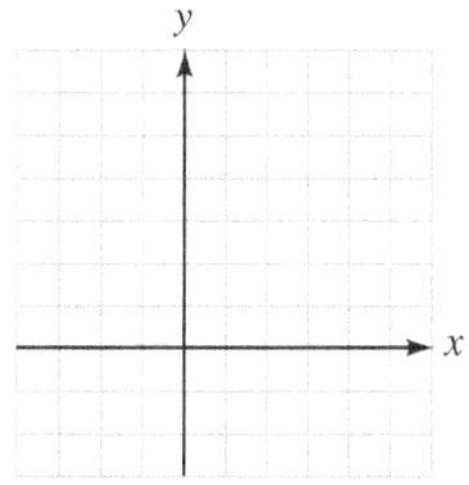

80. $y \leq 4 - x$

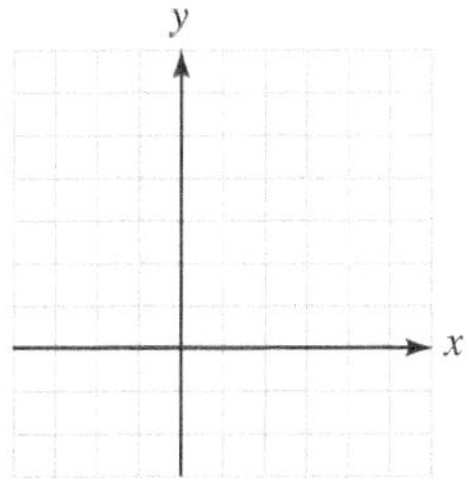

81. $-2 < x < 4$

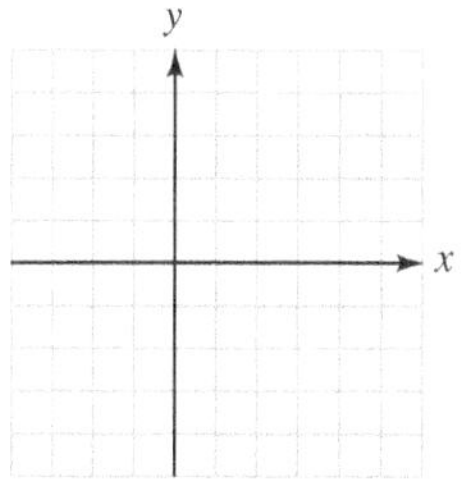

82. $y \leq -2$ or $y > 1$

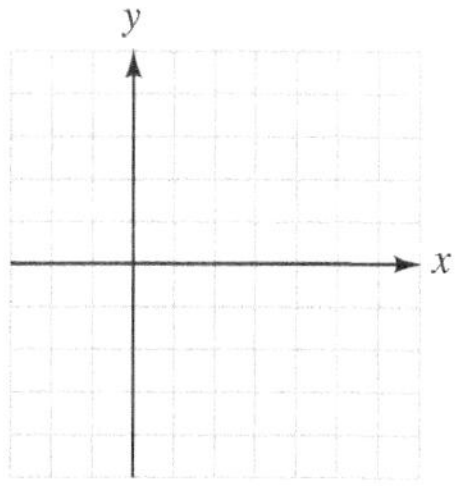

3 Test

Graph each equation.

1. $y = \frac{x}{2} + 1$

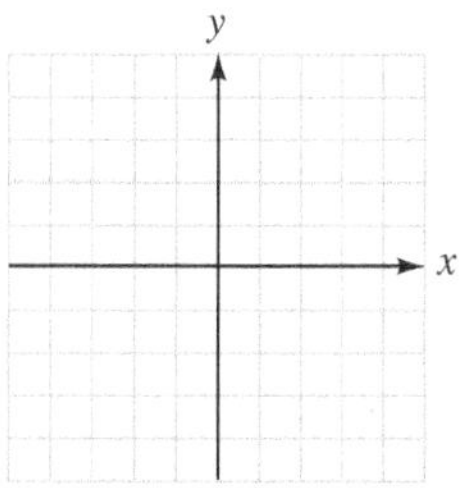

2. $2(x + 1) - y = 4$

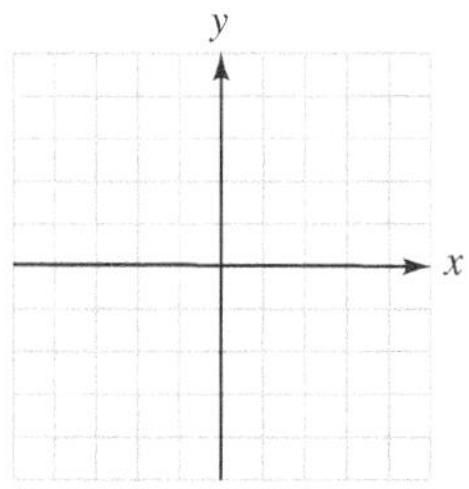

3. $x = 1$

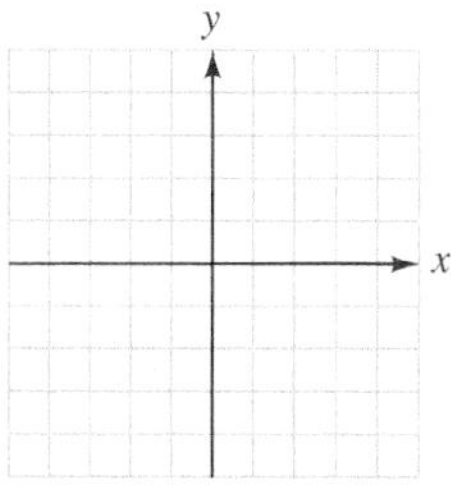

4. $2y = 8$

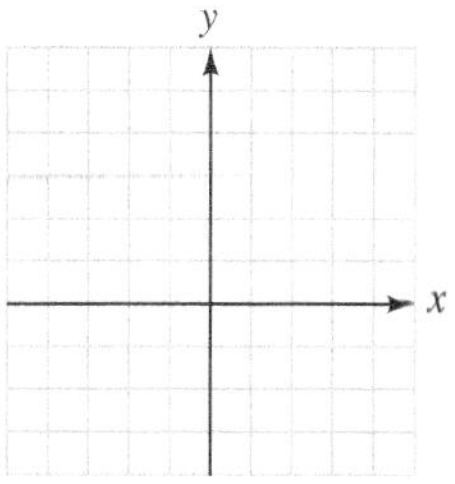

Unless otherwise noted, all content on this page is © Cengage Learning.

5. Find the slope of the line passing through $(-1, 3)$ and $(5, 8)$.

6. Find the slope of the line passing through $(-1, 3)$ and $(3, -1)$.

7. Find the slope of the line determined by $3x - 4y = 5$.

8. Find the y-intercept of the line determined by $2y - 7(x + 5) = 7$.

9. Find the slope of the line $y = 5$, if any.

10. Find the slope of the line $x = -2$, if any.

11. If two lines are parallel, their slopes are _____.

12. If two lines are nonvertical perpendicular lines, the product of their slopes is ___.

Determine whether lines with the given slopes are parallel, perpendicular, or neither.

13. 0.5 and -2

14. $\frac{20}{5}$ and 4

15. If a ramp rises 3 feet over a run of 12 feet, find the slope of the ramp.

16. If a business had sales of \$50,000 the second year in business and \$100,000 in sales the fifth year, find the annual dollar growth in sales.

17. Find the slope of a line parallel to a line with a slope of 2.

18. Find the slope of a line perpendicular to a line with a slope of 2.

19. Write the equation of a line that has a slope of 7 and passes through the point $(-2, 5)$. Give the result in point-slope form.

20. Write the equation of a line that has a slope of $\frac{3}{4}$ and a y-intercept of $(0, -5)$. Give the result in slope-intercept form.

21. Write the equation of a line that is perpendicular to the x-axis and passes through $(-7, 10)$.

22. Write the equation of a line that passes through $(3, -5)$ and is perpendicular to the line with the equation $y = \frac{1}{3}x + 11$. Give the result in slope-intercept form.

Suppose that $f(x) = 3x - 2$ and find each value.

23. $f(2)$

24. $f(-3)$

Graph the function and state its domain and range.

25. $f(x) = -|x| + 4$

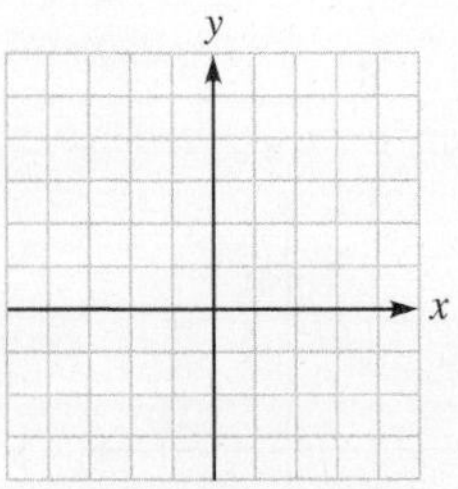

Determine whether each graph represents a function.

26.

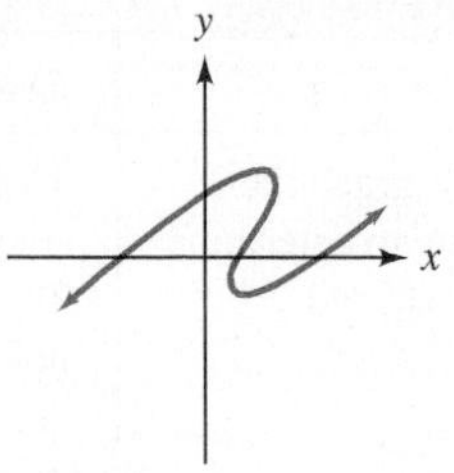

27.

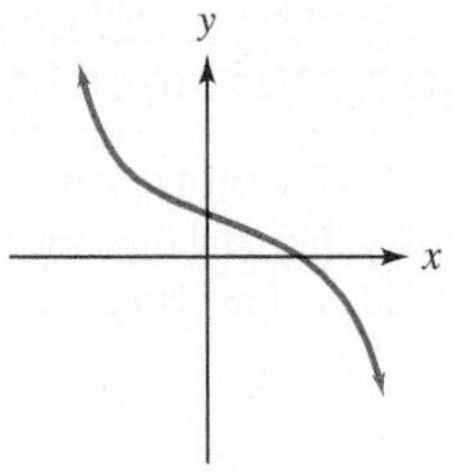

Graph each inequality.

28. $3x + 2y \geq 6$

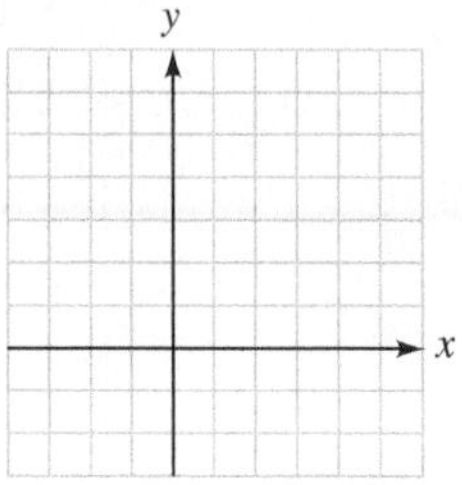

29. $-2 \leq y < 5$

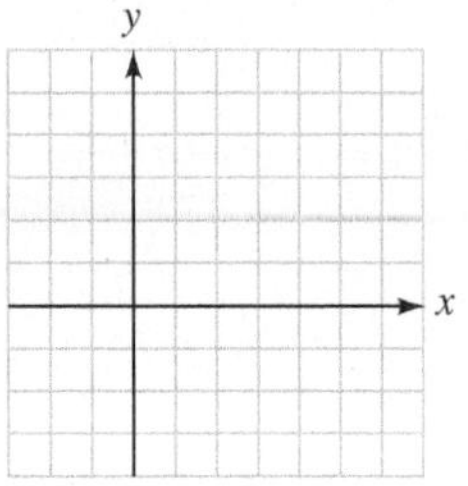

30. Find the equation of the boundary line, then write the inequality whose graph is shown.

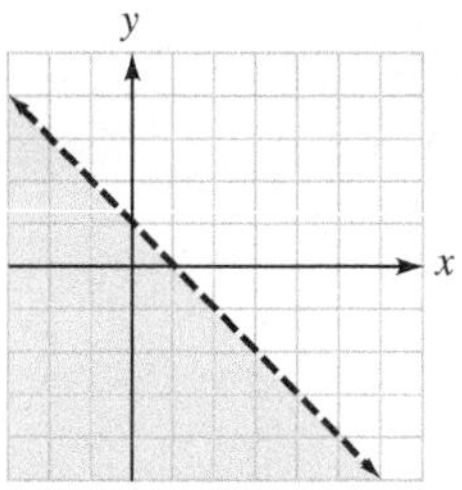

Unless otherwise noted, all content on this page is © Cengage Learning.

Polynomials

4

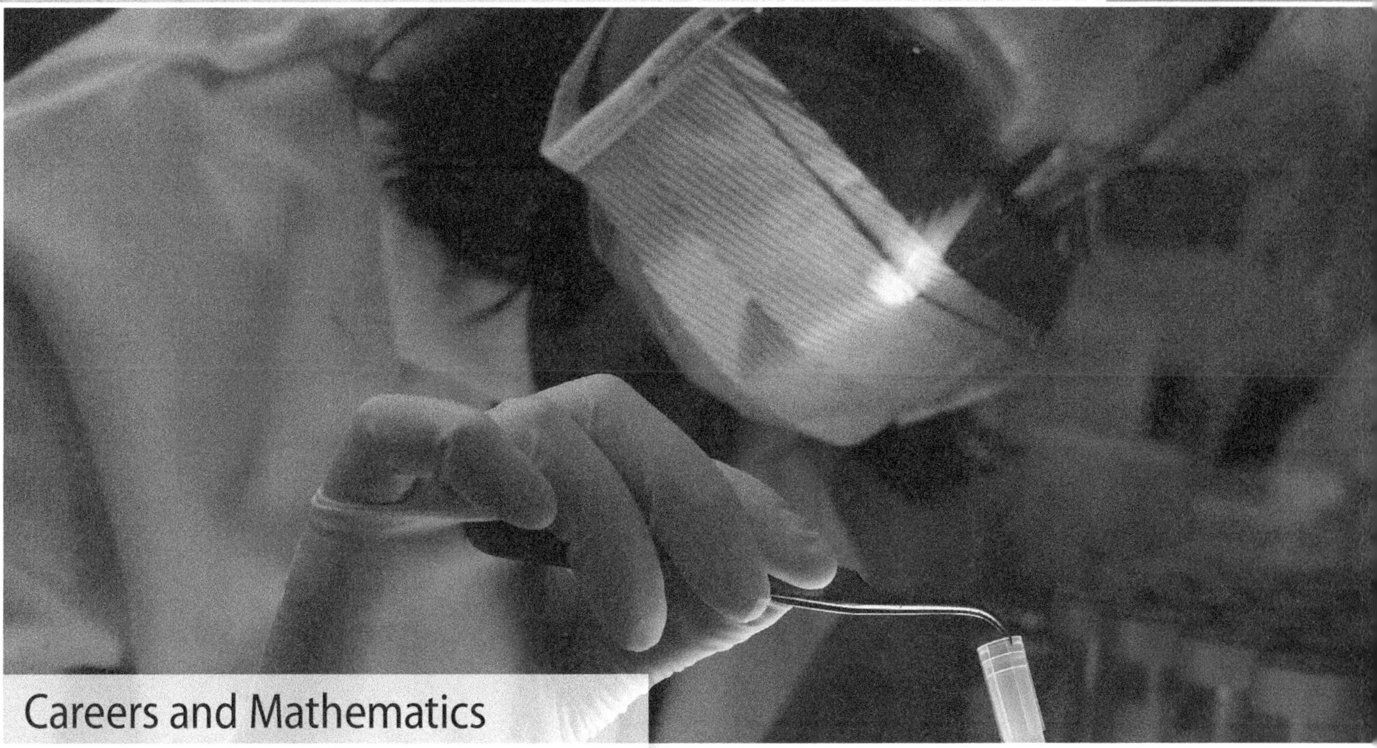

© emimkulivev/Shutterstock.com

Careers and Mathematics

MEDICAL SCIENTISTS

Medical scientists research human diseases to improve human health. Most conduct biomedical research to gain knowledge of the life processes of living organisms, including viruses, bacteria, and other infectious agents. They study biological systems to understand the causes of disease and develop treatments. Medical scientists try to identify changes in cells or chromosomes that signal the development of medical problems, such as various types of cancer.

Job Outlook:
Employment of medical scientists is expected to increase 36 percent by 2020. This is faster than the average for all occupations.

Annual Earnings:
$44,830–$142,800

For More Information:
http://www.bls.gov/ooh/life-physical-and-social-science/medical-scientists.htm

For a Sample Application:
See Problem 73 in Section 4.3.

REACH FOR SUCCESS
4.1 Natural-Number Exponents
4.2 Zero and Negative-Integer Exponents
4.3 Scientific Notation
4.4 Polynomials
4.5 Adding and Subtracting Polynomials
4.6 Multiplying Polynomials
4.7 Dividing Polynomials by Monomials
4.8 Dividing Polynomials by Polynomials
4.9 Synthetic Division
■ *Projects*
REACH FOR SUCCESS EXTENSION
CHAPTER REVIEW
CHAPTER TEST
CUMULATIVE REVIEW

In this chapter

In this chapter, we will recognize patterns for integer exponents and express very large and small numbers in scientific notation. We then discuss special algebraic expressions, called polynomials, and show how to add, subtract, multiply, and divide them.

Reach for Success Setting Course Goals

Two professional football teams do not show up on a Sunday afternoon as if they were playing a game of sandlot football. Instead, before the game they've studied films, they've practiced the plays—they have a "game plan."

In this exercise you will set some realistic goals for this course and then establish a "game plan" that will help you achieve your goals.

Melinda Nagy/Shutterstock.com

We're going to assume one of your goals is to be successful in this course.	The grade I am willing to work to achieve is a/an ______.
Most college classes recommend 2–3 hours of outside time per week for every 1 hour in class if students want to be successful. With mathematics, you may find it takes more time.	Considering other course commitments as well as any work and family commitments, state the number of hours *outside of class* you realistically believe you can devote each week to this one course. ______________
To help yourself be successful in this course, • complete all homework and other assignments on time. • take good notes in class. • ask questions in class. • study on a daily basis. • participate in a study group. • utilize the mathematics or tutor center. • attend instructor reviews (if available) or office hours. • access online tutoring services (if available).	List at least three things you are willing to add to your "game plan" and commit to, that will support your success in this course. 1. ______________ 2. ______________ 3. ______________ Can you think of anything else you can do to improve your performance in this class? ______________
Consider your answers to the questions above. Is your plan sufficient to achieve your goal?	YES ______ NO ______ Please explain. ______________

A Successful Study Strategy . . .

Set a goal, outline a plan of action, and stick to the plan to achieve that goal.

At the end of the chapter you will find an additional exercise to guide you through a successful semester.

Section 4.1 Natural-Number Exponents

Objectives

1. Write an exponential expression without exponents.
2. Write a repeated multiplication expression using exponents.
3. Simplify an expression by using the product rule for exponents.
4. Simplify an expression by using the power rules for exponents.
5. Simplify an expression by using the quotient rule for exponents.

Vocabulary

base exponent power

Getting Ready

Evaluate each expression.

1. 2^3	**2.** 3^2	**3.** $3(2)$	**4.** $2(3)$
5. $2^3 + 2^2$	**6.** $2^3 \cdot 2^2$	**7.** $3^3 - 3^2$	**8.** $\frac{3^3}{3^2}$

In this section, we will revisit the topic of exponents. This time we will develop the basic properties used to manipulate exponential expressions.

1 Write an exponential expression without exponents.

We have used natural-number exponents to indicate repeated multiplication. For example,

$$2^5 = 2 \cdot 2 \cdot 2 \cdot 2 \cdot 2 = 32 \qquad (-7)^3 = (-7)(-7)(-7) = -343$$
$$x^4 = x \cdot x \cdot x \cdot x \qquad -y^5 = -1 \cdot y \cdot y \cdot y \cdot y \cdot y$$

These examples suggest a definition for x^n, where n is a natural number.

NATURAL-NUMBER EXPONENTS

If n is a natural number, then

$$x^n = \overbrace{x \cdot x \cdot x \cdot \cdots \cdot x}^{n \text{ factors of } x}$$

In the exponential expression x^n, x is called the **base** and n is called the **exponent**. The entire expression is called a **power** of x.

$$\text{Base} \rightarrow x^n \leftarrow \text{Exponent}$$

If an exponent is a natural number, it tells how many times its base is to be used as a factor. An exponent of 1 indicates that its base is to be used one time as a factor, an exponent of 2 indicates that its base is to be used two times as a factor, and so on.

$$3^1 = 3 \qquad (-y)^1 = -y \qquad (-4z)^2 = (-4z)(-4z) \qquad (t^2)^3 = t^2 \cdot t^2 \cdot t^2$$

EXAMPLE 1 Find each value to show that **a.** $(-2)^4$ and **b.** -2^4 have different values.

Solution We find each power and show that the results are different.

a. $(-2)^4 = (-2)(-2)(-2)(-2)$
$= 16$

b. $-2^4 = -(2^4)$
$= -(2 \cdot 2 \cdot 2 \cdot 2)$
$= -16$

Since $16 \neq -16$, it follows that $(-2)^4 \neq -2^4$.
Note that in part a the base is -2 but in part b the base is 2.

SELF CHECK 1 Show that $(-4)^3$ and -4^3 have the same value.

COMMENT There is a pattern regarding even and odd exponents. If the exponent of a base is even, the result is positive. If the exponent of a base is odd, the result will be the same sign as the original base.

EXAMPLE 2 Write each expression without exponents.

a. r^3 **b.** $(-2s)^4$ **c.** $\left(\frac{1}{3}ab\right)^5$

Solution **a.** $r^3 = r \cdot r \cdot r$

b. $(-2s)^4 = (-2s)(-2s)(-2s)(-2s)$

c. $\left(\frac{1}{3}ab\right)^5 = \left(\frac{1}{3}ab\right)\left(\frac{1}{3}ab\right)\left(\frac{1}{3}ab\right)\left(\frac{1}{3}ab\right)\left(\frac{1}{3}ab\right)$

SELF CHECK 2 Write each expression without exponents. **a.** x^4
b. $(3z)^4$ **c.** $\left(-\frac{1}{2}xy\right)^3$

2 Write a repeated multiplication expression using exponents.

Many expressions can be written more compactly by using exponents.

EXAMPLE 3 Write each expression using one exponent.
a. $3 \cdot 3 \cdot 3 \cdot 3 \cdot 3$ **b.** $(5z)(5z)(5z)$

Solution **a.** Since 3 is used as a factor five times,

$$3 \cdot 3 \cdot 3 \cdot 3 \cdot 3 = 3^5$$

b. Since $5z$ is used as a factor three times,

$$(5z)(5z)(5z) = (5z)^3$$

SELF CHECK 3 Write each expression using one exponent.
a. $7 \cdot 7 \cdot 7 \cdot 7$ **b.** $\left(\frac{1}{3}xy\right)\left(\frac{1}{3}xy\right)$

3 Simplify an expression by using the product rule for exponents.

To develop a pattern for multiplying exponential expressions with the same base, we consider the product $x^2 \cdot x^3$. Since the expression x^2 means that x is to be used as a factor two times and the expression x^3 means that x is to be used as a factor three times, we have

$$x^2x^3 = \overbrace{x \cdot x}^{\text{2 factors of } x} \cdot \overbrace{x \cdot x \cdot x}^{\text{3 factors of } x}$$
$$= \overbrace{x \cdot x \cdot x \cdot x \cdot x}^{\text{5 factors of } x}$$
$$= x^5$$

In general,

$$x^m \cdot x^n = \overbrace{x \cdot x \cdot x \cdot \cdots \cdot x}^{m \text{ factors of } x} \cdot \overbrace{x \cdot x \cdot x \cdot \cdots \cdot x}^{n \text{ factors of } x}$$
$$= \overbrace{x \cdot x \cdot x \cdot x \cdot x \cdot x \cdot \cdots \cdot x \cdot x \cdot x}^{m + n \text{ factors of } x}$$
$$= x^{m+n}$$

This discussion suggests the following pattern: *To multiply two exponential expressions with the same base, keep the base and add the exponents.*

PRODUCT RULE FOR EXPONENTS

If m and n are natural numbers, then

$$x^mx^n = x^{m+n}$$

EXAMPLE 4 Simplify each expression.

a. $x^3x^4 = x^{3+4}$ Keep the base and add the exponents.

$= x^7$ $3 + 4 = 7$

b. $y^2y^4y = (y^2y^4)y$ Use the associative property of multiplication to group y^2 and y^4 together.

$= (y^{2+4})y$ Keep the base and add the exponents.

$= y^6y$ $2 + 4 = 6$

$= y^{6+1}$ Keep the base and add the exponents: $y = y^1$.

$= y^7$ $6 + 1 = 7$

SELF CHECK 4 Simplify each expression. **a.** zz^3 **b.** $x^2x^3x^6$

EXAMPLE 5 Simplify: $(2y^3)(3y^2)$

Solution

$(2y^3)(3y^2) = 2(3)y^3y^2$ Use the commutative and associative properties of multiplication to group the coefficients together and the variables together.

$= 6y^{3+2}$ Multiply the coefficients. Keep the base and add the exponents.

$= 6y^5$ $3 + 2 = 5$

SELF CHECK 5 Simplify: $(4x)(-3x^2)$

COMMENT The product rule for exponents applies only to exponential expressions with the same base. An expression such as x^2y^3 cannot be simplified, because x^2 and y^3 have different bases.

4 Simplify an expression by using the power rules for exponents.

To find another pattern of exponents, we consider the expression $(x^3)^4$, which can be written as $x^3 \cdot x^3 \cdot x^3 \cdot x^3$. Because each of the four factors of x^3 contains three factors of x, there are $4 \cdot 3$ (or 12) factors of x. Thus, the expression can be written as x^{12}.

$$(x^3)^4 = x^3 \cdot x^3 \cdot x^3 \cdot x^3$$
$$= \overbrace{\underbrace{x \cdot x \cdot x}_{x^3} \cdot \underbrace{x \cdot x \cdot x}_{x^3} \cdot \underbrace{x \cdot x \cdot x}_{x^3} \cdot \underbrace{x \cdot x \cdot x}_{x^3}}^{\text{12 factors of } x}$$
$$= x^{12}$$

In general,

$$(x^m)^n = \overbrace{x^m \cdot x^m \cdot x^m \cdot \cdots \cdot x^m}^{n \text{ factors of } x^m}$$
$$= \overbrace{x \cdot x \cdot x \cdot x \cdot x \cdot x \cdot x \cdot \cdots \cdot x}^{m \cdot n \text{ factors of } x}$$
$$= x^{m \cdot n}$$

The previous discussion suggests the following pattern: *To raise an exponential expression to a power, keep the base and multiply the exponents.*

POWER RULE FOR EXPONENTS

If m and n are natural numbers, then

$$(x^m)^n = x^{mn}$$

EXAMPLE 6 Simplify:

a. $(2^3)^7 = 2^{3 \cdot 7}$ Keep the base and multiply the exponents.
$= 2^{21}$ $3 \cdot 7 = 21$

b. $(z^7)^7 = z^{7 \cdot 7}$ Keep the base and multiply the exponents.
$= z^{49}$ $7 \cdot 7 = 49$

SELF CHECK 6 Simplify:
a. $(y^5)^2$ **b.** $(u^x)^y$

In the next example, the product and power rules of exponents are both used.

EXAMPLE 7 Simplify:

a. $(x^2x^5)^2 = (x^7)^2 = x^{14}$

b. $(y^6y^2)^3 = (y^8)^3 = y^{24}$

c. $(z^2)^4(z^3)^3 = z^8z^9 = z^{17}$

d. $(x^3)^2(x^5x^2)^3 = x^6(x^7)^3 = x^6x^{21} = x^{27}$

SELF CHECK 7 Simplify:
a. $(a^4a^3)^3$ **b.** $(k^9k^4)^2$ **c.** $(a^3)^3(a^4)^2$ **d.** $(y^5)^3(y^4y)^5$

To find more patterns for exponents, we consider the expressions $(2x)^3$ and $\left(\frac{2}{x}\right)^3$.

$$(2x)^3 = (2x)(2x)(2x) \qquad \left(\frac{2}{x}\right)^3 = \left(\frac{2}{x}\right)\left(\frac{2}{x}\right)\left(\frac{2}{x}\right) \quad (x \neq 0)$$

$$= (2 \cdot 2 \cdot 2)(x \cdot x \cdot x) \qquad = \frac{2 \cdot 2 \cdot 2}{x \cdot x \cdot x}$$

$$= 2^3x^3 \qquad = \frac{2^3}{x^3}$$

$$= 8x^3 \qquad = \frac{8}{x^3}$$

These examples suggest the following patterns: *To raise a product to a power, we raise each factor of the product to that power, and to raise a quotient to a power, we raise both the numerator and denominator to that power.*

PRODUCT TO A POWER RULE FOR EXPONENTS

If n is a natural number, then

$$(xy)^n = x^ny^n$$

QUOTIENT TO A POWER RULE FOR EXPONENTS

If n is a natural number, and if $y \neq 0$, then

$$\left(\frac{x}{y}\right)^n = \frac{x^n}{y^n}$$

EXAMPLE 8 Simplify. Assume no division by zero.

a. $(ab)^4 = a^4b^4$

b. $(3c)^3 = 3^3c^3$
$= 27c^3$

c. $(x^2y^3)^5 = (x^2)^5(y^3)^5$
$= x^{10}y^{15}$

d. $(-2x^3y)^2 = (-2)^2(x^3)^2y^2$
$= 4x^6y^2$

e. $\left(\frac{4}{k}\right)^3 = \frac{4^3}{k^3}$
$= \frac{64}{k^3}$

f. $\left(\frac{3x^2}{2y^3}\right)^5 = \frac{3^5(x^2)^5}{2^5(y^3)^5}$
$= \frac{243x^{10}}{32y^{15}}$

SELF CHECK 8 Simplify. Assume no division by zero.

a. $(cd)^5$ **b.** $(4x)^3$ **c.** $(a^5b)^3$

d. $(3x^2y)^2$ **e.** $\left(\frac{z}{8}\right)^3$ **f.** $\left(\frac{2x^3}{3y^2}\right)^4$

5 Simplify an expression by using the quotient rule for exponents.

To find a pattern for dividing exponential expressions, we consider the fraction $\frac{4^5}{4^2}$, where the exponent in the numerator is greater than the exponent in the denominator. We can simplify the fraction as follows:

$$\frac{4^5}{4^2} = \frac{4 \cdot 4 \cdot 4 \cdot 4 \cdot 4}{4 \cdot 4}$$

$$= \frac{\overset{1}{\cancel{4}} \cdot \overset{1}{\cancel{4}} \cdot 4 \cdot 4 \cdot 4}{\underset{1}{\cancel{4}} \cdot \underset{1}{\cancel{4}}}$$

$$= 4^3$$

The result of 4^3 has a base of 4 and an exponent of $5 - 2$ (or 3). This suggests that *to divide exponential expressions with the same base, we keep the base and subtract the exponents.*

QUOTIENT RULE FOR EXPONENTS

If m and n are natural numbers, $m > n$ and $x \neq 0$, then

$$\frac{x^m}{x^n} = x^{m-n}$$

EXAMPLE 9 Simplify. Assume no division by zero.

Solution

a. $\frac{x^4}{x^3} = x^{4-3}$
$= x^1$
$= x$

b. $\frac{8y^2y^6}{6y^3} = \frac{8y^8}{6y^3}$
$= \frac{8y^{8-3}}{6}$
$= \frac{4y^5}{3}$

c. $\frac{a^3a^5a^7}{a^4a} = \frac{a^{15}}{a^5}$
$= a^{15-5}$
$= a^{10}$

d. $\frac{(a^3b^4)^2}{ab^5} = \frac{a^6b^8}{ab^5}$
$= a^{6-1}b^{8-5}$
$= a^5b^3$

SELF CHECK 9 Simplify. Assume no division by zero.

a. $\frac{a^5}{a^3}$ **b.** $\frac{6b^2b^3}{4b^4}$ **c.** $\frac{a^4a^7a}{a^3a^5}$ **d.** $\frac{(x^2y^3)^2}{x^3y^4}$

SELF CHECK ANSWERS

1. both are -64 **2. a.** $x \cdot x \cdot x \cdot x$ **b.** $(3z)(3z)(3z)(3z)$ **c.** $\left(-\frac{1}{2}xy\right)\left(-\frac{1}{2}xy\right)\left(-\frac{1}{2}xy\right)$ **3. a.** 7^4 **b.** $\left(\frac{1}{3}xy\right)^2$ **4. a.** z^4 **b.** x^{11} **5.** $-12x^3$ **6. a.** y^{10} **b.** u^{xy} **7. a.** a^{21} **b.** k^{26} **c.** a^{17} **d.** y^{40} **8. a.** c^5d^5 **b.** $64x^3$ **c.** $a^{15}b^3$ **d.** $9x^4y^2$ **e.** $\frac{z^3}{512}$ **f.** $\frac{16x^{12}}{81y^8}$ **9. a.** a^2 **b.** $\frac{3b}{2}$ **c.** a^4 **d.** xy^2

NOW TRY THIS

Simplify each expression. Assume no division by zero.

1. If $x^{1/2}$ has meaning, find $(x^{1/2})^2$.
2. $-3^2(x^2 - 2^2)$
3. **a.** $x^{p+1}x^p$ **b.** $(x^{p+1})^2$ **c.** $\frac{x^{2p+1}}{x^p}$

4.1 Exercises

WARM-UPS *Perform the operations.*

1. $4 \cdot 4 \cdot 4$
2. $2 \cdot 2 \cdot 2 \cdot 2$
3. $(-7)(-7)(-7)$
4. $(-3)(-3)(-3)(-3)$

Evaluate each expression.

5. 6^2
6. $(-6)^2$
7. $3(4)^2$
8. $(3 \cdot 4)^2$

REVIEW

9. Graph the real numbers $-3, 0, 2$, and $-\frac{3}{2}$ on a number line.

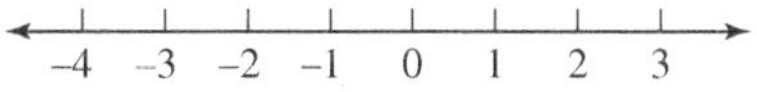

10. Graph the real numbers $-2 < x \leq 3$ on a number line.

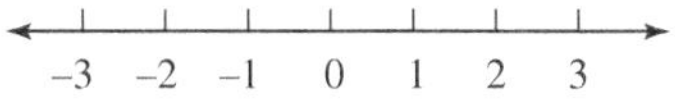

Write each algebraic expression as an English phrase.

11. $3(x + y)$
12. $3x + y$

Write each English phrase as an algebraic expression.

13. Three greater than the absolute value of twice x
14. The sum of the numbers y and z decreased by the sum of their squares

VOCABULARY AND CONCEPTS *Fill in the blanks. Assume no division by zero.*

15. The ____ of the exponential expression $(-5)^3$ is __. The exponent is _.
16. The base of the exponential expression -5^3 is _. The _______ is 3.
17. $(4y)^3$ means ___________.
18. Write $(-5x)(-5x)(-5x)(-5x)$ as a power. ______
19. $y^5 =$ _________
20. $x^m x^n =$ ____
21. $(xy)^n =$ ___
22. $\left(\frac{a}{b}\right)^n =$ __
23. $(a^b)^c =$ ___
24. $\frac{x^m}{x^n} =$ ____
25. The area of the square is $s \cdot s$. Why do you think the symbol s^2 is called "s squared"?

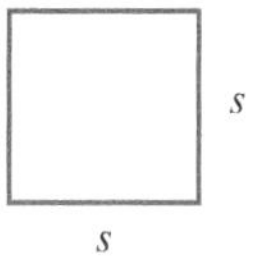

26. The volume of the cube is $s \cdot s \cdot s$. Why do you think the symbol s^3 is called "s cubed"?

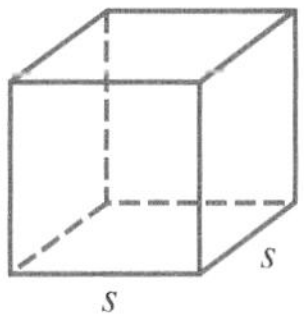

Identify the base and the exponent in each expression.

27. x^4
28. 4^x
29. 7^2
30. $(-7)^2$
31. $(2y)^3$
32. $(-3x)^2$
33. $-x^4$
34. $(-x)^4$
35. x
36. $(xy)^3$
37. $2x^3$
38. $-3y^6$

GUIDED PRACTICE *Evaluate each expression.* SEE EXAMPLE 1. (OBJECTIVE 1)

39. -5^2
40. $(-5)^2$
41. $2^2 + 3^2$
42. $2^3 - 2^2$
43. $-3(6^2 - 2^3)$
44. $2(4^3 + 3^2)$
45. $(-4)^3 + (-3)^2$
46. $-4^3 - 3^2$

Write each expression without using exponents. SEE EXAMPLE 2. (OBJECTIVE 1)

47. 5^3
48. -4^5
49. $-5x^6$
50. $3x^3$
51. $-2y^4$
52. $(-2y)^4$
53. $(3t)^5$
54. a^3b^2

Write each expression using one exponent. SEE EXAMPLE 3. (OBJECTIVE 2)

55. $2 \cdot 2 \cdot 2$
56. $5 \cdot 5$
57. $x \cdot x \cdot x \cdot x$
58. $y \cdot y \cdot y \cdot y \cdot y \cdot y$
59. $(2x)(2x)(2x)$
60. $(-4y)(-4y)$
61. $-4 \cdot t \cdot t \cdot t \cdot t$
62. $-8 \cdot a \cdot a \cdot a \cdot a \cdot a$

Simplify. SEE EXAMPLE 4. (OBJECTIVE 3)

63. x^4x^3
64. y^5y^2
65. x^5x^5
66. yy^3
67. $a^3a^4a^5$
68. $b^2b^3b^5$
69. $y^3(y^2y^4)$
70. $(y^4y)y^6$

Simplify. SEE EXAMPLE 5. (OBJECTIVE 3)

71. $4x^2(3x^5)$
72. $-2y(y^3)$
73. $(5x^4)(-2x)$
74. $(-3y^6)(-6y^5)$

Simplify. SEE EXAMPLE 6. (OBJECTIVE 4)

75. $(3^2)^4$
76. $(4^3)^3$
77. $(y^5)^3$
78. $(b^3)^6$

Unless otherwise noted, all content on this page is © Cengage Learning.

Simplify. SEE EXAMPLE 7. (OBJECTIVE 4)

79. $(x^2x^3)^5$
80. $(y^3y^4)^4$
81. $(a^2a^7)^3$
82. $(q^3q^3)^5$
83. $(x^5)^2(x^7)^3$
84. $(y^3y)^2(y^2)^2$
85. $(r^3r^2)^4(r^3r^5)^2$
86. $(yy^3)^3(y^2y^3)^4(y^3y^3)^2$

Simplify. Assume no division by 0. SEE EXAMPLE 8. (OBJECTIVE 4)

87. $(xy)^3$
88. $(uv^2)^4$
89. $(r^3s^2)^2$
90. $(a^3b^2)^3$
91. $(4ab^2)^2$
92. $(3x^2y)^3$
93. $(-2r^2s^3t)^3$
94. $(-3x^2y^4z)^2$
95. $\left(\frac{a}{b}\right)^3$
96. $\left(\frac{r^2}{s}\right)^4$
97. $\left(\frac{2x}{3y^2}\right)^4$
98. $\left(\frac{4u^2}{5v^4}\right)^3$

Simplify. Assume no division by 0. SEE EXAMPLE 9. (OBJECTIVE 5)

99. $\frac{x^5}{x^3}$
100. $\frac{a^6}{a^3}$
101. $\frac{y^3y^4}{yy^2}$
102. $\frac{b^4b^5}{b^2b^3}$
103. $\frac{12a^2a^3a^4}{4(a^4)^2}$
104. $\frac{16(aa^2)^3}{2a^2a^3}$
105. $\frac{(ab^2)^3}{(ab)^2}$
106. $\frac{(m^3n^4)^3}{(mn^2)^3}$

ADDITIONAL PRACTICE *Simplify. Assume no division by 0.*

107. tt^2
108. w^3w^5
109. $6x^3(-x^2)(-x^4)$
110. $-2x(-x^2)(-3x)$
111. $(-2a^5)^3$
112. $(-3b^7)^4$
113. $(3zz^2z^3)^5$
114. $(4t^3t^6t^2)^2$
115. $(s^3)^3(s^2)^2(s^5)^4$
116. $(s^2)^3(s^3)^2(s^4)^4$
117. $\left(\frac{-2a}{b}\right)^5$
118. $\left(\frac{2t}{3}\right)^4$
119. $\left(\frac{b^2}{3a}\right)^3$
120. $\left(\frac{a^3b}{c^4}\right)^5$
121. $\frac{17(x^4y^3)^8}{34(x^5y^2)^4}$
122. $\frac{35(r^3s^2)^2}{49r^2s^3}$
123. $\left(\frac{y^3y}{2yy^2}\right)^3$
124. $\left(\frac{3t^3t^4t^5}{4t^2t^6}\right)^3$
125. $\left(\frac{-2r^3r^3}{3r^4r}\right)^3$
126. $\left(\frac{-6y^4y^5}{5y^3y^5}\right)^2$
127. $\frac{20(r^4s^3)^4}{6(rs^3)^3}$
128. $\frac{15(x^2y^5)^5}{21(x^3y)^2}$

APPLICATIONS

129. Bouncing balls When a certain ball is dropped, it always rebounds to one-half of its previous height. If the ball is dropped from a height of 32 feet, explain why the expression $32\left(\frac{1}{2}\right)^4$ represents the height of the ball on the fourth bounce. Find the height of the fourth bounce.

130. Having babies The probability that a couple will have n baby boys in a row is given by the formula $\left(\frac{1}{2}\right)^n$. Find the probability that a couple will have four baby boys in a row.

131. Investing If an investment of $1,000 doubles every seven years, find the value of the investment after 28 years.

If P dollars are invested at a rate r, compounded annually, it will grow to A dollars in t years according to the formula

$$A = P(1 + r)^t$$

132. Compound interest How much will be in an account at the end of 2 years if $12,000 is invested at 5%, compounded annually?

133. Compound interest How much will be in an account at the end of 30 years if $8,000 is invested at 6%, compounded annually?

134. Investing Guess the answer to the following question. Then use a calculator to find the correct answer. Were you close?

If the value of 1¢ is to double every day, what will the penny be worth after 31 days?

WRITING ABOUT MATH

135. Describe how you would multiply two exponential expressions with like bases.

136. Describe how you would divide two exponential expressions with like bases.

SOMETHING TO THINK ABOUT

137. Is the operation of raising to a power commutative? That is, is $a^b = b^a$? Explain.

138. Is the operation of raising to a power associative? That is, is $(a^b)^c = a^{(b^c)}$? Explain.

Section 4.2 Zero and Negative-Integer Exponents

Objectives

1. Simplify an expression containing an exponent of zero.
2. Simplify an expression containing a negative-integer exponent.
3. Simplify an expression containing a variable exponent.

Getting Ready

Simplify by dividing out common factors. Assume that $y \neq 0, x \neq 0$.

1. $\dfrac{3 \cdot 3 \cdot 3}{3 \cdot 3 \cdot 3 \cdot 3}$ **2.** $\dfrac{2yy}{2yyy}$ **3.** $\dfrac{3xx}{3xx}$ **4.** $\dfrac{xxy}{xxxyy}$

In the previous section, we discussed natural-number exponents. We now continue the discussion to include 0 and negative-integer exponents.

1 Simplify an expression containing an exponent of zero.

When we discussed the quotient rule for exponents in the previous section, the exponent in the numerator was always greater than the exponent in the denominator. We now consider what happens when the exponents are equal.

If we apply the quotient rule to the fraction $\frac{5^3}{5^3}$, where the exponents in the numerator and denominator are equal, we obtain 5^0. However, because any nonzero number divided by itself equals 1, we also obtain 1.

$$\frac{5^3}{5^3} = 5^{3-3} = 5^0 \qquad \frac{5^3}{5^3} = \frac{\overset{1}{\cancel{5}} \cdot \overset{1}{\cancel{5}} \cdot \overset{1}{\cancel{5}}}{\underset{1}{\cancel{5}} \cdot \underset{1}{\cancel{5}} \cdot \underset{1}{\cancel{5}}} = 1$$

These are equal.

For this reason, we define 5^0 to be equal to 1. In general, the following is true.

ZERO EXPONENTS

If x is any nonzero real number, then

$$x^0 = 1$$

EXAMPLE 1 Write each expression without exponents.

a. $\left(\dfrac{1}{13}\right)^0 = 1$

b. $\dfrac{x^5}{x^5} = x^{5-5} \quad (x \neq 0)$
$= x^0$
$= 1$

c. $3x^0 = 3(1)$
$= 3$

d. $(3x)^0 = 1$

e. $\frac{6^n}{6^n} = 6^{n-n}$
$= 6^0$
$= 1$

f. $\frac{y^m}{y^m} = y^{m-m} \quad (y \neq 0)$
$= y^0$
$= 1$

Parts c and d illustrate that $3x^0 \neq (3x)^0$.

SELF CHECK 1 Write each expression without exponents. **a.** $(-0.115)^0$ **b.** $\frac{4^2}{4^2}$ **c.** $-2x^0$ **d.** $(-9y)^0$ **e.** $\frac{-2^x}{-2^x}$ **f.** $\frac{x^m}{x^m} \quad (x \neq 0)$

2 Simplify an expression containing a negative-integer exponent.

If we apply the quotient rule to $\frac{6^2}{6^5}$, where the exponent in the numerator is less than the exponent in the denominator, we obtain 6^{-3}. However, by dividing out two factors of 6, we also obtain $\frac{1}{6^3}$.

$$\frac{6^2}{6^5} = 6^{2-5} = \mathbf{6^{-3}} \qquad \frac{6^2}{6^5} = \frac{\overset{1}{\cancel{6}} \cdot \overset{1}{\cancel{6}}}{\underset{1}{\cancel{6}} \cdot \underset{1}{\cancel{6}} \cdot 6 \cdot 6 \cdot 6} = \mathbf{\frac{1}{6^3}}$$

These are equal.

For these reasons, we define 6^{-3} to be $\frac{1}{6^3}$. In general, the following is true.

NEGATIVE EXPONENTS If x is any nonzero number and n is a natural number, then

$$x^{-n} = \frac{1}{x^n} \quad \text{and} \quad \frac{1}{x^{-n}} = x^n$$

COMMENT A negative exponent can be viewed as "taking the reciprocal."

EXAMPLE 2 Express each quantity without negative exponents or parentheses. Assume no variables are 0.

a. $3^{-5} = \frac{1}{3^5}$
$= \frac{1}{243}$

b. $x^{-4} = \frac{1}{x^4}$

c. $(2x)^{-2} = \frac{1}{(2x)^2}$
$= \frac{1}{4x^2}$

d. $2x^{-2} = 2\left(\frac{1}{x^2}\right)$
$= \frac{2}{x^2}$

e. $(-3a)^{-4} = \frac{1}{(-3a)^4}$
$= \frac{1}{81a^4}$

f. $\frac{1}{(3x)^{-2}} = \frac{(3x)^2}{1}$
$= 9x^2$

SELF CHECK 2 Write each expression without negative exponents or parentheses. Assume no variables are 0. **a.** 4^{-3} **b.** a^{-5} **c.** $(3y)^{-3}$ **d.** $-5x^{-3}$ **e.** $(-2x)^{-3}$ **f.** $\frac{1}{(2x)^{-3}}$

Because of the definitions of negative and zero exponents, the product, power, and quotient rules are true for all integer exponents.

PROPERTIES OF INTEGER EXPONENTS

If m and n are integers and no base is 0 $(x \neq 0, y \neq 0)$, then

$$x^m x^n = x^{m+n} \qquad (x^m)^n = x^{mn} \qquad (xy)^n = x^n y^n \qquad \left(\frac{x}{y}\right)^n = \frac{x^n}{y^n}$$

$$x^0 = 1 \qquad x^{-n} = \frac{1}{x^n} \qquad \frac{1}{x^{-n}} = x^n \qquad \frac{x^m}{x^n} = x^{m-n}$$

EXAMPLE 3 Simplify and write the result without negative exponents. Assume no variables are 0.

COMMENT If m and n are natural numbers, $m < n$ and $x \neq 0$, then $\frac{x^m}{x^n} = \frac{1}{x^{n-m}}$.

a. $(x^{-3})^2 = x^{-6}$
$= \frac{1}{x^6}$

b. $\frac{x^3}{x^7} = x^{3-7}$
$= x^{-4}$
$= \frac{1}{x^4}$

c. $\frac{y^{-4}y^{-3}}{y^{-20}} = \frac{y^{-7}}{y^{-20}}$
$= y^{-7-(-20)}$
$= y^{-7+20}$
$= y^{13}$

d. $\frac{12a^3b^4}{4a^5b^2} = 3a^{3-5}b^{4-2}$
$= 3a^{-2}b^2$
$= \frac{3b^2}{a^2}$

e. $\left(-\frac{x^3y^2}{xy^{-3}}\right)^{-2} = (-x^{3-1}y^{2-(-3)})^{-2}$
$= (-x^2y^5)^{-2}$
$= \frac{1}{(-x^2y^5)^2}$
$= \frac{1}{x^4y^{10}}$

SELF CHECK 3 Simplify and write the result without negative exponents. Assume no variables are 0.

a. $(x^4)^{-3}$ **b.** $\frac{a^4}{a^8}$ **c.** $\frac{a^{-4}a^{-5}}{a^{-3}}$ **d.** $\frac{20x^5y^3}{5x^3y^6}$ **e.** $\left(\frac{x^4y^5}{x^{-7}y^9}\right)^{-3}$

3 Simplify an expression containing a variable exponent.

The properties of exponents are also true when the exponents are algebraic expressions.

EXAMPLE 4 Simplify. Assume no base is 0.

a. $x^{2m}x^{3m} = x^{2m+3m}$
$= x^{5m}$

b. $\frac{y^{2m}}{y^{4m}} = y^{2m-4m}$
$= y^{-2m}$
$= \frac{1}{y^{2m}}$

c. $a^{2m-1}a^{2m} = a^{2m-1+2m}$
$= a^{4m-1}$

d. $(b^{m+1})^2 = b^{(m+1)2}$
$= b^{2m+2}$

SELF CHECK 4 Simplify. Assume no base is 0.

a. $z^{3n}z^{2n}$ **b.** $\frac{z^{3n}}{z^{5n}}$ **c.** $x^{3m+2}x^m$ **d.** $(x^{m+2})^3$

Accent on technology

▸ Finding Present Value

To find out how much money P (called the *present value*) must be invested at an annual rate i (expressed as a decimal) to have \$$A$ in n years, we use the formula $P = A(1 + i)^{-n}$. To find out how much we must invest at 6% to have \$50,000 in 10 years, we substitute 50,000 for A, 0.06 (6%) for i, and 10 for n to get

$$P = A(1 + i)^{-n}$$
$$P = \mathbf{50{,}000}(1 + \mathbf{0.06})^{-\mathbf{10}}$$

To evaluate P with a calculator, we enter these numbers and press these keys:

(1 + .06) yˣ 10 +/− × 50000 — Using a calculator with a yˣ and a +/− key.

50000 (1 + .06) ∧ (−) 10 ENTER — Using a TI84 graphing calculator.

Either way, we see that we must invest \$27,919.74 to have \$50,000 in 10 years.

For instructions regarding the use of a Casio graphing calculator, please refer to the Casio Keystroke Guide in the back of the book.

SELF CHECK ANSWERS

1. a. 1 **b.** 1 **c.** −2 **d.** 1 **e.** 1 **f.** 1 **2. a.** $\frac{1}{64}$ **b.** $\frac{1}{a^5}$ **c.** $\frac{1}{27y^3}$ **d.** $-\frac{5}{x^3}$ **e.** $-\frac{1}{8x^3}$ **f.** $8x^3$ **3. a.** $\frac{1}{x^{12}}$ **b.** $\frac{1}{a^4}$ **c.** $\frac{1}{a^6}$ **d.** $\frac{4x^2}{y^3}$ **e.** $\frac{y^{12}}{x^{33}}$ **4. a.** z^{5n} **b.** $\frac{1}{z^{2n}}$ **c.** x^{4m+2} **d.** x^{3m+6}

NOW TRY THIS

Simplify each expression. Write your answer with positive exponents only. Assume no variables are 0.

1. $-2(x^2y^5)^0$

2. $-3x^{-2}$

3. $9^2 - 9^0$

4. Explain why the instructions above include the statement "Assume no variables are 0."

4.2 Exercises

WARM-UPS *Identify the base in each expression.*

1. $3x^{-2}$

2. $(3x)^{-2}$

3. $(5x)^0$

4. $5x^0$

Simplify by dividing out common factors. Assume no variable is 0.

5. $\dfrac{3 \cdot a \cdot a}{3 \cdot a \cdot a \cdot a}$

6. $\dfrac{-2 \cdot x \cdot x \cdot x}{-2 \cdot x \cdot x}$

7. $\dfrac{a \cdot a \cdot b}{a \cdot b \cdot b}$

8. $\dfrac{x \cdot x \cdot y \cdot y}{x \cdot x \cdot x \cdot y \cdot y}$

REVIEW

9. If $a = -2$ and $b = 3$, evaluate $\dfrac{3a^2 + 4b + 8}{a + 2b^2}$.

10. Evaluate: $|-3 + 5 \cdot 2|$

Solve each equation.

11. $5\left(x - \dfrac{1}{2}\right) = \dfrac{7}{2}$

12. $\dfrac{5(2 - x)}{6} = \dfrac{x + 6}{2}$

13. Solve $P = L + \dfrac{s}{f}i$ for s.

14. Solve $P = L + \dfrac{s}{f}i$ for i.

VOCABULARY AND CONCEPTS *Fill in the blanks.*

15. If x is any nonzero real number, then $x^0 =$ _. If x is any nonzero real number, then $x^{-n} =$ _.

16. Since $\dfrac{6^4}{6^4} = 6^{4-4} = 6^0$ and $\dfrac{6^4}{6^4} = 1$, we define 6^0 to be _.

17. Since $\frac{8^3}{8^5} = 8^{3-5} = 8^{-2}$ and $\frac{8^3}{8^5} = \frac{8\cdot8\cdot8}{8\cdot8\cdot8\cdot8\cdot8} = \frac{1}{8^2}$, we define 8^{-2} to be ___.

18. The amount P that must be deposited now to have A dollars in the future is called the __________.

GUIDED PRACTICE *Write each expression without exponents. Assume no variable is 0. SEE EXAMPLE 1. (OBJECTIVE 1)*

19. 9^0

20. $\frac{a^6}{a^6}$

21. $5x^0$

22. $(5x)^0$

23. $\left(\frac{a^2b^3}{ab^4}\right)^0$

24. $\frac{2}{3}\left(\frac{xyz}{x^2y}\right)^0$

25. $\frac{8^y}{8^y}$

26. $\frac{a^n}{a^n}$

27. $(-x)^0$

28. $-x^0$

29. $\frac{x^0 - 5x^0}{2x^0}$

30. $\frac{4a^0 + 2a^0}{3a^0}$

Simplify each expression by writing it as an expression without negative exponents or parentheses. Assume no variables are 0. SEE EXAMPLE 2. (OBJECTIVE 2)

31. 5^{-4}

32. 9^{-2}

33. a^{-5}

34. y^{-3}

35. $(2y)^{-4}$

36. $(-3x)^{-1}$

37. $(-5p)^{-3}$

38. $(-4z)^{-2}$

39. $(y^2y^4)^{-2}$

40. $(x^3x^2)^{-3}$

41. $-5x^{-4}$

42. $-7y^{-2}$

Simplify and write the result without negative exponents. Assume no variables are 0. SEE EXAMPLE 3. (OBJECTIVE 2)

43. $\frac{y^4}{y^5}$

44. $\frac{x^5}{x^9}$

45. $\frac{a^4}{a^9}$

46. $\frac{z^5}{z^8}$

47. $\frac{x^{-2}x^{-3}}{x^{-10}}$

48. $\frac{a^{-4}a^{-2}}{a^{-12}}$

49. $\frac{15a^3b^8}{3a^4b^4}$

50. $\frac{14b^5c^4}{21b^3c^5}$

51. $(a^{-6})^4$

52. $(-b^{-3})^9$

53. $\left(-\frac{b^5}{b^{-3}}\right)^{-4}$

54. $\left(\frac{a^4}{a^{-3}}\right)^3$

Simplify each expression. Assume no base is 0. SEE EXAMPLE 4. (OBJECTIVE 3)

55. $x^{2m}x^m$

56. $y^{5m}y^{4m}$

57. $\frac{x^{3n}}{x^{6n}}$

58. $\frac{y^m}{y^{7m}}$

59. $y^{3m+2}y^{-m}$

60. $x^{m+1}x^m$

61. $(x^{n+4})^3$

62. $(y^{m-3})^4$

63. $u^{2m}v^{3n}u^{3m}v^{-3n}$

64. $r^{2m}s^{-3}r^{3m}s^3$

65. $(y^{2-n})^{-4}$

66. $(x^{3-4n})^{-2}$

ADDITIONAL PRACTICE *Simplify each expression and write the result without using parentheses or negative exponents. Assume no variable base is 0.*

67. $2^5\cdot2^{-2}$

68. $10^2\cdot10^{-4}\cdot10^5$

69. $5^{-2}\cdot5^5\cdot5^{-3}$

70. $3^{-4}\cdot3^5\cdot3^{-3}$

71. $\frac{8^5\cdot8^{-3}}{8}$

72. $\frac{6^2\cdot6^{-3}}{6^{-2}}$

73. $\frac{2^5\cdot2^7}{2^6\cdot2^{-3}}$

74. $\frac{5^{-2}\cdot5^{-4}}{5^{-6}}$

75. a^{-9}

76. c^{-4}

77. $\frac{y^{3m}}{y^{2m}}$

78. $\frac{z^{4m}}{z^{2m}}$

79. $(4t)^{-3}$

80. $(-6r)^{-2}$

81. $(ab^2)^{-3}$

82. $(m^2n^3)^{-2}$

83. $(x^2y)^{-2}$

84. $(x^{-1}y^2)^{-3}$

85. $\frac{b^0b^3}{b^{-3}b^4}$

86. $\frac{(r^2)^3}{(r^3)^4}$

87. $(m^3n^4)^{-3}$

88. $(c^2d^3)^{-2}$

89. $\frac{x^{12}x^{-7}}{x^3x^4}$

90. $\frac{(b^3)^4}{(b^5)^4}$

91. $(ab^2)^{-2}$

92. $(x^2y)^{-3}$

93. $(-2x^3y^{-2})^{-5}$

94. $(-3u^{-2}v^3)^{-3}$

95. $(a^{-2}b^{-3})^{-4}$

96. $(y^{-3}z^5)^{-6}$

97. $\left(\frac{b^5}{b^{-2}}\right)^{-2}$

98. $\left(\frac{b^{-2}}{b^3}\right)^{-3}$

99. $\left(\frac{6a^2b^3}{2ab^2}\right)^{-2}$

100. $\left(\frac{15r^2s^{-2}t}{3r^{-3}s^3}\right)^{-3}$

101. $\left(\frac{18a^2b^3c^{-4}}{3a^{-1}b^2c}\right)^{-3}$

102. $\left(\frac{21x^{-2}y^2z^{-2}}{7x^3y^{-1}}\right)^{-2}$

103. $\left(\frac{-3r^4r^{-3}}{r^{-3}r^7}\right)^3$

104. $\left(\frac{12y^3z^{-2}}{3y^{-4}z^3}\right)^2$

105. $\left(\frac{14u^{-2}v^3}{21u^{-3}v}\right)^4$

106. $\frac{(17x^5y^{-5}z)^{-3}}{(17x^{-5}y^3z^2)^{-4}}$

107. $\frac{x^{3n}}{x^{6n}}$

108. $(y^2)^{m+1}$

APPLICATIONS *For 109–111, see Accent on Technology.*

109. Present value How much money must be invested at 7% to have \$100,000 in 40 years?

110. Present value How much money must be invested at 9% to have \$100,000 in 40 years?

111. Present value How much must be invested at 4% annual interest to have \$1,000,000 in 60 years?

112. Biology During bacterial reproduction, the time required for a population to double is called the **generation time**. If b bacteria are introduced into a medium, then after the generation time has elapsed, there will be $2b$ bacteria. After n generations, there will be $b \cdot 2^n$ bacteria. Give the meaning of this expression when $n = 0$.

WRITING ABOUT MATH

113. Explain how you would help a friend understand that 2^{-3} is not equal to -8.

114. Describe how you would verify on a calculator that

$$2^{-3} = \frac{1}{2^3}$$

SOMETHING TO THINK ABOUT

115. If a positive number x is raised to a negative power, is the result greater than, equal to, or less than x? Explore the possibilities.

116. We know that $x^{-n} = \frac{1}{x^n}$. Is it also true that $x^n = \frac{1}{x^{-n}}$? Explain.

Section 4.3 Scientific Notation

Objectives

1. Convert a number from standard notation to scientific notation.
2. Convert a number from scientific notation to standard notation.
3. Use scientific notation to simplify an expression.

Vocabulary

standard notation　　scientific notation

Getting Ready

Evaluate each expression.

1. 10^2	**2.** 10^3	**3.** 10^1	**4.** 10^{-2}
5. $5(10^2)$	**6.** $8(10^3)$	**7.** $3(10^1)$	**8.** $7(10^{-2})$

We now use exponents to express very large and very small numbers that are written in **standard notation** in a compact form called **scientific notation**. In science, almost all large and small numbers are written in scientific notation.

1 Convert a number from standard notation to scientific notation.

Scientists often deal with extremely large and extremely small numbers. For example,

- The distance from Earth to the Sun is approximately 150,000,000 kilometers.
- Ultraviolet light emitted from a mercury arc has a wavelength of approximately 0.000025 centimeter.

The large number of zeros in these numbers written in standard notation makes them difficult to read and hard to remember. Scientific notation provides a compact way of writing large and small numbers.

SCIENTIFIC NOTATION A number is written in scientific notation if it is written as the product of a number between 1 (including 1) and 10 and an integer power of 10.

Each of the following numbers is written in scientific notation.

3.67×10^6 $\quad$ 2.24×10^{-4} $\quad$ 9.875×10^{22}

Every number that is written in scientific notation has the following form:

$$\underbrace{\square.\square}_{\text{A number between 1 and 10.}} \times 10^{\overbrace{\square}^{\text{An integer exponent.}}}$$

EXAMPLE 1 Write 150,000,000 in scientific notation.

Solution We note that 1.5 lies between 1 and 10. To obtain 150,000,000, the decimal point in 1.5 must be moved eight places to the right. Because multiplying a number by 10 moves the decimal point one place to the right, we can accomplish this by multiplying 1.5 by 10 eight times.

1.5 0 0 0 0 0 0 0
8 places to the right.

150,000,000 written in scientific notation is 1.5×10^8.

SELF CHECK 1 Write 93,000,000 in scientific notation.

EXAMPLE 2 Write 0.000025 in scientific notation.

Solution We note that 2.5 is between 1 and 10. To obtain 0.000025, the decimal point in 2.5 must be moved five places to the left. We can accomplish this by dividing 2.5 by 10^5, which is equivalent to multiplying 2.5 by $\frac{1}{10^5}$ (or by 10^{-5}).

0 0 0 0 2.5
5 places to the left.

In scientific notation, 0.000025 is written 2.5×10^{-5}.

SELF CHECK 2 Write 0.0012 in scientific notation.

EXAMPLE 3 Write **a.** 235,000 and **b.** 0.00000235 in scientific notation.

Solution
a. $235{,}000 = 2.35 \times 10^5$, because 2.35 is between 1 and 10 and the decimal point must be moved 5 places to the right.
b. $0.00000235 = 2.35 \times 10^{-6}$, because 2.35 is between 1 and 10 and the decimal point must be moved 6 places to the left.

SELF CHECK 3 Write the following in scientific notation: **a.** 17,500 **b.** 0.657

Perspective

THE METRIC SYSTEM

A common metric unit of length is the kilometer, which is 1,000 meters. Because 1,000 is 10^3, we can write 1 km = 10^3 m. Similarly, 1 centimeter is one-hundredth of a meter: 1 cm = 10^{-2} m. In the metric system, prefixes such as *kilo* and *centi* refer to powers of 10. Other prefixes are used in the metric system, as shown in the table.

To appreciate the magnitudes involved, consider these facts: Light, which travels 186,000 miles every second, will travel about one foot in one nanosecond. The distance to the nearest star (except for the Sun) is 43 petameters, and the diameter of an atom is about 10 nanometers. To measure some quantities, however, even these units are inadequate. The Sun, for example, radiates 5×10^{26} watts. That's a lot of light bulbs!

Prefix	Symbol	Meaning	
peta	P	$10^{15} =$	1,000,000,000,000,000.
tera	T	$10^{12} =$	1,000,000,000,000.
giga	G	$10^9 =$	1,000,000,000.
mega	M	$10^6 =$	1,000,000.
kilo	k	$10^3 =$	1,000.
hecta	h	$10^2 =$	100.
deca	da	$10^1 =$	10.
deci	d	$10^{-1} =$	0.1
centi	c	$10^{-2} =$	0.01
milli	m	$10^{-3} =$	0.001
micro	μ	$10^{-6} =$	0.000 001
nano	n	$10^{-9} =$	0.000 000 001
pico	p	$10^{-12} =$	0.000 000 000 001

EXAMPLE 4 Write 432.0×10^{-5} in scientific notation.

Solution The number 432.0×10^{-5} is not written in scientific notation, because 432.0 is not a number between 1 and 10. To write the number in scientific notation, we proceed as follows:

$$\mathbf{432.0 \times 10^{-5}} = \mathbf{4.32 \times 10^2} \times 10^{-5} \quad \text{Write 432.0 in scientific notation.}$$
$$= 4.32 \times 10^{-3} \quad 10^2 \times 10^{-5} = 10^{-3}$$

SELF CHECK 4 Write 85×10^{-3} in scientific notation.

2 Convert a number from scientific notation to standard notation.

We can convert a number written in scientific notation to standard notation by reversing the process of converting standard notation to scientific notation. To convert a number to standard notation, move the decimal point the number of places indicated by the exponent. If the exponent is positive, this represents a large number and the decimal point will move to the right. If the exponent is negative, this represents a small number and the decimal point will move to the left. For example, to write 9.3×10^7 in standard notation, we move the decimal point seven places to the right. Since we already have one number to the right of the decimal, we will need to insert 6 zeros for place value.

$$9.3 \times 10^7 = 9.3 \times 10{,}000{,}000$$
$$= 93{,}000{,}000 \quad \text{Move the decimal point 7 places to the right.}$$

EXAMPLE 5 Write **a.** 3.4×10^5 and **b.** 2.1×10^{-4} in standard notation.

Solution **a.** $3.4 \times 10^5 = 3.4 \times 100{,}000$

$= 340{,}000$ Move the decimal point 5 places to the right.

b. $2.1 \times 10^{-4} = 2.1 \times \frac{1}{10^4}$

$= 2.1 \times \frac{1}{10{,}000}$

$= 0.00021$ Move the decimal point 4 places to the left.

SELF CHECK 5 Write the following in standard notation: **a.** 4.76×10^5 **b.** 9.8×10^{-3}

Each of the following numbers is written in both scientific and standard notation. In each case, the exponent gives the number of places that the decimal point moves, and the sign of the exponent indicates the direction that it moves.

$2.37 \times 10^6 = 2\,3\,7\,0\,0\,0\,0.$ Move the decimal point 6 places to the right.

$8.375 \times 10^{-3} = 0.0\,0\,8\,3\,7\,5$ Move the decimal point 3 places to the left.

$9.77 \times 10^0 = 9.77$ No movement of the decimal point.

3 Use scientific notation to simplify an expression.

Another advantage of scientific notation becomes apparent when we simplify fractions such as

$$\frac{(0.0032)(25{,}000)}{0.00040}$$

that contain very large or very small numbers. Although we can simplify this fraction by using arithmetic or a calculator, scientific notation provides an alternative way. First, we write each number in scientific notation; then we do the arithmetic on the numbers and the exponential expressions separately. Finally, we write the result in standard form, if desired.

$$\frac{(0.0032)(25{,}000)}{0.00040} = \frac{(3.2 \times 10^{-3})(2.5 \times 10^4)}{4.0 \times 10^{-4}}$$
$$= \frac{(3.2)(2.5)}{4.0} \times \frac{10^{-3}10^4}{10^{-4}}$$
$$= \frac{8.0}{4.0} \times 10^{-3+4-(-4)}$$
$$= 2.0 \times 10^5$$
$$= 200{,}000$$

EXAMPLE 6 **SPEED OF LIGHT** In a vacuum, light travels 1 meter in approximately 0.000000003 second. How long does it take for light to travel 500 kilometers?

Solution Since 1 kilometer = 1,000 meters, the length of time for light to travel 500 kilometers ($500 \cdot 1{,}000$ meters) is given by

$$\begin{aligned}(0.000000003)(500)(1{,}000) &= (3 \times 10^{-9})(5 \times 10^{2})(1 \times 10^{3}) \\ &= 3(5) \times 10^{-9+2+3} \\ &= \mathbf{15} \times 10^{-4} \\ &= \mathbf{1.5 \times 10^{1}} \times 10^{-4} \\ &= 1.5 \times 10^{-3} \\ &= 0.0015\end{aligned}$$

Light travels 500 kilometers in approximately 0.0015 second (or 1.5 millisecond).

SELF CHECK 6 How long does it take for light to travel 700 kilometers?

Accent on technology

▸ Finding Powers of Decimals

To find the value of $(453.46)^5$, we can use a calculator and enter these numbers and press these keys:

453.46 y^x 5 = Using a calculator with a y^x key.

453.46 ^ 5 ENTER Using a TI84 graphing calculator.

Either way, we have $(453.46)^5 = 1.917321395 \times 10^{13}$. Since this number is too large to show on the display, the calculator gives the result as `1.917321395 E13`.

For instructions regarding the use of a Casio graphing calculator, please refer to the Casio Keystroke Guide in the back of the book.

SELF CHECK ANSWERS

1. 9.3×10^7 **2.** 1.2×10^{-3} **3. a.** 1.75×10^4 **b.** 6.57×10^{-1} **4.** 8.5×10^{-2} **5. a.** 476,000 **b.** 0.0098 **6.** Light travels 700 kilometers in approximately 0.0021 second (or 2.1 milliseconds).

NOW TRY THIS

1. Write the result shown on the graphing calculator screen in

```
56900000*2570000
0000
            1.46233E18
```

a. scientific notation
b. standard notation

2. Write the result shown on the graphing calculator screen in

```
.000000562*.0000
903
           5.07486E⁻11
```

a. scientific notation
b. standard notation

3. There were approximately 1.45728×10^7 inches of wiring in one space shuttle. How many miles of wiring is this? (*Hint*: Recall that 5,280 feet = 1 mile.)

Unless otherwise noted, all content on this page is © Cengage Learning.

4.3 Exercises

WARM-UPS *Identify the number between 1 and 10 to be used to write the number in scientific notation.*

1. 39,000,000
2. 0.000000028
3. 83700
4. 921,400,000,000
5. 0.0000000001052
6. 0.000625

REVIEW

7. If $a = -1$, find the value of $-3a^{33}$.

8. Evaluate $\dfrac{3a^2 - 2b}{2a + 2b}$ if $a = 4$ and $b = 3$.

Determine which property of real numbers justifies each statement.

9. $a \cdot c = c \cdot a$

10. $7(u + 3) = 7u + 7 \cdot 3$

Solve each equation.

11. $6(x - 5) + 9 = 3x$

12. $8(3x - 5) - 4(2x + 3) = 12$

VOCABULARY AND CONCEPTS *Fill in the blanks.*

13. A number is written in ______________ when it is written as the product of a number between 1 (including 1) and 10 and an integer power of 10.

14. The number 125,000 is written in _______ notation.

GUIDED PRACTICE *Write each number in scientific notation.* SEE EXAMPLES 1–3. (OBJECTIVE 1)

15. 450,000
16. 4,750
17. 1,700,000
18. 290,000
19. 0.0059
20. 0.00083
21. 0.00000275
22. 0.000000055

Write each number in scientific notation. SEE EXAMPLE 4. (OBJECTIVE 1)

23. 42.5×10^2
24. 0.07×10^4
25. 0.37×10^{-4}
26. 25.2×10^{-3}

Write each number in standard notation. SEE EXAMPLE 5. (OBJECTIVE 2)

27. 2.3×10^2
28. 4.25×10^4
29. 8.12×10^5
30. 1.2×10^3
31. 1.15×10^{-3}
32. 8.16×10^{-5}
33. 9.76×10^{-4}
34. 6.52×10^{-3}

Use scientific notation to simplify each expression. Give all answers in standard notation. SEE EXAMPLE 6. (OBJECTIVE 3)

35. $(3.4 \times 10^2)(2.1 \times 10^3)$

36. $(4.1 \times 10^{-3})(3.4 \times 10^4)$

37. $\dfrac{9.3 \times 10^2}{3.1 \times 10^{-2}}$

38. $\dfrac{7.2 \times 10^6}{1.2 \times 10^8}$

39. $\dfrac{96{,}000}{(12{,}000)(0.00004)}$

40. $\dfrac{(0.48)(14{,}400{,}000)}{96{,}000{,}000}$

41. $\dfrac{2{,}475}{(132{,}000{,}000)(0.25)}$

42. $\dfrac{147{,}000{,}000{,}000{,}000}{25(0.000049)}$

ADDITIONAL PRACTICE *Write each number in scientific notation.*

43. 0.0000051
44. 0.04
45. 863,000,000
46. 514,000

47. $\dfrac{2.4 \times 10^2}{6 \times 10^{23}}$

48. $\dfrac{1.98 \times 10^2}{6 \times 10^{23}}$

Write each number in standard notation.

49. 37×10^7
50. 0.07×10^3
51. 0.32×10^{-4}
52. 617×10^{-2}

Determine which number of each pair is the larger.

53. 37.2 or 3.72×10^2
54. 37.2 or 3.72×10^{-1}
55. 3.72×10^3 or 4.72×10^3
56. 3.72×10^3 or 4.72×10^2
57. 3.72×10^{-1} or 4.72×10^{-2}
58. 3.72×10^{-3} or 2.72×10^{-2}

APPLICATIONS

59. Distance to Alpha Centauri The distance from Earth to the nearest star outside our solar system is approximately 25,700,000,000,000 miles. Write this number in scientific notation.

60. Speed of sound The speed of sound in air is 33,100 centimeters per second. Write this number in scientific notation.

61. Distance to Mars The distance from Mars to the Sun is approximately 1.14×10^8 miles. Write this number in standard notation.

62. Distance to Venus The distance from Venus to the Sun is approximately 6.7×10^7 miles. Write this number in standard notation.

63. Length of one meter One meter is approximately 0.00622 mile. Write this number in scientific notation.

64. Angstroms One angstrom is 1×10^{-7} millimeter. Write this number in standard notation.

65. Distance between Mercury and the Sun The distance from Mercury to the Sun is approximately 3.6×10^7 miles. Use scientific notation to express this distance in feet. (*Hint:* 5,280 feet = 1 mile.)

66. Oil reserves Recently, Venezuela was believed to have crude oil reserves of about 2.965×10^{11} barrels. A barrel contains 42 gallons of oil. Use scientific notation to express its oil reserves in gallons.

67. National debt The U.S. national debt in September 2011 was approximately $1,645.12 billion. Write this number in scientific notation, and then translate that to standard notation. (*Hint*: 1 billion = 1.0×10^9.)

68. National debt If the U.S. national debt continued to grow at a rate of $600,000.00 per day, calculate the debt in September 2012. Compare this debt to that of today. Has it increased or decreased?

69. Speed of sound The speed of sound in air is approximately 3.3×10^4 centimeters per second. Use scientific notation to express this speed in kilometers per second. (*Hint*: 100 centimeters = 1 meter and 1,000 meters = 1 kilometer.)

70. Light year One light year is approximately 5.87×10^{12} miles. Use scientific notation to express this distance in feet. (*Hint:* 5,280 feet = 1 mile.)

71. Wavelengths Some common types of electromagnetic waves are given in the table. List the wavelengths in order from shortest to longest.

Type	Use	Wavelength (m)
Visible light	Lighting	9.3×10^{-6}
Infrared	Photography	3.7×10^{-5}
X-rays	Medical	2.3×10^{-11}

72. Wavelengths More common types of electromagnetic waves are given in the table. List the wavelengths in order from longest to shortest.

Type	Use	Wavelength (m)
Radio waves	Communication	3.0×10^2
Microwaves	Cooking	1.1×10^{-2}
Ultraviolet	Sun lamp	6.1×10^{-8}

The bulk of the surface area of the red blood cell shown in the illustration is contained on its top and bottom. That area is $2\pi r^2$, twice the area of one circle. If there are N discs, their total surface area T will be N times the surface area of a single disc: $T = 2N\pi r^2$.

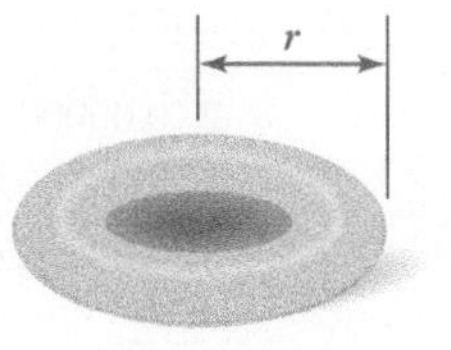

73. Red blood cells The red cells in human blood pick up oxygen in the lungs and carry it to all parts of the body. Each cell is a tiny circular disc with a radius of about 0.00015 in. Because the amount of oxygen carried depends on the surface area of the cells, and the cells are so tiny, a great number are needed—about 25 trillion in an average adult. Write these two numbers in scientific notation.

74. Red blood cells Find the total surface area of all the red blood cells in the body of an average adult. See Exercise 73.

WRITING ABOUT MATH

75. In what situations would scientific notation be more convenient than standard notation?

76. To multiply a number by a power of 10, we move the decimal point. Which way, and how far? Explain.

SOMETHING TO THINK ABOUT

77. Two positive numbers are written in scientific notation. How could you decide which is larger, without converting either to standard notation?

78. The product $1 \cdot 2 \cdot 3 \cdot 4 \cdot 5$, or 120, is called **5 factorial**, written 5!. Similarly, the number $6! = 6 \cdot 5 \cdot 4 \cdot 3 \cdot 2 \cdot 1 = 720$. Factorials get large very quickly. Calculate 30!, and write the number in standard notation. How large a factorial can you compute with a calculator?

Unless otherwise noted, all content on this page is © Cengage Learning.

Section 4.4

Polynomials

Objectives

1. Determine whether an expression is a polynomial.
2. Classify a polynomial as a monomial, binomial, or trinomial, if applicable.
3. Find the degree of a polynomial.
4. Evaluate a polynomial.
5. Evaluate a polynomial function.
6. Graph a linear, quadratic, and cubic polynomial function.

Vocabulary

polynomial
monomial
binomial
trinomial
degree of a monomial
degree of a polynomial
descending powers of a variable
ascending powers of a variable
polynomial function
quadratic function
parabola
cubic function

Getting Ready

Write each expression using exponents.

1. $2xxyyy$

2. $3xyyy$

3. $2xx + 3yy$

4. $xxx + yyy$

5. $(3xxy)(2xyy)$

6. $(5xyzzz)(xyz)$

7. $3(5xy)\left(\frac{1}{3}xy\right)$

8. $(xy)(xz)(yz)(xyz)$

In algebra, exponential expressions may be combined to form **polynomials**. In this section, we will introduce the topic of polynomials and graph some basic polynomial functions.

1 Determine whether an expression is a polynomial.

Recall that expressions such as

$$3x \qquad 4y^2 \qquad -8x^2y^3 \qquad \text{and} \qquad 25$$

with constant and/or variable factors are called *algebraic terms.* The coefficients of the first three of these terms are 3, 4, and −8, respectively. Because $25 = 25x^0$, 25 is referred to as a constant.

POLYNOMIALS

A polynomial is an algebraic expression that is a single term or the sum of several terms containing whole-number exponents on the variables.

Here are some examples of polynomials:

$$8xy^2t \qquad 3x + 2 \qquad 4y^2 - 2y + 3 \qquad 3a - 4b - 4c + 8d$$

COMMENT The expression $2x^3 - 3y^{-2}$ is not a polynomial, because the second term contains a negative exponent on a variable base.

EXAMPLE 1 Determine whether each expression is a polynomial.

a. $x^2 + 2x + 1$ — A polynomial.

b. $3x^{-1} - 2x - 3$ — No. The first term has a negative exponent on a variable base.

c. $\frac{1}{2}x^3 - 2.3x + 5$ — A polynomial.

d. $-2x + 3x^{1/2}$ — No. The second term has a fractional exponent on a variable base.

SELF CHECK 1 Determine whether each expression is a polynomial.

a. $3x^{-4} + 2x^2 - 3$ **b.** $7.5x^3 - 4x^2 - 3x$

c. $6x^2 - 9x^{1/3}$ **d.** $\frac{2}{3}x^4 + \frac{7}{9}x^3 - 7x$

2 Classify a polynomial as a monomial, binomial, or trinomial, if applicable.

A polynomial with one term is called a **monomial**. A polynomial with two terms is called a **binomial**. A polynomial with three terms is called a **trinomial**. Here are some examples.

Monomials	Binomials	Trinomials
$5x^2y$	$3u^3 - 4u^2$	$-5t^2 + 4t + 3$
$-6x$	$18a^2b + 4ab$	$27x^3 - 6x - 2$
29	$-29z^{17} - 1$	$-32r^6 + 7y^3 - z$

EXAMPLE 2 Classify each polynomial as a monomial, a binomial, or a trinomial, if applicable.

a. $5x^4 + 3x$ — Since the polynomial has two terms, it is a binomial.

b. $7x^4 - 5x^3 - 2$ — Since the polynomial has three terms, it is a trinomial.

c. $-5x^2y^3$ — Since the polynomial has one term, it is a monomial.

d. $9x^5 - 5x^2 + 8x - 7$ — Since the polynomial has four terms, it has no special name. It is none of these.

SELF CHECK 2 Classify each polynomial as a monomial, a binomial, or a trinomial, if applicable.

a. $5x$ **b.** $-5x^2 + 2x - 5$

c. $16x^2 - 9y^2$ **d.** $x^9 + 7x^4 - x^2 + 6x - 1$

3 Find the degree of a polynomial.

The monomial $7x^6$ is called a monomial of sixth degree or a monomial of degree 6, because the variable x occurs as a factor six times. The monomial $3x^3y^4$ is a monomial of the seventh degree, because the variables x and y occur as factors a total of seven times. Other examples are

$-2x^3$ is a monomial of degree 3.

$47x^2y^3$ is a monomial of degree 5.

$18x^4y^2z^8$ is a monomial of degree 14.

8 is a monomial of degree 0, because $8 = 8x^0$.

These examples illustrate the following definition.

DEGREE OF A MONOMIAL

If a is a nonzero coefficient, the **degree of the monomial** ax^n is n.

The degree of a monomial with several variables is the sum of the exponents on those variables.

COMMENT Note that the degree of ax^n is not defined when $a = 0$. Since $ax^n = 0$ when $a = 0$, the constant 0 has no defined degree.

Because each term of a polynomial is a monomial, we define the degree of a polynomial by considering the degree of each of its terms.

DEGREE OF A POLYNOMIAL

The **degree of a polynomial** is the degree of its term with largest degree.

For example,

- $x^2 + 2x$ is a binomial of degree 2, because the degree of its first term is 2 and the degree of its other term is less than 2.
- $3x^3y^2 + \mathbf{4x^4y^4} - 3x^3$ is a trinomial of degree 8, because the degree of its second term is 8 and the degree of each of its other terms is less than 8.
- $25x^4y^3z^7 - \mathbf{15xy^8z^{10}} - \mathbf{32x^8y^8z^3} + 4$ is a polynomial of degree 19, because its second and third terms are of degree 19. Its other terms have degrees less than 19.

EXAMPLE 3 Find the degree of each polynomial.

a. $-4x^3 - 5x^2 + 3x$ 3, the degree of the first term because it has largest degree.

b. $5x^4y^2 + 7xy^2 - 16x^3y^5$ 8, the degree of the last term.

c. $-17a^2b^3c^4 + 12a^3b^4c$ 9, the degree of the first term.

SELF CHECK 3 Find the degree of each polynomial.

a. $15p^3q^4 - 25p^4q^2$ **b.** $-14rs^3t^4 + 12r^3s^3t^3$

c. $16mn^6 - 9m^2n^2 + 3m^3n^3$

If the polynomial contains a single variable, we usually write it with its exponents in **descending order** where the term with the highest degree is listed first, followed by the term with the next highest degree, and so on. If we reverse the order, the polynomial is said to be written with its exponents in **ascending order**.

4 Evaluate a polynomial.

When a number is substituted for the variable in a polynomial, the polynomial takes on a numerical value. Finding that value is called *evaluating the polynomial*.

EXAMPLE 4 Evaluate the polynomial $3x^2 + 2$ when

a. $x = 0$ **b.** $x = 2$ **c.** $x = -3$ **d.** $x = -\frac{1}{5}$.

Solution

a.
$$\begin{aligned} 3x^2 + 2 &= 3(\mathbf{0})^2 + 2 \\ &= 3(0) + 2 \\ &= 0 + 2 \\ &= 2 \end{aligned}$$

b.
$$\begin{aligned} 3x^2 + 2 &= 3(\mathbf{2})^2 + 2 \\ &= 3(4) + 2 \\ &= 12 + 2 \\ &= 14 \end{aligned}$$

c. $3x^2 + 2 = 3(-3)^2 + 2$
$= 3(9) + 2$
$= 27 + 2$
$= 29$

d. $3x^2 + 2 = 3\left(-\frac{1}{5}\right)^2 + 2$
$= 3\left(\frac{1}{25}\right) + 2$
$= \frac{3}{25} + \frac{50}{25}$
$= \frac{53}{25}$

SELF CHECK 4 Evaluate $3x^2 + x - 2$ when **a.** $x = 2$ **b.** $x = -1$ **c.** $x = -2$ **d.** $x = \frac{1}{2}$

When we evaluate a polynomial for several values of its variable, we often write the results in a table.

EXAMPLE 5 Evaluate the polynomial $x^3 + 1$ for the following values and write the results in a table.
a. $x = -2$ **b.** $x = -1$ **c.** $x = 0$ **d.** $x = 1$ **e.** $x = 2$

Solution

	x	$x^3 + 1$	
a.	-2	-7	$x^3 + 1 = (-2)^3 + 1 = -7$
b.	-1	0	$x^3 + 1 = (-1)^3 + 1 = 0$
c.	0	1	$x^3 + 1 = (0)^3 + 1 = 1$
d.	1	2	$x^3 + 1 = (1)^3 + 1 = 2$
e.	2	9	$x^3 + 1 = (2)^3 + 1 = 9$

SELF CHECK 5 Complete the following table.

x	$-x^3 + 1$
-2	
-1	
0	
1	
2	

5 Evaluate a polynomial function.

Since the right sides of the functions $f(x) = 2x - 3$, $f(x) = x^2$, and $f(t) = -16t^2 + 64t$ are polynomials, they are called **polynomial functions**. We can evaluate these functions at specific values of the variable by evaluating the polynomial on the right side.

EXAMPLE 6 Given $f(x) = 2x - 3$, find $f(-2)$.

Solution To find $f(-2)$, we substitute -2 for x and evaluate the function.

$$\begin{aligned} f(x) &= 2x - 3 \\ f(-2) &= 2(-2) - 3 \\ &= -4 - 3 \\ &= -7 \end{aligned}$$

Thus, $f(-2) = -7$.

SELF CHECK 6 Given $f(x) = 2x - 3$, find $f(5)$.

Accent on technology

▸ Height of a Rocket

The height h (in feet) of a toy rocket launched straight up into the air with an initial velocity of 64 feet per second is given by the polynomial function

$$h = f(t) = -16t^2 + 64t$$

In this case, the height h is the dependent variable, and the time t is the independent variable because the height depends on the time elapsed. To find the height of the rocket 3.5 seconds after launch, we substitute 3.5 for t and evaluate h.

$$h = -16t^2 + 64t$$
$$h = -16(\mathbf{3.5})^2 + 64(\mathbf{3.5})$$

To evaluate h with a calculator, we enter these numbers and press these keys:

16 +/− × 3.5 x² + (64 × 3.5) = — Using a calculator with a +/− key.

(−) 16 × 3.5 x² + 64 × 3.5 ENTER — Using a TI84 graphing calculator.

Either way, the display reads 28. After 3.5 seconds, the rocket will be 28 feet above the ground.

For instructions regarding the use of a Casio graphing calculator, please refer to the Casio Keystroke Guide in the back of the book.

6 Graph a linear, quadratic, and cubic polynomial function.

We can graph polynomial functions as we graphed equations in Section 3.2. We make a table of values, plot points, and draw the line or curve that passes through those points.

In the next example, we graph the function $f(x) = 2x - 3$. Since its graph is a line, recall that it is a linear function.

COMMENT The ordered pair (x, y) can be written as $(x, f(x))$.

EXAMPLE 7 Graph: $f(x) = 2x - 3$

Solution We substitute numbers for x, compute the corresponding values of $f(x)$, and list the results in a table, as in Figure 4-1. We then plot the pairs $(x, f(x))$ and draw a line through the points, as shown in the figure. From the graph, we can see that x can be any value. This confirms that the domain is the set of real numbers $\mathbb{R}$. We also can see that $f(x)$ can be any value. This confirms the range is also the set of real numbers $\mathbb{R}$.

$f(x) = 2x - 3$

x	$f(x)$	$(x, f(x))$
−3	−9	(−3, −9)
−2	−7	(−2, −7)
−1	−5	(−1, −5)
0	−3	(0, −3)
1	−1	(1, −1)
2	1	(2, 1)
3	3	(3, 3)

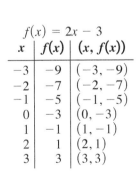

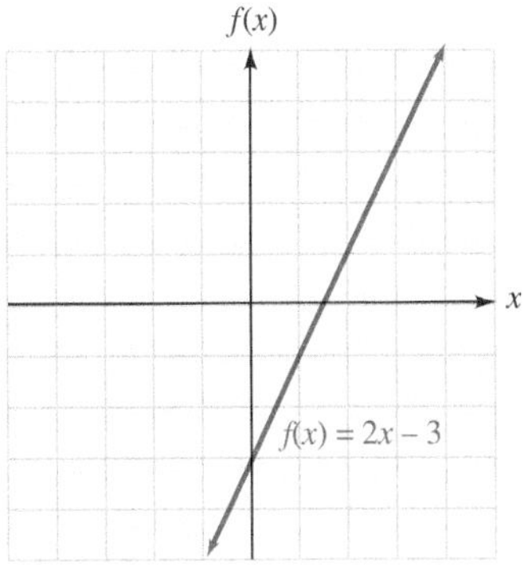

Figure 4-1

SELF CHECK 7 Graph: $f(x) = \frac{1}{2}x + 3$

Unless otherwise noted, all content on this page is © Cengage Learning.

In the next example, we graph the function $f(x) = x^2$, referred to as the squaring function. Since the polynomial on the right side is of second degree, we call this function a **quadratic function**.

EXAMPLE 8 Graph $f(x) = x^2$. State the domain and the range.

Solution We substitute numbers for x, compute the corresponding values of $f(x)$, and list the results in a table, as in Figure 4-2. We then plot the pairs $(x, f(x))$ and draw a smooth curve through the points, as shown in the figure. This curve is called a **parabola**. From the graph, we can see that x can be any value. This confirms that the domain is the set of real numbers $\mathbb{R}$. We can also see that $f(x)$ is always a positive number or 0. This confirms that the range is $\{f(x) \mid f(x) \text{ is a real number and } f(x) \geq 0\}$. In interval notation, this is $[0, \infty)$.

Amalie Noether
1882–1935
Albert Einstein described Noether as the most creative female mathematical genius since the beginning of higher education for women. Her work was in the area of abstract algebra. Although she received a doctoral degree in mathematics, she was denied a mathematics position in Germany because she was a woman.

$f(x) = x^2$

x	$f(x)$	$(x, f(x))$
−3	9	(−3, 9)
−2	4	(−2, 4)
−1	1	(−1, 1)
0	0	(0, 0)
1	1	(1, 1)
2	4	(2, 4)
3	9	(3, 9)

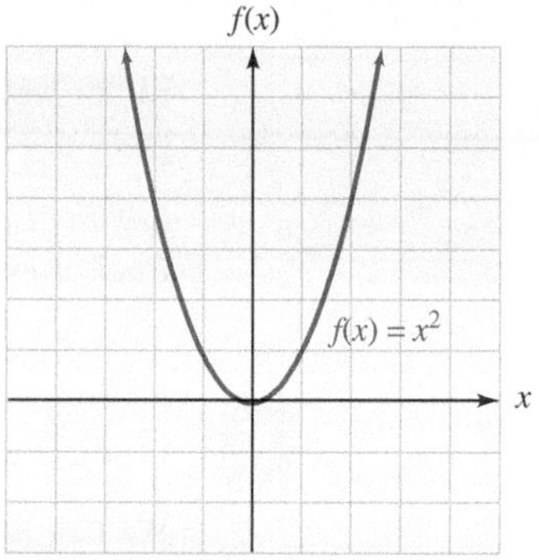

Figure 4-2

SELF CHECK 8 Graph $f(x) = x^2 - 3$. State the domain and the range.

In the next example, we graph the function $f(x) = x^3$, referred to as the cubing function. Since the polynomial on the right side is of third degree, we call this function a **cubic function**.

EXAMPLE 9 Graph $f(x) = x^3$. State the domain and the range.

Solution We substitute numbers for x, compute the corresponding values of $f(x)$, and list the results in a table, as in Figure 4-3. We then plot the pairs $(x, f(x))$ and draw a smooth curve through the points, as shown in the figure. From the figure, we can see that the domain and the range are both $\mathbb{R}$.

$f(x) = x^3$

x	$f(x)$	$(x, f(x))$
−2	−8	(−2, −8)
−1	−1	(−1, −1)
0	0	(0, 0)
1	1	(1, 1)
2	8	(2, 8)

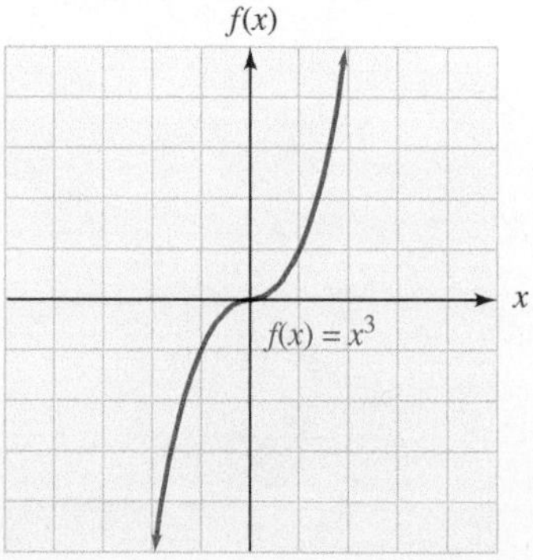

Figure 4-3

SELF CHECK 9 Graph $f(x) = x^3 + 3$. State the domain and the range.

Unless otherwise noted, all content on this page is © Cengage Learning.

Accent on technology

▸ Graphing Polynomial Functions

It is possible to use a graphing calculator to generate tables and graphs for polynomial functions. For example, Figure 4-4 shows calculator tables and the graphs of $f(x) = 2x - 3$, $f(x) = x^2$, and $f(x) = x^3$. (*Note*: Although you are graphing the function $f(x)$, the calculator will represent it as y.)

X	Y_1
−3	−9
−2	−7
−1	−5
0	−3
1	−1
2	1
3	3

X = −3

X	Y_1
−3	9
−2	4
−1	1
0	0
1	1
2	4
3	9

X = −3

X	Y_1
−3	−27
−2	−8
−1	−1
0	0
1	1
2	8
3	27

X = −3

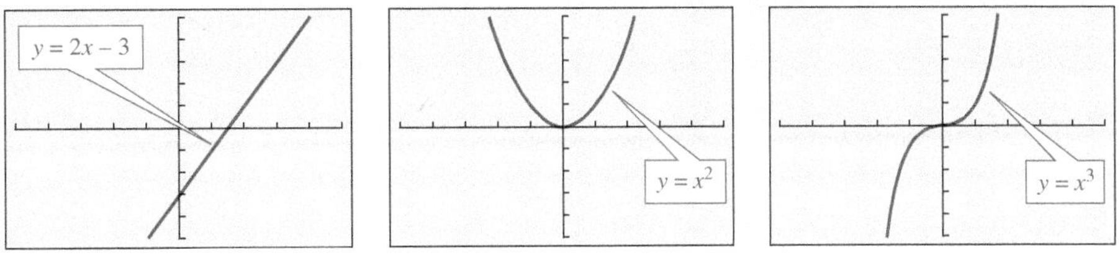

(a) (b) (c)

Figure 4-4

For instructions regarding the use of a Casio graphing calculator, please refer to the Casio Keystroke Guide in the back of the book.

EXAMPLE 10 Graph $f(x) = x^2 - 2x$. State the domain and the range.

Solution We substitute numbers for x, compute the corresponding values of $f(x)$, and list the results in a table, as in Figure 4-5. We then plot the pairs $(x, f(x))$ and draw a smooth curve through the points, as shown in the figure. From the graph, we can see the domain is $\mathbb{R}$ and the range is $[-1, \infty)$.

$f(x) = x^2 - 2x$

x	$f(x)$	$(x, f(x))$
−2	8	(−2, 8)
−1	3	(−1, 3)
0	0	(0, 0)
1	−1	(1, −1)
2	0	(2, 0)
3	3	(3, 3)
4	8	(4, 8)

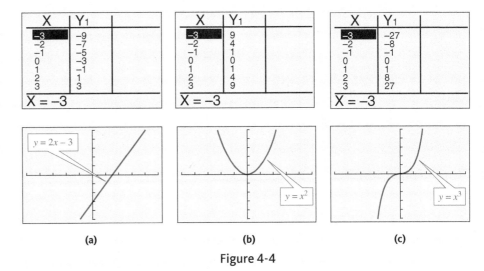

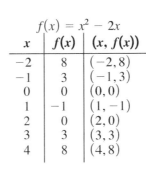

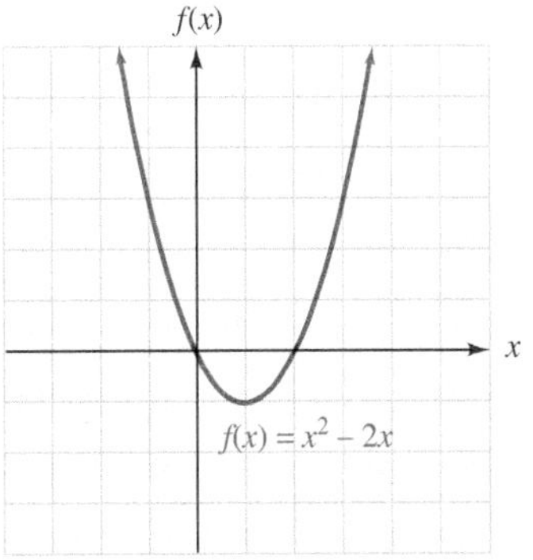

Figure 4-5

SELF CHECK 10 Graph $f(x) = x^2 - 2x$. State the domain and the range.

Unless otherwise noted, all content on this page is © Cengage Learning.

Everyday connections
NBA Salaries

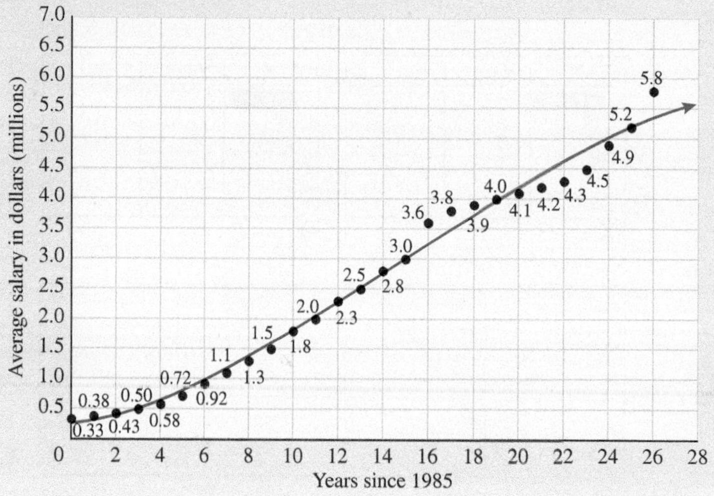

$$f(t) = 0.2457 + 0.0563t + 0.0127t^2 - 0.0002802t^3$$

The polynomial function shown above models average player salary in the National Basketball Association during the time period 1985–2011, where t equals the number of years since 1985. The black dots represent average salaries and points on the red graph represent predicted salaries. Use the graph to answer the following questions.

1. **a.** What was the average player salary in 2002?

   b. What was the average player salary predicted by the function f in 2002?
2. What is the predicted average salary for a player in 2017?
3. Does this function yield a realistic prediction of the average player salary in 2022?

SELF CHECK ANSWERS

1. a. no **b.** yes **c.** no **d.** yes **2. a.** monomial **b.** trinomial **c.** binomial **d.** none of these **3. a.** 7 **b.** 9 **c.** 7 **4. a.** 12 **b.** 0 **c.** 8 **d.** $-\frac{3}{4}$ **5.** 9, 2, 1, 0, −7 **6.** 7

7.

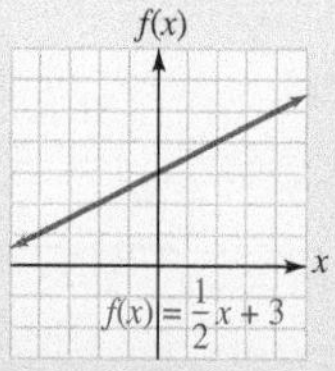

8.

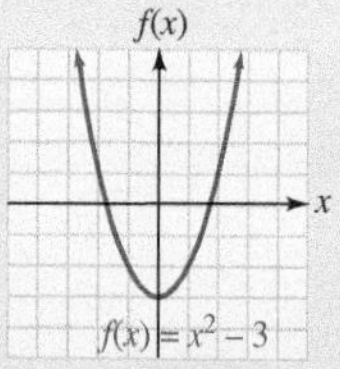

D: ℝ, R: $[-3, \infty)$

9.

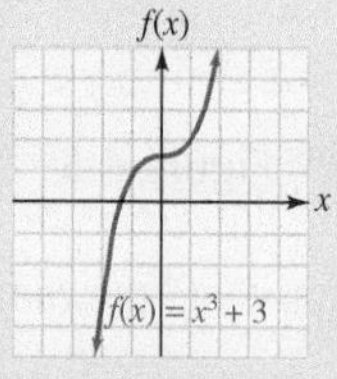

D: ℝ, R: ℝ

10.

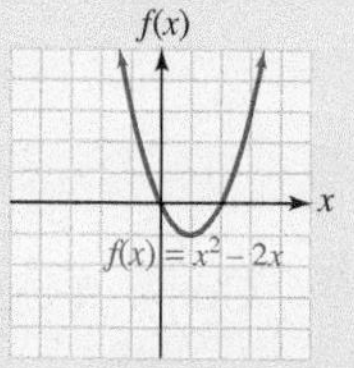

D: ℝ, R: $[-1, \infty)$

Unless otherwise noted, all content on this page is © Cengage Learning.

NOW TRY THIS

1. Classify each polynomial and state its degree.
   a. $9x^2 - 6x^8$
   b. 1
2. If $f(x) = -3x^2 - 2x$,
   a. find $f(-4)$.
   b. find $f(3p)$.
3. Use a graphing calculator to graph $f(x) = \frac{1}{2}x^3 - x^2 + 3x + 1$.

4.4 Exercises

WARM-UPS *Write each expression using exponents.*

1. $5 \cdot a \cdot a + 2 \cdot b \cdot b \cdot b$
2. $7 \cdot x \cdot x \cdot x + 8 \cdot y \cdot y$
3. $4 \cdot a \cdot a \cdot a \cdot b \cdot b$
4. $9 \cdot x \cdot x \cdot y \cdot y \cdot y \cdot y$
5. $x \cdot x \cdot x + 3 \cdot x \cdot x + x$
6. $5 \cdot a + 4 \cdot a \cdot a + a \cdot a \cdot a$
7. $a \cdot a \cdot a + b \cdot b \cdot b$
8. $2 \cdot x \cdot x + 5 \cdot y \cdot y$

REVIEW *Solve each equation.*

9. $5(u - 5) + 9 = 2(u + 4)$
10. $8(3a - 5) - 12 = 4(2a + 3)$

Solve each inequality and graph the solution set.

11. $-4(3y + 2) \le 28$
12. $-5 < 3t + 4 \le 13$

Write each expression without using parentheses or negative exponents. Assume no variable is zero.

13. $(x^3x^5)^4$
14. $(b^4)^3(b^2)^3$
15. $\left(\dfrac{y^2y^5}{y^4}\right)^3$
16. $\left(\dfrac{2t^3}{t}\right)^{-4}$

VOCABULARY AND CONCEPTS *Fill in the blanks.*

17. An expression such as $3t^4$ with a constant and/or a variable is called an ________ term.
18. A ________ is an algebraic expression that is one term, or the sum of two or more terms, containing whole-number exponents on the variables.
19. A ________ is a polynomial with one term. A ________ is a polynomial with two terms. A ________ is a polynomial with three terms.
20. If $a \neq 0$, the ______ of ax^n is n.
21. The degree of a monomial with several variables is the ____ of the exponents on those variables.
22. The graph of a _____ function is a line.
23. A function of the form $y = f(x)$ where $f(x)$ is a polynomial is called a ________ function.
24. The function $f(x) = x^2$ is called the squaring or ________ function.
25. The function $f(x) = x^3$ is called the cubing or _____ function.
26. The graph of a quadratic function is called a ________.
27. The polynomial $8x^5 - 3x^3 + 6x^2 - 1$ is written with its exponents in ________ order. Its degree is _.
28. The polynomial $-2x + x^2 - 5x^3 + 7x^4$ is written with its exponents in ________ order. Its degree is _.
29. Any equation in x and y where each input value x determines exactly one output value y is called a ________.
30. $f(x)$ is read as _____.

GUIDED PRACTICE *Determine whether each expression is a polynomial.* SEE EXAMPLE 1. (OBJECTIVE 1)

31. $x^4 + 3x^2 - 5$
32. $2x^{-2} - 3x + 7$
33. $3x^{1/2} - 4$
34. $0.5x^5 - 0.25x^2$

Classify each polynomial as a monomial, a binomial, a trinomial, or none of these. SEE EXAMPLE 2. (OBJECTIVE 2)

35. $3x + 7$
36. $5x^2 + 2x$
37. $3y^2 + 4y + 3$
38. $3xy$
39. $3z^2$
40. $3x^4 - 2x^3 + 3x - 1$
41. $5t - 32$
42. $-8x^3y^2z^5$

Give the degree of each polynomial. SEE EXAMPLE 3. (OBJECTIVE 3)

43. $4x^7 - 5x^2$
44. $3x^5 - 4x^2$
45. $-2x^2 + 3x^3$
46. $-6x^4 + 3x^2 - 5x$
47. $3x^2y^3 + 5x^3y^5$
48. $-2x^2y^3 + 4x^3y^2z$
49. $-5r^2s^2t - 3r^3st^2 + 3$
50. $8r^3s^2t^4 - 3r^5t^6 + 2$

Evaluate $-x^2 - 4$ for each value. SEE EXAMPLE 4. (OBJECTIVE 4)

51. $x = 0$ **52.** $x = 1$

53. $x = -1$ **54.** $x = -2$

Complete each table. SEE EXAMPLE 5. (OBJECTIVE 4)

55.

x	$x^2 - 3$
-2	
-1	
0	
1	
2	

56.

x	$-x^2 + 3$
-2	
-1	
0	
1	
2	

57.

x	$x^3 + 2$
-2	
-1	
0	
1	
2	

58.

x	$-x^3 + 2$
-2	
-1	
0	
1	
2	

If $f(x) = 5x + 1$, find each value. SEE EXAMPLE 6. (OBJECTIVE 5)

59. $f(-3)$ **60.** $f(4)$

61. $f\left(-\frac{1}{2}\right)$ **62.** $f\left(\frac{2}{5}\right)$

Graph each polynomial function. State the domain and range. SEE EXAMPLES 7–10. (OBJECTIVE 6)

63. $f(x) = x^2 - 1$

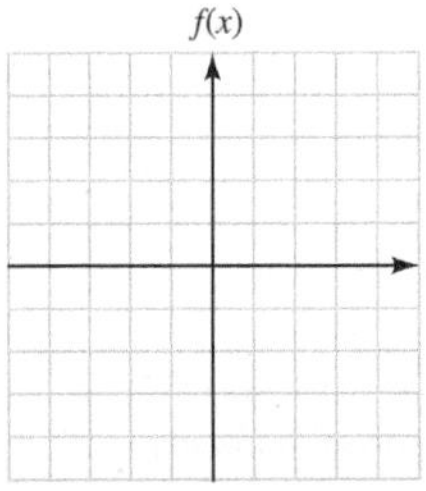

64. $f(x) = x^2 + 2$

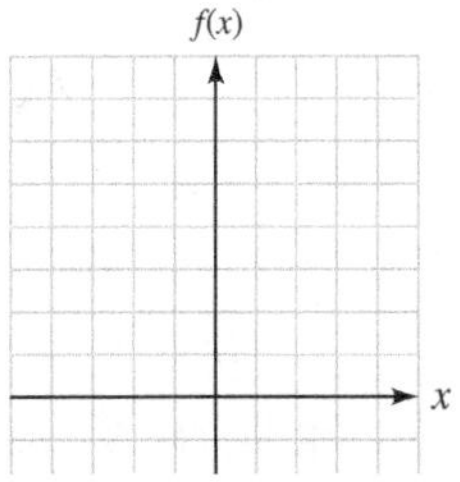

65. $f(x) = x^3 + 2$

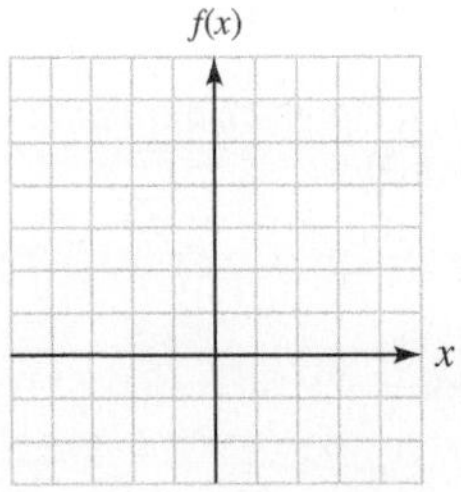

66. $f(x) = x^3 - 2$

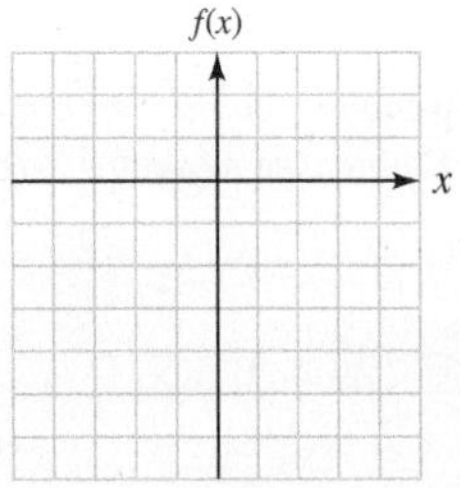

ADDITIONAL PRACTICE *Evaluate $5x - 3$ for each value.*

67. $x = 2$ **68.** $x = 0$

69. $x = -1$ **70.** $x = -2$

Classify each polynomial as a monomial, a binomial, a trinomial, or none of these.

71. $s^2 - 23s + 31$ **72.** $12x^3 - 12x^2 + 36x - 3$

73. $3x^5 - 2x^4 - 3x^3 + 17$ **74.** x^3

75. $\frac{1}{2}x^3 + 3$ **76.** $x^3 - 1$

Give an example of a polynomial that is . . .

77. a binomial **78.** a monomial

79. a trinomial **80.** not a monomial, a binomial, or a trinomial

81. of degree 3 **82.** of degree 1

83. of degree 0 **84.** of no defined degree

Give the degree of each polynomial.

85. $x^{12} + 3x^2y^3z^4$ **86.** 17^2x

87. 38 **88.** -25

If $f(x) = x^2 - 2x + 3$, find each value.

89. $f(5)$ **90.** $f(3)$

91. $f(-2)$ **92.** $f(-1)$

93. $f(0.5)$ **94.** $f(1.2)$

APPLICATIONS *Use a calculator to help solve each.*

95. Height of a rocket See the Accent on Technology section on page 255. Find the height of the rocket 2 seconds after launch.

96. Height of a rocket Again referring to page 255, make a table of values to find the rocket's height at various times. For what values of t will the height of the rocket be 0?

97. Computing revenue The revenue r (in dollars) that a manufacturer of desk chairs receives is given by the polynomial function

$$r = f(d) = -0.08d^2 + 100d$$

where d is the number of chairs manufactured. Find the revenue received when 815 chairs are manufactured.

98. Falling balloons Some students threw balloons filled with water from a dormitory window. The height h (in feet) of the balloons t seconds after being thrown is given by the polynomial function

$$h = f(t) = -16t^2 + 12t + 20$$

How far above the ground is a balloon 1.5 seconds after being thrown?

99. Stopping distance The number of feet that a car travels before stopping depends on the driver's reaction time and the braking distance. For one driver, the stopping distance d is given by the function $d = f(v) = 0.04v^2 + 0.9v$, where v is the velocity of the car. Find the stopping distance when the driver is traveling at 30 mph.

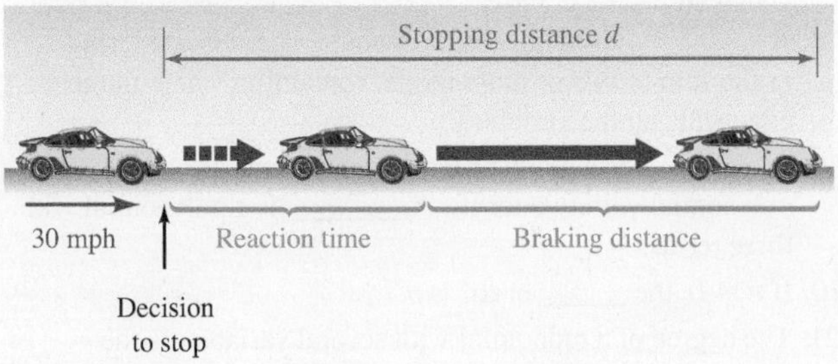

Unless otherwise noted, all content on this page is © Cengage Learning.

100. Stopping distance Find the stopping distance of the car discussed in Exercise 99 when the driver is going 70 mph.

WRITING ABOUT MATH

101. Describe how to determine the degree of a polynomial.

102. Describe how to classify a polynomial as a monomial, a binomial, a trinomial, or none of these.

SOMETHING TO THINK ABOUT

103. Find a polynomial whose value will be 1 if you substitute $\frac{3}{2}$ for x.

104. Graph the function $f(x) = -x^2$. What do you discover?

Section 4.5 Adding and Subtracting Polynomials

Objectives

1. Add two or more monomials.
2. Subtract two monomials.
3. Add two polynomials.
4. Subtract two polynomials.
5. Simplify an expression using the order of operations and combining like terms.
6. Solve an application requiring operations with polynomials.

Vocabulary

subtrahend minuend

Getting Ready

Combine like terms and simplify, if possible.

1. $3x + 2x$	**2.** $5y - 3y$	**3.** $19x + 6x$	**4.** $8z - 3z$
5. $9r + 3r$	**6.** $4r - 3s$	**7.** $7r - 7r$	**8.** $17r - 17r^2$

In this section, we will discuss how to add and subtract polynomials.

1 Add two or more monomials.

Recall that like terms have the same variables with the same exponents. For example,

$3xyz^2$ and $-2xyz^2$ are like terms. The variables and their exponents are the same.

$\frac{1}{2}ab^2c$ and $\frac{1}{3}a^2bc^2$ are unlike terms. The variables and their corresponding exponents are NOT the same.

Also recall that to combine like terms, we add (or subtract) their coefficients and keep the same variables with the same exponents. For example,

$$2y + 5y = (2 + 5)y \qquad -3x^2 + 7x^2 = (-3 + 7)x^2$$
$$= 7y \qquad = 4x^2$$

Likewise,

$$4x^3y^2 + 9x^3y^2 = 13x^3y^2 \qquad 4r^2s^3t^4 + 7r^2s^3t^4 = 11r^2s^3t^4$$

EXAMPLE 1 Perform the following additions.

a. $5xy^3 + 7xy^3 = 12xy^3$

b. $-7x^2y^2 + 6x^2y^2 + 3x^2y^2 = -x^2y^2 + 3x^2y^2$
$= 2x^2y^2$

c. $(2x^2)^2 + 81x^4 = 4x^4 + 81x^4$ $(2x^2)^2 = (2x^2)(2x^2) = 4x^4$
$= 85x^4$

SELF CHECK 1 Perform the following additions. **a.** $6a^3b^2 + 5a^3b^2$ **b.** $-2pq^2 + 5pq^2 + 8pq^2$ **c.** $27x^6 + (2x^2)^3$

2 Subtract two monomials.

To subtract one monomial from another, we add the opposite of the monomial that is to be subtracted. In symbols, $x - y = x + (-y)$.

EXAMPLE 2 Find each difference.

a. $8x^2 - 3x^2 = 8x^2 + (-3x^2)$
$= 5x^2$

b. $6x^3y^2 - 9x^3y^2 = 6x^3y^2 + (-9x^3y^2)$
$= -3x^3y^2$

c. $-3r^2st^3 - 5r^2st^3 = -3r^2st^3 + (-5r^2st^3)$
$= -8r^2st^3$

SELF CHECK 2 Find each difference. **a.** $12m^3 - 7m^3$ **b.** $-4p^3q^2 - 8p^3q^2$ **c.** $-8x^3y^2z - 12x^3y^2z$

3 Add two polynomials.

Because of the distributive property, we can remove parentheses enclosing several terms when the sign preceding the parentheses is +.

$$\begin{aligned} +(3x^2 + 3x - 2) &= +1(3x^2 + 3x - 2) \\ &= 1(3x^2) + 1(3x) + 1(-2) \\ &= 3x^2 + 3x + (-2) \\ &= 3x^2 + 3x - 2 \end{aligned}$$

We can add polynomials by removing parentheses, if necessary, and then combining any like terms that are contained within the polynomials.

EXAMPLE 3 Add: $(3x^2 - 3x + 2) + (2x^2 + 7x - 4)$

Solution $(3x^2 - 3x + 2) + (2x^2 + 7x - 4)$

$= 3x^2 - 3x + 2 + 2x^2 + 7x - 4$ Remove parentheses.

$= 3x^2 + 2x^2 - 3x + 7x + 2 - 4$ Use the commutative property of addition.

$= 5x^2 + 4x - 2$ Combine like terms.

SELF CHECK 3 Add: $(2a^2 - a + 4) + (5a^2 + 6a - 5)$

Additions such as Example 3 often are written with like terms aligned vertically. We then can add the polynomials column by column.

$$\begin{array}{r} 3x^2 - 3x + 2 \\ + \underline{2x^2 + 7x - 4} \\ 5x^2 + 4x - 2 \end{array}$$

EXAMPLE 4 Add:

$$\begin{array}{r} 4x^2y + 8x^2y^2 - 3x^2y^3 \\ + \underline{3x^2y - 8x^2y^2 + 8x^2y^3} \\ 7x^2y \qquad\qquad + 5x^2y^3 \end{array}$$

SELF CHECK 4 Add:

$$\begin{array}{r} 4pq^2 + 6pq^3 - 7pq^4 \\ + \underline{2pq^2 - 8pq^3 + 9pq^4} \end{array}$$

4 Subtract two polynomials.

We can remove parentheses enclosing several terms when the sign preceding the parentheses is negative by distributing a -1 to each term *within the parentheses.*

$$\begin{aligned} -(3x^2 + 3x - 2) &= \mathbf{-1}(3x^2 + 3x - 2) \\ &= \mathbf{-1}(3x^2) + (\mathbf{-1})(3x) + (\mathbf{-1})(-2) \\ &= -3x^2 + (-3x) + 2 \\ &= -3x^2 - 3x + 2 \end{aligned}$$

This suggests that the way to subtract polynomials is to remove parentheses by changing the sign of all terms being subtracted and then combine like terms.

EXAMPLE 5 Subtract:

a. $(3x - 4) - (5x + 7) = 3x - 4 - 5x - 7$
$= -2x - 11$

b. $(3x^2 - 4x - 6) - (2x^2 - 6x + 12) = 3x^2 - 4x - 6 - 2x^2 + 6x - 12$
$= x^2 + 2x - 18$

c. $(-4rt^3 + 2r^2t^2) - (-3rt^3 + 2r^2t^2) = -4rt^3 + 2r^2t^2 + 3rt^3 - 2r^2t^2$
$= -rt^3$

SELF CHECK 5 Subtract: **a.** $(4x - 1) - (x - 5)$ **b.** $(6x^2 - 5x + 4) - (-3x^2 + 6x + 1)$ **c.** $(-2a^2b + 5ab^2) - (-5a^2b - 7ab^2)$

To subtract polynomials in vertical form, we add the negative of the **subtrahend** (the bottom polynomial) to the **minuend** (the top polynomial) to obtain the difference.

EXAMPLE 6 Subtract $(3x^2y - 2xy^2)$ from $(2x^2y + 4xy^2)$.

Solution We write the subtraction in vertical form, change the signs of the terms of the subtrahend, and add.

$$\begin{array}{rr} & 2x^2y + 4xy^2 \\ - & \underline{3x^2y - 2xy^2} \end{array} \quad \rightarrow \quad \begin{array}{rr} & 2x^2y + 4xy^2 \\ + & \underline{-\,3x^2y + 2xy^2} \\ & -\,x^2y + 6xy^2 \end{array}$$

In horizontal form, the result is the same.

$$\begin{aligned} 2x^2y + 4xy^2 - (3x^2y - 2xy^2) &= 2x^2y + 4xy^2 - 3x^2y + 2xy^2 \\ &= -x^2y + 6xy^2 \end{aligned}$$

SELF CHECK 6 Subtract:

$$\begin{array}{rr} & 5p^2q - 6pq + 7q \\ - & \underline{2p^2q + 2pq - 8q} \end{array}$$

EXAMPLE 7 Subtract $(6xy^2 + 4x^2y^2 - x^3y^2)$ from $(-2xy^2 - 3x^3y^2)$.

Solution

$$\begin{array}{rr} & -2xy^2 \qquad\qquad - 3x^3y^2 \\ - & \underline{6xy^2 + 4x^2y^2 - x^3y^2} \end{array} \quad \rightarrow \quad \begin{array}{rr} & -2xy^2 \qquad\qquad - 3x^3y^2 \\ + & \underline{-6xy^2 - 4x^2y^2 + x^3y^2} \\ & -8xy^2 - 4x^2y^2 - 2x^3y^2 \end{array}$$

COMMENT Be careful of the order of subtration. "Subtract x from y" means $y - x$.

In horizontal form, the result is the same.

$$\begin{aligned} &-2xy^2 - 3x^3y^2 - (6xy^2 + 4x^2y^2 - x^3y^2) \\ &= -2xy^2 - 3x^3y^2 - 6xy^2 - 4x^2y^2 + x^3y^2 \\ &= -8xy^2 - 4x^2y^2 - 2x^3y^2 \end{aligned}$$

SELF CHECK 7 Subtract $(-2pq^2 - 2p^2q^2 + 3p^3q^2)$ from $(5pq^2 + 3p^2q^2 - p^3q^2)$.

5 Simplify an expression using the order of operations and combining like terms.

Because of the distributive property, we can remove parentheses enclosing several terms that are multiplied by a monomial by multiplying every term within the parentheses by that monomial. For example, to add $3(2x + 5)$ and $2(4x - 3)$, we proceed as follows:

$$\begin{aligned} 3(2x + 5) + 2(4x - 3) &= 6x + 15 + 8x - 6 \\ &= 6x + 8x + 15 - 6 && \text{Use the commutative property of addition.} \\ &= 14x + 9 && \text{Combine like terms.} \end{aligned}$$

EXAMPLE 8 Simplify:

a. $$\begin{aligned} 3(x^2 + 4x) + 2(x^2 - 4) &= 3x^2 + 12x + 2x^2 - 8 \\ &= 5x^2 + 12x - 8 \end{aligned}$$

b. $$\begin{aligned} 8(y^2 - 2y + 3) - 4(2y^2 + y - 3) &= 8y^2 - 16y + 24 - 8y^2 - 4y + 12 \\ &= -20y + 36 \end{aligned}$$

c. $-4(x^2y^2 - x^2y + 3x) - (x^2y^2 - 2x) + 3(x^2y^2 + 2x^2y)$
$= -4x^2y^2 + 4x^2y - 12x - x^2y^2 + 2x + 3x^2y^2 + 6x^2y$
$= -2x^2y^2 + 10x^2y - 10x$

SELF CHECK 8 Simplify: **a.** $2(a^3 - 3a) + 5(a^3 + 2a)$
b. $5(m^2 + 6m - 9) - (8m^2 - 2m + 5)$
c. $5(x^2y + 2x^2) - (x^2y - 3x^2)$

6 Solve an application requiring operations with polynomials.

EXAMPLE 9 **PROPERTY VALUES** A house purchased for \$95,000 is expected to appreciate according to the formula $y_1 = 2{,}500x + 95{,}000$, where y_1 is the value of the house after x years. A second house purchased for \$125,000 is expected to appreciate according to the formula $y_2 = 4{,}500x + 125{,}000$. Find one formula that will give the value of both properties after x years.

Solution The value of the first house after x years is given by the polynomial $2{,}500x + 95{,}000$. The value of the second house after x years is given by the polynomial $4{,}500x + 125{,}000$. The value of both houses will be the sum of these two polynomials.

$$2{,}500x + 95{,}000 + 4{,}500x + 125{,}000 = 7{,}000x + 220{,}000$$

The total value y of the properties is given by $y = 7{,}000x + 220{,}000$.

SELF CHECK 9 Find the total values of the properties after 20 years.

SELF CHECK ANSWERS

1. a. $11a^3b^2$ **b.** $11pq^2$ **c.** $35x^6$ **2. a.** $5m^3$ **b.** $-12p^3q^2$ **c.** $-20x^3y^2z$ **3.** $7a^2 + 5a - 1$ **4.** $6pq^2 - 2pq^3 + 2pq^4$ **5. a.** $3x + 4$ **b.** $9x^2 - 11x + 3$ **c.** $3a^2b + 12ab^2$ **6.** $3p^2q - 8pq + 15q$ **7.** $7pq^2 + 5p^2q^2 - 4p^3q^2$ **8. a.** $7a^3 + 4a$ **b.** $-3m^2 + 32m - 50$ **c.** $4x^2y + 13x^2$ **9.** \$360,000

NOW TRY THIS

1. If the lengths of the sides of a triangle represent consecutive even integers, find a polynomial that represents the perimeter of the triangle.
2. If the length of a rectangle is $(15x - 3)$ ft and the width is $(8x + 17)$ ft, find a polynomial that represents the perimeter.
3. If the length of one side of a rectangle is represented by the polynomial $(4x - 18)$ cm, and the perimeter is $(12x - 36)$ cm, find a polynomial that represents the width.

4.5 Exercises

WARM-UPS *Simplify.*

1. $x(7 + 2)$ **2.** $y(3 + 9)$
3. $a(16 - 4)$ **4.** $b(12 - 8)$

Determine whether the terms are like or unlike.

5. $6x^2, 6x$ **6.** $5a, 7b$
7. $4x^3, 5x^3$ **8.** $-2a^2b, 6a^2b$

REVIEW *Let a = 3, b = −2, c = −1, and d = 2. Evaluate each expression.*

9. $ab + cd$
10. $ac - bd$
11. $a(b - c)$
12. $d(b + a)$
13. Solve the inequality $-4(2x - 9) \geq 12$ and graph the solution set.
14. The kinetic energy of a moving object is given by the formula

$$K = \frac{mv^2}{2}$$

Solve the formula for m.

VOCABULARY AND CONCEPTS *Fill in the blanks.*

15. A ________ is a polynomial with one term.
16. If two polynomials are subtracted in vertical form, the bottom polynomial is called the ________, and the top polynomial is called the ________.
17. To add like monomials, add the numerical ________ and keep the ________.
18. $a - b = a +$ ___
19. To add two polynomials, combine any ________ contained in the polynomials.
20. To subtract polynomials, use the distributive property to remove parentheses and combine ________.

If the terms are like terms, add them. If they are unlike terms, state unlike terms.

21. $3y, 4y$
22. $3x^2, 5x^2$
23. $3x, 3y$
24. $3x^2, 6x$
25. $3x^3, 4x^3, 6x^3$
26. $-2y^4, -6y^4, 10y^4$
27. $-5x^3y^2, 13x^3y^2$
28. $23, 12x$
29. $15x^4y^2, -9x^4y^2, 4x^4y^2$
30. $32x^5y^3, -21x^5y^3, -11x^5y^3$
31. $-x^2y, xy, 3xy^2$
32. $4x^3y^2z, -6x^3y^2z, 2x^3y^2z$

GUIDED PRACTICE *Simplify.* SEE EXAMPLE 1. (OBJECTIVE 1)

33. $4y + 5y$
34. $3t + 6t$
35. $15x^2 + 10x^2$
36. $25r^4 + 15r^4$
37. $-7t^6 + 3(t^2)^3$
38. $-6(p^2)^4 + 10p^8$
39. $26x^2y^4 + 3x^2y^4$
40. $-16a^4b^2 + 10a^4b^2$

Simplify. SEE EXAMPLE 2. (OBJECTIVE 2)

41. $-18a - 3a$
42. $20b - 15b$
43. $32u^3 - 16u^3$
44. $25xy^2 - 7xy^2$
45. $18x^5y^2 - 11x^5y^2$
46. $17x^6y - 22x^6y$
47. $-14ab^3 - 6ab^3$
48. $17m^2n - 20m^2n$

Add. SEE EXAMPLE 3. (OBJECTIVE 3)

49. $(3x + 7) + (4x - 3)$
50. $(5y - 8) + (2y + 6)$
51. $(6y^2 - 2y + 5) + (2y^2 + 5y - 8)$
52. $(3x^2 - 3x - 2) + (3x^2 + 4x - 3)$

Perform the operation. SEE EXAMPLE 4. (OBJECTIVE 3)

53.
$$\begin{array}{r} 3x^2 + 4x + 5 \\ +\ \underline{2x^2 - 3x + 6} \end{array}$$

54.
$$\begin{array}{r} 2x^3 + 2x^2 - 3x + 5 \\ +\ \underline{3x^3 - 4x^2 - x - 7} \end{array}$$

55.
$$\begin{array}{r} 2x^3 - 3x^2 + 4x - 7 \\ +\ \underline{-9x^3 - 4x^2 - 5x + 6} \end{array}$$

56.
$$\begin{array}{r} -3x^3 + 4x^2 - 4x + 9 \\ +\ \underline{2x^3 \qquad\quad + 9x - 3} \end{array}$$

Subtract. SEE EXAMPLE 5. (OBJECTIVE 4)

57. $(4a + 3) - (2a - 4)$
58. $(5b - 7) - (3b + 5)$
59. $(2a^2 - 6a + 3) - (-3a^2 - 4a + 5)$
60. $(4b^2 + 5b - 1) - (-4b^2 - 3b - 6)$

Perform the operation. SEE EXAMPLE 6. (OBJECTIVE 4)

61.
$$\begin{array}{r} 3x^2 + 4x - 5 \\ -\ \underline{-2x^2 - 2x + 3} \end{array}$$

62.
$$\begin{array}{r} 3y^2 - 4y + 7 \\ -\ \underline{6y^2 - 6y - 13} \end{array}$$

63.
$$\begin{array}{r} 4x^3 + 4x^2 - 3x + 10 \\ -\ \underline{5x^3 - 2x^2 - 4x - 4} \end{array}$$

64.
$$\begin{array}{r} 3x^3 + 4x^2 + 7x + 12 \\ -\ \underline{-4x^3 + 6x^2 + 9x - 3} \end{array}$$

Perform the operation. SEE EXAMPLE 7. (OBJECTIVE 4)

65. Subtract $(8x + 2y)$ from $(-3x - 7y)$.
66. Subtract $(2x + 5y)$ from $(5x - 8y)$.
67. Subtract $(4x^2 - 3x + 2)$ from $(2x^2 - 3x + 1)$.
68. Subtract $(-4a + b)$ from $(6a^2 + 5a - b)$.

Simplify. SEE EXAMPLE 8. (OBJECTIVE 5)

69. $2(x + 3) + 4(x - 2)$
70. $5(y - 3) - 7(y + 4)$
71. $2(x^2 - 5x - 4) - 3(x^2 - 5x - 4) + 6(x^2 - 5x - 4)$
72. $7(x^2 + 3x + 1) + 9(x^2 + 3x + 1) - 5(x^2 + 3x + 1)$

ADDITIONAL PRACTICE *Perform the operations.*

73. $3rst + 4rst + 7rst$

74. $-2ab + 7ab - 3ab$

75. $-4a^2bc + 5a^2bc - 7a^2bc$

76. $(3x)^2 - 4x^2 + 10x^2$

77. $-3x^3y^6 + 2(xy^2)^3 - (3x)^3y^6$

78. $(-3x^2y)^4 + (4x^4y^2)^2 - 2x^8y^4$

79. $5x^5y^{10} - (2xy^2)^5 + (3x)^5y^{10}$

80. $5(x + y) + 7(x + y)$

81. $-8(x - y) + 11(x - y)$

82. $(4c^2 + 3c - 2) + (3c^2 + 4c + 2)$

83. $(-3z^2 - 4z + 7) + (2z^2 + 2z - 1) - (2z^2 - 3z + 7)$

84.
$$\begin{array}{r} -3x^2y + 4xy + 25y^2 \\ +\ \underline{\ 5x^2y - 3xy - 12y^2} \end{array}$$

85.
$$\begin{array}{r} -6x^3z - 4x^2z^2 + 7z^3 \\ +\ \underline{-7x^3z + 9x^2z^2 - 21z^3} \end{array}$$

86.
$$\begin{array}{r} -2x^2y^2 - 4xy + 12y^2 \\ -\ \underline{10x^2y^2 + 9xy - 24y^2} \end{array}$$

87.
$$\begin{array}{r} 25x^3 - 45x^2z + 31xz^2 \\ -\ \underline{12x^3 + 27x^2z - 17xz^2} \end{array}$$

88. $2(a^2b^2 - ab) - 3(ab + 2ab^2) + (b^2 - ab + a^2b^2)$

89. $3(xy^2 + y^2) - 2(xy^2 - 4y^2 + y^3) + 2(y^3 + y^2)$

90. $-4(x^2y^2 + xy^3 + xy^2z) - 2(x^2y^2 - 4xy^2z) - 2(8xy^3 - y)$

91. Find the sum when $(x^2 + x - 3)$ is added to the sum of $(2x^2 - 3x + 4)$ and $(3x^2 - 2)$.

92. Find the difference when $(t^3 - 2t^2 + 2)$ is subtracted from the sum of $(3t^3 + t^2)$ and $(-t^3 + 6t - 3)$.

93. Find the difference when $(-3z^3 - 4z + 7)$ is subtracted from the sum of $(2z^2 + 3z - 7)$ and $(-4z^3 - 2z - 3)$.

94. Find the sum when $(3x^2 + 4x - 7)$ is added to the sum of $(-2x^2 - 7x + 1)$ and $(-4x^2 + 8x - 1)$.

APPLICATIONS *Consider the following information: If a house was purchased for \$105,000 and is expected to appreciate \$900 per year, its value y after x years is given by the formula* $y = 900x + 105,000$. *SEE EXAMPLE 9. (OBJECTIVE 6)*

95. Value of a house Find the expected value of the house in 10 years.

96. Value of a house A second house was purchased for \$120,000 and was expected to appreciate \$1,000 per year. Find a polynomial equation that will give the value of the house in x years.

97. Value of a house Find one polynomial equation that will give the combined value y of both houses after x years.

98. Value of two houses Find the value of the two houses after 25 years.

Consider the following information: A business bought two computers, one for \$6,600 and the other for \$9,200. The first computer is expected to depreciate \$1,100 per year and the second \$1,700 per year.

99. Value of a computer Write a polynomial equation that will give the value of

a. the first computer after x years.

b. the second computer after x years.

100. Value of two computers

a. Find one polynomial equation that will give the value of both computers after x years.

b. Find the value of the computers after 3 years.

WRITING ABOUT MATH

101. How do you recognize like terms?

102. How do you add like terms?

SOMETHING TO THINK ABOUT *Let* $P(x) = 3x - 5$. *Find each value.*

103. $P(x + h) + P(x)$

104. $P(x + h) - P(x)$

105. If $P(x) = x^{23} + 5x^2 + 73$ and $Q(x) = x^{23} + 4x^2 + 73$, find $P(7) - Q(7)$.

106. If two numbers written in scientific notation have the same power of 10, they can be added as similar terms:

$$\mathbf{2} \times 10^3 + \mathbf{3} \times 10^3 = \mathbf{5} \times 10^3$$

Without converting to standard form, how could you add

$$2 \times 10^3 + 3 \times 10^4$$

Section 4.6 Multiplying Polynomials

Objectives

1. Multiply two or more monomials.
2. Multiply a polynomial by a monomial.
3. Multiply a binomial by a binomial.
4. Multiply a polynomial by a binomial.
5. Solve an equation that simplifies to a linear equation.
6. Solve an application involving multiplication of polynomials.

Vocabulary

FOIL method   squaring a binomial   conjugate binomials

Getting Ready

Simplify.

1. $(2x)(3)$ **2.** $(3xxx)(x)$ **3.** $5x^2 \cdot x$ **4.** $8x^2x^3$

Use the distributive property to remove parentheses.

5. $3(x + 5)$ **6.** $-2(x + 5)$ **7.** $4(y - 3)$ **8.** $-2(y^2 - 3)$

We now discuss how to multiply polynomials by beginning with a review of multiplying two monomials. We will then introduce multiplying polynomials with more than one term. The section concludes with a discussion of how these techniques can be used to solve linear equations.

1 Multiply two or more monomials.

We have previously multiplied monomials by other monomials. For example, to multiply $4x^2$ by $-2x^3$, we use the commutative and associative properties of multiplication to group the numerical factors together and the variable factors together. Then we multiply the numerical factors and multiply the variable factors.

$$\begin{aligned} 4x^2(-2x^3) &= 4(-2)x^2x^3 \\ &= -8x^5 \end{aligned}$$

This example suggests the following strategy.

MULTIPLYING MONOMIALS

To multiply two simplified monomials, multiply the numerical factors and then multiply the variable factors.

EXAMPLE 1 Multiply: **a.** $3x^5(2x^5)$ **b.** $-2a^2b^3(5ab^2)$

Solution **a.** $3x^5(2x^5) = 3(2)x^5x^5$

$= 6x^{10}$

b. $-2a^2b^3(5ab^2) = -2(5)a^2ab^3b^2$

$= -10a^3b^5$

SELF CHECK 1 Multiply: **a.** $(5a^2b^3)(6a^3b^4)$ **b.** $(-15p^3q^2)(5p^3q^2)$

2 Multiply a polynomial by a monomial.

To find the product of a monomial and a polynomial with more than one term, we use the distributive property. To multiply $2x + 4$ by $5x$, for example, we proceed as follows:

$5x(2x + 4) = 5x \cdot 2x + 5x \cdot 4$ Use the distributive property to remove parentheses.

$= 10x^2 + 20x$ Multiply the monomials $5x \cdot 2x = 10x^2$ and $5x \cdot 4 = 20x$.

This example suggests the following process.

MULTIPLYING POLYNOMIALS BY MONOMIALS

To multiply a polynomial with more than one term by a monomial, use the distributive property to remove parentheses and simplify.

EXAMPLE 2 Multiply: **a.** $3a^2(3a^2 - 5a)$ **b.** $-2xz^2(2x - 3z + 2z^2)$

Solution **a.** $3a^2(3a^2 - 5a) = 3a^2 \cdot 3a^2 - 3a^2 \cdot 5a$ Use the distributive property to remove parentheses.

$= 9a^4 - 15a^3$ Multiply.

b. $-2xz^2(2x - 3z + 2z^2)$

$= -2xz^2 \cdot 2x + (-2xz^2) \cdot (-3z) + (-2xz^2) \cdot 2z^2$ Use the distributive property to remove parentheses.

$= -4x^2z^2 + 6xz^3 + (-4xz^4)$ Multiply.

$= -4x^2z^2 + 6xz^3 - 4xz^4$

Recall that subtracting is equivalent to adding the opposite.

SELF CHECK 2 Multiply:

a. $2p^3(3p^2 - 5p)$

b. $-5a^2b(3a + 2b - 4ab)$

3 Multiply a binomial by a binomial.

To multiply two binomials, we must use the distributive property more than once. For example, to multiply $(2a - 4)$ by $(3a + 5)$, we proceed as follows.

$(2a - 4)(3a + 5) = (2a - 4) \cdot 3a + (2a - 4) \cdot 5$ Use the distributive property to remove parentheses.

$= 3a(2a - 4) + 5(2a - 4)$ Use the commutative property of multiplication.

$$= 3a \cdot 2a + 3a \cdot (-4) + 5 \cdot 2a + 5 \cdot (-4)$$ Use the distributive property.

$$= 6a^2 - 12a + 10a - 20$$ Do the multiplications.

$$= 6a^2 - 2a - 20$$ Combine like terms.

This example suggests the following strategy.

MULTIPLYING TWO BINOMIALS

To multiply two binomials, multiply each term of one binomial by each term of the other binomial and combine like terms.

EXAMPLE 3 Find each product.
(Using the FOIL method)

a. $(3x + 4)(2x - 3) = 3x(2x) + 3x(-3) + 4(2x) + 4(-3)$

$= 6x^2 - 9x + 8x - 12$

$= 6x^2 - x - 12$

b. $(2y - 7)(5y - 4) = 2y(5y) + 2y(-4) + (-7)(5y) + (-7)(-4)$

$= 10y^2 - 8y - 35y + 28$

$= 10y^2 - 43y + 28$

c. $(2r - 3s)(2r + t) = 2r(2r) + 2r(t) - 3s(2r) - 3s(t)$

$= 4r^2 + 2rt - 6sr - 3st$

$= 4r^2 + 2rt - 6rs - 3st$

SELF CHECK 3 Find each product.

a. $(5x + 3)(6x + 7)$
b. $(2a - 1)(3a + 2)$
c. $(5y - 2z)(2y + 3z)$

To multiply binomials, we can apply the distributive property using a mnemonic device, called the **FOIL method**. FOIL is an acronym for **F**irst terms, **O**uter terms, **I**nner terms, and **L**ast terms. To use this method to multiply $(2a - 4)$ by $(3a + 5)$, we

1. multiply the **F**irst terms $2a$ and $3a$ to obtain $6a^2$,
2. multiply the **O**uter terms $2a$ and 5 to obtain $10a$,
3. multiply the **I**nner terms -4 and $3a$ to obtain $-12a$, and
4. multiply the **L**ast terms -4 and 5 to obtain -20.

Then we simplify the resulting polynomial, if possible.

COMMENT FOIL is simply a mnemonic for applying the distributive property to multiply two binomials in a given order.

F L

$(2a - 4)(3a + 5) = 2a(3a) + 2a(5) + (-4)(3a) + (-4)(5)$

I

O

$= 6a^2 + 10a - 12a - 20$ Simplify.

$= 6a^2 - 2a - 20$ Combine like terms.

EXAMPLE 4 Simplify each expression.

a. $3(2x - 3)(x + 1)$

$= 3(2x^2 + 2x - 3x - 3)$ Multiply the binomials.

$= 3(2x^2 - x - 3)$ Combine like terms.

$= 6x^2 - 3x - 9$ Use the distributive property to remove parentheses.

b. $(x + 1)(x - 2) - 3x(x + 3)$

$= x^2 - 2x + x - 2 - 3x^2 - 9x$ Use the distributive property to remove parentheses.

$= -2x^2 - 10x - 2$ Combine like terms.

SELF CHECK 4 Simplify each expression.
a. $-3(6x - 5)(2x - 3)$
b. $(x + 3)(2x - 1) + 2x(x - 1)$

The products discussed in Example 5 are sometimes called special products. These include **squaring a binomial** and multiplying conjugate binomials. Binomials that have the same terms, but with opposite signs between the terms, are called **conjugate binomials**.

EXAMPLE 5 Find each product.

a. $(x + y)^2 = (x + y)(x + y)$ Square the binomial.

$= x^2 + xy + xy + y^2$ Distribute.

$= x^2 + 2xy + y^2$ Combine like terms.

The square of the sum of two quantities has three terms: *the square of the first quantity, plus twice the product of the quantities, plus the square of the second quantity.*

COMMENT Note that

$$(x + y)^2 \neq x^2 + y^2$$

and

$$(x - y)^2 \neq x^2 - y^2$$

b. $(x - y)^2 = (x - y)(x - y)$ Square the binomial.

$= x^2 - xy - xy + y^2$ Distribute.

$= x^2 - 2xy + y^2$ Combine like terms.

The square of the difference of two quantities has three terms: *the square of the first quantity, minus twice the product of the quantities, plus the square of the second quantity.*

c. $(x + y)(x - y) = x^2 - xy + xy - y^2$ Distribute.

$= x^2 - y^2$ Combine like terms.

The product of the sum and the difference of two quantities is a binomial. *It is the product of the first quantities minus the product of the second quantities.*

SELF CHECK 5 Find each product.
a. $(p + 2)^2$
b. $(p - 2)^2$
c. $(p + 2q)(p - 2q)$

Because the products discussed in Example 5 occur so often, it may be helpful to recognize their forms.

However, you should multiply these as binomials by applying the distributive property until you discover the pattern for yourself.

SPECIAL PRODUCTS

$$(x + y)^2 = x^2 + 2xy + y^2$$
$$(x - y)^2 = x^2 - 2xy + y^2$$
$$(x + y)(x - y) = x^2 - y^2$$

4 Multiply a polynomial by a binomial.

We must use the distributive property more than once to multiply a polynomial by a binomial. For example, to multiply $(3x^2 + 3x - 5)$ by $(2x + 3)$, we proceed as follows:

$$\begin{aligned}(2x + 3)(3x^2 + 3x - 5) &= (2x + 3)3x^2 + (2x + 3)3x + (2x + 3)(-5)\\ &= 3x^2(2x + 3) + 3x(2x + 3) - 5(2x + 3)\\ &= 6x^3 + 9x^2 + 6x^2 + 9x - 10x - 15\\ &= 6x^3 + 15x^2 - x - 15\end{aligned}$$

This example suggests the following process.

MULTIPLYING POLYNOMIALS

To multiply one polynomial by another, multiply each term of one polynomial by each term of the other polynomial and combine like terms.

It is often convenient to organize the work vertically.

EXAMPLE 6 **a.** Multiply:

$$\begin{array}{lrrrr} & & 3a^2 & - 4a & + 7 \\ & \times & & 2a & + 5 \\ \hline 5(3a^2 - 4a + 7) \rightarrow & & 15a^2 & - 20a & + 35 \\ 2a(3a^2 - 4a + 7) \rightarrow & 6a^3 & - 8a^2 & + 14a & \\ \hline & 6a^3 & + 7a^2 & - 6a & + 35 \end{array}$$

b. Multiply:

$$\begin{array}{lrrrrr} & & & 3y^2 & - 5y & + 4 \\ & \times & & - 4y^2 & & - 3 \\ \hline -3(3y^2 - 5y + 4) \rightarrow & & & - 9y^2 & + 15y & - 12 \\ -4y^2(3y^2 - 5y + 4) \rightarrow & -12y^4 & + 20y^3 & - 16y^2 & & \\ \hline & -12y^4 & + 20y^3 & - 25y^2 & + 15y & - 12 \end{array}$$

SELF CHECK 6 Multiply:

a. $(3x + 2)(2x^2 - 4x + 5)$

b. $(-2x^2 + 3)(2x^2 - 4x - 1)$

COMMENT An expression can be simplified by combining its like terms. An equation (two quantities set equal) can be solved. Remember that

Expressions are to be simplified.

Equations are to be solved.

5 Solve an equation that simplifies to a linear equation.

To solve an equation such as $(x + 2)(x + 3) = x(x + 7)$, we can use the distributive property to remove the parentheses on the left and the right sides and proceed as follows:

$$(x + 2)(x + 3) = x(x + 7)$$

$$x^2 + 3x + 2x + 6 = x^2 + 7x$$ Use the distributive property to remove parentheses.

$$x^2 + 5x + 6 = x^2 + 7x$$ Combine like terms.

$$5x + 6 = 7x$$ Subtract x^2 from both sides.

$$6 = 2x$$ Subtract $5x$ from both sides.

$$3 = x$$ Divide both sides by 2.

Check: $(x + 2)(x + 3) = x(x + 7)$

$$(3 + 2)(3 + 3) \stackrel{?}{=} 3(3 + 7)$$ Replace x with 3.

$$5(6) \stackrel{?}{=} 3(10)$$ Do the additions within parentheses.

$$30 = 30$$

Since the answer checks, the solution is 3.

EXAMPLE 7 Solve: $(x + 5)(x + 4) = (x + 9)(x + 10)$

Solution We remove parentheses on both sides of the equation and proceed as follows:

$$(x + 5)(x + 4) = (x + 9)(x + 10)$$

$$x^2 + 4x + 5x + 20 = x^2 + 10x + 9x + 90$$ Use the distributive property to remove parentheses.

$$x^2 + 9x + 20 = x^2 + 19x + 90$$ Combine like terms.

$$9x + 20 = 19x + 90$$ Subtract x^2 from both sides.

$$20 = 10x + 90$$ Subtract $9x$ from both sides.

$$-70 = 10x$$ Subtract 90 from both sides.

$$-7 = x$$ Divide both sides by 10.

Check: $(x + 5)(x + 4) = (x + 9)(x + 10)$

$$(-7 + 5)(-7 + 4) \stackrel{?}{=} (-7 + 9)(-7 + 10)$$ Replace x with -7.

$$(-2)(-3) \stackrel{?}{=} (2)(3)$$ Do the additions within parentheses.

$$6 = 6$$

Since the result checks, the solution is -7.

SELF CHECK 7 Solve: $(x + 2)(x - 4) = (x + 6)(x - 3)$

6 Solve an application involving multiplication of polynomials.

EXAMPLE 8 **DIMENSIONS OF A PAINTING** A square painting is surrounded by a border 2 inches wide. If the area of the border is 96 square inches, find the dimensions of the painting.

Figure 4-6
© Shutterstock.com/Olga Lyubkina

Solution

Analyze the problem: Refer to Figure 4-6, which shows a square painting surrounded by a border 2 inches wide. We can let x represent the length in inches of each side of the square painting. The outer rectangle is also a square, and one length is $(x + 2 + 2)$ or $(x + 4)$ inches.

Form an equation: We know that the area of the border is 96 square inches, the area of the larger square is $(x + 4)(x + 4)$, and the area of the painting is $x \cdot x$. If we subtract the area of the painting from the area of the larger square, the difference is 96 (the area of the border).

The area of the large square	minus	the area of the square painting	equals	the area of the border.
$(x + 4)(x + 4)$	$-$	$x \cdot x$	$=$	96

Solve the equation:

$$(x + 4)(x + 4) - x^2 = 96$$

$$x^2 + 8x + 16 - x^2 = 96 \quad \text{Use the distributive property to remove parentheses.}$$

$$8x + 16 = 96 \quad \text{Combine like terms.}$$

$$8x = 80 \quad \text{Subtract 16 from both sides.}$$

$$x = 10 \quad \text{Divide both sides by 8.}$$

State the conclusion: The dimensions of the painting are 10 inches by 10 inches.

Check the result: Check the result.

SELF CHECK 8 If the area of the border is 112 square inches, find the dimensions of the painting.

SELF CHECK ANSWERS

1. a. $30a^5b^7$ **b.** $-75p^6q^4$ **2. a.** $6p^5 - 10p^4$ **b.** $-15a^3b - 10a^2b^2 + 20a^3b^2$ **3. a.** $30x^2 + 53x + 21$ **b.** $6a^2 + a - 2$ **c.** $10y^2 + 11yz - 6z^2$ **4. a.** $-36x^2 + 84x - 45$ **b.** $4x^2 + 3x - 3$ **5. a.** $p^2 + 4p + 4$ **b.** $p^2 - 4p + 4$ **c.** $p^2 - 4q^2$ **6. a.** $6x^3 - 8x^2 + 7x + 10$ **b.** $-4x^4 + 8x^3 + 8x^2 - 12x - 3$ **7.** 2 **8.** The dimensions are 12 inches by 12 inches.

NOW TRY THIS

Simplify or solve as appropriate.

1. $-\frac{1}{2}x(8x^2 - 16x + 2)$
2. $(2x - 3)(4x^2 + 6x + 9)$
3. $(x - 2)(x + 5) = (x - 1)(x + 8)$
4. Find a representation of the area of a square with one side represented by $(3x + 5)$ ft.

4.6 Exercises

WARM-UPS *Simplify.*

1. $3x^2(5x)$
2. $6y(2y^2)$
3. $-4x(2y)$
4. $7x(-3y)$

Use the distributive property to remove parentheses.

5. $-5(x - 4)$
6. $-3(y^2 + 2)$
7. $2(4x^2 - 9)$
8. $7(z^2 + z)$

REVIEW *For 9–12, determine which property of real numbers justifies each statement.*

9. $4x + 5x^2 = 5x^2 + 4x$
10. $(x + 3) + y = x + (3 + y)$
11. $3(ab) = (ab)3$
12. $a + 0 = a$
13. Solve: $\frac{5}{3}(5y + 6) - 10 = 0$
14. Solve: $F = \frac{GMm}{d^2}$ for m

VOCABULARY AND CONCEPTS *Fill in the blanks.*

15. A polynomial with one term is called a ________.
16. A binomial is a polynomial with ___ terms. The binomials $(a + b)$ and $(a - b)$ are called ________ binomials.
17. Products in the form $(a + b)^2$, $(a - b)^2$, or $(a + b)(a - b)$ are called ______________.
18. In the acronym FOIL, F stands for _________, O stands for _________, I stands for _________ and L stands for _________.

Consider the product $(2x + 5)(3x - 4)$.

19. The product of the first terms is ___.
20. The product of the outer terms is ___.
21. The product of the inner terms is ___.
22. The product of the last terms is ___.

GUIDED PRACTICE *Multiply.* SEE EXAMPLE 1. (OBJECTIVE 1)

23. $(3x^2)(4x^3)$
24. $(-2a^3)(3a^2)$
25. $(-5t^3)(2t^4)$
26. $(-6a^2)(-3a^5)$
27. $(2x^2y^3)(3x^3y^2)$
28. $(-x^3y^6z)(x^2y^2z^7)$
29. $(3b^2)(-2b)(4b^3)$
30. $(3y)(2y^2)(-y^4)$

Multiply. SEE EXAMPLE 2. (OBJECTIVE 2)

31. $3(x + 4)$
32. $-3(a - 2)$
33. $-4(t + 7)$
34. $6(s^2 - 3)$
35. $3x(x - 2)$
36. $-5y(y + 3)$
37. $-2x^2(3x^2 - x)$
38. $4b^3(2 - 2b)$
39. $3xy(x + y)$
40. $-4x^2(3x^2 - x)$
41. $-6x^2(2x^2 + 3x - 5)$
42. $3y^3(2y^2 - 7y - 8)$
43. $\frac{1}{4}x^2(8x^5 - 4)$
44. $\frac{4}{3}a^2b(6a - 5b)$
45. $-\frac{2}{3}r^2t^2(9r - 3t)$
46. $-\frac{4}{5}p^2q(10p + 15q)$

Find each product. SEE EXAMPLE 3. (OBJECTIVE 3)

47. $(a + 4)(a + 5)$
48. $(y - 3)(y + 5)$
49. $(3x - 2)(x + 4)$
50. $(t + 4)(2t - 3)$
51. $(4a - 2)(2a - 3)$
52. $(2b - 1)(3b + 4)$
53. $(3x - 5)(2x + 1)$
54. $(5y - 7)(2y - 5)$
55. $(2s + 3t)(3s - t)$
56. $(3a - 2b)(4a + b)$
57. $(u + v)(u + 2t)$
58. $(x - 5y)(a + 2y)$

Simplify each expression. SEE EXAMPLE 4. (OBJECTIVE 3)

59. $2(x - 4)(x + 1)$
60. $-3(2x + 3y)(3x - 4y)$
61. $3a(a + b)(a - b)$
62. $-2r(r + s)(r + s)$
63. $(3x - 2y)(x + y)$
64. $(2a - 3b)(3a - 2b)$
65. $(2x - 3)(x + 1) - 5x(x + 2)$
66. $(x + 2)(3x - 1) + 3x(x - 2)$

Find each product. SEE EXAMPLE 5. (OBJECTIVE 3)

67. $(x + 5)^2$
68. $(y - 9)^2$
69. $(x - 4)^2$
70. $(a + 3)^2$
71. $(4t + 3)^2$
72. $(3t - 2)^2$
73. $(x \quad 2y)^2$
74. $(3a + 2b)^2$
75. $(r + 4)(r - 4)$
76. $(y + 6)(y - 6)$
77. $(4x + 5)(4x - 5)$
78. $(5z + 1)(5z - 1)$

Multiply. SEE EXAMPLE 6. (OBJECTIVE 4)

79. $(2x + 3)(x^2 + 4x - 1)$ **80.** $(3x - 2)(2x^2 - x + 2)$

81. $(4t + 3)(t^2 + 2t + 3)$

82. $(3x + y)(2x^2 - 3xy + y^2)$

83. $\begin{array}{r} 4x + 3 \\ \underline{x + 2} \end{array}$ **84.** $\begin{array}{r} 5r + 6 \\ \underline{2r - 1} \end{array}$

85. $\begin{array}{r} 4x - 2y \\ \underline{3x + 5y} \end{array}$ **86.** $\begin{array}{r} x^2 + x + 1 \\ \underline{x - 1} \end{array}$

Solve each equation. SEE EXAMPLE 7. (OBJECTIVE 5)

87. $(s - 4)(s + 1) = s^2 + 5$

88. $(y - 5)(y - 2) = y^2 - 4$

89. $z(z + 2) = (z + 4)(z - 4)$

90. $(z + 3)(z - 3) = z(z - 3)$

91. $(x + 4)(x - 4) = (x - 2)(x + 6)$

92. $(y - 1)(y + 6) = (y - 3)(y - 2) + 8$

93. $(a - 3)^2 = (a + 3)^2$

94. $(b + 2)^2 = (b - 1)^2$

ADDITIONAL PRACTICE *Simplify.*

95. $(x^2y^5)(x^2z^5)(-3y^2z^3)$ **96.** $(-r^4st^2)(2r^2st)(rst)$

97. $(x + 3)(2x - 3)$ **98.** $(2x + 3)(2x - 5)$

99. $(t - 3)(t - 3)$ **100.** $(z - 5)(z - 5)$

101. $(-2r - 3s)(2r + 7s)$ **102.** $(2a - 3b)^2$

103. $(3x - 2)^2$ **104.** $(x - 2y)(x^2 + 2xy + 4y^2)$

105. $(3x - y)(x^2 + 3xy - y^2)$

106. $(xyz^3)(xy^2z^2)^3$

Simplify or solve as appropriate.

107. $3xy(x + y) - 2x(xy - x)$

108. $(a + b)(a - b) - (a + b)(a + b)$

109. $(2x - 1)(2x + 1) = x(4x + 1)$

110. $7s^2 + (s - 3)(2s + 1) = (3s - 1)^2$

111. $(x + 2)^2 = (x - 2)^2$

112. $(2s - 3)(s + 2) = (2s + 1)(s - 3)$

113. $(x + y)(x - y) + x(x + y)$

114. $(x - 3)^2 - (x + 3)^2$

115. $(3x + 4)(2x - 2) - (2x + 1)(x + 3)$

116. $4 + (2y - 3)^2 = (2y - 1)(2y + 3)$

117. $3y(y + 2) = 3(y + 1)(y - 1)$

118. $(b + 2)(b - 2) + 2b(b + 1)$

APPLICATIONS SEE EXAMPLE 8. (OBJECTIVE 6)

119. Millstones The radius of one millstone in the illustration is 3 meters greater than the radius of the other, and their areas differ by 15π square meters. Find the radius of the larger millstone.

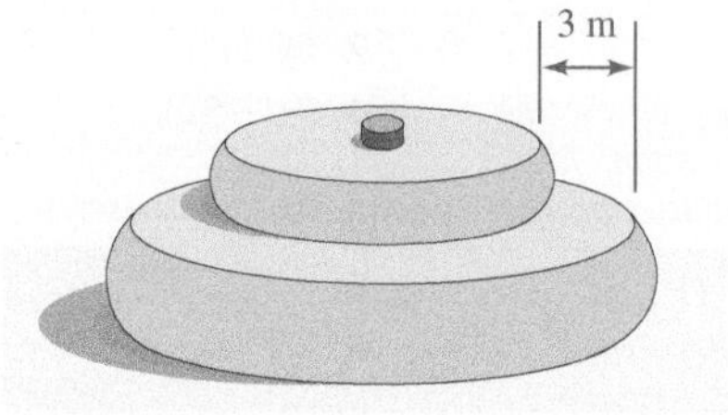

120. Bookbinding Two square sheets of cardboard used for making book covers differ in area by 44 square inches. An edge of the larger square is 2 inches greater than an edge of the smaller square. Find the length of an edge of the smaller square.

121. Baseball In major league baseball, the distance between bases is 30 feet greater than it is in softball. The bases in major league baseball mark the corners of a square that has an area 4,500 square feet greater than for softball. Find the distance between the bases in baseball.

122. Pulley designs The radius of one pulley in the illustration is 1 inch greater than the radius of the second pulley, and their areas differ by 4π square inches. Find the radius of the smaller pulley.

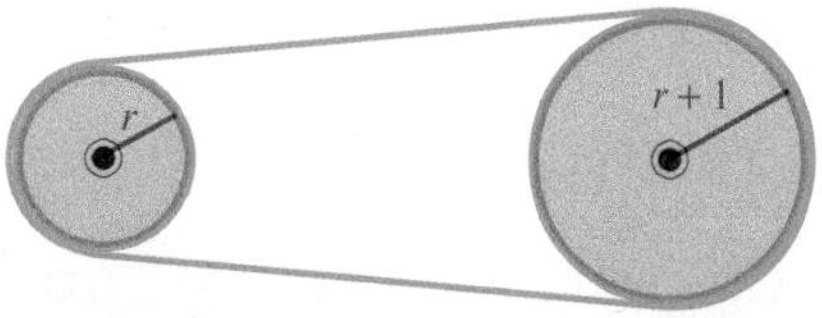

WRITING ABOUT MATH

123. Describe the steps involved in finding the product of a binomial and its conjugate.

124. Writing the expression $(x + y)^2$ as $x^2 + y^2$ illustrates a common error. Explain.

SOMETHING TO THINK ABOUT

125. The area of the square in the illustration is the total of the areas of the four smaller regions. The picture illustrates the product $(x + y)^2$. Explain.

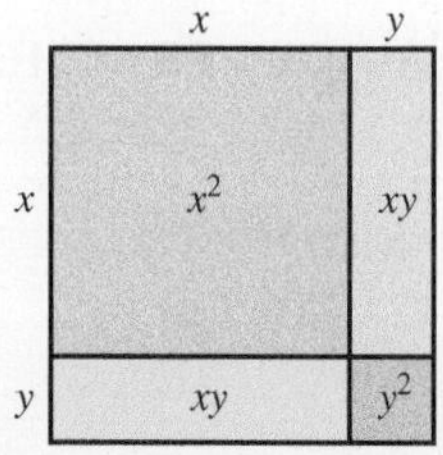

126. The illustration represents the product of two binomials. Explain.

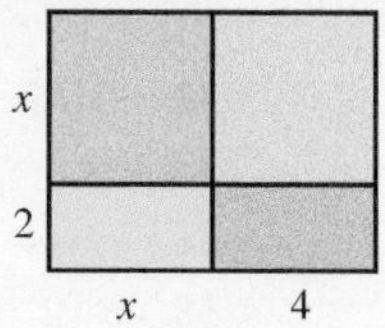

Unless otherwise noted, all content on this page is © Cengage Learning.

Section 4.7 Dividing Polynomials by Monomials

Objectives

1. Divide a monomial by a monomial.
2. Divide a polynomial by a monomial.
3. Solve a formula for a specified variable.

Getting Ready

Simplify each fraction.

1. $\dfrac{4x^2y^3}{2xy}$ 2. $\dfrac{9xyz}{9xz}$ 3. $\dfrac{15x^2y}{10x}$ 4. $\dfrac{6x^2y}{6xy^2}$

5. $\dfrac{(2x^2)(5y^2)}{10xy}$ 6. $\dfrac{(5x^3y)(6xy^3)}{10x^4y^4}$

In this section, we will show how to divide polynomials by monomials. We will discuss how to divide polynomials by polynomials and introduce all vocabulary associated with long division in the next section.

1 Divide a monomial by a monomial.

We have seen that dividing by a number is equivalent to multiplying by its reciprocal. For example, dividing the number 8 by 2 gives the same answer as multiplying 8 by $\frac{1}{2}$.

$$\frac{8}{2} = 4 \quad \text{and} \quad 8 \cdot \frac{1}{2} = 4$$

In general, the following is true.

DIVISION

$$\frac{a}{b} = a \cdot \frac{1}{b} \quad (b \neq 0)$$

Recall that to simplify a fraction, we write both its numerator and its denominator as the product of several factors and then divide out all common factors. For example,

$$\frac{20}{25} = \frac{4 \cdot 5}{5 \cdot 5} \qquad \text{Factor: } 20 = 4 \cdot 5 \text{ and } 25 = 5 \cdot 5$$

$$= \frac{4 \cdot \overset{1}{\cancel{5}}}{\underset{1}{\cancel{5}} \cdot 5} \qquad \text{Divide out the common factor of 5.}$$

$$= \frac{4}{5}$$

We can use the same method to simplify algebraic fractions that contain variables. We must assume, however, that no variable is 0.

$$\frac{3p^2q}{6pq^3} = \frac{3\cdot p\cdot p\cdot q}{2\cdot 3\cdot p\cdot q\cdot q\cdot q}$$ Factor: $p^2 = p\cdot p, 6 = 2\cdot 3$, and $q^3 = q\cdot q\cdot q$.

$$= \frac{\overset{1}{\cancel{3}}\cdot \overset{1}{\cancel{p}}\cdot p\cdot \overset{1}{\cancel{q}}}{2\cdot \underset{1}{\cancel{3}}\cdot \underset{1}{\cancel{p}}\cdot \underset{1}{\cancel{q}}\cdot q\cdot q}$$ Divide out the common factors of 3, p, and q.

$$= \frac{p}{2q^2}$$

To divide monomials, we can either use the previous method or use the rules of exponents.

COMMENT In all examples and exercises in this section, we will assume that no variables are 0 to avoid the possibility of division by 0. Recall that division by 0 is undefined.

EXAMPLE 1 Simplify: **a.** $\frac{x^2y}{xy^2}$ **b.** $\frac{-8a^3b^2}{4ab^3}$

Solution

Using Fractions

a. $$\frac{x^2y}{xy^2} = \frac{x\cdot x\cdot y}{x\cdot y\cdot y}$$

$$= \frac{\overset{1}{\cancel{x}}\cdot x\cdot \overset{1}{\cancel{y}}}{\underset{1}{\cancel{x}}\cdot y\cdot \underset{1}{\cancel{y}}}$$

$$= \frac{x}{y}$$

Using the Properties of Exponents

$$\frac{x^2y}{xy^2} = x^{2-1}y^{1-2}$$

$$= x^1y^{-1}$$

$$= x\cdot\frac{1}{y}$$

$$= \frac{x}{y}$$

Using Fractions

b. $$\frac{-8a^3b^2}{4ab^3} = \frac{-2\cdot 4\cdot a\cdot a\cdot a\cdot b\cdot b}{4\cdot a\cdot b\cdot b\cdot b}$$

$$= \frac{-2\cdot \overset{1}{\cancel{4}}\cdot \overset{1}{\cancel{a}}\cdot a\cdot a\cdot \overset{1}{\cancel{b}}\cdot \overset{1}{\cancel{b}}}{\underset{1}{\cancel{4}}\cdot \underset{1}{\cancel{a}}\cdot \underset{1}{\cancel{b}}\cdot \underset{1}{\cancel{b}}\cdot b}$$

$$= \frac{-2a^2}{b}$$

Using the Properties of Exponents

$$\frac{-8a^3b^2}{4ab^3} = \frac{(-1)2^3a^3b^2}{2^2ab^3}$$

$$= (-1)2^{3-2}a^{3-1}b^{2-3}$$

$$= (-1)2^1a^2b^{-1}$$

$$= -2a^2\cdot\frac{1}{b}$$

$$= \frac{-2a^2}{b}$$

SELF CHECK 1 Simplify: **a.** $\frac{r^5s^3}{rs^4}$ **b.** $\frac{-5p^2q^3}{10pq^4}$

2 Divide a polynomial by a monomial.

In Chapter 1, we saw that

$$\frac{a}{d} + \frac{b}{d} = \frac{a+b}{d}$$

Since this is true, we also have

$$\frac{a+b}{d} = \frac{a}{d} + \frac{b}{d}$$

This suggests that, to divide a polynomial by a monomial, we can divide each term of the polynomial in the numerator by the monomial in the denominator.

EXAMPLE 2 Simplify: $\dfrac{9x + 6y}{3xy}$

Solution

$$\frac{9x + 6y}{3xy} = \frac{9x}{3xy} + \frac{6y}{3xy} \quad \text{Divide each term in the numerator by the monomial.}$$

$$= \frac{3}{y} + \frac{2}{x} \quad \text{Simplify each fraction.}$$

SELF CHECK 2 Simplify: $\dfrac{4a - 8b}{4ab}$

EXAMPLE 3 Simplify: $\dfrac{6x^2y^2 + 4x^2y - 2xy}{2xy}$

Solution

COMMENT Remember that any nonzero value divided by itself is 1.

$$\frac{6x^2y^2 + 4x^2y - 2xy}{2xy}$$

$$= \frac{6x^2y^2}{2xy} + \frac{4x^2y}{2xy} - \frac{2xy}{2xy} \quad \text{Divide each term in the numerator by the monomial.}$$

$$= 3xy + 2x - 1 \quad \text{Simplify each fraction.}$$

SELF CHECK 3 Simplify: $\dfrac{9a^2b - 6ab^2 + 3ab}{3ab}$

EXAMPLE 4 Simplify: $\dfrac{12a^3b^2 - 4a^2b + a}{6a^2b^2}$

Solution

$$\frac{12a^3b^2 - 4a^2b + a}{6a^2b^2}$$

$$= \frac{12a^3b^2}{6a^2b^2} - \frac{4a^2b}{6a^2b^2} + \frac{a}{6a^2b^2} \quad \text{Divide each term in the numerator by the monomial.}$$

$$= 2a - \frac{2}{3b} + \frac{1}{6ab^2} \quad \text{Simplify each fraction.}$$

SELF CHECK 4 Simplify: $\dfrac{14p^3q + pq^2 - p}{7p^2q}$

EXAMPLE 5 Simplify: $\dfrac{(x - y)^2 - (x + y)^2}{xy}$

Solution

$$\frac{(x - y)^2 - (x + y)^2}{xy}$$

$$= \frac{x^2 - 2xy + y^2 - (x^2 + 2xy + y^2)}{xy} \quad \text{Square the binomials in the numerator.}$$

$$= \frac{x^2 - 2xy + y^2 - x^2 - 2xy - y^2}{xy} \quad \text{Use the distributive property to remove parentheses.}$$

$$= \frac{-4xy}{xy} \quad \text{Combine like terms.}$$

$$= -4 \quad \text{Simplify.}$$

SELF CHECK 5 Simplify: $\dfrac{(x + y)^2 - (x - y)^2}{2xy}$

3 Solve a formula for a specified variable.

The cross-sectional area of the trapezoidal drainage ditch shown in Figure 4-7 is given by the formula $A = \frac{1}{2}h(B + b)$, where B and b represent its bases and h represents its height. To solve the formula for b, we proceed as follows.

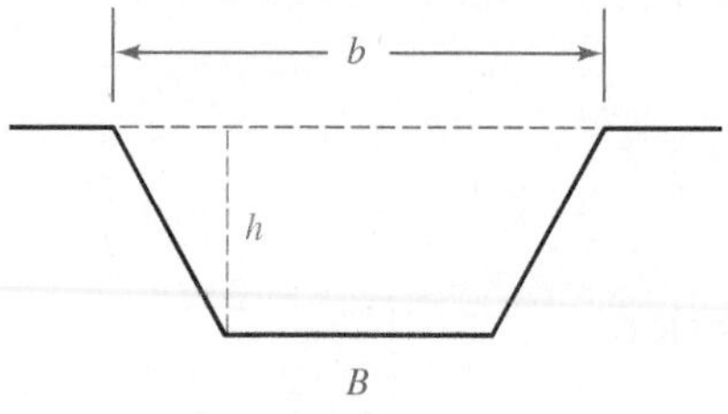

Figure 4-7

$$A = \frac{1}{2}h(B + b)$$

$$2(A) = 2\left[\frac{1}{2}h(B + b)\right] \quad \text{Multiply both sides by 2 to clear fractions.}$$

$$2A = h(B + b) \quad \text{Simplify.}$$

$$2A = hB + hb \quad \text{Use the distributive property to remove parentheses.}$$

$$2A - hB = hB - hB + hb \quad \text{Subtract } hB \text{ from both sides.}$$

$$2A - hB = hb \quad \text{Combine like terms.}$$

$$\frac{2A - hB}{h} = \frac{hb}{h} \quad \text{Divide both sides by } h.$$

$$\frac{2A - hB}{h} = b$$

EXAMPLE 6 Another student worked the previous problem in a different way and got a result of $b = \frac{2A}{h} - B$. Is this also correct?

Solution To show that this result is correct, we must show that $\frac{2A - hB}{h} = \frac{2A}{h} - B$. We can do this by dividing $(2A - hB)$ by h.

$$\frac{2A - hB}{h} = \frac{2A}{h} - \frac{hB}{h} \quad \text{Divide each term in the numerator by the monomial.}$$

$$= \frac{2A}{h} - B \quad \text{Simplify: } \frac{hB}{h} = B.$$

The results are the same.

SELF CHECK 6 Suppose another student got $2A - B$. Is this result correct?

Unless otherwise noted, all content on this page is © Cengage Learning.

SELF CHECK ANSWERS

1. a. $\frac{r^4}{s}$ **b.** $-\frac{p}{2q}$ **2.** $\frac{1}{b} - \frac{2}{a}$ **3.** $3a - 2b + 1$ **4.** $2p + \frac{q}{7p} - \frac{1}{7pq}$ **5.** 2 **6.** no

NOW TRY THIS

Perform each division.

1. $\dfrac{6 - 2i}{3}$

2. a. $\dfrac{2x^{p+1}}{6x^{p-1}}$ b. $\dfrac{x^{m-1}}{x^{1-m}}$

3. $\dfrac{(x + 3)^4 - (x + 3)^2}{(x + 3)^2}$

4.7 Exercises

WARM-UPS *Simplify each fraction.*

1. $\dfrac{3}{21}$
2. $\dfrac{27}{81}$
3. $\dfrac{-64}{72}$
4. $\dfrac{-125}{50}$
5. $\dfrac{70}{420}$
6. $\dfrac{-3{,}612}{-3{,}612}$
7. $\dfrac{8{,}423}{-8{,}423}$
8. $\dfrac{-288}{-112}$

REVIEW *Identify each polynomial as a monomial, a binomial, a trinomial, or none of these.*

9. $5a^2b + 2ab^2$
10. $-3x^3y$
11. $-2x^3 + 3x^2 - 4x + 12$
12. $17t^2 - 15t + 27$
13. Find the degree of the trinomial $3x^2 - 2x + 4$.
14. What is the numerical coefficient of the second term of the trinomial $-7t^2 - 5t + 17$?

VOCABULARY AND CONCEPTS *Fill in the blanks.*

15. A __________ is an algebraic expression in which the exponents on the variables are whole numbers.
16. A _________ is a polynomial with one algebraic term.
17. Any nonzero value divided by itself is _.
18. Division by _ is undefined.
19. $\dfrac{1}{b} \cdot a = \dfrac{}{\_}$
20. $\dfrac{15x - 6y}{6xy} = \dfrac{15x}{\_\_} - \dfrac{6y}{6xy}$

GUIDED PRACTICE *In all fractions, assume that no denominators are 0. Perform each division by simplifying each fraction. Write all answers without using negative or zero exponents.* SEE EXAMPLE 1. (OBJECTIVE 1)

21. $\dfrac{xy}{yz}$
22. $\dfrac{a^2b}{ab^2}$
23. $\dfrac{r^3s^2}{rs^3}$
24. $\dfrac{y^4z^3}{y^2z^2}$
25. $\dfrac{8x^3y^2}{4xy^3}$
26. $\dfrac{-3y^3z}{6yz^2}$
27. $\dfrac{12u^5v}{-4u^2v^3}$
28. $\dfrac{16rst^2}{-8rst^3}$

Simplify. SEE EXAMPLE 2. (OBJECTIVE 2)

29. $\dfrac{6x + 9y}{3xy}$
30. $\dfrac{8x + 12y}{4xy}$
31. $\dfrac{xy + 6}{3y}$
32. $\dfrac{ab + 10}{2b}$
33. $\dfrac{5x - 10y}{25xy}$
34. $\dfrac{2x - 32}{16x}$
35. $\dfrac{3x^2 + 6y^3}{3x^2y^2}$
36. $\dfrac{4a^2 - 9b^2}{12ab}$

Simplify. SEE EXAMPLES 3–4. (OBJECTIVE 2)

37. $\dfrac{4x - 2y + 8z}{4xy}$
38. $\dfrac{5a^2 + 10b^2 - 15ab}{5ab}$
39. $\dfrac{12x^3y^2 - 8x^2y - 4x}{4xy}$
40. $\dfrac{12a^2b^2 - 8a^2b - 4ab}{4ab}$

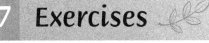

41. $\dfrac{-25x^2y + 30xy^2 - 5xy}{-5xy}$

42. $\dfrac{-30a^2b^2 - 15a^2b - 10ab^2}{-10ab}$

43. $\dfrac{15a^3b^2 - 10a^2b^3}{5a^2b^2}$

44. $\dfrac{9a^4b^3 - 16a^3b^4}{12a^2b}$

Simplify each numerator and perform the division. **SEE EXAMPLE 5.** *(OBJECTIVE 2)*

45. $\dfrac{5x(4x - 2y)}{2y}$

46. $\dfrac{9y^2(x^2 - 3xy)}{3x^2}$

47. $\dfrac{(-2x)^3 + (3x^2)^2}{6x^2}$

48. $\dfrac{(-3x^2y)^3 + (3xy^2)^3}{27x^3y^4}$

49. $\dfrac{4x^2y^2 - 2(x^2y^2 + xy)}{2xy}$

50. $\dfrac{-5a^3b - 5a(ab^2 - a^2b)}{10a^2b^2}$

51. $\dfrac{(a + b)^2 - (a - b)^2}{2ab}$

52. $\dfrac{(x - y)^2 + (x + y)^2}{2x^2y^2}$

Determine whether the two formulas are the same. **SEE EXAMPLE 6.** *(OBJECTIVE 3)*

53. $l = \dfrac{P - 2w}{2}$ and $l = \dfrac{P}{2} - w.$

54. $r = \dfrac{G + 2b}{2b}$ and $r = \dfrac{G}{2b} + b.$

55. Phone bills On a phone bill, the following formulas are given to compute the average cost per minute of x minutes of phone usage. Are they equivalent?

$$C = \frac{0.15x + 12}{x} \quad \text{and} \quad C = 0.15 + \frac{12}{x}$$

56. Electric bills On an electric bill, the following formulas are given to compute the average cost of x kwh of electricity. Are they equivalent?

$$C = \frac{0.08x + 5}{x} \quad \text{and} \quad C = 0.08x + \frac{5}{x}$$

ADDITIONAL PRACTICE *In all fractions, assume that no denominators are 0. Simplify each expression.*

57. $\dfrac{120}{160}$

58. $\dfrac{-90}{360}$

59. $\dfrac{5{,}880}{2{,}660}$

60. $\dfrac{-762}{366}$

61. $\dfrac{-16r^3y^2}{-4r^2y^4}$

62. $\dfrac{35xyz^2}{-7x^2yz}$

63. $\dfrac{-65rs^2t}{15r^2s^3t}$

64. $\dfrac{112u^3z^6}{-42u^3z^6}$

65. $\dfrac{x^2x^3}{xy^6}$

66. $\dfrac{(xy)^2}{x^2y^3}$

67. $\dfrac{(a^3b^4)^3}{ab^4}$

68. $\dfrac{(a^2b^3)^3}{a^6b^6}$

69. $\dfrac{12a + 2b}{6ab}$

70. $\dfrac{5ab + 30a^2}{10a}$

71. $\dfrac{16x - 8y}{24xy}$

72. $\dfrac{30xy^2 - 24x^2y + 12xy}{6xy}$

73. $\dfrac{2x(8x - 3y)}{4x}$

74. $\dfrac{3a^2b - 6(ab + a^2b^2)}{3ab}$

75. $\dfrac{8x^3 + 16x^2 - 4x}{4x}$

76. $\dfrac{12y^4 - 9y^3 + 6y^2}{-6y^2}$

77. $\dfrac{-(3x^3y^4)^3}{-(9x^4y^5)^2}$

78. $\dfrac{-15b^5 + 12b^3 - 18b^2 - 3b}{3b}$

79. $\dfrac{(a^2a^3)^4}{(a^4)^3}$

80. $\dfrac{(t^{-3}t^5)}{(t^2)^{-3}}$

81. $\dfrac{(3x - y)(2x - 3y)}{6xy}$

82. $\dfrac{(2m - n)(3m - 2n)}{-3m^2n^2}$

WRITING ABOUT MATH

83. Describe how you would simplify the fraction

$$\frac{4x^2y + 8xy^2}{4xy}$$

84. A student incorrectly attempts to simplify the fraction $\frac{3x + 5}{x + 5}$ as follows:

$$\frac{3x + 5}{x + 5} = \frac{3x + 5}{x + 5} = 3$$

How would you explain the error?

SOMETHING TO THINK ABOUT

85. If $x = 501$, evaluate $\dfrac{x^{500} - x^{499}}{x^{499}}$.

86. An exercise reads as follows:

$$\textit{Simplify:} \quad \frac{3x^3y + 6xy^2}{3xy^3}$$

It contains a misprint: one mistyped letter or digit. The correct answer is $\frac{x^2}{y} + 2$. Correct the exercise.

Section 4.8 Dividing Polynomials by Polynomials

Objectives

1. Divide a polynomial by a binomial.
2. Divide a polynomial by a binomial by first writing exponents in descending order.
3. Divide a polynomial with one or more missing terms by a binomial.

Vocabulary

divisor
dividend
quotient
remainder

Getting Ready

Divide.

1. $12\overline{)156}$ **2.** $17\overline{)357}$ **3.** $13\overline{)247}$ **4.** $19\overline{)247}$

We now complete our work of operations on polynomials by considering how to divide one polynomial by another.

1 Divide a polynomial by a binomial.

To divide one polynomial by another, we use a method similar to long division in arithmetic. Recall that the parts of a division problem are defined as

$$\begin{array}{r} \textit{quotient} + \textit{remainder} \\ \textit{divisor}\overline{)\textit{dividend}} \end{array}$$

Recall that division by zero is undefined. Therefore, the divisor cannot be 0. We must exclude any value of the variable that will result in a divisor of zero.

EXAMPLE 1 Divide $(x^2 + 5x + 6)$ by $(x + 2)$. Assume no division by 0.

Solution Here the **divisor** is $x + 2$ and the **dividend** is $x^2 + 5x + 6$. We proceed as follows:

Step 1:

$$\begin{array}{r} x \\ x + 2\overline{)x^2 + 5x + 6} \end{array}$$

How many times does x divide x^2? $\frac{x^2}{x} = x$
Write x above the division symbol.

Step 2:

$$\begin{array}{r} x \\ x + 2\overline{)x^2 + 5x + 6} \\ \underline{x^2 + 2x} \end{array}$$

Multiply each item in the divisor by x.
Write the product under $x^2 + 5x$ and draw a line.

Step 3:

$$\begin{array}{r} x \\ x + 2\overline{)x^2 + 5x + 6} \\ \underline{-x^2 - 2x} \\ 3x + 6 \end{array}$$

Subtract $(x^2 + 2x)$ from $(x^2 + 5x)$ by adding the negative of $(x^2 + 2x)$ to $(x^2 + 5x)$.

Bring down the 6.

Step 4:

$$\begin{array}{r} x + 3 \\ x + 2\overline{)x^2 + 5x + 6} \\ -x^2 - 2x \\ \hline 3x + 6 \end{array}$$

How many times does x divide $3x$? $\frac{3x}{x} = +3$
Write $+3$ above the division symbol.

Step 5:

$$\begin{array}{r} x + 3 \\ x + 2\overline{)x^2 + 5x + 6} \\ -x^2 - 2x \\ \hline 3x + 6 \\ -3x - 6 \\ \hline \end{array}$$

Multiply each term in the divisor by 3. Write the product under the $3x + 6$ and draw a line.

Step 6:

$$\begin{array}{r} x + 3 \\ x + 2\overline{)x^2 + 5x + 6} \\ -x^2 - 2x \\ \hline 3x + 6 \\ -3x - 6 \\ \hline 0 \end{array}$$

Subtract $(3x + 6)$ from $(3x + 6)$ by adding the negative of $(3x + 6)$.

The **quotient** is $x + 3$, and the **remainder** is 0.

Step 7: Check by verifying that $x + 2$ times $x + 3$ is $x^2 + 5x + 6$.

$$\begin{aligned} (x + 2)(x + 3) &= x^2 + 3x + 2x + 6 \\ &= x^2 + 5x + 6 \end{aligned}$$

SELF CHECK 1 Divide: $(x^2 + 7x + 12)$ by $(x + 3)$ Assume no division by 0.

We need to consider division problems with a remainder other than 0 such as in the next example.

EXAMPLE 2 Divide: $\dfrac{6x^2 - 7x - 2}{2x - 1}$ Assume no division by 0.

Solution Here the divisor is $2x - 1$ and the dividend is $6x^2 - 7x - 2$.

Step 1:

$$\begin{array}{r} 3x \\ 2x - 1\overline{)6x^2 - 7x - 2} \end{array}$$

How many times does $2x$ divide $6x^2$? $\frac{6x^2}{2x} = 3x$
Write $3x$ above the division symbol.

Step 2:

$$\begin{array}{r} 3x \\ 2x - 1\overline{)6x^2 - 7x - 2} \\ 6x^2 - 3x \\ \hline \end{array}$$

Multiply each term in the divisor by $3x$.
Write the product under $6x^2 - 7x$ and draw a line.

Step 3:

$$\begin{array}{r} 3x \\ 2x - 1\overline{)6x^2 - 7x - 2} \\ -6x^2 + 3x \\ \hline -4x - 2 \end{array}$$

Subtract $(6x^2 - 3x)$ from $(6x^2 - 7x)$ by adding the negative of $(6x^2 - 3x)$ to $(6x^2 - 7x)$.

Bring down the -2.

Step 4:

$$\begin{array}{r} 3x - 2 \\ 2x - 1\overline{)6x^2 - 7x - 2} \\ -6x^2 + 3x \\ \hline -4x - 2 \end{array}$$

How many times does $2x$ divide $-4x$? $\frac{-4x}{2x} = -2$
Write -2 above the division symbol.

Step 5:

$$\begin{array}{r} 3x - 2 \\ 2x - 1\overline{)6x^2 - 7x - 2} \\ -6x^2 + 3x \\ \hline -4x - 2 \\ +4x + 2 \\ \hline \end{array}$$

Multiply each term in the divisor by -2.
Write the product under $-4x - 2$ and draw a line.

Step 6:

$$\begin{array}{r} 3x - 2 \\ 2x - 1\overline{\big)6x^2 - 7x - 2} \\ \underline{-\ 6x^2 + 3x} \\ -4x - 2 \\ \underline{+4x - 2} \\ -4 \end{array}$$

Subtract $(-4x + 2)$ from $(-4x - 2)$ by adding the negative of $(-4x + 2)$.

COMMENT The division process ends when the degree of the remainder is less than the degree of the divisor.

Here the quotient is $3x - 2$, and the remainder is -4. It is common to write the answer in quotient $+ \frac{\text{remainder}}{\text{divisor}}$ form:

$$3x - 2 + \frac{-4}{2x - 1}$$

where the fraction $\frac{-4}{2x - 1}$ is formed by dividing the remainder by the divisor.

Step 7: To check the answer, we multiply $(3x - 2)$ by $(2x - 1)$ and add (-4). The result should be the dividend.

$$\begin{aligned}(2x - 1)(3x - 2) - 4 &= 6x^2 - 4x - 3x + 2 - 4 \\ &= 6x^2 - 7x - 2\end{aligned}$$

SELF CHECK 2 Divide: $\frac{8x^2 + 6x - 3}{2x + 3}$ Assume no division by 0.

Division by a binomial may include more than one variable as illustrated in the next example.

EXAMPLE 3

Divide: $\frac{6x^2 - xy - y^2}{3x + y}$ Assume no division by 0.

Solution Here the divisor is $3x + y$ and the dividend is $6x^2 - xy - y^2$.

Step 1:

$$\begin{array}{r} 2x \\ 3x + y\overline{\big)6x^2 - xy - y^2} \end{array}$$

How many times does $3x$ divide $6x^2$? $\frac{6x^2}{3x} = 2x$
Write $2x$ above the division symbol.

Step 2:

$$\begin{array}{r} 2x \\ 3x + y\overline{\big)6x^2 - xy - y^2} \\ \underline{6x^2 + 2xy} \end{array}$$

Multiply each term in the divisor by $2x$.
Write the product under $6x^2 - xy$ and draw a line.

Step 3:

$$\begin{array}{r} 2x \\ 3x + y\overline{\big)6x^2 - xy - y^2} \\ \underline{-6x^2 - 2xy} \\ -3xy - y^2 \end{array}$$

Subtract $(6x^2 + 2xy)$ from $(6x^2 - xy)$ by adding the negative of $(6x^2 + 2xy)$ to $(6x^2 - xy)$.

Bring down the $-y^2$.

Step 4:

$$\begin{array}{r} 2x - y \\ 3x + y\overline{\big)6x^2 - xy - y^2} \\ \underline{-6x^2 - 2xy} \\ -3xy - y^2 \end{array}$$

How many times does $3x$ divide $-3xy$? $\frac{-3xy}{3x} = -y$
Write $-y$ above the division symbol.

Step 5:

$$\begin{array}{r} 2x - y \\ 3x + y\overline{\big)6x^2 - xy - y^2} \\ \underline{-6x^2 - 2xy} \\ -3xy - y^2 \\ \underline{+3xy + y^2} \end{array}$$

Multiply each term in the divisor by $-y$. Write the product under the $-3x - y^2$ and draw a line.

Step 6:

$$\begin{array}{r} 2x - y \\ 3x + y\overline{)6x^2 - xy - y^2} \\ \underline{-6x^2 - 2xy} \\ -3xy - y^2 \\ \underline{+3xy + y^2} \\ 0 \end{array}$$

Subtract $(-3xy - y^2)$ from $(-3xy - y^2)$ by adding the negative of $(-3xy - y^2)$.

The quotient is $2x - y$ and the remainder is 0.

SELF CHECK 3 Divide $(6x^2 - xy - y^2)$ by $(2x - y)$. Assume no division by 0.

2 Divide a polynomial by a binomial by first writing exponents in descending order.

The division method works best when exponents of the terms in the divisor and the dividend are written in descending order. This means that the term involving the highest power of x appears first, the term involving the second-highest power of x appears second, and so on. For example, the terms in

$$3x^3 + 2x^2 - 7x + 5 \qquad 5 = 5x^0$$

have their exponents written in descending order.

If the powers in the dividend or divisor are not in descending order, we can use the commutative property of addition to write them that way.

EXAMPLE 4 Divide: $\dfrac{4x^2 + 2x^3 + 12 - 2x}{x + 3}$ Assume no division by 0.

Solution We write the dividend so that the exponents are in descending order and divide.

$$\begin{array}{r} 2x^2 - 2x + 4 \\ x + 3\overline{)2x^3 + 4x^2 - 2x + 12} \\ \underline{-2x^3 - 6x^2} \\ -2x^2 - 2x \\ \underline{+2x^2 + 6x} \\ +4x + 12 \\ \underline{-4x - 12} \\ 0 \end{array}$$

Check: $(x + 3)(2x^2 - 2x + 4) = 2x^3 - 2x^2 + 4x + 6x^2 - 6x + 12$
$= 2x^3 + 4x^2 - 2x + 12$

SELF CHECK 4 Divide: $\dfrac{x^2 - 10x + 6x^3 + 4}{2x - 1}$ Assume no division by 0.

3 Divide a polynomial with one or more missing terms by a binomial.

When we write the terms of a dividend in descending powers of x, we may notice that some powers of x are missing. For example, if the dividend is $3x^4 - 7x^2 - 3x + 15$, the term involving x^3 is missing. When this happens, we should either write the term with a coefficient of 0 or leave a blank space for it. In this case, we would write the dividend as

$$3x^4 + \mathbf{0x^3} - 7x^2 - 3x + 15 \qquad \text{or} \qquad 3x^4 \qquad -7x^2 - 3x + 15$$

EXAMPLE 5 Divide: $\frac{x^2 - 4}{x + 2}$ Assume no division by 0.

Solution Since $x^2 - 4$ does not have a term involving x, we must either include the term $0x$ or leave a space for it.

$$\begin{array}{r} x - 2 \\ x + 2\overline{)x^2 + 0x - 4} \\ \underline{-x^2 - 2x} \\ -2x - 4 \\ \underline{+2x + 4} \\ 0 \end{array}$$

Check: $(x + 2)(x - 2) = x^2 - 2x + 2x - 4$
$= x^2 - 4$

SELF CHECK 5 Divide: $\frac{x^2 - 9}{x - 3}$ Assume no division by 0.

SELF CHECK ANSWERS

1. $x + 4$ **2.** $4x - 3 + \frac{6}{2x + 3}$ **3.** $3x + y$ **4.** $3x^2 + 2x - 4$ **5.** $x + 3$

NOW TRY THIS

Assume no division by 0.

1. Identify the missing term(s): $8x^3 - 7x + 2x^5 - x^2$
2. Perform the division: $\frac{x^2 + 3x - 5}{x + 3}$
3. $(8x^2 - 2x + 3) \div (1 + 2x)$
4. The area of a rectangle is represented by $(3x^2 + 17x - 6)$ m^2 and the width is represented by $(3x - 1)$ m. Find a polynomial representation of the length.

4.8 Exercises

WARM-UPS *Divide and give the answer in quotient +* $\frac{\text{remainder}}{\text{divisor}}$ *form. Assume no division by 0.*

1. $16\overline{)384}$ **2.** $26\overline{)806}$
3. $19\overline{)271}$ **4.** $15\overline{)241}$

If each of the expressions is a divisor, determine what value must be excluded. Write your answer as x ≠ value.

5. $x + 2$ **6.** $3 - x$
7. $2x - 7$ **8.** $4x + 5$

REVIEW

9. List the composite numbers between 20 and 30.

10. Graph the set of prime numbers between 10 and 20 on a number line.

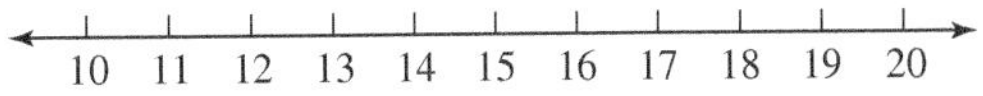

Let a = −2 and b = 3. Evaluate each expression.

11. $|a - b|$ **12.** $|a + b|$
13. $-|a^2 - b^2|$ **14.** $a - |-b|$

Simplify each expression.

15. $4(3x^2 - 5x + 2) + 3(2x^2 + 6x - 3)$
16. $-2(y^3 + 2y^2 - y) - 3(3y^3 + y)$

VOCABULARY AND CONCEPTS *Fill in the blanks.*

17. In the long division $x + 1\overline{)x^2 + 2x + 1}$, $x + 1$ is called the ______, and $x^2 + 2x + 1$ is called the ______.

Unless otherwise noted, all content on this page is © Cengage Learning.

18. The answer to a division problem is called the _______.

19. If a division does not come out even, the leftover part is called a ________.

20. The exponents in $2x^4 + 3x^3 + 4x^2 - 7x - 2$ are said to be written in _________ order.

Write each polynomial with the powers in descending order.

21. $8x^2 + 2x^3 - 5 + 3x$

22. $5x^2 + 7x^3 - 3x - 9$

23. $7x + 6x^3 - 4x^2 + 5x^4$

24. $7x^5 + x^3 - x^2 + 2x^4$

Identify the missing terms in each polynomial.

25. $5x^4 + 2x^2 - 1$

26. $-3x^5 - 2x^3 + 4x - 6$

GUIDED PRACTICE *Perform each division. Assume no division by 0.* SEE EXAMPLE 1. (OBJECTIVE 1)

27. Divide $(x^2 + 5x + 4)$ by $(x + 1)$.

28. Divide $(y^2 + 13y + 12)$ by $(y + 1)$.

29. $x + 5\overline{)x^2 + 7x + 10}$

30. $x + 6\overline{)x^2 + 5x - 6}$

31. $\dfrac{x^2 - 5x + 6}{x - 2}$

32. $\dfrac{z^2 - 7z + 12}{z - 3}$

33. $a - 7\overline{)a^2 - 11a + 28}$

34. $t - 6\overline{)t^2 - 3t - 18}$

Perform each division. Assume no division by 0. SEE EXAMPLE 2. (OBJECTIVE 1)

35. $\dfrac{8a^2 - 10a + 3}{4a - 3}$

36. $\dfrac{9a^2 - 9a - 4}{3a - 4}$

37. $\dfrac{3b^2 + 11b + 6}{3b + 2}$

38. $\dfrac{8a^2 + 2a - 3}{2a - 1}$

39. $\dfrac{2x^2 + 5x + 2}{2x + 3}$

40. $\dfrac{3x^2 - 8x + 3}{3x - 2}$

41. $\dfrac{4x^2 + 6x - 1}{2x + 1}$

42. $\dfrac{6x^2 - 11x + 2}{3x - 1}$

Perform each division. Assume no division by 0. SEE EXAMPLE 3. (OBJECTIVE 1)

43. Divide $(a^2 + 2ab + b^2)$ by $(a + b)$.

44. Divide $(a^2 - 2ab + b^2)$ by $(a - b)$.

45. $x + 2y\overline{)2x^2 + 3xy - 2y^2}$

46. $x + 3y\overline{)2x^2 + 5xy - 3y^2}$

47. $\dfrac{2x^2 - 7xy + 3y^2}{2x - y}$

48. $\dfrac{3x^2 + 5xy - 2y^2}{x + 2y}$

49. $\dfrac{12a^2 - ab - b^2}{3a - b}$

50. $\dfrac{2m^2 + 7mn - 4n^2}{2m - n}$

Write the powers of x in descending order (if necessary) and perform each division. Assume no division by 0. SEE EXAMPLE 4. (OBJECTIVE 2)

51. $5x + 3\overline{)11x + 10x^2 + 3}$

52. $2x - 7\overline{)-x - 21 + 2x^2}$

53. $4 + 2x\overline{)-10x - 28 + 2x^2}$

54. $1 + 3x\overline{)9x^2 + 1 + 6x}$

Perform each division. Assume no division by 0. SEE EXAMPLE 5. (OBJECTIVE 3)

55. $\dfrac{x^2 - 16}{x - 4}$

56. $\dfrac{x^2 - 25}{x + 5}$

57. $\dfrac{4x^2 - 9}{2x + 3}$

58. $\dfrac{25x^2 - 16}{5x - 4}$

59. $\dfrac{x^3 - 8}{x - 2}$

60. $\dfrac{x^3 + 27}{x + 3}$

ADDITIONAL PRACTICE *Perform each division. If there is a remainder, leave the answer in quotient + $\frac{\text{remainder}}{\text{divisor}}$ form. Assume no division by 0.*

61. $2x + 3\overline{)2x^3 + 7x^2 + 4x - 3}$

62. $2x - 1\overline{)2x^3 - 3x^2 + 5x - 2}$

63. $\dfrac{x^3 + 3x^2 + 3x + 1}{x + 1}$

64. $\dfrac{x^3 + 6x^2 + 12x + 8}{x + 2}$

65. $3x - 4\overline{)15x^3 - 23x^2 + 16x}$

66. $2y + 3\overline{)21y^2 + 6y^3 - 20}$

67. $3x + 2\overline{)6x^3 + 10x^2 + 7x + 2}$

68. $4x + 3\overline{)4x^3 - 5x^2 - 2x + 3}$

69. $\dfrac{2x^3 + 7x^2 + 4x - 4}{2x + 3}$

70. $\dfrac{6x^3 + x^2 + 2x - 2}{3x - 1}$

71. $\dfrac{2x^3 + 4x^2 - 2x + 3}{x - 2}$

72. $3x - 2y\overline{)-10y^2 + 13xy + 3x^2}$

73. $2x - y\overline{)xy - 2y^2 + 6x^2}$

74. $2x + y\overline{)2x^3 + 3x^2y + 3xy^2 + y^3}$

WRITING ABOUT MATH

75. Distinguish among *dividend, divisor, quotient,* and *remainder.*

76. How would you check the results of a division?

SOMETHING TO THINK ABOUT

77. Find the error in the following work.

$$\begin{array}{r} x + 1 \\ x - 2\overline{)x^2 + 3x - 2} \\ \underline{x^2 - 2x} \\ x - 2 \\ \underline{x - 2} \\ 0 \end{array}$$

78. Find the error in the following work.

$$\begin{array}{r} 3x \\ x+2\overline{)3x^2+10x+7} \\ \underline{3x^2+6x} \\ 4x+7 \end{array} = 3x + \frac{4x+7}{x+2}$$

79. Divide: $\dfrac{a^3+a}{a+3}$

80. Divide: $\dfrac{x^3+y^3}{x+y}$

Section 4.9 Synthetic Division

Objectives

1. Divide a polynomial by a binomial of the form $(x - r)$ using synthetic division.
2. Apply the remainder theorem to evaluate a polynomial.
3. Apply the factor theorem to determine whether a specific value is a zero of a polynomial.

Vocabulary

synthetic division
remainder theorem
factor theorem
zero of a polynomial

Getting Ready

Divide each polynomial $P(x)$ by $x - 2$ and find $P(2)$.

1. $P(x) = x^2 - x - 1$
2. $P(x) = x^2 + x + 3$

1 Divide a polynomial by a binomial of the form $(x - r)$ using synthetic division.

There is a method, called **synthetic division**, that we can use to divide a polynomial by a binomial of the form $(x - r)$. To see how it works, we consider the division of $(4x^3 - 5x^2 - 11x + 20)$ by $(x - 2)$.

$$\begin{array}{r} 4x^2 + 3x - 5 \\ x-2\overline{)4x^3 - 5x^2 - 11x + 20} \\ -\underline{4x^3 - 8x^2} \\ 3x^2 - 11x \\ -\underline{3x^2 - 6x} \\ -5x + 20 \\ -\underline{-5x + 10} \\ 10 \end{array} \text{(remainder)} \qquad \begin{array}{r} 4 \quad 3 \;-5 \\ 1-2\overline{)4 - 5 - 11 \quad 20} \\ -\underline{4 - 8} \\ 3 - 11 \\ -\underline{3 - 6} \\ -5 \quad 20 \\ -\underline{-5 \quad 10} \\ 10 \end{array} \text{(remainder)}$$

On the left is the long division, and on the right is the same division with the variables and their exponents removed and the numbers in the quotient moved to the left. The various powers of x can be remembered without actually writing them, because the exponents of the terms in the divisor, dividend, and quotient were written in descending order.

We can further shorten the version on the right. The numbers printed in color on the previous page need not be written, because they are duplicates of the numbers above them. We can write the long division as shown on the left, and shorten the process further by compressing the work vertically and eliminating the 1 (the coefficient of x in the divisor) as shown on the right.

$$\begin{array}{rrrrrr}
 & 4 & 3 & -5 & \\
1-2) & 4 & -5 & -11 & 20 \\
 - & -8 & & & \\
 & & 3 & & \\
 & - & -6 & & \\
 & & - & 5 & \\
 & & & - & 10 \\
 & & & & 10
\end{array}
\qquad\qquad
\begin{array}{rrrrr}
 & 4 & 3 & -5 & \\
-2) & 4 & -5 & -11 & 20 \\
- & & -8 & -6 & 10 \\
\hline
 & & 3 & -5 & 10
\end{array}$$

If we write the 4 in the quotient on the bottom line, the bottom line gives the coefficients of the quotient and the remainder. If we eliminate the top line, the division appears as follows:

$$\begin{array}{r|rrrr}
-2 & 4 & -5 & -11 & 20 \\
- & & -8 & -6 & 10 \\
\hline
 & 4 & 3 & -5 & 10
\end{array}$$

The bottom line was obtained by subtracting the middle line from the top line. If we replace the -2 in the divisor by $+2$, the division process will reverse the signs of every entry in the middle line, and then the bottom line can be obtained by addition. This gives the final form of the synthetic division.

$$\begin{array}{r|rrrrl}
+2 & 4 & -5 & -11 & 20 & \text{The coefficients of the dividend} \\
+ & & 8 & 6 & -10 & \\
\hline
 & 4 & 3 & -5 & 10 & \text{The coefficients of the quotient and the remainder}
\end{array}$$

Thus,

$$\frac{4x^3 - 5x^2 - 11x + 20}{x - 2} = 4x^2 + 3x - 5 + \frac{10}{x - 2}$$

EXAMPLE 1 Use synthetic division to divide $(6x^2 + 5x - 2)$ by $(x - 5)$. Assume no division by 0.

Solution We write the coefficients in the dividend and the 5 in the divisor in the following form:

$$\begin{array}{r|rrr}
5 & 6 & 5 & -2 \\
 & & & \\
\hline
\end{array}$$

Then we follow these steps:

$$\begin{array}{r|rrrl}
5 & 6 & 5 & -2 & \text{Begin by bringing down the 6.} \\
 & \downarrow & & & \\
\hline
 & 6 & & &
\end{array}$$

$$\begin{array}{r|rrrl}
5 & 6 & 5 & -2 & \text{Multiply 5 by 6 to obtain 30.} \\
 & & 30 & & \\
\hline
 & 6 & & &
\end{array}$$

$$\begin{array}{r|rrrl}
5 & 6 & 5 & -2 & \text{Add 5 and 30 to obtain 35.} \\
 & & 30 & & \\
\hline
 & 6 & 35 & &
\end{array}$$

$$\begin{array}{r|rrrl}
5 & 6 & 5 & -2 & \text{Multiply 35 by 5 to obtain 175.} \\
 & & 30 & 175 & \\
\hline
 & 6 & 35 & &
\end{array}$$

COMMENT When using synthetic division the quotient will be one degree less than the dividend.

$$\begin{array}{r|rrr} 5 & 6 & 5 & -2 \\ & & 30 & 175 \\ \hline & 6 & 35 & 173 \end{array}$$ Add -2 and 175 to obtain 173.

The numbers 6 and 35 represent the quotient $(6x + 35)$, and 173 is the remainder. Thus,

$$\frac{6x^2 + 5x - 2}{x - 5} = 6x + 35 + \frac{173}{x - 5}$$

SELF CHECK 1 Use synthetic division to divide $(6x^2 - 5x + 2)$ by $(x - 5)$. Assume no division by 0.

EXAMPLE 2

Use synthetic division to divide $(5x^3 + x^2 - 3)$ by $(x - 2)$. Assume no division by 0.

Solution We begin by writing

$$\begin{array}{r|rrrr} 2 & 5 & 1 & 0 & -3 \\ \hline \end{array}$$ Write 0 for the coefficient of x, the missing term.

and complete the division as follows:

$$\begin{array}{r|rrrr} 2 & 5 & 1 & 0 & -3 \\ & & 10 & & \\ \hline & 5 & 11 & & \end{array} \qquad \begin{array}{r|rrrr} 2 & 5 & 1 & 0 & -3 \\ & & 10 & 22 & \\ \hline & 5 & 11 & 22 & \end{array} \qquad \begin{array}{r|rrrr} 2 & 5 & 1 & 0 & -3 \\ & & 10 & 22 & 44 \\ \hline & 5 & 11 & 22 & 41 \end{array}$$

The numbers 5, 11, and 22 represent the quotient $5x^2 + 11x + 22$ and 41 is the remainder. Thus,

$$\frac{5x^3 + x^2 - 3}{x - 2} = 5x^2 + 11x + 22 + \frac{41}{x - 2}$$

SELF CHECK 2 Use synthetic division to divide $(5x^3 - x^2 + 3)$ by $(x - 2)$. Assume no division by 0.

EXAMPLE 3

Use synthetic division to divide $(5x^2 + 6x^3 + 2 - 4x)$ by $(x + 2)$. Assume no division by 0.

Solution First, we write the dividend with the exponents in descending order.

$$6x^3 + 5x^2 - 4x + 2$$

Then we write the divisor in $(x - r)$ form: $x - (-2)$. Using synthetic division, we begin by writing

$$\begin{array}{r|rrrr} -2 & 6 & 5 & -4 & 2 \\ & & & & \\ \hline \end{array}$$

and complete the division.

$$\begin{array}{r|rrrr} -2 & 6 & 5 & -4 & 2 \\ & & -12 & 14 & -20 \\ \hline & 6 & -7 & 10 & -18 \end{array}$$

Thus,

$$\frac{5x^2 + 6x^3 + 2 - 4x}{x + 2} = 6x^2 - 7x + 10 + \frac{-18}{x + 2}$$

SELF CHECK 3 Divide $(2x - 4x^2 + 3x^3 - 3)$ by $(x - 1)$. Assume no division by 0.

2 Apply the remainder theorem to evaluate a polynomial.

Synthetic division not only provides a method for simplifying some divisions, it can also be used to evaluate a polynomial by applying the **remainder theorem.**

REMAINDER THEOREM If a polynomial $P(x)$ is divided by $(x - r)$, the remainder is $P(r)$.

We illustrate the remainder theorem in the next example.

EXAMPLE 4 Let $P(x) = 2x^3 - 3x^2 - 2x + 1$. Find

a. $P(3)$. **b.** the remainder when $P(x)$ is divided by $(x - 3)$.

Solution **a.**

$$\begin{aligned} P(3) &= 2(3)^3 - 3(3)^2 - 2(3) + 1 \quad \text{Substitute 3 for } x. \\ &= 2(27) - 3(9) - 6 + 1 \\ &= 54 - 27 - 6 + 1 \\ &= 22 \end{aligned}$$

b. We can use the following synthetic division to find the remainder when $P(x) = 2x^3 - 3x^2 - 2x + 1$ is divided by $x - 3$.

$$\begin{array}{r|rrrr} 3 & 2 & -3 & -2 & 1 \\ & & 6 & 9 & 21 \\ \hline & 2 & 3 & 7 & 22 \end{array} \quad \text{The remainder is 22.}$$

The results of parts **a** and **b** illustrate that when $P(x)$ is divided by $(x - 3)$, the remainder is $P(3)$.

SELF CHECK 4 Use the polynomial of Example 4 to find:
a. $P(2)$. **b.** the remainder when the polynomial is divided by $(x - 2)$.

COMMENT It is often more efficient to find $P(r)$ by using synthetic division than by substituting r for x in $P(x)$. This is especially true if r is a fraction or decimal.

3 Apply the factor theorem to determine whether a specific value is a zero of a polynomial.

Recall that if two quantities are multiplied, each is called a *factor* of the product. Thus, $(x - 2)$ is one factor of $6x - 12$, because $6(x - 2) = 6x - 12$. The **factor theorem** enables us to find one factor of a polynomial if the remainder of a certain division is 0.

FACTOR THEOREM If $P(x)$ is a polynomial in x, then

$$P(r) = 0 \quad \text{if and only if} \quad (x - r) \text{ is a factor of } P(x).$$

If $P(x)$ is a polynomial in x and if $P(r) = 0$, r is called a **zero of the polynomial**.

EXAMPLE 5 Let $P(x) = 3x^3 - 5x^2 + 3x - 10$. Show that

a. $P(2) = 0$. **b.** $(x - 2)$ is a factor of $P(x)$.

Solution **a.**

$$\begin{aligned} P(2) &= 3(2)^3 - 5(2)^2 + 3(2) - 10 \\ &= 3(8) - 5(4) + 6 - 10 \\ &= 24 - 20 + 6 - 10 \\ &= 0 \end{aligned}$$

$$P(2) = 0$$

Therefore, 2 is a zero of the polynomial.

b. Divide $P(x)$ by $(x - 2)$.

$$\begin{array}{r|rrrr} 2 & 3 & -5 & 3 & -10 \\ & & 6 & 2 & 10 \\ \hline & 3 & 1 & 5 & 0 \end{array}$$

Because the remainder is 0, the numbers 3, 1, and 5 in the synthetic division represent the quotient $(3x^2 + x + 5)$. Thus,

$$\underbrace{(x - 2)}_{\text{Divisor}} \cdot \underbrace{(3x^2 + x + 5)}_{\text{quotient}} + \underbrace{0}_{\text{remainder}} = \underbrace{3x^3 - 5x^2 + 3x - 10}_{\text{the dividend, } P(x)}$$

or

$$(x - 2)(3x^2 + x + 5) = 3x^3 - 5x^2 + 3x - 10$$

Thus, $(x - 2)$ is a factor of $P(x)$.

SELF CHECK 5 Use the polynomial $P(x) = x^2 - 5x + 6$ and show that $P(2) = 0$ and that $(x - 2)$ is a factor of $P(x)$.

The result in Example 5 is true because the remainder, $P(2)$, is 0. If the remainder had not been 0, then $(x - 2)$ would not have been a factor of $P(x)$.

Accent on technology

▸ Approximating Zeros of Polynomials

COMMENT The x-coordinates of the x-intercepts of the graph represent what we call the real zeros of a polynomial.

We can use a graphing calculator to approximate the real zeros of a polynomial function $f(x)$. For example, to find the real zeros of $f(x) = 2x^3 - 6x^2 + 7x - 21$, we graph the function as in Figure 4-7 on the next page.

It appears from the figure that the function f has a zero at $x = 3$. To be certain, find $f(3)$.

$$f(3) = 2(3)^3 - 6(3)^2 + 7(3) - 21 = 0 \quad \text{Substitute 3 for } x.$$

From the factor theorem, we know that if $P(3) = 0$, then $(x - 3)$ is a factor of the polynomial. To find the other factors, we can synthetically divide by 3.

$$\begin{array}{r|rrrr} 3 & 2 & -6 & 7 & -21 \\ & & 6 & 0 & 21 \\ \hline & 2 & 0 & 7 & 0 \end{array}$$

Thus, $f(x) = (x - 3)(2x^2 + 7)$. Since $(2x^2 + 7)$ cannot be factored over the real numbers, we can conclude that 3 is the only real zero of the polynomial function.

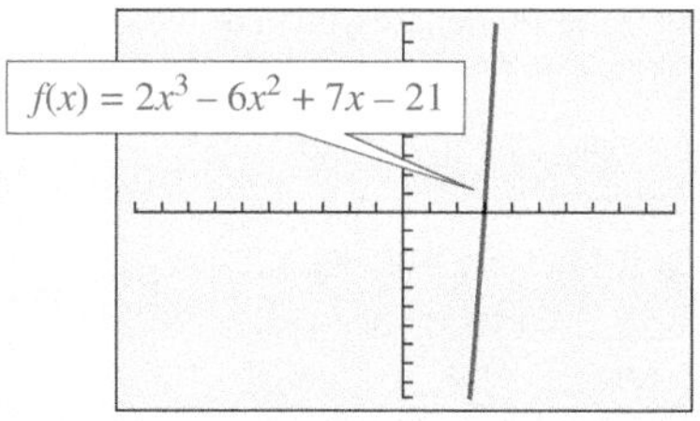

Figure 4-7

For instructions regarding the use of a Casio graphing calculator, please refer to the Casio Keystroke Guide in the back of the book.

Unless otherwise noted, all content on this page is © Cengage Learning.

SELF CHECK ANSWERS

1. $6x + 25 + \frac{127}{x - 5}$ **2.** $5x^2 + 9x + 18 + \frac{39}{x - 2}$ **3.** $3x^2 - x + 1 + \frac{-2}{x - 1}$ **4.** both are 1
5. $(x - 2)(x - 3) = x^2 - 5x + 6$

NOW TRY THIS

1. Given that $(x + 2)$ is a factor of $(2x^3 - 3x^2 - 11x + 6)$, find the remaining factors.
2. Solve $2x^3 - 3x^2 - 11x + 6 = 0$ without a calculator.
3. Find the x-intercepts for $f(x) = 2x^3 - 3x^2 - 11x + 6$.

4.9 Exercises

For Exercise Set 4.9, *assume no division by 0.*

WARM-UPS *Use long division to find each remainder.*

1. $(2x^2 + 3x - 18) \div (x + 4)$
2. $(4x^2 + 15x + 8) \div (x + 3)$

Evaluate.

3. $2x^2 + 3x - 18$ when $x = -4$
4. $4x^2 + 15x + 8$ when $x = -3$

REVIEW *Let* $f(x) = 3x^2 + 2x - 1$ *and find each value.*

5. $f(-1)$
6. $f(3)$
7. $f(2a)$
8. $f(-2t)$

Remove parentheses and simplify.

9. $5(x^2 - 3x + 2) + 4(3x^2 - x + 1)$
10. $-2(3y^3 - 2y + 7) - 3(y^2 + 2y - 4) + 4(y^3 + 2y - 1)$

VOCABULARY AND CONCEPTS *Fill in the blanks.*

11. In order to use synthetic division, the divisor must be in the form ____.
12. If the degree of the leading term of the dividend is 5 and the divisor is $(x - r)$, the degree of the leading term of the quotient will be _.
13. The remainder theorem states that if a polynomial $P(x)$ is divided by $(x - r)$, the remainder is ____.
14. The factor theorem states that if $P(x)$ is a polynomial in x, then $P(r) = 0$ if and only if ____ is a factor of $P(x)$.

GUIDED PRACTICE *Use synthetic division to perform each division.* SEE EXAMPLE 1. (OBJECTIVE 1)

15. $(x^2 - 13x + 30) \div (x - 9)$
16. $(x^2 - 6x + 5) \div (x - 5)$
17. $(x^2 + 4x - 10) \div (x + 5)$
18. $(x^2 + x - 6) \div (x - 2)$

Use synthetic division to perform each division. SEE EXAMPLE 2. (OBJECTIVE 1)

19. $(2x^3 - 5x - 6) \div (x - 2)$
20. $(x^3 - x + 6) \div (x - 2)$
21. $(4x^3 - 2x + 36) \div (x + 2)$
22. $(2x^3 - 5x^2 - 9) \div (x - 3)$

Use synthetic division to perform each division. SEE EXAMPLE 3. (OBJECTIVE 1)

23. $(5x^2 + 6x^3 + 4) \div (x + 1)$
24. $(-13x^2 + 2x^3 - 5 + 23x) \div (x - 4)$
25. $(4x^2 + 3x^3 + 8) \div (x + 2)$
26. $(4 - 3x^2 + x) \div (x - 4)$

Let $P(x) = 2x^3 - 4x^2 + 2x - 1$. *Evaluate* $P(x)$ *for the given value by using the remainder theorem. Then evaluate by substituting the value of x into the polynomial and simplifying.* SEE EXAMPLE 4. (OBJECTIVE 2)

27. $P(1)$
28. $P(2)$
29. $P(-3)$
30. $P(-1)$
31. $P(3)$
32. $P(-4)$
33. $P(-5)$
34. $P(4)$

Use the remainder theorem and synthetic division to find $P(r)$. (OBJECTIVE 2)

35. $P(x) = x^3 - 4x^2 + x - 2$; $r = 2$
36. $P(x) = x^3 - 3x^2 + x + 1$; $r = 1$
37. $P(x) = x^4 - 2x^3 + x^2 - 3x + 2$; $r = -2$
38. $P(x) = 2x^3 + x + 2$; $r = 3$

Use the factor theorem and determine whether the first expression is a factor of P(x). Factor, if possible. SEE EXAMPLE 5. (OBJECTIVE 3)

39. $x - 3$; $P(x) = x^3 - 3x^2 + 5x - 15$
40. $x + 1$; $P(x) = x^3 + 2x^2 - 2x - 3$

41. $x - 6$; $P(x) = 2x^3 - 8x^2 + x - 36$
42. $x + 2$; $P(x) = 3x^2 - 7x + 4$

ADDITIONAL PRACTICE *Divide using synthetic division.*

43. $(3x^3 - 10x^2 + 5x - 6) \div (x - 3)$
44. $(2x^3 - 9x^2 + 10x - 3) \div (x - 3)$
45. $(x^2 - 5x + 14) \div (x + 2)$
46. $(5x^2 - 8x - 2) \div (x - 2)$
47. $(x^2 + 8 + 6x) \div (x + 4)$
48. $(x^2 - 15 - 2x) \div (x + 3)$
49. $(x^2 + 13x + 42) \div (x + 6)$
50. $(4x^3 + 5x^2 - 1) \div (x + 2)$

Use the remainder theorem and synthetic division to find P(r).

51. $P(x) = x^3 + x^2 + 1$; $r = -2$
52. $P(x) = x^5 + 3x^4 - x^2 + 1$; $r = -1$
53. $P(x) = 2x^3 - x^2 + 4x - 5$; $r = \frac{1}{2}$
54. $P(x) = 3x^3 + x^2 - 8x + 4$; $r = \frac{2}{3}$

Use the factor theorem and determine whether the first expression is a factor of P(x). Factor, if possible.

55. x; $P(x) = 7x^3 - 5x^2 - 8x$
56. x; $P(x) = 5x^3 + 6x + 2$
57. $x - 5$; $P(x) = x^3 - 125$
58. $x - 2$; $P(x) = x^3 + 8$

Evaluate P(x) for the given value using the remainder theorem.

59. $P(x) = 3x^5 + 1$; $c = -\frac{1}{2}$
60. $P(x) = 5x^7 - 7x^4 + x^2 + 1$; $c = 2$

Let $Q(x) = x^4 - 3x^3 + 2x^2 + x - 3$. *Evaluate Q(x) for the given value by using the remainder theorem.*

61. $Q(-1)$
62. $Q(0)$
63. $Q(2)$
64. $Q(-2)$

Use a calculator to work each.

65. Find 2^6 by using synthetic division to evaluate the polynomial $P(x) = x^6$ at $x = 2$. Then check the answer by evaluating 2^6 with a calculator.
66. Find $(-3)^5$ by using synthetic division to evaluate the polynomial $P(x) = x^5$ at $x = -3$. Then check the answer by evaluating $(-3)^5$ with a calculator.

WRITING ABOUT MATH

67. If you are given $P(x)$, explain how to use synthetic division to calculate $P(a)$.
68. Explain the factor theorem.

SOMETHING TO THINK ABOUT *Suppose that* $P(x) = x^{100} - x^{99} + x^{98} - x^{97} + \cdots + x^2 - x + 1$.

69. Find the remainder when $P(x)$ is divided by $(x - 1)$.
70. Find the remainder when $P(x)$ is divided by $(x + 1)$.

Projects

PROJECT 1

Let $f(x) = 3x^2 + 3x - 2$, $g(x) = 2x^2 - 5$, and $t(x) = x + 2$. Perform each operation.

a. $f(x) + g(x)$
b. $g(x) - t(x)$
c. $f(x) \cdot g(x)$
d. $\dfrac{f(x)}{t(x)}$
e. $\dfrac{[g(x)]^2}{t(x)}$

PROJECT 2

To discover a pattern in the behavio
sider the polynomial $2x^2 - 3x$ –
polynomial at $x = 1$ and $x = 3$. Th
mial by $(x - 1)$ and again by $(x$ –

a. What do you notice about the rema
divisions?
b. Try others. Does the pattern hold wh
the polynomial at $x = -2$?
c. Does the pattern hold for other polyno
ment and report your conclusions.

Reach for Success

EXTENSION OF STUDY STRATEGIES

Let's consider how you can increase test preparedness *prior* to an exam. What would you do differently if you only knew then what you know now?

There is a known learning pattern called the Forgetting Curve, first recognized in 1885 by Hermann Ebbinghaus. The basic premise is that by reviewing material from a one-hour lecture for ten minutes the next day you are firing the synapses that make connections in your brain. Then, reviewing the material just a few minutes every day after that significantly increases your ability to recall that information. When you repeat this information, you send a message to your brain, "Here it is again; it must be important! I need to know this."

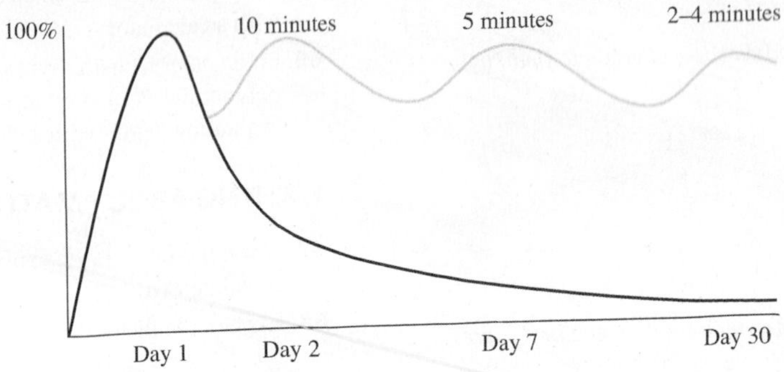

According to the graph, by Day 7, how much time would you need to spend to remember the material?____________

By Day 30, how much time would you need to review the material?____________

Can you think of factors that might influence the speed of forgetting?____________

____

…el overwhelmed by this amount of review? Does it seem to take too much time?____________

…y sessions rarely help with exam performance. Would it surprise you to know that it actually takes *less* …daily than it does to re… …the night before a test?____________

…l share … …or.

… …naterial, its connectedness to a previously learned topic, … …u can remember the material. While studying mathematics, … …p increase the connections to other material and make it easier

Unless otherwise noted, all content on this page is © Cengage Learning.

4 Review

SECTION 4.1 Natural-Number Exponents

DEFINITIONS AND CONCEPTS	EXAMPLES
If n is a natural number, then $x^n = \overbrace{x \cdot x \cdot x \cdot \cdots \cdot x}^{n \text{ factors of } x}$	$x^5 = \overbrace{x \cdot x \cdot x \cdot x \cdot x}^{5 \text{ factors of } x}$ $\quad x^7 = \overbrace{x \cdot x \cdot x \cdot x \cdot x \cdot x \cdot x}^{7 \text{ factors of } x}$
If m and n are integers, then	
$x^m x^n = x^{m+n}$	$x^2 \cdot x^7 = x^{2+7} = x^9$
$(x^m)^n = x^{m \cdot n}$	$(x^2)^7 = x^{2 \cdot 7} = x^{14}$
$(xy)^n = x^n y^n$	$(xy)^3 = x^3 y^3$
$\left(\frac{x}{y}\right)^n = \frac{x^n}{y^n} \quad (y \neq 0)$	$\left(\frac{x}{y}\right)^3 = \frac{x^3}{y^3} \quad (y \neq 0)$
$\frac{x^m}{x^n} = x^{m-n} \quad (x \neq 0)$	$\frac{x^7}{x^2} = x^{7-2} = x^5 \quad (x \neq 0)$

REVIEW EXERCISES

Write each expression without exponents.

1. $(-3x)^4$
2. $\left(\frac{1}{2}pq\right)^3$

Evaluate each expression.

3. 5^3
4. 3^5
5. $(-6)^2$
6. -6^2
7. $3^2 + 2^2$
8. $(3 + 2)^2$

Perform the operations and simplify.

9. x^4x^6
10. x^2x^7
11. $(y^7)^3$
12. $(x^{21})^2$
13. $(ab)^3$
14. $(3x)^4$
15. $b^3b^4b^5$
16. $-z^2(z^3y^2)$
17. $(16s)^2s$
18. $-3y(y^5)$
19. $(2x^4y^2)^3$
20. $(5x^3y)^2$
21. $\frac{x^7}{x^3}$
22. $\left(\frac{x^2y}{xy^2}\right)^2$
23. $\frac{8(y^2x)^2}{4(yx^2)^2}$
24. $\frac{(5y^2z^3)^3}{25(yz)^5}$

SECTION 4.2 Zero and Negative-Integer Exponents

DEFINITIONS AND CONCEPTS	EXAMPLES
$x^0 = 1 \quad (x \neq 0)$	$(2x)^0 = 1 \quad (x \neq 0)$
$x^{-n} = \frac{1}{x^n} \quad (x \neq 0)$	$x^{-3} = \frac{1}{x^3} \quad (x \neq 0)$
$\frac{1}{x^{-n}} = x^n \quad (x \neq 0)$	$\frac{1}{x^{-3}} = x^3 \quad (x \neq 0)$

REVIEW EXERCISES

Write each expression without negative exponents or parentheses.

25. x^0
26. $(3x^2y^2)^0$
27. $(3x^0)^2$
28. $(3x^2y^0)^2$
29. x^{-3}
30. $x^{-2}x^3$
31. y^4y^{-3}
32. $\frac{x^3}{x^{-7}}$
33. $(x^{-3}x^4)^{-2}$
34. $(a^{-2}b)^{-3}$
35. $\left(\frac{x^2}{x}\right)^{-5}$
36. $\left(\frac{15z^4}{5z^3}\right)^{-2}$

SECTION 4.3 Scientific Notation

DEFINITIONS AND CONCEPTS	EXAMPLES
A number is written in scientific notation if it is written as the product of a number between 1 (including 1) and 10 and an integer power of 10.	4,582,000,000 is written as 4.582×10^9 in scientific notation. 0.00035 is written as 3.5×10^{-4} in scientific notation.

REVIEW EXERCISES

Write each number in scientific notation.

37. 728
38. 6,230
39. 0.0275
40. 0.00942
41. 7.73
42. 753×10^3
43. 0.018×10^{-2}
44. 600×10^2

Write each number in standard notation.

45. 3.87×10^4
46. 7.98×10^{-5}
47. 2.68×10^0
48. 5.76×10^1
49. 739×10^{-2}
50. 0.437×10^{-3}
51. $\dfrac{(0.00012)(0.00004)}{0.00000016}$
52. $\dfrac{(4{,}800)(20{,}000)}{600{,}000}$

SECTION 4.4 Polynomials

DEFINITIONS AND CONCEPTS	EXAMPLES
A polynomial is an algebraic expression that is one term or the sum of terms containing whole-number exponents on the variables.	Polynomials: $9xy$, $5x^2 + 9x - 1$, and $11x - 5y$
If a is a nonzero coefficient, the degree of the monomial ax^n is n. The degree of a polynomial is the same as the degree of its term with largest degree.	Find the degree of each term and the degree of the polynomial $8x^2 - 5x + 3$. The degree of the first term is 2. The degree of the second term is 1. The degree of the third term is 0. The degree of the polynomial is 2.
When a number is substituted for the variable in a polynomial, the polynomial takes on a numerical value.	Evaluate $5x - 4$ when $x = -3$. $5x - 4 = 5(-3) - 4$ Substitute -3 for x. $= -15 - 4$ Simplify. $= -19$
Finding a function value for a polynomial uses the same process as evaluating a polynomial for a specified value.	If $f(x) = x^2 - 8x + 3$, find $f(-3)$. $f(x) = x^2 - 8x + 3$ $f(-3) = (-3)^2 - 8(-3) + 3$ Substitute -3 for x. $= 9 + 24 + 3$ Simplify. $= 36$ Since the result is 36, $f(-3) = 36$.
To graph a polynomial function, create a table of values, plot the ordered pairs $(x, f(x))$, and draw a smooth curve through those points. Determine the domain and the range from the graph.	Graph the polynomial function $f(x) = x^2 - 8x + 3$ and state the domain and range:

x	$f(x) = x^2 - 8x + 3$	$(x, f(x))$
-1	$f(-1) = (-1)^2 - 8(-1) + 3 = 12$	$(-1, 12)$
0	$f(0) = (0)^2 - 8(0) + 3 = 3$	$(0, 3)$
1	$f(1) = (1)^2 - 8(1) + 3 = -4$	$(1, -4)$
2	$f(2) = (2)^2 - 8(2) + 3 = -9$	$(2, -9)$
4	$f(4) = (4)^2 - 8(4) + 3 = -13$	$(4, -13)$
5	$f(5) = (5)^2 - 8(5) + 3 = -12$	$(5, -12)$

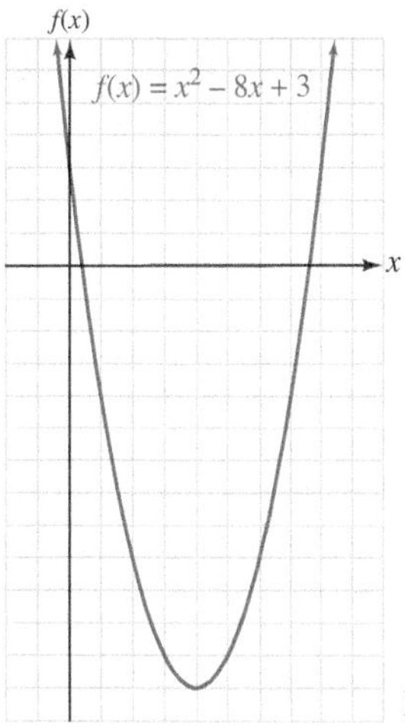

D: ℝ, R: [−13, ∞)

REVIEW EXERCISES

Find the degree of each polynomial and classify it as a monomial, a binomial, or a trinomial.

53. $29x^8$

54. $5^3x + x^2$

55. $-3x^5 + x - 1$

56. $9xy + 21x^3y^2$

Evaluate $3x + 2$ *for each value of* x.

57. $x = 7$

58. $x = -4$

59. $x = -2$

60. $x = \dfrac{2}{3}$

Evaluate $5x^4 - x$ *for each value of* x.

61. $x = 3$

62. $x = 0$

63. $x = -2$

64. $x = -0.3$

If $f(x) = x^2 - 4$, *find each value.*

65. $f(0)$

66. $f(-4)$

67. $f(-2)$

68. $f\left(\dfrac{1}{2}\right)$

Graph each polynomial function. State the domain and range.

69. $f(x) = x^2 - 5$

70. $f(x) = x^3 - 2$

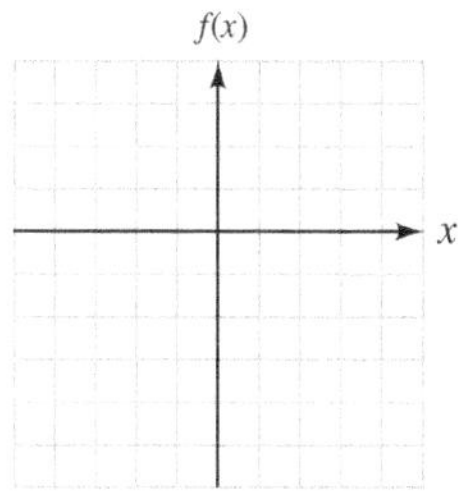

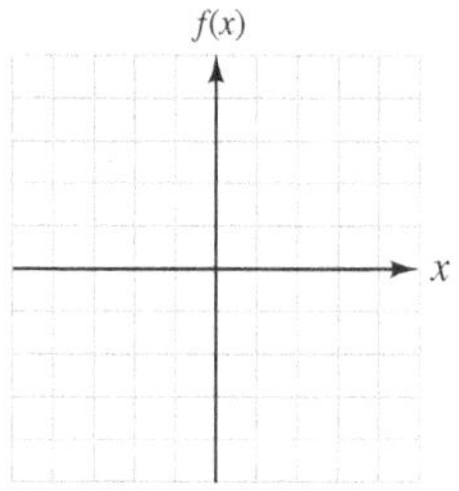

SECTION 4.5 Adding and Subtracting Polynomials

DEFINITIONS AND CONCEPTS	EXAMPLES
We can add polynomials by removing parentheses, if necessary, and then combining any like terms that are contained within the polynomials.	$(8x^3 - 6x + 13) + (9x - 7)$ $= 8x^3 - 6x + 13 + 9x - 7$ Remove parentheses. $= 8x^3 + 3x + 6$ Combine like terms.
We can subtract polynomials by dropping the negative sign and the parentheses, and *changing the sign of every term within the second set of parentheses.*	$(8x^3 - 6x + 13) - (9x - 7)$ $= 8x^3 - 6x + 13 - 9x + 7$ Change the sign of each term in the second set of parentheses. $= 8x^3 - 15x + 20$ Combine like terms.

REVIEW EXERCISES

Simplify each expression, if possible.

71. $6x - 4x + x$

72. $5x + 4y$

73. $(xy)^2 + 3x^2y^2$

74. $-2x^2yz + 3yx^2z$

75. $(3x^2 + 2x) + (5x^2 - 8x)$

76. $(7a^2 + 2a - 5) - (3a^2 - 2a + 1)$

77. $3(9x^2 + 3x + 7) - 2(11x^2 - 5x + 9)$

78. $4(4x^3 + 2x^2 - 3x - 8) - 5(2x^3 - 3x + 8)$

Unless otherwise noted, all content on this page is © Cengage Learning.

SECTION 4.6 Multiplying Polynomials

DEFINITIONS AND CONCEPTS	EXAMPLES
To multiply two monomials, first multiply the numerical factors and then multiply the variable factors using the properties of exponents.	$(5x^2y^3)(4xy^2)$ $= 5(4)x^2xy^3y^2$ Use the commutative property of multiplication. $= 20x^3y^5$ Use multiplication and the properties of exponents.
To multiply a polynomial with more than one term by a monomial, multiply each term of the polynomial by the monomial and simplify.	$4x(3x^2 + 2x)$ $= \mathbf{4x} \cdot 3x^2 + \mathbf{4x} \cdot 2x$ Use the distributive property. $= 12x^3 + 8x^2$ Multiply.
To multiply two binomials, use the distributive property.	$(2x - 5)(x + 3)$ $= \mathbf{2x}(x) + \mathbf{2x}(3) + (-5)(x) + (-5)(3)$ $= 2x^2 + 6x - 5x - 15$ $= 2x^2 + x - 15$
Special products: Squaring a binomial: $(x + y)^2 = (x + y)(x + y) = x^2 + 2xy + y^2$ $(x - y)^2 = (x - y)(x - y) = x^2 - 2xy + y^2$ Multiplying conjugate binomials: $(x + y)(x - y) = x^2 - y^2$	$(x + 7)^2 = (x + 7)(x + 7)$ $= x^2 + 14x + 49$ $(x - 7)^2 = (x - 7)(x - 7)$ $= x^2 - 14x + 49$ $(2x + 3)(2x - 3) = 4x^2 - 6x + 6x - 9$ $= 4x^2 - 9$
To multiply one polynomial by another, multiply each term of one polynomial by each term of the other polynomial, and simplify.	$(x + 2)(4x^2 - x + 3)$ $4x^3 - x^2 + 3x + 8x^2 - 2x + 6$ $4x^3 + 7x^2 + x + 6$
To solve an equation that simplifies to a linear equation, use the distributive property and order of operations and proceed to solve.	Solve: $(x + 3)(x - 5) = x^2 - 3(x + 1)$ $x^2 - 5x + 3x - 15 = x^2 - 3x - 3$ $x^2 - 2x - 15 = x^2 - 3x - 3$ $-2x - 15 = -3x - 3$ $x - 15 = -3$ $x = 12$

REVIEW EXERCISES

Find each product.

79. $(4x^3y^5)(3x^2y)$ **80.** $(xyz^3)(x^3z)^2$

Find each product.

81. $4(2x + 3)$ **82.** $3(2x + 4)$

83. $x^2(3x^2 - 5)$ **84.** $2y^2(y^2 + 5y)$

85. $-x^2y(y^2 - xy)$ **86.** $-3xy(xy - x)$

Find each product.

87. $(x + 5)(x + 4)$ **88.** $(2x + 1)(x - 1)$

89. $(3a - 3)(2a + 2)$ **90.** $6(a - 1)(a + 1)$

91. $(a - b)(2a + b)$ **92.** $(3x - y)(2x + y)$

Find each product.

93. $(x + 6)(x + 6)$ **94.** $(x + 5)(x - 5)$

95. $(y - 7)(y + 7)$ **96.** $(x + 4)^2$

97. $(x - 3)^2$ **98.** $(y - 2)^2$

99. $(3y + 2)^2$ **100.** $(y^2 + 1)(y^2 - 1)$

Find each product.

101. $(3x + 1)(x^2 + 2x + 1)$

102. $(2a - 3)(4a^2 + 6a + 9)$

Solve each equation.

103. $x^2 + 3 = x(x + 3)$

104. $x^2 + x = (x + 1)(x + 2)$

105. $(x + 2)(x - 5) = (x - 4)(x - 1)$

106. $(x - 1)(x - 2) = (x - 3)(x + 1)$

107. $x^2 + x(x + 2) = x(2x + 1) + 1$

108. $(x + 5)(3x + 1) = x^2 + (2x - 1)(x - 5)$

SECTION 4.7 Dividing Polynomials by Monomials

DEFINITIONS AND CONCEPTS	EXAMPLES
To divide a polynomial by a monomial, divide each term in the numerator by the monomial in the denominator.	Divide: $\dfrac{12x^6 - 8x^4 + 2x}{2x}$ Assume no division by 0. $\dfrac{12x^6 - 8x^4 + 2x}{2x}$ $= \dfrac{12x^6}{2x} - \dfrac{8x^4}{2x} + \dfrac{2x}{2x}$ Divide each term in the numerator by the monomial in the denominator. $= 6x^5 - 4x^3 + 1$

REVIEW EXERCISES

Perform each division. Assume no variable is 0.

109. $\dfrac{3x + 6y}{2xy}$

110. $\dfrac{21x^2y^2 - 7xy}{7xy}$

111. $\dfrac{15a^2bc + 20ab^2c - 25abc^2}{-5abc}$

112. $\dfrac{(x + y)^2 + (x - y)^2}{-2xy}$

SECTION 4.8 Dividing Polynomials by Polynomials

DEFINITIONS AND CONCEPTS	EXAMPLES
Use long division to divide one polynomial by another. Answers are written in $quotient + \frac{remainder}{divisor}$ form.	Divide: $\dfrac{6x^2 - 3x + 5}{x - 3}$ Assume no division by 0. $$\begin{array}{r} 6x + 15 \\ x - 3\overline{)6x^2 - 3x + 5} \\ \underline{-6x^2 + 18x} \\ 15x + 5 \\ \underline{-15x + 45} \\ 50 \end{array}$$ The result is $6x + 15 + \dfrac{50}{x - 3}$.

REVIEW EXERCISES

Perform each division. Assume no division by 0.

113. $x + 2\overline{)x^2 + 3x + 5}$

114. $x - 1\overline{)x^2 - 6x + 5}$

115. $x - 4\overline{)3x^2 - 11x - 4}$

116. $3x - 1\overline{)3x^2 + 14x - 2}$

117. $2x - 1\overline{)6x^3 + x^2 + 1}$

118. $3x + 1\overline{)-13x - 4 + 9x^3}$

SECTION 4.9 Synthetic Division

DEFINITIONS AND CONCEPTS	EXAMPLES
Dividing a polynomial by a binomial of the form $(x - r)$ using synthetic division: Write the coefficients of the polynomial in the dividend and r in the divisor. Use multiplication and addition to complete the division.	Use synthetic division to perform the division $(5x^2 - 8x - 2) \div (x - 2)$ $\begin{array}{r\|rrr} 2 & 5 & -8 & -2 \\ & & 10 & 4 \\ \hline & \mathbf{5} & \mathbf{2} & \mathbf{2} \end{array}$ Thus, $(5x^2 - 8x - 2) \div (x - 2) = \mathbf{5x + 2} + \frac{\mathbf{2}}{x - 2}$.
Remainder theorem: If a polynomial $P(x)$ is divided by $(x - r)$, then the remainder is $P(r)$.	Given $P(x) = 8x^3 - 2x^2 - 9$, find the remainder when $P(x)$ is divided by $x + 2$. $\begin{array}{r\|rrrr} -2 & 8 & -2 & 0 & -9 \\ & & -16 & 36 & -72 \\ \hline & 8 & -18 & 36 & \mathbf{-81} \end{array}$ Insert a 0 for the missing term. The remainder is -81. Thus, $P(-2) = -81$.
Factor theorem: If $P(x)$ is divided by $(x - r)$, then $P(r) = 0$, if and only if $(x - r)$ is a factor of $P(x)$.	Determine whether $(x + 4)$ is a factor of $(x^4 + 4x^3 + 9x^2 + 37x + 4)$. $\begin{array}{r\|rrrrr} -4 & 1 & 4 & 9 & 37 & 4 \\ & & -4 & 0 & -36 & -4 \\ \hline & 1 & 0 & 9 & 1 & 0 \end{array}$ Since the remainder is 0, $(x + 4)$ is a factor of $(x^4 + 4x^3 + 9x^2 + 37x + 4)$.

REVIEW EXERCISES

Use the factor theorem and synthetic division to determine whether the first expression is a factor of P(x). Factor, if possible.

119. $x - 3$; $P(x) = x^3 - 7x^2 + 9x + 9$

120. $x + 5$; $P(x) = x^3 + 4x^2 - 5x + 5$ (*Hint*: Write $(x + 5)$ as $x - (-5)$.)

4 Test

1. Use exponents to rewrite $2xxxyyyy$.

2. Evaluate: $3^2 + 5^3$

Write each expression as an expression containing only one exponent.

3. $y^3(y^5y)$

4. $(-3b^2)(2b^3)(-b^2)$

5. $(2x^3)^5(x^2)^3$

6. $(2rr^2r^3)^3$

Simplify each expression. Write answers without using parentheses or negative exponents. Assume no variable is 0.

7. $-7x^0$

8. $5y^{-6}y^3$

9. $\dfrac{y^2}{yy^{-2}}$

10. $\left(\dfrac{a^2b^{-1}}{4a^3b^{-2}}\right)^{-3}$

11. Write 540,000 in scientific notation.

12. Write 0.0025 in scientific notation.

13. Write 7.4×10^3 in standard notation.

14. Write 6.7×10^{-4} in standard notation.

15. Classify $3x^2 + 2$ as a monomial, a binomial, or a trinomial.

16. Find the degree of the polynomial $3x^2y^3z^4 + 2x^3y^2z - 5x^2y^3z^5$.

17. Evaluate $x^2 + x - 2$ when $x = -2$.

Unless otherwise noted, all content on this page is © Cengage Learning.

18. Graph the polynomial function $f(x) = x^2 + 2$. State the domain and range.

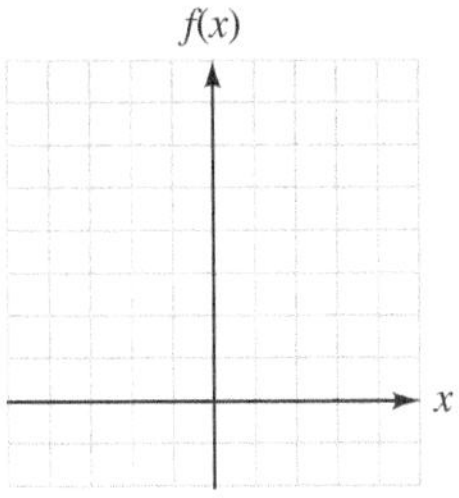

19. Simplify: $-6(x - y) + 2(x + y) - 3(x + 2y)$

20. Simplify: $-2(x^2 + 3x - 1) - 3(x^2 - x + 2) + 5(x^2 + 2)$

21.
$$\begin{array}{r} 3x^3 + 4x^2 - x - 7 \\ + \underline{2x^3 - 2x^2 + 3x + 2} \end{array}$$

22.
$$\begin{array}{r} 2x^2 - 7x + 3 \\ - \underline{3x^2 - 2x - 1} \end{array}$$

Find each product.

23. $(-2x^3)(2x^2y)$

24. $(-5x^4)(4x^5y)$

25. $(2x - 5)(3x + 4)$

26. $(2x - 3)(x^2 - 2x + 4)$

Simplify each expression. Assume no division by 0.

27. Simplify: $\dfrac{8x^2y^3z^4}{16x^3y^2z^4}$

28. Simplify: $\dfrac{6a^2 - 12b^2}{24ab}$

29. Divide: $2x + 3\overline{)2x^2 - x - 6}$

30. Solve: $(a + 2)^2 = (a - 3)^2$

31. Find the remainder: $\dfrac{x^3 - 4x^2 + 5x + 3}{x + 1}$

32. Use synthetic division to find the remainder when $(4x^3 + 3x^2 + 2x - 1)$ is divided by $(x - 2)$.

Cumulative Review

Evaluate each expression. Let $x = 2$ and $y = -5$.

1. $4 + 5x$

2. $3y^2 - 4$

3. $\dfrac{3x - y}{xy}$

4. $\dfrac{x^2 - y^2}{x + y}$

Solve each equation.

5. $\dfrac{4}{5}x + 6 = 18$

6. $x - 2 = \dfrac{x + 2}{3}$

7. $2(5x + 2) = 3(3x - 2)$

8. $4(y + 1) = -2(4 - y)$

Graph the solution of each inequality.

9. $3x - 4 > 2$

10. $7x - 9 < 5$

11. $-2 < -x + 3 < 5$

12. $0 \leq \dfrac{4 - x}{3} \leq 2$

Solve each formula for the indicated variable.

13. $A = p + prt$ for r

14. $A = \dfrac{1}{2}bh$ for h

Graph each equation.

15. $3x - 4y = 12$

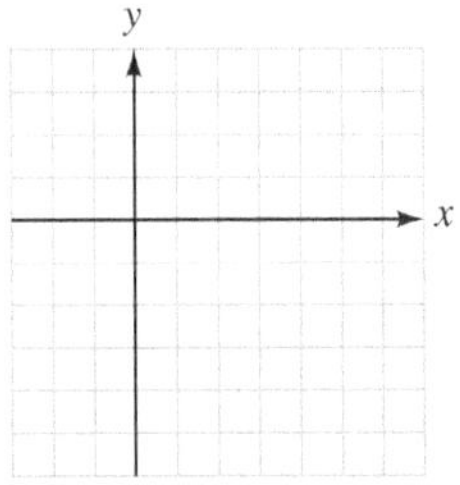

16. $y - 2 = \dfrac{1}{2}(x - 4)$

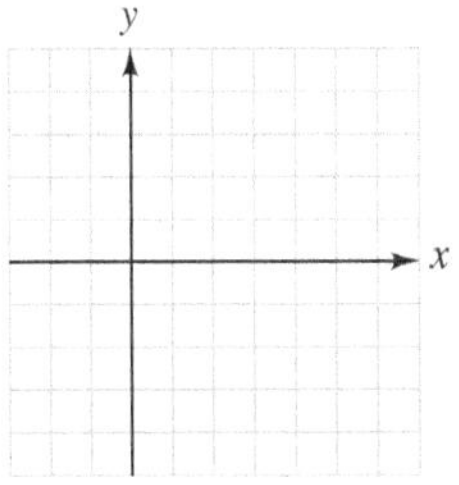

Let $f(x) = 5x - 2$ and find each value.

17. $f(4)$

18. $f(-1)$

19. $f(-2)$

20. $f\left(\dfrac{1}{5}\right)$

Write each expression as an expression using only one exponent. Assume no division by 0.

21. $y^4(y^2y^8)$

22. $\dfrac{x^5y^3}{x^4y^4}$

23. $\dfrac{a^4b^{-3}}{a^{-3}b^3}$

24. $\left(\dfrac{-x^{-2}y^3}{x^{-3}y^2}\right)^2$

Unless otherwise noted, all content on this page is © Cengage Learning.

Perform each operation. Assume no division by 0.

25. $(3x^2 + 2x - 7) - (2x^2 - 2x + 7)$

26. $(4x - 5)(3x + 2)$

27. $(x - 2)(x^2 + 2x + 4)$

28. $(2x^2 - 5x - 3) \div (x - 3)(x \neq 3)$

29. Astronomy The parsec, a unit of distance used in astronomy, is 3×10^{16} meters. The distance from Earth to Betelgeuse, a star in the constellation Orion, is 1.6×10^2 parsecs. Use scientific notation to express this distance in meters.

30. Surface area The total surface area A of a box with dimensions l, w, and d is given by the formula

$$A = 2lw + 2wd + 2ld$$

If $A = 202$, $l = 9$, and $w = 5$, find d.

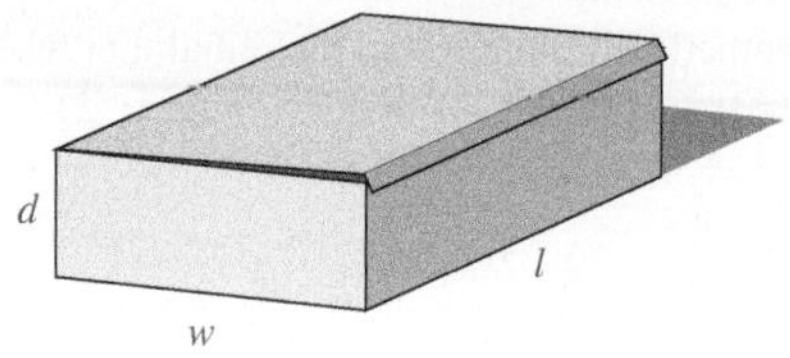

31. Concentric circles The area of the ring between the two concentric circles of radius r and R is given by the formula

$$A = \pi(R + r)(R - r)$$

If $r = 3$ and $R = 17$, find A to the nearest tenth.

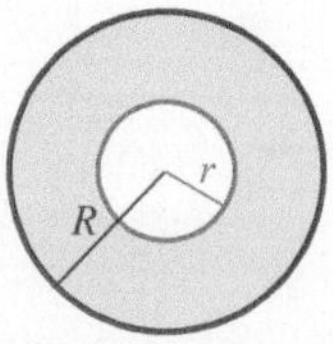

32. Employee discounts Employees at an appliance store can purchase merchandise at 25% less than the regular price. An employee buys a TV set for \$414.72, including 8% sales tax. Find the regular price of the TV.

Unless otherwise noted, all content on this page is © Cengage Learning.

Factoring Polynomials and Solving Equations by Factoring

5

Sergey Nivens/Shutterstock.com

Careers and Mathematics

ATHLETES, COACHES, AND SCOUTS

We are a nation of sports fans and sports players. Some of those who participate in amateur sports dream of becoming paid professional athletes, coaches, or sports officials, but very few beat the long odds of making a full-time living from professional sports. Nearly 42% of athletes and coaches work part time. Education and training for coaches and athletes vary greatly by the level and type of sport.

Job Outlook:
People who are state-certified to teach academic subjects in addition to physical education will have the best prospects for obtaining coaching and instructor jobs. Employment of athletes, coaches, and related workers is expected to increase faster than average for all occupations through 2020.
Annual Earnings:
Athletes: $28,340
Coaches and scouts: $26,950
For More Information:
http://www.bls.gov/oco/ocos251.htm
For a Sample Application:
See Problem 21 in Section 5.8.

REACH FOR SUCCESS
- **5.1** Factoring Out the Greatest Common Factor; Factoring by Grouping
- **5.2** Factoring the Difference of Two Squares
- **5.3** Factoring Trinomials with a Leading Coefficient of 1
- **5.4** Factoring Trinomials with a Leading Coefficient Other Than 1
- **5.5** Factoring the Sum and Difference of Two Cubes
- **5.6** Summary of Factoring Techniques
- **5.7** Solving Equations by Factoring
- **5.8** Solving Applications
- ■ ***Projects***

REACH FOR SUCCESS EXTENSION
CHAPTER REVIEW
CHAPTER TEST

In this chapter

In this chapter, we will reverse the operation of multiplying polynomials and show which polynomials were used to find a given product. We will use this skill to solve many equations and applications.

Reach for Success Organizing Your Time

© iStockphoto.com/Ryan Balderas

Life during your college years should be fun but it also requires time-management skills if you want to be academically successful. Although you previously examined a typical 24-hour day and a week, looking at a longer period will help you identify potential time conflicts. The popularity of electronic calendars has increased significantly. Regardless of the format, it is important to find a calendar that works for you.

Populate a two-week calendar with all the graded assignments from your syllabus in each of your classes.

Begin by filling in the dates in the chart. Then

1. Record all exams in the boxes.
2. Include due dates for any papers.
3. Add any labs/homework assignments.
4. Record employment hours including any scheduled overtime.

Sun _____	Mon _____	Tue _____	Wed _____	Thu _____	Fri _____	Sat _____
Sun _____	Mon _____	Tue _____	Wed _____	Thu _____	Fri _____	Sat _____

Do you know of any other time requirements for this month? If so, record them in your chart. Be sure to include any personal appointments and life events (e.g., birthdays, weddings).

Your calendar might look like this:

Sun 1	Mon 2 English outline due	Tue 3 Dentist 1 p.m. Work	Wed 4 Work	Thu 5 Mathematics test	Fri 6 RJ's school play	Sat 7 Volunteer work Wedding anniversary
Sun 8 Sociology group project meeting 2 p.m.	Mon 9 Work Kenley's Dr. appt. 2:30 p.m.	Tue 10 Turn in history paper Math lab due	Wed 11 Read history chapters 13–15	Thu 12 Sociology group project due Work	Fri 13 English paper due Fred's birthday	Sat 14 Work

A Successful Study Strategy . . .

Use a planner to schedule your entire semester. By seeing even a week at a time, you will be able to avoid staying up all night to finish a paper or prepare for an exam. All-nighters rarely benefit anyone!

At the end of the chapter you will find an additional exercise to guide you in planning for a successful semester.

Section 5.1 Factoring Out the Greatest Common Factor; Factoring by Grouping

Objectives

1. Identify the greatest common factor of two or more monomials.
2. Factor a polynomial containing a greatest common factor.
3. Factor a polynomial containing a negative greatest common factor.
4. Factor a polynomial containing a binomial greatest common factor.
5. Factor a four-term polynomial using grouping.

Vocabulary

fundamental theorem of arithmetic | greatest common factor (GCF) | factoring by grouping

Getting Ready

Simplify each expression.

1. $5(x + 3)$
2. $7(y - 8)$
3. $x(3x - 2)$
4. $y(5y + 9)$
5. $3(x + y) + a(x + y)$
6. $x(y + 1) + 5(y + 1)$
7. $5(x + 1) - y(x + 1)$
8. $x(x + 2) - y(x + 2)$

In this chapter, we will reverse the operation of multiplication and show how to find the factors of a known product. The process of finding the individual factors of a product is called *factoring*. We will limit our discussion of factoring polynomials to those that factor using only rational numbers.

1 Identify the greatest common factor of two or more monomials.

Recall that a natural number greater than 1 whose only factors are 1 and the number itself is called a prime number.

The prime numbers less than 50 are

$$2, 3, 5, 7, 11, 13, 17, 19, 23, 29, 31, 37, 41, 43, \text{ and } 47$$

A natural number is said to be in prime-factored form if it is written as the product of factors that are prime numbers.

To find the prime-factored form of a natural number, we can use a factoring tree. For example, to find the prime-factored form of 60, we proceed as follows:

Solution 1

1. Start with 60.
2. Factor 60 as $6 \cdot 10$.
3. Factor 6 and 10.

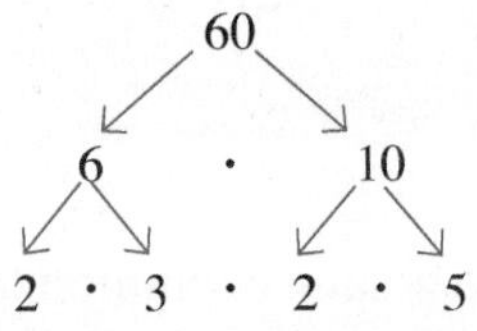

Solution 2

1. Start with 60.
2. Factor 60 as $4 \cdot 15$.
3. Factor 4 and 15.

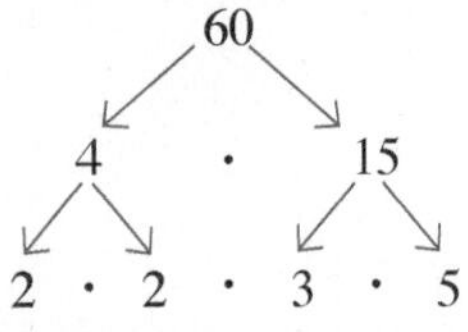

We stop when only prime numbers appear. In either case, the prime factorization of 60 is $2 \cdot 2 \cdot 3 \cdot 5$. Thus, the prime-factored form of 60 is $2^2 \cdot 3 \cdot 5$. This illustrates the **fundamental theorem of arithmetic**, which states that there is exactly one prime factorization for any natural number greater than 1.

COMMENT We found the GCF, 6, by using the prime factorizations of 42, 60, and 90.

The largest natural number that divides each group of natural numbers is called their **greatest common factor (GCF)**. The GCF of 42, 60, and 90 is 6, because 6 is the largest natural number that divides each of these numbers:

$$\frac{42}{6} = 7 \qquad \frac{60}{6} = 10 \qquad \text{and} \qquad \frac{90}{6} = 15$$

$$42 = 2 \cdot 3 \cdot 7$$
$$60 = 2 \cdot 2 \cdot 3 \cdot 5$$
$$90 = 2 \cdot 3 \cdot 3 \cdot 5$$

Algebraic monomials also can have a greatest common factor. The right sides of the equations show the prime factorizations of $6a^2b^3$, $4a^3b^2$, and $18a^2b$.

$$6a^2b^3 = 2 \cdot 3 \cdot a \cdot a \cdot b \cdot b \cdot b$$
$$4a^3b^2 = 2 \cdot 2 \cdot a \cdot a \cdot a \cdot b \cdot b$$
$$18a^2b = 2 \cdot 3 \cdot 3 \cdot a \cdot a \cdot b$$

Since all three of these monomials have one factor of 2, two factors of a, and one factor of b, the GCF is

$$2 \cdot a \cdot a \cdot b \qquad \text{or} \qquad 2a^2b$$

Perspective

Much of the mathematics that we have inherited from earlier times is the result of teamwork. In a battle early in the 12th century, control of the Spanish city of Toledo was taken from the Mohammedans, who had ruled there for four centuries. Libraries in this great city contained many books written in Arabic, full of knowledge that was unknown in Europe.

The Archbishop of Toledo wanted to share this knowledge with the rest of the world. He knew that these books should be translated into Latin, the universal language of scholarship. But what European scholar could read Arabic? The citizens of Toledo knew both Arabic and Spanish, and most scholars of Europe could read Spanish.

Teamwork saved the day. A citizen of Toledo read the Arabic text aloud, in Spanish. The scholars listened to the Spanish version and wrote it down in Latin. One of these scholars was an Englishman, Robert of Chester. It was he who translated al-Khowarazmi's book, *Ihm al-jabr wa'l muqabalah*, the beginning of the subject we now know as algebra.

EXAMPLE 1 Find the GCF of $10x^3y^2$, $60x^2y$, and $30xy^2$.

Solution Find the prime factorization of each of the three monomials.

$$10x^3y^2 = 2 \cdot 5 \cdot x \cdot x \cdot x \cdot y \cdot y$$
$$60x^2y = 2 \cdot 2 \cdot 3 \cdot 5 \cdot x \cdot x \cdot y$$
$$30xy^2 = 2 \cdot 3 \cdot 5 \cdot x \cdot y \cdot y$$

List each common factor the least number of times it appears in any one monomial: 2, 5, x, and y. Find the product of the factors in the list:

$$2 \cdot 5 \cdot x \cdot y = 10xy$$

SELF CHECK 1 Find the GCF of $20a^2b^3$, $12ab^4$, and $8a^3b^2$.

The process of factoring will be used to convert an addition or subtraction problem to a multiplication problem. This technique will then be used to solve equations later in this chapter and to simplify rational expressions in the next chapter.

2 Factor a polynomial containing a greatest common factor.

Recall that the distributive property provides a way to multiply a polynomial by a monomial. For example,

$$3x^2(2x - 3y) = 3x^2 \cdot 2x + 3x^2 \cdot (-3y)$$
$$= 6x^3 - 9x^2y$$

To reverse this process and factor the product $6x^3 - 9x^2y$, we can find the GCF of each term (which is $3x^2$) and then use the distributive property.

$$6x^3 - 9x^2y = 3x^2 \cdot 2x + 3x^2 \cdot (-3y)$$
$$= 3x^2(2x - 3y)$$

This process is called factoring out the greatest common factor.

FINDING THE GREATEST COMMON FACTOR (GCF)

1. Identify the number of terms.
2. Find the prime factorization of each term.
3. List each common factor the least number of times it appears in any one term.
4. Find the product of the factors found in the list to obtain the GCF.

EXAMPLE 2 Factor: $12y^2 + 20y$

Solution To find the GCF, we find the prime factorization of $12y^2$ and $20y$.

$$\left.\begin{aligned} 12y^2 &= 2 \cdot 2 \cdot 3 \cdot y \cdot y \\ 20y &= 2 \cdot 2 \cdot 5 \cdot y \end{aligned}\right\} \quad \text{GCF} = 4y$$

We can use the distributive property to factor out the GCF of $4y$.

$$12y^2 + 20y = 4y \cdot 3y + 4y \cdot 5$$
$$= 4y(3y + 5)$$

Check by verifying that $4y(3y + 5) = 12y^2 + 20y$.

SELF CHECK 2 Factor: $15x^3 - 20x^2$

EXAMPLE 3 Factor: $35a^3b^2 - 14a^2b^3$

Solution To find the GCF, we find the prime factorization of $35a^3b^2$ and $-14a^2b^3$.

$$\left.\begin{aligned} 35a^3b^2 &= 5 \cdot \mathbf{7 \cdot a \cdot a} \cdot a \cdot \mathbf{b \cdot b} \\ -14a^2b^3 &= -2 \cdot \mathbf{7 \cdot a \cdot a \cdot b \cdot b} \cdot b \end{aligned}\right\} \quad \text{GCF} = 7a^2b^2$$

We factor out the GCF of $7a^2b^2$.

$$\begin{aligned} 35a^3b^2 - 14a^2b^3 &= \mathbf{7a^2b^2} \cdot 5a + \mathbf{7a^2b^2} \cdot (-2b) \\ &= \mathbf{7a^2b^2}(5a - 2b) \end{aligned}$$

Check: $7a^2b^2(5a - 2b) = 35a^3b^2 - 14a^2b^3$

SELF CHECK 3 Factor: $40x^2y^3 + 15x^3y^2$

EXAMPLE 4 Factor: $a^2b^2 - ab$

Solution We factor out the GCF, which is ab.

$$\begin{aligned} a^2b^2 - ab &= \mathbf{ab} \cdot ab + \mathbf{ab}(-1) \\ &= \mathbf{ab}(ab - 1) \end{aligned}$$

Check: $ab(ab - 1) = a^2b^2 - ab$

COMMENT The last term of $a^2b^2 - ab$ has an implied coefficient of -1. When ab is factored out, we must write the coefficient of -1.

SELF CHECK 4 Factor: $x^3y^5 - x^2y^3$

EXAMPLE 5 Factor: $12x^3y^2z + 6x^2yz - 3xz$

Solution We factor out the GCF, which is $3xz$.

$$\begin{aligned} 12x^3y^2z + 6x^2yz - 3xz &= \mathbf{3xz} \cdot 4x^2y^2 + \mathbf{3xz} \cdot 2xy + \mathbf{3xz}(-1) \\ &= \mathbf{3xz}(4x^2y^2 + 2xy - 1) \end{aligned}$$

Check: $3xz(4x^2y^2 + 2xy - 1) = 12x^3y^2z + 6x^2yz - 3xz$

SELF CHECK 5 Factor: $6ab^2c - 12a^2bc + 3ab$

3 Factor a polynomial containing a negative greatest common factor.

It is often useful to factor -1 out of a polynomial, especially if the leading coefficient is negative.

EXAMPLE 6 Factor -1 out of $-a^3 + 2a^2 - 4$.

Solution $-a^3 + 2a^2 - 4$

$= (\mathbf{-1})a^3 + (\mathbf{-1})(-2a^2) + (\mathbf{-1})4$ Write each term with a factor of -1.

$= \mathbf{-1}(a^3 - 2a^2 + 4)$ Factor out the GCF, -1.

$= -(a^3 - 2a^2 + 4)$

Check: $-(a^3 - 2a^2 + 4) = -a^3 + 2a^2 - 4$

SELF CHECK 6 Factor -1 out of $-b^4 - 3b^2 + 2$.

EXAMPLE 7 Factor out the negative of the GCF: $-18a^2b + 6ab^2 - 12a^2b^2$

Solution The GCF is $6ab$. To factor out its negative, we factor out $-6ab$.

$$\begin{aligned}-18a^2b + 6ab^2 - 12a^2b^2 &= (\mathbf{-6ab})3a + (\mathbf{-6ab})(-b) + (\mathbf{-6ab})2ab \\ &= \mathbf{-6ab}(3a - b + 2ab)\end{aligned}$$

Check: $-6ab(3a - b + 2ab) = -18a^2b + 6ab^2 - 12a^2b^2$

SELF CHECK 7 Factor out the negative of the GCF: $-25xy^2 - 15x^2y + 30x^2y^2$

Sometimes a polynomial does not contain a GCF that is a monomial, but rather a polynomial. The next objective will discuss greatest common factors that are binomials or a product of a monomial and a binomial.

4 Factor a polynomial containing a binomial greatest common factor.

If the GCF of several terms is a polynomial, we can factor out the common polynomial factor. For example, since $(a + b)$ is a common factor of $(a + b)x$ and $(a + b)y$, we can factor out the $(a + b)$.

$$(\mathbf{a + b})x + (\mathbf{a + b})y = (\mathbf{a + b})(x + y)$$

We can check by verifying that $(a + b)(x + y) = (a + b)x + (a + b)y$.

EXAMPLE 8 Factor $(a + 3)$ out of $(a + 3) + (a + 3)^2$.

Solution Recall that $a + 3$ is equal to $(a + 3)^1$ and that $(a + 3)^2$ is equal to $(a + 3)(a + 3)$. We can factor out $(a + 3)$ and simplify.

$$\begin{aligned}(a + 3) + (a + 3)^2 &= (\mathbf{a + 3})\cdot 1 + (\mathbf{a + 3})\cdot(a + 3) \\ &= (\mathbf{a + 3})[1 + (a + 3)] && \text{Factor out } a + 3\text{, the GCF.} \\ &= (a + 3)(a + 4) && \text{Combine like terms.}\end{aligned}$$

SELF CHECK 8 Factor out $(y + 2)$: $(y + 2)^2 - 3(y + 2)$

EXAMPLE 9 Factor: $6a^2b^2(x + 2y) - 9ab(x + 2y)$

Solution The GCF of $6a^2b^2$ and $9ab$ is $3ab$. We can factor out this GCF as well as $(x + 2y)$.

$$\begin{aligned}&6a^2b^2(x + 2y) - 9ab(x + 2y) \\ &= \mathbf{3ab}\cdot 2ab(\mathbf{x + 2y}) - \mathbf{3ab}\cdot 3(\mathbf{x + 2y}) \\ &= \mathbf{3ab(x + 2y)}(2ab - 3) && \text{Factor out } 3ab(x + 2y)\text{, the GCF.}\end{aligned}$$

SELF CHECK 9 Factor: $4p^3q^2(2a + b) + 8p^2q^3(2a + b)$

5 Factor a four-term polynomial using grouping.

Suppose we want to factor

$$ax + ay + cx + cy$$

Although no factor is common to all four terms, there is a common factor of a in $ax + ay$ and a common factor of c in $cx + cy$. In this case, we group the first two terms and group the last two terms. We can factor out the a from the first two terms and the c from the last two terms to obtain

$$ax + ay + cx + cy = a(x + y) + c(x + y)$$
$$= (x + y)(a + c) \quad \text{Factor out } (x + y).$$

We can check the result by multiplication.

$$(x + y)(a + c) = ax + cx + ay + cy$$
$$= ax + ay + cx + cy$$

Thus, $ax + ay + cx + cy$ factors as $(x + y)(a + c)$. This type of factoring is called **factoring by grouping**.

EXAMPLE 10 Factor: $2c + 2d - cd - d^2$

Solution $2c + 2d - cd - d^2 = 2(c + d) - d(c + d)$ Factor out 2 from $(2c + 2d)$ and $-d$ from $(-cd - d^2)$.

$= (c + d)(2 - d)$ Factor out $(c + d)$.

Check: $(c + d)(2 - d) = 2c - cd + 2d - d^2$
$= 2c + 2d - cd - d^2$

SELF CHECK 10 Factor: $3a + 3b - ac - bc$

EXAMPLE 11 Factor: $5x^3 - 10x^2 + 4x - 8$

Solution $5x^3 - 10x^2 + 4x - 8 = 5x^2(x - 2) + 4(x - 2)$ Factor out $5x^2$ from $(5x^3 - 10x^2)$ and 4 from $(4x - 8)$.

$= (x - 2)(5x^2 + 4)$ Factor out $(x - 2)$.

SELF CHECK 11 Factor: $6y^3 + 3y^2 - 14y - 7$

COMMENT When factoring expressions, the final result must be a product. Expressions such as $2(c + d) - d(c + d)$ and $x(xy - a) - 1(xy - a)$ are *not* in factored form.

EXAMPLE 12 Factor: **a.** $a(c - d) + b(d - c)$ **b.** $ac + bd - ad - bc$

Solution **a.** $a(c - d) + b(d - c) = a(c - d) - b(-d + c)$ Factor -1 from $(d - c)$.

$= a(c - d) - b(c - d)$ $\quad -d + c = c - d$

$= (c - d)(a - b)$ Factor out $(c - d)$.

b. In this example, we cannot factor anything from the first two terms or the last two terms. However, if we rearrange the terms, we can factor by grouping.

$ac + bd - ad - bc = ac - ad + bd - bc$ $\quad bd - ad = -ad + bd$

$= a(c - d) + b(d - c)$ Factor a from $(ac - ad)$ and b from $(bd - bc)$.

$= a(c - d) - b(c - d)$ $\quad d - c = -1(c - d)$

$= (c - d)(a - b)$ Factor out $(c - d)$.

SELF CHECK 12 Factor: **a.** $2(x - y) - z(y - x)$
b. $ax - by - ay + bx$

COMMENT In Example 12(b), we also could have factored the polynomial if we had rearranged the terms as $ac - bc - ad + bd$.

SELF CHECK ANSWERS

1. $4ab^2$ **2.** $5x^2(3x - 4)$ **3.** $5x^2y^2(8y + 3x)$ **4.** $x^2y^3(xy^2 - 1)$ **5.** $3ab(2bc - 4ac + 1)$ **6.** $-(b^4 + 3b^2 - 2)$ **7.** $-5xy(5y + 3x - 6xy)$ **8.** $(y + 2)(y - 1)$ **9.** $4p^2q^2(2a + b)(p + 2q)$ **10.** $(a + b)(3 - c)$ **11.** $(2y + 1)(3y^2 - 7)$ **12. a.** $(x - y)(2 - z)$ **b.** $(a + b)(x - y)$

NOW TRY THIS

Factor.

1. $(x + y)(x^2 - 3) + (x + y)$
2. **a.** $x^{2n} + x^n$
   b. $x^3 + x^{-1}$
3. Which of the following is equivalent to $\frac{3 - x}{x + 2}$? There may be more than one answer.
   a. $\frac{-(x - 3)}{x + 2}$ **b.** $\frac{x - 3}{x + 2}$ **c.** $\frac{-x + 3}{x + 2}$ **d.** $-\frac{x - 3}{x + 2}$

5.1 Exercises

WARM-UPS *Find the greatest common factor.*

1. 3, 6, and 9
2. 2, 6, 10
3. 4, 16, 32
4. 5, 15, 20
5. 4, 6, 10
6. 6, 12, 21
7. 12, 18, 24
8. 30, 45, 60

REVIEW *Solve each equation and check all solutions.*

9. $3x - 2(x + 1) = 5$
10. $5(y - 1) + 1 = y$
11. $\frac{2x - 7}{5} = 3$
12. $2x - \frac{x}{2} = 5x$

VOCABULARY AND CONCEPTS *Fill in the blanks.*

13. If a natural number is written as the product of prime numbers, it is written in ____________ form.
14. The ____________________________ states that each natural number greater than 1 has exactly one prime factorization.
15. The GCF of several natural numbers is the ______ number that divides each of the numbers.
16. To find the prime factorization of a natural number, you can use a ___________.
17. To factor a four-term polynomial, it is often necessary to factor by _______.
18. Check the results of a factoring problem by ___________.

Find the prime factorization of each number.

19. 12
20. 24
21. 40
22. 98
23. 225
24. 144
25. 288
26. 968

GUIDED PRACTICE *Find the GCF of the given monomials.* SEE EXAMPLE 1. (OBJECTIVE 1)

27. $3x, 6xy$
28. $5xy^2, 10xy$
29. $5x^2, 10x$
30. $8y^3, 12y^2x$
31. $4ab, 18b$
32. $7a^2b, 14ab^2$
33. $6x^2y^2, 12xyz, 18xy^2z^3$
34. $4a^3b^2c, 12ab^2c^2, 20ab^2c^2$

Complete each factorization. SEE EXAMPLE 2. (OBJECTIVE 2)

35. $9a + 15 = 3(\quad + 5)$
36. $3t - 27 = 3(t - \quad)$
37. $4a + 12 = \quad(a + 3)$
38. $5b - 15 = \quad(b - 3)$
39. $8x + 12 = \quad(2x + 3)$
40. $6x - 15 = \quad(2x - 5)$
41. $4y^2 + 8y - 2xy = 2y(2y + \quad - \quad)$
42. $3b^2 - 9b - 6ab = 3b(\quad - \quad - 2a)$
43. $3x^2 - 6xy + 9xy^2 = \quad(\quad - 2y + 3y^2)$
44. $10x^3 + 8xy - 4xy^2 = 2x(\quad + 4y - \quad)$
45. $r^4 + r^2 = r^2(\quad + 1)$
46. $a^3 - a^2 = \quad(a - 1)$

Factor each polynomial by factoring out the GCF. SEE EXAMPLE 2. (OBJECTIVE 2)

47. $3x + 6$
48. $2y - 10$
49. $4x - 8$
50. $4t + 12$
51. $6x^2 - 9x$
52. $15a^2 + 3a$
53. $4b^3 - 10b^2$
54. $8y^3 - 12y^2$

Factor each polynomial by factoring out the GCF. SEE EXAMPLE 3. (OBJECTIVE 2)

55. $t^3 + 2t^2$ **56.** $b^3 - 3b^2$

57. $10x^2y^3 + 15xy^4$ **58.** $16a^4b^2 - 24a^2b^3$

Factor each polynomial by factoring out the GCF. SEE EXAMPLE 4. (OBJECTIVE 2)

59. $a^3b^3z^3 - a^2b^3z^2$ **60.** $r^3s^6t^9 + r^2s^2t^2$

61. $24x^2y^3z^4 + 8xy^2z^3$ **62.** $3x^2y^3 - 9x^4y^3z$

Factor each polynomial by factoring out the GCF. SEE EXAMPLE 5. (OBJECTIVE 2)

63. $3x + 3y - 6z$ **64.** $2x - 4y + 8z$

65. $ab + ac - ad$ **66.** $4y^2 + 8y - 2xy$

67. $rs - rt + ru$ **68.** $3x^2 - 6xy + 9xy^2$

69. $a^2b^2x^2 + a^3b^2x^2 - a^3b^3x^3$

70. $4x^2y^2z^2 - 6xy^2z^2 + 12xyz^2$

Factor out −1 from each polynomial. SEE EXAMPLE 6. (OBJECTIVE 3)

71. $-x - 2$ **72.** $-y + 3$

73. $-a - b$ **74.** $-x - 2y$

75. $-2x + 5y$ **76.** $-3x + 8z$

77. $-3ab - 5ac + 9bc$ **78.** $-6yz + 12xz - 5xy$

Factor out the negative of the GCF. SEE EXAMPLE 7. (OBJECTIVE 3)

79. $-3x^2y - 6xy^2$

80. $-4a^2b^2 + 6ab^2$

81. $-4a^2b^2c^2 + 14a^2b^2c - 10ab^2c^2$

82. $-25x^4y^3z^2 + 30x^2y^3z^4$

Complete each factorization. (OBJECTIVE 4)

83. $a(x + y) + b(x + y) = (x + y)$____

84. $x(a + b) + p(a + b) =$ ____$(x + p)$

85. $p(m - n) - q(m - n) =$ ____$(p - q)$

86. $(r - s)p - (r - s)q = (r - s)$____

87. $3(r - 2s) - x(r - 2s) =$ ____$(3 - x)$

88. $x(a + 2b) + y(a + 2b) = (a + 2b)$____

89. $(x + 3)(x + 1) - y(x + 1) = (x + 1)$____

90. $x(x^2 + 2) - y(x^2 + 2) =$ ____$(x - y)$

Factor each expression. SEE EXAMPLE 8. (OBJECTIVE 4)

91. $x(y + 1) - 5(y + 1)$

92. $3(x + y) - a(x + y)$

93. $(3x - y)(x^2 - 2) + (x^2 - 2)$

94. $(x - 5y)(a + 2) - (x - 5y)$

95. $(x + y)^2 + b(x + y)$ **96.** $(a - b)c + (a - b)d$

97. $(x - 3)^2 + (x - 3)$

98. $(3t + 5)^2 - (3t + 5)$

Factor each expression. SEE EXAMPLE 9. (OBJECTIVE 4)

99. $5a(2a - 1) - 10b(2a - 1)$

100. $3x^2(r + 3s) - 6y^2(r + 3s)$

101. $3x(c - 3d) + 6y(c - 3d)$

102. $9a^2b^2(3x - 2y) - 6ab(3x - 2y)$

Factor each polynomial by grouping. SEE EXAMPLE 10. (OBJECTIVE 5)

103. $2x + 2y + ax + ay$ **104.** $bx + bz + 5x + 5z$

105. $9p - 9q + mp - mq$ **106.** $7r + 7s - kr - ks$

Factor each expression. SEE EXAMPLE 11. (OBJECTIVE 5)

107. $9x^3 + 3x^2 + 12x + 4$ **108.** $6y^3 - 12y^2 - 5y + 10$

109. $8a^3 - 2a^2 - 4a + 1$ **110.** $7b^3 + 14b^2 + 2b + 4$

Factor each expression. SEE EXAMPLE 12. (OBJECTIVE 5)

111. $ax + bx - a - b$ **112.** $mp - np - m + n$

113. $x(a - b) + y(b - a)$ **114.** $p(m - n) - q(n - m)$

ADDITIONAL PRACTICE *Factor each expression completely.*

115. $r^4 + r^2$ **116.** $a^3 + a^2$

117. $12uvw^3 - 18uv^2w^2$

118. $14xyz - 16x^2y^2z$

119. $-14a^6b^6 + 49a^2b^3 - 21ab$

120. $-5a^2b^3c + 15a^3b^4c^2 - 25a^4b^3c$

121. $3x(a + b + c) - 2y(a + b + c)$

122. $2m(a - 2b + 3c) - 21xy(a - 2b + 3c)$

123. $14x^2y(r + 2s - t) - 21xy(r + 2s - t)$

124. $5xy^3(2x - y + 3z) + 25xy^2(2x - y + 3z)$

125. $3tv - 9tw + uv - 3uw$

126. $ce - 2cf + 3de - 6df$

127. $-4abc - 4ac^2 + 2bc + 2c^2$

128. $2x^3z - 4x^2z + 32xz - 64z$

129. $ax^3 + bx^3 + 2ax^2y + 2bx^2y$

130. $4a^2b + 12a^2 - 8ab - 24a$

131. $y^3 - 3y^2 - 5y + 15$

132. $3ab + 9a - 2b - 6$

133. $2r - bs - 2s + br$

134. $xy + 7 + y + 7x$

135. $ar^2 - brs + ars - br^2$

136. $a^2bc + a^2c + abc + ac$

WRITING ABOUT MATH

137. When we add $5x$ and $7x$, we combine like terms: $5x + 7x = 12x$. Explain how this is related to factoring out a common factor.

138. Explain how you would factor $x(a - b) + y(b - a)$.

SOMETHING TO THINK ABOUT

139. Think of two positive integers. Divide their product by their greatest common factor. Why do you think the result is called the least common multiple of the two integers? (*Hint*: The multiples of an integer such as 5 are 5, 10, 15, 20, 25, 30, and so on.)

140. Two integers are called *relatively prime* if their greatest common factor is 1. For example, 6 and 25 are relatively prime, but 6 and 15 are not. If the greatest common factor of three integers is 1, must any two of them be relatively prime? Explain.

Section 5.2 Factoring the Difference of Two Squares

Objectives

1. Factor the difference of two squares.
2. Completely factor a polynomial.

Vocabulary

difference of two squares | sum of two squares | prime polynomial

Getting Ready

Multiply the binomials.

1. $(a + b)(a - b)$ **2.** $(2r + s)(2r - s)$

3. $(3x + 2y)(3x - 2y)$ **4.** $(4x^2 + 3)(4x^2 - 3)$

Whenever we multiply binomial conjugates, binomials of the form $(x + y)$ and $(x - y)$, we obtain a binomial of the form $x^2 - y^2$.

$$(x + y)(x - y) = x^2 - xy + xy - y^2$$
$$= x^2 - y^2$$

In this section, we will show how to reverse the multiplication process and factor binomials such as $x^2 - y^2$ into binomial conjugates.

1 Factor the difference of two squares.

The binomial $x^2 - y^2$ is called the **difference of two squares**, because x^2 is the square of x and y^2 is the square of y. The difference of the squares of two quantities always factors into binomial conjugates.

FACTORING THE DIFFERENCE OF TWO SQUARES

$$x^2 - y^2 = (x + y)(x - y)$$

COMMENT The factorization of $x^2 - y^2$ also can be expressed as $(x - y)(x + y)$.

To factor $x^2 - 9$, we note that it can be written in the form $x^2 - 3^2$.

$$x^2 - 3^2 = (x + 3)(x - 3)$$

We can check by verifying that $(x + 3)(x - 3) = x^2 - 9$.

To factor the difference of two squares, it is helpful to know the integers that are perfect squares. The number 400, for example, is a perfect square, because $20^2 = 400$. The integer squares less than 400 are

1, 4, 9, 16, 25, 36, 49, 64, 81, 100, 121, 144, 169, 196, 225, 256, 289, 324, 361

Expressions containing variables such as x^4y^2 are also perfect squares, because they can be written as the square of a quantity:

$$x^4y^2 = (x^2y)^2$$

EXAMPLE 1 Factor: $25x^2 - 49$

Solution We can write $25x^2 - 49$ in the form $(5x)^2 - 7^2$.

$$\begin{aligned} 25x^2 - 49 &= (5x)^2 - 7^2 \\ &= (5x + 7)(5x - 7) \end{aligned}$$

We can check by multiplying $(5x + 7)$ and $(5x - 7)$.

$$\begin{aligned} (5x + 7)(5x - 7) &= 25x^2 - 35x + 35x - 49 \\ &= 25x^2 - 49 \end{aligned}$$

SELF CHECK 1 Factor: $16a^2 - 81$

EXAMPLE 2 Factor: $4y^4 - 25z^2$

Solution We can write $4y^4 - 25z^2$ in the form $(2y^2)^2 - (5z)^2$.

$$\begin{aligned} 4y^4 - 25z^2 &= (2y^2)^2 - (5z)^2 \\ &= (2y^2 + 5z)(2y^2 - 5z) \end{aligned}$$

Check by multiplication.

SELF CHECK 2 Factor: $9m^2 - 64n^4$

2 Completely factor a polynomial.

We often must factor out a greatest common factor before factoring the difference of two squares. To factor $8x^2 - 32$, for example, we factor out the GCF of 8 and then factor the resulting difference of two squares.

$$\begin{aligned} 8x^2 - 32 &= 8(x^2 - 4) && \text{Factor out 8, the GCF.} \\ &= 8(x^2 - 2^2) && \text{Write 4 as } 2^2. \\ &= 8(x + 2)(x - 2) && \text{Factor the difference of two squares.} \end{aligned}$$

We can check by multiplication:

$$8(x+2)(x-2) = 8(x^2-4)$$
$$= 8x^2 - 32$$

EXAMPLE 3 Factor completely: $2a^2x^3y - 8b^2xy$

Solution We factor out the GCF of $2xy$ and then factor the resulting difference of two squares.

$2a^2x^3y - 8b^2xy$	
$= \mathbf{2xy} \cdot a^2x^2 - \mathbf{2xy} \cdot 4b^2$	The GCF is $2xy$.
$= \mathbf{2xy}(a^2x^2 - 4b^2)$	Factor out $2xy$.
$= 2xy[(ax)^2 - (2b)^2]$	Write a^2x^2 as $(ax)^2$ and $4b^2$ as $(2b)^2$.
$= 2xy(ax+2b)(ax-2b)$	Factor the difference of two squares.

Check by multiplication.

SELF CHECK 3 Factor completely: $2p^2q^2s - 18r^2s$

Sometimes we must factor a difference of two squares more than once to completely factor a polynomial. For example, the binomial $625a^4 - 81b^4$ can be written in the form $(25a^2)^2 - (9b^2)^2$, which factors as

$$625a^4 - 81b^4 = (25a^2)^2 - (9b^2)^2$$
$$= (25a^2 + 9b^2)(\mathbf{25a^2 - 9b^2})$$

Since the factor $25a^2 - 9b^2$ can be written in the form $(5a)^2 - (3b)^2$, it is the difference of two squares and can be factored as $(5a+3b)(5a-3b)$. Thus,

$$625a^4 - 81b^4 = (25a^2 + 9b^2)(5a+3b)(5a-3b)$$

The binomial $25a^2 + 9b^2$ is the **sum of two squares**, because it can be written in the form $(5a)^2 + (3b)^2$. If we are limited to rational coefficients, binomials that are the sum of two squares cannot be factored unless they contain a GCF. Polynomials that do not factor are called **prime polynomials**.

EXAMPLE 4 Factor completely: $2x^4y - 32y$

Solution

$2x^4y - 32y = \mathbf{2y} \cdot x^4 - \mathbf{2y} \cdot 16$	The GCF is $2y$.
$= \mathbf{2y}(x^4 - 16)$	Factor out $2y$.
$= 2y(x^2+4)(\mathbf{x^2 - 4})$	Factor $x^4 - 16$.
$= 2y(x^2+4)(\mathbf{x+2})(\mathbf{x-2})$	Factor $x^2 - 4$. Note that $x^2 + 4$ is the sum of two squares and does not factor using rational coefficients.

Check by multiplication.

SELF CHECK 4 Factor completely: $48a^5 - 3ab^4$

Example 5 requires the techniques of factoring out a common factor, factoring by grouping, and factoring the difference of two squares.

EXAMPLE 5 Factor completely: $2x^3 - 8x + 2yx^2 - 8y$

Solution $2x^3 - 8x + 2yx^2 - 8y = 2(x^3 - 4x + yx^2 - 4y)$ Factor out 2, the GCF.

$= 2[x(x^2 - 4) + y(x^2 - 4)]$ Factor out x from $(x^3 - 4x)$ and y from $(yx^2 - 4y)$.

$= 2[(x^2 - 4)(x + y)]$ Factor out $x^2 - 4$.

$= 2(x + 2)(x - 2)(x + y)$ Factor $x^2 - 4$.

Check by multiplication.

SELF CHECK 5 Factor completely: $3a^3 - 12a + 3a^2b - 12b$

SELF CHECK ANSWERS
1. $(4a + 9)(4a - 9)$ **2.** $(3m + 8n^2)(3m - 8n^2)$ **3.** $2s(pq + 3r)(pq - 3r)$ **4.** $3a(4a^2 + b^2)(2a + b)(2a - b)$ **5.** $3(a + 2)(a - 2)(a + b)$

NOW TRY THIS

Factor completely.

1. a. $x^2 - \frac{1}{9}$
   b. $2x^2 - 0.72$
   c. $16 - x^2$
2. $(x + y)^2 - 25$
3. $x^{2n} - 9$

5.2 Exercises

WARM-UPS *Complete each factorization.*

1. $x^2 - 9 = (x + 3)$
2. $y^2 - 36 = (y + 6)$
3. $z^2 - 4 = \quad (z - 2)$
4. $p^2 - q^2 = (p + q)$
5. $25 - t^2 = (5 + t)$
6. $36 - r^2 = \quad (6 - r)$
7. $81 - y^2 = \quad (9 - y)$
8. $49 - y^4 = (7 + y^2)$
9. $4m^2 - 9n^2 = (2m + 3n)$
10. $16x^4 - 49y^2 = (4x^2 + 7y)$
11. $25y^6 - 64x^4 = \quad (5y^3 - 8x^2)$
12. $16p^2 - 25q^2 = \quad (4p - 5q)$

REVIEW

13. In the study of the flow of fluids, Bernoulli's law is given by the following equation. Solve it for p.
$$\frac{p}{w} + \frac{v^2}{2g} + h = k$$
14. Solve Bernoulli's law for h. (See Exercise 13.)

VOCABULARY AND CONCEPTS *Fill in the blanks.*

15. A binomial of the form $a^2 - b^2$ is called the ______________ ______.
16. A binomial of the form _______ is called the sum of two squares.
17. A polynomial that cannot be factored over the rational numbers is said to be a _____ polynomial.
18. The ____ of two squares cannot be factored unless it has a GCF.

GUIDED PRACTICE *Factor each polynomial.* SEE EXAMPLE 1. *(OBJECTIVE 1)*

19. $x^2 - 36$
20. $x^2 - 25$
21. $y^2 - 49$
22. $y^2 - 81$

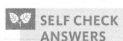

23. $4y^2 - 49$

24. $16z^2 - 9$

25. $25x^4 - 81$

26. $64x^4 - 9$

Factor each polynomial. SEE EXAMPLE 2. (OBJECTIVE 1)

27. $9x^2 - y^2$

28. $4x^2 - z^2$

29. $25t^2 - 36u^2$

30. $49u^2 - 64v^2$

31. $100a^2 - 49b^2$

32. $36a^2 - 121b^2$

33. $x^4 - 9y^2$

34. $121a^2 - 144b^2$

Factor each polynomial completely. SEE EXAMPLE 3. (OBJECTIVE 2)

35. $8x^2 - 32y^2$

36. $2a^2 - 200b^2$

37. $2a^2 - 8y^2$

38. $3x^2 - 27y^2$

39. $100x^2 - 16y^2$

40. $45u^2 - 20v^2$

41. $x^3 - xy^2$

42. $a^2b - b^3$

43. $4a^2x - 9b^2x$

44. $4b^2y - 16c^2y$

45. $3m^3 - 3mn^2$

46. $5x^3 - 5xy^2$

Factor each polynomial completely. SEE EXAMPLE 4. (OBJECTIVE 2)

47. $a^4 - 16$
48. $b^4 - 256$
49. $a^4 - b^4$
50. $m^4 - 16n^4$

Factor each polynomial completely. SEE EXAMPLE 5. (OBJECTIVE 2)

51. $2x^4 - 2y^4$
52. $a^5 - ab^4$
53. $a^4b - b^5$
54. $m^5 - 16mn^4$
55. $2x^4y - 512y^5$
56. $2x^8y^2 - 32y^6$
57. $a^3 - 9a + 3a^2 - 27$
58. $2x^3 - 18x - 6x^2 + 54$

ADDITIONAL PRACTICE

Factor each polynomial completely. If a polynomial is prime, so indicate.

59. $49y^2 - 225z^4$
60. $196x^4 - 169y^2$
61. $4x^4 - x^2y^2$
62. $9xy^2 - 4xy^4$
63. $x^4 - 81$
64. $y^4 - 625$
65. $16y^8 - 81z^4$
66. $x^8 - y^8$
67. $a^2 + b^2$
68. $36a^2 + 49b^2$
69. $x^8y^8 - 1$
70. $a^3 - 49a + 2a^2 - 98$
71. $2a^3b - 242ab^3$
72. $25x^2 + 36y^2$
73. $3a^{10} - 3a^2b^4$
74. $2p^{10}q - 32p^2q^5$
75. $3a^8 - 243a^4b^8$
76. $a^6b^2 - a^2b^6c^4$
77. $a^2b^7 - 625a^2b^3$
78. $16x^3y^4z - 81x^3y^4z^5$
79. $3a^5y + 6ay^5$
80. $81r^4 - 256s^4$
81. $144a^4 + 169b^4$
82. $2m^3n^2 - 32mn^2 + 8m^2 - 128$
83. $2x^9y + 2xy^9$
84. $25(a - b)^2 - 16$
85. $49(y + 1)^2 - x^2$
86. $b^3 - 25b - 2b^2 + 50$
87. $y^3 - 16y - 3y^2 + 48$
88. $243r^5s - 48rs^5$
89. $50c^4d^2 - 8c^2d^4$

WRITING ABOUT MATH

90. Explain how to factor the difference of two squares.
91. Explain why $x^4 - y^4$ is not completely factored as $(x^2 + y^2)(x^2 - y^2)$.

SOMETHING TO THINK ABOUT

92. It is easy to multiply 399 by 401 without a calculator: The product is $400^2 - 1$, or 159,999. Explain.
93. Use the method in the previous exercise to find $498 \cdot 502$ without a calculator.

Section 5.3 Factoring Trinomials with a Leading Coefficient of 1

Objectives

1. Factor a trinomial of the form $ax^2 + bx + c$, $a = 1$, by trial and error.
2. Factor a trinomial containing a negative greatest common factor.
3. Identify a prime trinomial.
4. Factor a polynomial completely.
5. Factor a trinomial of the form $ax^2 + bx + c$, $a = 1$, by grouping (*ac* method).
6. Factor a perfect-square trinomial.

Vocabulary

key number perfect-square trinomial

Getting Ready

Multiply the binomials.

1. $(x + 6)(x + 6)$ **2.** $(y - 7)(y - 7)$ **3.** $(a - 3)(a - 3)$

4. $(x + 4)(x + 5)$ **5.** $(r - 2)(r - 5)$ **6.** $(m + 3)(m - 7)$

7. $(a - 3b)(a + 4b)$ **8.** $(u - 3v)(u - 5v)$ **9.** $(x + 4y)(x - 6y)$

We now discuss how to factor trinomials of the form $x^2 + bx + c$, where the coefficient of x^2 is 1 and there are no common factors.

1 Factor a trinomial of the form $ax^2 + bx + c$, $a = 1$ by trial and error.

The product of two binomials is often a trinomial. For example,

$$(x + 3)(x + 3) = x^2 + 6x + 9 \quad \text{and} \quad (x - 3y)(x + 4y) = x^2 + xy - 12y^2$$

For this reason, we should not be surprised that many trinomials factor into the product of two binomials. To develop a method for factoring trinomials, we multiply $(x + a)$ and $(x + b)$.

$(x + a)(x + b) = x^2 + bx + ax + ab$ Use the distributive property.

$= x^2 + ax + bx + ab$ Apply the commutative property of addition.

$= x^2 + (a + b)x + ab$ Factor the GCF, x, out of $ax + bx$.

From the result, we can see that

- the first term is the product of x and x.
- the coefficient of the middle term is the sum of a and b, and
- the last term is the product of a and b.

We can use these facts to factor trinomials with leading coefficients of 1.

EXAMPLE 1 Factor: $x^2 + 5x + 6$

Solution To factor this trinomial, we will write it as the product of two binomials. Since the first term of the trinomial is x^2, the first term of each binomial factor must be x because $x \cdot x = x^2$. To fill in the following blanks, we must find two integers whose product is $+6$ and whose sum is $+5$.

$$x^2 + 5x + 6 = (x \quad)(x \quad)$$

The positive factorizations of 6 and the sums of the factors are shown in the following table.

Product of the factors	Sum of the factors
$1(6) = 6$	$1 + 6 = 7$
$2(3) = 6$	$2 + 3 = 5$

The last row contains the integers $+2$ and $+3$, whose product is $+6$ and whose sum is $+5$. So we can fill in the blanks with $+2$ and $+3$.

$$x^2 + 5x + 6 = (x + 2)(x + 3)$$

To check the result, we verify that $(x + 2)$ times $(x + 3)$ is $x^2 + 5x + 6$.

$$(x + 2)(x + 3) = x^2 + 3x + 2x + 2 \cdot 3$$
$$= x^2 + 5x + 6$$

SELF CHECK 1 Factor: $y^2 + 5y + 4$

COMMENT In Example 1, the factors can be written in either order due to the commutative property of multiplication. An equivalent factorization is $x^2 + 5x + 6 = (x + 3)(x + 2)$.

EXAMPLE 2 Factor: $y^2 - 7y + 12$

Solution Since the first term of the trinomial is y^2, the first term of each binomial factor must be y. To fill in the following blanks, we must find two integers whose product is $+12$ and whose sum is -7.

$$y^2 - 7y + 12 = (y \quad)(y \quad)$$

The factorizations of 12 and the sums of the factors are shown in the table.

Product of the factors	Sum of the factors
$1(12) = 12$	$1 + 12 = 13$
$2(6) = 12$	$2 + 6 = 8$
$3(4) = 12$	$3 + 4 = 7$
$-1(-12) = 12$	$-1 + (-12) = -13$
$-2(-6) = 12$	$-2 + (-6) = -8$
$-3(-4) = 12$	$-3 + (-4) = -7$

The last row contains the integers -3 and -4, whose product is $+12$ and whose sum is -7. So we can fill in the blanks with -3 and -4.

$$y^2 - 7y + 12 = (y - 3)(y - 4)$$

To check the result, we verify that $(y - 3)$ times $(y - 4)$ is $y^2 - 7y + 12$.

$$\begin{aligned}(y - 3)(y - 4) &= y^2 - 3y - 4y + 12\\ &= y^2 - 7y + 12\end{aligned}$$

SELF CHECK 2 Factor: $p^2 - 5p + 6$

EXAMPLE 3 Factor: $a^2 + 2a - 15$

Solution Since the first term is a^2, the first term of each binomial factor must be a. To fill in the blanks, we must find two integers whose product is -15 and whose sum is $+2$.

$$a^2 + 2a - 15 = (a \quad)(a \quad)$$

The factorizations of -15 and the sums of the factors are shown in the table.

Product of the factors	Sum of the factors
$1(-15) = -15$	$1 + (-15) = -14$
$3(-5) = -15$	$3 + (-5) = -2$
$5(-3) = -15$	$5 + (-3) = 2$
$15(-1) = -15$	$15 + (-1) = 14$

The third row contains the integers $+5$ and -3, whose product is -15 and whose sum is $+2$. So we can fill in the blanks with $+5$ and -3.

$$a^2 + 2a - 15 = (a + 5)(a - 3)$$

Check: $(a + 5)(a - 3) = a^2 - 3a + 5a - 15$
$= a^2 + 2a - 15$

SELF CHECK 3 Factor: $p^2 + 3p - 18$

EXAMPLE 4 Factor: $z^2 - 4z - 21$

Solution Since the first term is z^2, the first term of each binomial factor must be z. To fill in the blanks, we must find two integers whose product is -21 and whose sum is -4.

$$z^2 - 4z - 21 = (z \quad)(z \quad)$$

The factorizations of -21 and the sums of the factors are shown in the table.

Product of the factors	Sum of the factors
$1(-21) = -21$	$1 + (-21) = -20$
$3(-7) = -21$	$3 + (-7) = -4$
$7(-3) = -21$	$7 + (-3) = 4$
$21(-1) = -21$	$21 + (-1) = 20$

The second row contains the integers $+3$ and -7, whose product is -21 and whose sum is -4. So we can fill in the blanks with $+3$ and -7.

$$z^2 - 4z - 21 = (z + 3)(z - 7)$$

Check: $(z + 3)(z - 7) = z^2 - 7z + 3z - 21$
$= z^2 - 4z - 21$

SELF CHECK 4 Factor: $q^2 - 2q - 24$

COMMENT When factoring a trinomial written in descending order, if the last term is positive the binomial factors will have the same sign as the middle term. If the last term is negative the binomial factors will have different signs.

The next example is a trinomial containing two variables.

EXAMPLE 5 Factor: $x^2 + xy - 6y^2$

Solution Since the first term is x^2, the first term of each binomial factor must be x. Since the last term is $-6y^2$, the second term of each binomial factor has a factor of y. To fill in the blanks, we must find coefficients whose product is -6 that will give a middle coefficient of 1.

$$x^2 + xy - 6y^2 = (x \quad y)(x \quad y)$$

The factorizations of -6 and the sums of the factors are shown in the table.

Product of the factors	Sum of the factors
$1(-6) = -6$	$1 + (-6) = -5$
$2(-3) = -6$	$2 + (-3) = -1$
$3(-2) = -6$	$3 + (-2) = 1$
$6(-1) = -6$	$6 + (-1) = 5$

The third row contains the integers 3 and -2. These are the only integers whose product is -6 and will give the correct middle coefficient of 1. So we can fill in the blanks with 3 and -2.

$$x^2 + xy - 6y^2 = (x + 3y)(x - 2y)$$

Check:
$$\begin{aligned}(x + 3y)(x - 2y) &= x^2 - 2xy + 3xy - 6y^2 \\ &= x^2 + xy - 6y^2\end{aligned}$$

SELF CHECK 5 Factor: $a^2 + ab - 12b^2$

2 Factor a trinomial containing a negative greatest common factor.

When the coefficient of the first term is -1, we begin by factoring out -1.

EXAMPLE 6 Factor: $-x^2 + 2x + 15$

Solution We factor out -1 and then factor the trinomial.

$$\begin{aligned}-x^2 + 2x + 15 &= -(x^2 - 2x - 15) && \text{Factor out } -1. \\ &= -(x - 5)(x + 3) && \text{Factor } x^2 - 2x - 15.\end{aligned}$$

Check:
$$\begin{aligned}-(x - 5)(x + 3) &= -(x^2 + 3x - 5x - 15) \\ &= -(x^2 - 2x - 15) \\ &= -x^2 + 2x + 15\end{aligned}$$

COMMENT In Example 6, it is not necessary to factor out the -1, but by doing so, it usually will be easier to factor the remaining trinomial.

SELF CHECK 6 Factor: $-x^2 + 11x - 18$

3 Identify a prime trinomial.

If a trinomial cannot be factored using only rational coefficients, it is called a prime polynomial over the set of rational numbers.

EXAMPLE 7 Factor: $x^2 + 2x + 3$

Solution To factor the trinomial, we must find two integers whose product is +3 and whose sum is 2. The possible factorizations of 3 and the sums of the factors are shown in the table.

Product of the factors	Sum of the factors
$1(3) = 3$	$1 + 3 = 4$
$-1(-3) = 3$	$-1 + (-3) = -4$

Since two integers whose product is +3 and whose sum is +2 do not exist, $x^2 + 2x + 3$ cannot be factored. It is a prime trinomial.

SELF CHECK 7 Factor: $x^2 - 4x + 6$

4 Factor a polynomial completely.

To *factor* an expression means to factor the expression *completely*. The following examples require more than one type of factoring.

EXAMPLE 8 Factor: $-3ax^2 + 9a - 6ax$

Solution We first write the trinomial in descending powers of x and factor out the common factor of $-3a$.

$$\begin{aligned} -3ax^2 + 9a - 6ax &= -3ax^2 - 6ax + 9a \\ &= -3a\,(\mathbf{x^2 + 2x - 3}) \end{aligned}$$

Then we factor the trinomial $x^2 + 2x - 3$.

$$-3ax^2 + 9a - 6ax = -3a\,(\mathbf{x + 3})(\mathbf{x - 1})$$

Check:
$$\begin{aligned} -3a(x + 3)(x - 1) &= -3a(x^2 + 2x - 3) \\ &= -3ax^2 - 6ax + 9a \\ &= -3ax^2 + 9a - 6ax \end{aligned}$$

SELF CHECK 8 Factor: $-2pq^2 + 6p - 4pq$

We have factored four-term polynomials by grouping two terms and two terms. The next example will require grouping three terms.

EXAMPLE 9 Factor: $m^2 - 2mn + n^2 - 64a^2$

Solution We group the first three terms together and factor the resulting trinomial.

$$\begin{aligned} m^2 - 2mn + n^2 - 64a^2 &= (m - n)(m - n) - 64a^2 \\ &= (m - n)^2 - (8a)^2 \end{aligned}$$

Then we factor the resulting difference of two squares.

$$\begin{aligned} m^2 - 2mn + n^2 - 64a^2 &= (m - n)^2 - (8a)^2 \\ &= (m - n + 8a)(m - n - 8a) \end{aligned}$$

SELF CHECK 9 Factor: $p^2 + 4pq + 4q^2 - 25y^2$

5 Factor a trinomial of the form $ax^2 + bx + c$, $a = 1$, by grouping (*ac* method).

Carl Friedrich Gauss

1777–1855

Many people consider Gauss to be the greatest mathematician of all time. He made contributions in the areas of number theory, solutions of equations, geometry of curved surfaces, and statistics. For his efforts, he has earned the title "Prince of the Mathematicians."

An alternative way of factoring trinomials of the form $ax^2 + bx + c$, $a = 1$, uses the technique of factoring by grouping, sometimes referred to as the *ac method*. For example, to factor $x^2 + x - 12$ by grouping, we proceed as follows:

1. Determine the values of a and c $(a = 1$ and $c = -12)$ and find ac:

   $$(1)(-12) = -12$$

   This number is called the **key number**.

2. Find two factors of the key number -12 whose sum is $b = 1$. Two such factors are $+4$ and -3.

   $$+4(-3) = -12 \qquad \text{and} \qquad +4 + (-3) = 1$$

3. Use the factors $+4$ and -3 as the coefficients of two terms whose sum is x. Write these two terms to replace x.

   $$x^2 + x - 12 = x^2 \mathbf{+ 4x - 3x} - 12 \qquad x = +4x - 3x$$

4. Factor the right side of the previous equation by grouping.

   $$x^2 + 4x - 3x - 12 = x(\mathbf{x + 4}) - 3(\mathbf{x + 4})$$ Factor x out of $(x^2 + 4x)$ and -3 out of $(-3x - 12)$.

   $$= (\mathbf{x + 4})(x - 3)$$ Factor out $(x + 4)$.

Check this factorization by multiplication.

EXAMPLE 10 Factor $y^2 + 7y + 10$ by grouping.

Solution We note that this equation is in the form $y^2 + by + c$, with $a = 1$, $b = 7$, and $c = 10$. First, we determine the key number ac:

$$ac = 1(10) = 10$$

Then we find two factors of 10 whose sum is $b = 7$. Two such factors are $+2$ and $+5$. We use these factors as the coefficients of two terms whose sum is $7y$.

$$y^2 \mathbf{+ 7y} + 10 = y^2 \mathbf{+ 2y + 5y} + 10 \qquad 7y = +2y + 5y$$

Finally, we factor the right side of the previous equation by grouping.

$$y^2 + 2y + 5y + 10 = y(\mathbf{y + 2}) + 5(\mathbf{y + 2})$$ Factor out y from $(y^2 + 2y)$ and factor out 5 from $(5y + 10)$.

$$= (\mathbf{y + 2})(y + 5)$$ Factor out $(y + 2)$.

SELF CHECK 10 Use grouping to factor $p^2 - 7p + 12$.

EXAMPLE 11 Factor: $z^2 - 4z - 21$

Solution This is the trinomial of Example 4. To factor it by grouping, we note that the trinomial is in the form $z^2 + bz + c$, with $a = 1$, $b = -4$, and $c = -21$. First, we determine the key number ac:

$$ac = 1(-21) = -21$$

Then we find two factors of -21 whose sum is $b = -4$. Two such factors are $+3$ and -7. We use these factors as the coefficients of two terms whose sum is $-4z$. We write these two terms to replace $-4z$.

$$z^2 \mathbf{- 4z} - 21 = z^2 \mathbf{+ 3z - 7z} - 21 \qquad -4z = +3z - 7z$$

Unless otherwise noted, all content on this page is © Cengage Learning.

Finally, we factor the right side of the previous equation by grouping.

$$z^2 + 3z - 7z - 21 = z(z + 3) - 7(z + 3)$$ Factor out z from $(z^2 + 3z)$ and factor out -7 from $(-7z - 21)$.

$$= (z + 3)(z - 7)$$ Factor out $(z + 3)$.

SELF CHECK 11 Use grouping to factor $a^2 + 2a - 15$. This is the trinomial of Example 3.

6 Factor a perfect-square trinomial.

We have discussed the following special-product relationships used to square binomials.

1. $(x + y)^2 = x^2 + 2xy + y^2$

2. $(x - y)^2 = x^2 - 2xy + y^2$

These relationships can be used in reverse order to factor special trinomials called **perfect-square trinomials**.

PERFECT-SQUARE TRINOMIALS

1. $x^2 + 2xy + y^2 = (x + y)^2$

2. $x^2 - 2xy + y^2 = (x - y)^2$

In words, Formula 1 states that *if a trinomial is the square of one quantity, plus twice the product of the two quantities, plus the square of the second quantity, it factors into the square of the sum of the quantities.*

Formula 2 states that *if a trinomial is the square of one quantity, minus twice the product of the two quantities, plus the square of the second quantity, it factors into the square of the difference of the quantities.*

The trinomials on the left sides of the previous equations are perfect-square trinomials because they are the results of squaring a binomial. Although we can factor perfect-square trinomials by using the techniques discussed earlier in this section, we usually can factor them by inspecting their terms. For example, $x^2 + 8x + 16$ is a perfect-square trinomial, because

- The first term x^2 is the square of x.
- The last term 16 is the square of 4.
- The middle term $8x$ is twice the product of x and 4.

Thus,

$$x^2 + 8x + 16 = x^2 + 2(x)(4) + 4^2$$
$$= (x + 4)^2$$

EXAMPLE 12 Factor: $x^2 - 10x + 25$

Solution $x^2 - 10x + 25$ is a perfect-square trinomial, because

- The first term x^2 is the square of x.
- The last term 25 is the square of 5.
- The middle term $-10x$ is the negative of twice the product of x and 5.

Thus,

$$x^2 - 10x + 25 = x^2 - 2(x)(5) + 5^2$$
$$= (x - 5)^2$$

SELF CHECK 12 Factor: $x^2 + 10x + 25$

SELF CHECK ANSWERS

1. $(y + 1)(y + 4)$ **2.** $(p - 3)(p - 2)$ **3.** $(p + 6)(p - 3)$ **4.** $(q + 4)(q - 6)$ **5.** $(a - 3b)(a + 4b)$ **6.** $-(x - 9)(x - 2)$ **7.** prime **8.** $-2p(q + 3)(q - 1)$ **9.** $(p + 2q + 5y)(p + 2q - 5y)$ **10.** $(p - 4)(p - 3)$ **11.** $(a + 5)(a - 3)$ **12.** $(x + 5)^2$

NOW TRY THIS

Factor.

1. $18 + 3x - x^2$
2. $x^2 + \frac{2}{5}x + \frac{1}{25}$
3. $x^{2n} + x^n - 2$

5.3 Exercises

WARM-UPS *Find two numbers whose product is (a) and whose sum is (b).*

1. (a) 4 (b) 5
2. (a) 6 (b) −5
3. (a) −6 (b) 1
4. (a) −6 (b) −1
5. (a) 4 (b) −5
6. (a) −12 (b) 4
7. (a) −18 (b) 7
8. (a) 24 (b) 11

REVIEW *Graph the solution of each inequality on a number line.*

9. $x - 3 > 5$
10. $x + 4 \le 3$
11. $-3x - 5 \ge 4$
12. $2x - 3 < 7$
13. $\frac{3(x - 1)}{4} < 12$
14. $\frac{-2(x + 3)}{3} \ge 9$
15. $-2 < x \le 4$
16. $-5 \le x + 1 < 5$

VOCABULARY AND CONCEPTS *Complete each relationship for a perfect-square trinomial.*

17. $x^2 + 2xy + y^2 =$ ________
18. $x^2 - 2xy + y^2 =$ ________

Complete each factorization.

19. $x^2 + 5x + 6 = (x + 2)(x + \quad)$
20. $x^2 - 5x + 6 = (x \quad 2)(x \quad 3)$
21. $x^2 + x - 6 = (x \quad 2)(x + \quad)$
22. $x^2 - x - 6 = (x \quad 3)(x + \quad)$
23. $x^2 + 5x - 6 = (x + \quad)(x - \quad)$
24. $x^2 - 7x + 6 = (x - \quad)(x - \quad)$
25. $y^2 + 6y + 8 = (y + \quad)(y + \quad)$
26. $z^2 - 3z - 10 = (z + \quad)(z - \quad)$
27. $x^2 - xy - 2y^2 = (x + \quad)(x - \quad)$
28. $a^2 + ab - 6b^2 = (a + \quad)(a - \quad)$

GUIDED PRACTICE *Factor.* SEE EXAMPLE 1. (OBJECTIVE 1)

29. $x^2 + 5x + 4$
30. $y^2 + 4y + 3$
31. $z^2 + 6z + 8$
32. $x^2 + 7x + 10$
33. $x^2 + 8x + 15$
34. $z^2 + 8z + 16$
35. $x^2 + 12x + 20$
36. $y^2 + 9y + 20$

Factor. SEE EXAMPLE 2. (OBJECTIVE 1)

37. $t^2 - 9t + 14$
38. $c^2 - 9c + 8$
39. $x^2 - 8x + 12$
40. $p^2 - 10p + 16$
41. $x^2 - 9x + 20$
42. $t^2 - 3t + 2$
43. $r^2 - 8r + 7$
44. $y^2 - 9y + 18$

Factor. SEE EXAMPLE 3. (OBJECTIVE 1)

45. $q^2 + 8q - 9$
46. $x^2 + 5x - 24$
47. $s^2 + 11s - 26$
48. $b^2 + 6b - 7$
49. $c^2 + 4c - 5$
50. $x^2 + 8x - 20$
51. $y^2 + 4y - 12$
52. $a^2 + 9a - 36$

Factor. SEE EXAMPLE 4. (OBJECTIVE 1)

53. $b^2 - 5b - 6$

54. $t^2 - 5t - 50$

55. $a^2 - 10a - 39$

56. $a^2 - 4a - 5$

57. $m^2 - 3m - 10$

58. $y^2 - 2y - 35$

59. $x^2 - 3x - 40$

60. $x^2 - 6x - 16$

Factor. SEE EXAMPLE 5. (OBJECTIVE 1)

61. $m^2 + 5mn - 14n^2$

62. $m^2 - mn - 12n^2$

63. $a^2 - 4ab - 12b^2$

64. $a^2 + 7ab - 18b^2$

65. $a^2 + 10ab + 9b^2$

66. $u^2 + 2uv - 15v^2$

67. $m^2 - 11mn + 10n^2$

68. $x^2 + 6xy + 9y^2$

Factor each trinomial. Factor out -1 *first.* SEE EXAMPLE 6. (OBJECTIVE 2)

69. $-x^2 - 7x - 10$

70. $-x^2 + 9x - 20$

71. $-y^2 - 2y + 15$

72. $-y^2 - 3y + 18$

73. $-t^2 - 4t + 32$

74. $-t^2 - t + 30$

75. $-r^2 + 14r - 40$

76. $-r^2 + 14r - 45$

Factor each trinomial. If prime, so indicate. SEE EXAMPLE 7. (OBJECTIVE 3)

77. $a^2 + 3a + 10$

78. $v^2 + 9v + 15$

79. $r^2 - 9r - 12$

80. $b^2 + 6b - 18$

Factor. SEE EXAMPLE 8. (OBJECTIVE 4)

81. $2x^2 + 20x + 42$

82. $-2b^2 + 20b - 18$

83. $3y^3 - 21y^2 + 18y$

84. $-5a^3 + 25a^2 - 30a$

85. $3z^2 - 15tz + 12t^2$

86. $5m^2 + 45mn - 50n^2$

87. $-4x^2y - 4x^3 + 24xy^2$

88. $3x^2y^3 + 3x^3y^2 - 6xy^4$

Factor. SEE EXAMPLE 9. (OBJECTIVE 4)

89. $x^2 + 4x + 4 - y^2$

90. $p^2 - 2p + 1 - q^2$

91. $b^2 - 6b + 9 - c^2$

92. $m^2 + 8m + 16 - n^2$

Factor. SEE EXAMPLES 10–11. (OBJECTIVE 5)

93. $x^2 + 6x + 5$

94. $y^2 + 4y + 3$

95. $t^2 - 9t + 14$

96. $c^2 - 9c + 8$

97. $a^2 + 6a - 16$

98. $x^2 + 5x - 24$

99. $y^2 - y - 30$

100. $a^2 - 4a - 32$

Factor. SEE EXAMPLE 12. (OBJECTIVE 6)

101. $x^2 + 6x + 9$

102. $x^2 + 10x + 25$

103. $y^2 - 8y + 16$

104. $z^2 - 2z + 1$

105. $u^2 - 18u + 81$

106. $v^2 - 14v + 49$

107. $x^2 + 4xy + 4y^2$

108. $a^2 + 12ab + 36b^2$

ADDITIONAL PRACTICE

Factor. Write each trinomial in descending powers of one variable, if necessary. If a polynomial is prime, so indicate.

109. $4 - 5x + x^2$

110. $y^2 + 7 + 8y$

111. $10y + 9 + y^2$

112. $x^2 - 13 - 12x$

113. $-r^2 + 2s^2 + rs$

114. $u^2 - 3v^2 + 2uv$

115. $x^2 - 9x - 27$

116. $-a^2 + 5b^2 + 4ab$

117. $-a^2 - 6ab - 5b^2$

118. $3y^3 + 6y^2 + 3y$

119. $12xy + 4x^2y - 72y$

120. $y^2 + 2yz + z^2$

121. $r^2 - 2rs + 4s^2$

122. $r^2 + 24r + 144$

123. $r^2 - 10rs + 25s^2$

124. $a^2 + 2ab + b^2 - 4$

125. $a^2 + 6a + 9 - b^2$

126. $5y^3 + 10y^2 + 5y$

127. $t^2 + 18t + 81$

128. $c^2 - a^2 + 8a - 16$

WRITING ABOUT MATH

129. Explain how you would write a trinomial in descending order.

130. Explain how to use the distributive property to check the factoring of a trinomial.

SOMETHING TO THINK ABOUT

131. Two students factor $2x^2 + 20x + 42$ and get two different answers: $(2x + 6)(x + 7)$ and $(x + 3)(2x + 14)$. Do both answers check? Why don't they agree? Is either completely correct?

132. Find the error:

$x = y$	
$x^2 = xy$	Multiply both sides by x.
$x^2 - y^2 = xy - y^2$	Subtract y^2 from both sides.
$(x + y)(x - y) = y(x - y)$	Factor.
$x + y = y$	Divide both sides by $(x - y)$.
$y + y = y$	Substitute y for its equal, x.
$2y = y$	Combine like terms.
$2 = 1$	Divide both sides by y.

Section 5.4 Factoring Trinomials with a Leading Coefficient Other Than 1

Objectives

1. Factor a trinomial of the form $ax^2 + bx + c$ using trial and error.
2. Factor a trinomial of the form $ax^2 + bx + c$ by grouping (ac method).
3. Factor a polynomial involving a perfect-square trinomial.

Getting Ready

Multiply and combine like terms.

1. $(2x + 1)(3x + 2)$
2. $(3y - 2)(2y - 5)$
3. $(4t - 3)(2t + 3)$
4. $(2r + 5)(2r - 3)$
5. $(2m - 3)(3m - 2)$
6. $(4a + 3)(4a + 1)$

In the previous section, we saw how to factor trinomials whose leading coefficients are 1. We now show how to factor trinomials whose leading coefficients are other than 1.

1 Factor a trinomial of the form $ax^2 + bx + c$ using trial and error.

In the previous section, we only considered factors of c. We must now consider more combinations of factors when we factor trinomials with leading coefficients other than 1.

EXAMPLE 1 Factor: $2x^2 + 5x + 3$

Solution Since the first term is $2x^2$, the first terms of the binomial factors must be $2x$ and x. To fill in the blanks, we must find two factors of $+3$ that will give a middle term of $+5x$.

$$(2x \quad)(x \quad)$$

Since the sign of each term of the trinomial is +, we need to consider only positive factors of the last term (3). Since the positive factors of 3 are 1 and 3, there are two possible factorizations.

$$(2x + 1)(x + 3) \quad \text{or} \quad (2x + 3)(x + 1)$$

The first possibility is incorrect, because it gives a middle term of $7x$. The second possibility is correct, because it gives a middle term of $5x$. Thus,

$$2x^2 + 5x + 3 = (2x + 3)(x + 1)$$

Check by multiplication.

SELF CHECK 1 Factor: $3x^2 + 7x + 2$

EXAMPLE 2 Factor: $6x^2 - 17x + 5$

Solution Since the first term is $6x^2$, the first terms of the binomial factors must be $6x$ and x or $3x$ and $2x$. To fill in the blanks, we must find two factors of $+5$ that will give a middle term of $-17x$.

$$(6x \quad)(x \quad) \quad \text{or} \quad (3x \quad)(2x \quad)$$

Since the sign of the third term is $+$ and the sign of the middle term is $-$, we need to consider only negative factors of the last term (5). Since the negative factors of 5 are -1 and -5, there are four possible factorizations.

	$(6x - 1)(x - 5)$	$(6x - 5)(x - 1)$
The one to choose ⟶	$(3x - 1)(2x - 5)$	$(3x - 5)(2x - 1)$

Only the possibility printed in red gives the correct middle term of $-17x$. Thus,

$$6x^2 - 17x + 5 = (3x - 1)(2x - 5)$$

Check by multiplication.

SELF CHECK 2 Factor: $6x^2 - 7x + 2$

EXAMPLE 3 Factor: $3y^2 - 5y - 12$

Solution Since the sign of the third term of $3y^2 - 5y - 12$ is $-$, the signs between the binomial factors will be opposites. Because the first term is $3y^2$, the first terms of the binomial factors must be $3y$ and y.

Since $1(-12), 2(-6), 3(-4), 12(-1), 6(-2)$, and $4(-3)$ all give a product of -12, there are 12 possible combinations to consider.

	$(3y + 1)(y - 12)$	$(3y - 12)(y + 1)$
	$(3y + 2)(y - 6)$	$(3y - 6)(y + 2)$
	$(3y + 3)(y - 4)$	$(3y - 4)(y + 3)$
	$(3y + 12)(y - 1)$	$(3y - 1)(y + 12)$
	$(3y + 6)(y - 2)$	$(3y - 2)(y + 6)$
The one to choose ⟶	$(3x + 4)(y - 3)$	$(3y - 3)(y + 4)$

The combinations printed in blue cannot work, because one of the factors has a common factor. This implies that $3y^2 - 5y - 12$ would have a common factor, which it does not.

After mentally trying the remaining factors, we see that only $(3y + 4)(y - 3)$ gives the correct middle term of $-5x$. Thus,

$$3y^2 - 5y - 12 = (3y + 4)(y - 3)$$

Check by multiplication.

SELF CHECK 3 Factor: $5a^2 - 7a - 6$

EXAMPLE 4 Factor: $6b^2 + 7b - 20$

Solution Since the first term is $6b^2$, the first terms of the binomial factors must be $6b$ and b or $3b$ and $2b$. To fill in the blanks, we must find two factors of -20 that will give a middle term of $+7b$.

$$(6b \quad)(b \quad) \qquad \text{or} \qquad (3b \quad)(2b \quad)$$

Since the sign of the third term is $-$, the signs inside the binomial factors will be different. Because the factors of the last term (20) are 1, 2, 4, 5, 10, and 20, there are many possible combinations for the last terms. We must try to find a combination that will give a last term of -20 and a sum of the products of the outer terms and inner terms of $+7b$.

If we choose factors of $6b$ and b for the first terms and -5 and 4 for the last terms, we have

$$(6b - 5)(b + 4)$$

$$\begin{array}{r} -5b \\ \underline{24b} \\ 19b \end{array}$$

which gives an incorrect middle term of $19b$.

If we choose factors of $3b$ and $2b$ for the first terms and -4 and $+5$ for the last terms, we have

$$(3b - 4)(2b + 5)$$

$$\begin{array}{r} -8b \\ \underline{15b} \\ 7b \end{array}$$

which gives the correct middle term of $+7b$ and the correct last term of -20. Thus,

$$6b^2 + 7b - 20 = (3b - 4)(2b + 5)$$

Check by multiplication.

SELF CHECK 4 Factor: $4x^2 + 4x - 3$

Although, the next example has two variables, the process remains the same. When two variables are involved, first write the polynomial in descending powers of one of the variables. Then, proceed as previously discussed.

EXAMPLE 5 Factor: $2x^2 + 7xy + 6y^2$

Solution Notice the polynomial is written in descending powers of x. Since the first term is $2x^2$, the first terms of the binomial factors must be $2x$ and x. To fill in the blanks, we must find two factors of $6y^2$ that will give a middle term of $+7xy$.

$$(2x \quad)(x \quad)$$

Since the sign of each term is $+$, the signs inside the binomial factors will be $+$. The possible factors of the last term, $6y^2$, are

$$y \text{ and } 6y \qquad \text{or} \qquad 3y \text{ and } 2y$$

We must try to find a combination that will give a last term of $+6y^2$ and a middle term of $+7xy$.

If we choose y and $6y$ to be the factors of the last term, we have

$$(2x + y)(x + 6y)$$

$$\begin{array}{r} xy \\ \underline{12xy} \\ 13xy \end{array}$$

which gives an incorrect middle term of $13xy$.

If we choose $3y$ and $2y$ to be the factors of the last term, we have

$$(2x + 3y)(x + 2y)$$

$$\begin{array}{r} 3xy \\ \underline{4xy} \\ 7xy \end{array}$$

which gives a correct middle term of $7xy$. Thus,

$$2x^2 + 7xy + 6y^2 = (2x + 3y)(x + 2y)$$

Check by multiplication.

SELF CHECK 5 Factor: $4x^2 + 8xy + 3y^2$

Because some guesswork is often necessary, it is difficult to give specific rules for factoring trinomials. However, the following hints are often helpful.

FACTORING GENERAL TRINOMIALS USING TRIAL AND ERROR

1. Write the trinomial in descending powers of one variable.
2. Factor out any GCF (including -1 if that is necessary to make the coefficient of the first term positive).
3. If the sign of the third term is $+$, the signs between the terms of the binomial factors are the same as the sign of the middle term. If the sign of the third term is $-$, the signs between the terms of the binomial factors are opposites.
4. Try combinations of first terms and last terms until you find one that works, or until you exhaust all the possibilities. If no combination works, the trinomial is prime.
5. Check the factorization by multiplication.

EXAMPLE 6 Factor: $2x^2y - 8x^3 + 3xy^2$

Solution **Step 1:** Write the trinomial in descending powers of x.

$$-8x^3 + 2x^2y + 3xy^2$$

Step 2: Factor out the negative of the GCF, which is $-x$.

$$-8x^3 + 2x^2y + 3xy^2 = -x(8x^2 - 2xy - 3y^2)$$

Step 3: Because the sign of the third term of the trinomial factor is $-$, the signs within its binomial factors will be opposites.

Step 4: Find the binomial factors of the trinomial.

$$\begin{aligned} -8x^3 + 2x^2y + 3xy^2 &= -x(8x^2 - 2xy - 3y^2) \\ &= -x(2x + y)(4x - 3y) \end{aligned}$$

Step 5: Check by multiplication.

$$\begin{aligned} -x(2x + y)(4x - 3y) &= -x(8x^2 - 6xy + 4xy - 3y^2) \\ &= -x(8x^2 - 2xy - 3y^2) \\ &= -8x^3 + 2x^2y + 3xy^2 \\ &= 2x^2y - 8x^3 + 3xy^2 \end{aligned}$$

SELF CHECK 6 Factor: $12y - 2y^3 - 2y^2$

When there are a larger number of possible combinations, some people prefer to use an alternative to trial and error, the *ac* method. A visual of the *ac* method, referred to as the Box method, will also be discussed.

2 Factor a trinomial of the form $ax^2 + bx + c$ by grouping (*ac* method).

Another way to factor trinomials of the form $ax^2 + bx + c$ uses the grouping (*ac* method), first discussed in the previous section. For example, to factor $6x^2 - 17x + 5$ (Example 2) by grouping, we note that $a = 6$, $b = -17$, and $c = 5$ and proceed as follows:

1. Determine the product ac: $6(+5) = 30$. This is the *key number.*
2. Find two factors of the key number 30 whose sum is -17. Two such factors are -15 and -2.

   $$-15(-2) = 30 \quad \text{and} \quad -15 + (-2) = -17$$

3. Use -15 and -2 as coefficients of two terms to be placed between $6x^2$ and 5 to replace $-17x$.

   $$6x^2 \mathbf{- 17x} + 5 = 6x^2 \mathbf{- 15x - 2x} + 5$$

4. Factor the right side of the previous equation by grouping.

   $$6x^2 - 15x - 2x + 5 = 3x(\mathbf{2x - 5}) - 1(\mathbf{2x - 5})$$ Factor the GCF, $3x$, from $(6x^2 - 15x)$ and -1 from $(-2x + 5)$.

   $$= (\mathbf{2x - 5})(3x - 1)$$ Factor out the GCF, $(2x - 5)$.

   Verify this factorization by multiplication.

EXAMPLE 7 Factor $4y^2 + 12y + 5$ by grouping.

Solution To factor this trinomial by grouping, we note that it is written in the form $ay^2 + by + c$, with $a = 4$, $b = 12$, and $c = 5$. Since $a = 4$ and $c = 5$, we have $ac = 20$.

We now find two factors of 20 whose sum is 12. Two such factors are 10 and 2. We use these factors as coefficients of two terms to be placed between $4y^2$ and 5 to replace $+12y$.

$$4y^2 \mathbf{+ 12y} + 5 = 4y^2 \mathbf{+ 10y + 2y} + 5$$

Finally, we factor the right side of the previous equation by grouping.

$$4y^2 + 10y + 2y + 5 = 2y(2y + 5) + (2y + 5)$$ Factor the GCF, $2y$, from $(4y^2 + 10y)$.

$$= 2y(\mathbf{2y + 5}) + 1 \cdot (\mathbf{2y + 5})$$ Write the coefficient of 1.

$$= (\mathbf{2y + 5})(2y + 1)$$ Factor out the GCF, $(2y + 5)$.

Check by multiplication.

SELF CHECK 7 Use grouping to factor $2p^2 - 7p + 3$.

EXAMPLE 8 Factor: $6b^2 + 7b - 20$

Solution This is the trinomial of Example 4. Since $a = 6$ and $c = -20$ in the trinomial, $ac = -120$. We now find two factors of -120 whose sum is $+7$. Two such factors are 15 and -8. We use these factors as coefficients of two terms to be placed between $6b^2$ and -20 to replace $+7b$.

$$6b^2 + 7b - 20 = 6b^2 + \mathbf{15b} - \mathbf{8b} - 20$$

Finally, we factor the right side of the previous equation by grouping.

$$6b^2 + 15b - 8b - 20 = 3b(\mathbf{2b+5}) - 4(\mathbf{2b+5})$$ Factor the GCF, $3b$, from $(6b^2 + 15b)$ and -4 from $(-8b - 20)$.

$$= (\mathbf{2b+5})(3b - 4)$$ Factor out the GCF, $(2b + 5)$.

Check by multiplication.

COMMENT When using the grouping method, if no pair of factors of ac produces the desired value b, the trinomial is prime over the rationals.

SELF CHECK 8 Factor: $3y^2 - 4y - 4$

Another option for factoring general trinomials is to use a *box method*, which is a visual of the *ac* method. To illustrate, we will factor $6x^2 + 5x - 4$.

1. Draw a 2×2 box.

2. In the top left corner, write the first term of $6x^2$ and in the bottom right corner write the last term, -4.

$6x^2$	
	-4

3. Multiply these two terms to get $-24x^2$. Find two factors of $-24x^2$ that will add to give the middle term of the trinomial, $5x$. These will be $8x$ and $-3x$. Write these two values in the remaining boxes. Their placement does not matter.

$6x^2$	$-3x$
$8x$	-4

4. Factor the greatest common factor from each row and each column.

	$2x$	-1
$3x$	$6x^2$	$-3x$
4	$8x$	-4

The sums of the factors of the rows and columns are the factors of the original trinomial, $6x^2 + 5x - 4$.

$$6x^2 + 5x - 4 = (3x + 4)(2x - 1)$$

3 Factor a polynomial involving a perfect-square trinomial.

As before, we can factor perfect-square trinomials by inspection.

EXAMPLE 9 Factor: $4x^2 - 20x + 25$

Solution $4x^2 - 20x + 25$ is a perfect-square trinomial, because

- The first term $4x^2$ is the square of $2x$: $(2x)^2 = 4x^2$.
- The last term 25 is the square of 5: $5^2 = 25$.
- The middle term $-20x$ is the negative of twice the product of $2x$ and 5.

Thus,

$$4x^2 - 20x + 25 = (2x)^2 - 2(2x)(5) + 5^2$$
$$= (2x - 5)^2$$

Check by multiplication.

COMMENT It is recommended that you factor a perfect-square trinomial as you would any other trinomial until you discover the pattern for yourself.

SELF CHECK 9 Factor: $9x^2 - 12x + 4$

The next examples combine several factoring techniques. Although factoring a polynomial with four terms typically involves grouping two terms and two terms, there are times we need to group three terms. It may be helpful to look for a perfect-square trinomial in either the first three or last three terms.

EXAMPLE 10 Factor: $4x^2 - 4xy + y^2 - 9$

Solution

$4x^2 - 4xy + y^2 - 9$

$= (4x^2 - 4xy + y^2) - 9$ — Group the first three terms.

$= (2x - y)(2x - y) - 9$ — Factor the perfect-square trinomial.

$= (2x - y)^2 - 9$ — Write $(2x - y)(2x - y)$ as $(2x - y)^2$.

$= [(2x - y) + 3][(2x - y) - 3]$ — Factor the difference of two squares.

$= (2x - y + 3)(2x - y - 3)$ — Simplify each factor.

Check by multiplication.

SELF CHECK 10 Factor: $x^2 + 4x + 4 - y^2$

EXAMPLE 11 Factor: $9 - 4x^2 - 4xy - y^2$

Solution

$9 - 4x^2 - 4xy - y^2 = 9 - (4x^2 + 4xy + y^2)$ — Factor -1 from the last three terms.

$= 9 - (2x + y)(2x + y)$ — Factor the perfect-square trinomial.

$= 9 - (2x + y)^2$ — Write $(2x + y)(2x + y)$ as $(2x + y)^2$.

$= [3 + (2x + y)][3 - (2x + y)]$ — Factor the difference of two squares.

$= (3 + 2x + y)(3 - 2x - y)$ — Simplify each factor.

Check by multiplication.

SELF CHECK 11 Factor: $16 - a^2 - 2ab - b^2$

SELF CHECK ANSWERS

1. $(3x + 1)(x + 2)$ 2. $(3x - 2)(2x - 1)$ 3. $(5a + 3)(a - 2)$ 4. $(2x + 3)(2x - 1)$ 5. $(2x + 3y)(2x + y)$ 6. $-2y(y + 3)(y - 2)$ 7. $(2p - 1)(p - 3)$ 8. $(3y + 2)(y - 2)$ 9. $(3x - 2)^2$ 10. $(x + 2 + y)(x + 2 - y)$ 11. $(4 + a + b)(4 - a - b)$

NOW TRY THIS

1. If the area of a rectangle can be expressed by the polynomial $(35x^2 - 31x + 6)$ cm^2, find polynomial expressions for the length and width.
2. Factor, if possible:
   a. $-8x^2 - 15 + 22x$
   b. $15x^2 - 28x - 12$

5.4 Exercises

WARM-UPS *Fill in the blanks.*

1. $6x^2 + 7x + 2 = (2x + 1)(3x + \square)$
2. $3t^2 + t - 2 = (3t - \square)(t + 1)$
3. $6x^2 + x - 2 = (3x + \square)(2x - \square)$
4. $6x^2 + 5x + 1 = (\square x + 1)(3x + 1)$
5. $15x^2 - 7x - 4 = (5x - \square)(3x + 1)$
6. $2x^2 + 5x + 3 = (2x + \square)(x + 1)$

REVIEW

7. The nth term l of an arithmetic sequence is

$$l = f + (n - 1)d$$

where f is the first term and d is the common difference. Remove the parentheses and solve for n.

8. The sum S of n consecutive terms of an arithmetic sequence is

$$S = \frac{n}{2}(f + l)$$

where f is the first term and l is the nth term. Solve for f.

VOCABULARY AND CONCEPTS *Fill in the blanks.*

9. If the sign of the first and third terms of a trinomial are +, the signs within the binomial factors are _________ the sign of the middle term.
10. If the sign of the first term of a trinomial is + and the sign of the third term is −, the signs within the binomial factors are _________.
11. An alternative method to trial and error is the _________ or grouping.
12. A visual of the ac method is the _________.

Complete each factorization.

13. $6x^2 + 5x - 1 = (x \square 1)(6x \square 1)$
14. $6x^2 + x - 1 = (2x \square 1)(3x \square 1)$
15. $4x^2 + 4x - 3 = (2x + \square)(2x - \square)$
16. $4x^2 - x - 3 = (4x + \square)(x - \square)$
17. $12x^2 - 7xy + y^2 = (3x - \square)(4x - \square)$
18. $6x^2 + 5xy - 6y^2 = (2x + \square)(3x - \square)$

GUIDED PRACTICE *Factor.* SEE EXAMPLE 1. (OBJECTIVE 1)

19. $3a^2 + 7a + 2$
20. $6y^2 + 7y + 2$
21. $3a^2 + 10a + 3$
22. $2b^2 + 13b + 15$
23. $5t^2 + 13t + 6$
24. $16y^2 + 10y + 1$
25. $16x^2 + 16x + 3$
26. $4z^2 + 13z + 3$

Factor. SEE EXAMPLE 2. (OBJECTIVE 1)

27. $5y^2 - 23y + 12$
28. $10x^2 - 9x + 2$
29. $2y^2 - 7y + 3$
30. $7z^2 - 26z + 15$
31. $16m^2 - 14m + 3$
32. $4t^2 - 4t + 1$
33. $6x^2 - 7x + 2$
34. $20x^2 - 23x + 6$

Factor. SEE EXAMPLE 3. (OBJECTIVE 1)

35. $3a^2 - 4a - 4$
36. $2x^2 - 3x - 2$
37. $12y^2 - y - 1$
38. $8u^2 - 2u - 15$
39. $12y^2 - 5y - 2$
40. $10x^2 - x - 2$
41. $10y^2 - 3y - 1$
42. $14x^2 - 3x - 2$

Factor each trinomial. SEE EXAMPLE 4. (OBJECTIVE 1)

43. $8q^2 + 10q - 3$
44. $2m^2 + 5m - 12$
45. $10x^2 + 21x - 10$
46. $4y^2 + 5y - 6$
47. $30x^2 - 23x - 14$
48. $18x^2 - 15x - 25$
49. $8x^2 - 14x - 15$
50. $6x^2 - 7x - 20$

Factor. SEE EXAMPLE 5. (OBJECTIVE 1)

51. $2x^2 + 3xy + y^2$
52. $5m^2 + 8mn + 3n^2$
53. $3x^2 - 4xy + y^2$
54. $2b^2 - 5bc + 2c^2$
55. $10p^2 - 11pq - 6q^2$
56. $2u^2 + 3uv - 2v^2$
57. $6p^2 - pq - 2q^2$
58. $8r^2 - 10rs - 25s^2$

Write the terms of each trinomial in descending powers of one variable. Then factor. SEE EXAMPLE 6. (OBJECTIVE 1)

59. $15 + 8a^2 - 26a$
60. $16 - 40a + 25a^2$
61. $12x^2 + 10y^2 - 23xy$
62. $3ab + 20a^2 - 2b^2$

Factor. SEE EXAMPLES 7–8. (OBJECTIVE 2)

63. $6x^2 + 7x + 2$
64. $6m^2 + 19m + 3$
65. $-26x + 6x^2 - 20$
66. $-42 + 9a^2 - 3a$

Factor. SEE EXAMPLE 9. (OBJECTIVE 3)

67. $9x^2 - 12x + 4$
68. $9x^2 + 6x + 1$
69. $25x^2 + 30x + 9$
70. $16y^2 - 24y + 9$
71. $9a^2 + 24a + 16$
72. $4x^2 - 4x + 1$
73. $16x^2 - 40x + 25$
74. $25x^2 - 20x + 4$

Factor. SEE EXAMPLES 10–11. (OBJECTIVE 3)

75. $16x^2 + 8xy + y^2 - 9$
76. $9x^2 - 6x + 1 - d^2$
77. $9 - a^2 - 4ab - 4b^2$
78. $16m^2 - 24m - n^2 + 9$

ADDITIONAL PRACTICE

Factor. If the polynomial is prime, so indicate.

79. $-12y^2 - 12 + 25y$
80. $-12t^2 + 1 + 4t$
81. $5x^2 + 2 + x$
82. $25 + 2u^2 + 3u$
83. $4x^2 + 10x - 6$
84. $9x^2 + 21x - 18$
85. $y^3 + 13y^2 + 12y$
86. $6x^3 - 15x^2 - 9x$
87. $9y^3 + 3y^2 - 6y$
88. $30r^5 + 63r^4 - 30r^3$
89. $6s^5 - 26s^4 - 20s^3$
90. $4a^2 - 15ab + 9b^2$
91. $12x^2 + 5xy - 3y^2$
92. $b^2 + 4a^2 + 16ab$
93. $3b^2 + 3a^2 - ab$
94. $4a^2 - 4ab - 8b^2$
95. $25x^2 + 20xy + 4y^2$
96. $6r^2 + rs - 2s^2$
97. $-16x^4y^3 + 30x^3y^4 + 4x^2y^5$
98. $9p^2 + 1 + 6p - q^2$
99. $24a^2 + 14ab + 2b^2$
100. $9a^2 + 6ac - c^2$

WRITING ABOUT MATH

101. Describe an organized approach to finding all of the possibilities when you attempt to factor $12x^2 - 4x + 9$.

102. Explain how to determine whether a trinomial is prime.

SOMETHING TO THINK ABOUT

103. For what values of b will the trinomial $6x^2 + bx + 6$ be factorable?

104. For what values of b will the trinomial $5y^2 - by - 3$ be factorable?

Section 5.5 Factoring the Sum and Difference of Two Cubes

Objectives

1. Factor the sum of two cubes.
2. Factor the difference of two cubes.
3. Factor a polynomial involving the sum or difference of two cubes.

Vocabulary

sum of two cubes

difference of two cubes

Getting Ready

Find each product.

1. $(x - 3)(x^2 + 3x + 9)$
2. $(x + 2)(x^2 - 2x + 4)$
3. $(y + 4)(y^2 - 4y + 16)$
4. $(r - 5)(r^2 + 5r + 25)$
5. $(a - b)(a^2 + ab + b^2)$
6. $(a + b)(a^2 - ab + b^2)$

Recall that the difference of the squares of two quantities factors into the product of two binomials. One binomial is the sum of the quantities, and the other is the difference of the quantities.

$$x^2 - y^2 = (x + y)(x - y) \qquad \text{or} \qquad \mathrm{F}^2 - \mathrm{L}^2 = (\mathrm{F} + \mathrm{L})(\mathrm{F} - \mathrm{L})$$

In this section, we will discuss formulas for factoring the *sum of two cubes* and the *difference of two cubes.*

1 Factor the sum of two cubes.

To discover the pattern for factoring the sum of two cubes, we find the following product:

$$(x + y)(x^2 - xy + y^2) = x^3 - x^2y + xy^2 + x^2y - xy^2 + y^3 \quad \text{Use the distributive property.}$$

$$= x^3 + y^3 \quad \text{Combine like terms.}$$

This result justifies the formula for factoring the **sum of two cubes**.

FACTORING THE SUM OF TWO CUBES

$$x^3 + y^3 = (x + y)(x^2 - xy + y^2)$$

To factor the sum of two cubes, it is helpful to know the cubes of the numbers from 1 to 10:

1, 8, 27, 64, 125, 216, 343, 512, 729, 1,000

Expressions containing variables such as x^6y^3 are also perfect cubes, because they can be written as the cube of a quantity:

$$x^6y^3 = (x^2y)^3$$

EXAMPLE 1 Factor: $x^3 + 8$

Solution The binomial $x^3 + 8$ is the sum of two cubes, because

$$x^3 + 8 = x^3 + 2^3$$

Thus, $x^3 + 8$ factors as $(x + 2)$ times the trinomial $(x^2 - 2 \cdot x + 2^2)$.

$$\begin{aligned} \mathbf{x^3 + 8} &= \mathbf{x^3 + 2^3} \\ &= \mathbf{(x + 2)(x^2 - x \cdot 2 + 2^2)} \\ &= (x + 2)(x^2 - 2x + 4) \end{aligned}$$

To check, we can use the distributive property and combine like terms.

$$\begin{aligned} (x + 2)(x^2 - 2x + 4) &= x^3 - 2x^2 + 4x + 2x^2 - 4x + 8 \\ &= x^3 - 8 \end{aligned}$$

SELF CHECK 1 Factor: $p^3 + 64$

EXAMPLE 2 Factor: $8b^3 + 27c^3$

Solution The binomial $8b^3 + 27c^3$ is the sum of two cubes, because

$$8b^3 + 27c^3 = (2b)^3 + (3c)^3$$

Thus, the binomial $8b^3 + 27c^3$ factors as $(2b + 3c)$ times the trinomial $(2b)^2 - (2b)(3c) + (3c)^2$.

$$\begin{aligned} \mathbf{8b^3 + 27c^3} &= \mathbf{(2b)^3 + (3c)^3} \\ &= \mathbf{(2b + 3c)[(2b)^2 - (2b)(3c) + (3c)^2]} \\ &= (2b + 3c)(4b^2 - 6bc + 9c^2) \end{aligned}$$

To check, we can use the distributive property and combine like terms.

$$\begin{aligned} &(2b + 3c)(4b^2 - 6bc + 9c^2) \\ &\quad = 8b^3 - 12b^2c + 18bc^2 + 12b^2c - 18bc^2 + 27c^3 \\ &\quad = 8b^3 + 27c^3 \end{aligned}$$

SELF CHECK 2 Factor: $1{,}000p^3 + q^3$

2 Factor the difference of two cubes.

To discover the pattern for factoring the difference of two cubes, we find the following product:

$$\begin{aligned} (x - y)(x^2 + xy + y^2) &= x^3 + x^2y + xy^2 - x^2y - xy^2 - y^3 && \text{Use the distributive property.} \\ &= x^3 - y^3 && \text{Combine like terms.} \end{aligned}$$

This result justifies the formula for factoring the **difference of two cubes**.

FACTORING THE DIFFERENCE OF TWO CUBES

$$x^3 - y^3 = (x - y)(x^2 + xy + y^2)$$

EXAMPLE 3 Factor: $a^3 - 64b^3$

Solution The binomial $a^3 - 64b^3$ is the difference of two cubes.

$$a^3 - 64b^3 = a^3 - (4b)^3$$

Thus, its factors are the difference $a - 4b$ and the trinomial $a^2 + a(4b) + (4b)^2$.

$$\begin{aligned} a^3 - 64b^3 &= a^3 - (4b)^3 \\ &= (a - 4b)[a^2 + a(4b) + (4b)^2] \\ &= (a - 4b)(a^2 + 4ab + 16b^2) \end{aligned}$$

To check, we can use the distributive property and combine like terms.

$$\begin{aligned} &(a - 4b)(a^2 + 4ab + 16b^2) \\ &\quad = a^3 + 4a^2b + 16ab^2 - 4a^2b - 16ab^2 - 64b^3 \\ &\quad = a^3 - 64b^3 \end{aligned}$$

SELF CHECK 3 Factor: $27p^3 - 8$

3 Factor a polynomial involving the sum or difference of two cubes.

Sometimes we must factor out a greatest common factor before factoring a sum or difference of two cubes.

EXAMPLE 4 Factor: $-2t^5 + 128t^2$

Solution

$-2t^5 + 128t^2 = -2t^2(t^3 - 64)$	Factor the GCF, $-2t^2$, from $(-2t^5 + 128t^2)$.
$= -2t^2(t^3 - 4^3)$	Write $t^3 - 64$ as $t^3 - 4^3$.
$= -2t^2(t - 4)(t^2 + 4t + 16)$	Factor $t^3 - 4^3$.

We can check by multiplication.

SELF CHECK 4 Factor: $-3p^4 + 81p$

If a binomial is both the difference of two squares and the difference of two cubes, we will factor the difference of two squares first.

EXAMPLE 5 Factor: $x^6 - 64$

Solution If we consider the polynomial to be the difference of two squares, we can factor it as follows:

$$\begin{aligned} x^6 - 64 &= (x^3)^2 - 8^2 \\ &= (x^3 + 8)(x^3 - 8) \end{aligned}$$

Because $x^3 + 8$ is the sum of two cubes and $x^3 - 8$ is the difference of two cubes, each of these binomials can be factored.

$$\begin{aligned} x^6 - 64 &= (x^3 + 8)(x^3 - 8) \\ &= (x + 2)(x^2 - 2x + 4)(x - 2)(x^2 + 2x + 4) \end{aligned}$$

We can check by multiplication.

SELF CHECK 5 Factor: $a^6 - 1$

SELF CHECK ANSWERS

1. $(p + 4)(p^2 - 4p + 16)$ 2. $(10p + q)(100p^2 - 10pq + q^2)$ 3. $(3p - 2)(9p^2 + 6p + 4)$ 4. $-3p(p - 3)(p^2 + 3p + 9)$ 5. $(a + 1)(a^2 - a + 1)(a - 1)(a^2 + a + 1)$

NOW TRY THIS

Factor.

1. $x^3 - \frac{1}{8}$
2. $x^3 - y^{12}$
3. $64x^3 - 8$
4. $x^3(x^2 - 9) - 8(x^2 - 9)$

5.5 Exercises

WARM-UPS *Write each expression as a cube.*

1. 8
2. x^6
3. -27
4. 125
5. y^{12}
6. $27x^3$
7. $-y^9$
8. $-64y^6$

Square each expression.

9. 3
10. x^2
11. $4y^2$
12. $3x$
13. $-2x$
14. $-3y^2$
15. $-x^5$
16. $-5y^3$

REVIEW

17. The length of one fermi is 1×10^{-13} centimeter, approximately the radius of a proton. Express this number in standard notation.
18. In the 14th century, the Black Plague killed about 25,000,000 people, which was 25% of the population of Europe. Find the population at that time, expressed in scientific notation.

VOCABULARY AND CONCEPTS *Fill in the blanks.*

19. A polynomial in the form of $a^3 + b^3$ is called a ______________.
20. A polynomial in the form of $a^3 - b^3$ is called a ______________.

Complete each formula.

21. $x^3 + y^3 = (x + y)$
22. $x^3 - y^3 = (x - y)$

GUIDED PRACTICE *Factor.* SEE EXAMPLE 1. (OBJECTIVE 1)

23. $a^3 + 8$
24. $b^3 + 125$
25. $125x^3 + 8$
26. $8 + x^3$
27. $y^3 + 1$
28. $1 + 8x^3$
29. $125 + a^3$
30. $64 + b^3$

Factor. SEE EXAMPLE 2. (OBJECTIVE 1)

31. $m^3 + n^3$
32. $a^3 + 8b^3$
33. $x^3 + y^3$
34. $27x^3 + y^3$
35. $8u^3 + w^3$
36. $x^3y^3 + 1$

Factor. SEE EXAMPLE 3. (OBJECTIVE 2)

37. $x^3 - y^3$
38. $b^3 - 27$
39. $x^3 - 8$
40. $a^3 - 64$
41. $s^3 - t^3$
42. $27 - y^3$
43. $125p^3 - q^3$
44. $x^3 - 27y^3$
45. $27a^3 - b^3$
46. $64x^3 - 27$

Factor. SEE EXAMPLE 4. (OBJECTIVE 3)

47. $2x^3 + 54$
48. $5x^3 - 5$
49. $-x^3 + 216$
50. $-x^3 - 125$
51. $64m^3x - 8n^3x$
52. $16r^4 + 128rs^3$
53. $x^4y + 216xy^4$
54. $16a^5 - 54a^2b^3$

Factor each polynomial completely. Factor a difference of two squares first. SEE EXAMPLE 5. (OBJECTIVE 3)

55. $x^6 - 1$
56. $x^6 - y^6$
57. $x^{12} - y^6$
58. $a^{12} - 64$

ADDITIONAL PRACTICE *Factor.*

59. $y^3 + 8$
60. $x^3 - y^9$
61. $27x^3 + 125$
62. $81r^4s^2 - 24rs^5$
63. $64 - z^3$
64. $27x^3 - 125y^3$
65. $64x^3 + 27y^3$
66. $216a^4b^4 - 1{,}000ab^7$
67. $3(x^3 + y^3) - z(x^3 + y^3)$
68. $y^7z - yz^4$
69. $x(27y^3 - z^3) + 5(27y^3 - z^3)$
70. $(m^3 + 8n^3) + (m^3x + 8n^3x)$
71. $(a^3x + b^3x) - (a^3y + b^3y)$
72. $(a^4 + 27a) - (a^3b + 27b)$
73. $y^3(y^2 - 1) - 27(y^2 - 1)$
74. $z^3(y^2 - 4) + 8(y^2 - 4)$

WRITING ABOUT MATH

75. Explain how to factor $a^3 + b^3$.
76. Explain the difference between $x^3 - y^3$ and $(x - y)^3$.

SOMETHING TO THINK ABOUT

77. Let $a = 11$ and $b = 7$. Use a calculator to verify that
$a^3 - b^3 = (a - b)(a^2 + ab + b^2)$

78. Let $p = 5$ and $q = -2$. Use a calculator to verify that
$p^3 + q^3 = (p + q)(p^2 - pq + q^2)$

Section 5.6 Summary of Factoring Techniques

Objectives

1 Factor a polynomial.

Getting Ready

Multiply.

1. $3ax(x + a)$	**2.** $(x + 3y)(x - 3y)$
3. $(x - 2)(x^2 + 2x + 4)$	**4.** $2(x + 2)(x - 2)$
5. $(x - 5)(x + 2)$	**6.** $(2x - 3)(3x - 2)$
7. $2(3x - 1)(x - 2)$	**8.** $(a + b)(x + y)(x - y)$

In this section, we will discuss ways to approach any factoring problem.

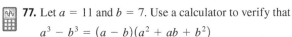

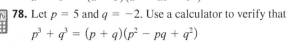

1 Factor a polynomial.

Suppose we want to factor the trinomial

$$x^4y + 7x^3y - 18x^2y$$

Since it is written in descending powers of x, we begin by identifying the problem type. The first type we look for is one that contains a common factor. Because the trinomial has the greatest common factor of x^2y, we factor it out first:

$$x^4y + 7x^3y - 18x^2y = x^2y(x^2 + 7x - 18)$$

We can factor the remaining trinomial $x^2 + 7x - 18$ as $(x + 9)(x - 2)$. Thus,

$$\begin{aligned} x^4y + 7x^3y - 18x^2y &= x^2y(x^2 + 7x - 18) \\ &= x^2y(x + 9)(x - 2) \end{aligned}$$

To identify the type of factoring problem, we follow these steps.

FACTORING A POLYNOMIAL

1. Write the polynomial in descending powers of one variable.
2. Factor out all common factors.
3. If an expression has two terms, check to see if the problem type is
   a. the difference of two squares: $x^2 - y^2 = (x + y)(x - y)$
   b. the sum of two cubes: $x^3 + y^3 = (x + y)(x^2 - xy + y^2)$
   c. the difference of two cubes: $x^3 - y^3 = (x - y)(x^2 + xy + y^2)$
4. If an expression has three terms, check to see if it is a perfect-square trinomial:

   $$\begin{aligned} x^2 + 2xy + y^2 &= (x + y)(x + y) \\ x^2 - 2xy + y^2 &= (x - y)(x - y) \end{aligned}$$

   If the trinomial is not a perfect-square trinomial, attempt to factor the trinomial as a general trinomial.
5. If an expression has four terms, try to factor the expression by grouping. It may be necessary to rearrange the terms.
6. Continue factoring until each nonmonomial factor is prime.
7. If the polynomial does not factor, the polynomial is *prime* over the set of rational numbers.
8. Check the results by multiplying.

EXAMPLE 1 Factor: $x^5y^2 - xy^6$

Solution Since the polynomial is written in descending powers of x, we begin by factoring out the greatest common factor of xy^2.

$$x^5y^2 - xy^6 = xy^2(x^4 - y^4)$$

The expression $x^4 - y^4$ has two terms. We check to see whether it is the difference of two squares, which it is. As the difference of two squares, it factors as $(x^2 + y^2)(x^2 - y^2)$.

$$\begin{aligned} x^5y^2 - xy^6 &= xy^2(x^4 - y^4) \\ &= xy^2(x^2 + y^2)(x^2 - y^2) \end{aligned}$$

The binomial $x^2 + y^2$ is the sum of two squares and cannot be factored. However, $x^2 - y^2$ is the difference of two squares and factors as $(x + y)(x - y)$.

$$\begin{aligned} x^5y^2 - xy^6 &= xy^2(x^4 - y^4) \\ &= xy^2(x^2 + y^2)(x^2 - y^2) \\ &= xy^2(x^2 + y^2)(x + y)(x - y) \end{aligned}$$

Since each individual factor is prime, the given expression is in completely factored form.

SELF CHECK 1 Factor: $-a^5b + ab^5$

EXAMPLE 2 Factor: $x^6 - x^4y^2 - x^3y^3 + xy^5$

Solution Factor out the greatest common factor of x.

$$x^6 - x^4y^2 - x^3y^3 + xy^5 = x(x^5 - x^3y^2 - x^2y^3 + y^5)$$

Since $x^5 - x^3y^2 - x^2y^3 + y^5$ has four terms, we try factoring it by grouping:

$$\begin{aligned} x^6 - x^4y^2 - x^3y^3 + xy^5 &= x(x^5 - x^3y^2 - x^2y^3 + y^5) \\ &= x[x^3(x^2 - y^2) - y^3(x^2 - y^2)] \\ &= x(x^2 - y^2)(x^3 - y^3) \quad \text{Factor out } x^2 - y^2. \end{aligned}$$

Finally, we factor the difference of two squares and the difference of two cubes:

$$\begin{aligned} x^6 - x^4y^2 - x^3y^3 + xy^5 &= x(x^2 - y^2)(x^3 - y^3) \\ &= x(x + y)(x - y)(x - y)(x^2 + xy + y^2) \\ &= x(x + y)(x - y)^2(x^2 + xy + y^2) \quad (x - y)(x - y) = (x - y)^2 \end{aligned}$$

Since each factor is prime, the given expression is in completely factored form.

SELF CHECK 2 Factor: $2a^5 - 2a^2b^3 - 8a^3 + 8b^3$

SELF CHECK ANSWERS

1. $-ab(a^2 + b^2)(a + b)(a - b)$ **2.** $2(a + 2)(a - 2)(a - b)(a^2 + ab + b^2)$

NOW TRY THIS

Factor.

1. $4x^2 + 16$

2. $ax^2 + bx^2 - 36a - 36b$

3. $9x^2 - 9x$

4. $64 - x^6$

5.6 Exercises

WARM-UPS *Indicate which factoring technique you would use first, if any.*

1. $3x^2 - 9x$

2. $49 - 16a^2$

3. $125 + r^3s^3$

4. $ax + ay - x - y$

5. $x^2 + 36$

6. $16x^2 - 24$

7. $25r^2 - s^4$

8. $8a^3 - 27b^3$

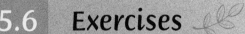

REVIEW *Solve each equation, if possible.*

9. $2(t + 5) + t = 3(t + 2) + 4$
10. $5 + 3(2x - 1) = 2(4 + 3x) - 24$
11. $6 + 2(t + 3) = t + 3$
12. $4m - 3 = -2(m + 1) - 3$

VOCABULARY AND CONCEPTS *Fill in the blanks.*

13. The first step in any factoring problem is to factor out all common ______, if possible.
14. If a polynomial has two terms, check to see if it is the ____________________, the sum of two cubes, or the ________ of two cubes.
15. If a polynomial has three terms, try to factor it as the product of two ________.
16. If a polynomial has four or more terms, try factoring by _______.

PRACTICE *Factor.*

17. $6x + 3$
18. $x^2 - 36$
19. $x^2 + 10x + 9$
20. $a^3 - 27$
21. $8t^2 - 6t - 9$
22. $4x^2 - 25$
23. $t^2 - 2t + 1$
24. $6p^2 - 3p - 2$
25. $2x^2 - 50$
26. $t^4 - 16$
27. $x^2 + 7x + 1$
28. $10r^2 - 13r - 4$
29. $-2x^5 + 128x^2$
30. $49 - 28z + 4z^2$
31. $14t^3 - 40t^2 + 6t^4$
32. $6x^2 + 7x - 20$
33. $6x^2 - x - 16$
34. $30a^4 + 5a^3 - 200a^2$
35. $6a^3 + 35a^2 - 6a$
36. $21t^3 - 10t^2 + t$
37. $16x^2 - 40x^3 + 25x^4$
38. $25a^2 - 60a + 36$
39. $-84x^2 - 147x - 12x^3$
40. $x^3 - 5x^2 - 25x + 125$
41. $8x^6 - 8$
42. $16x^2 + 64$
43. $5x^3 - 5x^5 + 25x^2$
44. $12y^3 - 27y$
45. $9x^2 + 12x + 16$
46. $70p^4q^3 - 35p^4q^2 + 49p^5q^2$
47. $2ab^2 + 8ab - 24a$
48. $2x^2y - 4xy^2$
49. $-8p^3q^7 - 4p^2q^3$
50. $8m^2n^3 - 24mn^4$
51. $4a^2 - 4ab + b^2 - 9$
52. $3rs + 6r^2 - 18s^2$
53. $8a^3 - b^3$
54. $ac + ad + bc + bd$
55. $x^2y^2 - 2x^2 - y^2 + 2$
56. $a^2c + a^2d^2 + bc + bd^2$
57. $a^2 + 2ab + b^2 - y^2$
58. $2x^3 + 54y^3$
59. $a^2(x - 3) - b^2(x - 3)$
60. $5x^3y^3z^4 + 25x^2y^3z^2 - 35x^3y^2z^5$
61. $8p^6 - 27q^6$
62. $3c^2 - 11cd - 4d^2$
63. $125p^3 - 64y^3$
64. $8a^2x^3y - 2b^2xy$
65. $-16x^4y^2z + 24x^5y^3z^4 - 15x^2y^3z^7$
66. $2ac + 4ad + bc + 2bd$
67. $81p^4 - 16q^4$
68. $4x^2 + 9y^2$
69. $54x^3 + 250y^6$
70. $4x^2 + 4x + 1 - y^2$
71. $x^5 - x^3y^2 + x^2y^3 - y^5$
72. $a^3x^3 - a^3y^3 + b^3x^3 - b^3y^3$
73. $2a^2c - 2b^2c + 4a^2d - 4b^2d$
74. $3a^2x^2 + 6a^2x + 3a^2 - 3b^2$

WRITING ABOUT MATH

75. Explain how to identify the type of factoring required to factor a polynomial.
76. Which factoring technique do you find most difficult? Why?

SOMETHING TO THINK ABOUT

77. Write $x^6 - y^6$ as $(x^3)^2 - (y^3)^2$, factor it as the difference of two squares, and show that you get

$$(x + y)(x^2 - xy + y^2)(x - y)(x^2 + xy + y^2)$$

Write $x^6 - y^6$ as $(x^2)^3 - (y^2)^3$, factor it as the difference of two cubes, and show that you get

$$(x + y)(x - y)(x^4 + x^2y^2 + y^4)$$

78. Verify that the results of Exercise 77 agree by showing the parts in color agree. Which do you think is completely factored?

Section 5.7 Solving Equations by Factoring

Objectives

1. Solve a quadratic equation in one variable using the zero-factor property.
2. Solve a higher-order polynomial equation in one variable.

Vocabulary

quadratic equation
quadratic form
zero-factor property
higher-order polynomial equation

Getting Ready

Solve each equation.

1. $x + 3 = 4$ **2.** $y - 8 = 5$ **3.** $3x - 2 = 7$ **4.** $5y + 9 = 19$

In this section, we will learn how to use factoring to solve many equations that contain second-degree polynomials in one variable. These equations are called *quadratic equations.*

Equations such as

$$3x + 2 = 0 \quad \text{and} \quad 9x - 6 = 0$$

that contain first-degree polynomials are *linear equations.* Equations such as

$$9x^2 - 6x = 0 \quad \text{and} \quad 3x^2 + 4x - 7 = 0$$

that contain second-degree polynomials are called **quadratic equations**.

QUADRATIC EQUATIONS

A quadratic equation in one variable is an equation of the form

$ax^2 + bx + c = 0$ (This is called **quadratic form**.)

where a, b, and c are real numbers, and $a \neq 0$.

1 Solve a quadratic equation in one variable using the zero-factor property.

Many quadratic equations can be solved by factoring. For example, to solve the quadratic equation

$$x^2 + 5x - 6 = 0$$

which is already in quadratic form, we begin by factoring the trinomial and writing the equation as

(1) $$(x + 6)(x - 1) = 0$$

This equation indicates that the product of two quantities is 0. However, if the product of two quantities is 0, then at least one of those quantities must be 0. This fact is called the **zero-factor property**.

ZERO-FACTOR PROPERTY

Suppose a and b represent two real numbers.

If $ab = 0$, then $a = 0$ or $b = 0$.

COMMENT In mathematics, when we use the word "or" it is understood to mean one or the other or both. Thus, if $a \cdot b = 0$, both a and b could be equal to 0.

By applying the zero-factor property to Equation 1, we have

$$x + 6 = 0 \qquad \text{or} \qquad x - 1 = 0$$

We can solve each of these linear equations to get

$$x = -6 \qquad \text{or} \qquad x = 1$$

To check, we substitute −6 for x, and then 1 for x in the original equation and simplify.

For x = −6	***For x = 1***
$x^2 + 5x - 6 = 0$	$x^2 + 5x - 6 = 0$
$(-6)^2 + 5(-6) - 6 \stackrel{?}{=} 0$	$(1)^2 + 5(1) - 6 \stackrel{?}{=} 0$
$36 - 30 - 6 \stackrel{?}{=} 0$	$1 + 5 - 6 \stackrel{?}{=} 0$
$6 - 6 \stackrel{?}{=} 0$	$6 - 6 \stackrel{?}{=} 0$
$0 = 0$	$0 = 0$

Both solutions check.

The quadratic equations $9x^2 - 6x = 0$ and $4x^2 - 25 = 0$ are each missing a term. The first equation is missing the constant term, and the second equation is missing the term involving x. These types of equations often can be solved by factoring.

EXAMPLE 1 Solve: $9x^2 - 6x = 0$

Solution We begin by factoring the left side of the equation.

$$9x^2 - 6x = 0$$

$$3x(3x - 2) = 0 \qquad \text{Factor out the common factor of } 3x.$$

By the zero-factor property, we have

$$3x = 0 \qquad \text{or} \qquad 3x - 2 = 0$$

Solve each linear equation.

$$3x = 0 \qquad \text{or} \qquad 3x - 2 = 0$$

$$3x = 2 \qquad \text{Add 2 to both sides.}$$

$$\frac{3x}{3} = \frac{0}{3} \qquad \frac{3x}{3} = \frac{2}{3} \qquad \text{Divide both sides of each equation by 3.}$$

$$x = 0 \qquad x = \frac{2}{3}$$

Check: We substitute these results for x in the original equation and simplify.

For $x = 0$

$$9x^2 - 6x = 0$$
$$9(0)^2 - 6(0) \stackrel{?}{=} 0$$
$$0 - 0 \stackrel{?}{=} 0$$
$$0 = 0$$

For $x = \frac{2}{3}$

$$9x^2 - 6x = 0$$
$$9\left(\frac{2}{3}\right)^2 - 6\left(\frac{2}{3}\right) \stackrel{?}{=} 0$$
$$9\left(\frac{4}{9}\right) - 6\left(\frac{2}{3}\right) \stackrel{?}{=} 0$$
$$4 - 4 \stackrel{?}{=} 0$$
$$0 = 0$$

Both solutions check.

SELF CHECK 1 Solve: $5y^2 + 10y = 0$

EXAMPLE 2 Solve: $4x^2 - 25 = 0$

Solution We proceed as follows:

$$4x^2 - 25 = 0$$
$$(2x + 5)(2x - 5) = 0 \quad \text{Factor } 4x^2 - 25.$$
$$2x + 5 = 0 \quad \text{or} \quad 2x - 5 = 0 \quad \text{Apply the zero-factor property.}$$
$$2x = -5 \quad | \quad 2x = 5 \quad \text{Isolate the variable term.}$$
$$x = -\frac{5}{2} \quad | \quad x = \frac{5}{2} \quad \text{Divide both sides of each equation by 2.}$$

Check each solution.

For $x = -\frac{5}{2}$

$$4x^2 - 25 = 0$$
$$4\left(-\frac{5}{2}\right)^2 - 25 \stackrel{?}{=} 0$$
$$4\left(\frac{25}{4}\right) - 25 \stackrel{?}{=} 0$$
$$0 = 0$$

For $x = \frac{5}{2}$

$$4x^2 - 25 = 0$$
$$4\left(\frac{5}{2}\right)^2 - 25 \stackrel{?}{=} 0$$
$$4\left(\frac{25}{4}\right) - 25 \stackrel{?}{=} 0$$
$$0 = 0$$

Both solutions check.

SELF CHECK 2 Solve: $9p^2 - 64 = 0$

In the next example, we solve an equation whose polynomial is a trinomial.

EXAMPLE 3 Solve: $x^2 - 3x - 18 = 0$

Solution

$$x^2 - 3x - 18 = 0$$
$$(x + 3)(x - 6) = 0 \quad \text{Factor } x^2 - 3x - 18.$$
$$x + 3 = 0 \quad \text{or} \quad x - 6 = 0 \quad \text{Apply the zero-factor property.}$$
$$x = -3 \quad | \quad x = 6 \quad \text{Solve each linear equation.}$$

Check each solution.

SELF CHECK 3 Solve: $x^2 + 3x - 18 = 0$

EXAMPLE 4 Solve: $2x^2 + 3x = 2$

Solution We write the equation in the form $ax^2 + bx + c = 0$ and solve for x.

COMMENT To apply the zero-factor property, the equation must be set equal to 0 prior to factoring.

$$2x^2 + 3x = 2$$

$$2x^2 + 3x - 2 = 0 \quad \text{Subtract 2 from both sides.}$$

$$(2x - 1)(x + 2) = 0 \quad \text{Factor } 2x^2 + 3x - 2.$$

$$2x - 1 = 0 \quad \text{or} \quad x + 2 = 0 \quad \text{Apply the zero-factor property.}$$

$$2x = 1 \quad \Big| \quad x = -2 \quad \text{Solve each linear equation.}$$

$$x = \frac{1}{2}$$

Check each solution.

SELF CHECK 4 Solve: $3x^2 - 5x = 2$

Some equations must be simplified before we write them in quadratic form. Many times this requires the distributive property as illustrated in the next example.

EXAMPLE 5 Solve: $(x - 2)(x - 6) = -3$

Solution We must write the equation in the form $ax^2 + bx + c = 0$ before we can solve for x. We first multiply the binomials and then set the result equal to zero.

$$(x - 2)(x - 6) = -3$$

$$x^2 - 6x - 2x + 12 = -3 \quad \text{Multiply.}$$

$$x^2 - 8x + 12 = -3 \quad \text{Combine like terms.}$$

$$x^2 - 8x + 15 = 0 \quad \text{Add 3 to both sides.}$$

$$(x - 3)(x - 5) = 0 \quad \text{Factor the trinomial.}$$

$$(x - 3) = 0 \quad \text{or} \quad x - 5 = 0 \quad \text{Apply the zero-factor property.}$$

$$x = 3 \quad \Big| \quad x = 5 \quad \text{Solve each linear equation.}$$

Check each solution.

SELF CHECK 5 Solve: $(x + 1)(x - 5) = 7$

2 Solve a higher-order polynomial equation in one variable.

A **higher-order polynomial equation** is any equation in one variable with a degree of 3 or larger.

EXAMPLE 6 Solve: $x^3 - 2x^2 - 63x = 0$

Solution We begin by completely factoring the left side.

$$x^3 - 2x^2 - 63x = 0$$

$$x(x^2 - 2x - 63) = 0 \quad \text{Factor out } x\text{, the GCF.}$$

$$x(x + 7)(x - 9) = 0 \quad \text{Factor the trinomial.}$$

$$x = 0 \quad \text{or} \quad x + 7 = 0 \quad \text{or} \quad x - 9 = 0 \quad \text{Set each factor equal to 0.}$$

$$\Big| \quad x = -7 \quad \Big| \quad x = 9 \quad \text{Solve each linear equation.}$$

Check each solution.

SELF CHECK 6 Solve: $x^3 - x^2 - 2x = 0$

As with quadratic equations, higher-order equations in one variable must be set equal to zero to solve by factoring.

EXAMPLE 7 Solve: $6x^3 + 12x = 17x^2$

Solution To set the equation equal to 0, we subtract $17x^2$ from both sides. Then we proceed as follows:

$$6x^3 + 12x = 17x^2$$

$$6x^3 - 17x^2 + 12x = 0 \qquad \text{Subtract } 17x^2 \text{ from both sides.}$$

$$x(6x^2 - 17x + 12) = 0 \qquad \text{Factor out } x\text{, the GCF.}$$

$$x(2x - 3)(3x - 4) = 0 \qquad \text{Factor } 6x^2 - 17x + 12.$$

$$x = 0 \quad \text{or} \quad 2x - 3 = 0 \quad \text{or} \quad 3x - 4 = 0 \qquad \text{Set each factor equal to 0.}$$

$$2x = 3 \qquad 3x = 4 \qquad \text{Solve the linear equations.}$$

$$x = \frac{3}{2} \qquad x = \frac{4}{3}$$

Check each solution.

SELF CHECK 7 Solve: $6x^3 + 7x^2 = 5x$

Everyday connections
Selling Calendars

© Shutterstock.com/HomeStudio

A bookshop is selling calendars at a price of \$4 each. At this price, the store can sell 12 calendars per day. The manager estimates that for each \$1 increase in the selling price, the store will sell 3 fewer calendars per day. Each calendar costs the store \$2. We can represent the store's total daily profit from calendar sales by the function $p(x) = -3x^2 + 30x - 48$, where x represents the selling price, in dollars, of a calendar. Find the selling price at which the profit $p(x)$ equals zero.

SELF CHECK ANSWERS

1. 0, −2 **2.** $\frac{8}{3}, -\frac{8}{3}$ **3.** 3, −6 **4.** $2, -\frac{1}{3}$ **5.** 6, −2 **6.** 0, 2, −1 **7.** $0, \frac{1}{2}, -\frac{5}{3}$

NOW TRY THIS

Solve each equation.

1. $8x^2 - 8 = 0$

2. $x^2 = 3x$

3. $x(x + 10) = -25$

5.7 Exercises

WARM-UPS *Solve.*

1. $(x-8)(x-7)=0$
2. $(x+9)(x-2)=0$
3. $(x-2)(x+3)=0$
4. $(x-3)(x-2)=0$
5. $(x-4)(x+1)=0$
6. $(x+5)(x+2)=0$
7. $(2x-5)(3x+6)=0$
8. $(3x-4)(x+1)=0$

REVIEW *Simplify each expression and write all results without using negative exponents.*

9. $u^3u^2u^4$
10. $\frac{y^6}{y^8}$
11. $\frac{a^3b^4}{a^2b^5}$
12. $(-2x^6)^0$

VOCABULARY AND CONCEPTS *Fill in the blanks.*

13. An equation of the form $ax^2+bx+c=0$, where $a \neq 0$, is called a ________ equation.
14. The property "If $ab=0$, then $a=$ _ or $b=$ _" is called the ________ property.
15. A quadratic equation contains a ______-degree polynomial in one variable.
16. If the product of three factors is 0, then at least one of the numbers must be _.

GUIDED PRACTICE *Solve.* SEE EXAMPLE 1. (OBJECTIVE 1)

17. $x^2+7x=0$
18. $x^2-12x=0$
19. $x^2-2x+1=0$
20. $x^2+x-20=0$
21. $x^2-3x=0$
22. $x^2+5x=0$
23. $5x^2+7x=0$
24. $2x^2-5x=0$
25. $x^2-7x=0$
26. $2x^2+10x=0$
27. $3x^2+8x=0$
28. $5x^2-x=0$

Solve. SEE EXAMPLE 2. (OBJECTIVE 1)

29. $x^2-25=0$
30. $x^2-36=0$
31. $9y^2-4=0$
32. $16z^2-25=0$

Solve. SEE EXAMPLE 3. (OBJECTIVE 1)

33. $x^2-13x+12=0$
34. $x^2+7x+6=0$
35. $x^2-2x-15=0$
36. $x^2+x-20=0$
37. $x^2-3x-18=0$
38. $x^2+3x-10=0$
39. $x^2-x-20=0$
40. $x^2-10x+24=0$

Solve. SEE EXAMPLE 4. (OBJECTIVE 1)

41. $6x^2+x=2$
42. $12x^2+5x=3$
43. $2x^2-5x=-2$
44. $5p^2-6p=-1$
45. $x^2=49$
46. $z^2=25$
47. $4x^2=81$
48. $9y^2=64$

Solve. SEE EXAMPLE 5. (OBJECTIVE 1)

49. $x(6x+5)=6$
50. $x(2x-3)=14$
51. $(x+1)(8x+1)=18x$
52. $4x(3x+2)=x+12$

Solve. SEE EXAMPLE 6. (OBJECTIVE 2)

53. $(x+4)(x-5)(x-7)=0$
54. $(x+2)(x+3)(x-4)=0$
55. $(x-1)(x^2+5x+6)=0$
56. $(x-2)(x^2-8x+7)=0$
57. $x^3+3x^2+2x=0$
58. $x^3-7x^2+10x=0$
59. $x^3-27x-6x^2=0$
60. $x^3-22x-9x^2=0$

Solve. SEE EXAMPLE 7. (OBJECTIVE 2)

61. $6x^3+20x^2=-6x$
62. $2x^3-2x^2=4x$
63. $x^3+7x^2=x^2-9x$
64. $x^3+10x^2=2x^2-16x$

ADDITIONAL PRACTICE *Solve.*

65. $x^2-4x=0$
66. $15x^2-20x=0$
67. $9x^2+5x=0$
68. $5x^2+x=0$
69. $(x+3)(x^2+2x-15)=0$
70. $(x+4)(x^2-2x-15)$
71. $x^2-4x-21=0$
72. $x^2+2x-15=0$
73. $2y-8=-y^2$
74. $-3y+18=y^2$
75. $(p^2-81)(p+2)=0$
76. $(4q^2-49)(q-7)=0$
77. $15x^2-2=7x$
78. $8x^2+10x=3$
79. $x^2+8-9x=0$
80. $45+x^2-14x=0$
81. $a^2+8a=-15$
82. $a^2-a=56$
83. $3x^2-8x=3$
84. $2x^2-11x=21$
85. $2x^2+x-3=0$
86. $6q^2-5q+1=0$

87. $14m^2 + 23m + 3 = 0$

88. $35n^2 - 34n + 8 = 0$

89. $(x + 2)(x^2 + x - 20) = 0$

90. $(x + 1)(x^2 - 9x + 8) = 0$

91. $z^2 - 81 = 0$

92. $x^2 - 16 = 0$

93. $4x^2 - 1 = 0$

94. $9y^2 - 1 = 0$

95. $x^3 + 1.3x^2 - 0.3x = 0$

96. $2.4x^3 - x^2 - 0.4x = 0$

WRITING ABOUT MATH

97. If the product of several numbers is 0, at least one of the numbers is 0. Explain why.

98. Explain the error in this solution.

$$5x^2 + 2x = 10$$
$$x(5x + 2) = 10$$
$$x = 10 \quad \text{or} \quad 5x + 2 = 10$$
$$5x = 8$$
$$x = \frac{8}{5}$$

SOMETHING TO THINK ABOUT

99. Solve in two ways: $3a^2 + 9a - 2a - 6 = 0$

100. Solve in two ways: $p^2 - 2p + p - 2 = 0$

Section 5.8 Solving Applications

Objectives

1. Solve an integer application using a quadratic equation.
2. Solve a motion application using a quadratic equation.
3. Solve a geometric application using a quadratic equation.

Getting Ready

1. One side of a square is s inches long. Find an expression that represents its area.
2. The length of a rectangle is 4 centimeters more than twice the width. If w represents the width, find an expression that represents the length.
3. If x represents the smaller of two consecutive integers, find an expression that represents their product.
4. The length of a rectangle is 3 inches greater than the width. If w represents the width of the rectangle, find an expression that represents the area.

Finally, we can use the methods for solving quadratic equations discussed in the previous section to solve applications.

1 Solve an integer application using a quadratic equation.

EXAMPLE 1 One integer is 5 less than another and their product is 84. Find the integers.

Analyze the problem We are asked to find two integers. Let x represent the larger number. Then $x - 5$ represents the smaller number.

Form an equation We know that the product of the integers is 84. Since a product refers to multiplication, we can form the equation $x(x - 5) = 84$.

Solve the equation — To solve the equation, we proceed as follows.

$$x(x - 5) = 84$$

$$x^2 - 5x = 84 \quad \text{Use the distributive property to remove parentheses.}$$

$$x^2 - 5x - 84 = 0 \quad \text{Subtract 84 from both sides.}$$

$$(x - 12)(x + 7) = 0 \quad \text{Factor.}$$

$$x - 12 = 0 \quad \text{or} \quad x + 7 = 0 \quad \text{Apply the zero-factor property.}$$

$$x = 12 \quad \bigg| \quad x = -7 \quad \text{Solve each linear equation.}$$

We have two different values for the first integer,

$$x = 12 \quad \text{or} \quad x = -7$$

and two different values for the second integer,

$$x - 5 = 7 \quad \text{or} \quad x - 5 = -12$$

State the conclusion — There are two pairs of integers: 12 and 7, and -7 and -12.

Check the result — The number 7 is five less than 12 and $12 \cdot 7 = 84$. The number -12 is five less than -7 and $-7 \cdot -12 = 84$. Both pairs of integers check.

SELF CHECK 1 One integer is 7 more than another and their product is 60. Find the integers.

COMMENT In Example 1, we could have let x represent the smaller number, in which case the larger number would be described as $(x + 5)$. The results would be the same.

2 Solve a motion application using a quadratic equation.

EXAMPLE 2 **FLYING OBJECTS** If an object is launched straight up into the air with an initial velocity of 112 feet per second, its height after t seconds is given by the formula

$$h = 112t - 16t^2$$

where h represents the height of the object in feet. After this object has been launched, in how many seconds will it hit the ground?

Analyze the problem — We are asked to find the number of seconds it will take for an object to hit the ground. When the object is launched, it will go up and then come down. When it hits the ground, its height will be 0. So, we let $h = 0$.

Form an equation — If we substitute 0 for h in the formula $h = 112t - 16t^2$, the new equation will be $0 = 112t - 16t^2$ and we will solve for t.

$$h = 112t - 16t^2$$

$$0 = 112t - 16t^2$$

Solve the equation — We solve the equation as follows.

$$0 = 112t - 16t^2$$

$$0 = 16t(7 - t) \quad \text{Factor out } 16t\text{, the GCF.}$$

$$16t = 0 \quad \text{or} \quad 7 - t = 0 \quad \text{Set each factor equal to 0.}$$

$$t = 0 \quad \bigg| \quad t = 7 \quad \text{Solve each linear equation.}$$

When $t = 0$, the object's height above the ground is 0 feet, because it has not been released. When $t = 7$, the height is again 0 feet. The object has hit the ground.

State the conclusion — The object hits the ground in 7 seconds.

Check the result When $t = 7$,

$$h = 112(7) - 16(7)^2$$
$$= 184 - 16(49)$$
$$= 0$$

Since the height is 0 feet, the object has hit the ground after 7 seconds.

SELF CHECK 2 If this object is launched with an initial velocity of 96 feet per second, how many seconds will it take to hit the ground?

3 Solve a geometric application using a quadratic equation.

Recall that the area of a rectangle is given by the formula

$$A = lw$$

where A represents the area, l the length, and w the width of the rectangle. The perimeter of a rectangle is given by the formula

$$P = 2l + 2w$$

where P represents the perimeter of the rectangle, l the length, and w the width.

EXAMPLE 3 **RECTANGLES** Assume that the rectangle in Figure 5-1 has an area of 52 square centimeters and that its length is 1 centimeter more than 3 times its width. Find the perimeter of the rectangle.

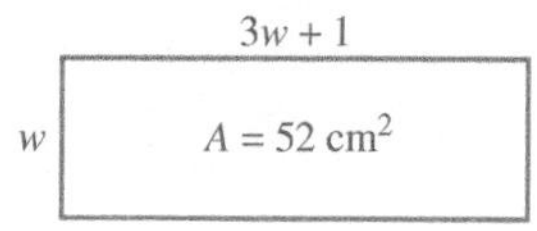

Figure 5-1

Analyze the problem We are asked to find the perimeter of the rectangle. To do so, we must know both the length and the width. If we let w represent the width of the rectangle in centimeters, then $(3w + 1)$ represents its length in centimeters.

Form and solve an equation We are given that the area of the rectangle is 52 square centimeters. We can use this fact to find the values of the width and length and then find the perimeter. To find the width, we can substitute 52 for A and $(3w + 1)$ for l in the formula $A = lw$ and solve for w.

$$A = lw$$
$$52 = (3w + 1)w$$
$$52 = 3w^2 + w \quad \text{Use the distributive property to remove parentheses.}$$
$$0 = 3w^2 + w - 52 \quad \text{Subtract 52 from both sides.}$$
$$0 = (3w + 13)(w - 4) \quad \text{Factor.}$$
$$3w + 13 = 0 \quad \text{or} \quad w - 4 = 0 \quad \text{Apply the zero-factor property.}$$
$$3w = -13 \qquad w = 4 \quad \text{Solve each linear equation.}$$
$$w = -\frac{13}{3}$$

Because the width of a rectangle cannot be negative, we discard the result $w = -\frac{13}{3}$. Thus, the width of the rectangle is 4 centimeters, and the length is given by

$$3w + 1 = 3(4) + 1$$
$$= 12 + 1$$
$$= 13$$

Unless otherwise noted, all content on this page is © Cengage Learning.

The dimensions of the rectangle are 4 centimeters by 13 centimeters. We find the perimeter by substituting 13 for l and 4 for w in the formula for the perimeter.

$$\begin{aligned} P &= 2l + 2w \\ &= 2(13) + 2(4) \\ &= 26 + 8 \\ &= 34 \end{aligned}$$

State the conclusion: The perimeter of the rectangle is 34 centimeters.

Check the result: A rectangle with dimensions of 13 centimeters by 4 centimeters does have an area of 52 square centimeters, and the length is 1 centimeter more than 3 times the width. A rectangle with these dimensions has a perimeter of 34 centimeters.

SELF CHECK 3 If the rectangle has an area of 80 square centimeters, use the same relationship for the length and the width to find the perimeter.

EXAMPLE 4

TRIANGLES The triangle in Figure 5-2 has an area of 10 square centimeters and a height that is 3 centimeters less than twice the length of its base. Find the length of the base and the height of the triangle.

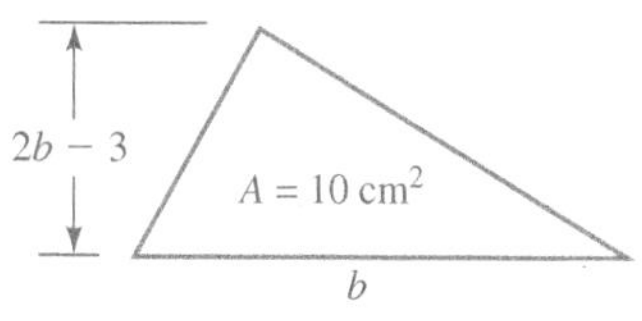

Figure 5-2

Analyze the problem: We are asked to find the length of the base and the height of the triangle, so we will let b represent the length of the base of the triangle in centimeters. Then $(2b - 3)$ represents the height in centimeters.

Form and solve an equation: Because the area is 10 square centimeters, we can substitute 10 for A and $(2b - 3)$ for h in the formula $A = \frac{1}{2}bh$ and solve for b.

$$A = \frac{1}{2}bh$$

$$10 = \frac{1}{2}b(2b - 3)$$

$$20 = b(2b - 3) \quad \text{Multiply both sides by 2 to clear fractions.}$$

$$20 = 2b^2 - 3b \quad \text{Use the distributive property to remove parentheses.}$$

$$0 = 2b^2 - 3b - 20 \quad \text{Subtract 20 from both sides.}$$

$$0 = (2b + 5)(b - 4) \quad \text{Factor.}$$

$$2b + 5 = 0 \quad \text{or} \quad b - 4 = 0 \quad \text{Set both factors equal to 0.}$$

$$2b = -5 \quad\quad b = 4 \quad \text{Solve each linear equation.}$$

$$b = -\frac{5}{2}$$

Because a triangle cannot have a negative number for the length of its base, we discard the result $b = -\frac{5}{2}$. The length of the base of the triangle is 4 centimeters.

State the conclusion: Its height is $2(4) - 3$, or 5 centimeters.

Check the result: If the base of the triangle has a length of 4 centimeters and the height of the triangle is 5 centimeters, its height is 3 centimeters less than twice the length of its base. Its area is 10 square centimeters.

$$A = \frac{1}{2}bh = \frac{1}{2}(4)(5) = 2(5) = 10$$

Unless otherwise noted, all content on this page is © Cengage Learning.

SELF CHECK 4 If the triangle in the figure had an area of 52 square centimeters, use the same relationship for the height and base to find the length of the base and the height of the triangle.

SELF CHECK ANSWERS

1. 5 and 12; −5 and −12 **2.** It will hit the ground after 6 seconds. **3.** The perimeter of the rectangle is 42 cm. **4.** The length of the base is 8 cm and the height is 13 cm.

NOW TRY THIS

A cell phone is shaped like a rectangle and its longer edge is one inch shorter than twice its shorter edge. If the area of the phone is 10 square inches, find the dimensions of the phone.

5.8 Exercises

WARM-UPS *Write the formula for . . .*

1. the area of a rectangle.
2. the area of a triangle.
3. the area of a square.
4. the volume of a rectangular solid.
5. the perimeter of a rectangle.
6. the perimeter of a square.

REVIEW *Solve each equation.*

7. $-2(5x + 2) = 3(2 - 3x)$
8. $4(3a - 2) - 12 = 2a$
9. **Rectangles** A rectangle is 5 times as long as it is wide, and its perimeter is 132 feet. Find its area.
10. **Investing** A woman invested $15,000, part at 7% simple annual interest and part at 8% annual interest. If she receives $1,100 interest per year, how much did she invest at 7%?

VOCABULARY AND CONCEPTS *Fill in the blanks.*

11. The first step in the problem-solving process is to ______ the problem.
12. The last step in the problem-solving process is to ________ _____.

APPLICATIONS

13. **Integer** One integer is 4 more than another. Their product is 32. Find the integers.
14. **Integer** One integer is 5 less than 4 times another. Their product is 21. Find the integers.
15. **Integer** If 4 is added to the square of an integer, the result is 5 less than 10 times that integer. Find the integer(s).
16. **Integer** If 5 times the square of an integer is added to 3 times the integer, the result is 2. Find the integer.

An object has been launched straight up into the air. The formula $h = vt - 16t^2$ gives the height h of the object above the ground after t seconds when it is launched upward with an initial velocity v.

17. **Time of flight** After how many seconds will an object hit the ground if it was launched with a velocity of 144 feet per second?
18. **Time of flight** After how many seconds will an object hit the ground if it was launched with a velocity of 160 feet per second?
19. **Ballistics** If a cannonball is fired with an upward velocity of 224 feet per second, at what times will it be at a height of 640 feet?
20. **Ballistics** A cannonball's initial upward velocity is 128 feet per second. At what times will it be 192 feet above the ground?
21. **Exhibition diving** At a resort, tourists watch swimmers dive from a cliff to the water 64 feet below. A diver's height h above the water t seconds after diving is given by $h = -16t^2 + 64$. How long does a dive last?
22. **Forensic medicine** The kinetic energy E of a moving object is given by $E = \frac{1}{2}mv^2$, where m is the mass of the object (in kilograms) and v is the object's velocity (in meters per second). Kinetic energy is measured in joules. By the damage done to a victim, a police pathologist determines that the energy of a 3-kilogram mass at impact was 54 joules. Find the velocity at impact.

In Exercises 23–24, note that in the triangle $y^2 = h^2 + x^2$. (Pythagorean theorem)

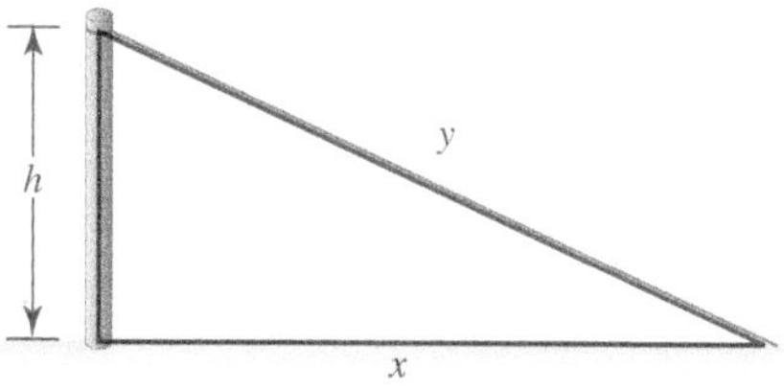

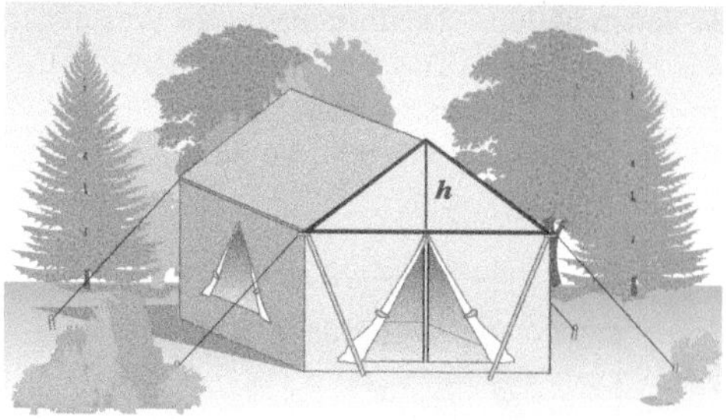

23. Ropes courses A camper slides down the cable of a high-adventure ropes course to the ground as shown in the illustration. At what height did the camper start his slide?

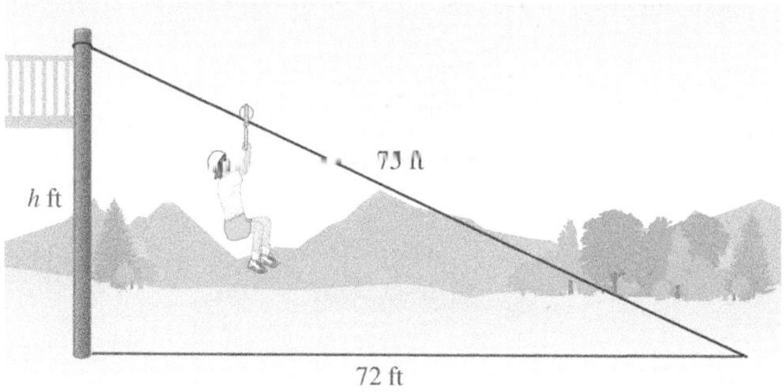

24. Ropes courses If the pole and the landing area discussed in Exercise 23 are 48 feet apart and the high end of the cable is 36 feet, how long is the cable?

25. Insulation The area of the rectangular slab of foam insulation is 36 square meters. Find the dimensions of the slab.

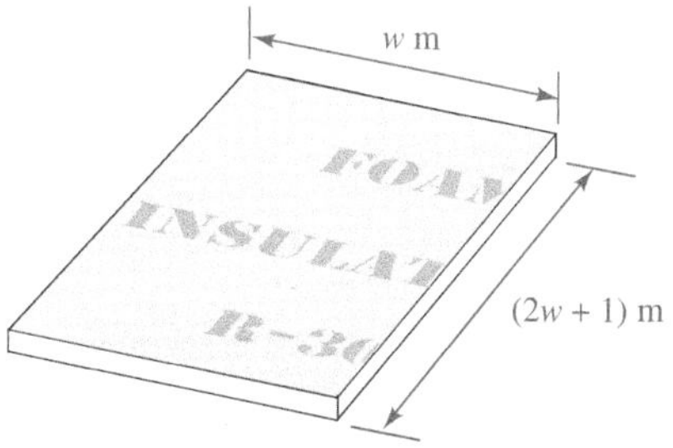

26. Shipping pallets The length of a rectangular shipping pallet is 2 feet less than 3 times its width. Its area is 21 square feet. Find the dimensions of the pallet.

27. Carpentry A rectangular room containing 143 square feet is 2 feet longer than it is wide. How long a crown molding is needed to trim the perimeter of the ceiling?

28. Designing tents The length of the base of the triangular sheet of canvas above the door of the tent shown is 2 feet more than twice its height. The area is 30 square feet. Find the height and the length of the base of the triangle.

29. Dimensions of a triangle The height of a triangle is 2 inches less than 5 times the length of its base. The area is 36 square inches. Find the length of the base and the height of the triangle.

30. Area of a triangle The base of a triangle is numerically 3 less than its area, and the height is numerically 6 less than its area. Find the area of the triangle.

31. Area of a triangle The length of the base and the height of a triangle are numerically equal. Their sum is 6 less than the number of units in the area of the triangle. Find the area of the triangle.

32. Dimensions of a parallelogram The formula for the area of a parallelogram is $A = bh$. The area of the parallelogram in the illustration is 200 square centimeters. If its base is twice its height, how long is the base?

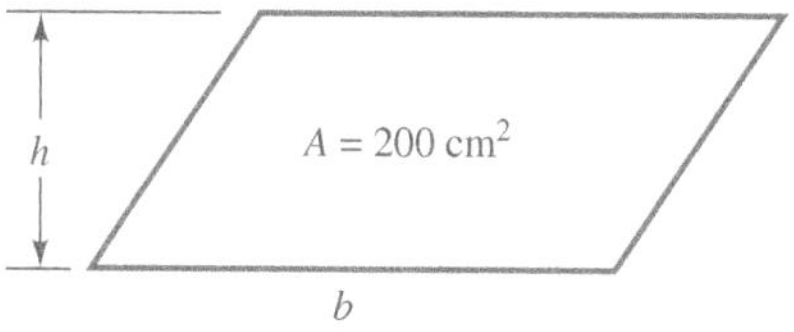

33. Swimming pool borders The owners of the rectangular swimming pool want to surround the pool with a crushed-stone border of uniform width. They have enough stone to cover 74 square meters. How wide should they make the border? (*Hint:* The area of the larger rectangle minus the area of the smaller is the area of the border.)

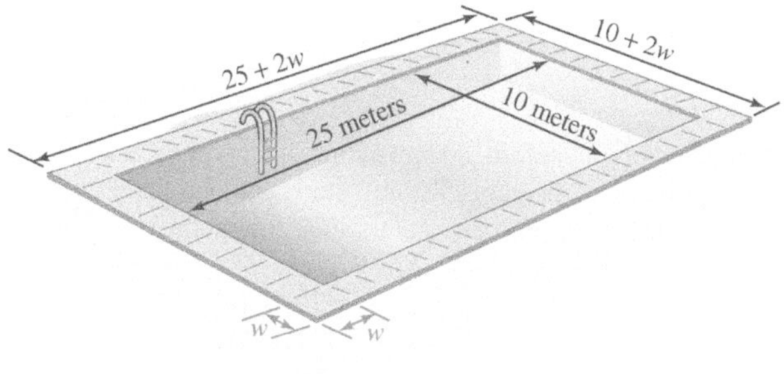

Unless otherwise noted, all content on this page is © Cengage Learning.

34. House construction The formula for the area of a trapezoid is $A = \frac{h(B + b)}{2}$. The area of the trapezoidal truss in the illustration is 24 square meters. Find the height of the trapezoid if one base is 8 meters and the other base is the same as the height.

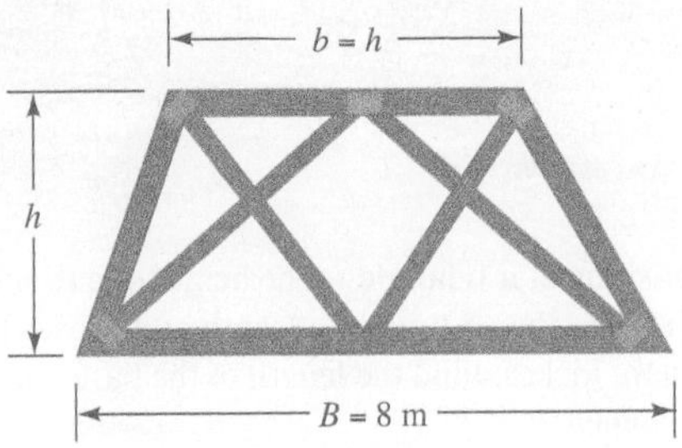

35. Volume of a solid The volume of a rectangular solid is given by the formula $V = lwh$, where l is the length, w is the width, and h is the height. The volume of the rectangular solid in the illustration is 210 cubic centimeters. Find the width of the rectangular solid if its length is 10 centimeters and its height is 1 centimeter longer than twice its width.

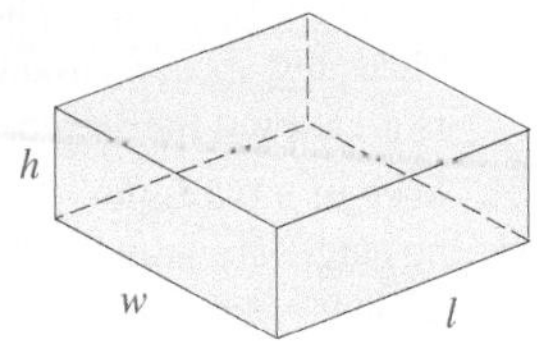

36. Volume of a pyramid The volume of a pyramid is given by the formula $V = \frac{Bh}{3}$, where B is the area of its base and h is its height. The volume of the pyramid in the illustration is 192 cubic centimeters. Find the dimensions of its rectangular base if one edge of the base is 2 centimeters longer than the other, and the height of the pyramid is 12 centimeters.

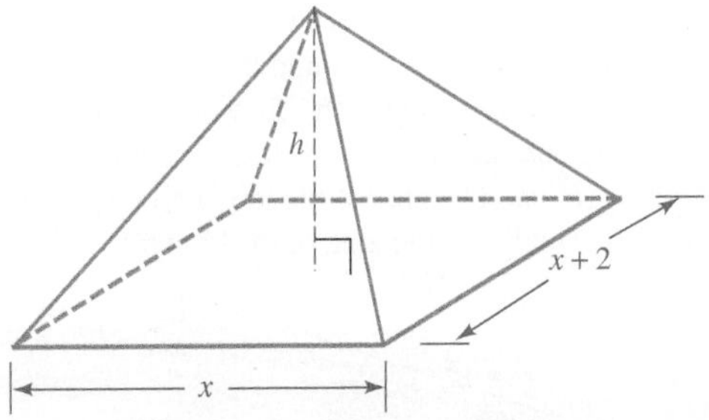

37. Volume of a pyramid The volume of a pyramid is 84 cubic centimeters. Its height is 9 centimeters, and one side of its rectangular base is 3 centimeters shorter than the other. Find the dimensions of its base. (See Exercise 36.)

38. Volume of a solid The volume of a rectangular solid is 180 cubic centimeters. Its height is 3 centimeters, and its width is 4 centimeters shorter than its length. Find the sum of its length and width. (See Exercise 35.)

39. Sewage treatment In one step in waste treatment, sewage is exposed to air by placing it in circular aeration pools. One sewage processing plant has two such pools, with diameters of 38 and 44 meters. Find the combined area of the pools.

40. Sewage treatment To meet new clean-water standards, the plant in Exercise 39 must double its capacity by building another pool. Find the radius of the circular pool that the engineering department should specify to double the plant's capacity.

In Exercises 41–42, $a^2 + b^2 = c^2$.

41. Tornado damage The tree shown below was blown down in a tornado. Find x and the height of the tree when it was standing.

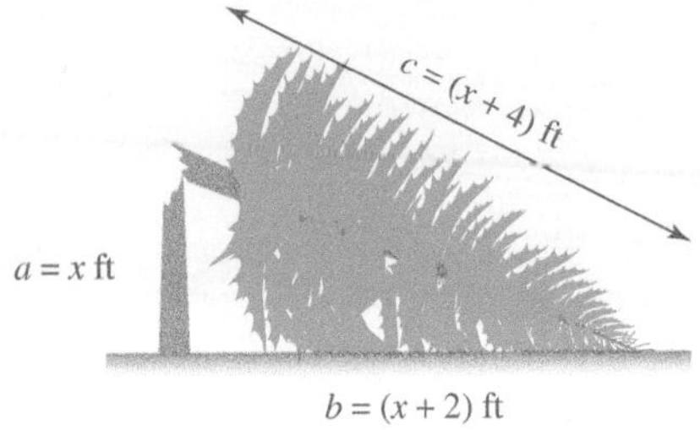

42. Car repairs To work under a car, a mechanic drives it up steel ramps like the ones shown below. Find the length of each side of the ramp.

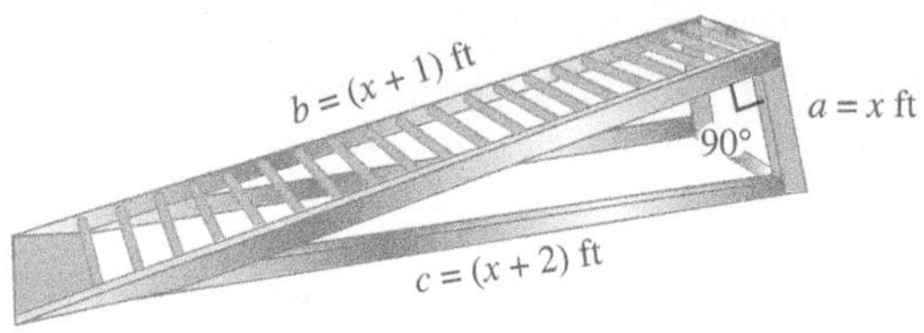

WRITING ABOUT MATH

43. Explain the steps you would use to set up and solve an application problem.

44. Explain how you should check the solution to an application.

SOMETHING TO THINK ABOUT

45. Here is an easy-sounding problem:

> *The length of a rectangle is 2 feet greater than the width, and the area is 18 square feet. Find the width of the rectangle.*

Set up the equation. Can you solve it? Why not?

46. Does the equation in Exercise 45 have a solution, even if you can't find it? If it does, find an estimate of the solution.

Unless otherwise noted, all content on this page is © Cengage Learning.

Projects

Because the length of each side of the largest square in Figure 5-3 is $x + y$, its area is $(x + y)^2$. This area is also the sum of four smaller areas, which illustrates the factorization

$$x^2 + 2xy + y^2 = (x + y)^2$$

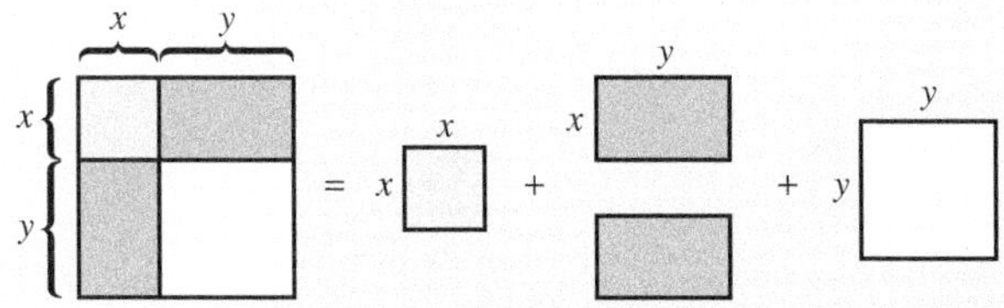

Figure 5-3

What factorization is illustrated by each of the following figures?

a.

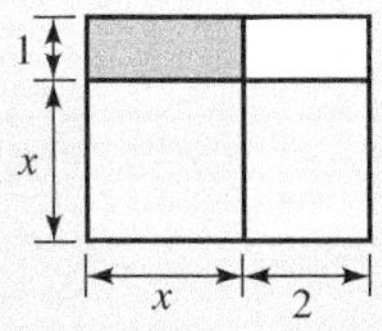

b.

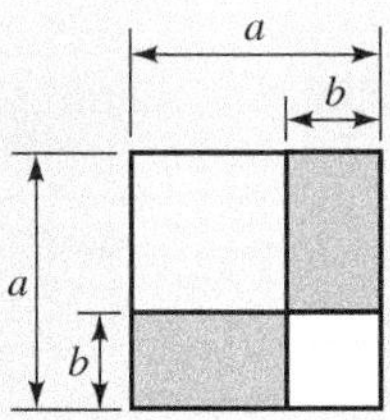

c.

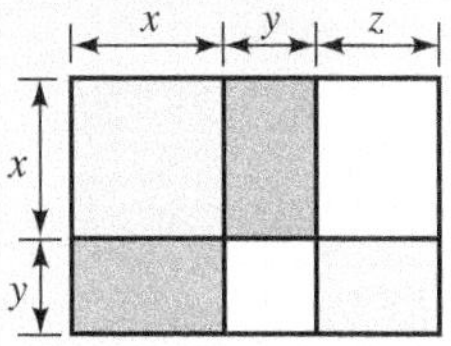

d.

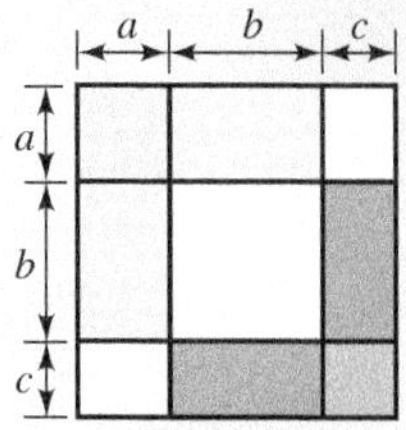

e. Factor the expression

$$a^2 + ac + 2a + ab + bc + 2b$$

and draw a figure that illustrates the factorization.

f. Verify the factorization

$$x^3 + 3x^2y + 3xy^2 + y^3 = (x + y)^3$$

Hint. Expand the right side:

$$(x + y)^3 = (x + y)(x + y)(x + y)$$

Then draw a figure that illustrates the factorization.

Unless otherwise noted, all content on this page is © Cengage Learning.

Reach for Success

EXTENSION OF ORGANIZING YOUR TIME

Now that you have created a calendar of all your responsibilities, let's look at how to establish *priorities*. This will allow you to avoid potential conflicts that could create unnecessary stress. Planning is the key to academic success!

Establish *your* criteria for dealing with scheduling conflicts. You cannot make a decision without knowing what criteria to use.

You might consider:	What criteria do you use to establish priorities?
• Financial: Is it going to cost a lot to do it?	1. ______
• Time: Do you have the time to do it?	2. ______
• Importance: Are there reasons to do it?	3. ______
• Risk: What happens if you do not do it or if you are late?	4. ______
• Ability to complete: Do you have the skills needed?	

Now, look at the two-week calendar below. Do you see any potential scheduling conflicts? ______

______

Sun 1	Mon 2	Tue 3	Wed 4	Thu 5	Fri 6	Sat 7
	English outline due	Dentist 1 p.m. Work	Work	Mathematics test	RJ's school play	Volunteer work Wedding anniversary
Sun 8	**Mon 9**	**Tue 10**	**Wed 11**	**Thu 12**	**Fri 13**	**Sat 14**
Sociology group project meeting 2 p.m.	Work Kenley's Dr. appt. 2:30 p.m.	Turn in history paper Math lab due	Read history chapters 13–15	Sociology group project due Work	English paper due Fred's birthday	Work

If you have to work on Tuesday and Wednesday the 3rd and 4th, when do you study for the mathematics test due on the 5th?

Your plan: ______

______

As you can see, the second week is busy with many assignments. To avoid time conflicts, consider completing some of the assignments the first week rather than waiting until the second week. What other strategies could you use?

______

______

______

Most instructors are willing to help you with study skills in addition to mathematical skills. Do not hesitate to ask them!

5 Review

SECTION 5.1 Factoring Out the Greatest Common Factor; Factoring by Grouping

DEFINITIONS AND CONCEPTS	EXAMPLES
A natural number is in **prime-factored form** if it is written as the product of prime-number factors.	$42 = 6 \cdot 7 = 2 \cdot 3 \cdot 7$ $56 = 8 \cdot 7 = 2 \cdot 4 \cdot 7 = 2 \cdot 2 \cdot 2 \cdot 7 = 2^3 \cdot 7$
The **greatest common factor (GCF)** of several monomials is found by taking each common prime factor the fewest number of times it appears in any one monomial.	Find the GCF of $12x^3y$, $42x^2y^2$, and $32x^2y^3$. $12x^3y = 2 \cdot 2 \cdot 3 \cdot x \cdot x \cdot x \cdot y$ $42x^2y^2 = 2 \cdot 3 \cdot 7 \cdot x \cdot x \cdot y \cdot y$ $32x^2y^3 = 2 \cdot 2 \cdot 2 \cdot 2 \cdot 2 \cdot x \cdot x \cdot y \cdot y \cdot y$ GCF $= 2x^2y$
If the leading coefficient of a polynomial is negative, it is often useful to factor out -1.	Factor completely: $-x^3 + 3x^2 - 5$ $-x^3 + 3x^2 - 5$ $= (-1)x^3 + (-1)(-3x^2) + (-1)5$ $= -1(x^3 - 3x^2 + 5)$ Factor out -1. $= -(x^3 - 3x^2 + 5)$ The coefficient of 1 need not be written.
If a polynomial has four terms, consider factoring it by grouping.	Factor completely: $x^2 + xy + 3x + 3y$ Factor x from $(x^2 + xy)$ and 3 from $(3x + 3y)$ and proceed as follows: $x^2 + xy + 3x + 3y = x(x + y) + 3(x + y)$ $= (x + y)(x + 3)$ Factor out $(x + y)$.

REVIEW EXERCISES

Find the prime factorization of each number.

1. 24
2. 45
3. 96
4. 102
5. 87
6. 99
7. 2,050
8. 4,096

Factor.

9. $4x + 12y$
10. $5ax^2 + 15a$
11. $7x^2 + 14x$
12. $9x^2 - 3x$
13. $2x^3 + 4x^2 - 8x$
14. $-ax - ay + az$
15. $ax + ay - a$
16. $x^2yz + xy^2z$
17. $(a - b)x + (a - b)y$
18. $(x + y)^2 + (x + y)$
19. $2x^2(x + 2) + 6x(x + 2)$
20. $5x(a + b)^2 - 10x(a + b)$
21. $3p + 9q + ap + 3aq$
22. $ar - 2as + 7r - 14s$
23. $x^2 + ax + bx + ab$
24. $xy + 2x - 2y - 4$
25. $xa + yb + ya + xb$
26. $x^3 - 4x^2 + 3x - 12$

SECTION 5.2 Factoring the Difference of Two Squares

DEFINITIONS AND CONCEPTS	EXAMPLES
To factor the difference of two squares, use the pattern $x^2 - y^2 = (x + y)(x - y)$	$x^2 - 36 = x^2 - 6^2 = (x + 6)(x - 6)$
If we limited to rational coefficients, binomials that are the sum of two squares cannot be factored over the real numbers unless they contain a GCF.	$9x^2 + 36 = 9(x^2 + 4)$ Factor out 9, the GCF. $(x^2 + 4)$ does not factor.

REVIEW EXERCISES

Factor.

27. $x^2 - 25$
28. $x^2y^2 - 16$
29. $(x + 2)^2 - y^2$
30. $z^2 - (x + y)^2$
31. $2x^2y + 18y^3$
32. $(x + y)^2 - z^2$

SECTIONS 5.3–5.4 Factoring Trinomials

DEFINITIONS AND CONCEPTS	EXAMPLES
Factor trinomials using these steps (trial and error):	Factor completely: $12 - x^2 - x$
1. Write the trinomial with the exponents of one variable in descending order.	**1.** We will begin by writing the exponents of x in descending order. $12 - x^2 - x = -x^2 - x + 12$
2. Factor out any greatest common factor (including -1 if that is necessary to make the coefficient of the first term positive).	**2.** Factor out -1 to get $= -(x^2 + x - 12)$
3. If the sign of the third term is +, the signs between the terms of the binomial factors are the same as the sign of the trinomial's second term. If the sign of the third term is −, the signs between the terms of the binomials are opposites.	**3.** Since the sign of the third term is −, the signs between the binomials are opposites.
4. Try various combinations of first terms and last terms until you find the one that works. If none work, the trinomial is prime.	**4.** We find the combination that works. $= -(x + 4)(x - 3)$
5. Check by multiplication.	**5.** Since $-(x + 4)(x - 3) = 12 - x^2 - x$, the factorization is correct.
Factor trinomials by grouping (ac method) using these steps: **1.** Write the trinomial in $ax^2 + bx + c$ form. **2.** Find the key number ac. **3.** Find two factors of the key number whose sum is b. **4.** Use the factors as the coefficients of two terms whose sum is bx to replace bx in the trinomial. **5.** Factor the polynomial by grouping.	$2x^2 - x - 10$ $a = 2, b = -1, c = -10$ Determine the key number $ac = 2(-10) = -20$. $= 2x^2 - 5x + 4x - 10$ Replace $-x$ with $-5x + 4x$. (The two factors whose product is -20 and difference is -1 are -5 and 4.) $= x(2x - 5) + 2(2x - 5)$ Factor by grouping. $= (2x - 5)(x + 2)$

REVIEW EXERCISES

Factor.

33. $x^2 + 7x + 10$

34. $x^2 - 8x + 15$

35. $x^2 + 2x - 24$

36. $x^2 - 4x - 12$

37. $2x^2 - 5x - 3$

38. $3x^2 - 14x - 5$

39. $15x^2 + x - 2$

40. $6x^2 + 3x - 3$

41. $6x^3 + 17x^2 - 3x$

42. $4x^3 - 5x^2 - 6x$

43. $12x - 4x^3 - 2x^2$

44. $-4a^3 + 4a^2b + 24ab^2$

SECTION 5.5 Factoring the Sum and Difference of Two Cubes

DEFINITIONS AND CONCEPTS	EXAMPLES
The sum and difference of two cubes factor according to the patterns	
$x^3 + y^3 = (x + y)(x^2 - xy + y^2)$	$x^3 + 64 = x^3 + 4^3 = (x + 4)(x^2 - x \cdot 4 + 4^2)$ $= (x + 4)(x^2 - 4x + 16)$
$x^3 - y^3 = (x - y)(x^2 + xy + y^2)$	$x^3 - 64 = x^3 - 4^3 = (x - 4)(x^2 - x(-4) + (-4)^2)$ $= (x - 4)(x^2 + 4x + 16)$

REVIEW EXERCISES

Factor.

45. $c^3 - 125$

46. $d^3 + 8$

47. $2x^3 + 54$

48. $2ab^4 - 2ab$

SECTION 5.6 Summary of Factoring Techniques

DEFINITIONS AND CONCEPTS	EXAMPLES
Factoring polynomials:	Factor: $-2x^6 + 2y^6$
1. Write the polynomial in descending powers of one variable.	
2. Factor out all common factors.	Factor out the common factor of -2. $-2x^6 + 2y^6 = -2 \cdot x^6 - (-2)y^6 = -2(x^6 - y^6)$
3. If an expression has two terms, check to see if it is **a.** the **difference of two squares**: $a^2 - b^2 = (a + b)(a - b)$ **b.** the **sum of two cubes**: $a^3 + b^3 = (a + b)(a^2 - ab + b^2)$ **c.** the **difference of two cubes**: $a^3 - b^3 = (a - b)(a^2 + ab + b^2)$	Identify $x^6 - y^6$ as the difference of two squares and factor it: $-2(x^6 - y^6) = -2[(x^3)^2 - (y^3)^2] = -2(x^3 + y^3)(x^3 - y^3)$ Then identify $(x^3 + y^3)$ as the sum of two cubes and $(x^3 - y^3)$ as the difference of two cubes and factor each binomial: $-2(x^6 - y^6) = -2[(x^3)^2 - (y^3)^2] = -2(x^3 + y^3)(x^3 - y^3)$ $= -2(x + y)(x^2 - xy + y^2)(x - y)(x^2 + xy + y^2)$

4. If an expression has three terms, check to see if it is a **perfect-square trinomial square**: $a^2 + 2ab + b^2 = (a + b)(a + b)$ $a^2 - 2ab + b^2 = (a - b)(a - b)$ If the trinomial is not a perfect-square trinomial, attempt to factor it as a **general trinomial**.	Factor: $4x^2 - 12x + 9$ This is a perfect-square trinomial because it has the form $(2x)^2 - 2(2x)(3) + (3)^2$. It factors as $(2x - 3)^2$.
5. If an expression has four or more terms, factor it by **grouping**. **6.** Continue factoring until each individual factor is prime, except possibly a monomial factor. **7.** If the polynomial does not factor, the polynomial is **prime** over the set of rational numbers. **8.** Check the results by multiplying.	Factor: $ax - bx + ay - by$ Since the expression has four terms, use factoring by grouping: $ax - bx + ay - by = x(a - b) + y(a - b)$ $= (a - b)(x + y)$

REVIEW EXERCISES

Factor.

49. $3x^2y - xy^2 - 6xy + 2y^2$

50. $5x^2 - 5x - 30$

51. $2a^2x + 2abx + a^3 + a^2b$

52. $2x^2 - 8x - 11$

53. $x^2 - 9 + ax + 3a$

54. $10x^3 - 80y^3$

SECTION 5.7 Solving Equations by Factoring

DEFINITIONS AND CONCEPTS	EXAMPLES
A **quadratic equation** is an equation of the form $ax^2 + bx + c = 0$, where a, b, and c are real numbers and $a \neq 0$.	$2x^2 + 5x = 8$ and $x^2 - 5x = 0$ are quadratic equations.
Zero-factor property: If a and b represent two real numbers and if $ab = 0$, then $a = 0$ or $b = 0$. The zero-factor property can be extended to any number of factors.	To solve the quadratic equation $x^2 - 3x = 4$, proceed as follows: $x^2 - 3x = 4$ $x^2 - 3x - 4 = 0$ Subtract 4 from both sides to write the equation in quadratic form. $(x + 1)(x - 4) = 0$ Factor $x^2 - 3x - 4$. $x + 1 = 0$ or $x - 4 = 0$ Apply the zero-factor property. $x = -1$ | $x = 4$ Solve each linear equation.

REVIEW EXERCISES

Solve.

55. $x^2 + 5x = 0$

56. $2x^2 - 6x = 0$

57. $3x^2 = 2x$

58. $5x^2 + 25x = 0$

59. $y^2 - 49 = 0$

60. $x^2 - 25 = 0$

61. $a^2 - 9a + 20 = 0$

62. $(x - 1)(x + 4) = 6$

63. $2x - x^2 + 24 = 0$

64. $16 + x^2 - 10x = 0$

65. $2x^2 - 5x - 3 = 0$

66. $2x^2 + x - 3 = 0$

67. $16x^2 = 9$

68. $9x^2 = 4$

69. $x^3 - 7x^2 + 12x = 0$

70. $x^3 + 5x^2 + 6x = 0$

71. $2x^3 + 5x^2 = 3x$

72. $3x^3 - 2x = x^2$

SECTION 5.8 Solving Applications

DEFINITIONS AND CONCEPTS	EXAMPLES
Use the methods for solving quadratic equations discussed in Section 5.7 to solve applications.	Assume that the area of a rectangle is 240 square inches and that its length is 4 inches less than twice its width. Find the perimeter of the rectangle.
1. Analyze the problem.	Let w represent the width of the rectangle in inches. Then $(2w - 4)$ represents its length. We can find the length and width by substituting into the formula for the area: $A = l \cdot w$.
2. Form an equation.	$240 = (2w - 4)w$
3. Solve the equation.	$240 = 2w^2 - 4w$ Use the distributive property to remove parentheses. $0 = 2w^2 - 4w - 240$ Subtract 240 from both sides. $0 = w^2 - 2w - 120$ Divide both sides by 2. $0 = (w - 12)(w + 10)$ Factor. $w - 12 = 0$ or $w + 10 = 0$ Apply the zero-factor property. $w = 12$ | $w = -10$ Solve each linear equation. Because the width cannot be negative, we discard the result $w = -10$. Thus, the width of the rectangle is 12 inches, and the length is given by $2w - 4 = 2(12) - 4$ $= 24 - 4$ $= 20$ The dimensions of the rectangle are 12 in. by 20 in. We find the perimeter by substituting 20 for l and 12 for w in the formula for perimeter. $P = 2l + 2w = 2(20) + 2(12) = 40 + 24 = 64$
4. State the conclusion.	The perimeter of the rectangle is 64 inches.
5. Check the result.	A rectangle with dimensions 12 in. by 20 in. does have an area of 240 square inches, and the length is 4 inches less than twice the width.

REVIEW EXERCISES

73. Numbers The sum of two numbers is 12, and their product is 35. Find the numbers.

74. Numbers If 3 times the square of a positive number is added to 5 times the number, the result is 2. Find the number.

75. Dimensions of a rectangle A rectangle is 2 feet longer than it is wide, and its area is 48 square feet. Find its dimensions.

76. Gardening A rectangular flower bed is 3 feet longer than twice its width, and its area is 27 square feet. Find its dimensions.

77. Geometry A rectangle is 3 feet longer than it is wide. Its area is numerically equal to its perimeter. Find its dimensions.

78. Geometry A triangle has a height 1 foot longer than its base. If its area is 36 square feet, find its height.

5 Test

1. Find the prime factorization of 120.
2. Find the prime factorization of 108.

Factor.

3. $60ab^2c^3 + 30a^3b^2c - 25a$
4. $3x^2(a + b) - 6xy(a + b)$
5. $ax + ay + bx + by$
6. $x^2 - 64$
7. $2a^2 - 32b^2$
8. $16x^4 - 81y^4$
9. $x^2 + 5x - 6$
10. $x^2 - 9x - 22$
11. $-x^2 - 10xy - 9y^2$
12. $6x^2 - 30xy + 24y^2$
13. $3x^2 + 13x + 4$
14. $2a^2 + 5a - 12$
15. $2x^2 + 3x - 1$
16. $12 - 25x + 12x^2$
17. $12a^2 + 6ab - 36b^2$
18. $x^3 - 8y^3$
19. $216 + 8a^3$
20. $x^9z^3 - y^3z^6$

Solve each equation.

21. $x^2 = -10x$
22. $2x^2 + 5x + 3 = 0$
23. $16y^2 - 64 = 0$
24. $-3(y - 6) + 2 = y^2 + 2$
25. $10x^2 - 13x = 9$
26. $10x^2 - x = 9$
27. $10x^2 + 43x = 9$
28. $10x^2 - 89x = 9$
29. **Cannon fire** A cannonball is fired straight up into the air with a velocity of 192 feet per second. In how many seconds will it hit the ground? (Its height above the ground is given by the formula $h = vt - 16t^2$, where v is the velocity and t is the time in seconds.)
30. **Base of a triangle** The base of a triangle with an area of 40 square meters is 2 meters longer than it is high. Find the base of the triangle.

Rational Expressions and Equations; Proportion and Variation

6

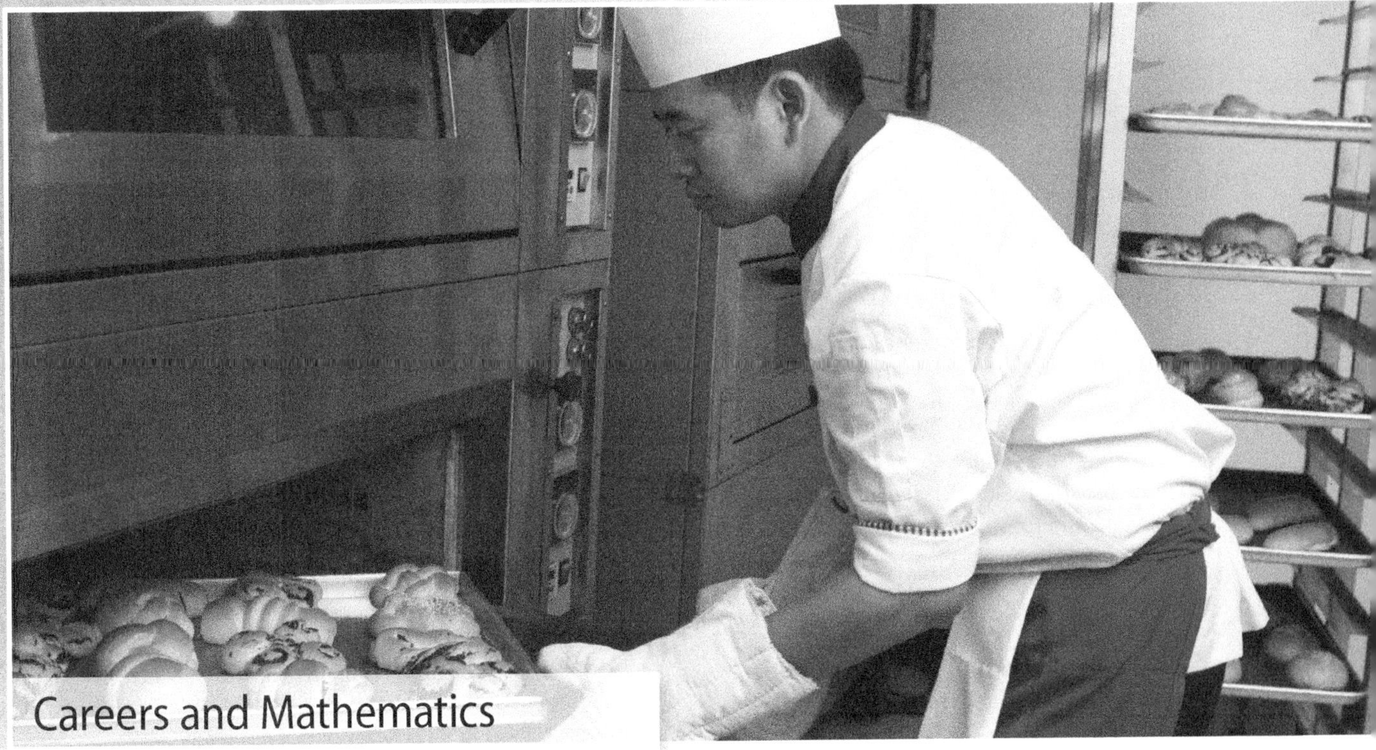

© OtnaYdur/Shutterstock.com

Careers and Mathematics

FOOD-PROCESSING OCCUPATIONS; BAKER

Bakers mix and bake ingredients in accordance to recipes to produce varying quantities of breads, pastries, and other baked goods. Bakers commonly are employed in grocery stores and specialty shops that produce small quantities of baked goods. In manufacturing, bakers produce goods in large quantities. Bakers make up 21% of all food-processing workers. Training varies widely among the food-processing occupations. Bakers often start as apprentices or as trainees.

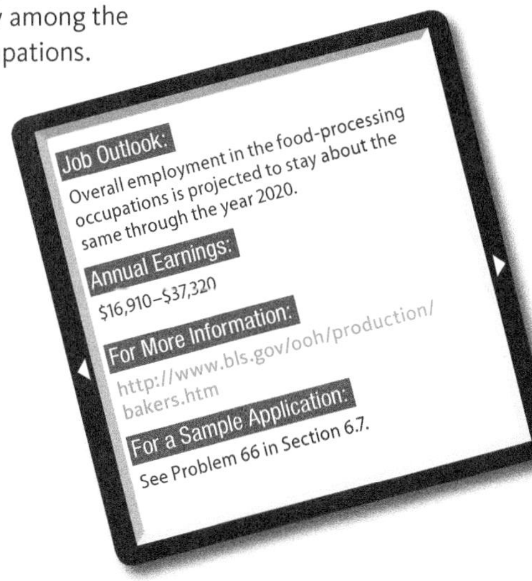
Job Outlook:
Overall employment in the food-processing occupations is projected to stay about the same through the year 2020.
Annual Earnings:
$16,910–$37,320
For More Information:
http://www.bls.gov/ooh/production/bakers.htm
For a Sample Application:
See Problem 66 in Section 6.7.

REACH FOR SUCCESS
6.1 Simplifying Rational Expressions
6.2 Multiplying and Dividing Rational Expressions
6.3 Adding and Subtracting Rational Expressions
6.4 Simplifying Complex Fractions
6.5 Solving Equations That Contain Rational Expressions
6.6 Solving Applications Whose Models Contain Rational Expressions
6.7 Proportion and Variation
- *Projects*

REACH FOR SUCCESS EXTENSION
CHAPTER REVIEW
CHAPTER TEST
CUMULATIVE REVIEW

In this chapter

In Chapter 6, we will discuss rational expressions, the fractions of algebra. After learning how to simplify, add, subtract, multiply, and divide them, we will solve equations and applications that involve rational expressions. We then will conclude by discussing proportion and variation.

Reach for Success Preparing for a Test

Preparation is necessary to be successful in any event. You have practiced and worked out the kinks. You are well rested, and you are motivated. Now is the time. Are you ready?

When preparing for class tests, what can you do to increase your confidence and feel less anxious?

© bikeriderlondon/Shutterstock.com

To know what you need to spend time reviewing, you might want to organize and categorize your notes according to:

- what you clearly understand.
- what is a little unclear.
- what you do not understand at all.

If you have note cards, move the cards that you understand to the back of the deck. If you do not have note cards, highlight the topics in your notebook that you do not understand. Continue to work on these. Seek help.

List the objectives that are giving you the most difficulty.

1. ______
2. ______
3. ______
4. ______
5. ______
6. ______
7. ______
8. ______

How many questions will be on the exam? ______

Will vocabulary be tested? ______

Will the formulas be provided? ______

Determine how much time you should allow yourself for each question on a test if you have 60 minutes to take a 20-question test. ______

Write down the formulas that you will need for the exam.

______

Use the Chapter Review at the end of the chapter to prepare for the exam. After completing the review, take the Chapter Test without notes, grade it, and mark the objectives you missed.

If the test includes completion, multiple choice, true/false, essay, or matching, be certain to practice those formats.

Did your instructor provide a written or computerized review for the test? ______ If so, work it as if it were the test—in a single sitting and without notes. How did you do? ______

Note any objectives you missed.

______

If your instructor did not have a review, make up your own practice test, alone or with a group. Then, take it as described above.

How will you further review any objectives you missed?

______

A Successful Study Strategy . . .

Begin your review early and practice every day to increase your confidence and thus reduce anxiety.

At the end of the chapter you will find an additional exercise to guide you in planning for a successful semester.

Section 6.1 Simplifying Rational Expressions

Objectives

1. Identify all values of a variable for which a rational expression is undefined.
2. Find the domain of a rational function.
3. Write a rational expression in simplest form.
4. Simplify a rational expression containing factors that are negatives.

Vocabulary

rational expression   simplest form

Getting Ready

Simplify.

1. $\frac{12}{16}$ 2. $\frac{16}{8}$ 3. $\frac{25}{55}$ 4. $\frac{36}{72}$

Fractions such as $\frac{1}{2}$ and $\frac{3}{4}$ that are the quotient of two integers are *rational numbers*. Expressions such as

$$\frac{a}{a+2} \quad \text{and} \quad \frac{5x^2+3}{x^2+x-12}$$

where the denominators and/or numerators are polynomials, are called **rational expressions**. Since rational expressions indicate division, we must exclude any values of the variable that will make the denominator equal to 0. For example, a cannot be -2 in the rational expression

$$\frac{a}{a+2}$$

because the denominator will be 0:

$$\frac{a}{a+2} = \frac{-2}{-2+2} = \frac{-2}{0}$$

When the denominator of a rational expression is 0, we say that the expression is undefined.

1 Identify all values of a variable for which a rational expression is undefined.

EXAMPLE 1 Identify all values of x such that the following rational expression is undefined.

$$\frac{5x^2+3}{x^2+x-12}$$

Solution To find the values of x that make the rational expression undefined, we set its denominator equal to 0 and solve for x.

$$x^2 + x - 12 = 0$$

$$(x + 4)(x - 3) = 0 \quad \text{Factor the trinomial.}$$

$$x + 4 = 0 \quad \text{or} \quad x - 3 = 0 \quad \text{Apply the zero-factor property.}$$

$$x = -4 \quad | \quad x = 3 \quad \text{Solve each equation.}$$

We can check by substituting 3 and -4 for x and verifying that these values make the denominator of the rational expression equal to 0.

For x = 3

$$\frac{5x^2 + 3}{x^2 + x - 12} = \frac{5(3)^2 + 3}{3^2 + 3 - 12} = \frac{5(9) + 3}{9 + 3 - 12} = \frac{45 + 3}{12 - 12} = \frac{48}{0}$$

For x = −4

$$\frac{5x^2 + 3}{x^2 + x - 12} = \frac{5(-4)^2 + 3}{(-4)^2 + (-4) - 12} = \frac{5(16) + 3}{16 - 4 - 12} = \frac{80 + 3}{12 - 12} = \frac{83}{0}$$

Since the denominator is 0 when $x = 3$ or $x = -4$, the rational expression $\frac{5x^2 + 3}{x^2 + x - 12}$ is undefined at these values.

SELF CHECK 1 Find all values of x such that the following rational expression is undefined.

$$\frac{3x^2 - 2}{x^2 - 2x - 3}$$

2 Find the domain of a rational function.

The same process is used to find the domain of a rational function. Recall that the domain is the set of all values that can be substituted for the variable. We learned that the domain for the linear function $f(x) = x + 3$ was $\mathbb{R}$, all real numbers.

EXAMPLE 2 Find the domain for the rational function:

$$f(x) = \frac{5x^2 + 3}{x^2 + x - 12}$$

Solution We follow the steps in Example 1 to find the values that would make the expression undefined. Those values are 3 and -4. To write the domain, we must identify all values that can be substituted. Therefore, in set-builder notation, the domain of the rational function

$$f(x) = \frac{5x^2 + 3}{x^2 + x - 12}$$

is $\{x | x \in \mathbb{R}, x \neq 3, -4\}$. This is read as "the set of all values of x such that x is a real number where $x \neq 3$ and $x \neq -4$." In interval notation, $(-\infty, -4)\cup(-4, 3)\cup(3, \infty)$.

SELF CHECK 2 Find the domain for the rational function:

$$f(x) = \frac{6x + 1}{x^2 + 5x - 6}$$

3 Write a rational expression in simplest form.

We have seen that a fraction can be simplified by dividing out common factors shared by its numerator and denominator. For example,

$$\frac{18}{30} = \frac{3 \cdot 6}{5 \cdot 6} = \frac{3 \cdot \overset{1}{\cancel{6}}}{5 \cdot \underset{1}{\cancel{6}}} = \frac{3}{5} \qquad -\frac{6}{15} = -\frac{3 \cdot 2}{3 \cdot 5} = -\frac{\overset{1}{\cancel{3}} \cdot 2}{\underset{1}{\cancel{3}} \cdot 5} = -\frac{2}{5}$$

These examples illustrate the *fundamental property of fractions*, first discussed in Chapter 1.

THE FUNDAMENTAL PROPERTY OF FRACTIONS

If a, b, and x are real numbers, then

$$\frac{a \cdot x}{b \cdot x} = \frac{a}{b} \quad (b \neq 0 \text{ and } x \neq 0)$$

Since rational expressions are fractions, we can use the fundamental property of fractions to simplify rational expressions. We factor the numerator and denominator of the rational expression and divide out all common factors. When all common factors have been divided out, we say that the rational expression has been written in **simplest form**.

EXAMPLE 3 Simplify $\frac{21x^2y}{14xy^2}$. Assume that the denominator is not 0.

Solution We will factor the numerator and the denominator and then divide out any common factors, if possible.

$$\frac{21x^2y}{14xy^2} = \frac{3 \cdot 7 \cdot x \cdot x \cdot y}{2 \cdot 7 \cdot x \cdot y \cdot y} \quad \text{Factor the numerator and denominator.}$$

$$= \frac{3 \cdot \overset{1}{\cancel{7}} \cdot \overset{1}{\cancel{x}} \cdot x \cdot \overset{1}{\cancel{y}}}{2 \cdot \underset{1}{\cancel{7}} \cdot \underset{1}{\cancel{x}} \cdot y \cdot \underset{1}{\cancel{y}}} \quad \text{Divide out the common factors of 7, } x\text{, and } y.$$

$$= \frac{3x}{2y}$$

This rational expression also can be simplified by using the rules of exponents.

$$\frac{21x^2y}{14xy^2} = \frac{3 \cdot 7}{2 \cdot 7}x^{2-1}y^{1-2} \quad \frac{x^2}{x} = x^{2-1}; \frac{y}{y^2} = y^{1-2}$$

$$= \frac{3}{2}xy^{-1} \quad 2 - 1 = 1; 1 - 2 = -1$$

$$= \frac{3}{2} \cdot \frac{x}{y} \quad y^{-1} = \frac{1}{y}$$

$$= \frac{3x}{2y} \quad \text{Multiply.}$$

SELF CHECK 3 Simplify $\frac{32a^3b^2}{24ab^4}$. Assume that the denominator is not 0.

EXAMPLE 4 Simplify $\frac{x^2 + 3x}{3x + 9}$. Assume that the denominator is not 0.

Solution We will factor the numerator and the denominator and then divide out any common factors, if possible.

$$\frac{x^2 + 3x}{3x + 9} = \frac{x(x + 3)}{3(x + 3)} \quad \text{Factor the numerator and the denominator.}$$

$$= \frac{x\overset{1}{\cancel{(x + 3)}}}{3\underset{1}{\cancel{(x + 3)}}} \quad \text{Divide out the common factor of } x + 3.$$

$$= \frac{x}{3}$$

SELF CHECK 4 Simplify $\frac{x^2 - 5x}{5x - 25}$. Assume that the denominator is not 0.

Perspective

The fraction $\frac{8}{4}$ is equal to 2, because $4 \cdot 2 = 8$. The expression $\frac{8}{0}$ is undefined, because there is no number x for which $0 \cdot x = 8$. The expression $\frac{0}{0}$ presents a different problem, however, because $\frac{0}{0}$ seems to equal any number. For example, $\frac{0}{0} = 17$, because $0 \cdot 17 = 0$. Similarly, $\frac{0}{0} = \pi$, because $0 \cdot \pi = 0$. Since "no answer" and "any answer" are both unacceptable, division by 0 is not allowed.

Although $\frac{0}{0}$ represents many numbers, there is often one best answer. In the 17th century, mathematicians such as Sir Isaac Newton (1642–1727) and Gottfried Wilhelm von Leibniz (1646–1716) began to look more closely at expressions related to the fraction $\frac{0}{0}$. One of these expressions, called a **derivative**, is the foundation of **calculus**, an important area of mathematics discovered independently by both Newton and Leibniz. They discovered that under certain conditions, there was one best answer. Expressions related to $\frac{0}{0}$ are called **indeterminate forms**.

Any number divided by 1 remains unchanged. For example,

$$\frac{37}{1} = 37, \qquad \frac{5x}{1} = 5x, \qquad \text{and} \qquad \frac{3x + y}{1} = 3x + y$$

In general, for any real number a, the following is true.

DIVISION BY 1

$$\frac{a}{1} = a$$

EXAMPLE 5 Simplify $\frac{x^3 + x^2}{1 + x}$. Assume that the denominator is not 0.

Solution We will factor the numerator and then divide out any common factors, if possible.

$$\frac{x^3 + x^2}{1 + x} = \frac{x^2(x + 1)}{1 + x} \quad \text{Factor the numerator.}$$

$$= \frac{x^2\overset{1}{\cancel{(x + 1)}}}{\underset{1}{\cancel{1 + x}}} \quad \text{Divide out the common factor of } x + 1.$$

$$= \frac{x^2}{1}$$

$$= x^2 \quad \text{Denominators of 1 need not be written.}$$

SELF CHECK 5 Simplify $\frac{x^2 - x}{x - 1}$. Assume that the denominator is not 0.

EXAMPLE 6 Simplify $\frac{x^2 + 13x + 12}{x^2 - 144}$. Assume that the denominator is not 0.

Solution We will factor the numerator and the denominator and then divide out any common factors, if possible.

$$\frac{x^2 + 13x + 12}{x^2 - 144} = \frac{(x + 1)(x + 12)}{(x + 12)(x - 12)} \quad \text{Factor the numerator and denominator.}$$

$$= \frac{(x + 1)\overset{1}{\cancel{(x + 12)}}}{\underset{1}{\cancel{(x + 12)}}(x - 12)} \quad \text{Divide out the common factor of } (x + 12).$$

$$= \frac{x + 1}{x - 12}$$

SELF CHECK 6 Simplify $\frac{x^2 - 9}{x^3 - 3x^2}$. Assume that the denominator is not 0.

COMMENT Remember that only *factors* common to the *entire numerator* and *entire denominator* can be divided out. *Terms* that are common to the numerator and denominator *cannot* be divided out.

$$\frac{5 + 8}{5} = \frac{13}{5} \qquad \cancel{\frac{5 + 8}{5} = \frac{\overset{1}{\cancel{5}} + 8}{\underset{1}{\cancel{5}}} = \frac{1 + 8}{1} = 9}$$

EXAMPLE 7 Simplify $\frac{5(x + 3) - 5}{7(x + 3) - 7}$. Assume that the denominator is not 0.

Solution We cannot divide out $(x + 3)$, because it is not a factor of the entire numerator, nor is it a factor of the entire denominator. Instead, we simplify the numerator and denominator, factor them, and divide out all common factors, if any.

$$\frac{5(x + 3) - 5}{7(x + 3) - 7} = \frac{5x + 15 - 5}{7x + 21 - 7} \quad \text{Use the distributive property.}$$

$$= \frac{5x + 10}{7x + 14} \quad \text{Combine like terms.}$$

$$= \frac{5(x + 2)}{7(x + 2)} \quad \text{Factor the numerator and denominator.}$$

$$= \frac{5\overset{1}{\cancel{(x + 2)}}}{7\underset{1}{\cancel{(x + 2)}}} \quad \text{Divide out the common factor of } (x + 2).$$

$$= \frac{5}{7}$$

SELF CHECK 7 Simplify $\frac{4(x - 2) + 4}{3(x - 2) + 3}$. Assume that the denominator is not 0.

EXAMPLE 8 Simplify: $\dfrac{x(x+3)-3(x-1)}{x^2+3}$

Solution Since the denominator x^2+3 is always positive, there are no restrictions on x. To simplify the fraction, we will simplify the numerator and then divide out any common factors, if possible.

$$\frac{x(x+3)-3(x-1)}{x^2+3} = \frac{x^2+3x-3x+3}{x^2+3} \quad \text{Use the distributive property.}$$

$$= \frac{x^2+3}{x^2+3} \quad \text{Combine like terms.}$$

$$= \frac{\overset{1}{\cancel{(x^2+3)}}}{\underset{1}{\cancel{(x^2+3)}}} \quad \text{Divide out the common factor of } (x^2+3).$$

$$= 1$$

SELF CHECK 8 Simplify: $\dfrac{a(a+2)-2(a-1)}{a^2+2}$

Sometimes rational expressions do not simplify. For example, to attempt to simplify

$$\frac{x^2+x-2}{x^2+x}$$

we factor the numerator and denominator.

$$\frac{x^2+x-2}{x^2+x} = \frac{(x+2)(x-1)}{x(x+1)}$$

Because there are no factors common to the numerator and denominator, this rational expression is already in simplest form.

EXAMPLE 9 Simplify $\dfrac{x^3+8}{x^2+ax+2x+2a}$. Assume that the denominator is not 0.

Solution We will factor the numerator and the denominator and then divide out any common factors, if possible.

$$\frac{x^3+8}{x^2+ax+2x+2a} = \frac{(x+2)(x^2-2x+4)}{x(x+a)+2(x+a)} \quad \text{Factor the numerator and begin to factor the denominator.}$$

$$= \frac{(x+2)(x^2-2x+4)}{(x+a)(x+2)} \quad \text{Finish factoring the denominator.}$$

$$= \frac{\overset{1}{\cancel{(x+2)}}(x^2-2x+4)}{(x+a)\underset{1}{\cancel{(x+2)}}} \quad \text{Divide out the common factor of } (x+2).$$

$$= \frac{x^2-2x+4}{x+a}$$

SELF CHECK 9 Simplify $\dfrac{ab+3a-2b-6}{a^3-8}$. Assume that the denominator is not 0.

4 Simplify a rational expression containing factors that are negatives.

If the terms of two polynomials are the same, except for signs, the polynomials are called *negatives* or *opposites* of each other. For example,

$x - y$ and $y - x$ are negatives (opposites),

$2a - 1$ and $1 - 2a$ are negatives (opposites), and

$3x^2 - 2x + 5$ and $-3x^2 + 2x - 5$ are negatives (opposites).

Example 10 shows why the quotient of two polynomials that are negatives is always -1.

EXAMPLE 10 Simplify **a.** $\dfrac{x - y}{y - x}$ **b.** $\dfrac{2a - 1}{1 - 2a}$. Assume that no denominators are 0.

Solution We can rearrange terms in each numerator, factor out -1, and proceed as follows:

a.
$$\frac{x - y}{y - x} = \frac{-y + x}{y - x} = \frac{-(y - x)}{y - x} = \frac{-\overset{1}{\cancel{(y - x)}}}{\underset{1}{\cancel{y - x}}} = -1$$

b.
$$\frac{2a - 1}{1 - 2a} = \frac{-1 + 2a}{1 - 2a} = \frac{-(1 - 2a)}{1 - 2a} = \frac{-\overset{1}{\cancel{(1 - 2a)}}}{\underset{1}{\cancel{1 - 2a}}} = -1$$

SELF CHECK 10 Simplify $\dfrac{3p - 2q}{2q - 3p}$. Assume that the denominator is not 0.

The previous example suggests this important result.

DIVISION OF NEGATIVES

The quotient of any nonzero expression and its negative is -1. In symbols, we have

if $a \neq b$, then $\dfrac{a - b}{b - a} = -1$.

SELF CHECK ANSWERS

1. $3, -1$ **2.** $\{x | x \in \mathbb{R}, x \neq -6, 1\}$; $(-\infty, -6) \cup (-6, 1) \cup (1, \infty)$ **3.** $\frac{4a^2}{3b^2}$ **4.** $\frac{x}{5}$ **5.** x **6.** $\frac{x + 3}{x^2}$ **7.** $\frac{4}{3}$ **8.** 1 **9.** $\frac{b + 3}{a^2 + 2a + 4}$ **10.** -1

NOW TRY THIS

1. Evaluate $\dfrac{x - 3}{x + 4}$ for **a.** $x = 3$ **b.** $x = 0$ **c.** $x = -4$

2. Simplify $\dfrac{4x + 20}{4x - 12}$. Assume $x \neq 3$.

3. a. Find all value(s) of x for which $\dfrac{x + 1}{9x^2 - x}$ is undefined.

b. Find the domain of $f(x) = \dfrac{x + 1}{9x^2 - x}$.

6.1 Exercises

WARM-UPS *Simplify each rational expression.*

1. $\frac{14}{21}$
2. $\frac{34}{17}$
3. $\frac{12}{16}$
4. $\frac{100}{25}$
5. $\frac{15}{35}$
6. $\frac{28}{35}$
7. $\frac{-18}{54}$
8. $-\frac{20}{12}$

REVIEW

9. State the associative property of addition.
10. State the distributive property.
11. What is the additive identity?
12. What is the multiplicative identity?
13. Find the additive inverse of $-\frac{7}{5}$.
14. Find the multiplicative inverse of $-\frac{7}{5}$.

VOCABULARY AND CONCEPTS *Fill in the blanks.*

15. In a fraction, the part above the fraction bar is called the ________.
16. In a fraction, the part below the fraction bar is called the __________.
17. The denominator of a fraction cannot be _.
18. A fraction that has polynomials in its numerator and denominator is called a ______ expression.
19. $x - 2$ and $2 - x$ are called _______ of each other.
20. To *simplify* a rational expression means to write it in _____ terms.
21. The fundamental property of fractions states that $\frac{ac}{bc} =$ _.
22. Any number x divided by 1 is _.
23. To simplify a rational expression, we _____ the numerator and denominator and divide out _______ factors.
24. A rational expression cannot be simplified when it is written in __________.

GUIDED PRACTICE *Find all values of the variable for which the following rational expressions are undefined.* SEE EXAMPLE 1. (OBJECTIVE 1)

25. $\frac{4y + 1}{y + 4}$
26. $\frac{5x - 2}{x - 6}$
27. $\frac{3x - 13}{x^2 - x - 2}$
28. $\frac{3p^2 + 7p}{8p^2 + 2p - 1}$
29. $\frac{5x + 2}{(x + 7)(2x - 1)}$
30. $\frac{6x^2 + 7}{(3x + 2)(4x - 5)}$
31. $\frac{2x^2 + 1}{3x^2 + x}$
32. $\frac{5x - 4}{x^2 - 3x}$

Find the domain. SEE EXAMPLE 2. (OBJECTIVE 2)

33. $f(x) = \frac{4x^2 + 3x}{5x - 2}$
34. $f(x) = \frac{12x - 7}{6x + 5}$
35. $f(x) = \frac{2m^2 + 5m}{2m^2 - m - 3}$
36. $f(x) = \frac{5q^2 - 3}{6q^2 - q - 2}$

Write each expression in simplest form. If it is already in simplest form, so indicate. Assume that no denominators are 0. SEE EXAMPLE 3. (OBJECTIVE 3)

37. $\frac{4x}{2}$
38. $\frac{2x}{4}$
39. $\frac{-25y^2}{5y}$
40. $\frac{-6x}{18}$
41. $\frac{6x^2y}{6xy^2}$
42. $\frac{x^2y^3}{x^2y^4}$
43. $\frac{2x^2}{3y}$
44. $\frac{7y^2}{5x^2}$

Write each expression in simplest form. If it is already in simplest form, so indicate. Assume that no denominators are 0. SEE EXAMPLE 4. (OBJECTIVE 3)

45. $\frac{x^2 + 7x}{2x + 14}$
46. $\frac{a^2 - 10a}{4a - 40}$
47. $\frac{3x + 15}{x^2 - 25}$
48. $\frac{x^2 + 3x}{2x + 6}$
49. $\frac{3x + 6}{2x + 1}$
50. $\frac{x^2 + 6x}{5x^2 + 6}$
51. $\frac{10x - 5}{18x - 9}$
52. $\frac{2x^2 - 2x}{5x - 5}$

Write each expression in simplest form. If it is already in simplest form, so indicate. Assume that no denominators are 0. SEE EXAMPLE 5. (OBJECTIVE 3)

53. $\frac{x + 3}{3(x + 3)}$
54. $\frac{2(x + 7)}{x + 7}$
55. $\frac{5x + 35}{x + 7}$
56. $\frac{x + x}{2}$

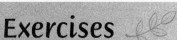

57. $\dfrac{5x^2 + 10x}{x - 2}$

58. $\dfrac{8y^2 - 4y}{y - 4}$

59. $\dfrac{3y^2 + 12y}{4 + y}$

60. $\dfrac{6 + x}{5x^2 + 30x}$

Write each expression in simplest form. If it is already in simplest form, so indicate. Assume that no denominators are 0. SEE EXAMPLE 6. (OBJECTIVE 3)

61. $\dfrac{x^2 + 3x + 2}{x^2 + x - 2}$

62. $\dfrac{x^2 + x - 6}{x^2 - x - 2}$

63. $\dfrac{x^2 - 8x + 15}{x^2 - x - 6}$

64. $\dfrac{x^2 - 6x - 7}{x^2 + 8x + 7}$

65. $\dfrac{2x^2 - 8x}{x^2 - 6x + 8}$

66. $\dfrac{3y^2 - 15y}{y^2 - 3y - 10}$

67. $\dfrac{2a^3 - 16}{2a^2 + 4a + 8}$

68. $\dfrac{3y^3 + 81}{y^2 - 3y + 9}$

69. $\dfrac{x^2 - 4x + 4}{x^2 - 4}$

70. $\dfrac{x^2 - 81}{x^2 - 18x + 81}$

71. $\dfrac{30x^2 - 14x - 8}{3x^2 + 4x + 1}$

72. $\dfrac{x^3 - x}{5x^3 - 5x}$

Write each expression in simplest form. If it is already in simplest form, so indicate. Assume that no denominators are 0. SEE EXAMPLES 7–8. (OBJECTIVE 3)

73. $\dfrac{4(x + 3) + 4}{3(x + 2) + 6}$

74. $\dfrac{x^2 - 3(2x - 3)}{x^2 - 9}$

75. $\dfrac{x^2 + 5x + 4}{2(x + 3) - (x + 2)}$

76. $\dfrac{x^2 - 9}{(2x + 3) - (x + 6)}$

Write each expression in simplest form. If it is already in simplest form, so indicate. Assume that no denominators are 0. SEE EXAMPLE 9. (OBJECTIVE 3)

77. $\dfrac{x^3 + 1}{ax + a + x + 1}$

78. $\dfrac{x^3 - 8}{ax + x - 2a - 2}$

79. $\dfrac{ab + b + 2a + 2}{ab + a + b + 1}$

80. $\dfrac{xy + 2y + 3x + 6}{x^2 + 5x + 6}$

Write each expression in simplest form. If it is already in simplest form, so indicate. Assume that no denominators are 0. SEE EXAMPLE 10. (OBJECTIVE 4)

81. $\dfrac{x - y}{y - x}$

82. $\dfrac{d - c}{c - d}$

83. $\dfrac{6x - 3y}{3y - 6x}$

84. $\dfrac{3c - 4d}{4c - 3d}$

ADDITIONAL PRACTICE

Write each expression in simplest form. If it is already in simplest form, so indicate. Assume that no denominators are 0.

85. $\dfrac{45}{9a}$

86. $\dfrac{48}{16y}$

87. $\dfrac{15x^2y}{5xy^2}$

88. $\dfrac{12xz}{4xz^2}$

89. $\dfrac{x^2 + 3x + 2}{x^3 + x^2}$

90. $\dfrac{x^2 - 8x + 16}{x^2 - 16}$

91. $\dfrac{3x + 3y}{x^2 + xy}$

92. $\dfrac{xy + 2x^2}{2xy + y^2}$

93. $\dfrac{3y + xy}{3x + xy}$

94. $\dfrac{6x^2 - 13x + 6}{3x^2 + x - 2}$

95. $\dfrac{xz - 2x}{yz - 2y}$

96. $\dfrac{x^2 - 2x - 15}{x^2 + 2x - 15}$

97. $\dfrac{15x - 3x^2}{25y - 5xy}$

98. $\dfrac{x^3 + 1}{x^2 - x + 1}$

99. $\dfrac{4 + 2(x - 5)}{3x - 5(x - 2)}$

100. $\dfrac{x^3 - 1}{x^2 + x + 1}$

101. $\dfrac{x^2 + 4x - 77}{x^2 - 4x - 21}$

102. $\dfrac{x^2 - 10x + 25}{25 - x^2}$

103. $\dfrac{xy + 3y + 3x + 9}{x^2 - 9}$

104. $\dfrac{ab + b^2 + 2a + 2b}{a^2 + 2a + ab + 2b}$

105. $\dfrac{2x^2 - 8}{x^2 - 3x + 2}$

106. $\dfrac{3x^2 - 27}{x^2 + 3x - 18}$

107. $\dfrac{a + b - c}{c - a - b}$

108. $\dfrac{x - y - z}{z + y - x}$

109. $\dfrac{6a - 6b + 6c}{9a - 9b + 9c}$

110. $\dfrac{3a - 3b - 6}{2a - 2b - 4}$

WRITING ABOUT MATH

111. Explain why $\dfrac{x - 7}{7 - x} = -1$.

112. Exercise 99 has two possible answers: $\dfrac{x - 3}{5 - x}$ and $-\dfrac{x - 3}{x - 5}$. Why is either answer correct?

SOMETHING TO THINK ABOUT

113. Find two different-looking but correct answers for the following problem.

Simplify: $\dfrac{y^2 + 5(2y + 5)}{25 - y^2}$

Section 6.2 Multiplying and Dividing Rational Expressions

Objectives

1. Multiply two rational expressions and write the result in simplest form.
2. Multiply a rational expression by a polynomial and write the result in simplest form.
3. Divide two rational expressions and write the result in simplest form.
4. Divide a rational expression by a polynomial and write the result in simplest form.
5. Perform combined operations on three or more rational expressions.

Getting Ready

Multiply or divide the fractions and simplify.

1. $\frac{3}{7}\cdot\frac{14}{9}$ **2.** $\frac{21}{15}\cdot\frac{10}{3}$ **3.** $\frac{19}{38}\cdot 6$ **4.** $42\cdot\frac{3}{21}$

5. $\frac{4}{9}\div\frac{8}{45}$ **6.** $\frac{11}{7}\div\frac{22}{14}$ **7.** $\frac{75}{12}\div\frac{50}{6}$ **8.** $\frac{13}{5}\div\frac{26}{20}$

Just like arithmetic fractions, rational expressions can be multiplied, divided, added, and subtracted. In this section, we will show how to multiply and divide rational expressions.

1 Multiply two rational expressions and write the result in simplest form.

Recall that to multiply fractions, we multiply their numerators and multiply their denominators. For example, to find the product of $\frac{4}{7}$ and $\frac{3}{5}$, we proceed as follows.

$$\frac{4}{7}\cdot\frac{3}{5} = \frac{4\cdot 3}{7\cdot 5} \quad \text{Multiply the numerators and multiply the denominators.}$$

$$= \frac{12}{35} \quad \text{Simplify.}$$

This suggests the rule for multiplying rational expressions.

MULTIPLYING RATIONAL EXPRESSIONS

If a, b, c, and d are polynomials, then

$$\frac{a}{b}\cdot\frac{c}{d} = \frac{ac}{bd} \quad \text{provided no denominators are 0.}$$

EXAMPLE 1 Multiply. Assume that no denominators are 0.

a. $\frac{1}{3}\cdot\frac{2}{5}$ **b.** $\frac{7}{9}\cdot\frac{-5}{3x}$ **c.** $\frac{x^2}{2}\cdot\frac{3}{y^2}$ **d.** $\frac{q+1}{q}\cdot\frac{q-1}{q-2}$

Solution We will multiply the numerators, multiply the denominators, and then simplify, if possible.

a. $\frac{1}{3} \cdot \frac{2}{5} = \frac{1 \cdot 2}{3 \cdot 5}$
$= \frac{2}{15}$

b. $\frac{7}{9} \cdot \frac{-5}{3x} = \frac{7(-5)}{9 \cdot 3x}$
$= \frac{-35}{27x}$

c. $\frac{x^2}{2} \cdot \frac{3}{y^2} = \frac{x^2 \cdot 3}{2 \cdot y^2}$
$= \frac{3x^2}{2y^2}$

d. $\frac{q+1}{q} \cdot \frac{q-1}{q-2} = \frac{(q+1)(q-1)}{q(q-2)}$

SELF CHECK 1 Multiply. Assume that no denominators are 0.

a. $\frac{4}{5} \cdot \frac{7}{9}$ **b.** $\frac{-5}{3} \cdot \frac{2x}{3}$ **c.** $\frac{x}{y^2} \cdot \frac{5x}{4}$ **d.** $\frac{y+1}{9y} \cdot \frac{y+2}{y-1}$

EXAMPLE 2

Multiply $\frac{35x^2y}{7y^2z} \cdot \frac{z}{5xy}$. Assume that no denominators are 0.

Solution We will multiply the numerators, multiply the denominators, and then simplify, if possible.

$\frac{35x^2y}{7y^2z} \cdot \frac{z}{5xy} = \frac{35x^2yz}{35y^3zx}$ Multiply the numerators and multiply the denominators.

$= \frac{x}{y^2}$ Simplify using properties of exponents.

SELF CHECK 2 Multiply $\frac{a^2b^2}{2a} \cdot \frac{9a^3}{3b^3}$. Assume that no denominators are 0.

EXAMPLE 3

Multiply $\frac{x^2 - x}{2x + 4} \cdot \frac{x + 2}{x}$. Assume that no denominators are 0.

Solution We will multiply the numerators, multiply the denominators, and then simplify.

$\frac{x^2 - x}{2x + 4} \cdot \frac{x + 2}{x} = \frac{(x^2 - x)(x + 2)}{(2x + 4)(x)}$ Multiply the numerators and multiply the denominators.

$= \frac{x(x-1)(x+2)}{2(x+2)x}$ Factor.

$= \frac{\overset{1}{\cancel{x}}(x-1)\overset{1}{\cancel{(x+2)}}}{2\underset{1}{\cancel{(x+2)}}\underset{1}{\cancel{x}}}$ Divide out common factors.

$= \frac{x-1}{2}$

SELF CHECK 3 Multiply $\frac{x^2 + x}{3x + 6} \cdot \frac{x + 2}{x + 1}$. Assume that no denominators are 0.

Recall from arithmetic that $\frac{2}{3} \cdot \frac{9}{8}$ could be simplified before multiplying:

$$\frac{\overset{1}{\cancel{2}}}{\underset{1}{\cancel{3}}} \cdot \frac{\overset{3}{\cancel{9}}}{\underset{4}{\cancel{8}}} = \frac{3}{4}$$

We could approach Example 3 in the same way. Factor each expression that is not in completely factored form and divide out common factors.

$$\frac{x^2 - x}{2x + 4} \cdot \frac{x + 2}{x} = \frac{x(x - 1)}{2(x + 2)} \cdot \frac{x + 2}{x}$$
$$= \frac{\overset{1}{\cancel{x}}(x - 1)}{2\underset{1}{\cancel{(x + 2)}}} \cdot \frac{\overset{1}{\cancel{x + 2}}}{\underset{1}{\cancel{x}}}$$
$$= \frac{x - 1}{2}$$

We will work Example 4 using this method.

EXAMPLE 4 Multiply $\frac{x^2 - 3x}{x^2 - x - 6}$ and $\frac{x^2 + x - 2}{x^2 - x}$. Assume that no denominators are 0.

Solution We will factor the numerators and denominators, simplify, and then multiply.

$$\frac{x^2 - 3x}{x^2 - x - 6} \cdot \frac{x^2 + x - 2}{x^2 - x}$$
$$= \frac{x(x - 3)}{(x - 3)(x + 2)} \cdot \frac{(x + 2)(x - 1)}{x(x - 1)} \quad \text{Factor where possible.}$$
$$= \frac{\overset{1}{\cancel{x}}\overset{1}{\cancel{(x - 3)}}}{\underset{1}{\cancel{(x - 3)}}\underset{1}{\cancel{(x + 2)}}} \cdot \frac{\overset{1}{\cancel{(x + 2)}}\overset{1}{\cancel{(x - 1)}}}{\underset{1}{\cancel{x}}\underset{1}{\cancel{(x - 1)}}} \quad \text{Divide out common factors.}$$
$$= 1$$

SELF CHECK 4 Multiply $\frac{a^2 + a}{a^2 - 4} \cdot \frac{a^2 - a - 2}{a^2 + 2a + 1}$. Assume that no denominators are 0.

2 Multiply a rational expression by a polynomial and write the result in simplest form.

Since any number divided by 1 remains unchanged, we can write any polynomial as a rational expression by writing it with a denominator of 1.

EXAMPLE 5 Multiply $\frac{x^2 + x}{x^2 + 8x + 7} \cdot (x + 7)$. Assume that the denominator is not 0.

Solution We will write $x + 7$ as $\frac{x + 7}{1}$, multiply the numerators, multiply the denominators, and then simplify.

$$\frac{x^2 + x}{x^2 + 8x + 7} \cdot (x + 7) = \frac{x^2 + x}{x^2 + 8x + 7} \cdot \frac{x + 7}{1} \quad \text{Write } x + 7 \text{ as a fraction with a denominator of 1.}$$
$$= \frac{x(x + 1)(x + 7)}{(x + 1)(x + 7)1} \quad \text{Multiply the fractions and factor where possible.}$$
$$= \frac{x\overset{1}{\cancel{(x + 1)}}\overset{1}{\cancel{(x + 7)}}}{1\underset{1}{\cancel{(x + 1)}}\underset{1}{\cancel{(x + 7)}}} \quad \text{Divide out all common factors.}$$
$$= x$$

SELF CHECK 5 Multiply $(a - 7) \cdot \frac{a^2 - a}{a^2 - 8a + 7}$. Assume that the denominator is not 0.

3 Divide two rational expressions and write the result in simplest form.

Recall that division by a nonzero number is equivalent to multiplying by the reciprocal of that number. Thus, to divide two fractions, we can invert the *divisor* (the fraction following the ÷ sign) and multiply. For example, to divide $\frac{4}{7}$ by $\frac{3}{5}$, we proceed as follows:

$$\frac{4}{7} \div \frac{3}{5} = \frac{4}{7} \cdot \frac{5}{3} \quad \text{Invert } \tfrac{3}{5} \text{ and write as multiplication.}$$

$$= \frac{20}{21} \quad \text{Multiply the numerators and multiply the denominators.}$$

This suggests the rule for dividing rational expressions.

DIVIDING RATIONAL EXPRESSIONS

If a, b, c, and d are polynomials, then

$$\frac{a}{b} \div \frac{c}{d} = \frac{a}{b} \cdot \frac{d}{c} = \frac{ad}{bc} \quad \text{provided no denominators are equal to 0.}$$

EXAMPLE 6 Divide, assuming that no denominators are 0.

a. $\frac{7}{13} \div \frac{21}{26}$ **b.** $\frac{-9x}{35y} \div \frac{15x^2}{14}$

Solution We will change each division to a multiplication and then multiply the resulting rational expressions.

a. $\frac{7}{13} \div \frac{21}{26} = \frac{7}{13} \cdot \frac{26}{21}$ Invert the divisor and multiply.

$= \frac{\overset{1}{\cancel{7}}}{\underset{1}{\cancel{13}}} \cdot \frac{\overset{2}{\cancel{26}}}{\underset{3}{\cancel{21}}}$ Divide out common factors.

$= \frac{2}{3}$

b. $\frac{-9x}{35y} \div \frac{15x^2}{14} = \frac{-9x}{35y} \cdot \frac{14}{15x^2}$ Invert the divisor and multiply.

$= \frac{\overset{-3}{\cancel{-9x}}}{\underset{5}{\cancel{35y}}} \cdot \frac{\overset{2}{\cancel{14}}}{\underset{5x}{\cancel{15x^2}}}$ Divide out common factors.

$= -\frac{6}{25xy}$ Multiply the remaining factors.

SELF CHECK 6 Divide. Assume that no denominators are 0.

a. $\frac{15}{14} \div \frac{25}{7}$ **b.** $\frac{-8a}{3b} \div \frac{16a^2}{9b^2}$

EXAMPLE 7 Divide $\frac{x^2+x}{3x-15} \div \frac{x^2+2x+1}{6x-30}$. Assume that no denominators are 0.

Solution We invert the divisor and write the division as multiplication and then multiply the resulting rational expressions.

$$\frac{x^2+x}{3x-15} \div \frac{x^2+2x+1}{6x-30}$$

$$= \frac{x^2+x}{3x-15} \cdot \frac{6x-30}{x^2+2x+1} \quad \text{Invert the divisor and multiply.}$$

$$= \frac{x(x+1)}{3(x-5)} \cdot \frac{6(x-5)}{(x+1)(x+1)} \quad \text{Factor where possible.}$$

$$= \frac{x\cancel{(x+1)}^1}{\cancel{3}_1\cancel{(x-5)}_1} \cdot \frac{^2\cancel{6}\cancel{(x-5)}^1}{\cancel{(x+1)}_1(x+1)} \quad \text{Divide out all common factors.}$$

$$= \frac{2x}{x+1} \quad \text{Multiply the fractions.}$$

SELF CHECK 7 Divide $\frac{a^2-1}{a^2+4a+3} \div \frac{a-1}{a^2+2a-3}$. Assume that no denominators are 0.

4 Divide a rational expression by a polynomial and write the result in simplest form.

To divide a rational expression by a polynomial, we write the polynomial as a rational expression with a denominator of 1 and then divide the expressions.

EXAMPLE 8 Divide $\frac{2x^2-3x-2}{2x+1} \div (4-x^2)$. Assume that no denominators are 0.

Solution We will write $4-x^2$ as $\frac{4-x^2}{1}$, invert the divisor, and then multiply the resulting rational expressions.

$$\frac{2x^2-3x-2}{2x+1} \div (4-x^2)$$

$$= \frac{2x^2-3x-2}{2x+1} \div \frac{4-x^2}{1} \quad \text{Write } 4-x^2 \text{ as a fraction with a denominator of 1.}$$

$$= \frac{2x^2-3x-2}{2x+1} \cdot \frac{1}{4-x^2} \quad \text{Invert the divisor and multiply.}$$

$$= \frac{(2x+1)(x-2)}{2x+1} \cdot \frac{1}{(2+x)(2-x)} \quad \text{Factor where possible.}$$

$$= \frac{\cancel{(2x+1)}^1\cancel{(x-2)}^{-1}}{\cancel{2x+1}_1} \cdot \frac{1}{(2+x)\cancel{(2-x)}_1} \quad \text{Divide out common factors: } \frac{x-2}{2-x} = -1$$

$$= \frac{-1}{2+x} \quad \text{Multiply.}$$

$$\text{OR } -\frac{1}{2+x} \quad \frac{-a}{b} = -\frac{a}{b}$$

SELF CHECK 8 Divide $(b-a) \div \frac{a^2-b^2}{a^2+ab}$. Assume that no denominators are 0.

5 Perform combined operations on three or more rational expressions.

Unless parentheses indicate otherwise, we will perform multiplications and divisions in order from left to right.

EXAMPLE 9 Simplify $\frac{x^2 - x - 6}{x - 2} \div \frac{x^2 - 4x}{x^2 - x - 2} \cdot \frac{x - 4}{x^2 + x}$. Assume that no denominators are 0.

Solution Since there are no parentheses to indicate otherwise, we perform the division first.

$$\frac{x^2 - x - 6}{x - 2} \div \frac{x^2 - 4x}{x^2 - x - 2} \cdot \frac{x - 4}{x^2 + x}$$

$$= \frac{x^2 - x - 6}{x - 2} \cdot \frac{x^2 - x - 2}{x^2 - 4x} \cdot \frac{x - 4}{x^2 + x}$$ Invert the divisor and multiply.

$$= \frac{(x + 2)(x - 3)(x + 1)(x - 2)(x - 4)}{(x - 2)x(x - 4)x(x + 1)}$$ Factor where possible and multiply.

$$= \frac{(x + 2)(x - 3)\overset{1}{\cancel{(x + 1)}}\overset{1}{\cancel{(x - 2)}}\overset{1}{\cancel{(x - 4)}}}{\underset{1}{\cancel{(x - 2)}}x\underset{1}{\cancel{(x - 4)}}x\underset{1}{\cancel{(x + 1)}}}$$ Divide out all common factors.

$$= \frac{(x + 2)(x - 3)}{x^2}$$

SELF CHECK 9 Simplify $\frac{a^2 + ab}{ab - b^2} \cdot \frac{a^2 - b^2}{a^2 + ab} \div \frac{a + b}{b}$. Assume that no denominators are 0.

EXAMPLE 10 Simplify $\frac{x^2 + 6x + 9}{x^2 - 2x}\left(\frac{x^2 - 4}{x^2 + 3x} \div \frac{x + 2}{x}\right)$. Assume that no denominators are 0.

Solution We perform the division within the parentheses first.

$$\frac{x^2 + 6x + 9}{x^2 - 2x}\left(\frac{x^2 - 4}{x^2 + 3x} \div \frac{x + 2}{x}\right)$$

$$= \frac{x^2 + 6x + 9}{x^2 - 2x}\left(\frac{x^2 - 4}{x^2 + 3x} \cdot \frac{x}{x + 2}\right)$$ Invert the divisor and multiply.

$$= \frac{(x + 3)(x + 3)}{x(x - 2)}\left[\frac{(x - 2)(x + 2)}{x(x + 3)} \cdot \frac{x}{x + 2}\right]$$ Factor where possible.

$$= \frac{(x + 3)(x + 3)}{x(x - 2)}\left[\frac{(x - 2)\overset{1}{\cancel{(x + 2)}}}{\underset{1}{\cancel{x}}(x + 3)} \cdot \frac{\overset{1}{\cancel{x}}}{\underset{1}{\cancel{x + 2}}}\right]$$ Divide out all common factors within the brackets.

$$= \frac{(x + 3)(x + 3)}{x(x - 2)} \cdot \frac{x - 2}{x + 3}$$ Remove brackets.

$$= \frac{\overset{1}{\cancel{(x + 3)}}(x + 3)}{x\underset{1}{\cancel{(x - 2)}}} \cdot \frac{\overset{1}{\cancel{x - 2}}}{\underset{1}{\cancel{x + 3}}}$$ Divide out all common factors.

$$= \frac{x + 3}{x}$$

SELF CHECK 10 Simplify $\frac{x^2 - 2x}{x^2 + 6x + 9} \div \left(\frac{x^2 - 4}{x^2 + 3x} \cdot \frac{x}{x + 2}\right)$. Assume that no denominators are 0.

SELF CHECK ANSWERS

1. a. $\frac{28}{45}$ **b.** $\frac{-10x}{9}$ **c.** $\frac{5x^2}{4y^2}$ **d.** $\frac{(y+1)(y+2)}{9y(y-1)}$ **2.** $\frac{3a^4}{2b}$ **3.** $\frac{x}{3}$ **4.** $\frac{a}{a+2}$ **5.** a **6. a.** $\frac{3}{10}$ **b.** $-\frac{3b}{2a}$ **7.** $a-1$ **8.** $-a$ **9.** 1 **10.** $\frac{x}{x+3}$

NOW TRY THIS

Simplify. Assume no division by 0.

1. $(x^2-4x-12)\cdot\dfrac{(x+6)^2}{x^2-36}$
2. $\dfrac{x^2-9}{x-2}\div\dfrac{9-x^2}{3x-6}$
3. $\dfrac{\frac{1}{2}}{\frac{3}{4}}$
4. $\dfrac{\frac{3}{5}-\frac{2}{3}}{\frac{7}{3}+\frac{2}{5}}$

6.2 Exercises

WARM-UPS *Fill in the boxes. Assume no denominator is 0.*

1. $\frac{5}{2}\cdot\frac{3}{\square}=\frac{3}{2}$

2. $\frac{7}{3}\cdot\frac{5}{\square}=\frac{35}{18}$

3. $\frac{x}{2}\cdot\frac{3}{\square}=\frac{3}{2}$

4. $\frac{2\square}{5}\cdot\frac{7}{x+1}=\frac{14}{5}$

5. $\frac{9}{\square}\div\frac{3}{2}=\frac{3}{2}$

6. $\frac{3}{7}\div\frac{\square}{\square}=1$

7. $\frac{\square}{2}\div\frac{x}{3}=\frac{3x}{2}$

8. $\frac{1}{\square}\div\frac{x}{1(x+1)}=\frac{1}{x}$

REVIEW *Simplify each expression. Write all answers without using negative exponents. Assume that no denominators are 0.*

9. $2x^3y^2(-3x^2y^4z)$

10. $\dfrac{8x^4y^5}{-2x^3y^2}$

11. $(5y)^{-3}$

12. $(a^{-2}a)^{-3}$

13. $\dfrac{x^{3m}}{x^{4m}}$

14. $(3x^2y^3)^0$

Perform the operations and simplify.

15. $-4(y^3-4y^2+3y-2)+6(-2y^2+4)-4(-2y^3-y)$

16. $y-5\overline{)5y^3-3y^2+4y-1}$ $(y\neq 5)$

VOCABULARY AND CONCEPTS *Fill in the blanks.*

17. To multiply fractions, we multiply their ________ and multiply their __________.

18. Unless parentheses indicate otherwise, do multiplications and divisions in order from __ to ___.

19. $\frac{a}{b}\cdot\frac{c}{d}=$ ___

20. To write a polynomial in fractional form, we insert a denominator of _.

21. $\frac{a}{b}\div\frac{c}{d}=\frac{a}{b}\cdot$ _

22. To divide two fractions, invert the _____ and _______.

GUIDED PRACTICE *Perform the multiplication. Assume that no denominators are 0.* SEE EXAMPLE 1. (OBJECTIVE 1)

23. $\frac{5}{7}\cdot\frac{9}{13}$

24. $\frac{3}{5}\cdot\frac{11}{7}$

25. $\frac{5x}{y}\cdot\frac{4x}{3y^2}$

26. $\frac{-8x^2}{z^4}\cdot\frac{2x}{5}$

27. $\frac{z+7}{7}\cdot\frac{z+2}{z}$

28. $\frac{a-3}{a}\cdot\frac{a+3}{5}$

29. $\frac{-3a}{a+2}\cdot\frac{a-1}{5}$

30. $\frac{-b}{b-1}\cdot\frac{b+2}{b-2}$

Perform the multiplication. Assume that no denominator is 0. Simplify the answers. SEE EXAMPLE 2. (OBJECTIVE 1)

31. $\frac{2y}{z}\cdot\frac{z}{3}$

32. $\frac{5x}{y}\cdot\frac{4}{x}$

33. $\frac{5y}{7}\cdot\frac{7x}{5z}$

34. $\frac{4x}{3y}\cdot\frac{3y}{7x}$

35. $\frac{x}{2x^2}\cdot\frac{-28xy}{7}$

36. $\frac{-2xy}{x^2}\cdot\frac{3xy}{2}$

37. $\frac{ab^2}{a^2b}\cdot\frac{b^2c^2}{abc}\cdot\frac{abc^2}{a^3c^2}$

38. $\frac{x^3y}{z}\cdot\frac{xz^3}{x^2y^2}\cdot\frac{yz}{xyz}$

Perform the multiplication. Assume that no denominators are 0. Simplify the answers, if possible. SEE EXAMPLE 3. (OBJECTIVE 1)

39. $\frac{x-2}{2}\cdot\frac{2x}{x-2}$

40. $\frac{y+7}{y}\cdot\frac{3y}{y+7}$

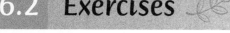

41. $\dfrac{5y-5}{y-1} \cdot \dfrac{y}{10y^2}$

42. $\dfrac{x-7}{4x+8} \cdot \dfrac{x+2}{x^2-49}$

43. $\dfrac{3y-12}{y+8} \cdot \dfrac{y^2+8y}{y-4}$

44. $\dfrac{y^2+3y}{9} \cdot \dfrac{3x}{y+3}$

45. $\dfrac{5z-10}{z+2} \cdot \dfrac{3}{3z-6}$

46. $\dfrac{4x-8}{4x-4} \cdot \dfrac{x^2-x}{x}$

Perform the multiplication. Assume that no denominators are 0. Simplify the answers, if possible. SEE EXAMPLE 4. (OBJECTIVE 1)

47. $\dfrac{z^2+4z-5}{5z-5} \cdot \dfrac{5z}{z+5}$

48. $\dfrac{x^2+x-6}{5x} \cdot \dfrac{5x-10}{x+3}$

49. $\dfrac{(x+1)^2}{x+1} \cdot \dfrac{x+2}{x+1}$

50. $\dfrac{(y-3)^2}{y-3} \cdot \dfrac{y-3}{y-3}$

51. $\dfrac{m^2-2m-3}{2m+4} \cdot \dfrac{m^2-4}{m^2+3m+2}$

52. $\dfrac{p^2-p-6}{3p-9} \cdot \dfrac{p^2-9}{p^2+6p+9}$

53. $\dfrac{abc^2}{a+1} \cdot \dfrac{c}{a^2b^2} \cdot \dfrac{a^2+a}{ac}$

54. $\dfrac{x^3yz^2}{4x+8} \cdot \dfrac{x^2-4}{2x^2y^2z^2} \cdot \dfrac{8yz}{x-2}$

Perform the multiplication. Assume that no denominators are 0. Simplify the answers when possible. SEE EXAMPLE 5. (OBJECTIVE 2)

55. $\dfrac{x-5}{2x-8} \cdot (x-4)$

56. $\dfrac{x^2+x}{x^2+4x+3} \cdot (x+3)$

57. $(5x-10) \cdot \dfrac{x^2+2x}{x^2-4}$

58. $(6x-8) \cdot \dfrac{x-2}{9x-12}$

Perform each division. Assume that no denominators are 0. Simplify answers when possible. SEE EXAMPLE 6. (OBJECTIVE 3)

59. $\dfrac{2}{3} \div \dfrac{1}{2}$

60. $\dfrac{3}{4} \div \dfrac{1}{3}$

61. $\dfrac{21}{14} \div \dfrac{5}{2}$

62. $\dfrac{14}{3} \div \dfrac{10}{3}$

63. $\dfrac{3x}{2} \div \dfrac{x}{2}$

64. $\dfrac{y}{6} \div \dfrac{2}{3y}$

65. $\dfrac{x^2y}{3xy} \div \dfrac{xy^2}{6y}$

66. $\dfrac{2xz}{z} \div \dfrac{4x^2}{z^2}$

Perform each division. Assume that no denominators are 0. Simplify answers when possible. SEE EXAMPLE 7. (OBJECTIVE 3)

67. $\dfrac{x+2}{3x} \div \dfrac{x+2}{2}$

68. $\dfrac{z-3}{3z} \div \dfrac{z+3}{z}$

69. $\dfrac{x^2-4}{3x+6} \div \dfrac{x-2}{x+2}$

70. $\dfrac{x^2-16}{3x+12} \div \dfrac{x-4}{x+4}$

71. $\dfrac{y(y+2)}{y^2(y-3)} \div \dfrac{y^2(y+2)}{(y-3)^2}$

72. $\dfrac{(z-3)^2}{4z^2} \div \dfrac{z-3}{8z}$

73. $\dfrac{x^2-x-6}{2x^2+9x+10} \div \dfrac{x^2-25}{2x^2+15x+25}$

74. $\dfrac{x^2-2x-35}{3x^2+27x} \div \dfrac{x^2+7x+10}{6x^2+12x}$

75. $\dfrac{5x^2+13x-6}{x+3} \div \dfrac{5x^2-17x+6}{x-2}$

76. $\dfrac{2x^2+8x-42}{x-3} \div \dfrac{2x^2+14x}{x^2+5x}$

77. $\dfrac{x^2+7xy+12y^2}{x^2+2xy-8y^2} \cdot \dfrac{x^2-xy-2y^2}{x^2+4xy+3y^2}$

78. $\dfrac{m^2+9mn+20n^2}{m^2-25n^2} \cdot \dfrac{m^2-9mn+20n^2}{m^2-16n^2}$

Perform the operations. Assume that no denominators are 0. Simplify answers when possible. SEE EXAMPLE 8. (OBJECTIVE 4)

79. $\dfrac{3x+9}{x+1} \div (x+3)$

80. $\dfrac{2x-5}{6x+1} \div (16x-40)$

81. $(3x+9) \div \dfrac{x^2-9}{6x}$

82. $(x-2) \div \dfrac{6x-12}{x+2}$

Perform the operations. Assume that no denominators are 0. Simplify answers when possible. SEE EXAMPLE 9. (OBJECTIVE 5)

83. $\dfrac{x}{3} \cdot \dfrac{9}{4} \div \dfrac{x^2}{6}$

84. $\dfrac{y^3}{3y} \cdot \dfrac{3y^2}{4} \div \dfrac{15}{20}$

85. $\dfrac{y^2}{2} \div \dfrac{4}{y} \cdot \dfrac{y^2}{8}$

86. $\dfrac{x^2}{18} \div \dfrac{x^3}{6} \div \dfrac{12}{x^2}$

87. $\dfrac{x^2-4}{2x+6} \div \dfrac{x+2}{4} \cdot \dfrac{x+3}{x-2}$

88. $\dfrac{2}{3x-3} \div \dfrac{2x+2}{x-1} \cdot \dfrac{5}{x+1}$

89. $\dfrac{x^2+x-6}{x^2-4} \cdot \dfrac{x^2+2x}{x-2} \div \dfrac{x^2+3x}{x+2}$

90. $\dfrac{x^2-x-6}{x^2+6x-7} \cdot \dfrac{x^2+x-2}{x^2+2x} \div \dfrac{x^2+7x}{x^2-3x}$

Perform the operations. Assume that no denominators are 0. Simplify answers when possible. SEE EXAMPLE 10. (OBJECTIVE 5)

91. $\dfrac{x^2-1}{x^2-9}\left(\dfrac{x+3}{x+2} \div \dfrac{5}{x+2}\right)$

92. $\dfrac{x^2 - 5x}{x + 1}\left(\dfrac{x + 1}{x^2 + 3x} \div \dfrac{x - 5}{x - 3}\right)$

93. $\dfrac{x - x^2}{x^2 - 4}\left(\dfrac{2x + 4}{x + 2} \div \dfrac{5}{x + 2}\right)$

94. $\dfrac{2}{3x - 3} \div \left(\dfrac{2x + 2}{x - 1} \cdot \dfrac{5}{x + 1}\right)$

ADDITIONAL PRACTICE *Perform the indicated operation(s). Assume that no denominators are 0. Simplify answers when possible.*

95. $\dfrac{8z}{2x} \cdot \dfrac{16x}{3x}$

96. $\dfrac{2x^2y}{3xy} \cdot \dfrac{3xy^2}{2}$

97. $\dfrac{2x^2z}{z} \cdot \dfrac{5x}{z}$

98. $\dfrac{10r^2st^3}{6rs^2} \cdot \dfrac{3r^3t}{2rst} \cdot \dfrac{2s^3t^4}{5s^2t^3}$

99. $\dfrac{3a^3b}{25cd^3} \cdot \dfrac{-5cd^2}{6ab} \cdot \dfrac{10abc^2}{2bc^2d}$

100. $\dfrac{3y}{8} \div \dfrac{2y}{4y}$

101. $\dfrac{3x}{y} \div \dfrac{2x}{4}$

102. $\dfrac{4x}{3x} \div \dfrac{2y}{9y}$

103. $\dfrac{14}{7y} \div \dfrac{10}{5z}$

104. $\dfrac{y - 9}{y + 9} \cdot \dfrac{y}{9}$

105. $\dfrac{(x + 7)^2}{x + 7} \div \dfrac{(x - 3)^2}{x + 7}$

106. $\dfrac{x^2 - 1}{3x - 3} \div \dfrac{x + 1}{3}$

107. $\dfrac{x^2 - 16}{x - 4} \div \dfrac{3x + 12}{x}$

108. $\dfrac{3x^2 + 5x + 2}{x^2 - 9} \cdot \dfrac{x - 3}{x^2 - 4} \cdot \dfrac{x^2 + 5x + 6}{6x + 4}$

109. $\dfrac{a^2 - ab + b^2}{a^3 + b^3} \cdot \dfrac{ac + ad + bc + bd}{c^2 - d^2}$

110. $\dfrac{ab + 4a + 2b + 8}{b^2 + 4b + 16} \div \dfrac{b^2 - 16}{b^3 - 64}$

111. $\dfrac{xw - xz + wy - yz}{x^2 + 2xy + y^2} \cdot \dfrac{x^3 - y^3}{z^2 - w^2}$

112. $\dfrac{s^3 - r^3}{r^2 + rs + s^2} \div \dfrac{pr - ps - qr + qs}{q^2 - p^2}$

113. $\dfrac{p^3 - p^2q + pq^2}{mp - mq + np - nq} \div \dfrac{q^3 + p^3}{q^2 - p^2}$

114. $\dfrac{x^2 - y^2}{x^4 - x^3} \div \dfrac{x - y}{x^2} \div \dfrac{x^2 + 2xy + y^2}{x + y}$

WRITING ABOUT MATH

115. Explain how to multiply two fractions and how to simplify the result.

116. Explain why any mathematical expression can be written as a fraction.

117. To divide fractions, you must first know how to multiply fractions. Explain.

118. Explain how to do the division $\frac{a}{b} \div \frac{c}{d} \div \frac{e}{f}$.

SOMETHING TO THINK ABOUT

119. Let x equal a number of your choosing. Without simplifying first, use a calculator to evaluate

$$\frac{x^2 + x - 6}{x^2 + 3x} \cdot \frac{x^2}{x - 2}$$

Try again, with a different value of x. If you were to simplify the expression, what do you think you would get?

120. Simplify the expression in Exercise 119 to determine whether your answer was correct.

Section 6.3 Adding and Subtracting Rational Expressions

Objectives

1. Add two rational expressions with like denominators and write the answer in simplest form.
2. Subtract two rational expressions with like denominators and write the answer in simplest form.
3. Find the least common denominator (LCD) of two or more polynomials and use it to write equivalent rational expressions.
4. Add two rational expressions with unlike denominators and write the answer in simplest form.
5. Subtract two rational expressions with unlike denominators and write the answer in simplest form.

Vocabulary

least common denominator (LCD)

Getting Ready

Add or subtract the fractions and simplify.

1. $\frac{1}{5} + \frac{3}{5}$
2. $\frac{3}{7} + \frac{4}{7}$
3. $\frac{3}{8} + \frac{4}{8}$
4. $\frac{18}{19} + \frac{20}{19}$
5. $\frac{5}{9} - \frac{4}{9}$
6. $\frac{7}{12} - \frac{1}{12}$
7. $\frac{7}{13} - \frac{9}{13}$
8. $\frac{20}{10} - \frac{7}{10}$

In this section, we will discuss how to add and subtract rational expressions.

1 Add two rational expressions with like denominators and write the answer in simplest form.

To add rational expressions with a common denominator, we follow the same process we use to add arithmetic fractions; add their numerators and keep the common denominator. For example,

$$\frac{2x}{7} + \frac{3x}{7} = \frac{2x + 3x}{7}$$ Add the numerators and keep the common denominator.

$$= \frac{5x}{7}$$ $2x + 3x = 5x$

In general, we have the following result.

ADDING RATIONAL EXPRESSIONS WITH LIKE DENOMINATORS

If a, b, and d represent polynomials, then

$$\frac{a}{d} + \frac{b}{d} = \frac{a + b}{d}$$ provided the denominator is not equal to 0.

EXAMPLE 1 Perform each addition. Assume that no denominators are 0.

Solution In each part, we will add the numerators and keep the common denominator.

a. $$\frac{xy}{8z} + \frac{3xy}{8z} = \frac{xy + 3xy}{8z}$$ Add the numerators and keep the common denominator.

$$= \frac{4xy}{8z}$$ Combine like terms.

$$= \frac{xy}{2z}$$ $\frac{4xy}{8z} = \frac{4 \cdot xy}{4 \cdot 2z} = \frac{xy}{2z}$, because $\frac{4}{4} = 1$

b. $$\frac{3x + y}{5x} + \frac{x + y}{5x} = \frac{3x + y + x + y}{5x}$$ Add the numerators and keep the common denominator.

$$= \frac{4x + 2y}{5x}$$ Combine like terms.

SELF CHECK 1 Perform each addition. Assume no denominators are 0.

a. $\frac{x}{7} + \frac{y}{7}$ **b.** $\frac{3x}{7y} + \frac{4x}{7y}$

COMMENT After adding two fractions, simplify the result if possible.

EXAMPLE 2 Add: $\frac{3x + 21}{5x + 10} + \frac{8x + 1}{5x + 10} \quad (x \neq -2)$

Solution Since the rational expressions have the same denominator, we add their numerators and keep the common denominator.

$$\frac{3x + 21}{5x + 10} + \frac{8x + 1}{5x + 10} = \frac{3x + 21 + 8x + 1}{5x + 10} \quad \text{Add the fractions.}$$

$$= \frac{11x + 22}{5x + 10} \quad \text{Combine like terms.}$$

$$= \frac{11\overset{1}{\cancel{(x + 2)}}}{5\underset{1}{\cancel{(x + 2)}}} \quad \text{Factor and divide out the common factor of } (x + 2).$$

$$= \frac{11}{5}$$

SELF CHECK 2 Add: $\frac{x + 4}{6x - 12} + \frac{x - 8}{6x - 12} \quad (x \neq 2)$

2 Subtract two rational expressions with like denominators and write the answer in simplest form.

To subtract rational expressions with a common denominator, we subtract their numerators and keep the common denominator.

SUBTRACTING RATIONAL EXPRESSIONS WITH LIKE DENOMINATORS

If a, b, and d represent polynomials, then

$$\frac{a}{d} - \frac{b}{d} = \frac{a - b}{d} \quad \text{provided the denominator is not equal to 0.}$$

EXAMPLE 3 Subtract, assuming no divisions by zero.

a. $\frac{5x}{3} - \frac{2x}{3}$ **b.** $\frac{5x + 1}{x - 3} - \frac{4x - 2}{x - 3}$

Solution In each part, the rational expressions have the same denominator. To subtract them, we subtract their numerators and keep the common denominator.

a. $$\frac{5x}{3} - \frac{2x}{3} = \frac{5x - 2x}{3} \quad \text{Subtract the numerators and keep the common denominator.}$$

$$= \frac{3x}{3} \quad \text{Combine like terms.}$$

$$= \frac{x}{1} \quad \text{Divide out the common factor.}$$

$$= x$$

b. $\frac{5x+1}{x-3} - \frac{4x-2}{x-3} = \frac{(5x+1)-(4x-2)}{x-3}$ Subtract the numerators and keep the common denominator.

$= \frac{5x+1-4x+2}{x-3}$ Use the distributive property to remove parentheses.

$= \frac{x+3}{x-3}$ Combine like terms.

SELF CHECK 3 Subtract: **a.** $\frac{9y}{4} - \frac{5y}{4}$ **b.** $\frac{2y+1}{y+5} - \frac{y-4}{y+5}$ $(y \neq -5)$

To add and/or subtract three or more rational expressions, we follow the order of operations.

EXAMPLE 4 Simplify: $\frac{3x+1}{x-7} - \frac{5x+2}{x-7} + \frac{2x+1}{x-7}$ $(x \neq 7)$

Solution This example involves both addition and subtraction of rational expressions. Unless parentheses indicate otherwise, we do additions and subtractions from left to right.

$\frac{3x+1}{x-7} - \frac{5x+2}{x-7} + \frac{2x+1}{x-7}$

$= \frac{(3x+1)-(5x+2)+(2x+1)}{x-7}$ Combine the numerators and keep the common denominator.

$= \frac{3x+1-5x-2+2x+1}{x-7}$ Use the distributive property to remove parentheses.

$= \frac{0}{x-7}$ Combine like terms.

$= 0$ Simplify.

SELF CHECK 4 Simplify: $\frac{2a-3}{a-5} + \frac{3a+2}{a-5} - \frac{24}{a-5}$ $(a \neq 5)$

Example 4 is a reminder that if the numerator of a rational expression is 0 and the denominator is not, the value of the expression is 0.

3 Find the least common denominator (LCD) of two or more polynomials and use it to write equivalent rational expressions.

Since the denominators of the fractions in the addition $\frac{4}{7} + \frac{3}{5}$ are different, we cannot add the fractions in their present form.

four-sevenths + three-fifths

↑ Different denominators ↑

To add these fractions, we need to find a common denominator. The smallest common denominator (called the **least** or **lowest common denominator**) is the easiest one to use.

LEAST COMMON DENOMINATOR

The least common denominator (LCD) for a set of fractions is the smallest number that each denominator will divide exactly.

We now review the method of writing two fractions using the LCD. In the addition $\frac{4}{7} + \frac{3}{5}$, the denominators are 7 and 5. The smallest number that 7 and 5 will divide exactly is 35. This is the LCD. We now build each fraction into a fraction with a denominator of 35.

$$\frac{4}{7} + \frac{3}{5} = \frac{4 \cdot 5}{7 \cdot 5} + \frac{3 \cdot 7}{5 \cdot 7}$$ Multiply numerator and denominator of $\frac{4}{7}$ by 5, and multiply numerator and denominator of $\frac{3}{5}$ by 7.

$$= \frac{20}{35} + \frac{21}{35}$$ Do the multiplications.

Now that the fractions have a common denominator, we can add them.

$$\frac{20}{35} + \frac{21}{35} = \frac{20 + 21}{35} = \frac{41}{35}$$

EXAMPLE 5 Write each rational expression as an equivalent expression with a denominator of $30y \quad (y \neq 0)$.

a. $\frac{1}{2y}$ **b.** $\frac{3y}{5}$ **c.** $\frac{7x}{10y}$

Solution To build each rational expression into an expression with a denominator of $30y$, we multiply the numerator and denominator by what it takes to make the denominator $30y$.

a. $\frac{1}{2y} = \frac{1 \cdot \mathbf{15}}{2y \cdot \mathbf{15}} = \frac{15}{30y}$

b. $\frac{3y}{5} = \frac{3y \cdot \mathbf{6y}}{5 \cdot \mathbf{6y}} = \frac{18y^2}{30y}$

c. $\frac{7x}{10y} = \frac{7x \cdot \mathbf{3}}{10y \cdot \mathbf{3}} = \frac{21x}{30y}$

SELF CHECK 5 Write each rational expression as a rational expression with a denominator of $30ab$ $(a, b \neq 0)$.

a. $\frac{1}{5b}$ **b.** $\frac{8b}{3}$ **c.** $\frac{5a}{6b}$

There is a process that we can use to find the least common denominator of several rational expressions.

FINDING THE LEAST COMMON DENOMINATOR (LCD)

1. List the different denominators that appear in the rational expressions.
2. Completely factor each denominator.
3. Form a product using each different factor obtained in Step 2. Use each different factor the *greatest* number of times it appears in any *one* factorization. The product formed by multiplying these factors is the LCD.

EXAMPLE 6 Find the LCD of $\frac{5a}{24b}$, $\frac{11a}{18b}$, and $\frac{35a}{36b}$ $\quad (b \neq 0)$.

Solution We list and factor each denominator into the product of prime numbers.

$$24b = 2 \cdot 2 \cdot 2 \cdot 3 \cdot b = 2^3 \cdot 3 \cdot b$$

$$18b = 2 \cdot 3 \cdot 3 \cdot b = 2 \cdot 3^2 \cdot b$$

$$36b = 2 \cdot 2 \cdot 3 \cdot 3 \cdot b = 2^2 \cdot 3^2 \cdot b$$

We then form a product with factors of 2, 3, and b. To find the LCD, we use each of these factors the *greatest* number of times it appears in any one factorization. We use 2 three times, because it appears three times as a factor of 24. We use 3 twice, because it occurs twice as a factor of 18 and 36. We use b once because it occurs once in each factor of $24b$, $18b$, and $36b$.

$$\begin{aligned} \text{LCD} &= 2\cdot 2\cdot 2\cdot 3\cdot 3\cdot b \\ &= 8\cdot 9\cdot b \\ &= 72b \end{aligned}$$

SELF CHECK 6 Find the LCD of $\frac{3y}{28z}$, $\frac{7xy}{12z}$, and $\frac{5x}{21z}$ $(z \neq 0)$.

4 Add two rational expressions with unlike denominators and write the answer in simplest form.

The process for adding and subtracting rational expressions with different denominators is the same as the process for adding and subtracting arithmetic expressions with different numerical denominators.

For example, to add $\frac{4x}{7}$ and $\frac{3x}{5}$, we first find the LCD of 7 and 5, which is 35. We then build the rational expressions so that each one has a denominator of 35. Finally, we add the results.

$$\frac{4x}{7} + \frac{3x}{5} = \frac{4x\cdot 5}{7\cdot 5} + \frac{3x\cdot 7}{5\cdot 7}$$ Multiply numerator and denominator of $\frac{4x}{7}$ by 5 and numerator and denominator of $\frac{3x}{5}$ by 7.

$$= \frac{20x}{35} + \frac{21x}{35}$$ Do the multiplications.

$$= \frac{41x}{35}$$ Add the numerators and keep the common denominator.

The following steps summarize how to add rational expressions that have unlike denominators.

ADDING RATIONAL EXPRESSIONS WITH UNLIKE DENOMINATORS

To add rational expressions with unlike denominators:

1. Find the LCD.
2. Write each rational expression as an equivalent expression with a denominator that is the LCD.
3. Add the resulting fractions.
4. Simplify the result, if possible.

EXAMPLE 7 Add: $\frac{5a}{24b}$, $\frac{11a}{18b}$, and $\frac{35a}{36b}$ $(b \neq 0)$

Solution In Example 6, we saw that the LCD of these rational expressions is $2\cdot 2\cdot 2\cdot 3\cdot 3\cdot b = 72b$. To add the rational expressions, we first factor each denominator:

$$\frac{5a}{24b} + \frac{11a}{18b} + \frac{35a}{36b} = \frac{5a}{2\cdot 2\cdot 2\cdot 3\cdot b} + \frac{11a}{2\cdot 3\cdot 3\cdot b} + \frac{35a}{2\cdot 2\cdot 3\cdot 3\cdot b}$$

In each resulting expression, we multiply the numerator and the denominator by whatever it takes to build the denominator to the lowest common denominator of $2\cdot 2\cdot 2\cdot 3\cdot 3\cdot b$.

$$= \frac{5a \cdot 3}{2 \cdot 2 \cdot 2 \cdot 3 \cdot b \cdot 3} + \frac{11a \cdot 2 \cdot 2}{2 \cdot 3 \cdot 3 \cdot b \cdot 2 \cdot 2} + \frac{35a \cdot 2}{2 \cdot 2 \cdot 3 \cdot 3 \cdot b \cdot 2}$$

$$= \frac{15a + 44a + 70a}{72b}$$ Do the multiplications.

$$= \frac{129a}{72b}$$ Add the fractions.

$$= \frac{43a}{24b}$$ Simplify.

SELF CHECK 7 Add: $\frac{3x}{28z}, \frac{5x}{21z},$ and $\frac{7x}{12z}$ $(z \neq 0)$

EXAMPLE 8 Add: $\frac{x+1}{x-2} + \frac{x+3}{x-1}$ $(x \neq 1, 2)$

Solution We first find the LCD.

$$\left.\begin{matrix} x-2 \\ x-1 \end{matrix}\right\} \quad \text{LCD} = (x-2)(x-1)$$

Then build the rational expressions so that each one has a denominator of $(x-2)(x-1)$.

$$\frac{x+1}{x-2} + \frac{x+3}{x-1} = \frac{(x+1)(x-1)}{(x-2)(x-1)} + \frac{(x-2)(x+3)}{(x-2)(x-1)}$$ Multiply the numerator and denominator of $\frac{(x+1)}{(x-2)}$ by $\frac{(x-1)}{(x-1)}$ and $\frac{(x+3)}{(x-1)}$ by $\frac{(x-2)}{(x-2)}$.

$$= \frac{x^2-1}{(x-2)(x-1)} + \frac{x^2+x-6}{(x-2)(x-1)}$$ Do the multiplications.

$$= \frac{2x^2+x-7}{(x-2)(x-1)}$$ Add the fractions.

This is in simplified form.

SELF CHECK 8 Add: $\frac{x+5}{x+1} + \frac{x-4}{x-3}$ $(x \neq -1, 3)$

EXAMPLE 9 Add: $\frac{1}{x} + \frac{x}{y}$ $(x, y \neq 0)$

Solution By inspection, the LCD is xy.

COMMENT Building fractions means to multiply each fraction by 1 in the form necessary to obtain an equivalent fraction with a new denominator of the LCD.

$$\frac{1}{x} + \frac{x}{y} = \frac{1(y)}{x(y)} + \frac{(x)x}{(x)y}$$ Build the fractions.

$$= \frac{y}{xy} + \frac{x^2}{xy}$$ Do the multiplications.

$$= \frac{y+x^2}{xy}$$ Add the fractions.

SELF CHECK 9 Add: $\frac{a}{b} + \frac{3}{a}$ $(a, b \neq 0)$

5 Subtract two rational expressions with unlike denominators and write the answer in simplest form.

To subtract rational expressions with unlike denominators, we first write them as expressions with the same denominator and then subtract the numerators.

EXAMPLE 10 Subtract: $\frac{x}{x+1} - \frac{3}{x}$ $(x \neq 0, -1)$

Solution Because x and $x + 1$ represent different values and have no common factors, the least common denominator (LCD) is their product, $(x + 1)x$.

$$\frac{x}{x+1} - \frac{3}{x} = \frac{x(x)}{(x+1)x} - \frac{3(x+1)}{x(x+1)}$$ Build the fractions to obtain the common denominator.

$$= \frac{x(x) - 3(x+1)}{(x+1)x}$$ Subtract the numerators and keep the common denominator.

$$= \frac{x^2 - 3x - 3}{(x+1)x}$$ Do the multiplication in the numerator.

SELF CHECK 10 Subtract: $\frac{a}{a-1} - \frac{5}{a}$ $(a \neq 0, 1)$

EXAMPLE 11 Subtract: $\frac{a}{a-1} - \frac{2}{a^2-1}$ $(a \neq 1, -1)$

Solution To find the LCD, we factor the denominators, where possible.

$$\left.\begin{aligned} a - 1 &= \mathbf{a - 1} \\ a^2 - 1 &= (a+1)(\mathbf{a - 1}) \end{aligned}\right\} \quad \text{LCD} = (a+1)(a-1)$$

After finding the LCD, we proceed as follows:

$$\frac{a}{a-1} - \frac{2}{a^2-1}$$

$$= \frac{a}{(a-1)} - \frac{2}{(a+1)(a-1)}$$ Factor the denominator.

$$= \frac{a(\mathbf{a+1})}{(a-1)(\mathbf{a+1})} - \frac{2}{(a+1)(a-1)}$$ Build the first fraction.

$$= \frac{a(a+1) - 2}{(a-1)(a+1)}$$ Subtract the numerators and keep the common denominator.

$$= \frac{a^2 + a - 2}{(a-1)(a+1)}$$ Use the distributive property to remove parentheses.

$$= \frac{(a+2)\overset{1}{\cancel{(a-1)}}}{\underset{1}{\cancel{(a-1)}}(a+1)}$$ Factor and divide out the common factor of $a - 1$.

$$= \frac{a+2}{a+1}$$ Simplify.

SELF CHECK 11 Subtract: $\frac{b}{b+1} - \frac{3}{b^2-1}$ $(b \neq 1, -1)$

EXAMPLE 12 Subtract: $\frac{3}{x-y} - \frac{x}{y-x} \quad (x \neq y)$

Solution We note that the second denominator is the negative of the first, so we can multiply both the numerator and the denominator of the second fraction by -1 to obtain

$$\frac{3}{x-y} - \frac{x}{y-x} = \frac{3}{x-y} - \frac{-1x}{-1(y-x)}$$ Multiply both the numerator and the denominator by -1.

$$= \frac{3}{x-y} - \frac{-x}{-y+x}$$ Use the distributive property to remove parentheses.

$$= \frac{3}{x-y} - \frac{-x}{x-y}$$ Apply the commutative property of addition.

$$= \frac{3-(-x)}{x-y}$$ Subtract the numerators and keep the common denominator.

$$= \frac{3+x}{x-y}$$

SELF CHECK 12 Subtract: $\frac{5}{a-b} - \frac{2}{b-a} \quad (a \neq b)$

EXAMPLE 13 Perform the operations $\frac{3}{x^2-y^2} + \frac{2}{x-y} - \frac{1}{x+y}$. Assume that no denominator is 0.

Solution Find the least common denominator.

$$\left.\begin{aligned} x^2 - y^2 &= (x-y)(x+y) \\ x - y &= x - y \\ x + y &= x + y \end{aligned}\right\}$$ Factor each denominator, where possible.

Since the least common denominator is $(x-y)(x+y)$, we build each fraction into a new fraction with that common denominator.

$$\frac{3}{x^2-y^2} + \frac{2}{x-y} - \frac{1}{x+y}$$

$$= \frac{3}{(x-y)(x+y)} + \frac{2}{x-y} - \frac{1}{x+y}$$ Factor.

$$= \frac{3}{(x-y)(x+y)} + \frac{2(x+y)}{(x-y)(x+y)} - \frac{1(x-y)}{(x+y)(x-y)}$$ Build each fraction.

$$= \frac{3+2(x+y)-1(x-y)}{(x-y)(x+y)}$$ Combine the numerators and keep the common denominator.

$$= \frac{3+2x+2y-x+y}{(x-y)(x+y)}$$ Use the distributive property to remove parentheses.

$$= \frac{3+x+3y}{(x-y)(x+y)}$$ Combine like terms.

SELF CHECK 13 Perform the operations: $\frac{5}{a^2-b^2} - \frac{3}{a+b} + \frac{4}{a-b} \quad (a \neq b, -b)$

SELF CHECK ANSWERS

1. a. $\frac{x+y}{7}$ b. $\frac{x}{y}$ 2. $\frac{1}{3}$ 3. a. y b. 1 4. 5 5. a. $\frac{6a}{30ab}$ b. $\frac{80ab^2}{30ab}$ c. $\frac{25a^2}{30ab}$ 6. $84z$ 7. $\frac{13x}{14z}$ 8. $\frac{2x^2-x-19}{(x+1)(x-3)}$ 9. $\frac{a^2+3b}{ab}$ 10. $\frac{a^2-5a+5}{a(a-1)}$ 11. $\frac{b^2-b-3}{(b+1)(b-1)}$ 12. $\frac{7}{a-b}$ 13. $\frac{a+7b+5}{(a+b)(a-b)}$

NOW TRY THIS

Simplify each expression. Assume no denominators are 0.

1. $\frac{5}{x+3}-\frac{2}{x-3}$
2. $\left(\frac{2}{3x}-1\right)\div\left(\frac{4}{9x}-x\right)$
3. $x^{-1}+x^{-2}$

6.3 Exercises

WARM-UPS *Determine whether the expressions are equal.*

1. $\frac{2}{3}, \frac{12}{18}$
2. $\frac{3}{8}, \frac{15}{40}$
3. $\frac{6}{11}, \frac{24}{42}$
4. $\frac{5}{10}, \frac{15}{30}$
5. $\frac{x}{3}, \frac{3x}{9}$
6. $\frac{5}{3}, \frac{5x}{3y}$ $(y \neq 0)$
7. $\frac{5}{3}, \frac{5x}{3x}$ $(x \neq 0)$
8. $\frac{4y}{20}, \frac{y}{5}$

REVIEW *Write each number in prime-factored form.*

9. 81
10. 64
11. 136
12. 242
13. 102
14. 315
15. 144
16. 217

VOCABULARY AND CONCEPTS *Fill in the blanks.*

17. The ____ for a set of rational expressions is the smallest number that each denominator divides exactly.
18. When we multiply the numerator and denominator of a rational expression by some number to get a common denominator, we say that we are _______ the fraction.
19. To add two rational expressions with like denominators, we add their _________ and keep the ________________.
20. To subtract two rational expressions with _____ denominators, we need to find a common denominator.

GUIDED PRACTICE *Perform the operations. Simplify answers, if possible. Assume that no denominators are 0.* SEE EXAMPLES 1–2. (OBJECTIVE 1)

21. $\frac{1}{8a}+\frac{1}{8a}$
22. $\frac{3}{4y}+\frac{3}{4y}$
23. $\frac{2x}{y}+\frac{2x}{y}$
24. $\frac{8y}{7x}+\frac{6y}{7x}$
25. $\frac{4y-1}{y-4}+\frac{5y+3}{y-4}$
26. $\frac{2x+3}{x+4}+\frac{3x-1}{x+4}$
27. $\frac{3x-5}{x-2}+\frac{6x-13}{x-2}$
28. $\frac{8x-7}{x+3}+\frac{2x+37}{x+3}$

Perform the operations. Simplify answers, if possible. Assume that no denominators are 0. SEE EXAMPLE 3. (OBJECTIVE 2)

29. $\frac{35}{72}-\frac{44}{72}$
30. $\frac{5}{48}-\frac{11}{48}$
31. $\frac{9y}{3x}-\frac{6y}{3x}$
32. $\frac{9y}{x}-\frac{5y}{x}$
33. $\frac{4y-5}{2y}-\frac{3}{2y}$
34. $\frac{6x+3}{5x}-\frac{3x}{5x}$
35. $\frac{6x-5}{3xy}-\frac{3x-5}{3xy}$
36. $\frac{7x+7}{5y}-\frac{2x+7}{5y}$
37. $\frac{3y-2}{y+3}-\frac{2y-5}{y+3}$
38. $\frac{5x+8}{x+5}-\frac{3x-2}{x+5}$
39. $\frac{5y+3}{y-4}-\frac{4y-1}{y-4}$
40. $\frac{2x-1}{x-3}-\frac{x+2}{x-3}$

Perform the operations. Simplify answers, if possible. Assume that no denominators are 0. SEE EXAMPLE 4. (OBJECTIVES 1–2)

41. $\frac{11x}{12} + \frac{7x}{12} - \frac{2x}{12}$

42. $\frac{13y}{32} + \frac{13y}{32} - \frac{10y}{32}$

43. $\frac{3x + 1}{x - 2} + \frac{5x + 2}{x - 2} - \frac{2x + 1}{x - 2}$

44. $\frac{2y - 3}{y + 2} - \frac{y - 3}{y + 2} + \frac{3y + 6}{y + 2}$

45. $\frac{4b + 5}{b + 1} - \frac{6b - 2}{b + 1} + \frac{b - 7}{b + 1}$

46. $\frac{7a + 1}{a - 1} + \frac{a + 1}{a - 1} - \frac{10a}{a - 1}$

47. $\frac{x + 1}{x - 2} - \frac{2(x - 3)}{x - 2} + \frac{3(x + 1)}{x - 2}$

48. $\frac{3xy}{x - y} - \frac{x(3y - x)}{x - y} - \frac{x(x - y)}{x - y}$

Build each fraction into an equivalent fraction with the indicated denominator. Assume that no denominators are 0. SEE EXAMPLE 5. (OBJECTIVE 3)

49. $\frac{21}{8}$; 32

50. $\frac{7}{x}$; xy

51. $\frac{8}{x}$; x^2y

52. $\frac{7}{y}$; xy^2

53. $\frac{4x}{x + 3}$; $(x + 3)^2$

54. $\frac{5y}{y - 2}$; $(y - 2)^2$

55. $\frac{2y}{x}$; $x^2 + x$

56. $\frac{3x}{y}$; $y^2 - y$

57. $\frac{z}{z - 1}$; $z^2 - 1$

58. $\frac{z}{z + 3}$; $z^2 - 9$

59. $\frac{2}{x + 1}$; $x^2 + 3x + 2$

60. $\frac{3}{x - 1}$; $x^2 + x - 2$

Several denominators are given. Find the LCD. SEE EXAMPLE 6. (OBJECTIVE 3)

61. $2x, 6x$

62. $10y, 15y$

63. $3x, 6y, 9xy$

64. $2x^2, 6y, 3xy$

65. $x^2 - 4, x + 2$

66. $y^2 - 9, y - 3$

67. $x^2 + 6x, x + 6, x$

68. $xy^2 - xy, xy, y - 1$

Perform the operations. Simplify answers, if possible. Assume that no denominators are 0. SEE EXAMPLE 7. (OBJECTIVE 4)

69. $\frac{4x}{3y} + \frac{2x}{y}$

70. $\frac{5x}{2z} + \frac{7x}{6z}$

71. $\frac{x + 2}{2x} + \frac{x - 1}{3x}$

72. $\frac{x + 3}{x^2} + \frac{x + 5}{2x}$

Perform the operations. Simplify answers, if possible. Assume that no denominators are 0. SEE EXAMPLE 8. (OBJECTIVE 4)

73. $\frac{x + 1}{x - 1} + \frac{x - 1}{x + 1}$

74. $\frac{2x}{x + 2} + \frac{x + 1}{x - 3}$

75. $\frac{x}{5x + 2} + \frac{x - 1}{x + 2}$

76. $\frac{2x - 1}{3x + 2} + \frac{x + 4}{2x + 3}$

Perform the operations. Simplify answers, if possible. Assume that no denominators are 0. SEE EXAMPLE 9. (OBJECTIVE 4)

77. $\frac{x - 2}{x} + \frac{y + 2}{y}$

78. $\frac{a + 2}{b} + \frac{b - 2}{a}$

79. $\frac{3y}{x} + \frac{x + 1}{y - 1}$

80. $\frac{a}{b} + \frac{2b - 1}{a + 2}$

Perform the operations. Simplify answers, if possible. Assume that no denominators are 0. SEE EXAMPLE 10. (OBJECTIVE 5)

81. $\frac{5}{x} - \frac{x + 2}{x + 1}$

82. $\frac{x + 2}{x + 1} - \frac{5}{x}$

83. $\frac{2x + 3}{x + 5} - \frac{x - 1}{x + 2}$

84. $\frac{x - 1}{x + 5} - \frac{2x + 3}{x + 2}$

Perform the operations. Simplify answers, if possible. Assume that no denominators are 0. SEE EXAMPLE 11. (OBJECTIVE 5)

85. $\frac{x}{x - 2} + \frac{4 + 2x}{x^2 - 4}$

86. $\frac{y}{y + 3} - \frac{2y - 6}{y^2 - 9}$

87. $\frac{x + 1}{x + 2} - \frac{x^2 + 1}{x^2 - x - 6}$

88. $\frac{x + 1}{2x + 4} - \frac{x^2}{2x^2 - 8}$

Perform the operations. Simplify answers, if possible. Assume that no denominators are 0. SEE EXAMPLE 12. (OBJECTIVE 5)

89. $\dfrac{y+3}{y-1} - \dfrac{y+4}{1-y}$

90. $\dfrac{2x+2}{x-2} - \dfrac{2x}{2-x}$

91. $\dfrac{x+5}{2x-y} - \dfrac{x-1}{y-2x}$

92. $\dfrac{2a-b}{a-2b} - \dfrac{a+3b}{2b-a}$

Perform the operations. Simplify answers, if possible. Assume that no denominators are 0. SEE EXAMPLE 13. (OBJECTIVES 4–5)

93. $\dfrac{2x}{x^2-3x+2} + \dfrac{2x}{x-1} - \dfrac{x}{x-2}$

94. $\dfrac{4a}{a-2} - \dfrac{3a}{a-3} + \dfrac{4a}{a^2-5a+6}$

95. $\dfrac{a}{a-1} - \dfrac{2}{a+2} + \dfrac{3(a-2)}{a^2+a-2}$

96. $\dfrac{2x}{x-1} + \dfrac{3x}{x+1} - \dfrac{x+3}{x^2-1}$

ADDITIONAL PRACTICE *Several denominators are given. Find the LCD.*

97. x^2-x-6, x^2-9

98. x^2-4x-5, x^2-25

99. $\dfrac{15x}{6y} - \dfrac{7x}{8}$

Perform the operations. Assume that no denominators are 0.

100. $\dfrac{2x}{y} - \dfrac{x}{y}$

101. $\dfrac{1}{2} + \dfrac{2}{3}$

102. $\dfrac{2y}{9} + \dfrac{y}{3}$

103. $\dfrac{2}{3} - \dfrac{5}{6}$

104. $\dfrac{8a}{15} - \dfrac{5a}{12}$

105. $\dfrac{2y}{5x} - \dfrac{y}{2}$

106. $\dfrac{x}{x+1} + \dfrac{x-1}{x}$

107. $\dfrac{x+2}{x} + \dfrac{x-5}{x+2}$

108. $\dfrac{x+5}{xy} - \dfrac{x-1}{x^2y}$

109. $\dfrac{y-7}{y^2} - \dfrac{y+7}{2y}$

110. $\dfrac{x}{3y} + \dfrac{2x}{3y} - \dfrac{x}{3y}$

111. $\dfrac{5y}{8x} + \dfrac{4y}{8x} - \dfrac{y}{8x}$

112. $\dfrac{5r^2}{2r} - \dfrac{r^2}{2r}$

113. $\dfrac{3x}{y+2} - \dfrac{3y}{y+2} + \dfrac{x+y}{y+2}$

114. $\dfrac{3y}{x-5} + \dfrac{x}{x-5} - \dfrac{y-x}{x-5}$

115. $\dfrac{-a}{3a^2-27} + \dfrac{1}{3a+9}$

116. $\dfrac{d}{d^2+6d+5} - \dfrac{d}{d^2+5d+4}$

117. $14 + \dfrac{10}{y^2}$

118. $\dfrac{2}{x} - 3x$

WRITING ABOUT MATH

119. Explain how to add rational expressions with the same denominator.

120. Explain how to subtract rational expressions with the same denominator.

121. Explain how to find a lowest common denominator.

122. Explain how to add two rational expressions with different denominators.

SOMETHING TO THINK ABOUT

123. Find the error:

$$\frac{2x+3}{x+5} - \frac{x+2}{x+5} = \frac{2x+3-x+2}{x+5}$$
$$= \frac{x+5}{x+5}$$
$$= 1$$

124. Find the error:

$$\frac{5x-4}{y} + \frac{x}{y} = \frac{5x-4+x}{y+y}$$
$$= \frac{6x-4}{2y}$$
$$= \frac{3x-2}{y}$$

Show that each formula is true.

125. $\dfrac{a}{b} + \dfrac{c}{d} = \dfrac{ad+bc}{bd}$

126. $\dfrac{a}{b} - \dfrac{c}{d} = \dfrac{ad-bc}{bd}$

Section 6.4 Simplifying Complex Fractions

Objectives

1. Simplify a complex fraction.
2. Simplify a fraction containing terms with negative exponents.

Vocabulary

complex fraction

Getting Ready

Use the distributive property to remove parentheses, and simplify.

1. $3\left(1 + \frac{1}{3}\right)$ **2.** $10\left(\frac{1}{5} - 2\right)$ **3.** $4\left(\frac{3}{2} + \frac{1}{4}\right)$ **4.** $14\left(\frac{3}{7} - 1\right)$

5. $x\left(\frac{3}{x} + 3\right)$ **6.** $y\left(\frac{2}{y} - 1\right)$ **7.** $4x\left(3 - \frac{1}{2x}\right)$ **8.** $6xy\left(\frac{1}{2x} + \frac{1}{3y}\right)$

In this section, we will consider fractions that contain fractions. These complicated fractions are called *complex fractions.*

1 Simplify a complex fraction.

Fractions such as

$$\frac{\frac{1}{3}}{4}, \quad \frac{\frac{5}{3}}{\frac{2}{9}}, \quad \frac{x + \frac{1}{2}}{3 - x}, \quad \text{and} \quad \frac{\frac{x+1}{2}}{x + \frac{1}{x}}$$

that contain fractions in their numerators and/or denominators are called **complex fractions**. Complex fractions should be simplified. For example, we can simplify

$$\frac{\frac{5x}{3}}{\frac{2y}{9}}$$

by doing the division:

$$\frac{\frac{5x}{3}}{\frac{2y}{9}} = \frac{5x}{3} \div \frac{2y}{9} = \frac{5x}{3} \cdot \frac{9}{2y} = \frac{5x \cdot 3 \cdot \overset{1}{\cancel{3}}}{\underset{1}{\cancel{3}} \cdot 2y} = \frac{15x}{2y}$$

There are two methods that we can use to simplify complex fractions.

SIMPLIFYING COMPLEX FRACTIONS

METHOD 1

Write the numerator and the denominator of the complex fraction as single fractions. Then divide the fractions and simplify.

METHOD 2

Multiply the numerator and denominator of the complex fraction by the LCD of the fractions in its numerator and denominator. Then simplify the results, if possible.

Hypatia
AD 370–415
Hypatia is the earliest known woman in the history of mathematics. She was a professor at the University of Alexandria. Because of her scientific beliefs, she was considered to be a heretic. At the age of 45, she was attacked by a mob and murdered for her beliefs.

Using Method 1 to simplify $\dfrac{\dfrac{3x}{5}+1}{2-\dfrac{x}{5}}$ (assuming no division by 0), we proceed as follows:

$$\frac{\frac{3x}{5}+1}{2-\frac{x}{5}} = \frac{\frac{3x}{5}+\frac{5}{5}}{\frac{10}{5}-\frac{x}{5}}$$ Write 1 as $\frac{5}{5}$ and 2 as $\frac{10}{5}$.

$$= \frac{\frac{3x+5}{5}}{\frac{10-x}{5}}$$ Add the fractions in the numerator and subtract the fractions in the denominator.

$$= \frac{3x+5}{5} \div \frac{10-x}{5}$$ Write the complex fraction as an equivalent division problem.

$$= \frac{3x+5}{5} \cdot \frac{5}{10-x}$$ Invert the divisor and multiply.

$$= \frac{(3x+5)\overset{1}{\cancel{5}}}{\underset{1}{\cancel{5}}(10-x)}$$ Multiply the fractions.

$$= \frac{3x+5}{10-x}$$ Divide out the common factor of 5: $\frac{5}{5}=1$.

To use Method 2, we first determine that the LCD of the fractions in the numerator and denominator is 5. We then multiply both the numerator and denominator by 5.

$$\frac{\frac{3x}{5}+1}{2-\frac{x}{5}} = \frac{5\left(\frac{3x}{5}+1\right)}{5\left(2-\frac{x}{5}\right)}$$ Multiply both numerator and denominator by 5.

$$= \frac{5\cdot\frac{3x}{5}+5\cdot 1}{5\cdot 2-5\cdot\frac{x}{5}}$$ Use the distributive property to remove parentheses.

$$= \frac{3x+5}{10-x}$$ Simplify.

With practice, you will be able to see which method is easier to understand in any given situation.

Unless otherwise noted, all content on this page is © Cengage Learning.

EXAMPLE 1 Simplify $\dfrac{\frac{x}{3}}{\frac{y}{3}}$. Assume that no denominators are 0.

Solution We will simplify the complex fraction using both methods.

Method 1

$$\frac{\frac{x}{3}}{\frac{y}{3}} = \frac{x}{3} \div \frac{y}{3}$$
$$= \frac{x}{3} \cdot \frac{3}{y}$$
$$= \frac{3x}{3y}$$
$$= \frac{x}{y}$$

Method 2

$$\frac{\frac{x}{3}}{\frac{y}{3}} = \frac{3\left(\frac{x}{3}\right)}{3\left(\frac{y}{3}\right)}$$
$$= \frac{\frac{x}{1}}{\frac{y}{1}}$$
$$= \frac{x}{y}$$

SELF CHECK 1 Simplify $\dfrac{\frac{a}{4}}{\frac{5}{b}}$. Assume no denominator is 0.

EXAMPLE 2 Simplify $\dfrac{\frac{x}{x+1}}{\frac{y}{x}}$. Assume no denominator is 0.

Solution We will simplify the complex fraction using both methods.

Method 1

$$\frac{\frac{x}{x+1}}{\frac{y}{x}} = \frac{x}{x+1} \div \frac{y}{x}$$
$$= \frac{x}{x+1} \cdot \frac{x}{y}$$
$$= \frac{x^2}{y(x+1)}$$

Method 2

$$\frac{\frac{x}{x+1}}{\frac{y}{x}} = \frac{x(x+1)\left(\frac{x}{x+1}\right)}{x(x+1)\left(\frac{y}{x}\right)}$$
$$= \frac{\frac{x^2}{1}}{\frac{y(x+1)}{1}}$$
$$= \frac{x^2}{y(x+1)}$$

SELF CHECK 2 Simplify $\dfrac{\frac{x}{y}}{\frac{x}{y+1}}$. Assume no denominator is 0.

EXAMPLE 3 Simplify $\dfrac{1 + \frac{1}{x}}{1 - \frac{1}{x}}$. Assume no denominator is 0.

Solution We will simplify the complex fraction using both methods.

Method 1

$$\frac{1 + \frac{1}{x}}{1 - \frac{1}{x}} = \frac{\frac{x}{x} + \frac{1}{x}}{\frac{x}{x} - \frac{1}{x}}$$

$$= \frac{\frac{x + 1}{x}}{\frac{x - 1}{x}}$$

$$= \frac{x + 1}{x} \div \frac{x - 1}{x}$$

$$= \frac{x + 1}{x} \cdot \frac{x}{x - 1}$$

$$= \frac{(x + 1)x}{x(x - 1)}$$

$$= \frac{x + 1}{x - 1}$$

Method 2

$$\frac{1 + \frac{1}{x}}{1 - \frac{1}{x}} = \frac{x\left(1 + \frac{1}{x}\right)}{x\left(1 - \frac{1}{x}\right)}$$

$$= \frac{x + 1}{x - 1}$$

SELF CHECK 3 Simplify $\dfrac{\frac{1}{x} + 1}{\frac{1}{x} - 1}$. Assume no denominator is 0.

EXAMPLE 4 Simplify $\dfrac{1}{1 + \frac{1}{x + 1}}$. Assume no denominator is 0.

Solution We will simplify this complex fraction by using Method 2 only.

$$\frac{1}{1 + \frac{1}{x + 1}} = \frac{(x + 1) \cdot 1}{(x + 1)\left(1 + \frac{1}{x + 1}\right)}$$ Multiply the numerator and denominator of the complex fraction by $(x + 1)$.

$$= \frac{x + 1}{(x + 1)1 + 1}$$ Use the distributive property.

$$= \frac{x + 1}{x + 2}$$ Simplify.

SELF CHECK 4 Simplify $\dfrac{2}{\frac{1}{x + 2} - 2}$. Assume no denominator is 0.

2 Simplify a fraction containing terms with negative exponents.

Many fractions with terms containing negative exponents are complex fractions as the next example illustrates.

EXAMPLE 5 Simplify $\dfrac{x^{-1} + y^{-2}}{x^{-2} - y^{-1}}$. Assume no denominator is 0.

Solution We will write each expression using positive exponents and then simplify the complex fraction using Method 2:

$$\frac{x^{-1} + y^{-2}}{x^{-2} - y^{-1}} = \frac{\frac{1}{x} + \frac{1}{y^2}}{\frac{1}{x^2} - \frac{1}{y}}$$ Write without negative exponents.

$$= \frac{x^2y^2\left(\frac{1}{x} + \frac{1}{y^2}\right)}{x^2y^2\left(\frac{1}{x^2} - \frac{1}{y}\right)}$$ Multiply the numerator and denominator of the complex fraction by x^2y^2, the LCD.

$$= \frac{xy^2 + x^2}{y^2 - x^2y}$$ Distribute.

$$= \frac{x(y^2 + x)}{y(y - x^2)}$$ Factor the numerator and denominator.

The result cannot be simplified. Therefore either of the last two steps is a correct answer.

SELF CHECK 5 Simplify $\dfrac{x^{-2} - y^{-1}}{x^{-1} + y^{-2}}$. Assume no denominator is 0.

SELF CHECK ANSWERS **1.** $\frac{ab}{20}$ **2.** $\frac{y+1}{y}$ **3.** $\frac{1+x}{1-x}$ **4.** $\frac{2(x+2)}{-2x-3}$ **5.** $\frac{y(y-x^2)}{x(y^2+x)}$

NOW TRY THIS

Simplify each complex fraction. Assume no division by 0.

1. $\dfrac{\frac{a}{y^2}}{\frac{b}{x^3}}$

2. $\dfrac{\frac{x}{x+2} + \frac{5}{x}}{\frac{1}{3x} + \frac{x}{2x+4}}$

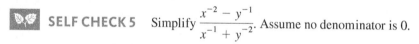

6.4 Exercises

WARM-UPS *Simplify each complex fraction.*

1. $\dfrac{\frac{2}{3}}{\frac{1}{2}}$

2. $\dfrac{\frac{2}{3}}{\frac{3}{4}}$

3. $\dfrac{\frac{6}{7}}{\frac{8}{21}}$

4. $\dfrac{\frac{4}{5}}{\frac{32}{15}}$

5. $\dfrac{\frac{7}{8}}{\frac{49}{4}}$

6. $\dfrac{2}{\frac{1}{2}}$

7. $\dfrac{\frac{1}{2}}{2}$

8. $\dfrac{1 + \frac{1}{2}}{\frac{1}{2}}$

9. $\dfrac{\frac{2}{3} + 1}{\frac{1}{3} + 1}$

10. $\dfrac{\frac{5}{4} - 3}{\frac{3}{4} - 3}$

11. $\dfrac{\frac{1}{2} + \frac{3}{4}}{\frac{3}{2} + \frac{1}{4}}$

12. $\dfrac{\frac{2}{3} - \frac{5}{2}}{\frac{2}{3} - \frac{3}{2}}$

REVIEW *Write each expression as an expression involving only one exponent. Assume no variable is zero.*

13. t^5t^2t

14. $(a^0a^2)^3$

15. $-2r(r^3)^2$

16. $(b^4)^2(b^6)^0$

Write each expression without parentheses or negative exponents.

17. $\left(\dfrac{3r}{4r^3}\right)^{-4}$

18. $\left(\dfrac{12y^{-3}}{3y^2}\right)^{-3}$

19. $\left(\dfrac{6r^{-2}}{2r^3}\right)^{-2}$

20. $\left(\dfrac{4x^3}{5x^{-3}}\right)^{-2}$

VOCABULARY AND CONCEPTS *Fill in the blanks.*

21. If a fraction contains a fraction in its numerator and/or denominator, it is called a ____________.

22. The denominator of the complex fraction $\dfrac{\frac{3}{x} + \frac{x}{y}}{\frac{1}{x} + 2}$ is _____.

23. In Method 1, we write the numerator and denominator of a complex fraction as _____ fractions and then _____.

24. In Method 2, we multiply the numerator and denominator of the complex fraction by the ____ of the fractions in its numerator and denominator.

GUIDED PRACTICE *Simplify each complex fraction. Assume no division by 0.* SEE EXAMPLE 1. (OBJECTIVE 1)

25. $\dfrac{\frac{2x}{y}}{\frac{4}{xy}}$

26. $\dfrac{\frac{y}{x}}{\frac{x}{xy}}$

27. $\dfrac{\frac{5t^2}{9x^2}}{\frac{3t}{x^2t}}$

28. $\dfrac{\frac{4w^2}{5t}}{\frac{w}{15t}}$

Simplify each complex fraction. Assume no division by 0. SEE EXAMPLE 2. (OBJECTIVE 1)

29. $\dfrac{\frac{a}{b}}{\frac{a}{a+1}}$

30. $\dfrac{\frac{x}{y-1}}{\frac{x}{y}}$

31. $\dfrac{\frac{x}{y-1}}{\frac{x}{y+1}}$

32. $\dfrac{\frac{x+y}{x-y}}{\frac{x}{x+y}}$

Simplify each complex fraction. Assume no division by 0. SEE EXAMPLE 3. (OBJECTIVE 1)

33. $\dfrac{\frac{1}{y} + 3}{\frac{3}{y} - 2}$

34. $\dfrac{2 + \frac{1}{x}}{2 - \frac{3}{x}}$

35. $\dfrac{5 + \frac{3}{x}}{3 + \frac{2}{x}}$

36. $\dfrac{\frac{3}{x} - 3}{\frac{9}{x} - 3}$

37. $\dfrac{\frac{2}{a+2} + 1}{\frac{3}{a+2}}$

38. $\dfrac{3 - \frac{2}{m-3}}{\frac{4}{m-3}}$

39. $\dfrac{\frac{3}{x} + \frac{4}{x+1}}{\frac{2}{x+1} - \frac{3}{x}}$

40. $\dfrac{\frac{5}{y-3} - \frac{2}{y}}{\frac{1}{y} + \frac{2}{y-3}}$

41. $\dfrac{\frac{3y}{x} - y}{y - \frac{y}{x}}$

42. $\dfrac{\frac{y}{x} + 3y}{y + \frac{2y}{x}}$

43. $\dfrac{1}{\frac{1}{x}+\frac{1}{y}}$

44. $\dfrac{1}{\frac{b}{a}-\frac{a}{b}}$

45. $\dfrac{\frac{2}{x}}{\frac{2}{y}-\frac{4}{x}}$

46. $\dfrac{\frac{2y}{3}}{\frac{2y}{3}-\frac{8}{y}}$

47. $\dfrac{\frac{3}{x}+\frac{2x}{y}}{\frac{4}{x}}$

48. $\dfrac{\frac{4}{a}-\frac{a}{b}}{\frac{b}{a}}$

Simplify each complex fraction. Assume no division by 0. SEE EXAMPLE 4. *(OBJECTIVE 1)*

49. $\dfrac{\frac{1}{x+1}}{1+\frac{1}{x+1}}$

50. $\dfrac{\frac{1}{x-1}}{1-\frac{1}{x-1}}$

51. $\dfrac{\frac{x}{x+2}}{\frac{x}{x+2}+x}$

52. $\dfrac{\frac{2}{x-2}}{\frac{2}{x-2}-1}$

53. $\dfrac{\frac{2}{x}-\frac{3}{x+1}}{\frac{2}{x+1}-\frac{3}{x}}$

54. $\dfrac{\frac{5}{y}+\frac{4}{y+1}}{\frac{4}{y}-\frac{5}{y+1}}$

55. $\dfrac{\frac{m}{m+2}-\frac{2}{m-1}}{\frac{3}{m+2}+\frac{m}{m-1}}$

56. $\dfrac{\frac{2a}{a-3}+\frac{1}{a-2}}{\frac{a}{a-2}-\frac{3}{a-3}}$

Simplify each complex fraction. Assume no division by 0. SEE EXAMPLE 5. *(OBJECTIVE 2)*

57. $\dfrac{x^{-2}+1}{x^{-1}+1}$

58. $\dfrac{3x^{-1}+2}{3x^{-1}-1}$

59. $\dfrac{y^{-2}+1}{y^{-2}-1}$

60. $\dfrac{1+x^{-1}}{x^{-1}-1}$

61. $\dfrac{a^{-2}+a}{a+1}$

62. $\dfrac{t-t^{-2}}{1-t^{-1}}$

63. $\dfrac{2x^{-1}+4x^{-2}}{2x^{-2}+x^{-1}}$

64. $\dfrac{x^{-2}-3x^{-3}}{3x^{-2}-9x^{-3}}$

ADDITIONAL PRACTICE

Simplify each complex fraction. Assume no division by 0.

65. $\dfrac{\frac{y}{x-1}}{\frac{y}{x}}$

66. $\dfrac{\frac{a}{b}}{\frac{a-1}{b}}$

67. $\dfrac{3+\frac{3}{x-1}}{3-\frac{3}{x}}$

68. $\dfrac{2-\frac{2}{x+1}}{2+\frac{2}{x}}$

69. $\dfrac{\frac{2}{x+2}}{\frac{3}{x-3}+\frac{1}{x}}$

70. $\dfrac{\frac{1}{x-1}-\frac{4}{x}}{\frac{3}{x+1}}$

71. $\dfrac{\frac{1}{x}+\frac{2}{x+1}}{\frac{2}{x-1}-\frac{1}{x}}$

72. $\dfrac{\frac{3}{x+1}-\frac{2}{x-1}}{\frac{1}{x+2}+\frac{2}{x-1}}$

73. $\dfrac{\frac{1}{y^2+y}-\frac{1}{xy+x}}{\frac{1}{xy+x}-\frac{1}{y^2+y}}$

74. $\dfrac{\frac{2}{b^2-1}-\frac{3}{ab-a}}{\frac{3}{ab-a}-\frac{2}{b^2-1}}$

75. $\dfrac{1-25y^{-2}}{1+10y^{-1}+25y^{-2}}$

76. $\dfrac{1-9x^{-2}}{1-6x^{-1}+9x^{-2}}$

WRITING ABOUT MATH

77. Explain how to use Method 1 to simplify

$$\dfrac{1+\frac{1}{x}}{3-\frac{1}{x}}$$

78. Explain how to use Method 2 to simplify the expression in Exercise 77.

SOMETHING TO THINK ABOUT

79. Simplify each complex fraction:

$$\frac{1}{1+1},\ \frac{1}{1+\frac{1}{2}},\ \frac{1}{1+\frac{1}{1+\frac{1}{2}}},\ \frac{1}{1+\frac{1}{1+\frac{1}{1+\frac{1}{2}}}}$$

80. In Exercise 79, what is the pattern in the numerators and denominators of the four answers? What would be the next answer?

Section 6.5 Solving Equations That Contain Rational Expressions

Objectives

1. Solve an equation that contains one or more rational expressions.
2. Identify extraneous solutions.

Vocabulary

extraneous solution

Getting Ready

Simplify.

1. $3\left(x + \frac{1}{3}\right)$
2. $8\left(x - \frac{1}{8}\right)$
3. $x\left(\frac{3}{x} + 2\right)$
4. $3y\left(\frac{1}{3} - \frac{2}{y}\right)$
5. $6x\left(\frac{5}{2x} + \frac{2}{3x}\right)$
6. $9x\left(\frac{7}{9} + \frac{2}{3x}\right)$
7. $(y - 1)\left(\frac{1}{y - 1} + 1\right)$
8. $(x + 2)\left(3 - \frac{1}{x + 2}\right)$

We will now use our knowledge of rational expressions to solve equations that contain rational expressions with variables in their denominators. To do so, we will use new equation-solving methods that sometimes lead to false solutions. For this reason, it is important to check all apparent answers.

1 Solve an equation that contains one or more rational expressions.

To solve equations containing rational expressions, we use the same process we did with equations containing fractions. To clear the fractions, we multiply both sides of the equation by the LCD of the rational expressions that appear in the equation. To review this process, we will solve an equation containing only numerical denominators.

$$\frac{x}{3} + 1 = \frac{x}{6}$$

$$\mathbf{6}\left(\frac{x}{3} + 1\right) = \mathbf{6}\left(\frac{x}{6}\right) \quad \text{Multiply both sides of the equation by 6, the LCD, to clear fractions.}$$

We then use the distributive property to remove parentheses, simplify, and solve the resulting equation for x.

$$\mathbf{6}\cdot\frac{x}{3} + \mathbf{6}\cdot 1 = \mathbf{6}\cdot\frac{x}{6}$$

$$2x + 6 = x$$

$$x + 6 = 0 \quad \text{Subtract } x \text{ from both sides.}$$

$$x = -6 \quad \text{Subtract 6 from both sides.}$$

Check: $\frac{x}{3} + 1 = \frac{x}{6}$

$$\frac{-6}{3} + 1 \stackrel{?}{=} \frac{-6}{6} \quad \text{Substitute } -6 \text{ for } x.$$

$$-2 + 1 \stackrel{?}{=} -1 \quad \text{Simplify.}$$

$$-1 = -1$$

Because -6 satisfies the original equation, it is the solution.

EXAMPLE 1 Solve: $\frac{4}{x} + 1 = \frac{6}{x}$

Solution Note that $x = 0$ is a restricted value since it creates division by 0. To clear the equation of rational expressions, we multiply both sides of the equation by the LCD of $\frac{4}{x}$, 1, and $\frac{6}{x}$, which is x.

$$\frac{4}{x} + 1 = \frac{6}{x}$$

$$x\left(\frac{4}{x} + 1\right) = x\left(\frac{6}{x}\right) \quad \text{Multiply both sides by } x \text{, the LCD.}$$

$$x \cdot \frac{4}{x} + x \cdot 1 = x \cdot \frac{6}{x} \quad \text{Use the distributive property.}$$

$$4 + x = 6 \quad \text{Simplify.}$$

$$x = 2 \quad \text{Subtract 4 from both sides.}$$

Check: $\frac{4}{x} + 1 = \frac{6}{x}$

$$\frac{4}{2} + 1 \stackrel{?}{=} \frac{6}{2} \quad \text{Substitute 2 for } x.$$

$$2 + 1 \stackrel{?}{=} 3 \quad \text{Simplify.}$$

$$3 = 3$$

Because 2 satisfies the original equation, it is the solution.

SELF CHECK 1 Solve: $\frac{6}{x} - 1 = \frac{3}{x}$

2 Identify extraneous solutions.

If we multiply both sides of an equation by an expression that involves a variable, as we did in Example 1, we must check the apparent solutions. The next example shows why.

EXAMPLE 2 Solve: $\frac{x+3}{x-1} = \frac{4}{x-1}$

Solution Find the restricted value of the denominators. In this example $x \neq 1$. To clear the equation of rational expressions, we multiply both sides by $(x - 1)$, the LCD of the fractions contained in the equation.

$$\frac{x+3}{x-1} = \frac{4}{x-1}$$

$$(x-1)\frac{x+3}{x-1} = (x-1)\frac{4}{x-1} \quad \text{Multiply both sides by } (x-1) \text{, the LCD.}$$

$$x + 3 = 4 \quad \text{Simplify.}$$
$$x = 1 \quad \text{Subtract 3 from both sides.}$$

COMMENT Whenever a restricted (excluded) value is a possible solution, it will be extraneous.

Because both sides were multiplied by an expression containing a variable, we must check to see if the apparent solution is a value that must be excluded. If we replace x with 1 in the original equation, both denominators will become 0. Therefore, 1 is not a solution. Such false solutions are often called **extraneous solutions**. Because 1 does not satisfy the original equation, there is no solution. The solution set of the equation is $\varnothing$.

SELF CHECK 2 Solve: $\dfrac{x+5}{x-2} = \dfrac{7}{x-2}$

The next two examples suggest the steps to follow when solving equations that contain rational expressions.

SOLVING EQUATIONS CONTAINING RATIONAL EXPRESSIONS

1. Find any restrictions on the variable. Remember that the denominator of a fraction cannot be 0.
2. Multiply both sides of the equation by the LCD of the rational expressions appearing in the equation to clear the equation of fractions.
3. Solve the resulting equation.
4. If an apparent solution of an equation is a restricted value, that value must be excluded. Check all solutions for extraneous roots.

EXAMPLE 3 Solve: $\dfrac{3x+1}{x+1} - 2 = \dfrac{3(x-3)}{x+1}$

Solution Since the denominator $x + 1$ cannot be 0, $x \neq -1$. To clear the equation of rational expressions, we multiply both sides by $(x + 1)$, the LCD of the rational expressions contained in the equation. We then can solve the resulting equation.

$$\frac{3x+1}{x+1} - 2 = \frac{3(x-3)}{x+1}$$

$$(x+1)\left[\frac{3x+1}{x+1} - 2\right] = (x+1)\left[\frac{3(x-3)}{x+1}\right] \quad \text{Multiply both sides by } (x+1), \text{ the LCD.}$$

$$3x + 1 + (x+1)(-2) = 3(x-3) \quad \text{Use the distributive property to remove brackets.}$$

$$3x + 1 - 2x - 2 = 3x - 9 \quad \text{Use the distributive property to remove parentheses.}$$

$$x - 1 = 3x - 9 \quad \text{Combine like terms.}$$

$$-2x = -8 \quad \text{On both sides, subtract } 3x \text{ and add 1.}$$

$$x = 4 \quad \text{Divide both sides by } -2.$$

The apparent solution 4 is not an excluded value. We will check the solution to verify our work.

Check: $\dfrac{3x+1}{x+1} - 2 = \dfrac{3(x-3)}{x+1}$

$$\frac{3(4)+1}{4+1} - 2 \stackrel{?}{=} \frac{3(4-3)}{4+1} \quad \text{Substitute 4 for } x.$$

$$\frac{13}{5} - \frac{10}{5} \stackrel{?}{=} \frac{3(1)}{5}$$

$$\frac{3}{5} = \frac{3}{5}$$

Because 4 satisfies the original equation, it is the solution.

SELF CHECK 3 Solve: $\frac{12}{x+1} - 5 = \frac{2}{x+1}$

To solve an equation with rational expressions, we often will have to factor a denominator to determine the least common denominator.

EXAMPLE 4 Solve: $1 = \frac{3}{x-2} - \frac{12}{x^2-4}$

Solution To find the LCD and any restricted values of x, we must factor the second denominator.

$$1 = \frac{3}{x-2} - \frac{12}{x^2-4}$$

$$1 = \frac{3}{x-2} - \frac{12}{(x+2)(x-2)} \quad \text{Factor } x^2 - 4.$$

Since $x + 2$ and $x - 2$ cannot be 0, $x \neq -2, 2$.

To clear the equation of rational expressions, we multiply both sides by $(x + 2)(x - 2)$, the LCD of the fractions contained in the equation.

$$(x+2)(x-2)(1) = (x+2)(x-2)\left[\frac{3}{x-2} - \frac{12}{(x+2)(x-2)}\right]$$

Multiply both sides by $(x + 2)(x - 2)$.

$$(x+2)(x-2) = (x+2)(x-2)\frac{3}{x-2} + (x+2)(x-2)\left(\frac{-12}{(x+2)(x-2)}\right)$$

Use the distributive property to remove brackets.

$$(x+2)(x-2) = (x+2)(3) - 12 \quad \text{Simplify.}$$

$$x^2 - 4 = 3x + 6 - 12 \quad \text{Use the distributive property to remove parentheses.}$$

$$x^2 - 4 = 3x - 6 \quad \text{Simplify.}$$

$$x^2 - 3x - 4 + 6 = 0 \quad \text{Subtract } 3x \text{ and add 6 to both sides.}$$

$$x^2 - 3x + 2 = 0 \quad \text{Combine like terms.}$$

$$(x-2)(x-1) = 0 \quad \text{Factor the left side.}$$

$$(x-2) = 0 \quad \text{or} \quad (x-1) = 0 \quad \text{Apply the zero-factor property.}$$

$$x = 2 \quad \Big| \quad x = 1 \quad \text{Solve each equation.}$$

Because 2 is an excluded value, it is an extraneous solution. Verify that 1 is the solution of the given equation.

SELF CHECK 4 Solve: $\frac{x-4}{x-3} + \frac{x-2}{x-3} = x - 3$

EXAMPLE 5 Solve: $\frac{4}{5} + y = \frac{4y-50}{5y-25}$

Solution Since $5y - 25$ cannot be 0, $y \neq 5$. Thus, 5 is a restricted value.

$$\frac{4}{5} + y = \frac{4y-50}{5y-25}$$

$$\frac{4}{5} + y = \frac{4y - 50}{5(y - 5)}$$ Factor $5y - 25$.

$$5(y - 5)\left[\frac{4}{5} + y\right] = 5(y - 5)\left[\frac{4y - 50}{5(y - 5)}\right]$$ Multiply both sides by $5(y - 5)$, the LCD.

$$4(y - 5) + 5y(y - 5) = 4y - 50$$ Use the distributive property to remove brackets.

$$4y - 20 + 5y^2 - 25y = 4y - 50$$ Use the distributive property to remove parentheses.

$$5y^2 - 25y - 20 = -50$$ Subtract $4y$ from both sides and rearrange terms.

$$5y^2 - 25y + 30 = 0$$ Add 50 to both sides.

$$y^2 - 5y + 6 = 0$$ Divide both sides by 5.

$$(y - 3)(y - 2) = 0$$ Factor $y^2 - 5y + 6$.

$$y - 3 = 0 \quad \text{or} \quad y - 2 = 0$$ Apply the zero-factor property.

$$y = 3 \quad \Big| \quad y = 2$$

Verify that 3 and 2 both satisfy the original equation.

SELF CHECK 5 Solve: $\frac{x - 6}{3x - 9} - \frac{1}{3} = \frac{x}{2}$

Many formulas are equations that contain rational expressions. The formula $\frac{1}{r} = \frac{1}{r_1} + \frac{1}{r_2}$ is used in electronics to calculate parallel resistances. To solve the formula for r, we eliminate the denominators by multiplying both sides by the LCD, which is rr_1r_2.

$$\frac{1}{r} = \frac{1}{r_1} + \frac{1}{r_2}$$

$$rr_1r_2\left(\frac{1}{r}\right) = rr_1r_2\left(\frac{1}{r_1} + \frac{1}{r_2}\right)$$ Multiply both sides by rr_1r_2 the LCD.

$$r_1r_2 = rr_2 + rr_1$$ Use the distributive property to remove parentheses.

$$r_1r_2 = r(r_2 + r_1)$$ Factor out r, the GCF.

$$\frac{r_1r_2}{r_2 + r_1} = r$$ Divide both sides by $r_2 + r_1$.

or

$$r = \frac{r_1r_2}{r_2 + r_1}$$

SELF CHECK ANSWERS **1.** 3 **2.** $\varnothing$, 2 is extraneous **3.** 1 **4.** 5; 3 is extraneous **5.** 1, 2

NOW TRY THIS

1. Solve: $\frac{6}{x} = x - 5$
2. Solve: $\frac{x - 2}{(x + 3)^2} - \frac{5}{x + 3} + 1 = 0$
3. Explain how to identify extraneous solutions.

6.5 Exercises

WARM-UPS *Solve each equation and check the solution.*

1. $\frac{y}{3} + 6 = \frac{4y}{3}$
2. $\frac{2y}{5} - 8 = \frac{4y}{5}$
3. $\frac{z - 3}{2} = z + 2$
4. $\frac{b + 2}{3} = b - 2$
5. $\frac{5(x + 1)}{8} = x + 1$
6. $\frac{3(x - 1)}{2} + 2 = x$

Indicate the LCD you will use to clear the fractions. Do not solve. Assume no denominators are zero.

7. $\frac{x - 3}{x + 5} = \frac{x}{2}$
8. $\frac{1}{x - 1} = \frac{8}{x}$
9. $\frac{y}{y - 1} + 5 = \frac{y + 1}{y + 2}$
10. $\frac{5x - 8}{3x} + 3x = \frac{x}{15}$
11. $\frac{5}{x^2 - 9} + \frac{3x}{x - 3} = \frac{4}{x + 3}$
12. $\frac{7y}{3y + 6} - \frac{2}{y + 2} = 5$

REVIEW *Factor each expression.*

13. $x^2 + 8x$
14. $x^3 - 27$
15. $2x^2 + x - 3$
16. $6a^2 - 5a - 6$
17. $4x^2 + 10x - 6$
18. $x^4 - 81$

VOCABULARY AND CONCEPTS *Fill in the blanks.*

19. False solutions that result from multiplying both sides of an equation by a variable are called ________ solutions.
20. If you multiply both sides of an equation by an expression that involves a variable, you must _____ the solution.
21. To clear an equation of rational expressions, we multiply both sides by the ____ of the expressions in the equation.
22. To clear the equation $\frac{x}{x - 2} - \frac{x}{x - 1} = 5$ of fractions, we multiply both sides by ___________.

GUIDED PRACTICE *Solve each equation and check the solution.* SEE EXAMPLE 1. (OBJECTIVE 1)

23. $\frac{3}{x} + 2 = 3$
24. $\frac{2}{x} + 9 = 11$
25. $\frac{x}{x + 2} + 3 = \frac{2x}{x + 2}$
26. $\frac{11}{b} + \frac{13}{b} = 12$
27. $\frac{2}{y + 1} + 5 = \frac{12}{y + 1}$
28. $\frac{1}{t - 3} = \frac{-2}{t - 3} + 1$
29. $\frac{1}{x - 1} + \frac{3}{x - 1} = 1$
30. $\frac{3}{p + 6} - 2 = \frac{7}{p + 6}$

Solve each equation and check the solution. Identify any extraneous values. SEE EXAMPLE 2. (OBJECTIVE 2)

31. $\frac{a^2}{a + 2} - \frac{4}{a + 2} = a$
32. $\frac{z^2}{z + 1} + 2 = \frac{1}{z + 1}$
33. $\frac{x}{x - 5} - \frac{5}{x - 5} = 3$
34. $\frac{3}{y - 2} + 1 = \frac{3}{y - 2}$

Solve each equation and check the solution. Identify any extraneous values. SEE EXAMPLE 3. (OBJECTIVE 2)

35. $\frac{2x + 1}{x + 5} - 1 = \frac{3x - 2}{x + 5}$
36. $\frac{3x + 1}{x + 3} + \frac{x - 3}{x + 3} = 2$
37. $\frac{x - 4}{x - 3} + \frac{x - 2}{x - 3} = x - 3$
38. $\frac{4x}{x - 1} + 1 = \frac{x + 3}{x - 1}$

Solve each equation and check the solution. Identify any extraneous values. SEE EXAMPLE 4. (OBJECTIVE 2)

39. $\frac{v}{v + 2} + \frac{1}{v - 1} = 1$
40. $\frac{b + 2}{b + 3} + 1 = \frac{-7}{b - 5}$
41. $\frac{u}{u - 1} + \frac{1}{u} = \frac{u^2 + 1}{u^2 - u}$
42. $\frac{3}{x - 2} + \frac{1}{x} = \frac{2(3x + 2)}{x^2 - 2x}$
43. $\frac{5}{x} + \frac{3}{x + 2} = \frac{-6}{x(x + 2)}$
44. $\frac{x - 3}{x - 2} - \frac{1}{x} = \frac{x - 3}{x}$
45. $\frac{-5}{s^2 + s - 2} + \frac{3}{s + 2} = \frac{1}{s - 1}$
46. $\frac{n}{n^2 - 9} + \frac{n + 8}{n + 3} = \frac{n - 8}{n - 3}$

Solve each equation and check the solution. Identify any extraneous values. SEE EXAMPLE 5. (OBJECTIVE 2)

47. $y + \frac{3}{4} = \frac{3y - 50}{4y - 24}$
48. $y + \frac{2}{3} = \frac{2y - 12}{3y - 9}$
49. $\frac{3}{5x - 20} + \frac{4}{5} = \frac{3}{5x - 20} - \frac{x}{5}$
50. $\frac{x}{x - 1} - \frac{12}{x^2 - x} = \frac{-1}{x - 1}$

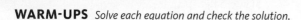

51. $\frac{7}{q^2 - q - 2} + \frac{1}{q + 1} = \frac{3}{q - 2}$

52. $\frac{x - 3}{4x - 4} + \frac{1}{9} = \frac{x - 5}{6x - 6}$

53. $\frac{3y}{3y - 6} + \frac{8}{y^2 - 4} = \frac{2y}{2y + 4}$

54. $1 - \frac{3}{b} = \frac{-8b}{b^2 + 3b}$

ADDITIONAL PRACTICE *Solve each equation.*

55. $\frac{c - 4}{4} = \frac{c + 4}{8}$

56. $\frac{x}{2} + 4 = \frac{3x}{2}$

57. $\frac{x}{5} - \frac{x}{3} = -8$

58. $\frac{3a}{2} + \frac{a}{3} = -22$

59. $\frac{x + 2}{2} - 3x = x + 8$

60. $\frac{3x - 1}{6} - \frac{x + 3}{2} = \frac{3x + 4}{3}$

61. $\frac{3r}{2} - \frac{3}{r} = \frac{3r}{2} + 3$

62. $\frac{2p}{3} - \frac{1}{p} = \frac{2p - 1}{3}$

63. $\frac{1}{3} + \frac{2}{x - 3} = 1$

64. $\frac{3}{5} + \frac{7}{x + 2} = 2$

65. $\frac{z - 4}{z - 3} = \frac{z + 2}{z + 1}$

66. $\frac{a + 2}{a + 8} = \frac{a - 3}{a - 2}$

67. $\frac{x - 2}{x - 3} + \frac{x - 1}{x^2 - 8x + 15} = 1$

68. $\frac{x + 1}{x + 2} + \frac{1}{x^2 + x - 2} = 1$

69. $\frac{1}{a} + \frac{1}{b} = 1$ for a

70. $\frac{1}{a} - \frac{1}{b} = 1$ for b

71. $\frac{a}{b} + \frac{c}{d} = 1$ for b

72. $\frac{a}{b} - \frac{c}{d} = 1$ for a

73. Solve the formula $\frac{1}{r} = \frac{1}{r_1} + \frac{1}{r_2}$ for r_1.

74. Solve the formula $\frac{1}{r} = \frac{1}{r_1} + \frac{1}{r_2}$ for r_2.

APPLICATIONS

75. Optics The focal length f of a lens is given by the formula

$$\frac{1}{f} = \frac{1}{d_1} + \frac{1}{d_2}$$

where d_1 is the distance from the object to the lens and d_2 is the distance from the lens to the image. Solve the formula for f.

76. Solve the formula in Exercise 75 for d_1.

WRITING ABOUT MATH

77. Explain how you would decide what to do first when you solve an equation that involves fractions.

78. Explain why it is important to check your solutions to an equation that contains fractions with variables in the denominator.

SOMETHING TO THINK ABOUT

79. What numbers are equal to their own reciprocals?

80. Solve: $x^{-2} + x^{-1} = 0$.

Section 6.6 Solving Applications Whose Models Contain Rational Expressions

Objectives

1. Solve an application using a rational equation.

Getting Ready

1. If it takes 5 hours to fill a pool, what part could be filled in 1 hour?
2. $\$x$ is invested at 5% annual interest. Write an expression for the interest earned in one year.
3. Write an expression for the amount of an investment that earns $\$y$ interest in one year at 5%.
4. Express how long it takes to travel y miles at 52 mph.

In this section, we will consider applications whose solutions depend on solving equations containing rational expressions.

1 Solve an application using a rational equation.

EXAMPLE 1 **NUMBERS** If the same number is added to both the numerator and denominator of the fraction $\frac{3}{5}$, the result is $\frac{4}{5}$. Find the number.

Analyze the problem: We are asked to find a number. We will let n represent the unknown number.

Form an equation: If we add the number n to both the numerator and denominator of the fraction $\frac{3}{5}$, we will get $\frac{4}{5}$. This gives the equation

$$\frac{3+n}{5+n} = \frac{4}{5}$$

Solve the equation: To solve the equation, we proceed as follows:

$$\frac{3+n}{5+n} = \frac{4}{5}$$

$$5(5+n)\frac{3+n}{5+n} = 5(5+n)\frac{4}{5} \quad \text{Multiply both sides by } 5(5+n) \text{, the LCD.}$$

$$5(3+n) = (5+n)4 \quad \text{Use the distributive property.}$$

$$15 + 5n = 20 + 4n \quad \text{Use the distributive property to remove parentheses.}$$

$$15 + n = 20 \quad \text{Subtract } 4n \text{ from both sides.}$$

$$n = 5 \quad \text{Subtract 15 from both sides.}$$

State the conclusion: The number is 5.

Check the result: Add 5 to both the numerator and denominator of $\frac{3}{5}$ and get

$$\frac{3+5}{5+5} = \frac{8}{10} = \frac{4}{5}$$

The result checks.

SELF CHECK 1 If the same number is subtracted from both the numerator and denominator of the fraction $\frac{4}{5}$, the result is $\frac{5}{6}$. Find the number.

As we did in Example 1, it is important to state the conclusion after solving an application.

EXAMPLE 2 **FILLING AN OIL TANK** An inlet pipe can fill an oil tank in 7 days, and a second inlet pipe can fill the same tank in 9 days. If both pipes are used, how long will it take to fill the tank?

Analyze the problem

We are asked to find how long it will take to fill the tank, so we let x represent the number of days it will take to fill the tank.

Form an equation

The key is to note what each pipe can do in 1 day. If you add what the first pipe can do in 1 day to what the second pipe can do in 1 day, the sum is what they can do together in 1 day. Since the first pipe can fill the tank in 7 days, it can do $\frac{1}{7}$ of the job in 1 day. Since the second pipe can fill the tank in 9 days, it can do $\frac{1}{9}$ of the job in 1 day. If it takes x days for both pipes to fill the tank, together they can do $\frac{1}{x}$ of the job in 1 day. This gives the equation

What the first inlet pipe can do in 1 day	plus	what the second inlet pipe can do in 1 day	equals	what they can do together in 1 day.
$\frac{1}{7}$	$+$	$\frac{1}{9}$	$=$	$\frac{1}{x}$

Solve the equation

To solve the equation, we proceed as follows:

$$\frac{1}{7} + \frac{1}{9} = \frac{1}{x}$$

$$\mathbf{63x}\left(\frac{1}{7} + \frac{1}{9}\right) = \mathbf{63x}\left(\frac{1}{x}\right) \quad \text{Multiply both sides by } 63x\text{, the LCD.}$$

$$9x + 7x = 63 \quad \text{Use the distributive property to remove parentheses and simplify.}$$

$$16x = 63 \quad \text{Combine like terms.}$$

$$x = \frac{63}{16} \quad \text{Divide both sides by 16.}$$

State the conclusion

It will take $\frac{63}{16}$ or $3\frac{15}{16}$ days for both inlet pipes to fill the tank.

Check the result

In $\frac{63}{16}$ days, the first pipe fills $\frac{1}{7}\left(\frac{63}{16}\right)$ of the tank, and the second pipe fills $\frac{1}{9}\left(\frac{63}{16}\right)$ of the tank. The sum of these efforts, $\frac{9}{16} + \frac{7}{16}$, is equal to one full tank.

SELF CHECK 2 If an inlet pipe can fill the oil tank in 6 days, and a second inlet pipe can fill the same tank in 11 days, how long will it take to fill the tank if both pipes are used?

EXAMPLE 3 **TRACK AND FIELD** A coach can run 10 miles in the same amount of time that her best student athlete can run 12 miles. If the student can run 1 mph faster than the coach, how fast can the student run?

Analyze the problem

We are asked to find how fast the student can run. Since we know that the student runs 1 mph faster than the coach, we will let r represent the rate of the coach and $r + 1$ represent the rate of the student. In this case, we want to find the rate of the student, which is $(r + 1)$.

Form an equation

This is a uniform motion problem, based on the formula $d = rt$, where d is the distance traveled, r is the rate, and t is the time. If we solve this formula for t, we obtain

$$t = \frac{d}{r}$$

If the coach runs 10 miles at some unknown rate of r mph, it will take $\frac{10}{r}$ hours. If the student runs 12 miles at some unknown rate of $(r + 1)$ mph, it will take $\frac{12}{r+1}$ hours. We can organize the information of the problem as in Table 6-1.

TABLE 6-1

	d	=	r	$\cdot$	t
Student	12		$r + 1$		$\frac{12}{r+1}$
Coach	10		r		$\frac{10}{r}$

Because the times are given to be equal, we know that $\frac{12}{r+1} = \frac{10}{r}$. This gives the equation

The time it takes the student to run 12 miles	equals	the time it takes the coach to run 10 miles.
$\frac{12}{r+1}$	$=$	$\frac{10}{r}$

Solve the equation We can solve the equation as follows:

COMMENT This example could have been set up with r representing the student's rate and $r - 1$ the coach's rate.

$$\frac{12}{r+1} = \frac{10}{r}$$

$$r(r+1)\frac{12}{r+1} = r(r+1)\frac{10}{r} \quad \text{Multiply both sides by } r(r+1)\text{, the LCD.}$$

$$12r = 10(r+1) \quad \text{Simplify.}$$

$$12r = 10r + 10 \quad \text{Use the distributive property to remove parentheses.}$$

$$2r = 10 \quad \text{Subtract } 10r \text{ from both sides.}$$

$$r = 5 \quad \text{Divide both sides by 2.}$$

State the conclusion The coach can run 5 mph. The student, running 1 mph faster, can run 6 mph.

Check the result Verify that this result checks.

SELF CHECK 3 The coach runs 8 miles in the same amount of time that her student can run 12 miles. If the student can run 2 mph faster than the coach, how fast can the student run?

EXAMPLE 4 **COMPARING INVESTMENTS** At one bank, a sum of money invested for one year will earn \$96 interest. If invested in bonds, that same money would earn \$120, because the interest rate paid by the bonds is 1% greater than that paid by the bank. Find the bank's rate of interest.

Analyze the problem We are asked to find the bank's rate of interest, so we can let r represent the bank's rate. If the interest on the bonds is 1% greater, then the bonds' interest rate will be $r + 0.01$.

Form an equation This is an interest problem that is based on the formula $i = pr$, where i is the interest, p is the principal (the amount invested), and r is the annual rate of interest.

If we solve this formula for p, we obtain

$$p = \frac{i}{r}$$

If an investment at a bank earns \$96 interest at some unknown rate r, the principal invested is $\frac{96}{r}$. If an investment in bonds earns \$120 interest at some unknown rate $(r + 0.01)$, the principal invested is $\frac{120}{r + 0.01}$. We can organize the information of the problem as in Table 6-2.

TABLE 6-2

	Interest	=	Principal	·	Rate
Bank	96		$\frac{96}{r}$		r
Bonds	120		$\frac{120}{r + 0.01}$		$r + 0.01$

Because the same principal would be invested in either account, we can set up the following equation:

$$\frac{96}{r} = \frac{120}{r + 0.01}$$

Solve the equation We can solve the equation as follows:

$$\frac{96}{r} = \frac{120}{r + 0.01}$$

$$\boldsymbol{r(r + 0.01)} \cdot \frac{96}{r} = \frac{120}{r + 0.01} \cdot \boldsymbol{r(r + 0.01)}$$ Multiply both sides by $r(r + 0.01)$, the LCD.

$$96(r + 0.01) = 120r$$ Use the distributive property.

$$96r + 0.96 = 120r$$ Use the distributive property to remove parentheses.

$$0.96 = 24r$$ Subtract $96r$ from both sides.

$$0.04 = r$$ Divide both sides by 24.

State the conclusion The bank's interest rate is 0.04 or 4%. The bonds pay 5% interest, a rate 1% greater than that paid by the bank.

Check the results Verify that these rates check.

SELF CHECK 4 Find the bank's rate of interest if the sum of money will earn \$90 interest. The same money would earn \$120 in bonds, because the interest rate is 1% greater than that paid by the bank.

SELF CHECK ANSWERS

1. The number is –1. **2.** It will take $\frac{66}{17}$ or $3\frac{15}{17}$ days for both inlet pipes to fill the tank. **3.** The student can run 6 mph. **4.** The bank's annual interest rate is 3%. The bonds pay 4% annual interest.

NOW TRY THIS

Chris can clean a house in 3 hours and Cheryl can clean the house in 2 hours. Their son, Tyler, can scatter toys all over the house in 4 hours. If Tyler starts scattering toys at the same time Chris and Cheryl start cleaning, when (if ever) will the house be clean?

6.6 Exercises

WARM-UPS

1. Write the formula that relates the principal p that is invested, the earned interest i, and the rate r for 1 year.

2. Write the formula that relates the distance d traveled at a speed r, for a time t.

3. Write the formula that relates the cost C of purchasing q items that cost $\$d$ each.

4. Write the formula that relates the value v of a mixture of n pounds costing $\$p$ per pound.

REVIEW *Solve each equation.*

5. $x^2 - 5x - 6 = 0$

6. $x^2 - 81 = 0$

7. $(y - 3)(y^2 + 5y + 4) = 0$

8. $(x^2 - 1)(x^2 - 4) = 0$

9. $y^3 - y^2 = 0$

10. $3b^3 - 27b = 0$

11. $4(y - 3) = -y^2$

12. $6t^3 + 35t^2 = 6t$

VOCABULARY AND CONCEPTS

13. List the five steps used in problem solving.

14. Write 3% as a decimal.

APPLICATIONS *Solve and verify your answer. SEE EXAMPLE 1. (OBJECTIVE 1)*

15. Numbers If the denominator of $\frac{3}{4}$ is increased by a number and the numerator is doubled, the result is 1. Find the number.

16. Numbers If a number is added to the numerator of $\frac{7}{8}$ and the same number is subtracted from the denominator, the result is 2. Find the number.

17. Numbers If a number is added to the numerator of $\frac{3}{4}$ and twice as much is added to the denominator, the result is $\frac{4}{7}$. Find the number.

18. Numbers If a number is added to the numerator of $\frac{4}{9}$ and twice as much is subtracted from the denominator, the result is –9. Find the number.

Solve and verify your answer. SEE EXAMPLE 2. (OBJECTIVE 1)

19. Grading papers It takes a teacher 60 minutes to grade a set of quizzes and takes her aide twice as long to do the same amount of grading. How long will it take them to grade a set of quizzes if they work together?

20. Printing schedules It takes a printer 12 hours to print the class schedules for all of the students in a college. A faster printer can do the job in 9 hours. How long will it take to do the job if both printers are used?

21. Filling a pool An inlet pipe can fill an empty swimming pool in 5 hours, and another inlet pipe can fill the pool in 4 hours. How long will it take both pipes to fill the pool?

22. Roofing a house A homeowner estimates that it will take 7 days to roof his house. A professional roofer estimates that he could roof the house in 4 days. How long will it take if the homeowner helps the roofer?

Solve and verify your answer. SEE EXAMPLE 3. (OBJECTIVE 1)

23. Flying speeds On average, a Canada goose can fly 10 mph faster than a Great Blue heron. Find their flying speeds if a goose can fly 180 miles in the same time it takes a heron to fly 120 miles.

24. Touring A tourist can bicycle 28 miles in the same time as he can walk 8 miles. If he can ride 10 mph faster than he can walk, how much time should he allow to walk a 30-mile trail? (*Hint*: How fast can he walk?)

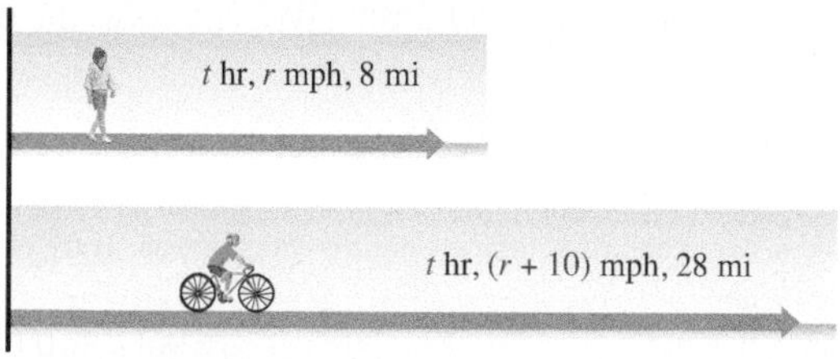

25. Comparing travel A plane can fly 300 miles in the same time as it takes a car to go 120 miles. If the car travels 90 mph slower than the plane, find the speed of the plane.

26. Wind speed A plane can fly 300 miles downwind in the same amount of time as it can travel 210 miles upwind. Find the velocity of the wind if the plane can fly 255 mph in still air.

Solve and verify your answer. SEE EXAMPLE 4. (OBJECTIVE 1)

27. Comparing investments Two certificates of deposit pay interest at rates that differ by 1%. Money invested for one year in the first CD earns \$175 interest. The same principal invested in the other CD earns \$200. Find the two rates of interest.

28. Comparing interest rates Two bond funds pay interest at rates that differ by 2%. Money invested for one year in the first fund earns \$315 interest. The same amount invested in the other fund earns \$385. Find the lower rate of interest.

29. Comparing interest rates Two mutual funds pay interest at rates that differ by 4%. Money invested for one year in the first fund earns \$300 interest. The same amount invested in the other fund earns \$60. Find the higher rate of interest.

30. Comparing interest rates Two banks pay interest at rates that differ by 1%. Money invested for one year in the first account earns \$105 interest. The same amount invested in the other account earns \$125. Find the two rates of interest.

Unless otherwise noted, all content on this page is © Cengage Learning.

ADDITIONAL PRACTICE *Solve and verify your answer.*

31. Filling a pool One inlet pipe can fill an empty pool in 4 hours, and a drain can empty the pool in 8 hours. How long will it take the pipe to fill the pool if the drain is left open?

32. Sewage treatment A sludge pool is filled by two inlet pipes. One pipe can fill the pool in 15 days and the other pipe can fill it in 21 days. However, if no sewage is added, waste removal will empty the pool in 36 days. How long will it take the two inlet pipes to fill an empty pool?

33. Sales A bookstore can purchase several calculators for a total cost of $120. If each calculator cost $1 less, the bookstore could purchase 10 additional calculators at the same total cost. How many calculators can be purchased at the regular price?

34. Furnace repairs A repairman purchased several furnace-blower motors for a total cost of $210. If his cost per motor had been $5 less, he could have purchased 1 additional motor. How many motors did he buy at the regular rate?

35. Boating A boat that can travel 18 mph in still water can travel 22 miles downstream in the same amount of time that it can travel 14 miles upstream. Find the speed of the current in the river.

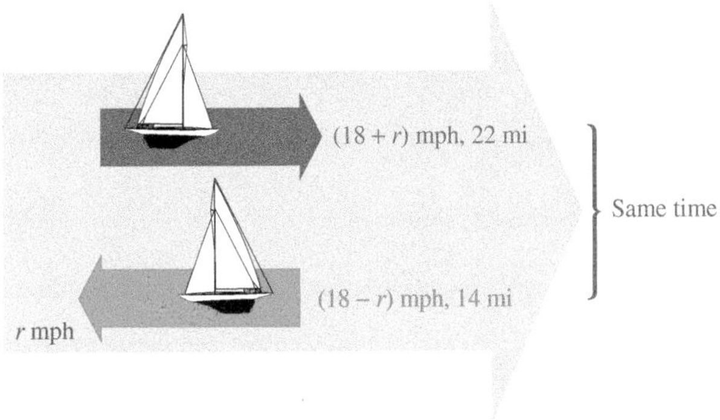

36. Conveyor belts The diagram shows how apples are processed for market. Although the second conveyor belt is shorter, an apple spends the same time on each belt because the second belt moves 1 ft/sec slower than the first. Find the speed of each belt.

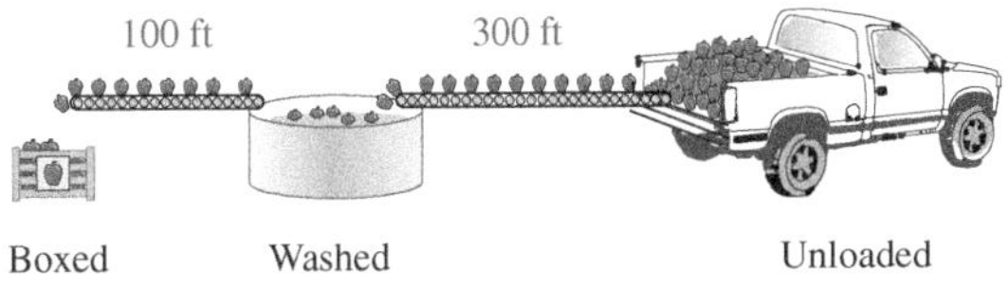

37. Numbers The sum of a number and its reciprocal is $\frac{13}{6}$. Find the numbers.

38. Numbers The sum of the reciprocals of two consecutive even integers is $\frac{7}{24}$. Find the integers.

39. Road Rallies The London to Tashkent (Uzbekistan) Road Rally takes roughly 2 weeks to complete. One leg of the race, from Nurburgring, Germany, to Wroclaw, Poland, covers 512 miles and the leg from Krakow, Poland, to Kiev, Ukraine, covers 528 miles. If the speed from Germany to Poland is 20 mph faster than that from Poland to Ukraine and the difference in times is 4 hours, find the rates of travel for the 2 legs. (*Hint*: Add the 4 hours to the expression for the faster rate.)

40. Pony Express The Pony Express carried mail from St. Joseph, MO, to Sacramento, CA, between 1860 and 1861. Pony Express riders stopped only long enough to change horses and these were saddled and waiting. The distance covered was 1,800 miles. A car travels 52 mph faster than the horse. If it took the Pony Express riders 7.5 times as long to make the trip as by car, find how long it took the Pony Express to make the trip.

41. Sharing costs Some office workers bought a $60 gift for their boss. If there had been five more employees to contribute, everyone's cost would have been $2 less. How many workers contributed to the gift?

42. Sales A dealer bought some radios for a total of $1,200. She gave away 6 radios as gifts, sold each of the rest for $10 more than she paid for each radio, and broke even. How many radios did she buy?

43. River tours A river boat tour begins by going 60 miles upstream against a 5 mph current. Then the boat turns around and returns with the current. What still-water speed should the captain use to complete the tour in 5 hours?

44. Travel time A company president flew 680 miles in a corporate jet but returned in a smaller plane that could fly only half as fast. If the total travel time was 6 hours, find the speeds of the planes.

WRITING ABOUT MATH

45. The key to solving shared work problems is to ask, "How much of the job could be done in 1 unit of time?" Explain.

46. It is difficult to check the solution of a shared work problem. Explain how you could decide if the answer is at least reasonable.

SOMETHING TO THINK ABOUT

47. Create a problem, involving either investment income or shared work, that can be solved by an equation that contains rational expressions.

48. Solve the problem you created in Exercise 47.

Unless otherwise noted, all content on this page is © Cengage Learning.

Section 6.7 Proportion and Variation

Objectives

1. Solve a proportion.
2. Solve an application involving a proportion.
3. Use similar triangles to determine a length of one of the sides.
4. Solve an application involving direct variation.
5. Solve an application involving inverse variation.
6. Solve an application involving joint variation.
7. Solve an application involving combined variation.

Vocabulary

ratio	means	inverse variation
unit cost	similar triangles	joint variation
rates	direct variation	combined variation
proportion	constant of proportionality (constant of variation)	
extremes		

Getting Ready

Solve each equation.

1. $30k = 70$
2. $\dfrac{k}{4{,}000^2} = 90$

Classify each function as a linear function or a rational function.

3. $f(x) = 3x$
4. $f(x) = \dfrac{3}{x} \quad (x > 0)$

In this section, we will discuss ratio and proportion. Then we will use these skills to solve variation problems.

1 Solve a proportion.

The comparison of two numbers is often called a **ratio.** For example, the fraction $\frac{2}{3}$ can be read as "the ratio of 2 to 3." Some more examples of ratios are

$$\frac{4x}{7y} \text{ (the ratio of } 4x \text{ to } 7y) \qquad \text{and} \qquad \frac{x-2}{3x} \text{ (the ratio of } (x-2) \text{ to } 3x)$$

Ratios often are used to express **unit costs**, such as the cost per pound of ground round steak.

$$\begin{array}{r} \text{The cost of a package of ground round} \rightarrow \\ \text{The weight of the package} \rightarrow \end{array} \frac{\$18.75}{5 \text{ lb}} = \$3.75 \text{ per lb} \leftarrow \text{The cost per pound}$$

Ratios also are used to express **rates**, such as an average rate of speed.

$$\begin{array}{r} \text{A distance traveled} \rightarrow \\ \text{in a period of time} \rightarrow \end{array} \frac{372 \text{ miles}}{6 \text{ hours}} = 62 \text{ mph} \leftarrow \text{The average rate of speed}$$

An equation indicating that two ratios are equal is called a **proportion**. Two examples of proportions are

$$\frac{1}{4} = \frac{2}{8} \quad \text{and} \quad \frac{4}{7} = \frac{12}{21}$$

In the proportion $\frac{a}{b} = \frac{c}{d}$, a and d are called the **extremes** and b and c are called the **means**.

To develop a fundamental property of proportions, we suppose that

$$\frac{a}{b} = \frac{c}{d}$$

is a proportion and multiply both sides of the equation by bd, the LCD, to obtain

$$bd\left(\frac{a}{b}\right) = bd\left(\frac{c}{d}\right)$$

$$\frac{\cancel{b}da}{\cancel{b}} = \frac{b\cancel{d}c}{\cancel{d}}$$

$$ad = bc$$

Thus, if $\frac{a}{b} = \frac{c}{d}$, then $ad = bc$. This shows that in a proportion, the *product of the extremes equals the product of the means.*

EXAMPLE 1 Solve: $\dfrac{x + 1}{x} = \dfrac{x}{x + 2}$ Assume no denominators are 0.

Solution We will use the fact that *in a proportion, the product of the extremes is equal to the product of the means.*

$$\frac{x + 1}{x} = \frac{x}{x + 2}$$

$(x + 1)(x + 2) = x \cdot x$ — In a proportion, the product of the extremes equals the product of the means.

$x^2 + 3x + 2 = x^2$ — Multiply.

$3x + 2 = 0$ — Subtract x^2 from both sides.

$x = -\dfrac{2}{3}$ — Subtract 2 from both sides and divide by 3.

Thus, the solution is $-\frac{2}{3}$.

SELF CHECK 1 Solve: $\dfrac{x - 1}{x} = \dfrac{x}{x + 3}$ Assume no denominators are 0.

EXAMPLE 2 Solve: $\dfrac{5a + 2}{2a} = \dfrac{18}{a + 4}$ Assume no denominators are 0.

Solution We will use the fact that *in a proportion, the product of the extremes is equal to the product of the means.*

$$\frac{5a + 2}{2a} = \frac{18}{a + 4}$$

$(5a + 2)(a + 4) = 2a(18)$ — In a proportion, the product of the extremes equals the product of the means.

$5a^2 + 22a + 8 = 36a$ — Multiply.

$5a^2 - 14a + 8 = 0$ — Subtract $36a$ from both sides because this is a quadratic equation.

$(5a - 4)(a - 2) = 0$ — Factor.

$$\begin{aligned} 5a - 4 &= 0 \quad \text{or} \quad & a - 2 &= 0 \quad && \text{Set each factor equal to 0.} \\ 5a &= 4 & a &= 2 && \text{Solve each linear equation.} \\ a &= \frac{4}{5} \end{aligned}$$

Thus, the solutions are $\frac{4}{5}$ and 2.

SELF CHECK 2 Solve: $\frac{3x + 1}{12} = \frac{x}{x + 2}$ Assume no denominators are 0.

2 Solve an application involving a proportion.

EXAMPLE 3 **GROCERY SHOPPING** If 7 pears cost \$2.73, how much will 11 pears cost?

Solution We can let c represent the cost in dollars of 11 pears. The price per pear of 7 pears is $\frac{\$2.73}{7}$, and the price per pear of 11 pears is $\frac{\$c}{11}$. Since these costs are equal, we can set up and solve the following proportion.

$$\begin{aligned} \frac{2.73}{7} &= \frac{c}{11} \\ 11(2.73) &= 7c && \text{In a proportion, the product of the extremes is equal to the product of the means.} \\ 30.03 &= 7c && \text{Multiply.} \\ \frac{30.03}{7} &= c && \text{Divide both sides by 7.} \\ c &= 4.29 && \text{Simplify.} \end{aligned}$$

Eleven pears will cost \$4.29.

SELF CHECK 3 How much will 28 pears cost?

Everyday connections
Highest and Longest Jumpers

A kangaroo rat can jump 45 times its body length, the longest jumper of all mammals. Its body length averages 4.5 inches. The longest jump by a human is 29 feet 4 inches (Mike Powell (6 feet 2 inches) of USA in 1991).

Nicholas Taffs/Shutterstock.com

Herbert Kratky/Shutterstock.com

A flea measures 0.029 inch high and can jump up 4.06 inches. The highest jump by a human is 8 feet 5 inches (Javier Sotomayor (6 feet 5 inches) of Cuba in 1991).

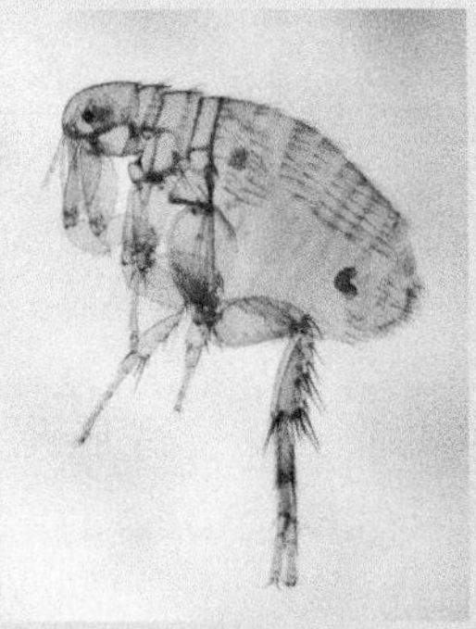

Sergejus Byckovskis/Shutterstock.com

Natursports/Shutterstock.com

Sources: http://www.sports-reference.com/olympics/athletes/po/mike-powell-1.html

http://www.iaaf.org/athletes/biographies/country=cubathcode=2718/index.html

(continued)

1. If the height of a 6 foot 2 inch man were proportional to that of a kangaroo rat, how far could the man jump?

2. If the distance the kangaroo rat could jump were proportional to the distance the man could jump, how far would the kangaroo rat be able to jump?

3. If the height of a 6 foot 5 inch man were proportional to that of a flea, how high could the man jump?

4. If the height a flea could jump were proportional to the height a man could jump, how high would the flea be able to jump?

3 Use similar triangles to determine a length of one of the sides.

If two angles of one triangle have the same measure as two angles of a second triangle, the triangles will have the same shape. In this case, we call the triangles **similar triangles**. Here are some facts about similar triangles.

SIMILAR TRIANGLES

If two triangles are similar, then

1. the three angles of the first triangle have the same measures, respectively, as the three angles of the second triangle.
2. the lengths of all corresponding sides are in proportion.

The triangles shown in Figure 6-1 are similar because the measures of their corresponding angles are the same. Therefore, the measures of their corresponding sides are in proportion.

$$\frac{2}{4} = \frac{x}{2x}, \quad \frac{x}{2x} = \frac{1}{2}, \quad \frac{1}{2} = \frac{2}{4}$$

COMMENT It is also true that if the measures of the corresponding sides of two triangles are in proportion, the triangles are similar.

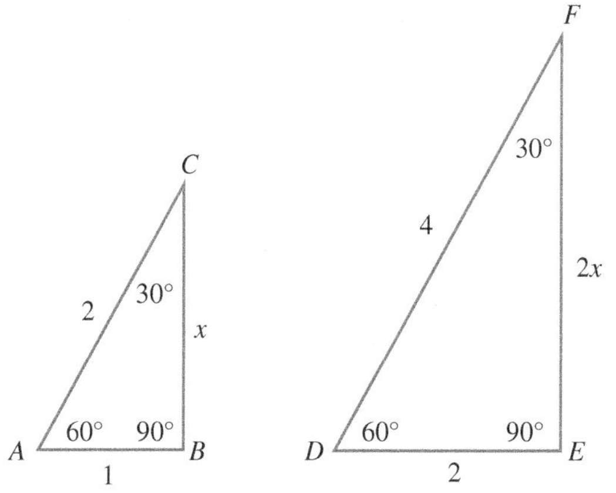

Figure 6-1

The properties of similar triangles often enable us to find the lengths of the sides of a triangle indirectly. For example, on a sunny day we can find the height of a tree and stay safely on the ground.

EXAMPLE 4 **HEIGHT OF A TREE** A tree casts a shadow of 29 feet at the same time as a vertical yardstick casts a shadow of 2.5 feet. Find the height of the tree.

Solution We will let h represent the height of the tree in feet. Refer to Figure 6-2 on the next page, which illustrates the triangles determined by the tree and its shadow, and the yardstick and its shadow. Because the triangles have the same shape, they are similar, and the

Unless otherwise noted, all content on this page is © Cengage Learning.

measures of their corresponding sides are in proportion. We can find the value of h by setting up and solving the following proportion.

$$\frac{h}{3} = \frac{29}{2.5}$$

$2.5h = 3(29)$ In a proportion, the product of the extremes is equal to the product of the means.

$2.5h = 87$ Simplify.

$h = 34.8$ Divide both sides by 2.5.

The tree is about 35 feet tall.

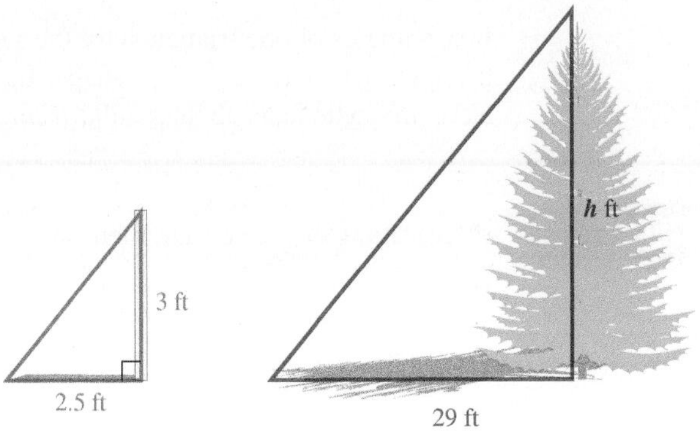

Figure 6-2

SELF CHECK 4 Find the height of the tree if its shadow is 25 feet long.

4 Solve an application involving direct variation.

To introduce **direct variation**, we consider the formula

$$C = \pi D$$

for the circumference of a circle, where C is the circumference, D is the diameter, and $\pi \approx 3.14159$. If we double the diameter of a circle, we determine another circle with a larger circumference C_1 such that

$$C_1 = \pi(2D) = 2\pi D = 2C$$

Thus, doubling the diameter results in doubling the circumference. Likewise, if we triple the diameter, we triple the circumference.

COMMENT In this section, we will assume that k is a positive number.

In this formula, we say that the variables C and D *vary directly*, or that they are *directly proportional*. This is because as one variable gets larger, so does the other, and in a predictable way. In this example, the constant π is called the **constant of variation** or the *constant of proportionality*.

DIRECT VARIATION The words "y varies directly with x" or "y is directly proportional to x" mean that $y = kx$ for some nonzero constant k. The constant k is called the constant of variation or the constant of proportionality.

Since the relationship for direct variation ($y = kx$) defines a linear function, its graph is always a line with a y-intercept at the origin. The graph of $y = kx$ appears in Figure 6-3 for three positive values of k.

Unless otherwise noted, all content on this page is © Cengage Learning.

One real-world example of direct variation is Hooke's law from physics. Hooke's law states that the distance a spring will stretch varies directly with the force that is applied to it.

If d represents a distance and f represents a force, Hooke's law is expressed mathematically as

$$d = kf$$

where k is the constant of variation. If the spring stretches 10 inches when a weight of 6 pounds is attached, the value of k can be found as follows:

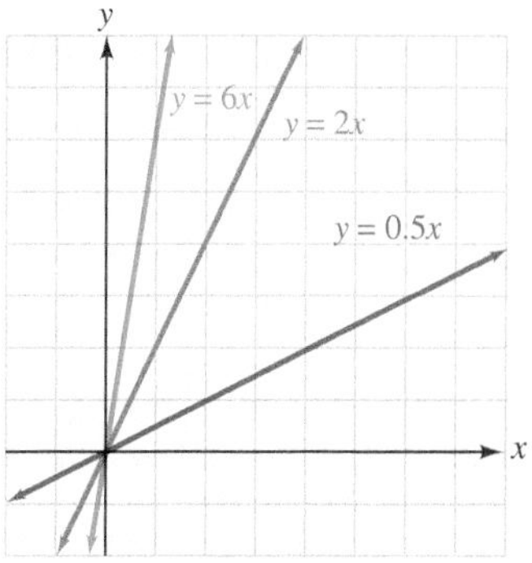

Figure 6-3

$$d = kf$$

$$10 = k(6) \quad \text{Substitute 10 for } d \text{ and 6 for } f.$$

$$\frac{5}{3} = k \quad \text{Divide both sides by 6 and simplify.}$$

To find the force required to stretch the spring a distance of 35 inches, we solve the equation $d = kf$ for f, with $d = 35$ and $k = \frac{5}{3}$.

$$d = kf$$

$$35 = \frac{5}{3}f \quad \text{Substitute 35 for } d \text{ and } \tfrac{5}{3} \text{ for } k.$$

$$105 = 5f \quad \text{Multiply both sides by 3.}$$

$$21 = f \quad \text{Divide both sides by 5.}$$

The force required to stretch the spring a distance of 35 inches is 21 pounds.

EXAMPLE 5 **DIRECT VARIATION** The distance traveled in a given time is directly proportional to the speed. If a car travels 70 miles at 30 mph, how far will it travel in the same time at 45 mph?

Solution The words *distance is directly proportional to speed* can be expressed by the equation

$$(1) \quad d = ks$$

where d is distance, k is the constant of variation, and s is the speed. To find the value of k, we substitute 70 for d and 30 for s, and solve.

$$d = ks$$

$$70 = k(30)$$

$$k = \frac{7}{3}$$

To find the distance traveled at 45 mph, we substitute $\frac{7}{3}$ for k and 45 for s in Equation 1 and simplify.

$$d = ks$$

$$d = \frac{7}{3}(45)$$

$$= 105$$

In the time it took to travel 70 miles at 30 mph, the car could travel 105 miles at 45 mph.

SELF CHECK 5 How far will the car travel in the same time at 60 mph?

Unless otherwise noted, all content on this page is © Cengage Learning.

5 Solve an application involving inverse variation.

In the formula $w = \frac{12}{l}$, w gets smaller as l gets larger, and w gets larger as l gets smaller. This is an example of **inverse variation**. Since these variables vary in opposite directions in a predictable way, we say that the variables *vary inversely*, or that they are *inversely proportional.* The constant 12 is the constant of variation.

INVERSE VARIATION

The words "y varies inversely with x" or "y is inversely proportional to x" mean that $y = \frac{k}{x}$ for some nonzero constant k. The constant k is called the constant of variation.

The formula for inverse variation $\left(y = \frac{k}{x}\right)$ defines a rational function. The graph of $y = \frac{k}{x}$ appears in Figure 6-4 for three positive values of k.

Because of gravity, an object in space is attracted to Earth. The force of this attraction varies inversely with the square of the distance from the object to Earth's center.

If f represents the force and d represents the distance, this information can be expressed by the equation

$$f = \frac{k}{d^2}$$

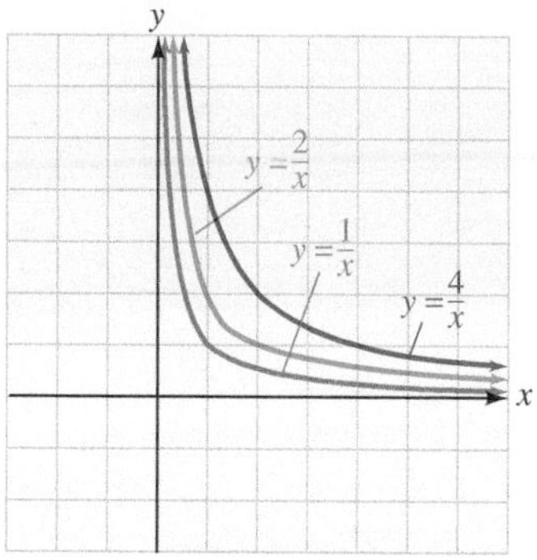

Figure 6-4

If we know that an object 4,000 miles from Earth's center is attracted to Earth with a force of 90 pounds, we can find the value of k.

$$f = \frac{k}{d^2}$$

$$\mathbf{90} = \frac{k}{\mathbf{4{,}000}^2} \quad \text{Substitute 90 for } f \text{ and 4,000 for } d.$$

$$k = 90(4{,}000^2)$$

$$= 1.44 \times 10^9$$

To find the force of attraction when the object is 5,000 miles from Earth's center, we proceed as follows:

$$f = \frac{k}{d^2}$$

$$f = \frac{\mathbf{1.44 \times 10^9}}{\mathbf{5{,}000}^2} \quad \text{Substitute } 1.44 \times 10^9 \text{ for } k \text{ and 5,000 for } d.$$

$$= 57.6$$

The object will be attracted to Earth with a force of 57.6 pounds when it is 5,000 miles from Earth's center.

EXAMPLE 6 **LIGHT INTENSITY** The intensity I of light received from a light source varies inversely with the square of the distance from the source. If the intensity of a light source 4 feet from an object is 8 candelas, find the intensity at a distance of 2 feet.

Solution The words *intensity varies inversely with the square of the distance d* can be expressed by the equation

$$I = \frac{k}{d^2}$$

Unless otherwise noted, all content on this page is © Cengage Learning.

To find the value of k, we substitute 8 for I and 4 for d and solve.

$$I = \frac{k}{d^2}$$

$$8 = \frac{k}{4^2}$$

$$128 = k$$

To find the intensity when the object is 2 feet from the light source, we substitute 2 for d and 128 for k and simplify.

$$I = \frac{k}{d^2}$$

$$I = \frac{\mathbf{128}}{\mathbf{2}^2}$$

$$= 32$$

The intensity at 2 feet is 32 candelas.

SELF CHECK 6 Find the intensity at a distance of 8 feet.

6 Solve an application involving joint variation.

There are times when one variable varies with the product of several variables. For example, the area of a triangle varies directly with the product of its base and height:

$$A = \frac{1}{2}bh$$

Such variation is called **joint variation**.

JOINT VARIATION

If one variable varies directly with the product of two or more variables, the relationship is called joint variation. If y varies jointly with x and z, then $y = kxz$. The nonzero constant k is called the constant of variation.

EXAMPLE 7 **VOLUME OF A CONE** The volume V of a cone varies jointly with its height h and the area of its base b. If $V = 6\text{ cm}^3$ when $h = 3$ cm and $B = 6\text{ cm}^2$, find V when $h = 2$ cm and $B = 8\text{ cm}^2$.

Solution The words *V varies jointly with h and B* can be expressed by the equation

$V = khB$ The relationship also can be read as "V is directly proportional to the product of h and B."

We can find the value of k by substituting 6 for V, 3 for h, and 6 for B.

$$V = khB$$

$$6 = k(3)(6)$$

$$6 = k(18)$$

$\frac{1}{3} = k$ Divide both sides by 18; $\frac{6}{18} = \frac{1}{3}$.

To find V when $h = 2$ and $B = 8$, we substitute these values into the formula $V = \frac{1}{3}hB$.

$$V = \frac{1}{3}hB$$

$$V = \left(\frac{1}{3}\right)(2)(8)$$

$$= \frac{16}{3}$$

The volume is $5\frac{1}{3}$ cm^3.

SELF CHECK 7 Find V when $B = 10$ cm.

7 Solve an application involving combined variation.

Many applied problems involve a combination of direct and inverse variation. Such variation is called **combined variation**.

EXAMPLE 8 **BUILDING HIGHWAYS** The time it takes to build a highway varies directly with the length of the road, but inversely with the number of workers. If it takes 100 workers 4 weeks to build 2 miles of highway, how long will it take 80 workers to build 10 miles of highway?

Solution We can let t represent the time in weeks, l represent the length in miles, and w represent the number of workers. The relationship among these variables can be expressed by the equation

$$t = \frac{kl}{w}$$

We substitute 4 for t, 100 for w, and 2 for l to find the value of k:

$$4 = \frac{k(2)}{100}$$

$$400 = 2k \quad \text{Multiply both sides by 100.}$$

$$200 = k \quad \text{Divide both sides by 2.}$$

We now substitute 80 for w, 10 for l, and 200 for k in the equation $t = \frac{kl}{w}$ and simplify:

$$t = \frac{kl}{w}$$

$$t = \frac{200(10)}{80}$$

$$= 25$$

It will take 25 weeks for 80 workers to build 10 miles of highway.

SELF CHECK 8 How long will it take 60 workers to build 6 miles of highway?

SELF CHECK ANSWERS **1.** $\frac{3}{2}$ **2.** $\frac{2}{3}$, 1 **3.** \$10.92 **4.** 30 ft **5.** 140 mi **6.** 2 candelas **7.** $6\frac{2}{3}$ cm^3 **8.** 20 weeks

NOW TRY THIS

1. Solve $\frac{5x + 2}{6x + 3} = \frac{2x}{2x + 1}$ as a proportion.
2. Solve $\frac{5x + 2}{6x + 3} = \frac{2x}{2x + 1}$ by multiplying by the LCD.

   Explain to a classmate why you did not get the same answer as in Problem 1.

The time T, in hours, required for a satellite to complete an orbit around Earth varies directly as the radius r of the orbit measured from the center of Earth and inversely as the velocity in miles per hour. The radius of Earth is approximately 4,000 miles.

3. The space shuttle made one orbit around Earth in 1.5 hours at a rate of 17,000 mph. Its altitude above Earth's surface was about 200 miles. Find the constant of variation rounded to 2 decimal places.
4. The Pentagon's Defense Satellite Communications System (DSCS) orbits at 23,500 miles above Earth's surface at a speed of approximately 6,955 mph. Using the same constant of variation, how long is its orbit? What can you conclude about the satellite?

6.7 Exercises

WARM-UPS *Which expressions are proportions?*

1. $\frac{3}{5} = \frac{6}{10}$
2. $\frac{1}{2} = \frac{1}{3}$
3. $\frac{1}{2} = \frac{1}{4}$
4. $\frac{1}{x} = \frac{2}{2x}$

Find the value of k.

5. $t = \frac{kl}{w}$ when $t = 20$, $l = 4$, and $w = 40$
6. $a = kbc$ when $a = 12$, $b = 2$, and $c = 3$

REVIEW *Simplify each expression.*

7. $(x^4 \cdot x^{-5})^2$
8. $\left(\frac{a^3a^5}{a^{-2}}\right)^3$
9. $\frac{3y^0 - 5y^0}{y^0}$
10. $\left(\frac{2r^{-2}r^{-3}}{4r^{-5}}\right)^{-3}$
11. Write 470,000 in scientific notation.
12. Write 0.000047 in scientific notation.
13. Write 2.5×10^{-3} in standard notation.
14. Write 2.5×10^4 in standard notation.

VOCABULARY AND CONCEPTS *Fill in the blanks.*

15. Ratios are used to express ________ and ____.
16. An equation stating that two ratios are equal is called a ________.
17. In a proportion, the product of the ________ is equal to the product of the ______.
18. If two angles of one triangle have the same measures as two angles of a second triangle, the triangles are ______.
19. The equation $y = kx$ indicates ______ variation.
20. The equation $y = \frac{k}{x}$ indicates ______ variation.
21. Inverse variation is represented by a ______ function.
22. The relationship for direct variation $(y = kx)$ is represented by a _____ function through the origin.
23. The equation $y = kxz$ indicates ____ variation and k represents the ____________________.
24. The equation $y = \frac{kx}{z}$ indicates ________ variation.

Determine whether the graph represents direct variation, inverse variation, or neither.

25.

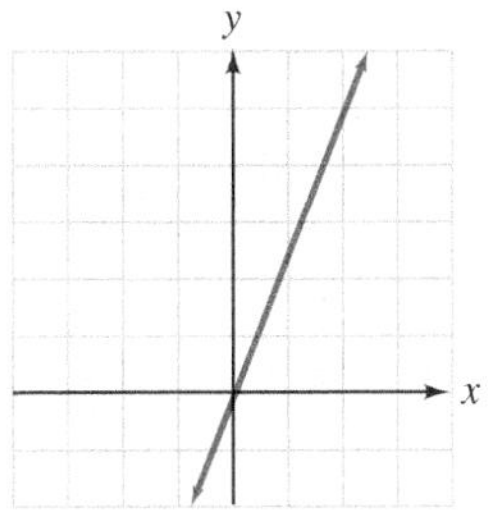

26.

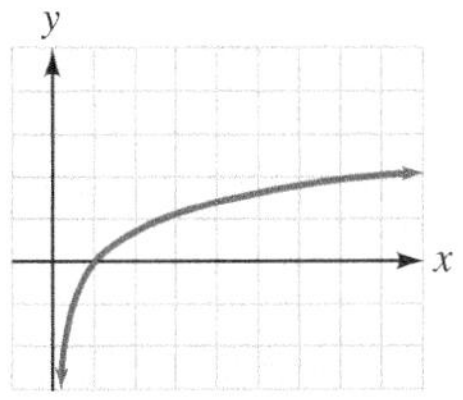

27.

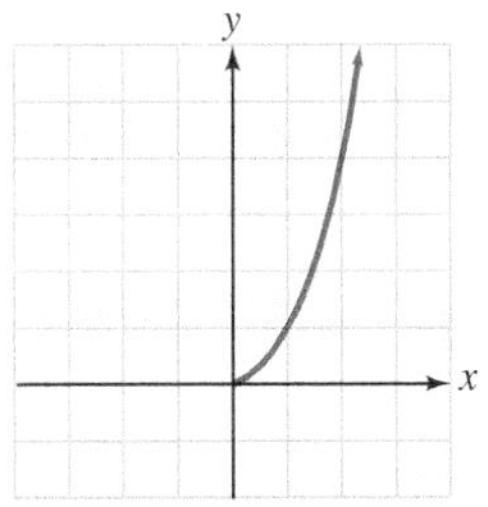

28.

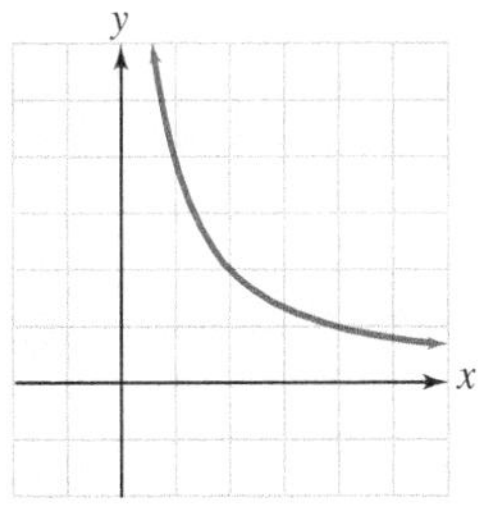

Unless otherwise noted, all content on this page is © Cengage Learning.

GUIDED PRACTICE *Solve each proportion for the variable. SEE EXAMPLE 1. (OBJECTIVE 1)*

29. $\frac{x}{8} = \frac{28}{32}$

30. $\frac{4}{y} = \frac{6}{27}$

31. $\frac{r-2}{3} = \frac{r}{5}$

32. $\frac{5}{n} = \frac{7}{n+2}$

33. $\frac{x+2}{x-2} = \frac{7}{3}$

34. $\frac{4}{x+3} = \frac{3}{5}$

35. $\frac{5}{5z+3} = \frac{2z}{2z^2+6}$

36. $\frac{z+2}{z+6} = \frac{z-4}{z-2}$

Solve each proportion. SEE EXAMPLE 2. (OBJECTIVE 1)

37. $\frac{1}{x+3} = \frac{-2x}{x+5}$

38. $\frac{x-1}{x+1} = \frac{2}{3x}$

39. $\frac{a-4}{a+2} = \frac{a-5}{a+1}$

40. $\frac{9t+6}{t(t+3)} = \frac{7}{t+3}$

Express each sentence as a formula. SEE EXAMPLES 5–8. (OBJECTIVES 4–7)

41. A varies directly with the square of p.

42. p varies directly with q.

43. a varies inversely with the square of b.

44. v varies inversely with the cube of r.

45. B varies jointly with m and n.

46. C varies jointly with x, y, and z.

47. X varies directly with w and inversely with q.

48. P varies directly with x and inversely with the square of q.

ADDITIONAL PRACTICE *Solve each proportion.*

49. $\frac{2}{c} = \frac{c-3}{2}$

50. $\frac{x}{7} = \frac{7}{x}$

51. $\frac{9}{5x} = \frac{3x}{15}$

52. $\frac{2}{x+6} = \frac{-2x}{5}$

53. $\frac{a+6}{a+4} = \frac{a-3}{a-4}$

54. $\frac{x+4}{5} = \frac{3(x-2)}{3}$

Express each sentence as a formula.

55. P varies directly with the square of a, and inversely with the cube of j.

56. M varies inversely with the cube of n, and jointly with x and the square of z.

Express each formula in words. In each formula, k is the constant of variation.

57. $L = kmn$

58. $P = \frac{km}{n}$

59. $E = kab^2$

60. $U = krs^2t$

61. $X = \frac{kx^2}{y^2}$

62. $Z = \frac{kw}{xy}$

63. $R = \frac{kL}{d^2}$

64. $e = \frac{kPL}{A}$

APPLICATIONS *Use an equation to solve. SEE EXAMPLE 3. (OBJECTIVE 2)*

65. Selling shirts Consider the following ad. How much will 5 shirts cost?

66. Cooking A recipe requires four 16-ounce bottles of catsup to make two gallons of spaghetti sauce. How many bottles are needed to make ten gallons of sauce?

67. Gas consumption A car gets 38 mpg. How much gas will be needed to go 323 miles?

68. Model railroading An HO-scale model railroad engine is 9 inches long. The HO scale is 87 feet to 1 foot. How long is a real engine?

69. Hobbies Standard dollhouse scale is 1 inch to 1 foot. Heidi's dollhouse is 32 inches wide. How wide would it be if it were a real house?

70. Staffing A school board has determined that there should be 3 teachers for every 50 students. How many teachers are needed for an enrollment of 2,700 students?

71. Drafting In a scale drawing, a 280-foot antenna tower is drawn 7 inches high. The building next to it is drawn 2 inches high. How tall is the actual building?

72. Mixing fuel The instructions on a can of oil intended to be added to lawnmower gasoline read:

Recommended	Gasoline	Oil
50 to 1	6 gal	16 oz

Are these instructions correct? (*Hint*: There are 128 ounces in 1 gallon.)

73. Recommended dosage The recommended child's dose of the sedative hydroxine is 0.006 gram per kilogram of body mass. Find the dosage for a 30-kg child.

74. Body mass The proper dose of the antibiotic cephalexin in children is 0.025 gram per kilogram of body mass. Find the mass of a child receiving a $1\frac{1}{8}$-gram dose.

Unless otherwise noted, all content on this page is © Cengage Learning.

Use similar triangles to solve. SEE EXAMPLE 4. (OBJECTIVE 3)

75. Height of a tree A tree casts a shadow of 28 feet at the same time as a 6-foot man casts a shadow of 4 feet. Find the height of the tree.

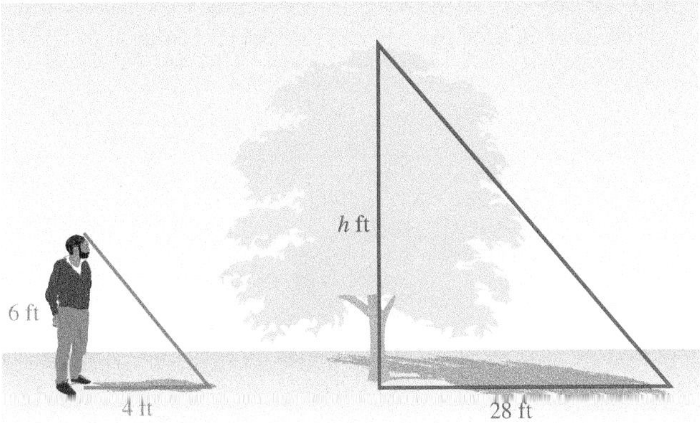

76. Height of a flagpole A man places a mirror on the ground and sees the reflection of the top of a flagpole. The two triangles in the illustration are similar. Find the height, h, of the flagpole.

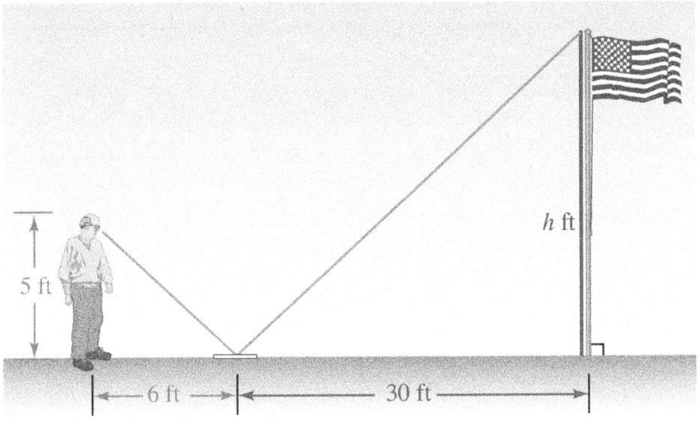

77. Width of a river Use the dimensions in the illustration to find w, the width of the river. The two triangles in the illustration are similar.

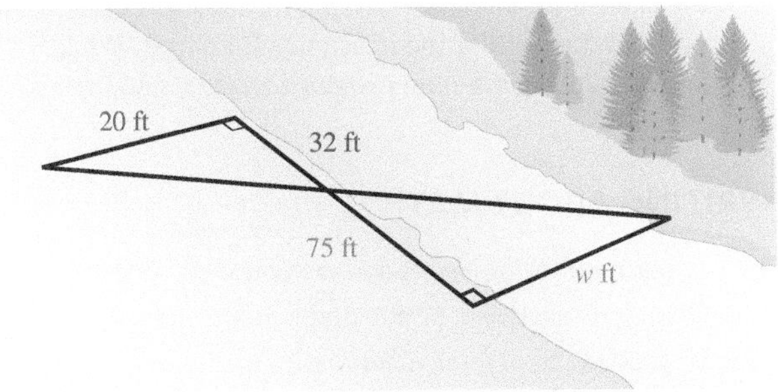

78. Flight paths An airplane ascends 150 feet at a constant rate as it flies a horizontal distance of 1,000 feet. How much altitude will it gain as it flies a horizontal distance of 1 mile? (*Hint*: 5,280 feet = 1 mile.)

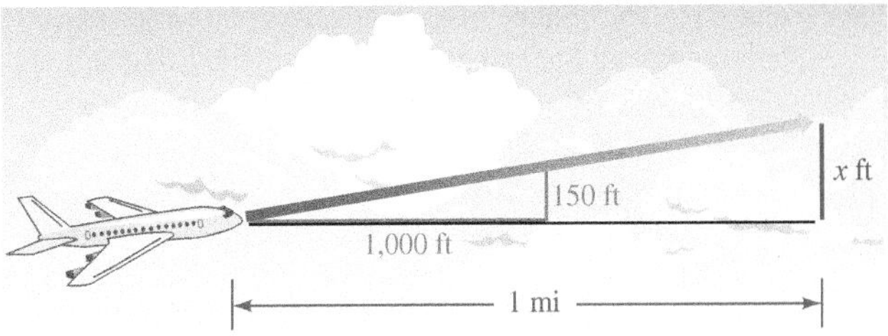

79. Flight paths An airplane descends 1,350 feet at a constant rate as it flies a horizontal distance of 1 mile. How much altitude is lost as it flies a horizontal distance of 5 miles?

80. Ski runs A ski course with $\frac{1}{2}$ mile of horizontal run falls 100 feet in every 300 feet of run. Find the height of the hill.

Solve each variation. SEE EXAMPLE 5. (OBJECTIVE 4)

81. Area of a circle The area of a circle varies directly with the square of its radius. The constant of variation is π. Find the area of a circle with a radius of 8 inches.

82. Falling objects An object in free fall travels a distance s that is directly proportional to the square of the time t. If an object falls 1,024 feet in 8 seconds, how far will it fall in 10 seconds?

83. Finding distance The distance that a car can travel is directly proportional to the number of gallons of gasoline it consumes. If a car can travel 288 miles on 12 gallons of gasoline, how far can it travel on a full tank of 18 gallons?

84. Farming A farmer's harvest in bushels varies directly with the number of acres planted. If 8 acres can produce 144 bushels, how many acres are required to produce 1,152 bushes?

Solve each variation. SEE EXAMPLE 6. (OBJECTIVE 5)

85. Farming The length of time that a given number of bushels of corn will last when feeding cattle varies inversely with the number of animals. If x bushels will feed 25 cows for 10 days, how long will the feed last for 10 cows?

86. Geometry For a fixed area, the length of a rectangle is inversely proportional to its width. A rectangle has a width of 18 feet and a length of 12 feet. If the length is increased to 16 feet, find the width.

87. Gas pressure Under constant temperature, the volume occupied by a gas is inversely proportional to the pressure applied. If the gas occupies a volume of 20 cubic inches under a pressure of 6 pounds per square inch, find the volume when the gas is subjected to a pressure of 10 pounds per square inch.

88. Value of a boat The value of a boat usually varies inversely with its age. If a boat is worth $7,000 when it is 3 years old, how much will it be worth when it is 7 years old?

Solve each variation. SEE EXAMPLE 7. (OBJECTIVE 6)

89. Geometry The area of a rectangle varies jointly with its length and width. If both the length and the width are tripled, by what factor is the area multiplied?

Unless otherwise noted, all content on this page is © Cengage Learning.

90. Geometry The volume of a rectangular solid varies jointly with its length, width, and height. If the length is doubled, the width is tripled, and the height is doubled, by what factor is the volume multiplied?

91. Costs of a trucking company The costs incurred by a trucking company vary jointly with the number of trucks in service and the number of hours each is used. When 4 trucks are used for 6 hours each, the costs are \$1,800. Find the costs of using 10 trucks, each for 12 hours.

92. Storing oil The number of gallons of oil that can be stored in a cylindrical tank varies jointly with the height of the tank and the square of the radius of its base. The constant of proportionality is 23.5. Find the number of gallons that can be stored in the cylindrical tank shown in the illustration.

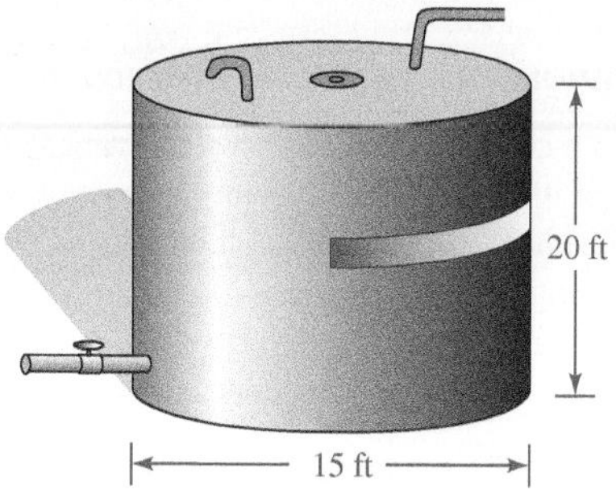

Solve each variation. SEE EXAMPLE 8. (OBJECTIVE 7)

93. Building construction The deflection of a beam is inversely proportional to its width and the cube of its depth. If the deflection of a 4-inch-by-4-inch beam is 1.1 inches, find the deflection of a 2-inch-by-8-inch beam positioned as in the illustration.

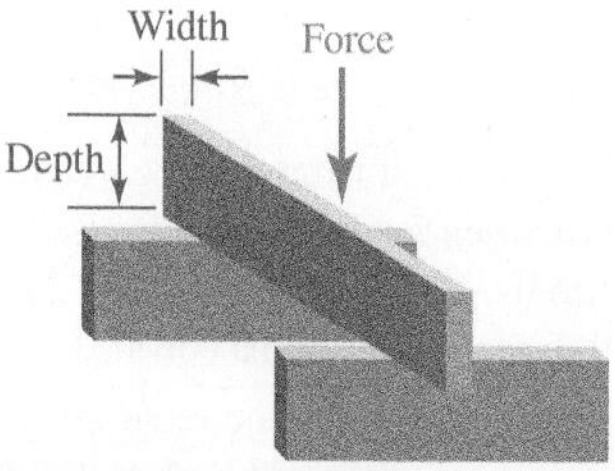

94. Building construction Find the deflection of the beam in Exercise 93 when the beam is positioned as in the illustration.

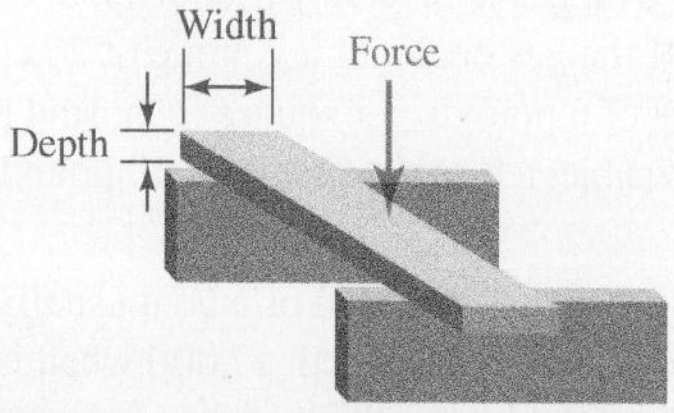

95. Gas pressure The pressure of a certain amount of gas is directly proportional to the temperature (measured on the Kelvin scale) and inversely proportional to the volume. A sample of gas at a pressure of 1 atmosphere occupies a volume of 1 cubic meter at a temperature of 273 Kelvin. When heated, the gas expands to twice its volume, but the pressure remains constant. To what temperature is it heated?

96. Tension A yo-yo, twirled at the end of a string, is kept in its circular path by the tension of the string. The tension T is directly proportional to the square of the speed s and inversely proportional to the radius r of the circle. In the illustration, the tension is 32 pounds when the speed is 8 feet/second and the radius is 6 feet. Find the tension when the speed is 4 feet/second and the radius is 3 feet.

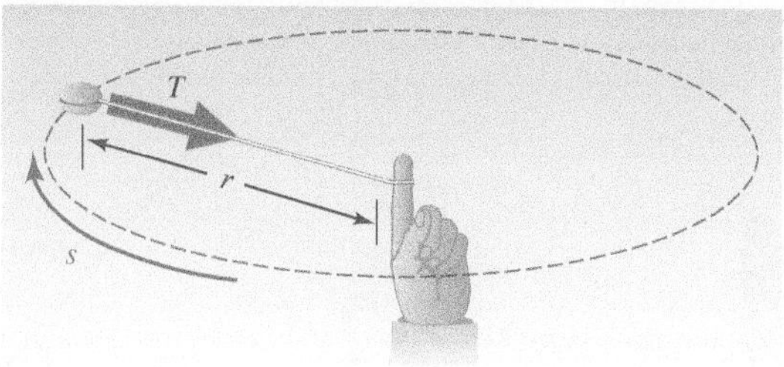

Solve each variation.

97. Organ pipes The frequency of vibration of air in an organ pipe is inversely proportional to the length l of the pipe. If a pipe 2 feet long vibrates 256 times per second, how many times per second will a 6-foot pipe vibrate?

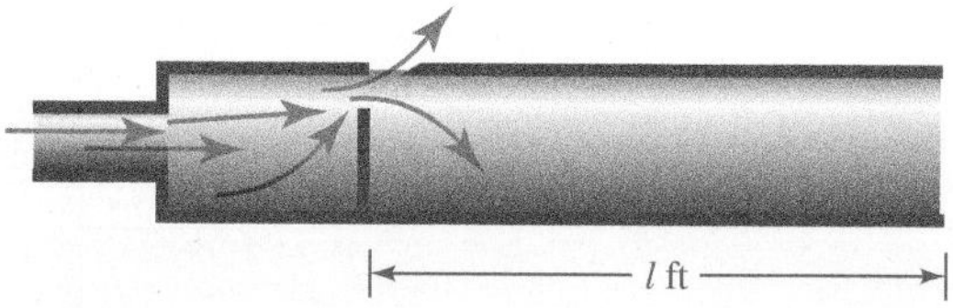

98. Finding the constant of variation A quantity l varies jointly with x and y and inversely with z. If the value of l is 24 when $x = 12$, $y = 4$, and $z = 8$, find k.

99. Electronics The voltage (in volts) measured across a resistor is directly proportional to the current (in amperes) flowing through the resistor. The constant of variation is the **resistance** (in ohms). If 6 volts is measured across a resistor carrying a current of 2 amperes, find the resistance.

100. Electronics The power (in watts) lost in a resistor (in the form of heat) is directly proportional to the square of the current (in amperes) passing through it. The constant of proportionality is the resistance (in ohms). What power is lost in a 5-ohm resistor carrying a 3-ampere current?

WRITING ABOUT MATH

101. Explain the terms *means* and *extremes*.

102. Distinguish between a *ratio* and a *proportion*.

103. Explain the term *joint variation*.

104. Explain why the equation $\frac{y}{x} = k$ indicates that y varies directly with x.

SOMETHING TO THINK ABOUT

105. As temperature increases on the Fahrenheit scale, it also increases on the Celsius scale. Is this direct variation? Explain.

106. As the cost of a purchase (less than \$5) increases, the amount of change received from a five-dollar bill decreases. Is this inverse variation? Explain.

Unless otherwise noted, all content on this page is © Cengage Learning.

Projects

PROJECT 1

If the sides of two similar triangles are in the ratio of 1 to 1, the triangles are said to be congruent. Congruent triangles have the same shape and the same size (area).

a. Draw several triangles with sides of length 1, 1.5, and 2 inches. Are the triangles all congruent? What general rule could you make?

b. Draw several triangles with the dimensions shown in the illustration on the right. Are the triangles all congruent? What general rule could you make?

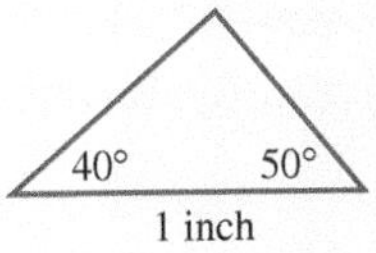

c. Draw several triangles with the dimensions shown in the illustration on the right. Are the triangles all congruent? What general rule could you make?

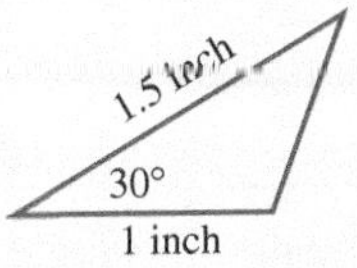

PROJECT 2

If y is equal to a polynomial divided by a polynomial, we call the resulting function a rational function. The simplest of these functions is defined by the equation $f(x) = \frac{1}{x}$. Since the denominator of this fraction cannot be 0, the domain of the function is the set of real numbers, except 0. Since a fraction with a numerator of 1 cannot be 0, the range is also the set of real numbers, except 0.

Construct the graph of this function by making a table of values containing at least eight ordered pairs, plotting the ordered pairs, and joining the points with two curves. Then graph each of the following rational functions and find each one's domain and range.

a. $f(x) = \frac{2}{x}$ **b.** $f(x) = -\frac{1}{x}$

c. $f(x) = -\frac{1}{x+1}$ **d.** $f(x) = \frac{2}{x-1}$

PROJECT 3

Suppose that the cost of telephone service is \$6 per month plus 5¢ per call. If n represents the number of calls made one month, the cost C of phone service that month will be given by $C = 0.05n + 6$. If we divide the total cost C by the number of calls n, we will obtain the average cost per call, which we will denote as $\bar{c}$.

$$(1)\quad \bar{c} = \frac{C}{n} = \frac{0.05n + 6}{n}$$

$\bar{c}$ is the average cost per call, C is the total monthly cost, and n is the number of phone calls made that month.

Use the rational function in Equation 1 to find the average monthly cost per call when

a. 5 calls were made **b.** 25 calls were made.

Assume that a phone company charges \$15 a month and 5¢ per phone call and answer the following questions.

a. Write a function that will give the cost C per month for making n phone calls.

b. Write a function that will give the average cost $\bar{c}$ per call during the month.

c. Find the cost if 45 phone calls were made during the month.

d. Find the average cost per call if 45 calls were made that month.

Reach for Success

EXTENSION OF PREPARING FOR A TEST

Now that you have a few strategies for preparing for a test, consider how to be emotionally and physically ready. *Emotional intelligence* is defined as the ability to identify one's emotions and then manage those emotions. A good start to physical fitness is a healthy lifestyle that includes proper nutrition, exercise, and sleep. If you feel good physically you will more likely feel good emotionally. Feeling emotionally and physically strong should contribute to improved test performance.

On a scale of 1 to 10, how would you rate your emotional readiness for the next test if 10 represents emotionally strong? 1 2 3 4 5 6 7 8 9 10	What strategies might you employ to improve your emotional readiness? 1. ________ 2. ________ 3. ________
On a scale of 1 to 10, how would you rate your physical health if 10 represents healthy and well? 1 2 3 4 5 6 7 8 9 10	What strategies might you employ to improve your physical health? 1. ________ 2. ________ 3. ________

Talk with your professor or talk with your advisor for additional strategies for successful test preparation.

6 Review

SECTION 6.1 Simplifying Rational Expressions

DEFINITIONS AND CONCEPTS	EXAMPLES
Fractions that are the quotient of two integers are *rational numbers*.	Rational numbers: $\frac{2}{13}$ $\frac{15}{28}$
Fractions that are the quotient of two polynomials are *rational expressions*.	Rational expressions: $\frac{x}{x-5}$ $\frac{5x-2}{x^2+x-3}$
$\frac{a}{0}$ is undefined. To find the values of the variable that make a rational expression undefined, we set the denominator equal to 0 and solve. The domain is the set of all values that can be substituted for the variable.	x cannot be 2 in the rational expression $\frac{x+1}{x-2}$ because 2 will cause the denominator to be 0. In interval notation, $(-\infty, 2) \cup (2, \infty)$. $x - 2 = 0$ Set the denominator equal to 0. $x = 2$ Solve. The domain of $f(x) = \frac{x-1}{x-2}$, is $\{x \mid x \in \mathbb{R}, x \neq 2\}$.
If b and c are not 0, then $\frac{a \cdot c}{b \cdot c} = \frac{a}{b}$	$\frac{22xy^3}{33x^2y} = \frac{2 \cdot \overset{1}{\cancel{11}} \cdot \overset{1}{\cancel{x}} \cdot \overset{1}{\cancel{y}} \cdot y \cdot y}{3 \cdot \underset{1}{\cancel{11}} \cdot \underset{1}{\cancel{x}} \cdot x \cdot \underset{1}{\cancel{y}}}$ Factor the numerator and denominator. $= \frac{2y^2}{3x}$ Divide out the common factors of 11, x, and y.
$\frac{a}{1} = a$	$\frac{2x^3 - 2x^2}{x-1} = \frac{2x^2\overset{1}{\cancel{(x-1)}}}{\underset{1}{\cancel{x-1}}}$ Factor the numerator. $= 2x^2$ Divide out the common factor of $x - 1$.
$\frac{a-b}{b-a} = -1$	Because $x - 3$ and $3 - x$ are opposites (additive inverses), their quotient is -1. $\frac{x-3}{3-x} = -1$

REVIEW EXERCISES

For what values of x is the rational expression undefined?

1. $\frac{x+5}{(x+4)(x-2)}$
2. $\frac{x-3}{x^2+x-6}$

Find the domain.

3. $f(x) = \frac{x}{x-7}$
4. $f(x) = \frac{2x-5}{x^2-5x}$

Write each expression in lowest terms.

5. $-\frac{51}{153}$
6. $\frac{105}{45}$
7. $\frac{3x^2}{6x^3}$
8. $\frac{5xy^2}{2x^2y^2}$
9. $\frac{x^2}{x^2+x}$
10. $\frac{x+2}{x^2+2x}$
11. $\frac{12x^2y}{4x^2y^2}$
12. $\frac{8x^2y}{2x(4xy)}$
13. $\frac{7a-5}{5-7a}$
14. $\frac{x^2-x-56}{x^2-5x-24}$
15. $\frac{2x^2-16x}{2x^2-18x+16}$
16. $\frac{a^2+2a+ab+2b}{a^2+2ab+b^2}$

SECTION 6.2 Multiplying and Dividing Rational Expressions

DEFINITIONS AND CONCEPTS	EXAMPLES
$\frac{a}{b} \cdot \frac{c}{d} = \frac{a \cdot c}{b \cdot d}$ $(b, d \neq 0)$	$\frac{x+1}{x+4} \cdot \frac{x^2-16}{x^2-x-2} = \frac{(x+1)(x^2-16)}{(x+4)(x^2-x-2)}$ Multiply the numerators and multiply the denominators. $= \frac{\overset{1}{\cancel{(x+1)}}\overset{1}{\cancel{(x+4)}}(x-4)}{\underset{1}{\cancel{(x+4)}}\underset{1}{\cancel{(x+1)}}(x-2)}$ Factor and divide out the common factors $= \frac{x-4}{x-2}$

$\frac{a}{b} \div \frac{c}{d} = \frac{a}{b} \cdot \frac{d}{c} \quad (b, c, d \neq 0)$	$\frac{6x^2}{2x+8} \div \frac{3x}{x^2-2x-24}$	
	$= \frac{6x^2}{2x+8} \cdot \frac{x^2-2x-24}{3x}$	Invert the denominator and multiply.
	$= \frac{6x^2}{2(x+4)} \cdot \frac{(x-6)(x+4)}{3x}$	Factor.
	$= \frac{\overset{1}{\cancel{3}}\,\overset{x}{\cancel{6x^2}}}{\cancel{2}\cancel{(x+4)}_{1}} \cdot \frac{(x-6)\overset{1}{\cancel{(x+4)}}}{\cancel{3x}}$	Divide out the common factors.
	$= \frac{x(x-6)}{1}$	
	$= x(x-6)$	Denominators of 1 need not be written.

REVIEW EXERCISES

Perform each multiplication and simplify.

17. $\frac{5x^2y}{3x} \cdot \frac{6x}{15y^2}$

18. $\frac{3x}{x^2-x} \cdot \frac{2x-2}{x^2}$

19. $\frac{x^2+3x+2}{x^2+2x} \cdot \frac{x}{x+1}$

20. $\frac{x^2+x}{3x-15} \cdot \frac{6x-30}{x^2+2x+1}$

Perform each division and simplify.

21. $\frac{9xy}{14x^2} \div \frac{18y^2}{21xy}$

22. $\frac{x^2+5x}{x^2+4x-5} \div \frac{x^2}{x-1}$

23. $\frac{x^2-x-6}{2x-1} \div \frac{x^2-2x-3}{2x^2+x-1}$

24. $\frac{x^2-3x}{x^2-x-6} \div \frac{x^2-x}{x^2+x-2}$

25. $\frac{x^2+4x+4}{x^2+x-6}\left(\frac{x-2}{x-1} \div \frac{x+2}{x^2+2x-3}\right)$

SECTION 6.3 Adding and Subtracting Rational Expressions

DEFINITIONS AND CONCEPTS	EXAMPLES	
$\frac{a}{d} + \frac{b}{d} = \frac{a+b}{d} \quad (d \neq 0)$	$\frac{4x+3}{2x} + \frac{x+1}{2x} = \frac{4x+3+x+1}{2x}$	Add the numerators and keep the common denominator.
	$= \frac{5x+4}{2x}$	Combine like terms.
$\frac{a}{d} - \frac{b}{d} = \frac{a-b}{d} \quad (d \neq 0)$	$\frac{4x+3}{2x} - \frac{x+1}{2x} = \frac{4x+3-(x+1)}{2x}$	Subtract the numerators and keep the common denominator.
	$= \frac{4x+3-x-1}{2x}$	Use the distributive property to remove the parentheses.
	$= \frac{3x+2}{2x}$	Combine like terms.

The least common denominator (LCD) for a set of fractions is the smallest number that each denominator will divide exactly.

The LCD of 6 and 9 is 18.
The LCD of 10 and 15 is 30.

Finding the Least Common Denominator (LCD):

1. List the different denominators that appear in the rational expressions.
2. Completely factor each denominator.
3. Form a product using each different factor obtained in Step 2. Use each different factor the *greatest* number of times it appears in any one factorization. The product formed by multiplying these factors is the LCD.

Find the LCD of $\frac{2x}{3y}, \frac{4x}{15y}$, and $\frac{5x}{18y}$.

$$3y = 3 \cdot y$$
$$15y = 3 \cdot 5 \cdot y$$
$$18y = 2 \cdot 3 \cdot 3 \cdot y$$

Form a product with factors of 2, 3 and y. We use 2 one time, because it appears only once as a factor of 18. We use 3 two times because it appears twice as a factor of 18. We use 5 one time, because it appears only once as a factor of 15. We use y once because it only occurs once in each factor of $3y$, $15y$, and $18y$.

$$\text{LCD} = 2 \cdot 3 \cdot 3 \cdot 5 \cdot y = 90y$$

To add or subtract rational expressions with unlike denominators, first find the LCD of the expressions. Then express each fraction in equivalent form with this LCD. Finally, add or subtract the expressions. Simplify, if possible.

To perform the subtraction $\frac{x+1}{x^2-9} - \frac{2}{x+3}$, we factor $x^2 - 9$ and find that the LCD is $(x+3)(x-3)$. Then, we proceed as follows:

$$\frac{x+1}{x^2-9} - \frac{2}{x+3}$$

$$= \frac{x+1}{(x+3)(x-3)} - \frac{2(x-3)}{(x+3)(x-3)}$$ Build the second fraction and factor the denominator of the first fraction.

$$= \frac{x+1-2(x-3)}{(x+3)(x-3)}$$ Subtract the numerators and keep the common denominator.

$$= \frac{x+1-2x+6}{(x+3)(x-3)}$$ Distribute.

$$= \frac{-x+7}{(x+3)(x-3)}$$ Combine like terms.

REVIEW EXERCISES

Perform each operation. Simplify all answers.

26. $\frac{5x}{x+3} + \frac{x-8}{x+3}$

27. $\frac{y}{x-y} - \frac{x}{x-y}$

28. $\frac{x}{x-1} + \frac{1}{x}$

29. $\frac{1}{7} - \frac{1}{x}$

30. $\frac{3}{x+1} - \frac{2}{x}$

31. $\frac{x+2}{2x} - \frac{2-x}{x^2}$

32. $\frac{x}{x+2} + \frac{3}{x} - \frac{4}{x^2+2x}$

33. $\frac{2}{x-1} - \frac{3}{x+1} + \frac{x-5}{x^2-1}$

SECTION 6.4 Simplifying Complex Fractions

DEFINITIONS AND CONCEPTS

To simplify a complex fraction, use either of these methods:

Method 1

Write the numerator and denominator of the complex fraction as single fractions, do the division of the fractions, and simplify.

EXAMPLES

Method 1

$$\frac{\frac{2x}{x+1} + \frac{3}{x}}{\frac{1}{x} + \frac{2}{x+1}}$$

$$= \frac{\frac{2x \cdot x}{x(x+1)} + \frac{3(x+1)}{x(x+1)}}{\frac{1(x+1)}{x(x+1)} + \frac{2x}{x(x+1)}}$$ Build each fraction.

$$= \frac{\frac{2x^2}{x(x+1)} + \frac{3x+3}{x(x+1)}}{\frac{x+1}{x(x+1)} + \frac{2x}{x(x+1)}}$$ Simplify wherever possible.

$$= \frac{2x^2+3x+3}{x(x+1)} \div \frac{x+1+2x}{x(x+1)}$$ Add the fractions and write the resulting complex fraction as an equivalent division problem.

$$= \frac{2x^2+3x+3}{x(x+1)} \div \frac{3x+1}{x(x+1)}$$ Simplify the numerators.

$$= \frac{2x^2+3x+3}{x(x+1)} \cdot \frac{x(x+1)}{3x+1}$$ Invert the divisor and multiply.

$$= \frac{2x^2+3x+3}{3x+1}$$ Multiply the fractions and simplify.

Method 2

Multiply both the numerator and the denominator of the complex fraction by the LCD of the fractions that appear in the numerator and the denominator, then simplify.

Method 2

$$\frac{\frac{2x}{x+1}+\frac{3}{x}}{\frac{1}{x}+\frac{2}{x+1}}$$

$$=\frac{x(x+1)\left(\frac{2x}{x+1}+\frac{3}{x}\right)}{x(x+1)\left(\frac{1}{x}+\frac{2}{x+1}\right)}$$ Multiply the numerator and the denominator by the LCD of $x(x + 1)$.

$$=\frac{x(x+1)\left(\frac{2x}{x+1}\right)+x(x+1)\left(\frac{3}{x}\right)}{x(x+1)\left(\frac{1}{x}\right)+x(x+1)\left(\frac{2}{x+1}\right)}$$ Distribute.

$$=\frac{x\cdot 2x+3(x+1)}{x+1+2x}$$ Do the multiplication.

$$=\frac{2x^2+3x+3}{3x+1}$$ Simplify.

REVIEW EXERCISES

Simplify each complex fraction.

34. $\dfrac{\frac{9}{4}}{\frac{4}{9}}$

35. $\dfrac{\frac{3}{2}+1}{\frac{2}{3}+1}$

36. $\dfrac{\frac{1}{x}+1}{\frac{1}{x}-1}$

37. $\dfrac{2+\frac{7}{x}}{3-\frac{1}{x^2}}$

38. $\dfrac{\frac{2}{x-1}+\frac{x-1}{x+1}}{\frac{1}{x^2-1}}$

39. $\dfrac{\frac{a}{b}+c}{\frac{b}{a}+c}$

SECTION 6.5 Solving Equations That Contain Rational Expressions

DEFINITIONS AND CONCEPTS

1. Find any restrictions on the variable. Remember that the denominator of a fraction cannot be 0.
2. Multiply both sides of the equation by the LCD of the rational expressions appearing in the equation to clear the equation of fractions.
3. Solve the resulting equation.
4. If an apparent solution of an equation is a restricted value, that value must be excluded. Check all solutions for extraneous roots.

EXAMPLES

Solve: $\frac{1}{x+1}+\frac{2}{x}=\frac{x}{x^2+x}$

$$x(x+1)\left(\frac{1}{x+1}+\frac{2}{x}\right)=x(x+1)\left[\frac{x}{x(x+1)}\right]$$ Multiply both sides by $x(x + 1)$, the LCD.

$$x(x+1)\left(\frac{1}{x+1}\right)+x(x+1)\left(\frac{2}{x}\right)=x(x+1)\left[\frac{x}{x(x+1)}\right]$$ Use the distributive property.

$$x+2(x+1)=x$$ Multiply each term by $x(x + 1)$.

$$x+2x+2=x$$ Distribute.

$$3x+2=x$$ Combine like terms.

$$2x=-2$$ Subtract x and 2 from both sides.

$$x=-1$$ Divide both sides by 2.

When -1 is substituted for x, the denominators become 0. Since division by 0 is undefined, $x = -1$ is extraneous. Since there is no solution, the solution set is $\varnothing$.

REVIEW EXERCISES

For Exercises 40–45 solve each equation and check all answers.

40. $\frac{4}{x} = \frac{6}{x-3}$

41. $\frac{7}{x+3} = \frac{5}{x+1}$

42. $\frac{2}{3x} + \frac{1}{x} = \frac{5}{9}$

43. $\frac{2x}{x+4} = \frac{3}{x-1}$

44. $\frac{2}{x-1} + \frac{3}{x+4} = \frac{-5}{x^2+3x-4}$

45. $\frac{4}{x+2} - \frac{3}{x+3} = \frac{6}{x^2+5x+6}$

46. Solve for r_1: $\frac{1}{r} = \frac{1}{r_1} + \frac{1}{r_2}$

47. The efficiency E of a Carnot engine is given by the formula

$$E = 1 - \frac{T_2}{T_1}$$

Solve the formula for T_1.

48. Nuclear medicine Radioactive tracers are used for diagnostic work in nuclear medicine. The effective half-life H of a radioactive material in a biological organism is given by the formula

$$H = \frac{RB}{R+B}$$

where R is the radioactive half-life and B is the biological half-life of the tracer. Solve the formula for R.

SECTION 6.6 Solving Applications Whose Models Contain Rational Expressions

DEFINITIONS AND CONCEPTS	EXAMPLES
1. Analyze the problem. **2.** Form an equation. **3.** Solve the equation. **4.** State the conclusion. **5.** Check the result.	An inlet pipe can fill a pond in 4 days, and a second inlet pipe can fill the same pond in 3 days. If both pipes are used, how long will it take to fill the pond? Let x represent the number of days it takes to fill the pond.

What the first inlet pipe can do in 1 day	plus	what the second inlet pipe can do in 1 day	equals	what they can do together in 1 day.
$\frac{1}{4}$	$+$	$\frac{1}{3}$	$=$	$\frac{1}{x}$

To solve the equation, we proceed as follows:

$$\frac{1}{4} + \frac{1}{3} = \frac{1}{x}$$

$$12x\left(\frac{1}{4} + \frac{1}{3}\right) = 12x\left(\frac{1}{x}\right)$$ Multiply both sides by $12x$.

$$3x + 4x = 12$$ Use the distributive property to remove parentheses and simplify.

$$7x = 12$$ Combine like terms.

$$x = \frac{12}{7}$$ Divide both sides by 7.

It will take $\frac{12}{7}$ or $1\frac{5}{7}$ days for both inlet pipes to fill the pond.

REVIEW EXERCISES

49. Pumping a basement If one pump can empty a flooded basement in 18 hours and a second pump can empty the basement in 20 hours, how long will it take to empty the basement when both pumps are used?

50. Painting houses If a homeowner can paint a house in 12 days and a professional painter can paint it in 8 days, how long will it take if they work together?

51. Jogging A jogger can bicycle 30 miles in the same time as he can jog 10 miles. If he can ride 10 mph faster than he can jog, how fast can he jog?

52. Wind speed A plane can fly 400 miles downwind in the same amount of time as it can travel 320 miles upwind. If the plane can fly at 360 mph in still air, find the velocity of the wind.

SECTION 6.7 Proportion and Variation

DEFINITIONS AND CONCEPTS	EXAMPLES

Solving a proportion:

In a proportion, the product of the extremes is equal to the product of the means.

To solve $\frac{x-1}{x+1} = \frac{2}{3x}$, proceed as follows:

$3x(x-1) = 2(x+1)$ — The product of the extremes is equal to the product of the means.

$3x^2 - 3x = 2x + 2$ — Use the distributive property to remove parentheses.

$3x^2 - 5x - 2 = 0$ — Subtract $2x$ and 2 from both sides.

$(3x+1)(x-2) = 0$ — Factor.

$3x + 1 = 0$ or $x - 2 = 0$ — Set each factor equal to 0.

$x = -\frac{1}{3}$ | $x = 2$ — Solve each linear equation.

The solutions are $-\frac{1}{3}, 2$.

Using similar triangles to find the length of a missing side:

If two angles of one triangle have the same measure as two angles of a second triangle, the triangles are similar.

The measure of corresponding sides of similar triangles are in proportion.

The two triangles below are similar. Find the value of x.

$\frac{x}{5} = \frac{8}{20}$ — Corresponding sides are in proportion.

$20x = 40$ — The product of the extremes is equal to the product of the means.

$x = 2$ — Solve for x.

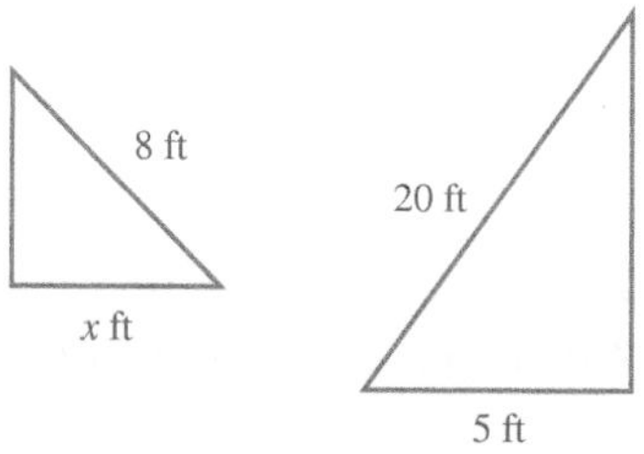

The length of the unknown side is 2 ft.

Direct variation:

$y = kx$ (k is a constant)

Express each sentence as a formula:
The distance, d, a car travels is directly proportional to the time, t, it has been traveling.

$d = kt$

Inverse variation:

$y = \frac{k}{x}$ (k is a constant)

The temperature, T, of the coffee in the mug varies inversely to the time, t, it has been sitting on the counter.

$T = \frac{k}{t}$

Joint variation:

$y = kxz$ (k is a constant)

The interest, I, on the money in a bank account is jointly proportional to the principle, P, and the interest rate, r.

$I = kPr$

Combined variation:

$y = \frac{kx}{z}$ (k is a constant)

The pressure, P, of the gas varies directly as the temperature, T, and inversely as the volume, V.

$P = \frac{kT}{V}$

REVIEW EXERCISES

Solve each proportion.

53. $\frac{x+1}{8} = \frac{4x-2}{24}$

54. $\frac{1}{x+6} = \frac{x+10}{12}$

55. Find the height of a tree if it casts a 44-foot shadow when a 4-foot tree casts a $2\frac{1}{2}$-foot shadow.

Unless otherwise noted, all content on this page is © Cengage Learning.

56. Assume that x varies directly with y. If $x = 18$ when $y = 3$, find the value of x when $y = 9$.

57. Assume that x varies inversely with y. If $x = 24$ when $y = 3$, find the value of y when $x = 12$.

58. Assume that x varies jointly with y and z. Find the constant of variation if $x = 24$ when $y = 3$ and $z = 4$.

59. Assume that x varies directly with t and inversely with y. Find the constant of variation if $x = 2$ when $t = 8$ and $y = 64$.

60. **Taxes** The property taxes in a city vary directly with the assessed valuation. If a tax of $1,575 is levied on a house assessed at $90,000, find the tax on a building assessed at $312,000.

6 Test

Assume no division by 0.

1. Simplify: $\dfrac{27x^3y^2}{45xy^3}$

2. Simplify: $\dfrac{2x^2 - x - 3}{4x^2 - 9}$

3. Simplify: $\dfrac{3(x+2) - 3}{2x - 4 - (x - 5)}$

4. Multiply and simplify: $\dfrac{12x^2y}{15xyz} \cdot \dfrac{25y^2z}{16xt}$

5. Multiply and simplify: $\dfrac{x^2 + 3x + 2}{3x + 9} \cdot \dfrac{x + 3}{x^2 - 4}$

6. Divide and simplify: $\dfrac{7ab^2}{24ac} \div \dfrac{21a^2b^3}{40abc^2}$

7. Divide and simplify: $\dfrac{x^2 - x}{3x^2 + 6x} \div \dfrac{3x - 3}{3x^3 + 6x^2}$

8. Simplify: $\dfrac{x^2 + xy}{x - y} \cdot \dfrac{x^2 - y^2}{x^2 - 2x} \div \dfrac{x^2 + 2xy + y^2}{x^2 - 4}$

9. Add: $\dfrac{6x + 5}{x - 2} + \dfrac{3x - 7}{x - 2}$

10. Subtract: $\dfrac{3y + 7}{2y + 3} - \dfrac{3(y - 2)}{2y + 3}$

11. Add: $\dfrac{x + 1}{x} + \dfrac{x - 1}{x + 1}$

12. Subtract: $\dfrac{5x}{x - 2} - 3$

13. Simplify: $\dfrac{\frac{9x^3}{xy^2}}{\frac{6y^2}{x^3y^4}}$

14. Simplify: $\dfrac{1 + \frac{y}{x}}{\frac{y}{x} - 1}$

15. Solve for x: $\dfrac{x}{10} - \dfrac{1}{2} = \dfrac{x}{5}$

16. Solve for x: $\dfrac{1}{x + 2} + \dfrac{1}{x - 2} = \dfrac{4}{x^2 - 4}$

17. Solve for x: $\dfrac{7}{x + 4} - \dfrac{1}{2} = \dfrac{3}{x + 4}$

18. Solve for B: $H = \dfrac{RB}{R + B}$

19. **Cleaning highways** One highway worker could pick up all the trash on a strip of highway in 7 hours, and his helper could pick up the trash in 9 hours. How long will it take them if they work together?

20. **Boating** A boat can motor 28 miles downstream in the same amount of time as it can motor 18 miles upstream. Find the speed of the current if the boat can motor at 23 mph in still water.

21. **Flight path** A plane drops 575 feet as it flies a horizontal distance of $\frac{1}{2}$ mile. How much altitude will it lose as it flies a horizontal distance of 7 miles?

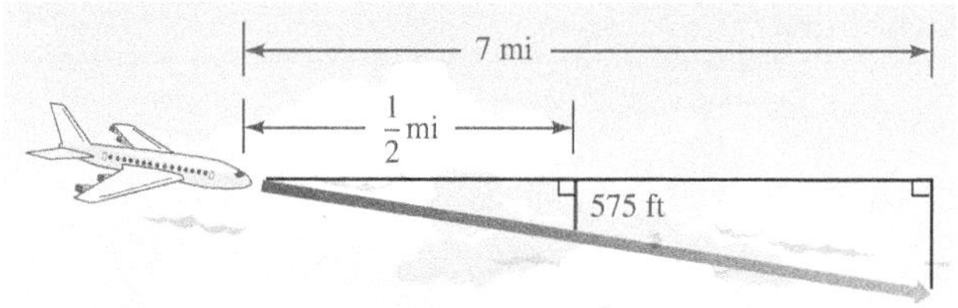

22. Solve for y: $\dfrac{y}{y - 1} = \dfrac{y - 2}{y}$

23. Find the height of a tree that casts a shadow of 12 feet when a vertical yardstick casts a shadow of 2 feet.

24. Solve the proportion: $\dfrac{3}{x - 2} = \dfrac{x + 3}{2x}$

25. V varies inversely with t. If $V = 55$ when $t = 20$, find the value of t when $V = 75$.

26. Given $f(x) = \frac{x + 6}{x - 3}$,

a. find all values of the variable for which $f(x)$ is undefined.

b. write the domain of $f(x)$ in interval notation.

Unless otherwise noted, all content on this page is © Cengage Learning.

Cumulative Review

1. x^3x^6
2. $(x^2)^5$
3. $\frac{x^5}{x^2}$
4. $(6x^4)^0$
5. $(3x^2 - 2x) + (6x^3 - 3x^2 - 1)$
6. $(4x^3 - 2x) - (2x^3 - 2x^2 - 3x + 1)$
7. $3(5x^2 - 4x + 3) + 2(-x^2 + 2x - 4)$
8. $4(3x^2 - 4x - 1) - 2(-2x^2 + 4x - 3)$

Perform each operation. Assume no division by 0.

9. $(2x^4y^3)(-7x^5y)$
10. $-5x^2(7x^3 - 2x^2 - 2)$
11. $(5x + 2)(4x + 1)$
12. $(5x - 4y)(3x + 2y)$
13. $x + 3\overline{)x^2 + 7x + 12}$
14. $2x - 3\overline{)2x^3 - x^2 - x - 3}$

Factor each expression.

15. $4xy^2 - 12x^2y^3$
16. $3(a + b) + x(a + b)$
17. $2a + 2b + ab + b^2$
18. $25p^4 - 16q^2$
19. $x^2 - 5x - 14$
20. $x^2 - xy - 6y^2$
21. $6a^2 - 7a - 20$
22. $8m^2 - 10mn - 3n^2$
23. $p^3 - 27q^3$
24. $8r^3 + 64s^3$

Solve each equation.

25. $\frac{4}{5}x + 6 = 18$
26. $5 - \frac{x + 2}{3} = 7 - x$
27. $6x^2 - x - 2 = 0$
28. $5x^2 = 10x$
29. $x^2 + 6x + 5 = 0$
30. $2y^2 + 5y - 12 = 0$

Solve each inequality and graph the solution set.

31. $4x - 7 > 1$
32. $7x - 9 < 5$
33. $-2 < -x + 3 < 5$
34. $0 \le \frac{4 - x}{3} \le 2$

Graph each equation.

35. $4x - 3y = 12$

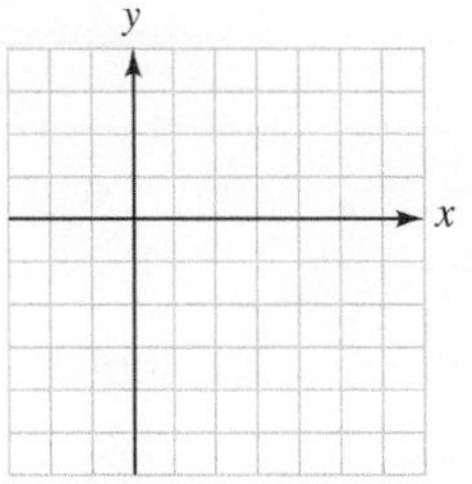

36. $3x + 4y = 4y + 12$

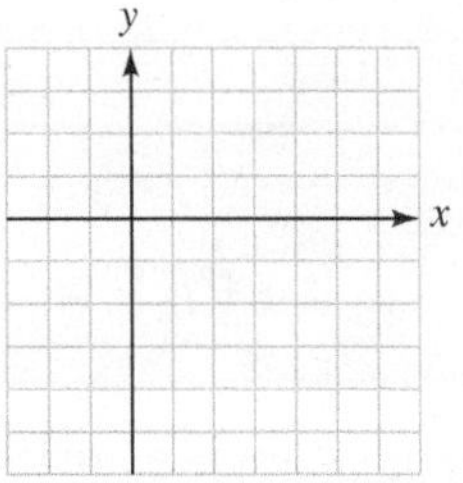

If $f(x) = 2x^2 - 3$, find each value.

37. $f(-1)$
38. $f(3)$
39. $f(-2)$
40. $f(2x)$

Simplify each fraction.

41. $\frac{x^2 + 3x + 2}{x^2 - 4}$
42. $\frac{x^2 + 2x - 15}{x^2 + 3x - 10}$

Perform the operation(s) and simplify when possible. Assume no division by 0.

43. $\frac{x^2 + x - 6}{5x - 5} \cdot \frac{5x - 10}{x + 3}$
44. $\frac{p^2 - p - 6}{3p - 9} \div \frac{p^2 + 6p + 9}{p^2 - 9}$
45. $\frac{3x}{x + 2} + \frac{5x}{x + 2} - \frac{7x - 2}{x + 2}$
46. $\frac{x - 1}{x + 1} + \frac{x + 1}{x - 1}$
47. $\frac{a + 1}{2a + 4} - \frac{a^2}{2a^2 - 8}$
48. $\frac{\frac{1}{x} + \frac{1}{y}}{\frac{1}{x} - \frac{1}{y}}$

Divide using synthetic division.

49. $(x^2 + 9x + 20) \div (x + 5)$
50. $(2x^2 + 4x - x^3 + 3) \div (x - 1)$

Unless otherwise noted, all content on this page is © Cengage Learning.

Transitioning to Intermediate Algebra

7

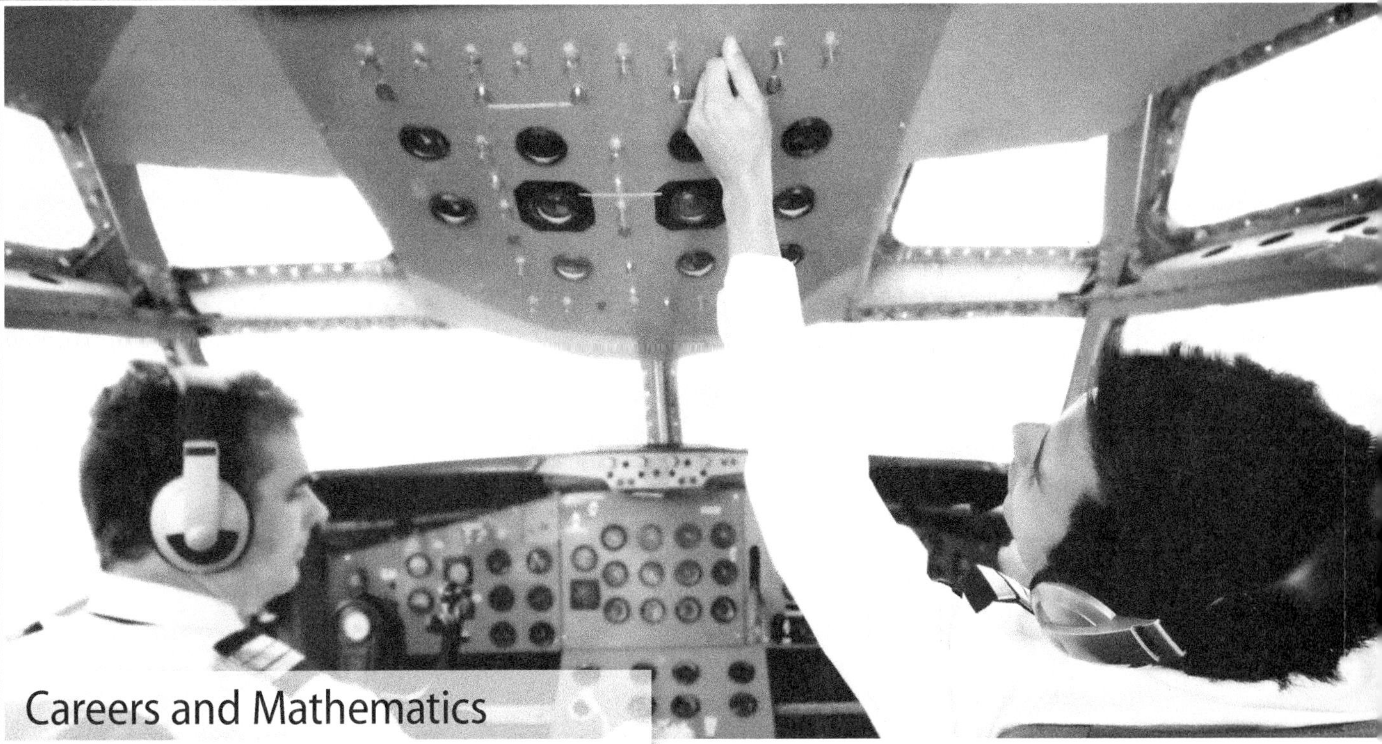

© Andresr/Shutterstock.com

Careers and Mathematics

AIRLINE AND COMMERCIAL PILOTS

Pilots are highly trained professionals who fly either airplanes or helicopters. Airline pilots, copilots, and flight engineers are commercial pilots who work as flight instructors at local airports or for large businesses that fly company cargo. Some fly executives in their own airplanes. Every pilot who is paid to transport passengers or cargo must have a commercial pilot's license with an instrument rating issued by the FAA.

Job Outlook:
Job opportunities for pilots are projected to increase 11%, about as fast as average for all occupations through 2020.
Annual Earnings:
$20,000–$166,400
For More Information:
http://www.bls.gov/oco/ocos107.htm
For a Sample Application:
See Project 2 at the end of this chapter.

REACH FOR SUCCESS

7.1 Review of Solving Linear Equations and Inequalities in One Variable

7.2 Review of Graphing Linear Equations, Finding the Slopes of Lines, and Writing Equations of Lines

7.3 Review of Functions

7.4 Review of Factoring and Solving Quadratic Equations

7.5 Review of Rational Expressions and Solving Rational Equations

7.6 Solving Equations in One Variable Containing an Absolute Value Expression

7.7 Solving Inequalities in One Variable Containing an Absolute Value Expression

■ ***Projects***

REACH FOR SUCCESS EXTENSION

CHAPTER REVIEW

CHAPTER TEST

In this chapter

In this chapter, we will review many concepts covered in the first six chapters to prepare you for the intermediate algebra level. If you have difficulty with any topic in this chapter, you might review the sections in the text in which we first discussed that topic.

Reach for Success Accessing Academic Resources

When a baseball pitcher starts to struggle with a certain type of pitch, he may fall behind in the ball/strike count. Often, a suggestion from his manager or pitching coach can give him the guidance he needs to correct it.

If you do not feel you are understanding as well as you should, or would like, then perhaps it is time to seek help. Using the varied resources available to you can improve your chances of a successful semester.

© Dennis Ku/Shutterstock.com

Obtain a hard copy or view an online version of your college catalog. Where did you find it?

____

Search for Academic Resources. List the resources that could help with your mathematics class.

____

____

Does your college have a mathematics lab or tutoring center? ______

If so, where is it located? ______________

What are the hours of operation?
M: ________ T: ________ W: ________
Th: ________ F: ________ S/Su: ________

Write down at least two days and times that you could go to the center to get help for an hour or more.

____

If not, what other resources are available? ______________

In addition to, or in place of, on-campus services, some colleges offer online tutoring.

Does your college offer online tutoring services for mathematics? ________

If so, what is the process to get enrolled? ______________

____

If not, have you tried the software that accompanies your textbook? ________

Colleges usually offer a wide variety of resources to help students succeed.

Are there other services at your college that might support your performance in this class?

If so, what are they? ______________

Does your college offer a College Success or Study Skills course? ______________

If so, would you consider enrolling? ________ Explain.

____

A Successful Study Strategy . . .

Know where to go to get help with your mathematics. In addition to those listed above, remember that your instructor may be your best resource.

At the end of the chapter you will find an additional exercise to guide you in planning for a successful semester.

Section 7.1 Review of Solving Linear Equations and Inequalities in One Variable

Objectives

1. Solve a linear equation in one variable.
2. Solve a linear equation in one variable that results in an identity or a contradiction.
3. Solve a formula for a specified variable.
4. Solve an application using a linear equation in one variable.
5. Solve a linear inequality in one variable.
6. Solve a compound inequality.

Vocabulary

equation	interval	absolute inequalities
solutions	open interval	conditional inequalities
conditional equation	interval notation	trichometry property
identity	closed interval	transitive property
contradiction	half-open interval	linear inequality
empty set	unbounded interval	

Getting Ready

State whether the given value of x will make each statement true.

1. $x + 3 = 5;\ 2$ **2.** $x - 5 = 3;\ 8$

3. $\frac{3x}{5} = 6;\ 5$ **4.** $2x + 3 = x - 4;\ -7$

In this section, we will review how to solve equations and inequalities.

1 Solve a linear equation in one variable.

Recall that an **equation** is a statement indicating that two mathematical expressions are equal. The set of numbers that satisfy an equation is called its *solution set,* and the elements in the solution set are called **solutions** or *roots* of the equation. Finding the solution set of an equation is called *solving the equation.*

To solve a linear equation, we will use the following two properties of equality to replace the equation with simpler equivalent equations that have the same solution set. We continue this process until we have isolated the variable on one side of the equal sign.

1. Addition Property of Equality: If any quantity is added to (or subtracted from) both sides of an equation, a new equation is formed that is equivalent to the original equation.
2. Multiplication Property of Equality: If both sides of an equation are multiplied (or divided) by the same nonzero constant, a new equation is formed that is equivalent to the original equation.

EXAMPLE 1 Solve: $3(2x - 1) = 2x + 9$

Solution We can use the distributive property to remove parentheses and then isolate x on the left side of the equation.

$$3(2x - 1) = 2x + 9$$

$$6x - 3 = 2x + 9$$ Use the distributive property to remove parentheses.

$$6x - 3 - \mathbf{2x} = 2x + 9 - \mathbf{2x}$$ To eliminate $2x$ from the right side, subtract $2x$ from both sides.

$$4x - 3 = 9$$ Combine like terms.

$$4x - 3 + \mathbf{3} = 9 + \mathbf{3}$$ To undo the subtraction by 3, add 3 to both sides.

$$4x = 12$$ Combine like terms.

$$x = 3$$ To undo the multiplication by 4, divide both sides by 4.

Check: We substitute 3 for x in the original equation to determine whether it satisfies the equation.

$$3(2x - 1) = 2x + 9$$

$$3(2 \cdot \mathbf{3} - 1) \stackrel{?}{=} 2 \cdot \mathbf{3} + 9$$

$$3(5) \stackrel{?}{=} 6 + 9$$ On the left side, perform the operations in parentheses first.

$$15 = 15$$

Since 3 satisfies the original equation, it is a solution. The solution set of the equation is {3}.

SELF CHECK 1 Solve: $2(3x - 2) = 3x - 13$

To solve more complicated linear equations, we will follow these steps.

SOLVING LINEAR EQUATIONS

1. If an equation contains fractions, multiply both sides of the equation by their least common denominator (LCD) to clear the fractions.
2. Use the distributive property to remove all grouping symbols and combine like terms.
3. Use the addition property to move all variables to one side of the equation (isolate the variable term). Combine like terms, if necessary.
4. Use the multiplication property to make the coefficient of the variable equal to 1 (isolate the variable).
5. Check the result by replacing the variable in the original equation with the possible solution and verifying that the number satisfies the equation.

EXAMPLE 2 Solve: $\frac{5}{3}(x - 3) = \frac{3}{2}(x - 2) + 2$

Solution **Step 1:** Since 6 is the smallest number that can be divided by both 2 and 3, we multiply both sides of the equation by 6, the LCD, to clear the fractions:

$$\frac{5}{3}(x - 3) = \frac{3}{2}(x - 2) + 2$$

$$\mathbf{6}\left[\frac{5}{3}(x - 3)\right] = \mathbf{6}\left[\frac{3}{2}(x - 2) + 2\right]$$ Multiply both sides of the equation by 6, the LCD.

$$10(x - 3) = 9(x - 2) + 12$$ $6 \cdot \frac{5}{3} = 10, 6 \cdot \frac{3}{2} = 9$, and $6 \cdot 2 = 12$

Step 2: We use the distributive property to remove parentheses and then combine like terms.

$10x - 30 = 9x - 18 + 12$ Use the distributive property to remove parentheses.

$10x - 30 = 9x - 6$ Combine like terms.

Step 3: We isolate the variable term by subtracting $9x$ and adding 30 to both sides.

$10x - 30 \mathbf{- 9x + 30} = 9x - 6 \mathbf{- 9x + 30}$

$x = 24$ Combine like terms.

Step 4: Since 1 is the coefficient of x in the above equation, Step 4 is unnecessary.

Step 5: We check by substituting 24 for x in the original equation and simplifying:

$$\frac{5}{3}(x - 3) = \frac{3}{2}(x - 2) + 2$$

$$\frac{5}{3}(\mathbf{24} - 3) \stackrel{?}{=} \frac{3}{2}(\mathbf{24} - 2) + 2$$

$$\frac{5}{3}(21) \stackrel{?}{=} \frac{3}{2}(22) + 2$$

$$5(7) \stackrel{?}{=} 33 + 2$$

$$35 = 35$$

Since 24 satisfies the equation, it is a solution. The solution set of the equation is {24}.

SELF CHECK 2 Solve: $\frac{2}{3}(x - 2) = \frac{5}{2}(x - 1) + 3$

2 Solve a linear equation in one variable that results in an identity or a contradiction.

The equations discussed so far have been **conditional equations**. For these equations, some number x is a solution and all others are not. An **identity** is an equation that is satisfied by every number x for which both sides of the equation are defined.

EXAMPLE 3 Solve: $2(x - 1) + 4 = 4(1 + x) - (2x + 2)$

Solution

$2(x - 1) + 4 = 4(1 + x) - (2x + 2)$

$2x - 2 + 4 = 4 + 4x - 2x - 2$ Use the distributive property to remove parentheses.

$2x + 2 = 2x + 2$ Combine like terms.

$2 = 2$ Subtract $2x$ from both sides.

Since $2 = 2$, the equation is true for every number x. Since every number x satisfies this equation, it is an identity. The solution set of the equation is the set of real numbers denoted by $\mathbb{R}$.

SELF CHECK 3 Solve: $3(x + 1) - (20 + x) = 5(x - 1) - 3(x + 4)$

A **contradiction** is an equation that has no solution.

EXAMPLE 4 Solve: $\frac{x-1}{3} + 4x = \frac{3}{2} + \frac{13x-2}{3}$

Solution

$$\frac{x-1}{3} + 4x = \frac{3}{2} + \frac{13x-2}{3}$$

$$6\left(\frac{x-1}{3} + 4x\right) = 6\left(\frac{3}{2} + \frac{13x-2}{3}\right)$$ To eliminate the fractions, multiply both sides by 6.

$$2(x-1) + 6(4x) = 9 + 2(13x-2)$$ Use the distributive property to remove the outer parentheses.

$$2x - 2 + 24x = 9 + 26x - 4$$ Use the distributive property to remove parentheses.

$$26x - 2 = 26x + 5$$ Combine like terms.

$$-2 = 5$$ Subtract $26x$ from both sides.

COMMENT If all the variable terms are eliminated when solving a linear equation, the results require interpretation. A false statement indicates no solution and a true statement indicates all real numbers.

Since $-2 = 5$ is false, no number x satisfies the equation. The solution set of the equation is $\varnothing$, called the **empty set**.

SELF CHECK 4 Solve: $\frac{x-2}{3} - 3 = \frac{1}{5} + \frac{x+1}{3}$

3 Solve a formula for a specified variable.

To solve a formula for a variable means to isolate that variable on one side of the equal sign and place all other quantities on the other side.

EXAMPLE 5 **WAGES AND COMMISSIONS** A sales clerk earns \$200 per week plus a 5% commission on the value of the merchandise she sells. What dollar volume must she sell each week to earn \$250, \$300, and \$350 in three successive weeks?

Solution The weekly earnings e are computed using the formula

$$e = 200 + 0.05v$$

where v represents the value in dollars of the merchandise sold. To find the value of v for the three values of e, we first solve the equation for v.

$$e = 200 + 0.05v$$

$$e - 200 = 0.05v$$ Subtract 200 from both sides.

$$\frac{e-200}{0.05} = v$$ Divide both sides by 0.05.

We can now substitute \$250, \$300, and \$350 for e and compute v.

$$v = \frac{e-200}{0.05} \qquad v = \frac{e-200}{0.05} \qquad v = \frac{e-200}{0.05}$$

$$v = \frac{\mathbf{250} - 200}{0.05} \qquad v = \frac{\mathbf{300} - 200}{0.05} \qquad v = \frac{\mathbf{350} - 200}{0.05}$$

$$v = 1{,}000 \qquad v = 2{,}000 \qquad v = 3{,}000$$

She must sell \$1,000 worth of merchandise the first week, \$2,000 worth in the second week, and \$3,000 worth in the third week.

SELF CHECK 5 What dollar volume must the clerk sell to earn \$600?

4 Solve an application using a linear equation in one variable.

EXAMPLE 6 **BUILDING A DOG RUN** A man has 28 meters of fencing to make a rectangular dog run. He wants the dog run to be 6 meters longer than it is wide. Find its dimensions.

Analyze the problem: To find the dimensions, we need to find both the length and the width. If w is chosen to represent the width in meters of the dog run, then $w + 6$ represents its length. (See Figure 7-1.)

Form an equation: The perimeter P of a rectangle is the distance around it. Because the dog run is a rectangle, opposite sides have the same length. The perimeter can be expressed as $2w + 2(w + 6)$ or as 28.

Figure 7-1

Two widths	plus	two lengths	equals	the perimeter.
$2 \cdot w$	$+$	$2 \cdot (w + 6)$	$=$	28

Solve the equation: We can solve this equation as follows:

$$2w + 2(w + 6) = 28$$
$$2w + 2w + 12 = 28 \quad \text{Use the distributive property to remove parentheses.}$$
$$4w + 12 = 28 \quad \text{Combine like terms.}$$
$$4w = 16 \quad \text{Subtract 12 from both sides.}$$
$$w = 4 \quad \text{Divide both sides by 4. This is the width in meters.}$$
$$w + 6 = 10 \quad \text{Add 6 to the width to find the length.}$$

State the conclusion: The dimensions of the dog run are 4 meters by 10 meters.

Check the result: If the dog run has a width of 4 meters and a length of 10 meters, its length is 6 meters longer than its width, and the perimeter is $2(4) + 2(10) = 28$.

SELF CHECK 6 If the man has 60 meters of fencing and wants the pen to be 10 meters longer than it is wide, find the dimensions.

5 Solve a linear inequality in one variable.

Recall that *inequalities* are statements indicating that two quantities might be unequal.

- $a < b$ means "a is less than b."
- $a > b$ means "a is greater than b."
- $a \leq b$ means "a is less than or equal to b."
- $a \geq b$ means "a is greater than or equal to b."

In Chapter 1, we saw that many inequalities can be graphed as regions on the number line, called **intervals**. For example, the graph of the inequality $-4 < x < 2$ is shown in Figure 7-2(a). Since neither endpoint is included, we say that the graph is an **open interval**.

Unless otherwise noted, all content on this page is © Cengage Learning.

In **interval notation**, this interval is denoted as $(-4, 2)$, where the parentheses indicate that the endpoints are not included.

The graph of the inequality $-2 \le x \le 5$ is shown in Figure 7-2(b). Since both endpoints are included, we say that the graph is a **closed interval**. This interval is denoted as $[-2, 5]$, where the brackets indicate that the endpoints are included.

Since one endpoint is included and one is not in the interval shown in Figure 7-2(c), we call the interval a **half-open interval**. This interval is denoted as $[-10, 10)$. Since the interval shown in Figure 7-2(d) extends forever in one direction, it is called an **unbounded interval**. This interval is denoted as $[-6, \infty)$, where the symbol ∞ is read as "infinity."

COMMENT Interval notation requires values to be written with the smaller number listed first.

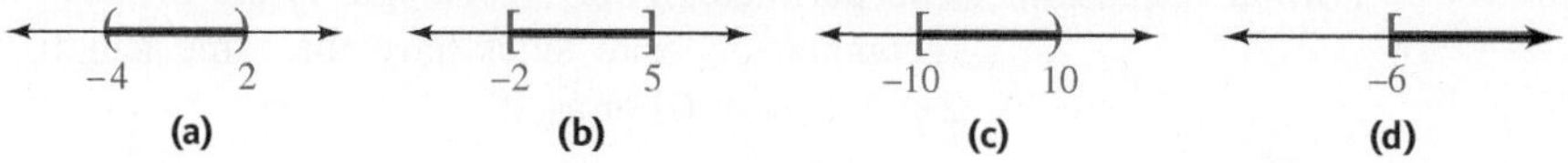

Figure 7-2

If a and b are real numbers, Table 7-1 shows the different types of intervals that can occur.

TABLE 7-1

Kind of interval	Inequality	Graph	Interval
Open interval	$a < x < b$	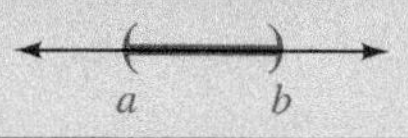	(a, b)
Half-open interval	$a \le x < b$	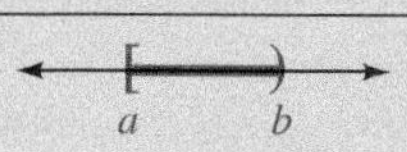	$[a, b)$
	$a < x \le b$	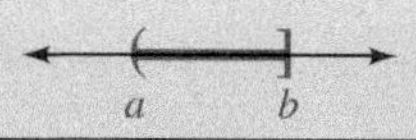	$(a, b]$
Closed interval	$a \le x \le b$	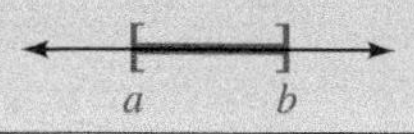	$[a, b]$
Unbounded interval	$x > a$	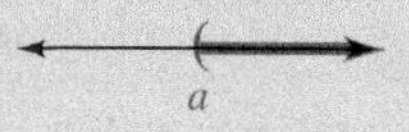	(a, ∞)
	$x \ge a$	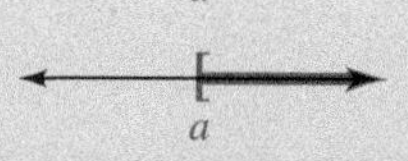	$[a, \infty)$
	$x < a$	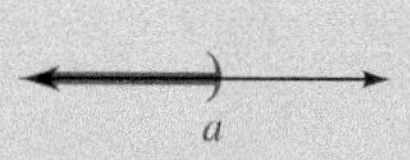	$(-\infty, a)$
	$x \le a$	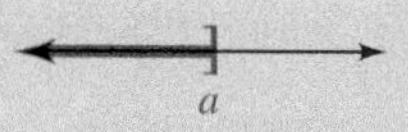	$(-\infty, a]$
	$-\infty < x < \infty$	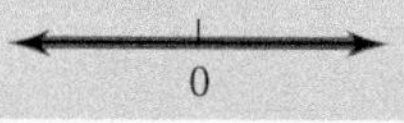	$(-\infty, \infty)$

Inequalities such as $x + 1 > x$, which are true for all numbers x, are called **absolute inequalities**. Note that solving this inequality results in $1 > 0$, which is true for all values of x. Inequalities such as $3x + 2 < 8$, which are true for some numbers x, but not all numbers x, are called **conditional inequalities**.

If a and b are two real numbers, then $a < b$, $a = b$, or $a > b$. This property, called the **trichotomy property**, indicates that one and only one of three statements is true about any two real numbers. Either

- the first number is less than the second,
- the first number is equal to the second,
- or the first number is greater than the second.

If a, b, and c are real numbers with $a < b$ and $b < c$, then $a < c$. This property, called the **transitive property**, indicates that if we have three numbers and the first number is less than the second and the second number is less than the third, then the first number is less than the third.

To solve an inequality, we use the following properties of inequalities.

Unless otherwise noted, all content on this page is © Cengage Learning.

PROPERTIES OF INEQUALITIES

1. Any real number can be added to (or subtracted from) both sides of an inequality to produce another inequality with the same direction as the original inequality.
2. If both sides of an inequality are multiplied (or divided) by a positive number, another inequality results with the same direction as the original inequality.
3. If both sides of an inequality are multiplied (or divided) by a negative number, another inequality results, but with the opposite direction from the original inequality.

Property 1 indicates that any number can be added to both sides of a true inequality to obtain another true inequality with the same direction. For example, if 4 is added to both sides of the inequality $3 < 12$, we have

$$3 + 4 < 12 + 4$$
$$7 < 16$$

is a true statement and the $<$ symbol remains a $<$ symbol. Adding 4 to both sides does not change the direction (sometimes called the *order*) of the inequality.

Subtracting 4 from both sides of $3 < 12$ does not change the direction of the inequality either since it also results in a true statement.

$$3 - 4 < 12 - 4$$
$$-1 < 8$$

Property 2 indicates that both sides of a true inequality can be multiplied by any positive number to obtain another true inequality with the same direction. For example, if both sides of the true inequality $-4 < 6$ are multiplied by 2, we obtain

$$2(-4) < 2(6)$$
$$-8 < 12$$

and the $<$ symbol remains a $<$ symbol. Multiplying both sides by 2 does not change the direction of the inequality.

Dividing both sides by 2 does not change the direction of the inequality either.

$$\frac{-4}{2} < \frac{6}{2}$$
$$-2 < 3$$

Property 3 indicates that if both sides of a true inequality are multiplied by any negative number, another true inequality results, but with the opposite direction. For example, if both sides of the true inequality $-4 < 6$ are multiplied by -2, we obtain

$$-4 < 6$$
$$-2(-4) > -2(6)$$
$$8 > -12$$

and the $<$ symbol becomes a $>$ symbol. Multiplying both sides by -2 reverses the direction of the inequality.

Dividing both sides by -2 also reverses the direction of the inequality.

$$-4 < 6$$
$$\frac{-4}{-2} > \frac{6}{-2}$$
$$2 > -3$$

COMMENT We must remember to reverse the inequality symbol every time we multiply or divide both sides by a negative number.

A **linear inequality** in one variable is any inequality that can be expressed in the form

$$ax + c < 0 \qquad ax + c > 0 \qquad ax + c \le 0 \qquad \text{or} \qquad ax + c \ge 0 \qquad (a \ne 0)$$

We can solve linear inequalities by applying the same steps that we use for solving linear equations, with one exception. If we multiply or divide both sides by a *negative* number, we must reverse the direction of the inequality.

EXAMPLE 7 Solve, then graph the solution set: **a.** $3(2x - 9) < 9$ **b.** $-4(3x + 2) \le 16$

Solution **a.** We solve the inequality as if it were an equation:

$$3(2x - 9) < 9$$

$6x - 27 < 9$ Use the distributive property to remove parentheses.

$6x < 36$ Add 27 to both sides.

$x < 6$ Divide both sides by 6.

The solution set is the interval $(-\infty, 6)$. The graph of the solution set is shown in Figure 7-3(a).

b. We solve the inequality as if it were an equation:

$$-4(3x + 2) \le 16$$

$-12x - 8 \le 16$ Use the distributive property to remove parentheses.

$-12x \le 24$ Add 8 to both sides.

$x \ge -2$ Divide both sides by -12 and reverse the $\le$ symbol.

The solution set is the interval $[-2, \infty)$. The graph of the solution set is shown in Figure 7-3(b).

(a) **(b)**

Figure 7-3

SELF CHECK 7 Solve, then graph the solution set:
a. $5(x - 2) \ge 5$ **b.** $-3(2x + 1) > 9$

EXAMPLE 8 Solve, then graph the solution set: $\frac{2}{3}(x + 2) > \frac{4}{5}(x - 3)$

Solution

$$\frac{2}{3}(x + 2) > \frac{4}{5}(x - 3)$$

$15 \cdot \frac{2}{3}(x + 2) > 15 \cdot \frac{4}{5}(x - 3)$ To clear the fractions, multiply both sides by 15, the LCD.

$10(x + 2) > 12(x - 3)$ $15 \cdot \frac{2}{3} = 10$ and $15 \cdot \frac{4}{5} = 12$

$10x + 20 > 12x - 36$ Use the distributive property to remove parentheses.

$-2x + 20 > -36$ Subtract $12x$ from both sides.

$-2x > -56$ Subtract 20 from both sides.

$x < 28$ Divide both sides by -2 and reverse the $>$ symbol.

The solution set is the interval $(-\infty, 28)$. Its graph is shown in Figure 7-4.

COMMENT The statement $287 > x$ is equivalent to $x < 28$.

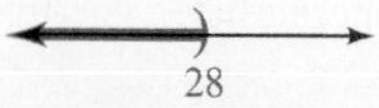

Figure 7-4

SELF CHECK 8 Solve, then graph the solution set: $\frac{1}{2}(x - 1) \le \frac{2}{3}(x + 1)$

Unless otherwise noted, all content on this page is © Cengage Learning.

6 Solve a compound inequality.

To say that x is between -3 and 8, we write a double inequality:

$-3 < x < 8$ Read as "-3 is less than x and x is less than 8."

This double inequality contains two different linear inequalities:

$-3 < x$ and $x < 8$

The word *and* indicates that these two inequalities are true at the same time.

DOUBLE INEQUALITIES The double inequality $c < x < d$ is equivalent to $c < x$ and $x < d$.

EXAMPLE 9 Solve, then graph the solution set: $-3 \le 2x + 5 < 7$

Solution This inequality means that $2x + 5$ is between -3 and 7 or possibly equal to -3. We can solve it by isolating x between the inequality symbols:

$-3 \le 2x + 5 < 7$

$-8 \le 2x < 2$ Subtract 5 from all three parts.

$-4 \le x < 1$ Divide all three parts by 2.

Figure 7-5

The solution set is the interval $[-4, 1)$. Its graph is shown in Figure 7-5.

SELF CHECK 9 Solve, then graph the solution set: $-3 < 2x - 5 \le 9$

EXAMPLE 10 Solve, then graph the solution set: $x + 3 < 2x - 1 < 4x - 3$

Solution Since it is impossible to isolate x between the inequality symbols, we solve each of its linear inequalities separately.

$x + 3 < 2x - 1$	and	$2x - 1 < 4x - 3$
$4 < x$		$2 < 2x$
		$1 < x$

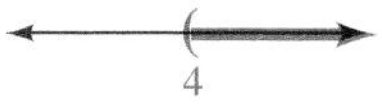

Figure 7-6

Only those values of x where $x > 4$ and $x > 1$ are in the solution set. Since all numbers greater than 4 are also greater than 1, the solutions are the values x where $x > 4$. The solution set is the interval $(4, \infty)$. The graph is shown in Figure 7-6.

SELF CHECK 10 Solve, then graph the solution set: $x - 5 \le 3x - 1 \le 5x + 5$

EXAMPLE 11 Solve the compound inequality, then graph the solution set: $x \le -3$ or $x \ge 8$

Solution The solution set of $x \le -3$ or $x \ge 8$ is the union of two intervals:

$(-\infty, -3] \cup [8, \infty)$

Its graph is shown in Figure 7-7.

The word *or* in the statement $x \le -3$ or $x \ge 8$ indicates that only one of the inequalities needs to be true to make the statement true.

Figure 7-7

SELF CHECK 11 Solve, then graph the solution set: $x < -2$ or $x > 4$

Unless otherwise noted, all content on this page is © Cengage Learning.

SELF CHECK ANSWERS

1. −3 **2.** −1 **3.** $\mathbb{R}$ **4.** $\varnothing$ **5.** \$8,000 **6.** The dimensions will be 10 meters by 20 meters.

7. a. $[3, \infty)$

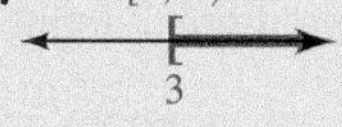

b. $(-\infty, -2)$

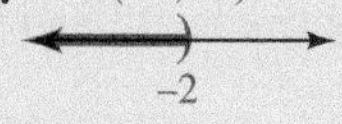

8. $[-7, \infty)$

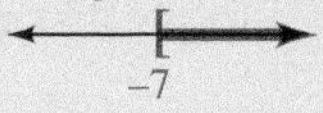

9. $(1, 7]$

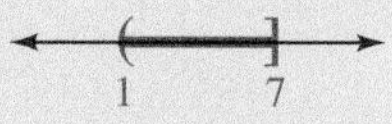

10. $[-2, \infty)$

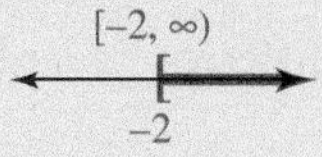

11. $(-\infty, -2) \cup (4, \infty)$

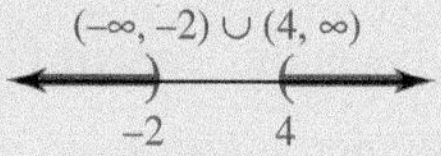

NOW TRY THIS

Solve each inequality, and express the result as a graph and in interval notation, if appropriate.

1. $6x - 2(x + 1) < 10$ or $-5x < -10$

2. $4x \geq 2(x - 6)$ and $-5x \geq 30$

7.1 Exercises

WARM-UPS *Solve each equation or inequality.*

1. $2x + 4 = 6$

2. $3x - 4 = 8$

3. $2x < 4$

4. $3x + 1 \geq 10$

5. $3x < -12$

6. $\frac{1}{2}x \geq -4$

REVIEW *Simplify each expression. Assume no variable is 0.*

7. $\left(\frac{t^3t^5t^{-6}}{t^2t^{-4}}\right)^{-3}$

8. $\left(\frac{a^{-2}b^3a^5b^{-2}}{a^6b^{-5}}\right)^{-4}$

9. Baking A man invested \$1,068 in baking equipment to make pies. Each pie requires \$3.50 in ingredients. If he can sell all the pies he can make for \$9.50 each, how many pies will he have to make to earn a profit?

10. Investing A woman invested \$15,000, part at 5% annual interest and the rest at 4%. If she earned \$700 in income over a one-year period, how much did she invest at 5%?

VOCABULARY AND CONCEPTS *Fill in the blanks.*

11. An _______ is a statement indicating that two mathematical expressions are equal.

12. If any quantity is _____ to both sides of an equation, a new equation is formed that is equivalent to the original equation.

13. If both sides of an equation are ________ (or ______) by the same nonzero number, a new equation is formed that is equivalent to the original equation.

14. The symbol $>$ is read as "___________."

15. An ______ is an equation that is true for all values of its variable.

16. An open interval has no ________.

17. A __________ is an equation that is true for no values of its variable.

18. The symbol $\leq$ is read as "_________ or equal to."

19. A ________ interval has one endpoint.

20. The transitive property states that if $a < b$ and $b < c$, then _____.

21. If both sides of an inequality are multiplied by a _______ number, a new relationship is formed that has the opposite direction from the original inequality.

22. If both sides of an inequality are multiplied by a _______ number, a new inequality is formed that has the same direction as the first.

GUIDED PRACTICE *Solve each equation.* SEE EXAMPLE 1. *(OBJECTIVE 1)*

23. $2x + 1 = 13$

24. $2x - 4 = 16$

25. $3(x + 1) = 15$

26. $-2(x + 5) = 30$

Unless otherwise noted, all content on this page is © Cengage Learning.

27. $2r - 5 = 1 - r$

28. $5s - 13 = s - 1$

29. $3(2y - 4) - 6 = 3y$

30. $2x + (2x - 3) = 5$

31. $5(5 - a) = 37 - 2a$

32. $4a + 17 = 7(a + 2)$

33. $4(y + 1) = -2(4 - y)$

34. $5(r + 4) = -2(r - 3)$

Solve each equation. SEE EXAMPLE 2. (OBJECTIVE 1)

35. $\frac{x}{2} - \frac{x}{3} = 4$

36. $\frac{x}{2} + \frac{x}{3} = 10$

37. $\frac{x}{6} + 1 = \frac{x}{3}$

38. $\frac{3}{2}(y + 4) = \frac{20 - y}{2}$

39. $\frac{a + 1}{3} + \frac{a - 1}{5} = \frac{2}{15}$

40. $\frac{a + 1}{4} + \frac{2a - 3}{4} = \frac{a}{2} - 2$

41. $\frac{2z + 3}{3} + \frac{3z - 4}{6} = \frac{z - 2}{2}$

42. $\frac{y - 8}{5} + 2 = \frac{2}{5} - \frac{y}{3}$

Solve each equation and indicate whether it is an identity or a contradiction. SEE EXAMPLES 3–4. (OBJECTIVE 2)

43. $4(2 - 3t) + 6t = -6t + 8$

44. $3x - 6 = -3x + 6(x - 2)$

45. $3(x - 4) + 6 = -2(x + 4) + 5x$

46. $2(x - 3) = \frac{3}{2}(x - 4) + \frac{x}{2}$

Solve each formula for the indicated variable. SEE EXAMPLE 5. (OBJECTIVE 3)

47. $V = \frac{1}{3}Bh$ for B

48. $A = \frac{1}{2}bh$ for b

49. $P = 2l + 2w$ for w

50. $P = 2l + 2w$ for l

51. $z = \frac{x - \mu}{\sigma}$ for x

52. $z = \frac{x - \mu}{\sigma}$ for μ

53. $y = mx + b$ for x

54. $y = mx + b$ for m

Solve each inequality. State the result in interval notation and graph the solution set. SEE EXAMPLE 7. (OBJECTIVE 5)

55. $5x - 3 > 7$

56. $7x - 9 < 5$

57. $6x + 5 \le -13$

58. $3x + 2 \ge 17$

59. $-3x - 1 \le 5$

60. $-2x + 6 \ge 16$

61. $-3(a + 2) > 2(a + 1)$

62. $-4(y - 1) < y + 8$

Solve each inequality. State the result in interval notation. SEE EXAMPLE 8. (OBJECTIVE 5)

63. $\frac{1}{2}y + 2 \ge \frac{1}{3}y - 4$

64. $\frac{1}{4}x - \frac{1}{3} \le x + 2$

65. $\frac{1}{5}y - \frac{8}{5} + 2 \le \frac{2}{5} - \frac{1}{3}y$

66. $\frac{1}{2}x - \frac{1}{3}x \ge 10$

67. $\frac{x + 1}{4} - \frac{2x - 4}{2} > \frac{3 - 2x}{4}$

68. $\frac{1}{3}x - 1 \ge \frac{1}{5}x + 1$

69. $\frac{3}{2}(x + 4) > 10 - \frac{x}{2}$

70. $\frac{3}{4}(x - 2) < x - 2$

Solve each inequality. State the result in interval notation and graph the solution set. SEE EXAMPLE 9. (OBJECTIVE 6)

71. $-2 < -b + 3 < 5$

72. $4 < -t - 2 < 9$

73. $15 > 2x - 7 > 9$

74. $25 > 3x - 2 > 7$

75. $-6 < -3(x - 4) \le 24$

76. $-4 \le -2(x + 8) < 8$

77. $0 \ge \frac{1}{2}x - 4 > 6$

78. $-2 \le \frac{5 - 3x}{2} \le 2$

Solve each inequality. State the result in interval notation, where appropriate. SEE EXAMPLE 10. (OBJECTIVE 6)

79. $x - 1 \le 2x + 5 < 3x$

80. $x + 2 < 3x - 1 < 4x + 1$

81. $x + 5 \le 2x - 3 \le 5x - 1$

82. $x + 1 < 2x + 2 < 3x + 3$

Solve each inequality. State the result in interval notation and graph the solution set. SEE EXAMPLE 11. (OBJECTIVE 6)

83. $3x + 2 < 8$ or $2x - 3 > 11$

84. $3x + 4 < -2$ or $3x + 4 > 10$

85. $-4(x + 2) \ge 12$ or $3x + 8 < 11$

86. $x < 3$ or $x > -3$

ADDITIONAL PRACTICE

Solve each equation or inequality. If an inequality, give the answer as a graph and in interval notation, where appropriate.

87. $x < -3$ and $x > 3$

88. $8(3a - 5) - 4(2a + 3) = 12$

89. $y(y + 2) = (y + 1)^2 - 1$

90. $x(x - 3) = (x - 1)^2 - (5 + x)$

91. $8 - 9y \ge -y$

92. $4 - 3x \le x$

93. $5(x - 2) \ge 0$ and $-3x < 9$

94. $2(a - 5) - (3a + 1) = 0$

95. $-6 \le \frac{1}{3}a + 1 < 0$

96. $0 \le \frac{4 - x}{3} \le 2$

97. $P = L + \frac{s}{f}i$ for s

98. $P = L + \frac{s}{f}i$ for f

APPLICATIONS

Solve each application. SEE EXAMPLE 6. (OBJECTIVE 4)

99. Finding dimensions The rectangular garden below is twice as long as it is wide. Find its dimensions.

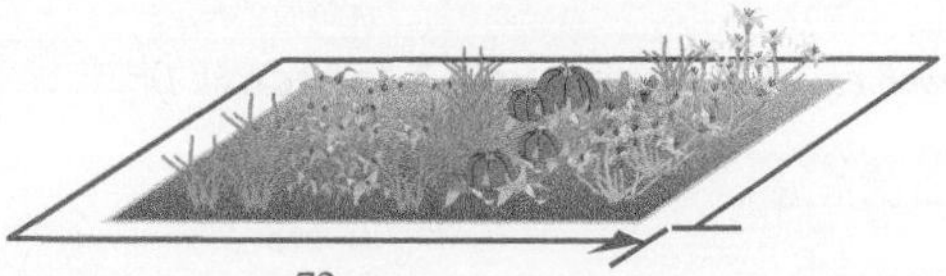

100. Fencing pastures A farmer has 624 feet of fencing to enclose the rectangular pasture shown below. Because a river runs along one side, fencing will be needed on only three sides. Find the dimensions of the pasture if its length is double its width and parallel to the river.

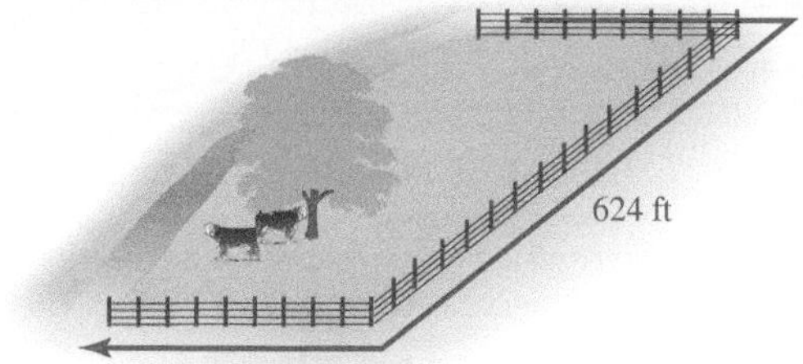

101. Fencing pens A man has 150 feet of fencing to build the pen shown below. If one end is a square, find the outside dimensions of the entire pen.

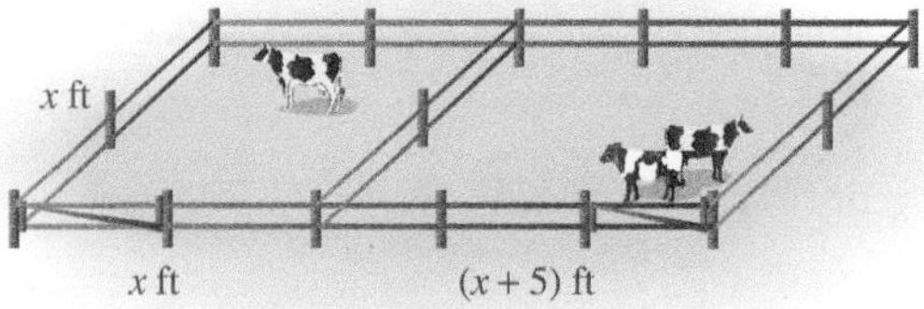

102. Enclosing swimming pools A woman wants to enclose the swimming pool shown in the illustration and have a walkway of uniform width all the way around. How wide will the walkway be if the woman uses 180 feet of fencing?

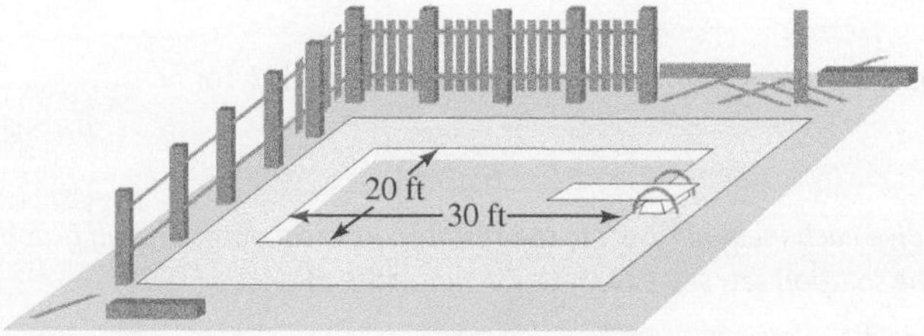

Unless otherwise noted, all content on this page is © Cengage Learning.

Solve each application.

103. Cutting boards The carpenter below saws a board into two pieces. He wants one piece to be 1 foot longer than twice the length of the shorter piece. Find the length of each piece.

104. Cutting beams A 30-foot steel beam is to be cut into two pieces. The longer piece is to be 6 feet more than 2 times as long as the shorter piece. Find the length of each piece.

105. Investing If a woman invests $10,000 at 8% annual interest, how much more must she invest at 9% so that her annual income will exceed $1,250?

106. Finding profit The wholesale cost of a radio is $27. A store owner knows that for the radio to sell, it must be priced under $42. If p is the profit, express the possible profit as an inequality.

107. Downloading music A student can afford to spend up to $275 on a smartphone and downloaded music. If the phone costs $210 and the music downloads are $1.10 each, find the greatest number of songs the student can download.

108. Grades A student has scores of 70, 77, and 85 on three exams. What score is needed on a fourth exam to make the student's average 80 or better?

WRITING ABOUT MATH

109. Explain the difference between a conditional equation, an identity, and a contradiction.

110. The techniques for solving linear equations and linear inequalities are similar, yet different. Explain.

SOMETHING TO THINK ABOUT

111. Find the error.

$$4(x + 3) = 16$$
$$4x + 3 = 16$$
$$4x = 13$$
$$x = \frac{13}{4}$$

112. Which of these relations is transitive?

a. $=$ **b.** $\le$ **c.** $\not\ge$ **d.** $\ne$

Section 7.2 Review of Graphing Linear Equations, Finding the Slopes of Lines, and Writing Equations of Lines

Objectives

1. Graph a linear equation in two variables by plotting points and by using the intercepts.
2. Graph a horizontal line and a vertical line.
3. Find the midpoint of a line segment between two points.
4. Find the slope of a line from a graph passing through two points and from an equation.
5. Determine whether two lines are parallel, perpendicular, or neither.
6. Write an equation of a line using the point-slope form and slope-intercept form.
7. Write an equation of a line modeling real-world data.

Unless otherwise noted, all content on this page is © Cengage Learning.

Vocabulary

coordinates (x, y)	rise	negative reciprocals
y-intercept	run	point-slope form
x-intercept	slope-intercept form	scattergram
midpoint	parallel lines	
slope	perpendicular lines	

Getting Ready

1. State the coordinates of each point.

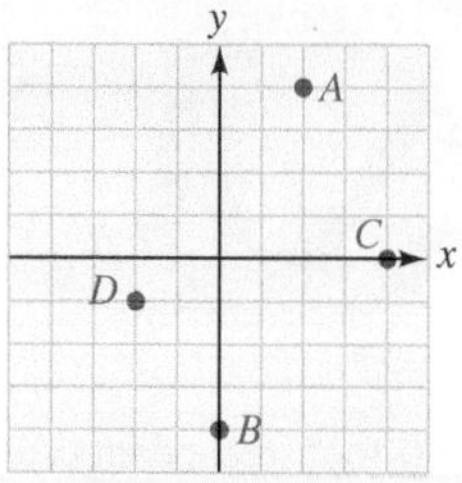

2. In the equation $2x + y = 5$, find the value of y when x has the following values.

a. $x = 2$ **b.** $x = -2$

c. $x = 0$ **d.** $x = \frac{3}{2}$

Simplify the following.

3. $\frac{7-2}{6+4}$ **4.** $\frac{-8-(-8)}{-3-4}$ **5.** $\frac{-1-6}{3-5}$ **6.** $\frac{-2-4}{-5-(-5)}$

In this section, we will review graphing linear equations, finding the slopes of lines, and writing equations of lines.

Recall that a linear equation is any equation in the form $Ax + By = C$, where A and B are not both zero. We will review several methods of graphing a linear equation: plotting points, using the intercepts, and using the slope and y-intercept.

1 Graph a linear equation in two variables by plotting points and by using the intercepts.

The *graph of an equation* in the variables x and y is the set of all points on a rectangular coordinate system with **coordinates (x, y)** that satisfy the equation. To graph the equation of a line, we may use any two ordered pairs that satisfy the equation or we may choose to use the x- and y-intercepts.

To graph the equation using intercepts, we first review the definitions of the x- and y-intercepts.

INTERCEPTS OF A LINE

The **y-intercept** of a line is the point $(0, b)$ where the line intersects the y-axis. To find the value of b, substitute 0 for x in the equation and solve for y.

The **x-intercept** of a line is the point $(a, 0)$ where the line intersects the x-axis. To find the value of a, substitute 0 for y in the equation and solve for x.

In Example 1, we will review both options.

EXAMPLE 1 Graph the equation $3x + 2y = 12$ by **a.** plotting points, and **b.** using the intercepts.

Solution **a.** To graph an equation, we can select values for either x or y, substitute them in the equation, and solve for the other variable. For example, if $x = 2$, then

$$3x + 2y = 12$$
$$3(2) + 2y = 12 \quad \text{Substitute 2 for } x.$$
$$6 + 2y = 12 \quad \text{Simplify.}$$
$$2y = 6 \quad \text{Subtract 6 from both sides.}$$
$$y = 3 \quad \text{Divide both sides by 2.}$$

One ordered pair that satisfies the equation is $(2, 3)$. If $y = 9$, we have

Unless otherwise noted, all content on this page is © Cengage Learning.

$$
\begin{aligned}
3x + 2y &= 12 \\
3x + 2(\mathbf{9}) &= 12 && \text{Substitute 9 for } y. \\
3x + 18 &= 12 && \text{Simplify.} \\
3x &= -6 && \text{Subtract 18 from both sides.} \\
x &= -2 && \text{Divide both sides by 3.}
\end{aligned}
$$

A second ordered pair that satisfies the equation is $(-2, 9)$.

The pairs $(2, 3)$ and $(-2, 9)$ and one other pair (as a check) that satisfies the equation are shown in the table in Figure 7-8. We plot each pair on a rectangular coordinate system and join the points to obtain the line shown in the figure. This line is the graph of the equation.

$3x + 2y = 12$

x	y	(x, y)
2	3	$(2, 3)$
-2	9	$(-2, 9)$
6	-3	$(6, -3)$

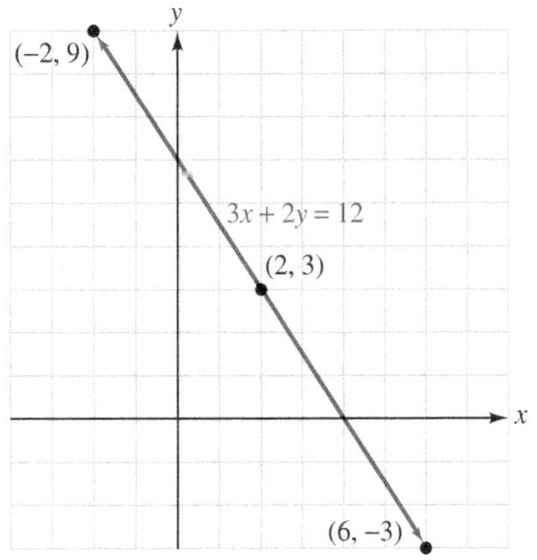

Figure 7-8

b. To find the y-intercept we substitute 0 for x and solve for y.

$$
\begin{aligned}
3x + 2y &= 12 \\
3(\mathbf{0}) + 2y &= 12 && \text{Substitute 0 for } x. \\
2y &= 12 && \text{Simplify.} \\
y &= 6 && \text{Divide both sides by 2.}
\end{aligned}
$$

The y-intercept is the point $(0, 6)$.

To find the x-intercept we substitute 0 for y and solve for x.

$$
\begin{aligned}
3x + 2y &= 12 \\
3x + 2(\mathbf{0}) &= 12 && \text{Substitute 0 for } y. \\
3x &= 12 && \text{Simplify.} \\
x &= 4 && \text{Divide both sides by 3.}
\end{aligned}
$$

The x-intercept is the point $(4, 0)$.

We use the intercepts to graph the line shown in Figure 7-9.

$3x + 2y = 12$

x	y	(x, y)
0	6	$(0, 6)$
4	0	$(4, 0)$

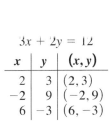

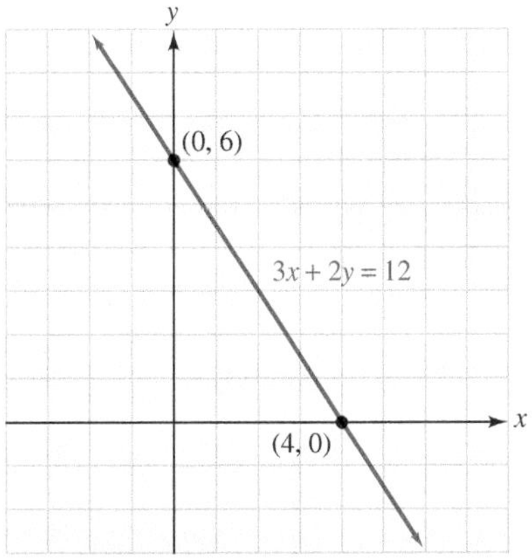

Figure 7-9

Unless otherwise noted, all content on this page is © Cengage Learning.

We can see that the graphs of the line are the same regardless of the method used.

SELF CHECK 1 Graph $4x - 2y = 12$ by **a.** plotting points, and **b.** using intercepts.

2 Graph a horizontal line and a vertical line.

Recall that the general form of a linear equation is $Ax + By = C$. When $A = 0$, the equation becomes $By = C$. This indicates a horizontal line. When $B = 0$, the equation becomes $Ax = C$, which indicates a vertical line. We will review these forms in the next example.

EXAMPLE 2 Graph: **a.** $y = 3$ **b.** $x = -2$

Solution **a.** Since the equation $y = 3$ does not contain an x term, the numbers chosen for x have no effect on y. The value of y is always 3.

After plotting the pairs (x, y) shown in Figure 7-10 and joining them with a straight line, we see that the graph is a horizontal line, parallel to the x-axis, with a y-intercept of $(0, 3)$. The line has no x-intercept.

b. Since the equation $x = -2$ does not contain a y term, the value of y can be any number. The value of x is always -2.

After plotting the pairs (x, y) shown in Figure 7-10 and joining them with a straight line, we see that the graph is a vertical line, parallel to the y-axis with an x-intercept of $(-2, 0)$. The line has no y-intercept.

$y = 3$

x	y	(x, y)
-3	3	$(-3, 3)$
0	3	$(0, 3)$
2	3	$(2, 3)$
4	3	$(4, 3)$

$x = -2$

x	y	(x, y)
-2	-2	$(-2, -2)$
-2	0	$(-2, 0)$
-2	2	$(-2, 2)$
-2	6	$(-2, 6)$

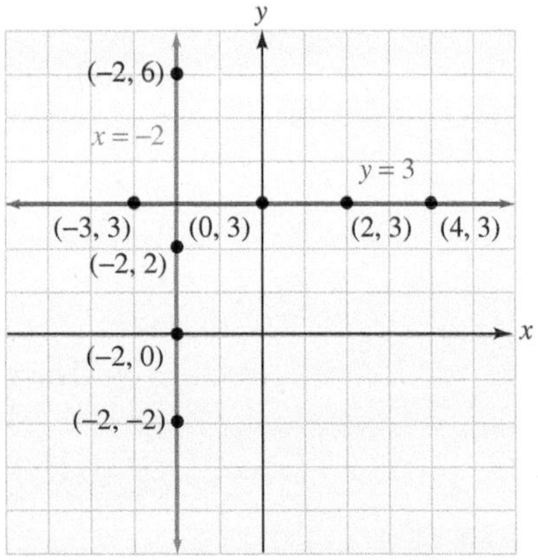

Figure 7-10

SELF CHECK 2 Graph: **a.** $x = 4$ **b.** $y = -2$

The results of Example 2 suggest the following facts.

EQUATIONS OF HORIZONTAL AND VERTICAL LINES

If a and b are real numbers, then

the graph of $y = b$ is a horizontal line with y-intercept at $(0, b)$. If $b = 0$, the line is the x-axis.

the graph of $x = a$ is a vertical line with x-intercept at $(a, 0)$. If $a = 0$, the line is the y-axis.

Unless otherwise noted, all content on this page is © Cengage Learning.

3 Find the midpoint of a line segment between two points.

If point M in Figure 7-11 lies midway between points $P(x_1, y_1)$ and $Q(x_2, y_2)$, point M is called the **midpoint** of segment PQ. To find the coordinates of M, we average the x-coordinates and average the y-coordinates of P and Q.

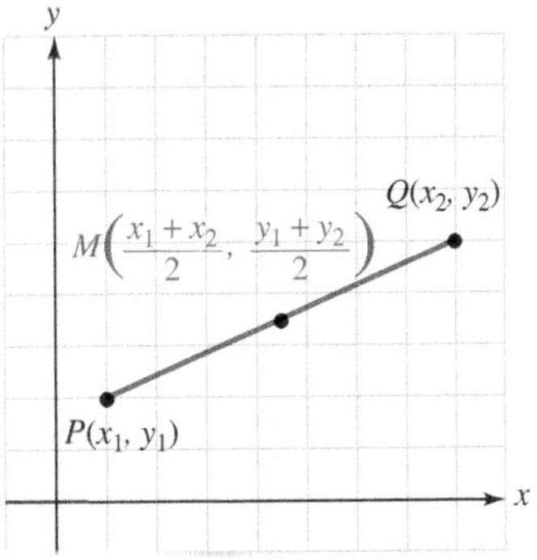

Figure 7-11

THE MIDPOINT FORMULA

The **midpoint** of the line segment $P(x_1, y_1)$ and $Q(x_2, y_2)$ is the point M with coordinates of

$$\left(\frac{x_1 + x_2}{2}, \frac{y_1 + y_2}{2}\right)$$

EXAMPLE 3 Find the midpoint of the segment joining $(-2, 4)$ and $(3, -5)$.

Solution To find the midpoint, we average the x-coordinates and the y-coordinates to obtain

$$\frac{x_1 + x_2}{2} = \frac{-2 + 3}{2} = \frac{1}{2} \quad \text{and} \quad \frac{y_1 + y_2}{2} = \frac{4 + (-5)}{2} = -\frac{1}{2}$$

The midpoint of the segment is the point $\left(\frac{1}{2}, -\frac{1}{2}\right)$.

SELF CHECK 3 Find the midpoint of $(5, -3)$ and $(-2, 5)$.

4 Find the slope of a line from a graph passing through two points and from an equation.

Knowing the slant (steepness) of a line is useful when working with linear equations. A measure of this slant is called the slope of the line.

Recall that the **slope** of a line is calculated as the $\frac{\text{rise}}{\text{run}}$, where the **rise** is the change in the vertical direction as we move from one point to another and the **run** is the change in the horizontal direction. To find the slope of a graphed line, we use this definition. However, if we are given two points through which we want to find the slope, we use the following relationship.

SLOPE OF A NONVERTICAL LINE

The slope m of the nonvertical line passing through points (x_1, y_1) and (x_2, y_2) is

$$m = \frac{\text{change in } y}{\text{change in } x} = \frac{y_2 - y_1}{x_2 - x_1}$$

Unless otherwise noted, all content on this page is © Cengage Learning.

EXAMPLE 4 Use the graph shown in Figure 7-12 to find the slope of the line.

Solution Using the concept of $\frac{\text{rise}}{\text{run}}$, we begin at the upper point and count how far we move to align horizontally with the lower point. We move down (negative direction) a distance of 8 units. So this rise is -8.

From this point we move to the right until we reach the bottom point. We move a distance of 5 in the positive direction, so the run is 5. The slope of the line is $\frac{-8}{5}$, or $-\frac{8}{5}$. We obtain the same result if we start with the lower point and move to the upper point.

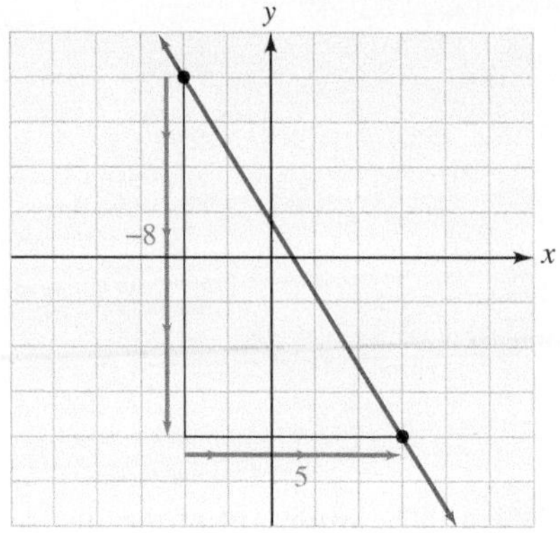

Figure 7-12

SELF CHECK 4 Use the graph to find the slope of the line.

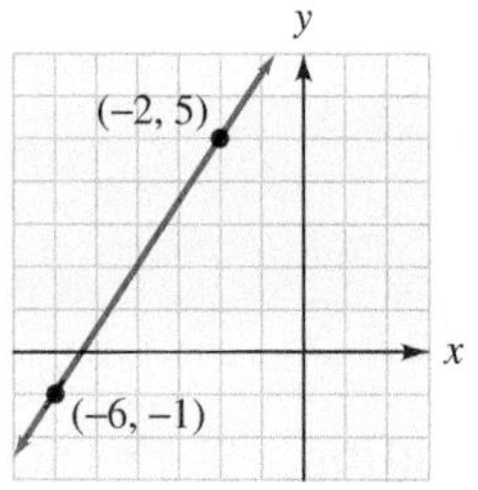

EXAMPLE 5 Find the slope of the line passing through the points $(-5, 2)$ and $(3, -4)$.

Solution We let $(x_1, y_1) = (-5, 2)$ and $(x_2, y_2) = (3, -4)$ then

$$m = \frac{\text{change in } y}{\text{change in } x}$$

$$= \frac{y_2 - y_1}{x_2 - x_1} \quad \text{This is the formula for slope.}$$

$$= \frac{-4 - 2}{3 - (-5)} \quad \text{Substitute } -4 \text{ for } y_2,\ 2 \text{ for } y_1,\ 3 \text{ for } x_2, \text{ and } -5 \text{ for } x_1.$$

$$= \frac{-6}{8} \quad \text{Simplify.}$$

$$= -\frac{3}{4}$$

The slope is $-\frac{3}{4}$. We obtain the same result if we let $(x_1, y_1) = (3, -4)$ and $(x_2, y_2) = (-5, 2)$.

SELF CHECK 5 Find the slope of the line passing through the points $(-7, -9)$ and $(2, 1)$.

Unless otherwise noted, all content on this page is © Cengage Learning.

To find the slope of a line from a given equation, we could graph the equation and count squares on the resulting line graph to determine the rise and the run; this could be difficult if we do not have integer coordinates. Another method is to find the x- and y-intercepts of the graph and use the slope formula. However, recall that the **slope-intercept form** of a line is $y = mx + b$, where m is the slope and b is the y-coordinate of the y-intercept. This form allows us to identify the slope as the coefficient of x.

EXAMPLE 6 Find the slope of the line determined by $3x - 4y = 12$.

Solution We must solve for y to write the equation in slope-intercept form.

$$3x - 4y = 12$$

$$-4y = -3x + 12 \quad \text{Subtract } 3x \text{ from both sides.}$$

$$y = \frac{-3x + 12}{-4} \quad \text{Divide both sides by } -4.$$

$$y = \frac{-3x}{-4} + \frac{12}{-4} \quad \text{Divide each term in the numerator by } -4.$$

$$y = \frac{3}{4}x - 3 \quad \text{Simplify.}$$

The coefficient of the x-term is $\frac{3}{4}$. Therefore, the slope of the line is $\frac{3}{4}$.

SELF CHECK 6 Find the slope of the line determined by $2x + 5y = 12$.

COMMENT When calculating slope, always subtract the y-values and the x-values in the same order.

$$m = \frac{y_2 - y_1}{x_2 - x_1} \quad \text{or}$$

$$m = \frac{y_1 - y_2}{x_1 - x_2}$$

Grace Murray Hopper
1906–1992

Grace Hopper graduated from Vassar College in 1928 and obtained a Master's degree from Yale in 1930. In 1943, she entered the U.S. Naval Reserve. While in the Navy, she became a programmer of the Mark I, the world's first large computer. She is credited for first using the word "bug" to refer to a computer problem. The first bug was actually a moth that flew into one of the relays of the Mark II. From then on, locating computer problems was called "debugging" the system.

We need to identify the slope of a horizontal line and a vertical line.

If $P(x_1, y_1)$ and $Q(x_2, y_2)$ are points on the horizontal line shown in Figure 7-13, then $y_1 = y_2$, and the numerator of the fraction

$$\frac{y_2 - y_1}{x_2 - x_1} \quad \text{On a horizontal line, } x_2 \neq x_1.$$

is 0. Thus, the value of the fraction is 0, and the slope of the horizontal line is 0.

If $P(x_1, y_1)$ and $Q(x_2, y_2)$ are two points on the vertical line shown in Figure 7-13, then $x_1 = x_2$, and the denominator of the fraction

$$\frac{y_2 - y_1}{x_2 - x_1} \quad \text{On a vertical line, } y_2 \neq y_1.$$

is 0. Since the denominator cannot be 0, a vertical line has an undefined slope.

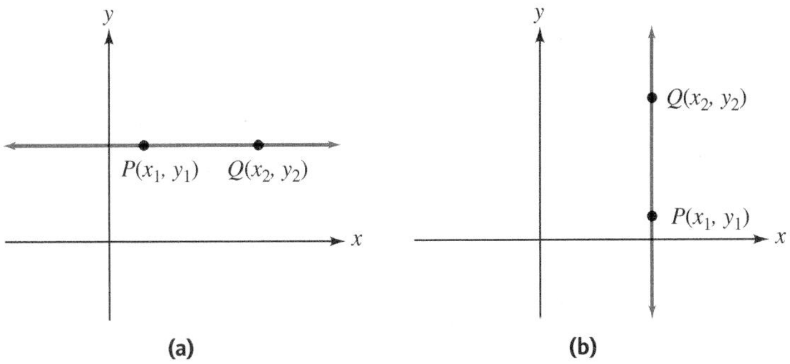

Figure 7-13

Unless otherwise noted, all content on this page is © Cengage Learning.

SLOPES OF HORIZONTAL AND VERTICAL LINES

All horizontal lines (lines with equations of the form $y = b$) have a slope of 0.

All vertical lines (lines with equations of the form $x = a$) have an undefined slope.

If a line rises as we follow it from left to right, as in Figure 7-14(a), its slope is positive. If a line drops as we follow it from left to right, as in Figure 7-14(b), its slope is negative. If a line is horizontal, as in Figure 7-14(c), its slope is 0. If a line is vertical, as in Figure 7-14(d), it has an undefined slope.

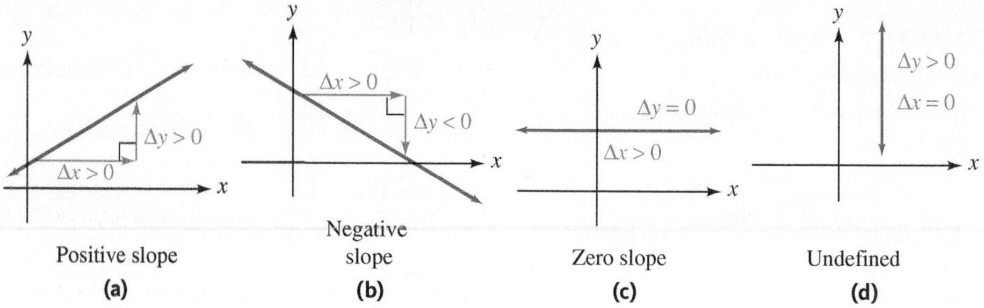

Figure 7-14

5 Determine whether two lines are parallel, perpendicular, or neither.

Recall that **parallel lines** have the same slope and that nonvertical **perpendicular lines** have slopes that are negative reciprocals.

Because a horizontal line is perpendicular to a vertical line, a line with a slope of 0 is perpendicular to a line with an undefined slope.

Two real numbers a and b are called **negative reciprocals** if $ab = -1$. For example,

$$-\frac{4}{3} \quad \text{and} \quad \frac{3}{4}$$

are negative reciprocals, because $-\frac{4}{3}\left(\frac{3}{4}\right) = -1$.

SLOPES OF PARALLEL LINES

Nonvertical parallel lines have the same slope, and lines having the same slope are parallel.

Since vertical lines are parallel, lines with undefined slope are parallel.

SLOPES OF PERPENDICULAR LINES

If two nonvertical lines are perpendicular, their slopes are negative reciprocals.

If the slopes of two lines are negative reciprocals, the lines are perpendicular.

EXAMPLE 7 Determine whether the given lines are parallel, perpendicular, or neither.

Solution **a.** $y = 5x - 2; \quad y = 5x + 9$

Comparing to $y = mx + b$, the slopes of both lines are 5; therefore the lines are parallel.

b. $y = \frac{2}{3}x - 6; \quad 3x + 2y = 8$

Unless otherwise noted, all content on this page is © Cengage Learning.

To find the slope of the second line, we first solve for y.

$$3x + 2y = 8$$
$$2y = -3x + 8$$
$$y = -\frac{3}{2}x + 4$$

The slope of the first line is $\frac{2}{3}$ and the slope of the second line is $-\frac{3}{2}$. Because $\frac{2}{3}$ and $-\frac{3}{2}$ are negative reciprocals, the lines are perpendicular.

c. $4x + 2y = 9; \quad -2x - 4y = 9$

We must solve both equations for y.

$$4x + 2y = 9$$
$$2y = -4x + 9$$
$$y = -2x + \frac{9}{2} \quad \text{The slope is } -2.$$
$$-2x - 4y = 9$$
$$-4y = 2x + 9$$
$$y = -\frac{1}{2}x - \frac{9}{4} \quad \text{The slope is } -\frac{1}{2}.$$

The slopes are not the same and not negative reciprocals. Therefore, the lines are neither parallel nor perpendicular.

SELF CHECK 7 Determine whether the given lines are parallel, perpendicular, or neither.

$3x - 9y = 6; \quad 12x + 4y = 6$

We now apply the concept of slope to write the equation of a line passing through two fixed points. We also will use slope as an aid in graphing lines.

6 Write an equation of a line using the point-slope form and slope-intercept form.

Recall from our work earlier in the text that there are two methods for writing an equation of a line given enough information to do so. We can use either the **point-slope form** or the **slope-intercept form** of an equation of a line.

POINT-SLOPE FORM The point-slope equation of the line passing through $P(x_1, y_1)$ and with slope m is

$$y - y_1 = m(x - x_1)$$

SLOPE-INTERCEPT FORM The slope-intercept equation of a line with slope m and y-intercept $(0, b)$ is

$$y = mx + b$$

EXAMPLE 8 Write an equation of the line with a slope of $-\frac{2}{3}$ and passing through $(-4, 5)$ using **a.** the point-slope form, and **b.** the slope-intercept form.

Solution **a.** For the point-slope form we substitute $-\frac{2}{3}$ for m, -4 for x_1, and 5 for y_1 and simplify.

$$y - y_1 = m(x - x_1) \quad \text{Point-slope form}$$

$$y - 5 = -\frac{2}{3}[x - (-4)] \quad \text{Substitute } -\tfrac{2}{3} \text{ for } m, -4 \text{ for } x_1, \text{ and 5 for } y_1.$$

$$y - 5 = -\frac{2}{3}(x + 4) \quad -(-4) = 4$$

The point-slope equation of the line is $y - 5 = -\frac{2}{3}(x + 4)$.

b. In the slope-intercept form, we substitute $-\frac{2}{3}$ for m, -4 for x, and 5 for y. Then we solve for b.

$$y = mx + b \quad \text{Slope-intercept form}$$

$$5 = -\frac{2}{3}(-4) + b$$

$$5 = \frac{8}{3} + b \quad \text{Multiply.}$$

$$\frac{7}{3} = b \quad \text{Subtract } \tfrac{8}{3} \text{ from both sides.}$$

We have solved for b, but now we must write the equation.

$$y = -\frac{2}{3}x + \frac{7}{3}$$

We generally leave an equation written in slope-intercept form because it is very useful when answering questions regarding real-world situations.

SELF CHECK 8 Write an equation of the line with slope of $\frac{5}{4}$ and passing through $(0, 5)$ using both the point-slope form and the slope-intercept form.

In Example 8(a), the answer $y - 5 = -\frac{2}{3}(x + 4)$ is not typically left in this form. To write the equation in general form $(Ax + By = C)$, we would clear fractions first.

$$y - 5 = -\frac{2}{3}(x + 4)$$

$$3[y - 5] = 3\left[-\frac{2}{3}(x + 4)\right] \quad \text{Multiply both sides by 3, the LCD.}$$

$$3y - 15 = -2(x + 4) \quad \text{Use the distributive property to remove brackets.}$$

$$3y - 15 = -2x - 8 \quad \text{Use the distributive property to remove parentheses.}$$

$$3y + 2x - 15 = -8 \quad \text{Add } 2x \text{ to both sides.}$$

$$3y + 2x = 7 \quad \text{Add 15 to both sides.}$$

$$2x + 3y = 7 \quad \text{Use the commutative property of addition.}$$

EXAMPLE 9 Write an equation of the line passing through $(-5, 4)$ and $(8, -6)$ in slope-intercept form.

Solution First we find the slope of the line.

$$m = \frac{y_2 - y_1}{x_2 - x_1}$$

$$= \frac{-6 - 4}{8 - (-5)} \quad \text{Substitute } -6 \text{ for } y_2, 4 \text{ for } y_1, 8 \text{ for } x_2, \text{ and } -5 \text{ for } x_1.$$

$$= -\frac{10}{13}$$

Because the line passes through both points, we can choose either one and substitute its coordinates into the slope-intercept form. If we choose $(-5, 4)$, we substitute -5 for x, 4 for y, and $-\frac{10}{13}$ for m and proceed as follows.

$$y = mx + b \quad \text{Slope-intercept form}$$

$$4 = -\frac{10}{13}(-5) + b \quad \text{Substitute } -\tfrac{10}{13} \text{ for } m, -5 \text{ for } x, \text{ and 4 for } y.$$

$$4 = \frac{50}{13} + b \quad \text{Multiply.}$$

$$\frac{2}{13} = b \quad \text{Subtract } \tfrac{50}{3} \text{ from both sides.}$$

The equation of the line is $y = -\frac{10}{13}x + \frac{2}{13}$.

SELF CHECK 9 Write an equation of the line passing through the points $(-2, 5)$ and $(4, -3)$ in slope-intercept form.

We now review the slope- and y-intercept method of graphing a line.

It is a straightforward approach to graph a linear equation when it is written in slope-intercept form. For example, to graph $y = \frac{4}{3}x - 2$, we note that $b = -2$ so the y-intercept is $(0, b) = (0, -2)$. (See Figure 7-15.)

Because the slope of the line is $\frac{\Delta y}{\Delta x} = \frac{4}{3}$, we locate another point on the line by starting at the point $(0, -2)$ and counting 3 units to the right and 4 units up. The line joining the two points is the graph of the equation.

We have reviewed parallel and perpendicular lines and writing equations of lines. We now combine these skills to write equations of lines given that they are parallel or perpendicular to another line.

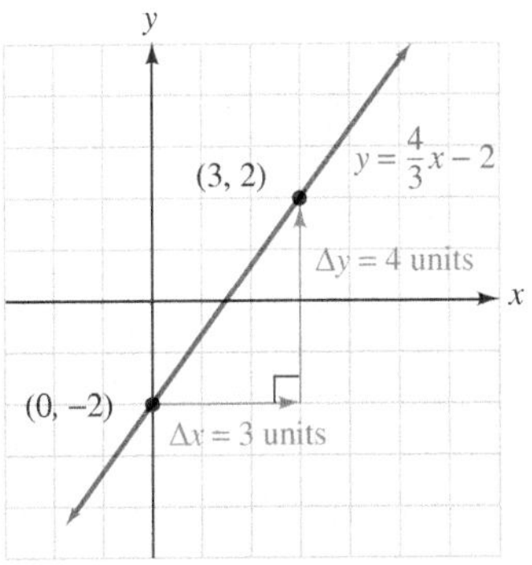

Figure 7-15

EXAMPLE 10 Write an equation of the line passing through $(-2, 5)$ and parallel to the line $y = 8x - 3$.

Solution The equation is solved for y, and the slope of the line given by $y = 8x - 3$ is the coefficient of x, which is 8. Since the desired equation is to have a graph that is parallel to the graph of $y = 8x - 3$, its slope also must be 8.

Unless otherwise noted, all content on this page is © Cengage Learning.

We will use the point-slope form, so we substitute -2 for x_1, 5 for y_1, and 8 for m and simplify.

$y - y_1 = m(x - x_1)$

$y - 5 = 8[x - (-2)]$ Substitute.

$y - 5 = 8(x + 2)$ $-(-2) = 2$

$y - 5 = 8x + 16$ Use the distributive property to remove parentheses.

$y = 8x + 21$ Add 5 to both sides.

The equation is $y = 8x + 21$.

SELF CHECK 10 Write an equation of the line that is parallel to the line $y = 6x + 8$ and passes through the origin.

EXAMPLE 11 Write an equation of the line passing through $(-2, 5)$ and perpendicular to the line $2x + 3y = 9$.

Solution To write this equation, we determine the slope by solving the equation for y.

$2x + 3y = 9$

$3y = -2x + 9$ Subtract $2x$ from both sides.

$y = -\frac{2}{3}x + 3$ Divide both sides by 3.

The slope of this line is $-\frac{2}{3}$. We need to know the slope of a line perpendicular to this line. We find the negative reciprocal, $\frac{3}{2}$, and use this as the slope for the perpendicular line. We substitute $\frac{3}{2}$ for m, -2 for x, and 5 for y and solve for b.

$y = mx + b$ Slope-intercept form

$5 = \frac{3}{2}(-2) + b$ Substitute $\frac{3}{2}$ for m, -2 for x, and 5 for y.

$5 = -3 + b$ Multiply.

$8 = b$ Add 3 to both sides.

The equation is $y = \frac{3}{2}x + 8$.

SELF CHECK 11 Write the equation of the line that is perpendicular to the line $2x + 5y = 12$ and passes through $(2, 4)$.

We summarize the various forms for the equation of a line in Table 7-2.

TABLE 7-2

Point-slope form of a linear equation	$y - y_1 = m(x - x_1)$ The slope is m, and the line passes through (x_1, y_1).
Slope-intercept form of a linear equation	$y = mx + b$ The slope is m, and the y-intercept is $(0, b)$.
General form of a linear equation	$Ax + By = C$ A and B cannot both be 0.
A horizontal line	$y = b$ The slope is 0, and the y-intercept is $(0, b)$.
A vertical line	$x = a$ The slope is undefined and the x-intercept is $(a, 0)$.

7 Write an equation of a line modeling real-world data.

COMMENT When the size of data is large, we can insert a ⧹⧹ (break) symbol on the x- or y-axis to indicate that the scale does not begin until the first value is listed.

We will now use many skills reviewed in this section to write an equation of a line modeling real-world data. We can then use the model to predict outcomes.

In statistics, the process of using one variable to predict another is called *regression.* For example, if we know a man's height, we can make a good prediction about his weight, because taller men usually weigh more than shorter men.

Figure 7-16(a) shows the result of sampling ten men at random and finding their heights and weights. The graph of the ordered pairs (h, w) is called a **scattergram**.

Man	Height in inches	Weight in pounds
1	66	140
2	68	150
3	68	165
4	70	180
5	70	165
6	71	175
7	72	200
8	74	190
9	75	210
10	75	215

(a)

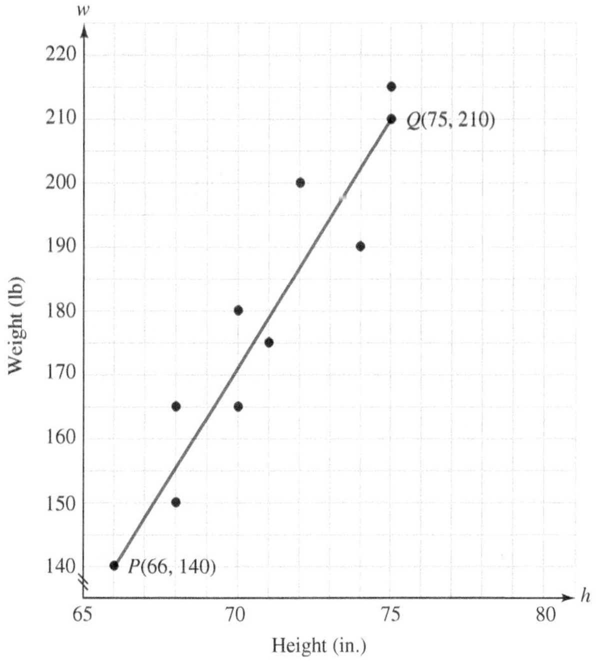

(b)

Figure 7-16

To write a prediction equation (sometimes called a *regression equation*), we must find the equation of the line that comes closer to all of the points in the scattergram than any other possible line. There are exact methods to find this equation, but we can only approximate it here.

To write an approximation of the regression equation, we place a straightedge on the scattergram shown in Figure 7-16(b) and draw the line joining two points that seems to best fit all the points. In the figure, line PQ is drawn, where point P has coordinates of $(66, 140)$ and point Q has coordinates of $(75, 210)$.

Our approximation of the regression equation will be the equation of the line passing through points P and Q. To find the equation of this line, we first find its slope.

$$\begin{aligned} m &= \frac{y_2 - y_1}{x_2 - x_1} \\ &= \frac{\mathbf{210} - \mathbf{140}}{\mathbf{75} - \mathbf{66}} \\ &= \frac{70}{9} \end{aligned}$$

We can then use point-slope form to find its equation.

Unless otherwise noted, all content on this page is © Cengage Learning.

$$y - y_1 = m(x - x_1)$$

$$y - 140 = \frac{70}{9}(x - 66) \quad \text{Choose } (66, 140) \text{ for } (x_1, y_1).$$

$$y = \frac{70}{9}x - \frac{4{,}620}{9} + 140 \quad \text{Use distributive property to remove parentheses and add 140 to both sides.}$$

$$y = \frac{70}{9}x - \frac{1{,}120}{3} \quad \text{Simplify.}$$

Using variables w and h to describe our data, our approximation of the regression equation is $w = \frac{70}{9}h - \frac{1{,}120}{3}$, where w represents weight and h represents height.

EXAMPLE 12 Using the model $w = \frac{70}{9}h - \frac{1{,}120}{3}$ (where w represents weight and h represents height), answer the following.

a. How much would you expect a 6′1″ man to weigh?
b. What does the slope represent in this model?
c. What is the w-intercept in this model? Is it reasonable?

Solution **a.** Because our data is given in inches, we convert 6′1″ to 73 inches. To predict the weight of a man this tall we substitute 73 for h in the model and simplify.

$$w = \frac{70}{9}h - \frac{1{,}120}{3}$$

$$w = \frac{70}{9}(73) - \frac{1{,}120}{3}$$

$$w \approx 194.4$$

We would predict that a 73-inch-tall man chosen at random will weigh about 194 pounds.

b. The slope is defined to be $\frac{\text{rise}}{\text{run}}$. In this model, for each inch in height increase, there is an increase of approximately $\frac{70}{9}$ pounds, or about 7.8 pounds.

c. The w-intercept is $\left(0, -\frac{1{,}120}{3}\right)$, or approximately $(0, -373)$. This is not reasonable because a person cannot weigh a negative amount.

SELF CHECK 12 A caterer charged \$4,625 for an event for 100 people and \$2,375 for an event for 50 people.

a. Write a model that relates the total cost and the number of people attending an event.
b. What does the slope represent in this model?
c. What is the c-intercept in this model and what does it represent?
d. How much would an event cost for 75 people?
e. Explain why the cost for 100 persons is not twice as much as for 50 people.

SELF CHECK ANSWERS

1. a.

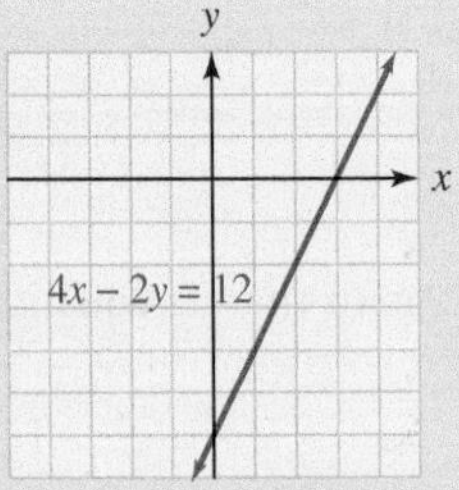

b.

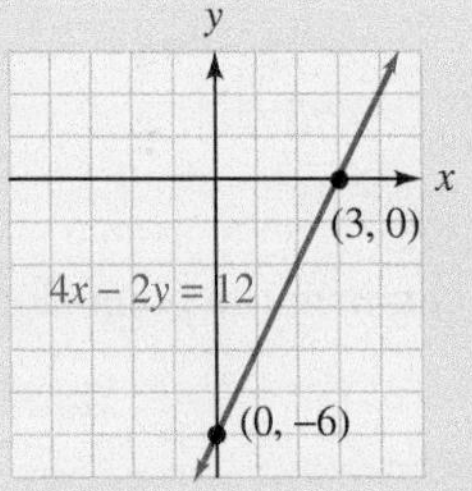

2.

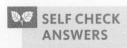

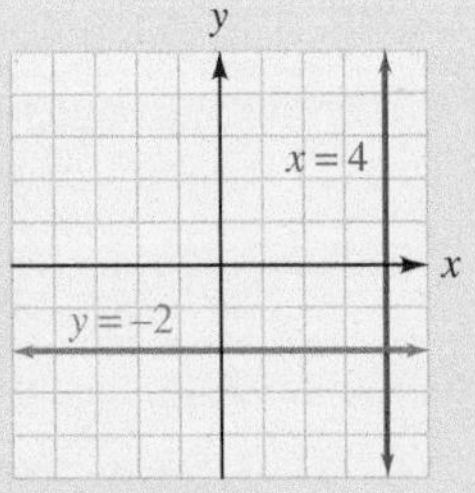

Unless otherwise noted, all content on this page is © Cengage Learning.

3. $\left(\frac{3}{2}, 1\right)$ **4.** $\frac{3}{2}$ **5.** $\frac{10}{9}$ **6.** $-\frac{2}{5}$ **7.** perpendicular **8.** $y - 5 = \frac{5}{4}x$; $y = \frac{5}{4}x + 5$ **9.** $y = -\frac{4}{3}x + \frac{7}{3}$ **10.** $y = 6x$ **11.** $y = \frac{5}{2}x - 1$ **12. a.** $c = 45p + 125$, where p represents the number of people at an event and c represents the total cost of the event **b.** The slope represents the cost per person, $45. **c.** The c-intercept represents the caterer's base cost, $125.00 **d.** $3,500 **e.** The total cost includes a $125 base cost regardless of the number of people attending.

NOW TRY THIS

1. Find the center of a circle with a diameter having endpoints at $(-5, 2)$ and $(-9, -7)$.

2. Find the midpoint of the segment joining
   a. $(p - 2, p)$ and $(4 - p, 5p - 2)$
   b. $(p + 5, 3p - 1)$ and $(6p, p + 9)$

3. Which of the following graphs could be the graph of $y = -px - q$, where $p > 0$ and $q > 0$?

a.
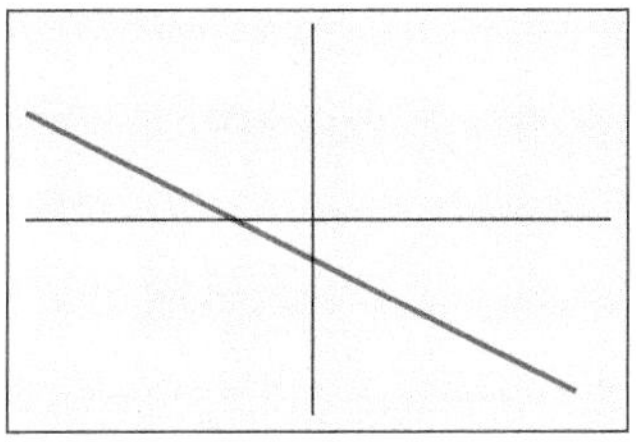

b.
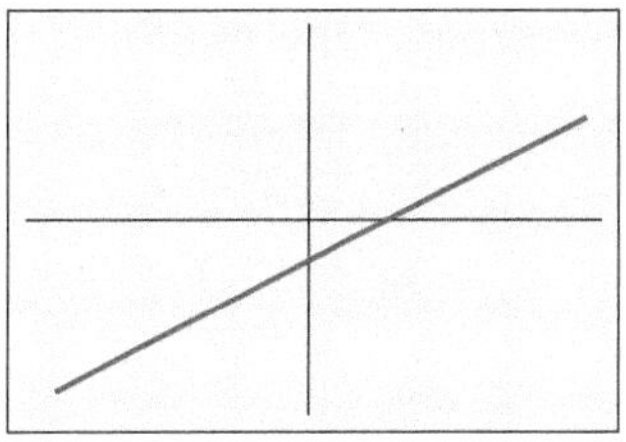

c.
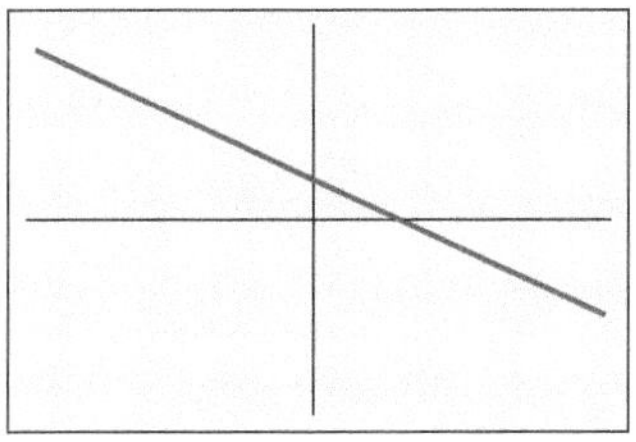

d.
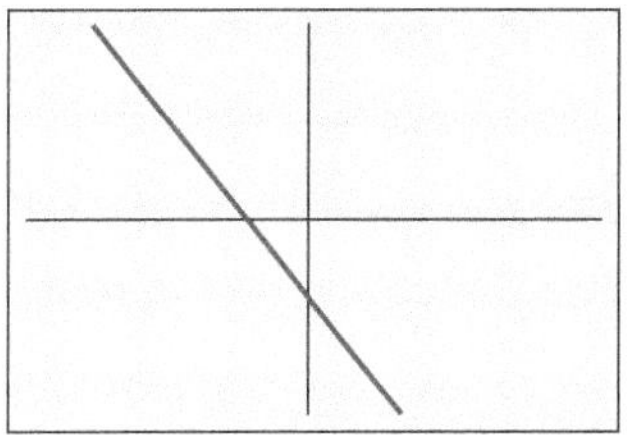

4. Write an equation of a line in slope-intercept form (if possible) with the given information.
   a. parallel to $y = -3$ through $(4, 6)$
   b. perpendicular to $x = 4$ through $(0, 0)$
   c. parallel to $3x = 2y + 8$ through $(-4, -6)$

7.2 Exercises

WARM-UPS *Determine whether the point (5, −2) is a solution to the given equations.*

1. $2x - 4y = 2$
2. $4x - 5y = 30$
3. $3x + 5y = 5$
4. $x + 2y = 9$

Determine whether the slope of the line in each graph is positive, negative, 0, or undefined.

5.
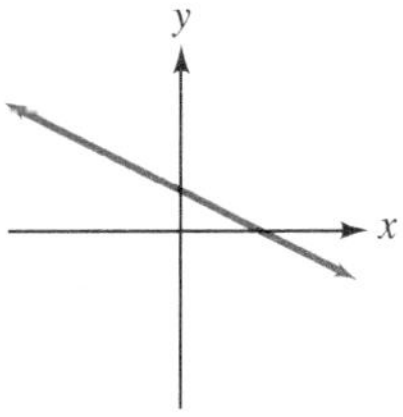

6.
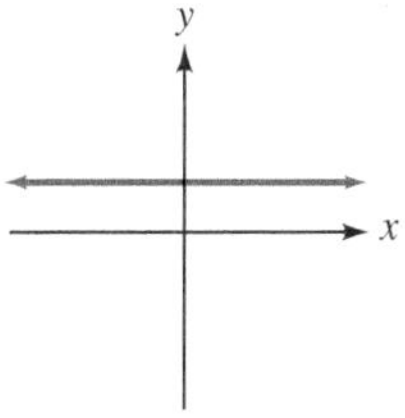

Unless otherwise noted, all content on this page is © Cengage Learning.

7.

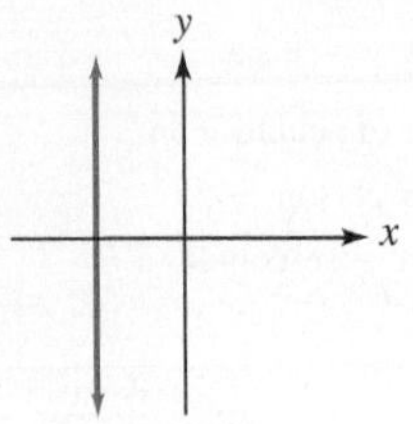

8.

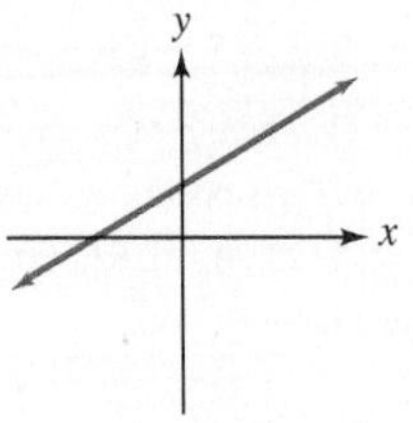

Determine whether the equations are in point-slope, slope-intercept, or general form.

9. $5x - 2y = 8$

10. $y - 6 = 5(x + 4)$

11. $y = -\frac{4}{7}x + 1$

12. $6x + 8y = 12$

REVIEW *Simplify each expression. Write all answers without negative exponents. Assume no variable is 0.*

13. $(x^3y^2)^3$

14. $\left(\frac{x^5}{x^3}\right)^3$

15. $(x^{-3}y^2)^{-4}$

16. $\left(\frac{3x^2y^3}{8}\right)^0$

Solve each equation.

17. $3(x + 2) + x = 5x$

18. $12b + 6(3 - b) = b + 3$

19. $\frac{5(2 - x)}{3} - 1 = x + 5$

20. $\frac{r - 1}{3} = \frac{r + 2}{6} + 2$

VOCABULARY AND CONCEPTS *Fill in the blanks.*

21. The y-intercept of a line is the point where the line intersects the _____.

22. The x-intercept of a line is the point where the line intersects the _____.

23. The graph of any equation of the form $x = a$, where a is a constant, is a ______ line.

24. The graph of any equation of the form $y = b$, where b is a constant, is a ________ line.

25. The midpoint of a segment with endpoints at (a, b) and (c, d) has coordinates of __________.

26. Slope is defined as the change in _ divided by the change in _.

27. A slope is a rate of ______.

28. The formula to compute slope is $m =$ _____.

29. The change in y (denoted as Δy) is the ___ of the line between two points.

30. The change in x (denoted as Δx) is the ___ of the line between two points.

31. The slope of a ________ line is 0.

32. The slope of a ______ line has an undefined slope.

33. The slopes of nonvertical ___________ lines are negative ________.

34. ______ lines have the same slope.

GUIDED PRACTICE *Graph each equation using a table of values.* SEE EXAMPLE 1. (OBJECTIVE 1)

35. $x + y = 4$

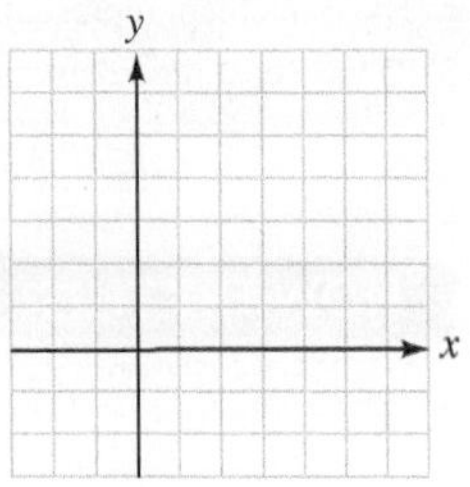

36. $x - y = 2$

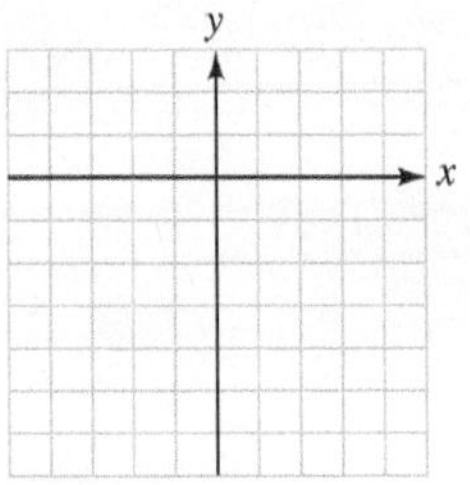

37. $2x - y = 3$

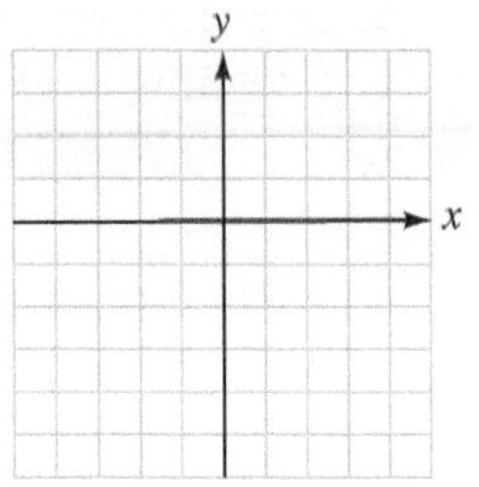

38. $x + 2y = 5$

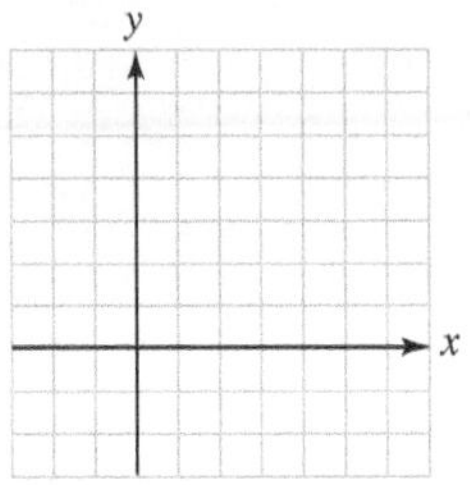

Graph each equation using x- and y-intercepts. SEE EXAMPLE 1. (OBJECTIVE 1)

39. $3x + 4y = 12$

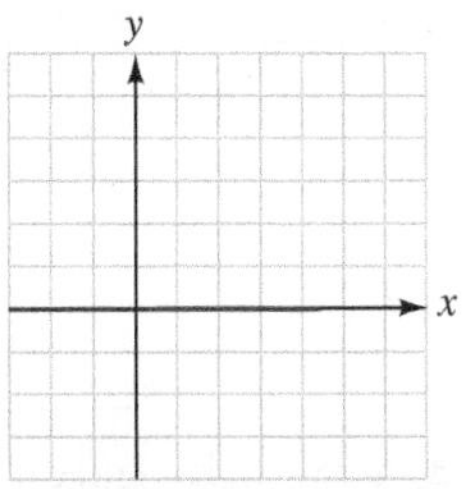

40. $4x - 3y = 12$

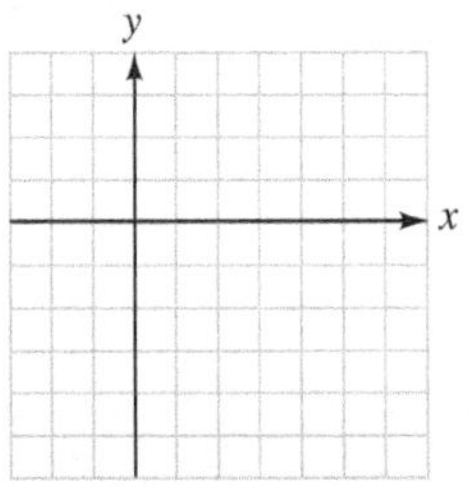

41. $y = -3x + 2$

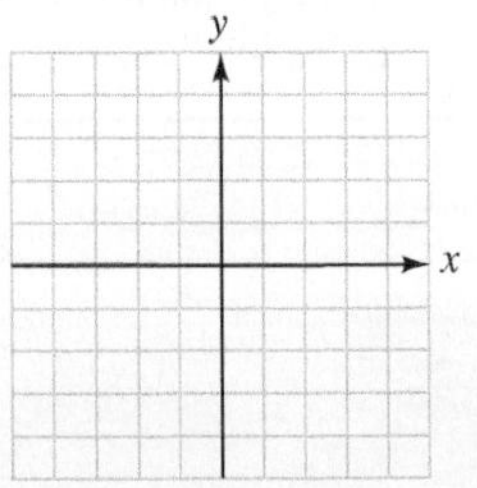

42. $y = 2x - 3$

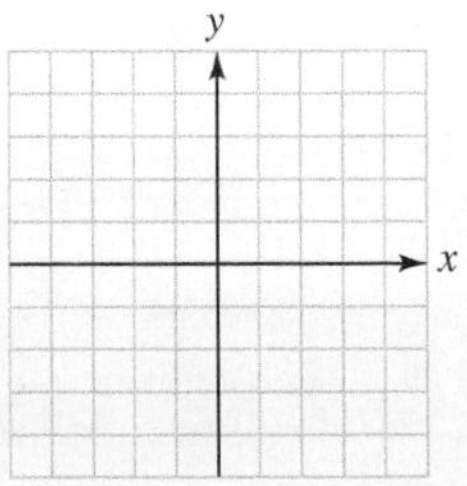

Graph each equation. SEE EXAMPLE 2. (OBJECTIVE 2)

43. $x = 3$

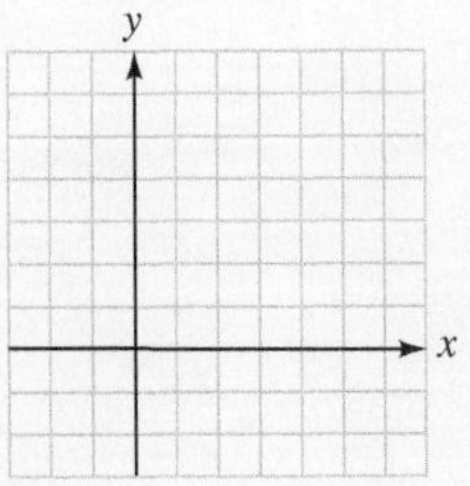

44. $y = -4$

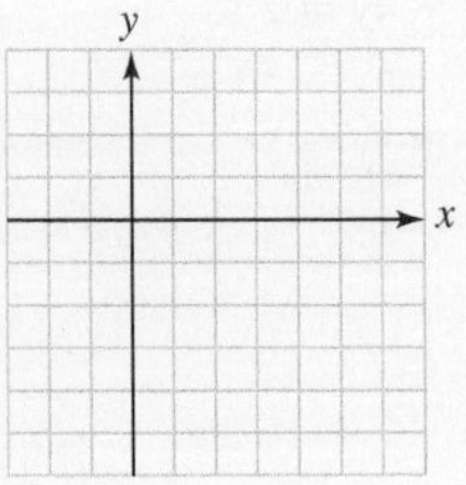

Unless otherwise noted, all content on this page is © Cengage Learning.

45. $-3y + 2 = 5$

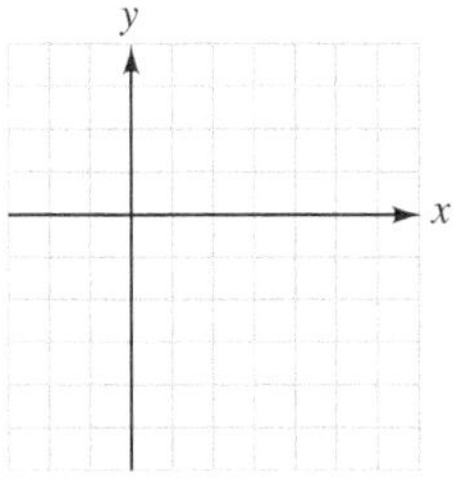

46. $-2x + 3 = 11$

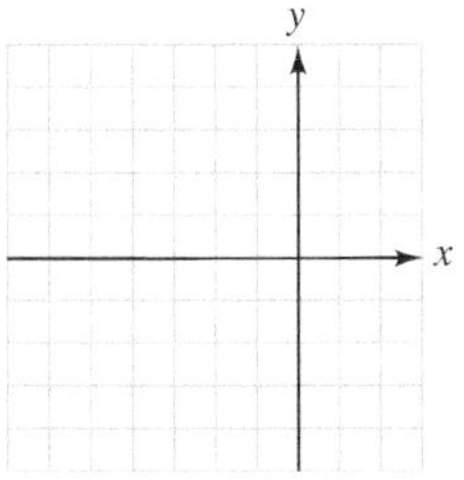

Find the midpoint of the segment joining the given points. SEE EXAMPLE 3. (OBJECTIVE 3)

47. $(0, 0), (6, 8)$
48. $(10, 4), (2, -2)$
49. $(-2, -8), (3, 4)$
50. $(-5, -2), (7, 3)$

Find the slope of the line. SEE EXAMPLE 4. (OBJECTIVE 4)

51.

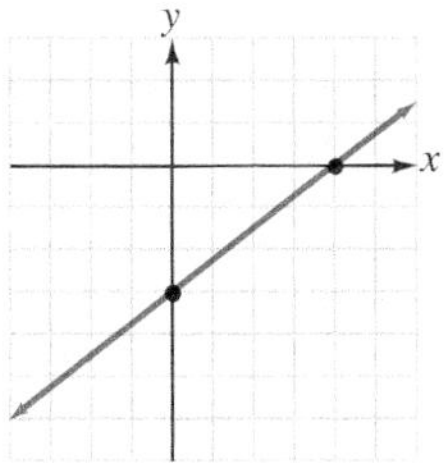

52.

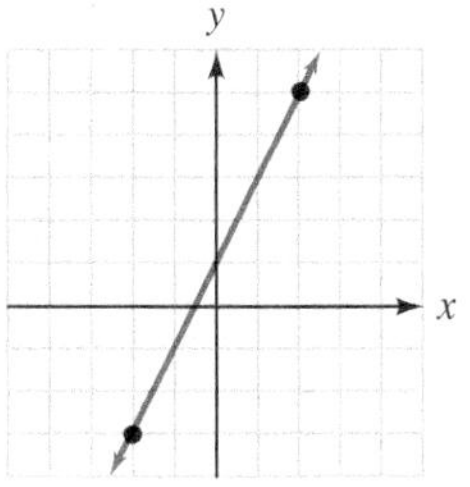

53.

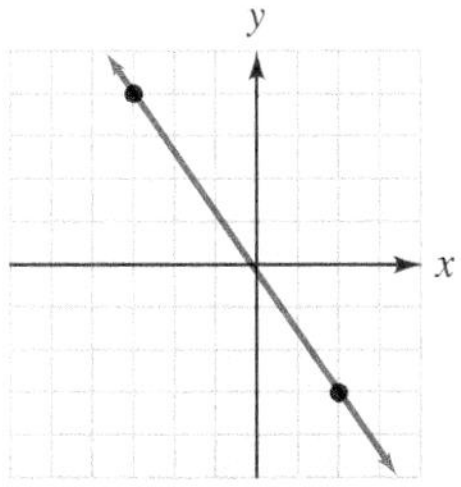

54.

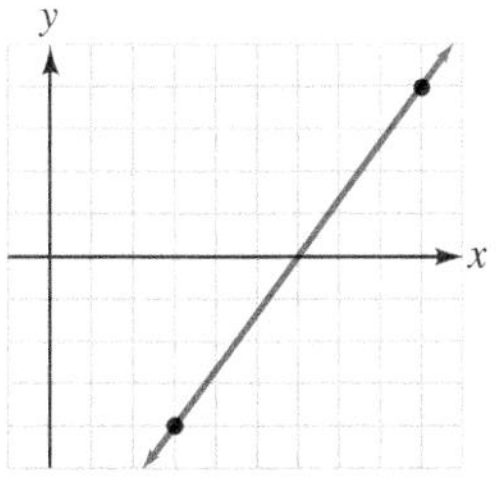

Find the slope of the line that passes through the given points, if possible. SEE EXAMPLE 5. (OBJECTIVE 4)

55. $(0, 0), (3, 9)$
56. $(9, 6), (0, 0)$
57. $(-1, 8), (6, 1)$
58. $(-5, -8), (3, 8)$
59. $(7, 5), (-9, 5)$
60. $(0, -8), (-5, 0)$
61. $(-7, -5), (-7, -2)$
62. $(3, -5), (3, 14)$

Find the slope of the line determined by each equation. SEE EXAMPLE 6. (OBJECTIVE 4)

63. $3x + 2y = 12$
64. $2x - y = 6$
65. $3x = 4y - 2$
66. $x = y$
67. $y = \dfrac{x - 4}{2}$
68. $x = \dfrac{3 - y}{4}$
69. $4y = 3(y + 2)$
70. $x + y = \dfrac{2 - 3y}{3}$

Determine whether the graphs of each pair of equations are parallel, perpendicular, or neither parallel nor perpendicular. SEE EXAMPLE 7. (OBJECTIVE 5)

71. $y = 3x + 4, y = 3x - 7$
72. $y = 4x - 13, y = \frac{1}{4}x + 13$
73. $x + y = 2, y = x + 5$
74. $x = y + 2, y = x + 3$
75. $y = 3x + 7, 2y = 6x - 9$
76. $2x + 3y = 9, 3x - 2y = 5$
77. $x = 3y + 4, y = -3x + 7$
78. $3x + 6y = 1, y = \frac{1}{2}x$

Use point-slope form to write the equation of the line with the given properties. SEE EXAMPLE 8. (OBJECTIVE 6)

79. $m = 5$, passing through $(0, 7)$
80. $m = -8$, passing through $(0, -2)$
81. $m = -3$, passing through $(2, 0)$
82. $m = 4$, passing through $(-5, 0)$

Use slope-intercept form to write the equation of the line with the given properties. SEE EXAMPLE 8. (OBJECTIVE 6)

83. $m = 3, b = 17$
84. $m = -2, b = 11$
85. $m = -7$, passing through $(7, 5)$
86. $m = 3$, passing through $(-2, -5)$

Use slope-intercept form to write the equation of the line passing through the two given points. SEE EXAMPLE 9. (OBJECTIVE 6)

87. $P(0, 0), Q(4, 4)$
88. $P(-5, -5), Q(0, 0)$
89. $P(3, 4), Q(0, -3)$
90. $P(4, 0), Q(6, -8)$

Write the equation of the line that passes through the given point and is parallel to the given line. Write the answer in slope-intercept form. SEE EXAMPLE 10. (OBJECTIVE 6)

91. $(0, 0)$, parallel to $y = 4x - 7$
92. $(0, 0)$, parallel to $x = -3y - 12$
93. $(2, 5)$, parallel to $4x - y = 7$
94. $(-6, 3)$, parallel to $y + 3x = -12$

Write the equation of the line that passes through the given point and is perpendicular to the given line. Write the answer in slope-intercept form. SEE EXAMPLE 11. (OBJECTIVE 6)

95. $(0, 0)$, perpendicular to $y = 4x - 7$
96. $(0, 0)$, perpendicular to $x = -3y - 12$
97. $(2, 5)$, perpendicular to $4x - y = 7$
98. $(-6, 3)$, perpendicular to $y + 3x = -12$

Do as indicated. SEE EXAMPLE 12. (OBJECTIVE 7)

A party planner charges \$6,775 for a formal party for 100 people and \$3,525 for 50 people.

99. Write an equation modeling this data.
100. What does the slope represent in this model?

Unless otherwise noted, all content on this page is © Cengage Learning.

A guttering company charges \$1,359.80 for installing 340 feet of guttering and \$839 for 200 feet of guttering.

101. Write an equation modeling this data.

102. What does the y-intercept represent in this model?

ADDITIONAL PRACTICE *Graph each equation.*

103. $3y = 6x - 9$

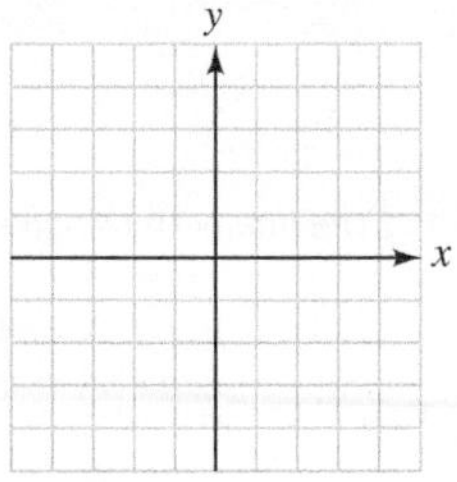

104. $-2y - 3x + 9 = 0$

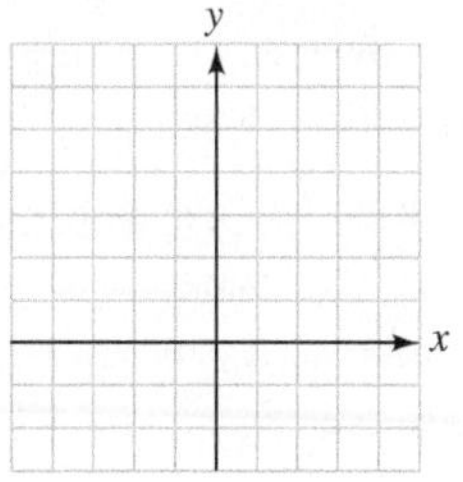

State the slope and the y-intercept of the line determined by the given equation.

105. $3x - 2y = 8$

106. $-2(x + 3y) = 5$

107. $-2x + 4y = 12$

108. $5(2x - 3y) = 4$

Write each equation in slope-intercept form to find the slope and the y-intercept. Then use the slope and y-intercept to graph the line.

109. $y + 1 = x$

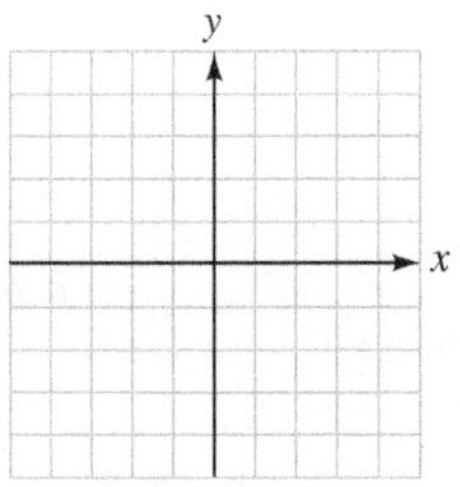

110. $x + y = 2$

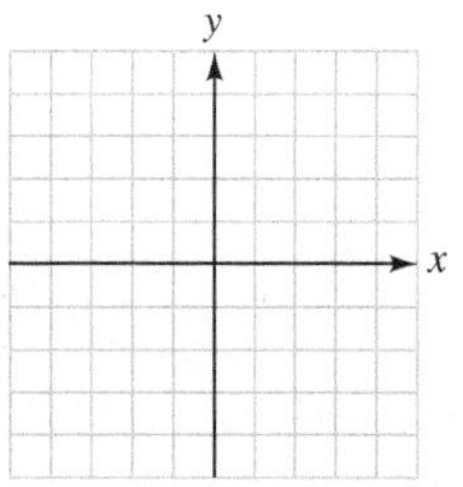

111. $x = \frac{3}{2}y - 3$

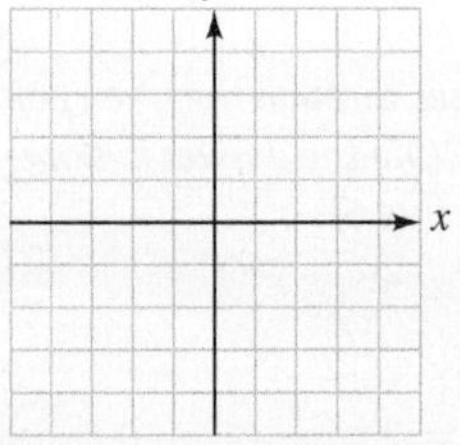

112. $x = -\frac{4}{5}y + 2$

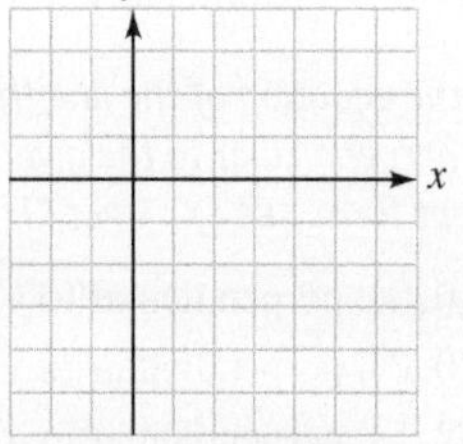

Write an equation of the line in slope-intercept form.

113.

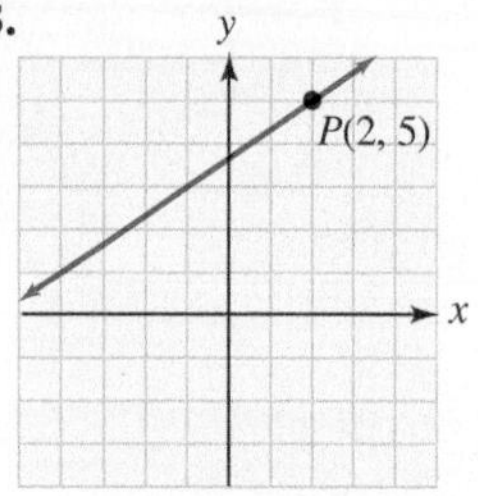

114.

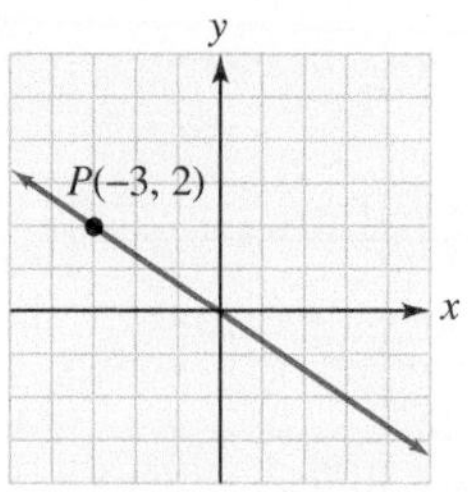

115.

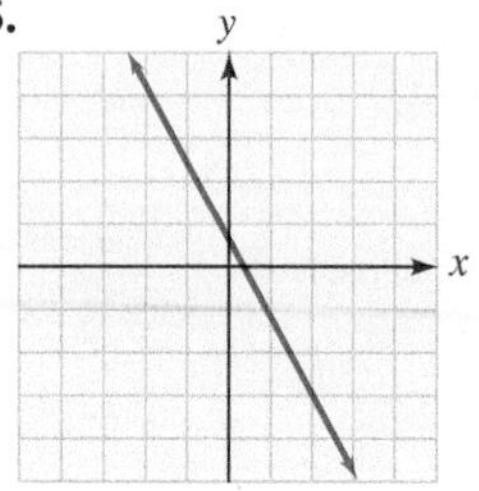

116.

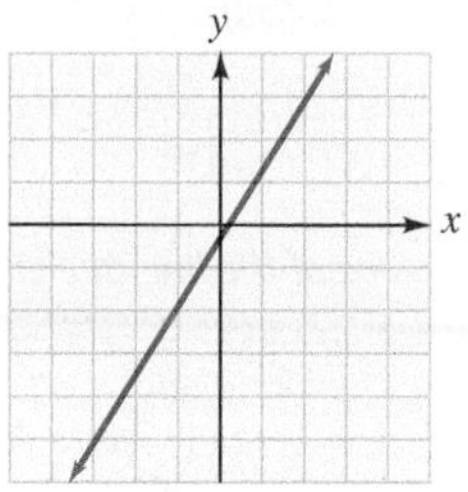

Write the equation of the line that passes through the given point and is parallel or perpendicular to the given line. Write the answer in slope-intercept form.

117. $(4, -2)$, parallel to $x = \frac{5}{4}y - 2$

118. $(1, -5)$, parallel to $x = -\frac{3}{4}y + 5$

119. $(4, -2)$, perpendicular to $x = \frac{5}{4}y - 2$

120. $(1, -5)$, perpendicular to $x = -\frac{3}{4}y + 5$

Find the midpoint of the segment joining the given points.

121. $(a - b, b), (a + b, 3b)$

122. $(2, -3), (4, -8)$

APPLICATIONS *For problems involving depreciation or appreciation, assume straight-line depreciation or appreciation.*

123. House appreciation A house purchased for \$125,000 is expected to appreciate according to the formula $y = 7{,}500x + 125{,}000$, where y is the value of the house after x years. Find the value of the house 5 years later and 10 years later.

124. Demand equations The number of TVs that consumers buy depends on price. The higher the price, the fewer people will buy. The equation that relates price to the number of TVs sold at that price is called a **demand equation**.

For a 25-inch TV, this equation is $p = -\frac{1}{10}q + 170$, where p is the price and q is the number of TVs sold at that price. How many TVs will be sold at a price of \$150?

Unless otherwise noted, all content on this page is © Cengage Learning.

125. Supply equations The number of TVs that manufacturers produce depends on price. The higher the price, the more TVs manufacturers will produce. The equation that relates price to the number of TVs produced at that price is called a **supply equation**.

For a 25-inch TV, the supply equation is $p = \frac{1}{10}q + 130$, where p is the price and q is the number of TVs produced for sale at that price. How many TVs will be produced if the price is \$150?

126. Crime prevention The number n of incidents of daytime burglaries requiring police response appears to be related to d, the money spent on salaries for residential patrols, by the equation

$$n = 430 - 0.005d$$

What expenditure would reduce the number of incidents to 350?

127. Grade of a road Find the slope of the road.

(*Hint*: 1 mi = 5,280 ft.)

128. Slope of a roof Find the slope of the roof.

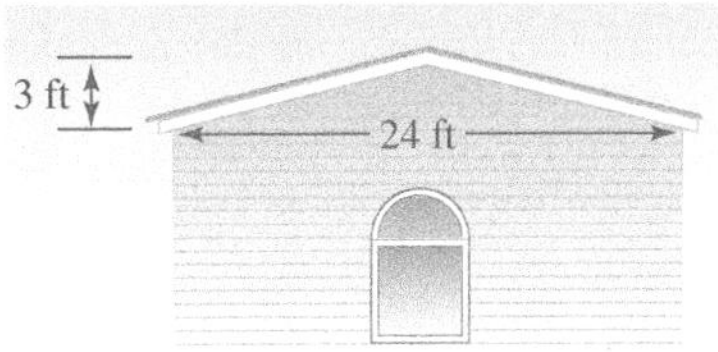

129. Physical fitness Find the slope of the treadmill for each setting listed in the table.

Height setting
2 in.
5 in.
8 in.

130. Value of a painting In 1987, the painting called *Irises* by Vincent van Gogh was auctioned from a starting value of \$15 million. It was sold 2 minutes later for \$49 million. Find the rate of increase in seconds.

131. Global warming The following line graphs estimate the global temperature rise between the years of 2010 and 2040. Find the average rate of temperature change (the slope) of Model A: Status quo.

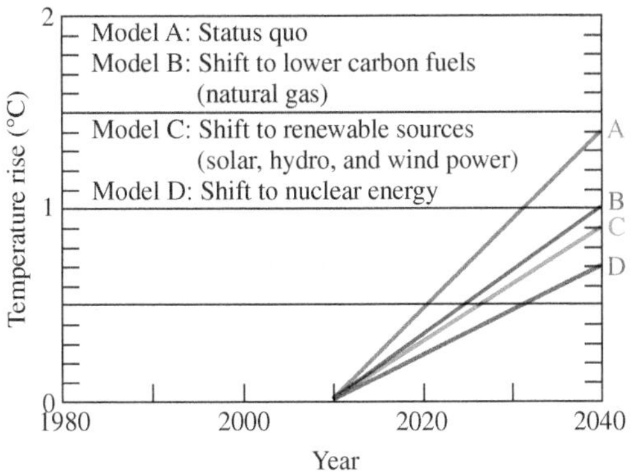

132. Predicting burglaries A police department knows that city growth and the number of burglaries are related by a linear equation. City records show that 575 burglaries were reported in a year when the local population was 77,000, and 675 were reported in a year when the population was 87,000. How many burglaries can be expected when the population reaches 110,000?

133. Wheelchair ramps The illustration shows two designs for a ramp to make a platform wheelchair accessible.

a. Find the slope of the ramp shown in design 1.

b. Find the slope of each part of the ramp shown in design 2.

c. Give one advantage and one disadvantage of each design.

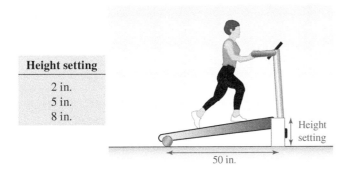

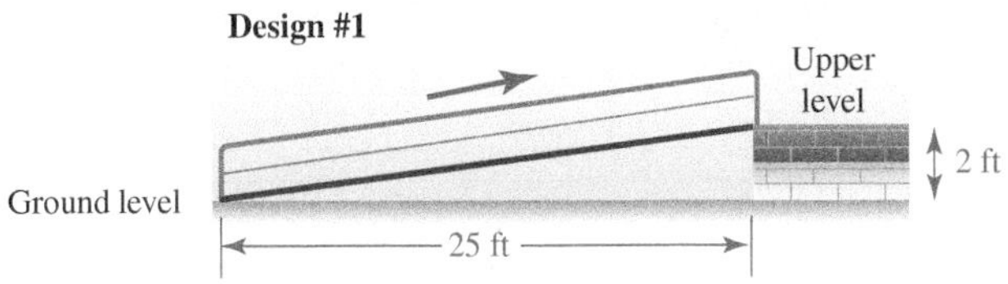

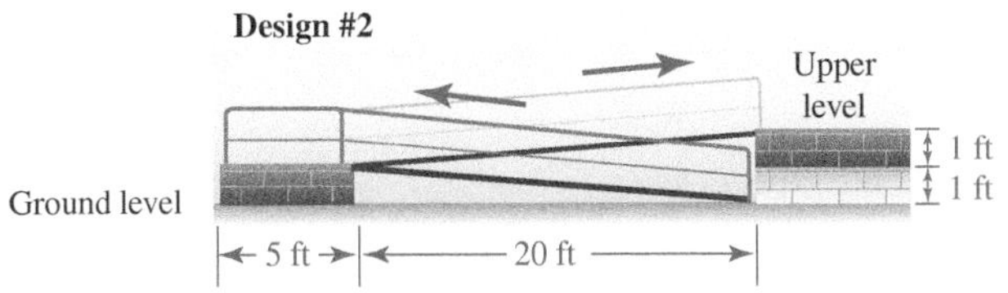

Unless otherwise noted, all content on this page is © Cengage Learning.

134. Car repair A garage charges a fixed amount, plus an hourly rate, to service a car. Use the information in the table to find the hourly rate.

A-1 Car Repair Typical charges	
2 hours	\$143
5 hours	\$320

135. Depreciation equations A business purchased the computer shown. It will be depreciated over a 5-year period, when it will probably be worth \$200. Find the depreciation equation.

136. Jackson Pollock's *No. 5, 1948* sold for \$140 million in 2006. If its estimated value in 2011 was \$156.8 million, find the straight-line appreciation equation.

137. Real estate listings Use the information given in the following description of the property to write a straight-line appreciation equation for the price of the house.

Vacation Home
\$122,000
Only 2 years old

- Great investment property!
- Expected to appreciate \$4,000/yr

Sq ft: 1,635	**Fam rm:** yes	**Den:** no
Bdrm: 3	**Ba:** 1.5	**Gar:** enclosed
A/C: yes	**Firepl:** yes	**Kit:** built-ins

138. Depreciation equations Find the depreciation equation for the value of the TV in the want ad in the illustration.

For Sale: 3-year-old 65-inch TV, \$1,750 new. Asking \$800. Call 715-5588. Ask for Joe.

WRITING ABOUT MATH

139. Explain how to graph a line using the intercept method.

140. Explain why a vertical line has an undefined slope.

141. Explain how to determine from their slopes whether two lines are parallel, perpendicular, or neither.

142. Explain how to find the equation of a line passing through two given points.

SOMETHING TO THINK ABOUT

143. If the line $y = ax + b$ passes through only quadrants I and II, what can be known about a and b?

144. What are the coordinates of the three points that divide the segment joining $P(a, b)$ and $Q(c, d)$ into four equal parts?

145. Find the slope of the line $Ax + By = C$.

146. The line passing through points $(1, 3)$ and $(-2, 7)$ is perpendicular to the line passing through points $(4, b)$ and $(8, -1)$. Find the value of b.

Unless otherwise noted, all content on this page is © Cengage Learning.

Section 7.3 Review of Functions

Objectives

1. Find the domain and range of a set of ordered pairs and determine whether the set is a function.
2. Find the domain and range from a graph and determine whether the graph represents a function.
3. Use function notation to evaluate a function at a given value.
4. Find the domain of a function given its equation.
5. Graph a function and determine the domain and range.
6. Write a linear function modeling real world data.

Vocabulary

relation	function	independent variable
domain	vertical line test	linear function
range	dependent variable	base year

Getting Ready

If $y = \frac{3}{2}x - 2$, *find the value of y for each value of x.*

1. $x = 2$ **2.** $x = 6$ **3.** $x = -12$ **4.** $x = -\frac{1}{2}$

In this section, we will review *relations* and *functions.*

1 Find the domain and range of a set of ordered pairs and determine whether the set is a function.

Table 7-3 shows the number of women serving in the U.S. House of Representatives for several recent sessions of Congress.

TABLE 7-3

Women in the U.S. House of Representatives

Session of Congress	107th	108th	109th	110th	111th	112th
Number of Female Representatives	59	59	68	71	76	73

We can display the data in the table as a set of ordered pairs, where the first component (or *input*) represents the session of Congress and the second component (or *output*) represents the number of women serving in that session.

$$(107, 59) \quad (108, 59) \quad (109, 68) \quad (110, 71) \quad (111, 76) \quad (112, 73)$$

Sets of ordered pairs like this are called **relations**. The set of all *first components* $\{107, 108, 109, 110, 111, 112\}$ is called the **domain** of the relation, and the set of all *second components* $\{59, 68, 71, 76, 73\}$ is called the **range** of the relation. Although 59 occurs twice as an output value, we list it only once in the range.

When each first component in a relation determines exactly one second component, the relation is called a **function**.

FUNCTIONS

A function is any set of ordered pairs (a relation) in which each first component (or input value) determines exactly one second component (or output value).

EXAMPLE 1 Find the domain and range of the relation $\{(3, 2), (5, -7), (-8, 2), (-9, -12)\}$ and determine whether it represents a function.

Solution Because the set of first components represents the domain, the domain is $\{3, 5, -8, -9\}$. Because the set of second components represents the range, the range is $\{2, -7, -12\}$. In this relation

the first component of 3 determines a second component of 2

the first component of 5 determines a second component of -7

the first component of -8 determines a second component of 2

the first component of -9 determines a second component of -12

Since each component in the domain determines exactly one component in the range, this relation is a function.

SELF CHECK 1 Find the domain and range of the relation $\{(-5, 6), (-12, 4), (8, 6), (5, 4)\}$ and determine whether it defines a function.

2 Find the domain and range from a graph and determine whether the graph represents a function.

A **vertical line test** can be used to determine whether the graph of an equation represents a function. If every vertical line that intersects a graph does so exactly once, the graph represents a function, because every number x determines a single value of y. If any vertical line that intersects a graph does so more than once, the graph cannot represent a function, because to one number x there would correspond more than one value of y.

The graph in Figure 7-17(a) represents a function, because every vertical line that intersects the graph does so exactly once. The graph in Figure 7-17(b) does not represent a function, because some vertical lines intersect the graph more than once.

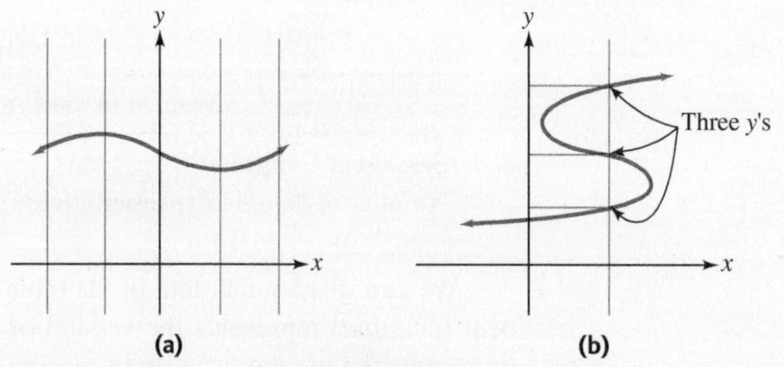

Figure 7-17

Unless otherwise noted, all content on this page is © Cengage Learning.

EXAMPLE 2 Find the domain and range of the relation determined by each graph and then state whether the graph defines a function.

a.

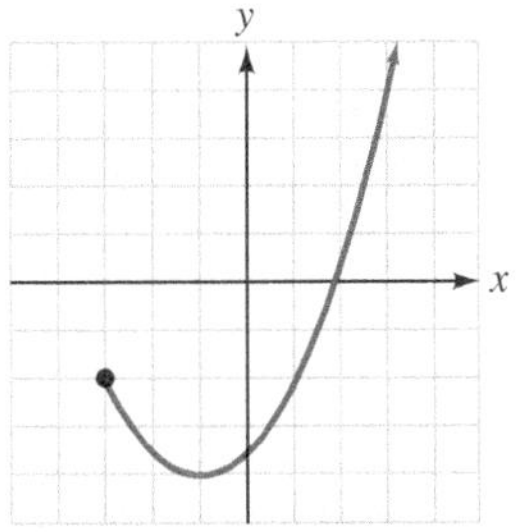

b.

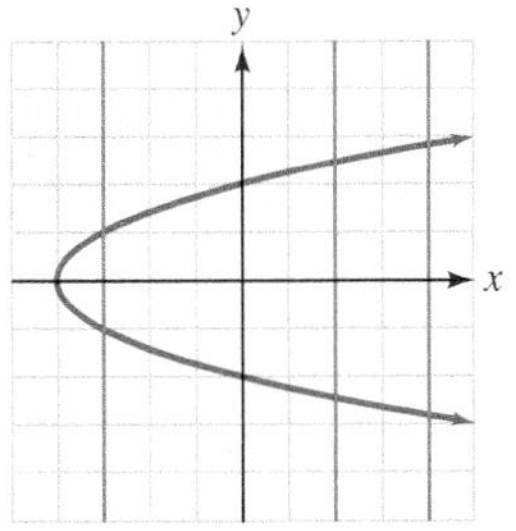

Solution

a. To find the domain, we look at the leftmost point on the graph and identify -3 as its x-coordinate. Since the graph continues forever to the right, there is no rightmost point. Therefore, the domain is $[-3, \infty)$.

To find the range, we look for the lowest point on the graph and identify -4 as its y-coordinate. Since the graph continues upward forever, there is no highest point. Therefore, the range is $[-4, \infty)$.

Since every vertical line that intersects the graph will do so exactly once, the vertical line test indicates that the graph is a function.

b. To find the domain, we note that the x-coordinate of the leftmost point is -4 and that there is no rightmost point. Therefore, the domain is $[-4, \infty)$.

To find the range, we note that there is no lowest point or highest point. Therefore, the range is $(-\infty, \infty)$.

Since many vertical lines that intersect the graph will do so more than once, the vertical line test indicates that the graph is not a function.

SELF CHECK 2 Find the domain and range of the relation determined by the graph and then tell whether the graph defines a function.

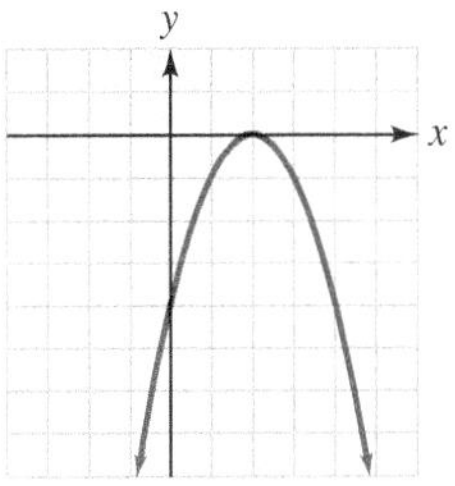

3 Use function notation to evaluate a function at a given value.

In Chapter 3, we introduced the following special notation, which is used to denote functions.

FUNCTION NOTATION

The notation $y = f(x)$ denotes that the variable y is a function of x.

COMMENT The notation $f(x)$ does not mean "f times x."

The notation $y = f(x)$ is read as "y equals f of x." Note that y and $f(x)$ are two notations for the same quantity. Thus, the equations $y = 4x + 3$ and $f(x) = 4x + 3$ are equivalent.

The notation $y = f(x)$ provides a way of denoting the value of y (the **dependent variable**) that corresponds to some number x (the **independent variable**).

Unless otherwise noted, all content on this page is © Cengage Learning.

EXAMPLE 3 Let $f(x) = 4x + 3$. Find: **a.** $f(3)$ **b.** $f(-1)$ **c.** $f(0)$
d. the value of x for which $f(x) = 7$

Solution **a.** We replace x with 3:

$$f(x) = 4x + 3$$
$$f(3) = 4(3) + 3$$
$$= 12 + 3$$
$$= 15$$

b. We replace x with -1:

$$f(x) = 4x + 3$$
$$f(-1) = 4(-1) + 3$$
$$= -4 + 3$$
$$= -1$$

c. We replace x with 0:

$$f(x) = 4x + 3$$
$$f(0) = 4(0) + 3$$
$$= 3$$

d. We replace $f(x)$ with 7.

$$f(x) = 4x + 3$$
$$7 = 4x + 3$$
$$4 = 4x \quad \text{Subtract 3 from both sides.}$$
$$x = 1 \quad \text{Divide both sides by 4.}$$

SELF CHECK 3 If $f(x) = -2x - 1$, find: **a.** $f(2)$ **b.** $f(-3)$ **c.** $f(0)$
d. the value of x for which $f(x) = -7$

We can think of a function as a machine that takes some input x and turns it into some output $f(x)$, as shown in Figure 7-18. The machine shown in Figure 7-19 turns the input number 2 into the output value -3 and turns the input number 6 into the output value -11. The set of numbers that we can put into the machine is the domain of the function, and the set of numbers that comes out is the range.

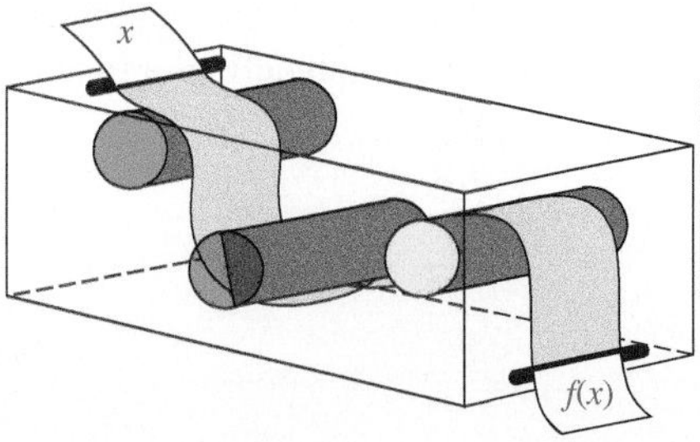

Figure 7-18

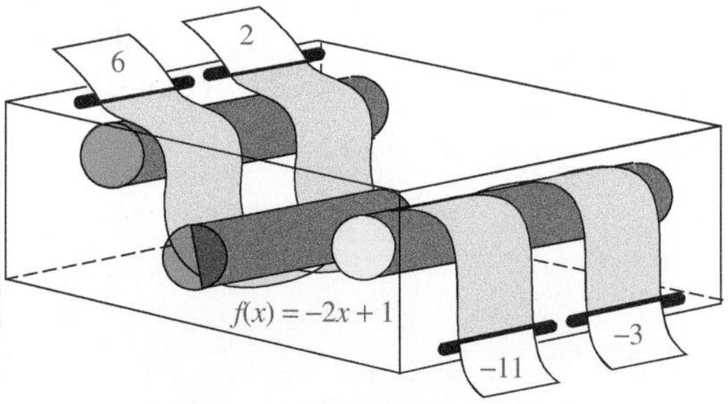

Figure 7-19

The letter f used in the notation $y = f(x)$ represents the word *function*. However, other letters can be used to represent functions. The notations $y = g(x)$ and $y = h(x)$ also denote functions involving the independent variable x.

EXAMPLE 4 Let $f(x) = 4x - 1$. Find: **a.** $f(3) + f(2)$ **b.** $f(a) - f(b)$

Solution **a.** We find $f(3)$ and $f(2)$ separately.

$$f(x) = 4x - 1 \qquad f(x) = 4x - 1$$
$$f(3) = 4(3) - 1 \qquad f(2) = 4(2) - 1$$
$$= 12 - 1 \qquad = 8 - 1$$
$$= 11 \qquad = 7$$

We then add the results to obtain $f(3) + f(2) = 11 + 7 = 18$.

b. We find $f(a)$ and $f(b)$ separately.

$$f(x) = 4x - 1 \qquad f(x) = 4x - 1$$
$$f(a) = 4a - 1 \qquad f(b) = 4b - 1$$

Unless otherwise noted, all content on this page is © Cengage Learning.

We then subtract the results to obtain

$$\begin{aligned} f(a) - f(b) &= (4a - 1) - (4b - 1) \\ &= 4a - 1 - 4b + 1 \\ &= 4a - 4b \end{aligned}$$

SELF CHECK 4 Let $g(x) = -2x + 3$. Find: **a.** $g(-2) + g(3)$ **b.** $g(a) - g(b)$

4 Find the domain of a function given its equation.

The domain of a function that is defined by an equation is the set of all numbers that are permissible replacements for its variable.

EXAMPLE 5 Find the domain of the functions defined by

a. $f(x) = x^2 + 8x - 3$ **b.** $f(x) = \dfrac{1}{x - 2}$

Solution **a.** Since any real number can be substituted for x in the function $f(x) = x^2 + 8x - 3$ to obtain a single value y, the domain is $(-\infty, \infty)$.

b. The number 2 cannot be substituted for x in the function $f(x) = \frac{1}{x - 2}$, because that would make the denominator 0. However, any real number, except 2, can be substituted for x to obtain a single value y. Therefore, the domain is the set of all real numbers except 2. This is the interval $(-\infty, 2) \cup (2, \infty)$.

SELF CHECK 5 Find the domain of the function defined by

a. $g(x) = -9x + 2$ **b.** $h(x) = \dfrac{2}{3x-1}$

5 Graph a function and determine the domain and range.

The *graph of a function* is the graph of the ordered pairs $(x, f(x))$ that define the function. To illustrate, we graph the function $f(x) = -2x + 1$ as in Figure 7-20. Since every real number x determines a corresponding value of y, the domain is the interval $(-\infty, \infty)$. Since the values of y can be any real number, the range is the interval $(-\infty, \infty)$. This graph is that of a linear function.

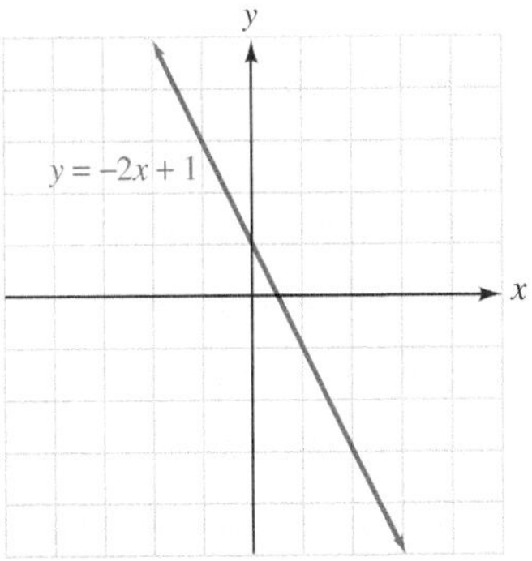

Figure 7-20

LINEAR FUNCTIONS

A **linear function** is a function defined by an equation that can be written in the form

$$f(x) = mx + b \quad \text{or} \quad y = mx + b$$

where m is the slope of the line graph and $(0, b)$ is the y-intercept.

Unless otherwise noted, all content on this page is © Cengage Learning.

EXAMPLE 6 Graph the function $f(x) = \left|\frac{3}{2}x - 3\right| + 2$ and determine the domain and range.

Solution To graph the function, we select values for x, solve for $f(x)$, and plot the ordered pairs. The graph appears in Figure 7-21.

x	$f(x)$
−2	8
0	5
2	2
3	3.5
4	5

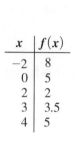

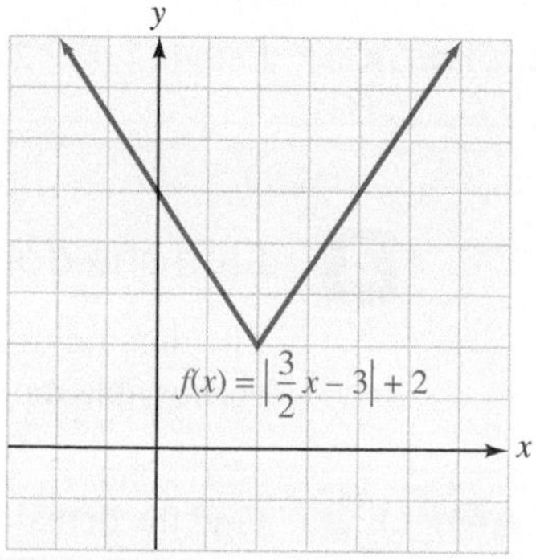

Figure 7-21

From the graph, we can see that the domain is the interval $(-\infty, \infty)$ and the range is the interval $[2, \infty)$.

SELF CHECK 6 Graph $f(x) = \left|\frac{3}{4}x\right| + 2$ and determine the domain and range.

6 Write a linear function modeling real-world data.

Writing a linear function from data is the same process as writing an equation of a line given two points.

EXAMPLE 7 **NEWSPAPER DECLINES** The number of daily newspapers dropped from 1,611 in 1990 to about 1,383 in 2009. Write a linear function to model this data and use the function to estimate the number of daily newspapers in 2020.

Solution Because 1990 and 2009 represent years, they have no real meaning in the context of the problem. Therefore, we designate 1990 as our **base year**, or zero. We represent 2009 by 19, because it is 19 away from 1990. Our ordered pairs are thus $(0, 1{,}611)$, which is also the y-intercept, and $(19, 1{,}383)$. To write the linear function, we must first find the slope of the line between these two points.

$$m = \frac{1{,}383 - 1{,}611}{19 - 0} = \frac{-228}{19} = -12$$

The linear function is $f(x) = -12x + 1{,}611$, where x represents the number of years since 1990.

Because $2020 - 1990 = 30$, to estimate the number of daily newspapers in 2020, we need to find $f(30)$.

$$\begin{aligned} f(x) &= -12x + 1{,}611 \\ f(\mathbf{30}) &= -12(\mathbf{30}) + 1{,}611 \\ &= -360 + 1{,}611 \\ &= 1{,}251 \end{aligned}$$

There will be approximately 1,251 daily newspapers in 2020.

Unless otherwise noted, all content on this page is © Cengage Learning.

SELF CHECK 7 The life expectancy of a man in 1995 was 72.54 years. By 2011 it had increased to 75.68 years. Write a linear function that models this data and use it to estimate the life expectancy of a man in 2019.

SELF CHECK ANSWERS

1. D: $\{-5, -12, 8, 5\}$; R: $\{6, 4\}$, it is a function **2.** D: $(-\infty, \infty)$; R: $(-\infty, 0]$, yes **3. a.** -5 **b.** 5 **c.** -1 **d.** 3 **4. a.** 4 **b.** $-2a + 2b$ **5. a.** $(-\infty, \infty)$ **b.** $\left(-\infty, \frac{1}{3}\right) \cup \left(\frac{1}{3}, \infty\right)$ **6.** D: $(-\infty, \infty)$; R: $[2, \infty)$

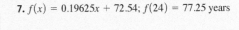

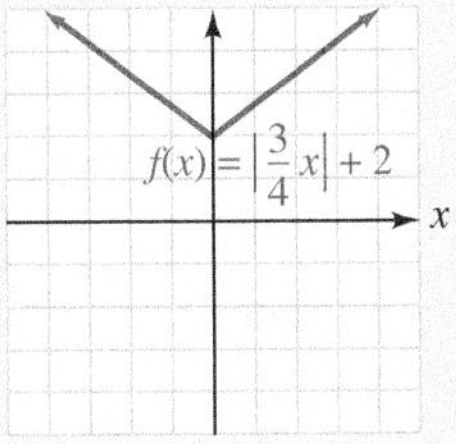

7. $f(x) = 0.19625x + 72.54$; $f(24) = 77.25$ years

NOW TRY THIS

Given $f(x) = 3x - 4$, *find*

1. $f(x - 2)$
2. $f(x + h)$
3. $f(x + h) - f(x)$

7.3 Exercises

WARM-UPS *Fill in the blanks.*

1. The denominator of a fraction can never be _.

2. In the equation $y = mx + b$, m is the ____ of the line.

3. In the equation $y = mx + b$, b is the y-coordinate of the _________ of the graph.

4. In the equation $y = 5x - 9$, the independent variable is _.

5. In the equation $y = -2x + 4$, the dependent variable is _.

6. In the equation $y = 3x + 7$, any substitution for x is called an ____ value.

REVIEW *Solve each equation.*

7. $\dfrac{y + 2}{2} = 4(y + 2)$

8. $\dfrac{3z - 1}{6} - \dfrac{3z + 4}{3} = \dfrac{z + 3}{2}$

9. $\dfrac{2a}{3} + \dfrac{1}{2} = \dfrac{6a - 1}{6}$

10. $\dfrac{2x + 3}{5} - \dfrac{3x - 1}{3} = \dfrac{x - 1}{15}$

VOCABULARY AND CONCEPTS *Fill in the blanks.*

11. Any set of ordered pairs defines a ______.

12. A _______ is a correspondence between a set of input values and a set of output values, where each ____ value determines exactly one _____ value.

13. In a function, the set of all inputs is called the ______ of the function.

14. In a function, the set of all outputs is called the _____ of the function.

15. To decide whether a graph determines a function, use the ______________.

16. A linear function is any function that can be written in the form ______________.

17. If a vertical line intersects a graph more than once, the graph ______ represent a function.

18. The notation $f(3)$ is the value of _ when $x = 3$.

19. The domain is the set of all ___ elements of a relation.

20. If every vertical line intersects a graph just one time, the graph represents a _______.

GUIDED PRACTICE *Find the domain and range of each relation and determine if it is a function.* SEE EXAMPLE 1. (OBJECTIVE 1)

21. $\{(3, -2), (5, 0), (-4, -5), (0, 0)\}$

22. $\{(9, 2), (3, 3), (-6, 2), (2, 3)\}$

23. $\{(-2, 3), (6, 8), (-2, 5), (5, 4)\}$

24. $\{(3, -2), (5, 2), (4, 5), (3, 0)\}$

Unless otherwise noted, all content on this page is © Cengage Learning.

State the domain and range of each graph in interval notation and determine whether the relation represents a function. **SEE EXAMPLE 2. (OBJECTIVE 2)**

25.

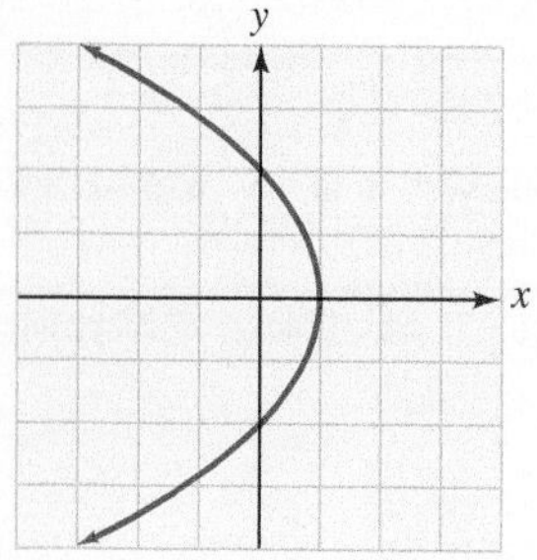

26.

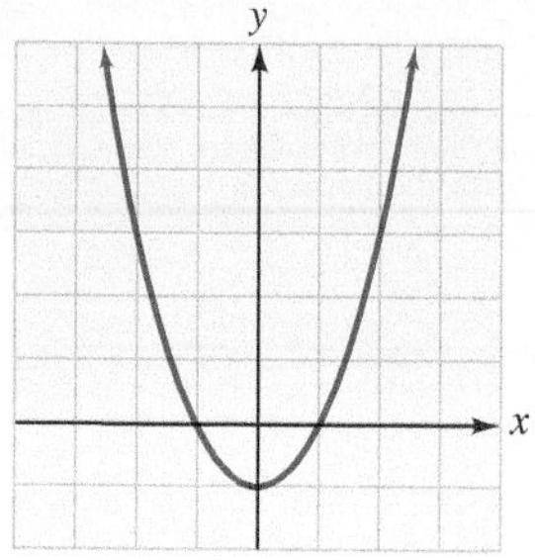

27.

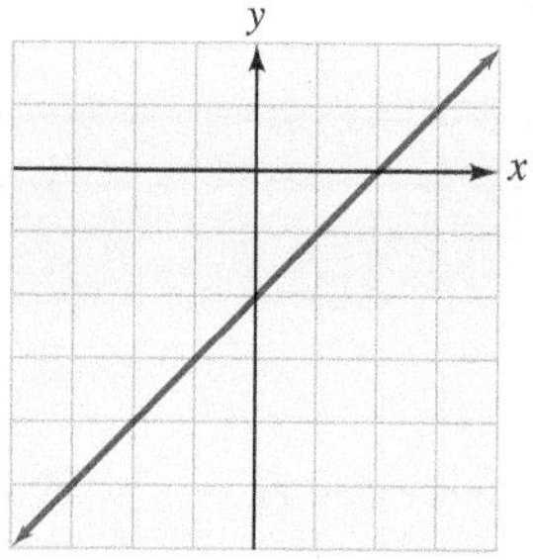

28.

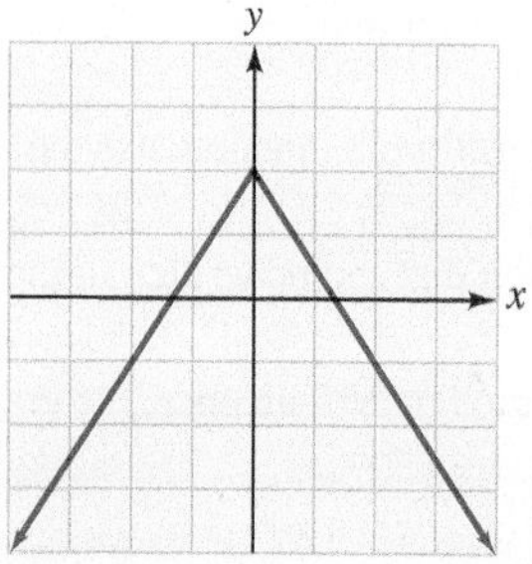

Find $f(3)$, $f(-1)$, and all values of x for which $f(x) = 0$. **SEE EXAMPLE 3. (OBJECTIVE 3)**

29. $f(x) = 3x$ **30.** $f(x) = -4x$

31. $f(x) = 2x - 3$ **32.** $f(x) = 3x - 5$

Find $f(-2)$, $f(2)$, and $f(3)$. **SEE EXAMPLE 3. (OBJECTIVE 3)**

33. $f(x) = x^2$ **34.** $f(x) = 7x - 2$

35. $f(x) = x^3 - 1$ **36.** $f(x) = x^3$

37. $f(x) = |x| + 2$ **38.** $f(x) = |x| - 5$

39. $f(x) = x^2 - 2$ **40.** $f(x) = x^2 + 3$

State each value given that $f(x) = 2x + 1$. **SEE EXAMPLE 4. (OBJECTIVE 3)**

41. $f(3) + f(2)$ **42.** $f(1) - f(-1)$

43. $f(b) - f(a)$ **44.** $f(b) + f(a)$

Write the domain of each function in interval notation. **SEE EXAMPLE 5. (OBJECTIVE 4)**

45. $f(x) = \dfrac{1}{x - 4}$ **46.** $f(x) = \dfrac{5}{x + 1}$

47. $f(x) = \dfrac{1}{x + 3}$ **48.** $f(x) = \dfrac{3}{x - 4}$

49. $f(x) = \dfrac{x}{x^2 + 2}$ **50.** $f(x) = \dfrac{x}{x - 3}$

51. $f(x) = \dfrac{x}{2x - 1}$

52. $f(x) = \dfrac{x + 2}{5x + 9}$

Sketch the graph of each function and state the domain and range in interval notation. **SEE EXAMPLE 6. (OBJECTIVE 5)**

53. $2x - 3y = 6$

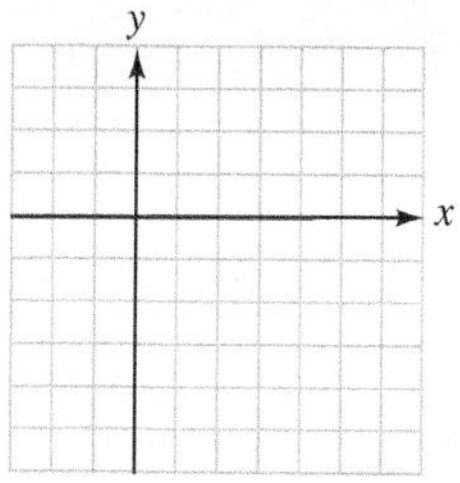

54. $3x + 2y = -6$

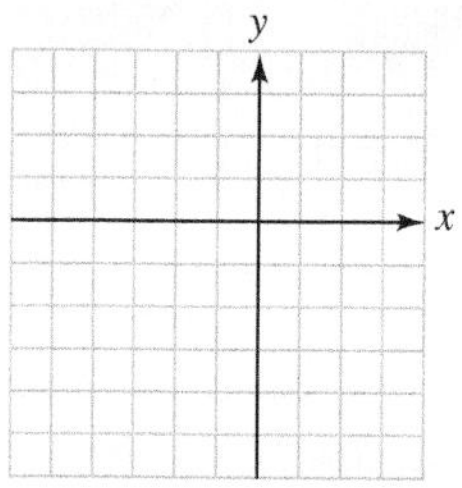

55. $f(x) = x^2 - 2x - 3$

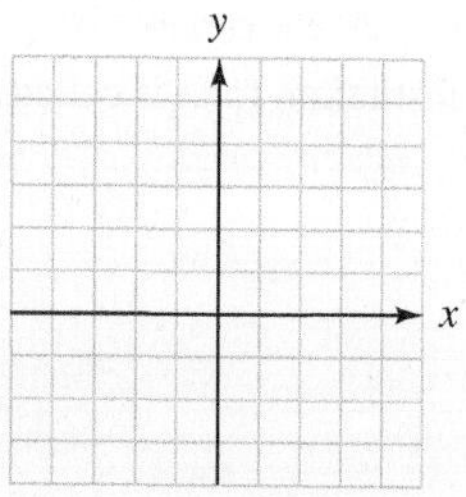

56. $f(x) = -|x| + 2$

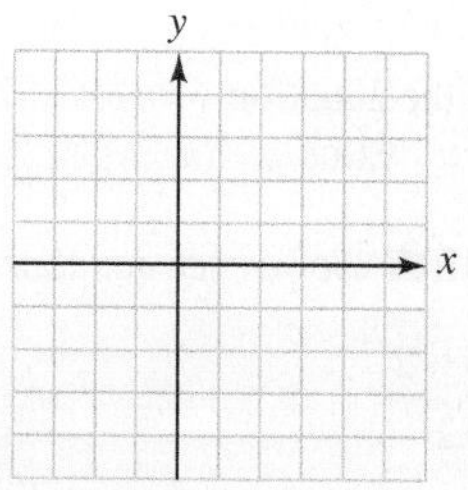

ADDITIONAL PRACTICE *For each function, find $f(2)$, $f(3)$, and $f(-1)$.*

57. $f(x) = (x + 1)^2$ **58.** $f(x) = (x - 3)^2$

59. $f(x) = 2x^2 - x$ **60.** $f(x) = 5x^2 + 2x$

61. $f(x) = 7 + 5x$ **62.** $f(x) = 3 + 3x$

63. $f(x) = 9 - 2x$ **64.** $f(x) = 12 + 3x$

Unless otherwise noted, all content on this page is © Cengage Learning.

Find each value given that $f(x) = 2x + 1$.

65. $f(b) - f(0)$

66. $f(b) - f(1)$

67. $f(0) + f\left(-\frac{1}{2}\right)$

68. $f(a) + f(2a)$

Find $g(w)$ and $g(w + 1)$.

69. $g(x) = 2x$

70. $g(x) = -3x$

71. $g(x) = 3x - 5$

72. $g(x) = 2x - 7$

Determine whether each equation defines a linear function.

73. $y = 3x^2 + 2$

74. $y = \frac{x - 3}{2}$

75. $x = 3y - 4$

76. $x = \frac{8}{y}$

APPLICATIONS *For Exercises 77–80, write a linear function to model the given data and make the indicated estimations.* SEE EXAMPLE 7. (OBJECTIVE 6)

77. College tuition The average tuition for a full-time undergraduate student in a public college in 2002 was \$4,790. In 2012 it was \$8,240. Using the linear function model, estimate the cost of tuition in 2018.

78. Wheat harvest While the number of acres of wheat harvested in Texas and Colorado has increased, nationwide it has decreased from 55,699,000 acres in 2008 to 47,637,000 in 2010. Using the linear function model, estimate the number of acres of wheat harvested in 2017.

79. Violent crime In 2000 there were approximately 507 violent crimes per 100,000 people. In 2009 violent crimes dropped to about 429 per 100,000 people. Using the linear function model, estimate the number of violent crimes in 2021 for a city with a population of 400,000.

80. Student enrollment The number of full-time male students between 22 and 24 grew from 1,355,000 to 1,700,000 from 2005 to 2010. Using the linear function model, estimate the number of full-time male students between 22 and 24 in 2017.

81. Ballistics A bullet shot straight up is s feet above the ground after t seconds, where $f(t) = -16t^2 + 256t$. Find the height of the bullet 3 seconds after it is shot.

82. Artillery fire A mortar shell is s feet above the ground after t seconds, where $f(t) = -16t^2 + 512t + 64$. Find the height of the shell 20 seconds after it is fired.

83. Dolphins See the illustration. The height h in feet reached by a dolphin t seconds after breaking the surface of the water is given by

$$f(t) = -16t^2 + 32t$$

How far above the water will the dolphin be 1.5 seconds after a jump?

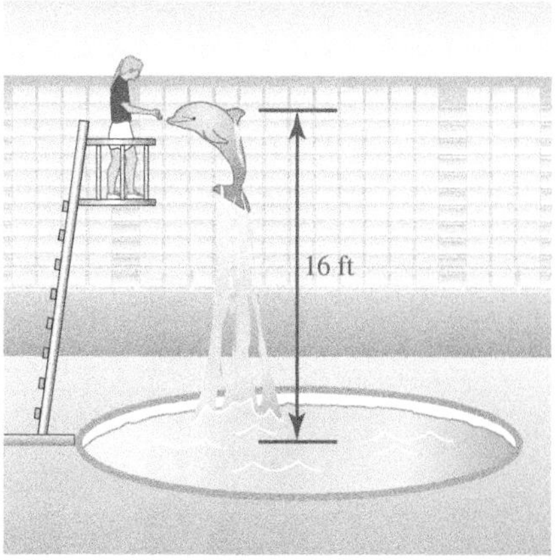

84. Housing A housing contractor lists the following costs.

Fees and permits	\$14,000
Cost per square foot	\$102

a. Write a linear function that describes the cost of building a house with s square feet.

b. Find the cost to build a house having 1,800 square feet.

85. Conversion from degrees Celsius to degrees Fahrenheit The temperature in degrees Fahrenheit that is equivalent to a temperature in degrees Celsius is given by the function $F(C) = \frac{9}{5}C + 32$. Find the Fahrenheit temperature that is equivalent to 25°C.

86. Conversion from degrees Fahrenheit to degrees Celsius The temperature in degrees Celsius that is equivalent to a temperature in degrees Fahrenheit is given by the function $C(F) = \frac{5}{9}F - \frac{160}{9}$. Find the Celsius temperature that is equivalent to 14°F.

87. Selling DVD players An electronics firm manufactures portable DVD players, receiving \$120 for each unit it makes. If x represents the number of units produced, the income received is determined by the *revenue function* $R(x) = 120x$. The manufacturer has fixed costs of \$12,000 per month and variable costs of \$57.50 for each unit manufactured. Thus, the *cost function* is $C(x) = 57.50x + 12{,}000$. How many DVD players must the company sell for revenue to equal cost?

88. Selling tires A tire company manufactures premium tires, receiving \$130 for each tire it makes. If the manufacturer has fixed costs of \$15,512.50 per month and variable costs of \$93.50 for each tire manufactured, how many tires must the company sell for revenue to equal cost?

WRITING ABOUT MATH

89. Explain the concepts of function, domain, and range.

90. Explain why the constant function is a special case of a linear function.

SOMETHING TO THINK ABOUT *Let $f(x) = 2x + 1$ and $g(x) = x^2$. Assume that $f(x) \neq 0$ and $g(x) \neq 0$.*

91. Is $f(x) + g(x)$ equal to $g(x) + f(x)$?

92. Is $f(x) - g(x)$ equal to $g(x) - f(x)$?

Unless otherwise noted, all content on this page is © Cengage Learning.

Section 7.4 Review of Factoring and Solving Quadratic Equations

Objectives

1. Factor a polynomial by factoring out the greatest common factor (GCF).
2. Factor a polynomial with four terms or six terms by grouping.
3. Factor a difference of two squares.
4. Factor a trinomial by trial and error and by grouping (*ac* method).
5. Factor a sum and difference of two cubes.
6. Solve a quadratic equation by factoring.

Vocabulary

ac method
key number
quadratic equation
zero-factor property
quadratic form

Getting Ready

Perform each multiplication.

1. $3x^2y(2x - y)$
2. $(x + 2)(x - 2)$
3. $(x + 2)(x - 3)$
4. $(2x + 3)(3x - 1)$
5. $(x - 3)(x^2 + 3x + 9)$
6. $(x + 2)(x^2 - 2x + 4)$

In this section, we will review the types of factoring as discussed in Chapter 5 and apply those techniques to solve a quadratic equation.

1 Factor a polynomial by factoring out the greatest common factor (GCF).

EXAMPLE 1 Factor out the greatest common factor (GCF): $3xy^2z^3 + 6xz^2 - 9xyz^4$

Solution We begin by factoring each monomial:

$$3xy^2z^3 = \mathbf{3} \cdot \mathbf{x} \cdot y \cdot y \cdot \mathbf{z} \cdot \mathbf{z} \cdot z$$
$$6xz^2 = \mathbf{3} \cdot 2 \cdot \mathbf{x} \cdot \mathbf{z} \cdot \mathbf{z}$$
$$-9xyz^4 = -3 \cdot \mathbf{3} \cdot \mathbf{x} \cdot y \cdot \mathbf{z} \cdot \mathbf{z} \cdot z \cdot z$$

Since each term has one factor of 3, one factor of x, and two factors of z and there are no other common factors, $3xz^2$ is the GCF of the three terms. We can use the distributive property to factor it out.

$$3xy^2z^3 + 6xz^2 - 9xyz^4 = \mathbf{3xz^2} \cdot y^2z + \mathbf{3xz^2} \cdot 2 + \mathbf{3xz^2}(-3yz^2)$$
$$= \mathbf{3xz^2}(y^2z + 2 - 3yz^2)$$

SELF CHECK 1 Factor out the greatest common factor: $4ab^3 - 6a^2b^2$

EXAMPLE 2 Factor out the negative of the greatest common factor: $-6u^2v^3 + 8u^3v^2$

Solution Since the GCF of the two terms is $2u^2v^2$, the negative of the GCF is $-2u^2v^2$. To factor it out, we proceed as follows:

$$\begin{aligned} -6u^2v^3 + 8u^3v^2 &= -2u^2v^2 \cdot 3v + 2u^2v^2 \cdot 4u \\ &= -2u^2v^2 \cdot 3v + (-2u^2v^2)(-4u) \\ &= -2u^2v^2(3v - 4u) \end{aligned}$$

SELF CHECK 2 Factor out the negative of the greatest common factor: $-3p^3q + 6p^2q^2$

EXAMPLE 3 Factor x^{2n} from $x^{4n} + x^{3n} + x^{2n}$.

Solution We can write the trinomial in the form

$$x^{2n} \cdot x^{2n} + x^{2n} \cdot x^n + x^{2n} \cdot 1$$

and then factor out x^{2n}.

$$\begin{aligned} x^{4n} + x^{3n} + x^{2n} &= x^{2n} \cdot x^{2n} + x^{2n} \cdot x^n + x^{2n} \cdot 1 \\ &= x^{2n}(x^{2n} + x^n + 1) \end{aligned}$$

SELF CHECK 3 Factor $2a^n$ from $6a^{2n} - 4a^{n+1}$.

2 Factor a polynomial with four terms or six terms by grouping.

Although there is no factor common to all four terms of $ac + ad + bc + bd$, there is a factor of a in the first two terms and a factor of b in the last two terms. We can factor out these common factors.

$$ac + ad + bc + bd = a(c + d) + b(c + d)$$

We can now factor out $c + d$ on the right side.

$$ac + ad + bc + bd = (c + d)(a + b)$$

The grouping in this type of problem is not always unique. For example, if we write the expression $ac + ad + bc + bd$ in the form $ac + bc + ad + bd$ and factor c from the first two terms and d from the last two terms, we obtain the result with the factors in reverse order.

$$\begin{aligned} ac + bc + ad + bd &= c(a + b) + d(a + b) \\ &= (a + b)(c + d) \end{aligned}$$

EXAMPLE 4 Factor: $3ax^2 + 3bx^2 + a + 5bx + 5ax + b$

Solution Although there is no factor common to all six terms, $3x^2$ can be factored out of the first two terms, and $5x$ can be factored out of the fourth and fifth terms to obtain

COMMENT Some examples may require applying the commutative property of addition.

$$\mathbf{3ax^2 + 3bx^2} + a + \mathbf{5bx + 5ax} + b = \mathbf{3x^2(a + b)} + a + \mathbf{5x(b + a)} + b$$

This result can be written in the form

$$3ax^2 + 3bx^2 + a + 5bx + 5ax + b = 3x^2(a + b) + 5x(a + b) + 1(a + b)$$

Since $a + b$ is common to all three terms, it can be factored out.

$$3ax^2 + 3bx^2 + a + 5bx + 5ax + b = (a + b)(3x^2 + 5x + 1)$$

SELF CHECK 4 Factor: $2px^2 + 2qx^2 + p + 7qx + 7px + q$

EXAMPLE 5 Factor: $8x^3 + 4x^2 - 6x - 3$

Solution There is no common factor to all four terms, but $4x^2$ can be factored out of the first two terms and -3 can be factored out of the last two terms to obtain

$$8x^3 + 4x^2 - 6x - 3 = 4x^2(2x + 1) - 3(2x + 1)$$

Since $2x + 1$ is common to both terms, it can be factored out as a GCF.

$$= 4x^2(2x + 1) - 3(2x + 1)$$
$$= (2x + 1)(4x^2 - 3)$$

SELF CHECK 5 Factor: $2x^3 - 8x^2 + 5x - 20$

3 Factor a difference of two squares.

Recall the formula for factoring the difference of two squares.

FACTORING THE DIFFERENCE OF TWO SQUARES

$$x^2 - y^2 = (x + y)(x - y)$$

If we think of the difference of two squares as the square of a **F**irst quantity minus the square of a **L**ast quantity, we have the formula

$$F^2 - L^2 = (F + L)(F - L)$$

In words, we say, *to factor the square of a* **F***irst quantity minus the square of a* **L***ast quantity, we multiply the* **F***irst plus the* **L***ast by the* **F***irst minus the* **L***ast.*

To factor $49x^2 - 16$, for example, we write $49x^2 - 16$ in the form $(7x)^2 - (4)^2$ and use the formula for factoring the difference of two squares:

$$49x^2 - 16 = (7x)^2 - (4)^2$$
$$= (7x + 4)(7x - 4)$$

We can verify this result by multiplying $7x + 4$ and $7x - 4$ and observing that the result is $49x^2 - 16$.

COMMENT Expressions such as $(7x)^2 + (4)^2$ are the sum of two squares and cannot be factored in the real number system. The binomial $49x^2 + 16$ is prime.

EXAMPLE 6 Factor: $(x + y)^4 - z^4$

Solution This expression is the difference of two squares that can be factored:

$$(x + y)^4 - z^4 = [(x + y)^2]^2 - (z^2)^2$$
$$= [(x + y)^2 + z^2][(x + y)^2 - z^2]$$

The factor $(x + y)^2 + z^2$ is the sum of two squares and is prime. The factor $(x + y)^2 - z^2$ is the difference of two squares and can be factored as $(x + y + z)(x + y - z)$. Thus,

$$(x + y)^4 - z^4 = [(x + y)^2 + z^2][(x + y)^2 - z^2]$$
$$= [(x + y)^2 + z^2](x + y + z)(x + y - z)$$

SELF CHECK 6 Factor: $a^4 - (b + c)^4$

EXAMPLE 7 Factor: $2x^4y - 32y$

Solution We factor out the GCF of $2y$ and proceed as follows:

$$\begin{aligned} 2x^4y - 32y &= 2y(x^4 - 16) && \text{Factor out the GCF of } 2y. \\ &= 2y(x^2 + 4)(x^2 - 4) && \text{Factor } x^4 - 16. \\ &= 2y(x^2 + 4)(x + 2)(x - 2) && \text{Factor } x^2 - 4. \end{aligned}$$

SELF CHECK 7 Factor: $3ap^4 - 243a$

4 Factor a trinomial by trial and error and by grouping (*ac* method).

To factor a trinomial of the form $x^2 + bx + c$ by trial and error, we follow these steps:

FACTORING TRINOMIALS

1. Write the trinomial in descending powers of one variable.
2. Factor out any GCF, including -1.
3. List the factorizations of the third term of the trinomial.
4. Select the factorization in which the sum of the factors is the coefficient of the middle term.

EXAMPLE 8 Factor: **a.** $x^2 - 6x + 8$ **b.** $30x - 4xy - 2xy^2$

Solution **a.** Since this trinomial is already written in descending powers of x and there are no common factors, we can move to Step 3 and list the possible factorizations of the third term, which is 8.

The one to choose ↓

$8(1)$ $4(2)$ $-8(-1)$ $\mathbf{-4(-2)}$

In this trinomial, the coefficient of the middle term is -6, and the only factorization where the sum of the factors is -6 is $-4(-2)$. Thus,

$$x^2 - 6x + 8 = (x - 4)(x - 2)$$

Because of the commutative property of multiplication, the order of the factors is not important. We can verify this result by multiplication.

b. We begin by writing the trinomial in descending powers of y:

$$-2xy^2 - 4xy + 30x = -2xy^2 - 4xy + 30x$$

Since each term in the trinomial has a common factor of $-2x$, it can be factored out as a GCF.

$$-2xy^2 - 4xy + 30x = -2x(y^2 + 2y - 15)$$

To factor $y^2 + 2y - 15$, we list the factors of -15 and find the pair whose sum is 2.

The one to choose ↓

$15(-1)$ $\mathbf{5(-3)}$ $1(-15)$ $3(-5)$

The only factorization where the sum of the factors is 2 (the coefficient of the middle term of $y^2 + 2y - 15$) is $5(-3)$. Thus,

$$30x - 4xy - 2xy^2 = -2x(y^2 + 2y - 15)$$
$$= -2x(y + 5)(y - 3)$$

Verify this result by multiplication.

SELF CHECK 8 Factor: **a.** $x^2 + 5x + 6$ **b.** $16a - 2ap^2 - 4ap$

There are usually more combinations of factors to consider when factoring trinomials with leading coefficients other than 1. For example, to factor $5x^2 + 7x + 2$, we must find two binomials of the form $ax + b$ and $cx + d$ such that

$$5x^2 + 7x + 2 = (ax + b)(cx + d)$$

Since the first term of the trinomial $5x^2 + 7x + 2$ is $5x^2$, the first terms of the binomial factors must be $5x$ and x.

$$5x^2 + 7x + 2 = \overbrace{(5x + b)(x}^{5x^2} + d)$$

The product of the last terms must be $+2$, and the sum of the products of the outer and inner terms must be $+7x$. We must find two numbers whose product is $+2$ that will give a middle term of $+7x$. However, the factors of 2 must be multiplied by the factors of $5x^2$ to produce this middle term of $+7x$.

$$5x^2 + 7x + 2 = (5x + b)(x + d) \qquad \text{last terms: } 2, \quad \text{O} + \text{I} = 7x$$

Since $2(1)$and $(-2)(-1)$ both give a product of 2, there are four combinations to consider:

$$(5x + 2)(x + 1) \qquad (5x - 2)(x - 1) \qquad (5x + 1)(x + 2) \qquad (5x - 1)(x - 2)$$

Of these combinations, only the first gives the correct middle term of $7x$.

(1) $5x^2 + 7x + 2 = (5x + 2)(x + 1)$

We can verify this result by multiplication.

Pierre de Fermat
1601–1665
Pierre de Fermat shares the honor with Descartes for discovering analytic geometry, and with Pascal for developing the theory of probability. But to Fermat alone goes credit for founding number theory. He is probably most famous for a theorem called *Fermat's last theorem*. It states that if n represents a number greater than 2, there are no whole numbers, a, b, and c, that satisfy the equation

$$a^n + b^n = c^n$$

If a trinomial has the form $ax^2 + bx + c$, with integer coefficients and $a \neq 0$, we can test to see if it is factorable. If the value of $b^2 - 4ac$ is a perfect square, the trinomial can be factored using only integers. If the value is not a perfect square, the trinomial is prime and cannot be factored using only integers.

For example, $5x^2 + 7x + 2$ is a trinomial in the form $ax^2 + bx + c$ with

$$a = 5, \qquad b = 7, \qquad \text{and} \qquad c = 2$$

For this trinomial, the value of $b^2 - 4ac$ is

$$b^2 - 4ac = 7^2 - 4(5)(2) = 49 - 40 = 9$$

Since 9 is a perfect square, the trinomial is factorable. Its factorization is shown in Equation 1.

TEST FOR FACTORABILITY

A trinomial of the form $ax^2 + bx + c$, with integer coefficients and $a \neq 0$, will factor into two binomials with integer coefficients if the value of $b^2 - 4ac$ is a perfect square.

If $b^2 - 4ac = 0$, the factors will be the same.

If $b^2 - 4ac$ is not a perfect square, the trinomial is prime.

Unless otherwise noted, all content on this page is © Cengage Learning.

EXAMPLE 9 Factor: $3p^2 - 4p - 4$

Solution In this trinomial, $a = 3$, $b = -4$, and $c = -4$. To see whether it factors, we evaluate $b^2 - 4ac$.

$$b^2 - 4ac = (-4)^2 - 4(3)(-4) = 16 + 48 = 64$$

Since 64 is a perfect square, the trinomial is factorable.

To factor the trinomial, we note that the first terms of the binomial factors must be $3p$ and p to give the first term of $3p^2$.

$3p^2$

$$3p^2 - 4p - 4 = (3p + ?)(p + ?)$$

The product of the last terms must be -4, and the sum of the products of the outer terms and the inner terms must be $-4p$. We must find two numbers whose product is -4 that will give a middle term of $-4p$. However, the factors of -4 must be multiplied by the factors of $3p^2$ to produce this middle term of $-4p$.

-4

$$3p^2 - 4p - 4 = (3p + ?)(p + ?)$$

$O + I = -4p$

Since $1(-4)$, $-1(4)$, and $-2(2)$ all give a product of -4, there are six combinations to consider:

$$(3p + 1)(p - 4) \qquad (3p - 4)(p + 1) \qquad (3p - 1)(p + 4)$$
$$(3p + 4)(p - 1) \qquad (3p - 2)(p + 2) \qquad (3p + 2)(p - 2)$$

Of these combinations, only the one in red gives the required middle term of $-4p$. Thus,

$$3p^2 - 4p - 4 = (3p + 2)(p - 2)$$

SELF CHECK 9 Factor, if possible: $2m^2 - 3m - 9$

Recall the following steps for factoring trinomials by trial and error.

FACTORING A GENERAL TRINOMIAL

1. Write the trinomial in descending powers of one variable.
2. Factor out any GCF (including -1 if that is necessary to make the coefficient of the first term positive).
3. Test the trinomial for factorability.
4. Factor the remaining trinomial.
   a. If the sign of the third term is $+$, the signs between the terms of the binomial factors are the same as the sign of the middle term. If the sign of the third term is $-$, the signs between the terms of the binomial factors are opposite.
   b. Try combinations of the factors of the first term and last term until you find one where the sum of the product of the outer and inner terms is the middle term of the trinomial. If no combination works, the trinomial is prime.
5. Check the factorization by multiplication, including any GCF found in Step 2.

EXAMPLE 10 Factor: $24y + 10xy - 6x^2y$

Solution We write the trinomial in descending powers of x and factor out $-2y$:

$$\begin{aligned} 24y + 10xy - 6x^2y &= -6x^2y + 10xy + 24y \\ &= -2y(3x^2 - 5x - 12) \end{aligned}$$

In the trinomial $3x^2 - 5x - 12$, $a = 3$, $b = -5$, and $c = -12$. Thus,

$$b^2 - 4ac = (-5)^2 - 4(3)(-12) = 25 + 144 = 169$$

Since 169 is a perfect square, the trinomial will factor.

Since the sign of the third term of $3x^2 - 5x - 12$ is $-$, the signs between the binomial factors will be opposite. Because the first term is $3x^2$, the first terms of the binomial factors must be $3x$ and x.

$$-2y(3x^2 - 5x - 12) = -2y(\overbrace{3x \quad)(x}^{3x^2} \quad)$$

The product of the last terms must be -12, and the sum of the outer terms and the inner terms must be $-5x$.

$$-2y(3x^2 - 5x - 12) = -2y(3x \quad ?)(x \quad ?)$$

(last terms: -12; $\text{O} + \text{I} = -5x$)

Since $1(-12), 2(-6), 3(-4), 12(-1), 6(-2)$, and $4(-3)$ all give a product of -12, there are 12 combinations to consider.

$(3x + 1)(x - 12)$	$(3x - 12)(x + 1)$	$(3x + 2)(x - 6)$
$(3x - 6)(x + 2)$	$(3x + 3)(x - 4)$	$(3x - 4)(x + 3)$
$(3x + 12)(x - 1)$	$(3x - 1)(x + 12)$	$(3x + 6)(x - 2)$
$(3x - 2)(x + 6)$	$(3x + 4)(x - 3)$ ↑ The one to choose	$(3x - 3)(x + 4)$

The six combinations marked in blue cannot work because one of the factors has a common factor. This implies that $3x^2 - 5x - 12$ would have a common factor, which it does not.

After trying the remaining combinations, we find that only $(3x + 4)(x - 3)$ gives the proper middle term of $-5x$. Thus,

$$\begin{aligned} 24y + 10xy - 6x^2y &= -2y(3x^2 - 5x - 12) \\ &= -2y(3x + 4)(x - 3) \end{aligned}$$

SELF CHECK 10 Factor: $18a - 6ap^2 + 3ap$

EXAMPLE 11 Factor: $x^{2n} + x^n - 2$

Solution Since the first term is x^{2n}, the first terms of the factors must be x^n and x^n.

$$x^{2n} + x^n - 2 = (\overbrace{x^n \quad)(x^n}^{x^{2n}} \quad)$$

Since the third term is -2, the last terms of the factors must have opposite signs, have a product of -2, and give a middle term of x^n. The only combination that works is

$$x^{2n} + x^n - 2 = (x^n + 2)(x^n - 1)$$

SELF CHECK 11 Factor: $a^{2n} + 2a^n - 3$

Many times a four-term polynomial is factored by grouping the first two and last two terms. However, there are some four-term polynomials that are factored by grouping the first three terms and factoring those as a trinomial, as illustrated in the next example.

EXAMPLE 12 Factor: $x^2 + 6x + 9 - z^2$

Solution We group the first three terms together and factor the trinomial to obtain

$$\begin{aligned} x^2 + 6x + 9 - z^2 &= (x + 3)(x + 3) - z^2 \\ &= (x + 3)^2 - z^2 \end{aligned}$$

We can now factor the difference of two squares to get

$$x^2 + 6x + 9 - z^2 = (x + 3 + z)(x + 3 - z)$$

SELF CHECK 12 Factor: $y^2 + 4y + 4 - t^2$

Many of these trinomials could have been factored using a different method. Factoring by grouping, also called the ***ac* method**, can be used to factor trinomials of the form $ax^2 + bx + c$. For example, to factor the trinomial $6x^2 + 7x - 3$ where $a = 6$, $b = 7$, and $c = -3$, we proceed as follows:

1. Find the product ac: $6(-3) = -18$. This number is called the **key number**.
2. Find the two factors of the key number -18 whose sum is $b = 7$. These are 9 and -2.

   $9(-2) = -18$ and $9 + (-2) = 7$
3. Use the factors 9 and -2 as coefficients of terms, the sum of which is $+7x$. We replace $+7x$ with $9x$ and $-2x$.

   $$6x^2 + 7x - 3 = 6x^2 + 9x - 2x - 3$$
4. Factor by grouping:

   $$\begin{aligned} 6x^2 + 9x - 2x - 3 &= 3x(2x + 3) - (2x + 3) \\ &= (2x + 3)(3x - 1) \quad \text{Factor out } 2x + 3. \end{aligned}$$

   We can verify this factorization by multiplication.
   For additional examples see Chapter 5.

5 Factor a sum and difference of two cubes.

Recall the following formula for factoring the sum of two cubes.

FACTORING THE SUM OF TWO CUBES

$$x^3 + y^3 = (x + y)(x^2 - xy + y^2)$$

If we think of the sum of two cubes as the cube of a **F**irst quantity plus the cube of a **L**ast quantity, we have the formula

$$F^3 + L^3 = (F + L)(F^2 - FL + L^2)$$

In words, we say, *to factor the cube of a* **First** *quantity plus the cube of a* **Last** *quantity, we multiply the* **First** *plus the* **Last** *by*

- *the* **First** *squared*
- *minus the* **First** *times the* **Last**
- *plus the* **Last** *squared.*

EXAMPLE 13 Factor: **a.** $x^3 + 8$ **b.** $8b^3 + 27c^3$

Solution **a.** The binomial $x^3 + 8$ is the sum of two cubes, because

$$x^3 + 8 = x^3 + 2^3$$

Thus, $x^3 + 8$ factors as $(x + 2)$ times the trinomial $x^2 - 2x + 2^2$.

$$\begin{aligned} F^3 + L^3 &= (F + L)(F^2 - F\ L + L^2) \\ x^3 + 2^3 &= (x + 2)(x^2 - x \cdot 2 + 2^2) \\ &= (x + 2)(x^2 - 2x + 4) \end{aligned}$$

b. The binomial $8b^3 + 27c^3$ is the sum of two cubes, because

$$8b^3 + 27c^3 = (2b)^3 + (3c)^3$$

Thus, $8b^3 + 27c^3$ factors as $(2b + 3c)$ times the trinomial $(2b)^2 - (2b)(3c) + (3c)^2$.

$$\begin{aligned} F^3 + L^3 &= (F + L)\ (F^2 - F\ L + L^2) \\ (2b)^3 + (3c)^3 &= (2b + 3c)[(2b)^2 - (2b)(3c) + (3c)^2] \\ &= (2b + 3c)(4b^2 - 6bc + 9c^2) \end{aligned}$$

SELF CHECK 13 Factor: **a.** $p^3 + 64$
b. $27p^3 + 125q^3$

Recall the following formula for factoring the difference of two cubes.

FACTORING THE DIFFERENCE OF TWO CUBES

$$x^3 - y^3 = (x - y)(x^2 + xy + y^2)$$

If we think of the difference of two cubes as the cube of a **First** quantity minus the cube of a **Last** quantity, we have the formula

$$F^3 - L^3 = (F - L)(F^2 + FL + L^2)$$

In words, we say, *to factor the cube of a* **First** *quantity minus the cube of a* **Last** *quantity, we multiply the* **First** *minus the* **Last** *by*

- *the* **First** *squared*
- *plus the* **First** *times the* **Last**
- *plus the* **Last** *squared.*

EXAMPLE 14 Factor: **a.** $a^3 - 64b^3$ **b.** $-2t^5 + 128t^2$

Solution **a.** The binomial $a^3 - 64b^3$ is the difference of two cubes.

$$a^3 - 64b^3 = a^3 - (4b)^3$$

Thus, its factors are the difference $a - 4b$ and the trinomial $a^2 + a(4b) + (4b)^2$.

$$\begin{array}{l} F^3 - L^3 = (F - L)(F^2 + F\,L + L^2) \\ \downarrow \quad \downarrow \quad\quad \downarrow \quad \downarrow \quad \downarrow \quad \downarrow \downarrow \quad \downarrow \\ a^3 - (4b)^3 = (a - 4b)[a^2 + a(4b) + (4b)^2] \\ \qquad\qquad = (a - 4b)(a^2 + 4ab + 16b^2) \end{array}$$

b. $-2t^5 + 128t^2 = -2t^2(t^3 - 64)$ Factor out $-2t^2$.

$= -2t^2(t - 4)(t^2 + 4t + 16)$ Factor $t^3 - 64$.

SELF CHECK 14 Factor:
a. $27p^3 - 8$
b. $-3p^4 + 81p$

EXAMPLE 15 Factor: $x^6 - 64$

Solution The binomial $x^6 - 64$ is both the difference of two squares and the difference of two cubes. We will factor the difference of two squares first.

$$\begin{aligned} x^6 - 64 &= (x^3)^2 - 8^2 \\ &= (x^3 + 8)(x^3 - 8) \end{aligned}$$

Since $x^3 + 8$ is the sum of two cubes and $x^3 - 8$ is the difference of two cubes, each of these binomials can be factored.

$$\begin{aligned} x^6 - 64 &= (x^3 + 8)(x^3 - 8) \\ &= (x + 2)(x^2 - 2x + 4)(x - 2)(x^2 + 2x + 4) \end{aligned}$$

SELF CHECK 15 Factor: $a^6 - 1$

6 Solve a quadratic equation by factoring.

Recall that an equation such as $3x^2 + 4x - 7 = 0$ or $-5y^2 + 3y + 8 = 0$ is called a *quadratic* (or *second-degree*) equation.

QUADRATIC EQUATIONS

A **quadratic equation** is any equation that can be written in the form

$$ax^2 + bx + c = 0$$

where a, b, and c are real numbers and $a \neq 0$.

Many quadratic equations can be solved by factoring and then applying the **zero-factor property**.

ZERO-FACTOR PROPERTY

Assume a and b are real numbers.

If $ab = 0$, then $a = 0$ or $b = 0$.

The zero-factor property and its extensions state that *if the product of two or more factors is* 0, *then at least one of the factors must be* 0.

Many equations that do not appear to be quadratic can be put into **quadratic form** $(ax^2 + bx + c = 0)$ and then solved by factoring.

EXAMPLE 16 Solve: $x = \frac{6}{5} - \frac{6}{5}x^2$

Solution We write the equation in quadratic form and then solve by factoring.

$$x = \frac{6}{5} - \frac{6}{5}x^2$$

$$5x = 6 - 6x^2$$ Multiply both sides by 5 to clear fractions.

$$6x^2 + 5x - 6 = 0$$ Add $6x^2$ to both sides and subtract 6 from both sides.

$$(3x - 2)(2x + 3) = 0$$ Factor the trinomial.

$$3x - 2 = 0 \quad \text{or} \quad 2x + 3 = 0$$ Set each factor equal to 0.

$$3x = 2 \qquad 2x = -3$$

$$x = \frac{2}{3} \qquad x = -\frac{3}{2}$$

Verify that both solutions check.

SELF CHECK 16 Solve: $x = \frac{6}{7}x^2 - \frac{3}{7}$

COMMENT To solve a quadratic equation by factoring, be sure to set the equation equal to 0 before applying the zero-factor property.

SELF CHECK ANSWERS

1. $2ab^2(2b - 3a)$ **2.** $-3p^2q(p - 2q)$ **3.** $2a^n(3a^n - 2a)$ **4.** $(p + q)(2x^2 + 7x + 1)$ **5.** $(2x^2 + 5)(x - 4)$ **6.** $[a^2 + (b + c)^2](a + b + c)(a - b - c)$ **7.** $3a(p^2 + 9)(p + 3)(p - 3)$ **8. a.** $(x + 3)(x + 2)$ **b.** $-2a(p + 4)(p - 2)$ **9.** $(2m + 3)(m - 3)$ **10.** $-3a(2p + 3)(p - 2)$ **11.** $(a^n + 3)(a^n - 1)$ **12.** $(y + 2 + t)(y + 2 - t)$ **13. a.** $(p + 4)(p^2 - 4p + 16)$ **b.** $(3p + 5q)(9p^2 - 15pq + 25q^2)$ **14. a.** $(3p - 2)(9p^2 + 6p + 4)$ **b.** $-3p(p - 3)(p^2 + 3p + 9)$ **15.** $(a + 1)(a^2 - a + 1)(a - 1)(a^2 + a + 1)$ **16.** $\frac{3}{2}, -\frac{1}{3}$

NOW TRY THIS

Factor completely.

1. $x^2 - x + \frac{1}{4}$

2. $3x^{2/3} - x^{1/3} - 2$

3. $3(\tan x)^2 - \tan x - 2$

4. Compare the results of factoring $x^6 - 1$ as a difference of two squares to factoring $x^6 - 1$ as a difference of two cubes. What conclusions can be drawn?

7.4 Exercises

WARM-UPS *Complete each factorization.*

1. $8x^3 - 12x^5 = 4x^3(\_\_\_\_\_\_)$
2. $3x^2 - 6x = \_\_(x - 2)$
3. $x^2 - 1 = (\_\_\_\_\_)(x - 1)$
4. $25x^2 - 9 = (\_\_\_\_\_\_)(5x - 3)$
5. $x^2 + 5x - 6 = (x - 1)(\_\_\_\_\_)$
6. $x^2 - 5x + 6 = (\_\_\_\_\_)(x - 3)$

REVIEW

7. The speed of sound is approximately 1.1×10^3 ft/sec. Express this number in standard notation.

8. The time t (in hours) it takes to complete a job is given by the following equation. Find the value of t.

$$\frac{t}{10} + \frac{t}{5} = 1$$

Solve each equation.

9. $\frac{2}{3}(5t - 3) = 38$
10. $2q^2 - 9 = q(q + 3) + q^2$

VOCABULARY AND CONCEPTS *Fill in the blanks.*

11. Factoring out a common monomial is based on the distributive property, which is $a(b + c) =$ ______.

12. A polynomial with 4 or 6 terms is usually factored by ______.

13. If a polynomial cannot be factored over the real numbers, it is said to be _____.

14. Factoring the difference of two squares is based on the formula $F^2 - L^2 =$ __________.

15. The two methods that can be used to factor a trinomial are _________ and the *ac* (grouping) method.

16. $x^3 + y^3 = (x + y)$_________

17. $x^3 - y^3 = (x - y)$_________

18. To be factored completely all factors of a polynomial, except a monomial, must be _____.

GUIDED PRACTICE *Factor out the GCF from each polynomial.* SEE EXAMPLE 1. (OBJECTIVE 1)

19. $2x + 8$
20. $3y - 9$
21. $2x^2 - 6x$
22. $3y^3 + 3y^2$

Factor out the negative GCF from each polynomial. SEE EXAMPLE 2. (OBJECTIVE 1)

23. $-3a + 6$
24. $-6b + 12$
25. $-6x^2 - 3xy$
26. $-15y^3 - 25y^2$

Factor out the designated common factor. SEE EXAMPLE 3. (OBJECTIVE 1)

27. x^2 from $x^{n+2} + x^{n+3}$
28. y^3 from $y^{n+3} + y^{n+5}$
29. y^n from $2y^{n+2} - 3y^{n+3}$
30. x^n from $4x^{n+3} - 5x^{n+5}$

Factor each polynomial by grouping. SEE EXAMPLE 4. (OBJECTIVE 2)

31. $ax + bx + ay + by$
32. $ar - br + as - bs$
33. $x^2 + yx + 2x + 2y$
34. $2c + 2d - cd - d^2 + cx + dx$

Factor each polynomial by grouping. SEE EXAMPLE 5. (OBJECTIVE 2)

35. $8a^3 - 2a^2 + 12a - 3$
36. $3b^3 + 2b^2 - 15b - 10$
37. $5x^3 + x^2 + 5x + 1$
38. $y^3 - y^2 + y - 1$

Factor each binomial completely. SEE EXAMPLES 6 AND 7. (OBJECTIVE 3)

39. $x^2 - 4$
40. $(x + y)^2 - 9$
41. $9y^2 - 64$
42. $16x^4 - 81y^2$
43. $x^4 - y^4$
44. $16a^4 - 81b^4$
45. $2x^2 - 288$
46. $8x^2 - 72$

Factor each trinomial completely, if possible. If it cannot be factored, state that the polynomial is prime. SEE EXAMPLE 8. (OBJECTIVE 4)

47. $x^2 + 5x + 6$
48. $y^2 + 7y + 6$
49. $a^2 + 5a - 52$
50. $b^2 + 9b - 38$

Test each trinomial for factorability and factor it completely, if possible. If it cannot be factored, state that the polynomial is prime. SEE EXAMPLE 9. (OBJECTIVE 4)

51. $2x^2 - 11x + 5$
52. $3c^2 - 5c - 12$
53. $6y^2 + 7y + 2$
54. $6x^2 - 11x + 3$

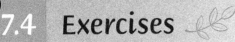

Factor each trinomial completely, if possible. If it cannot be factored, state that the polynomial is prime. SEE EXAMPLE 10. (OBJECTIVE 4)

55. $3x^4 - 10x^3 + 3x^2$

56. $6y^5 + 7y^4 + 2y^3$

57. $-8x^3 - 18x + 24x^2$

58. $-6x^2 - 4x - 9x^3$

Factor each trinomial completely, if possible. If it cannot be factored, state that the polynomial is prime. SEE EXAMPLE 11. (OBJECTIVE 4)

59. $x^{2n} + 2x^n + 1$

60. $b^{2n} - b^n - 6$

61. $6x^{2n} + 7x^n - 3$

62. $x^{4n} - 2x^{2n} + 1$

63. $x^{4n} + 2x^{2n}y^{2n} + y^{4n}$

64. $12y^{4n} + 10y^{2n} + 2$

65. $y^{6n} + 2y^{3n}z + z^2$

66. $2a^{6n} - 3a^{3n} - 2$

Factor each expression completely. SEE EXAMPLE 12. (OBJECTIVE 4)

67. $x^2 + 4x + 4 - y^2$

68. $x^2 - 6x + 9 - 4y^2$

69. $x^2 + 2x + 1 - 9z^2$

70. $x^2 + 10x + 25 - 16z^2$

Factor each polynomial completely. SEE EXAMPLE 13. (OBJECTIVE 5)

71. $y^3 + 1$

72. $b^3 + 125$

73. $8 + x^3$

74. $2x^3 + 54$

Factor each polynomial completely. SEE EXAMPLE 14. (OBJECTIVE 5)

75. $a^3 - 64$

76. $x^3 - 8$

77. $27 - y^3$

78. $2x^3 - 2$

Factor each polynomial completely. Factor a difference of two squares first. SEE EXAMPLE 15. (OBJECTIVE 5)

79. $x^6 - 1$

80. $x^6 - y^6$

81. $x^{12} - y^6$

82. $a^{12} - 64$

Solve. SEE EXAMPLE 16. (OBJECTIVE 6)

83. $\frac{3a^2}{2} = \frac{1}{2} - a$

84. $x^2 = \frac{1}{6}(5x + 4)$

85. $\frac{1}{2}x^2 - \frac{5}{4}x = -\frac{1}{2}$

86. $x^2 + \frac{7}{3}x = -\frac{5}{4}$

ADDITIONAL PRACTICE *Factor each polynomial completely.*

87. $27z^3 + 12z^2 + 3z$

88. $25t^6 - 10t^3 + 5t^2$

89. $15x^2y - 10x^2y^2$

90. $63x^3y^2 + 81x^2y^4$

91. $13ab^2c^3 - 26a^3b^2c$

92. $4x^2yz^2 + 4xy^2z^2$

93. $24s^3 - 12s^2t + 6st^2$

94. $18y^2z^2 + 12y^2z^3 - 24y^4z^3$

95. $81a^4 - 49b^2$

96. $64r^6 - 121s^2$

97. $(x - y)^2 - z^2$

98. $a^2 - (b + c)^2$

99. $-a^2 + 4a + 32$

100. $-x^2 - 2x + 15$

101. $2x^2 + 8x - 42$

102. $2y^2 + 4y - 48$

103. $-3x^2 + 15x - 18$

104. $-3y^2 - 24y + 60$

105. $a^2 - 3ab - 4b^2$

106. $b^2 + 2bc - 80c^2$

107. $8a^2 + 6a - 9$

108. $15b^2 + 4b - 4$

109. $5x^2 + 4x + 1$

110. $6z^2 + 17z + 12$

111. $8x^2 - 10x + 3$

112. $4a^2 + 20a + 3$

113. $2y^2 + yt - 6t^2$

114. $3x^2 - 10xy - 8y^2$

115. $-3a^2 + ab + 2b^2$

116. $-2x^2 + 3xy + 5y^2$

117. $a^2b^2 - 13ab^2 + 22b^2$

118. $a^2b^2x^2 - 18a^2b^2x + 81a^2b^2$

119. $8u^3 + 27$

120. $27x^3 + y^3$

121. $a^3 + 8b^3$

122. $64x^3 + 27y^3$

123. $s^3 - t^3$

124. $x^3 - 27y^3$

125. $27a^3 - b^3$

126. $27x^3 - 125y^3$

127. $2x^3 - 32x$

128. $3x^3 - 243x$

129. $x^4 + 8x^2 + 15$

130. $x^4 + 11x^2 + 24$

131. $y^4 - 13y^2 + 30$

132. $y^4 - 13y^2 + 42$

133. $a^4 - 13a^2 + 36$

134. $b^4 - 17b^2 + 16$

135. $-x^3 + 216$

136. $-x^3 - 125$

137. $a^2 - b^2 + a + b$

138. $2x + y + 4x^2 - y^2$

139. $24m^5n - 3m^2n^4$

140. $z^3(y^2 - 4) + 8(y^2 - 4)$

141. $4c^2 - a^2 - 6ab - 9b^2$

142. $c^2 - 4a^2 + 4ab - b^2$

143. $x^4y + 216xy^4$

144. $16a^5 - 54a^2b^3$

Solve.

145. $x\left(3x + \frac{22}{5}\right) = 1$

146. $x\left(\frac{x}{11} - \frac{1}{7}\right) = \frac{6}{77}$

WRITING ABOUT MATH

147. Explain how you would factor -1 from a trinomial.

148. Explain how you would test the polynomial $ax^2 + bx + c$ for factorability.

149. Explain how to factor $a^3 + b^3$.

150. Explain the difference between $x^3 - y^3$ and $(x - y)^3$.

SOMETHING TO THINK ABOUT

151. Because it is the difference of two squares, $x^2 - q^2$ always factors. Does the test for factorability predict this?

152. The polynomial $ax^2 + ax + a$ factors because a is a common factor. Does the test for factorability predict this? Is there something wrong with the test? Explain.

153. Factor: $x^{32} - y^{32}$

154. Find the error.

$$x = y$$
$$x^2 = xy$$
$$x^2 - y^2 = xy - y^2$$
$$(x + y)(x - y) = y(x - y)$$
$$\frac{(x + y)(x - y)}{x - y} = \frac{y(x - y)}{x - y}$$
$$x + y = y$$
$$y + y = y$$
$$2y = y$$
$$\frac{2y}{y} = \frac{y}{y}$$
$$2 = 1$$

Section 7.5 Review of Rational Expressions and Solving Rational Equations

Objectives

1. Find all values of a variable for which a rational expression is undefined.
2. Simplify a rational expression.
3. Multiply and divide two rational expressions.
4. Add and subtract two rational expressions.
5. Simplify a complex fraction.
6. Solve a rational equation.

Vocabulary

rational expression, rational equation, complex fraction

Getting Ready

Perform each operation.

1. $\frac{2}{3} \cdot \frac{5}{2}$ **2.** $\frac{2}{3} \div \frac{5}{2}$ **3.** $\frac{2}{3} + \frac{5}{2}$ **4.** $\frac{2}{3} - \frac{5}{2}$

1 Find all values of a variable for which a rational expression is undefined.

Recall that a **rational expression** is a fraction that indicates the quotient of two polynomials. Because rational expressions indicate division, we must exclude any values of the variable that cause the denominator to be equal to 0. When the denominator is 0, the expression is undefined.

EXAMPLE 1 Find all values of x for which the following rational expression is undefined.

$$\frac{x^2 + 2x - 5}{3x^2 + 11x - 4}$$

Solution To find the values of x that will make the expression undefined, we set the denominator equal to 0 and solve for x.

$$3x^2 + 11x - 4 = 0$$

$$(3x - 1)(x + 4) = 0 \quad \text{Factor the trinomial.}$$

$$3x - 1 = 0 \quad \text{or} \quad x + 4 = 0 \quad \text{Set each factor equal to 0.}$$

$$3x = 1 \quad \Big| \quad x = -4$$

$$x = \frac{1}{3}$$

If $x = \frac{1}{3}$ or $x = -4$, the denominator of the rational expression will be 0 and, thus, undefined.

SELF CHECK 1 Find all values of x such that the following rational expression is undefined.

$$\frac{x^2 + 5}{2x^2 + 5x + 2}$$

COMMENT Values that make an expression undefined are sometimes called excluded, restricted, or undefined values.

Note: For the remainder of this section, we will assume that no denominator is 0.

2 Simplify a rational expression.

To manipulate rational expressions, we use the same rules as we use to simplify, multiply, divide, add, and subtract arithmetic fractions.

EXAMPLE 2 Simplify: $\dfrac{2x^2 + 11x + 12}{3x^2 + 11x - 4}$

Solution We factor the numerator and denominator and divide out all common factors.

$$\frac{2x^2 + 11x + 12}{3x^2 + 11x - 4} = \frac{(2x + 3)\overset{1}{\cancel{(x + 4)}}}{(3x - 1)\underset{1}{\cancel{(x + 4)}}}$$

$$= \frac{2x + 3}{3x - 1} \qquad \frac{x + 4}{x + 4} = 1$$

COMMENT The expression $\frac{2x + 3}{3x - 1}$ is in simplest form. The x in the numerator is a *factor* of the first term only. It is not a factor of the entire numerator. Likewise, the x in the denominator is not a factor of the entire denominator.

SELF CHECK 2 Simplify: $\dfrac{2x^2 + 5x + 2}{3x^2 + 5x - 2}$

EXAMPLE 3 Simplify: $\dfrac{3x^2 - 10xy - 8y^2}{4y^2 - xy}$

Solution We factor the numerator and denominator and proceed as follows.

$$\frac{3x^2 - 10xy - 8y^2}{4y^2 - xy} = \frac{(3x + 2y)\overset{-1}{\cancel{(x - 4y)}}}{y\underset{1}{\cancel{(4y - x)}}}$$

Because $x - 4y$ and $4y - x$ are negatives, their quotient is -1.

$$= \frac{-(3x + 2y)}{y}$$

$$= \frac{-3x - 2y}{y}$$

SELF CHECK 3 Simplify: $\dfrac{-2a^2 - ab + 3b^2}{a^2 - ab}$

3 Multiply and divide two rational expressions.

To multiply two rational expressions, we multiply the numerators and multiply the denominators.

EXAMPLE 4 Multiply: $\dfrac{6x^2 + 5x - 4}{2x^2 + 5x + 3} \cdot \dfrac{8x^2 + 6x - 9}{12x^2 + 7x - 12}$

Solution We multiply the rational expressions, factor each polynomial, and simplify.

$$\frac{6x^2 + 5x - 4}{2x^2 + 5x + 3} \cdot \frac{8x^2 + 6x - 9}{12x^2 + 7x - 12}$$

$$= \frac{(6x^2 + 5x - 4)(8x^2 + 6x - 9)}{(2x^2 + 5x + 3)(12x^2 + 7x - 12)}$$

Multiply the numerators and multiply the denominators.

$$= \frac{(3x + 4)(2x - 1)(4x - 3)(2x + 3)}{(2x + 3)(x + 1)(3x + 4)(4x - 3)}$$

Factor the polynomials.

$$= \frac{\overset{1}{\cancel{(3x + 4)}}(2x - 1)\overset{1}{\cancel{(4x - 3)}}\overset{1}{\cancel{(2x + 3)}}}{\underset{1}{\cancel{(2x + 3)}}(x + 1)\underset{1}{\cancel{(3x + 4)}}\underset{1}{\cancel{(4x - 3)}}}$$

Divide out the common factors.

$$= \frac{2x - 1}{x + 1}$$

SELF CHECK 4 Multiply: $\dfrac{2x^2 + 5x + 3}{3x^2 + 5x + 2} \cdot \dfrac{2x^2 - 5x + 3}{4x^2 - 9}$

In Example 4, we would obtain the same answer if we factored first and divided out the common factors before we multiplied.

To divide two rational expressions, we invert the divisor and multiply.

EXAMPLE 5 Divide: $\dfrac{x^3+8}{x+1} \div \dfrac{x^2-2x+4}{2x^2-2}$

Solution Using the rule for division of fractions, we invert the divisor and multiply.

$$\begin{aligned}\frac{x^3+8}{x+1} \div \frac{x^2-2x+4}{2x^2-2} &= \frac{x^3+8}{x+1}\cdot\frac{2x^2-2}{x^2-2x+4}\\ &= \frac{(x^3+8)(2x^2-2)}{(x+1)(x^2-2x+4)}\\ &= \frac{(x+2)\overset{1}{\cancel{(x^2-2x+4)}}2\overset{1}{\cancel{(x+1)}}(x-1)}{\underset{1}{\cancel{(x+1)}}\underset{1}{\cancel{(x^2-2x+4)}}} \qquad \begin{aligned}2x^2-2 &= 2(x^2-1)\\ &= 2(x+1)(x-1)\end{aligned}\\ &= 2(x+2)(x-1)\end{aligned}$$

SELF CHECK 5 Divide: $\dfrac{x^3+27}{x^2-4} \div \dfrac{x^2-3x+9}{x+2}$

EXAMPLE 6 Simplify: $\dfrac{x^2+2x-3}{6x^2+5x+1} \div \dfrac{2x^2-2}{2x^2-5x-3}\cdot\dfrac{6x^2+4x-2}{x^2-2x-3}$

Solution We write the division as a multiplication by inverting the divisor. Since multiplications and divisions are performed from left to right, only the middle rational expression should be inverted. Finally, we multiply the rational expressions, factor each polynomial, and divide out the common factors.

$$\begin{aligned}\frac{x^2+2x-3}{6x^2+5x+1} \div \frac{2x^2-2}{2x^2-5x-3}\cdot\frac{6x^2+4x-2}{x^2-2x-3} &= \frac{x^2+2x-3}{6x^2+5x+1}\cdot\frac{2x^2-5x-3}{2x^2-2}\cdot\frac{6x^2+4x-2}{x^2-2x-3}\\ &= \frac{(x^2+2x-3)(2x^2-5x-3)(6x^2+4x-2)}{(6x^2+5x+1)(2x^2-2)(x^2-2x-3)}\\ &= \frac{(x+3)\overset{1}{\cancel{(x-1)}}\overset{1}{\cancel{(2x+1)}}\overset{1}{\cancel{(x-3)}}\overset{1}{\cancel{2}}(3x-1)\overset{1}{\cancel{(x+1)}}}{(3x+1)\underset{1}{\cancel{(2x+1)}}\underset{1}{\cancel{2}}(x+1)\underset{1}{\cancel{(x-1)}}\underset{1}{\cancel{(x-3)}}\underset{1}{\cancel{(x+1)}}}\\ &= \frac{(x+3)(3x-1)}{(3x+1)(x+1)}\end{aligned}$$

SELF CHECK 6 $\dfrac{2x^2-2x-4}{x^2+2x-8}\cdot\dfrac{3x^2+15x}{x+1} \div \dfrac{4x^2-100}{x^2-x-20}$

4 Add and subtract two rational expressions.

To add or subtract rational expressions with like denominators, we add or subtract the numerators and keep the same denominator. Whenever possible, we must simplify the result.

EXAMPLE 7 Simplify: $\dfrac{4x}{x+2}+\dfrac{7x}{x+2}$

Solution

$$\frac{4x}{x+2}+\frac{7x}{x+2}=\frac{4x+7x}{x+2}$$
$$=\frac{11x}{x+2}$$

SELF CHECK 7 Simplify: $\dfrac{4a}{a+3}+\dfrac{2a}{a+3}$

To add or subtract rational expressions with unlike denominators, we must find the least common denominator and then write each fraction in an equivalent form using the least common denominator.

EXAMPLE 8 Simplify: $\dfrac{4x}{x+2}-\dfrac{7x}{x-2}$

Solution The least common denominator of $x+2$ and $x-2$ is $(x+2)(x-2)$. To write each fraction in an equivalent form with this common denominator we proceed as follows.

COMMENT The − sign between the fractions applies to both terms of $7x^2+14x$.

$$\frac{4x}{x+2}-\frac{7x}{x-2}=\frac{4x(x-2)}{(x+2)(x-2)}-\frac{(x+2)7x}{(x+2)(x-2)}$$ $\frac{x-2}{x-2}=1;\frac{x+2}{x+2}=1$

$$=\frac{(4x^2-8x)-(7x^2+14x)}{(x+2)(x-2)}$$ Use the distributive property to remove parentheses.

$$=\frac{4x^2-8x-7x^2-14x}{(x+2)(x-2)}$$ Subtract the numerators and keep the denominator.

$$=\frac{-3x^2-22x}{(x+2)(x-2)}$$ Combine like terms.

SELF CHECK 8 Simplify: $\dfrac{3a}{a+3}-\dfrac{2a}{a-3}$

EXAMPLE 9 Add: $\dfrac{x}{x^2-2x+1}+\dfrac{3}{x^2-1}$

Solution We factor each denominator to find the LCD.

$$x^2-2x+1=(x-1)(x-1)=(x-1)^2$$
$$x^2-1=(x+1)(x-1)$$

We take the highest power of each factor to form the LCD of $(x-1)^2(x+1)$.

We now write each rational expression with its denominator in factored form and write each rational expression with an LCD of $(x-1)^2(x+1)$. Finally, we add them.

$$\frac{x}{x^2-2x+1}+\frac{3}{x^2-1}=\frac{x}{(x-1)(x-1)}+\frac{3}{(x+1)(x-1)}$$
$$=\frac{x(x+1)}{(x-1)(x-1)(x+1)}+\frac{3(x-1)}{(x+1)(x-1)(x-1)}$$
$$=\frac{x^2+x+3x-3}{(x-1)(x-1)(x+1)}$$
$$=\frac{x^2+4x-3}{(x-1)^2(x+1)}$$ This result does not simplify.

SELF CHECK 9 Add: $\dfrac{3}{a^2 + a} + \dfrac{2}{a^2 - 1}$

5 Simplify a complex fraction.

Recall that a **complex fraction** is a fraction with a rational expression in its numerator and/or its denominator. Examples of complex fractions are

$$\frac{\frac{3}{5}}{\frac{6}{7}}, \qquad \frac{\frac{x+2}{3}}{x-4}, \qquad \text{and} \qquad \frac{\frac{3x^2-2}{2x}}{3x - \frac{2}{y}}$$

We will demonstrate two methods for simplifying complex fractions by simplifying $\dfrac{\frac{3a}{b}}{\frac{6ac}{b^2}}$.

Method 1: We write the complex fraction as a division and proceed as follows.

$$\frac{\frac{3a}{b}}{\frac{6ac}{b^2}} = \frac{3a}{b} \div \frac{6ac}{b^2}$$

$$= \frac{3a}{b} \cdot \frac{b^2}{6ac} \qquad \text{Invert the divisor and multiply.}$$

$$= \frac{b}{2c} \qquad \text{Multiply the fractions and simplify.}$$

Method 2: We multiply the numerator and denominator by the LCD of $\frac{3a}{b}$ and $\frac{6ac}{b^2}$, b^2, and simplify.

$$\frac{\frac{3a}{b}}{\frac{6ac}{b^2}} = \frac{\frac{3a}{b} \cdot \mathbf{b^2}}{\frac{6ac}{b^2} \cdot \mathbf{b^2}} \qquad \frac{b^2}{b^2} = 1$$

$$= \frac{\frac{3ab^2}{b}}{\frac{6ab^2c}{b^2}}$$

$$= \frac{3ab}{6ac} \qquad \text{Simplify the fractions in the numerator and denominator.}$$

$$= \frac{b}{2c} \qquad \text{Divide out the common factor of } 3a.$$

EXAMPLE 10 Simplify: $\dfrac{\frac{1}{x} + \frac{1}{y}}{\frac{1}{x} - \frac{1}{y}}$

Solution **Method 1:** We add the rational expressions in the numerator and in the denominator and proceed as follows.

$$\frac{\frac{1}{x} + \frac{1}{y}}{\frac{1}{x} - \frac{1}{y}} = \frac{\frac{1 \cdot y}{x \cdot y} + \frac{x \cdot 1}{x \cdot y}}{\frac{1 \cdot y}{x \cdot y} - \frac{x \cdot 1}{x \cdot y}}$$

$$= \frac{\frac{y + x}{xy}}{\frac{y - x}{xy}}$$

$$= \frac{y + x}{xy} \div \frac{y - x}{xy}$$

$$= \frac{y + x}{xy} \cdot \frac{xy}{y - x} \quad \text{Invert the divisor and multiply.}$$

$$= \frac{y + x}{y - x} \quad \text{Multiply and then divide out the factors of } x \text{ and } y.$$

Norbert Wiener
1894–1964
A child prodigy, Norbert Wiener received his PhD from Harvard University at the age of 19. As a professor of mathematics at MIT, Wiener analyzed the nature of information and communication, and created a new field called cybernetics. Without this study, modern computers would not exist.

Method 2: We multiply the numerator and denominator by xy (the LCD of the rational expressions appearing in the complex fraction) and simplify.

$$\frac{\frac{1}{x} + \frac{1}{y}}{\frac{1}{x} - \frac{1}{y}} = \frac{xy\left(\frac{1}{x} + \frac{1}{y}\right)}{xy\left(\frac{1}{x} - \frac{1}{y}\right)} \quad \frac{xy}{xy} = 1$$

$$= \frac{\frac{xy}{x} + \frac{xy}{y}}{\frac{xy}{x} - \frac{xy}{y}} \quad \text{Use the distributive property to remove parentheses.}$$

$$= \frac{y + x}{y - x} \quad \text{Simplify the fractions.}$$

SELF CHECK 10 Simplify: $\dfrac{\frac{1}{x} - \frac{1}{y}}{\frac{1}{x} + \frac{1}{y}}$

EXAMPLE 11 Simplify: $\dfrac{x^{-1} + y^{-1}}{x^{-2} - y^{-2}}$

Solution **Method 1:** We proceed as follows.

$$\frac{x^{-1} + y^{-1}}{x^{-2} - y^{-2}} = \frac{\frac{1}{x} + \frac{1}{y}}{\frac{1}{x^2} - \frac{1}{y^2}} \quad \text{Write the fraction without using negative exponents.}$$

Unless otherwise noted, all content on this page is © Cengage Learning.

$$= \frac{\frac{y}{xy} + \frac{x}{xy}}{\frac{y^2}{x^2y^2} - \frac{x^2}{x^2y^2}}$$ Obtain a common denominator for the numerator and for the denominator.

$$= \frac{\frac{y + x}{xy}}{\frac{y^2 - x^2}{x^2y^2}}$$ Add the fractions in the numerator and denominator.

$$= \frac{y + x}{xy} \div \frac{y^2 - x^2}{x^2y^2}$$ Write the complex fraction as a division.

$$= \frac{y + x}{xy} \cdot \frac{x^2y^2}{(y - x)(y + x)}$$ Invert the divisor, multiply, and factor.

$$= \frac{(y + x)x^2y^2}{xy(y - x)(y + x)}$$ Multiply the numerators and the denominators.

$$= \frac{xy}{y - x}$$ Divide out the common factors of x, y, and $y + x$ in the numerator and denominator.

Method 2: We multiply both numerator and denominator by x^2y^2, the LCD of the rational expressions, and proceed as follows:

$$\frac{x^{-1} + y^{-1}}{x^{-2} - y^{-2}} = \frac{\frac{1}{x} + \frac{1}{y}}{\frac{1}{x^2} - \frac{1}{y^2}}$$ Write the fraction without negative exponents.

$$= \frac{x^2y^2\left(\frac{1}{x} + \frac{1}{y}\right)}{x^2y^2\left(\frac{1}{x^2} - \frac{1}{y^2}\right)}$$ $\frac{x^2y^2}{x^2y^2} = 1$

$$= \frac{xy^2 + x^2y}{y^2 - x^2}$$ Use the distributive property to remove parentheses, and simplify.

$$= \frac{xy(y + x)}{(y + x)(y - x)}$$ Factor the numerator and denominator.

$$= \frac{xy}{y - x}$$ Divide out the common factor $y + x$.

SELF CHECK 11 Simplify: $\frac{x^{-1} - y^{-1}}{x^{-2}}$

COMMENT $x^{-1} + y^{-1}$ means $\frac{1}{x} + \frac{1}{y}$, and $(x + y)^{-1}$ means $\frac{1}{x + y}$.

EXAMPLE 12 Simplify: $\dfrac{\dfrac{2x}{1-\frac{1}{x}}+3}{3-\dfrac{2}{x}}$

Solution We begin by multiplying the numerator and denominator of

$$\frac{2x}{1-\dfrac{1}{x}}$$

by x. This will eliminate the complex fraction in the numerator of the original fraction.

$$\frac{\dfrac{2x}{1-\frac{1}{x}}+3}{3-\dfrac{2}{x}}=\frac{\dfrac{x2x}{x\left(1-\frac{1}{x}\right)}+3}{3-\dfrac{2}{x}} \qquad \frac{x}{x}=1$$

$$=\frac{\dfrac{2x^2}{x-1}+3}{3-\dfrac{2}{x}}$$

We then multiply the numerator and denominator of the previous complex fraction by $x(x-1)$, the LCD of $\frac{2x^2}{x-1}$, 3, and $\frac{2}{x}$, and simplify:

$$\frac{\dfrac{2x}{1-\frac{1}{x}}+3}{3-\dfrac{2}{x}}=\frac{x(x-1)\left(\dfrac{2x^2}{x-1}+3\right)}{x(x-1)\left(3-\dfrac{2}{x}\right)} \qquad \frac{x(x-1)}{x(x-1)}=1$$

$$=\frac{2x^3+3x(x-1)}{3x(x-1)-2(x-1)} \qquad \text{Use the distributive property.}$$

$$=\frac{2x^3+3x^2-3x}{3x^2-5x+2} \qquad \text{Use the distributive property to remove parentheses.}$$

This result does not simplify.

SELF CHECK 12 Simplify: $\dfrac{\dfrac{3}{1-\frac{2}{x}}+1}{2-\dfrac{1}{x}}$

Perspective

Each of the complex fractions in the list

$$1+\frac{1}{2},\ 1+\frac{1}{1+\frac{1}{2}},\ 1+\frac{1}{1+\frac{1}{1+\frac{1}{2}}},\ 1+\frac{1}{1+\frac{1}{1+\frac{1}{1+\frac{1}{2}}}},\ldots$$

can be simplified by using the value of the expression preceding it. For example, to simplify the second expression in the list, replace $1+\frac{1}{2}$ with $\frac{3}{2}$.

$$1+\frac{1}{1+\frac{1}{2}}=1+\frac{1}{\frac{3}{2}}=1+\frac{2}{3}=\frac{5}{3}$$

To simplify the third expression, replace $1+\frac{1}{1+\frac{1}{2}}$ with $\frac{5}{3}$:

$$1+\frac{1}{1+\frac{1}{1+\frac{1}{2}}}=1+\frac{1}{\frac{5}{3}}=1+\frac{3}{5}=\frac{8}{5}$$

Can you show that the expressions in the list simplify to the fractions $\frac{3}{2}, \frac{5}{3}, \frac{8}{5}, \frac{13}{8}, \frac{21}{13}, \frac{34}{21}, \ldots$?

Do you see a pattern, and can you predict the next fraction?

Use a calculator to write each of these fractions as a decimal. The values produced get closer and closer to the irrational number 1.61803398875 . . . , which is known as the **golden ratio**. This number often appears in the architecture of the ancient Greeks and Egyptians. The width of the stairs in front of the Greek Parthenon (Illustration 1), divided by the building's height, is the golden ratio. The height of the triangular face of the Great Pyramid of Cheops (Illustration 2), divided by the pyramid's width, is also the golden ratio.

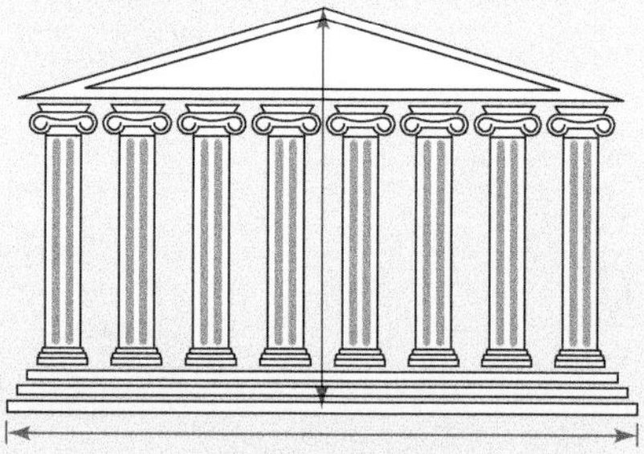

ILLUSTRATION 1

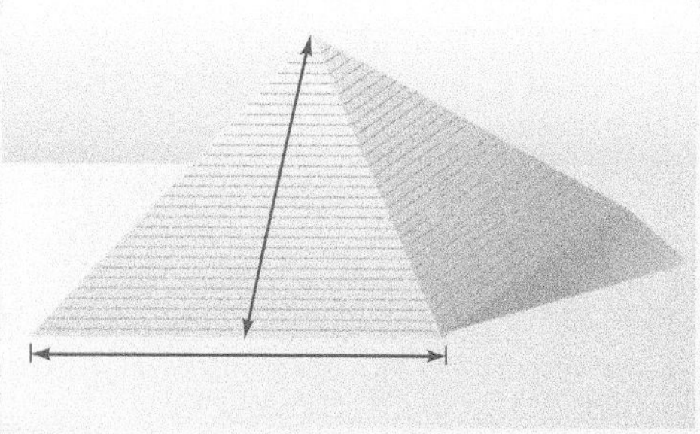

ILLUSTRATION 2

We now show how to solve equations that contain rational expressions.

6 Solve a rational equation.

If an equation contains one or more rational expressions, it is called a **rational equation**. Some examples are

$$\frac{3}{5}+\frac{7}{x+2}=2,\qquad \frac{x+3}{x-3}=\frac{2}{x^2-4},\qquad\text{and}\qquad \frac{-x^2+10}{x^2-1}+\frac{3x}{x-1}=\frac{2x}{x+1}$$

To solve a rational equation, we can multiply both sides of the equation by a nonzero expression *to clear the equation of fractions.*

EXAMPLE 13 Solve: $\dfrac{-x^2+10}{x^2-1}+\dfrac{3x}{x-1}=\dfrac{2x}{x+1}$

Solution We first note that x cannot be 1 or -1, because this would give a 0 in a denominator. If $x \neq 1$ and $x \neq -1$, we can clear the equation of fractions by multiplying both sides by the LCD of the three rational expressions, $(x+1)(x-1)$, and proceed as follows:

Unless otherwise noted, all content on this page is © Cengage Learning.

$$\frac{-x^2 + 10}{x^2 - 1} + \frac{3x}{x - 1} = \frac{2x}{x + 1}$$

$$\frac{-x^2 + 10}{(x + 1)(x - 1)} + \frac{3x}{x - 1} = \frac{2x}{x + 1}$$ Factor $x^2 - 1$.

$$(x + 1)(x - 1)\left[\frac{-x^2 + 10}{(x + 1)(x - 1)} + \frac{3x}{x - 1} = \frac{2x}{x + 1}\right](x + 1)(x - 1)$$ Multiply both sides by the LCD, $(x + 1)(x - 1)$ to clear fractions.

$$\frac{(x + 1)(x - 1)(-x^2 + 10)}{(x + 1)(x - 1)} + \frac{3x(x + 1)(x - 1)}{x - 1} = \frac{2x(x + 1)(x - 1)}{x + 1}$$ Use the distributive property

$$-x^2 + 10 + 3x(x + 1) = 2x(x - 1)$$ Divide out common factors.

$$-x^2 + 10 + 3x^2 + 3x = 2x^2 - 2x$$ Use the distributive property to remove parentheses.

$$2x^2 + 10 + 3x = 2x^2 - 2x$$ Combine like terms.

$$10 + 3x = -2x$$ Subtract $2x^2$ from both sides.

$$10 + 5x = 0$$ Add $2x$ to both sides.

$$5x = -10$$ Subtract 10 from both sides.

$$x = -2$$ Divide both sides by 5.

The solution is -2. Verify that -2 is a solution of the original equation.

SELF CHECK 13 Solve: $\dfrac{2x^2}{x^2 - 4} = \dfrac{3}{x + 2} + \dfrac{2x}{x - 2}$

SELF CHECK ANSWERS

1. $-\frac{1}{2}, -2$ 2. $\frac{2x + 1}{3x - 1}$ 3. $\frac{-2a - 3b}{a}$ 4. $\frac{x - 1}{3x + 2}$ 5. $\frac{x + 3}{x - 2}$ 6. $\frac{3x}{2}$ 7. $\frac{6a}{a + 3}$ 8. $\frac{a^2 - 15a}{(a + 3)(a - 3)}$ 9. $\frac{5a - 3}{a(a + 1)(a - 1)}$ 10. $\frac{y - x}{y + x}$ 11. $\frac{xy - x^2}{y}$ 12. $\frac{4x^2 - 2x}{2x^2 - 5x + 2}$ 13. $\frac{6}{7}$

NOW TRY THIS

Simplify.

1. $5x(x - 2)^{-1} - 6(x + 3)^{-1}$
2. $8(x - 2)^{-2} - 30(x - 2)^{-1} + 7$

Solve.

3. $\dfrac{2(x - 5)}{x - 2} = \dfrac{6x + 12}{4 - x^2}$

7.5 Exercises

Assume no division by 0.

WARM-UPS *Simplify each rational expression by dividing out common factors.*

1. $\dfrac{6(x + 9)}{3(x + 9)}$
2. $\dfrac{x^2(7 - x)}{x(7 - x)}$
3. $\dfrac{x + 5}{5 + x}$
4. $\dfrac{3 + x}{x + 3}$
5. $\dfrac{x - 5}{5 - x}$
6. $\dfrac{3 - x}{x - 3}$
7. $\dfrac{x + 2}{(x - 1)(x + 2)}$
8. $\dfrac{(x - 1)(x + 1)}{x - 1}$

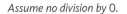

REVIEW *Graph each interval.*

9. $(-\infty, -4) \cup [5, \infty)$ **10.** $(4, 8]$

Solve each equation.

11. $x^2 - 5x - 6 = 0$ **12.** $x^2 + 9x + 20 = 0$

13. $2x^2 - 5x - 3 = 0$ **14.** $5x^2 + 23x + 12 = 0$

15. $a^4 - 13a^2 + 36 = 0$ **16.** $x^4 - 10x^2 + 9 = 0$

17. $P = 2l + 2w$ for w **18.** $S = \dfrac{a - lr}{1 - r}$ for a

VOCABULARY AND CONCEPTS *Fill in the blanks.*

19. $\dfrac{ax}{bx} = \_$ $(b \neq 0, x \neq 0)$

20. $\dfrac{a}{b} \cdot \dfrac{c}{d} = \_$ $(b \neq 0, d \neq 0)$

21. $\dfrac{a}{b} \div \dfrac{c}{d} = \_$ $(b \neq 0, c \neq 0, d \neq 0)$

22. $\dfrac{a}{b} + \dfrac{c}{b} = \_\_\_\_$ $(b \neq 0)$

GUIDED PRACTICE *Find all values of the variable that will make the following undefined.* SEE EXAMPLE 1. (OBJECTIVE 1).

23. $\dfrac{3x - 5}{2x}$ **24.** $\dfrac{9x + 4}{7x}$

25. $\dfrac{x + 5}{x - 2}$ **26.** $\dfrac{3x^2 + 4}{x + 6}$

27. $\dfrac{x^2 - 5}{3x + 2}$ **28.** $\dfrac{x^2 + x - 9}{5x - 4}$

29. $\dfrac{2x^2 + x - 9}{2x^2 - 5x - 3}$ **30.** $\dfrac{x^2 - 9x + 10}{5x^2 + 23x + 12}$

For the remainder of the exercises, assume no denominator is 0. Simplify each rational expression. SEE EXAMPLE 2. (OBJECTIVE 2)

31. $\dfrac{3x^2 - 12}{x^2 - x - 2}$ **32.** $\dfrac{x^2 + 2x - 15}{x^2 - 25}$

33. $\dfrac{x^2 + 2x + 1}{x^2 + 4x + 3}$ **34.** $\dfrac{6x^2 + x - 2}{8x^2 + 2x - 3}$

35. $\dfrac{4x^2 + 24x + 32}{16x^2 + 8x - 48}$ **36.** $\dfrac{a^2 - 4}{5a^2 + 20a + 20}$

37. $\dfrac{x^3 + 8}{x^2 - 2x + 4}$ **38.** $\dfrac{x^2 + 3x + 9}{x^3 - 27}$

Simplify each rational expression. SEE EXAMPLE 3. (OBJECTIVE 2)

39. $\dfrac{y - x}{x^2 - y^2}$ **40.** $\dfrac{3m - 6n}{6n - 3m}$

41. $\dfrac{x - y}{x^2 - y^2}$ **42.** $\dfrac{2x^2 + 2x - 12}{x^2 - 4}$

43. $\dfrac{3x^2 - 3y^2}{x^2 + 2xy + y^2}$ **44.** $\dfrac{5x - 2y}{5x^2 - 7xy + 2y^2}$

45. $\dfrac{ax + by - ay - bx}{b^2 - a^2}$ **46.** $\dfrac{x^2 - 2xy}{2y^2 + x - 2y - xy}$

Perform the operation and simplify. SEE EXAMPLE 4. (OBJECTIVE 3)

47. $\dfrac{x^2 + 2x + 1}{x} \cdot \dfrac{x^2 - x}{x^2 - 1}$

48. $\dfrac{x^2 - x - 6}{x^2 - 4} \cdot \dfrac{x^2 - x - 2}{9 - x^2}$

49. $\dfrac{a + 6}{a^2 - 16} \cdot \dfrac{3a - 12}{3a + 18}$

50. $\dfrac{2x^2 - x - 3}{x^2 - 1} \cdot \dfrac{x^2 + x - 2}{2x^2 + x - 6}$

51. $\dfrac{9x^2 + 3x - 20}{3x^2 - 7x + 4} \cdot \dfrac{3x^2 - 5x + 2}{9x^2 + 18x + 5}$

52. $\dfrac{3t^2 - t - 2}{6t^2 - 5t - 6} \cdot \dfrac{4t^2 - 9}{2t^2 + 5t + 3}$

53. $\dfrac{2p^2 - 5p - 3}{p^2 - 9} \cdot \dfrac{2p^2 + 5p - 3}{2p^2 + 5p + 2}$

54. $\dfrac{x^2 - y^2}{2x^2 + 2xy + x + y} \cdot \dfrac{2x^2 - 5x - 3}{yx - 3y - x^2 + 3x}$

Perform the operation and simplify. SEE EXAMPLE 5. (OBJECTIVE 3)

55. $\dfrac{x^2 - 16}{x^2 - 25} \div \dfrac{x + 4}{x - 5}$

56. $\dfrac{a^2 - 9}{a^2 - 49} \div \dfrac{a + 3}{a + 7}$

57. $\dfrac{a^2 + 2a - 35}{12x} \div \dfrac{ax - 3x}{a^2 + 4a - 21}$

58. $\dfrac{x^2 - 4}{2b - bx} \div \dfrac{x^2 + 4x + 4}{2b + bx}$

59. $\dfrac{x^2 - 6x + 9}{4 - x^2} \div \dfrac{x^2 - 9}{x^2 - 8x + 12}$

60. $\dfrac{2x^2 - 7x - 4}{20 - x - x^2} \div \dfrac{2x^2 - 9x - 5}{x^2 - 25}$

61. $(2x^2 - 15x + 25) \div \dfrac{2x^2 - 3x - 5}{x + 1}$

62. $(x^2 - 6x + 9) \div \dfrac{x^2 - 9}{x + 3}$

Perform the operations and simplify. SEE EXAMPLE 6. (OBJECTIVE 3)

63. $\frac{6a^2-7a-3}{a^2-1} \div \frac{4a^2-12a+9}{a^2-1} \cdot \frac{2a^2-a-3}{3a^2-2a-1}$

64. $\frac{x^2-x-12}{x^2+x-2} \div \frac{x^2-6x+8}{x^2-3x-10} \cdot \frac{x^2-3x+2}{x^2-2x-15}$

65. $\frac{4x^2-10x+6}{x^4-3x^3} \div \frac{2x-3}{2x^3} \cdot \frac{x-3}{2x-2}$

66. $\frac{2}{x-1} \div \frac{x^2-1}{2x} \div \frac{x}{x^2+2x+1}$

67. $(x+3) \cdot \frac{x^2}{2x-6} \div \frac{2}{x-3}$

68. $(4x^2-9) \div \frac{2x^2+7x+6}{x+2} \div (2x+3)$

69. $\frac{2x^2+5x-3}{x^2+2x-3} \div \left(\frac{x^2+2x-35}{x^2-6x+5} \div \frac{x^2-9x+14}{2x^2-5x+2}\right)$

70. $\frac{x^2-4}{x^2-x-6} \div \left(\frac{x^2-x-2}{x^2-8x+15} \cdot \frac{x^2-3x-10}{x^2+3x+2}\right)$

Perform the operation(s) and simplify. SEE EXAMPLE 7. (OBJECTIVE 4)

71. $\frac{x}{x+4} + \frac{5}{x+4}$

72. $\frac{3}{a+6} - \frac{a}{a+6}$

73. $\frac{5x}{x+1} + \frac{3}{x+1} - \frac{2x}{x+1}$

74. $\frac{4}{a+4} - \frac{2a}{a+4} + \frac{3a}{a+4}$

75. $\frac{3x}{2x+2} + \frac{x+4}{2x+2}$

76. $\frac{4y}{y-4} - \frac{16}{y-4}$

77. $\frac{3(x^2+x)}{x^2-5x+6} + \frac{-3(x^2-x)}{x^2-5x+6}$

78. $\frac{2x+4}{x^2+13x+12} - \frac{x+3}{x^2+13x+12}$

Perform the operation and simplify. SEE EXAMPLE 8. (OBJECTIVE 4)

79. $\frac{a+b}{3} + \frac{a-b}{7}$

80. $\frac{x-y}{2} + \frac{x+y}{3}$

81. $\frac{a}{2} + \frac{2a}{5}$

82. $\frac{b}{6} + \frac{3a}{4}$

83. $\frac{3}{4x} + \frac{2}{3x}$

84. $\frac{2}{5a} + \frac{3}{2b}$

85. $x + \frac{1}{x}$

86. $2 - \frac{1}{x+1}$

87. $\frac{3}{x+2} + \frac{5}{x-4}$

88. $\frac{2}{a+4} - \frac{6}{a+3}$

89. $\frac{7}{x+3} + \frac{4x}{x+6}$

90. $\frac{x+2}{x+5} - \frac{x-3}{x+7}$

Perform the operation(s) and simplify. SEE EXAMPLE 9. (OBJECTIVE 4)

91. $1 + x - \frac{x}{x-5}$

92. $2 - x + \frac{3}{x-9}$

93. $\frac{8}{x^2-9} + \frac{2}{x-3} - \frac{6}{x}$

94. $\frac{x}{x^2-4} - \frac{x}{x+2} + \frac{2}{x}$

95. $\frac{3}{x+1} - \frac{2}{x-1} + \frac{x+3}{x^2-1}$

96. $\frac{2}{x-2} + \frac{3}{x+2} - \frac{x-1}{x^2-4}$

97. $\frac{x}{x^2+5x+6} + \frac{x}{x^2-4}$

98. $\frac{x}{3x^2-2x-1} + \frac{4}{3x^2+10x+3}$

Simplify each complex fraction. SEE EXAMPLE 10. (OBJECTIVE 5)

99. $\frac{\frac{1}{a}+\frac{1}{b}}{\frac{1}{a}}$

100. $\frac{\frac{1}{b}}{\frac{1}{a}-\frac{1}{b}}$

101. $\frac{\frac{y}{x}-\frac{x}{y}}{\frac{1}{x}+\frac{1}{y}}$

102. $\frac{\frac{y}{x}-\frac{x}{y}}{\frac{1}{y}-\frac{1}{x}}$

103. $\frac{\frac{1}{a}-\frac{1}{b}}{\frac{a}{b}-\frac{b}{a}}$

104. $\frac{\frac{1}{a}+\frac{1}{b}}{\frac{a}{b}-\frac{b}{a}}$

105. $\frac{\frac{1}{a+1}+1}{\frac{3}{a-1}+1}$

106. $\frac{2+\frac{3}{x+1}}{\frac{1}{x}+x+x^2}$

Simplify each complex fraction. SEE EXAMPLE 11. (OBJECTIVE 5)

107. $\frac{x^{-1}+y^{-1}}{x^{-1}-y^{-1}}$

108. $\frac{(x+y)^{-1}}{x^{-1}+y^{-1}}$

109. $\frac{x-y^{-2}}{y-x^{-2}}$

110. $\frac{x^{-2}-y^{-2}}{x^{-1}-y^{-1}}$

Simplify each complex fraction. SEE EXAMPLE 12. (OBJECTIVE 5)

111. $\frac{1+\frac{a}{b}}{1-\frac{a}{1-\frac{a}{b}}}$

112. $\frac{1+\frac{2}{1+\frac{a}{b}}}{1-\frac{a}{b}}$

113. $a + \dfrac{a}{1 + \dfrac{a}{a + 1}}$

114. $b + \dfrac{b}{1 - \dfrac{b + 1}{b}}$

Solve. SEE EXAMPLE 13. (OBJECTIVE 6)

115. $\dfrac{3}{y} - \dfrac{1}{12} = \dfrac{7}{4y}$

116. $\dfrac{2}{x + 5} - \dfrac{1}{6} = \dfrac{1}{x + 4}$

117. $\dfrac{y^2}{y + 2} + 4 = \dfrac{y + 6}{y + 2} + 3$

118. $\dfrac{7}{x + 9} - \dfrac{x + 2}{2} = \dfrac{x + 4}{x + 9}$

ADDITIONAL PRACTICE *Simplify each expression.*

119. $\dfrac{a^{-2}b^2}{x^{-1}y} \cdot \dfrac{a^4b^4}{x^2y^3}$

120. $\dfrac{(a^3)^2}{b^{-1}} \div \dfrac{(a^3)^{-2}}{b^{-1}}$

121. $\dfrac{m^2 - n^2}{2x^2 + 3x - 2} \cdot \dfrac{2x^2 + 5x - 3}{n^2 - m^2}$

122. $\dfrac{x^2 + 3x + xy + 3y}{x^2 - 9} \cdot \dfrac{x - 3}{x + 3}$

123. $\dfrac{x^3 + y^3}{x^3 - y^3} \div \dfrac{x^2 - xy + y^2}{x^2 + xy + y^2}$

124. $\dfrac{2x^2 + 3xy + y^2}{y^2 - x^2} \div \dfrac{6x^2 + 5xy + y^2}{2x^2 - xy - y^2}$

125. $\dfrac{x + 8}{x - 3} - \dfrac{x - 14}{3 - x}$

126. $\dfrac{3 - x}{2 - x} + \dfrac{x - 1}{x - 2}$

127. $\dfrac{x - 2}{x^2 - 3x} + \dfrac{2x - 1}{x^2 + 3x} - \dfrac{2}{x^2 - 9}$

128. $\dfrac{2}{x - 1} - \dfrac{2x}{x^2 - 1} - \dfrac{x}{x^2 + 2x + 1}$

129. $\dfrac{x + y}{x^{-1} + y^{-1}}$

130. $\dfrac{x - y}{x^{-1} - y^{-1}}$

Solve.

131. $\dfrac{3}{x} + \dfrac{4}{x + 1} = 2$

132. $\dfrac{6}{x - 2} + 1 = \dfrac{12}{x - 1}$

WRITING ABOUT MATH

133. Explain how to simplify a rational expression.

134. Explain how to multiply two rational expressions.

135. Explain how to divide two rational expressions.

136. Explain how to add two rational expressions.

SOMETHING TO THINK ABOUT

137. A student compared his answer, $\frac{a - 3b}{2b - a}$, with the answer, $\frac{3b - a}{a - 2b}$, in the back of the text. Is the student's answer correct?

138. Another student shows this work:

$$\frac{3x^2 + 6}{3y} = \frac{\cancel{3}x^2 + \overset{2}{\cancel{6}}}{\cancel{3}y} = \frac{x^2 + 2}{y}$$

Is the student's work correct?

139. In which parts can you divide out the 4's?

a. $\dfrac{4x}{4y}$ **b.** $\dfrac{4x}{x + 4}$ **c.** $\dfrac{4 + x}{4 + y}$ **d.** $\dfrac{4x}{4 + 4y}$

140. In which parts can you divide out the 3's?

a. $\dfrac{3x + 3y}{3z}$ **b.** $\dfrac{3(x + y)}{3x + y}$ **c.** $\dfrac{x + 3}{3y}$ **d.** $\dfrac{3x + 3y}{3a - 3b}$

Section 7.6 Solving Equations in One Variable Containing an Absolute Value Expression

Objectives

1. Solve an equation containing a single absolute value expression.
2. Solve an equation where two absolute value expressions are equal.

Getting Ready

Simplify each expression.

1. $-(-6)$ **2.** $-(-5)$ **3.** $-(x-2)$ **4.** $-(2-\pi)$

In this section, we will review the definition of absolute value and show how to solve equations that contain absolute values. Recall the following definition of the absolute value of x.

ABSOLUTE VALUE

If $x \geq 0$, then $|x| = x$.

If $x < 0$, then $|x| = -x$

This definition associates a nonnegative real number with any real number.

- If $x \geq 0$, then x is its own absolute value.
- If $x < 0$, then $-x$ (which is positive) is the absolute value.

Either way, $|x|$ is positive or 0:

$$|x| \geq 0 \quad \text{for all real numbers } x$$

Let us consider how to find each of the following absolute values.

a. $|9|$: Since $9 \geq 0$, 9 is its own absolute value: $|9| = 9$.

b. $|-5|$: Since $-5 < 0$, the negative of -5 is the absolute value:

$$|-5| = -(-5) = 5$$

c. $|0|$: Since $0 \geq 0$, 0 is its own absolute value: $|0| = 0$.

d. $-|-10|$: $-|-10| = -(10) = -10$

e. $|2-\pi|$: Since $\pi \approx 3.14$, it follows that $2 - \pi < 0$. Thus,

$$|2-\pi| = -(2-\pi) = \pi - 2$$

COMMENT The placement of a – sign in an expression containing an absolute value symbol is important. For example, $|-19| = 19$, but $-|19| = -19$.

1 Solve an equation containing a single absolute value expression.

Now that we have reviewed finding the absolute value, we will consider how to solve an equation containing an absolute value expression. In the equation $|x| = 5$, x can be either 5 or -5, because

$$|5| = 5 \qquad \text{and} \qquad |-5| = 5$$

Thus, if $|x| = 5$, then $x = 5$ or $x = -5$. In general, the following is true.

ABSOLUTE VALUE EQUATIONS

If $k > 0$, then

$$|x| = k \qquad \text{is equivalent to} \qquad x = k \text{ or } x = -k$$

The absolute value of x can be interpreted as the distance on the number line from a point to the origin. The solutions of $|x| = k$ are represented by the two points that lie exactly k units from the origin. (See Figure 7-22 on the next page.)

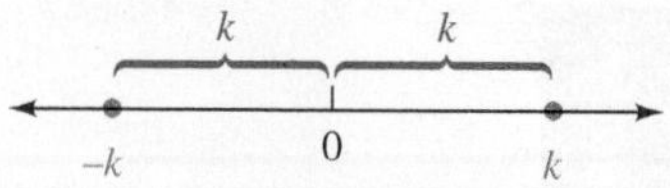

Figure 7-22

The equation $|x - 3| = 7$ indicates that a point on the number line with a coordinate of $x - 3$ is 7 units from the origin. Thus, $x - 3$ can be either 7 or -7.

$$x - 3 = 7 \quad \text{or} \quad x - 3 = -7$$
$$x = 10 \quad \Big| \quad x = -4$$

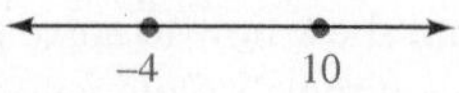

Figure 7-23

The solutions of $|x - 3| = 7$ are 10 and -4. (See Figure 7-23). If either of these numbers is substituted for x in the equation, it results in a true statement:

$$\begin{aligned} |x - 3| &= 7 \\ |\mathbf{10} - 3| &\stackrel{?}{=} 7 \\ |7| &\stackrel{?}{=} 7 \\ 7 &= 7 \end{aligned} \qquad \begin{aligned} |x - 3| &= 7 \\ |\mathbf{-4} - 3| &\stackrel{?}{=} 7 \\ |-7| &\stackrel{?}{=} 7 \\ 7 &= 7 \end{aligned}$$

EXAMPLE 1 Solve: $|3x - 2| = 5$

Solution We can write $|3x - 2| = 5$ as

$$3x - 2 = 5 \quad \text{or} \quad 3x - 2 = -5$$

and solve each equation for x:

$$\begin{aligned} 3x - 2 &= 5 \\ 3x &= 7 \\ x &= \frac{7}{3} \end{aligned} \quad \text{or} \quad \begin{aligned} 3x - 2 &= -5 \\ 3x &= -3 \\ x &= -1 \end{aligned}$$

Verify that both solutions check.

SELF CHECK 1 Solve: $|2x + 3| = 5$

To solve more complicated equations with a term involving an absolute value, we must isolate the absolute value before attempting to solve for the variable.

EXAMPLE 2 Solve: $\left|\frac{2}{3}x + 3\right| + 4 = 10$

Solution We first isolate the absolute value on the left side.

$$\left|\frac{2}{3}x + 3\right| + 4 = 10$$

(1) $$\left|\frac{2}{3}x + 3\right| = 6 \qquad \text{Subtract 4 from both sides.}$$

We can now write Equation 1 as

$$\frac{2}{3}x + 3 = 6 \quad \text{or} \quad \frac{2}{3}x + 3 = -6$$

Unless otherwise noted, all content on this page is © Cengage Learning.

and solve each equation for x:

$$\frac{2}{3}x + 3 = 6 \quad \text{or} \quad \frac{2}{3}x + 3 = -6$$

$$\frac{2}{3}x = 3 \qquad \frac{2}{3}x = -9$$ Subtract 3 from both sides of each equation.

$$2x = 9 \qquad 2x = -27$$ To clear fractions, multiply both sides of each equation by 3.

$$x = \frac{9}{2} \qquad x = -\frac{27}{2}$$ Divide both sides of each equation by 2.

Verify that both solutions check.

SELF CHECK 2 Solve: $\left|\frac{3}{2}x - 3\right| + 1 = 7$

EXAMPLE 3 Solve: $\left|7x + \frac{1}{2}\right| = -4$

Solution Since the absolute value of a number cannot be negative, no value of x can make the equation $\left|7x + \frac{1}{2}\right| = -4$ true. Since this equation has no solutions, its solution set is $\varnothing$.

SELF CHECK 3 Solve: $-|3x + 2| = 4$

EXAMPLE 4 Solve: $\left|\frac{1}{2}x - 5\right| - 4 = -4$

Solution We first isolate the absolute value on the left side.

$$\left|\frac{1}{2}x - 5\right| - 4 = -4$$

$$\left|\frac{1}{2}x - 5\right| = 0$$ Add 4 to both sides.

Since 0 is the only number whose absolute value is 0, the binomial $\frac{1}{2}x - 5$ must be 0, and we have

$$\frac{1}{2}x - 5 = 0$$

$$\frac{1}{2}x = 5$$ Add 5 to both sides.

$$x = 10$$ Multiply both sides by 2.

Verify that 10 satisfies the original equation.

SELF CHECK 4 Solve: $\left|\frac{2}{3}x - 4\right| + 2 = 2$

Accent on technology

▸ Solving Absolute Value Equations

We can solve absolute value equations with a graphing calculator. For example, to solve $|2x - 3| = 9$, we graph the equations $y_1 = |2x - 3|$ and $y_2 = 9$ on the same coordinate system, as shown in Figure 7-24. To enter the equation $y_1 = |2x - 3|$ on a TI-84 Plus calculator, press these keys.

Y = MATH (▸) ENTER (abs( ) 2 X, Γ, Θ, η − 3)

The equation $|2x - 3| = 9$ will be true for all x-coordinates of points that lie on both graphs. By using the INTERSECT feature, we can see that the graphs intersect when $x = -3$ and $x = 6$. These are the solutions of the equation.

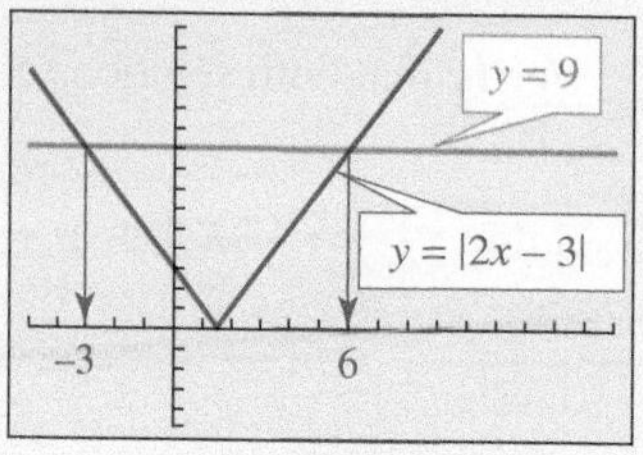

Figure 7-24

For instructions regarding the use of a Casio graphing calculator, please refer to the Casio Keystroke Guide in the back of the the book.

2 Solve an equation where two absolute value expressions are equal.

The equation $|a| = |b|$ is true when $a = b$ or when $a = -b$. For example,

$$|3| = |3| \qquad |3| = |-3|$$
$$3 = 3 \qquad 3 = 3$$

Thus, we have the following result.

EQUATIONS WITH TWO ABSOLUTE VALUES

If a and b represent algebraic expressions, the equation $|a| = |b|$ is equivalent to the pair of equations

$$a = b \quad \text{or} \quad a = -b$$

EXAMPLE 5 Solve: $|5x + 3| = |3x + 25|$

Solution This equation is true when $5x + 3 = 3x + 25$, or when $5x + 3 = -(3x + 25)$. We solve each equation for x.

$$5x + 3 = 3x + 25 \quad \text{or} \quad 5x + 3 = -(3x + 25)$$
$$2x = 22 \qquad\qquad 5x + 3 = -3x - 25$$
$$x = 11 \qquad\qquad 8x = -28$$
$$x = -\frac{28}{8}$$
$$x = -\frac{7}{2}$$

Verify that both solutions check.

SELF CHECK 5 Solve: $|4x - 3| = |2x + 5|$

Unless otherwise noted, all content on this page is © Cengage Learning.

SELF CHECK ANSWERS **1.** 1, −4 **2.** 6, −2 **3.** ∅ **4.** 6 **5.** 4, $-\frac{1}{3}$

NOW TRY THIS

Express each of the following as an absolute value equation.

1. $x = 2$ or $x = -2$
2. $2x - 3 = 7$ or $2x - 3 = -7$
3. $x + 4 = 6$ or $x + 2 = -8$

7.6 Exercises

WARM-UPS *Find each absolute value.*

1. $|8|$ **2.** $|16|$
3. $|-5|$ **4.** $|-28|$
5. $|-6|$ **6.** $|-15|$
7. $-|5|$ **8.** $-|4|$
9. $-|-20|$ **10.** $-|-25|$
11. $|2\pi - 4|$ **12.** $|\pi - 4|$

Select the smaller of the two numbers.

13. $|2|, |5|$ **14.** $|6|, |3|$
15. $|5|, |-8|$ **16.** $|-6|, |2|$
17. $|-2|, |10|$ **18.** $|-6|, -|6|$
19. $|-3|, -|-4|$ **20.** $|-3|, |-2|$

REVIEW *Solve each equation.*

21. $3(2a - 1) = 2a$ **22.** $\frac{t}{6} - \frac{t}{3} = -1$

23. $\frac{5x}{2} - 1 = \frac{x}{3} + 12$ **24.** $4b - \frac{b+9}{2} = \frac{b+2}{5} - \frac{8}{5}$

VOCABULARY AND CONCEPTS *Fill in the blanks.*

25. If $x \geq 0$, then $|x| =$ _.
26. If $x < 0$, then $|x| =$ __.
27. $|x| \geq$ _ for all real numbers x.
28. If $k > 0$, then $|x| = k$ is equivalent to __________.
29. If $|a| = |b|$, then $a = b$ or ______.
30. If $k > 0$, the equation $|x| = k$ has ___ solutions.

GUIDED PRACTICE *Solve each equation.* SEE EXAMPLE 1. *(OBJECTIVE 1)*

31. $|x| = 8$ **32.** $|x| = 9$
33. $|x| = 12$ **34.** $|x| = 4.23$
35. $|x - 3| = 6$ **36.** $|x + 4| = 8$
37. $|x + 5| = 9$ **38.** $|x - 4| = 10$
39. $|2x - 3| = 5$ **40.** $|4x - 4| = 20$
41. $|3x + 2| = 16$ **42.** $|5x - 3| = 22$

Solve each equation. SEE EXAMPLE 2. *(OBJECTIVE 1)*

43. $|x + 3| + 7 = 10$ **44.** $|2 - x| + 3 = 5$

45. $|0.3x - 3| - 2 = 7$ **46.** $|0.1x + 8| - 1 = 1$

Solve each equation. SEE EXAMPLE 3. *(OBJECTIVE 1)*

47. $|x - 21| = -8$ **48.** $\left|\frac{7}{2}x + 3\right| = -5$

49. $|6x + 2| + 8 = 5$ **50.** $|4x - 7| + 2 = 0$

Solve each equation. SEE EXAMPLE 4. *(OBJECTIVE 1)*

51. $\left|\frac{3}{5}x - 4\right| - 2 = -2$

52. $\left|\frac{3}{4}x + 2\right| + 4 = 4$

53. $\left|\frac{5}{4}x + 1\right| - 2 = 5$

54. $\left|\frac{2}{5}x - 3\right| + 6 = 9$

Solve each equation. SEE EXAMPLE 5. *(OBJECTIVE 2)*

55. $|2x + 1| = |3x + 3|$ **56.** $|3x + 1| = |x - 5|$

57. $|3x - 1| = |x + 5|$ **58.** $|2 - x| = |3x + 2|$

59. $|4x + 3| = |9 - 2x|$ **60.** $\left|\frac{x}{2} + 2\right| = \left|\frac{x}{2} - 2\right|$

61. $|5x - 7| = |4x + 1|$ **62.** $|7x + 12| = |x - 6|$

ADDITIONAL PRACTICE *Solve each equation.*

63. $|3x + 24| = 0$

64. $|2x + 10| = 0$

65. $\left|\dfrac{x}{2} - 1\right| = 3$

66. $\left|\dfrac{4x - 64}{4}\right| = 32$

67. $|3 - 4x| = 5$

68. $|8 - 5x| = 18$

69. $\left|x + \dfrac{1}{3}\right| = |x - 3|$

70. $\left|x - \dfrac{1}{4}\right| = |x + 4|$

71. $|3x + 7| = -|8x - 2|$

72. $-|17x + 13| = |3x - 14|$

73. $\left|\dfrac{3x + 48}{3}\right| = 12$

74. $\left|\dfrac{x}{2} + 2\right| = 4$

75. $\left|\dfrac{8x + 1}{7}\right| + 6 = 1$

76. $\left|\dfrac{9x + 6}{3}\right| = 8$

Use a graphing calculator to solve each equation. Give the result to the nearest tenth.

77. $|0.75x + 0.12| = 12.3$

78. $|-0.47x - 1.75| = 5.1$

WRITING ABOUT MATH

79. Explain how to find the absolute value of a number.

80. Explain why the equation $|x| + 5 = 0$ has no solution.

SOMETHING TO THINK ABOUT

81. For what values of k does $|x| + k = 0$ have exactly two solutions?

82. For what value of k does $|x| + k = 0$ have exactly one solution?

83. Construct several examples to show that $|a \cdot b| = |a| \cdot |b|$.

84. Construct several examples to show that $\left|\dfrac{a}{b}\right| = \dfrac{|a|}{|b|}$.

85. Construct several examples to show that $|a + b| \neq |a| + |b|$.

86. Construct several examples to show that $|a - b| \neq |a| - |b|$.

87. For what value of x is $x = |1 - x|$?

88. For what value of x is $x = |x + 1|$?

Section 7.7 Solving Inequalities in One Variable Containing an Absolute Value Expression

Objectives

1 Solve an inequality in one variable containing one absolute value expression.

Getting Ready

Solve each inequality.

1. $2x + 3 > 5$ **2.** $-3x - 1 < 5$ **3.** $2x - 5 \leq 9$

In this section, we will show how to solve inequalities that have an expression containing an absolute value.

1 Solve an inequality in one variable containing one absolute value expression.

The inequality $|x| < 5$ indicates that a point with coordinate x is *less than* 5 units from the origin. (See Figure 7-25.) Thus, x is between -5 and 5, and

$$|x| < 5 \qquad \text{is equivalent to} \qquad -5 < x < 5$$

The solution to the inequality $|x| < k \quad (k > 0)$ includes the coordinates of the points on the number line that are *less than* k units from the origin. (See Figure 7-26.)

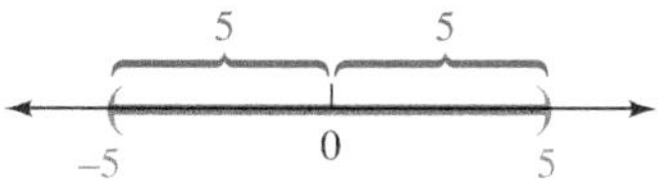

Figure 7-25

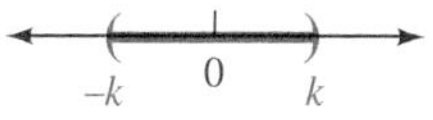

Figure 7-26

We have the following facts.

ABSOLUTE VALUE INEQUALITIES (<, ≤)

$|x| < k \qquad \text{is equivalent to} \qquad -k < x < k \quad (k > 0)$

$|x| \leq k \qquad \text{is equivalent to} \qquad -k \leq x \leq k \quad (k \geq 0)$

To solve an inequality containing an absolute value expression, we first isolate the absolute value expression on one side of the inequality. Then we use one of the previous properties, provided k is nonnegative, to write the inequality as a double inequality.

EXAMPLE 1 Solve, then graph the solution set: $|2x - 3| - 2 < 7$

Solution To isolate the absolute value term, we first add 2 to both sides of the inequality to obtain

$$|2x - 3| < 9$$

We can then write the inequality as the double inequality

$$-9 < 2x - 3 < 9$$

and solve for x by isolating x between the inequality symbols:

$-9 < 2x - 3 < 9$

$-6 < 2x < 12$ Add 3 to all three parts.

$-3 < x < 6$ Divide all three parts by 2.

Figure 7-27

Any number between -3 and 6, not including either -3 or 6, is in the solution set. This is the interval $(-3, 6)$. The graph is shown in Figure 7-27.

SELF CHECK 1 Solve, then graph the solution set: $|3x + 1| < 5$

EXAMPLE 2 Solve, then graph the solution set: $|3x + 2| \leq 5$

Solution Because the absolute value is already isolated, we write the expression as the double inequality

$$-5 \leq 3x + 2 \leq 5$$

Unless otherwise noted, all content on this page is © Cengage Learning.

and solve for x:

$$-5 \le 3x + 2 \le 5$$

$$-7 \le 3x \le 3 \quad \text{Subtract 2 from all three parts.}$$

$$-\frac{7}{3} \le x \le 1 \quad \text{Divide all three parts by 3.}$$

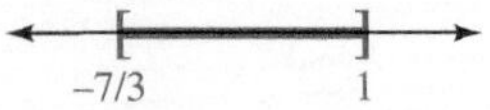

Figure 7-28

The solution set is the interval $\left[-\frac{7}{3}, 1\right]$. The graph is shown in Figure 7-28.

SELF CHECK 2 Solve, then graph the solution set: $|2x - 3| \le 5$

The inequality $|x| > 5$ can be interpreted to mean that a point with coordinate x is *more than* 5 units from the origin. (See Figure 7-29.)

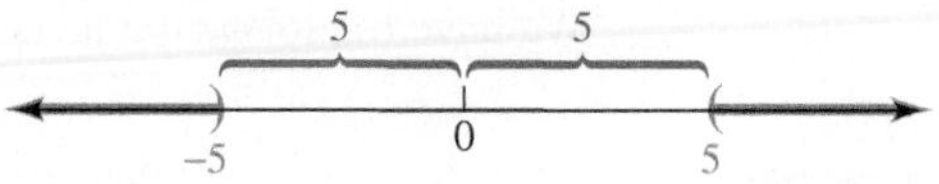

Figure 7-29

Thus, $x < -5$ or $x > 5$.

In general, the inequality $|x| > k \quad (k > 0)$ can be interpreted to mean that a point with coordinate x is *more than* k units from the origin. (See Figure 7-30.)

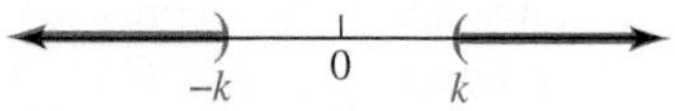

Figure 7-30

Thus,

$|x| > k$ is equivalent to $x < -k$ or $x > k$

The *or* indicates union, an either/or situation. It is necessary for x to satisfy only one of the two conditions to be in the solution set.

To summarize,

ABSOLUTE VALUE INEQUALITIES (>, ≥)

If k is a nonnegative constant, then

$|x| > k$ is equivalent to $x < -k$ or $x > k$

$|x| \ge k$ is equivalent to $x \le -k$ or $x \ge k$

EXAMPLE 3 Solve, then graph the solution set: $|5x - 10| > 20$

Solution We write the inequality as two separate inequalities

$$5x - 10 < -20 \quad \text{or} \quad 5x - 10 > 20$$

and solve each one for x:

$5x - 10 < -20$	or	$5x - 10 > 20$	
$5x < -10$		$5x > 30$	Add 10 to both sides.
$x < -2$		$x > 6$	Divide both sides by 5.

Thus, x is either less than -2 or greater than 6.

$$x < -2 \quad \text{or} \quad x > 6$$

Unless otherwise noted, all content on this page is © Cengage Learning.

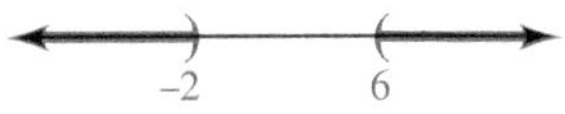

Figure 7-31

This is the union of two intervals $(-\infty, -2) \cup (6, \infty)$. The graph is shown in Figure 7-31.

SELF CHECK 3 Solve, then graph the solution set: $|3x - 2| > 4$

EXAMPLE 4 Solve, then graph the solution set: $\left|\dfrac{3-x}{5}\right| \geq 6$

Solution We write the inequality as two separate inequalities

$$\frac{3-x}{5} \leq -6 \quad \text{or} \quad \frac{3-x}{5} \geq 6$$

and solve each one for x:

$$\frac{3-x}{5} \leq -6 \quad \text{or} \quad \frac{3-x}{5} \geq 6$$

$3 - x \leq -30$	$3 - x \geq 30$	Multiply both sides by 5 to clear fractions.
$-x \leq -33$	$-x \geq 27$	Subtract 3 from both sides.
$x \geq 33$	$x \leq -27$	Divide both sides by −1 and reverse the direction of the inequality symbol.

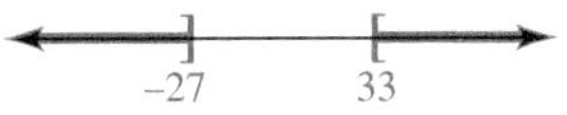

Figure 7-32

The solution set is $(-\infty, -27] \cup [33, \infty)$. The graph is shown in Figure 7-32.

SELF CHECK 4 Solve, then graph the solution set: $\left|\dfrac{4-x}{3}\right| \geq 2$

Everyday Connections

The Shortest Distance Between Two Points

Perhaps one of the most famous examples of an inequality involving absolute values is the so-called Triangle Inequality describing one of the Propositions in Euclid's *Elements*. In layman's terms, the Triangle Inequality provides the basis for the rule of thumb, "The shortest distance between two points is a straight line." Using mathematical terminology, we say that given any triangle formed by the points A, B, and C, the length of any one side of the triangle must be shorter than the sum of the other two sides. Using mathematical notation for the triangle in the illustration, we write $|x + y| < |x| + |y|$.

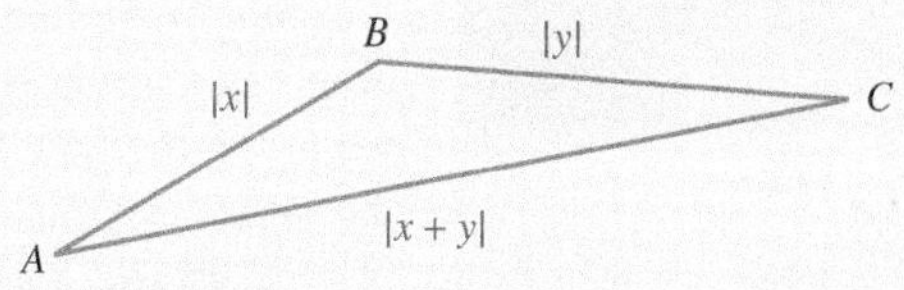

Use the Triangle Inequality to determine whether the given numbers represent the lengths of the sides of a triangle.

1. 8, 14, 20
2. 10, 11, 22

Unless otherwise noted, all content on this page is © Cengage Learning.

EXAMPLE 5 Solve, then graph the solution set: $\left|\frac{2}{3}x - 2\right| - 3 > 6$

Solution We begin by adding 3 to both sides to isolate the absolute value on the left side. We then proceed as follows:

$$\left|\frac{2}{3}x - 2\right| - 3 > 6$$

$$\left|\frac{2}{3}x - 2\right| > 9 \qquad \text{Add 3 to both sides.}$$

$$\frac{2}{3}x - 2 < -9 \quad \text{or} \quad \frac{2}{3}x - 2 > 9$$

$$\frac{2}{3}x < -7 \quad \Big| \quad \frac{2}{3}x > 11 \qquad \text{Add 2 to both sides.}$$

$$2x < -21 \quad \Big| \quad 2x > 33 \qquad \text{Multiply both sides by 3.}$$

$$x < -\frac{21}{2} \quad \Big| \quad x > \frac{33}{2} \qquad \text{Divide both sides by 2.}$$

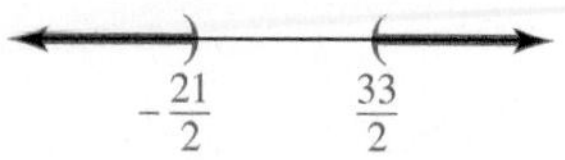

Figure 7-33

The solution set is $\left(-\infty, -\frac{21}{2}\right) \cup \left(\frac{33}{2}, \infty\right)$, whose graph appears in Figure 7-33.

SELF CHECK 5 Solve, then graph the solution set: $\left|\frac{3}{2}x + 1\right| - 2 > 1$

EXAMPLE 6 Solve, then graph the solution set: $|3x - 5| \geq -2$

Solution Since the absolute value of any number is nonnegative, and since any nonnegative number is larger than -2, the inequality is true for all x. The solution set is $(-\infty, \infty)$, whose graph appears in Figure 7-34.

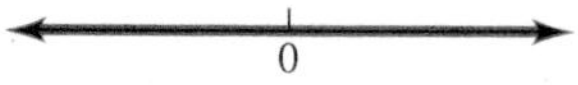

Figure 7-34

SELF CHECK 6 Solve, then graph the solution set: $|2x + 3| > -5$

Accent on technology

▸ Solving Absolute Value Inequalities

We can solve many absolute value inequalities by a graphing method. For example, to solve $|2x - 3| < 9$, we graph the equations $y_1 = |2x - 3|$ and $y_2 = 9$ on the same coordinate system. If we use window settings of $[-5, 15]$ for x and $[-5, 15]$ for y, we see the graph shown in Figure 7-35.

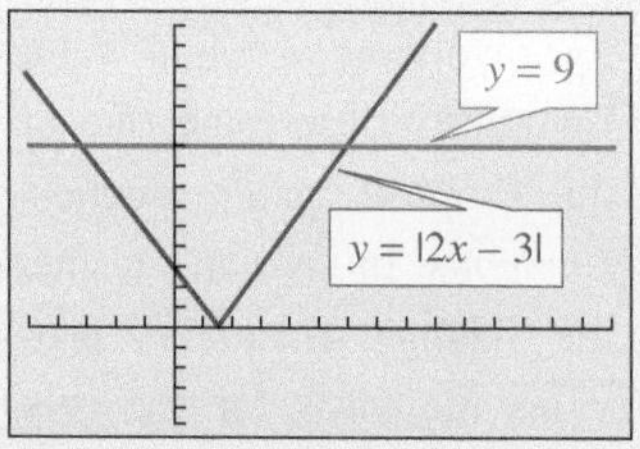

Figure 7-35

The inequality $|2x - 3| < 9$ will be true for all x-coordinates of points that lie on the graph of $y = |2x - 3|$ and below the graph of $y = 9$. By using the TRACE feature, we can see that these values of x are in the interval $(-3, 6)$.

The inequality $|2x - 3| > 9$ will be true for all x-coordinates of points that lie on the graph of $y = |2x - 3|$ and above the graph of $y = 9$. By using the TRACE feature, we can see that these values of x are in the union of two intervals $(-\infty, -3) \cup (6, \infty)$.

For instructions regarding the use of a Casio graphing calculator, please refer to the Casio Keystroke Guide in the back of the book.

Unless otherwise noted, all content on this page is © Cengage Learning.

SELF CHECK ANSWERS

1. $\left(-2, \frac{4}{3}\right)$ (−2, 4/3)

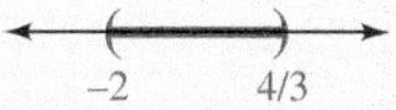

2. $[-1, 4]$ [−1, 4]

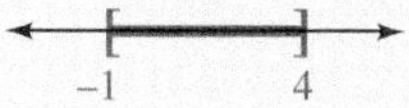

3. $\left(-\infty, -\frac{2}{3}\right) \cup (2, \infty)$ (−∞, −2/3) ∪ (2, ∞)

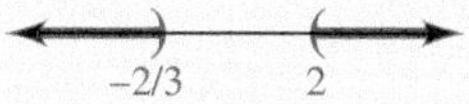

4. $(-\infty, -2] \cup [10, \infty)$ (−∞, −2] ∪ [10, ∞)

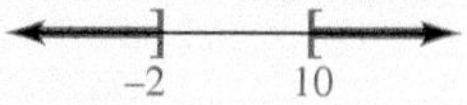

5. $\left(-\infty, -\frac{8}{3}\right) \cup \left(\frac{4}{3}, \infty\right)$ (−∞, −8/3) ∪ (4/3, ∞)

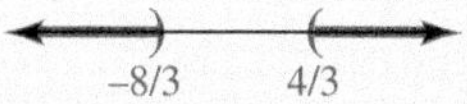

6. $(-\infty, \infty)$ (−∞, ∞)

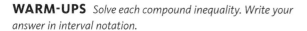

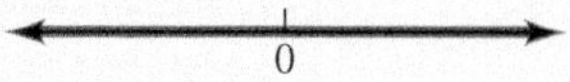

NOW TRY THIS

1. Express each of the following as an absolute value inequality.
   a. $-5 \leq x \leq 5$
   b. $3x - 5 < -7$ or $3x - 5 > 7$
2. Solve the inequality $|x + 5| \leq 2x - 9$. Write the solution in interval notation and graph it.

7.7 Exercises

WARM-UPS *Solve each compound inequality. Write your answer in interval notation.*

1. $x > -8$ and $x < 8$
2. $x < -8$ or $x > 8$
3. $x < -1$ or $x > 1$
4. $x \leq 1$ and $x \geq -1$
5. $x + 1 \leq 2$ and $x + 1 \geq -2$
6. $x + 1 \geq 2$ or $x + 1 \leq -2$
7. $-5 < 2x + 3 < 5$
8. $2x + 3 \leq -5$ or $2x + 3 \geq 5$

REVIEW *Solve each formula for the given variable.*

9. $A = p + prt$ for t
10. $A = p + prt$ for r
11. $S = 2lw + 4wh$ for l
12. $V = \frac{1}{3}Bh$ for B

VOCABULARY AND CONCEPTS *Fill in the blanks.*

13. If $k > 0$, then $|x| < k$ is equivalent to __________.
14. If $k > 0$, then ______ is equivalent to $-k \leq x \leq k$.
15. If k is a nonnegative constant, then $|x| > k$ is equivalent to ____________.
16. If k is a nonnegative constant, then ______ is equivalent to $x \leq -k$ or $x \geq k$.

GUIDED PRACTICE *Solve each inequality. Write the solution set in interval notation and graph it.* SEE EXAMPLE 1. (OBJECTIVE 1)

17. $|x| < 8$
18. $|x| < 7$
19. $|2x| < 8$
20. $|3x| < 27$
21. $|x + 1| < 2$
22. $|3 - 2x| < 7$
23. $|3x - 2| < 10$
24. $|2x - 3| < 6$

Solve each inequality. Write the solution set in interval notation and graph it. SEE EXAMPLE 2. (OBJECTIVE 1)

25. $|x + 9| \leq 12$
26. $|x - 8| \leq 12$
27. $|4x - 1| \leq 7$
28. $|4 - 3x| \leq 13$

Unless otherwise noted, all content on this page is © Cengage Learning.

Solve each inequality. Write the solution set in interval notation and graph it. SEE EXAMPLE 3. (OBJECTIVE 1)

29. $|5x| > 5$

30. $|7x| > 7$

31. $|x - 12| > 24$

32. $|3x + 2| > 14$

Solve each inequality. Write the solution set in interval notation and graph it. SEE EXAMPLE 4. (OBJECTIVE 1)

33. $|x + 5| \geq 7$

34. $|2x - 5| \geq 25$

35. $|2 - 3x| \geq 8$

36. $|-1 - 2x| \geq 5$

Solve each inequality. Write the solution set in interval notation and graph it. SEE EXAMPLE 5. (OBJECTIVE 1)

37. $\left|\frac{1}{3}x + 7\right| + 5 > 6$

38. $\left|\frac{1}{2}x - 3\right| - 4 < 2$

39. $|3x + 1| + 2 < 6$

40. $-|5x - 1| + 2 < 0$

Solve each inequality. Write the solution set in interval notation, if possible, and graph it. SEE EXAMPLE 6. (OBJECTIVE 1)

41. $-2|3x - 4| < 16$

42. $|7x + 2| > -8$

43. $|5x - 1| + 4 \leq 0$

44. $|3x + 2| \leq -3$

ADDITIONAL PRACTICE *Solve each inequality. Write the solution set in interval notation, if possible, and graph it.*

45. $|2x + 1| + 2 \leq 2$

46. $\left|\frac{x - 5}{10}\right| \leq 0$

47. $|4x + 3| > 0$

48. $\left|3\left(\frac{x + 4}{4}\right)\right| > 0$

49. $3|2x + 5| \geq 9$

50. $\left|\frac{7}{3}x - \frac{3}{5}\right| \geq 1$

51. $\left|\frac{3}{5}x + \frac{7}{3}\right| < 2$

52. $\left|\frac{x - 2}{3}\right| \leq 4$

53. $-|3x + 1| < -8$

54. $-|2x - 3| < -7$

55. $\left|\frac{x - 2}{3}\right| > 4$

56. $|5x - 12| < -5$

57. $\left|\frac{1}{6}x + 6\right| + 2 < 2$

58. $3\left|\frac{1}{3}(x - 2)\right| + 2 \leq 3$

59. $\left|\frac{1}{7}x + 1\right| \leq 0$

60. $\left|\frac{3}{5}x - 2\right| + 3 \leq 3$

61. $\left|\frac{1}{5}x - 5\right| + 4 > 4$

62. $|8x - 3| > 0$

63. $|3x - 2| + 2 \geq 0$

64. $|4x + 3| > -5$

Use a calculator to solve each inequality. Give each answer in interval notation. Round to the nearest tenth.

65. $|0.5x + 0.7| < 2.6$

66. $|1.25x - 0.75| < 3.15$

67. $|2.15x - 3.05| > 3.8$

68. $|-3.57x + 0.12| > 2.75$

WRITING ABOUT MATH

69. Explain how parentheses and brackets are used when graphing inequalities.

70. If $k > 0$, explain the difference between the solution sets of $|x| < k$ and $|x| > k$.

SOMETHING TO THINK ABOUT

71. Under what conditions is $|x| + |y| > |x + y|$?

72. Under what conditions is $|x| + |y| = |x + y|$?

Projects

PROJECT 1

The expression $1 + x + x^2 + x^3$ is a polynomial of degree 3. The polynomial $1 + x + x^2 + x^3 + x^4$ has the same pattern, but one more term. Its degree is 4. As the pattern continues and more terms are added, the degree of the polynomial increases. If there were no end to the number of terms, the "polynomial" would have infinitely many terms, and no defined degree:

$$1 + x + x^2 + x^3 + x^4 + x^5 + x^6 + \cdots$$

Such "unending polynomials," called **power series**, are studied in calculus. However, this particular series is the result of a division of polynomials:

- Consider the division $\frac{1}{1 - x}$. Find the quotient by filling in more steps of this long division:

Step 1:

$$\begin{array}{r} 1 \\ 1 - x \overline{)\,1 + 0x + 0x^2 +} \\ \underline{1 - x} \\ x \end{array}$$

Step 2:

$$\begin{array}{r} 1 \\ 1 - x \overline{)\,1 + 0x + 0x^2 +} \\ \underline{1 - x} \\ x + 0x^2 \\ \underline{x - x^2} \\ x^2 \end{array}$$

To determine how the fraction $\frac{1}{1 - x}$ and the series $1 + x + x^2 + x^3 + x^4 + x^5 + x^6 + \cdots$ could be equal, try this experiment.

- Let $x = \frac{1}{2}$ and evaluate $\frac{1}{1 - x}$.
- Again, let $x = \frac{1}{2}$ and evaluate the series. Because you cannot add infinitely many numbers, just add the first 3 or 4 or 5 terms and see if you find a pattern. Use a calculator to complete this table:

Polynomial	Value at $x = \frac{1}{2}$
$1 + x + x^2$	
$1 + x + x^2 + x^3$	
$1 + x + x^2 + x^3 + x^4$	
$1 + x + x^2 + x^3 + x^4 + x^5$	
$1 + x + x^2 + x^3 + x^4 + x^5 + x^6$	

What number do the values in the second column seem to be approaching? That number is called the *sum* of the series.

- Explain why the nonterminating decimal 1.1111111 . . . represents the infinite series

$$1 + \left(\frac{1}{10}\right) + \left(\frac{1}{10}\right)^2 + \left(\frac{1}{10}\right)^3 + \left(\frac{1}{10}\right)^4 + \left(\frac{1}{10}\right)^5 + \left(\frac{1}{10}\right)^6 + \cdots$$

- Using the fraction $\frac{1}{1 - x}$, explain why $1.11111\ldots = \frac{10}{9}$.
- Verify that $\frac{10}{9} = 1.11111\ldots$ by dividing 10 by 9.

PROJECT 2

In order for a plane to fly, the proper amount of lift must be provided by the wings. Two factors that determine lift are controlled by the pilot. One is the speed (or velocity) of the plane, and the other is the *angle of attack*, which is the angle between the direction the plane is aimed and the direction it is actually moving, as shown in the illustration.

For one particular plane weighing 2,050 pounds, the lift, velocity, and angle of attack are related by the equation

$$L = (0.017a + 0.023)V^2$$

(Continued)

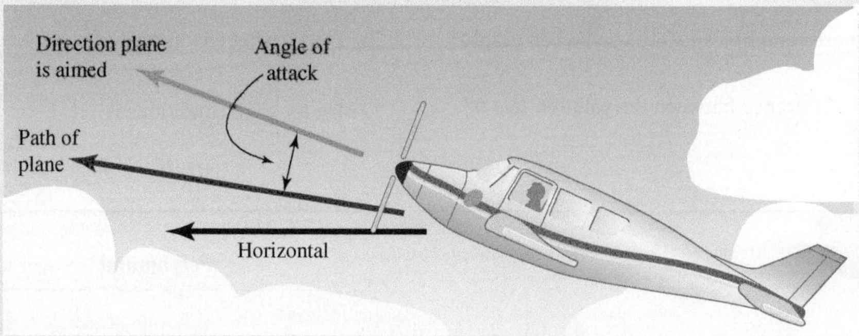

where L is the lift in pounds, a is the angle of attack in degrees, and V is the velocity in feet per second. To support the plane, the lift must equal the plane's weight.

a. Find the correct angle of attack when the velocity of the plane is 88.64 mph. (*Hint*: You must change the velocity to units of feet per second.)

b. As the angle of attack approaches 17°, the plane begins to stall. With more cargo on the return trip, the same plane weighs 2,325 pounds. If the pilot allows the velocity to drop to 80 feet per second (about 55 mph), will the plane stall?

Unless otherwise noted, all content on this page is © Cengage Learning.

Reach for Success

EXTENSION OF ACCESSING ACADEMIC RESOURCES

Now that you know where to go to get mathematics help, let's move on and look at some other resources.

An advisor can discuss your workload and course load to help you find a good balance. An advisor can also explain the consequences of failing a course or falling below a 2.0 GPA.	Do you have any issues for which an advisor would be good resource? If so, list them. ______ ______ ______
Some colleges assign students to a specific academic advisor while others offer "drop in" service to students.	Were you assigned an academic advisor?______ If so, what is your advisor's name?______ What is his/her phone number?______ Where is his/her office? ______ Have you met with your advisor this semester?______ If not, where is the advising office located?______ What is the telephone number?______ Have you met with any advisor this semester?______
The following may not apply to all students.	
Financial aid officers can provide information on obtaining money for college expenses through scholarships, grants, and loans.	Could you benefit from speaking with a financial aid officer? ______ Write the office phone number here. ______ Where is the office located? ______
The Office of Veteran Affairs is a valuable resource for all students who have served in the armed forces at any time.	Could you benefit from speaking with a Veterans' Affairs officer? ______ Write the office phone number here. ______ Where is the office located? ______
Most colleges offer personal counseling on a short-term basis.	Could you benefit from speaking with a personal counselor? ______ Write the phone number of the counseling office. ______ Where is it located?______

Taking advantage of one or more of these resources may help guide you on the path to success.

7 Review

SECTION 7.1 Review of Solving Linear Equations and Inequalities in One Variable

DEFINITIONS AND CONCEPTS	EXAMPLES
Addition and multiplication properties of equality If a and b are real numbers and $a = b$, then $a + c = b + c \quad a - c = b - c$ $ac = bc \quad (c \neq 0) \quad \frac{a}{c} = \frac{b}{c} \quad (c \neq 0)$	If $a = b$, then $a + 3 = b + 3 \quad a - 5 = b - 5$ $4a = 4b \quad \frac{a}{7} = \frac{b}{7}$
Solving linear equations: **To solve a linear equation in one variable, follow these steps:** **1.** If the equation contains fractions, multiply both sides of the equation by a number that will clear the fractions. **2.** Use the distributive property to remove all sets of parentheses and combine like terms. **3.** Use the addition and subtraction properties to move all variables to one side of the equation (isolate the variable term) and all numbers on the other side. Combine like terms, if necessary. **4.** Use the multiplication and division properties to make the coefficient of the variable equal to 1 (isolate the variable). **5.** Check the result by replacing the variable in the original equation with the possible solution and verifying that the number satisfies the equation.	To solve $\frac{x-2}{5} - x = \frac{8}{5} - x + 2$, we first clear the fractions by multiplying both sides by 5 and proceeding as follows: $5\left(\frac{x-2}{5} - x\right) = 5\left(\frac{8}{5} - x + 2\right)$ $x - 2 - 5x = 8 - 5x + 10$ Use the distributive property to remove parentheses. $-4x - 2 = 18 - 5x$ Combine like terms. $x = 20$ Add $5x$ and 2 to both sides. Verify that 20 satisfies the original equation.
An **identity** is an equation that is satisfied by every number x for which both sides of the equation are defined. The solution of an identity is the set of *all real numbers* and is denoted by $\mathbb{R}$.	Solve. $-5(2x + 3) + 8x - 9 = 6x - 8(x + 3)$ $-10x - 15 + 8x - 9 = 6x - 8x - 24$ Use the distributive property to remove parentheses. $-2x - 24 = -2x - 24$ Combine like terms. Since the left side of the equation is the same as the right side, every number x will satisfy the equation. The equation is an identity and its solution set is all real numbers, $\mathbb{R}$.
A **contradiction** is an equation that has no solution. Its solution set is the **empty set**, denoted by $\varnothing$.	Solve. $5(2x + 3) - 8x - 9 = 8x - 6(x + 3)$ $10x + 15 - 8x - 9 = 8x - 6x - 18$ Use the distributive property to remove parentheses. $2x + 6 = 2x - 18$ Combine like terms. $6 = -18$ Subtract $2x$ from both sides. Because $6 \neq -18$, there is no number that will satisfy the equation. Therefore, it is a contradiction and the solution set is the empty set, $\varnothing$.

To solve a formula for an indicated variable means to isolate that variable on one side of the equation.	To solve $ax + by = c$ for y, we proceed as follows: $ax + by = c$ $by = c - ax$ Subtract ax from both sides. $\frac{by}{b} = \frac{c - ax}{b}$ Divide both sides by b. $y = \frac{c - ax}{b}$ $\frac{b}{b} = 1$
Solving linear inequalities: **Trichotomy property:** $a < b, a = b$, or $a > b$	For the values x and 3, either $x < 3, x = 3$, or $x > 3$.
Transitive properties: If $a < b$ and $b < c$, then $a < c$. If $a > b$ and $b > c$, then $a > c$.	If $-2 < 5$ and $5 < 10$, then $-2 < 10$. If $20 > 7$ and $7 > -5$, then $20 > -5$.
Properties of inequality: If a and b are real numbers and $a < b$, then $a + c < b + c$ $a - c < b - c$ $ac < bc \quad (c > 0)$ $ac > bc \quad (c < 0)$ $\frac{a}{c} < \frac{b}{c} \quad (c > 0)$ $\frac{a}{c} > \frac{b}{c} \quad (c < 0)$	To solve the linear inequality $3(2x + 6) < 18$, use the same steps as for solving equations. $3(2x + 6) < 18$ $6x + 18 < 18$ Use the distributive property to remove parentheses. $6x < 0$ Subtract 18 from both sides. $x < 0$ Divide both sides by 6. The solution set is $\{x \mid x < 0\}$ or, in interval notation, $(-\infty, 0)$. The graph is shown. The parenthesis at 0 indicates that 0 is not included in the solution set.  To solve $-4(3x - 4) \leq -8$, proceed as follows: $-4(3x - 4) \leq -8$ $-12x + 16 \leq -8$ Use the distributive property to remove parentheses. $-12x \leq -24$ Subtract 16 from both sides. $x \geq 2$ Divide both sides by -12 and reverse the inequality symbol. The solution set is $\{x \mid x \geq 2\}$ or, in interval notation, $[2, \infty)$. The graph is shown. The bracket at 2 indicates that 2 is included in the solution set. 
Compound inequalities: If we cannot isolate the variable between the inequality symbols, we will need to write the inequality as two separate statements: $c < x < d$ is equivalent to $c < x$ and $x < d$.	To solve the inequality $-7 \leq 2x - 5 < 3$, isolate x between the inequality symbols. $-7 \leq 2x - 5 < 3$ $-2 \leq 2x < 8$ Add 5 to all three parts. $-1 \leq x < 4$ Divide all three parts by 2. The solution set is $\{x \mid -1 \leq x < 4\}$ or, in interval notation, $[-1, 4)$. The graph is shown. 

Unless otherwise noted, all content on this page is © Cengage Learning.

REVIEW EXERCISES

Solve and check each equation.

1. $3(y - 1) = 24$
2. $2(x + 7) = 42$
3. $13(x - 9) - 2 = 7x - 5$
4. $\frac{8(x - 5)}{3} = 2(x - 4)$
5. $2(x - 3) - x = 5x - 4(x + 2)$
6. $2(x + 7) - 8 = 4x - 2(x - 3)$

Solve for the indicated variable.

7. $V = \frac{1}{3}\pi r^2 h$ for h
8. $V = \frac{1}{6}ab(x + y)$ for x

9. **Carpentry** A carpenter wants to cut a 15-foot rafter so that one piece is 4 times as long as the other. Where should he cut the board?

10. **Geometry** A rectangle is 5 feet shorter than twice the width. If the perimeter of the rectangle is 110 feet, find its area.

Solve each inequality. Give each solution set in interval notation and graph it.

11. $\frac{1}{3}y - 2 \geq \frac{1}{2}y + 2$
12. $\frac{7}{4}(x + 3) < \frac{3}{8}(x - 3)$
13. $3 < 3x + 4 < 10$
14. $4x > 3x + 2 > x - 3$

SECTION 7.2 Review of Graphing Linear Equations, Finding the Slopes of Lines, and Writing Equations of Lines

DEFINITIONS AND CONCEPTS	EXAMPLES
Graphing a linear equation: $Ax + By = C$ (A and B are not both 0.) **To graph a linear equation by plotting points:** Find two ordered pairs (x, y) that satisfy the equation. Plot each pair on the rectangular coordinate system. Draw a line passing through the two points. Use a third point as a check.	The equation $x + y = -2$ is written in general form. To graph it, find three ordered pairs that satisfy the equation. x | y | (x, y) 0 | −2 | $(0, -2)$ 2 | −4 | $(2, -4)$ 3 | −5 | $(3, -5)$ 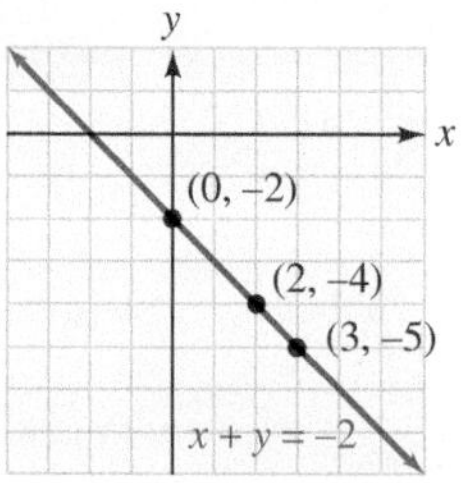 Then plot the points and draw a line passing through them.
The **y-intercept** of a line is the point where the line intersects the y-axis.	The y-intercept of the graph above is the point with coordinates $(0, -2)$.
The **x-intercept** of a line is the point where the line intersects the x-axis.	The x-intercept of the graph above is the point with coordinates $(-2, 0)$.
To graph a linear equation by the intercept method: 1. Find the x-intercept by setting $y = 0$. 2. Find the y-intercept by setting $x = 0$.	To graph $2x - 4y = 12$, find the intercepts. 1. Let $y = 0$. $2x - 4(0) = 12$; $2x = 12$; $x = 6$ The x-intercept is $(6, 0)$. 2. Let $x = 0$. $2(0) - 4y = 12$; $-4y = 12$; $y = -3$ The y-intercept is $(0, -3)$.

Unless otherwise noted, all content on this page is © Cengage Learning.

3. Plot the intercepts and draw a line passing through them.

3. Plot the intercepts and draw the line shown in the figure.

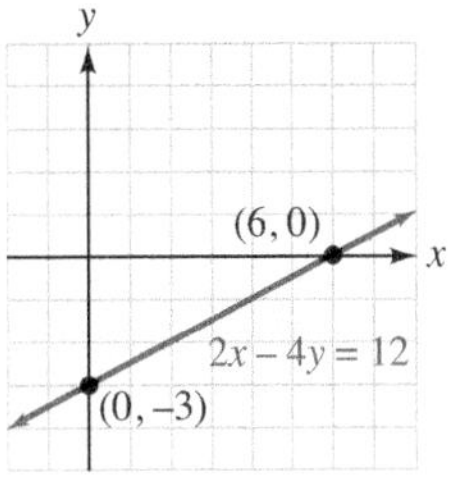

4. Find one additional point as a check.

4. As a check, let $x = 2$.

$$2(2) - 4y = 12$$
$$4 - 4y = 12$$
$$-4y = 8$$
$$y = -2$$

A third point is $(2, -2)$.
The point $(2, -2)$ lies on the line.

Graph of a horizontal line:

$y = b$ y-intercept at $(0, b)$

Graph of a vertical line:

$x = a$ x-intercept at $(a, 0)$

The graph of $y = -3$ is a horizontal line passing through $(0, -3)$.

The graph of $x = 2$ is a vertical line passing through $(2, 0)$.

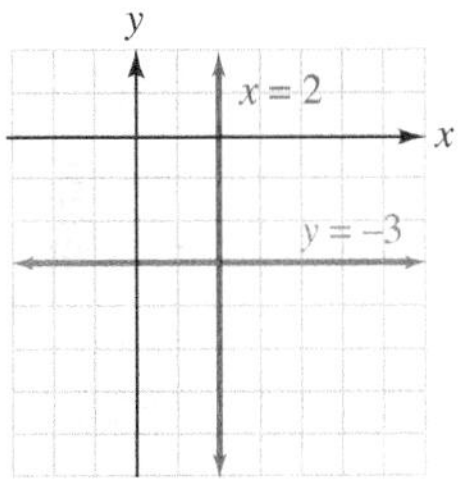

Midpoint formula:
If $P(x_1, y_1)$ and $Q(x_2, y_2)$, the midpoint of segment PQ is

$$\left(\frac{x_1 + x_2}{2}, \frac{y_1 + y_2}{2}\right)$$

To find the midpoint of the line segment joining $P(3, -2)$ and $Q(2, -5)$, find the mean of the x-coordinates and the mean of the y-coordinates:

$$\frac{x_1 + x_2}{2} = \frac{3 + 2}{2} = \frac{5}{2}$$

$$\frac{y_1 + y_2}{2} = \frac{-2 + (-5)}{2} = -\frac{7}{2}$$

The midpoint of segment PQ is the point $\left(\frac{5}{2}, -\frac{7}{2}\right)$.

Slope of a nonvertical line:
If $x_2 \neq x_1$,

$$m = \frac{\Delta y}{\Delta x} = \frac{y_2 - y_1}{x_2 - x_1}$$

The slope of a line passing through $(-1, 4)$ and $(5, -3)$ is given by

$$m = \frac{y_2 - y_1}{x_2 - x_1} = \frac{-3 - 4}{5 - (-1)} = \frac{-7}{6}$$

Horizontal lines have a slope of 0.

$y = 5$ is a horizontal line with slope 0.

Vertical lines have an undefined slope.

$x = 5$ is a vertical line with an undefined slope.

Parallel lines have the same slope.

Two lines with slopes $\frac{3}{2}$ and $\frac{3}{2}$ are parallel if they have different y-intercepts.

The slopes of two nonvertical perpendicular lines are negative reciprocals.

Two lines with slopes 5 and $-\frac{1}{5}$ are perpendicular.

Equations of a line:
Point-slope form:

$$y - y_1 = m(x - x_1)$$

Write the point-slope form of a line passing through $(-3, 5)$ with slope $\frac{1}{2}$.

$y - y_1 = m(x - x_1)$ This is point slope form.

$y - 5 = \frac{1}{2}[x - (-3)]$ Substitute.

$y - 5 = \frac{1}{2}(x + 3)$

Unless otherwise noted, all content on this page is © Cengage Learning.

Slope-intercept form: $y = mx + b$	Write the slope-intercept form of a line with slope $\frac{1}{2}$ and y-intercept $(0, -4)$. $y = mx + b$ This is slope-intercept form. $y = \frac{1}{2}x - 4$ Substitute. $y = \frac{1}{2}x - 4$

REVIEW EXERCISES

Graph each equation.

15. $x + y = 4$

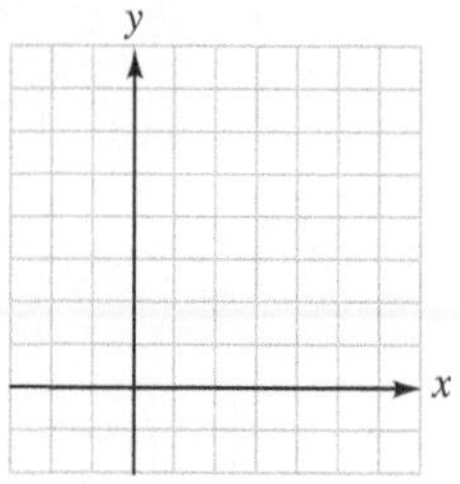

16. $2x - y = 8$

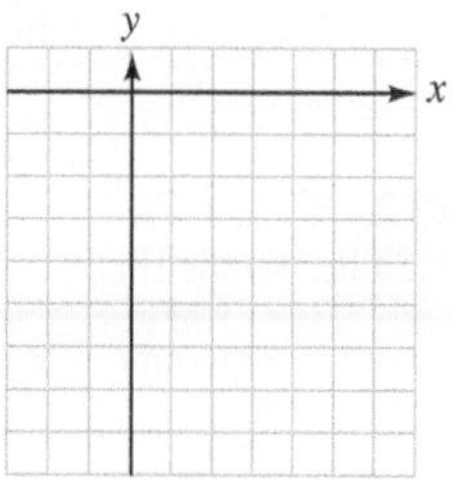

17. $y = 3x + 4$

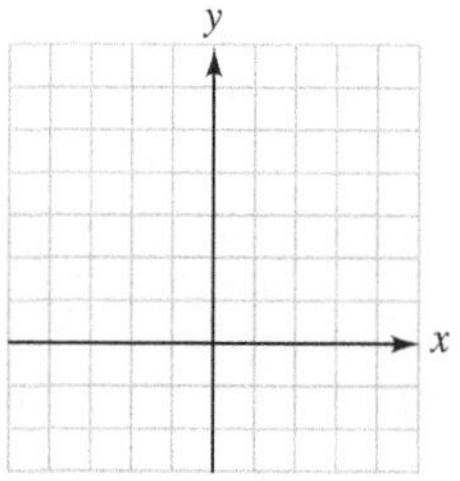

18. $x = 4 - 2y$

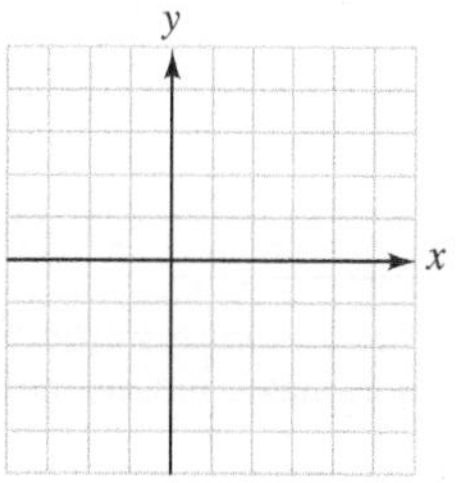

19. $y = 4$

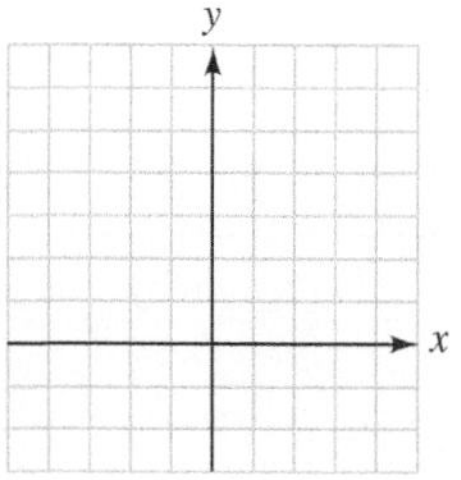

20. $x = -2$

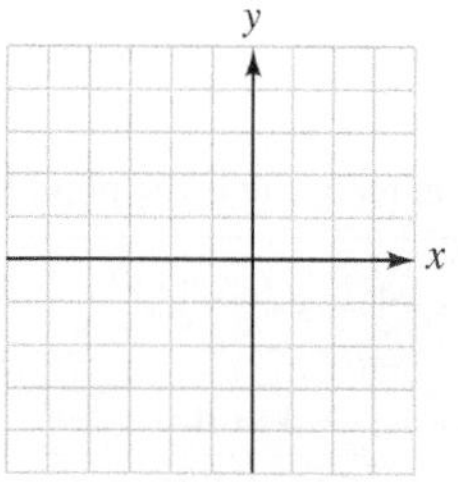

21. $2(x + 3) = x + 2$

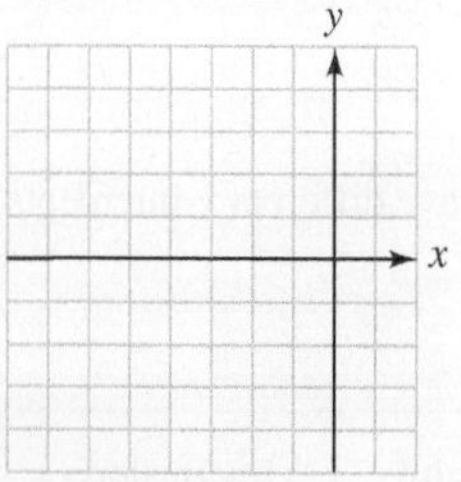

22. $3y = 2(y - 1)$

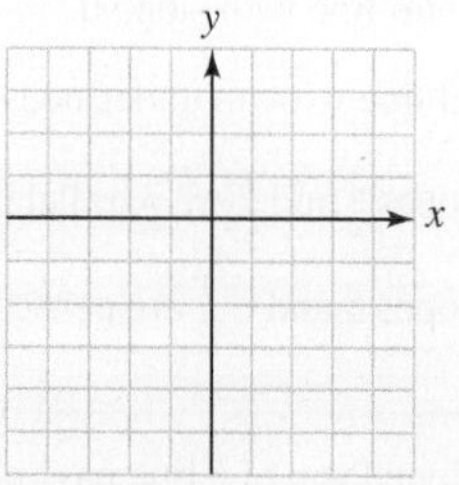

23. Find the midpoint of the line segment joining $(-3, 5)$ and $(6, 11)$.

Find the slope of the line passing through points P and Q, if possible.

24. $P(3, 5)$ and $Q(5, 7)$ **25.** $P(-3, -2)$ and $Q(6, 12)$

26. $P(-3, 4)$ and $Q(-5, -6)$ **27.** $P(5, 4)$ and $Q(-6, -9)$

28. $P(-3, 4)$ and $Q(1, 4)$ **29.** $P(-5, -4)$ and $Q(-5, 8)$

Find the slope of the graph of each equation, if one exists.

30. $3x - 2y = 18$ **31.** $x + 2y = 8$

32. $2(x + 3) = 10$ **33.** $3y + 2 = 9$

Determine whether the lines with the given slopes are parallel, perpendicular, or neither.

34. $m_1 = 4, m_2 = -0.25$ **35.** $m_1 = 0.5, m_2 = \frac{1}{2}$

36. $m_1 = 0.5, m_2 = -\frac{1}{2}$ **37.** $m_1 = 5, m_2 = \frac{-1}{5}$

38. Sales growth If the sales of a new business were $66,000 in its first year and $120,000 in its fourth year, find the rate of growth in sales per year.

Write the equation of the line with the given properties. Write the equation in slope-intercept form.

39. slope of 3; passing through $P(8, 5)$

40. passing through $(-2, 5)$ and $(6, -7)$

41. passing through $(3, 5)$; parallel to the graph of $3x - 2y = 7$

42. passing through $(-3, -5)$; perpendicular to the graph of $y = -\frac{1}{4}x + 3$

43. Graph the line $y = \frac{2}{3}x + 1$ using the slope-intercept method.

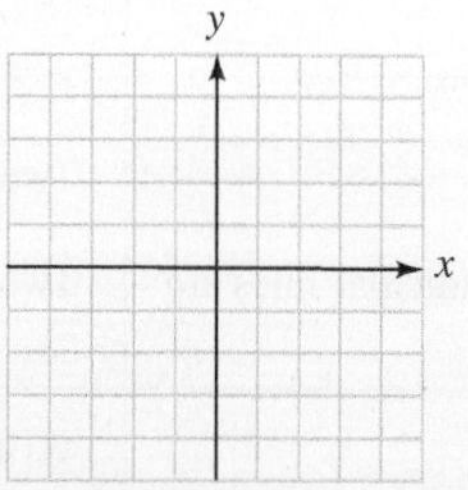

44. Are the lines represented by $2x + 3y = 8$ and $3x - 2y = 10$ parallel or perpendicular?

45. Depreciation A business purchased a copy machine for $8,700 and will depreciate it on a straight-line basis over the next 5 years. At the end of its useful life, it will be sold as scrap for $100. Find its depreciation equation.

Unless otherwise noted, all content on this page is © Cengage Learning.

SECTION 7.3 Review of Functions

DEFINITIONS AND CONCEPTS	EXAMPLES
A **relation** is any set of ordered pairs. A **function** is any set of ordered pairs (a relation) in which each first component (input value) determines exactly one second component (output value). The **domain** is the set of input values. The **range** is the set of output values.	Find the domain and range of the relation $\{(2, -1), (6, 3), (2, 5)\}$ and determine whether it defines a function. **Domain:** $\{2, 6\}$ because 2 and 6 are the first components in the ordered pairs. **Range:** $\{-1, 3, 5\}$ because -1, 3, and 5 are the second components in the ordered pairs. The relation does not define a function because the first component 2 determines two different second components.
The **vertical line test** can be used to determine whether a graph represents a function.	Find the domain and range of the function represented by the graph and determine whether it is a function. 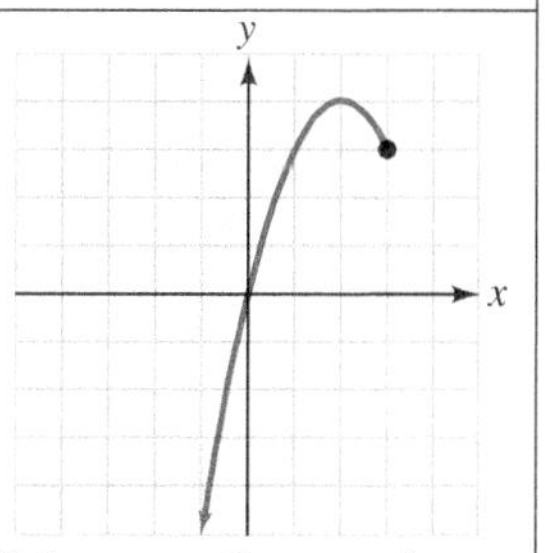 Since the graph extends forever to the left, and stops at $x = 3$ on the right, the domain is $(-\infty, 3]$. Since the graph extends forever downward and ends at $y = 4$, the domain is $(-\infty, 4]$. Since every vertical line that intersects the graph will do so exactly once, the vertical line test indicates that the graph is a function.
$f(k)$ represents the value of $f(x)$ when $x = k$.	If $f(x) = -7x + 4$, find $f(2)$. $f(x) = -7x + 4$ $f(2) = -7(2) + 4$ Substitute 2 for x. $= -14 + 4$ $= -10$ In ordered pair form, we can write $(2, -10)$.
The domain of a function of x that is defined by an equation is the set of all numbers that are permissible replacements for x.	Find the domain of $f(x) = \frac{-5}{x + 7}$. The number -7 cannot be substituted for x in the function, because that would make the denominator 0. However, any real number, except -7, can be substituted for x to obtain a single value y. Therefore, the domain is the set of all real numbers except -7. This is the union of two intervals $(-\infty, -7) \cup (-7, \infty)$.
Find the domain and range of a function by graphing.	Graph $f(x) = \|x + 1\| - 2$ by plotting points. The domain is $(-\infty, \infty)$ and the range is $[-2, \infty)$. 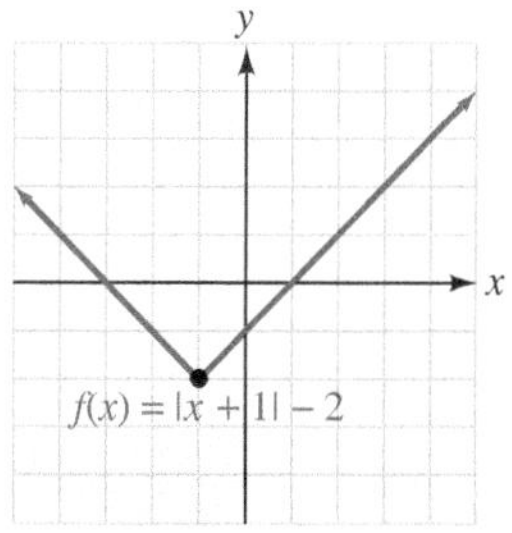
Write a linear function modeling real-world data by following the same procedure as writing an equation of a line given two points.	A watch that sold for $2,500 in 2001 was worth $3,000 in 2011. Write a linear function modeling this data. We let 2001 be our base year, giving us the 2 points $(0, 2{,}500)$ and $(10, 3{,}000)$. Find the slope. $\frac{3{,}000 - 2{,}500}{10 - 0} = \frac{500}{10} = 50$ The slope is 50 and the y-intercept is $(0, 2{,}500)$. The linear function is $f(x) = 50x + 2{,}500$.

Unless otherwise noted, all content on this page is © Cengage Learning.

REVIEW EXERCISES

Determine whether each equation determines y to be a function of x.

46. $y = 6x - 4$
47. $y = 4 - x$
48. $y^2 = x$
49. $|y| = x^2$

Assume that $f(x) = 3x + 4$ and $g(x) = x^2 - 3$ and find each value.

50. $f(-3)$
51. $g(8)$
52. $g(-2)$
53. $f(5)$

Find the domain of the function.

54. $f(x) = \dfrac{4}{2 - x}$

55. $f(x) = \dfrac{7}{x - 3}$

56. $f(x) = 7$

Find the domain and range of each function by graphing.

57. $f(x) = 6x + 2$
58. $f(x) = 3x - 10$
59. $f(x) = x^2 - 1$

Use the vertical line test to determine whether each graph represents a function.

60.

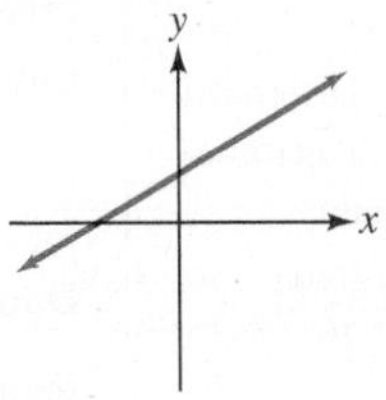

61.

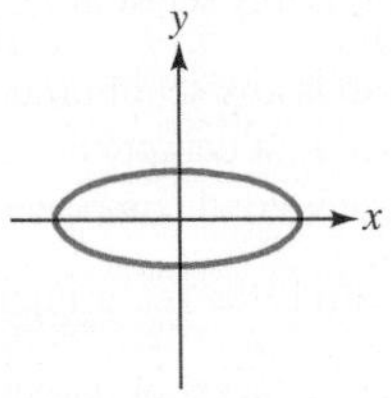

62.

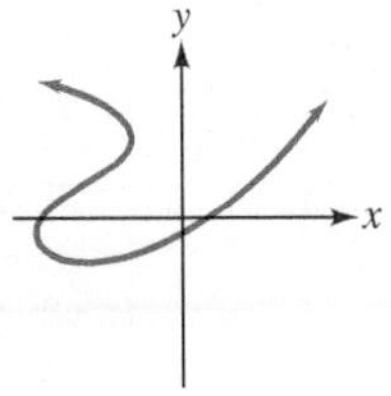

63.

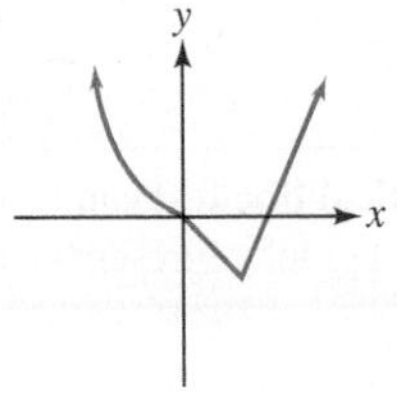

SECTION 7.4 Review of Factoring and Solving Quadratic Equations

DEFINITIONS AND CONCEPTS	EXAMPLES
GCF and factoring by grouping: Always factor out all common factors as the first step in a factoring problem.	To factor $36x^3 - 6x^2 + 12x$, use the distributive property to factor out the greatest common factor of $6x$. $6x(6x^2 - x + 2)$ Factor the common term.
If an expression has four or more terms, try to factor the expression by grouping.	Factor. $25x^3 - 10x^2 + 20x - 8$ $= (25x^3 - 10x^2) + (20x - 8)$ Group the first two terms and the last two terms. $= 5x^2(5x - 2) + 4(5x - 2)$ Factor each grouping. $= (5x - 2)(5x^2 + 4)$ Factor out $5x - 2$.
Difference of two squares: $x^2 - y^2 = (x + y)(x - y)$	Factor. $64x^2 - 25$ $= (8x)^2 - (5)^2$ Write each term as a perfect square. $= (8x + 5)(8x - 5)$ Factor.
Sum of two cubes: $x^3 + y^3 = (x + y)(x^2 - xy + y^2)$	Factor. $64x^3 + 125$ $= (4x)^3 + (5)^3$ Write each term as a cube. $= (4x + 5)[(4x)^2 - (4x)(5) + (5)^2]$ Factor. $= (4x + 5)(16x^2 - 20x + 25)$ Simplify.

Unless otherwise noted, all content on this page is © Cengage Learning.

Difference of two cubes: $x^3 - y^3 = (x - y)(x^2 + xy + y^2)$	Factor. $27a^3 - 1$ $= (3a)^3 - (1)^3$ Write each term as a cube. $= (3a - 1)[(3a)^2 + (3a)(1) + (1)^2]$ Factor. $= (3a - 1)(9a^2 + 3a + 1)$ Simplify.
Factoring a trinomial of the form $x^2 + bx + c$ by trial and error: **1.** Write the trinomial in descending powers of one variable. **2.** Factor out any GCF, including −1. **3.** List the factorizations of the third term of the trinomial. **4.** Select the factorization in which the sum of the factors is the coefficient of the middle term.	To factor $x^2 - 8x + 12$, first note that the trinomial is written in descending powers of x. Then note that the factors of +12 are 1 and 12 −1 and −12 2 and 6 −2 and −6 3 and 4 −3 and −4 Since −2 and −6 are the only pair whose sum is −8, the factorization is $(x - 2)(x - 6)$.
Test for factorability: A trinomial of the form $ax^2 + bx + c \quad (a \neq 0)$ will factor with integer coefficients if $b^2 - 4ac$ is a perfect square.	To determine whether $5x^2 - 8x + 3$ is factorable with integer coefficients, note that $a = 5$, $b = -8$, and $c = 3$. Calculate $b^2 - 4ac$. $b^2 - 4ac = (-8)^2 - 4(5)(3) = 64 - 60 = 4$ Since 4 is a perfect square, $5x^2 - 8x + 3$ is factorable with integer coefficients.
Factoring a trinomial with a leading coefficient other than 1 by trial and error: **1.** Write the trinomial in descending powers of one variable. **2.** Factor out any GCF (including −1 if that is necessary to make the coefficient of the first term positive). **3.** Test the trinomial for factorability. **4.** Factor the remaining trinomials. **a.** If the sign of the third term of a trinomial is +, the signs between the terms of each binomial factor are the same as the sign of the middle term of the trinomial. If the sign of the third term is −, the signs between the terms of the binomials are opposites. **b.** Try combinations of the factors of the first term and last term until you find the one where the sum of the product of the outer and inner terms is the middle term of the trinomial. If no combination works, the trinomial is prime. **5.** Check the factorization by multiplication including any GCF found in Step 2.	To factor $5x^2 - 8x + 3$, follow the steps shown on the left. **1.** The polynomial is already written in descending powers of x. **2.** In this polynomial, there are no common factors. **3.** We have previously seen that $5x^2 - 8x + 3$ is factorable. **4. a.** Since the last term is + and the middle term −, the signs in both sets of parentheses will be −. The factors of the first term, $5x^2$, are $5x$ and $1x$. The factors of the last term, 3, are 3 and 1. **b.** The possible combinations are $(5x - 3)(x - 1)$ and $(5x - 1)(x - 3)$ The pair that will give a middle term of $-8x$ is $(5x - 3)(x - 1)$ **5.** Verify the factorization.
Solving quadratic equations by factoring: A quadratic equation is any equation that can be written in the form $ax^2 + bx + c = 0$, where a, b, and c are real numbers and $a \neq 0$. **Zero-factor property:** If $xy = 0$, then $x = 0$ or $y = 0$.	To solve $7x^2 = 2\left(x + \frac{5}{2}\right)$, proceed as follows: $7x^2 = 2x + 5$ Use the distributive property to remove parentheses. $7x^2 - 2x - 5 = 0$ Subtract $2x$ and 5 from both sides to write the equation in quadratic form. $(7x + 5)(x - 1) = 0$ Factor the trinomial. $7x + 5 = 0$ or $x - 1 = 0$ Apply the zero-factor property. $x = -\frac{5}{7}$ | $x = 1$ Solve each equation. Verify that the solutions are $-\frac{5}{7}$ and 1.

REVIEW EXERCISES

Factor each polynomial.

64. $3x + 6$

65. $5x^2y^3 - 10xy^2$

66. $-8x^2y^3z^4 - 12x^4y^3z^2$

67. $12a^6b^4c^2 + 15a^2b^4c^6$

68. $x^3 + 2x^2 + 3x + 6$

69. $ac + bc + 3a + 3b$

70. Factor x^n from $x^{2n} + x^n$.

71. Factor y^{2n} from $y^{2n} - y^{3n}$.

Factor each polynomial.

72. $x^4 + 4y + 4x^2 + x^2y$

73. $a^5 + b^2c + a^2c + a^3b^2$

74. $z^2 + 25$

75. $y^2 - 121$

76. $2x^4 - 98$

77. $3x^6 - 300x^2$

78. $y^2 + 11y + 10$

79. $z^2 - 11z + 30$

80. $-2x^2 + 5x - 2$

81. $-y^2 + 5y + 24$

82. $y^3 + y^2 - 2y$

83. $2a^4 + 4a^3 - 6a^2$

84. $15x^2 - 57xy - 12y^2$

85. $30x^2 + 65xy + 10y^2$

86. $x^2 + 4x + 4 - 4p^4$

87. $y^2 + 3y + 2 + 2x + xy$

Factor each polynomial.

88. $x^3 + 64$

89. $8y^3 - 512$

Solve.

90. $2x^2 - 18 = 0$

91. $2x^2 + 9x = 5$

SECTION 7.5 Review of Rational Expressions and Solving Rational Equations

DEFINITIONS AND CONCEPTS	EXAMPLES
Finding values of a variable for which a rational expression is undefined: Because the denominator of a fraction cannot be equal to 0, set the denominator equal to 0 and solve for the variable to find the restricted values.	To find the values of the variable for which the rational expression $\frac{x^2+6x-1}{x^2-5x+6}$ is undefined, set the denominator, $x^2 - 5x + 6$, equal to 0 and solve. $x^2 - 5x + 6 = 0$ $(x-3)(x-2) = 0$ Factor the trinomial. $x - 3 = 0$ or $x - 2 = 0$ Set each factor equal to 0. $x = 3$ | $x = 2$ Solve each linear equation. The values of x that will make $\frac{x^2+6x-1}{x^2-5x+6}$ undefined are 3 and 2.
Simplifying a rational expression: To simplify a rational expression, factor the numerator and denominator and divide out all factors common to the numerator and denominator. $\frac{ak}{bk} = \frac{a}{b}$ $(b \neq 0 \text{ and } k \neq 0)$	To simplify $\frac{x^2-5x+6}{x^2-9}$, factor the numerator and the denominator and divide out any resulting common factors: $\frac{x^2-5x+6}{x^2-9} = \frac{\overset{1}{\cancel{(x-3)}}(x-2)}{(x+3)\underset{1}{\cancel{(x-3)}}} = \frac{x-2}{x+3}$
Multiplying rational expressions: To multiply two rational expressions, follow the same procedure as multiplying two fractions. $\frac{a}{b} \cdot \frac{c}{d} = \frac{ac}{bd}$ $(b \neq 0, d \neq 0)$ Simplify the result, if possible.	$\frac{a+6}{ax+4a-4x-16} \cdot \frac{3a-12}{3a+18}$ $= \frac{(a+6)(3a-12)}{(ax+4a-4x-16)(3a+18)}$ Multiply the numerators and multiply the denominators. $= \frac{\cancel{(a+6)}\cancel{3}\cancel{(a-4)}}{\cancel{(a-4)}(x+4)\cancel{3}\cancel{(a+6)}}$ Factor. $= \frac{1}{x+4}$ Divide all common factors.

Dividing rational expressions: To divide two rational expressions, follow the same procedure as dividing two fractions. $\frac{a}{b} \div \frac{c}{d} = \frac{a}{b} \cdot \frac{d}{c} \quad (b \neq 0, d \neq 0, c \neq 0)$ Simplify the result, if possible.	$\frac{x^2 - 4}{2b - bx} \div \frac{x^2 + 4x + 4}{2b + bx}$ $= \frac{x^2 - 4}{2b - bx} \cdot \frac{2b + bx}{x^2 + 4x + 4}$ Invert the divisor and multiply. $= \frac{(x^2 - 4)(2b + bx)}{(2b - bx)(x^2 + 4x + 4)}$ Multiply the numerators and multiply the denominators. $= \frac{(x - 2)\cancel{(x + 2)}\cancel{b}\cancel{(2 + x)}}{\cancel{b}(2 - x)\cancel{(x + 2)}\cancel{(x + 2)}}$ Factor. $= \frac{x - 2}{2 - x}$ Divide out all common factors. $= -1$ $x - 2$ and $2 - x$ are opposites.
Adding and subtracting rational expressions with the same denominator: To add or subtract two rational expressions with like denominators, add the numerators and keep the denominator. Simplify, if possible. $\frac{a}{b} + \frac{c}{b} = \frac{a + c}{b} \quad (b \neq 0)$ $\frac{a}{b} - \frac{c}{b} = \frac{a - c}{b} \quad (b \neq 0)$	$\frac{2x + 4}{x^2 + 13x + 12} - \frac{x + 3}{x^2 + 13x + 12}$ $= \frac{2x + 4 - x - 3}{x^2 + 13x + 12}$ Subtract the numerators and keep the common denominator. $= \frac{x + 1}{x^2 + 13x + 12}$ Combine like terms. $= \frac{\cancel{x + 1}}{(x + 12)\cancel{(x + 1)}}$ Factor the denominator. $= \frac{1}{x + 12}$ Divide out the common factor $x + 1$.
Finding the LCD of two rational expressions: To find the LCD of the denominators of two fractions, factor each denominator and use each factor the greatest number of times that it appears in any one denominator. The product of these factors is the LCD.	To find the LCD of $\frac{2a}{a^2 - 2a - 8}$ and $\frac{3}{a^2 - 5a + 4}$, factor each denominator to obtain $\frac{2a}{(a - 4)(a + 2)} \qquad \frac{3}{(a - 4)(a - 1)}$ The LCD is $(a - 4)(a + 2)(a - 1)$.
Adding and subtracting rational expressions with different denominators. To add or subtract two rational expressions with different denominators, find the LCD of the two expressions and write each fraction as an equivalent fraction with the new denominator. Add the numerators and keep the denominator. Simplify, if possible.	$\frac{2a}{a^2 - 2a - 8} + \frac{3}{a^2 - 5a + 4}$ $= \frac{2a}{(a - 4)(a + 2)} + \frac{3}{(a - 4)(a - 1)}$ Factor each denominator. Write each fraction with the LCD of $(a - 4)(a + 2)(a - 1)$. $= \frac{2a(a - 1)}{(a - 4)(a + 2)(a - 1)} + \frac{3(a + 2)}{(a - 4)(a - 1)(a + 2)}$ $= \frac{2a^2 - 2a}{(a - 4)(a + 2)(a - 1)} + \frac{3a + 6}{(a - 4)(a - 1)(a + 2)}$ $= \frac{2a^2 + a + 6}{(a - 4)(a + 2)(a - 1)}$ This expression cannot be simplified.

Solving rational equations: To solve a rational equation, we multiply both sides of the equation by the LCD of the rational expressions in the equation to clear it of fractions. Because this method can produce *extraneous solutions,* it is necessary to check all possible solutions.	To solve $1 = \frac{3}{x-2} - \frac{12}{x^2-4}$, first determine that x cannot be 2 or -2. Then multiply both sides by the LCD of $(x+2)(x-2)$. $(x+2)(x-2)(1) = \left(\frac{3}{x-2} - \frac{12}{(x+2)(x-2)}\right)(x+2)(x-2)$ $(x+2)(x-2) = 3(x+2) - 12$ $x^2 - 4 = 3x + 6 - 12$ Use the distributive property to remove parentheses. $x^2 - 4 = 3x - 6$ Combine like terms. $x^2 - 3x + 2 = 0$ Add $-3x$ and 6 to both sides to write in quadratic form. $(x-2)(x-1) = 0$ Factor. $x - 2 = 0$ or $x - 1 = 0$ Set each factor equal to 0. $x = 2$ | $x = 1$ Solve each linear equation. Since 2 is a restricted value, it is extraneous. The only solution is 1.

REVIEW EXERCISES

Find all values of the variable for which the following are undefined.

92. $\frac{5x^2 + 2x + 1}{5x^2 + 14x - 3}$

93. $\frac{x^2 + x + 1}{2x^2 - 3x - 2}$

Perform the operations and simplify. Assume no division by 0.

94. $\frac{x^2 + 4x + 4}{x^2 - x - 6} \cdot \frac{x^2 - 9}{x^2 + 5x + 6}$

95. $\frac{x^3 - 64}{x^2 + 4x + 16} \div \frac{x^2 - 16}{x + 4}$

96. $\frac{5y}{x - y} - \frac{3y + 1}{x - y}$

97. $\frac{3x - 1}{x^2 + 2} + \frac{3(x - 2)}{x^2 + 2}$

98. $\frac{3}{x + 2} + \frac{2}{x + 3}$

99. $\frac{4x}{x - 4} - \frac{3}{x + 3}$

100. $\frac{x^2 + 3x + 2}{x^2 - x - 6} \cdot \frac{3x^2 - 3x}{x^2 - 3x - 4} \div \frac{x^2 + 3x + 2}{x^2 - 2x - 8}$

101. $\frac{x^2 - x - 6}{x^2 - 3x - 10} \div \frac{x^2 - x}{x^2 - 5x} \cdot \frac{x^2 - 4x + 3}{x^2 - 6x + 9}$

102. $\frac{2x}{x + 1} + \frac{3x}{x + 2} + \frac{4x}{x^2 + 3x + 2}$

103. $\frac{5x}{x - 3} + \frac{5}{x^2 - 5x + 6} + \frac{x + 3}{x - 2}$

104. $\frac{3(x + 2)}{x^2 - 1} - \frac{2}{x + 1} + \frac{4(x + 3)}{x^2 - 2x + 1}$

105. $\frac{x}{x^2 + 4x + 4} + \frac{2x}{x^2 - 4} - \frac{x^2 - 4}{x - 2}$

Simplify each complex fraction.

106. $\frac{\frac{1}{x} + \frac{2}{y}}{\frac{2}{x} - \frac{1}{y}}$

107. $\frac{x^{-1} - y^{-1}}{x^{-1} + y^{-1}}$

Solve.

108. $\frac{2}{x + 3} - \frac{x}{3 - x} = -\frac{10}{x^2 - 9}$

109. $\frac{x}{x - 2} + \frac{7}{x + 5} = \frac{14}{x^2 + 3x - 10}$

SECTION 7.6 Solving Equations in One Variable Containing an Absolute Value Expression

DEFINITIONS AND CONCEPTS	EXAMPLES
If $x \geq 0$, $\|x\| = x$.	$\|4\| = 4$
If $x < 0$, $\|x\| = -x$.	$\|-4\| = -(-4) = 4$

If $k > 0$, $|x| = k$ is equivalent to $x = k$ or $x = -k$.

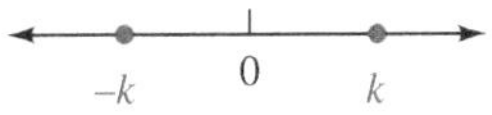

When solving an absolute value equation, first isolate the absolute value.

To solve the equation $|2x - 3| = 1$, write $|2x - 3| = 1$ as

$$2x - 3 = 1 \quad \text{or} \quad 2x - 3 = -1$$

and solve each equation for x:

$$\begin{array}{rcl} 2x - 3 = 1 & \text{or} & 2x - 3 = -1 \\ 2x = 4 & & 2x = 2 \\ x = 2 & & x = 1 \end{array}$$

Verify that both solutions check.

$|a| = |b|$ is equivalent to $a = b$ or $a = -b$.

To solve the equation $|4x - 3| = |2x + 15|$, note that the equation is true when $4x - 3 = 2x + 15$ or when $4x - 3 = -(2x + 15)$.

Then solve each equation for x.

$$\begin{array}{rcl} 4x - 3 = 2x + 15 & \text{or} & 4x - 3 = -(2x + 15) \\ 4x = 2x + 18 & & 4x - 3 = -2x - 15 \\ 2x = 18 & & 6x = -12 \\ x = 9 & & x = -2 \end{array}$$

Verify that both solutions check.

REVIEW EXERCISES

Solve and check each equation.

110. $|4x + 2| = 10$

111. $\left|\frac{3}{2}x - 4\right| = 0$

112. $|3x + 2| + 7 = 2$

113. $|3x - 4| = |4x - 3|$

SECTION 7.7 Solving Inequalities in One Variable Containing an Absolute Value Expression

DEFINITIONS AND CONCEPTS | **EXAMPLES**

If $k > 0$, $|x| < k$ is equivalent to $-k < x < k$.

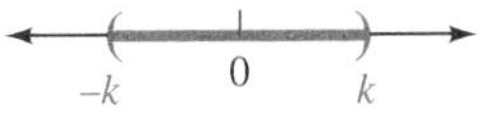

To solve the inequality $|2x - 3| < 1$, note that $|2x - 3| < 1$ is equivalent to $-1 < 2x - 3 < 1$, an inequality that we solve by isolating the variable between the inequality symbols.

$$\begin{array}{ll} -1 < 2x - 3 < 1 & \\ 2 < 2x < 4 & \text{Add 3 to all three parts.} \\ 1 < x < 2 & \text{Divide all three parts by 2.} \end{array}$$

The solution set contains all numbers between 1 and 2, not including either 1 or 2. This is the interval $(1, 2)$, whose graph is shown below.

$|x| > k$ is equivalent to $x < -k$ or $x > k$.

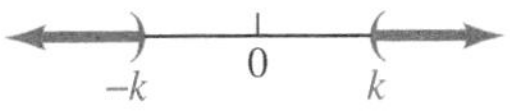

To solve the inequality $|5x - 5| > 15$, write the inequality as two separate inequalities and solve each one. Since $|5x - 5| > 15$ is equivalent to

$$5x - 5 < -15 \quad \text{or} \quad 5x - 5 > 15$$

we have

$$\begin{array}{rclll} 5x - 5 < -15 & \text{or} & 5x - 5 > 15 & \\ 5x < -10 & & 5x > 20 & \text{Add 5 to both sides.} \\ x < -2 & & x > 4 & \text{Divide both sides by 5.} \end{array}$$

Unless otherwise noted, all content on this page is © Cengage Learning.

Thus, x is either less than -2 or greater than 4.

$$x < -2 \quad \text{or} \quad x > 4$$

This is the union of two intervals $(-\infty, -2) \cup (4, \infty)$, whose graph appears below.

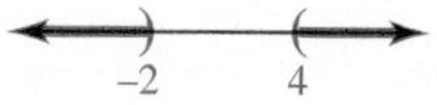

REVIEW EXERCISES

Solve each inequality. Give each solution in interval notation and graph it.

114. $|2x + 7| < 3$

115. $|3x - 8| \geq 4$

116. $\left|\frac{3}{2}x - 14\right| \geq 0$

117. $\left|\frac{3}{2}x - 14\right| < 0$

7 Test

Solve each equation.

1. $9(x + 4) + 4 = 4(x - 5)$ **2.** $\frac{y - 1}{5} + 2 = \frac{2y - 3}{3}$

3. $7 - 2(x - 3) = 8 - 2x + 5$

4. Solve $P = L + \frac{s}{f}i$ for i.

5. Cutting pipe A 20-foot pipe is to be cut into three pieces. One piece is to be twice as long as another, and the third piece is to be six times as long as the shortest. Find the length of the longest piece.

6. Geometry The length of a rectangle with a perimeter of 26 centimeters is 5 centimeters longer than it is wide. Find its area.

Solve. Express the solution in interval notation and as a graph, where appropriate.

7. $-2(2x + 3) \geq 14$

8. $-2 < \frac{x - 4}{3} < 4$

9. $|2x + 3| - 1 = 10$

10. $|3x + 4| = |x + 12|$

11. $|x + 3| \leq 4$

12. $|2x - 4| > 22$

13. Graph the equation $2x - 5y = 10$.

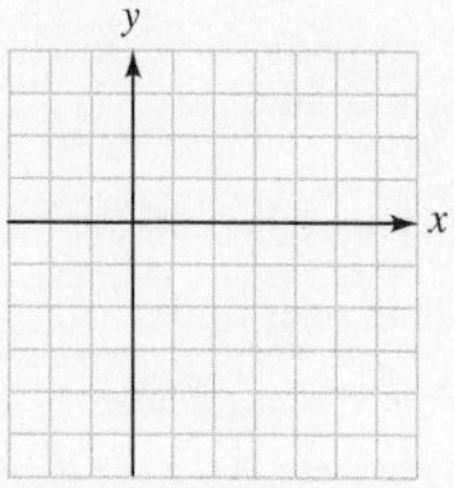

14. Find the coordinates of the midpoint of the line segment shown on the graph.

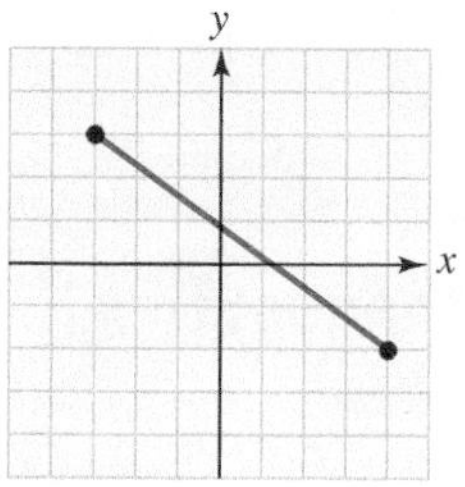

15. Find the x- and y-intercepts of the graph of $y = \frac{x - 3}{5}$.

16. Is the graph of $x - 7 = 0$ a horizontal or a vertical line?

Find the slope of each line, if possible.

17. the line through $(-2, 4)$ and $(6, 8)$

18. the graph of $2x - 3y = 8$

19. the graph of $x = 12$

20. the graph of $y = 12$

21. Write the equation of the line with slope of $\frac{2}{3}$ that passes through $(4, -5)$. Give the answer in slope-intercept form.

22. Write the equation of the line that passes through $(-2, 6)$ and $(-4, -10)$. Give the answer in slope-intercept form.

23. Find the slope and the y-intercept of the graph of $-2(x - 3) = 3(2y + 5)$.

24. Determine whether the graphs of $4x - y = 12$ and $y = \frac{1}{4}x + 3$ are parallel, perpendicular, or neither.

25. Determine whether the graphs of $y = -\frac{2}{3}x + 4$ and $2y = 3x - 3$ are parallel, perpendicular, or neither.

Unless otherwise noted, all content on this page is © Cengage Learning.

26. Write the equation of the line that passes through the origin and is parallel to the graph of $y = \frac{3}{2}x - 7$.

27. Find the domain and range of the function $f(x) = x^3$ by graphing.

28. Find the domain and range of the function $f(x) = |x|$ by graphing.

Let $f(x) = 3x + 1$ *and* $g(x) = x^2 - 2$. *Find each value.*

29. $f(3)$

30. $g(0)$

31. $f(a)$

32. $g(-x)$

Determine whether each graph represents a function.

33.

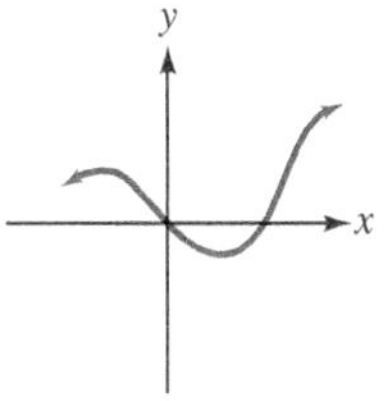

34.

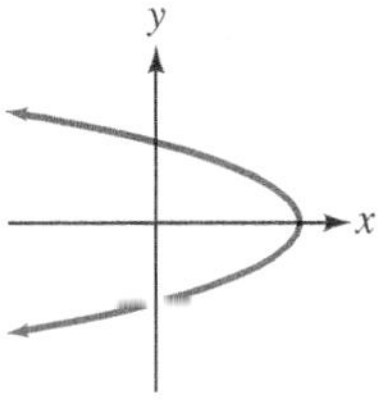

Factor each polynomial.

35. $3xy^2 + 6x^2y$

36. $ax - xy + ay - y^2$

37. $x^2 - 49$

38. $b^3 + 125$

39. $b^2 - 3b + 5$

40. $x^2 + 8x + 15$

41. $6b^2 + b - 2$

42. $6u^2 + 9u - 6$

43. $x^2 + 6x + 9 - y^2$

Perform the operations and simplify, if necessary. Write all answers without negative exponents.

44. $\frac{u^2 + 5u + 6}{u^2 - 4} \cdot \frac{u^2 - 5u + 6}{u^2 - 9}$

45. $\frac{x^3 + y^3}{4} \div \frac{x^2 - xy + y^2}{2x + 2y}$

46. $\frac{x + 2}{x + 1} - \frac{x + 1}{x + 2}$

Simplify each complex fraction.

47. $\dfrac{\frac{2u^2w^3}{v^2}}{\frac{4uw^4}{uv}}$

48. $\dfrac{\frac{x}{y} + \frac{1}{2}}{\frac{x}{2} - \frac{1}{y}}$

Solve.

49. $3x^2 = x$

50. $\frac{a - 1}{a + 3} - \frac{1 - 2a}{3 - a} = \frac{2 - a}{a - 3}$

Unless otherwise noted, all content on this page is © Cengage Learning.

Solving Systems of Linear Equations and Inequalities

8

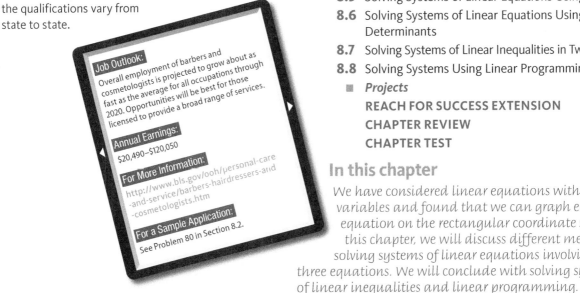

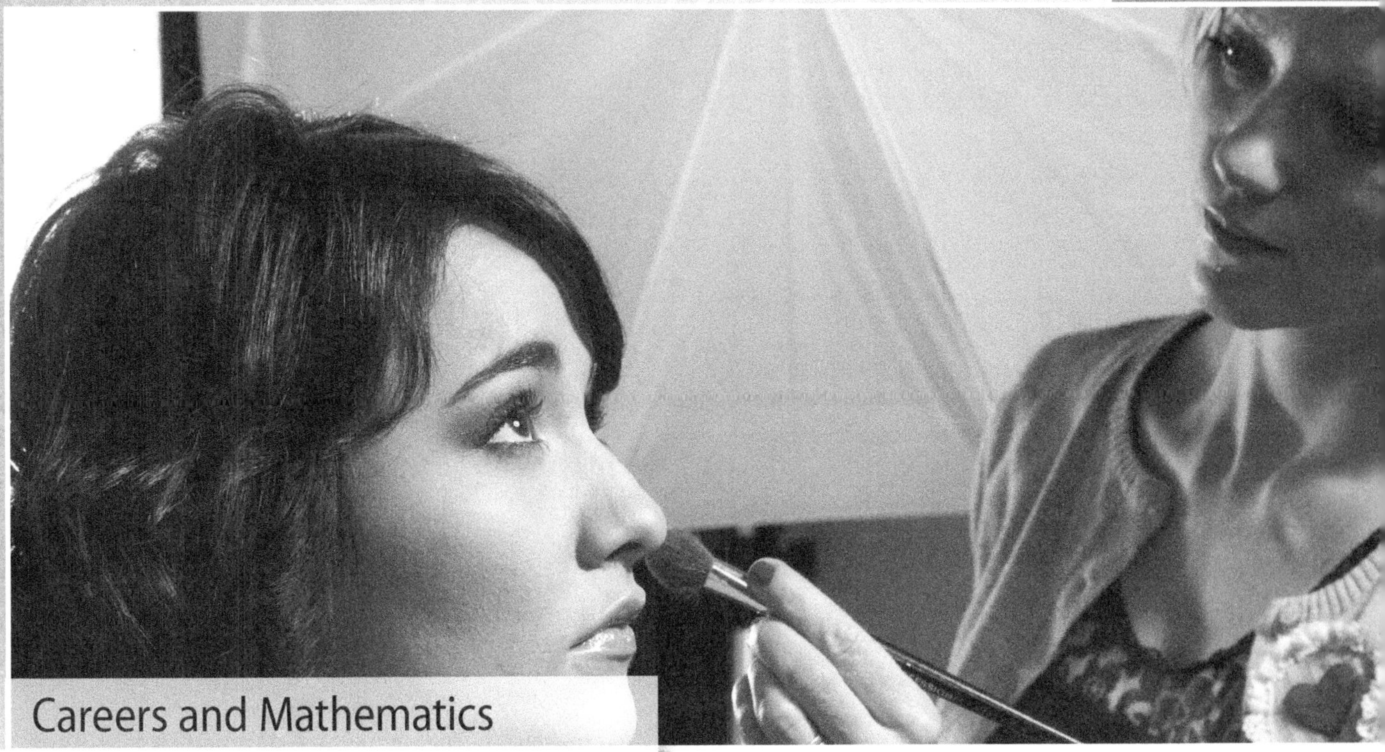

© Danie Nel/Shutterstock.com

Careers and Mathematics

COSMETOLOGISTS

Cosmetologists offer services such as hair care, make-up, and manicures and pedicures. Make-up artists' skills can range from providing a personal service for special occasions to those required for the unique needs of the television, movie, and photography industries.

Almost half of all cosmetologists are self-employed and all are required to be licensed, although the qualifications vary from state to state.

Job Outlook:
Overall employment of barbers and cosmetologists is projected to grow about as fast as the average for all occupations through 2020. Opportunities will be best for those licensed to provide a broad range of services.

Annual Earnings:
\$20,490–\$120,050

For More Information:
http://www.bls.gov/ooh/personal-care-and-service/barbers-hairdressers-and-cosmetologists.htm

For a Sample Application:
See Problem 80 in Section 8.2.

REACH FOR SUCCESS
- 8.1 Solving Systems of Linear Equations by Graphing
- 8.2 Solving Systems of Linear Equations by Substitution and Elimination
- 8.3 Solving Applications of Systems of Linear Equations in Two Variables
- 8.4 Solving Systems of Three Linear Equations in Three Variables
- 8.5 Solving Systems of Linear Equations Using Matrices
- 8.6 Solving Systems of Linear Equations Using Determinants
- 8.7 Solving Systems of Linear Inequalities in Two Variables
- 8.8 Solving Systems Using Linear Programming
- *Projects*

REACH FOR SUCCESS EXTENSION
CHAPTER REVIEW
CHAPTER TEST

In this chapter

We have considered linear equations with two variables and found that we can graph each equation on the rectangular coordinate system. In this chapter, we will discuss different methods for solving systems of linear equations involving two or three equations. We will conclude with solving systems of linear inequalities and linear programming.

Reach for Success Examining Test Results

Now that you have your exam score, what are you going to do with it?

Regardless of whether your score is high or low, it is important that you take the time to find out *why* you missed any of the questions so that you might do better the next time.

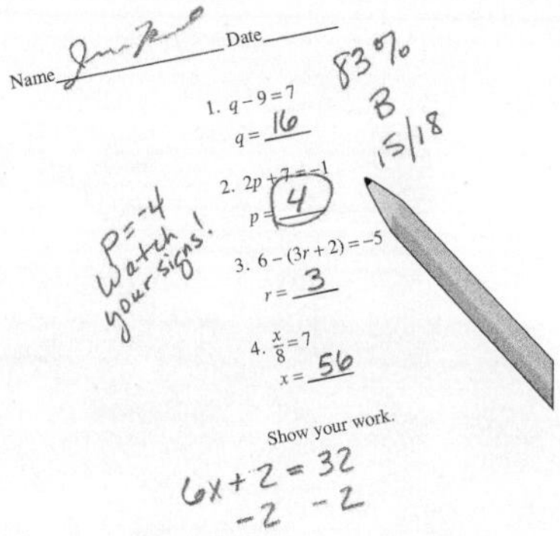

Look at different types of errors that can be made on a test. First, consider the *careless errors*. These are generally sign errors, arithmetic errors, miscopying from one line to another, and misreading the directions.

Go over your test and identify any problems for which you lost points for careless errors.

How many points did you lose? ____________

By reviewing your work prior to submitting your test, you may be able to avoid losing points to careless errors.

Now look at *conceptual errors*. These errors reflect a lack of understanding the material. (These will require you to answer some tough questions and to be totally truthful with yourself!)

Review your test and identify any problems for which you lost points because you did not understand the material.

How many points did you lose? ____________

Now, think back to the days prior to the test.

Were you in class the day these topics were covered? ______

Did you do *all* the homework assigned for the topics? ______

Did you ask questions in class? ____________

Did you go to your professor or the mathematics/tutor center for additional help? ____________

Next, look at *test-taking strategies*. These include writing down formulas at the start of the test, using all the allotted time, and previewing questions to work from easier to more difficult.

Did you write down formulas at the start of the test? ______

Did you preview to work questions you understood first? ____________

How much of the allotted time did you use? ____________

Set up an appointment with your instructor to discuss your exam errors. This is especially critical if your instructor does not allow you to keep the exam.

Describe the result of the consultation with your instructor. ____

To separate anxiety interference from conceptual errors, rework the problems by covering up your test and sliding a blank sheet of paper down as you progress. What was the result of covering the problems and reworking? ______ ____

If you now get them correct, math anxiety might be interfering and you should consult with your instructor.

If you still miss them, it might be a lack of understanding and you should seek tutoring.

A Successful Study Strategy . . .

Review each test in detail. Consider test-taking as an opportunity to show what you know. This positive attitude can improve performance.

At the end of the chapter you will find an additional exercise to guide you in planning for a successful semester.

Unless otherwise noted, all content on this page is © Cengage Learning.

Section 8.1 Solving Systems of Linear Equations by Graphing

Objectives

1. Solve a system of two linear equations by graphing.
2. Recognize that an inconsistent system has no solution.
3. Express the infinitely many solutions of a dependent system as a general ordered pair.
4. Use a system of linear equations in two variables to solve a linear equation in one variable graphically.

Vocabulary

system of equations
consistent system
equivalent systems
inconsistent system
distinct lines
independent equations
dependent equations

Getting Ready

Let $y = -3x + 2$. Find the value of y when

1. $x = 0$.
2. $x = 3$.
3. $x = -3$.
4. $x = -\frac{1}{3}$.

Find five pairs of numbers

5. with a sum of 12.
6. with a difference of 3.

In the pair of linear equations

$$\begin{cases} x + 2y = 4 \\ 2x - y = 3 \end{cases}$$

(called a **system of equations**) there are infinitely many ordered pairs (x, y) that satisfy the first equation and infinitely many ordered pairs (x, y) that satisfy the second equation. However, there is only one ordered pair (x, y) that satisfies both equations at the same time. The process of finding this ordered pair is called *solving the system.*

1 Solve a system of two linear equations by graphing.

In general, we follow these steps.

THE GRAPHING METHOD

1. On a single set of coordinate axes, carefully graph each equation.
2. Find the coordinates of the point where the graphs intersect, if applicable.
3. Check the solution in both of the original equations, if applicable.
4. If the graphs have no point in common, the system has no solution.
5. If the graphs of the equations coincide (are the same), the system has infinitely many solutions that can be expressed as a general ordered pair.

When a system of equations has a solution (as in Example 1), the system is called a **consistent system**.

EXAMPLE 1 Solve the system by graphing: $\begin{cases} x + 2y = 4 \\ 2x - y = 3 \end{cases}$

Solution We graph both equations on one set of coordinates axes, as shown in Figure 8-1.

$x + 2y = 4$

x	y	(x, y)
4	0	$(4, 0)$
0	2	$(0, 2)$
−2	3	$(-2, 3)$

$2x - y = 3$

x	y	(x, y)
1	−1	$(1, -1)$
0	−3	$(0, -3)$
−1	−5	$(-1, -5)$

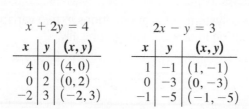

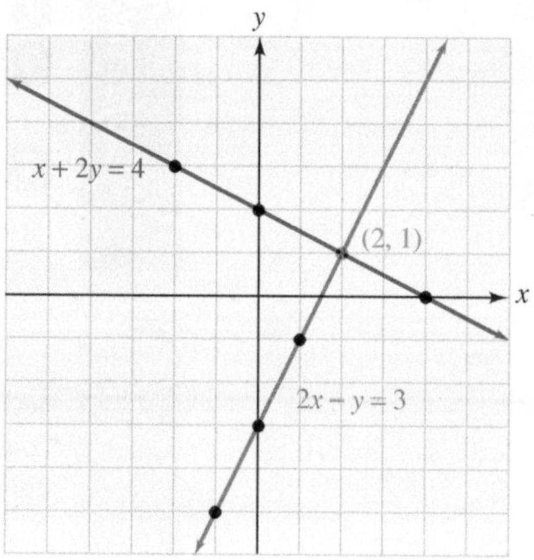

Figure 8-1

Although infinitely many ordered pairs (x, y) satisfy $x + 2y = 4$, and infinitely many ordered pairs (x, y) satisfy $2x - y = 3$, only the coordinates of the point where the graphs intersect satisfy both equations. Since the intersection point has coordinates of $(2, 1)$, the solution is the ordered pair $(2, 1)$, or $x = 2$ and $y = 1$.

When we check the solution, we substitute 2 for x and 1 for y in both equations and verify that $(2, 1)$ satisfies each one as shown below.

$$\begin{aligned} x + 2y &= 4 \\ \mathbf{2} + 2(\mathbf{1}) &= 4 \\ 2 + 2 &\stackrel{?}{=} 4 \\ 4 &= 4 \end{aligned} \qquad \begin{aligned} 2x - y &= 3 \\ 2(\mathbf{2}) - \mathbf{1} &= 3 \\ 4 - 1 &\stackrel{?}{=} 3 \\ 3 &= 3 \end{aligned}$$

SELF CHECK 1 Solve the system by graphing: $\begin{cases} 2x + y = 4 \\ x - 3y = -5 \end{cases}$

If the equations in two systems are equivalent, the systems are called **equivalent systems**. In Example 2, we solve a more difficult system by writing it as a simpler equivalent system.

EXAMPLE 2 Solve the system by graphing: $\begin{cases} \frac{3}{2}x - y = \frac{5}{2} \\ x + \frac{1}{2}y = 4 \end{cases}$

Solution We multiply both sides of $\frac{3}{2}x - y = \frac{5}{2}$ by 2 to clear the fractions and obtain the equation $3x - 2y = 5$. We multiply both sides of $x + \frac{1}{2}y = 4$ by 2 to clear the fraction and obtain the equation $2x + y = 8$. This will result in a new system

$$\begin{cases} 3x - 2y = 5 \\ 2x + y = 8 \end{cases}$$

which is equivalent to the original system but has no fractions. If we graph each equation in the new system, as in Figure 8-2, we see that the coordinates of the point where the two lines intersect are $(3, 2)$.

Unless otherwise noted, all content on this page is © Cengage Learning.

$3x - 2y = 5$

x	y	(x, y)
0	$-\frac{5}{2}$	$\left(0, -\frac{5}{2}\right)$
$\frac{5}{3}$	0	$\left(\frac{5}{3}, 0\right)$

$2x + y = 8$

x	y	(x, y)
4	0	$(4, 0)$
1	6	$(1, 6)$

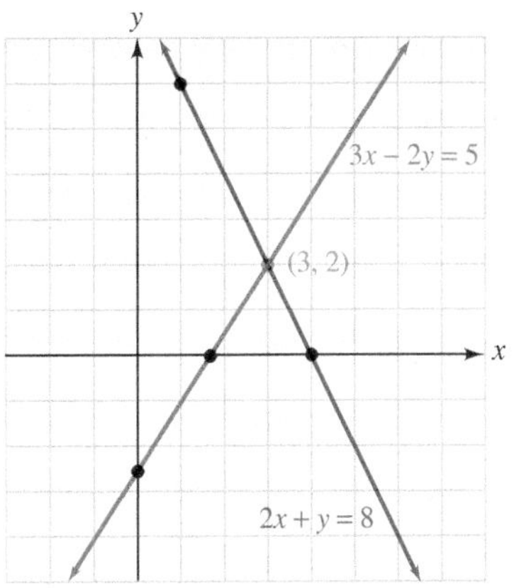

Figure 8-2

To check the solution, $(3, 2)$, we substitute 3 for x and 2 for y in the two original equations.

$$\frac{3}{2}x - y = \frac{5}{2} \qquad\qquad x + \frac{1}{2}y = 4$$

$$\frac{3}{2}(\mathbf{3}) - \mathbf{2} \stackrel{?}{=} \frac{5}{2} \qquad\qquad \mathbf{3} + \frac{1}{2}(\mathbf{2}) \stackrel{?}{=} 4$$

$$\frac{9}{2} - 2 \stackrel{?}{=} \frac{5}{2} \qquad\qquad 3 + 1 \stackrel{?}{=} 4$$

$$\frac{5}{2} = \frac{5}{2} \qquad\qquad 4 = 4$$

SELF CHECK 2 Solve the system by graphing: $\begin{cases} \frac{5}{2}x - y = 2 \\ x + \frac{1}{3}y = 3 \end{cases}$

2 Recognize that an inconsistent system has no solution.

When a system has no solution (as in Example 3), it is called an **inconsistent system**. Since there is no solution, the solution set is $\varnothing$.

EXAMPLE 3 Solve the system by graphing: $\begin{cases} 2x + 3y = 6 \\ 4x + 6y = 24 \end{cases}$

Solution We graph both equations on one set of coordinate axes, as shown in Figure 8-3 on the next page. In this example, the graphs appear to be parallel. We can show that this is true by writing each equation in slope-intercept form.

$$2x + 3y = 6 \qquad\qquad 4x + 6y = 24$$

$$3y = -2x + 6 \qquad\qquad 6y = -4x + 24$$

$$y = -\frac{2}{3}x + 2 \qquad\qquad y = -\frac{2}{3}x + 4$$

The slope of both lines is $-\frac{2}{3}$, but their y-intercepts, $(2, 0)$ and $(4, 0)$, are different, thus they are called **distinct lines**. The slopes are equal and the y-intercepts are different, therefore the lines are parallel.

Since the graphs are parallel lines, the lines do not intersect, and the system does not have a solution. The system is an *inconsistent system* and its solution set is $\varnothing$.

Unless otherwise noted, all content on this page is © Cengage Learning.

$2x + 3y = 6$

x	y	(x, y)
3	0	(3, 0)
0	2	(0, 2)
−3	4	(−3, 4)

$4x + 6y = 24$

x	y	(x, y)
6	0	(6, 0)
0	4	(0, 4)
−3	6	(−3, 6)

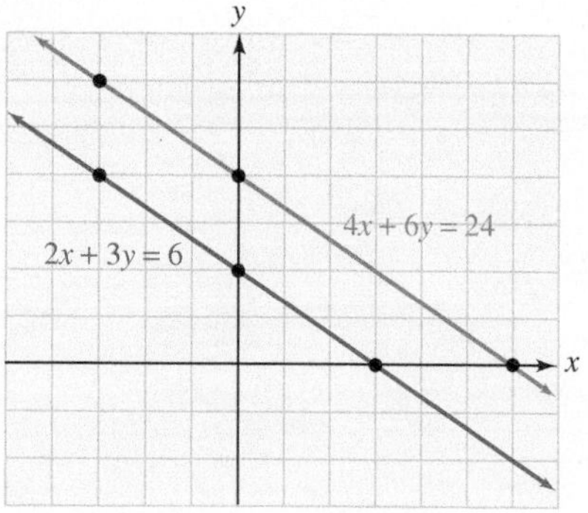

Figure 8-3

SELF CHECK 3 Solve the system by graphing: $\begin{cases} 2x + 3y = 6 \\ y = -\frac{2}{3}x - 3 \end{cases}$

3 Express the infinitely many solutions of a dependent system as a general ordered pair.

When the equations of a system have different graphs (as in Examples 1, 2, and 3), the equations are called **independent equations**. Two equations with the same graph are called **dependent equations**, as in Example 4.

EXAMPLE 4 Solve the system by graphing: $\begin{cases} 2y - x = 4 \\ 2x + 8 = 4y \end{cases}$

Solution We graph each equation on one set of coordinate axes, as shown in Figure 8-4. Since the graphs are the same line (coincide), the system has infinitely many solutions. In fact, any ordered pair (x, y) that satisfies one equation satisfies the other.

$2y - x = 4$

x	y	(x, y)
−4	0	(−4, 0)
0	2	(0, 2)

$2x + 8 = 4y$

x	y	(x, y)
−4	0	(−4, 0)
0	2	(0, 2)

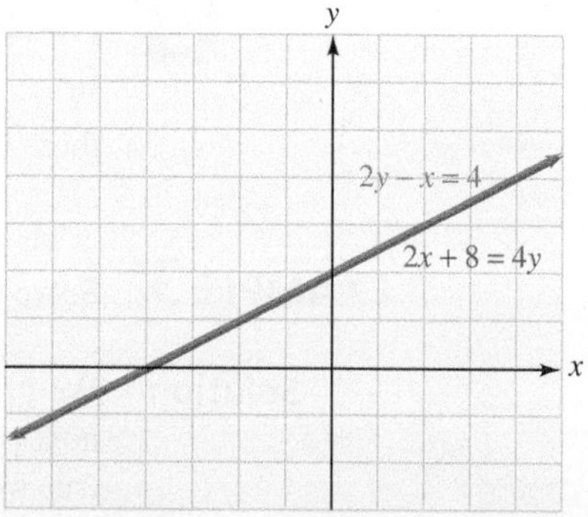

Figure 8-4

To find a general ordered pair solution, we can solve either equation of the system for either variable. If we solve one equation for y, we will obtain an expression for y in terms of x,

$$\begin{aligned} 2y - x &= 4 \\ 2y &= x + 4 \\ y &= \frac{1}{2}x + 2 \end{aligned}$$

Substituting $\frac{1}{2}x + 2$ for y in the ordered pair (x, y), we obtain the general solution $\left(x, \frac{1}{2}x + 2\right)$. Since x is the independent variable, we can substitute any value for x to obtain a value for y to generate infinitely many ordered pair solutions.

Unless otherwise noted, all content on this page is © Cengage Learning.

SELF CHECK 4 Solve the system by graphing: $\begin{cases} 2x - y = 4 \\ x = \frac{1}{2}y + 2 \end{cases}$

We summarize the possibilities that can occur when we graph two linear equations, each with two variables.

SOLVING SYSTEMS OF TWO LINEAR EQUATIONS

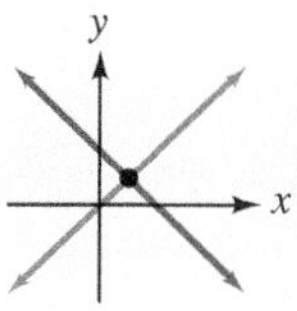

If the lines are distinct and intersect, the equations are independent and the system is consistent. **One solution exists**, expressed as (x, y).

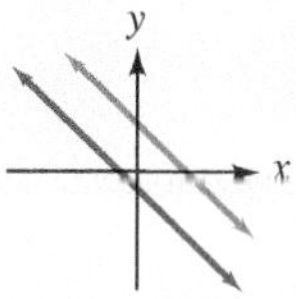

If the lines are distinct and parallel, the equations are independent and the system is inconsistent. **No solution exists**, expressed as $\varnothing$.

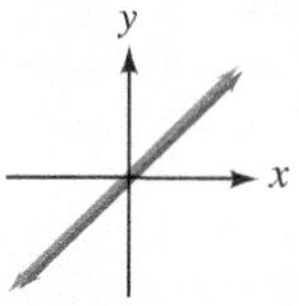

If the lines coincide (are the same), the equations are dependent and the system is consistent. **Infinitely many solutions exist**, expressed as a general ordered pair such as $(x, ax + b)$.

Accent on technology

▸ Solving Systems by Graphing

To solve the system

$$\begin{cases} 3x + 2y = 12 \\ 2x - 3y = 12 \end{cases}$$

using a graphing calculator, we first solve each equation for y so we can enter them into a graphing calculator. After solving for y, we obtain the following equivalent system. The graph is shown in Figure 8-5.

$$\begin{cases} y = -\frac{3}{2}x + 6 \\ y = \frac{2}{3}x - 4 \end{cases}$$

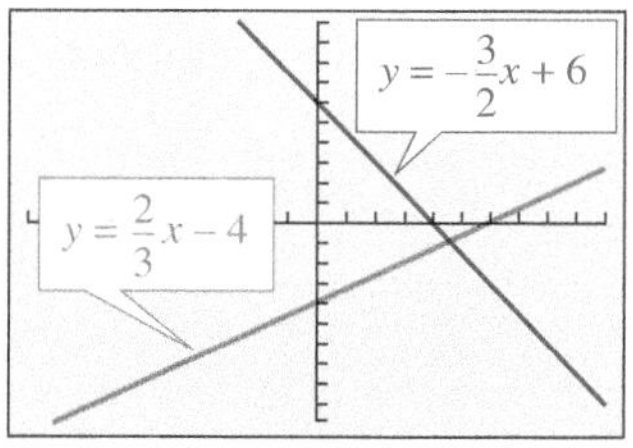

Figure 8-5

COMMENT The intersect method will work only if the x-coordinate of the intersection point is visible on the screen. If it is not, you may need to adjust the viewing window.

We can solve this system using the INTERSECT feature found in the CALC menu on a TI-84 graphing calculator. From the graph, press **2nd TRACE** (CALC) and select "5: intersect" by using the down arrow to highlight and then press **ENTER** or by pressing 5 on the keyboard to obtain Figure 8-6(a) on the next page. We direct the calculator to look for the intersection by selecting a point on the first line and pressing **ENTER** and a point on the second line and pressing **ENTER**. We can move the cursor closer to the intersection point using the **TRACE** key and then press **ENTER** again. We obtain Figure 8-6(b). From the figure, we see that the solution is approximately $(4.6153846, -0.9230769)$.

(Continued)

Unless otherwise noted, all content on this page is © Cengage Learning.

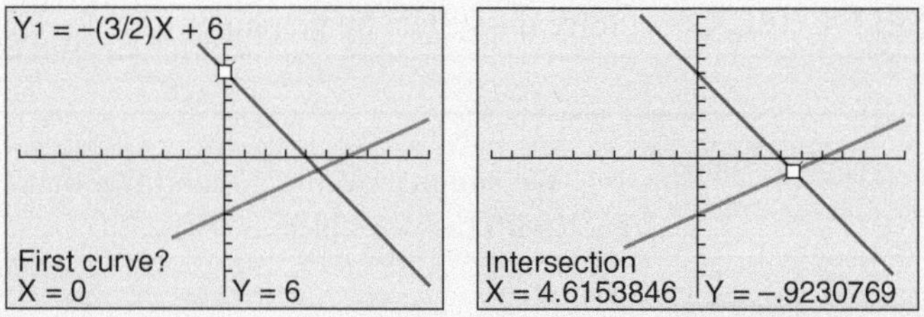

(a) (b)

Figure 8-6

Refer to the tear-out card for the method for converting decimal answers to fractions and verify that the exact solution is $x = \frac{60}{13}$ and $y = -\frac{12}{13}$ or $\left(\frac{60}{13}, \frac{-12}{13}\right)$.

For instructions regarding the use of a Casio graphing calculator, please refer to the Casio Keystroke Guide in the back of the book.

4 Use a system of linear equations in two variables to solve a linear equation in one variable graphically.

The graphing method discussed in this section can be used to solve equations in one variable.

EXAMPLE 5 Solve $2x + 4 = -2$ graphically.

Solution To solve $2x + 4 = -2$ graphically, we can set the left and right sides of the equation equal to y. The graphs of $y = 2x + 4$ and $y = -2$ are shown in Figure 8-7. To solve $2x + 4 = -2$, we need to find the value of x that makes $2x + 4$ equal to -2. The point of intersection of the graphs is $(-3, -2)$. This indicates that if x is -3, the expression $2x + 4$ equals -2. So the solution of $2x + 4 = -2$ is -3. Check this result by substituting 3 for x in the original equation.

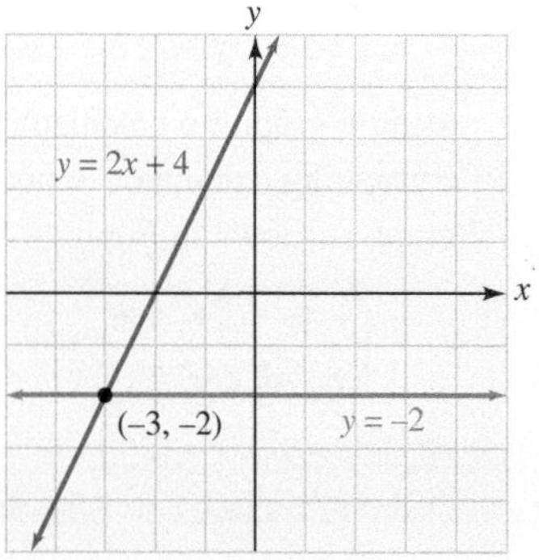

Figure 8-7

SELF CHECK 5 Solve $2x + 4 = 2$ graphically.

Accent on technology

▸ Solving Equations Graphically

To solve $2(x - 3) + 3 = 7$ with a TI-84 graphing calculator, we graph the left and right sides of the equation in the same window by entering

$$Y_1 = 2(x - 3) + 3$$
$$Y_2 = 7$$

Figure 8-8(a) shows the graphs, generated using settings of $[-10, 10]$ for x and for y.

The coordinates of the point of intersection of the graphs can be determined using the INTERSECT feature outlined in the previous Accent on Technology.

Unless otherwise noted, all content on this page is © Cengage Learning.

COMMENT Notice that this is an equation in one variable. Although the intersection will display both an x-value and a y-value, only the value of x is the solution.

In Figure 8-8(b), we see that the point of intersection is $(5, 7)$, which indicates that 5 is the solution of $2(x - 3) + 3 = 7$.

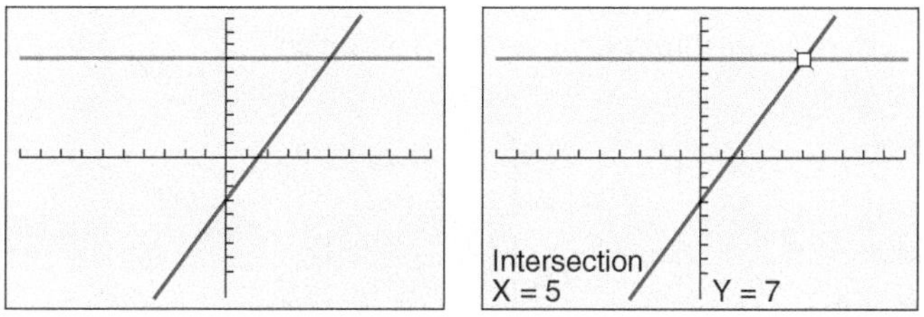

(a) (b)

Figure 8-8

For instructions regarding the use of a Casio graphing calculator, please refer to the Casio Keystroke Guide in the back of the book.

SELF CHECK ANSWERS

1. $(1, 2)$

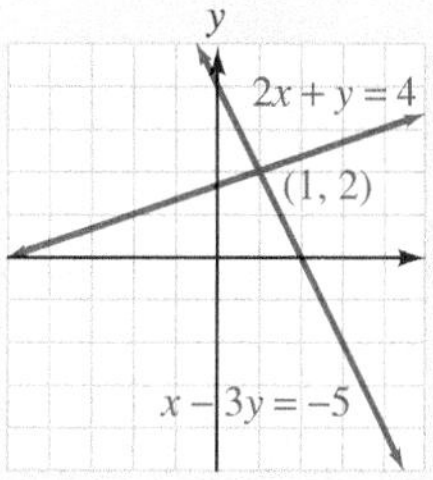

2. $(2, 3)$

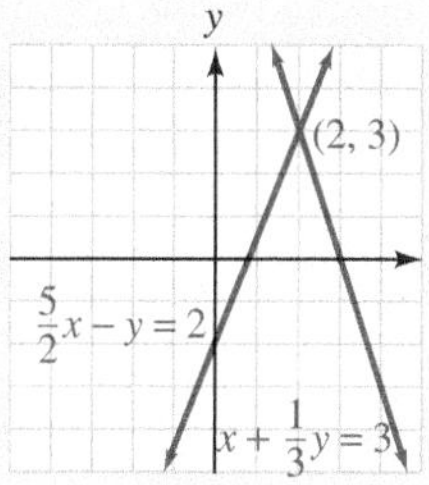

3. $\varnothing$

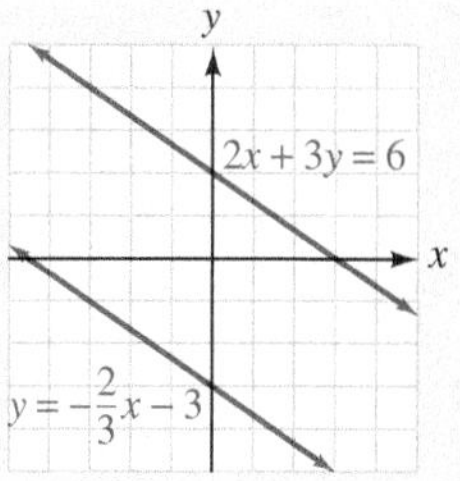

4. $(x, 2x - 4)$

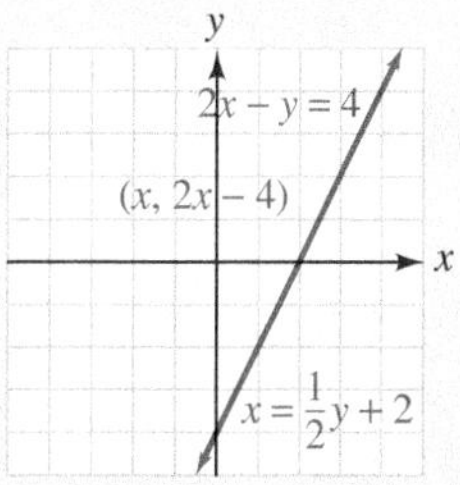

5. -1

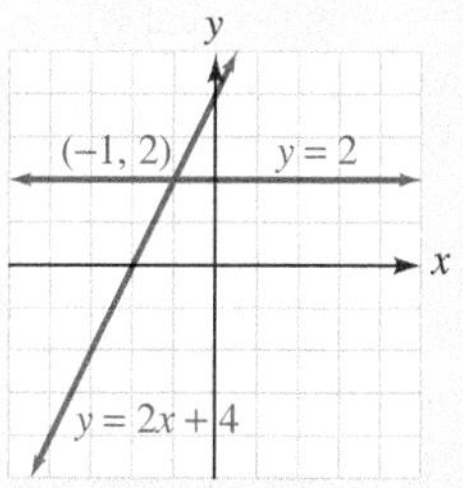

NOW TRY THIS

Solve each system by graphing.

a. $\begin{cases} y = x^2 - 4 \\ y = 2x - 1 \end{cases}$ **b.** $\begin{cases} y = -|x + 1| + 4 \\ y = -\frac{1}{3}x + 1 \end{cases}$ **c.** $\begin{cases} y = |x - 2| \\ y = x^2 \end{cases}$

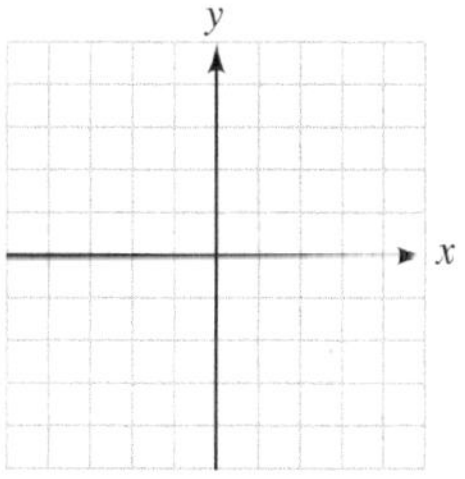

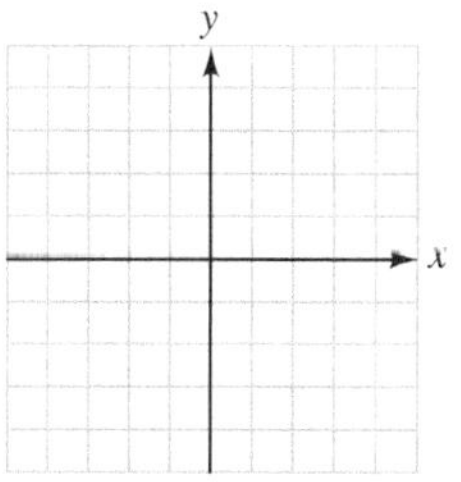

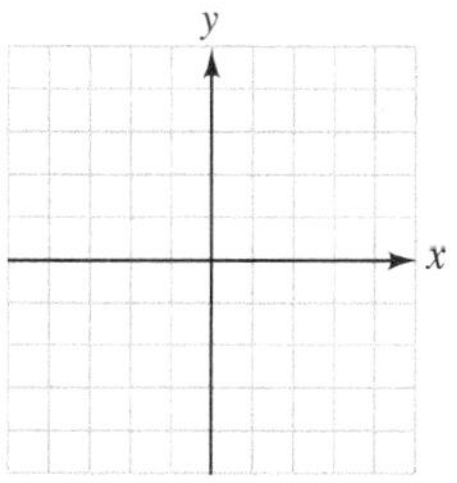

Unless otherwise noted, all content on this page is © Cengage Learning.

8.1 Exercises

WARM-UPS *Determine whether the ordered pair is a solution of the system of equations.*

1. $(1, 3);\ \begin{cases} y = 3x \\ y = \frac{1}{3}x + \frac{8}{3} \end{cases}$

2. $(-1, 2);\ \begin{cases} y = 2x + 4 \\ y = x + 2 \end{cases}$

3. $(2, -1);\ \begin{cases} 2x - 4y = 8 \\ 4x + y = 9 \end{cases}$

4. $(-4, 3);\ \begin{cases} 4x - y = -19 \\ 3x + 2y = -6 \end{cases}$

REVIEW *Write each number in scientific notation.*

5. 850,000,000

6. 0.000000479

7. 239×10^3

8. 465×10^{-4}

VOCABULARY AND CONCEPTS *Fill in the blanks.*

9. If two or more equations are considered at the same time, they are called a ______ of equations.

10. When a system of equations has one or more solutions, it is called a ________ system.

11. If a system has no solution, it is called an __________ system. Its solution set is ∅.

12. If two equations have different graphs, they are called __________ equations.

13. Two equations with the same graph are called ________ equations.

14. If the equations in two systems are equivalent, the systems are called ________ systems.

GUIDED PRACTICE *Solve each system by graphing.* SEE EXAMPLE 1. (OBJECTIVE 1)

15. $\begin{cases} x + y = 6 \\ x - y = 2 \end{cases}$

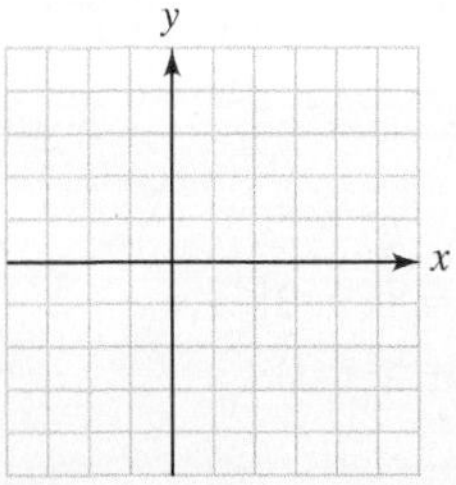

16. $\begin{cases} x - y = 4 \\ 2x + y = 5 \end{cases}$

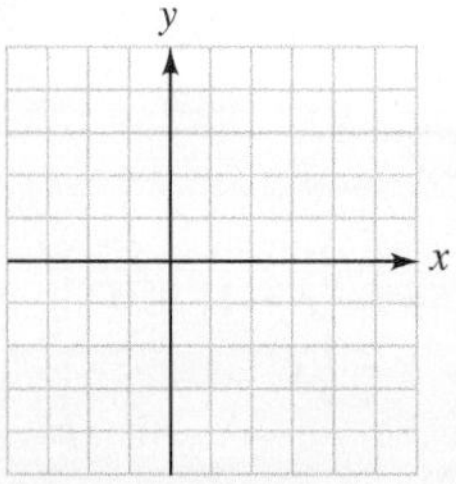

17. $\begin{cases} 2x + y = 1 \\ x - 2y = -7 \end{cases}$

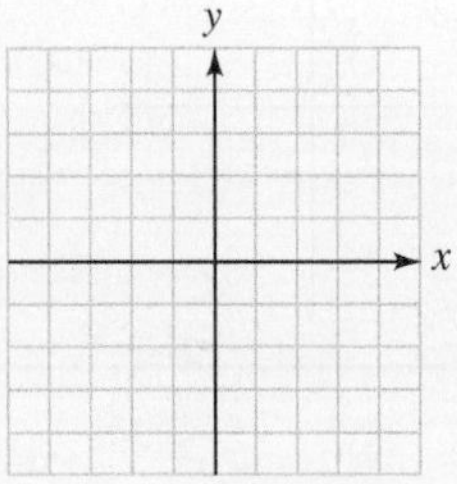

18. $\begin{cases} 3x - y = -3 \\ 2x + y = -7 \end{cases}$

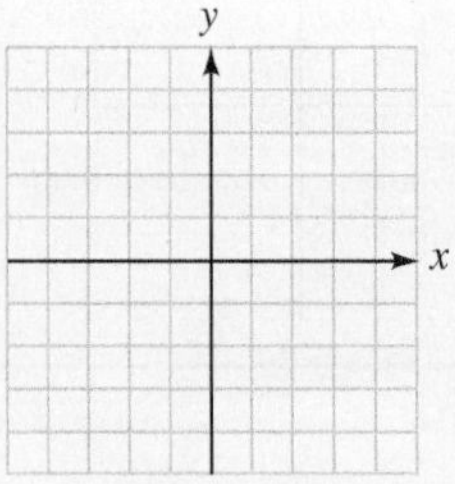

19. $\begin{cases} 2x + 3y = 0 \\ 2x + y = 4 \end{cases}$

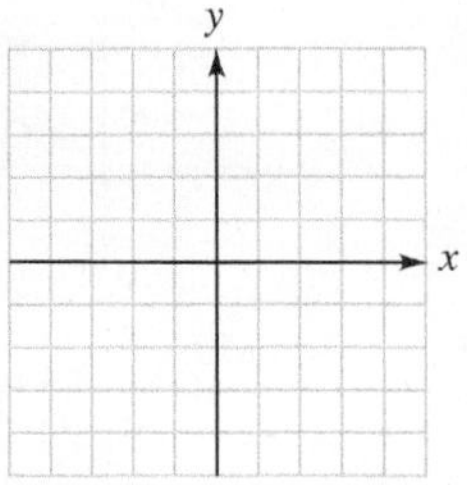

20. $\begin{cases} 3x - 2y = 0 \\ 2x + 3y = 0 \end{cases}$

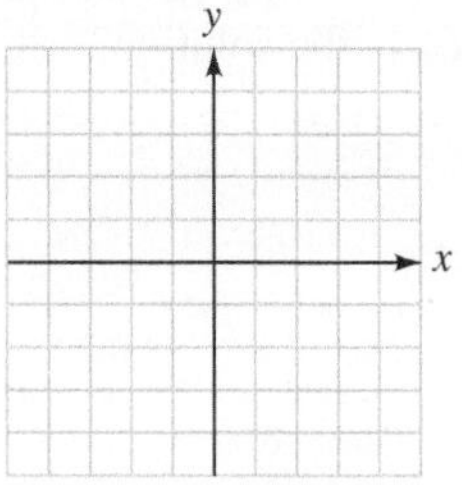

21. $\begin{cases} y = 3 \\ x = 2 \end{cases}$

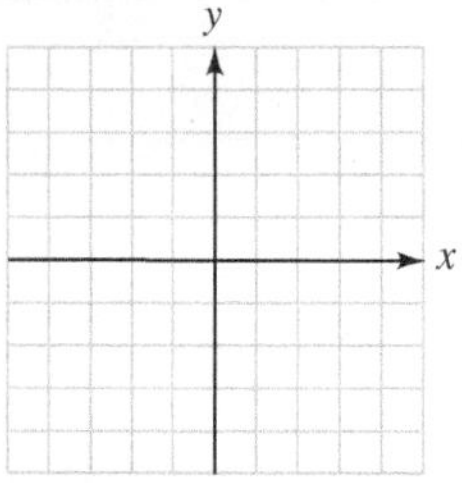

22. $\begin{cases} 2x + 3y = -15 \\ 2x + y = -9 \end{cases}$

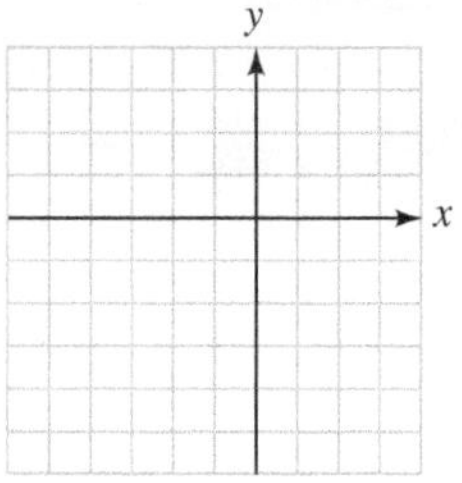

Solve each system by graphing. SEE EXAMPLE 2. (OBJECTIVE 1)

23. $\begin{cases} x = 2 \\ y = \frac{4 - x}{2} \end{cases}$

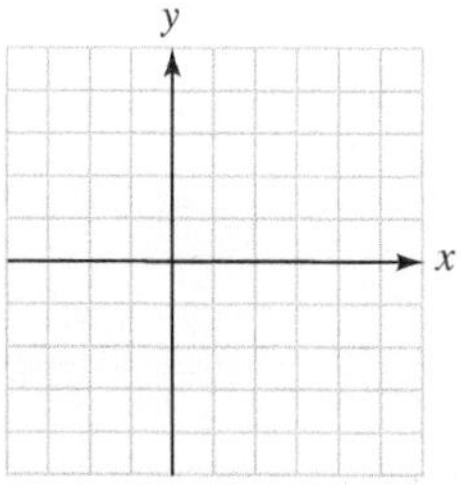

24. $\begin{cases} y = -2 \\ x = \frac{4 + 3y}{2} \end{cases}$

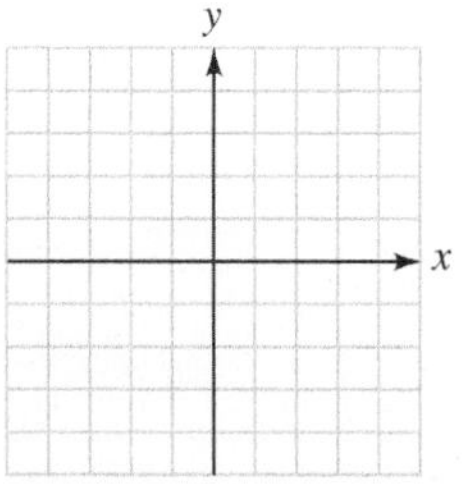

25. $\begin{cases} \frac{5}{2}x + y = \frac{1}{2} \\ 2x - \frac{3}{2}y = 5 \end{cases}$

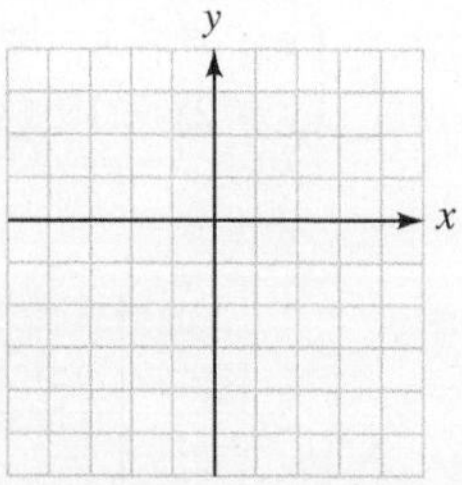

26. $\begin{cases} x = \frac{5y - 4}{2} \\ x - \frac{5}{3}y + \frac{1}{3} = 0 \end{cases}$

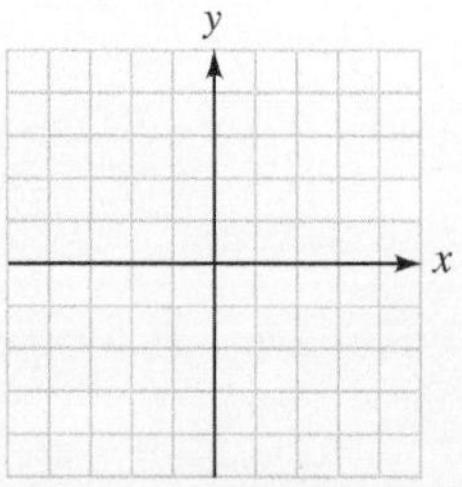

Unless otherwise noted, all content on this page is © Cengage Learning.

Solve each system by graphing. State whether the system is inconsistent or if the equations are dependent. Write each solution set. SEE EXAMPLE 3. (OBJECTIVE 2)

27. $\begin{cases} 3x = 5 - 2y \\ 3x + 2y = 7 \end{cases}$

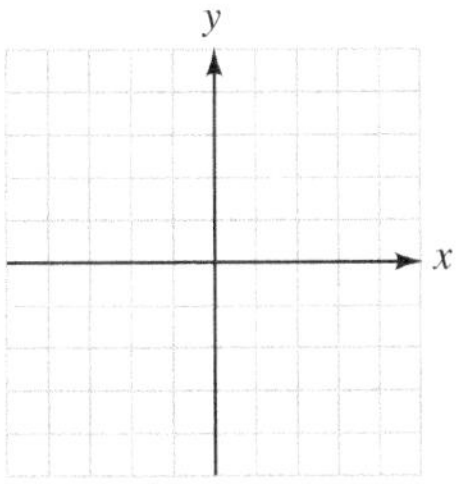

28. $\begin{cases} y = x \\ x - y = 7 \end{cases}$

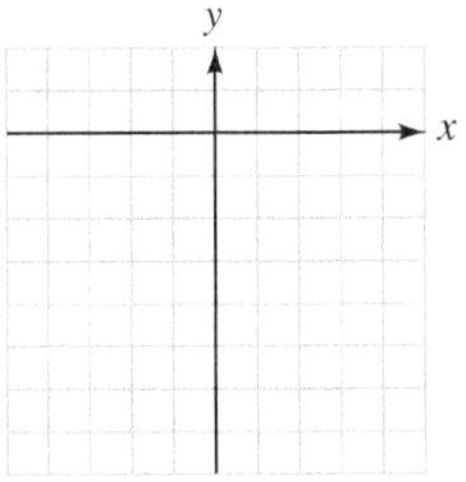

29. $\begin{cases} x = 2y - 8 \\ y = \frac{1}{2}x - 5 \end{cases}$

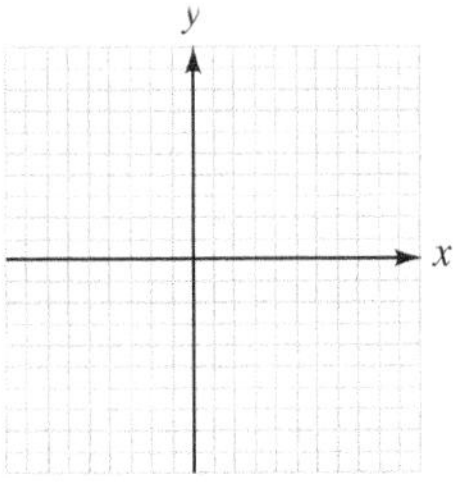

30. $\begin{cases} 8x - 2y = 7 \\ y = 4x + 3 \end{cases}$

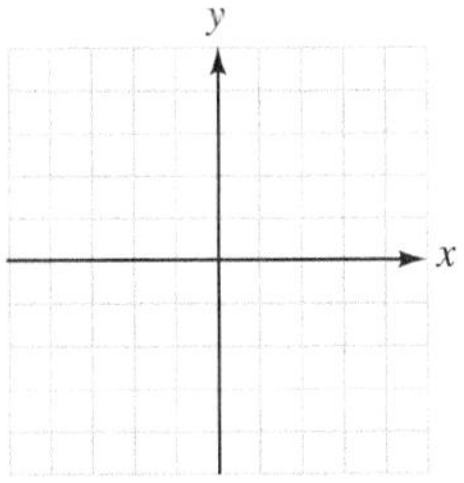

Solve each system by graphing. State whether the system is inconsistent or if the equations are dependent. Write each solution set. SEE EXAMPLE 4. (OBJECTIVE 3)

31. $\begin{cases} x = 3 - 2y \\ 2x + 4y = 6 \end{cases}$

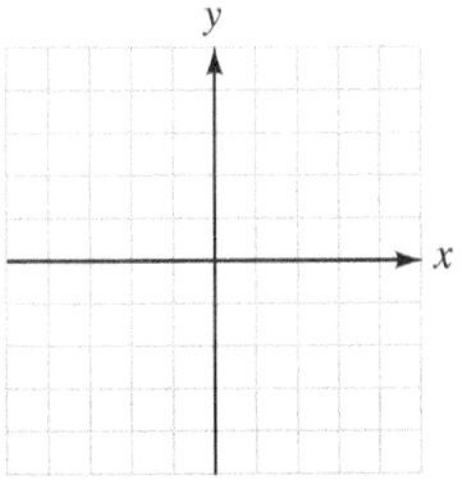

32. $\begin{cases} x = y \\ y - x = 0 \end{cases}$

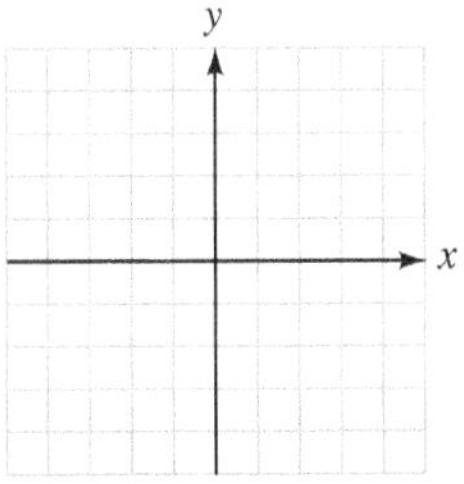

33. $\begin{cases} 6x + 3y = 9 \\ y + 2x = 3 \end{cases}$

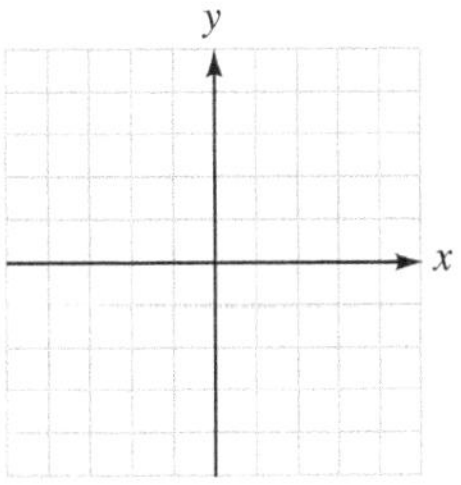

34. $\begin{cases} x - 3y = 9 \\ 3x - 27 = 9y \end{cases}$

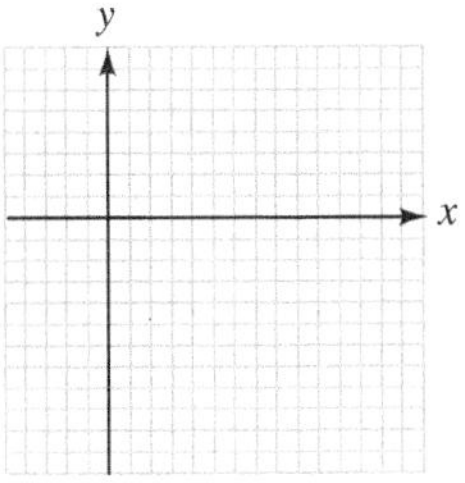

Solve each equation graphically. SEE EXAMPLE 5. (OBJECTIVE 4).

35. $\frac{3}{2}a + 4 = -5$

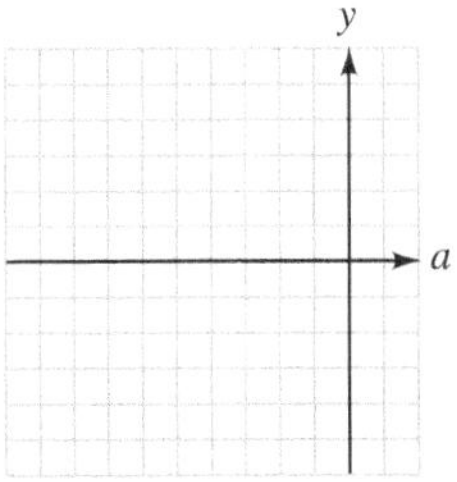

36. $-8(a - 1) = -8$

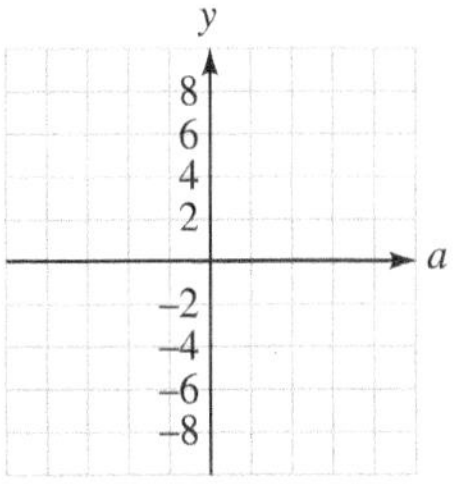

37. $a + 2 = -a - 1$

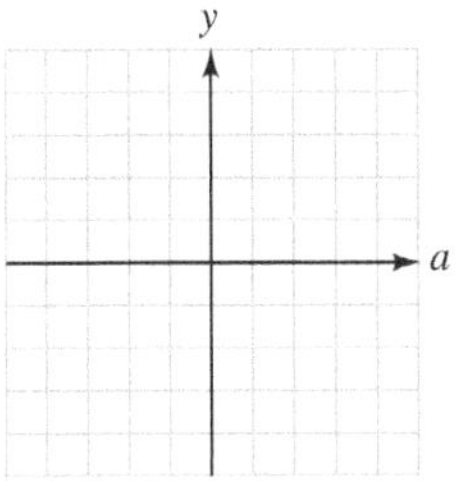

38. $2(a - 3) = -3a - 1$

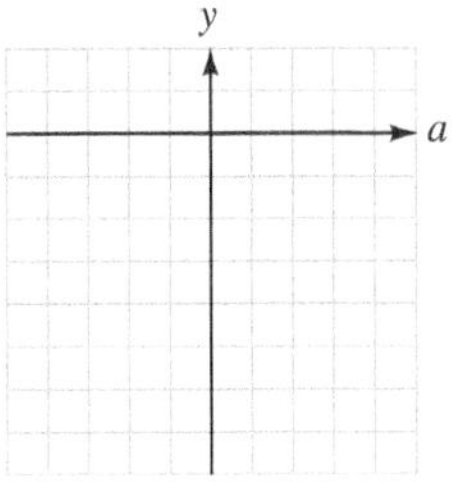

ADDITIONAL PRACTICE *Determine whether the following systems will have one solution, no solution, or infinitely many solutions.*

39. $\begin{cases} y = 5x \\ y = 5x + 3 \end{cases}$

40. $\begin{cases} y = 4x \\ y = 3x + x \end{cases}$

41. $\begin{cases} y = 6x + 1 \\ y = -6x + 1 \end{cases}$

42. $\begin{cases} y = -3x + 2 \\ -3x = y \end{cases}$

Solve each system by graphing.

43. $\begin{cases} x = \dfrac{11 - 2y}{3} \\ y = \dfrac{11 - 6x}{4} \end{cases}$

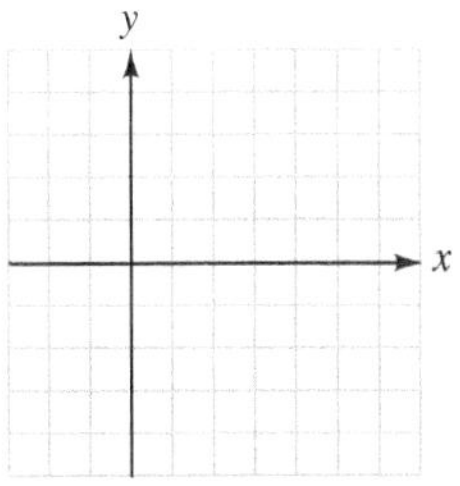

44. $\begin{cases} x = \dfrac{1 - 3y}{4} \\ y = \dfrac{12 + 3x}{2} \end{cases}$

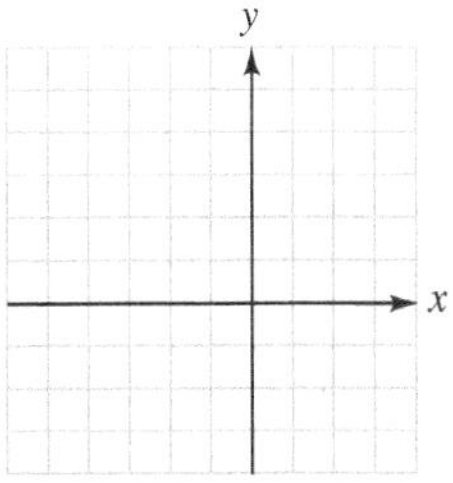

45. $\begin{cases} \frac{5}{2}x + 3y = 6 \\ y = \dfrac{24 - 10x}{12} \end{cases}$

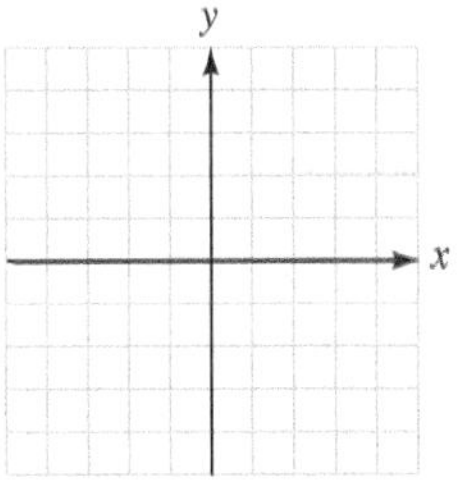

46. $\begin{cases} 2x = 5y - 11 \\ 3x = 2y \end{cases}$

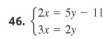

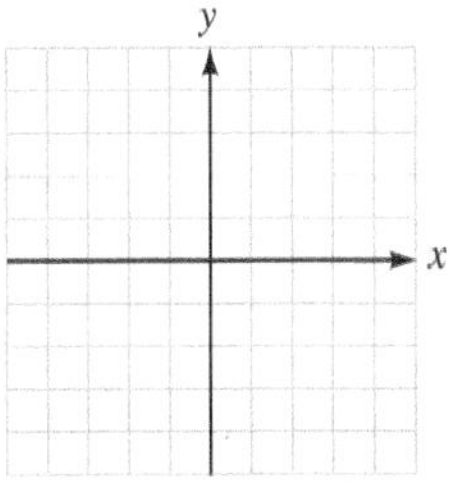

Unless otherwise noted, all content on this page is © Cengage Learning.

Use a graphing calculator to solve each system. Give the exact answer.

47. $\begin{cases} x = 13 - 4y \\ 3x = 4 + 2y \end{cases}$

48. $\begin{cases} 3x = 7 - 2y \\ 2x = 2 + 4y \end{cases}$

49. $\begin{cases} x = -\frac{3}{2}y \\ x = \frac{3}{2}y - 2 \end{cases}$

50. $\begin{cases} x = \frac{3y - 1}{4} \\ y = \frac{4 - 8x}{3} \end{cases}$

Use a graphing calculator to solve each system. Give all answers to the nearest hundredth.

51. $\begin{cases} y = 3.2x - 1.5 \\ y = -2.7x - 3.7 \end{cases}$

52. $\begin{cases} y = -0.45x + 5 \\ y = 5.55x - 13.7 \end{cases}$

53. $\begin{cases} 1.7x + 2.3y = 3.2 \\ y = 0.25x + 8.95 \end{cases}$

54. $\begin{cases} 2.75x = 12.9y - 3.79 \\ 7.1x - y = 35.76 \end{cases}$

APPLICATIONS

55. Retailing The cost of manufacturing one type of camera and the revenue from the sale of those cameras are shown in the illustration.

a. From the illustration, find the cost of manufacturing 15,000 cameras.

b. From the illustration, find the revenue obtained by selling 20,000 cameras.

c. For what number of cameras sold will the revenue equal the cost?

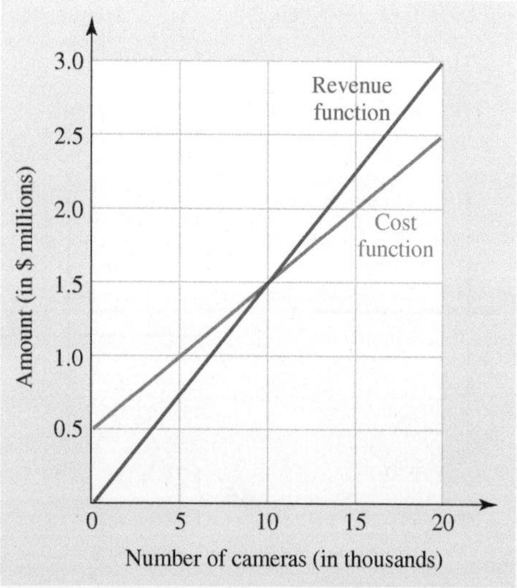

56. Food service

a. Estimate the point of intersection of the two graphs shown in the illustration. Express your answer in the form (year, number of meals).

b. What information about dining out does the point of intersection provide?

Number of Take-Out and On-Premise Meals Purchased at Commercial Restaurants per Person Annually

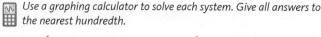

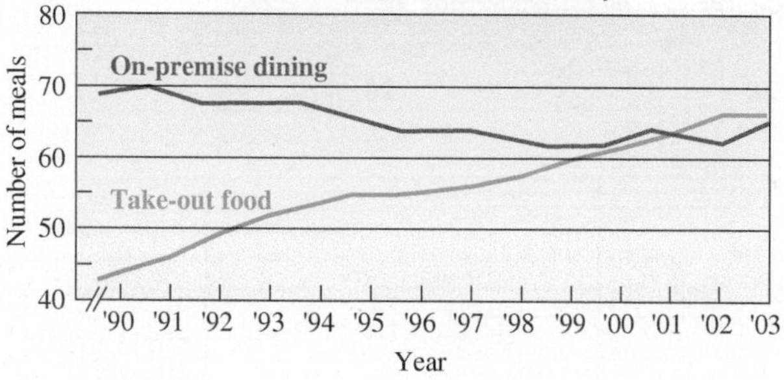

57. Navigation Two ships are sailing on the same coordinate system. One ship is following a course described by $2x + 3y = 6$, and the other is following a course described by $2x - 3y = 9$.

a. Is there a possibility of a collision?

b. Find the coordinates of the danger point.

c. Is a collision a certainty?

58. Navigation Two airplanes are flying at the same altitude and in the same coordinate system. One plane is following a course described by $y = \frac{2}{5}x - 2$, and the other is following a course described by $x = \frac{5y + 7}{2}$. Is there a possibility of a collision?

59. Car repairs Smith Chevrolet charges \$50 per hour for labor on car repairs. Lopez Ford charges a diagnosis fee of \$30 plus \$40 per hour for labor. If the labor on an engine repair costs the same at either shop, how long does the repair take?

60. Landscaping A landscaper installs some trees and shrubs at a bank. He installs 25 plants for a total cost of \$1,500. How many trees (t) and how many shrubs (s) did he install if each tree costs \$100 and each shrub costs \$50?

61. Cost and revenue The function $C(x) = 200x + 400$ gives the cost for a college to offer x sections of an introductory class in CPR (cardiopulmonary resuscitation). The function $R(x) = 280x$ gives the amount of revenue the college brings in when offering x sections of CPR. Find the point where the cost equals the revenue by graphing each function on the same coordinate system.

62. In Exercise 61, how many sections does the college need to offer to make a profit on the CPR training course?

WRITING ABOUT MATH

63. Explain how to solve a system of two equations in two variables.

64. Can a system of two equations in two variables have exactly two solutions? Why or why not?

SOMETHING TO THINK ABOUT

65. Form an independent system of equations with a solution of $(-5, 2)$.

66. Form a dependent system of equations with a solution of $(-5, 2)$.

Unless otherwise noted, all content on this page is © Cengage Learning.

Section 8.2 Solving Systems of Linear Equations by Substitution and Elimination

Objectives

1. Solve a system of two linear equations by substitution.
2. Solve a system of two linear equations by elimination (addition).
3. Identify an inconsistent linear system.
4. Identify a dependent linear system and express the answer as a general ordered pair.
5. Write a repeating decimal in fractional form.
6. Solve an application by setting up and solving a system of two linear equations.

Getting Ready

Remove parentheses by using the distributive property.

1. $3(2x - 7)$ **2.** $-4(3x + 5)$

Substitute $x - 3$ for y and remove parentheses by using the distributive property.

3. $3y$ **4.** $-2(y + 2)$

The graphing method discussed in the previous section provides a way to visualize the process of solving systems of equations. However, it can be difficult to use to solve systems of higher order, such as three equations, each with three variables. In this section, we will discuss algebraic methods that enable us to solve such systems.

1 Solve a system of two linear equations by substitution.

To solve a system of two linear equations (each with two variables) by substitution, we use the following steps.

THE SUBSTITUTION METHOD

1. If necessary, solve one equation for one of its variables, preferably a variable with a coefficient of 1.
2. Substitute the resulting expression for the variable obtained in Step 1 into the other equation and solve that equation.
3. Find the value of the other variable by substituting the value of the variable found in Step 2 into any equation containing both variables.
4. State the solution.
5. Check the solution in both of the original equations.

EXAMPLE 1 Solve the system by substitution: $\begin{cases} 4x + y = 13 \\ -2x + 3y = -17 \end{cases}$

Solution **Step 1:** We solve the first equation for y, because y has a coefficient of 1 and no fractions are introduced.

$$4x + y = 13$$
$$(1) \quad y = -4x + 13 \quad \text{Subtract } 4x \text{ from both sides.}$$

Step 2: We then substitute $-4x + 13$ for y in the second equation of the system and solve for x.

$$\begin{aligned} -2x + 3y &= -17 \\ -2x + 3(\mathbf{-4x + 13}) &= -17 && \text{Substitute.} \\ -2x - 12x + 39 &= -17 && \text{Use the distributive property to remove parentheses.} \\ -14x &= -56 && \text{Combine like terms and subtract 39 from both sides.} \\ x &= 4 && \text{Divide both sides by } -14. \end{aligned}$$

Step 3: To find the value of y, we substitute 4 for x in Equation 1 and simplify:

$$\begin{aligned} y &= -4x + 13 \\ y &= -4(\mathbf{4}) + 13 && \text{Substitute.} \\ &= -3 \end{aligned}$$

Step 4: The solution is $x = 4$ and $y = -3$, or $(4, -3)$. The graphs of these two equations would intersect at the point with coordinates $(4, -3)$.

Step 5: To verify that this solution satisfies both equations, we substitute $x = 4$ and $y = -3$ into each equation in the original system and simplify.

$$\begin{aligned} 4x + y &= 13 \\ 4(\mathbf{4}) + (\mathbf{-3}) &\stackrel{?}{=} 13 \\ 16 - 3 &\stackrel{?}{=} 13 \\ 13 &= 13 \end{aligned} \qquad \begin{aligned} -2x + 3y &= -17 \\ -2(\mathbf{4}) + 3(\mathbf{-3}) &\stackrel{?}{=} -17 \\ -8 - 9 &\stackrel{?}{=} -17 \\ -17 &= -17 \end{aligned}$$

Since the ordered pair $(4, -3)$ satisfies both equations of the system, the solution checks.

SELF CHECK 1 Solve the system by substitution: $\begin{cases} x + 3y = 9 \\ 2x - y = -10 \end{cases}$

EXAMPLE 2 Solve the system by substitution: $\begin{cases} \frac{4}{3}x + \frac{1}{2}y = -\frac{2}{3} \\ \frac{1}{2}x + \frac{2}{3}y = \frac{5}{3} \end{cases}$

Solution First we find an equivalent system without fractions by multiplying both sides of each equation by 6, the LCD.

$$\begin{matrix} (1) \\ (2) \end{matrix} \quad \begin{cases} 8x + 3y = -4 \\ 3x + 4y = 10 \end{cases}$$

Because no variable in either equation has a coefficient of 1, it is impossible to avoid fractions when solving for a variable. We solve Equation 2 for x.

$$\begin{aligned} 3x + 4y &= 10 \\ 3x &= -4y + 10 && \text{Subtract } 4y \text{ from both sides.} \\ (3) \quad x &= -\frac{4}{3}y + \frac{10}{3} && \text{Divide both sides by 3.} \end{aligned}$$

We then substitute $-\frac{4}{3}y + \frac{10}{3}$ for x in Equation 1 and solve for y.

$$\begin{aligned} 8x + 3y &= -4 & \\ 8\left(-\frac{4}{3}y + \frac{10}{3}\right) + 3y &= -4 & \text{Substitute.} \\ -\frac{32}{3}y + \frac{80}{3} + 3y &= -4 & \text{Use the distributive property to remove parentheses.} \\ -32y + 80 + 9y &= -12 & \text{Multiply both sides by 3 to clear fractions.} \\ -23y &= -92 & \text{Combine like terms and subtract 80 from both sides.} \\ y &= 4 & \text{Divide both sides by } -23. \end{aligned}$$

We can find the value of x by substituting 4 for y in Equation 3 and simplifying:

$$\begin{aligned} x &= -\frac{4}{3}y + \frac{10}{3} & \\ &= -\frac{4}{3}(4) + \frac{10}{3} & \text{Substitute.} \\ &= -\frac{6}{3} & -\tfrac{16}{3} + \tfrac{10}{3} = -\tfrac{6}{3} \\ &= -2 & \end{aligned}$$

The solution is the ordered pair $(-2, 4)$. Verify that this solution checks in both of the equations in the original system.

SELF CHECK 2 Solve the system by substitution: $\begin{cases} \frac{2}{3}x + \frac{1}{2}y = 1 \\ \frac{1}{3}x - \frac{3}{2}y = 4 \end{cases}$

2 Solve a system of two linear equations by elimination (addition).

Sometimes using the substitution method can be cumbersome, especially if none of the coefficients are 1. We have another algebraic method of solving systems that may be more efficient. In the elimination (addition) method, we combine the equations of the system in a way that will eliminate the terms involving one of the variables.

THE ELIMINATION (ADDITION) METHOD

1. If necessary, write both equations of the system in $Ax + By = C$ form.
2. If necessary, multiply the terms of one or both of the equations by constants chosen to make the coefficients of one of the variables opposites.
3. Add the equations and solve the resulting equation, if possible.
4. Substitute the value obtained in Step 3 into either of the original equations and solve for the remaining variable, if applicable.
5. State the solution obtained in Steps 3 and 4.
6. Check the solution in both equations of the original system.

EXAMPLE 3 Solve the system by elimination: $\begin{cases} 4x + y = 13 \\ -2x + 3y = -17 \end{cases}$

Solution This is the same system we solved in Example 1. Since both equations are already written in $Ax + By = C$ form, Step 1 is unnecessary.

Step 2: To solve the system by elimination, we multiply the second equation by 2 to make the coefficients of x opposites.

$$\begin{cases} 4x + y = 13 \\ -4x + 6y = -34 \end{cases}$$

Step 3: When we add the equations, the terms involving x are eliminated because $4x$ and $-4x$ are additive inverses and their sum is 0. We will obtain

$$7y = -21$$

$$y = -3 \quad \text{Divide both sides by 7.}$$

Step 4: To find the value of x, we substitute -3 for y in either of the original equations and solve. If we use the first equation, we have

$$4x + y = 13$$

$$4x + (-3) = 13 \quad \text{Substitute.}$$

$$4x = 16 \quad \text{Add 3 to both sides.}$$

$$x = 4 \quad \text{Divide both sides by 4.}$$

Step 5: The solution is $x = 4$ and $y = -3$, or $(4, -3)$.

Step 6: The check was completed in Example 1.

SELF CHECK 3 Solve the system by elimination: $\begin{cases} 3x + 2y = 0 \\ 2x - y = -7 \end{cases}$

EXAMPLE 4 Solve the system by elimination: $\begin{cases} \frac{4}{3}x + \frac{1}{2}y = -\frac{2}{3} \\ \frac{1}{2}x + \frac{2}{3}y = \frac{5}{3} \end{cases}$

Solution This is the same system we solved in Example 2. To solve it by elimination, we may find an equivalent system with no fractions by multiplying both sides of each equation by 6 to obtain

$$\begin{cases} 8x + 3y = -4 & (1) \\ 3x + 4y = 10 & (2) \end{cases}$$

We can solve for x by eliminating the terms involving y. To do so, we multiply both sides of Equation 1 by 4 and both sides of Equation 2 by -3 to obtain

$$\begin{cases} 32x + 12y = -16 \\ -9x - 12y = -30 \end{cases}$$

When these equations are added, the y-terms are eliminated and we obtain the result

$$23x = -46$$

$$x = -2 \quad \text{Divide both sides by 23.}$$

To find the value of y, we substitute -2 for x in either Equation 1 or Equation 2. If we substitute -2 for x in Equation 2, we have

$$3x + 4y = 10$$

$$3(-2) + 4y = 10 \quad \text{Substitute.}$$

$$-6 + 4y = 10 \quad \text{Simplify.}$$

$$4y = 16 \quad \text{Add 6 to both sides.}$$

$$y = 4 \quad \text{Divide both sides by 4.}$$

The solution is $(-2, 4)$. The check was completed in Example 2.

SELF CHECK 4 Solve the system by elimination: $\begin{cases} \frac{4}{3}x + \frac{1}{2}y = -3 \\ \frac{1}{2}x + \frac{2}{3}y = -\frac{1}{6} \end{cases}$

3 Identify an inconsistent linear system.

In the next two examples, we encounter the situation in which both variables are simultaneously eliminated but with different interpretations.

EXAMPLE 5 Solve the system using any method: $\begin{cases} y = 2x + 4 \\ 8x - 4y = 7 \end{cases}$

Solution Because the first equation is already solved for y, we will use the substitution method.

$$8x - 4y = 7$$

$$8x - 4(2x + 4) = 7 \quad \text{Substitute } 2x + 4 \text{ for } y \text{ in the second equation.}$$

We then solve this equation for x:

$$8x - 8x - 16 = 7 \quad \text{Use the distributive property to remove parentheses.}$$

$$-16 = 7 \quad \text{Combine like terms.}$$

This false statement indicates that the equations in the system are independent (they represent distinct lines), but that the system is inconsistent. If the equations of this system were graphed, the graphs would be parallel. Since the system has no point of intersection, its solution set is $\varnothing$.

SELF CHECK 5 Solve the system using any method: $\begin{cases} x = -\frac{5}{2}y + 5 \\ y = -\frac{2}{5}x + 5 \end{cases}$

4 Identify a dependent linear system and express the answer as a general ordered pair.

In the next example, both variables are eliminated but the result is different.

EXAMPLE 6 Solve the system using any method: $\begin{cases} 4x + 6y = 12 \\ -2x - 3y = -6 \end{cases}$

Solution Since the equations are written in $Ax + By = C$ form, we will use the elimination (addition) method. To eliminate the x-terms when adding the equations, we multiply both sides of the second equation by 2 to obtain

$$\begin{cases} 4x + 6y = 12 \\ -4x - 6y = -12 \end{cases}$$

After adding the left and right sides, we have

$$0x + 0y = 0$$

$$0 = 0$$

Here, both the x- and y-terms are eliminated. The true statement $0 = 0$ indicates that the equations in this system are dependent and that the system is consistent.

Note that the equations of the system are equivalent, because when the second equation is multiplied by -2, it becomes the first equation. The line graphs of these equations would coincide. Since any ordered pair that satisfies one of the equations also satisfies the other, there are infinitely many solutions.

COMMENT When we solved a linear equation in one variable, a result of $0 = 0$ indicated a solution of all real numbers. When we solve a system of linear equations in two variables a result of $0 = 0$ indicates a solution of an infinite number of ordered pairs that could be written in the form $(x, ax + b)$.

To find a general solution, we can solve either equation of the system to obtain $y = -\frac{2}{3}x + 2$. Substituting $-\frac{2}{3}x + 2$ for y in the ordered pair (x, y), we obtain the general solution $\left(x, -\frac{2}{3}x + 2\right)$.

SELF CHECK 6 Solve the system using any method: $\begin{cases} x = -\frac{5}{2}y + 5 \\ y = -\frac{2}{5}x + 2 \end{cases}$

5 Write a repeating decimal in fractional form.

We have seen that to write a fraction in decimal form, we divide its numerator by its denominator. The result is often a repeating decimal. By using systems of equations, we can write a repeating decimal in fractional form.

EXAMPLE 7 Write $0.2\overline{54}$ as a fraction.

Solution To write $0.2\overline{54}$ as a fraction, we note that the decimal has a repeating block of two digits and then form an equation by setting x equal to the decimal.

(1) $x = 0.2545454\ldots$

We then form another equation by multiplying both sides of Equation 1 by 10^2.

(2) $100x = 25.4545454\ldots \quad 10^2 = 100$

We can *subtract* each side of Equation 1 from the corresponding side of Equation 2 to obtain

$$\begin{aligned} 100x &= 25.4\,54\,54\,54\ldots \\ x &= 0.2\,54\,54\,54\ldots \\ \hline 99x &= 25.2 \end{aligned}$$

Finally, we solve $99x = 25.2$ for x and simplify the fraction.

$$x = \frac{25.2}{99} = \frac{25.2 \cdot \mathbf{10}}{99 \cdot \mathbf{10}} = \frac{252}{990} = \frac{18 \cdot 14}{18 \cdot 55} = \frac{14}{55}$$

We can use a calculator to verify that the decimal representation of $\frac{14}{55}$ is $0.2\overline{54}$.

SELF CHECK 7 Write $0.3\overline{72}$ as a fraction.

The key step in solving Example 7 was multiplying both sides of Equation 1 by 10^2. If there had been n digits in the repeating block of the decimal, we would have multiplied both sides of Equation 1 by 10^n.

6 Solve an application by setting up and solving a system of two linear equations.

To solve applications using two variables, we follow the same problem-solving strategy discussed in Chapter 2, except that we use two variables and form two equations instead of one.

EXAMPLE 8 **RETAIL SALES** A store advertises two types of cell phones, one selling for \$67 and the other for \$100. If the receipts from the sale of 36 phones totaled \$2,940, how many of each type were sold?

Analyze the problem — We can let x represent the number of phones sold for \$67 and let y represent the number of phones sold for \$100.

Form two equations — Because a total of 36 phones were sold, we can form the equation

The number of lower-priced phones	plus	the number of higher-priced phones	equals	the total number of phones.
x	$+$	y	$=$	36

We know that the receipts for the sale of x of the \$67 phones will be \$$67x$ and that the receipts for the sale of y of the \$100 phones will be \$$100y$. Since the sum of these receipts is \$2,940, the second equation is

The value of the lower-priced phones	plus	the value of the higher-priced phones	equals	the total receipts.
$67x$	$+$	$100y$	$=$	2,940

Solve the system — To find out how many of each type of phone were sold, we must solve the system

$$\begin{aligned} &(1) \quad \\ &(2) \quad \end{aligned} \begin{cases} x + y = 36 \\ 67x + 100y = 2{,}940 \end{cases}$$

We multiply both sides of Equation 1 by -100, add the resulting equation to Equation 2, and solve for x:

$$\begin{aligned} -100x - 100y &= -3{,}600 \\ 67x + 100y &= 2{,}940 \\ \hline -33x \qquad\quad &= -660 \\ x &= 20 \quad \text{Divide both sides by } -33. \end{aligned}$$

To find the value of y, we substitute 20 for x in Equation 1 and solve.

$$\begin{aligned} x + y &= 36 \\ 20 + y &= 36 \quad \text{Substitute.} \\ y &= 16 \quad \text{Subtract 20 from both sides.} \end{aligned}$$

State the conclusion — The store sold 20 of the \$67 phones and 16 of the \$100 phones.

Check the result — If 20 of one type were sold and 16 of the other type were sold, a total of 36 phones were sold.

Since the value of the lower-priced phones is $20(\$67) = \$1{,}340$ and the value of the higher-priced phones is $16(\$100) = \$1{,}600$, the total receipts are \$2,940.

SELF CHECK 8 If the sale of the 36 phones totaled \$2,775, how many of each type were sold?

EXAMPLE 9 **MIXING SOLUTIONS** How many ounces of a 5% saline solution and how many ounces of a 20% saline solution must be mixed together to obtain 50 ounces of a 15% saline solution?

Analyze the problem To find how many ounces of each type of saline solution should be mixed, we can let x represent the number of ounces of the 5% saline solution and y represent the number of ounces of the 20% saline solution. (See Figure 8-9.)

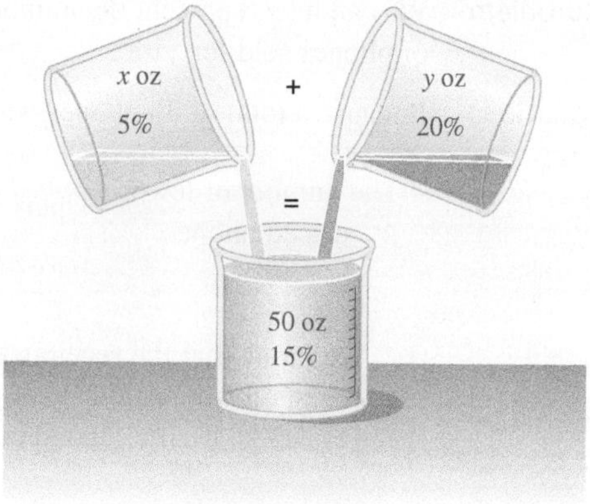

Figure 8-9

Form two equations Because a total of 50 ounces are needed, one of the equations will be

The number of ounces of 5% saline solution	plus	the number of ounces of 20% saline solution	equals	the total number of ounces in the mixture.
x	$+$	y	$=$	50

The amount of salt in the x ounces of 5% saline solution is $0.05x$, and the amount of salt in the y ounces of 20% saline solution is $0.20y$. The amount of salt in the 50 ounces of 15% saline solution will be $0.15(50)$. This gives the equation

The salt in the 5% solution	plus	the salt in the 20% solution	equals	the salt in the mixture.
$0.05x$	$+$	$0.20y$	$=$	$0.15(50)$

Solve the system To find out how many ounces of each are needed, we solve the following system:

$$\begin{cases} x + y = 50 & (1) \\ 0.05x + 0.20y = 7.5 \quad 0.15(50) = 7.5 & (2) \end{cases}$$

To solve this system by substitution, we can solve Equation 1 for y

$$x + y = 50$$
$$(3) \quad y = 50 - x \quad \text{Subtract } x \text{ from both sides.}$$

and then substitute $50 - x$ for y in Equation 2.

$$0.05x + 0.20y = 7.5$$
$$0.05x + 0.20(\mathbf{50 - x}) = 7.5 \quad \text{Substitute.}$$
$$5x + 20(50 - x) = 750 \quad \text{Multiply both sides by 100.}$$
$$5x + 1{,}000 - 20x = 750 \quad \text{Use the distributive property to remove parentheses.}$$
$$-15x = -250 \quad \text{Combine like terms and subtract 1,000 from both sides.}$$
$$x = \frac{-250}{-15} \quad \text{Divide both sides by } -15.$$
$$x = \frac{50}{3} \quad \text{Simplify.}$$

Unless otherwise noted, all content on this page is © Cengage Learning.

To find the value of y, we substitute $\frac{50}{3}$ for x in Equation 3:

$$\begin{aligned} y &= 50 - x \\ &= 50 - \mathbf{\frac{50}{3}} \quad \text{Substitute.} \\ &= \frac{100}{3} \end{aligned}$$

State the conclusion — To obtain 50 ounces of a 15% saline solution, we must mix $\frac{50}{3}\left(16\frac{2}{3}\right)$ ounces of the 5% saline solution with $\frac{100}{3}\left(33\frac{1}{3}\right)$ ounces of the 20% saline solution.

Check the result — We note that $16\frac{2}{3}$ ounces of solution plus $33\frac{1}{3}$ ounces of solution equals the required 50 ounces of solution. We also note that 5% of $16\frac{2}{3} \approx 0.83$, and 20% of $33\frac{1}{3} \approx 6.67$, giving a total of 7.5, which is 15% of 50.

SELF CHECK 9 How many ounces of a 10% saline solution and how many ounces of a 15% saline solution must be mixed together to obtain 50 ounces of a 12% saline solution?

Systems of equations can be used to solve many applications. Sometimes the equations may involve more than two variables.

In manufacturing, running a machine involves both *setup costs* and *unit costs*. Setup costs include the cost of preparing a machine to do a certain job. Unit costs depend on the number of items to be manufactured, including costs of raw materials and labor.

Suppose that a certain machine has a setup cost of \$600 and a unit cost of \$3 per item. If x items will be manufactured using this machine, the cost will be

$$\text{Cost} = 600 + 3x \quad \text{Cost} = \text{setup cost} + \text{unit cost} \cdot \text{the number of items}$$

Furthermore, suppose that a larger and more efficient machine has a setup cost of \$800 and a unit cost of \$2 per item. The cost of manufacturing x items using this machine is

$$\text{Cost on larger machine} = 800 + 2x$$

The *break point* is the number of units x that need to be manufactured to make the cost of running either machine the same. The break point can be found by setting the two costs equal to each other and solving for x.

$$\begin{aligned} 600 + 3x &= 800 + 2x \\ x &= 200 \quad \text{Subtract 600 and } 2x \text{ from both sides.} \end{aligned}$$

The break point is 200 units, because the cost using either machine is \$1,200 when $x = 200$.

$$\begin{aligned} \text{Cost on small machine} &= 600 + 3x \\ &= 600 + 3(\mathbf{200}) \\ &= 600 + 600 \\ &= \$1{,}200 \end{aligned} \qquad \begin{aligned} \text{Cost on large machine} &= 800 + 2x \\ &= 800 + 2(\mathbf{200}) \\ &= 800 + 400 \\ &= \$1{,}200 \end{aligned}$$

EXAMPLE 10 **BREAK-POINT ANALYSIS** One machine has a setup cost of \$400 and a unit cost of \$1.50, and another machine has a setup cost of \$500 and a unit cost of \$1.25. Find the break point.

Analyze the problem — The cost of manufacturing x units using machine 1 is x times \$1.50, plus the setup cost of \$400. The cost of manufacturing x units using machine 2 is x times \$1.25, plus the setup cost of \$500. The break point occurs when the costs are equal.

Form two equations The cost C_1 using machine 1 is

The cost of using machine 1	equals	the cost of manufacturing x units at \$1.50 per unit	plus	the setup cost.
C_1	$=$	$1.5x$	$+$	400

The cost C_2 using machine 2 is

The cost of using machine 2	equals	the cost of manufacturing x units at \$1.25 per unit	plus	the setup cost.
C_2	$=$	$1.25x$	$+$	500

Solve the system To find the break point, we must solve the system $\begin{cases} C_1 = 1.5x + 400 \\ C_2 = 1.25x + 500 \end{cases}$. Since the break point occurs when $C_1 = C_2$, we can substitute $1.5x + 400$ for C_2 to obtain

$$1.5x + 400 = 1.25x + 500$$

$$1.5x = 1.25x + 100 \quad \text{Subtract 400 from both sides.}$$

$$0.25x = 100 \quad \text{Subtract } 1.25x \text{ from both sides.}$$

$$x = 400 \quad \text{Divide both sides by 0.25.}$$

State the conclusion The break point is 400 units.

Check the result For 400 units, the cost using machine 1 is $400 + 1.5(400) = 400 + 600 = \$1{,}000$. The cost using machine 2 is $500 + 1.25(400) = 500 + 500 = \$1{,}000$. Since the costs are equal, the break point is 400 units.

SELF CHECK 10 One machine has a setup cost of \$300 and a unit cost of \$1.50 and another machine has a setup cost of \$700 and a unit cost of \$1.25. Find the break point.

Some applications include geometric figures. A *parallelogram* is a four-sided figure with its opposite sides parallel. (See Figure 8-10(a).) Here are some important facts about parallelograms.

1. Opposite sides of a parallelogram have the same measure.
2. Opposite angles of a parallelogram have the same measure.
3. Consecutive angles of a parallelogram are supplementary.
4. A diagonal of a parallelogram (see Figure 8-10(b)) divides the parallelogram into two *congruent triangles*—triangles with the same shape and same area.
5. In Figure 8-10(b), $\angle 1$ and $\angle 2$, and $\angle 3$ and $\angle 4$, are called pairs of *alternate interior angles.* When a diagonal intersects two parallel sides of a parallelogram, all pairs of alternate interior angles have the same measure.

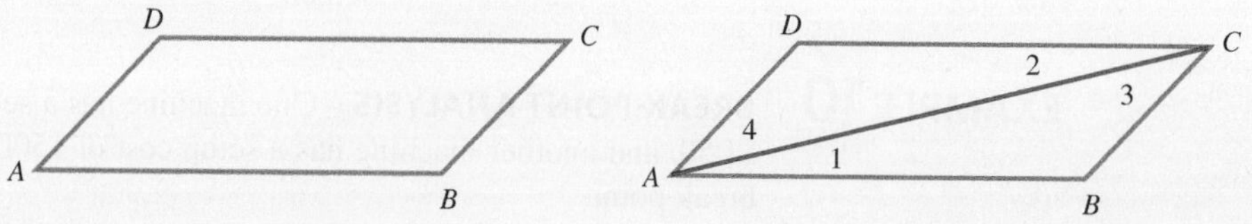

(a) (b)

Figure 8-10

Unless otherwise noted, all content on this page is © Cengage Learning.

EXAMPLE 11 **PARALLELOGRAMS** Refer to the parallelogram shown in Figure 8-11 and find the values of x and y.

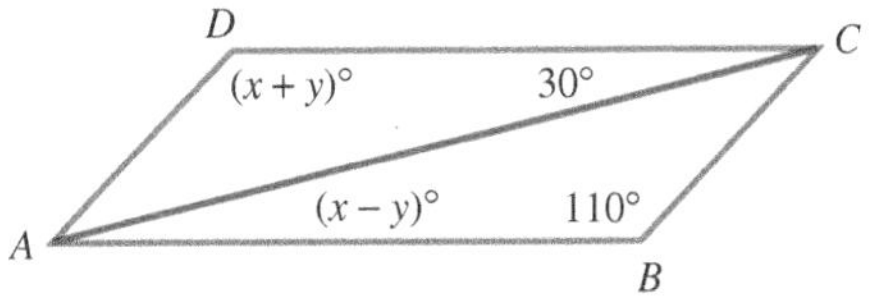

Figure 8-11

Solution Since diagonal AC intersects two parallel sides, the alternate interior angles that are formed have the same measure. Thus, $(x - y)° = 30°$. Since opposite angles of a parallelogram have the same measure, we know that $(x + y)° = 110°$. We can form the following system of equations and solve it by addition.

$$\begin{aligned} &(1) \quad \begin{cases} x - y = 30 \\ x + y = 110 \end{cases} \\ &(2) \end{aligned}$$

$$2x = 140 \quad \text{Add Equations 1 and 2 to eliminate } y.$$

$$x = 70 \quad \text{Divide both sides by 2.}$$

We can substitute 70 for x in Equation 2 and solve for y.

$$x + y = 110$$

$$70 + y = 110 \quad \text{Substitute 70 for } x.$$

$$y = 40 \quad \text{Subtract 70 from both sides.}$$

Thus, $x = 70$ and $y = 40$.

SELF CHECK 11 If the angle at B is 112° and the other angle symbol (DCA) is 42°, find the value of x and y.

SELF CHECK ANSWERS

1. $(-3, 4)$ **2.** $(3, -2)$ **3.** $(-2, 3)$ **4.** $(-3, 2)$ **5.** $\varnothing$ **6.** $\left(x, -\frac{2}{5}x + 2\right)$ **7.** $\frac{41}{110}$ **8.** The store sold 25 of the \$67 phones and 11 of the \$100 phones. **9.** There should be 20 ounces of 15% saline solution and 30 ounces of the 10% saline solution. **10.** The break point is 1,600 units. **11.** $x = 77$ and $y = 35$

NOW TRY THIS

Use substitution to solve the system: $\begin{cases} t = r + 1 \\ r + s + t = 0 \\ r + s = -2 \end{cases}$

8.2 Exercises

WARM-UPS *Substitute $x - 2$ for y and solve for x.*

1. $3x - 4y = 7$

2. $5x + 3y = 18$

Add the left and right sides of the equations in each system and write the result.

3. $\begin{cases} 4x - 2y = 6 \\ 3x + 2y = 8 \end{cases}$

4. $\begin{cases} 5x + 3y = 10 \\ -5x - 7y = -8 \end{cases}$

Unless otherwise noted, all content on this page is © Cengage Learning.

REVIEW *Simplify each expression. Write all answers without using negative exponents.*

5. $(a^3b^4)^2(ab^2)^3$

6. $\left(\dfrac{a^2b^3c^4d}{ab^2c^3d^4}\right)^{-3}$

7. $\left(\dfrac{-3x^3y^4}{x^{-5}y^3}\right)^{-4}$

8. $\dfrac{5t^0 - 2t^0 + 3}{4t^0 + t^0}$

VOCABULARY AND CONCEPTS *Fill in the blanks.*

9. Running a machine involves both ____ costs and ___ costs.

10. The _____ point is the number of units that need to be manufactured to make the cost the same on either of two machines.

11. A ___________ is a four-sided figure with both pairs of opposite sides parallel.

12. _______ sides of a parallelogram have the same length.

13. _______ angles of a parallelogram have the same measure.

14. _________ angles of a parallelogram are supplementary.

GUIDED PRACTICE *Solve each system by substitution.* SEE EXAMPLE 1. (OBJECTIVE 1)

15. $\begin{cases} y = 3x \\ x + y = 8 \end{cases}$

16. $\begin{cases} y = x + 2 \\ x + 2y = 16 \end{cases}$

17. $\begin{cases} x - y = 2 \\ 2x + y = 13 \end{cases}$

18. $\begin{cases} x - y = -4 \\ 3x - 2y = -5 \end{cases}$

Solve each system by substitution. SEE EXAMPLE 2. (OBJECTIVE 1)

19. $\begin{cases} \dfrac{x}{2} + \dfrac{y}{2} = 6 \\ \dfrac{x}{2} - \dfrac{y}{2} = -2 \end{cases}$

20. $\begin{cases} \dfrac{x}{2} - \dfrac{y}{3} = -4 \\ \dfrac{x}{2} + \dfrac{y}{9} = 0 \end{cases}$

21. $\begin{cases} x = \dfrac{2}{3}y \\ y = 4x + 5 \end{cases}$

22. $\begin{cases} \dfrac{3x}{5} + \dfrac{5y}{3} = 2 \\ \dfrac{6x}{5} - \dfrac{5y}{3} = 1 \end{cases}$

Solve each system by elimination (addition). SEE EXAMPLE 3. (OBJECTIVE 2)

23. $\begin{cases} 2x - y = -5 \\ 2x + y = -3 \end{cases}$

24. $\begin{cases} x + 3y = 7 \\ x - 3y = -11 \end{cases}$

25. $\begin{cases} 3x + 5y = 2 \\ 4x - 3y = 22 \end{cases}$

26. $\begin{cases} 4x - 3y = 9 \\ 3x + 2y = 11 \end{cases}$

27. $\begin{cases} 5x - 2y = 19 \\ 3x + 4y = 1 \end{cases}$

28. $\begin{cases} 2y - 3x = -13 \\ 3x - 17 = 4y \end{cases}$

29. $\begin{cases} 4x + 6y = 5 \\ 8x - 9y = 3 \end{cases}$

30. $\begin{cases} 4x + 9y = 8 \\ 2x - 6y = -3 \end{cases}$

Solve each system by elimination (addition). SEE EXAMPLE 4. (OBJECTIVE 2)

31. $\begin{cases} \frac{5}{6}x + \frac{2}{3}y = \frac{7}{6} \\ \frac{10}{7}x - \frac{4}{9}y = \frac{17}{21} \end{cases}$

32. $\begin{cases} \frac{3x}{2} - \frac{2y}{3} = 0 \\ \frac{3x}{4} + \frac{4y}{3} = \frac{5}{2} \end{cases}$

33. $\begin{cases} \dfrac{3}{4}x + \dfrac{2}{3}y = 7 \\ \dfrac{3}{5}x - \dfrac{1}{2}y = 18 \end{cases}$

34. $\begin{cases} \dfrac{2}{3}x - \dfrac{1}{4}y = -8 \\ \dfrac{1}{2}x - \dfrac{3}{8}y = -9 \end{cases}$

Solve each system using any method. SEE EXAMPLE 5. (OBJECTIVE 3)

35. $\begin{cases} 3x = 2y - 4 \\ 6x - 4y = -4 \end{cases}$

36. $\begin{cases} x = \dfrac{3}{2}y + 5 \\ 2x - 3y = 8 \end{cases}$

37. $\begin{cases} x = 5y + 2 \\ 3x = 15y + 10 \end{cases}$

38. $\begin{cases} 3x - 6y = 8 \\ -6x + 12y = -10 \end{cases}$

Solve each system using any method. SEE EXAMPLE 6. (OBJECTIVE 4)

39. $\begin{cases} 12x = 4y + 6 \\ 6x - 2y = 3 \end{cases}$

40. $\begin{cases} 8x - 4y = 16 \\ 2x - 4 = y \end{cases}$

41. $\begin{cases} y - 2x = 6 \\ 4x + 12 = 2y \end{cases}$

42. $\begin{cases} 6x - 2y = 7 \\ 12x - 4y = 14 \end{cases}$

Write each repeating decimal as a fraction. Simplify the answer when possible. SEE EXAMPLE 7. (OBJECTIVE 5)

43. $0.\overline{6}$

44. $0.\overline{29}$

45. $-0.34\overline{89}$

46. $-2.3\overline{47}$

ADDITIONAL PRACTICE *Solve each system using any method.*

47. $\begin{cases} y = 2x \\ x + y = 6 \end{cases}$

48. $\begin{cases} y = -x \\ 2x + y = 4 \end{cases}$

49. $\begin{cases} 3x - 4y = 9 \\ x + 2y = 8 \end{cases}$

50. $\begin{cases} 3x - 2y = -10 \\ 6x + 5y = 25 \end{cases}$

51. $\begin{cases} x - y = 6 \\ x + y = 2 \end{cases}$

52. $\begin{cases} 4x + 2y = 4 \\ 2x + y = 5 \end{cases}$

53. $\begin{cases} 2x + 2y = -1 \\ 3x + 4y = 0 \end{cases}$

54. $\begin{cases} 5x + 3y = -7 \\ 3x - 3y = 7 \end{cases}$

55. $\begin{cases} \dfrac{2}{5}x - \dfrac{1}{6}y = \dfrac{7}{10} \\ \dfrac{3}{4}x - \dfrac{2}{3}y = \dfrac{19}{8} \end{cases}$

56. $\begin{cases} \dfrac{5}{6}x - y = \dfrac{-13}{6} \\ \dfrac{3}{2}x + y = \dfrac{-19}{2} \end{cases}$

57. $\begin{cases} 3x + 5y = -14 \\ 2x - 3y = 16 \end{cases}$

58. $\begin{cases} 2x + 3y = 8 \\ 6y + 4x = 16 \end{cases}$

Solve each system.

59. $\begin{cases} \dfrac{1}{x} + \dfrac{1}{y} = \dfrac{5}{6} \\ \dfrac{1}{x} - \dfrac{1}{y} = \dfrac{1}{6} \end{cases}$

60. $\begin{cases} \dfrac{1}{x} + \dfrac{1}{y} = \dfrac{9}{20} \\ \dfrac{1}{x} - \dfrac{1}{y} = \dfrac{1}{20} \end{cases}$

61. $\begin{cases} \dfrac{1}{x} + \dfrac{2}{y} = -1 \\ \dfrac{2}{x} - \dfrac{1}{y} = -7 \end{cases}$

62. $\begin{cases} \dfrac{3}{x} - \dfrac{2}{y} = -30 \\ \dfrac{2}{x} - \dfrac{3}{y} = -30 \end{cases}$

APPLICATIONS *Use a system of two linear equations to solve.*
SEE EXAMPLES 8–11. (OBJECTIVE 6)

63. Merchandising A pair of shoes and a sweater cost \$110. If the sweater cost \$20 more than the shoes, how much did the sweater cost?

64. Merchandising A sporting goods salesperson sells 2 fishing reels and 5 rods for \$270. The next day, the salesperson sells 4 reels and 2 rods for \$220. How much does each cost?

65. Electronics Two resistors in the voltage divider circuit in the illustration have a total resistance of 1,375 ohms. To provide the required voltage, R_1 must be 125 ohms greater than R_2. Find both resistances.

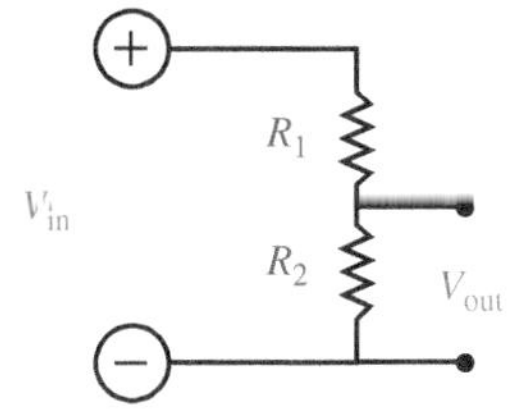

66. Stowing baggage A small aircraft can carry 950 pounds of baggage, distributed between two storage compartments. On one flight, the plane is fully loaded, with 150 pounds more baggage in one compartment than the other. How much is stowed in each compartment?

67. Geometry The rectangular field in the illustration is surrounded by 72 meters of fencing. If the field is partitioned as shown, a total of 88 meters of fencing is required. Find the dimensions of the field.

68. Geometry In a right triangle, one acute angle is 15° greater than two times the other acute angle. Find the difference between the angles.

69. Investment income Part of \$8,000 was invested at 10% interest and the rest at 12%. If the annual income from these investments was \$900, how much was invested at each rate?

70. Investment income Part of \$12,000 was invested at 6% interest and the rest at 7.5%. If the annual income from these investments was \$810, how much was invested at each rate?

71. Mixing solutions How many ounces of the two alcohol solutions in the illustration must be mixed to obtain 100 ounces of a 12.2% solution?

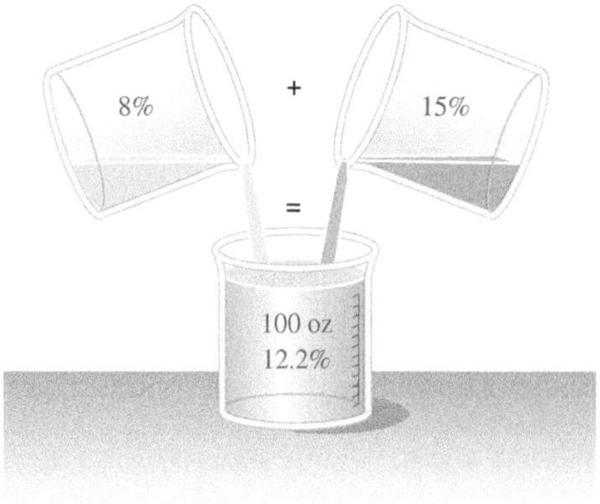

72. Mixing candy How many pounds each of candy shown in the illustration must be mixed to obtain 60 pounds of candy that is worth \$3 per pound?

73. Travel A car travels 50 miles in the same time that a plane travels 180 miles. The speed of the plane is 143 mph faster than the speed of the car. Find the speed of the car.

74. Travel A car and a truck leave Rockford at the same time, heading in opposite directions. When they are 350 miles apart, the car has gone 70 miles farther than the truck. How far has the car traveled?

75. Making bicycles A bicycle manufacturer builds racing bikes and mountain bikes, with the per-unit manufacturing costs shown in the table. The company has budgeted \$15,900 for labor and \$13,075 for materials. How many bicycles of each type can be built?

Model	Cost of materials	Cost of labor
Racing	\$55	\$60
Mountain	\$70	\$90

76. Farming A farmer keeps some animals on a strict diet. Each animal is to receive 15 grams of protein and 7.5 grams of carbohydrates. The farmer uses two food mixes with nutrients shown in the illustration. How many grams of each mix should be used to provide the correct nutrients for each animal?

Mix	Protein	Carbohydrates
Mix A	12%	9%
Mix B	15%	5%

Unless otherwise noted, all content on this page is © Cengage Learning.

77. Milling brass plates Two machines can mill a brass plate. One machine has a setup cost of \$300 and a cost per plate of \$2. The other machine has a setup cost of \$500 and a cost per plate of \$1. Find the break point.

78. Printing books A printer has two presses. One has a setup cost of \$210 and can print the pages of a certain book for \$5.98. The other press has a setup cost of \$350 and can print the pages of the same book for \$5.95. Find the break point.

79. Managing a computer store The manager of a computer store knows that his fixed costs are \$8,925 per month and that his unit cost is \$850 for every computer sold. If he can sell all the computers he can get for \$1,275 each, how many computers must he sell each month to break even?

80. Managing a makeup studio A makeup studio specializing in photo shoots has fixed costs of \$23,600 per month. The owner estimates that the cost for each indoor shoot is \$910. This cost covers labor and makeup. If her studio can book as many indoor shoots as she wants at a price of \$1500 each, how many shoots will it take each month to break even?

81. Running a small business A person invests \$18,375 to set up a small business that produces a piece of computer software that will sell for \$29.95. If each piece can be produced for \$5.45, how many pieces must be sold to break even?

82. Running a record company Three people invest \$35,000 each to start a record company that will produce reissues of classic jazz. Each release will be a set of 3 CDs that will retail for \$15 per disc. If each set can be produced for \$18.95, how many sets must be sold for the investors to make a profit?

Break-point analysis *A paint manufacturer can choose between two processes for manufacturing house paint, with monthly costs shown in the table. Assume that the paint sells for \$18 per gallon.*

Process	Fixed costs	Unit cost (per gallon)
A	\$32,500	\$13
B	\$80,600	\$ 5

83. For process A, how many gallons must be sold for the manufacturer to break even?

84. For process B, how many gallons must be sold for the manufacturer to break even?

85. If expected sales are 6,000 gallons per month, which process should the company use?

86. If expected sales are 7,000 gallons per month, which process should the company use?

Making water pumps *A manufacturer of automobile water pumps is considering retooling for one of two manufacturing processes, with monthly fixed costs and unit costs as indicated in the table. Each water pump can be sold for \$50.*

Process	Fixed costs	Unit cost (per gallon)
A	\$12,390	\$29
B	\$20,460	\$17

87. For process A, how many water pumps must be sold for the manufacturer to break even?

88. For process B, how many water pumps must be sold for the manufacturer to break even?

89. If expected sales are 550 per month, which process should be used?

90. If expected sales are 600 per month, which process should be used?

91. If expected sales are 650 per month, which process should be used?

92. At what monthly sales level is process B better?

93. Geometry If two angles are supplementary, their sum is 180°. If the difference between two supplementary angles is 110°, find the measure of each angle.

94. Geometry If two angles are complementary, their sum is 90°. If one of two complementary angles is 16° greater than the other, find the measure of each angle.

95. Find the values of x and y in the parallelogram.

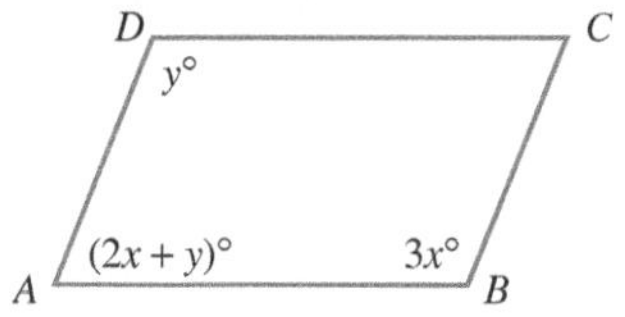

96. Find the values of x and y in the parallelogram.

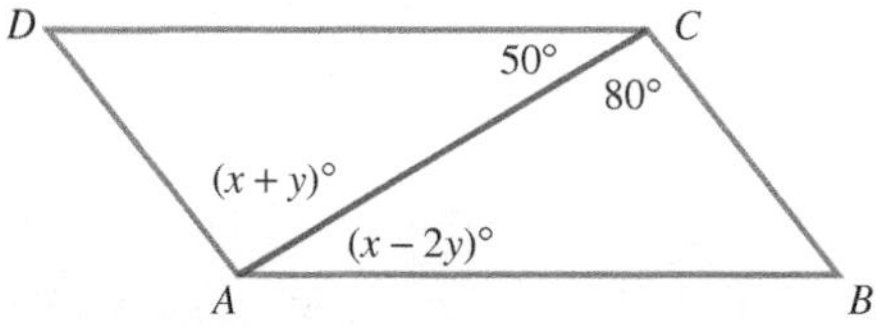

97. Physical therapy To rehabilitate her knee, an athlete does leg extensions. Her goal is to regain a full 90° range of motion in this exercise. Use the information in the illustration to determine her current range of motion in degrees.

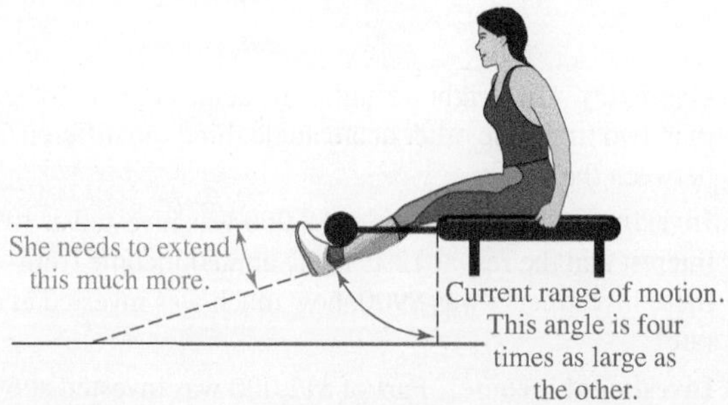

98. The Marine Corps The Marine Corps War Memorial in Arlington, Virginia, portrays the raising of the U.S. flag on Iwo Jima during World War II. Find the measure of the two

Unless otherwise noted, all content on this page is © Cengage Learning.

angles shown in the illustration if the measure of one of the angles is 15° less than twice the other.

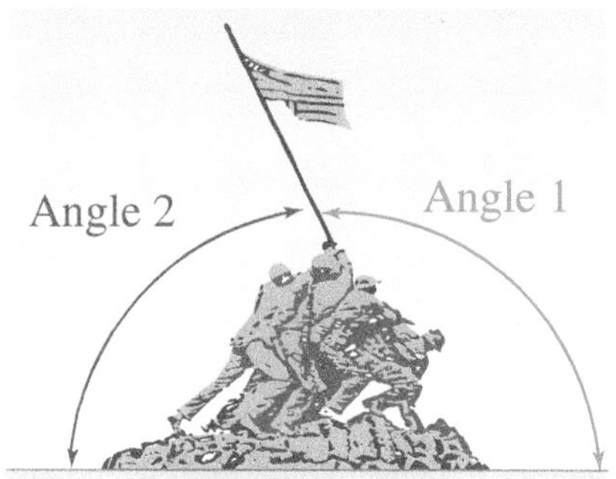

99. Radio frequencies In a radio, an inductor and a capacitor are used in a resonant circuit to select a wanted radio station at a frequency f and reject all others. The inductance L and the capacitance C determine the inductance reactance X_L and the capacitive reactance X_C of that circuit, where

$$X_L = 2\pi fL \qquad \text{and} \qquad X_C = \frac{1}{2\pi fC}$$

The radio station selected will be at the frequency f where $X_L = X_C$. Write a formula for f^2 in terms of L and C.

100. Choosing salary plans A sales clerk can choose from two salary options:

1. a straight 7% commission
2. \$150 + 2% commission

How much would the clerk have to sell for each plan to produce the same monthly paycheck?

WRITING ABOUT MATH

101. Which method would you use to solve the following system? Why?

$$\begin{cases} y = 3x + 1 \\ 3x + 2y = 12 \end{cases}$$

102. Which method would you use to solve the following system? Why?

$$\begin{cases} 2x + 4y = 9 \\ 3x - 5y = 20 \end{cases}$$

SOMETHING TO THINK ABOUT

103. Under what conditions will a system of two equations in two variables be inconsistent?

104. Under what conditions will the equations of a system of two equations in two variables be dependent?

Section 8.3 Solving Applications of Systems of Linear Equations in Two Variables

Objective

1 Solve an application using a system of linear equations.

Getting Ready

Let x and y represent two numbers. Use an algebraic expression to denote each phrase.

1. The sum of x and y
2. The difference when y is subtracted from x
3. The product of x and y
4. The quotient x divided by y
5. Give the formula for the area of a rectangle.
6. Give the formula for the perimeter of a rectangle.

We have previously set up equations involving one variable to solve applications. In this section, we consider ways to solve word problems by using equations in two variables.

Unless otherwise noted, all content on this page is © Cengage Learning.

1 Solve an application using a system of linear equations.

The following steps are helpful when solving applications involving two unknown quantities.

PROBLEM SOLVING

1. Read the problem and *analyze* the facts. Identify the unknowns by asking yourself "What am I asked to find?"
   a. Select different variables to represent two unknown quantities.
   b. Write a sentence to define each variable.
2. Form two equations involving each of the two variables. This will create a system of two equations in two variables. (This may require reading the problem several times to understand the given facts. What information is given? Is there a formula that applies to this situation? Will a sketch, chart, or diagram help you visualize the facts of the problem?)
3. Solve the system using the most convenient method: graphing, substitution, or elimination.
4. State the conclusion as a sentence.
5. Check the solution in the words of the problem.

EXAMPLE 1 **FARMING** A farmer raises wheat and soybeans on 215 acres. If he wants to plant 31 more acres in wheat than in soybeans, how many acres of each should he plant?

Analyze the problem The farmer plants two fields, one in wheat and one in soybeans. We are asked to find how many acres of each he should plant. So, we let w represent the number of acres of wheat and s represent the number of acres of soybeans.

Form two equations We know that the *number of acres* of wheat planted plus the *number of acres* of soybeans planted will equal a total of 215 *acres*. So we can form the equation

The number of acres planted in wheat	plus	the number of acres planted in soybeans	equals	the total number of acres.
w	$+$	s	$=$	215

Since the farmer wants to plant 31 more acres in wheat than in soybeans, we can form the equation

The number of acres planted in wheat	minus	the number of acres planted in soybeans	equals	31 acres.
w	$-$	s	$=$	31

Solve the system We can now solve the system

$$\begin{aligned} &(1) \quad \begin{cases} w + s = 215 \\ w - s = 31 \end{cases} \\ &(2) \end{aligned}$$

by the elimination method.

$$\begin{aligned} w + s &= 215 \\ w - s &= \ \ 31 \\ \hline 2w \quad &= 246 \\ w \quad &= 123 \quad \text{Divide both sides by 2.} \end{aligned}$$

To find s, we substitute 123 for w in Equation 1.

$$\begin{aligned} w + s &= 215 \\ \mathbf{123} + s &= 215 \quad \text{Substitute 123 for } w. \\ s &= 92 \quad \text{Subtract 123 from both sides.} \end{aligned}$$

State the conclusion The farmer should plant 123 acres of wheat and 92 acres of soybeans.

Check the result The total acreage planted is $123 + 92$, or 215 acres. The area planted in wheat is 31 acres greater than that planted in soybeans, because $123 - 92 = 31$. The answers check.

SELF CHECK 1 A farmer raises wheat and soybeans on 163 acres. If he wants to plant 29 more acres in wheat than in soybeans, how many acres of each should he plant?

EXAMPLE 2

LAWN CARE An installer of underground irrigation systems wants to cut a 20-foot length of plastic tubing into two pieces. The longer piece is to be 2 feet longer than twice the shorter piece. Find the length of each piece.

Analyze the problem Refer to Figure 8-12, which shows the pipe. We need to find the length of each pipe, so we let s represent the *length* of the shorter piece in feet and l represent the *length* of the longer piece in feet.

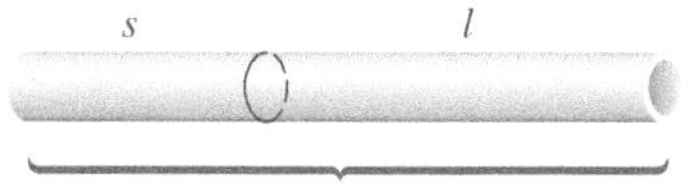

Figure 8-12

Form two equations Since the length of the plastic tube is 20 feet, we can form the equation

The length of the shorter piece	plus	the length of the longer piece	equals	the total length of the pipe.
s	$+$	l	$=$	20

Since the longer piece is 2 feet longer than twice the shorter piece, we can form the equation

The length of the longer piece	equals	2	times	the length of the shorter piece	plus	2 feet.
l	$=$	2	$\cdot$	s	$+$	2

Solve the system We can use the substitution method to solve the system

$$\begin{cases} s + l = 20 \\ l = 2s + 2 \end{cases}$$

$$\begin{aligned} s + (2s + 2) &= 20 && \text{Substitute } 2s + 2 \text{ for } l \text{ in the first equation.} \\ 3s + 2 &= 20 && \text{Remove parentheses and combine like terms.} \\ 3s &= 18 && \text{Subtract 2 from both sides.} \\ s &= 6 && \text{Divide both sides by 3.} \end{aligned}$$

State the conclusion The shorter piece should be 6 feet long. To find the length of the longer piece, we substitute 6 for s in the first equation and solve for l.

$$\begin{aligned} s + l &= 20 \\ 6 + l &= 20 && \text{Substitute.} \\ l &= 14 && \text{Subtract 6 from both sides.} \end{aligned}$$

The longer piece should be 14 feet long.

Check the result The sum of 6 and 14 is 20 and 14 is 2 more than twice 6. The answers check.

Unless otherwise noted, all content on this page is © Cengage Learning.

SELF CHECK 2 If the installer has a 24-foot length of plastic tubing, find the length of each piece if the longer piece is 3 feet longer than twice the shorter piece.

EXAMPLE 3 **GARDENING** Tom has 150 feet of fencing to enclose a rectangular garden. If the length is to be 5 feet less than 3 times the width, find the area of the garden.

Analyze the problem To find the area of a rectangle, we need to know its length and width, so we can let l represent the length of the garden in feet and w represent the width in feet. See Figure 8-13.

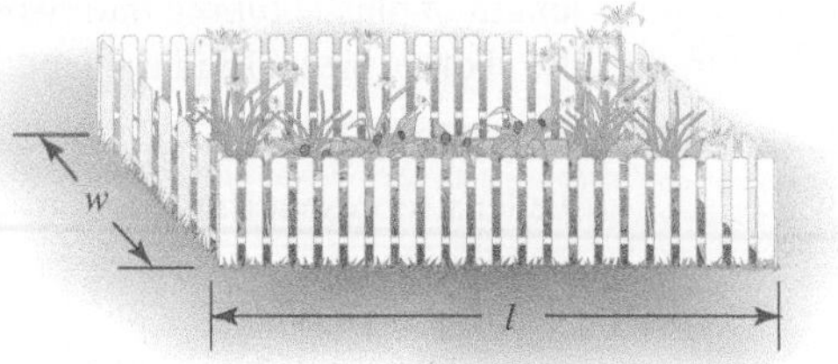

Figure 8-13

Form two equations Since the perimeter of the rectangle is 150 feet, and this is two lengths plus two widths, we can form the equation:

2	times	the length of the garden	plus	2	times	the width of the garden	equals	the perimeter of the garden.
2	$\cdot$	l	+	2	$\cdot$	w	=	150

Since the length is 5 feet less than 3 times the width, we can form the equation

The length of the garden	equals	3	times	the width of the garden	minus	5 feet.
l	=	3	$\cdot$	w	−	5

Solve the system Because one variable is already isolated, we can use the substitution method to solve this system.

$$\begin{cases} 2l + 2w = 150 \\ l = 3w - 5 \end{cases}$$

$$\begin{aligned} 2(3w - 5) + 2w &= 150 && \text{Substitute } 3w - 5 \text{ for } l \text{ in the first equation.} \\ 6w - 10 + 2w &= 150 && \text{Use the distributive property.} \\ 8w - 10 &= 150 && \text{Combine like terms.} \\ 8w &= 160 && \text{Add 10 to both sides.} \\ w &= 20 && \text{Divide both sides by 8.} \end{aligned}$$

The width of the garden is 20 feet. To find the length, we substitute 20 for w in the second equation and simplify.

$$\begin{aligned} l &= 3w - 5 \\ &= 3(20) - 5 && \text{Substitute 20 for } w. \\ &= 60 - 5 \\ &= 55 \end{aligned}$$

The length of the garden is 55 feet.

Unless otherwise noted, all content on this page is © Cengage Learning.

Although we have found the length and the width of the garden, we are asked to find the area. Since the dimensions of the rectangle are 55 feet by 20 feet, and the area of a rectangle is given by the formula

$$A = l \cdot w \quad \text{Area = length times width}$$

we have

$$A = 55 \cdot 20$$
$$= 1{,}100$$

State the conclusion The garden covers an area of 1,100 square feet.

Check the result Because the dimensions of the garden are 55 feet by 20 feet, the perimeter is

$$P = 2l + 2w$$
$$= 2(\mathbf{55}) + 2(\mathbf{20}) \quad \text{Substitute for } l \text{ and } w.$$
$$= 110 + 40$$
$$= 150$$

It is also true that 55 feet is 5 feet less than 3 times 20 feet. The answers check.

SELF CHECK 3 Tom has 160 feet of fencing to enclose a rectangular garden. If the length is to be 5 feet more than twice the width, find the area of the garden.

Everyday connections
Paralympic Medals

The International Olympic Committee sets the specifications for all medals awarded in the Olympics and Paralympics. They must be at least 60 millimeters in diameter and at least 3 millimeters thick. Gold medals must be 92.5% pure silver and plated with at least 6 grams of gold. The gold medal for the 2010 Vancouver Winter Paralympics was 100 millimeters in diameter, 6 millimeters thick, and because each medal was unique, each weighed between 500 and 576 grams.

Photo by Adrian Pang, available under a Creative Commons Attribution license.

When the medal was awarded in early 2010, 6 grams of gold and 481 grams of silver were worth $478.75. In late 2011, when the price of silver had doubled and gold increased by 150%, the same medal was worth $850.40.

1. What was the price of one gram of gold and one gram of silver in 2010?

2. What was the price of one gram of gold and one gram of silver in 2011?

Source: http://www.olympic.org/Documents/Reference_documents_Factsheets/Winter_Games_Medals_FACTSHEET_EN.pdf

EXAMPLE 4 **INVESTING** Terri and Juan earned $650 in interest from a one-year investment of $15,000. If Terri invested some of the money at 4% annual interest and Juan invested the rest at 5% annual interest, how much did each invest?

Analyze the problem We are asked to find how much money Terri and Juan invested. We can let x represent the amount invested by Terri and y represent the amount of money invested by Juan. We are told that Terri invested an unknown part of the $15,000 at 4% interest and Juan invested the rest at 5% interest. Together, these investments earned $650 in interest.

Form two equations Because the total investment is $15,000, we have

The amount invested by Terri	plus	the amount invested by Juan	equals	the total amount invested.
x	$+$	y	$=$	15,000

Since the interest on x dollars invested at 4% is $0.04x$, the interest on y dollars invested at 5% is $0.05y$, and the combined income is $650, we have

The interest on the 4% investment	plus	the interest on the 5% investment	equals	the total interest earned.
$0.04x$	$+$	$0.05y$	$=$	650

Thus, we have the system

$$\begin{aligned} &(1) \quad \begin{cases} x + y = 15{,}000 \\ 0.04x + 0.05y = 650 \end{cases} \\ &(2) \end{aligned}$$

Solve the system To solve the system, we use the elimination method.

$$\begin{aligned} -4x - 4y &= -60{,}000 && \text{Multiply both sides of Equation 1 by } -4. \\ 4x + 5y &= 65{,}000 && \text{Multiply both sides of Equation 2 by 100 to clear the decimals.} \\ \hline y &= 5{,}000 && \text{Add the equations together.} \end{aligned}$$

To find the value of x, we substitute 5,000 for y in Equation 1 and simplify.

$$\begin{aligned} x + y &= 15{,}000 \\ x + \mathbf{5{,}000} &= 15{,}000 && \text{Substitute 5,000 for } y. \\ x &= 10{,}000 && \text{Subtract 5,000 from both sides.} \end{aligned}$$

State the conclusion Terri invested $10,000, and Juan invested $5,000.

Check the result

$$\begin{aligned} \$10{,}000 + \$5{,}000 &= \$15{,}000 && \text{The two investments total \$15,000.} \\ 0.04(\$10{,}000) &= \$400 && \text{Terri earned \$400.} \\ 0.05(\$5{,}000) &= \$250 && \text{Juan earned \$250.} \end{aligned}$$

The combined interest is $400 + $250 = $650. The answers check.

SELF CHECK 4 Terri and Juan earned $800 from a one-year investment of $20,000. If Terri invested some of the money at 3% and Juan invested the rest at 5%, how much did each invest? How much interest did each earn?

All of the previous examples could be solved using one equation in one variable as you did in Chapter 2. In Examples 5–7, using two variables to create a system of linear equations will be more efficient.

EXAMPLE 5 **BOATING** A boat traveled 30 kilometers downstream in 3 hours and made the return trip in 5 hours. Find the speed of the boat in still water.

Analyze the problem We are asked to find the speed of the boat, so we let s represent the speed of the boat in km/hr in still water. Recall from earlier problems that when traveling upstream or downstream, the current affects that speed. Therefore, we let c represent the speed of the current, in km/hr.

Form two equations Traveling downstream, the rate of the boat will be the speed of the boat in still water, s, in km/hr plus the speed of the current, c, in km/hr. Thus, the rate of the boat going downstream is $(s + c)$.

Traveling upstream, the rate of the boat will be the speed of the boat in still water, s, minus the speed of the current, c. Thus, the rate of the boat going upstream is $(s - c)$.

We can organize the information of the problem as in Table 8-1.

TABLE 8-1

	Distance	=	Rate	·	Time
Downstream	30		$s + c$		3
Upstream	30		$s - c$		5

Because $d = r \cdot t$, the information in the table produces two equations in two variables.

$$\begin{cases} 30 = 3(s + c) \\ 30 = 5(s - c) \end{cases}$$

After distributing and writing the equations in $ax + by = c$ form, we have

$$\begin{cases} 3s + 3c = 30 \\ 5s - 5c = 30 \end{cases}$$

Solve the system

To solve this system by elimination, we multiply the first equation by 5, the second equation by 3, add the equations, and solve for s.

$$\begin{aligned} 15s + 15c &= 150 \\ 15s - 15c &= 90 \\ \hline 30s \quad &= 240 \\ s &= 8 \quad \text{Divide both sides by 30.} \end{aligned}$$

State the conclusion

The speed of the boat in still water is 8 kilometers per hour.

Check the result

If $s = 8$,

$$\begin{aligned} 3(8) + 3c &= 30 \\ 24 + 3c &= 30 \\ 3c &= 6 \\ c &= 2 \end{aligned}$$

The speed of the current is 2 kilometers per hour. In 3 hours, the boat will travel downstream $3(8 + 2) = 30$ kilometers. In 5 hours, the boat will return upstream $5(8 - 2) = 30$ kilometers. The answers check.

SELF CHECK 5 A boat traveled 36 kilometers downstream in 3 hours and made the return trip in 4 hours. Find the speed of the boat in still water.

EXAMPLE 6

MEDICAL TECHNOLOGY A laboratory technician has one batch of antiseptic that is 40% alcohol and a second batch that is 60% alcohol. She needs to make 8 liters of solution that is 55% alcohol. How many liters of each batch should she use?

Analyze the problem

We need to know how many liters of each type of alcohol she should use, so we can let x represent the number of liters to be used from batch 1 and let y represent the number of liters to be used from batch 2.

Form two equations

Some 60% alcohol solution must be added to some 40% alcohol solution to make a 55% alcohol solution. We can organize the information of the problem as in Table 8-2 on page 574.

TABLE 8-2

	Fractional part that is alcohol	·	Number of liters of solution	=	Number of liters of alcohol
Batch 1	0.40		x		$0.40x$
Batch 2	0.60		y		$0.60y$
Mixture	0.55		8		$0.55(8)$

The information in Table 8-2 provides the system.

$$\begin{cases} x + y = 8 \\ 0.40x + 0.60y = 0.55(8) \end{cases}$$

The *number of liters* of batch 1 plus the *number of liters* of batch 2 equals the *total number of liters* in the mixture.

The *amount of alcohol* in batch 1 plus the *amount of alcohol* in batch 2 equals the total *amount of alcohol* in the mixture.

Solve the system We can use substitution to solve this system. Solve the first equation for x.

$$x = -y + 8$$

$0.40(-y + 8) + 0.60y = 4.4$ Substitute the expression for x in the second equation and $0.55(8) = 4.4$.

$-0.40y + 3.2 + 0.60y = 4.4$ Use the distributive property.

$0.20y + 3.2 = 4.4$ Combine like terms.

$0.20y = 1.2$ Subtract 3.2 from both sides.

$y = 6$ Divide both sides by 0.20.

To find the value of x, we substitute 6 for y in the first equation and simplify:

$x + y = 8$

$x + \mathbf{6} = 8$ Substitute 6 for y.

$x = 2$ Subtract 6 from both sides.

State the conclusion The technician should use 2 liters of the 40% alcohol solution and 6 liters of the 60% alcohol solution.

Check the result Mixing 2 liters of batch 1 with 6 liters of batch 2 produces 8 liters.

The amount of alcohol in each is $0.40(2) + 0.60(6) = 0.55(8)$.

$$0.8 + 3.6 = 4.40$$

$$4.4 = 4.4$$

The answers check.

SELF CHECK 6 What if the technician wanted 10 liters of solution that is 46% alcohol. How many liters of each batch should she use?

EXAMPLE 7 **MANUFACTURING** The set-up cost of a machine that mills brass plates is $750. After set-up, it costs $0.25 to mill each plate. Management is considering the use of a larger machine that can produce the same plates at a cost of $0.20 per plate. If the set-up cost of the larger machine is $1,200, how many plates would the company have to produce to make the switch worthwhile?

Analyze the problem We can let p represent the number of brass plates produced. Then, we will let c represent the total cost of milling p plates (set-up cost plus cost per plate).

Form two equations To determine whether the switch is worthwhile, we need to know if the larger machine can produce the plates cheaper than the old machine and if so, when that occurs. We begin by finding the number of plates that will cost the same to produce on either machine (called the *break-even point*).

If we call the machine currently being used machine 1, and the larger machine 2, we can form the two equations

The cost of making p plates on machine 1	equals	the set-up cost of machine 1	plus	the cost per plate on machine 1	times	the number of plates p to be made.
c_1	$=$	750	$+$	0.25	$\cdot$	p

The cost of making p plates on machine 2	equals	the set-up cost of machine 2	plus	the cost per plate on machine 2	times	the number of plates p to be made.
c_2	$=$	1,200	$+$	0.20	$\cdot$	p

Solve the system Since the costs at the break-even point are equal $(c_1 = c_2)$, we can use the substitution method to solve the system

$$\begin{cases} c_1 = 750 + 0.25p \\ c_2 = 1{,}200 + 0.20p \end{cases}$$

$$750 + 0.25p = 1{,}200 + 0.20p \quad \text{Substitute } 750 + 0.25p \text{ for } c_2 \text{ in the second equation.}$$

$$0.25p = 450 + 0.20p \quad \text{Subtract 750 from both sides.}$$

$$0.05p = 450 \quad \text{Subtract } 0.20p \text{ from both sides.}$$

$$p = 9{,}000 \quad \text{Divide both sides by 0.05.}$$

State the conclusion If 9,000 plates are milled, the cost will be the same on either machine. If more than 9,000 plates are milled, the cost will be cheaper on the larger machine, because it mills the plates less expensively than the smaller machine.

Check the solution Figure 8-14 verifies that the break-even point occurs when 9,000 plates are produced. It also interprets the solution graphically.

Current machine
$c_1 = 750 + 0.25p$

p	c
0	750
1,000	1,000
5,000	2,000

Larger machine
$c_2 = 1{,}200 + 0.20p$

p	c
0	1,200
4,000	2,000
12,000	3,600

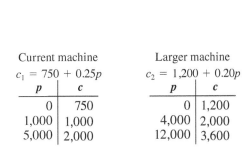

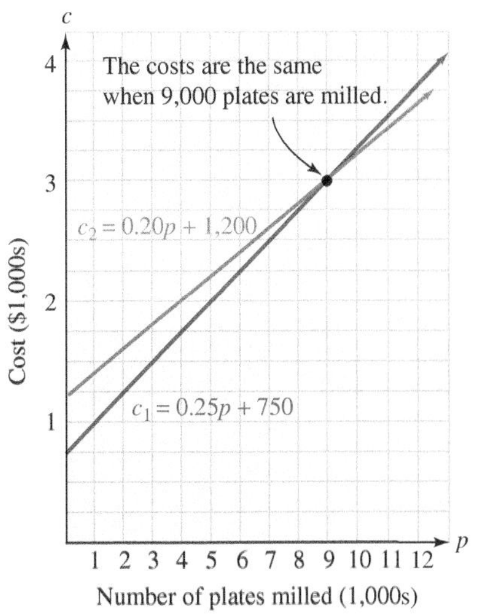

Figure 8-14

SELF CHECK 7 If the set-up cost of the larger machine is $1,500, how many plates would the company have to produce to make the switch worthwhile?

Unless otherwise noted, all content on this page is © Cengage Learning.

SELF CHECK ANSWERS

1. The farmer should plant 96 acres of wheat and 67 acres of soybeans. **2.** The shorter piece should be 7 feet long. The longer piece should be 17 feet long. **3.** The garden covers an area of 1,375 square feet. **4.** Terri and Juan each invested \$10,000. Terri earned \$300 in interest while Juan earned \$500. **5.** The speed of the boat in still water is 10.5 km/hr. **6.** The technician should use 7 liters of the 40% alcohol solution and 3 liters of the 60% alcohol solution. **7.** If 15,000 plates are milled, the cost will be the same. If more than 15,000 plates are milled, the cost will be cheaper on the larger machine.

NOW TRY THIS

A chemist has 20 ml of a 30% alcohol solution. How much pure alcohol must she add so that the resulting solution contains 50% alcohol?

8.3 Exercises

WARM-UPS *For each of the following, identify the two variables needed to set up a system of equations and write a description of what each variable represents.*

1. Prince's Pizza sold a total of 52 pizzas and calzones.

2. The length of a garden is 8 feet longer than twice its width.

3. Danielle and Kinley invested a total of \$15,000.

4. Tony spent \$24 on a total purchase of 6 hot dogs and drinks.

5. A nurse needs 25 milliliters of a 15% saline solution but only has 9% saline solution and 20% saline solution available.

6. The cost of 6 shirts and 3 pairs of jeans is \$123.

7. The theater department sold a total of 273 adult tickets and student tickets.

8. A candy store owner wants to mix a total of 28 pounds of cashews and Brazil nuts.

Set up an equation for each of the statements in Exercises 1–8 using the variables you identified. State, in words, what the equation means.

9. Exercise 1

10. Exercise 2

11. Exercise 3

12. Exercise 4

13. Exercise 5

14. Exercise 6

15. Exercise 7

16. Exercise 8

REVIEW *Graph.*

17. $x < 4$

18. $x \geq -3$

19. $-1 < x \leq 2$

20. $-2 \leq x \leq 0$

Write each product using exponents.

21. $9 \cdot 9 \cdot 9 \cdot 9 \cdot a$

22. $5(\pi)(r)(r)$

23. $x \cdot x \cdot x \cdot y \cdot y \cdot y \cdot y$

24. $(-5)(-5)$

VOCABULARY AND CONCEPTS *Fill in the blanks.*

25. A _______ is a letter that represents a number.

26. An _______ is a statement indicating that two quantities are equal.

27. $\begin{cases} a + 2b = 9 \\ a = 4b + 3 \end{cases}$ is a ______ of linear equations.

28. A _______ of a system of two linear equations satisfies both equations simultaneously.

GUIDED PRACTICE *Use two equations in two variables to solve each application.* SEE EXAMPLE 1. (OBJECTIVE 1)

29. Government The salaries of the President and Vice President of the United States total $592,600 a year. If the President makes $207,400 more than the Vice President, find each of their salaries.

30. Splitting the lottery Chayla and Lena pool their resources to buy several lottery tickets. They win $250,000! They agree that Lena should get $50,000 more than Chayla, because she gave most of the money. How much will Chayla get?

31. Figuring inheritances In his will, a man left his older son $5,000 more than twice as much as he left his younger son. If the estate is worth $742,250, how much did the younger son get?

32. Causes of death The number of American women that died from cancer is approximately seven times the number that died from diabetes. If the number of deaths from these two causes was 308,000, how many American women died from each cause?

Use two equations in two variables to solve each application. SEE EXAMPLE 2. (OBJECTIVE 1)

33. Cutting pipe A plumber wants to cut the pipe shown in the illustration into two pieces so that one piece is 5 feet longer than the other. How long should each piece be?

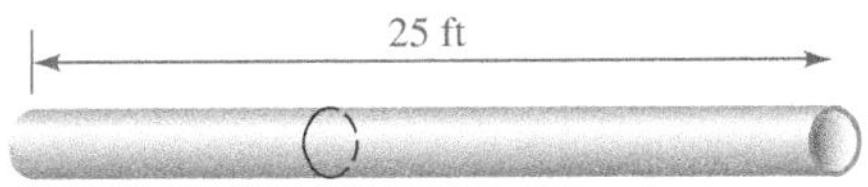

34. Cutting lumber A carpenter wants to cut a 20-foot board into two pieces so that one piece is 4 times as long as the other. How long should each piece be?

35. Geometry The perimeter of the rectangle shown in the illustration is 110 feet. Find its dimensions.

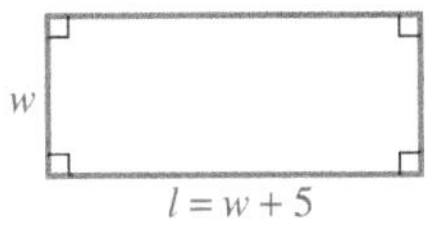

36. Geometry A rectangle is 3 times as long as it is wide, and its perimeter is 80 centimeters. Find its dimensions.

Use two equations in two variables to solve each application. SEE EXAMPLE 3. (OBJECTIVE 1)

37. Geometry The length of a rectangle is 3 feet less than twice its width. If its perimeter is 48 feet, find its area.

38. Geometry A 50-meter path surrounds a rectangular garden. The width of the garden is two-thirds its length. Find its area.

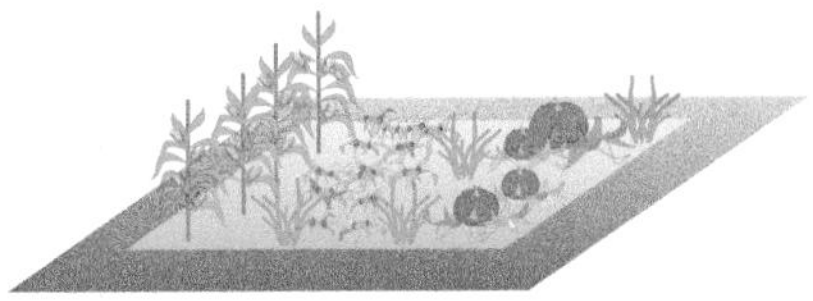

39. At the movies At an IMAX theater, the giant rectangular movie screen has a width 26 feet less than its length. If its perimeter is 332 feet, find the area of the screen.

40. Geometry The length of a residential lot is 20 feet longer than twice the width. If its perimeter is 340 feet, find its area.

Use two equations in two variables to solve each application. SEE EXAMPLE 4. (OBJECTIVE 1)

41. Investing money Bill invested some money at 5% annual interest, and Janette invested some at 7%. If their combined interest was $310 on a total investment of $5,000, how much did Bill invest?

42. Investing money Peter invested some money at 6% annual interest, and Martha invested some at 12%. If their combined investment was $6,000 and their combined interest was $540, how much money did Martha invest?

43. Buying tickets Students can buy tickets to a basketball game for $1. The admission for nonstudents is $2. If 350 tickets are sold and the total receipts are $450, how many student tickets are sold?

44. Buying tickets If receipts for the movie advertised in the illustration were $720 for an audience of 190 people, how many senior citizens attended?

Use two equations in two variables to solve each application. SEE EXAMPLE 5. (OBJECTIVE 1)

45. Boating A boat can travel 24 miles downstream in 2 hours and can make the return trip in 3 hours. Find the speed of the boat in still water.

Unless otherwise noted, all content on this page is © Cengage Learning.

46. **Aviation** With the wind, a plane can fly 3,000 miles in 5 hours. Against the same wind, the trip takes 6 hours. Find the airspeed of the plane (the speed in still air).

47. **Aviation** An airplane can fly downwind a distance of 600 miles in 2 hours. However, the return trip against the same wind takes 3 hours. Find the speed of the wind.

48. **Finding the speed of a current** It takes a motorboat 4 hours to travel 56 miles down a river, and it takes 3 hours longer to make the return trip. Find the speed of the current.

Use two equations in two variables to solve each application.
SEE EXAMPLE 6. (OBJECTIVE 1)

49. **Mixing chemicals** A chemist has one solution that is 40% alcohol and another that is 55% alcohol. How much of each must she use to make 15 liters of a solution that is 50% alcohol?

50. **Mixing pharmaceuticals** A nurse has a solution that is 25% alcohol and another that is 50% alcohol. How much of each must he use to make 20 liters of a solution that is 40% alcohol?

51. **Mixing nuts** A merchant wants to mix the peanuts with the cashews shown in the illustration to get 48 pounds of mixed nuts to sell at $4 per pound. How many pounds of each should the merchant use?

52. **Mixing peanuts and candy** A merchant wants to mix peanuts worth $3 per pound with jelly beans worth $1.50 per pound to make 30 pounds of a mixture worth $2.10 per pound. How many pounds of each should he use?

Use two equations in two variables to solve each application.
SEE EXAMPLE 7. (OBJECTIVE 1)

53. **Choosing a furnace** A high-efficiency 90+ furnace costs $2,250 and costs an average of $412 per year to operate in Rockford, IL. An 80+ furnace costs only $1,715 but costs $466 per year to operate. Find the break-even point.

54. **Choosing a furnace** See Exercise 53. If you intended to live in a house for seven years, which furnace would you choose?

55. **Making tires** A company has two molds to form tires. One mold has a set-up cost of $600 and the other a set-up cost of $1,100. The cost to make each tire on the first machine is $15, and the cost per tire on the second machine is $13. Find the break-even point.

56. **Making tires** See Exercise 55. If you planned a production run of 500 tires, which mold would you use?

ADDITIONAL PRACTICE *Use two equations in two variables to solve each application.*

57. **Buying contact lens cleaner** Two bottles of contact lens cleaner and three bottles of soaking solution cost $29.40, and three bottles of cleaner and two bottles of soaking solution cost $28.60. Find the cost of each.

58. **Buying clothes** Two pairs of shoes and four pairs of socks cost $109, and three pairs of shoes and five pairs of socks cost $160. Find the cost of a pair of socks.

59. **Raising livestock** A rancher raises five times as many cows as horses. If he has 168 animals, how many cows does he have?

60. **Grass seed mixture** A landscaper used 100 pounds of grass seed containing twice as much bluegrass as rye. He added 15 more pounds of bluegrass to the mixture before seeding a lawn. How many pounds of bluegrass did he use?

61. **TV programming** The producer of a 30-minute documentary about World War I divided it into two parts. Four times as much program time was devoted to the causes of the war as to the outcome. How long was each part of the documentary?

62. **Selling ice cream** At a store, ice cream cones cost $1.90 and sundaes cost $2.65. One day, the receipts for a total of 148 cones and sundaes were $328.45. How many cones were sold?

63. **Integers** One integer is three times another, and their sum is 112. Find the integers.

64. **Integers** The sum of two integers is 38, and their difference is 12. Find the integers.

65. **Integers** Three times one integer plus another integer is 29. If the first integer plus twice the second is 18, find the integers.

66. **Integers** Twice one integer plus another integer is 21. If the first integer plus 3 times the second is 33, find the integers.

67. **Investing money** An investment of $950 at one rate of interest and $1,200 at a higher rate together generate an annual income of $88.50. If the investment rates differ by 2%, find the lower rate.

68. **Motion** A man drives for a while at 45 mph. Realizing that he is running late, he increases his speed to 60 mph and completes his 405-mile trip in 8 hours. How long does he drive at 45 mph?

69. **Selling radios** An electronics store put two types of car radios on sale. One model sold for $87, and the other sold for $119. During the sale, the receipts for 25 radios sold were $2,495. How many of the less expensive radios were sold?

70. **Buying baseball equipment** One catcher's mitt and ten outfielder's gloves cost $239.50. How much does each cost if one catcher's mitt and five outfielder's gloves cost $134.50?

Unless otherwise noted, all content on this page is © Cengage Learning.

71. Buying painting supplies Two partial receipts for paint supplies appear in the illustration. How much did each gallon of paint and each brush cost?

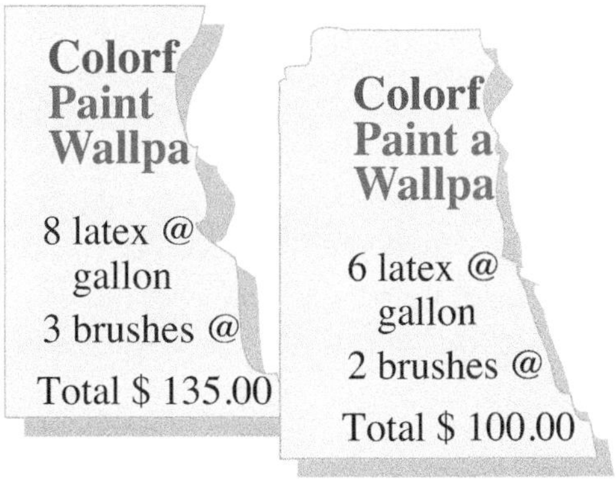

72. Equilibrium price The number of canoes sold at a marina depends on price. As the price gets higher, fewer canoes will be sold. The equation that relates the price of a canoe to the number sold is called a **demand equation**. Suppose that the demand equation for canoes is

$$p = -\frac{1}{2}q + 1{,}300$$

where p is the price and q is the number sold at that price.

The number of canoes produced also depends on price. As the price gets higher, more canoes will be manufactured. The equation that relates the number of canoes produced to the price is called a **supply equation**. Suppose that the supply equation for canoes is

$$p = \frac{1}{3}q + \frac{1{,}400}{3}$$

where p is the price and q is the number produced at that price. The equilibrium price is the price at which supply equals demand. Find the equilibrium price.

WRITING ABOUT MATH

73. Which problem in the preceding set did you find the hardest? Why?

74. Which problem in the preceding set did you find the easiest? Why?

SOMETHING TO THINK ABOUT

75. In the illustration below, how many nails will balance one nut?

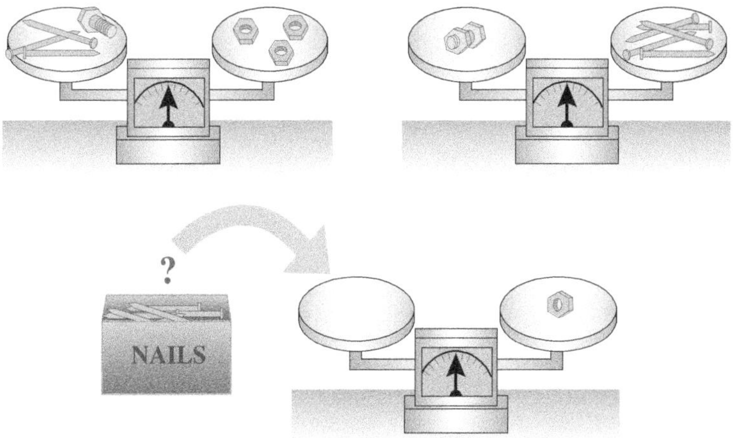

Section 8.4 Solving Systems of Three Linear Equations in Three Variables

Objectives

1. Solve a system of three linear equations in three variables.
2. Identify an inconsistent system.
3. Identify a dependent system and express the solution as an ordered triple in terms of one of the variables.
4. Solve an application by setting up and solving a system of three linear equations in three variables.

Unless otherwise noted, all content on this page is © Cengage Learning.

Vocabulary

ordered triple plane

Getting Ready

Determine whether the equation $x + 2y + 3z = 6$ is satisfied by the following values.

1. $x = 1, y = 1, z = 1$
2. $x = -2, y = 1, z = 2$
3. $x = 2, y = -2, z = -1$
4. $x = 2, y = 2, z = 0$

We now extend the definition of a linear equation to include equations of the form $ax + by + cz = d$ where a, b, c, and d are numerical values (d is a constant). The solution of a system of three linear equations with three variables is an **ordered triple** of numbers if the equations are independent and the system consistent. For example, the solution of the system

$$\begin{cases} 2x + 3y + 4z = 20 \\ 3x + 4y + 2z = 17 \\ 3x + 2y + 3z = 16 \end{cases} \qquad (x, y, z), \text{ which equals } (1, 2, 3)$$

is the ordered triple $(1, 2, 3)$. Each equation is satisfied if $x = 1$, $y = 2$, and $z = 3$.

$$\begin{aligned} 2x + 3y + 4z &= 20 \\ 2(1) + 3(2) + 4(3) &\stackrel{?}{=} 20 \\ 2 + 6 + 12 &\stackrel{?}{=} 20 \\ 20 &= 20 \end{aligned} \qquad \begin{aligned} 3x + 4y + 2z &= 17 \\ 3(1) + 4(2) + 2(3) &\stackrel{?}{=} 17 \\ 3 + 8 + 6 &\stackrel{?}{=} 17 \\ 17 &= 17 \end{aligned} \qquad \begin{aligned} 3x + 2y + 3z &= 16 \\ 3(1) + 2(2) + 3(3) &\stackrel{?}{=} 16 \\ 3 + 4 + 9 &\stackrel{?}{=} 16 \\ 16 &= 16 \end{aligned}$$

The graph of an equation of the form $ax + by + cz = d$ is a flat surface called a **plane**. A system of three linear equations in three variables is consistent or inconsistent, depending on how the three planes corresponding to the three equations intersect. Figure 8-15 illustrates some of the possibilities.

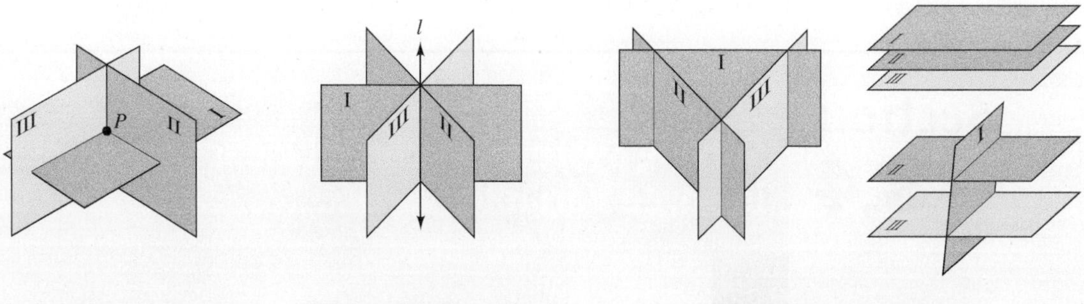

The three planes intersect at a single point P: One solution. A consistent system.

(a)

The three planes have a line l in common: Infinitely many solutions. A consistent system.

(b)

The three planes have no point in common: No solutions. An inconsistent system.

(c)

Figure 8-15

1 Solve a system of three linear equations in three variables.

To solve a system of three linear equations in three variables by elimination, we follow these steps.

Unless otherwise noted, all content on this page is © Cengage Learning.

SOLVING THREE LINEAR EQUATIONS IN THREE VARIABLES

1. Write all equations in the system in $ax + by + cz = d$ form.
2. Select any two equations and eliminate a variable.
3. Select a different pair of equations and eliminate the same variable.
4. Solve the resulting pair of two equations in two variables.
5. To find the value of the third variable, substitute the values of the two variables found in Step 4 into any equation containing all three variables and solve the equation.
6. Check the solution in all three of the original equations.

EXAMPLE 1 Solve the system: $\begin{cases} 2x + y + 4z = 12 \\ x + 2y + 2z = 9 \\ 3x - 3y - 2z = 1 \end{cases}$

Solution We are given the system

COMMENT To keep track of the equations it is helpful to number them.

$$\begin{aligned} &(1) \quad \\ &(2) \quad \\ &(3) \quad \end{aligned}\begin{cases} 2x + y + 4z = 12 \\ x + 2y + 2z = 9 \\ 3x - 3y - 2z = 1 \end{cases}$$

If we select Equations 2 and 3 and add them, the variable z is eliminated:

$$\begin{array}{lrcl} (2) & x + 2y + 2z &=& 9 \\ (3) & 3x - 3y - 2z &=& 1 \\ \hline (4) & 4x - y &=& 10 \end{array}$$

We now select a different pair of equations, 1 and 3 (we could have used Equations 1 and 2) and eliminate z again. If each side of Equation 3 is multiplied by 2 and the resulting equation is added to Equation 1, z is again eliminated:

$$\begin{array}{lrcl} (1) & 2x + y + 4z &=& 12 \\ & 6x - 6y - 4z &=& 2 \\ \hline (5) & 8x - 5y &=& 14 \end{array}$$

Equations 4 and 5 form a system of two equations in two variables:

$$\begin{aligned} &(4) \quad \\ &(5) \quad \end{aligned}\begin{cases} 4x - y = 10 \\ 8x - 5y = 14 \end{cases}$$

To solve this system, we multiply Equation 4 by -5 and add the resulting equation to Equation 5 to eliminate y:

$$\begin{array}{lrcl} & -20x + 5y &=& -50 \\ (5) & 8x - 5y &=& 14 \\ \hline & -12x &=& -36 \end{array}$$

$x = 3$ Divide both sides by -12.

To find the value of y, we substitute 3 for x in any equation containing only x and y (such as Equation 5) and solve.

(5) $8x - 5y = 14$

$8(3) - 5y = 14$ Substitute.

$24 - 5y = 14$ Simplify.

$-5y = -10$ Subtract 24 from both sides.

$y = 2$ Divide both sides by -5.

To find the value of z, we substitute 3 for x and 2 for y in an equation containing x, y, and z (such as Equation 1) and solve.

$$\begin{aligned} (1) \qquad 2x + y + 4z &= 12 & \\ 2(3) + 2 + 4z &= 12 & \text{Substitute.} \\ 8 + 4z &= 12 & \text{Simplify.} \\ 4z &= 4 & \text{Subtract 8 from both sides.} \\ z &= 1 & \text{Divide both sides by 4.} \end{aligned}$$

The solution of the system is $(3, 2, 1)$. Verify that these values satisfy each equation in the original system.

SELF CHECK 1 Solve the system: $\begin{cases} 2x + y + 4z = 16 \\ x + 2y + 2z = 11 \\ 3x - 3y - 2z = -9 \end{cases}$

2 Identify an inconsistent system.

The next example is an inconsistent system and has no solution.

EXAMPLE 2 Solve the system: $\begin{cases} 2x + y - 3z = -3 \\ 3x - 2y + 4z = 2 \\ 4x + 2y - 6z = -7 \end{cases}$

Solution We are given the system of equations

$$\begin{matrix} (1) \\ (2) \\ (3) \end{matrix} \quad \begin{cases} 2x + y - 3z = -3 \\ 3x - 2y + 4z = 2 \\ 4x + 2y - 6z = -7 \end{cases}$$

We can multiply Equation 1 by 2 and add the resulting equation to Equation 2 to eliminate y.

$$\begin{aligned} & 4x + 2y - 6z = -6 \\ (2) \quad & 3x - 2y + 4z = 2 \\ \hline (4) \quad & 7x \qquad\quad - 2z = -4 \end{aligned}$$

We now add Equations 2 and 3 to again eliminate y.

$$\begin{aligned} (2) \quad & 3x - 2y + 4z = 2 \\ (3) \quad & 4x + 2y - 6z = -7 \\ \hline (5) \quad & 7x \qquad\quad - 2z = -5 \end{aligned}$$

Equations 4 and 5 form the system

$$\begin{matrix} (4) \\ (5) \end{matrix} \quad \begin{cases} 7x - 2z = -4 \\ 7x - 2z = -5 \end{cases}$$

Since $7x - 2z$ cannot equal both -4 and -5, this system is inconsistent; it has no solution. Its solution set is $\varnothing$.

SELF CHECK 2 Solve the system: $\begin{cases} 2x + y - 3z = 8 \\ 3x - 2y + 4z = 10 \\ 4x + 2y - 6z = -5 \end{cases}$

3 Identify a dependent system and express the solution as an ordered triple in terms of one of the variables.

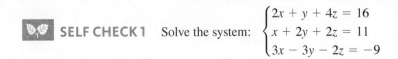

When the equations in a system of two equations in two variables were dependent, the system had infinitely many solutions. This is not always true for systems of three equations in three variables. In fact, a system can have dependent equations and still be inconsistent. Figure 8-16 illustrates the different possibilities.

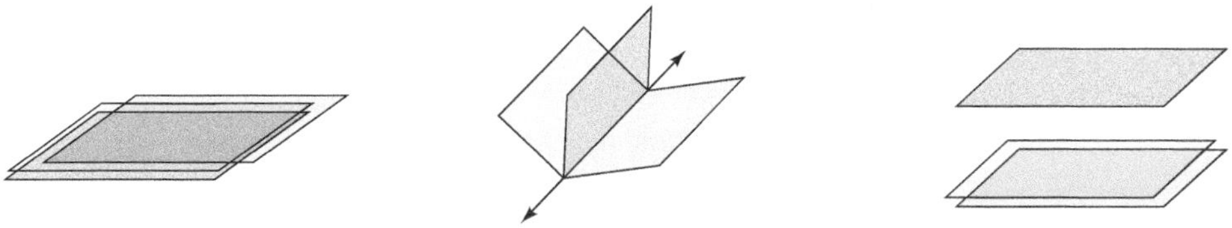

When three planes coincide, the equations are dependent, and there are infinitely many solutions.

(a)

When three planes intersect in a common line, the equations are dependent, and there are infinitely many solutions.

(b)

When two planes coincide and are parallel to a third plane, the system is inconsistent, and there are no solutions.

(c)

Figure 8-16

EXAMPLE 3 Solve the system: $\begin{cases} 3x - 2y + z = -1 \\ 2x + y - z = 5 \\ 5x - y = 4 \end{cases}$

Solution We can add the first two equations to obtain

$$\begin{aligned} (1) \quad & 3x - 2y + z = -1 \\ (2) \quad & 2x + y - z = 5 \\ (4) \quad & 5x - y = 4 \end{aligned}$$

Since Equation 4 is the same as the third equation of the system, the equations of the system are dependent, and there will be infinitely many solutions.

To write the general solution to this system, we can solve Equation 4 for y to obtain

$$\begin{aligned} 5x - y &= 4 \\ -y &= -5x + 4 && \text{Subtract } 5x \text{ from both sides.} \\ y &= 5x - 4 && \text{Multiply both sides by } -1. \end{aligned}$$

We then can substitute $5x - 4$ for y in the first equation of the system and solve for z to obtain

$$\begin{aligned} 3x - 2y + z &= -1 \\ 3x - 2(5x - 4) + z &= -1 && \text{Substitute.} \\ 3x - 10x + 8 + z &= -1 && \text{Use the distributive property to remove parentheses.} \\ -7x + 8 + z &= -1 && \text{Combine like terms.} \\ z &= 7x - 9 && \text{Add } 7x \text{ and } -8 \text{ to both sides.} \end{aligned}$$

COMMENT The solution in Example 3 is called a *general* solution. If we had eliminated different variables, we could have expressed the general solution in terms of y or in terms of z.

Since we have found the values of y and z in terms of x, every solution to the system has the form $(x, 5x - 4, 7x - 9)$, where x can be any real number. For example,

If $x = 1$, a solution is $(1, 1, -2)$. $5(1) - 4 = 1$, and $7(1) - 9 = -2$

If $x = 2$, a solution is $(2, 6, 5)$. $5(2) - 4 = 6$, and $7(2) - 9 = 5$

If $x = 3$, a solution is $(3, 11, 12)$. $5(3) - 4 = 11$, and $7(3) - 9 = 12$

This system has infinitely many solutions of the form $(x, 5x - 4, 7x - 9)$.

SELF CHECK 3 Solve the system: $\begin{cases} 3x + 2y + z = -1 \\ 2x - y - z = 5 \\ 5x + y = 4 \end{cases}$

Unless otherwise noted, all content on this page is © Cengage Learning.

4 Solve an application by setting up and solving a system of three linear equations in three variables.

EXAMPLE 4 **MANUFACTURING HAMMERS** A company makes three types of hammers—good, better, and best. The cost of making each type of hammer is \$4, \$6, and \$7, respectively, and the hammers sell for \$6, \$9, and \$12. Each day, the cost of making 100 hammers is \$520, and the daily revenue from their sale is \$810. How many of each type are manufactured?

Analyze the problem: We need to know the number of each type of hammer manufactured, so we will let x represent the number of good hammers, y represent the number of better hammers, and z represent the number of best hammers.

Form three equations: Since x represents the number of good hammers, y represents the number of better hammers, and z represents the number of best hammers, we know that

The total number of hammers is $x + y + z$.
The cost of making good hammers is \$$4x$ (\$4 times x hammers).
The cost of making better hammers is \$$6y$ (\$6 times y hammers).
The cost of making best hammers is \$$7z$ (\$7 times z hammers).
The revenue received by selling good hammers is \$$6x$ (\$6 times x hammers).
The revenue received by selling better hammers is \$$9y$ (\$9 times y hammers).
The revenue received by selling best hammers is \$$12z$ (\$12 times z hammers).

The information leads to three equations:

The number of good hammers	plus	the number of better hammers	plus	the number of best hammers	equals	the total number of hammers.
x	$+$	y	$+$	z	$=$	100
The cost of making good hammers	plus	the cost of making better hammers	plus	the cost of making best hammers	equals	the total cost.
$4x$	$+$	$6y$	$+$	$7z$	$=$	520
The revenue from the good hammers	plus	the revenue from the better hammers	plus	the revenue from the best hammers	equals	the total revenue.
$6x$	$+$	$9y$	$+$	$12z$	$=$	810

Solve the system: These three equations give the following system:

$$\begin{cases} x + y + z = 100 & (1) \\ 4x + 6y + 7z = 520 & (2) \\ 6x + 9y + 12z = 810 & (3) \end{cases}$$

that we can solve as follows:

If we multiply Equation 1 by -7 and add the result to Equation 2, we obtain

$$\begin{aligned} -7x - 7y - 7z &= -700 \\ 4x + 6y + 7z &= 520 \\ \hline -3x - y \qquad &= -180 \quad (4) \end{aligned}$$

If we multiply Equation 1 by -12 and add the result to Equation 3, we obtain

$$\begin{array}{lrcl} & -12x - 12y - 12z & = & -1{,}200 \\ & 6x + 9y + 12z & = & 810 \\ \hline (5) & -6x - 3y & = & -390 \end{array}$$

If we multiply Equation 4 by -3 and add it to Equation 5, we obtain

$$\begin{array}{rcl} 9x + 3y & = & 540 \\ -6x - 3y & = & -390 \\ \hline 3x & = & 150 \end{array}$$

$x = 50$ Divide both sides by 3.

To find y, we substitute 50 for x in Equation 4:

$$-3x - y = -180$$
$$-3(\mathbf{50}) - y = -180 \quad \text{Substitute.}$$
$$-y = -30 \quad \text{Add 150 to both sides.}$$
$$y = 30 \quad \text{Divide both sides by } -1$$

To find z, we substitute 50 for x and 30 for y in Equation 1:

$$x + y + z = 100$$
$$\mathbf{50} + \mathbf{30} + z = 100$$
$$z = 20 \quad \text{Subtract 80 from both sides.}$$

State the conclusion Each day, the company makes 50 good hammers, 30 better hammers, and 20 best hammers. This totals 100. The cost of producing the hammers is $\$4(50) + \$6(30) + \$7(20)$ or \$520 and the revenue from the sale of the hammers is $\$6(50) + \$9(30) + \$12(20)$ or \$810.

Check the result Check the solution in each equation in the original system.

SELF CHECK 4 If the cost of the 100 hammers is \$560 and the daily revenue is \$870, how many of each type are manufactured?

EXAMPLE 5

CURVE FITTING The equation of the parabola shown in Figure 8-17 is of the form $y = ax^2 + bx + c$. Find the equation of the parabola.

Solution Since the parabola passes through the points shown in the figure, each pair of coordinates satisfies the equation $y = ax^2 + bx + c$. For example, $(-1, 5)$ substituted in $y = ax^2 + bx + c$ becomes $5 = a(-1)^2 + b(-1) + c$ or simplified as $a - b + c = 5$. If we substitute the x- and y-values of each point into the equation and simplify, we obtain the following system.

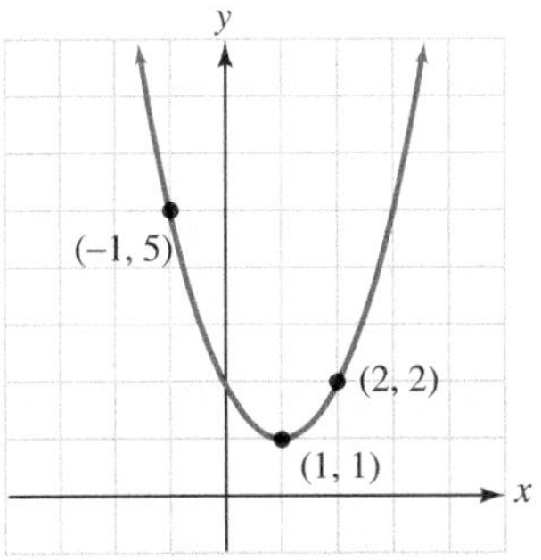

Figure 8-17

$$\begin{array}{ll} (1) & \\ (2) & \\ (3) & \end{array} \begin{cases} a - b + c = 5 & \text{for point } (-1, 5) \\ a + b + c = 1 & \text{for point } (1, 1) \\ 4a + 2b + c = 2 & \text{for point } (2, 2) \end{cases}$$

If we add Equations 1 and 2, we obtain $2a + 2c = 6$. If we multiply Equation 1 by 2 and add the result to Equation 3, we get $6a + 3c = 12$. We can then divide both sides of $2a + 2c = 6$ by 2 and divide both sides of $6a + 3c = 12$ by 3 to obtain the system

$$\begin{array}{l} (4) \\ (5) \end{array} \begin{cases} a + c = 3 \\ 2a + c = 4 \end{cases}$$

If we multiply Equation 4 by -1 and add the result to Equation 5, we get $a = 1$. To find the value of c, we can substitute 1 for a in Equation 4 and find that $c = 2$. To find the value of b, we can substitute 1 for a and 2 for c in Equation 2 and find that $b = -2$.

Unless otherwise noted, all content on this page is © Cengage Learning.

After we substitute these values of a, b, and c into the equation $y = ax^2 + bx + c$ we have the equation of the parabola.

$$y = ax^2 + bx + c$$
$$y = 1x^2 + (-2)x + 2$$
$$y = x^2 - 2x + 2$$

SELF CHECK 5 If a parabola passes through the points $(1, -10)$, $(-2, 8)$, and $(5, -6)$, find the equation of the parabola.

SELF CHECK ANSWERS **1.** $(1, 2, 3)$ **2.** $\varnothing$ **3.** infinitely many solutions of the form $(x, 4 - 5x, -9 + 7x)$ **4.** The company makes 30 good hammers, 50 better hammers, and 20 best hammers. **5.** $y = x^2 - 5x - 6$

NOW TRY THIS

The manager of a coffee bar wants to mix some Peruvian Organic coffee worth \$15 per pound with some Colombian coffee worth \$10 per pound and Indian Malabar coffee worth \$18 per pound to obtain 50 pounds of a blend that he can sell for \$17.50 per pound. He wants to use 10 fewer pounds of the Indian Malabar than Peruvian Organic. How many pounds of each should he use? (*Hint*: This problem is based on the formula $V = np$, where V represents value, n represents the number of pounds, and p represents the price per pound.)

8.4 Exercises

WARM-UPS *Is the ordered triple a solution of the system?*

1. $(1, 1, 1)$; $\begin{cases} 2x + y - 3z = 0 \\ 3x - 2y + 4z = 5 \\ 4x + 2y - 6z = 0 \end{cases}$

2. $(2, 1, 1)$; $\begin{cases} 3x + 2y - z = 5 \\ 2x - 3y + 2z = 4 \\ 4x - 2y + 3z = 10 \end{cases}$

REVIEW *Consider the line passing through $(-1, -5)$ and $(2, 3)$.*

3. Find the slope of the line.

4. Write the equation of the line in slope-intercept form.

Find each value if $f(x) = 2x^2 + 1$.

5. $f(0)$ **6.** $f(-2)$

7. $f(3s)$ **8.** $f(4t)$

VOCABULARY AND CONCEPTS *Fill in the blanks.*

9. The graph of the equation $2x + 3y + 4z = 5$ is a flat surface called a _____.

10. When three planes coincide, the equations of the system are _______, and there are _______ many solutions.

11. When three planes intersect in a line, the system will have _______ many solutions.

12. When three planes are parallel, the system will have __ solutions.

Determine whether the ordered triple is a solution of the given system.

13. $(1, 2, 1)$, $\begin{cases} x + y - z = 2 \\ 2x + y + 3z = 7 \\ x - \frac{1}{3}y + \frac{2}{3}z = 1 \end{cases}$

14. $(-3, 2, -1)$, $\begin{cases} 2x + 2y + 3z = 5 \\ 3x + y = -6 \\ x + y + 2z = 1 \end{cases}$

GUIDED PRACTICE *Solve each system.* SEE EXAMPLE 1. *(OBJECTIVE 1)*

15. $\begin{cases} x + y + z = 4 \\ 2x + y - z = 1 \\ 2x - 3y + z = 1 \end{cases}$

16. $\begin{cases} x + y + z = 4 \\ x - y + z = 2 \\ x - y - z = 0 \end{cases}$

17. $\begin{cases} 4x + 3z = 4 \\ 2y - 6z = -1 \\ 8x + 4y + 3z = 9 \end{cases}$

18. $\begin{cases} 2x + 3y + 2z = 1 \\ 2x - 3y + 2z = -1 \\ 4x + 3y - 2z = 4 \end{cases}$

Solve each system. **SEE EXAMPLE 2.** *(OBJECTIVE 2)*

19. $\begin{cases} 3a - 2b - c = 4 \\ 6a - 4b - 2c = 10 \\ a + 3b + c = 2 \end{cases}$

20. $\begin{cases} 2x + y - z = 1 \\ x + 2y + 2z = 2 \\ 4x + 5y + 3z = 3 \end{cases}$

Solve each system. **SEE EXAMPLE 3.** *(OBJECTIVE 3)*

21. $\begin{cases} 7x - 2y - z = 1 \\ 9x - 6y + z = 7 \\ x - 2y + z = 3 \end{cases}$

22. $\begin{cases} x - 2y + 3z = 9 \\ -x + 3y = -4 \\ 2x - 5y + 3z = 13 \end{cases}$

23. $\begin{cases} x + 2y + z = 1 \\ 3x + y + 3z = 3 \\ -2x + y - 2z = -2 \end{cases}$

24. $\begin{cases} -x + y - 2z = -5 \\ 3x + 2y = 4 \\ 4x - 4y + 8z = 20 \end{cases}$

ADDITIONAL PRACTICE *Solve each system.*

25. $\begin{cases} a + 3b + c = 10 \\ 3a + b + c = 12 \\ a + b + 3c = 8 \end{cases}$

26. $\begin{cases} 3x - y - 2z = 12 \\ x + y + 6z = 8 \\ 2x - 2y - z = 11 \end{cases}$

27. $\begin{cases} x + \frac{1}{3}y + z = 13 \\ \frac{1}{2}x - y + \frac{1}{3}z = -2 \\ x + \frac{1}{2}y - \frac{1}{3}z = 2 \end{cases}$

28. $\begin{cases} x - \frac{1}{5}y - z = 9 \\ \frac{1}{4}x + \frac{1}{5}y - \frac{1}{2}z = 5 \\ 2x + y + \frac{1}{6}z = 12 \end{cases}$

29. $\begin{cases} x - 3y + z = 1 \\ 2x - y - 2z = 2 \\ x + 2y - 3z = -1 \end{cases}$

30. $\begin{cases} 2x + y - 3z = 5 \\ x - 2y + 4z = 9 \\ 4x + 2y - 6z = 1 \end{cases}$

31. $\begin{cases} 2x + 3y + 4z = 6 \\ 2x - 3y - 4z = -4 \\ 4x + 6y + 8z = 12 \end{cases}$

32. $\begin{cases} x - 3y + 4z = 2 \\ 2x + y + 2z = 3 \\ 4x - 5y + 10z = 7 \end{cases}$

33. $\begin{cases} 2x + 2y + 3z = 10 \\ 3x + y - z = 0 \\ x + y + 2z = 6 \end{cases}$

34. $\begin{cases} x - y + z = 4 \\ x + 2y - z = -1 \\ x + y - 3z = -2 \end{cases}$

APPLICATIONS *Use a system of three linear equations to solve.* **SEE EXAMPLES 4–5.** *(OBJECTIVE 4)*

35. Making statues An artist makes three types of ceramic statues at a monthly cost of \$650 for 180 statues. The manufacturing costs for the three types are \$5, \$4, and \$3. If the statues sell for \$20, \$12, and \$9, respectively, how many of each type should be made to produce \$2,100 in monthly revenue?

36. Manufacturing footballs A factory manufactures three types of footballs at a monthly cost of \$2,425 for 1,125 footballs. The manufacturing costs for the three types of footballs are \$4, \$3, and \$2. These footballs sell for \$16, \$12, and \$10, respectively. How many of each type are manufactured if the monthly profit is \$9,275? (*Hint*: Profit = Income − Cost.)

37. Curve fitting Find the equation of the parabola shown in the illustration.

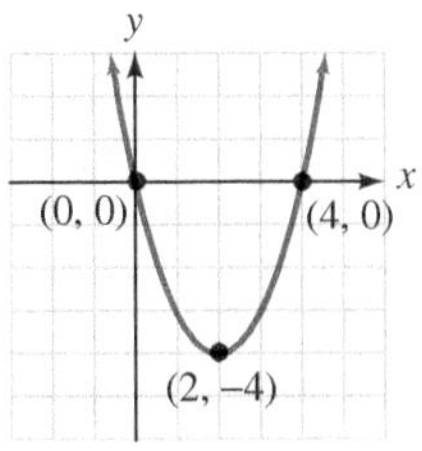

38. Curve fitting Find the equation of the parabola shown in the illustration.

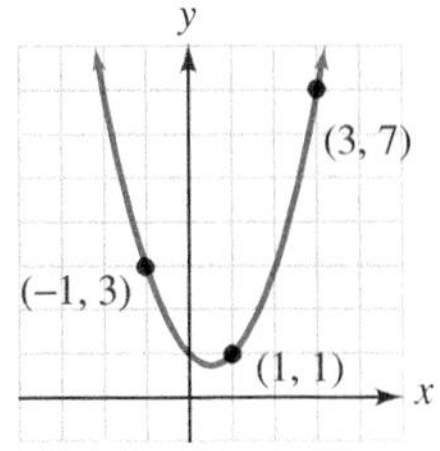

39. Integers The sum of three integers is 19. The third integer is twice the second, and the second integer is 5 more than the first. Find the integers.

40. Integers The sum of three integers is 48. If the first integer is doubled, the sum is 60. If the second integer is doubled, the sum is 63. Find the integers.

41. Geometry The sum of the angles in any triangle is 180°. In triangle *ABC*, $\angle A$ is 80° less than the sum of $\angle B$ and $\angle C$, and $\angle C$ is 50° less than twice $\angle B$. Find the measure of each angle.

42. Geometry The sum of the angles of any four-sided figure is 360°. In the quadrilateral, $\angle A = \angle B$, $\angle C$ is 20° greater than $\angle A$, and $\angle D = 40°$. Find the measure of each angle.

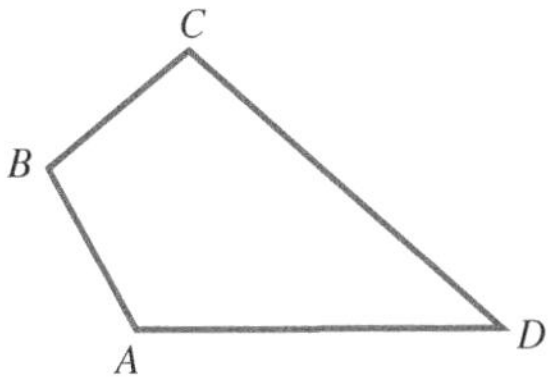

43. Nutritional planning One unit of each of three foods contains the nutrients shown in the table. How many units of each must be used to provide exactly 11 grams of fat, 6 grams of carbohydrates, and 10 grams of protein?

Food	Fat	Carbohydrates	Protein
A	1	1	2
B	2	1	1
C	2	1	2

44. Nutritional planning One unit of each of three foods contains the nutrients shown in the table. How many units of each must be used to provide exactly 14 grams of fat, 9 grams of carbohydrates, and 9 grams of protein?

Food	Fat	Carbohydrates	Protein
A	2	1	2
B	3	2	1
C	1	1	2

Unless otherwise noted, all content on this page is © Cengage Learning.

45. Concert tickets Tickets for a concert cost \$5, \$3, and \$2. Twice as many \$5 tickets were sold as \$2 tickets. The receipts for 750 tickets were \$2,625. How many of each price ticket were sold?

46. Mixing nuts The owner of a candy store mixed some peanuts worth \$3 per pound, some cashews worth \$9 per pound, and some Brazil nuts worth \$9 per pound to get 50 pounds of a mixture that would sell for \$6 per pound. She used 15 fewer pounds of cashews than peanuts. How many pounds of each did she use?

47. Chainsaw sculpting A north woods sculptor carves three types of statues with a chainsaw. The times required for carving, sanding, and painting a totem pole, a bear, and a deer are shown in the table. How many of each should be produced to use all available labor hours?

	Totem pole	Bear	Deer	Time available
Carving	2 hours	2 hours	1 hour	14 hours
Sanding	1 hour	2 hours	2 hours	15 hours
Painting	3 hours	2 hours	2 hours	21 hours

48. Making clothing A clothing manufacturer makes coats, shirts, and slacks. The times required for cutting, sewing, and packaging each item are shown in the table. How many of each should be made to use all available labor hours?

	Coats	Shirts	Slacks	Time available
Cutting	20 min	15 min	10 min	115 hr
Sewing	60 min	30 min	24 min	280 hr
Packaging	5 min	12 min	6 min	65 hr

49. Earth's atmosphere Use the information in the graph to determine what percent of Earth's atmosphere is nitrogen, is oxygen, and is other gases.

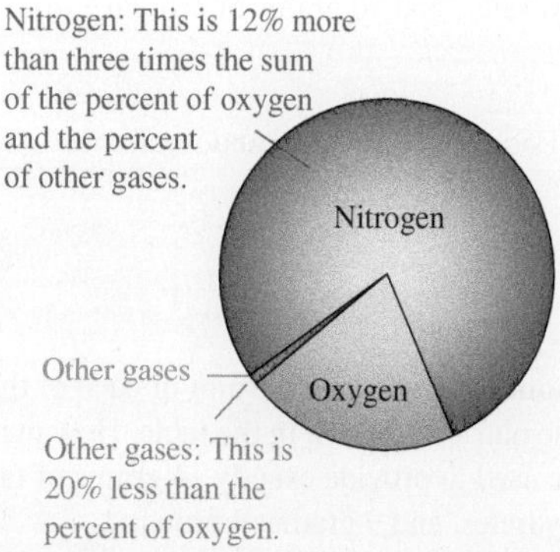
Nitrogen: This is 12% more than three times the sum of the percent of oxygen and the percent of other gases.
Nitrogen
Other gases
Oxygen
Other gases: This is 20% less than the percent of oxygen.

50. NFL records Jerry Rice, who played with the San Francisco 49ers and the Oakland Raiders, holds the all-time record for touchdown passes caught (197). Here are interesting facts about three quarterbacks who helped him achieve this record.

- He caught 17 more TD passes from Steve Young (SF) than he did from Joe Montana.
- He caught 50 more TD passes from Joe Montana (SF) than he did from Rich Gannon (Oak).
- He caught a total of 171 TD passes from Young, Montana, and Gannon.

Determine the number of touchdown passes Rice caught from Young, from Montana, and from Gannon.

The equation of a circle is of the form $x^2 + y^2 + cx + dy + e = 0$.

51. Curve fitting Find the equation of the circle shown in the illustration.

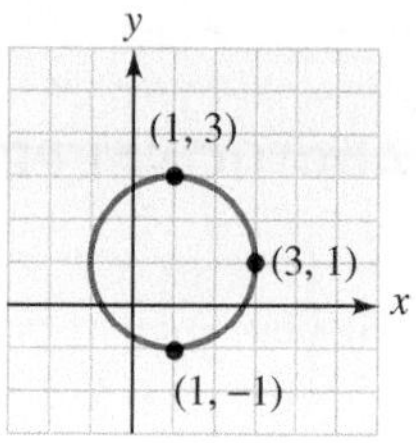
y
(1, 3)
(3, 1)
x
(1, −1)

52. Curve fitting Find the equation of the circle shown in the illustration.

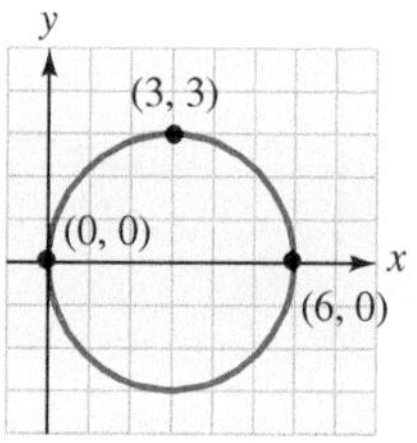
y
(3, 3)
(0, 0)
x
(6, 0)

WRITING ABOUT MATH

53. What makes a system of three equations in three variables inconsistent?

54. What makes the equations of a system of three equations in three variables dependent?

SOMETHING TO THINK ABOUT

55. Solve the system:

$$\begin{cases} x + y + z + w = 3 \\ x - y - z - w = -1 \\ x + y - z - w = 1 \\ x + y - z + w = 3 \end{cases}$$

56. Solve the system:

$$\begin{cases} 2x + y + z + w = 3 \\ x - 2y - z + w = -3 \\ x - y - 2z - w = -3 \\ x + y - z + 2w = 4 \end{cases}$$

Unless otherwise noted, all content on this page is © Cengage Learning.

Section 8.5 Solving Systems of Linear Equations Using Matrices

Objectives

1. Solve a system of linear equations with the same number of equations as variables using Gaussian elimination.
2. Solve a system with more linear equations than variables using row operations on a matrix.
3. Solve a system with fewer linear equations than variables using row operations on a matrix.

Vocabulary

matrix	augmented matrix	triangular form of a matrix
element of a matrix	coefficient matrix	back substitution
square matrix	Gaussian elimination	

Getting Ready

Multiply the first row by 2 and add the result to the second row.

1. $\begin{matrix} 2 & 3 & 5 \\ 1 & 2 & 3 \end{matrix}$

2. $\begin{matrix} -1 & 0 & 4 \\ 2 & 3 & -1 \end{matrix}$

Multiply the first row by −1 and add the result to the second row.

3. $\begin{matrix} 2 & 3 & 5 \\ 1 & 2 & 3 \end{matrix}$

4. $\begin{matrix} -1 & 0 & 4 \\ 2 & 3 & -1 \end{matrix}$

In this section, we will discuss an alternative method used for solving systems of linear equations. This method involves the use of *matrices.*

1 Solve a system of linear equations with the same number of equations as variables using Gaussian elimination.

MATRIX

A **matrix** is any rectangular array of numbers.

Some examples of matrices are

$$A = \begin{bmatrix} 1 & 2 & 3 \\ 4 & 5 & 6 \end{bmatrix} \quad B = \begin{bmatrix} 1 & 2 \\ 3 & 4 \\ 5 & 6 \end{bmatrix} \quad C = \begin{bmatrix} 2 & 4 & 6 \\ 8 & 10 & 12 \\ 14 & 16 & 18 \end{bmatrix}$$

The numbers in each matrix are called **elements**. Because matrix A has two rows and three columns, it is called a 2×3 matrix (read "2 by 3" matrix). Matrix B is a 3×2 matrix, because the matrix has three rows and two columns. Matrix C is a 3×3 matrix (three rows and three columns).

Arthur Cayley
1821–1895
Cayley taught mathematics at Cambridge University. When he refused to take religious vows, he was fired and became a lawyer. After 14 years, he returned to mathematics and to Cambridge. Cayley was a major force in developing the theory of matrices.

Any matrix with the same number of rows and columns, like matrix C, is called a **square matrix**.

To use matrices to solve systems of linear equations, we consider the system

$$\begin{cases} x - 2y - z = 6 \\ 2x + 2y - z = 1 \\ -x - y + 2z = 1 \end{cases}$$

which can be represented by the following matrix, called an **augmented matrix**:

$$\left[\begin{array}{rrr|r} 1 & -2 & -1 & 6 \\ 2 & 2 & -1 & 1 \\ -1 & -1 & 2 & 1 \end{array}\right]$$

The first three columns of the augmented matrix form a 3×3 matrix called a **coefficient matrix**. It is determined by the coefficients of x, y, and z in the equations of the system. The 3×1 matrix to the right of the dashed line is determined by the constants in the equations.

Coefficient matrix ***Column of constants***

$$\begin{bmatrix} 1 & -2 & 1 \\ 2 & 2 & -1 \\ -1 & -1 & 2 \end{bmatrix} \qquad \begin{bmatrix} 6 \\ 1 \\ 1 \end{bmatrix}$$

Each row of the augmented matrix represents one equation of the system:

$$\left[\begin{array}{rrr|r} 1 & -2 & -1 & 6 \\ 2 & 2 & -1 & 1 \\ -1 & -1 & 2 & 1 \end{array}\right] \begin{array}{c} \leftrightarrow \\ \leftrightarrow \\ \leftrightarrow \end{array} \begin{cases} x - 2y - z = 6 \\ 2x + 2y - z = 1 \\ -x - y + 2z = 1 \end{cases}$$

To solve a 3×3 system of equations by **Gaussian elimination**, we transform an augmented matrix into the following matrix that has all 0's below its main diagonal, which is formed by the elements a, e, and h.

$$\left[\begin{array}{ccc|c} a & b & c & d \\ 0 & e & f & g \\ 0 & 0 & h & i \end{array}\right] \quad (a, b, c, \ldots, i \text{ are real numbers})$$

We often can write a matrix in this form, called **triangular form**, by using the following operations.

ELEMENTARY ROW OPERATIONS

1. Any two rows of a matrix can be interchanged.
2. Any row of a matrix can be multiplied by a nonzero constant.
3. Any row of a matrix can be changed by adding a nonzero constant multiple of another row to it.

- A type 1 row operation corresponds to interchanging two equations of a system.
- A type 2 row operation corresponds to multiplying both sides of an equation by a nonzero constant.
- A type 3 row operation corresponds to adding a nonzero multiple of one equation to another.

None of these operations will change the solution of the given system of equations.

After we have written the matrix in triangular form, we can solve the corresponding system of equations by a process called **back substitution**, as shown in Example 1.

Unless otherwise noted, all content on this page is © Cengage Learning.

EXAMPLE 1 Solve the system using matrices: $\begin{cases} x - 2y - z = 6 \\ 2x + 2y - z = 1 \\ -x - y + 2z = 1 \end{cases}$

Solution We can represent the system with the following augmented matrix:

$$\left[\begin{array}{rrr|r} 1 & -2 & -1 & 6 \\ 2 & 2 & -1 & 1 \\ -1 & -1 & 2 & 1 \end{array}\right]$$

To get 0's under the 1 in the first column, we multiply row 1 of the augmented matrix by -2 and add it to row 2 to get a new row 2. We then add row 1 to row 3 to obtain a new row 3.

$$\left[\begin{array}{rrr|r} 1 & -2 & -1 & 6 \\ 0 & 6 & 1 & -11 \\ 0 & -3 & 1 & 7 \end{array}\right] \quad \begin{array}{r} \\ -2R1 + R2 \to R2 \\ R1 + R3 \to R3 \end{array}$$

To get a 0 under the 6 in the second column of the previous matrix, we multiply row 2 by $\frac{1}{2}$ and add it to row 3.

$$\left[\begin{array}{rrr|r} 1 & -2 & 1 & 6 \\ 0 & 6 & 1 & -11 \\ 0 & 0 & \frac{3}{2} & \frac{3}{2} \end{array}\right] \quad \begin{array}{r} \\ \\ \frac{1}{2}R2 + R3 \to R3 \end{array}$$

Finally, to clear the fraction in the third row, third column, we multiply row 3 by $\frac{2}{3}$, which is the reciprocal of $\frac{3}{2}$.

$$\left[\begin{array}{rrr|r} 1 & -2 & 1 & 6 \\ 0 & 6 & 1 & -11 \\ 0 & 0 & 1 & 1 \end{array}\right] \quad \begin{array}{r} \\ \\ \frac{2}{3}R3 \to R3 \end{array}$$

The final matrix represents the system of equations

$$\begin{array}{l} (1) \\ (2) \\ (3) \end{array} \quad \begin{cases} x - 2y - z = 6 \\ 0x + 6y + z = -11 \\ 0x + 0y + z = 1 \end{cases}$$

From Equation 3, we can see that $z = 1$. Now we begin the back substitution process. To find the value of y, we substitute 1 for z in Equation 2 and solve.

$$\begin{array}{lrcll} (2) & 6y + z & = & -11 & \\ & 6y + 1 & = & -11 & \text{Substitute 1 for } z. \\ & 6y & = & -12 & \text{Subtract 1 from both sides.} \\ & y & = & -2 & \text{Divide both sides by 6.} \end{array}$$

Thus, $y = -2$. To find the value of x, we substitute 1 for z and -2 for y in Equation 1 and solve.

$$\begin{array}{lrcll} (1) & x - 2y - z & = & 6 & \\ & x - 2(-2) - 1 & = & 6 & \text{Substitute 1 for } z \text{ and } -2 \text{ for } y. \\ & x + 3 & = & 6 & \text{Simplify.} \\ & x & = & 3 & \text{Subtract 3 from both sides.} \end{array}$$

Thus, $x = 3$. The solution of the given system is $(3, -2, 1)$. Verify that this ordered triple satisfies each equation of the original system.

SELF CHECK 1 Solve the system using matrices: $\begin{cases} x - 2y - z = 2 \\ 2x + 2y - z = -5 \\ -x - y + 2z = 7 \end{cases}$

2 Solve a system with more linear equations than variables using row operations on a matrix.

We can use matrices to solve systems that have more equations than variables.

EXAMPLE 2 Solve the system using matrices: $\begin{cases} x + y = -1 \\ 2x - y = 7 \\ -x + 2y = -8 \end{cases}$

Solution This system can be represented by the following augmented matrix:

$$\left[\begin{array}{cc|c} 1 & 1 & -1 \\ 2 & -1 & 7 \\ -1 & 2 & -8 \end{array}\right]$$

To obtain 0's under the 1 in the first column, we multiply row 1 by -2 and add it to row 2. Then we can add row 1 to row 3.

$$\left[\begin{array}{cc|c} 1 & 1 & -1 \\ 0 & \mathbf{-3} & 9 \\ 0 & 3 & -9 \end{array}\right] \quad \begin{array}{l} \\ -2R1 + R2 \rightarrow R2 \\ R1 + R3 \rightarrow R3 \end{array}$$

To get a 0 under the -3 in the second column, we can add row 2 to row 3.

$$\left[\begin{array}{cc|c} 1 & 1 & -1 \\ 0 & -3 & 9 \\ 0 & 0 & 0 \end{array}\right] \quad R2 + R3 \rightarrow R3$$

Finally, to get a 1 in the second row, second column, we multiply row 2 by $-\frac{1}{3}$.

$$\left[\begin{array}{cc|c} 1 & 1 & -1 \\ 0 & 1 & -3 \\ 0 & 0 & 0 \end{array}\right] \quad -\tfrac{1}{3}R2 \rightarrow R2$$

The final matrix represents the system

$$\begin{cases} x + y = -1 \\ 0x + y = -3 \\ 0x + 0y = 0 \end{cases}$$

The third equation can be discarded, because $0x + 0y = 0$ for all x and y. From the second equation, we can read that $y = -3$. To find the value of x, we substitute -3 for y in the first equation and solve.

$$x + y = -1$$
$$x + (\mathbf{-3}) = -1 \quad \text{Substitute } -3 \text{ for } y.$$
$$x = 2 \quad \text{Add 3 to both sides.}$$

The solution is $(2, -3)$. Verify that this solution satisfies all three equations of the original system.

SELF CHECK 2 Solve the system using matrices: $\begin{cases} x + y = 1 \\ 2x - y = 8 \\ -x + 2y = -7 \end{cases}$

COMMENT If the last row of the final matrix in Example 2 had been representative of the form $0x + 0y = k$, where $k \neq 0$, the system would not have a solution. No values of x and y could make the expression $0x + 0y$ equal to a nonzero constant k.

3 Solve a system with fewer linear equations than variables using row operations on a matrix.

We also can solve many systems that have more variables than equations.

EXAMPLE 3 Solve the system using matrices: $\begin{cases} x + y - 2z = -1 \\ 2x - y + z = -3 \end{cases}$

Solution This system can be represented by the following augmented matrix.

$$\left[\begin{array}{ccc|c} 1 & 1 & -2 & -1 \\ 2 & -1 & 1 & -3 \end{array}\right]$$

To obtain a 0 under the 1 in the first column, we multiply row 1 by -2 and add it to row 2.

$$\left[\begin{array}{ccc|c} 1 & 1 & -2 & -1 \\ 0 & -3 & 5 & -1 \end{array}\right] \quad -2R1 + R2 \rightarrow R2$$

Then to get a 1 in the second row, second column, we multiply row 2 by $-\frac{1}{3}$.

$$\left[\begin{array}{ccc|c} 1 & 1 & -2 & -1 \\ 0 & 1 & -\frac{5}{3} & \frac{1}{3} \end{array}\right] \quad -\tfrac{1}{3}R2 \rightarrow R2$$

The final matrix represents the system

$$\begin{cases} x + y - 2z = -1 \\ y - \frac{5}{3}z = \frac{1}{3} \end{cases}$$

We add $\frac{5}{3}z$ to both sides of the second equation to obtain

$$y = \frac{1}{3} + \frac{5}{3}z$$

We have not found a specific value for y. However, we have found y in terms of z.

To find a value of x in terms of z, we substitute $\frac{1}{3} + \frac{5}{3}z$ for y in the first equation and simplify to obtain

$$x + y - 2z = -1$$

$$x + \frac{1}{3} + \frac{5}{3}z - 2z = -1 \quad \text{Substitute.}$$

$$x + \frac{1}{3} - \frac{1}{3}z = -1 \quad \text{Combine like terms.}$$

$$x - \frac{1}{3}z = -\frac{4}{3} \quad \text{Subtract } \tfrac{1}{3} \text{ from both sides.}$$

$$x = -\frac{4}{3} + \frac{1}{3}z \quad \text{Add } \tfrac{1}{3}z \text{ to both sides.}$$

A solution of this system must have the form

$$\left(-\frac{4}{3} + \frac{1}{3}z, \frac{1}{3} + \frac{5}{3}z, z\right) \quad \text{This solution is a general solution of the system.}$$

for all values of z.

SELF CHECK 3 Solve the system using matrices: $\begin{cases} x + y - 2z = 11 \\ 2x - y + z = -2 \end{cases}$

Everyday connections
Staffing

Matrices with the same number of rows and columns can be added. We simply add their corresponding elements. For example,

$$\begin{bmatrix} 2 & 3 & -4 \\ -1 & 2 & 5 \end{bmatrix} + \begin{bmatrix} 3 & -1 & 0 \\ 4 & 3 & 2 \end{bmatrix}$$
$$= \begin{bmatrix} 2+3 & 3+(-1) & -4+0 \\ -1+4 & 2+3 & 5+2 \end{bmatrix}$$
$$= \begin{bmatrix} 5 & 2 & -4 \\ 3 & 5 & 7 \end{bmatrix}$$

To multiply a matrix by a constant, we multiply each element of the matrix by the constant. For example,

$$5 \cdot \begin{bmatrix} 2 & 3 & -4 \\ -1 & 2 & 5 \end{bmatrix}$$
$$= \begin{bmatrix} 5 \cdot 2 & 5 \cdot 3 & 5 \cdot (-4) \\ 5 \cdot (-1) & 5 \cdot 2 & 5 \cdot 5 \end{bmatrix}$$
$$= \begin{bmatrix} 10 & 15 & -20 \\ -5 & 10 & 25 \end{bmatrix}$$

Since matrices provide a good way to store information in computers, they often are used in applied problems. For example, suppose there are 66 security officers employed at either the downtown office or the suburban office:

Downtown Office

	Male	Female
Day shift	12	18
Night shift	3	0

Suburban Office

	Male	Female
Day shift	14	12
Night shift	5	2

The information about the employees is contained in the following matrices.

$$D = \begin{bmatrix} 12 & 18 \\ 3 & 0 \end{bmatrix} \quad \text{and} \quad S = \begin{bmatrix} 14 & 12 \\ 5 & 2 \end{bmatrix}$$

The entry in the first row-first column in matrix D gives the information that 12 males work the day shift at the downtown office. Company management can add the matrices D and S to find corporate-wide totals:

$$D + S = \begin{bmatrix} 12 & 18 \\ 3 & 0 \end{bmatrix} + \begin{bmatrix} 14 & 12 \\ 5 & 2 \end{bmatrix}$$
$$= \begin{bmatrix} 26 & 30 \\ 8 & 2 \end{bmatrix}$$

We interpret the total to mean:

	Male	Female
Day shift	26	30
Night shift	8	2

If one-third of the force in each category at the downtown location retires, the downtown staff would be reduced to $\frac{2}{3}D$ people. We can compute $\frac{2}{3}D$ by multiplying each entry by $\frac{2}{3}$.

$$\frac{2}{3}D = \frac{2}{3}\begin{bmatrix} 12 & 18 \\ 3 & 0 \end{bmatrix}$$
$$= \begin{bmatrix} 8 & 12 \\ 2 & 0 \end{bmatrix}$$

After retirements, downtown staff would be

	Male	Female
Day shift	8	12
Night shift	2	0

SELF CHECK ANSWERS

1. $(1, -2, 3)$ **2.** $(3, -2)$ **3.** $\left(3 + \frac{1}{3}z, 8 + \frac{5}{3}z, z\right)$

NOW TRY THIS

A toy company builds authentic models of a compact car, a sedan, and a truck. The times required for preparation, assembly, and post-production are given below. Use matrices to help determine how many of each should be made in order to use all available labor hours.

	Compact	Sedan	Truck	Total labor hours
Preparation	1 hr	1 hr	2 hrs	60 hrs
Assembly	2 hrs	3 hrs	4 hrs	130 hrs
Post-production	2 hrs	2 hrs	3 hrs	100 hrs

8.5 Exercises

WARM-UPS *Write the system of equations that correspond to the following.*

1. $\left[\begin{array}{cc|c} 2 & -1 & 3 \\ 1 & 5 & 7 \end{array}\right]$

2. $\left[\begin{array}{cc|c} 1 & -3 & 5 \\ 0 & 4 & -1 \end{array}\right]$

3. $\left[\begin{array}{ccc|c} 3 & -2 & 4 & 1 \\ 5 & 2 & -3 & -7 \\ -1 & 9 & 8 & 0 \end{array}\right]$

4. $\left[\begin{array}{ccc|c} 1 & 5 & -1 & 6 \\ 0 & 4 & -2 & 3 \\ -1 & -1 & 0 & 8 \end{array}\right]$

REVIEW *Write each number in scientific notation.*

5. 470,000,000

6. 0.0000089

7. 75×10^4

8. 0.63×10^4

VOCABULARY AND CONCEPTS *Fill in the blanks.*

9. A ______ is a rectangular array of numbers.

10. The numbers in a matrix are called its ________.

11. A 3×4 matrix has _ rows and 4 _______.

12. A ______ matrix has the same number of rows as columns.

13. An __________ matrix of a system of equations includes the __________ matrix and the column of constants.

14. If a matrix has all 0's below its main diagonal, it is written in _________ form.

Consider the system $\begin{cases} 2x - 3y = 9 \\ 4x + 2y = 2 \end{cases}$

15. Find the coefficient matrix.

16. Find the augmented matrix.

Determine whether each matrix is in triangular form.

17. $\begin{bmatrix} 5 & 3 & -1 \\ 0 & 7 & 6 \\ 0 & 0 & 2 \end{bmatrix}$

18. $\begin{bmatrix} 7 & -2 & 3 \\ 0 & 1 & 4 \\ 0 & 6 & 0 \end{bmatrix}$

Use a row operation on the first matrix to find the missing number in the second matrix.

19. $\begin{bmatrix} 3 & 1 & 2 \\ 4 & 5 & 7 \end{bmatrix}$ $\begin{bmatrix} 3 & 1 & 2 \\ 1 & 4 & \square \end{bmatrix}$

20. $\begin{bmatrix} -1 & 3 & 2 \\ 1 & -2 & 3 \end{bmatrix}$ $\begin{bmatrix} -1 & 3 & 2 \\ \square & 1 & 5 \end{bmatrix}$

21. $\begin{bmatrix} -5 & -1 & 3 \\ -2 & 3 & 1 \end{bmatrix}$ $\begin{bmatrix} -5 & -1 & 3 \\ -4 & 6 & \square \end{bmatrix}$

22. $\begin{bmatrix} 2 & 1 & -3 \\ 2 & 6 & 1 \end{bmatrix}$ $\begin{bmatrix} 6 & 3 & \square \\ 2 & 6 & 1 \end{bmatrix}$

GUIDED PRACTICE *Use matrices to solve each system of equations. Give a general solution if necessary.* SEE EXAMPLE 1. (OBJECTIVE 1)

23. $\begin{cases} 2x - 3y = 3 \\ 2x - y = 5 \end{cases}$

24. $\begin{cases} x + y = 3 \\ x - y = -1 \end{cases}$

25. $\begin{cases} x + 2y = -2 \\ 3x - y = 8 \end{cases}$

26. $\begin{cases} 2x - 3y = 16 \\ -4x + y = -22 \end{cases}$

27. $\begin{cases} x + y + z = 6 \\ x + 2y + z = 8 \\ x + y + 2z = 9 \end{cases}$

28. $\begin{cases} x - y + z = 2 \\ x + 2y - z = 6 \\ 2x - y - z = 3 \end{cases}$

29. $\begin{cases} x - y = 1 \\ y + z = 1 \\ x + z = 2 \end{cases}$

30. $\begin{cases} x + z = 1 \\ x + y = 2 \\ 2x + y + z = 3 \end{cases}$

Use matrices to solve each system of equations. SEE EXAMPLE 2. (OBJECTIVE 2)

31. $\begin{cases} x + y = 3 \\ 3x - y = 1 \\ 2x + y = 4 \end{cases}$

32. $\begin{cases} x - y = -5 \\ 2x + 3y = 5 \\ x + y = 1 \end{cases}$

33. $\begin{cases} 2x + y = 7 \\ x - y = 2 \\ -x + 3y = -2 \end{cases}$

34. $\begin{cases} 3x - y = 2 \\ -6x + 3y = 0 \\ -x + 2y = -4 \end{cases}$

Use matrices to solve each system of equations. Give a general solution. SEE EXAMPLE 3. (OBJECTIVE 3)

35. $\begin{cases} x + 3y + 2z = 4 \\ -x - 2y - z = -2 \end{cases}$

36. $\begin{cases} 2x - 4y + 3z = 6 \\ -4x + 6y + 4z = -6 \end{cases}$

37. $\begin{cases} -3x - 2y + 2z = 0 \\ -x + y - z = -5 \end{cases}$

38. $\begin{cases} 4x - 3y + 5z = 0 \\ 4x + 3y - z = 0 \end{cases}$

ADDITIONAL PRACTICE *Use matrices to solve each system of equations.*

39. $\begin{cases} 5x - 4y = 8 \\ 10x + 3y = -6 \end{cases}$

40. $\begin{cases} 5x - 4y = 10 \\ x - 7y = 2 \end{cases}$

41. $\begin{cases} 5a = 24 + 2b \\ 5b = 3a + 16 \end{cases}$

42. $\begin{cases} 3m = 2n + 16 \\ 2m = -5n - 2 \end{cases}$

43. $\begin{cases} 3a + b - 3c = 5 \\ a - 2b + 4c = 10 \\ a + b + c = 13 \end{cases}$

44. $\begin{cases} 2a + b - 3c = -1 \\ 3a - 2b - c = -5 \\ a - 3b - 2c = -12 \end{cases}$

45. $\begin{cases} x + 2y + 2z = 2 \\ 2x + y - z = 1 \\ 4x + 5y + 3z = 3 \end{cases}$

46. $\begin{cases} x + 2y - z = 3 \\ 2x - y + 2z = 6 \\ x - 3y + 3z = 4 \end{cases}$

47. $\begin{cases} 2x - y = 4 \\ x + 3y = 2 \\ -x - 4y = -2 \end{cases}$

48. $\begin{cases} 3x - 2y = 5 \\ x + 2y = 7 \\ -3x - y = -11 \end{cases}$

49. $\begin{cases} x + 3y = 7 \\ x + y = 3 \\ 3x + y = 5 \end{cases}$

50. $\begin{cases} x + y = 3 \\ x - 2y = -3 \\ x - y = 1 \end{cases}$

51. $\begin{cases} 5x - 2y = 4 \\ 2x - 4y = -8 \end{cases}$

52. $\begin{cases} 2x - y = -1 \\ x - 2y = 1 \end{cases}$

53. $\begin{cases} 2x + y = -4 \\ 6x + 3y = 1 \end{cases}$

54. $\begin{cases} x - 4y = 6 \\ -3x + 12y = 10 \end{cases}$

55. $\begin{cases} 3x - 2y + 4z = 4 \\ x + y + z = 3 \\ 6x - 2y - 3z = 10 \end{cases}$

56. $\begin{cases} 2x + y - z = -6 \\ x - y - z = -2 \\ -4x + 3y + z = 6 \end{cases}$

57. $\begin{cases} 3x - y = 9 \\ -6x + 2y = -18 \end{cases}$

58. $\begin{cases} x - y = 1 \\ -3x + 3y = -3 \end{cases}$

59. $\begin{cases} x + y + z = 6 \\ x - y + z = 2 \end{cases}$

60. $\begin{cases} x - y = 0 \\ y + z = 3 \\ x + z = 3 \end{cases}$

61. $\begin{cases} x + 2y + z = 1 \\ 2x - y + 2z = 2 \\ 3x + y + 3z = 3 \end{cases}$

62. $\begin{cases} x - 2y + 3z = 9 \\ -x + 3y = -4 \\ 2x - 5y + 3z = 13 \end{cases}$

63. $\begin{cases} 2x + y + 3z = 3 \\ -2x - y + z = 5 \\ 4x - 2y + 2z = 2 \end{cases}$

64. $\begin{cases} 3x + 2y + z = 8 \\ 6x - y + 2z = 16 \\ -9x + y - z = -20 \end{cases}$

APPLICATIONS

65. Piggy banks When a child breaks open her piggy bank, she finds a total of 64 coins, consisting of nickels, dimes, and quarters. The total value of the coins is \$6. If the nickels were dimes, and the dimes were nickels, the value of the coins would be \$5. How many nickels, dimes, and quarters were in the piggy bank?

66. Theater seating The illustration shows the cash receipts and the ticket prices from two sold-out performances of a play. Find the number of seats in each of the three sections of the 800-seat theater.

Sunday Ticket Receipts

Matinee	\$13,000
Evening	\$23,000

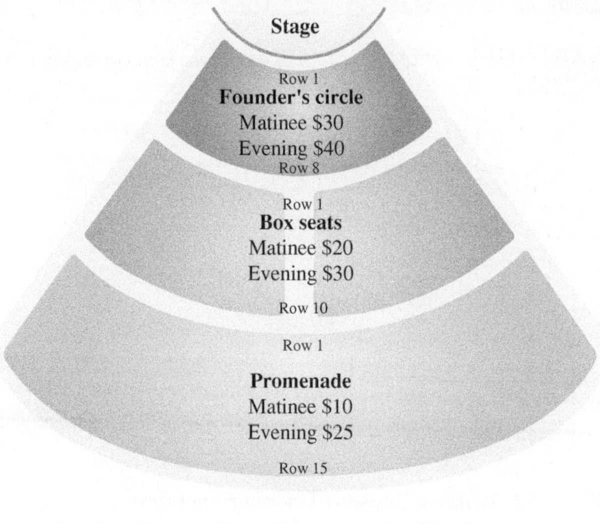

Remember that the equation of a parabola is of the form $y = ax^2 + bx + c$.

67. Curve fitting Find the equation of the parabola passing through the points $(0, 1)$, $(1, 2)$, and $(-1, 4)$.

68. Curve fitting Find the equation of the parabola passing through the points $(0, 1)$, $(1, 1)$, and $(-1, -1)$.

69. Physical therapy After an elbow injury, a volleyball player has restricted movement of her arm. Her range of motion (the measure of $\angle 1$) is 28° less than the measure of $\angle 2$. Find the measure of each angle.

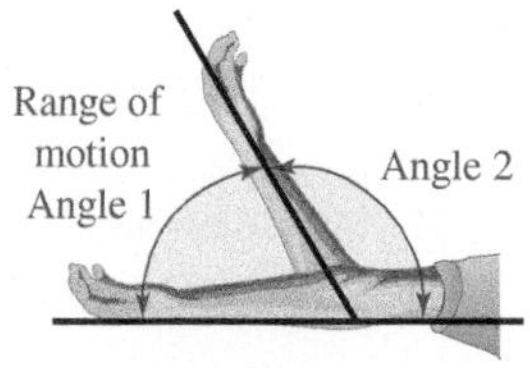

70. Cats and dogs In 2011, there were approximately 164 million dogs and cats in the United States. If there were 8 million more cats than dogs, how many dogs and cats were there?

Remember these facts from geometry. Then solve each problem using two variables.

Two angles whose measures add up to 90° are complementary.
Two angles whose measures add up to 180° are supplementary.
The sum of the measures of the interior angles in a triangle is 180°.

71. Geometry One angle is 28° larger than its complement. Find the measure of each angle.

72. Geometry One angle is 46° larger than its supplement. Find the measure of each angle.

Unless otherwise noted, all content on this page is © Cengage Learning.

73. Geometry In the triangle below, $\angle B$ is 25° more than $\angle A$, and $\angle C$ is 5° less than twice $\angle A$. Find the measure of each angle in the triangle.

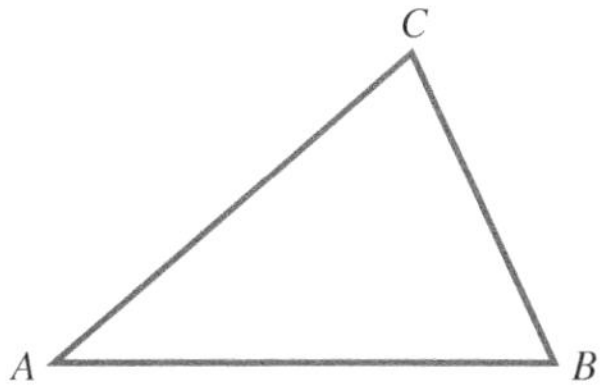

74. Geometry In the triangle below, $\angle A$ is 10° less than $\angle B$, and $\angle B$ is 10° less than $\angle C$. Find the measure of each angle in the triangle.

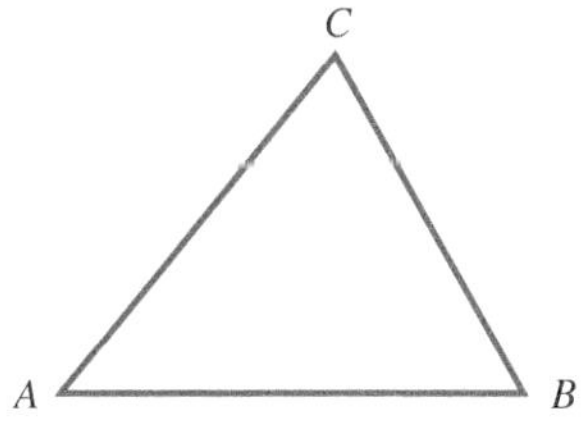

WRITING ABOUT MATH

75. Explain how to check the solution of a system of equations.

76. Explain how to perform a type 3 row operation.

SOMETHING TO THINK ABOUT

77. If the system represented by

$$\left[\begin{array}{ccc|c} 1 & 1 & 0 & 1 \\ 0 & 0 & 1 & 2 \\ 0 & 0 & 0 & k \end{array}\right]$$

has no solution, what do you know about k?

78. Is it possible for a system with fewer equations than variables to have no solution? Illustrate.

Section 8.6 Solving Systems of Linear Equations Using Determinants

Objectives

1. Find the determinant of a 2×2 and a 3×3 matrix without a calculator.
2. Solve a system of linear equations using Cramer's rule.

Vocabulary

determinant | minors | Cramer's rule

Getting Ready

Find each value.

1. $3(-4) - 2(5)$

2. $5(2) - 3(-4)$

3. $2(2 - 5) - 3(5 - 2) + 2(4 - 3)$

4. $-3(5 - 2) + 2(3 + 1) - 2(5 + 1)$

Unless otherwise noted, all content on this page is © Cengage Learning.

We now discuss a final method for solving systems of linear equations. This method involves *determinants*, an idea related to the concept of matrices.

1 Find the determinant of a 2 × 2 and a 3 × 3 matrix without a calculator.

If a matrix A has the same number of rows as columns, it is called a *square matrix.* To each square matrix A, there is associated a number called its *determinant,* represented by the symbol $|A|$.

VALUE OF A 2 × 2 DETERMINANT

If a, b, c, and d are real numbers, the **determinant** of the matrix $\begin{bmatrix} a & b \\ c & d \end{bmatrix}$ is

$$\begin{vmatrix} a & b \\ c & d \end{vmatrix} = ad - bc$$

COMMENT Note that if A is a matrix, $|A|$ represents the determinant of A. If A is a number, $|A|$ represents the absolute value of A.

The determinant of a 2 × 2 matrix is the number that is equal to the product of the numbers on the major diagonal

$$\begin{vmatrix} a & b \\ c & d \end{vmatrix}$$

minus the product of the numbers on the other diagonal

$$\begin{vmatrix} a & b \\ c & d \end{vmatrix}$$

EXAMPLE 1 Evaluate the determinants: **a.** $\begin{vmatrix} 3 & 2 \\ 6 & 9 \end{vmatrix}$ **b.** $\begin{vmatrix} -5 & \frac{1}{2} \\ -1 & 0 \end{vmatrix}$

Solution **a.** $\begin{vmatrix} 3 & 2 \\ 6 & 9 \end{vmatrix} = 3(9) - 2(6)$

$= 27 - 12$

$= 15$

b. $\begin{vmatrix} -5 & \frac{1}{2} \\ -1 & 0 \end{vmatrix} = -5(0) - \frac{1}{2}(-1)$

$= 0 + \frac{1}{2}$

$= \frac{1}{2}$

SELF CHECK 1 Evaluate the determinant: $\begin{vmatrix} 4 & -3 \\ 2 & 1 \end{vmatrix}$

A 3 × 3 determinant can be evaluated by expanding by **minors.**

VALUE OF A 3 × 3 DETERMINANT

Minor of a_1 ↓   Minor of b_1 ↓   Minor of c_1 ↓

$$\begin{vmatrix} a_1 & b_1 & c_1 \\ a_2 & b_2 & c_2 \\ a_3 & b_3 & c_3 \end{vmatrix} = a_1\begin{vmatrix} b_2 & c_2 \\ b_3 & c_3 \end{vmatrix} - b_1\begin{vmatrix} a_2 & c_2 \\ a_3 & c_3 \end{vmatrix} + c_1\begin{vmatrix} a_2 & b_2 \\ a_3 & b_3 \end{vmatrix}$$

To find the minor of a_1, we find the determinant formed by crossing out the elements of the matrix that are in the same row and column as a_1:

$$\begin{vmatrix} a_1 & b_1 & c_1 \\ a_2 & b_2 & c_2 \\ a_3 & b_3 & c_3 \end{vmatrix} \qquad \text{The minor of } a_1 \text{ is } \begin{vmatrix} b_2 & c_2 \\ b_3 & c_3 \end{vmatrix}.$$

To find the minor of b_1, we cross out the elements of the matrix that are in the same row and column as b_1:

$$\begin{vmatrix} a_1 & b_1 & c_1 \\ a_2 & b_2 & c_2 \\ a_3 & b_3 & c_3 \end{vmatrix} \qquad \text{The minor of } b_1 \text{ is } \begin{vmatrix} a_2 & c_2 \\ a_3 & c_3 \end{vmatrix}.$$

To find the minor of c_1, we cross out the elements of the matrix that are in the same row and column as c_1:

$$\begin{vmatrix} a_1 & b_1 & c_1 \\ a_2 & b_2 & c_2 \\ a_3 & b_3 & c_3 \end{vmatrix} \qquad \text{The minor of } c_1 \text{ is } \begin{vmatrix} a_2 & b_2 \\ a_3 & b_3 \end{vmatrix}.$$

We can evaluate a 3×3 determinant by expanding it along any row or column. To determine the signs between the terms of the expansion of a 3×3 determinant, we use the following array of signs.

ARRAY OF SIGNS FOR A 3 × 3 DETERMINANT

$$\begin{matrix} + & - & + \\ - & + & - \\ + & - & + \end{matrix}$$

The pattern above (alternating signs within the rows and columns beginning with a positive in row 1, column 1) holds true for any square matrix.

EXAMPLE 2 Evaluate the determinant: $\begin{vmatrix} 1 & 3 & -2 \\ 2 & 1 & 3 \\ 1 & 2 & 3 \end{vmatrix}$

Solution

Minor of 1 ↓ Minor of 3 ↓ Minor of −2 ↓

$$\begin{vmatrix} 1 & 3 & -2 \\ 2 & 1 & 3 \\ 1 & 2 & 3 \end{vmatrix} = 1\begin{vmatrix} 1 & 3 \\ 2 & 3 \end{vmatrix} - 3\begin{vmatrix} 2 & 3 \\ 1 & 3 \end{vmatrix} + (-2)\begin{vmatrix} 2 & 1 \\ 1 & 2 \end{vmatrix}$$

$$= 1(3 - 6) - 3(6 - 3) - 2(4 - 1)$$

$$= -3 - 9 - 6$$

$$= -18$$

SELF CHECK 2 Evaluate the determinant: $\begin{vmatrix} 2 & -1 & 3 \\ 1 & 2 & -2 \\ 3 & 1 & 1 \end{vmatrix}$

Perspective

LEWIS CARROLL

One of the more amusing historical anecdotes concerning matrices and determinants involves the English mathematician Charles Dodgson, also known as Lewis Carroll. The anecdote describes how England's Queen Victoria so enjoyed reading Carroll's book *Alice in Wonderland* that she requested a copy of his next publication. To her great surprise, she received an autographed copy of a mathematics text titled *An Elementary Treatise on Determinants.* The story was repeated as fact so often that Carroll finally included an explicit disclaimer in his book *Symbolic Logic*, insisting that the incident never actually occurred.

© Morphart Creations Inc./Shutterstock.com

Evaluate each determinant.

1. $\begin{vmatrix} 3 & 4 \\ 2 & 1 \end{vmatrix}$ **2.** $\begin{vmatrix} 3 & 4 & 2 \\ 1 & -1 & 5 \\ 1 & 2 & -2 \end{vmatrix}$

Source: http://mathworld.wolfram.com/Determinant.html

EXAMPLE 3 Evaluate $\begin{vmatrix} 1 & 3 & -2 \\ 2 & 1 & 3 \\ 1 & 2 & 3 \end{vmatrix}$ by expanding on the middle column.

Solution This is the determinant of Example 2. To expand it along the middle column, we use the signs of the middle column of the array of signs:

Minor of 3 ↓ Minor of 1 ↓ Minor of 2 ↓

$$\begin{vmatrix} 1 & 3 & -2 \\ 2 & 1 & 3 \\ 1 & 2 & 3 \end{vmatrix} = -3\begin{vmatrix} 2 & 3 \\ 1 & 3 \end{vmatrix} + 1\begin{vmatrix} 1 & -2 \\ 1 & 3 \end{vmatrix} - 2\begin{vmatrix} 1 & -2 \\ 2 & 3 \end{vmatrix}$$

$$= -3(6 - 3) + 1[3 - (-2)] - 2[3 - (-4)]$$

$$= -3(3) + 1(5) - 2(7)$$

$$= -9 + 5 - 14$$

$$= -18$$

As expected, we get the same value as in Example 2.

SELF CHECK 3 Evaluate: $\begin{vmatrix} 2 & -1 & 3 \\ 1 & 2 & 2 \\ 3 & 1 & 1 \end{vmatrix}$

Accent on technology

▸ Evaluating Determinants

To use a TI-84 graphing calculator to evaluate the determinant in Example 3, $\begin{vmatrix} 1 & 3 & -2 \\ 2 & 1 & 3 \\ 1 & 2 & 3 \end{vmatrix}$, we first enter the matrix by pressing the **MATRX** key, selecting EDIT, and pressing the **ENTER** key. We then enter the dimensions and the elements of the matrix to obtain Figure 8-18(a). We then press **2ND QUIT** to clear the screen. We then press **MATRX**, select MATH(1 det), and press **ENTER** to obtain Figure 8-18(b). Next, press **MATRX** (NAMES) and press **ENTER** to obtain Figure 8-18(c). Press **ENTER** again to find the determinant. The value of the determinant is −18.

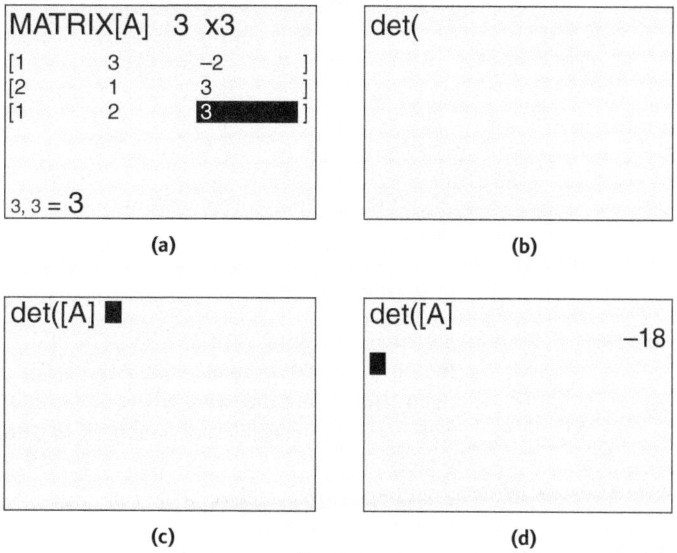

Figure 8-18

For instructions regarding the use of a Casio graphing calculator, please refer to the Casio Keystroke Guide in the back of the book.

2 Solve a system of linear equations using Cramer's rule.

The method of using determinants to solve systems of equations is called **Cramer's rule**, named after the 18th-century mathematician Gabriel Cramer. To develop Cramer's rule, we consider the system

$$\begin{cases} ax + by = e \\ cx + dy = f \end{cases}$$

where x and y are variables and a, b, c, d, e, and f are numerical values.

If we multiply both sides of the first equation by d and multiply both sides of the second equation by $-b$, we can add the equations and eliminate y:

$$\begin{array}{rcl} adx + bdy &=& ed \\ -bcx - bdy &=& -bf \\ \hline adx - bcx \qquad &=& ed - bf \end{array}$$

Gabriel Cramer

1704–1752

Although other mathematicians had worked with determinants, it was the work of Cramer that popularized them.

To solve for x, we use the distributive property to write $adx - bcx$ as $(ad - bc)x$ on the left side and divide each side by $ad - bc$:

$$(ad - bc)x = ed - bf$$

$$x = \frac{ed - bf}{ad - bc} \quad (ad - bc) \neq 0$$

We can find y in a similar manner. After eliminating the variable x, we have

$$y = \frac{af - ec}{ad - bc} \quad (ad \quad bc) \neq 0$$

Determinants provide a way of remembering these formulas. Note that the denominator for both x and y is

$$\begin{vmatrix} a & b \\ c & d \end{vmatrix} = ad - bc$$

Unless otherwise noted, all content on this page is © Cengage Learning.

The numerators can be expressed as determinants also:

$$x = \frac{ed - bf}{ad - bc} = \frac{\begin{vmatrix} e & b \\ f & d \end{vmatrix}}{\begin{vmatrix} a & b \\ c & d \end{vmatrix}} \quad \text{and} \quad x = \frac{af - ec}{ad - bc} = \frac{\begin{vmatrix} a & e \\ c & f \end{vmatrix}}{\begin{vmatrix} a & b \\ c & d \end{vmatrix}}$$

If we compare these formulas with the original system

$$\begin{cases} ax + by = e \\ cx + dy = f \end{cases}$$

we note that in the expressions for x and y above, the denominator determinant is formed by using the coefficients $a, b, c,$ and d of the variables in the equations. The numerator determinants are the same as the denominator determinant, except that the column of coefficients of the variable for which we are solving is replaced with the column of constants e and f.

CRAMER'S RULE FOR TWO LINEAR EQUATIONS IN TWO VARIABLES

The solution of the system $\begin{cases} ax + by = e \\ cx + dy = f \end{cases}$ is given by

$$x = \frac{D_x}{D} \quad \text{and} \quad y = \frac{D_y}{D}$$

where $D = \begin{vmatrix} a & b \\ c & d \end{vmatrix}$, $D_x = \begin{vmatrix} e & b \\ f & d \end{vmatrix}$, and $D_y = \begin{vmatrix} a & e \\ c & f \end{vmatrix}$.

If $D \neq 0$, the system is consistent and the equations are independent.

If $D = 0$ and D_x or D_y is nonzero, the system is inconsistent.

EXAMPLE 4 Use Cramer's rule to solve $\begin{cases} 4x - 3y = 6 \\ -2x + 5y = 4 \end{cases}$.

Solution The value of x is the quotient of two determinants. The denominator determinant is made up of the coefficients of x and y:

$$D = \begin{vmatrix} 4 & -3 \\ -2 & 5 \end{vmatrix}$$

To solve for x, we form the numerator determinant from the denominator determinant by replacing its first column (the coefficients of x) with the column of constants (6 and 4).

To solve for y, we form the numerator determinant from the denominator determinant by replacing the second column (the coefficients of y) with the column of constants (6 and 4).

To find the values of x and y, we evaluate each determinant:

$$x = \frac{D_x}{D} = \frac{\begin{vmatrix} 6 & -3 \\ 4 & 5 \end{vmatrix}}{\begin{vmatrix} 4 & -3 \\ -2 & 5 \end{vmatrix}} = \frac{6(5) - (-3)(4)}{4(5) - (-3)(-2)} = \frac{30 + 12}{20 - 6} = \frac{42}{14} = 3$$

$$y = \frac{D_y}{D} = \frac{\begin{vmatrix} 4 & 6 \\ -2 & 4 \end{vmatrix}}{\begin{vmatrix} 4 & -3 \\ -2 & 5 \end{vmatrix}} = \frac{4(4) - 6(-2)}{4(5) - (-3)(-2)} = \frac{16 + 12}{20 - 6} = \frac{28}{14} = 2$$

The solution of this system is $(3, 2)$. Verify that $x = 3$ and $y = 2$ satisfy each equation in the given system.

SELF CHECK 4 Solve the system: $\begin{cases} 2x - 3y = -16 \\ 3x + 5y = 14 \end{cases}$

EXAMPLE 5

Use Cramer's rule to solve $\begin{cases} 7x = 8 - 4y \\ 2y = 3 - \frac{7}{2}x \end{cases}$.

Solution We multiply both sides of the second equation by 2 to eliminate the fraction and write the system in the form

$$\begin{cases} 7x + 4y = 8 \\ 7x + 4y = 6 \end{cases}$$

When we attempt to use Cramer's rule to solve this system for x, we obtain

$$x = \frac{D_x}{D} = \frac{\begin{vmatrix} 8 & 4 \\ 6 & 4 \end{vmatrix}}{\begin{vmatrix} 7 & 4 \\ 7 & 4 \end{vmatrix}} = \frac{8}{0}, \text{ which is undefined.}$$

This system is inconsistent because the denominator determinant is 0 and the numerator determinant is not 0. Since this system has no solution, its solution set is $\varnothing$.

We can see directly from the system that it is inconsistent. For any values of x and y, it is impossible that 7 times x plus 4 times y could be both 8 and 6.

SELF CHECK 5 Solve the system: $\begin{cases} 3x = 8 - 4y \\ y = \frac{5}{2} - \frac{3}{4}x \end{cases}$

EXAMPLE 6

Use Cramer's rule to solve $\begin{cases} y = -3x + 1 \\ 6x + 2y = 2 \end{cases}$.

Solution We first write the system in the form

$$\begin{cases} 3x + y = 1 \\ 6x + 2y = 2 \end{cases}$$

When we attempt to use Cramer's rule to solve this system for x, we obtain

$$x = \frac{D_x}{D} = \frac{\begin{vmatrix} 1 & 1 \\ 2 & 2 \end{vmatrix}}{\begin{vmatrix} 3 & 1 \\ 6 & 2 \end{vmatrix}} = \frac{0}{0}, \text{ which is indeterminate.}$$

When we attempt to use Cramer's rule to solve this system for y, we obtain

$$y = \frac{D_y}{D} = \frac{\begin{vmatrix} 3 & 1 \\ 6 & 2 \end{vmatrix}}{\begin{vmatrix} 3 & 1 \\ 6 & 2 \end{vmatrix}} = \frac{0}{0}, \text{ which is indeterminate.}$$

This system is consistent and its equations are dependent because every determinant is 0. Every solution of one equation is also a solution of the other. The solution is the general ordered pair $(x, -3x + 1)$.

SELF CHECK 6 Solve the system: $\begin{cases} y = 2x - 5 \\ 10x - 5y = 25 \end{cases}$

CRAMER'S RULE FOR THREE LINEAR EQUATIONS IN THREE VARIABLES

The solution of the system $\begin{cases} ax + by + cz = j \\ dx + ey + fz = k \\ gx + hy + iz = l \end{cases}$ is given by

$$x = \frac{D_x}{D}, \quad y = \frac{D_y}{D}, \quad \text{and} \quad z = \frac{D_z}{D}$$

where

$$D = \begin{vmatrix} a & b & c \\ d & e & f \\ g & h & i \end{vmatrix} \qquad D_x = \begin{vmatrix} \mathbf{j} & b & c \\ \mathbf{k} & e & f \\ \mathbf{l} & h & i \end{vmatrix}$$

$$D_y = \begin{vmatrix} a & \mathbf{j} & c \\ d & \mathbf{k} & f \\ g & \mathbf{l} & i \end{vmatrix} \qquad D_z = \begin{vmatrix} a & b & \mathbf{j} \\ d & e & \mathbf{k} \\ g & h & \mathbf{l} \end{vmatrix}$$

If $D \neq 0$, the system is consistent and the equations are independent.

If $D = 0$ and D_x or D_y or D_z is nonzero, the system is inconsistent.

If the determinant of the coefficient, D, is 0, the system is inconsistent or has infinitely many solutions. Use another method to solve.

EXAMPLE 7 Use Cramer's rule to solve $\begin{cases} 2x + y + 4z = 12 \\ x + 2y + 2z = 9 \\ 3x - 3y - 2z = 1 \end{cases}$.

Solution The denominator determinant is the determinant formed by the coefficients of the variables. To form the numerator determinants, we substitute the column of constants for the coefficients of the variable to be found. We form the quotients for $x, y,$ and z and evaluate the determinants:

$$x = \frac{D_x}{D} = \frac{\begin{vmatrix} 12 & 1 & 4 \\ 9 & 2 & 2 \\ 1 & -3 & -2 \end{vmatrix}}{\begin{vmatrix} 2 & 1 & 4 \\ 1 & 2 & 2 \\ 3 & -3 & -2 \end{vmatrix}} = \frac{12\begin{vmatrix} 2 & 2 \\ -3 & -2 \end{vmatrix} - 1\begin{vmatrix} 9 & 2 \\ 1 & -2 \end{vmatrix} + 4\begin{vmatrix} 9 & 2 \\ 1 & -3 \end{vmatrix}}{2\begin{vmatrix} 2 & 2 \\ -3 & -2 \end{vmatrix} - 1\begin{vmatrix} 1 & 2 \\ 3 & -2 \end{vmatrix} + 4\begin{vmatrix} 1 & 2 \\ 3 & -3 \end{vmatrix}} = \frac{12(2) - (-20) + 4(-29)}{2(2) - (-8) + 4(-9)} = \frac{-72}{-24} = 3$$

$$y = \frac{D_y}{D} = \frac{\begin{vmatrix} 2 & 12 & 4 \\ 1 & 9 & 2 \\ 3 & 1 & -2 \end{vmatrix}}{\begin{vmatrix} 2 & 1 & 4 \\ 1 & 2 & 2 \\ 3 & -3 & -2 \end{vmatrix}} = \frac{2\begin{vmatrix} 9 & 2 \\ 1 & -2 \end{vmatrix} - 12\begin{vmatrix} 1 & 2 \\ 3 & -2 \end{vmatrix} + 4\begin{vmatrix} 1 & 9 \\ 3 & 1 \end{vmatrix}}{-24} = \frac{2(-20) - 12(-8) + 4(-26)}{-24} = \frac{-48}{-24} = 2$$

$$z = \frac{D_z}{D} = \frac{\begin{vmatrix} 2 & 1 & 12 \\ 1 & 2 & 9 \\ 3 & -3 & 1 \end{vmatrix}}{\begin{vmatrix} 2 & 1 & 4 \\ 1 & 2 & 2 \\ 3 & -3 & -2 \end{vmatrix}} = \frac{2\begin{vmatrix} 2 & 9 \\ -3 & 1 \end{vmatrix} - 1\begin{vmatrix} 1 & 9 \\ 3 & 1 \end{vmatrix} + 12\begin{vmatrix} 1 & 2 \\ 3 & -3 \end{vmatrix}}{-24} = \frac{2(29) - 1(-26) + 12(-9)}{-24} = \frac{-24}{-24} = 1$$

The solution of this system is $(3, 2, 1)$.

SELF CHECK 7 Solve the system: $\begin{cases} x + y + 2z = 6 \\ 2x - y + z = 9 \\ x + y - 2z = -6 \end{cases}$

SELF CHECK ANSWERS **1.** 10 **2.** 0 **3.** −20 **4.** $(-2, 4)$ **5.** ∅ **6.** $(x, 2x - 5)$ **7.** $(2, -2, 3)$

NOW TRY THIS

Solve for x.

1. $\begin{vmatrix} x & 2 \\ x & 3 \end{vmatrix} = 4$

2. $\begin{vmatrix} 2x & 3 \\ -5x & -3 \end{vmatrix} = 10(x - 1)$

3. $\begin{vmatrix} x + 4 & 3 \\ 2x - 5 & 2 \end{vmatrix} = 2x + 5$

4. $\begin{vmatrix} 2 & x & -1 \\ 1 & 2x & 4 \\ -4 & x & 1 \end{vmatrix} = 30$

8.6 Exercises

WARM-UPS *Evaluate ad − bc for the following.*

1. $a = 4, b = 3, c = 2, d = -1$

2. $a = 2, b = 0, c = -1, d = 4$

3. $a = 0, b = 8, c = 2, d = 5$

When using Cramer's rule to solve the system $\begin{cases} 2x - y = -21 \\ 4x + 5y = 7 \end{cases}$

4. Set up the denominator determinant for x.

5. Set up the numerator determinant for x.

6. Set up the numerator determinant for y.

REVIEW *Solve each equation.*

7. $5(x - 2) - (3 - x) = x + 2$

8. $\frac{3}{7}x = 2(x + 11)$

9. $\frac{5}{3}(5x + 6) - 10 = 0$

10. $5 - 3(2x - 1) = 2(4 + 3x) - 24$

VOCABULARY AND CONCEPTS *Fill in the blanks.*

11. A determinant is a ______ that is associated with a ______ matrix.

12. The value of $\begin{vmatrix} a & b \\ c & d \end{vmatrix}$ is ______.

13. The minor of b_1 in $\begin{vmatrix} a_1 & b_1 & c_1 \\ a_2 & b_2 & c_2 \\ a_3 & b_3 & c_3 \end{vmatrix}$ is ________.

14. We can evaluate a determinant by expanding it along any ___ or ______.

15. The method of solving a system of linear equations using determinants is called __________.

16. The setup for the denominator determinant for the value of x in the system $\begin{cases} 5x + 3y = 6 \\ 4x - 2y = 7 \end{cases}$ is ________.

17. If $D \neq 0$, then the system is ________ and the equations are ________.

18. If the denominator determinant for y in a system of equations is zero, the equations of the system are ________ or the system is _________.

GUIDED PRACTICE *Evaluate each determinant.* SEE EXAMPLE 1. (OBJECTIVE 1)

19. $\begin{vmatrix} 4 & 2 \\ -3 & 2 \end{vmatrix}$

20. $\begin{vmatrix} 3 & -2 \\ -2 & 4 \end{vmatrix}$

21. $\begin{vmatrix} -2 & 5 \\ 1 & -3 \end{vmatrix}$

22. $\begin{vmatrix} -1 & -2 \\ -3 & -4 \end{vmatrix}$

Evaluate each determinant. SEE EXAMPLES 2–3. (OBJECTIVE 1)

23. $\begin{vmatrix} 1 & 1 & 1 \\ 1 & 0 & 2 \\ 0 & 2 & 0 \end{vmatrix}$

24. $\begin{vmatrix} 1 & 2 & 0 \\ 0 & 1 & 2 \\ 0 & 0 & 1 \end{vmatrix}$

25. $\begin{vmatrix} -1 & 2 & 1 \\ 2 & 1 & -3 \\ 1 & 1 & 1 \end{vmatrix}$

26. $\begin{vmatrix} 1 & 2 & 3 \\ 1 & 2 & 3 \\ 1 & 2 & 3 \end{vmatrix}$

27. $\begin{vmatrix} 1 & -2 & 3 \\ -2 & 1 & 1 \\ -3 & -2 & 1 \end{vmatrix}$

28. $\begin{vmatrix} 1 & 1 & 2 \\ 2 & 1 & -2 \\ 3 & 1 & 3 \end{vmatrix}$

29. $\begin{vmatrix} 2 & 4 & 6 \\ 1 & 3 & 5 \\ 9 & 8 & 7 \end{vmatrix}$

30. $\begin{vmatrix} 1 & 4 & 7 \\ 2 & 5 & 8 \\ 3 & 6 & 9 \end{vmatrix}$

Use Cramer's rule to solve each system, if possible. If the equations of the system are dependent, give a general solution. SEE EXAMPLE 4. (OBJECTIVE 2)

31. $\begin{cases} 4x + 3y = 5 \\ 3x - 2y = -9 \end{cases}$

32. $\begin{cases} 3x - y = -3 \\ 2x + y = -7 \end{cases}$

33. $\begin{cases} x + y = 6 \\ x - y = 2 \end{cases}$

34. $\begin{cases} x - y = 4 \\ 2x + y = 5 \end{cases}$

Use Cramer's rule to solve each system, if possible. If the equations of the system are dependent, give a general solution. SEE EXAMPLE 5. (OBJECTIVE 2)

35. $\begin{cases} 4x = 2y - 5 \\ y = \dfrac{4x - 5}{2} \end{cases}$

36. $\begin{cases} y = \dfrac{11 - 3x}{2} \\ x = \dfrac{11 - 4y}{6} \end{cases}$

37. $\begin{cases} 2x = \dfrac{y + 5}{2} \\ y = 4x - 6 \end{cases}$

38. $\begin{cases} y = \dfrac{12 + 2x}{3} \\ x = \dfrac{3}{2}y + 3 \end{cases}$

Use Cramer's rule to solve each system, if possible. If the equations of the system are dependent, give a general solution. SEE EXAMPLE 6. (OBJECTIVE 2)

39. $\begin{cases} 2x + 3y = 9 \\ y = -\dfrac{2}{3}x + 3 \end{cases}$

40. $\begin{cases} x = \dfrac{12 - 6y}{5} \\ y = \dfrac{24 - 10x}{12} \end{cases}$

41. $\begin{cases} 4x - 3y = 6 \\ y = \dfrac{4x - 6}{3} \end{cases}$

42. $\begin{cases} 2x + 3y = 12 \\ x = \dfrac{12 - 3y}{2} \end{cases}$

Use Cramer's rule to solve each system, if possible. If the equations of the system are dependent, give a general solution. SEE EXAMPLE 7. (OBJECTIVE 2)

43. $\begin{cases} x + y + z = 4 \\ x + y - z = 0 \\ x - y + z = 2 \end{cases}$

44. $\begin{cases} x + y + z = 4 \\ x - y + z = 2 \\ x - y - z = 0 \end{cases}$

45. $\begin{cases} x + y + 2z = 7 \\ x + 2y + z = 8 \\ 2x + y + z = 9 \end{cases}$

46. $\begin{cases} x + 2y + 2z = 10 \\ 2x + y + 2z = 9 \\ 2x + 2y + z = 1 \end{cases}$

47. $\begin{cases} 2x + y - z = 1 \\ x + 2y + 2z = 2 \\ 4x + 5y + 3z = 3 \end{cases}$

48. $\begin{cases} 2x - y + 4z + 2 = 0 \\ 5x + 8y + 7z = -8 \\ x + 3y + z + 3 = 0 \end{cases}$

49. $\begin{cases} 2x + 3y + 4z = 6 \\ 2x - 3y - 4z = -4 \\ 4x + 6y + 8z = 12 \end{cases}$

50. $\begin{cases} x - 3y + 4z - 2 = 0 \\ 2x + y + 2z - 3 = 0 \\ 4x - 5y + 10z - 7 = 0 \end{cases}$

ADDITIONAL PRACTICE *Evaluate each determinant.*

51. $\begin{vmatrix} 2a & b \\ b & 2a \end{vmatrix}$

52. $\begin{vmatrix} y & x \\ x + y & x - y \end{vmatrix}$

53. $\begin{vmatrix} a & 2a & -a \\ 2 & -1 & 3 \\ 1 & 2 & -3 \end{vmatrix}$

54. $\begin{vmatrix} 1 & 2b & -3 \\ 2 & -b & 2 \\ 1 & 3b & 1 \end{vmatrix}$

55. $\begin{vmatrix} 1 & a & b \\ 1 & 2a & 2b \\ 1 & 3a & 3b \end{vmatrix}$

56. $\begin{vmatrix} a & b & c \\ 0 & b & c \\ 0 & 0 & c \end{vmatrix}$

Use a graphing calculator to evaluate each determinant.

57. $\begin{bmatrix} 2 & -3 & 4 \\ -1 & 2 & 4 \\ 3 & -3 & 1 \end{bmatrix}$

58. $\begin{bmatrix} -3 & 2 & -5 \\ 3 & -2 & 6 \\ 1 & -3 & 4 \end{bmatrix}$

59. $\begin{bmatrix} 2 & 1 & -3 \\ -2 & 2 & 4 \\ 1 & -2 & 2 \end{bmatrix}$

60. $\begin{bmatrix} 4 & 2 & -3 \\ 2 & -5 & 6 \\ 2 & 5 & -2 \end{bmatrix}$

Use Cramer's rule to solve each system.

61. $\begin{cases} 2x + 3y = 0 \\ 4x - 6y = -4 \end{cases}$

62. $\begin{cases} 4x - 3y = -1 \\ 8x + 3y = 4 \end{cases}$

63. $\begin{cases} y = \dfrac{-2x + 1}{3} \\ 3x - 2y = 8 \end{cases}$

64. $\begin{cases} 2x + 3y = -1 \\ x = \dfrac{y - 9}{4} \end{cases}$

65. $\begin{cases} x = \dfrac{5y - 4}{2} \\ y = \dfrac{3x - 1}{5} \end{cases}$

66. $\begin{cases} y = \dfrac{1 - 5x}{2} \\ x = \dfrac{3y + 10}{4} \end{cases}$

67. $\begin{cases} 2x + y + z = 5 \\ x - 2y + 3z = 10 \\ x + y - 4z = -3 \end{cases}$

68. $\begin{cases} 3x + 2y - z = -8 \\ 2x - y + 7z = 10 \\ 2x + 2y - 3z = -10 \end{cases}$

69. $\begin{cases} 4x + 3z = 4 \\ 2y - 6z = -1 \\ 8x + 4y + 3z = 9 \end{cases}$

70. $\begin{cases} \frac{1}{2}x + y + z + \frac{3}{2} = 0 \\ x + \frac{1}{2}y + z - \frac{1}{2} = 0 \\ x + y + \frac{1}{2}z + \frac{1}{2} = 0 \end{cases}$

71. $\begin{cases} x + y = 1 \\ \frac{1}{2}y + z = \frac{5}{2} \\ x - z = -3 \end{cases}$

72. $\begin{cases} 3x + 4y + 14z = 7 \\ -\frac{1}{2}x - y + 2z = \frac{3}{2} \\ x + \frac{3}{2}y + \frac{5}{2}z = 1 \end{cases}$

Solve each equation.

73. $\begin{vmatrix} x & 2 \\ -3 & 1 \end{vmatrix} = 2$

74. $\begin{vmatrix} x & -x \\ 2 & -3 \end{vmatrix} = -5$

75. $\begin{vmatrix} x & -2 \\ 3 & 1 \end{vmatrix} = \begin{vmatrix} 4 & 2 \\ x & 3 \end{vmatrix}$

76. $\begin{vmatrix} x & 3 \\ x & 2 \end{vmatrix} = \begin{vmatrix} 3 & 2 \\ 1 & 1 \end{vmatrix}$

APPLICATIONS

77. Signaling A system of sending signals uses two flags held in various positions to represent letters of the alphabet. The illustration shows how the letter U is signaled. Find the measures x and y, if y is to be 30° more than x.

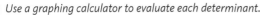

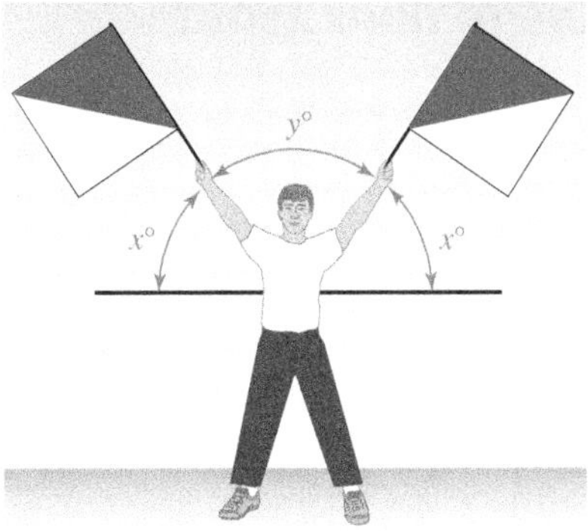

78. Inventories The table shows an end-of-the-year inventory report for a warehouse that supplies electronics stores. If the warehouse stocks two models of cordless telephones, one valued at \$67 and the other at \$100, how many of each model of phone did the warehouse have at the time of the inventory?

Item	Number	Merchandise value
Television	800	\$1,005,450
Radios	200	\$15,785
Cordless phones	360	\$29,400

79. Investing A student wants to average a 6.6% return by investing \$20,000 in the three stocks listed in the table. Because HiTech is considered to be a high-risk investment, he wants to invest three times as much in SaveTel and HiGas combined as he invests in HiTech. How much should he invest in each stock?

Stock	Rate of return
HiTech	10%
SaveTel	5%
HiGas	6%

80. Investing A woman wants to average a $7\frac{1}{3}\%$ return by investing \$30,000 in three certificates of deposit. She wants to invest five times as much in the 8% CD as in the 6% CD. How much should she invest in each CD?

Type of CD	Rate of return
12 month	6%
24 month	7%
36 month	8%

WRITING ABOUT MATH

81. Explain how to find the minor of an element of a determinant.

82. Explain how to find the value of x when solving a system of linear equations by Cramer's rule.

Unless otherwise noted, all content on this page is © Cengage Learning.

SOMETHING TO THINK ABOUT

83. Show that

$$\begin{vmatrix} x & y & 1 \\ -2 & 3 & 1 \\ 3 & 5 & 1 \end{vmatrix} = 0$$

is the equation of the line passing through $(-2, 3)$ and $(3, 5)$.

84. Show that

$$\frac{1}{2}\begin{vmatrix} 0 & 0 & 1 \\ 3 & 0 & 1 \\ 0 & 4 & 1 \end{vmatrix}$$

is the area of the triangle with vertices at $(0, 0)$, $(3, 0)$, and $(0, 4)$.

Determinants with more than 3 rows and 3 columns can be evaluated by expanding them by minors. The sign array for a 4 × 4 determinant is

$$\begin{matrix} + & - & + & - \\ - & + & - & + \\ + & - & + & - \\ - & + & - & + \end{matrix}$$

Evaluate each determinant.

85. $\begin{vmatrix} 1 & 0 & 2 & 1 \\ 2 & 1 & 1 & 3 \\ 1 & 1 & 1 & 1 \\ 2 & 1 & 1 & 1 \end{vmatrix}$

86. $\begin{vmatrix} 1 & 2 & -1 & 1 \\ -2 & 1 & 3 & -1 \\ 0 & 1 & 1 & 2 \\ 2 & 0 & 3 & 1 \end{vmatrix}$

Section 8.7 Solving Systems of Linear Inequalities in Two Variables

Objectives

1. Determine whether an ordered pair is a solution to a given linear inequality.
2. Graph a linear inequality in one or two variables.
3. Solve an application involving a linear inequality in two variables.
4. Graph the solution set of a system of linear inequalities in one or two variables.
5. Solve an application using a system of linear inequalities.

Vocabulary

linear inequality in two variables

half-plane

test point

doubly shaded region

Getting Ready

Graph $y = \frac{1}{3}x + 3$ *and determine whether the given point lies on the line, above the line, or below the line.*

1. $(0, 0)$ **2.** $(0, 4)$ **3.** $(2, 2)$ **4.** $(6, 5)$

5. $(-3, 2)$ **6.** $(6, 8)$ **7.** $(-6, 0)$ **8.** $(-9, 5)$

In this section, we will discuss how to solve linear inequalities in two variables graphically. Then we will show how to solve systems of inequalities. We conclude this section by using these skills to solve applications.

1 Determine whether an ordered pair is a solution to a given linear inequality.

A **linear inequality in two variables** is an inequality that can be written in one of the following forms:

$$Ax + By > C \qquad Ax + By < C \qquad Ax + By \geq C \qquad Ax + By \leq C$$

where A, B, and C are real numbers and A and B are not both 0. Some examples of linear inequalities are

$$2x - y > -3 \qquad y < 3 \qquad x + 47 \geq 6 \qquad x \leq -2$$

An ordered pair (x, y) is a solution of an inequality in two variables if a true statement results when the values of x and y are substituted into the inequality.

EXAMPLE 1 Determine whether each ordered pair is a solution of $y \geq x - 5$.

a. $(4, 2)$ **b.** $(0, -6)$ **c.** $(5, 0)$

Solution **a.** To determine whether $(4, 2)$ is a solution, we substitute 4 for x and 2 for y.

$$y \geq x - 5$$
$$2 \geq 4 - 5$$
$$2 \geq -1$$

Since $2 \geq -1$ is a true inequality, $(4, 2)$ is a solution.

b. To determine whether $(0, -6)$ is a solution, we substitute 0 for x and -6 for y.

$$y \geq x - 5$$
$$-6 \geq 0 - 5$$
$$-6 \geq -5$$

Since $-6 \geq -5$ is a false inequality, $(0, -6)$ is not a solution.

c. To determine whether $(5, 0)$ is a solution, we substitute 5 for x and 0 for y.

$$y \geq x - 5$$
$$0 \geq 5 - 5$$
$$0 \geq 0$$

Since $0 \geq 0$ is a true inequality, $(5, 0)$ is a solution.

SELF CHECK 1 Determine whether each ordered pair is a solution of $y < x + 4$.

a. $(2, 8)$ **b.** $(3, -4)$ **c.** $(-4, 0)$

2 Graph a linear inequality in one or two variables.

The graph of $y = x - 5$ is a line consisting of the points whose coordinates satisfy the equation. The graph of the inequality $y \geq x - 5$ is not a line but rather an area bounded by a line, called a **half-plane**. The half-plane consists of the points whose coordinates satisfy the inequality. The line serves as the *boundary* of the half-plane.

EXAMPLE 2 Graph the inequality: $y \geq x - 5$

Solution Because equality is included in the original inequality, we begin by graphing the equation $y = x - 5$ with a *solid* line, as in Figure 8-19(a). Because the graph of $y \geq x - 5$ also indicates that y can be greater than $x - 5$, the coordinates of points other than those shown in Figure 8-19(a) satisfy the inequality. To determine which side of the line to shade, we select a **test point** and substitute its value into the inequality. A convenient test point is the origin.

$$y \geq x - 5$$
$$\mathbf{0} \geq \mathbf{0} - 5 \quad \text{Substitute 0 for } x \text{ and 0 for } y.$$
$$0 \geq -5$$

Because $0 \geq -5$ is true, the coordinates of the origin satisfy the original inequality. In fact, the coordinates of every point on the same side of the line as the origin satisfy the inequality. The graph of $y \geq x - 5$ is the half-plane that is shaded in Figure 8-19(b).

$y = x - 5$

x	y	(x, y)
0	−5	(0, −5)
5	0	(5, 0)

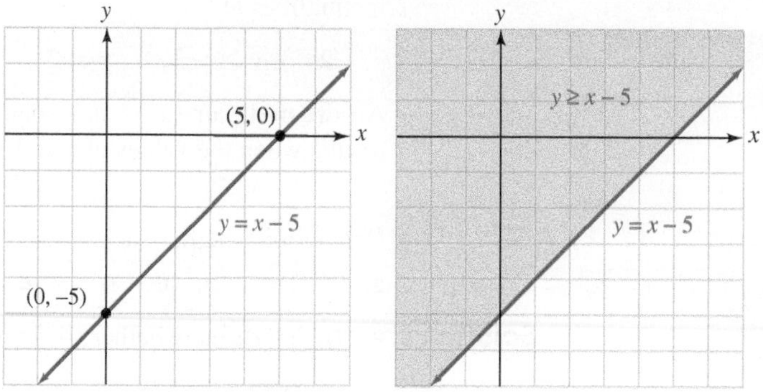

(a) (b)

Figure 8-19

SELF CHECK 2 Graph: $y \geq -x - 2$

EXAMPLE 3 Graph: $x + 2y < 6$

Solution We find the boundary by graphing the equation $x + 2y = 6$. Since the symbol $<$ does not include equality, the points on the graph of $x + 2y = 6$ will not be a part of the graph. To indicate this, we draw the boundary line as a *dashed* line. See Figure 8-20.

To determine which half-plane to shade, we substitute the coordinates of a test point that lies on one side of the boundary line into $x + 2y < 6$. The origin is a convenient choice.

$$x + 2y < 6$$
$$\mathbf{0} + 2(\mathbf{0}) < 6 \quad \text{Substitute 0 for } x \text{ and 0 for } y.$$
$$0 < 6$$

Since $0 < 6$ is true, we shade the side of the line that includes the origin. The graph is shown in Figure 8-20.

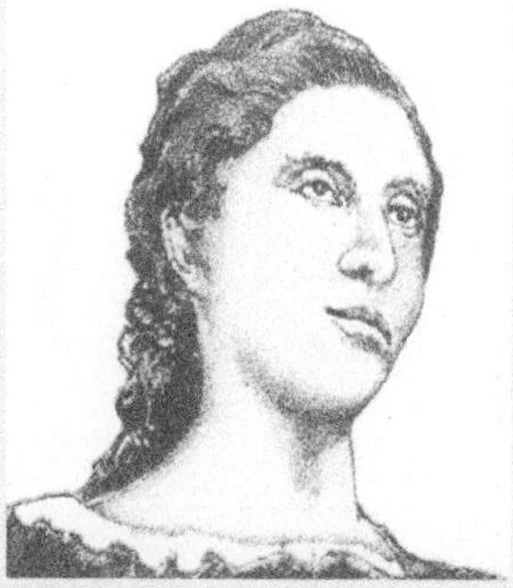

Sophie Germain
1776–1831
Sophie Germain was 13 years old during the French Revolution. Because of dangers caused by the insurrection in Paris, she was kept indoors and spent most of her time reading about mathematics in her father's library. Since interest in mathematics was considered inappropriate for a woman at that time, much of her work was written under the pen name of M. LeBlanc.

$x + 2y = 6$

x	y	(x, y)
0	3	(0, 3)
6	0	(6, 0)
4	1	(4, 1)

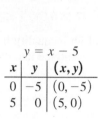

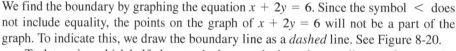

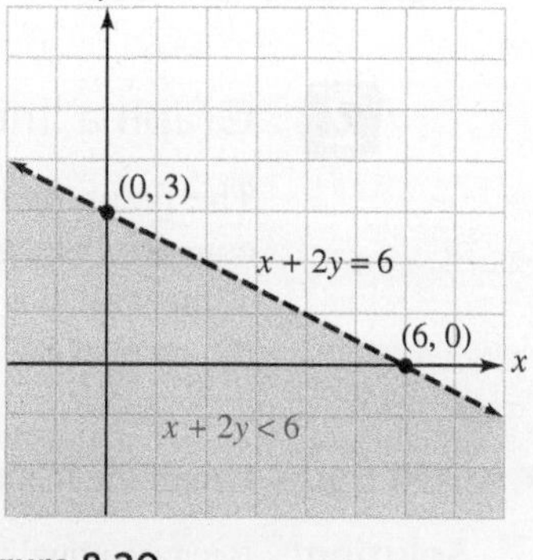

Figure 8-20

SELF CHECK 3 Graph: $-2x + y > -4$

Unless otherwise noted, all content on this page is © Cengage Learning.

COMMENT The decision to use a dashed line or solid line is determined by the inequality symbol. If the symbol is $<$ or $>$, the line is dashed. If it is $\leq$ or $\geq$, the line is solid.

Alternatively, the decision to shade above or below the boundary can be determined by the direction of the inequality after it is solved for y. If $y < mx + b$, shade below the boundary and if $y > mx + b$, shade above it.

EXAMPLE 4 Graph: $2x - y < 0$

Solution To find the boundary line, we graph the equation $2x - y = 0$. Since the symbol $<$ does not include equality, the points on the boundary are not a part of the graph of $2x - y < 0$. To show this, we draw the boundary as a dashed line. See Figure 8-21(a).

To determine which half-plane to shade, we substitute the coordinates of a test point that lies on one side of the boundary into $2x - y < 0$. Because the origin lies on the line, we must choose a different test point. Point $T(0, 2)$, for example, is below the boundary line. See Figure 8-21(a). To see if point $T(2, 0)$ satisfies $2x - y < 0$, we substitute 2 for x and 0 for y in the inequality.

$$\begin{aligned} 2x - y &< 0 \\ 2(2) - 0 &< 0 \\ 4 &< 0 \end{aligned}$$

Since $4 < 0$ is false, the coordinates of point T do not satisfy the inequality, and point T is not on the side of the line we want to shade. Instead, we shade the other side of the boundary line. The graph of the solution set of $2x - y < 0$ is shown in Figure 8-21(b).

$2x - y = 0$

x	y	(x, y)
0	0	(0, 0)
−1	−2	(−1, −2)
3	6	(3, 6)

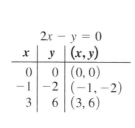

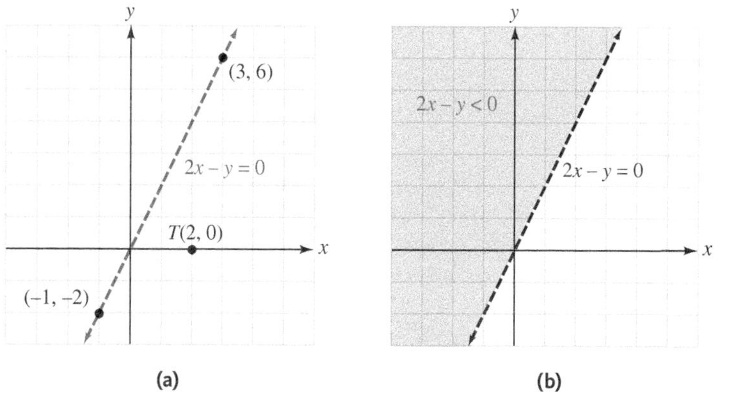

Figure 8-21

SELF CHECK 4 Graph: $3x - y > 0$

3 Solve an application involving a linear inequality in two variables.

EXAMPLE 5 **EARNING MONEY** Chen has two part-time jobs, one paying \$5 per hour and the other paying \$6 per hour. He must earn at least \$120 per week to pay his expenses while attending college. Write an inequality and graph it to show the various ways he can schedule his time to achieve his goal.

Solution If we let x represent the number of hours Chen works on the first job and y the number of hours he works on the second job, we have

Unless otherwise noted, all content on this page is © Cengage Learning.

The hourly rate on the first job	times	the hours worked on the first job	plus	the hourly rate on the second job	times	the hours worked on the second job	is at least	$120.
$5	·	x	+	$6	·	y	$\geq$	$120

The graph of the inequality $5x + 6y \geq 120$ is shown in Figure 8-22. Since Chen cannot work a negative number of hours, the graph in the figure has no meaning when either x or y is negative. Any point in the shaded region indicates a possible way Chen can schedule his time and earn $120 or more per week. For example, if he works 20 hours on the first job and 10 hours on the second job, he will earn

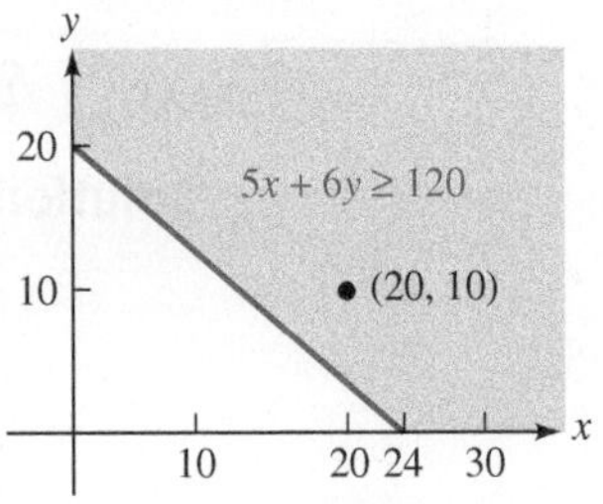

Figure 8-22

$$\begin{aligned} \$5(20) + \$6(10) &= \$100 + \$60 \\ &= \$160 \end{aligned}$$

SELF CHECK 5 If Chen's expenses are $150 per week and his two jobs pay $9 and $10 per hour, write an inequality and graph it to show how he can schedule his time to achieve his goal.

4 Graph the solution set of a system of linear inequalities in one or two variables.

We have seen that the graph of a linear inequality in two variables is a half-plane. Therefore, we would expect the graph of a system of two linear inequalities to contain two half-planes. For example, to solve the system

$$\begin{cases} x + y \geq 1 \\ x - y \geq 1 \end{cases}$$

we graph each inequality and then superimpose the graphs on one set of coordinate axes.

The graph of $x + y \geq 1$ includes the graph of the equation $x + y = 1$ and all points above it. Because the boundary line is included, we draw it with a solid line. See Figure 8-23(a).

The graph of $x - y \geq 1$ includes the graph of the equation $x - y = 1$ and all points below it. Because the boundary line is included, we draw it with a solid line. See Figure 8-23(b).

$x + y = 1$

x	y	(x, y)
0	1	(0, 1)
1	0	(1, 0)
2	−1	(2, −1)

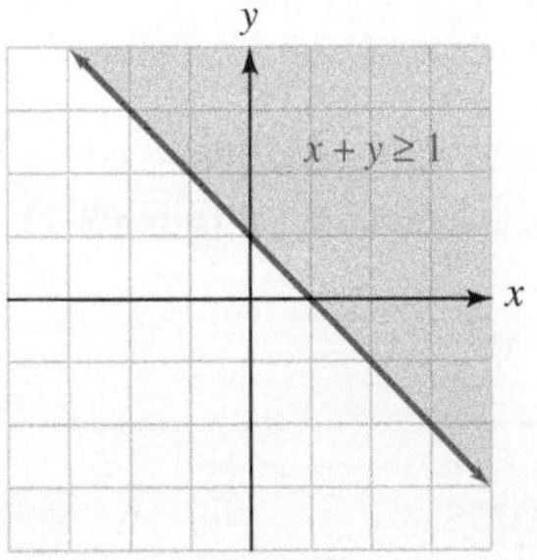

(a)

$x - y = 1$

x	y	(x, y)
0	−1	(0, −1)
1	0	(1, 0)
2	1	(2, 1)

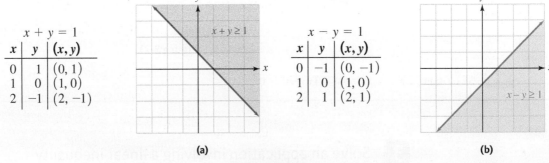

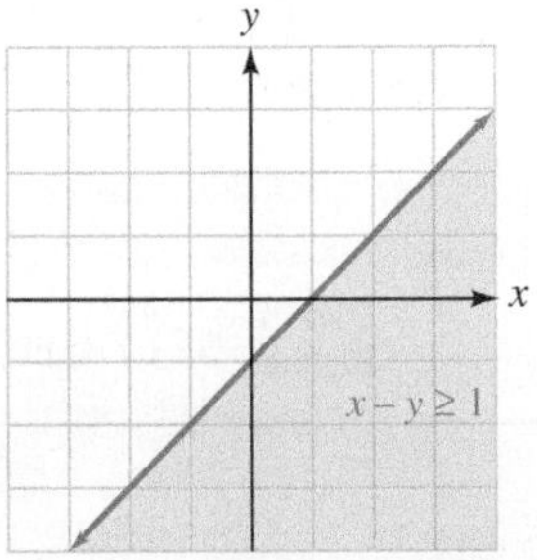

(b)

Figure 8-23

In Figure 8-24(a), we show the result when the graphs are superimposed on one coordinate system. The area that is shaded twice represents the set of solutions of the given system. Any point in the **doubly shaded region** has coordinates that satisfy both of the inequalities.

Figure 8-24(a) is our "working" graph. Just as we solved compound inequalities in one variable, we shade the various regions of the graph and then interpret the results. The solution is only that part of the graph that is doubly shaded, so we create a new graph, Figure 8-24(b), with only the solution.

Unless otherwise noted, all content on this page is © Cengage Learning.

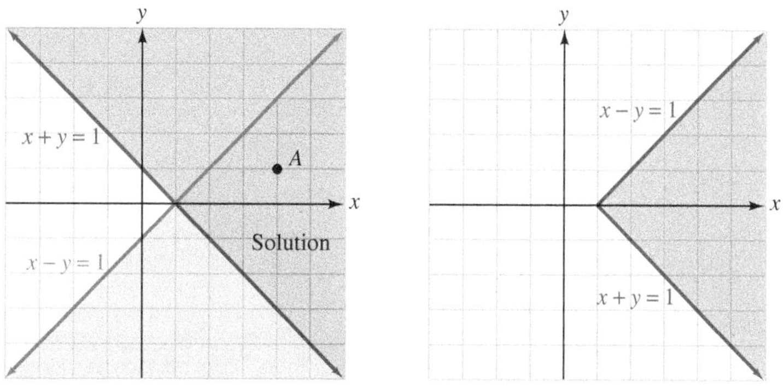

(a) (b)

Figure 8-24

To see that this is true, we select a test point, such as point A, that lies in the doubly shaded region and show that its coordinates (4, 1) satisfy both inequalities.

$x + y \geq 1$	$x - y \geq 1$
$4 + 1 \geq 1$	$4 - 1 \geq 1$
$5 \geq 1$	$3 \geq 1$

Since the coordinates of point A satisfy both inequalities, point A is a solution and therefore all points in the doubly shaded region are solutions.

In general, to solve systems of linear inequalities, we will take the following steps.

SOLVING SYSTEMS OF INEQUALITIES

1. Graph each inequality in the system on the same coordinate axes using solid or dashed lines as appropriate.
2. Find the region where the graphs overlap.
3. Select a test point from the region to verify the solution.
4. Graph only the solution set, if there is one.

EXAMPLE 6 Graph the solution set: $\begin{cases} 2x + y < 4 \\ -2x + y > 2 \end{cases}$

Solution We graph each inequality on one set of coordinate axes, as in Figure 8-25(a).

- The graph of $2x + y < 4$ includes all points below the line $2x + y = 4$. Since the boundary is not included, we draw it as a dashed line.
- The graph of $-2x + y > 2$ includes all points above the line $-2x + y = 2$. Since the boundary is not included, we draw it as a dashed line.

The area that is shaded twice represents the set of solutions of the given system. We create a new graph that reflects only the solution, Figure 8-25(b).

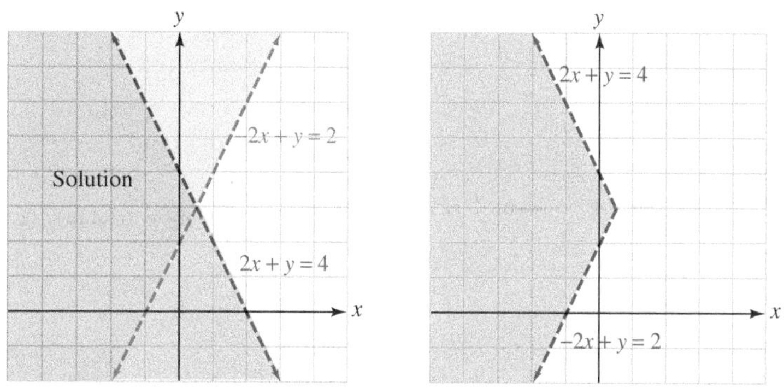

(a) (b)

Figure 8-25

Unless otherwise noted, all content on this page is © Cengage Learning.

Select a test point in the doubly shaded region and show that it satisfies both inequalities.

SELF CHECK 6 Graph the solution set: $\begin{cases} x + 3y \le 6 \\ -x + 3y < 6 \end{cases}$

EXAMPLE 7 Graph the solution set: $\begin{cases} x \le 2 \\ y > 3 \end{cases}$

Solution We graph each inequality on one set of coordinate axes, as in Figure 8-26(a).

- The graph of $x \le 2$ includes all points on the line $x = 2$ and all points to the left of the line. Since the boundary line is included, we draw it as a solid line.
- The graph $y > 3$ includes all points above the line $y = 3$. Since the boundary is not included, we draw it as a dashed line.

The area that is shaded twice represents the set of solutions of the given system. Pick a point in the doubly shaded region and show that this is truc. Graph the solution on a new set of axes as in Figure 8-26(b).

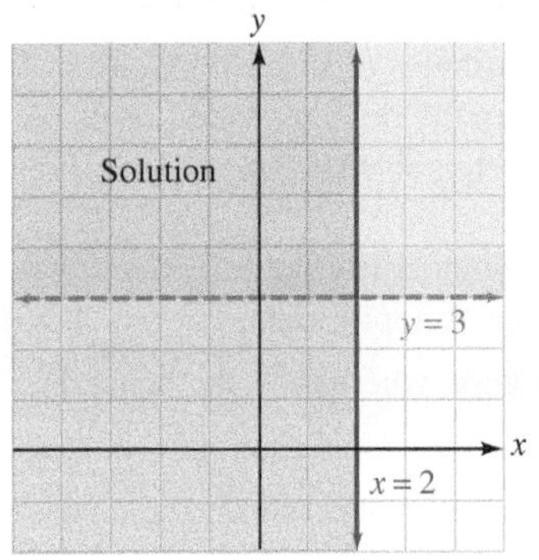

(a)

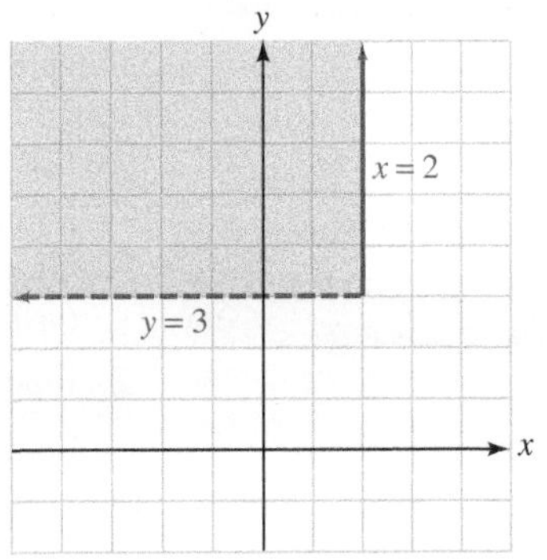

(b)

Figure 8-26

SELF CHECK 7 Graph the solution set: $\begin{cases} y \ge 1 \\ x > 2 \end{cases}$

EXAMPLE 8 Graph the solution set: $\begin{cases} y < 3x - 1 \\ y \ge 3x + 1 \end{cases}$

Solution We graph each inequality, as in Figure 8-27.

- The graph of $y < 3x - 1$ includes all of the points below the dashed line $y = 3x - 1$.
- The graph of $y \ge 3x + 1$ includes all of the points on and above the solid line $y = 3x + 1$.

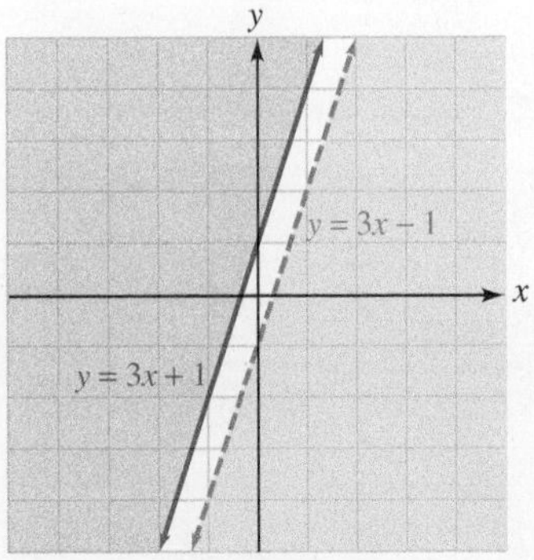

Figure 8-27

Unless otherwise noted, all content on this page is © Cengage Learning.

Since the graphs of these inequalities do not intersect, the solution set is $\varnothing$.

SELF CHECK 8 Graph the solution set: $\begin{cases} y \geq -\frac{1}{2} + 1 \\ y \leq -\frac{1}{2}x - 1 \end{cases}$

5 Solve an application using a system of linear inequalities.

EXAMPLE 9 **LANDSCAPING** A man budgets from \$300 to \$600 for trees and shrubs to landscape his yard. After shopping around, he finds that good trees cost \$150 and mature shrubs cost \$75. What combinations of trees and shrubs can he afford to buy?

Analyze the problem The man wants to spend *at least* \$300 but *not more than* \$600 for trees and shrubs.

Form two inequalities We can let x represent the number of trees purchased and y the number of shrubs purchased. We can then form the following system of inequalities.

The cost of a tree	times	the number of trees purchased	plus	the cost of a shrub	times	the number of shrubs purchased	should be at least	\$300.
\$150	$\cdot$	x	$+$	\$75	$\cdot$	y	$\geq$	\$300

The cost of a tree	times	the number of trees purchased	plus	the cost of a shrub	times	the number of shrubs purchased	should not be more than	\$600.
\$150	$\cdot$	x	$+$	\$75	$\cdot$	y	$\leq$	\$600

Solve the system We graph the system

$$\begin{cases} 150x + 75y \geq 300 \\ 150x + 75y \leq 600 \end{cases}$$

as in Figure 8-28. The coordinates of each point shown in the graph give a possible combination of the number of trees (x) and the number of shrubs (y) that can be purchased. These possibilities are

$(0, 4), (0, 5), (0, 6), (0, 7), (0, 8)$
$(1, 2), (1, 3), (1, 4), (1, 5), (1, 6)$
$(2, 0), (2, 1), (2, 2), (2, 3), (2, 4)$
$(3, 0), (3, 1), (3, 2), (4, 0)$

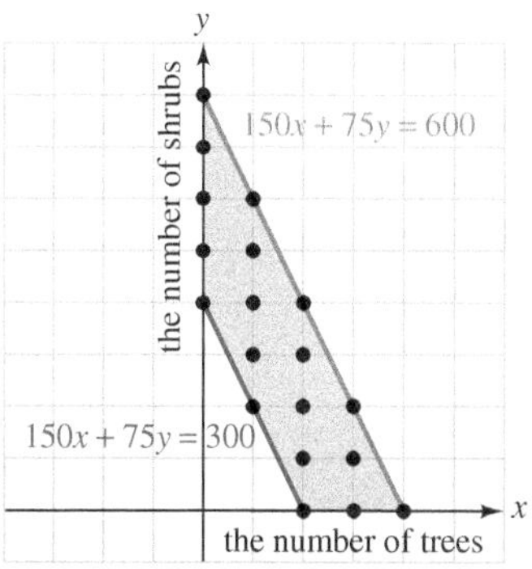

Figure 8-28

Only these points can be used, because the man cannot buy part of a tree or part of a shrub.

SELF CHECK 9 From the graph, what is the maximum number of trees he may purchase? At what geometric point does this occur?

Unless otherwise noted, all content on this page is © Cengage Learning.

SELF CHECK ANSWERS

1. a. no **b.** yes **c.** no

2.

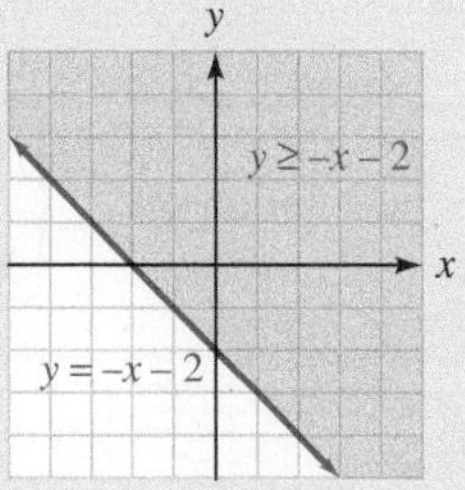

3.

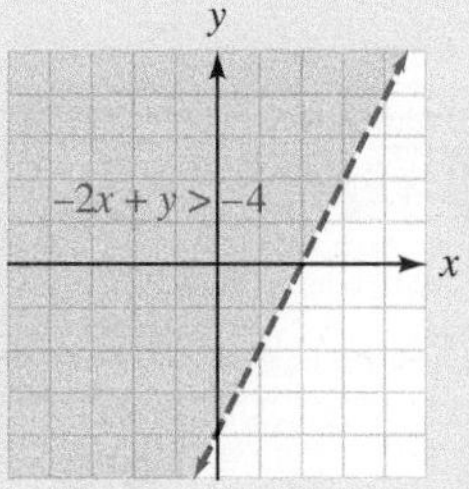

4.

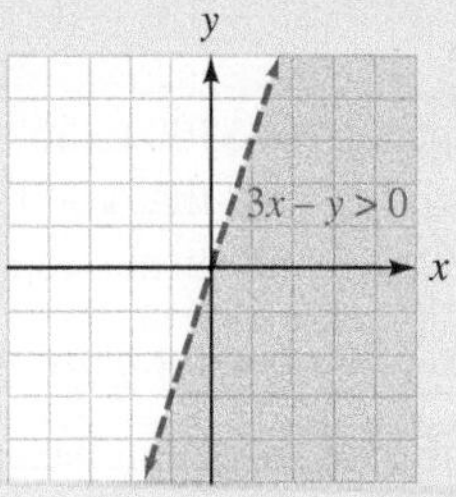

5.

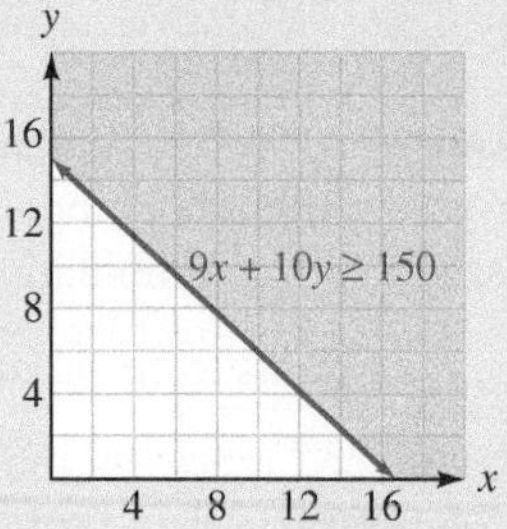

6.

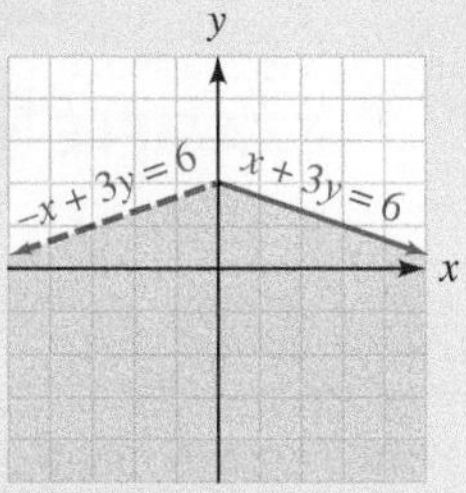

7.

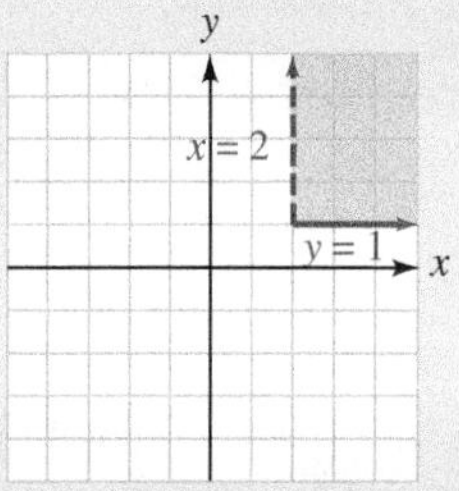

8. $\varnothing$

9. Four trees is the maximum number he may purchase and occurs at one of the x-interecepts, (4, 0). He would not be able to purchase any shrubs.

NOW TRY THIS

Solve each system by graphing.

1. $\begin{cases} y \geq x \\ x < -y + 2 \end{cases}$

2. $\begin{cases} x - y > 4 \\ y < x + 5 \end{cases}$

3. $\begin{cases} x - y > 0 \\ x \geq 3 \\ -y > 2 \end{cases}$

Unless otherwise noted, all content on this page is © Cengage Learning.

8.7 Exercises

WARM-UPS *Determine whether the following coordinates satisfy $y > 3x + 2$.*

1. $(0, 0)$ **2.** $(5, 5)$
3. $(-2, 4)$ **4.** $(-3, -6)$

Determine whether the graph of each inequality should be shaded above or below the boundary.

5. $y > x - 2$ **6.** $y \leq x + 4$
7. $x - y > -1$ **8.** $x - y < 5$

REVIEW

9. Solve: $4x - 6 = 18$.
10. Solve: $2(x - 4) \leq -12$.
11. Solve: $A = P + Prt$ for t.
12. Does the graph of $y = -3x$ pass through the origin?

Simplify each expression.

13. $7x + 4(x - 6)$
14. $5a - 2(6 - a)$
15. $3(y - x) + 5y + 8x$
16. $3p + 2(q - p) + q$

VOCABULARY AND CONCEPTS *Fill in the blanks.*

17. $2x - y \leq 4$ is a linear ________ in x and y.
18. The symbol $\leq$ means ________ or ______.
19. In the accompanying graph, the line $2x - y = 4$ is the ________ of the graph $2x - y \leq 4$.
20. In the accompanying graph, the line $2x - y = 4$ divides the rectangular coordinate system into two ____ ______.

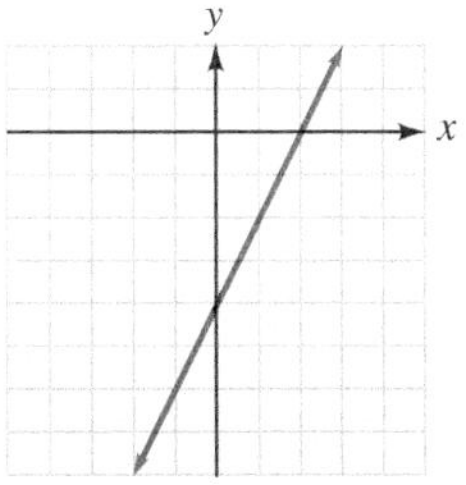

21. $\begin{cases} x + y > 5 \\ x - y < 3 \end{cases}$ is a system of linear ________.

22. The ________ of a system of linear inequalities are all the ordered pairs that make all of the inequalities of the system true at the same time.
23. Any point in the ____________ region of the graph of the solution of a system of two linear inequalities has coordinates that satisfy both of the inequalities of the system.
24. To graph a linear inequality such as $2x + y > 4$, first graph the boundary with a dashed line. Then pick a test ____ to determine which half-plane to shade.
25. Determine whether the graph of each linear inequality includes the boundary line.
a. $x - y > 5$ **b.** $4x + 3y \leq 12$
26. If a false statement results when the coordinates of a test point are substituted into a linear inequality, which half-plane should be shaded to represent the solution of the inequality?

GUIDED PRACTICE *Determine whether each ordered pair is a solution of the given inequality.* SEE EXAMPLE 1. (OBJECTIVE 1)

27. Determine whether each ordered pair is a solution of $5x - 3y \geq 0$.
a. $(1, 1)$ **b.** $(-2, -3)$
c. $(0, 0)$ **d.** $\left(\frac{1}{5}, \frac{4}{3}\right)$
28. Determine whether each ordered pair is a solution of $x + 3y < -20$.
a. $(3, -9)$ **b.** $(0, 0)$
c. $(2, 1)$ **d.** $\left(-\frac{1}{2}, -8\right)$
29. Determine whether each ordered pair is a solution of $x + y > 4$.
a. $(0, 4)$ **b.** $(1, 5)$
c. $\left(-1, \frac{1}{2}\right)$ **d.** $\left(-\frac{3}{4}, 7\right)$
30. Determine whether each ordered pair is a solution of $x - 2y < -6$.
a. $(4, 2)$ **b.** $(0, 2)$
c. $(-1, 5)$ **d.** $\left(\frac{3}{4}, 6\right)$

Graph each inequality. SEE EXAMPLE 2. (OBJECTIVE 2)

31. $y \leq x + 2$

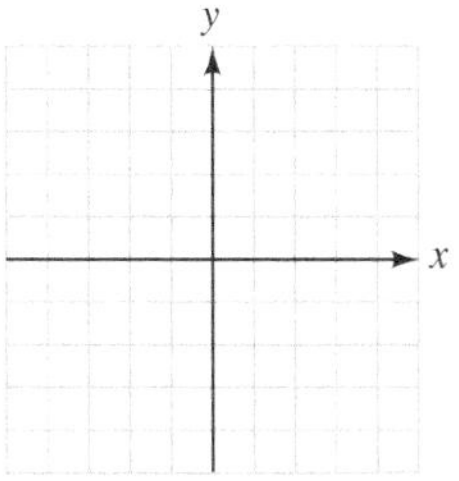

32. $y \leq -x + 1$

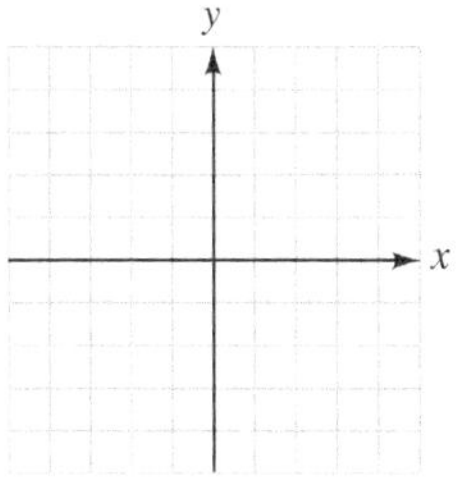

33. $y \leq 4x$

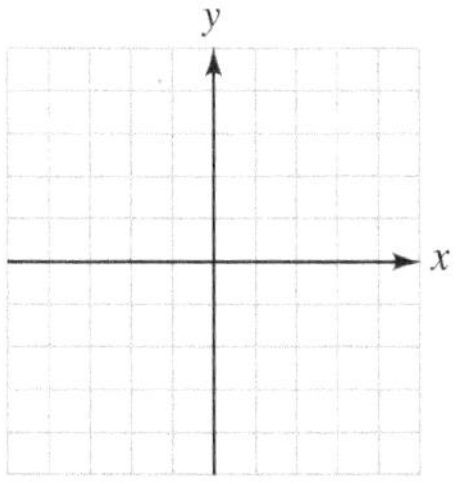

34. $y \geq 3 - x$

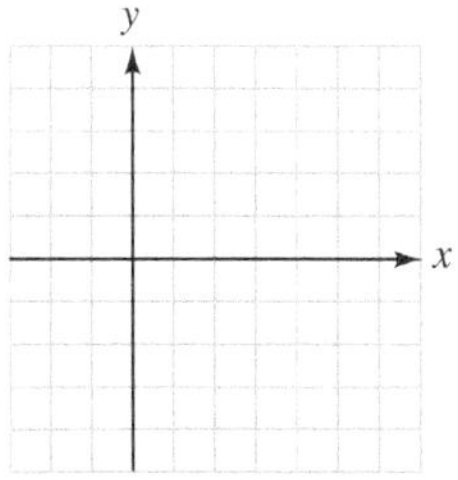

Graph each inequality. SEE EXAMPLE 3. (OBJECTIVE 2)

35. $y > x - 3$

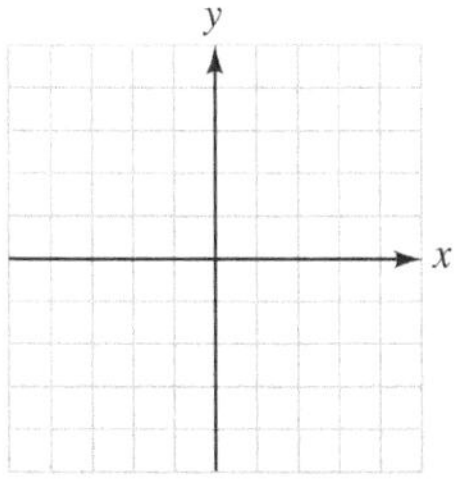

36. $y + 2x < 0$

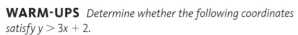

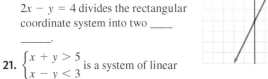

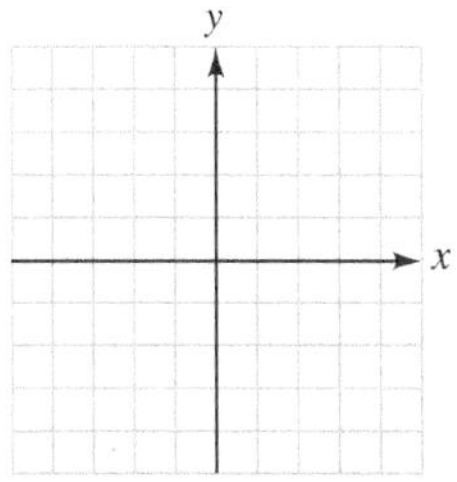

Unless otherwise noted, all content on this page is © Cengage Learning.

37. $y > 2x - 4$

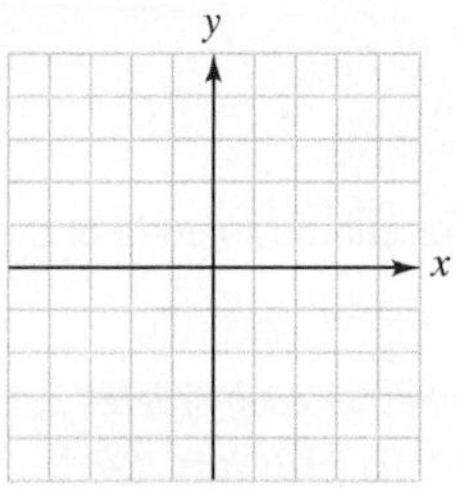

38. $y < 2 - x$

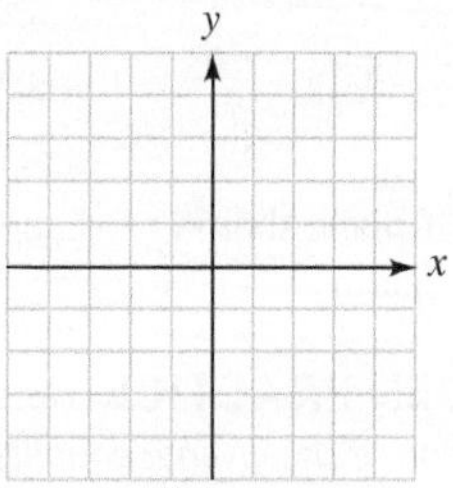

Graph each inequality. SEE EXAMPLE 4. (OBJECTIVE 2)

39. $2x - y \le 4$

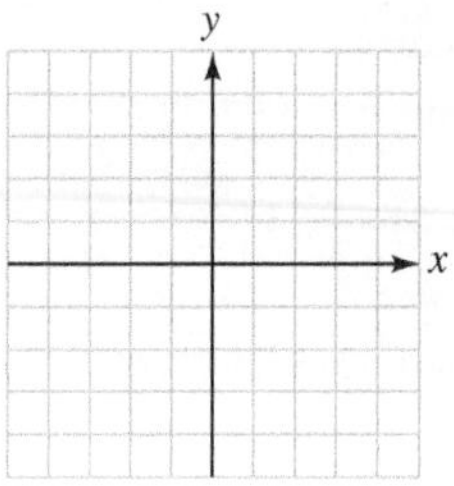

40. $3x - 4y > 12$

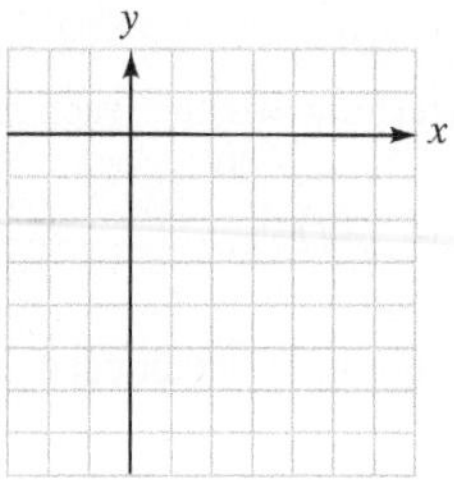

41. $x - 2y \le 4$

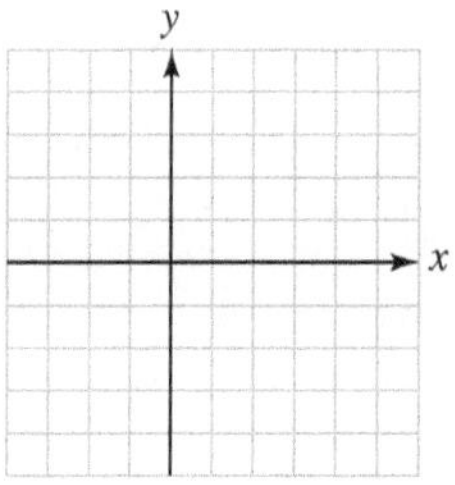

42. $7x - 2y < 21$

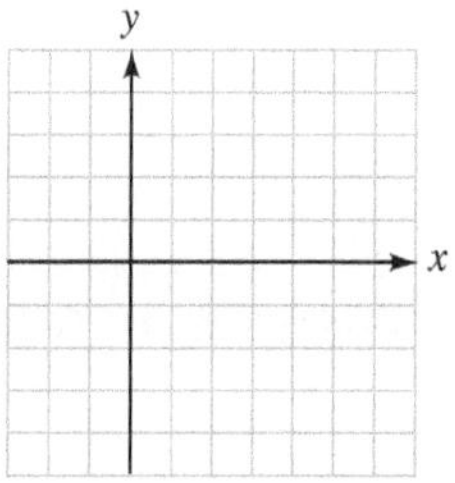

Graph the solution set of each system of inequalities, when possible. SEE EXAMPLE 6. (OBJECTIVE 4)

43. $\begin{cases} x + 2y \le 3 \\ 2x - y \ge 1 \end{cases}$

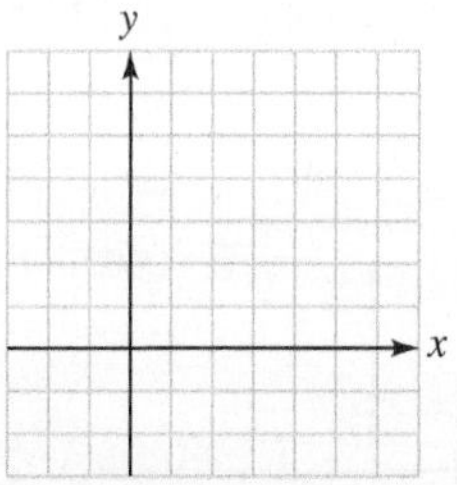

44. $\begin{cases} 2x + y \ge 3 \\ x - 2y \le -1 \end{cases}$

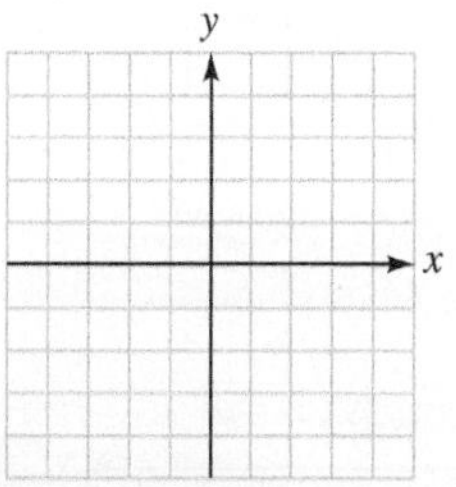

45. $\begin{cases} x + y < -1 \\ x - y > -1 \end{cases}$

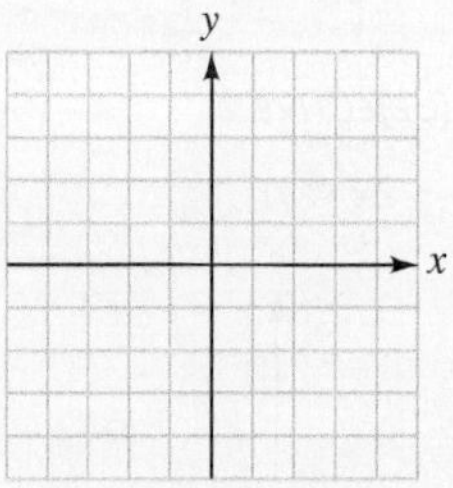

46. $\begin{cases} x + y > 2 \\ x - y < -2 \end{cases}$

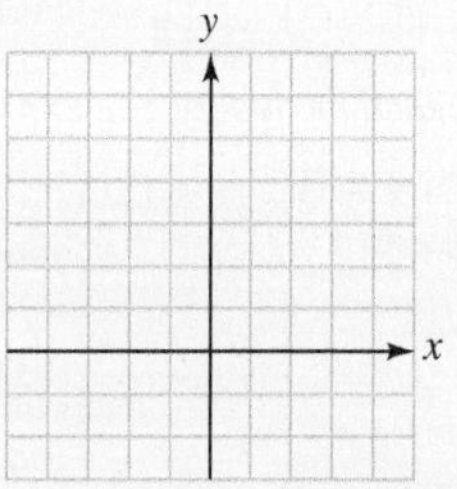

Graph the solution set of each system of inequalities, when possible. SEE EXAMPLE 7. (OBJECTIVE 4)

47. $\begin{cases} x > 2 \\ y \le 3 \end{cases}$

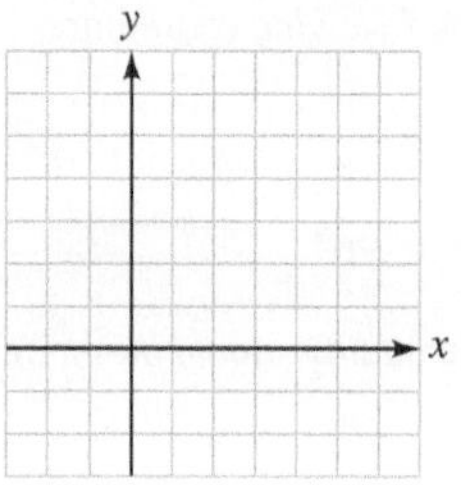

48. $\begin{cases} x \ge -1 \\ y > -2 \end{cases}$

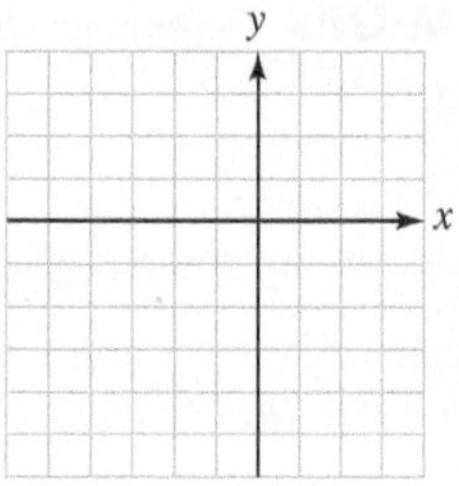

49. $\begin{cases} x \le 0 \\ y < 0 \end{cases}$

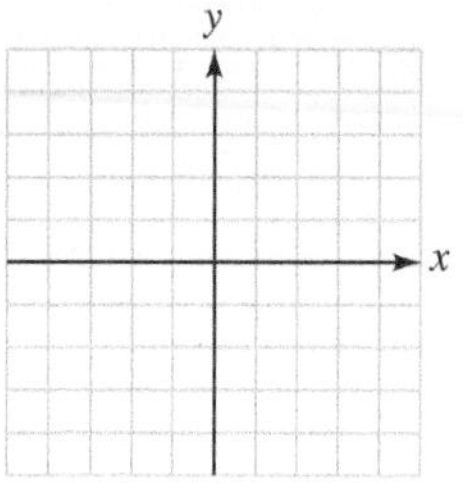

50. $\begin{cases} x < -2 \\ y \ge 3 \end{cases}$

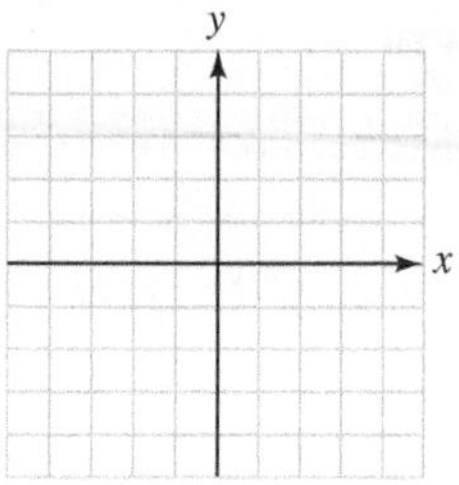

Graph the solution set of each system of inequalities, when possible. If not possible, state $\varnothing$. SEE EXAMPLE 8. (OBJECTIVE 4)

51. $\begin{cases} x + y < 1 \\ x + y > 3 \end{cases}$

52. $\begin{cases} y < 2x - 1 \\ 2x - y < -4 \end{cases}$

53. $\begin{cases} y \le -\frac{4}{3}x - 2 \\ 4x + 3y > 15 \end{cases}$

54. $\begin{cases} 3x + y < -2 \\ y > 3(1 - x) \end{cases}$

ADDITIONAL PRACTICE *Graph the solution.*

55. $y < 3x$

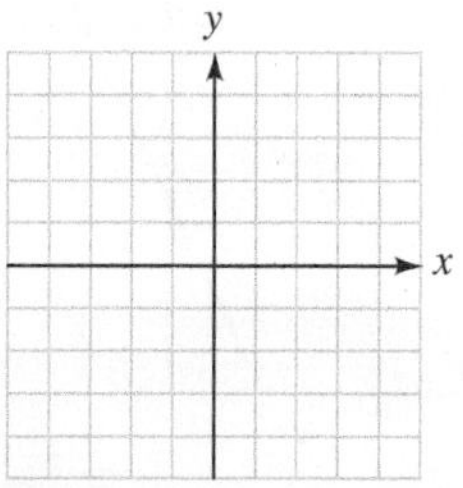

56. $3x + 2y \ge 12$

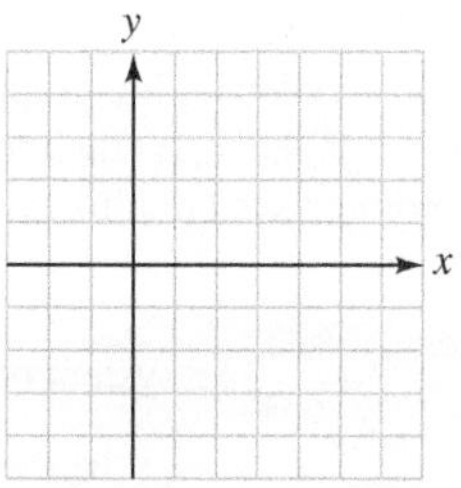

57. $\begin{cases} 2x - y < 4 \\ x + y \ge -1 \end{cases}$

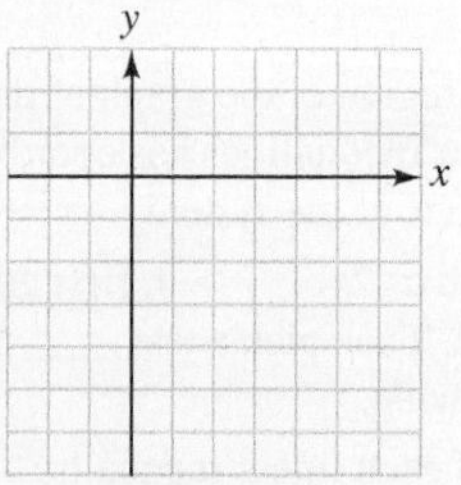

58. $\begin{cases} x - y \ge 5 \\ x + 2y < -4 \end{cases}$

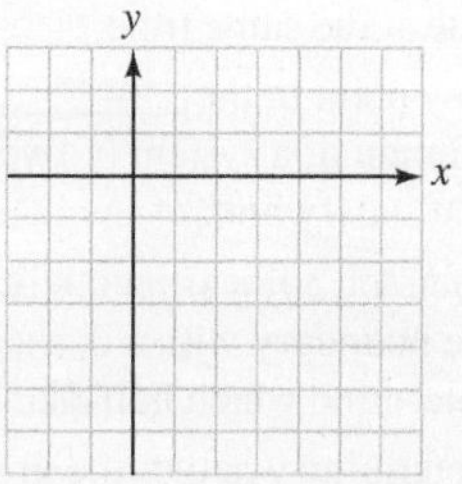

Unless otherwise noted, all content on this page is © Cengage Learning.

59. $y < 2 - 3x$

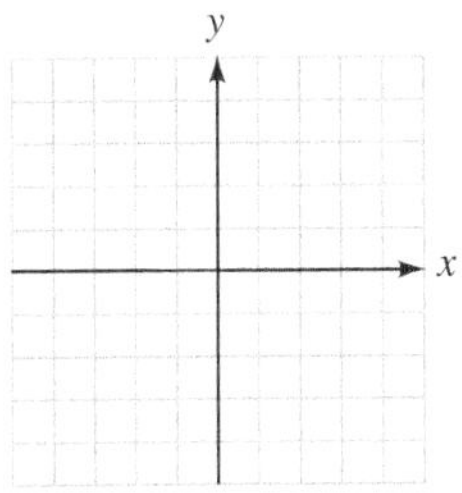

60. $y \geq 5 - 2x$

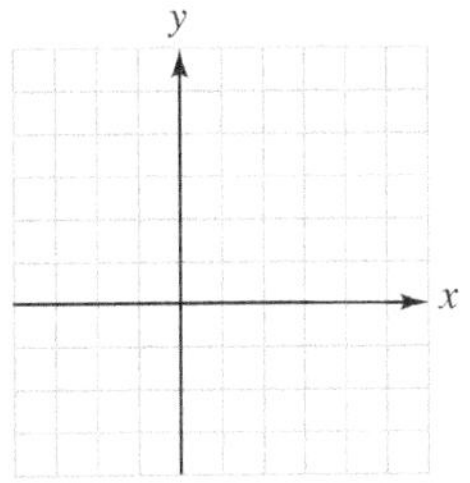

61. $x < 2$

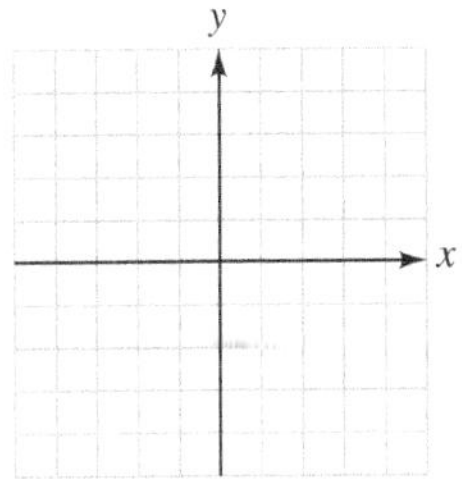

62. $2y - x < 8$

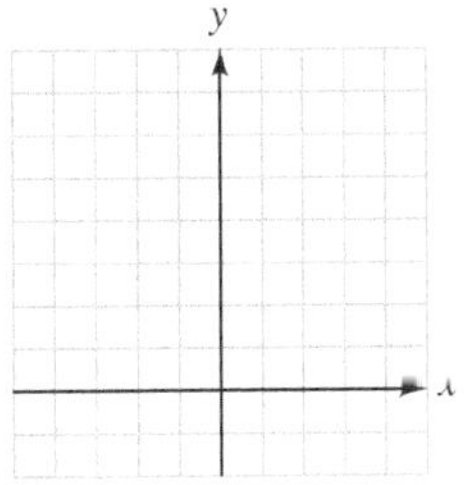

63. $y + 9x \geq 3$

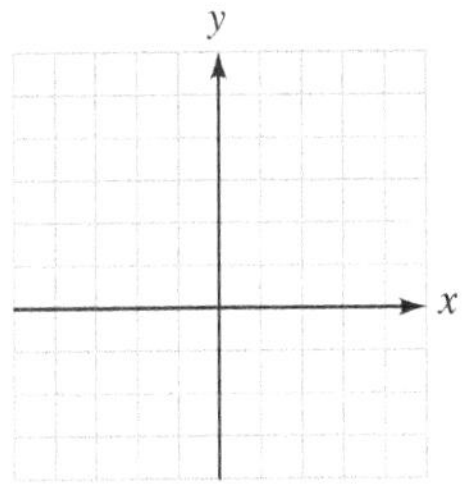

64. $y > -3$

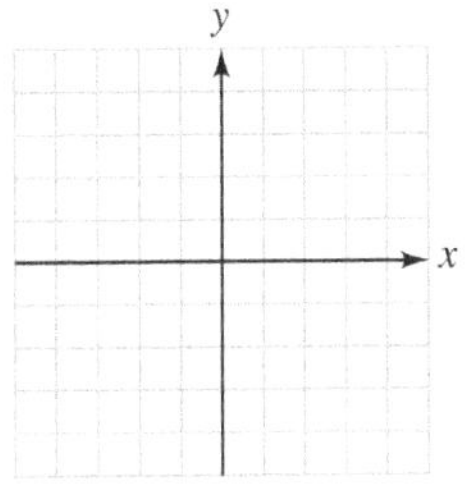

65. $4x + 3y \leq 12$

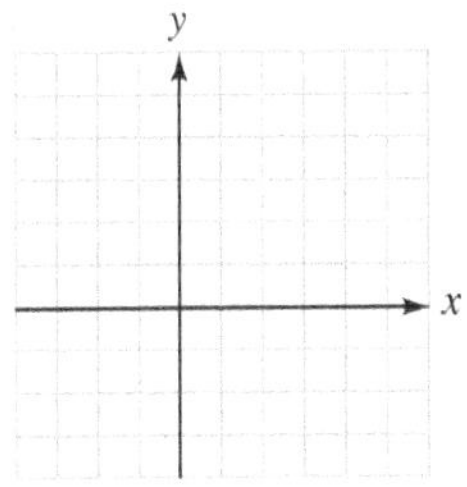

66. $5x + 4y \geq 20$

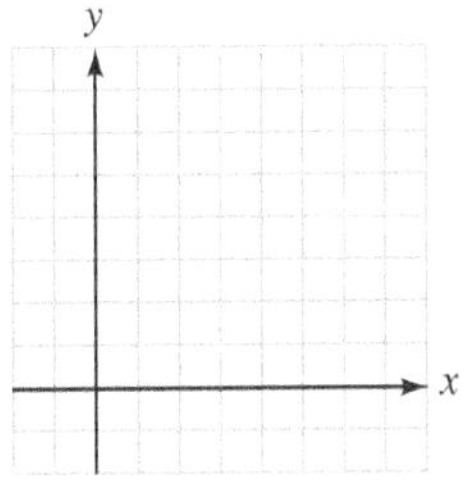

67. $y \leq 1$

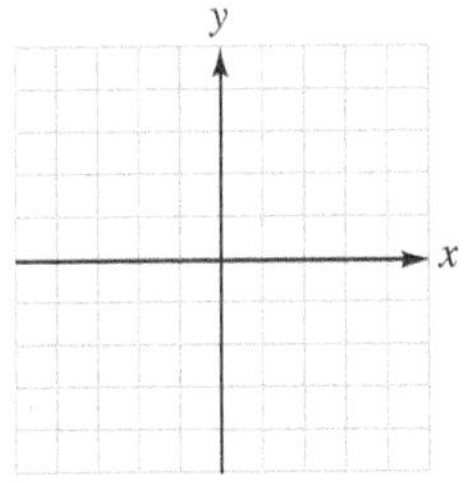

68. $x \geq -4$

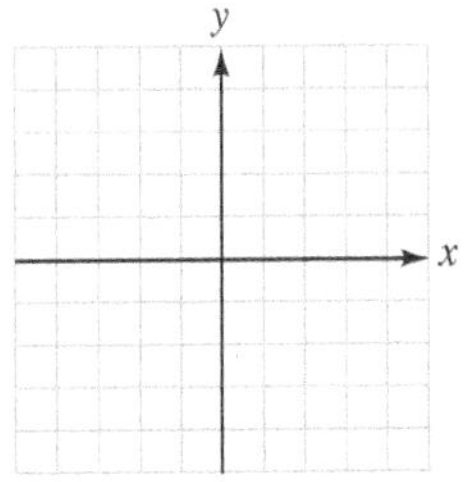

69. $\begin{cases} 3x + 4y > -7 \\ 2x - 3y \geq 1 \end{cases}$

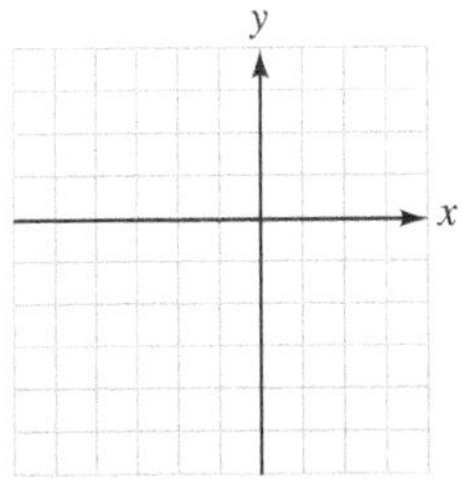

70. $\begin{cases} 3x + y \leq 1 \\ 4x - y > -8 \end{cases}$

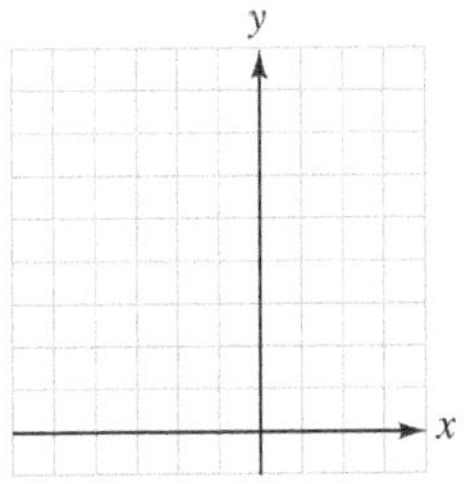

71. $\begin{cases} 2x - 4y > -6 \\ 3x + y \geq 5 \end{cases}$

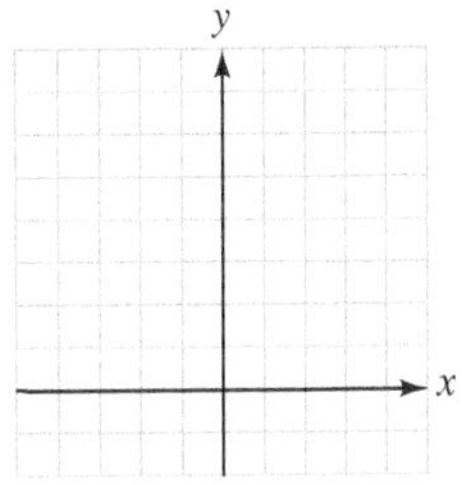

72. $\begin{cases} 2x - 3y < 0 \\ 2x + 3y \geq 12 \end{cases}$

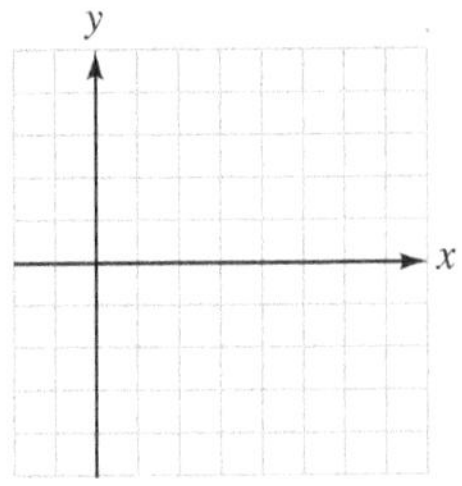

73. $\begin{cases} \frac{x}{2} + \frac{y}{3} \geq 2 \\ \frac{x}{2} - \frac{y}{2} < -1 \end{cases}$

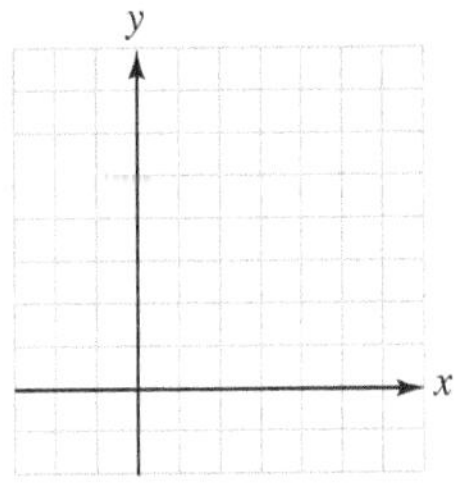

74. $\begin{cases} \frac{x}{3} - \frac{y}{2} < -3 \\ \frac{x}{3} + \frac{y}{2} > -1 \end{cases}$

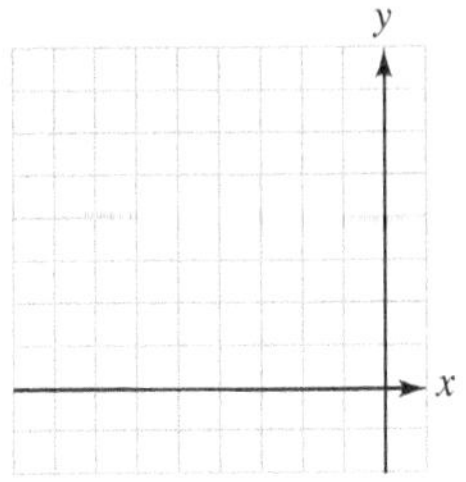

APPLICATIONS *Graph each inequality for nonnegative values of x and y. Then give some ordered pairs that satisfy the inequality.* *SEE EXAMPLE 5. (OBJECTIVE 3)*

75. Production planning It costs a bakery \$3 to make a cake and \$4 to make a pie. Production costs cannot exceed \$120 per day. Find an inequality that shows the possible combinations of cakes, x, and pies, y, that can be made, and graph it in the illustration.

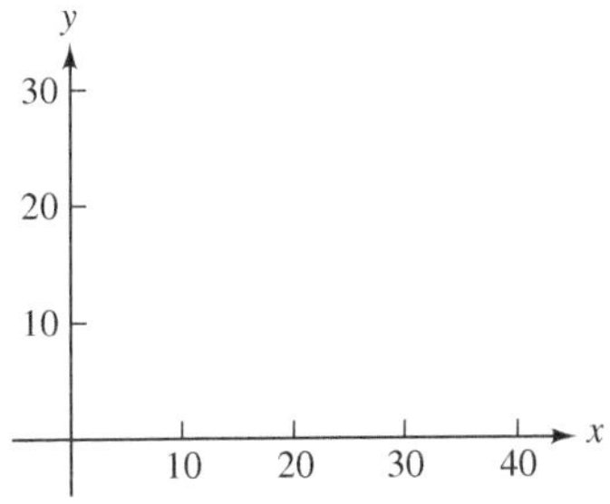

76. Hiring babysitters Tomiko has a choice of two babysitters. Sitter 1 charges \$6 per hour, and sitter 2 charges \$7 per hour. Tomiko can afford no more than \$42 per week for sitters. Find an inequality that shows the possible ways that she can hire sitter 1 (x) and sitter 2 (y), and graph it in the illustration.

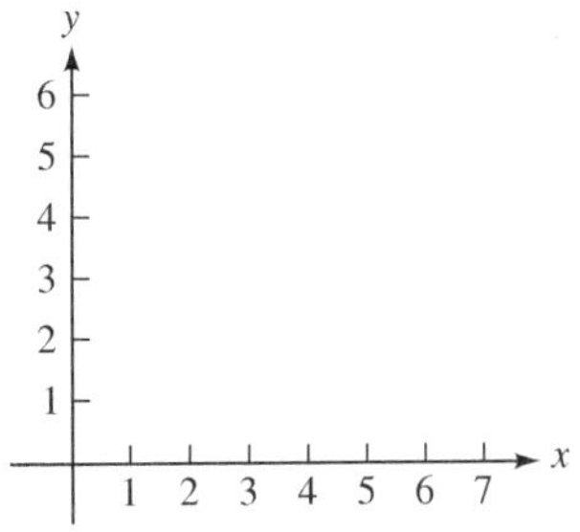

Unless otherwise noted, all content on this page is © Cengage Learning.

77. Inventory A clothing store advertises that it maintains an inventory of at least $4,400 worth of men's jackets. A leather jacket costs $100, and a nylon jacket costs $88. Find an inequality that shows the possible ways that leather jackets, x, and nylon jackets, y, can be stocked, and graph it in the illustration.

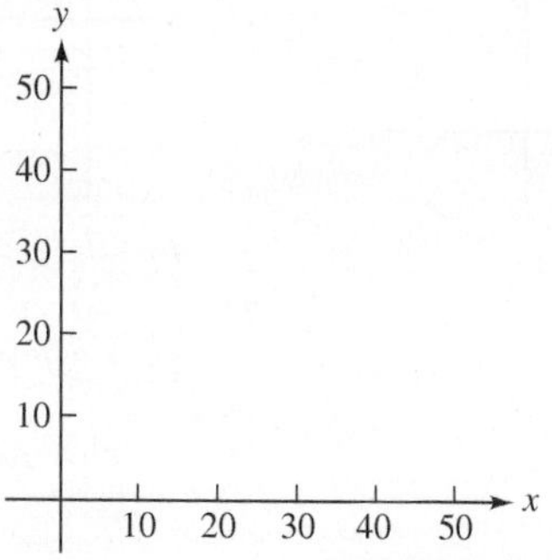

78. Making sporting goods To keep up with demand, a sporting goods manufacturer allocates at least 2,400 units of time per day to make baseballs and footballs. It takes 20 units of time to make a baseball and 30 units of time to make a football. Find an inequality that shows the possible ways to schedule the time to make baseballs, x, and footballs, y, and graph it in the illustration.

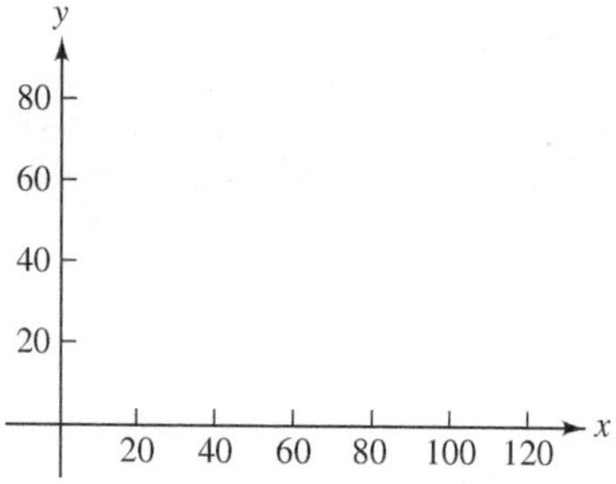

79. Investing Robert has up to $8,000 to invest in two companies. Stock in Robotronics sells for $40 per share, and stock in Macrocorp sells for $50 per share. Find an inequality that shows the possible ways that he can buy shares of Robotronics, x, and Macrocorp, y, and graph it in the illustration.

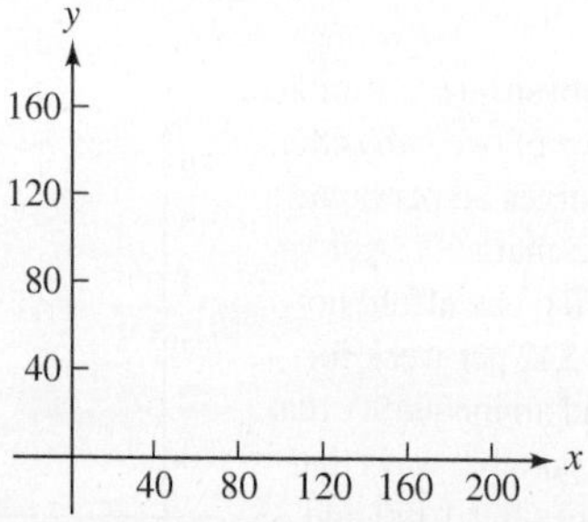

80. Buying tickets Tickets to the Rockford Rox baseball games cost $6 for reserved seats and $4 for general admission. Nightly receipts must be at least $10,200 to meet expenses. Find an inequality that shows the possible ways that the Rox can sell reserved seats, x, and general admission tickets, y, and graph it in the illustration.

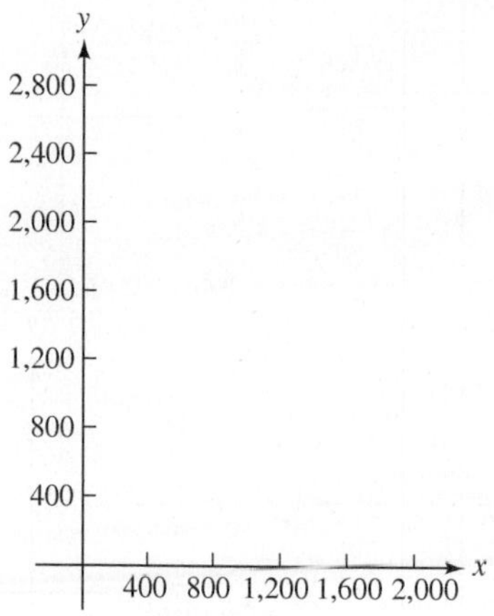

Graph each system of inequalities and give two possible solutions to each problem. SEE EXAMPLE 9. (OBJECTIVE 5)

81. Buying CDs Melodic Music has compact discs on sale for either $10 or $15. A customer wants to spend at least $30 but no more than $60 on CDs. Find a system of inequalities whose graph will show the possible combinations of $10 CDs, x, and $15 CDs, y, that the customer can buy, and graph it in the illustration.

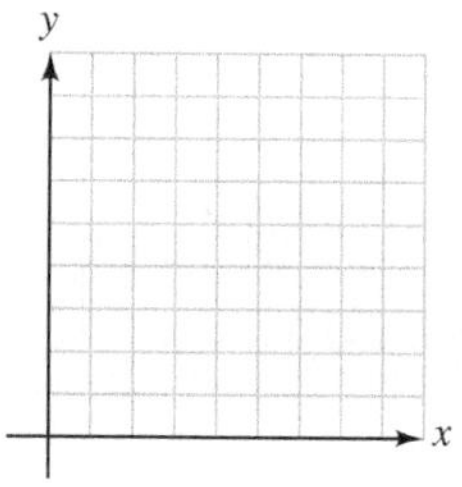

82. Buying boats Dry Boatworks wholesales aluminum boats for $800 and fiberglass boats for $600. Northland Marina wants to order at least $2,400 but no more than $4,800 worth of boats. Find a system of inequalities whose graph will show the possible combinations of aluminum boats, x, and fiberglass boats, y, that can be ordered, and graph it in the illustration.

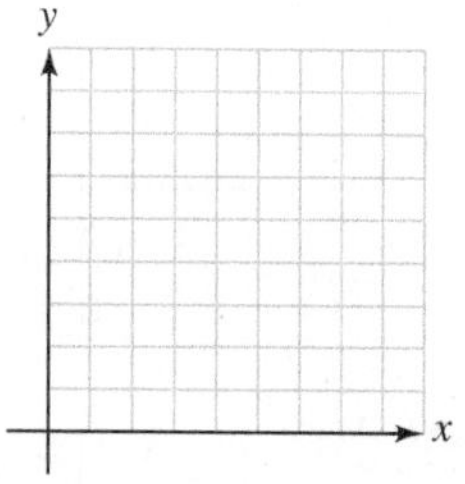

83. Buying furniture A distributor wholesales desk chairs for $150 and side chairs for $100. Best Furniture wants to order no more than $900 worth of chairs and wants to order more side chairs than desk chairs. Find a system of inequalities whose graph will show the possible combinations of desk chairs, x, and side chairs, y, that can be ordered, and graph it in the illustration.

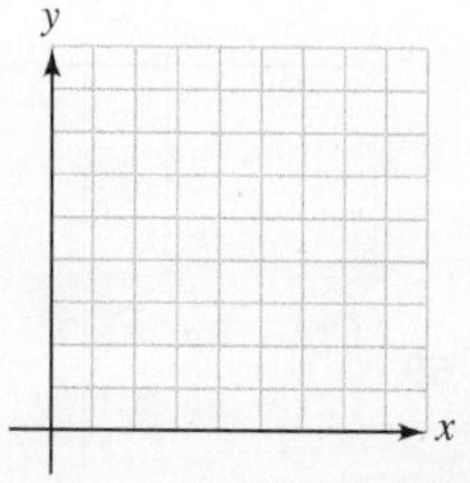

Unless otherwise noted, all content on this page is © Cengage Learning.

84. Ordering furnace equipment J. Bolden Heating Company wants to order no more than \$2,000 worth of electronic air cleaners and humidifiers from a wholesaler that charges \$500 for air cleaners and \$200 for humidifiers. Bolden wants more humidifiers than air cleaners. Find a system of inequalities whose graph will show the possible combinations of air cleaners, x, and humidifiers, y, that can be ordered, and graph it in the illustration.

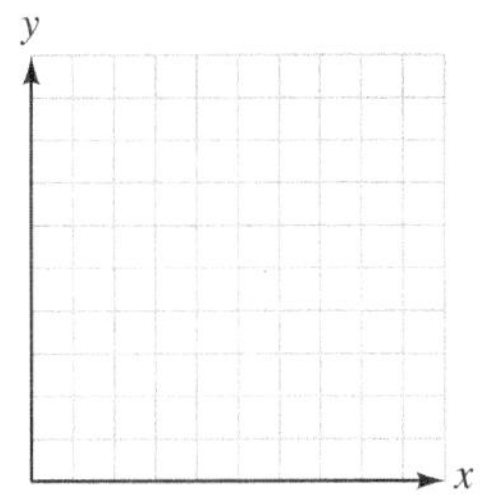

WRITING ABOUT MATH

85. Explain how to find the boundary for the graph of an inequality.

86. Explain how to decide which side of the boundary line to shade.

87. Explain how to use graphing to solve a system of inequalities.

88. Explain when a system of inequalities will have no solutions.

SOMETHING TO THINK ABOUT

89. What are some limitations of the graphing method for solving inequalities?

90. Graph $y = 3x + 1$, $y < 3x + 1$, and $y > 3x + 1$. What do you discover?

91. Can a system of inequalities have
   a. no solutions?
   b. exactly one solution?
   c. infinitely many solutions?

92. Find a system of two inequalities that has a solution of $(2, 0)$ but no solutions of the form (x, y) where $y < 0$.

Section 8.8 Solving Systems Using Linear Programming

Objectives

1 Find the maximum and minimum value of an equation in the form $P = ax + by$, subject to specific constraints.

2 Solve an application using linear programming.

Vocabulary

linear programming
objective function
constraints
feasibility region

Getting Ready

Evaluate $2x + 3y$ for each pair of coordinates.

1. $(0, 0)$ **2.** $(3, 0)$ **3.** $(2, 2)$ **4.** $(0, 4)$

Unless otherwise noted, all content on this page is © Cengage Learning.

We now use our knowledge of solving systems of inequalities in two variables to solve linear programming applications.

1 Find the maximum and minimum value of an equation in the form $P = ax + by$, subject to specific constraints.

Linear programming is a mathematical technique used to find the optimal allocation of resources in the military, business, telecommunications, and other fields. It got its start during World War II when it became necessary to move huge quantities of people, materials, and supplies as efficiently and economically as possible.

To solve a linear program, we maximize (or minimize) a function (called the **objective function**) subject to given conditions on its variables. These conditions (called **constraints**) are usually given as a system of linear inequalities. For example, suppose that the annual profit (in millions of dollars) earned by a business is given by the equation $P = y + 2x$, where the profit is determined by the sale of two different items, x and y, and are subject to the following constraints:

$$\begin{cases} 3x + y \leq 120 \\ x + y \leq 60 \\ x \geq 0 \\ y \geq 0 \end{cases}$$

To find the maximum profit P that can be earned by the business, we solve the system of inequalities as shown in Figure 8-29(a) and find the coordinates of each *corner point* of the region R, called a **feasibility region**. We can then write the profit equation

$$P = y + 2x \qquad \text{in the form} \qquad y = -2x + P$$

The equation $y = -2x + P$ is the equation of a set of parallel lines, each with a slope of -2 and a y-intercept of P. To find the line that passes through region R and provides the maximum value of P, we refer to Figure 8-29(b) and locate the line with the greatest y-intercept. Since line l has the greatest y-intercept and intersects region R at the corner point $(30, 30)$, the maximum value of P (subject to the given constraints) is

$$\begin{aligned} P &= y + 2x \\ &= 30 + 2(30) \\ &= 90 \end{aligned}$$

Thus, the maximum profit P that can be earned is \$90 million. This profit occurs when $x = 30$ and $y = 30$.

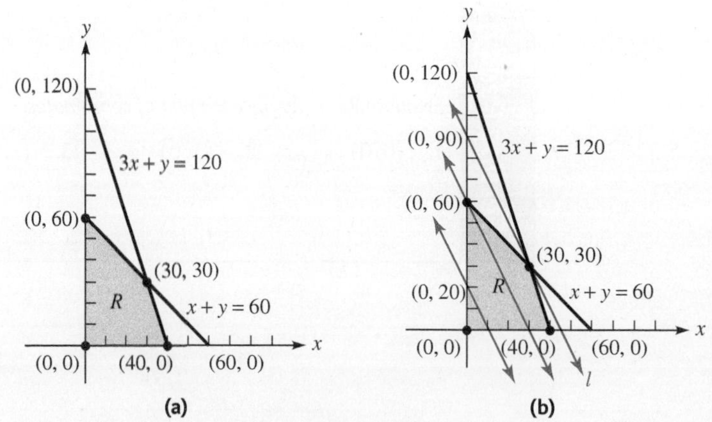

Figure 8-29

Unless otherwise noted, all content on this page is © Cengage Learning.

Perspective

HOW TO SOLVE IT

As a young student, George Polya (1888–1985) enjoyed mathematics and understood the solutions presented by his teachers. However, Polya had questions still asked by mathematics students today: "Yes, the solution works, but how is it possible to come up with such a solution? How could I discover such things by myself?" These questions still concerned him years later when, as Professor of Mathematics at Stanford University, he developed an approach to teaching mathematics that was very popular with faculty and students. His book, *How to Solve It*, became a bestseller.

George Polya
1888–1985

Polya's problem-solving approach involves four steps.

- *Understand the problem.* What is the unknown? What information is known? What are the conditions?
- *Devise a plan.* Have you seen anything like it before? Do you know any related problems you have solved before? If you can't solve the proposed problem, can you solve a similar but easier problem?
- *Carry out the plan.* Check each step. Can you explain why each step is correct?
- *Look back.* Examine the solution. Can you check the result? Can you use the result, or the method, to solve any other problem?

The preceding discussion illustrates the following important fact.

MAXIMUM OR MINIMUM OF AN OBJECTIVE FUNCTION

If a linear function, subject to the constraints of a system of linear inequalities in two variables, attains a maximum or a minimum value, that value will occur at a corner point or along an entire edge of the region R that represents the solution of the system.

EXAMPLE 1 If $P = 2x + 3y$, find the maximum value of P subject to the following constraints:

$$\begin{cases} x + y \le 4 \\ 2x + y \le 6 \\ x \ge 0 \\ y \ge 0 \end{cases}$$

Solution We solve the system of inequalities to find the feasibility region R shown in Figure 8-30. The coordinates of its corner points are $(0, 0)$, $(3, 0)$, $(0, 4)$, and $(2, 2)$.

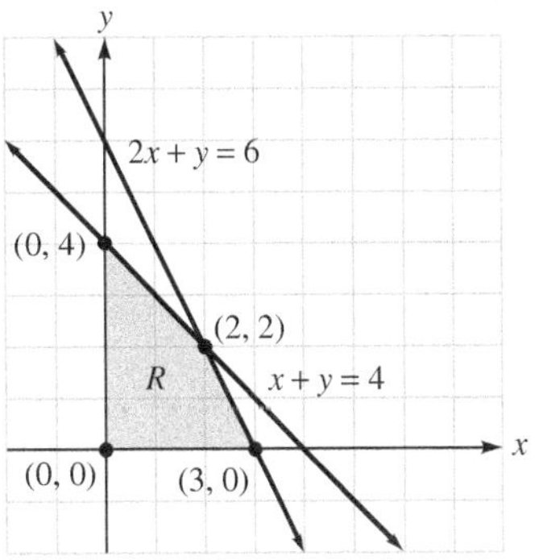

Figure 8-30

Since the maximum value of P will occur at a corner of R, we substitute the coordinates of each corner point into the objective function $P = 2x + 3y$ and find the one that gives the maximum value of P.

Unless otherwise noted, all content on this page is © Cengage Learning.

Point	$P = 2x + 3y$
$(0, 0)$	$P = 2(0) + 3(0) = 0$
$(3, 0)$	$P = 2(3) + 3(0) = 6$
$(2, 2)$	$P = 2(2) + 3(2) = 10$
$(0, 4)$	$P = 2(0) + 3(4) = 12$

The maximum value $P = 12$ occurs when $x = 0$ and $y = 4$.

SELF CHECK 1 Find the maximum value of $P = 4x + 3y$, subject to the constraints of Example 1.

EXAMPLE 2 If $P = 3x + 2y$, find the minimum value of P subject to the following constraints:

$$\begin{cases} x + y \geq 1 \\ x - y \leq 1 \\ x - y \geq 0 \\ x \leq 2 \end{cases}$$

Solution We refer to the feasibility region shown in Figure 8-31 with corner points at $\left(\frac{1}{2}, \frac{1}{2}\right)$, $(2, 2)$, $(2, 1)$, and $(1, 0)$.

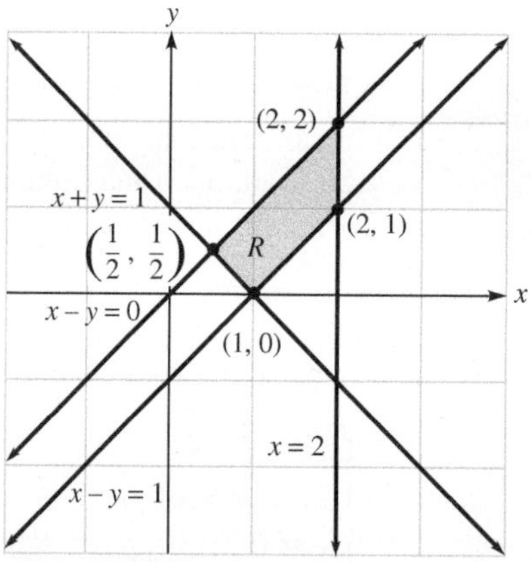

Figure 8-31

Since the minimum value of P occurs at a corner point of region R, we substitute the coordinates of each corner point into the objective function $P = 3x + 2y$ and find the one that gives the minimum value of P.

Point	$P = 3x + 2y$
$\left(\frac{1}{2}, \frac{1}{2}\right)$	$P = 3\left(\frac{1}{2}\right) + 2\left(\frac{1}{2}\right) = \frac{5}{2}$
$(2, 2)$	$P = 3(2) + 2(2) = 10$
$(2, 1)$	$P = 3(2) + 2(1) = 8$
$(1, 0)$	$P = 3(1) + 2(0) = 3$

The minimum value $P = \frac{5}{2}$ occurs when $x = \frac{1}{2}$ and $y = \frac{1}{2}$.

SELF CHECK 2 Find the minimum value of $P = 2x + y$, subject to the constraints of Example 2.

Unless otherwise noted, all content on this page is © Cengage Learning.

2 Solve an application using linear programming.

Linear programming applications can be complex and involve hundreds of variables. In this section, we will consider a few simpler situations. Since they involve only two variables, we can solve them using graphical methods.

EXAMPLE 3 **INCOME** An accountant prepares tax returns for individuals and for small businesses. On average, each individual return requires 3 hours of her time and 1 hour of computer time. Each business return requires 4 hours of her time and 2 hours of computer time. Because of other business considerations, her time is limited to 240 hours, and the computer time is limited to 100 hours. If she earns a profit of \$80 on each individual return and a profit of \$150 on each business return, how many returns of each type should she prepare to maximize her profit?

Solution First, we organize the given information into a table.

	Individual tax return	Business tax return	Time available
Accountant's time	3	4	240 hours
Computer time	1	2	100 hours
Profit	\$80	\$150	

Then we solve using the following steps.

Find the objective function

Suppose that x represents the number of individual returns to be completed and y represents the number of business returns to be completed. Because each of the x individual returns will earn an \$80 profit, and each of the y business returns will earn a \$150 profit, the total profit is given by the equation

$$P = 80x + 150y$$

Find the feasibility region

The number of individual returns and business returns cannot be negative, thus we know that $x \geq 0$ and $y \geq 0$.

Each of the x individual returns will take 3 hours of her time, each of the y business returns will take 4 hours of her time, and the total number of hours she will work will be $(3x + 4y)$ hours. This amount must be less than or equal to her available time, which is 240 hours. Thus, the inequality $3x + 4y \leq 240$ is a constraint on the accountant's time.

Each of the x individual returns will take 1 hour of computer time, each of the y business returns will take 2 hours of computer time, and the total number of hours of computer time will be $(x + 2y)$ hours. This amount must be less than or equal to the available computer time, which is 100 hours. Thus, the inequality $x + 2y \leq 100$ is a constraint on the computer time.

We have the following constraints on the values of x and y.

$$\begin{cases} x \geq 0 & \text{The number of individual returns is nonnegative.} \\ y \geq 0 & \text{The number of business returns is nonnegative.} \\ 3x + 4y \leq 240 & \text{The accountant's time must be less than or equal to 240 hours.} \\ x + 2y \leq 100 & \text{The computer time must be less than or equal to 100 hours.} \end{cases}$$

To find the feasibility region, we graph each of the constraints to find region R, as in Figure 8-32 on the next page. The four corner points of this region have coordinates of $(0, 0)$, $(80, 0)$, $(40, 30)$, and $(0, 50)$.

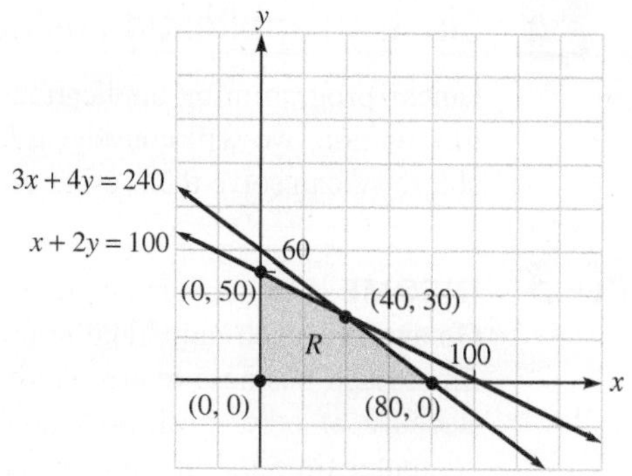

Figure 8-32

Find the maximum profit

To find the maximum profit, we substitute the coordinates of each corner point into the objective function $P = 80x + 150y$.

Point	$P = 80x + 150y$
(0, 0)	$P = 80(0) + 150(0) = 0$
(80, 0)	$P = 80(80) + 150(0) = 6{,}400$
(40, 30)	$P = 80(40) + 150(30) = 7{,}700$
(0, 50)	$P = 80(0) + 150(50) = 7{,}500$

From the table, we can see that the accountant will earn a maximum profit of \$7,700 if she prepares 40 individual returns and 30 business returns.

SELF CHECK 3 The accountant can make \$7,500 when $x = 0$ and $y = 50$. Interpret the meaning of these values.

EXAMPLE 4 **SUPPLEMENTS** Healthtab and Robust are two diet supplements. Each Healthtab tablet costs 50¢ and contains 3 units of calcium, 20 units of Vitamin C, and 40 units of iron. Each Robust tablet costs 60¢ and contains 4 units of calcium, 40 units of Vitamin C, and 30 units of iron. At least 24 units of calcium, 200 units of Vitamin C, and 120 units of iron are required for the daily needs of one patient. How many tablets of each supplement should be taken daily for a minimum cost? Find the daily minimum cost.

Solution First, we organize the given information into a table.

	Healthtab	Robust	Amount required
Calcium	3	4	24
Vitamin C	20	40	200
Iron	40	30	120
Cost	50¢	60¢	

Find the objective function

We let x represent the number of Healthtab tablets to be taken daily and y the corresponding number of Robust tablets. Because each of the x Healthtab tablets will cost 50¢, and each of the y Robust tablets will cost 60¢, the total cost will be given by the equation

$$C = 0.50x + 0.60y \quad 50¢ = \$0.50 \text{ and } 60¢ = \$0.60$$

Find the feasibility region

Since there are requirements for calcium, Vitamin C, and iron, there is a constraint for each. Note that neither x nor y can be negative.

Unless otherwise noted, all content on this page is © Cengage Learning.

$$\begin{cases} 3x + 4y \geq 24 & \text{The amount of calcium must be greater than or equal to 24 units.} \\ 20x + 40y \geq 200 & \text{The amount of Vitamin C must be greater than or equal to 200 units.} \\ 40x + 30y \geq 120 & \text{The amount of iron must be greater than or equal to 120 units.} \\ x \geq 0, y \geq 0 & \text{The number of tablets taken must be greater than or equal to 0.} \end{cases}$$

We graph the inequalities to find the feasibility region and the coordinates of its corner points, as in Figure 8-33.

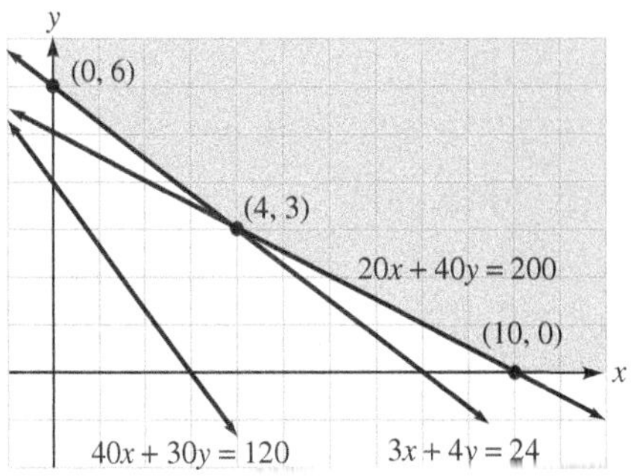

Figure 8-33

Find the minimum cost

In this case, the feasibility region is not bounded on all sides. The coordinates of the corner points are $(0, 6)$, $(4, 3)$, and $(10, 0)$. To find the minimum cost, we substitute each pair of coordinates into the objective function.

Point	$C = 0.50x + 0.60y$
$(0, 6)$	$C = 0.50(\mathbf{0}) + 0.60(\mathbf{6}) = 3.60$
$(4, 3)$	$C = 0.50(\mathbf{4}) + 0.60(\mathbf{3}) = 3.80$
$(10, 0)$	$C = 0.50(\mathbf{10}) + 0.60(\mathbf{0}) = 5.00$

A minimum cost will occur if no Healthtab and 6 Robust tablets are taken daily. The minimum daily cost is \$3.60.

SELF CHECK 4 If the minimum daily cost is chosen, determine the number of units of calcium, Vitamin C, and iron that the patient will receive.

EXAMPLE 5 **PRODUCTION SCHEDULES** A television program director must schedule comedy skits and musical numbers for prime-time variety shows. Each comedy skit requires 2 hours of rehearsal time, costs \$3,000, and brings in \$20,000 from the show's sponsors. Each musical number requires 1 hour of rehearsal time, costs \$6,000, and generates \$12,000. If 250 hours are available for rehearsal, and \$600,000 is budgeted for comedy and music, how many segments of each type should be produced to maximize income? Find the maximum income.

Solution First, we organize the given information into a table.

	Comedy	Musical	Available
Rehearsal time (hours)	2	1	250
Cost (in \$1,000s)	3	6	600
Generated income (in \$1,000s)	20	12	

Find the objective function

We let x represent the number of comedy skits and y the number of musical numbers to be scheduled. Because each of the x comedy skits generates \$20 thousand, the income generated by the comedy skits is \$$20x$ thousand. The musical numbers produce \$$12y$ thousand. The objective function to be maximized is

$$V = 20x + 12y$$

Unless otherwise noted, all content on this page is © Cengage Learning.

Find the feasibility region

Since there are limits on rehearsal time and budget, there is a constraint for each. Note that neither x nor y can be negative.

$$\begin{cases} 2x + y \le 250 & \text{The total rehearsal time must be less than or equal to 250 hours.} \\ 3x + 6y \le 600 & \text{The total cost must be less than or equal to \$600 thousand.} \\ x \ge 0, y \ge 0 & \text{The numbers of skits and musical numbers must be greater than or equal to 0.} \end{cases}$$

We graph the inequalities to find the feasibility region shown in Figure 8-34 and find the coordinates of each corner point.

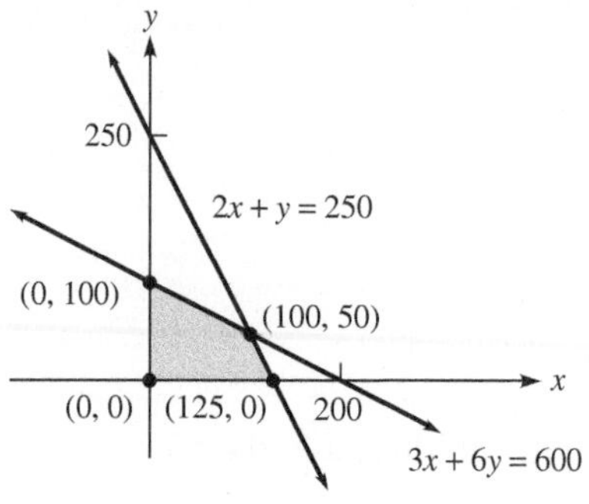

Figure 8-34

Find the maximum income

The coordinates of the corner points of the feasible region are $(0, 0)$, $(0, 100)$, $(100, 50)$, and $(125, 0)$. To find the maximum income, we substitute each pair of coordinates into the objective function.

Corner point	$V = 20x + 12y$
$(0, 0)$	$V = 20(\mathbf{0}) + 12(\mathbf{0}) = 0$
$(0, 100)$	$V = 20(\mathbf{0}) + 12(\mathbf{100}) = 1{,}200$
$(100, 50)$	$V = 20(\mathbf{100}) + 12(\mathbf{50}) = 2{,}600$
$(125, 0)$	$V = 20(\mathbf{125}) + 12(\mathbf{0}) = 2{,}500$

Maximum income will occur if 100 comedy skits and 50 musical numbers are scheduled. The maximum income will be 2,600 thousand dollars, or \$2,600,000.

SELF CHECK 5 If a survey reveals the public wants only to watch comedy skits, what is the maximum income the station will make?

Everyday connections
Fuel Oil Production

Crude Oil, Gasoline, and Natural Gas Futures

Heating Oil NY Harbor	\$3.0962
Gasoline NY Harbor	\$3.1129

Suppose that the refining process at United States refineries requires the production of at least three gallons of gasoline for each gallon of fuel oil. To meet the anticipated demands of winter, at least 2.10 million gallons of fuel oil will need to be produced per day. The demand for gasoline is not more than 9 million gallons a day. The wholesale price of gasoline is \$3.1129 per gallon and the wholesale price of fuel oil is \$3.0962/gal.

1. How much of each should be produced in order to maximize revenue?

2. What is the maximum revenue?

Source: http://www.wtrg.com/#Crude

Unless otherwise noted, all content on this page is © Cengage Learning.

SELF CHECK ANSWERS

1. 14 **2.** $\frac{3}{2}$ **3.** She can make \$7,500 by doing no individual returns and 50 business returns. **4.** The patient will receive 24 units of calcium, 240 units of Vitamin C, and 180 units of iron. **5.** The maximum income will be \$2,500,000.

NOW TRY THIS

The FDA set the standard that fat calories should be limited to between 20% and 35% of the total daily intake, carbohydrates between 45% and 65%, and protein between 10% and 35%. Due to diabetes, Dale must limit his carbohydrate intake to the minimum (45%) and consume between 1,600 and 2,000 calories per day. Let x represent the number of fat calories and y the number of protein calories.

1. Between what numbers of calories can Dale consume daily in carbohydrates?
2. Between what numbers of calories does this leave for fats (x) and protein (y) combined?
3. Between what numbers of calories can he allot to fats?
4. Between what numbers of calories can be allotted to proteins?
5. What are the corner points in this graph?
6. Which one will minimize fat but maximize protein?

8.8 Exercises

WARM-UPS *Evaluate $P = 2x + 5y$ when*

1. $x = 0, y = 4$

2. $x = 2, y = 0$

Solve each system of linear equations.

3. $\begin{cases} x + y = 5 \\ 3x + 2y = 12 \end{cases}$

4. $\begin{cases} x + 3y = 6 \\ 4x + 3y = 12 \end{cases}$

REVIEW *Consider the line passing through* (3, 5) *and* (−1, −4).

5. Find the slope of the line.

6. Write the equation of the line in general form.

7. Write the equation of the line in slope-intercept form.

8. Write the equation of the line that passes through the origin and is parallel to the line.

VOCABULARY AND CONCEPTS *Fill in the blanks.*

9. In a linear program, the inequalities are called _________.

10. Ordered pairs that satisfy the constraints of a linear program are called _______ solutions.

11. The function to be maximized (or minimized) in a linear program is called the _______ function.

12. The objective function of a linear program attains a maximum (or minimum), subject to the constraints, at a ______ or along an ____ of the feasibility region.

GUIDED PRACTICE *Maximize P subject to the following constraints.* SEE EXAMPLE 1. (OBJECTIVE 1)

13. $P = 2x + 3y$
$$\begin{cases} x \ge 0 \\ y \ge 0 \\ x + y \le 4 \end{cases}$$

14. $P = 3x + 2y$
$$\begin{cases} x \ge 0 \\ y \ge 0 \\ x + y \le 4 \end{cases}$$

15. $P = y + \frac{1}{2}x$
$$\begin{cases} x \ge 0 \\ y \ge 0 \\ 2y - x \le 1 \\ y - 2x \ge -2 \end{cases}$$

16. $P = 4y - x$
$$\begin{cases} x \le 2 \\ y \ge 0 \\ x + y \ge 1 \\ 2y - x \le 1 \end{cases}$$

Minimize P subject to the following constraints. SEE EXAMPLE 2. (OBJECTIVE 1)

17. $P = 5x + 12y$
$$\begin{cases} x \ge 0 \\ y \ge 0 \\ x + y \le 4 \end{cases}$$

18. $P = 3x + 6y$
$$\begin{cases} x \ge 0 \\ y \ge 0 \\ x + y \le 4 \end{cases}$$

19. $P = 3y + x$
$$\begin{cases} x \geq 0 \\ y \geq 0 \\ 2y - x \leq 1 \\ y - 2x \geq -2 \end{cases}$$

20. $P = 5y + x$
$$\begin{cases} x \leq 2 \\ y \geq 0 \\ x + y \geq 1 \\ 2y - x \leq 1 \end{cases}$$

ADDITIONAL PRACTICE *Maximize P subject to the following constraints.*

21. $P = 2x + y$
$$\begin{cases} y \geq 0 \\ y - x \leq 2 \\ 2x + 3y \leq 6 \\ 3x + y \leq 3 \end{cases}$$

22. $P = x - 2y$
$$\begin{cases} x + y \leq 5 \\ y \leq 3 \\ x \leq 2 \\ x \geq 0 \\ y \geq 0 \end{cases}$$

23. $P = 3x - 2y$
$$\begin{cases} x \leq 1 \\ x \geq -1 \\ y - x \leq 1 \\ x - y \leq 1 \end{cases}$$

24. $P = x - y$
$$\begin{cases} 5x + 4y \leq 20 \\ y \leq 5 \\ x \geq 0 \\ y \geq 0 \end{cases}$$

Minimize P subject to the following constraints.

25. $P = 6x + 2y$
$$\begin{cases} y \geq 0 \\ y - x \leq 2 \\ 2x + 3y \leq 6 \\ 3x + y \leq 3 \end{cases}$$

26. $P = 2y - x$
$$\begin{cases} x \geq 0 \\ y \geq 0 \\ x + y \leq 5 \\ x + 2y \geq 2 \end{cases}$$

27. $P = 2x - 2y$
$$\begin{cases} x \leq 1 \\ x \geq -1 \\ y - x \leq 1 \\ x - y \leq 1 \end{cases}$$

28. $P = y - 2x$
$$\begin{cases} x + 2y \leq 4 \\ 2x + y \leq 4 \\ x + 2y \geq 2 \\ 2x + y \geq 2 \end{cases}$$

APPLICATIONS *Use an objective function and a system of inequalities that describe the constraints in each problem. Use a graph of the feasibility region, showing the corner points to find the maximum or minimum value of the objective function.* SEE EXAMPLES 3–5. (OBJECTIVE 2)

29. Furniture Two woodworkers, Tom and Carlos, bring in $100 for making a table and $80 for making a chair. On average, Tom must work 3 hours and Carlos 2 hours to make a chair. Tom must work 2 hours and Carlos 6 hours to make a table. If neither wants to work more than 42 hours per week, how many tables and how many chairs should they make each week to maximize their income? Find the maximum income.

	Table	Chair	Time available
Income ($)	100	80	
Tom's time (hr)	2	3	42
Carlos's time (hr)	6	2	42

30. Crafts Two artists, Nina and Rob, make yard ornaments. They bring in $80 for each wooden snowman they make and $64 for each wooden Santa Claus. On average, Nina must work 4 hours and Rob 2 hours to make a snowman. Nina must work 3 hours and Rob 4 hours to make a Santa Claus. If neither wants to work more than 20 hours per week, how many of each ornament should they make each week to maximize their income? Find the maximum income.

	Snowman	Santa Claus	Time available
Income ($)	80	64	
Nina's time (hr)	4	3	20
Rob's time (hr)	2	4	20

31. Inventories An electronics store manager stocks from 20 to 30 IBM-compatible computers and from 30 to 50 Macintosh computers. There is room in the store to stock up to 60 computers. The manager receives a commission of $50 on the sale of each IBM-compatible computer and $40 on the sale of each Macintosh computer. If the manager can sell all of the computers, how many should she stock to maximize her commissions? Find the maximum commission.

Inventory	IBM	Macintosh
Minimum	20	30
Maximum	30	50
Commission	$50	$40

32. Diet A diet requires at least 16 units of Vitamin C and at least 34 units of Vitamin B complex. Two food supplements are available that provide these nutrients in the amounts and costs shown in the table. How much of each should be used to minimize the cost?

Supplement	Vitamin C	Vitamin B	Cost
A	3 units/g	2 units/g	3¢/g
B	2 units/g	6 units/g	4¢/g

33. Production Manufacturing DVD players and TVs requires the use of the electronics, assembly, and finishing departments of a factory, according to the following schedule:

	Hours for DVD player	Hours for TV	Hours available per week
Electronics	3	4	180
Assembly	2	3	120
Finishing	2	1	60

Each DVD player has a profit of \$40, and each TV has a profit of \$32. How many DVD players and TVs should be manufactured weekly to maximize profit? Find the maximum profit.

34. Production A company manufactures one type of computer chip that runs at 1.66 GHz and another that runs at 2.66 GHz. The company can make a maximum of 50 fast chips per day and a maximum of 100 slow chips per day. It takes 6 hours to make a fast chip and 3 hours to make a slow chip, and the company's employees can provide up to 360 hours of labor per day. If the company makes a profit of \$20 on each 2.66-GHz chip and \$27 on each 1.66-GHz chip, how many of each type should be manufactured to earn the maximum profit?

35. Financial planning A stockbroker has \$200,000 to invest in stocks and bonds. She wants to invest at least \$100,000 in stocks and at least \$50,000 in bonds. If stocks have an annual yield of 9% and bonds have an annual yield of 7%, how much should she invest in each to maximize her income? Find the maximum return.

36. Production A small country exports soybeans and flowers. Soybeans require 8 workers per acre, flowers require 12 workers per acre, and 100,000 workers are available. Government contracts require that there be at least 3 times as many acres of soybeans as flowers planted. It costs \$250 per acre to plant soybeans and \$300 per acre to plant flowers, and there is a budget of \$3 million. If the profit from soybeans is \$1,600 per acre and the profit from flowers is \$2,000 per acre, how many acres of each crop should be planted to maximize profit? Find the maximum profit.

WRITING ABOUT MATH

37. What is meant by the constraints of a linear program?

38. What is meant by a feasible solution of a linear program?

SOMETHING TO THINK ABOUT

39. Try to construct a linear programming problem. What difficulties do you encounter?

40. Try to construct a linear programming problem that will have a maximum at every point along an edge of the feasibility region.

Projects

PROJECT 1

The number of units of a product that will be produced depends on the unit price of the product. As the unit price gets higher, the product will be produced in greater quantity, because the producer will make more money on each item. The *supply* of the product will grow, and we say that supply *is a function of* (or *depends on*) the unit price. Furthermore, as the price rises, fewer consumers will buy the product, and the *demand* will decrease. The demand for the product is also a function of the unit price.

In this project, we will assume that both supply and demand are *linear* functions of the unit price. Thus, the graph of supply (the y-coordinate) versus price (the x-coordinate) is a line with positive slope. The graph of the demand function is a line with negative slope. Because these two lines cannot be parallel, they must intersect. The price at which supply equals demand is called the *market price:* At this price, the same number of units of the product will be sold as are manufactured.

You work for Soda Pop Inc. and have the task of analyzing the sales figures for the past year. You have been provided with the following supply and demand functions. (Supply and demand are measured in cases per week; p, the price per case, is measured in dollars.)

The demand for soda is $D(p) = 19{,}000 - 2{,}200p$.
The supply of soda is $S(p) = 3{,}000 + 1{,}080p$.

Both functions are true for values of p from \$3.50 to \$5.75.

Graph both functions on the same set of coordinate axes, being sure to label each graph, and include any other important information. Then write a report for your supervisor that answers the following questions.

a. Explain why producers will be able to sell all of the soda they make when the price is \$3.50 per case. How much money will the producers take in from these sales?

b. How much money will producers take in from sales when the price is \$5.75 per case? How much soda will not be sold?

c. Find the market price for soda (to the nearest cent). How many cases per week will be sold at this price? How much money will the producers take in from sales at the market price?

d. Explain why prices always tend toward the market price. That is, explain why the unit price will rise if the demand is greater than the supply, and why the unit price will fall if supply is greater than demand.

(Continued)

PROJECT 2

Goodstuff Produce Company has two large water canals that feed the irrigation ditches on its fruit farm. One of these canals runs directly north and south, and the other runs directly east and west. The canals cross at the center of the farm property (the origin) and divide the farm into four quadrants. The company is interested in digging some new irrigation ditches in a portion of the northeast quadrant. You have been hired to plan the layout of the new system.

Your design is to make use of ditch Z, which is already present. This ditch runs from a point 300 meters north of the origin to a point 400 meters east of the origin. The owners of Goodstuff want two new ditches.

- Ditch A is to begin at a point 100 meters north of the origin and follow a line that travels 3 meters north for every 7 meters it travels east until it intersects ditch Z.
- Ditch B is to run from the origin to ditch Z in such a way that it exactly bisects the area in the northeast quadrant that is south of both ditch Z and ditch A.

You are to provide the equations of the lines that the three ditches follow, as well as the exact location of the gates that will be installed where the ditches intersect one another. Be sure to provide explanations and organized work that will clearly display the desired information and assure the owners of Goodstuff that they will get exactly what they want.

Reach for Success REVIEWING YOUR GAME PLAN

In an activity in a previous chapter, you set goals for this course. It is now time to check your progress toward achieving them.

A. Fill in the blanks from your *original goals* for this course.

The grade I am willing to work to achieve is a/an _____.

Considering other course commitments as well as any work and family commitments, state the number of hours *outside of class* you realistically believe you can devote each week to this one course. _____

List at least three things you are willing to add to your game plan that will support your success in this course.

1. ______________________________
2. ______________________________
3. ______________________________

Can you think of anything else you can do to improve your performance in this class? ______________________________

____

B. Considering your performance in this course to date, please answer these questions.

What is your approximate grade in this class now? _____ Is this the grade you want to earn? _____ If this is your goal grade, or a higher one, congratulations! It appears you stayed with your game plan and should continue with this plan for the remainder of the semester.

If the grade is lower than what you want, let's look at some factors that may have contributed to this.

List the items on your game plan that you have done to be successful in this course.

____

Congratulations on following through with these items!

If there are items on your game plan that you have not done, list these and consider why you have not been able to follow through with your plan. Some reasons might be work schedule, unreliable child care, other course demands, content more difficult than anticipated, motivational issues, or personal issues.

____

C. Now, let's determine if there are changes you can make now to turn the semester around.

Is it necessary to modify the grade you would be working toward? _____ What is that new grade? _____

Of the items listed in Part B above, consider and list any changes you could make now to improve your chance of success in this course before the end of the semester.

____

____

D. Is your success in this course a high enough priority to make the changes you listed above? _____

8 Review

SECTION 8.1 Solving Systems of Linear Equations by Graphing

DEFINITIONS AND CONCEPTS

1. On a single set of coordinate axes, carefully graph each equation.
2. Find the coordinates of the point where the graphs intersect, if applicable.
3. Check the solution in both of the original equations, if applicable.
4. If the graphs have no point in common, the system has no solution.
5. If the graphs of the equations coincide (are the same), the system has infinitely many solutions that can be expressed as a general ordered pair.

In a graph of two equations, each with two variables:

If the lines are distinct (are different) and intersect, the equations are *independent* and the system is *consistent.* **One solution exists**, expressed as (x, y).

If the lines are distinct and parallel, the equations are *independent* and the system is *inconsistent.* **No solution exists**, expressed as $\varnothing$.

If the lines coincide (are the same), the equations are *dependent* and the system is *consistent.* **Infinitely many solutions exist**, expressed as $(x, ax + b)$.

EXAMPLES

Solve each system by graphing.

a. $\begin{cases} x + y = 6 \\ x - y = 2 \end{cases}$

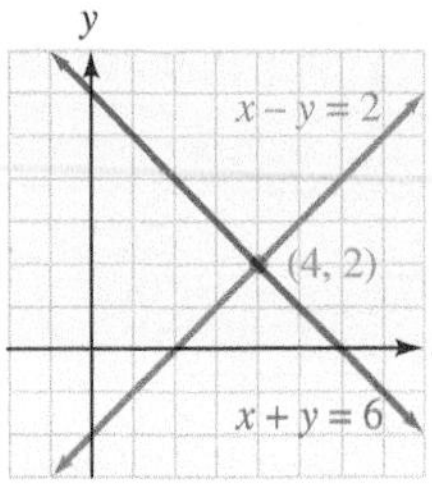

The solution is $(4, 2)$.

b. $\begin{cases} 2x + y = 6 \\ y = -2x - 3 \end{cases}$

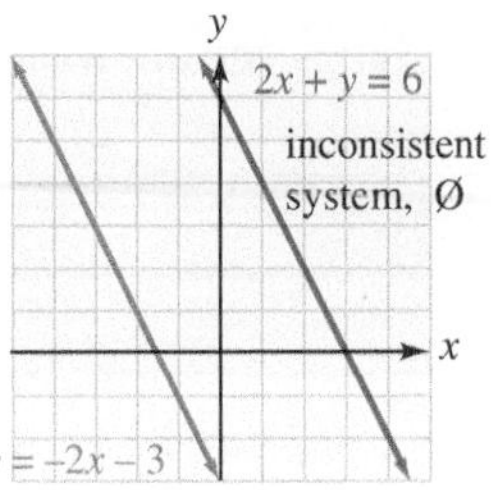

The solution set is $\varnothing$.

c. $\begin{cases} x - 3y = 6 \\ y = \dfrac{1}{3}x - 2 \end{cases}$

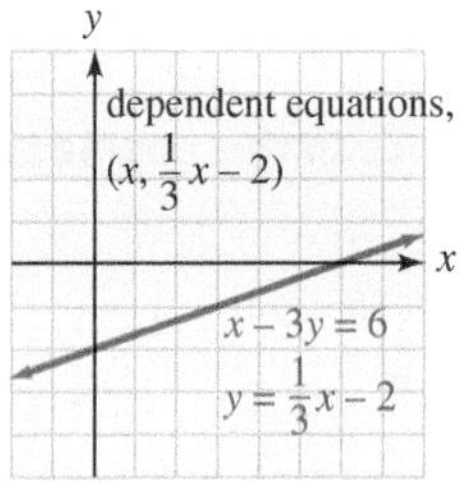

The equations of the system are dependent, and infinitely many solutions exist. A general solution is $\left(x, \frac{1}{3}x - 2\right)$.

REVIEW EXERCISES

Solve each system by the graphing method.

1. $\begin{cases} 2x + y = 11 \\ -x + 2y = 7 \end{cases}$

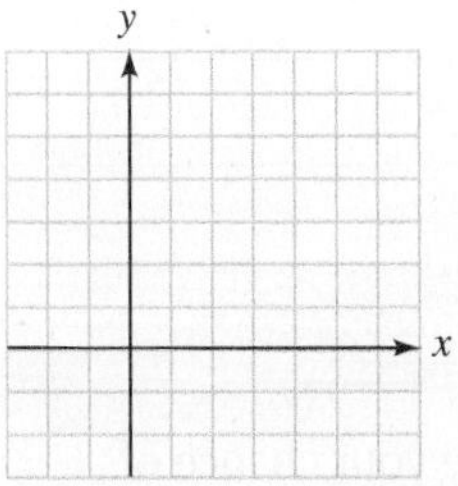

2. $\begin{cases} 3x + 2y = 0 \\ 2x - 3y = -13 \end{cases}$

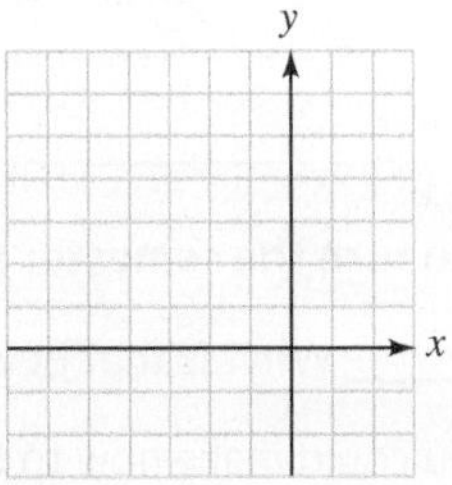

3. $\begin{cases} \frac{1}{2}x + \frac{1}{3}y = 2 \\ y = 6 - \frac{3}{2}x \end{cases}$

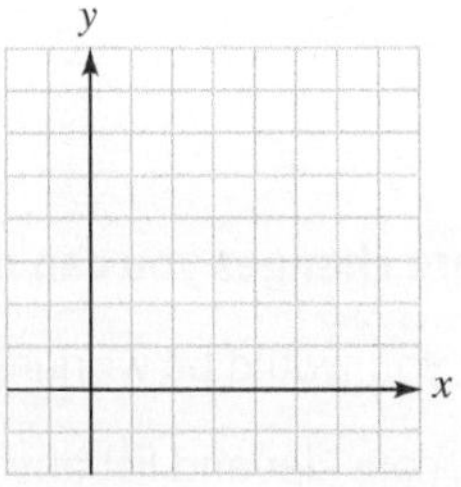

4. $\begin{cases} \frac{1}{3}x - \frac{1}{2}y = 1 \\ 6x - 9y = 2 \end{cases}$

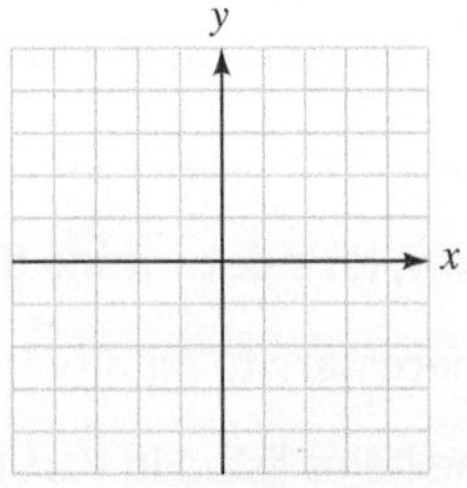

Unless otherwise noted, all content on this page is © Cengage Learning.

SECTION 8.2 Solving Systems of Linear Equations by Substitution and Elimination

DEFINITIONS AND CONCEPTS	EXAMPLES
1. If necessary, solve one equation for one of its variables, preferably a variable with a coefficient of 1. 2. Substitute the resulting expression for the variable obtained in Step 1 into the other equation and solve that equation. 3. Find the value of the other variable by substituting the value of the variable found in Step 2 into any equation containing both variables. 4. State the solution. 5. Check the solution in both of the original equations.	Solve by substitution: $\begin{cases} x = 3y - 9 \\ 2x - y = 2 \end{cases}$ Since the first equation is already solved for x, we will substitute its right side for x in the second equation. $2(\mathbf{3y - 9}) - y = 2$ $6y - 18 - y = 2$ Use the distributive property to remove parentheses. $5y - 18 = 2$ Combine like terms. $5y = 20$ Add 18 to both sides. $y = 4$ Divide both sides by 5. To find x, we can substitute 4 for y in the first equation. $x = 3(\mathbf{4}) - 9$ $x = 3$ The solution is $(3, 4)$.
1. If necessary, write both equations of the system in $Ax + By = C$ form. 2. If necessary, multiply the terms of one or both of the equations by constants chosen to make the coefficients of one of the variables opposites. 3. Add the equations and solve the resulting equation, if possible. 4. Substitute the value obtained in Step 3 into either of the original equations and solve for the remaining variable, if applicable. 5. State the solution obtained in Steps 3 and 4. 6. Check the solution in both equations of the original system.	Solve by elimination: $\begin{cases} 2x - 3y = 8 \\ x + 2y = 4 \end{cases}$ To eliminate x, we multiply the second equation by -2 and add the result to the first equation. $2x - 3y = 8$ $-2x - 4y = -8$ $-7y = 0$ $y = 0$ Divide both sides by 7. To find x, we can substitute 0 for y in the first equation. $2x - 3(\mathbf{0}) = 8$ Substitute. $2x = 8$ Simplify. $x = 4$ Divide both sides by 2. The solution is $(4, 0)$.

REVIEW EXERCISES

Solve each system by substitution.

5. $\begin{cases} y = 2x + 5 \\ 3x + 4y = 9 \end{cases}$

6. $\begin{cases} y = 3x + 5 \\ 3x - y = -5 \end{cases}$

7. $\begin{cases} x + 2y = 11 \\ 2x - y = 2 \end{cases}$

8. $\begin{cases} 2x + 3y = -2 \\ 3x + 5y = -2 \end{cases}$

Solve each system by elimination.

9. $\begin{cases} x - y = 1 \\ 5x + 2y = -16 \end{cases}$

10. $\begin{cases} 3x + 2y = 1 \\ 2x - 3y = 5 \end{cases}$

11. $\begin{cases} 2x + 6 = -3y \\ -2x = 3y + 6 \end{cases}$

12. $\begin{cases} y = \dfrac{2x - 3}{2} \\ x = \dfrac{2y + 7}{2} \end{cases}$

SECTION 8.3 Solving Applications of Systems of Linear Equations in Two Variables

DEFINITIONS AND CONCEPTS	EXAMPLES
Systems of equations are useful in solving many types of applications.	**Boating** A boat traveled 30 kilometers downstream in 3 hours and traveled 12 kilometers in 3 hours against the current. Find the speed of the boat in still water. *Analyze the problem* We can let s represent the speed of the boat in still water in km/hr and let c represent the speed of the current in km/hr. *Form two equations* The rate of speed of the boat while going downstream is $(s + c)$. The rate of the boat while going upstream is $(s - c)$. Because $d = r \cdot t$, the information gives two equations in two variables. $\begin{cases} 30 = 3(s + c) \\ 12 = 3(s - c) \end{cases}$ After removing parentheses and rearranging terms, we have (1) (2) $\begin{cases} 3s + 3c = 30 \\ 3s - 3c = 12 \end{cases}$ *Solve the system* To solve this system by elimination (addition), we add the equations, and solve for s. $3s + 3c = 30$ $\underline{3s - 3c = 12}$ $6s = 42$ $s = 7$ Divide both sides by 6. *State the conclusion* The speed of the boat in still water is 7 kilometers per hour. *Check the result* If $s = 7$, $3(7) + 3c = 30$ $21 + 3c = 30$ $3c = 9$ $c = 3$ The speed of the current is 3 km/hr. In 3 hours, the boat will travel downstream $3(7 + 3) = 30$ km. In 3 hours, the boat will travel against the current $3(7 - 3) = 12$ km. The answers check.

REVIEW EXERCISES

13. Integers One number is 3 times another, and their sum is 84. Find the numbers.

14. Geometry The length of a rectangle is 3 times its width, and its perimeter is 24 feet. Find its dimensions.

15. Buying grapefruit A grapefruit costs 15 cents more than an orange. Together, they cost 85 cents. Find the cost of a grapefruit.

16. Utility bills A man's electric bill for January was \$23 less than his gas bill. The two utilities cost him a total of \$109. Find the amount of his gas bill.

17. Buying groceries Two gallons of milk and 3 dozen eggs cost \$6.80. Three gallons of milk and 2 dozen eggs cost \$7.35. How much does each gallon of milk cost?

18. Investing money Carlos invested part of \$4,500 in a 4% certificate of deposit account and the rest in a 3% passbook account. If the total annual interest from both accounts is \$160, how much did he invest at 3%?

19. Boating It takes a boat 4 hours to travel 56 miles down a river and 3 hours longer to make the return trip. Find the speed of the current.

20. Medical technology A laboratory technician has one batch of solution that is 10% saline and a second batch that is 60% saline. He would like to make 50 milliliters of solution that is 30% saline. How many liters of each batch should he use?

SECTION 8.4 Solving Systems of Three Linear Equations in Three Variables

DEFINITIONS AND CONCEPTS	EXAMPLES
Strategy for solving three linear equations in three variables: **1.** Write all equations in the system in $ax + by + cz = d$ form. **2.** Select any two equations and eliminate a variable. **3.** Select a different pair of equations and eliminate the same variable. **4.** Solve the resulting pair of two equations in two variables. **5.** To find the value of the third variable, substitute the values of the two variables found in Step 4 into any equation containing all three variables and solve the equation. **6.** Check the solution in all three of the original equations.	To solve the system $\begin{cases} x + y + z = 4 \\ x - 2y - z = -9 \\ 2x - y + 2z = -1 \end{cases}$, we can add the first and second equations to obtain Equation 1: $\begin{aligned} x + y + z &= 4 \\ x - 2y - z &= -9 \\ \hline (1)\quad 2x - y \quad &= -5 \end{aligned}$ We now multiply the second equation by 2 and add it to the third equation to obtain Equation 2: $\begin{aligned} 2x - 4y - 2z &= -18 \\ 2x - y + 2z &= -1 \\ \hline (2)\quad 4x - 5y \quad &= -19 \end{aligned}$ To solve the system $\begin{cases} 2x - y = -5 \\ 4x - 5y = -19 \end{cases}$, formed by Equations 1 and 2, we can multiply the first equation by -2 and add the result to the second equation to eliminate x. $\begin{aligned} -4x + 2y &= 10 \\ 4x - 5y &= -19 \\ \hline -3y &= -9 \\ y &= 3 \end{aligned}$ Divide both sides by -3. We can substitute 3 into either equation of the system to find x. $2x - y = -5$ This is the first equation of the system. $2x - \mathbf{3} = -5$ Substitute 3 for y. $2x = -2$ Add 3 to both sides. $x = -1$ Divide both sides by 2. We now can substitute -1 for x and 3 for y into any of the equations in the original system and solve for z: $x + y + z = 4$ This is the first equation of the original system. $\mathbf{-1} + \mathbf{3} + z = 4$ Substitute. $2 + z = 4$ Simplify. $z = 2$ Subtract 2 from both sides. The solution is $(-1, 3, 2)$.

REVIEW EXERCISES

Solve each system.

21. $\begin{cases} x + y + z = 6 \\ x - y - z = -4 \\ -x + y - z = 2 \end{cases}$

22. $\begin{cases} 2x + 3y + z = -5 \\ -x + 2y - z = -6 \\ 3x + y + 2z = 4 \end{cases}$

SECTION 8.5 Solving Systems of Linear Equations Using Matrices

DEFINITIONS AND CONCEPTS	EXAMPLES
A **matrix** is any rectangular array of numbers. Systems of linear equations can be solved using matrices and the method of **Gaussian elimination** and back substitution. A matrix with m rows and n columns is called an $m \times n$ matrix.	To use matrices to solve the system $$\begin{cases} x + y + z = 4 \\ 2x - y + 2z = -1 \\ x - 2y - z = -9 \end{cases}$$ we can represent it with the following augmented matrix: $$\left[\begin{array}{ccc\|c} 1 & 1 & 1 & 4 \\ 2 & -1 & 2 & -1 \\ 1 & -2 & -1 & -9 \end{array}\right]$$ To get 0's under the 1 in the first column, we multiply row 1 of the augmented matrix by -2 and add it to row 2 to get a new row 2. We then multiply row 1 by -1 and add it to row 3 to get a new row 3. $$\left[\begin{array}{ccc\|c} 1 & 1 & 1 & 4 \\ 0 & \mathbf{-3} & 0 & -9 \\ 0 & -3 & -2 & -13 \end{array}\right] \quad \begin{array}{l} \\ -2R1 + R2 \rightarrow R2 \\ -1R1 + R3 \rightarrow R3 \end{array}$$ To get a 0 under the -3 in the second column of the previous matrix, we multiply row 2 by -1 and add it to row 3.
$$\left[\begin{array}{ccc\|c} a & b & c & d \\ 0 & e & f & g \\ 0 & 0 & h & i \end{array}\right]$$ ($a, b, c, \ldots i$ are real numbers) This matrix is in triangular form.	$$\left[\begin{array}{ccc\|c} 1 & 1 & 1 & 4 \\ 0 & -3 & 0 & -9 \\ 0 & 0 & \mathbf{-2} & -4 \end{array}\right] \quad -1R2 + R3 \rightarrow R3$$ Finally, to obtain a 1 in the third row, third column, we multiply row 3 by $-\frac{1}{2}$. $$\left[\begin{array}{ccc\|c} 1 & 1 & 1 & 4 \\ 0 & -3 & 0 & -9 \\ 0 & 0 & 1 & 2 \end{array}\right] \quad -\tfrac{1}{2}R3 \rightarrow R3$$ The final matrix represents the system $$\begin{array}{l} (1) \\ (2) \\ (3) \end{array} \begin{cases} x + y + z = 4 \\ 0x - 3y + 0z = -9 \\ 0x + 0y + z = 2 \end{cases}$$ From Equation 3, we see that $z = 2$. From Equation 2, we see that $y = 3$. To find the value of x, we use back substitution to substitute 2 for z and 3 for y in Equation 1 and solve for x: (1) $x + y + z = 4$ $x + \mathbf{3} + \mathbf{2} = 4$ Substitute. $x + 5 = 4$ Simplify. $x = -1$ Subtract 5 from both sides. Thus, $x = -1$. The solution of the given system is $(-1, 3, 2)$. Verify that this ordered triple satisfies each equation of the original system.

REVIEW EXERCISES

Solve each system by using matrices.

23. $\begin{cases} -x + 5y + 2z = -6 \\ x - 5y - z = 6 \\ 3x - 15y - 6z = 18 \end{cases}$

24. $\begin{cases} x + y + z = 6 \\ 2x - y + z = 1 \\ 4x + y - z = 5 \end{cases}$

25. $\begin{cases} x + y = 3 \\ x - 2y = -3 \\ 2x + y = 4 \end{cases}$

26. $\begin{cases} x - 3y + z = 4 \\ 2x - 5y + 3z = 6 \end{cases}$

SECTION 8.6 Solving Systems of Linear Equations Using Determinants

DEFINITIONS AND CONCEPTS	EXAMPLES
A **determinant of a square matrix** is a number. $\begin{vmatrix} a & b \\ c & d \end{vmatrix} = ad - bc$	Find the determinant: $\begin{vmatrix} 8 & -3 \\ 2 & -1 \end{vmatrix}$ $\begin{vmatrix} 8 & -3 \\ 2 & -1 \end{vmatrix} = 8(-1) - (-3)(2) = -8 + 6 = -2$
$\begin{vmatrix} a_1 & b_1 & c_1 \\ a_2 & b_2 & c_2 \\ a_3 & b_3 & c_3 \end{vmatrix} = a_1\begin{vmatrix} b_2 & c_2 \\ b_3 & c_3 \end{vmatrix} - b_1\begin{vmatrix} a_2 & c_2 \\ a_3 & c_3 \end{vmatrix} + c_1\begin{vmatrix} a_2 & b_2 \\ a_3 & b_3 \end{vmatrix}$	To evaluate the determinant $\begin{vmatrix} 1 & 3 & -2 \\ 1 & -2 & -1 \\ 2 & -1 & 3 \end{vmatrix}$, we can expand by minors: Minor of 1 ↓, Minor of 3 ↓, Minor of −2 ↓ $\begin{vmatrix} 1 & 3 & -2 \\ 1 & -2 & -1 \\ 2 & -1 & 3 \end{vmatrix} = 1\begin{vmatrix} -2 & -1 \\ -1 & 3 \end{vmatrix} - 3\begin{vmatrix} 1 & -1 \\ 2 & 3 \end{vmatrix} + (-2)\begin{vmatrix} 1 & -2 \\ 2 & -1 \end{vmatrix}$ $= 1(-6 - 1) - 3(3 + 2) - 2(-1 + 4)$ $= -7 - 15 - 6$ $= -28$
Cramer's rule for two linear equations in two variables: The solution of the system $\begin{cases} ax + by = e \\ cx + dy = f \end{cases}$ is given by $x = \dfrac{D_x}{D} = \dfrac{\begin{vmatrix} e & b \\ f & d \end{vmatrix}}{\begin{vmatrix} a & b \\ c & d \end{vmatrix}}$ and $y = \dfrac{D_y}{D} = \dfrac{\begin{vmatrix} a & e \\ c & f \end{vmatrix}}{\begin{vmatrix} a & b \\ c & d \end{vmatrix}}$ If $D \neq 0$, the system is consistent and the equations are independent. If $D = 0$ and D_x or D_y is nonzero, the system is inconsistent. If every determinant is 0, the system is consistent but the equations are dependent.	Solve using Cramer's rule: $\begin{cases} 2x - 4y = -14 \\ 3x + y = -7 \end{cases}$ $x = \dfrac{D_x}{D} = \dfrac{\begin{vmatrix} -14 & -4 \\ -7 & 1 \end{vmatrix}}{\begin{vmatrix} 2 & -4 \\ 3 & 1 \end{vmatrix}} = \dfrac{-14 - 28}{2 - (-12)} = \dfrac{-42}{14} = -3$ $y = \dfrac{D_y}{D} = \dfrac{\begin{vmatrix} 2 & -14 \\ 3 & -7 \end{vmatrix}}{\begin{vmatrix} 2 & -4 \\ 3 & 1 \end{vmatrix}} = \dfrac{-14 - (-42)}{2 - (-12)} = \dfrac{28}{14} = 2$ The solution is $(-3, 2)$.
Cramer's rule for three linear equations in three variables: The solution of the system $\begin{cases} ax + by + cz = j \\ dx + ey + fz = k \\ gx + hy + iz = l \end{cases}$ is given by $x = \frac{D_x}{D}$, $y = \frac{D_y}{D}$, and $z = \frac{D_z}{D}$ where $D = \begin{vmatrix} a & b & c \\ d & e & f \\ g & h & i \end{vmatrix}$ $D_x = \begin{vmatrix} j & b & c \\ k & e & f \\ l & h & i \end{vmatrix}$ $D_y = \begin{vmatrix} a & j & c \\ d & k & f \\ g & l & i \end{vmatrix}$ $D_z = \begin{vmatrix} a & b & j \\ d & e & k \\ g & h & l \end{vmatrix}$	To use Cramer's rule to solve $\begin{cases} x + y + z = 4 \\ 2x - y + 2z = -1 \\ x - 2y - z = -9 \end{cases}$, we can find D, D_x, D_y, and D_z and substitute these values into the formulas $x = \dfrac{D_x}{D}$, $y = \dfrac{D_y}{D}$, and $z = \dfrac{D_z}{D}$ After forming and evaluating the determinants, we will obtain $D = 6$, $D_x = -6$, $D_y = 18$, and $D_z = 12$ and we have $x = \dfrac{D_x}{D} = \dfrac{-6}{6} = -1$, $y = \dfrac{D_y}{D} = \dfrac{18}{6} = 3$, $z = \dfrac{D_z}{D} = \dfrac{12}{6} = 2$ The solution of this system is $(-1, 3, 2)$.

If $D \neq 0$, the system is consistent and the equations are independent.

If $D = 0$ and D_x or D_y or D_z is nonzero, the system is inconsistent.

If every determinant is 0, the system is consistent but the equations are dependent.

REVIEW EXERCISES

Evaluate each determinant.

27. $\begin{vmatrix} 3 & 1 \\ -2 & 4 \end{vmatrix}$

28. $\begin{vmatrix} -4 & 5 \\ 6 & -2 \end{vmatrix}$

29. $\begin{vmatrix} -1 & 2 & -1 \\ 2 & -1 & 3 \\ 1 & -2 & 2 \end{vmatrix}$

30. $\begin{vmatrix} 3 & -2 & 2 \\ 1 & -2 & -2 \\ 2 & 1 & -1 \end{vmatrix}$

Use Cramer's rule to solve each system.

31. $\begin{cases} 3x + 4y = 10 \\ 2x - 3y = 1 \end{cases}$

32. $\begin{cases} 2x - 5y = -17 \\ 3x + 2y = 3 \end{cases}$

33. $\begin{cases} x + 2y + z = 0 \\ 2x + y + z = 3 \\ x + y + 2z = 5 \end{cases}$

34. $\begin{cases} x + y + z = 1 \\ -x + y - z = 3 \\ -x - y + z = 5 \end{cases}$

SECTION 8.7 Solving Systems of Linear Inequalities in Two Variables

DEFINITIONS AND CONCEPTS	EXAMPLES
1. Graph each inequality in the system on the same coordinate axes using solid or dashed lines as appropriate. **2.** Find the region where the graphs overlap. **3.** Select a test point from the region to verify the solution. **4.** Graph only the solution set, if there is one. If a given inequality is $<$ or $>$, the boundary line is dashed. If a given inequality is $\leq$ or $\geq$, the boundary line is solid.	Graph the solution set: $\begin{cases} x + y < 4 \\ 2x - y \geq 6 \end{cases}$ 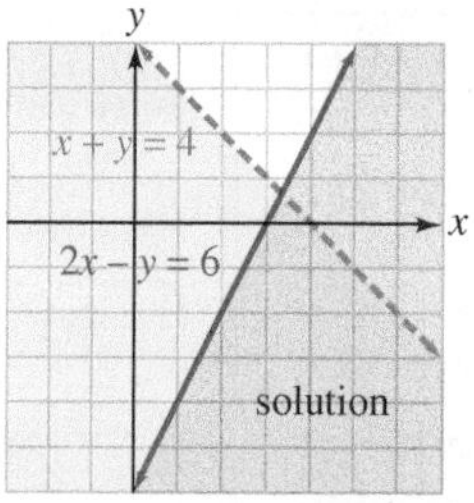 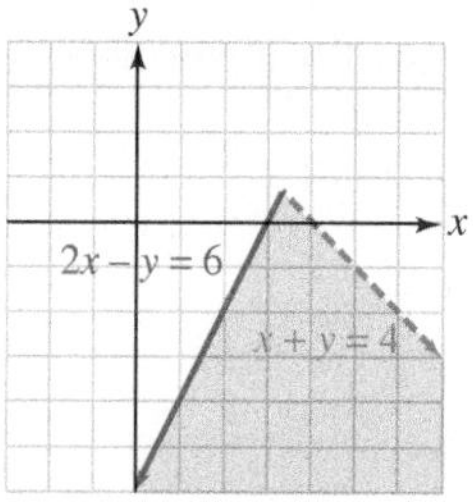 The graph of $x + y < 4$ includes all points below the line $x + y = 4$. Since the boundary is not included, we draw it as a dashed line. The graph of $2x - y \geq 6$ includes all points below the line $2x - y = 6$. Since the boundary is included, we draw it as a solid line. The solution is graphed above on the right.

REVIEW EXERCISES

Graph each inequality.

35. $y \geq x + 2$

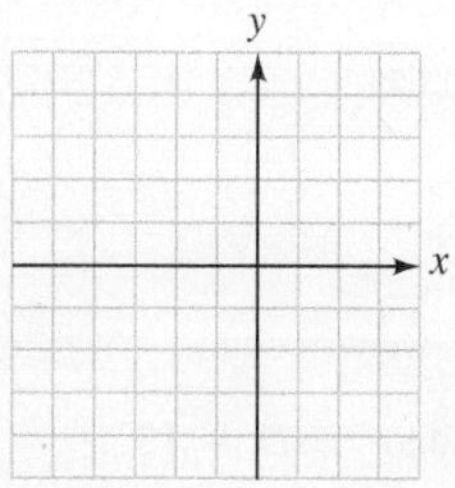

36. $x < 3$

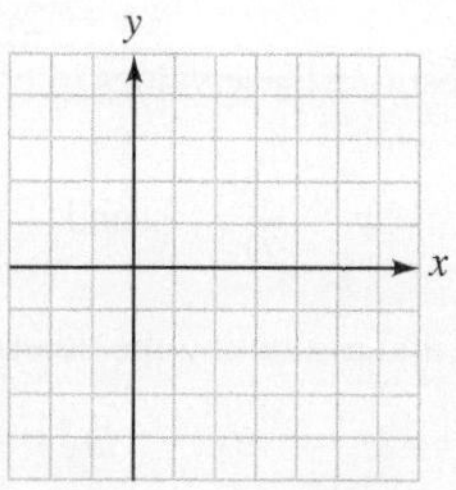

Solve each system of inequalities.

37. $\begin{cases} 5x + 3y < 15 \\ 3x - y > 3 \end{cases}$

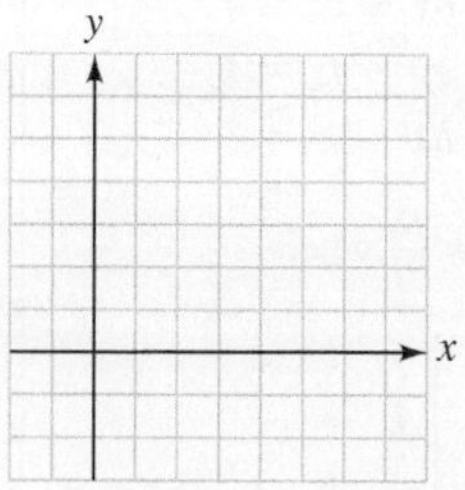

38. $\begin{cases} 5x - 3y \geq 5 \\ 3x + 2y \geq 3 \end{cases}$

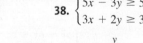

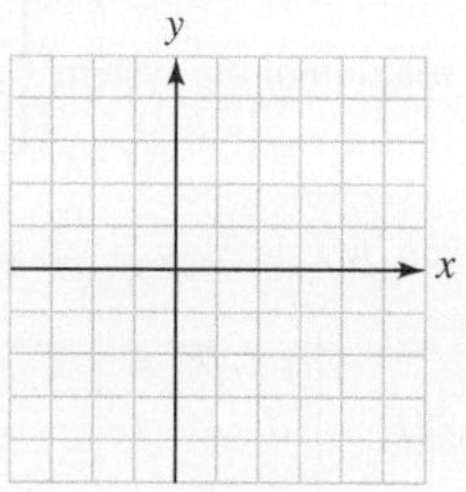

Unless otherwise noted, all content on this page is © Cengage Learning.

39. $\begin{cases} x \geq 3y \\ y < 3x \end{cases}$

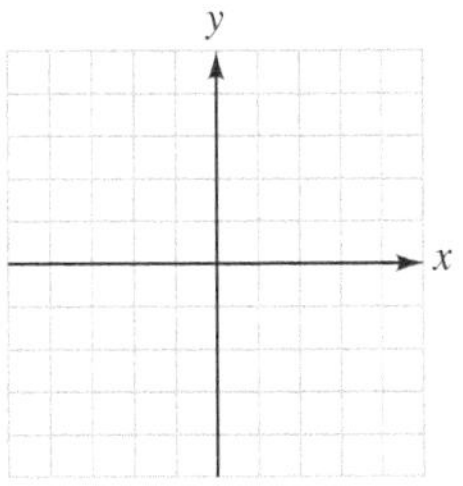

40. $\begin{cases} x \geq 0 \\ x \leq 3 \end{cases}$

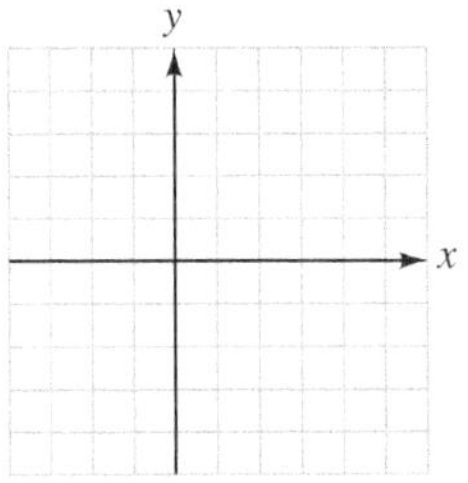

41. Shopping A mother wants to spend at least \$40 but no more than \$60 on her child's school uniform. If shirts sell for \$10 and pants sell for \$20, find a system of inequalities that describe the possible numbers of shirts, x, and pants, y, that she can buy. Graph the system and give two possible solutions.

SECTION 8.8 Solving Systems Using Linear Programming

DEFINITIONS AND CONCEPTS

If a linear function, subject to the constraints of a system of linear inequalities in two variables, attains a maximum or a minimum value, that value will occur at a corner or along an entire edge of the region R that represents the solution of the system.

EXAMPLES

To solve a linear program, we maximize (or minimize) a function (called the objective function) subject to given conditions on its variables, called constraints.

To find the maximum value P of the function $P = 3x + 2y$ subject to the constraints

$$\begin{cases} x - y \leq 4 \\ x + y \leq 6 \\ x \geq 0 \\ y \geq 0 \end{cases}$$

we solve the system of inequalities to find the feasibility region R shown in the figure below. The coordinates of its corner points are $(0, 0)$, $(4, 0)$, $(0, 6)$, and $(5, 1)$.

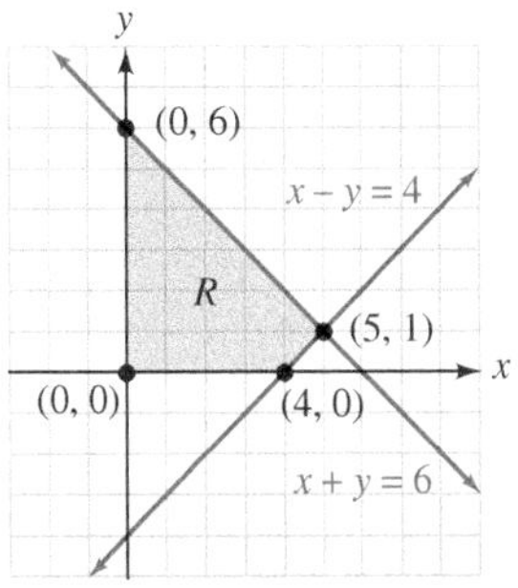

Since the maximum value of P will occur at a corner of R, we substitute the coordinates of each corner point into the objective function $P = 3x + 2y$ and find the one that gives the maximum value of P.

Point	$P = 3x + 2y$
$(0, 0)$	$P = 3(\mathbf{0}) + 2(\mathbf{0}) = 0$
$(4, 0)$	$P = 3(\mathbf{4}) + 2(\mathbf{0}) = 12$
$(5, 1)$	$P = 3(\mathbf{5}) + 2(\mathbf{1}) = 17$
$(0, 6)$	$P = 3(\mathbf{0}) + 2(\mathbf{6}) = 12$

The maximum value $P = 17$ occurs when $x = 5$ and $y = 1$.

Unless otherwise noted, all content on this page is © Cengage Learning.

REVIEW EXERCISES

42. Maximize $P = 3x + y$ subject to $\begin{cases} x \geq 0 \\ y \geq 0 \\ x + y \leq 4 \end{cases}$

43. Fertilizer A company manufactures fertilizers X and Y. Each 50-pound bag requires three ingredients, which are available in the limited quantities shown in the table.

Ingredient	Number of pounds in fertilizer X	Number of pounds in fertilizer Y	Total number of pounds available
Nitrogen	6	10	20,000
Phosphorus	8	6	16,400
Potash	6	4	12,000

The profit on each bag of fertilizer X is \$6, and on each bag of Y, \$5. How many bags of each should be produced to maximize profit?

8 Test

1. Solve $\begin{cases} 2x + y = 5 \\ y = 2x - 3 \end{cases}$ by graphing.

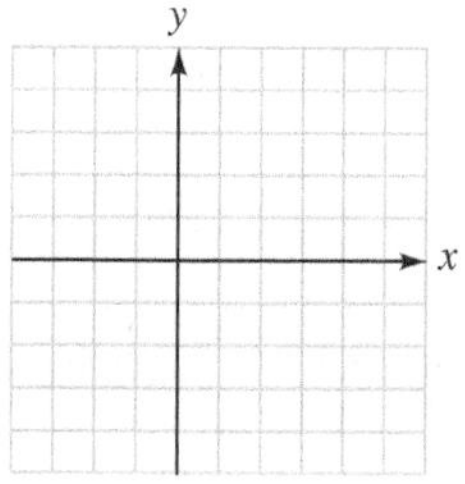

2. Use substitution to solve: $\begin{cases} 2x - 4y = 14 \\ x = -2y + 7 \end{cases}$

3. Use elimination to solve: $\begin{cases} 2x + 3y = -5 \\ 3x - 2y = 12 \end{cases}$

4. Use any method to solve: $\begin{cases} \frac{x}{2} - \frac{y}{4} = -4 \\ x + y = -2 \end{cases}$

Consider the system $\begin{cases} 3(x + y) = x - 3 \\ -y = \frac{2x + 3}{3} \end{cases}$.

5. Are the equations of the system dependent or independent?

6. Is the system consistent or inconsistent?

Use an elementary row operation to find the missing number in the second matrix.

7. $\begin{bmatrix} 1 & 3 & -2 \\ 4 & -2 & -1 \end{bmatrix}, \begin{bmatrix} 1 & 3 & -2 \\ -1 & -17 & \square \end{bmatrix}$

8. $\begin{bmatrix} -1 & 3 & 6 \\ 3 & -2 & 4 \end{bmatrix}, \begin{bmatrix} -1 & 3 & 6 \\ 5 & -8 & \square \end{bmatrix}$

Consider the system $\begin{cases} x + y + z = 4 \\ x + y - z = 6 \\ 2x - 3y + z = -1 \end{cases}$.

9. Write the augmented matrix that represents the system.

10. Solve for x.

11. Solve for y.

12. Solve for z.

Use matrices to solve each system.

13. $\begin{cases} x + y = 4 \\ 2x - y = 2 \end{cases}$

14. $\begin{cases} x + y = 2 \\ x - y = -4 \\ 2x + y = 1 \end{cases}$

Evaluate each determinant.

15. $\begin{vmatrix} 2 & -3 \\ 4 & 5 \end{vmatrix}$

16. $\begin{vmatrix} -3 & -4 \\ -2 & 3 \end{vmatrix}$

17. $\begin{vmatrix} 1 & 2 & 0 \\ 2 & 0 & 3 \\ 1 & -2 & 2 \end{vmatrix}$

18. $\begin{vmatrix} 1 & -2 & 3 \\ 3 & 1 & -2 \\ 2 & -4 & 6 \end{vmatrix}$

Consider the system $\begin{cases} x - y = -6 \\ 3x + y = -6 \end{cases}$, *which is to be solved with Cramer's rule.*

19. When solving for x, what is the numerator determinant? (Do not evaluate it.)

20. When solving for y, what is the denominator determinant? (Do not evaluate it.)

21. Solve the system for x.

22. Solve the system for y.

Use a system of equations in two variables to solve each application.

23. Numbers The sum of two numbers is −18. One number is 2 greater than 3 times the other. Find the product of the numbers.

24. Water parks A father paid \$119 for his family of 7 to spend the day at Magic Waters water park. How many adult tickets did he buy?

Admission	
Adult ticket	\$21
Child ticket	\$14

Unless otherwise noted, all content on this page is © Cengage Learning.

25. Investing A woman invested some money at 3% annual interest and some at 4% annual interest. The interest on the combined investment of $10,000 was $340 for one year. How much was invested at 4%?

26. Kayaking A kayaker can paddle 8 miles down a river in 2 hours and make the return trip in 4 hours. Find the speed of the current in the river.

Solve each system of inequalities by graphing.

27. $\begin{cases} x + y < 3 \\ x - y < 1 \end{cases}$

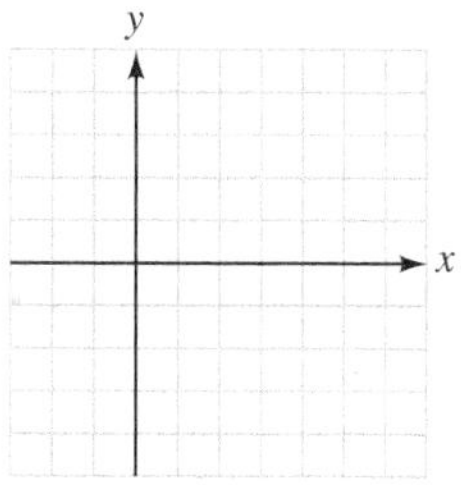

28. $\begin{cases} 2x + 3y \le 6 \\ x \ge 2 \end{cases}$

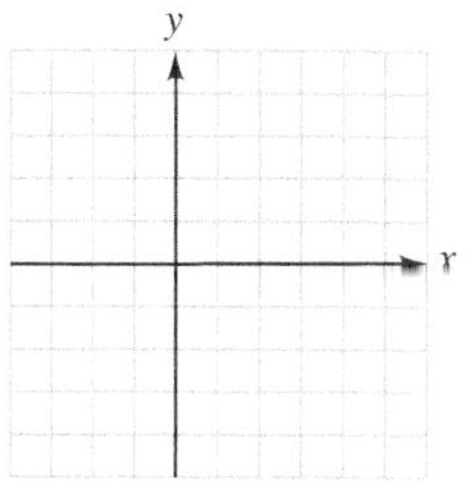

Solve each system.

29. $\begin{cases} 2x - 3y \ge 6 \\ y \le -x + 1 \end{cases}$

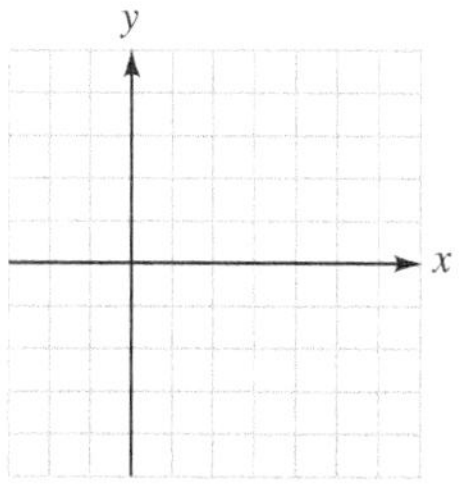

30. Maximize $P = 3x - y$ subject to $\begin{cases} y \ge 1 \\ y \le 2 \\ y \le 3x + 1 \\ x \le 1 \end{cases}$

Unless otherwise noted, all content on this page is © Cengage Learning.

Radicals and Rational Exponents

9

© Mircea BEZERGHEANU/Shutterstock.com

Careers and Mathematics

PHOTOGRAPHERS

Photographers produce and preserve images that paint a picture, tell a story, or record an event. They use either a traditional camera that records images on silver halide film that is developed into prints or a digital camera that electronically records images.

More than half of all photographers are self-employed. Employers usually seek applicants with a "good eye," imagination, and creativity. Entry-level positions in photojournalism generally require a college degree in journalism or photography.

Job Outlook:
Employment of photographers is expected to increase about as fast as the average for all occupations through 2020. However, photographers can expect intense competition for job openings because the work is attractive to many people.

Annual Earnings:
$18,380–$37,370

For More Information:
http://www.bls.gov/oco/ocos264.htm

For a Sample Application:
See Problem 109 in Section 9.5.

REACH FOR SUCCESS
9.1 Radical Expressions
9.2 Applications of the Pythagorean Theorem and the Distance Formula
9.3 Rational Exponents
9.4 Simplifying and Combining Radical Expressions
9.5 Multiplying Radical Expressions and Rationalizing
9.6 Radical Equations
9.7 Complex Numbers
■ *Projects*
REACH FOR SUCCESS EXTENSION
CHAPTER REVIEW
CHAPTER TEST

In this chapter

In this chapter, we will reverse the squaring process and learn how to find square roots of numbers. We also will learn how to find other roots of numbers, simplify radical expressions, solve radical equations, and simplify and perform operations on complex numbers.

Reach for Success Taking Notes in Class

You're trying to take notes in class but your instructor is talking faster than you can write. You need to know what is important and how you can reflect this in your notes.

When reading, most students use a highlighter to identify what is important. In a mathematics class, consider using two different colored inks to clarify the procedures.

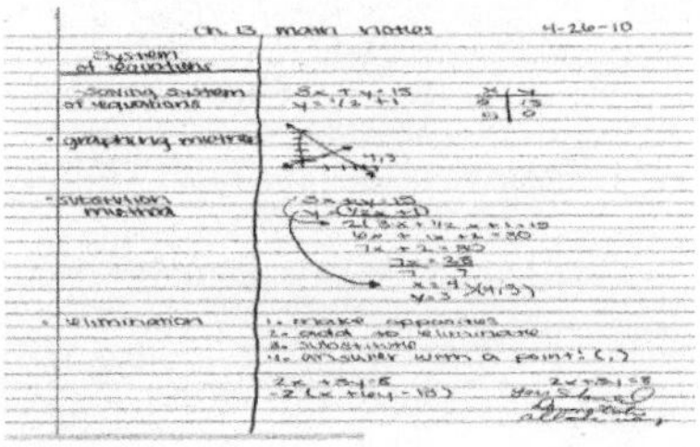

A professor speaks an average of 150 words per minute but you can only write at a rate of 35 words per minute and can think at a much faster rate of 350 words per minute.

What strategies can you use to take effective notes and help you stay focused during class?

1. ________________________________

2. ________________________________

In your reading-intensive courses, notes are generally presented from left to right, whereas in mathematics, problems can be worked vertically. This can pose a challenge in identifying what is important in each problem.

We use a red color change to emphasize any processes we use to complete the exercise. We use black to show the results of the processes.

In each blank at the right, write why you think a different color is used.

Solve the equation:	Read the Instructions
$5x - 9 = 3(x + 3)$	Original example
$5x - 9 = 3(x + 3)$	______________ (Why the red arrows?)
$5x - 9 = 3x + 9$	
$-3x \quad -3x$	______________ (Why the red − 3x?)
$2x - 9 = 9$	
$+9 \quad +9$	______________ (Why the red + 9?)
$\frac{2x}{2} = \frac{18}{2}$	______________ (Why the red division by 2?)
$x = 9$	

Try this problem. Solve the equation using a pencil and a red pen appropriately.

The steps have been identified to help you.

$6x + 3 = -5(x - 2)$	Original example
$6x + 3 = -5(x - 2)$	Insert red arrows where necessary.
______________	Write the result of distributing.
______________	Add 5x to each side of the equation. (Show your work in red.)
______________	Write the result of the addition.
______________	Subtract 3 from each side of the equation. (Show your work in red.)
______________	Write the result of the subtraction.
______________	Divide each term by 11. (Show your work in red.)
______________	Write the result of the division.

A Successful Study Strategy . . .

Use at least two colors of pens when taking notes in your mathematics class. Reviewing your notes is more effective because they have been "highlighted."

At the end of the chapter you will find an additional exercise to guide you in planning for a successful college experience.

Unless otherwise noted, all content on this page is © Cengage Learning.

Section 9.1 Radical Expressions

Objectives

1. Simplify a perfect-square root.
2. Simplify a perfect-square root expression.
3. Simplify a perfect-cube root.
4. Simplify a perfect *n*th root.
5. Find the domain of a square-root function and a cube-root function.
6. Use a square root to solve an application.

Vocabulary

square root	integer squares	index
radical sign	cube root	square-root function
radicand	odd root	cube-root function
principal square root	even root	standard deviation

Getting Ready

Find each power.

1. 0^2 **2.** 4^2 **3.** $(-4)^2$ **4.** -4^2

5. $\left(\frac{2}{5}\right)^3$ **6.** $\left(-\frac{3}{4}\right)^4$ **7.** $(7xy)^2$ **8.** $(7xy)^3$

In this section, we will discuss perfect square roots and other perfect roots of algebraic expressions. We also will consider their related functions.

1 Simplify a perfect-square root.

When solving an equation, we often must find what number must be squared to obtain a second number a. If such a number can be found, it is called a **square root** of a. For example,

- 0 is a square root of 0, because $0^2 = 0$.
- 4 is a square root of 16, because $4^2 = 16$.
- -4 is a square root of 16, because $(-4)^2 = 16$.
- $7xy$ is a square root of $49x^2y^2$, because $(7xy)^2 = 49x^2y^2$.
- $-7xy$ is a square root of $49x^2y^2$, because $(-7xy)^2 = 49x^2y^2$.

All positive numbers have two real-number square roots: one that is positive and one that is negative.

EXAMPLE 1 Find the two square roots of 121.

Solution The two square roots of 121 are 11 and -11, because

$$11^2 = 121 \qquad \text{and} \qquad (-11)^2 = 121$$

SELF CHECK 1 Find the two square roots of 144.

To express square roots, we use the symbol $\sqrt{}$, called a **radical sign**. For example,

$\sqrt{121} = 11$ Read as "The positive square root of 121 is 11."

$-\sqrt{121} = -11$ Read as "The negative square root of 121 is -11."

The number under the radical sign is called a **radicand**.

COMMENT The **principal square root** of a positive number is always positive. Although 5 and -5 are both square roots of 25, only 5 is the principal square root. The radical expression $\sqrt{25}$ represents 5. The radical expression $-\sqrt{25}$ represents -5.

SQUARE ROOT OF *a*

If $a > 0$, $\sqrt{a}$ is the positive number, the principal square root of a, whose square is a. In symbols,

$$(\sqrt{a})^2 = a$$

If $a = 0$, $\sqrt{a} = \sqrt{0} = 0$. The principal square root of 0 is 0.

If $a < 0$, $\sqrt{a}$ is not a real number.

Because of the previous definition, the square root of any number squared is that number. For example,

$$(\sqrt{10})^2 = \sqrt{10} \cdot \sqrt{10} = 10 \qquad (\sqrt{a})^2 = \sqrt{a} \cdot \sqrt{a} = a$$

EXAMPLE 2 Simplify each radical.

a. $\sqrt{1} = 1$ **b.** $\sqrt{81} = 9$

c. $-\sqrt{81} = -9$ **d.** $-\sqrt{225} = -15$

e. $\sqrt{\frac{1}{4}} = \frac{1}{2}$ **f.** $-\sqrt{\frac{16}{121}} = -\frac{4}{11}$

g. $\sqrt{0.04} = 0.2$ **h.** $-\sqrt{0.0009} = -0.03$

SELF CHECK 2 Simplify: **a.** $-\sqrt{49}$ **b.** $\sqrt{\frac{25}{49}}$ **c.** $\sqrt{0.0036}$

Numbers such as 1, 4, 9, 16, 49, and 1,600 are called **integer squares**, because each one is the square of an integer. The square root of every integer square is an integer.

$$\sqrt{1} = 1 \quad \sqrt{4} = 2 \quad \sqrt{9} = 3 \quad \sqrt{16} = 4 \quad \sqrt{49} = 7 \quad \sqrt{1{,}600} = 40$$

Perspective

CALCULATING SQUARE ROOTS

The Bakhshali manuscript is an early mathematical manuscript that was discovered in India in the late 19th century. Mathematical historians estimate that the manuscript was written sometime around A.D. 400. One section of the manuscript presents a procedure for calculating square roots using arithmetic. Specifically, we can use the formula

$$\sqrt{Q} = A + \frac{b}{2A} - \left(\frac{b^2}{4A(2A^2 + b)}\right)$$

where A^2 = a perfect square close to the number Q, and $b = Q - A^2$.

For example, if we want to compute an approximation of $\sqrt{21}$, we can choose $A^2 = 16$. Thus, $A = 4$ and $b = 21 - 16 = 5$. So we obtain

$$\sqrt{21} = 4 + \frac{5}{(2)(4)} - \left(\frac{5^2}{(4)(4)((2)(4)^2 + 5)}\right)$$

$$= 4 + \frac{5}{8} - \left(\frac{25}{(16)(37)}\right)$$

$$= 4 + \frac{5}{8} - \frac{25}{592} \approx 4.58277027$$

Using the square root key on a calculator, we see that, to nine decimal places, $\sqrt{21} = 4.582575695$. Therefore, the formula gives an answer that is correct to three decimal places.

1. Use the formula to approximate $\sqrt{105}$. How accurate is your answer?

2. Use the formula to approximate $\sqrt{627}$. How accurate is your answer?

Source: http://www.gap-system.org/~history/HistTopics/Bakhshali_manuscript.html

Accent on technology

▸ Approximating Square Roots

The square roots of many positive integers are not rational numbers. For example, $\sqrt{11}$ is an *irrational number*. To find an approximate value of $\sqrt{11}$ with a calculator, we enter these numbers and press these keys.

11 **2ND** $\sqrt{\ }$ Using a scientific calculator

2ND X² ($\sqrt{\ }$) 11 **ENTER** Using a graphing calculator

Either way, we will see that

$$\sqrt{11} \approx 3.31662479$$

For instructions regarding the use of a Casio graphing calculator, please refer to the Casio Keystroke Guide in the back of the book.

Square roots of negative numbers are not real numbers. For example, $\sqrt{-9}$ is not a real number, because no real number squared equals -9. Square roots of negative numbers come from a set called *imaginary numbers*, which we will discuss later in this chapter.

2 Simplify a perfect-square root expression.

If $x \neq 0$, the positive number x^2 has x and $-x$ for its two square roots. To denote the principal square root of x^2, we must know whether x is positive or negative.

If $x > 0$, we can write

$\sqrt{x^2} = x$ $\sqrt{x^2}$ represents the positive square root of x^2, which is x.

If x is negative, then $-x > 0$, and we can write

$\sqrt{x^2} = -x$ $\sqrt{x^2}$ represents the positive square root of x^2, which is $-x$.

If we do not know whether x is positive or negative, we must use absolute value symbols to guarantee that $\sqrt{x^2}$ is positive.

DEFINITION OF $\sqrt{x^2}$

If x can be any real number, then

$$\sqrt{x^2} = |x|$$

EXAMPLE 3 Simplify each expression. Assume that x can be any real number.

a. $\sqrt{16x^2} = \sqrt{(4x)^2}$ Write $16x^2$ as $(4x)^2$.

$= |4x|$ Because $16x^2 = (|4x|)^2$. Since x could be negative, absolute value symbols are needed.

$= 4|x|$ Since 4 is a positive constant in the product $4x$, we write it outside the absolute value symbols.

b. $\sqrt{x^2 + 2x + 1}$

$= \sqrt{(x + 1)^2}$ Factor $x^2 + 2x + 1$.

$= |x + 1|$ Because $(x + 1)$ can be negative, absolute value symbols are needed.

c. $\sqrt{x^4} = x^2$ Because $x^4 = (x^2)^2$, and $x^2 \geq 0$, no absolute value symbols are needed.

SELF CHECK 3 Simplify: **a.** $\sqrt{25a^2}$ **b.** $\sqrt{x^2 + 4x + 4}$ **c.** $\sqrt{16a^4}$

3 Simplify a perfect-cube root.

The **cube root of *x*** is any number whose cube is x. For example,

4 is a cube root of 64, because $4^3 = 64$.
$3x^2y$ is a cube root of $27x^6y^3$, because $(3x^2y)^3 = 27x^6y^3$.
$-2y$ is a cube root of $-8y^3$, because $(-2y)^3 = -8y^3$.

CUBE ROOT OF *a*

The cube root of a is denoted as $\sqrt[3]{a}$ and is the number whose cube is a. In symbols,

$$(\sqrt[3]{a})^3 = a$$

If a is any real number, then

$$\sqrt[3]{a^3} = a$$

We note that 64 has two real-number square roots, 8 and -8. However, 64 has only one real-number cube root, 4, because 4 is the only real number whose cube is 64. *Since every real number has exactly one real cube root, it is unnecessary to use absolute value symbols when simplifying cube roots.*

EXAMPLE 4 Simplify each radical.

a. $\sqrt[3]{125} = 5$ Because $5^3 = 5 \cdot 5 \cdot 5 = 125$

b. $\sqrt[3]{\frac{1}{8}} = \frac{1}{2}$ Because $\left(\frac{1}{2}\right)^3 = \frac{1}{2} \cdot \frac{1}{2} \cdot \frac{1}{2} = \frac{1}{8}$

c. $\sqrt[3]{-27x^3} = -3x$ Because $(-3x)^3 = (-3x)(-3x)(-3x) = -27x^3$

d. $\sqrt[3]{-\frac{8a^3}{27b^3}} = -\frac{2a}{3b}$ $\left(-\frac{2a}{3b}\right)^3 = \left(-\frac{2a}{3b}\right)\left(-\frac{2a}{3b}\right)\left(-\frac{2a}{3b}\right) = -\frac{8a^3}{27b^3}$

e. $\sqrt[3]{0.216x^3y^6} = 0.6xy^2$ $(0.6xy^2)^3 = (0.6xy^2)(0.6xy^2)(0.6xy^2) = 0.216x^3y^6$

SELF CHECK 4 Simplify: **a.** $\sqrt[3]{1{,}000}$ **b.** $\sqrt[3]{\frac{1}{27}}$ **c.** $\sqrt[3]{125a^3}$ **d.** $\sqrt[3]{-64x^3}$

COMMENT The previous examples suggest that if a can be factored into three equal factors, any one of those factors is a cube root of a.

4 Simplify a perfect *n*th root.

Just as there are square roots and cube roots, there are fourth roots, fifth roots, sixth roots, and so on.

In the radical $\sqrt[n]{x}$, n is called the **index** (or **order**) of the radical. When the index is 2, the radical is a square root, and we usually do not write the index.

$$\sqrt[2]{x} = \sqrt{x}$$

COMMENT When n is an even number greater than 1 and $x < 0$, $\sqrt[n]{x}$ is not a real number. For example, $\sqrt[4]{-81}$ is not a real number, because no real number raised to the 4th power is -81. However, when n is odd, $\sqrt[n]{x}$ is a real number.

When n is an odd natural number greater than 1, $\sqrt[n]{x}$ represents an **odd root**. Since every real number has only one real nth root when n is odd, we do not need to use absolute value symbols when finding odd roots. For example,

$\sqrt[5]{243} = 3$ because $3^5 = 243$

$\sqrt[7]{-128x^7} = -2x$ because $(-2x)^7 = -128x^7$

When n is an even natural number greater than 1, $\sqrt[n]{x}$ represents an **even root**. In this case, there will be one positive and one negative real nth root. For example, the two real sixth roots of 729 are 3 and -3, because $3^6 = 729$ and $(-3)^6 = 729$. When finding even roots, we use absolute value symbols to guarantee that the principal nth root is positive.

$\sqrt[4]{(-3)^4} = |-3| = 3$ — $3^4 = (-3)^4$. We also could simplify this as follows: $\sqrt[4]{(-3)^4} = \sqrt[4]{81} = 3$

$\sqrt[6]{729x^6} = |3x| = 3|x|$ — $(3|x|)^6 = 729x^6$. The absolute value symbols guarantee that the sixth root is positive.

EXAMPLE 5 Simplify each radical.

a. $\sqrt[4]{625} = 5$, because $5^4 = 625$ — Read $\sqrt[4]{625}$ as "the fourth root of 625."

b. $\sqrt[5]{-32} = -2$, because $(-2)^5 = -32$ — Read $\sqrt[5]{-32}$ as "the fifth root of -32."

c. $\sqrt[6]{\frac{1}{64}} = \frac{1}{2}$, because $\left(\frac{1}{2}\right)^6 = \frac{1}{64}$ — Read $\sqrt[6]{\frac{1}{64}}$ as "the sixth root of $\frac{1}{64}$."

d. $\sqrt[7]{10^7} = 10$, because $10^7 = 10^7$ — Read $\sqrt[7]{10^7}$ as "the seventh root of 10^7."

SELF CHECK 5 Simplify: **a.** $\sqrt[4]{\frac{1}{81}}$ **b.** $\sqrt[5]{10^5}$

When finding the nth root of an nth power, we can use the following rules.

DEFINITION OF $\sqrt[n]{a^n}$

If n is an odd natural number greater than 1, then $\sqrt[n]{a^n} = a$.
If n is an even natural number, then $\sqrt[n]{a^n} = |a|$.

EXAMPLE 6 Simplify each radical. Assume that x can be any real number.

Solution

a. $\sqrt[5]{x^5} = x$ — Since n is odd, absolute value symbols are not needed.

b. $\sqrt[4]{16x^4} = |2x| = 2|x|$ — Since n is even and x can be negative, absolute value symbols are needed to guarantee that the result is positive.

c. $\sqrt[6]{(x+4)^6} = |x+4|$ — Absolute value symbols are needed to guarantee that the result is positive.

d. $\sqrt[3]{(x+1)^3} = x + 1$ — Since n is odd, absolute value symbols are not needed.

e. $\sqrt{(x^2 + 6x + 9)^2} = \sqrt{[(x+3)^2]^2}$ — Factor $x^2 + 6x + 9$.

$= \sqrt{(x+3)^4}$

$= (x+3)^2$ — Since $(x+3)^2$ is always positive, absolute value symbols are not needed.

SELF CHECK 6 Simplify. Assume that a can be any real number. **a.** $\sqrt[4]{16a^4}$ **b.** $\sqrt[5]{(a+5)^5}$ **c.** $\sqrt{(x^2+10x+25)^2}$

We summarize the possibilities for $\sqrt[n]{x}$ as follows:

DEFINITION FOR $\sqrt[n]{x}$

Assume n is a natural number greater than 1 and x is a real number.

If $x > 0$, then $\sqrt[n]{x}$ is the positive number such that $\left(\sqrt[n]{x}\right)^n = x$.

If $x = 0$, then $\sqrt[n]{x} = 0$.

If $x < 0$ $\begin{cases} \text{and n is odd, then } \sqrt[n]{x} \text{ is the real number such that } \left(\sqrt[n]{x}\right)^n = x. \\ \text{and n is even, then } \sqrt[n]{x} \text{ is not a real number.} \end{cases}$

5 Find the domain of a square-root function and a cube-root function.

Since there is one principal square root for every nonnegative real number x, the equation $f(x) = \sqrt{x}$ determines a function, called the **square-root function**.

EXAMPLE 7 Consider the function $f(x) = \sqrt{x}$.

a. Find the domain. **b.** Graph the function. **c.** Find the range.

Solution

a. To find the domain, we note that $x \geq 0$ in the function because the radicand must be nonnegative. Thus, the domain is the set of nonnegative real numbers. In interval notation, the domain is $[0, \infty)$.

b. We can create a table of values and plot points to obtain the graph shown in Figure 9-1(a). If we use a graphing calculator, we can choose window settings of $[-1, 9]$ for x and $[-2, 5]$ for y to see the graph shown in Figure 9-1(b).

$f(x) = \sqrt{x}$

x	$f(x)$	$(x, f(x))$
0	0	(0, 0)
1	1	(1, 1)
4	2	(4, 2)
9	3	(9, 3)

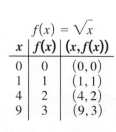

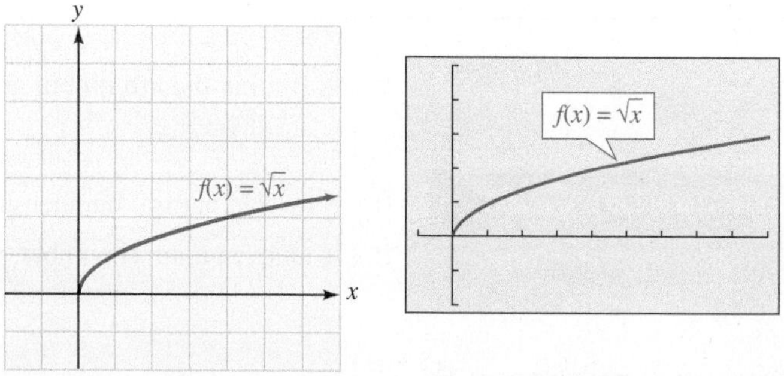

(a) (b)

Figure 9-1

c. From either graph, we can conclude that the range of the function is the set of nonnegative real numbers, which is the interval $[0, \infty)$. The graph also confirms that the domain is the interval $[0, \infty)$.

SELF CHECK 7 Graph $f(x) = \sqrt{x} + 2$ and compare it to the graph of $f(x) = \sqrt{x}$. Find the domain and the range.

Unless otherwise noted, all content on this page is © Cengage Learning.

The graphs of many functions are translations or reflections of the square-root function. For example, if $k > 0$,

- The graph of $f(x) = \sqrt{x} + k$ is the graph of $f(x) = \sqrt{x}$ translated k units upward.
- The graph of $f(x) = \sqrt{x} - k$ is the graph of $f(x) = \sqrt{x}$ translated k units downward.
- The graph of $f(x) = \sqrt{x + k}$ is the graph of $f(x) = \sqrt{x}$ translated k units to the left.
- The graph of $f(x) = \sqrt{x - k}$ is the graph of $f(x) = \sqrt{x}$ translated k units to the right.
- The graph of $f(x) = -\sqrt{x}$ is the graph of $f(x) = \sqrt{x}$ reflected about the x-axis.

The equation $f(x) = \sqrt[3]{x}$ defines the **cube-root function**. From the graph shown in Figure 9-2(a), we can see that the domain and range of the function $f(x) = \sqrt[3]{x}$ are the set of real numbers, $\mathbb{R}$, and in interval notation $(-\infty, \infty)$. Note that the graph of $f(x) = \sqrt[3]{x}$ passes the vertical line test. Figures 9-2(b) and 9-2(c) show several translations of the cube-root function.

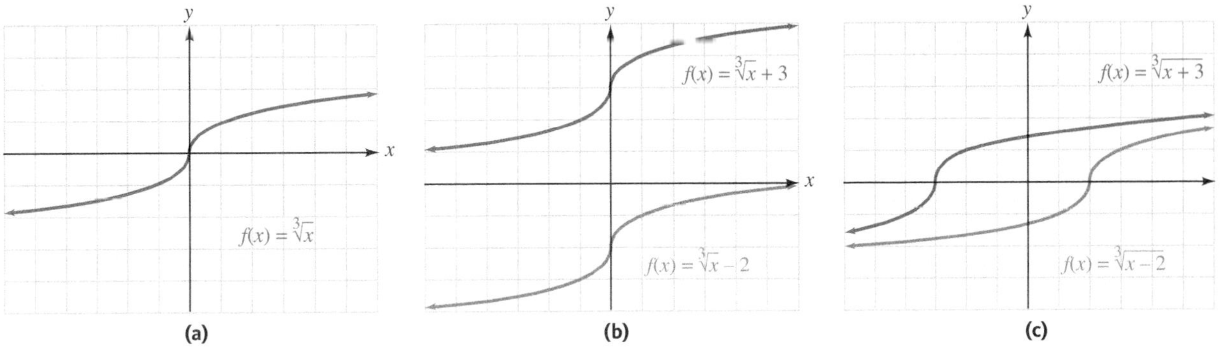

Figure 9-2

EXAMPLE 8 Consider the function $f(x) = \sqrt[3]{x + 3} + 1$.

a. Find the domain. **b.** Graph the function. **c.** Find the range.

Solution **a.** To find the domain, we note that the radicand is not restricted because the function is a cube root. Therefore, the domain is all real numbers, $\mathbb{R}$, and represented by the interval $(-\infty, \infty)$.

b. This graph will be the same shape as $f(x) = \sqrt[3]{x}$, translated 3 units to the left and 1 unit upward. (See Figure 9-3(a).) We can confirm this graph by using a graphing calculator. If we select window settings of $[-10, 5]$ for x and $[-5, 5]$ for y, we will obtain the graph shown in Figure 9-3(b).

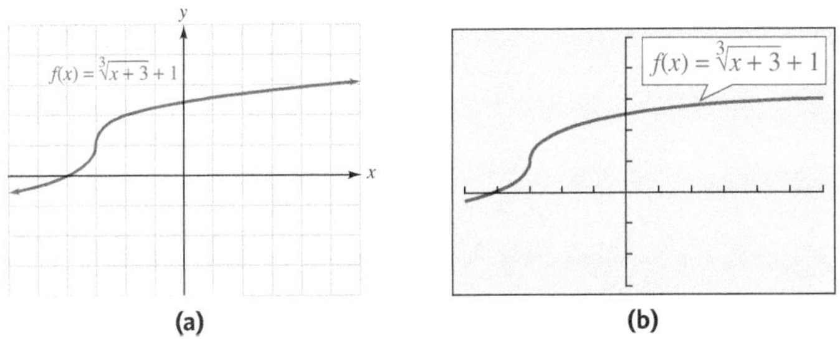

Figure 9-3

c. From either graph, we can conclude that the range is the interval $(-\infty, \infty)$. The graph also confirms that the domain is the interval $(-\infty, \infty)$.

SELF CHECK 8 Graph $f(x) = -\sqrt[3]{x - 2}$ and find the domain and range.

Unless otherwise noted, all content on this page is © Cengage Learning.

6 Use a square root to solve an application.

In the real world, there are many models that involve square root relationships.

EXAMPLE 9 **PERIOD OF A PENDULUM** The *period of a pendulum* is the time required for the pendulum to swing back and forth to complete one cycle. (See Figure 9-4.) The period t (in seconds) is a function of the pendulum's length l in feet, which is defined by the formula

$$t = f(l) = 2\pi\sqrt{\frac{l}{32}}$$

Find the period of a pendulum that is 5 feet long.

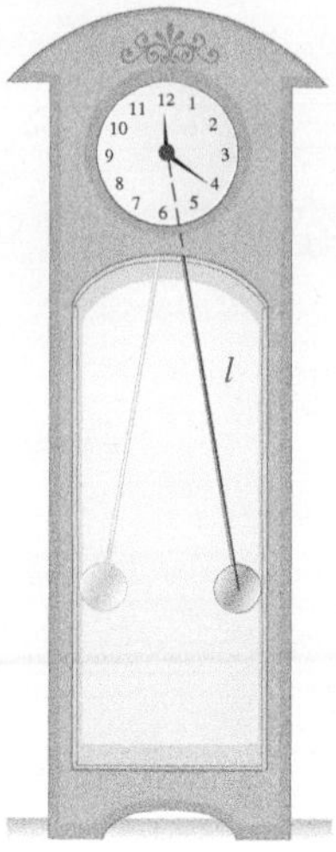

Figure 9-4

Solution We substitute 5 for l in the formula and simplify.

$$t = 2\pi\sqrt{\frac{l}{32}}$$

$$t = 2\pi\sqrt{\frac{5}{32}} \quad \text{Substitute.}$$

$$\approx 2.483647066 \quad \text{Use a calculator.}$$

To the nearest tenth, the period is 2.5 seconds.

SELF CHECK 9 To the nearest hundredth, find the period of a pendulum that is 3 feet long.

Accent on technology

▸ Finding the Period of a Pendulum

To solve Example 9 with a graphing calculator with window settings of $[-2, 10]$ for x and $[-2, 10]$ for y, we graph the function $f(x) = 2\pi\sqrt{\frac{x}{32}}$, as in Figure 9-5(a). We use the value feature of the calculator to find the value of y when $x = 5$. From the graph, press **2nd** **TRACE** (CALC) **ENTER**. Enter 5 **ENTER**. The period is given by the y-value shown on the screen.

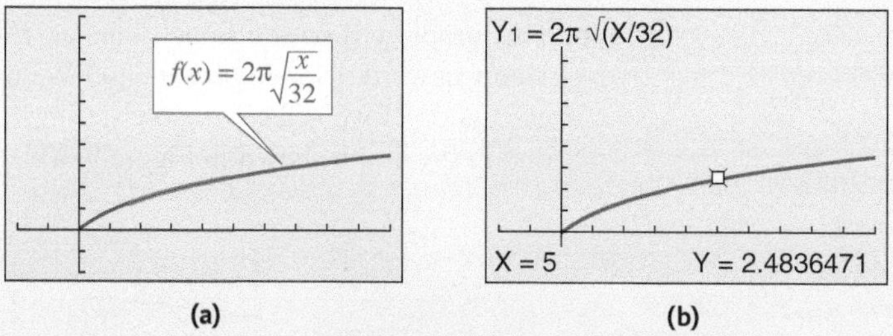

Figure 9-5

For instructions regarding the use of a Casio graphing calculator, please refer to the Casio Keystroke Guide in the back of the book.

In statistics, the **standard deviation** of a data set is a measure of how tightly the data points are grouped around the mean (average) of the data set. The smaller the value, the more tightly the data points are grouped around the mean.

To see how to compute the standard deviation of a distribution, we consider the distribution 4, 5, 5, 8, 13 and construct the following table.

Unless otherwise noted, all content on this page is © Cengage Learning.

Original terms	Mean of the distribution	Differences (original term minus mean)	Squares of the differences from the mean
4	7	−3	9
5	7	−2	4
5	7	−2	4
8	7	1	1
13	7	6	36

The population *standard deviation* of the distribution is the positive square root of the mean of the numbers shown in column 4 of the table.

$$\text{Standard deviation} = \sqrt{\frac{\text{sum of the squares of the differences from the mean}}{\text{number of differences}}}$$

$$= \sqrt{\frac{9 + 4 + 4 + 1 + 36}{5}}$$

$$= \sqrt{\frac{54}{5}}$$

$$\approx 3.286335345 \quad \text{Use a calculator.}$$

To the nearest hundredth, the standard deviation of the given distribution is 3.29.

The symbol for the population standard deviation is σ, the lowercase Greek letter *sigma*.

EXAMPLE 10 Which of the following distributions is more tightly grouped around the mean?

a. 3, 5, 7, 8, 12 **b.** 1, 4, 6, 11

Solution We compute the standard deviation of each distribution.

a.

Original terms	Mean of the distribution	Differences (original term minus mean)	Squares of the differences from the mean
3	7	−4	16
5	7	−2	4
7	7	0	0
8	7	1	1
12	7	5	25

$$\sigma = \sqrt{\frac{16 + 4 + 0 + 1 + 25}{5}} = \sqrt{\frac{46}{5}} \approx 3.03$$

b.

Original terms	Mean of the distribution	Differences (original term minus mean)	Squares of the differences from the mean
1	5.5	−4.5	20.25
4	5.5	−1.5	2.25
6	5.5	0.5	0.25
11	5.5	5.5	30.25

$$\sigma = \sqrt{\frac{20.25 + 2.25 + 0.25 + 30.25}{4}} = \sqrt{\frac{53}{4}} \approx 3.64$$

Since the standard deviation for the first distribution is less than the standard deviation for the second, the first distribution is more tightly grouped around the mean.

SELF CHECK 10 Which of the following distributions is more tightly grouped around the mean?
a. 4, 6, 8, 10, 11 **b.** 3, 5, 7, 9

SELF CHECK ANSWERS

1. 12, −12 **2. a.** −7 **b.** $\frac{5}{7}$ **c.** 0.06 **3. a.** $5|a|$ **b.** $|x + 2|$ **c.** $4a^2$
4. a. 10 **b.** $\frac{1}{3}$ **c.** $5a$ **d.** $-4x$ **5. a.** $\frac{1}{3}$ **b.** 10 **6. a.** $2|a|$ **b.** $a + 5$ **c.** $(x + 5)^2$
7. It is 2 units higher.

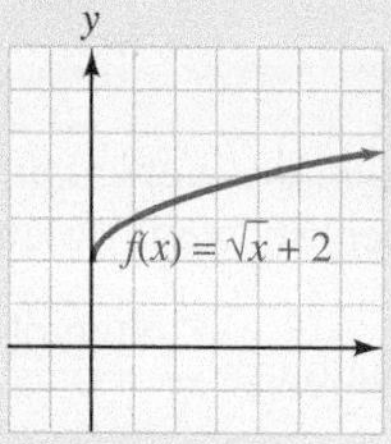

D: $[0, \infty)$
R: $[2, \infty)$

8.

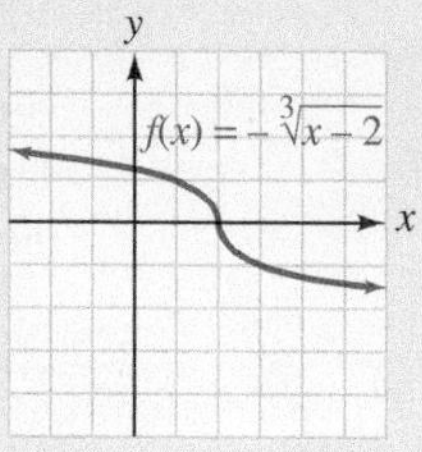

D: $(-\infty, \infty)$
R: $(-\infty, \infty)$

9. 1.92 sec **10.** Distribution **b** is more tightly grouped around the mean.

NOW TRY THIS

1. Simplify: $\sqrt{100x^{100}}$
2. Without using a calculator, between which two integers will the value of each expression be found?
   a. $\sqrt{8}$ b. $\sqrt{54}$ c. $\sqrt[3]{54}$
3. Given that $2^{-1} = \frac{1}{2}$, $2^0 = 1$, $2^1 = 2$, $2^2 = 4$, and $2^3 = 8$, between which two integers would you expect the value of each expression to be found?
   a. $2^{1/2}$ b. $2^{3/2}$ c. $2^{5/2}$

9.1 Exercises

WARM-UPS *Fill in the blanks.*

1. $(\_\_\_)^2 = 25$
2. $(\_\_\_)^2 = 81$
3. $(\_\_\_)^2 = 4x^2$
4. $(\_\_\_)^2 = 36a^2$
5. $(\_\_\_)^3 = 8$
6. $(\_\_\_)^3 = -27$
7. $(\_\_\_)^4 = 81$
8. $(\_\_\_)^4 = 16$

REVIEW *Simplify each rational expression. Assume no denominators are zero.*

9. $\dfrac{x^2 + 7x + 10}{x^2 - 4}$
10. $\dfrac{a^3 - b^3}{b^2 - a^2}$

Perform the operations. Assume no denominators are zero.

11. $\dfrac{x^2 - x - 6}{x^2 - 2x - 3} \cdot \dfrac{x^2 - 1}{x^2 + x - 2}$

12. $\dfrac{x^2 - 3x - 4}{x^2 - 5x + 6} \div \dfrac{x^2 - 2x - 3}{x^2 - x - 2}$

13. $\dfrac{3}{m + 1} + \dfrac{3m}{m - 1}$
14. $\dfrac{2x + 3}{3x - 1} - \dfrac{x - 4}{2x + 1}$

VOCABULARY AND CONCEPTS *Fill in the blanks.*

15. $5x^2$ is the square root of $25x^4$, because _____ $= 25x^4$ and 6 is a square root of 36 because _______.
16. The numbers 1, 4, 9, 16, 25, . . . are called ____________.
17. The principal square root of x $(x > 0)$ is the ______ square root of x.
18. The graph of $f(x) = \sqrt{x} + 3$ is the graph of $f(x) = \sqrt{x}$ translated _ units __.
19. The graph of $f(x) = \sqrt{x + 5}$ is the graph of $f(x) = \sqrt{x}$ translated _ units to the ___.
20. When n is an odd natural number greater than 1, $\sqrt[n]{x}$ represents an ___ root.
21. Given the radical $\sqrt[a]{b}$, the symbol $\sqrt{}$ is the _____ sign, a is the _____, and b is the _______.
22. The square-root function $f(x) = \sqrt{x}$ has the domain _____, while the cube-root function $f(x) = \sqrt[3]{x}$ has the domain _____.
23. $\sqrt{x^2} =$ __
24. $\left(\sqrt[3]{x}\right)^3 =$ _
25. $\sqrt[3]{x^3} =$ _
26. $\sqrt{0} =$ _
27. When n is a positive ____ number, $\sqrt[n]{x}$ represents an even root.

Unless otherwise noted, all content on this page is © Cengage Learning.

28. The _______ deviation of a set of numbers is a measure of how tightly the data points are grouped around the mean of the data set.

Identify the radicand in each expression.

29. $-\sqrt{7y^2}$

30. $6\sqrt{ab}$

31. $ab^2\sqrt{a^2 + b^3}$

32. $\frac{1}{2}x\sqrt{\frac{x}{y}}$

GUIDED PRACTICE *Find the two square roots of each number.* SEE EXAMPLE 1. (OBJECTIVE 1)

33. 100

34. 25

35. 49

36. 169

Find each square root, if possible. SEE EXAMPLE 2. (OBJECTIVE 1)

37. $\sqrt{121}$

38. $\sqrt{144}$

39. $-\sqrt{64}$

40. $-\sqrt{1}$

41. $-\sqrt{\frac{25}{49}}$

42. $\sqrt{\frac{49}{81}}$

43. $\sqrt{-4}$

44. $\sqrt{-121}$

45. $\sqrt{0.16}$

46. $\sqrt{0.25}$

Find each square root. Assume that all variables are unrestricted, and use absolute value symbols when necessary. SEE EXAMPLE 3. (OBJECTIVE 2)

47. $\sqrt{25x^2}$

48. $\sqrt{16b^2}$

49. $\sqrt{64y^4}$

50. $\sqrt{16y^4}$

51. $\sqrt{(x + 3)^2}$

52. $\sqrt{(a + 6)^2}$

53. $\sqrt{a^2 + 6a + 9}$

54. $\sqrt{x^2 + 14x + 49}$

Simplify each cube root. SEE EXAMPLE 4. (OBJECTIVE 3)

55. $\sqrt[3]{1}$

56. $\sqrt[3]{512}$

57. $\sqrt[3]{-27}$

58. $\sqrt[3]{-8}$

59. $\sqrt[3]{-\frac{64}{27}}$

60. $\sqrt[3]{\frac{125}{216}}$

61. $\sqrt[3]{0.064}$

62. $\sqrt[3]{0.001}$

63. $\sqrt[3]{125y^3}$

64. $\sqrt[3]{-27x^6}$

65. $\sqrt[3]{-1{,}000p^3q^3}$

66. $\sqrt[3]{343a^6b^3}$

Simplify each radical, if possible. SEE EXAMPLE 5. (OBJECTIVE 4)

67. $-\sqrt[5]{243}$

68. $\sqrt[5]{-32}$

69. $-\sqrt[4]{16}$

70. $-\sqrt[4]{625}$

71. $\sqrt[6]{64}$

72. $\sqrt[6]{729}$

73. $\sqrt[4]{\frac{16}{625}}$

74. $\sqrt[5]{-\frac{243}{32}}$

75. $\sqrt[4]{-256}$

76. $\sqrt[6]{-729}$

77. $-\sqrt[5]{-\frac{1}{32}}$

78. $-\sqrt[4]{\frac{81}{256}}$

Simplify each radical. Assume that all variables are unrestricted, and use absolute value symbols where necessary. SEE EXAMPLE 6. (OBJECTIVE 4)

79. $\sqrt[4]{81x^4}$

80. $\sqrt[6]{64x^6}$

81. $\sqrt[3]{8a^3}$

82. $\sqrt[5]{32a^5}$

83. $\sqrt[4]{\frac{1}{16}x^4}$

84. $\sqrt[4]{\frac{1}{81}x^8}$

85. $\sqrt[4]{x^{12}}$

86. $\sqrt[8]{x^{24}}$

87. $\sqrt[5]{-x^5}$

88. $\sqrt[3]{-x^6}$

89. $\sqrt[3]{-27a^6}$

90. $\sqrt[5]{-32x^5}$

Graph each function and state the domain and range. SEE EXAMPLE 7. (OBJECTIVE 5)

91. $f(x) = \sqrt{x + 4}$

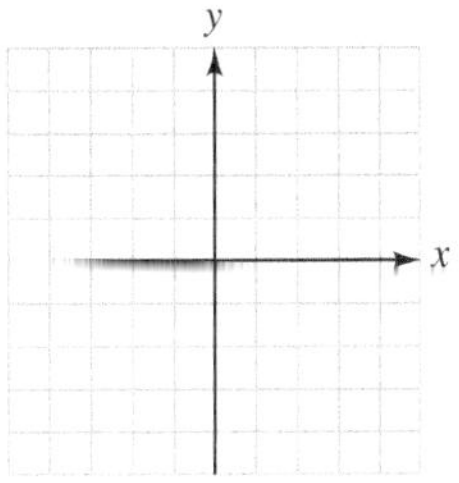

92. $f(x) = -\sqrt{x - 2}$

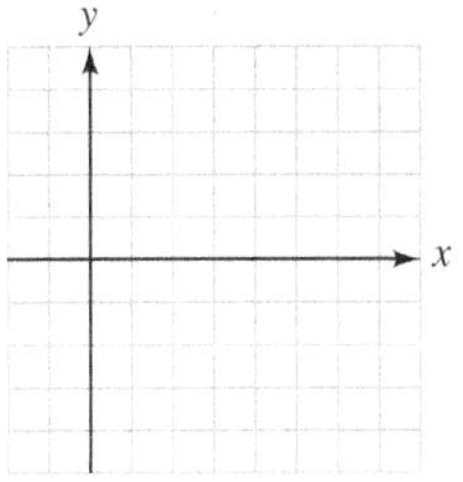

93. $f(x) = -\sqrt{x} - 3$

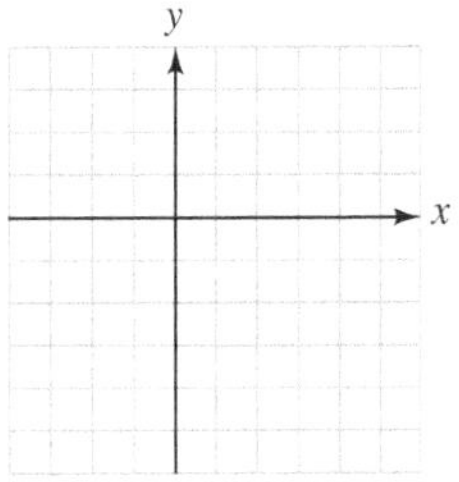

94. $f(x) = \sqrt{x + 1} + 2$

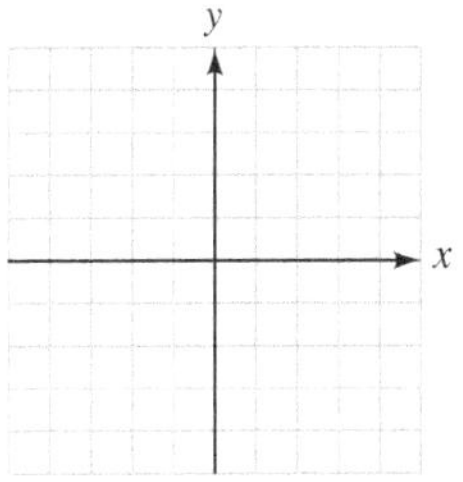

Graph each function and state the domain and range. SEE EXAMPLE 8. (OBJECTIVE 5)

95. $f(x) = \sqrt[3]{x} - 1$

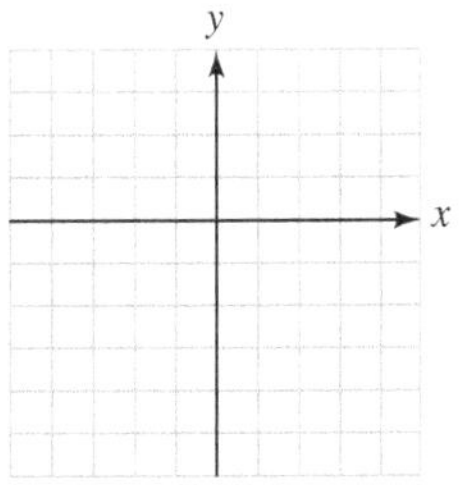

96. $f(x) = \sqrt[3]{x + 1} + 2$

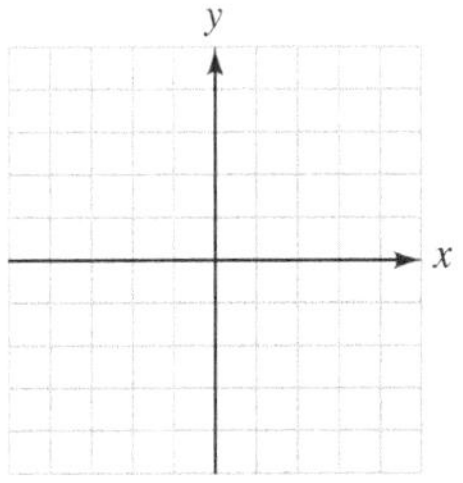

97. $f(x) = -\sqrt[3]{x - 1} - 2$

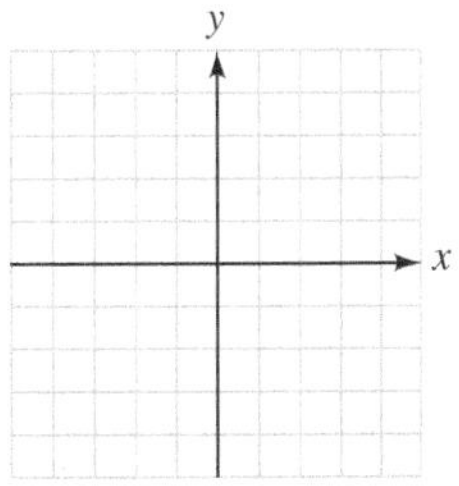

98. $f(x) = -\sqrt[3]{x} - 1$

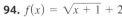

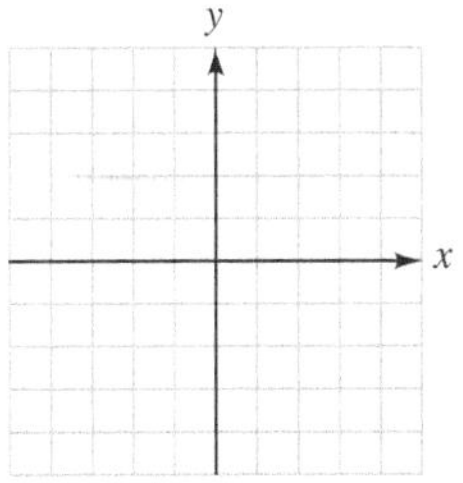

Unless otherwise noted, all content on this page is © Cengage Learning.

ADDITIONAL PRACTICE *Simplify each radical. Assume that all variables are unrestricted, and use absolute value symbols where necessary.*

99. $\sqrt{(-4)^2}$

100. $\sqrt{(-9)^2}$

101. $\sqrt{-36}$

102. $-\sqrt{-4}$

103. $\sqrt{(-5b)^2}$

104. $\sqrt{(-8c)^2}$

105. $\sqrt{t^2 + 24t + 144}$

106. $\sqrt{m^2 + 30m + 225}$

107. $\sqrt[3]{-\frac{1}{8}m^6n^3}$

108. $\sqrt[3]{\frac{27}{1{,}000}a^6b^6}$

109. $\sqrt[3]{0.008z^9}$

110. $\sqrt[3]{0.064s^9t^6}$

111. $\sqrt[25]{(x + 2)^{25}}$

112. $\sqrt[44]{(x + 4)^{44}}$

113. $\sqrt[8]{0.00000001x^{16}y^8}$

114. $\sqrt[5]{0.00032x^{10}y^5}$

Use a calculator to find each square root. Give the answer to four decimal places.

115. $\sqrt{12}$

116. $\sqrt{340}$

117. $\sqrt{679.25}$

118. $\sqrt{0.0063}$

APPLICATIONS *For Exercises 119–122, use a calculator to solve.* SEE EXAMPLES 9–10. (OBJECTIVE 6)

119. Period of a pendulum The longest pendulum in a working clock is 13 feet. Find the period of the pendulum.

120. Period of a pendulum A small clock has a pendulum that is two inches in length. Find the period of the pendulum.

121. Statistics Find the standard deviation of the following distribution to the nearest hundredth: 2, 5, 5, 6, 7

122. Statistics Find the standard deviation of the following distribution to the nearest hundredth: 3, 6, 7, 9, 11, 12

123. Statistics In statistics, the formula

$$s_{\bar{x}} = \frac{s}{\sqrt{N}}$$

gives an estimate of the standard error of the mean. Find $s_{\bar{x}}$ to four decimal places when $s = 65$ and $N = 30$.

124. Statistics In statistics, the formula

$$\sigma_{\bar{x}} = \frac{\sigma}{\sqrt{N}}$$

gives the standard deviation of means of samples of size N. Find $\sigma_{\bar{x}}$ to four decimal places when $\sigma = 12.7$ and $N = 32$.

125. Radius of a circle The radius r of a circle is given by the formula $r = \sqrt{\frac{A}{\pi}}$, where A is its area. Find the radius of a circle whose area is 9π square units.

126. Diagonal of a baseball diamond The diagonal d of a square is given by the formula $d = \sqrt{2s^2}$, where s is the length of each side. Find the diagonal of the baseball diamond.

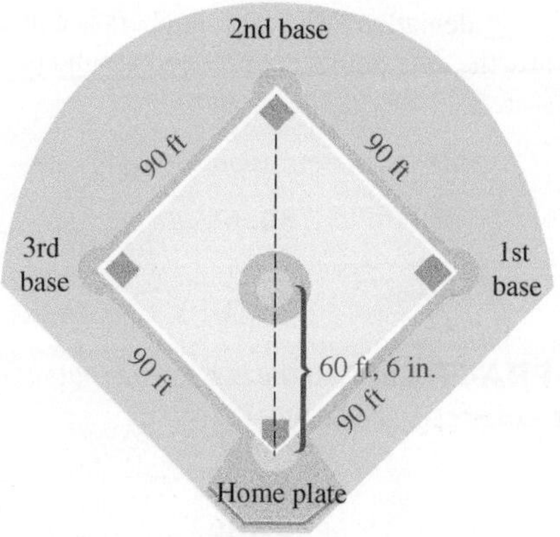

127. Falling objects The time t (in seconds) that it will take for an object to fall a distance of s feet is given by the formula

$$t = \frac{\sqrt{s}}{4}$$

If a stone is dropped down a 256-foot well, how long will it take it to hit bottom?

128. Law enforcement Police sometimes use the formula $s = k\sqrt{l}$ to estimate the speed s (in mph) of a car involved in an accident. In this formula, l is the length of the skid in feet, and k is a constant depending on the condition of the pavement. For wet pavement, $k \approx 3.24$. How fast was a car going if its skid was 400 feet on wet pavement?

129. Electronics When the resistance in a circuit is 18 ohms, the current I (measured in amperes) and the power P (measured in watts) are related by the formula

$$I = \sqrt{\frac{P}{18}}$$

Find the current used by an electrical appliance that is rated at 980 watts.

130. Medicine The approximate pulse rate p (in beats per minute) of an adult who is t inches tall is given by the formula

$$p = \frac{590}{\sqrt{t}}$$

Find the approximate pulse rate of an adult who is 71 inches tall.

WRITING ABOUT MATH

131. If x is any real number, then $\sqrt{x^2} = x$ is not correct. Explain.

132. If x is any real number, then $\sqrt[3]{x^3} = |x|$ is not correct. Explain.

SOMETHING TO THINK ABOUT

133. Is $\sqrt{x^2 - 4x + 4} = x - 2$? What are the exceptions?

134. When is $\sqrt{x^2} \neq x$?

Unless otherwise noted, all content on this page is © Cengage Learning.

Section 9.2

Applications of the Pythagorean Theorem and the Distance Formula

Objectives

1 Apply the Pythagorean theorem to find the length of one side of a right triangle.

2 Find the distance between two points on the coordinate plane.

Vocabulary

leg
hypotenuse
Pythagorean theorem
distance formula

Getting Ready

Evaluate each expression.

1. $3^2 + 4^2$
2. $5^2 + 12^2$
3. $(5 - 2)^2 + (2 + 1)^2$
4. $(111 - 21)^2 + (60 - 4)^2$

In this section, we will discuss the Pythagorean theorem, a theorem that shows the relationship of the sides of a right triangle. We will then use this theorem to develop a formula to calculate the distance between two points on the coordinate plane.

1 Apply the Pythagorean theorem to find the length of one side of a right triangle.

If we know the lengths of the two **legs** of a right triangle, we can find the length of the **hypotenuse** (the side opposite the 90° angle) by using the **Pythagorean theorem**. In fact, if we know the lengths of any two sides of a right triangle, we can find the length of the third side.

PYTHAGOREAN THEOREM

If a and b are the lengths of two legs of a right triangle and c is the length of the hypotenuse, then

$$a^2 + b^2 = c^2$$

In any right triangle, the square of the length of the hypotenuse is equal to the sum of the squares of the lengths of the two legs.

Suppose the right triangle shown in Figure 9-6 has legs of length 3 and 4 units. To find the length of the hypotenuse, we can use the Pythagorean theorem.

$$a^2 + b^2 = c^2$$
$$3^2 + 4^2 = c^2 \quad \text{Substitute.}$$
$$9 + 16 = c^2 \quad \text{Simplify.}$$
$$25 = c^2 \quad \text{Add.}$$

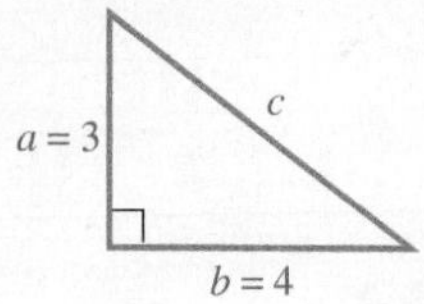

Figure 9-6

To find the value of c, we ask "what number when squared is equal to 25?" There are two such numbers: the positive square root of 25 and the negative square root of 25. Since c represents the length of the hypotenuse and cannot be negative, it follows that c is the positive square root of 25.

$$\sqrt{25} = c \quad \text{Recall that the radical symbol } \sqrt{\ } \text{ represents the positive, or principal, square root of a number.}$$
$$5 = c$$

The length of the hypotenuse is 5 units.

EXAMPLE 1 **FIGHTING FIRES** To fight a forest fire, the forestry department plans to clear a rectangular fire break around the fire, as shown in Figure 9-7. Crews are equipped with mobile communications with a 3,000-yard range. Can crews at points A and B remain in radio contact?

Solution Points A, B, and C form a right triangle. The lengths of its sides are represented as a, b, and c in yards, where a is opposite point A, b is opposite point B, and c is opposite point C. To find the distance c, we can use the Pythagorean theorem, substituting 2,400 for a and 1,000 for b and solving for c.

$$a^2 + b^2 = c^2$$
$$\mathbf{2{,}400}^2 + \mathbf{1{,}000}^2 = c^2 \quad \text{Substitute.}$$
$$5{,}760{,}000 + 1{,}000{,}000 = c^2 \quad \text{Square each value.}$$
$$6{,}760{,}000 = c^2 \quad \text{Add.}$$
$$\sqrt{6{,}760{,}000} = c \quad \text{Since } c \text{ represents a length, it must be the positive square root of 6,760,000.}$$
$$2{,}600 = c \quad \text{Use a calculator to find the square root.}$$

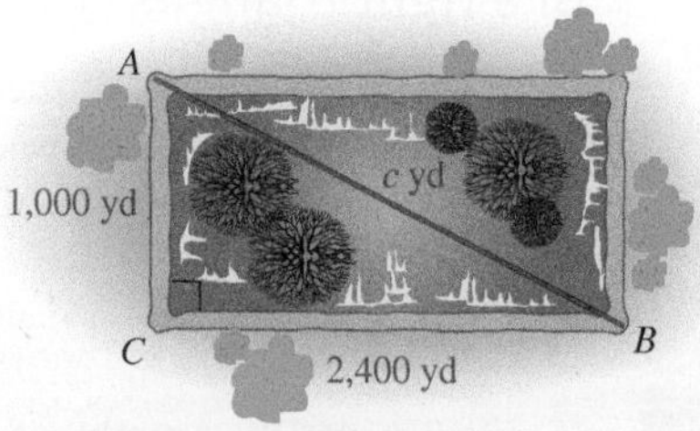

Figure 9-7

The two crews are 2,600 yards apart. Because this distance is less than the 3,000-yard range of the radios, they can communicate.

SELF CHECK 1 Can the crews communicate if $b = 1{,}850$ yards?

Unless otherwise noted, all content on this page is © Cengage Learning.

Perspective

Pythagoras was a teacher. Although it was unusual at that time, his classes were coeducational. He and his followers formed a secret society with two rules: Membership was for life, and members could not reveal the secrets they knew.

Much of their teaching was good mathematics, but some ideas were strange. To them, numbers were sacred. Because beans were used as counters to represent numbers, Pythagoreans refused to eat beans. They also believed that the *only* numbers were the whole numbers.

To them, fractions were not numbers; $\frac{2}{3}$ was just a way of comparing the whole numbers 2 and 3. They believed that whole numbers were the building blocks of the universe. The basic Pythagorean doctrine was, "All things are numbers," and they meant *whole* numbers.

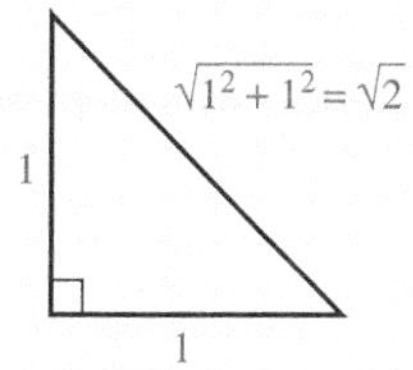

The Pythagorean theorem was an important discovery of the Pythagorean school, yet it caused some controversy. The right triangle in the illustration has two legs of length 1. By the Pythagorean theorem, the length of the hypotenuse is $\sqrt{2}$. One of their own group, Hippasus of Metapontum, discovered that $\sqrt{2}$ is an irrational number: There are *no* whole numbers a and b that make the fraction $\frac{a}{b}$ exactly equal to $\sqrt{2}$. This discovery was not appreciated by the other Pythagoreans. How could everything in the universe be described with whole numbers, when the side of this simple triangle couldn't? The Pythagoreans had a choice. Either expand their beliefs, or cling to the old. According to legend, the group was at sea at the time of the discovery. Rather than upset the system, they threw Hippasus overboard.

2 Find the distance between two points on the coordinate plane.

We can use the Pythagorean theorem to develop a formula to find the distance between any two points that are graphed on a rectangular coordinate system.

To find the distance d between points P and Q shown in Figure 9-8, we construct the right triangle PRQ. Because line segment RQ is vertical, point R will have the same x-coordinate as point Q. Because line segment PR is horizontal, point R will have the same y-coordinate as point P. The distance between P and R is $|x_2 - x_1|$, and the distance between R and Q is $|y_2 - y_1|$. We apply the Pythagorean theorem to the right triangle PRQ to get

$$(PQ)^2 = (PR)^2 + (RQ)^2$$ Read PQ as "the length of segment PQ."

$$d^2 = |x_2 - x_1|^2 + |y_2 - y_1|^2$$ Substitute the value of each expression.

$$d^2 = (x_2 - x_1)^2 + (y_2 - y_1)^2$$ $|x_2 - x_1|^2 = (x_2 - x_1)^2$ and $|y_2 - y_1|^2 = (y_2 - y_1)^2$

(1) $$d = \sqrt{(x_2 - x_1)^2 + (y_2 - y_1)^2}$$ Since d represents a length, it must be the positive square root of $(x_2 - x_1)^2 + (y_2 - y_1)^2$.

Equation 1 is called the **distance formula**.

Pythagoras of Samos
569?–475? B.C.

Pythagoras is thought to be the world's first pure mathematician. Although he is famous for the theorem that bears his name, he is often called "the father of music," because a society he led discovered some of the fundamentals of musical harmony. This secret society had numerology as its religion. The society is also credited with the discovery of irrational numbers.

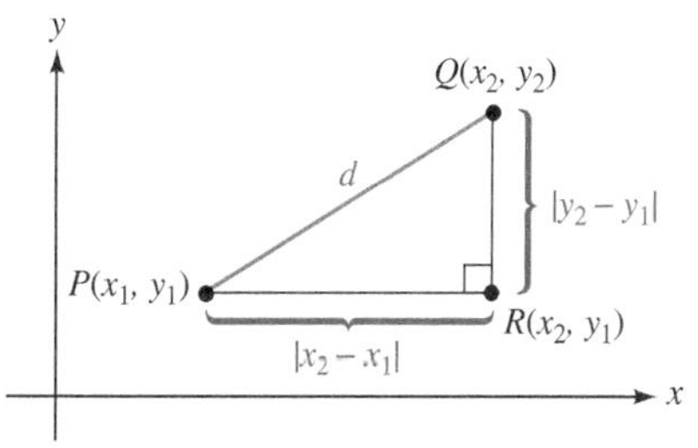

Figure 9-8

DISTANCE FORMULA

The distance d between two points (x_1, y_1) and (x_2, y_2) is given by the formula

$$d = \sqrt{(x_2 - x_1)^2 + (y_2 - y_1)^2}$$

Unless otherwise noted, all content on this page is © Cengage Learning.

EXAMPLE 2 Find the distance between the points $(-2, 3)$ and $(4, -5)$.

Solution To find the distance, we can use the distance formula by substituting 4 for x_2, -2 for x_1, -5 for y_2, and 3 for y_1.

COMMENT Recall that a square root symbol is a grouping symbol and that any radicand must be simplified first.

$$\sqrt{a^2 + b^2} \neq \sqrt{a} + \sqrt{b}$$

$$
\begin{aligned}
d &= \sqrt{(x_2 - x_1)^2 + (y_2 - y_1)^2} \\
&= \sqrt{[4 - (-2)]^2 + (-5 - 3)^2} && \text{Substitute.} \\
&= \sqrt{(4 + 2)^2 + (-5 - 3)^2} && \text{Simplify.} \\
&= \sqrt{6^2 + (-8)^2} && \text{Simplify.} \\
&= \sqrt{36 + 64} && \text{Square each value.} \\
&= \sqrt{100} && \text{Add.} \\
&= 10 && \text{Take the square root.}
\end{aligned}
$$

The distance between the two points is 10 units.

SELF CHECK 2 Find the distance between $P(-2, -2)$ and $Q(3, 10)$.

EXAMPLE 3 **BUILDING A FREEWAY** In a city, streets run north and south, and avenues run east and west. Streets are 850 feet apart and avenues are 850 feet apart. The city plans to construct a straight freeway from the intersection of 25th Street and 8th Avenue to the intersection of 115th Street and 64th Avenue. How long will the freeway be?

Solution We can represent the roads by the coordinate system in Figure 9-9, where the units on each axis represent 850 feet. We represent the end of the freeway at 25th Street and 8th Avenue by the point $(x_1, y_1) = (25, 8)$. The other end is $(x_2, y_2) = (115, 64)$.

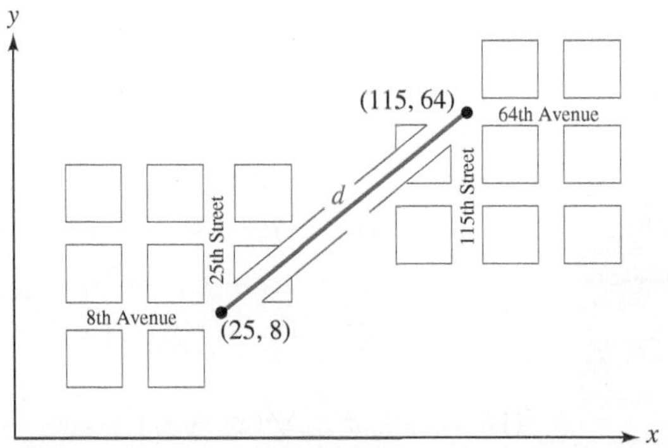

Figure 9-9

We can use the distance formula to find the number of units between the two designated points.

$$
\begin{aligned}
d &= \sqrt{(x_2 - x_1)^2 + (y_2 - y_1)^2} \\
d &= \sqrt{(115 - 25)^2 + (64 - 8)^2} && \text{Substitute.} \\
&= \sqrt{90^2 + 56^2} && \text{Subtract.} \\
&= \sqrt{8{,}100 + 3{,}136} && \text{Square each value.} \\
&= \sqrt{11{,}236} && \text{Add.} \\
&= 106 && \text{Use a calculator to find the square root.}
\end{aligned}
$$

There are approximately 106 units between the two designated points, and because each unit is 850 feet, the length of the freeway is $106(850) = 90{,}100$ feet. Since 5,280 feet $=$ 1 mile, we can divide 90,100 by 5,280 to convert 90,100 feet to 17.064394 miles. Thus, the freeway will be about 17 miles long.

Unless otherwise noted, all content on this page is © Cengage Learning.

SELF CHECK 3 Find the diagonal distance in feet from the intersection of 25th Street and 8th Avenue to the intersection of 28th Street and 12th Avenue.

SELF CHECK ANSWERS **1.** no **2.** 13 units **3.** 4,250 ft

NOW TRY THIS

1. Determine whether the points $(4, -2), (-2, -4)$, and $(-4, 2)$ are the vertices of an *isosceles triangle* (two equal sides), an *equilateral triangle* (three equal sides), or neither. Is the triangle a right triangle?
2. Find the distance between points with coordinates of $(2x + 1, x + 1)$ and $(2x - 3, x - 2)$.

9.2 Exercises

WARM-UPS *Evaluate each expression.*

1. $\sqrt{625}$ **2.** $\sqrt{289}$

3. $\sqrt{7^2 + 24^2}$ **4.** $\sqrt{15^2 + 8^2}$

5. $\sqrt{5^2 - 3^2}$ **6.** $\sqrt{5^2 - 4^2}$

REVIEW *Find each product.*

7. $(2x + 5)(3x - 4)$ **8.** $(6y - 7)(4y + 5)$

9. $(4a - 3b)(5a - 2b)$ **10.** $(4r - 3)(2r^2 + 3r - 4)$

VOCABULARY AND CONCEPTS *Fill in the blanks.*

11. In a right triangle, the side opposite the 90° angle is called the ________.

12. In a right triangle, the two shorter sides are called ___.

13. If a and b are the lengths of two legs of a right triangle and c is the length of the hypotenuse, then __________.

14. In any right triangle, the square of the length of the hypotenuse is equal to the ___ of the squares of the lengths of the two ___. This fact is known as the ______________.

15. If $x^2 = 25$ and x is positive, we can conclude that x is the ______ square root of 25. Thus, $x = 5$.

16. The formula for finding the distance between two points on a rectangular coordinate system is $d =$ ____________________.

GUIDED PRACTICE *The lengths of two sides of the right triangle ABC shown in the illustration are given. Find the length of the third side.* SEE EXAMPLE 1. (OBJECTIVE 1)

17. $a = 6$ ft and $b = 8$ ft

18. $a = 10$ cm and $c = 26$ cm

19. $b = 18$ m and $c = 82$ m

20. $b = 7$ ft and $c = 25$ ft

21. $a = 14$ in. and $c = 50$ in.

22. $a = 8$ cm and $b = 15$ cm

23. $a = \sqrt{8}$ mi and $b = \sqrt{8}$ mi

24. $a = \sqrt{11}$ ft and $b = \sqrt{38}$ ft

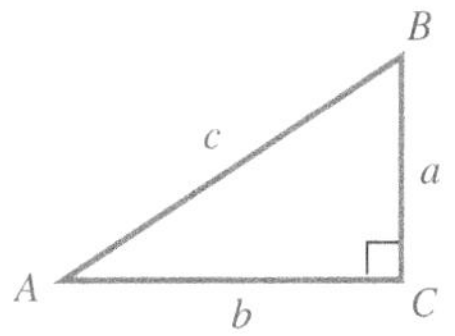

Find the distance between the given points. If an answer is not exact, use a calculator and give an approximation to the nearest tenth. SEE EXAMPLE 2. (OBJECTIVE 2)

25. $(0, 0), (3, -4)$ **26.** $(0, 0), (-6, 8)$

27. $(10, 8), (-2, 3)$ **28.** $(5, 9), (8, 13)$

29. $(-7, 12), (-1, 4)$ **30.** $(-5, -2), (7, 3)$

31. $(12, -6), (-3, 2)$ **32.** $(-14, -9), (10, -2)$

33. $(-3, 5), (-5, -5)$ **34.** $(2, -3), (4, -8)$

35. $(-9, 3), (4, 7)$ **36.** $(-1, -3), (-5, 8)$

ADDITIONAL PRACTICE *In Exercises 37–40, use a calculator to approximate each value to the nearest foot. The baseball diamond is a square, 90 feet on a side.*

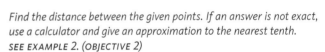

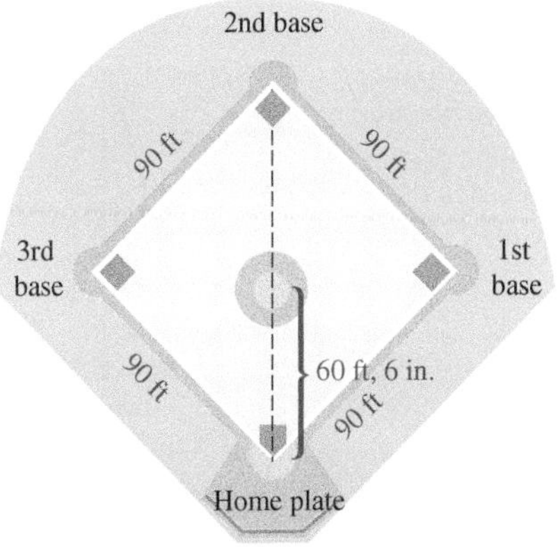

Unless otherwise noted, all content on this page is © Cengage Learning.

37. Baseball How far must a catcher throw the ball to throw out a runner stealing second base?

38. Baseball In baseball, the pitcher's mound is 60 feet, 6 inches from home plate. How far from the mound is second base?

39. Baseball If the third baseman fields a ground ball 10 feet directly behind third base along the third base line, how far must he throw the ball to throw a runner out at first base?

40. Baseball The shortstop fields a grounder at a point one-third of the way from second base to third base. How far will he have to throw the ball to make an out at first base?

For Exercises 41–44, approximate each answer to the nearest tenth.

41. Geometry Find the length of the diagonal of one of the faces of the cube.

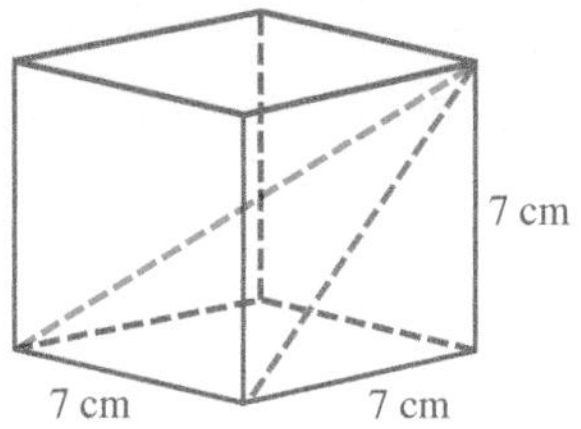

42. Geometry Find the length of the diagonal of the cube shown in the illustration in Exercise 41.

43. Supporting a weight A weight placed on the tight wire pulls the center down 1 foot. By how much is the wire stretched? Round the answer to the nearest hundredth of a foot.

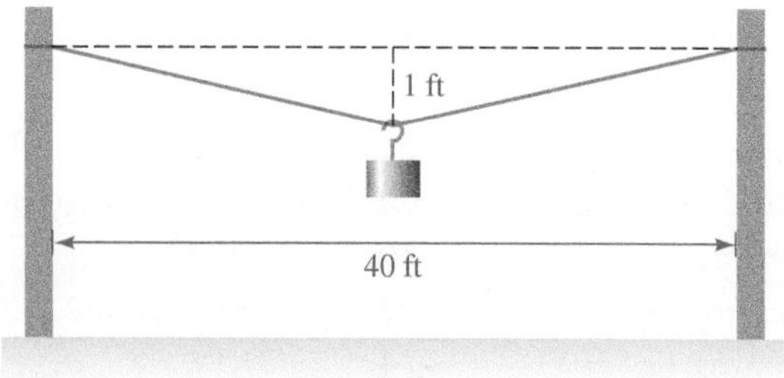

44. Supporting a weight If the weight in Exercise 43 pulls the center down 2 feet, by how much would the wire stretch? Round the answer to the nearest tenth of a foot.

APPLICATIONS *Solve each application.* SEE EXAMPLE 1. (OBJECTIVE 1)

45. Sailing Refer to the sailboat in the illustration. How long must a rope be to fasten the top of the mast to the bow?

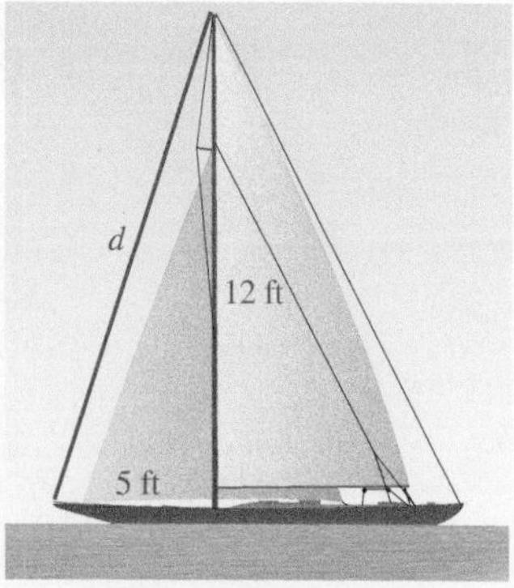

46. Carpentry The gable end of the roof shown is divided in half by a vertical brace. Find the distance from an eave to the peak.

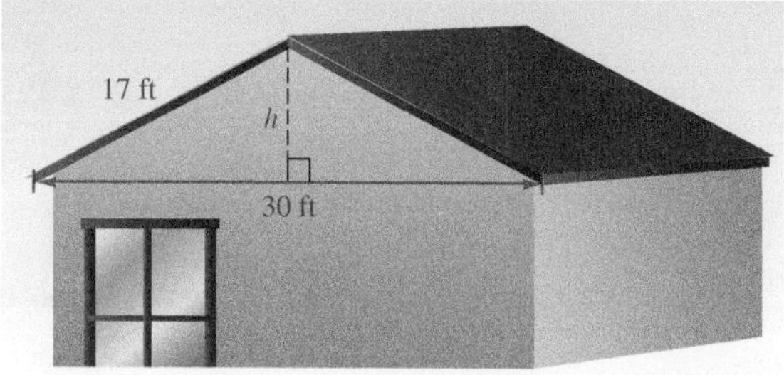

47. Reach of a ladder The base of the 37-foot ladder in the illustration is 9 feet from the wall. Will the top reach a window ledge that is 35 feet above the ground?

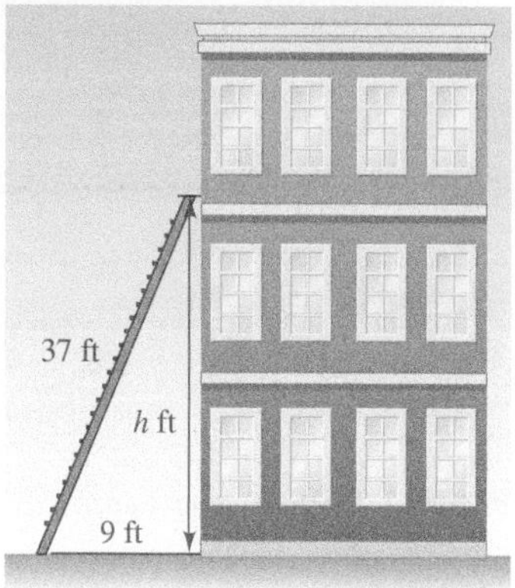

48. Geometry The side, s, of a square with area A square feet is given by the formula $s = \sqrt{A}$. Find the perimeter of a square with an area of 49 square feet.

Solve each application. SEE EXAMPLE 3. (OBJECTIVE 2)

49. Telephone service The telephone cable in the illustration currently runs from A to B to C to D. How much cable is required to run from A to D directly?

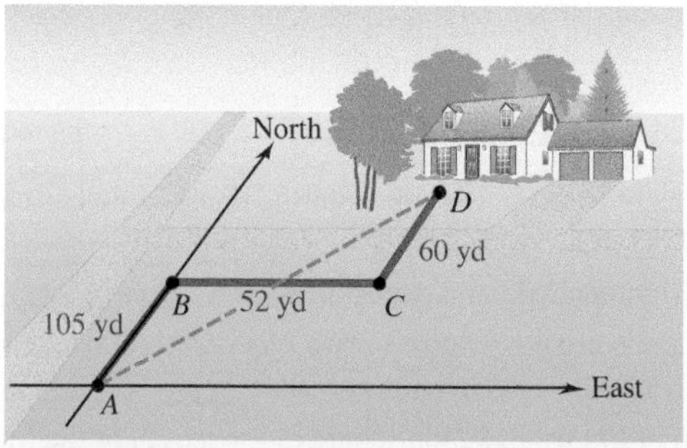

50. Electric service The power company routes its lines as shown in the illustration. How much wire could be saved by going directly from A to E?

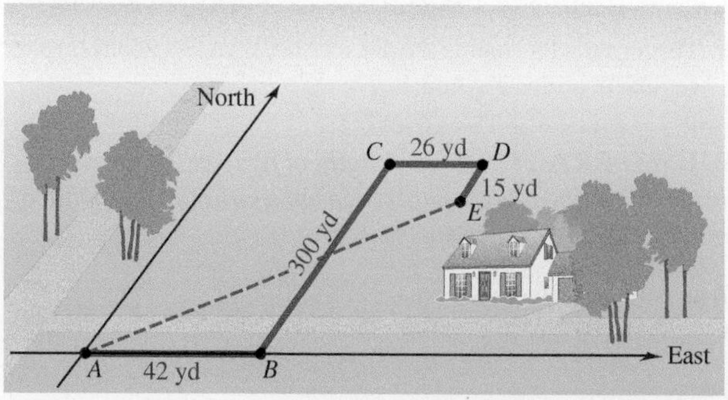

Unless otherwise noted, all content on this page is © Cengage Learning.

51. Surface area of a cube The total surface area, A, of a cube is related to its volume, V, by the formula $A = 6\sqrt[3]{V^2}$. Find the surface area of a cube with a volume of 8 cubic centimeters.

52. Area of many cubes A grain of table salt is a cube with a volume of approximately 6×10^{-6} cubic in., and there are about 1.5 million grains of salt in one cup. Find the total surface area of the salt in one cup. (See Exercise 51.)

Use a calculator to help answer each question.

53. Packing a tennis racket The diagonal d of a rectangular box with dimensions $a \times b \times c$ is given by

$$d = \sqrt{a^2 + b^2 + c^2}$$

Will the racket fit in the shipping carton?

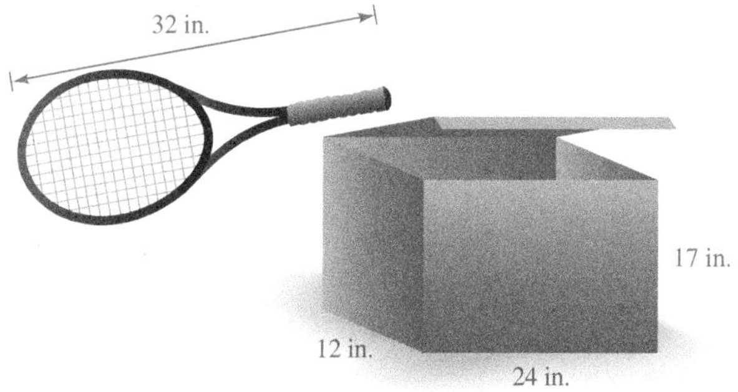

54. Packing a tennis racket Will the racket in Exercise 53 fit in a carton with dimensions that are 15 in. · 15 in. · 17 in.?

55. Shipping packages A delivery service won't accept a package for shipping if any dimension exceeds 21 inches. An archaeologist wants to ship a 36-inch femur bone. Will it fit in a 3-inch-tall box that has a 21-inch-square base?

56. Shipping packages Can the archaeologist in Exercise 55 ship the femur bone in a cubical box 21 inches on an edge?

57. Geometry Show that the point $(5, 1)$ is equidistant from points $(7, 0)$ and $(3, 0)$.

58. Geometry Show that a triangle with vertices at $(2, 3)$, $(-3, 4)$, and $(1, -2)$ is a right triangle. (*Hint*: If the Pythagorean theorem holds, the triangle is a right triangle.)

59. Geometry Show that a triangle with vertices at $(-2, 4)$, $(2, 8)$, and $(6, 4)$ is isosceles.

60. Geometry Show that a triangle with vertices at $(-2, 13)$, $(-8, 9)$, and $(-2, 5)$ is isosceles.

WRITING ABOUT MATH

61. State the Pythagorean theorem.

62. Explain the distance formula.

SOMETHING TO THINK ABOUT

63. Body mass The formula

$$I = \frac{703w}{h^2}$$

(where w is weight in pounds and h is height in inches) can be used to estimate body mass index, I. The scale shown in the table can be used to judge a person's risk of heart attack. A girl weighing 104 pounds is 54.1 inches tall. Find her estimated body mass index.

20–26	normal
27–29	higher risk
30 and above	very high risk

64. What is the risk of a heart attack for a man who is 6 feet tall and weighs 210 pounds?

65. Bowling The velocity, v, of an object after it has fallen d feet is given by the equation $v^2 = 64d$. If an inexperienced bowler lofts the ball 4 feet, with what velocity does it strike the alley?

66. Bowling Using the formula from Exercise 65, find the velocity if the bowler lofts the ball 3 feet. Round to the nearest tenth.

Unless otherwise noted, all content on this page is © Cengage Learning.

Section 9.3 Rational Exponents

Objectives

1. Simplify an expression that contains a positive rational exponent with a numerator of 1.
2. Simplify an expression that contains a positive rational exponent with a numerator other than 1.
3. Simplify an expression that contains a negative rational exponent.
4. Simplify an expression that contains rational exponents by applying the properties of exponents.
5. Simplify a radical expression by first writing it as an expression with a rational exponent.

Getting Ready

Simplify each expression. Assume no variable is zero.

1. x^3x^4	**2.** $(a^3)^4$	**3.** $\dfrac{a^8}{a^4}$	**4.** a^0
5. x^{-4}	**6.** $\dfrac{1}{x^{-3}}$	**7.** $\left(\dfrac{b^2}{c^3}\right)^3$	**8.** $(a^2a^3)^2$

1 Simplify an expression that contains a positive rational exponent with a numerator of 1.

We have seen that positive integer exponents indicate the number of times that a base is to be used as a factor in a product. For example, x^5 means that x is to be used as a factor five times.

$$x^5 = \overbrace{x \cdot x \cdot x \cdot x \cdot x}^{5 \text{ factors of } x}$$

Furthermore, we recall the following properties of exponents.

RULES OF EXPONENTS

If there are no divisions by 0, then for all integers m and n,

1. $x^mx^n = x^{m+n}$	**2.** $(x^m)^n = x^{mn}$	**3.** $(xy)^n = x^ny^n$
4. $\left(\dfrac{x}{y}\right)^n = \dfrac{x^n}{y^n}$	**5.** $x^0 = 1 \quad (x \neq 0)$	**6.** $x^{-n} = \dfrac{1}{x^n}$
7. $\dfrac{x^m}{x^n} = x^{m-n}$	**8.** $\left(\dfrac{x}{y}\right)^{-n} = \left(\dfrac{y}{x}\right)^n$	**9.** $\dfrac{1}{x^{-n}} = x^n$

To show how to raise bases to rational powers, we consider the expression $10^{1/2}$. Since rational exponents must obey the same rules as integer exponents, the square of $10^{1/2}$ is equal to 10.

$$\begin{aligned}(10^{1/2})^2 &= 10^{(1/2)2} && \text{Keep the base and multiply the exponents.}\\ &= 10^1 && \tfrac{1}{2}\cdot 2 = 1\\ &= 10 && 10^1 = 10\end{aligned}$$

However, we have seen that

$$(\sqrt{10})^2 = 10$$

Since $(10^{1/2})^2$ and $(\sqrt{10})^2$ both equal 10, we define $10^{1/2}$ to be $\sqrt{10}$. Likewise, we define

$$10^{1/3} \text{ to be } \sqrt[3]{10} \qquad \text{and} \qquad 10^{1/4} \text{ to be } \sqrt[4]{10}$$

RATIONAL EXPONENTS

If n is a natural number greater than 1, and $\sqrt[n]{x}$ is a real number, then

$$x^{1/n} = \sqrt[n]{x}$$

EXAMPLE 1 Simplify each expression. Assume all variables represent nonnegative values.

a. $9^{1/2}$ **b.** $-\left(\frac{16}{9}\right)^{1/2}$ **c.** $(-64)^{1/3}$ **d.** $16^{1/4}$ **e.** $\left(\frac{1}{32}\right)^{1/5}$ **f.** $0^{1/8}$
g. $-(32x^5)^{1/5}$ **h.** $(xyz)^{1/4}$

Solution

a. $9^{1/2} = \sqrt{9} = 3$

b. $-\left(\frac{16}{9}\right)^{1/2} = -\sqrt{\frac{16}{9}} = -\frac{4}{3}$

c. $(-64)^{1/3} = \sqrt[3]{-64} = -4$

d. $16^{1/4} = \sqrt[4]{16} = 2$

e. $\left(\frac{1}{32}\right)^{1/5} = \sqrt[5]{\frac{1}{32}} = \frac{1}{2}$

f. $0^{1/8} = \sqrt[8]{0} = 0$

g. $-(32x^5)^{1/5} = -\sqrt[5]{32x^5} = -2x$

h. $(xyz)^{1/4} = \sqrt[4]{xyz}$

SELF CHECK 1 Simplify each expression. Assume that all variables represent nonnegative values.

a. $16^{1/2}$ **b.** $\left(\frac{27}{8}\right)^{1/3}$ **c.** $-(16x^4)^{1/4}$

EXAMPLE 2 Write each radical using a rational exponent. **a.** $\sqrt[4]{5xyz}$ **b.** $\sqrt[5]{xy^2}$

Solution **a.** $\sqrt[4]{5xyz} = (5xyz)^{1/4}$ **b.** $\sqrt[5]{xy^2} = (xy^2)^{1/5}$

SELF CHECK 2 Write each radical using a rational exponent.
a. $\sqrt[6]{4ab}$ **b.** $\sqrt[3]{a^2b^4}$

As with radicals, when n is even in the expression $x^{1/n}(n > 1)$, there are two real nth roots and we must use absolute value symbols to guarantee that the simplified result is positive.

When n is odd, there is only one real nth root, and we do not need to use absolute value symbols.

When n is even and x is negative, the expression $x^{1/n}$ is not a real number.

EXAMPLE 3 Assume that all variables can be any real number, and simplify each expression using absolute value symbols when necessary.

a. $(-27x^3)^{1/3} = -3x$ — $-27x^3 = (-3x)^3$. Since n is odd, no absolute value symbols are needed.

b. $(49x^2)^{1/2} = |7x| = 7|x|$ — $49x^2 = (|7x|)^2$. Since $7x$ can be negative, absolute value symbols are needed.

c. $(256a^8)^{1/8} = 2|a|$ — $256a^8 = (2|a|)^8$. Since a can be any real number, $2a$ can be negative. Thus, absolute value symbols are needed.

d. $[(y+1)^2]^{1/2} = |y+1|$ — $(y+1)^2 = |y+1|^2$. Since y can be any real number, $y+1$ can be negative, and the absolute value symbols are needed.

e. $(25b^4)^{1/2} = 5b^2$ — $25b^4 = (5b^2)^2$. Since $b^2 \geq 0$, no absolute value symbols are needed.

f. $(-256x^4)^{1/4}$ is not a real number. — No real number raised to the 4th power is $-256x^4$.

SELF CHECK 3 Simplify each expression using absolute value symbols when necessary.

a. $(625a^4)^{1/4}$ **b.** $(b^4)^{1/2}$ **c.** $(-8x^3)^{1/3}$ **d.** $(64x^2)^{1/2}$
e. $(-9x^4)^{1/2}$

We summarize the cases as follows.

SUMMARY OF THE DEFINITIONS OF $x^{1/n}$

Assume n is a natural number greater than 1 and x is a real number.

If $x > 0$, then $x^{1/n}$ is the positive number such that $(x^{1/n})^n = x$.

If $x = 0$, then $x^{1/n} = 0$.

If $x < 0$ $\begin{cases} \text{and } n \text{ is odd, then } x^{1/n} \text{ is the real number such that } (x^{1/n})^n = x. \\ \text{and } n \text{ is even, then } x^{1/n} \text{ is not a real number.} \end{cases}$

2 Simplify an expression that contains a positive rational exponent with a numerator other than 1.

We can extend the definition of $x^{1/n}$ to include rational exponents with numerators other than 1. For example, since $4^{3/2}$ can be written as $(4^{1/2})^3$, we have

$$4^{3/2} = (4^{1/2})^3 = (\sqrt{4})^3 = 2^3 = 8$$

Thus, we can simplify $4^{3/2}$ by cubing the square root of 4. We can also simplify $4^{3/2}$ by taking the square root of 4 cubed.

$$4^{3/2} = (4^3)^{1/2} = 64^{1/2} = \sqrt{64} = 8$$

In general, we have the following strategy.

CHANGING FROM RATIONAL EXPONENTS TO RADICALS

If m and n are positive integers, $x \geq 0$, and $\frac{m}{n}$ is in simplified form, then

$$x^{m/n} = (\sqrt[n]{x})^m = \sqrt[n]{x^m}$$

We can interpret $x^{m/n}$ in two ways:

1. $x^{m/n}$ means the mth power of the nth root of x.
2. $x^{m/n}$ means the nth root of the mth power of x.

EXAMPLE 4 Simplify each expression. **a.** $27^{2/3}$ **b.** $\left(\frac{1}{16}\right)^{3/4}$ **c.** $(-8x^3)^{4/3}$

Solution

a. $27^{2/3} = \left(\sqrt[3]{27}\right)^2 = 3^2 = 9$ or $27^{2/3} = \sqrt[3]{27^2} = \sqrt[3]{729} = 9$

b. $\left(\frac{1}{16}\right)^{3/4} = \left(\sqrt[4]{\frac{1}{16}}\right)^3 = \left(\frac{1}{2}\right)^3 = \frac{1}{8}$ or $\left(\frac{1}{16}\right)^{3/4} = \sqrt[4]{\left(\frac{1}{16}\right)^3} = \sqrt[4]{\frac{1}{4{,}096}} = \frac{1}{8}$

c. $(-8x^3)^{4/3} = \left(\sqrt[3]{-8x^3}\right)^4 = (-2x)^4 = 16x^4$ or $(-8x^3)^{3/4} = \sqrt[3]{(-8x^3)^4} = \sqrt[3]{4{,}096x^{12}} = 16x^4$

COMMENT To avoid large numbers, it is usually better to find the root of the base first, as shown in Example 4.

SELF CHECK 4 Simplify: **a.** $16^{3/2}$ **b.** $(-27x^6)^{2/3}$

Accent on technology

▸ Rational Exponents

We can evaluate expressions containing rational exponents using the exponential key y^x or x^y on a scientific calculator. For example, to evaluate $10^{2/3}$, we enter

10 y^x (2 ÷ 3) = **4.641588834**

Note that parentheses were used when entering the power. Without them, the calculator would interpret the entry as $10^2 \div 3$.

To evaluate the exponential expression using a graphing calculator, we use the ∧ key, which raises a base to a power. Again, we use parentheses when entering the power.

10 ∧ (2 ÷ 3) **ENTER** **10∧(2/3)** **4.641588834**

To the nearest hundredth, $10^{2/3} \approx 4.64$.

For instructions regarding the use of a Casio graphing calculator, please refer to the Casio Keystroke Guide in the back of the book.

3 Simplify an expression that contains a negative rational exponent.

To be consistent with the definition of negative integer exponents, we define $x^{-m/n}$ as follows.

DEFINITION OF $x^{-m/n}$

If m and n are natural numbers, $\frac{m}{n}$ is in simplified form, and $x^{1/n}$ is a real number $(x \neq 0)$, then

$$x^{-m/n} = \frac{1}{x^{m/n}} \quad \text{and} \quad \frac{1}{x^{-m/n}} = x^{m/n}$$

EXAMPLE 5 Write each expression without negative exponents.

a. $64^{-1/2}$ **b.** $(16x^{-4})^{-1/4}$ **c.** $(-32x^5)^{-2/5}$ **d.** $(-16)^{-3/4}$

Solution

a. $64^{-1/2} = \frac{1}{64^{1/2}}$

$= \frac{1}{8}$

b. $(16x^{-4})^{-1/4} = 16^{-1/4}(x^{-4})^{-1/4}$

$= \frac{1}{16^{1/4}}x$

$= \frac{1}{2}x$ or $\frac{x}{2}$

c. $(-32x^5)^{-2/5} = \frac{1}{(-32x^5)^{2/5}} \quad (x \neq 0)$

$= \frac{1}{[(-32x^5)^{1/5}]^2}$

$= \frac{1}{(-2x)^2}$

$= \frac{1}{4x^2}$

d. $(-16)^{-3/4}$ is not a real number, because $(-16)^{1/4}$ is not a real number.

SELF CHECK 5 Write each expression without negative exponents.

a. $25^{-3/2}$ **b.** $(-27a^{-3})^{-2/3}$ **c.** $(-81)^{-3/4}$

COMMENT A base of 0 raised to a negative power is undefined, because 0^{-2} would equal $\frac{1}{0^2}$, which is undefined since we cannot divide by 0.

4 Simplify an expression that contains rational exponents by applying the properties of exponents.

We can use the properties of exponents to simplify many expressions with rational exponents.

EXAMPLE 6 Write all answers without negative exponents. Assume that all variables represent positive numbers. Thus, no absolute value symbols are necessary.

a. $5^{2/7}5^{3/7} = 5^{2/7+3/7}$ Apply the rule $x^m x^n = x^{m+n}$.

$= 5^{5/7}$ Add: $\frac{2}{7} + \frac{3}{7} = \frac{5}{7}$

b. $(5^{2/7})^3 = 5^{(2/7)(3)}$ Apply the rule $(x^m)^n = x^{mn}$.

$= 5^{6/7}$ Multiply: $\frac{2}{7}(3) = \frac{2}{7}\left(\frac{3}{1}\right) = \frac{6}{7}$

c. $(a^{2/3}b^{1/2})^6 = (a^{2/3})^6(b^{1/2})^6$ Apply the rule $(xy)^n = x^n y^n$.

$= a^{12/3}b^{6/2}$ Apply the rule $(x^m)^n = x^{mn}$ twice.

$= a^4b^3$ Simplify the exponents.

d. $\frac{a^{8/3}a^{1/3}}{a^2} = a^{8/3+1/3-2}$ Apply the rules $x^m x^n = x^{m+n}$ and $\frac{x^m}{x^n} = x^{m-n}$.

$= a^{8/3+1/3-6/3}$ $2 = \frac{6}{3}$

$= a^{3/3}$ $\frac{8}{3} + \frac{1}{3} - \frac{6}{3} = \frac{3}{3}$

$= a$ $\frac{3}{3} = 1$

SELF CHECK 6 Simplify: **a.** $(x^{1/3}y^{3/2})^6$ **b.** $\frac{x^{5/3}x^{2/3}}{x^{1/3}}$

EXAMPLE 7 Assume that all variables represent positive numbers and perform the operations.

COMMENT Note that $a + a^{7/5} \neq a^{1+7/5}$. The expression $a + a^{7/5}$ cannot be simplified, because a and $a^{7/5}$ are not like terms.

a. $a^{4/5}(a^{1/5} + a^{3/5}) = a^{4/5}a^{1/5} + a^{4/5}a^{3/5}$ Use the distributive property to remove parentheses.

$= a^{4/5+1/5} + a^{4/5+3/5}$ Apply the rule $x^m x^n = x^{m+n}$.

$= a^{5/5} + a^{7/5}$ Simplify the exponents.

$= a + a^{7/5}$

b. $x^{1/2}(x^{-1/2} + x^{1/2}) = x^{1/2}x^{-1/2} + x^{1/2}x^{1/2}$ Use the distributive property to remove parentheses.

$= x^{1/2-1/2} + x^{1/2+1/2}$ Apply the rule $x^m x^n = x^{m+n}$.

$= x^0 + x^1$ Simplify.

$= 1 + x$ $x^0 = 1$

c. $(x^{2/3} + 1)(x^{2/3} - 1) = x^{4/3} - x^{2/3} + x^{2/3} - 1$ Use the distributive property to remove parentheses.

$= x^{4/3} - 1$ Combine like terms.

d. $(x^{1/2} + y^{1/2})^2 = (x^{1/2} + y^{1/2})(x^{1/2} + y^{1/2})$ $(x + y)^2 = (x + y)(x + y)$

$= x + 2x^{1/2}y^{1/2} + y$ Use the distributive property to remove parentheses.

SELF CHECK 7 Assume that all variables represent positive numbers and perform the operations.

a. $p^{1/5}(p^{4/5} + p^{2/5})$ **b.** $(p^{2/3} + q^{1/3})(p^{2/3} - q^{1/3})$

5 Simplify a radical expression by first writing it as an expression with a rational exponent.

We can simplify many radical expressions by using the following steps.

USING RATIONAL EXPONENTS TO SIMPLIFY RADICALS

1. Write the radical expression as an exponential expression with rational exponents.
2. Simplify the rational exponents.
3. Write the exponential expression as a radical.

EXAMPLE 8 Simplify. Assume variables represent positive values.

a. $\sqrt[4]{3^2}$ **b.** $\sqrt[8]{x^6}$ **c.** $\sqrt[9]{27x^6y^3}$

Solution

a. $\sqrt[4]{3^2} = 3^{2/4}$ Apply the rule $\sqrt[n]{x^m} = x^{m/n}$.

$= 3^{1/2}$ $\frac{2}{4} = \frac{1}{2}$

$= \sqrt{3}$ Write using radical notation.

b. $\sqrt[8]{x^6} = x^{6/8}$ Apply the rule $\sqrt[n]{x^m} = x^{m/n}$.

$= x^{3/4}$ $\frac{6}{8} = \frac{3}{4}$

$= \sqrt[4]{x^3}$ Write using radical notation.

c. $\sqrt[9]{27x^6y^3} = (3^3x^6y^3)^{1/9}$ Write 27 as 3^3 and write the radical as an exponential expression.

$= 3^{3/9}x^{6/9}y^{3/9}$ Raise each factor to the $\frac{1}{9}$ power by multiplying the fractional exponents.

$= 3^{1/3}x^{2/3}y^{1/3}$ Simplify each fractional exponent.

$= (3x^2y)^{1/3}$ Apply the rule $(xy)^n = x^ny^n$.

$= \sqrt[3]{3x^2y}$ Write using radical notation.

SELF CHECK 8 Simplify. Assume variables represent positive values.

a. $\sqrt[6]{3^3}$ b. $\sqrt[10]{x^8}$ c. $\sqrt[4]{49x^2y^2}$

SELF CHECK ANSWERS

1. a. 4 b. $\frac{3}{2}$ c. $-2x$ 2. a. $(4ab)^{1/6}$ b. $(a^2b^4)^{1/3}$ 3. a. $5|a|$ b. b^2 c. $-2x$ d. $8|x|$ e. not a real number 4. a. 64 b. $9x^4$ 5. a. $\frac{1}{125}$ b. $\frac{1}{9}a^2$ or $\frac{a^2}{9}$ c. not a real number 6. a. x^2y^9 b. x^2 7. a. $p + p^{3/5}$ b. $p^{4/3} - q^{2/3}$ 8. a. $\sqrt{3}$ b. $\sqrt[5]{x^4}$ c. $\sqrt{7xy}$

NOW TRY THIS

1. Evaluate each exponential expression, if possible.
   a. $-64^{-2/3}$ b. $-64^{-3/2}$ c. $(-64)^{-2/3}$
   d. $(-64)^{-3/2}$
2. Simplify each expression and then write the answer using radical notation. Assume all variables represent positive numbers.
   a. $x^{a/4} \cdot x^{a/2}$ b. $\dfrac{x^{n/m}}{x^{(n-1)/m}}$
3. Simplify $\dfrac{\sqrt{5}}{\sqrt[3]{5}}$ and write the answer using radical notation.

9.3 Exercises

Assume no division by zero.

WARM-UPS *Simplify.*

1. -7^2
2. -9^2
3. 5^3
4. 2^3
5. 4^{-2}
6. 10^{-3}
7. $\sqrt{4^3}$
8. $\sqrt{2^4}$
9. $\sqrt[3]{8^2}$
10. $\sqrt[3]{3^2 \cdot 3}$

REVIEW *Solve each inequality.*

11. $8x - 7 \le 1$
12. $4(6 - 2y) < -16$
13. $\frac{4}{5}(r - 3) > \frac{2}{3}(r + 2)$
14. $-4 < 2x - 4 \le 8$
15. **Mixing solutions** How much water must be added to 5 pints of a 20% alcohol solution to dilute it to a 15% solution?
16. **Selling apples** A grocer bought some boxes of apples for \$70. However, 4 boxes were spoiled. The grocer sold the remaining boxes at a profit of \$2 each. How many boxes did the grocer sell if she managed to break even?

VOCABULARY AND CONCEPTS *Fill in the blanks.*

17. $a^4 =$ ________
18. $a^ma^n =$ ___
19. $(a^m)^n =$ ___
20. $(ab)^n =$ ___
21. $\left(\frac{a}{b}\right)^n = \frac{\_\_\_}{\_\_\_}$
22. $a^0 =$ _, provided $a \ne$ _.
23. $a^{-n} = \frac{\_\_\_}{\_\_\_}$, provided $a \ne$ _.
24. $\frac{a^m}{a^n} =$ ____, provided $a \ne 0$.
25. $\left(\frac{a}{b}\right)^{-n} = \frac{\_\_\_}{\_\_\_}$
26. $x^{1/n} =$ ___
27. $(x^n)^{1/n} =$ ___, provided n is even.
28. $x^{m/n} = \sqrt[n]{x^m} =$ ______

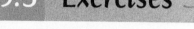

GUIDED PRACTICE *Change each expression into radical notation. (OBJECTIVE 1)*

29. $5^{1/4}$
30. $26^{1/2}$
31. $8^{1/5}$
32. $29^{1/8}$
33. $(13a)^{1/7}$
34. $(4ab)^{1/6}$
35. $\left(\frac{1}{2}x^3y\right)^{1/4}$
36. $\left(\frac{3}{4}a^2b^2\right)^{1/5}$
37. $(6a^3b)^{1/4}$
38. $(5pq^2)^{1/3}$
39. $(x^2 + y^2)^{1/2}$
40. $(a^4 + b^4)^{1/4}$

Simplify each expression, if possible. SEE EXAMPLE 1. (OBJECTIVE 1)

41. $49^{1/2}$
42. $64^{1/2}$
43. $27^{1/3}$
44. $125^{1/3}$
45. $\left(\frac{1}{9}\right)^{1/2}$
46. $\left(\frac{1}{16}\right)^{1/2}$
47. $\left(\frac{1}{8}\right)^{1/3}$
48. $\left(\frac{1}{16}\right)^{1/4}$
49. $-81^{1/4}$
50. $-125^{1/3}$
51. $(-25)^{1/2}$
52. $(-216)^{1/2}$

Change each radical to an exponential expression. SEE EXAMPLE 2. (OBJECTIVE 1)

53. $\sqrt{7}$
54. $\sqrt[3]{12}$
55. $\sqrt[4]{3a}$
56. $\sqrt[5]{8xz}$
57. $5\sqrt[7]{b}$
58. $4\sqrt[3]{p}$
59. $\sqrt[6]{\frac{1}{7}abc}$
60. $\sqrt[7]{\frac{3}{8}p^2q}$
61. $\sqrt[5]{\frac{1}{2}mn}$
62. $\sqrt[8]{\frac{2}{7}p^2q}$
63. $\sqrt[3]{x^2 + y^2}$
64. $\sqrt[4]{a^4 - b^4}$

Simplify each expression. Assume that all variables can be any real number, and use absolute value symbols if necessary. SEE EXAMPLE 3. (OBJECTIVE 1)

65. $(25y^2)^{1/2}$
66. $(16x^4)^{1/4}$
67. $(243x^5)^{1/5}$
68. $(-27x^3)^{1/3}$
69. $[(x + 1)^4]^{1/4}$
70. $[(x + 5)^3]^{1/3}$
71. $(-81x^{12})^{1/4}$
72. $(-16x^4)^{1/2}$

Simplify each expression, if possible. SEE EXAMPLE 4. (OBJECTIVE 2)

73. $25^{3/2}$
74. $27^{2/3}$
75. $81^{3/4}$
76. $100^{3/2}$
77. $(-27x^3)^{4/3}$
78. $(-1{,}000x^{12})^{2/3}$
79. $\left(\frac{1}{8}\right)^{2/3}$
80. $\left(\frac{4}{9}\right)^{3/2}$

Write each expression without using negative exponents. Assume that all variables represent positive numbers. SEE EXAMPLE 5. (OBJECTIVE 3)

81. $4^{-1/2}$
82. $8^{-1/3}$
83. $4^{-3/2}$
84. $25^{-5/2}$
85. $(16x^2)^{-3/2}$
86. $(81c^4)^{-3/2}$
87. $(-27y^3)^{-2/3}$
88. $(-8z^9)^{-2/3}$
89. $\left(\frac{1}{4}\right)^{-3/2}$
90. $\left(\frac{4}{25}\right)^{-3/2}$
91. $\left(\frac{27}{8}\right)^{-4/3}$
92. $\left(\frac{25}{49}\right)^{-3/2}$

Perform the operations. Write answers without negative exponents. Assume that all variables represent positive numbers. SEE EXAMPLE 6. (OBJECTIVE 4)

93. $5^{4/9}5^{4/9}$
94. $4^{2/5}4^{2/5}$
95. $(4^{1/5})^3$
96. $(3^{1/3})^5$
97. $6^{-2/3}6^{-4/3}$
98. $5^{1/3}5^{-5/3}$
99. $\frac{9^{4/5}}{9^{3/5}}$
100. $\frac{7^{2/3}}{7^{1/2}}$
101. $\frac{7^{1/2}}{7^0}$
102. $\frac{3^{4/3}3^{1/3}}{3^{2/3}}$
103. $\frac{2^{5/6}2^{1/3}}{2^{1/2}}$
104. $\frac{5^{1/3}5^{1/2}}{5^{1/3}}$
105. $(x^{1/4} \cdot x^{3/2})^8$
106. $(a^{1/3} \cdot a^{3/4})^{12}$
107. $(a^{2/3})^{1/3}$
108. $(t^{4/5})^{10}$

Perform the operations. Write answers without negative exponents. Assume that all variables represent positive numbers. SEE EXAMPLE 7. (OBJECTIVE 4)

109. $y^{1/3}(y^{2/3} + y^{5/3})$
110. $y^{2/5}(y^{-2/5} + y^{3/5})$
111. $x^{3/5}(x^{7/5} - x^{2/5} + 1)$
112. $(x^{1/2} + 2)(x^{1/2} - 2)$
113. $(x^{1/2} + y^{1/2})(x^{1/2} - y^{1/2})$
114. $(x^{2/3} - x)(x^{2/3} + x)$
115. $(x^{2/3} + y^{2/3})^2$
116. $(a^{3/2} - b^{3/2})^2$

Use rational exponents to simplify each radical. Assume that all variables represent positive numbers. SEE EXAMPLE 8. (OBJECTIVE 5)

117. $\sqrt[6]{p^3}$
118. $\sqrt[8]{q^2}$
119. $\sqrt[4]{25b^2}$
120. $\sqrt[9]{-8x^6}$

ADDITIONAL PRACTICE *Simplify each expression, if possible. Assume all variables represent positive numbers. Write answers without negative exponents.*

121. $16^{1/4}$
122. $625^{1/4}$
123. $32^{1/5}$
124. $0^{1/5}$
125. $0^{1/3}$
126. $(-243)^{1/5}$
127. $(-27)^{1/3}$
128. $(-125)^{1/3}$
129. $(25x^4)^{3/2}$
130. $(27a^3b^3)^{2/3}$
131. $\left(\frac{8x^3}{27}\right)^{2/3}$
132. $\left(\frac{27}{64y^6}\right)^{2/3}$
133. $(-32p^5)^{-2/5}$
134. $(16q^6)^{-5/2}$
135. $\left(-\frac{8x^3}{27}\right)^{-1/3}$
136. $\left(\frac{16}{81y^4}\right)^{-3/4}$
137. $(a^{1/2}b^{1/3})^{3/2}$
138. $(a^{3/5}b^{3/2})^{2/3}$

139. $(mn^{-2/3})^{-3/5}$

140. $(r^{-2}s^3)^{1/3}$

141. $\dfrac{(4x^3y)^{1/2}}{(9xy)^{1/2}}$

142. $\dfrac{(27x^3y)^{1/3}}{(8xy^2)^{2/3}}$

143. $(27x^{-3})^{-1/3}$

144. $(16a^{-2})^{-1/2}$

145. $x^{4/3}(x^{2/3} + 3x^{5/3} - 4)$

146. $(x^{1/3} + x^2)(x^{1/3} - x^2)$

147. $(x^{-1/2} - x^{1/2})^2$

148. $(a^{1/2} - b^{2/3})^2$

Use a calculator to evaluate each expression. Round to the nearest hundredth.

149. $15^{1/3}$

150. $50.5^{1/4}$

151. $1.045^{1/5}$

152. $(-1{,}000)^{2/5}$

Use a calculator to evaluate each expression. Round to the nearest hundredth.

153. $17^{-1/2}$

154. $2.45^{-2/3}$

155. $(-0.25)^{-1/5}$

156. $(-17.1)^{-3/7}$

WRITING ABOUT MATH

157. Explain how you would decide whether $a^{1/n}$ is a real number.

158. The expression $(a^{1/2} + b^{1/2})^2$ is not equal to $a + b$. Explain.

SOMETHING TO THINK ABOUT

159. The fraction $\frac{2}{4}$ is equal to $\frac{1}{2}$. Is $16^{2/4}$ equal to $16^{1/2}$? Explain.

160. How would you evaluate an expression with a mixed-number exponent? For example, what is $8^{1\frac{1}{3}}$? What is $25^{2\frac{1}{2}}$? Explain.

Section 9.4 Simplifying and Combining Radical Expressions

Objectives

1. Simplify a radical expression by applying the properties of radicals.
2. Add and subtract two or more radical expressions.
3. Find the length of a side of a 30°–60°–90° triangle and a 45°–45°–90° triangle.

Vocabulary

like (similar) radicals   altitude

Getting Ready

Simplify each radical. Assume that all variables represent positive numbers.

1. $\sqrt{225}$
2. $\sqrt{576}$
3. $\sqrt[3]{125}$
4. $\sqrt[3]{343}$
5. $\sqrt{16x^4}$
6. $\sqrt{\dfrac{64}{121}x^6}$
7. $\sqrt[3]{27a^3b^9}$
8. $\sqrt[3]{-8a^{12}}$

In this section, we will introduce the multiplication and division properties of radicals and use them to simplify radical expressions. Then we will add and subtract radical expressions and apply these techniques to find the lengths of the sides of special right triangles.

1 Simplify a radical expression by applying the properties of radicals.

Many properties of exponents have counterparts in radical notation. Because $a^{1/n}b^{1/n} = (ab)^{1/n}$, we have

$$(1) \quad \sqrt[n]{a}\sqrt[n]{b} = \sqrt[n]{ab}$$

For example,

$$\sqrt{5}\sqrt{5} = \sqrt{5 \cdot 5} = \sqrt{5^2} = 5$$
$$\sqrt[3]{7x}\sqrt[3]{49x^2} = \sqrt[3]{7x \cdot 7^2x^2} = \sqrt[3]{7^3 \cdot x^3} = 7x$$
$$\sqrt[4]{2x^3}\sqrt[4]{8x} = \sqrt[4]{2x^3 \cdot 2^3x} = \sqrt[4]{2^4 \cdot x^4} = 2x \quad (x > 0)$$

If we write Equation 1 in a different order, we have the following rule.

MULTIPLICATION PROPERTY OF RADICALS

If $\sqrt[n]{a}$ and $\sqrt[n]{b}$ are real numbers, then

$$\sqrt[n]{ab} = \sqrt[n]{a}\sqrt[n]{b}$$

If all radicals represent real numbers, *the nth root of the product of two numbers is equal to the product of their nth roots.*

COMMENT The multiplication property of radicals applies to the *n*th root of the product of two numbers. There is no such property for sums or differences. A radical symbol is a grouping symbol. Thus, any addition or subtraction within the radicand must be completed first.

A second property of radicals involves quotients. Because

$$\frac{a^{1/n}}{b^{1/n}} = \left(\frac{a}{b}\right)^{1/n} \quad (b \neq 0)$$

it follows that

$$(2) \quad \frac{\sqrt[n]{a}}{\sqrt[n]{b}} = \sqrt[n]{\frac{a}{b}} \quad (b \neq 0)$$

For example,

$$\frac{\sqrt{8x^3}}{\sqrt{2x}} = \sqrt{\frac{8x^3}{2x}} = \sqrt{4x^2} = 2x \quad (x > 0)$$
$$\frac{\sqrt[3]{54x^5}}{\sqrt[3]{2x^2}} = \sqrt[3]{\frac{54x^5}{2x^2}} = \sqrt[3]{27x^3} = 3x \quad (x \neq 0)$$

If we write Equation 2 in a different order, we have the following rule.

Carl Friedrich Gauss
1777–1855
Many people consider Gauss to be the greatest mathematician of all time. He made contributions in the areas of number theory, solutions of equations, geometry of curved surfaces, and statistics. For his efforts, he earned the title "Prince of the Mathematicians."

DIVISION PROPERTY OF RADICALS

If $\sqrt[n]{a}$ and $\sqrt[n]{b}$ are real numbers, then

$$\sqrt[n]{\frac{a}{b}} = \frac{\sqrt[n]{a}}{\sqrt[n]{b}} \quad (b \neq 0)$$

Unless otherwise noted, all content on this page is © Cengage Learning.

As long as all radicals represent real numbers, *the nth root of the quotient of two numbers is equal to the quotient of their nth roots.*

A radical expression is said to be in simplified form when each of the following statements is true.

SIMPLIFIED FORM OF A RADICAL EXPRESSION

A radical expression is in simplified form when

1. Each numerical factor in the radicand, in prime factored form, appears to a power that is less than the index of the radical.
2. Each variable factor in the radicand appears to a power that is less than the index of the radical.
3. The radicand contains no fractions or negative numbers.
4. No radicals appear in the denominator of a fraction.

EXAMPLE 1 Simplify: **a.** $\sqrt{12}$ **b.** $\sqrt{98}$ **c.** $\sqrt[3]{54}$

COMMENT Use a factoring tree.

a.

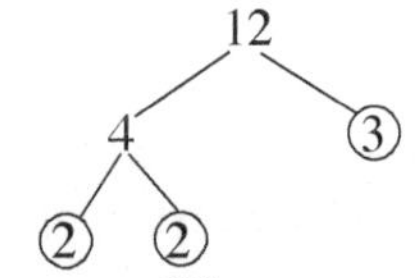

b.

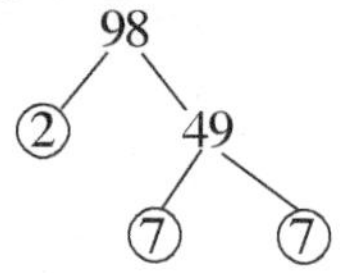

In parts **a** and **b**, each factor pair produces a single number outside the radical.

c.

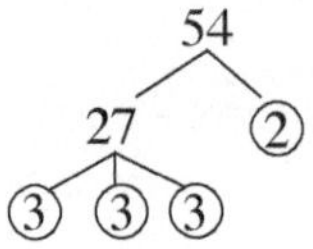

In part **c**, each factor triple produces a single number outside the radical.

Solution **a.** Recall that squares of integers, such as 1, 4, 9, 16, 25, and 36, are *perfect squares.* To simplify $\sqrt{12}$, we factor 12 so that one factor is the largest perfect square that divides 12. Since 4 is the largest perfect-square factor of 12, we write 12 as $4 \cdot 3$, use the multiplication property of radicals, and simplify.

$$\begin{aligned}\sqrt{12} &= \sqrt{\mathbf{4} \cdot \mathbf{3}} && \text{Write 12 as } 4 \cdot 3.\\ &= \sqrt{4}\sqrt{3} && \sqrt{4 \cdot 3} = \sqrt{4}\sqrt{3}\\ &= 2\sqrt{3} && \sqrt{4} = 2\end{aligned}$$

b. Since the largest perfect-square factor of 98 is 49, we have

$$\begin{aligned}\sqrt{98} &= \sqrt{\mathbf{49} \cdot \mathbf{2}} && \text{Write 98 as } 49 \cdot 2.\\ &= \sqrt{49}\sqrt{2} && \sqrt{49 \cdot 2} = \sqrt{49}\sqrt{2}\\ &= 7\sqrt{2} && \sqrt{49} = 7\end{aligned}$$

c. Numbers that are cubes of integers, such as 1, 8, 27, 64, 125, and 216, are called *perfect cubes.* Since the largest perfect-cube factor of 54 is 27, we have

$$\begin{aligned}\sqrt[3]{54} &= \sqrt[3]{\mathbf{27} \cdot \mathbf{2}} && \text{Write 54 as } 27 \cdot 2.\\ &= \sqrt[3]{27}\sqrt[3]{2} && \sqrt[3]{27 \cdot 2} = \sqrt[3]{27}\sqrt[3]{2}\\ &= 3\sqrt[3]{2} && \sqrt[3]{27} = 3\end{aligned}$$

SELF CHECK 1 Simplify: **a.** $\sqrt{20}$ **b.** $\sqrt[3]{24}$

EXAMPLE 2 Simplify: **a.** $\sqrt{\dfrac{15}{49x^2}}$ $(x > 0)$ **b.** $\sqrt[3]{\dfrac{10x^2}{27y^6}}$ $(y \neq 0)$

Solution **a.** We can write the square root of the quotient as the quotient of the square roots and simplify the denominator. Since $x > 0$, we have

$$\begin{aligned}\sqrt{\frac{15}{49x^2}} &= \frac{\sqrt{15}}{\sqrt{49x^2}}\\ &= \frac{\sqrt{15}}{7x}\end{aligned}$$

b. We can write the cube root of the quotient as the quotient of two cube roots. Since $y \neq 0$, we have

$$\sqrt[3]{\frac{10x^2}{27y^6}} = \frac{\sqrt[3]{10x^2}}{\sqrt[3]{27y^6}}$$
$$= \frac{\sqrt[3]{10x^2}}{3y^2}$$

SELF CHECK 2 Simplify: **a.** $\sqrt{\frac{11}{36a^2}}$ $(a > 0)$ **b.** $\sqrt[3]{\frac{8a^2}{125y^3}}$ $(y \neq 0)$

EXAMPLE 3 Simplify each expression. Assume that all variables represent positive numbers.

a. $\sqrt{128a^5}$ **b.** $\sqrt[3]{24x^5}$ **c.** $\frac{\sqrt{45xy^2}}{\sqrt{5x}}$ **d.** $\frac{\sqrt[3]{-432x^5}}{\sqrt[3]{8x}}$

Solution **a.** We can write $128a^5$ as $64a^4 \cdot 2a$ and use the multiplication property of radicals.

$\sqrt{128a^5} = \sqrt{64a^4 \cdot 2a}$ $\quad$ $64a^4$ is the largest perfect square that divides $128a^5$.
$= \sqrt{64a^4}\sqrt{2a}$ $\quad$ Use the multiplication property of radicals.
$= 8a^2\sqrt{2a}$ $\quad$ $\sqrt{64a^4} = 8a^2$

COMMENT Be sure to include the index of radicals in each step to avoid errors.

b. We can write $24x^5$ as $8x^3 \cdot 3x^2$ and use the multiplication property of radicals.

$\sqrt[3]{24x^5} = \sqrt[3]{8x^3 \cdot 3x^2}$ $\quad$ $8x^3$ is the largest perfect cube that divides $24x^5$.
$= \sqrt[3]{8x^3}\sqrt[3]{3x^2}$ $\quad$ Use the multiplication property of radicals.
$= 2x\sqrt[3]{3x^2}$ $\quad$ $\sqrt[3]{8x^3} = 2x$

c. We can write the quotient of the square roots as the square root of a quotient.

$\frac{\sqrt{45xy^2}}{\sqrt{5x}} = \sqrt{\frac{45xy^2}{5x}}$ $\quad$ Use the quotient property of radicals.
$= \sqrt{9y^2}$ $\quad$ Simplify the fraction.
$= 3y$

d. We can write the quotient of the cube roots as the cube root of a quotient.

$\frac{\sqrt[3]{-432x^5}}{\sqrt[3]{8x}} = \sqrt[3]{\frac{-432x^5}{8x}}$ $\quad$ Use the quotient property of radicals.
$= \sqrt[3]{-54x^4}$ $\quad$ Simplify the fraction.
$= \sqrt[3]{-27x^3 \cdot 2x}$ $\quad$ $-27x^3$ is the largest perfect cube that divides $-54x^4$.
$= \sqrt[3]{-27x^3}\sqrt[3]{2x}$ $\quad$ Use the multiplication property of radicals.
$= -3x\sqrt[3]{2x}$

SELF CHECK 3 Simplify each expression. Assume that all variables represent positive numbers.

a. $\sqrt{98b^3}$ **b.** $\sqrt[3]{54y^5}$ **c.** $\frac{\sqrt{50ab^2}}{\sqrt{2a}}$

To simplify more complicated radicals, we can use the prime factorization of the radicand to find its perfect-square factors. For example, to simplify $\sqrt{3{,}168x^5y^7}$, we first find the prime factorization of $3{,}168x^5y^7$.

$$3{,}168x^5y^7 = 2^5 \cdot 3^2 \cdot 11 \cdot x^5 \cdot y^7$$

Then we have

$$\begin{aligned}\sqrt{3{,}168x^5y^7} &= \sqrt{2^4\cdot 3^2\cdot x^4\cdot y^6\cdot 2\cdot 11\cdot x\cdot y}\\ &= \sqrt{2^4\cdot 3^2\cdot x^4\cdot y^6}\sqrt{2\cdot 11\cdot x\cdot y} \quad \text{Write each perfect square under the left radical and each nonperfect square under the right radical.}\\ &= 2^2\cdot 3x^2y^3\sqrt{22xy}\\ &= 12x^2y^3\sqrt{22xy}\end{aligned}$$

2 Add and subtract two or more radical expressions.

Radical expressions with the same index and the same radicand are called **like** or **similar radicals**. For example, $3\sqrt{2}$ and $5\sqrt{2}$ are like radicals. However,

$3\sqrt{5}$ and $5\sqrt{2}$ are not like radicals, because the radicands are different.

$3\sqrt{5}$ and $2\sqrt[3]{5}$ are not like radicals, because the indexes are different.

We often can combine like terms. For example, to simplify the expression $3\sqrt{2} + 2\sqrt{2}$, we use the distributive property to factor out $\sqrt{2}$ and simplify.

$$\begin{aligned}3\sqrt{2} + 2\sqrt{2} &= (3 + 2)\sqrt{2}\\ &= 5\sqrt{2}\end{aligned}$$

Radicals with the same index but different radicands can often be written as like radicals. For example, to simplify the expression $\sqrt{27} - \sqrt{12}$, we simplify both radicals and combine the resulting like radicals.

$$\begin{aligned}\sqrt{27} - \sqrt{12} &= \sqrt{9\cdot 3} - \sqrt{4\cdot 3}\\ &= \sqrt{9}\sqrt{3} - \sqrt{4}\sqrt{3} \quad \text{Apply the multiplication property of radicals.}\\ &= 3\sqrt{3} - 2\sqrt{3} \quad \text{Simplify.}\\ &= (3 - 2)\sqrt{3} \quad \text{Factor out } \sqrt{3}\text{, the GCF.}\\ &= \sqrt{3}\end{aligned}$$

As the previous examples suggest, we can use the following procedure to add or subtract radicals.

ADDING AND SUBTRACTING RADICALS

To add or subtract radicals, simplify each radical and combine all like radicals by adding the coefficients and keeping the common radical.

EXAMPLE 4 Simplify: $2\sqrt{12} - 3\sqrt{48} + 3\sqrt{3}$

Solution We simplify each radical separately and combine like radicals.

$$\begin{aligned}2\sqrt{12} - 3\sqrt{48} + 3\sqrt{3} &= 2\sqrt{4\cdot 3} - 3\sqrt{16\cdot 3} + 3\sqrt{3}\\ &= 2\sqrt{4}\sqrt{3} - 3\sqrt{16}\sqrt{3} + 3\sqrt{3}\\ &= 2(2)\sqrt{3} - 3(4)\sqrt{3} + 3\sqrt{3}\\ &= 4\sqrt{3} - 12\sqrt{3} + 3\sqrt{3}\\ &= (4 - 12 + 3)\sqrt{3}\\ &= -5\sqrt{3}\end{aligned}$$

SELF CHECK 4 Simplify: $3\sqrt{75} - 2\sqrt{12} + 2\sqrt{48}$

EXAMPLE 5 Simplify: $\sqrt[3]{16} - \sqrt[3]{54} + \sqrt[3]{24}$

Solution We simplify each radical separately and combine like radicals.

$$\begin{aligned}\sqrt[3]{16} - \sqrt[3]{54} + \sqrt[3]{24} &= \sqrt[3]{8 \cdot 2} - \sqrt[3]{27 \cdot 2} + \sqrt[3]{8 \cdot 3}\\ &= \sqrt[3]{8}\sqrt[3]{2} - \sqrt[3]{27}\sqrt[3]{2} + \sqrt[3]{8}\sqrt[3]{3}\\ &= 2\sqrt[3]{2} - 3\sqrt[3]{2} + 2\sqrt[3]{3}\\ &= -\sqrt[3]{2} + 2\sqrt[3]{3}\end{aligned}$$

COMMENT We cannot combine $-\sqrt[3]{2}$ and $2\sqrt[3]{3}$, because the radicals have different radicands.

SELF CHECK 5 Simplify: $\sqrt[3]{24} - \sqrt[3]{16} + \sqrt[3]{54}$

Remember to include the index of radicals in each step.

EXAMPLE 6 Simplify: $\sqrt[3]{16x^4} + \sqrt[3]{54x^4} - \sqrt[3]{-128x^4}$

Solution We simplify each radical separately, factor out $\sqrt[3]{2x}$, and combine like radicals.

$$\begin{aligned}&\sqrt[3]{16x^4} + \sqrt[3]{54x^4} - \sqrt[3]{-128x^4}\\ &= \sqrt[3]{8x^3 \cdot 2x} + \sqrt[3]{27x^3 \cdot 2x} - \sqrt[3]{-64x^3 \cdot 2x}\\ &= \sqrt[3]{8x^3}\sqrt[3]{2x} + \sqrt[3]{27x^3}\sqrt[3]{2x} - \sqrt[3]{-64x^3}\sqrt[3]{2x}\\ &= 2x\sqrt[3]{2x} + 3x\sqrt[3]{2x} + 4x\sqrt[3]{2x}\\ &= (2x + 3x + 4x)\sqrt[3]{2x}\\ &= 9x\sqrt[3]{2x}\end{aligned}$$

SELF CHECK 6 Simplify: $\sqrt{32x^3} + \sqrt{50x^3} - \sqrt{18x^3}$ $(x > 0)$

3 Find the length of a side of a 30°–60°–90° triangle and a 45°–45°–90° triangle.

An isosceles right triangle is a right triangle with two legs of equal length. If we know the length of one leg of an isosceles right triangle, we can use the Pythagorean theorem to find the length of the hypotenuse. Since the triangle shown in Figure 9-10 is a right triangle, we have

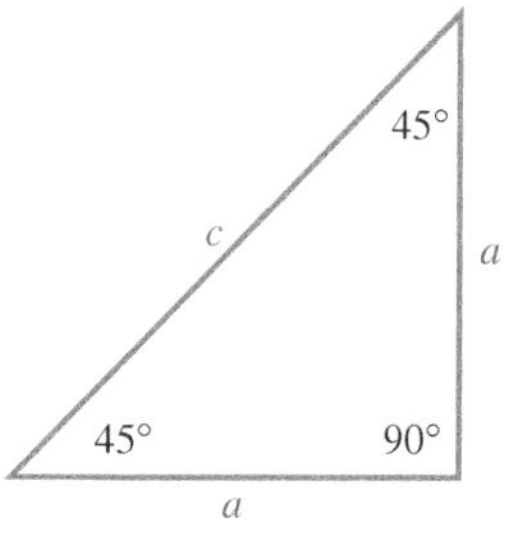

Figure 9-10

$c^2 = a^2 + a^2$	Use the Pythagorean theorem.
$c^2 = 2a^2$	Combine like terms.
$c = \sqrt{2a^2}$	Since c represents a length, take the positive square root.
$c = a\sqrt{2}$	$\sqrt{2a^2} = \sqrt{2}\sqrt{a^2} = \sqrt{2}a = a\sqrt{2}$. No absolute value symbols are needed, because a is positive.

Thus, *in an isosceles right triangle, the length of the hypotenuse is the length of one leg times* $\sqrt{2}$.

EXAMPLE 7 **GEOMETRY** If one leg of the isosceles right triangle shown in Figure 9-10 is 10 feet long, find the length of the hypotenuse.

Solution Since the length of the hypotenuse is the length of a leg times $\sqrt{2}$, we have

$$c = 10\sqrt{2}$$

The length of the hypotenuse is $10\sqrt{2}$ feet. To two decimal places, the length is 14.14 feet.

SELF CHECK 7 Find the length of the hypotenuse of an isosceles right triangle if one leg is 12 meters long.

Unless otherwise noted, all content on this page is © Cengage Learning.

If the length of the hypotenuse of an isosceles right triangle is known, we can use the Pythagorean theorem to find the length of each leg.

EXAMPLE 8 **GEOMETRY** Approximate the length of each leg of the isosceles right triangle shown in Figure 9-11 to two decimal places.

Solution We use the Pythagorean theorem.

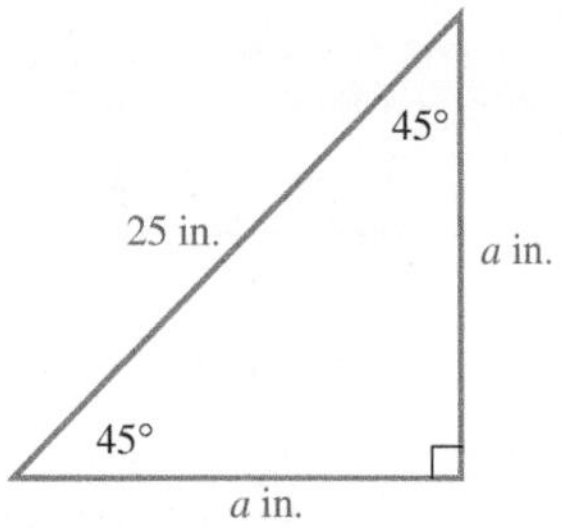

Figure 9-11

$$c^2 = a^2 + a^2$$

$$25^2 = 2a^2 \quad \text{Substitute 25 for } c \text{ and combine like terms.}$$

$$\frac{625}{2} = a^2 \quad \text{Square 25 and divide both sides by 2.}$$

$$\sqrt{\frac{625}{2}} = a \quad \text{Since } a \text{ represents a length, take the positive square root.}$$

$$a \approx 17.67766953 \quad \text{Use a calculator.}$$

To two decimal places, the length is 17.68 inches.

SELF CHECK 8 If the length of the hypotenuse of an isosceles right triangle is 48 inches, approximate the length of each leg to two decimal places.

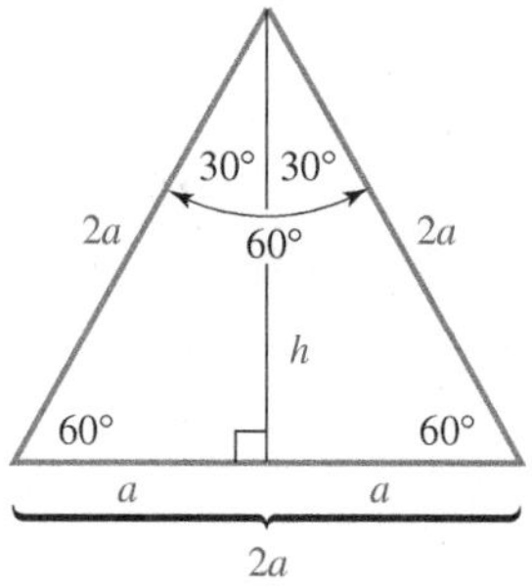

Figure 9-12

COMMENT In summary, the lengths of the sides are as follows:

$30°-60°-90°$; $s, s\sqrt{3}, 2s$

$45°-45°-90°$; $s, s, s\sqrt{2}$

From geometry, we know that an *equilateral triangle* is a triangle with three sides of equal length and three 60° angles. An **altitude** is the perpendicular distance from a vertex of a triangle to the opposite side. If an altitude is drawn upon the base of an equilateral triangle, as shown in Figure 9-12, it bisects the base and divides the triangle into two 30°–60°–90° triangles. We can see that the shortest leg of each 30°–60°–90° triangle is a units long. Recall that in any triangle, the side opposite the smallest angle is the shortest side and the side opposite the largest angle is the longest. Thus,

The length of the shorter leg of a 30°–60°–90° triangle is half as long as its hypotenuse.

We can find the length of the altitude, h, by using the Pythagorean theorem.

$$a^2 + h^2 = (2a)^2$$

$$a^2 + h^2 = 4a^2 \quad (2a)^2 = (2a)(2a) = 4a^2$$

$$h^2 = 3a^2 \quad \text{Subtract } a^2 \text{ from both sides.}$$

$$h = \sqrt{3a^2} \quad \text{Since } h \text{ represents a length, take the positive square root.}$$

$$h = a\sqrt{3} \quad \sqrt{3a^2} = \sqrt{3}\sqrt{a^2} = a\sqrt{3}. \text{ No absolute value symbols are needed, because } a \text{ is positive.}$$

Thus,

The length of the longer leg is the length of the shorter leg times $\sqrt{3}$ and the length of the hypotenuse is twice the length of the shorter leg.

EXAMPLE 9 **GEOMETRY** Find the length of the hypotenuse and the longer leg of the right triangle shown in Figure 9-13.

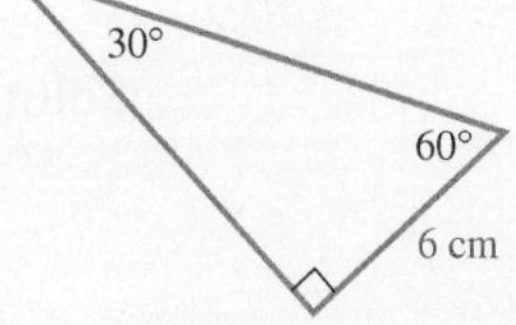

Figure 9-13

Solution The known length of 6 cm is opposite the 30° angle. Since the shorter leg of the triangle is half as long as its hypotenuse, the hypotenuse is 12 centimeters long.

Since the length of the longer leg is the length of the shorter leg times $\sqrt{3}$, the longer leg is $6\sqrt{3}$ (about 10.39) centimeters long.

Unless otherwise noted, all content on this page is © Cengage Learning.

SELF CHECK 9 Find the length of the hypotenuse and the longer leg of a 30°–60°–90° right triangle if the shorter leg is 8 centimeters long.

EXAMPLE 10 **GEOMETRY** Find the length of each leg of the triangle shown in Figure 9-14.

Solution Since the shorter leg of a 30°–60°–90° triangle is half as long as its hypotenuse, the shorter leg is $\frac{9}{2}$ centimeters long.

Since the length of the longer leg is the length of the shorter leg times $\sqrt{3}$, the longer leg is $\frac{9}{2}\sqrt{3}$ (or about 7.79) centimeters long.

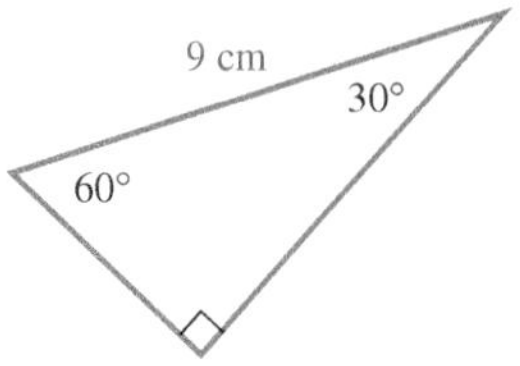

Figure 9-14

SELF CHECK 10 Find the length of the longer leg of a right triangle if its hypotenuse is 10 cm long.

SELF CHECK ANSWERS

1. a. $2\sqrt{5}$ **b.** $2\sqrt[3]{3}$ **2. a.** $\frac{\sqrt{11}}{6a}$ **b.** $\frac{2\sqrt[3]{a^2}}{5y}$ **3. a.** $7b\sqrt{2b}$ **b.** $3y\sqrt[3]{2y^2}$ **c.** $5b$ **4.** $19\sqrt{3}$ **5.** $2\sqrt[3]{3} + \sqrt[3]{2}$ **6.** $6x\sqrt{2x}$ **7.** $12\sqrt{2}$ m **8.** To two decimal places, the length is 33.94 inches. **9.** 16 cm, $8\sqrt{3}$ cm **10.** $5\sqrt{3}$ cm

NOW TRY THIS

Simplify.

1. $\sqrt[3]{8x - 8} + \sqrt[3]{x - 1}$

2. $\frac{5\sqrt[3]{32}}{2}$

9.4 Exercises

WARM-UPS *Factor such that one factor is the largest perfect square.*

1. 28

2. 24

3. 18

4. 50

Factor such that one factor is the largest perfect cube.

5. 16

6. 40

7. 54

8. 108

Combine like terms.

9. $12x^2 + 12x^2$

10. $20y^3 - 17y^3$

REVIEW *Perform each operation.*

11. $7x^3y(-3x^4y^{-5})$

12. $-2a^2b^{-2}(4a^{-2}b^4 - 2a^2b + 3a^3b^2)$

13. $(3t + 2)^2$

14. $(5r - 3s)(5r + 2s)$

15. $2p - 5\overline{)6p^2 - 7p - 25}$

16. $3m + n\overline{)6m^3 - m^2n + 2mn^2 + n^3}$

VOCABULARY AND CONCEPTS *Fill in the blanks. Assume a and b are positive numbers and n is a natural number greater than 1.*

17. $\sqrt[n]{ab} =$ ______

18. $\sqrt[n]{\frac{a}{b}} =$ ______

Unless otherwise noted, all content on this page is © Cengage Learning.

19. If two radicals have the same index and the same radicand, they are called ___ radicals.

20. The perpendicular distance from a vertex of a triangle to the opposite side is called an ______.

GUIDED PRACTICE *Simplify each expression. Assume that all variables represent positive numbers.* (OBJECTIVE 1)

21. $\sqrt{6}\sqrt{6}$

22. $\sqrt{11}\sqrt{11}$

23. $\sqrt{t}\sqrt{t}$

24. $-\sqrt{z}\sqrt{z}$

25. $\sqrt[3]{5x^2}\sqrt[3]{25x}$

26. $\sqrt[4]{25a}\sqrt[4]{25a^3}$

27. $\dfrac{\sqrt{500}}{\sqrt{5}}$

28. $\dfrac{\sqrt{128}}{\sqrt{2}}$

29. $\dfrac{\sqrt{98x^3}}{\sqrt{2x}}$

30. $\dfrac{\sqrt{125a^5}}{\sqrt{5a}}$

31. $\dfrac{\sqrt{180ab^4}}{\sqrt{5ab^2}}$

32. $\dfrac{\sqrt{112ab^3}}{\sqrt{7ab}}$

33. $\dfrac{\sqrt[3]{48}}{\sqrt[3]{6}}$

34. $\dfrac{\sqrt[3]{64}}{\sqrt[3]{8}}$

35. $\dfrac{\sqrt[3]{189a^4}}{\sqrt[3]{7a}}$

36. $\dfrac{\sqrt[3]{243x^7}}{\sqrt[3]{9x}}$

Simplify. SEE EXAMPLE 1. (OBJECTIVE 1)

37. $\sqrt{20}$

38. $\sqrt{8}$

39. $-\sqrt{200}$

40. $-\sqrt{250}$

41. $\sqrt[3]{80}$

42. $\sqrt[3]{270}$

43. $\sqrt[3]{-81}$

44. $\sqrt[3]{-72}$

45. $\sqrt[4]{32}$

46. $\sqrt[4]{48}$

47. $\sqrt[5]{96}$

48. $\sqrt[7]{256}$

Simplify. Assume no denominators are 0. SEE EXAMPLE 2. (OBJECTIVE 1)

49. $\sqrt{\dfrac{5}{49a^2}}$

50. $\sqrt{\dfrac{11}{36x^2}}$

51. $\sqrt[3]{\dfrac{7a^3}{64}}$

52. $\sqrt[3]{\dfrac{4b^3}{125}}$

53. $\sqrt[4]{\dfrac{3p^4}{10{,}000q^4}}$

54. $\sqrt[5]{\dfrac{4r^5}{243s^{10}}}$

55. $\sqrt[5]{\dfrac{3m^{15}}{32n^{10}}}$

56. $\sqrt[6]{\dfrac{5a^6}{64b^{12}}}$

Simplify. Assume that all variables represent positive numbers. SEE EXAMPLE 3. (OBJECTIVE 1)

57. $\sqrt{63y^3}$

58. $\sqrt{125x^3}$

59. $\sqrt{48a^5}$

60. $\sqrt{160b^5}$

61. $-\sqrt{112a^3}$

62. $\sqrt{147a^5}$

63. $\sqrt{175a^2b^3}$

64. $\sqrt{128a^3b^5}$

65. $\dfrac{-\sqrt{150ab^4}}{\sqrt{2a}}$

66. $\dfrac{\sqrt{360xy^3}}{\sqrt{3xy}}$

67. $\dfrac{\sqrt[3]{-108x^6y}}{2y^3}$

68. $\dfrac{-\sqrt[3]{-243a^4b}}{3ab^3}$

69. $\sqrt[3]{16x^{12}y^3}$

70. $\sqrt[3]{40a^3b^6}$

71. $\sqrt{\dfrac{z^2}{16x^2}}$

72. $\sqrt{\dfrac{b^4}{64a^8}}$

Simplify. SEE EXAMPLE 4. (OBJECTIVE 2)

73. $\sqrt{3} + \sqrt{27}$

74. $\sqrt{8} + \sqrt{32}$

75. $\sqrt{2} - \sqrt{8}$

76. $\sqrt{20} - \sqrt{125}$

77. $\sqrt{98} - \sqrt{50}$

78. $\sqrt{72} - \sqrt{200}$

79. $3\sqrt{125} + 4\sqrt{45}$

80. $2\sqrt{72} + 5\sqrt{32}$

81. $3\sqrt{54} + 2\sqrt{28} - 4\sqrt{96}$

82. $4\sqrt{300} - 2\sqrt{108} + 3\sqrt{200}$

Simplify. SEE EXAMPLE 5. (OBJECTIVE 2)

83. $\sqrt[3]{24} + \sqrt[3]{81}$

84. $\sqrt[3]{16} + \sqrt[3]{128}$

85. $\sqrt[3]{32} - \sqrt[3]{108}$

86. $\sqrt[3]{80} - \sqrt[3]{10{,}000}$

87. $2\sqrt[3]{125} - 5\sqrt[3]{64}$

88. $\sqrt[3]{81} - \sqrt[3]{24}$

89. $2\sqrt[3]{16} - \sqrt[3]{54} - 3\sqrt[3]{128}$

90. $\sqrt[3]{250} - 4\sqrt[3]{5} + \sqrt[3]{16}$

Simplify. All variables represent positive numbers. SEE EXAMPLE 6. (OBJECTIVE 2)

91. $\sqrt[3]{81x^5} - \sqrt[3]{24x^5}$

92. $\sqrt[3]{16x^4} - \sqrt[3]{54x^4}$

93. $\sqrt{25yz^2} + \sqrt{9yz^2}$

94. $\sqrt{36xy^2} + \sqrt{49xy^2}$

95. $\sqrt{y^5} - \sqrt{9y^5} - \sqrt{25y^5}$

96. $\sqrt{8y^7} + \sqrt{32y^7} - \sqrt{2y^7}$

97. $3\sqrt[3]{-2x^4} - \sqrt[3]{54x^4}$

98. $2\sqrt[3]{320a^5} - 2\sqrt[3]{-135a^5}$

Use the figure for Exercises 99–106. Find the lengths of the remaining sides of the triangle. SEE EXAMPLE 7. (OBJECTIVE 3)

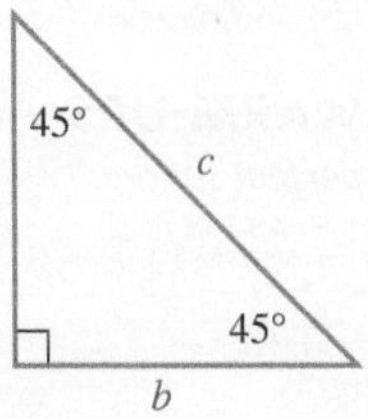

99. $b = \dfrac{2}{3}$

100. $b = \dfrac{3}{10}$

101. $a = 5\sqrt{2}$

102. $a = 12\sqrt{2}$

Find the lengths of the remaining sides of the triangle. SEE EXAMPLE 8. (OBJECTIVE 3)

103. $c = 5\sqrt{2}$

104. $c = 8\sqrt{2}$

105. $c = 7\sqrt{2}$

106. $c = 16\sqrt{2}$

Unless otherwise noted, all content on this page is © Cengage Learning.

Use the figure for Exercises 107–114. Find the lengths of the remaining sides of the triangle. SEE EXAMPLE 9. (OBJECTIVE 3)

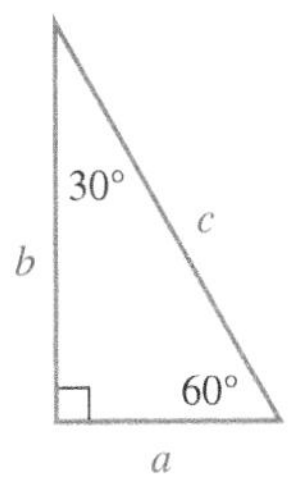

107. $a = 5$

108. $a = 8$

109. $b = 9\sqrt{3}$

110. $b = 18\sqrt{3}$

Find the lengths of the remaining sides of the triangle. SEE EXAMPLE 10. (OBJECTIVE 3)

111. $c = 24$

112. $c = 8$

113. $c = 15$

114. $c = 25$

ADDITIONAL PRACTICE *Simplify. Assume that all variables represent positive numbers.*

115. $4\sqrt{2x} + 6\sqrt{2x}$

116. $\sqrt{25y^2z} - \sqrt{16y^2z}$

117. $\sqrt[4]{32x^{12}y^4}$

118. $\sqrt[5]{64x^{10}y^5}$

119. $\sqrt[4]{\dfrac{5x}{16z^4}}$

120. $\sqrt[3]{\dfrac{11a^2}{125b^6}}$

121. $\sqrt{98} - \sqrt{50} - \sqrt{72}$

122. $\sqrt{20} + \sqrt{125} - \sqrt{80}$

123. $3\sqrt[3]{27} + 12\sqrt[3]{216}$

124. $14\sqrt[4]{32} - 15\sqrt[4]{162}$

125. $23\sqrt[4]{768} + \sqrt[4]{48}$

126. $3\sqrt[4]{512} + 2\sqrt[4]{32}$

127. $4\sqrt[4]{243} - \sqrt[4]{48}$

128. $\sqrt[4]{48} - \sqrt[4]{243} - \sqrt[4]{768}$

129. $6\sqrt[3]{5y} + 3\sqrt[3]{5y}$

130. $8\sqrt[5]{7a^2} - 7\sqrt[5]{7a^2}$

131. $10\sqrt[6]{12xyz} - \sqrt[6]{12xyz}$

132. $3\sqrt[4]{x^4y} - 2\sqrt[4]{x^4y}$

133. $\sqrt[5]{x^6y^2} + \sqrt[5]{32x^6y^2} + \sqrt[5]{x^6y^2}$

134. $\sqrt[3]{xy^4} + \sqrt[3]{8xy^4} - \sqrt[3]{27xy^4}$

135. $\sqrt{x^2 + 2x + 1} + \sqrt{x^2 + 2x + 1}$

136. $\sqrt{4x^2 + 12x + 9} + \sqrt{9x^2 + 6x + 1}$

Find the lengths of the remaining sides in each triangle. Approximate each answer to two decimal places.

137.

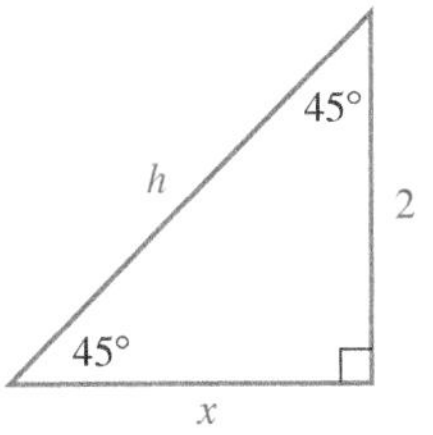

138.

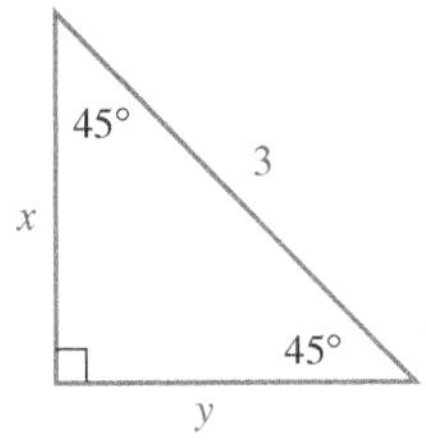

139.

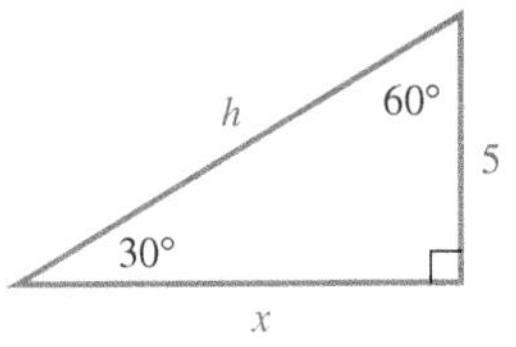

140.

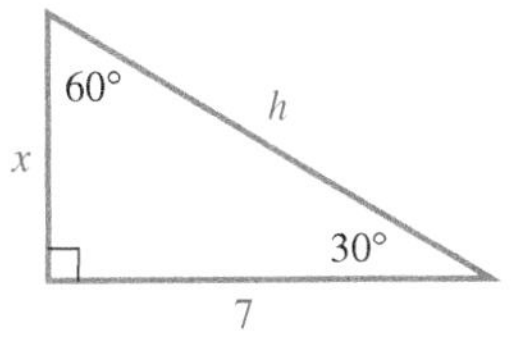

141.

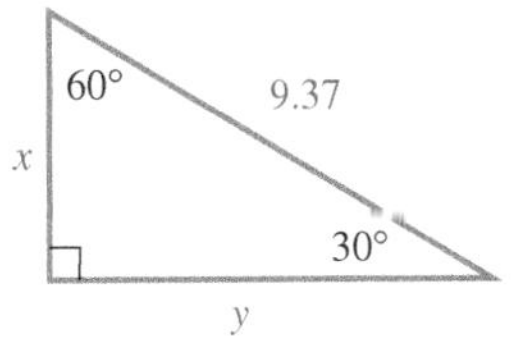

142.

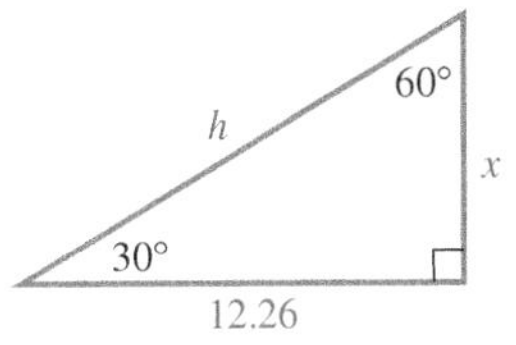

143.

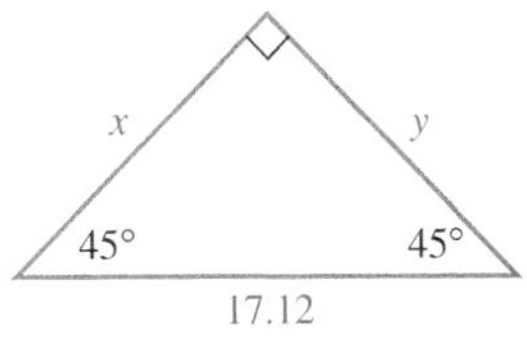

144.

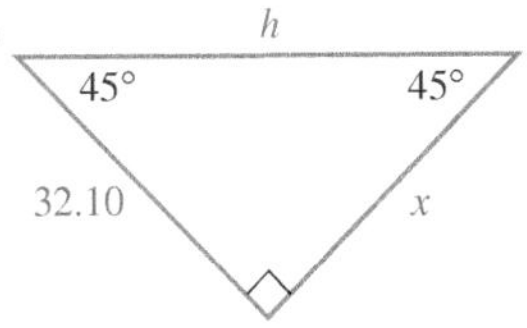

APPLICATIONS *Find the exact answer and then approximate to the nearest hundredth.*

145. Hardware The sides of a regular hexagonal nut are 10 millimeters long. Find the height h in millimeters of the nut.

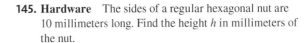

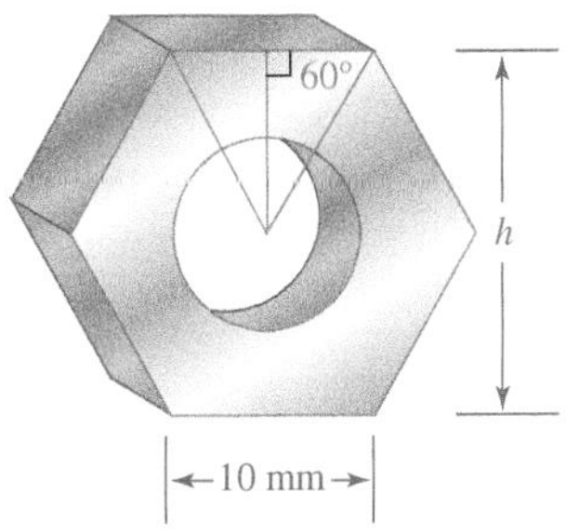

Unless otherwise noted, all content on this page is © Cengage Learning.

146. Ironing boards Find the height h, in inches, of the ironing board shown in the illustration.

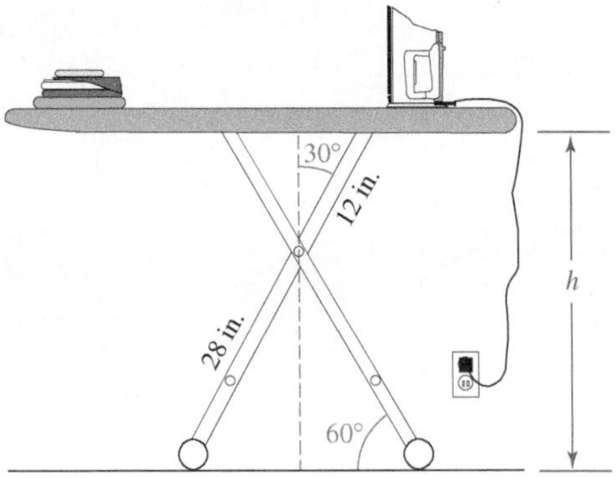

WRITING ABOUT MATH

147. Explain how to recognize like radicals.

148. Explain how to combine like radicals.

SOMETHING TO THINK ABOUT

149. Find the sum.

$$\sqrt{3} + \sqrt{3^2} + \sqrt{3^3} + \sqrt{3^4} + \sqrt{3^5}$$

150. Can you find any numbers a and b such that

$$\sqrt{a + b} = \sqrt{a} + \sqrt{b}$$

Section 9.5 Multiplying Radical Expressions and Rationalizing

Objectives

1. Multiply two radical expressions.
2. Rationalize the denominator of a fraction that contains a radical expression.
3. Rationalize the numerator of a fraction that contains a radical expression.
4. Solve an application containing a radical expression.

Vocabulary

rationalize denominators   conjugates   rationalize numerators

Getting Ready

Simplify.

1. a^3a^4 **2.** $\frac{b^5}{b^2}$ **3.** $a(a - 2)$ **4.** $3b^2(2b + 3)$

5. $(a + 2)(a - 5)$ **6.** $(2a + 3b)(2a - 3b)$

We now learn how to multiply radical expressions and rationalize the denominators or numerators of radical expressions. Then we will use these skills to solve applications.

1 Multiply two radical expressions.

Radical expressions with the same index can be multiplied and divided.

Unless otherwise noted, all content on this page is © Cengage Learning.

EXAMPLE 1 Multiply: **a.** $(3\sqrt{6})(2\sqrt{3})$ **b.** $(\sqrt[3]{3a})(\sqrt[3]{9a^4})$

Solution We use the commutative and associative properties of multiplication to multiply the coefficients and the radicals separately. Then we simplify any radicals in the product, if possible.

a.
$$\begin{aligned}(3\sqrt{6})(2\sqrt{3}) &= 3(2)\sqrt{6}\sqrt{3} && \text{Multiply the coefficients and multiply the radicals.}\\ &= 6\sqrt{18} && 3(2) = 6 \text{ and } \sqrt{6}\sqrt{3} = \sqrt{18}\\ &= 6\sqrt{9}\sqrt{2} && \sqrt{18} = \sqrt{9\cdot 2} = \sqrt{9}\sqrt{2}\\ &= 6(3)\sqrt{2} && \text{Simplify.}\\ &= 18\sqrt{2}\end{aligned}$$

b.
$$\begin{aligned}(\sqrt[3]{3a})(\sqrt[3]{9a^4}) &= \sqrt[3]{27a^5} && \text{Multiply the radicals.}\\ &= \sqrt[3]{27a^3\cdot a^2} && \text{Factor } 27a^5.\\ &= \sqrt[3]{27a^3}\sqrt[3]{a^2} && \sqrt[3]{ab} = \sqrt[3]{a}\sqrt[3]{b}\\ &= 3a\sqrt[3]{a^2} && \sqrt[3]{27a^3} = 3a\end{aligned}$$

SELF CHECK 1 Multiply: **a.** $(-2\sqrt{8})(5\sqrt{6})$ **b.** $(\sqrt[3]{4a^2})(\sqrt[3]{6a^2})$

To multiply a radical expression with two or more terms by a radical expression, we use the distributive property to remove parentheses and then simplify each resulting term, if possible.

EXAMPLE 2 Multiply: $3\sqrt{3}(4\sqrt{8} - 5\sqrt{10})$

Solution
$$\begin{aligned}&3\sqrt{3}(4\sqrt{8} - 5\sqrt{10})\\ &= 3\sqrt{3}\cdot 4\sqrt{8} - 3\sqrt{3}\cdot 5\sqrt{10} && \text{Use the distributive property to remove parentheses.}\\ &= 12\sqrt{24} - 15\sqrt{30} && \text{Multiply the coefficients and multiply the radicals.}\\ &= 12\sqrt{4}\sqrt{6} - 15\sqrt{30} && \text{Use the multiplication property of radicals.}\\ &= 12(2)\sqrt{6} - 15\sqrt{30} && \text{Simplify.}\\ &= 24\sqrt{6} - 15\sqrt{30}\end{aligned}$$

SELF CHECK 2 Multiply: $4\sqrt{2}(3\sqrt{5} - 2\sqrt{8})$

To multiply two radical expressions, each with two or more terms, we use the distributive property as we did when we multiplied two polynomials. Then we simplify each resulting term, if possible.

EXAMPLE 3 Multiply: $(\sqrt{7} + \sqrt{2})(\sqrt{7} - 3\sqrt{2})$

Solution
$$\begin{aligned}&(\sqrt{7} + \sqrt{2})(\sqrt{7} - 3\sqrt{2})\\ &= \sqrt{7}\sqrt{7} - 3\sqrt{7}\sqrt{2} + \sqrt{2}\sqrt{7} - 3\sqrt{2}\sqrt{2}\\ &= 7 - 3\sqrt{14} + \sqrt{14} - 3(2)\\ &= 7 - 2\sqrt{14} - 6\\ &= 1 - 2\sqrt{14}\end{aligned}$$

SELF CHECK 3 Multiply: $(\sqrt{5} + 2\sqrt{3})(\sqrt{5} - \sqrt{3})$

EXAMPLE 4 Multiply: $(\sqrt{3x} - \sqrt{5})(\sqrt{2x} + \sqrt{10})$

Solution $(\sqrt{3x} - \sqrt{5})(\sqrt{2x} + \sqrt{10})$

$$= \sqrt{3x}\sqrt{2x} + \sqrt{3x}\sqrt{10} - \sqrt{5}\sqrt{2x} - \sqrt{5}\sqrt{10}$$
$$= \sqrt{6x^2} + \sqrt{30x} - \sqrt{10x} - \sqrt{50}$$
$$= \sqrt{6}\sqrt{x^2} + \sqrt{30x} - \sqrt{10x} - \sqrt{25}\sqrt{2}$$
$$= \sqrt{6}x + \sqrt{30x} - \sqrt{10x} - 5\sqrt{2}$$

COMMENT Note that x in the answer is not under the radical in the first term, but it is under the radical in the second and third terms.

SELF CHECK 4 Multiply: $(\sqrt{x} + 1)(\sqrt{x} - 3)$

COMMENT It is important to draw radical signs so they completely cover the radicand, but no more than the radicand. For example $\sqrt{6x}$ and $\sqrt{6}x$ are not the same expressions. To avoid confusion, we can use the commutative property of multiplication and write an expression such as $\sqrt{6}x$ in the form $x\sqrt{6}$.

2 Rationalize the denominator of a fraction that contains a radical expression.

To divide radical expressions, we **rationalize the denominator** of a fraction to write the denominator with a rational number. For example, to divide $\sqrt{70}$ by $\sqrt{3}$ we write the division as the fraction

$$\frac{\sqrt{70}}{\sqrt{3}}$$

To rationalize the radical in the denominator, we multiply the numerator and the denominator by a number that will result in a perfect square under the radical. Because $3 \cdot 3 = 9$ and 9 is a perfect square, $\sqrt{3}$ is such a number.

$$\frac{\sqrt{70}}{\sqrt{3}} = \frac{\sqrt{70} \cdot \sqrt{3}}{\sqrt{3} \cdot \sqrt{3}} \quad \text{Multiply numerator and denominator by } \sqrt{3}.$$

$$= \frac{\sqrt{210}}{3} \quad \text{Multiply the radicals.}$$

Since there is no radical in the denominator and $\sqrt{210}$ cannot be simplified, the expression $\frac{\sqrt{210}}{3}$ is in simplest form, and the division is complete.

EXAMPLE 5 Rationalize each denominator.

a. $\sqrt{\frac{20}{7}}$ **b.** $\frac{4}{\sqrt[3]{2}}$

Solution **a.** We first write the square root of the quotient as the quotient of two square roots.

$$\sqrt{\frac{20}{7}} = \frac{\sqrt{20}}{\sqrt{7}} \quad \text{Recall, } \sqrt[n]{\frac{a}{b}} = \frac{\sqrt[n]{a}}{\sqrt[n]{b}} \quad b \neq 0.$$

Because the denominator is a square root, we must then multiply the numerator and the denominator by a number that will result in a rational number in the denominator.

Such a number is $\sqrt{7}$.

$$\frac{\sqrt{20}}{\sqrt{7}} = \frac{\sqrt{20}\cdot\sqrt{7}}{\sqrt{7}\cdot\sqrt{7}} \quad \text{Multiply numerator and denominator by } \sqrt{7}.$$

$$= \frac{\sqrt{140}}{7} \quad \text{Multiply the radicals.}$$

$$= \frac{2\sqrt{35}}{7} \quad \text{Simplify } \sqrt{140}\text{: } \sqrt{140} = \sqrt{4\cdot 35} = \sqrt{4}\sqrt{35} = 2\sqrt{35}$$

Another approach is to simplify as follows:

$$\sqrt{\frac{20}{7}} = \sqrt{\frac{20\cdot 7}{7\cdot 7}}$$

$$= \sqrt{\frac{4\cdot 5\cdot 7}{49}}$$

$$= \frac{\sqrt{4}\sqrt{35}}{\sqrt{49}}$$

$$= \frac{2\sqrt{35}}{7}$$

b. Since the denominator is a cube root, we multiply the numerator and the denominator by a cube root of a number that will result in a perfect cube under the radical sign. Since 8 is a perfect cube, and $8 = 2\cdot 4$, then $\sqrt[3]{4}$ is such a number.

$$\frac{4}{\sqrt[3]{2}} = \frac{4\cdot\sqrt[3]{4}}{\sqrt[3]{2}\cdot\sqrt[3]{4}} \quad \text{Multiply numerator and denominator by } \sqrt[3]{4}.$$

$$= \frac{4\sqrt[3]{4}}{\sqrt[3]{8}} \quad \text{Multiply the radicals in the denominator.}$$

$$= \frac{4\sqrt[3]{4}}{2} \quad \text{Simplify.}$$

$$= 2\sqrt[3]{4}$$

Here is another approach:

$$\frac{4}{\sqrt[3]{2}} = \frac{4}{\sqrt[3]{2}}\cdot\frac{\sqrt[3]{2^2}}{\sqrt[3]{2^2}}$$

$$= \frac{4\sqrt[3]{4}}{\sqrt[3]{2^3}}$$

$$= \frac{4\sqrt[3]{4}}{2}$$

$$= 2\sqrt[3]{4}$$

SELF CHECK 5 Rationalize each denominator. **a.** $\sqrt{\frac{8}{5}}$ **b.** $\frac{5}{\sqrt[4]{3}}$

EXAMPLE 6 Rationalize the denominator: $\frac{\sqrt[3]{5}}{\sqrt[3]{18}}$

Solution We multiply the numerator and the denominator by a number that will result in a perfect cube under the radical sign in the denominator.

Since 216 is the smallest perfect cube that is divisible by 18 $(216 \div 18 = 12)$, multiplying the numerator and the denominator by $\sqrt[3]{12}$ will result in the smallest possible perfect cube under the radical in the denominator.

$$\frac{\sqrt[3]{5}}{\sqrt[3]{18}} = \frac{\sqrt[3]{5} \cdot \sqrt[3]{12}}{\sqrt[3]{18} \cdot \sqrt[3]{12}} \quad \text{Multiply numerator and denominator by } \sqrt[3]{12}.$$

$$= \frac{\sqrt[3]{60}}{\sqrt[3]{216}} \quad \text{Multiply the radicals.}$$

$$= \frac{\sqrt[3]{60}}{6} \quad \text{Simplify.}$$

There is another approach. Since $18 = 2 \cdot 3^2$, we need factors of $2^2 \cdot 3$ to obtain a perfect cube $2^3 \cdot 3^3$.

$$\frac{\sqrt[3]{5}}{\sqrt[3]{18}} = \frac{\sqrt[3]{5}\sqrt[3]{2^2 \cdot 3}}{\sqrt[3]{2 \cdot 3^2}\sqrt[3]{2^2 \cdot 3}}$$

$$= \frac{\sqrt[3]{5}\sqrt[3]{12}}{\sqrt[3]{2^3 \cdot 3^3}}$$

$$= \frac{\sqrt[3]{60}}{2 \cdot 3}$$

$$= \frac{\sqrt[3]{60}}{6}$$

SELF CHECK 6 Rationalize the denominator. $\frac{\sqrt[3]{2}}{\sqrt[3]{9}}$

EXAMPLE 7 Rationalize the denominator of $\frac{\sqrt{5xy^2}}{\sqrt{xy^3}}$ (x and y are positive numbers).

Solution

Method 1

$$\frac{\sqrt{5xy^2}}{\sqrt{xy^3}} = \sqrt{\frac{5xy^2}{xy^3}}$$

$$= \sqrt{\frac{5}{y}}$$

$$= \frac{\sqrt{5}}{\sqrt{y}}$$

$$= \frac{\sqrt{5}\sqrt{y}}{\sqrt{y}\sqrt{y}}$$

$$= \frac{\sqrt{5y}}{y}$$

Method 2

$$\frac{\sqrt{5xy^2}}{\sqrt{xy^3}} = \sqrt{\frac{5xy^2}{xy^3}}$$

$$= \sqrt{\frac{5}{y}}$$

$$= \sqrt{\frac{5 \cdot y}{y \cdot y}}$$

$$= \frac{\sqrt{5y}}{\sqrt{y^2}}$$

$$= \frac{\sqrt{5y}}{y}$$

COMMENT When possible, simplify prior to rationalizing.

SELF CHECK 7 Rationalize the denominator. $\frac{\sqrt{4ab^3}}{\sqrt{2a^2b^2}}$ $(a > 0, b > 0)$

To rationalize the denominator of a fraction with square roots in a binomial denominator, we can multiply the numerator and denominator by the **conjugate** of the denominator.

CONJUGATES The conjugate of $(a + b)$ is $(a - b)$, and the conjugate of $(a - b)$ is $(a + b)$.

If we multiply an expression such as $\left(5 + \sqrt{2}\right)$ by its conjugate $\left(5 - \sqrt{2}\right)$, we will obtain an expression without any radical terms.

$$\begin{aligned}\left(5 + \sqrt{2}\right)\left(5 - \sqrt{2}\right) &= 25 - 5\sqrt{2} + 5\sqrt{2} - 2 \\ &= 23\end{aligned}$$

EXAMPLE 8 Rationalize the denominator. $\dfrac{1}{\sqrt{2} + 1}$

Solution We multiply the numerator and denominator of the fraction by $\left(\sqrt{2} - 1\right)$, which is the conjugate of the denominator.

$$\begin{aligned}\frac{1}{\sqrt{2} + 1} &= \frac{1\left(\sqrt{2} - 1\right)}{\left(\sqrt{2} + 1\right)\left(\sqrt{2} - 1\right)} && \frac{\sqrt{2} - 1}{\sqrt{2} - 1} = 1 \\ &= \frac{\sqrt{2} - 1}{\left(\sqrt{2}\right)^2 - \sqrt{2} + \sqrt{2} - 1} && \text{Multiply.} \\ &= \frac{\sqrt{2} - 1}{2 - 1} && \text{Simplify.} \\ &= \sqrt{2} - 1 && \frac{\sqrt{2} - 1}{2 - 1} = \frac{\sqrt{2} - 1}{1} = \sqrt{2} - 1\end{aligned}$$

SELF CHECK 8 Rationalize the denominator. $\dfrac{2}{\sqrt{3} + 1}$

EXAMPLE 9 Rationalize the denominator. $\dfrac{\sqrt{x} + \sqrt{2}}{\sqrt{x} - \sqrt{2}}$ $(x > 0, x \neq 2)$

Solution We multiply the numerator and denominator by $\left(\sqrt{x} + \sqrt{2}\right)$, which is the conjugate of the denominator, and simplify.

$$\begin{aligned}\frac{\sqrt{x} + \sqrt{2}}{\sqrt{x} - \sqrt{2}} &= \frac{\left(\sqrt{x} + \sqrt{2}\right)\left(\sqrt{x} + \sqrt{2}\right)}{\left(\sqrt{x} - \sqrt{2}\right)\left(\sqrt{x} + \sqrt{2}\right)} \\ &= \frac{x + \sqrt{2x} + \sqrt{2x} + 2}{\left(\sqrt{x}\right)^2 + \sqrt{2x} - \sqrt{2x} - \left(\sqrt{2}\right)^2} && \text{Use the distributive property.} \\ &= \frac{x + 2\sqrt{2x} + 2}{x - 2} && \text{Combine like terms.}\end{aligned}$$

SELF CHECK 9 Rationalize the denominator. $\dfrac{\sqrt{x} - \sqrt{2}}{\sqrt{x} + \sqrt{2}}$ $(x > 0)$

3 Rationalize the numerator of a fraction that contains a radical expression.

In calculus, we sometimes have to **rationalize a numerator** by multiplying the numerator and denominator of the fraction by the conjugate of the numerator. The process is similar to rationalizing the denominator, but the denominator may still contain a radical expression.

EXAMPLE 10 Rationalize the numerator. $\dfrac{\sqrt{x}-3}{\sqrt{x}}$ $(x > 0)$

Solution We multiply the numerator and denominator by $\left(\sqrt{x}+3\right)$, which is the conjugate of the numerator.

$$\frac{\sqrt{x}-3}{\sqrt{x}} = \frac{\left(\sqrt{x}-3\right)\left(\sqrt{x}+3\right)}{\sqrt{x}\left(\sqrt{x}+3\right)}$$

$$= \frac{x+3\sqrt{x}-3\sqrt{x}-9}{x+3\sqrt{x}}$$

$$= \frac{x-9}{x+3\sqrt{x}}$$

SELF CHECK 10 Rationalize the numerator. $\dfrac{\sqrt{x}+3}{\sqrt{x}}$ $(x > 0)$

4 Solve an application containing a radical expression.

EXAMPLE 11 **PHOTOGRAPHY** Many camera lenses (see Figure 9-15) have an adjustable opening called the *aperture*, which controls the amount of light passing through the lens. The *f-number* of a lens is its *focal length* divided by the diameter of its circular aperture.

© Brocorwin/Shutterstock.com

Figure 9-15

$$f\text{-number} = \frac{f}{d} \qquad f \text{ is the focal length, and } d \text{ is the diameter of the aperture.}$$

A lens with a focal length of 12 centimeters and an aperture with a diameter of 6 centimeters has an f-number of $\frac{12}{6}$ and is an $f/2$ lens. If the area of the aperture is reduced to admit half as much light, the f-number of the lens will change. Find the new f-number.

Solution We first find the area of the aperture when its diameter is 6 centimeters.

$A = \pi r^2$ This is the formula for the area of a circle.

$A = \pi(3)^2$ Since a radius is half the diameter, substitute 3 for r.

$A = 9\pi$

When the size of the aperture is reduced to admit half as much light, the area of the aperture will be $\frac{9\pi}{2}$ square centimeters. To find the diameter of a circle with this area, we proceed as follows:

$A = \pi r^2$ This is the formula for the area of a circle.

$\dfrac{9\pi}{2} = \pi\left(\dfrac{d}{2}\right)^2$ Substitute $\frac{9\pi}{2}$ for A and $\frac{d}{2}$ for r.

$\dfrac{9\pi}{2} = \dfrac{\pi d^2}{4}$ $\left(\frac{d}{2}\right)^2 = \frac{d^2}{4}$

$\dfrac{4}{\pi}\left(\dfrac{9\pi}{2}\right) = \dfrac{4}{\pi}\left(\dfrac{\pi d^2}{4}\right)$ Multiply both sides by $\frac{4}{\pi}$, the reciprocal of the coefficient of d^2.

$18 = d^2$ Simplify.

$d = 3\sqrt{2}$ $\sqrt{18} = \sqrt{9}\sqrt{2} = 3\sqrt{2}$

Since the focal length of the lens is still 12 centimeters and the diameter is now $3\sqrt{2}$ centimeters, the new f-number of the lens is

$$f\text{-number} = \frac{f}{d} = \frac{12}{3\sqrt{2}} \qquad \text{Substitute 12 for } f \text{ and } 3\sqrt{2} \text{ for } d.$$

$$= \frac{4}{\sqrt{2}} \quad \text{Simplify.}$$

$$= \frac{4\sqrt{2}}{2} \quad \text{Rationalize the denominator.}$$

$$= 2\sqrt{2} \quad \text{Simplify.}$$

$$\approx 2.828427125 \quad \text{Use a calculator.}$$

The lens is now an $f/2.8$ lens.

SELF CHECK 11 If the area of the aperture is enlarged to admit twice as much light, find the f-number.

SELF CHECK ANSWERS

1. a. $-40\sqrt{3}$ **b.** $2a^2\sqrt[3]{3a}$ **2.** $12\sqrt{10}$ 32 **3.** $1+\sqrt{15}$ **4.** $x-2\sqrt{x}-3$ **5. a.** $\frac{2\sqrt{10}}{3}$ **b.** $\frac{5\sqrt[4]{27}}{3}$ **6.** $\frac{\sqrt[3]{6}}{3}$ **7.** $\frac{\sqrt{2ab}}{a}$ **8.** $\sqrt{3}-1$ **9.** $\frac{x-2\sqrt{2x}+2}{x-2}$ **10.** $\frac{x-9}{x-3\sqrt{x}}$ **11.** The lens is now an $f/1.4$ lens.

NOW TRY THIS

Find the domain of each of the following. State your answer in interval notation.

1. $f(x)=\frac{3x-2}{\sqrt{x}+1}$

2. $g(x)=\frac{3x-2}{\sqrt{x+1}}$

3. $h(x)=\frac{\sqrt{x-2}}{\sqrt{x}+1}$

9.5 Exercises

WARM-UPS *Simplify each expression. Assume that all variables represent positive numbers.*

1. $\sqrt{5}\sqrt{5}$

2. $\sqrt{11}\sqrt{11}$

3. $\sqrt{xy}\sqrt{xy}$

4. $\sqrt{ab}\sqrt{ab}$

5. $\sqrt[3]{5}\sqrt[3]{5}\sqrt[3]{5}$

6. $\sqrt[3]{7}\sqrt[3]{7}\sqrt[3]{7}$

7. $\sqrt[3]{a^2b}\sqrt[3]{ab^2}$

8. $\sqrt[3]{xy^2}\sqrt[3]{x^2y}$

REVIEW *Solve.*

9. $\frac{2}{3-a}=1$

10. $5(s-4)=-5(s-4)$

11. $\frac{8}{b-2}+\frac{3}{2-b}=-\frac{1}{b}$

12. $\frac{2}{x-2}+\frac{1}{x+1}=\frac{1}{(x+1)(x-2)}$

VOCABULARY AND CONCEPTS *Fill in the blanks.*

13. To multiply $2\sqrt{7}$ by $3\sqrt{5}$, we multiply _ by 3 and then multiply ____ by ____.

14. To multiply $2\sqrt{5}(3\sqrt{8}+\sqrt{3})$, we use the ________ property to remove parentheses and simplify each resulting term.

15. The conjugate of $(\sqrt{x}+1)$ is ________.

16. To multiply $(\sqrt{3}+\sqrt{2})(\sqrt{3}-2\sqrt{2})$, we use the ________________.

17. To rationalize the denominator of $\frac{1}{\sqrt{3}-1}$, multiply both the numerator and denominator by the ________ of the denominator, $\sqrt{3}+1$.

18. To rationalize the numerator of $\frac{\sqrt{5}+2}{\sqrt{5}-2}$, multiply both the numerator and denominator by ________.

GUIDED PRACTICE *Simplify. All variables represent positive values.* SEE EXAMPLE 1. (OBJECTIVE 1)

19. $\sqrt{2}\sqrt{8}$

20. $\sqrt{3}\sqrt{27}$

21. $\sqrt{5}\sqrt{10}$

22. $\sqrt{7}\sqrt{35}$

23. $2\sqrt{3}\sqrt{6}$

24. $3\sqrt{11}\sqrt{33}$

25. $\sqrt[3]{5}\sqrt[3]{25}$

26. $\sqrt[3]{7}\sqrt[3]{49}$

27. $\sqrt[3]{5r^2s}\sqrt[3]{2r}$

28. $\sqrt[3]{3xy^2}\sqrt[3]{9x^3}$

29. $\sqrt{ab^3}\sqrt{ab}$

30. $\sqrt{8x}\sqrt{2x^3y}$

31. $(2\sqrt{3x^3y})(5\sqrt{6xy^2})$

32. $(7\sqrt{18a})(3\sqrt{3a^3b})$

33. $\sqrt{x(x+3)}\sqrt{x^3(x+3)}$
34. $\sqrt{y^2(x+y)}\sqrt{(x+y)^3}$

Simplify. SEE EXAMPLE 2. (OBJECTIVE 1)

35. $3\sqrt{5}(4-\sqrt{5})$
36. $2\sqrt{7}(3\sqrt{7}-1)$
37. $3\sqrt{2}(4\sqrt{3}+2\sqrt{7})$
38. $-2\sqrt{3}(3\sqrt{5}-6\sqrt{7})$

Simplify. SEE EXAMPLE 3. (OBJECTIVE 1)

39. $(\sqrt{2}+1)(\sqrt{2}-3)$
40. $(2\sqrt{3}+1)(\sqrt{3}-1)$
41. $(3\sqrt{2}+2)(\sqrt{2}+1)$
42. $(4\sqrt{3}-3)(\sqrt{3}+2)$

Simplify. All variables represent positive values. SEE EXAMPLE 4. (OBJECTIVE 1)

43. $(4\sqrt{x}+3)(2\sqrt{x}-5)$
44. $(5\sqrt{y}-4)(2\sqrt{y}-5)$
45. $(\sqrt{3a}+\sqrt{6})(\sqrt{2a}-\sqrt{3})$
46. $(\sqrt{7y}-\sqrt{5})(\sqrt{14y}+\sqrt{10})$

Rationalize each denominator. SEE EXAMPLE 5. (OBJECTIVE 2)

47. $\sqrt{\frac{1}{7}}$
48. $\sqrt{\frac{5}{3}}$
49. $\sqrt{\frac{2}{3}}$
50. $\sqrt{\frac{3}{2}}$
51. $\frac{\sqrt{24}}{\sqrt{5}}$
52. $\frac{\sqrt{8}}{\sqrt{3}}$
53. $\frac{\sqrt{7}}{\sqrt{12}}$
54. $\frac{\sqrt{5}}{\sqrt{32}}$

Rationalize each denominator. SEE EXAMPLE 6. (OBJECTIVE 2)

55. $\frac{3}{\sqrt[3]{16}}$
56. $\frac{2}{\sqrt[3]{6}}$
57. $\frac{\sqrt[3]{7}}{\sqrt[3]{72}}$
58. $\frac{\sqrt[3]{9}}{\sqrt[3]{54}}$

Rationalize each denominator. All variables represent positive values. SEE EXAMPLE 7. (OBJECTIVE 2)

59. $\frac{\sqrt{8x^2y}}{\sqrt{xy}}$
60. $\frac{\sqrt{9xy}}{\sqrt{3x^2y}}$
61. $\frac{\sqrt{10xy^2}}{\sqrt{2xy^3}}$
62. $\frac{\sqrt{5ab^2c}}{\sqrt{10abc}}$

Rationalize each denominator. SEE EXAMPLE 8. (OBJECTIVE 2)

63. $\frac{1}{\sqrt{2}-1}$
64. $\frac{3}{\sqrt{3}-1}$
65. $\frac{\sqrt{2}}{\sqrt{5}+3}$
66. $\frac{\sqrt{3}}{\sqrt{3}-2}$
67. $\frac{\sqrt{3}+1}{\sqrt{3}-1}$
68. $\frac{\sqrt{2}-1}{\sqrt{2}+1}$
69. $\frac{\sqrt{7}-\sqrt{2}}{\sqrt{2}+\sqrt{7}}$
70. $\frac{\sqrt{3}+\sqrt{2}}{\sqrt{3}-\sqrt{2}}$

Rationalize each denominator. All variables represent positive values. SEE EXAMPLE 9. (OBJECTIVE 2)

71. $\frac{2}{\sqrt{x}+1}$
72. $\frac{3}{\sqrt{x}-2}$
73. $\frac{x}{\sqrt{x}-4}$
74. $\frac{2x}{\sqrt{x}+1}$
75. $\frac{\sqrt{x}-\sqrt{y}}{\sqrt{x}+\sqrt{y}}$
76. $\frac{\sqrt{x}+\sqrt{y}}{\sqrt{x}-\sqrt{y}}$

Rationalize each numerator. All variables represent positive values. SEE EXAMPLE 10. (OBJECTIVE 3)

77. $\frac{\sqrt{3}+1}{2}$
78. $\frac{\sqrt{5}-1}{2}$
79. $\frac{\sqrt{x}+3}{x}$
80. $\frac{2+\sqrt{x}}{5x}$

ADDITIONAL PRACTICE *Simplify. All variables represent positive values.*

81. $(3\sqrt[3]{9})(2\sqrt[3]{3})$
82. $(2\sqrt[3]{16})(-\sqrt[3]{4})$
83. $\sqrt{5ab}\sqrt{5a}$
84. $\sqrt{15rs^2}\sqrt{10r}$
85. $\sqrt[3]{a^5b}\sqrt[3]{16ab^5}$
86. $\sqrt[3]{3x^4y}\sqrt[3]{18x}$
87. $\sqrt[3]{6x^2(y+z)^2}\sqrt[3]{18x(y+z)}$
88. $\sqrt[3]{9x^2y(z+1)^2}\sqrt[3]{6xy^2(z+1)}$
89. $-2\sqrt{5x}(4\sqrt{2x}-3\sqrt{3})$
90. $3\sqrt{7t}(2\sqrt{7t}+3\sqrt{3t^2})$
91. $(\sqrt{5z}+\sqrt{3})(\sqrt{5z}+\sqrt{3})$
92. $(\sqrt{5x}+\sqrt{3})(\sqrt{5x}-\sqrt{3})$
93. $(3\sqrt{2r}-2)^2$
94. $(2\sqrt{3t}+5)^2$
95. $-2(\sqrt{3x}+\sqrt{3})^2$
96. $3(\sqrt{5x}-\sqrt{3})^2$

Rationalize each denominator. All variables represent positive values.

97. $\frac{3}{\sqrt[3]{9}}$
98. $\frac{2}{\sqrt[3]{a}}$
99. $\frac{1}{\sqrt[4]{4}}$
100. $\frac{4}{\sqrt[4]{32}}$
101. $\frac{1}{\sqrt[5]{16}}$
102. $\frac{1}{\sqrt[5]{2}}$

103. $\dfrac{\sqrt[3]{4a^2}}{\sqrt[3]{2ab}}$

104. $\dfrac{\sqrt[3]{9x}}{\sqrt[3]{3xy}}$

105. $\dfrac{2z - 1}{\sqrt{2z} - 1}$

106. $\dfrac{3t - 1}{\sqrt{3t} + 1}$

Rationalize each numerator. All variables represent positive values.

107. $\dfrac{\sqrt{x} + \sqrt{y}}{\sqrt{x}}$

108. $\dfrac{\sqrt{x} - \sqrt{y}}{\sqrt{x} + \sqrt{y}}$

APPLICATIONS *Solve each application. SEE EXAMPLE 11. (OBJECTIVE 4)*

109. **Photography** We have seen that a lens with a focal length of 12 centimeters and an aperture $3\sqrt{2}$ centimeters in diameter is an $f/2.8$ lens. Find the f-number if the area of the aperture is again cut in half.

110. **Photography** A lens with a focal length of 12 centimeters and an aperture 3 centimeters in diameter is an $f/4$ lens. Find the f-number if the area of the aperture is cut in half.

111. **Targets** The radius r of the target is given by the formula

$$r = \sqrt{\frac{A}{\pi}}$$

where A is the area. Write the formula in a form in which the denominator is not part of the radicand.

112. **Pulse rates** The approximate pulse rate (in beats per minute) of an adult who is t inches tall is given by the function

$$p(t) = \frac{590}{\sqrt{t}}$$

Write the formula in a form in which the denominator is a rational expression.

113. If the hypotenuse of an isosceles right triangle is 8 cm, find the length of each leg.

114. The hypotenuse of a 45°–45°–90° triangle is 14 m. Find the length of each leg.

115. The longer leg of a 30°–60°–90° triangle is 6 ft. Find the length of remaining sides.

116. The altitude of an equilateral triangle is 24 mm. Find the lengths of the sides of the triangle.

WRITING ABOUT MATH

117. Explain how to simplify a fraction with the monomial denominator $\sqrt[3]{3}$.

118. Explain how to simplify a fraction with the monomial denominator $\sqrt[3]{9}$.

SOMETHING TO THINK ABOUT *Assume that x is a rational number.*

119. Write the numerator of $\dfrac{\sqrt{x} - 3}{4}$ as a rational number.

120. Rationalize the numerator: $\dfrac{2\sqrt{3x} + 4}{\sqrt{3x} - 1}$

Section 9.6 Radical Equations

Objectives

1. Solve a radical equation containing one radical.
2. Solve a radical equation containing two radicals.
3. Solve a radical equation containing three radicals.
4. Solve a formula containing a radical for a specified variable.

Unless otherwise noted, all content on this page is © Cengage Learning.

Vocabulary

power rule

Getting Ready

Simplify.

1. $(\sqrt{a})^2$ **2.** $(\sqrt{5x})^2$ **3.** $(\sqrt{x+4})^2$ **4.** $(\sqrt[3]{y-3})^3$

In this section, we will solve equations that contain radicals. To do so, we will use the **power rule**. Then, we will use this rule to solve an application.

1 Solve a radical equation containing one radical.

THE POWER RULE

If x, y, and n are real numbers and $x = y$, then

$$x^n = y^n$$

When we raise both sides of an equation to the same power, the resulting equation might not be equivalent to the original equation. For example, if we square both sides of the equation

(1) $x = 3$ With a solution set of $\{3\}$

we obtain the equation

(2) $x^2 = 9$ With a solution set of $\{3, -3\}$

Equations 1 and 2 are not equivalent, because they have different solution sets, and the solution -3 of Equation 2 does not satisfy Equation 1. Since raising both sides of an equation to the same power can produce an equation with roots (extraneous) that do not satisfy the original equation, we must check each possible solution in the original equation.

EXAMPLE 1 Solve: $\sqrt{x+3} = 4$

Solution Because the radical term is isolated, to eliminate the radical, we apply the power rule by squaring both sides of the equation, and proceed as follows:

$$\sqrt{x+3} = 4$$

$$(\sqrt{x+3})^2 = (4)^2 \quad \text{Square both sides of the equation.}$$

$$x + 3 = 16 \quad \text{Simplify.}$$

$$x = 13 \quad \text{Subtract 3 from both sides.}$$

To check the apparent solution of 13, we can substitute 13 for x and determine whether it satisfies the original equation.

$$\sqrt{x+3} = 4$$
$$\sqrt{13+3} \stackrel{?}{=} 4 \quad \text{Substitute 13 for } x.$$
$$\sqrt{16} \stackrel{?}{=} 4$$
$$4 = 4$$

Since 13 satisfies the original equation, it is a solution.

SELF CHECK 1 Solve: $\sqrt{a-2} = 3$

To solve an equation containing radicals, we follow these steps.

SOLVING AN EQUATION CONTAINING RADICALS

1. Isolate one radical expression on one side of the equation.
2. Raise both sides of the equation to the power that is the same as the index of the radical.
3. Solve the resulting equation. If it still contains a radical, go back to Step 1.
4. Check the possible solutions to eliminate the extraneous ones.

EXAMPLE 2

HEIGHT OF A BRIDGE The distance d (in feet) that an object will fall in t seconds is given by the formula

$$t = \sqrt{\frac{d}{16}}$$

To find the height of a bridge, a man drops a stone into the water. (See Figure 9-16.) If it takes the stone 3 seconds to hit the water, how far above the river is the bridge?

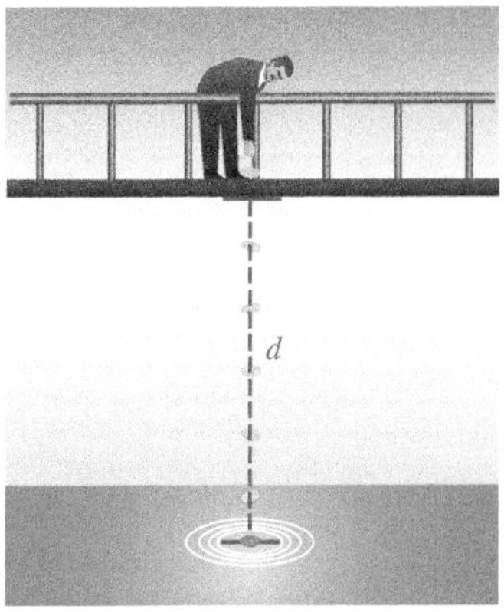

Figure 9-16

Solution We substitute 3 for t in the formula and solve for d.

$$t = \sqrt{\frac{d}{16}}$$
$$3 = \sqrt{\frac{d}{16}}$$
$$9 = \frac{d}{16} \quad \text{Apply the power rule (square both sides) to eliminate the square root.}$$
$$144 = d \quad \text{Multiply both sides by 16.}$$

The bridge is 144 feet above the river.

SELF CHECK 2 How high is the bridge if it takes 4 seconds for the stone to hit the water?

COMMENT Be certain to isolate the radical term prior to applying the power rule.

EXAMPLE 3

Solve: $\sqrt{3x+1} + 1 = x$

Solution We first subtract 1 from both sides to isolate the radical. Then, to eliminate the radical, we square both sides of the equation and proceed as follows:

Unless otherwise noted, all content on this page is © Cengage Learning.

$$\sqrt{3x+1}+1=x$$

$$\sqrt{3x+1}=x-1 \quad \text{Subtract 1 from both sides.}$$

$$\left(\sqrt{3x+1}\right)^2=(x-1)^2 \quad \text{Apply the power rule.}$$

$$3x+1=x^2-2x+1 \quad \text{Simplify.}$$

$$0=x^2-5x \quad \text{Subtract } 3x \text{ and 1 from both sides to write the equation in quadratic form.}$$

$$0=x(x-5) \quad \text{Factor.}$$

$$x=0 \quad \text{or} \quad x-5=0 \quad \text{Set each factor equal to 0.}$$

$$x=0 \quad \Big| \quad x=5$$

We must check each possible solution to see whether it satisfies the original equation.

Check:

$$\sqrt{3x+1}+1=x \qquad \sqrt{3x+1}+1=x$$

$$\sqrt{3(\mathbf{0})+1}+1\stackrel{?}{=}\mathbf{0} \qquad \sqrt{3(\mathbf{5})+1}+1\stackrel{?}{=}\mathbf{5}$$

$$\sqrt{1}+1\stackrel{?}{=}0 \qquad \sqrt{16}+1\stackrel{?}{=}5$$

$$2\neq 0 \qquad 5=5$$

Since 0 does not check, it is extraneous and must be discarded. The only solution of the original equation is 5.

SELF CHECK 3 Solve: $\sqrt{4x+1}+1=x$

Accent on technology

▸Solving Equations Containing Radicals

To find solutions for $\sqrt{3x+1}+1=x$ with a graphing calculator, we graph the functions $f(x)=\sqrt{3x+1}+1$ and $g(x)=x$, and then adjust the window settings to $[-5,10]$ for x and $[-2,8]$ for y as in Figure 9-17(a).

We can find the exact x-coordinate of the intersection point by using the INTERSECT command found in the CALC menu. We see that $x=5$, as in Figure 9-17(b).

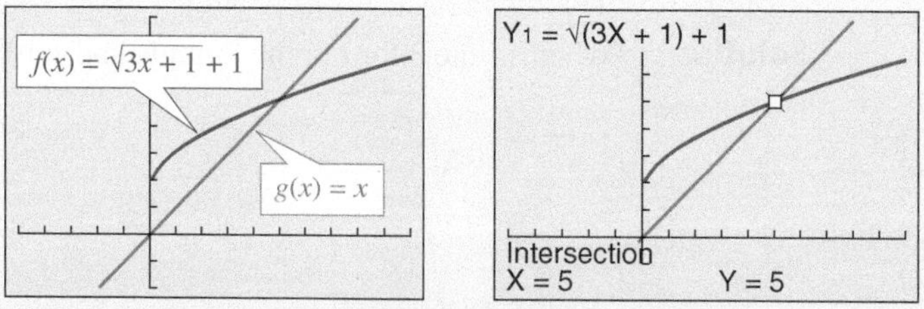

(a) (b)

Figure 9-17

For instructions regarding the use of a Casio graphing calculator, please refer to the Casio Keystroke Guide in the back of the book.

EXAMPLE 4 Solve: $\sqrt[3]{x^3+7}=x+1$

Solution To eliminate the radical, we cube both sides of the equation and proceed as follows:

$$\sqrt[3]{x^3+7}=x+1$$

$$\left(\sqrt[3]{x^3+7}\right)^3=(x+1)^3 \quad \text{Cube both sides to eliminate the cube root.}$$

$$x^3+7=x^3+3x^2+3x+1 \quad \text{Simplify.}$$

$$0=3x^2+3x-6 \quad \text{Subtract } x^3 \text{ and 7 from both sides to write the equation in quadratic form.}$$

$$0=x^2+x-2 \quad \text{Divide both sides by 3.}$$

Unless otherwise noted, all content on this page is © Cengage Learning.

$$0 = (x + 2)(x - 1)$$ Factor the trinomial.

$$x + 2 = 0 \quad \text{or} \quad x - 1 = 0$$ Set each factor equal to 0.

$$x = -2 \quad\quad x = 1$$

We check each possible solution to see whether it satisfies the original equation.

Check:

$$\begin{aligned} \sqrt[3]{x^3 + 7} &= x + 1 \\ \sqrt[3]{(-2)^3 + 7} &\stackrel{?}{=} -2 + 1 \\ \sqrt[3]{-8 + 7} &\stackrel{?}{=} -1 \\ \sqrt[3]{-1} &\stackrel{?}{=} -1 \\ -1 &= -1 \end{aligned} \qquad \begin{aligned} \sqrt[3]{x^3 + 7} &= x + 1 \\ \sqrt[3]{1 + 7} &\stackrel{?}{=} 1 + 1 \\ \sqrt[3]{8} &\stackrel{?}{=} 2 \\ 2 &= 2 \end{aligned}$$

Both solutions satisfy the original equation; thus the solutions are −2 and 1.

SELF CHECK 4 Solve: $\sqrt[3]{x^3 + 8} = x + 2$

2 Solve a radical equation containing two radicals.

When more than one radical appears in an equation, it is often necessary to apply the power rule more than once.

EXAMPLE 5 Solve: $\sqrt{x} + \sqrt{x + 2} = 2$

Solution We subtract $\sqrt{x}$ from both sides to isolate one radical on one side of the equation. To remove one of the radicals, we square both sides of the equation.

$$\sqrt{x} + \sqrt{x + 2} = 2$$

$$\sqrt{x + 2} = 2 - \sqrt{x}$$ Subtract $\sqrt{x}$ from both sides.

$$\left(\sqrt{x + 2}\right)^2 = \left(2 - \sqrt{x}\right)^2$$ Square both sides to eliminate the square root on the left.

$$x + 2 = 4 - 4\sqrt{x} + x$$ $\left(2 - \sqrt{x}\right)\left(2 - \sqrt{x}\right) = 4 - 2\sqrt{x} - 2\sqrt{x} + x = 4 - 4\sqrt{x} + x$

$$2 = 4 - 4\sqrt{x}$$ Subtract x from both sides.

$$-2 = -4\sqrt{x}$$ Subtract 4 from both sides.

$$\frac{1}{2} = \sqrt{x}$$ Divide both sides by −4.

$$\frac{1}{4} = x$$ Square both sides to eliminate the square root on the right side.

Check:

$$\begin{aligned} \sqrt{x} + \sqrt{x + 2} &= 2 \\ \sqrt{\frac{1}{4}} + \sqrt{\frac{1}{4} + 2} &\stackrel{?}{=} 2 \\ \frac{1}{2} + \sqrt{\frac{9}{4}} &\stackrel{?}{=} 2 \\ \frac{1}{2} + \frac{3}{2} &\stackrel{?}{=} 2 \\ 2 &= 2 \end{aligned}$$

The solution checks.

SELF CHECK 5 Solve: $\sqrt{a} + \sqrt{a + 3} = 3$

3 Solve a radical equation containing three radicals.

EXAMPLE 6 Solve: $\sqrt{x+2}+\sqrt{2x}=\sqrt{18-x}$

Solution In this case, one radical is already isolated on one side of the equation, so we begin by squaring both sides. Then we proceed as follows.

$$\sqrt{x+2}+\sqrt{2x}=\sqrt{18-x}$$

$$\left(\sqrt{x+2}+\sqrt{2x}\right)^2=\left(\sqrt{18-x}\right)^2$$ Square both sides to eliminate the square root on the right side.

$$\left(\sqrt{x+2}+\sqrt{2x}\right)\left(\sqrt{x+2}+\sqrt{2x}\right)=18-x$$ Simplify.

$$\left(\sqrt{x+2}\right)^2+\left(\sqrt{x+2}\right)\left(\sqrt{2x}\right)+\left(\sqrt{2x}\right)\left(\sqrt{x+2}\right)+\left(\sqrt{2x}\right)^2=18-x$$ Multiply.

$$x+2+\sqrt{2x(x+2)}+\sqrt{2x(x+2)}+2x=18-x$$ Simplify.

$$3x+2+2\sqrt{2x(x+2)}=18-x$$ Combine like terms.

$$2\sqrt{2x(x+2)}=16-4x$$ Subtract $3x$ and 2 from both sides.

$$\sqrt{2x(x+2)}=8-2x$$ Divide both sides by 2.

$$\left(\sqrt{(2x)(x+2)}\right)^2=(8-2x)^2$$ Square both sides to eliminate the square root on the left side.

$$2x(x+2)=64-32x+4x^2$$

$$2x^2+4x=64-32x+4x^2$$ Use the distributive property to remove parentheses.

$$0=2x^2-36x+64$$ Write the equation in quadratic form.

$$0=x^2-18x+32$$ Divide both sides by 2.

$$0=(x-16)(x-2)$$ Factor the trinomial.

$$x-16=0 \quad \text{or} \quad x-2=0$$ Set each factor equal to 0.

$$x=16 \quad | \quad x=2$$

Verify that 2 satisfies the equation, but 16 does not. Thus, the only solution is 2.

SELF CHECK 6 Solve: $\sqrt{3x+4}+\sqrt{x+9}=\sqrt{x+25}$

4 Solve a formula containing a radical for a specified variable.

To *solve a formula for a variable* means to isolate that variable on one side of the equation, with all other quantities on the other side.

EXAMPLE 7 **DEPRECIATION RATES** Some office equipment that is now worth V dollars originally cost C dollars 3 years ago. The rate r at which it has depreciated is given by

$$r=1-\sqrt[3]{\frac{V}{C}}$$

Solve the formula for C.

Solution We begin by isolating the cube root on the right side of the equation.

$$r=1-\sqrt[3]{\frac{V}{C}}$$

$$r-1=-\sqrt[3]{\frac{V}{C}}$$ Subtract 1 from both sides.

$$(r-1)^3=\left(-\sqrt[3]{\frac{V}{C}}\right)^3$$ To eliminate the radical, cube both sides.

$$(r - 1)^3 = -\frac{V}{C}$$ Simplify the right side.

$$C(r - 1)^3 = -V$$ Multiply both sides by C.

$$C = -\frac{V}{(r - 1)^3}$$ Divide both sides by $(r - 1)^3$.

SELF CHECK 7 A formula used in statistics to determine the size of a sample to obtain a desired degree of accuracy is

$$E = z_0\sqrt{\frac{pq}{n}}$$

Solve the formula for n.

SELF CHECK ANSWERS **1.** 11 **2.** 256 ft **3.** 6; 0 is extraneous **4.** 0, −2 **5.** 1 **6.** 0, $-\frac{196}{3}$ is extraneous **7.** $n = \frac{z_0^2 pq}{E^2}$

NOW TRY THIS

Solve.

1. $x^{1/3} = 2$

2. $x^{2/3} = 4$

3. $(x + 1)^{-1/2} = 3$

9.6 Exercises

WARM-UPS *Simplify.*

1. $(x^{1/2})^2$

2. $(a^{1/3})^3$

3. $[(x + 3)^{1/2}]^2$

4. $[(x - 4)^{1/3}]^3$

5. $(x + 7)^2$

6. $(x - 6)^2$

REVIEW *If $f(x) = 3x^2 - 4x + 2$, find each quantity.*

7. $f(-2)$

8. $f(-3)$

9. $f(3)$

10. $f\left(\frac{1}{2}\right)$

VOCABULARY AND CONCEPTS *Fill in the blanks.*

11. If x, y, and n are real numbers and $x = y$, then ______, called the ________.

12. When solving equations containing radicals, first _____ one radical on one side of the equation.

13. To solve the equation $\sqrt{x + 4} = 5$, we first _____ both sides.

14. To solve the equation $\sqrt[3]{x + 4} = 2$, we first ____ both sides.

15. Squaring both sides of an equation can introduce ________ solutions.

16. Always remember to _____ the solutions of an equation containing radicals to eliminate any ________ solutions.

GUIDED PRACTICE *Solve each equation.* SEE EXAMPLE 1. (OBJECTIVE 1)

17. $\sqrt{4x + 5} = 3$

18. $\sqrt{7x - 10} = 12$

19. $\sqrt{4x - 8} + 4 = 10$

20. $\sqrt{6x + 13} - 2 = 5$

21. $\sqrt[3]{7n - 1} = 3$

22. $\sqrt[3]{9p + 10} = 4$

23. $x = \frac{\sqrt{12x - 5}}{2}$

24. $x = \frac{\sqrt{16x - 12}}{2}$

Solve each equation. Identify any extraneous solution. SEE EXAMPLE 3. (OBJECTIVE 1)

25. $\sqrt{2y - 5} + 4 = y$

26. $\sqrt{3x + 13} - 3 = x$

27. $\sqrt{-5x + 24} = 6 - x$

28. $\sqrt{-x + 2} = x - 2$

Solve each equation. SEE EXAMPLE 4. (OBJECTIVE 1)

29. $\sqrt[3]{x^3 - 1} = x - 1$

30. $\sqrt[3]{x^3 - 7} = x - 1$

31. $\sqrt[3]{x^3 + 56} - 2 = x$

32. $\sqrt[3]{x^3 - 19} = x - 3$

Solve each equation. Identify any extraneous solution. SEE EXAMPLE 5. (OBJECTIVE 2)

33. $2\sqrt{4x+1} = \sqrt{x+4}$
34. $\sqrt{3(x+4)} = \sqrt{5x-12}$
35. $\sqrt{x+5} = \sqrt{7-x}$
36. $\sqrt{10-x} = \sqrt{2x+4}$
37. $1 + \sqrt{z} = \sqrt{z+3}$
38. $\sqrt{x} + 5 = \sqrt{x+35}$
39. $\sqrt{4s+1} - \sqrt{6s} = -1$
40. $\sqrt{y+7} + 3 = \sqrt{y+4}$

Solve each equation. Identify any extraneous solution. SEE EXAMPLE 6. (OBJECTIVE 3)

41. $\sqrt{y} + \sqrt{5} = \sqrt{y+15}$
42. $\sqrt{x+1} + \sqrt{3x} = \sqrt{5x+1}$
43. $\sqrt{3x} - \sqrt{x+1} = \sqrt{x-2}$
44. $\sqrt{x+2} + \sqrt{2x-3} = \sqrt{11-x}$

Solve each formula for the indicated variable. SEE EXAMPLE 7. (OBJECTIVE 4)

45. $v = \sqrt{2gh}$ for g
46. $d = 1.4\sqrt{h}$ for h
47. $T = 2\pi\sqrt{\frac{l}{32}}$ for l
48. $d = \sqrt[3]{\frac{12V}{\pi}}$ for V
49. $r = \sqrt[3]{\frac{A}{P}} - 1$ for A
50. $r = \sqrt[3]{\frac{A}{P}} - 1$ for P
51. $L_A = L_B\sqrt{1 - \frac{v^2}{c^2}}$ for v^2
52. $R_1 = \sqrt{\frac{A}{\pi} - R_2^{\,2}}$ for A

ADDITIONAL PRACTICE *Solve each equation. Identify any extraneous solution.*

53. $\sqrt{y+2} = 4 - y$
54. $\sqrt{22y+86} = y + 9$
55. $2\sqrt{x} = \sqrt{5x-16}$
56. $3\sqrt{x} = \sqrt{3x+12}$
57. $\sqrt{2y+1} = 1 - 2\sqrt{y}$
58. $\sqrt{u} + 3 = \sqrt{u-3}$
59. $\sqrt{2x+5} + \sqrt{x+2} = 5$
60. $\sqrt{2x+5} + \sqrt{2x+1} + 4 = 0$
61. $5r + 4 = \sqrt{5r+20} + 4r$
62. $\sqrt{x}\sqrt{x+16} = 15$
63. $\sqrt{x}\sqrt{x+6} = 4$
64. $\frac{6}{\sqrt{x+5}} = \sqrt{x}$
65. $\sqrt[4]{x^4 + 4x^2 - 4} = -x$
66. $\sqrt[4]{8x-8} + 2 = 0$
67. $2 + \sqrt{u} = \sqrt{2u+7}$
68. $\sqrt[4]{12t+4} + 2 = 0$
69. $u = \sqrt[4]{u^4 - 6u^2 + 24}$
70. $\sqrt{6t+1} - 3\sqrt{t} = -1$
71. $\sqrt{x-5} - \sqrt{x+3} = 4$
72. $\sqrt[4]{10p+1} = \sqrt[4]{11p-7}$
73. $\sqrt{x+8} - \sqrt{x-4} = -2$
74. $\sqrt[4]{10y+2} = 2\sqrt[4]{2}$
75. $\sqrt{z-1} + \sqrt{z+2} = 3$
76. $\sqrt{16v+1} + \sqrt{8v+1} = 12$
77. $\sqrt{\sqrt{a} + \sqrt{a+8}} = 2$
78. $\sqrt{\sqrt{2y} - \sqrt{y-1}} = 1$
79. $\frac{\sqrt{2x}}{\sqrt{x+2}} = \sqrt{x-1}$
80. $\sqrt{8-x} - \sqrt{3x-8} = \sqrt{x-4}$

APPLICATIONS *Solve each application.* SEE EXAMPLE 2. (OBJECTIVE 1)

81. Highway design A curve banked at 8° will accommodate traffic traveling s mph if the radius of the curve is r feet, according to the formula $s = 1.45\sqrt{r}$. If engineers expect 65-mph traffic, what radius should they specify?

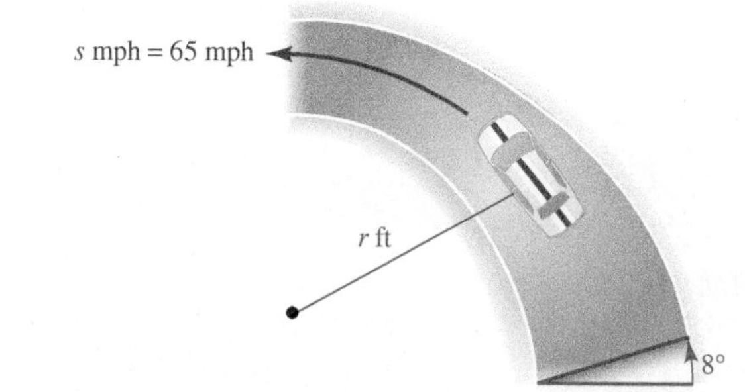

82. Horizon distance The higher a lookout tower is built, the farther an observer can see. That distance d (called the *horizon distance,* measured in miles) is related to the height h of the observer (measured in feet) by the formula $d = 1.4\sqrt{h}$. How tall must a lookout tower be to see the edge of the forest, 25 miles away?

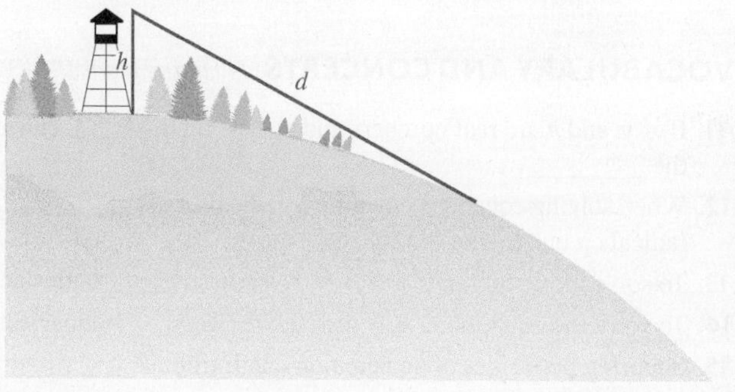

83. Generating power The power generated by a windmill is related to the velocity of the wind by the formula

$$v = \sqrt[3]{\frac{P}{0.02}}$$

where P is the power (in watts) and v is the velocity of the wind (in mph). Find the speed of the wind when the windmill is generating 500 watts of power.

84. Carpentry During construction, carpenters often brace walls as shown in the illustration, where the length of the brace is given by the formula

$$l = \sqrt{f^2 + h^2}$$

If a carpenter nails a 10-ft brace to the wall 6 feet above the floor, how far from the base of the wall should he nail the brace to the floor?

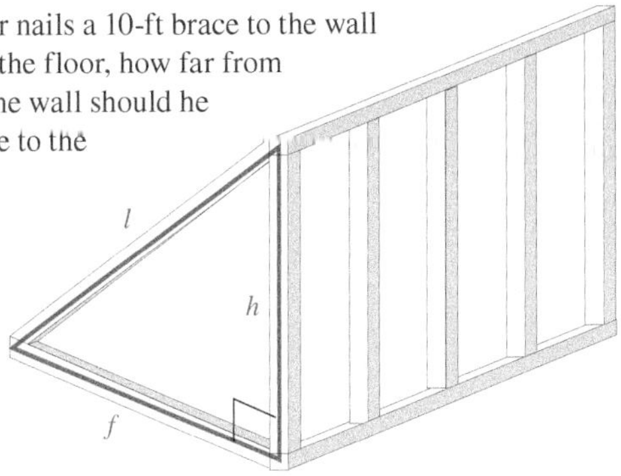

Use a graphing calculator.

85. Depreciation The formula

$$r = 1 - \sqrt[n]{\frac{S}{C}}$$

gives the annual depreciation rate r of a car that had an original cost of C dollars, a useful life of n years, and a trade-in value of S dollars. Find the annual depreciation rate of a car that cost \$22,000 and was sold 15 years later as salvage for \$900. Give the result to the nearest percent.

86. Savings accounts The interest rate r earned by a savings account after n compoundings is given by the formula

$$\sqrt[n]{\frac{V}{P}} - 1 = r$$

where V is the current value and P is the original principal. What interest rate r was paid on an account in which a deposit of \$1,000 grew to \$1,338.23 after 5 compoundings?

87. Marketing The number of wrenches that will be produced at a given price can be predicted by the formula $s = \sqrt{5x}$, where s is the supply (in thousands) and x is the price (in dollars). If the demand, d, for wrenches can be predicted by the formula $d = \sqrt{100 - 3x^2}$, find the equilibrium price.

88. Marketing The number of footballs that will be produced at a given price can be predicted by the formula $s = \sqrt{23x}$, where s is the supply (in thousands) and x is the price (in dollars). If the demand, d, for footballs can be predicted by the formula $d = \sqrt{312 - 2x^2}$, find the equilibrium price.

89. Medicine The resistance R to blood flow through an artery can be found using the formula

$$r = \sqrt[4]{\frac{8kl}{\pi R}}$$

where r is the radius of the artery, k is the viscosity of blood, and l is the length of the artery. Solve the formula for R.

90. Generating power The power P generated by a windmill is given by the formula

$$s = \sqrt[3]{\frac{P}{0.02}}$$

where s is the speed of the wind. Solve the formula for P.

WRITING ABOUT MATH

91. If both sides of an equation are raised to the same power, the resulting equation might not be equivalent to the original equation. Explain.

92. Explain why you must check each apparent solution of a radical equation.

SOMETHING TO THINK ABOUT

93. Solve: $\sqrt[3]{2x} = \sqrt{x}$

94. Solve: $\sqrt[4]{x} = \sqrt{\frac{x}{4}}$

Unless otherwise noted, all content on this page is © Cengage Learning.

Section 9.7 Complex Numbers

Objectives

1. Simplify an imaginary number.
2. Write a complex number in $a + bi$ form.
3. Determine whether two complex numbers are equal.
4. Simplify an expression containing complex numbers.
5. Rationalize the denominator of a fraction that contains a complex number.
6. Find a specified power of i and apply this pattern to simplifying an expression.
7. Find the absolute value of a complex number.

Vocabulary

imaginary number
complex number
complex conjugates
absolute value of a complex number

Getting Ready

Perform the following operations.

1. $(3x + 5) + (4x - 5)$

2. $(3x + 5) - (4x - 5)$

3. $(3x + 5)(4x - 5)$

4. $(3x + 5)(3x - 5)$

We have seen that square roots of negative numbers are not real numbers. However, there is a set of numbers, called the *complex numbers*, in which negative numbers do have square roots. In this section, we will discuss this broader set of numbers.

1 Simplify an imaginary number.

Consider the number $\sqrt{-3}$. Since no real number squared is -3, $\sqrt{-3}$ is not a real number. For years, people believed that numbers such as

$$\sqrt{-1}, \quad \sqrt{-3}, \quad \sqrt{-4}, \quad \text{and} \quad \sqrt{-9}$$

were nonsense. In the 17th century, René Descartes (1596–1650) called them **imaginary numbers**. Today, imaginary numbers have many important uses, such as describing the behavior of alternating current in electronics.

The imaginary number $\sqrt{-1}$ often is denoted by the letter i:

$$i = \sqrt{-1}$$

Because i represents the square root of -1, it follows that

$$i^2 = -1$$

Perspective

The Pythagoreans (ca. 500 B.C.) understood the universe as a harmony of whole numbers. They did not classify fractions as numbers, and were upset that $\sqrt{2}$ was not the ratio of whole numbers. For 2,000 years, little progress was made in the understanding of the various kinds of numbers.

The father of algebra, François Vieta (1540–1603), understood the whole numbers, fractions, and certain irrational numbers. But he was unable to accept negative numbers, and certainly not imaginary numbers.

René Descartes (1596–1650) thought these numbers to be nothing more than figments of his imagination, so he called them *imaginary numbers*. Leonhard Euler (1707–1783) used the letter i for $\sqrt{-1}$; Augustin Cauchy (1789–1857) used the term *conjugate*; and Carl Gauss (1777–1855) first used the word *complex*.

Today, we accept complex numbers without question, but it took many centuries and the work of many mathematicians to make them respectable.

If we assume that multiplication of imaginary numbers is commutative and associative, then

$$\begin{aligned}(2i)^2 &= 2^2i^2\\ &= 4(-1) \quad i^2 = -1\\ &= -4\end{aligned}$$

Since $(2i)^2 = -4$, $2i$ is a square root of -4, and we can write

$$\sqrt{-4} = 2i$$

This result also can be obtained by using the multiplication property of radicals:

$$\sqrt{-4} = \sqrt{4(-1)} = \sqrt{4}\sqrt{-1} = 2i$$

EXAMPLE 1 Simplify:

a. $\sqrt{-25}$ **b.** $\sqrt{\frac{-100}{49}}$

Solution We can use the multiplication property of radicals to simplify any imaginary number. For example,

a. $\sqrt{-25} = \sqrt{25(-1)} = \sqrt{25}\sqrt{-1} = 5i$

b. $\sqrt{\frac{-100}{49}} = \sqrt{\frac{100}{49}(-1)} = \frac{\sqrt{100}}{\sqrt{49}}\sqrt{-1} = \frac{10}{7}i$

SELF CHECK 1 Simplify: **a.** $\sqrt{-36}$ **b.** $\sqrt{\frac{-4}{9}}$

These examples illustrate the following rule.

PROPERTIES OF RADICALS

If at least one of a and b is a nonnegative real number, then

$$\sqrt{ab} = \sqrt{a}\sqrt{b} \quad \text{and} \quad \sqrt{\frac{a}{b}} = \frac{\sqrt{a}}{\sqrt{b}} \quad (b \neq 0)$$

COMMENT If a and b are both negative, then $\sqrt{ab} \neq \sqrt{a}\sqrt{b}$. Recall, the simplified form of a radical expression does not contain a negative radicand. Therefore the $\sqrt{-16}$ and the $\sqrt{-4}$ must be simplified prior to multiplying. For example, if $a = -16$ and $b = -4$, we have

$$\sqrt{(-16)}\sqrt{(-4)} = (4i)(2i) = 8i^2 = 8(-1) = -8$$

The imaginary numbers are a subset of a set of numbers called the *complex numbers.*

COMPLEX NUMBERS

A **complex number** is any number that can be written in the standard form $a + bi$, where a and b are real numbers and $i = \sqrt{-1}$.

In the complex number $a + bi$, a is called the *real part*, and b is called the *imaginary part.*

If $b = 0$, the complex number $a + bi$ is a real number. If $b \neq 0$ and $a = 0$, the complex number $0 + bi$ (or just bi) is an imaginary number.

Any imaginary number can be expressed in bi form. For example,

COMMENT The expression $\sqrt{3}i$ is sometimes written as $i\sqrt{3}$ to make it clear that i is not part of the radicand.

$$\sqrt{-1} = i$$
$$\sqrt{-9} = \sqrt{9(-1)} = \sqrt{9}\sqrt{-1} = 3i$$
$$\sqrt{-3} = \sqrt{3(-1)} = \sqrt{3}\sqrt{-1} = \sqrt{3}i$$

The relationship among the real numbers, the imaginary numbers, and the complex numbers is shown in Figure 9-18.

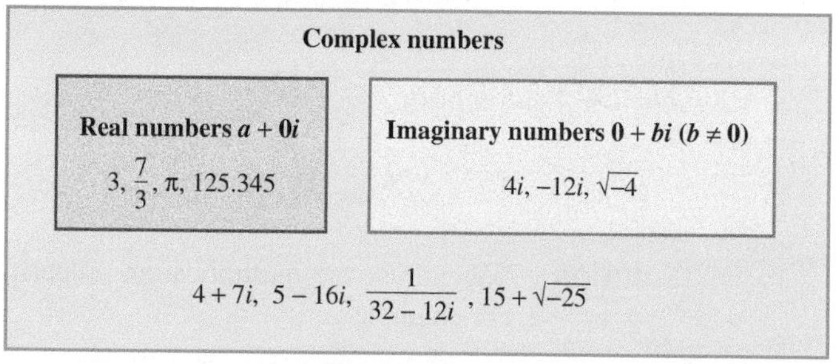

Figure 9-18

2 Write a complex number in $a + bi$ form.

The next example shows how to write complex numbers in $(a + bi)$ form. It is common to use $(a - bi)$ as a substitute for $a + (-b)i$.

EXAMPLE 2 Write each number in $a + bi$ form.

a. $7 = 7 + 0i$

b. $3i = 0 + 3i$

c. $4 - \sqrt{-16} = 4 - \sqrt{-1(16)}$
$= 4 - \sqrt{16}\sqrt{-1}$
$= 4 - 4i$

d. $5 + \sqrt{-11} = 5 + \sqrt{-1(11)}$
$= 5 + \sqrt{11}\sqrt{-1}$
$= 5 + \sqrt{11}i$

SELF CHECK 2 Write $3 - \sqrt{-25}$ in $a + bi$ form.

Unless otherwise noted, all content on this page is © Cengage Learning.

3 Determine whether two complex numbers are equal.

EQUALITY OF COMPLEX NUMBERS

The complex numbers $a + bi$ and $c + di$ are equal if and only if

$$a = c \quad \text{and} \quad b = d$$

Because of the preceding definition, complex numbers are equal when their real parts are equal and their imaginary parts are equal.

EXAMPLE 3

a. $2 + 3i = \sqrt{4} + \frac{6}{2}i$ because $2 = \sqrt{4}$ and $3 = \frac{6}{2}$.

b. $4 - 5i = \frac{12}{3} - \sqrt{25}i$ because $4 = \frac{12}{3}$ and $-5 = -\sqrt{25}$.

c. $x + yi = 4 + 7i$ if and only if $x = 4$ and $y = 7$.

SELF CHECK 3 Is $2 + 3i = \sqrt{4} - \frac{6}{2}i$?

4 Simplify an expression containing complex numbers.

Now that we can simplify a complex number, we will move to adding, subtracting, and multiplying them.

ADDITION AND SUBTRACTION OF COMPLEX NUMBERS

Complex numbers are added and subtracted as if they were binomials:

$$(a + bi) + (c + di) = (a + c) + (b + d)i$$
$$(a + bi) - (c + di) = (a + bi) + (-c - di) = (a - c) + (b - d)i$$

The preceding definition suggests that when adding or subtracting two complex numbers, we add or subtract the real parts and then add or subtract the imaginary parts.

EXAMPLE 4 Perform the operations.

a. $(8 + 4i) + (12 + 8i) = 8 + 4i + 12 + 8i$
$= 20 + 12i$

b. $(7 - 4i) + (9 + 2i) = 7 - 4i + 9 + 2i$
$= 16 - 2i$

c. $(-6 + i) - (3 - 4i) = -6 + i - 3 + 4i$
$= -9 + 5i$

d. $(2 - 4i) - (-4 + 3i) = 2 - 4i + 4 - 3i$
$= 6 - 7i$

SELF CHECK 4 Perform the operations.

a. $(3 - 5i) + (-2 + 7i)$ **b.** $(3 - 5i) - (-2 + 7i)$

To multiply a complex number by an imaginary number, we use the distributive property to remove parentheses and simplify. For example,

$$\begin{aligned} -5i(4-8i) &= -5i(4) - (-5i)8i && \text{Use the distributive property to remove parentheses.} \\ &= -20i + 40i^2 && \text{Simplify.} \\ &= -20i + 40(-1) && \text{Remember that } i^2 = -1. \\ &= -40 - 20i \end{aligned}$$

To multiply two complex numbers, we use the following definition.

MULTIPLYING COMPLEX NUMBERS

Complex numbers are multiplied as if they were binomials, with $i^2 = -1$:

$$\begin{aligned} (a+bi)(c+di) &= ac + adi + bci + bdi^2 \\ &= ac + adi + bci + bd(-1) \\ &= (ac - bd) + (ad + bc)i \end{aligned}$$

EXAMPLE 5 Multiply the complex numbers.

a. $$\begin{aligned} 2i(7-3i) &= 14i - 6i^2 && \text{Use the distributive property to remove parentheses.} \\ &= 14i + 6 && i^2 = -1 \\ &= 6 + 14i \end{aligned}$$

b. $$\begin{aligned} (2+3i)(3-2i) &= 6 - 4i + 9i - 6i^2 && \text{Use the distributive property to remove parentheses.} \\ &= 6 + 5i + 6 && i^2 = -1, \text{ combine } -4i \text{ and } 9i. \\ &= 12 + 5i \end{aligned}$$

c. $$\begin{aligned} (3+i)(1+2i) &= 3 + 6i + i + 2i^2 && \text{Use the distributive property to remove parentheses.} \\ &= 3 + 7i - 2 && i^2 = -1, \text{ combine } 6i \text{ and } i. \\ &= 1 + 7i \end{aligned}$$

d. $$\begin{aligned} (-4+2i)(2+i) &= -8 - 4i + 4i + 2i^2 && \text{Use the distributive property to remove parentheses.} \\ &= -8 - 2 && i^2 = -1, \text{ combine } -4i \text{ and } 4i. \\ &= -10 \end{aligned}$$

SELF CHECK 5 Multiply: **a.** $3(4-5i)$ **b.** $(-2+3i)(3-2i)$

COMPLEX CONJUGATES

The complex numbers $(a+bi)$ and $(a-bi)$ are called **complex conjugates**.

For example,

$(3+4i)$ and $(3-4i)$ are complex conjugates.

$(5-7i)$ and $(5+7i)$ are complex conjugates.

EXAMPLE 6 Find the product of $(3 + i)$ and its complex conjugate.

Solution The complex conjugate of $(3 + i)$ is $(3 - i)$. We can find the product as follows:

$$\begin{aligned}(3 + i)(3 - i) &= 9 - 3i + 3i - i^2 && \text{Use the distributive property.}\\ &= 9 - i^2 && \text{Combine like terms.}\\ &= 9 - (-1) && i^2 = -1\\ &= 10\end{aligned}$$

SELF CHECK 6 Multiply: $(2 + 3i)(2 - 3i)$

The product of the complex number $a + bi$ and its complex conjugate $a - bi$ is the real number $a^2 + b^2$, as the following work shows:

$$\begin{aligned}(a + bi)(a - bi) &= a^2 - abi + abi - b^2i^2 && \text{Use the distributive property to remove parentheses.}\\ &= a^2 - b^2(-1) && i^2 = -1\\ &= a^2 + b^2\end{aligned}$$

5 Rationalize the denominator of a fraction that contains a complex number.

If $b \neq 0$, the complex number $a + bi$ contains the square root $i = \sqrt{-1}$. Since a square root cannot remain in the denominator of a fraction, we often have to rationalize a denominator when dividing complex numbers.

EXAMPLE 7 Divide and write the result in $a + bi$ form: $\dfrac{1}{3 + i}$

Solution We can rationalize the denominator by multiplying the numerator and the denominator by the complex conjugate of the denominator.

$$\begin{aligned}\frac{1}{3 + i} &= \frac{1}{3 + i} \cdot \frac{3 - i}{3 - i} && \text{Multiply the numerator and denominator by the complex conjugate of the denominator.}\\ &= \frac{3 - i}{9 - 3i + 3i - i^2} && \text{Multiply.}\\ &= \frac{3 - i}{9 - (-1)} && i^2 = -1\\ &= \frac{3 - i}{10} && \text{Subtract.}\\ &= \frac{3}{10} - \frac{1}{10}i && \text{Write the result in } a + bi \text{ form.}\end{aligned}$$

SELF CHECK 7 Divide and write the result in $a + bi$ form. $\dfrac{1}{5 - i}$

EXAMPLE 8 Write $\frac{3-i}{2+i}$ in $a + bi$ form.

Solution We multiply the numerator and the denominator of the fraction by the complex conjugate of the denominator.

$$\frac{3-i}{2+i} = \frac{3-i}{2+i} \cdot \frac{2-i}{2-i} \qquad \frac{2-i}{2-i} = 1$$

$$= \frac{6-3i-2i+i^2}{4-2i+2i-i^2} \qquad \text{Multiply the numerators and multiply the denominators.}$$

$$= \frac{6-5i+(-1)}{4-(-1)} \qquad \text{Combine like terms and substitute } i^2 \text{ with } -1.$$

$$= \frac{5-5i}{4-(-1)} \qquad \text{Subtract.}$$

$$= \frac{5(1-i)}{5} \qquad \text{Factor out 5, the GCF.}$$

$$= 1 - i \qquad \text{Simplify.}$$

SELF CHECK 8 Rationalize the denominator: $\frac{2+i}{5-i}$

EXAMPLE 9 Write $\frac{4+\sqrt{-16}}{2+\sqrt{-4}}$ in $a + bi$ form.

Solution

$$\frac{4+\sqrt{-16}}{2+\sqrt{-4}} = \frac{4+4i}{2+2i} \qquad \text{Write each number in } a + bi \text{ form.}$$

$$= \frac{2\overset{1}{\cancel{(2+2i)}}}{\underset{1}{\cancel{2+2i}}} \qquad \text{Factor out 2, the GCF, and simplify.}$$

$$= 2 + 0i$$

SELF CHECK 9 Divide: $\frac{6+\sqrt{-81}}{2+\sqrt{-9}}$

COMMENT To avoid mistakes, always write complex numbers in $a + bi$ form prior to performing any operations with complex numbers.

6 Find a specified power of *i* and apply this pattern to simplifying an expression.

The powers of i produce an interesting pattern:

$$i = \sqrt{-1} = i \qquad i^5 = i^4 i = 1i = i$$

$$i^2 = \left(\sqrt{-1}\right)^2 = -1 \qquad i^6 = i^4 i^2 = 1(-1) = -1$$

$$i^3 = i^2 i = -1i = -i \qquad i^7 = i^4 i^3 = 1(-i) = -i$$

$$i^4 = i^2 i^2 = (-1)(-1) = 1 \qquad i^8 = i^4 i^4 = (1)(1) = 1$$

The pattern continues: $i, -1, -i, 1, \ldots$. When simplifying a specified power of i the calculation is easier if we can write the expression in terms of i^4 since $i^4 = 1$. If there is a remainder when the exponent is divided by 4, use the pattern $i, -1, -i, 1$ to simplify the expression as illustrated in the next example.

EXAMPLE 10 Simplify: i^{29}

Solution We note that 29 divided by 4 gives a quotient of 7 and a remainder of 1. Thus, $29 = 4 \cdot 7 + 1$, and

$$\begin{aligned} i^{29} &= i^{4\cdot 7+1} && 29 = 4\cdot 7 + 1 \\ &= (i^4)^7 \cdot i && i^{4\cdot 7+1} = i^{4\cdot 7} \cdot i^1 = (i^4)^7 \cdot i \\ &= 1^7 \cdot i && i^4 = 1 \\ &= i \end{aligned}$$

SELF CHECK 10 Simplify: i^{31}

The results of Example 10 illustrate the following process.

POWERS OF i

If n is a natural number that has a remainder of r when divided by 4, then

$$i^n = i^r.$$

When n is divisible by 4, the remainder r is 0 and $i^0 = 1$.

EXAMPLE 11 Simplify: i^{55}

Solution We divide 55 by 4 and get a remainder of 3. Therefore,

$$i^{55} = i^3 = -i$$

SELF CHECK 11 Simplify: i^{62}

EXAMPLE 12 Simplify each expression. If a denominator has a factor of i, rationalize the denominator. Write the result in $a + bi$ form.

a. $$\begin{aligned} 2i^2 + 4i^3 &= 2(-1) + 4(-i) \\ &= -2 - 4i \end{aligned}$$

b. $$\begin{aligned} \frac{3}{2i} &= \frac{3}{2i} \cdot \frac{i}{i} && \frac{i}{i} = 1 \\ &= \frac{3i}{2i^2} \\ &= \frac{3i}{2(-1)} \\ &= \frac{3i}{-2} \\ &= 0 - \frac{3}{2}i \end{aligned}$$

c. $$\begin{aligned} -\frac{5}{i} &= -\frac{5}{i} \cdot \frac{i}{i} && \frac{i}{i} = 1 \\ &= -\frac{5(i)}{i^2} \\ &= -\frac{5i}{-1} \\ &= 5i \\ &= 0 + 5i \end{aligned}$$

d. $$\begin{aligned} \frac{6}{i^3} &= \frac{6i}{i^3 i} && \frac{i}{i} = 1 \\ &= \frac{6i}{i^4} \\ &= \frac{6i}{1} \\ &= 6i \\ &= 0 + 6i \end{aligned}$$

SELF CHECK 12 Simplify and write the result in $a + bi$ form. **a.** $3i^3 - 2i^2$ **b.** $\frac{2}{3i}$

7 Find the absolute value of a complex number.

ABSOLUTE VALUE OF A COMPLEX NUMBER

The **absolute value** of the complex number $a + bi$ is $\sqrt{a^2 + b^2}$. In symbols,

$$|a + bi| = \sqrt{a^2 + b^2}$$

EXAMPLE 13 Find each absolute value.

a.
$$\begin{aligned} |3 + 4i| &= \sqrt{3^2 + 4^2} \\ &= \sqrt{9 + 16} \\ &= \sqrt{25} \\ &= 5 \end{aligned}$$

b.
$$\begin{aligned} |3 - 4i| &= \sqrt{3^2 + (-4)^2} \\ &= \sqrt{9 + 16} \\ &= \sqrt{25} \\ &= 5 \end{aligned}$$

c.
$$\begin{aligned} |-5 - 12i| &= \sqrt{(-5)^2 + (-12)^2} \\ &= \sqrt{25 + 144} \\ &= \sqrt{169} \\ &= 13 \end{aligned}$$

d.
$$\begin{aligned} |a + 0i| &= \sqrt{a^2 + 0^2} \\ &= \sqrt{a^2} \\ &= |a| \end{aligned}$$

SELF CHECK 13 Evaluate: $|5 + 12i|$

SELF CHECK ANSWERS

1. a. $6i$ **b.** $\frac{2}{3}i$ **2.** $3 - 5i$ **3.** no **4. a.** $1 + 2i$ **b.** $5 - 12i$ **5. a.** $12 - 15i$ **b.** $13i$ **6.** 13 **7.** $\frac{5}{26} + \frac{1}{26}i$ **8.** $\frac{9}{26} + \frac{7}{26}i$ **9.** $3 + 0i$ **10.** $-i$ **11.** -1 **12. a.** $2 - 3i$ **b.** $0 - \frac{2}{3}i$ **13.** 13

NOW TRY THIS

1. Simplify: $-\sqrt{-8}\sqrt{-2}$
2. Evaluate $3x^2 - 2x - 4$ for $x = 2 - 3i$.

9.7 Exercises

WARM-UPS *Identify each number as a real number or not a real number.*

1. $\sqrt{-9}$
2. $-\sqrt{25}$
3. $\sqrt{a}$ if $a \geq 0$
4. $\sqrt{x}$ if $x < 0$
5. $-\sqrt{-1}$
6. $\sqrt{-1}$

Find the conjugate for each expression.

7. $3 + \sqrt{5}$
8. $-7 - \sqrt{2}$

Combine like terms.

9. $5 + 3x + 7 + 4x$
10. $-3 - 4x + 9 - 6x$

REVIEW *Perform each operation.*

11. $\frac{x^2 - x - 6}{9 - x^2} \cdot \frac{x^2 + x - 6}{x^2 - 4}$ **12.** $\frac{3x + 4}{x - 2} + \frac{x - 4}{x + 2}$

13. Wind speed A plane that can fly 200 mph in still air makes a 330-mile flight with a tail wind and returns, flying into the same wind. Find the speed of the wind if the total flying time is $3\frac{1}{3}$ hours.

14. Finding rates A student drove a distance of 135 miles at an average speed of 50 mph. How much faster would he have to drive on the return trip to save 30 minutes of driving time?

VOCABULARY AND CONCEPTS *Fill in the blanks.*

15. $\sqrt{-1} =$ ___ **16.** $i^6 =$ ___

17. $i^7 =$ ___ **18.** $i^8 =$ ___

19. $\sqrt{-1}, \sqrt{-3}, \sqrt{-4}$ are examples of ________ numbers.

20. $\sqrt{ab} =$ _______, provided a and b are not both negative.

21. $\sqrt{\frac{a}{b}} =$ _____ $(b \neq 0)$, provided a and b are not both negative.

22. $3 + 5i$, $2 - 7i$, and $5 - \frac{1}{2}i$ are examples of _______ numbers.

23. The real part of $5 + 7i$ is __. The imaginary part is __.

24. $a + bi = c + di$ if and only if $a =$ __ and $b =$ __.

25. $a + bi$ and $a - bi$ are called complex _________.

26. $|a + bi| =$ ________

GUIDED PRACTICE *Write each imaginary number in simplified form.* SEE EXAMPLE 1. (OBJECTIVE 1)

27. $\sqrt{-25}$ **28.** $\sqrt{-16}$

29. $\sqrt{-121}$ **30.** $\sqrt{-81}$

31. $\sqrt{-7}$ **32.** $\sqrt{-29}$

33. $\sqrt{-8}$ **34.** $\sqrt{-24}$

Write each number in $a + bi$ *form.* SEE EXAMPLE 2. (OBJECTIVE 2)

35. 9 **36.** $4i$

37. $9 + \sqrt{-9}$ **38.** $16 + \sqrt{-36}$

Determine whether the complex numbers are equal. SEE EXAMPLE 3. (OBJECTIVE 3)

39. $3 + 7i, \sqrt{9} + (5 + 2)i$ **40.** $\sqrt{4} + \sqrt{25}i, 2 - (-5)i$

41. $\sqrt{4} + \sqrt{-4}, 2 - 2i$ **42.** $\sqrt{-9} - i, 4i$

Simplify each expression. Write all answers in standard form. SEE EXAMPLE 4. (OBJECTIVE 4)

43. $(4 + 7i) + (3 - 5i)$ **44.** $(5 + 3i) - (6 - 9i)$

45. $(7 - 3i) - (4 + 2i)$ **46.** $(8 + 3i) + (-7 - 2i)$

47. $(8 + 5i) + (7 + 2i)$ **48.** $(-6 - 8i) - (-9 - 3i)$

49. $(1 + i) - 2i + (5 - 7i)$ **50.** $(-9 + i) - 5i + (2 + 7i)$

Simplify each expression. Write all answers in standard form. SEE EXAMPLE 5. (OBJECTIVE 4)

51. $6i(3 - 4i)$ **52.** $-4i(3 + 4i)$

53. $-5i(5 - 5i)$ **54.** $2i(7 + 2i)$

55. $(2 + i)(3 - i)$ **56.** $(4 - i)(2 + i)$

57. $(5 + 2i)(4 - 6i)$ **58.** $(3 - 2i)(4 - 3i)$

59. $(4 + \sqrt{3}i)(3 - \sqrt{3}i)$ **60.** $(5 + \sqrt{3}i)(2 - \sqrt{3}i)$

61. $(4 + 3i)^2$ **62.** $(3 - 2i)^2$

Find the product of the complex number and its complex conjugate. SEE EXAMPLE 6. (OBJECTIVE 4)

63. $6 - 5i$ **64.** $7 - 2i$

65. $3 + 4i$ **66.** $5 + 2i$

Divide and write each expression in standard form. SEE EXAMPLE 7. (OBJECTIVE 5)

67. $\frac{5}{2 - i}$ **68.** $\frac{26}{3 - 2i}$

69. $\frac{10}{5 - i}$ **70.** $\frac{4}{2 + i}$

Divide and write each expression in standard form. SEE EXAMPLE 8. (OBJECTIVE 5)

71. $\frac{3 - 2i}{3 + 2i}$ **72.** $\frac{2 + 3i}{2 - 3i}$

73. $\frac{3 + 2i}{3 + i}$ **74.** $\frac{2 - 5i}{2 + 5i}$

Divide and write each expression in standard form. SEE EXAMPLE 9. (OBJECTIVE 5)

75. $\frac{-12}{7 - \sqrt{-1}}$ **76.** $\frac{4}{3 + \sqrt{-1}}$

77. $\frac{7}{5 - \sqrt{-9}}$ **78.** $\frac{3}{2 + \sqrt{-25}}$

Simplify each expression. SEE EXAMPLES 10–11. (OBJECTIVE 6)

79. i^5 **80.** i^7

81. i^{10} **82.** i^{12}

83. i^{37} **84.** i^{19}

85. i^{27} **86.** i^{22}

87. i^{64} **88.** i^{82}

89. i^{97} **90.** i^{200}

Simplify each expression. SEE EXAMPLE 12. (OBJECTIVES 5–6)

91. $5i^3 + 2i^2$ **92.** $4i^2 - 3i^3$

93. $\frac{1}{i}$ **94.** $\frac{1}{i^3}$

95. $\frac{4}{5i^3}$ **96.** $\frac{2}{3i}$

97. $\frac{3i}{8\sqrt{-9}}$ **98.** $\frac{5i^3}{2\sqrt{-4}}$

99. $\frac{-3}{5i^5}$ **100.** $\frac{-4}{6i^7}$

Find each value. SEE EXAMPLE 13. (OBJECTIVE 7)

101. $|6 + 8i|$ **102.** $|12 + 5i|$

103. $|12 - 5i|$ **104.** $|3 - 4i|$

105. $|5 + 7i|$ **106.** $|6 - 5i|$

107. $\left|\frac{3}{5} - \frac{4}{5}i\right|$ **108.** $\left|\frac{5}{13} + \frac{12}{13}i\right|$

ADDITIONAL PRACTICE *Are the two numbers equal?*

109. $8 + 5i, 2^3 + \sqrt{25}i^3$

110. $4 - 7i, -4i^2 + 7i^3$

Simplify each expression. Write the answer in standard form.

111. $(8 - \sqrt{-1})(-2 - \sqrt{-16})$

112. $(-1 + \sqrt{-4})(2 + \sqrt{-9})$

113. $(2 + 3i)^2$ **114.** $(1 - 3i)^2$

115. $\frac{5i}{6 + 2i}$ **116.** $\frac{-4i}{2 - 6i}$

117. $\frac{\sqrt{5} - \sqrt{3}i}{\sqrt{5} + \sqrt{3}i}$

118. $\frac{\sqrt{3} + \sqrt{2}i}{\sqrt{3} - \sqrt{2}i}$

119. $\left(\frac{i}{3 + 2i}\right)^2$

120. $\left(\frac{5 + i}{2 + i}\right)^2$

121. $(5 + 3i) - (3 - 5i) + \sqrt{-1}$

122. $(8 + 7i) - (-7 - \sqrt{-64}) + (3 - i)$

123. $(-8 - \sqrt{3}i) - (7 - 3\sqrt{3}i)$

124. $(4 + 3\sqrt{5}i) + (-6 - 4\sqrt{5}i)$

125. $(2 + i)(2 - i)(1 + i)$

126. $(3 + 2i)(3 - 2i)(i + 1)$

127. $(3 + i)[(3 - 2i) + (2 + i)]$

128. $(2 - 3i)[(5 - 2i) - (2i + 1)]$

APPLICATIONS *In electronics, the formula $V = IR$ is called Ohm's law. It gives the relationship in a circuit between the voltage V (in volts), the current I (in amperes), and the resistance R (in ohms).*

129. Electronics Find V when $I = 2 - 3i$ amperes and $R = 2 + i$ ohms.

130. Electronics Find R when $I = 3 - 2i$ amperes and $V = 18 + i$ volts.

In electronics, the formula $Z = \frac{V}{I}$ is used to find the impedance Z of a circuit, where V is the voltage and I is the current.

131. Electronics Find the impedance of a circuit when the voltage is $1.7 + 0.5i$ and the current is $0.5i$.

132. Electronics Find the impedance of a circuit when the voltage is $1.6 - 0.4i$ and the current is $-0.2i$.

WRITING ABOUT MATH

133. Determine how to decide whether two complex numbers are equal.

134. Define the complex conjugate of a complex number.

135. Show that $1 - 5i$ is a solution of $x^2 - 2x + 26 = 0$.

136. Show that $3 - 2i$ is a solution of $x^2 - 6x + 13 = 0$.

137. Show that i is a solution of $x^4 - 3x^2 - 4 = 0$.

138. Show that $2 + i$ is *not* a solution of $x^2 + x + 1 = 0$.

SOMETHING TO THINK ABOUT

139. Rationalize the numerator: $\frac{3 - i}{2}$

140. Rationalize the numerator: $\frac{2 + 3i}{2 - 3i}$

Projects

PROJECT 1

The size of a TV screen is measured along the diagonal of its screen, as shown in the illustrations. The screen of a traditional TV has an aspect ratio of 4:3. This means that the ratio of the width of the screen to its height is $\frac{4}{3}$. The screen of a wide-screen set has an aspect ratio of 16:9. This means that the ratio of the width of the screen to its height is $\frac{16}{9}$.

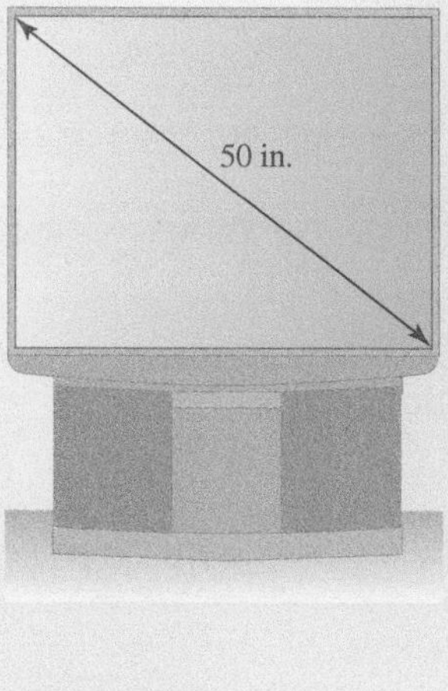

Unless otherwise noted, all content on this page is © Cengage Learning.

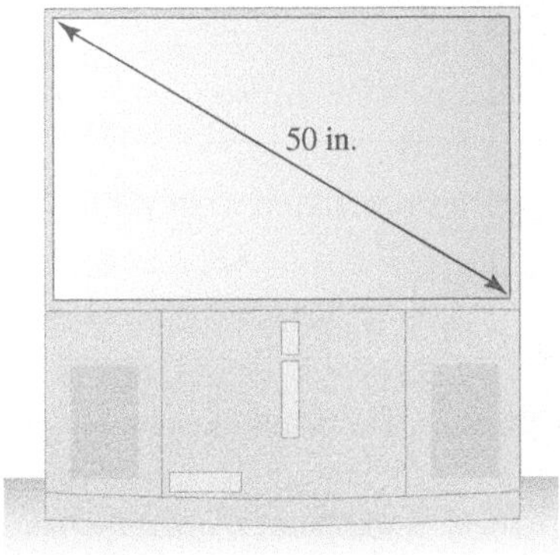

a. Find the width and height of the traditional-screen set shown in the illustration on the previous page. $\left(Hint: \frac{4}{3} = \frac{4x}{3x}.\right)$

b. Find the viewing area of the traditional-screen set in square inches.

c. Find the width and height of the wide-screen set shown in the illustration above.

d. Find the viewing area of the wide-screen set in square inches.

e. Which set has the larger viewing area? Give the answer as a percent.

PROJECT 2

Tom and Brian arrange to have a bicycle race. Each leaves his own house at the same time and rides to the other's house, whereupon the winner of the race calls his own house and leaves a message for the loser. A map of the race is shown in the illustration. Brian stays on the highway, averaging 21 mph. Tom knows that he and Brian are evenly matched when biking on the highway, so he cuts across country for the first part of his trip, averaging 15 mph. When Tom reaches the highway at point A, he turns right and follows the highway, averaging 21 mph.

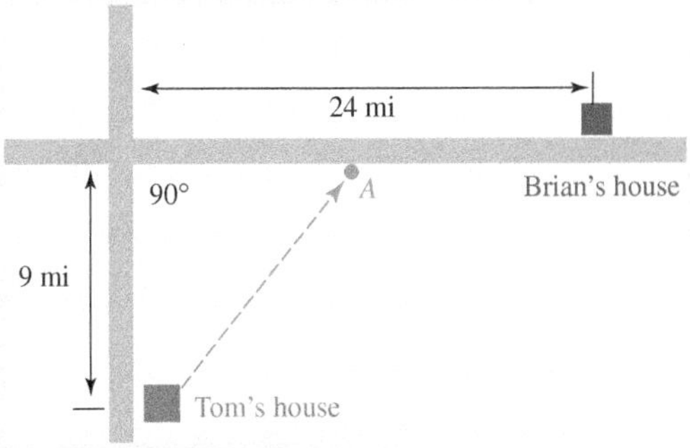

Tom and Brian never meet during the race and, amazingly, the race is a tie. Each of them calls the other at exactly the same moment!

a. How long (to the nearest second) did it take each person to complete the race?

b. How far from the intersection of the two highways is point A? (*Hint*: Set the travel times for Brian and Tom equal to each other. You may find two answers, but only one of them matches all of the information.)

c. Show that if Tom had started straight across country for Brian's house (in order to minimize the distance he had to travel), he would have lost the race. By how much time (to the nearest second) would he have lost? Then show that if Tom had biked across country to a point 9 miles from the intersection of the two highways, he would have won the race. By how much time (to the nearest second) would he have won?

Unless otherwise noted, all content on this page is © Cengage Learning.

Reach for Success

EXTENSION OF TAKING NOTES IN CLASS

You've learned to use two colors to take notes in class. In the previous exercise we listed *what* we did in each step. In this excercise, we will list both *what* steps we took and *why*.

To make an example easier to understand, we explain the steps we take, and why we take them, to find the answer. This serves two purposes:

- to use the vocabulary necessary for success in your mathematics course; and
- to help you understand the mathematical processes.

Find the slope and the y-intercept for the line with equation

$5x + 3y = 12$

In the blanks below list the steps taken to complete the exercise. The explanation of *why* has already been listed for you.

$5x + 3y = 12x$	Original exercise.
$\begin{array}{r} -5x \quad -5x \\ \hline \end{array}$	____________________ (to isolate the y-term)
$\dfrac{3y}{3} = -\dfrac{5x}{3} + \dfrac{12}{3}$	____________________ (to obtain a leading coefficient of 1 on y)
	____________________
$y = -\frac{5}{3}x + 4$	This is in $y = mx + b$ form. (slope/y-intercept form)
	____________________
The slope is $-\frac{5}{3}$ The y-intercept is (0, 4)	The slope is the coefficient of the x-term, and the y-intercept is a point that must be expressed as an ordered pair.

For each step taken in answering the question, state

- what steps were taken,
- explain why we took those steps.

Find the slope of the line perpendicular to the line $3x + y = 15$

$3x + y = 15$	Step: ____________________
$\begin{array}{r} -3x \quad -3x \\ \hline \end{array}$	(Why?) ____________________
$y = -3x + 15$	
The slope of the line is _________.	(Why?) ____________________
The slope of a line perpendicular to this line is _________.	(Why?) ____________________

Fill in the last two blanks of the exercise in the second column and explain how you arrived at those answers in the third column.

You can use the techniques to reinforce your understanding of the material, even if your instructor does not use these methods when explaining problems in class.

9 Review

SECTION 9.1 Radical Expressions

DEFINITIONS AND CONCEPTS	EXAMPLES
Simplifying radicals: Assume n is a natural number greater than 1 and x is a real number.	
If $x > 0$, then $\sqrt[n]{x}$ is the positive number such that $(\sqrt[n]{x})^n = x$.	$\sqrt{25} = 5$ because $5^2 = 25$ (the principal square root).
If $x = 0$, then $\sqrt[n]{x} = 0$.	$\sqrt[4]{0} = 0$ because $0^4 = 0$.
If $x < 0$, and n is odd, $\sqrt[n]{x}$ is the real number such that $(\sqrt[n]{x})^n = x$.	$\sqrt[3]{-64} = -4$ because $(-4)^3 = -64$.
If $x < 0$, and n is even, $\sqrt[n]{x}$ is not a real number.	$\sqrt{-4}$ is not a real number because there is no real number when squared that equals -4.
If n is an even natural number, $\sqrt[n]{a^n} = \lvert a \rvert$ If n is an odd natural number, greater than 1, $\sqrt[n]{a^n} = a$	Simplify. $\sqrt{64x^2} = 8\lvert x \rvert$ Absolute value bars are necessary because x could be a negative number. $\sqrt[3]{8x^3} = -2x$ Absolute value bars are not necessary because the index is odd.
Finding the domain of a radical function: If $f(x) = \sqrt[n]{x}$, then the domain of $f(x)$ is $(-\infty, \infty)$ when n is odd. If n is even, find the domain by setting the radicand greater than or equal to 0.	To find the domain of $g(x) = \sqrt[3]{x - 9}$, we note that in a cube root the radicand can be any real number. Therefore, x can be any real number, and the domain is $(-\infty, \infty)$. To find the domain of $f(x) = \sqrt{x + 4}$, set the radicand to be greater than or equal to 0, and solve for x. $x + 4 \geq 0$ The radicand must be ≥ 0. $x \geq -4$ Subtract 4 from each side. The domain is $[-4, \infty)$.
Standard deviation of a data set: Standard deviation $= \sqrt{\dfrac{\text{sum of the squares of the differences from the mean}}{\text{number of differences}}}$	Find the standard deviation of the data set 1, 3, 4, 8.

Original terms	Mean	Difference	Square of the differences
1	4	−3	9
3	4	−1	1
4	4	0	0
8	4	4	16

The sum of the squares of the differences is 26 and there are 4 values in the data set.

The standard deviation is $\sqrt{\dfrac{26}{4}} \approx 2.549509757$.

To the nearest hundredth, the standard deviation is 2.55.

REVIEW EXERCISES

Simplify each radical. Assume that x can be any number.

1. $\sqrt{81}$ **2.** $-\sqrt{169}$

3. $-\sqrt{36}$ **4.** $\sqrt{225}$

5. $\sqrt[3]{-27}$ **6.** $-\sqrt[3]{216}$

7. $\sqrt[4]{625}$ **8.** $\sqrt[5]{-32}$

9. $\sqrt{25x^2}$ **10.** $\sqrt{x^2 + 6x + 9}$

11. $\sqrt[3]{27a^6b^3}$ **12.** $\sqrt[4]{256x^8y^4}$

Graph each function and state the domain and range.

13. $f(x) = \sqrt{x + 2}$

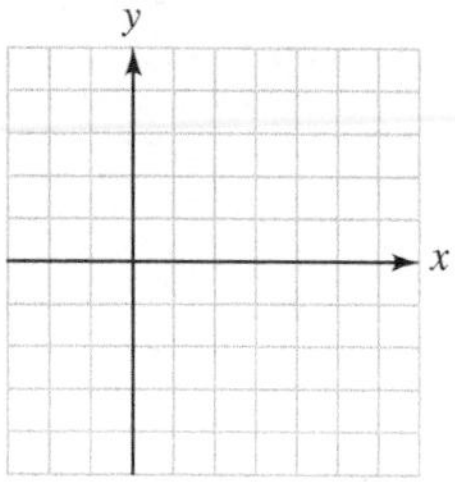

14. $f(x) = -\sqrt{x - 1}$

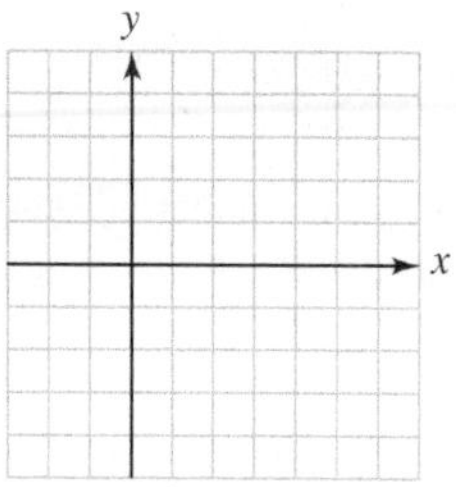

15. $f(x) = -\sqrt{x} + 2$

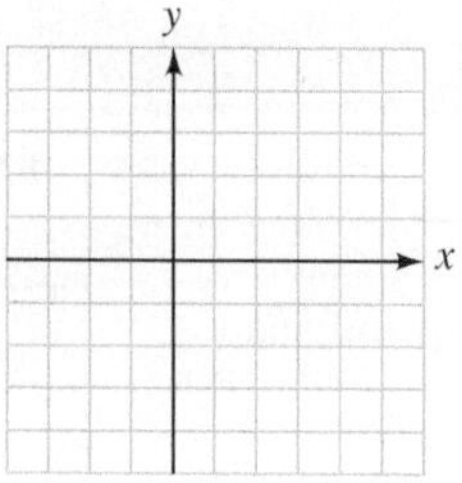

16. $f(x) = -\sqrt[3]{x} + 3$

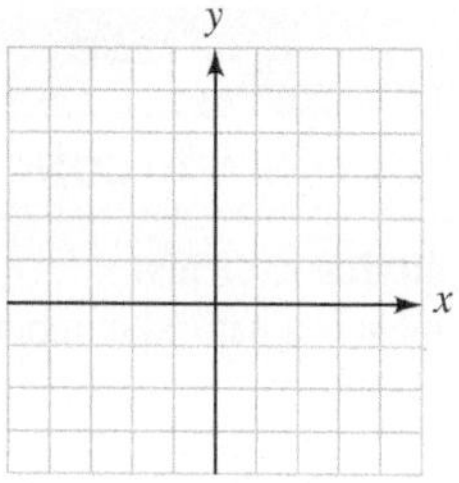

Consider the distribution 4, 8, 12, 16, 20.

17. Find the mean of the distribution.

18. Find the standard deviation.

SECTION 9.2 Applications of the Pythagorean Theorem and the Distance Formula

DEFINITIONS AND CONCEPTS	EXAMPLES
The Pythagorean theorem: If a and b are the lengths of the legs of a right triangle and c is the length of the hypotenuse, then $a^2 + b^2 = c^2$	To find the length of the hypotenuse of a right triangle with legs of length 9 ft and 12 ft, proceed as follows: $a^2 + b^2 = c^2$ Use the Pythagorean theorem. $(\mathbf{9})^2 + (\mathbf{12})^2 = c^2$ Substitute the values. $81 + 144 = c^2$ Square each value. $225 = c^2$ Add. $\sqrt{225} = c$ Since c is a length, take the positive square root. $15 = c$ $\sqrt{225} = 15$ The hypotenuse is 15 ft.
The distance formula: The distance between two points, (x_1, y_1) and (x_2, y_2), on a coordinate plane is $d = \sqrt{(x_2 - x_1)^2 + (y_2 - y_1)^2}$	To find the distance between $(8, -1)$ and $(5, 3)$, use the distance formula: $d = \sqrt{(\mathbf{x_2} - \mathbf{x_1})^2 + (\mathbf{y_2} - \mathbf{y_1})^2}$ Use the distance formula. $= \sqrt{(\mathbf{5} - \mathbf{8})^2 + [\mathbf{3} - (\mathbf{-1})]^2}$ Substitute the values. $= \sqrt{(-3)^2 + 4^2}$ Subtract. $= \sqrt{9 + 16}$ Square each value. $= \sqrt{25}$ Add. $= 5$ Simplify. The distance between the points is 5 units.

Unless otherwise noted, all content on this page is © Cengage Learning.

REVIEW EXERCISES

In Exercises 19–20, the horizon distance d (measured in miles) is related to the height h (measured in feet) of the observer by the formula $d = 1.4\sqrt{h}$.

19. View from a submarine A submarine's periscope extends 4.7 feet above the surface. About how far away is the horizon?

20. View from a submarine About how far out of the water must a submarine periscope extend to provide a 4-mile horizon?

21. Sailing A technique called *tacking* allows a sailboat to make progress into the wind. A sailboat follows the course in the illustration. Find d, the distance the boat advances into the wind.

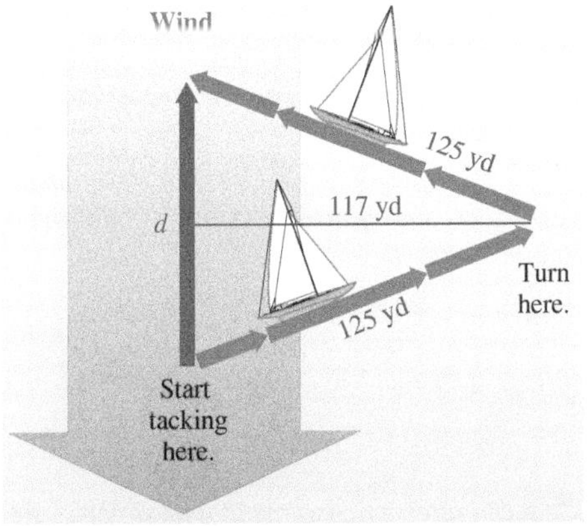

22. Communications Some campers 3,900 yards from a highway are talking to truckers on a citizen's band radio with an 8,900-yard range. Over what length of highway can these conversations take place?

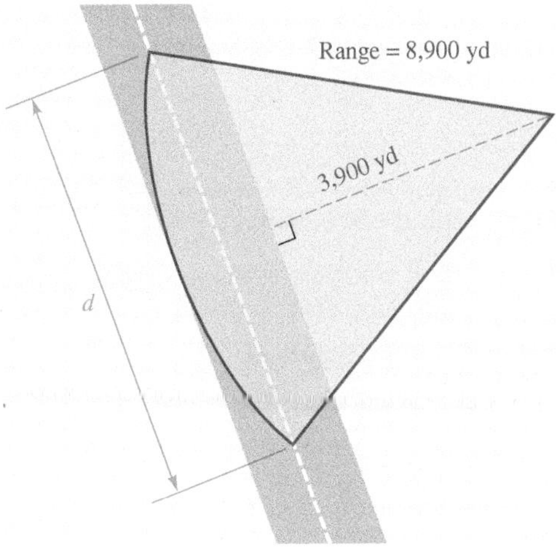

23. Find the distance between points $(1, 5)$ and $(9, -1)$.

24. Find the distance between points $(-4, 6)$ and $(-2, 8)$. Give the result to the nearest hundredth.

SECTION 9.3 Rational Exponents

DEFINITIONS AND CONCEPTS	EXAMPLES
Rational exponents with a numerator of 1: If n is a natural number $(n > 1)$ and $\sqrt[n]{x}$ is a real number, then $x^{1/n} = \sqrt[n]{x}$.	Simplify: $16^{1/2} = \sqrt{16} = 4$
If n is an even natural number, $(x^n)^{1/n} = \lvert x \rvert$.	$(25x^2)^{1/2} = \sqrt{25x^2} = 5\lvert x \rvert$
Assume n is a natural number $(n > 1)$ and x is a real number.	
If $x > 0$, then $x^{1/n}$ is the positive number such that $(x^{1/n})^n = x$.	$64^{1/2} = \sqrt{64} = 8$
If $x = 0$, then $x^{1/n} = 0$.	$0^{1/4} = \sqrt[4]{0} = 0$
If $x < 0$ and n is odd, then $x^{1/n}$ is the real number such that $(x^{1/n})^n = x$.	$(-27)^{1/3} = \sqrt[3]{-27} = -3$
If $x < 0$ and n is even, then $x^{1/n}$ is not a real number.	$(-64)^{1/2} = \sqrt{-64}$ is not a real number.
Assume m and n are positive integers and $x > 0$. **Rational exponents with a numerator other than 1:** $x^{m/n} = \sqrt[n]{x^m} = (\sqrt[n]{x})^m$	Simplify: $64^{2/3} = \sqrt[3]{(64)^2} = \sqrt[3]{4{,}096} = 16$; $64^{2/3} = (\sqrt[3]{64})^2 = 4^2 = 16$

Unless otherwise noted, all content on this page is © Cengage Learning.

Negative rational exponents: $x^{-m/n} = \frac{1}{x^{m/n}}$ $\frac{1}{x^{-m/n}} = x^{m/n}$ **Simplifying expressions with rational exponents:** Apply the properties of exponents.	Simplify: $25^{-1/2} = \frac{1}{25^{1/2}} = \frac{1}{\sqrt{25}} = \frac{1}{5}$ $\frac{1}{16^{-3/2}} = 16^{3/2} = (\sqrt{16})^3 = 4^3 = 64$ $\frac{x^{3/4} \cdot x^{2/3}}{x^{7/6}} = x^{3/4+2/3-7/6}$ Use the rules $x^m \cdot x^n = x^{m+n}$ and $\frac{x^m}{x^n} = x^{m-n}$. $= x^{1/4}$ $\frac{3}{4} + \frac{2}{3} - \frac{7}{6} = \frac{1}{4}$ $= \sqrt[4]{x}$ Write in radical notation.
Simplifying radical expressions: **1.** Write the radical expression as an exponential expression with rational exponents. **2.** Simplify the rational exponents. **3.** Write the exponential expression as a radical.	$\sqrt[4]{81x^2} = (81x^2)^{1/4}$ Use the rule $\sqrt[n]{x} = x^{1/n}$. $= (3^4x^2)^{1/4}$ $81 = 3^4$ $= 3x^{1/2}$ Use the rule $(xy)^m = x^m \cdot y^m$. $= 3\sqrt{x}$ Write in radical notation.

REVIEW EXERCISES

Simplify each expression, if possible. Assume that all variables represent positive numbers.

25. $81^{1/2}$
26. $-49^{1/2}$
27. $9^{3/2}$
28. $16^{3/2}$
29. $(-27)^{1/3}$
30. $-8^{2/3}$
31. $8^{-2/3}$
32. $64^{-1/3}$
33. $-49^{5/2}$
34. $\frac{1}{16^{5/2}}$
35. $\left(\frac{27}{125}\right)^{-2/3}$
36. $\left(\frac{4}{9}\right)^{-3/2}$
37. $(27x^3y)^{1/3}$
38. $(16x^2y^4)^{1/4}$
39. $(25x^3y^4)^{3/2}$
40. $(8u^2v^3)^{-2/3}$

Perform the multiplications. Assume that all variables represent positive numbers and write all answers without negative exponents.

41. $5^{1/4}5^{1/2}$
42. $a^{5/9}a^{2/9}$
43. $u^{1/2}(u^{1/2} - u^{-1/2})$
44. $v^{2/3}(v^{1/3} + v^{4/3})$
45. $(x^{1/2} + y^{1/2})^2$
46. $(a^{2/3} + b^{2/3})(a^{2/3} - b^{2/3})$

Simplify each expression. Assume that all variables are positive.

47. $\sqrt[6]{5^2}$
48. $\sqrt[8]{x^4}$
49. $\sqrt[9]{27a^3b^6}$
50. $\sqrt[4]{25a^2b^2}$

SECTION 9.4 Simplifying and Combining Radical Expressions

DEFINITIONS AND CONCEPTS	EXAMPLES
Properties of radicals: Assume n is a natural number greater than 1. $\sqrt[n]{ab} = \sqrt[n]{a}\sqrt[n]{b}$ if a and b are not both negative $\sqrt[n]{\frac{a}{b}} = \frac{\sqrt[n]{a}}{\sqrt[n]{b}}$ $(b \neq 0)$	Simplify: $\sqrt{24} = \sqrt{4 \cdot 6} = \sqrt{4}\sqrt{6} = 2\sqrt{6}$ $\sqrt[3]{24} = \sqrt[3]{8 \cdot 3} = \sqrt[3]{8}\sqrt[3]{3} = 2\sqrt[3]{3}$ $\sqrt{\frac{21}{64x^6}} = \frac{\sqrt{21}}{\sqrt{64x^6}} = \frac{\sqrt{21}}{8x^3}$ $\sqrt[3]{\frac{21}{64x^6}} = \frac{\sqrt[3]{21}}{\sqrt[3]{64x^6}} = \frac{\sqrt[3]{21}}{4x^2}$
Adding and subtracting radical expressions: Like radicals, radical expressions with the same index and the same radicand, can be combined by addition and subtraction. Radicals that are not alike often can be simplified to radicals that are alike and then combined.	Add: $9\sqrt{2} + 4\sqrt{2} = (9 + 4)\sqrt{2} = 13\sqrt{2}$ $\sqrt{2} + \sqrt{18} = \sqrt{2} + \sqrt{9}\sqrt{2} = \sqrt{2} + 3\sqrt{2} = 4\sqrt{2}$

Special right triangles: In a 45°–45°–90° triangle, the length of the hypotenuse is the length of one leg times $\sqrt{2}$.	If each leg of an isosceles triangle is 17 cm, the hypotenuse measures $17\sqrt{2}$ cm.
The shorter leg of a 30°–60°–90° triangle is half as long as the hypotenuse. The longer leg is the length of the shorter leg times $\sqrt{3}$.	If the shorter leg of a 30°–60°–90° triangle is 12 units, the hypotenuse is 24 cm and the longer leg is $12\sqrt{3}$ units.

REVIEW EXERCISES

Simplify each expression. Assume that all variables represent positive numbers.

51. $\sqrt{176}$
52. $\sqrt[3]{250}$
53. $\sqrt[4]{32}$
54. $\sqrt[5]{96}$
55. $\sqrt{8x^3}$
56. $\sqrt{24x^6y^5}$
57. $\sqrt[3]{16x^5y^4}$
58. $\sqrt[3]{54x^7y^3}$
59. $\dfrac{\sqrt{32x^3}}{\sqrt{2x}}$
60. $\dfrac{\sqrt[3]{16x^5}}{\sqrt[3]{2x^2}}$
61. $\sqrt[3]{\dfrac{2a^2b}{27x^3}}$
62. $\sqrt{\dfrac{17xy}{64a^4}}$

Simplify and combine like radicals. Assume that all variables represent positive numbers.

63. $\sqrt{12} + \sqrt{27}$
64. $\sqrt{18} - \sqrt{32}$
65. $2\sqrt[3]{3} - \sqrt[3]{24}$
66. $\sqrt[4]{32} + 2\sqrt[4]{162}$
67. $2x\sqrt{8} + 2\sqrt{200x^2} + \sqrt{50x^2}$
68. $3\sqrt{27a^3} - 2a\sqrt{3a} + 5\sqrt{75a^3}$
69. $\sqrt[3]{54} - 3\sqrt[3]{16} + 4\sqrt[3]{128}$
70. $2\sqrt[4]{32x^5} + 4\sqrt[4]{162x^5} - 5x\sqrt[4]{512x}$
71. **Geometry** Find the length of the hypotenuse of an isosceles right triangle whose legs measure 7 meters.
72. **Geometry** The hypotenuse of a 30°–60°–90° triangle measures $12\sqrt{3}$ centimeters. Find the length of each leg.

Find the value of x to two decimal places.

73.

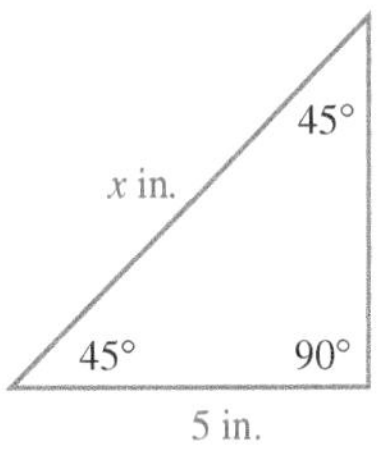

74.

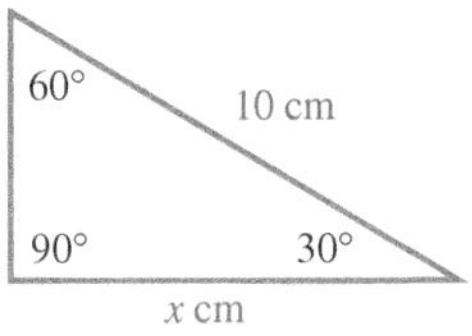

SECTION 9.5 Multiplying Radical Expressions and Rationalizing

DEFINITIONS AND CONCEPTS	EXAMPLES
Multiplying radical expressions: If two radicals have the same index, they can be multiplied.	Multiply: $\sqrt{3x}\sqrt{6x} = \sqrt{18x^2}\quad (x > 0)$ $= \sqrt{9x^2}\sqrt{2}$ $= 3x\sqrt{2}$ $(5 - 2\sqrt{3})(7 + 6\sqrt{3})$ $= 35 + 30\sqrt{3} - 14\sqrt{3} - 12\sqrt{9}$ Distribute. $= 35 + 16\sqrt{3} - 12(3)$ Combine like terms. $= 35 + 16\sqrt{3} - 36$ Multiply. $= -1 + 16\sqrt{3}$ Simplify.
Rationalizing the denominator: To eliminate a single radical in the denominator, we multiply the numerator and the denominator by a number that will result in a perfect square (or cube, or fourth power, etc.) under the radical in the denominator determined by the index.	Rationalize: $\sqrt{\dfrac{5}{8}} = \dfrac{\sqrt{5}}{\sqrt{8}}$ $\qquad \sqrt{\dfrac{a}{b}} = \dfrac{\sqrt{a}}{\sqrt{b}}$ $= \dfrac{\sqrt{5}}{\sqrt{8}} \cdot \dfrac{\sqrt{2}}{\sqrt{2}}$ Multiply by 1: $\dfrac{\sqrt{2}}{\sqrt{2}} = 1$ $= \dfrac{\sqrt{10}}{\sqrt{16}}$ Multiply radicals. $= \dfrac{\sqrt{10}}{4}$ Simplify.

Unless otherwise noted, all content on this page is © Cengage Learning.

To rationalize a fraction whose denominator has two terms with one or both containing square roots, we multiply its numerator and denominator by the *conjugate* of its denominator.	Rationalize: $\frac{5}{4-\sqrt{2}} \cdot \frac{4+\sqrt{2}}{4+\sqrt{2}}$ Multiply the numerator and the denominator by the conjugate of the denominator. $= \frac{20+5\sqrt{2}}{16-2}$ Multiply. $= \frac{20+5\sqrt{2}}{14}$ Simplify.

REVIEW EXERCISES

Simplify each expression. Assume that all variables represent positive numbers.

75. $(5\sqrt{3})(2\sqrt{5})$
76. $2\sqrt{6}\sqrt{216}$
77. $\sqrt{5x}\sqrt{4x}$
78. $\sqrt[3]{3x^2}\sqrt[3]{9x^4}$
79. $-\sqrt[3]{2x^2}\sqrt[3]{4x}$
80. $-\sqrt[4]{256x^5y^{11}}\sqrt[4]{625x^9y^3}$
81. $\sqrt{3}(\sqrt{27}-4)$
82. $\sqrt{5}(\sqrt{5}-6)$
83. $\sqrt{5}(\sqrt{2}-1)$
84. $\sqrt{3}(\sqrt{3}+\sqrt{2})$
85. $(\sqrt{5}+1)(\sqrt{5}-1)$
86. $(\sqrt{3}+\sqrt{2})(\sqrt{3}+\sqrt{2})$
87. $(\sqrt{x}+\sqrt{y})(\sqrt{x}-\sqrt{y})$
88. $(2\sqrt{u}+3)(3\sqrt{u}-4)$

Rationalize each denominator.

89. $\frac{1}{\sqrt{3}}$
90. $\frac{\sqrt{3}}{\sqrt{5}}$
91. $\frac{x}{\sqrt{xy}}$
92. $\frac{\sqrt[3]{uv}}{\sqrt[3]{u^5v^7}}$
93. $\frac{2}{\sqrt{2}-1}$
94. $\frac{\sqrt{2}}{\sqrt{3}-1}$
95. $\frac{2x-32}{\sqrt{x}+4}$
96. $\frac{\sqrt{a}+1}{\sqrt{a}-1}$

Rationalize each numerator. All variables represent positive numbers.

97. $\frac{\sqrt{3}}{5}$
98. $\frac{\sqrt[3]{9}}{3}$
99. $\frac{3-\sqrt{x}}{2}$
100. $\frac{\sqrt{a}-\sqrt{b}}{\sqrt{a}}$

SECTION 9.6 Radical Equations

DEFINITIONS AND CONCEPTS	EXAMPLES
The power rule: Assume x, y, and n are real numbers. If $x = y$, then $x^n = y^n$. When we raise both sides of an equation to the same power it can lead to extraneous solutions. Be sure to check all possible solutions. Prior to applying the power rule be certain to isolate one radical.	To solve $\sqrt{x+4} = x-2$, proceed as follows: $(\sqrt{x+4})^2 = (x-2)^2$ Square both sides of the equation to eliminate the square root. $x+4 = x^2-4x+4$ Square the binomial. $0 = x^2-5x$ Subtract x and 4 from both sides to write the equation in quadratic form. $0 = x(x-5)$ Factor x^2-5x. $x = 0$ or $x-5=0$ Set each factor equal to 0. $x = 0$ | $x = 5$ **Check:** $\sqrt{x+4} = x-2$; $\sqrt{x+4} = x-2$ $\sqrt{\mathbf{0}+4} \stackrel{?}{=} \mathbf{0}-2$; $\sqrt{\mathbf{5}+4} \stackrel{?}{=} \mathbf{5}-2$ $\sqrt{4} \stackrel{?}{=} -2$; $\sqrt{9} \stackrel{?}{=} 3$ $2 \neq -2$; $3 = 3$ Since 0 does not check, it is extraneous and must be discarded. The only solution of the original equation is 5.

REVIEW EXERCISES

Solve each equation.

101. $\sqrt{3y - 11} = \sqrt{y + 7}$ **102.** $u = \sqrt{25u - 144}$

103. $\sqrt{2x + 6} + 1 = x$ **104.** $\sqrt{z + 1} + \sqrt{z} = 2$

105. $\sqrt{2x + 5} - \sqrt{2x} = 1$ **106.** $\sqrt[3]{x^3 + 8} = x + 2$

SECTION 9.7 Complex Numbers

DEFINITIONS AND CONCEPTS	EXAMPLES
Simplifying imaginary numbers: $\sqrt{-1}$ is defined as the **imaginary number** i.	Simplify: $\sqrt{-12} = \sqrt{-4}\sqrt{3}$ Write -12 as $-4(3)$. $= \sqrt{-1}\sqrt{4}\sqrt{3}$ Write -4 as the product of -1 and 4. $= i(2)\sqrt{3}$ $\sqrt{-1} = i$ $= 2\sqrt{3}i$
Operations with complex numbers: If a, b, c, and d are real numbers and $i^2 = -1$, $a + bi = c + di$ if and only if $a = c$ and $b = d$ $(a + bi) + (c + di) = (a + c) + (b + d)i$ $(a + bi) - (c + di) = (a - c) + (b - d)i$ $(a + bi)(c + di) = (ac - bd) + (ad + bc)i$	$3 + 5i = \sqrt{9} + \frac{10}{2}i$ because $\sqrt{9} = 3$ and $\frac{10}{2} = 5$. Perform the indicated operations: $(6 + 5i) + (2 + 9i) = (6 + 2) + (5 + 9)i = 8 + 14i$ $(6 + 5i) - (2 + 9i) = (6 - 2) + (5 - 9)i = 4 - 4i$ $(6 - 5i)(2 + 9i) = 12 + 54i - 10i - 45i^2$ $= 12 + 44i - 45(-1)$ $= 12 + 44i + 45$ $= 57 + 44i$
Dividing complex numbers: To divide complex numbers, write the division as a fraction, and rationalize the denominator.	$\frac{6}{3 + i} \cdot \frac{3 - i}{3 - i} = \frac{18 - 6i}{9 - i^2}$ Multiply the numerator and the denominator by the conjugate of the denominator. Multiply the fractions. $= \frac{18 - 6i}{9 + 1}$ $i^2 = -1$ $= \frac{18 - 6i}{10}$ Add. $= \frac{18}{10} - \frac{6}{10}i$ Write in $a + bi$ form. $= \frac{9}{5} - \frac{3}{5}i$ Simplify each fraction.
Powers of i $i = \sqrt{-1}$ $i^2 = -1$ $i^3 = -i$ $i^4 = 1$ If n is a natural number that has a remainder of r when divided by 4, then $i^n = i^r$. When n is divisible by 4, the remainder r is 0 and $i^0 = 1$.	Simplify: $i^{81} = (i^4)^{20} \cdot i$ $= (i^1)^{20} \cdot i$ $= i$

Absolute value of a complex number: $\lvert a + bi \rvert = \sqrt{a^2 + b^2}$	Find the absolute value: $\lvert 4 + 8i \rvert = \sqrt{4^2 + 8^2}$ $\quad \lvert a + bi \rvert = \sqrt{a^2 + b^2}$ $= \sqrt{16 + 64}$ Square each value. $= \sqrt{80}$ Add. $= \sqrt{16 \cdot 5}$ Factor: $80 = 16(5)$ $= 4\sqrt{5}$ Simplify: $\sqrt{16 \cdot 5} = \sqrt{16}\sqrt{5} = 4\sqrt{5}$

REVIEW EXERCISES

Perform the operations and give all answers in a + bi form.

107. $(6 + 3i) + (4 - 15i)$

108. $(-5 - 22i) - (-7 + 28i)$

109. $\left(-32 + \sqrt{-144}\right) - \left(64 + \sqrt{-81}\right)$

110. $\left(-8 + \sqrt{-8}\right) + \left(6 - \sqrt{-32}\right)$

111. $(2 - 7i)(-3 + 4i)$

112. $(-4 + 5i)(3 + 2i)$

113. $\left(5 - \sqrt{-27}\right)\left(-6 + \sqrt{-12}\right)$

114. $\left(2 + \sqrt{-128}\right)\left(3 - \sqrt{-98}\right)$

115. $\dfrac{3}{4i}$

116. $\dfrac{-2}{5i^3}$

117. $\dfrac{6}{2 + i}$

118. $\dfrac{7}{3 - i}$

119. $\dfrac{4 + i}{4 - i}$

120. $\dfrac{3 - i}{3 + i}$

121. $\dfrac{3}{5 + \sqrt{-4}}$

122. $\dfrac{2}{3 - \sqrt{-9}}$

Simplify.

123. $\lvert 9 + 12i \rvert$

124. $\lvert 24 - 10i \rvert$

125. i^{39}

126. i^{52}

9 Test

Find each root.

1. $\sqrt{36}$

2. $\sqrt{125}$

3. $\sqrt{4x^2}$

4. $\sqrt[3]{8x^3}$

Graph each function and find the domain and range.

5. $f(x) = \sqrt{x - 2}$

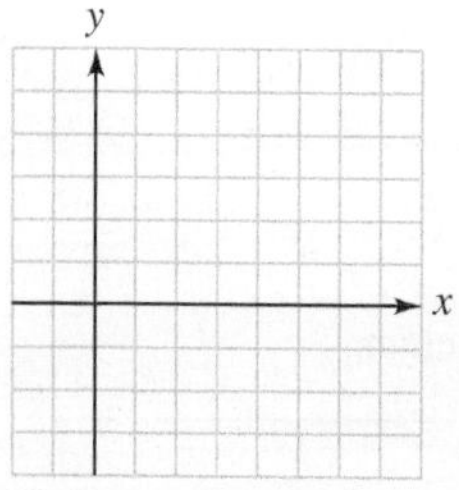

6. $f(x) = \sqrt[3]{x} + 3$

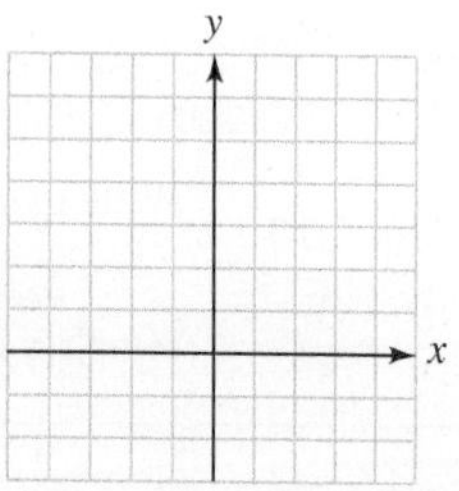

Use a calculator.

7. Shipping crates The diagonal brace on the shipping crate shown in the illustration is 53 inches. Find the height, *h*, of the crate.

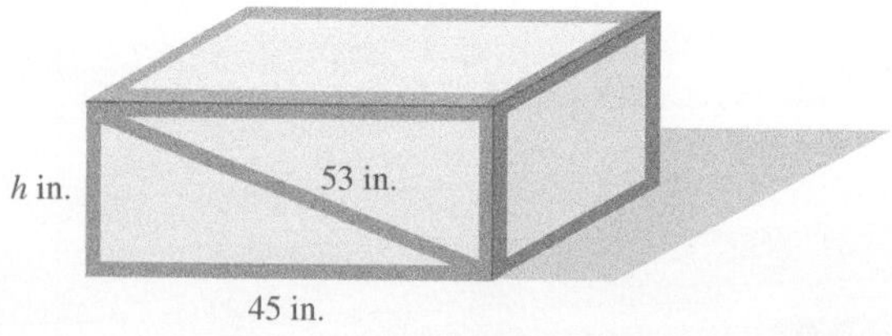

Unless otherwise noted, all content on this page is © Cengage Learning.

8. Pendulums The 2-meter pendulum rises 0.1 meter at the extremes of its swing. Find the width w of the swing in meters.

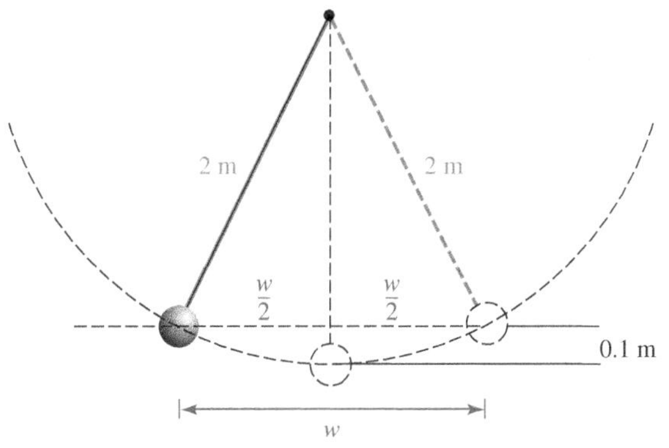

Find the distance between the points.

9. $(6, 8), (0, 0)$

10. $(-2, 5), (22, 12)$

Simplify each expression. Assume that all variables represent positive numbers, and write answers without using negative exponents.

11. $81^{1/4}$

12. $-64^{2/3}$

13. $36^{-3/2}$

14. $\left(-\frac{8}{27}\right)^{-2/3}$

15. $\frac{2^{5/3}2^{1/6}}{2^{1/2}}$

16. $\frac{(8x^3y)^{1/2}(8xy^5)^{1/2}}{(x^3y^6)^{1/3}}$

Simplify each expression. Assume that all variables represent positive numbers.

17. $\sqrt{108}$

18. $\sqrt{250x^3y^5}$

19. $\frac{\sqrt[3]{240x^{10}y^4}}{\sqrt[3]{6xy}}$

20. $\sqrt{\frac{3a^5}{48a^7}}$

Simplify each expression. Assume that the variables are unrestricted.

21. $\sqrt{45x^2}$

22. $\sqrt{48x^6}$

23. $\sqrt[3]{16x^9}$

24. $\sqrt{18x^4y^9}$

Simplify. Assume that all variables represent positive numbers.

25. $\sqrt{12} - \sqrt{27}$

26. $2\sqrt[3]{40} - \sqrt[3]{5,000} + 4\sqrt[3]{625}$

27. $2\sqrt{48y^5} - 3y\sqrt{12y^3}$

28. $\sqrt[4]{768z^5} + z\sqrt[4]{48z}$

Perform each operation and simplify, if possible. All variables represent positive numbers.

29. $-2\sqrt{xy}\left(3\sqrt{x} + \sqrt{xy^3}\right)$

30. $\left(3\sqrt{2} + \sqrt{3}\right)\left(2\sqrt{2} - 3\sqrt{3}\right)$

Rationalize each denominator.

31. $\frac{1}{\sqrt{5}}$

32. $\frac{3t - 1}{\sqrt{3t} - 1}$

Rationalize each numerator.

33. $\frac{\sqrt{3}}{\sqrt{7}}$

34. $\frac{\sqrt{a} + \sqrt{b}}{\sqrt{a} - \sqrt{b}}$

Solve.

35. $\sqrt[3]{6n + 4} - 4 = 0$

36. $1 - \sqrt{u} = \sqrt{u - 3}$

Perform the operations. Give all answers in $a + bi$ form.

37. $(2 + 4i) + (-3 + 7i)$

38. $\left(3 - \sqrt{-9}\right) - \left(-1 + \sqrt{-16}\right)$

39. $2i(3 - 4i)$

40. $(3 + 2i)(-4 - i)$

41. $\frac{1}{\sqrt{2}i}$

42. $\frac{2 + i}{3 - i}$

Quadratic and Other Nonlinear Functions and Inequalities

10

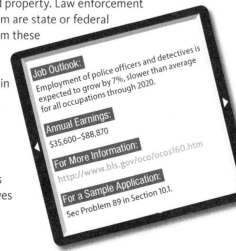

LightField Studios/Shutterstock.com

Careers and Mathematics

POLICE OFFICERS AND DETECTIVES

People depend on police officers and detectives to protect their lives and property. Law enforcement officers, some of whom are state or federal special agents, perform these duties in a variety of ways. Uniformed police officers maintain regular patrols and respond to calls for service, and some serve as undercover officers. Police work can be very dangerous and stressful. Detectives collect evidence for criminal cases.

Job Outlook:
Employment of police officers and detectives is expected to grow by 7%, slower than average for all occupations through 2020.

Annual Earnings:
$35,600–$88,870

For More Information:
http://www.bls.gov/oco/ocos160.htm

For a Sample Application:
See Problem 89 in Section 10.1.

REACH FOR SUCCESS

10.1 Solving Quadratic Equations Using the Square-Root Property and by Completing the Square

10.2 Solving Quadratic Equations Using the Quadratic Formula

10.3 The Discriminant and Equations That Can Be Written in Quadratic Form

10.4 Graphs of Quadratic Functions

10.5 Graphs of Other Nonlinear Functions

10.6 Solving Quadratic and Other Nonlinear Inequalities

■ *Projects*

REACH FOR SUCCESS EXTENSION

CHAPTER REVIEW

CHAPTER TEST

CUMULATIVE REVIEW

In this chapter

In this chapter, we will discuss general methods for solving equations in quadratic form, and we will consider the graphs of quadratic functions. Finally, we will discuss graphs of other nonlinear functions and solve quadratic and other nonlinear inequalities.

Reach for Success Learning Beyond Remembering

Can you name all five of the Great Lakes shown in the photo? Being able to do so means that you have learned and stored this information in your brain in such a way that you can access it any time you need it.

In this exercise, you will explore Bloom's Taxonomy of learning at the first three levels.

Provided by the SeaWiFS Project, NASA/Goddard Space Flight Center, and ORBIMAGE

If you couldn't name all five, does this list help you select the correct names?

Cumberland	*Erie*	*Huron*	*Michigan*
Ontario	*Victoria*	*Salt Lake*	*Superior*

If you could not name the five lakes without this list, you have learned only to *recognize* their names. The information is stored in your brain, but in a way that requires a "prompt" to be retrieved.

The triangle is an updated illustration of learning levels as inspired by psychologist Benjamin Bloom in 1956.

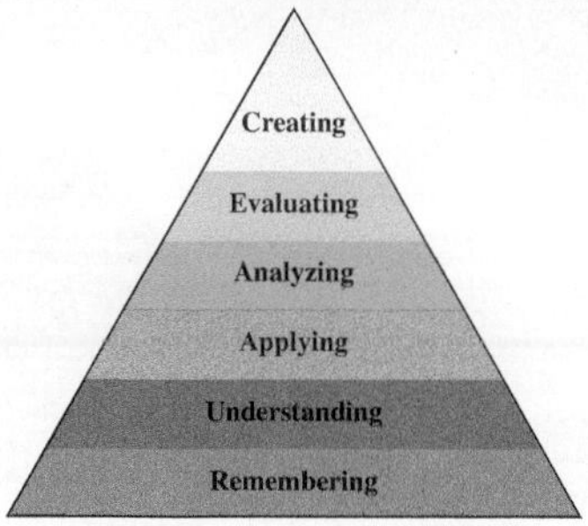

Ideally, you would like to learn material beyond recognition and remembering, to understanding, applying, analyzing, evaluating, and ultimately creating.

To help you *recall* the names of the Great Lakes, you can use the mnemonic device HOMES (Huron, Ontario, Michigan, Erie, Superior). Vocabulary and formulas in mathematics are necessary and creating your own mnemonic devices may make it easier to remember them.

Recalling the names of the Great Lakes is still at the *Remembering* level of Bloom's Taxonomy. "Explain why they are called the Great Lakes" is at the *Understanding* level. "Demonstrate the size of the Great Lakes in terms of one of the states" is at the *Applying* level.

Below are three questions you might see on a test covering the topic of slope. For each, identify the level of Bloom's Taxonomy corresponding to the degree of learning required: Remembering, Understanding, or Applying.

1. Given a table of data, find the slope of the line it represents. ______________
2. Write the formula used to calculate the slope of a line between two points. ______________
3. Given a graph of the national debt between 2008 and 2012, interpret how steeply it grew. ______

Do an Internet search for the levels of Bloom's Taxonomy. Check your answers to the previous three statements against the descriptions you find.

In your own words, explain the difference between

1. the Remembering level and the Understanding level.

____

2. the Understanding level and the Applying level.

____

It is impossible for a textbook to provide an example of every variation of a mathematics problem. Therefore it is important that you *learn* the skills in such a way that you can understand what you've learned and recall and apply it to a new situation. Many of the *Now Try This* exercises in this book are designed to help you transition to higher levels of learning.

A Successful Study Strategy . . .

Practicing your mathematics every day will help you learn at levels beyond remembering.

At the end of the chapter you will find an additional exercise to guide you to a successful semester.

Unless otherwise noted, all content on this page is © Cengage Learning.

Section 10.1 Solving Quadratic Equations Using the Square-Root Property and by Completing the Square

Objectives

1. Solve a quadratic equation by factoring.
2. Solve a quadratic equation by applying the square-root property.
3. Solve a quadratic equation by completing the square.
4. Solve an application requiring the use of the square-root property.

Vocabulary

square-root property
extracting the roots
completing the square
compound interest

Getting Ready

Factor each expression.

1. $x^2 - 25$
2. $b^2 - 81$
3. $6x^2 + x - 2$
4. $4x^2 - 4x - 3$

We begin this section by reviewing how to solve quadratic equations by factoring. We will then discuss how to solve these equations by applying the **square-root property**, sometimes called ***extracting the roots***, by completing the square, and then use these skills to solve applications.

1 Solve a quadratic equation by factoring.

Recall that a *quadratic equation* is an equation of the form $ax^2 + bx + c = 0 \quad (a \neq 0)$, where a, b, and c are real numbers. As a review, we will solve the first two examples by factoring.

EXAMPLE 1 Solve: $x^2 = 9$

Solution To solve this quadratic equation by factoring, we proceed as follows:

$$x^2 = 9$$

$$x^2 - 9 = 0 \qquad \text{Subtract 9 from both sides.}$$

$$(x + 3)(x - 3) = 0 \qquad \text{Factor the binomial.}$$

$$x + 3 = 0 \quad \text{or} \quad x - 3 = 0 \qquad \text{Set each factor equal to 0.}$$

$$x = -3 \quad \bigg| \quad x = 3 \qquad \text{Solve each linear equation.}$$

Check: ***For x = −3*** ***For x = 3***

$$x^2 = 9 \qquad x^2 = 9$$
$$(-3)^2 \stackrel{?}{=} 9 \qquad (3)^2 \stackrel{?}{=} 9$$
$$9 = 9 \qquad 9 = 9$$

Since both results check, the solutions are 3 and −3.

SELF CHECK 1 Solve: $p^2 = 64$

EXAMPLE 2 Solve: $6x^2 - 7x - 3 = 0$

Solution To solve this quadratic equation by factoring, we proceed as follows:

$$6x^2 - 7x - 3 = 0$$
$$(2x - 3)(3x + 1) = 0 \qquad \text{Factor.}$$
$$2x - 3 = 0 \quad \text{or} \quad 3x + 1 = 0 \qquad \text{Set each factor equal to 0.}$$
$$x = \frac{3}{2} \quad \Big| \quad x = -\frac{1}{3} \qquad \text{Solve each linear equation.}$$

Check: ***For*** $x = \frac{3}{2}$ ***For*** $x = -\frac{1}{3}$

$$6x^2 - 7x - 3 = 0 \qquad 6x^2 - 7x - 3 = 0$$
$$6\left(\frac{3}{2}\right)^2 - 7\left(\frac{3}{2}\right) - 3 \stackrel{?}{=} 0 \qquad 6\left(-\frac{1}{3}\right)^2 - 7\left(-\frac{1}{3}\right) - 3 \stackrel{?}{=} 0$$
$$6\left(\frac{9}{4}\right) - 7\left(\frac{3}{2}\right) - 3 \stackrel{?}{=} 0 \qquad 6\left(\frac{1}{9}\right) - 7\left(-\frac{1}{3}\right) - 3 \stackrel{?}{=} 0$$
$$\frac{27}{2} - \frac{21}{2} - \frac{6}{2} \stackrel{?}{=} 0 \qquad \frac{2}{3} + \frac{7}{3} - \frac{9}{3} \stackrel{?}{=} 0$$
$$0 = 0 \qquad 0 = 0$$

Since both results check, the solutions are $\frac{3}{2}$ and $-\frac{1}{3}$.

SELF CHECK 2 Solve: $6m^2 - 5m + 1 = 0$

Unfortunately, many quadratic expressions do not factor easily. For example, it would be difficult to solve $2x^2 + 4x + 1 = 0$ by factoring, because $2x^2 + 4x + 1$ cannot be factored by using only integers.

2 Solve a quadratic equation by applying the square-root property.

To develop general methods for solving all quadratic equations, we first solve $x^2 = c$ by a method similar to the one used in Example 1.

$$x^2 = c$$
$$x^2 - c = 0 \qquad \text{Subtract } c \text{ from both sides.}$$
$$x^2 - \left(\sqrt{c}\right)^2 = 0 \qquad c = \left(\sqrt{c}\right)^2$$
$$\left(x + \sqrt{c}\right)\left(x - \sqrt{c}\right) = 0 \qquad \text{Factor the difference of two squares.}$$
$$x + \sqrt{c} = 0 \quad \text{or} \quad x - \sqrt{c} = 0 \qquad \text{Set each factor equal to 0.}$$
$$x = -\sqrt{c} \quad \Big| \quad x = \sqrt{c} \qquad \text{Solve each linear equation.}$$

The two solutions of $x^2 = c$ are $x = \sqrt{c}$ and $x = -\sqrt{c}$.

THE SQUARE-ROOT PROPERTY (EXTRACTING THE ROOTS)

The equation $x^2 = c$ has two solutions. They are

$$x = \sqrt{c} \quad \text{or} \quad x = -\sqrt{c}$$

We often use the symbol $\pm\sqrt{c}$ to represent the two solutions $\sqrt{c}$ and $-\sqrt{c}$. The symbol $\pm\sqrt{c}$ is read as "the positive or negative square root of c."

EXAMPLE 3 Use the square-root property to solve $x^2 - 12 = 0$.

Solution We can write the equation as $x^2 = 12$ and use the square-root property.

$$x^2 - 12 = 0$$

$$x^2 = 12 \qquad \text{Add 12 to both sides.}$$

$$x = \sqrt{12} \quad \text{or} \quad x = -\sqrt{12} \qquad \text{Use the square-root property.}$$

$$x = 2\sqrt{3} \quad \Big| \quad x = -2\sqrt{3} \qquad \sqrt{12} = \sqrt{4}\sqrt{3} = 2\sqrt{3}$$

The solutions can be written as $\pm 2\sqrt{3}$. Verify that each one satisfies the equation.

SELF CHECK 3 Use the square-root property to solve $x^2 - 18 = 0$.

EXAMPLE 4 Use the square-root property to solve $(x - 3)^2 = 16$.

Solution We can use the square-root property.

$$(x - 3)^2 = 16$$

$$x - 3 = \sqrt{16} \quad \text{or} \quad x - 3 = -\sqrt{16} \qquad \text{Use the square-root property.}$$

$$x - 3 = 4 \quad \Big| \quad x - 3 = -4 \qquad \sqrt{16} = 4 \text{ and } -\sqrt{16} = -4$$

$$x = 3 + 4 \quad \Big| \quad x = 3 - 4 \qquad \text{Add 3 to both sides.}$$

$$x = 7 \quad \Big| \quad x = -1 \qquad \text{Simplify.}$$

Verify that each solution satisfies the equation.

SELF CHECK 4 Use the square-root property to solve $(x + 2)^2 = 9$.

Because we have learned to work with radicals and complex numbers, we can now solve any quadratic equation.

EXAMPLE 5 Use the square-root property to solve $9x^2 + 25 = 0$.

Solution We can write the equation as $x^2 = -\frac{25}{9}$ and use the square-root property.

$$9x^2 + 25 = 0$$

$$x^2 = -\frac{25}{9} \qquad \text{Subtract 25 from both sides and divide both sides by 9.}$$

$$x = \sqrt{-\frac{25}{9}} \quad \text{or} \quad x = -\sqrt{-\frac{25}{9}} \qquad \text{Use the square-root property.}$$

$$x = \sqrt{\frac{25}{9}}\sqrt{-1} \quad \Big| \quad x = -\sqrt{\frac{25}{9}}\sqrt{-1} \qquad \sqrt{-\tfrac{25}{9}} = \sqrt{\tfrac{25}{9}(-1)} = \sqrt{\tfrac{25}{9}}\sqrt{-1}$$

$$x = \frac{5}{3}i \quad \Big| \quad x = -\frac{5}{3}i \qquad \sqrt{\tfrac{25}{9}} = \tfrac{5}{3};\ \sqrt{-1} = i$$

Check:

$9x^2 + 25 = 0$	$9x^2 + 25 = 0$
$9\left(\frac{5}{3}i\right)^2 + 25 \stackrel{?}{=} 0$	$9\left(-\frac{5}{3}i\right)^2 + 25 \stackrel{?}{=} 0$
$9\left(\frac{25}{9}\right)i^2 + 25 \stackrel{?}{=} 0$	$9\left(\frac{25}{9}\right)i^2 + 25 \stackrel{?}{=} 0$
$25(-1) + 25 \stackrel{?}{=} 0$	$25(-1) + 25 \stackrel{?}{=} 0$
$0 = 0$	$0 = 0$

Since both results check, the solutions are $\pm\frac{5}{3}i$.

SELF CHECK 5 Use the square-root property to solve $4x^2 + 36 = 0$.

3 Solve a quadratic equation by completing the square.

All quadratic equations can be solved by a method called **completing the square**. This method is based on the special products

$$x^2 + 2ax + a^2 = (x + a)^2 \quad \text{and} \quad x^2 - 2ax + a^2 = (x - a)^2$$

Recall that the trinomials $x^2 + 2ax + a^2$ and $x^2 - 2ax + a^2$ are both *perfect-square trinomials*, because both factor as the square of a binomial. In each case, the coefficient of the first term is 1 and if we take one-half of the coefficient of x in the middle term and square it, we obtain the third term.

$$\left[\frac{1}{2}(2a)\right]^2 = a^2 \qquad \tfrac{1}{2}(2a) = \left(\tfrac{1}{2}\cdot 2\right)a = a$$

$$\left[\frac{1}{2}(-2a)\right]^2 = (-a)^2 = a^2 \qquad \tfrac{1}{2}(-2a) = \left(\tfrac{1}{2}\cdot -2\right)a = -a$$

EXAMPLE 6 Find the number that when added to each binomial results in a perfect-square trinomial: **a.** $x^2 + 10x$ **b.** $x^2 - 6x$ **c.** $x^2 - 11x$

Solution

a. To make $x^2 + 10x$ a perfect-square trinomial, we first find one-half of 10 to obtain 5 and square 5 to get 25.

$$\left[\frac{1}{2}(10)\right]^2 = (5)^2 = 25$$

If we add 25 to $x^2 + 10x$ we obtain $x^2 + 10x + 25$. This is a perfect-square trinomial because $x^2 + 10x + 25 = (x + 5)^2$.

b. To make $x^2 - 6x$ a perfect-square trinomial, we first find one-half of -6 to obtain -3 and square -3 to get 9.

$$\left[\frac{1}{2}(-6)\right]^2 = (-3)^2 = 9$$

If we add 9 to $x^2 - 6x$ we obtain $x^2 - 6x + 9$. This is a perfect-square trinomial because $x^2 - 6x + 9 = (x - 3)^2$.

c. To make $x^2 - 11x$ a perfect-square trinomial, we first find one-half of -11 to obtain $-\frac{11}{2}$ and square $-\frac{11}{2}$ to get $\frac{121}{4}$.

$$\left[\frac{1}{2}(-11)\right]^2 = \left(-\frac{11}{2}\right)^2 = \frac{121}{4}$$

If we add $\frac{121}{4}$ to $x^2 - 11x$ we obtain $x^2 - 11x + \frac{121}{4}$. This is a perfect-square trinomial because $x^2 - 11x + \frac{121}{4} = \left(x - \frac{11}{2}\right)^2$.

SELF CHECK 6 Find the number that when added to $a^2 - 5a$ results in a perfect-square trinomial.

To see geometrically why completing the square works on $x^2 + 10x$, we refer to Figure 10-1(a), which shows a polygon with an area of $x^2 + 10x$. To turn the polygon into a square, we can divide the area of $10x$ into two areas of $5x$ and then reassemble the polygon as shown in Figure 10-1(b). To fill in the missing corner, we must add a square with an area of $5^2 = 25$. Thus, we complete the square.

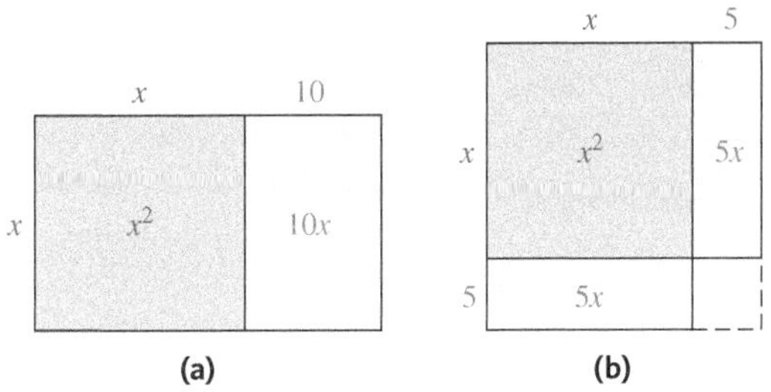

Figure 10-1

To solve an equation of the form $ax^2 + bx + c = 0$ $(a \neq 0)$ by completing the square, we use the following steps.

COMPLETING THE SQUARE

1. Make sure that the coefficient of x^2 is 1. If not, divide both sides of the equation by the coefficient of x^2.
2. If necessary, add a number to both sides of the equation to place the constant term on the right side of the equal sign.
3. Complete the square:
   a. Find one-half of the coefficient of x and square it.
   b. Add the square to both sides of the equation.
4. Factor the trinomial square on one side of the equation and combine like terms on the other side.
5. Solve the resulting equation by using the square-root property.

EXAMPLE 7 Solve $x^2 + 8x + 7 = 0$ by completing the square.

Solution **Step 1** In this example, the coefficient of x^2 is already 1.

Step 2 We add -7 to both sides to place the constant on the right side of the equal sign.

$$x^2 + 8x + 7 = 0$$
$$x^2 + 8x = -7$$

Step 3 The coefficient of x is 8, one-half of 8 is 4, and $4^2 = 16$. To complete the square, we add 16 to both sides.

$$x^2 + 8x + \mathbf{16} = -7 + \mathbf{16}$$

Unless otherwise noted, all content on this page is © Cengage Learning.

Step 4 Since the left side of the equation is a perfect-square trinomial, we can factor it as $(x + 4)^2$ and simplify on the right side to obtain

$$(x + 4)^2 = 9$$

Step 5 We can then solve the resulting equation by using the square-root property.

$$(x + 4)^2 = 9$$

$$x + 4 = \sqrt{9} \quad \text{or} \quad x + 4 = -\sqrt{9}$$
$$x + 4 = 3 \qquad\qquad x + 4 = -3$$
$$x = -1 \qquad\qquad x = -7$$

After checking both results, we see that the solutions are -1 and -7.

SELF CHECK 7 Solve $a^2 + 5a + 4 = 0$ by completing the square.

Although Example 7 was solved by completing the square, we could solve it by factoring the trinomial and applying the zero factor property. The results would be the same.

EXAMPLE 8 Solve $6x^2 + 5x - 6 = 0$ by completing the square.

Solution **Step 1** To make the coefficient of x^2 equal to 1, we divide both sides by 6.

$$6x^2 + 5x - 6 = 0$$

$$\frac{6x^2}{6} + \frac{5}{6}x - \frac{6}{6} = \frac{0}{6} \quad \text{Divide both sides by 6.}$$

$$x^2 + \frac{5}{6}x - 1 = 0 \quad \text{Simplify.}$$

Step 2 We add 1 to both sides to place the constant on the right side.

$$x^2 + \frac{5}{6}x \qquad = 1$$

Step 3 The coefficient of x is $\frac{5}{6}$, one-half of $\frac{5}{6}$ is $\frac{5}{12}$, and $\left(\frac{5}{12}\right)^2 = \frac{25}{144}$. To complete the square, we add $\frac{25}{144}$ to both sides.

$$x^2 + \frac{5}{6}x + \mathbf{\frac{25}{144}} = 1 + \mathbf{\frac{25}{144}}$$

Step 4 Since the left side of the equation is a perfect-square trinomial, we can factor it as $\left(x + \frac{5}{12}\right)^2$ and simplify the right side to obtain

$$\left(x + \frac{5}{12}\right)^2 = \frac{169}{144} \qquad 1 + \frac{25}{144} = \frac{144}{144} + \frac{25}{144} = \frac{169}{144}$$

Step 5 We can solve this equation by using the square-root property.

$$x + \frac{5}{12} = \sqrt{\frac{169}{144}} \quad \text{or} \quad x + \frac{5}{12} = -\sqrt{\frac{169}{144}} \qquad \text{Apply the square-root property.}$$

$$x + \frac{5}{12} = \frac{13}{12} \qquad x + \frac{5}{12} = -\frac{13}{12} \qquad \sqrt{\frac{169}{144}} = \frac{13}{12}$$

$$x = -\frac{5}{12} + \frac{13}{12} \qquad x = -\frac{5}{12} - \frac{13}{12} \qquad \text{Subtract } \frac{5}{12} \text{ from both sides.}$$

$$x = \frac{2}{3} \qquad x = -\frac{3}{2} \qquad \text{Add and simplify each fraction.}$$

After checking both results, we see that the solutions are $\frac{2}{3}$ and $-\frac{3}{2}$. Note that this equation could be solved by factoring.

SELF CHECK 8 Solve $6p^2 - 5p - 6 = 0$ by completing the square.

EXAMPLE 9 Solve $2x^2 + 4x + 1 = 0$ by completing the square.

Solution

$$2x^2 + 4x + 1 = 0$$

$$x^2 + 2x + \frac{1}{2} = \frac{0}{2}$$ Divide both sides by 2 to make the coefficient of x^2 equal to 1.

$$x^2 + 2x = -\frac{1}{2}$$ Subtract $\frac{1}{2}$ from both sides.

$$x^2 + 2x + \mathbf{1} = -\frac{1}{2} + \mathbf{1}$$ Square half the coefficient of x and add it to both sides: $\left(\frac{1}{2} \cdot 2\right)^2 = 1^2 = 1$

$$(x + 1)^2 = \frac{1}{2}$$ Factor on one side and combine like terms on the other.

$$x + 1 = \sqrt{\frac{1}{2}} \quad \text{or} \quad x + 1 = -\sqrt{\frac{1}{2}}$$ Use the square-root property.

$$x + 1 = \frac{\sqrt{2}}{2} \quad \Big| \quad x + 1 = -\frac{\sqrt{2}}{2}$$ $\sqrt{\frac{1}{2}} = \frac{1}{\sqrt{2}} = \frac{1 \cdot \sqrt{2}}{\sqrt{2}\sqrt{2}} = \frac{\sqrt{2}}{2}$

$$x = -1 + \frac{\sqrt{2}}{2} \quad \Big| \quad x = -1 - \frac{\sqrt{2}}{2}$$

These solutions can be written as $-1 \pm \frac{\sqrt{2}}{2}$.

COMMENT Note that $-1 \pm \frac{\sqrt{2}}{2}$ can be written as $-1 \pm \frac{\sqrt{2}}{2} = \frac{-2}{2} \pm \frac{\sqrt{2}}{2}$ $= \frac{-2 \pm \sqrt{2}}{2}$

SELF CHECK 9 Solve $3x^2 + 6x + 1 = 0$ by completing the square.

Accent on technology

▸ Checking Solutions of Quadratic Equations

We can use a graphing calculator to check the solutions of the equation $2x^2 + 4x + 1 = 0$ found in Example 9. Using a TI-84 Plus calculator, we first find an approximation of the decimal value of $-1 + \frac{\sqrt{2}}{2}$ by pressing these keys:

−1 **+ 2ND X²(√)2) ÷ 2 ENTER**

We will obtain the screen shown in Figure 10-2(a) on the next page. We can now store this decimal value in the calculator by pressing these keys:

STO *X, T, θ, n* ENTER

Finally, we enter $2x^2 + 4x + 1$ by pressing

2 ***X, T, θ, n*** **X²** (²) + 4 ***X, T, θ, n*** + 1

After pressing **ENTER** one more time, we will obtain the screen shown in Figure 10-2(b). The 0 on the screen confirms that $-1 + \frac{\sqrt{2}}{2}$ satisfies the equation $2x^2 + 4x + 1 = 0$ and is a solution.

(continued)

We can check the other solution in a similar way.

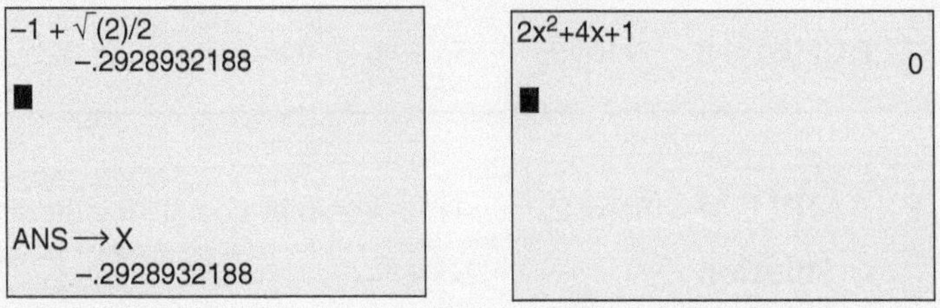

(a) (b)

Figure 10-2

For instructions regarding the use of a Casio graphing calculator, please refer to the Casio Keystroke Guide in the back of the book.

In the next example, the solutions are complex numbers.

EXAMPLE 10 Solve $3x^2 + 2x + 2 = 0$ by completing the square.

Solution

$$3x^2 + 2x + 2 = 0$$

$$x^2 + \frac{2}{3}x + \frac{2}{3} = \frac{0}{3}$$ Divide both sides by 3 to make the coefficient of x^2 equal to 1.

$$x^2 + \frac{2}{3}x = -\frac{2}{3}$$ Subtract $\frac{2}{3}$ from both sides.

$$x^2 + \frac{2}{3}x + \mathbf{\frac{1}{9}} = -\frac{2}{3} + \mathbf{\frac{1}{9}}$$ Square half the coefficient of x and add it to both sides. $\left(\frac{1}{2}\cdot\frac{2}{3}\right)^2 = \left(\frac{1}{3}\right)^2 = \frac{1}{9}$

$$\left(x + \frac{1}{3}\right)^2 = -\frac{5}{9}$$ Factor on one side and combine terms on the other: $\frac{1}{9} - \frac{2}{3} = \frac{1}{9} - \frac{6}{9} = -\frac{5}{9}$.

$$x + \frac{1}{3} = \sqrt{-\frac{5}{9}} \quad \text{or} \quad x + \frac{1}{3} = -\sqrt{-\frac{5}{9}}$$ Use the square-root property.

$$x + \frac{1}{3} = \sqrt{\frac{5}{9}}\sqrt{-1} \qquad x + \frac{1}{3} = -\sqrt{\frac{5}{9}}\sqrt{-1}$$ $\sqrt{-\frac{5}{9}} = \sqrt{\frac{5}{9}(-1)} = \sqrt{\frac{5}{9}}\sqrt{-1}$

$$x + \frac{1}{3} = \frac{\sqrt{5}}{3}i \qquad x + \frac{1}{3} = -\frac{\sqrt{5}}{3}i$$ $\sqrt{\frac{5}{9}} = \frac{\sqrt{5}}{\sqrt{9}} = \frac{\sqrt{5}}{3}$

$$x = -\frac{1}{3} + \frac{\sqrt{5}}{3}i \qquad x = -\frac{1}{3} - \frac{\sqrt{5}}{3}i$$ Subtract $\frac{1}{3}$ from both sides.

These solutions are written as $x = -\frac{1}{3} \pm \frac{\sqrt{5}}{3}i$.

SELF CHECK 10 Solve $x^2 + 4x + 6 = 0$ by completing the square.

4 Solve an application requiring the use of the square-root property.

Many applications involving equations containing squared terms can be solved using the square-root property.

Unless otherwise noted, all content on this page is © Cengage Learning.

EXAMPLE 11 **DVDs** A DVD used for recording movies has a surface area of 17.72 square inches on one side. Find the radius of a disc.

Solution The formula for the area of a circular disc is $A = \pi r^2$. We can find the radius of a disc by substituting 17.72 for A and solving for r.

$$A = \pi r^2$$

$$17.72 = \pi r^2 \quad \text{Substitute 17.72 for } A.$$

$$\frac{17.72}{\pi} = r^2 \quad \text{Divide both sides by } \pi.$$

$$r = \sqrt{\frac{17.72}{\pi}} \quad \text{or} \quad r = -\sqrt{\frac{17.72}{\pi}} \quad \text{Use the square-root property.}$$

Since the radius of a disc cannot be negative, we will discard the negative result. Thus, the radius of a disc is $\sqrt{\frac{17.72}{\pi}}$ inches or, to the nearest hundredth, 2.37 inches.

SELF CHECK 11 The surface area of the DVD is 114.31 square centimeters. Find the radius of the disc.

When you deposit money in a bank account, it earns interest. If you leave the money in the account, the earned interest is deposited back into the account and also earns interest. When this is the case, the account is earning **compound interest**. There is a formula we can use to compute the amount in an account at any time t.

FORMULA FOR COMPOUND INTEREST

If P dollars is deposited in an account and interest is paid once a year at an annual rate r, the amount A in the account after t years is given by the formula

$$A = P(1 + r)^t$$

EXAMPLE 12 **SAVING MONEY** A woman invests \$10,000 in an account. Find the annual interest rate if the account is worth \$11,025 in 2 years.

Solution We substitute 11,025 for A, 10,000 for P, and 2 for t in the compound interest formula and solve for r.

$$A = P(1 + r)^t$$

$$\mathbf{11{,}025} = \mathbf{10{,}000}(1 + r)^{\mathbf{2}} \quad \text{Substitute.}$$

$$\frac{11{,}025}{10{,}000} = (1 + r)^2 \quad \text{Divide both sides by 10,000.}$$

$$1.1025 = (1 + r)^2 \quad \frac{11{,}025}{10{,}000} = 1.1025$$

$$1 + r = 1.05 \quad \text{or} \quad 1 + r = -1.05 \quad \text{Use the square-root property: } \sqrt{1.1025} = 1.05$$

$$r = 0.05 \qquad\qquad r = -2.05 \quad \text{Subtract 1 from both sides.}$$

Since an interest rate cannot be negative, we must discard the result of -2.05. Thus, the annual interest rate is 0.05, or 5%.

We can check this result by substituting 0.05 for r, 10,000 for P, and 2 for t in the formula and confirming that the deposit of \$10,000 will grow to \$11,025 in 2 years.

$$A = P(1 + r)^t = \mathbf{10{,}000}(1 + \mathbf{0.05})^{\mathbf{2}} = 10{,}000(1.1025) = 11{,}025$$

SELF CHECK 12 Find the annual interest rate if the account is worth \$10,920.25 in 2 years.

SELF CHECK ANSWERS

1. $8, -8$ **2.** $\frac{1}{3}, \frac{1}{2}$ **3.** $\pm 3\sqrt{2}$ **4.** $1, -5$ **5.** $\pm 3i$ **6.** $\frac{25}{4}$ **7.** $-1, -4$ **8.** $-\frac{2}{3}, \frac{3}{2}$ **9.** $-1 \pm \frac{\sqrt{6}}{3}$ **10.** $-2 \pm \sqrt{2}i$ **11.** The radius, to the nearest hundredth, is 6.03 cm. **12.** The interest rate is 4.5%.

NOW TRY THIS

1. Solve using the square-root property.
   a. $(3x + 5)^2 = 18$
   b. $(x + 6)^2 = 0$
2. Solve $x^2 - 2\sqrt{2}x + 1 = 0$ by completing the square.

10.1 Exercises

WARM-UPS *Multiply.*

1. $(x + 5)(x - 5)$
2. $(x - 10)(x + 10)$
3. $(x + 2)^2$
4. $(x - 5)^2$
5. $(x - 7)^2$
6. $(x + 4)^2$

REVIEW *Solve each equation or inequality.*

7. $\frac{t + 9}{2} + \frac{t + 2}{5} = \frac{8}{5} + 4t$
8. $\frac{1 - 5x}{2x} + 4 = \frac{x + 3}{x}$
9. $3(t - 3) + 3t \le 2(t + 1) + t + 1$
10. $-2(y + 4) - 3y + 8 \ge 3(2y - 3) - y$

VOCABULARY AND CONCEPTS *Fill in the blanks.*

11. The square-root property, also called extracting the roots, states that the solutions of $x^2 = c$ are _______ and _________.
12. To complete the square on x in $x^2 + 6x = 17$, find one-half of _, square it to obtain _, and add _ to both sides of the equation.
13. The symbol $\pm$ is read as _______________.
14. The formula for annual compound interest is ____________.

GUIDED PRACTICE *Use factoring to solve each equation.* SEE EXAMPLE 1. (OBJECTIVE 1)

15. $x^2 - 81 = 0$
16. $y^2 - 121 = 0$
17. $2y^2 - 72 = 0$
18. $4y^2 - 64 = 0$

Use factoring to solve each equation. SEE EXAMPLE 2. (OBJECTIVE 1)

19. $y^2 + 7y + 12 = 0$
20. $x^2 + 9x + 20 = 0$
21. $6s^2 + 11s - 10 = 0$
22. $3x^2 + 10x - 8 = 0$

Use the square-root property to solve each equation. SEE EXAMPLE 3. (OBJECTIVE 2)

23. $x^2 = 36$
24. $x^2 = 144$
25. $z^2 = 5$
26. $u^2 = 24$

Use the square-root property to solve each equation. SEE EXAMPLE 4. (OBJECTIVE 2)

27. $(y + 3)^2 = 9$
28. $(y - 1)^2 = 4$
29. $(x - 2)^2 - 5 = 0$
30. $(x - 4)^2 - 7 = 0$

Use the square-root property to solve each equation. SEE EXAMPLE 5. (OBJECTIVE 2)

31. $16p^2 + 49 = 0$
32. $q^2 + 25 = 0$
33. $4m^2 + 81 = 0$
34. $9n^2 + 64 = 0$

Find the number that when added to the binomial will make it a perfect-square trinomial. SEE EXAMPLE 6. (OBJECTIVE 3)

35. $x^2 + 4x$
36. $x^2 - 6x$
37. $x^2 - 3x$
38. $x^2 + 7x$

Use completing the square to solve each equation. SEE EXAMPLE 7. (OBJECTIVE 3)

39. $x^2 + 4x - 12 = 0$
40. $x^2 + 6x + 5 = 0$
41. $x^2 - 6x + 8 = 0$
42. $x^2 + 8x + 15 = 0$

Use completing the square to solve each equation. SEE EXAMPLE 8. (OBJECTIVE 3)

43. $6x^2 + 11x + 3 = 0$
44. $6x^2 + x - 2 = 0$
45. $6x^2 - 7x - 5 = 0$
46. $4x^2 - x - 3 = 0$
47. $9 - 6r = 8r^2$
48. $11m - 10 = 3m^2$
49. $x + 1 = 2x^2$
50. $-2 = 2x^2 - 5x$

Use completing the square to solve each equation. SEE EXAMPLE 9. (OBJECTIVE 3)

51. $\frac{7x + 1}{5} = -x^2$
52. $\frac{3x^2}{8} = \frac{1}{8} - x$

53. $3x^2 - 6x + 1 = 0$

54. $3x^2 + 9x + 5 = 0$

Use completing the square to solve each equation. SEE EXAMPLE 10. *(OBJECTIVE 3)*

55. $p^2 + 2p + 2 = 0$

56. $x^2 - 8x + 17 = 0$

57. $y^2 + 8y + 18 = 0$

58. $t^2 + t + 3 = 0$

ADDITIONAL PRACTICE *Solve using any method.*

59. $9x - 8 = x^2$

60. $5t - 6 = t^2$

61. $3x^2 - 16 = 0$

62. $5x^2 - 49 = 0$

63. $(s - 7)^2 - 9 = 0$

64. $(t + 4)^2 = 16$

65. $(x + 5)^2 - 3 = 0$

66. $(x + 3)^2 - 7 = 0$

67. $2z^2 - 5z + 2 = 0$

68. $2x^2 - x - 1 = 0$

69. $3m^2 - 2m - 3 = 0$

70. $4p^2 + 2p + 3 = 0$

71. $5x^2 + 15x = 0$

72. $5x^2 + 11x = 0$

73. $x^2 + 5x + 4 = 0$

74. $x^2 - 11x + 30 = 0$

75. $x^2 - 9x - 10 = 0$

76. $x^2 - 7x + 10 = 0$

77. $2x^2 - x + 8 = 0$

78. $4x^2 + 2x + 5 = 0$

Solve for the indicated variable. Assume that all variables represent positive numbers. Express all radicals in simplified form.

79. $2d^2 = 3h$ for d

80. $2x^2 = d^2$ for d

81. $E = mc^2$ for c

82. $S = \frac{1}{2}gt^2$ for t

Find all values of x that will make f(x) = 0.

83. $f(x) = 2x^2 + x - 5$

84. $f(x) = 3x^2 - 2x - 4$

85. $f(x) = x^2 + x - 3$

86. $f(x) = x^2 + 2x - 4$

APPLICATIONS *Solve each application.* SEE EXAMPLES 11–12. *(OBJECTIVE 4)*

87. Falling object The distance s (in feet) that an object will fall in t seconds is given by the formula $s = 16t^2$. How long will it take an object to fall 256 feet?

88. Pendulum The time (in seconds) it takes a pendulum to swing back and forth to complete one cycle is related to its length l (in feet) by the formula:

$$l = \frac{32t^2}{4\pi^2}$$

How long will it take a 5-foot pendulum to swing through one cycle? Give the result to the nearest hundredth.

89. Law enforcement To estimate the speed s (in mph) of a car involved in an accident, police often use the formula $s^2 = 10.5l$, where l is the length of any skid mark. Approximately how fast was a car going that was involved in an accident and left skid marks of 495 feet?

90. Medicine The approximate pulse rate (in beats per minute) of an adult who is t inches tall is given by the formula

$$p^2 = \frac{348{,}100}{t}$$

Find the pulse rate of an adult who is 64 inches tall.

91. Saving money A student invests \$8,500 in a savings account drawing interest that is compounded annually. Find the annual rate if the money grows to \$9,193.60 in 2 years.

92. Saving money A woman invests \$12,500 in a savings account drawing interest that is compounded annually. Find the annual rate if the money grows to \$14,045 in 2 years.

93. Flags In 1912, an order by President Taft fixed the width and length of the U.S. flag in the ratio 1 to 1.9. If 100 square feet of cloth are to be used to make a U.S. flag, estimate its dimensions to the nearest $\frac{1}{4}$ foot.

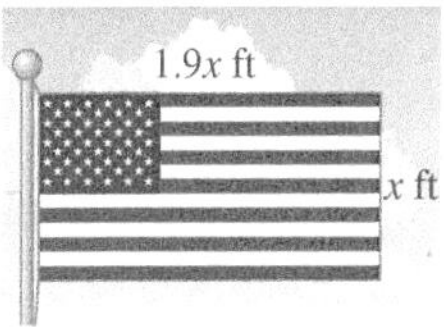

94. Accidents The height h (in feet) of an object that is dropped from a height of s feet is given by the formula $h = s - 16t^2$, where t is the time the object has been falling. A 5-foot-tall woman on a sidewalk looks directly overhead and sees a window washer drop a bottle from 4 stories up. How long does she have to get out of the way? (A story is 10 feet.)

WRITING ABOUT MATH

95. Explain why we did not round to the nearest tenth in Exercise 94.

96. Explain why a cannot be 0 in the quadratic equation $ax^2 + bx + c = 0$.

SOMETHING TO THINK ABOUT

97. What number must be added to $x^2 + \sqrt{3}x$ to make it a perfect-square trinomial?

98. Solve $x^2 + \sqrt{3}x - \frac{1}{4} = 0$ by completing the square.

Unless otherwise noted, all content on this page is © Cengage Learning.

Section 10.2 Solving Quadratic Equations Using the Quadratic Formula

Objectives

1. Solve a quadratic equation using the quadratic formula.
2. Solve a formula for a specified variable using the quadratic formula.
3. Solve an application involving a quadratic equation.

Vocabulary

quadratic formula

Getting Ready

Add a number to each binomial to complete the square. Then write the resulting trinomial as the square of a binomial.

1. $x^2 + 12x$

2. $x^2 - 7x$

Evaluate $\sqrt{b^2 - 4ac}$ for the following values.

3. $a = 6, b = 1, c = -2$

4. $a = 4, b = -4, c = -3$

Solving quadratic equations by completing the square is often tedious. Fortunately, there is a more direct way. In this section, we will develop a formula, called the **quadratic formula**, that we can use to solve quadratic equations primarily with arithmetic. To develop this formula, we will use the skills we learned in the last section and complete the square.

1 Solve a quadratic equation using the quadratic formula.

To develop a formula to solve quadratic equations, we will solve the general quadratic equation $ax^2 + bx + c = 0$ $(a \neq 0)$ by completing the square.

$$ax^2 + bx + c = 0$$

$$\frac{ax^2}{a} + \frac{bx}{a} + \frac{c}{a} = \frac{0}{a}$$ To make the coefficient of x^2 equal to 1, we divide both sides by a.

$$x^2 + \frac{bx}{a} = -\frac{c}{a}$$ $\frac{0}{a} = 0$; subtract $\frac{c}{a}$ from both sides.

$$x^2 + \frac{bx}{a} + \left(\frac{b}{2a}\right)^2 = \left(\frac{b}{2a}\right)^2 - \frac{c}{a}$$ Complete the square on x by adding $\left(\frac{b}{2a}\right)^2$ to both sides.

$$x^2 + \frac{b}{a}x + \frac{b^2}{4a^2} = \frac{b^2}{4a^2} - \frac{4ac}{4aa}$$ Simplify and obtain a common denominator on the right side.

$$\left(x + \frac{b}{2a}\right)^2 = \frac{b^2 - 4ac}{4a^2}$$ Factor the left side and add the fractions on the right side.

(1) $$x + \frac{b}{2a} = \pm\sqrt{\frac{b^2 - 4ac}{4a^2}}$$ Use the square-root property.

We can solve Equation 1 by writing it as two separate equations.

$$x + \frac{b}{2a} = \sqrt{\frac{b^2 - 4ac}{4a^2}} \quad \text{or} \quad x + \frac{b}{2a} = -\sqrt{\frac{b^2 - 4ac}{4a^2}}$$

$$x + \frac{b}{2a} = \frac{\sqrt{b^2 - 4ac}}{2a} \qquad x + \frac{b}{2a} = -\frac{\sqrt{b^2 - 4ac}}{2a}$$

$$x = -\frac{b}{2a} + \frac{\sqrt{b^2 - 4ac}}{2a} \qquad x = -\frac{b}{2a} - \frac{\sqrt{b^2 - 4ac}}{2a}$$

$$x = \frac{-b + \sqrt{b^2 - 4ac}}{2a} \qquad x = \frac{-b - \sqrt{b^2 - 4ac}}{2a}$$

These two solutions represent the *quadratic formula*.

THE QUADRATIC FORMULA

The solutions of $ax^2 + bx + c = 0$ $(a \neq 0)$ are obtained by the formula

$$x = \frac{-b \pm \sqrt{b^2 - 4ac}}{2a}$$

COMMENT Be sure to draw the fraction bar under both parts of the numerator, and be sure to draw the radical sign exactly over $b^2 - 4ac$.

EXAMPLE 1 Use the quadratic formula to solve $2x^2 - 3x - 5 = 0$.

Solution In this equation $a = 2$, $b = -3$, and $c = -5$.

$$x = \frac{-b \pm \sqrt{b^2 - 4ac}}{2a}$$

$$= \frac{-(-3) \pm \sqrt{(-3)^2 - 4(2)(-5)}}{2(2)} \quad \text{Substitute 2 for } a\text{, } -3 \text{ for } b\text{, and } -5 \text{ for } c.$$

$$= \frac{3 \pm \sqrt{9 + 40}}{4} \quad \text{Simplify.}$$

$$= \frac{3 \pm \sqrt{49}}{4} \quad \text{Add.}$$

$$= \frac{3 \pm 7}{4} \quad \text{Simplify the radical.}$$

$$x = \frac{3 + 7}{4} \quad \text{or} \quad x = \frac{3 - 7}{4}$$

$$x = \frac{10}{4} \qquad x = \frac{-4}{4}$$

$$x = \frac{5}{2} \qquad x = -1$$

After checking the results, we see that the solutions are $\frac{5}{2}$ and -1. Note that this equation can be solved by factoring.

SELF CHECK 1 Use the quadratic formula to solve $3x^2 - 5x - 2 = 0$.

It is important to write the quadratic equation in quadratic form $ax^2 + bx + c = 0$, prior to identifying the values of a, b and c.

EXAMPLE 2 Use the quadratic formula to solve $2x^2 + 1 = -4x$.

Solution We begin by writing the equation in $ax^2 + bx + c = 0$ form (called *standard form*) before identifying a, b, and c.

$$2x^2 + 4x + 1 = 0$$

In this equation, $a = 2$, $b = 4$, and $c = 1$.

$$x = \frac{-b \pm \sqrt{b^2 - 4ac}}{2a}$$

$$= \frac{-4 \pm \sqrt{4^2 - 4(2)(1)}}{2(2)} \quad \text{Substitute 2 for } a\text{, 4 for } b\text{, and 1 for } c.$$

$$= \frac{-4 \pm \sqrt{16 - 8}}{4} \quad \text{Simplify.}$$

$$= \frac{-4 \pm \sqrt{8}}{4} \quad \text{Subtract.}$$

$$= \frac{-4 \pm 2\sqrt{2}}{4} \quad \sqrt{8} = \sqrt{4 \cdot 2} = \sqrt{4}\sqrt{2} = 2\sqrt{2}$$

$$= \frac{-2 \pm \sqrt{2}}{2} \quad \frac{-4 \pm 2\sqrt{2}}{4} = \frac{2(-2 \pm \sqrt{2})}{4} = \frac{-2 \pm \sqrt{2}}{2}$$

Note that these solutions can be written as $-1 \pm \frac{\sqrt{2}}{2}$.

SELF CHECK 2 Use the quadratic formula to solve $3x^2 - 2x - 3 = 0$.

In the next example, the solutions are complex numbers.

EXAMPLE 3 Use the quadratic formula to solve $x^2 + x = -1$.

Solution We begin by writing the equation in standard form before identifying a, b, and c.

$$x^2 + x + 1 = 0$$

In this equation, $a = 1$, $b = 1$, and $c = 1$.

$$x = \frac{-b \pm \sqrt{b^2 - 4ac}}{2a}$$

$$= \frac{-1 \pm \sqrt{1^2 - 4(1)(1)}}{2(1)} \quad \text{Substitute 1 for } a\text{, 1 for } b\text{, and 1 for } c.$$

$$= \frac{-1 \pm \sqrt{1 - 4}}{2} \quad \text{Simplify the expression under the radical.}$$

$$= \frac{-1 \pm \sqrt{-3}}{2} \quad \text{Subtract.}$$

$$= \frac{-1 \pm \sqrt{3}i}{2} \quad \text{Simplify the radical expression.}$$

Note that these solutions should be written in $a + bi$ form as $-\frac{1}{2} \pm \frac{\sqrt{3}}{2}i$.

SELF CHECK 3 Use the quadratic formula to solve $a^2 + 2a = -3$.

Perspective

THE FIBONACCI SEQUENCE AND THE GOLDEN RATIO

Perhaps one of the most intriguing examples of how a mathematical idea can represent natural phenomena is the *Fibonacci Sequence*, a list of whole numbers that is generated by a very simple rule. This sequence was first developed by the Italian mathematician Leonardo da Pisa, more commonly known as Fibonacci. The Fibonacci Sequence is the following list of numbers

$$1, 1, 2, 3, 5, 8, 13, 21, \ldots$$

where each successive number in the list is obtained by adding the two preceding numbers. Although Fibonacci originally developed this sequence to solve a mathematical puzzle, subsequent study of the numbers in this sequence has uncovered many examples in the natural world in which this sequence emerges. For example, the arrangement of the seeds on the face of a sunflower, the hibernation periods of certain insects, and the branching patterns of many plants all give rise to Fibonacci numbers.

Among the many special properties of these numbers is the fact that, as we generate more and more numbers in the list, the ratio of successive numbers approaches a constant value. This value is designated by the symbol ϕ and often is referred to as the "Golden Ratio." One way to calculate the value of ϕ is to solve the quadratic equation $\phi^2 - \phi - 1 = 0$.

Leonardo Fibonacci

1. Using the quadratic formula, find the exact value of ϕ.

2. Using a calculator, find a decimal approximation of ϕ, correct to three decimal places.

Note: Negative solutions are not applicable in this context.

2 Solve a formula for a specified variable using the quadratic formula.

EXAMPLE 4 An object thrown straight up with an initial velocity of v_0 feet per second will reach a height of s feet in t seconds according to the formula $s = -16t^2 + v_0t$. Solve the formula for t.

Solution We begin by writing the equation in standard form:

$$s = -16t^2 + v_0t$$

$$16t^2 - v_0t + s = 0$$

In this equation, $a = 16$, $b = -v_0$, and $c = s$. We can use the quadratic formula to solve for t.

$$t = \frac{-b \pm \sqrt{b^2 - 4ac}}{2a}$$

$$t = \frac{-(-v_0) \pm \sqrt{(-v_0)^2 - 4(16)(s)}}{2(16)} \quad \text{Substitute into the quadratic formula.}$$

$$t = \frac{v_0 \pm \sqrt{v_0{}^2 - 64s}}{32} \quad \text{Simplify.}$$

SELF CHECK 4 Solve the formula $s = -16t^2 + 128t$ for t.

Unless otherwise noted, all content on this page is © Cengage Learning.

3 Solve an application involving a quadratic equation.

EXAMPLE 5 **DIMENSIONS OF A RECTANGLE** Find the dimensions of the rectangle shown in Figure 10-3, given that its area is 253 cm^2.

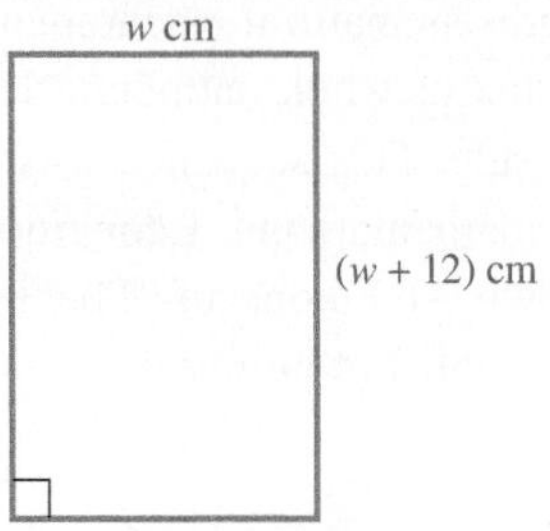

Figure 10-3

Solution If we let w represent the width of the rectangle in cm, then $(w + 12)$ represents its length in cm. Since the area of the rectangle is 253 square centimeters, we can form the equation

$$w(w + 12) = 253 \quad \text{Area of a rectangle} = \text{width} \cdot \text{length}$$

and solve it as follows:

$$w(w + 12) = 253$$
$$w^2 + 12w = 253 \quad \text{Use the distributive property to remove parentheses.}$$
$$w^2 + 12w - 253 = 0 \quad \text{Write the equation in quadratic form.}$$

Solution by factoring

$$(w - 11)(w + 23) = 0$$
$$w - 11 = 0 \quad \text{or} \quad w + 23 = 0$$
$$w = 11 \quad \bigg| \quad w = -23$$

Solution by quadratic formula

$$w = \frac{-12 \pm \sqrt{12^2 - 4(1)(-253)}}{2(1)}$$
$$= \frac{-12 \pm \sqrt{144 + 1{,}012}}{2}$$
$$= \frac{-12 \pm \sqrt{1{,}156}}{2}$$
$$= \frac{-12 \pm 34}{2}$$
$$w = 11 \quad \text{or} \quad w = -23$$

Since the rectangle cannot have a negative width, we discard the solution of -23. Thus, the only solution is $w = 11$. Since the rectangle is 11 centimeters wide and $(11 + 12)$ centimeters long, its dimensions are 11 centimeters by 23 centimeters.

Check: 23 is 12 more than 11, and the area of a rectangle with dimensions of 23 centimeters by 11 centimeters is 253 square centimeters.

SELF CHECK 5 Find the dimensions of the rectangle shown given that its area is 189 square centimeters.

SELF CHECK ANSWERS

1. $2, -\frac{1}{3}$ **2.** $\frac{1}{3} \pm \frac{\sqrt{10}}{3}$ **3.** $-1 \pm \sqrt{2}i$ **4.** $t = 4 \pm \frac{\sqrt{256 - s}}{4}$ **5.** The dimensions are 9 cm by 21 cm.

Unless otherwise noted, all content on this page is © Cengage Learning.

NOW TRY THIS

1. The length of a rectangular garden is 1 ft less than 3 times the width. If the area is 44 square feet, find the length of the garden.
2. The product of 2 consecutive integers is 90. Find the numbers.
3. Solve $x^3 - 8 = 0$. (*Hint*: Recall how to factor the difference of cubes.)
4. Graph $y = x^3 - 8$ and identify the x-intercept(s).

10.2 Exercises

WARM-UPS *Identify a, b, and c in each quadratic equation.*

1. $5x^2 - 3x = 4$

2. $-2x^2 + x = 5$

REVIEW *Solve for the indicated variable.*

3. $Ax + By = C$ for y

4. $R = \frac{kL}{d^2}$ for L

Simplify each radical.

5. $\sqrt{45}$

6. $\sqrt{288}$

7. $\frac{5}{\sqrt{5}}$

8. $\frac{1}{2 - \sqrt{3}}$

VOCABULARY AND CONCEPTS *Fill in the blanks.*

9. In the quadratic equation $7x^2 - 4x - 9 = 0$, $a =$ _, $b =$ ___, and $c =$ ___.

10. The solutions of $ax^2 + bx + c = 0$ $(a \neq 0)$ are given by the quadratic formula, which is $x =$ __________.

GUIDED PRACTICE *Solve each equation using the quadratic formula.* SEE EXAMPLE 1. (OBJECTIVE 1)

11. $x^2 + 6x + 5 = 0$

12. $x^2 - 3x + 2 = 0$

13. $x^2 - 5x - 14 = 0$

14. $x^2 - 2x - 35 = 0$

15. $x^2 + 9x = -20$

16. $y^2 - 18y = -81$

17. $2x^2 - x - 3 = 0$

18. $3x^2 - 10x + 8 = 0$

Solve each equation using the quadratic formula. SEE EXAMPLE 2. (OBJECTIVE 1)

19. $15x^2 - 14x = 8$

20. $4x^2 = -5x + 6$

21. $8u = -4u^2 - 3$

22. $4t + 3 = 4t^2$

23. $5x^2 + 5x + 1 = 0$

24. $4w^2 + 6w + 1 = 0$

25. $5x^2 + 2x - 1 = 0$

26. $7x^2 + 4x - 1 = 0$

Solve each equation using the quadratic formula. SEE EXAMPLE 3. (OBJECTIVE 1)

27. $x^2 + 2x + 2 = 0$

28. $5x^2 = 2x - 1$

29. $x^2 + 5x + 7 = 0$

30. $x^2 + 3x + 3 = 0$

31. $3x^2 - 4x = -2$

32. $2x^2 + 3x = -3$

33. $3x^2 - 2x = -3$

34. $3x^2 + 2x + 1 = 0$

Solve each formula for the indicated variable. SEE EXAMPLE 4. (OBJECTIVE 2)

35. $C = \frac{N^2 - N}{2}$, for N
(The formula for a selection sort in data processing)

36. $A = 2\pi r^2 + 2\pi hr$, for r
(The formula for the surface area of a right circular cylinder)

37. $x^2 - kx = -ay$ for x

38. $xy^2 + 3xy + 7 = 0$ for y

ADDITIONAL PRACTICE *Solve each equation using any method.*

39. $6x^2 - 5x - 4 = 0$

40. $2x^2 + 5x - 3 = 0$

41. $\frac{x^2}{2} + \frac{5}{2}x = -1$

42. $-3x = \frac{x^2}{2} + 2$

43. $3x^2 - 2 = 2x$

44. $-9x = 2 - 3x^2$

45. $16y^2 + 8y - 3 = 0$

46. $16x^2 + 16x + 3 = 0$

Find all x-values that will make $f(x) = 0$.

47. $f(x) = 4x^2 + 4x - 19$

48. $f(x) = 9x^2 + 12x - 8$

49. $f(x) = 3x^2 + 2x + 2$

50. $f(x) = 5x^2 + 2x + 1$

Use the quadratic formula and a calculator to solve each equation. Give all answers to the nearest hundredth.

51. $0.7x^2 - 3.5x - 25 = 0$

52. $-4.5x^2 + 0.2x + 3.75 = 0$

Note that a and b are the solutions to the equation $(x - a)(x - b) = 0$.

53. Find a quadratic equation that has a solution set of $\{3, 5\}$.

54. Find a quadratic equation that has a solution set of $\{-2, 7\}$.

55. Find a third-degree equation that has a solution set of $\{2, 3, -4\}$.

56. Find a fourth-degree equation that has a solution set of $\{2, -2, 5, -5\}$.

APPLICATIONS *Solve. SEE EXAMPLE 5. (OBJECTIVE 3)*

57. Dimensions of a rectangle The rectangle has an area of 96 square feet. Find its dimensions.

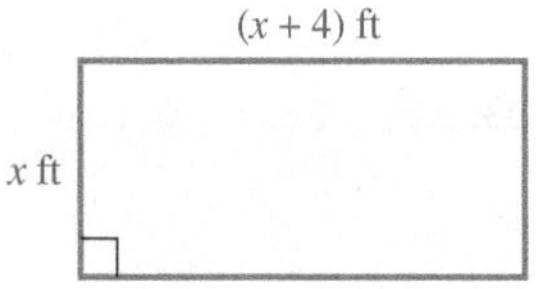

58. Dimensions of a window The area of the window is 77 square feet. Find its dimensions.

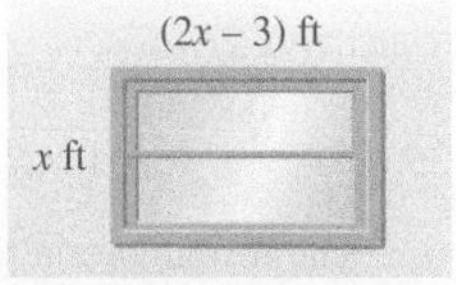

59. Side of a square The area of a square is numerically equal to its perimeter. Find the length of each side of the square.

60. Perimeter of a rectangle A rectangle is 2 inches longer than it is wide. Numerically, its area exceeds its perimeter by 11. Find the perimeter.

Solve.

61. Base of a triangle The height of a triangle is 5 centimeters longer than three times its base. Find the base of the triangle if its area is 6 square centimeters.

62. Height of a triangle The height of a triangle is 4 meters longer than twice its base. Find the height if the area of the triangle is 15 square meters.

63. Integer The product of two consecutive even integers is 168. Find the integers. (*Hint*: If one even integer is x, the next consecutive even integer is $x + 2$.)

64. Integer The product of two consecutive odd integers is 143. Find the integers. (*Hint*: If one odd integer is x, the next consecutive odd integer is $x + 2$.)

65. Integer The sum of the squares of two consecutive integers is 85. Find the integers. (*Hint*: If one integer is x, the next consecutive positive integer is $x + 1$.)

66. Integer The sum of the squares of three consecutive integers is 77. Find the integers. (*Hint*: If one integer is x, the next consecutive positive integer is $x + 1$, and the third is $x + 2$.)

67. Rates A woman drives her snowmobile 150 miles at the rate of r mph. She could have gone the same distance in 2 hours less time if she had increased her speed by 20 mph. Find r.

68. Rates Jeff bicycles 160 miles at the rate of r mph. The same trip would have taken 2 hours longer if he had decreased his speed by 4 mph. Find r.

69. Concert tickets Tickets to a concert cost \$4, and the projected attendance is 300 people. It is further projected that for every 10¢ increase in ticket price, the average attendance will decrease by 5. At what ticket price will the nightly receipts be \$1,248?

70. Bus fares A bus company has 3,000 passengers daily, paying a \$1.25 fare. For each 25¢ increase in fare, the company estimates that it will lose 80 passengers. What is the smallest increase in fare that will produce a \$4,970 daily revenue?

71. Profit The *Gazette's* profit is \$20 per year for each of its 3,000 subscribers. Management estimates that the profit per subscriber will increase by 1¢ for each additional subscriber over the current 3,000. How many subscribers will bring a total profit of \$120,000?

72. Interest rates A woman invests \$1,000 in a mutual fund for which interest is compounded annually at a rate r. After one year, she deposits an additional \$2,000. After two years, the balance A in the account is

$$A = 1{,}000(1 + r)^2 + 2{,}000(1 + r)$$

If this amount is \$3,368.10, find r.

73. Framing a picture The frame around the picture in the illustration has a constant width. To the nearest hundredth, how wide is the frame if its area equals the area of the picture?

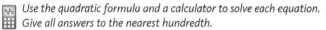

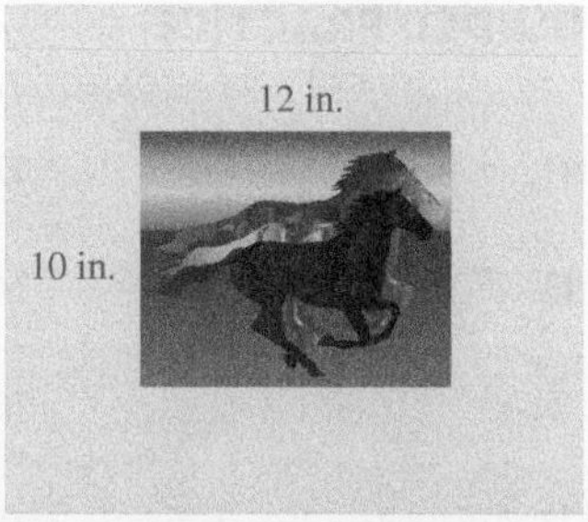

74. Metal fabrication A box with no top is to be made by cutting a 2-inch square from each corner of the square sheet of metal shown in the illustration. After bending up the sides, the volume of the box is to be 200 cubic inches. How large should the piece of metal be?

Unless otherwise noted, all content on this page is © Cengage Learning.

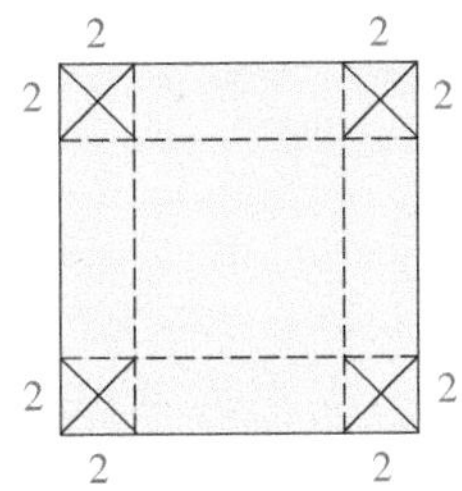

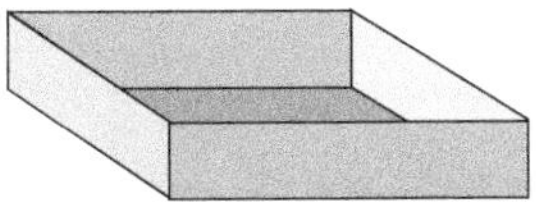

Use a calculator.

75. Labor force The labor force participation rate P (in percent) for workers ages 16 and older from 1966 to 2008 is approximated by the quadratic equation

$$P = -0.0072x^2 + 0.4904x + 58.2714$$

where $x = 0$ corresponds to the year 1966, $x = 1$ corresponds to 1967, and so on. (Thus, $0 \le x \le 42$.) In what year in this range were 65% of the workers ages 16 and older part of the workforce?

76. Space program The yearly budget B (in billions of dollars) for the National Aeronautics and Space Administration (NASA) is approximated by the quadratic equation

$$B = 0.0518x^2 - 0.2122x + 14.1112$$

where x is the number of years since 1997 and $0 \le x \le 12$. In what year does the model indicate that NASA's budget was about \$17 billion?

77. Chemistry A weak acid (0.1 M concentration) breaks down into free cations (the hydrogen ion, H^+) and anions (A^-). When this acid dissociates, the following equilibrium equation is established:

$$\frac{[H^+][A^-]}{[HA]} = 4 \times 10^{-4}$$

where $[H^+]$, the hydrogen ion concentration, is equal to $[A^-]$, the anion concentration. $[HA]$ is the concentration of the undissociated acid itself. Find $[H^+]$ at equilibrium. (*Hint*: If $[H^+] = x$, then $[HA] = 0.1 - x$.)

78. Chemistry A saturated solution of hydrogen sulfide (0.1 M concentration) dissociates into cation $[H^+]$ and anion $[HS^-]$, where $[H^+] = [HS^-]$. When this solution dissociates, the following equilibrium equation is established:

$$\frac{[H^+][HS^-]}{[HHS]} = 1.0 \times 10^{-7}$$

Find $[H^+]$. (*Hint*: If $[H^+] = x$, then $[HHS] = 0.1 - x$.)

WRITING ABOUT MATH

79. Explain why $x = -b \pm \frac{\sqrt{b^2 - 4ac}}{2a}$ is not a correct statement of the quadratic formula.

80. Explain why $x = \frac{b \pm \sqrt{b^2 - 4ac}}{2a}$ is not a correct statement of the quadratic formula.

SOMETHING TO THINK ABOUT *All of the equations we have solved so far have had rational-number coefficients. However, the quadratic formula can be used to solve quadratic equations with irrational or even imaginary coefficients. Try solving each of the following equations.*

81. $x^2 + 2\sqrt{2}x - 6 = 0$

82. $\sqrt{2}x^2 + x - \sqrt{2} = 0$

83. $x^2 - 3ix - 2 = 0$

84. $ix^2 + 3x - 2i = 0$

Section 10.3 The Discriminant and Equations That Can Be Written in Quadratic Form

Objectives

1. Use the discriminant to determine the type of solutions to a given quadratic equation.
2. Solve an equation that can be written in quadratic form.
3. Verify the solutions of a quadratic equation by showing that the sum of the solutions is $-\frac{b}{a}$ and the product is $\frac{c}{a}$.

Unless otherwise noted, all content on this page is © Cengage Learning.

Vocabulary

discriminant

Getting Ready

Evaluate $b^2 - 4ac$ for the following values.

1. $a = 2, b = 3$, and $c = -1$
2. $a = -2, b = 4$, and $c = -3$

Recall that we can test a trinomial of the form $ax^2 + bx + c$ $(a \neq 0)$ for factorability by evaluating $b^2 - 4ac$. If $b^2 - 4ac$ is a perfect square, the trinomial is factorable. We recognize that $b^2 - 4ac$ is the radicand in the quadratic formula. We can use that part of the quadratic formula to determine the type of solutions that a quadratic equation will have without solving it first.

1 Use the discriminant to determine the type of solutions to a given quadratic equation.

Suppose that the coefficients a, b, and c in the equation $ax^2 + bx + c = 0$ $(a \neq 0)$ are real numbers. Then the solutions of the equation are given by the quadratic formula

$$x = \frac{-b \pm \sqrt{b^2 - 4ac}}{2a} \quad (a \neq 0)$$

The value of $b^2 - 4ac$, called the **discriminant**, determines the type of solutions for any quadratic equation. Note that $b^2 - 4ac$ is a radicand; thus if $b^2 - 4ac \geq 0$, the solutions are real numbers and if $b^2 - 4ac < 0$, the solutions are nonreal complex numbers.

THE DISCRIMINANT

Given $ax^2 + bx + c = 0$ $(a \neq 0)$ and a, b, and c are real numbers,

if $b^2 - 4ac > 0$, there are two unequal real solutions.
if $b^2 - 4ac = 0$, there are two equal real solutions (called a *double root*).
if $b^2 - 4ac < 0$, the two solutions are complex conjugates.

If we further restrict a, b, and c to the rational numbers and the discriminant

is a perfect square greater than 0, there are two unequal rational solutions.
is positive but not a perfect square, the solutions are irrational and unequal.

EXAMPLE 1 Determine the type of solutions for the equations.

a. $x^2 + x + 1 = 0$ **b.** $3x^2 + 5x + 2 = 0$

Solution **a.** We calculate the discriminant for $x^2 + x + 1 = 0$.

$$\begin{aligned} b^2 - 4ac &= 1^2 - 4(1)(1) \quad a = 1, b = 1, \text{and } c = 1 \\ &= -3 \end{aligned}$$

Since $b^2 - 4ac < 0$, the solutions will be two complex conjugates.

b. We calculate the discriminant for $3x^2 + 5x + 2 = 0$.

$$\begin{aligned} b^2 - 4ac &= 5^2 - 4(3)(2) \quad a = 3, b = 5, \text{ and } c = 2 \\ &= 25 - 24 \\ &= 1 \end{aligned}$$

Since $b^2 - 4ac > 0$ and $b^2 - 4ac$ is a perfect square, there will be two unequal rational solutions.

SELF CHECK 1 Determine the type of solutions.

a. $x^2 + x - 1 = 0$

b. $4x^2 - 10x + 25 = 0$

EXAMPLE 2 What value of k will make the solutions of the equation $kx^2 - 12x + 9 = 0$ equal?

Solution We calculate the discriminant:

$$\begin{aligned} b^2 - 4ac &= (-12)^2 - 4(k)(9) \quad a = k, b = -12, \text{ and } c = 9 \\ &= 144 - 36k \\ &= -36k + 144 \end{aligned}$$

Since the solutions are to be equal, we let $-36k + 144 = 0$ and solve for k.

$$\begin{aligned} -36k + 144 &= 0 \\ -36k &= -144 \quad \text{Subtract 144 from both sides.} \\ k &= 4 \quad \text{Divide both sides by } -36. \end{aligned}$$

If $k = 4$, the solutions will be equal. Verify this by solving $4x^2 - 12x + 9 = 0$ and showing that the solutions are equal.

SELF CHECK 2 Find the value of k that will make the solutions of $kx^2 - 20x + 25 = 0$ equal.

2 Solve an equation that can be written in quadratic form.

Many equations that are not quadratic can be written in quadratic form $(ax^2 + bx + c = 0)$ and then solved using the techniques discussed in previous sections. For example, an inspection of the equation $x^4 - 5x^2 + 4 = 0$ shows that

$$x^4 - 5x^2 + 4 = 0$$

The leading term x^4 is the square of x^2, the variable part of the middle term: $x^4 = (x^2)^2$

The last term is a constant.

To solve the equation $x^4 - 5x^2 + 4 = 0$, we can write the equation in a different form and proceed as follows:

$$\begin{aligned} x^4 - 5x^2 + 4 &= 0 \\ (x^2)^2 - 5(x^2) + 4 &= 0 \end{aligned}$$

If we substitute u for each x^2, we will obtain a quadratic equation with the variable u that we can solve by factoring.

$$\begin{aligned} u^2 - 5u + 4 &= 0 \quad \text{Let } x^2 = u. \\ (u - 4)(u - 1) &= 0 \quad \text{Factor } u^2 - 5u + 4. \end{aligned}$$

$$u - 4 = 0 \quad \text{or} \quad u - 1 = 0 \quad \text{Set each factor equal to 0.}$$

$$u = 4 \quad | \quad u = 1$$

Since $u = x^2$, it follows that $x^2 = 4$ or $x^2 = 1$. Thus,

$$x^2 = 4 \quad \text{or} \quad x^2 = 1$$
$$x = 2 \quad \text{or} \quad x = -2 \quad \Big| \quad x = 1 \quad \text{or} \quad x = -1$$

This equation has four solutions: $1, -1, 2$, and -2. Verify that each one satisfies the original equation. Note that this equation can be solved by factoring.

EXAMPLE 3 Solve: $x - 7\sqrt{x} + 12 = 0$

Solution We examine the leading term and middle term.

The leading term x is the square of $\sqrt{x}$, the variable part of the middle term: $x = (\sqrt{x})^2$

$$x - 7\sqrt{x} + 12 = 0$$

If we write x as $(\sqrt{x})^2$, the equation takes the form

$$(\sqrt{x})^2 - 7\sqrt{x} + 12 = 0$$

and it is said to be *quadratic in* $\sqrt{x}$. We can solve this equation by letting $u = \sqrt{x}$ and factoring.

$u^2 - 7u + 12 = 0$	Replace each $\sqrt{x}$ with u.
$(u - 3)(u - 4) = 0$	Factor $u^2 - 7u + 12$.
$u - 3 = 0$ or $u - 4 = 0$	Set each factor equal to 0.
$u = 3$ | $u = 4$	

To find the value of x, we undo the substitutions by replacing each u with $\sqrt{x}$. Then we solve the radical equations by squaring both sides.

$$\sqrt{x} = 3 \quad \text{or} \quad \sqrt{x} = 4$$
$$x = 9 \quad \Big| \quad x = 16$$

The solutions are 9 and 16. Verify that both satisfy the original equation.

SELF CHECK 3 Solve: $x - 5\sqrt{x} + 6 = 0$

EXAMPLE 4 Solve: $2m^{2/3} - 2 = 3m^{1/3}$

Solution First we subtract $3m^{1/3}$ from both sides to write the equation in quadratic form with descending powers of m.

$$2m^{2/3} - 3m^{1/3} - 2 = 0$$

Then we write the equation in the form

$$2(m^{1/3})^2 - 3m^{1/3} - 2 = 0 \quad (m^{1/3})^2 = m^{2/3}$$

If we substitute u for $m^{1/3}$, this equation can be written in a form that can be solved by factoring.

$2u^2 - 3u - 2 = 0$	Replace each $m^{1/3}$ with u.
$(2u + 1)(u - 2) = 0$	Factor $2u^2 - 3u - 2$.
$2u + 1 = 0$ or $u - 2 = 0$	Set each factor equal to 0.
$u = -\frac{1}{2}$ | $u = 2$	

To find the value of m, we undo the substitutions by replacing each u with $m^{1/3}$ and solve each resulting equation by cubing both sides.

$$m^{1/3} = -\frac{1}{2} \quad \text{or} \quad m^{1/3} = 2$$

$$(m^{1/3})^3 = \left(-\frac{1}{2}\right)^3 \quad \Big| \quad (m^{1/3})^3 = (2)^3 \qquad \text{Cube both sides.}$$

$$m = -\frac{1}{8} \quad \Big| \quad m = 8 \qquad \text{Simplify.}$$

The solutions are $-\frac{1}{8}$ and 8. Verify that both satisfy the original equation.

SELF CHECK 4 Solve: $a^{2/3} = -3a^{1/3} + 10$

EXAMPLE 5 Solve: $\frac{24}{x} + \frac{12}{x+1} = 11$

Solution Since the denominator cannot be 0, x cannot be 0 or -1. If either 0 or -1 appears as a possible solution, it is extraneous and must be discarded.

$$\frac{24}{x} + \frac{12}{x+1} = 11$$

$$x(x+1)\left(\frac{24}{x} + \frac{12}{x+1}\right) = x(x+1)11 \qquad \text{Multiply both sides by } x(x+1) \text{, the LCD.}$$

$$24(x+1) + 12x = (x^2 + x)11 \qquad \text{Simplify.}$$

$$24x + 24 + 12x = 11x^2 + 11x \qquad \text{Use the distributive property to remove parentheses.}$$

$$36x + 24 = 11x^2 + 11x \qquad \text{Combine like terms.}$$

$$0 = 11x^2 - 25x - 24 \qquad \text{Write the equation in quadratic form.}$$

$$0 = (11x + 8)(x - 3) \qquad \text{Factor } 11x^2 - 25x - 24.$$

$$11x + 8 = 0 \quad \text{or} \quad x - 3 = 0 \qquad \text{Set each factor equal to 0.}$$

$$x = -\frac{8}{11} \quad \Big| \quad x = 3$$

Verify that $-\frac{8}{11}$ and 3 satisfy the original equation.

SELF CHECK 5 Solve: $\frac{12}{x} + \frac{6}{x+3} = 5$

EXAMPLE 6 Solve: $15a^{-2} - 8a^{-1} + 1 = 0$

Solution First we write the equation in the form

$$15(a^{-1})^2 - 8a^{-1} + 1 = 0 \qquad (a^{-1})^2 = a^{-2}$$

If we substitute u for a^{-1}, this equation can be written in a form that can be solved by factoring.

$$15u^2 - 8u + 1 = 0 \qquad \text{Replace each } a^{-1} \text{ with } u.$$

$$(5u - 1)(3u - 1) = 0 \qquad \text{Factor } 15u^2 - 8u + 1.$$

$$5u - 1 = 0 \quad \text{or} \quad 3u - 1 = 0 \qquad \text{Set each factor equal to 0.}$$

$$u = \frac{1}{5} \quad \Big| \quad u = \frac{1}{3}$$

To find the value of a, we undo the substitutions by replacing each u with a^{-1} and solve each resulting equation.

$$a^{-1} = \frac{1}{5} \qquad a^{-1} = \frac{1}{3}$$

$$\frac{1}{a} = \frac{1}{5} \qquad \frac{1}{a} = \frac{1}{3} \qquad a^{-1} = \frac{1}{a}$$

$$5 = a \qquad 3 = a \qquad \text{Solve the proportions.}$$

The solutions are 5 and 3. Verify that both satisfy the original equation.

SELF CHECK 6 Solve: $28c^{-2} - 3c^{-1} - 1 = 0$

EXAMPLE 7 Solve the formula $s = 16t^2 - 32$ for t.

Solution We proceed as follows:

$$s = 16t^2 - 32$$

$$s + 32 = 16t^2 \qquad \text{Add 32 to both sides.}$$

$$\frac{s + 32}{16} = t^2 \qquad \text{Divide both sides by 16.}$$

$$t^2 = \frac{s + 32}{16} \qquad \text{Write } t^2 \text{ on the left side.}$$

$$t = \pm\sqrt{\frac{s + 32}{16}} \qquad \text{Apply the square-root property.}$$

$$t = \pm\frac{\sqrt{s + 32}}{\sqrt{16}} \qquad \sqrt{\frac{a}{b}} = \frac{\sqrt{a}}{\sqrt{b}}$$

$$t = \pm\frac{\sqrt{s + 32}}{4}$$

SELF CHECK 7 Solve $a^2 + b^2 = c^2$ for a.

3 Verify the solutions of a quadratic equation by showing that the sum of the solutions is $-\frac{b}{a}$ and the product is $\frac{c}{a}$.

SOLUTIONS OF A QUADRATIC EQUATION

If r_1 and r_2 are the solutions of the quadratic equation $ax^2 + bx + c = 0$, with $a \neq 0$, then

$$r_1 + r_2 = -\frac{b}{a} \quad \text{and} \quad r_1 r_2 = \frac{c}{a}$$

Proof We note that the solutions to the equation are given by the quadratic formula

$$r_1 = \frac{-b + \sqrt{b^2 - 4ac}}{2a} \quad \text{and} \quad r_2 = \frac{-b - \sqrt{b^2 - 4ac}}{2a}$$

Thus,

$$r_1 + r_2 = \frac{-b + \sqrt{b^2 - 4ac}}{2a} + \frac{-b - \sqrt{b^2 - 4ac}}{2a}$$

$$= \frac{-b + \sqrt{b^2 - 4ac} - b - \sqrt{b^2 - 4ac}}{2a}$$ Keep the denominator and add the numerators.

$$= -\frac{2b}{2a}$$ Combine like terms.

$$= -\frac{b}{a}$$ Simplify.

and

$$r_1 r_2 = \frac{-b + \sqrt{b^2 - 4ac}}{2a} \cdot \frac{-b - \sqrt{b^2 - 4ac}}{2a}$$

$$= \frac{b^2 - (b^2 - 4ac)}{4a^2}$$ Multiply the numerators and multiply the denominators.

$$= \frac{b^2 - b^2 + 4ac}{4a^2}$$ Use the distributive property to remove parentheses.

$$= \frac{4ac}{4a^2}$$ Combine like terms.

$$= \frac{c}{a}$$ Simplify.

It can also be shown that if

$$r_1 + r_2 = -\frac{b}{a} \quad \text{and} \quad r_1 r_2 = \frac{c}{a}$$

then r_1 and r_2 are solutions of $ax^2 + bx + c = 0$. We can use this fact to check the solutions of quadratic equations.

EXAMPLE 8 Show that $\frac{3}{2}$ and $-\frac{1}{3}$ are solutions of $6x^2 - 7x - 3 = 0$.

Solution Since $a = 6$, $b = -7$, and $c = -3$, we have

$$-\frac{b}{a} = -\frac{-7}{6} = \frac{7}{6} \quad \text{and} \quad \frac{c}{a} = \frac{-3}{6} = -\frac{1}{2}$$

Since $\frac{3}{2} + \left(-\frac{1}{3}\right) = \frac{7}{6}$ and $\left(\frac{3}{2}\right)\left(-\frac{1}{3}\right) = -\frac{1}{2}$, these numbers are solutions. Solve the equation to verify that the roots are $\frac{3}{2}$ and $-\frac{1}{3}$.

SELF CHECK 8 Are $-\frac{3}{2}$ and $\frac{1}{3}$ solutions of $6x^2 + 7x - 3 = 0$?

SELF CHECK ANSWERS

1. a. two irrational and unequal **b.** two complex conjugates **2.** 4 **3.** 4, 9 **4.** 8, −125 **5.** 3, $-\frac{12}{5}$ **6.** −7, 4 **7.** $a = \pm\sqrt{c^2 - b^2}$ **8.** yes

NOW TRY THIS

1. Solve: $x - 3\sqrt{x} - 4 = 0$
2. Without substituting, show that $(3 + 5i)$ and $(3 - 5i)$ are solutions of $x^2 - 6x + 34 = 0$.
3. Find the discriminant of $\sqrt{2}x^2 - \sqrt{65}x - 2\sqrt{2} = 0$.

10.3 Exercises

WARM-UPS *Find $b^2 - 4ac$ when*

1. $a = 1, b = 1, c = 1$
2. $a = 2, b = 1, c = 1$

Are the following numbers solutions of $x^2 - 2x + 2 = 0$?

3. $1 - i$
4. $-1 - i$
5. $-1 + i$
6. $1 + i$

REVIEW *Solve each equation. Assume no division by 0.*

7. $\frac{1}{4} + \frac{1}{t} = \frac{1}{2t}$
8. $\frac{p-3}{3p} + \frac{1}{2p} = \frac{1}{4}$
9. Find the slope of the line passing through $(-3, -2)$ and $(4, 1)$.
10. Write an equation of the line passing through $(-3, -2)$ and $(4, 1)$ in general form.

VOCABULARY AND CONCEPTS *Consider the equation $ax^2 + bx + c = 0$ $(a \neq 0)$, and fill in the blanks.*

11. The discriminant is ________.
12. If $b^2 - 4ac < 0$, the solutions of the equation are two complex ________.
13. If $b^2 - 4ac$ is a nonzero perfect square, the solutions are ______ numbers and ______.
14. If r_1 and r_2 are the solutions of the equation, then $r_1 + r_2 =$ ___ and $r_1 r_2 =$ _.

GUIDED PRACTICE *Use the discriminant to determine what type of solutions exist for each quadratic equation. Do not solve the equation.* SEE EXAMPLE 1. (OBJECTIVE 1)

15. $9x^2 + 6x + 1 = 0$
16. $6x^2 - 5x - 6 = 0$
17. $10x^2 + 2x + 3 = 0$
18. $3x^2 + 10x - 2 = 0$
19. $6x^2 = 8x - 1$
20. $9x^2 = 12x - 4$
21. $x(2x - 3) = 20$
22. $x(x - 3) = -10$

Find the values of k that will make the solutions of each given quadratic equation equal. SEE EXAMPLE 2. (OBJECTIVE 1)

23. $x^2 + kx + 16 = 0$
24. $kx^2 - 12x + 4 = 0$
25. $4x^2 + 9 = kx$
26. $9x^2 - kx + 25 = 0$

Solve each equation. (OBJECTIVE 2)

27. $x^4 - 17x^2 + 16 = 0$
28. $x^4 - 10x^2 + 9 = 0$
29. $x^4 - 3x^2 = -2$
30. $x^4 - 19x^2 = -48$
31. $x^4 = 6x^2 - 5$
32. $x^4 = 8x^2 - 7$
33. $2x^4 - 10x^2 = -8$
34. $3x^4 + 12 = 15x^2$

Solve each equation. SEE EXAMPLE 3. (OBJECTIVE 2)

35. $x - 7\sqrt{x} + 10 = 0$
36. $x - 5\sqrt{x} + 4 = 0$
37. $2x - \sqrt{x} = 3$
38. $3x - 4 = -4\sqrt{x}$
39. $2x + x^{1/2} - 3 = 0$
40. $2x - x^{1/2} - 1 = 0$
41. $3x + 5x^{1/2} + 2 = 0$
42. $3x - 4x^{1/2} + 1 = 0$

Solve each equation. SEE EXAMPLE 4. (OBJECTIVE 2)

43. $x^{2/3} + 5x^{1/3} + 6 = 0$
44. $x^{2/3} - 2x^{1/3} - 3 = 0$
45. $3m^{2/3} - m^{1/3} - 2 = 0$
46. $4m^{2/3} - 8m^{1/3} + 3 = 0$

Solve each equation. SEE EXAMPLE 5. (OBJECTIVE 2)

47. $x + 10 + \frac{9}{x} = 0$
48. $x - 4 + \frac{3}{x} = 0$
49. $x + 3 = \frac{28}{x}$
50. $x + \frac{15}{x} = 8$
51. $\frac{1}{x-1} + \frac{3}{x+1} = 2$
52. $\frac{6}{x-2} - \frac{12}{x-1} = -1$
53. $\frac{1}{x+2} + \frac{24}{x+3} = 13$
54. $\frac{3}{x} + \frac{4}{x+1} = 2$

Solve each equation. SEE EXAMPLE 6. (OBJECTIVE 2)

55. $x^{-4} - 2x^{-2} + 1 = 0$
56. $4x^{-4} + 1 = 5x^{-2}$
57. $8a^{-2} - 10a^{-1} - 3 = 0$
58. $2y^{-2} - 5y^{-1} = 3$

Solve each equation for the indicated variable. SEE EXAMPLE 7. (OBJECTIVE 2)

59. $x^2 + y^2 = r^2$ for x
60. $x^2 + y^2 = r^2$ for y
61. $xy^2 + 5xy + 3 = 0$ for y
62. $kx = ay - x^2$ for x

Solve each equation and verify that the sum of the solutions is $-\frac{b}{a}$ and that the product of the solutions is $\frac{c}{a}$. SEE EXAMPLE 8. (OBJECTIVE 3)

63. $12x^2 - 5x - 2 = 0$

64. $8x^2 - 2x - 3 = 0$

65. $2x^2 + 5x + 1 = 0$

66. $3x^2 + 9x + 1 = 0$

67. $3x^2 - 2x + 4 = 0$

68. $2x^2 - x + 4 = 0$

69. $x^2 + 2x + 5 = 0$

70. $x^2 - 4x + 13 = 0$

ADDITIONAL PRACTICE

71. Use the discriminant to determine whether the solutions of $1{,}492x^2 + 1{,}776x - 1{,}984 = 0$ are real numbers.

72. Use the discriminant to determine whether the solutions of $1{,}776x^2 - 1{,}492x + 1{,}984 = 0$ are real numbers.

Solve using any method.

73. $4x - 5\sqrt{x} - 9 = 0$

74. $9x - 5\sqrt{x} = 4$

75. $3x^{2/3} - x^{1/3} - 2 = 0$

76. $4x^{2/3} + 4x^{1/3} + 1 = 0$

77. $2x^4 + 24 = 26x^2$

78. $4x^4 = -9 + 13x^2$

79. $x^4 - 24 = -2x^2$

80. $x^4 - 7x^2 = 18$

81. $4(2x - 1)^2 - 3(2x - 1) - 1 = 0$

82. $4(x^2 - 1)^2 + 13(x^2 - 1) + 9 = 0$

83. $x^{-2/3} - 2x^{-1/3} - 3 = 0$

84. $4x^{-1} - 5x^{-1/2} - 9 = 0$

85. $x + \frac{2}{x - 2} = 0$

86. $x + \frac{x + 5}{x - 3} = 0$

87. $8(m + 1)^{-2} - 30(m + 1)^{-1} + 7 = 0$

88. $2(p - 2)^{-2} + 3(p - 2)^{-1} - 5 = 0$

89. $I = \frac{k}{d^2}$ for d

90. $V = \frac{1}{3}\pi r^2 h$ for r

91. $\sigma = \sqrt{\frac{\Sigma x^2}{N} - \mu^2}$ for μ^2

92. $\sigma = \sqrt{\frac{\Sigma x^2}{N} - \mu^2}$ for N

Find the values of k that will make the solutions of each given quadratic equation equal.

93. $(k - 1)x^2 + (k - 1)x + 1 = 0$

94. $(k + 3)x^2 + 2kx + 4 = 0$

95. $(k + 4)x^2 + 2kx + 9 = 0$

96. $(k + 15)x^2 + (k - 30)x + 4 = 0$

97. Determine k such that the solutions of $3x^2 + 4x = k$ are complex numbers.

98. Determine k such that the solutions of $kx^2 - 4x = 7$ are complex numbers.

WRITING ABOUT MATH

99. Describe how to predict what type of solutions the equation $3x^2 - 4x + 5 = 0$ will have.

100. How is the discriminant related to the quadratic formula?

SOMETHING TO THINK ABOUT

101. Can a quadratic equation with integer coefficients have one real and one complex solution? Why?

102. Can a quadratic equation with complex coefficients have one real and one complex solution? Why?

Section 10.4 Graphs of Quadratic Functions

Objectives

1. Graph a quadratic function of the form $f(x) = ax^2$.
2. Use a vertical translation of $f(x) = ax^2$ to graph $f(x) = ax^2 + k$.
3. Use a horizontal translation of $f(x) = ax^2$ to graph $f(x) = a(x - h)^2$.
4. Use both a vertical and horizontal translation of $f(x) = ax^2$ to graph $f(x) = a(x - h)^2 + k$.
5. Graph a quadratic function written in standard form by writing it in the form $f(x) = a(x - h)^2 + k$.
6. Find the vertex of a parabola given $f(x) = ax^2 + bx + c$ using $\left(-\frac{b}{2a}, f\left(-\frac{b}{2a}\right)\right)$.
7. Graph a quadratic function written in standard form by finding the vertex, the axis of symmetry, and the x- and y-intercepts.
8. Solve an application using a quadratic function.

Vocabulary

quadratic functions
parabola
vertex
axis of symmetry
variance

Getting Ready

If $f(x) = 3x^2 + x - 2$, find each value.

1. $f(0)$ **2.** $f(1)$ **3.** $f(-1)$ **4.** $f(-2)$

If $x = -\frac{b}{2a}$, find the value of x when a and b have the following values.

5. $a = 3$ and $b = -6$ **6.** $a = 5$ and $b = -40$

In this section, we consider graphs of second-degree polynomial functions, called **quadratic functions**.

The graph shown in Figure 10-4 shows the height (in relation to time) of a toy rocket launched straight up into the air.

From the graph, we can see that the height of the rocket 2 seconds after it was launched is about 128 feet and that the height of the rocket 5 seconds after it was launched is 80 feet.

The parabola shown in Figure 10-4 is the graph of a quadratic function.

QUADRATIC FUNCTION

A quadratic function is a second-degree polynomial function of the form

$$f(x) = ax^2 + bx + c \quad (a \neq 0)$$

where a, b, and c are real numbers.

We begin the discussion of graphing quadratic functions by considering the graph of $f(x) = ax^2 + bx + c$, where $b = 0$ and $c = 0$.

COMMENT Note that the graph describes the height of the rocket, not the path of the rocket. The rocket goes straight up and comes straight down.

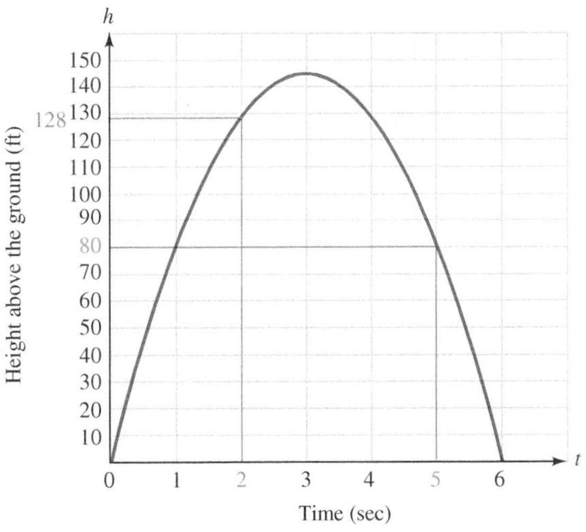

Figure 10-4

1 Graph a quadratic function of the form $f(x) = ax^2$.

EXAMPLE 1 Graph: **a.** $f(x) = x^2$ **b.** $g(x) = 3x^2$ **c.** $h(x) = \frac{1}{3}x^2$

Solution We create a table of ordered pairs that satisfy each equation, plot each point, and join them with a smooth curve, as in Figure 10-5. We note that the graph of $h(x) = \frac{1}{3}x^2$ is wider than the graph of $f(x) = x^2$, and that the graph of $g(x) = 3x^2$ is narrower than the graph of $f(x) = x^2$. In the function $f(x) = ax^2$, the smaller the value of $|a|$, the wider the graph.

a. $f(x) = x^2$

x	$f(x)$	$(x, f(x))$
-2	4	$(-2, 4)$
-1	1	$(-1, 1)$
0	0	$(0, 0)$
1	1	$(1, 1)$
2	4	$(2, 4)$

b. $g(x) = 3x^2$

x	$g(x)$	$(x, g(x))$
-2	12	$(-2, 12)$
-1	3	$(-1, 3)$
0	0	$(0, 0)$
1	3	$(1, 3)$
2	12	$(2, 12)$

c. $h(x) = \frac{1}{3}x^2$

x	$h(x)$	$(x, h(x))$
-2	$\frac{4}{3}$	$\left(-2, \frac{4}{3}\right)$
-1	$\frac{1}{3}$	$\left(-1, \frac{1}{3}\right)$
0	0	$(0, 0)$
1	$\frac{1}{3}$	$\left(1, \frac{1}{3}\right)$
2	$\frac{4}{3}$	$\left(2, \frac{4}{3}\right)$

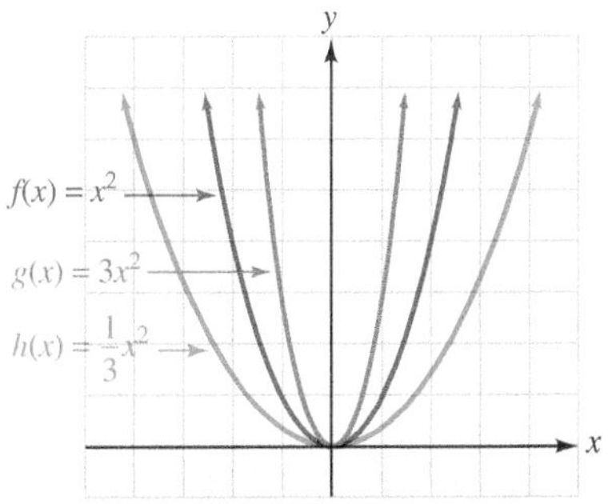

Figure 10-5

SELF CHECK 1 On the same set of coordinate axes, graph each function.
a. $f(x) = 2x^2$ **b.** $f(x) = \frac{1}{2}x^2$

If we consider the graph of $f(x) = -3x^2$, we will see that it opens downward and has the same shape as the graph of $g(x) = 3x^2$. This is called a *reflection*.

EXAMPLE 2 Graph: $f(x) = -3x^2$

Solution We create a table of ordered pairs that satisfy the equation, plot each point, and join them with a smooth curve, as in Figure 10-6 on the next page.

Unless otherwise noted, all content on this page is © Cengage Learning.

$f(x) = -3x^2$

x	$f(x)$	$(x, f(x))$
−2	−12	(−2, −12)
−1	−3	(−1, −3)
0	0	(0, 0)
1	−3	(1, −3)
2	−12	(2, −12)

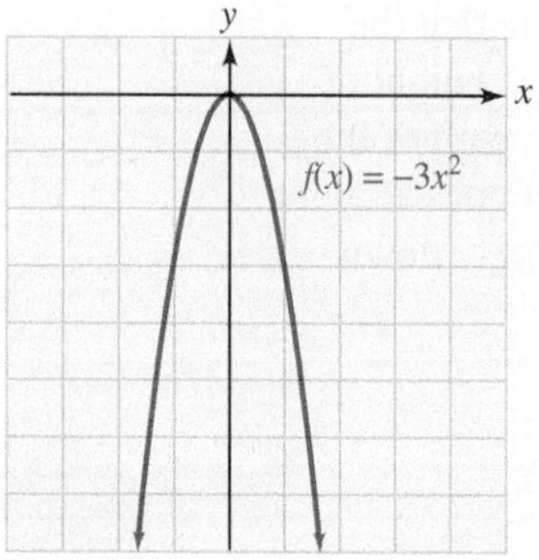

Figure 10-6

SELF CHECK 2 Graph: $f(x) = -\frac{1}{3}x^2$

The graphs of quadratic functions are called **parabolas**. They open upward when $a > 0$ and downward when $a < 0$. The lowest point (*minimum*) of a parabola that opens upward, or the highest point (*maximum*) of a parabola that opens downward, is called the **vertex** of the parabola. The vertex of the parabola shown in Figure 10-6 is the point $(0, 0)$.

The vertical line, called an **axis of symmetry**, that passes through the vertex divides the parabola into two congruent halves. The axis of symmetry of the parabola shown in Figure 10-6 is the y-axis, written as the equation $x = 0$.

2 Use a vertical translation of $f(x) = ax^2$ to graph $f(x) = ax^2 + k$.

Now let us consider a vertical translation of $f(x) = ax^2$ to graph $f(x) = ax^2 + k$. This means the graph of $f(x) = ax^2$ will retain its shape, but will be shifted k units upward or downward. In this case the vertex will shift to the point $(0, k)$. The axis of symmetry of the parabola is a vertical line written as the equation $x = 0$.

EXAMPLE 3 Graph: **a.** $f(x) = 2x^2$ **b.** $g(x) = 2x^2 + 3$ **c.** $h(x) = 2x^2 - 3$

Solution We create a table of ordered pairs that satisfy each equation, plot each point, and join them with a smooth curve, as in Figure 10-7. We note that the graph of $g(x) = 2x^2 + 3$ is identical to the graph of $f(x) = 2x^2$, except that it has been translated 3 units upward. The graph of $h(x) = 2x^2 - 3$ is identical to the graph of $f(x) = 2x^2$, except that it has been translated 3 units downward.

a. $f(x) = 2x^2$

x	$f(x)$	$(x, f(x))$
−2	8	(−2, 8)
−1	2	(−1, 2)
0	0	(0, 0)
1	2	(1, 2)
2	8	(2, 8)

b. $g(x) = 2x^2 + 3$

x	$g(x)$	$(x, g(x))$
−2	11	(−2, 11)
−1	5	(−1, 5)
0	3	(0, 3)
1	5	(1, 5)
2	11	(2, 11)

c. $h(x) = 2x^2 - 3$

x	$h(x)$	$(x, h(x))$
−2	5	(−2, 5)
−1	−1	(−1, −1)
0	−3	(0, −3)
1	−1	(1, −1)
2	5	(2, 5)

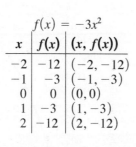

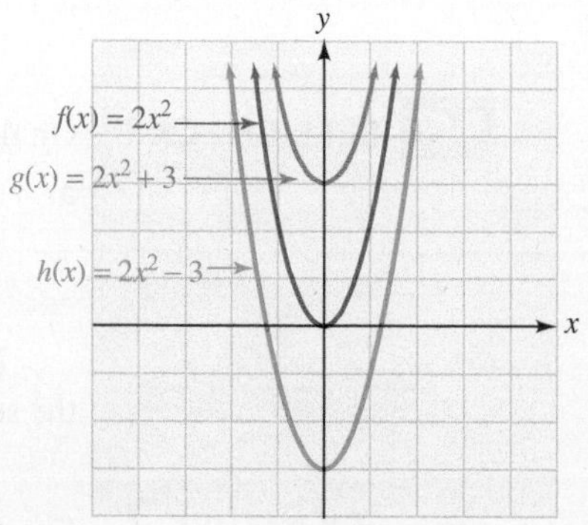

Figure 10-7

Unless otherwise noted, all content on this page is © Cengage Learning.

SELF CHECK 3 On the same set of coordinate axes, graph each function and tell how it differs from the graph of $f(x) = x^2$.

a. $f(x) = x^2 + 1$
b. $f(x) = x^2 - 5$

The results of Example 3 confirm the following facts.

VERTICAL TRANSLATIONS OF GRAPHS

If $y = f(x)$ is a function and k is a positive number, then

- The graph of $y = f(x) + k$ is identical to the graph of $y = f(x)$, except that it is translated k units upward.
- The graph of $y = f(x) - k$ is identical to the graph of $y = f(x)$, except that it is translated k units downward.

3 Use a horizontal translation of $f(x) = ax^2$ to graph $f(x) = a(x - h)^2$.

Now let us consider horizontal translations of $f(x) = ax^2$ to graph $f(x) = a(x - h)^2$. This means the graph of $f(x) = ax^2$ will retain its shape but will be shifted h units to the right or left. In this case the vertex will be shifted to the point $(h, 0)$. The axis of symmetry of the parabola is a vertical line written as the equation $x = h$.

EXAMPLE 4 Graph: **a.** $f(x) = 2x^2$ **b.** $g(x) = 2(x - 3)^2$ **c.** $h(x) = 2(x + 3)^2$

Solution We create a table of ordered pairs that satisfy each equation, plot each point, and join them with a smooth curve, as in Figure 10-8. We note that the graph of $g(x) = 2(x - 3)^2$ is identical to the graph of $f(x) = 2x^2$, except that it has been translated 3 units to the right. The graph of $h(x) = 2(x + 3)^2$ is identical to the graph of $f(x) = 2x^2$, except that it has been translated 3 units to the left.

a. $f(x) = 2x^2$

x	$f(x)$	$(x, f(x))$
-2	8	$(-2, 8)$
-1	2	$(-1, 2)$
0	0	$(0, 0)$
1	2	$(1, 2)$
2	8	$(2, 8)$

b. $g(x) = 2(x - 3)^2$

x	$g(x)$	$(x, g(x))$
1	8	$(1, 8)$
2	2	$(2, 2)$
3	0	$(3, 0)$
4	2	$(4, 2)$
5	8	$(5, 8)$

c. $h(x) = 2(x + 3)^2$

x	$h(x)$	$(x, h(x))$
-5	8	$(-5, 8)$
-4	2	$(-4, 2)$
-3	0	$(-3, 0)$
-2	2	$(-2, 2)$
-1	8	$(-1, 8)$

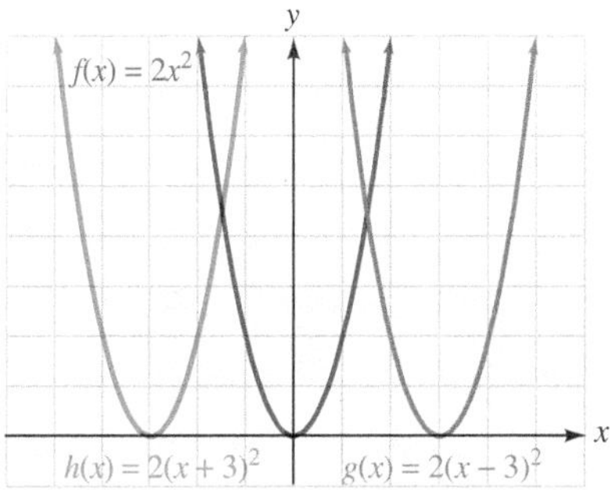

Figure 10-8

SELF CHECK 4 On the same set of coordinate axes, graph each function and tell how it differs from the graph of $f(x) = x^2$.

a. $f(x) = (x - 2)^2$
b. $f(x) = (x + 5)^2$

The results of Example 4 confirm the following facts.

Unless otherwise noted, all content on this page is © Cengage Learning.

HORIZONTAL TRANSLATIONS OF GRAPHS

If $y = f(x)$ is a function and h is a positive number, then

- The graph of $y = f(x - h)$ is identical to the graph of $y = f(x)$, except that it is translated h units to the right.
- The graph of $y = f(x + h)$ is identical to the graph of $y = f(x)$, except that it is translated h units to the left.

4 Use both a vertical and horizontal translation of $f(x) = ax^2$ to graph $f(x) = a(x - h)^2 + k$.

We will now consider both the vertical and horizontal translation of $f(x) = ax^2$ to graph $f(x) = a(x - h)^2 + k$. This means the graph of $f(x) = ax^2$ will retain its shape but will be shifted h units to the right or left and k units upward or downward. In this case the vertex will be shifted to the point (h, k) and the axis of symmetry of the parabola will be a vertical line written as the equation $x = h$.

EXAMPLE 5 Graph: $f(x) = 2(x - 3)^2 - 4$

Solution The graph of $f(x) = 2(x - 3)^2 - 4$ is identical to the graph of $g(x) = 2(x - 3)^2$, except that it has been translated 4 units downward. The graph of $g(x) = 2(x - 3)^2$ is identical to the graph of $h(x) = 2x^2$, except that it has been translated 3 units to the right. Thus, to graph $f(x) = 2(x - 3)^2 - 4$, we can graph $h(x) = 2x^2$ and shift it 3 units to the right and then 4 units downward, as shown in Figure 10-9.

The vertex of the graph is the point $(3, -4)$, and the axis of symmetry is the line $x = 3$.

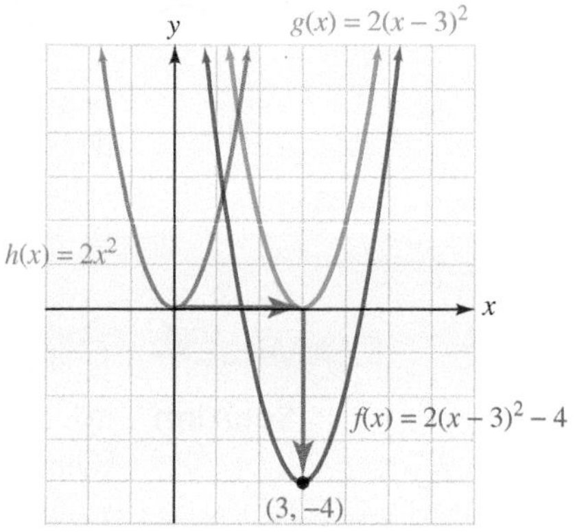

Figure 10-9

SELF CHECK 5 Graph: $f(x) = 2(x + 3)^2 + 1$

The results of Example 5 confirm the following facts.

VERTEX AND AXIS OF SYMMETRY OF A PARABOLA

The graph of the function

$$f(x) = a(x - h)^2 + k \quad (a \neq 0)$$

is a parabola with vertex at (h, k). (See Figure 10-10.)

The parabola opens upward when $a > 0$ and downward when $a < 0$. The axis of symmetry is the vertical line $x = h$.

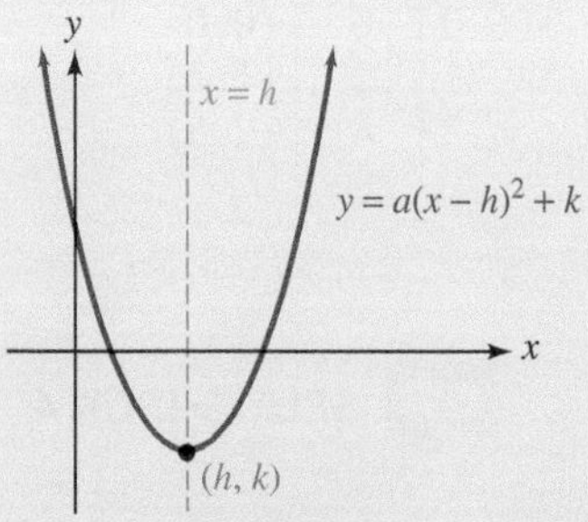

Figure 10-10

Unless otherwise noted, all content on this page is © Cengage Learning.

5 Graph a quadratic function written in standard form by writing it in the form $f(x) = a(x - h)^2 + k$.

To graph functions of the form $f(x) = ax^2 + bx + c$, we can complete the square to write the function in the form $f(x) = a(x - h)^2 + k$.

EXAMPLE 6 Graph: $f(x) = 2x^2 - 4x - 1$

Solution We complete the square on x to write the function in the form $f(x) = a(x - h)^2 + k$.

$f(x) = 2x^2 - 4x - 1$

$f(x) = 2(x^2 - 2x) - 1$ — Factor 2 from $(2x^2 - 4x)$.

$f(x) = 2(x^2 - 2x + \mathbf{1}) - 1 - \mathbf{2}$ — Complete the square on x. Since this adds 2 to the right side, we also subtract 2 from the right side.

(1) $f(x) = 2(x - 1)^2 - 3$ — Factor $(x^2 - 2x + 1)$ and combine like terms.

From Equation 1, we can see that the vertex will be at the point $(1, -3)$. We can plot the vertex and a few points on either side of the vertex and draw the graph, which appears in Figure 10-11.

COMMENT Note that this is the graph of $f(x) = 2x^2$ translated one unit to the right and three units downward.

$f(x) = 2x^2 - 4x - 1$

x	$f(x)$	$(x, f(x))$
−1	5	(−1, 5)
0	−1	(0, −1)
1	−3	(1, −3)
2	−1	(2, −1)
3	5	(3, 5)

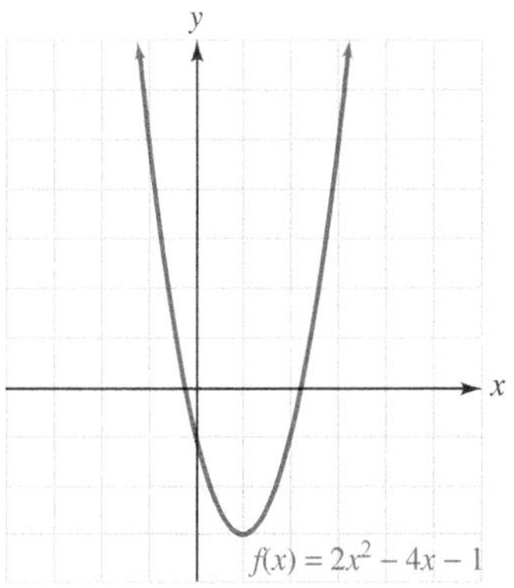

Figure 10-11

SELF CHECK 6 Graph: $f(x) = 2x^2 - 4x + 1$

6 Find the vertex of a parabola given $f(x) = ax^2 + bx + c$ using $\left(-\frac{b}{2a}, f\left(-\frac{b}{2a}\right)\right)$.

We can derive a formula for the vertex of the graph of $f(x) = ax^2 + bx + c$ by completing the square in the same manner as we did in Example 6.

$f(x) = ax^2 + bx + c$

$f(x) = a\left(x^2 + \frac{b}{a}x\right) + c$ — Factor a from the first two terms.

$f(x) = a\left(x^2 + \frac{b}{a}x + \frac{b^2}{4a^2}\right) + c - \frac{b^2}{4a}$ — Complete the square on x. Since this adds $\frac{a \cdot b^2}{4a^2} = \frac{b^2}{4a}$ to the right side, we also subtract $\frac{b^2}{4a}$ from the right side.

$f(x) = a\left[\left(x + \frac{b}{2a}\right)^2\right] + \frac{4ac - b^2}{4a}$ — Factor $x^2 + \frac{b}{a}x + \frac{b^2}{4a^2}$ and combine like terms.

$f(x) = a\left[x - \left(-\frac{b}{2a}\right)\right]^2 + \frac{4ac - b^2}{4a}$ — Compare to $f(x) = a(x - h)^2 + k$.

$$\uparrow h \qquad \uparrow k$$

Unless otherwise noted, all content on this page is © Cengage Learning.

The x-coordinate of the vertex is $-\frac{b}{2a}$. The y-coordinate of the vertex is $\frac{4ac - b^2}{4a}$. We can also find the y-coordinate of the vertex by substituting the x-coordinate, $-\frac{b}{2a}$, for x in the quadratic function and simplifying.

FORMULA FOR THE VERTEX OF A PARABOLA

The vertex of the graph of the quadratic function $f(x) = ax^2 + bx + c$ is

$$\left(-\frac{b}{2a}, f\left(-\frac{b}{2a}\right)\right)$$

and the axis of symmetry of the parabola is the vertical line $x = -\frac{b}{2a}$.

EXAMPLE 7 Find the vertex of the graph of $f(x) = 2x^2 - 4x - 1$.

Solution The function is written in $f(x) = ax^2 + bx + c$ form, where $a = 2$, $b = -4$, and $c = -1$. We can find the x-coordinate of the vertex by evaluating $-\frac{b}{2a}$.

$$-\frac{b}{2a} = -\frac{-4}{2(2)} = -\frac{-4}{4} = 1$$

We can find the y-coordinate by evaluating $f\left(-\frac{b}{2a}\right)$.

COMMENT This is the same as finding $f(1)$.

$$f\left(-\frac{b}{2a}\right) = f(1) = 2(1)^2 - 4(1) - 1 = -3$$

The vertex is the point $(1, -3)$. This agrees with the result we obtained in Example 6 by completing the square.

SELF CHECK 7 Find the vertex of the graph of $f(x) = 3x^2 - 12x + 8$.

7 Graph a quadratic function written in standard form by finding the vertex, the axis of symmetry, and the *x*- and *y*-intercepts.

Much can be determined about the graph of $f(x) = ax^2 + bx + c$ from the coefficients a, b, and c. This information is summarized as follows:

GRAPHING A QUADRATIC FUNCTION $f(x) = ax^2 + bx + c$

1. Determine whether the parabola opens upward or downward.

   If $a > 0$, the parabola opens upward.

   If $a < 0$, the parabola opens downward.
2. Find the vertex and axis of symmetry.
   - **a.** The x-coordinate of the vertex of the parabola is $-\frac{b}{2a}$.
   - **b.** To find the y-coordinate of the vertex, substitute $-\frac{b}{2a}$ for x and find $f\left(-\frac{b}{2a}\right)$.
   - **c.** The axis of symmetry is the vertical line passing through the vertex. The axis of symmetry is the equation $x = -\frac{b}{2a}$.
3. Find the intercepts.
   - **a.** The y-intercept is determined by the value of $f(x)$ when $x = 0$. The y-intercept is $(0, c)$.
   - **b.** The x-intercepts (if any) are determined by the values of x that make $f(x) = 0$. To find them, solve the quadratic equation $ax^2 + bx + c = 0$.
4. Plot another point using the axis of symmetry.
5. Draw a smooth curve through the points.

EXAMPLE 8 Graph: $f(x) = -2x^2 - 8x - 8$

Solution **Step 1** *Determine whether the parabola opens upward or downward.* The function is in the form $f(x) = ax^2 + bx + c$, with $a = -2$, $b = -8$, and $c = -8$. Since $a < 0$, the parabola opens downward.

Step 2 *Find the vertex and draw the axis of symmetry.* To find the coordinates of the vertex, we evaluate $-\frac{b}{2a}$ by substituting -2 for a and -8 for b.

$$x = -\frac{b}{2a} = -\frac{-8}{2(-2)} = -2$$

We then find $f(-2)$.

$$f\left(-\frac{b}{2a}\right) = f(-2) = -2(-2)^2 - 8(-2) - 8 = -8 + 16 - 8 = 0$$

The vertex of the parabola is the point $(-2, 0)$. The axis of symmetry is the line $x = -2$.

Step 3 *Find the x- and y-intercepts.* Since $c = -8$, the y-intercept of the parabola is $(0, -8)$. The point $(-4, -8)$, two units to the left of the axis of symmetry, must also be on the graph. We plot both points in black on the graph.

To find the x-intercepts, we set $f(x)$ equal to 0 and solve the resulting quadratic equation.

$$\begin{aligned} f(x) &= -2x^2 - 8x - 8 \\ 0 &= -2x^2 - 8x - 8 && \text{Set } f(x) = 0. \\ 0 &= x^2 + 4x + 4 && \text{Divide both sides by } -2. \\ 0 &= (x + 2)(x + 2) && \text{Factor the trinomial.} \end{aligned}$$

$$x + 2 = 0 \quad \text{or} \quad x + 2 = 0 \qquad \text{Set each factor equal to 0.}$$
$$x = -2 \qquad\qquad x = -2$$

Since the solutions are the same, the graph has only one x-intercept: $(-2, 0)$. This point is the vertex of the parabola and has already been plotted.

Step 4 *Plot another point.* Finally, we find another point on the parabola. If $x = -3$, then $f(-3) = -2$. We plot $(-3, -2)$ and use symmetry to determine that $(-1, -2)$ is also on the graph. Both points are in black.

Step 5 *Draw a smooth curve through the points,* as shown in Figure 10-12.

$f(x) = -2x^2 - 8x - 8$

x	$f(x)$	$(x, f(x))$
-3	-2	$(-3, -2)$

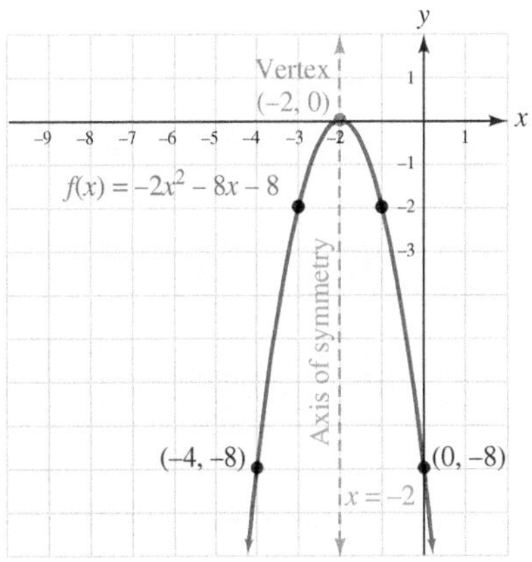

Figure 10-12

SELF CHECK 8 Graph: $f(x) = 2x^2 - 8x + 8$

Unless otherwise noted, all content on this page is © Cengage Learning.

Everyday connections
Astronaut Training

Some astronauts train for weightlessness in space travel by flying in C9B aircraft that follows a series of maneuvers to produce a "parabolic flight." A typical training mission may last up to 3 hours with the astronaut experiencing 40 to 50 weightlessness sessions. The aircraft altitude is modeled by the equation

$$f(x) = -\frac{1{,}600}{169}x^2 + \frac{8{,}000}{13}x + 24{,}000$$

where x represents time and $f(x)$ represents altitude.

The flight never flies below 24,000 feet during the training.

1. Find the vertex of the parabola and explain its meaning in the context of this problem.
2. How long does one flight last?
3. Weightlessness occurs for the middle 25 seconds. At what altitude will weightlessness first occur? Round your answer to the nearest foot.

http://jsc-aircraft-ops.jsc.nasa.gov/Reduced_Gravity/about.html

Accent on technology

▸ Graphing Quadratic Functions

To find the coordinates of the vertex of a graph we use the MINIMUM (or MAXIMUM) command found in the CALC menu. We first graph the function $f(x) = 0.7x^2 + 2x - 3.5$ as in Figure 10-13(a). We then select 3 in the CALC menu, enter -3 for a left guess, and press **ENTER**. We then enter 0 for a right guess and press **ENTER**. After pressing **ENTER** again, we will obtain the minimum value $(-1.42857, -4.928571)$, as shown in Figure 10-13(b).

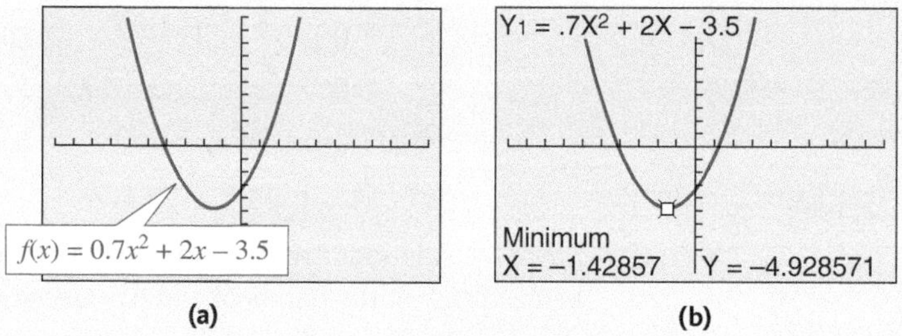

Figure 10-13

The solutions of the quadratic equation $0.7x^2 + 2x - 3.5 = 0$ are the values of x that will make $f(x) = 0$ in the function $f(x) = 0.7x^2 + 2x - 3.5$. To approximate these values, we solve the equation by using the ZERO command found in the CALC menu. From the graph in Figure 10-14(a), we select 2 in the CALC menu to obtain Figure 10-14(b). We enter -5 for a left guess and press **ENTER**. We then enter -2 for a right guess and press **ENTER**. After pressing **ENTER** again, we will obtain Figure 10-14(c). We can find the second solution from the graph as Figure 10-15(a) by selecting 2 in the

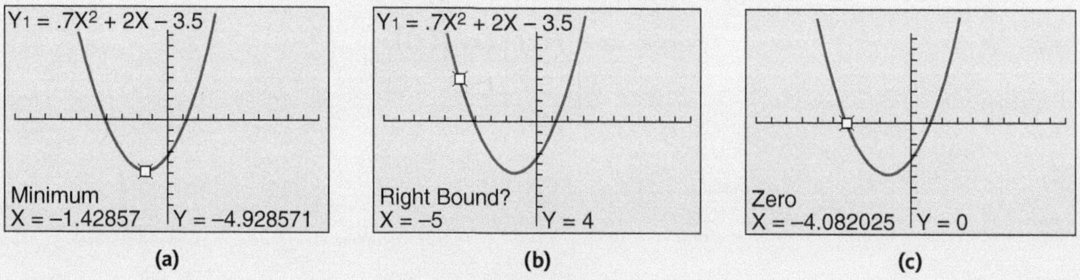

Figure 10-14

(continued)

Unless otherwise noted, all content on this page is © Cengage Learning.

CALC menu and enter 0 for a left guess and press **ENTER** to obtain Figure 10-15(b). By entering 3 and then **ENTER** we obtain Figure 10-15(c). The solutions, to three decimal points, are -4.082 and 1.225.

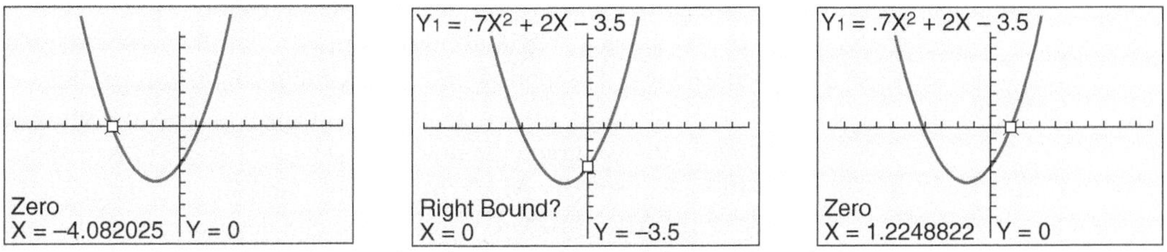

(a) (b) (c)

Figure 10-15

For instructions regarding the use of a Casio graphing calculator, please refer to the Casio Keystroke Guide in the back of the book.

8 Solve an application using a quadratic function.

EXAMPLE 9 **BALLISTICS** The ball shown in Figure 10-16(a) is thrown straight up by a Major League Baseball pitcher with a velocity of 128 feet per second. The function $s = h(t) = -16t^2 + 128t$ gives the relation between t (the time measured in seconds) and s (the number of feet the ball is above the ground). How long will it take the ball to reach its maximum height, and what is that height?

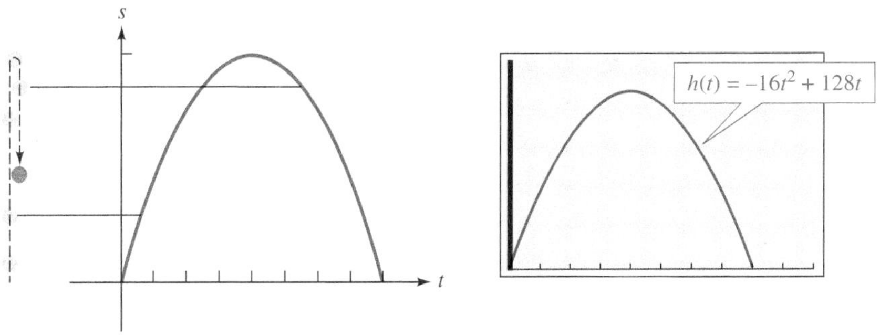

(a) (b)

Figure 10-16

Solution The graph of $s = h(t) = -16t^2 + 128t$ is a parabola. Since the coefficient of t^2 is negative, it opens downward. The time it takes the ball to reach its maximum height is given by the t-coordinate of its vertex, and the maximum height of the ball is given by the s-coordinate of the vertex. To find the vertex, we find its t-coordinate and s-coordinate. To find the t-coordinate, we compute

$$-\frac{b}{2a} = -\frac{128}{2(-16)} \quad b = 128 \text{ and } a = -16$$

$$= -\frac{128}{-32}$$

$$= 4$$

Unless otherwise noted, all content on this page is © Cengage Learning.

To find the s-coordinate, we substitute 4 for t in $h(t) = -16t^2 + 128t$.

$$\begin{aligned} h(4) &= -16(4)^2 + 128(4) \\ &= -256 + 512 \\ &= 256 \end{aligned}$$

Since $t = 4$ and $s = 256$ are the coordinates of the vertex, the ball will reach a maximum height in 4 seconds and that maximum height will be 256 feet.

To solve this with a graphing calculator with window settings of $[0, 10]$ for x and $[0, 300]$ for y, we graph the function $h(t) = -16t^2 + 128t$ to obtain the graph in Figure 10-16(b). By using the MAXIMUM command found under the CALC menu in the same way as the MINIMUM command, we can determine that the ball reaches a height of 256 feet in 4 seconds.

SELF CHECK 9 If the ball shown is thrown straight up with a velocity of 96 feet per second, how long will it take the ball to reach its maximum height and what is that height?

EXAMPLE 10

MAXIMIZING AREA A man wants to build the rectangular pen shown in Figure 10-17(a) to house his dog. If he uses one side of his barn, find the maximum area that he can enclose with 80 feet of fencing.

Solution If we use w to represent the width of the pen in feet, the length is represented by $(80 - 2w)$ feet. Since the area A of the pen is the product of its length and width, we have

$$\begin{aligned} A &= (80 - 2w)w \\ &= 80w - 2w^2 \\ &= -2w^2 + 80w \end{aligned}$$

Since the graph of $A = -2w^2 + 80w$ is a parabola opening downward, the maximum area will be given by the second-coordinate of the vertex of the graph. To find the vertex, we find its first coordinate by letting $b = 80$ and $a = -2$ and computing $-\frac{b}{2a}$.

$$-\frac{b}{2a} = -\frac{80}{2(-2)} = 20$$

We can then find the second coordinate of the vertex by substituting 20 into the function $A = -2w^2 + 80w$.

$$\begin{aligned} A &= -2w^2 + 80w \\ &= -2(20)^2 + 80(20) \\ &= -2(400) + 1{,}600 \\ &= -800 + 1{,}600 \\ &= 800 \end{aligned}$$

Thus, the coordinates of the vertex of the graph of the quadratic function are $(20, 800)$, and the maximum area is 800 square feet. This occurs when the width is 20 feet.

To solve this using a graphing calculator with window settings of $[0, 50]$ for x and $[0, 1{,}000]$ for y, we graph the function $A = -2w^2 + 80w$ to obtain the graph in Figure 10-17(b). By using the MAXIMUM command, we can determine that the maximum area is 800 square feet when the width is 20 feet.

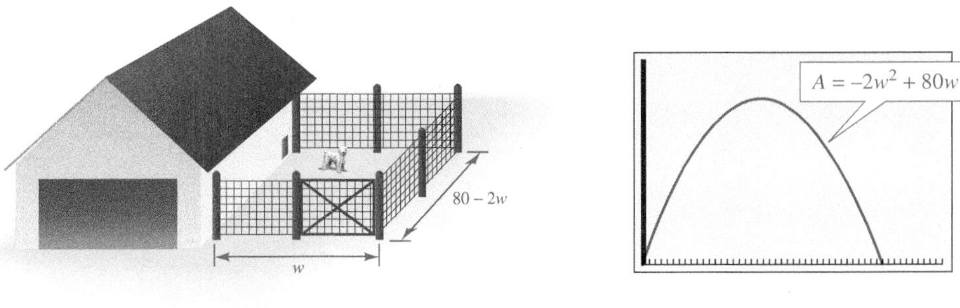

(a) (b)

Figure 10-17

SELF CHECK 10 Find the maximum area the man can enclose with 130 feet of fencing.

In statistics, the square of the standard deviation is called the **variance**, a way of looking at the variability of scores.

EXAMPLE 11 **VARIANCE** If p is the chance that a person selected at random has the H1N1 virus, then $(1 - p)$ is the chance that the person does not have the H1N1 virus. If 100 people are randomly sampled, we know from statistics that the variance of this type of sample distribution will be $100p(1 - p)$. Find the value of p that will maximize the variance.

Solution The variance is given by the function

$$v(p) = 100p(1 - p) \quad \text{or} \quad v(p) = -100p^2 + 100p$$

$$a = -100, b = 100, c = 0$$

$$\frac{-b}{2a} = \frac{-100}{2(-100)} = \frac{1}{2}$$

A value of $\frac{1}{2}$ or 0.5 will give maximum variance.

In this setting, all values of p are between 0 and 1, including 0 and 1.

We can use window settings of $[0, 1]$ for x when graphing the function $v(p) = -100p^2 + 100p$ on a graphing calculator. If we also use window settings of $[0, 30]$ for y, we will obtain the graph shown in Figure 10-18(a). After using the MAXIMUM command to obtain Figure 10-18(b), we can see that a value of 0.5 will yield the maximum variance.

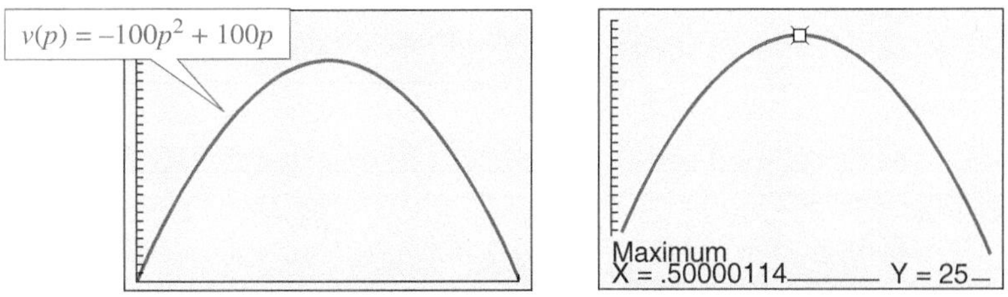

(a) (b)

Figure 10-18

SELF CHECK 11 What is the maximum variance in this example?

Unless otherwise noted, all content on this page is © Cengage Learning.

SELF CHECK ANSWERS

1.
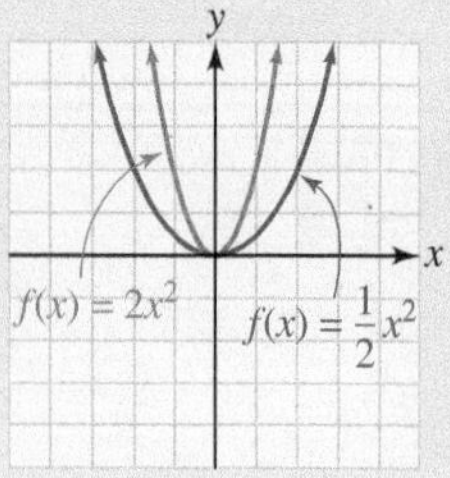

2.
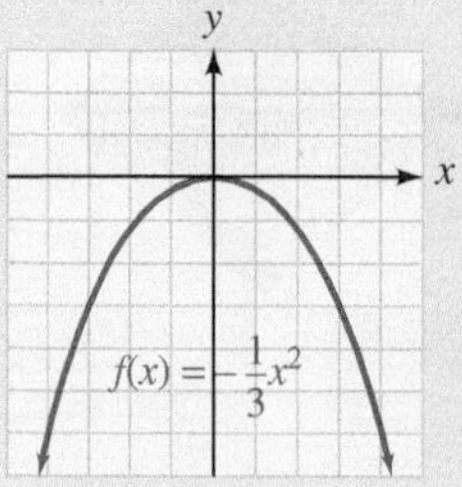

3.
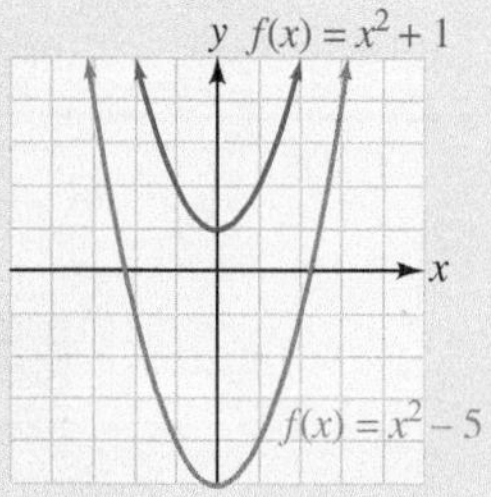

a. shifted 1 unit upward
b. shifted 5 units downward

4.
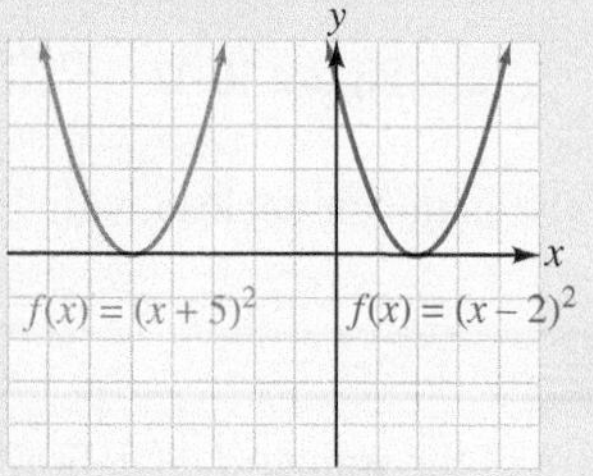

a. shifted 2 units to the right
b. shifted 5 units to the left

5.
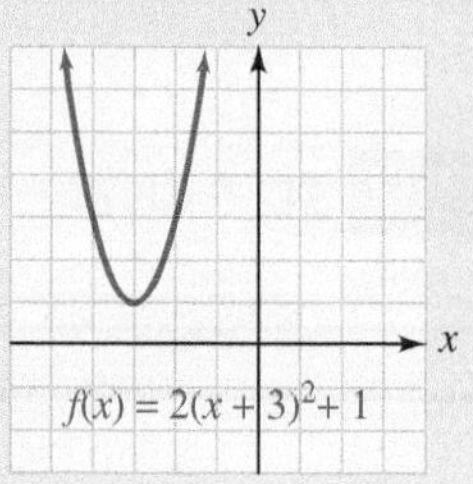

6.
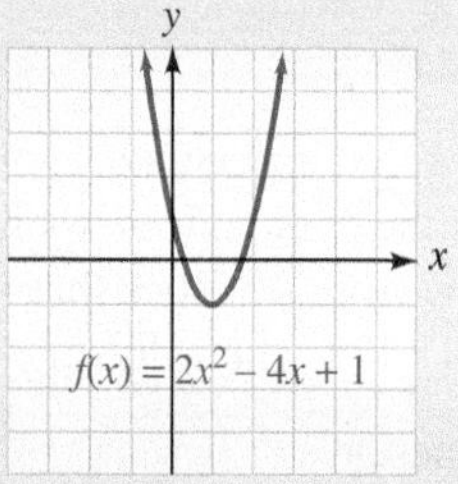

7. $(2, -4)$ **8.**
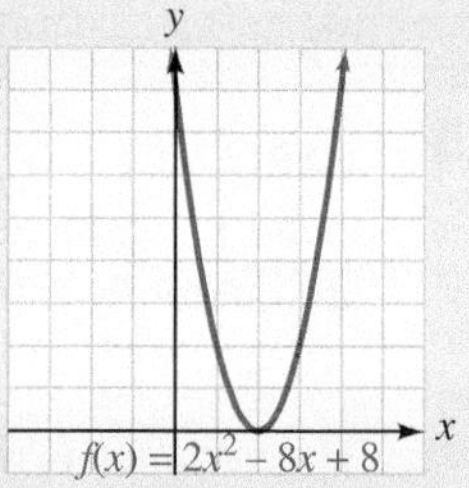

9. It will take 3 sec for the ball to reach its maximum height of 144 ft. **10.** The maximum area he can enclose is 2,112.5 ft^2. **11.** The maximum variance is 25.

NOW TRY THIS

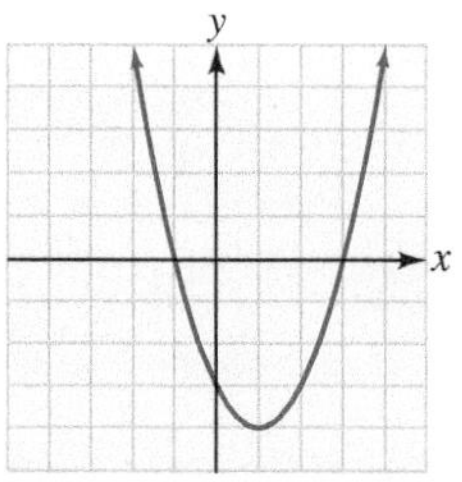

Answer the following questions about the graph of $f(x)$.

1. What is the vertex?
2. What is the axis of symmetry?
3. What is the domain?
4. What is the range?
5. For what values of x will $f(x) = 0$?
6. For what values of x will y be positive? $(f(x) > 0)$
7. For what values of x will y be negative? $(f(x) < 0)$

10.4 Exercises

WARM-UPS *Find the equation of a vertical line passing through the given point.*

1. $(3, 5)$
2. $(-2, 7)$
3. $(-1, 0)$
4. $(6, 4)$

Solve.

5. $f(x) = x^2 - 9$
6. $f(x) = x^2 - 2x - 3$

REVIEW *Find the value of x.*

7.
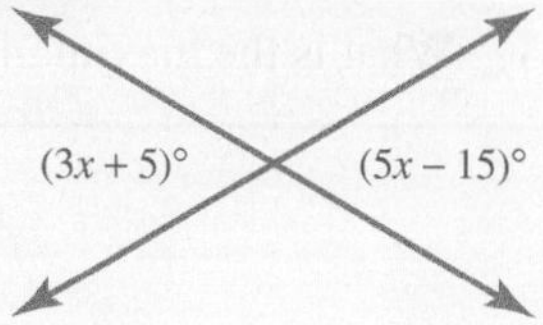

Unless otherwise noted, all content on this page is © Cengage Learning.

8. Lines r and s are parallel.

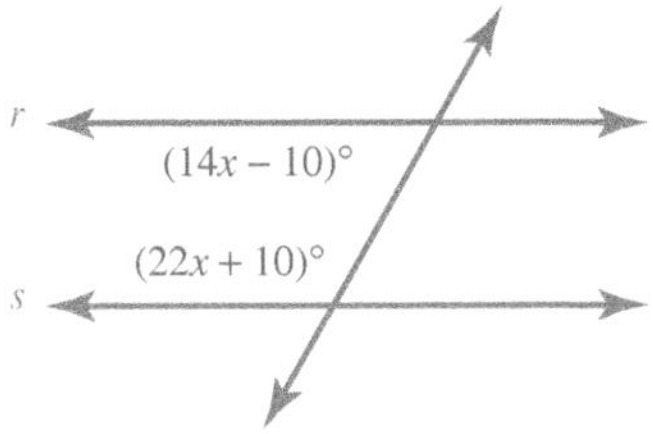

9. Travel Madison and St. Louis are 385 miles apart. One train leaves Madison and heads toward St. Louis at the rate of 30 mph. Three hours later, a second train leaves Madison, bound for St. Louis. If the second train travels at the rate of 55 mph, in how many hours will the faster train overtake the slower train?

10. Investing A woman invests \$25,000, some at 7% annual interest and the rest at 8%. If the annual income from both investments is \$1,900, how much is invested at the higher rate?

VOCABULARY AND CONCEPTS *Fill in the blanks.*

11. A quadratic function is a second-degree polynomial function that can be written in the form ____________________, where _______.

12. The graphs of quadratic functions are called ________.

13. The highest (________) or lowest (________) point on a parabola is called the _____.

14. A vertical line that divides a parabola into two congruent halves is called an ___ of symmetry.

15. The graph of $y = f(x) + k$ $(k > 0)$ is identical to the graph of $y = f(x)$, except that it is translated k units ______.

16. The graph of $y = f(x) - k$ $(k > 0)$ is a vertical translation of the graph of $y = f(x)$, k units ________.

17. The graph of $y = f(x - h)$ $(h > 0)$ is a horizontal translation of the graph of $y = f(x)$, h units ________.

18. The graph of $y = f(x + h)$ $(h > 0)$ is identical to the graph of $y = f(x)$, except that it is translated h units ________.

19. The graph of $y = f(x) = ax^2 + bx + c$ $(a \neq 0)$ opens _______ when $a > 0$.

20. In statistics, the square of the standard deviation is called the _______.

GUIDED PRACTICE *Graph each function.* SEE EXAMPLES 1–2. (OBJECTIVE 1)

21. $f(x) = x^2$

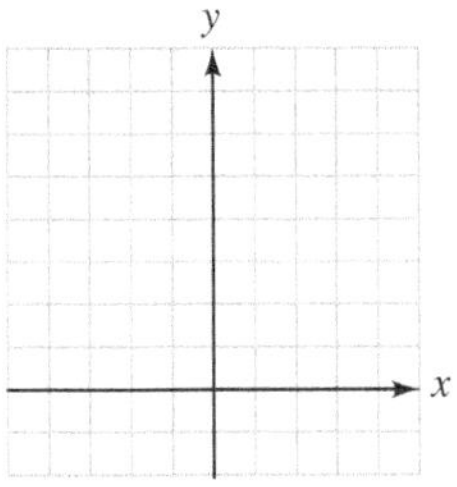

22. $f(x) = -x^2$

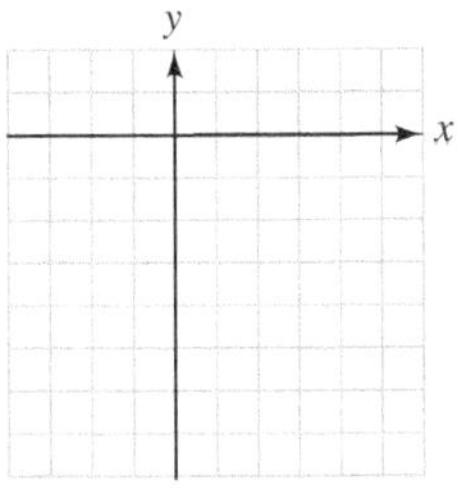

23. $f(x) = -2x^2$

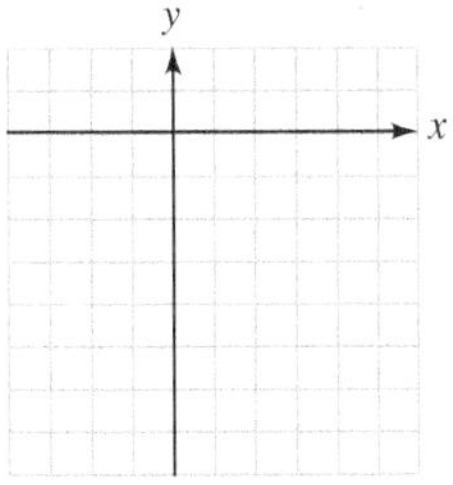

24. $f(x) = 3x^2$

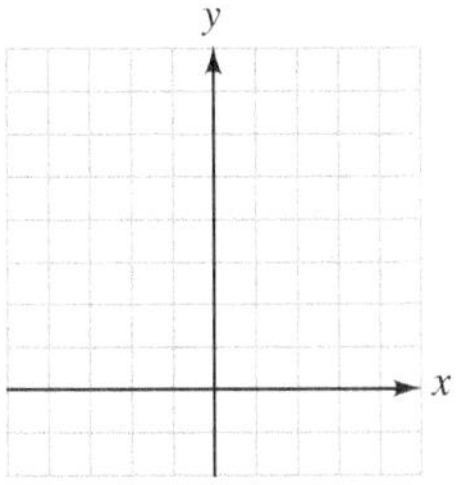

Graph each function. SEE EXAMPLE 3. (OBJECTIVE 2)

25. $f(x) = x^2 + 2$

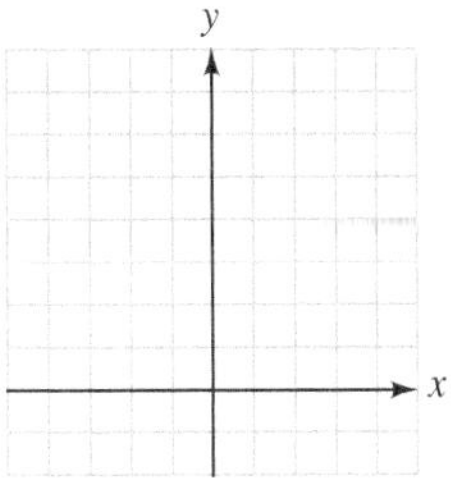

26. $f(x) = x^2 - 3$

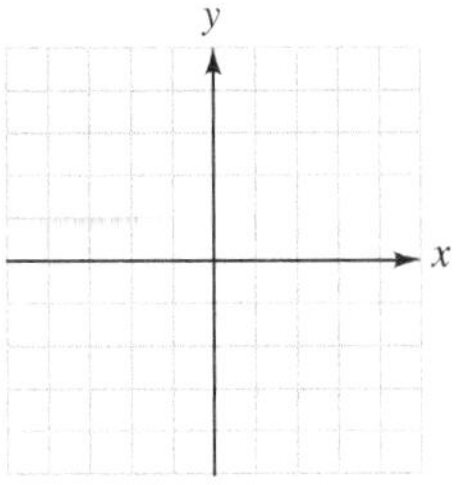

27. $f(x) = x^2 - 1$

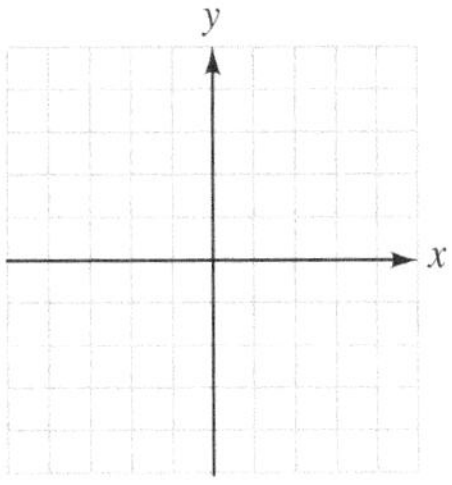

28. $f(x) = x^2 + 4$

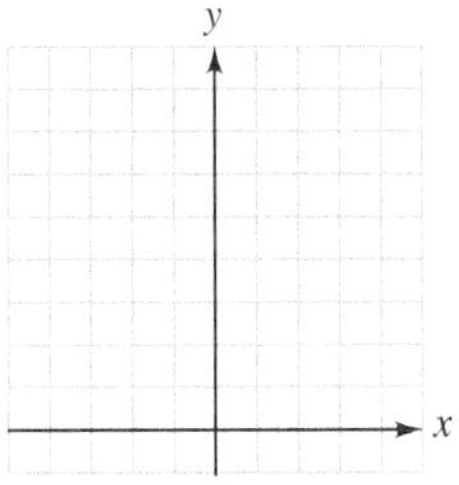

Graph each function. SEE EXAMPLE 4. (OBJECTIVE 3)

29. $f(x) = -(x - 2)^2$

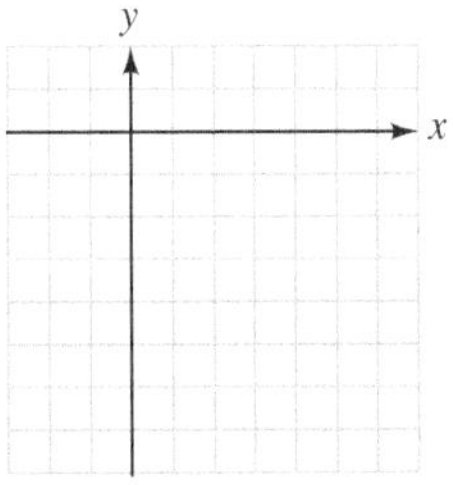

30. $f(x) = (x + 2)^2$

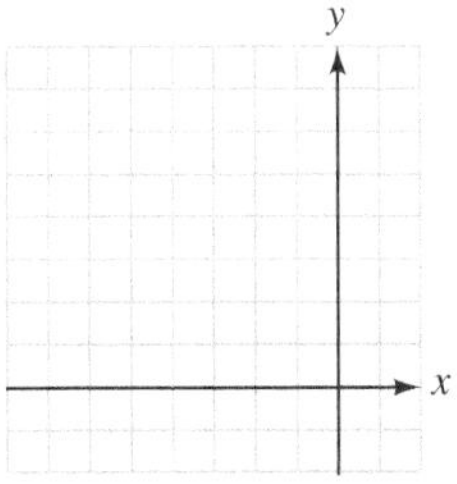

31. $f(x) = -(x - 3)^2$

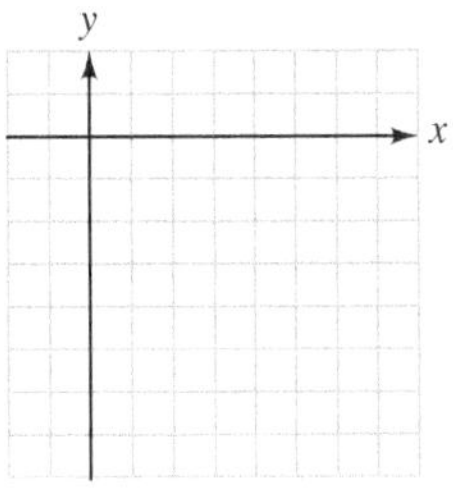

32. $f(x) = (x + 3)^2$

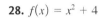

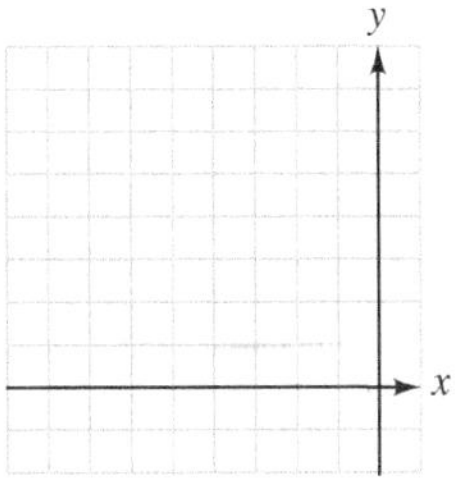

Unless otherwise noted, all content on this page is © Cengage Learning.

Graph each function. SEE EXAMPLE 5. (OBJECTIVE 4)

33. $f(x) = (x - 3)^2 + 2$

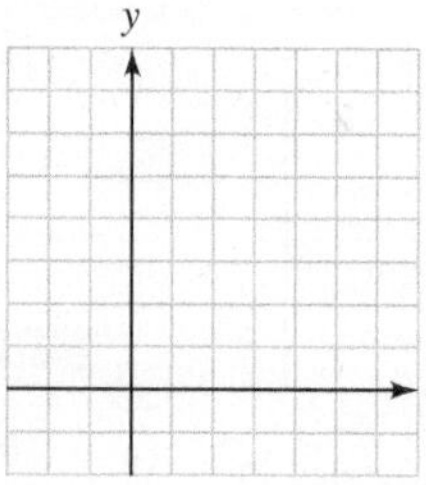

34. $f(x) = (x + 1)^2 - 2$

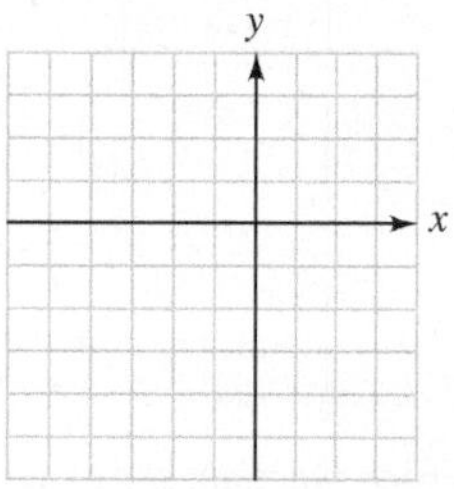

35. $y = (x - 4)^2 + 5$

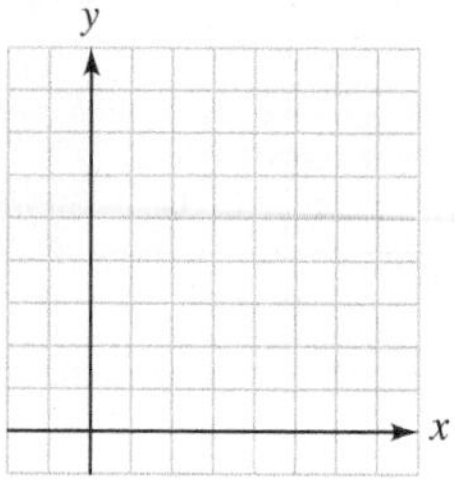

36. $y = 4(x + 5)^2 - 6$

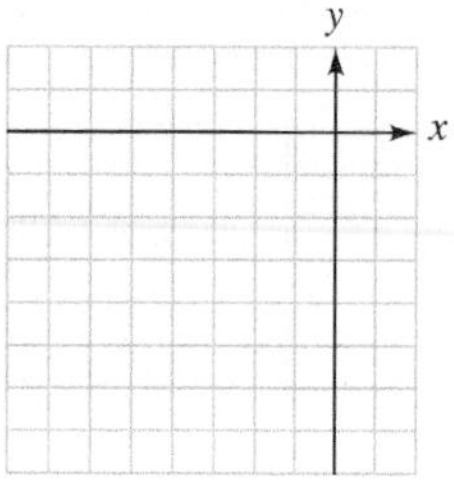

Find the coordinates of the vertex and the axis of symmetry of the graph of each equation. If necessary, complete the square on x to write the equation in the form $y = a(x - h)^2 + k$. ***Do not graph the equation.*** SEE EXAMPLE 6. (OBJECTIVE 5)

37. $y = (x - 1)^2 + 2$ **38.** $y = 2(x - 2)^2 - 1$

39. $y = 2(x + 3)^2 - 4$ **40.** $y = -3(x + 1)^2 + 3$

41. $y = -3x^2$ **42.** $y = 3x^2 - 3$

43. $y = 2x^2 - 4x$ **44.** $y = 3x^2 + 6x$

Use $\left(-\frac{b}{2a}, f\left(-\frac{b}{2a}\right)\right)$ to find the vertex of each function. SEE EXAMPLE 7. (OBJECTIVE 6)

45. $f(x) = 3x^2 + 6x + 1$ **46.** $f(x) = 5x^2 - 10x$

47. $f(x) = -3x^2 + 12x + 4$ **48.** $f(x) = 3x^2 - 8x + 5$

Find the coordinates of the vertex and the axis of symmetry of the graph of each equation. Use $\left(-\frac{b}{2a}, f\left(-\frac{b}{2a}\right)\right)$ to find the vertex and graph the equation. SEE EXAMPLE 8. (OBJECTIVE 7)

49. $f(x) = -2x^2 + 4x + 1$

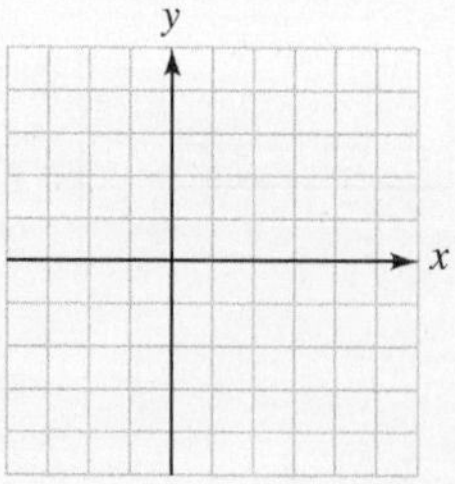

50. $f(x) = 3x^2 - 12x + 9$

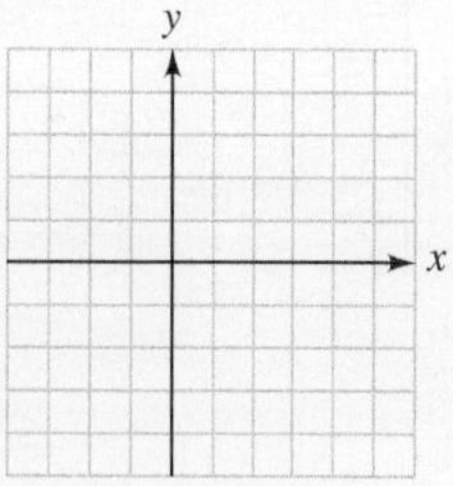

51. $f(x) = 3x^2 - 12x + 10$

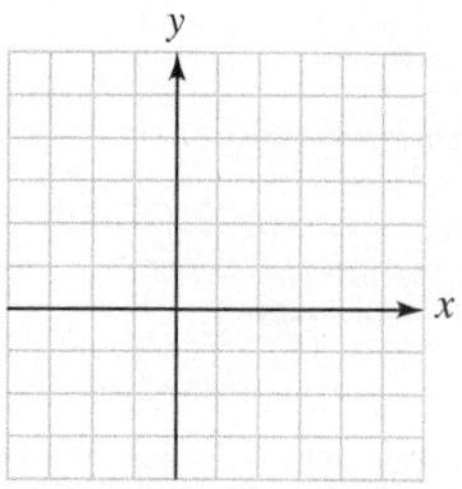

52. $f(x) = -2x^2 + 4x + 3$

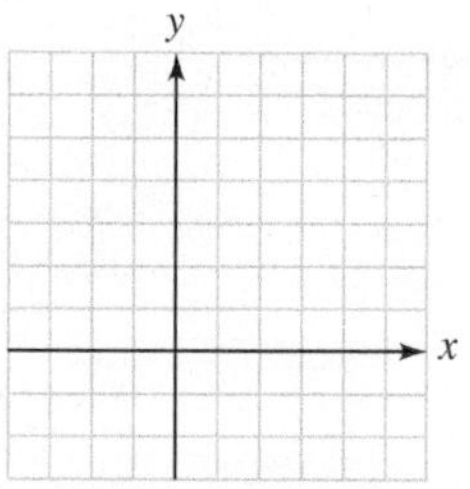

ADDITIONAL PRACTICE *Find the vertex and axis of symmetry using any method. Do not graph.*

53. $y = -4x^2 + 16x + 5$ **54.** $y = 5x^2 + 20x + 25$

55. $y - 7 = 6x^2 - 5x$ **56.** $y - 7 = 4x^2 + 10x$

Graph each function.

57. $f(x) = x^2 + x - 6$

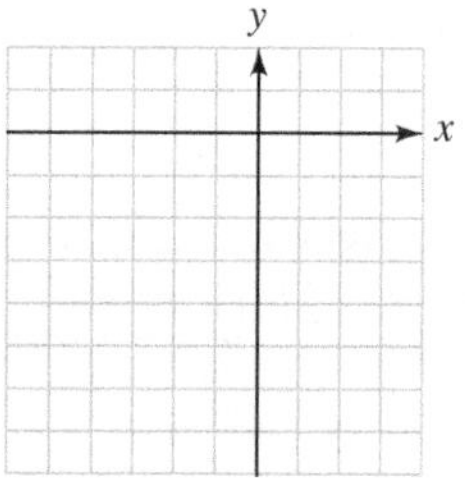

58. $f(x) = x^2 - x - 6$

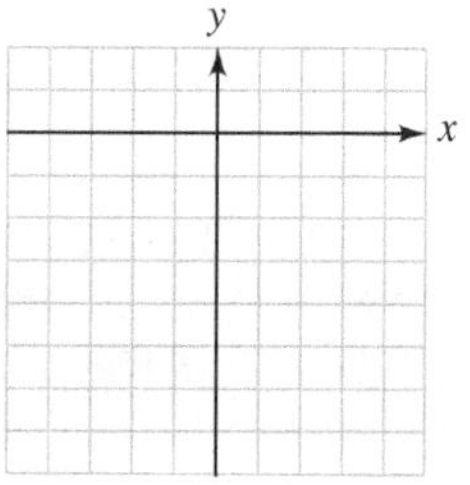

Use a graphing calculator to find the coordinates of the vertex of the graph of each quadratic function. Give results to the nearest hundredth.

59. $y = 2x^2 - x + 1$ **60.** $y = x^2 + 5x - 6$

61. $y = 7 + x - x^2$ **62.** $y = 2x^2 - 3x + 2$

Use a graphing calculator to solve each equation. If a result is not exact, give the result to the nearest hundredth.

63. $x^2 + 9x - 10 = 0$ **64.** $2x^2 - 5x - 3 = 0$

65. $0.5x^2 - 0.7x - 3 = 0$ **66.** $2x^2 - 0.5x - 2 = 0$

67. The equation $y - 3 = (x + 7)^2$ represents a quadratic function whose graph is a parabola. Find its vertex.

68. Show that $y = ax^2$, where $a \neq 0$, represents a quadratic function whose vertex is at the origin.

APPLICATIONS *Solve. Use a graphing calculator if necessary.* SEE EXAMPLES 9–11. (OBJECTIVE 8)

69. Ballistics If a ball is thrown straight up with an initial velocity of 48 feet per second, its height s after t seconds is given by the equation $s = 48t - 16t^2$. Find the maximum height attained by the ball and the time it takes for the ball to reach that height.

Unless otherwise noted, all content on this page is © Cengage Learning.

70. Ballistics From the top of the building, a ball is thrown straight up with an initial velocity of 32 feet per second. The equation $s = -16t^2 + 32t + 48$ gives the height s of the ball t seconds after it is thrown. Find the maximum height reached by the ball and the time it takes for the ball to hit the ground. (*Hint*: Let $s = 0$ and solve for t.)

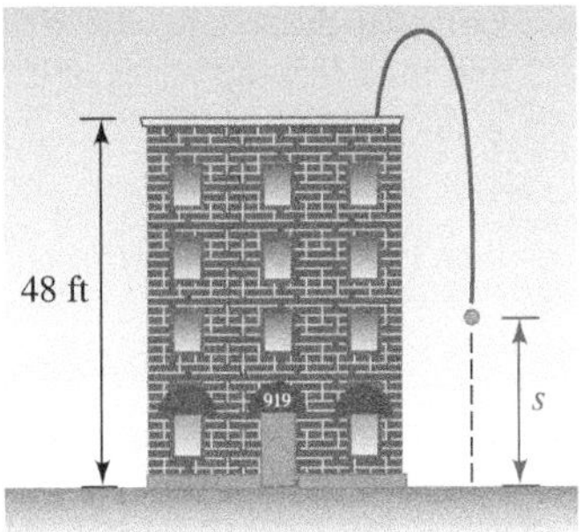

71. Maximizing area Find the dimensions of the rectangle of maximum area that can be constructed with 400 feet of fencing. Find the maximum area.

72. Fencing a field A farmer wants to fence in three sides of a rectangular field with 1,000 feet of fencing. The other side of the rectangle will be a river. If the enclosed area is to be maximum, find the dimensions of the field.

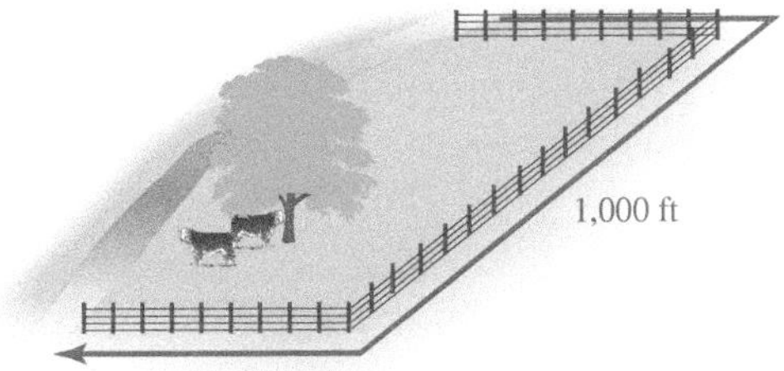

73. Finding the variance If p is the chance that a person sampled at random has high blood pressure, $1 - p$ is the chance that the person does not. If 50 people are sampled at random, the variance of the sample will be $50p(1 - p)$. What two values of p will give a variance of 9.375?

74. Finding the variance If p is the chance that a person sampled at random smokes, then $1 - p$ is the chance that the person doesn't. If 75 people are sampled at random, the variance of the sample will be $75p(1 - p)$. What two values of p will give a variance of 12?

75. Police investigations A police officer seals off the scene of a car collision using a roll of yellow police tape that is 300 feet long. What dimensions should be used to seal off the maximum rectangular area around the collision? What is the maximum area?

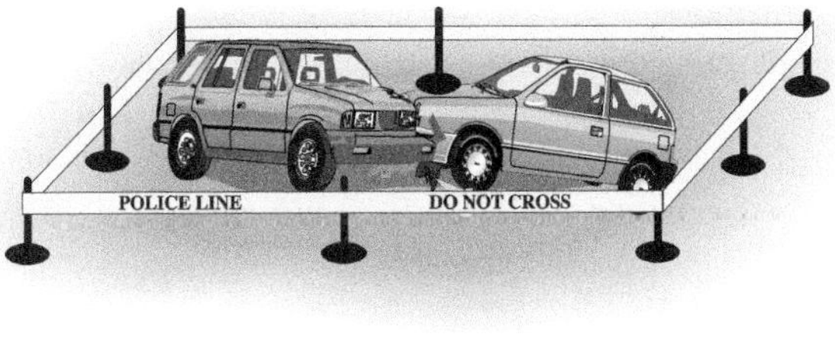

76. Operating costs The cost C in dollars of operating a certain concrete-cutting machine is related to the number of minutes n the machine is run by the function

$$C(n) = 2.2n^2 - 66n + 655$$

For what number of minutes is the cost of running the machine a minimum? What is the minimum cost?

77. Water usage The height (in feet) of the water level in a reservoir over a 1-year period is modeled by the function

$$H(t) = 3.3t^2 - 59.4t + 281.3$$

How low did the water level get that year?

78. School enrollment The total annual enrollment (in millions) in U.S. elementary and secondary schools for the years 1975–1996 is given by the function

$$E(x) = 0.058x^2 - 1.162x + 50.604$$

For this period, what was the lowest enrollment?

79. Maximizing revenue The revenue R received for selling x stereos is given by the equation

$$R = -\frac{x^2}{1{,}000} + 10x$$

Find the number of stereos that must be sold to obtain the maximum revenue.

80. Maximizing revenue In Exercise 79, find the maximum revenue.

81. Maximizing revenue The revenue received for selling x radios is given by the formula

$$R = -\frac{x^2}{728} + 9x$$

How many radios must be sold to obtain the maximum revenue? Find the maximum revenue.

82. Maximizing revenue The revenue received for selling x stereos is given by the formula

$$R = -\frac{x^2}{5} + 80x - 1{,}000$$

How many stereos must be sold to obtain the maximum revenue? Find the maximum revenue.

83. Maximizing revenue When priced at \$30 each, a toy has annual sales of 4,000 units. The manufacturer estimates that each \$1 increase in cost will decrease sales by 100 units. Find the unit price that will maximize total revenue. (*Hint*: Total revenue = price · the number of units sold.)

84. Maximizing revenue When priced at \$57, one type of camera has annual sales of 525 units. For each \$1 the camera is reduced in price, management expects to sell an additional 75 cameras. Find the unit price that will maximize total revenue. (*Hint*: Total revenue = price · the number of units sold.)

WRITING ABOUT MATH

85. The graph of $y = ax^2 + bx + c$ $(a \neq 0)$ passes the vertical line test. Explain why this shows that the equation defines a function.

86. The graph of $x = y^2 - 2y$ is a parabola. Explain why its graph does not represent a function.

Unless otherwise noted, all content on this page is © Cengage Learning.

SOMETHING TO THINK ABOUT

87. Can you use a graphing calculator to find solutions of the equation $x^2 + x + 1 = 0$? What is the problem? How do you interpret the result?

88. Complete the square on x in the equation $y = ax^2 + bx + c$ and show that the vertex of the parabolic graph is the point with coordinates of

$$\left(-\frac{b}{2a}, \frac{4ac - b^2}{4a}\right)$$

Section 10.5 Graphs of Other Nonlinear Functions

Objectives

1. Graph a vertical and horizontal translation of the cubic and absolute value functions.
2. Graph a reflection about the x-axis of the cubic and absolute value functions.
3. Interpret the graph of a rational function.
4. Find the domain and range of a rational function.

Vocabulary

cubic function
absolute value function
reflection
paraboloid
asymptote
horizontal asymptote
vertical asymptote

Getting Ready

State the slope and the y-intercept of each linear function.

1. $f(x) = 2x - 3$

2. $f(x) = -3x + 4$

Find the value of f(x) when $x = 2$ and $x = -1$.

3. $f(x) = 5x - 4$

4. $f(x) = \frac{1}{2}x + 3$

In the previous section, we discussed quadratic functions and horizontal and vertical translations. In this section, we will extend the discussion to include two other nonlinear functions, the cubic and absolute value functions. We will discuss reflections of these functions about the x-axis and graph rational functions.

Before we graph translations of the cubic and absolute value functions, we will review their basic graphs.

To graph the function $f(x) = x^3$, called the **cubic function**, we substitute values for x in the equation and compute the corresponding values of $f(x)$. For example, if $x = -2$, we have

$$\begin{aligned} f(x) &= x^3 \\ f(-2) &= (-2)^3 \quad \text{Substitute } -2 \text{ for } x. \\ &= -8 \end{aligned}$$

The ordered pair $(-2, -8)$ satisfies the equation and will lie on the graph. We list this pair and others that satisfy the equation in the table shown in Figure 10-19. We plot the points and draw a smooth curve through them to obtain the graph.

$f(x) = x^3$

x	$f(x)$	$(x, f(x))$
-2	-8	$(-2, -8)$
-1	-1	$(-1, -1)$
0	0	$(0, 0)$
1	1	$(1, 1)$
2	8	$(2, 8)$

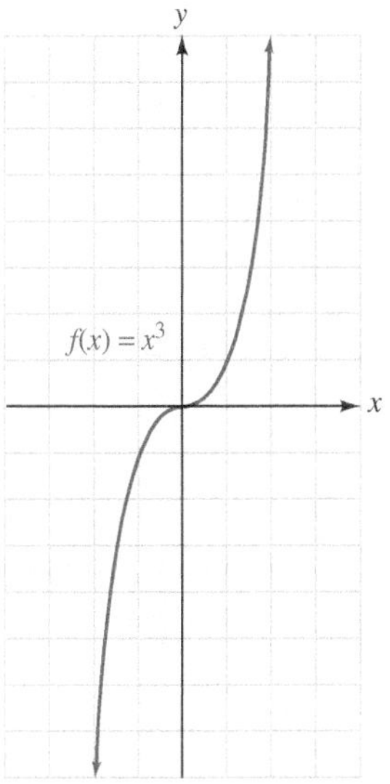

Figure 10-19

From the graph, we can see that the domain of the cubic function is the set of all real numbers, which is the interval $(-\infty, \infty)$, or $\mathbb{R}$. We can also see that the range is the interval $(-\infty, \infty)$.

To graph the function $f(x) = |x|$, called the **absolute value function**, substitute values for x in the equation and compute the corresponding values of y. For example, if $x = -3$, we have

$$\begin{aligned} f(x) &= |x| \\ f(\mathbf{-3}) &= |\mathbf{-3}| \quad \text{Substitute } -3 \text{ for } x. \\ &= 3 \end{aligned}$$

The ordered pair $(-3, 3)$ satisfies the equation and will lie on the graph. We list this pair and others that satisfy the equation in the table shown in Figure 10-20. We plot the points and draw a V-shaped graph through them.

From the graph, we see that the domain of the absolute value function is the set of real numbers, which is the interval $(-\infty, \infty)$. We can also see that the range is the set of non-negative real numbers, which is the interval $[0, \infty)$.

$f(x) = |x|$

x	$f(x)$	$(x, f(x))$
-3	3	$(-3, 3)$
-2	2	$(-2, 2)$
-1	1	$(-1, 1)$
0	0	$(0, 0)$
1	1	$(1, 1)$
2	2	$(2, 2)$
3	3	$(3, 3)$

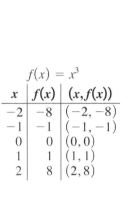

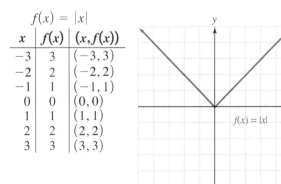

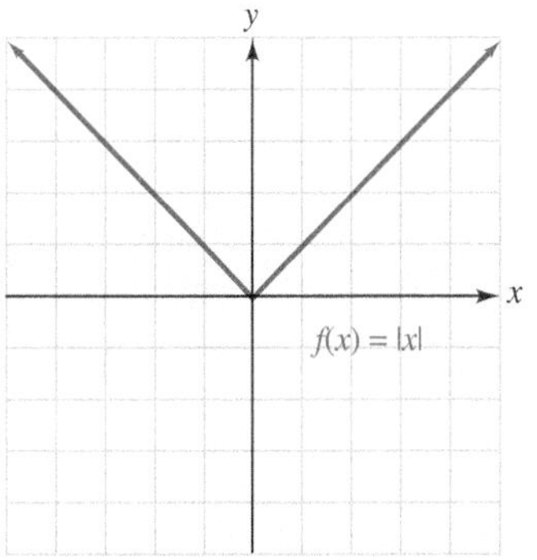

Figure 10-20

Unless otherwise noted, all content on this page is © Cengage Learning.

Accent on technology

▸ Graphing Functions

We can graph nonlinear functions with a graphing calculator. For example, to graph $f(x) = x^3$ in a standard window of $[-10, 10]$ for x and $[-10, 10]$ for y, we enter the function by using the **Y=** key and typing x ^ 3. We then press the **GRAPH** key to obtain the graph in Figure 10-21(a). To graph $f(x) = |x|$, we enter the function by selecting "abs" from the MATH menu and pressing the **GRAPH** key to obtain the graph in Figure 10-21(b).

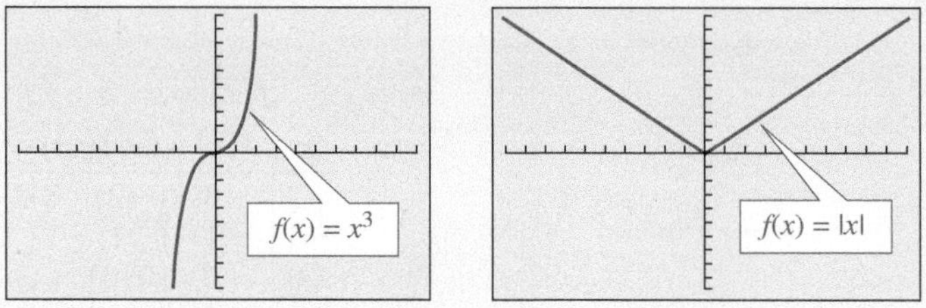

The cubic function
(a)

The absolute value function
(b)

Figure 10-21

When using a graphing calculator, we must be sure that the viewing window does not show a misleading graph. For example, if we graph $f(x) = |x|$ in the window $[0, 10]$ for x and $[0, 10]$ for y, we will obtain a misleading graph that looks like a line. (See Figure 10-22.) The complete graph is the V-shaped graph shown in Figure 10-21(b).

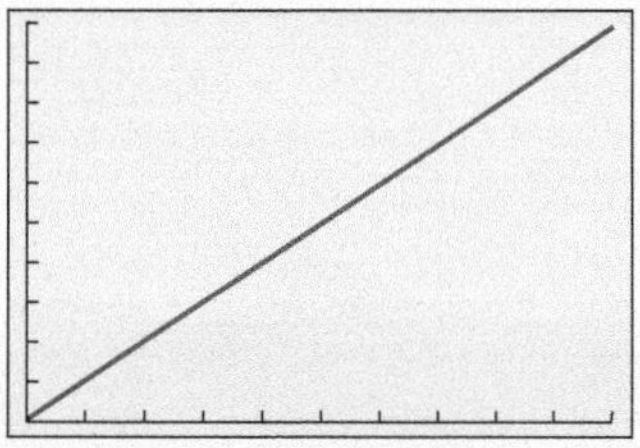

Figure 10-22

For instructions regarding the use of a Casio graphing calculator, please refer to the Casio Keystroke Guide in the back of the book.

1 Graph a vertical and horizontal translation of the cubic and absolute value functions.

Recall that the graph of the quadratic function $f(x) = x^2$ is a parabola whose vertex is at $(0, 0)$. Figure 10-23 shows the graph of $f(x) = x^2 + k$ for three different values of k. If $k = 0$, we obtain the graph of $f(x) = x^2$. If $k = 3$, we obtain the graph of $f(x) = x^2 + 3$, which is identical to the graph of $f(x) = x^2$, except that it is shifted 3 units upward. If $k = -4$, we obtain the graph of $f(x) = x^2 - 4$, which is identical to the graph of $f(x) = x^2$, except that it is shifted 4 units downward. These shifts are called vertical translations, as we learned in the previous section.

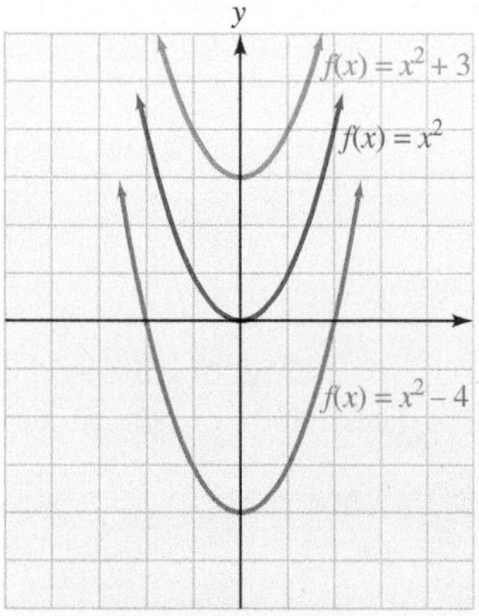

Figure 10-23

In general, we can make these observations.

Unless otherwise noted, all content on this page is © Cengage Learning.

VERTICAL TRANSLATIONS

If f is a function and k is a positive number, then

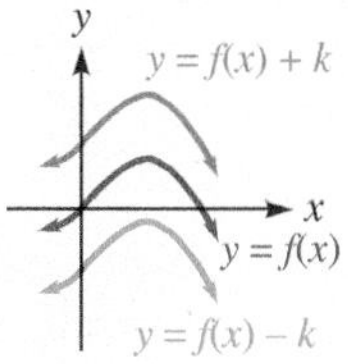

- The graph of $y = f(x) + k$ is identical to the graph of $y = f(x)$, except that it is translated k units upward.
- The graph of $y = f(x) - k$ is identical to the graph of $y = f(x)$, except that it is translated k units downward.

EXAMPLE 1 Graph: $f(x) = |x| + 2$

Solution The graph of $f(x) = |x| + 2$ will be the same V-shaped graph as $f(x) = |x|$, except that it is shifted 2 units upward. The graph appears in Figure 10-24.

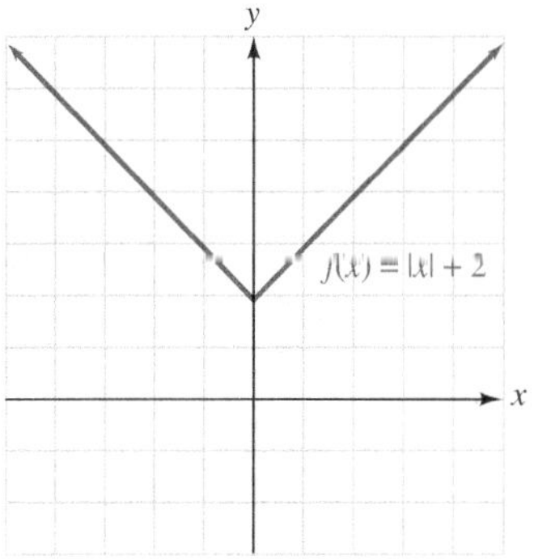

Figure 10-24

SELF CHECK 1 Graph: $f(x) = |x| - 3$

Figure 10-25 shows the graph of $f(x) = (x + h)^2$ for three different values of h. If $h = 0$, we obtain the graph of $f(x) = x^2$. The graph of $f(x) = (x - 3)^2$ is identical to the graph of $f(x) = x^2$, except that it is shifted 3 units to the right. The graph of $f(x) = (x + 2)^2$ is identical to the graph of $f(x) = x^2$, except that it is shifted 2 units to the left. These shifts are called horizontal translations, as we learned in the previous section.

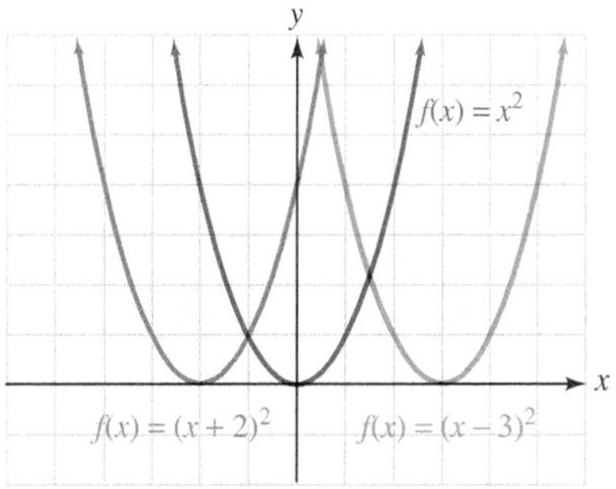

Figure 10-25

In general, we can make these observations.

HORIZONTAL TRANSLATIONS

If f is a function and k is a positive number, then

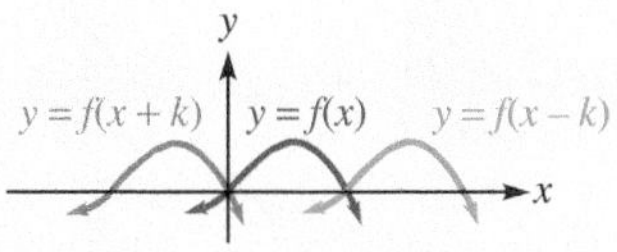

- The graph of $y = f(x - k)$ is identical to the graph of $y = f(x)$, except that it is translated k units to the right.
- The graph of $y = f(x + k)$ is identical to the graph of $y = f(x)$, except that it is translated k units to the left.

Unless otherwise noted, all content on this page is © Cengage Learning.

EXAMPLE 2 Graph: $f(x) = |x - 2|$

Solution The graph of $f(x) = |x - 2|$ will be the same shape as the graph of $f(x) = |x|$, except that it is shifted 2 units to the right. The graph appears in Figure 10-26.

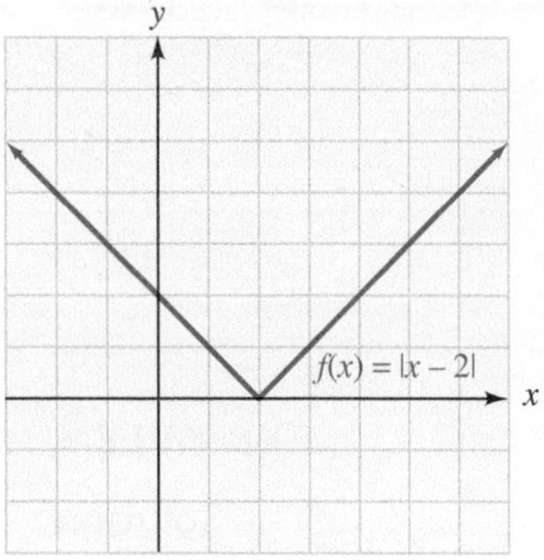

Figure 10-26

SELF CHECK 2 Graph: $f(x) = |x + 3|$

EXAMPLE 3 Graph: $f(x) = (x - 3)^3 + 2$

Solution We can graph this function by translating the graph of $f(x) = x^3$ to the right 3 units and then upward 2 units, as shown in Figure 10-27.

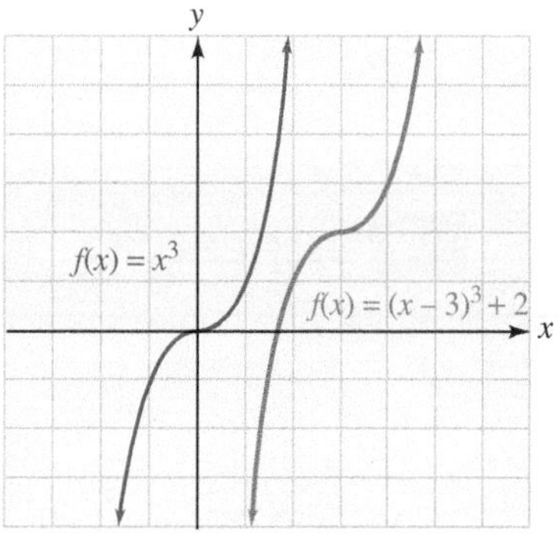

Figure 10-27

SELF CHECK 3 Graph: $f(x) = (x + 2)^3 - 3$

2 Graph a reflection about the *x*-axis of the cubic and absolute value functions.

We now consider the graph of $f(x) = -|x|$. To graph this function, we create a table of values, plot each point, and draw the graph, as in Figure 10-28.

$f(x) = -|x|$

x	$f(x)$	$(x, f(x))$
−3	−3	(−3, −3)
−2	−2	(−2, −2)
−1	−1	(−1, −1)
0	0	(0, 0)
1	−1	(1, −1)
2	−2	(2, −2)
3	−3	(3, −3)

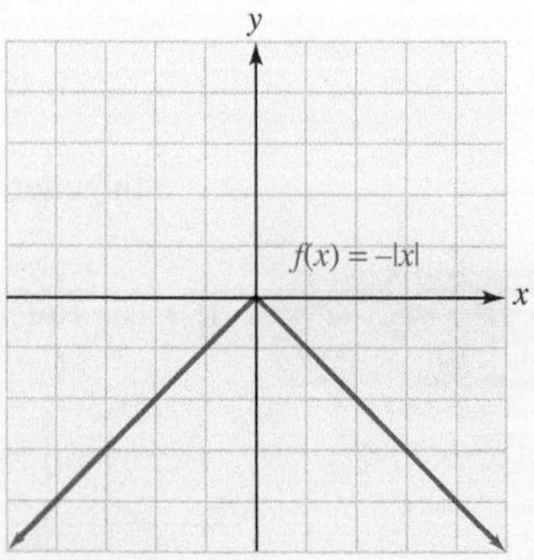

Figure 10-28

Unless otherwise noted, all content on this page is © Cengage Learning.

As we can see from the graph, its shape is the same as the graph of $f(x) = |x|$, except that it has been flipped upside down. We say that the graph of $f(x) = -|x|$ is a **reflection** about the x-axis. In general, we can make the following statement.

REFLECTIONS ABOUT THE x-AXIS

The graph of $y = -f(x)$ is identical to the graph of $y = f(x)$, except that it is reflected about the x-axis.

EXAMPLE 4 Graph the functions: **a.** $f(x) = -|x - 1| + 3$ **b.** $f(x) = -(x + 1)^3 + 2$

Solution **a.** We graph this function by translating the graph of $f(x) = -|x|$ to the right 1 unit and upward 3 units, as shown in Figure 10-29(a).

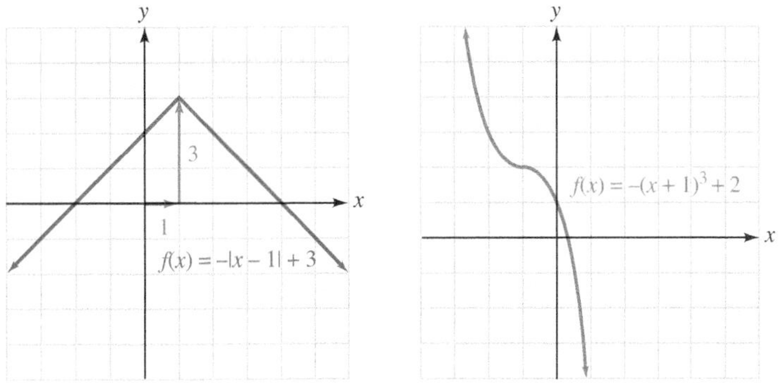

(a) (b)

Figure 10-29

b. We graph this function by translating the graph of $f(x) = -x^3$ to the left 1 unit and upward 2 units, as shown in Figure 10-29(b).

SELF CHECK 4 Graph: **a.** $f(x) = -|x + 2| - 3$ **b.** $f(x) = -(x - 1)^3 - 2$

Perspective

GRAPHS IN SPACE

In an xy-coordinate system, graphs of equations containing the two variables x and y are lines or curves. Other equations have more than two variables, and graphing them often requires some ingenuity and perhaps the aid of a computer. Graphs of equations with the three variables x, y, and z are viewed in a three-dimensional coordinate system with three axes. The coordinates of points in a three-dimensional coordinate system are ordered triples (x, y, z). For example, the points $P(2, 3, 4)$ and $Q(-1, 2, 3)$ are plotted in Illustration 1 on the next page.

Graphs of equations in three variables are not lines or curves, but flat planes or curved surfaces. Only the simplest of these equations can be conveniently graphed by hand; a computer provides the best images of others. The graph in Illustration 2 is called a **paraboloid**; it is the three-dimensional version of a parabola. Illustration 3 models a portion of the vibrating surface of a drum head.

(continued)

Unless otherwise noted, all content on this page is © Cengage Learning.

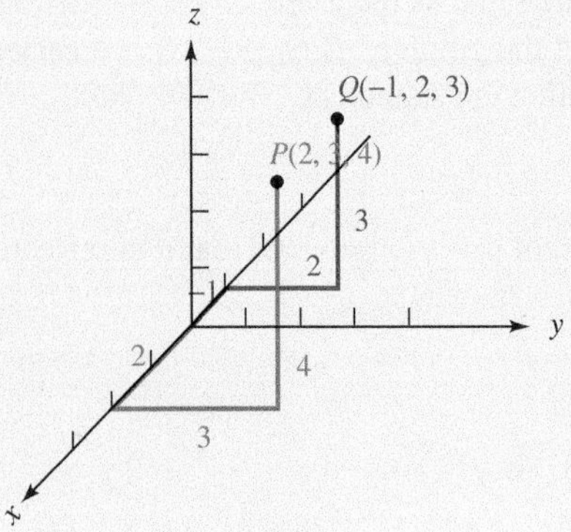

Illustration 1

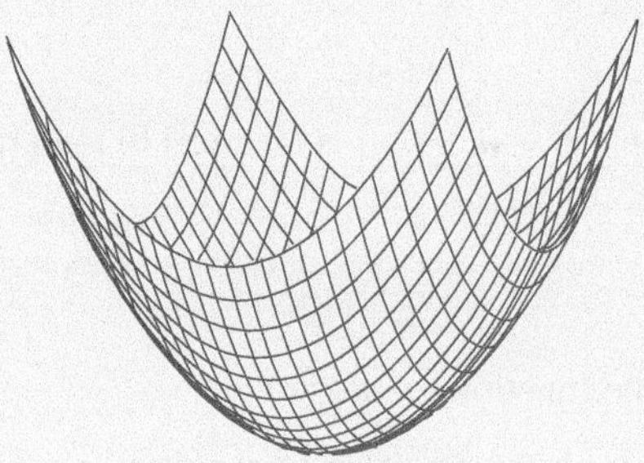

Illustration 2

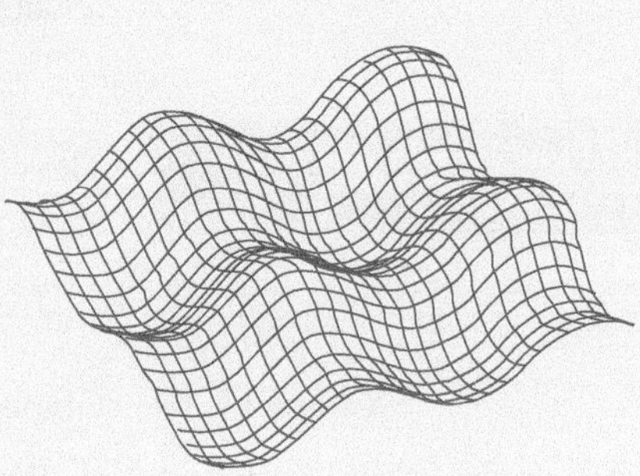

Illustration 3

3 Interpret the graph of a rational function.

Rational expressions often define functions. For example, if the cost of subscribing to an online information network is \$6 per month plus \$1.50 per hour of access time, the average (mean) hourly cost of the service is the total monthly cost, divided by the number of hours of access time:

$$\bar{c} = \frac{C}{n} = \frac{1.50n + 6}{n}$$ $\bar{c}$ is the mean hourly cost, C is the total monthly cost, and n is the number of hours the service is used.

The function

$$(1) \quad \bar{c} = f(n) = \frac{1.50n + 6}{n} \quad (n > 0)$$

gives the mean hourly cost of using the information network for n hours per month.

Figure 10-30 shows the graph of the rational function $\bar{c} = f(n) = \frac{1.50n + 6}{n} \quad (n > 0)$. Since $n > 0$, the domain of this function is the interval $(0, \infty)$.

From the graph, we can see that the mean hourly cost decreases as the number of hours of access time increases. Since the cost of each extra hour of access time is \$1.50, the mean hourly cost can approach \$1.50 but never drop below it. Thus, the graph of the function approaches the line $y = 1.5$ as n increases without bound.

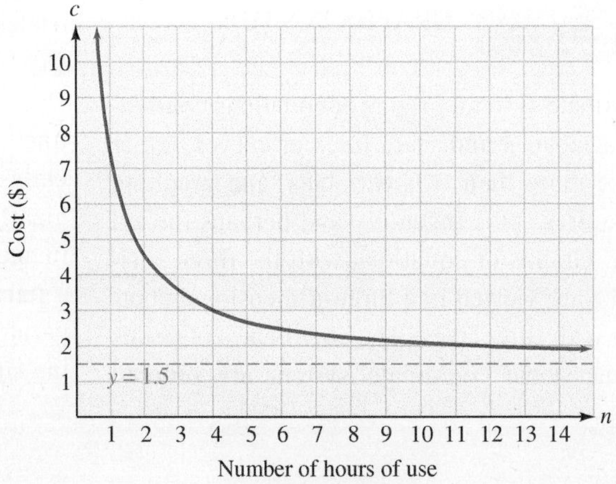

Figure 10-30

When a graph approaches a line as the dependent variable gets large, we call the line an **asymptote**. The line $y = 1.5$ is a **horizontal asymptote** of the graph.

Unless otherwise noted, all content on this page is © Cengage Learning.

As n gets smaller and approaches 0, the graph approaches the y-axis but never touches it. The y-axis is a **vertical asymptote** of the graph.

EXAMPLE 5 Find the mean hourly cost when the network described above is used for **a.** 3 hours **b.** 70.4 hours.

Solution **a.** To find the mean hourly cost for 3 hours of access time, we substitute 3 for n in Equation 1 and simplify:

$$\bar{c} = f(\mathbf{3}) = \frac{1.50(\mathbf{3}) + 6}{\mathbf{3}} = 3.5$$

The mean hourly cost for 3 hours of access time is \$3.50.

b. To find the mean hourly cost for 70.4 hours of access time, we substitute 70.4 for n in Equation 1 and simplify:

$$\bar{c} = f(\mathbf{70.4}) = \frac{1.50(\mathbf{70.4}) + 6}{\mathbf{70.4}} = 1.585227273$$

The mean hourly cost for 70.4 hours of access time is approximately \$1.59.

SELF CHECK 5 Find the mean hourly cost when the network is used for 5 hours.

4 Find the domain and range of a rational function.

Since division by 0 is undefined, any values that make the denominator 0 in a rational function must be excluded from the domain of the function.

EXAMPLE 6 Find the domain and range of $f(x) = \dfrac{3x + 2}{x^2 + x - 6}$.

Solution From the set of real numbers, we must exclude any values of x that make the denominator 0. To find these values, we set the denominator $x^2 + x - 6$ equal to 0 and solve for x.

$$x^2 + x - 6 = 0$$

$$(x + 3)(x - 2) = 0 \quad \text{Factor.}$$

$$x + 3 = 0 \quad \text{or} \quad x - 2 = 0 \quad \text{Set each factor equal to 0.}$$

$$x = -3 \quad \Big| \quad x = 2 \quad \text{Solve each linear equation.}$$

Thus, the domain of the function is the set of all real numbers except -3 and 2. In interval notation, the domain is $(-\infty, -3) \cup (-3, 2) \cup (2, \infty)$.

To find the range, we graph the function. We can find the domain and range of the function in Example 6 by looking at its graph. If we use window settings of $[-10, 10]$ for x and $[-10, 10]$ for y and graph the function

$$f(x) = \frac{3x + 2}{x^2 + x - 6}$$

we will obtain the graph in Figure 10-31.

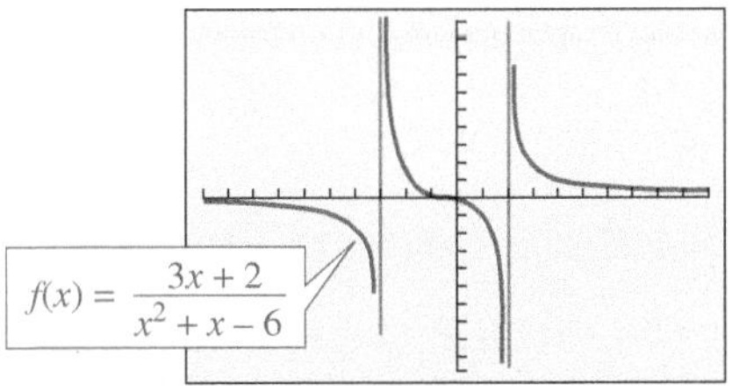

Figure 10-31

Unless otherwise noted, all content on this page is © Cengage Learning.

From the figure, we can see that

- As x approaches -3 from the left, the values of y decrease, and the graph approaches the vertical line $x = -3$.
- As x approaches -3 from the right, the values of y increase, and the graph approaches the vertical line $x = -3$.

From the figure, we also can see that

- As x approaches 2 from the left, the values of y decrease, and the graph approaches the vertical line $x = 2$.
- As x approaches 2 from the right, the values of y increase, and the graph approaches the vertical line $x = 2$.

The lines $x = -3$ and $x = 2$ are vertical asymptotes. Although the vertical lines in the graph appear to be the graphs of $x = -3$ and $x = 2$, they are not. Graphing calculators draw graphs by connecting dots whose x-coordinates are close together. Often, when two such points straddle a vertical asymptote and their y-coordinates are far apart, the calculator draws a line between them anyway, producing what appears to be a vertical asymptote. If you set your calculator to dot mode instead of connected mode, the vertical lines will not appear.

From Figure 10-31, we also can see that

- As x increases to the right of 2, the values of y decrease and approach the value $y = 0$.
- As x decreases to the left of -3, the values of y increase and approach the value $y = 0$.

The line $y = 0$ (the x-axis) is a horizontal asymptote. Graphing calculators do not draw lines that appear to be horizontal asymptotes. From the graph, we can see that all real numbers x, except -3 and 2, give a value of y. This confirms that the domain of the function is $(-\infty, -3) \cup (-3, 2) \cup (2, \infty)$. We also can see that y can be any value. Thus, the range is $(-\infty, \infty)$.

SELF CHECK 6 Find the domain and range of $f(x) = \dfrac{x+1}{x-2}$.

SELF CHECK ANSWERS

1.

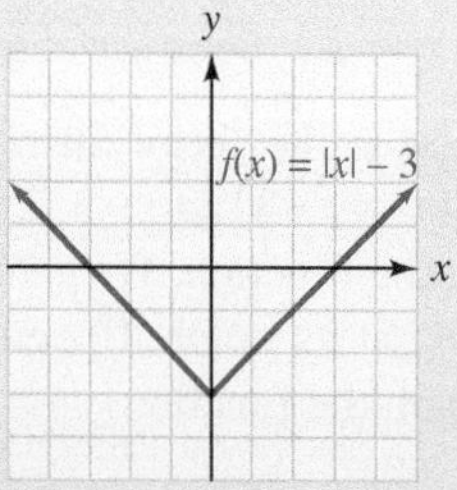

2.

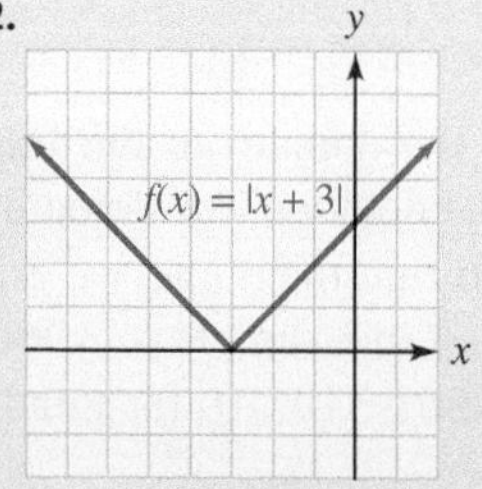

3.

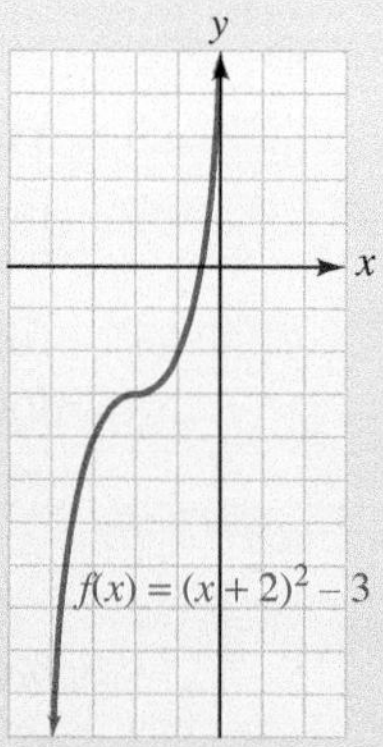

Unless otherwise noted, all content on this page is © Cengage Learning.

4. a.

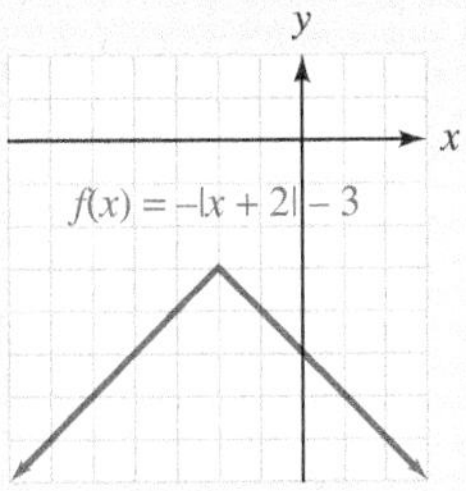

b.

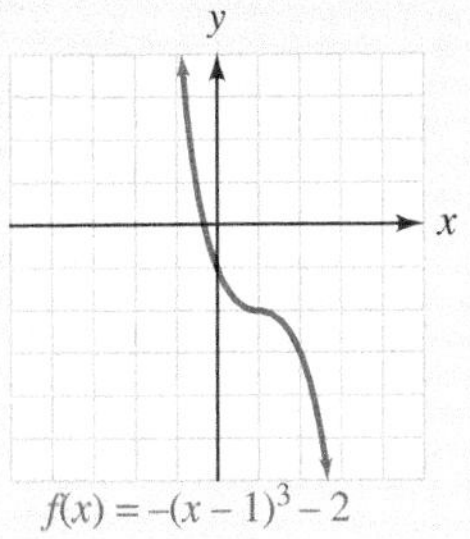

5. \$2.70 **6.** D: $(-\infty, 2) \cup (2, \infty)$; R: $(-\infty, 1) \cup (1, \infty)$

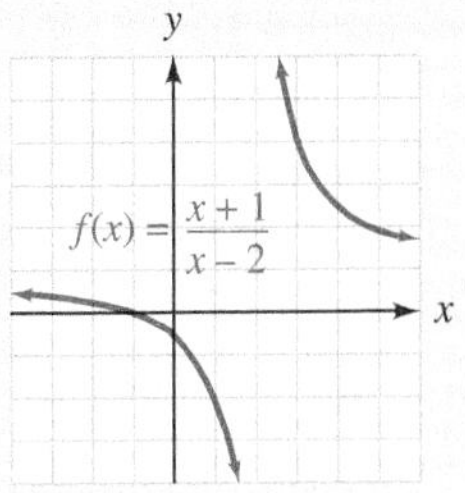

NOW TRY THIS

1. Given the graph of $f(x)$ below, sketch a graph of each translation or reflection.

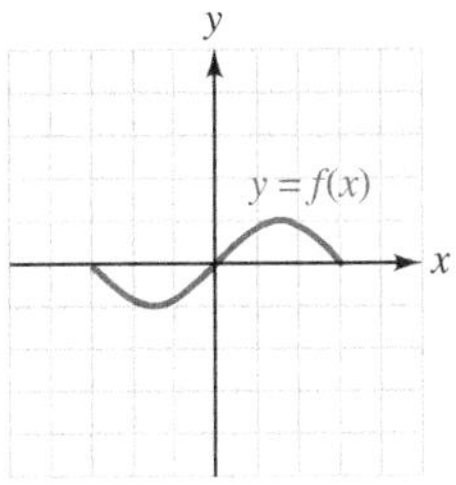

a. $f(x) + 2$ **b.** $-f(x)$ **c.** $f(x - 1)$

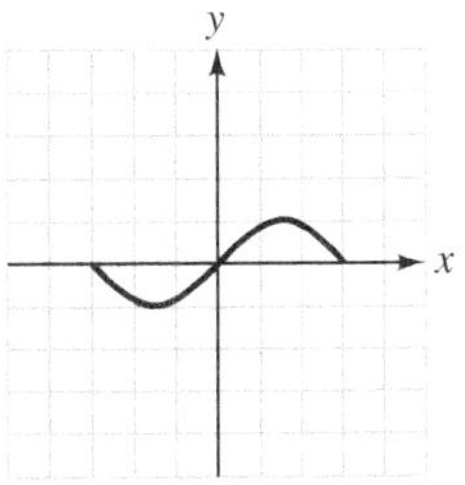

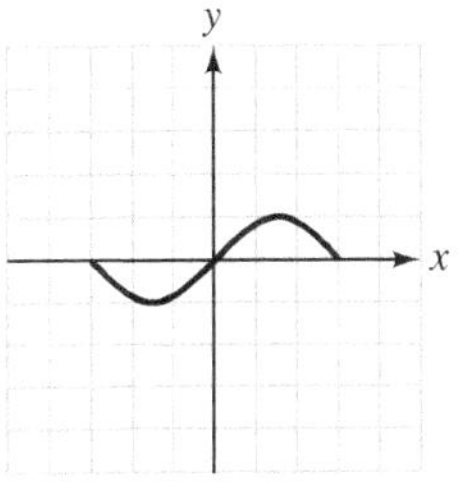

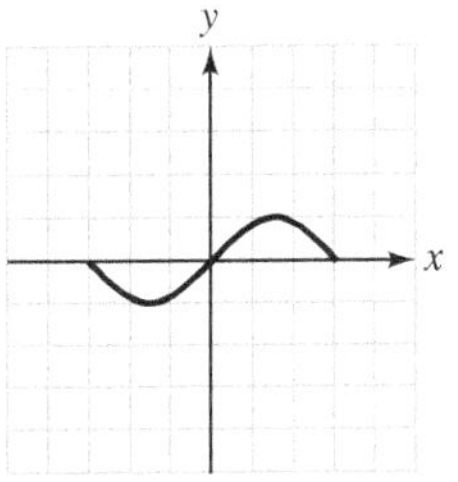

10.5 Exercises

WARM-UPS *Describe the graph as each of these relate to* $f(x) = x^2$.

1. $f(x) = x^2 + 3$

2. $f(x) = x^2 - 4$

3. $f(x) = (x + 3)^2$

4. $f(x) = (x - 4)^2$

REVIEW

5. List the prime numbers between 30 and 40.

6. State the associative property of addition.

7. State the commutative property of multiplication.

8. What is the additive identity element?

9. What is the multiplicative identity element?

10. Find the multiplicative inverse of $\frac{5}{3}$.

Unless otherwise noted, all content on this page is © Cengage Learning.

VOCABULARY AND CONCEPTS *Fill in the blanks.*

11. The function $f(x) = x^3$ is called the _____ function.

12. The function $f(x) = |x|$ is called the ____________ function.

13. Shifting the graph of an equation upward or downward is called a ______ translation.

14. Shifting the graph of an equation to the left or to the right is called a _________ translation.

15. The graph of $f(x) = x^3 + 5$ is the same as the graph of $f(x) = x^3$, except that it is shifted _ units ______.

16. The graph of $f(x) = x^3 - 2$ is the same as the graph of $f(x) = x^3$, except that it is shifted _ units _________.

17. The graph of $f(x) = |x - 5|^3$ is the same as the graph of $f(x) = |x^3|$, except that it is shifted _ units _________.

18. The graph of $f(x) = |x + 4|^3$ is the same as the graph of $f(x) = |x^3|$, except that it is shifted _ units ________.

19. The graph of $y = -f(x)$ is identical to the graph of $y = f(x)$, except that it is reflected about the _____.

20. If a function is the quotient of two polynomials, it is called a ______ function.

21. If a graph approaches a vertical line but never touches it, the line is called a vertical _________.

22. If a graph approaches a horizontal line, the line is called a horizontal _________.

GUIDED PRACTICE *Sketch each graph using a translation of the graph of $f(x) = x^3$ or $f(x) = |x|$.* SEE EXAMPLE 1. (OBJECTIVE 1)

23. $f(x) = x^3 - 3$

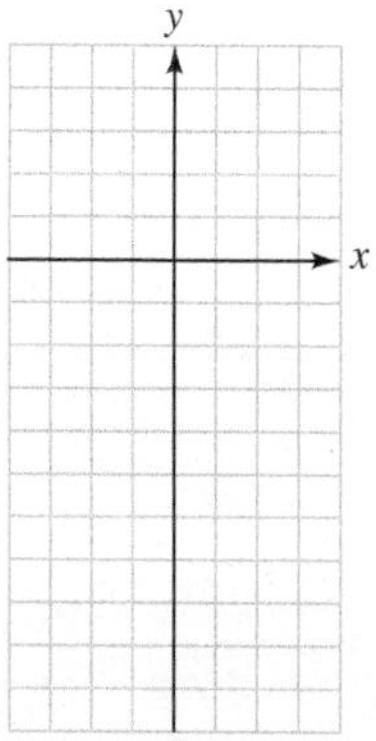

24. $f(x) = x^3 + 2$

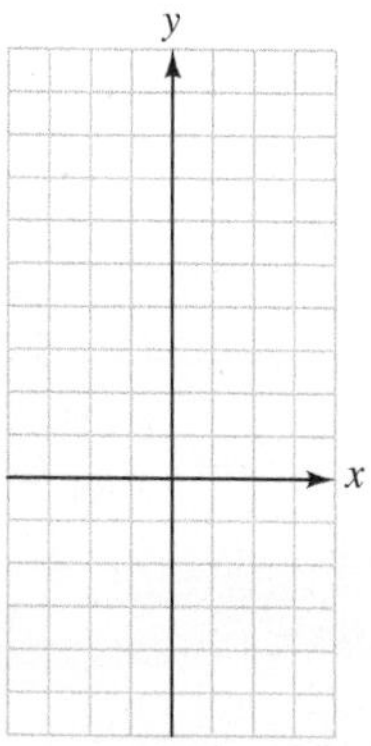

25. $f(x) = |x| - 2$

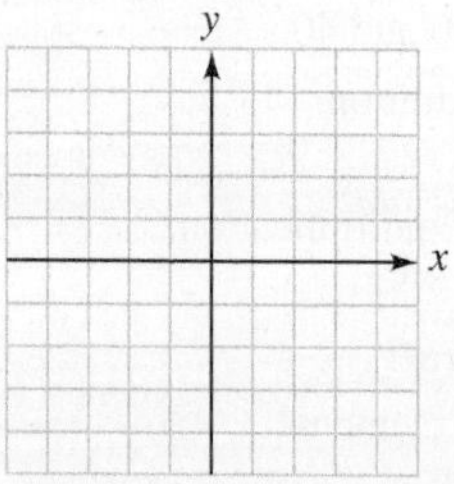

26. $f(x) = |x| + 1$

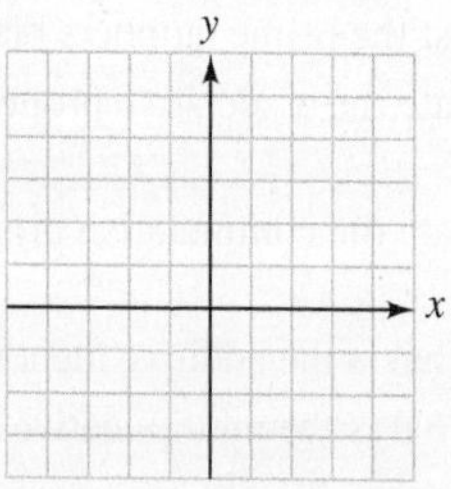

Sketch each graph using a translation of the graph of $f(x) = x^3$ or $f(x) = |x|$. SEE EXAMPLE 2. (OBJECTIVE 1)

27. $f(x) = (x - 1)^3$

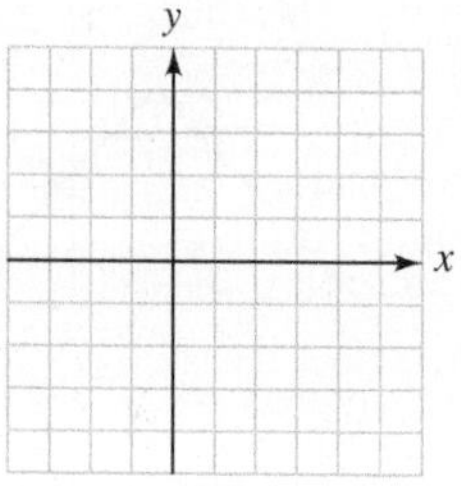

28. $f(x) = (x + 1)^3$

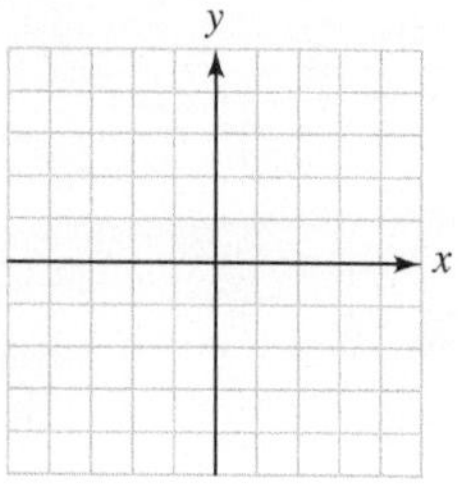

29. $f(x) = |x - 1|$

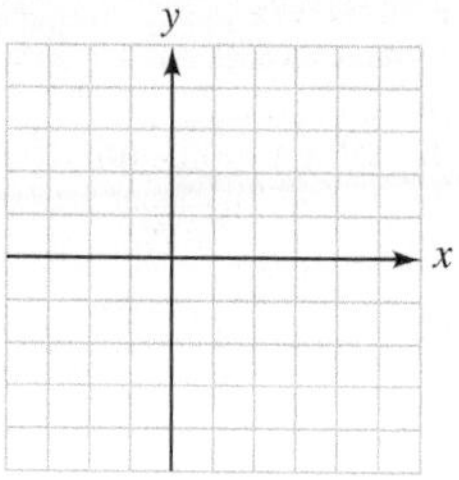

30. $f(x) = |x + 2|$

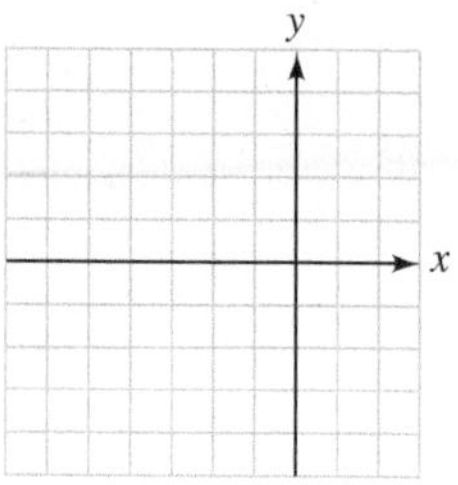

Sketch each graph using a translation of the graph of $f(x) = x^3$ or $f(x) = |x|$. SEE EXAMPLE 3. (OBJECTIVE 1)

31. $f(x) = |x - 2| - 1$

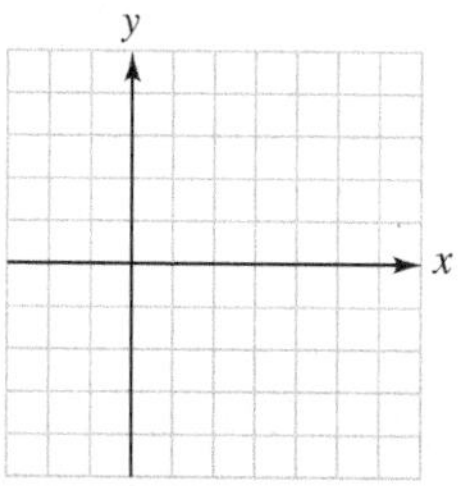

32. $f(x) = |x + 4| + 3$

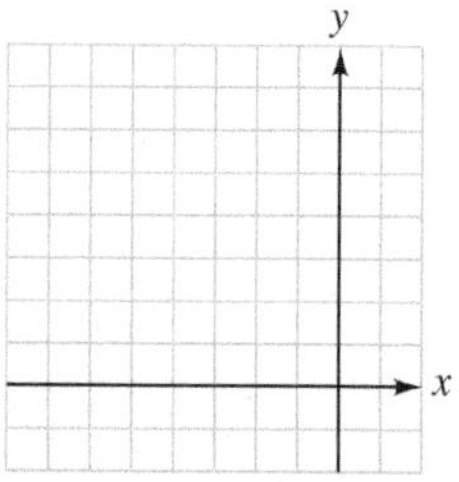

33. $f(x) = (x + 1)^3 - 2$

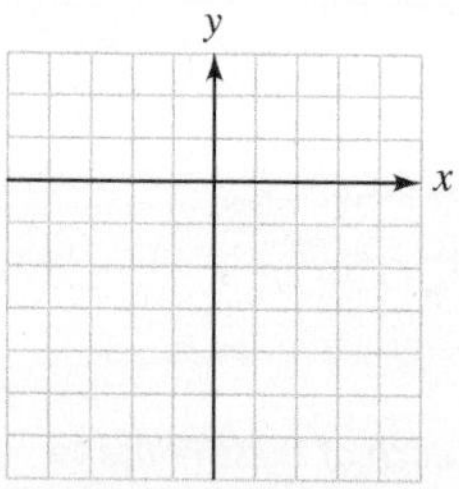

34. $f(x) = (x - 4)^3 + 1$

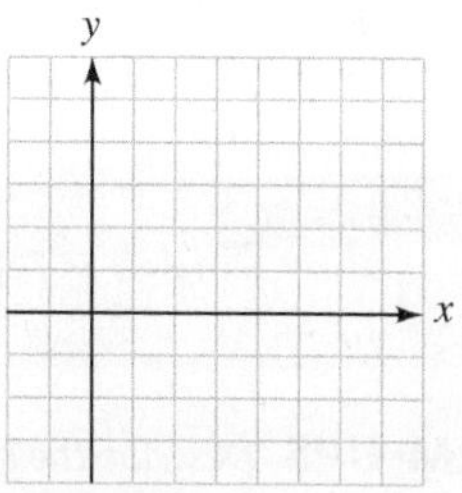

Graph each function. SEE EXAMPLE 4. (OBJECTIVE 2)

35. $f(x) = -x^2$

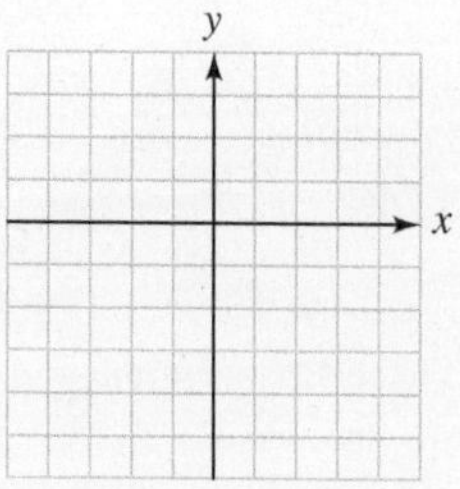

36. $f(x) = -(x - 2)^2 - 3$

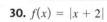

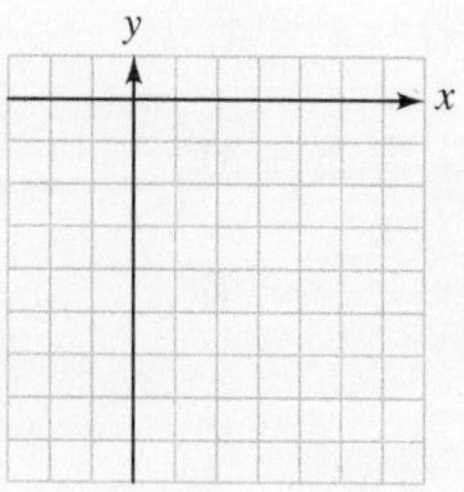

Unless otherwise noted, all content on this page is © Cengage Learning.

37. $f(x) = -|x + 1| - 2$

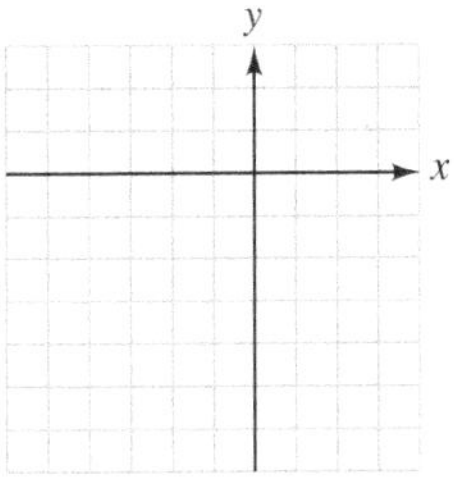

38. $f(x) = -|x - 2| - 3$

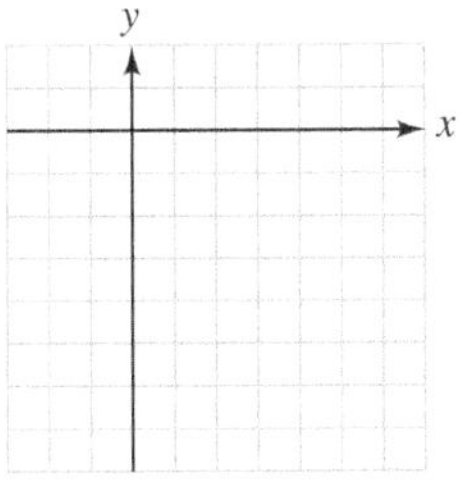

39. $f(x) = -x^3 + 2$

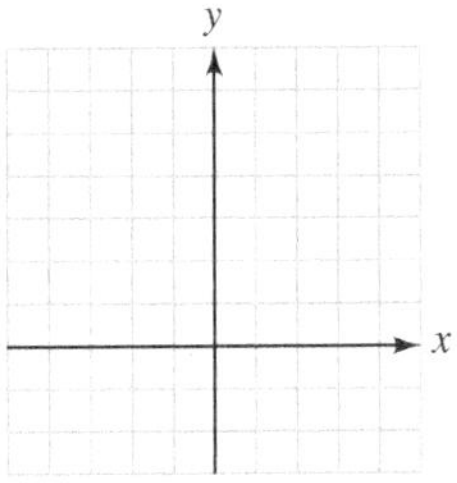

40. $f(x) = -x^3 - 3$

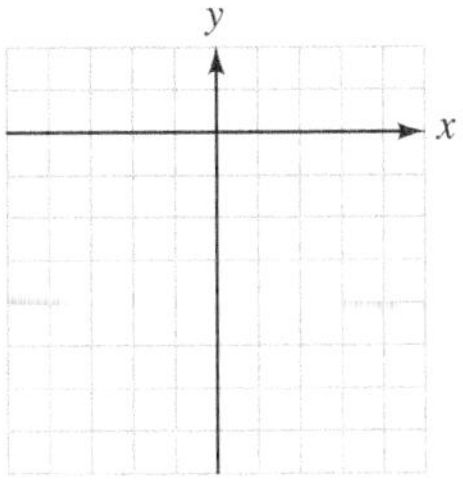

41. $f(x) = -(x + 1)^3 - 4$

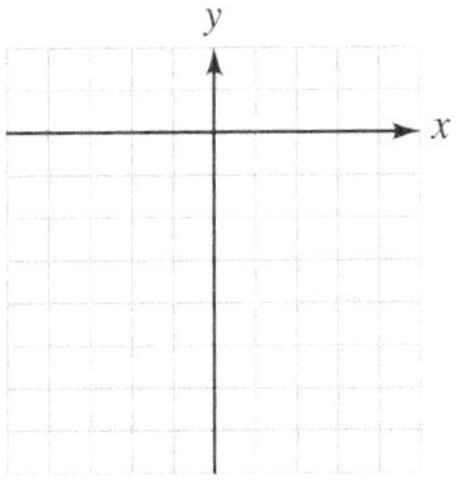

42. $f(x) = -(x - 2)^3 - 1$

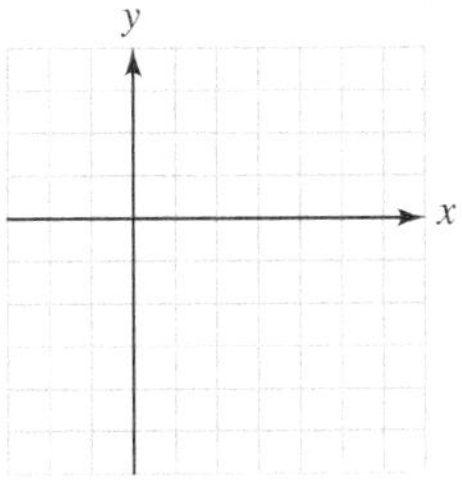

The time t it takes to travel 600 miles is a function of the mean rate of speed r: $t = f(r) = \frac{600}{r}$. *Find t for each value of r.* SEE EXAMPLE 5. (OBJECTIVE 3)

43. 30 mph
44. 40 mph
45. 50 mph
46. 60 mph

Find the domain of each rational function and write it in interval notation. Use a graphing calculator to graph each rational function to verify the domain and find the range. SEE EXAMPLE 6. (OBJECTIVE 4)

47. $f(x) = \dfrac{x}{x - 2}$

48. $f(x) = \dfrac{x + 2}{x}$

49. $f(x) = \dfrac{x + 1}{x^2 - 4}$

50. $f(x) = \dfrac{x - 2}{x^2 - 3x - 4}$

ADDITIONAL PRACTICE *Graph each function.*

51. $f(x) = |x| - 5$

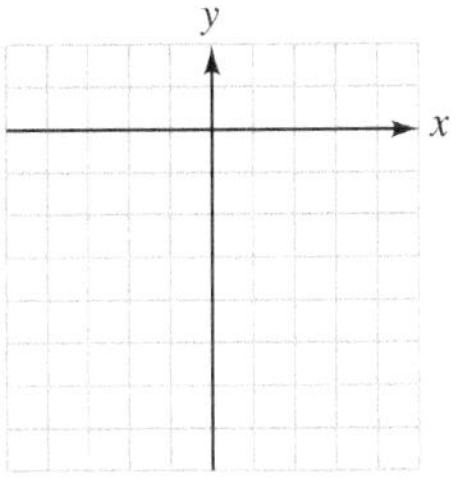

52. $f(x) = x^3 + 4$

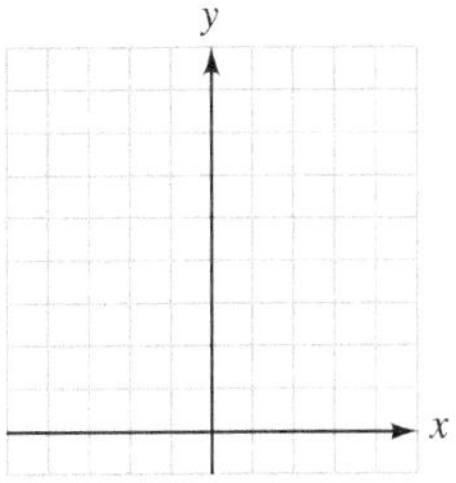

53. $f(x) = (x - 1)^3$

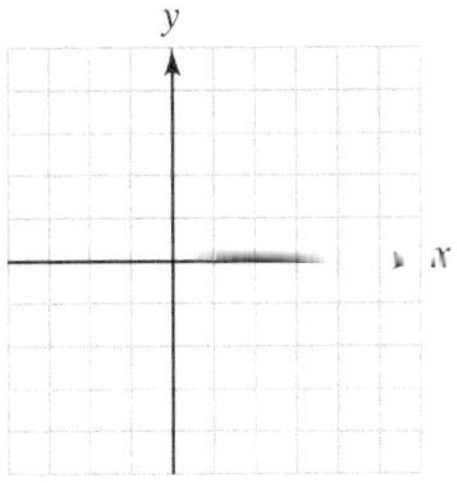

54. $f(x) = |x + 4|$

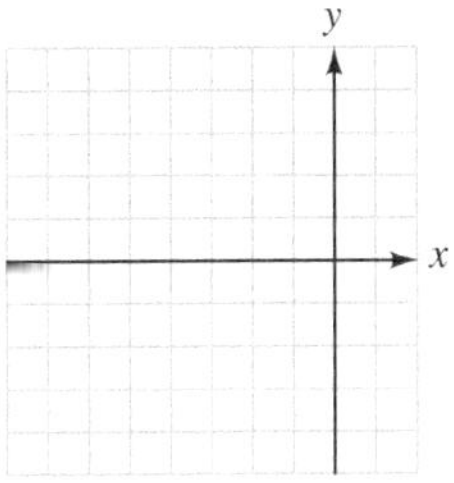

Use a graphing calculator to graph each function, using values of $[-4, 4]$ *for x and* $[-4, 4]$ *for y. The graph is not what it appears to be. Select a better viewing window to find the true graph.*

55. $f(x) = x^2 + 8$
56. $f(x) = x^3 - 8$

57. $f(x) = |x + 5|$
58. $f(x) = |x - 5|$

59. $f(x) = (x - 6)^2$
60. $f(x) = (x + 9)^2$

61. $f(x) = x^3 + 8$
62. $f(x) = x^3 - 12$

Suppose the cost (in dollars) of removing p% of the pollution in a river is given by the function $c = f(p) = \frac{50{,}000p}{100 - p}$ $(0 \le p < 100)$. *Find the cost of removing each percent of pollution.*

63. 20%
64. 30%
65. 50%
66. 80%

Unless otherwise noted, all content on this page is © Cengage Learning.

APPLICATIONS *A service club wants to publish a directory of its members. Some investigation shows that the cost of typesetting and photography will be \$700, and the cost of printing each directory will be \$1.25.*

67. Find a function that gives the total cost C of printing x directories.

68. Find a function that gives the mean cost per directory $\bar{c}$ of printing x directories.

69. Find the total cost of printing 250 directories.

70. Find the mean cost per directory if 250 directories are printed.

71. Find the mean cost per directory if 300 directories are printed.

72. Find the mean cost per directory if 500 directories are printed.

An electric company charges \$15 per month plus 11.4¢ for each kilowatt hour (kwh) of electricity used.

73. Find a function that gives the total cost C of n kwh of electricity.

74. Find a function that gives the mean cost per kwh $\bar{c}$ when using n kwh.

75. Find the total cost for using 1,850 kwh.

76. Find the mean cost per kwh when 1,850 kwh are used.

77. Find the mean cost per kwh when 2,500 kwh are used.

78. Find the mean cost per kwh when 1,000 kwh are used.

Assume that a person buys a horse for \$5,000 and plans to pay \$350 per month to board the horse.

79. Find a function that will give the total cost of owning the horse for x months.

80. Find a function that will give the mean cost per month $\bar{c}$ after owning the horse for x months.

81. Find the total cost of owning the horse for 10 years.

82. Find the mean cost per month if the horse is owned for 10 years.

WRITING ABOUT MATH

83. Explain how to graph an equation by plotting points.

84. Explain how the graphs of $y = (x - 4)^2 - 3$ and $y = x^2$ are related.

SOMETHING TO THINK ABOUT

85. Can a rational function have two horizontal asymptotes? Explain.

86. Use a graphing calculator to investigate the positioning of the vertical asymptotes of a rational function by graphing $y = \frac{x}{x - k}$ for several values of k. What do you observe?

Section 10.6 Solving Quadratic and Other Nonlinear Inequalities

Objectives

1. Solve a quadratic inequality.
2. Solve a rational inequality.
3. Solve a nonlinear inequality by graphing.

Vocabulary

quadratic inequality | critical values | critical points

Getting Ready

Factor each trinomial.

1. $x^2 + 2x - 15$

2. $x^2 - 3x + 2$

We have previously solved linear inequalities. In this section, we will discuss how to solve quadratic and rational inequalities.

1 Solve a quadratic inequality.

Quadratic inequalities in one variable, say x, are inequalities that can be written in one of the following forms, where $a \neq 0$:

$$ax^2 + bx + c < 0 \qquad ax^2 + bx + c > 0$$
$$ax^2 + bx + c \leq 0 \qquad ax^2 + bx + c \geq 0$$

To solve one of these inequalities, we must find its solution set. There are two algebraic methods for finding these methods.

For example, to solve

$$x^2 + x - 6 < 0$$

we must find the values of x that make the inequality true.

Method 1 To find these values, we can factor the trinomial to obtain

$$(x + 3)(x - 2) < 0$$

Since the product of $x + 3$ and $x - 2$ is to be less than 0, the values of the factors must be opposite in sign. This will happen when one of the factors is positive and the other is negative.

To keep track of the sign of $x + 3$, we can construct the following graph.

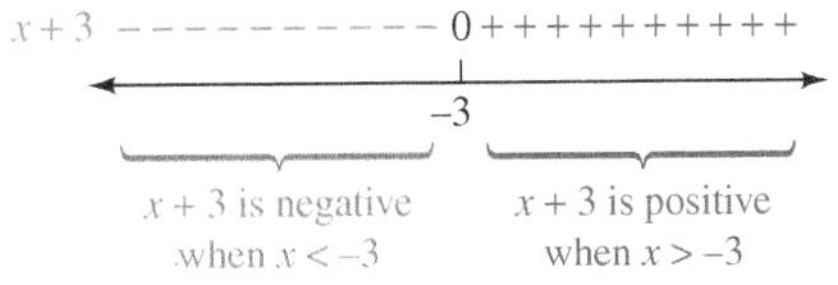

To keep track of the sign of $x - 2$, we can construct the following graph.

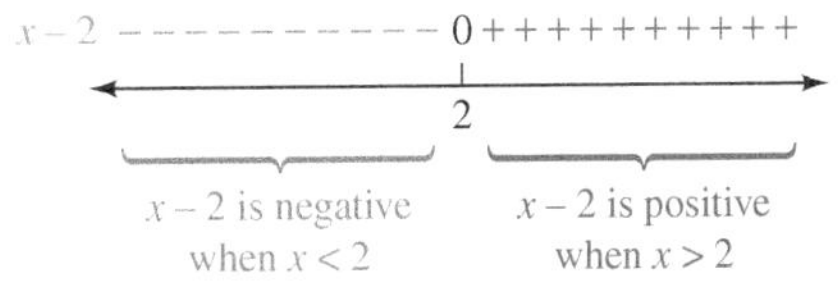

We can merge these graphs as shown in Figure 10-32 and note where the signs of the factors are opposite. This occurs in the interval $(-3, 2)$. Therefore, the product $(x + 3)(x - 2)$ will be less than 0 when

$$-3 < x < 2$$

The graph of the solution set is shown on the number line in Figure 10-32.

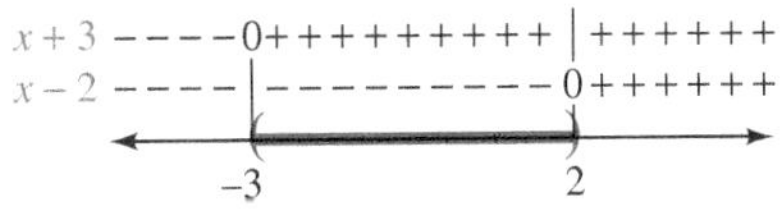

Figure 10-32

Method 2 Another way to solve the inequality $x^2 + x - 6 < 0$ is to first solve its related quadratic equation $x^2 + x - 6 = 0$. The solutions to this equation are sometimes called **critical values** and they establish points on the number line, called **critical points**.

Unless otherwise noted, all content on this page is © Cengage Learning.

$$x^2 + x - 6 = 0$$
$$(x + 3)(x - 2) = 0$$
$$x + 3 = 0 \quad \text{or} \quad x - 2 = 0$$
$$x = -3 \quad | \quad x = 2$$

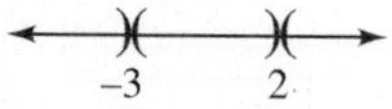

The graphs of these critical values establish the three intervals shown on the number line. To determine which intervals are solutions, we create a table of intervals, test a number in each, and determine whether it satisfies the inequality, as shown in Table 10-1.

TABLE 10-1

Interval	Test value	Inequality $x^2 + x - 6 < 0$	Result
$(-\infty, -3)$	-6	$(-6)^2 - (-6) - 6 \overset{?}{<} 0$ $36 < 0$ false	The numbers in this interval are not solutions.
$(-3, 2)$	0	$(0)^2 - (0) - 6 \overset{?}{<} 0$ $-6 < 0$ true	The numbers in this interval are solutions.
$(2, \infty)$	5	$(5)^2 - (5) - 6 \overset{?}{<} 0$ $14 < 0$ false	The numbers in this interval are not solutions.

Figure 10-33

The solution set is the same result as using Method 1, the interval $(-3, 2)$, as shown in Figure 10-33.

EXAMPLE 1 Solve: $x^2 + 2x - 3 \geq 0$

Solution **Method 1** We can factor the trinomial to get $(x - 1)(x + 3)$ and construct the sign chart shown in Figure 10-34.

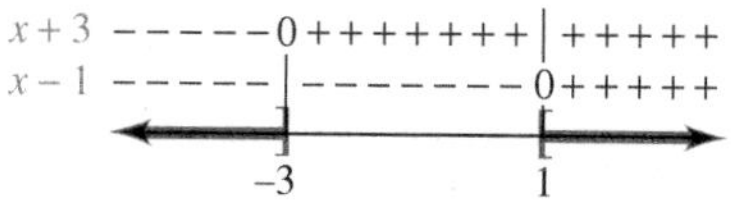

Figure 10-34

- $x - 1$ is 0 when $x = 1$, is positive when $x > 1$, and is negative when $x < 1$.
- $x + 3$ is 0 when $x = -3$, is positive when $x > -3$, and is negative when $x < -3$.

The product of $x - 1$ and $x + 3$ will be greater than 0, a positive value, when the signs of the binomial factors are the same. This occurs in the intervals $(-\infty, -3)$ and $(1, \infty)$. The numbers -3 and 1 are also included, because they make the product equal to 0. Thus, the solution set is the union of two intervals

$$(-\infty, -3] \cup [1, \infty)$$

The graph of the solution set is shown on the number line in Figure 10-34.

Method 2 We can obtain the same result by solving the related quadratic equation $x^2 + 2x - 3 = 0$ and establishing critical points on the number line.

$$x^2 + 2x - 3 = 0$$
$$(x + 3)(x - 1) = 0$$
$$x + 3 = 0 \quad \text{or} \quad x - 1 = 0$$
$$x = -3 \quad | \quad x = 1$$

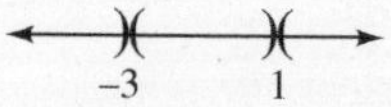

The graphs of these critical values establish the three intervals shown on the number line. To determine which intervals are solutions, we test an interval and see whether it satisfies the inequality, as shown in Table 10-2.

Unless otherwise noted, all content on this page is © Cengage Learning.

TABLE 10-2

Interval	Test value	Inequality $x^2 + 2x - 3 \geq 0$	Result
$(-\infty, -3)$	-5	$(-5)^2 + 2(-5) - 3 \overset{?}{\geq} 0$ $12 \geq 0$ true	The numbers in this interval are solutions.
$(-3, 1)$	0	$(0)^2 + 2(0) - 3 \overset{?}{\geq} 0$ $-3 \geq 0$ false	The numbers in this interval are not solutions.
$(1, \infty)$	4	$(4)^2 + 2(4) - 3 \overset{?}{\geq} 0$ $21 \geq 0$ true	The numbers in this interval are solutions.

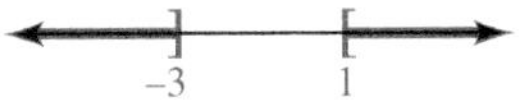

Figure 10-35

From the table, we see that numbers in the intervals $(-\infty, -3)$ and $(1, \infty)$ satisfy the inequality. Because this quadratic inequality includes equality, the critical values of $x = -3$ and $x = 1$ also satisfy the inequality. Thus, the solution set is the union of two intervals: $(-\infty, -3] \cup [1, \infty)$, as shown in Figure 10-35.

SELF CHECK 1 Solve $x^2 + 2x - 15 > 0$ and graph its solution set.

2 Solve a rational inequality.

Making a sign chart is useful for solving many inequalities that are neither linear nor quadratic.

EXAMPLE 2 Solve: $\frac{1}{x} < 6$

Solution We subtract 6 from both sides to make the right side equal to 0, find a common denominator, and add the fractions.

$$\frac{1}{x} < 6$$

$$\frac{1}{x} - 6 < 0 \quad \text{Subtract 6 from both sides.}$$

$$\frac{1}{x} - \frac{6x}{x} < 0 \quad \text{Write each fraction using the common denominator.}$$

$$\frac{1 - 6x}{x} < 0 \quad \text{Write as a single fraction.}$$

To use Method 1, we make a sign chart, as in Figure 10-36.

- The denominator x is 0 when $x = 0$, is positive when $x > 0$, and is negative when $x < 0$.
- The numerator $1 - 6x$ is 0 when $x = \frac{1}{6}$, is positive when $x < \frac{1}{6}$, and is negative when $x > \frac{1}{6}$.

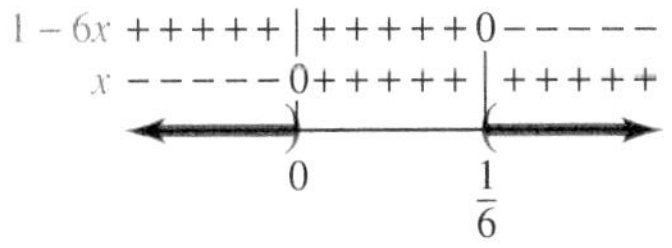

Figure 10-36

COMMENT Since we do not know whether x is positive, 0, or negative, multiplying both sides of the inequality $\frac{1}{x} < 6$ by x is a three-case situation:

- If $x > 0$, then $1 < 6x$.
- If $x = 0$, then the fraction $\frac{1}{x}$ is undefined.
- If $x < 0$, then $1 > 6x$.

If you multiply both sides by x and solve $1 < 6x$, you are only considering one case and will get only part of the answer.

Unless otherwise noted, all content on this page is © Cengage Learning.

The fraction $\frac{1 - 6x}{x}$ will be less than 0 when the numerator and denominator are opposite in sign. This occurs in the union of two intervals:

$$(-\infty, 0) \cup \left(\frac{1}{6}, \infty\right)$$

The graph of this union is shown in Figure 10-36.

To solve this inequality by Method 2, we can find the critical values by finding the values of x that make the numerator of $\frac{1 - 6x}{x}$ equal to 0 and the values of x that make the denominator equal to 0 and use test values in each interval to determine the solution. The critical values are the numbers $\frac{1}{6}$ and 0 and the solution is $(-\infty, 0) \cup \left(\frac{1}{6}, \infty\right)$.

SELF CHECK 2 Solve: $\frac{3}{x} > 5$

EXAMPLE 3 Solve: $\frac{x^2 - 3x + 2}{x - 3} \geq 0$

Solution We write the fraction with the numerator in factored form.

$$\frac{(x - 2)(x - 1)}{x - 3} \geq 0$$

To keep track of the signs of the binomials, we construct the sign chart shown in Figure 10-37. The fraction will be positive in the intervals where all factors are positive, or where two factors are negative. The numbers 1 and 2 are included, because they make the numerator (and thus the fraction) equal to 0. The number 3 is not included, because it creates 0 in the denominator.

The solution is the union of two intervals $[1, 2] \cup (3, \infty)$. The graph appears in Figure 10-37.

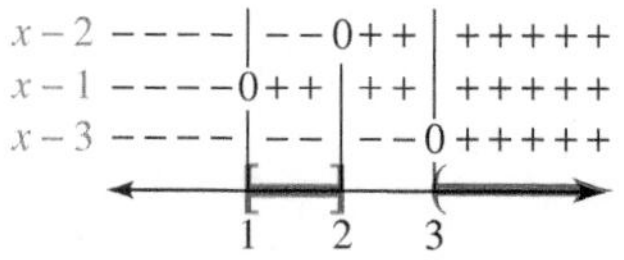

Figure 10-37

To solve this inequality by Method 2, we can find the critical values by finding the values of x that make the numerator of $\frac{(x - 2)(x - 1)}{x - 3}$ equal to 0 and the values of x that make the denominator equal to 0. The critical values are the numbers 1, 2, and 3 and the solution is $[1, 2] \cup (3, \infty)$.

SELF CHECK 3 Solve: $\frac{x + 2}{x^2 - 2x - 3} > 0$

EXAMPLE 4 Solve: $\frac{3}{x - 1} < \frac{2}{x}$

Solution We subtract $\frac{2}{x}$ from both sides to get 0 on the right side and proceed as follows:

$$\frac{3}{x - 1} < \frac{2}{x}$$

$$\frac{3}{x - 1} - \frac{2}{x} < 0 \quad \text{Subtract } \frac{2}{x} \text{ from both sides.}$$

Unless otherwise noted, all content on this page is © Cengage Learning.

$$\frac{3x}{(x-1)x} - \frac{2(x-1)}{x(x-1)} < 0$$ Write each fraction with the common denominator, $x(x-1)$.

$$\frac{3x - 2x + 2}{x(x-1)} < 0$$ Keep the denominator and subtract the numerators.

$$\frac{x+2}{x(x-1)} < 0$$ Combine like terms.

We can keep track of the signs of the three factors with the sign chart shown in Figure 10-38. The fraction will be negative in the intervals with either one or three negative factors. The numbers 0 and 1 are not included, because they give a 0 in the denominator, and the number -2 is not included, because it does not satisfy the inequality.

The solution is the union of two intervals $(-\infty, -2) \cup (0, 1)$, as shown in Figure 10-38.

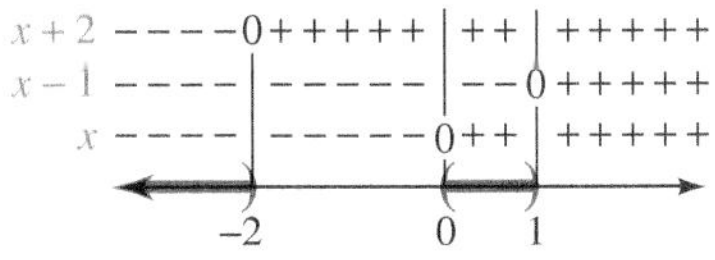

Figure 10-38

To solve this inequality by Method 2, we can find the critical values by finding the values of x that make the numerator of $\frac{x+2}{x(x-1)}$ equal to 0 and the values of x that make the denominator equal to 0 and use test values in each interval to determine the solution. The critical values are the numbers $-2, 0,$ and 1 and the solution is $(-\infty, -2) \cup (0, 1)$.

SELF CHECK 4 Solve: $\frac{2}{x+1} > \frac{1}{x}$

Accent on technology

▸ Solving Inequalities

To approximate the solutions of $x^2 + 2x - 3 \geq 0$ (Example 1) by graphing, we can use window settings of $[-10, 10]$ for x and $[-10, 10]$ for y and graph the quadratic function $y = x^2 + 2x - 3$, as in Figure 10-39. The solutions of the inequality will be those numbers x for which the graph of $y = x^2 + 2x - 3$ lies above or on the x-axis. We can trace to find that this interval is $(-\infty, -3] \cup [1, \infty)$.

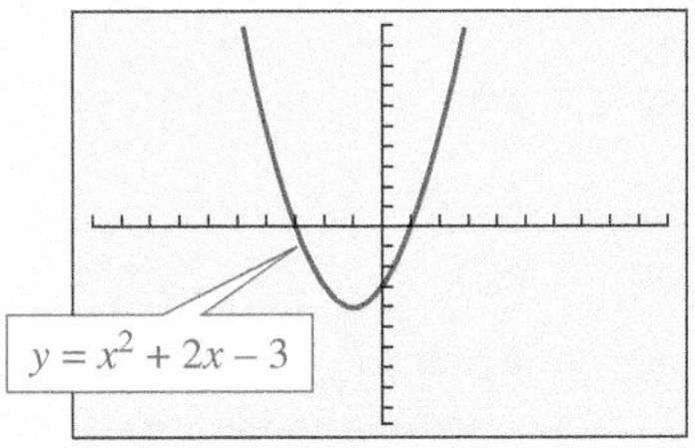

Figure 10-39

To approximate the solutions of $\frac{3}{x-1} < \frac{2}{x}$ (Example 4), we first write the inequality in the form

$$\frac{3}{x-1} - \frac{2}{x} < 0$$

Then we use window settings of $[-5, 5]$ for x and $[-3, 3]$ for y and graph the function $y = \frac{3}{x-1} - \frac{2}{x}$, as in Figure 10-40(a) on the next page. The solutions of the inequality will be those numbers x for which the graph lies below the x-axis.

(continued)

COMMENT Graphing calculators cannot determine whether a critical value is included in a solution set. You must make that determination yourself.

Unless otherwise noted, all content on this page is © Cengage Learning.

COMMENT We could continue tracing using our original graph. As you trace back to positive values of y, you will come to $x = 0$ $y =$. This means that $x = 0$ is a vertical asymptote. Keep tracing to your right and you will see that as x becomes larger the values of y are negative. They remain negative until you pass $x = 1$ (another vertical asymptote) at which time the y-values are positive. Therefore the solution is $(-\infty, -2) \cup (0, 1)$.

We can trace to see that the graph is below the x-axis when x is less than -2. Since we cannot see the graph in the interval $0 < x < 1$, we redraw the graph using window settings of $[-1, 2]$ for x and $[-25, 10]$ for y. See Figure 10-40(b).

We can now see that the graph is below the x-axis in the interval $(0, 1)$. Thus, the solution of the inequality is the union of two intervals:

$$(-\infty, -2) \cup (0, 1)$$

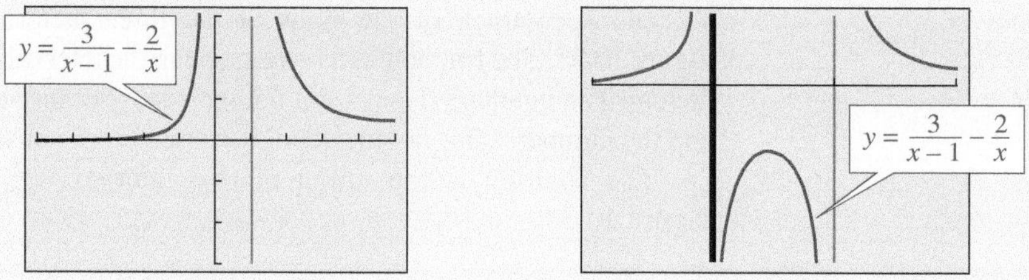

(a) (b)

Figure 10-40

For instructions regarding the use of a Casio graphing calculator, please refer to the Casio Keystroke Guide in the back of the book.

3 Solve a nonlinear inequality by graphing.

We now consider the graphs of nonlinear inequalities in two variables.

EXAMPLE 5 Solve: $y < -x^2 + 4$

Solution The graph of $y = -x^2 + 4$ is the parabolic boundary separating the region representing $y < -x^2 + 4$ and the region representing $y > -x^2 + 4$.

We graph $y = -x^2 + 4$ as a dashed parabola, because there is no equality symbol in the original inequality. Since the coordinates of the origin satisfy the inequality $y < -x^2 + 4$, the point $(0, 0)$ is in the graph. The complete graph is shown in Figure 10-41.

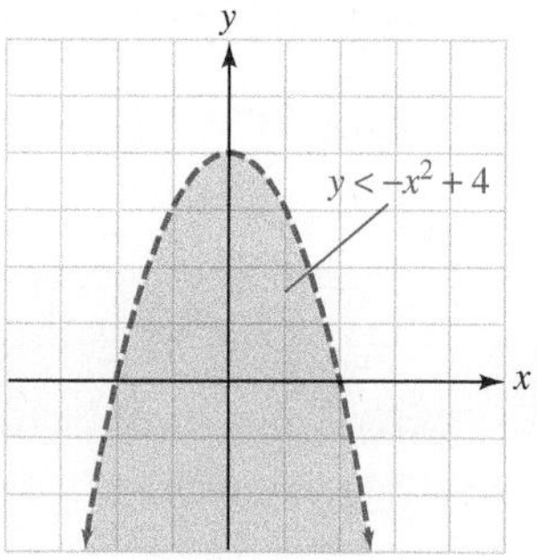

Figure 10-41

SELF CHECK 5 Solve: $y \geq -x^2 + 4$

EXAMPLE 6 Solve: $x \leq |y|$

Solution We first graph $x = |y|$ as in Figure 10-42(a), using a solid line because the symbol in the original inequality is $\leq$. Since the origin is on the graph, we cannot use it as a test point. However, another point, such as $(1, 0)$, will do. We substitute 1 for x and 0 for y into the inequality to obtain

$$x \leq |y|$$
$$1 \leq |0|$$
$$1 \leq 0$$

Since $1 \leq 0$ is a false statement, the point $(1, 0)$ does not satisfy the inequality and is not part of the graph. Thus, the graph of $x \leq |y|$ is to the left of the boundary.

Unless otherwise noted, all content on this page is © Cengage Learning.

The complete graph is shown in Figure 10-42(b).

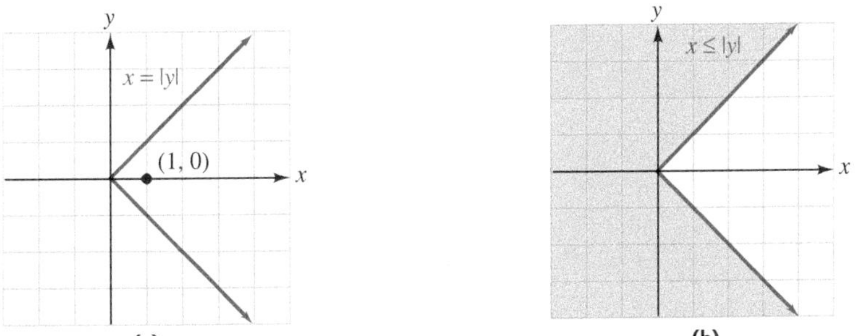

Figure 10-42

SELF CHECK 6 Solve: $x \geq -|y|$

SELF CHECK ANSWERS

1. $(-\infty, -5) \cup (3, \infty)$ 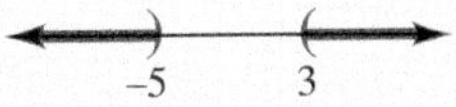

2. $\left(0, \frac{3}{5}\right)$ 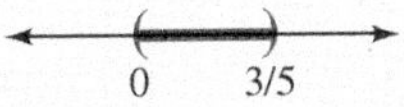

3. $(-2, -1) \cup (3, \infty)$ 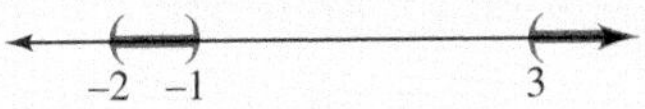

4. $(-1, 0) \cup (1, \infty)$ 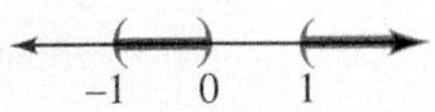

5. 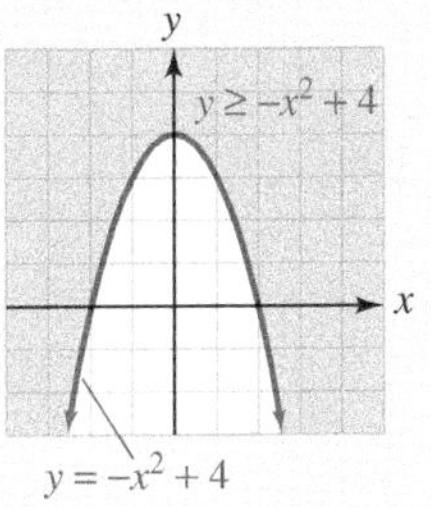

6. 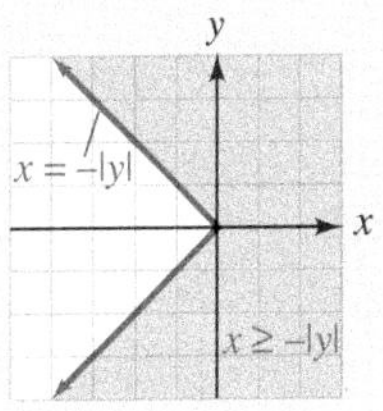

NOW TRY THIS

Find the domain of each of the following.

1. $f(x) = \sqrt{x - 6}$
2. $g(x) \geq x^2 - 5x - 6$
3. $h(x) = \sqrt{x^2 - 5x - 6}$
4. $k(x) = \dfrac{3x - 5}{\sqrt{x^2 - 5x - 6}}$

10.6 Exercises

WARM-UPS *Determine the value of x where $x - 2$ is*

1. 0.

2. positive.

3. negative.

Determine the value of x where $x + 3$ is

4. 0.

5. positive.

6. negative.

Unless otherwise noted, all content on this page is © Cengage Learning.

7. Multiply both sides of the equation $\frac{1}{x} < 2$ by x when x is positive and write the resulting inequality.

8. Multiply both sides of the equation $\frac{1}{x} < 2$ by x when x is negative and write the resulting inequality.

REVIEW *Write each statement as an equation.*

9. y varies directly with x.

10. y varies inversely with t.

11. t varies jointly with x and y.

12. d varies directly with t but inversely with u^2.

Find the slope of the graph of each equation.

13. $y = 3x - 4$

14. $\frac{2x - y}{5} = 8$

VOCABULARY AND CONCEPTS *Fill in the blanks.*

15. When $x > 3$, the binomial $x - 3$ is ______ than zero.

16. When $x < 3$, the binomial $x - 3$ is ___ than zero.

17. The inequality $x^2 + 9x + 18 \leq 0$ is an example of a ________ inequality.

18. The inequality $\frac{x+3}{x-2} > 0$ is an example of a ______ inequality.

19. If $x = 0$, the fraction $\frac{1}{x}$ is ________.

20. To solve $x^2 + 2x - 3 < 0$, we can find the solutions of the related equation $x^2 + 2x - 3 = 0$. The solutions are called ___________. They establish points on a number line that separate the line into ________.

21. To keep track of the signs of factors in a product or quotient, we can use a ____ chart.

22. The inequality $|x + 3| < 0$ will be graphed with a ______ line.

GUIDED PRACTICE *Solve each inequality. State each result in interval notation and graph the solution set.* SEE EXAMPLE 1. (OBJECTIVE 1)

23. $x^2 - 5x + 4 < 0$

24. $x^2 - 3x - 4 > 0$

25. $x^2 - 8x + 15 > 0$

26. $x^2 + 2x - 8 < 0$

27. $x^2 + x - 12 \leq 0$

28. $x^2 + 7x + 12 \geq 0$

29. $x^2 + 2x \geq 15$

30. $x^2 - 8x \leq -15$

31. $x^2 + 8x < -16$

32. $x^2 + 6x \geq -9$

33. $x^2 \geq 9$

34. $x^2 \geq 16$

Solve each inequality. State each result in interval notation and graph the solution set. SEE EXAMPLE 2. (OBJECTIVE 2)

35. $\frac{1}{x} < 2$

36. $\frac{1}{x} > 3$

37. $\frac{4}{x} \geq 2$

38. $-\frac{6}{x} < 12$

Solve each inequality. State each result in interval notation and graph the solution set. SEE EXAMPLE 3. (OBJECTIVE 2)

39. $\frac{x^2 - x - 12}{x - 1} < 0$

40. $\frac{x^2 + x - 6}{x - 4} \geq 0$

41. $\frac{x^2 + x - 20}{x + 2} \geq 0$

42. $\frac{x^2 - 10x + 25}{x + 5} < 0$

43. $\frac{x^2 - 4x + 4}{x + 4} < 0$

44. $\frac{2x^2 - 5x + 2}{x + 2} > 0$

45. $\frac{6x^2 - 5x + 1}{2x + 1} > 0$

46. $\frac{6x^2 + 11x + 3}{3x - 1} < 0$

Solve each inequality. State each result in interval notation and graph the solution set. SEE EXAMPLE 4. (OBJECTIVE 2)

47. $\frac{3}{x - 2} < \frac{4}{x}$

48. $\frac{-6}{x + 1} \geq \frac{1}{x}$

49. $\frac{-5}{x+2} \geq \frac{4}{2-x}$

50. $\frac{-6}{x-3} < \frac{5}{3-x}$

51. $\frac{7}{x-3} \geq \frac{2}{x+4}$

52. $\frac{-5}{x-4} < \frac{3}{x+1}$

53. $(x+2)^2 > 0$

54. $(x-3)^2 < 0$

Graph each inequality. SEE EXAMPLE 5. (OBJECTIVE 3)

55. $y < x^2 + 1$

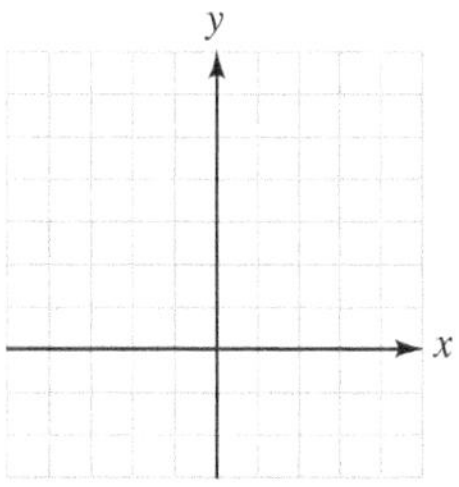

56. $y > x^2 - 3$

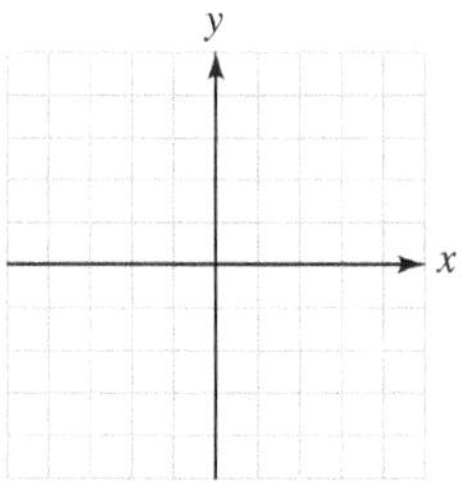

57. $y \leq x^2 + 5x + 6$

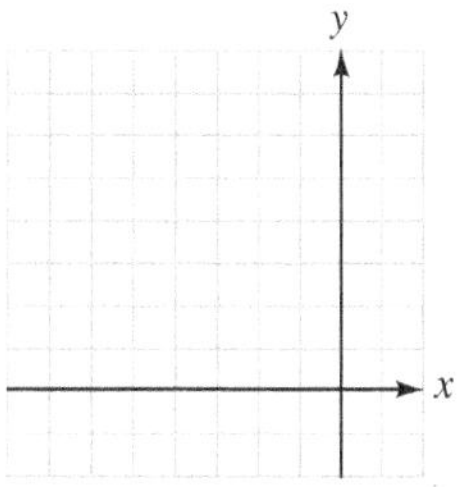

58. $y \geq x^2 + 5x + 4$

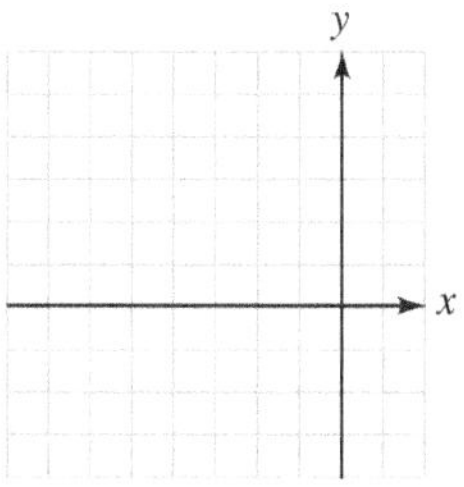

59. $y \geq (x-1)^2$

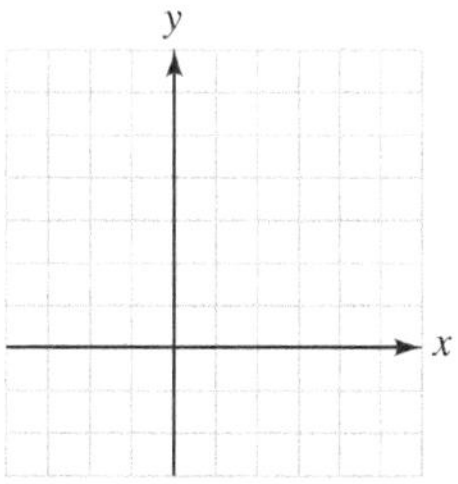

60. $y \leq (x+2)^2$

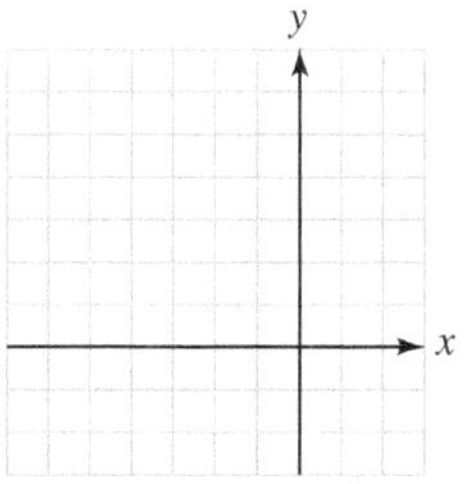

61. $-x^2 - y + 6 > -x$

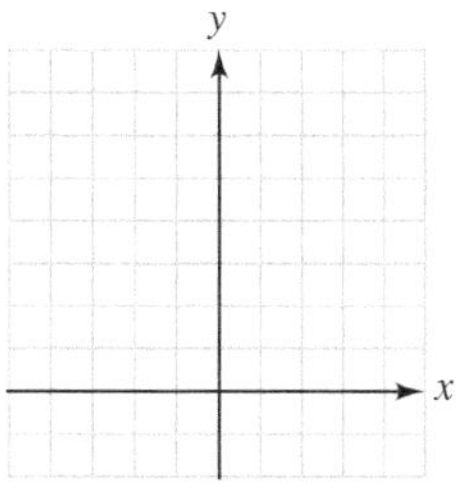

62. $y > (x+3)(x-2)$

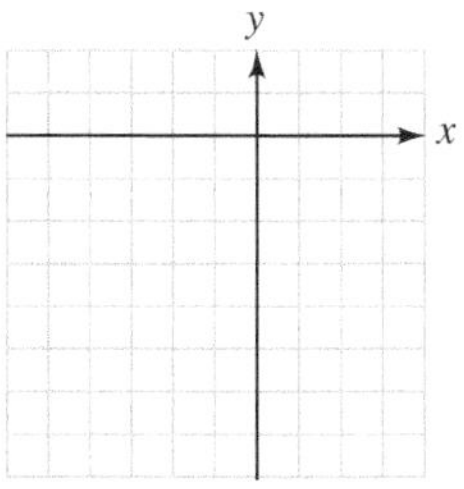

Graph each inequality. SEE EXAMPLE 6. (OBJECTIVE 3)

63. $y < |x+4|$

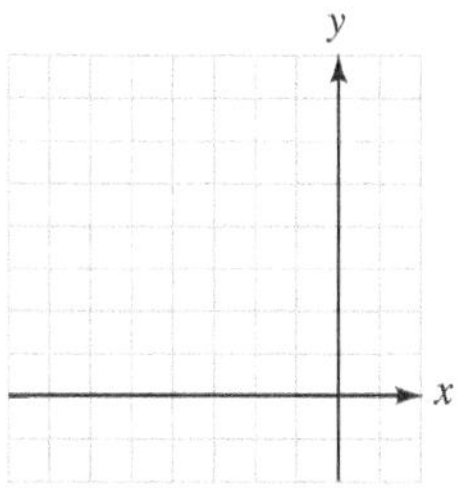

64. $y \geq |x-3|$

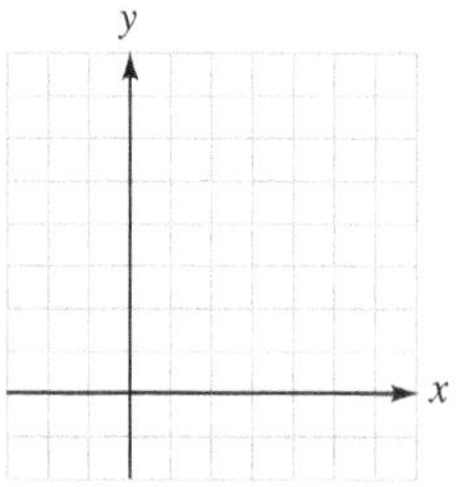

65. $y \leq -|x| + 2$

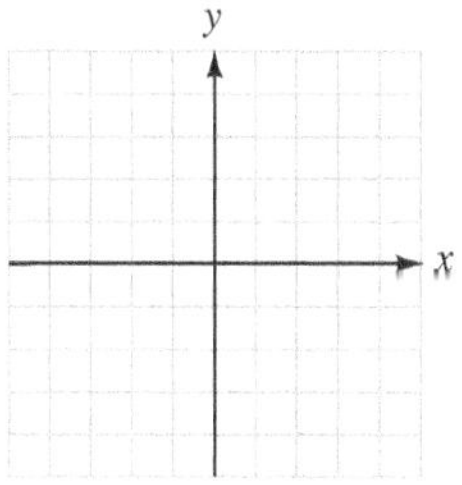

66. $y > |x| - 2$

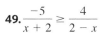

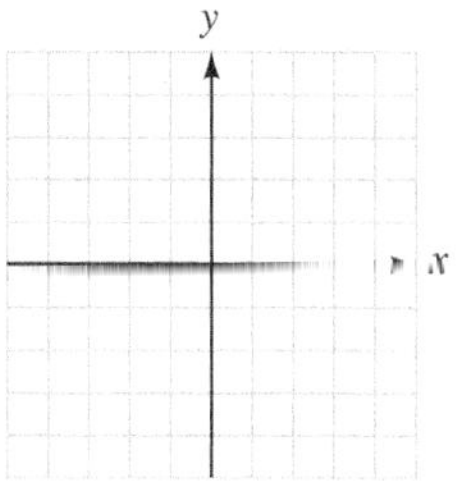

ADDITIONAL PRACTICE

Solve each inequality. State each result in interval notation and graph the solution set.

67. $2x^2 - 50 < 0$

68. $3x^2 - 243 < 0$

69. $-\frac{5}{x} < 3$

70. $\frac{4}{x} \geq 8$

71. $\frac{x}{x+4} \leq \frac{1}{x+1}$

72. $\frac{x}{x+9} \geq \frac{1}{x+1}$

73. $\frac{x}{x+16} > \frac{1}{x+1}$

74. $\frac{x}{x+25} < \frac{1}{x+1}$

Use a graphing calculator to solve each inequality. State the answer in interval notation.

75. $x^2 - 2x - 3 < 0$

76. $x^2 + x - 6 > 0$

77. $\frac{x+3}{x-2} \geq 0$

78. $\frac{3}{x} \leq 2$

Unless otherwise noted, all content on this page is © Cengage Learning.

WRITING ABOUT MATH

79. Explain why $(x - 4)(x + 5)$ will be positive only when the signs of $x - 4$ and $x + 5$ are the same.

80. Explain how to find the graph of $y \geq x^2$.

SOMETHING TO THINK ABOUT

81. Under what conditions will the fraction $\frac{(x-1)(x+4)}{(x+2)(x+1)}$ be positive?

82. Under what conditions will the fraction $\frac{(x-1)(x+4)}{(x+2)(x+1)}$ be negative?

Projects

PROJECT 1

Ballistics is the study of how projectiles fly. The general formula for the height above the ground of an object thrown straight up or down is given by the function

$$h(t) = -16t^2 + v_0t + h_0$$

where h is the object's height (in feet) above the ground t seconds after it is thrown. The initial velocity v_0 is the velocity with which the object is thrown, measured in feet per second. The initial height h_0 is the object's height (in feet) above the ground when it is thrown. (If $v_0 > 0$, the object is thrown upward; if $v_0 < 0$, the object is thrown downward.)

This formula takes into account the force of gravity, but disregards the force of air resistance. It is much more accurate for a smooth, dense ball than for a crumpled piece of paper.

One act in the Bungling Brothers Circus is Amazing Glendo's cannonball-catching act. A cannon fires a ball vertically into the air; Glendo, standing on a platform above the cannon, uses his catlike reflexes to catch the ball as it passes by on its way toward the roof of the big top. As the balls fly past, they are within Glendo's reach only during a two-foot interval of their upward path.

As an investigator for the company that insures the circus, you have been asked to find answers to the following questions. The answers will determine whether or not Bungling Brothers' insurance policy will be renewed.

a. In the first part of the act, cannonballs are fired from the end of a six-foot cannon with an initial velocity of 80 feet per second. Glendo catches one ball between 40 and 42 feet above the ground. Then he lowers his platform and catches another ball between 25 and 27 feet above the ground.

i. Show that if Glendo missed a cannonball, it would hit the roof of the 56-foot-tall big top. How long would it take for a ball to hit the big top? To prevent this from happening, a special net near the roof catches and holds any missed cannonballs.

ii. Find (to the nearest thousandth of a second) how long the cannonballs are within Glendo's reach for each of his catches. Which catch is easier? Why does your answer make sense? Your company is willing to insure against injuries to Glendo if he has at least 0.025 second to make each catch. Should the insurance be offered?

b. For Glendo's grand finale, the special net at the roof of the big top is removed, making Glendo's catch more significant to the people in the audience, who worry that if Glendo misses, the tent will collapse around them. To make it even more dramatic, Glendo's arms are tied to restrict his reach to a one-foot interval of the ball's flight, and he stands on a platform just under the peak of the big top, so that his catch is made at the very last instant (between 54 and 55 feet above the ground). For this part of the act, however, Glendo has the cannon charged with less gunpowder, so that the muzzle velocity of the cannon is 56 feet per second. Show work to prove that Glendo's big finale is in fact his easiest catch, and that even if he misses, the big top is never in any danger of collapsing, so insurance should be offered against injury to the audience.

PROJECT 2

The center of Sterlington is the intersection of Main Street (running east–west) and Due North Road (running north–south). The recreation area for the townspeople is Robin Park, a few blocks from there. The park is bounded on the south by Main Street and on every other side by Parabolic Boulevard, named for its distinctive shape. In fact, if Main Street and Due North Road were used as the axes of a rectangular coordinate system, Parabolic Boulevard would have the equation $y = -(x - 4)^2 + 5$, where each unit on the axes is 100 yards.

The city council has recently begun to consider whether or not to put two walkways through the park. (See Illustration 1.) The walkways would run from two points on Main Street and converge at the northernmost point of the park, dividing the area of the park exactly into thirds.

The city council is pleased with the aesthetics of this arrangement but needs to know two important facts.

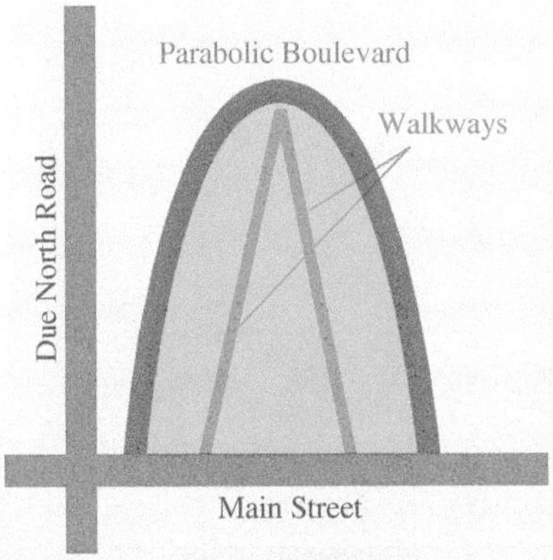

Illustration 1

a. For planning purposes, they need to know exactly where on Main Street the walkways would begin.
b. To budget for the construction, they need to know how long the walkways will be.

Provide answers for the city council, along with explanations and work to show that your answers are correct. Use the formula shown in Illustration 2, credited to Archimedes (287–212 B.C.), for the area under a parabola but above a line perpendicular to the axis of symmetry of the parabola.

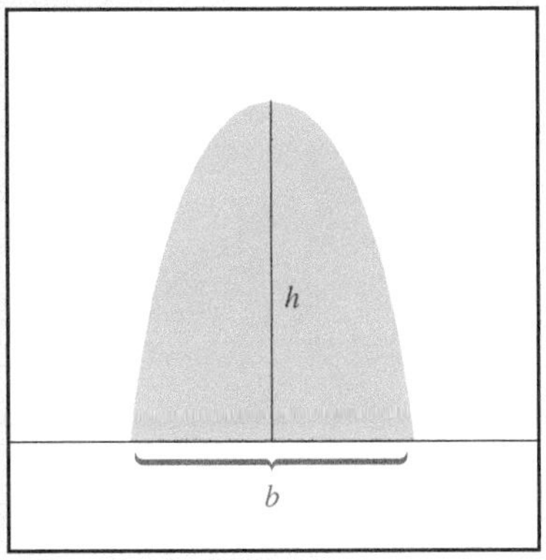

Shaded area $= \frac{2}{3} \cdot b \cdot h$

Illustration 2

Unless otherwise noted, all content on this page is © Cengage Learning.

Reach for Success

EXTRA PRACTICE EXERCISE LEARNING BEYOND REMEMBERING

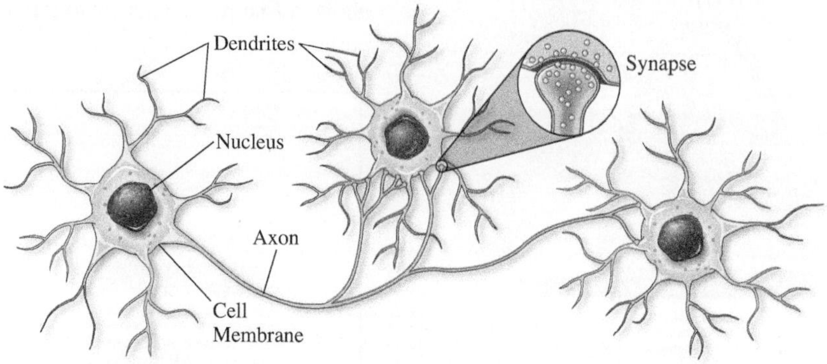

The brain cells shown in the figure were not connected at one time. No matter your age, when you first hear new information (for example, the formula for finding the slope of a line between two points), your brain starts to grow **dendrites**. Your instructor gives an example and explains the process. When you have to try a problem on your own, your brain connects the dendrites in this cell to other brain cells that have information you need (understanding coordinates of a point, evaluating a variable, and subtracting, dividing, and manipulating signed numbers). This is accomplished by moving the new information through the **axon** to **synapses** at its end. Once in the synapses, the information is "fired" to bridge the gap between this and other cells' dendrites. The more connections that are made (working problems in class), the more deeply you have embedded that information in your brain. Experts say "you must fire it to wire it," meaning that you must continue to make these connections (do many exercises until you've solidified your understanding) or risk not having enough connections when you need that information (for a test). The phrase "use it or lose it" is more accurate than we ever imagined!

Which of the following are true statements? Circle any/all that apply.

1. To learn new material you must be able to connect it to something you already know.
2. You must practice skills in order to "wire" them.
3. You can forget information if you have not "fired" enough synapses or have not "fired" them recently.
4. If you continue to learn new things, you can grow dendrites at any age.

Your ability to move from the remembering to the higher levels of learning depends on the number of new dendrites and connections you're able to make.

Unless otherwise noted, all content on this page is © Cengage Learning.

10 Review

SECTION 10.1 Solving Quadratic Equations Using the Square-Root Property and by Completing the Square

DEFINITIONS AND CONCEPTS	EXAMPLES
Square-root property: The equation $x^2 = c$ has two solutions: $x = \sqrt{c}$ and $x = -\sqrt{c}$	To solve $x^2 - 28 = 0$, proceed as follows: $x^2 = 28$ Add 28 to both sides. $x = \sqrt{28}$ or $x = -\sqrt{28}$ Use the square-root property. $x = 2\sqrt{7}$ | $x = -2\sqrt{7}$ $\sqrt{28} = \sqrt{4}\sqrt{7} = 2\sqrt{7}$ The solutions are $2\sqrt{7}$ and $-2\sqrt{7}$, or $\pm 2\sqrt{7}$.
Completing the square: 1. Be certain that the coefficient of x^2 is 1. If not, divide both sides of the equation by the coefficient of x^2. 2. If necessary, move the constant term to one side of the equal sign. 3. Complete the square: a. Find one-half of the coefficient of x and square it. b. Add the square to both sides of the equation. 4. Factor the trinomial square on one side of the equation and combine like terms on the other side. 5. Solve the resulting equation by using the square-root property.	To solve $2x^2 - 12x + 24 = 0$, we divide both sides of the equation by 2 to obtain 1 as the coefficient of x^2. 1. $\frac{2x^2}{2} - \frac{12x}{2} + \frac{24}{2} = \frac{0}{2}$ Divide both sides by 2. $x^2 - 6x + 12 = 0$ Simplify. 2. $x^2 - 6x = -12$ Subtract 12 from both sides. 3. Since $\left[\frac{1}{2}(-6)\right]^2 = (-3)^2 = 9$, we complete the square by adding 9 to both sides. $x^2 - 6x + \mathbf{9} = -12 + \mathbf{9}$ 4. $(x - 3)^2 = -3$ Factor the left side and combine terms on the right side. 5. $x - 3 = \sqrt{-3}$ or $x - 3 = -\sqrt{-3}$ Use the square-root property. $x - 3 = \sqrt{3}i$ | $x - 3 = -\sqrt{3}i$ Simplify $\sqrt{-3}$. $x = 3 + \sqrt{3}i$ | $x = 3 - \sqrt{3}i$ Add 3 to both sides. The solutions are $3 \pm \sqrt{3}i$.

REVIEW EXERCISES

Solve each equation by factoring or by using the square-root property.

1. $12x^2 + x - 6 = 0$
2. $6x^2 + 17x + 5 = 0$
3. $15x^2 + 2x - 8 = 0$
4. $(x + 3)^2 = 16$

Solve each equation by completing the square.

5. $x^2 + 8x + 12 = 0$
6. $2x^2 - 9x + 7 = 0$
7. $2x^2 - x - 5 = 0$

SECTION 10.2 Solving Quadratic Equations Using the Quadratic Formula

DEFINITIONS AND CONCEPTS	EXAMPLES
Quadratic formula: The solutions of $ax^2 + bx + c = 0 \quad (a \neq 0)$ are represented by the formula	To solve $2x^2 - 5x + 4 = 0$, note that in the equation $a = 2$, $b = -5$, and $c = 4$, and substitute these values into the quadratic formula. $x = \frac{-b \pm \sqrt{b^2 - 4ac}}{2a}$

$x = \dfrac{-b \pm \sqrt{b^2 - 4ac}}{2a}$	$= \dfrac{-(-5) \pm \sqrt{(-5)^2 - 4(2)(4)}}{2(2)}$	Substitute the values.
	$= \dfrac{5 \pm \sqrt{25 - 32}}{4}$	Simplify.
	$= \dfrac{5 \pm \sqrt{-7}}{4}$	Add.
	$= \dfrac{5 \pm i\sqrt{7}}{4}$	$\sqrt{-7} = i\sqrt{7}$
	The solutions are $\frac{5}{4} \pm \frac{\sqrt{7}}{4}i$.	

REVIEW EXERCISES

Solve each equation by using the quadratic formula.

8. $x^2 - 5x - 6 = 0$ **9.** $x^2 = 7x$

10. $2x^2 + 13x - 7 = 0$ **11.** $5x^2 - 2x - 3 = 0$

12. $2x^2 = x + 2$ **13.** $x^2 + x + 2 = 0$

14. Dimensions of a rectangle A rectangle is 2 centimeters longer than it is wide. If both the length and width are doubled, its area is increased by 72 square centimeters. Find the dimensions of the original rectangle.

15. Dimensions of a rectangle A rectangle is 1 foot longer than it is wide. If the length is tripled and the width is doubled, its area is increased by 30 square feet. Find the dimensions of the original rectangle.

16. Ballistics If a rocket is launched straight up into the air with an initial velocity of 112 feet per second, its height after t seconds is given by the formula $h = 112t - 16t^2$, where h represents the height of the rocket in feet. After launch, how long will it be before it hits the ground?

17. Ballistics What is the maximum height of the rocket discussed in Exercise 16?

SECTION 10.3 The Discriminant and Equations That Can Be Written in Quadratic Form

DEFINITIONS AND CONCEPTS	EXAMPLES
The discriminant is $b^2 - 4ac$, where a, b, and c are real numbers and $a \neq 0$. If $b^2 - 4ac > 0$, the equation $ax^2 + bx + c = 0$ has two real unequal solutions. If $b^2 - 4ac = 0$, the equation $ax^2 + bx + c = 0$ has two real equal solutions (called a *double root*). If $b^2 - 4ac < 0$, the equation $ax^2 + bx + c = 0$ has two solutions that are complex conjugates. Further, if a, b, and c are rational numbers and the discriminant: is a perfect square greater than 0, there are two unequal rational solutions. is positive but not a perfect square, the solutions are irrational and unequal.	To determine the type of solutions for the equation $x^2 - x + 9 = 0$, we calculate the discriminant. $b^2 - 4ac = (-1)^2 - 4(1)(9)$ $a = 1, b = -1$, and $c = 9$. $= 1 - 36$ $= -35$ Since $b^2 - 4ac < 0$, the two solutions are complex conjugates. To determine the type of solutions for the equation $2x^2 + 4x - 5 = 0$, we calculate the discriminant. $b^2 - 4ac = 4^2 - 4(2)(-5)$ $a = 2, b = 4$, and $c = -5$. $= 16 + 40$ $= 56$ Since $b^2 - 4ac > 0$, the solutions are unequal and irrational.
Solving equations quadratic in form: Use u substitution to write a quadratic equation.	To solve the equation $x^4 - 3x^2 - 4 = 0$, we can proceed as follows: $x^4 - 3x^2 - 4 = 0$ $(x^2)^2 - 3(x^2) - 4 = 0$ $u^2 - 3u - 4 = 0$ Substitute u for x^2. $(u - 4)(u + 1) = 0$ Factor $u^2 - 3u - 4$. $u - 4 = 0$ or $u + 1 = 0$ Set each factor equal to 0. $u = 4$ | $u = -1$

	Since $u = x^2$, it follows that $x^2 = 4$ or $x^2 = -1$. Thus, $x^2 = 4$ or $x^2 = -1$; $x = 2$ or $x = -2$ | $x = i$ or $x = -i$
Verifying solutions: If r_1 and r_2 are solutions of $ax^2 + bx + c = 0$ $(a \neq 0)$, then $r_1 + r_2 = -\frac{b}{a}$ and $r_1 r_2 = \frac{c}{a}$	To verify that $\frac{5}{3}$ and 2 are the solutions of $3x^2 - 11x + 10 = 0$, make the following calculations: $-\frac{b}{a} = -\frac{-11}{3} = \frac{11}{3}$ and $\frac{c}{a} = \frac{10}{3}$. Since $\frac{5}{3} + 2 = \frac{5}{3} + \frac{6}{3} = \frac{11}{3}$ and $\frac{5}{3}(2) = \frac{5}{3} \cdot \frac{2}{1} = \frac{10}{3}$, $\frac{5}{3}$ and 2 are the solutions of $3x^2 - 11x + 10 = 0$.

REVIEW EXERCISES

Use the discriminant to determine the types of solutions that exist for each equation.

18. $2x^2 + 5x - 4 = 0$

19. $4x^2 - 5x + 7 = 0$

20. Find the values of k that will make the solutions of $(k - 8)x^2 + (k + 16)x = -49$ equal.

21. Find the values of k such that the solutions of $3x^2 + 4x = k + 1$ will be real numbers.

Solve each equation.

22. $x - 8x^{1/2} + 7 = 0$

23. $a^{2/3} + a^{1/3} - 6 = 0$

24. $\frac{1}{x + 1} - \frac{1}{x} = -\frac{1}{x + 1}$

25. $\frac{6}{x + 2} + \frac{6}{x + 1} = 5$

26. Find the sum of the solutions of the equation $3x^2 - 14x + 3 = 0$.

27. Find the product of the solutions of the equation $3x^2 - 14x + 3 = 0$.

SECTION 10.4 Graphs of Quadratic Functions

DEFINITIONS AND CONCEPTS	EXAMPLES
Graphing quadratic functions: If f is a function and k and h positive numbers, then: The graph of $f(x) + k$ is identical to the graph of $f(x)$, except that it is translated k units upward. The graph of $f(x) - k$ is identical to the graph of $f(x)$, except that it is translated k units downward. The graph of $f(x - h)$ is identical to the graph of $f(x)$, except that it is translated h units to the right. The graph of $f(x + h)$ is identical to the graph of $f(x)$, except that it is translated h units to the left.	Graph each of the following. **a.** $f(x) = 4x^2$ **b.** $f(x) = 4x^2 - 3$ **c.** $f(x) = 4(x - 3)^2$ 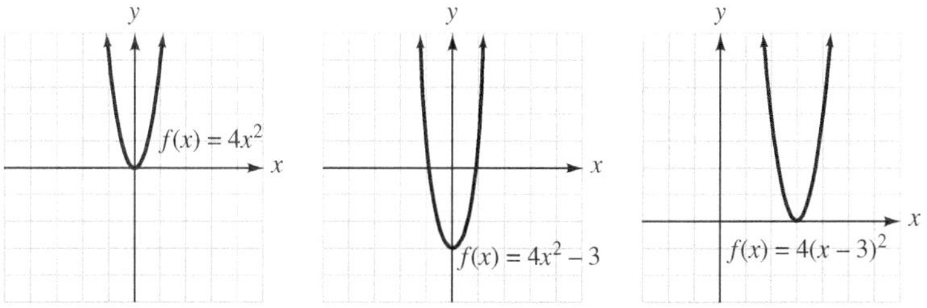
Finding the vertex of a parabola: If $a \neq 0$, the graph of $y = a(x - h)^2 + k$ is a parabola with vertex at (h, k). It opens upward when $a > 0$ and downward when $a < 0$.	The vertex of the graph of $y = 2(x - 3)^2 - 5$ is $(3, -5)$. Since $2 > 0$, the graph will open upward.

Unless otherwise noted, all content on this page is © Cengage Learning.

The coordinates of the vertex of the graph of

$$f(x) = ax^2 + bc + c \quad (a \neq 0)$$

are $\left(-\frac{b}{2a}, f\left(-\frac{b}{2a}\right)\right)$.

The axis of symmetry is the vertical line $x = -\frac{b}{2a}$.

Find the axis of symmetry and the vertex of the graph of $f(x) = 2x^2 - 4x - 5$.

$$x\text{-coordinate} = -\frac{b}{2a} = -\frac{-4}{2(2)} = -\frac{-4}{4} = \frac{4}{4} = 1$$

The axis of symmetry is $x = 1$.
To find the y-coordinate of the vertex, substitute 1 for x in $f(x) = 2x^2 - 4x - 5$.

$$\begin{aligned} f(x) &= 2x^2 - 4x - 5 \\ f(1) &= 2(1)^2 - 4(1) - 5 \\ &= 2 - 4 - 5 \\ &= -7 \end{aligned}$$

The vertex is $(1, -7)$.

REVIEW EXERCISES

Graph each function and state the coordinates of the vertex of the resulting parabola.

28. $y = 2x^2 - 3$

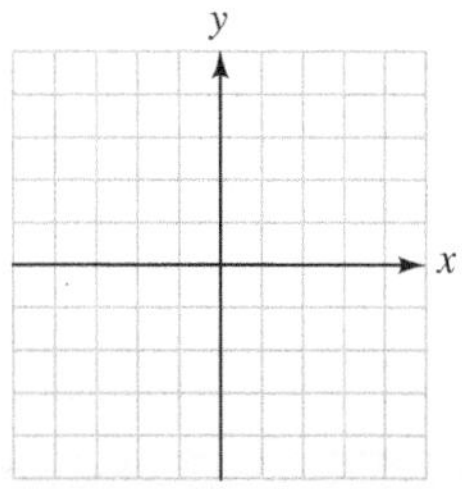

29. $y = -2x^2 - 1$

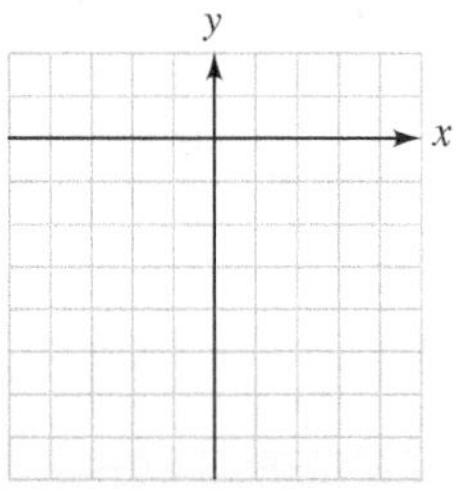

30. $y = -4(x - 2)^2 + 1$

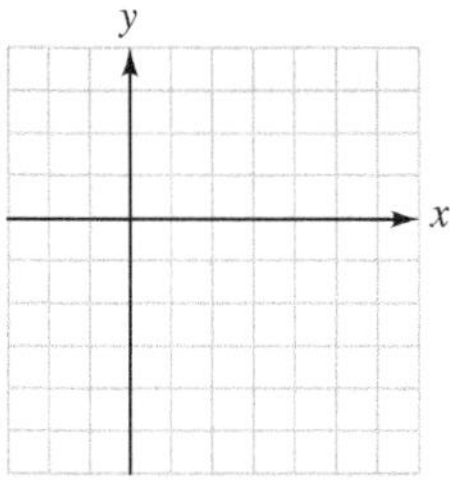

31. $y = 5x^2 + 10x - 1$

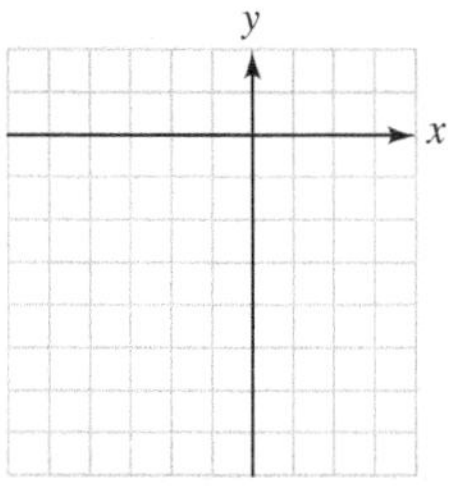

32. Find the vertex of the graph and the axis of symmetry of $f(x) = 4x^2 - 16x - 3$.

SECTION 10.5 Graphs of Other Nonlinear Functions

DEFINITIONS AND CONCEPTS	EXAMPLES
Graphs of nonlinear equations are not lines.	
Vertical translations: If f is a function and k is a positive number, then	
• The graph of $f(x) + k$ is identical to the graph of $f(x)$, except that it is translated k units upward.	The graph of $f(x) = \|x\| + 4$ will be the same shape as the graph of $f(x) = \|x\|$, but shifted 4 units upward.
• The graph of $f(x) - k$ is identical to the graph of $f(x)$, except that it is translated k units downward.	The graph of $f(x) = \|x\| - 3$ will be the same shape as the graph of $f(x) = \|x\|$, but shifted 3 units downward.
Horizontal translations: If f is a function and k is a positive number, then	
• The graph of $f(x - k)$ is identical to the graph of $f(x)$, except that it is translated k units to the right.	The graph of $f(x) = (x - 3)^3$ will be the same shape as the graph of $f(x) = x^3$, but shifted 3 units to the right.
• The graph of $f(x + k)$ is identical to the graph of $f(x)$, except that it is translated k units to the left.	The graph of $f(x) = (x + 4)^3$ will be the same shape as the graph of $f(x) = x^3$, but shifted 4 units to the left.
Reflections: The graph of $y = -f(x)$ is the graph of $f(x)$ reflected about the x-axis.	The graph of $f(x) = -x^3$ is the graph of $f(x) = x^3$ reflected about the x-axis.

Unless otherwise noted, all content on this page is © Cengage Learning.

Finding the domain of a rational function: The domain of a rational function is all values for which the function is defined.	Since the denominator of the fraction in $f(x) = \frac{x+1}{x-2}$ cannot be 0, the domain of $f(x)$ is all real numbers except 2. In interval notation, this is $(-\infty, 2) \cup (2, \infty)$.
Vertical asymptotes occur where a rational function is not defined.	In the graph of $f(x) = \frac{x+1}{x-2}$, there will be a vertical asymptote at $x = 2$ because the function is not defined when x is 2. 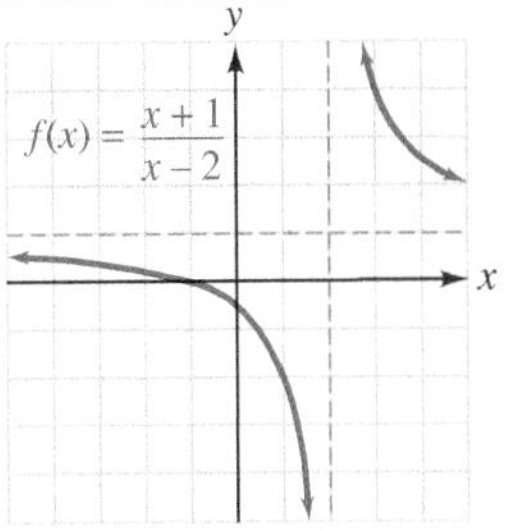
We may obtain the range from the graph. A horizontal asymptote occurs when a value is excluded from the range.	From the graph, we see that the range is $(-\infty, 1) \cup (1, \infty)$. There will be a horizontal asymptote at $y = 1$.

REVIEW EXERCISES

Graph each function.

33. $f(x) = |x| - 3$

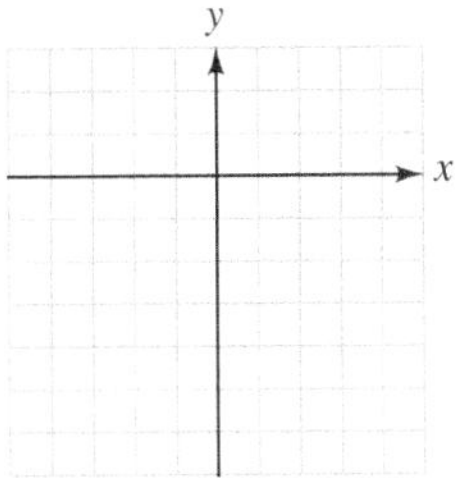

34. $f(x) = |x| - 4$

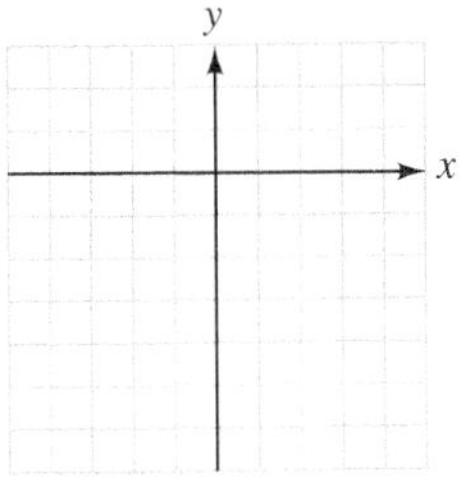

35. $f(x) = (x - 2)^3$

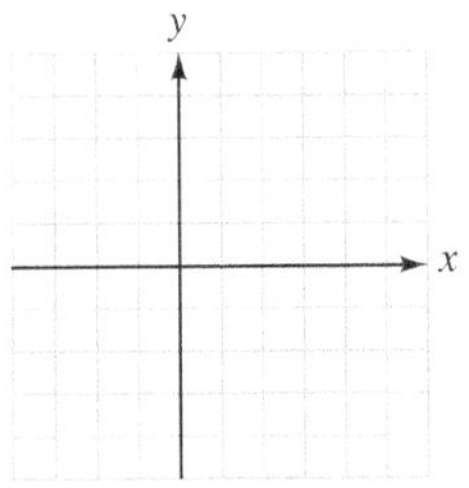

36. $f(x) = (x + 4)^3 - 3$

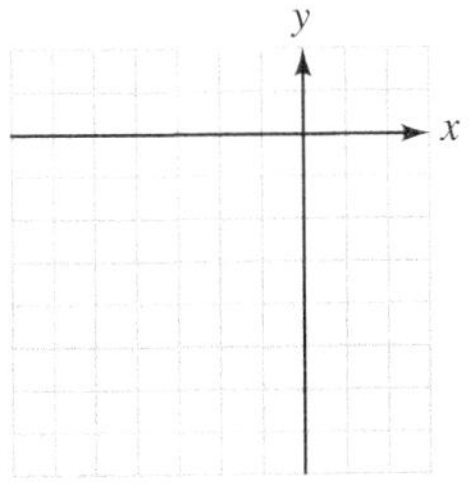

37. $f(x) = -x^3 - 2$

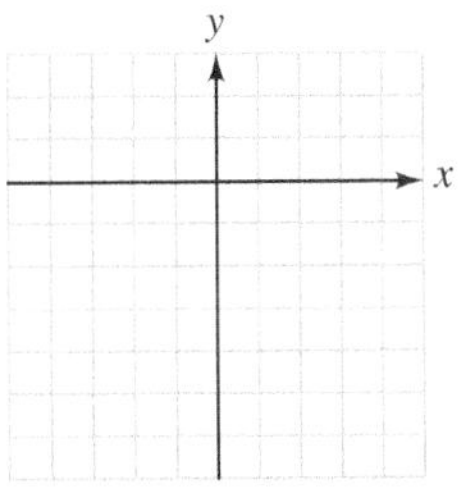

38. $f(x) = -|x - 1| + 2$

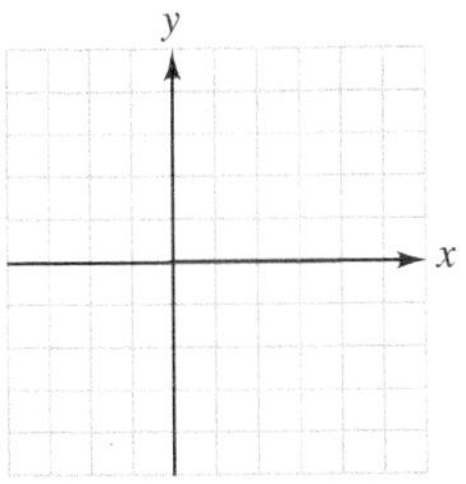

Use a graphing calculator to graph each function. Compare the results in Review Exercises 33–38.

39. $f(x) = x^2 - 3$ **40.** $f(x) = |x| - 4$

41. $f(x) = (x - 2)^3$ **42.** $f(x) = |x + 4| - 3$

43. $f(x) = -x^3 - 2$ **44.** $f(x) = -|x - 1| + 2$

Use a graphing calculator to graph each rational function and find its domain and range.

45. $f(x) = \dfrac{2}{x - 2}$ **46.** $f(x) = \dfrac{x}{x + 3}$

Unless otherwise noted, all content on this page is © Cengage Learning.

SECTION 10.6 Solving Quadratic and Other Nonlinear Inequalities

<table>
<tr><th>DEFINITIONS AND CONCEPTS</th><th>EXAMPLES</th></tr>
<tr><td>Solving quadratic inequalities:
To solve a quadratic inequality in one variable,

Method 1: Make a sign chart and determine which intervals are solutions.
or
Method 2: Create a table of intervals and test each.</td><td>To solve $x^2 - 2x - 8 > 0$ by method 1, factor the trinomial to get $(x - 4)(x + 2)$ and construct the chart shown below. The critical values will occur at 4 and -2.

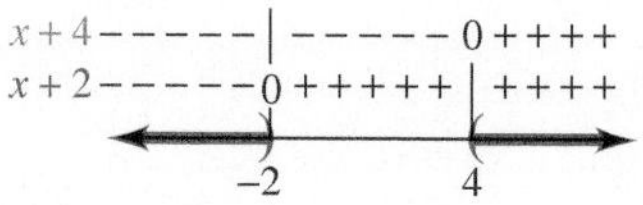

$x - 4$ is positive when $x > 4$, and is negative when $x < 4$.
$x + 2$ is positive when $x > -2$, and is negative when $x < -2$.

The product of $x - 4$ and $x + 2$ will be greater than 0 when the signs of the binomial factors are the same. This occurs in the intervals $(-\infty, -2)$ and $(4, \infty)$. The numbers -2 and 4 are not included, because equality is not indicated in the original inequality. Thus, the solution set in interval notation is

$(-\infty, -2) \cup (4, \infty)$

The graph of the solution set is shown on the number line below.

</td></tr>
<tr><td>Graphing rational inequalities:
To solve inequalities with rational expressions, get 0 on the right side, add the fractions, and then factor the numerator and denominator. Use a sign chart to determine the solution.</td><td>To solve $\frac{1}{x} \geq -4$, add 4 to both sides to make the right side equal to 0 and proceed as follows:

$\frac{1}{x} + 4 \geq 0$
$\frac{1}{x} + \frac{4x}{x} \geq 0$ Write each fraction with a common denominator.
$\frac{1 + 4x}{x} \geq 0$ Add.

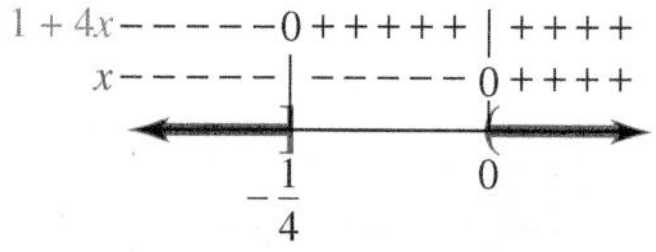

Finally, make a sign chart, as shown.

The denominator x is undefined when $x = 0$, positive when $x > 0$, and negative when $x < 0$. The numerator $1 + 4x$ is 0 when $x = -\frac{1}{4}$, positive when $x > -\frac{1}{4}$, and negative when $x < -\frac{1}{4}$. The critical values occur when the numerator or denominator is 0: 0 and $-\frac{1}{4}$.

The fraction $\frac{1 + 4x}{x}$ will be greater than or equal to 0 when the numerator and denominator are the same sign and when the numerator is 0. This occurs in the interval

$\left(-\infty, -\frac{1}{4}\right] \cup (0, \infty)$

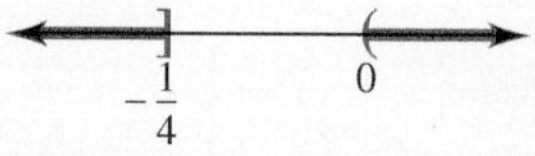

The graph of this interval is shown.</td></tr>
</table>

Unless otherwise noted, all content on this page is © Cengage Learning.

Solving nonlinear inequalities by graphing:
To solve a nonlinear inequality, first graph the equation. Then determine which region represents the solution to the inequality.

To graph $x > |y|$, first graph $x = |y|$ as in the illustration below using a dashed line, because equality is not indicated in the original inequality.

Since the origin is on the graph, we cannot use it as a test point. We select another point, such as $(1, 0)$. We substitute 1 for x and 0 for y into the inequality to get

$$x > |y|$$
$$\mathbf{1} > |\mathbf{0}|$$
$$1 > 0$$

Since $1 > 0$ is a true statement, the point $(1, 0)$ satisfies the inequality and is part of the graph. Thus, the solution to $x > |y|$ is to the right of the boundary.

The complete solution is shown.

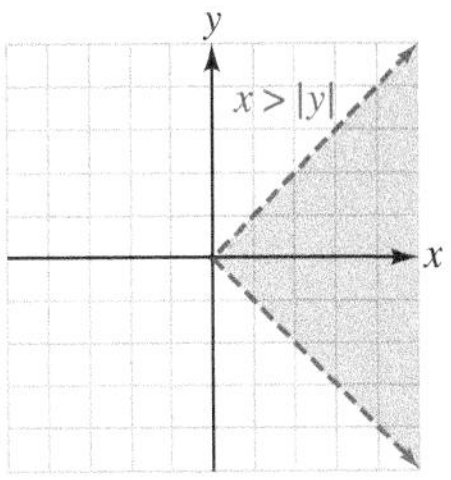

REVIEW EXERCISES

Solve each inequality. State each result in interval notation and graph the solution set.

47. $x^2 + 2x - 35 > 0$ **48.** $x^2 + 7x - 18 < 0$

49. $\frac{3}{x} \leq 5$ **50.** $\frac{2x^2 - x - 28}{x - 1} > 0$

Use a graphing calculator to solve each inequality. Compare the results with Review Exercises 47–50.

51. $x^2 + 2x - 35 > 0$ **52.** $x^2 + 7x - 18 < 0$

53. $\frac{3}{x} \leq 5$ **54.** $\frac{2x^2 - x - 28}{x - 1} > 0$

Solve each inequality.

55. $f(x) < \frac{1}{2}x^2 - 1$ **56.** $y \geq -|x|$

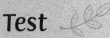

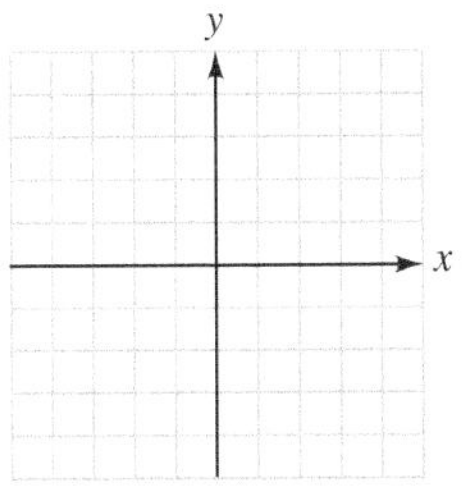

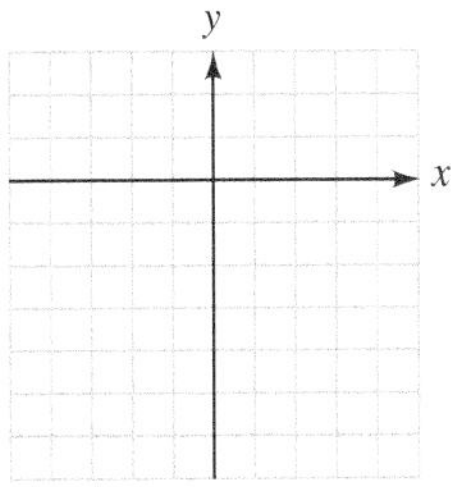

10 Test

Solve each equation by factoring.

1. $x^2 - 3x - 28 = 0$ **2.** $x(6x + 19) = -15$

Solve each equation using the square root method.

3. $x^2 - 144 = 0$ **4.** $(x - 5)^2 = -6$

Solve each equation by completing the square.

5. $x^2 + 6x + 7 = 0$ **6.** $x^2 - 5x - 3 = 0$

Solve each equation by the quadratic formula.

7. $2x^2 + 5x + 1 = 0$ **8.** $x^2 - 2x + 6 = 0$

9. Determine whether the solutions of $4x^2 + 3x + 10 = 0$ are real or nonreal numbers.

10. For what value(s) of k are the solutions of $4x^2 - 2kx + k - 1 = 0$ equal?

11. One leg of a right triangle is 14 inches longer than the other, and the hypotenuse is 26 inches. Find the length of the shorter leg.

12. Solve: $2y - 3y^{1/2} + 1 = 0$

Unless otherwise noted, all content on this page is © Cengage Learning.

13. Graph $f(x) = \frac{1}{2}x^2 - 4$ and state the coordinates of its vertex.

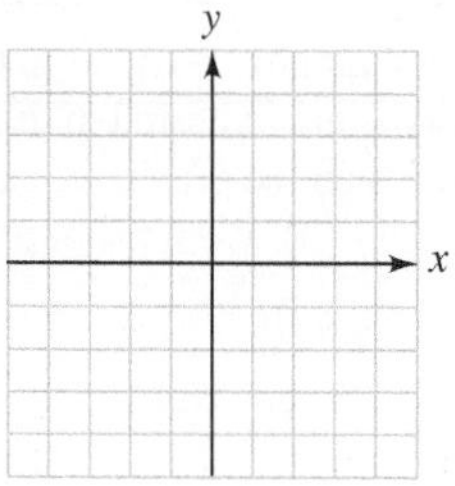

14. State the vertex and axis of symmetry of the graph of $f(x) = -3x^2 + 12x - 5$.

Graph each function.

15. $f(x) = (x - 3)^2 + 1$

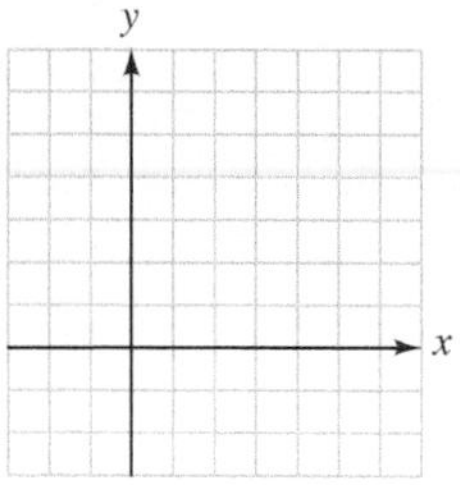

16. $f(x) = x^3$

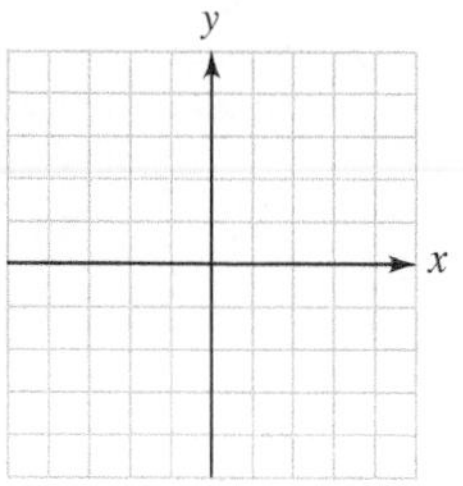

17. $f(x) = |x| - 2$

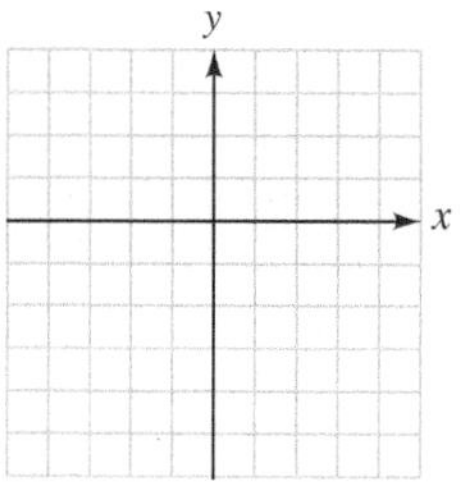

18. $f(x) = (x + 1)^3 - 3$

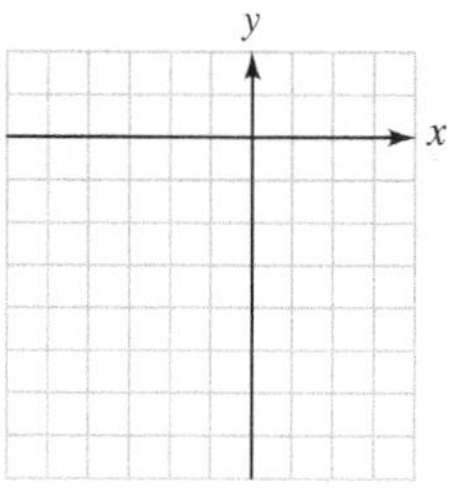

19. $f(x) = -|x - 2| + 1$

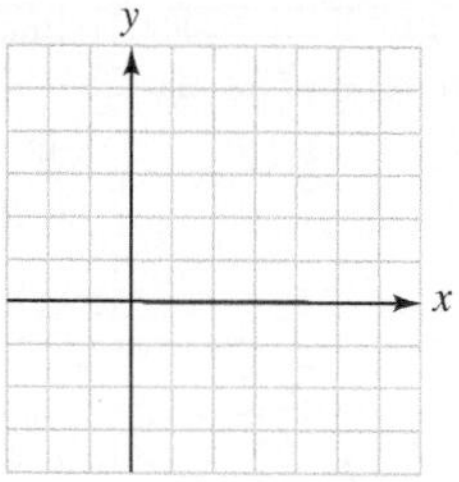

20. $f(x) \le -x^2 + 3$

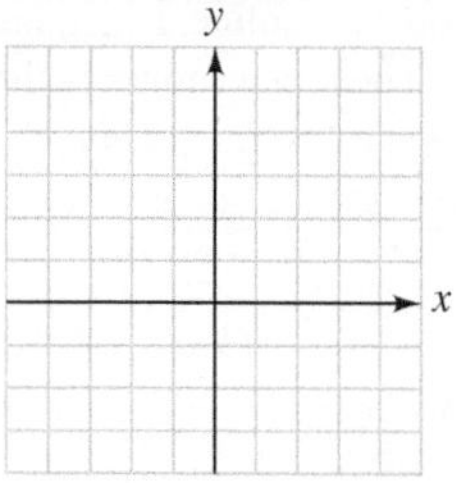

Solve each inequality and graph the solution set.

21. $x^2 - 2x - 8 > 0$

22. $x^2 + 5x - 6 \le 0$

23. $\dfrac{x - 2}{x + 3} \le 0$

24. $|y| \le x$

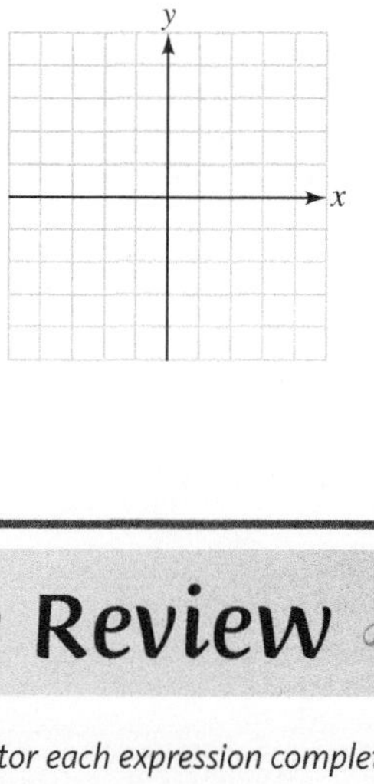

25. $f(x) < |x - 2| + 3$

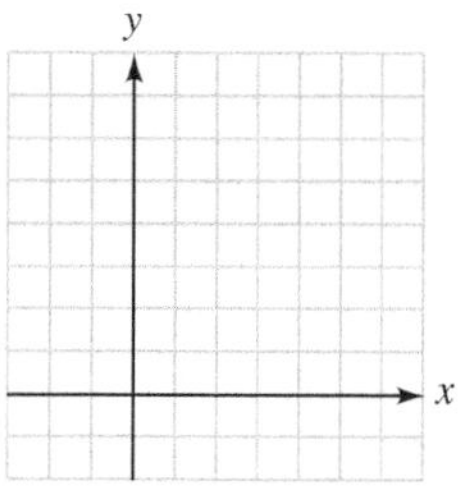

Cumulative Review

Find the domain and range of each function.

1. $f(x) = 5x^2 - 2$

2. $f(x) = -|x - 4|$

Write the equation of the line with the given properties.

3. $m = 4$, passing through $(-3, -5)$

4. parallel to the graph of $2x + 3y = 6$ and passing through $(0, -2)$

Perform each operation.

5. $(2a^2 + 4a - 7) - 2(3a^2 - 4a)$

6. $(4x - 3)(5x + 2)$

Factor each expression completely using only integers.

7. $x^4 - 16y^4$

8. $15x^2 - 2x - 8$

Solve each equation.

9. $x^2 - 8x - 9 = 0$

10. $6a^3 - 2a = a^2$

Simplify each expression. Assume that all variables represent positive numbers.

11. $\sqrt{36a^2b^4}$

12. $\sqrt{48t^3}$

13. $\sqrt[3]{-64y^6}$

14. $\sqrt[3]{\dfrac{128x^4}{2x}}$

Unless otherwise noted, all content on this page is © Cengage Learning.

15. $8^{-1/3}$

16. $27^{2/3}$

17. $\dfrac{y^{2/3}y^{5/3}}{y^{1/3}}$

18. $\dfrac{x^{5/3}x^{1/2}}{x^{3/4}}$

Graph each function and give the domain and the range.

19. $f(x) = \sqrt{x-2}$

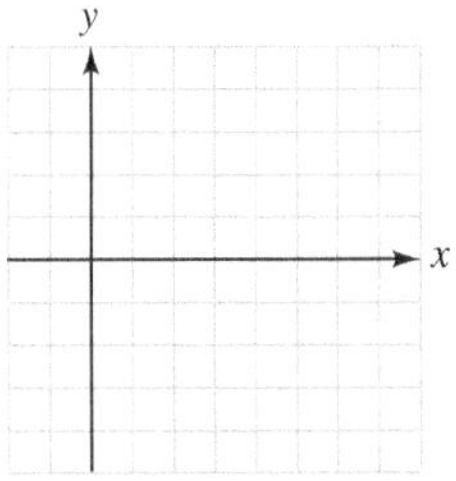

20. $f(x) = -\sqrt{x+2}$

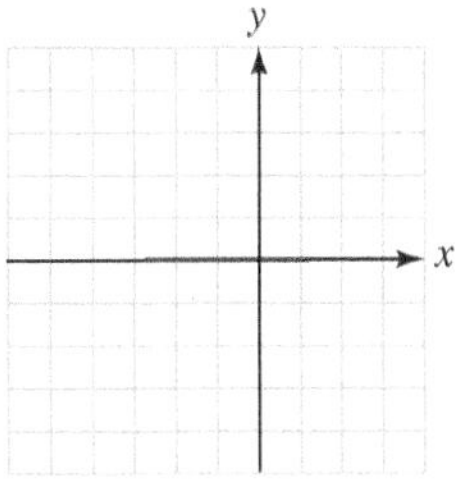

Perform the operations.

21. $\left(x^{1/2} - x^{1/3}\right)\left(x^{2/3} + x^{1/3}\right)$

22. $\left(x^{-1/2} + x^{1/2}\right)^2$

Simplify each statement. Assume no division by 0.

23. $\sqrt{50} - \sqrt{8} + \sqrt{32}$

24. $-3\sqrt[4]{32} - 2\sqrt[4]{162} + 5\sqrt[4]{48}$

25. $3\sqrt{2}(2\sqrt{3} - 4\sqrt{12})$

26. $\dfrac{5}{\sqrt[3]{x}}$

27. $\dfrac{\sqrt{x}+2}{\sqrt{x}-1}$

28. $\sqrt[6]{x^3y^3}$

Solve each equation.

29. $5\sqrt{x+2} = x + 8$

30. $\sqrt{x} + \sqrt{x+2} = 2$

31. Find the length of the hypotenuse of the right triangle shown in Illustration 1.

32. Find the length of the hypotenuse of the right triangle shown in Illustration 2.

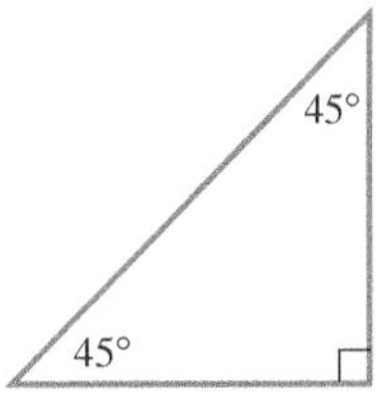

Illustration 1

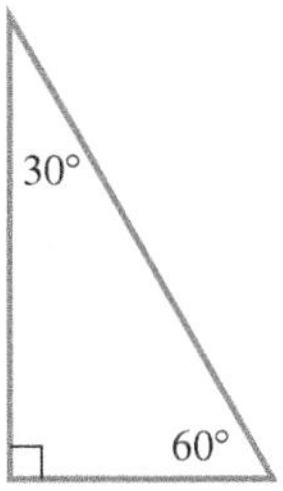

Illustration 2

33. Find the distance between $(-2, 6)$ and $(4, 14)$.

34. Solve $x^2 + 25$ using the square root method.

35. Use the method of completing the square to solve $2x^2 + x - 3 = 0$.

36. Use the quadratic formula to solve $3x^2 + 4x - 1 = 0$.

37. Graph $f(x) = \frac{1}{2}x^2 + 5$ and state the coordinates of its vertex.

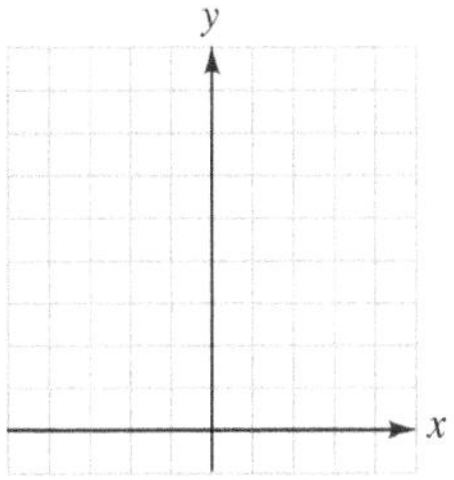

38. Graph $f(x) \leq -x^2 + 3$ and find the coordinates of its vertex.

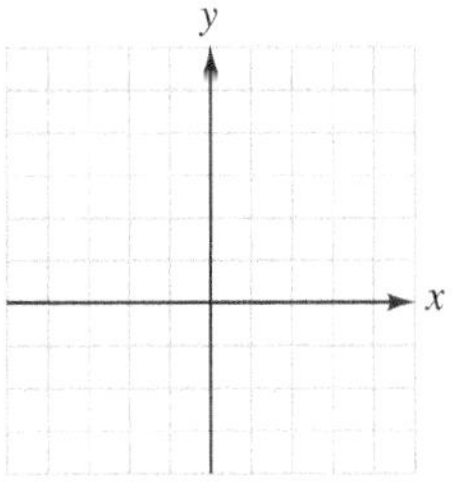

Write each expression as a real number or as a complex number in $a + bi$ form.

39. $(5 + 7i) + (8 - 10i)$

40. $(7 - 4i) - (12 + 3i)$

41. $(6 + 5i)(6 - 5i)$

42. $(3 + i)(3 - 3i)$

43. $(3 - 2i) - (4 + i)^2$

44. $\dfrac{5}{3 - i}$

45. $|3 + 2i|$

46. $|5 - 6i|$

47. For what values of k will the solutions of $2x^2 + 4x = k$ be equal?

48. Solve: $a - 7a^{1/2} + 12 = 0$

Solve each inequality and graph the solution set on the number line.

49. $x^2 - x - 6 > 0$

50. $\dfrac{x-3}{x+2} \leq 0$

51. Graph $f(x) = -|x - 5| - 2$.

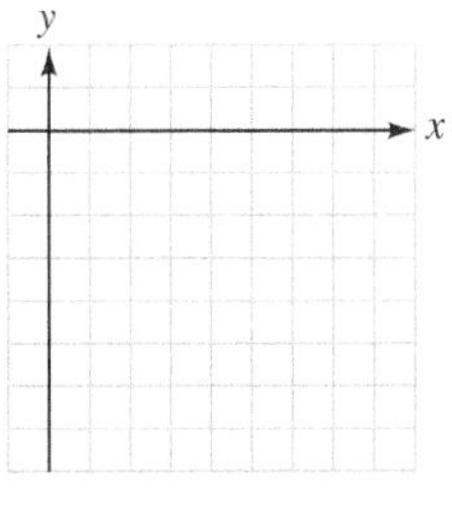

Unless otherwise noted, all content on this page is © Cengage Learning.

52. Graph $f(x) = (x - 1)^3 + 4$.

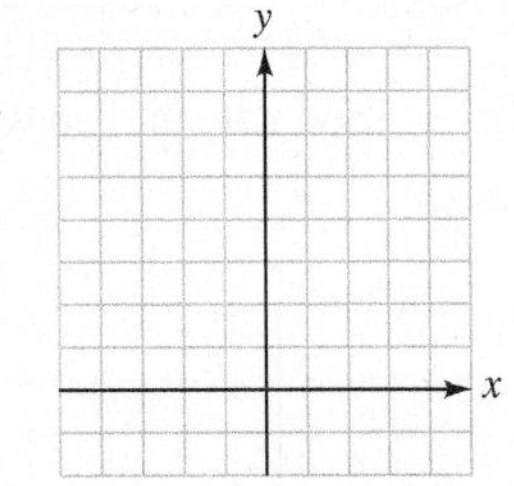

Solve.

53. $\frac{5}{x} - 10 > 0$

54. $\frac{x - 3}{2} \geq \frac{x + 3}{x}$

Simplify. Assume no division by 0.

55. $\frac{x^2 - 9}{x^2 - 3x} \div \frac{x^2 + 9x + 18}{x}$

56. $\frac{x + 5}{x + 7} - \frac{x - 3}{x - 4}$

57. $\dfrac{\frac{1}{x} + 4}{\frac{2}{x^2} + 3}$

Solve.

58. $\frac{x - 2}{x - 3} - \frac{1}{x} = \frac{1}{x - 3}$

Unless otherwise noted, all content on this page is © Cengage Learning.

Algebra, Composition, and Inverses of Functions; Exponential and Logarithmic Functions

11

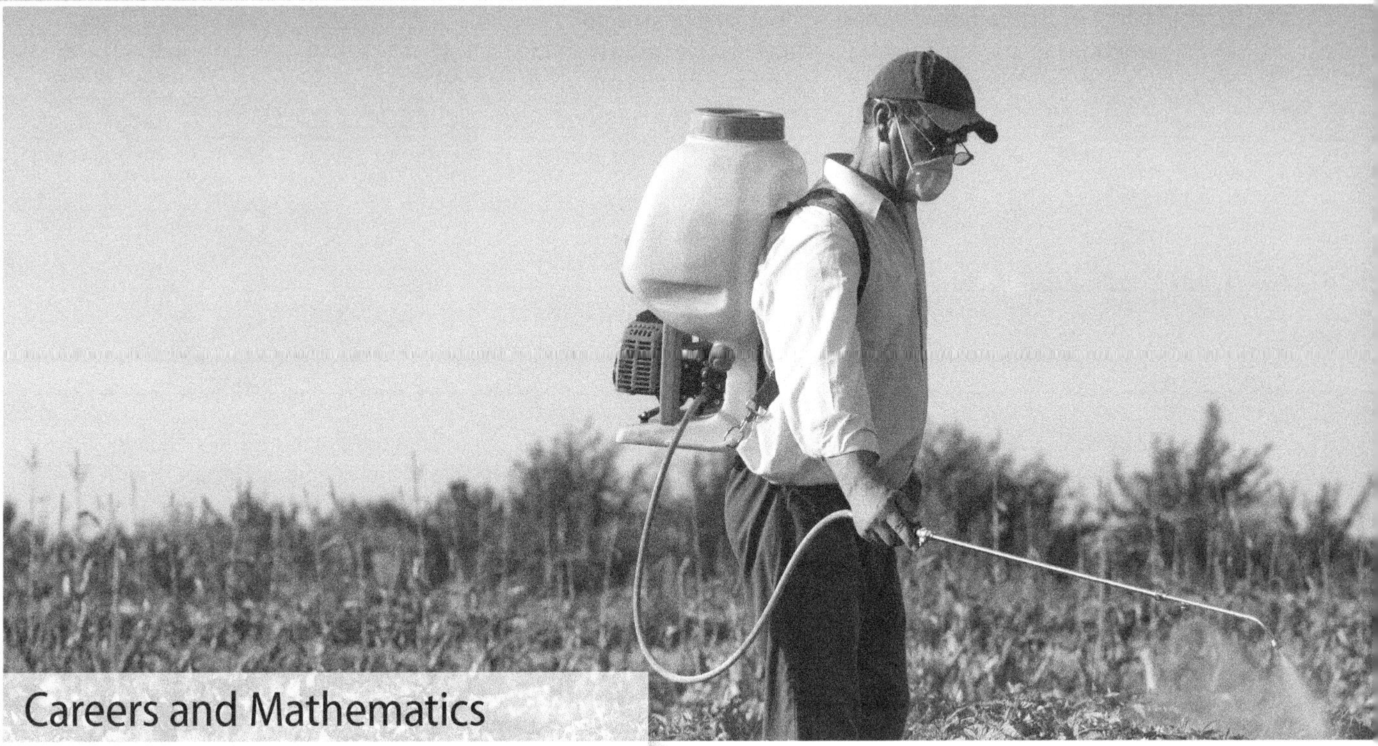

© Fotokostic/Shutterstock.com

Careers and Mathematics

PEST-CONTROL WORKERS

Few people welcome roaches, rats, mice, spiders, termites, fleas, ants, and bees into their homes. It is the job of pest-control workers to locate, identify, destroy, control, and repel these pests.

Roughly 86 percent of pest control workers are employed in services related to the building industry, primarily in states with warmer climates. About 7 percent are self-employed.

Both federal and state laws require pest-control workers to be certified.

Job Outlook:
Employment of pest control workers is expected to grow 26% between now and 2020, which is faster than the average for all occupations.

Yearly Earnings:
$20,340–$46,930

For More Information:
http://www.bls.gov/oco/ocos254.htm

For a Sample Application:
See Problem 89 in Section 11.8.

REACH FOR SUCCESS
11.1 Algebra and Composition of Functions
11.2 Inverses of Functions
11.3 Exponential Functions
11.4 Base-*e* Exponential Functions
11.5 Logarithmic Functions
11.6 Natural Logarithms
11.7 Properties of Logarithms
11.8 Exponential and Logarithmic Equations
■ *Projects*
REACH FOR SUCCESS EXTENSION
CHAPTER REVIEW
CHAPTER TEST

In this chapter

In this chapter, we introduce algebraic operations, composition, and inverses of functions. We will discuss two functions that are important in many applications. Exponential functions are used to compute compound interest, find radioactive decay, and model population growth. Logarithmic functions are used to measure acidity of solutions, drug dosage, intensity of an amplifier, magnitude of earthquakes, and safe noise levels in factories.

Reach for Success Planning for Your Future

The various courses and services that are needed to attain your academic and career goals should fit together like the pieces of a puzzle. Planning for your future involves more than concentrating on a single piece, such as a single course. It involves all of the pieces necessary to create the picture in your particular puzzle.

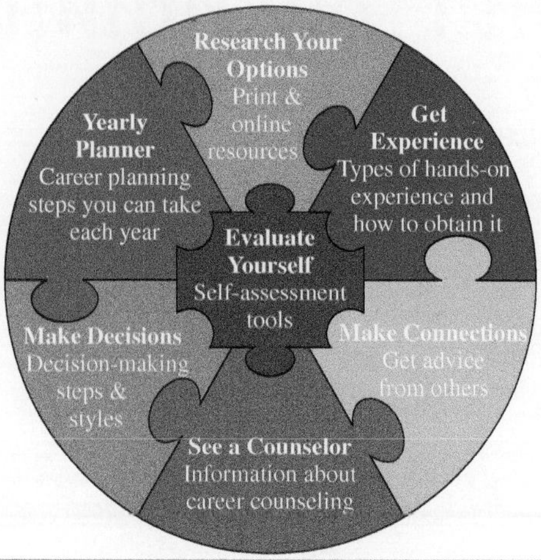

SHORT-TERM EDUCATIONAL PLANNING

This includes ensuring there is sufficient funding to attend college and then selecting the appropriate courses for next semester.

Where can you find your grade point average (GPA)? What factors can affect your GPA?

____

____

Check with your financial aid office to find the requirements for both merit-based and need-based financial aid.

____

____

List the importance of a high GPA.

1. ______________________________

2. ______________________________

Taking the entire required mathematics sequence consecutively improves your chance of success by more than 50%.

Familiarize yourself with the mathematics sequence required to meet graduation requirements at your college or transfer school for your major. What courses must you take?

____

____

To be successful, you need to start in the right course and the right format for you. Assessment tests will typically place you in the right course. The format of the course is usually your choice and varies by college.

Some format possibilities include computer-based lecture, mini-sessions, express, weekend, learning communities, and hybrid, blended, or fully online.

What formats are available at your college for the mathematics courses you need?

____

____

____

Which format is a good fit for you? ______________

Explain. ______________________________

A Successful Study Strategy . . .

Take all of your required courses that have prerequisites without interruption in the semester sequence.

At the end of the chapter you will find an additional exercise to guide you in planning for a successful college experience.

Unless otherwise noted, all content on this page is © Cengage Learning.

Section 11.1 Algebra and Composition of Functions

Objectives

1. Find the sum, difference, product, and quotient of two functions.
2. Find the composition of two functions.
3. Find the difference quotient of a function.
4. Solve an application requiring the composition of two functions.

Vocabulary

composition | composite functions | difference quotient

Getting Ready

Simplify.

1. $2x + 1 + x - 2$

2. $2x + 1 - (x - 2)$

3. $(2x + 1)(x - 2)$

Throughout the text, we have talked about functions. In this section, we will show how to add, subtract, multiply, and divide them, processes that are sometimes referred to as the algebra of functions. We also will show how to find the composition of two functions.

1 Find the sum, difference, product, and quotient of two functions.

We now consider how functions can be added, subtracted, multiplied, and divided.

OPERATIONS ON FUNCTIONS

If the domains and ranges of functions f and g are subsets of the real numbers,

the *sum* of f and g, denoted as $f + g$, is defined by

$$(f + g)(x) = f(x) + g(x)$$

the *difference* of f and g, denoted as $f - g$, is defined by

$$(f - g)(x) = f(x) - g(x)$$

the *product* of f and g, denoted as $f \cdot g$, is defined by

$$(f \cdot g)(x) = f(x)g(x)$$

the *quotient* of f and g, denoted as f/g, is defined by

$$(f/g)(x) = \frac{f(x)}{g(x)} \quad (g(x) \neq 0)$$

The domain of each of these functions is the set of real numbers x that are in the domain of both f and g. In the case of the quotient, there is the further restriction that $g(x) \neq 0$.

EXAMPLE 1 Let $f(x) = 2x^2 + 1$ and $g(x) = 5x - 3$. Find each function and its domain.

a. $f + g$ **b.** $f - g$

Solution **a.**
$$\begin{aligned}(f + g)(x) &= f(x) + g(x)\\ &= (2x^2 + 1) + (5x - 3)\\ &= 2x^2 + 5x - 2\end{aligned}$$

The domain of $f + g$ is the set of real numbers that are in the domain of both f and g. Since the domain of both f and g is the interval $(-\infty, \infty)$, the domain of $f + g$ is also the interval $(-\infty, \infty)$.

b.
$$\begin{aligned}(f - g)(x) &= f(x) - g(x)\\ &= (2x^2 + 1) - (5x - 3)\\ &= 2x^2 + 1 - 5x + 3 && \text{Use the distributive property to remove parentheses.}\\ &= 2x^2 - 5x + 4 && \text{Combine like terms.}\end{aligned}$$

Since the domain of both f and g is $(-\infty, \infty)$, the domain of $f - g$ is also the interval $(-\infty, \infty)$.

SELF CHECK 1 Let $f(x) = 3x - 2$ and $g(x) = 2x^2 + 3x$. Find each function and its domain.

a. $f + g$ **b.** $f - g$

EXAMPLE 2 Let $f(x) = 2x^2 + 1$ and $g(x) = 5x - 3$. Find each function and its domain.

a. $f \cdot g$ **b.** f/g

Solution **a.**
$$\begin{aligned}(f \cdot g)(x) &= f(x)g(x)\\ &= (2x^2 + 1)(5x - 3)\\ &= 10x^3 - 6x^2 + 5x - 3 && \text{Multiply.}\end{aligned}$$

The domain of $f \cdot g$ is the set of real numbers that are in the domain of both f and g. Since the domain of both f and g is the interval $(-\infty, \infty)$, the domain of $f \cdot g$ is also the interval $(-\infty, \infty)$.

b.
$$\begin{aligned}(f/g)(x) &= \frac{f(x)}{g(x)}\\ &= \frac{2x^2 + 1}{5x - 3}\end{aligned}$$

Since the denominator of the fraction cannot be 0, $x \neq \frac{3}{5}$. The domain of f/g is the union of two intervals $\left(-\infty, \frac{3}{5}\right) \cup \left(\frac{3}{5}, \infty\right)$.

SELF CHECK 2 Let $f(x) = 2x^2 - 3$ and $g(x) = x - 1$. Find each function and its domain.

a. $f \cdot g$ **b.** f/g

2 Find the composition of two functions.

We have seen that a function can be represented by a machine: We put in a number from the domain, and a number from the range comes out. For example, if we put the number 2 into the machine shown in Figure 11-1(a), the number $f(2) = 5(2) - 2 = 8$ comes out. In general, if we put x into the machine shown in Figure 11-1(b), the value $f(x)$ comes out.

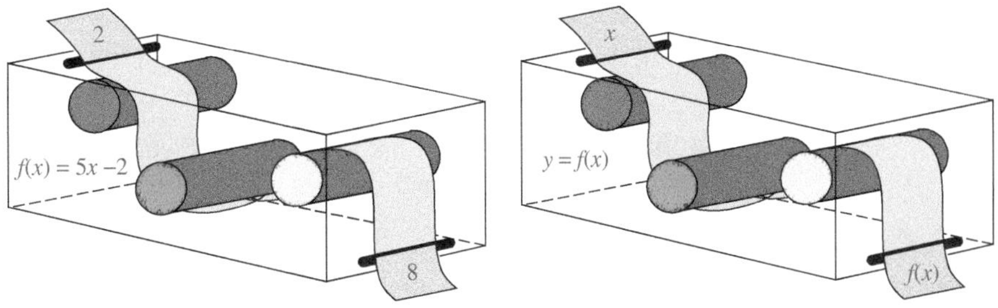

(a) (b)

Figure 11-1

Often one quantity is a function of a second quantity that depends, in turn, on a third quantity. For example, the cost of a car trip is a function of the gasoline consumed. The amount of gasoline consumed, in turn, is a function of the number of miles driven. Such chains of dependence can be analyzed mathematically as **compositions of functions**.

The function machines shown in Figure 11-2 illustrate the composition of functions f and g. When we put a number x into the function g, $g(x)$ comes out. The value $g(x)$ goes into function f, which transforms $g(x)$ into $f(g(x))$. This two-step process defines a new function, called a **composite function**. If the function machines for g and f were connected to make a single machine, that machine would be named $f \circ g$, read as "f composition g" or "f of g."

To be in the domain of the composite function $f \circ g$, a number x has to be in the domain of g. Also, the output of g must be in the domain of f. Thus, the domain of $f \circ g$ consists of those numbers x that are in the domain of g, and for which $g(x)$ is in the domain of f.

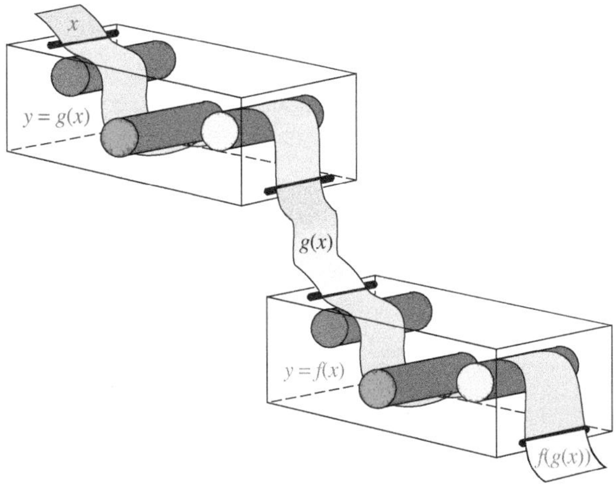

Figure 11-2

COMPOSITE FUNCTIONS The **composite function** $f \circ g$ is defined by

$$(f \circ g)(x) = f(g(x))$$

COMMENT Note that in this example, $(f \circ g)(x) \neq (g \circ f)(x)$. This shows that the composition of functions is not commutative.

For example, if $f(x) = 4x - 5$ and $g(x) = 3x + 2$, then

$$\begin{aligned} (f \circ g)(x) &= f(g(x)) \\ &= f(3x + 2) \\ &= 4(3x + 2) - 5 \\ &= 12x + 8 - 5 \\ &= 12x + 3 \end{aligned} \qquad \begin{aligned} (g \circ f)(x) &= g(f(x)) \\ &= g(4x - 5) \\ &= 3(4x - 5) + 2 \\ &= 12x - 15 + 2 \\ &= 12x - 13 \end{aligned}$$

Unless otherwise noted, all content on this page is © Cengage Learning.

EXAMPLE 3 Let $f(x) = 2x + 1$ and $g(x) = x - 4$. Find

a. $(f \circ g)(9)$ **b.** $(f \circ g)(x)$ **c.** $(g \circ f)(-2)$

Solution **a.** $(f \circ g)(9)$ means $f(g(9))$. In Figure 11-3(a), function g receives the number 9, subtracts 4, and releases the number 5. The 5 then goes into the f function, which doubles 5 and adds 1. The final result, 11, is the output of the composite function $f \circ g$:

$$(f \circ g)(9) = f(\mathbf{g(9)}) = f(\mathbf{5}) = 2(\mathbf{5}) + 1 = 11$$

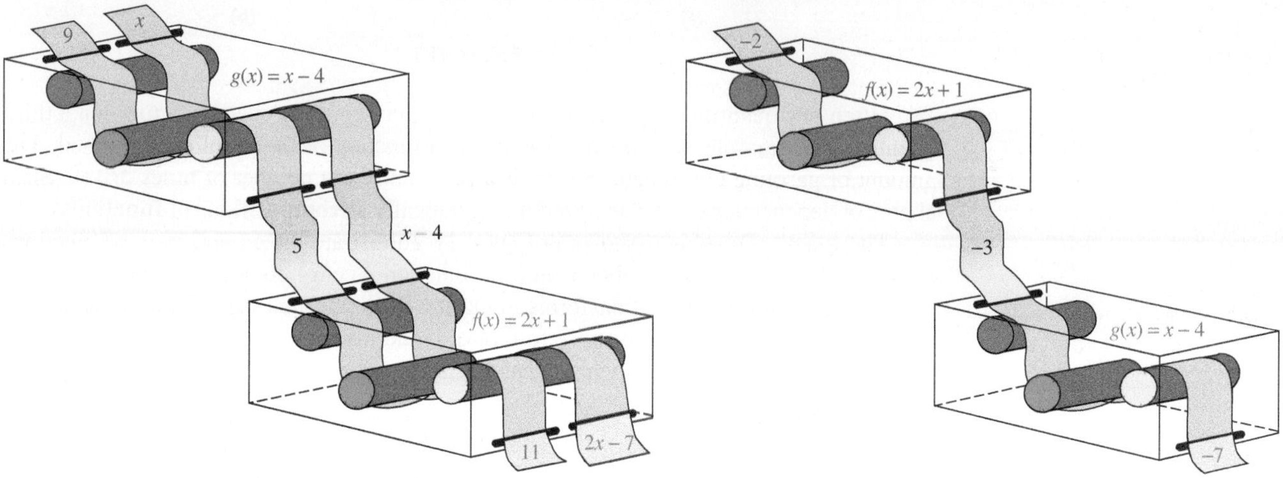

(a) (b)

Figure 11-3

b. $(f \circ g)(x)$ means $f(g(x))$. In Figure 11-3(a), function g receives the number x, subtracts 4, and releases the number $(x - 4)$. The $(x - 4)$ then goes into the f function, which doubles $(x - 4)$ and adds 1. The final result, $(2x - 7)$, is the output of the composite function $f \circ g$.

$$(f \circ g)(x) = f(\mathbf{g(x)}) = f(\mathbf{x - 4}) = 2(\mathbf{x - 4}) + 1 = 2x - 7$$

c. $(g \circ f)(-2)$ means $g(f(-2))$. In Figure 11-3(b), function f receives the number -2, doubles it and adds 1, and releases -3 into the g function. Function g subtracts 4 from -3 and releases a final result of -7. Thus,

$$(g \circ f)(-2) = g(\mathbf{f(-2)}) = g(\mathbf{-3}) = \mathbf{-3} - 4 = -7$$

SELF CHECK 3 Let $f(x) = 3x + 2$ and $g(x) = 9x - 5$. Find

a. $(f \circ g)(2)$ **b.** $(g \circ f)(x)$ **c.** $(g \circ f)(-4)$

3 Find the difference quotient of a function.

An important function in calculus, called the **difference quotient**, represents the slope of a line at a given point on the graph of a function. The difference quotient is defined as follows:

$$\frac{f(x + h) - f(x)}{h}, \quad h \neq 0$$

Unless otherwise noted, all content on this page is © Cengage Learning.

EXAMPLE 4 If $f(x) = x^2 - 4$, evaluate the difference quotient.

Solution First, we evaluate $f(x + h)$.

$$f(x) = x^2 - 4$$
$$f(x + h) = (x + h)^2 - 4 \quad \text{Substitute } x + h \text{ for } h.$$
$$= x^2 + 2xh + h^2 - 4 \quad (x + h)^2 = x^2 + 2hx + h^2$$

Then we note that $f(x) = x^2 - 4$. We can now substitute the values of $f(x + h)$ and $f(x)$ into the difference quotient and simplify.

$$\frac{f(x + h) - f(x)}{h} = \frac{(x^2 + 2xh + h^2 - 4) - (x^2 - 4)}{h}$$
$$= \frac{x^2 + 2xh + h^2 - 4 - x^2 + 4}{h} \quad \text{Use the distributive property to remove parentheses.}$$
$$= \frac{2xh + h^2}{h} \quad \text{Combine like terms.}$$
$$= \frac{h(2x + h)}{h} \quad \text{Factor out } h\text{, the GCF, in the numerator.}$$
$$= 2x + h \quad \text{Divide out } h; \frac{h}{h} = 1.$$

The difference quotient for this function simplifies as $2x + h$.

SELF CHECK 4 If $f(x) = 3x^2 - 4$, evaluate the difference quotient.

4 Solve an application requiring the composition of two functions.

EXAMPLE 5 **TEMPERATURE CHANGE** A laboratory sample is removed from a cooler at a temperature of 15° Fahrenheit. Technicians are warming the sample at a controlled rate of 3° F per hour. Express the sample's Celsius temperature as a function of the time, t (in hours), since it was removed from refrigeration.

Solution The temperature of the sample is 15° F when $t = 0$. Because it warms at 3° F per hour, it warms $(3t)°$ after t hours. The Fahrenheit temperature after t hours is given by the function

$$F(t) = 3t + 15$$

The Celsius temperature is a function of the Fahrenheit temperature, given by the formula

$$C(F) = \frac{5}{9}(F - 32)$$

To express the sample's Celsius temperature as a function of time, we find the composition function $C \circ F$.

$$(C \circ F)(t) = C(F(t))$$
$$= \frac{5}{9}(F(t) - 32)$$
$$= \frac{5}{9}[(3t + 15) - 32] \quad \text{Substitute } 3t + 15 \text{ for } F(t).$$
$$= \frac{5}{9}(3t - 17) \quad \text{Combine like terms.}$$

$= \frac{15}{9}t - \frac{85}{9}$ Use the distributive property to remove parentheses.

$= \frac{5}{3}t - \frac{85}{9}$ Simplify.

SELF CHECK 5 If the sample temperature is 51° Fahrenheit and technicians are cooling the sample at 3° F per hour, express the sample's Celsius temperature as a function of the time since it was put into refrigeration.

SELF CHECK ANSWERS **1. a.** $2x^2 + 6x - 2, (-\infty, \infty)$ **b.** $-2x^2 - 2, (-\infty, \infty)$ **2. a.** $2x^3 - 2x^2 - 3x + 3, (-\infty, \infty)$ **b.** $\frac{2x^2 - 3}{x - 1}, (-\infty, 1) \cup (1, \infty)$ **3. a.** 41 **b.** $27x + 13$ **c.** -95 **4.** $6x + 3h$ **5.** $(C \circ F)(t) = \frac{95}{9} - \frac{5}{3}t$

NOW TRY THIS

Given $f(x) = 3x - 2$ *and* $g(x) = x^2 - 5x + 1$, *find*

1. $(g \circ f)(-1)$
2. $(g \circ f)(x)$
3. $(f \circ f)(x)$

11.1 Exercises

Assume no denominators are 0.

WARM-UPS *If* $f(x) = 2x + 3$ *and* $g(x) = 4x^2 - 2$, *find the following.*

1. $f(-3)$
2. $g(-2)$
3. $g(x + 1)$
4. $f(x + 1)$

Simplify.

5. $(2x + 3) + (4x^2 - 2)$
6. $(4x^2 - 2) - (2x + 3)$
7. $(2x + 3) - (4x^2 - 2)$
8. $(2x + 3)(4x^2 - 2)$

REVIEW *Simplify each expression.*

9. $\frac{5x^2 - 13x - 6}{9 - x^2}$
10. $\frac{2x^3 + 14x^2}{3 + 2x - x^2} \cdot \frac{x^2 - 3x}{x}$
11. $\frac{8 + 2x - x^2}{12 + x - 3x^2} \div \frac{3x^2 + 5x - 2}{3x - 1}$
12. $\frac{x - 1}{1 + \frac{x}{x - 2}}$

VOCABULARY AND CONCEPTS *Fill in the blanks.*

13. $(f + g)(x) =$ ________
14. $(f - g)(x) =$ ________
15. $(f \cdot g)(x) =$ ________
16. $(f/g)(x) = \frac{\quad}{\quad}$ $(g(x) \neq 0)$
17. In Exercises 13–15, the domain of each function is the set of real numbers x that are in the ______ of both f and g.
18. The ________ of functions f and g is denoted by $(f \circ g)(x)$ or $f \circ g$.
19. $(f \circ g)(x) =$ ______
20. The difference quotient is defined as ________.
21. In calculus, the difference quotient represents the slope of a line at a ________ on a graph.
22. In the difference quotient, $h \neq$ _.

GUIDED PRACTICE *Let* $f(x) = 3x$ *and* $g(x) = 4x$. *Find each function and its domain.* SEE EXAMPLE 1. (OBJECTIVE 1)

23. $f + g$
24. $f - g$
25. $g - f$
26. $g + f$

Let $f(x) = 3x$ *and* $g(x) = 4x$. *Find each function and its domain.* SEE EXAMPLE 2. (OBJECTIVE 1)

27. $f \cdot g$
28. f/g
29. g/f
30. $g \cdot f$

Let $f(x) = 2x + 1$ and $g(x) = x - 3$. Find each function and its domain. SEE EXAMPLES 1–2. *(OBJECTIVE 1)*

31. $f + g$
32. $f - g$
33. $g - f$
34. $g + f$
35. $f \cdot g$
36. f/g
37. g/f
38. $g \cdot f$

Let $f(x) = 2x + 1$ and $g(x) = x^2 - 1$. Find each value. SEE EXAMPLE 3. *(OBJECTIVE 2)*

39. $(f \circ g)(4)$ **40.** $(g \circ f)(4)$
41. $(g \circ f)(-3)$ **42.** $(f \circ g)(-3)$
43. $(f \circ g)(0)$ **44.** $(g \circ f)(0)$
45. $(f \circ g)\left(\frac{1}{4}\right)$ **46.** $(g \circ f)\left(\frac{1}{4}\right)$
47. $(f \circ g)(x)$ **48.** $(g \circ f)(x)$
49. $(g \circ f)(3x)$ **50.** $(f \circ g)(3x)$

Find $\frac{f(x + h) - f(x)}{h}$. SEE EXAMPLE 4. *(OBJECTIVE 3)*

51. $f(x) = 4x + 5$ **52.** $f(x) = 5x - 1$
53. $f(x) = x^2$ **54.** $f(x) = x^2 - 1$
55. $f(x) = 2x^2 - 1$ **56.** $f(x) = 3x^2$
57. $f(x) = x^2 + x$ **58.** $f(x) = x^2 - x$
59. $f(x) = x^2 + 3x - 4$ **60.** $f(x) = x^2 - 4x + 3$
61. $f(x) = 2x^2 + 3x - 7$ **62.** $f(x) = 3x^2 - 2x + 4$

ADDITIONAL PRACTICE

Let $f(x) = 3x - 2$ and $g(x) = 2x^2 + 1$. Find each function and its domain.

63. $f - g$
64. $f + g$
65. f/g
66. $f \cdot g$

Let $f(x) = x^2 - 1$ and $g(x) = x^2 - 4$. Find each function and its domain.

67. $f - g$
68. $f + g$
69. g/f
70. $g \cdot f$

Let $f(x) = 3x - 2$ and $g(x) = x^2 + x$. Find each value.

71. $(f \circ g)(4)$ **72.** $(g \circ f)(4)$
73. $(g \circ f)(-2)$ **74.** $(f \circ g)(-2)$
75. $(g \circ f)(0)$ **76.** $(f \circ g)(0)$
77. $(g \circ f)(x)$ **78.** $(f \circ g)(x)$

For problems, 79–90, find $\frac{f(x) - f(a)}{x - a}$ $(x \neq a)$.

79. $f(x) = 4x - 7$ **80.** $f(x) = 6x + 5$
81. $f(x) = x^2$ **82.** $f(x) = x^2 - 1$
83. $f(x) = 2x^2 - 1$ **84.** $f(x) = 3x^2$
85. $f(x) = x^2 + x$ **86.** $f(x) = x^2 - x$
87. $f(x) = x^2 + 3x - 4$ **88.** $f(x) = x^2 - 4x + 3$
89. $f(x) = 2x^2 + 3x - 7$ **90.** $f(x) = 3x^2 - 2x + 4$

91. If $f(x) = x + 1$ and $g(x) = 2x - 5$, show that $(f \circ g)(x) \neq (g \circ f)(x)$.

92. If $f(x) = x^2 + 1$ and $g(x) = 3x^2 - 2$, show that $(f \circ g)(x) \neq (g \circ f)(x)$.

93. If $f(x) = x^2 + 2x - 3$, find $f(a)$, $f(h)$, and $f(a + h)$. Then show that $f(a + h) \neq f(a) + f(h)$.

94. If $g(x) = 2x^2 + 10$, find $g(a)$, $g(h)$, and $g(a + h)$. Then show that $g(a + h) \neq g(a) + g(h)$.

95. If $f(x) = x^3 - 1$, find $\frac{f(x + h) - f(x)}{h}$.

96. If $f(x) = x^3 + 2$, find $\frac{f(x + h) - f(x)}{h}$.

APPLICATIONS

Solve each application. SEE EXAMPLE 5. *(OBJECTIVE 4)*

97. Alloys A molten alloy must be cooled slowly to control crystallization. When removed from the furnace, its temperature is 2,700° F, and it will be cooled at 200° F per hour. Express the Celsius temperature as a function of the number of hours t since cooling began.

98. Weather forecasting A high-pressure area promises increasingly warmer weather for the next 48 hours. The temperature is now 34° Celsius and will rise 1° C every 6 hours. Express the Fahrenheit temperature as a function of the number of hours from now.

WRITING ABOUT MATH

99. Explain how to find the domain of f/g.

100. Explain why the difference quotient represents the slope of a line passing through $(x, f(x))$ and $(x + h, f(x + h))$.

SOMETHING TO THINK ABOUT

101. Is composition of functions associative? Choose functions f, g, and h and determine whether $[f \circ (g \circ h)](x) = [(f \circ g) \circ h](x)$.

102. Choose functions f, g, and h and determine whether $f \circ (g + h) = f \circ g + f \circ h$.

Section 11.2 Inverses of Functions

Objectives

1. Determine whether a function is one-to-one.
2. Apply the horizontal line test to determine whether the graph of a function is one-to-one.
3. Find the inverse of a function.

Vocabulary

one-to-one function
horizontal line test
inverse of a function
identity function

Getting Ready

Solve each equation for y.

1. $x = 3y + 2$

2. $x = \frac{3}{2}y + 5$

We already know that real numbers have inverses. For example, the additive inverse of 3 is -3, because $3 + (-3) = 0$. The multiplicative inverse of 3 is $\frac{1}{3}$, because $3\left(\frac{1}{3}\right) = 1$. In a similar way, functions have inverses. After discussing one-to-one functions, we will learn how to find the inverse of a function.

1 Determine whether a function is one-to-one.

Recall that for each input into a function, there is a single output. For some functions, different inputs have the same output, as shown in Figure 11-4(a). For other functions, different inputs have different outputs, as shown in Figure 11-4(b).

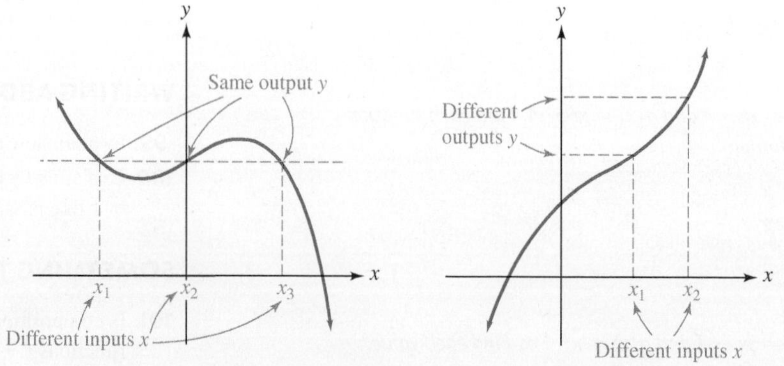

Not a one-to-one function
(a)

A one-to-one function
(b)

Figure 11-4

Unless otherwise noted, all content on this page is © Cengage Learning.

When every output of a function corresponds to exactly one input, we say that the function is *one-to-one.*

ONE-TO-ONE FUNCTIONS

A function is called **one-to-one** if each input value of x in the domain determines a different output value of y in the range.

EXAMPLE 1 Determine whether the following are one-to-one. **a.** $f(x) = x^2$ **b.** $f(x) = x^3$

Solution **a.** The function $f(x) = x^2$ is not one-to-one, because different input values x can determine the same output value y. For example, inputs of 3 and -3 produce the same output value of 9.

$$f(3) = 3^2 = 9 \qquad \text{and} \qquad f(-3) = (-3)^2 = 9$$

b. The function $f(x) = x^3$ is one-to-one, because different input values x determine different output values of y for all x. This is because different numbers have different cubes.

SELF CHECK 1 Determine whether $f(x) = 2x + 3$ is one-to-one.

2 Apply the horizontal line test to determine whether the graph of a function is one-to-one.

Similar to the vertical line test used to determine if a graph represents a function, a **horizontal line test** can be used to determine whether the graph of a function represents a one-to-one function. If every horizontal line that intersects the graph of a function does so only once, the function is one-to-one. Otherwise, the function is not one-to-one. See Figure 11-5.

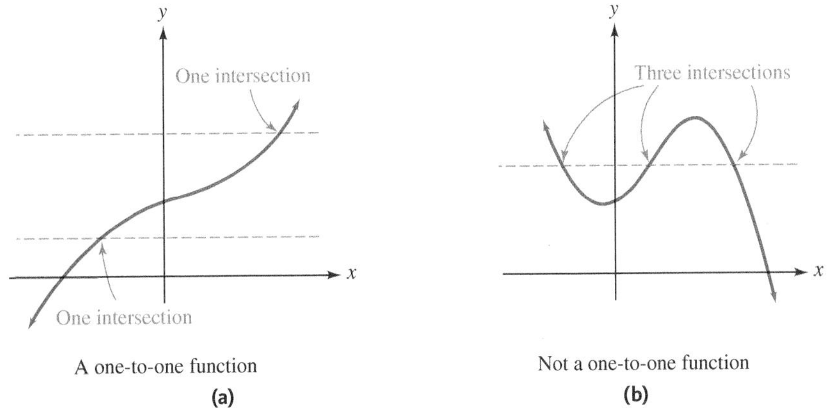

Figure 11-5

EXAMPLE 2 The graphs in Figure 11-6 on the next page represent functions. Use the horizontal line test to determine whether the graphs represent one-to-one functions.

Solution **a.** Because many horizontal lines intersect the graph shown in Figure 11-6(a) twice, the graph does not represent a one-to-one function.

b. Because each horizontal line that intersects the graph in Figure 11-6(b) does so exactly once, the graph does represent a one-to-one function.

Unless otherwise noted, all content on this page is © Cengage Learning.

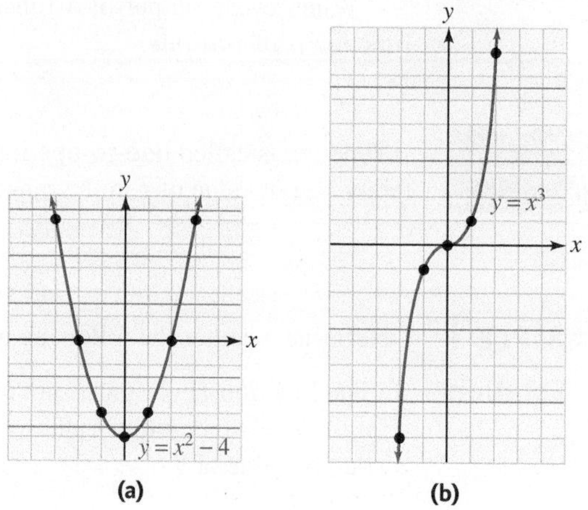

Figure 11-6

SELF CHECK 2 Does the following graph represent a one-to-one function?

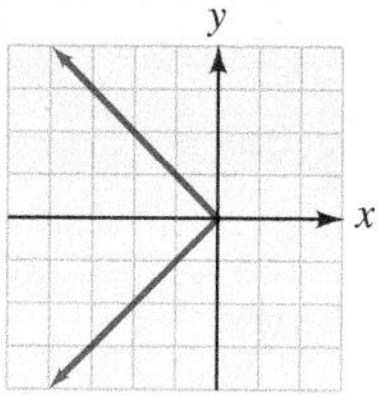

COMMENT Use the vertical line test to determine whether a graph represents a function. If it does, use the horizontal line test to determine whether the function is one to-one.

3 Find the inverse of a function.

The function defined by $C = \frac{5}{9}(F - 32)$ is the formula that we use to convert degrees Fahrenheit to degrees Celsius. If we substitute a Fahrenheit reading into the formula, we obtain a Celsius reading. For example, if we substitute 41° for F we obtain a Celsius reading of 5°.

$$C = \frac{5}{9}(F - 32)$$

$$= \frac{5}{9}(\mathbf{41} - 32) \quad \text{Substitute 41 for } F.$$

$$= \frac{5}{9}(9)$$

$$= 5$$

If we want to find a Fahrenheit reading from a Celsius reading, we need a formula into which we can substitute a Celsius reading and have a Fahrenheit reading result. Such a formula is $F = \frac{9}{5}C + 32$, which takes the Celsius reading of 5° and turns it back into a Fahrenheit reading of 41°.

$$F = \frac{9}{5}C + 32$$

$$= \frac{9}{5}(\mathbf{5}) + 32 \quad \text{Substitute 5 for } C.$$

$$= 41$$

Unless otherwise noted, all content on this page is © Cengage Learning.

The functions defined by these two formulas do opposite things. The first turns 41° F into 5° C, and the second turns 5° C back into 41° F. For this reason, we say that the functions are *inverses* of each other.

If f is the function determined by the table shown in Figure 11-7(a), it turns the number 1 into 10, 2 into 20, and 3 into 30. Since the inverse of f must turn 10 back into 1, 20 back into 2, and 30 back into 3, it consists of the ordered pairs shown in Figure 11-7(b).

Function f

x	y
1	10
2	20
3	30

↑ Domain ↑ Range

(a)

Inverse of f

x	y
10	1
20	2
30	3

↑ Domain ↑ Range

(b)

Note that the inverse of f is also a function.

Figure 11-7

We note that the domain of f and the range of its inverse is {1, 2, 3}. The range of f and the domain of its inverse is {10, 20, 30}.

This example suggests that to form the **inverse of a function** f, we simply interchange the coordinates of each ordered pair that determines f. When the inverse of a function is also a function, we call it *f inverse* and denote it with the symbol f^{-1}.

COMMENT The symbol $f^{-1}(x)$ is read as "the inverse of $f(x)$" or just "f inverse." The -1 in the notation $f^{-1}(x)$ is not an exponent.

FINDING THE INVERSE OF A ONE-TO-ONE FUNCTION

We find the inverse of a function as follows:

1. Replace $f(x)$ with y.
2. Interchange the variables x and y.
3. Solve the resulting equation for y. If the inverse is also a function, use the notation $f^{-1}(x)$.

EXAMPLE 3 If $f(x) = 4x + 2$, find the inverse of f and determine whether it is a function.

Solution Because $f(x) = 4x + 2$ is a linear equation, its graph is a line. The horizontal line test indicates that the function is one-to-one. Therefore, its inverse is also a function. To find the inverse, we replace $f(x)$ with y and interchange the positions of x and y.

$$f(x) = 4x + 2$$

$$y = 4x + 2 \quad \text{Replace } f(x) \text{ with } y.$$

$$x = 4y + 2 \quad \text{Interchange the variables } x \text{ and } y.$$

Then we solve the equation for y.

$$x = 4y + 2$$

$$x - 2 = 4y \quad \text{Subtract 2 from both sides.}$$

$$y = \frac{x - 2}{4} \quad \text{Divide both sides by 4 and write } y \text{ on the left side.}$$

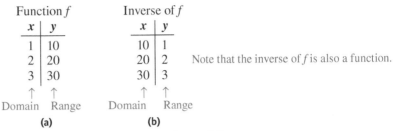

Expressing the inverse in function notation, we have

$$f^{-1}(x) = \frac{x-2}{4}$$

SELF CHECK 3 If $f(x) = -5x - 3$, find the inverse of f and determine whether it is a function.

COMMENT If the inverse of a function is also a function, we write the inverse function using function notation.

To emphasize an important relationship between a function and its inverse, we substitute some number x, such as $x = 3$, into the function $f(x) = 4x + 2$ of Example 3. The corresponding value of y produced is

$$f(3) = 4(3) + 2 = 14$$

If we substitute 14 into the inverse function, f^{-1}, the corresponding value of y that is produced is

$$f^{-1}(14) = \frac{14-2}{4} = 3$$

Thus, the function f turns 3 into 14, and the inverse function f^{-1} turns 14 back into 3. In general, *the composition of a function and its inverse is the **identity function**.*

Recall that in the real number system, 0 is called the *additive identity* because $0 + x = x$ and 1 is called the *multiplicative identity* because $1 \cdot x = x$. There is an identity for functions as well. The identity function is defined by the equation $I(x) = x$. Under this function, the value that corresponds to any real number x is x itself. If f is any function, the composition of f with the identity function is the function f:

$$(f \circ I)(x) = (I \circ f)(x) = f(x)$$

We can show this as follows:

$(f \circ I)(x)$ means $f(I(x))$. Because $I(x) = x$, we have

$$(f \circ I)(x) = f(I(x)) = f(x)$$

$(I \circ f)(x)$ means $I(f(x))$. Because I passes any number through unchanged, we have $I(f(x)) = f(x)$ and

$$(I \circ f)(x) = I(f(x)) = f(x)$$

To prove that $f(x) = 4x + 2$ and $f^{-1}(x) = \frac{x-2}{4}$ are inverse functions, we must show that their composition (in both directions) is the identity function:

$$\begin{aligned} (f \circ f^{-1})(x) &= f(f^{-1}(x)) \\ &= f\left(\frac{x-2}{4}\right) \\ &= 4\left(\frac{x-2}{4}\right) + 2 \\ &= x - 2 + 2 \\ &= x \end{aligned} \qquad \begin{aligned} (f^{-1} \circ f)(x) &= f^{-1}(f(x)) \\ &= f^{-1}(4x + 2) \\ &= \frac{4x + 2 - 2}{4} \\ &= \frac{4x}{4} \\ &= x \end{aligned}$$

Thus, $(f \circ f^{-1})(x) = (f^{-1} \circ f)(x) = x$, which is the identity function $I(x)$.

EXAMPLE 4 The set of all pairs (x, y) determined by $3x + 2y = 6$ is a function. Find its inverse function, and graph the function and its inverse on one coordinate system.

Solution To find the inverse function of $3x + 2y = 6$, we interchange x and y to obtain

$$3y + 2x = 6$$

and then solve the equation for y.

$$3y + 2x = 6$$

$$3y = -2x + 6 \quad \text{Subtract } 2x \text{ from both sides.}$$

$$y = -\frac{2}{3}x + 2 \quad \text{Divide both sides by 3.}$$

Since the resulting equation represents a function, we write it in inverse function notation as

$$f^{-1}(x) = -\frac{2}{3}x + 2$$

The graphs of $3x + 2y = 6$ and $f^{-1}(x) = -\frac{2}{3}x + 2$ appear in Figure 11-8.

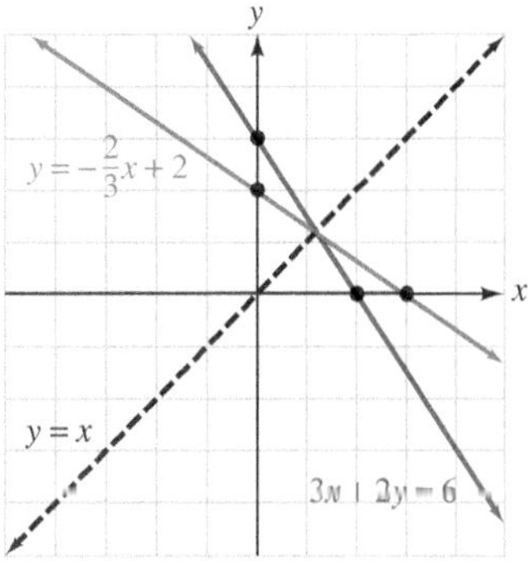

Figure 11-8

COMMENT The dashed line $y = x$ shown in Figure 11-8 is included to illustrate the symmetry of the function and its inverse.

SELF CHECK 4 Find the inverse of the function defined by $2x - 3y = 6$. Graph the function and its inverse on one coordinate system.

In Example 4, the graph of $3x + 2y = 6$ and $f^{-1}(x) = -\frac{2}{3}x + 2$ are symmetric about the line $y = x$. In general, *any function and its inverse are symmetric about the line* $y = x$, because when the coordinates (a, b) satisfy an equation, the coordinates (b, a) will satisfy its inverse.

In each example so far, the inverse of a function has been another function. This is not always true, as the following example will show.

EXAMPLE 5 Find the inverse of the function determined by $f(x) = x^2$.

Solution

$y = x^2$ Replace $f(x)$ with y.

$x = y^2$ Interchange x and y.

$y = \pm\sqrt{x}$ Use the square-root property and write y on the left side.

When the inverse $y = \pm\sqrt{x}$ is graphed as in Figure 11-9, we see that the graph does not pass the vertical line test. Thus, it is not a function.

The graph of $y = x^2$ is also shown in the figure. As expected, the graphs of $y = x^2$ and $y = \pm\sqrt{x}$ are symmetric about the line $y = x$.

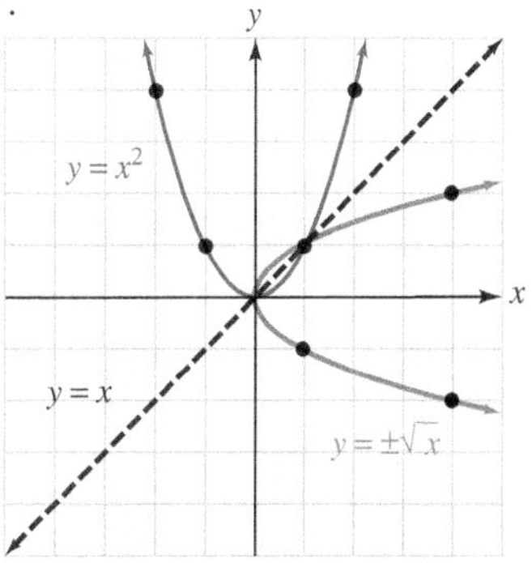

Figure 11-9

SELF CHECK 5 Find the inverse of the function determined by $f(x) = 4x^2$.

EXAMPLE 6 Find the inverse of $f(x) = x^3$.

Solution To find the inverse, we proceed as follows:

$y = x^3$ Replace $f(x)$ with y.

$x = y^3$ Interchange the variables x and y.

$\sqrt[3]{x} = y$ Take the cube root of both sides.

Unless otherwise noted, all content on this page is © Cengage Learning.

We note that to each number x there corresponds one real cube root. Thus, $y = \sqrt[3]{x}$ represents a function. In function notation, we have

$$f^{-1}(x) = \sqrt[3]{x}$$

SELF CHECK 6 Find the inverse of $f(x) = x^5$.

If a function is not one-to-one, we may be able to restrict its domain to create a one-to-one function.

EXAMPLE 7 Find the inverse of $f(x) = x^2\ (x \geq 0)$. Then determine whether the inverse is a function. Graph the function and its inverse.

Solution The domain is restricted to $x \geq 0$ so that $f(x)$ is one-to-one. The inverse of the function $f(x) = x^2$ with $x \geq 0$ is

$y = x^2$ with $x \geq 0$ — Replace $f(x)$ with y.

$x = y^2$ with $y \geq 0$ — Interchange the variables x and y.

$y = \pm\sqrt{x}$ with $y \geq 0$ — Solve for y and write y on the left side.

Considering the restriction $y \geq 0$, the equation can be written more simply as

$y = \sqrt{x}$ — This is the inverse of $f(x) = x^2$.

In this equation, each number x gives only one value of y. Thus, the inverse is a function, which we can write as

$$f^{-1}(x) = \sqrt{x}$$

The graphs of the two functions appear in Figure 11-10. The line $y = x$ is included so that we can see that the graphs are symmetric about the line $y = x$.

$y = x^2$ and $x \geq 0$

x	y	(x, y)
0	0	(0, 0)
1	1	(1, 1)
2	4	(2, 4)
3	9	(3, 9)

$x = y^2$ and $y \geq 0$

x	y	(x, y)
0	0	(0, 0)
1	1	(1, 1)
4	2	(4, 2)
9	3	(9, 3)

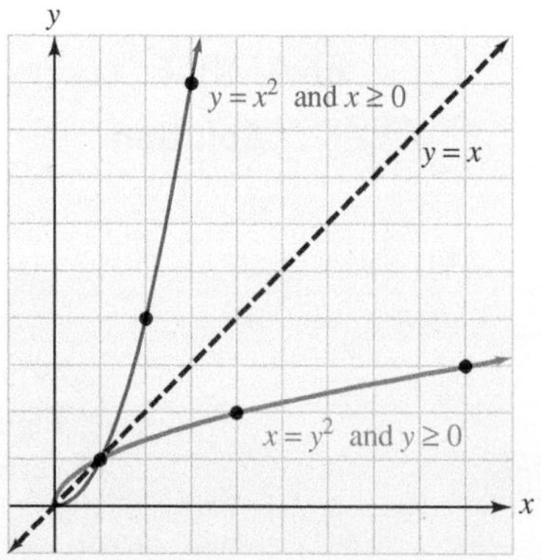

Figure 11-10

SELF CHECK 7 Find the inverse of $f(x) = x^2 - 8\quad (x \geq 0)$. Graph the function and its inverse.

Unless otherwise noted, all content on this page is © Cengage Learning.

SELF CHECK ANSWERS

1. yes **2.** no **3.** $f^{-1}(x) = -\frac{1}{5}x - \frac{3}{5}$; yes **4.** $f^{-1}(x) = \frac{3}{2}x + 3$

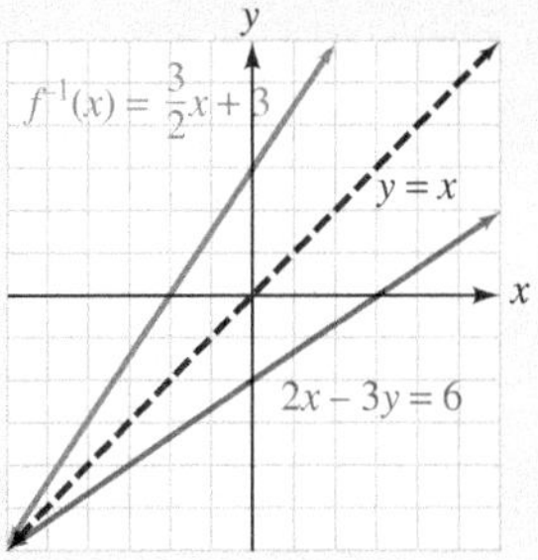

5. $y = \pm\frac{\sqrt{x}}{2}$ **6.** $f^{-1}(x) = \sqrt[5]{x}$ **7.** $f^{-1}(x) = \sqrt{x + 8}$

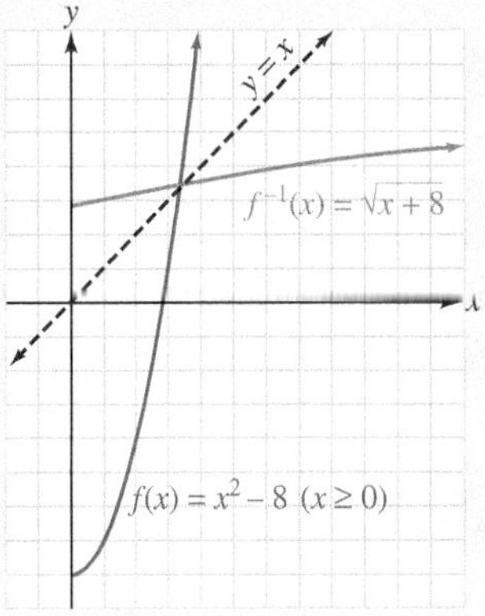

NOW TRY THIS

1. Find the inverse of $f(x) = \frac{1}{x}$.
2. Given $f(x) = \frac{5x + 1}{3x - 2}$, find $f^{-1}(x)$.

11.2 Exercises

WARM-UPS *Given $f(x) = 2x + 1$ and $g(x) = \frac{1}{2}x - \frac{1}{2}$, find the following and write the results as ordered pairs.*

1. $f(0)$ **2.** $g(1)$
3. $f(3)$ **4.** $g(7)$
5. $f(-1)$ **6.** $g(-1)$

REVIEW *Write each complex number in a + bi form or find each value.*

7. $7 - \sqrt{-49}$ **8.** $(6 - 4i) + (2 + 7i)$

9. $(3 + 4i)(2 - 3i)$ **10.** $\frac{6 + 7i}{3 - 4i}$

11. $|6 - 8i|$ **12.** $\left|\frac{2 + i}{3 - i}\right|$

VOCABULARY AND CONCEPTS *Fill in the blanks.*

13. A function is called ________ if each input determines a different output.

14. If every ________ line that intersects the graph of a function does so only once, the function is one-to-one.

15. If a one-to-one function turns an input of 2 into an output of 5, the inverse function will turn 5 into _.

16. The symbol $f^{-1}(x)$ is read as ___________ or _______.

17. $(f \circ f^{-1})(x) = (f^{-1} \circ f)(x) =$ _.

18. The graphs of a function and its inverse are symmetrical about the line _____.

GUIDED PRACTICE *Determine whether each function is one-to-one.* SEE EXAMPLE 1. (OBJECTIVE 1)

19. $f(x) = 5x$ **20.** $f(x) = |x|$
21. $f(x) = x^4$ **22.** $f(x) = x^3 - 2$

Unless otherwise noted, all content on this page is © Cengage Learning.

Each graph represents a function. Use the horizontal line test to determine whether the function is one-to-one. SEE EXAMPLE 2. (OBJECTIVE 2)

23.

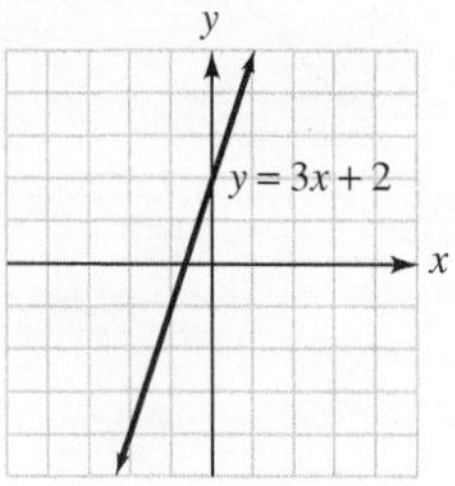

24.

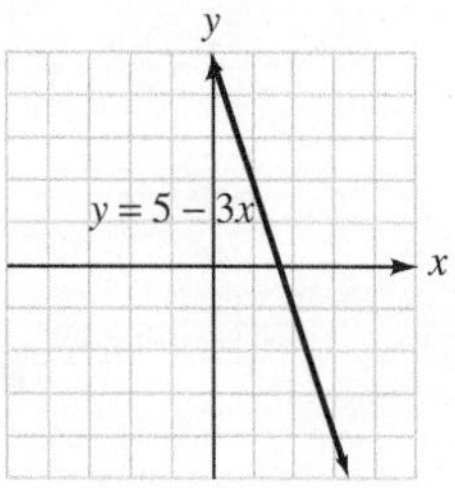

25.

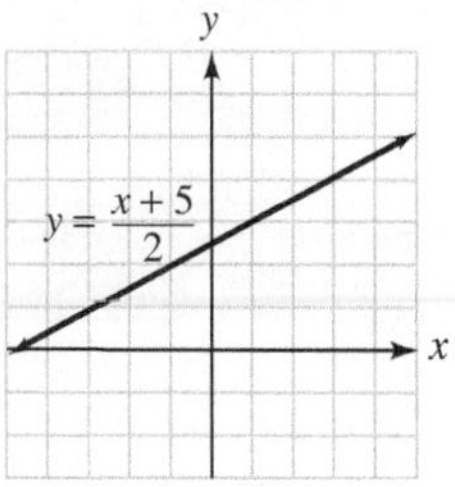

26.

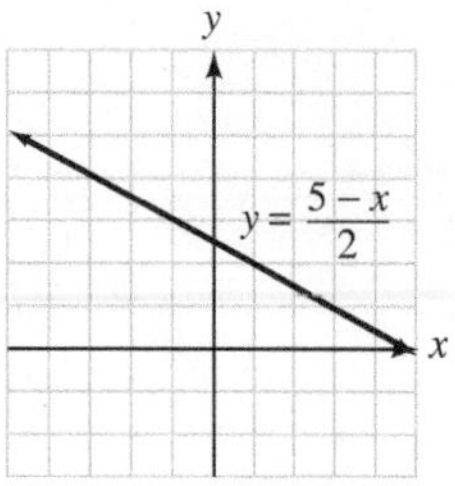

27.

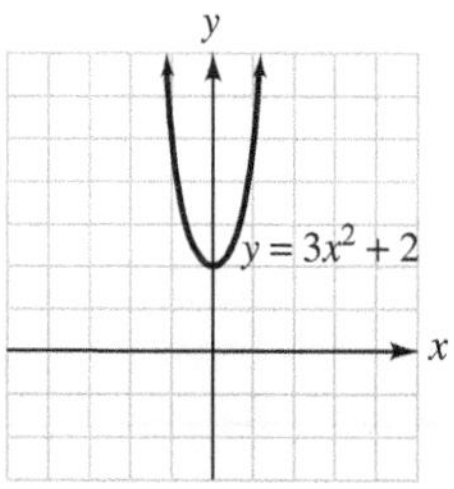

28.

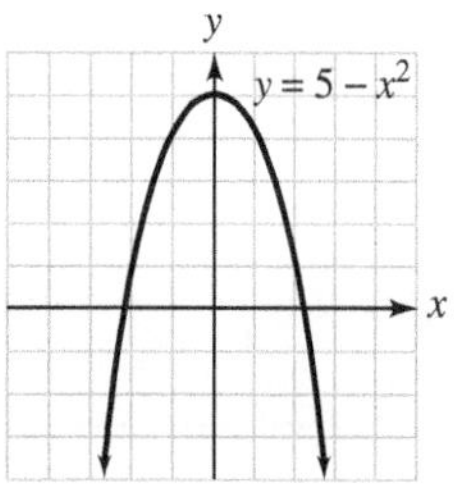

29.

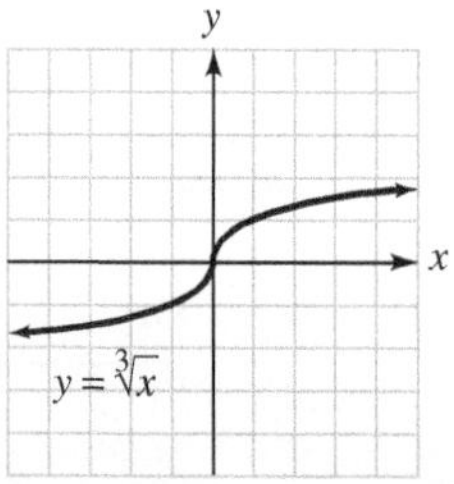

30.

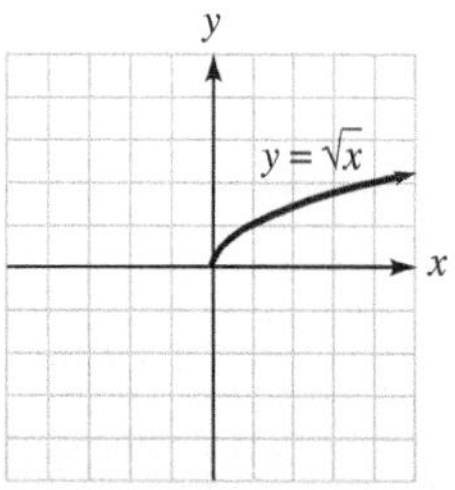

Find the inverse of each set of ordered pairs (x, y) and determine whether the inverse is a function. (OBJECTIVE 3)

31. $\{(4,3),(3,2),(2,1)\}$

32. $\{(4,1),(5,1),(6,1),(7,1)\}$

33. $\{(1,2),(2,3),(1,3),(1,5)\}$

34. $\{(-1,-1),(0,0),(1,1),(2,2)\}$

Find the inverse of each function and express it in the form $y = f^{-1}(x)$. Verify each result by showing that $(f \circ f^{-1})(x) = (f^{-1} \circ f)(x) = x$. SEE EXAMPLE 3. (OBJECTIVE 3)

35. $f(x) = 4x + 1$

36. $y + 1 = 5x$

37. $x + 4 = 5y$

38. $x = 3y + 1$

Find the inverse of each function. Then graph the function and its inverse on one coordinate system. Draw the line of symmetry on the graph. SEE EXAMPLE 4. (OBJECTIVE 3)

39. $y = 4x + 3$

40. $x = 3y - 1$

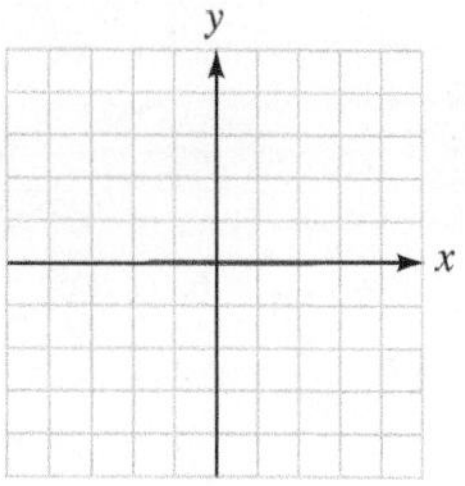

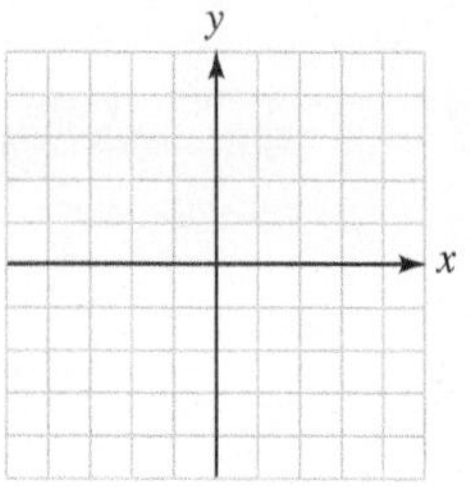

41. $x = \dfrac{y - 2}{3}$

42. $y = \dfrac{x + 3}{4}$

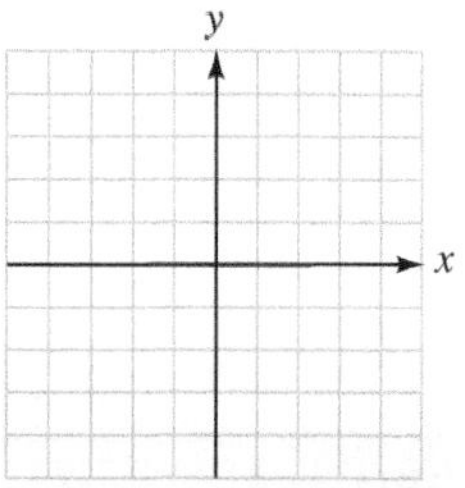

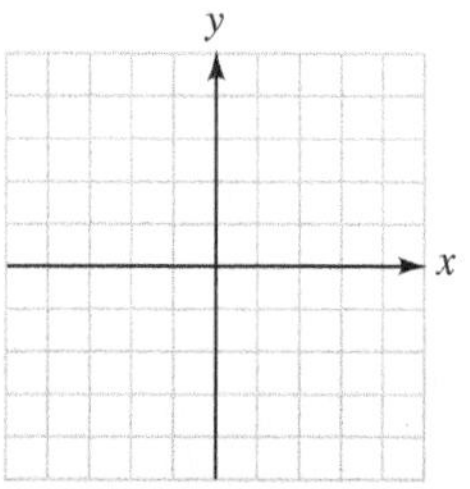

Find the inverse of each function and determine whether it is a function. If it is a function, express it in function notation. SEE EXAMPLES 5–6. (OBJECTIVE 3)

43. $y = x^2 + 9$

44. $y = x^2 + 5$

45. $y = x^3$

46. $xy = 4$

Graph each equation and its inverse on one set of coordinate axes. Draw the line of symmetry. SEE EXAMPLE 7. (OBJECTIVE 3)

47. $y = x^2 + 1, x \geq 0$

48. $y = \dfrac{1}{4}x^2 - 3, x \geq 0$

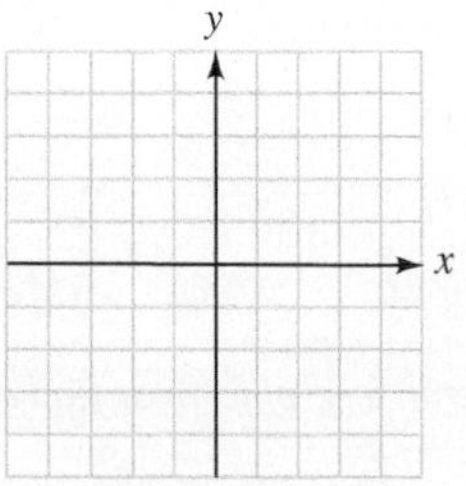

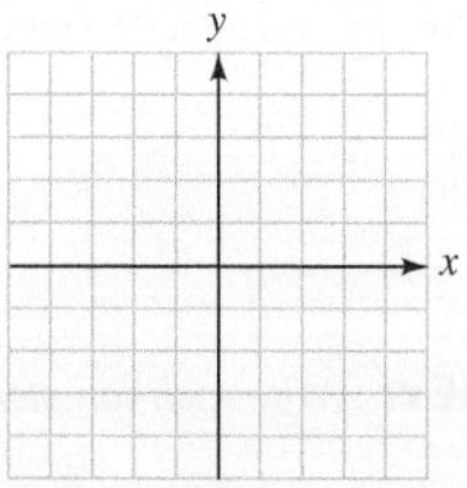

49. $y = \sqrt{x}, x \geq 0$

50. $y = -\sqrt{x}, x \geq 0$

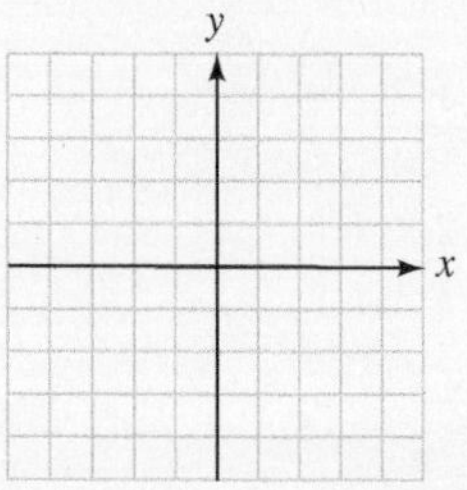

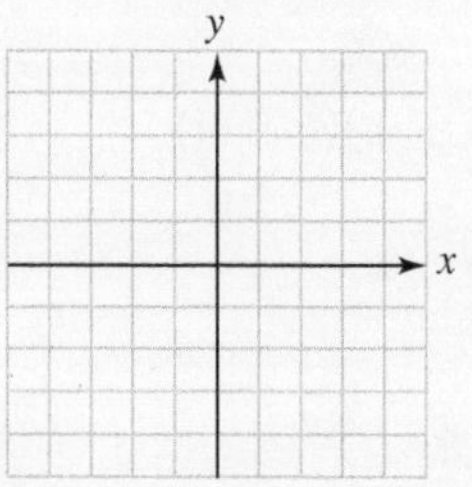

Unless otherwise noted, all content on this page is © Cengage Learning.

ADDITIONAL PRACTICE *Find the inverse of each function. Do not rationalize denominators.*

51. $\{(1, 1), (2, 4), (3, 9), (4, 16)\}$

52. $\{(1, 1), (2, 1), (3, 1), (4, 1)\}$

53. $y = -\sqrt[3]{x}$

54. $y = \sqrt[3]{x}$

55. $f(x) = 2x^3 - 3$

56. $f(x) = \dfrac{3}{x^3} - 1$

57. $f(x) = \dfrac{x - 7}{3}$

58. $f(x) = \dfrac{2x + 6}{3}$

59. $4x - 5y = 20$

60. $3x + 5y = 15$

Graph each equation and its inverse on one set of coordinate axes. Draw the line of symmetry.

61. $3x - y = 5$

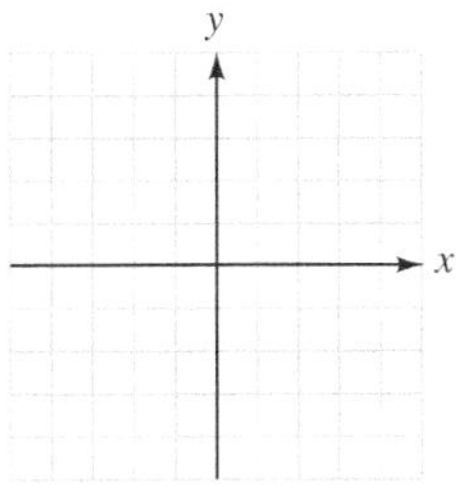

62. $2x + 3y = 9$

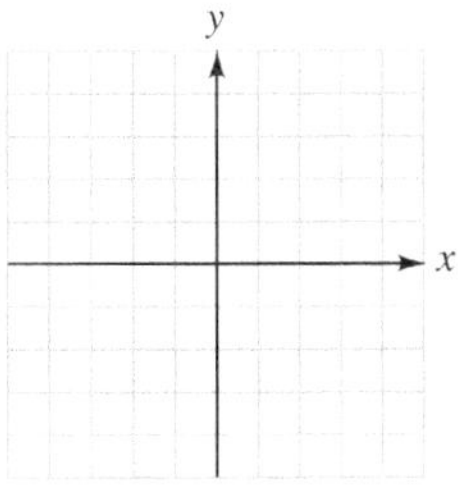

63. $3(x + y) = 2x + 4$

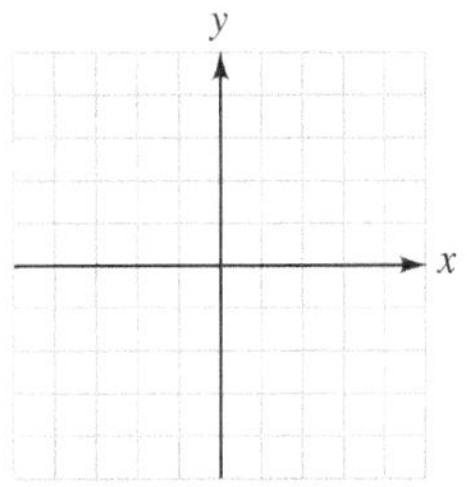

64. $-4(y - 1) + x = 2$

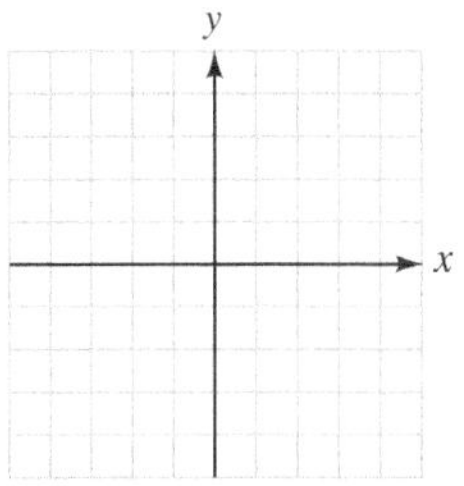

WRITING ABOUT MATH

65. Explain the purpose of the vertical line test.

66. Explain the purpose of the horizontal line test.

SOMETHING TO THINK ABOUT

67. Find the inverse of $y = \dfrac{x + 1}{x - 1}$.

68. Using the functions of Exercise 67, show that $(f \circ f^{-1})(x) = x$.

Section 11.3 Exponential Functions

Objectives

1. Graph an exponential function.
2. Graph a translation of an exponential function.
3. Evaluate an application containing an exponential function.

Vocabulary

exponential function	compound interest	periodic interest rate
increasing function	present value	compounding period
decreasing function	future value	

Unless otherwise noted, all content on this page is © Cengage Learning.

Getting Ready

Find each value.

1. 2^3 **2.** $25^{1/2}$ **3.** 5^{-2} **4.** $\left(\frac{3}{2}\right)^{-3}$

The graph in Figure 11-11 shows the balance in an investment account in which \$10,000 was invested in 2000 at 9% annual interest, compounded monthly. The graph shows that in the year 2010, the value of the account was \$25,000, and in the year 2030, the value will be approximately \$147,000. The curve shown in Figure 11-11 is the graph of a function called an *exponential function*, the topic of this section.

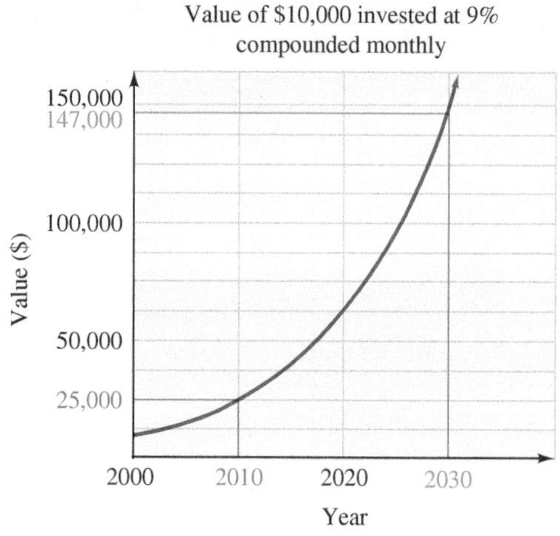

Figure 11-11

1 Graph an exponential function.

If $b > 0$ and $b \neq 1$, the function $f(x) = b^x$ is called an **exponential function**. Since x can be any real number, its domain is the set of real numbers. This is the interval $(-\infty, \infty)$. Since b is positive, the value of $f(x)$ is positive and the range is the set of positive numbers. This is the interval $(0, \infty)$.

Since $b \neq 1$, an exponential function cannot be the constant function $f(x) = 1^x$, in which $f(x) = 1$ for every real number x.

EXPONENTIAL FUNCTIONS

An exponential function with base b is defined by the equation

$$f(x) = b^x \ (b > 0, b \neq 1, \text{ and } x \text{ is a real number})$$

The domain of any exponential function is the interval $(-\infty, \infty)$. The range is the interval $(0, \infty)$.

Since the domain and range of $f(x) = b^x$ are subsets of the real numbers, we can graph exponential functions on a rectangular coordinate system.

Unless otherwise noted, all content on this page is © Cengage Learning.

EXAMPLE 1 Graph: $f(x) = 2^x$

Solution To graph $f(x) = 2^x$, we find several points (x, y) whose coordinates satisfy the equation, plot the points, and join them with a smooth curve, as shown in Figure 11-12.

$f(x) = 2^x$

x	$f(x)$	$(x, f(x))$
-1	$\frac{1}{2}$	$\left(-1, \frac{1}{2}\right)$
0	1	$(0, 1)$
1	2	$(1, 2)$
2	4	$(2, 4)$
3	8	$(3, 8)$

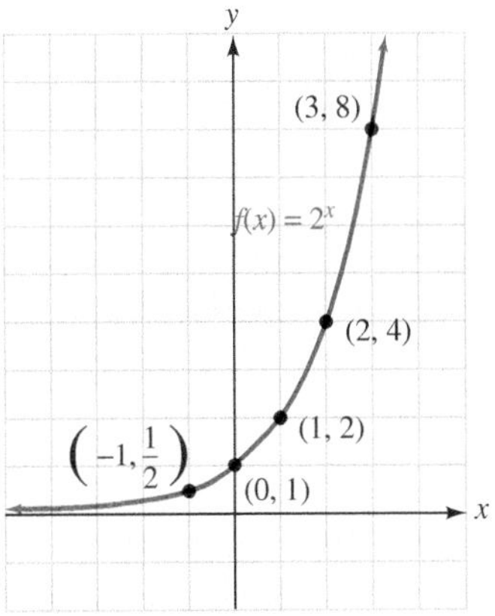

Figure 11-12

By looking at the graph, we can verify that the domain is the interval $(-\infty, \infty)$ and that the range is the interval $(0, \infty)$.

Note that as x decreases, the values of $f(x)$ decrease and approach 0, but will never be 0. Thus, the x-axis is the horizontal asymptote of the graph.

Also note that the graph of $f(x) = 2^x$ passes through the points $(0, 1)$ and $(1, 2)$.

SELF CHECK 1 Graph: $f(x) = 4^x$

In Example 1, as the values of x increase the values of $f(x)$ increase. When the graph of a function rises as we move to the right, we call the function an **increasing function**. When $b > 1$, the larger the value of b, the steeper the curve.

EXAMPLE 2 Graph: $f(x) = \left(\frac{1}{2}\right)^x$

Solution We find and plot pairs (x, y) that satisfy the equation. The graph of $f(x) = \left(\frac{1}{2}\right)^x$ appears in Figure 11-13.

$f(x) = \left(\frac{1}{2}\right)^x$

x	$f(x)$	$(x, f(x))$
-2	4	$(-2, 4)$
-1	2	$(-1, 2)$
0	1	$(0, 1)$
1	$\frac{1}{2}$	$\left(1, \frac{1}{2}\right)$

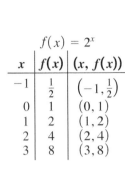

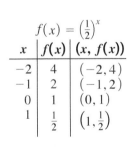

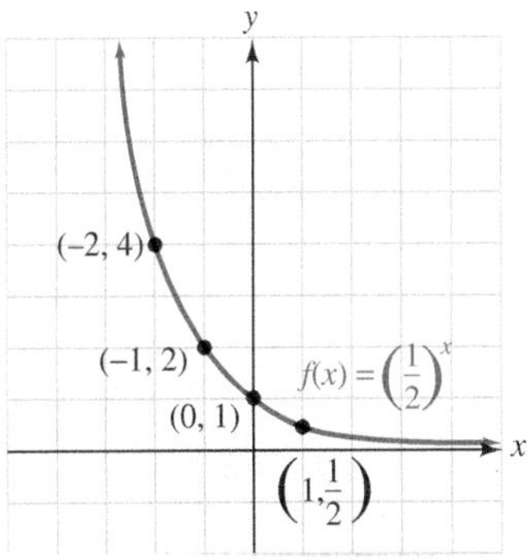

Figure 11-13

By looking at the graph, we can see that the domain is the interval $(-\infty, \infty)$ and that the range is the interval $(0, \infty)$.

In this case, as x increases, the values of $f(x)$ decrease and approach 0. The x-axis is a horizontal asymptote. Note that the graph of $f(x) = \left(\frac{1}{2}\right)^x$ passes through the points $(0, 1)$ and $\left(1, \frac{1}{2}\right)$.

Unless otherwise noted, all content on this page is © Cengage Learning.

SELF CHECK 2 Graph: $f(x) = \left(\frac{1}{4}\right)^x$

In Example 2, as the values of x increase the values of $f(x)$ decrease. When the graph of a function drops as we move to the right, we call the function a **decreasing function**. When $0 < b < 1$, the smaller the value of b, the steeper the curve.

Examples 1 and 2 illustrate the following properties of exponential functions.

PROPERTIES OF EXPONENTIAL FUNCTIONS

The domain of the exponential function $f(x) = b^x$ is the interval $(-\infty, \infty)$.
The range is the interval $(0, \infty)$.
The graph has a y-intercept of $(0, 1)$.
The x-axis is an asymptote of the graph.
The graph of $f(x) = b^x$ passes through the point $(1, b)$.
If $b > 1$, then $f(x) = b^x$ is an increasing function.

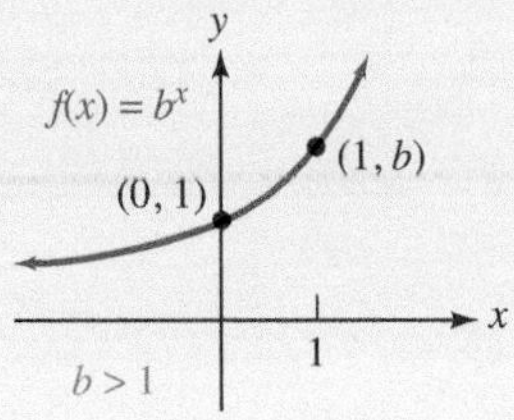

Increasing function

If $0 < b < 1$, then $f(x) = b^x$ is a decreasing function.

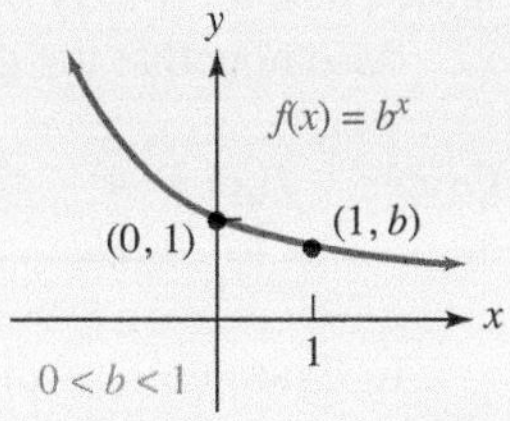

Decreasing function

EXAMPLE 3

From the graph of $f(x) = b^x$ shown in Figure 11-14, find the value of b.

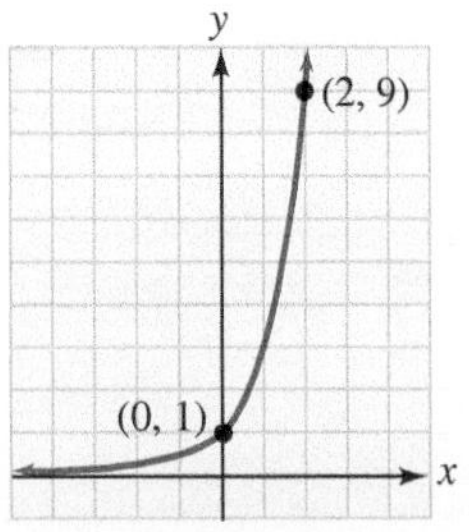

Figure 11-14

Solution We first note that the graph passes through $(0, 1)$. Since the point $(2, 9)$ is on the graph, we substitute 9 for y and 2 for x in the equation $y = b^x$ to obtain

$$y = b^x$$
$$9 = b^2$$
$$3 = b \quad \text{Because 3 is the positive number whose square is 9}$$

The base b is 3.

SELF CHECK 3 From the graph of $f(x) = b^x$ shown to the right, find the value of b.

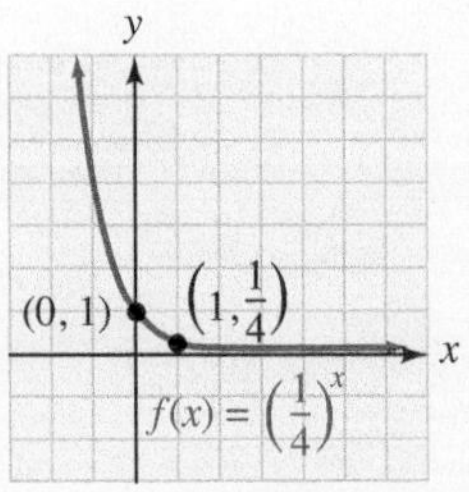

Unless otherwise noted, all content on this page is © Cengage Learning.

Accent on technology

▸ Graphing One-To-One Exponential Functions

To use a graphing calculator to graph $f(x) = \left(\frac{2}{3}\right)^x$ and $f(x) = \left(\frac{3}{2}\right)^x$, we enter the right sides of the equations. The screen will show the following equations.

$\backslash Y_1 = (2/3)$^X

$\backslash Y_2 = (3/2)$^X

If we use window settings of $[-10, 10]$ for x and $[-2, 10]$ for y and press **GRAPH**, we will obtain the graph shown in Figure 11-15.

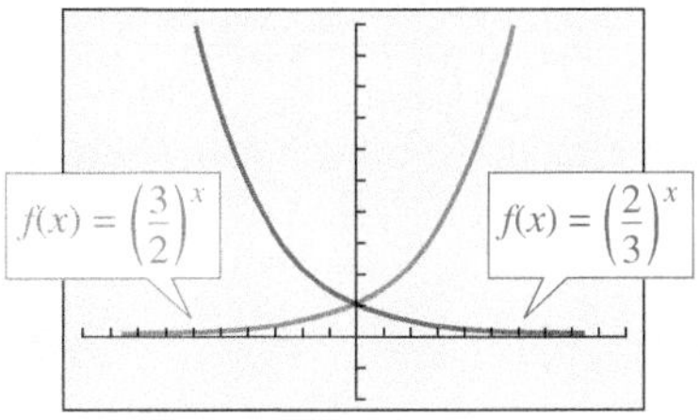

Figure 11-15

The graph of $f(x) = \left(\frac{2}{3}\right)^x$ passes through the points $(0, 1)$ and $\left(1, \frac{2}{3}\right)$. Since $\frac{2}{3} < 1$, the function is decreasing.

The graph of $f(x) = \left(\frac{3}{2}\right)^x$ passes through the points $(0, 1)$ and $\left(1, \frac{3}{2}\right)$. Since $\frac{3}{2} > 1$, the function is increasing.

Since both graphs pass the horizontal line test, each function is one-to-one.

For instructions regarding the use of a Casio graphing calculator, please refer to the Casio Keystroke Guide in the back of the book.

2 Graph a translation of an exponential function.

We have seen that when $k > 0$ the graph of

$y = f(x) + k$ is the graph of $y = f(x)$ translated k units upward.

$y = f(x) - k$ is the graph of $y = f(x)$ translated k units downward.

$y = f(x - k)$ is the graph of $y = f(x)$ translated k units to the right.

$y = f(x + k)$ is the graph of $y = f(x)$ translated k units to the left.

EXAMPLE 4 On one set of axes, graph $f(x) = 2^x$ and $f(x) = 2^x + 3$.

Solution The graph of $f(x) = 2^x + 3$ is identical to the graph of $f(x) = 2^x$, except that it is translated 3 units upward. (See Figure 11-16.)

$f(x) = 2^x$

x	$f(x)$	$(x, f(x))$
-4	$\frac{1}{16}$	$\left(-4, \frac{1}{16}\right)$
0	1	$(0, 1)$
2	4	$(2, 4)$

$f(x) = 2^x + 3$

x	$f(x)$	$(x, f(x))$
-4	$3\frac{1}{16}$	$\left(-4, 3\frac{1}{16}\right)$
0	4	$(0, 4)$
2	7	$(2, 7)$

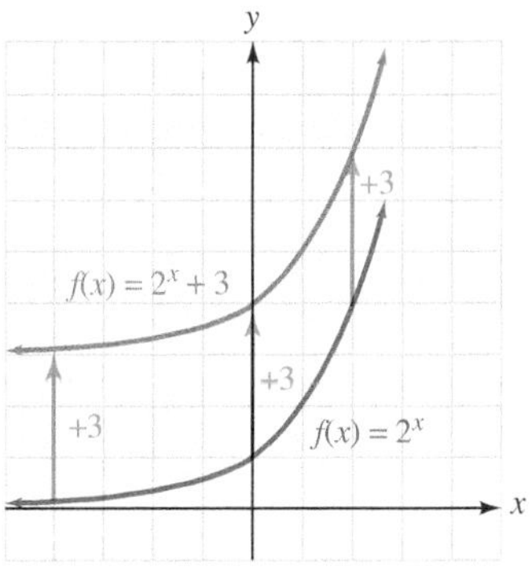

Figure 11-16

SELF CHECK 4 On one set of axes, graph $f(x) = 4^x$ and $f(x) = 4^x - 3$.

Unless otherwise noted, all content on this page is © Cengage Learning.

EXAMPLE 5 On one set of axes, graph $f(x) = 2^x$ and $f(x) = 2^{(x+3)}$.

Solution The graph of $f(x) = 2^{(x+3)}$ is identical to the graph of $f(x) = 2^x$, except that it is translated 3 units to the left. (See Figure 11-17.)

$f(x) = 2^x$

x	$f(x)$	$(x, f(x))$
0	1	(0, 1)
2	4	(2, 4)
4	16	(4, 16)

$f(x) = 2^{(x+3)}$

x	$f(x)$	$(x, f(x))$
0	8	(0, 8)
−1	4	(−1, 4)
−2	2	(−2, 2)

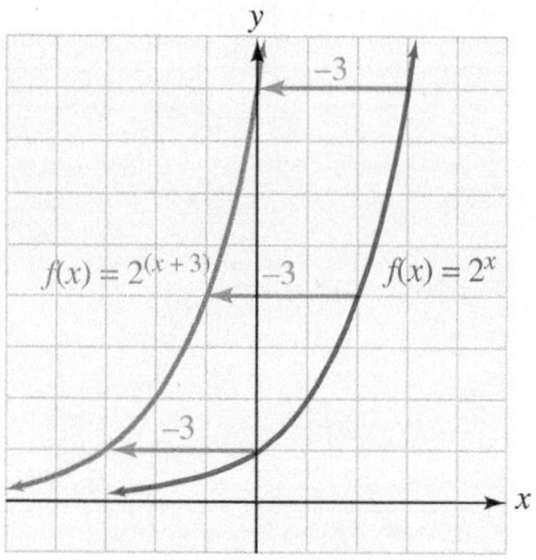

Figure 11-17

SELF CHECK 5 On one set of axes, graph $f(x) = 4^x$ and $f(x) = 4^{(x-3)}$.

The graphs of $f(x) = kb^x$ and $f(x) = b^{kx}$ are vertical and horizontal stretchings, respectively, of the graph of $f(x) = b^x$. To graph these functions, we can plot several points and join them with a smooth curve or use a graphing calculator.

Accent on technology

▸ Graphing Exponential Functions

To use a TI-84 graphing calculator to graph the exponential function $f(x) = 3(2^{x/3})$, we enter the right side of the equation. The display will show the equation

\Y$_1$ = 3(2^(X/3))

If we use window settings of $[-10, 10]$ for x and $[-2, 18]$ for y and press **GRAPH**, we will obtain the graph shown in Figure 11-18.

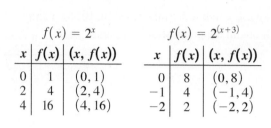

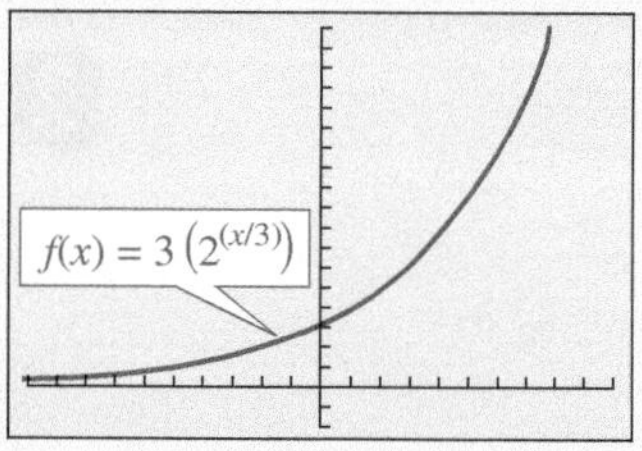

Figure 11-18

For instructions regarding the use of a Casio graphing calculator, please refer to the Casio Keystroke Guide in the back of the book.

3 Evaluate an application containing an exponential function.

EXAMPLE 6 **CELL PHONE GROWTH** The worldwide cellular telephone industry continues to experience exponential growth. The exponential function $S(n) = 1.6(1.173)^n$ approximates the number of cellular telephone subscribers in billions from 2004 to 2011, where n is the number of years since 2004.

a. How many subscribers were there in 2004?

b. How many subscribers were there in 2011?

Solution **a.** To find the number of subscribers in 2004, we substitute 0 for n in the function and find $S(0)$.

$$S(n) = 1.6(1.173)^n$$

$$S(\mathbf{0}) = 1.6(1.173)^{\mathbf{0}} \quad \text{Substitute 0 for } n.$$

$$= 1.6 \cdot 1 \quad (1.173)^0 = 1$$

$$= 1.6$$

Unless otherwise noted, all content on this page is © Cengage Learning.

In 2004, there were approximately 1.6 billion cellular telephone subscribers in the world.

b. To find the number of subscribers in 2011, we substitute 7 for n in the function and find $S(7)$.

$$S(n) = 1.6(1.173)^n$$
$$S(7) = 1.6(1.173)^7 \quad \text{Substitute 7 for } n.$$
$$\approx 4.88884274 \quad \text{Use a calculator to find an approximation.}$$

In 2011, there were approximately 4.89 billion cellular telephone subscribers.

SELF CHECK 6 If the cellular telephone industry continues to grow at the same rate, approximate the number of subscribers in 2020.

Calculating compound interest is another application that involves an exponential function.

If we deposit $\$P$ in an account paying an annual simple interest rate r, we can find the amount A in the account at the end of t years by using the formula $A = P + Prt$ or $A = P(1 + rt)$.

Suppose that we deposit \$500 in an account that pays interest every six months. Then $P = 500$, and after six months $\left(\frac{1}{2} \text{ year}\right)$, the amount in the account will be

$$A = 500(1 + rt)$$
$$= 500\left(1 + r\cdot\frac{1}{2}\right) \quad \text{Substitute } \tfrac{1}{2} \text{ for } t.$$
$$= 500\left(1 + \frac{r}{2}\right)$$

The account will begin the second six-month period with a value of $\$500\left(1 + \frac{r}{2}\right)$. After the second six-month period, the amount will be

$$A = P(1 + rt)$$
$$A = \left[500\left(1 + \frac{r}{2}\right)\right]\left(1 + r\cdot\frac{1}{2}\right) \quad \text{Substitute } 500\left(1 + \tfrac{r}{2}\right) \text{ for } P \text{ and } \tfrac{1}{2} \text{ for } t.$$
$$= 500\left(1 + \frac{r}{2}\right)\left(1 + \frac{r}{2}\right)$$
$$= 500\left(1 + \frac{r}{2}\right)^2$$

At the end of the third six-month period, the amount in the account will be

$$A = 500\left(1 + \frac{r}{2}\right)^3$$

In this discussion, the earned interest is deposited back in the account and also earns interest. When this is the case, we say that the account is earning **compound interest**.

FORMULA FOR COMPOUND INTEREST

If $\$P$ is deposited in an account and interest is paid k times a year at an annual rate r, the amount A in the account after t years is given by

$$A = P\left(1 + \frac{r}{k}\right)^{kt}$$

EXAMPLE 7 **SAVING FOR COLLEGE** To save for college, parents invest \$12,000 for their newborn child in a mutual fund that should average a 10% annual return. If the quarterly interest is reinvested, how much will be available in 18 years?

Solution We substitute 12,000 for P, 0.10 for r, and 18 for t into the formula for compound interest and calculate A. Since interest is paid quarterly, $k = 4$.

$$A = P\left(1 + \frac{r}{k}\right)^{kt}$$

$$A = \mathbf{12{,}000}\left(1 + \frac{\mathbf{0.10}}{\mathbf{4}}\right)^{\mathbf{4(18)}}$$

$$= 12{,}000(1 + 0.025)^{72}$$

$$= 12{,}000(1.025)^{72}$$

$$\approx 71{,}006.74 \qquad \text{Use a calculator.}$$

In 18 years, the account should be worth \$71,006.74.

SELF CHECK 7 How much would be available if the parents invest \$20,000?

In business applications, the initial amount of money deposited is the **present value** (PV). The amount to which the money will grow is called the **future value** (FV). The interest rate used for each compounding period is the **periodic interest rate** (i), and the number of times interest is compounded is the number of **compounding periods** (n). Using these definitions, we have an alternative formula for compound interest.

FORMULA FOR FUTURE VALUE

$$FV = PV(1 + i)^n \quad \text{where } i = \frac{r}{k} \text{ and } n = kt$$

This formula appears on business calculators. To use this formula to solve Example 7, we proceed as follows:

$$FV = PV(1 + i)^n$$

$$FV = \mathbf{12{,}000}(1 + \mathbf{0.025})^{\mathbf{72}} \qquad i = \tfrac{0.10}{4} = 0.025 \text{ and } n = 4(18) = 72$$

$$\approx \$71{,}006.74$$

Accent on technology

▸ Solving Investment Problems

Suppose \$1 is deposited in an account earning 6% annual interest, compounded monthly. To use a graphing calculator to estimate how much money will be in the account in 100 years, we can substitute 1 for P, 0.06 for r, and 12 for k into the formula

$$A = P\left(1 + \frac{r}{k}\right)^{kt}$$

$$A = \mathbf{1}\left(1 + \frac{\mathbf{0.06}}{\mathbf{12}}\right)^{\mathbf{12}t}$$

and simplify to obtain

$$A = (1.005)^{12t}$$

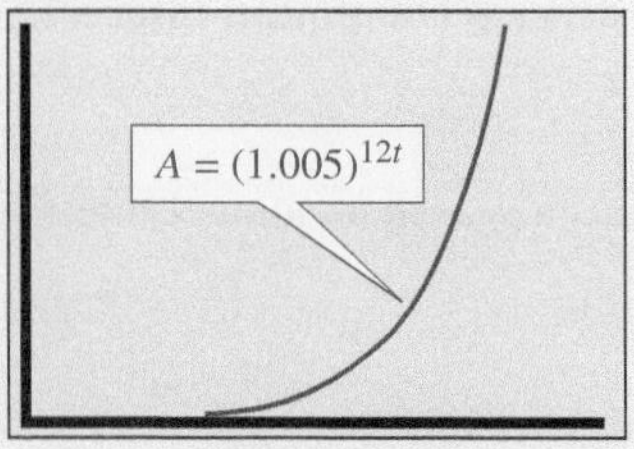

Figure 11-19

We now graph $A = (1.005)^{12t}$ using a TI-84 calculator with window settings of $[0, 120]$ for t and $[0, 400]$ for A to obtain the graph shown in Figure 11-19. From the graph, we can see that the money grows slowly in the early years and rapidly in the later years.

Unless otherwise noted, all content on this page is © Cengage Learning.

We can use the VALUE feature under the CALC menu to determine that exactly $397.44 will be in the account in 100 years.

For instructions regarding the use of a Casio graphing calculator, please refer to the Casio Keystroke Guide in the back of the book.

SELF CHECK ANSWERS

1.

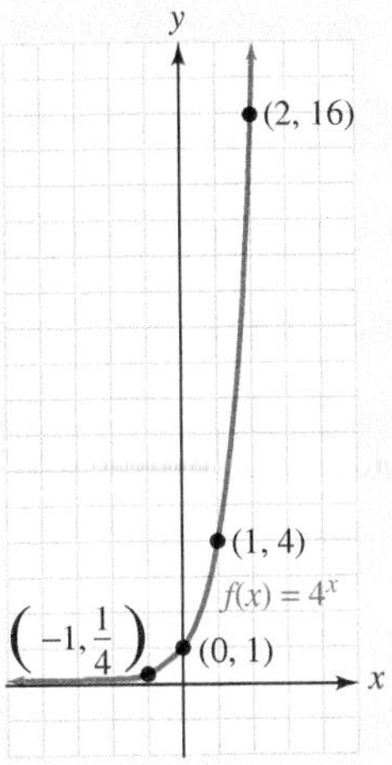

2.

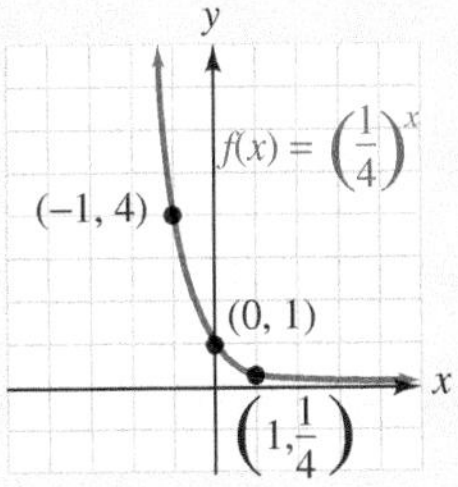

3. $\frac{1}{4}$

4.

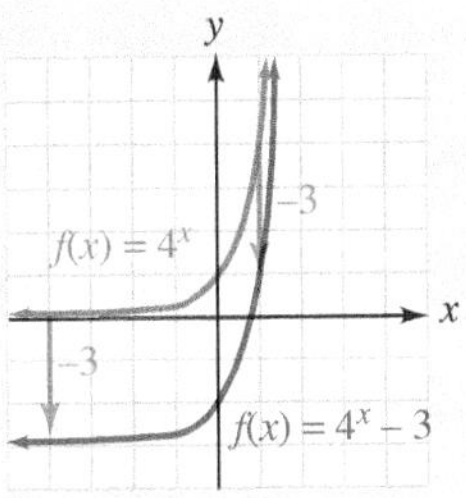

5.

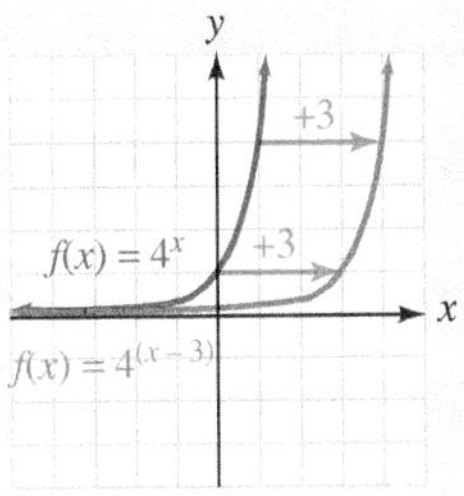

6. ≈ 20.55 billion **7.** $118,344.56

NOW TRY THIS

If you were given $1 on May 1, $2 on May 2, $4 on May 3, $8 on May 4, and double the previous day's earnings each day after, how much would you earn on

a. May 10?
b. May 15?
c. May 21?

Try to find an equation to model this situation.

11.3 Exercises

WARM-UPS *If $x = 2$, evaluate each expression.*

1. 3^x **2.** 7^x

3. $3(6^x)$ **4.** $5^{(x-1)}$

If $x = -2$, evaluate each expression.

5. 3^x **6.** 7^x

7. $3(6^x)$ **8.** $5^{(x-1)}$

REVIEW *In the illustration, lines r and s are parallel.*

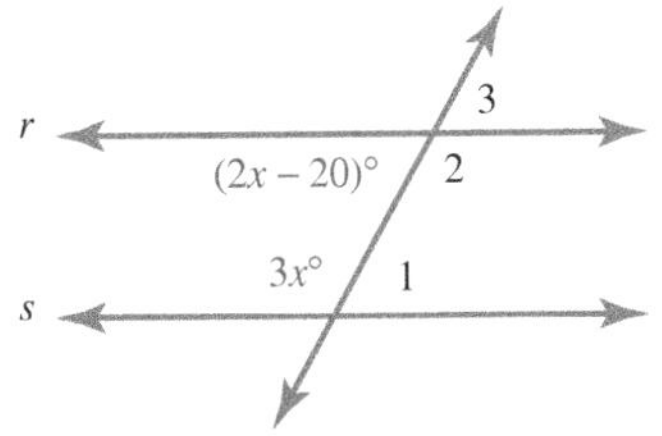

Unless otherwise noted, all content on this page is © Cengage Learning.

9. Find the value of x.

10. Find the measure of $\angle 1$.

11. Find the measure of $\angle 2$.

12. Find the measure of $\angle 3$.

VOCABULARY AND CONCEPTS *Fill in the blanks.*

13. If $b > 0$ and $b \neq 1$, $f(x) = b^x$ is called an ________ function.

14. The ______ of an exponential function is $(-\infty, \infty)$.

15. The range of an exponential function is the interval ______.

16. The graph of $f(x) = 3^x$ passes through the points $(0, \_)$ and $(1, \_)$.

17. If $b > 1$, then $f(x) = b^x$ is an ________ function.

18. If $0 < b < 1$, then $f(x) = b^x$ is a ________ function.

19. A formula for compound interest is $A =$ ________.

20. An alternative formula for compound interest is $FV =$ ________, where PV stands for present value, i stands for the ________________, and n stands for the number of ________________.

21. The graph of $f(x) = b^x$ passes through the point $(1, \_)$.

22. The _ axis is an asymptote of the graph $f(x) = b^x$.

23. $f(x) = b^{(x+1)}$ is the graph of $f(x) = b^x$ translated 1 unit to the ___.

24. $f(x) = b^x + 1$ is the graph of $f(x) = b^x$ translated 1 unit ______.

GUIDED PRACTICE *Graph each exponential function. Check your work with a graphing calculator.* SEE EXAMPLE 1. (OBJECTIVE 1)

25. $f(x) = 3^x$

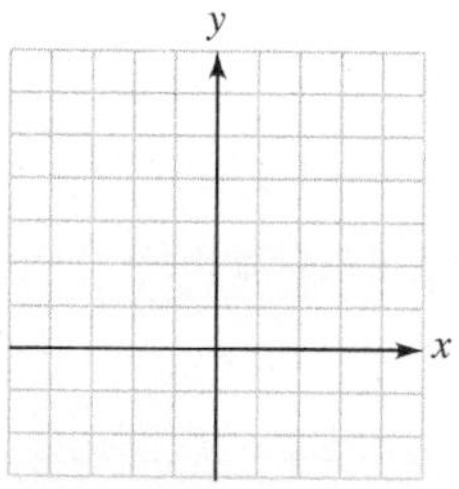

26. $f(x) = 5^x$

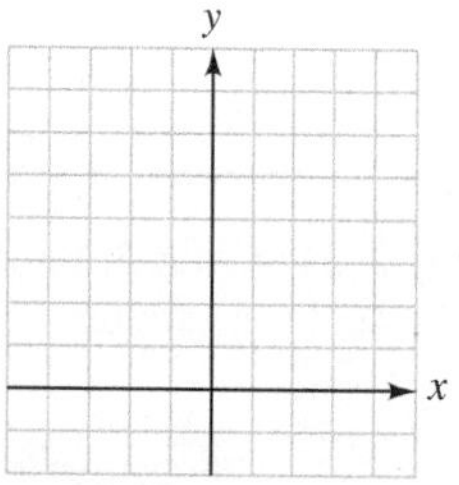

27. $f(x) = \left(\frac{5}{2}\right)^x$

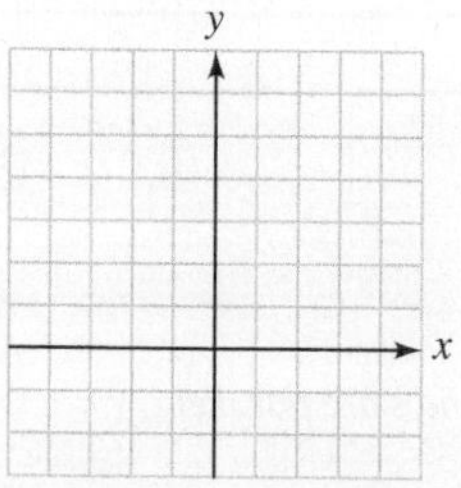

28. $f(x) = \left(\frac{4}{3}\right)^x$

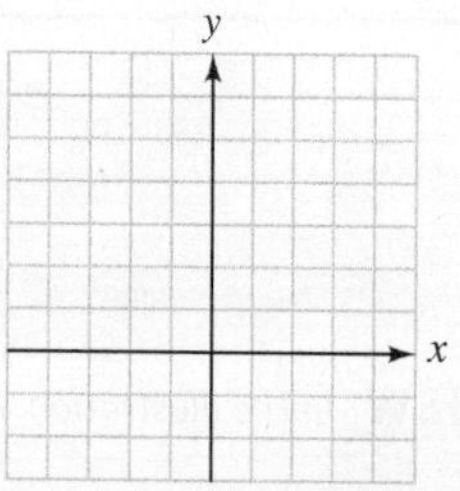

Graph each exponential function. Check your work with a graphing calculator. SEE EXAMPLE 2. (OBJECTIVE 1)

29. $f(x) = \left(\frac{1}{3}\right)^x$

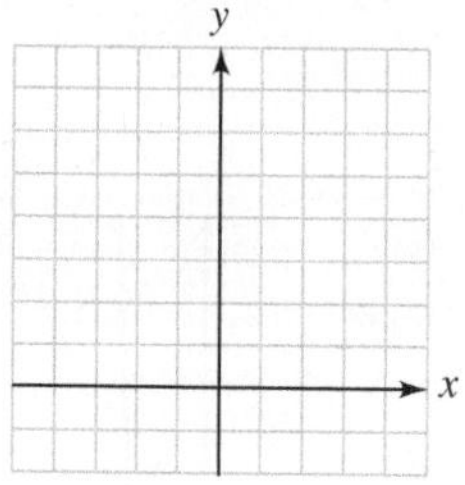

30. $f(x) = \left(\frac{1}{5}\right)^x$

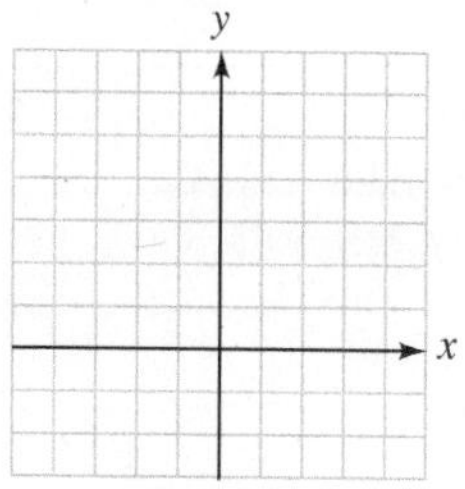

31. $f(x) = \left(\frac{2}{5}\right)^x$

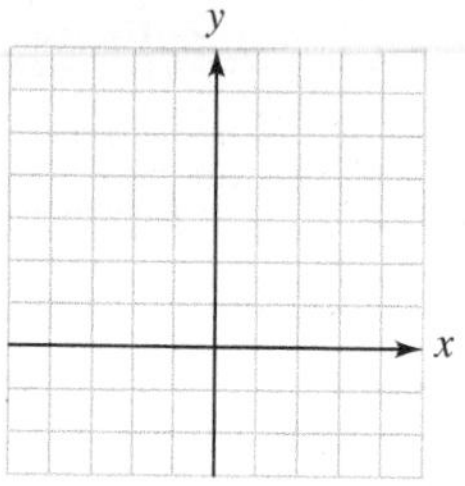

32. $f(x) = \left(\frac{3}{4}\right)^x$

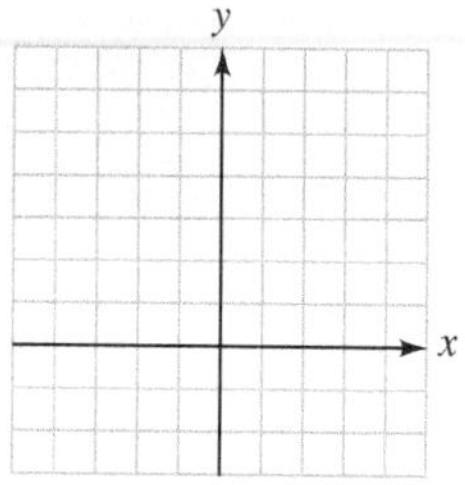

Find the value of b that would cause the graph of $f(x) = b^x$ to look like the graph indicated. SEE EXAMPLE 3. (OBJECTIVE 1)

33.

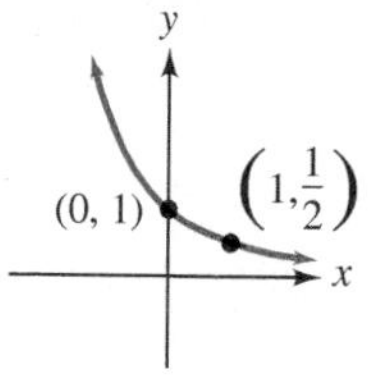

34.

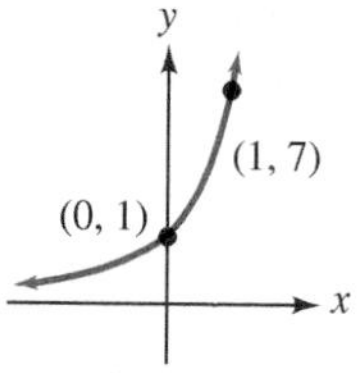

35.

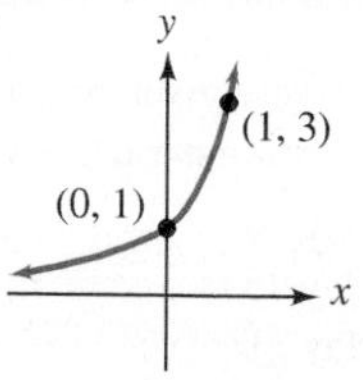

36.

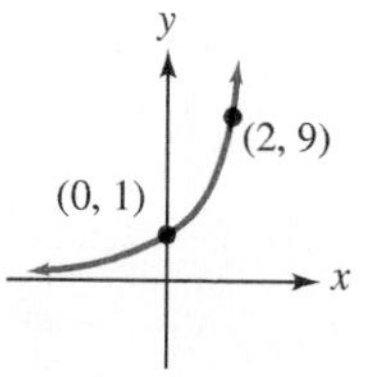

Graph each exponential function. Check your work with a graphing calculator. SEE EXAMPLES 4–5. (OBJECTIVE 2)

37. $f(x) = 3^x - 2$

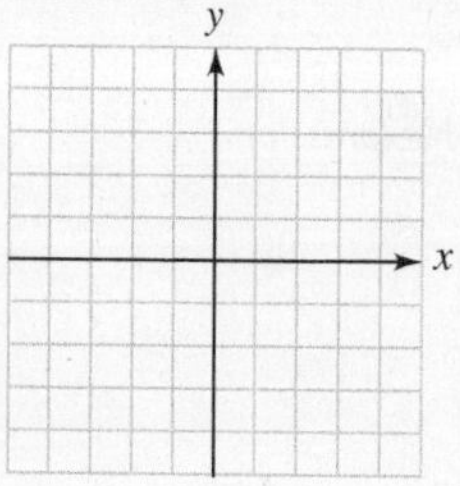

38. $f(x) = 2^x + 1$

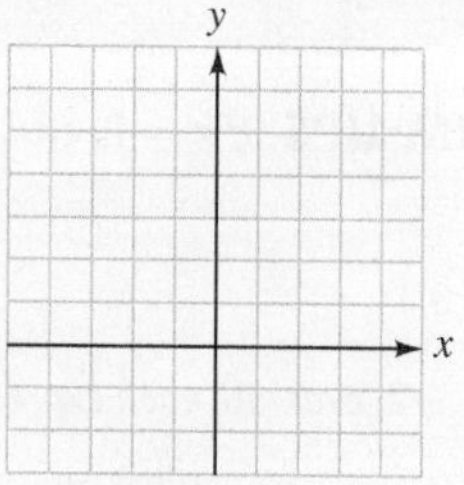

Unless otherwise noted, all content on this page is © Cengage Learning.

39. $f(x) = 3^{(x-1)}$

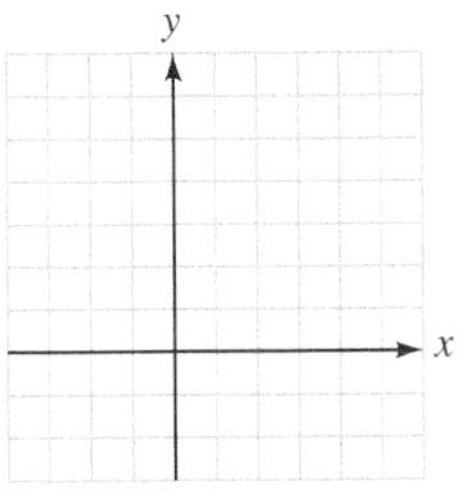

40. $f(x) = 2^{(x+1)}$

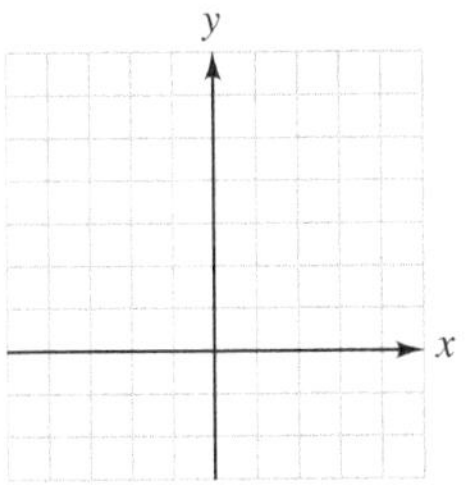

ADDITIONAL PRACTICE *Find the value of b that would cause the graph of $f(x) = b^x$ to look like the graph indicated.*

41.

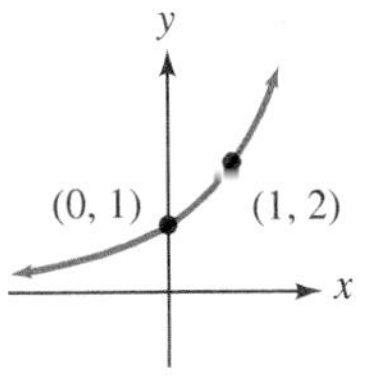

42.

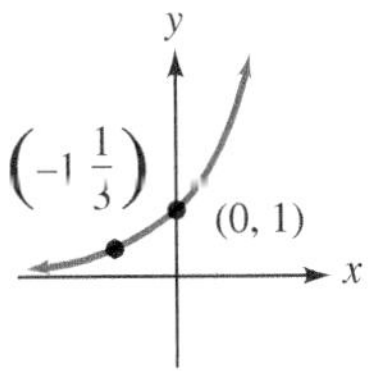

Use a graphing calculator to graph each function. Determine whether the function is an increasing or a decreasing function.

43. $f(x) = \frac{1}{2}(3^{x/2})$

44. $f(x) = -3(2^{x/3})$

45. $f(x) = 2(3^{-x/2})$

46. $f(x) = -\frac{1}{4}(2^{-x/2})$

APPLICATIONS *Evaluate each application.* SEE EXAMPLES 6–7. *(OBJECTIVE 3)*

47. Cell phone usage Refer to Example 6 and state the number of cell phone users in 2005.

48. Cell phone usage Refer to Example 6 and state the number of cell phone users in 2010.

49. Radioactive decay A radioactive material decays according to the formula $A = A_0\left(\frac{2}{3}\right)^t$, where A_0 is the initial amount present and t is measured in years. Find an expression for the amount present in 8 years.

50. Bacteria cultures A colony of 6 million bacteria is growing in a culture medium. The population P after t hours is given by the formula $P = (6 \times 10^6)(2.3)^t$. Find the population after 4 hours.

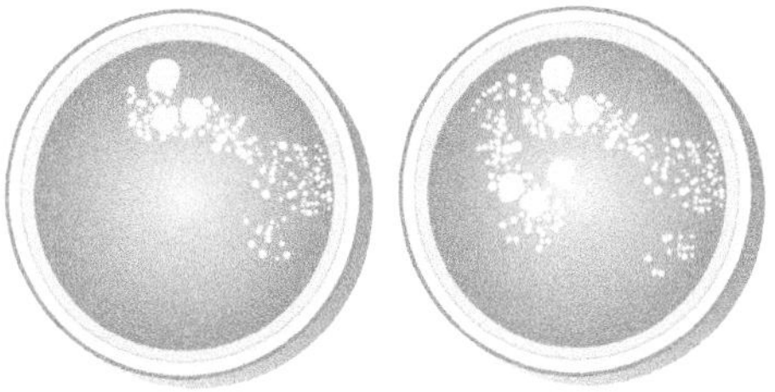

Assume that there are no deposits or withdrawals.

51. Compound interest An initial deposit of \$10,000 earns 8% interest, compounded quarterly. How much will be in the account after 10 years?

52. Compound interest An initial deposit of \$10,000 earns 8% interest, compounded monthly. How much will be in the account after 10 years?

53. Comparing interest rates How much more interest could \$100,000 earn in 10 years, compounded quarterly, if the annual interest rate were $4\frac{1}{2}$% instead of 4%?

54. Comparing savings plans Which institution in the two ads provides the better investment?

Fidelity Savings & Loan
Earn 5.25%
compounded monthly

Union Trust
Money Market Account
paying 5.35%
compounded annually

55. Compound interest If \$1 had been invested on July 4, 1776, at 5% interest, compounded annually, what would it be worth on July 4, 2076?

56. Frequency of compounding \$10,000 is invested in each of two accounts, both paying 6% annual interest. In the first account, interest compounds quarterly, and in the second account, interest compounds daily. Find the difference between the accounts after 20 years.

57. Discharging a battery The charge remaining in a battery decreases as the battery discharges. The charge C (in coulombs) after t days is given by the formula $C = (3 \times 10^{-4})(0.7)^t$. Find the charge after 5 days.

58. Town population The population of North Rivers is decreasing exponentially according to the formula $P = 3{,}745(0.93)^t$, where t is measured in years from the present date. Find the population in 6 years, 9 months.

Unless otherwise noted, all content on this page is © Cengage Learning.

59. Salvage value A small business purchases a computer for \$4,700. It is expected that its value each year will be 75% of its value in the preceding year. If the business disposes of the computer after 5 years, find its salvage value (the value after 5 years).

60. Louisiana Purchase In 1803, the United States acquired territory from France in the Louisiana Purchase. The country doubled its territory by adding 827,000 square miles of land for \$15 million. If the value of the land has appreciated at the rate of 6% each year, approximate what one square mile of land would be worth in 2023?

WRITING ABOUT MATH

61. If the world population is increasing exponentially, why is there cause for concern?

62. How do the graphs of $y = b^x$ differ when $b > 1$ and $0 < b < 1$?

SOMETHING TO THINK ABOUT

63. In the definition of the exponential function, b could not equal 0. Why not?

64. In the definition of the exponential function, b could not be negative. Why not?

Section 11.4 Base-*e* Exponential Functions

Objectives

1. Compute the continuously compounded interest of an investment given the principal, rate, and duration.
2. Graph an exponential function with a base of *e*.
3. Solve an application involving exponential growth or decay.

Vocabulary

continuous compound interest — annual growth rate — exponential function with a base of *e*

Getting Ready

Evaluate $\left(1 + \frac{1}{n}\right)^n$ for the following values. Round each answer to the nearest hundredth.

1. $n = 1$ **2.** $n = 2$ **3.** $n = 4$ **4.** $n = 10$

In this section, we will discuss special exponential functions with a base of *e*, an important irrational number.

1 Compute the continuously compounded interest of an investment given the principal, rate, and duration.

If a bank pays interest twice a year, we say that interest is *compounded semiannually.* If it pays interest four times a year, we say that interest is *compounded quarterly.* If it pays interest continuously (infinitely many times in a year), we say that interest is *compounded continuously.*

To develop the formula for **continuous compound interest**, we start with the formula

$$A = P\left(1 + \frac{r}{k}\right)^{kt} \quad \text{This is the formula for compound interest.}$$

and substitute rn for k. Since r and k are positive numbers, so is n.

$$A = P\left(1 + \frac{r}{rn}\right)^{rnt}$$

We can then simplify the fraction $\frac{r}{rn}$ and use the commutative property of multiplication to change the order of the exponents.

$$A = P\left(1 + \frac{1}{n}\right)^{nrt}$$

Finally, we can use a property of exponents to write this formula as

(1) $$A = P\left[\left(1 + \frac{1}{n}\right)^{n}\right]^{rt}$$ Use the property $a^{mn} = (a^m)^n$.

To find the value of $\left(1 + \frac{1}{n}\right)^n$, we use a calculator to evaluate it for several values of n, as shown in Table 11-1.

Leonhard Euler
1707–1783
Euler first used the letter *i* to represent $\sqrt{-1}$, the letter *e* for the base of natural logarithms, and the symbol Σ for summation. Euler was one of the most prolific mathematicians of all time, contributing to almost all areas of mathematics. Much of his work was accomplished after he became blind.

TABLE 11-1

n	$\left(1 + \frac{1}{n}\right)^n$
1	2
2	2.25
4	2.44140625 . . .
12	2.61303529 . . .
365	2.71456748 . . .
1,000	2.71692393 . . .
100,000	2.71826823 . . .
1,000,000	2.71828046 . . .

The results suggest that as n gets larger, the value of $\left(1 + \frac{1}{n}\right)^n$ approaches an irrational number with a value of 2.71828. . . . This number is called *e*, which has the following approximate value.

$$e \approx \mathbf{2.718281828459}$$

In continuous compound interest, k (the number of compoundings) is infinitely large. Since k, r, and n are all positive and $k = rn$, as k gets very large (approaches infinity), then so does n. Therefore, we can replace $\left(1 + \frac{1}{n}\right)^n$ in Equation 1 with e to obtain

$$A = Pe^{rt}$$

FORMULA FOR EXPONENTIAL GROWTH

If a quantity P increases or decreases at an annual rate r, compounded continuously, then the amount A after t years is given by

$$A = Pe^{rt}$$

If time is measured in years, then r is called the **annual growth rate**. If r is negative, the "growth" represents a decrease, commonly referred to as *decay*.

To compute the amount to which \$12,000 will grow if invested for 18 years at 10% annual interest, compounded continuously, we substitute 12,000 for P, 0.10 for r, and 18 for t in the formula for exponential growth:

$$\begin{aligned} A &= Pe^{rt} \\ &= \mathbf{12{,}000}e^{\mathbf{0.10}(18)} \\ &= 12{,}000e^{1.8} \\ &\approx 72{,}595.76957 \quad \text{Use a calculator.} \end{aligned}$$

Unless otherwise noted, all content on this page is © Cengage Learning.

After 18 years, the account will contain \$72,595.77. This is \$1,589.03 more than the result in Example 7 in the previous section, where interest was compounded quarterly.

EXAMPLE 1 **CONTINUOUS COMPOUND INTEREST** If \$25,000 accumulates interest at an annual rate of 8%, compounded continuously, find the balance in the account in 50 years.

Solution We substitute 25,000 for P, 0.08 for r, and 50 for t.

$$A = Pe^{rt}$$
$$A = \mathbf{25{,}000}e^{(\mathbf{0.08})(\mathbf{50})}$$
$$= 25{,}000e^{4}$$
$$\approx 1{,}364{,}953.751 \quad \text{Use a calculator.}$$

In 50 years, the balance will be \$1,364,953.75—over one million dollars.

SELF CHECK 1 Find the balance in 60 years.

2 Graph an exponential function with a base of *e*.

The function $f(x) = e^x$ is called the **exponential function** with a base of e. To graph the function $f(x) = e^x$, we plot several points and join them with a smooth curve, as shown in Figure 11-20.

EXAMPLE 2 Graph: $f(x) = e^x$

Solution

$f(x) = e^x$

x	$f(x)$	$(x, f(x))$
−2	0.1	(−2, 0.1)
−1	0.4	(−1, 0.4)
0	1	(0, 1)
1	2.7	(1, 2.7)
2	7.4	(2, 7.4)

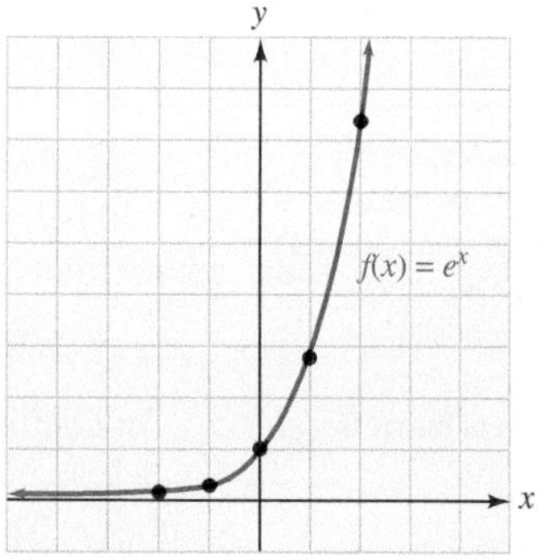

Figure 11-20

SELF CHECK 2 Graph: $f(x) = 3e^x$

Accent on technology

Translations of the Exponential Function

Figure 11-21(a) shows the graphs of $f(x) = e^x$, $f(x) = e^x + 5$, and $f(x) = e^x - 3$. To graph these functions on a TI-84 graphing calculator with window settings of $[-3, 6]$ for x and $[-5, 15]$ for y, we enter the right sides of the equations after the symbols $Y_1 =$, $Y_2 =$, and $Y_3 =$. To enter e^x, press the keys **2nd LN** (e^x) **X, t, θ, n)**. The display will show

$Y_1 = e$^(x)
$Y_2 = e$^$(x) + 5$
$Y_3 = e$^$(x) - 3$

Unless otherwise noted, all content on this page is © Cengage Learning.

After graphing these functions, we can see that the graph of $f(x) = e^x + 5$ is 5 units above the graph of $f(x) = e^x$, and that the graph of $f(x) = e^x - 3$ is 3 units below the graph of $f(x) = e^x$.

Figure 11-21(b) shows the calculator graphs of $f(x) = e^x$, $f(x) = e^{x+5}$, and $f(x) = e^{x-3}$. To graph these functions with window settings of $[-7, 10]$ for x and $[-5, 15]$ for y, we enter the right sides of the equations after the symbols $Y_1 =$, $Y_2 =$, $Y_3 =$. The display will show

$$Y_1 = e^\wedge(x)$$
$$Y_2 = e^\wedge(x + 5)$$
$$Y_3 = e^\wedge(x - 3)$$

After graphing these functions, we can see that the graph of $f(x) = e^{x+5}$ is 5 units to the left of the graph of $f(x) = e^x$, and that the graph of $f(x) = e^{x-3}$ is 3 units to the right of the graph of $f(x) = e^x$.

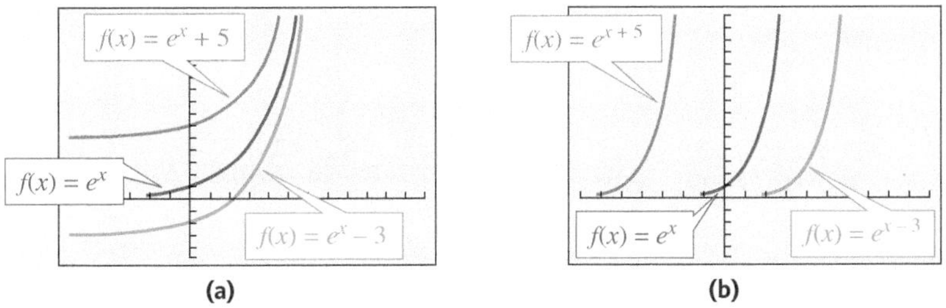

Figure 11-21

For instructions regarding the use of a Casio graphing calculator, please refer to the Casio Keystroke Guide in the back of the book.

Accent on technology

▸ Graphing Exponential Functions with a Base of *e*

Figure 11-22 shows the calculator graph of $f(x) = 3e^{-x/2}$. To graph this function with window settings of $[-7, 10]$ for x and $[-5, 15]$ for y, we enter the right side of the equation after the symbol $Y_1 =$. The display will show the equation

$$Y_1 = 3(e^\wedge(-x/2))$$

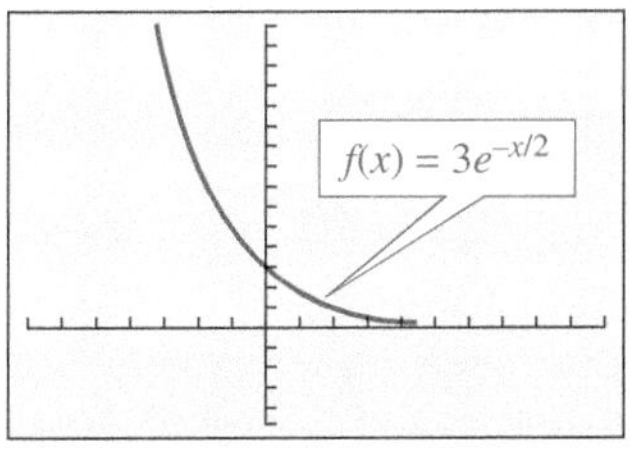

Figure 11-22

The graph has a y-intercept of $(0, 3)$.

For instructions regarding the use of a Casio graphing calculator, please refer to the Casio Keystroke Guide in the back of the book.

3 Solve an application involving exponential growth or decay.

An equation based on the exponential function provides a model for population growth. In the Malthusian model for population growth, the future or past population of a colony is related to the present population by the formula $A = Pe^{rt}$.

EXAMPLE 3 **CITY PLANNING** The population of a city is currently 15,000, but changing economic conditions are causing the population to decrease by 2% each year. If this trend continues, estimate the population in 30 years.

Solution Since the population is decreasing by 2% each year, the annual growth rate is -2%, or -0.02. We can substitute -0.02 for r, 30 for t, and 15,000 for P in the formula for exponential growth and find the value of A.

Unless otherwise noted, all content on this page is © Cengage Learning.

$$A = Pe^{rt}$$
$$A = \mathbf{15{,}000}e^{-\mathbf{0.02(30)}}$$
$$= 15{,}000e^{-0.6}$$
$$\approx 8{,}232.174541$$

In 30 years, city planners expect a population of approximately 8,232.

SELF CHECK 3 Estimate the population in 50 years.

The English economist Thomas Robert Malthus (1766–1834) pioneered in population study. He believed that poverty and starvation were unavoidable, because the human population tends to grow exponentially, but the food supply tends to grow linearly.

EXAMPLE 4 **POPULATION GROWTH** Suppose that a country with a population of 1,000 people is growing exponentially according to the formula

$$P = 1{,}000e^{0.02t}$$

where t is in years and 0.02 represents a 2% growth rate. Furthermore, assume that the food supply measured in adequate food per day per person is growing linearly according to the formula

$$y = 30.625x + 2{,}000$$

where x is in years. In how many years will the population outstrip the food supply?

Solution We can use a graphing calculator, with window settings of $[0, 100]$ for x and $[0, 10{,}000]$ for y. After graphing the functions, we obtain Figure 11-23(a).

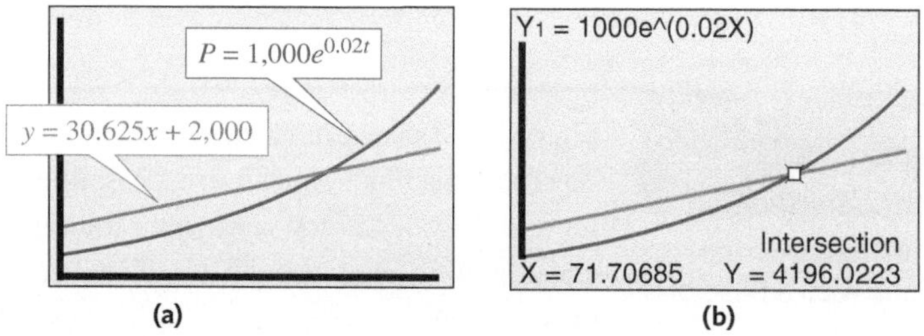

Figure 11-23

If we use the INTERSECT feature in the CALC menu, we obtain Figure 11-23(b). The food supply will be adequate for about 71.7 years when the population will be about 4,196.

SELF CHECK 4 Suppose that the population grows at a 3% rate and the food supply growth remains the same. For how many years will the food supply be adequate?

The atomic structure of a radioactive material changes as the material emits radiation. The amount of radioactive material that is present decays exponentially according to the following formula.

RADIOACTIVE DECAY FORMULA

The amount A of radioactive material present at a time t is given by the formula

$$A = A_0e^{kt}$$

where A_0 is the amount that was present at $t = 0$ and k is a negative number.

Unless otherwise noted, all content on this page is © Cengage Learning.

EXAMPLE 5 **RADIOACTIVE DECAY** The radioactive material radon-22 decays according to the formula $A = A_0e^{-0.181t}$, where t is expressed in days. How much radon-22 will be left if a sample of 100 grams decays for 20 days?

Solution To find the number of grams of radon-22 that will be left, we substitute 100 for A_0 and 20 for t and simplify.

COMMENT Note that the radioactive decay formula is the same as the formula for exponential growth, except for the variables.

$$\begin{aligned} A &= A_0e^{-0.181t} \\ A &= \mathbf{100}e^{-0.181(\mathbf{20})} \\ &= 100e^{-3.62} && -0.181 \cdot 20 = -3.62 \\ &= 100(0.0267826765) && e^{-3.62} \approx 0.0267826765 \\ &\approx 2.678267649 && \text{Multiply.} \end{aligned}$$

To the nearest hundredth, 2.68 grams of radon-22 will be left in 20 days.

SELF CHECK 5 To the nearest hundredth, how much radon-22 will be left in 30 days?

SELF CHECK ANSWERS

1. \$3,037,760.44 **2.**

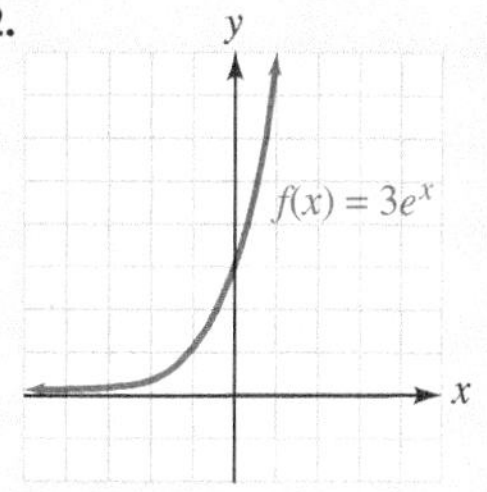

3. approximately 5,518 **4.** about 38.6 years **5.** 0.44 gram

NOW TRY THIS

Give all answers to the nearest milligram (mg).

The half-life (how much time has passed when half of the drug remains) of ibuprofen is approximately 1.8 hours. This translates to $k \approx -0.39$. (We will learn how to compute this value in Section 11.8.)

1. If a woman takes two 200 mg tablets, how many mg of ibuprofen are in her system 4 hours after taking the medication?
2. If she takes 2 more tablets 4 hours after taking the first dose, how many mg are in her system 6 hours after the first dose?
3. If she doesn't take any more medication, how many mg are in her system 12 hours after the first dose? 18 hours?

11.4 Exercises

WARM-UPS *Use a calculator to find each value to the nearest hundredth.*

1. e^0
2. e^1
3. $2e^3$
4. $-3e^2$
5. $e^{-1.5}$
6. $e^{-3.7}$

REVIEW *Simplify each expression. Assume that all variables represent positive numbers.*

7. $\sqrt{320x^7}$
8. $\sqrt[3]{-81x^4y^5}$
9. $5\sqrt{98y^3} - 4y\sqrt{18y}$
10. $\sqrt[4]{48z^5} + \sqrt[4]{768z^5}$

Unless otherwise noted, all content on this page is © Cengage Learning.

VOCABULARY AND CONCEPTS *Fill in the blanks.*

11. To two decimal places, the value of e is ____.

12. The formula for continuous compound interest is $A =$ ____.

13. Since $e > 1$, the base-e exponential function is a(n) ________ function.

14. The graph of the exponential function $y = e^x$ passes through the points $(0, 1)$ and _____.

15. The Malthusian population growth formula is _______.

16. The Malthusian theory is pessimistic, because ________ grows exponentially, but food supplies grow ______.

GUIDED PRACTICE *Graph each function. Check your work with a graphing calculator. Compare each graph to the graph of $f(x) = e^x$. SEE EXAMPLE 2. (OBJECTIVE 2)*

17. $f(x) = e^x + 1$

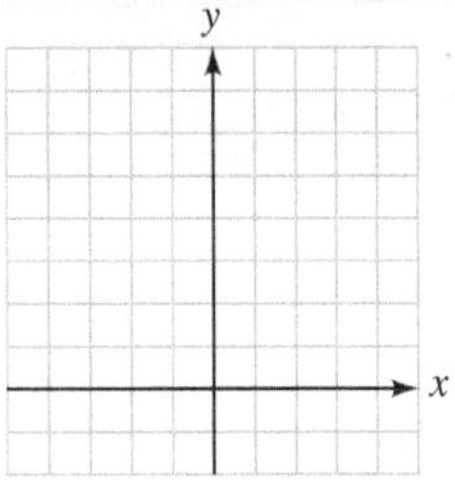

18. $f(x) = e^x - 2$

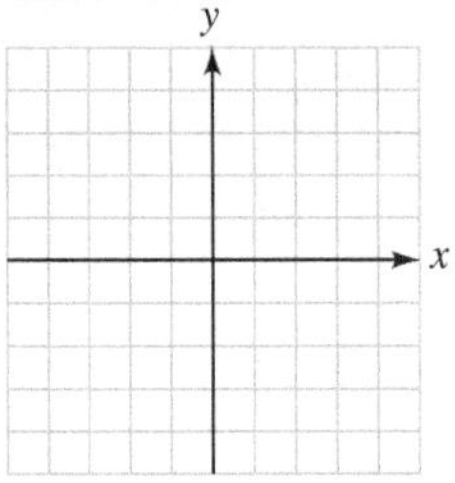

19. $f(x) = e^{x+3}$

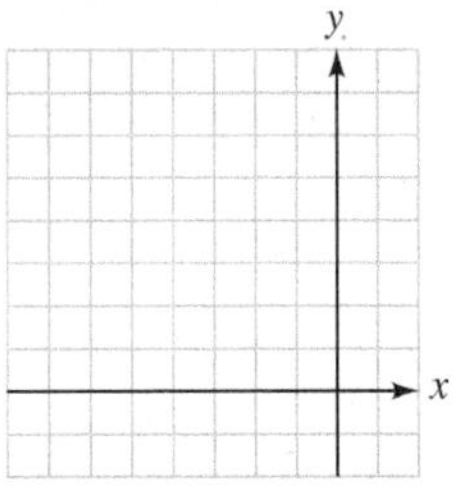

20. $f(x) = e^{x-5}$

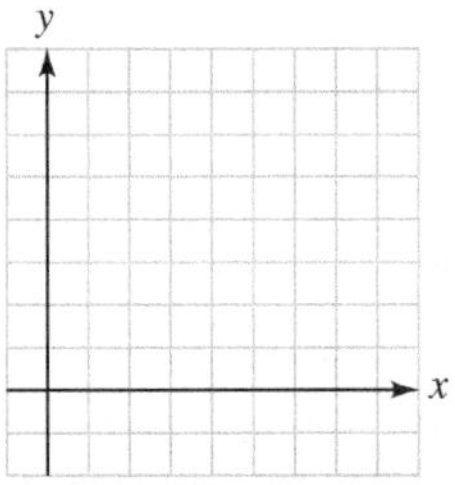

Determine whether the graph of $f(x) = e^x$, where x is a natural number, could look like the graph shown here. (OBJECTIVE 2)

21.

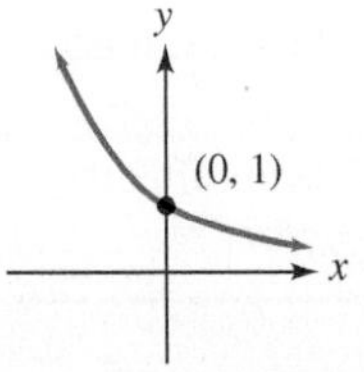

22.

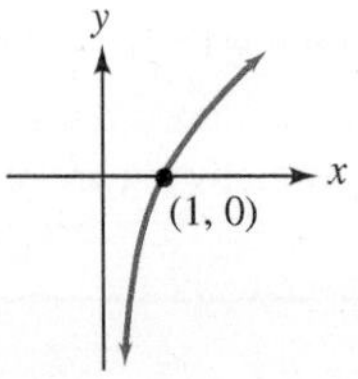

23.

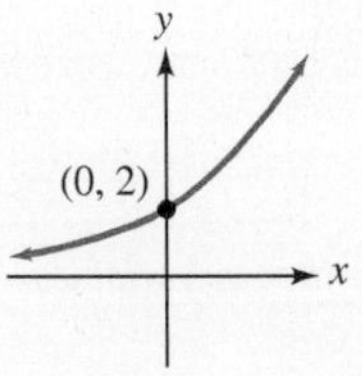

24.

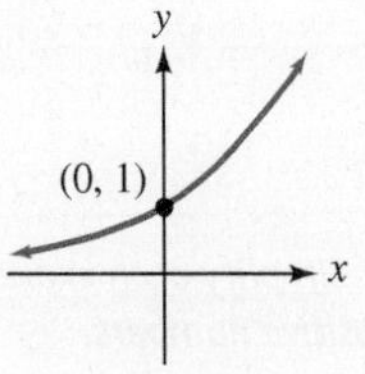

ADDITIONAL PRACTICE *Graph each function.*

25. $f(x) = -e^x$

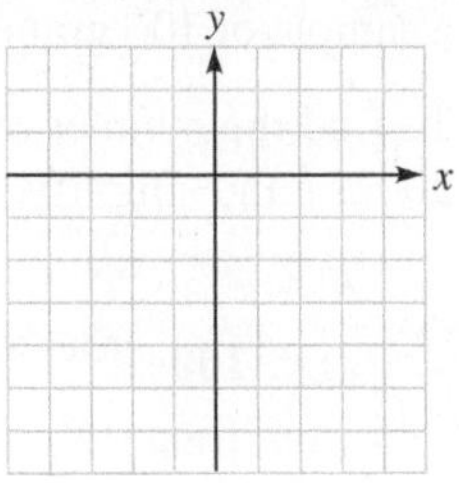

26. $f(x) = -e^x + 1$

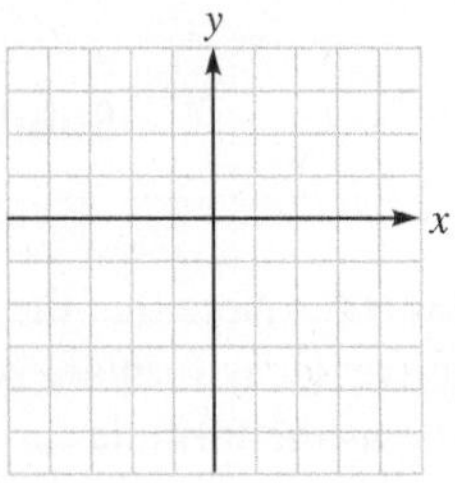

27. $f(x) = 2e^x$

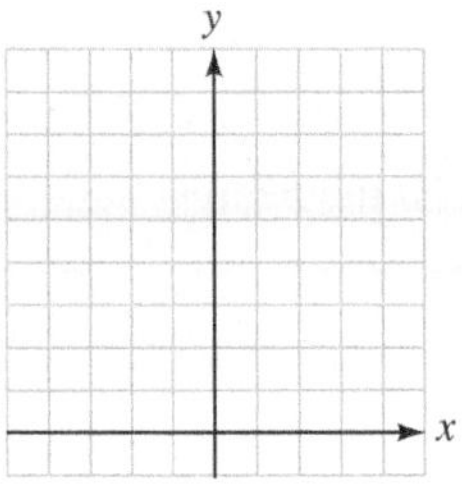

28. $f(x) = \frac{1}{2}e^x$

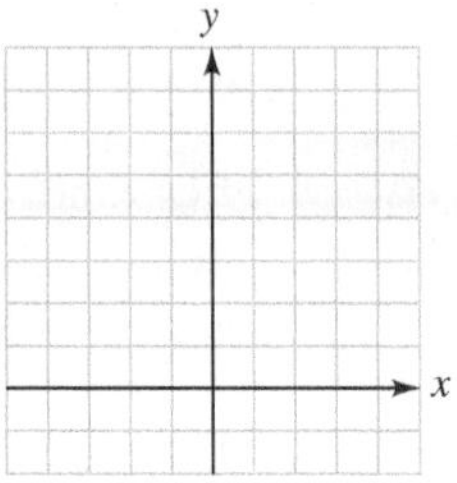

APPLICATIONS *Use a calculator to help solve. Assume that there are no deposits or withdrawals. SEE EXAMPLE 1. (OBJECTIVE 1)*

29. Continuous compound interest An investment of $9,000 earns 4% interest, compounded continuously. What will the investment be worth in 20 years?

30. Continuous compound interest An investment of $6,000 earns 7% interest, compounded continuously. What will the investment be worth in 35 years?

31. Determining the initial deposit An account now contains $25,000 and has been accumulating 5% annual interest, compounded continuously, for 15 years. Find the initial deposit.

32. Determining a previous balance An account now contains $8,000 and has been accumulating 8% annual interest, compounded continuously. How much was in the account 6 years ago?

Use a calculator to help solve. SEE EXAMPLE 3. (OBJECTIVE 3)

33. World population growth Earth's population is approximately 6 billion people and is growing at an annual rate of 1.9%. Assuming a Malthusian growth model, find the world population in 30 years.

34. World population growth Earth's population is approximately 6 billion people and is growing at an annual rate of 1.9%. Assuming a Malthusian growth model, find the world population in 40 years.

35. World population growth Assuming a Malthusian growth model and an annual growth rate of 1.9%, by what factor will Earth's current population increase in 50 years? (See Exercise 33.)

36. Population growth The growth of a population is modeled by

$$P = 173e^{0.03t}$$

How large will the population be when $t = 30$?

Unless otherwise noted, all content on this page is © Cengage Learning.

Use a calculator to help solve. SEE EXAMPLE 4. (OBJECTIVE 3)

37. In Example 4, suppose that better farming methods change the formula for food growth to $y = 31x + 2{,}000$. How long will the food supply be adequate?

38. In Example 4, suppose that a birth-control program changed the formula for population growth to $P = 1{,}000e^{0.01t}$. How long will the food supply be adequate?

Use a calculator to help solve. Round each answer to the nearest hundredth. SEE EXAMPLE 5. (OBJECTIVE 3)

39. Radioactive decay The radioactive material iodine-131 decays according to the formula $A = A_0e^{-0.087t}$, where t is expressed in days. To the nearest hundredth, how much iodine-131 will be left if a sample of 75 grams decays for 40 days?

40. Radioactive decay The radioactive material strontium-89 decays according to the formula $A = A_0e^{-0.013t}$, where t is expressed in days. To the nearest hundredth, how much strontium-89 will be left if a sample of 25 grams decays for 45 days?

41. Radioactive decay The radioactive material tin-126 decays according to the formula $A = A_0e^{-0.00000693t}$, where t is expressed in years. To the nearest hundredth, how much tin-126 will be left if a sample of 2,500 grams decays for 100 years?

42. Radioactive decay The radioactive material plutonium-239 decays according to the formula $A = A_0e^{-0.0000284t}$, where t is expressed in years. To the nearest hundredth, how much plutonium-239 will be left if a sample of 75 grams decays for 50,000 years?

Use a graphing calculator to solve.

43. Compounding methods An initial deposit of \$5,000 grows at an annual rate of 8.5% for 5 years. Compute the final balances resulting from continuous compounding and annual compounding.

44. Compounding methods An initial deposit of \$30,000 grows at an annual rate of 8% for 20 years. Compute the final balances resulting from continuous compounding and annual compounding.

45. Population decline The decline of a population is modeled by

$$P = 8{,}000e^{-0.008t}$$

How large will the population be when $t = 50$?

46. Epidemics The spread of foot and mouth disease through a herd of cattle can be modeled by the formula

$$P = P_0e^{0.27t} \text{ (}t\text{ is in days)}$$

If a rancher does not act quickly to treat two cases, how many cattle will have the disease in 10 days?

47. Medicine The concentration, x, of a certain drug in an organ after t minutes is given by $x = 0.08(1 - e^{-0.1t})$. Find the concentration of the drug after 30 minutes.

48. Medicine Refer to Exercise 47. Find the initial concentration of the drug.

49. Skydiving Before her parachute opens, a skydiver's velocity v in meters per second is given by $v = 50(1 - e^{-0.2t})$. Find the initial velocity.

50. Skydiving Refer to Exercise 49 and find the velocity after 20 seconds.

51. Free-falling objects After t seconds, a certain falling object has a velocity v given by $v = 50(1 - e^{-0.3t})$. Which is falling faster after 2 seconds, this object or the skydiver in Exercise 49?

52. Alcohol absorption In one individual, the blood alcohol level t minutes after drinking two shots of whiskey is given by $P = 0.3(1 - e^{-0.05t})$. Find the blood alcohol level after 15 minutes.

53. Depreciation A camping trailer originally purchased for \$4,570 is continuously losing value at the rate of 6% per year. Find its value when it is $6\frac{1}{2}$ years old.

54. Depreciation A boat purchased for \$7,500 has been continuously decreasing in value at the rate of 2% each year. It is now 8 years, 3 months old. Find its value.

WRITING ABOUT MATH

55. Explain why the graph of $f(x) = e^x - 5$ is 5 units below the graph of $f(x) = e^x$.

56. Explain why the graph of $f(x) = e^{(x+5)}$ is 5 units to the left of the graph of $f(x) = e^x$.

SOMETHING TO THINK ABOUT

57. The value of e can be calculated to any degree of accuracy by adding the first several terms of the following list:

$$1, 1, \frac{1}{2}, \frac{1}{2 \cdot 3}, \frac{1}{2 \cdot 3 \cdot 4}, \frac{1}{2 \cdot 3 \cdot 4 \cdot 5}, \ldots$$

The more terms that are added, the closer the sum will be to e. Add the first six numbers in the preceding list. To how many decimal places is the sum accurate?

58. Graph the function defined by the equation

$$y = f(x) = \frac{e^x + e^{-x}}{2}$$

from $x = -2$ to $x = 2$. The graph will look like a parabola, but it is not. The graph, called a **catenary**, is important in the design of power distribution networks, because it represents the shape of a uniform flexible cable whose ends are suspended from the same height. The function is called the **hyperbolic cosine function**.

59. If $e^{t+10} = ke^t$, find k.

60. If $e^{5t} = k^t$, find k.

Section 11.5 Logarithmic Functions

Objectives

1. Write a logarithmic function as an exponential function and write an exponential function as a logarithmic function.
2. Graph a logarithmic function.
3. Graph a vertical and horizontal translation of a logarithmic function.
4. Evaluate a common logarithm.
5. Solve an application involving a logarithm.

Vocabulary

logarithmic function
logarithm
common logarithm
decibel
Richter scale

Getting Ready

Find each value.

1. 7^0 **2.** 5^2 **3.** 5^{-2} **4.** $16^{1/2}$

In this section, we consider the inverse function of an exponential function $f(x) = b^x$. The inverse function is called a *logarithmic function*. These functions can be used to solve applications from fields such as electronics, seismology, and business.

1 Write a logarithmic function as an exponential function and write an exponential function as a logarithmic function.

Since the exponential function $y = b^x$ is one-to-one, it has an inverse function defined by the equation $x = b^y$. To express this inverse function in the form $y = f^{-1}(x)$, we must solve the equation $x = b^y$ for y. To do this, we need the following definition.

LOGARITHMIC FUNCTIONS

If $b > 0$ and $b \neq 1$, the **logarithmic function with base b** is defined by

$$y = \log_b x \quad \text{if and only if} \quad x = b^y$$

COMMENT Since the domain of the logarithmic function is the set of positive numbers, the logarithm of 0 or the logarithm of a negative number is not defined in the real-number system.

The domain of the logarithmic function is the interval $(0, \infty)$. The range is the interval $(-\infty, \infty)$.

Since the function $y = \log_b x$ is the inverse of the one-to-one exponential function $y = b^x$, the logarithmic function is also one-to-one.

The previous definition guarantees that any pair (x, y) that satisfies the equation $y = \log_b x$ also satisfies the equation $x = b^y$.

$\log_4 1 = 0$	because	$1 = 4^0$
$\log_5 25 = 2$	because	$25 = 5^2$
$\log_5 \frac{1}{25} = -2$	because	$\frac{1}{25} = 5^{-2}$
$\log_{16} 4 = \frac{1}{2}$	because	$4 = 16^{1/2}$
$\log_2 \frac{1}{8} = -3$	because	$\frac{1}{8} = 2^{-3}$
$\log_b x = y$	because	$x = b^y$

COMMENT Since $b^y = x$ is equivalent to $y = \log_b x$, then $b^{\log_b x} = x$ by substitution.

In each of these examples, the **logarithm** of a number is an exponent. In fact,

$\log_b x$ is the exponent to which b is raised to get x.

In equation form, we write

$$b^{\log_b x} = x$$

EXAMPLE 1

Find the value of y in each equation. **a.** $\log_6 1 = y$ **b.** $\log_3 27 = y$ **c.** $\log_5 \frac{1}{5} = y$ **d.** $\log_3 81 = x$

Solution

a. We can write the equation $\log_6 1 = y$ as the equivalent exponential equation $6^y = 1$. Since $6^0 = 1$, it follows that $y = 0$. Thus,

$$\log_6 1 = 0$$

b. $\log_3 27 = y$ is equivalent to $3^y = 27$. Since $3^3 = 27$, it follows that $3^y = 3^3$, and $y = 3$. Thus,

$$\log_3 27 = 3$$

c. $\log_5 \frac{1}{5} = y$ is equivalent to $5^y = \frac{1}{5}$. Since $5^{-1} = \frac{1}{5}$, it follows that $5^y = 5^{-1}$, and $y = -1$. Thus,

$$\log_5 \frac{1}{5} = -1$$

d. $\log_3 81 = x$ is equivalent to $3^x = 81$. Because $3^4 = 81$, it follows that $3^x = 3^4$. Thus, $x = 4$.

SELF CHECK 1 Find the value of y in each equation. **a.** $\log_3 9 = y$ **b.** $\log_2 64 = y$ **c.** $\log_5 \frac{1}{125} = y$ **d.** $\log_2 32 = x$

EXAMPLE 2

Find the value of x in each equation. **a.** $\log_x 125 = 3$ **b.** $\log_4 x = 3$

Solution

a. $\log_x 125 = 3$ is equivalent to $x^3 = 125$. Because $5^3 = 125$, it follows that $x^3 = 5^3$. Thus, $x = 5$.

b. $\log_4 x = 3$ is equivalent to $4^3 = x$. Because $4^3 = 64$, it follows that $x = 64$.

SELF CHECK 2 Find the value of x in each equation. **a.** $\log_x 8 = 3$ **b.** $\log_5 x = 2$

EXAMPLE 3

Find the value of x in each equation.

a. $\log_{1/3} x = 2$ **b.** $\log_{1/3} x = -2$ **c.** $\log_{1/3} \frac{1}{27} = x$

Solution **a.** $\log_{1/3} x = 2$ is equivalent to $\left(\frac{1}{3}\right)^2 = x$. Thus, $x = \frac{1}{9}$.

b. $\log_{1/3} x = -2$ is equivalent to $\left(\frac{1}{3}\right)^{-2} = x$. Thus,

$$x = \left(\frac{1}{3}\right)^{-2} = 3^2 = 9$$

c. $\log_{1/3} \frac{1}{27} = x$ is equivalent to $\left(\frac{1}{3}\right)^x = \frac{1}{27}$. Because $\left(\frac{1}{3}\right)^3 = \frac{1}{27}$, it follows that $x = 3$.

SELF CHECK 3 Find the value of x in each equation. **a.** $\log_{1/4} x = 3$ **b.** $\log_{1/4} x = -2$

2 Graph a logarithmic function.

To graph the logarithmic function $f(x) = \log_2 x$, we calculate and plot several points with coordinates (x, y) that satisfy the equation $x = 2^y$. After joining these points with a smooth curve, we have the graph shown in Figure 11-24(a).

COMMENT Substituting values for y and solving for x may be easier.

To graph $f(x) = \log_{1/2} x$, we calculate and plot several points with coordinates (x, y) that satisfy the equation $x = \left(\frac{1}{2}\right)^y$. After joining these points with a smooth curve, we have the graph shown in Figure 11-24(b).

$f(x) = \log_2 x$

x	$f(x)$	$(x, f(x))$
$\frac{1}{4}$	-2	$\left(\frac{1}{4}, -2\right)$
$\frac{1}{2}$	-1	$\left(\frac{1}{2}, -1\right)$
1	0	$(1, 0)$
2	1	$(2, 1)$
4	2	$(4, 2)$
8	3	$(8, 3)$

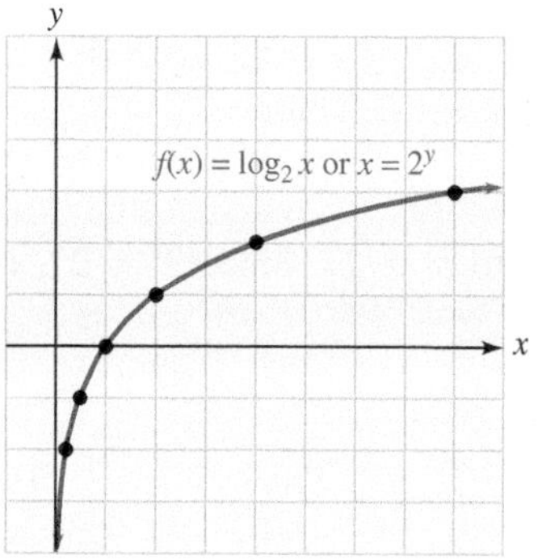

(a)

$f(x) = \log_{1/2} x$

x	$f(x)$	$(x, f(x))$
$\frac{1}{4}$	2	$\left(\frac{1}{4}, 2\right)$
$\frac{1}{2}$	1	$\left(\frac{1}{2}, 1\right)$
1	0	$(1, 0)$
2	-1	$(2, -1)$
4	-2	$(4, -2)$
8	-3	$(8, -3)$

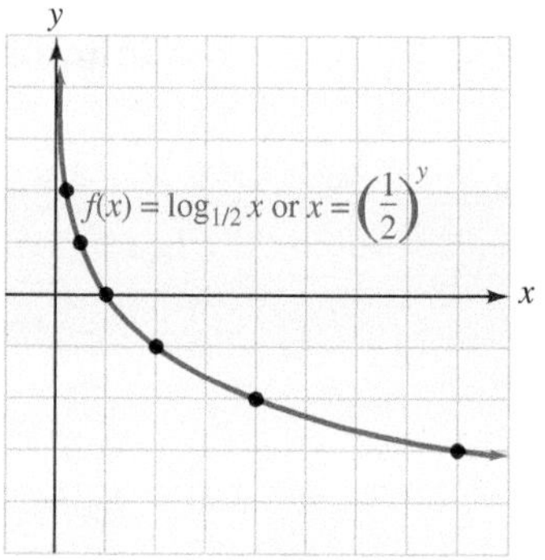

(b)

Figure 11-24

EXAMPLE 4 Graph the function defined by $f(x) = \log_3 x$.

Solution To graph the logarithmic function, we calculate and plot several points with coordinates (x, y) that satisfy the equation $x = 3^y$. (See Figure 11-25.)

x	$f(x)$	$(x, f(x))$
1	0	$(1, 0)$
3	1	$(3, 1)$
9	2	$(9, 2)$

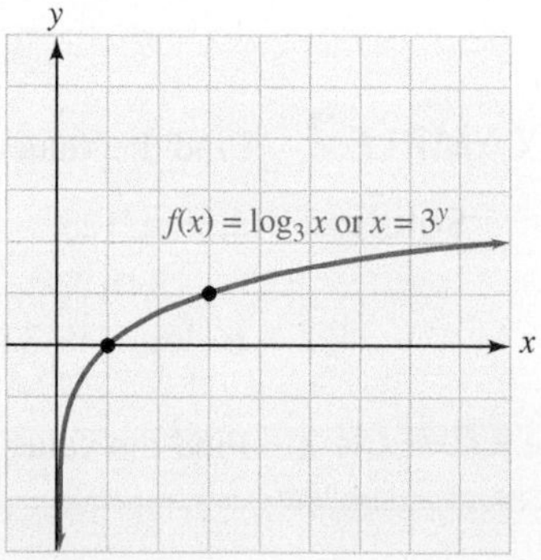

Figure 11-25

SELF CHECK 4 Graph the function defined by $f(x) = \log_4 x$.

Unless otherwise noted, all content on this page is © Cengage Learning.

The graphs of all logarithmic functions $(y = \log_b x)$ are similar to those in Figure 11-26. If $b > 1$, the logarithmic function is increasing, as in Figure 11-26(a). If $0 < b < 1$, the logarithmic function is decreasing, as in Figure 11-26(b).

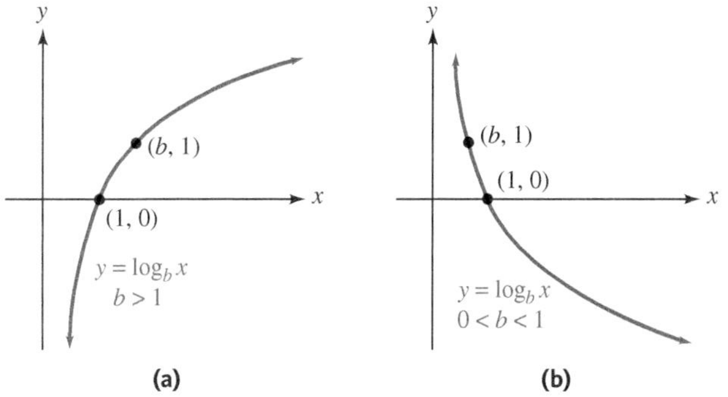

Figure 11-26

The graph of $y = \log_b x$ has the following properties.

1. It passes through the point $(1, 0)$.
2. It passes through the point $(b, 1)$.
3. The y-axis is an asymptote.
4. The domain is $(0, \infty)$ and the range is $(-\infty, \infty)$.

The exponential and logarithmic functions are inverses of each other and, therefore, have symmetry about the line $y = x$. The graphs $y = \log_b x$ and $y = b^x$ are shown in Figure 11-27(a) when $b > 1$, and in Figure 11-27(b) when $0 < b < 1$.

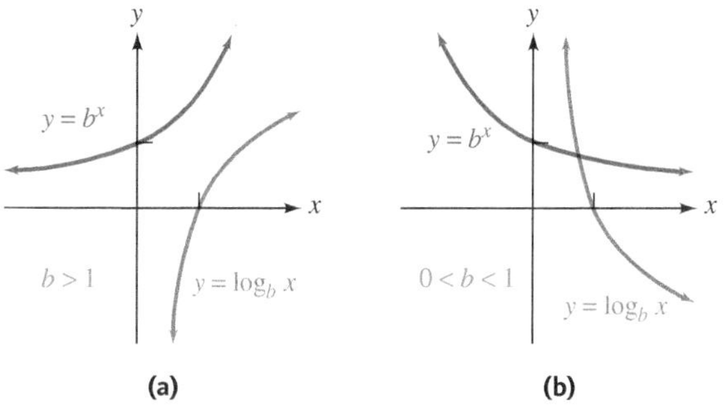

Figure 11-27

3 Graph a vertical and horizontal translation of a logarithmic function.

The graphs of many functions involving logarithms are translations of the basic logarithmic graphs.

EXAMPLE 5 Graph the function defined by $f(x) = 3 + \log_2 x$.

Solution The graph of $f(x) = 3 + \log_2 x$ is identical to the graph of $f(x) = \log_2 x$, except that it is translated 3 units upward. (See Figure 11-28.)

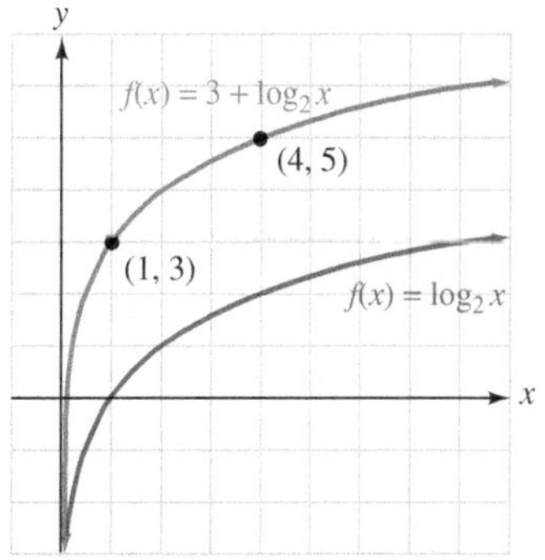

Figure 11-28

Unless otherwise noted, all content on this page is © Cengage Learning.

SELF CHECK 5 Graph: $f(x) = \log_3 x - 2$

EXAMPLE 6 Graph: $f(x) = \log_{1/2}(x - 1)$

Solution The graph of $f(x) = \log_{1/2}(x - 1)$ is identical to the graph of $f(x) = \log_{1/2} x$, except that it is translated 1 unit to the right. (See Figure 11-29.)

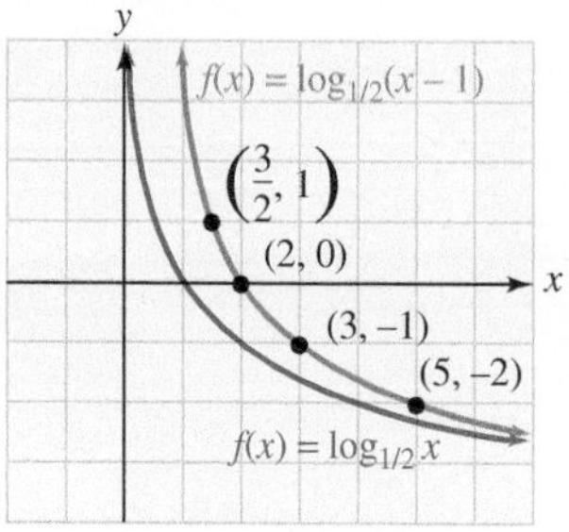

Figure 11-29

SELF CHECK 6 Graph: $f(x) = \log_{1/3}(x + 2)$

Accent on technology

▸ Graphing Logarithmic Functions

Graphing calculators can graph logarithmic functions directly only if the base of the logarithmic function is 10 or e. To use a TI-84 calculator to graph $f(x) = -2 + \log_{10}\left(\frac{1}{2}x\right)$, we enter the right side of the equation after the symbol $Y_1 =$. The display will show the equation

$$Y_1 = -2 + \log(1/2{*}x)$$

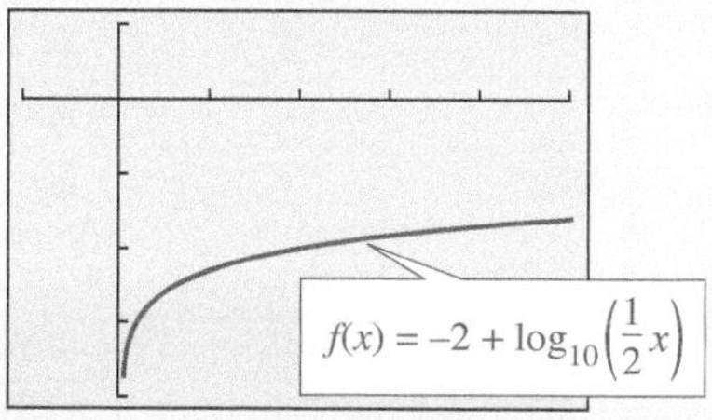

Figure 11-30

If we use window settings of $[-1, 5]$ for x and $[-4, 1]$ for y and press **GRAPH**, we will obtain the graph shown in Figure 11-30.

For instructions regarding the use of a Casio graphing calculator, please refer to the Casio Keystroke Guide in the back of the book.

4 Evaluate a common logarithm.

For computational purposes and in many applications, we will use base-10 logarithms (also called **common logarithms**). When the base b is not indicated in the notation $\log x$, we assume that $b = 10$:

$$\textbf{log } x \textbf{ means } \log_{10} x$$

Because base-10 logarithms appear so often, it is a good idea to become familiar with the following base-10 logarithms:

$\log_{10} \frac{1}{100} = -2$ because $10^{-2} = \frac{1}{100}$

$\log_{10} \frac{1}{10} = -1$ because $10^{-1} = \frac{1}{10}$

$\log_{10} 1 = 0$ because $10^0 = 1$

$\log_{10} 10 = 1$ because $10^1 = 10$

$\log_{10} 100 = 2$ because $10^2 = 100$

$\log_{10} 1{,}000 = 3$ because $10^3 = 1{,}000$

In general, we have

$$\log_{10} 10^x = x$$

Unless otherwise noted, all content on this page is © Cengage Learning.

Accent on technology

▸ Finding Base-10 (Common) Logarithms

Before calculators, extensive tables provided logarithms of numbers. Today, logarithms can be computed quickly with a calculator. For example, to find log 32.58 with a scientific calculator, we enter these numbers and press these keys:

32.58 **LOG**

The display will read **1.51295108**. To four decimal places, $\log 32.58 = 1.5130$.

To use a TI-84 graphing calculator, we enter these numbers and press these keys:

LOG 32.58 **)** **ENTER**

The display will read **LOG (32.58)**
1.51295108.

To four decimal places, $\log 32.58 = 1.5130$.

For instructions regarding the use of a Casio graphing calculator, please refer to the Casio Keystroke Guide in the back of the book.

EXAMPLE 7 Find the value of x in the equation $\log x = 0.3568$. Round to four decimal places.

Solution The equation $\log x = 0.3568$ is equivalent to $10^{0.3568} = x$. To find the value of x with a calculator, we can enter these numbers and press these keys:

Scientific Calculator	*Graphing Calculator*
10 y^x 0.3568 =	10 ∧ 0.3568 **ENTER**

Either way, the display will read **2.274049951**. To four decimal places,

$$x = 2.2740$$

SELF CHECK 7 Find the value of x: $\log x = 2.7$. Round to four decimal places.

5 Solve an application involving a logarithm.

Common logarithms are used in electrical engineering to express the voltage gain (or loss) of an electronic device such as an amplifier. The unit of gain (or loss), called the **decibel**, is defined by a logarithmic relation.

DECIBEL VOLTAGE GAIN

If E_O is the output voltage of a device and E_I is the input voltage, the decibel voltage gain is given by

$$\text{dB gain} = 20 \log \frac{E_O}{E_I}$$

If the input to an amplifier is 0.4 volt and the output is 50 volts, we can find the decibel voltage gain by substituting 0.4 for E_I and 50 for E_O into the formula for dB gain:

$$\text{dB gain} = 20 \log \frac{E_O}{E_I}$$

$$\text{dB gain} = 20 \log \frac{\mathbf{50}}{\mathbf{0.4}}$$
$$= 20 \log 125$$
$$\approx 42 \qquad \text{Use a calculator.}$$

The amplifier provides a 42-decibel voltage gain.

In seismology, the study of earthquakes, common logarithms are used to measure the magnitude (ground motion) of earthquakes on the **Richter scale**. The magnitude of an earthquake is given by the following logarithmic function.

RICHTER SCALE

If R is the magnitude of an earthquake, A is the amplitude (measured in micrometers), and P is the period (the time of one oscillation of Earth's surface, measured in seconds), then

$$R = \log \frac{A}{P}$$

EXAMPLE 8

MEASURING EARTHQUAKES Find the measure on the Richter scale of an earthquake with an amplitude of 10,000 micrometers (1 centimeter) and a period of 0.1 second.

Solution We substitute 10,000 for A and 0.1 for P in the Richter scale formula and simplify.

$$R = \log \frac{A}{P}$$
$$R = \log \frac{\mathbf{10{,}000}}{\mathbf{0.1}}$$
$$= \log 100{,}000$$
$$= 5$$

The earthquake measures 5 on the Richter scale.

SELF CHECK 8 Find the measure on the Richter scale of an earthquake with an amplitude of 3,162,277 micrometers and a period of 0.1 second.

SELF CHECK ANSWERS

1. a. 2 **b.** 6 **c.** −3 **d.** 5 **2. a.** 2 **b.** 25 **3. a.** $\frac{1}{64}$ **b.** 16 **4.**

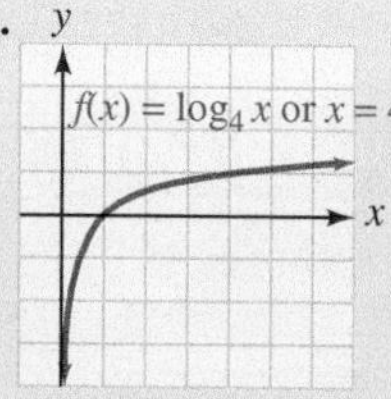

5.

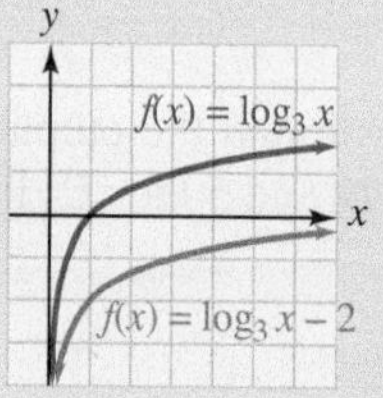

6.

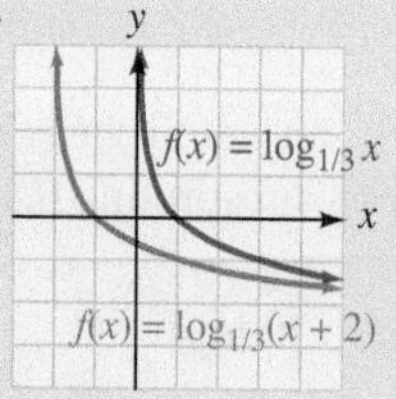

7. 501.1872 **8.** It measures 7.5 on the Richter scale.

Unless otherwise noted, all content on this page is © Cengage Learning.

NOW TRY THIS

In 1989, an earthquake measuring 7.1 on the Richter scale rocked San Francisco.

1. Write a logarithmic equation to describe the earthquake.
2. Write the equation in exponential form.

In 1964, an earthquake measuring 9.1 on the Richter scale devastated Juneau, Alaska.

3. Write the equation in exponential form.
4. Given that each earthquake had the same period, the amplitude in Juneau was how many times greater than that in San Francisco?

11.5 Exercises

WARM-UPS *Fill in the blanks.*

1. $2^{\square} = 8$
2. $3^{\square} = 9$
3. $\square^3 = 125$
4. $\square^3 = 27$
5. $5^{\square} = 25$
6. $\square^4 = 81$
7. $2^{\square} = 32$
8. $\square^{-3} = \frac{1}{27}$
9. $\square^{-2} = \frac{1}{4}$

REVIEW *Solve each equation.*

10. $\sqrt[3]{6x + 4} = 4$
11. $\sqrt{3x - 4} = \sqrt{-7x + 2}$
12. $\sqrt{a + 1} - 1 = 3a$
13. $3 - \sqrt{t - 3} = \sqrt{t}$

VOCABULARY AND CONCEPTS *Fill in the blanks.*

14. The equation $y = \log_b x$ is equivalent to _____ .
15. The domain of the logarithmic function is the interval ______.
16. The _____ of the logarithmic function is the interval $(-\infty, \infty)$.
17. $b^{\log_b x} =$ _.
18. Because an exponential function is one-to-one, it has an ______ function that is called a _________ function.
19. $\log_b x$ is the _______ to which b is raised to obtain x.
20. The y-axis is an ________ to the graph of $f(x) = \log_b x$.
21. The graph of $f(x) = \log_b x$ passes through the points _____ and _____.
22. A logarithm with a base of 10 is called a _______ logarithm and $\log_{10} 10^x =$ _.
23. The decibel voltage gain is found using the equation dB gain = _______.
24. The magnitude of an earthquake is measured by the formula $R =$ ____.

GUIDED PRACTICE *Write each equation in exponential form.* (*OBJECTIVE 1*)

25. $\log_4 64 = 3$
26. $\log_5 5 = 1$
27. $\log_{1/2} \frac{1}{8} = 3$
28. $\log_{1/3} 1 = 0$
29. $\log_4 \frac{1}{64} = -3$
30. $\log_6 \frac{1}{36} = -2$
31. $\log_{1/2} \frac{1}{8} = 3$
32. $\log_{1/5} 25 = -2$

Write each equation in logarithmic form. (*OBJECTIVE 1*)

33. $7^2 = 49$
34. $10^4 = 10{,}000$
35. $6^{-2} = \frac{1}{36}$
36. $5^{-3} = \frac{1}{125}$
37. $\left(\frac{1}{2}\right)^{-5} = 32$
38. $\left(\frac{1}{3}\right)^{-3} = 27$
39. $x^y = z$
40. $m^n = p$

Find each value of x. SEE EXAMPLE 1. (*OBJECTIVE 1*)

41. $\log_2 32 = x$
42. $\log_3 9 = x$
43. $\log_5 125 = x$
44. $\log_6 216 = x$

Find each value of x. SEE EXAMPLE 2. (*OBJECTIVE 1*)

45. $\log_7 x = 2$
46. $\log_5 x = 0$
47. $\log_6 x = 1$
48. $\log_2 x = 4$
49. $\log_{25} x = \frac{1}{2}$
50. $\log_4 x = \frac{1}{2}$
51. $\log_5 x = -2$
52. $\log_{27} x = -\frac{1}{3}$
53. $\log_x 5^3 = 3$
54. $\log_x 5 = 1$
55. $\log_x \frac{9}{4} = 2$
56. $\log_x \frac{\sqrt{3}}{3} = \frac{1}{2}$

Find each value of x. SEE EXAMPLE 3. (OBJECTIVE 1)

57. $\log_{1/2}\frac{1}{8} = x$

58. $\log_{1/3}\frac{1}{81} = x$

59. $\log_{1/4}x = 3$

60. $\log_{1/3}x = 4$

Graph each function. Determine whether each function is an increasing or decreasing function. SEE EXAMPLE 4. (OBJECTIVE 2)

61. $f(x) = \log_3 x$

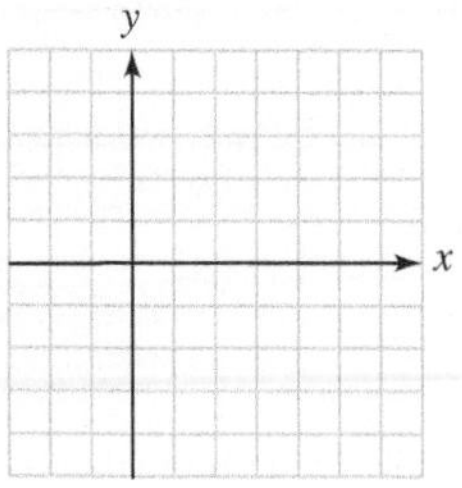

62. $f(x) = \log_{1/3} x$

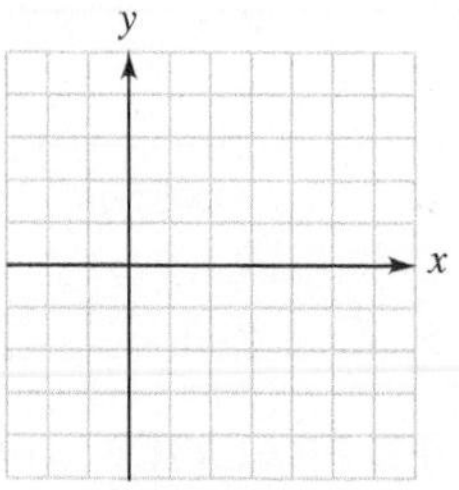

63. $f(x) = \log_{1/2} x$

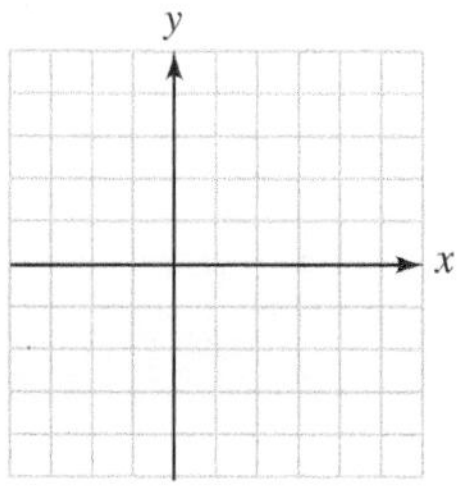

64. $f(x) = \log_4 x$

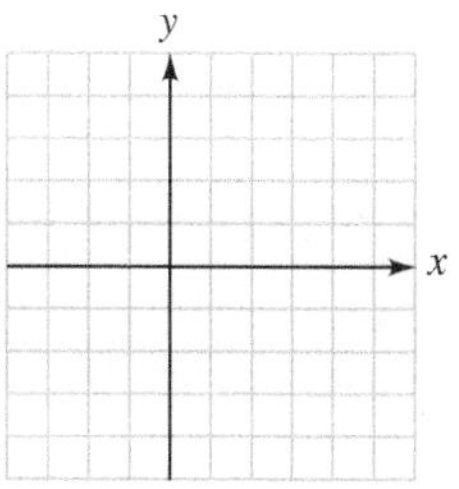

Graph each pair of inverse functions on a single coordinate system. (OBJECTIVE 2)

65. $f(x) = 2^x$
$g(x) = \log_2 x$

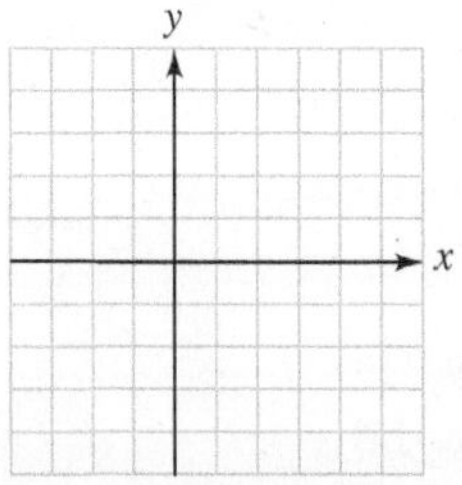

66. $f(x) = \left(\frac{1}{2}\right)^x$
$g(x) = \log_{1/2} x$

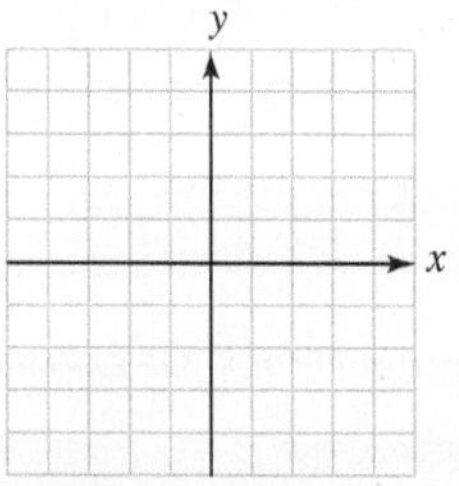

67. $f(x) = \left(\frac{1}{4}\right)^x$
$g(x) = \log_{1/4} x$

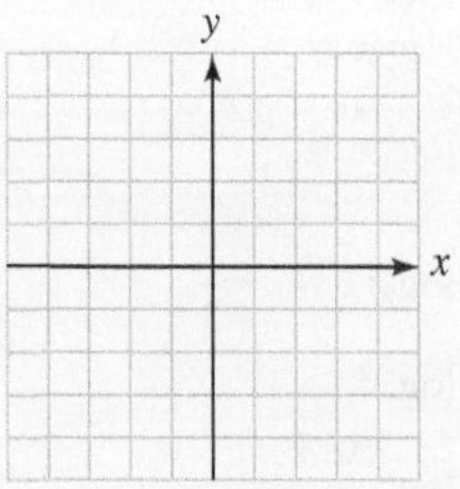

68. $f(x) = 4^x$
$g(x) = \log_4 x$

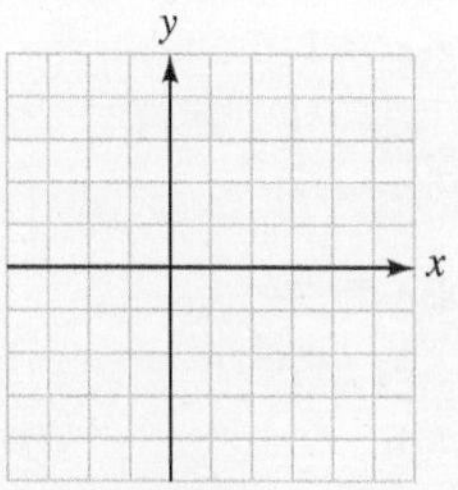

Graph each function. SEE EXAMPLES 5–6. (OBJECTIVE 3)

69. $f(x) = 3 + \log_3 x$

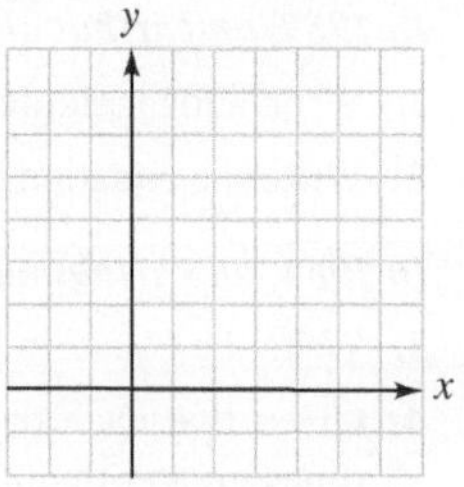

70. $f(x) = \log_{1/3} x - 1$

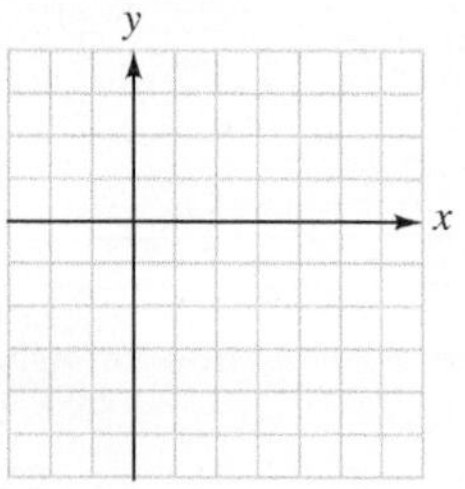

71. $f(x) = \log_{1/2}(x - 2)$

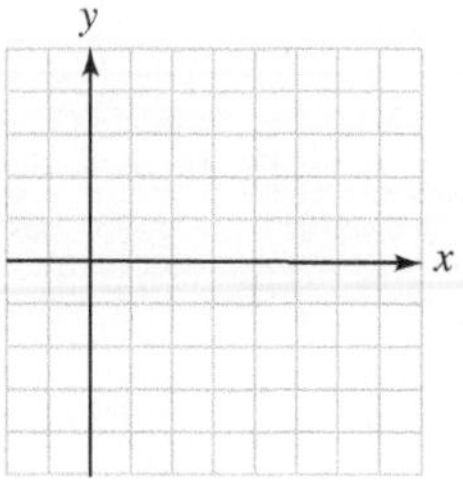

72. $f(x) = \log_4(x + 2)$

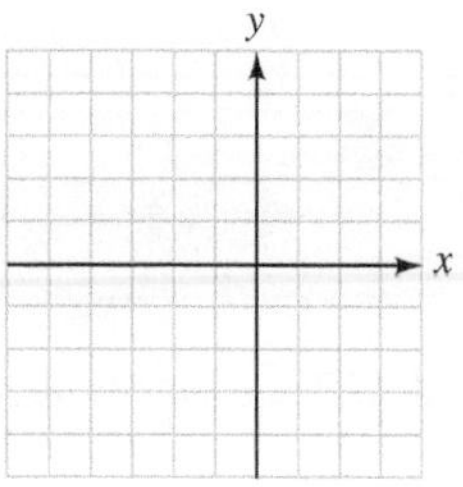

Use a calculator to find each value. Give answers to four decimal places. (OBJECTIVE 4)

73. $\log 8.25$

74. $\log 0.77$

75. $\log 0.00867$

76. $\log 375.876$

Use a calculator to find each value of y. If an answer is not exact, give the answer to two decimal places. SEE EXAMPLE 7. (OBJECTIVE 4)

77. $\log y = 4.24$

78. $\log y = 0.926$

79. $\log y = -3.71$

80. $\log y = -0.28$

ADDITIONAL PRACTICE *Find each value of x.*

81. $\log_{36} x = -\frac{1}{2}$

82. $\log_{27} x = -\frac{1}{3}$

83. $\log_{1/2} 8 = x$

84. $\log_{1/2} 16 = x$

85. $\log_{1/2} x = -3$

86. $\log_{1/4} x = -2$

87. $\log_4 2 = x$

88. $\log_{125} 5 = x$

89. $\log_{100}\frac{1}{1{,}000} = x$

90. $\log_{5/2}\frac{4}{25} = x$

91. $\log_{27} 9 = x$

92. $\log_{12} x = 0$

93. $\log_{2\sqrt{2}} x = 2$

94. $\log_4 8 = x$

95. $\log_x \frac{1}{64} = -3$

96. $\log_x \frac{1}{100} = -2$

97. $2^{\log_2 4} = x$

98. $3^{\log_3 5} = x$

99. $x^{\log_4 6} = 6$

100. $x^{\log_3 8} = 8$

101. $\log 10^3 = x$

102. $\log 10^{-2} = x$

103. $10^{\log x} = 100$

104. $10^{\log x} = \frac{1}{10}$

Unless otherwise noted, all content on this page is © Cengage Learning.

Use a calculator to find each value of y. If an answer is not exact, give the answer to two decimal places.

105. $\log y = 1.4023$

106. $\log y = 2.6490$

107. $\log y = \log 8$

108. $\log y = \log 7$

Find the value of b, if any, that would cause the graph of $f(x) = \log_b x$ to look like the graph indicated.

109.

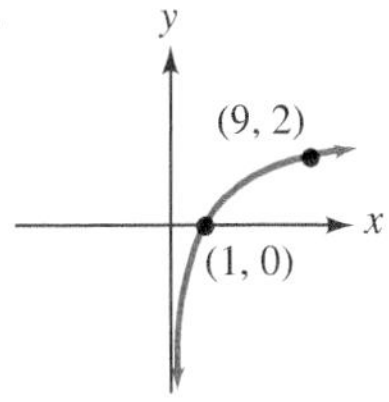

110.

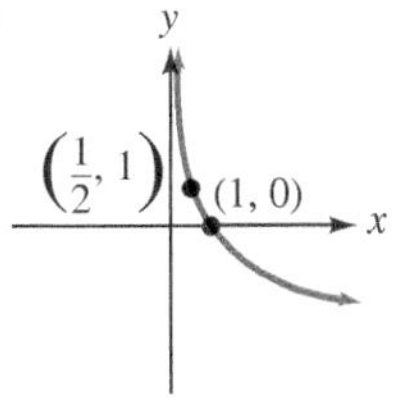

111.

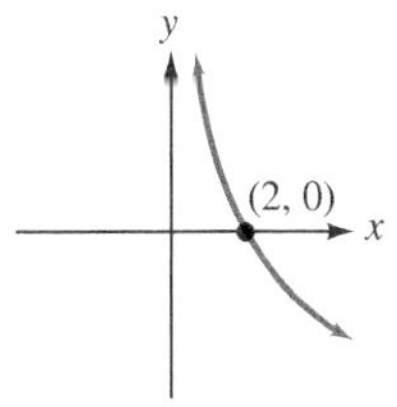

112.

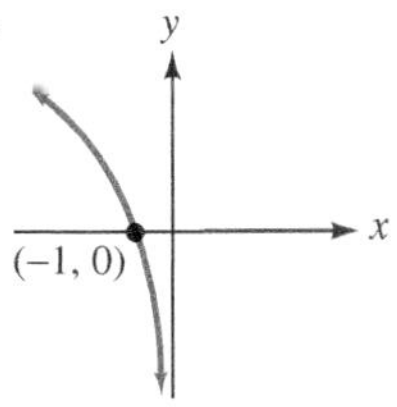

Use a calculator to help solve. If an answer is not exact, round to the nearest tenth. SEE EXAMPLE 8. *(OBJECTIVE 5)*

113. Earthquakes An earthquake has an amplitude of 5,000 micrometers and a period of 0.2 second. Find its measure on the Richter scale.

114. Earthquakes The period of an earthquake with amplitude of 80,000 micrometers is 0.08 second. Find its measure on the Richter scale.

115. Earthquakes An earthquake has a period of $\frac{1}{4}$ second and an amplitude of 2,500 micrometers. Find its measure on the Richter scale.

116. Earthquakes By what factor must the amplitude of an earthquake change to increase its magnitude by 1 point on the Richter scale? Assume that the period remains constant.

Use a calculator to help solve. If an answer is not exact, round to the nearest tenth.

117. Depreciation Business equipment is often depreciated using the double declining-balance method. In this method, a piece of equipment with a life expectancy of N years, costing $\$C$, will depreciate to a value of $\$V$ in n years, where n is given by the formula

$$n = \frac{\log V - \log C}{\log\left(1 - \frac{2}{N}\right)}$$

A computer that cost $17,000 has a life expectancy of 5 years. If it has depreciated to a value of $2,000, how old is it?

118. Depreciation See Exercise 117. A printer worth $470 when new had a life expectancy of 12 years. If it is now worth $189, how old is it?

119. Time for money to grow If $\$P$ is invested at the end of each year in an annuity earning annual interest at a rate r, the amount in the account will be $\$A$ after n years, where

$$n = \frac{\log\left(\frac{Ar}{P} + 1\right)}{\log(1 + r)}$$

If $1,000 is invested each year in an annuity earning 12% annual interest, how long will it take for the account to be worth $20,000?

120. Time for money to grow If $5,000 is invested each year in an annuity earning 8% annual interest, how long will it take for the account to be worth $50,000? (See Exercise 119.)

APPLICATIONS

Use a calculator to help solve. If an answer is not exact, round to the nearest tenth. (OBJECTIVE 5)

121. Finding the gain of an amplifier Find the dB gain of an amplifier if the input voltage is 0.71 volt when the output voltage is 20 volts.

122. Finding the gain of an amplifier Find the dB gain of an amplifier if the output voltage is 2.8 volts when the input voltage is 0.05 volt.

123. dB gain of an amplifier Find the dB gain of the amplifier.

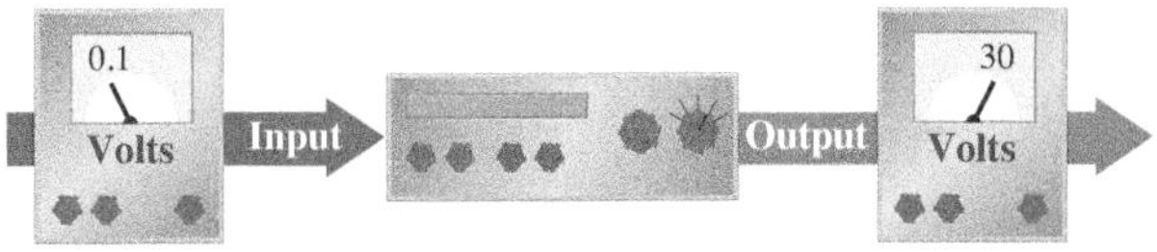

124. dB gain of an amplifier An amplifier produces an output of 80 volts when driven by an input of 0.12 volts. Find the amplifier's dB gain.

WRITING ABOUT MATH

125. Describe the appearance of the graph of $y = f(x) = \log_b x$ when $0 < b < 1$ and when $b > 1$.

126. Explain why it is impossible to find the logarithm of a negative number.

SOMETHING TO THINK ABOUT

127. Graph $f(x) = -\log_3 x$. How does the graph compare to the graph of $f(x) = \log_3 x$?

128. Find a logarithmic function that passes through the points $(1, 0)$ and $(5, 1)$.

129. Explain why an earthquake measuring 7 on the Richter scale is much worse than an earthquake measuring 6.

Unless otherwise noted, all content on this page is © Cengage Learning.

Section 11.6 Natural Logarithms

Objectives

1. Evaluate a natural logarithm.
2. Solve a logarithmic equation with a calculator.
3. Graph a natural logarithmic function.
4. Solve an application involving a natural logarithm.

Vocabulary

natural logarithm

Getting Ready

Evaluate each expression.

1. $\log_4 16$ **2.** $\log_2 \frac{1}{8}$ **3.** $\log_5 5$ **4.** $\log_7 1$

In this section, we will discuss special logarithmic functions with a base of e. They play an important role in advanced mathematics and applications in other scientific fields.

1 Evaluate a natural logarithm.

John Napier
1550–1617
Napier is famous for his work with natural logarithms. In fact, natural logarithms are often called *Napierian logarithms*. He also invented a device, called *Napier's rods*, that did multiplications mechanically. His device was a forerunner of modern-day computers.

We have seen the importance of base-e exponential functions in mathematical models of events in nature. Base-e logarithms are just as important. They are called **natural logarithms** or Napierian logarithms, after John Napier (1550–1617), and usually are written as $\ln x$, rather than $\log_e x$:

$$\ln x \quad \text{means} \quad \log_e x$$

As with all logarithmic functions, the domain of $f(x) = \ln x$ is the interval $(0, \infty)$, and the range is the interval $(-\infty, \infty)$.

We have seen that the logarithm of a number is an exponent. For natural logarithms,

$\ln x$ is the exponent to which e is raised to obtain x.

In equation form, we write

$$e^{\ln x} = x$$

To find the base-e logarithms of numbers, we can use a calculator.

Unless otherwise noted, all content on this page is © Cengage Learning.

Accent on technology

▸ Evaluating Logarithms

To use a scientific calculator to find the value of ln 9.87, we enter these numbers and press these keys:

9.87 **LN**

The display will read **2.289499853**. To four decimal places, ln 9.87 = 2.2895.

To use a TI-84 graphing calculator, we enter

LN 9.87 **) ENTER**

The display will read **LN (9.87)** **2.289499853**.

For instructions regarding the use of a Casio graphing calculator, please refer to the Casio Keystroke Guide in the back of the book.

EXAMPLE 1 Use a calculator to find each value. **a.** ln 17.32 **b.** $\ln(\log 0.05)$

Solution **a.** We can enter these numbers and press these keys:

Scientific Calculator	*Graphing Calculator*
17.32 **LN**	**LN** 17.32 **) ENTER**

Either way, the result is 2.851861903.

b. We can enter these numbers and press these keys:

Scientific Calculator	*Graphing Calculator*
0.05 **LOG LN**	**LN (LOG** 0.05 **)) ENTER**

Either way, we obtain an error, because log 0.05 is a negative number. Because the domain of $\ln x$ is $(0, \infty)$, we cannot take the logarithm of a negative number.

SELF CHECK 1 Find each value to four decimal places.

a. $\ln \pi$ **b.** $\ln\left(\log \frac{1}{2}\right)$

2 Solve a logarithmic equation with a calculator.

EXAMPLE 2 Find the value of x to four decimal places.

a. $\ln x = 1.335$ **b.** $\ln x = \log 5.5$

Solution **a.** The equation $\ln x = 1.335$ is equivalent to $e^{1.335} = x$. Enter $e^{1.335}$ into the calculator and the display will read 3.799995946. To four decimal places,

$$x = 3.8000$$

b. The equation $\ln x = \log 5.5$ is equivalent to $e^{\log 5.5} = x$. Enter $e^{\log 5.5}$ into the calculator and the display will read 2.096695826. To four decimal places,

$$x = 2.0967$$

SELF CHECK 2 Find the value of x to four decimal places.

a. $\ln x = 2.5437$ **b.** $\log x = \ln 5$

3 Graph a natural logarithmic function.

The equation $y = \ln x$ is equivalent to the equation $x = e^y$. To graph $f(x) = \ln x$, we can plot points that satisfy the equation $x = e^y$ and join them with a smooth curve, as shown in Figure 11-31(a). Figure 11-31(b) shows the calculator graph.

$f(x) = \ln x$

x	$f(x)$	$(x, f(x))$
$\frac{1}{e} \approx 0.4$	-1	$(0.4, -1)$
1	0	$(1, 0)$
$e \approx 2.7$	1	$(2.7, 1)$
$e^2 \approx 7.4$	2	$(7.4, 2)$

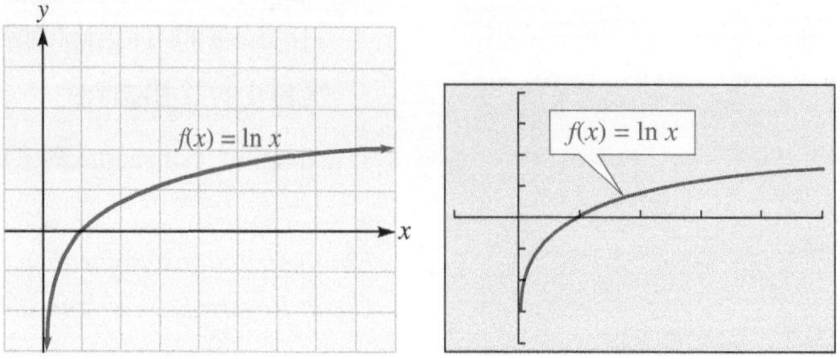

(a) (b)

Figure 11-31

EXAMPLE 3 Graph: $f(x) = \ln x - 1$

Solution The graph of $f(x) = \ln x - 1$ is 1 unit below the graph of $f(x) = \ln x$. (See Figure 11-32.)

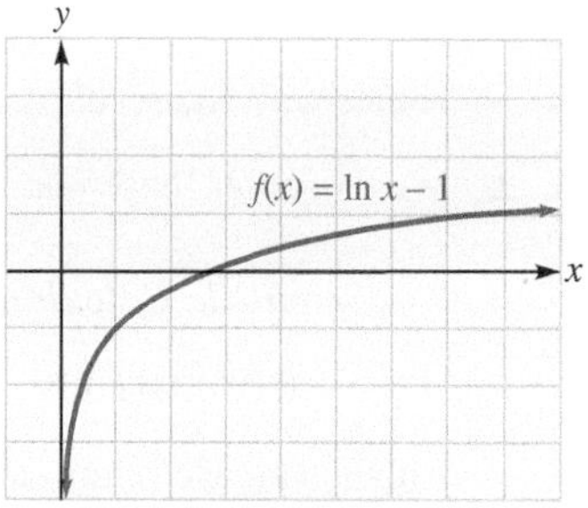

Figure 11-32

SELF CHECK 3 Graph: $f(x) = \ln x + 1$

Accent on technology

▸ Graphing Logarithmic Functions

Many graphs of logarithmic functions involve translations of the graph of $f(x) = \ln x$. For example, Figure 11-33 shows calculator graphs of the functions $f(x) = \ln x$, $f(x) = \ln x + 2$, and $f(x) = \ln x - 3$.

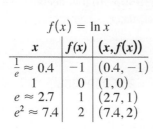

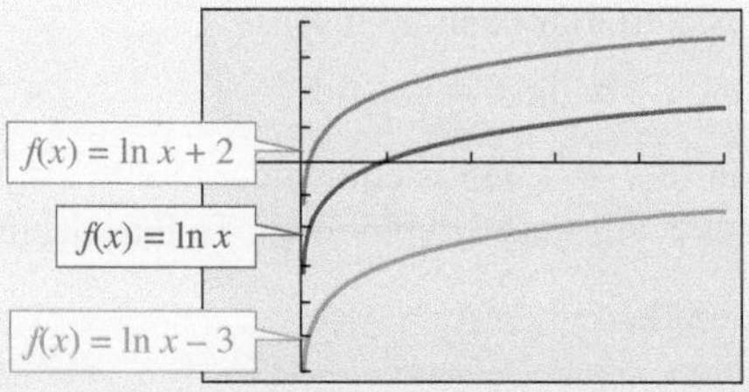

The graph of $f(x) = \ln x + 2$ is 2 units above the graph of $f(x) = \ln x$.

The graph of $f(x) = \ln x - 3$ is 3 units below the graph of $f(x) = \ln x$.

Figure 11-33

Unless otherwise noted, all content on this page is © Cengage Learning.

Figure 11-34 shows the calculator graphs of the functions $f(x) = \ln x$, $f(x) = \ln(x - 2)$, and $f(x) = \ln(x + 2)$.

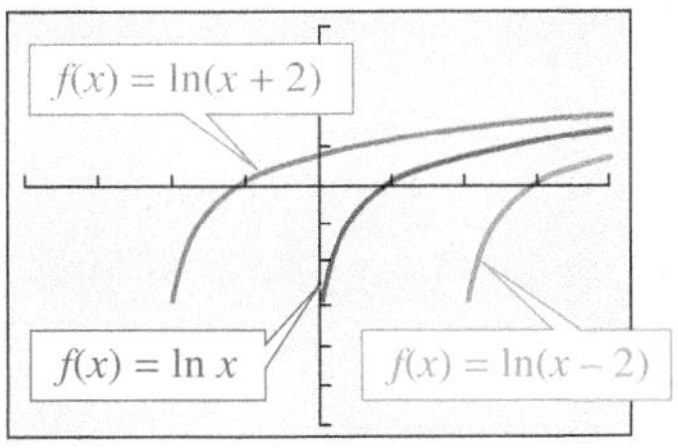

The graph of $f(x) = \ln(x - 2)$ is 2 units to the right of the graph of $f(x) = \ln x$.

The graph of $f(x) = \ln(x + 2)$ is 2 units to the left of the graph of $f(x) = \ln x$.

Figure 11-34

For instructions regarding the use of a Casio graphing calculator, please refer to the Casio Keystroke Guide in the back of the book.

4 Solve an application involving a natural logarithm.

If a population grows exponentially at a certain annual rate, the time required for the population to double is called the *doubling time*.

EXAMPLE 4 **DOUBLING TIME** If the Earth's population continues to grow at the approximate rate of 2% per year, how long will it take for its population to double?

Solution In Section 11.4, we learned that the formula for population growth is $A = Pe^{rt}$, where P is the population at time $t = 0$, A is the population after t years, and r is the annual rate of growth, compounded continuously. Since we want to find out how long it takes for the population to double, we can substitute $2P$ for A and 0.02 for r in the formula and proceed as follows.

$$A = Pe^{rt}$$

$$2P = Pe^{0.02t} \quad \text{Substitute } 2P \text{ for } A \text{ and } 0.02 \text{ for } r.$$

$$2 = e^{0.02t} \quad \text{Divide both sides by } P.$$

$$\ln 2 = 0.02t \quad \ln 2 \text{ is the exponent to which } e \text{ is raised to obtain 2.}$$

$$34.65735903 \approx t \quad \text{Divide both sides by 0.02 and simplify.}$$

The population will double in about 35 years.

SELF CHECK 4 If the world population's annual growth rate could be reduced to 1.5% per year, what would be the doubling time? Give the result to the nearest year.

By solving the formula $A = Pe^{rt}$ for t, we can obtain a simpler formula for finding the doubling time.

$$A = Pe^{rt}$$

$$2P = Pe^{rt} \quad \text{Substitute } 2P \text{ for } A.$$

$$2 = e^{rt} \quad \text{Divide both sides by } P.$$

$$\ln 2 = rt \quad \ln 2 \text{ is the exponent to which } e \text{ is raised to get 2.}$$

$$\frac{\ln 2}{r} = t \quad \text{Divide both sides by } r.$$

This result gives a specific formula for finding the doubling time.

Unless otherwise noted, all content on this page is © Cengage Learning.

FORMULA FOR DOUBLING TIME

If r is the annual rate (compounded continuously) and t is the time required for a population to double, then

$$t = \frac{\ln 2}{r}$$

EXAMPLE 5 **DOUBLING TIME** How long will it take \$1,000 to double at an annual rate of 8%, compounded continuously?

Solution We can substitute 0.08 for r and simplify:

$$t = \frac{\ln 2}{r}$$

$$t = \frac{\ln 2}{\mathbf{0.08}}$$

$$\approx 8.664339757$$

It will take about $8\frac{2}{3}$ years for the money to double.

COMMENT To find the doubling time, you can use either the method in Example 4 or the method in Example 5.

SELF CHECK 5 How long will it take at 9%, compounded continuously?

SELF CHECK ANSWERS

1. a. 1.1447 **b.** no value **2. a.** 12.7267 **b.** 40.6853 **3.**

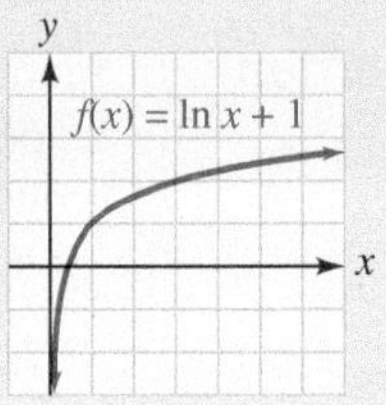

4. 46 years **5.** about 7.7 years

NOW TRY THIS

Between 2000 and 2007, McKinney, TX, was ranked overall as the fastest growing city in the United States.

1. The population more than doubled from 54,369 to 115,620 during this period. To the nearest tenth, what was the average growth rate?
2. If the growth rate slowed to half that above and was estimated to remain steady for the next 10 years, project the population in 2017.
3. Discuss with another student (or in a group) the difficulties a city might face with such a growth rate.

11.6 Exercises

WARM-UPS *Write each logarithmic equation as an exponential equation.*

1. $\log_3 9 = 2$

2. $\log_a b = c$

Write each exponential equation as a logarithmic equation.

3. $2^3 = 8$

4. $x^y = z$

REVIEW *Write the equation of the required line.*

5. having a slope of -7 and a y-intercept of $(0, 2)$

6. passing through the point $(3, 2)$ and perpendicular to the line $y = \frac{2}{3}x - 12$

Unless otherwise noted, all content on this page is © Cengage Learning.

7. parallel to the line $3x + 2y = 9$ and passing through the point $(-3, 5)$
8. vertical line through the point $(4, 1)$
9. horizontal line through the point $(-3, 6)$

Simplify each expression. Assume no denominators are 0.

10. $\dfrac{3x - 5}{9x^2 - 25}$

11. $\dfrac{x + 1}{x} + \dfrac{x - 1}{x + 1}$

12. $\dfrac{x^2 + 3x + 2}{3x + 12} \cdot \dfrac{x + 4}{x^2 - 4}$

13. $\dfrac{1 + \frac{y}{x}}{\frac{y}{x} - 1}$

VOCABULARY AND CONCEPTS *Fill in the blanks.*

14. The expression ____ means ___ and is called a ______ logarithm.
15. The domain of the function $f(x) = \ln x$ is the interval ______ and the range is the interval ________.
16. The graph of $f(x) = \ln x$ has the _____ as an asymptote.
17. In the expression $\log x$, the base is understood to be __.
18. In the expression $\ln x$, the base is understood to be _.
19. If a population grows exponentially at a rate r, the time it will take the population to double is given by the formula

$t = \dfrac{\quad}{\quad}$.

20. The logarithm of a negative number is ________.

GUIDED PRACTICE *Use a calculator to find each value, if possible. Express all answers to four decimal places.* SEE EXAMPLE 1. (OBJECTIVE 1)

21. $\ln 25.25$
22. $\ln 0.523$
23. $\ln 9.89$
24. $\ln 0.00725$

Use a calculator to find the values of y, if possible. Express all answers to four decimal places. SEE EXAMPLE 2. (OBJECTIVE 2)

25. $\ln y = 2.3015$
26. $\ln y = 1.548$
27. $\ln y = 3.17$
28. $\ln y = 0.837$

Use a graphing calculator to graph each function. SEE EXAMPLE 3. (OBJECTIVE 3)

29. $y = -\ln x$
30. $y = \ln x^2$
31. $y = \ln(-x)$
32. $y = \ln\left(\dfrac{1}{2}x\right)$

Determine whether the graph could represent the graph of $y = \ln x$.

33.

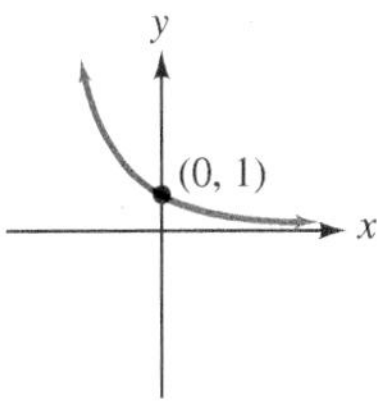

34.

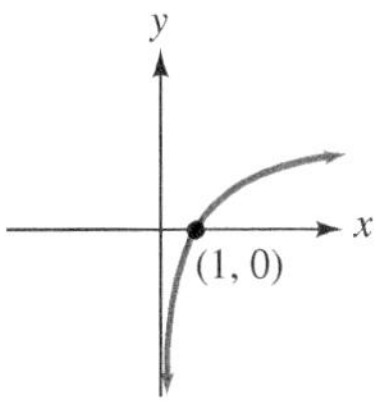

35.

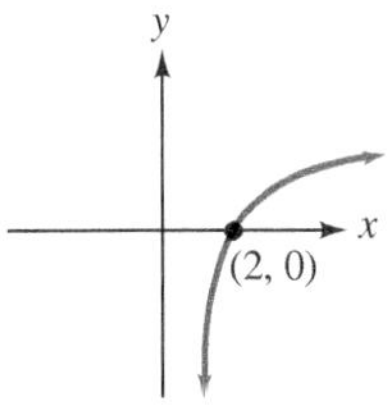

36.

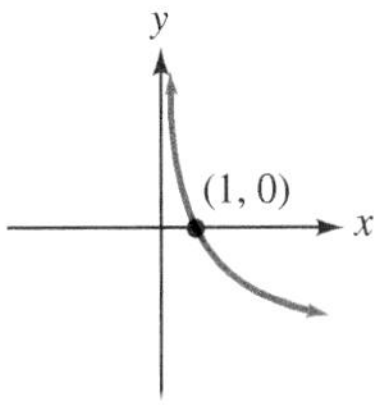

ADDITIONAL PRACTICE *Use a calculator to find each value, if possible. Express all answers to four decimal places.*

37. $\log(\ln 3)$
38. $\ln(\log 28.8)$
39. $\ln(\log 0.7)$
40. $\log(\ln 0.2)$

Use a calculator to find each value of y, if possible. Express all answers to four decimal places.

41. $\ln y = -4.72$
42. $\ln y = -0.48$
43. $\log y = \ln 4$
44. $\ln y = \log 5$

APPLICATIONS *Use a calculator to solve. Round each answer to the nearest tenth.* SEE EXAMPLES 4–5. (OBJECTIVE 4)

45. **Population growth** See the ad. How long will it take the population of River City to double?

46. **Population growth** A population growing at an annual rate r will triple in a time t given by the formula

$$t = \frac{\ln 3}{r}$$

How long will it take the population of a town growing at the rate of 8% per year to triple?

47. **Doubling money** How long will it take \$1,000 to double if it is invested at an annual rate of 5%, compounded continuously?
48. **Tripling money** Find the length of time for \$25,000 to triple if invested at 6% annual interest, compounded continuously. (See Exercise 46.)

Unless otherwise noted, all content on this page is © Cengage Learning.

49. Making Jell-O After the contents of a package of Jell-O are combined with boiling water, the mixture is placed in a refrigerator whose temperature remains a constant 38° F. Estimate the number of hours t that it will take for the Jell-O to cool to 50° F using the formula

$$t = -\frac{1}{0.9}\ln\frac{50 - T_r}{212 - T_r}$$

where T_r is the temperature of the refrigerator.

50. Forensic medicine To estimate the number of hours t that a murder victim had been dead, a coroner used the formula

$$t = \frac{1}{0.25}\ln\frac{98.6 - T_s}{82 - T_s}$$

where T_s is the temperature of the surroundings where the body was found. If the crime took place in an apartment where the thermostat was set at 72° F, approximately how long ago did the murder occur?

WRITING ABOUT MATH

51. The time it takes money to double at an annual rate r, compounded continuously, is given by the formula $t = (\ln 2)/r$. Explain why money doubles more quickly as the rate increases.

52. The time it takes money to triple at an annual rate r, compounded continuously, is given by the formula $t = (\ln 3)/r$. Explain why money triples less quickly as the rate decreases.

SOMETHING TO THINK ABOUT

53. Use the formula $P = P_0e^{rt}$ to verify that P will be three times as large as P_0 when $t = \frac{\ln 3}{r}$.

54. Use the formula $P = P_0e^{rt}$ to verify that P will be four times as large as P_0 when $t = \frac{\ln 4}{r}$.

55. Find a formula to find how long it will take a sum of money to become five times as large.

56. Use a graphing calculator to graph

$$f(x) = \frac{1}{1 + e^{-2x}}$$

and discuss the graph.

Section 11.7 Properties of Logarithms

Objectives

1. Simplify a logarithmic expression by applying a property or properties of logarithms.
2. Expand a logarithmic expression.
3. Write a logarithmic expression as a single logarithm.
4. Evaluate a logarithm by applying properties of logarithms.
5. Apply the change-of-base formula.
6. Solve an application using one or more logarithmic properties.

Vocabulary

change-of-base formula   pH of a solution

Getting Ready

Simplify each expression. Assume $x \neq 0$.

1. x^mx^n **2.** x^0 **3.** $(x^m)^n$ **4.** $\frac{x^m}{x^n}$

In this section, we will consider many properties of logarithms. We will then use these properties to solve applications.

1 Simplify a logarithmic expression by applying a property or properties of logarithms.

Since logarithms are exponents, the properties of exponents have counterparts in the theory of logarithms. We begin with four basic properties.

PROPERTIES OF LOGARITHMS

If b is a positive number and $b \neq 1$, then

1. $\log_b 1 = 0$ **2.** $\log_b b = 1$

3. $\log_b b^x = x$ **4.** $b^{\log_b x} = x \quad (x > 0)$

Properties 1 through 4 follow directly from the definition of a logarithm.

1. $\log_b 1 = 0$, because $b^0 = 1$.

2. $\log_b b = 1$, because $b^1 = b$.

3. $\log_b b^x = x$, because $b^x = b^x$.

4. $b^{\log_b x} = x$, because $\log_b x$ is the exponent to which b is raised to get x.

Properties 3 and 4 also indicate that the composition of the exponential and logarithmic functions with the same base (in both directions) is the identity function. This is expected, because the exponential and logarithmic functions with the same base are inverse functions.

EXAMPLE 1 Simplify each expression. **a.** $\log_5 1$ **b.** $\log_3 3$ **c.** $\log_7 7^3$ **d.** $b^{\log_b 7}$

Solution

a. By Property 1, $\log_5 1 = 0$, because $5^0 = 1$.

b. By Property 2, $\log_3 3 = 1$, because $3^1 = 3$.

c. By Property 3, $\log_7 7^3 = 3$, because $7^3 = 7^3$.

d. By Property 4, $b^{\log_b 7} = 7$, because $\log_b 7$ is the power to which b is raised to get 7.

SELF CHECK 1 Simplify. **a.** $\log_4 1$ **b.** $\log_5 5$ **c.** $\log_2 2^4$ **d.** $5^{\log_5 2}$

The next two properties state that

The logarithm of a product is the sum of the logarithms.

The logarithm of a quotient is the difference of the logarithms.

THE PRODUCT AND QUOTIENT PROPERTIES OF LOGARITHMS

If M, N, and b are positive numbers and $b \neq 1$, then

5. $\log_b MN = \log_b M + \log_b N$ **6.** $\log_b \frac{M}{N} = \log_b M - \log_b N$

Proof To prove the product property of logarithms, we let $x = \log_b M$ and $y = \log_b N$. We use the definition of logarithms to write each equation in exponential form.

$$M = b^x \quad \text{and} \quad N = b^y$$

Then $MN = b^x b^y$ and a property of exponents gives

$$MN = b^{x+y} \qquad b^x b^y = b^{x+y}\text{: Keep the base and add the exponents.}$$

We write this exponential equation in logarithmic form as

$$\log_b MN = x + y$$

Substituting the values of x and y completes the proof.

$$\log_b MN = \log_b M + \log_b N$$

The proof of the quotient property of logarithms is similar.

COMMENT By the product property of logarithms, the logarithm of a *product* is equal to the *sum* of the logarithms. The logarithm of a sum or a difference usually does not simplify. In general,

$$\log_b (M + N) \neq \log_b M + \log_b N$$

$$\log_b (M - N) \neq \log_b M - \log_b N$$

By the quotient property of logarithms, the logarithm of a *quotient* is equal to the *difference* of the logarithms. The logarithm of a quotient is not the quotient of the logarithms:

$$\log_b \frac{M}{N} \neq \frac{\log_b M}{\log_b N}$$

Accent
on technology

▸ Verifying Properties
 of Logarithms

We can use a calculator to illustrate the product property of logarithms by showing that

$$\ln[(3.7)(15.9)] = \ln 3.7 + \ln 15.9$$

We calculate the left and right sides of the equation separately and compare the results. To use a calculator to find $\ln[(3.7)(15.9)]$, we enter these numbers and press these keys:

3.7 **×** 15.9 **=** **LN** Using a scientific calculator

LN 3.7 **×** 15.9 **)** **ENTER** Using a TI-84 graphing calculator

The display will read **4.074651929** .

To find $\ln 3.7 + \ln 15.9$, we enter these numbers and press these keys:

3.7 **LN** **+** 15.9 **LN** **=** Using a scientific calculator

LN 3.7 **)** **+** **LN** 15.9 **)** **ENTER** Using a TI-84 graphing calculator

The display will read **4.074651929** . Since the left and right sides are equal, the equation is true.

For instructions regarding the use of a Casio graphing calculator, please refer to the Casio Keystroke Guide in the back of the book.

The power rule of logarithms states that

> *The logarithm of an expression to a power is the power times the logarithm of the expression.*

THE POWER RULE
OF LOGARITHMS

If M, p, and b are positive numbers and $b \neq 1$, then

7. $\log_b M^p = p \log_b M$

Proof To prove the power rule, we let $x = \log_b M$, write the expression in exponential form, and raise both sides to the pth power:

$$M = b^x$$
$$(M)^p = (b^x)^p \quad \text{Raise both sides to the } p\text{th power.}$$
$$M^p = b^{px} \quad \text{Keep the base and multiply the exponents.}$$

Using the definition of logarithms gives

$$\log_b M^p = px$$

Substituting the value for x completes the proof.

$$\log_b M^p = p \log_b M$$

The logarithmic property of equality states that

If the logarithms of two numbers are equal, the numbers are equal.

THE LOGARITHMIC PROPERTY OF EQUALITY

If x, y, and b are positive numbers and $b \neq 1$, then

8. If $\log_b x = \log_b y$, then $x = y$.

The logarithmic property of equality follows from the fact that the logarithmic function is a one-to-one function. It will be important in the next section when we solve logarithmic equations.

2 Expand a logarithmic expression.

We can use the properties of logarithms to write a logarithm as the sum or difference of several logarithms.

EXAMPLE 2 Assume that b, x, y, and z are positive numbers and $b \neq 1$. Write each expression in terms of the logarithms of x, y, and z.

a. $\log_b (xyz)$ **b.** $\log_b \frac{xy}{z}$

Solution

a. $\log_b (xyz) = \log_b x + \log_b y + \log_b z$ The log of a product is the sum of the logs.

b. $\log_b \frac{xy}{z} = \log_b(xy) - \log_b z$ The log of a quotient is the difference of the logs.

$= (\log_b x + \log_b y) - \log_b z$ The log of a product is the sum of the logs.

$= \log_b x + \log_b y - \log_b z$ Remove parentheses.

SELF CHECK 2 Write $\log_b \frac{x}{yz}$ in terms of the logarithms of x, y, and z.

EXAMPLE 3 Assume that b, x, y, and z are positive numbers and $b \neq 1$. Write each expression in terms of the logarithms of x, y, and z.

a. $\log_b(x^2y^3z)$ **b.** $\log_b \frac{\sqrt{x}}{y^3z}$

Solution **a.** $\log_b(x^2y^3z) = \log_b x^2 + \log_b y^3 + \log_b z$ The log of a product is the sum of the logs.

$= 2\log_b x + 3\log_b y + \log_b z$ The log of an expression to a power is the power times the log of the expression.

b. $\log_b \frac{\sqrt{x}}{y^3z} = \log_b \sqrt{x} - \log_b(y^3z)$ The log of a quotient is the difference of the logs.

$= \log_b x^{1/2} - (\log_b y^3 + \log_b z)$ $\sqrt{x} = x^{1/2}$. The log of a product is the sum of the logs.

$= \frac{1}{2}\log_b x - (3\log_b y + \log_b z)$ The log of a power is the power times the log.

$= \frac{1}{2}\log_b x - 3\log_b y - \log_b z$ Use the distributive property to remove parentheses.

SELF CHECK 3 Write $\log_b \sqrt[4]{\frac{x^3y}{z}}$ in terms of the logarithms of x, y, and z.

3 Write a logarithmic expression as a single logarithm.

We can use the properties of logarithms to combine several logarithms into one logarithm.

EXAMPLE 4 Assume that b, x, y, and z are positive numbers and $b \neq 1$. Write each expression as one logarithm.

a. $3\log_b x + \frac{1}{2}\log_b y$ **b.** $\frac{1}{2}\log_b(x-2) - \log_b y + 3\log_b z$

Solution **a.** $3\log_b x + \frac{1}{2}\log_b y = \log_b x^3 + \log_b y^{1/2}$ A power times a log is the log of the power.

$= \log_b(x^3y^{1/2})$ The sum of two logs is the log of a product.

$= \log_b\left(x^3\sqrt{y}\right)$ $y^{1/2} = \sqrt{y}$

b. $\frac{1}{2}\log_b(x-2) - \log_b y + 3\log_b z$

$= \log_b(x-2)^{1/2} - \log_b y + \log_b z^3$ A power times a log is the log of the power.

$= \log_b \frac{(x-2)^{1/2}}{y} + \log_b z^3$ The difference of two logs is the log of the quotient.

$= \log_b \frac{z^3\sqrt{x-2}}{y}$ The sum of two logs is the log of a product.

SELF CHECK 4 Write the expression as one logarithm.

$2\log_b x + \frac{1}{2}\log_b y - 2\log_b(x-y)$

We summarize the properties of logarithms.

PROPERTIES OF LOGARITHMS

If b, M, and N are positive numbers and $b \neq 1$, then

1. $\log_b 1 = 0$
2. $\log_b b = 1$
3. $\log_b b^x = x$
4. $b^{\log_b x} = x$
5. $\log_b MN = \log_b M + \log_b N$
6. $\log_b \frac{M}{N} = \log_b M - \log_b N$
7. $\log_b M^p = p \log_b M$
8. If $\log_b x = \log_b y$, then $x = y$.

4 Evaluate a logarithm by applying properties of logarithms.

EXAMPLE 5 Given that $\log 2 \approx 0.3010$ and $\log 3 \approx 0.4771$, find approximations without using a calculator for the following:

a. $\log 6$ **b.** $\log 9$ **c.** $\log 18$ **d.** $\log 2.5$

Solution

a.
$$\log 6 = \log(2 \cdot 3)$$
$= \log 2 + \log 3$ The log of a product is the sum of the logs.
$\approx 0.3010 + 0.4771$ Substitute the approximate value of each logarithm.
≈ 0.7781

b.
$$\log 9 = \log(3^2)$$
$= 2 \log 3$ The log of a power is the power times the log.
$\approx 2(0.4771)$ Substitute the approximate value of log 3.
≈ 0.9542

c.
$$\log 18 = \log(2 \cdot 3^2)$$
$= \log 2 + \log 3^2$ The log of a product is the sum of the logs.
$= \log 2 + 2 \log 3$ The log of a power is the power times the log.
$\approx 0.3010 + 2(0.4771)$ Substitute the approximate value of each logarithm.
≈ 1.2552

d.
$$\log 2.5 = \log\left(\frac{5}{2}\right)$$
$= \log 5 - \log 2$ The log of a quotient is the difference of the logs.
$= \log \frac{10}{2} - \log 2$ Write 5 as $\frac{10}{2}$.
$= \log 10 - \log 2 - \log 2$ The log of a quotient is the difference of the logs.
$= 1 - 2 \log 2$ $\log_{10} 10 = 1$
$\approx 1 - 2(0.3010)$ Substitute the approximate value of log 2.
≈ 0.3980

SELF CHECK 5 Use the values given in Example 5 and approximate:

a. $\log 1.5$ **b.** $\log 0.2$

5 Apply the change-of-base formula.

If we know the base-a logarithm of a number, we can find its logarithm to some other base b with a formula called the **change-of-base formula**.

CHANGE-OF-BASE FORMULA

If a, b, and x are real numbers, $b > 0$, and $b \neq 1$, then

$$\log_b x = \frac{\log_a x}{\log_a b}$$

Proof To prove this formula, we begin with the equation $\log_b x = y$.

$y = \log_b x$	
$x = b^y$	Write the equation in exponential form.
$\mathbf{\log_a x = \log_a b^y}$	Take the base-a logarithm of both sides.
$\log_a x = y \log_a b$	The log of a power is the power times the log.
$y = \dfrac{\log_a x}{\log_a b}$	Divide both sides by $\log_a b$.
$\log_b x = \dfrac{\log_a x}{\log_a b}$	Refer to the first equation and substitute $\log_b x$ for y.

If we know logarithms to base a (for example, $a = 10$), we can find the logarithm of x to a new base b. We simply divide the base-a logarithm of x by the base-a logarithm of b.

EXAMPLE 6 Approximate to 4 decimal places $\log_4 9$ using base-10 logarithms.

Solution We substitute 4 for b, 10 for a, and 9 for x into the change-of-base formula.

COMMENT $\frac{\log_a x}{\log_a b}$ means that one logarithm is to be divided by the other.

$$\log_b \mathbf{x} = \frac{\log_a \mathbf{x}}{\log_a \mathbf{b}}$$

$$\log_4 \mathbf{9} = \frac{\log_{10} \mathbf{9}}{\log_{10} \mathbf{4}}$$

$$\approx 1.584962501$$

To four decimal places, $\log_4 9 = 1.5850$.

SELF CHECK 6 Approximate $\log_5 3$ to four decimal places using base-10 logarithms.

COMMENT It does not matter what base you choose when applying the change-of-base formula. You could use base-e (natural logarithm) and obtain the same result. In the example above, $\log_4 \mathbf{9} = \frac{\ln 9}{\ln 4} \approx 1.584962501$.

6 Solve an application using one or more logarithmic properties.

Common logarithms are used to express the acidity of solutions. The more acidic a solution, the greater the concentration of hydrogen ions. This concentration is indicated by the pH scale, or hydrogen ion index. The **pH of a solution** is defined by the following equation.

pH OF A SOLUTION

If $[H^+]$ is the hydrogen ion concentration in gram-ions per liter, then

$$\text{pH} = -\log[H^+]$$

EXAMPLE 7 **FINDING THE pH OF A SOLUTION** Find the pH of pure water, which has a hydrogen ion concentration of 10^{-7} gram-ions per liter.

Solution Since pure water has approximately 10^{-7} gram-ions per liter, its pH is

$$\begin{aligned} pH &= -\log[H^+] \\ pH &= -\log \mathbf{10^{-7}} \\ &= -(-7)\log 10 && \text{The log of a power is the power times the log.} \\ &= -(-7)(1) && \log 10 = 1 \\ &= 7 \end{aligned}$$

SELF CHECK 7 Find the pH of black coffee, which has a hydrogen ion concentration of 10^{-5} gram-ions per liter.

EXAMPLE 8 **FINDING THE HYDROGEN ION CONCENTRATION** Find the hydrogen ion concentration of seawater if its pH is 8.5.

Solution To find its hydrogen ion concentration, we substitute 8.5 for the pH and find $[H^+]$.

$$\begin{aligned} 8.5 &= -\log[H^+] \\ -8.5 &= \log[H^+] && \text{Multiply both sides by } -1. \\ [H^+] &= 10^{-8.5} && \text{Write the equation in exponential form.} \end{aligned}$$

We can use a calculator to find that

$$[H^+] \approx 3.2 \times 10^{-9} \text{ gram-ions per liter}$$

SELF CHECK 8 Find the hydrogen ion concentration of healthy skin if its pH is 4.5.

In physiology, experiments suggest that the relationship between the loudness and the intensity of sound is logarithmic and has been named the Weber–Fechner law.

WEBER–FECHNER LAW If L is the apparent loudness of a sound, I is the actual intensity, and k is a constant, then

$$L = k \ln I$$

EXAMPLE 9 **WEBER–FECHNER LAW** Find the increase in intensity that will cause the apparent loudness of a sound to double.

Solution If the original loudness L_O is caused by an actual intensity I_O, then

(1) $$L_O = k \ln I_O$$

To double the apparent loudness, we multiply both sides of Equation 1 by 2 and use the power rule of logarithms:

$$\begin{aligned} \mathbf{2}\, L_O &= \mathbf{2}\, k \ln I_O \\ &= k \ln(I_O)^2 \end{aligned}$$

To double the loudness of a sound, the intensity must be squared.

SELF CHECK 9 What decrease in intensity will cause a sound to be half as loud?

SELF CHECK ANSWERS

1. a. 0 **b.** 1 **c.** 4 **d.** 2 **2.** $\log_b x - \log_b y - \log_b z$ **3.** $\frac{1}{4}(3\log_b x + \log_b y - \log_b z)$ **4.** $\log_b \frac{x^2\sqrt{y}}{(x-y)^2}$ **5. a.** 0.1761 **b.** −0.6990 **6.** 0.6826 **7.** 5 **8.** $[H^+] \approx 3.2 \times 10^{-5}$ gram-ions per liter **9.** the square root of the intensity

NOW TRY THIS

Evaluate.

1. $\log 5 + \log 20$

2. $\log_3 24 - \log_3 8$

3. $7\log_2 \frac{1}{2} - \log_2 \frac{1}{8}$

11.7 Exercises

WARM-UPS *Find the value of x in each equation.*

1. $\log_5 25 = x$

2. $\log_x 5 = 1$

3. $\log_7 x = 3$

4. $\log_2 x = -2$

5. $\log_9 x = \frac{1}{2}$

6. $\log_x 4 = 2$

7. $\log_{1/3} x = 2$

8. $\log_{16} 4 = x$

9. $\log_x \frac{1}{4} = -2$

REVIEW *Consider the line that passes through (−2, 3) and (4, −4).*

10. Find the slope of the line.

11. Find the distance between the points.

12. Find the midpoint of the segment.

13. Write an equation of the line.

VOCABULARY AND CONCEPTS *Fill in the blanks. Assume all variables are positive numbers and b ≠ 1.*

14. $\log_b 1 =$ _

15. $\log_b b =$ _

16. $\log_b MN = \log_b$ _ $+ \log_b$ _

17. $b^{\log_b x} =$ _

18. If $\log_b x = \log_b y$, then _ = _

19. $\log_b \frac{M}{N} = \log_b M$ _ $\log_b N$

20. $\log_b x^p = p \cdot \log_b$ _

21. $\log_b b^x =$ _

22. $\log_b(A + B)$ ___ $\log_b A + \log_b B$

23. $\log_b A + \log_b B$ ___ $\log_b AB$

24. The change-of-base formula states that $\log_b x =$ _____.

GUIDED PRACTICE *Simplify each expression.* SEE EXAMPLE 1. (OBJECTIVE 1)

25. $\log_3 1 =$ _

26. $\log_7 7 =$ _

27. $\log_6 6^2 =$ _

28. $7^{\log_7 9} =$ _

29. $5^{\log_5 10} =$ _

30. $\log_4 4^3 =$ _

31. $\log_2 2 =$ _

32. $\log_4 1 =$ _

Use a calculator to verify each equation. (OBJECTIVE 1)

33. $\log[(2.5)(3.7)] = \log 2.5 + \log 3.7$

34. $\ln \frac{11.3}{6.1} = \ln 11.3 - \ln 6.1$

35. $\ln(2.25)^4 = 4\ln 2.25$

36. $\log 45.37 = \frac{\ln 45.37}{\ln 10}$

Assume that x, y, z, and b are positive numbers (b ≠ 1). Use the properties of logarithms to write each expression in terms of the logarithms of x, y, and z. SEE EXAMPLE 2. (OBJECTIVE 2)

37. $\log_b 7xy$

38. $\log_b 4xz$

39. $\log_b \frac{5x}{y}$

40. $\log_b \frac{x}{yz}$

Assume that x, y, z, and b are positive numbers (b ≠ 1). Use the properties of logarithms to write each expression in terms of the logarithms of x, y, and z. SEE EXAMPLE 3. (OBJECTIVE 2)

41. $\log_b x^3y^2$

42. $\log_b xy^2z^3$

43. $\log_b \frac{x^3\sqrt{z}}{y^5}$

44. $\log_b \sqrt{xy}$

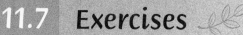

Assume that x, y, z, and b are positive numbers ($b \neq 1$). Use the properties of logarithms to write each expression as the logarithm of a single quantity. SEE EXAMPLE 4. (OBJECTIVE 3)

45. $\log_b(x - 3) - 5 \log_b x$

46. $\log_b x + \log_b(x + 2) - \log_b 8$

47. $7 \log_b x + \frac{1}{2} \log_b z$

48. $-2 \log_b x - 3 \log_b y + \log_b z$

Assume that $\log 4 \approx 0.6021$, $\log 7 \approx 0.8451$, and $\log 9 \approx 0.9542$. *Use these values and the properties of logarithms to approximate each value. Do not use a calculator.* SEE EXAMPLE 5. (OBJECTIVE 4)

49. $\log 28$ **50.** $\log \frac{7}{9}$

51. $\log 2.25$ **52.** $\log 1.75$

53. $\log \frac{49}{36}$ **54.** $\log \frac{4}{63}$

55. $\log 252$ **56.** $\log 49$

Use a calculator and the change-of-base formula to find each logarithm to four decimal places. SEE EXAMPLE 6. (OBJECTIVE 5)

57. $\log_3 7$ **58.** $\log_7 3$

59. $\log_{1/3} 3$ **60.** $\log_{1/2} 6$

61. $\log_3 8$ **62.** $\log_5 10$

63. $\log_{\sqrt{2}} \sqrt{5}$ **64.** $\log_{\pi} e$

ADDITIONAL PRACTICE *Simplify each expression.*

65. $\log_7 1 =$ _ **66.** $\log_9 9 =$ _

67. $\log_3 3^7 =$ _ **68.** $5^{\log_5 8} =$ _

69. $8^{\log_8 10} =$ __ **70.** $\log_4 4^2 =$ _

71. $\log_9 9 =$ _ **72.** $\log_3 1 =$ _

Assume that x, y, z, and b are positive numbers ($b \neq 1$). Use the properties of logarithms to write each expression in terms of the logarithms of x, y, and z.

73. $\log_b\left(\frac{xy}{z}\right)^{1/3}$

74. $\log_b \frac{x^4\sqrt{y}}{z^2}$

75. $\log_b \frac{\sqrt[3]{x}}{\sqrt[4]{yz}}$

76. $\log_b \sqrt[4]{\frac{x^3y^2}{z^4}}$

Assume that x, y, z, and b are positive numbers ($b \neq 1$). Use the properties of logarithms to write each expression as the logarithm of a single quantity. Simplify, if possible.

77. $3 \log_b(x + 2) - 4 \log_b y + \frac{1}{2} \log_b(x + 1)$

78. $3 \log_b(x + 1) - 2 \log_b(x + 2) + \log_b x$

79. $\log_b\left(\frac{x}{z} + x\right) - \log_b\left(\frac{y}{z} + y\right)$

80. $\log_b(xy + y^2) - \log_b(xz + yz) + \log_b z$

Assume that $\log 4 \approx 0.6021$, $\log 7 \approx 0.8451$, *and* $\log 9 \approx 0.9542$. *Use these values and the properties of logarithms to approximate each value. Do not use a calculator.*

81. $\log 112$ **82.** $\log 324$

83. $\log \frac{144}{49}$ **84.** $\log \frac{324}{63}$

Use a calculator to verify each equation.

85. $\log \sqrt{24.3} = \frac{1}{2} \log 24.3$ **86.** $\ln 8.75 = \frac{\log 8.75}{\log e}$

Determine whether each statement is true. If a statement is false, explain why.

87. $\log_b 0 = 1$

88. $\log_b(x + y) \neq \log_b x + \log_b y$

89. $\log_b xy = (\log_b x)(\log_b y)$

90. $\log_b ab = \log_b a + 1$

91. $\log_7 7^7 = 7$

92. $7^{\log_7 7} = 7$

93. $\frac{\log_b A}{\log_b B} = \log_b A - \log_b B$

94. $\log_b(A - B) = \frac{\log_b A}{\log_b B}$

95. $3 \log_b \sqrt[3]{a} = \log_b a$

96. $\frac{1}{3} \log_b a^3 = \log_b a$

97. $\log_b \frac{1}{a} = -\log_b a$

98. $\log_b 2 = \log_2 b$

APPLICATIONS *Use a calculator to find each value.*
SEE EXAMPLES 7–9. (OBJECTIVE 6)

99. pH of a solution Find the pH of a solution with a hydrogen ion concentration of 1.7×10^{-5} gram-ions per liter.

100. Hydrogen ion concentration Find the hydrogen ion concentration of a saturated solution of calcium hydroxide whose pH is 13.2.

101. Aquariums To test for safe pH levels in a fresh-water aquarium, a test strip is compared with the scale shown in the illustration. Find the corresponding range in the hydrogen ion concentration.

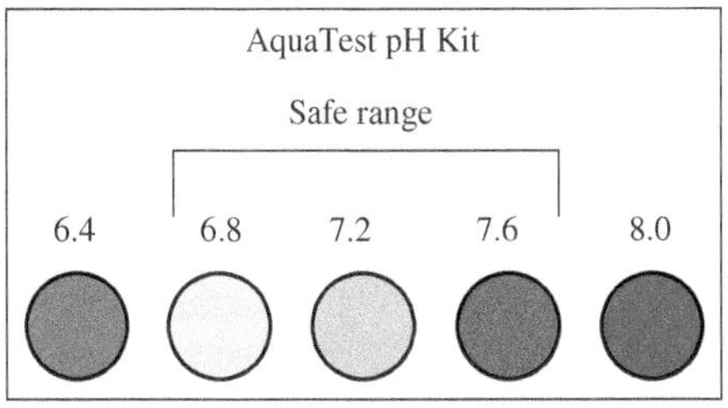

Unless otherwise noted, all content on this page is © Cengage Learning.

102. pH of pickles The hydrogen ion concentration of sour pickles is 6.31×10^{-4}. Find the pH.

103. Change in loudness If the intensity of a sound is doubled, find the apparent change in loudness.

104. Change in loudness If the intensity of a sound is tripled, find the apparent change in loudness.

105. Change in intensity What change in intensity of sound will cause an apparent tripling of the loudness?

106. Change in intensity What increase in the intensity of a sound will cause the apparent loudness to be multiplied by 4?

WRITING ABOUT MATH

107. Explain why ln(log 0.9) is undefined in the real-number system.

108. Explain why $\log_b(\ln 1)$ is undefined in the real-number system.

SOMETHING TO THINK ABOUT

109. Show that $\ln(e^x) = x$.

110. If $\log_b 3x = 1 + \log_b x$, find the value of x.

111. Show that $\log_{b^2} x = \frac{1}{2}\log_b x$.

112. Show that $e^{x \ln a} = a^x$.

Section 11.8 Exponential and Logarithmic Equations

Objectives

1. Solve an exponential equation.
2. Solve a logarithmic equation.
3. Solve an application involving an exponential or logarithmic equation.

Vocabulary

exponential equation   logarithmic equation   half-life

Getting Ready

Write each expression without using exponents.

1. $\log x^2$
2. $\log x^{1/2}$
3. $\log x^0$
4. $\log a^b + b \log a$

An **exponential equation** is an equation that contains a variable in one of its exponents. Some examples of exponential equations are

$$3^x = 5, \quad 6^{x-3} = 2^x, \quad \text{and} \quad 3^{2x+1} - 10(3^x) + 3 = 0$$

A **logarithmic equation** is an equation with logarithmic expressions that contain a variable. Some examples of logarithmic equations are

$$\log(2x) = 25, \quad \ln x - \ln(x - 12) = 24, \quad \text{and} \quad \log x = \log\frac{1}{x} + 4$$

In this section, we will learn how to solve many of these equations.

1 Solve an exponential equation.

One technique we use to solve exponential equations is to apply one of the properties:

if $a = b$, then $\log a = \log b$, or
if $a = b$, then $\ln a = \ln b$.

EXAMPLE 1 Solve: $4^x = 7$

Solution Since logarithms of equal numbers are equal, we can take the common or natural logarithm of both sides of the equation and obtain a new equation. We will use the common logarithm. The power rule of logarithms then provides a process of writing the variable x as a coefficient instead of an exponent.

$$4^x = 7$$

$$\mathbf{log}(4^x) = \mathbf{log}(7) \quad \text{Take the common logarithm of both sides.}$$

$$x \log 4 = \log 7 \quad \text{The log of a power is the power times the log.}$$

$$(1) \quad x = \frac{\log 7}{\log 4} \quad \text{Divide both sides by (log 4).}$$

$$\approx 1.403677461 \quad \text{Use a calculator.}$$

To four decimal places, $x = 1.4037$.

SELF CHECK 1 Solve $5^x = 4$. Give the result to four decimal places.

COMMENT The right side of Equation 1 calls for a division, not a subtraction.

$$\frac{\log 7}{\log 4} \quad \text{means} \quad (\log 7) \div (\log 4)$$

It is the expression $\log\left(\frac{7}{4}\right)$ that means $\log 7 - \log 4$.

EXAMPLE 2 Solve: $73 = 1.6(1.03)^t$

Solution We divide both sides by 1.6 to isolate the exponential expression. Thus, we obtain

$$\frac{73}{1.6} = 1.03^t$$

and solve the equation. If $a = b$ then $b = a$.

$$1.03^t = \frac{73}{1.6}$$

$$\mathbf{log}(1.03^t) = \mathbf{log}\left(\frac{73}{1.6}\right) \quad \text{Take the common logarithm of both sides.}$$

$$t \log 1.03 = \log \frac{73}{1.6} \quad \text{The logarithm of a power is the power times the logarithm.}$$

$$t = \frac{\log \frac{73}{1.6}}{\log 1.03} \quad \text{Divide both sides by log 1.03.}$$

$$t \approx 129.2493444 \quad \text{Use a calculator.}$$

To four decimal places, $t = 129.2493$.

SELF CHECK 2 Solve $47 = 2.5(1.05)^t$. Give the result to 4 decimal places.

EXAMPLE 3 Solve: $6^{x-3} = 2^x$

Solution

$$6^{x-3} = 2^x$$

$$\log(6^{x-3}) = \log(2^x)$$ Take the common logarithm of both sides.

$$(x - 3)\log 6 = x \log 2$$ The log of a power is the power times the log.

$$x \log 6 - 3 \log 6 = x \log 2$$ Use the distributive property to remove parentheses.

$$x \log 6 - x \log 2 = 3 \log 6$$ Add (3 log 6) and subtract (x log 2) from both sides.

$$x(\log 6 - \log 2) = 3 \log 6$$ Factor out x, the GCF, on the left side.

$$x = \frac{3 \log 6}{\log 6 - \log 2}$$ Divide both sides by (log 6 − log 2).

$$x \approx 4.892789261$$ Use a calculator.

To four decimal places, $x = 4.8928$.

SELF CHECK 3 Solve to four decimal places: $5^{x-2} = 3^x$

Another technique we use to solve exponential equations is to write the expressions with the same base and then set their corresponding exponents equal.

If $a^n = a^m$, then $n = m$.

EXAMPLE 4 Solve: $2^{x^2+2x} = \frac{1}{2}$

Solution We can use the fact that both expressions involve a base of 2 to write the equation in a different form. Since $\frac{1}{2} = 2^{-1}$, we can write the equation in the form

$$2^{x^2+2x} = 2^{-1}$$

Equal quantities with equal bases have equal exponents. Therefore,

$$x^2 + 2x = -1$$

$$x^2 + 2x + 1 = 0$$ Add 1 to both sides.

$$(x + 1)(x + 1) = 0$$ Factor the trinomial.

$x + 1 = 0$ or $x + 1 = 0$ Set each factor equal to 0.

$x = -1$ | $x = -1$ -1 is a double root.

Verify that -1 satisfies the equation.

SELF CHECK 4 Solve: $3^{x^2-2x} = \frac{1}{3}$

Accent on technology

▸ Solving Exponential Equations

To use a TI-84 graphing calculator to approximate the solutions of $2^{x^2+2x} = \frac{1}{2}$ (see Example 4), we can graph the two corresponding functions

$$f(x) = 2^{x^2+2x} \quad \text{and} \quad f(x) = \frac{1}{2}$$

and find the intersection.

If we use window settings of $[-4, 4]$ for x and $[-2, 6]$ for y, we obtain the graph shown in Figure 11-35(a).

We use the intersect feature to find the point(s) of intersection as illustrated in Figure 11.35(b). Since $x = -1$ is the only intersection point, -1 is the only solution.

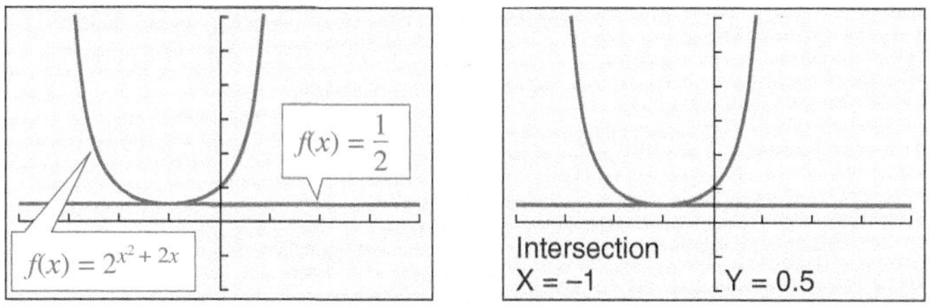

(a) (b)

Figure 11-35

For instructions regarding the use of a Casio graphing calculator, please refer to the Casio Keystroke Guide in the back of the book.

2 Solve a logarithmic equation.

In each of the following examples, we use the properties of logarithms to change a logarithmic equation into an algebraic equation. Recall that the logarithmic property of equality states that if $\log_b x = \log_b y$, then $x = y$. Be certain to check any apparent solutions.

EXAMPLE 5 Solve: $\log_b(3x + 2) - \log_b(2x - 3) = 0$

Solution

$\log_b(3x + 2) - \log_b(2x - 3) = 0$

$\log_b(3x + 2) = \log_b(2x - 3)$ Add $\log_b(2x - 3)$ to both sides.

$3x + 2 = 2x - 3$ If $\log_b r = \log_b s$, then $r = s$.

$x = -5$ Subtract $2x$ and 2 from both sides.

Check:

$\log_b(3x + 2) - \log_b(2x - 3) = 0$

$\log_b[3(-5) + 2] - \log_b[2(-5) - 3] \stackrel{?}{=} 0$

$\log_b(-13) - \log_b(-13) \stackrel{?}{=} 0$

Since we cannot take the logarithm of a negative number, the apparent solution of -5 must be discarded. Since this equation has no solution, its solution set is $\varnothing$.

COMMENT Example 5 illustrates that you must check the apparent solutions of all logarithmic equations.

SELF CHECK 5 Solve: $\log_b(5x + 2) - \log_b(7x - 2) = 0$

COMMENT To solve a logarithmic equation, consider whether writing the equation in exponential form would help you solve for the variable.

EXAMPLE 6 Solve: $\log x + \log(x - 3) = 1$

Solution

$\log x + \log(x - 3) = 1$

$\log[x(x - 3)] = 1$ The sum of two logs is the log of a product.

$x(x - 3) = 10^1$ Use the definition of logarithms to write the equation in exponential form.

$x^2 - 3x - 10 = 0$ Use the distributive property to remove parentheses and write in quadratic form.

$(x + 2)(x - 5) = 0$ Factor the trinomial.

$x + 2 = 0$ or $x - 5 = 0$ Set each factor equal to 0.

$x = -2$ | $x = 5$

Unless otherwise noted, all content on this page is © Cengage Learning.

Check: The number -2 is not a solution, because we cannot take the logarithm of a negative number. It is extraneous. We will check the remaining number, 5.

$$\log x + \log(x-3) = 1$$

$$\log 5 + \log(5-3) \stackrel{?}{=} 1 \quad \text{Substitute 5 for } x.$$

$$\log 5 + \log 2 \stackrel{?}{=} 1$$

$$\log 10 \stackrel{?}{=} 1 \quad \text{The sum of two logs is the log of a product.}$$

$$1 = 1 \quad \log 10 = 1$$

Since 5 satisfies the equation, it is a solution.

SELF CHECK 6 Solve: $\log x + \log(x+3) = 1$

EXAMPLE 7 Solve: $\dfrac{\log(5x-6)}{\log x} = 2$

Solution We know that the $\log x$ cannot equal 0, thus, we can multiply both sides of the equation by $\log x$ to obtain

$$\log(5x-6) = 2\log x$$

and apply the power rule of logarithms to obtain

$$\log(5x-6) = \log x^2$$

By Property 8 of logarithms, $5x - 6 = x^2$. Thus,

$$5x - 6 = x^2$$

$$0 = x^2 - 5x + 6 \quad \text{Write in quadratic form.}$$

$$0 = (x-3)(x-2) \quad \text{Factor the trinomial.}$$

$$x - 3 = 0 \quad \text{or} \quad x - 2 = 0 \quad \text{Set each factor equal to 0.}$$

$$x = 3 \quad \Big| \quad x = 2 \quad \text{Solve each equation.}$$

Verify that both 2 and 3 satisfy the equation.

SELF CHECK 7 Solve: $\dfrac{\log(5x+6)}{\log x} = 2$

Accent on technology

▸ Solving Logarithmic Equations

To use a graphing calculator to approximate the solutions of $\log x + \log(x-3) = 1$ (see Example 6), we can graph the corresponding functions

$$f(x) = \log(x) + \log(x-3) \quad \text{and} \quad f(x) = 1$$

If we use window settings of $[0, 10]$ for x and $[-2, 2]$ for y, we obtain the graph shown in Figure 11-36. Since the solution of the equation is the intersection, we can find the solution by using the INTERCEPT command. The solution is $x = 5$.

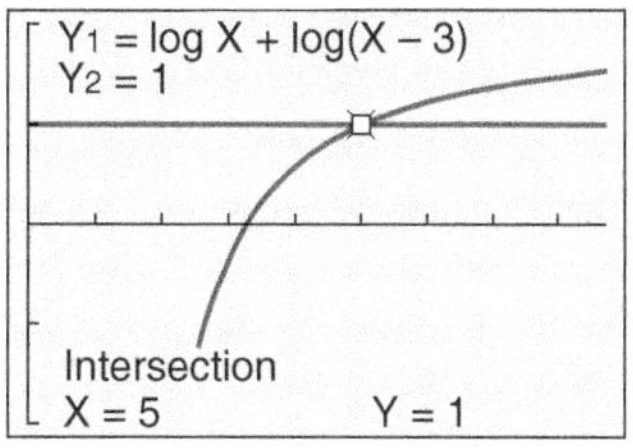

Figure 11-36

For instructions regarding the use of a Casio graphing calculator, please refer to the Casio Keystroke Guide in the back of the book.

3 Solve an application involving an exponential or logarithmic equation.

We have previously discussed that the amount A of radiation present in a radioactive material decays exponentially according to the formula $A = A_0e^{kt}$, where A_0 is the amount of radioactive material present at time $t = 0$, and k is a negative number.

Experiments have determined the time it takes for one-half of a sample of a radioactive element to decompose. That time is a constant, called the given material's **half-life**.

EXAMPLE 8 **HALF-LIFE OF RADON-22** In Example 5 of Section 11.4, we learned that radon-22 decays according to the formula $A = A_0e^{-0.181t}$, where t is expressed in days. Find the material's half-life to the nearest hundredth.

Solution At time $t = 0$, the amount of radioactive material is A_0. At the end of one half-life, the amount present will be $\frac{1}{2}A_0$. To find the material's half-life, we can substitute $\frac{1}{2}A_0$ for A in the formula and solve for t.

$$A = A_0e^{-0.181t}$$

$$\frac{1}{2}A_0 = A_0e^{-0.181t} \qquad \text{Substitute } \tfrac{1}{2}A_0 \text{ for } A.$$

$$\frac{1}{2} = e^{-0.181t} \qquad \text{Divide both sides by } A_0.$$

$$\ln\left(\frac{1}{2}\right) = \ln(e^{-0.181t}) \qquad \text{Take the natural logarithm of both sides.}$$

$$-0.6931471806 \approx -0.181t \ln e \qquad \text{Find } \ln\tfrac{1}{2}\text{, and use the property } \ln M^p = p \ln M.$$

$$3.829542434 \approx t \qquad \text{Divide both sides by } -0.181 \text{ and note that } \ln e = 1.$$

COMMENT We can divide both sides of the equation by A_0 because we know it is not zero.

To the nearest hundredth, the half-life of radon-22 is 3.83 days.

SELF CHECK 8 To the nearest hundredth, find the half-life of iodine-131 given that it decays according to the formula $A = A_0e^{-0.087t}$ where t is expressed in days.

When a living organism dies, the oxygen/carbon dioxide cycle common to all living things stops and carbon-14, a radioactive isotope with a half-life of 5,730 years, is no longer absorbed. By measuring the amount of carbon-14 present in an ancient object, archaeologists can estimate the object's age. The formula for radioactive decay is $A = A_0e^{kt}$, where A_0 is the original amount of carbon-14 present, t is the age of the object, and k is a negative number.

Unless otherwise noted, all content on this page is © Cengage Learning.

EXAMPLE 9 **CARBON-14 DATING** To the nearest hundred years, how old is a wooden statue that retains one-third of its original carbon-14 content?

Solution We can find the value of k in the formula $A = A_0e^{kt}$ by using the fact that after 5,730 years, half of the original amount of carbon-14 will remain.

$$A = A_0e^{kt}$$

$$\frac{1}{2}A_0 = A_0e^{k(5{,}730)} \quad \text{Substitute } \tfrac{1}{2}A_0 \text{ for } A \text{ and 5,730 for } t.$$

$$\frac{1}{2} = e^{5{,}730k} \quad \text{Divide both sides by } A_0.$$

$$\ln\left(\frac{1}{2}\right) = \ln(e^{5{,}730k}) \quad \text{Take the natural logarithm of both sides.}$$

$$-0.6931471806 \approx 5{,}730k \ln e \quad \text{Find } \ln \tfrac{1}{2}\text{, and use the property } \ln M^p = p \ln M.$$

$$-0.000120968094 \approx k \quad \text{Divide both sides by 5,730 and note that } \ln e = 1.$$

COMMENT Do not round any values until the last step.

Thus, the formula for radioactive decay for carbon-14 can be written as

$$A \approx A_0e^{-0.000120968094t}$$

Since $\frac{1}{3}$ of the original carbon-14 still remains, we can proceed as follows:

$$A \approx A_0e^{-0.000120968094t}$$

$$\frac{1}{3}A_0 \approx A_0e^{-0.000120968094t} \quad \text{Substitute } \tfrac{1}{3}A_0 \text{ for } A.$$

$$\frac{1}{3} \approx e^{-0.000120968094t} \quad \text{Divide both sides by } A_0.$$

$$\ln\left(\frac{1}{3}\right) \approx \ln(e^{-0.000120968094t}) \quad \text{Take the natural logarithm of both sides.}$$

$$-1.098612289 \approx (-0.000120968094t) \ln e \quad \text{Find } \ln \tfrac{1}{3}\text{, and use the property } \ln M^p = p \ln M.$$

$$9081.835155 \approx t \quad \text{Divide both sides by } -0.000120968094 \text{ and note that } \ln e = 1.$$

To the nearest one hundred years, the statue is 9,100 years old.

SELF CHECK 9 To the nearest hundred years, find the age of a statue that retains 25% of its original carbon-14 content.

Recall that when there is sufficient food and space, populations of living organisms tend to increase exponentially according to the Malthusian growth model.

MALTHUSIAN GROWTH MODEL

If P is the population at some time t, P_0 is the initial population at $t = 0$, and k is the rate of growth, then

$$P = P_0e^{kt}$$

COMMENT Note that this formula is the same as all exponential growth formulas, except for the variables.

EXAMPLE 10 **POPULATION GROWTH** The bacteria in a laboratory culture increased from an initial population of 500 to 1,500 in 3 hours. How long will it take for the population to reach 10,000?

Solution We substitute 500 for P_0, 1,500 for P, and 3 for t and simplify to find k:

$$P = P_0 e^{kt}$$

$$\mathbf{1{,}500} = \mathbf{500}(e^{k3}) \quad \text{Substitute 1,500 for } P\text{, 500 for } P_0\text{, and 3 for } t.$$

$$3 = e^{3k} \quad \text{Divide both sides by 500.}$$

$$3k = \ln 3 \quad \text{Write the equation in logarithmic form.}$$

$$k = \frac{\ln 3}{3} \quad \text{Divide both sides by 3.}$$

To find when the population will reach 10,000, we substitute 10,000 for P, 500 for P_0, and $\frac{\ln 3}{3}$ for k in the equation $P = P_0 e^{kt}$ and solve for t:

$$P = P_0 e^{kt}$$

$$\mathbf{10{,}000} = \mathbf{500}e^{[(\ln 3)/3]t}$$

$$20 = e^{[(\ln 3)/3]t} \quad \text{Divide both sides by 500.}$$

$$\left(\frac{\ln 3}{3}\right)t = \ln 20 \quad \text{Write the equation in logarithmic form.}$$

$$t = \frac{3 \ln 20}{\ln 3} \quad \text{Multiply both sides by } \tfrac{3}{\ln 3}.$$

$$\approx 8.180499084 \quad \text{Use a calculator.}$$

The culture will reach 10,000 bacteria in about 8 hours.

SELF CHECK 10 How long will it take to reach 20,000?

EXAMPLE 11 **GENERATION TIME** If a medium is inoculated with a bacterial culture that contains 1,000 cells per milliliter, how many generations will pass by the time the culture has grown to a population of 1 million cells per milliliter?

Solution During bacterial reproduction, the time required for a population to double is called the *generation time*. If b bacteria are introduced into a medium, then after the generation time of the organism has elapsed, there are $2b$ cells. After another generation, there are $2(2b)$, or $4b$ cells, and so on. After n generations, the number of cells present will be

(1) $$B = b \cdot 2^n$$

To find the number of generations that have passed while the population grows from b bacteria to B bacteria, we solve Equation 1 for n.

$$\mathbf{\log B} = \mathbf{\log(b \cdot 2^n)} \quad \text{Take the common logarithm of both sides.}$$

$$\log B = \log b + n \log 2 \quad \text{Apply the product and power rules of logarithms.}$$

$$\log B - \log b = n \log 2 \quad \text{Subtract } \log b \text{ from both sides.}$$

$$n = \frac{1}{\log 2}(\log B - \log b) \quad \text{Multiply both sides by } \tfrac{1}{\log 2}.$$

(2) $$n = \frac{1}{\log 2}\left(\log \frac{B}{b}\right) \quad \text{Use the quotient rule of logarithms.}$$

Equation 2 is a formula that gives the number of generations that will pass as the population grows from b bacteria to B bacteria.

To find the number of generations that have passed while a population of 1,000 cells per milliliter has grown to a population of 1 million cells per milliliter, we substitute 1,000 for b and 1,000,000 for B in Equation 2 on the previous page and solve for n.

$$n = \frac{1}{\log 2}\log\frac{1{,}000{,}000}{1{,}000}$$

$$= \frac{1}{\log 2}\log 1{,}000 \quad \text{Simplify.}$$

$$\approx 3.321928095(3) \quad \frac{1}{\log 2} \approx 3.321928095 \text{ and } \log 1{,}000 = 3.$$

$$\approx 9.965784285$$

Approximately 10 generations will have passed.

SELF CHECK 11 How many generations will pass by the time the culture has grown to a population of 10 million cells per milliliter?

Everyday connections
U.S. Population Growth

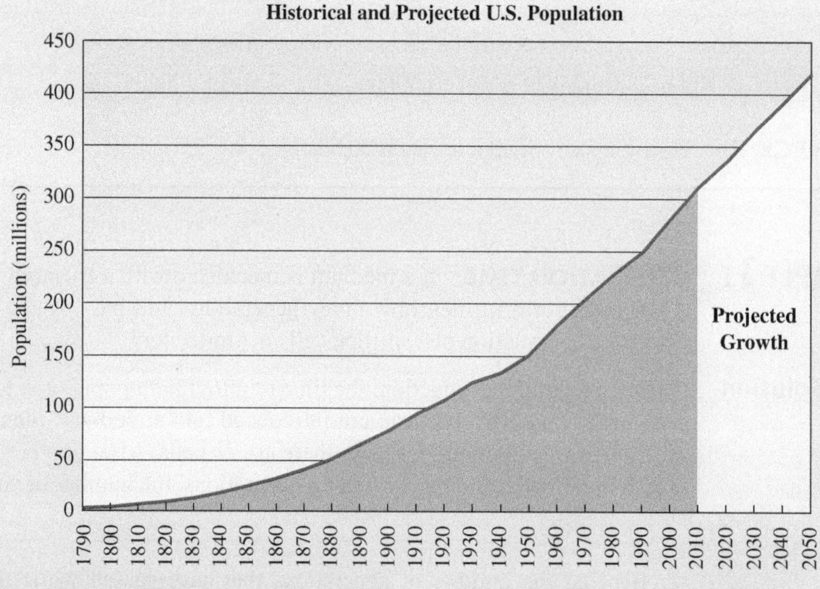

Population growth in the United States can be modeled by an exponential function of the form $P(t) = P_0 \cdot e^{rt}$, where $P_0 =$ the initial population during a given time interval, and t represents the number of years in that time interval.

1. Given that the United States population was approximately 200 million in 1970 and approximately 280 million in 2000, determine the growth rate of the population during this time period.
2. Given that the United States population was 280 million in 2000 and approximately 300 million in 2010, determine the growth rate of the population during this time period.
3. Using the population growth rate from Question 2, to the nearest year how long would it take for the United States population to double?

Source: http://www.npg.org/popfacts.htm

Unless otherwise noted, all content on this page is © Cengage Learning.

SELF CHECK ANSWERS **1.** 0.8614 **2.** 60.1321 **3.** 6.3013 **4.** 1 **5.** 2 **6.** 2; -5 is extraneous **7.** 6; -1 is extraneous **8.** 7.97 days **9.** about 11,500 years **10.** about 10 hours **11.** approximately 13 generations

NOW TRY THIS

Given $f(x) = \log_3(x + 2) + \log_3 x$,

1. solve $f(x) = 1$.
2. solve $f(x) = 2$.

11.8 Exercises

WARM-UPS *Write each logarithmic equation as an exponential equation.*

1. $\log x = 2$ **2.** $\log y = 1$

3. $\log(x + 1) = 1$

4. $\log(x - 2) = 3$

Simplify, using a calculator, and approximate to four decimal places.

5. $\frac{\log 4}{\log 3}$ **6.** $\log\left(\frac{4}{3}\right)$

7. $-\log\left(\frac{5}{7}\right)$ **8.** $-\frac{\log 5}{\log 7}$

REVIEW *Solve each equation.*

9. $4x^2 - 64x = 0$ **10.** $9y^2 - 49 = 0$

11. $3p^2 + 10p = 8$ **12.** $4t^2 + 1 = -6t$

VOCABULARY AND CONCEPTS *Fill in the blanks.*

13. An equation with a variable as an exponent is called a(n) __________ equation.

14. An equation with a logarithmic expression that contains a variable is a(n) __________ equation.

15. The formula for radioactive decay is $A =$ _____.

16. The _______ of a radioactive element is determined by how long it takes for half of a sample to decompose.

GUIDED PRACTICE *Solve each exponential equation. If an answer is not exact, give the answer to four decimal places.* SEE EXAMPLE 1. (OBJECTIVE 1)

17. $3^x = 8$ **18.** $7^x = 12$

19. $e^t = 50$ **20.** $e^{-t} = 0.25$

Solve each exponential equation. If an answer is not exact, give the answer to four decimal places. SEE EXAMPLE 2. (OBJECTIVE 1)

21. $7 = 4.3(1.01)^t$ **22.** $52 = 5.1(1.03)^t$

23. $5 = 2.1(1.04)^t$ **24.** $61 = 1.5(1.02)^t$

Solve each exponential equation. If an answer is not exact, give the answer to four decimal places. SEE EXAMPLE 3. (OBJECTIVE 1)

25. $6^{x-2} = 4$ **26.** $5^{x+1} = 3$

27. $4^{x+1} = 7^x$ **28.** $5^{x-3} = 3^{2x}$

Solve each exponential equation. Give the answer to four decimal places. SEE EXAMPLE 4. (OBJECTIVE 1)

29. $5^{x^2+2x} = 125$ **30.** $3^{x^2-3x} = 81$

31. $3^{x^2+4x} = \frac{1}{81}$ **32.** $7^{x^2+3x} = \frac{1}{49}$

Use a calculator to solve each equation, if possible. Give all answers to the nearest tenth. (OBJECTIVE 1)

33. $2^{x+1} = 7$ **34.** $3^{x-1} = 2^x$

35. $2^{x^2-2x} - 8 = 0$ **36.** $3^x - 10 = 3^{-x}$

Solve each logarithmic equation. Check all solutions. SEE EXAMPLE 5. (OBJECTIVE 2)

37. $\log 9x = \log 27$

38. $\log 5x = \log 45$

39. $\log(4x + 3) = \log(x + 9)$

40. $\log(x^2 + 4x) = \log(x^2 + 16)$

Solve each logarithmic equation. Check all solutions. SEE EXAMPLE 6. (OBJECTIVE 2)

41. $\log x^2 = 2$

42. $\log x^3 = 3$

43. $\log x + \log(x - 3) = 1$

44. $\log x + \log(x + 9) = 1$

45. $\log x + \log(x - 15) = 2$

46. $\log x + \log(x + 21) = 2$

47. $\log(x + 90) = 3 - \log x$

48. $\log(x - 90) = 3 - \log x$

Solve each logarithmic equation. Check all solutions. SEE EXAMPLE 7. (OBJECTIVE 2)

49. $\frac{\log(2x + 1)}{\log(x - 1)} = 2$

50. $\dfrac{\log(4x + 9)}{\log(2x - 3)} = 2$

51. $\dfrac{\log(3x + 4)}{\log x} = 2$

52. $\dfrac{\log(8x - 7)}{\log x} = 2$

Use a graphing calculator to solve each equation. If an answer is not exact, give all answers to the nearest tenth. (OBJECTIVE 2)

53. $\log x + \log(x - 15) = 2$

54. $\log x + \log(x + 3) = 1$

55. $\ln(2x + 5) - \ln 3 = \ln(x - 1)$

56. $\log(x^2 + 4x)^2 = 1$

ADDITIONAL PRACTICE *Solve each exponential equation. If an answer is not exact, give the answer to four decimal places.*

57. $5^x = 4^x$

58. $3^{2x} = 4^x$

59. $7^{x^2} = 10$

60. $8^{x^2} = 11$

61. $8^{x^2} = 9^x$

62. $5^{x^2} = 2^{5x}$

Solve each logarithmic equation. Check all solutions.

63. $\log(5 - 4x) - \log(x + 25) = 0$

64. $\log(3x + 5) - \log(2x + 6) = 0$

65. $\log(x - 6) - \log(x - 2) = \log\dfrac{5}{x}$

66. $\log(3 - 2x) - \log(x + 9) = 0$

Solve each equation. If the answer is not exact, round to four decimal places.

67. $4^{x+2} - 4^x = 15$ (*Hint*: $4^{x+2} = 4^x4^2$.)

68. $3^{x+3} + 3^x = 84$ (*Hint*: $3^{x+3} = 3^x3^3$.)

69. $\dfrac{\log(5x + 6)}{2} = \log x$

70. $\dfrac{1}{2}\log(4x + 5) = \log x$

71. $\log_3 x = \log_3\left(\dfrac{1}{x}\right) + 4$

72. $\log_5(7 + x) + \log_5(8 - x) - \log_5 2 = 2$

73. $2(3^x) = 6^{2x}$

74. $2(3^{x+1}) = 3(2^{x-1})$

75. $\log x^2 = (\log x)^2$

76. $\log(\log x) = 1$

77. $2\log_2 x = 3 + \log_2(x - 2)$

78. $2\log_3 x - \log_3(x - 4) = 2 + \log_3 2$

79. $\log(7y + 1) = 2\log(y + 3) - \log 2$

80. $2\log(y + 2) = \log(y + 2) - \log 12$

81. $\log\dfrac{4x + 1}{2x + 9} = 0$

82. $\log\dfrac{2 - 5x}{2(x + 8)} = 0$

APPLICATIONS *Solve each application.* SEE EXAMPLE 8. (OBJECTIVE 3)

83. Half-life To the nearest day, find the half-life of strontium-89, given that it decays according to the formula $A = A_0e^{-0.013t}$.

84. Half-life To the nearest thousand years, find the half-life of plutonium-239, given that it decays according to the formula $A = A_0e^{-0.0000284t}$.

85. Radioactive decay In two years, 20% of a radioactive element decays. Find its half-life.

86. Tritium decay The half-life of tritium is 12.4 years. How long will it take for 25% of a sample of tritium to decompose?

Solve each application. SEE EXAMPLE 9. (OBJECTIVE 3)

87. Carbon-14 dating The bone fragment shown in the illustration contains 60% of the carbon-14 that it is assumed to have had initially. How old is it?

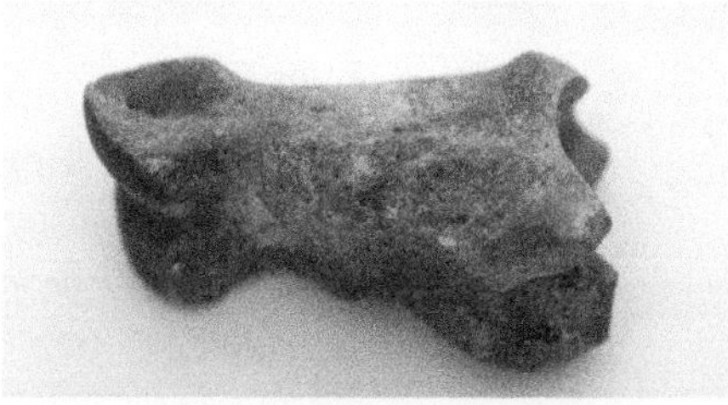
© Falk Kienas/Shutterstock.com

88. Carbon-14 dating Only 10% of the carbon-14 in a small wooden bowl remains. How old is the bowl?

Solve each application. SEE EXAMPLE 10. (OBJECTIVE 3)

89. Rodent control The rodent population in a city is currently estimated at 30,000. If it is expected to double every 5 years, when will the population reach 1 million?

90. Population growth The population of a city is expected to triple every 15 years. When can the city planners expect the present population of 140 persons to double?

91. Bacterial culture A bacterial culture doubles in size every 24 hours. By how much will it have increased in 36 hours?

92. Bacterial growth A bacterial culture grows according to the formula

$$P = P_0a^t$$

If it takes 5 days for the culture to triple in size, how long will it take to double in size?

Solve each application. SEE EXAMPLE 11. (OBJECTIVE 3)

93. Medicine If a medium is inoculated with a bacterial culture containing 500 cells per milliliter, how many generations will have passed by the time the culture contains 5×10^6 cells per milliliter?

94. Medicine If a medium is inoculated with a bacterial culture containing 800 cells per milliliter, how many generations will have passed by the time the culture contains 6×10^7 cells per milliliter?

Solve each application.

95. Thorium decay An isotope of thorium, ^{227}Th, has a half-life of 18.4 days. How long will it take for 80% of the sample to decompose?

96. Lead decay An isotope of lead, ^{201}Pb, has a half-life of 8.4 hours. How many hours ago was there 30% more of the substance?

97. Compound interest If \$500 is deposited in an account paying 8.5% annual interest, compounded semiannually, how long will it take for the account to increase to \$800?

98. Continuous compound interest In Exercise 97, how long will it take if the interest is compounded continuously?

99. Compound interest If \$1,300 is deposited in a savings account paying 9% interest, compounded quarterly, how long will it take the account to increase to \$2,100?

100. Compound interest A sum of \$5,000 deposited in an account grows to \$7,000 in 5 years. Assuming annual compounding, what interest rate is being paid?

101. Rule of seventy A rule of thumb for finding how long it takes an investment to double is called the **rule of seventy**. To apply the rule, divide 70 by the interest rate written as a percent. At 5%, it takes $\frac{70}{5} = 14$ years to double an investment. At 7%, it takes $\frac{70}{7} = 10$ years. Explain why this formula works.

102. Oceanography The intensity I of a light a distance x meters beneath the surface of a lake decreases exponentially. From the illustration, find the depth at which the intensity will be 20%.

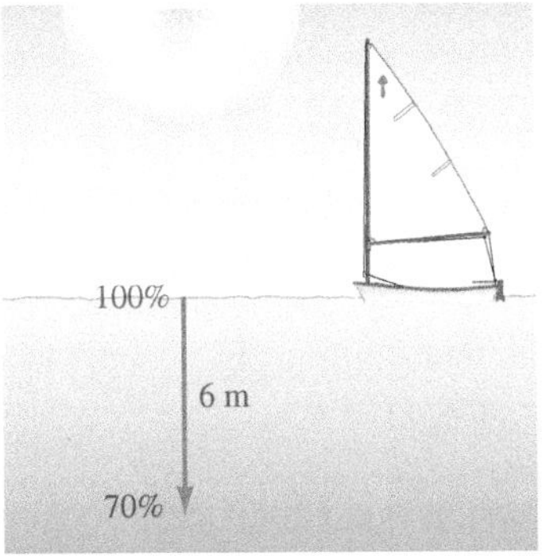

WRITING ABOUT MATH

103. Explain how to solve the equation $2^x = 7$.

104. Explain how to solve the equation $x^2 = 7$.

SOMETHING TO THINK ABOUT

105. Without solving the following equation, find the values of x that cannot be a solution:

$$\log(x - 3) - \log(x^2 + 2) = 0$$

106. Solve the equation $x^{\log x} = 10{,}000$.

Projects

PROJECT 1

When an object moves through air, it encounters air resistance. So far, all ballistics problems in this text have ignored air resistance. We now consider the case where an object's fall is affected by air resistance.

At relatively low velocities ($v < 200$ feet per second), the force resisting an object's motion is a constant multiple of the object's velocity:

$$\text{Resisting force} = f_r = bv$$

where b is a constant that depends on the size, shape, and texture of the object, and has units of kilograms per second. This is known as **Stokes' law of resistance**.

In a vacuum, the downward velocity of an object dropped with an initial velocity of 0 feet per second is

$$v(t) = 32t \quad \text{(no air resistance)}$$

t seconds after it is released. However, with air resistance, the velocity is given by the formula

$$v(t) = \frac{32m}{b}(1 - e^{-(b/m)t})$$

where m is the object's mass (in kilograms). There is also a formula for the distance an object falls (in feet) during the first t seconds after release, taking into account air resistance:

$$d(t) = \frac{32m}{b}t - \frac{32m^2}{b^2}(1 - e^{-(b/m)t})$$

Without air resistance, the formula would be

$$d(t) = 16t^2$$

a. Fearless Freda, a renowned skydiving daredevil, performs a practice dive from a hot-air balloon with an altitude of 5,000 feet. With her parachute on, Freda has a mass of 75 kg, so that $b = 15$ kg/sec. How far (to the nearest foot) will Freda fall in 5 seconds? Compare this with the answer you get by disregarding air resistance.

b. What downward velocity (to the nearest ft/sec) does Freda have after she has fallen for 2 seconds? For

(continued)

Unless otherwise noted, all content on this page is © Cengage Learning.

5 seconds? Compare these answers with the answers you get by disregarding air resistance.

c. Find Freda's downward velocity after falling for 20, 22, and 25 seconds. (Without air resistance, Freda would hit the ground in less than 18 seconds.) Note that Freda's velocity increases only slightly. This is because for a large enough velocity, the force of air resistance almost counteracts the force of gravity; after Freda has been falling for a few seconds, her velocity becomes nearly constant. The constant velocity that a falling object approaches is called the *terminal velocity.*

$$\text{Terminal velocity} = \frac{32m}{b}$$

Find Freda's terminal velocity for her practice dive.

d. In Freda's show, she dives from a hot-air balloon with an altitude of only 550 feet, and pulls her ripcord when her velocity is 100 feet per second. (She can't tell her speed, but she knows how long it takes to reach that speed.) It takes a fall of 80 more feet for the chute to open fully, but then the chute increases the force of air resistance, making $b = 80$. After that, Freda's velocity approaches the terminal velocity of an object with this new b-value.

To the nearest hundredth of a second, how long should Freda fall before she pulls the ripcord? To the nearest foot, how close is she to the ground when she pulls the ripcord? How close to the ground is she when the chute takes full effect? At what velocity will Freda hit the ground?

PROJECT 2

If an object at temperature T_0 is surrounded by a constant temperature T_s (for instance, an oven or a large amount of fluid that has a constant temperature), the temperature of the object will change with time t according to the formula

$$T(t) = T_s + (T_0 - T_s)e^{-kt}$$

This is **Newton's law of cooling and warming**. The number k is a constant that depends on how well the object absorbs and dispels heat.

In the course of brewing "yo ho! grog," the dread pirates of Hancock Isle have learned that it is important that their rather disgusting, soupy mash be heated slowly to allow all of the ingredients a chance to add their particular offensiveness to the mixture. However, after the mixture has simmered for several hours, it is equally important that the grog be cooled very quickly, so that it retains its potency. The kegs of grog are then stored in a cool spring.

By trial and error, the pirates have learned that by placing the mash pot into a tub of boiling water (100° C), they can heat the mash in the correct amount of time. They have also learned that they can cool the grog to the temperature of the spring by placing it in ice caves for 1 hour.

With a thermometer, you find that the pirates heat the mash from 20° C to 95° C and then cool the grog from 95° C to 7° C. Calculate how long the pirates cook the mash, and how cold the ice caves are. Assume that $k = 0.5$, and t is measured in hours.

Reach for Success

EXTENSION OF PLANNING FOR YOUR FUTURE

Now that you've considered short-term planning, let's move on and look at your future career goals.

LONG-TERM EDUCATIONAL PLANNING Take advantage of your college resources to help you identify your area of interest. Ultimately, you will need to declare a major.	Your college may have resources to help your choose a major. List a few of these resources. __________ __________
Have you chosen a major? __________ If so, consider careers in that field and list at least three. **1.** __________ **2.** __________ **3.** __________	If not, visit your career services area, see an advisor, or talk with a favorite professor. With whom did you consult? __________ What advice did you receive? __________ __________
What are potential employers looking for in your career interest? __________ __________	Circle True or False for the following statement. Most companies want employees with the problem-solving skills developed in mathematics courses. True False
There are certificate programs and associate's, bachelor's, master's, and doctoral degrees.	What type of certificate or degree is required for your field of interest? __________ What additional mathematics courses (if any) are required for your major? __________ __________

Proper planning can help you avoid taking classes that might not be required for your degree, thus saving you time and money.

11 Review

SECTION 11.1 Algebra and Composition of Functions

DEFINITIONS AND CONCEPTS	EXAMPLES
Operations with functions: $(f+g)(x) = f(x) + g(x)$ $(f-g)(x) = f(x) - g(x)$ $(f \cdot g)(x) = f(x)g(x)$ $(f/g)(x) = \dfrac{f(x)}{g(x)} \quad (g(x) \neq 0)$ The domain of each of these functions is the set of real numbers x that are in the domain of both f and g. In the case of the quotient, there is the further restriction that $g(x) \neq 0$.	Given $f(x) = 6x - 2$ and $g(x) = x^2 + 4$, find each function and its domain. **a.** $f+g$ **b.** $f-g$ **c.** $f \cdot g$ **d.** f/g **a.** $f+g = f(x) + g(x)$ $= (6x-2) + (x^2+4)$ $= x^2 + 6x + 2$ Combine like terms. D: $(-\infty, \infty)$ **b.** $f-g = (6x-2) - (x^2+4)$ $= 6x - 2 - x^2 - 4$ Use the distributive property to remove parentheses. $= -x^2 + 6x - 6$ Combine like terms. D: $(-\infty, \infty)$ **c.** $f \cdot g = (6x-2)(x^2+4)$ $= 6x^3 + 24x - 2x^2 - 8$ Multiply. $= 6x^3 - 2x^2 + 24x - 8$ Write exponents in descending order. D: $(-\infty, \infty)$ **d.** $\dfrac{g}{f} = \dfrac{x^2+4}{6x-2}$ D: $\left(-\infty, \frac{1}{3}\right) \cup \left(\frac{1}{3}, \infty\right)$
Composition of functions: $(f \circ g)(x) = f(g(x))$	Let $f(x) = 5x - 4$ and $g(x) = x^2 + 2$. Find **a.** $(f \circ g)(x)$ **b.** $(g \circ f)(x)$ **a.** $(f \circ g)(x)$ means $f(g(x))$. $f(g(x)) = f(x^2+2)$ $= 5(x^2+2) - 4$ Substitute (x^2+2) for x. $= 5x^2 + 10 - 4$ Multiply. $= 5x^2 + 6$ Add. **b.** $(g \circ f)(x)$ means $g(f(x))$. $g(f(x)) = g(5x-4)$ $= (5x-4)^2 + 2$ Substitute $(5x-4)$ for x. $= 25x^2 - 40x + 16 + 2$ Square the binomial. $= 25x^2 - 40x + 18$ Add.

The difference quotient: The difference quotient is defined as follows: $\dfrac{f(x+h)-f(x)}{h}, \quad h \neq 0$	To evaluate the difference quotient for $f(x) = x^2 + 2x - 6$, first evaluate $f(x + h)$. $f(x) = x^2 + 2x - 6$ $f(x+h) = (x+h)^2 + 2(x+h) - 6$ Substitute $(x + h)$ for h. $= x^2 + 2xh + h^2 + 2x + 2h - 6$ $(x+h)^2 = x^2 + 2hx + h^2$ Now substitute the values of $f(x + h)$ and $f(x)$ into the difference quotient and simplify. $\dfrac{f(x+h)-f(x)}{h}$ $= \dfrac{(x^2 + 2xh + h^2 + 2x + 2h - 6) - (x^2 + 2x - 6)}{h}$ $= \dfrac{x^2 + 2xh + h^2 + 2x + 2h - 6 - x^2 - 2x + 6}{h}$ Use the distributive property to remove parentheses. $= \dfrac{2xh + h^2 + 2h}{h}$ Combine like terms. $= \dfrac{h(2x + h + 2)}{h}$ Factor out h, the GCF, in the numerator. $= 2x + h + 2$ Divide out h. The difference quotient for this function is $2x + h + 2$.

REVIEW EXERCISES

Let $f(x) = 2x$ and $g(x) = x + 1$. Find each function or value.

For 1–4 only, find the domain.

1. $f + g$
2. $f - g$
3. $f \cdot g$
4. f/g
5. $(f \circ g)(3)$
6. $(g \circ f)(-2)$
7. $(f \circ g)(x)$
8. $(g \circ f)(x)$

SECTION 11.2 Inverses of Functions

DEFINITIONS AND CONCEPTS	EXAMPLES
Horizontal line test: If every horizontal line that intersects the graph of a function does so only once, the function is one-to-one.	Determine whether the graph of each function is one-to-one. **a.** 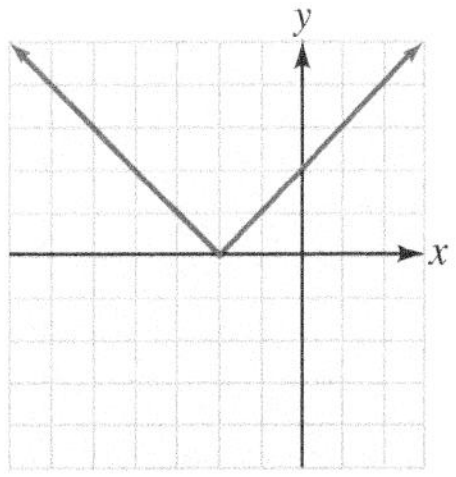 not one-to-one **b.** 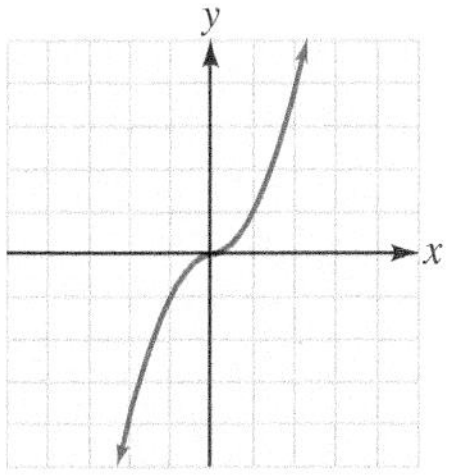 one-to-one **a.** The function is not one-to-one because a horizontal line will cross the graph more than once. **b.** The function is one-to-one because any horizontal line will cross the graph no more than once.

Unless otherwise noted, all content on this page is © Cengage Learning.

Finding the inverse of a one-to-one function:
We find the inverse of a function as follows:

1. Replace $f(x)$ with y.
2. Interchange the variables x and y.
3. Solve the resulting equation for y. If the inverse is also a function, use the notation $f^{-1}(x)$.

To find the inverse of $f(x) = x^3 - 5$, we proceed as follows:

$y = x^3 - 5$ Replace $f(x)$ with y.

$x = y^3 - 5$ Interchange the variables x and y.

$x + 5 = y^3$ Add 5 to both sides.

$\sqrt[3]{x + 5} = y$ Take the cube root of both sides.

Because there corresponds one real cube root for each x, $y = \sqrt[3]{x + 5}$ represents a function. In function notation, we describe the inverse as

$$f^{-1}(x) = \sqrt[3]{x + 5}$$

REVIEW EXERCISES

Graph each function and use the horizontal line test to decide whether the function is one-to-one.

9. $f(x) = 2(x - 3)$

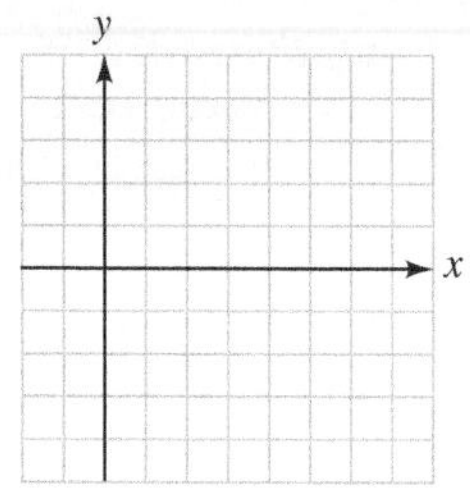

10. $f(x) = x(2x - 3)$

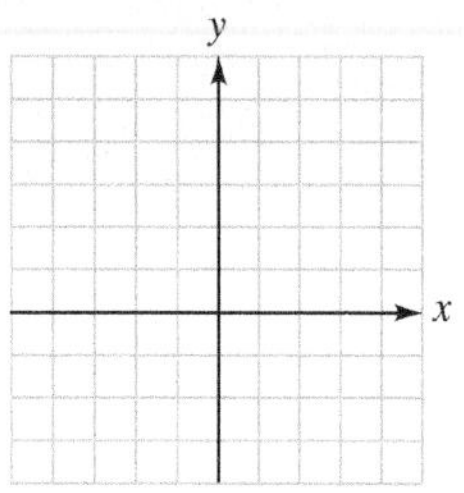

11. $f(x) = -3(x - 2)^2 + 5$

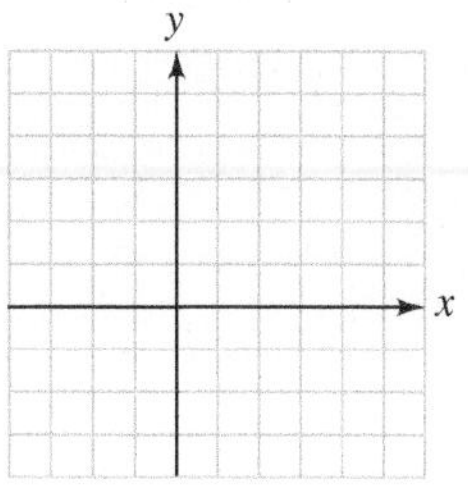

12. $f(x) = |x|$

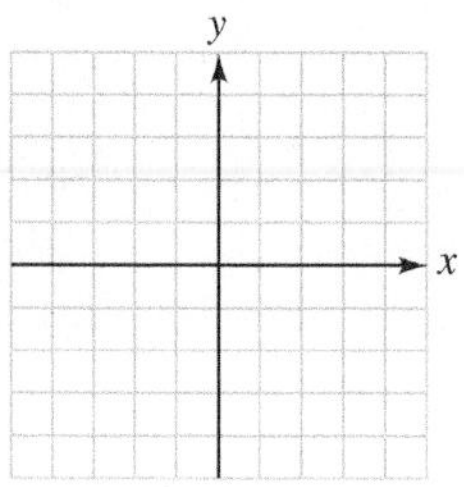

Find the inverse of each function. Do not rationalize the result.

13. $f(x) = 7x - 2$

14. $f(x) = 4x + 5$

15. $y = 2x^2 - 1 \quad (x \geq 0)$

SECTION 11.3 Exponential Functions

DEFINITIONS AND CONCEPTS | **EXAMPLES**

An exponential function with base b is defined by the equation

$f(x) = b^x \quad (b > 0, b \neq 1$ and x is a real number$)$

The domain is $(-\infty, \infty)$ and the range is $(0, \infty)$.

Graph $f(x) = 4^x$ and $f(x) = \left(\frac{1}{4}\right)^x$.

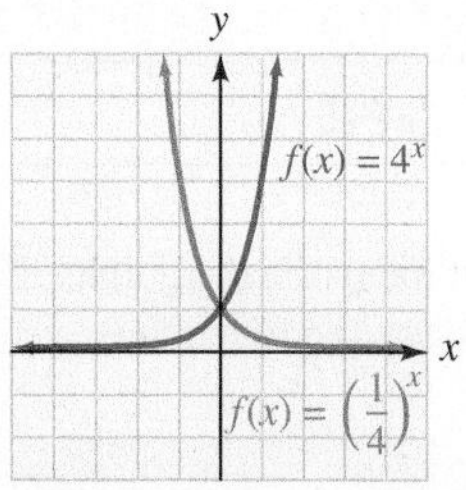

Increasing/decreasing functions:
The graph of $f(x) = b^x$ is an increasing function if $b > 1$ and decreasing if $0 < b < 1$.

The graph of $f(x) = 4^x$ is an increasing function because as the values of x increase the values of $f(x)$ also increase.

The graph of $f(x) = \left(\frac{1}{4}\right)^x$ is a decreasing function because as the values of x increase the values of $f(x)$ decrease.

Unless otherwise noted, all content on this page is © Cengage Learning.

Graphing translations:
Graphing a translation of an exponential function follows the same rules as for polynomial functions.

Graph $f(x) = 4^x - 2$ and $f(x) = 4^{x-2}$.

The graph of $f(x) = 4^x - 2$ is the same as the graph of $f(x) = 4^x$ but shifted 2 units downward.

The graph of $f(x) = 4^{x-2}$ is the same as the graph of $f(x) = 4^x$ but shifted 2 units to the right.

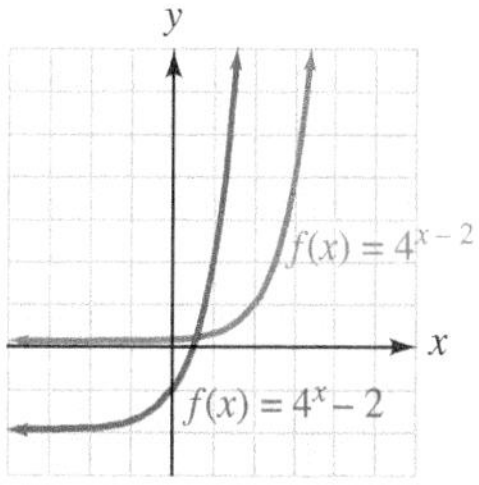

Compound interest:
The amount of money in an account with an initial deposit of $P at an annual interest rate of $r\%$, compounded k times a year for t years, can be found using the formula $A = P\left(1 + \frac{r}{k}\right)^{kt}$.

To find the balance in an account after 3 years when $5,000 is deposited at 6% interest, compounded quarterly, we note that

$$P = 5{,}000 \qquad r = 0.06 \qquad k = 4 \qquad t = 3$$

We can substitute these values into the formula for compound interest and proceed as follows:

$$A = P\left(1 + \frac{r}{k}\right)^{kt}$$

$$A = \mathbf{5{,}000}\left(1 + \frac{\mathbf{0.06}}{\mathbf{4}}\right)^{\mathbf{4(3)}}$$

$$A = 5{,}000(1 + 0.015)^{12}$$

$$A = 5{,}000(1.015)^{12}$$

$$A \approx 5{,}978.090857 \qquad \text{Use a calculator.}$$

There will be $5,978.09 in the account after 3 years.

REVIEW EXERCISES

16. Sketch the graph of an increasing exponential function.
17. Sketch the graph of a decreasing exponential function.

Graph the function defined by each equation.

18. $f(x) = 3^x$

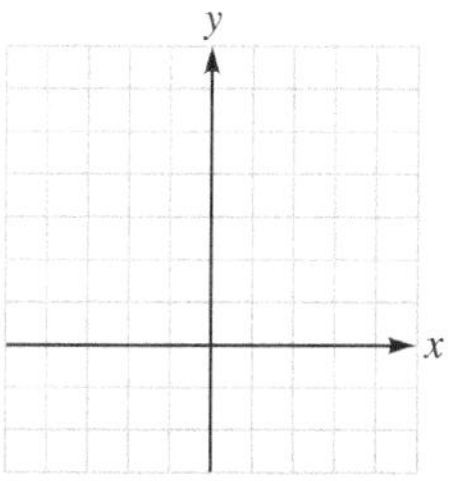

19. $f(x) = \left(\frac{1}{3}\right)^x$

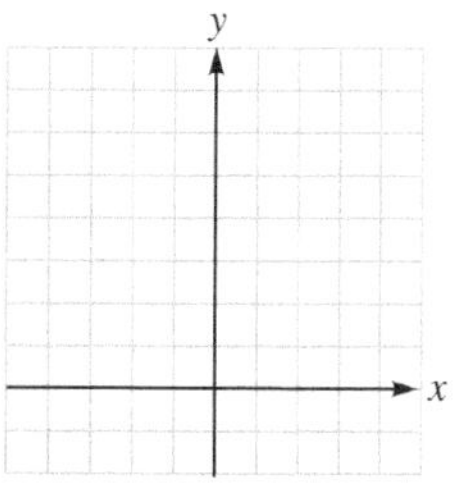

20. The graph of $f(x) = 6^x$ will pass through the points $(0, x)$ and $(1, y)$. Find the values of x and y.

21. Give the domain and range of the function $f(x) = 7^x$.

Graph each function by using a translation.

22. $f(x) = \left(\frac{1}{2}\right)^x - 2$

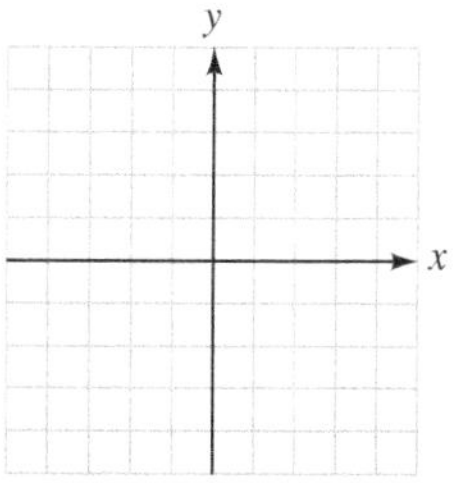

23. $f(x) = \left(\frac{1}{2}\right)^{x+2}$

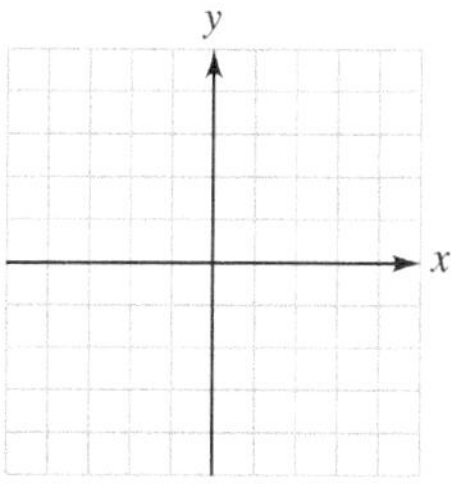

24. Savings How much will $25,000 become if it earns 5% per year for 35 years, compounded quarterly?

Unless otherwise noted, all content on this page is © Cengage Learning.

SECTION 11.4 Base-*e* Exponential Functions

DEFINITIONS AND CONCEPTS | **EXAMPLES**

$e \approx 2.71828182845904$

Continuous compound interest:
The amount of money A in an account with an initial deposit of \$$P$ at an annual interest rate of r%, compounded continuously for t years, can be found using the formula

$$A = Pe^{rt}$$

To find the balance in an account after 3 years when \$5,000 is deposited at 6% interest, compounded continuously, we note that

$$P = 5{,}000 \quad r = 0.06 \quad t = 3$$

We can substitute these values into the formula $A = Pe^{rt}$ and simplify to obtain

$$A = Pe^{rt}$$
$$A = \mathbf{5{,}000}e^{\mathbf{0.06(3)}}$$
$$A = 5{,}000e^{0.18}$$
$$A \approx 5{,}986.09 \quad \text{Use a calculator.}$$

There will be \$5,986.09 in the account.

Malthusian population growth:
The same formula for continuous compound interest is used for population growth, where A is the new population, P is the previous population, r is the growth rate, and t is the number of years.

$$A = Pe^{rt}$$

To find the population of a town in 5 years with a current population of 4,000 and a growth rate of 1.5%, we note that

$$P = 4{,}000 \quad r = 0.015 \quad t = 5$$

We can substitute these values into the formula $A = Pe^{rt}$ to obtain

$$A = Pe^{rt}$$
$$A = \mathbf{4{,}000}e^{\mathbf{0.015(5)}}$$
$$A = 4{,}000e^{0.075}$$
$$A \approx 4{,}311.54 \quad \text{Use a calculator.}$$

There will be about 4,312 people in 5 years.

Radioactive decay:
The formula $A = A_0e^{kt}$ can be used to determine the amount of material left (A) when an initial amount (A_0) has been decaying for t years at a rate of k percent per year. The value of k in this type of problem will be negative.

The radioactive material radon-22 decays with a rate of $k = -0.181$ for t days. To find how much radon-22 will be left if a sample of 100 grams decays for 30 days, we can substitute the values into the radioactive decay formula.

$$A = A_0e^{kt}$$
$$A = \mathbf{100}e^{\mathbf{-0.181(30)}}$$
$$A = 100e^{-5.43}$$
$$A \approx .4383095803 \quad \text{Use a calculator.}$$

There will be approximately 0.44 grams remaining.

REVIEW EXERCISES

25. If \$25,000 accumulates interest at an annual rate of 5%, compounded continuously, how much will be in the account in 35 years?

Graph each function.

26. $f(x) = e^x + 1$

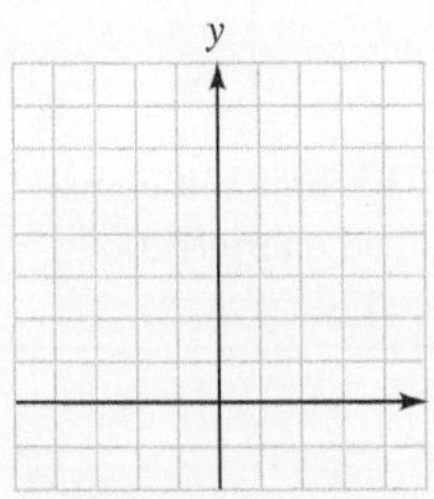

27. $f(x) = e^{x-3}$

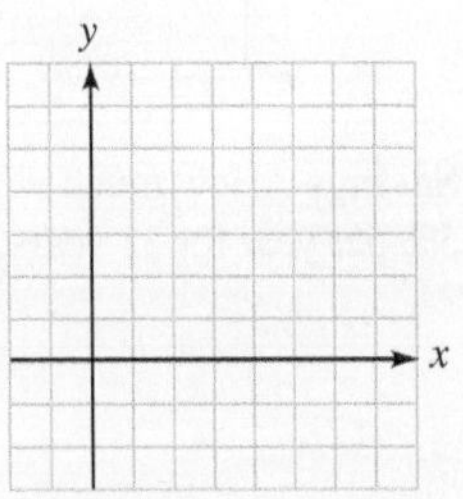

28. U.S. population The population of the United States is approximately 275,000,000 people. Find the population in 50 years if $k = 0.015$.

29. Radioactive decay The radioactive material strontium-90 decays according to the formula $A = A_0e^{-0.0244t}$, where t is expressed in years. To the nearest hundredth, how much of the material will remain if a sample of 50 grams decays for 20 years?

Unless otherwise noted, all content on this page is © Cengage Learning.

SECTION 11.5 Logarithmic Functions

DEFINITIONS AND CONCEPTS | **EXAMPLES**

If $b > 0$ and $b \neq 1$,

$$y = \log_b x \quad \text{if and only if} \quad x = b^y$$

The graph of $f(x) = \log_b x$ has the following properties.

1. It passes through the point $(1, 0)$.
2. It passes through the point $(b, 1)$.
3. The y-axis is an asymptote.
4. The domain is $(0, \infty)$ and the range is $(-\infty, \infty)$.

If $b > 1$, $y = \log_b x$ is an increasing function.
If $0 < b < 1$, $y = \log_b x$ is a decreasing function.

$y = \log_3 27$ is equivalent to $27 = 3^y$.

Since $\log_4 x = 2$ is equivalent to $x = 4^2$, we have $x = 16$.

Since $\log_2 32 = x$ is equivalent to $32 = 2^x$ or $2^5 = 2^x$, we have $x = 5$.

Since $\log_x 64 = 3$ is equivalent to $64 = x^3$ or $4^3 = x^3$, we have $x = 4$.

Graph $f(x) = \log_2 x$

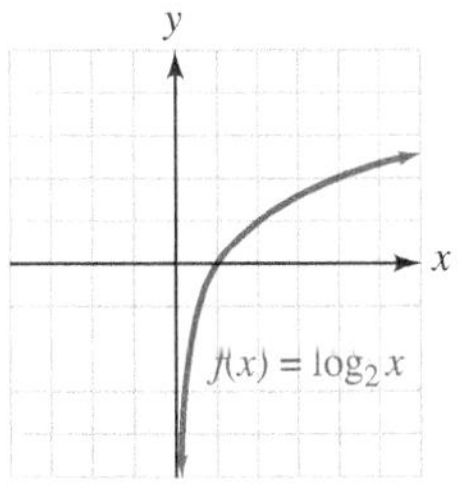

If E_O is the output voltage of a device and E_I is the input voltage, the decibel voltage gain is given by

$$\text{dB gain} = 20 \log \frac{E_O}{E_I}$$

If the input to an amplifier is 0.6 volt and the output is 40 volts, find the decibel voltage gain.

$$\text{dB gain} = 20 \log \frac{40}{0.6}$$
$$\approx 36.5\text{-decibel voltage gain}$$

If R is the magnitude of an earthquake, A is the amplitude (measured in micrometers), and P is the period (the time of one oscillation of Earth's surface, measured in seconds), then find the earthquake's measure.

$$R = \log \frac{A}{P}$$

An earthquake has an amplitude of 8,000 micrometers and a period of 0.02 second. Find its measure on the Richter scale.

$$R = \log \frac{8{,}000}{0.02}$$
$$\approx 5.6 \text{ on the Richter scale}$$

REVIEW EXERCISES

30. State the domain and range of the logarithmic function $y = \log_b x$.

31. Explain why the functions $y = b^x$ and $y = \log_b x$ are called inverse functions.

Find each value.

32. $\log_4 16$

33. $\log_{16} \frac{1}{4}$

34. $\log_\pi 1$

35. $\log_5 0.04$

36. $\log_b \sqrt{b}$

37. $\log_a \sqrt[3]{a}$

Find the value of x.

38. $\log_3 x = 4$

39. $\log_{\sqrt{3}} x = 4$

40. $\log_{\sqrt{3}} x = 6$

41. $\log_{0.1} 10 = x$

42. $\log_x 2 = -\frac{1}{3}$

43. $\log_x 16 = 4$

44. $\log_{0.25} x = -1$

45. $\log_{0.125} x = -\frac{1}{3}$

46. $\log_{\sqrt{2}} 32 = x$

47. $\log_{\sqrt{5}} x = -4$

48. $\log_{\sqrt{3}} 9\sqrt{3} = x$

49. $\log_{\sqrt{5}} 5\sqrt{5} = x$

Graph each function.

50. $f(x) = \log(x - 2)$

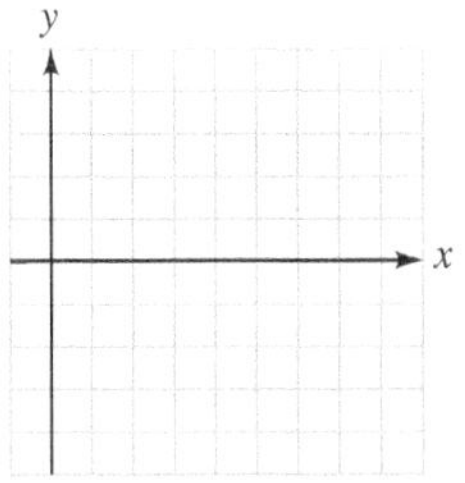

51. $f(x) = 3 + \log x$

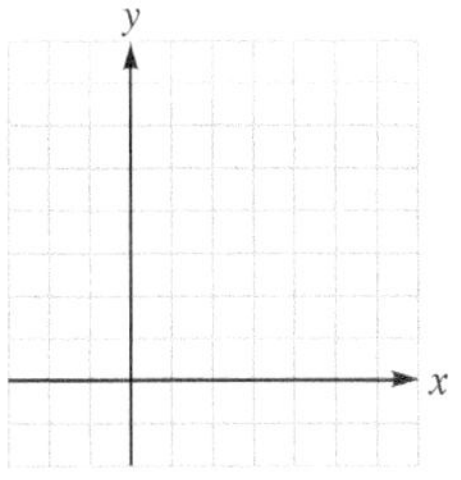

Graph each pair of equations on one set of coordinate axes.

52. $y = 4^x$ and $y = \log_4 x$

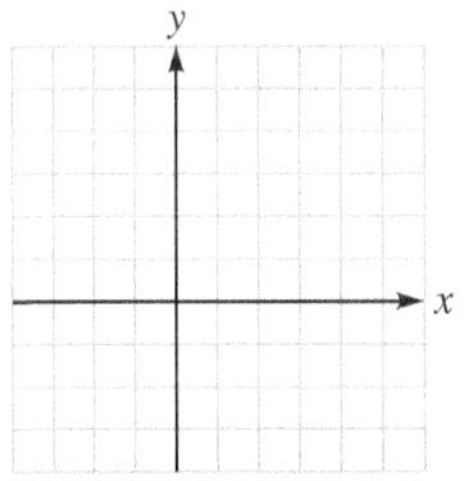

53. $y = \left(\frac{1}{3}\right)^x$ and $y = \log_{1/3} x$

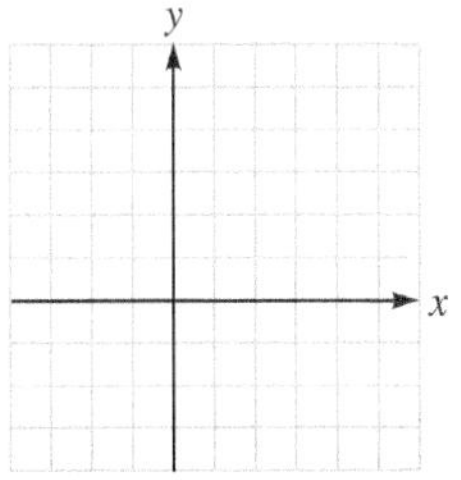

Unless otherwise noted, all content on this page is © Cengage Learning.

54. dB gain An amplifier has an output of 18 volts when the input is 0.04 volt. Find the dB gain.

55. Earthquakes An earthquake had a period of 0.3 second and an amplitude of 7,500 micrometers. Find its measure on the Richter scale.

SECTION 11.6 Natural Logarithms

DEFINITIONS AND CONCEPTS	EXAMPLES
$\ln x$ means $\log_e x$.	Use a calculator to approximate $\ln 5.89$ to 4 decimal places. $\ln 5.89 \approx 1.7733$
Population doubling time: $A = A_0e^{kt}$ or $t = \frac{\ln 2}{r}$	To find the time it takes for the population of a colony to double if the growth rate is 1.5% per year, substitute 0.015 into the formula for doubling time: $t = \frac{\ln 2}{r}$ $t = \frac{\ln 2}{\mathbf{0.015}}$ $t \approx 46.2098$ Use a calculator. The colony will double in population in about 46 years.

REVIEW EXERCISES

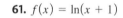

Use a calculator to find each value to four decimal places.

56. $\ln 362$

57. $\ln(\log 7.85)$

Find the value of x.

58. $\ln x = 2.336$

59. $\ln x = \log 8.8$

Graph each function.

60. $f(x) = 1 + \ln x$

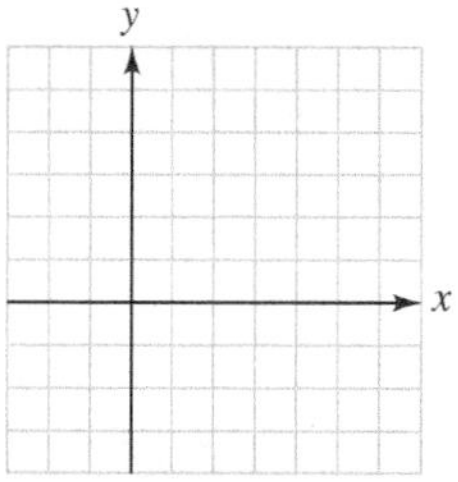

61. $f(x) = \ln(x + 1)$

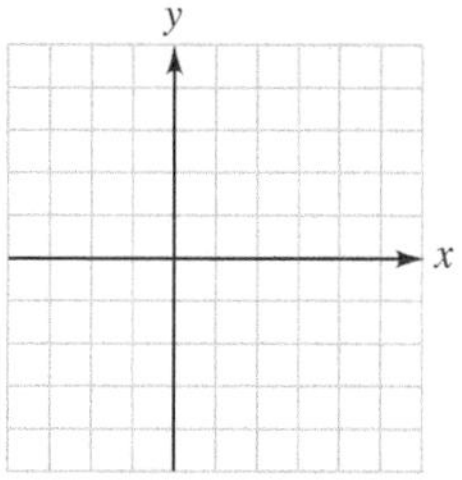

62. U.S. population How long will it take the population of the United States to double if the growth rate is 3% per year?

SECTION 11.7 Properties of Logarithms

DEFINITIONS AND CONCEPTS	EXAMPLES
Properties of logarithms: If all variables are positive numbers and $b \neq 1$, **1.** $\log_b 1 = 0$ **2.** $\log_b b = 1$ **3.** $\log_b b^x = x$ **4.** $b^{\log_b x} = x$ **5.** $\log_b MN = \log_b M + \log_b N$ **6.** $\log_b \frac{M}{N} = \log_b M - \log_b N$ **7.** $\log_b M^p = p \log_b M$ **8.** If $\log_b x = \log_b y$, then $x = y$.	 **1.** $\log_b 1 = 0$ because $b^0 = 1$. **2.** $\log_b b = 1$ because $b^1 = b$. **3.** $\log_b b^x = x$ because $b^x = b^x$. **4.** $b^{\log_b x} = x$ because $b^y = x$ and $y = \log_b x$. **5.** $\log_5 xy = \log_5 x + \log_5 y$ **6.** $\log_5 \frac{x}{y} = \log_5 x - \log_5 y$ **7.** $\log_5 x^3 = 3 \log_5 x$ **8.** If $\log_3 x = \log_3 9$, then $x = 9$.

Unless otherwise noted, all content on this page is © Cengage Learning.

Change-of-base formula:
If a, b, and x are real numbers, $b > 0$, and $b \neq 0$, then

$$\log_b x = \frac{\log_a x}{\log_a b}$$

To evaluate $\log_7 16$, substitute 7 and 16 into the change-of-base formula.

$$\log_7 \mathbf{16} = \frac{\log_{10} \mathbf{16}}{\log_{10} \mathbf{7}} \approx 1.4248 \quad \text{Use a calculator.}$$

or

$$\log_7 \mathbf{16} = \frac{\ln \mathbf{16}}{\ln \mathbf{7}} \approx 1.4248 \quad \text{Use a calculator.}$$

If $[H^+]$ is the hydrogen ion concentration in gram-ions per liter, then

$$pH = -\log [H^+]$$

Find the pH of a solution that has a hydrogen concentration of 10^{-4} grams per liter

$$\begin{aligned} pH &= -\log(10^{-4}) \\ &= 4 \end{aligned}$$

If L is the apparent loudness of a sound, I is the actual intensity, and k is a constant, then

$$L = k \ln I$$

Find the decrease in intensity that will cause the apparent loudness of a sound to be a fourth as loud.

$$\begin{aligned} \frac{1}{4} L_0 &= \frac{1}{4} k \ln I_0 \\ &= k \ln (I_0)^{\frac{1}{4}} \end{aligned}$$

To decrease the loudness of a sound by a fourth, the intensity is the fourth root.

REVIEW EXERCISES

Simplify each expression.

63. $\log_5 1$
64. $\log_9 9$
65. $\log_7 7^3$
66. $8^{\log_8 5}$
67. $\log 10^5$
68. $\ln 1$
69. $10^{\log_{10} 7}$
70. $e^{\ln 2}$
71. $\log_a a^6$
72. $\ln e^3$

Write each expression in terms of the logarithms of x, y, and z. Simplify, if possible.

73. $\log_b \dfrac{x^5 y^4}{z^2}$

74. $\log_b \sqrt{\dfrac{x}{yz^2}}$

Write each expression as the logarithm of one quantity.

75. $3 \log_b x - 5 \log_b y + 7 \log_b z$

76. $\dfrac{1}{2} \log_b x + 3 \log_b y - 7 \log_b z$

Assume that $\log a = 0.6$, $\log b = 0.36$, *and* $\log c = 2.4$. *Find the value of each expression.*

77. $\log abc$
78. $\log a^2 b$
79. $\log \dfrac{ac}{b}$
80. $\log \dfrac{a^2}{c^3 b^2}$

81. To four decimal places, approximate $\log_5 17$.

82. pH of grapefruit The pH of grapefruit juice is about 3.1. Find its hydrogen ion concentration.

83. Change in loudness Find the decrease in loudness if the intensity is cut in half.

SECTION 11.8 Exponential and Logarithmic Equations

DEFINITIONS AND CONCEPTS	EXAMPLES
Solving exponential and logarithmic equations: If $a^n = a^m$, then $n = m$	Solve each equation. **a.** $\begin{aligned} 9^x &= 3^{x-1} \\ (3^2)^x &= 3^{x-1} \\ 3^{2x} &= 3^{x-1} \\ 2x &= x - 1 \\ x &= -1 \end{aligned}$

If the bases of an exponential equation are not the same, take the logarithm of both sides. Then use the properties of logarithms to solve for the variable.

b. $9^x = 5$

$\ln(9^x) = \ln 5$ Take the natural logarithm of both sides.

$x \ln 9 = \ln 5$ The logarithm of a power is the power times the logarithm.

$x = \dfrac{\ln 5}{\ln 9}$ Divide both sides by ln 9.

$x \approx .7325$ Use a calculator.

Use the properties of logarithms to combine multiple logarithms into a single logarithm.

c. $\log_2(x + 1) + \log_2(x - 1) = 3$

$\log_2[(x + 1)(x - 1)] = 3$ Write as a single logarithm.

$\log_2(x^2 - 1) = 3$ Multiply inside the brackets.

$2^3 = x^2 - 1$ Write as an exponential expression.

$8 = x^2 - 1$ Simplify.

$9 = x^2$ Add 1 to both sides.

$x = 3, -3$ Take the square root of both sides.

Since 3 checks, it is a solution. Since -3 does not check, it is extraneous.

REVIEW EXERCISES

Solve each equation for x. If an answer is not exact, round to four decimal places.

84. $3^x = 7$

85. $5^{x+2} = 625$

86. $25 = 5.5(1.05)^t$

87. $4^{2t-1} = 64$

88. $2^x = 3^{x-1}$

89. $3^{x^2+4x} = \dfrac{1}{27}$

90. $\log x + \log(29 - x) = 2$

91. $\log_2 x + \log_2(x - 2) = 3$

92. $\log_2(x + 2) + \log_2(x - 1) = 2$

93. $\dfrac{\log(7x - 12)}{\log x} = 2$

94. $\log x + \log(x - 5) = \log 6$

95. $\log 3 - \log(x - 1) = -1$

96. $e^{x \ln 2} = 9$

97. $\ln x = \ln(x - 1)$

98. $\ln x = \ln(x - 1) + 1$

99. $\ln x = \log_{10} x$ (*Hint*: Use the change-of-base formula.)

100. Carbon-14 dating A wooden statue found in Egypt has a carbon-14 content that is two-thirds of that found in living wood. If the half-life of carbon-14 is 5,730 years, how old is the statue?

11 Test

Let f(x) = 4x and g(x) = x − 1. Find each function and its domain.

1. $g + f$

2. $f - g$

3. $g \cdot f$

4. g/f

Let f(x) = 4x and g(x) = x − 1. Find each value.

5. $(g \circ f)(1)$

6. $(f \circ g)(0)$

7. $(f \circ g)(-1)$

8. $(g \circ f)(-2)$

Let f(x) = 4x and g(x) = x − 1. Find each function.

9. $(f \circ g)(x)$

10. $(g \circ f)(x)$

Find the inverse of the function.

11. $3x + 2y = 12$

Graph each function.

12. $f(x) = 2^x + 1$

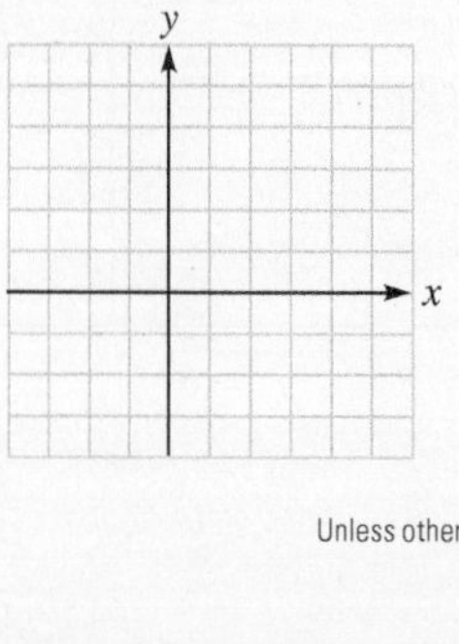

13. $f(x) = 2^{-x}$

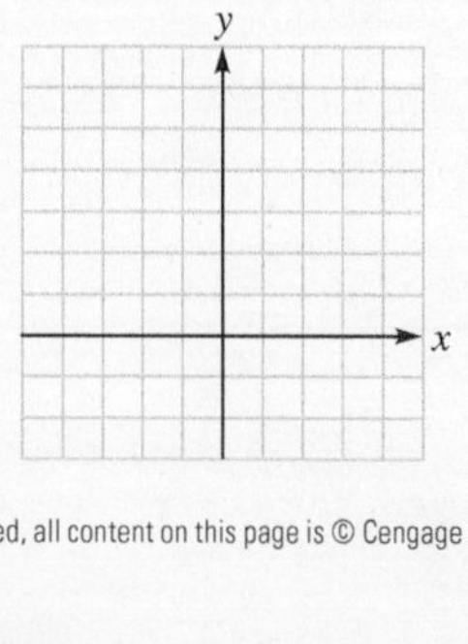

Unless otherwise noted, all content on this page is © Cengage Learning.

Solve each equation.

14. **Radioactive decay** A radioactive material decays according to the formula $A = A_0(2)^{-t}$. How much of a 3-gram sample will be left in 6 years?

15. **Investing** An initial deposit of $1,000 earns 6% interest, compounded twice a year. How much will be in the account in one year?

16. Graph the function $f(x) = e^x$.

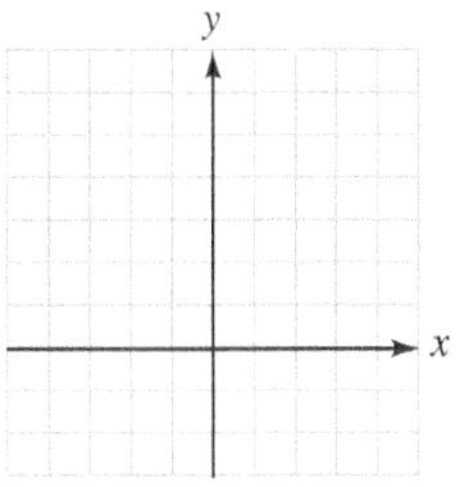

17. **Investing** An account contains $2,000 and has been earning 8% interest, compounded continuously. How much will be in the account in 10 years?

Find the value of x.

18. $\log_9 81 = x$

19. $\log_x 81 = 4$

20. $\log_3 x = -3$

21. $\log_x 49 = 2$

22. $\log_{3/2} \frac{9}{4} = x$

23. $\log_{2/3} x = -3$

Graph each function.

24. $f(x) = -\log_3 x$

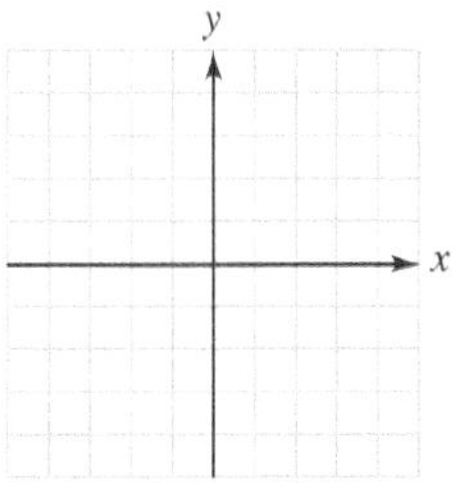

25. $f(x) = \ln x$

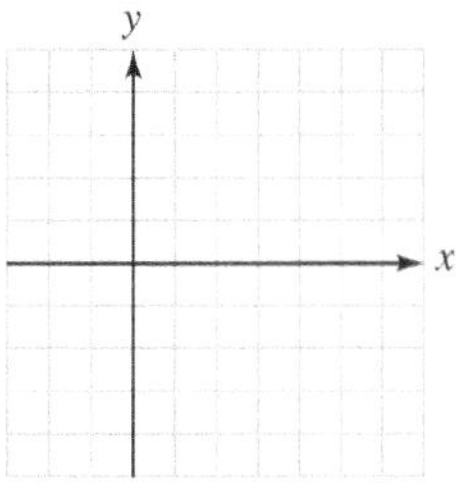

Write each expression in terms of the logarithms of a, b, and c.

26. $\log xy^3z^4$

27. $\ln\sqrt{\frac{a}{b^2c}}$

Write each expression as a logarithm of a single quantity.

28. $\frac{1}{2}\log(a + 2) + \log b - 3\log c$

29. $\frac{1}{3}(\log a - 2\log b) - \log c$

Assume that $\log 2 \approx 0.3010$ *and* $\log 3 \approx 0.3010$. *Find each value. Do not use a calculator.*

30. $\log 18$

31. $\log \frac{8}{3}$

Use the change-of-base formula to find each logarithm. Do not simplify the answer.

32. $\log_3 2$

33. $\log_\pi e$

Determine whether each statement is true. If a statement is not true, explain why.

34. $\log_a ab = 1 + \log_a b$

35. $\frac{\log a}{\log b} = \log a - \log b$

36. $\log a^{-3} = \frac{1}{3\log a}$

37. $\ln(-x) = -\ln x$

38. Find the pH of a solution with a hydrogen ion concentration of 3.7×10^{-7}. (*Hint*: $\text{pH} = -\log[\text{H}^+]$.)

39. Find the dB gain of an amplifier when $E_O = 60$ volts and $E_I = 0.3$ volt. (*Hint*: dB gain $= 20\log(E_O/E_I)$.)

Solve each equation.

40. $7^x = 4$

41. $3^{x-1} = 100^x$

Solve each equation.

42. $\log(3x + 1) = \log(5x - 7)$

43. $\log x + \log(x - 9) = 1$

Unless otherwise noted, all content on this page is © Cengage Learning.

Conic Sections, Systems of Equations and Inequalities, and More Graphing

12

moodboard/Cultura/Jupiterimages

Careers and Mathematics

WATER TRANSPORTATION OCCUPATIONS

The movement of cargo and passengers between nations depends on workers in the water transportation occupations, also known on commercial ships as merchant mariners. They operate and maintain deep-sea merchant ships, tugboats, ferries, and excursion vessels.

About 17 percent of water transportation employees work in inland waters, primarily the Mississippi River system; 23 percent on the deep seas and the Great Lakes; and another 24 percent in harbor operations.

Job Outlook:
Employment in this industry is projected to grow 20 percent through 2020. This is faster than average for all occupations.
Annual Earnings:
$24,890–$99,690
For More Information:
http://www.bls.gov/oco/ocos247.htm
For a Sample Application:
See Problem 38 in Section 12.3.

REACH FOR SUCCESS
12.1 The Circle and the Parabola
12.2 The Ellipse
12.3 The Hyperbola
12.4 Solving Systems of Equations and Inequalities Containing One or More Second-Degree Terms
12.5 Piecewise-Defined Functions and the Greatest Integer Function
■ *Projects*
REACH FOR SUCCESS EXTENSION
CHAPTER REVIEW
CHAPTER TEST
CUMULATIVE REVIEW

In this chapter

We have seen that the graphs of linear functions are straight lines, and that the graphs of quadratic functions are parabolas. In this chapter, we will discuss some special curves called conic sections. We will solve systems of equations and inequalities. Then we will discuss piecewise-defined functions and step functions.

Reach for Success Managing Your Stress

For life, it is useful to know how to relax. While taking an exam or studying, it is particularly important.

We will focus on relaxation techniques with an awareness that strong academic preparation will also reduce stress.

© Aleksandr Markin/Shutterstock.com

Although we may not be able to eliminate some of the causes of stress, we may be able to minimize the physical and mental responses to it in order to focus on the task at hand.

Identify any physical manifestations of stress you may be experiencing.

____

List any strategies you currently use to manage your stress.

____

Here are a few relaxation techniques you may use while studying or taking an exam:

- Releasing shoulder tension

Raise both shoulders simultaneously. Hold them up for a few seconds and then quickly drop them down again. Did you feel the release of the tension?

____

- Visualizing a calm environment

Visualize a place that brings you a sense of calmness. Where is this place? ______________________
Hold that thought . . . and try to work a mathematics problem.
Can you remain calm enough to stay focused on that problem? ______________________

- Clearing your mind

Take 4 to 5 seconds to slowly take a deep breath in through your nose. Hold it for 4 or 5 seconds, if possible. Slowly release through your mouth taking 4 to 5 seconds to do this. Wait 4 to 5 seconds and do it again. Repeat several times.

- Deep breathing

Try the breathing technique for one minute (4 or 5 full breaths). Do you notice a reduction in stress?

____

Laughter, good eating habits, adequate sleep, and daily exercise can all help to reduce your stress level.

A Successful Study Strategy . . .

Apply relaxation techniques to help reduce the physical aspects of stress in order to focus on studying and test taking.

At the end of the chapter you will find an additional exercise to guide you to a successful semester.

Unless otherwise noted, all content on this page is © Cengage Learning.

Section 12.1 The Circle and the Parabola

Objectives

1. Find the center and radius of a circle given an equation in standard form.
2. Find the center and radius of a circle given an equation in general form.
3. Write an equation of a circle in both standard and general form given the center and the radius.
4. Solve an application involving a circle.
5. Graph a parabola of the form $x = (y - k)^2 + h$.
6. Solve an application involving a parabola.

Vocabulary

conic section
circle
center of a circle
radius of a circle
standard form of the equation of a circle
point circle
general form of the equation of a circle
paraboloid

Getting Ready

Square each binomial.

1. $(x - 2)^2$ **2.** $(x + 4)^2$

What number must be added to each binomial to make it a perfect square trinomial?

3. $x^2 + 9x$ **4.** $x^2 - 12x$

In this chapter, we will introduce a group of curves called *conic sections.*

The graphs of second-degree equations in x and y represent figures that were investigated in the 17th century by René Descartes (1596–1650) and Blaise Pascal (1623–1662). Descartes discovered that graphs of second-degree equations fall into one of several categories: a pair of lines, a point, a circle, a parabola, an ellipse, a hyperbola, or no graph at all. Because all of these graphs can be formed by the intersection of a plane and a right-circular cone, they are called **conic sections**. See Figure 12-1 on the next page.

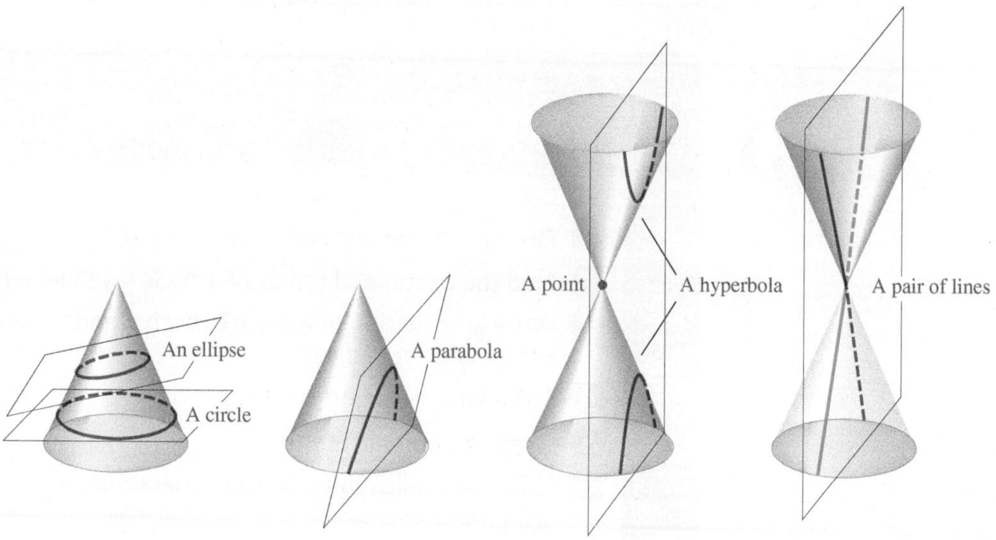

Figure 12-1

1 Find the center and radius of a circle given an equation in standard form.

A *circle* is one of the most common of the conic sections, generally seen in circular gears, pizza cutters, and Ferris wheels. Because of the circle's importance, we will begin the study of conic sections with the circle.

Every conic section can be represented by a second-degree equation in x and y. To find the form of an equation of a circle, we use the following definition.

THE CIRCLE

A **circle** is the set of all points in a plane that are a fixed distance from a point, called its **center**. The fixed distance is the **radius** of the circle.

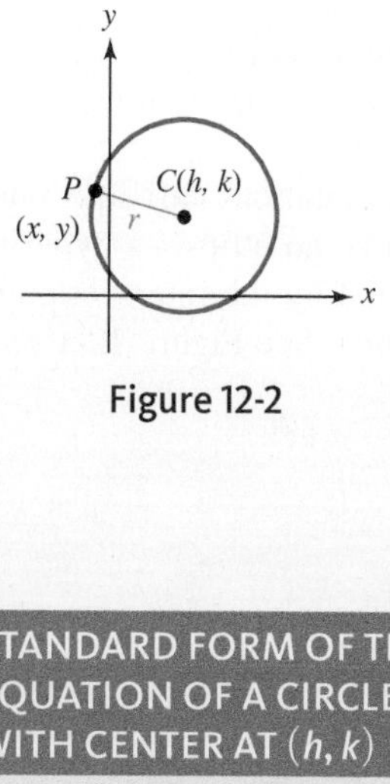

Figure 12-2

To develop the general equation of a circle, we must write the equation of a circle with a radius of r and with center at some point $C(h, k)$, as in Figure 12-2. This task is equivalent to finding all points $P(x, y)$ such that the length of line segment CP is r. We can use the distance formula to find r.

$$r = \sqrt{(x - h)^2 + (y - k)^2}$$

We then square both sides to obtain

$$(1) \quad r^2 = (x - h)^2 + (y - k)^2 \quad \text{or} \quad (x - h)^2 + (y - k)^2 = r^2$$

This equation is called the **standard form of the equation of a circle** with radius r and center at the point with coordinates (h, k).

STANDARD FORM OF THE EQUATION OF A CIRCLE WITH CENTER AT (h, k)

Any equation that can be written in the form

$$(x - h)^2 + (y - k)^2 = r^2$$

has a graph that is a circle with radius r and center at point (h, k).

If $r = 0$, the graph reduces to a single point called a **point circle**. If $r^2 < 0$, a circle does not exist. If both coordinates of the center are 0, the center of the circle is the origin.

Unless otherwise noted, all content on this page is © Cengage Learning.

STANDARD FORM OF THE EQUATION OF A CIRCLE WITH CENTER AT (0, 0)

Any equation that can be written in the form

$$x^2 + y^2 = r^2$$

has a graph that is a circle with radius r and center at the origin.

EXAMPLE 1 Find the center and the radius of each circle and then graph it.

a. $(x - 4)^2 + (y - 1)^2 = 9$ **b.** $x^2 + y^2 = 25$ **c.** $(x + 3)^2 + y^2 = 12$

Solution **a.** We can determine the center and the radius of a circle when its equation is written in standard form.

$$(x - \mathbf{4})^2 + (y - \mathbf{1})^2 = \mathbf{9}$$
$$\uparrow \qquad\qquad \uparrow \qquad \uparrow$$
$$(x - h)^2 + (y - k)^2 = r^2$$

$h = 4, k = 1$, and $r^2 = 9$. Since the radius of a circle must be positive, $r = 3$.

The center of the circle is $(h, k) = (4, 1)$ and the radius is 3.

To plot four points on the circle, we move up, down, left, and right 3 units from the center, as shown in Figure 12-3(a). Then we draw a circle through the points to obtain the graph of $(x - 4)^2 + (y - 1)^2 = 9$, as shown in Figure 12-3(b).

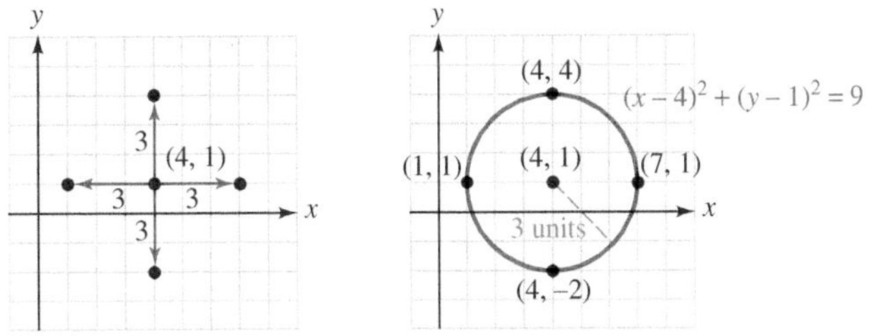

(a) (b)

Figure 12-3

b. To verify that the center of the circle is the origin, we can write $x^2 + y^2 = 25$ in the following way:

$$(x - \mathbf{0})^2 + (y - \mathbf{0})^2 = \mathbf{25}$$
$$\uparrow \qquad\qquad \uparrow \qquad \uparrow$$
$$h \qquad\qquad k \qquad r^2$$

$h = 0, k = 0$, and $r^2 = 25$. Since the radius of a circle must be positive, $r = 5$.

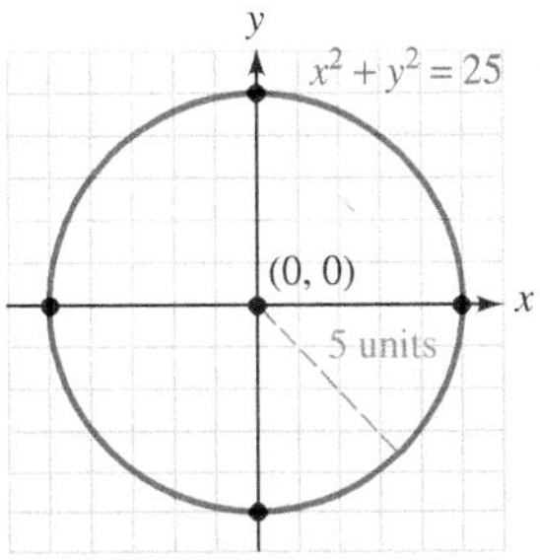

Figure 12-4

This illustrates that the center of the circle is at $(0, 0)$ and the radius is 5.

To plot four points on the circle, we move up, down, left, and right 5 units from the center. Then we draw a circle through the points to obtain the graph of $x^2 + y^2 = 25$, as shown in Figure 12-4.

c. To determine h in the equation $(x + 3)^2 + y^2 = 12$, it is helpful to write $x + 3$ as $x - (-3)$.

Standard form requires a minus symbol here.

$$\downarrow$$
$$[x - (\mathbf{-3})]^2 + (y - \mathbf{0})^2 = \mathbf{12}$$
$$\uparrow \qquad\qquad \uparrow \qquad \uparrow$$
$$h \qquad\qquad k \qquad r^2$$

$h = -3, k = 0$, and $r^2 = 12$

Unless otherwise noted, all content on this page is © Cengage Learning.

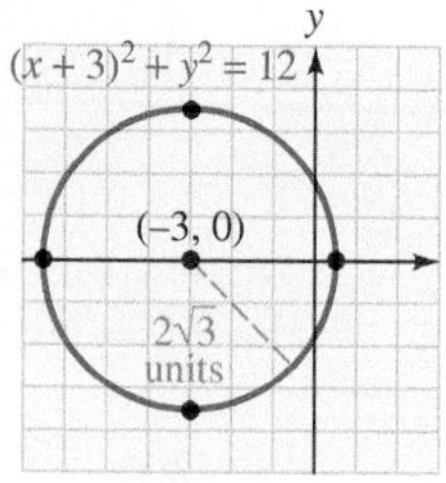

Figure 12-5

If $r^2 = 12$, by the square root property

$$r = \pm\sqrt{12} = \pm 2\sqrt{3}$$

Because the radius must be positive, $r = 2\sqrt{3}$. The center of the circle is at $(-3, 0)$ and the radius is $2\sqrt{3}$.

To plot four points on the circle, we move upward, downward, left, and right $2\sqrt{3} \approx 3.5$ units from the center. Then we draw a circle through the points to obtain the graph of $(x + 3)^2 + y^2 = 12$, as shown in Figure 12-5.

SELF CHECK 1 Find the center and the radius of each circle and then graph it.
a. $(x - 3)^2 + (y + 4)^2 = 4$ **b.** $x^2 + y^2 = 8$

2 Find the center and radius of a circle given an equation in general form.

Another important form of the equation of a circle is called the **general form of the equation of a circle**.

GENERAL FORM OF THE EQUATION OF A CIRCLE

The equation of any circle can be written in the form

$$x^2 + y^2 + Dx + Ey + F = 0$$

EXAMPLE 2 Graph: $x^2 + y^2 - 4x + 2y - 20 = 0$

Solution Since the equation matches the general form of the equation of a circle, we know that its graph will be a circle. To find its center and radius, we must complete the square on both x and y and write the equation in standard form.

$$x^2 + y^2 - 4x + 2y = 20 \quad \text{Add 20 to both sides.}$$
$$x^2 - 4x + y^2 + 2y = 20 \quad \text{Rearrange terms using the commutative property of addition.}$$

To complete the square on x and y, add 4 and 1 to both sides.

$$x^2 - 4x + 4 + y^2 + 2y + 1 = 20 + 4 + 1$$
$$(x - 2)^2 + (y + 1)^2 = 25 \quad \text{Factor } x^2 - 4x + 4 \text{ and } y^2 + 2y + 1.$$

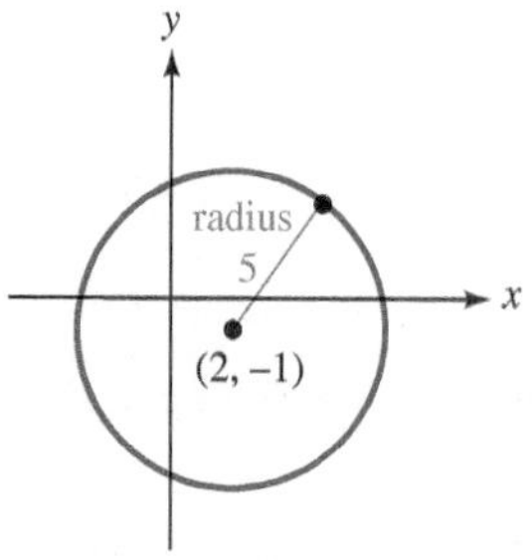

$x^2 + y^2 - 4x + 2y = 20$

Figure 12-6

We can now see that this result is the standard equation of a circle with a radius of 5 and center at $h = 2$ and $k = -1$. If we plot the center and draw a circle with a radius of 5 units, we will obtain the circle shown in Figure 12-6.

SELF CHECK 2 Write the equation $x^2 + y^2 + 2x - 4y - 11 = 0$ in standard form and graph it.

3 Write an equation of a circle in both standard and general form given the center and the radius.

EXAMPLE 3 Find the general form of the equation of the circle with radius 5 and center at $(3, 2)$.

Solution We substitute 5 for r, 3 for h, and 2 for k in the standard form of a circle and proceed as follows:

$$(x - h)^2 + (y - k)^2 = r^2$$
$$(x - 3)^2 + (y - 2)^2 = 5^2$$

Unless otherwise noted, all content on this page is © Cengage Learning.

$$x^2 - 6x + 9 + y^2 - 4y + 4 = 25 \quad (x-3)^2 = x^2 - 6x + 9;\ (y-2)^2 = y^2 - 4y + 4$$
$$x^2 + y^2 - 6x - 4y - 12 = 0 \quad \text{Subtract 25 from both sides and simplify.}$$

The general form of the equation is $x^2 + y^2 - 6x - 4y - 12 = 0$.

SELF CHECK 3 Find the general form of the equation of the circle with radius 6 and center at $(2, 3)$.

Accent on technology

▸ Graphing Circles

Since the graphs of circles fail the vertical line test, their equations do not represent functions. It is somewhat more difficult to use a graphing calculator to graph equations that are not functions. For example, to graph the circle described by $(x-1)^2 + (y-2)^2 = 4$, we must split the equation into two functions and graph each one separately. We begin by solving the equation for y.

$$(x-1)^2 + (y-2)^2 = 4$$
$$(y-2)^2 = 4 - (x-1)^2 \quad \text{Subtract } (x-1)^2 \text{ from both sides.}$$
$$y - 2 = \pm\sqrt{4 - (x-1)^2} \quad \text{Use the square-root property.}$$
$$y = 2 \pm \sqrt{4 - (x-1)^2} \quad \text{Add 2 to both sides.}$$

This equation defines two functions. If we use window settings of $[-3, 5]$ for x and $[-3, 5]$ for y and graph the functions

$$y = 2 + \sqrt{4 - (x-1)^2} \quad \text{and} \quad y = 2 - \sqrt{4 - (x-1)^2}$$

Using a TI-84 calculator, we see the distorted circle shown in Figure 12-7(a). To see a better representation, graphing calculators have a square feature, ZSquare, that displays an equal unit distance on both the x- and y-axes. After applying this feature (zoom 5), we obtain the circle shown in Figure 12-7(b).

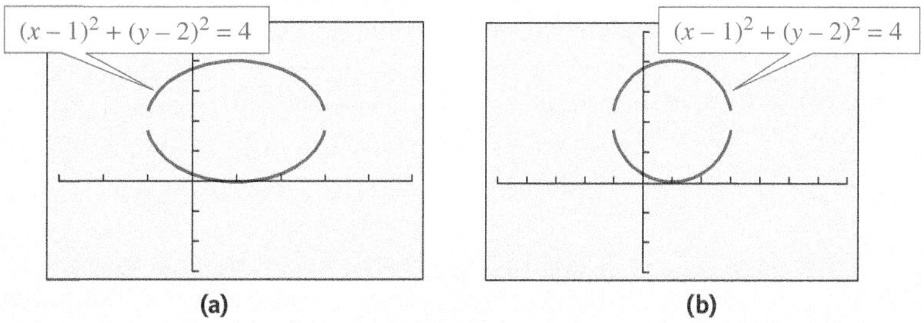

Figure 12-7

For instructions regarding the use of a Casio graphing calculator, please refer to the Casio Keystroke Guide in the back of the book.

4 Solve an application involving a circle.

EXAMPLE 4 **TV TRANSLATORS** The broadcast area of a TV station is bounded by the circle $x^2 + y^2 = 3{,}600$, where x and y are measured in miles. A translator station picks up the signal and retransmits it from the center of a circular area bounded by $(x+30)^2 + (y-40)^2 = 1{,}600$. Find the location of the translator and the greatest distance from the main transmitter that the signal can be received.

Solution The coverage of the TV station is bounded by $x^2 + y^2 = 60^2$, a circle centered at the origin with a radius of 60 miles, as shown in Figure 12-8 on the next page. Because the translator is at the center of the circle $(x+30)^2 + (y-40)^2 = 1{,}600$, it is located at $(-30, 40)$, a point 30 miles west and 40 miles north of the TV station. The radius of the translator's coverage is $\sqrt{1{,}600}$, or 40 miles.

Unless otherwise noted, all content on this page is © Cengage Learning.

As shown in the figure, the greatest distance of reception is the sum of A, the distance from the translator to the TV station, and 40 miles, the radius of the translator's coverage.

To find A, we use the distance formula to find the distance between $(x_1, y_1) = (-30, 40)$ and the origin, $(x_2, y_2) = (0, 0)$.

$$A = \sqrt{(x_1 - x_2)^2 + (y_1 - y_2)^2}$$
$$A = \sqrt{(-30 - 0)^2 + (40 - 0)^2}$$
$$= \sqrt{(-30)^2 + 40^2}$$
$$= \sqrt{2{,}500}$$
$$= 50$$

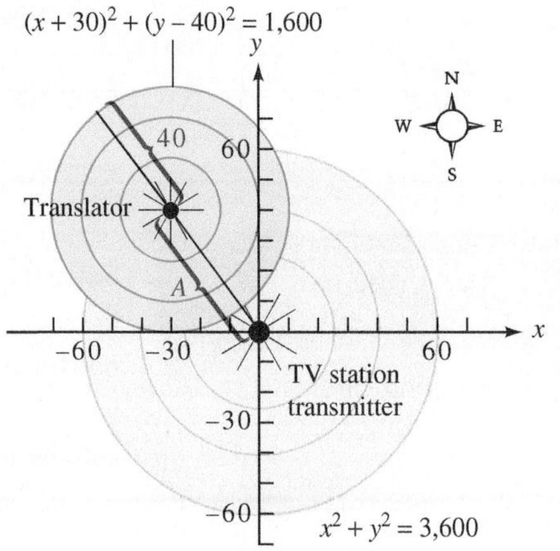

Figure 12-8

Since the translator is 50 miles from the TV station and it broadcasts the signal an additional 40 miles, the greatest reception distance is 50 + 40, or 90 miles.

SELF CHECK 4 If the circular area is bounded by $(x - 60)^2 + (y + 80)^2 = 1{,}600$, find the location of the translator.

5 Graph a parabola of the form $x = (y - k)^2 + h$.

Parabolas can be rotated to generate dish-shaped surfaces called **paraboloids**, as shown in Figure 12-9(a). Any light or sound placed at the *focus* of a paraboloid is reflected outward in parallel paths. This property makes parabolic surfaces ideal for flashlight and headlight reflectors. It also makes parabolic surfaces good antennas, because signals captured by such antennas are concentrated at the focus. Parabolic mirrors are capable of concentrating the rays of the Sun at a single point and thereby generating tremendous heat. This property is used in the design of solar furnaces.

Any object thrown upward and outward travels in a parabolic path, as shown in Figure 12-9(b). In architecture, many arches are parabolic in shape, because this provides strength. Cables that support suspension bridges hang in a form similar to that of a parabola. (See Figure 12-9(c).)

Parabolas

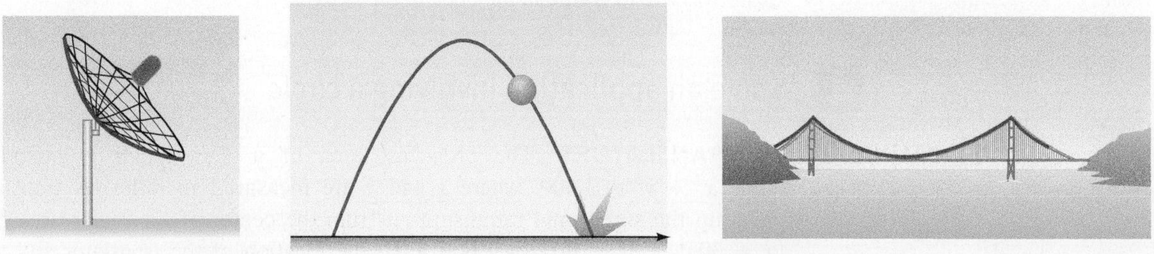

Figure 12-9

We have seen that equations of the form $y = a(x - h)^2 + k$, with $a \neq 0$, represent parabolas with the vertex at the point (h, k). They open upward when $a > 0$ and downward when $a < 0$.

Unless otherwise noted, all content on this page is © Cengage Learning.

Equations of the form $x = a(y - k)^2 + h \quad (a \neq 0)$, also represent parabolas with vertex at point (h, k). However, they open to the right when $a > 0$ and to the left when $a < 0$. Parabolas that open to the right or left do not represent functions, because their graphs fail the vertical line test.

Standard equations of many parabolas are summarized in the following table.

EQUATIONS OF PARABOLAS

Parabola opening	*Vertex at origin*	*Vertex at* (h, k)
Upward	$y = ax^2 \quad (a > 0)$	$y = a(x - h)^2 + k \quad (a > 0)$
Downward	$y = ax^2 \quad (a < 0)$	$y = a(x - h)^2 + k \quad (a < 0)$
Right	$x = ay^2 \quad (a > 0)$	$x = a(y - k)^2 + h \quad (a > 0)$
Left	$x = ay^2 \quad (a < 0)$	$x = a(y - k)^2 + h \quad (a < 0)$

EXAMPLE 5 Graph each equation. **a.** $x = \frac{1}{2}y^2$ **b.** $x = -2(y - 2)^2 + 3$

Solution **a.** We can create a table of ordered pairs that satisfy the equation, plot each pair, and draw the parabola, as in Figure 12-10(a). Because the equation is of the form $x = ay^2$ with $a > 0$, the parabola opens to the right and has its vertex at the origin.

b. We can create a table of ordered pairs that satisfy the equation, plot each pair, and draw the parabola, as in Figure 12-10(b). Because the equation is of the form $x = a(y - k)^2 + h \quad (a < 0)$, the parabola opens to the left and has its vertex at the point with coordinates $(3, 2)$.

$x = \frac{1}{2}y^2$

x	y	(x, y)
0	0	$(0, 0)$
2	2	$(2, 2)$
2	−2	$(2, -2)$
8	4	$(8, 4)$
8	−4	$(8, -4)$

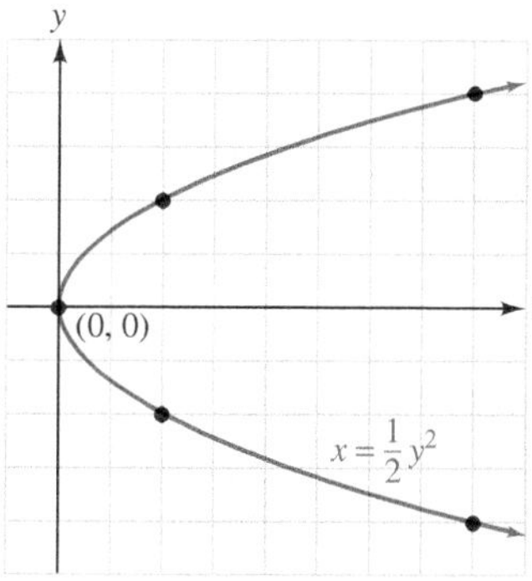

(a)

$x = -2(y - 2)^2 + 3$

x	y	(x, y)
−5	0	$(-5, 0)$
1	1	$(1, 1)$
3	2	$(3, 2)$
1	3	$(1, 3)$
−5	4	$(-5, 4)$

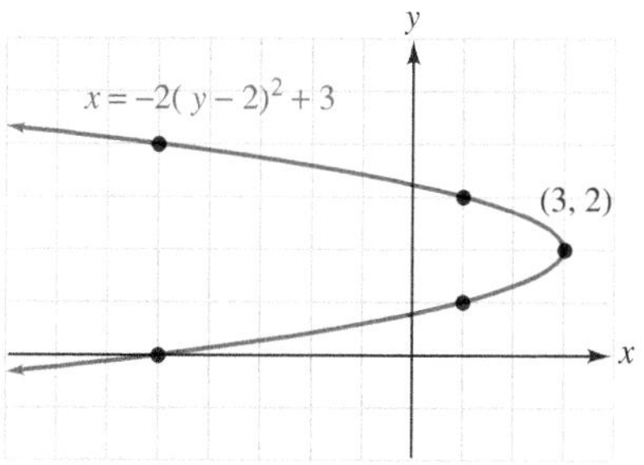

(b)

Figure 12-10

SELF CHECK 5 Graph: $x = \frac{1}{2}(y - 1)^2 - 2$

The general forms of the equations of a parabola are as follows:

GENERAL FORM OF THE EQUATION OF A PARABOLA THAT OPENS UPWARD OR DOWNWARD

The general form of the equation of a parabola that opens upward or downward is

$$y = ax^2 + bx + c \quad (a \neq 0)$$

If $a > 0$, the parabola opens upward. If $a < 0$, the parabola opens downward.

GENERAL FORM OF THE EQUATION OF A PARABOLA THAT OPENS LEFT OR RIGHT

The general form of the equation of a parabola that opens to the left or to the right is

$$x = ay^2 + by + c \quad (a \neq 0)$$

If $a > 0$, the parabola opens to the right. If $a < 0$, the parabola opens to the left.

Unless otherwise noted, all content on this page is © Cengage Learning.

EXAMPLE 6 Graph: $x = -2y^2 + 12y - 15$

Solution This equation is in the general form of a parabola that opens left or right. Since $a = -2$, the parabola opens to the left. To find the coordinates of its vertex, we write the equation in standard form by completing the square on y.

COMMENT Notice the similarity to writing an equation in $y = a(x - h)^2 + k$ form.

$$\begin{aligned} x &= -2y^2 + 12y - 15 \\ &= -2(y^2 - 6y) - 15 && \text{Factor out } -2 \text{ from } -2y^2 + 12y. \\ &= \mathbf{-2(y^2 - 6y + 9) - 15 + 18} && \text{Subtract and add 18; } -2(9) = -18. \\ &= -2(y - 3)^2 + 3 \end{aligned}$$

Because the equation is written in the form $x = a(y - k)^2 + h$, we can determine that the parabola has its vertex at $(3, 3)$. The graph is shown in Figure 12-11.

$x = -2y^2 + 12y - 15$

x	y	(x, y)
−5	1	$(-5, 1)$
1	2	$(1, 2)$
3	3	$(3, 3)$
1	4	$(1, 4)$
−5	5	$(-5, 5)$

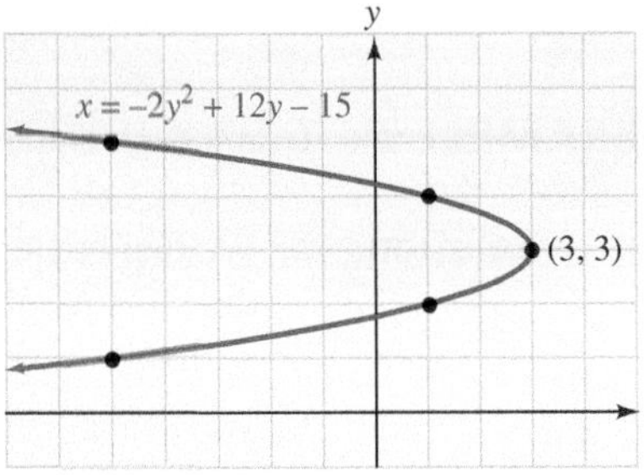

Figure 12-11

SELF CHECK 6 Graph: $x = 0.5y^2 - y - 1$

COMMENT To find the y-coordinate of the vertex for any horizontal parabola $(x = ay^2 + by + c)$, we could compute $-\frac{b}{2a}$. To find the x-coordinate, we could substitute the value of $-\frac{b}{2a}$ for y and find the value of x.

6 Solve an application involving a parabola.

EXAMPLE 7 **GATEWAY ARCH** The shape of the Gateway Arch in St. Louis is approximately a parabola 630 feet high and 630 feet wide, as shown in Figure 12-12(a). How high is the arch 100 feet from its foundation?

Solution We place the parabola in a coordinate system as in Figure 12-12(b), with ground level on the x-axis and the vertex of the parabola at the point $(h, k) = (0, 630)$.

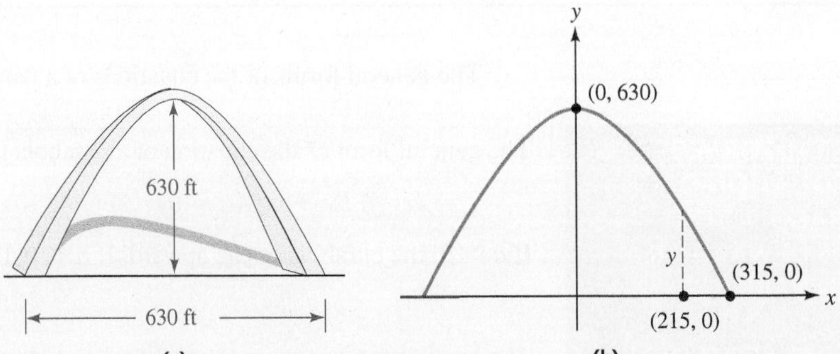

Figure 12-12

Unless otherwise noted, all content on this page is © Cengage Learning.

The equation of this downward-opening parabola has the form

$$\begin{aligned} y &= a(x - h)^2 + k && \text{With } a < 0 \\ &= a(x - \mathbf{0})^2 + \mathbf{630} && \text{Substitute } h = 0 \text{ and } k = 630. \\ &= ax^2 + 630 && \text{Simplify.} \end{aligned}$$

Because the Gateway Arch is 630 feet wide at its base, the parabola passes through the point $\left(\frac{630}{2}, 0\right)$, or $(315, 0)$. To find the value of a in the equation of the parabola, we proceed as follows:

$$\begin{aligned} y &= ax^2 + 630 && \\ \mathbf{0} &= a(\mathbf{315})^2 + 630 && \text{Substitute 315 for } x \text{ and 0 for } y. \\ \frac{-630}{315^2} &= a && \text{Subtract 630 from both sides and divide both sides by } 315^2. \\ -\frac{2}{315} &= a && \text{Simplify: } \frac{-630}{315^2} = \frac{-2}{315} \end{aligned}$$

The equation of the parabola that approximates the shape of the Gateway Arch is

$$y = -\frac{2}{315}x^2 + 630$$

To find the height of the arch at a point 100 feet from its foundation, we substitute 315 − 100, or 215, for x in the equation of the parabola and solve for y.

$$\begin{aligned} y &= -\frac{2}{315}x^2 + 630 \\ &= -\frac{2}{315}(\mathbf{215})^2 + 630 \\ &= 336.5079365 \end{aligned}$$

At a point 100 feet from the foundation, the height of the arch is about 337 feet.

SELF CHECK 7 How high is the arch 150 feet from its foundation?

SELF CHECK ANSWERS

1. a. $(3, -4), 2$

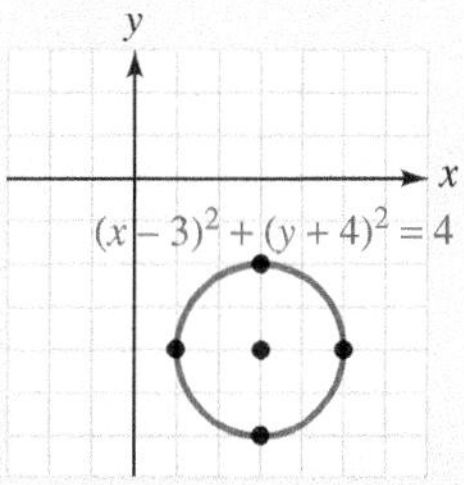

b. $(0, 0), 2\sqrt{2}$

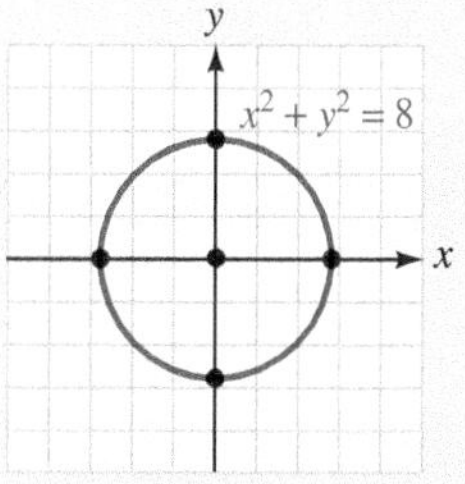

2. $(x + 1)^2 + (y - 2)^2 = 16$

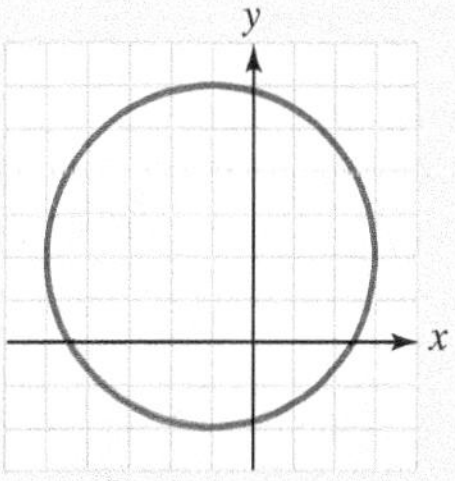

$x^2 + y^2 + 2x - 4y - 11 = 0$

3. $x^2 + y^2 - 4x - 6y - 23 = 0$

4. It is located at a point 60 miles east and 80 miles south of the station.

Unless otherwise noted, all content on this page is © Cengage Learning.

5.

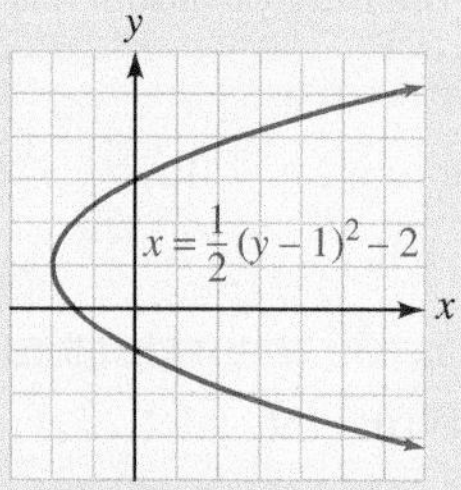

6.

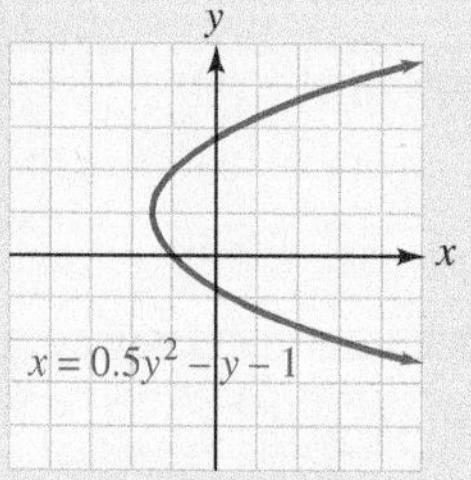

7. From a point 150 feet from the foundation, the height of the arch is about 457 feet.

NOW TRY THIS

The equation $4y^2 + (4x - 1)y + x^2 - 5x - 3 = 0$ is an equation for a parabola that is neither vertical nor horizontal. Use the quadratic formula to solve for y and then graph the equation with a calculator.

12.1 Exercises

WARM-UPS *Determine whether the graph of the parabola opens upward or downward.*

1. $y = -5x^2 + 4x + 1$
2. $y = 8x^2 - 7x - 1$
3. $y = x^2 - 6x - 7$
4. $y = -4x^2 - 7x + 1$

Factor.

5. $x^2 + 14x + 49$
6. $x^2 - 18x + 81$
7. $x^2 - 20x + 100$
8. $x^2 + 12x + 36$

REVIEW *Solve each equation.*

9. $|2x - 5| = 7$
10. $\left|\frac{4 - 3x}{5}\right| = 12$
11. $|3x + 4| = |5x - 2|$
12. $|6 - 4x| = |x + 2|$

VOCABULARY AND CONCEPTS *Fill in the blanks.*

13. A _____ section is determined by the intersection of a plane and a right-circular cone.
14. A _____ is the set of all points in a _____ that are a fixed distance from a given point. The fixed distance is called the _____ and the point is called the _____.
15. The equation of the circle $x^2 + (y - 3)^2 = 16$ is in _______ form with the center at _____ and a radius _.
16. The graph of the equation $x^2 + y^2 = 0$ is a _________.
17. The equation $x^2 + y^2 - 10x - 8y - 8 = 0$ is an equation of a _____ written in _______ form.
18. The graph of $y = ax^2$ $(a > 0)$ is a _______ with vertex at the _____ that opens _______.
19. The graph of $x = a(y - 2)^2 + 3$ $(a > 0)$ is a _______ with vertex at _____ that opens to the ____.
20. The graph of $x = a(y - 1)^2 - 3$ $(a < 0)$ is a _______ with vertex at _______ that opens to the ___.

GUIDED PRACTICE *Graph each equation and state the center and radius of the circle.* SEE EXAMPLE 1. (OBJECTIVE 1)

21. $x^2 + y^2 = 9$

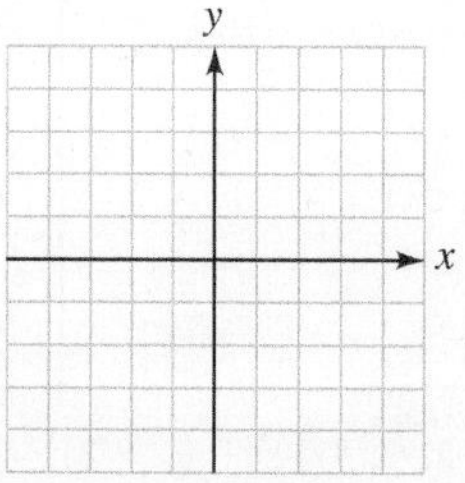

22. $x^2 + y^2 = 16$

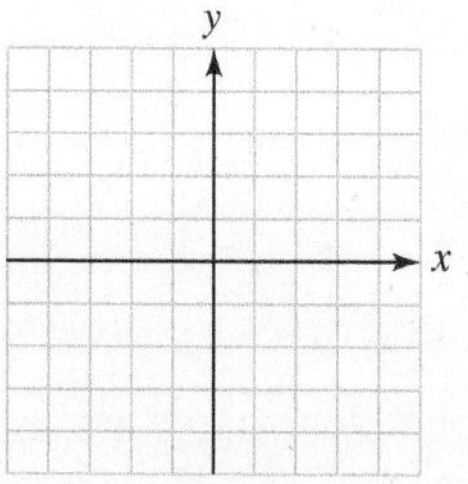

23. $(x - 2)^2 + y^2 = 9$

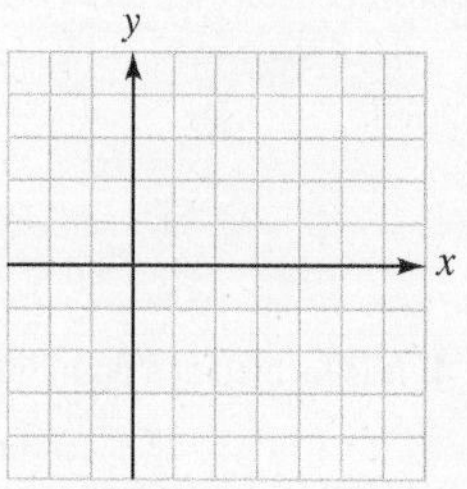

24. $x^2 + (y - 3)^2 = 4$

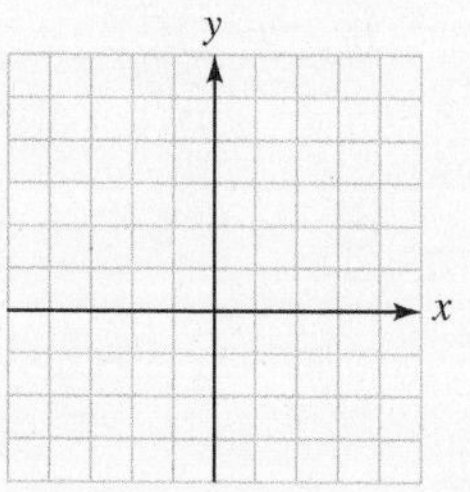

Unless otherwise noted, all content on this page is © Cengage Learning.

25. $(x-2)^2+(y-4)^2=4$

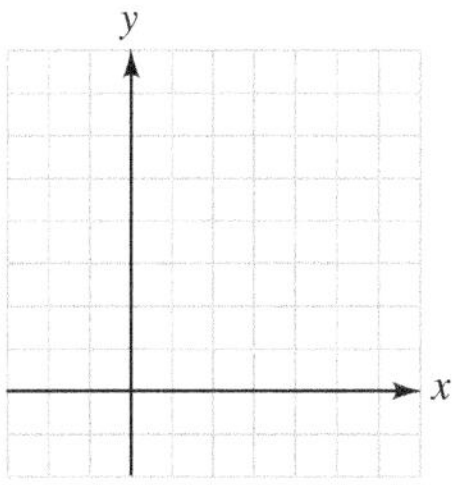

26. $(x-3)^2+(y-2)^2=4$

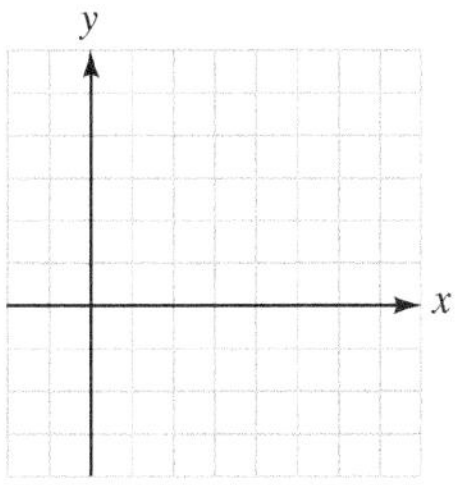

27. $(x+3)^2+(y-1)^2=16$

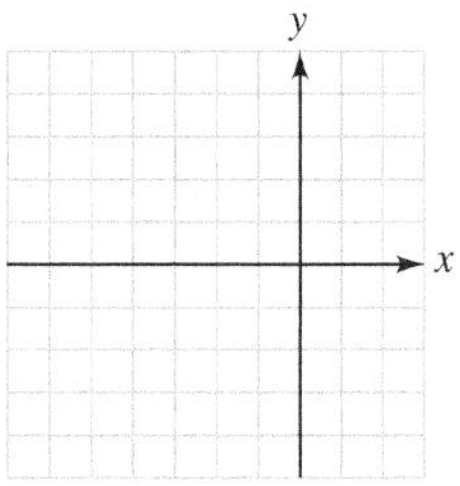

28. $(x-1)^2+(y+4)^2=9$

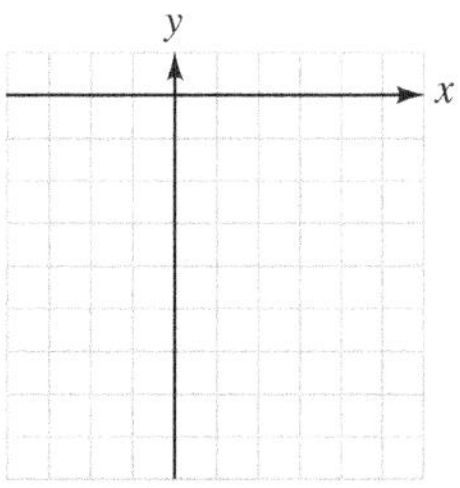

Graph each circle. Give the coordinates of the center and state the radius. SEE EXAMPLE 2. (OBJECTIVE 2)

29. $x^2+y^2+2x-8=0$

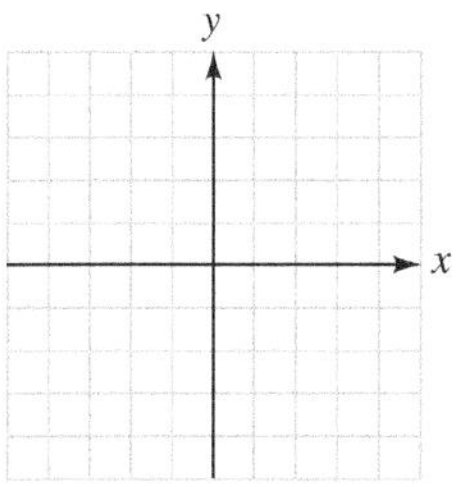

30. $x^2+y^2-4y=12$

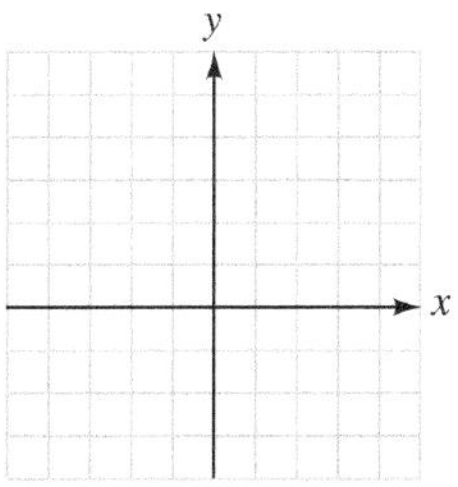

31. $9x^2+9y^2-12y=5$

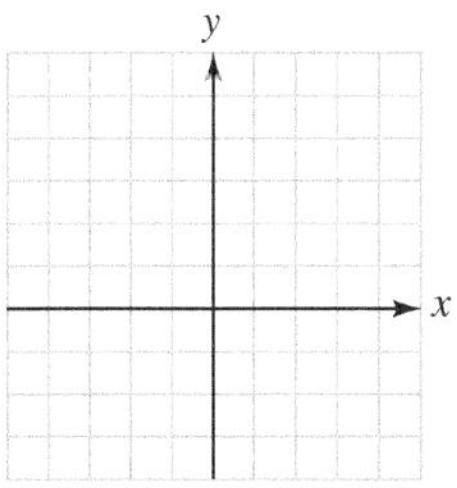

32. $4x^2+4y^2+4y=15$

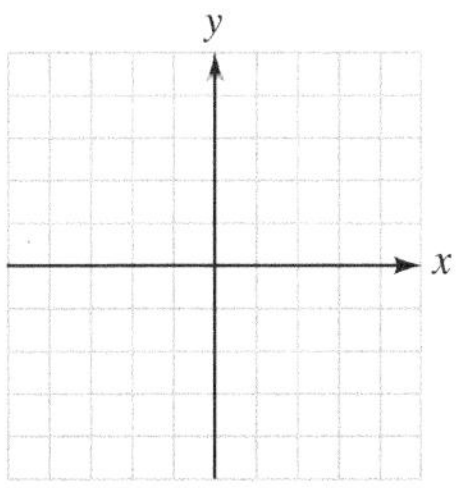

33. $x^2+y^2-2x+4y=-1$

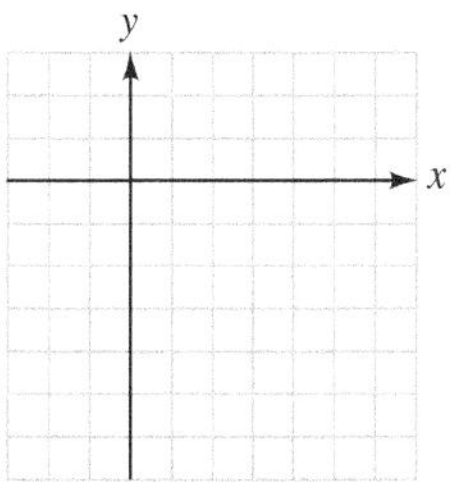

34. $x^2+y^2+4x+2y=4$

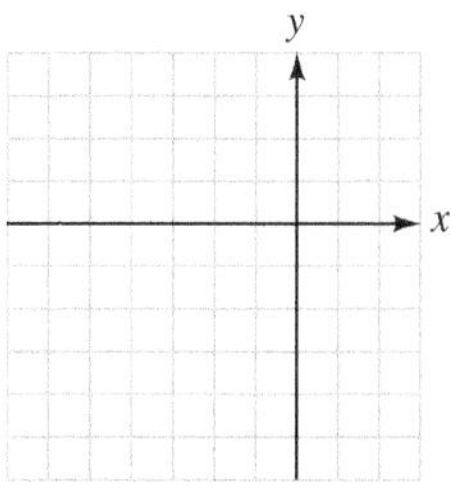

35. $x^2+y^2+6x-4y=-12$

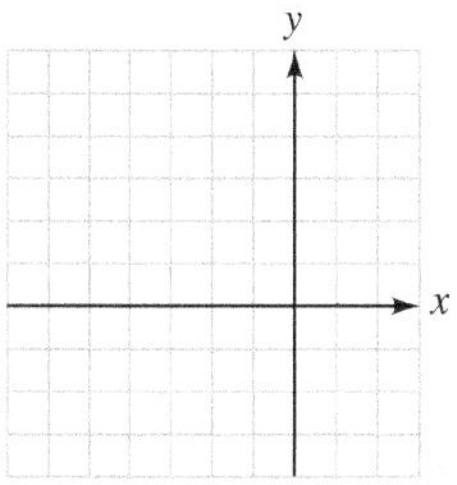

36. $x^2+y^2+8x+2y=-13$

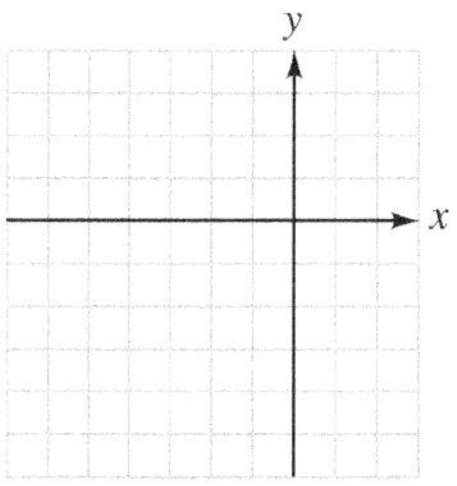

Write the equation of the circle with the following properties in standard form and in general form. SEE EXAMPLE 3. (OBJECTIVE 3)

37. center at origin; radius 3

38. center at origin; radius 4

39. center at $(2, 4)$; radius 7

40. center at $(5, 3)$; radius 2

41. center at $(-4, 7)$; radius 10

42. center at $(5, -4)$; radius 6

43. center at the origin; diameter $8\sqrt{5}$

44. center at the origin; diameter $4\sqrt{3}$

Unless otherwise noted, all content on this page is © Cengage Learning.

State the vertex of each parabola and graph the equation. SEE EXAMPLE 5. (OBJECTIVE 5)

45. $x = y^2$

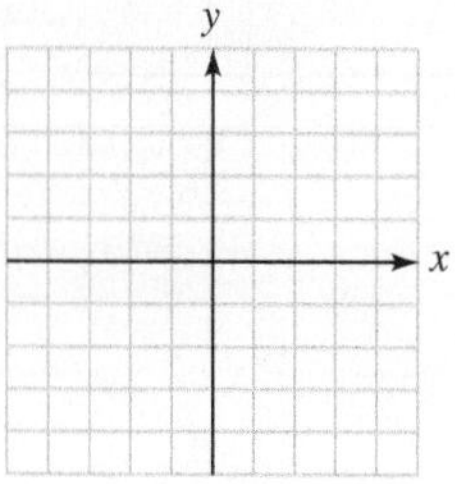

46. $x = -y^2 + 1$

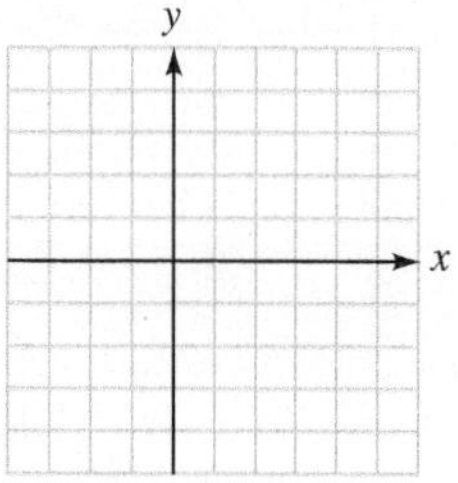

47. $x = -\frac{1}{4}y^2$

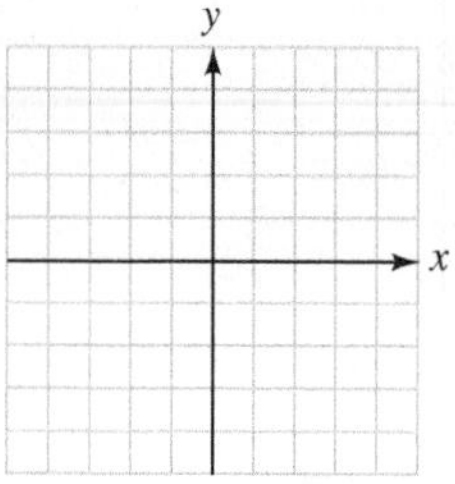

48. $x = 4y^2$

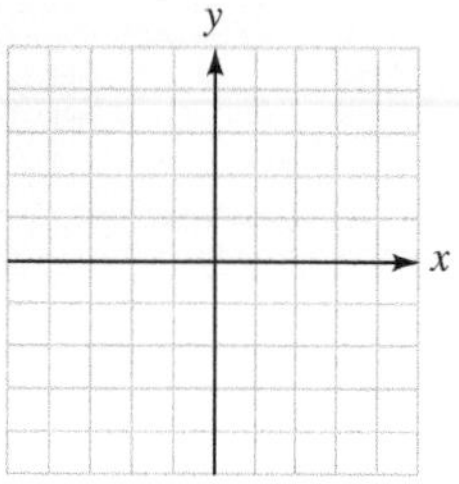

State the vertex of each parabola and graph the equation. SEE EXAMPLE 6. (OBJECTIVE 5)

49. $y^2 + 4x - 6y = -1$

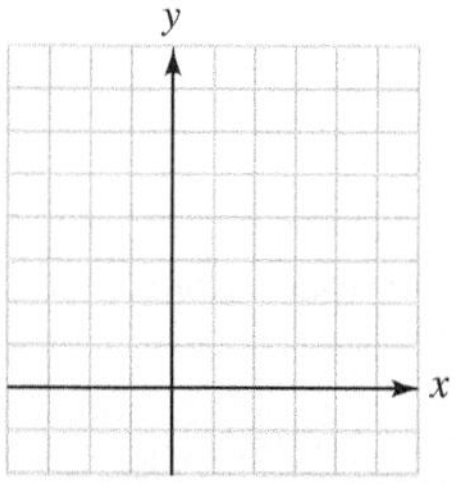

50. $x = \frac{1}{2}y^2 + 2y$

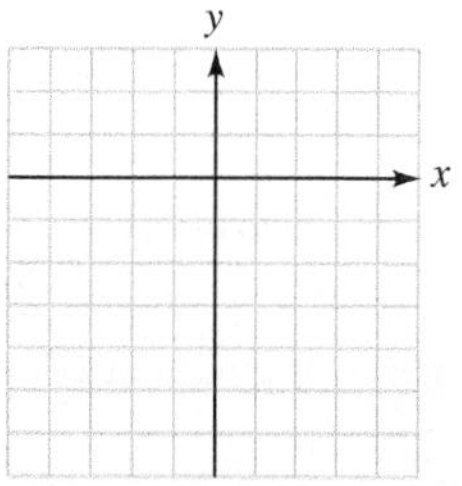

51. $y = -x^2 - x + 1$

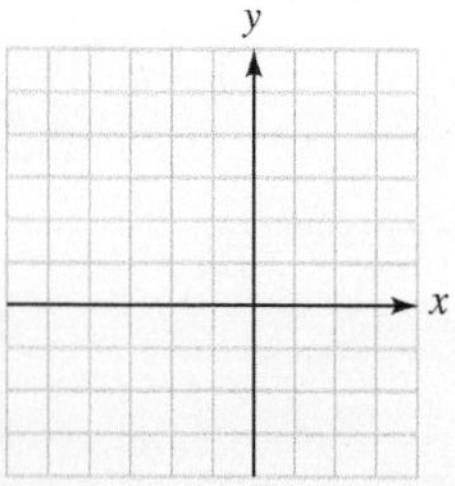

52. $x^2 - 2y - 2x = -7$

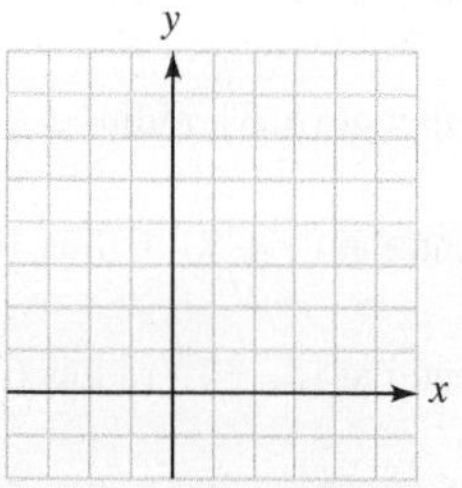

ADDITIONAL PRACTICE *Graph each equation.*

53. $x^2 + (y + 3)^2 = 1$

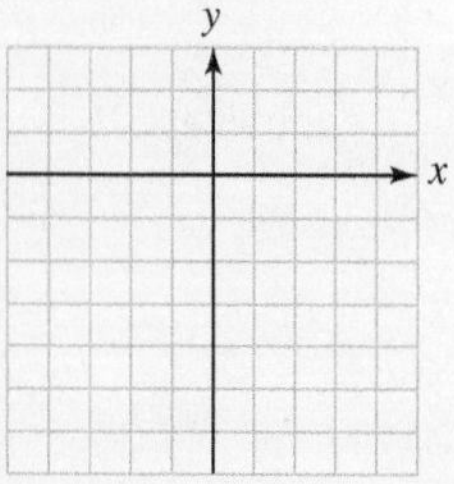

54. $(x + 4)^2 + y^2 = 1$

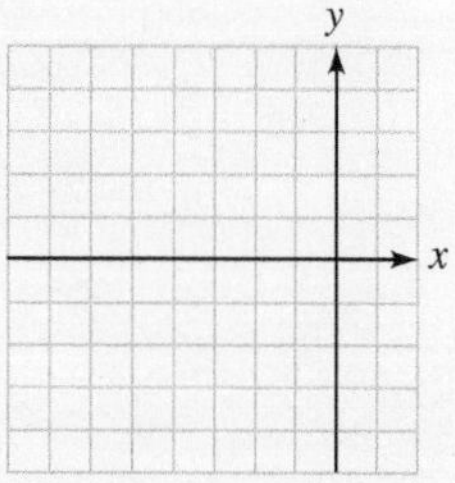

55. $y = 2(x - 1)^2 + 3$

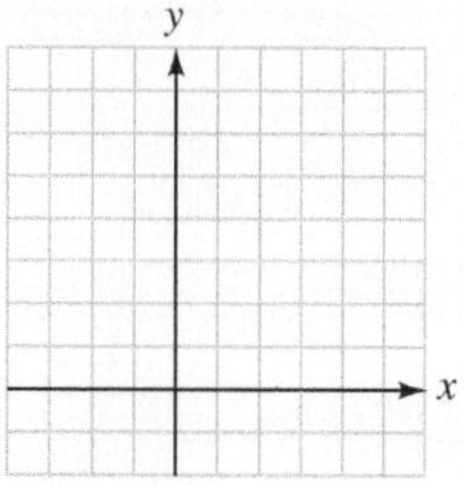

56. $y = -2(x + 1)^2 + 2$

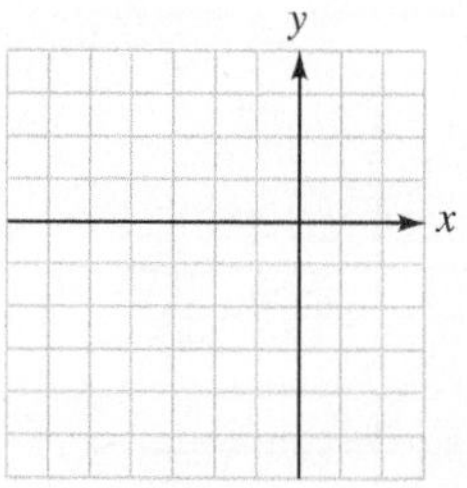

57. $y = x^2 + 4x + 5$

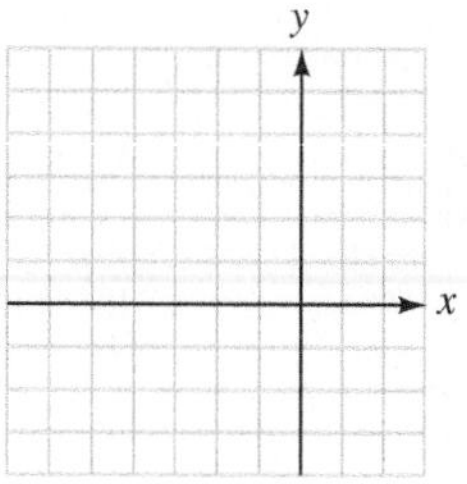

58. $y = -x^2 - 2x + 3$

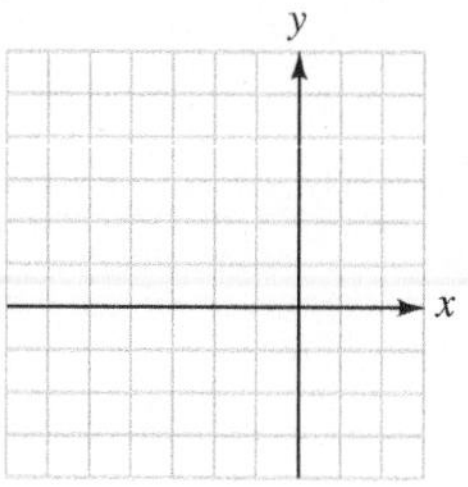

59. $y = 3(x + 1)^2 - 2$

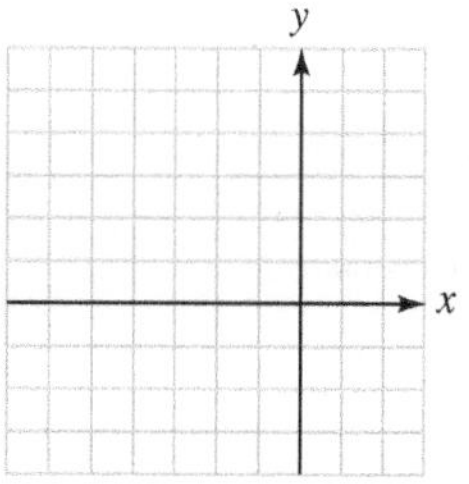

60. $y = -3(x - 1)^2 - 2$

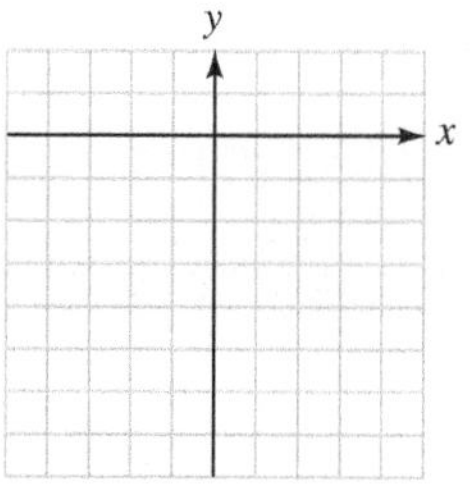

Use a graphing calculator to graph each equation.

61. $3x^2 + 3y^2 = 16$

62. $2x^2 + 2y^2 = 9$

63. $(x + 1)^2 + y^2 = 16$

64. $x^2 + (y - 2)^2 = 4$

65. $x = 2y^2$

66. $x = y^2 - 4$

67. $x^2 - 2x + y = 6$

68. $x = -2(y - 1)^2 + 2$

Unless otherwise noted, all content on this page is © Cengage Learning.

APPLICATIONS *Solve each application.* SEE EXAMPLE 4. (OBJECTIVE 4)

69. Meshing gears For design purposes, the large gear is the circle $x^2 + y^2 = 16$. The smaller gear is a circle centered at $(7, 0)$ and tangent (touching at 1 point) to the larger circle. Find the equation of the smaller gear.

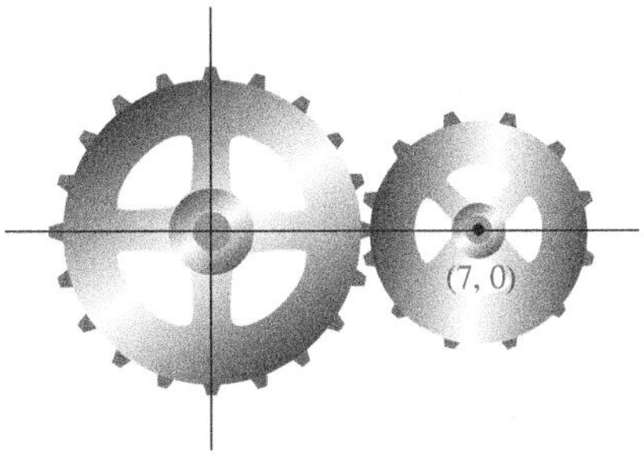

70. Width of a walkway The following walkway is bounded by the two circles $x^2 + y^2 = 2{,}500$ and $(x - 10)^2 + y^2 = 900$, measured in feet. Find the largest and the smallest width of the walkway.

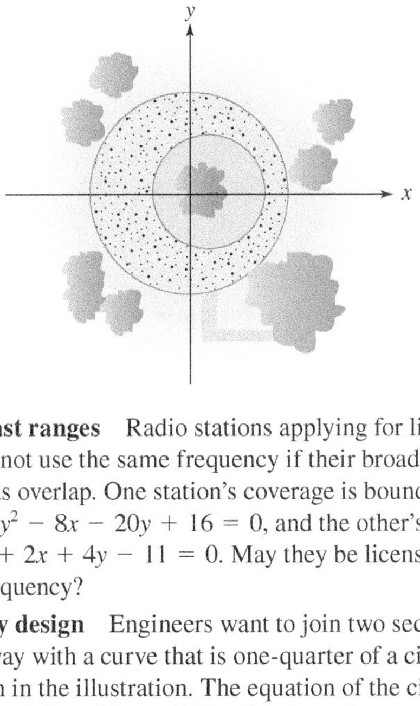

71. Broadcast ranges Radio stations applying for licensing may not use the same frequency if their broadcast areas overlap. One station's coverage is bounded by $x^2 + y^2 - 8x - 20y + 16 = 0$, and the other's by $x^2 + y^2 + 2x + 4y - 11 = 0$. May they be licensed for the same frequency?

72. Highway design Engineers want to join two sections of highway with a curve that is one-quarter of a circle as shown in the illustration. The equation of the circle is $x^2 + y^2 - 16x - 20y + 155 = 0$, where distances are measured in kilometers. Find the locations (relative to the center of town) of the intersections of the highway with State and with Main.

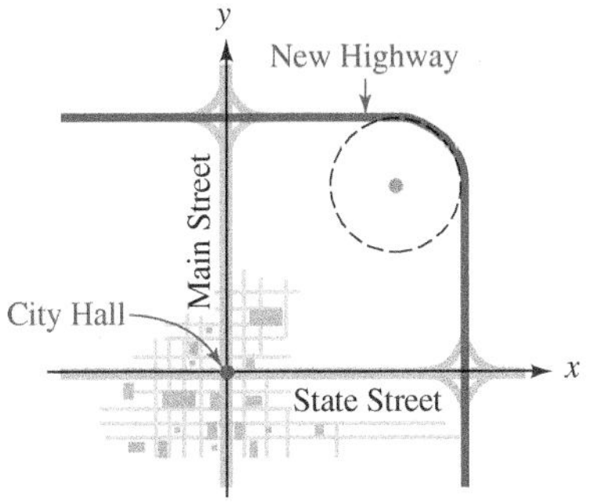

Solve each application. SEE EXAMPLE 7. (OBJECTIVE 6)

73. Projectiles The cannonball in the illustration follows the parabolic trajectory $y = 30x - x^2$. Where does it land?

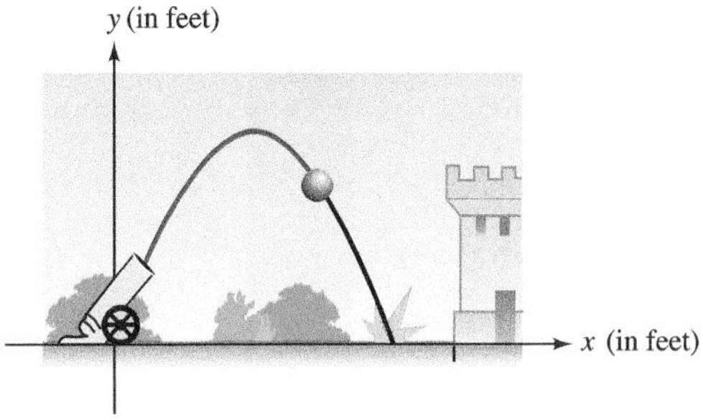

74. Projectiles In Exercise 73, how high does the cannonball rise?

75. Path of a comet If the path of a comet is given by the equation $2y^2 - 9x = 18$, how far is it from the Sun at the vertex of the orbit? Distances are measured in astronomical units (AU).

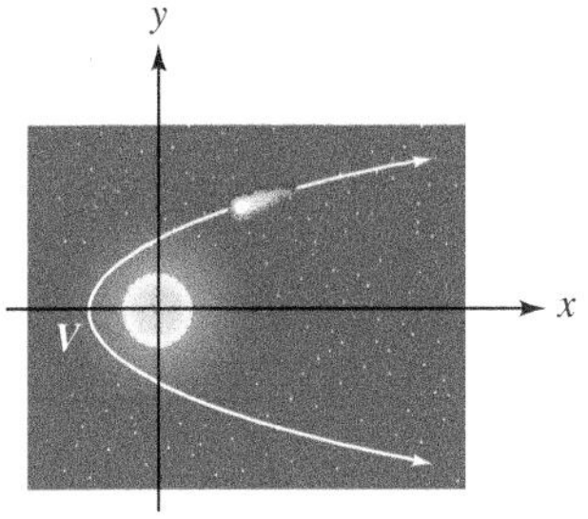

76. Satellite antennas The cross section of the satellite antenna is a parabola given by the equation $y = \frac{1}{16}x^2$, with distances measured in feet. If the dish is 8 feet wide, how deep is it?

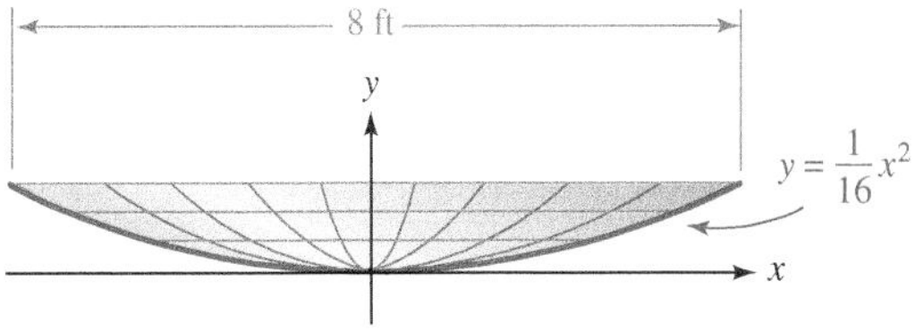

WRITING ABOUT MATH

77. Explain how to decide from its equation whether the graph of a parabola opens upward, downward, right, or left.

78. From the equation of a circle, explain how to determine the radius and the coordinates of the center.

SOMETHING TO THINK ABOUT

79. From the values of a, h, and k, explain how to determine the number of x-intercepts of the graph of $y = a(x - h)^2 + k$.

80. Under what conditions will the graph of $x = a(y - k)^2 + h$ have no y-intercepts?

Unless otherwise noted, all content on this page is © Cengage Learning.

Section 12.2 The Ellipse

Objectives

1. Graph an ellipse given an equation in standard form.
2. Graph an ellipse given an equation in general form.
3. Solve an application involving an ellipse.

Vocabulary

ellipse	foci	major axis
focus	vertices	minor axis

Getting Ready

Solve each equation for the indicated variable $(a \neq 0, b \neq 0)$.

1. $\frac{y^2}{b^2} = 1$ for y **2.** $\frac{x^2}{a^2} = 1$ for x

A third conic section is an oval-shaped curve called an *ellipse*. Ellipses can be nearly round or they can be long and narrow. In this section, we will learn how to construct ellipses and how to graph equations that represent them.

Ellipses have optical and acoustical properties that are useful in architecture and engineering. For example, many arches are portions of an ellipse, because the shape is pleasing to the eye. (See Figure 12-13(a).) The planets and many comets have elliptical orbits. (See Figure 12-13(b).) Gears are often cut into elliptical shapes to provide nonuniform motion. (See Figure 12-13(c).)

Ellipses

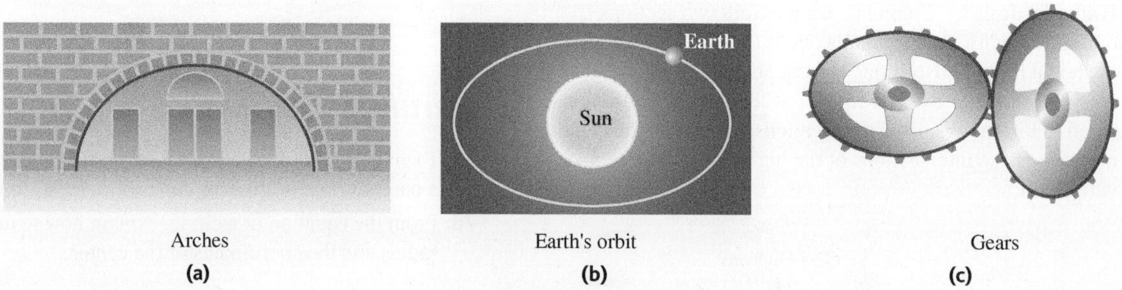

Figure 12-13

Unless otherwise noted, all content on this page is © Cengage Learning.

1 Graph an ellipse given an equation in standard form.

THE ELLIPSE

An **ellipse** is the set of all points P in the plane the sum of whose distances from two fixed points is a constant. See Figure 12-14, in which $d_1 + d_2$ is a constant.

Each of the two fixed points is called a **focus**. Midway between the **foci** is the *center* of the ellipse.

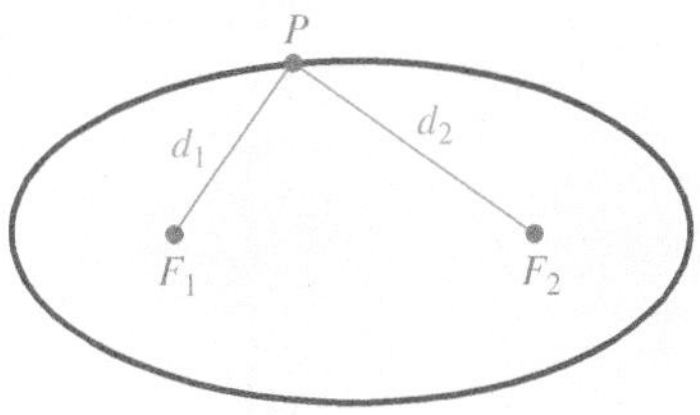

Figure 12-14

We can construct an ellipse by placing two thumbtacks fairly close together, as in Figure 12-15. We then tie each end of a piece of string to a thumbtack, catch the loop with the point of a pencil, and, while keeping the string taut, draw the ellipse.

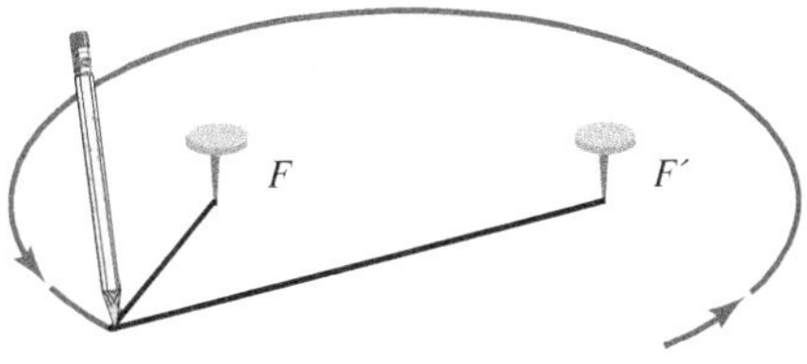

Figure 12-15

Sir Isaac Newton

1642–1727

Newton was an English scientist and mathematician. Because he was not a good farmer, he went to Cambridge University to become a preacher. When he had to leave Cambridge because of the plague, he made some of his most important discoveries. He is best known in mathematics for developing calculus and in physics for discovering the laws of motion. Newton probably contributed more to science and mathematics than anyone else in history.

Using this method, we can construct an ellipse of any specific size. For example, to construct an ellipse that is 10 inches wide and 6 inches high, we must find the length of string to use and the distance between thumbtacks.

To do this, we will let a represent the distance between the center and vertex V, as shown in Figure 12-16(a). We will also let c represent the distance between the center of the ellipse and either focus. When the pencil is at vertex V, the length of the string is $c + a + (a - c)$, or just $2a$. Because $2a$ is the 10-inch width of the ellipse, the string needs to be 10 inches long. The distance $2a$ is constant for any point on the ellipse, including point B shown in Figure 12-16(b).

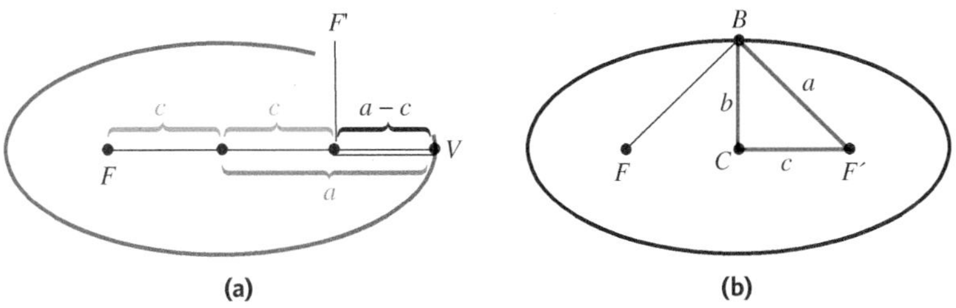

Figure 12-16

From right triangle BCF' in Figure 12-16(b) and the Pythagorean theorem, we can find the value of c as follows:

$$a^2 = b^2 + c^2 \qquad \text{or} \qquad c = \sqrt{a^2 - b^2}$$

Unless otherwise noted, all content on this page is © Cengage Learning.

Since distance b is one-half of the height of the ellipse, $b = 3$. Since $2a = 10$, $a = 5$. We can now substitute $a = 5$ and $b = 3$ into the formula to find the value of c:

$$\begin{aligned} c &= \sqrt{5^2 - 3^2} \\ &= \sqrt{25 - 9} \\ &= \sqrt{16} \\ &= 4 \end{aligned}$$

Since $c = 4$, the distance between the thumbtacks must be 8 inches. We can construct the ellipse by tying a 10-inch string to thumbtacks that are 8 inches apart.

To graph ellipses, we can create a table of ordered pairs that satisfy the equation, plot them, and join the points with a smooth curve.

EXAMPLE 1 Graph: $\frac{x^2}{36} + \frac{y^2}{9} = 1$

Solution We note that the equation can be written in the form

$$\frac{x^2}{6^2} + \frac{y^2}{3^2} = 1 \quad 36 = 6^2 \text{ and } 9 = 3^2$$

After creating a table of ordered pairs that satisfy the equation, plotting each of them, and joining the points with a curve, we obtain the ellipse shown in Figure 12-17.

We note that the center of the ellipse is the origin, the ellipse intersects the x-axis at points $(6, 0)$ and $(-6, 0)$, and the ellipse intersects the y-axis at points $(0, 3)$ and $(0, -3)$.

$\frac{x^2}{36} + \frac{y^2}{9} = 1$

x	y	(x, y)
-6	0	$(-6, 0)$
-4	± 2.2	$(-4, \pm 2.2)$
-2	± 2.8	$(-2, \pm 2.8)$
0	± 3	$(0, \pm 3)$
2	± 2.8	$(2, \pm 2.8)$
4	± 2.2	$(4, \pm 2.2)$
6	0	$(6, 0)$

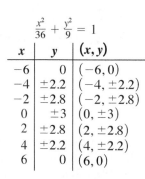

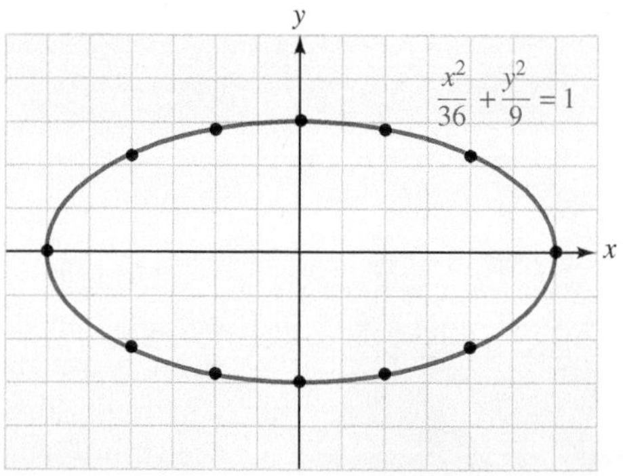

Figure 12-17

SELF CHECK 1 Graph: $\frac{x^2}{4} + \frac{y^2}{16} = 1$

Example 1 illustrates that the graph of

$$\frac{x^2}{a^2} + \frac{y^2}{b^2} = 1$$

is an ellipse centered at the origin. To find the x-intercepts of the graph, we can let $y = 0$ and solve for x.

$$\begin{aligned} \frac{x^2}{a^2} + \frac{0^2}{b^2} &= 1 \\ \frac{x^2}{a^2} + 0 &= 1 \\ x^2 &= a^2 \\ x = a \quad \text{or} \quad x &= -a \end{aligned}$$

The x-intercepts are $(a, 0)$ and $(-a, 0)$.

Unless otherwise noted, all content on this page is © Cengage Learning.

To find the y-intercepts, we let $x = 0$ and solve for y.

$$\frac{0^2}{a^2} + \frac{y^2}{b^2} = 1$$

$$0 + \frac{y^2}{b^2} = 1$$

$$y^2 = b^2$$

$$y = b \quad \text{or} \quad y = -b$$

The y-intercepts are $(0, b)$ and $(0, -b)$.

In general, we have the following results.

EQUATIONS OF AN ELLIPSE CENTERED AT THE ORIGIN

The equation of an ellipse centered at the origin, with x-intercepts at $V_1(a, 0)$ and $V_2(-a, 0)$ and with y-intercepts of $(0, b)$ and $(0, -b)$, is

$$\frac{x^2}{a^2} + \frac{y^2}{b^2} = 1 \quad (a > b > 0)$$ See Figure 12-18(a).

The equation of an ellipse centered at the origin, with y-intercepts at $V_1(0, a)$ and $V_2(0, -a)$ and x-intercepts of $(b, 0)$ and $(-b, 0)$, is

$$\frac{x^2}{b^2} + \frac{y^2}{a^2} = 1 \quad (a > b > 0)$$ See Figure 12-18(b).

In Figure 12-18, the points V_1 and V_2 are the **vertices** of the ellipse, the midpoint of segment V_1V_2 is the *center* of the ellipse, and the distance between the center and either vertex is a. The segment V_1V_2 is called the **major axis**, and the segment joining either $(0, b)$ and $(0, -b)$ or $(b, 0)$ and $(-b, 0)$ is called the **minor axis**.

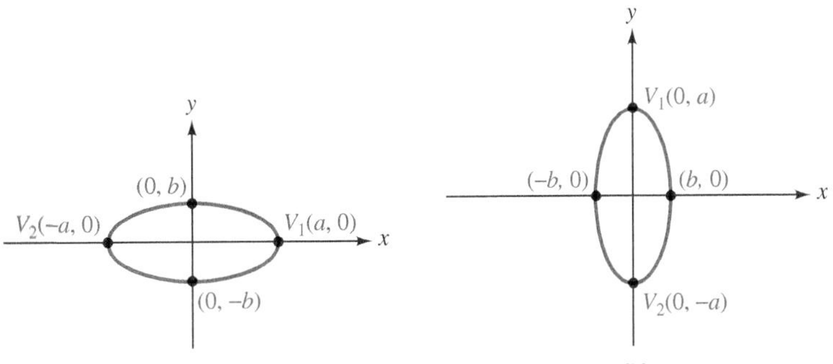

Figure 12-18

The equations for ellipses centered at (h, k) are as follows.

STANDARD EQUATION OF A HORIZONTAL ELLIPSE CENTERED AT (h, k)

The equation of a horizontal ellipse centered at (h, k), with major axis parallel to the x-axis, is

$$\frac{(x-h)^2}{a^2} + \frac{(y-k)^2}{b^2} = 1 \quad (a > b > 0)$$

STANDARD EQUATION OF A VERTICAL ELLIPSE CENTERED AT (h, k)

The equation of a vertical ellipse centered at (h, k), with major axis parallel to the y-axis, is

$$\frac{(x-h)^2}{b^2} + \frac{(y-k)^2}{a^2} = 1 \quad (a > b > 0)$$

Unless otherwise noted, all content on this page is © Cengage Learning.

COMMENT To determine whether an ellipse is horizontal or vertical, look at the denominators in its standard equation. If the largest denominator is associated with the x-term, the ellipse will be horizontal. If the largest denominator is associated with the y-term, the ellipse will be vertical.

EXAMPLE 2 Graph: $25(x-2)^2 + 16(y+3)^2 = 400$

Solution We first write the equation in standard form.

$$25(x-2)^2 + 16(y+3)^2 = 400$$

$$\frac{25(x-2)^2}{400} + \frac{16(y+3)^2}{400} = \frac{400}{400} \quad \text{Divide both sides by 400.}$$

$$\frac{(x-2)^2}{16} + \frac{(y+3)^2}{25} = 1 \quad \text{Simplify each fraction.}$$

$$\frac{(x-2)^2}{4^2} + \frac{[y-(-3)]^2}{5^2} = 1$$

Because $5 > 4$ this is the equation of a vertical ellipse centered at $(h, k) = (2, -3)$ with major axis parallel to the y-axis and with $b = 4$ and $a = 5$. We first plot the center, as shown in Figure 12-19. Since a is the distance from the center to a vertex, we can locate the vertices by counting 5 units above and 5 units below the center. The vertices are at points $(2, 2)$ and $(2, -8)$.

Since $b = 4$, we can locate two more points on the ellipse by counting 4 units to the left and 4 units to the right of the center. The points $(-2, -3)$ and $(6, -3)$ are also on the graph.

Using these four points as guides, we can draw the ellipse.

$$\frac{(x-2)^2}{16} + \frac{(y+3)^2}{25} = 1$$

x	y	(x, y)
2	2	$(2, 2)$
2	−8	$(2, -8)$
6	−3	$(6, -3)$
−2	−3	$(-2, -3)$

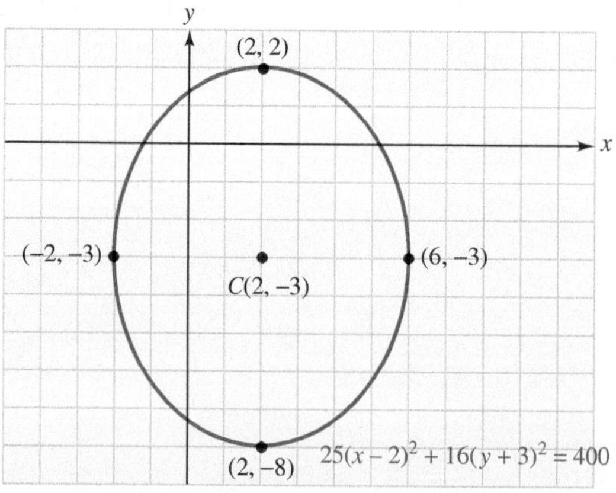

Figure 12-19

SELF CHECK 2 Graph: $16(x-1)^2 + 9(y+2)^2 = 144$

Accent on technology

▸ Graphing Ellipses

To use a graphing calculator to graph

$$\frac{(x+2)^2}{4} + \frac{(y-1)^2}{25} = 1$$

we first clear the fractions by multiplying both sides by 100 and solving for y.

$$25(x+2)^2 + 4(y-1)^2 = 100 \quad \text{Multiply both sides by 100.}$$

$$4(y-1)^2 = 100 - 25(x+2)^2 \quad \text{Subtract } 25(x+2)^2 \text{ from both sides.}$$

Unless otherwise noted, all content on this page is © Cengage Learning.

$$(y-1)^2 = \frac{100 - 25(x+2)^2}{4} \qquad \text{Divide both sides by 4.}$$

$$y - 1 = \pm\frac{\sqrt{100 - 25(x+2)^2}}{2} \qquad \text{Use the square-root property.}$$

$$y = 1 \pm \frac{\sqrt{100 - 25(x+2)^2}}{2} \qquad \text{Add 1 to both sides.}$$

If we use window settings $[-6, 6]$ for x and $[-6, 6]$ for y and graph the functions

$$y = 1 + \frac{\sqrt{100 - 25(x+2)^2}}{2} \quad \text{and} \quad y = 1 - \frac{\sqrt{100 - 25(x+2)^2}}{2}$$

On a TI-84 calculator, we will obtain the ellipse shown in Figure 12-20.

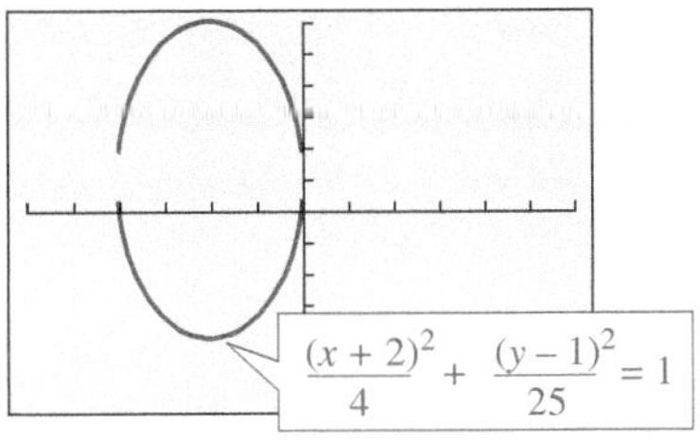

Figure 12-20

For instructions regarding the use of a Casio graphing calculator, please refer to the Casio Keystroke Guide in the back of the book.

2 Graph an ellipse given an equation in general form.

Another important form of the equation of an ellipse is called the *general form.*

GENERAL FORM OF THE EQUATION OF AN ELLIPSE

The equation of any ellipse can be written in the form

$$Ax^2 + Cy^2 + Dx + Ey + F = 0$$

COMMENT Notice that there is no B coefficient in the general form. It is reserved for an xy term that would rotate the ellipse off the x- and y-axis. These will not be considered in this course.

We can use completing the square to write the general form of the equation of an ellipse.

EXAMPLE 3 Write $4x^2 + 9y^2 - 16x - 18y - 11 = 0$ in standard form to show that the equation represents an ellipse. Then graph the equation.

Solution We write the equation in standard form by completing the square on x and y.

$$4x^2 + 9y^2 - 16x - 18y - 11 = 0$$

$$4x^2 + 9y^2 - 16x - 18y = 11 \qquad \text{Add 11 to both sides.}$$

$$4x^2 - 16x + 9y^2 - 18y = 11 \qquad \text{Use the commutative property to rearrange terms.}$$

Unless otherwise noted, all content on this page is © Cengage Learning.

$$4(x^2 - 4x) + 9(y^2 - 2y) = 11$$ Factor 4 from $4x^2 - 16x$ and factor 9 from $9y^2 - 18y$ to obtain coefficients of 1 for the squared terms.

$$4(x^2 - 4x + \mathbf{4}) + 9(y^2 - 2y + \mathbf{1}) = 11 + \mathbf{16} + \mathbf{9}$$ Complete the square to make $x^2 - 4x$ and $y^2 - 2y$ perfect trinomial squares. Since 16 and 9 are added to the left side, add 16 and 9 to the right side.

$$4(x - 2)^2 + 9(y - 1)^2 = 36$$ Factor $x^2 - 4x + 4$ and $y^2 - 2y + 1$.

$$\frac{(x - 2)^2}{9} + \frac{(y - 1)^2}{4} = 1$$ Divide both sides by 36.

Since this equation is of the form $\frac{(x-h)^2}{a^2} + \frac{(y-k)^2}{b^2} = 1$ $(a > b > 0)$, it represents an ellipse with $h = 2$, $k = 1$, $a = 3$, and $b = 2$. Its graph is shown in Figure 12-21.

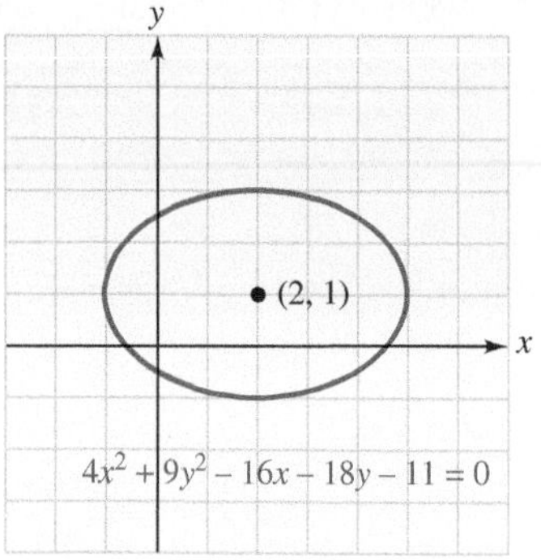

Figure 12-21

SELF CHECK 3 Graph: $4x^2 - 8x + 9y^2 - 36y = -4$

COMMENT To distinguish between an equation of an ellipse and an equation of a circle, look at the coefficients of the squared terms. If the coefficients are the same, the equation is the equation of a circle. If the coefficients are different, but both positive, the equation is the equation of an ellipse.

3 Solve an application involving an ellipse.

EXAMPLE 4 **LANDSCAPE DESIGN** A landscape architect is designing an elliptical pool that will fit in the center of a 20-by-30-foot rectangular garden, leaving at least 5 feet of space on all sides. Find the equation of the ellipse.

Solution We place the rectangular garden in a coordinate system, as in Figure 12-22. To maintain 5 feet of clearance at the ends of the ellipse, the vertices must be the points $V_1(10, 0)$ and $V_2(-10, 0)$. Similarly, the y-intercepts are the points $(0, 5)$ and $(0, -5)$.

The equation of the ellipse has the form

$$\frac{x^2}{a^2} + \frac{y^2}{b^2} = 1$$

with $a = 10$ and $b = 5$. Thus, the equation of the boundary of the pool is

$$\frac{x^2}{100} + \frac{y^2}{25} = 1$$

Unless otherwise noted, all content on this page is © Cengage Learning.

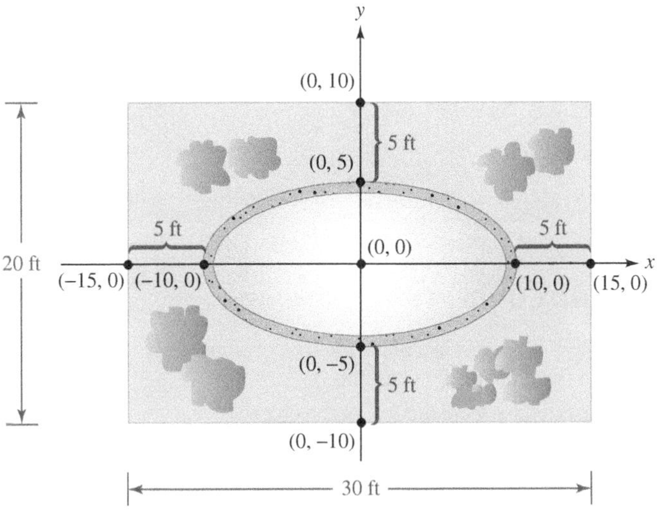

Figure 12-22

SELF CHECK 4 If the orientation of the garden is changed to 30 feet by 20 feet, find the equation of the pool.

SELF CHECK ANSWERS

1.

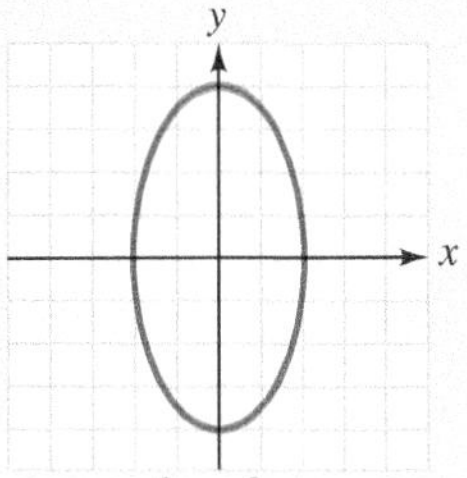

$\frac{x^2}{4} + \frac{y^2}{16} = 1$

2.

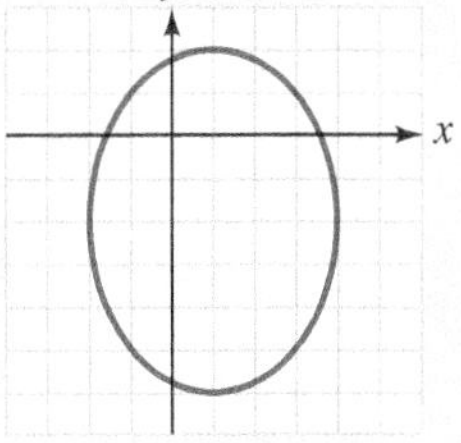

$\frac{(x-1)^2}{9} + \frac{(y+2)^2}{16} = 1$

3.

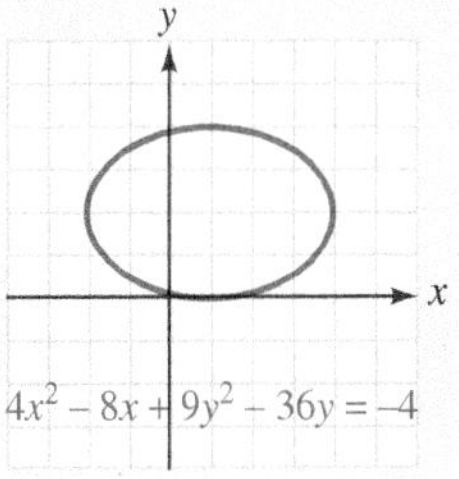

4. $\frac{x^2}{25} + \frac{y^2}{100} = 1$

NOW TRY THIS

1. If the vertices of an ellipse are $(5, -1)$ and $(5, 5)$ and the endpoints of the minor axis are $(3, 2)$ and $(7, 2)$, find the equation.
2. The eccentricity of an ellipse is $\frac{c}{a}$. If the center is located at $(4, -1)$ and the eccentricity is $\frac{2}{3}$, $a = 3$, find the equation of the horizontal ellipse.

12.2 Exercises

WARM-UPS *Evaluate when $x = 0$. Write as ordered pairs.*

1. $\frac{x^2}{4} + \frac{y^2}{9} = 1$

2. $\frac{x^2}{36} + \frac{y^2}{25} = 1$

Evaluate when $y = 0$. Write as ordered pairs.

3. $\frac{x^2}{4} + \frac{y^2}{9} = 1$

4. $\frac{x^2}{36} + \frac{y^2}{25} = 1$

Unless otherwise noted, all content on this page is © Cengage Learning.

REVIEW *Find each product.*

5. $5x^{-3}y^4(7x^3 + 2y^{-4})$

6. $(3x^{-3} - y^{-3})(3x^{-3} + y^{-3})$

Write each expression without using negative exponents.

7. $\dfrac{x^{-2} + y^{-2}}{x^{-2} - y^{-2}}$ **8.** $\dfrac{2x^{-3} - 2y^{-3}}{4x^{-3} + 4y^{-3}}$

VOCABULARY AND CONCEPTS *Fill in the blanks.*

9. An _____ is the set of all points in a plane the ___ of whose distances from two fixed points is a constant.

10. The fixed points in Exercise 9 are the ___ of the ellipse.

11. The midpoint of the line segment joining the foci of an ellipse is called the _____ of the ellipse.

12. The graph of $\frac{x^2}{a^2} + \frac{y^2}{b^2} = 1$ $(a > b > 0)$ has vertices at ______, y-intercepts at _______, and eccentricity of _.

13. The center of the ellipse with an equation of $\frac{x^2}{a^2} + \frac{y^2}{b^2} = 1$ is ____ with the ________ having a length of $2a$ and the minor axis having a length of __.

14. The center of the ellipse with an equation of $\frac{(x-h)^2}{a^2} + \frac{(y-k)^2}{b^2} = 1$ is the point ____ .

GUIDED PRACTICE *Graph each equation.* SEE EXAMPLE 1. *(OBJECTIVE 1)*

15. $\dfrac{x^2}{4} + \dfrac{y^2}{9} = 1$

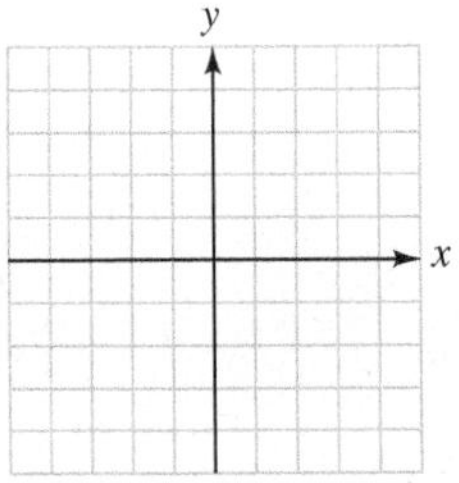

16. $x^2 + \dfrac{y^2}{9} = 1$

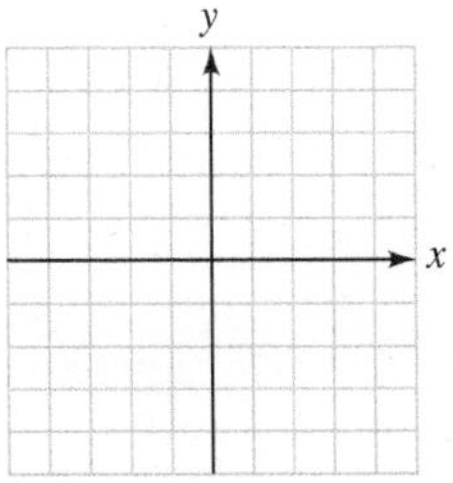

17. $\dfrac{(x-2)^2}{9} + \dfrac{(y-1)^2}{4} = 1$

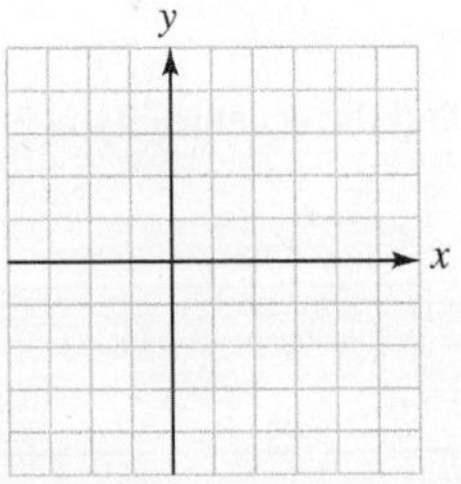

18. $\dfrac{(x-1)^2}{9} + \dfrac{(y-3)^2}{4} = 1$

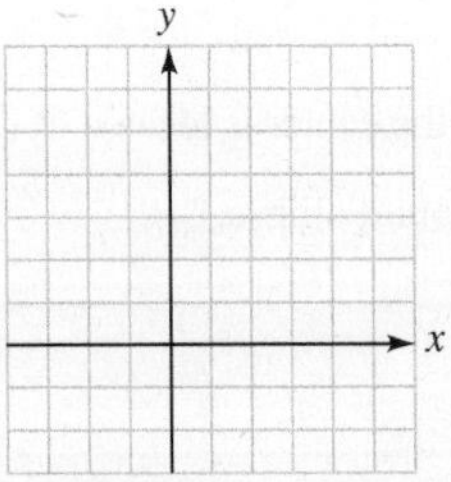

Graph each equation. SEE EXAMPLE 2. *(OBJECTIVE 1)*

19. $x^2 + 9y^2 = 9$

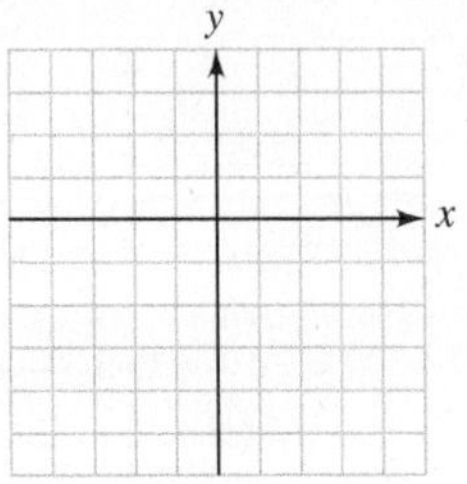

20. $25x^2 + 9y^2 = 225$

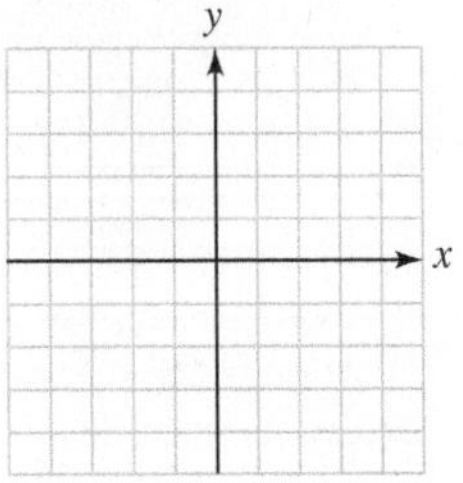

21. $(x+1)^2 + 4(y+2)^2 = 4$

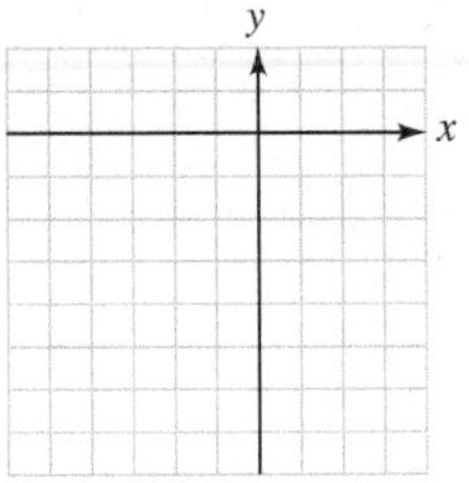

22. $9(x-5)^2 + (y+2)^2 = 9$

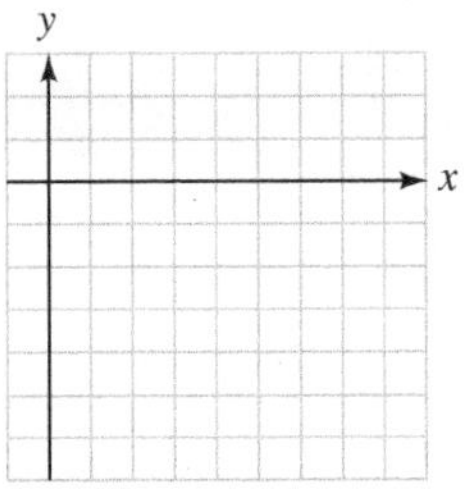

Write each equation in standard form and graph it. SEE EXAMPLE 3. *(OBJECTIVE 2)*

23. $x^2 + 4y^2 - 4x + 8y + 4 = 0$

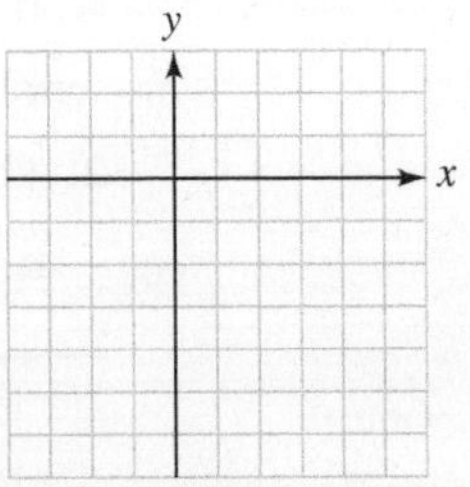

24. $x^2 + 4y^2 - 2x - 16y = -13$

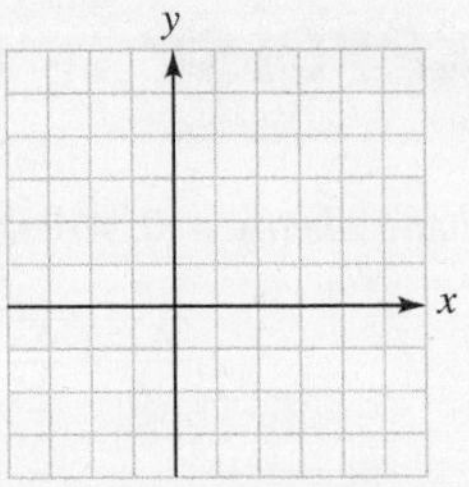

Unless otherwise noted, all content on this page is © Cengage Learning.

25. $9x^2 + 4y^2 - 18x + 16y = 11$

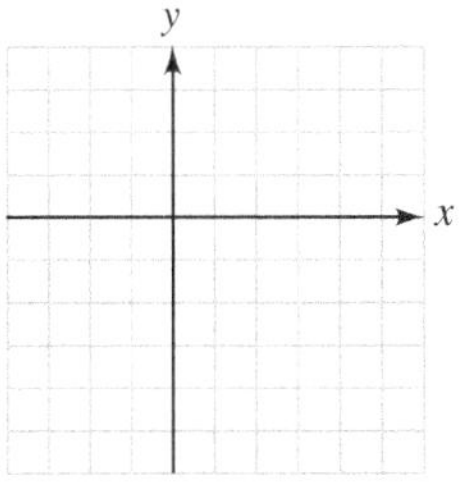

26. $16x^2 + 25y^2 - 160x - 200y + 400 = 0$

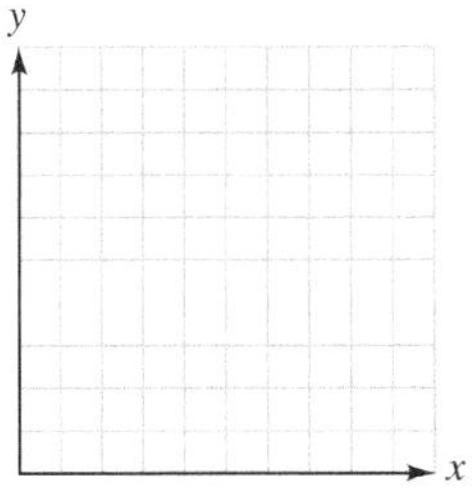

ADDITIONAL PRACTICE *Graph each equation.*

27. $\dfrac{x^2}{9} + \dfrac{y^2}{16} = 1$

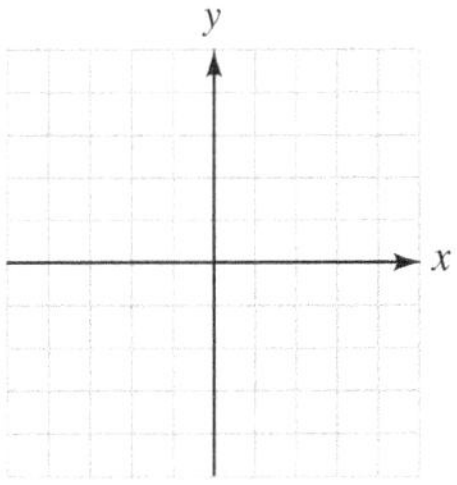

28. $\dfrac{x^2}{25} + \dfrac{y^2}{36} = 1$

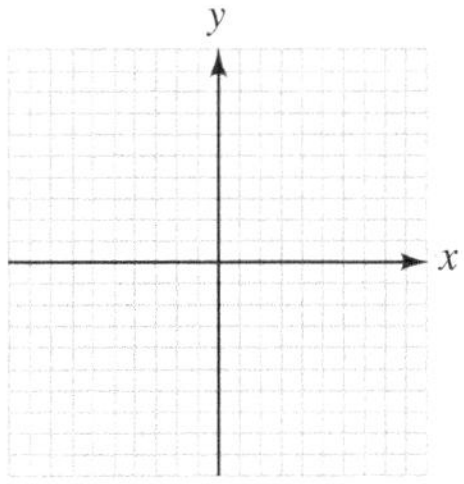

29. $\dfrac{(x-2)^2}{16} + \dfrac{y^2}{25} = 1$

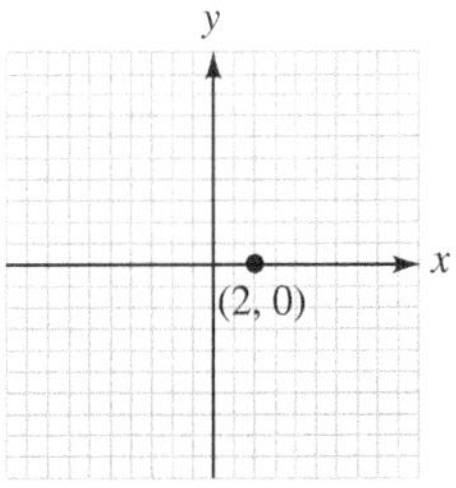

30. $\dfrac{x^2}{25} + \dfrac{(y+1)^2}{36} = 1$

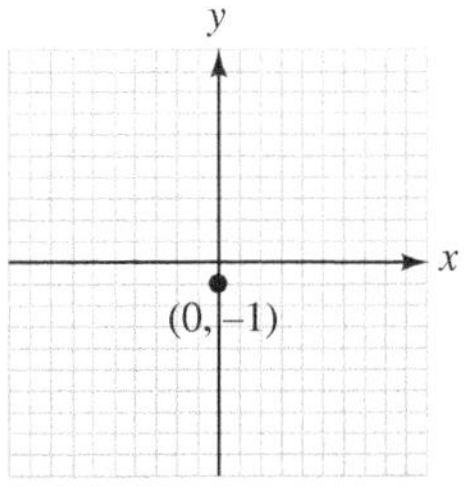

31. $16x^2 + 4y^2 = 64$

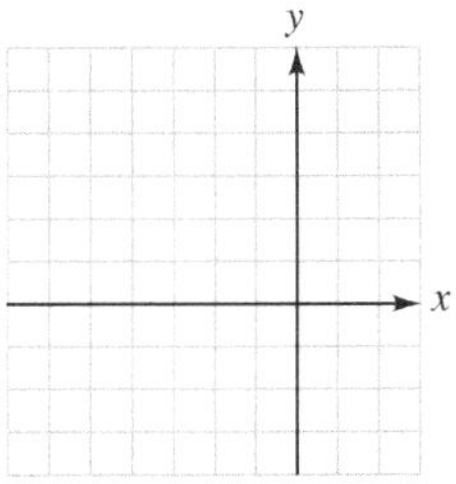

32. $4x^2 + 9y^2 = 36$

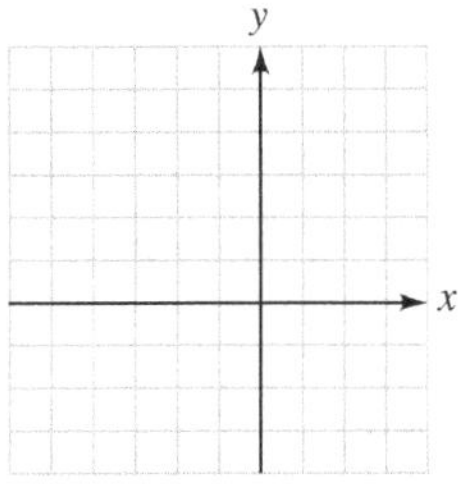

33. $25(x+1)^2 + 9y^2 = 225$

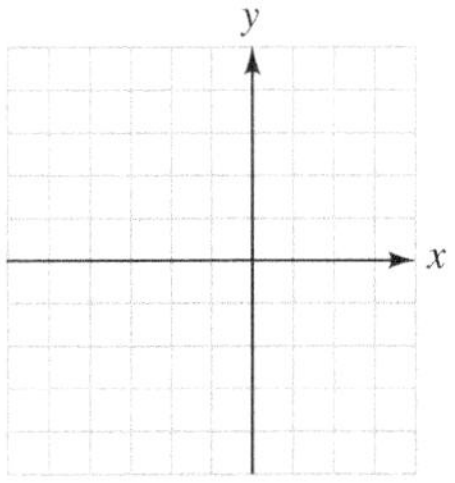

34. $4(x-6)^2 + 25(y-3)^2 = 100$

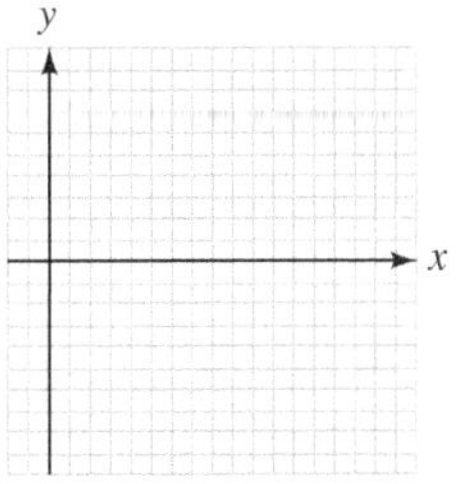

Use a graphing calculator to graph each equation.

35. $\dfrac{x^2}{9} + \dfrac{y^2}{4} = 1$

36. $x^2 + 16y^2 = 16$

37. $\dfrac{x^2}{4} + \dfrac{(y-1)^2}{9} = 1$

38. $\dfrac{(x+1)^2}{9} + \dfrac{(y-2)^2}{4} = 1$

APPLICATIONS *Solve each application.* SEE EXAMPLE 4. *(OBJECTIVE 3)*

39. Fitness equipment With elliptical cross-training equipment, the feet move through the natural elliptical pattern that one experiences when walking, jogging, or running. Write the equation of the elliptical pattern shown below.

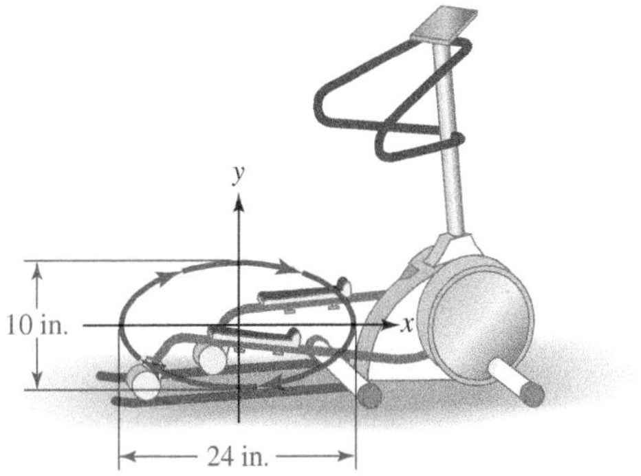

Unless otherwise noted, all content on this page is © Cengage Learning.

40. Pool tables Find the equation of the outer edge of the elliptical pool table shown below. Assume the red ball is at the focus.

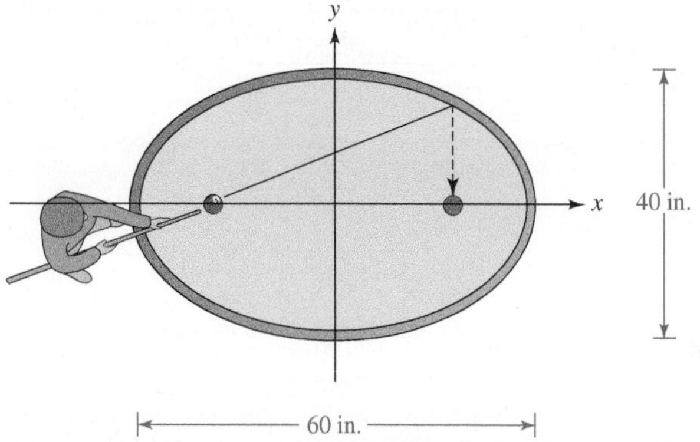

41. Designing an underpass The arch of the underpass is half of an ellipse. Find the equation of the arch.

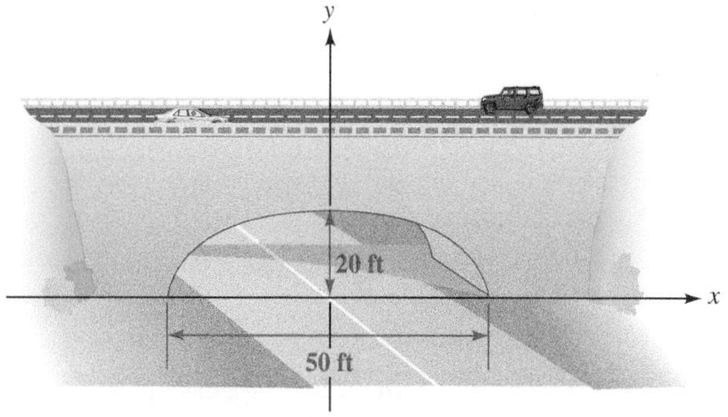

42. Calculating clearance Find the height of the elliptical arch in Exercise 41 at a point 10 feet from the center of the roadway.

43. Area of an ellipse The area A of the ellipse

$$\frac{x^2}{a^2} + \frac{y^2}{b^2} = 1$$

is given by $A = \pi ab$. Find the area of the ellipse $25x^2 + 4y^2 = 100$.

44. Area of a track The elliptical track is bounded by the ellipses $4x^2 + 9y^2 = 576$ and $9x^2 + 25y^2 = 900$. Find the area of the track. (See Exercise 43.)

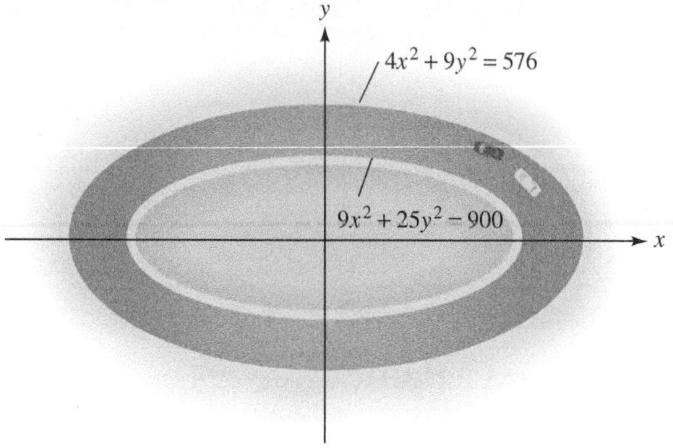

WRITING ABOUT MATH

45. Explain how to find the x- and the y-intercepts of the graph of the ellipse

$$\frac{x^2}{a^2} + \frac{y^2}{b^2} = 1$$

46. Explain the relationship between the center, focus, and vertex of an ellipse.

SOMETHING TO THINK ABOUT

47. What happens to the graph of $\frac{x^2}{a^2} + \frac{y^2}{b^2} = 1$ when $a = b$?

48. Explain why the graph of $x^2 + 4x + y^2 - 8y + 30 = 0$ does not exist.

Section 12.3 The Hyperbola

Objectives

1. Graph a hyperbola given an equation in standard form.
2. Graph a hyperbola given an equation in general form.
3. Graph a hyperbola of the form $xy = k$.
4. Solve an application involving a hyperbola.

Unless otherwise noted, all content on this page is © Cengage Learning.

Vocabulary

hyperbola | fundamental rectangle | asymptotes

Getting Ready

Find the value of y when $\frac{x^2}{25} - \frac{y^2}{9} = 1$ *and x is the given value. Give each result to the nearest tenth.*

1. $x = 6$ **2.** $x = -7$

The last conic section, the *hyperbola*, is a curve with two branches. In this section, we will graph equations that represent hyperbolas.

Hyperbolas are the basis of a navigational system known as LORAN (LOng RAnge Navigation). (See Figure 12-23.) They are also used to find the source of a distress signal, are the basis for the design of hypoid gears, and describe the paths of some comets.

Hyperbola

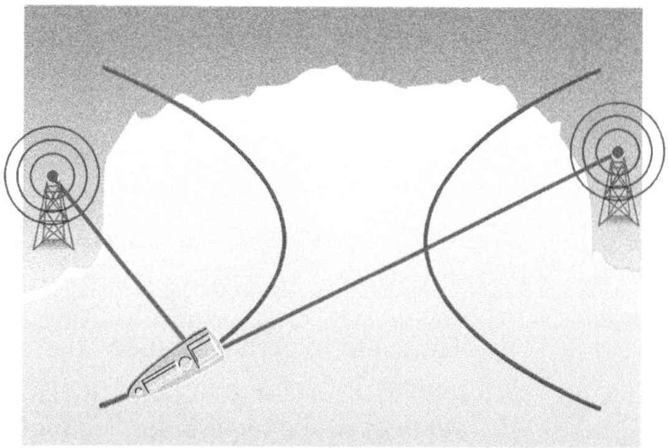

Figure 12-23

1 Graph a hyperbola given an equation in standard form.

THE HYPERBOLA

A **hyperbola** is the set of all points P in the plane for which the difference of the distances of each point from two fixed points, the foci, is a constant.

See Figure 12-24, in which $d_1 - d_2$ is a constant.

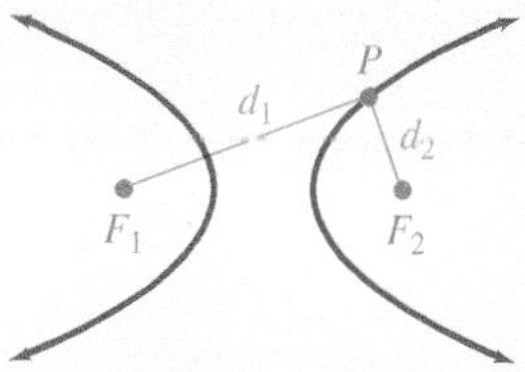

Figure 12-24

Unless otherwise noted, all content on this page is © Cengage Learning.

The graph of the standard form of the equation

$$\frac{x^2}{25} - \frac{y^2}{9} = 1$$

is a hyperbola. To graph the equation, we create a table of ordered pairs that satisfy the equation, plot each pair, and join the points with a smooth curve as in Figure 12-25.

$\frac{x^2}{25} - \frac{y^2}{9} = 1$

x	y	(x, y)
-7	± 2.9	$(-7, \pm 2.9)$
-6	± 2.0	$(-6, \pm 2.0)$
-5	0	$(-5, 0)$
5	0	$(5, 0)$
6	± 2.0	$(6, \pm 2.0)$
7	± 2.9	$(7, \pm 2.9)$

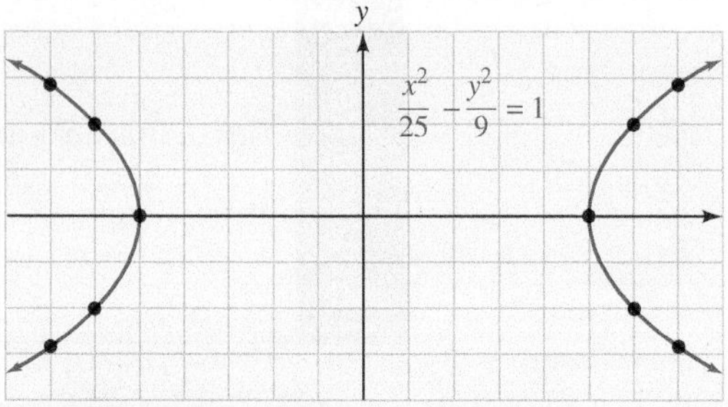

Figure 12-25

This graph is centered at the origin and intersects the x-axis at $(5, 0)$ and $(-5, 0)$. We also note that the graph does not intersect the y-axis.

It is possible to draw a hyperbola by plotting only 4 points. For example, if we want to graph the hyperbola with an equation of

$$\frac{x^2}{a^2} - \frac{y^2}{b^2} = 1$$

we first look at the x- and y-intercepts. To find the x-intercepts, we let $y = 0$ and solve for x:

$$\frac{x^2}{a^2} - \frac{0^2}{b^2} = 1$$

$$x^2 = a^2$$

$$x = \pm a$$

Thus, the hyperbola crosses the x-axis at the points $V_1(a, 0)$ and $V_2(-a, 0)$, called the **vertices** of the hyperbola. See Figure 12-26.

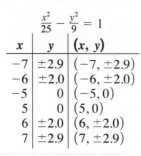

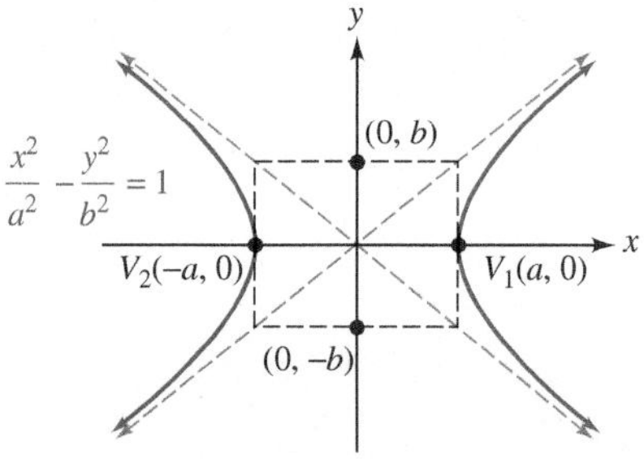

Figure 12-26

To find the y-intercepts, we let $x = 0$ and solve for y:

$$\frac{0^2}{a^2} - \frac{y^2}{b^2} = 1$$

$$y^2 = -b^2$$

$$y = \pm\sqrt{-b^2}$$

Since b^2 is always positive, $\sqrt{-b^2}$ is an imaginary number. This means that the hyperbola does not cross the y-axis.

We construct a rectangle, called the **fundamental rectangle**, whose sides pass horizontally through $\pm b$ on the y-axis and vertically through $\pm a$ on the x-axis. The extended diagonals of the rectangle will be **asymptotes** of the hyperbola, which can be useful when sketching the graph.

Unless otherwise noted, all content on this page is © Cengage Learning.

EQUATION OF A HYPERBOLA CENTERED AT THE ORIGIN WITH VERTICES ON THE x-AXIS

Any equation that can be written in the form

$$\frac{x^2}{a^2} - \frac{y^2}{b^2} = 1$$

has a graph that is a hyperbola centered at the origin. The x-intercepts are the vertices $V_1(a, 0)$ and $V_2(-a, 0)$. There are no y-intercepts.

The asymptotes of the hyperbola are the extended diagonals of the fundamental rectangle shown in Figure 12-27.

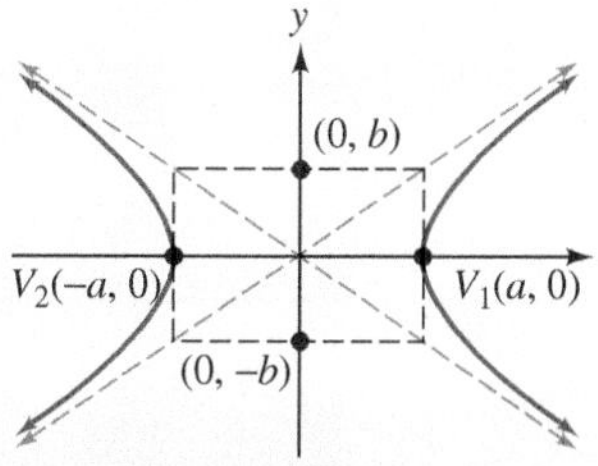

Figure 12-27

The branches of the hyperbola in previous discussions open to the left and to the right. It is possible for hyperbolas to have different orientations with respect to the x- and y-axes. For example, the branches of a hyperbola can open upward and downward. In that case, the following equation applies.

EQUATION OF A HYPERBOLA CENTERED AT THE ORIGIN WITH VERTICES ON THE y-AXIS

Any equation that can be written in the form

$$\frac{y^2}{a^2} - \frac{x^2}{b^2} = 1$$

has a graph that is a hyperbola centered at the origin, as in Figure 12-28. The y-intercepts are the vertices $V_1(0, a)$ and $V_2(0, -a)$. There are no x-intercepts.

The asymptotes of the hyperbola are the extended diagonals of the rectangle shown in Figure 12-28.

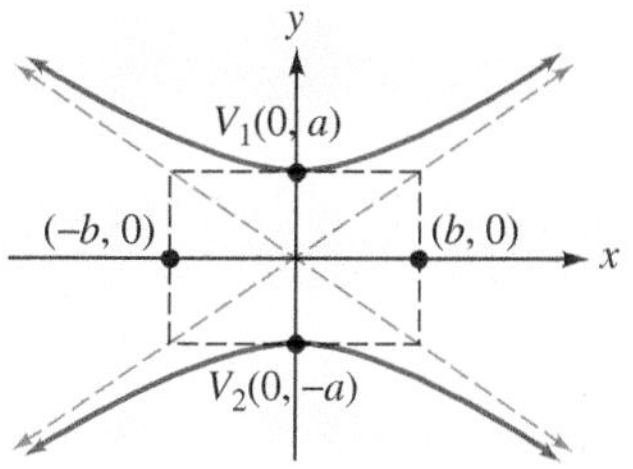

Figure 12-28

COMMENT To determine whether a hyperbola opens horizontally, as in Figure 12-27, or vertically, as in Figure 12-28, we look at the signs of the terms. If the term containing x^2 is positive, the hyperbola will open horizontally. If the term containing y^2 is positive, the hyperbola will open vertically.

EXAMPLE 1

Graph: $9y^2 - 4x^2 = 36$

Solution To write the equation in standard form, we divide both sides by 36 to obtain

$$\frac{9y^2}{36} - \frac{4x^2}{36} = 1$$

$$\frac{y^2}{4} - \frac{x^2}{9} = 1 \quad \text{Simplify each fraction.}$$

Because the term containing y^2 is positive, the hyperbola will open vertically. We can find the y-intercepts of the graph by letting $x = 0$ and solving for y:

$$\frac{y^2}{4} - \frac{0^2}{9} = 1$$

$$y^2 = 4$$

Thus, $y = \pm 2$, and the vertices of the hyperbola are $V_1(0, 2)$ and $V_2(0, -2)$. (See Figure 12-29.)

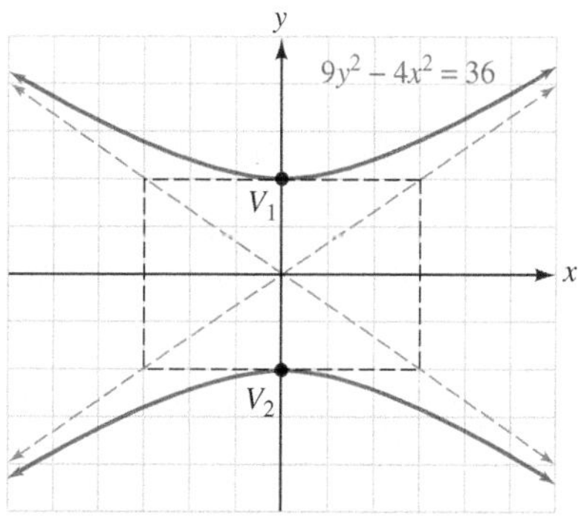

Figure 12-29

Unless otherwise noted, all content on this page is © Cengage Learning.

Since $\pm\sqrt{9} = \pm 3$, we can use the points $(3, 0)$ and $(-3, 0)$ on the x-axis to help draw the fundamental rectangle. We then draw its extended diagonals as asymptotes to help sketch the hyperbola.

SELF CHECK 1 Graph: $9x^2 - 4y^2 = 36$

Accent on technology

▸ Graphing Hyperbolas

To graph $\frac{x^2}{9} - \frac{y^2}{16} = 1$ using a graphing calculator, we follow the same procedure that we used for circles and ellipses. To write the equation as two functions, we solve for y to obtain $y = \pm\frac{\sqrt{16x^2 - 144}}{3}$. Then we graph the following two functions in a square window setting on a TI-84 calculator to obtain the graph of the hyperbola shown in Figure 12-30.

$$y = \frac{\sqrt{16x^2 - 144}}{3} \quad \text{and} \quad y = -\frac{\sqrt{16x^2 - 144}}{3}$$

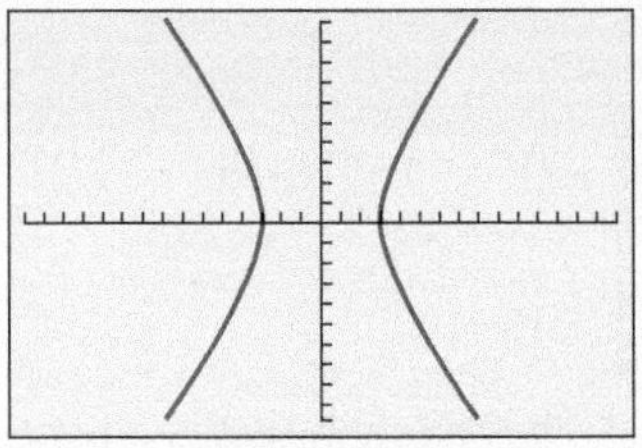

Figure 12-30

For instructions regarding the use of a Casio graphing calculator, please refer to the Casio Keystroke Guide in the back of the book.

If a hyperbola is centered at a point with coordinates (h, k), the following equations apply.

EQUATIONS OF HYPERBOLAS CENTERED AT (h, k)

Any equation that can be written in the form

$$\frac{(x-h)^2}{a^2} - \frac{(y-k)^2}{b^2} = 1$$

is a hyperbola centered at (h, k) that opens left and right (horizontally).

Any equation of the form

$$\frac{(y-k)^2}{a^2} - \frac{(x-h)^2}{b^2} = 1$$

is a hyperbola centered at (h, k) that opens upward and downward (vertically).

EXAMPLE 2 Graph: $\frac{(x-3)^2}{16} - \frac{(y+1)^2}{4} = 1$

Solution We write the equation in the form

$$\frac{(x-3)^2}{4^2} - \frac{[y-(-1)]^2}{2^2} = 1$$

to see that its graph will be a hyperbola centered at the point $(h, k) = (3, -1)$. Its vertices are located at $a = 4$ units to the right and left of center, at $(7, -1)$ and $(-1, -1)$. Since $b = 2$, we can count 2 units above and below center to locate points $(3, 1)$ and $(3, -3)$. With these points, we can draw the fundamental rectangle along

Unless otherwise noted, all content on this page is © Cengage Learning.

with its extended diagonals to locate the asymptotes. We can then sketch the hyperbola, as shown in Figure 12-31.

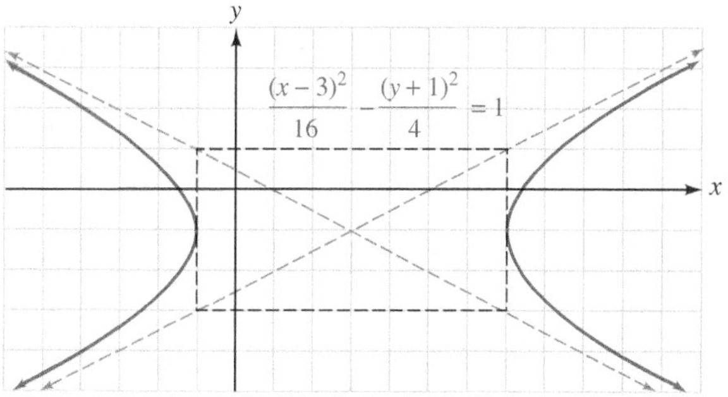

Figure 12-31

SELF CHECK 2 Graph: $\frac{(x+2)^2}{9} - \frac{(y-1)^2}{4} = 1$

2 Graph a hyperbola given an equation in general form.

Another important form of the equation of a hyperbola is called the *general form*.

GENERAL FORM OF THE EQUATION OF A HYPERBOLA

The equation of any hyperbola can be written in the form

$$Ax^2 - Cy^2 + Dx + Ey + F = 0$$

Notice that for a hyperbola the coefficient for either the x^2 or y^2 term is negative. Recall for an ellipse the coefficients were both positive. As we did with ellipses we will need to complete the square to identify critical information for graphing.

EXAMPLE 3 Write the equation $x^2 - y^2 - 2x + 4y - 12 = 0$ in standard form to show that the equation represents a hyperbola. Then graph it.

Solution We proceed as follows.

$$x^2 - y^2 - 2x + 4y - 12 = 0$$

$$x^2 - y^2 - 2x + 4y = 12 \quad \text{Add 12 to both sides.}$$

$$x^2 - 2x - y^2 + 4y = 12 \quad \text{Use the commutative property to rearrange the } x \text{ and } y \text{ terms.}$$

$$x^2 - 2x - 1(y^2 - 4y) = 12 \quad \text{Factor } -1 \text{ from } -y^2 + 4y.$$

We then complete the square on x and y to make perfect trinomial squares.

$$x^2 - 2x + 1 - (y^2 - 4y + 4) = 12 + 1 - 4$$

We then factor $x^2 - 2x + 1$ and $y^2 - 4y + 4$ to obtain

$$(x-1)^2 - (y-2)^2 = 9$$

$$\frac{(x-1)^2}{9} - \frac{(y-2)^2}{9} = 1 \quad \text{Divide both sides by 9.}$$

This is the equation of a hyperbola centered at $(1, 2)$. Its graph is shown in Figure 12-32.

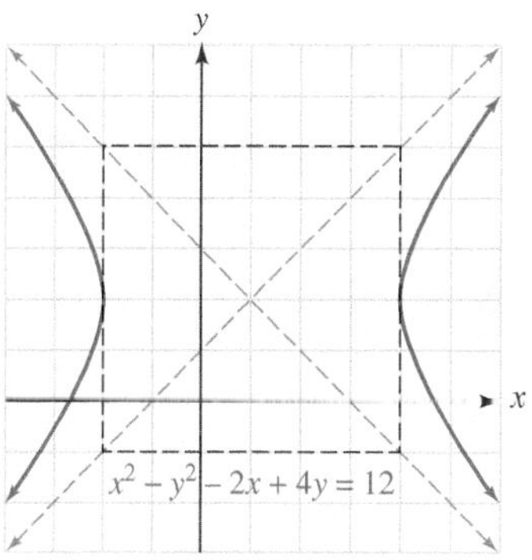

Figure 12-32

SELF CHECK 3 Graph: $x^2 - 4y^2 + 2x - 8y = 7$

Unless otherwise noted, all content on this page is © Cengage Learning.

Everyday connections
Focus on Conics

Satellite dishes and flashlight reflectors are familiar examples of a conic's ability to reflect a beam of light or to concentrate incoming satellite signals at one point. That property is shown in the following illustration.

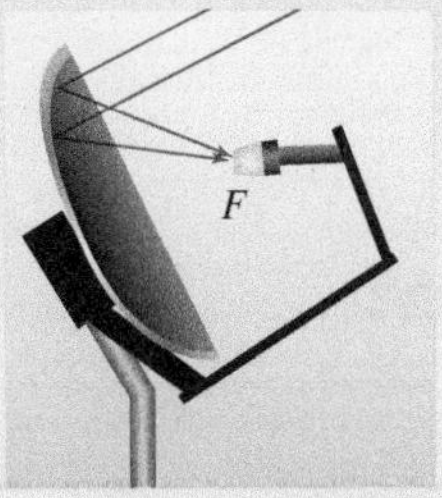

1. Another form of an equation for a parabola with the vertex at $(0, 0)$ is $x^2 = 4py$, where p is the distance from the vertex to the focus. Find the coordinate of the focus for the parabola defined by the equation $x^2 = y$.

An ellipse has two foci, the points marked in the following illustration. Any light or signal that starts at one focus will be reflected to the other. This property is the basis of whispering galleries, where a person standing at one focus can clearly hear another person speaking at the other focus.

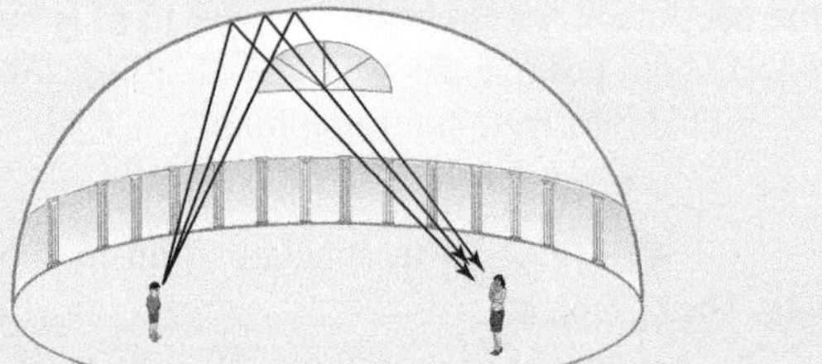

2. If the height of an elliptical rotunda is 32 feet and its width is 80 feet, how far from the center of the room should two people stand to witness the "whispering chamber" effect?

The focal property of the ellipse is also used in *lithotripsy*, a medical procedure for treating kidney stones. The patient is placed in an elliptical tank of water with the kidney stone at one focus. Shock waves from a small controlled explosion at the other focus are concentrated on the stone, pulverizing it.

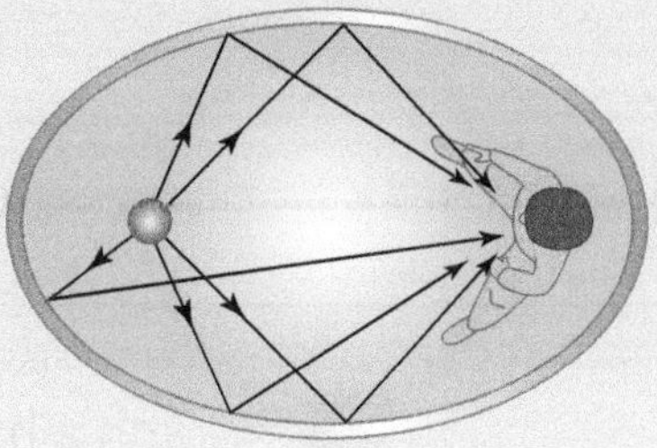

The hyperbola also has two foci, the two points labeled F in the following illustration. As in the ellipse, light aimed at one focus is reflected toward the other. Hyperbolic mirrors are used in some reflecting telescopes.

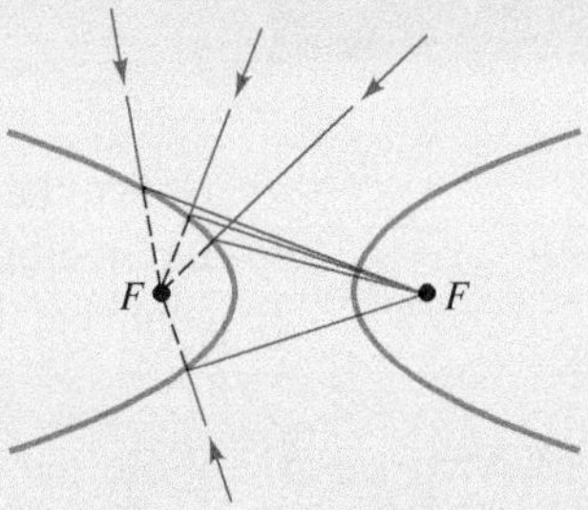

3 Graph a hyperbola of the form $xy = k$.

There is a special type of hyperbola (also centered at the origin) that does not intersect either the x- or the y-axis. These hyperbolas have equations of the form $xy = k$, where $k \neq 0$.

Unless otherwise noted, all content on this page is © Cengage Learning.

EXAMPLE 4 Graph: $xy = -8$

Solution We create a table of ordered pairs, plot each pair, and join the points with a smooth curve to obtain the hyperbola in Figure 12-33.

$xy = -8$

x	y	(x, y)
1	−8	(1, −8)
2	−4	(2, −4)
4	−2	(4, −2)
8	−1	(8, −1)
−1	8	(−1, 8)
−2	4	(−2, 4)
−4	2	(−4, 2)
−8	1	(−8, 1)

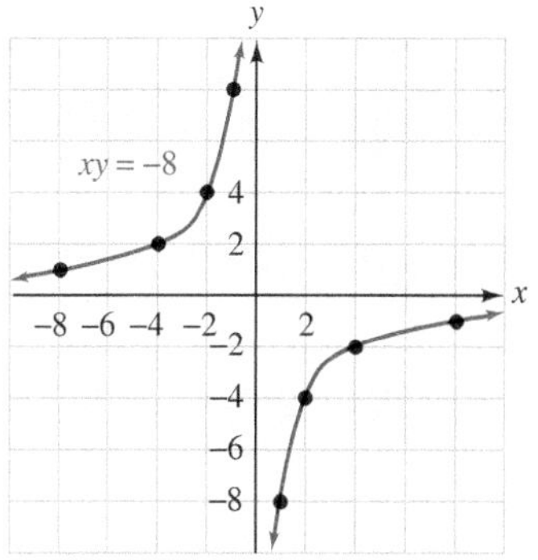

Figure 12-33

SELF CHECK 4 Graph: $xy = 6$

The result in Example 4 illustrates the following general equation.

EQUATIONS OF HYPERBOLAS OF THE FORM $xy = k$

Any equation of the form $xy = k$, where $k \neq 0$, has a graph that is a hyperbola that does not intersect either the x-axis or the y-axis.

4 Solve an application involving a hyperbola.

EXAMPLE 5 **ATOMIC STRUCTURE** In an experiment that led to the discovery of the atomic structure of matter, Lord Rutherford (1871–1937) shot high-energy alpha particles toward a thin sheet of gold. Because many were reflected, Rutherford showed the existence of the nucleus of a gold atom. The alpha particle in Figure 12-34 is repelled by the nucleus at the origin; it travels along the hyperbolic path given by $4x^2 - y^2 = 16$. How close does the particle come to the nucleus?

Solution To find the distance from the nucleus at the origin, we must find the coordinates of the vertex V. To do so, we write the equation of the particle's path in standard form:

$$4x^2 - y^2 = 16$$

$$\frac{4x^2}{16} - \frac{y^2}{16} = \frac{16}{16} \quad \text{Divide both sides by 16.}$$

$$\frac{x^2}{4} - \frac{y^2}{16} = 1 \quad \text{Simplify.}$$

$$\frac{x^2}{2^2} - \frac{y^2}{4^2} = 1 \quad \text{Write 4 as } 2^2 \text{ and 16 as } 4^2.$$

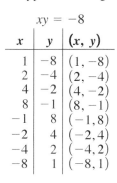

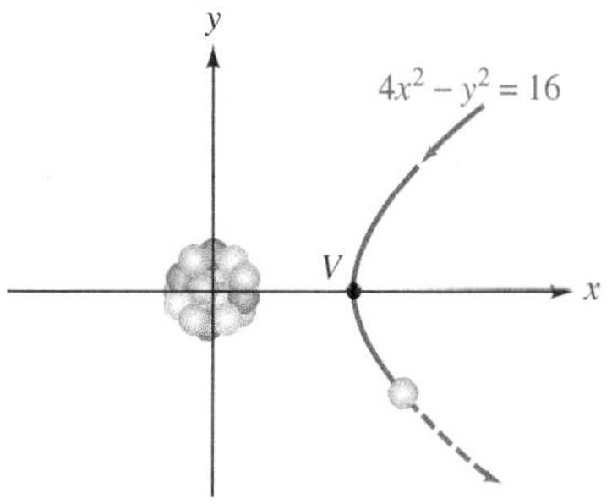

Figure 12-34

This equation is in the form

$$\frac{x^2}{a^2} - \frac{y^2}{b^2} = 1$$

Unless otherwise noted, all content on this page is © Cengage Learning.

with $a = 2$. Thus, the vertex of the path is $(2, 0)$. The particle is never closer than 2 units from the nucleus.

SELF CHECK 5 Identify the vertex for the other branch of the hyperbola.

SELF CHECK ANSWERS

1. 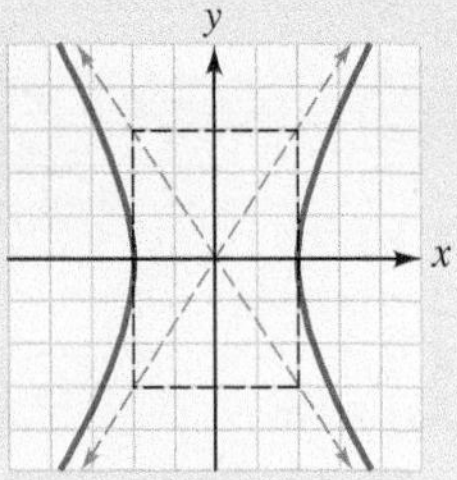

2. 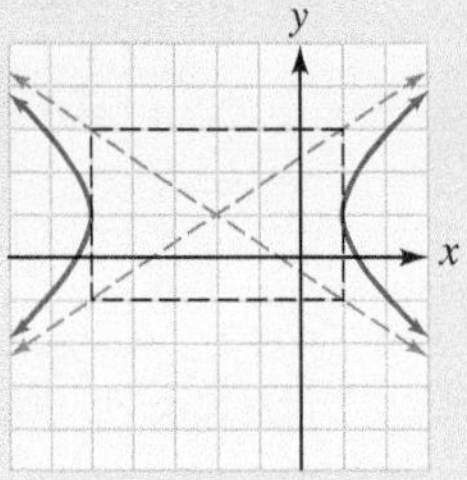

3. 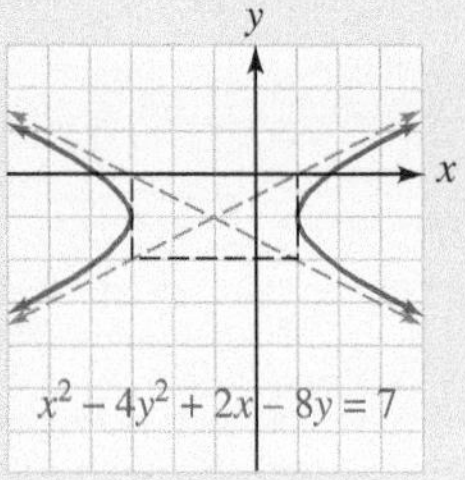

4. 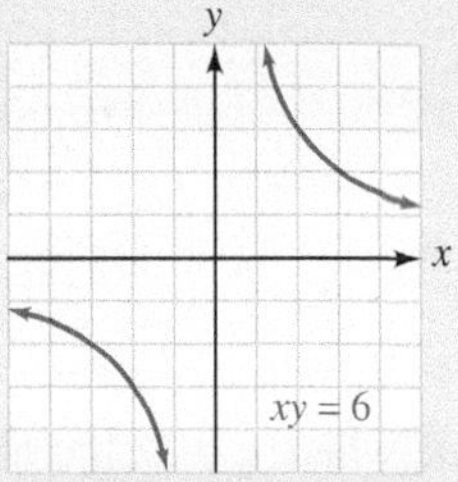

5. $(-2, 0)$

NOW TRY THIS

Given the equation $\frac{x^2}{a^2} - \frac{y^2}{b^2} = 1$*, the equations for the asymptotes are* $y = \frac{b}{a}x$*, and* $y = -\frac{b}{a}x$*.*

1. Find the equations for the asymptotes for the hyperbola described by the equation $\frac{x^2}{25} - \frac{y^2}{9} = 1$.
2. Find the equations for the asymptotes for the hyperbola described by the equation $\frac{y^2}{4} - \frac{x^2}{9} = 1$.
3. Find the equations for the asymptotes for the hyperbola described by the equation $\frac{(x+2)^2}{9} - \frac{(y-1)^2}{4} = 1$.

12.3 Exercises

WARM-UPS *Evaluate when y = 0. Write as ordered pairs.*

1. $\frac{x^2}{4} - \frac{y^2}{9} = 1$
2. $\frac{x^2}{36} - \frac{y^2}{16} = 1$

REVIEW *Factor each expression.*

3. $-5x^4 + 10x^3 - 15x^2$
4. $9a^2 - b^2$
5. $14a^2 - 15ab - 9b^2$
6. $27p^3 - 64q^3$

VOCABULARY AND CONCEPTS *Fill in the blanks.*

7. A ________ is the set of all points in a plane for which the ________ of the distances from two fixed points is a constant.
8. The fixed points in Exercise 7 are the ___ of the hyperbola.
9. The midpoint of the line segment joining the foci of a hyperbola is called the _____ of the hyperbola.

Unless otherwise noted, all content on this page is © Cengage Learning.

10. To graph a hyperbola, we locate its center and vertices, sketch the __________________, and sketch the asymptotes.

11. The hyperbolic graph of $\frac{x^2}{a^2} - \frac{y^2}{b^2} = 1$ has x-intercepts of _______. There are no __________.

12. The center of the hyperbola with an equation of $\frac{x^2}{a^2} - \frac{y^2}{b^2} = 1$ is the point _____. The center of the hyperbola with an equation of $\frac{(x-h)^2}{a^2} - \frac{(y-k)^2}{b^2} = 1$ is the point _____.

GUIDED PRACTICE *Graph each hyperbola.* SEE EXAMPLE 1. (OBJECTIVE 1)

13. $\frac{x^2}{9} - \frac{y^2}{4} = 1$

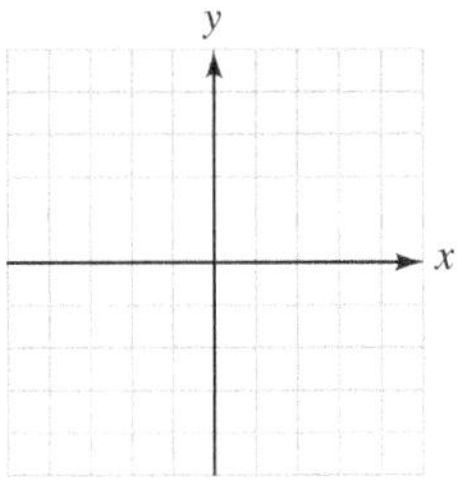

14. $\frac{x^2}{4} - \frac{y^2}{4} = 1$

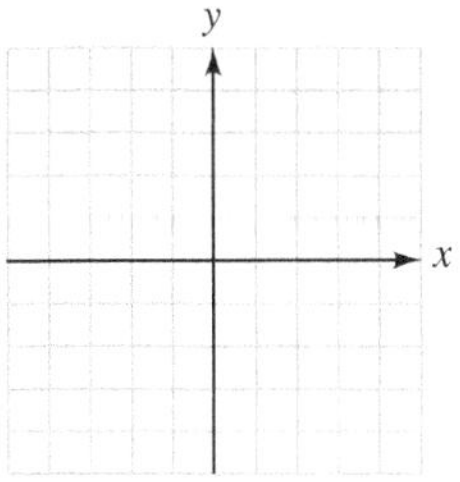

15. $\frac{y^2}{4} - \frac{x^2}{9} = 1$

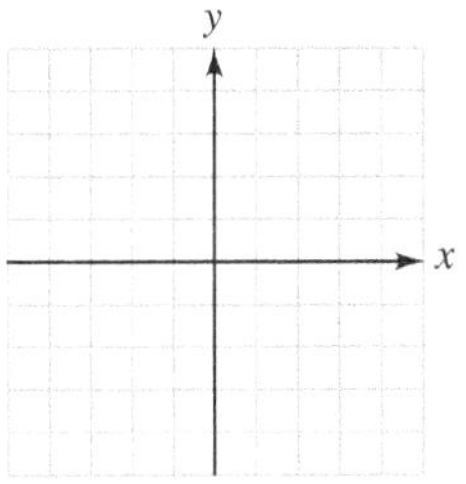

16. $\frac{y^2}{4} - \frac{x^2}{64} = 1$

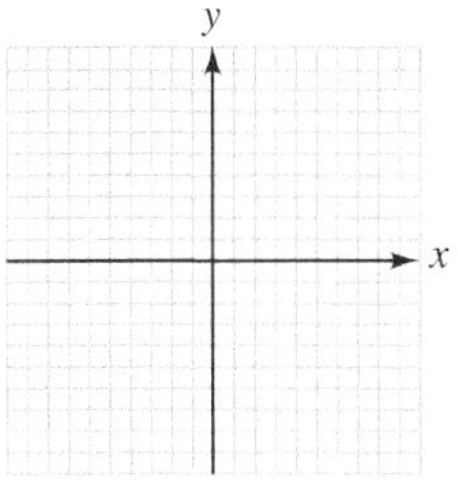

Graph each hyperbola. SEE EXAMPLE 2. (OBJECTIVE 1)

17. $\frac{(x-2)^2}{9} - \frac{y^2}{16} = 1$

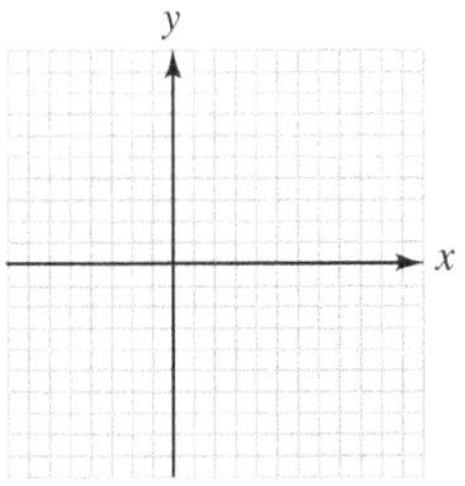

18. $\frac{(x+2)^2}{16} - \frac{(y-3)^2}{25} = 1$

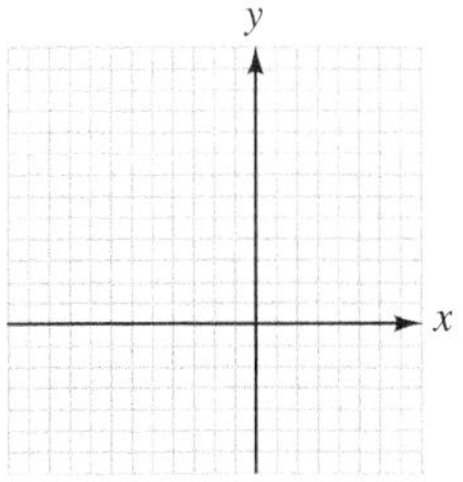

19. $\frac{(y+1)^2}{1} - \frac{(x-2)^2}{4} = 1$

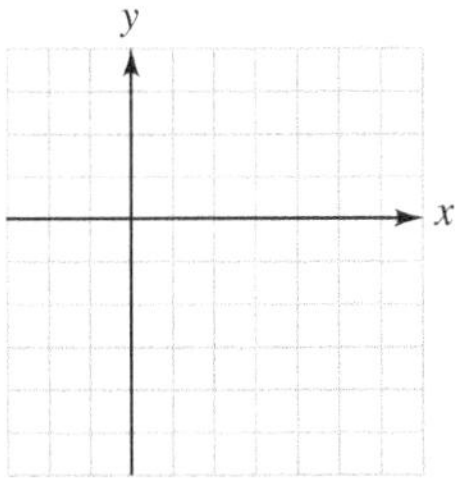

20. $\frac{(y-2)^2}{4} - \frac{(x+1)^2}{1} = 1$

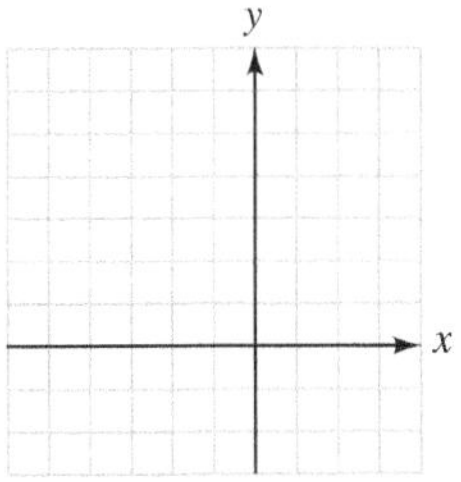

Write each equation in standard form and graph it. SEE EXAMPLE 3. (OBJECTIVE 2)

21. $4x^2 - y^2 + 8x - 4y = 4$

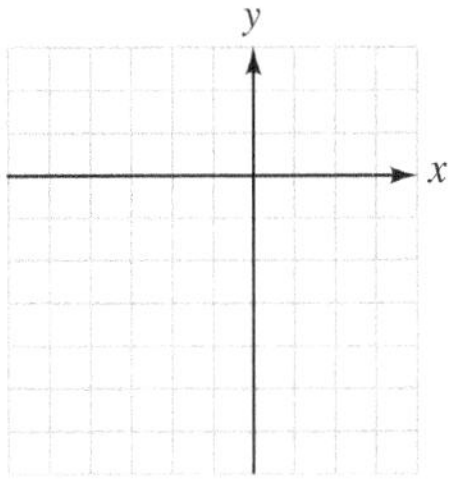

22. $x^2 - 9y^2 - 4x - 54y = 86$

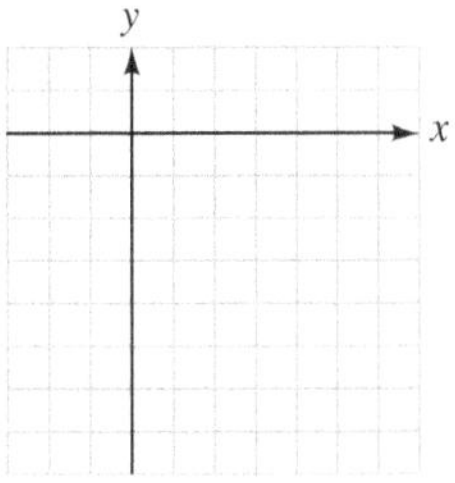

23. $4y^2 - x^2 + 8y + 4x = 4$

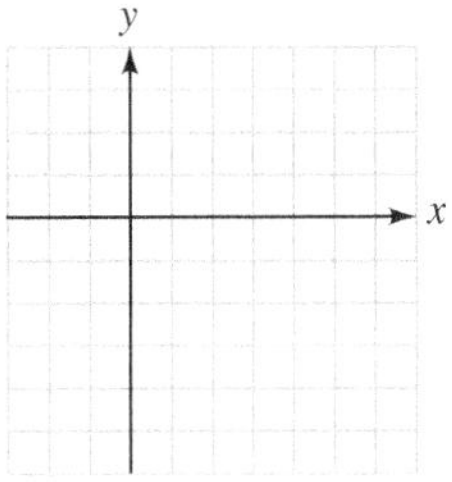

24. $y^2 - 4x^2 - 4y - 8x = 4$

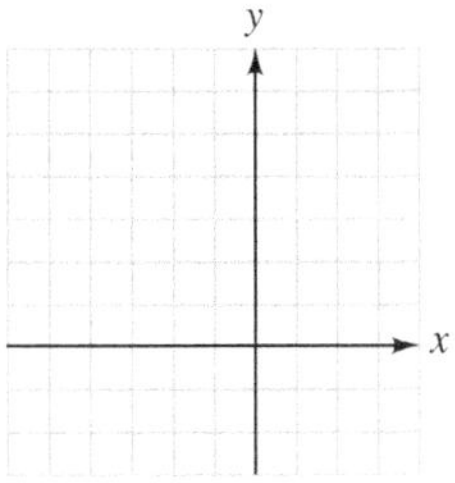

Graph each hyperbola. SEE EXAMPLE 4. (OBJECTIVE 3)

25. $xy = 10$

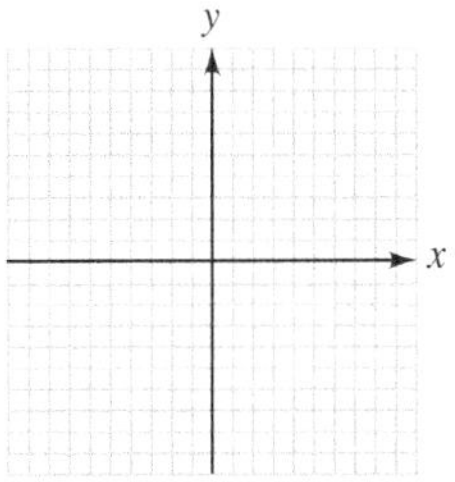

26. $xy = -10$

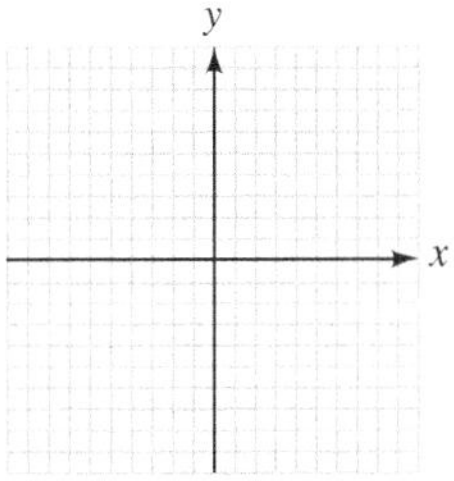

27. $xy = -12$

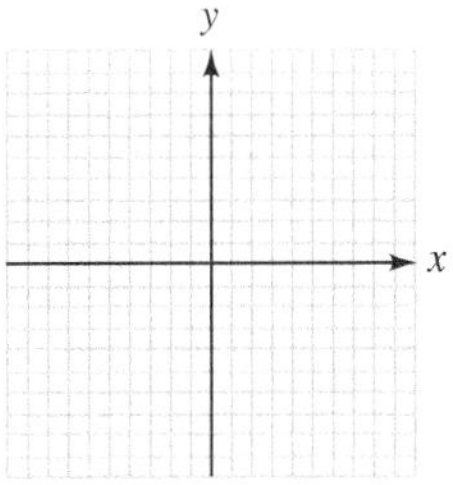

28. $xy = 6$

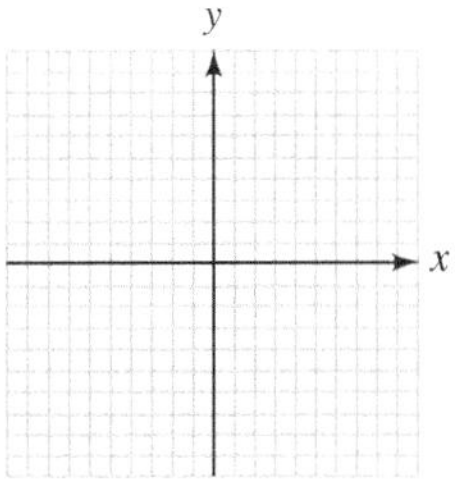

ADDITIONAL PRACTICE *Graph each equation.*

29. $25x^2 - y^2 = 25$

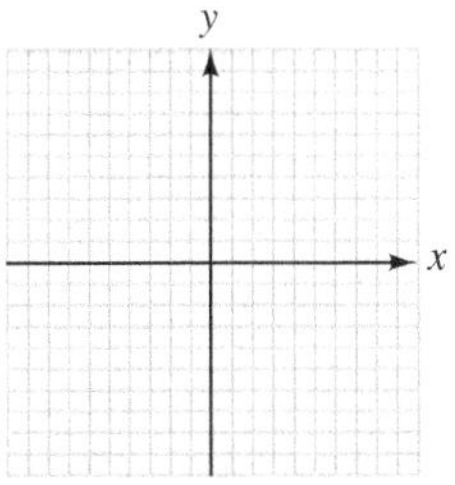

Unless otherwise noted, all content on this page is © Cengage Learning.

30. $9x^2 - 4y^2 = 36$

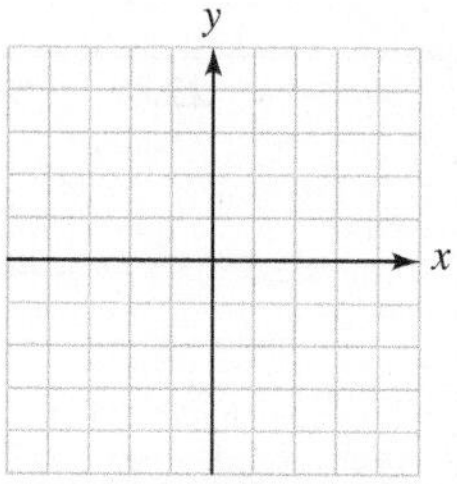

31. $4(x + 3)^2 - (y - 1)^2 = 4$

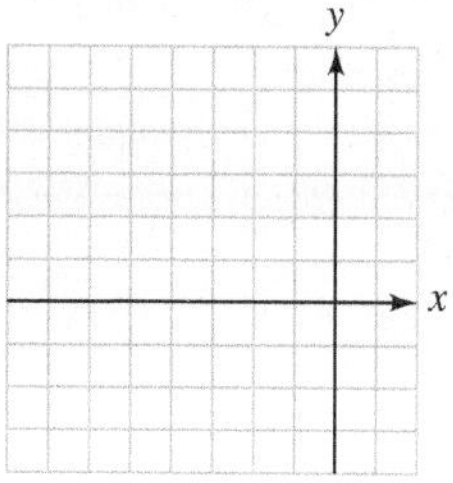

32. $(x + 5)^2 - 16y^2 = 16$

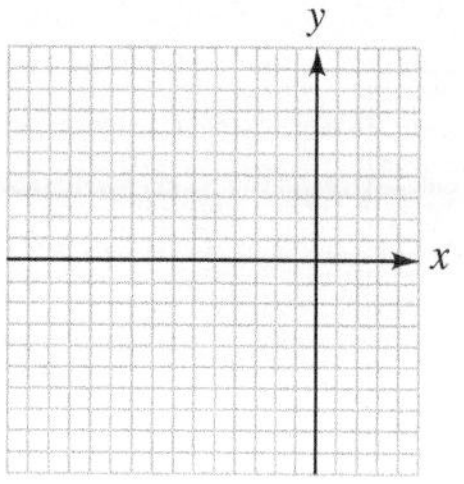

Use a graphing calculator to graph each equation.

33. $\frac{x^2}{9} - \frac{y^2}{4} = 1$

34. $y^2 - 16x^2 = 16$

35. $\frac{x^2}{4} - \frac{(y - 1)^2}{9} = 1$

36. $\frac{(y + 1)^2}{9} - \frac{(x - 2)^2}{4} = 1$

APPLICATIONS *Solve each application.* SEE EXAMPLE 5. (OBJECTIVE 4)

37. Alpha particles The particle in the illustration approaches the nucleus at the origin along the path $9y^2 - x^2 = 81$. How close does the particle come to the nucleus?

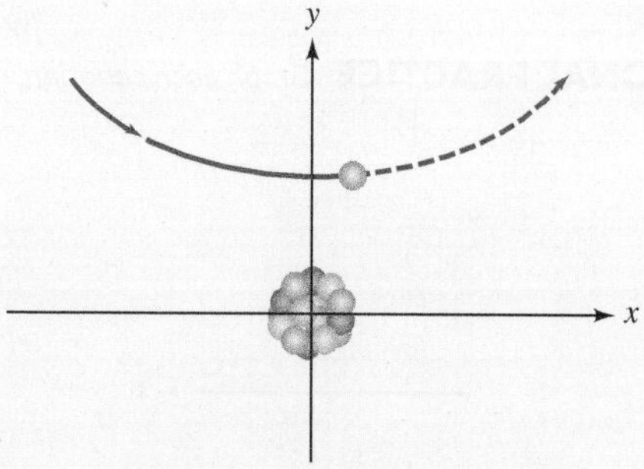

38. LORAN By determining the difference of the distances between the ship and two land-based radio transmitters, the LORAN system places the ship on the hyperbola $x^2 - 4y^2 = 576$. If the ship is also 5 miles out to sea, find its coordinates.

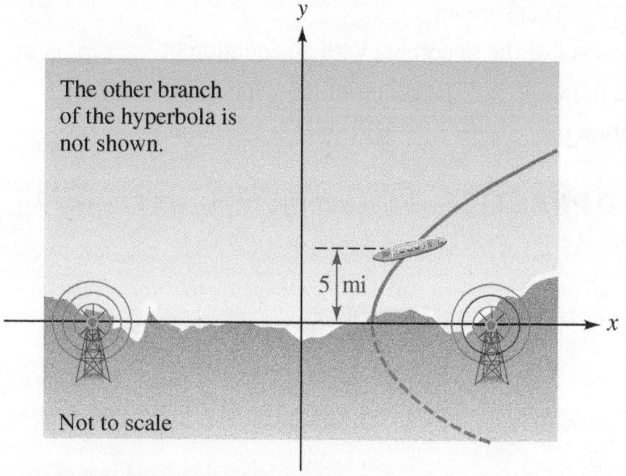

39. Electrostatic repulsion Two similarly charged particles are shot together for an almost head-on collision, as shown in the illustration. They repel each other and travel the two branches of the hyperbola given by $x^2 - 4y^2 = 4$. How close do they get?

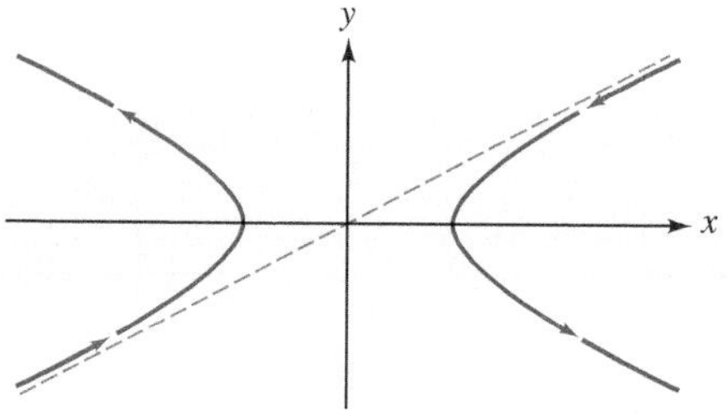

40. Sonic boom The position of the sonic boom caused by faster-than-sound aircraft is the hyperbola $y^2 - x^2 = 25$ in the coordinate system shown below. How wide is the hyperbola 5 units from its vertex?

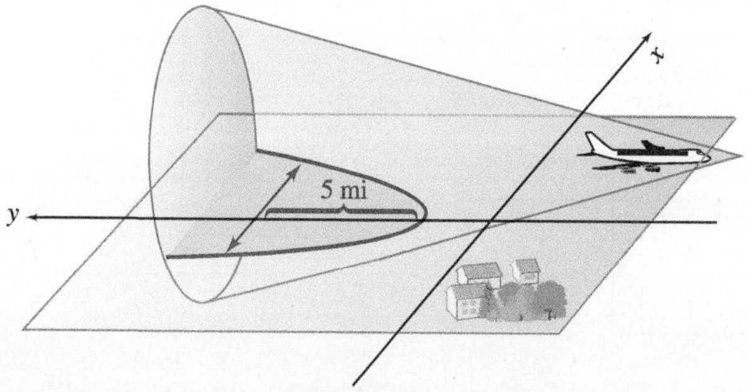

WRITING ABOUT MATH

41. Explain how to find the x- and the y-intercepts of the graph of the hyperbola

$$\frac{x^2}{a^2} - \frac{y^2}{b^2} = 1$$

42. Explain why the graph of the hyperbola

$$\frac{x^2}{a^2} - \frac{y^2}{b^2} = 1$$

has no y-intercept.

Unless otherwise noted, all content on this page is © Cengage Learning.

SOMETHING TO THINK ABOUT

43. Describe the fundamental rectangle of

$$\frac{x^2}{a^2} - \frac{y^2}{b^2} = 1$$

when $a = b$.

44. The hyperbolas $x^2 - y^2 = 1$ and $y^2 - x^2 = 1$ are called **conjugate hyperbolas**. Graph both on the same axes. What do they have in common?

Section 12.4 Solving Systems of Equations and Inequalities Containing One or More Second-Degree Terms

Objectives

1. Solve a system of equations containing one or more second-degree terms by graphing.
2. Solve a system of equations containing one or more second-degree terms by substitution.
3. Solve a system of equations containing one or more second-degree terms by elimination.
4. Solve a system of inequalities containing one or more second-degree terms.

Getting Ready

Add the left sides and the right sides of the following equations.

1. $\begin{array}{r} 3x^2 + 3y^2 = 12 \\ 4x^2 - 3y^2 = 32 \\ \hline \end{array}$

2. $\begin{array}{r} -12x^2 - 5y^2 = -17 \\ 12x^2 + 2y^2 = 25 \\ \hline \end{array}$

In this section, we discuss ways to solve systems of two equations in two variables where at least one of the equations is of second degree.

1 Solve a system of equations containing one or more second-degree terms by graphing.

EXAMPLE 1 Solve by graphing: $\begin{cases} x^2 + y^2 = 25 \\ 2x + y = 10 \end{cases}$

Solution The graph of $x^2 + y^2 = 25$ is a circle with center at the origin and radius of 5. The graph of $2x + y = 10$ is a line. Depending on whether the line is a secant (intersecting the circle at two points) or a tangent (intersecting the circle at one point) or does not intersect the circle at all, there are two, one, or no solutions to the system, respectively.

After graphing the circle and the line, as shown in Figure 12-35, we see that there are two intersection points $(3, 4)$ and $(5, 0)$. These are the solutions of the system.

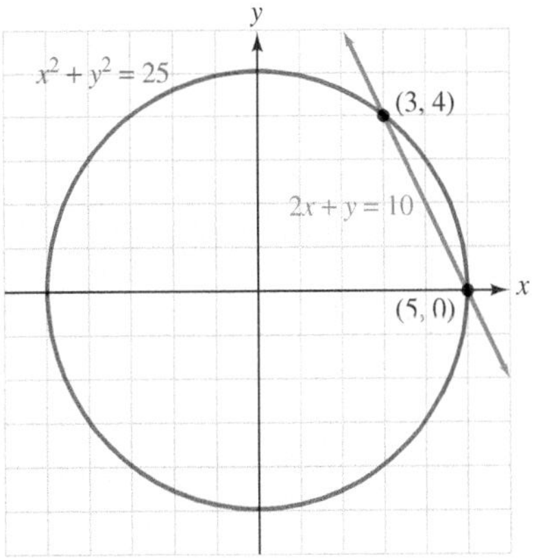

Figure 12-35

Unless otherwise noted, all content on this page is © Cengage Learning.

SELF CHECK 1 Solve by graphing: $\begin{cases} x^2 + y^2 = 13 \\ y = -\frac{1}{5}x + \frac{13}{5} \end{cases}$

Accent on technology

▸ Solving Systems of Equations

To solve Example 1 with a graphing calculator, we graph the circle and the line on one set of coordinate axes (see Figure 12-36(a)). We can find the coordinates of the intersection points of the graphs by using the INTERSECT feature found in the CALC menu. (See Figure 12-36(b) and Figure 12-36(c).)

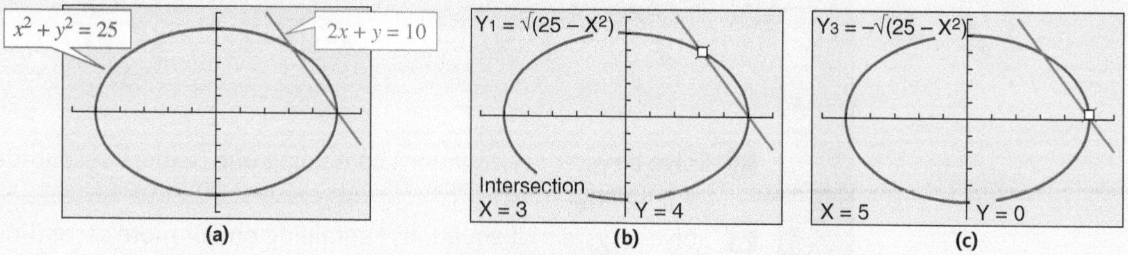

Figure 12-36

For instructions regarding the use of a Casio graphing calculator, please refer to the Casio Keystroke Guide in the back of the book.

2 Solve a system of equations containing one or more second-degree terms by substitution.

Algebraic methods also can be used to solve systems of equations. We will solve the system from Example 1 using substitution.

EXAMPLE 2 Solve using substitution: $\begin{cases} x^2 + y^2 = 25 \\ 2x + y = 10 \end{cases}$

Solution This system has one second-degree equation and one first-degree equation. The first equation is the equation of a circle and the second equation is the equation of a line. Since a line can intersect a circle in 0, 1, or 2 points, this system can have 0, 1, or 2 solutions.

We can solve the system by substitution. Solving the linear equation for y gives

$$2x + y = 10$$
$$y = -2x + 10$$

We can substitute $-2x + 10$ for y in the second-degree equation and solve the resulting quadratic equation for x:

$$x^2 + y^2 = 25$$
$$x^2 + (-2x + 10)^2 = 25$$
$$x^2 + 4x^2 - 40x + 100 = 25 \quad (-2x + 10)(-2x + 10) = 4x^2 - 40x + 100$$
$$5x^2 - 40x + 75 = 0 \quad \text{Combine like terms and subtract 25 from both sides.}$$
$$x^2 - 8x + 15 = 0 \quad \text{Divide both sides by 5.}$$
$$(x - 5)(x - 3) = 0 \quad \text{Factor } x^2 - 8x + 15.$$
$$x - 5 = 0 \quad \text{or} \quad x - 3 = 0 \quad \text{Set each factor equal to 0.}$$
$$x = 5 \quad \Big| \quad x = 3$$

Unless otherwise noted, all content on this page is © Cengage Learning.

If we substitute 5 for x in the original equation $x^2 + y^2 = 25$, we obtain $y = 0$. If we substitute 3 for x in the original equation, we obtain $y = 4$ or $y = -4$. We must check both solutions in the original second equation, $2x + y = 10$. The ordered pair $(3, -4)$ does not check and is, therefore, not a solution. There are two solutions, $(5, 0)$ and $(3, 4)$.

SELF CHECK 2 Solve by substitution: $\begin{cases} x^2 + y^2 = 13 \\ y = -\frac{1}{5}x + \frac{13}{5} \end{cases}$

EXAMPLE 3 Solve using substitution: $\begin{cases} 4x^2 + 9y^2 = 5 \\ y = x^2 \end{cases}$

Solution This system has two second-degree equations. The first is the equation of an ellipse and the second is the equation of a parabola. Since an ellipse and a parabola can intersect in 0, 1, 2, 3, or 4 points, this system can have 0, 1, 2, 3, or 4 solutions.

We can solve this system by substitution.

$$4x^2 + 9y^2 = 5$$

$$4y + 9y^2 = 5 \qquad \text{Substitute } y \text{ for } x^2.$$

$$9y^2 + 4y - 5 = 0 \qquad \text{Write in quadratic form.}$$

$$(9y - 5)(y + 1) = 0 \qquad \text{Factor } 9y^2 + 4y - 5.$$

$$9y - 5 = 0 \quad \text{or} \quad y + 1 = 0 \qquad \text{Set each factor equal to 0.}$$

$$y = \frac{5}{9} \quad \Big| \quad y = -1$$

Since $y = x^2$, the values of x are found by solving the equations

$$x^2 = \frac{5}{9} \quad \text{and} \quad x^2 = -1$$

Because $x^2 = -1$ has no real number solutions, this possibility is discarded. The solutions of $x^2 = \frac{5}{9}$ are

$$x = \frac{\sqrt{5}}{3} \quad \text{or} \quad x = -\frac{\sqrt{5}}{3}$$

The solutions of the system are $\left(\frac{\sqrt{5}}{3}, \frac{5}{9}\right)$ and $\left(-\frac{\sqrt{5}}{3}, \frac{5}{9}\right)$.

SELF CHECK 3 Solve using substitution: $\begin{cases} x^2 + y^2 = 20 \\ y = x^2 \end{cases}$

3 Solve a system of equations containing one or more second-degree terms by elimination.

EXAMPLE 4 Solve by elimination: $\begin{cases} 3x^2 + 2y^2 = 36 \\ 4x^2 - y^2 = 4 \end{cases}$

Solution This system has two second-degree equations. The first equation is the equation of an ellipse and the second equation is the equation of a hyperbola. Since an ellipse and a hyperbola can intersect in 0, 1, 2, 3, or 4 points, this system can have 0, 1, 2, 3, or 4 solutions.

Since both equations are in the form $ax^2 + by^2 = c$, we can solve the system by elimination. To do so, we can copy the first equation and multiply the second equation by 2 to obtain the equivalent system

$$\begin{cases} 3x^2 + 2y^2 = 36 \\ 8x^2 - 2y^2 = 8 \end{cases}$$

We add the equations to eliminate y and solve the resulting equation for x:

$$11x^2 = 44$$
$$x^2 = 4$$
$$x = 2 \quad \text{or} \quad x = -2$$

To find y, we substitute 2 for x and then -2 for x in the first equation and proceed as follows:

For $x = 2$	**For $x = -2$**
$3x^2 + 2y^2 = 36$	$3x^2 + 2y^2 = 36$
$3(2)^2 + 2y^2 = 36$	$3(-2)^2 + 2y^2 = 36$
$12 + 2y^2 = 36$	$12 + 2y^2 = 36$
$2y^2 = 24$	$2y^2 = 24$
$y^2 = 12$	$y^2 = 12$
$y = +\sqrt{12}$ or $y = -\sqrt{12}$	$y = +\sqrt{12}$ or $y = -\sqrt{12}$
$y = 2\sqrt{3}$ | $y = -2\sqrt{3}$	$y = 2\sqrt{3}$ | $y = -2\sqrt{3}$

The four solutions of this system are

$$(2, 2\sqrt{3}), (2, -2\sqrt{3}), (-2, 2\sqrt{3}), \text{ and } (-2, -2\sqrt{3})$$

SELF CHECK 4 Solve: $\begin{cases} x^2 + 4y^2 = 16 \\ x^2 - y^2 = 1 \end{cases}$

4 Solve a system of inequalities containing one or more second-degree terms.

EXAMPLE 5 Graph the solution set of the system: $\begin{cases} y < x^2 \\ y > \frac{x^2}{4} - 2 \end{cases}$

Solution The graph of $y = x^2$ is the parabola shown in Figure 12-37(a), which opens upward and has its vertex at the origin. Because equality is not included, the parabola is drawn with a dashed line. The points with coordinates that satisfy the inequality $y < x^2$ are those points below the parabola.

The graph of $y > \frac{x^2}{4} - 2$ is a parabola opening upward, with vertex at $(0, -2)$. However, this time the points with coordinates that satisfy the inequality are those points above the parabola. Because equality is not included, the parabola is drawn with a dashed line.

$y = x^2$

x	y	(x, y)
0	0	(0, 0)
1	1	(1, 1)
−1	1	(−1, 1)
2	4	(2, 4)
−2	4	(−2, 4)

$y = \frac{x^2}{4} - 2$

x	y	(x, y)
0	−2	(0, −2)
2	−1	(2, −1)
−2	−1	(−2, −1)
4	2	(4, 2)
−4	2	(−4, 2)

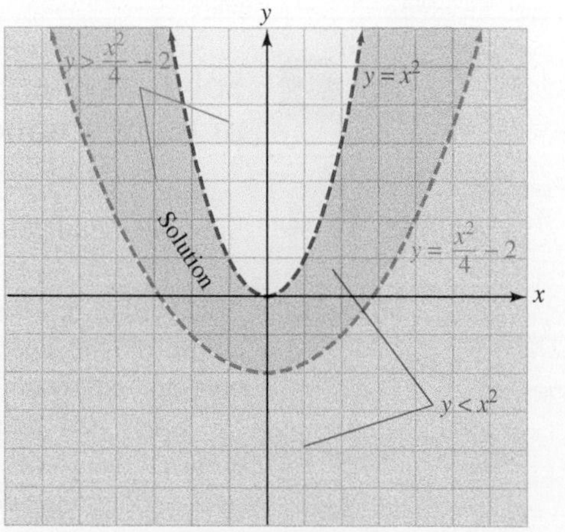

Figure 12-37 (a)

Unless otherwise noted, all content on this page is © Cengage Learning.

The graph of the solution set of the system will be the area between the parabolas as shown in Figure 12-37(b).

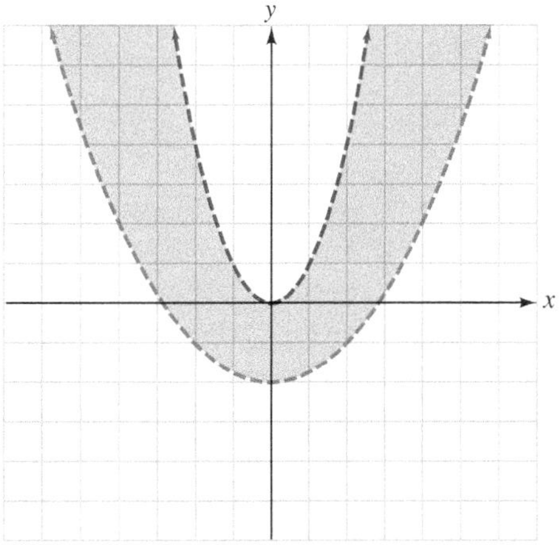

Figure 12-37 (b)

SELF CHECK 5 Graph the solution of the system: $\begin{cases} x^2 + y^2 \le 4 \\ y > x^2 - 2 \end{cases}$

SELF CHECK ANSWERS

1. (3, 2), (−2, 3)

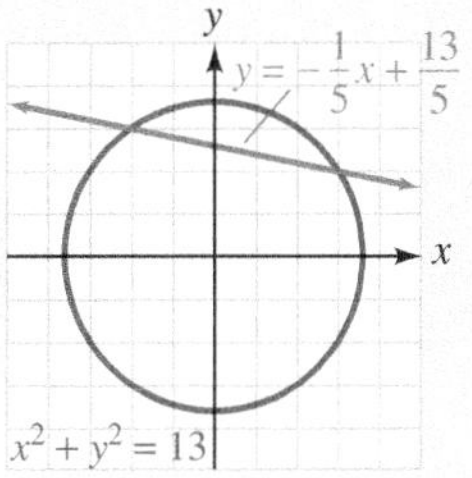

2. (3, 2), (−2, 3) **3.** (2, 4), (−2, 4) **4.** $(2, \sqrt{3}), (2, -\sqrt{3}), (-2, \sqrt{3}), (-2, -\sqrt{3})$ **5.**

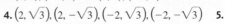

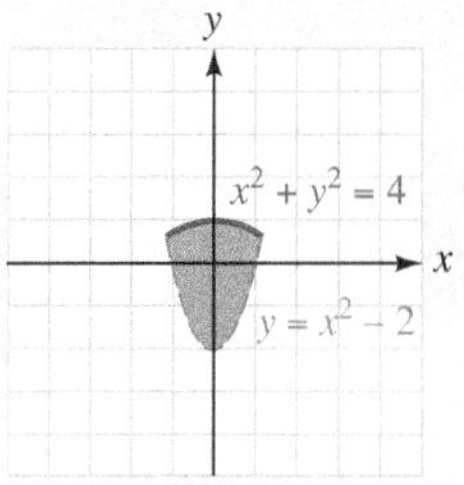

NOW TRY THIS

Solve by substitution: $\begin{cases} x^2 - y^2 = 4 \\ 9x^2 + 16y^2 = 144 \end{cases}$

12.4 Exercises

WARM-UPS *State the possible number of solutions of a system when the graphs of the equations are*

1. a line and a parabola.

2. a line and an ellipse.

3. a circle and a parabola.

4. a circle and an ellipse.

REVIEW *Simplify each radical expression. Assume that all variables represent positive numbers.*

5. $\sqrt{200y^2} - 7\sqrt{98y^2}$

6. $9b\sqrt{112b} - 4\sqrt{175b^3}$

7. $\dfrac{3t\sqrt{2t} + 2\sqrt{2t^3}}{\sqrt{18t} + \sqrt{2t}}$

8. $\sqrt[3]{\dfrac{x}{4}} - \sqrt[3]{\dfrac{x}{32}} + \sqrt[3]{\dfrac{x}{500}}$

Unless otherwise noted, all content on this page is © Cengage Learning.

VOCABULARY AND CONCEPTS *Fill in the blanks.*

9. We can solve systems of equations by _______, elimination (addition), or _________.

10. The two algebraic methods of solving a system of equations are _________ and _________.

11. A line can intersect an ellipse in at most ___ points.

12. A parabola can intersect a circle in as few as _ points.

13. A parabola can intersect an ellipse in at most ___ points.

14. An ellipse can intersect a hyperbola in at most ___ points.

15. A system of inequalities is solved by _______.

16. If an inequality contains the $\leq$ symbol, the boundary will be ____.

GUIDED PRACTICE *Solve each system by graphing.* SEE EXAMPLE 1. (OBJECTIVE 1)

17. $\begin{cases} 8x^2 + 32y^2 = 256 \\ x = 2y \end{cases}$

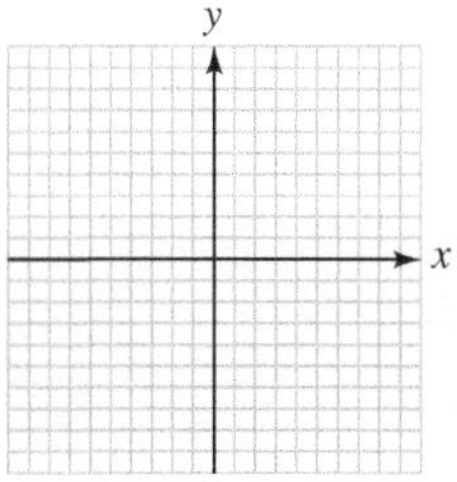

18. $\begin{cases} x^2 + y^2 = 2 \\ x + y = 2 \end{cases}$

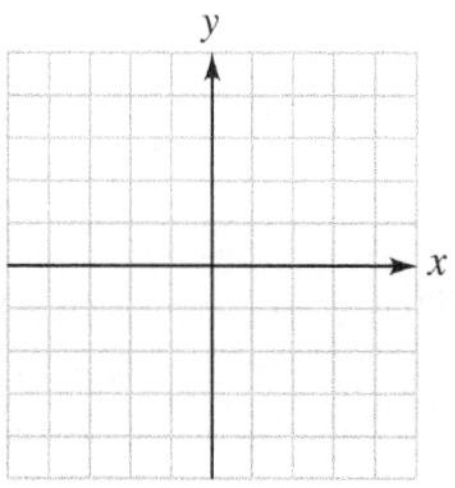

19. $\begin{cases} x^2 - 13 = -y^2 \\ y = 2x - 4 \end{cases}$

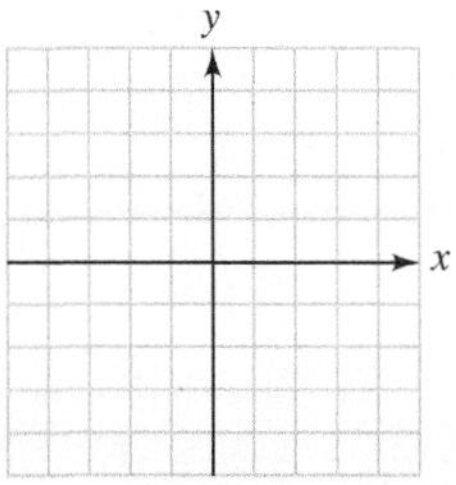

20. $\begin{cases} x^2 + y^2 = 5 \\ x + y = 3 \end{cases}$

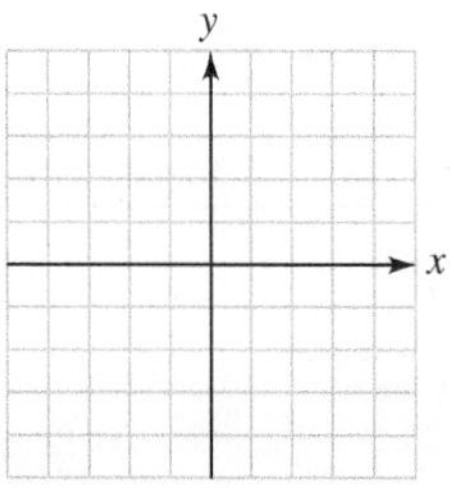

Solve each system by substitution. SEE EXAMPLE 2. (OBJECTIVE 2)

21. $\begin{cases} 25x^2 + 9y^2 = 225 \\ 5x + 3y = 15 \end{cases}$

22. $\begin{cases} x^2 - x - y = 2 \\ 4x - 3y = 0 \end{cases}$

23. $\begin{cases} x^2 + y^2 = 2 \\ x + y = 2 \end{cases}$

24. $\begin{cases} x^2 + y^2 = 5 \\ x + y = 3 \end{cases}$

Solve each system by substitution. SEE EXAMPLE 3. (OBJECTIVE 2)

25. $\begin{cases} x^2 + y^2 = 20 \\ y = x^2 \end{cases}$

26. $\begin{cases} x^2 + y^2 = 30 \\ y = x^2 \end{cases}$

27. $\begin{cases} 2x^2 + y^2 = 6 \\ x^2 - y^2 = 3 \end{cases}$

28. $\begin{cases} x^2 - y^2 = 1 \\ y = x^2 - 1 \end{cases}$

Solve each system by elimination. SEE EXAMPLE 4. (OBJECTIVE 3)

29. $\begin{cases} x^2 + y^2 = 13 \\ x^2 - y^2 = 5 \end{cases}$

30. $\begin{cases} x^2 - y^2 = -5 \\ 3x^2 + 2y^2 = 30 \end{cases}$

31. $\begin{cases} x^2 + y^2 = 25 \\ 2x^2 - 3y^2 = 5 \end{cases}$

32. $\begin{cases} y = x^2 - 4 \\ x^2 - y^2 = -16 \end{cases}$

Graph the solution set of each system of inequalities. SEE EXAMPLE 5. (OBJECTIVE 4)

33. $\begin{cases} 2x - y > 4 \\ y < -x^2 + 2 \end{cases}$

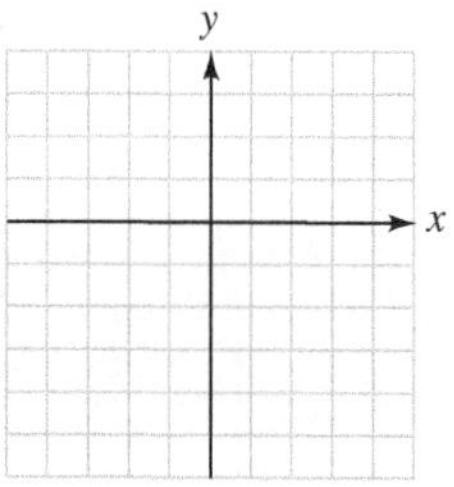

34. $\begin{cases} x \leq y^2 \\ y \geq x \end{cases}$

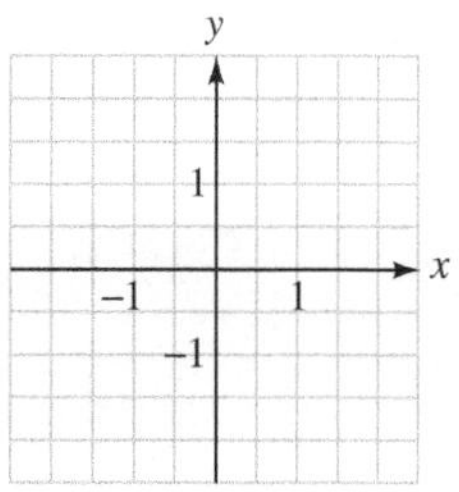

35. $\begin{cases} y > x^2 - 4 \\ y < -x^2 + 4 \end{cases}$

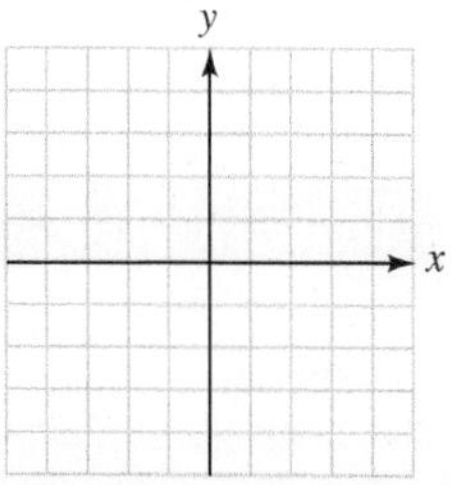

36. $\begin{cases} x \geq y^2 \\ y \geq x^2 \end{cases}$

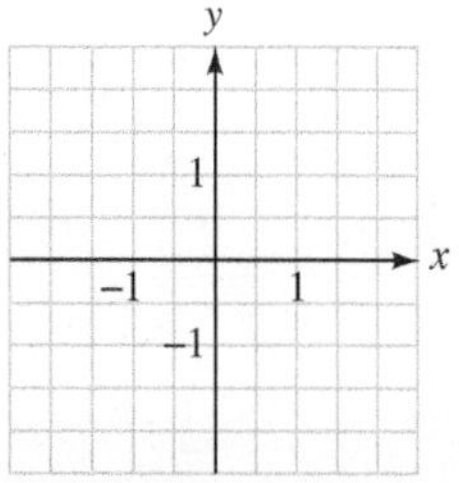

ADDITIONAL PRACTICE *Solve using any method (including a graphing calculator).*

37. $\begin{cases} 9x^2 - 7y^2 = 81 \\ x^2 + y^2 = 9 \end{cases}$

38. $\begin{cases} 6x^2 + 8y^2 = 182 \\ 8x^2 - 3y^2 = 24 \end{cases}$

39. $\begin{cases} x^2 + y^2 = 36 \\ 49x^2 + 36y^2 = 1{,}764 \end{cases}$

40. $\begin{cases} x^2 + y^2 = 10 \\ 2x^2 - 3y^2 = 5 \end{cases}$

Unless otherwise noted, all content on this page is © Cengage Learning.

41. $\begin{cases} x^2 + y^2 = 10 \\ y = 3x^2 \end{cases}$

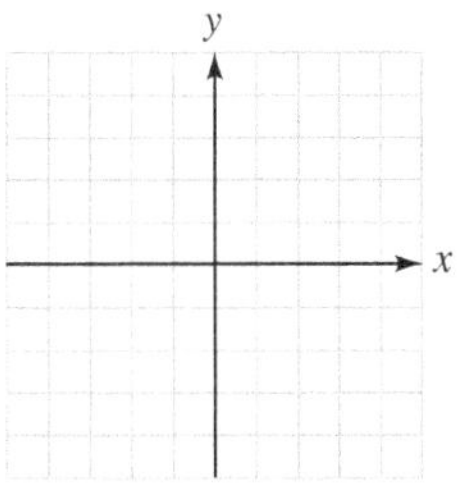

42. $\begin{cases} x^2 + y^2 = 13 \\ y = x^2 - 1 \end{cases}$

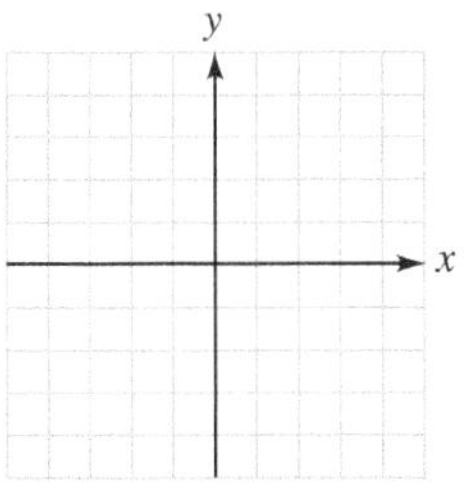

43. $\begin{cases} x^2 + y^2 = 25 \\ 12x^2 + 64y^2 = 768 \end{cases}$

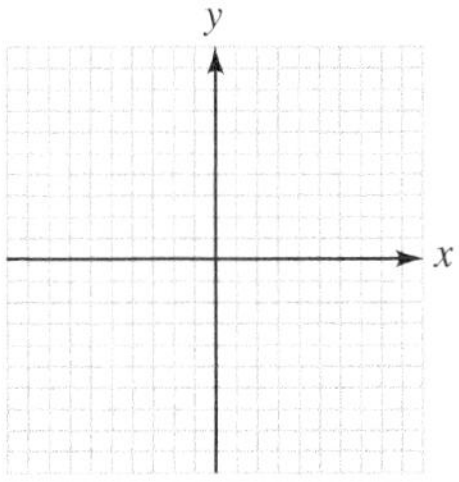

44. $\begin{cases} x^2 + y^2 = 20 \\ y = x^2 \end{cases}$

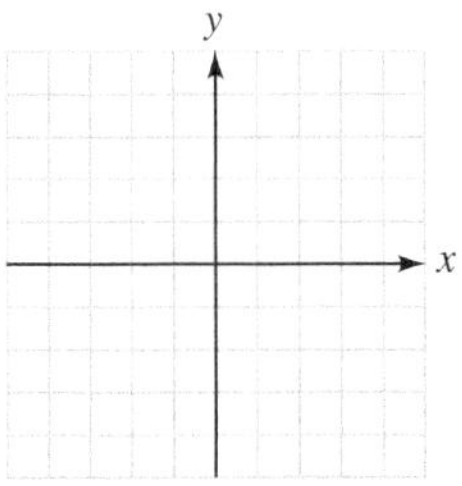

45. $\begin{cases} x^2 + y^2 = 5 \\ y = x + 1 \end{cases}$

46. $\begin{cases} x^2 - y^2 = 7 \\ 2x^2 + y^2 = 14 \end{cases}$

47. $\begin{cases} x^2 + y^2 = 15 \\ x^2 - y^2 = -9 \end{cases}$

48. $\begin{cases} xy = -\frac{9}{2} \\ 3x + 2y = 6 \end{cases}$

49. $\begin{cases} y^2 = 40 - x^2 \\ y = x^2 - 10 \end{cases}$

50. $\begin{cases} x^2 - 6x - y = -5 \\ x^2 - 6x + y = -5 \end{cases}$

51. $\begin{cases} \frac{1}{x} + \frac{2}{y} = 1 \\ \frac{2}{x} - \frac{1}{y} = \frac{1}{3} \end{cases}$

52. $\begin{cases} \frac{1}{x} + \frac{3}{y} = 4 \\ \frac{2}{x} - \frac{1}{y} = 7 \end{cases}$

53. $\begin{cases} 3y^2 = xy \\ 2x^2 + xy - 84 = 0 \end{cases}$

54. $\begin{cases} \frac{1}{x} + \frac{1}{y} = 5 \\ \frac{1}{x} - \frac{1}{y} = -3 \end{cases}$

55. $\begin{cases} xy = \frac{1}{6} \\ y + x = 5xy \end{cases}$

56. $\begin{cases} xy = \frac{1}{12} \\ y + x = 7xy \end{cases}$

APPLICATIONS *Use a graphing calculator to help solve each application.*

57. Integers The product of two integers is 32, and their sum is 12. Find the integers.

58. Numbers The sum of the squares of two numbers is 221, and the sum of the numbers is 9. Find the numbers.

59. Geometry The area of a rectangle is 63 square centimeters, and its perimeter is 32 centimeters. Find the dimensions of the rectangle.

60. Investing money Grant receives \$225 annual income from one investment. Jeff invested \$500 more than Grant, but at an annual rate of 1% less. Jeff's annual income is \$240. What is the amount and rate of Grant's investment?

61. Investing money Carol receives \$67.50 annual income from one investment. Juan invested \$150 more than Carol at an annual rate of $1\frac{1}{2}$% more. Juan's annual income is \$94.50. What is the amount and rate of Carol's investment? (*Hint*: There are two answers.)

62. Artillery The shell fired from the base of the hill follows the parabolic path $y = -\frac{1}{6}x^2 + 2x$ with distances measured in miles. The hill has a slope of $\frac{1}{3}$. How far from the gun is the point of impact? (*Hint*: Find the coordinates of the point and then the distance.)

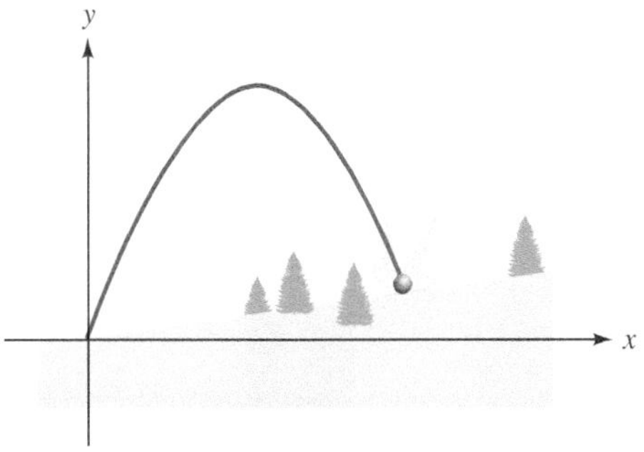

63. Driving rates Jim drove 306 miles. Jim's brother made the same trip at a speed 17 mph slower than Jim did and required an extra $1\frac{1}{2}$ hours. What was Jim's rate and time?

WRITING ABOUT MATH

64. Describe the benefits of the graphical method for solving a system of equations.

65. Describe the drawbacks of the graphical method.

SOMETHING TO THINK ABOUT

66. If the graphs of the two independent equations of a system are parabolas, how many solutions might the system have?

67. If the graphs of the two independent equations of a system are hyperbolas, how many solutions might the system have?

68. If the graphs of the two equations of a system are lines, how many solutions might the system have?

69. If the graphs of the two equations of a system are circles, how many solutions might the system have?

Unless otherwise noted, all content on this page is © Cengage Learning.

Section 12.5 Piecewise-Defined Functions and the Greatest Integer Function

Objectives

1. Graph a piecewise-defined function and determine the open intervals over which the function is increasing, decreasing, and constant.
2. Graph the greatest integer function.
3. Solve an application involving the greatest integer function or a piecewise-defined function.

Vocabulary

increasing
decreasing
piecewise-defined function
greatest integer function
step function

Getting Ready

1. Is $f(x) = x^2$ positive or negative when $x > 0$?
2. Is $f(x) = -x^2$ positive or negative when $x > 0$?
3. What is the largest integer that is less than 98.6?
4. What is the largest integer that is less than -2.7?

If the values of $f(x)$ increase as x increases on an open interval, we say that the function is **increasing** *on the interval* (see Figure 12-38(a)). If the values of $f(x)$ decrease as x increases on an open interval, we say that the function is **decreasing** *on the interval* (see Figure 12-38(b)). If the values of $f(x)$ remain constant as x increases on an open interval, we say that the function is *constant on the interval* (see Figure 12-38(c)).

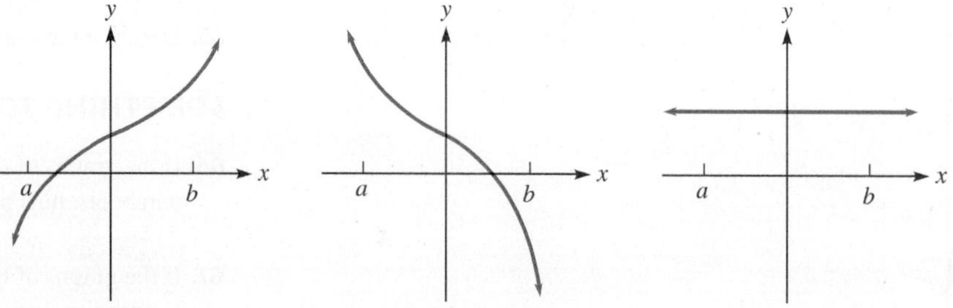

Increasing on (a, b) (a) Decreasing on (a, b) (b) Constant on (a, b) (c)

Figure 12-38

Some functions are defined by using different equations for different parts of their domains. Such functions are called **piecewise-defined functions**.

Unless otherwise noted, all content on this page is © Cengage Learning.

1 Graph a piecewise-defined function and determine the open intervals over which the function is increasing, decreasing, and constant.

A basic piecewise-defined function is the definition of $|x|$, which can be written in the form

$$f(x) = \begin{cases} x \text{ when } x \geq 0 \\ -x \text{ when } x < 0 \end{cases}$$

When x is in the interval $[0, \infty)$, we use the function $f(x) = x$ to evaluate. However, when x is in the interval $(-\infty, 0)$, we use the function $f(x) = -x$ to evaluate. The graph of the resulting piecewise function is shown in Figure 12-39.

For $x \geq 0$

x	y	(x, y)
0	0	$(0, 0)$
1	1	$(1, 1)$
2	2	$(2, 2)$
3	3	$(3, 3)$

For $x < 0$

x	y	(x, y)
−4	4	$(-4, 4)$
−3	3	$(-3, 3)$
−2	2	$(-2, 2)$
−1	1	$(-1, 1)$

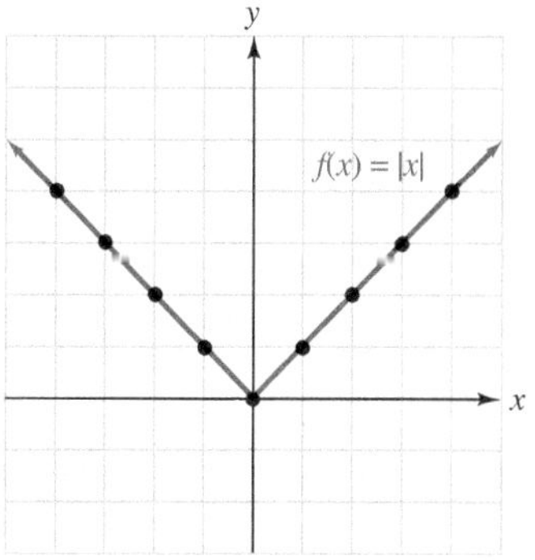

Figure 12-39

The function, shown in Figure 12-39, is decreasing on the interval $(-\infty, 0)$ and is increasing on the interval $(0, \infty)$.

EXAMPLE 1 Graph the piecewise-defined function given by

$$f(x) = \begin{cases} x^2 \text{ when } x \leq 0 \\ x \text{ when } 0 < x < 2 \\ -1 \text{ when } x \geq 2 \end{cases}$$

and determine where the function is increasing, decreasing, or constant.

Solution For each number x, we decide which of the three equations will be used to find the corresponding value of y:

- For numbers $x \leq 0$, $f(x)$ is determined by $f(x) = x^2$, and the graph is the left half of a parabola. See Figure 12-40. Since the values of $f(x)$ decrease on this graph as x increases, the function is decreasing on the interval $(-\infty, 0)$.
- For numbers $0 < x < 2$, $f(x)$ is determined by $f(x) = x$, and the graph is part of a line. Since the values of $f(x)$ increase on this graph as x increases, the function is increasing on the interval $(0, 2)$.
- For numbers $x \geq 2$, $f(x)$ is the constant -1, and the graph is part of a horizontal line. Since the values of $f(x)$ remain constant on this line, the function is constant on the interval $(2, \infty)$.

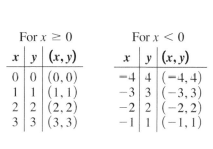

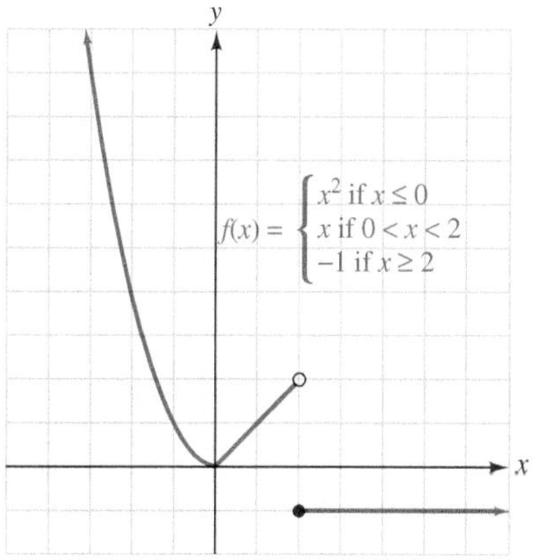

Figure 12-40

Unless otherwise noted, all content on this page is © Cengage Learning.

The use of solid and open circles on the graph indicates that $f(x) = -1$ when $x = 2$.

Since every number x determines exactly one value y, the domain of this function is the interval $(-\infty, \infty)$. The range is $\{-1\} \cup [0, \infty)$.

SELF CHECK 1 Graph: $f(x) = \begin{cases} 2x \text{ when } x \le 0 \\ \frac{1}{2}x \text{ when } x > 0 \end{cases}$ and determine where the function is increasing, decreasing, or constant.

2 Graph the greatest integer function.

The **greatest integer function** is important in computer applications. It is a function determined by the equation

$f(x) = [\![x]\!]$ Read as "y equals the greatest integer in x."

where the value of y that corresponds to x is the greatest integer that is less than or equal to x. For example,

$$[\![4.7]\!] = 4, \quad \left[\!\!\left[2\frac{1}{2}\right]\!\!\right] = 2, \quad [\![\pi]\!] = 3, \quad [\![-3.7]\!] = -4, \quad [\![-5.7]\!] = -6$$

COMMENT One way to help determine the greatest integer for a value that is not an integer is to visualize the number on a number line. The integer directly to the left of the number is the greatest integer.

EXAMPLE 2 Graph: $f(x) = [\![x]\!]$

Solution We list several intervals and the corresponding values of the greatest integer function:

$[0, 1)$ $y = [\![x]\!] = 0$ For numbers from 0 to 1, not including 1, the greatest integer in the interval is 0.

$[1, 2)$ $y = [\![x]\!] = 1$ For numbers from 1 to 2, not including 2, the greatest integer in the interval is 1.

$[2, 3)$ $y = [\![x]\!] = 2$ For numbers from 2 to 3, not including 3, the greatest integer in the interval is 2.

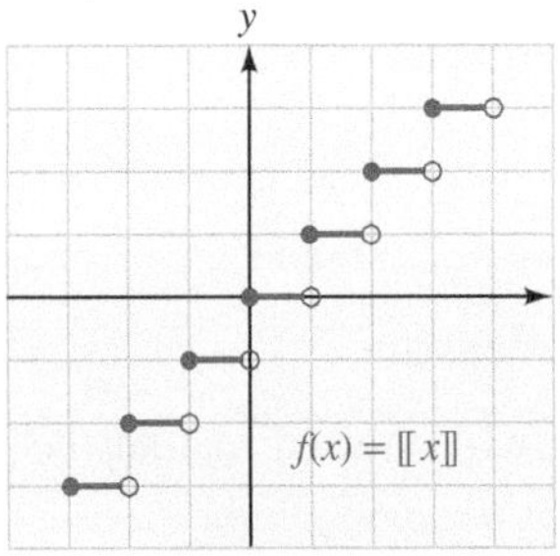

Figure 12-41

In each interval, the values of y are constant, but they jump by 1 at integer values of x. The graph is shown in Figure 12-41. From the graph, we see that the domain is $(-\infty, \infty)$, and the range is the set of integers $\{\ldots, -3, -2, -1, 0, 1, 2, 3, \ldots\}$.

SELF CHECK 2 Graph: $f(x) = [\![x]\!] + 1$

Since the greatest integer function contains a series of horizontal line segments, it is an example of a group of functions called **step functions**.

3 Solve an application involving the greatest integer function or a piecewise-defined function.

EXAMPLE 3 **PRINTING STATIONERY** To print stationery, a printer charges \$10 for setup charges, plus \$20 for each box. Any portion of a box counts as a full box. Graph this step function.

Unless otherwise noted, all content on this page is © Cengage Learning.

Solution If we order stationery and cancel the order before it is printed, the cost will be \$10. Thus, the ordered pair $(0, 10)$ will be on the graph.

If we purchase 1 box, the cost will be \$10 for setup plus \$20 for printing, for a total cost of \$30. Thus, the ordered pair $(1, 30)$ will be on the graph.

The cost of $1\frac{1}{2}$ boxes will be the same as the cost of 2 boxes, or \$50. Thus, the ordered pairs $(1.5, 50)$ and $(2, 50)$ will be on the graph. The complete graph is shown in Figure 12-42.

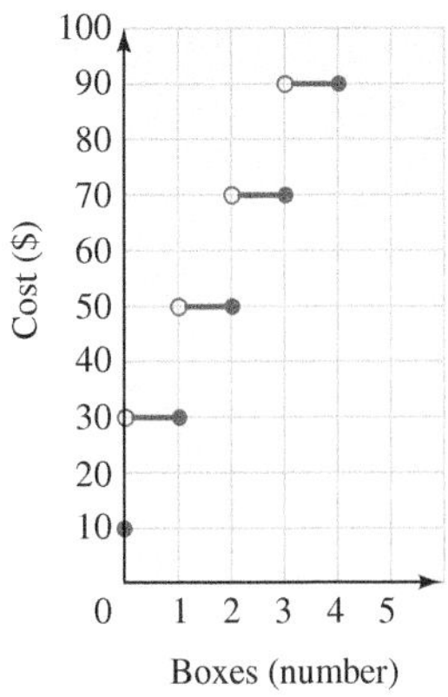

Figure 12-42

SELF CHECK 3 How much will $3\frac{1}{2}$ boxes cost?

EXAMPLE 4 **COMPUTING TAXES** The IRS taxes for a single person earning an adjusted gross income less than \$83,500 a year can be computed from the piecewise-defined function

$$f(x) = \begin{cases} 0.10x & \text{if } 0 \le x < 8{,}500 \\ 0.15(x - 8{,}500) + 850 & \text{if } 8{,}500 \le x < 34{,}500 \\ 0.25(x - 34{,}500) + 4{,}750 & \text{if } 34{,}500 \le x < 83{,}500 \end{cases}$$

Compute the taxes owed for an adjusted gross income of \$23,257.

Solution Because \$23,257 falls between \$8,500 and \$34,500, we use the function $f(x) = 0.15(x - 8{,}500) + 850$ to compute the taxes owed. We evaluate this function for $x = 23{,}257$.

$$\begin{aligned} f(x) &= 0.15(x - 8{,}500) + 850 \\ f(\mathbf{23{,}257}) &= 0.15(\mathbf{23{,}257} - 8{,}500) + 850 \\ &= 0.15(14{,}757) + 850 \\ &= 2{,}213.55 + 850 \\ &= 3{,}063.55 \end{aligned}$$

The taxes owed on an adjusted gross income of \$23,257 are \$3,063.55.

SELF CHECK 4 Compute the taxes owed on an adjusted gross income of \$82,368.

SELF CHECK ANSWERS

1.

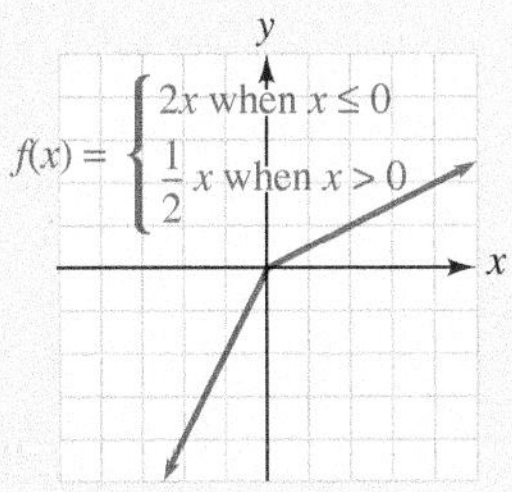

The function is increasing on the interval $(-\infty, \infty)$.

2.

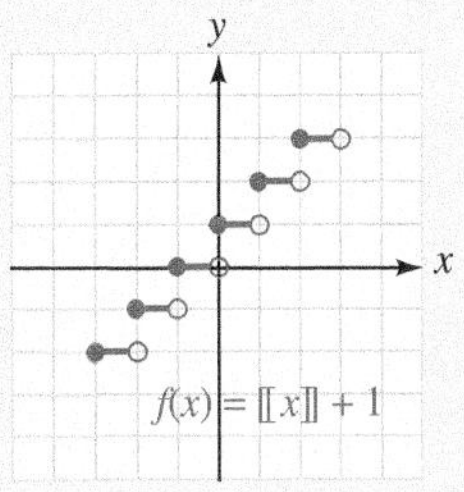

3. \$90 **4.** \$16,717.00

Unless otherwise noted, all content on this page is © Cengage Learning.

NOW TRY THIS

Graph the piecewise-defined function.

$$f(x) = \begin{cases} -x^2 \text{ if } x < 1 \\ [\![x]\!] \text{ if } 1 \le x < 4 \\ x - 1 \text{ if } x \ge 4 \end{cases}$$

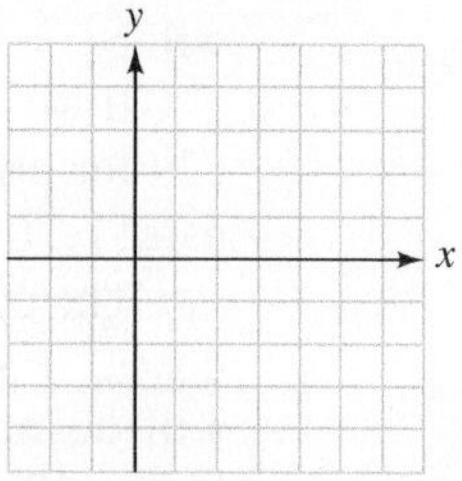

12.5 Exercises

WARM-UPS *For each function, evaluate f(1), f(2), and f(3). Determine if the function values, y, are increasing or decreasing.*

1. $f(x) = x^2$
2. $f(x) = |x|$

For each function, find f(−3), f(−2), and f(−1). Determine if the function values, y, are increasing or decreasing.

3. $f(x) = x^2$
4. $f(x) = |x|$

REVIEW *Find the value of x. Assume that lines r and s are parallel.*

5.

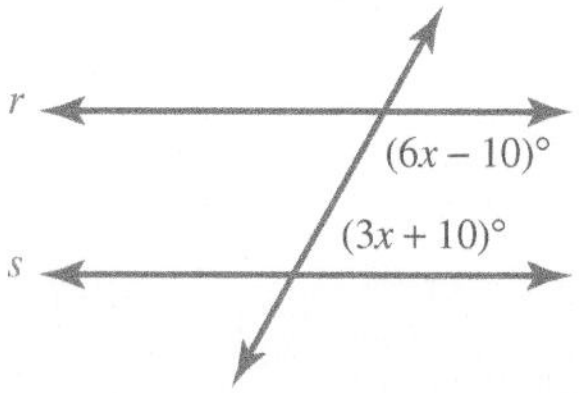

6.

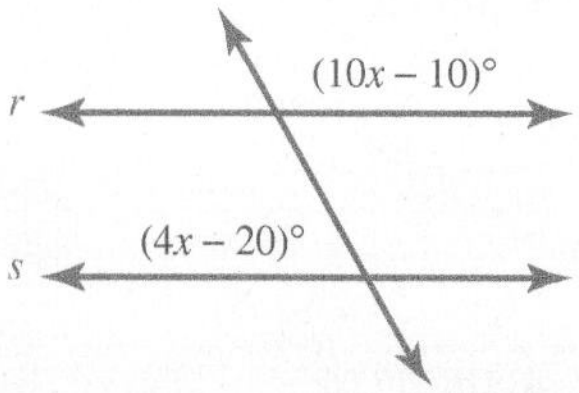

VOCABULARY AND CONCEPTS *Fill in the blanks.*

7. Piecewise-defined functions are defined by using different functions for different parts of their _______.
8. When the values $f(x)$ increase as the values of x increase over an interval, we say that the function is ________ over that interval.
9. In a _______ function, as the values of x increase the values of ____ are the same.
10. When the values of $f(x)$ decrease as the values of x _______ over an interval, we say that the function is decreasing over that interval.
11. When the graph of a function contains a series of horizontal line segments, the function is called a ____ function.
12. The function that gives the largest integer that is less than or equal to a number x is called the ____________ function.

GUIDED PRACTICE *Give the intervals on which each function is increasing, decreasing, or constant.* (OBJECTIVE 1)

13.

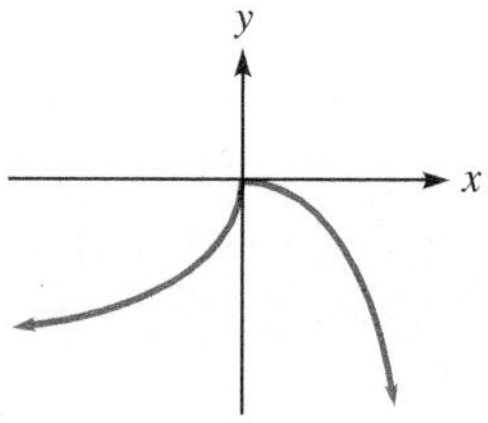

14.

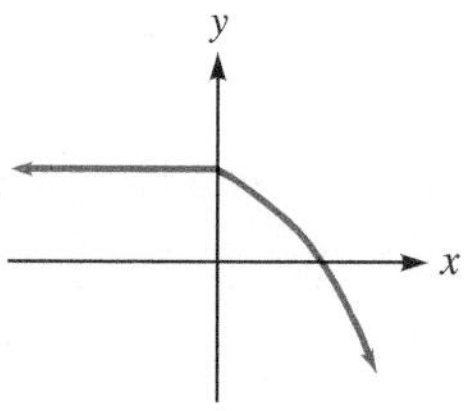

15.

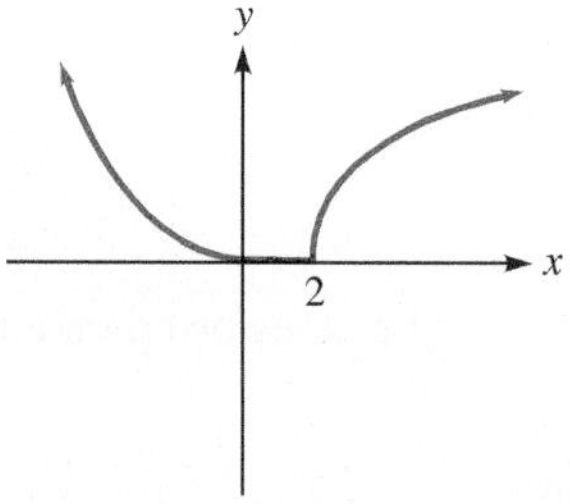

16.

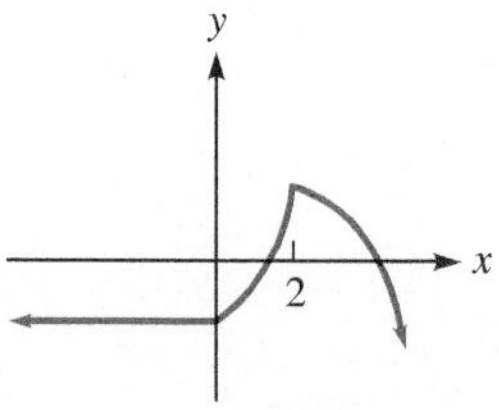

Graph each function and state the intervals on which f(x) is increasing, decreasing, or constant. SEE EXAMPLE 1. (OBJECTIVE 1)

17. $f(x) = \begin{cases} -1 \text{ when } x \le 0 \\ x \text{ when } x > 0 \end{cases}$

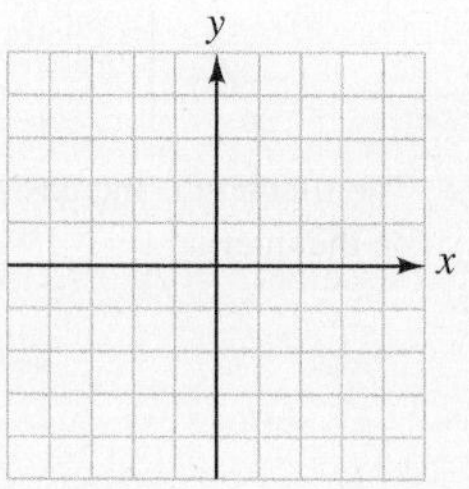

18. $f(x) = \begin{cases} -2 \text{ if } x \le 0 \\ x^2 \text{ if } x > 0 \end{cases}$

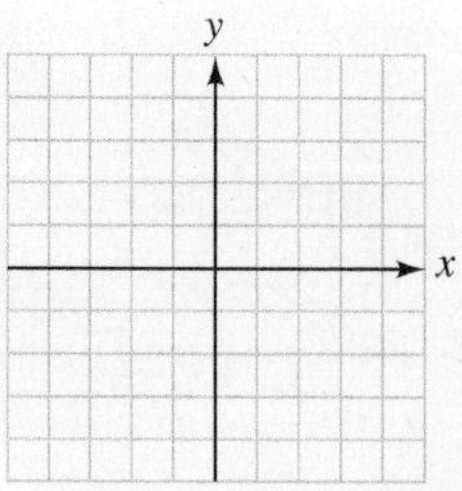

Unless otherwise noted, all content on this page is © Cengage Learning.

19. $f(x) = \begin{cases} -x \text{ if } x \le 0 \\ x \text{ if } 0 < x < 2 \\ -x \text{ if } x \ge 2 \end{cases}$

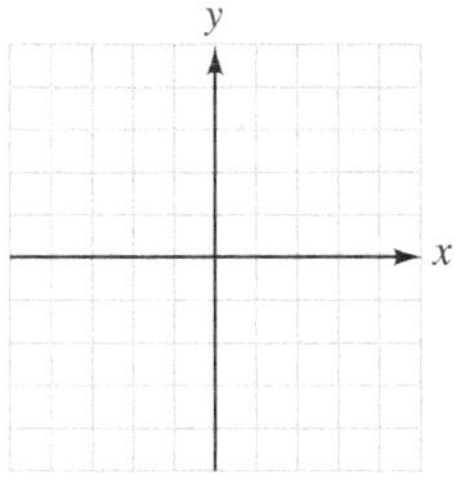

20. $f(x) = \begin{cases} -x \text{ if } x < 0 \\ x^2 \text{ if } 0 \le x \le 1 \\ 1 \text{ if } x > 1 \end{cases}$

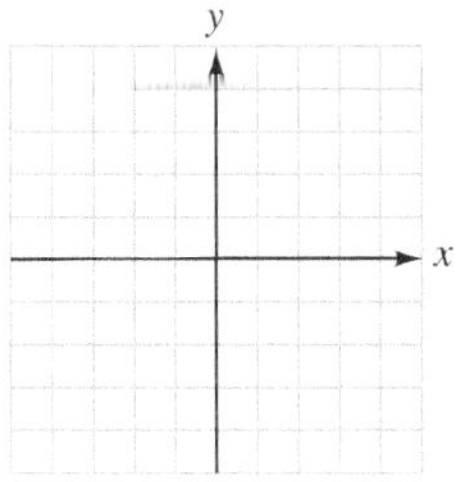

Graph each function. SEE EXAMPLE 2. (OBJECTIVE 2)

21. $f(x) = -[\![x]\!]$

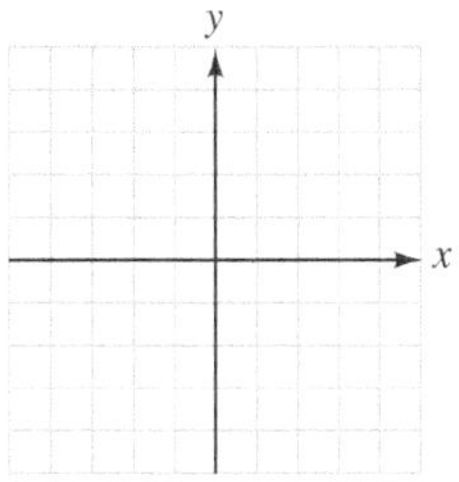

22. $f(x) = [\![x]\!] + 2$

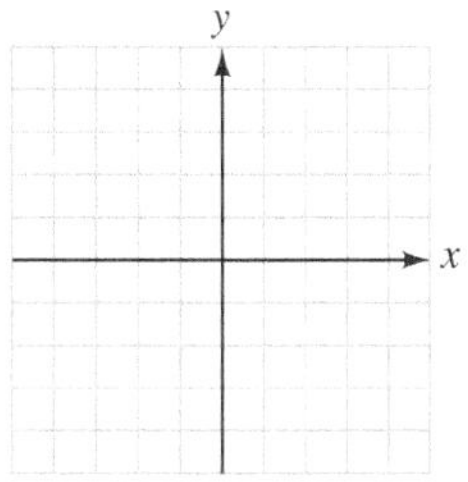

23. $f(x) = 2[\![x]\!]$

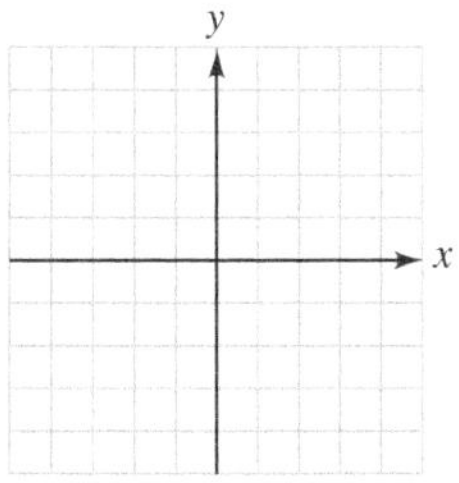

24. $f(x) = \left[\!\left[\frac{1}{2}x\right]\!\right]$

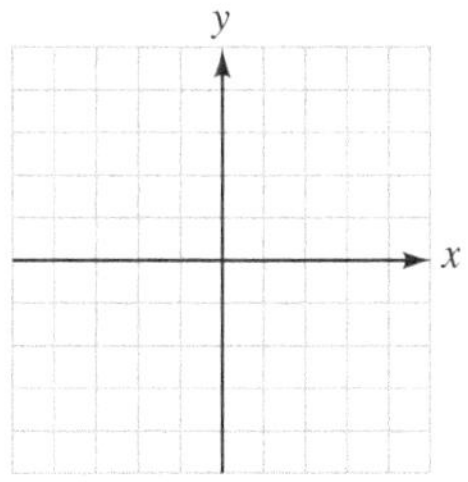

ADDITIONAL PRACTICE

25. Signum function Computer programmers use a function denoted by $f(x) = \operatorname{sgn} x$ that is defined in the following way:

$$f(x) = \begin{cases} -1 \text{ if } x < 0 \\ 0 \text{ if } x = 0 \\ 1 \text{ if } x > 0 \end{cases}$$

Graph this function.

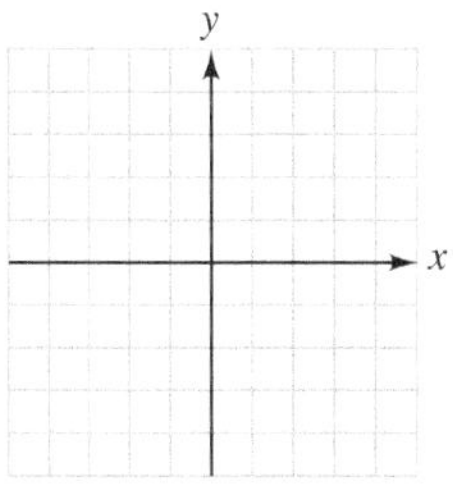

26. Heaviside unit step function This function, used in calculus, is defined by

$$f(x) = \begin{cases} 1 \text{ if } x > 0 \\ 0 \text{ if } x < 0 \end{cases}$$

Graph this function.

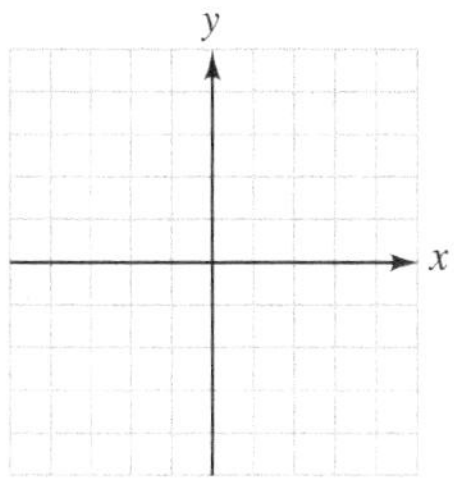

APPLICATIONS *Solve each application.* SEE EXAMPLES 3 AND 4. (OBJECTIVE 3)

27. Renting a ski jet A marina charges \$20 to rent a ski jet for 1 hour, plus \$5 for every extra hour (or portion of an hour). Graph the ordered pairs (h, c), where h represents the number of hours and c represents the cost. Find the cost if the ski is used for 2.5 hours.

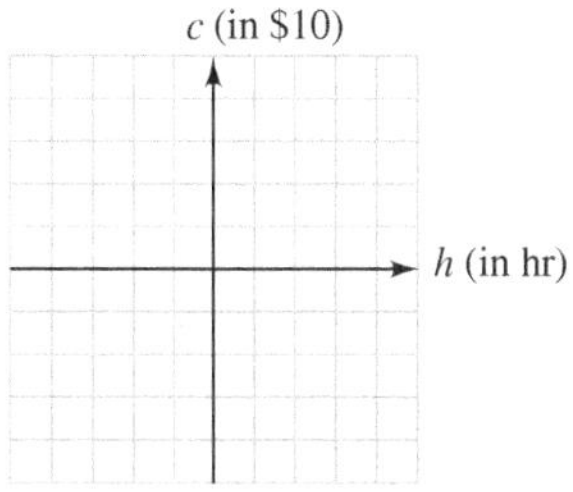

28. Riding in a taxi A cab company charges \$3 for a trip up to 1 mile, and \$2 for every extra mile (or portion of a mile). Graph the ordered pairs (m, c), where m represents the number of miles traveled and c represents the cost. Find the cost to ride $10\frac{1}{4}$ miles.

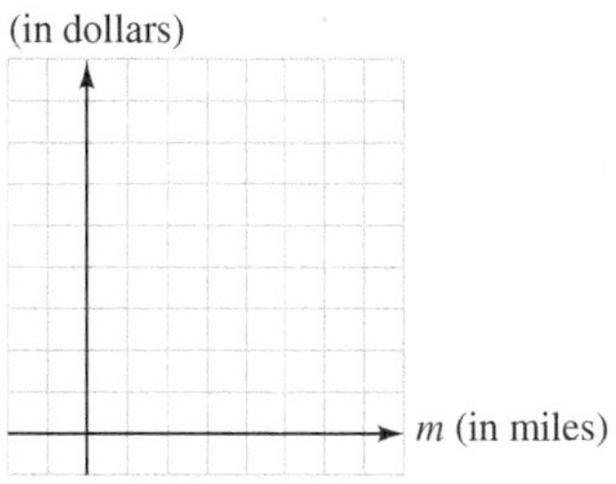

Unless otherwise noted, all content on this page is © Cengage Learning.

29. Information access Computer access to international data network A costs \$10 per day plus \$8 per hour or fraction of an hour. Network B charges \$15 per day, but only \$6 per hour or fraction of an hour. For each network, graph the ordered pairs (t, C), where t represents the connect time and C represents the total cost. Find the minimum daily usage at which it would be more economical to use network B.

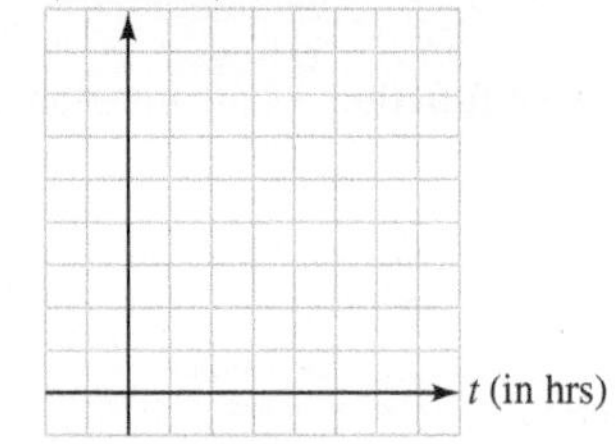

30. Royalties A publisher has agreed to pay the author of a novel 7% royalties on sales of the first 50,000 copies and 10% on sales thereafter. If the book sells for \$10, express the royalty income I as a function of s, the number of copies sold, and graph the function. (*Hint*: When sales are into the second 50,000 copies, how much was earned on the first 50,000?)

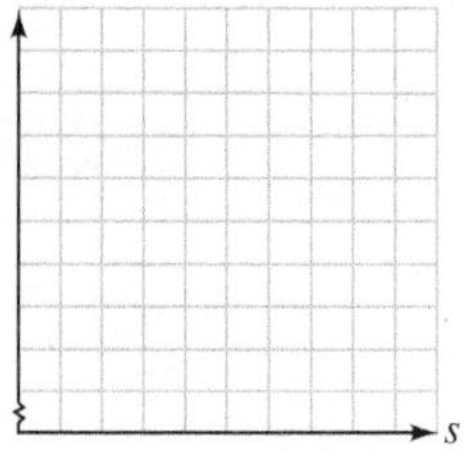

WRITING ABOUT MATH

31. Explain how to decide whether a function is increasing on the interval (a, b).

32. Describe the greatest integer function and why it might be called a step function.

SOMETHING TO THINK ABOUT

33. Find a piecewise-defined function that is increasing on the interval $(-\infty, 6)$ and decreasing on the interval $(6, \infty)$.

34. Find a piecewise-defined function that is constant on the interval $(-\infty, 0)$, increasing on the interval $(0, 3)$, and decreasing on the interval $(3, \infty)$.

Project

The zillionaire G. I. Luvmoney is known for his love of flowers. On his estate, he recently set aside a circular plot of land with a radius of 100 yards to be made into a flower garden. He has hired your landscape design firm to do the job. If Luvmoney is satisfied, he will hire your firm to do more lucrative jobs. Here is Luvmoney's plan.

The center of the circular plot of land is to be the origin of a rectangular coordinate system. You are to make 100 circles, all centered at the origin, with radii of 1 yard, 2 yards, 3 yards, and so on up to the outermost circle, which will have a radius of 100 yards. Inside the innermost circle, he wants a fountain with a circular walkway around it. In the ring between the first and second circle, he wants to plant his favorite kind of flower, in the next ring his second favorite, and so on, until you reach the edge of the circular plot. Luvmoney provides you with a list ranking his 99 favorite flowers.

The first thing he wants to know is the area of each ring, so that he will know how many of each plant to order. Then he wants a simple formula that will give the area of any ring just by substituting in the number of the ring.

He also wants a walkway to go through the garden in the form of a hyperbolic path, following the equation

$$x^2 - \frac{y^2}{9} = 1$$

Luvmoney wants to know the x- and y-coordinates of the points where the path will intersect the circles, so that those points can be marked with stakes to keep gardeners from planting flowers where the walkway will later be built. He wants a formula (or two) that will enable him to put in the number of a circle and get out the intersection points.

Finally, although cost has no importance for Luvmoney, his accountants will want an estimate of the total cost of all of the flowers.

You go back to your office with Luvmoney's list. You find that because the areas of the rings grow from the inside of the garden to the outside, and because of Luvmoney's ranking of flowers, a strange thing happens. The first ring of flowers will cost \$360, and the flowers in every ring after that will cost 110% as much as the flowers in the previous ring. That is, the second ring of flowers will cost $\$360(1.1) = \$396$, the third will cost \$435.60, and so on.

Answer all of Luvmoney's questions, and show work that will convince him that you are right.

Unless otherwise noted, all content on this page is © Cengage Learning.

Reach for Success MANAGING YOUR STRESS

Now that you have learned several relaxation techniques to help manage your stress, let's focus on the power of positive self-talk. Typically, if you *think* you cannot do something, it will usually become a self-fulfilling prophecy. With a "can-do" attitude, mathematics can be learned. However, with anything you want to do (and do well) it is going to require practice. Prior to practicing, you need to believe you can do it.

Would you describe your attitude toward mathematics as positive or negative? ______________ ______________	Do you know *why* you have this attitude? ______________ ______________ ______________
Ask yourself this, "Am I willing to do what it takes to learn mathematics this semester?" ______________ ______________	If so, what can you do this semester to become a stronger mathematics student? **1.** ______________ **2.** ______________ **3.** ______________ **4.** ______________
In society, people are embarrassed if they cannot read or write. Why do you think people accept not being able to do mathematics? ______________ ______________ ______________	What attitude do you see toward mathematics in those around you in your personal life? ______________ ______________ ______________ ______________ Find a reference to positive or negative attitudes toward mathematics in a movie or television program. ______________ ______________

Academic preparation will help manage your stress and help you keep a positive attitude toward the subject. Keep using all the relaxation techniques to manage your stress!

12 Review

SECTION 12.1 The Circle and the Parabola

DEFINITIONS AND CONCEPTS	EXAMPLES

DEFINITIONS AND CONCEPTS

Equations of a circle:

Standard forms:

$x^2 + y^2 = r^2$
center $(0, 0)$, radius r

EXAMPLES

To graph a circle, we need to know the center and the radius.
Graph each circle.

a. $x^2 + y^2 = 25$

$C(0, 0)$ Compare to the formula.

$r^2 = 25$ Compare to the formula.

$r = 5$

The center is $(0, 0)$ and the radius is 5 units.

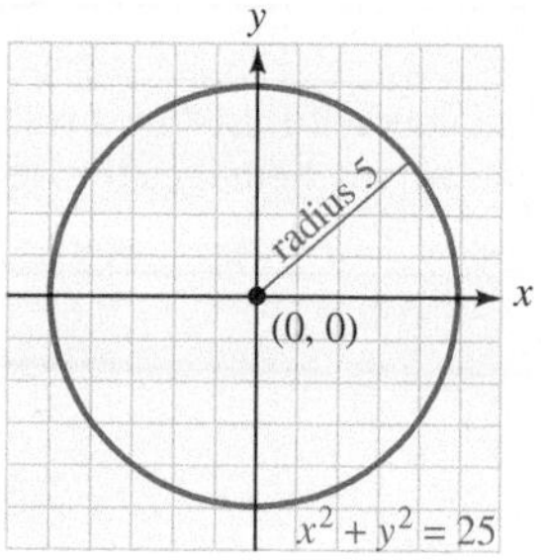

$(x - h)^2 + (y - k)^2 = r^2$
center (h, k), radius r

b. $(x - 3)^2 + (y + 4)^2 = 16$

$(x - 3)^2 + [y - (-4)]^2 = 16$ Write in standard form.

$C(3, -4)$ Comparing to the formula, $h = 3$ and $k = -4$.

$r^2 = 16$ Compare to the formula.

$r = 4$ Take the positive square root of 16.

The center is at $(3, -4)$ and the radius is 4 units.

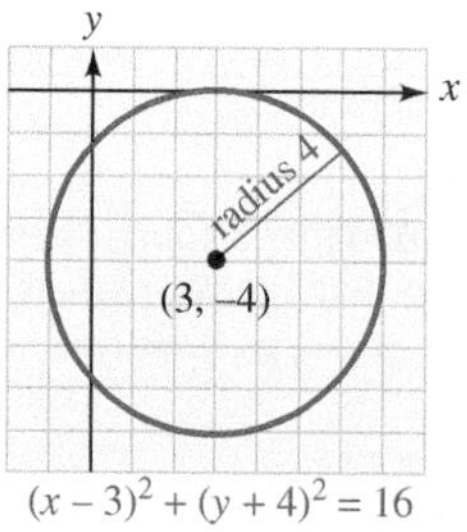

General form:

$x^2 + y^2 + Dx + Ey + F = 0$

c. $x^2 + y^2 + 2x + 6y + 6 = 0$
To write the equation in standard form, we can complete the square on both x and y.

$$x^2 + y^2 + 2x + 6y = -6$$ Subtract 6 from both sides.

$$x^2 + 2x + y^2 + 6y = -6$$ Rearrange the terms.

$$x^2 + 2x + \mathbf{1} + y^2 + 6y + \mathbf{9} = -6 + \mathbf{1} + \mathbf{9}$$ Complete the square on x and y.

$$(x + 1)^2 + (y + 3)^2 = 4$$ Factor and simplify.

Unless otherwise noted, all content on this page is © Cengage Learning.

Comparing to the standard form, we see that $C(-1, -3)$, $r^2 = 4$, and $r = 2$. The center is at $(-1, -3)$ with a radius of 2 units.

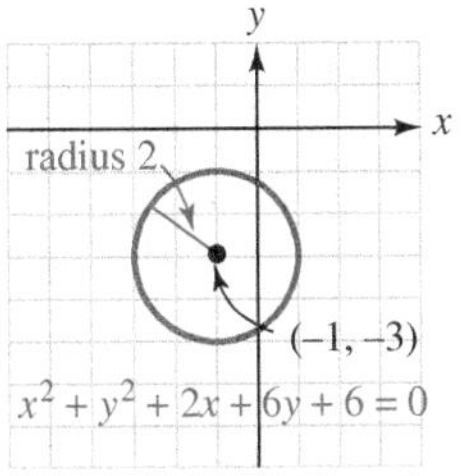

Equations of parabolas:

Parabola opening	Vertex at origin	
Upward	$y = ax^2$	$(a > 0)$
Downward	$y = ax^2$	$(a < 0)$
Right	$x = ay^2$	$(a > 0)$
Left	$x = ay^2$	$(a < 0)$

Parabola opening	Vertex at (h, k)	
Upward	$y = a(x - h)^2 + k$	$(a > 0)$
Downward	$y = a(x - h)^2 + k$	$(a < 0)$
Right	$x = a(y - k)^2 + h$	$(a > 0)$
Left	$x = a(y - k)^2 + h$	$(a < 0)$

Graph each parabola.

a. $x = y^2$
The parabola is horizontal and opens to the right because $a > 0$. To obtain the graph, we can plot several points and connect them with a smooth curve.

$x = y^2$

x	y	(x, y)
0	0	$(0, 0)$
4	2	$(4, 2)$
4	−2	$(4, -2)$
9	3	$(9, 3)$
9	−3	$(9, -3)$

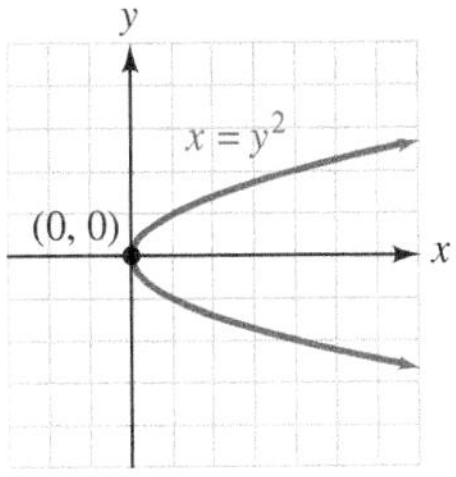

b. $x = -2(y - 1)^2 + 3$
The parabola is horizontal and opens to the left because $a < 0$. To obtain the graph, we can plot several points and connect them with a smooth curve.

$x = -2(y - 1)^2 + 3$

x	y	(x, y)
1	0	$(1, 0)$
1	2	$(1, 2)$
3	1	$(3, 1)$
−5	−1	$(-5, -1)$
−5	3	$(-5, 3)$

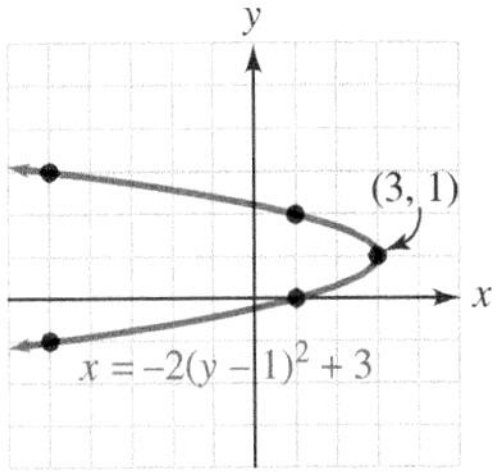

REVIEW EXERCISES

Graph each equation.

1. $(x - 1)^2 + (y + 2)^2 = 9$

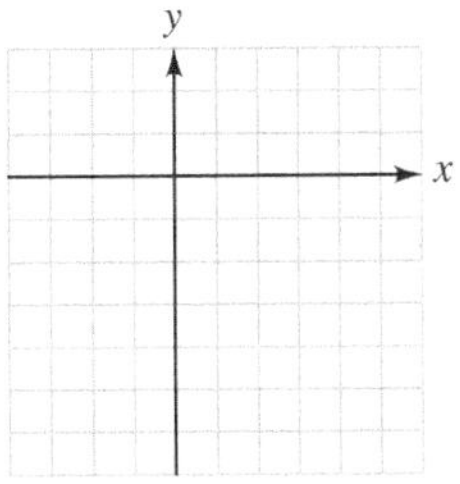

2. $x^2 + y^2 = 16$

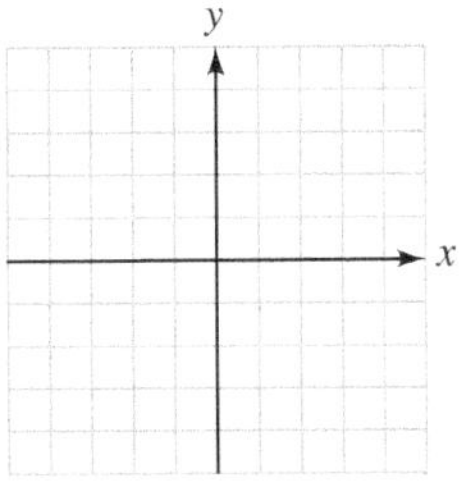

3. Write the equation in standard form and graph it.

$x^2 + y^2 + 4x - 2y = 4$

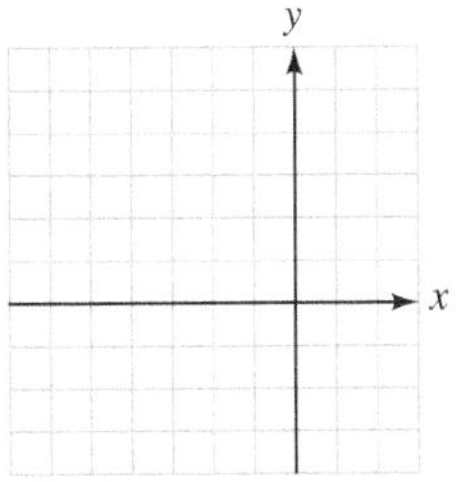

Unless otherwise noted, all content on this page is © Cengage Learning.

Graph each equation.

4. $x = -3(y-2)^2 + 5$

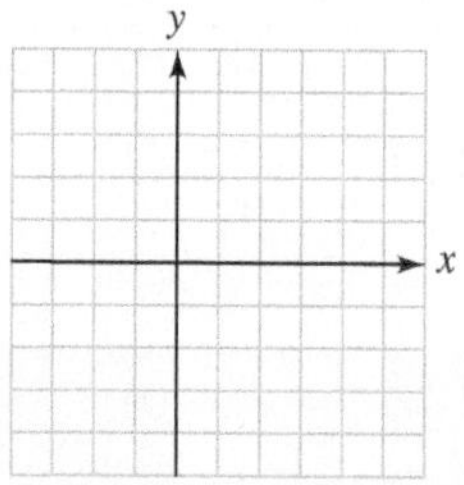

5. $x = 2(y+1)^2 - 2$

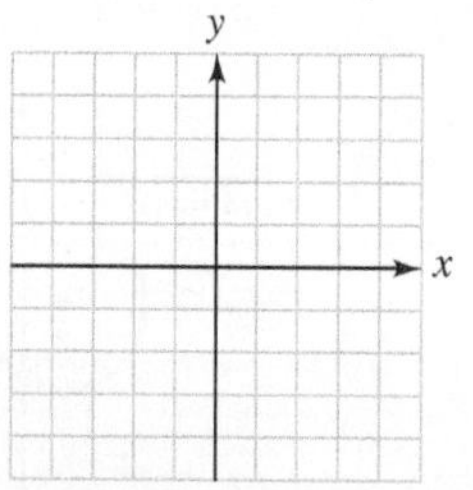

6. $y = -2(x-1)^2 + 3$

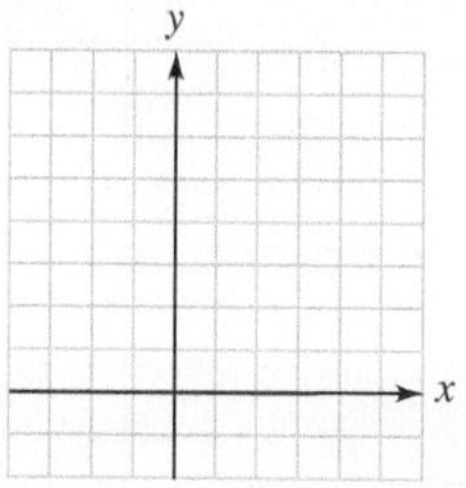

7. $y = (x+2)^2 - 3$

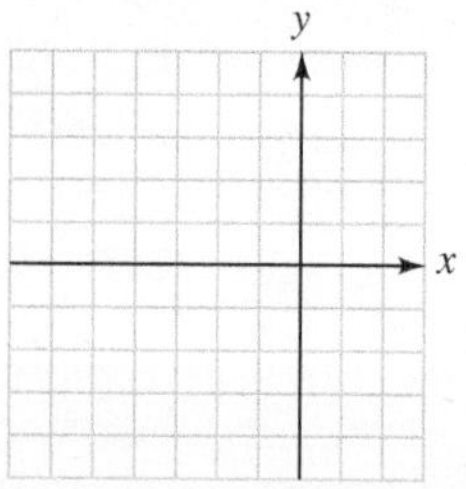

SECTION 12.2 The Ellipse

DEFINITIONS AND CONCEPTS

Equations of an ellipse:

Standard forms:

Center at $(0, 0)$

$$\frac{x^2}{a^2} + \frac{y^2}{b^2} = 1 \quad (a > b > 0)$$

$$\frac{x^2}{b^2} + \frac{y^2}{a^2} = 1 \quad (a > b > 0)$$

Center at (h, k)

$$\frac{(x-h)^2}{a^2} + \frac{(y-k)^2}{b^2} = 1$$

$$\frac{(x-h)^2}{b^2} + \frac{(y-k)^2}{a^2} = 1$$

In either case,

The length of the major axis is $2a$.
The length of the minor axis is $2b$.

General form:

$$Ax^2 + Cy^2 + Dx + Ey + F = 0$$

EXAMPLES

To graph an ellipse, we need to know the center and the endpoints of the major and minor axes.

Graph each ellipse:

a. $\frac{x^2}{9} + \frac{y^2}{16} = 1$

Comparing to the formula, the center is $(0, 0)$ and

$a^2 = 16 \qquad b^2 = 9$
$a = 4 \qquad b = 3$

The ellipse will be vertical because the larger denominator is associated with the y-term. The endpoints of the major axis will be 4 units above and below the center, $(0, 4)$ and $(0, -4)$. The endpoints of the minor axis will be 3 units to the left and right of center, $(3, 0)$ and $(-3, 0)$.

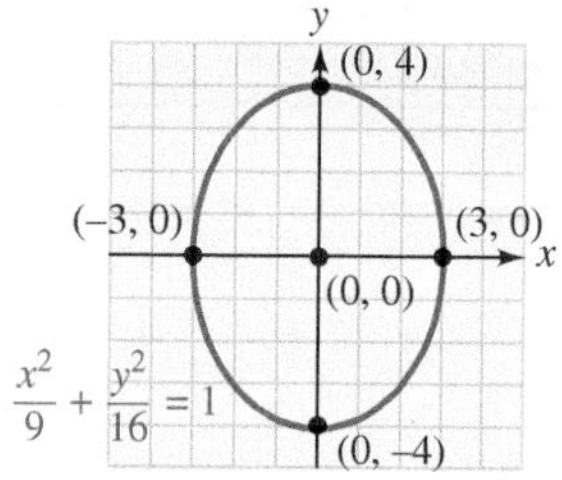

b. $\frac{(x-3)^2}{9} + \frac{(y-4)^2}{4} = 1$

From the formula, the center is $(3, 4)$ and

$a^2 = 9 \qquad b^2 = 4$
$a = 3 \qquad b = 2$

The ellipse will be horizontal because the larger denominator is associated with the x-term. The endpoints of the major axis will be 3 units to the left and right of the center, $(6, 4)$ and $(0, 4)$. The endpoints of the minor axis will be 2 units above and below the center, $(3, 6)$ and $(3, 2)$.

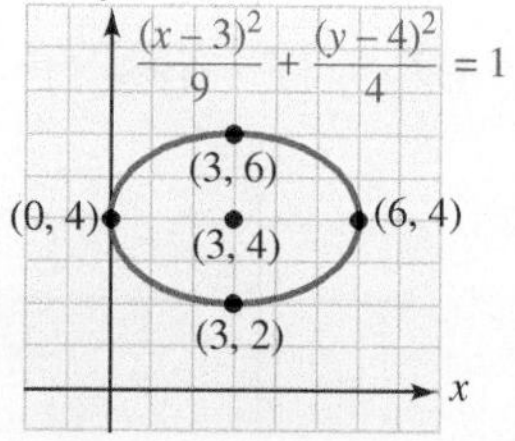

To find the standard equation of the ellipse with equation $4x^2 + y^2 - 8x - 2y - 11 = 0$, proceed as follows:

$4x^2 + y^2 - 8x - 2y = 11$	Add 11 to both sides.
$4x^2 - 8x + y^2 - 2y = 11$	Apply the commutative property of addition to rearrange terms.

Unless otherwise noted, all content on this page is © Cengage Learning.

Equation	Explanation
$4(x^2 - 2x) + y^2 - 2y = 11$	Factor to obtain a coefficient of 1 for the term involving x-squared.
$4(x^2 - 2x + \mathbf{1}) + (y^2 - 2y + \mathbf{1}) = 11 + \mathbf{4} + \mathbf{1}$	Complete the square on both x and y.
$4(x-1)^2 + (y-1)^2 = 16$	Factor and simplify.
$\frac{(x-1)^2}{4} + \frac{(y-1)^2}{16} = 1$	Divide both sides by 16.

REVIEW EXERCISES

Graph each ellipse.

8. $9x^2 + 16y^2 = 144$

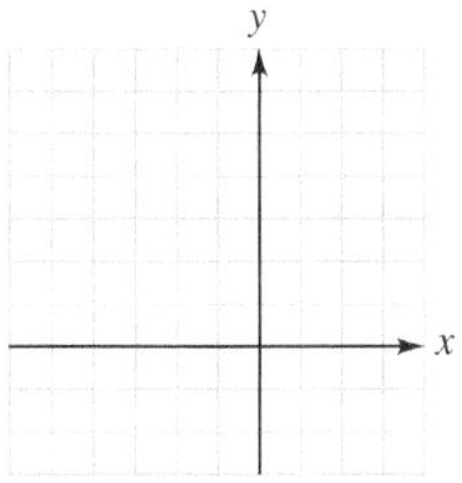

9. $\frac{(x-2)^2}{4} + \frac{(y-1)^2}{9} = 1$

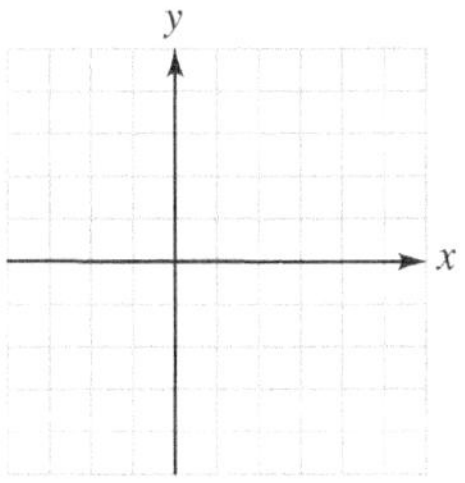

10. Write the equation in standard form and graph it.

$4x^2 + 9y^2 + 8x - 18y = 23$

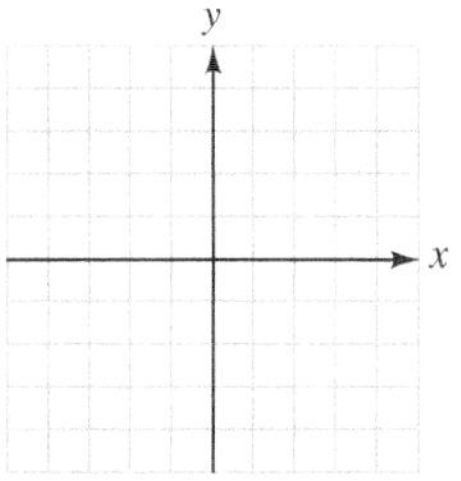

SECTION 12.3 The Hyperbola

DEFINITIONS AND CONCEPTS	EXAMPLES
Equations of a hyperbola:	To graph a hyperbola, we need to know where it is centered, the coordinates of the vertices, and the location of the asymptotes.

Equations of a hyperbola:

Standard forms:

Center at $(0, 0)$

$\frac{x^2}{a^2} - \frac{y^2}{b^2} = 1$ opens left or right $V_1(a, 0)$, $V_2(-a, 0)$

$\frac{y^2}{a^2} - \frac{x^2}{b^2} = 1$ opens upward or downward $V_1(0, a)$, $V_2(0, -a)$

EXAMPLES

To graph a hyperbola, we need to know where it is centered, the coordinates of the vertices, and the location of the asymptotes.

Graph: $9y^2 - 4x^2 = 36$

$\frac{9y^2}{36} - \frac{4x^2}{36} = 1$ Write the equation in standard form.

$\frac{y^2}{4} - \frac{x^2}{9} = 1$ Simplify each fraction.

From the previous equation, we can determine that $a = 2$ and $b = 3$. Because the y-term is positive, the hyperbola will be vertical and the vertices of the hyperbola are $V_1(0, 2)$ and $V_2(0, -2)$.

Since $b = 3$, we can use the points $(3, 0)$ and $(-3, 0)$ on the x-axis to help draw the fundamental rectangle. We then draw its extended diagonals to locate the asymptotes and sketch the hyperbola.

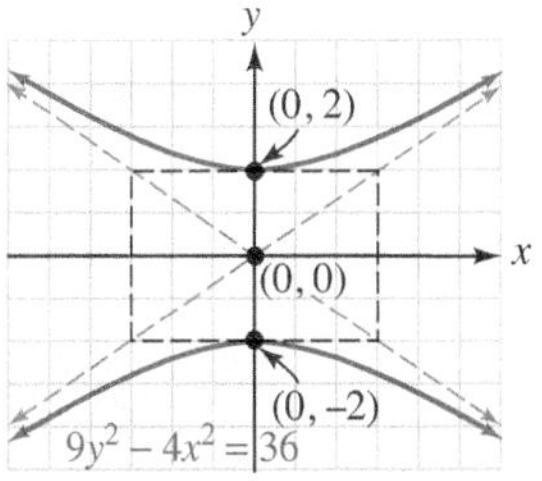

Unless otherwise noted, all content on this page is © Cengage Learning.

Center at (h, k) $\dfrac{(x-k)^2}{a^2} - \dfrac{(y-h)^2}{b^2} = 1$ opens left or right $\dfrac{(y-k)^2}{a^2} - \dfrac{(x-h)^2}{b^2} = 1$ opens upward or downward	Graph: $\dfrac{(x-3)^2}{4} - \dfrac{(y+1)^2}{4} = 1$ 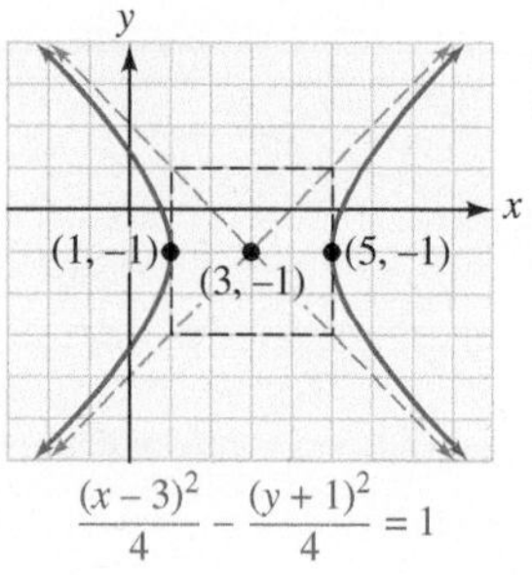 From the equation, we see that the hyperbola is centered at $(3, -1)$. Its vertices are located 2 units to the right and left of center, at $(5, -1)$ and $(1, -1)$. Since $b = 2$, we can count 2 units below and above center to locate points $(3, -3)$ and $(3, 1)$. With these points, we can draw the fundamental rectangle along with its extended diagonals to locate the asymptotes. Then we can sketch the hyperbola.
General form: $Ax^2 - Cy^2 + Dx + Ey + F = 0$	Write $x^2 - y^2 + 6x + 4y = 4$ in standard form and graph. $x^2 - y^2 + 6x + 4y = 4$ $x^2 + 6x - y^2 + 4y - 4 = 0$ $x^2 + 6x + 9 - (y^2 - 4y + 4) = 9$ $(x+3)^2 - (y-2)^2 = 9$ $\dfrac{(x+3)^2}{9} - \dfrac{(y-2)^2}{9} = 1$ Comparing to the standard equation, this is a horizontal hyperbola with centered at $(-3, 2)$. 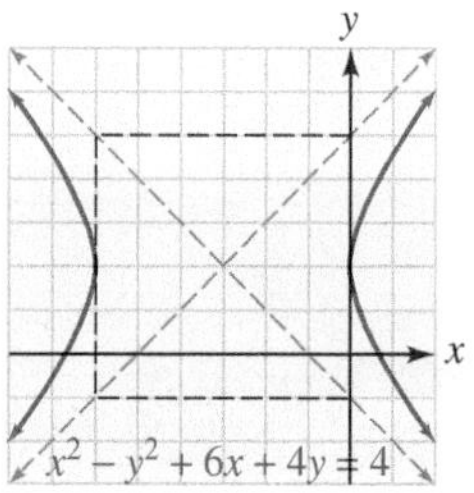
$xy = k, k \neq 0$	The equation $xy = 6$ is a hyperbola that does not intersect the x-axis or the y-axis. 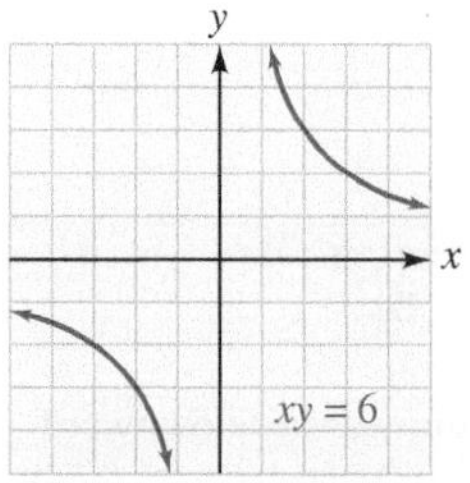

REVIEW EXERCISES

Graph each hyperbola.

11. $9x^2 - y^2 = -9$

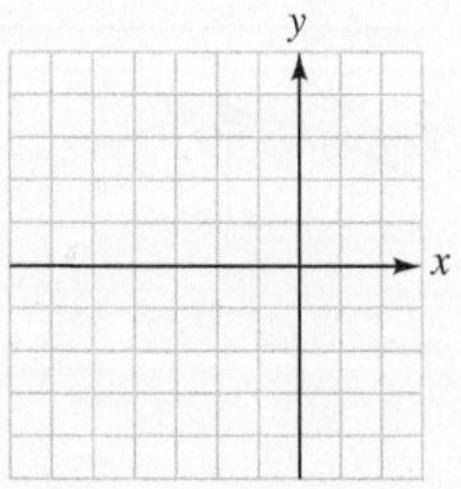

12. $xy = 9$

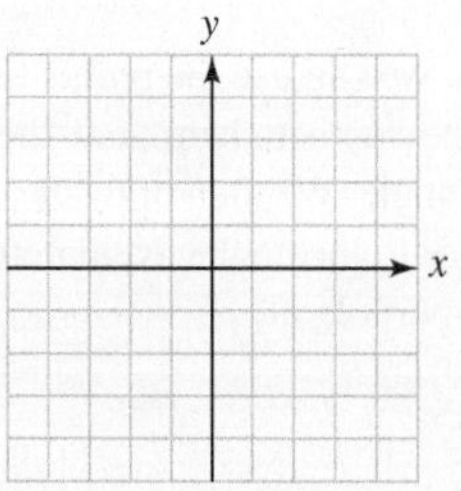

13. Determine whether the equation $2y^2 - 4x^2 + 8x - 8y = 8$ represents an ellipse or a hyperbola.

14. Write the equation in standard form and graph it.

$$9x^2 - 4y^2 - 18x - 8y = 31$$

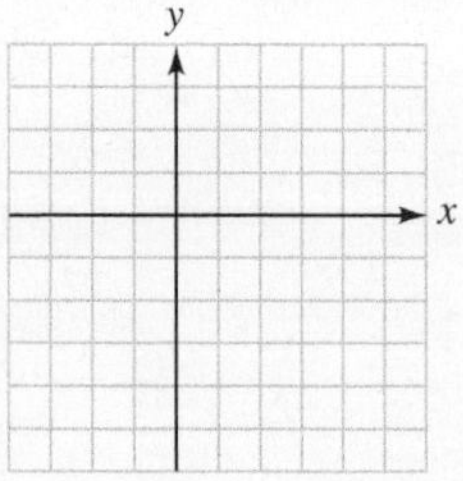

Unless otherwise noted, all content on this page is © Cengage Learning.

SECTION 12.4 Solving Systems of Equations and Inequalities Containing One or More Second-Degree Terms

DEFINITIONS AND CONCEPTS	EXAMPLES

Solve by graphing:
To solve a system of equations by graphing, graph each equation. The coordinates of the intersection points of the graphs will be the solutions of the system.

Solve the system by graphing.

$$\begin{cases} x^2 + y^2 = 4 \\ x - y = 2 \end{cases}$$

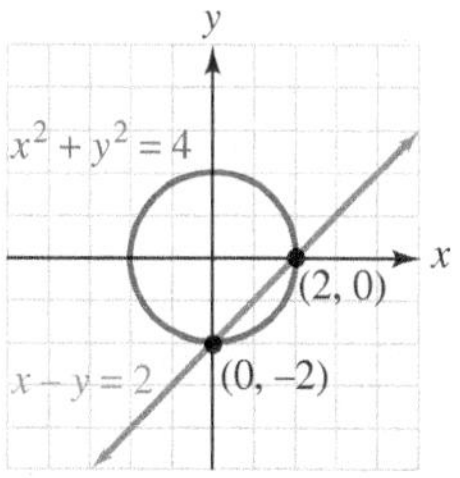

The equation $x^2 + y^2 = 4$ is that of a circle and $x - y = 2$ is that of a line. There is a possibility of zero, one, or two solutions.

After graphing the equations, we find that the two graphs intersect at the points $(2, 0)$, and $(0, -2)$. These are the solutions.

Solve by substitution:
Solve one equation for one variable, and substitute the result for that variable in the other equation.

Solve the system by substitution.

$$\begin{cases} x^2 + 9y^2 = 10 \\ y = x^2 \end{cases}$$

The equation $x^2 + 9y^2 = 10$ is that of an ellipse and $y = x^2$ is that of a parabola. We have the possibility of 0, 1, 2, 3, or 4 solutions. To find the solutions, proceed as follows:

$$x^2 + 9y^2 = 10$$

$$y + 9y^2 = 10 \quad \text{Substitute } y \text{ for } x^2.$$

$$9y^2 + y - 10 = 0 \quad \text{Write in quadratic form.}$$

$$(9y + 10)(y - 1) = 0 \quad \text{Factor } 9y^2 + y - 10.$$

$$9y + 10 = 0 \quad \text{or} \quad y - 1 = 0 \quad \text{Set each factor equal to 0.}$$

$$y = -\frac{10}{9} \quad \Big| \quad y = 1$$

Since $y = x^2$, the values of x can be found by solving the equations

$$x^2 = -\frac{10}{9} \quad \text{and} \quad x^2 = 1$$

Because $x^2 = -\frac{10}{9}$ has no real solutions, this possibility is discarded. The solutions of $x^2 = 1$ are

$$x = 1 \quad \text{or} \quad x = -1$$

The solutions of the system are $(1, 1)$ and $(-1, 1)$.

Solve by elimination (addition):
To solve a system of equations by elimination, add the equations to eliminate one of the variables. Then solve the resulting equation for the other variable.

If the original system does not contain opposite coefficients on one of the variables, multiply one or both of the equations by constants that will produce additive inverses.

Solve the system by elimination (addition).

$$\begin{cases} 4x^2 - y^2 = 1 \\ 4x^2 + y^2 = 1 \end{cases}$$

The equation $4x^2 - y^2 = 1$ is that of a hyperbola and the equation $4x^2 + y^2 = 1$ is that of an ellipse. We have the possibility of 0, 1, 2, 3, or 4 solutions.

If we add the equations $\begin{cases} 4x^2 - y^2 = 1 \\ 4x^2 + y^2 = 1 \end{cases}$, we have

$$8x^2 = 2$$

$$x^2 = \frac{1}{4}$$

$$x = \frac{1}{2}, -\frac{1}{2}$$

Unless otherwise noted, all content on this page is © Cengage Learning.

After substituting each value of x for y in the first equation, we have

$$4\left(\frac{1}{2}\right)^2 - y^2 = 1 \qquad 4\left(-\frac{1}{2}\right)^2 - y^2 = 1$$

$$4\left(\frac{1}{4}\right) - y^2 = 1 \qquad 4\left(\frac{1}{4}\right) - y^2 = 1$$

$$1 - y^2 = 1 \qquad 1 - y^2 = 1$$

$$-y^2 = 0 \qquad -y^2 = 0$$

$$y = 0 \qquad y = 0$$

The solutions are $\left(\frac{1}{2}, 0\right)$ and $\left(-\frac{1}{2}, 0\right)$.

Systems of inequalities are solved by graphing.

Solve: $\begin{cases} y \geq x^2 \\ y \leq -x^2 + 4 \end{cases}$

The graph of $y = x^2$ is the parabola that opens upward and has its vertex at the origin. The points with coordinates that satisfy the inequality $y \geq x^2$ are those points above the parabola and include the points on the parabola.

The graph of $y \leq -x^2 + 4$ is a parabola opening downward, with vertex at $(0, 4)$. The points with coordinates that satisfy the inequality are those points below the parabola and include the points on the parabola. The graph of the solution set of the system is the area between the parabolas.

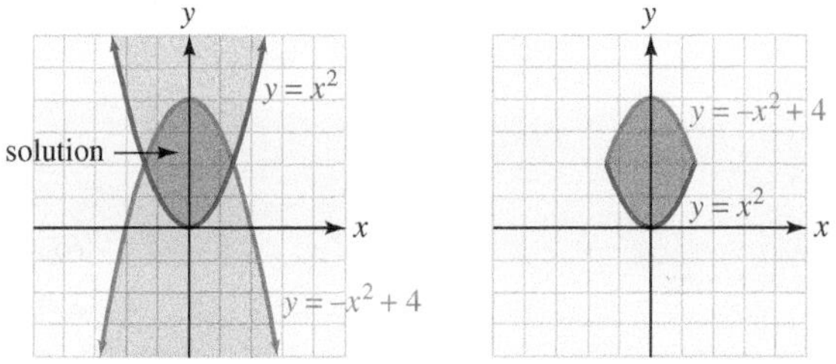

REVIEW EXERCISES

Solve each system of equations.

15. $\begin{cases} x^2 + y^2 = 13 \\ y = x^2 - 1 \end{cases}$

16. $\begin{cases} x^2 + y^2 = 20 \\ x^2 - y^2 = -12 \end{cases}$

17. Graph the solution set in the system: $\begin{cases} y \geq x^2 - 4 \\ y < x + 3 \end{cases}$

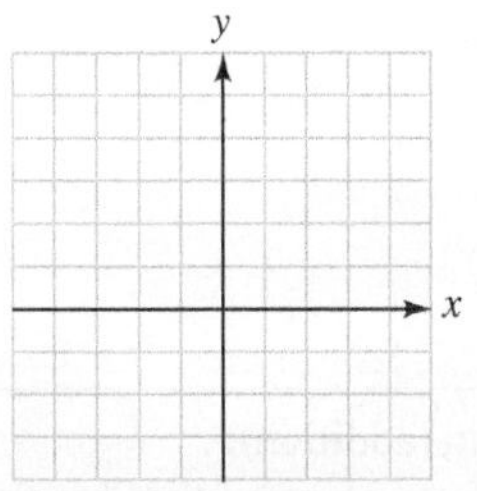

Unless otherwise noted, all content on this page is © Cengage Learning.

SECTION 12.5 Piecewise-Defined Functions and the Greatest Integer Function

DEFINITIONS AND CONCEPTS	EXAMPLES

Piecewise-defined functions:
A piecewise-defined function is a function that has different equations for different intervals of x.

Increasing, decreasing, and constant functions:
A function is increasing on the interval (a, b) if the values of $f(x)$ increase as x increases from a to b.

A function is decreasing on the interval (a, b) if the values of $f(x)$ decrease as x increases from a to b.

A function is constant on the interval (a, b) if the value of $f(x)$ is constant as x increases from a to b.

EXAMPLES

Graph the function

$$f(x) = \begin{cases} -x^2 & \text{if } x < 0 \\ -x & \text{if } 0 \le x < 3 \\ -1 & \text{if } x \ge 3 \end{cases}$$

and determine the intervals where the function is increasing, decreasing, and constant.

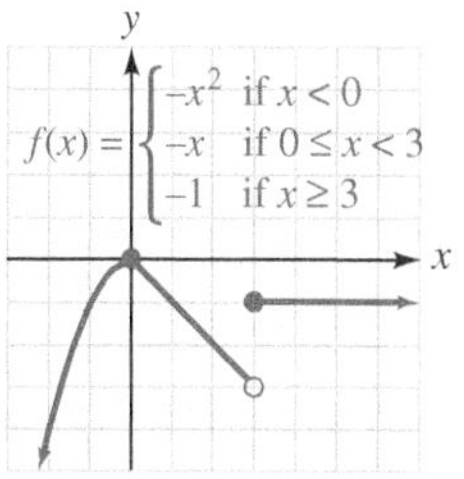

For each number x, we decide which of the three equations will be used to find the corresponding value of y:

- For numbers $x < 0$, $f(x)$ is determined by $f(x) = -x^2$, and the graph is the left half of a parabola. Since the values of $f(x)$ increase on this graph as x increases, the function is increasing on the interval $(-\infty, 0)$.
- For numbers $0 < x < 3$, $f(x)$ is determined by $f(x) = -x$, and the graph is part of a line. Since the values of $f(x)$ decrease on this graph as x increases, the function is decreasing on the interval $(0, 3)$.
- For numbers $x \ge 3$, $f(x)$ is the constant -1, and the graph is part of a horizontal line. Since the values of $f(x)$ remain constant on this line, the function is constant on the interval $(3, \infty)$.

Greatest integer function:
The function $f(x) = [\![x]\!]$ describes the greatest integer function where the value of y that corresponds to x is the greatest integer that is less than or equal to x. To find the greatest integer for a value that is not an integer, visualize the number on a number line and the integer directly to the left is the greatest integer.

EXAMPLES

Graph: $f(x) = [\![x]\!] - 1$

We list several intervals and the corresponding values of the greatest integer function:

$[0, 1)$	$f(x) = [\![x]\!] - 1$	For numbers from 0 to 1, not including 1, the greatest integer in the interval is 0 and then we subtract 1 and graph -1 in the interval.
$[1, 2)$	$f(x) = [\![x]\!] - 1$	For numbers from 1 to 2, not including 2, the greatest integer in the interval is 1 and then we subtract 1 and graph 0 in the interval.
$[2, 3)$	$f(x) = [\![x]\!] - 1$	For numbers from 2 to 3, not including 3, the greatest integer in the interval is 0 and then we subtract 1 and graph 1 in the interval.

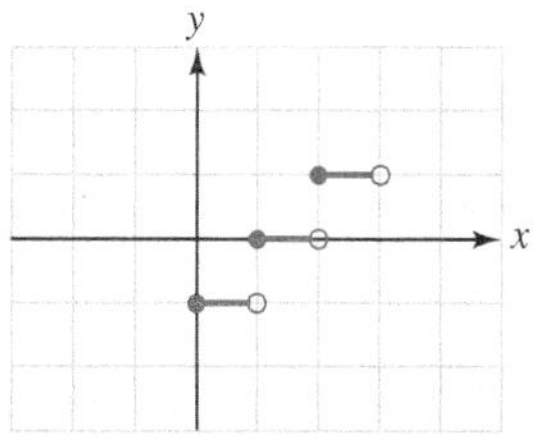

Unless otherwise noted, all content on this page is © Cengage Learning.

REVIEW EXERCISES

18. Determine the intervals where the function is increasing, decreasing, or constant.

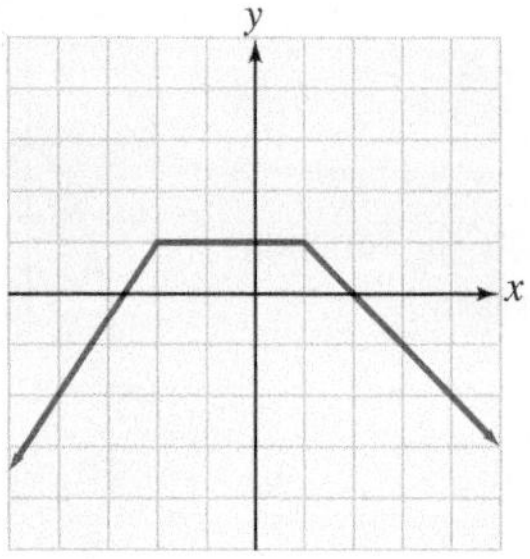

Graph each function.

19. $f(x) = \begin{cases} x \text{ if } x \leq 1 \\ -x^2 \text{ if } x > 1 \end{cases}$

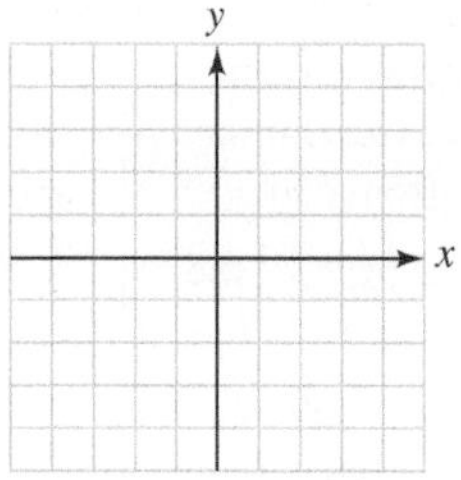

20. $f(x) = 3[\![x]\!]$

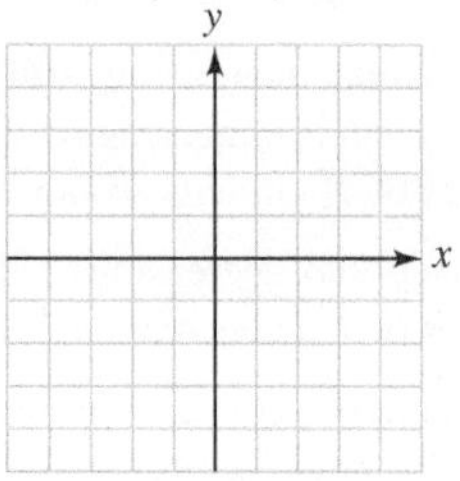

12 Test

1. Find the center and the radius of the circle $(x + 5)^2 + (y - 2)^2 = 9$.

2. Find the center and the radius of the circle $x^2 + y^2 + 8x - 4y = 5$.

Graph each equation.

3. $(x + 1)^2 + (y - 2)^2 = 9$

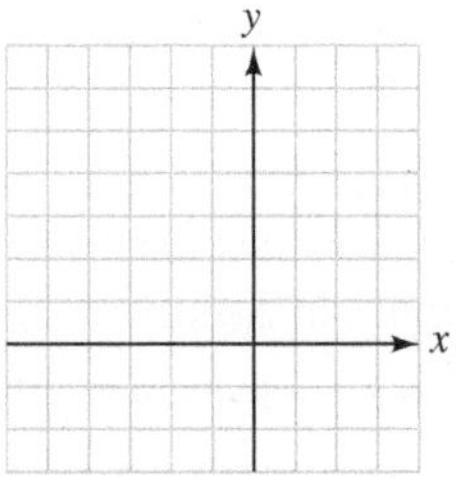

4. $x = (y - 2)^2 - 1$

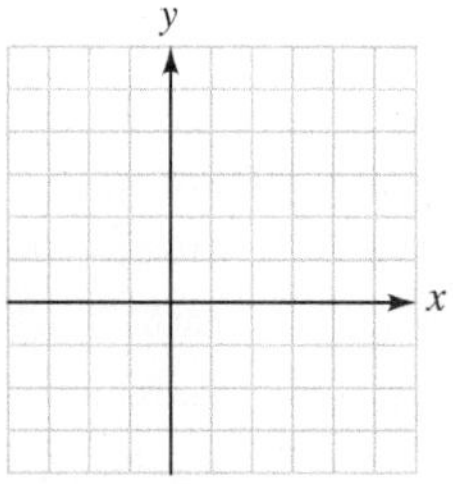

5. $9x^2 + 4y^2 = 36$

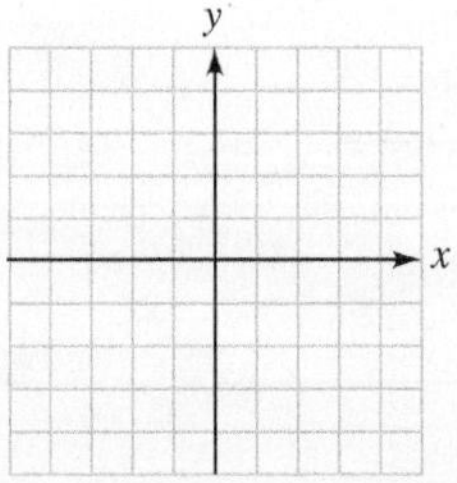

6. $\dfrac{(x - 2)^2}{9} - y^2 = 1$

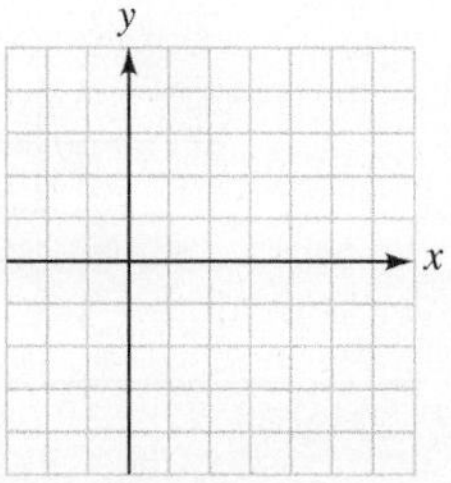

Write each equation in standard form and graph the equation.

7. $4x^2 + y^2 - 24x + 2y = -33$

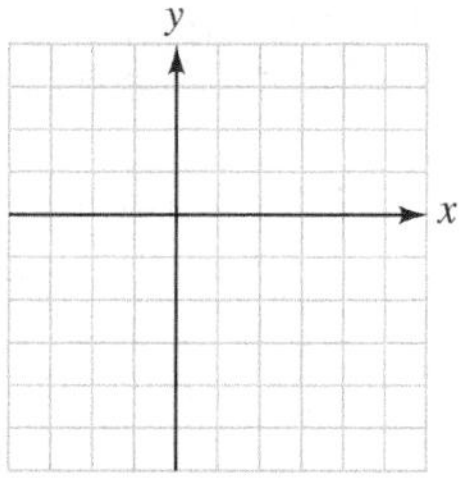

8. $x^2 + y^2 + 6x - 2y = -1$

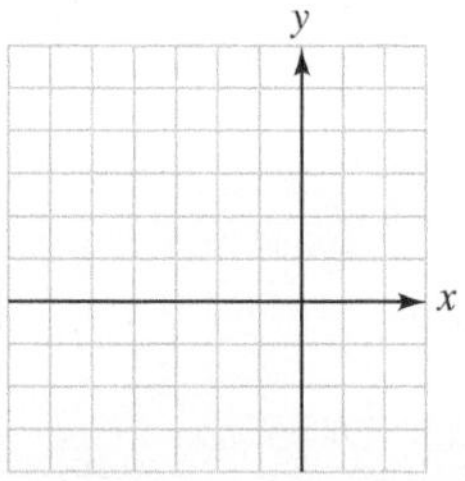

9. $x^2 - 9y^2 + 2x + 36y = 44$

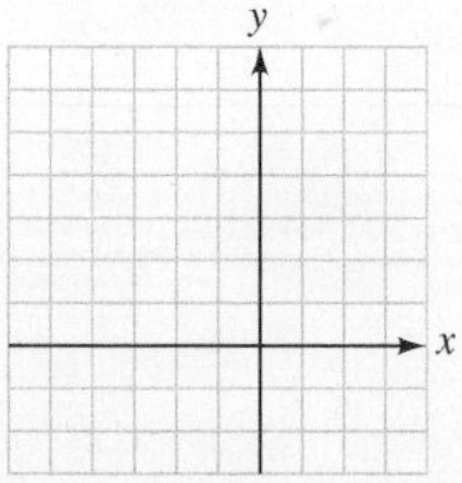

Unless otherwise noted, all content on this page is © Cengage Learning.

Solve each system.

10. $\begin{cases} 3x + 4y = 12 \\ 9x^2 + 16y^2 = 144 \end{cases}$

11. $\begin{cases} x^2 + y^2 = 25 \\ 4x^2 - 9y = 0 \end{cases}$

12. $\begin{cases} x - y < 3 \\ y \le x^2 - 6x + 7 \end{cases}$

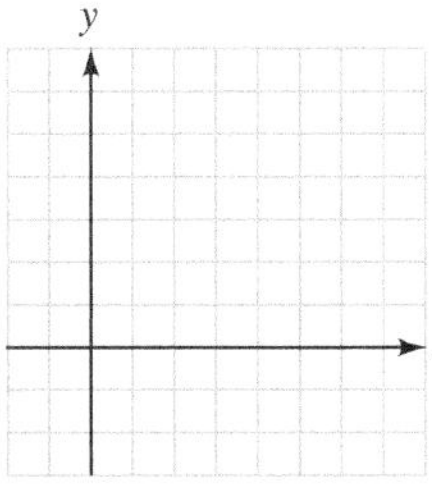

13. Determine the interval where the function is increasing, decreasing, or constant.

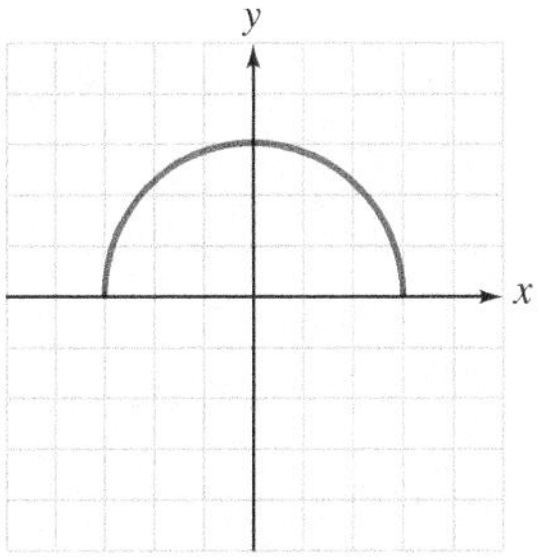

14. Graph: $f(x) = \begin{cases} -x^2, \text{ when } x < 0 \\ -x, \text{ when } x \ge 0 \end{cases}$

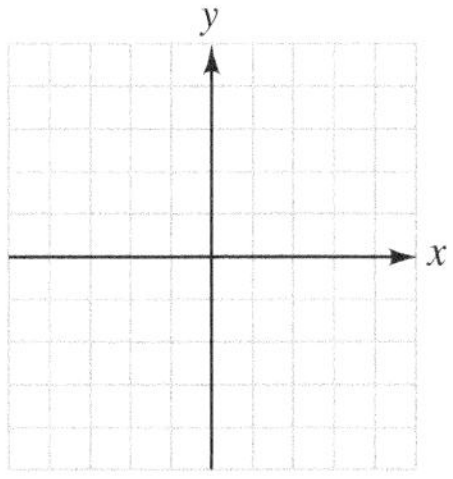

15. Graph: $f(x) = 2[\![x]\!]$

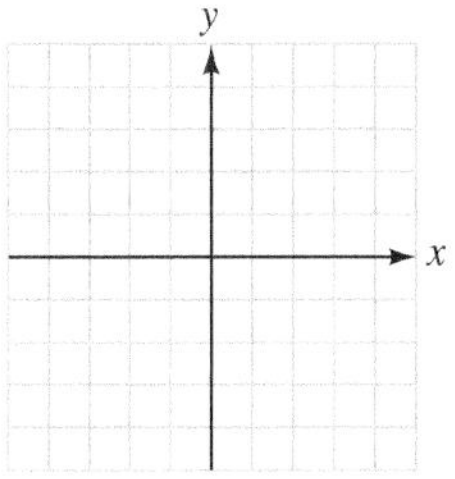

Cumulative Review

Perform the operations.

1. $(2x - 5y)(4x + y)$

2. $(a^n + 1)(a^n - 3)$

Simplify each fraction. Assume no division by 0.

3. $\dfrac{7a - 14}{a^2 - 7a + 10}$

4. $\dfrac{a^4 - 5a^2 + 4}{a^2 + 3a + 2}$

Perform the operations and simplify the result. Assume no division by 0.

5. $\dfrac{a^2 - a - 6}{a^2 - 4} \div \dfrac{a^2 - 9}{a^2 + a - 6}$

6. $\dfrac{2}{a - 2} + \dfrac{3}{a + 2} - \dfrac{a - 1}{a^2 - 4}$

Determine whether the graphs of the linear equations are parallel, perpendicular, or neither.

7. $3x - 4y = 12,\ y = \dfrac{3}{4}x - 5$

8. $y = 3x + 4,\ x = -3y + 4$

Write the equation of each line with the following properties.

9. $m = -6$, passing through $(0, 3)$

10. passing through $(8, -5)$ and $(-5, 4)$

Graph each inequality.

11. $2x - 3y < 6$

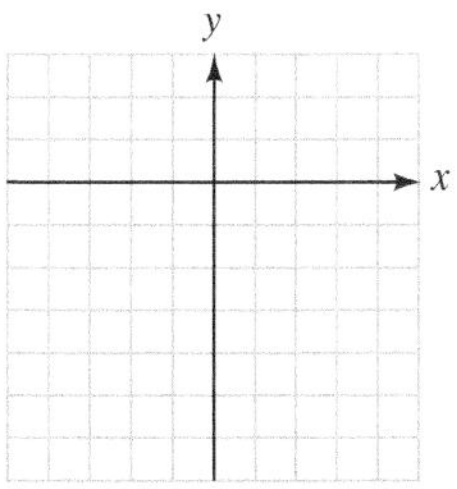

12. $y \ge x^2 - 4$

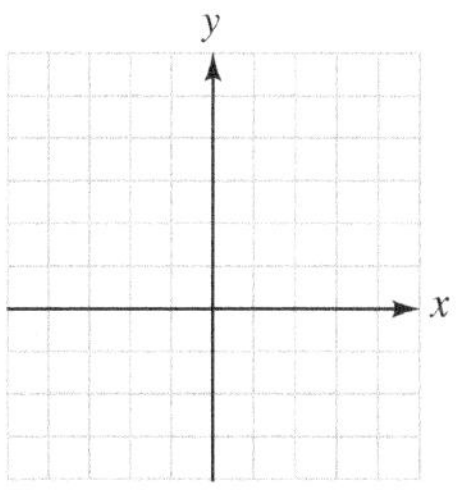

Simplify each expression.

13. $\sqrt{98} + \sqrt{8} - \sqrt{32}$

14. $12\sqrt[3]{648x^4} + 3\sqrt[3]{81x^4}$

Unless otherwise noted, all content on this page is © Cengage Learning.

Solve each equation.

15. $\sqrt{3a+1} = a - 1$

16. $x - \dfrac{2x}{x-5} = 1 - \dfrac{10}{x-5}$

17. $10a^2 + 21a - 10 = 0$

18. $2x^2 - 2x + 5 = 0$

19. If $f(x) = x^2 - 3$ and $g(x) = 4x + 1$, find $(f \circ g)(x)$.

20. Find the inverse function of $y = 2x^3 - 1$.

21. Graph: $y = \left(\dfrac{1}{2}\right)^x$

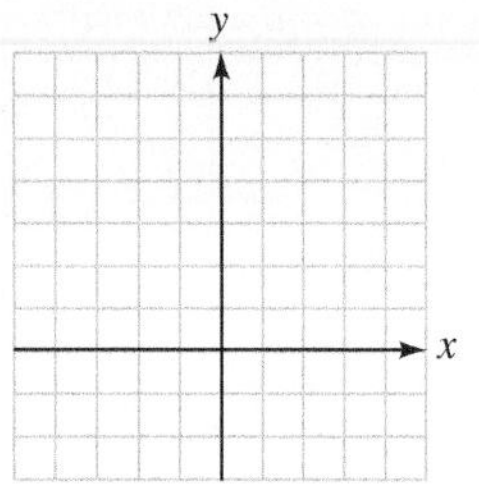

22. Write $y = \log_5 x$ as an exponential equation.

Solve each equation.

23. $2^{x+1} = 8^{x^2-1}$

24. $2 \log 5 + \log x - \log 4 = 2$

Graph each equation.

25. $x^2 + (y+1)^2 = 9$

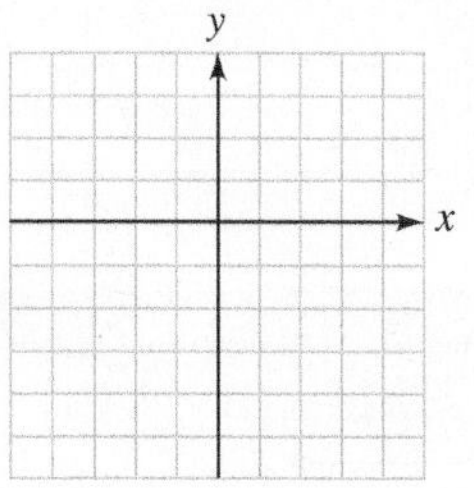

26. $x^2 - 9(y+1)^2 = 9$

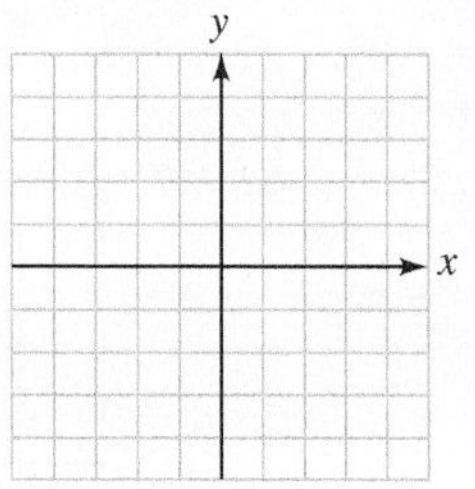

Solve each system.

27. $\begin{cases} x + 2y + 3z = 4 \\ 5x + 6y + 7z = 8 \\ 9x + 10y + 11z = 12 \end{cases}$

28. $\begin{cases} x^2 + y^2 = 5 \\ x^2 - y^2 = 3 \end{cases}$

29. $\begin{cases} x^2 + y^2 = 25 \\ 4x^2 - 9y = 0 \end{cases}$

30. Solve the system: $\begin{cases} y \geq x^2 \\ y < x + 3 \end{cases}$

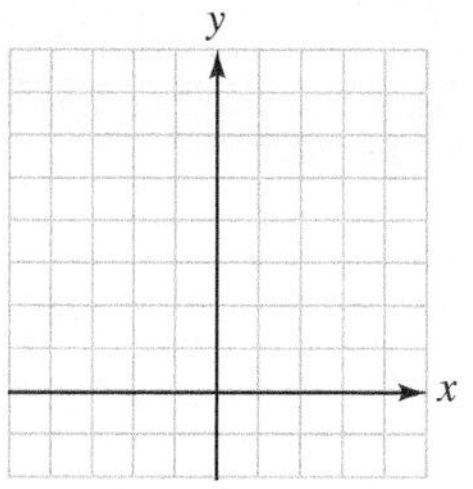

Unless otherwise noted, all content on this page is © Cengage Learning.

Miscellaneous Topics

13

© wavebreakmedia ltd./Shutterstock.com

Careers and Mathematics

FINANCIAL ANALYSTS

Financial analysts provide analysis and guidance to businesses to help them with their investment decisions. These specialists gather financial information, analyze it, and make recommendations to their clients. Financial analysts specialize in an industry, product, or geographical area. A bachelor's or graduate degree is required for financial analysts.

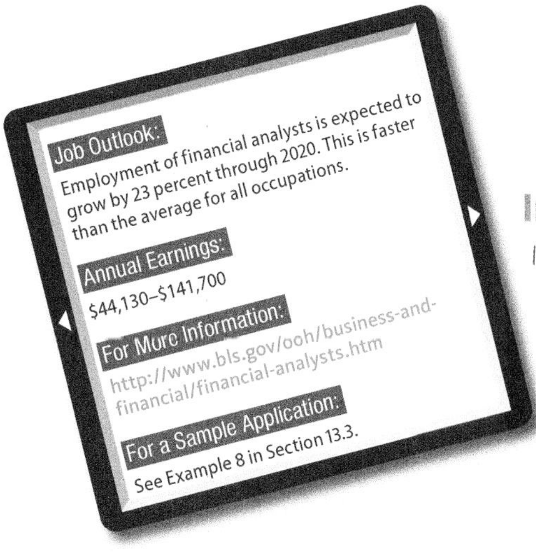
Job Outlook:
Employment of financial analysts is expected to grow by 23 percent through 2020. This is faster than the average for all occupations.
Annual Earnings:
$44,130–$141,700
For More Information:
http://www.bls.gov/ooh/business-and-financial/financial-analysts.htm
For a Sample Application:
See Example 8 in Section 13.3.

REACH FOR SUCCESS
13.1 The Binomial Theorem
13.2 Arithmetic Sequences and Series
13.3 Geometric Sequences and Series
13.4 Infinite Geometric Sequences and Series
13.5 Permutations and Combinations
13.6 Probability
■ *Projects*
REACH FOR SUCCESS EXTENSION
CHAPTER REVIEW
CHAPTER TEST
CUMULATIVE REVIEW

In this chapter

In this chapter, we introduce several topics that have applications in advanced mathematics and in many occupational areas. The binomial theorem, permutations, and combinations are used in statistics. Arithmetic and geometric sequences are used in the mathematics of finance.

Reach for Success Adjusting to College Life

Do you feel "connected" to your college? Do you want to be?

Becoming a part of the college culture can occur by accessing college resources, actively participating during class, working with study groups, or joining a club or organization.

Monkey Business Images/ Shutterstock.com

If you want to "fit in" at college, start with understanding why you selected this college.	Why did you choose to attend this college? Location? ___ Cost? ___Size? ___ College Mission? ___ Other? ________________
As a college student, you are considered an adult. With the privileges and freedom of college life come responsibilities and certain expected behavior.	List three differences between college and high school behavior. 1. ________________ 2. ________________ 3. ________________
Adjusting to college means being familiar with classroom and/or online etiquette.* For example, texting during class is considered inappropriate behavior by most professors.	List three other behaviors that might interfere with learning in the college classroom. 1. ________________ 2. ________________ 3. ________________
A form of social interaction is forming study groups for your classes. Do you belong to a study group for this mathematics class? ________________ If so, when do you meet ________________ and for how long? ________________	If not, exchange emails or phone numbers with at least one other student. Write that information here: ________________ Include times this study group is available to meet. ________________
There are usually opportunities for social engagement. Review a list of the organizations at your college and name two that may interest you. 1. ________________ 2. ________________	Do you have an area of interest for which there is no organization or chapter at your college? If so,use the internet to explore it. Write the name of the organization ________________ and explain its mission. ________________ Would a faculty member be interested in helping guide you in the creation of a local chapter at your college? If so, who? ________________

*Discussion boards are a technique some instructors use to help online students feel a part of the college community. Proper etiquette, for example, respectful dialogue, needs to be followed here as well.

A Successful Study Strategy . . .

Form study groups for your courses at the beginning of the semester. Being a part of this group or a larger one may help motivate you to do well in your courses.

At the end of the chapter you will find an additional exercise to guide you in planning for a successful college experience.

Unless otherwise noted, all content on this page is © Cengage Learning.

Section 13.1 The Binomial Theorem

Objectives

1. Raise a binomial to a power.
2. Complete a specified number of rows of Pascal's triangle.
3. Simplify an expression involving factorial notation.
4. Apply the binomial theorem to expand a binomial.
5. Find a specified term of a binomial expansion.

Vocabulary

Pascal's triangle | factorial notation | binomial theorem

Getting Ready

Raise each binomial to the indicated power.

1. $(x + 2)^2$
2. $(x - 3)^2$
3. $(x + 1)^3$
4. $(x - 2)^3$

We have seen how to square and cube binomials. In this section, we will learn how to raise binomials to higher powers without performing the multiplication and to find any term of a binomial expansion.

1 Raise a binomial to a power.

We have discussed how to raise binomials to positive-integer powers. For example, we know that

$$(a + b)^2 = a^2 + 2ab + b^2$$

and that

$$\begin{aligned}(a + b)^3 &= (a + b)(a + b)^2 \\ &= (a + b)(a^2 + 2ab + b^2) \\ &= a^3 + 2a^2b + ab^2 + a^2b + 2ab^2 + b^3 && \text{Use the distributive property to remove parentheses} \\ &= a^3 + 3a^2b + 3ab^2 + b^3 && \text{Combine like terms.}\end{aligned}$$

To show how to raise binomials to positive-integer powers without performing the actual multiplication, we consider the following binomial expansions:

$$\begin{aligned}(a + b)^0 &= 1 && \text{1 term} \\ (a + b)^1 &= a + b && \text{2 terms} \\ (a + b)^2 &= a^2 + 2ab + b^2 && \text{3 terms}\end{aligned}$$

Blaise Pascal
1623–1662
Pascal was torn between the fields of religion and mathematics. Each surfaced at times in his life to dominate his interest. In mathematics, Pascal made contributions to the study of conic sections, probability, and differential calculus. At the age of 19, he invented a calculating machine. He is best known for a triangular array of numbers that bears his name.

$(a + b)^3 = a^3 + 3a^2b + 3ab^2 + b^3$ 4 terms

$(a + b)^4 = a^4 + 4a^3b + 6a^2b^2 + 4ab^3 + b^4$ 5 terms

$(a + b)^5 = a^5 + 5a^4b + 10a^3b^2 + 10a^2b^3 + 5ab^4 + b^5$ 6 terms

$(a + b)^6 = a^6 + 6a^5b + 15a^4b^2 + 20a^3b^3 + 15a^2b^4 + 6ab^5 + b^6$ 7 terms

Several patterns appear in these expansions:

1. Each expansion has one more term than the power of the binomial.
2. The degree of each term in each expansion is equal to the exponent of the binomial that is being expanded. For example, in the expansion of $(a + b)^5$, the sum of the exponents in each term is 5:

$$(a + b)^5 = a^5 + 5\underbrace{a^4b}_{4+1=5} + 10\underbrace{a^3b^2}_{3+2=5} + 10\underbrace{a^2b^3}_{2+3=5} + 5\underbrace{ab^4}_{1+4=5} + b^5$$

3. The first term in each expansion is a raised to the power of the binomial, and the last term in each expansion is b raised to the power of the binomial.
4. The exponents on a decrease by 1 in each successive term. The exponents of b, beginning with $b^0 = 1$ in the first term, increase by 1 in each successive term. For example, the expansion of $(a + b)^4$ is

$$a^4b^0 + 4a^3b^1 + 6a^2b^2 + 4a^1b^3 + a^0b^4$$

Thus, the variables have the pattern

$$a^n, \quad a^{n-1}b, \quad a^{n-2}b^2, \quad \ldots, \quad ab^{n-1}, \quad b^n$$

2 Complete a specified number of rows of Pascal's triangle.

To see another pattern, we write the coefficients of each expansion in the following triangular array:

$$\begin{array}{ccccccccccccc l}
 & & & & & & 1 & & & & & & & \text{Row 0} \\
 & & & & & 1 & & 1 & & & & & & \text{Row 1} \\
 & & & & 1 & & 2 & & 1 & & & & & \text{Row 2} \\
 & & & 1 & & 3 & & 3 & & 1 & & & & \text{Row 3} \\
 & & 1 & & 4 & & 6 & & 4 & & 1 & & & \text{Row 4} \\
 & 1 & & 5 & & 10 & & 10 & & 5 & & 1 & & \text{Row 5} \\
1 & & 6 & & 15 & & 20 & & 15 & & 6 & & 1 & \text{Row 6}
\end{array}$$

In this array, called **Pascal's triangle**, each entry between the 1's is the sum of the closest pair of numbers in the line immediately above it. For example, the first 15 in the bottom row is the sum of the 5 and 10 immediately above it. Pascal's triangle continues with the same pattern forever. The next two lines are

$$\begin{array}{ccccccccccccccccc l}
 & 1 & & 7 & & 21 & & 35 & & 35 & & 21 & & 7 & & 1 & & \text{Row 7} \\
1 & & 8 & & 28 & & 56 & & 70 & & 56 & & 28 & & 8 & & 1 & \text{Row 8}
\end{array}$$

EXAMPLE 1 Expand: $(x + y)^5$

Solution The first term in the expansion is x^5, and the exponents of x decrease by 1 in each successive term. A y first appears in the second term, and the exponents on y increase by 1 in each successive term, concluding when the term y^5 is reached. Thus, the variables in the expansion are

$$x^5, \quad x^4y, \quad x^3y^2, \quad x^2y^3, \quad xy^4, \quad y^5$$

The coefficients of these variables are given in row 5 of Pascal's triangle.

Unless otherwise noted, all content on this page is © Cengage Learning.

1 5 10 10 5 1

Combining this information gives the following expansion:

$$(x + y)^5 = x^5 + 5x^4y + 10x^3y^2 + 10x^2y^3 + 5xy^4 + y^5$$

SELF CHECK 1 Expand: $(x + y)^4$

EXAMPLE 2 Expand: $(u - v)^4$

Solution We note that $(u - v)^4$ can be written in the form $[u + (-v)]^4$. The variables in this expansion are

$$u^4, \quad u^3(-v), \quad u^2(-v)^2, \quad u(-v)^3, \quad (-v)^4$$

and the coefficients are given in row 4 of Pascal's triangle.

1 4 6 4 1

Thus, the required expansion is

$$\begin{aligned}(u - v)^4 &= u^4 + 4u^3(-v) + 6u^2(-v)^2 + 4u(-v)^3 + (-v)^4 \\ &= u^4 - 4u^3v + 6u^2v^2 - 4uv^3 + v^4\end{aligned}$$

SELF CHECK 2 Expand: $(x - y)^5$

3 Simplify an expression involving factorial notation.

Although Pascal's triangle gives the coefficients of the terms in a binomial expansion, it can be a tedious way to expand a binomial for large powers. To develop a more efficient way, we introduce **factorial notation**.

FACTORIAL NOTATION

If n is a natural number, the symbol $n!$ (read as "n factorial" or as "factorial n") is defined as

$$n! = n(n - 1)(n - 2)(n - 3) \cdots (3)(2)(1)$$

Zero factorial is defined as

$$0! = 1$$

EXAMPLE 3 Write each expression without using factorial notation.

a. $2!$ **b.** $5!$ **c.** $-9!$ **d.** $(n - 2)!$ **e.** $4! \cdot 0!$

Solution

a. $2! = 2 \cdot 1 = 2$

b. $5! = 5 \cdot 4 \cdot 3 \cdot 2 \cdot 1 = 120$

c. $-9! = -1 \cdot 9 \cdot 8 \cdot 7 \cdot 6 \cdot 5 \cdot 4 \cdot 3 \cdot 2 \cdot 1 = -362{,}880$

d. $(n - 2)! = (n - 2)(n - 3)(n - 4) \cdot \cdots \cdot 3 \cdot 2 \cdot 1$

e. $4! \cdot 0! = (4 \cdot 3 \cdot 2 \cdot 1) \cdot 1 = 24$

COMMENT According to the previous definition, part **d** is meaningful only if $n - 2$ is a natural number.

SELF CHECK 3 Write each expression without using factorial notation.

a. $6!$ **b.** $x!$

Accent on technology

▸ Factorials

We can find factorials using a calculator. For example, to find 12! with a scientific calculator, we enter

12 **x!** (You may have to use a **2ND** or **SHIFT** key first.) **479001600**

To find 12! on a TI-84 graphing calculator, we enter

12 **MATH ▶ ▶ ▶ 4 ENTER**

12!
479001600

For instructions regarding the use of a Casio graphing calculator, please refer to the Casio Keystroke Guide in the back of the book.

To discover an important property of factorials, we note that

$$5 \cdot 4! = 5 \cdot 4 \cdot 3 \cdot 2 \cdot 1 = 5!$$
$$7 \cdot 6! = 7 \cdot 6 \cdot 5 \cdot 4 \cdot 3 \cdot 2 \cdot 1 = 7!$$
$$10 \cdot 9! = 10 \cdot 9 \cdot 8 \cdot 7 \cdot 6 \cdot 5 \cdot 4 \cdot 3 \cdot 2 \cdot 1 = 10!$$

These examples suggest the following property.

PROPERTY OF FACTORIALS

If n is a positive integer, then

$$n(n-1)! = n!$$

EXAMPLE 4 Simplify: **a.** $\frac{6!}{5!}$ **b.** $\frac{10!}{8!(10-8)!}$

Solution **a.** If we write 6! as $6 \cdot 5!$, we can simplify the fraction by removing the common factor 5! in the numerator and denominator.

$$\frac{6!}{5!} = \frac{6 \cdot 5!}{5!} = \frac{6 \cdot \cancel{5!}}{\cancel{5!}} = 6 \quad \text{Simplify: } \tfrac{5!}{5!} = 1$$

b. First, we subtract within the parentheses. Then we write 10! as $10 \cdot 9 \cdot 8!$ and simplify.

$$\frac{10!}{8!(10-8)!} = \frac{10!}{8! \cdot 2!} = \frac{10 \cdot 9 \cdot \cancel{8!}}{\cancel{8!} \cdot 2!} = \frac{5 \cdot 2 \cdot 9}{2 \cdot 1} = 45 \quad \text{Simplify: } \tfrac{8!}{8!} = 1. \text{ Factor 10 as } 5 \cdot 2 \text{ and simplify: } \tfrac{2}{2} = 1$$

SELF CHECK 4 Simplify: **a.** $\frac{4!}{3!}$ **b.** $\frac{7!}{5!(7-5)!}$

4 Apply the binomial theorem to expand a binomial.

Now that we have developed an understanding of factorial notation, we will use this notation to state the **binomial theorem**, which will provide an alternative method for expanding a binomial for large powers of n.

THE BINOMIAL THEOREM If n is any positive integer, then

$$(a+b)^n = a^n + \frac{n!}{1!(n-1)!}a^{n-1}b + \frac{n!}{2!(n-2)!}a^{n-2}b^2 + \frac{n!}{3!(n-3)!}a^{n-3}b^3 + \cdots + \frac{n!}{r!(n-r)!}a^{n-r}b^r + \cdots + b^n$$

In the binomial theorem, the exponents of the variables follow the familiar pattern:

- The sum of the exponents on a and b in each term is n,
- the exponents on a decrease in each subsequent term, and
- the exponents on b increase in each subsequent term.

Only the method of finding the coefficients is different. Except for the first and last terms, the numerator of each coefficient is $n!$. If the exponent of b in a particular term is r, the denominator of the coefficient of that term is $r!(n-r)!$.

EXAMPLE 5 Use the binomial theorem to expand $(a+b)^3$.

Solution We can substitute directly into the binomial theorem and simplify:

$$\begin{aligned}(a+b)^3 &= a^3 + \frac{3!}{1!(3-1)!}a^2b + \frac{3!}{2!(3-2)!}ab^2 + b^3\\ &= a^3 + \frac{3!}{1!\cdot 2!}a^2b + \frac{3!}{2!\cdot 1!}ab^2 + b^3\\ &= a^3 + \frac{3\cdot 2\cdot 1}{1\cdot 2\cdot 1}a^2b + \frac{3\cdot 2\cdot 1}{2\cdot 1\cdot 1}ab^2 + b^3\\ &= a^3 + 3a^2b + 3ab^2 + b^3\end{aligned}$$

SELF CHECK 5 Use the binomial theorem to expand $(a+b)^4$.

EXAMPLE 6 Use the binomial theorem to expand $(x-y)^4$.

Solution We can write $(x-y)^4$ in the form $[x+(-y)]^4$, substitute directly into the binomial theorem, and simplify:

COMMENT To expand $(x-y)^n$, the signs on the terms will alternate.

$$\begin{aligned}(x-y)^4 &= [x+(-y)]^4\\ &= x^4 + \frac{4!}{1!(4-1)!}x^3(-y) + \frac{4!}{2!(4-2)!}x^2(-y)^2\\ &\quad + \frac{4!}{3!(4-3)!}x(-y)^3 + (-y)^4\\ &= x^4 - \frac{4\cdot 3!}{1!\cdot 3!}x^3y + \frac{4\cdot 3\cdot 2!}{2!\cdot 2!}x^2y^2 - \frac{4\cdot 3!}{3!\cdot 1!}xy^3 + y^4\\ &= x^4 - 4x^3y + 6x^2y^2 - 4xy^3 + y^4 \quad \text{Note the alternating signs.}\end{aligned}$$

SELF CHECK 6 Use the binomial theorem to expand $(x-y)^3$.

EXAMPLE 7 Use the binomial theorem to expand $(3u-2v)^4$.

Solution We write $(3u-2v)^4$ in the form $[3u+(-2v)]^4$ and let $a = 3u$ and $b = -2v$. Then we can use the binomial theorem to expand $(a+b)^4$.

$$(a+b)^4 = a^4 + \frac{4!}{1!(4-1)!}a^3b + \frac{4!}{2!(4-2)!}a^2b^2 + \frac{4!}{3!(4-3)!}ab^3 + b^4$$
$$= a^4 + 4a^3b + 6a^2b^2 + 4ab^3 + b^4$$

Now we can substitute $3u$ for a and $-2v$ for b and simplify:

$$(3u-2v)^4 = (3u)^4 + 4(3u)^3(-2v) + 6(3u)^2(-2v)^2 + 4(3u)(-2v)^3 + (-2v)^4$$
$$= 81u^4 - 216u^3v + 216u^2v^2 - 96uv^3 + 16v^4$$

SELF CHECK 7 Use the binomial theorem to expand $(2a-3b)^3$.

5 Find a specified term of a binomial expansion.

To find the fourth term of the expansion of $(a+b)^9$, we could raise the binomial $a+b$ to the 9th power and look at the fourth term. However, this task would be tedious. By using the binomial theorem, we can construct the fourth term without finding the complete expansion of $(a+b)^9$.

EXAMPLE 8 Find the fourth term in the expansion of $(a+b)^9$.

Solution Since b^1 appears in the second term, b^2 appears in the third term, and so on, the exponent on b in the fourth term is 3. Since the exponent on b added to the exponent on a must equal 9, the exponent on a must be 6. Thus, the variables of the fourth term are

a^6b^3 The sum of the exponents must be 9.

We can find the coefficient of a^6b^3 by using the formula $\frac{n!}{r!(n-r)!}$, where n is the power of the expansion and r is the exponent of the second variable b.

To find the coefficient of the fourth term, we substitute 9 for n and 3 for r and simplify.

$$\frac{n!}{r!(n-r)!} = \frac{9!}{3!(9-3)!}$$

The complete fourth term is

$$\frac{9!}{3!(9-3)!}a^6b^3 = \frac{9\cdot 8\cdot 7\cdot 6!}{3\cdot 2\cdot 1\cdot 6!}a^6b^3$$
$$= 84a^6b^3$$

SELF CHECK 8 Find the third term of the expansion of $(a+b)^9$.

EXAMPLE 9 Find the sixth term in the expansion of $(x-y)^7$.

Solution We first find the sixth term of $[x+(-y)]^7$. In the sixth term, the exponent on $(-y)$ is 5. Thus, the variables in the sixth term are

$x^2(-y)^5$ The sum of the exponents must be 7.

The coefficient of these variables is

$$\frac{n!}{r!(n-r)!} = \frac{7!}{5!(7-5)!}$$

The complete sixth term is

$$\frac{7!}{5!(7-5)!}x^2(-y)^5 = -\frac{7\cdot6\cdot5!}{5!\cdot2\cdot1}x^2y^5$$
$$= -21x^2y^5$$

SELF CHECK 9 Find the fourth term of the expansion of $(a-b)^7$.

EXAMPLE 10 Find the fourth term of the expansion of $(2x-3y)^6$.

Solution We can let $a = 2x$ and $b = -3y$ and find the fourth term of the expansion of $(a+b)^6$:

$$\frac{6!}{3!(6-3)!}a^3b^3 = \frac{6\cdot5\cdot4\cdot3!}{3!\cdot3\cdot2\cdot1}a^3b^3$$
$$= 20a^3b^3$$

We can now substitute $2x$ for a and $-3y$ for b and simplify:

$$20a^3b^3 = 20(2x)^3(-3y)^3$$
$$= -4{,}320x^3y^3$$

The fourth term is $-4{,}320x^3y^3$.

SELF CHECK 10 Find the third term of the expansion of $(2a-3b)^6$.

SELF CHECK ANSWERS

1. $x^4+4x^3y+6x^2y^2+4xy^3+y^4$ **2.** $x^5-5x^4y+10x^3y^2-10x^2y^3+5xy^4-y^5$ **3. a.** 720 **b.** $x(x-1)(x-2)\cdot\cdots\cdot3\cdot2\cdot1$ **4. a.** 4 **b.** 21 **5.** $a^4+4a^3b+6a^2b^2+4ab^3+b^4$ **6.** $x^3-3x^2y+3xy^2-y^3$ **7.** $8a^3-36a^2b+54ab^2-27b^3$ **8.** $36a^7b^2$ **9.** $-35a^4b^3$ **10.** $2{,}160a^4b^2$

NOW TRY THIS

1. Expand $(1+2i)^7$ using any method.

Find the specified term of each expansion.

2. fourth term of $\left(5x+\frac{1}{4}\right)^6$
3. fifth term of $(3-2i)^8$

13.1 Exercises

WARM-UPS *Simplify.*

1. $6\cdot5\cdot4\cdot3\cdot2\cdot1$

2. $4\cdot3\cdot2\cdot1$

3. $\dfrac{6\cdot5\cdot4\cdot3\cdot2\cdot1}{4\cdot3\cdot2\cdot1}$

4. $\dfrac{7\cdot6\cdot5\cdot4\cdot3\cdot2\cdot1}{5\cdot4\cdot3\cdot2\cdot1}$

Square each binomial.

5. $(x+3y)^2$

6. $(2x+3y)^2$

7. $(x-3y)^2$

8. $(2x-3y)^2$

REVIEW *Find each value of x.*

9. $\log_9 81 = x$

10. $\log_x 100 = 2$

11. $\log_{25} x = \dfrac{1}{2}$

12. $\log_{1/2}\dfrac{1}{8} = x$

VOCABULARY AND CONCEPTS *Fill in the blanks.*

13. Every binomial expansion has ___ more term than the power of the binomial.

14. The first term in the expansion of $(a+b)^{20}$ is ___.

15. The triangular array that can be used to find the coefficients of a binomial expansion is called ______ triangle.
16. The symbol 5! is read as "__________."
17. $8 \cdot 7 \cdot 6 \cdot 5 \cdot 4 \cdot 3 \cdot 2 \cdot 1 =$ _ (Write your answer in factorial notation.)
18. $6! = 6 \cdot$ _
19. $0! =$ _
20. According to the binomial theorem, the third term of the expansion of $(a + b)^n$ is ______________.

GUIDED PRACTICE *Expand each expression using Pascal's triangle.* SEE EXAMPLES 1–2. (OBJECTIVES 1–2)

21. $(a + b)^3$
22. $(a + b)^4$
23. $(a - b)^4$
24. $(a - b)^3$

Evaluate. SEE EXAMPLE 3. (OBJECTIVE 3)

25. $4!$
26. $7!$
27. $-8!$
28. $-6!$
29. $3! + 4!$
30. $2!(3!)$
31. $3!(4!)$
32. $4! + 4!$

Simplify. SEE EXAMPLE 4. (OBJECTIVE 3)

33. $\dfrac{9!}{11!}$
34. $\dfrac{13!}{10!}$
35. $\dfrac{49!}{47!}$
36. $\dfrac{101!}{100!}$
37. $\dfrac{9!}{7! \cdot 0!}$
38. $\dfrac{7!}{5! \cdot 0!}$
39. $\dfrac{5!}{3!(5-3)!}$
40. $\dfrac{6!}{4!(6-4)!}$

Use the binomial theorem to expand each expression. SEE EXAMPLE 5. (OBJECTIVE 4)

41. $(x + y)^3$
42. $(x + y)^4$
43. $(a + b)^6$
44. $(a + b)^5$

Use the binomial theorem to expand each expression. SEE EXAMPLE 6. (OBJECTIVE 4)

45. $(a - b)^4$
46. $(x - y)^3$
47. $(x - 2y)^3$
48. $(2x - y)^4$

Use the binomial theorem to expand each expression. SEE EXAMPLE 7. (OBJECTIVE 4)

49. $(2x + y)^3$
50. $(x + 2y)^3$
51. $(2x + 3y)^3$
52. $(3x - 2y)^3$

Find the specified term of each expansion. SEE EXAMPLE 8. (OBJECTIVE 5)

53. $(a + b)^4$; third term
54. $(a + b)^3$; third term
55. $(x + y)^6$; fifth term
56. $(x + y)^7$; fifth term

Find the specified term of each expansion. SEE EXAMPLE 9. (OBJECTIVE 5)

57. $(x - y)^4$; fourth term
58. $(x - y)^5$; second term
59. $(x - y)^8$; third term
60. $(x - y)^9$; seventh term

Find the specified term of each expansion. SEE EXAMPLE 10. (OBJECTIVE 5)

61. $(4x + y)^5$; third term
62. $(x + 4y)^5$; fourth term
63. $(x - 3y)^4$; second term
64. $(3x - y)^5$; third term
65. $(2x - 5)^7$; fourth term
66. $(2x + 3)^6$; sixth term
67. $(2x - 3y)^5$; fifth term
68. $(3x - 2y)^4$; second term

ADDITIONAL PRACTICE *Evaluate each expression.*

69. $8(7!)$
70. $4!(5)$
71. $\dfrac{7!}{5!(7-5)!}$
72. $\dfrac{8!}{6!(8-6)!}$
73. $\dfrac{5!(8-5)!}{4!7!}$
74. $\dfrac{6!7!}{(8-3)!(7-4)!}$

Use a calculator to find each factorial.

75. $11!$
76. $13!$
77. $20!$
78. $55!$

Expand using any method.

79. $(3 + 2y)^4$
80. $(2x + 3)^4$
81. $\left(\dfrac{x}{2} - \dfrac{y}{3}\right)^3$
82. $\left(\dfrac{x}{3} + \dfrac{y}{2}\right)^3$
83. $(x - y)^5$
84. $(a - b)^6$
85. $\left(\dfrac{x}{2} + \dfrac{y}{3}\right)^4$
86. $\left(\dfrac{x}{3} - \dfrac{y}{2}\right)^4$
87. Without referring to the text, write the first ten rows of Pascal's triangle.
88. Find the sum of the numbers in each row of the first ten rows of Pascal's triangle. What is the pattern?

Find the specified term of each expansion.

89. $(\sqrt{2}x + \sqrt{3}y)^6$; third term

90. $(\sqrt{3}x + \sqrt{2}y)^5$; second term

91. $\left(\frac{x}{2} - \frac{y}{3}\right)^4$; second term

92. $\left(\frac{x}{3} + \frac{y}{2}\right)^5$; fourth term

93. $(x + 5)^6$; third term

94. $(x - 2)^4$; second term

95. $(a + b)^n$; fourth term

96. $(a + b)^n$; third term

97. $(a - b)^n$; fifth term

98. $(a - b)^n$; sixth term

99. $(a + b)^n$; rth term

100. $(a + b)^n$; $(r + 1)$th term

WRITING ABOUT MATH

101. Explain how to construct Pascal's triangle.

102. Explain how to find the variables of the terms in the expansion of $(r + s)^4$.

103. Explain how to find the coefficients in the expansion of $(x + y)^5$.

104. Explain why the signs alternate in the expansion of $(x - y)^9$.

SOMETHING TO THINK ABOUT

105. If we apply the pattern of the coefficients to the coefficient of the first term in a binomial expansion, the coefficient would be $\frac{n!}{0!(n-0)!}$. Show that this expression is 1.

106. If we apply the pattern of the coefficients to the coefficient of the last term in a binomial expansion, the coefficient would be $\frac{n!}{n!(n-n)!}$. Show that this expression is 1.

107. Find the sum of the numbers in the designated diagonal rows of Pascal's triangle shown in the illustration. What is the pattern?

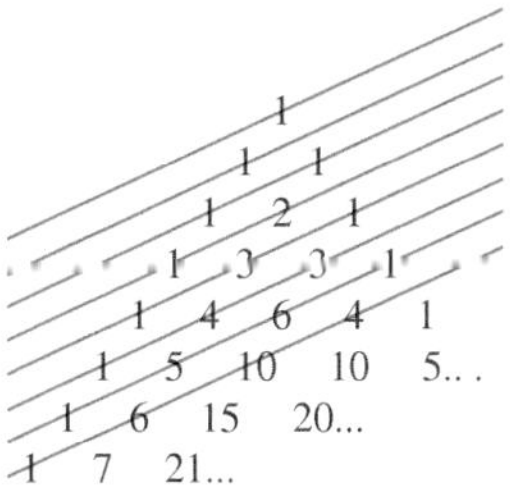

108. Find the constant term in the expansion of $\left(x + \frac{1}{x}\right)^{10}$.

109. Find the coefficient of a^5 in the expansion of $\left(a - \frac{1}{a}\right)^9$.

Section 13.2 Arithmetic Sequences and Series

Objectives

1. Find a specified term of a sequence given the general term.
2. Find a specified term of an arithmetic sequence given the first term and the common difference.
3. Find a specified term of an arithmetic sequence given the first term and one or more other terms.
4. Insert one or more arithmetic means between two numbers.
5. Find the sum of the first n terms of an arithmetic sequence.
6. Expand and find the sum of a series written in summation notation.

Vocabulary

sequence	general term	series
Fibonacci sequence	arithmetic sequence	arithmetic series
finite sequence	common difference	summation notation
infinite sequence	arithmetic means	index of summation

Unless otherwise noted, all content on this page is © Cengage Learning.

Getting Ready

Complete each table.

1.

n	$2n + 1$
1	
2	
3	
4	

2.

n	$3n - 5$
3	
4	
5	
6	

In the next two sections, we will discuss ordered lists of numbers called *sequences.* In this section, we will discuss sequences in general and then examine a special type of sequence called an *arithmetic sequence.*

A **sequence** is a function whose domain is the set of natural numbers. For example, the function $f(n) = 3n + 2$, where n is a natural number, is a sequence. Because a sequence is a function whose domain is the set of natural numbers, we can write its values as a list. If the natural numbers are substituted for n, the function $f(n) = 3n + 2$ generates the list

$$5, 8, 11, 14, 17, \ldots$$

It is common to call the list, as well as the function, a sequence. Each number in the list is called a *term* of the sequence. Other examples of sequences are

$1^3, 2^3, 3^3, 4^3, \ldots$	The ordered list of the cubes of the natural numbers
$4, 8, 12, 16, \ldots$	The ordered list of the positive multiples of 4
$2, 3, 5, 7, 11, \ldots$	The ordered list of prime numbers
$1, 1, 2, 3, 5, 8, 13, 21, \ldots$	The Fibonacci sequence

The **Fibonacci sequence** is named after the 12th-century mathematician Leonardo of Pisa—also known as Fibonacci. Beginning with the 2, each term of the sequence is the sum of the two preceding terms.

Finite sequences contain a finite number of terms and **infinite sequences** contain infinitely many terms. One example of each type of sequence is:

Finite sequence: 1, 5, 9, 13, 17, 21, 25

Infinite sequence: 3, 6, 9, 12, 15, . . . The ellipsis, . . . , indicates that the sequence goes on forever.

1 Find a specified term of a sequence given the general term.

In this section, we will use a_n (read as "a sub n") to denote the nth term of a sequence. For example, in the sequence 3, 6, 9, 12, 15, . . . , we have

1st term	2nd term	3rd term	4th term	5th term	
3,	6,	9,	12,	15,	. . .
↑	↑	↑	↑	↑	
a_1	a_2	a_3	a_4	a_5	

To describe all the terms of a sequence, we can write a formula for a_n, called the **general term** of the sequence. For the sequence 3, 6, 9, 12, 15, . . . , we note that $a_1 = 3 \cdot \mathbf{1}$, $a_2 = 3 \cdot \mathbf{2}$, $a_3 = 3 \cdot \mathbf{3}$, and so on. In general, the nth term of the sequence is found by multiplying n by 3.

$$a_n = 3n$$

We can use this formula to find any term of the sequence. For example, to find the 12th term, we substitute 12 for n.

$$a_{12} = 3(\mathbf{12}) = 36$$

EXAMPLE 1 Given an infinite sequence with $a_n = 2n - 3$, find:

a. the first four terms **b.** a_{50}

Solution **a.** To find the first four terms of the sequence, we substitute 1, 2, 3, and 4 for n in $a_n = 2n - 3$ and simplify.

$a_1 = 2(1) - 3 = -1$ Substitute 1 for n.

$a_2 = 2(2) - 3 = 1$ Substitute 2 for n.

$a_3 = 2(3) - 3 = 3$ Substitute 3 for n.

$a_4 = 2(4) - 3 = 5$ Substitute 4 for n.

The first four terms of the sequence are -1, 1, 3, and 5.

b. To find a_{50}, the 50th term of the sequence, we let $n = 50$:

$$a_{50} = 2(50) - 3 = 97$$

SELF CHECK 1 Given an infinite sequence with $a_n = 3n + 5$, find:

a. the first three terms **b.** a_{100}

2 Find a specified term of an arithmetic sequence given the first term and the common difference.

One common type of sequence is the arithmetic sequence.

ARITHMETIC SEQUENCE

An **arithmetic sequence** is a sequence of the form

$$a_1, \quad a_1 + d, \quad a_1 + 2d, \quad \ldots, \quad a_1 + (n - 1)d, \ldots$$

where a_1 is the *first term* and d is the **common difference**. The *nth term* is given by

$$a_n = a_1 + (n - 1)d$$

We note that the second term of an arithmetic sequence has an addend of $1d$, the third term has an addend of $2d$, the fourth term has an addend of $3d$, and the nth term has an addend of $(n - 1)d$. We also note that the difference between any two consecutive terms in an arithmetic sequence is d.

EXAMPLE 2 An arithmetic sequence has a first term of 5 and a common difference of 4.

a. Write the first six terms of the sequence.

b. Write the 25th term of the sequence.

Solution **a.** Since the first term is $a_1 = 5$ and the common difference is $d = 4$, the first six terms are

$$\underset{a_1}{\overset{5,}{\uparrow}} \quad \underset{a_2}{\overset{5+4,}{\uparrow}} \quad \underset{a_3}{\overset{5+2(4),}{\uparrow}} \quad \underset{a_4}{\overset{5+3(4),}{\uparrow}} \quad \underset{a_5}{\overset{5+4(4),}{\uparrow}} \quad \underset{a_6}{\overset{5+5(4)}{\uparrow}}$$

or

5, 9, 13, 17, 21, 25

b. The nth term is $a_n = a_1 + (n - 1)d$. Since we want the 25th term, we let $n = 25$:

$$\begin{aligned} a_n &= a_1 + (n - 1)d \\ a_{25} &= 5 + (25 - 1)4 \quad \text{Remember that } a_1 = 5 \text{ and } d = 4. \\ &= 5 + 24(4) \\ &= 5 + 96 \\ &= 101 \end{aligned}$$

SELF CHECK 2 **a.** Write the seventh term of the sequence in Example 2.
b. Write the 30th term of the sequence in Example 2.

3 Find a specified term of an arithmetic sequence given the first term and one or more other terms.

EXAMPLE 3 The first three terms of an arithmetic sequence are 3, 8, and 13. Find:

a. the 67th term **b.** the 100th term

Solution We first find d, the common difference. It is the difference between two successive terms:

$$d = 8 - 3 = 13 - 8 = 5$$

a. We substitute 3 for a_1, 67 for n, and 5 for d in the formula for the nth term and simplify:

$$\begin{aligned} a_n &= a_1 + (n - 1)d \\ a_{67} &= 3 + (67 - 1)5 \\ &= 3 + 66(5) \\ &= 333 \end{aligned}$$

b. We substitute 3 for a_1, 100 for n, and 5 for d in the formula for the nth term, and simplify:

$$\begin{aligned} a_n &= a_1 + (n - 1)d \\ a_{100} &= 3 + (100 - 1)5 \\ &= 3 + 99(5) \\ &= 498 \end{aligned}$$

SELF CHECK 3 Find the 50th term of the sequence in Example 3.

EXAMPLE 4 The first term of an arithmetic sequence is 12, and the 50th term is 3,099. Write the first six terms of the sequence.

Solution The key is to find the common difference. Because the 50th term of this sequence is 3,099, we can let $n = 50$ and solve the following equation for d:

$$\begin{aligned} a_n &= a_1 + (n - 1)d \\ a_{50} &= 12 + (50 - 1)d \\ 3{,}099 &= 12 + 49d \quad \text{Substitute 3,099 for } a_{50} \text{ and simplify.} \\ 3{,}087 &= 49d \quad \text{Subtract 12 from both sides.} \\ 63 &= d \quad \text{Divide both sides by 49.} \end{aligned}$$

The common difference is 63, and since the first term of the sequence is 12, its first six terms are

$$12, 75, 138, 201, 264, 327 \quad \text{Add 63 to a term to get the next term.}$$

SELF CHECK 4 The first term of an arithmetic sequence is 15, and the 12th term is 92. Write the first four terms of the sequence.

4 Insert one or more arithmetic means between two numbers.

If numbers are inserted between two numbers a and b to form an arithmetic sequence, the inserted numbers are called **arithmetic means** between a and b.

If a single number is inserted between the numbers a and b to form an arithmetic sequence, that number is called the *arithmetic mean* between a and b.

EXAMPLE 5 Insert two arithmetic means between 6 and 27.

Solution Because we want to insert two arithmetic means between 6 and 27, this implies that the first term is $a_1 = 6$ and 27 is the fourth term, a_4. We must find the common difference so that the terms

$$\underset{a_1}{\overset{\uparrow}{6,}} \quad \underset{a_2}{\overset{\uparrow}{6 + d,}} \quad \underset{a_3}{\overset{\uparrow}{6 + 2d,}} \quad \underset{a_4}{\overset{\uparrow}{27}}$$

form an arithmetic sequence. To find the value of d, we substitute 6 for a_1 and 4 for n into the formula for the nth term:

$$\begin{aligned} a_n &= a_1 + (n - 1)d \\ a_4 &= 6 + (4 - 1)d \\ 27 &= 6 + 3d && \text{Substitute 27 for } a_4 \text{ and simplify.} \\ 21 &= 3d && \text{Subtract 6 from both sides.} \\ 7 &= d && \text{Divide both sides by 3.} \end{aligned}$$

The common difference is 7. The two arithmetic means between 6 and 27 are

$$\begin{aligned} 6 + d &= 6 + 7 \\ &= 13 \end{aligned} \quad \text{or} \quad \begin{aligned} 6 + 2d &= 6 + 2(7) \\ &= 6 + 14 \\ &= 20 \end{aligned}$$

The numbers 6, 13, 20, and 27 are the first four terms of an arithmetic sequence.

SELF CHECK 5 Insert two arithmetic means between 8 and 44.

5 Find the sum of the first n terms of an arithmetic sequence.

We now consider a formula that calculates the sum of the first n terms of an arithmetic sequence. To develop this formula, we let S_n represent the sum of the first n terms of an arithmetic sequence:

$$S_n = a_1 + [a_1 + d] + [a_1 + 2d] + \cdots + [a_1 + (n - 1)d]$$

We write the same sum again, but in reverse order:

$$S_n = [a_1 + (n - 1)d] + [a_1 + (n - 2)d] + [a_1 + (n - 3)d] + \cdots + a_1$$

We add these two equations together, term by term, to obtain

$$\begin{aligned}2S_n &= [2a_1 + (n-1)d] + [2a_1 + d + dn - 2d] + [2a_1 + 2d + dn - 3d] \\ &\qquad + \cdots + [2a_1 + (n-1)d] \\ &= [2a_1 + (n-1)d] + [2a_1 + dn - d] + [2a_1 + dn - d] \\ &\qquad + \cdots + [2a_1 + (n-1)d] \\ 2S_n &= [2a_1 + (n-1)d] + [2a_1 + (n-1)d] + [2a_1 + (n-1)d] \\ &\qquad + \cdots + [2a_1 + (n-1)d]\end{aligned}$$

Because there are n equal terms on the right side of the preceding equation, we can write

$$\begin{aligned}2S_n &= n[2a_1 + (n-1)d] \\ 2S_n &= n[a_1 + \mathbf{a_1} + \mathbf{(n-1)d}] && 2a_1 = a_1 + a_1 \\ 2S_n &= n[a_1 + \mathbf{a_n}] && \text{Substitute } a_n \text{ for } a_1 + (n-1)d. \\ S_n &= \frac{n(a_1 + a_n)}{2}\end{aligned}$$

This reasoning establishes the following.

SUM OF THE FIRST n TERMS OF AN ARITHMETIC SEQUENCE

The sum of the first n terms of an arithmetic sequence is given by the formula

$$S_n = \frac{n(a_1 + a_n)}{2} \quad \text{with} \quad a_n = a_1 + (n-1)d$$

where a_1 is the first term, a_n is the nth term, and n is the number of terms in the sequence.

EXAMPLE 6 Find the sum of the first 40 terms of the arithmetic sequence 4, 10, 16, ...

Solution In order to find the sum, we will need to identify the common difference, d, and the 40th term, a_{40}. The common difference is $16 - 10 = 10 - 4 = 6$ and $a_{40} = 4 + (40 - 1)6 = 238$. Then we substitute into the formula for S_n:

$$\begin{aligned}S_n &= \frac{n(a_1 + a_{40})}{2} \\ S_{40} &= \frac{40(4 + 238)}{2} \\ &= 20(242) \\ &= 4{,}840\end{aligned}$$

The sum of the first 40 terms is 4,840.

SELF CHECK 6 Find the sum of the first 50 terms of the arithmetic sequence 3, 8, 13, ...

When the commas between the terms of a sequence are replaced by + signs, we call the indicated sum a **series**. The sum of the terms of an arithmetic sequence is called an **arithmetic series**. Some examples are

$4 + 8 + 12 + 16 + 20 + 24$	Since this series has a limited number of terms, it is a finite arithmetic series.
$5 + 8 + 11 + 14 + 17 + \cdots$	Since this series has infinitely many terms, it is an infinite arithmetic series.

6 Expand and find the sum of a series written in summation notation.

COMMENT This objective introduces summation notation for series, not just arithmetic series.

We can use a shorthand notation for indicating the sum of a finite number of consecutive terms in a series. This notation, called **summation notation**, involves the Greek letter Σ (sigma). The expression

$$\sum_{k=2}^{5} 3k \quad \text{Read as "the summation of } 3k \text{ from } k = 2 \text{ to } k = 5."$$

designates the sum of all terms obtained if we successively substitute the numbers 2, 3, 4, and 5 for k, called the **index of the summation**. Thus, we have

$$\begin{aligned} \sum_{k=2}^{5} 3k &= \underset{k=2}{3(2)} + \underset{k=3}{3(3)} + \underset{k=4}{3(4)} + \underset{k=5}{3(5)} \\ &= 6 + 9 + 12 + 15 \\ &= 42 \end{aligned}$$

EXAMPLE 7 Write the series associated with each summation.

a. $\sum_{k=1}^{4} k$ **b.** $\sum_{k=2}^{5} (k-1)^3$

Solution **a.** $\sum_{k=1}^{4} k = 1 + 2 + 3 + 4$

b. $$\begin{aligned} \sum_{k=2}^{5} (k-1)^3 &= (2-1)^3 + (3-1)^3 + (4-1)^3 + (5-1)^3 \\ &= 1^3 + 2^3 + 3^3 + 4^3 \\ &= 1 + 8 + 27 + 64 \end{aligned}$$

SELF CHECK 7 Write the series associated with the summation $\sum_{t=3}^{5} t^2$.

EXAMPLE 8 Find each sum. **a.** $\sum_{k=3}^{5} (2k+1)$ **b.** $\sum_{k=2}^{5} k^2$ **c.** $\sum_{k=1}^{3} (3k^2+3)$

Solution **a.** $$\begin{aligned} \sum_{k=3}^{5} (2k+1) &= [2(3)+1] + [2(4)+1] + [2(5)+1] \\ &= 7 + 9 + 11 \\ &= 27 \end{aligned}$$

b. $$\begin{aligned} \sum_{k=2}^{5} k^2 &= 2^2 + 3^2 + 4^2 + 5^2 \\ &= 4 + 9 + 16 + 25 \\ &= 54 \end{aligned}$$

c. $\sum_{k=1}^{3}(3k^2 + 3) = [3(1)^2 + 3] + [3(2)^2 + 3] + [3(3)^2 + 3]$

$= 6 + 15 + 30$

$= 51$

SELF CHECK 8 Evaluate: $\sum_{k=1}^{4}(2k^2 - 3)$

SELF CHECK ANSWERS **1. a.** 8, 11, 14 **b.** 305 **2. a.** 29 **b.** 121 **3.** 248 **4.** 15, 22, 29, 36 **5.** 20, 32 **6.** 6,275 **7.** 9 + 16 + 25 **8.** 48

NOW TRY THIS

Given $a_5 = 3a - 2$ and $a_9 = 11a + 10$ in an arithmetic sequence:

1. Find the first 5 terms.
2. Find the sum of the first 10 terms.

13.2 Exercises

WARM-UPS *Evaluate each algebraic expression for $n = 1$, $n = 2$, $n = 3$, and $n = 4$.*

1. $2n$ **2.** n^2

3. $n + 4$ **4.** $n - 3$

5. $5 + (n - 1)2$ **6.** $3 + (n - 1)7$

REVIEW *Perform the operations and simplify, if possible. Assume no division by 0.*

7. $5(3x^2 - 6x + 4) + 3(4x^2 + 6x - 2)$

8. $(4p + q)(2p^2 - 3pq + 2q^2)$

9. $\frac{3a + 4}{a - 2} + \frac{3a - 4}{a + 2}$

10. $2t - 3\overline{)8t^4 - 12t^3 + 8t^2 - 16t + 6}$

VOCABULARY AND CONCEPTS *Fill in the blanks.*

11. A _______ is a function whose domain is the set of natural numbers. If a sequence has a limited number of terms, it is called a _____ sequence. If it has an unlimited number of terms, it is an ______ sequence.

12. The sequence 1, 1, 2, 3, 5, 8, 13, 21, . . . is called the _______ sequence.

13. The sequence 3, 9, 15, 21, . . . is an example of an ________ sequence with a common ________ of 6 and a first term of 3.

14. The last term (or nth term) of an arithmetic sequence is given by the formula _______________.

15. If a number c is inserted between two numbers a and b to form an arithmetic sequence, then c is called the ___________ between a and b.

16. The sum of the first n terms of an arithmetic sequence is given by the formula $S_n =$ ________.

17. The indicated sum of the terms of an arithmetic sequence is called an arithmetic _____.

18. The symbol Σ is the Greek letter _____.

19. $\sum_{k=1}^{7} k$ means ________________.

20. In the expression $\sum_{k=1}^{5}(2k - 5)$, k is called the _____ of the summation.

GUIDED PRACTICE *Given the infinite sequence $a_n = 3n - 2$, find each value.* SEE EXAMPLE 1. (OBJECTIVE 1)

21. a_2 **22.** a_3

23. a_{30} **24.** a_{50}

Write the first five terms of each arithmetic sequence with the given properties. SEE EXAMPLE 2. (OBJECTIVE 2)

25. $a_1 = 3, d = 2$

26. $a_1 = -2, d = 3$

27. $a_1 = -5, d = -3$

28. $a_1 = 8, d = -5$

Write (a) the first five terms of each arithmetic sequence with the given properties; (b) the specified term of the sequence. SEE EXAMPLE 2. (OBJECTIVE 2)

29. $a_1 = 4, d = 3$; find a_{15}.

30. $a_1 = -2, d = 3$; find a_{20}.

31. $a_1 = -7, d = -2$; find a_{30}.

32. $a_1 = 10, d = -6$; find a_{12}.

Find the specific term given the first three terms of an arithmetic sequence. SEE EXAMPLE 3. (OBJECTIVE 3)

33. 2, 7, 12, . . . ; 50th term

34. 10, 7, 4, . . . ; 23rd term

35. $5, -1, -7, \ldots$; 49th term

36. $-2, 5, 12 \ldots$; 61st term

Write the first five terms of each arithmetic sequence with the given properties. SEE EXAMPLE 4. (OBJECTIVE 3)

37. $a_1 = 5$, fifth term is 29

38. $a_1 = 4$, sixth term is 39

39. $a_1 = -4$, sixth term is -39

40. $a_1 = -5$, fifth term is -37

Insert the specified number of arithmetic means. SEE EXAMPLE 5. (OBJECTIVE 4)

41. Insert three arithmetic means between 2 and 11.

42. Insert four arithmetic means between 5 and 25.

43. Insert four arithmetic means between 10 and 20.

44. Insert three arithmetic means between 20 and 30.

Find the sum of the first n terms of each arithmetic sequence. SEE EXAMPLE 6. (OBJECTIVE 5)

45. 1, 4, 7, . . . ; $n = 30$

46. 2, 6, 10, . . . ; $n = 28$

47. $-5, -1, 3, \ldots; n = 17$

48. $-7, -1, 5, \ldots; n = 15$

Write the series associated with each summation. SEE EXAMPLE 7. (OBJECTIVE 6)

49. $\sum_{k=1}^{4}(3k)$

50. $\sum_{k=1}^{3}(k-9)^2$

51. $\sum_{k=4}^{6}k^2$

52. $\sum_{k=3}^{5}(-2k)$

Find each sum. SEE EXAMPLE 8. (OBJECTIVE 6)

53. $\sum_{k=1}^{4}6k$

54. $\sum_{k=2}^{5}3k$

55. $\sum_{k=3}^{4}(k^2+3)$

56. $\sum_{k=2}^{6}(k^2+1)$

ADDITIONAL PRACTICE *Write the first five terms of each arithmetic sequence with the given properties.*

57. $d = 7$, sixth term is -83.

58. $d = 3$, seventh term is 12.

59. $d = -3$, seventh term is 16.

60. $d = -5$, seventh term is -12.

61. The 19th term is 131 and the 20th term is 138.

62. The 16th term is 70 and the 18th term is 78.

Find the specified value.

63. Find the 30th term of the arithmetic sequence with $a_1 = 7$ and $d = 12$.

64. Find the 55th term of the arithmetic sequence with $a_1 = -5$ and $d = 4$.

65. Find the 37th term of the arithmetic sequence with a second term of -4 and a third term of -9.

66. Find the 40th term of the arithmetic sequence with a second term of 6 and a fourth term of 16.

67. Find the first term of the arithmetic sequence with a common difference of 11 and whose 27th term is 263.

68. Find the common difference of the arithmetic sequence with a first term of -164 if its 36th term is -24.

69. Find the common difference of the arithmetic sequence with a first term of 40 if its 44th term is 556.

70. Find the first term of the arithmetic sequence with a common difference of -5 and whose 23rd term is -625.

71. Find the arithmetic mean between 10 and 19.

72. Find the arithmetic mean between 5 and 23.

73. Find the arithmetic mean between -4.5 and 7.

74. Find the arithmetic mean between -6.3 and -5.2.

Find the sum of the first n terms.

75. Second term is 7, third term is 12; $n = 12$.

76. Second term is 5, fourth term is 9; $n = 16$.

77. $f(n) = 2n + 1$, nth term is 31; n is a natural number.

78. $f(n) = 4n + 3$, nth term is 23; n is a natural number.

79. Find the sum of the first 50 natural numbers.

80. Find the sum of the first 100 natural numbers.

81. Find the sum of the first 50 odd natural numbers.

82. Find the sum of the first 50 even natural numbers.

Find each sum.

83. $\sum_{k=4}^{4}(2k+4)$

84. $\sum_{k=3}^{5}(3k^2-7)$

APPLICATIONS *Solve each application.*

85. Saving money Yasmeen puts $60 into a safety deposit box. Each month after that, she puts $50 more in the box. Write the first six terms of an arithmetic sequence that gives the monthly amounts in her savings, and find her savings after 10 years.

86. Installment loans Maria borrowed $10,000, interest-free, from her mother. She agreed to pay back the loan in monthly installments of $275. Write the first six terms of an arithmetic sequence that shows the balance due after each month, and find the balance due after 17 months.

87. Designing a patio Each row of bricks in the triangular patio is to have one more brick than the previous row, ending with the longest row of 150 bricks. How many bricks will be needed?

88. **Falling objects** The equation $s = 16t^2$ represents the distance s in feet that an object will fall in t seconds. After 1 second, the object has fallen 16 feet. After 2 seconds, the object has fallen 64 feet, and so on. Find the distance that the object will fall during the second and third seconds.

89. **Falling objects** Refer to Exercise 88. How far will the object fall during the 12th second?

90. **Interior angles** The sums of the angles of several polygons are given in the table. Assuming that the pattern continues, complete the table.

Figure	Number of sides	Sum of angles
Triangle	3	180°
Quadrilateral	4	360°
Pentagon	5	540°
Hexagon	6	720°
Octagon	8	
Dodecagon	12	

WRITING ABOUT MATH

91. Define an arithmetic sequence.

92. Develop the formula for finding the sum of the first n terms of an arithmetic sequence.

SOMETHING TO THINK ABOUT

93. Write the addends of the sum given by

$$\sum_{n=1}^{6}\left(\frac{1}{2}n + 1\right)$$

94. Find the sum of the sequence given in Exercise 93.

95. Show that the arithmetic mean between a and b is the average of a and b: $\frac{a+b}{2}$

96. Show that the sum of the two arithmetic means between a and b is $a + b$.

97. Show that $\sum_{k=1}^{5} 5k = 5\sum_{k=1}^{5} k$.

98. Show that $\sum_{k=3}^{6}(k^2 + 3k) = \sum_{k=3}^{6} k^2 + \sum_{k=3}^{6} 3k$.

99. Show that $\sum_{k=1}^{n} 3 = 3n$. (*Hint*: Consider 3 to be $3k^0$.)

100. Show that $\sum_{k=1}^{3}\frac{k^2}{k} \neq \frac{\sum_{k=1}^{3} k^2}{\sum_{k=1}^{3} k}$.

Section 13.3 Geometric Sequences and Series

Objectives

1. Find a specified term of a geometric sequence given the first term and the common ratio.
2. Find a specified term of a geometric sequence given the first term and one or more other terms.
3. Find one or more geometric means given two terms of a sequence.
4. Find the sum of the first n terms of a geometric sequence.
5. Solve an application involving a geometric sequence.

Vocabulary

geometric sequence   common ratio   geometric mean

Unless otherwise noted, all content on this page is © Cengage Learning.

Getting Ready

Complete each table.

1.

n	$5(2^n)$
1	
2	
3	

2.

n	$6(3^n)$
1	
2	
3	

In this section, we consider another common type of sequence called a *geometric sequence.*

1 Find a specified term of a geometric sequence given the first term and the common ratio.

Each term of a geometric sequence is found by multiplying the previous term by the same number.

GEOMETRIC SEQUENCE

A **geometric sequence** is a sequence of the form

$$a_1, \quad a_1r, \quad a_1r^2, \quad a_1r^3, \quad \dots \quad a_1r^{n-1}, \quad \dots$$

where a_1 is the *first term* and r is the **common ratio**. The *nth term* is given by

$$a_n = a_1r^{n-1}$$

We note that the second term of a geometric sequence has a factor of r^1, the third term has a factor of r^2, the fourth term has a factor of r^3, and the nth term has a factor of r^{n-1}. We also note that the quotient obtained when any term is divided by the previous term is r.

EXAMPLE 1 A geometric sequence has a first term of 5 and a common ratio of 3.

a. Write the first five terms of the sequence.

b. Find the ninth term.

Solution

a. Since the first term is $a_1 = 5$ and the common ratio is $r = 3$, the first five terms are

$$\underset{a_1}{\overset{5,}{\uparrow}} \quad \underset{a_2}{\overset{5(3),}{\uparrow}} \quad \underset{a_3}{\overset{5(3^2),}{\uparrow}} \quad \underset{a_4}{\overset{5(3^3),}{\uparrow}} \quad \underset{a_5}{\overset{5(3^4)}{\uparrow}}$$

Each term is found by multiplying the previous term by 3.

or

$$5, 15, 45, 135, 405$$

b. The nth term is a_1r^{n-1} where $a_1 = 5$ and $r = 3$. Because we want the ninth term, we let $n = 9$:

$$\begin{aligned} a_n &= a_1r^{n-1} \\ a_9 &= 5(3)^{9-1} \\ &= 5(3)^8 \\ &= 5(6{,}561) \\ &= 32{,}805 \end{aligned}$$

SELF CHECK 1 A geometric sequence has a first term of 3 and a common ratio of 4.

a. Write the first four terms.

b. Find the eighth term.

2 Find a specified term of a geometric sequence given the first term and one or more other terms.

EXAMPLE 2 The first three terms of a geometric sequence are 16, 4, and 1. Find the seventh term.

Solution We note that the first term a_1 is 16 and that

$$r = \frac{a_2}{a_1} = \frac{4}{16} = \frac{1}{4} \quad \text{Also note that } \frac{a_3}{a_2} = \frac{1}{4}.$$

We substitute 16 for a_1, $\frac{1}{4}$ for r, and 7 for n in the formula for the nth term and simplify:

$$\begin{aligned} a_n &= a_1 r^{n-1} \\ a_7 &= 16\left(\frac{1}{4}\right)^{7-1} \\ &= 16\left(\frac{1}{4}\right)^{6} \\ &= 16\left(\frac{1}{4{,}096}\right) \\ &= \frac{1}{256} \end{aligned}$$

SELF CHECK 2 Find the tenth term of the sequence in Example 2.

EXAMPLE 3 Write the first five terms of the geometric sequence whose first term is -3 and fourth term is -192.

Solution Here the first term is $a_1 = -3$ and $a_4 = -192$. We substitute -3 for a_1 and -192 for a_4 into the formula for the nth term of a geometric sequence and solve for r.

$$\begin{aligned} a_n &= a_1 r^{n-1} \\ -192 &= -3r^{4-1} \\ 64 &= r^3 && \text{Divide both sides by } -3. \\ 4 &= r && \text{Take the cube root of both sides.} \end{aligned}$$

The first five terms are $-3, -12, -48, -192, -768$.

SELF CHECK 3 Write the first five terms of the geometric sequence whose first term is -64 and sixth term is -2.

3 Find one or more geometric means given two terms of a sequence.

If numbers are inserted between two numbers a and b to form a geometric sequence, the inserted numbers are called **geometric means** between a and b.

If a single number is inserted between the numbers a and b to form a geometric sequence, that number is called the **geometric mean** between a and b.

EXAMPLE 4 Insert two geometric means between 7 and 1,512.

Solution Here the first term is $a_1 = 7$. Because we are inserting two geometric means between 7 and 1,512, the fourth term is $a_4 = 1{,}512$. To find the common ratio r so that the terms

$$\begin{array}{cccc} 7, & 7r, & 7r^2, & 1{,}512 \\ \uparrow & \uparrow & \uparrow & \uparrow \\ a_1 & a_2 & a_3 & a_4 \end{array}$$

form a geometric sequence, we substitute 4 for n and 7 for a_1 into the formula for the nth term of a geometric sequence and solve for r.

$$\begin{aligned} a_n &= a_1 r^{n-1} \\ a_4 &= 7r^{4-1} \\ 1{,}512 &= 7r^3 \\ 216 &= r^3 && \text{Divide both sides by 7.} \\ 6 &= r && \text{Take the cube root of both sides.} \end{aligned}$$

The two geometric means between 7 and 1,512 are

$$\mathbf{7r = 7(6) = 42} \qquad \text{and} \qquad \mathbf{7r^2 = 7(6)^2 = 7(36) = 252}$$

The numbers 7, 42, 252, and 1,512 are the first four terms of a geometric sequence.

SELF CHECK 4 Insert three positive geometric means between 1 and 16.

EXAMPLE 5 Find a geometric mean between 2 and 20.

Solution We want to find the middle term of the three-term geometric sequence

$$\begin{array}{ccc} 2, & 2r, & 20 \\ \uparrow & \uparrow & \uparrow \\ a_1 & a_2 & a_3 \end{array}$$

with $a_1 = 2$, $a_3 = 20$, and $n = 3$. To find the value of r, we substitute these values into the formula for the nth term of a geometric sequence:

$$\begin{aligned} a_n &= ar^{n-1} \\ a_3 &= 2r^{3-1} \\ 20 &= 2r^2 \\ 10 &= r^2 && \text{Divide both sides by 2.} \\ \pm\sqrt{10} &= r && \text{Take the square root of both sides.} \end{aligned}$$

Because r can be either $\sqrt{10}$ or $-\sqrt{10}$, there are two values for the geometric mean. They are

$$\mathbf{2r = 2\sqrt{10}} \qquad \text{and} \qquad \mathbf{2r = -2\sqrt{10}}$$

The numbers 2, $2\sqrt{10}$, 20 and 2, $-2\sqrt{10}$, 20 both form geometric sequences. The common ratio of the first sequence is $\sqrt{10}$, and the common ratio of the second sequence is $-\sqrt{10}$.

SELF CHECK 5 Find a geometric mean between 2 and 200.

4 Find the sum of the first n terms of a geometric sequence.

There is a formula that gives the sum of the first n terms of a geometric sequence. To develop this formula, we let S_n represent the sum of the first n terms of a geometric sequence.

(1) $S_n = a_1 + a_1r + a_1r^2 + a_1r^3 + \cdots + a_1r^{n-1}$ This is a geometric series.

We multiply both sides of Equation 1 by r to obtain

(2) $S_nr = a_1r + a_1r^2 + a_1r^3 + \cdots + a_1r^{n-1} + a_1r^n$

We now subtract Equation 2 from Equation 1 and solve for S_n:

$$S_n - S_nr = a_1 - a_1r^n$$

$$S_n(1 - r) = a_1 - a_1r^n \quad \text{Factor out } S_n \text{ from the left side.}$$

$$S_n = \frac{a_1 - a_1r^n}{1 - r} \quad \text{Divide both sides by } 1 - r.$$

This reasoning establishes the following formula.

SUM OF THE FIRST n TERMS OF A GEOMETRIC SEQUENCE

The sum of the first n terms of a geometric sequence is given by the formula

$$S_n = \frac{a_1 - a_1r^n}{1 - r} \quad (r \neq 1)$$

where S_n is the sum, a_1 is the first term, r is the common ratio, and n is the number of terms.

EXAMPLE 6 Find the sum of the first six terms of the geometric sequence 250, 50, 10, ...

Solution Here $a_1 = 250$, $r = \frac{1}{5}$, and $n = 6$. We substitute these values into the formula for the sum of the first n terms of a geometric sequence and simplify:

$$S_n = \frac{a_1 - a_1r^n}{1 - r}$$

$$S_n = \frac{250 - 250\left(\frac{1}{5}\right)^6}{1 - \frac{1}{5}}$$

$$= \frac{250 - 250\left(\frac{1}{15{,}625}\right)}{\frac{4}{5}}$$

$$= \frac{5}{4}\left(250 - \frac{250}{15{,}625}\right)$$

$$= \frac{5}{4}\left(\frac{3{,}906{,}000}{15{,}625}\right)$$

$$= 312.48$$

The sum of the first six terms is 312.48.

SELF CHECK 6 Find the sum of the first five terms of the geometric sequence 100, 20, 4, ...

5 Solve an application involving a geometric sequence.

EXAMPLE 7 **GROWTH OF A TOWN** The mayor of Eagle River (population 1,500) predicts a growth rate of 4% each year for the next 10 years. Find the expected population of Eagle River 10 years from now.

Solution Let P_0 be the initial population of Eagle River. After 1 year, there will be a different population, P_1. The initial population (P_0) plus the growth (the product of P_0 and the rate of growth, r) will equal this new population P_1:

$$P_1 = P_0 + P_0 r = P_0(1 + r)$$

The population after 2 years will be P_2, and

$$\begin{aligned} P_2 &= P_1 + P_1 r \\ &= \mathbf{P_1}(1 + r) && \text{Factor out } P_1. \\ &= \mathbf{P_0(1 + r)}(1 + r) && P_1 = P_0(1 + r). \\ &= P_0(1 + r)^2 \end{aligned}$$

The population after 3 years will be P_3, and

$$\begin{aligned} P_3 &= P_2 + P_2 r \\ &= \mathbf{P_2}(1 + r) && \text{Factor out } P_2. \\ &= \mathbf{P_0(1 + r)^2}(1 + r) && P_2 = P_0(1 + r)^2. \\ &= P_0(1 + r)^3 \end{aligned}$$

The yearly population figures

$$P_0, \quad P_1, \quad P_2, \quad P_3, \quad \ldots$$

or

$$P_0, \quad P_0(1 + r), \quad P_0(1 + r)^2, \quad P_0(1 + r)^3, \quad \ldots$$

form a geometric sequence with a first term of P_0 and a common ratio of $1 + r$. The population of Eagle River after 10 years is P_{10}, which is the 11th term of this sequence:

$$\begin{aligned} a_n &= ar^{n-1} \\ P_{10} = a_{11} &= P_0(1 + r)^{10} \\ &= 1{,}500(1 + 0.04)^{10} \\ &= 1{,}500(1.04)^{10} \\ &\approx 1{,}500(1.480244285) \\ &\approx 2{,}220 \end{aligned}$$

The expected population 10 years from now is 2,220 people.

SELF CHECK 7 Find the expected population of Eagle River 20 years from now.

EXAMPLE 8 **AMOUNT OF AN ANNUITY** An *annuity* is a sequence of equal payments made periodically over a length of time. The sum of the payments and the interest earned during the *term* of the annuity is called the *amount* of the annuity.

After a sales clerk works six months, her employer will begin an annuity for her and will contribute \$500 every six months to a fund that pays 8% annual interest. After she has been employed for two years, what will be the amount of the annuity?

Solution Because the payments are to be made semiannually, there will be four payments of \$500, each earning a rate of 4% per six-month period. These payments will occur at the end of 6 months, 12 months, 18 months, and 24 months. The first payment, to be made after 6 months, will earn interest for three interest periods. Thus, the amount of the first payment is $\$500(1.04)^3$. The amounts of each of the four payments after two years are shown in Table 13-1 on the next page.

The amount of the annuity is the sum of the amounts of the individual payments, a sum of \$2,123.23.

TABLE 13-1

Payment (at the end of period)	Amount of payment at the end of 2 years
1	$\$500(1.04)^3 = \$562.43$
2	$\$500(1.04)^2 = \$540.80$
3	$\$500(1.04)^1 = \$520.00$
4	$\$500 = \$500.00$
	$A_n = \$2{,}123.23$

SELF CHECK 8 If the interest rate is only 4% annual interest, what will be the amount of the annuity?

SELF CHECK ANSWERS

1. a. 3, 12, 48, 192 **b.** 49,152 **2.** $\frac{1}{16{,}384}$ **3.** $-64, -32, -16, -8, -4$ **4.** 2, 4, 8 **5.** 20, -20 **6.** 124.96 **7.** The expected population 20 years from now is 3,286. **8.** \$2,060.80

NOW TRY THIS

Given $a_1 = 8x - 12$ and $r = \frac{1}{2}$ in a geometric sequence:

1. Find the first 4 terms.
2. Find the sum of the first 4 terms by adding the terms.
3. Set up the formula for finding the sum of the first 4 terms and simplify it to show that the result is the same as the result obtained in Exercise 2.

13.3 Exercises

WARM-UPS *Simplify.*

1. $2(4^2)$
2. $2(4^3)$
3. $4(3)^{5-1}$
4. $4(3)^{7-1}$
5. $7\left(\frac{1}{3}\right)^{4-1}$
6. $7\left(\frac{1}{3}\right)^{5-1}$

REVIEW *Solve each inequality. Assume no division by 0.*

7. $x^2 - 4x - 5 \leq 0$
8. $a^2 - 9a + 20 \geq 0$
9. $\dfrac{x-4}{x+3} \geq 0$
10. $\dfrac{t^2 + t - 20}{t+2} < 0$

VOCABULARY AND CONCEPTS *Fill in the blanks.*

11. A sequence of the form $a_1, a_1r, a_1r^2, \ldots$ is called a ________ sequence.
12. The formula for the nth term of a geometric sequence is ________.
13. In a geometric sequence, r is called the ____________.
14. A number inserted between two numbers a and b to form a geometric sequence is called a geometric _____ between a and b.
15. The sum of the first n terms of a geometric sequence is given by the formula ____________.
16. In the formula for Exercise 15, a_1 is the ____ term of the sequence.

Find the next term in each geometric sequence.

17. 1, 3, 9, . . .

18. $1, \frac{1}{3}, \frac{1}{9}, \ldots$

Find the common ratio in each geometric sequence.

19. 0.2, 0.5, 1.25, . . .

20. $\sqrt{3}, 3, 3\sqrt{3}, \ldots$

Find the value of x in each geometric sequence.

21. $2, x, 18, 54, \ldots$

22. $3, x, \frac{1}{3}, \frac{1}{9}, \ldots$

GUIDED PRACTICE

Write the first five terms of each geometric sequence and find the eighth term. SEE EXAMPLE 1. (OBJECTIVE 1)

23. $a_1 = 4, r = 3$

24. $a_1 = -2, r = 2$

25. $a_1 = -5, r = \frac{1}{5}$

26. $a_1 = 8, r = \frac{1}{2}$

Given the first three terms of a geometric sequence, find the specified term. SEE EXAMPLE 2. (OBJECTIVE 2)

27. 1, 3, 9, . . . ; sixth term

28. 2, 6, 18, . . . ; fifth term

29. $2, 1, \frac{1}{2}, \ldots$; seventh term

30. $1, \frac{1}{3}, \frac{1}{9}, \ldots$; fifth term

Write the first five terms of each geometric sequence. SEE EXAMPLE 3. (OBJECTIVE 2)

31. $a_1 = 2, r > 0$, third term is 32.

32. $a_1 = 3$, fourth term is 24.

33. $a_1 = 2, r < 0$, third term is 50.

34. $a_1 = -81$, sixth term is $\frac{1}{3}$.

Insert the specified number of geometric means. SEE EXAMPLE 4. (OBJECTIVE 3)

35. Insert three positive geometric means between 2 and 162.

36. Insert four geometric means between 3 and 96.

37. Insert four geometric means between −4 and −12,500.

38. Insert three geometric means (two positive and one negative) between −64 and −1,024.

Insert the specified geometric mean between the given numbers. SEE EXAMPLE 5. (OBJECTIVE 3)

39. Find the negative geometric mean between 2 and 128.

40. Find the positive geometric mean between 3 and 243.

41. Find the positive geometric mean between 10 and 20.

42. Find the negative geometric mean between 5 and 15.

Find the sum of the first n terms of each geometric sequence. SEE EXAMPLE 6. (OBJECTIVE 4)

43. 120, 60, 30, . . . ; $n = 4$

44. 2, −6, 18, . . . ; $n = 6$

45. 2, −6, 18, . . . ; $n = 5$

46. 256, −64, 16, . . . ; $n = 5$

ADDITIONAL PRACTICE

Find the first five terms of the geometric sequence.

47. The second term is 10, and the third term is 50.

48. The third term is −27, and the fourth term is 81.

49. $a_1 = -64, r < 0$, fifth term is −4.

50. $a_1 = -64, r > 0$, fifth term is −4.

Find the indicated quantity.

51. Find the ninth term of the geometric sequence with $a_1 = 4$ and $r = 3$.

52. Find the 12th term of the geometric sequence with $a_1 = 64$ and $r = \frac{1}{2}$.

53. Find the first term of the geometric sequence with a common ratio of −3 and an eighth term of −81.

54. Find the first term of the geometric sequence with a common ratio of 2 and a tenth term of 384.

55. Find the common ratio of the geometric sequence with a first term of −8 and a sixth term of −1,944.

56. Find the common ratio of the geometric sequence with a first term of 12 and a sixth term of $\frac{3}{8}$.

57. Find a geometric mean, if possible, between −50 and 10.

58. Find a negative geometric mean, if possible, between −25 and −5.

Find the sum of the first n terms of each geometric sequence.

59. 3, −6, 12, . . . ; $n = 8$

60. 3, 6, 12, . . . ; $n = 8$

61. 3, 6, 12, . . . ; $n = 7$

62. 3, −6, 12, . . . ; $n = 7$

63. The second term is 1, and the third term is $\frac{1}{5}$; $n = 4$.

64. The second term is 1, and the third term is 4; $n = 5$.

65. The third term is −2, and the fourth term is 1; $n = 6$.

66. The third term is −3, and the fourth term is 1; $n = 5$.

APPLICATIONS

Use a calculator to help solve. SEE EXAMPLES 7–8. (OBJECTIVE 5)

67. Population growth The population of Union is predicted to increase by 6% each year. What will be the population of Union 5 years from now if its current population is 500?

68. Population decline The population of Bensonville is decreasing by 10% each year. If its current population is 98, what will be the population 8 years from now?

69. Declining savings John has \$10,000 in a safety deposit box. Each year he spends 12% of what is left in the box. How much will be in the box after 15 years?

70. Savings growth Lu Ling has \$5,000 in a savings account earning 12% annual interest. How much will be in her account 10 years from now? (Assume that Lu Ling makes no deposits or withdrawals.)

71. House appreciation A house appreciates by 6% each year. If the house is worth \$70,000 today, how much will it be worth 12 years from now?

72. Motorboat depreciation A motorboat that cost \$5,000 when new depreciates at a rate of 9% per year. How much will the boat be worth in 5 years?

73. Inscribed squares Each inscribed square in the illustration joins the midpoints of the next larger square. The area of the first square, the largest, is 1 square unit. Find the area of the 12th square.

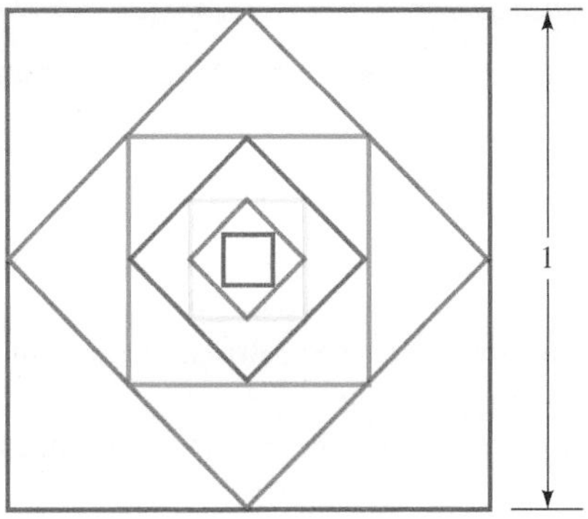

74. Genealogy The following family tree spans 3 generations and lists 7 people. How many names would be listed in a family tree that spans 10 generations?

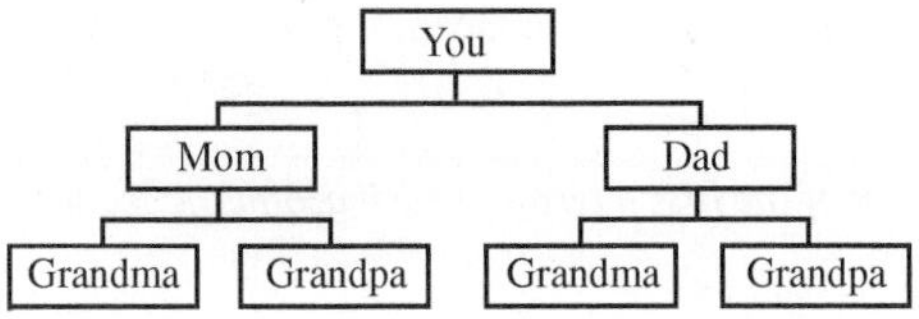

75. Annuities Find the amount of an annuity if \$1,000 is paid semiannually for two years at 6% annual interest. Assume that the first of the four payments is made immediately.

76. Annuities Note that the amounts shown in Table 13-1 form a geometric sequence. Verify the answer for Example 8 by using the formula for the sum of a geometric sequence.

WRITING ABOUT MATH

77. Define a geometric sequence.

78. Develop the formula for finding the sum of the first n terms of a geometric sequence.

SOMETHING TO THINK ABOUT

79. Show that the formula for the sum of the first n terms of a geometric sequence can be found by using the formula $S = \frac{a_1 - a_n r}{1 - r}$.

80. Show that the geometric mean between a and b is $\sqrt{ab}$.

81. If $a > b > 0$, which is larger: the arithmetic mean between a and b or the geometric mean between a and b?

82. Is there a geometric mean between -5 and 5?

83. Show that the formula for the sum of the first n terms of a geometric sequence can be written in the form

$$S_n = \frac{a_1 - a_n r}{1 - r} \quad \text{where } a_n = a_1 r^{n-1}$$

84. Show that the formula for the sum of the first n terms of a geometric sequence can be written in the form

$$S_n = \frac{a_1(1 - r^n)}{1 - r}$$

Section 13.4 Infinite Geometric Sequences and Series

Objectives

1. Find the nth partial sum of an infinite geometric series.
2. Determine whether the sum of an infinite geometric sequence exists.
3. Find the sum of an infinite geometric series, if possible.
4. Write a repeating decimal as a common fraction.

Unless otherwise noted, all content on this page is © Cengage Learning.

Vocabulary

infinite geometric series   partial sum

Getting Ready

Evaluate each expression.

1. $\dfrac{2}{1 - \frac{1}{2}}$ 2. $\dfrac{3}{1 - \frac{1}{3}}$ 3. $\dfrac{\frac{3}{2}}{1 - \frac{1}{2}}$ 4. $\dfrac{\frac{5}{3}}{1 - \frac{1}{3}}$

In this section, we will consider geometric sequences with infinitely many terms.

If we form the sum of the terms of an infinite geometric sequence, we obtain a series called an **infinite geometric series**. For example, if the common ratio r is 3, we have

Infinite geometric sequence	***Infinite geometric series***
2, 6, 18, 54, 162, . . .	$2 + 6 + 18 + 54 + 162 + \cdots$

Under certain conditions, we can find the sum of an infinite geometric series. To define this sum, we consider the geometric series

$$a_1 + a_1r + a_1r^2 + a_1r^3 + \cdots + a_1r^{n-1}$$

- The first **partial sum**, S_1, of the sequence is $S_1 = a_1$.
- The second partial sum, S_2, of the sequence is $S_2 = a_1 + a_1r$.
- The third partial sum, S_3, of the sequence is $S_3 = a_1 + a_1r + a_1r^2$.
- The nth partial sum, S_n, of the sequence is $S_n = a_1 + a_1r + a_1r^2 + \cdots + a_1r^{n-1}$.

1 Find the *n*th partial sum of an infinite geometric series.

EXAMPLE 1 Given the sequence 2, 6, 18, 54, 162, . . . find

a. S_4 and **b.** S_7

Solution We know $a_1 = 2$ and the common ratio is $18 \div 6 = 6 \div 2 = 3$ and we will substitute these values into the formula.

a.
$$\begin{aligned} S_n &= \frac{a_1 - a_1r^n}{1 - r} \\ S_4 &= \frac{2 - 2(3)^4}{1 - 3} \\ &= \frac{2 - 2(81)}{-2} \\ &= \frac{2 - 162}{-2} \\ &= \frac{-160}{-2} \\ &= 80 \end{aligned}$$

The 4th partial sum is 80.

b.
$$\begin{aligned} S_n &= \frac{a_1 - a_1r^n}{1 - r} \\ S_7 &= \frac{2 - 2(3)^7}{1 - 3} \\ &= \frac{2 - 2(2{,}187)}{-2} \\ &= \frac{2 - 4{,}374}{-2} \\ &= \frac{-4372}{-2} \\ &= 2{,}186 \end{aligned}$$

The 7th partial sum is 2,186.

SELF CHECK 1 Given the sequence 3, 6, 12, 24, . . . find **a.** S_8 and **b.** S_{15}

If the nth *partial sum*, S_n, approaches some number S_∞ as n approaches infinity, then S_∞ is called the *sum of the infinite geometric series.*

To develop a formula for finding the sum (if it exists) of an infinite geometric series, we consider the formula for the sum of the first n terms of a geometric sequence.

$$S_n = \frac{a_1 - a_1r^n}{1 - r} \quad (r \neq 1)$$

If $|r| < 1$ and a_1 is constant, then the term a_1r^n in the above formula approaches 0 as n becomes very large. For example,

$$a_1\left(\frac{1}{4}\right)^1 = \frac{1}{4}a_1, \quad a_1\left(\frac{1}{4}\right)^2 = \frac{1}{16}a_1, \quad a_1\left(\frac{1}{4}\right)^3 = \frac{1}{64}a_1$$

and so on. When n is very large, the value of a_1r^n is negligible, and the term a_1r^n in the above formula can be ignored. This reasoning justifies the following.

SUM OF AN INFINITE GEOMETRIC SERIES

If a_1 is the first term and r is the common ratio of an infinite geometric sequence, and if $|r| < 1$, the sum of the related geometric series is represented by the formula

$$S_\infty = \frac{a_1}{1 - r}$$

COMMENT Recall that $|r| < 1$ is equivalent to the inequality $-1 < r < 1$. This implies that an infinite geometric series will have a sum if r is between -1 and 1.

2 Determine whether the sum of an infinite geometric sequence exists.

To determine whether S_∞ can be calculated, we look at the ratio of the sequence. If $-1 < r < 1$, then the infinite geometric sequence has a sum. If $r < -1$ or $r > 1$, the sum cannot be calculated.

EXAMPLE 2 Determine whether the S_∞ of each infinite geometric sequence can be found.

a. 2, 4, 8, . . . **b.** 3, 1, $\frac{1}{3}$, . . .

Solution We must calculate the common ratio and determine if it falls between -1 and 1.

a. 2, 4, 8, . . .
The ratio of the sequence is $8 \div 4 = 4 \div 2 = 2$. Since 2 does not fall between -1 and 1, the sum of the terms of the infinite geometric sequence cannot be found.

b. 3, 1, $\frac{1}{3}$, . . .
The ratio of the sequence is $\frac{1}{3} \div 1 = 1 \div 3 = \frac{1}{3}$. Since $-1 < \frac{1}{3} < 1$, the sum of the terms of the infinite geometric sequence can be found.

SELF CHECK 2 Determine whether the S_∞ of each infinite geometric sequence can be found.
a. 8, 4, 2, . . . **b.** -5, 10, -20, . . .

In the next example, we will find the S_∞ of an infinite geometric series, if possible.

3 Find the sum of an infinite geometric series, if possible.

EXAMPLE 3 Find the sum of the infinite geometric series $125 + 25 + 5 + \cdots$.

Solution Here $a_1 = 125$ and $r = \frac{1}{5}$. Since $|r| = \left|\frac{1}{5}\right| = \frac{1}{5} < 1$, we can find the sum of the series. We substitute 125 for a_1 and $\frac{1}{5}$ for r in the formula $S_\infty = \frac{a_1}{1-r}$ and simplify:

$$S_\infty = \frac{a_1}{1-r} = \frac{125}{1-\frac{1}{5}} = \frac{125}{\frac{4}{5}} = \frac{5}{4}(125) = \frac{625}{4}$$

The sum of the series is $\frac{625}{4}$ or 156.25.

SELF CHECK 3 Find the sum of the infinite geometric series $100 + 20 + 4 + \cdots$.

EXAMPLE 4 Find the sum of the infinite geometric series $64 + (-4) + \frac{1}{4} + \cdots$.

Solution Here $a_1 = 64$ and $r = -\frac{1}{16}$. Since $|r| = \left|-\frac{1}{16}\right| = \frac{1}{16} < 1$, we can find the sum of the series. We substitute 64 for a_1 and $-\frac{1}{16}$ for r in the formula $S_\infty = \frac{a_1}{1-r}$ and simplify:

$$S_\infty = \frac{a_1}{1-r} = \frac{64}{1-\left(-\frac{1}{16}\right)} = \frac{64}{\frac{17}{16}} = \frac{16}{17}(64) = \frac{1{,}024}{17}$$

The sum of the terms of the infinite geometric sequence $64, -4, \frac{1}{4}, \ldots$ is $\frac{1{,}024}{17}$.

SELF CHECK 4 Find the sum of the infinite geometric series $81 + 27 + 9 + \cdots$.

EXAMPLE 5 Find the sum of the infinite geometric series $\frac{9}{2} + 6 + 8 + \cdots$, if possible.

Solution Here $a_1 = \frac{9}{2}$ and $r = \frac{4}{3}$. Since $|r| = \left|\frac{4}{3}\right| = \frac{4}{3} > 1$, there is no sum.

SELF CHECK 5 Find the sum of the infinite geometric series $\frac{5}{2} + 5 + 10 + \cdots$, if possible.

4 Write a repeating decimal as a common fraction.

We can use the sum of an infinite geometric series to write a repeating decimal as a common fraction.

EXAMPLE 6 Write $0.\overline{8}$ as a common fraction.

Solution The decimal 0.8 can be written as an infinite geometric series:

$$0.\overline{8} = 0.888\cdots = \frac{8}{10} + \frac{8}{100} + \frac{8}{1{,}000} + \cdots$$

where $a_1 = \frac{8}{10}$ and $r = \frac{1}{10}$. Because $|r| = \left|\frac{1}{10}\right| = \frac{1}{10} < 1$, we can find the sum as follows:

$$S_\infty = \frac{a_1}{1-r} = \frac{\frac{8}{10}}{1-\frac{1}{10}} = \frac{\frac{8}{10}}{\frac{9}{10}} = \frac{8}{9}$$

Thus, $0.\overline{8} = \frac{8}{9}$. Long division will verify that $\frac{8}{9} = 0.888\ldots$.

SELF CHECK 6 Write $0.\overline{5}$ as a common fraction.

EXAMPLE 7 Write $0.\overline{25}$ as a common fraction.

Solution The decimal $0.\overline{25}$ can be written as an infinite geometric series:

$$0.\overline{25} = 0.252525\ldots = \frac{25}{100} + \frac{25}{10{,}000} + \frac{25}{1{,}000{,}000} + \ldots$$

where $a_1 = \frac{25}{100}$ and $r = \frac{1}{100}$. Since $|r| = \left|\frac{1}{100}\right| = \frac{1}{100} < 1$, we can find the sum as follows:

$$S_\infty = \frac{a_1}{1-r} = \frac{\frac{25}{100}}{1-\frac{1}{100}} = \frac{\frac{25}{100}}{\frac{99}{100}} = \frac{25}{99}$$

Thus, $0.\overline{25} = \frac{25}{99}$. Long division will verify that this is true.

SELF CHECK 7 Write $0.\overline{15}$ as a common fraction.

SELF CHECK ANSWERS **1. a.** 765 **b.** 98,301 **2. a.** yes **b.** no **3.** 125 **4.** $\frac{243}{2}$ **5.** no sum because $r > 1$ **6.** $\frac{5}{9}$ **7.** $\frac{5}{33}$

NOW TRY THIS

Find each sum, if possible.

1. $\sum_{k=1}^{\infty} 3\left(\frac{2}{3}\right)^k$
2. $\sum_{k=1}^{\infty} \frac{1}{4}\left(\frac{3}{2}\right)^k$
3. $\sum_{k=1}^{10} 5\left(\frac{1}{2}k\right)$

13.4 Exercises

WARM-UPS *Find the common ratio in each infinite geometric sequence.*

1. $2, 6, 18, \ldots$

2. $100, 50, 25, \ldots$

3. $50, 10, 2, \ldots$

4. $\frac{1}{50}, \frac{1}{10}, \frac{1}{2}, \ldots$

5. $1, \frac{1}{5}, \frac{1}{25}, \ldots$

6. $\frac{2}{5}, \frac{1}{5}, \frac{1}{10}, \ldots$

REVIEW *Determine whether each equation determines y to be a function of x.*

7. $y = -2x^3 + 5$
8. $xy = 12$
9. $3x = y^2 + 4$
10. $x = |y|$

VOCABULARY AND CONCEPTS *Fill in the blanks.*

11. If a geometric sequence has infinitely many terms, it is called an ______ geometric sequence.
12. The third partial sum of the series $2 + 6 + 18 + 54 + \cdots$ is ________ $= 26$.
13. The formula for the sum of an infinite geometric series with $|r| < 1$ is ________.
14. Write $0.\overline{7}$ as an infinite geometric series.

____

GUIDED PRACTICE *Find the sum of each infinite geometric series.* SEE EXAMPLE 3. (OBJECTIVE 3)

15. $48 + 12 + 3 + \cdots$
16. $12 + 6 + 3 + \cdots$
17. $200 + 40 + 8 + \cdots$
18. $45 + 15 + 5 + \cdots$

Find the sum of each infinite geometric series. SEE EXAMPLE 4. (OBJECTIVE 3)

19. $12 + (-6) + 3 + \cdots$
20. $8 + (-4) + 2 + \cdots$
21. $-81 + 27 + (-9) + \cdots$
22. $-45 + 15 + (-5) + \cdots$

Find the sum of each infinite geometric series, if possible. SEE EXAMPLE 5. (OBJECTIVE 3)

23. $-3 + (-6) + (-12) + \cdots$
24. $4 + 12 + 36 + \cdots$
25. $\frac{3}{4} + \frac{3}{2} + 3 + \cdots$
26. $-\frac{5}{6} + \left(-\frac{5}{3}\right) + \left(-\frac{10}{3}\right) + \cdots$

Write each decimal as a common fraction. Then check the answer by using long division. SEE EXAMPLE 6. (OBJECTIVE 4)

27. $0.\overline{1}$
28. $0.\overline{2}$
29. $-0.\overline{3}$
30. $-0.\overline{4}$

Write each decimal as a common fraction. Then check the answer by using long division. SEE EXAMPLE 7. (OBJECTIVE 4)

31. $0.\overline{12}$
32. $0.\overline{21}$
33. $0.\overline{75}$
34. $0.\overline{57}$

ADDITIONAL PRACTICE *Find the sum of each infinite geometric series, if possible.*

35. $-54 + 18 + (-6) + \cdots$
36. $-112 + (-28) + (-7) + \cdots$
37. $-\frac{27}{2} + (-9) + (-6) + \cdots$
38. $\frac{18}{25} + \frac{6}{5} + 2 + \cdots$

APPLICATIONS *Solve each application.*

39. Controlling moths To reduce the population of a destructive moth, biologists release 1,000 sterilized male moths each day into the environment. If 80% of these moths alive one day survive until the next, then the population of sterile males is the sum of the infinite geometric sequence

$$1{,}000 + 1{,}000(0.8) + 1{,}000(0.8)^2 + 1{,}000(0.8)^3 + \cdots$$

Find the long-term population.

40. Controlling moths If mild weather increases the day-to-day survival rate of the sterile male moths in Exercise 41 to 90%, find the long-term population.

41. Bouncing balls On each bounce, the rubber ball in the illustration rebounds to a height one-half of that from which it fell. Find the total distance the ball travels.

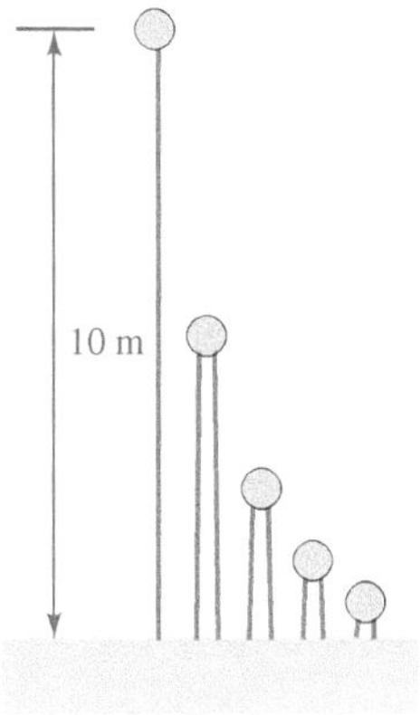

42. Bouncing balls A golf ball is dropped from a height of 12 feet. On each bounce, it returns to a height two-thirds of that from which it fell. Find the total distance the ball travels.

WRITING ABOUT MATH

43. Why must the absolute value of the common ratio be less than 1 before an infinite geometric series can have a sum?
44. Can an infinite arithmetic series have a sum?

SOMETHING TO THINK ABOUT

45. An infinite geometric series has a sum of 5 and a first term of 1. Find the common ratio.
46. An infinite geometric series has a common ratio of $-\frac{2}{3}$ and a sum of 9. Find the first term.
47. Show that $0.\overline{9} = 1$.
48. Show that $1.\overline{9} = 2$.
49. Does $0.999999 = 1$? Explain.
50. If $f(x) = 1 + x + x^2 + x^3 + x^4 + \cdots$, find $f\left(\frac{1}{2}\right)$ and $f\left(-\frac{1}{2}\right)$.

Unless otherwise noted, all content on this page is © Cengage Learning.

Section 13.5 Permutations and Combinations

Objectives

1. Use the multiplication principle to determine the number of ways one event can be followed by another.
2. Use permutations to find the number of n things taken r at a time if order is important.
3. Use combinations to find the number of n things taken r at a time.
4. Use combinations to find the coefficients of the terms of a binomial expansion.

Vocabulary

tree diagram
event
multiplication principle for events
permutation
combination

Getting Ready

Evaluate each expression.

1. $4 \cdot 3 \cdot 2 \cdot 1$
2. $5 \cdot 4 \cdot 3 \cdot 2 \cdot 1$
3. $\dfrac{6 \cdot 5 \cdot 4 \cdot 3 \cdot 2 \cdot 1}{4 \cdot 3 \cdot 2 \cdot 1}$
4. $\dfrac{8 \cdot 7 \cdot 6 \cdot 5 \cdot 4 \cdot 3 \cdot 2 \cdot 1}{2(5 \cdot 4 \cdot 3 \cdot 2 \cdot 1)}$

In this section, we will discuss methods of counting the different ways we can do something like arranging books on a shelf or selecting a committee. These kinds of problems are important in statistics, insurance, telecommunications, and other fields.

1 Use the multiplication principle to determine the number of ways one event can be followed by another.

Steven goes to the cafeteria for lunch. He has a choice of three different sandwiches (hamburger, hot dog, or ham and cheese) and four different beverages (cola, root beer, orange, or milk). How many different lunches can he choose?

He has three choices of sandwich, and for any one of these choices, he has four choices of drink. The different options are shown in the **tree diagram** in Figure 13-1.

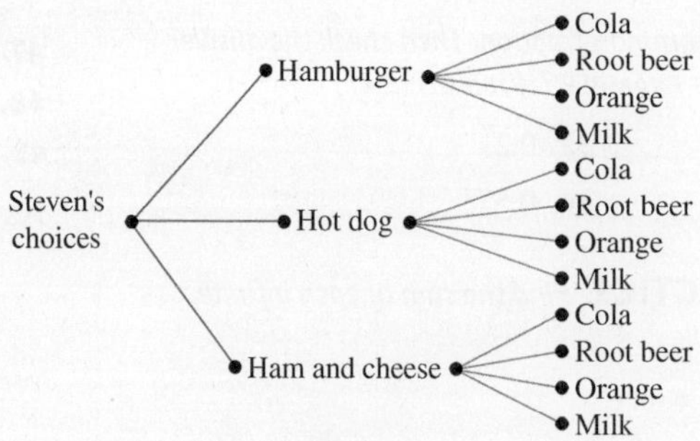

Figure 13-1

Unless otherwise noted, all content on this page is © Cengage Learning.

The tree diagram illustrates that he has a total of 12 different lunches from which to choose. One of the possibilities is a hamburger with a cola, and another is a hot dog with milk.

A situation that can have several different outcomes—such as choosing a sandwich—is called an **event**. Choosing a sandwich and choosing a beverage can be thought of as two events. The preceding example illustrates the **multiplication principle for events**.

MULTIPLICATION PRINCIPLE FOR EVENTS

Let E_1 and E_2 be two events. If E_1 can be done in a_1 ways, and if—after E_1 has occurred—E_2 can be done in a_2 ways, the event "E_1 followed by E_2" can be done in $a_1 \cdot a_2$ ways.

EXAMPLE 1 After dinner, Heidi plans to watch the evening news and then a situation comedy on TV. If there are choices of four news broadcasts and two comedies, in how many ways can she choose to watch television?

Solution Let E_1 be the event "watching the news" and E_2 be the event "watching a comedy." Because there are four ways to accomplish E_1 and two ways to accomplish E_2, the number of choices that she has is $4 \cdot 2 = 8$.

SELF CHECK 1 If Josie has 7 shirts and 5 pairs of pants, how many outfits could be created?

The multiplication principle can be extended to any number of events. In Example 2, we use it to complete the number of ways that we can arrange objects in a row.

EXAMPLE 2 In how many ways can we arrange five books on a shelf?

Solution We can fill the first space with any of the 5 books, the second space with any of the remaining 4 books, the third space with any of the remaining 3 books, the fourth space with any of the remaining 2 books, and the fifth space with the remaining 1 (or last) book. By the multiplication principle for events, the number of ways in which the books can be arranged is

$$5 \cdot 4 \cdot 3 \cdot 2 \cdot 1 = 120$$

SELF CHECK 2 In how many ways can 4 men line up in a row?

EXAMPLE 3 If Juan has six flags, each of a different color, to hang on a flagpole, how many different arrangements can he display by using four flags?

Solution Juan must find the number of arrangements of 4 flags when there are 6 flags from which to choose. He can hang any one of the 6 flags in the top position, any one of the remaining 5 flags in the second position, any one of the remaining 4 flags in the third position, and any one of the remaining 3 flags in the lowest position. By the multiplication principle for events, the total number of arrangements is

$$6 \cdot 5 \cdot 4 \cdot 3 = 360$$

SELF CHECK 3 How many different arrangements can he display if each uses three flags?

2 Use permutations to find the number of n things taken r at a time if order is important.

When computing the number of possible arrangements of objects such as books on a shelf or flags on a pole, we are finding the number of **permutations** of those objects. In these cases, *order is important.* A blue flag followed by a yellow flag is different than a yellow flag followed by a blue flag.

In Example 2, we found that the number of permutations of five books, using all five of them, is 120. In Example 3, we found that the number of permutations of six flags, using four of them, is 360.

The symbol $P(n, r)$, read as "the number of permutations of n things taken r at a time," is often used to express permutation situations. In Example 2, we found that $P(5, 5) = 120$. In Example 3, we found that $P(6, 4) = 360$.

EXAMPLE 4 If Sarah has seven flags, each of a different color, to hang on a flagpole, how many different arrangements can she display by using three flags?

Solution She must find $P(7, 3)$ (the number of permutations of 7 things taken 3 at a time). In the top position Sarah can hang any of the 7 flags, in the middle position any one of the remaining 6 flags, and in the bottom position any one of the remaining 5 flags. According to the multiplication principle for events,

$$P(7, 3) = 7 \cdot 6 \cdot 5 = 210$$

She can display 210 arrangements using only three of the seven flags.

SELF CHECK 4 How many different arrangements can Sarah display using four flags?

COMMENT Refer to the calculator tear out card for instructions on finding permutations and combinations using your calculator.

Although it is correct to write $P(7, 3) = 7 \cdot 6 \cdot 5$, there is an advantage in changing the form of this answer to obtain a formula for computing $P(7, 3)$:

$$\begin{aligned} P(7, 3) &= 7 \cdot 6 \cdot 5 \\ &= \frac{7 \cdot 6 \cdot 5 \cdot 4 \cdot 3 \cdot 2 \cdot 1}{4 \cdot 3 \cdot 2 \cdot 1} && \text{Multiply both the numerator and denominator by } 4 \cdot 3 \cdot 2 \cdot 1. \\ &= \frac{7!}{4!} \\ &= \frac{7!}{(7-3)!} \end{aligned}$$

The generalization of this process provides the following formula.

COMPUTING $P(n, r)$ The number of permutations of n things taken r at a time is given by the formula

$$P(n, r) = \frac{n!}{(n-r)!}$$

EXAMPLE 5 Compute: **a.** $P(8, 2)$ **b.** $P(7, 5)$ **c.** $P(n, n)$ **d.** $P(n, 0)$

Solution We substitute into the permutation formula $P(n, r) = \frac{n!}{(n-r)!}$.

a.
$$\begin{aligned} P(8, 2) &= \frac{8!}{(8-2)!} \\ &= \frac{8 \cdot 7 \cdot 6!}{6!} \\ &= 8 \cdot 7 \\ &= 56 \end{aligned}$$

b.
$$\begin{aligned} P(7, 5) &= \frac{7!}{(7-5)!} \\ &= \frac{7 \cdot 6 \cdot 5 \cdot 4 \cdot 3 \cdot 2!}{2!} \\ &= 7 \cdot 6 \cdot 5 \cdot 4 \cdot 3 \\ &= 2{,}520 \end{aligned}$$

c. $P(n,n) = \dfrac{n!}{(n-n)!} = \dfrac{n!}{0!} = \dfrac{n!}{1} = n!$

d. $P(n,0) = \dfrac{n!}{(n-0)!} = \dfrac{n!}{n!} = 1$

SELF CHECK 5 Compute: **a.** $P(10,6)$ **b.** $P(10,0)$

Parts **c** and **d** of Example 5 establish the following formulas.

FINDING $P(n, n)$ AND $P(n, 0)$

The number of permutations of n things taken n at a time and n things taken 0 at a time are given by the formulas

$$P(n,n) = n! \quad \text{and} \quad P(n,0) = 1$$

EXAMPLE 6 **TV PROGRAMMING** **a.** In how many ways can a TV executive arrange the Saturday night lineup of 6 programs if there are 15 programs from which to choose? **b.** If there are only 6 programs from which to choose?

Solution **a.** To find the number of permutations of 15 programs taken 6 at a time, we will use the formula $P(n,r) = \frac{n!}{(n-r)!}$ with $n = 15$ and $r = 6$.

$$P(15,6) = \frac{15!}{(15-6)!} = \frac{15 \cdot 14 \cdot 13 \cdot 12 \cdot 11 \cdot 10 \cdot 9!}{9!} = 15 \cdot 14 \cdot 13 \cdot 12 \cdot 11 \cdot 10 = 3{,}603{,}600$$

b. To find the number of permutations of 6 programs taken 6 at a time, we use the formula $P(n,n) = n!$ with $n = 6$.

$$P(6,6) = 6! = 720$$

SELF CHECK 6 How many ways can the executive select 6 programs if there are 20 programs from which to choose?

3 Use combinations to find the number of n things taken r at a time.

Suppose that a student must read 4 books from a reading list of 10 books. The order in which he reads them is not important. For the moment, however, let's assume that order is important and find the number of permutations of 10 things taken 4 at a time.

$$P(10,4) = \frac{10!}{(10-4)!} = \frac{10 \cdot 9 \cdot 8 \cdot 7 \cdot 6!}{6!} = 10 \cdot 9 \cdot 8 \cdot 7 = 5{,}040$$

If order is important, there are 5,040 ways of choosing 4 books when there are 10 books from which to choose. However, because the order in which the student reads the books does not matter, the previous result of 5,040 is too large. Since there are 24 (or 4!) ways of ordering the 4 books that are chosen, the result of 5,040 is exactly 24 (or 4!) times too large. Therefore, the number of choices that the student has is the number of permutations of 10 things taken 4 at a time, divided by 24:

$$\frac{P(10,4)}{24} = \frac{5{,}040}{24} = 210$$

The student has 210 ways of choosing 4 books to read from the list of 10 books.

In situations where *order is not important*, we are interested in **combinations**, not permutations. The symbols $C(n,r)$ and $\binom{n}{r}$ both mean the number of combinations of n things taken r at a time.

If a selection of r books is chosen from a total of n books, the number of possible selections is $C(n,r)$ and there are $r!$ arrangements of the r books in each selection. If we consider the selected books as an ordered grouping, the number of orderings is $P(n,r)$. Therefore, we have

(1) $r! \cdot C(n,r) = P(n,r)$

We can divide both sides of Equation 1 by $r!$ to obtain the formula for finding $C(n,r)$:

$$C(n,r) = \binom{n}{r} = \frac{P(n,r)}{r!} = \frac{n!}{r!(n-r)!}$$

FINDING $C(n,r)$

The number of combinations of n things taken r at a time is given by

$$C(n,r) = \frac{n!}{r!(n-r)!}$$

EXAMPLE 7 Compute: **a.** $C(8,5)$ **b.** $\binom{7}{2}$ **c.** $C(n,n)$ **d.** $C(n,0)$

Solution We will substitute into the combination formula $C(n,r) = \frac{n!}{r!(n-r)!}$.

a.
$$C(8,5) = \frac{8!}{5!(8-5)!} = \frac{8 \cdot 7 \cdot 6 \cdot 5!}{5! \cdot 3!} = 8 \cdot 7 = 56$$

b.
$$\binom{7}{2} = \frac{7!}{2!(7-2)!} = \frac{7 \cdot 6 \cdot 5!}{2 \cdot 1 \cdot 5!} = 21$$

c.
$$C(n,n) = \frac{n!}{n!(n-n)!} = \frac{n!}{n!(0!)} = \frac{n!}{n!(1)} = 1$$

d.
$$C(n,0) = \frac{n!}{0!(n-0)!} = \frac{n!}{0! \cdot n!} = \frac{1}{0!} = \frac{1}{1} = 1$$

The symbol $C(n,0)$ indicates that we choose 0 things from the available n things.

SELF CHECK 7 Compute: **a.** $C(9,6)$ **b.** $C(10,10)$

Parts **c** and **d** of Example 7 establish the following formulas.

FINDING $C(n, n)$ AND $C(n, 0)$

The number of combinations of n things taken n at a time is 1. The number of combinations of n things taken 0 at a time is 1.

$$C(n, n) = 1 \quad \text{and} \quad C(n, 0) = 1$$

EXAMPLE 8

SELECTING COMMITTEES If 15 students want to select a committee of 4 students, how many different committees are possible?

Solution Since the ordering of people on each possible committee is not important, we find the number of combinations of 15 people selected 4 at a time:

$$\begin{aligned} C(15, 4) &= \frac{15!}{4!(15-4)!} \\ &= \frac{15 \cdot 14 \cdot 13 \cdot 12 \cdot 11!}{4 \cdot 3 \cdot 2 \cdot 1 \cdot 11!} \\ &= \frac{15 \cdot 14 \cdot 13 \cdot 12}{4 \cdot 3 \cdot 2 \cdot 1} \\ &= 1{,}365 \end{aligned}$$

There are 1,365 possible committees.

SELF CHECK 8 In how many ways can 20 students select a committee of 5 students?

EXAMPLE 9

CONGRESS A committee in Congress consists of ten Democrats and eight Republicans. In how many ways can a subcommittee be chosen if it is to contain five Democrats and four Republicans?

Solution There are $C(10, 5)$ ways of choosing the 5 Democrats and $C(8, 4)$ ways of choosing the 4 Republicans. By the multiplication principle for events, there are $C(10, 5) \cdot C(8, 4)$ ways of choosing the subcommittee:

$$\begin{aligned} C(10, 5) \cdot C(8, 4) &= \frac{10!}{5!(10-5)!} \cdot \frac{8!}{4!(8-4)!} \\ &= \frac{10 \cdot 9 \cdot 8 \cdot 7 \cdot 6 \cdot 5!}{120 \cdot 5!} \cdot \frac{8 \cdot 7 \cdot 6 \cdot 5 \cdot 4!}{24 \cdot 4!} \\ &= \frac{10 \cdot 9 \cdot 8 \cdot 7 \cdot 6}{120} \cdot \frac{8 \cdot 7 \cdot 6 \cdot 5}{24} \\ &= 17{,}640 \end{aligned}$$

There are 17,640 possible subcommittees.

SELF CHECK 9 In how many ways can a subcommittee be chosen if it is to contain four members from each party?

4 Use combinations to find the coefficients of the terms of a binomial expansion.

We have seen that the expansion of $(x + y)^3$ is

$$(x + y)^3 = 1x^3 + 3x^2y + 3xy^2 + 1y^3$$

and that

$$\binom{3}{0} = 1, \quad \binom{3}{1} = 3, \quad \binom{3}{2} = 3, \quad \text{and} \quad \binom{3}{3} = 1$$

Putting these facts together provides the following way of writing the expansion of $(x + y)^3$.

$$(x + y)^3 = \binom{3}{0}x^3 + \binom{3}{1}x^2y + \binom{3}{2}xy^2 + \binom{3}{3}y^3$$

Likewise, we have

$$(x + y)^4 = \binom{4}{0}x^4 + \binom{4}{1}x^3y + \binom{4}{2}x^2y^2 + \binom{4}{3}xy^3 + \binom{4}{4}y^4$$

The generalization of this idea allows us to state the binomial theorem using combinatorial notation.

THE BINOMIAL THEOREM If n is any positive integer, then

$$(a + b)^n = \binom{n}{0}a^n + \binom{n}{1}a^{n-1}b + \binom{n}{2}a^{n-2}b^2 + \cdots + \binom{n}{r}a^{n-r}b^r + \cdots + \binom{n}{n}b^n$$

EXAMPLE 10 Use the combinatorial form of the binomial theorem to expand $(x + y)^6$.

Solution

$$\begin{aligned}(x + y)^6 &= \binom{6}{0}x^6 + \binom{6}{1}x^5y + \binom{6}{2}x^4y^2 + \binom{6}{3}x^3y^3 + \binom{6}{4}x^2y^4 + \binom{6}{5}xy^5 + \binom{6}{6}y^6 \\ &= x^6 + 6x^5y + 15x^4y^2 + 20x^3y^3 + 15x^2y^4 + 6xy^5 + y^6\end{aligned}$$

SELF CHECK 10 Use the combinatorial form of the binomial theorem to expand $(a + b)^2$.

EXAMPLE 11 Use the combinatorial form of the binomial theorem to expand $(2x - y)^3$.

Solution

$$\begin{aligned}(2x - y)^3 &= [2x + (-y)]^3 \\ &= \binom{3}{0}(2x)^3 + \binom{3}{1}(2x)^2(-y) + \binom{3}{2}(2x)(-y)^2 + \binom{3}{3}(-y)^3 \\ &= 1(2x)^3 + 3(4x^2)(-y) + 3(2x)(y^2) + 1(-y)^3 \\ &= 8x^3 - 12x^2y + 6xy^2 - y^3\end{aligned}$$

SELF CHECK 11 Use the combinatorial form of the binomial theorem to expand $(3a + b)^3$.

SELF CHECK ANSWERS

1. 35 **2.** 24 **3.** 120 **4.** 840 **5. a.** 151,200 **b.** 1 **6.** 27,907,200 **7. a.** 84 **b.** 1 **8.** 15,504 **9.** 14,700 **10.** $a^2 + 2ab + b^2$ **11.** $27a^3 + 27a^2b + 9ab^2 + b^3$

NOW TRY THIS

1. In Texas, license plate numbering for passenger cars has often been changed to accommodate a growing population and the fact that new plates must be issued every 8 years. Find the greatest number of license plates that could be issued during each of these periods. A or B represents a letter and 0 or 1 represents a number.

1975–1982	AAA-000	(I or O is not used)
1982–1990	000-AAA	(I or O is not used)
1990–1998	BBB-00B	(no vowels or Q used)
1998–2004	B00-BBB	(no vowels or Q used)
2004–2007	000-BBB	(no vowels or Q used)
2007–mid-2009	BBB-000	(no vowels or Q used)

2. Find the greatest number of license plates that could be distributed from 1975–mid-2009.

3. Beginning in mid-2009, Texas began issuing 7-digit license numbers of the form BB0-B000 (no vowels or Q is used). How many license plates can be issued using this form?

4. Personalized license plates can be purchased from an authorized third party with the following specifications. The plate can consist of 1–6 letters or numbers including no more than 2 symbols (dash, space, period, @ for heart, or * for the state icon). A letter or number must be in the first position. How many plates could be created within these parameters?

Source: http://www.licenseplateinfo.com/txchart/pass-tables.html

13.5 Exercises

WARM-UPS *Simplify.*

1. $\frac{7!}{3!}$
2. $\frac{6!}{4!}$
3. $\frac{5!}{1!}$
4. $\frac{5!}{0!}$
5. $\frac{8!}{3!5!}$
6. $\frac{9!}{5!4!}$

REVIEW *Find each value of x. Assume no division by 0.*

7. $|2x - 7| = 17$
8. $2x^2 - x = 15$
9. $\frac{3}{x - 5} = \frac{8}{x}$
10. $\frac{3}{x} = \frac{x - 2}{8}$

VOCABULARY AND CONCEPTS *Fill in the blanks.*

11. If an event E_1 can be done in p ways and, after it occurs, a second event E_2 can be done in q ways, the event E_1 followed by E_2 can be done in ___ ways.
12. We can use a ___ diagram to illustrate the multiplication principle.
13. A _________ is an arrangement of objects in which order matters.
14. The symbol ______ means the number of permutations of n things taken r at a time.
15. The formula for the number of permutations of n things taken r at a time is _____________.
16. $P(n, n) =$ _ and $P(n, 0) =$ _.
17. A _________ is an arrangement of objects in which order does not matter.
18. The symbol $C(n, r)$ or __ means the number of _________ of n things taken r at a time.
19. The formula for the number of combinations of n things taken r at a time is _____________.
20. $C(n, n) =$ _ and $C(n, 0) =$ _.

GUIDED PRACTICE *Evaluate using permutations.* SEE EXAMPLE 5. (OBJECTIVE 2)

21. $P(5, 5)$
22. $P(4, 4)$
23. $P(6, 2)$
24. $P(3, 2)$
25. $P(4, 2) \cdot P(5, 2)$
26. $P(n, n) \cdot P(n, 0)$
27. $\frac{P(n, n)}{P(n, 0)}$
28. $\frac{P(6, 2)}{P(5, 4)}$

Evaluate using combinations. SEE EXAMPLE 7. (OBJECTIVE 3)

29. $C(4, 3)$

30. $C(7, 5)$

31. $\binom{8}{5}$

32. $\binom{6}{4}$

33. $\binom{7}{4}\binom{7}{5}$

34. $\binom{9}{8}\binom{9}{7}$

35. $\frac{C(38, 37)}{C(19, 18)}$

36. $\frac{C(25, 23)}{C(40, 39)}$

Use the combinatorial form of the binomial theorem to expand each binomial. SEE EXAMPLE 10. (OBJECTIVE 4)

37. $(x + y)^4$

38. $(x + y)^8$

39. $(a + b)^5$

40. $(a + b)^7$

Use the combinatorial form of the binomial theorem to expand each binomial. SEE EXAMPLE 11. (OBJECTIVE 4)

41. $(2x + y)^3$

42. $(2x + 1)^4$

43. $(4x - y)^3$

44. $(x - 2y)^5$

ADDITIONAL PRACTICE *Evaluate.*

45. $C(12, 0) \cdot C(12, 12)$

46. $\frac{C(8, 0)}{C(8, 1)}$

47. $\frac{P(6, 2) \cdot P(7, 3)}{P(5, 1)}$

48. $\frac{P(8, 3)}{P(5, 3) \cdot P(4, 3)}$

49. $C(n, 2)$

50. $C(n, 3)$

Expand the binomial.

51. $(3x - 2)^4$

52. $(3 - x^2)^3$

Find the indicated term of the binomial expansion.

53. $(x - 5y)^5$; fourth term

54. $(2x - y)^5$; third term

55. $(x^2 - y^3)^4$; second term

56. $(x^3 - y^2)^4$; fourth term

APPLICATIONS *Solve each application.* SEE EXAMPLES 1–3. (OBJECTIVE 1)

57. Arranging an evening Kyoro plans to go to dinner and see a movie. In how many ways can she arrange her evening if she has a choice of five movies and seven restaurants?

58. Travel choices Paula has five ways to travel from New York to Chicago, three ways to travel from Chicago to Denver, and four ways to travel from Denver to Los Angeles. How many choices are available to Paula if she travels from New York to Los Angeles?

59. Arranging books In how many ways can seven books be placed on a shelf?

60. Lining up In how many ways can the people shown be placed in a line?

Solve each application. SEE EXAMPLES 4 AND 6. (OBJECTIVE 2)

61. Making license plates How many six-digit license plates can be manufactured? Note that there are ten choices—0, 1, 2, 3, 4, 5, 6, 7, 8, 9—for each digit.

62. Making license plates How many six-digit license plates can be manufactured if no digit can be repeated?

63. Making license plates How many six-digit license plates can be manufactured if no license can begin with 0 and if no digit can be repeated?

64. Making license plates How many license plates can be manufactured with two letters followed by four digits?

65. Phone numbers How many seven-digit phone numbers are available in area code 815 if no phone number can begin with 0 or 1?

66. Phone numbers How many ten-digit phone numbers are available if area codes 000 and 911 cannot be used and if no local number can begin with 0 or 1?

67. Arranging books In how many ways can four novels and five biographies be arranged on a shelf if the novels are placed on the left?

68. Making a ballot In how many ways can six candidates for mayor and four candidates for the county board be arranged on a ballot if all of the candidates for mayor must be placed on top?

69. Combination locks How many permutations does a combination lock have if each combination has three numbers, no two numbers of any combination are equal, and the lock has 25 numbers?

70. Combination locks How many permutations does a combination lock have if each combination has three numbers, no two numbers of any combination are equal, and the lock has 50 numbers?

71. Arranging appointments The receptionist at a dental office has only three appointment times available before Tuesday, and ten patients have toothaches. In how many ways can the receptionist fill those appointments?

72. Computers In many computers, a word consists of 32 *bits*—a string of thirty-two 1's and 0's. How many different words are possible?

73. Palindromes A palindrome is any word, such as *madam* or *radar*, that reads the same backward and forward. How many five-digit numerical palindromes (such as 13531) are there? (*Hint*: A leading 0 would be dropped.)

Unless otherwise noted, all content on this page is © Cengage Learning.

74. **Call letters** The call letters of U.S. commercial radio stations have 3 or 4 letters, and the first is always a W or a K. How many radio stations could this system support?

Solve each application. SEE EXAMPLES 8–9. (OBJECTIVE 3)

75. **Planning a picnic** A class of 14 students wants to pick a committee of 3 students to plan a picnic. How many committees are possible?
76. **Choosing books** Jeff must read 3 books from a reading list of 15 books. How many choices does he have?
77. **Forming committees** The number of three-person committees that can be formed from a group of people is ten. How many people are in the group?
78. **Forming committees** The number of three-person committees that can be formed from a group of people is 20. How many people are in the group?
79. **Winning a lottery** In one state lottery, anyone who picks the correct 6 numbers (in any order) wins. With the numbers 0 through 99 available, how many choices are possible?
80. **Taking a test** The instructions on a test read: "Answer any ten of the following fifteen questions. Then choose one of the remaining questions for homework and turn in its solution tomorrow." In how many ways can the questions be chosen?
81. **Forming a committee** In how many ways can we select a committee of two men and two women from a group containing three men and four women?
82. **Forming a committee** In how many ways can we select a committee of three women and two men from a group containing five women and three men?
83. **Choosing clothes** In how many ways can we select 2 shirts and 3 neckties from a group of 12 shirts and 10 neckties?
84. **Choosing clothes** In how many ways can we select five dresses and two coats from a wardrobe containing nine dresses and three coats?

WRITING ABOUT MATH

85. State the multiplication principle for events.
86. Explain why *permutation lock* would be a better name for a combination lock.

SOMETHING TO THINK ABOUT

87. How many ways could five people stand in line if two people insist on standing together?
88. How many ways could five people stand in line if two people refuse to stand next to each other?

Section 13.6 Probability

Objectives

1 Find a sample space for an experiment.
2 Find the probability of an event.

Vocabulary

probability experiment
sample space
event

Getting Ready

Answer each question.

1. What are the possible outcomes on one roll of one die?
2. What are the possible outcomes after flipping a single coin one time?

The **probability** that an event will occur is a measure of the likelihood of that event. A tossed coin, for example, can land in two ways, either heads or tails. Because one of these two equally likely outcomes is heads, we expect that out of several tosses, about half will be heads. We say that the probability of obtaining heads in a single toss of the coin is $\frac{1}{2}$.

If records show that out of 100 days with weather conditions like today's, 30 have received rain, the weather service will report, "There is a $\frac{30}{100}$ or 30% probability of rain today."

1 Find a sample space for an experiment.

Activities such as tossing a coin, rolling a die, drawing a card, and predicting rain are called **experiments**. For any experiment, a list of all possible outcomes is called a **sample space**. For example, the sample space S for the experiment of tossing two coins is the set

$$S = \{(\mathrm{H},\mathrm{H}), (\mathrm{H},\mathrm{T}), (\mathrm{T},\mathrm{H}), (\mathrm{T},\mathrm{T})\}$$ There are four possible outcomes.

where the ordered pair (H,T) represents the outcome "heads on the first coin and tails on the second coin."

EXAMPLE 1 List the sample space of the experiment "rolling two dice a single time."

Solution We can list ordered pairs and let the first number be the result on the first die and the second number the result on the second die. The sample space S is the following set of ordered pairs:

$$\begin{array}{llllll}(1,1) & (1,2) & (1,3) & (1,4) & (1,5) & (1,6)\\ (2,1) & (2,2) & (2,3) & (2,4) & (2,5) & (2,6)\\ (3,1) & (3,2) & (3,3) & (3,4) & (3,5) & (3,6)\\ (4,1) & (4,2) & (4,3) & (4,4) & (4,5) & (4,6)\\ (5,1) & (5,2) & (5,3) & (5,4) & (5,5) & (5,6)\\ (6,1) & (6,2) & (6,3) & (6,4) & (6,5) & (6,6)\end{array}$$

By counting, we see that the experiment has 36 equally likely possible outcomes.

SELF CHECK 1 List the sample space of the experiment "flipping one coin followed by rolling 1 die."

2 Find the probability of an event.

An **event** is a subset of the sample space of an experiment. For example, if E is the event "getting at least one heads" in the experiment of tossing two coins, then

$$E = \{(\mathrm{H},\mathrm{H}), (\mathrm{H},\mathrm{T}), (\mathrm{T},\mathrm{H})\}$$ There are 3 ways of getting at least one heads.

Because the outcome of getting at least one heads can occur in 3 out of 4 possible ways, we say that the *probability* of E is $\frac{3}{4}$, and we write

$$P(E) = P(\text{at least one heads}) = \frac{3}{4}$$

PROBABILITY OF AN EVENT

If a sample space of an experiment has n distinct and equally likely outcomes and E is an event that occurs in s of those ways, the *probability of E* is

$$P(E) = \frac{s}{n}$$

Since $0 \le s \le n$, it follows that $0 \le \frac{s}{n} \le 1$. This implies that all probabilities have a value from 0 to 1. If an event cannot happen, its probability is 0. If an event is certain to happen, its probability is 1.

EXAMPLE 2 Find the probability of the event "rolling a sum of 7 on one roll of two dice."

Solution In the sample space listed in Example 1, the following ordered pairs give a sum of 7:

$$\{(1, 6), (2, 5), (3, 4), (4, 3), (5, 2), (6, 1)\}$$

Since there are 6 ordered pairs whose numbers give a sum of 7 out of a total of 36 equally likely outcomes, we have

$$P(E) = P(\text{rolling a sum of 7}) = \frac{s}{n} = \frac{6}{36} = \frac{1}{6}$$

SELF CHECK 2 Find the probability of rolling a sum of 4.

COMMENT A standard playing deck of 52 cards has two red suits, hearts and diamonds, and two black suits, clubs and spades. Each suit has 13 cards, including the ace, face cards (king, queen, and jack), and numbers 2 through 10. We will refer to a standard deck of cards in many examples and exercises.

EXAMPLE 3 Find the probability of drawing an ace on one draw from a standard card deck.

Solution Since there are 4 aces in the deck, the number of favorable outcomes is $s = 4$. Since there are 52 cards in the deck, the total number of possible outcomes is $n = 52$. The probability of drawing an ace is the ratio of the number of favorable outcomes to the number of possible outcomes.

$$P(\text{an ace}) = \frac{s}{n} = \frac{4}{52} = \frac{1}{13}$$

The probability of drawing an ace is $\frac{1}{13}$.

SELF CHECK 3 Find the probability of drawing a red ace on one draw from a standard card deck.

Everyday connections
Winning the Lottery

© iStockphoto.com/David Birkbeck

Suppose a certain state lottery has the following Mega Millions design.

- A spinning machine contains 49 different balls. Each ball is labeled with exactly one number from the list $\{1, 2, 3, \ldots, 49\}$. The spinning machine ensures that each ball has an equal chance of being selected.
- Six different balls are drawn, without replacement, from the spinning machine.
- Each lottery ticket has six different numbers from the list $\{1, 2, 3, \ldots, 49\}$.
- The last number must match the sixth drawn number.

Suppose that a grand-prize-winning ticket consists of all six numbers drawn from the machine. Find the probability associated with the winning ticket.

EXAMPLE 4 Find the probability of drawing 5 cards, all hearts, from a standard card deck.

Solution The number of ways we can draw 5 hearts from the 13 hearts is $C(13, 5)$, the number of combinations of 13 things taken 5 at a time. The number of ways to draw 5 cards from the deck is $C(52, 5)$, the number of combinations of 52 things taken 5 at a time. The probability of drawing 5 hearts is the ratio of the number of favorable outcomes to the number of possible outcomes.

$$P(5\text{ hearts}) = \frac{s}{n} = \frac{C(13,5)}{C(52,5)}$$

$$P(5\text{ hearts}) = \frac{\frac{13!}{5!8!}}{\frac{52!}{5!47!}}$$

$$= \frac{13!}{\cancel{5!}8!} \cdot \frac{\cancel{5!}47!}{52!} \qquad \frac{5!}{5!} = 1$$

$$= \frac{13 \cdot 12 \cdot 11 \cdot 10 \cdot 9 \cdot 8!}{8!} \cdot \frac{47!}{52 \cdot 51 \cdot 50 \cdot 49 \cdot 48 \cdot 47!}$$

$$= \frac{13 \cdot 12 \cdot 11 \cdot 10 \cdot 9}{52 \cdot 51 \cdot 50 \cdot 49 \cdot 48}$$

$$= \frac{33}{66{,}640}$$

The probability of drawing 5 hearts is $\frac{33}{66{,}640}$ or $4.951980792 \times 10^{-4}$.

SELF CHECK 4 Find the probability of drawing 6 cards, all diamonds, from a standard card deck.

SELF CHECK ANSWERS

1. $S = \{(H,1), (H,2), (H,3), (H,4), (H,5), (H,6), (T,1), (T,2), (T,3), (T,4), (T,5), (T,6)\}$ **2.** $\frac{1}{12}$ **3.** $\frac{1}{26}$ **4.** $\frac{33}{391{,}510}$ or $8.428903476 \times 10^{-5}$

NOW TRY THIS

The notation $P(E)$ represents the probability that event E will occur. The notation $P(\overline{E})$ represents the probability that event E will *not* occur. Because either event E or event $\overline{E}$ must occur, $P(E) + P(\overline{E}) = 1$. Thus, $P(E) = 1 - P(\overline{E})$.

Sometimes it is easier to determine $P(E)$ by finding $P(\overline{E})$ and subtracting the result from 1.

1. In a family of 5 children, find the probability that at least one child is a girl.
2. In a clinical trial, the probability of experiencing a serious side effect is 0.15 while the probability of a minor side effect is 0.45, with no overlapping side effects. What is the probability that a person in the trial picked at random experiences no side effects?

13.6 Exercises

WARM-UPS *Write a combination to represent*

1. the number of ways of drawing a red card from a standard deck.
2. the number of ways of drawing a club card from a standard deck.

REVIEW *Solve each equation.*

3. $4^{3x} = \frac{1}{64}$
4. $6^{-x+2} = \frac{1}{36}$
5. $5^{x^2-2x} = 125$
6. $2^{x^2-3x} = 16$
7. $3^{x^2+4x} = \frac{1}{81}$
8. $7^{x^2+3x} = \frac{1}{49}$

VOCABULARY AND CONCEPTS *Fill in the blanks.*

9. An __________ is any activity for which the outcome is uncertain.
10. A list of all possible outcomes for an experiment is called a __________.
11. The probability of an event E is defined as $P(E) =$ _.
12. If an event is certain to happen, its probability is _.
13. If an event cannot happen, its probability is _.
14. All probability values are between _ and _, inclusive.
15. Fill in the blanks to find the probability of drawing a red face card from a standard deck.
    a. The number of red face cards is _.
    b. The number of cards in the deck is __.
    c. The probability is __ or __.

16. Fill in the blanks to find the probability of drawing 4 aces from a standard card deck.
   a. The number of ways to draw 4 aces from 4 aces is $C(4, 4) = \_$.
   b. The number of ways to draw 4 cards from 52 cards is $C(52, 4) = \_\_\_\_\_\_$.
   c. The probability is ______.

GUIDED PRACTICE *List the sample space of each experiment.* *SEE EXAMPLE 1. (OBJECTIVE 1)*

17. Rolling one die followed by tossing one coin

18. Tossing three coins

19. Selecting a letter of the alphabet

20. Picking a one-digit number

Two dice are rolled. Find the probability of each event. *SEE EXAMPLE 2. (OBJECTIVE 2)*

21. Rolling a sum of 5
22. Rolling a sum of 6
23. Rolling a sum less than 6
24. Rolling a sum greater than 9

Find the probability of each event. SEE EXAMPLE 3. (OBJECTIVE 2)

25. Drawing a diamond on one draw from a standard card deck
26. Drawing a face card from a standard deck
27. Drawing a red face card from a standard deck
28. Drawing the ace of spades from a standard deck

Find the probability of each event. SEE EXAMPLE 4. (OBJECTIVE 2)

29. Drawing 6 diamonds from a standard deck without replacing the cards after each draw
30. Drawing 5 aces from a standard deck without replacing the cards after each draw
31. Drawing 5 clubs from the black cards in a standard deck
32. Drawing 5 cards, all red, from a standard deck

ADDITIONAL PRACTICE *An ordinary die is rolled once. Find the probability of each event.*

33. Rolling a 2
34. Rolling a number greater than 4
35. Rolling a number larger than 1 but less than 6
36. Rolling an odd number

Balls numbered from 1 to 42 are placed in a container and stirred. If one is drawn at random, find the probability of each result.

37. The number is less than 20.
38. The number is less than 50.
39. The number is a prime number.
40. The number is less than 10 or greater than 40.

Refer to the following spinner. If the spinner is spun, find the probability of each event. Assume that the spinner never stops on a line.

41. The spinner stops on red.
42. The spinner stops on green.
43. The spinner stops on brown.
44. The spinner stops on yellow.

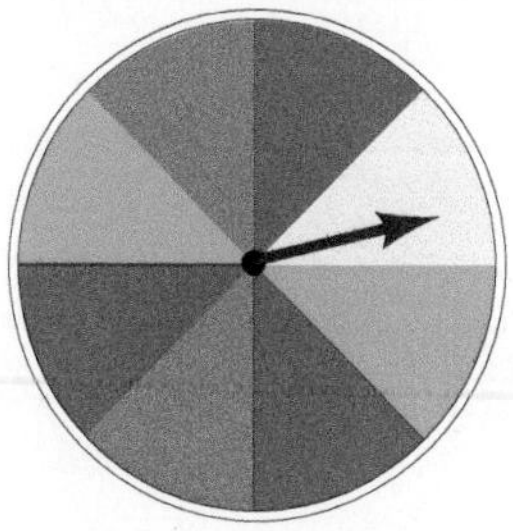

Find the probability of each event.

45. Drawing a face card from a standard deck followed by a 10 after replacing the first card
46. Drawing an ace from a standard deck followed by a 10 after replacing the first card
47. Rolling a sum of 7 on one roll of two dice
48. Drawing a red egg from a basket containing 5 red eggs and 7 blue eggs
49. Drawing a yellow egg from a basket containing 5 red eggs and 7 yellow eggs

Assume that the probability that a backup generator will fail a test is $\frac{1}{2}$ and that the college in question has 4 backup generators. After completing Exercise 50, find each probability in Exercises 51–56.

50. Construct a sample space for the test.

51. All generators will fail the test.
52. Exactly 1 generator will fail.
53. Exactly 2 generators will fail.
54. Exactly 3 generators will fail.
55. No generator will fail.
56. Find the sum of the probabilities in Exercises 51–55.

APPLICATIONS *Solve each application.*

57. **Quality control** In a batch of 10 tires, 2 are known to be defective. If 4 tires are chosen at random, find the probability that all 4 tires are good.
58. **Medicine** Out of a group of 9 patients treated with a new drug, 4 suffered a relapse. If 3 patients are selected at random from the group of 9, find the probability that none of the 3 patients suffered a relapse.

Unless otherwise noted, all content on this page is © Cengage Learning.

A survey of 282 people is taken to determine the opinions of doctors, teachers, and lawyers on a proposed piece of legislation, with the results shown in the table. A person is chosen at random from those surveyed. Refer to the table to find each probability.

59. The person favors the legislation.

60. A doctor opposes the legislation.

61. A person who opposes the legislation is a lawyer.

	Number that favor	Number that oppose	Number with no opinion	Total
Doctors	70	32	17	119
Teachers	83	24	10	117
Lawyers	23	15	8	46
Total	176	71	35	282

WRITING ABOUT MATH

62. Explain why all probability values range from 0 to 1.

63. Explain the concept of probability.

SOMETHING TO THINK ABOUT *If $P(A)$ represents the probability of event A, and $P(B|A)$ represents the probability that event B will occur after event A, then*

$$P(A \text{ and } B) = P(A) \cdot P(B|A)$$

64. In a school, 30% of the students are gifted in math and 10% are gifted in art and math. If a student is gifted in math, find the probability that the student is also gifted in art.

65. The probability that a person owns a luxury car is 0.2, and the probability that the owner of such a car also owns a second car is 0.7. Find the probability that a person chosen at random owns both a luxury car and a second car.

Projects

PROJECT 1

Baytown is building an auditorium. The city council has already decided on the layout shown in the illustration. Each of the sections A, B, C, D, E is to be 60 feet in length from front to back. The aisle widths cannot be changed due to fire regulations. The one thing left to decide is how many rows of seats to put in each section. Based on the following information regarding each section of the auditorium, help the council decide on a final plan.

Sections A and C each have four seats in the front row, five seats in the second row, six seats in the third row, and so on, adding one seat per row as we count from front to back.

Section B has eight seats in the front row and adds one seat per row as we count from front to back.

Sections D and E each have 28 seats in the front row and add two seats per row as we count from front to back.

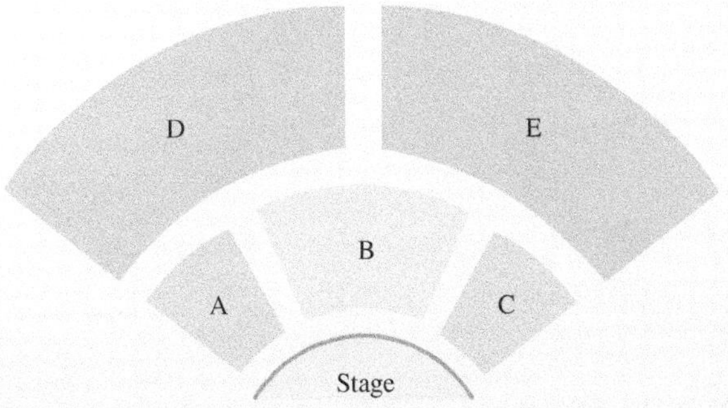

a. One plan calls for a distance of 36 inches, front to back, for each row of seats. Another plan allows for 40 inches (an extra four inches of legroom) for each row. How many seats will the auditorium have under each of these plans?

b. Another plan calls for the higher-priced seats (Sections A, B, and C) to have the extra room afforded by 40-inch rows, but for Sections D and E to have enough rows to make sure that the auditorium holds at least 2,700 seats. Determine how many rows Section D and E would have to contain for this to work. (This answer should be an integer.) How much space (to the nearest tenth of an inch) would be allotted for each row in Sections D and E?

PROJECT 2

Pascal's triangle contains a wealth of interesting patterns. You have seen two in Exercises 88 and 107 of Section 13.1. Here are a few more.

a. Find the hockey-stick pattern in the numbers in the illustration. What would be the missing number in the rightmost hockey stick? Does this pattern work for larger hockey sticks? Experiment.

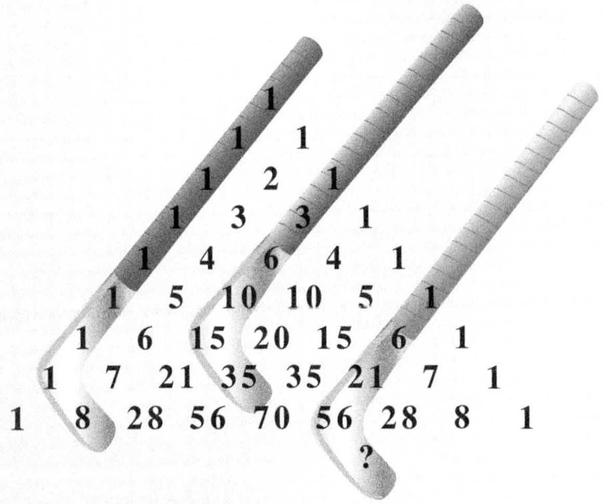

Unless otherwise noted, all content on this page is © Cengage Learning.

b. In Illustration 1, find the pattern in the sums of increasingly larger portions of Pascal's triangle. Find the sum of all of the numbers up to and including the row that begins 1 10 45 . . . .

c. In Illustration 2, find the pattern in the sums of the squares of the numbers in each row of the triangle. What is the sum of the squares of the numbers in the row that begins 1 10 45 . . . ? (*Hint*: Calculate $P(2, 1)$, $P(4, 2)$, $P(6, 3)$, ... Do these numbers appear elsewhere in the triangle?)

d. In 1653, Pascal described the triangle in *Treatise on the Arithmetic Triangle,* writing, "I have left out many more properties than I have included. It is remarkable how fertile in properties this triangle is. *Everyone can try his hand.*" Accept Pascal's invitation. Find some of the triangle's patterns for yourself and share your discoveries with your class. Illustration 3 is an idea to get you started.

				1
			1	1 1
		1	1 1	1 2 1
	1	1 1	1 2 1	1 3 3 1
1	1 1	1 2 1	1 3 3 1	1 4 6 4 1
= 1	= 3	= 7	= ?	= ?

ILLUSTRATION 1

$$1^2 = 1$$
$$1^2 + 1^2 = 2$$
$$1^2 + 2^2 + 1^2 = 6$$
$$1^2 + 3^2 + 3^2 + 1^2 = 20$$
$$1^2 + 4^2 + 6^2 + 4^2 + 1^2 = 70$$
$$1^2 + 5^2 + 10^2 + 10^2 + 5^2 + 1^2 = ?$$
$$1^2 + 6^2 + 15^2 + 20^2 + 15^2 + 6^2 + 1^2 = ?$$

ILLUSTRATION 2

1
1 1
1 2 1
1 3 3 1
1 4 6 4 1
1 5 10 10 5 1
1 6 15 20 15 6 1
1 7 21 35 35 21 7 1
1 8 28 56 70 56 28 8 1

ILLUSTRATION 3

Unless otherwise noted, all content on this page is © Cengage Learning.

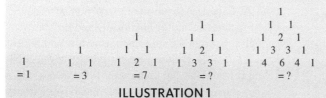

Reach for Success ADJUSTING TO COLLEGE LIFE

An excellent strategy to "fit in" at college and to feel connected is to enroll in classes in which two or more courses are "linked" or those in which a service component is included. To give back to the institution or community, consider volunteering.

Define learning communities by searching the internet. (You may find more than one variation of the definition.) __________ __________ __________	Does your college offer Learning Communities? Yes or No If so, identify one that may interest you. __________
Define Service Learning by searching the internet. __________ __________ __________	Does your college offer Service Learning? Yes or No If so, list a class that offers it as a component. __________ __________
What is your strongest subject? __________	Is there a club in that area? __________ If so, do they need tutors? __________
Are you bilingual? Yes or No	Does your college have an English as a second Language (ESL) program? Yes or No If so, does your college have a need for students willing to be speaking partners? Yes or No
Does your community offer any volunteer opportunities in which you are interested? __________	If so, what are they? __________ __________

Proper planning can help you avoid taking classes that might not be required for your degree, thus saving you time and money.

13 Review

SECTION 13.1 The Binomial Theorem

DEFINITIONS AND CONCEPTS	EXAMPLES
The symbol $n!$ **(factorial)** is defined as $n! = n(n-1)(n-2) \cdot \ldots \cdot (3)(2)(1)$ where n is a natural number. $0! = 1$ $n(n-1)! = n!$ (n is a natural number)	$7! = 7 \cdot 6 \cdot 5 \cdot 4 \cdot 3 \cdot 2 \cdot 1 = 5{,}040$ $\frac{5!}{2!3!} = \frac{5 \cdot 4 \cdot \cancel{3!}}{2 \cdot 1 \cdot \cancel{3!}} = \frac{20}{2} = 10$ $7 \cdot 6! = 7 \cdot \mathbf{6 \cdot 5 \cdot 4 \cdot 3 \cdot 2 \cdot 1} = 7!$
The binomial theorem: If n is any positive integer, then $(a+b)^n = a^n + \frac{n!}{1!(n-1)!}a^{n-1}b + \frac{n!}{2!(n-2)!}a^{n-2}b^2 + \cdots + \frac{n!}{r!(n-r)!}a^{n-r}b^r + \cdots + b^n$	$(x-2y)^4$ $= [x + (-2y)]^4$ $= x^4 + \frac{4!}{1!3!}x^3(-2y) + \frac{4!}{2!2!}x^2(-2y)^2 + \frac{4!}{3!1!}x(-2y)^3 + (-2y)^4$ $= x^4 + 4x^3(-2y) + 6x^2(4y^2) + 4x(-8y^3) + 16y^4$ $= x^4 - 8x^3y + 24x^2y^2 - 32xy^3 + 16y^4$
Specific term of a binomial expansion: The binomial theorem can be used to find a specific term of a binomial expansion. The coefficient of the variables can be found using the formula $\frac{n!}{r!(n-r)!}$ where n is the power of the expansion and r is the exponent of the second variable.	To find the third term of $(x+y)^5$, note that the exponent on y will be 2, because the exponent on y is 1 less than the number of the term. The power on x will be 3 because the sum of the powers must be 5. Thus, the variables will be x^3y^2. The coefficient of the variables can be found with the formula. $\frac{5!}{2!(5-2)!}x^3y^2$ Substitute the values into the formula. $\frac{5 \cdot 4 \cdot \cancel{3!}}{2 \cdot 1 \cdot \cancel{3!}}x^3y^2$ Expand the factorials. $10x^3y^2$ Simplify.

REVIEW EXERCISES

Evaluate each expression.

1. $(5!)(2!)$
2. $\frac{5!}{3!}$
3. $\frac{6!}{2!(6-2)!}$
4. $\frac{12!}{3!(12-3)!}$
5. $(n-n)!$
6. $\frac{10!}{6!}$

Use the binomial theorem to find each expansion.

7. $(x+y)^5$
8. $(x-y)^4$
9. $(4x-y)^3$
10. $(x+4y)^3$

Find the specified term in each expansion.

11. $(x+y)^5$; fifth term
12. $(x-y)^5$; fourth term
13. $(3x-4y)^3$; second term
14. $(4x+3y)^4$; third term

SECTION 13.2 Arithmetic Sequences and Series

DEFINITIONS AND CONCEPTS	EXAMPLES
An **arithmetic sequence** is a sequence of the form $a_1, a_1 + d, a_1 + 2d, \ldots, a_1 + (n - 1)d$ where a_1 is the first term. the nth term is $a_n = a_1 + (n - 1)d$. d is the common difference.	Find the first 5 terms of the sequence with $a_1 = 15$ and $d = -2$. $a_1 = 15$ $a_2 = 15 + (-2) = 13$ $a_3 = 15 + 2(-2) = 11$ $a_4 = 15 + 3(-2) = 9$ $a_5 = 15 + 4(-2) = 7$
Arithmetic means: If numbers are inserted between two given numbers a and b to form an arithmetic sequence, the inserted numbers are **arithmetic means** between a and b.	To insert two arithmetic means between 7 and 25, note that $a_1 = 7$ and $a_4 = 25$. Substitute these values into the formula for the nth term and solve for d. $a_n = a_1 + (n - 1)d$ $25 = 7 + (4 - 1)d$ Substitute values. $25 = 7 + 3d$ Simplify. $18 = 3d$ Subtract 7 from both sides. $6 = d$ Divide both sides by 3. The common difference is 6. The two arithmetic means are $7 + 6 = 13$ and $7 + 2(6) = 19$.
Sum of terms of arithmetic sequences: The sum of the first n terms of an arithmetic sequence is given by $S_n = \dfrac{n(a_1 + a_n)}{2}$ with $a_n = a_1 + (n - 1)d$ where a_1 is the first term, a_n is the nth term, and n is the number of terms in the sequence.	To find the sum of the first 12 terms of the sequence 4, 12, 20, 28, . . . , note that the common difference is 8 and substitute it into the formula for the nth term to find the twelfth term: $a_{12} = 4 + (12 - 1)8 = 4 + 88 = 92$ Then substitute into the formula for the sum of the first n terms: $S_{12} = \dfrac{12(4 + 92)}{2} = \dfrac{12(96)}{2} = \dfrac{1{,}152}{2} = 576$
Summation notation: $\sum_{k=1}^{n} f(k) = f(1) + f(2) + \cdots + f(n)$ Summation notation describes a series.	$\sum_{k=1}^{3}(3k - 1) = [3(1) - 1] + [3(2) - 1] + [3(3) - 1]$ $= (3 - 1) + (6 - 1) + (9 - 1)$ $= 2 + 5 + 8$ $= 15$

REVIEW EXERCISES

15. Find the eighth term of an arithmetic sequence whose first term is 7 and whose common difference is 5.

16. Write the first five terms of the arithmetic sequence whose ninth term is 242 and whose seventh term is 212.

17. Find two arithmetic means between 8 and 25.

18. Find the sum of the first 20 terms of the sequence 11, 18, 25, ...

19. Find the sum of the first ten terms of the sequence $9, 6\frac{1}{2}, 4, \ldots$.

Find each sum.

20. $\sum_{k=1}^{5} 4k$

21. $\sum_{k=1}^{4} (k^2 + 1)$

22. $\sum_{k=1}^{4} (3k - 4)$

23. $\sum_{k=10}^{10} 36k$

SECTION 13.3 Geometric Sequences and Series

DEFINITIONS AND CONCEPTS	EXAMPLES
Geometric sequences: A geometric sequence is a sequence of the form $a_1, a_1r, a_1r^2, a_1r^3, \dots, a_1r^{n-1}$ where a_1 is the first term. $a_n = a_1r^{n-1}$ is the nth term. r is the common ratio.	The first four terms of the geometric sequence with $a_1 = 8$ and $r = 2$ are as follows: $a_1 = 8$ $a_2 = 8(2) = 16$ $a_3 = 8(2)^2 = 32$ $a_4 = 8(2)^3 = 64$
Geometric means: If numbers are inserted between a and b to form a geometric sequence, the inserted numbers are geometric means between a and b.	To insert the positive geometric mean between 4 and 1, note that $a_1 = 4$ and $a_3 = 1$. Then substitute these values into the formula for the nth term and solve for r. $a_n = a_1r^{n-1}$ Formula for the nth term $a_3 = 4r^{3-1}$ Formula for the 3rd term $1 = 4r^2$ Substitute 1 for a_3. $\frac{1}{4} = r^2$ Divide both sides by 4. $\frac{1}{2} = r$ Take the positive square root of both sides. Since $r = \frac{1}{2}$, $a_2 = a_1r = 4\left(\frac{1}{2}\right) = 2$. The geometric mean between 4 and 1 is 2.
Sum of terms of a geometric sequence: The sum of the first n terms of a geometric sequence is given by $S_n = \frac{a_1 - a_1r^n}{1 - r} \quad (r \neq 1)$ where S_n is the sum, a_1 is the first term, r is the common ratio, and n is the number of terms in the sequence.	To find the sum of the first 4 terms of the geometric sequence defined by $a_n = 2(3)^n$, note that the first term and the common ratio are $a_1 = 2(3)^1 = 2(3) = 6$ and $r = 3$ Then substitute into the formula for the sum of the first n terms and simplify: $S_n = \frac{a_1 - a_1r^n}{1 - r}$ $S_4 = \frac{6 - 6(3)^4}{1 - 3}$ Substitute values into the formula. Thus, $S_4 = \frac{6 - 6(81)}{-2} = \frac{6 - 486}{-2} = 240$ The sum of the first 4 terms is 240.

REVIEW EXERCISES

24. Write the first five terms of the geometric sequence whose fourth term is 3 and whose fifth term is $\frac{3}{2}$.

25. Find the fifth term of a geometric sequence with a first term of $\frac{1}{16}$ and a common ratio of 2.

26. Find two geometric means between −6 and 384.

27. Find the sum of the first six terms of the sequence 240, 120, 60, ...

28. Find the sum of the first eight terms of the sequence $\frac{1}{8}, -\frac{1}{4}, \frac{1}{2}, \dots$.

29. Car depreciation A $5,000 car depreciates at the rate of 20% of the previous year's value. How much is the car worth after 5 years?

30. Stock appreciation The value of Mia's stock portfolio is expected to appreciate at the rate of 18% per year. How much will the portfolio be worth in 10 years if its current value is $25,700?

31. Planting corn A farmer planted 300 acres in corn this year. He intends to plant an additional 75 acres in corn in each successive year until he has 1,200 acres in corn. In how many years will that be?

32. Falling objects If an object is in free fall, the sequence 16, 48, 80, . . . represents the distance in feet that the object falls during the first second, during the second second, during the third second, and so on. How far will the object fall during the first 10 seconds?

SECTION 13.4 Infinite Geometric Sequences and Series

DEFINITIONS AND CONCEPTS	EXAMPLES
Sum of an infinite geometric series: If r is the common ratio of an infinite geometric series and if $\|r\| < 1$, the sum of the series is given by $S_\infty = \dfrac{a_1}{1-r}$ where a_1 is the first term and r is the common ratio.	To find the sum of the geometric series $8 + 4 + 2 + \cdots$, note that the first term is 8 and the common ratio is $r = \frac{1}{2}$. Since $\left\|\frac{1}{2}\right\| < 1$, the sum exists. To find it, proceed as follows: $S_\infty = \dfrac{a_1}{1-r}$ $S_\infty = \dfrac{8}{1-\frac{1}{2}}$ Substitute values into the formula. Thus, $S_\infty = \dfrac{8}{\frac{1}{2}} = 8 \cdot 2 = 16$ The sum is 16.
To write a repeating decimal as a common fraction, write the decimal as an infinite geometric series and find the sum.	Write $0.\overline{7}$ as a common fraction. Write the infinite geometric series $\frac{7}{10} + \frac{7}{100} + \frac{7}{1{,}000} + \cdots$ $S_\infty = \dfrac{\frac{7}{10}}{1-\frac{1}{10}} = \dfrac{7}{10} \div \dfrac{9}{10} = \dfrac{7}{10} \cdot \dfrac{10}{9} = \dfrac{7}{9}$

REVIEW EXERCISES

33. Find the sum of the infinite geometric series $100 + 20 + 4 + \cdots$.

34. Write the decimal $0.\overline{05}$ as a common fraction.

SECTION 13.5 Permutations and Combinations

DEFINITIONS AND CONCEPTS	EXAMPLES
Multiplication principle for events: If E_1 and E_2 are two events and if E_1 can be done in a_1 ways and E_2 can be done in a_2 ways, then the event "E_1 followed by E_2" can be done in $a_1 \cdot a_2$ ways.	If Danielle has 5 shirts, 4 pairs of jeans, and 3 pairs of shoes, she has a choice of $5 \cdot 4 \cdot 3 = 60$ different outfits to wear.
Formulas for permutations: $P(n, r) = \dfrac{n!}{(n-r)!}$ $P(n, n) = n!$ $P(n, 0) = 1$ If order matters, use the permutation formula.	$P(6, 4) = \dfrac{6!}{(6-4)!} = \dfrac{6 \cdot 5 \cdot 4 \cdot 3 \cdot 2 \cdot 1}{(2)!} = \dfrac{720}{2 \cdot 1} = 360$ $P(5, 5) = 5! = 5 \cdot 4 \cdot 3 \cdot 2 \cdot 1 = 120$ $P(5, 0) = 1$
Formulas for combinations: $C(n, r) = \dbinom{n}{r} = \dfrac{n!}{r!(n-r)!}$ $C(n, n) = \dbinom{n}{n} = 1 \quad C(n, 0) = \dbinom{n}{0} = 1$ If order does not matter, use the combination formula.	$C(12, 5) = \dfrac{12!}{5!(12-5)!} = \dfrac{12!}{5!(7)!}$ $= \dfrac{12 \cdot 11 \cdot 10 \cdot 9 \cdot 8 \cdot \cancel{7!}}{5 \cdot 4 \cdot 3 \cdot 2 \cdot 1 \cdot \cancel{7!}} = \dfrac{12 \cdot 11 \cdot 10 \cdot 9 \cdot 8}{5 \cdot 4 \cdot 3 \cdot 2 \cdot 1} = 792$ $C(5, 5) = 1 \quad C(5, 0) = 1$

The Binomial Theorem (Combinatorial Form): If n is any positive integer, then $$(a+b)^n = \binom{n}{0}a^n + \binom{n}{1}a^{n-1}b + \binom{n}{2}a^{n-2}b^2 + \cdots + \binom{n}{r}a^{n-r}b^r + \cdots + \binom{n}{n}b^n$$	Use the combinatorial form of the binomial theorem to expand $(x-y)^3$. $$(x-y)^3 = \binom{3}{0}x^3 - \binom{3}{1}x^2y + \binom{3}{2}xy^2 - \binom{3}{3}y^3$$ $$= 1x^3 - 3x^2y + 3xy^2 - 1y^3$$ $$= x^3 - 3x^2y + 3xy^2 - y^3$$

REVIEW EXERCISES

35. Planning a trip If there are 17 flights from New York to Chicago, and 8 flights from Chicago to San Francisco, in how many different ways could a passenger plan a trip from New York to San Francisco?

Evaluate each expression.

36. $P(7,7)$

37. $P(5,0)$

38. $P(8,6)$

39. $\dfrac{P(9,6)}{P(10,7)}$

Evaluate each expression.

40. $C(8,8)$

41. $C(7,0)$

42. $\binom{8}{6}$

43. $\binom{10}{4}$

44. $C(6,3) \cdot C(7,3)$

45. Use the combinatorial form of the binomial theorem to expand $(x+2)^3$.

46. Lining up In how many ways can five people be arranged in a line?

47. Lining up In how many ways can three men and five women be arranged in a line if the women are placed ahead of the men?

48. Choosing people In how many ways can we pick three people from a group of ten?

49. Forming committees In how many ways can we pick a committee of two Democrats and two Republicans from a group containing five Democrats and six Republicans?

SECTION 13.6 Probability

DEFINITIONS AND CONCEPTS	EXAMPLES
An event that cannot occur has a probability of 0. An event that is certain to occur has a probability of 1. All other events have probabilities between 0 and 1.	The probability of rolling a 7 with a single standard die is 0 because there is no 7 on a standard die. The probability of drawing a red card from a stack of red cards is 1 because all the cards are red.
Sample space: The sample space consists of all possibilities of outcomes for an experiment.	The sample space of rolling one standard die is the set of all possible outcomes: $\{1,2,3,4,5,6\}$
Probability of an event: If S is the *sample space* of an experiment with n distinct and equally likely outcomes, and E is an event that occurs in s of those ways, then the probability of E is $P(E) = \frac{s}{n}$	To find the probability of rolling a prime number with one roll of a standard die, first determine that there are 6 possible outcomes in the sample space: The set of possible outcomes is {1, 2, 3, 4, 5, 6}. Then determine that there are 3 possible favorable outcomes: The prime numbers on a face of a die are 2, 3, and 5. Thus, $P(\text{prime number}) = \frac{3}{6} = \frac{1}{2}$ The probability of rolling a prime number is $\frac{1}{2}$.

Unless otherwise noted, all content on this page is © Cengage Learning.

REVIEW EXERCISES

In Exercises 50–52, assume that a dart is randomly thrown at the colored chart.

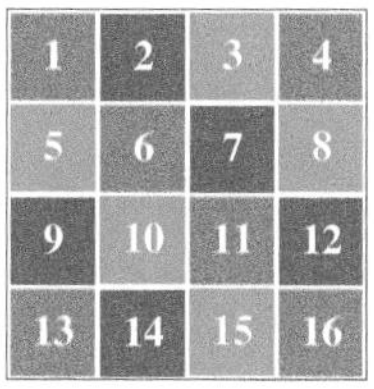

50. What is the probability that the dart lands in a blue area?
51. What is the probability that the dart lands in an even-numbered area?
52. What is the probability that the dart lands in an area whose number is greater than 2?
53. Find the probability of rolling a sum of 5 on one roll of two dice.
54. Find the probability of drawing a black card from a stack of red cards.
55. Find the probability of drawing a 10 from a standard deck of cards.
56. Find the probability of drawing a 5-card poker hand that has exactly 3 aces.
57. Find the probability of drawing 5 cards, all diamonds, from a standard card deck.

13 Test

1. Evaluate: $\frac{9!}{7!}$
2. Evaluate: $0!$
3. Find the second term in the expansion of $(x - y)^5$.
4. Find the third term in the expansion of $(x + 2y)^4$.
5. Find the tenth term of an arithmetic sequence with the first three terms of 3, 10, and 17.
6. Find the sum of the first 12 terms of the sequence $-2, 3, 8, \ldots$.
7. Find two arithmetic means between 2 and 98.
8. Evaluate: $\sum_{k=1}^{4}(3k - 4)$
9. Find the seventh term of the geometric sequence whose first three terms are $-\frac{1}{9}, -\frac{1}{3}$, and -1.
10. Find the sum of the first six terms of the sequence $\frac{1}{27}, \frac{1}{9}, \frac{1}{3}, \ldots$.
11. Find two geometric means between 3 and 648.
12. Find the sum of all of the terms of the infinite geometric sequence $9, 3, 1, \ldots$.
13. Find the sum of $5, 10, 20, \ldots$

Find the value of each expression.

14. $P(7, 3)$
15. $P(8, 8)$
16. $C(5, 3)$
17. $C(8, 3)$
18. $C(6, 0) \cdot P(6, 5)$
19. $\frac{P(6, 4)}{C(6, 4)}$
20. Expand $(x - 3)^4$.
21. **Selecting people** In how many ways can we select five people from a group of eight?
22. **Selecting committees** From a group of five men and four women, how many three-person committees can be created that will include two women?

Find each probability.

23. Rolling a 6 on one roll of a die
24. Drawing a jack or a queen from a standard card deck
25. Receiving 5 hearts for a 5-card poker hand
26. Tossing 2 heads in 5 tosses of a fair coin

Cumulative Review

1. Use graphing to solve: $\begin{cases} 2x + y = 5 \\ x - 2y = 0 \end{cases}$
2. Use substitution to solve: $\begin{cases} 3x + y = 4 \\ 2x - 3y = -1 \end{cases}$
3. Use elimination to solve: $\begin{cases} x + 2y = -2 \\ 2x - y = 6 \end{cases}$
4. Use any method to solve: $\begin{cases} \frac{x}{10} + \frac{y}{5} = \frac{1}{2} \\ \frac{x}{2} - \frac{y}{5} = \frac{13}{10} \end{cases}$
5. Evaluate: $\begin{vmatrix} 4 & -3 \\ 2 & -2 \end{vmatrix}$

Unless otherwise noted, all content on this page is © Cengage Learning.

6. Use Cramer's rule and solve for y only:
$$\begin{cases} 4x - 3y = -1 \\ 3x + 4y = -7 \end{cases}$$

7. Solve: $\begin{cases} x + y + z = 1 \\ 2x - y - z = -4 \\ x - 2y + z = 4 \end{cases}$

8. Solve for z only: $\begin{cases} x + 2y + 3z = 6 \\ 3x + 2y + z = 6 \\ 2x + 3y + z = 6 \end{cases}$

9. Solve by graphing: $\begin{cases} 3x - 2y < 6 \\ y < -x + 2 \end{cases}$

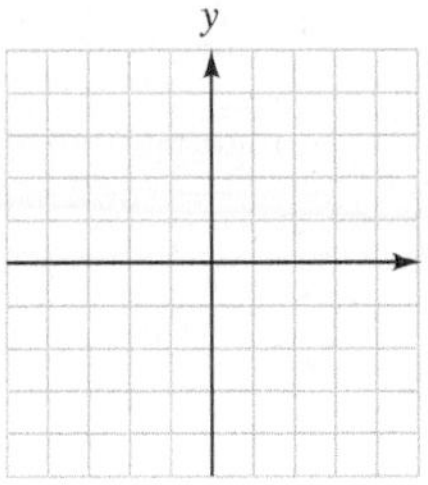

10. Solve by graphing: $\begin{cases} y < x + 2 \\ 3x + y \leq 6 \end{cases}$

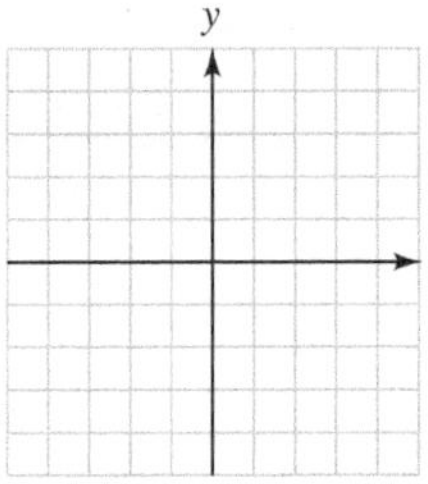

11. Graph: $f(x) = \left(\frac{1}{2}\right)^x$

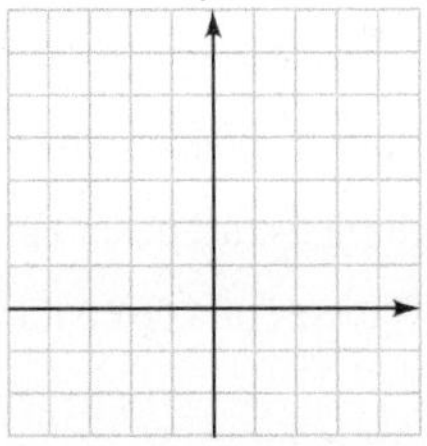

12. Write $y = \log_2 x$ as an exponential equation.

Find the value of x.

13. $\log_x 27 = 3$

14. $\log_5 125 = x$

15. $\log_2 x = -4$

16. $\log_5 x = 0$

17. Find the inverse of $y = \log_2 x$.

18. If $\log_{10} 10^x = y$, then y equals what quantity?

If $\log 7 = 0.8451$ *and* $\log 14 = 1.1461$, *evaluate each expression without using a calculator or tables.*

19. $\log 98$

20. $\log 2$

21. $\log 27$

22. $\log \frac{7}{5}$ (*Hint*: $\log 10 = 1$.)

23. Solve: $2^{x+5} = 3^x$

24. Solve: $\log 5 + \log x - \log 4 = 1$

Use a calculator for Exercises 25–27.

25. Boat depreciation How much will a \$9,000 boat be worth after 9 years if it depreciates 12% per year?

26. Find $\log_7 9$ to four decimal places.

27. Evaluate: $\dfrac{5!7!}{6!}$

28. Use the binomial theorem to expand $(3a - b)^4$.

29. Find the seventh term in the expansion of $(2x - y)^8$.

30. Find the 20th term of an arithmetic sequence with a first term of -11 and a common difference of 6.

31. Find the sum of the first 20 terms of an arithmetic sequence with a first term of 6 and a common difference of 3.

32. Insert two arithmetic means between -3 and 30.

33. Evaluate: $\displaystyle\sum_{k=1}^{4} 2k^2$

34. Evaluate: $\displaystyle\sum_{k=3}^{6} (4k - 5)$

35. Find the seventh term of a geometric sequence with a first term of $\frac{1}{27}$ and a common ratio of 3.

36. Find the sum of the first ten terms of the sequence $\frac{1}{64}, \frac{1}{32}, \frac{1}{16}, \ldots$.

37. Insert two geometric means between -3 and 192.

38. Find the sum of all the terms of the sequence $9, 3, 1, \ldots$.

39. Evaluate: $P(9, 3)$

40. Evaluate: $C(10, 7)$

41. Evaluate: $\dfrac{C(8, 4) \cdot C(8, 0)}{P(6, 2)}$

42. If $n > 1$, which is smaller: $P(n, n)$ or $C(n, n)$?

43. Lining up In how many ways can seven people stand in a line?

44. Forming a committee In how many ways can a committee of three people be chosen from a group containing nine people?

45. Cards Find the probability of drawing a red face card from a standard deck of cards.

Unless otherwise noted, all content on this page is © Cengage Learning.

GLOSSARY

absolute value The distance between a given number and 0 on a number line, usually denoted with | |

addition property of equality The property that states if a, b, and c are real numbers and $a = b$, then $a + c = b + c$

additive inverses A pair of numbers, a and b, are additive inverses if $a + b = 0$; also called **negatives** or **opposites**.

algebraic expression A combination of variables, numbers, and arithmetic operations; an algebraic expression does not contain an equality or inequality symbol.

algebraic term An expression that is a constant, variable, or a product of constants and variables; for example, 37, xyz, and $32t$ are terms.

algorithm A repeating series of steps or processes

altitude The length of the perpendicular line from a vertex of a triangle to its opposite side

angle A geometric figure consisting of two rays emanating from a common point

annual growth rate The rate r that a quantity P increases or decreases within one year; the variable r in the exponential growth formula $A = Pe^{rt}$

area The amount of surface enclosed by a two-dimensional geometric figure; expressed as unit^2

arithmetic means Numbers that are inserted between two elements in an arithmetic sequence to form a new arithmetic sequence

arithmetic sequence A sequence in which each successive term is equal to the sum of the preceding term and a constant; written $a, a + d, a + 2d, a + 3d, \ldots$

arithmetic series The indicated sum of the terms of an arithmetic sequence

ascending order An ordering of the terms of a polynomial such that a given variable's exponents occur in increasing order

associative properties The properties for addition and multiplication that state for real numbers a, b, and c, $(a + b) + c = a + (b + c)$ and $(ab)c = a(bc)$. *Note*: The associative property does not hold for subtraction or division.

asymptote A straight line that is approached by a graph

augmented matrix A matrix representing a system of linear equations written in standard form, with columns consisting of the coefficients and a column of the constant terms of the equations

axis of symmetry For the parabola $f(x) = ax^2 + bx + c$, a vertical line, $x = -\frac{b}{2a}$

back substitution The process of finding the value of a second (or third, etc.) variable after finding the first by substituting the first into an equation of two variables and, if necessary, substituting both values into an equation of three values to find the third, etc.

base In the expression x^y, x is the base and y is the exponent. The base, x, will be used as a factor y times.

base angles of an isosceles triangle The two congruent angles of an isosceles triangle opposite the two sides that are of the same length

binomial A polynomial with exactly two terms

binomial theorem The theorem used to expand a binomial

Cartesian coordinate system A grid composed of a horizontal axis and a vertical axis that allows us to identify each point in a plane with a unique ordered pair of numbers; also called the **rectangular coordinate system**

Cartesian plane The set of all points in the Cartesian coordinate system

center of a circle The point that is equidistant from all points on a circle

center of a hyperbola The midpoint of the segment joining the foci of a hyperbola

change-of-base formula A formula used to convert a logarithm from one base a to some other base b

circle The set of all points in a plane that are the same distance from a fixed point (center)

circumference The distance around a circle

closure properties The properties that state the sum, difference, product, or quotient of two real numbers is a real number

coefficient matrix A matrix consisting of the coefficients of the variables in a system of linear equations

combinations The number of ways to choose r things taken from a set of n things if order is not important

common difference The constant difference between successive terms in an arithmetic sequence, that is, the number d in the formula $a_n = a_1 + (n - 1)d$

common logarithm A logarithm with a base of 10

common ratio The constant quotient of two successive terms in a geometric sequence

commutative properties The properties for addition and multiplication that state for real numbers a and b, $a + b = b + a$ and $ab = ba$. *Note*: The commutative property does not hold for subtraction and division.

complementary angles Two angles whose measures sum to 90°

completing the square The process of forming a perfect-square trinomial from a binomial of the form $x^2 + bx$

complex fraction A fraction that contains a fraction in its numerator and/or denominator

complex number The sum of a real number and an imaginary number

composite function A new function formed when one function is evaluated in terms of another, denoted by $(f \circ g)(x)$

composite numbers A natural number greater than 1 with factors other than 1 and itself

composition The process of using the output of one function as the input of another function

compound inequality A single statement representing the intersection of two inequalities

compounding periods The number of times per year interest is compounded

compound interest The total interest earned on money deposited in an account where the interest for each compounding period is deposited back into the account, which increases the principle for the next compounding period

conditional equation An equation in one variable that has exactly one solution

cone A three-dimensional surface with a circular base whose cross sections (parallel to the base) are circles that decrease in diameter until it comes to a point

conic section A graph that can be formed by the intersection of a plane and a right-circular cone

conjugate An expression that contains the same two terms as another but with opposite signs between them; for example, $a + \sqrt{b}$ and $a - \sqrt{b}$

consistent system A system of equations with at least one solution

constant A term whose variable factor(s) have an exponent of 0

constraints Inequalities that limit the possible values of the variables of the objective function

contradiction An equation that is false for all values of its variables; its solution set is $\varnothing$.

coordinate The number that corresponds to a given point on a number line

coordinate plane The plane that contains the x-axis and y-axis

Cramer's rule A method using determinants to solve systems of linear equations

critical points Given a quadratic or rational inequality, the points on the number line corresponding to its critical values

critical values Given a quadratic inequality in standard form, the solutions to the quadratic equation $ax^2 + bx + c = 0$; given a rational inequality in standard form, the values where the denominator is equal to zero and the numerator is equal to zero

cube A rectangular solid whose faces are all congruent squares

cube root If $a = b^3$, then b is a cube root of a.

cube-root function The function $f(x) = \sqrt[3]{x}$

cubic function A polynomial function whose equation is of third degree; alternately, a function whose equation is of the form $f(x) = ax^3 + bx^2 + cx + d$

cubic units The units given to a volume measurement

cubing function The function $f(x) = x^3$

cylinder A three-dimensional surface whose cross sections (parallel to the base) are the same size and shape as the base of the cylinder

decibel A unit of measure that expresses the voltage gain (or loss) of an electronic device such as an amplifier

decreasing function A function whose value decreases as the independent value increases (graph drops as we move to the right)

degree A unit of measure for an angle

degree of a monomial The sum of the exponents of all variables in a monomial

degree of a polynomial The largest degree of a polynomial's terms

denominator The expression below the bar of a fraction

dependent equations A system of equations that has infinitely many solutions

dependent variable The variable in an equation of two variables whose value is determined by the independent variable (usually y in an equation involving x and y)

descending powers of a variable An ordering of the terms of a polynomial such that a given variable's exponents occur in decreasing order

determinant A calculated value from the elements in a square matrix: For a two-by-two matrix $\begin{bmatrix} a & b \\ c & d \end{bmatrix}$, the determinant is $ad - bc$. For a three-by-three determinant or larger, we use the method of expanding by minors.

diameter A line segment connecting two points on a circle and passing through the center; the length of such a line segment

difference The result of subtracting two expressions

difference of two cubes An expression of the form $a^3 - b^3$

difference of two squares An expression of the form $a^2 - b^2$

difference quotient The function defined by $\frac{f(x + h) - f(x)}{h}$

direct variation A relationship between two variables x and y of the form $y = kx$, where k is the constant of proportionality

discriminant The part of the quadratic formula, $b^2 - 4ac$, used to determine the number and nature of solutions to the quadratic equation

distributive property The property that states for real numbers a, b, and c, $a(b + c) = ab + ac$ and $(b + c)a = ab + ac$.

dividend In long division, the expression under the division symbol; in a fraction, the expression in the numerator

divisor In long division, the expression in front of the division symbol; in a fraction, the expression in the denominator

domain of a function The set of all permissible input values of a function

domain of a relation The set of all first elements (components) of a relation

double inequality Two inequalities written together to indicate a set of numbers that lie between two fixed values

doubly shaded region The area of a graph depicting the intersection of two half-planes

eccentricity A measure of the flatness of an ellipse

element of a matrix One of the entries in a matrix

elimination (addition) An algebraic method for solving a system of equations where one variable is eliminated when the equations are added

ellipse The set of all points in a plane the sum of whose distances from two fixed points (foci) is a constant

ellipses A symbol consisting of three dots (. . .) meaning "and so forth" in the same manner

empty set A set that has no elements, denoted by the symbol ∅ or { }

equal ratios Two (or more) ratios that represent equal values

equation A statement indicating that two quantities are equal

equivalent equations Two equations that have the same solution set

equivalent fractions Two fractions with different denominators describing the same value; for example, $\frac{3}{11}$ and $\frac{6}{22}$ are equivalent fractions.

equivalent systems Two systems of equations that have the same solution set

even integers The set of integers that can be divided exactly by 2. *Note*: 0 is an even integer.

even root If $b = \sqrt[n]{a}$, b is an even root if n is even.

event A situation that can have several different outcomes or a subset of the sample space of an experiment

exponent In the expression x^y, x is the base and y is the exponent. The exponent y states the number of times that the base x will be used as a factor.

exponential equation An equation that contains a variable in one of its exponents

exponential expression An expression of the form y^x, also called a **power of *x***

exponential function A function of the form $f(x) = ab^x$

exponential function (natural) A function of the form $f(t) = Pe^{kt}$

extraneous solution A solution to an equation that does not result in a true statement when substituted for the variable in the original equation

extremes In the proportion $\frac{a}{b} = \frac{c}{d}$, the numbers a and d are called the extremes.

factorial For a natural number n, the product of all the natural numbers less than or equal to n, denoted as $n!$ (*Exception:* $0!$ is defined as 1.)

factoring The process of finding the individual factors of a product

factoring tree A visual representation of factors of a number, usually used as a tool to express the number in prime-factored form

factors of a number The natural numbers that divide a given number; for example, the factors of 12 are 1, 2, 3, 4, 6, and 12.

factor theorem The theorem that states if $P(x)$ is a polynomial, then $P(r) = 0$ if and only if $(x - r)$ is a factor of $P(r)$

feasibility region The set of points that satisfy all of the constraints of a system of inequalities

Fibonacci sequence The sequence 1, 1, 2, 3, 5, 8, . . . with every successive term the sum of the two previous terms

finite sequence A sequence with a finite number of terms

foci of an ellipse The two fixed points used in the definition of the ellipse

foci of a hyperbola The two fixed points used in the definition of the hyperbola

FOIL method An acronym representing the order for multiplying the terms of two binomials

formula A literal equation in which the variables correspond to defined quantities

fulcrum The point on which a lever pivots

fundamental theorem of arithmetic The theorem that states there is exactly one prime factorization for any natural number greater than 1

future value The amount to which an invested amount of money will grow

Gaussian elimination A method of solving a system of linear equations by working with its associated augmented matrix

general form of the equation of a circle The equation of a circle written in the form $x^2 + y^2 + Dx + Ey + F = 0$

general form of the equation of an ellipse The equation of an ellipse written in the form $Ax^2 + Cy^2 + Dx + Ey + F = 0$

general form of the equation of a hyperbola The equation of a hyperbola written in the form $Ax^2 - Cy^2 + Dx + Ey + F = 0$

general form of a linear equation A linear equation written in the form $Ax + By = C$, A and B not both equal to 0.

geometric means Numbers that are inserted between two elements in a geometric sequence to form a new geometric sequence

geometric sequence A sequence in which each successive term is equal to the product of the preceding term and a constant, written $a_1, a_1r, a_1r^2, a_1r^3, \ldots$

geometric series The indicated sum of the terms of a geometric sequence.

graph of a point A point's location in the rectangular or Cartesian plane

greatest common factor (GCF) The largest expression that is a factor of each of a group of expressions

greatest integer function The function whose output for a given x is the greatest integer that is less than or equal to x, denoted by $[\![\ ]\!]$

grouping symbol A symbol (such as a radical sign or a fraction bar) or pair of symbols (such as parentheses or braces) that indicate that these operations should be computed before other operations

half-life The time it takes for one-half of a sample of a radioactive element to decompose

half-plane A subset of the coordinate plane consisting of all points on a given side of a boundary line

horizontal line A line parallel to the x-axis; a line with the equation $y = b$

horizontal-line test A test used to determine if a given function is one-to-one: If every horizontal line that intersects the graph of the function does so exactly once, then the graph is the graph of a one-to-one function.

horizontal translation A graph that is the same shape as a given graph, except that it is shifted horizontally

hyperbola The set of all points in a plane the difference of whose distances from two fixed points (foci) is a constant

hypotenuse The longest side of a right triangle, the side opposite the 90° angle

identity An equation that is true for all values of its variables; its solution set is $\mathbb{R}$.

identity elements The additive identity is 0, because for all real a, $a + 0 = 0 + a = a$. The multiplicative identity is 1, because for all real a, $a \cdot 1 = 1 \cdot a = a$.

identity function The function whose rule is $f(x) = x$

imaginary number The square root of a negative number

improper fraction A fraction whose numerator is greater than or equal to its denominator

inconsistent system A system of equations that has no solution; its solution set is $\varnothing$.

increasing function A function whose value increases as the independent variable increases (graph rises as we move to the right)

independent equations A system of two equations with two variables for which each equation's graph is different

independent variable The variable in an equation of two variables to which we assign input values (usually x in an equation involving x and y)

index If $b = \sqrt[n]{a}$, we say n is the index of the radical.

inequality A mathematical statement indicating that two quantities are not necessarily equal

inequality symbols A set of six symbols that are used to describe two expressions that are not equal:

- $\approx$ "is approximately equal to"
- $\neq$ "is not equal to"
- $<$ "is less than"
- $>$ "is greater than"
- $\leq$ "is less than or equal to"
- $\geq$ "is greater than or equal to"

infinite sequence A sequence with infinitely many terms

input value A value substituted for the independent variable

integers The set of numbers given by $\{\ldots, -4, -3, -2, -1, 0, 1, 2, 3, 4, \ldots\}$

integer square An integer that is the second power of another integer

intercept method The method of graphing a linear equation by first graphing the intercepts, and then connecting them by a straight line

intersection of two intervals A combination of two intervals that contains any point in the first interval that is also in the second interval

interval A set of all real numbers between two given real numbers a and b; it must be specified whether or not a and b are included in the interval.

interval notation A way of writing intervals using brackets or parentheses to distinguish between endpoints that are included in a given interval, and endpoints that are not included

inverse function The function, denoted by f^{-1}, obtained by reversing the coordinates of each ordered pair $(x, f(x))$; for example, if $f(2) = 3$, then $f^{-1}(3) = 2$.

irrational numbers The set of numbers that cannot be put in the form of a fraction with integer numerator and nonzero integer denominator; can be expressed only as nonterminating, nonrepeating decimals

isosceles triangle A triangle that has two sides of the same length

least (or lowest) common denominator The smallest expression that is exactly divisible by the denominators of a set of rational expressions

like radicals Radicals that have the same index and radicand

like signs Two numbers that are both positive or both negative

like terms Terms containing the same variables with the same exponents

linear equation in one variable An equation of the form $ax + b = c$, where a, b, and c are real numbers and $a \neq 0$

linear equation in two variables An equation of the form $Ax + By = C$, where A, B, and C are real numbers; A and B cannot both be 0.

linear function A function whose graph is a line; a function whose equation is of the form $f(x) = ax + b$

linear inequality A statement that can be written in one of the following forms: $y > ax + b$, $y \geq ax + b$, $y < ax + b$, or $y \leq ax + b$

linear inequality in two variables An inequality containing an expression of the form $Ax + By$ (A and B not both equal to zero)

linear programming A mathematical technique used to find the optimal allocation of resources

literal equation An equation with more than one variable, usually a formula

logarithm The exponent to which b is raised to obtain x, denoted by $\log_b x$, $b > 0$, $b \neq 1$

logarithmic equation An equation with logarithmic expressions that contain a variable $y = \log_b x$ if and only if $x = b^y$

logarithmic function The function $f(x) = \log_b x; b \neq 1, b > 0$

lowest terms A fraction written in such a way so that no integer greater than 1 divides both its numerator and its denominator; also called **simplest form**

major axis (transverse) of an ellipse The line segment passing through the foci joining the vertices of an ellipse

matrix A rectangular array of numbers

mean The sum of a collection of values divided by the number of values

means In the proportion $\frac{a}{b} = \frac{c}{d}$, the numbers b and c are called the means.

median The middle value of a collection of values, obtained by arranging them in increasing order, and either choosing the middle value, or the average of the two middle values

midpoint The point halfway between two given points

minor A number associated with an element of a square matrix used to find a determinant of next lower order, by crossing out the elements of the matrix that are in the same row and column as the element

minor axis of an ellipse The line segment passing through the center of an ellipse perpendicular to the major axis

minuend The first expression in the difference of two expressions; for example, in the expression $(3x^2 + 2x - 5) - (4x^2 - x + 5)$, the expression $(3x^2 + 2x - 5)$ is the minuend.

mixed number A rational number expressed as an integer plus a proper fraction

mode The value that occurs most often in a collection of values; there can be more than one mode for a collection of values.

monomial A polynomial with exactly one term

multiplication principle of events The method of determining the number of ways that one event can be followed by another

multiplication property of equality The property that states if a, b, and c are real numbers and $a = b$, then $ac = bc$.

multiplicative identity Is 1, because for all real numbers a, $a \cdot 1 = 1 \cdot a = a$

multiplicative inverse Two numbers whose product is 1; also called **reciprocals**

natural logarithm A logarithm with a base of e

natural numbers The set of numbers given by $\{1, 2, 3, 4, 5, \ldots\}$; also called **positive integers**

negative numbers The set of numbers less than zero

negatives A pair of numbers, a and b, represented by points that lie on opposite sides of the origin and at equal distances from the origin. If a and b are negatives, $a + b = 0$. Also called **opposites** or **additive inverses**

number line A line that is used to represent real numbers or sets of real numbers

numerator The expression above the bar of a fraction

numerical coefficient The number factor of a term; for example, the numerical coefficient of $8ab$ is 8.

objective function A function whose value is maximized or minimized

odd integers The set of integers that cannot be divided exactly by 2

odd root If $b = \sqrt[n]{a}$, b is an odd root if n is odd.

opposites A pair of numbers, a and b, represented by points that lie on opposite sides of the origin and at equal distances from the origin. If a and b are opposites, $a + b = 0$. Also called **negatives** or **additive inverses**

ordered pair A pair of real numbers, written as (x, y), that describes a unique point in the Cartesian plane

origin The point on a number line that represents the number zero; the point on the rectangular coordinate system that represents the point $(0, 0)$

output value The value of the dependent variable, determined by the choice of input value

parabola The graph of a quadratic function

partial sum The sum of finitely many consecutive terms of a series, starting at the first term

Pascal's triangle A triangular array of numbers in which 1's begin and end each row, and each entry between the 1's is the sum of the closest pair of numbers in the line immediately above it; the first 4 rows of Pascal's triangle are

$$\begin{array}{ccccccc} & & & 1 & & & \\ & & 1 & & 1 & & \\ & 1 & & 2 & & 1 & \\ 1 & & 3 & & 3 & & 1 \end{array}$$

percent The numerator of a fraction whose denominator is 100

perfect squares The exact second power of another integer or polynomial; for example, 9 is a perfect square (3^2) as is $a^2 + 2ab + b^2$, which is equal to $(a + b)^2$.

perfect trinomial square A trinomial of the form $a^2 + 2ab + b^2$ or of the form $a^2 - 2ab + b^2$

perimeter The distance around a non-circular geometric figure

periodic interest rate The interest rate used for each compounding period

permutations The number of ways to choose r things taken from a set of n things if order is important

perpendicular lines Two lines that meet at a 90° angle, two lines whose slopes are negative reciprocals of each other, or a horizontal line and a vertical line

piecewise-defined function A function defined by using different equations for different parts of its domain

plane The graph of an equation of the form $ax + by + cz = d$; a, b, c NOT all equal to 0.

point circle A circle with radius zero; a single point

point-slope form The equation of a line in the form $y - y_1 = m(x - x_1)$, where m is the slope and (x_1, y_1) is a point on the line

polynomial An algebraic expression that is a single term or the sum of several terms containing whole-number exponents on the variables

polynomial function A function whose equation is a polynomial; for example, $f(x) = x^2 - 3x + 6$

positive integers The set of numbers given by $\{1, 2, 3, 4, 5, \ldots\}$, that is, the set of integers that are greater than zero; also called the **natural numbers**

power of *x* An expression of the form x^y; also called an **exponential expression**

power rule The rule that states if $a = b$, then $a^n = b^n$ where n is a natural number

prime-factored form A natural number written as the product of factors that are prime numbers

prime number A natural number greater than 1 that can be divided exactly only by 1 and itself

prime polynomial A polynomial that does not factor over the rational numbers

principal square root The positive square root of a number

probability A measure of the likelihood of an event

product The result of multiplying two or more expressions

proper fraction A fraction whose numerator is less than its denominator

proportion A statement that two ratios are equal

pyramid A three-dimensional surface whose cross sections (parallel to the base) are polygons the same shape as the base but decrease in size until it comes to a point

Pythagorean theorem If the length of the hypotenuse of a right triangle is c and the lengths of the two legs are a and b, then $a^2 + b^2 = c^2$.

quadrants The four regions in the rectangular coordinate system formed by the x- and y-axes

quadratic equation An equation that can be written in the form $ax^2 + bx + c = 0$ where $a \neq 0$

quadratic formula A formula that produces the solutions to the general quadratic equation; $x = \frac{-b \pm \sqrt{b^2 - 4ac}}{2a}$

quadratic function A polynomial function whose equation in one variable is of second degree; alternately, a function whose equation is of the form $f(x) = ax^2 + bx + c$ $(a \neq 0)$

quadratic inequality An inequality involving a quadratic polynomial in one variable

quotient The result of dividing two expressions

radical sign A symbol used to represent the root of a number

radicand The expression under a radical sign

radius A segment drawn from the center of a circle to a point on the circle; the length of such a line segment

range of a function The set of all output values of a function

range of a relation The set of all second elements of a relation

rate A ratio used to compare two quantities of different units

ratio The comparison of two quantities by their indicated quotient

rational equation An equation that contains one or more rational expressions

rational expression The quotient of two polynomials, where the polynomial in the denominator cannot be equal to 0

rational function A function of the form $f(x) = \frac{p(x)}{q(x)}$ where p and q are polynomial functions of x and $q(x) \neq 0$

rationalizing the denominator The process of simplifying a fraction so that there are no radicals in the denominator

rationalizing the numerator The process of simplifying a fraction so that there are no radicals in the numerator

rational numbers The set of fractions that have an integer numerator and a nonzero integer denominator; alternately, any terminating or repeating decimal

real numbers The set that contains the rational numbers and the irrational numbers

reciprocal One number is the reciprocal of another if their product is 1; for example $\frac{3}{11}$ is the reciprocal of $\frac{11}{3}$. Also called **multiplicative inverses**

rectangular coordinate system A grid that allows us to identify each point in a plane with a unique pair of numbers; also called the **Cartesian coordinate system**

reflection A graph that is identical in shape to a given graph, except that it is flipped across a line (usually the x- or y-axis)

relation A set of ordered pairs

remainder The computation of $x \div y$ will yield an expression of the form $q + \frac{r}{y}$. The quantity r is called the remainder.

remainder theorem The theorem that states if a polynomial function $P(x)$ is divided by $(x - r)$, the remainder is $P(r)$

repeating decimal A decimal expression of a fraction that eventually falls into an infinitely repeating pattern

Richter scale A unit of measure that expresses magnitude (ground motion) of an earthquake

right angle An angle whose measure is 90°

right triangle A triangle that has an angle whose measure is 90°

root A number that makes an equation true when substituted for its variable; if there are several variables, then the set of numbers that make the equation true; also called a **solution**

roster method A method of describing a set by listing its elements within braces

sample space The set of all possible outcomes for an experiment

scientific notation The representation of a number as the product of a number between 1 and 10 (1 included), and an integer power of ten, for example, 6.02×10^{23}

sequence An ordered set of numbers that are defined by the position (1st, 2nd, 3rd, etc.) they hold

set A collection of objects whose members are listed or defined within braces

set-builder notation A method of describing a set that uses a variable (or variables) to represent the elements and a rule to determine the possible values of the variable; for example, we can describe the natural numbers as $\{x \mid x$ is an integer and $x > 0\}$.

similar triangles Two triangles that have the same shape, that is, two triangles whose corresponding angles have the same measure

simplest form The form of a fraction where no expression other than 1 divides both its numerator and its denominator

simultaneous solution Values for the variables in a system of equations that satisfy all of the equations

slope-intercept form The equation of the line in the form $y = mx + b$ where m is the slope and b is the y-coordinate of the y-intercept

solution A number that makes an equation true when substituted for its variable; if there are several variables, then the numbers that make the equation true; also called a **root**

solution of an inequality The numbers that make a given inequality true

solution set The set of all values that makes an equation true when substituted for the variable in the original equation

sphere A three-dimensional surface in which all points are equidistant from a fixed point (center); a ball shape

square matrix A matrix with the same number of rows as columns

square root If $a = b^2$, then b is a square root of a.

square-root function The function $f(x) = \sqrt{x}$

square-root property The property that states the equation $x^2 = c$ has two solutions: $\sqrt{c}$ and $-\sqrt{c}$

square units The units given to an area measurement

squaring function The function $f(x) = x^2$

standard deviation A measure of how closely a set of data is grouped about its mean

standard form of the equation of a circle The equation of a circle written in the form $(x - h)^2 + (y - k)^2 = r^2$, where (h, k) is the center and r is the radius

standard form of the equation of an ellipse The equation of an ellipse written in the form $\frac{(x-h)^2}{a^2} + \frac{(y-k)^2}{b^2} = 1$, where (h, k) is the center.

standard form of the equation of a hyperbola The equation of a hyperbola written in the form $\frac{(x-h)^2}{a^2} - \frac{(y-k)^2}{b^2} = 1$, where (h, k) is the center.

standard notation The representation of a given number as an integer part, followed by a decimal; for example, 212.3337012

step function A function whose graph is a series of horizontal line segments

straight angle An angle whose measure is 180°

subset A set, all of whose elements are included in a different set; for example, the set $\{1, 3, 5\}$ is a subset of the set $\{1, 2, 3, 4, 5\}$.

subtrahend The second expression in the difference of two expressions; for example, in the expression $(3x^2 + 2x - 5) - (4x^2 - x + 5)$, the expression $(4x^2 - x + 5)$ is the subtrahend.

sum The result of adding two expressions

summation notation A shorthand notation for indicating the sum of a number of consecutive terms in a series, denoted by Σ

sum of infinite geometric series The indicated sum of the terms of a geometric sequence for which $|r| < 1$

sum of two cubes An expression of the form $a^3 + b^3$

sum of two squares An expression of the form $a^2 + b^2$

supplementary angles Two angles whose measures sum to 180°

synthetic division A method used to divide a polynomial by a binomial of the form $x - r$

system of equations Two or more equations that are solved simultaneously

term An expression that is a number, a variable, or a product of numbers and variables; for example, 37, xyz, and $3x$ are terms.

terminating decimal A decimal expression for a fraction that contains a finite number of decimal places

tree diagram A way of visually representing all the possible outcomes of an event

triangular form of a matrix A matrix with all zeros below its main diagonal

trinomial A polynomial with exactly three terms

unit cost The ratio of an item's cost to its quantity; expressed with a denominator of 1

unlike signs Two numbers have unlike signs if one is positive and one is negative.

unlike terms Terms that are not like terms

variables Letters that are used to represent real numbers

vertex The lowest point of a parabola that opens upward, or the highest point of a parabola that opens downward

vertex angle of an isosceles triangle The angle in an isosceles triangle formed by the two sides that are of the same length

vertical line A line parallel to the y-axis; a line with the equation $x = b$

vertical-line test A method of determining whether the graph of an equation represents a function

vertical translation A graph that is the same shape as a given graph, only shifted vertically

vertices of a hyperbola The endpoints of the axis that passes through the foci

vertices of an ellipse The endpoints of the major axis

volume The amount of space enclosed by a three-dimensional geometric figure

whole numbers The set of numbers given by $\{0, 1, 2, 3, 4, 5, \ldots\}$; that is, the set of integers that are greater than or equal to zero

x-axis The horizontal number line in the rectangular coordinate system

x-coordinate The first number in an ordered pair

x-intercept The point $(a, 0)$ where a graph intersects the x-axis

y-axis The vertical number line in the rectangular coordinate system

y-coordinate The second number in an ordered pair

y-intercept The point $(0, b)$ where a graph intersects the y-axis

zero-factor property The property of real numbers that states if the product of two quantities is 0, then at least one of those quantities must be equal to 0

APPENDIX 1: Measurement Conversions

Every day we are confronted with the need to quantify things. We drive a number of miles (distance) to work or school. We fill our cars with a number of gallons (volume) of fuel. We purchase a quantity of deli lunch meats (weight).

In this appendix we will look at the two systems of measurement most commonly used in the United States: the American system and the metric system. We will examine the units of measure in each system for

Length (distance)   Volume (capacity)   Weight (mass)

A.1.1 Measurement in the American System

The American system continues to find wide usage in the United States, particularly in consumer applications such as retail and grocery stores. But even those areas are slowly converting to the metric system. As an example, milk is still sold by the gallon, but many carbonated beverages are now sold by the liter.

1 Identify the units of length in the American system and convert from one unit of length to another.

The standard units of length in the American system are the **inch**, **foot**, **yard**, and **mile**. They are related to one another in the following ways:

1 foot (ft) = 12 inches (in.)

1 yard (yd) = 3 feet = 36 inches

1 mile (mi) = 5,280 feet = 1,760 yards = 63,360 inches

To convert a measurement from one unit of length to another, we use a **unit conversion factor**, a special fraction that is equal to 1 (the unit). It is written with the desired conversion units in the numerator and the initial units in the denominator:

$$\frac{\text{the numerator is written in units that you wish to find}}{\text{the denominator is written in units that you wish to remove}}$$

For example, because there are three feet in one yard, the unit conversion for yards to feet is written as:

$$\frac{3 \text{ feet}}{1 \text{ yard}}$$

Read as "3 feet per yard," it is algebraically equal to 1. Therefore, to convert units of a 100-yard football field to feet, we multiply the initial quantity (100 yards) by the yards-to-feet conversion factor.

$$\begin{aligned} 100 \text{ yards} &= 100 \cancel{\text{yards}} \cdot \frac{3 \text{ feet}}{1 \cancel{\text{yard}}} \\ &= 100 \cdot 3 \\ &= 300 \text{ feet} \end{aligned}$$

To convert the units of measure among inches, feet, yards and miles, we use the following unit conversion factors.

To convert from	Unit conversion factor
Feet to inches	12 in. / 1 ft
Yards to feet	3 ft / 1 yd
Yards to inches	36 in. / 1 yd
Miles to feet	5,280 ft / 1 mi

To convert from	Unit conversion factor
Inches to feet	1 ft / 12 in.
Feet to yards	1 yd / 3 ft
Inches to yards	1 yd / 36 in.
Feet to miles	1 mi / 5,280 ft

EXAMPLE 1 Determine the height (length) of a 20-foot flagpole in yards.

Solution

$$\begin{aligned} 20 \text{ feet} &= 20 \cancel{\text{feet}} \cdot \frac{1 \text{ yard}}{3 \cancel{\text{feet}}} \\ &= \frac{20}{3} \\ &= 6\frac{2}{3} \text{ yards} \end{aligned}$$

SELF CHECK 1 Determine the height of a 25-foot tower in yards.

EXAMPLE 2 Express the length of a 1.5-mile horse race track in feet.

Solution

$$\begin{aligned} 1.5 \text{ miles} &= 1.5 \cancel{\text{miles}} \cdot \frac{5{,}280 \text{ feet}}{1 \cancel{\text{mile}}} \\ &= 1.5 \cdot 5{,}280 \text{ feet} \\ &= 7{,}920 \text{ feet} \end{aligned}$$

SELF CHECK 2 Express the length of a 2-mile horse race track in feet.

EXAMPLE 3 What is the height in miles, to the nearest tenth of a mile, of an airplane flying at 30,000 feet?

Solution

$$\begin{aligned} 30{,}000 \text{ feet} &= 30{,}000 \cancel{\text{feet}} \cdot \frac{1 \text{ mile}}{5{,}280 \cancel{\text{feet}}} \\ &= \frac{30{,}000 \text{ miles}}{5{,}280} \\ &= 5.7 \text{ miles} \end{aligned}$$

SELF CHECK 3 What is the height in miles, to the nearest tenth of a mile, of an airplane flying at 35,000 feet?

2 Identify the units of volume in the American system and convert from one unit of volume to another.

The standard units of volume (sometimes called capacity) in the American system are the fluid **ounce**, **cup**, **pint**, **quart**, and **gallon**. They are related to one another in the following ways:

1 cup (c) = 8 fluid ounces (fl oz)

1 pint (pt) = 2 cups

1 quart (qt) = 2 pints

1 gallon (gal) = 4 quarts

In the same manner as shown in converting units of length, we use a unit conversion factor to convert units of volume. To convert the units of measure among fluid ounces, cups, pints, quarts, and gallons, we use the following unit conversion factors.

To convert from	Unit conversion factor
Cups to fluid ounces	8 fl oz / 1 c
Pints to cups	2 c / 1 pt / 16 fl oz
Quarts to pints	2 pt / 1 qt / 32 fl oz
Gallons to quarts	4 qt / 1 gal / 128 fl oz

To convert from	Unit conversion factor
Fluid ounces to cups	1 c / 8 fl oz
Cups to pints	1 pt / 2 c / 16 fl oz
Pints to quarts	1 qt / 2 pt / 32 fl oz
Quarts to gallons	1 gal / 4 qt / 128 fl oz

EXAMPLE 4 How many quarts are in a 20-gallon automobile gasoline tank?

Solution

$$\begin{aligned} 20 \text{ gallons} &= 20 \cancel{\text{ gallons}} \cdot \frac{4 \text{ quarts}}{1 \cancel{\text{ gallon}}} \\ &= 20 \cdot 4 \text{ quarts} \\ &= 80 \text{ quarts} \end{aligned}$$

SELF CHECK 4 How many quarts are in a 30-gallon automobile gasoline tank?

EXAMPLE 5 How many fluid ounces are in 6 pints of ice cream?

Solution

$$\begin{aligned} 6 \text{ pints} &= 6 \cancel{\text{ pints}} \cdot \frac{2 \cancel{\text{ cups}}}{1 \cancel{\text{ pint}}} \cdot \frac{8 \text{ fluid ounces}}{1 \cancel{\text{ cup}}} \\ &= 6 \cdot 2 \cdot 8 \\ &= 96 \text{ fluid ounces} \end{aligned}$$

SELF CHECK 5 How many fluid ounces are in 4 pints of ice cream?

EXAMPLE 6 How many cups of milk are in a half-gallon container?

Solution

$$\begin{aligned} \frac{1}{2} \text{ gallon} &= \frac{1}{2} \cancel{\text{ gallon}} \cdot \frac{4 \cancel{\text{ quarts}}}{1 \cancel{\text{ gallon}}} \cdot \frac{2 \cancel{\text{ pints}}}{1 \cancel{\text{ quart}}} \cdot \frac{2 \text{ cups}}{1 \cancel{\text{ pint}}} \\ &= \frac{1}{2} \cdot 4 \cdot 2 \cdot 2 \\ &= 8 \text{ cups} \end{aligned}$$

SELF CHECK 6 How many cups of milk are in a $\frac{3}{4}$-gallon container?

3 Identify the units of weight in the American system and convert from one unit of weight to another.

The standard units of weight in the American system are the **ounce**, **pound**, and **ton**. They are related to one another in the following ways:

1 pound (lb) = 16 ounces (oz)

1 ton (T) = 2,000 pounds

In the same manner as shown in converting units of length and units of volume, we use a unit conversion factor to convert units of weight. To convert the units of measure among ounces, pounds, and tons, we use the following unit conversion factors.

To convert from	Unit conversion factor
Pounds to ounces	16 oz / 1 lb
Tons to pounds	2,000 lb / 1 T

To convert from	Unit conversion factor
Ounces to pounds	1 lb / 16 oz
Pounds to tons	1 T / 2,000 lb

EXAMPLE 7 How many pounds are in a 256-ounce bowling ball?

Solution

$$
\begin{aligned}
256 \text{ ounces} &= 256 \cancel{\text{ounces}} \cdot \frac{1 \text{ pound}}{16 \cancel{\text{ounces}}} \\
&= \frac{256 \text{ pounds}}{16} \\
&= 16 \text{ pounds}
\end{aligned}
$$

SELF CHECK 7 How many pounds are in a 192-ounce bowling ball?

EXAMPLE 8 How many pounds of material are in a 3-ton automobile?

Solution

$$
\begin{aligned}
3 \text{ tons} &= 3 \cancel{\text{tons}} \cdot \frac{2{,}000 \text{ pounds}}{1 \cancel{\text{ton}}} \\
&= 3 \cdot 2{,}000 \text{ pounds} \\
&= 6{,}000 \text{ pounds}
\end{aligned}
$$

SELF CHECK 8 How many pounds of material are in a 2-ton automobile?

EXAMPLE 9 How much does a 25-pound Atlantic salmon weigh in ounces?

Solution

$$
\begin{aligned}
25 \text{ pounds} &= 25 \cancel{\text{pounds}} \cdot \frac{16 \text{ ounces}}{1 \cancel{\text{pound}}} \\
&= 25 \cdot 16 \text{ ounces} \\
&= 400 \text{ ounces}
\end{aligned}
$$

SELF CHECK 9 How much does a 20-pound Atlantic salmon weigh in ounces?

A.1.2 Measurement in the Metric System

The metric system is widely used internationally. In the United States, it is primarily used in scientific, medical, and engineering applications, but its use is slowly expanding. The metric system, like the decimal numeration system, is based on the number 10. This makes the conversions from one metric unit to another much easier than in the American system.

1 Identify the units of length in the metric system and convert from one unit of length to another.

The primary unit of length in the metric system is the **meter**. It is slightly longer than one American yard, or about 39 inches. The other units of length are obtained by multiplying or dividing the meter by powers of 10. The names of the units are given by adding a prefix to the basic unit, the meter.

1 kilometer (km) = 1,000 meters

1 hectometer (hm) = 100 meters

1 decameter (dam) = 10 meters

1 meter (m) = primary metric unit of length

1 decimeter (dm) = 0.1 meter

1 centimeter (cm) = 0.01 meter

1 millimeter (mm) = 0.001 meter

For conversions in the metric system, we could use a unit conversion factor in the same manner as in the American system. Instead, we use a much simpler method of shifting the decimal point of a measure left or right to change to larger or smaller units. The following chart illustrates how to convert among the metric units of length.

	mm	cm	dm	m	dam	hm	km
1 meter equals	1,000.0	100.0	10.0	1.0	0.1	0.01	0.001
Example	5,250.0			5.25			
	Initial unit	3 places to the *right* >----------------------->		Larger desired unit	If desired unit is 3 places to the *right*, then move the initial decimal point 3 places to the *left*.		
	Smaller desired unit	3 places to the *left* <-----------------------<		Initial unit	If desired unit is 3 places to the *left*, then move the initial decimal point 3 places to the *right*.		

The conversion requires counting the number of places—and noting the direction—between the desired unit and the initial unit. For example, if the initial measurement is expressed in millimeters and the desired unit is meters, we note the number of places and the direction (in the chart) between millimeters and meters. Meter units are 3 places *to the right* of millimeter units, so the decimal point must be shifted 3 places *to the left*. In the chart, 5,250.0 millimeters is converted to meters by moving its decimal point 3 places *to the left*, resulting in 5.25 meters.

COMMENT To convert a measure in any unit to a *larger* unit, move the decimal point of the initial measure to the *left*. To convert a measure in any unit to a *smaller* unit, move the decimal point of the initial measure to the *right*.

EXAMPLE 10 How high, in centimeters, is a 3-meter ceiling in a room?

Solution There are 2 places between centimeters and meters. To convert from a larger unit to a smaller unit, we move the decimal point 2 places to the right.

3 meters = 300 centimeters

SELF CHECK 10 How high, in centimeters, is a 4-meter ceiling in a room?

EXAMPLE 11 How long is a 5-kilometer race in meters?

Solution There are 3 places between kilometers and meters. To convert from a larger unit to a smaller unit, we move the decimal point 3 places to the right.

5 kilometers = 5,000 meters

SELF CHECK 11 How long is a 10-kilometer race in meters?

EXAMPLE 12 How many 6-centimeter-long nails can be cut from an 18-meter length of nail wire?

Solution There are 2 places between centimeters and meters. To convert from a larger unit to a smaller unit, we move the decimal point 2 places to the right.

18 meters = 1,800 centimeters

There will be 1,800/6 = 300 nails cut from each nail wire.

SELF CHECK 12 How many 5-centimeter-long nails can be cut from an 18-meter length of nail wire?

2 Identify the units of volume in the metric system and convert from one unit of volume to another.

The primary unit of volume (capacity) in the metric system is the **liter**. It is defined as the volume of a cube whose sides are each 10 centimeters long. (A special unit of volume, used in medical fields, is the **cubic centimeter**. It is defined as the volume of a cube whose sides are each 1 centimeter long and it is equivalent to 1 millimeter.) In the same manner as metric units of length, we obtain the other units of volume by multiplying or dividing the liter by powers of 10. The names of the units are given by adding a prefix to the basic unit, the liter.

1 kiloliter (kl) = 1,000 liters

1 hectoliter (hl) = 100 liters

1 decaliter (dal) = 10 liters

1 liter (l) = primary metric unit of volume

1 deciliter (dl) = 0.1 liter

1 centiliter (cl) = 0.01 liter

1 milliliter (ml) = 0.001 liter = 1 cm^3 (cubic centimeter)

Conversions of metric volume are performed in the same manner as conversions of metric length, by moving the decimal point left or right. The following chart illustrates how to convert among metric units of volume.

	ml	cl	dl	l	dal	hl	kl
1 liter equals	1,000.0	100.0	10.0	1.0	0.1	0.01	0.001
Example		800.0		8.0			
		Initial unit	2 places to the *right* >---------->	Larger desired unit	If desired unit is 2 places to the *right*, then move the initial decimal point 2 places to the *left*.		
		Smaller desired unit	2 places to the *left* <----------<	Initial unit	If desired unit is 2 places to the *left*, then move the initial decimal point 2 places to the *right*.		

For example, to convert from centiliters to liters, we note that liter units are 2 places *to the right* of centiliter units, so we move the decimal point of the initial measure 2 places *to the left*. In the chart, 800.0 centiliters is converted to liters by moving its decimal point 2 places to the left, resulting in 8 liters.

COMMENT Remember that to convert a measure in any unit to a *larger* unit, move the decimal point of the initial measure to the *left*. To convert a measure in any unit to a *smaller* unit, move the decimal point of the initial measure to the *right*.

EXAMPLE 13 How many centiliters are in a 2-liter bottle of juice?

Solution There are 2 places between centiliters and liters. To convert from a larger unit to a smaller unit, we move the decimal point 2 places to the right.

2 liters = 200 centiliters

SELF CHECK 13 How many centiliters are in a 1.5-liter bottle of juice?

EXAMPLE 14 How many kiloliters are in a 4,000-centiliter gas tank?

Solution There are 5 places between kiloliters and centiliters. To convert from a smaller unit to a larger unit, we move the decimal point 5 places to the left.

4,000 centiliters = 0.04 kiloliter

SELF CHECK 14 How many kiloliters are in a 3,000-centiliter gas tank?

EXAMPLE 15 How many milliliters (or cubic centimeters) are in $\frac{1}{2}$ liter of saline solution?

Solution There are 3 places between milliliters and liters. To convert from a larger unit to a smaller unit, we move the decimal point 3 places to the right.

0.5 liter = 500 milliliters

SELF CHECK 15 How many milliliters (or cubic centimeters) are in $\frac{3}{4}$ liter of saline solution?

3 Identify the units of mass (or weight) in the metric system and convert from one unit of mass to another.

The primary unit of mass in the metric system is the **gram**. It is defined as the weight of water in a cube whose sides are each 1 centimeter long. The names of the units are given by adding a prefix to the basic unit, the gram.

1 kilogram (kg) = 1,000 grams

1 hectogram (hg) = 100 grams

1 decagram (dag) = 10 grams

1 gram (g) = primary metric unit of mass

1 decigram (dg) = 0.1 gram

1 centigram (cg) = 0.01 gram

1 milligram (mg) = 0.001 gram

In the same manner as the metric units of length and volume, we obtain the other units of mass by multiplying or dividing the gram by powers of 10. We convert metric mass in the same manner as metric length and volume, by moving the decimal point left or right. The following chart illustrates how to convert among the metric units of mass.

		mg	cg	dg	g	dag	hg	kg
1 gram equals		1,000.0	100.0	10.0	1.0	0.1	0.01	0.001
Example					4,600.0			4.60
	If desired unit is 3 places to the *right*, then move the initial decimal point 3 places to the *left*.				Initial unit	3 places to the *right* >-------------------->		Larger desired unit
	If desired unit is 3 places to the *left*, then move the initial decimal point 3 places to the *right*.				Smaller desired unit	3 places to the *left* <--------------------<		Initial unit

COMMENT Remember that to convert a measure in any unit to a *larger* unit, move the decimal point of the initial measure to the *left*. To convert a measure in any unit to a *smaller* unit, move the decimal point of the initial measure to the *right*.

EXAMPLE 16 Four raisins weigh 5 grams. How many milligrams do they weigh?

Solution There are 3 places between milligrams and grams. To convert from a larger unit to a smaller unit, we move the decimal point 3 places to the right.

1 gram = 1,000 milligrams

Thus, four raisins weigh 5 grams or 5,000 milligrams.

SELF CHECK 16 Four raisins weigh 6 grams. How many milligrams do they weigh?

EXAMPLE 17 How many grams in a 6-kilogram bowling ball?

Solution There are 3 places between kilograms and grams. To convert from a larger unit to a smaller unit, we move the decimal point 3 places to the right.

6 kilograms = 6,000 grams

SELF CHECK 17 How many grams in a 4-kilogram bowling ball?

EXAMPLE 18 How many hectograms are in 27 centigrams?

Solution There are 4 places between kilograms and grams. To convert from a smaller unit to a larger unit, we move the decimal point 4 places to the left.

27 centigrams = 0.0027 hectogram

SELF CHECK 18 How many hectograms are in 30 centigrams?

A.1.3 Conversions Between Metric and American Systems

Consider the following questions:

- Is a 4-mile race longer than a 5-kilometer run?
- Is a 4-millimeter closed-end wrench larger than a $\frac{1}{2}$-inch socket wrench?
- Is a 5.2-liter engine smaller than a 352-cubic inch Hemi?

Because we may need to understand what a given quantity in one system actually means in the other system, we will look at the methods used for such conversions.

COMMENT The labels on many consumer goods often will show both American and metric measures for the package content. A soft conversion lists the American system first, 18 oz (510 g) for example, whereas a hard conversion lists the metric system first, such as 500 ml (16.9 fl oz).

1 Convert units of length from one system to another.

The following table illustrates common conversions of length between the American and metric systems. These can be used to create unit conversion factors.

Conversions of Length	
1 in. ≈ 2.54 cm	1 cm ≈ 0.39 in.
1 ft ≈ 0.30 m	1 m ≈ 3.28 ft
1 yd ≈ 0.91 m	1 m ≈ 1.09 yd
1 mi ≈ 1.61 km	1 km ≈ 0.62 mi

For any conversion, the unit conversion factor will be written as

$$\frac{\text{the numerator is written in units that you wish to find}}{\text{the denominator is written in units that you wish to remove}}$$

For example, to convert a measure of 87 feet to meters, the process would be

$$\begin{aligned} 87\text{ feet} &\approx 87\,\cancel{\text{feet}}\cdot\frac{1\text{ meter}}{3.28\,\cancel{\text{feet}}} \\ &\approx \frac{87\text{ meters}}{3.28} \\ &\approx 26.524\text{ meters} \end{aligned}$$

EXAMPLE 19 A bluefin tuna is about 6.5 feet long. How long is that in meters?

Solution

$$\begin{aligned} 6.5\text{ feet} &\approx 6.5\,\cancel{\text{feet}}\cdot\frac{1\text{ meter}}{3.28\,\cancel{\text{feet}}} \\ &\approx \frac{6.5\text{ meters}}{3.28} \\ &\approx 2\text{ meters} \end{aligned}$$

SELF CHECK 19 A bluefin tuna is about 6 feet long. How long is that in meters?

Converting lengths is important when working with miles per hour and kilometers per hour. Fortunately, for practical purposes, most speedometers have the equivalents posted for us.

2 Convert units of volume (or capacity) from one system to another.

The following table illustrates common conversions of volume between the American and metric systems. These can be used to create unit conversion factors.

Conversions of Volume (Capacity)	
1 fl oz ≈ 29.57 ml	1 l ≈ 33.81 fl oz
1 pt ≈ 0.47 l	1 l ≈ 2.11 pt
1 qt ≈ 0.95 l	1 l ≈ 1.06 qt
1 gal ≈ 3.79 l	1 l ≈ 0.264 gal

We construct a unit conversion factor for volumes in the same manner as for length. Remember that for any conversion, the unit conversion factor will be written as

$$\frac{\text{the numerator is written in units that you wish to find}}{\text{the denominator is written in units that you wish to remove}}$$

For example, to convert a measure of 7 gallons to liters, the process would be

$$\begin{aligned} 7 \text{ gallons} &\approx 7 \cancel{\text{gallons}} \cdot \frac{1 \text{ liter}}{0.264 \cancel{\text{gallon}}} \\ &\approx \frac{7 \text{ liters}}{0.264} \\ &\approx 26.515 \text{ liters} \end{aligned}$$

EXAMPLE 20 A 2-liter soft drink is equivalent to how many gallons?

Solution

$$\begin{aligned} 2 \text{ liters} &\approx 2 \cancel{\text{liters}} \cdot \frac{1 \text{ gallon}}{3.79 \cancel{\text{liters}}} \\ &\approx \frac{2 \text{ gallons}}{3.79} \\ &\approx \frac{1}{2} \text{ gallon} \end{aligned}$$

SELF CHECK 20 A 3-liter bottle is equivalent to how many gallons?

3 Convert units of weight (or mass) from one system to another.

The following table illustrates common conversions of weight between the American and metric systems. These can be used to create unit conversion factors.

Conversions of Weight (Mass)	
1 oz ≈ 28.35 g	1 g ≈ 0.035 oz
1 lb ≈ 0.5 kg	1 kg ≈ 2.20 lb

We construct a unit conversion factor for weight in the same manner as for length and volume. For example, to convert a measure of 4 pounds to kilograms, the process would be

$$\begin{aligned} 4 \text{ pounds} &\approx 4 \cancel{\text{pounds}} \cdot \frac{1 \text{ kilogram}}{2.20 \cancel{\text{pounds}}} \\ &\approx \frac{4 \text{ kilograms}}{2.20} \\ &\approx 1.818 \text{ kilograms} \end{aligned}$$

EXAMPLE 21 A bluefin tuna weighs about 250 kilograms. How much is that in pounds?

Solution

$$250 \text{ kilograms} \approx 250 \cancel{\text{kilograms}} \cdot \frac{2.20 \text{ pounds}}{\cancel{\text{kilograms}}}$$
$$\approx 250 \cdot 2.20 \text{ pounds}$$
$$\approx 550 \text{ pounds}$$

SELF CHECK 21 A bluefin tuna weighs about 440 pounds. How much is that in kilograms?

SELF CHECK ANSWERS

1. $8\frac{1}{3}$ yd **2.** 10,560 ft **3.** 6.6 mi **4.** 120 qt **5.** 64 fl oz **6.** 12 c **7.** 12 lb **8.** 4,000 lb **9.** 320 oz **10.** 400 cm **11.** 10,000 m **12.** 360 nails **13.** 150 cl **14.** 0.03 kl **15.** 750 cm^3 **16.** 6,000 mg **17.** 4,000 g **18.** 0.003 hg **19.** ≈1.8 m **20.** ≈0.8 gallon **21.** ≈200 kg

A.1 Exercises

Make the appropriate conversions.

1. Find, in feet, the height of a 27.3-yard tree.

2. Find the length of an Olympic-size swimming pool (50 yards) in feet.

3. Find the height of Mt. McKinley to the nearest tenth of a mile if it is 20,320 feet high.

4. The Atlanta Motor Speedway Nascar Track is a 1.54-mile oval. A major 500-mile race is generally run there each fall. How may laps does the race take and how many feet are covered in each lap?

5. In the Space Shuttle approximately 1,214,400 feet of wiring is necessary to maintain control of, and communication with, the spacecraft. How many miles of wire is this?

6. Grass sod is sold in 1 foot × 1 yard rectangles. If a front yard is 85 feet long, how many rectangles will it take to make 1 row lengthwise?

7. FAA rules require pilots flying in unpressurized aircraft to use supplemental oxygen for flights above 12,500 feet. What is that altitude to the nearest tenth of a mile?

8. How many gallons of oil are there in a 120-quart container?

9. A gasoline-powered lawn mower holds $\frac{1}{2}$ gallon of gas. How many cups is that?

10. A bottle of sports drink contains 32 ounces and comes in a case of 4 bottles. How many gallons are there in a case?

11. A case of soft drinks contains 24 12-ounce cans. How many 32-ounce bottles would that fill?

12. A pint of premium ice cream costs about $3.50. If the price per gallon were consistent with the price per pint, how much would $\frac{1}{2}$ gallon cost?

13. A woman bought 5 gallons of shelled pecans. If a pint freezer bag truly holds a pint, how many freezer bags will she need to store the 5 gallons of pecans?

14. A recipe for custard calls for 3 pints of milk. How many cups is that?

15. The Endeavor, the last Space Shuttle flown, weighed 4.5 million pounds at liftoff. Two minutes into the flight, it had already lost half that weight. What was its weight at the 2-minute mark in tons?

16. How many 4-ounce packages can be obtained from a 17.2-pound chunk of tuna?

17. To be granted agricultural status for tax purposes in Texas, land must support at least 1,000 pounds per acre of hoofed livestock. The XIT Ranch in Texas was once the largest in the United States, with 3,500,000 acres. To maintain its agricultural tax status, how many tons of cattle would have to be raised on the ranch? If an average grown cow weighs approximately 1,650 pounds, how many head of cattle would be needed?

18. The European Grand Prix is a race of 308.883 kilometers with speeds up to 360 kilometers per hour. If the race is 57 laps, how long is each lap in meters?

19. A heavy-duty trash bag is 2 millimeters thick. How thick is this in meters?

20. A 2-liter bottle of a diet cola contains how many milliliters?

21. A quarter-bottle of champagne holds 18.75 centiliters while the Nebuchadnezzar bottle holds 15 liters. How many centiliters does the Nebuchadenezzar hold?

22. In medicine, 1 cubic centimeter is equivalent to 1 milliliter. If a dose calls for 150 cubic centimeters, how many liters is that?

23. One Troy ounce of gold weighs 31.1034768 grams. If Chris bought 5 Troy ounces of gold, how many kilograms would he have?

24. A baby kangaroo weighs just 2 grams at birth but an adult can weigh up to 90 kilograms. How many grams can an adult weigh?

25. Approximately how many feet are there in one kilometer?

26. If gasoline is 69.4¢ per liter, about how much is gas per gallon?

27. A recipe calls for 340.2 grams of butter to make 12 croissants. How many ounces of butter is in each croissant?

APPENDIX 2: Symmetries of Graphs

There are several ways that a graph can exhibit symmetry about the coordinate axes and the origin. It is often easier to draw graphs of equations if we first find the x- and y-intercepts and find any of the following symmetries of the graph:

1. ***y*-axis symmetry:** If the point $(-x, y)$ lies on a graph whenever the point (x, y) does, as in Figure I-1(a), we say that the graph is **symmetric about the *y*-axis**.
2. **Symmetry about the origin:** If the point $(-x, -y)$ lies on the graph whenever the point (x, y) does, as in Figure I-1(b), we say that the graph is **symmetric about the origin**.
3. ***x*-axis symmetry:** If the point $(x, -y)$ lies on the graph whenever the point (x, y) does, as in Figure I-1(c), we say that the graph is **symmetric about the *x*-axis**.

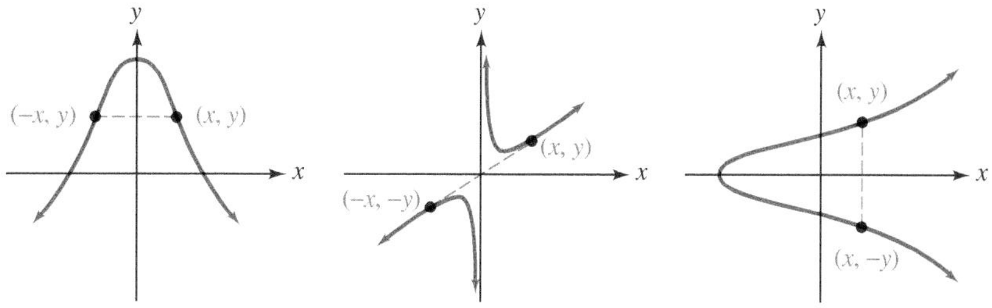

(a) (b) (c)

Figure I-1

Tests for Symmetry for Graphs in x and y

- To test a graph for y-axis symmetry, replace x with $-x$. If the new equation is equivalent to the original equation, the graph is symmetric about the y-axis. Symmetry about the y-axis will occur whenever x appears with only even exponents.
- To test a graph for symmetry about the origin, replace x with $-x$ and y with $-y$. If the resulting equation is equivalent to the original equation, the graph is symmetric about the origin.
- To test a graph for x-axis symmetry, replace y with $-y$. If the resulting equation is equivalent to the original equation, the graph is symmetric about the x-axis. The only function that is symmetric about the x-axis is $f(x) = 0$.

Unless otherwise noted, all content on this page is © Cengage Learning.

EXAMPLE 1 Find the intercepts and the symmetries of the graph of $y = f(x) = x^3 - 9x$. Then graph and state the domain and range.

Solution ***x*-intercepts:** To find the x-intercepts, we let $y = 0$ and solve for x.

$$y = x^3 - 9x$$

$$0 = x^3 - 9x \qquad \text{Substitute 0 for } y.$$

$$0 = x(x^2 - 9) \qquad \text{Factor out } x\text{, the GCF.}$$

$$0 = x(x + 3)(x - 3) \qquad \text{Factor } x^2 - 9.$$

$$x = 0 \quad \text{or} \quad x + 3 = 0 \quad \text{or} \quad x - 3 = 0 \qquad \text{Set each factor equal to 0.}$$

$$x = -3 \qquad\qquad x = 3$$

Since the x-coordinates of the x-intercepts are 0, −3, and 3, the graph intersects the x-axis at $(0, 0)$, $(-3, 0)$, and $(3, 0)$.

***y*-intercepts:** To find the y-intercepts, we let $x = 0$ and solve for y.

$$y = x^3 - 9x$$

$$y = 0^3 - 9(0) \qquad \text{Substitute 0 for } x.$$

$$y = 0$$

Since the y-coordinate of the y-intercept is 0, the graph intersects the y-axis at $(0, 0)$.

Symmetry: We test for symmetry *about the y-axis* by replacing x with $-x$, simplifying, and comparing the result to the original equation.

(1) $y = x^3 - 9x$ — This is the original equation.

$y = (-x)^3 - 9(-x)$ — Replace x with $-x$.

(2) $y = -x^3 + 9x$ — Simplify.

Because Equation 2 is not equivalent to Equation 1, the graph is not symmetric about the y-axis.

We test for symmetry *about the origin* by replacing x with $-x$ and y with $-y$, respectively, and comparing the result to the original equation.

(1) $y = x^3 - 9x$ — This is the original equation.

$-y = (-x)^3 - 9(-x)$ — Replace x with $-x$, and y with $-y$.

$-y = -x^3 + 9x$ — Simplify.

(3) $y = x^3 - 9x$ — Multiply both sides by −1 to solve for y.

Because Equation 3 is equivalent to Equation 1, the graph is symmetric about the origin.

Because the equation is the equation of a nonzero function, there is no symmetry *about the x-axis.*

To graph the equation, we plot the x-intercepts of $(-3, 0)$, $(0, 0)$, and $(3, 0)$ and the y-intercept of $(0, 0)$. We also plot other points for positive values of x and use the symmetry about the origin to draw the rest of the graph, as in Figure I-2(a). (Note that the scale on the x-axis is different from the scale on the y-axis.)

If we graph the equation with a graphing calculator, with window settings of $[-10, 10]$ for x and $[-10, 10]$ for y, we will obtain the graph shown in Figure I-2(b).

From the graph, we can see that the domain is the interval $(-\infty, \infty)$, and the range is the interval $(-\infty, \infty)$.

$f(x) = x^3 - 9x$

x	$f(x)$
0	0
1	-8
2	-10
3	0
4	28

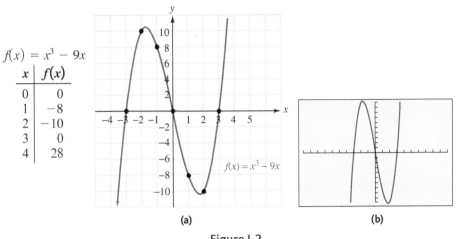

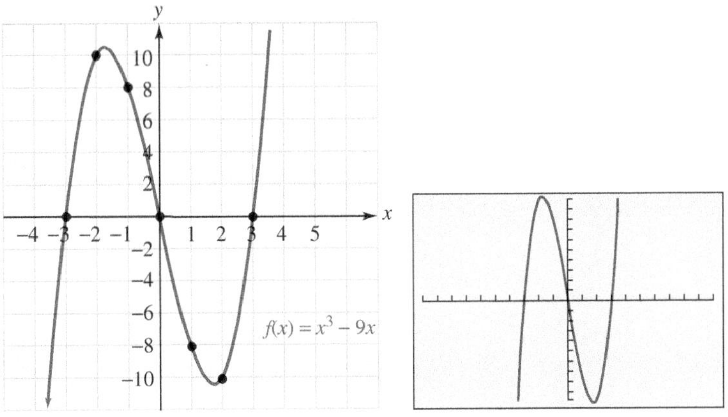

(a) (b)

Figure I-2

EXAMPLE 2 Graph the function $f(x) = |x| - 2$ and state the domain and range.

Solution ***x*-intercepts:** To find the x-intercepts, we let $y = 0$ and solve for x.

$$y = |x| - 2$$
$$\mathbf{0} = |x| - 2$$
$$2 = |x|$$
$$x = -2 \quad \text{or} \quad x = 2$$

Since -2 and 2 are solutions, the points $(-2, 0)$ and $(2, 0)$ are the x-intercepts, and the graph passes through $(-2, 0)$ and $(2, 0)$.

***y*-intercepts:** To find the y-intercepts, we let $x = 0$ and solve for y.

$$y = |x| - 2$$
$$y = |\mathbf{0}| - 2$$
$$y = -2$$

Since $y = -2$, $(0, -2)$ is the y-intercept, and the graph passes through the point $(0, -2)$.

Symmetry: To test for y-axis symmetry, we replace x with $-x$.

(4) $y = |x| - 2$ This is the original equation.

$y = |\mathbf{-x}| - 2$ Replace x with $-x$.

(5) $y = |x| - 2$ $|-x| = |x|$

Since Equation 5 is equivalent to Equation 4, the graph is symmetric about the y-axis. The graph has no other symmetries.

We plot the x- and y-intercepts and several other points (x, y), and use the y-axis symmetry to obtain the graph shown in Figure I-3(a) on the next page.

If we graph the equation with a graphing calculator, with window settings of $[-10, 10]$ for x and $[-10, 10]$ for y, we will obtain the graph shown in Figure I-3(b).

From the graph, we see that the domain is the interval $(-\infty, \infty)$, and the range is the interval $[-2, \infty)$.

Unless otherwise noted, all content on this page is © Cengage Learning.

$f(x) = |x| - 2$

x	$f(x)$
0	−2
1	−1
2	0
3	1
4	2

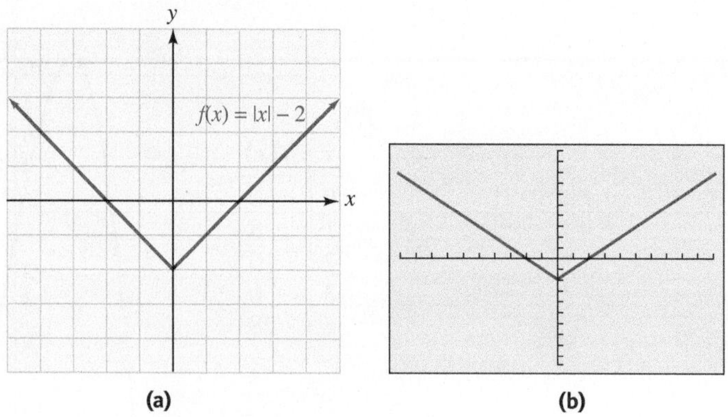

Figure I-3

A.2 Exercises

Find the symmetries of the graph of each equation. Do not draw the graph.

1. $y = x^2 - 1$

2. $y = x^3$

3. $y = x^5$

4. $y = x^4$

5. $y = -x^2 + 2$

6. $y = x^3 + 1$

7. $y = x^2 - x$

8. $y^2 = x + 7$

9. $y = -|x + 2|$

10. $y = |x| - 3$

11. $|y| = x$

12. $y = 2\sqrt{x}$

Graph each function and state its domain and range. Check each graph with a graphing calculator.

13. $f(x) = x^4 - 4$

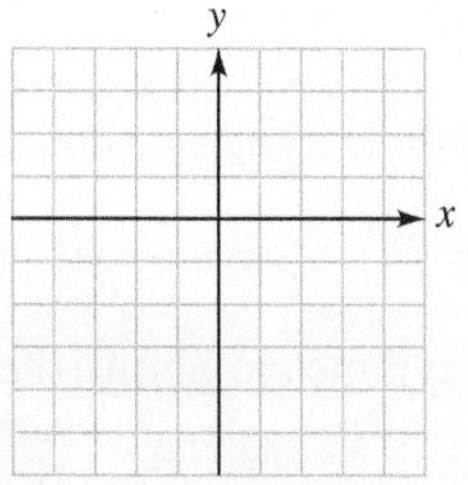

14. $f(x) = \frac{1}{2}x^4 - 1$

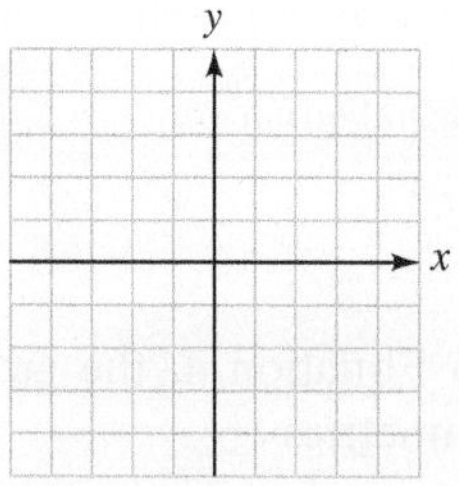

15. $f(x) = -x^3$

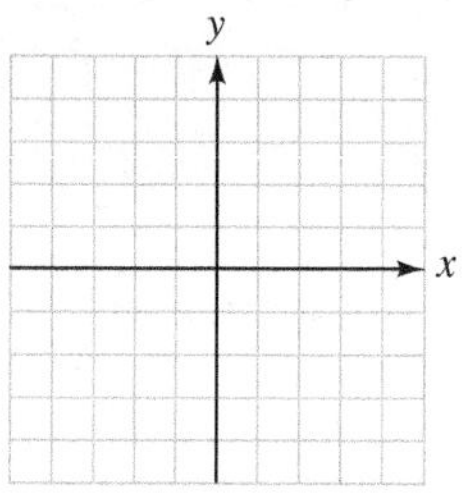

16. $f(x) = x^3 + 2$

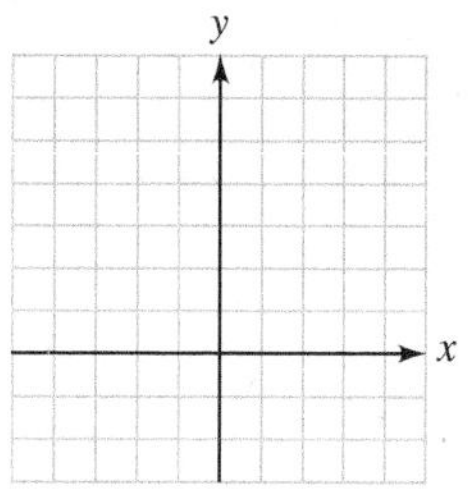

17. $f(x) = x^4 + x^2$

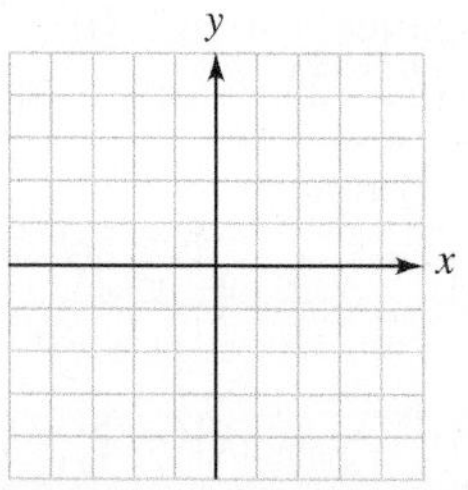

18. $f(x) = 3 - x^4$

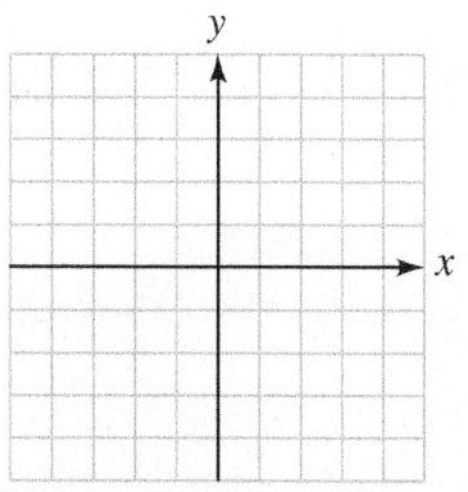

19. $f(x) = x^3 - x$

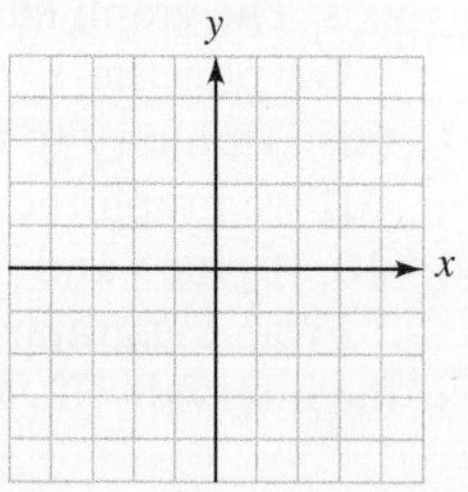

20. $f(x) = x^3 + x$

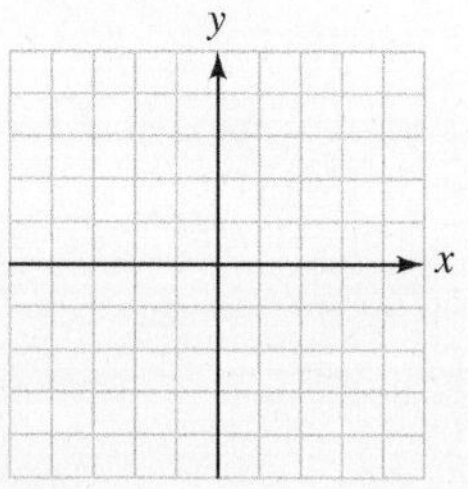

Unless otherwise noted, all content on this page is © Cengage Learning.

21. $f(x) = \frac{1}{2}|x| - 1$

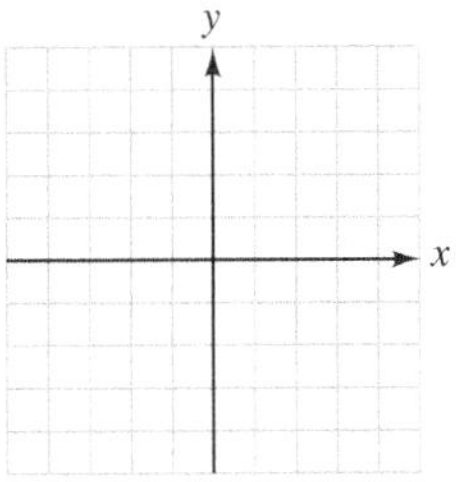

22. $f(x) = -|x| + 1$

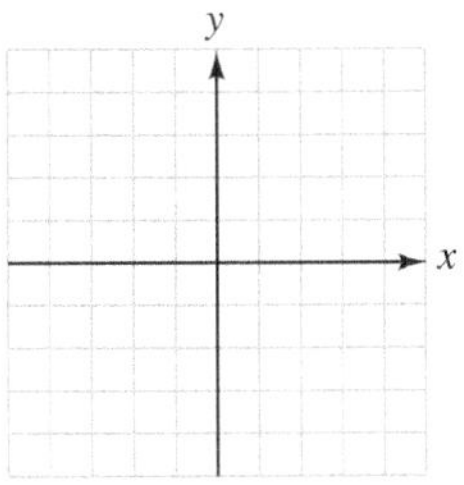

23. $f(x) = -|x + 2|$

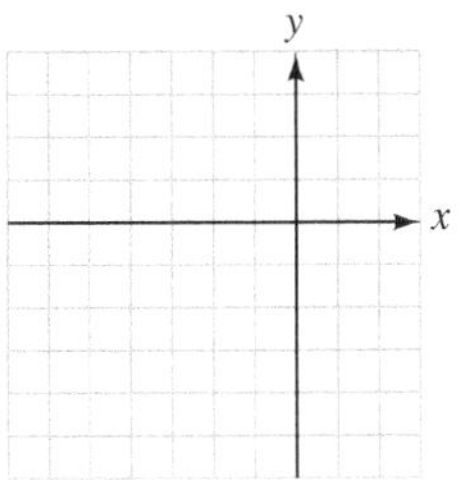

24. $f(x) = |x - 2|$

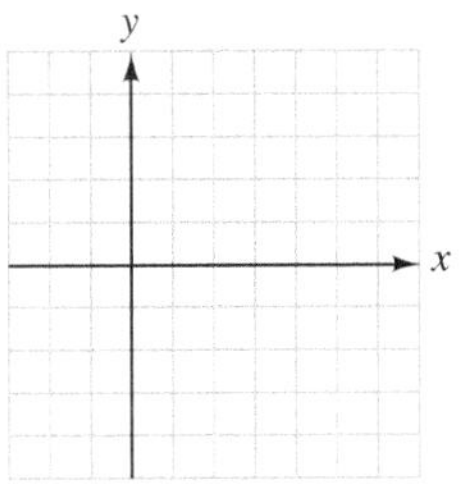

Unless otherwise noted, all content on this page is © Cengage Learning.

APPENDIX 3: Tables

TABLE A Powers and Roots

n	n^2	$\sqrt{n}$	n^3	$\sqrt[3]{n}$	n	n^2	$\sqrt{n}$	n^3	$\sqrt[3]{n}$
1	1	1.000	1	1.000	51	2,601	7.141	132,651	3.708
2	4	1.414	8	1.260	52	2,704	7.211	140,608	3.733
3	9	1.732	27	1.442	53	2,809	7.280	148,877	3.756
4	16	2.000	64	1.587	54	2,916	7.348	157,464	3.780
5	25	2.236	125	1.710	55	3,025	7.416	166,375	3.803
6	36	2.449	216	1.817	56	3,136	7.483	175,616	3.826
7	49	2.646	343	1.913	57	3,249	7.550	185,193	3.849
8	64	2.828	512	2.000	58	3,364	7.616	195,112	3.871
9	81	3.000	729	2.080	59	3,481	7.681	205,379	3.893
10	100	3.162	1,000	2.154	60	3,600	7.746	216,000	3.915
11	121	3.317	1,331	2.224	61	3,721	7.810	226,981	3.936
12	144	3.464	1,728	2.289	62	3,844	7.874	238,328	3.958
13	169	3.606	2,197	2.351	63	3,969	7.937	250,047	3.979
14	196	3.742	2,744	2.410	64	4,096	8.000	262,144	4.000
15	225	3.873	3,375	2.466	65	4,225	8.062	274,625	4.021
16	256	4.000	4,096	2.520	66	4,356	8.124	287,496	4.041
17	289	4.123	4,913	2.571	67	4,489	8.185	300,763	4.062
18	324	4.243	5,832	2.621	68	4,624	8.246	314,432	4.082
19	361	4.359	6,859	2.668	69	4,761	8.307	328,509	4.102
20	400	4.472	8,000	2.714	70	4,900	8.367	343,000	4.121
21	441	4.583	9,261	2.759	71	5,041	8.426	357,911	4.141
22	484	4.690	10,648	2.802	72	5,184	8.485	373,248	4.160
23	529	4.796	12,167	2.844	73	5,329	8.544	389,017	4.179
24	576	4.899	13,824	2.884	74	5,476	8.602	405,224	4.198
25	625	5.000	15,625	2.924	75	5,625	8.660	421,875	4.217
26	676	5.099	17,576	2.962	76	5,776	8.718	438,976	4.236
27	729	5.196	19,683	3.000	77	5,929	8.775	456,533	4.254
28	784	5.292	21,952	3.037	78	6,084	8.832	474,552	4.273
29	841	5.385	24,389	3.072	79	6,241	8.888	493,039	4.291
30	900	5.477	27,000	3.107	80	6,400	8.944	512,000	4.309
31	961	5.568	29,791	3.141	81	6,561	9.000	531,441	4.327
32	1,024	5.657	32,768	3.175	82	6,724	9.055	551,368	4.344
33	1,089	5.745	35,937	3.208	83	6,889	9.110	571,787	4.362
34	1,156	5.831	39,304	3.240	84	7,056	9.165	592,704	4.380
35	1,225	5.916	42,875	3.271	85	7,225	9.220	614,125	4.397
36	1,296	6.000	46,656	3.302	86	7,396	9.274	636,056	4.414
37	1,369	6.083	50,653	3.332	87	7,569	9.327	658,503	4.431
38	1,444	6.164	54,872	3.362	88	7,744	9.381	681,472	4.448
39	1,521	6.245	59,319	3.391	89	7,921	9.434	704,969	4.465
40	1,600	6.325	64,000	3.420	90	8,100	9.487	729,000	4.481
41	1,681	6.403	68,921	3.448	91	8,281	9.539	753,571	4.498
42	1,764	6.481	74,088	3.476	92	8,464	9.592	778,688	4.514
43	1,849	6.557	79,507	3.503	93	8,649	9.644	804,357	4.531
44	1,936	6.633	85,184	3.530	94	8,836	9.695	830,584	4.547
45	2,025	6.708	91,125	3.557	95	9,025	9.747	857,375	4.563
46	2,116	6.782	97,336	3.583	96	9,216	9.798	884,736	4.579
47	2,209	6.856	103,823	3.609	97	9,409	9.849	912,673	4.595
48	2,304	6.928	110,592	3.634	98	9,604	9.899	941,192	4.610
49	2,401	7.000	117,649	3.659	99	9,801	9.950	970,299	4.626
50	2,500	7.071	125,000	3.684	100	10,000	10.000	1,000,000	4.642

TABLE B Base-10 Logarithms

N	0	1	2	3	4	5	6	7	8	9
1.0	.0000	.0043	.0086	.0128	.0170	.0212	.0253	.0294	.0334	.0374
1.1	.0414	.0453	.0492	.0531	.0569	.0607	.0645	.0682	.0719	.0755
1.2	.0792	.0828	.0864	.0899	.0934	.0969	.1004	.1038	.1072	.1106
1.3	.1139	.1173	.1206	.1239	.1271	.1303	.1335	.1367	.1399	.1430
1.4	.1461	.1492	.1523	.1553	.1584	.1614	.1644	.1673	.1703	.1732
1.5	.1761	.1790	.1818	.1847	.1875	.1903	.1931	.1959	.1987	.2014
1.6	.2041	.2068	.2095	.2122	.2148	.2175	.2201	.2227	.2253	.2279
1.7	.2304	.2330	.2355	.2380	.2405	.2430	.2455	.2480	.2504	.2529
1.8	.2553	.2577	.2601	.2625	.2648	.2672	.2695	.2718	.2742	.2765
1.9	.2788	.2810	.2833	.2856	.2878	.2900	.2923	.2945	.2967	.2989
2.0	.3010	.3032	.3054	.3075	.3096	.3118	.3139	.3160	.3181	.3201
2.1	.3222	.3243	.3263	.3284	.3304	.3324	.3345	.3365	.3385	.3404
2.2	.3424	.3444	.3464	.3483	.3502	.3522	.3541	.3560	.3579	.3598
2.3	.3617	.3636	.3655	.3674	.3692	.3711	.3729	.3747	.3766	.3784
2.4	.3802	.3820	.3838	.3856	.3874	.3892	.3909	.3927	.3945	.3962
2.5	.3979	.3997	.4014	.4031	.4048	.4065	.4082	.4099	.4116	.4133
2.6	.4150	.4166	.4183	.4200	.4216	.4232	.4249	.4265	.4281	.4298
2.7	.4314	.4330	.4346	.4362	.4378	.4393	.4409	.4425	.4440	.4456
2.8	.4472	.4487	.4502	.4518	.4533	.4548	.4564	.4579	.4594	.4609
2.9	.4624	.4639	.4654	.4669	.4683	.4698	.4713	.4728	.4742	.4757
3.0	.4771	.4786	.4800	.4814	.4829	.4843	.4857	.4871	.4886	.4900
3.1	.4914	.4928	.4942	.4955	.4969	.4983	.4997	.5011	.5024	.5038
3.2	.5051	.5065	.5079	.5092	.5105	.5119	.5132	.5145	.5159	.5172
3.3	.5185	.5198	.5211	.5224	.5237	.5250	.5263	.5276	.5289	.5302
3.4	.5315	.5328	.5340	.5353	.5366	.5378	.5391	.5403	.5416	.5428
3.5	.5441	.5453	.5465	.5478	.5490	.5502	.5514	.5527	.5539	.5551
3.6	.5563	.5575	.5587	.5599	.5611	.5623	.5635	.5647	.5658	.5670
3.7	.5682	.5694	.5705	.5717	.5729	.5740	.5752	.5763	.5775	.5786
3.8	.5798	.5809	.5821	.5832	.5843	.5855	.5866	.5877	.5888	.5899
3.9	.5911	.5922	.5933	.5944	.5955	.5966	.5977	.5988	.5999	.6010
4.0	.6021	.6031	.6042	.6053	.6064	.6075	.6085	.6096	.6107	.6117
4.1	.6128	.6138	.6149	.6160	.6170	.6180	.6191	.6201	.6212	.6222
4.2	.6232	.6243	.6253	.6263	.6274	.6284	.6294	.6304	.6314	.6325
4.3	.6335	.6345	.6355	.6365	.6375	.6385	.6395	.6405	.6415	.6425
4.4	.6435	.6444	.6454	.6464	.6474	.6484	.6493	.6503	.6513	.6522
4.5	.6532	.6542	.6551	.6561	.6571	.6580	.6590	.6599	.6609	.6618
4.6	.6628	.6637	.6646	.6656	.6665	.6675	.6684	.6693	.6702	.6712
4.7	.6721	.6730	.6739	.6749	.6758	.6767	.6776	.6785	.6794	.6803
4.8	.6812	.6821	.6830	.6839	.6848	.6857	.6866	.6875	.6884	.6893
4.9	.6902	.6911	.6920	.6928	.6937	.6946	.6955	.6964	.6972	.6981
5.0	.6990	.6998	.7007	.7016	.7024	.7033	.7042	.7050	.7059	.7067
5.1	.7076	.7084	.7093	.7101	.7110	.7118	.7126	.7135	.7143	.7152
5.2	.7160	.7168	.7177	.7185	.7193	.7202	.7210	.7218	.7226	.7235
5.3	.7243	.7251	.7259	.7267	.7275	.7284	.7292	.7300	.7308	.7316
5.4	.7324	.7332	.7340	.7348	.7356	.7364	.7372	.7380	.7388	.7396

TABLE B (continued)

N	0	1	2	3	4	5	6	7	8	9
5.5	.7404	.7412	.7419	.7427	.7435	.7443	.7451	.7459	.7466	.7474
5.6	.7482	.7490	.7497	.7505	.7513	.7520	.7528	.7536	.7543	.7551
5.7	.7559	.7566	.7574	.7582	.7589	.7597	.7604	.7612	.7619	.7627
5.8	.7634	.7642	.7649	.7657	.7664	.7672	.7679	.7686	.7694	.7701
5.9	.7709	.7716	.7723	.7731	.7738	.7745	.7752	.7760	.7767	.7774
6.0	.7782	.7789	.7796	.7803	.7810	.7818	.7825	.7832	.7839	.7846
6.1	.7853	.7860	.7868	.7875	.7882	.7889	.7896	.7903	.7910	.7917
6.2	.7924	.7931	.7938	.7945	.7952	.7959	.7966	.7973	.7980	.7987
6.3	.7993	.8000	.8007	.8014	.8021	.8028	.8035	.8041	.8048	.8055
6.4	.8062	.8069	.8075	.8082	.8089	.8096	.8102	.8109	.8116	.8122
6.5	.8129	.8136	.8142	.8149	.8156	.8162	.8169	.8176	.8182	.8189
6.6	.8195	.8202	.8209	.8215	.8222	.8228	.8235	.8241	.8248	.8254
6.7	.8261	.8267	.8274	.8280	.8287	.8293	.8299	.8306	.8312	.8319
6.8	.8325	.8331	.8338	.8344	.8351	.8357	.8363	.8370	.8376	.8382
6.9	.8388	.8395	.8401	.8407	.8414	.8420	.8426	.8432	.8439	.8445
7.0	.8451	.8457	.8463	.8470	.8476	.8482	.8488	.8494	.8500	.8506
7.1	.8513	.8519	.8525	.8531	.8537	.8543	.8549	.8555	.8561	.8567
7.2	.8573	.8579	.8585	.8591	.8597	.8603	.8609	.8615	.8621	.8627
7.3	.8633	.8639	.8645	.8651	.8657	.8663	.8669	.8675	.8681	.8686
7.4	.8692	.8698	.8704	.8710	.8716	.8722	.8727	.8733	.8739	.8745
7.5	.8751	.8756	.8762	.8768	.8774	.8779	.8785	.8791	.8797	.8802
7.6	.8808	.8814	.8820	.8825	.8831	.8837	.8842	.8848	.8854	.8859
7.7	.8865	.8871	.8876	.8882	.8887	.8893	.8899	.8904	.8910	.8915
7.8	.8921	.8927	.8932	.8938	.8943	.8949	.8954	.8960	.8965	.8971
7.9	.8976	.8982	.8987	.8993	.8998	.9004	.9009	.9015	.9020	.9025
8.0	.9031	.9036	.9042	.9047	.9053	.9058	.9063	.9069	.9074	.9079
8.1	.9085	.9090	.9096	.9101	.9106	.9112	.9117	.9122	.9128	.9133
8.2	.9138	.9143	.9149	.9154	.9159	.9165	.9170	.9175	.9180	.9186
8.3	.9191	.9196	.9201	.9206	.9212	.9217	.9222	.9227	.9232	.9238
8.4	.9243	.9248	.9253	.9258	.9263	.9269	.9274	.9279	.9284	.9289
8.5	.9294	.9299	.9304	.9309	.9315	.9320	.9325	.9330	.9335	.9340
8.6	.9345	.9350	.9355	.9360	.9365	.9370	.9375	.9380	.9385	.9390
8.7	.9395	.9400	.9405	.9410	.9415	.9420	.9425	.9430	.9435	.9440
8.8	.9445	.9450	.9455	.9460	.9465	.9469	.9474	.9479	.9484	.9489
8.9	.9494	.9499	.9504	.9509	.9513	.9518	.9523	.9528	.9533	.9538
9.0	.9542	.9547	.9552	.9557	.9562	.9566	.9571	.9576	.9581	.9586
9.1	.9590	.9595	.9600	.9605	.9609	.9614	.9619	.9624	.9628	.9633
9.2	.9638	.9643	.9647	.9652	.9657	.9661	.9666	.9671	.9675	.9680
9.3	.9685	.9689	.9694	.9699	.9703	.9708	.9713	.9717	.9722	.9727
9.4	.9731	.9736	.9741	.9745	.9750	.9754	.9759	.9763	.9768	.9773
9.5	.9777	.9782	.9786	.9791	.9795	.9800	.9805	.9809	.9814	.9818
9.6	.9823	.9827	.9832	.9836	.9841	.9845	.9850	.9854	.9859	.9863
9.7	.9868	.9872	.9877	.9881	.9886	.9890	.9894	.9899	.9903	.9908
9.8	.9912	.9917	.9921	.9926	.9930	.9934	.9939	.9943	.9948	.9952
9.9	.9956	.9961	.9965	.9969	.9974	.9978	.9983	.9987	.9991	.9996

TABLE C Base-*e* Logarithms

N	0	1	2	3	4	5	6	7	8	9
1.0	.0000	.0100	.0198	.0296	.0392	.0488	.0583	.0677	.0770	.0862
1.1	.0953	.1044	.1133	.1222	.1310	.1398	.1484	.1570	.1655	.1740
1.2	.1823	.1906	.1989	.2070	.2151	.2231	.2311	.2390	.2469	.2546
1.3	.2624	.2700	.2776	.2852	.2927	.3001	.3075	.3148	.3221	.3293
1.4	.3365	.3436	.3507	.3577	.3646	.3716	.3784	.3853	.3920	.3988
1.5	.4055	.4121	.4187	.4253	.4318	.4383	.4447	.4511	.4574	.4637
1.6	.4700	.4762	.4824	.4886	.4947	.5008	.5068	.5128	.5188	.5247
1.7	.5306	.5365	.5423	.5481	.5539	.5596	.5653	.5710	.5766	.5822
1.8	.5878	.5933	.5988	.6043	.6098	.6152	.6206	.6259	.6313	.6366
1.9	.6419	.6471	.6523	.6575	.6627	.6678	.6729	.6780	.6831	.6881
2.0	.6931	.6981	.7031	.7080	.7129	.7178	.7227	.7275	.7324	.7372
2.1	.7419	.7467	.7514	.7561	.7608	.7655	.7701	.7747	.7793	.7839
2.2	.7885	.7930	.7975	.8020	.8065	.8109	.8154	.8198	.8242	.8286
2.3	.8329	.8372	.8416	.8459	.8502	.8544	.8587	.8629	.8671	.8713
2.4	.8755	.8796	.8838	.8879	.8920	.8961	.9002	.9042	.9083	.9123
2.5	.9163	.9203	.9243	.9282	.9322	.9361	.9400	.9439	.9478	.9517
2.6	.9555	.9594	.9632	.9670	.9708	.9746	.9783	.9821	.9858	.9895
2.7	.9933	.9969	1.0006	.0043	.0080	.0116	.0152	.0188	.0225	.0260
2.8	1.0296	.0332	.0367	.0403	.0438	.0473	.0508	.0543	.0578	.0613
2.9	.0647	.0682	.0716	.0750	.0784	.0818	.0852	.0886	.0919	.0953
3.0	1.0986	.1019	.1053	.1086	.1119	.1151	.1184	.1217	.1249	.1282
3.1	.1314	.1346	.1378	.1410	.1442	.1474	.1506	.1537	.1569	.1600
3.2	.1632	.1663	.1694	.1725	.1756	.1787	.1817	.1848	.1878	.1909
3.3	.1939	.1969	.2000	.2030	.2060	.2090	.2119	.2149	.2179	.2208
3.4	.2238	.2267	.2296	.2326	.2355	.2384	.2413	.2442	.2470	.2499
3.5	1.2528	.2556	.2585	.2613	.2641	.2669	.2698	.2726	.2754	.2782
3.6	.2809	.2837	.2865	.2892	.2920	.2947	.2975	.3002	.3029	.3056
3.7	.3083	.3110	.3137	.3164	.3191	.3218	.3244	.3271	.3297	.3324
3.8	.3350	.3376	.3403	.3429	.3455	.3481	.3507	.3533	.3558	.3584
3.9	.3610	.3635	.3661	.3686	.3712	.3737	.3762	.3788	.3813	.3838
4.0	1.3863	.3888	.3913	.3938	.3962	.3987	.4012	.4036	.4061	.4085
4.1	.4110	.4134	.4159	.4183	.4207	.4231	.4255	.4279	.4303	.4327
4.2	.4351	.4375	.4398	.4422	.4446	.4469	.4493	.4516	.4540	.4563
4.3	.4586	.4609	.4633	.4656	.4679	.4702	.4725	.4748	.4770	.4793
4.4	.4816	.4839	.4861	.4884	.4907	.4929	.4951	.4974	.4996	.5019
4.5	1.5041	.5063	.5085	.5107	.5129	.5151	.5173	.5195	.5217	.5239
4.6	.5261	.5282	.5304	.5326	.5347	.5369	.5390	.5412	.5433	.5454
4.7	.5476	.5497	.5518	.5539	.5560	.5581	.5602	.5623	.5644	.5665
4.8	.5686	.5707	.5728	.5748	.5769	.5790	.5810	.5831	.5851	.5872
4.9	.5892	.5913	.5933	.5953	.5974	.5994	.6014	.6034	.6054	.6074
5.0	1.6094	.6114	.6134	.6154	.6174	.6194	.6214	.6233	.6253	.6273
5.1	.6292	.6312	.6332	.6351	.6371	.6390	.6409	.6429	.6448	.6467
5.2	.6487	.6506	.6525	.6544	.6563	.6582	.6601	.6620	.6639	.6658
5.3	.6677	.6696	.6715	.6734	.6752	.6771	.6790	.6808	.6827	.6845
5.4	.6864	.6882	.6901	.6919	.6938	.6956	.6974	.6993	.7011	.7029

Use the properties of logarithms and ln 10 ≈ 2.3026 to find logarithms of numbers less than 1 or greater than 10.

TABLE C (*continued*)

N	0	1	2	3	4	5	6	7	8	9
5.5	1.7047	.7066	.7084	.7102	.7120	.7138	.7156	.7174	.7192	.7210
5.6	.7228	.7246	.7263	.7281	.7299	.7317	.7334	.7352	.7370	.7387
5.7	.7405	.7422	.7440	.7457	.7475	.7492	.7509	.7527	.7544	.7561
5.8	.7579	.7596	.7613	.7630	.7647	.7664	.7681	.7699	.7716	.7733
5.9	.7750	.7766	.7783	.7800	.7817	.7834	.7851	.7867	.7884	.7901
6.0	1.7918	.7934	.7951	.7967	.7984	.8001	.8017	.8034	.8050	.8066
6.1	.8083	.8099	.8116	.8132	.8148	.8165	.8181	.8197	.8213	.8229
6.2	.8245	.8262	.8278	.8294	.8310	.8326	.8342	.8358	.8374	.8390
6.3	.8405	.8421	.8437	.8453	.8469	.8485	.8500	.8516	.8532	.8547
6.4	.8563	.8579	.8594	.8610	.8625	.8641	.8656	.8672	.8687	.8703
6.5	1.8718	.8733	.8749	.8764	.8779	.8795	.8810	.8825	.8840	.8856
6.6	.8871	.8886	.8901	.8916	.8931	.8946	.8961	.8976	.8991	.9006
6.7	.9021	.9036	.9051	.9066	.9081	.9095	.9110	.9125	.9140	.9155
6.8	.9169	.9184	.9199	.9213	.9228	.9242	.9257	.9272	.9286	.9301
6.9	.9315	.9330	.9344	.9359	.9373	.9387	.9402	.9416	.9430	.9445
7.0	1.9459	.9473	.9488	.9502	.9516	.9530	.9544	.9559	.9573	.9587
7.1	.9601	.9615	.9629	.9643	.9657	.9671	.9685	.9699	.9713	.9727
7.2	.9741	.9755	.9769	.9782	.9796	.9810	.9824	.9838	.9851	.9865
7.3	.9879	.9892	.9906	.9920	.9933	.9947	.9961	.9974	.9988	2.0001
7.4	2.0015	.0028	.0042	.0055	.0069	.0082	.0096	.0109	.0122	.0136
7.5	2.0149	.0162	.0176	.0189	.0202	.0215	.0229	.0242	.0255	.0268
7.6	.0281	.0295	.0308	.0321	.0334	.0347	.0360	.0373	.0386	.0399
7.7	.0412	.0425	.0438	.0451	.0464	.0477	.0490	.0503	.0516	.0528
7.8	.0541	.0554	.0567	.0580	.0592	.0605	.0618	.0631	.0643	.0656
7.9	.0669	.0681	.0694	.0707	.0719	.0732	.0744	.0757	.0769	.0782
8.0	2.0794	.0807	.0819	.0832	.0844	.0857	.0869	.0882	.0894	.0906
8.1	.0919	.0931	.0943	.0956	.0968	.0980	.0992	.1005	.1017	.1029
8.2	.1041	.1054	.1066	.1078	.1090	.1102	.1114	.1126	.1138	.1150
8.3	.1163	.1175	.1187	.1199	.1211	.1223	.1235	.1247	.1258	.1270
8.4	.1282	.1294	.1306	.1318	.1330	.1342	.1353	.1365	.1377	.1389
8.5	2.1401	.1412	.1424	.1436	.1448	.1459	.1471	.1483	.1494	.1506
8.6	.1518	.1529	.1541	.1552	.1564	.1576	.1587	.1599	.1610	.1622
8.7	.1633	.1645	.1656	.1668	.1679	.1691	.1702	.1713	.1725	.1736
8.8	.1748	.1759	.1770	.1782	.1793	.1804	.1815	.1827	.1838	.1849
8.9	.1861	.1872	.1883	.1894	.1905	.1917	.1928	.1939	.1950	.1961
9.0	2.1972	.1983	.1994	.2006	.2017	.2028	.2039	.2050	.2061	.2072
9.1	.2083	.2094	.2105	.2116	.2127	.2138	.2148	.2159	.2170	.2181
9.2	.2192	.2203	.2214	.2225	.2235	.2246	.2257	.2268	.2279	.2289
9.3	.2300	.2311	.2322	.2332	.2343	.2354	.2364	.2375	.2386	.2396
9.4	.2407	.2418	.2428	.2439	.2450	.2460	.2471	.2481	.2492	.2502
9.5	2.2513	.2523	.2534	.2544	.2555	.2565	.2576	.2586	.2597	.2607
9.6	.2618	.2628	.2638	.2649	.2659	.2670	.2680	.2690	.2701	.2711
9.7	.2721	.2732	.2742	.2752	.2762	.2773	.2783	.2793	.2803	.2814
9.8	.2824	.2834	.2844	.2854	.2865	.2875	.2885	.2895	.2905	.2915
9.9	.2925	.2935	.2946	.2956	.2966	.2976	.2986	.2996	.3006	.3016

APPENDIX 4: Answers to Selected Exercises

Getting Ready (page 3)

1. 1, 2, 3, etc. **2.** $\frac{1}{2}, \frac{2}{3}$, etc. **3.** $-3, -21$, etc.

Exercises 1.1 (page 11)

11. -7 **13.** set **15.** whole **17.** integers **19.** subset **21.** rational **23.** real **25.** natural, prime **27.** odd **29.** $<$ **31.** variables **33.** 7 **35.** parenthesis, open **37.** distance, 6 **39.** 1, 2, 6, 9 **41.** 1, 2, 6, 9 **43.** $-3, -1, 0, 1, 2, 6, 9$ **45.** $-3, -\frac{1}{2}, -1, 0, 1, 2, \frac{5}{3}, \sqrt{7}$, 3.25, 6, 9 **47.** $-3, -1, 1, 9$ **49.** 6, 9 **51.** $<$ **53.** $>$ **55.** $>$ **57.** $>$

59. 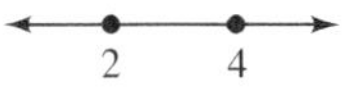

; 4, 4 **61.** 

; 11, 11

63. 

; –2, –2 **65.** 

; 8, 8

67. 

69. 

71. 

73. 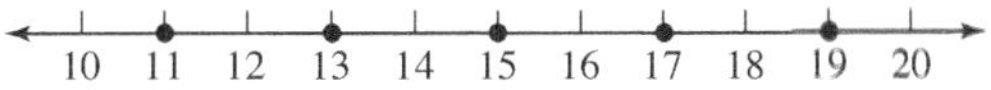

75. 36 **77.** 0 **79.** -23 **81.** 8 **83.** 9; natural, odd, composite, and whole number **85.** 0; even integer, whole number **87.** 24; natural, even, composite, and whole number **89.** 3; natural, odd, prime, and whole number **91.** $<$ **93.** $=$ **95.** $<$ **97.** $>$ **99.** $=$ **101.** $9 > 4$ **103.** $8 \le 8$ **105.** $3 + 4 = 7$ **107.** $\sqrt{2} \approx 1.41$ **109.** $7 \ge 3$ **111.** $0 < 6$ **113.** $8 < 3 + 8$ **115.** $10 - 4 > 6 - 2$ **117.** $3 \cdot 4 > 2 \cdot 3$ **119.** $\frac{24}{6} > \frac{12}{4}$

121. 

123. 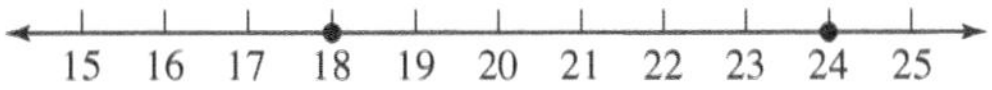

125. 

127. 2

Getting Ready (page 13)

1. 250 **2.** 148 **3.** 16,606 **4.** 105

Exercises 1.2 (page 25)

1. 3 **3.** 6 **5.** $\frac{3}{8}$ **7.** $\frac{9}{16}$ **9.** $\frac{11}{9}$ **11.** $\frac{1}{6}$ **13.** 5.72 **15.** 0.5 **17.** 5.17 **19.** true **21.** false **23.** false **25.** true **27.** $=$ **29.** $=$ **31.** numerator **33.** undefined **35.** prime **37.** improper **39.** 1 **41.** multiply **43.** numerators, denominator **45.** least common denominator, equivalent **47.** terminating, 2 **49.** divisor, dividend, quotient **51.** $2 \cdot 2 \cdot 2 \cdot 3$ **53.** $2 \cdot 2 \cdot 2 \cdot 2 \cdot 3$ **55.** $\frac{1}{2}$ **57.** $\frac{3}{4}$ **59.** $\frac{3}{2}$ **61.** $\frac{9}{8}$ **63.** $\frac{2}{15}$ **65.** $\frac{8}{5}$ **67.** 10 **69.** $\frac{20}{3}$ **71.** $\frac{4}{15}$ **73.** $\frac{5}{8}$ **75.** 24 **77.** $\frac{9}{10}$ **79.** $\frac{6}{5}$ **81.** $\frac{2}{17}$ **83.** $\frac{4}{21}$ **85.** $\frac{22}{35}$ **87.** $5\frac{1}{5}$ **89.** $1\frac{2}{3}$ **91.** $1\frac{1}{4}$ **93.** $\frac{5}{9}$ **95.** 0.6, terminating **97.** $0.4\overline{09}$, repeating **99.** 359.24 **101.** 44.785 **103.** 112.32 **105.** 4.55 **107.** 496.26; 496.258 **109.** 6,025.40; 6,025.398 **111.** $\frac{3}{2}$ **113.** $\frac{1}{4}$ **115.** $\frac{1}{4}$ **117.** $\frac{14}{5}$ **119.** $\frac{19}{15}$ **121.** $\frac{17}{12}$ **123.** $\frac{9}{4}$ **125.** $\frac{29}{3}$ **127.** 498.26 **129.** 3,337.52 **131.** 10.02 **133.** 55.21 **135.** $31\frac{1}{6}$ acres **137.** 65 yd **139.** \$68.45 million **141.** \$12,240 **143.** 13,475 **145.** \$20,944,000 **147.** \$2,201.95 **149.** \$1,170 **151.** the high-capacity boards **153.** 205,200 lb **155.** the high-efficiency furnace

Getting Ready (page 29)

1. 4 **2.** 9 **3.** 27 **4.** 8 **5.** $\frac{1}{4}$ **6.** $\frac{1}{27}$ **7.** $\frac{8}{125}$ **8.** $\frac{27}{1{,}000}$

Exercises 1.3 (page 36)

1. 32 **3.** 64 **5.** $\frac{8}{27}$ **7.** y **9.** $4x$

11. 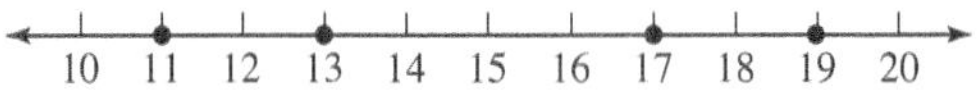

13. prime number **15.** exponent **17.** grouping **19.** perimeter, circumference **21.** $P = 4s$; units **23.** $P = 2l + 2w$; units **25.** $P = a + b + c$; units **27.** $P = a + b + c + d$; units **29.** $C = \pi D$ or $C = 2\pi r$; units **31.** $V = lwh$; cubic units **33.** $V = \frac{1}{3}Bh$; cubic units **35.** $V = \frac{4}{3}\pi r^3$; cubic units **37.** $6 \cdot 6$; 36 **39.** $\left(-\frac{1}{5}\right)\left(-\frac{1}{5}\right)\left(-\frac{1}{5}\right)\left(-\frac{1}{5}\right)$; $\frac{1}{625}$ **41.** $x \cdot x \cdot x$ **43.** $8 \cdot z \cdot z \cdot z \cdot z$ **45.** $4x \cdot 4x \cdot 4x$ **47.** $3 \cdot 6y \cdot 6y$ **49.** 36 **51.** 10,000 **53.** 80 **55.** 216 **57.** 11 **59.** 3 **61.** 13 **63.** 8 **65.** 17 **67.** 8 **69.** $\frac{1}{144}$ **71.** 11 **73.** 1 **75.** 1 **77.** 20 in. **79.** 15 m **81.** 36 m^2 **83.** 55 ft^2 **85.** approx. 88 m **87.** approx. 1,386 ft^2 **89.** 6 cm^3 **91.** approx. 905 m^3 **93.** approx. 1,056 cm^3 **95.** 36 **97.** 18 **99.** 36 **101.** 2 **103.** 16 **105.** 21 **107.** 11 **109.** 8 **111.** $\frac{8}{9}$ **113.** 493.039 **115.** 640.09 **117.** $(3 \cdot 8) + (5 \cdot 3)$ **119.** $(3 \cdot 8 + 5) \cdot 3$ **121.** 40,764.51 ft^3 **123.** $121\frac{3}{5}$ m **125.** 480 ft^3 **127.** 8 **131.** bigger

Getting Ready (page 39)

1. 17.52 **2.** 2.94 **3.** 2 **4.** 1 **5.** 96 **6.** 382

Exercises 1.4 (page 45)

1. 5 **3.** 3 **5.** 4 **7.** 2 **9.** 20 **11.** 24 **13.** arrows **15.** unlike **17.** subtract, greater

Unless otherwise noted, all content on this page is © Cengage Learning.

19. add, opposite **21.** 14 **23.** -9 **25.** $\frac{12}{35}$
27. 78 **29.** 4 **31.** 0.5 **33.** $\frac{5}{12}$ **35.** -34.58
37. 7 **39.** -1 **41.** 0 **43.** -19 **45.** 1 **47.** 2.2
49. 4 **51.** 12 **53.** 5 **55.** $\frac{1}{2}$ **57.** 4 **59.** -7
61. 11 **63.** 12 **65.** 1 **67.** 1 **69.** -1.52
71. -7.08 **73.** 3 **75.** 2.45 **77.** -8 **79.** -7
81. -8 **83.** 3 **85.** 1.3 **87.** $-8\frac{3}{4}$ **89.** -4.2
91. -6 **93.** $-\frac{29}{30}$ **95.** \$235 **97.** $+9$ **99.** $-4°$
101. 2,000 yr **103.** 1,325 m **105.** 4,000 ft **107.** 5°
109. 12,187 **111.** 700 **113.** \$422.66 **115.** \$83,425.57

Getting Ready (page 47)

1. 56 **2.** 54 **3.** 72 **4.** 63 **5.** 9 **6.** 6
7. 8 **8.** 8

Exercises 1.5 (page 52)

1. 3 **3.** 24 **5.** 2 **7.** 9 **9.** 9 **11.** 1,125 lb
13. -45 **15.** positive **17.** positive **19.** positive
21. a **23.** 0 **25.** 36 **27.** 56 **29.** -90 **31.** 448
33. -24 **35.** 25 **37.** -64 **39.** 90 **41.** 5
43. -3 **45.** -8 **47.** -10 **49.** -24 **51.** 4
53. -4 **55.** 2 **57.** -4 **59.** -20 **61.** -9
63. -16 **65.** 1 **67.** -4 **69.** 2 **71.** undefined
73. 9 **75.** 88 **77.** 3 **79.** -8 **81.** -96
83. -420 **85.** 49 **87.** -9 **89.** 5 **91.** 9
93. 5 **95.** undefined **97.** 20 **99.** $-\frac{11}{12}$ **101.** $-\frac{1}{6}$
103. $-\frac{7}{36}$ **105.** $-\frac{25}{144}$ **107.** \$5,100 **109.** $\frac{-18}{-3} = +6$
111. a. $-$ \$2,400 **b.** $-$ \$969 **c.** $-$ \$1,044 **d.** $-$ \$4,413
113. 2-point loss per day **115.** yes

Getting Ready (page 54)

1. sum **2.** product **3.** quotient **4.** difference
5. quotient **6.** difference **7.** product **8.** sum

Exercises 1.6 (page 59)

1. sum **3.** product **5.** quotient **7.** difference
9. 532 **11.** $\frac{1}{2}$ **13.** sum **15.** multiplication
17. algebraic **19.** term, coefficient **21.** $x + y$
23. $x - 3$ **25.** $(2x)y$ **27.** $3xy$ **29.** $\frac{y}{x}$ **31.** $\frac{3z}{4x}$
33. 3 **35.** 120 **37.** -5 **39.** 60 **41.** undefined
43. 5 **45.** 1; -7 **47.** 3; -1 **49.** 4; -3 **51.** 3; 9
53. 4; 5 **55.** $z + \frac{x}{y}$ **57.** $z - xy$ **59.** $\frac{xy}{x + z}$
61. $\frac{x - 4}{3y}$ **63.** the sum of y and 4 **65.** the product of x, y, and the sum of x and y **67.** the quotient obtained when the sum of x and 2 is divided by z **69.** the quotient obtained when y is divided by z **71.** the product of 2, x, and y
73. the quotient obtained when 5 is divided by the sum of x and y
75. $x + z$; 10 **77.** $y - z$; 2 **79.** $yz - 3$; 5 **81.** $\frac{xy}{z}$; 16
83. 19 and x **85.** x **87.** 3, x, y, and z **89.** 17, x, and z
91. 5, 1, and 8 **93.** x and y **95.** 75 **97.** x and y
99. $c + 6$ **101. a.** $h - 20$ **b.** $c + 20$ **103.** \$35,000n
105. $500 - x$ **107.** \$$(3d + 5)$ **109.** 49,995,000

Getting Ready (page 62)

1. 17 **2.** 17 **3.** 38.6 **4.** 38.6 **5.** 56 **6.** 56
7. 0 **8.** 1 **9.** 777 **10.** 777

Exercises 1.7 (page 67)

7. $x + y^2 \geq z$ **9.** $\geq$ **11.** positive **13.** real **15.** a
17. $(b + c)$ **19.** ac **21.** a **23.** element, multiplication
25. $a, \frac{1}{a}$, multiplicative **27.** 10 **29.** -24 **31.** 144
33. 3 **35.** Both are 12. **37.** Both are 29. **39.** Both are 60. **41.** Both are 175. **43.** $3x + 15$ **45.** $5z - 20$
47. $-6x - 2y$ **49.** $x^2 + 3x$ **51.** $-ax - bx$
53. $-4x^2 - 4x - 8$ **55.** $-5, \frac{1}{5}$ **57.** $-\frac{1}{3}, 3$ **59.** 0, none
61. $\frac{2}{3}, -\frac{3}{2}$ **63.** 0.2, -5 **65.** $-\frac{5}{4}, \frac{4}{5}$ **67.** $8x + 16$
69. y^3x **71.** $(y + x)z$ **73.** $x(yz)$ **75.** Both are 0.
77. Both are -6. **79.** $-6a - 24$ **81.** $-3x^2 + 3xa$
83. comm. prop. of add. **85.** comm. prop. of mult.
87. distrib. prop. **89.** comm. prop. of add. **91.** identity for mult. **93.** add. inverse **95.** identity for add.

Chapter Review (page 71)

1. 1, 2, 3, 4, 5 **2.** 2, 3, 5 **3.** 1, 3, 5 **4.** 4 **5.** $-6, 0, 5$
6. $-6, -\frac{2}{3}, 0, 2.6, 5$ **7.** 5 **8.** all of them **9.** $-6, 0$
10. 5 **11.** $\sqrt{2}, \pi$ **12.** $-6, -\frac{2}{3}$ **13.** $<$
14. $<$ **15.** $>$ **16.** $=$ **17.** 9 **18.** -8
19. 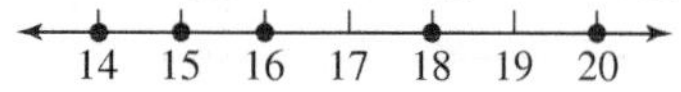

20. 

21. 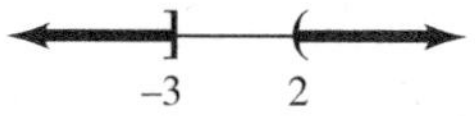

22. 

23. 5 **24.** 25 **25.** $\frac{5}{3}$ **26.** $\frac{8}{3}$ **27.** $\frac{1}{3}$ **28.** $\frac{1}{3}$
29. 1 **30.** $\frac{5}{2}$ **31.** $\frac{4}{3}$ **32.** $\frac{1}{3}$ **33.** $\frac{9}{20}$ **34.** $\frac{73}{63}$
35. $\frac{11}{21}$ **36.** $\frac{2}{15}$ **37.** $8\frac{11}{12}$ **38.** $2\frac{11}{12}$ **39.** 80.19
40. 20.99 **41.** 6.48 **42.** 3.7 **43.** 4.70 **44.** 26.36
45. 3.57 **46.** 3.75 **47.** $66\frac{3}{4}$ acres **48.** 6.85 hr
49. 85 **50.** 40.2 ft **51.** 81 **52.** $\frac{4}{9}$ **53.** 0.25
54. 33 **55.** 25 **56.** 49 **57.** $22\frac{3}{4}$ sq ft
58. 15,133.6 ft^3 **59.** 34 **60.** 14 **61.** 8 **62.** 7
63. 98 **64.** 38 **65.** 3 **66.** 15 **67.** 58
68. 4 **69.** 3 **70.** 3 **71.** 22 **72.** 1
73. 24 **74.** -33 **75.** -6.5 **76.** $\frac{1}{3}$ **77.** -12
78. 16 **79.** 1.2 **80.** -3.54 **81.** 19 **82.** 7
83. -5 **84.** -7 **85.** $-\frac{3}{10}$ **86.** 1 **87.** 1
88. $-\frac{1}{7}$ **89.** 40 **90.** 60 **91.** $\frac{1}{4}$ **92.** 1.3875
93. -35 **94.** -105 **95.** $-\frac{2}{3}$ **96.** -17.22
97. 3 **98.** 7 **99.** $\frac{7}{2}$ **100.** 6
101. -6 **102.** 18 **103.** 26 **104.** 7 **105.** 6
106. $\frac{3}{2}$ **107.** xz **108.** $x + 2y$ **109.** $2(x + y)$
110. $x - yz$ **111.** the product of 5, x, and z
112. 5 decreased by the product of y and z **113.** 4 less than the product of x and y **114.** the sum of x, y, and z, divided by twice their product **115.** 1 **116.** -2 **117.** 8
118. 4 **119.** 4 **120.** 6 **121.** 3 **122.** 6

Unless otherwise noted, all content on this page is © Cengage Learning.

123. 4 **124.** undefined **125.** −7 **126.** 39 **127.** undefined **128.** −2 **129.** 3 **130.** 7 **131.** 1 **132.** 9 **133.** closure prop. of add. **134.** comm. prop. of mult. **135.** assoc. prop. of add. **136.** distrib. prop. **137.** comm. prop. of add. **138.** assoc. prop. of mult. **139.** comm. prop. of add. **140.** identity for mult. **141.** add. inverse **142.** identity for add.

Chapter 1 Test (page 77)

1. 31, 37, 41, 43, 47 **2.** 2 **3.** 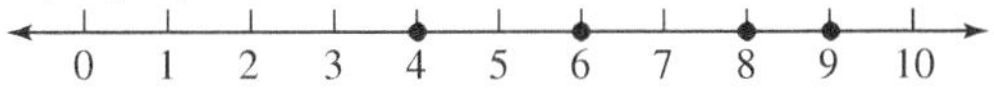

4. 

5. −17 **6.** 0 **7.** = **8.** < **9.** > **10.** = **11.** $\frac{13}{20}$ **12.** $\frac{4}{5}$ **13.** $\frac{3}{14}$ **14.** $\frac{9}{2}$ **15.** −1 **16.** $-\frac{1}{13}$ **17.** 33.3 **18.** 401.63 ft^2 **19.** 64 cm^2 **20.** 1,539 in^3 **21.** −2 **22.** −14 **23.** −4 **24.** 12 **25.** 5 **26.** −23 **27.** $\frac{xy}{x+y}$ **28.** $5y - (x + y)$ **29.** $24x + 14y$ **30.** $\$(12a + 8b)$ **31.** −5 **32.** 4 **33.** $3x + 6$ **34.** $-pr + pt$ **35.** 0 **36.** 5 **37.** comm. prop. of mult. **38.** distrib. prop. **39.** comm. prop. of add. **40.** mult. inverse prop.

Getting Ready (page 81)

1. −3 **2.** 7 **3.** x **4.** 1 **5.** $\frac{1}{x}$ **6.** 1 **7.** 4 **8.** 4 **9.** 3

Exercises 2.1 (page 91)

1. addition **3.** subtraction **5.** division **7.** multiplication **9.** $\frac{22}{15}$ **11.** $\frac{25}{27}$ **13.** 33 **15.** −317 **17.** equation, expression **19.** equivalent **21.** equal **23.** equal **25.** regular price **27.** 100 **29.** equation **31.** expression **33.** equation **35.** expression **37.** yes **39.** no **41.** yes **43.** yes **45.** yes **47.** yes **49.** 15 **51.** −14 **53.** 13 **55.** $\frac{1}{2}$ **57.** 3 **59.** −21 **61.** −4 **63.** $-\frac{7}{15}$ **65.** 18 **67.** 15 **69.** $\frac{1}{3}$ **71.** $-\frac{3}{2}$ **73.** 4 **75.** −11 **77.** −9 **79.** 27 **81.** $\frac{1}{8}$ **83.** 98 **85.** 85 **87.** 4,912 **89.** $9,345 **91.** $90 **93.** 80 **95.** 19 **97.** 520 **99.** 320 **101.** 150 **103.** 20% **105.** 3.3 **107.** −60 **109.** −28 **111.** −5.58 **113.** $\frac{5}{2}$ **115.** 36 **117.** −33 **119.** $-\frac{1}{3}$ **121.** $-\frac{1}{12}$ **123.** $-\frac{1}{5}$ **125.** 5 **127.** −3 **129.** $-\frac{1}{2}$ **131.** 6% **133.** 200% **135.** $270 **137.** 6% **139.** $260 **141.** 234 **143.** 55% **145.** 2,760 **147.** 370 **149.** $17,750 **151.** $2.22 **153.** $4.95 **155.** $145,149 **159.** about 3.16

Getting Ready (page 94)

1. 22 **2.** 36 **3.** 5 **4.** $\frac{13}{2}$ **5.** −1 **6.** −1 **7.** $\frac{7}{9}$ **8.** $-\frac{19}{3}$

Exercises 2.2 (page 100)

1. add 9 **3.** add 3 **5.** multiply by 3 **7.** 3 **9.** 50 cm **11.** 80.325 in.^2 **13.** cost **15.** percent **17.** percent of increase **19.** 1 **21.** −2 **23.** −1 **25.** 3 **27.** 3 **29.** −6 **31.** −3 **33.** −54 **35.** −16 **37.** 27 **39.** −33 **41.** 28 **43.** 3 **45.** 10 **47.** 4 **49.** 10 **51.** −4 **53.** 10 **55.** −4 **57.** 2 **59.** −5 **61.** $\frac{3}{2}$ **63.** 20 **65.** $\frac{15}{4}$ **67.** $\frac{16}{5}$ **69.** 11 **71.** −7 **73.** $-\frac{2}{3}$ **75.** 0 **77.** 6 **79.** $\frac{4}{3}$ **81.** $325 **83.** 6 days **85.** $50 **87.** 6% to 15% **89.** 5 **91.** 3 **93.** 30 min **95.** $7,400 **97.** no chance; he needs 112 **101.** $\frac{7x+4}{22} = \frac{1}{2}$

Getting Ready (page 102)

1. $3x + 4x$ **2.** $7x + 2x$ **3.** $8w - 3w$ **4.** $10y - 4y$ **5.** $7x$ **6.** $9x$ **7.** $5w$ **8.** $6y$

Exercises 2.3 (page 107)

1. expression **3.** equation **5.** equation **7.** identity **9.** contradiction **11.** 0 **13.** 2 **15.** $\frac{13}{56}$ **17.** $\frac{48}{35}$ **19.** variables, like, unlike, coefficient **21.** identity, contradiction **23.** $20x$ **25.** $3x^2$ **27.** unlike terms **29.** $2x + 12$ **31.** $7z - 15$ **33.** $12x + 121$ **35.** $3y - 15$ **37.** $7x + 2y$ **39.** −41 **41.** 3 **43.** −2 **45.** 3 **47.** 1 **49.** 2 **51.** 6 **53.** 12 **55.** −9 **57.** −20 **59.** 9 **61.** 16 **63.** 5 **65.** 4 **67.** identity, $\mathbb{R}$ **69.** contradiction, $\varnothing$ **71.** contradiction, $\varnothing$ **73.** identity, $\mathbb{R}$ **75.** expression, $x - 2y$ **77.** equation, 8 **79.** expression, $5x + 7$ **81.** equation, −20 **83.** equation, $\mathbb{R}$ **85.** equation, $\frac{1}{2}$ **87.** equation, 2 **89.** expression, $5x + 24$ **91.** equation, −3 **93.** expression, $0.7m + 22.16$ **95.** equation, 1 **97.** equation, 0 **99.** 0.9 **105.** 0

Getting Ready (page 109)

1. 3 **2.** −5 **3.** r **4.** $-a$ **5.** 7 **6.** 12 **7.** d **8.** s

Exercises 2.4 (page 113)

1. $a = -\frac{c}{b}$ **3.** $b = ac$ **5.** $3x - 4y$ **7.** $-a - 8$ **9.** literal **11.** isolate **13.** subtract **15.** $I = \frac{E}{R}$ **17.** $w = \frac{V}{lh}$ **19.** $y = x - 12$ **21.** $h = \frac{3V}{B}$ **23.** $h = \frac{3V}{\pi r^2}$ **25.** $x = 2y - 2$ **27.** $B = \frac{2}{5}A - 3$ or $B = \frac{2A - 15}{5}$ **29.** $q = \frac{2p - hr}{h}$ or $q = \frac{2p}{h} - r$ **31.** $r = \frac{G}{2b} + 1$ or $r = \frac{G + 2b}{2b}$ **33.** $t = \frac{d}{r}$; $t = 7$ **35.** $b = P - a - c$; $b = 16$ **37.** $y = \frac{5 - 3x}{2}$ **39.** $d = \frac{C}{\pi}$ **41.** $w = \frac{P - 2l}{2}$ **43.** $t = \frac{A - P}{Pr}$ **45.** $w = \frac{2gK}{v^2}$ **47.** $g = \frac{wv^2}{2K}$ **49.** $M = \frac{Fd^2}{Gm}$ **51.** $p = \frac{i}{rt}$; $p = 750$ **53.** $h = \frac{2K}{a + b}$; $h = 8$ **55.** $h = \frac{3V}{\pi r^2}$; $h = 3$ in. **57.** $I = \frac{E}{R}$; $I = 4$ amp **59.** $R = \frac{P}{I^2}$; $R = 13.78$ ohms **61.** $m = \frac{Fd^2}{GM}$ **63.** $D = \frac{L - 3.25r - 3.25R}{2}$; $D = 6$ ft **65.** $C \approx 0.1304T$, about 13% of taxable income **69.** 90,000,000,000 joules

Unless otherwise noted, all content on this page is © Cengage Learning.

Getting Ready (page 116)

1. $(x + x + 2)$ ft or $(2x + 2)$ ft **2.** $(x + 3x)$ ft or $4x$ ft
3. $P = 2l + 2w$

Exercises 2.5 (page 122)

1. $12 - x$ **3.** $0.07(18{,}000 - x)$ **5.** 200 cm^3
7. $8x + 2$ **9.** $-\frac{3}{2}$ **11.** \$1,488 **13.** $2l + 2w$
15. vertex **17.** degrees **19.** straight **21.** supplementary
23. 4 ft and 8 ft **25.** 5 ft, 10 ft, 15 ft **27.** 6 ft, 8 ft, 10 ft **29.** 14,000 hardcovers, 196,000 paperbacks
31. 10° **33.** 47° **35.** 159° **37.** 87° **39.** 44°
41. 130° **43.** 19 cm by 26 cm **45.** 12 in. by 18 in.
47. 19 ft **49.** 20° **51.** \$6,000 **53.** \$4,500 at 9% and \$19,500 at 14% **55.** \$3,750 in each account **57.** \$5,000
59. 6% and 7%

Getting Ready (page 124)

1. 60 mi **2.** 385 mi **3.** 5.6 gal **4.** 9 lb

Exercises 2.6 (page 130)

1. $60h$ mi **3.** 5 oz **5.** -9 **7.** 2 **9.** 2
11. 6 **13.** $d = rt$ **15.** $v = pn$ **17.** 3 hr **19.** 3.5 days
21. 6.5 hr **23.** $\frac{2}{7}$ hr or approximately 17 min
25. 10 hr **27.** 7.5 hr **29.** 65 mph and 45 mph
31. 4 hr **33.** 2.5 liters **35.** 30 gal **37.** 50 gal
39. 7.5 oz **41.** 40 lb lemon drops and 60 lb jelly beans **43.** \$1.20 **45.** 45 lb **47.** \$5.60

Getting Ready (page 132)

1. 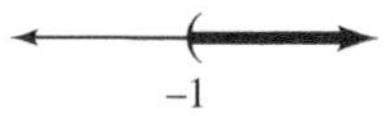

2. 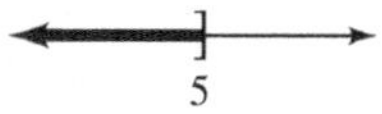

3. 

4. 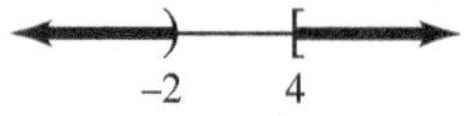

Exercises 2.7 (page 137)

1. same **3.** reverses **5.** same **7.** $5x^2 - 2y^2$
9. $-x + 14$ **11.** is less than; is greater than **13.** double inequality **15.** inequality
17. $x > 3$; 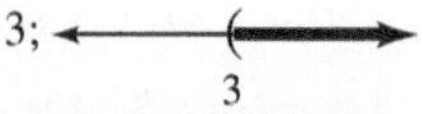

19. $x \leq -1$; 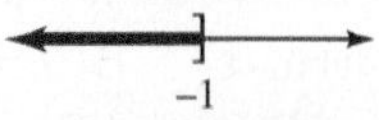

21. $x \leq 4$; 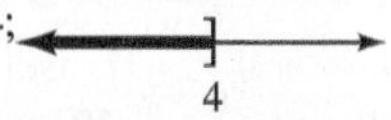

23. $x > -3$; 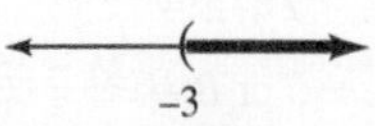

25. $x > -1$; 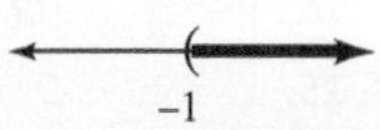

27. $x \geq 3$; 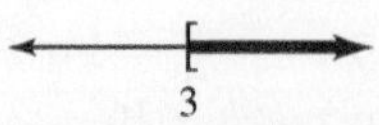

29. $x \geq -10$; 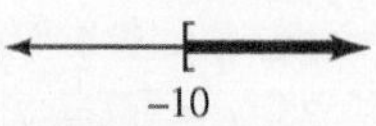

31. $x > -3$; 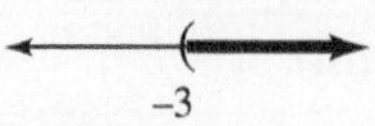

33. $x < -4$; 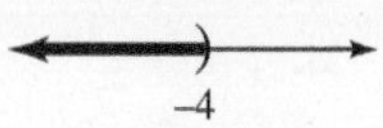

35. $x < -2$; 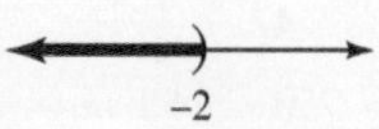

37. $x > 3$; 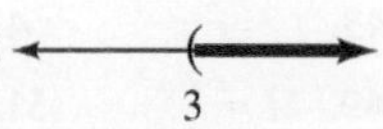

39. $x \geq 10$; 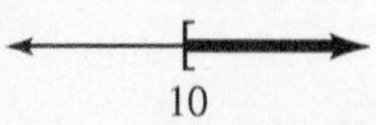

41. $7 < x < 10$; 

43. $-3 < x \leq 5$; 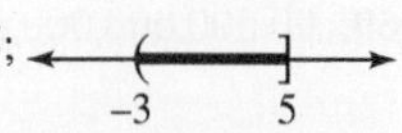

45. $-10 \leq x \leq 0$; 

47. $-4 < x < 1$; 

49. $x > -1$; 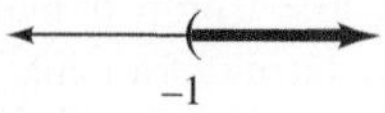

51. $x \geq 2$; 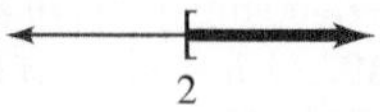

53. $x \leq 20$; 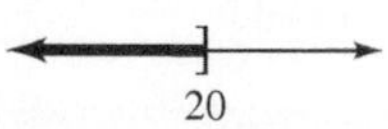

55. $x > -7$; 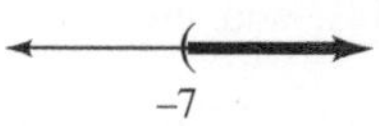

57. $x \geq 4$; 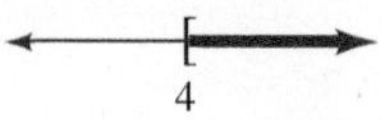

59. $-5 < x < -3$; 

61. $-6 \leq x \leq 10$; 

63. $2 \leq x < 3$; 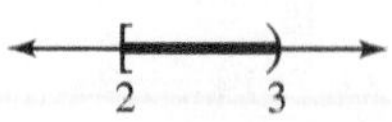

65. $-1 \leq x < 2$; 

67. $-6 < x < 14$; 

69. $s \geq 98$ **71.** $s \geq 17$ cm **73.** $r \geq 27$ mpg
75. 0.1 mi $\leq x \leq$ 2.5 mi **77.** 3.3 mi $< x <$ 4.1 mi
79. $73.4° < F < 78.8°$ **81.** 37.052 in. $< C <$ 38.308 in.
83. 68.18 kg $< w <$ 86.36 kg **85.** 5 ft $< w <$ 9 ft

Chapter Review (page 141)

1. expression **2.** equation **3.** no **4.** yes **5.** -1
6. -13 **7.** 4 **8.** -15 **9.** -1 **10.** 0 **11.** $-\frac{1}{3}$
12. $-\frac{1}{2}$ **13.** \$105.40 **14.** \$97.70 **15.** 5 **16.** -2
17. $-\frac{1}{2}$ **18.** $\frac{3}{2}$ **19.** 18 **20.** -35 **21.** $-\frac{1}{2}$ **22.** 6
23. 245 **24.** 1,300 **25.** 37% **26.** 12.5% **27.** 3
28. 2 **29.** 1 **30.** 1 **31.** 1 **32.** -2 **33.** 2
34. 7 **35.** -2 **36.** -1 **37.** 5 **38.** 3 **39.** -6
40. -4 **41.** 8 **42.** 30 **43.** 15 **44.** 4 **45.** $\frac{21}{2}$
46. 44 **47.** \$320 **48.** 6.5% **49.** 96.4%
50. 52.8% **51.** $14x$ **52.** $19a$ **53.** $5b$ **54.** $-2x$
55. $-2y$ **56.** not like terms **57.** $9x$ **58.** $6 - 7x$
59. 2 **60.** -8 **61.** 7 **62.** 13 **63.** -3
64. -41 **65.** 9 **66.** -7 **67.** 7 **68.** 4
69. 14 **70.** 39 **71.** identity, $\mathbb{R}$ **72.** contradiction, $\varnothing$
73. identity, $\mathbb{R}$ **74.** contradiction, $\varnothing$ **75.** $R = \frac{E}{I}$
76. $t = \frac{i}{pr}$ **77.** $R = \frac{P}{I^2}$ **78.** $B = \frac{3V}{h}$
79. $c = p - a - b$ **80.** $m = \frac{y - b}{x}$ **81.** $h = \frac{V}{\pi r^2}$
82. $B = \frac{2}{3}A - 4$ or $B = \frac{2A - 12}{3}$ **83.** $G = \frac{Fd^2}{Mm}$
84. $m = \frac{RT}{PV}$ **85.** 5 ft from one end **86.** 15° **87.** 45°
88. 78° **89.** 105° **90.** 13 in. **91.** \$16,000 at 7%, \$11,000 at 9% **92.** 30 min **93.** 2 hr **94.** 9 hr
95. 24 liters **96.** 1 liter **97.** 10 lb of each
98. 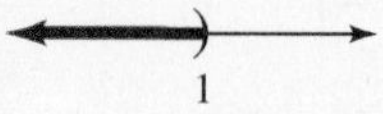

99. 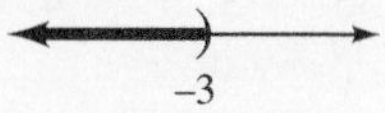

100. 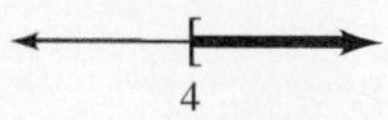

101. 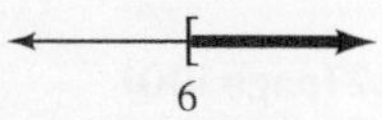

102. 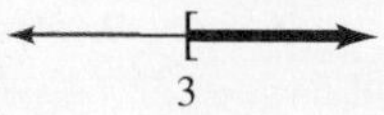

103. 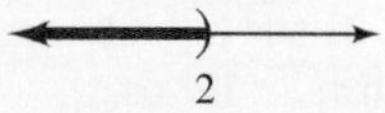

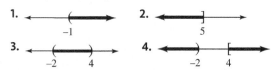

Unless otherwise noted, all content on this page is © Cengage Learning.

104. 

105. 

106. 6 ft $< l \le$ 20 ft

Chapter 2 Test (page 146)

1. solution **2.** not a solution **3.** −36 **4.** 47
5. −12 **6.** −7 **7.** −2 **8.** 1 **9.** 7 **10.** −3
11. $6x - 15$ **12.** $8x - 10$ **13.** −18 **14.** $-36x + 13$
15. $\mathbb{R}$ **16.** $\varnothing$ **17.** −2 **18.** 0 **19.** $t = \frac{d}{r}$
20. $h = \frac{S - 2\pi r^2}{2\pi r}$ **21.** $h = \frac{A}{2\pi r}$ **22.** $y = \frac{-x + 5}{2}$
23. 75° **24.** 101° **25.** \$6,000 at 6%, \$4,000 at 5%
26. $\frac{3}{5}$ hr or 36 min **27.** $7\frac{1}{2}$ liters **28.** 40 lb

29. 

30. 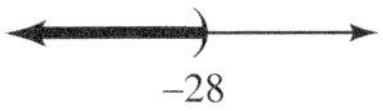

31. 

32. 

Cumulative Review (page 147)

1. integer, rational, real **2.** rational, real

3. 

4. 

5. 0 **6.** $\frac{7}{10}$ **7.** $8\frac{1}{10}$
8. 35.65 **9.** 45 **10.** −2 **11.** 16 **12.** 0
13. 9.9 **14.** 5,275 **15.** 5 **16.** 37, y
17. $-5x + 7y$ **18.** $x - 5$ **19.** $-2x^2y^3$ **20.** $-9x + 3$
21. −9 **22.** 41 **23.** $\frac{7}{4}$ **24.** $\mathbb{R}$ **25.** $h = \frac{2A}{b + B}$
26. $x = \frac{y - b}{m}$ **27.** \$22,814.56 **28.** \$900 **29.** \$22,690
30. 125 lb **31.** no **32.** 7.3 ft and 10.7 ft **33.** 540 kWh
34. 185 ft **35.** −9 **36.** 1 **37.** 280 **38.** −564

39. 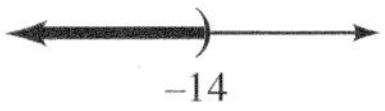

40. 

Getting Ready (page 151)

1. 

2. 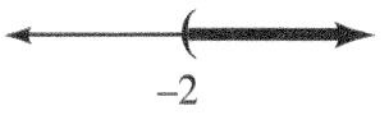

3. 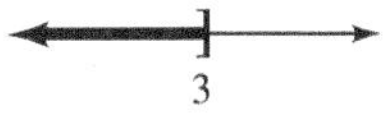

4. 

Exercises 3.1 (page 158)

1. III **3.** IV **5.** 15 **7.** 8 **9.** 7 **11.** −49
13. ordered pair **15.** origin **17.** rectangular, Cartesian
19. coordinates **21.** no **23.** origin, left, up **25.** II

27. 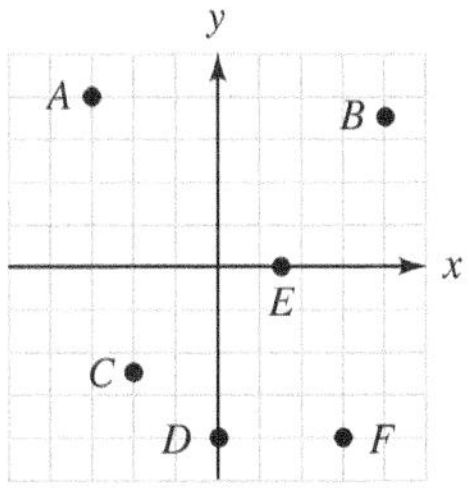

29. 2, −2, 4, 3, 0 **31.** 10 min before the workout, her heart rate was 60 beats per min. **33.** 150 beats per min
35. approximately 5 min and 50 min after starting
37. 10 beats per min faster after cooldown **39.** \$2
41. \$7

43. 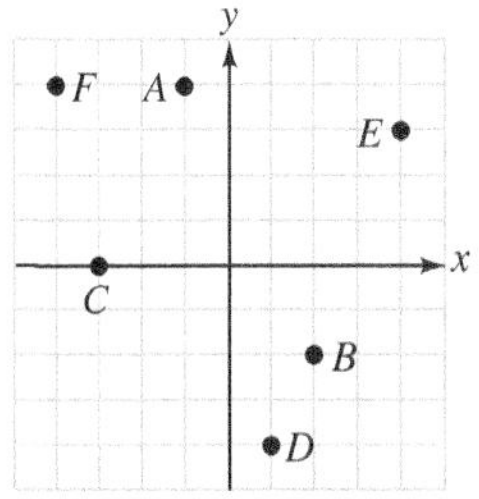

$A(-1, 4), B(2, -2), C(-3, 0),$
$D(1, -4), E(4, 3), F(-4, 4)$

45. 15¢; 85¢ **47.** 10¢
49. Carbondale $(5\frac{1}{2}, \text{J})$, Champaign $(6\frac{1}{2}, \text{D})$, Chicago (8, B), Peoria $(5\frac{1}{2}, \text{C})$, Rockford $(5\frac{1}{2}, \text{A})$
51. a. 60°; 4 ft **b.** 30°; 4 ft

53. 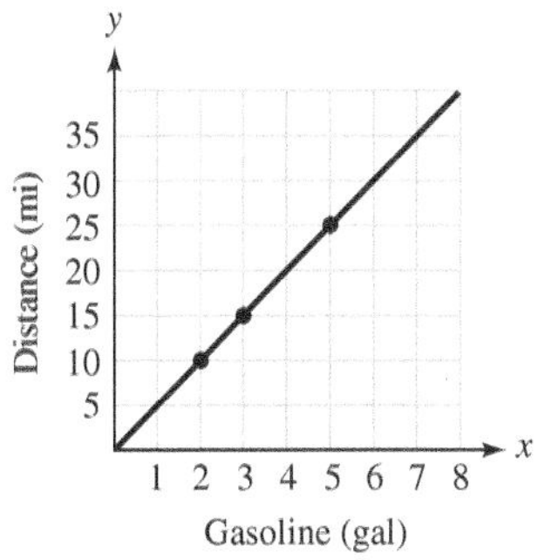

a. 30 mi **b.** 8 gal **c.** 32.5 mi

55. 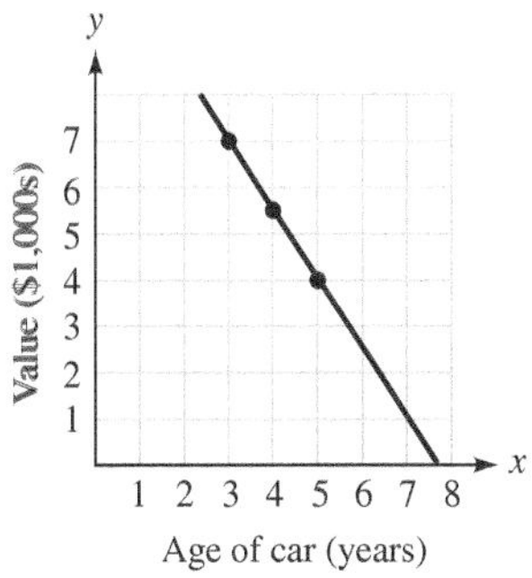

a. A 3-yr-old car is worth \$7,000. **b.** \$1,000 **c.** 6 yr

Getting Ready (page 161)

1. 1 **2.** 5 **3.** −3 **4.** 2

Exercises 3.2 (page 171)

1. 2 **3.** 5 **5.** vertical (except $x = 0$) **7.** −18
9. an expression **11.** 1.25 **13.** 0.1 **15.** linear, two
17. independent, dependent **19.** linear
21. y-intercept **23.** yes **25.** no
27. $-3, -2, -5, -7; (0, -3), (1, -2), (-2, -5), (-4, -7)$
29. $0, -2, -6, 4; (0, 0), (1, -2), (3, -6), (-2, 4)$

Unless otherwise noted, all content on this page is © Cengage Learning.

31.
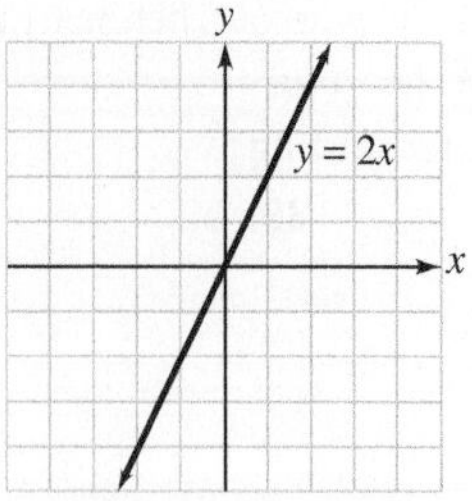

33.
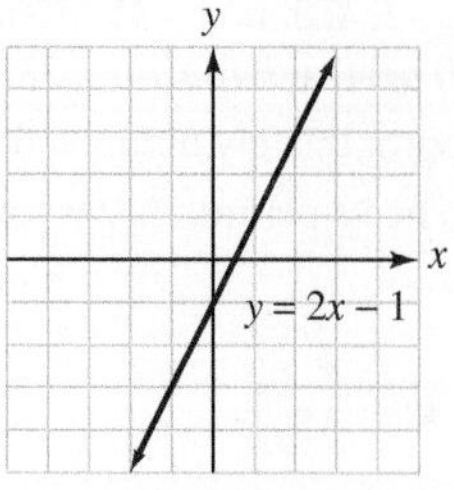

35.
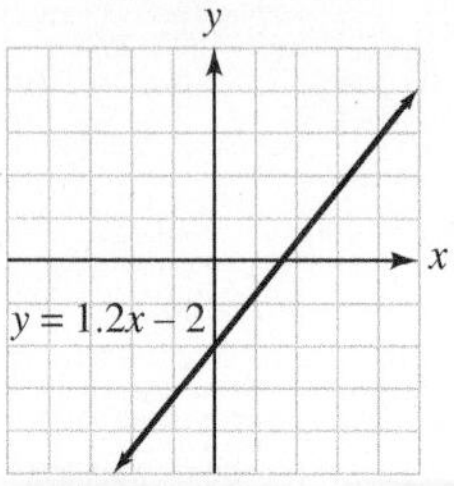

37.
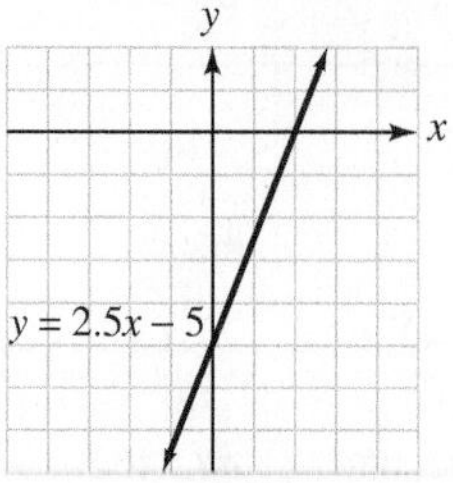

39.
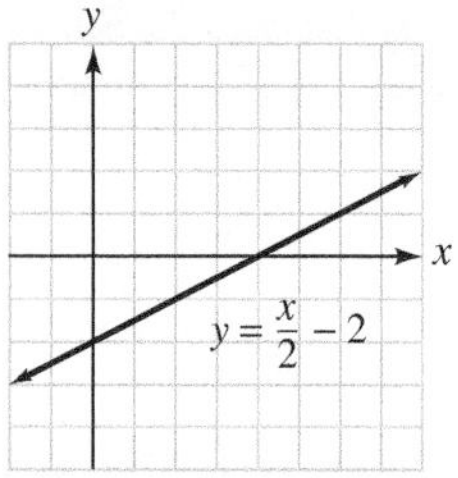

41.
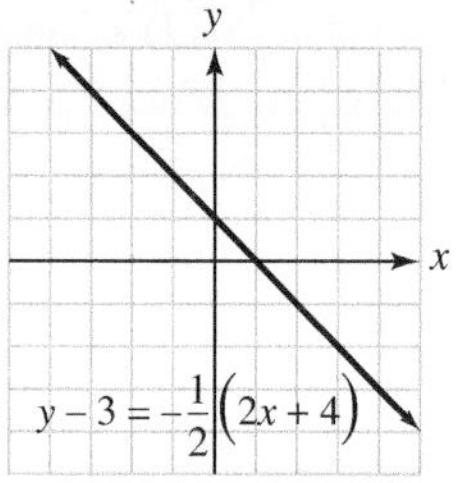

43.
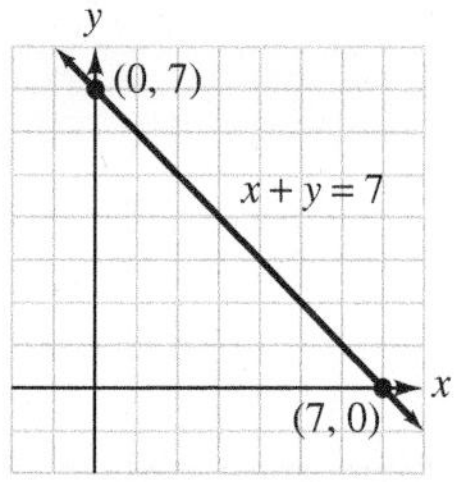

45.
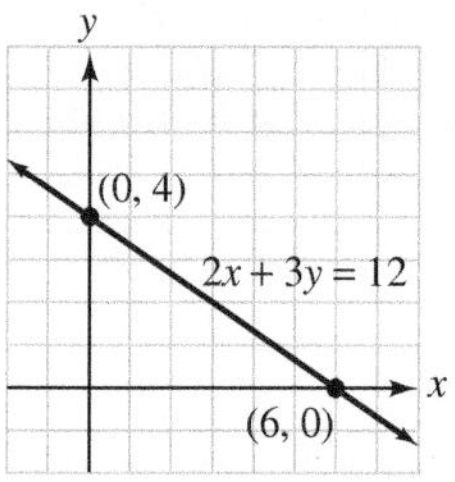

47.
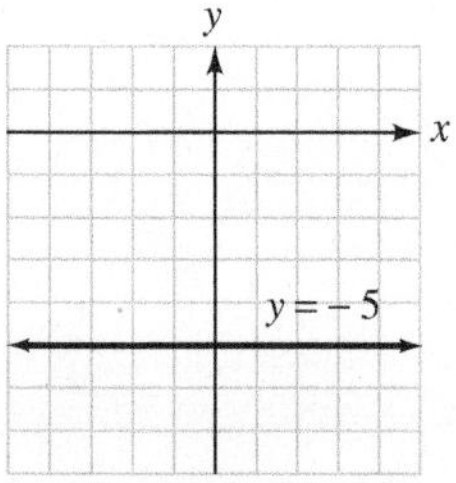

49.
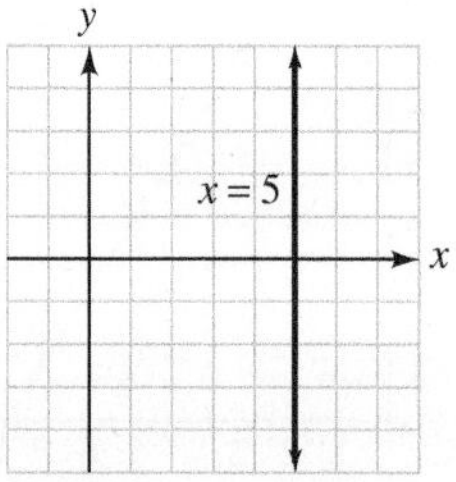

51.
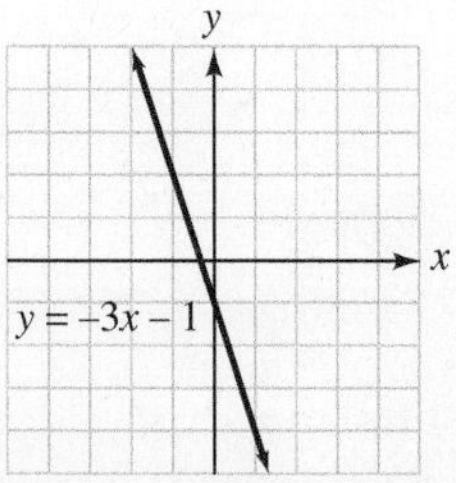

53.
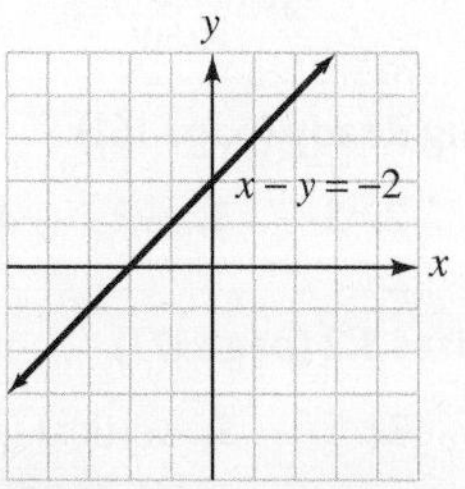

55.
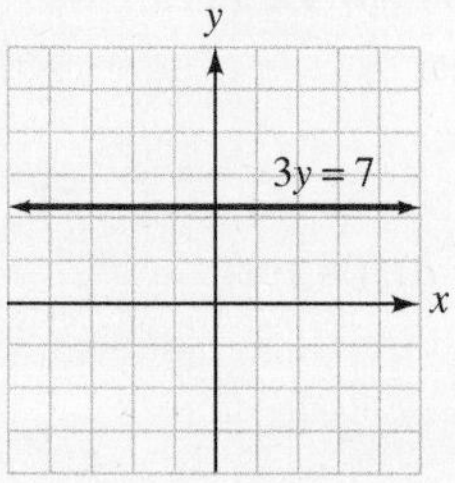

57.
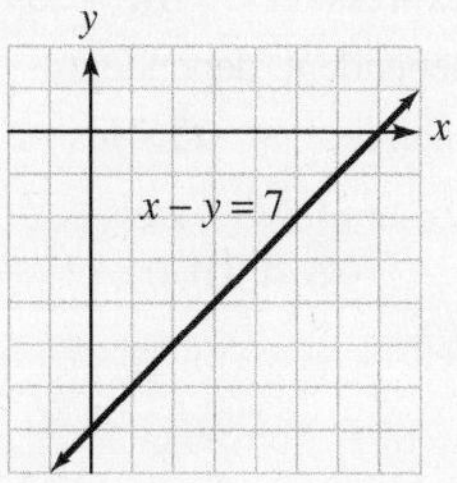

59.
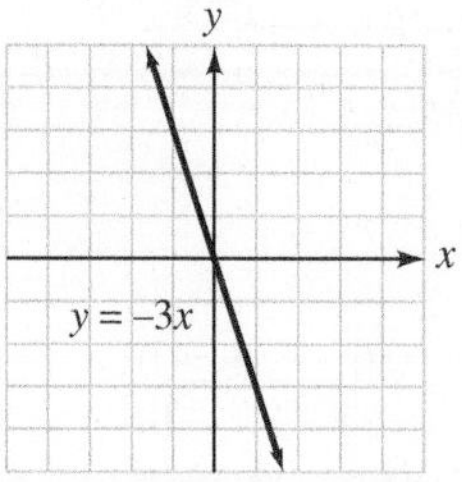

61.
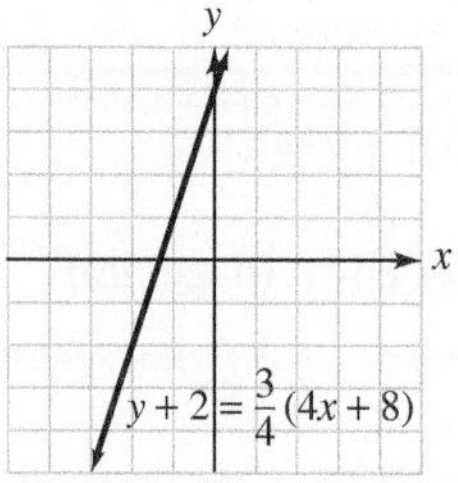

63.
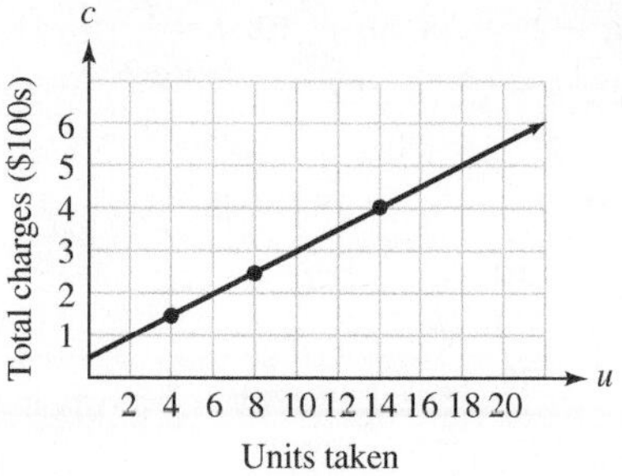

a. $c = 50 + 25u$ **b.** 150, 250, 400; $(4, 150), (8, 250), (14, 400)$ **c.** The service fee is \$50. **d.** \$850

65.
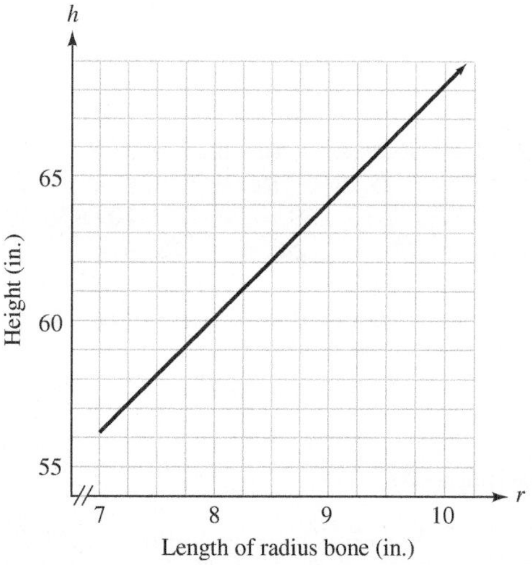

a. 56.2, 62.1, 64.0; $(7, 56.2), (8.5, 62.1), (9, 64.0)$ **b.** taller the woman is. **c.** 68 in. **73.** $(6, 6)$

75. $\left(-\frac{1}{2}, \frac{5}{2}\right)$ **77.** $(7, 6)$ **79.** $\left(\frac{2x-1}{2}, -\frac{1}{2}\right)$

Getting Ready (page 175)

1. 1 **2.** −5 **3.** 0 **4.** undefined

Exercises 3.3 (page 185)

1. $\frac{2}{5}$ **3.** undefined **5.** $6a - 12$ **7.** $-2z - 4w$
9. $a - 5b$ **11.** y, x **13.** rise, run, rise, run
15. hypotenuse **17.** perpendicular
19. increasing, decreasing **21.** $\frac{2}{3}$ **23.** $\frac{4}{3}$ **25.** $-\frac{7}{8}$
27. 3 **29.** 3 **31.** 1 **33.** $-\frac{1}{3}$ **35.** $\frac{13}{5}$
37. $-\frac{3}{2}$ **39.** $\frac{2}{5}$ **41.** $\frac{1}{2}$ **43.** 5 **45.** 0
47. undefined **49.** 0 **51.** undefined
53. 0 **55.** undefined **57.** perpendicular

Unless otherwise noted, all content on this page is © Cengage Learning.

59. neither **61.** parallel **63.** perpendicular
65. $-\frac{3}{5}$ **67.** -1 **69.** negative **71.** positive
73. undefined **75.** not the same line
77. not the same line **79.** same line **81.** parallel
83. perpendicular **85.** neither **87.** $y = 0, m = 0$
89. $\frac{1}{220}$ **91.** $\frac{1}{5}$ **93.** 3.5 students per yr
95. \$642.86 per year **99.** 4

Getting Ready (page 188)

1. $y = -2x + 12$ **2.** $y = 3x - 7$ **3.** $y = -2x + \frac{9}{2}$
4. $y = \frac{5}{4}x - 3$

Exercises 3.4 (page 192)

1. $2, (3, 5)$ **3.** $\frac{7}{8}, (-4, -5)$ **5.** 6 **7.** -1
9. 7,950 **11.** $y - y_1 = m(x - x_1)$ **13.** $(1, 2), 2, 3$
15. $y - 0 = 3(x - 0)$ **17.** $y - (-2) = -7[x - (-1)]$
19. $y - 3 = 2[x - (-5)]$ **21.** $y - 5 = -\frac{6}{7}(x - 6)$
23. $y = -5x + 7$ **25.** $y = 5x + 7$ **27.** $y = -3x + 6$
29. $y = \frac{1}{3}x - 4$ **31.** $y = \frac{2}{3}x + \frac{11}{3}$ **33.** $y = -\frac{1}{2}x - 1$
35.

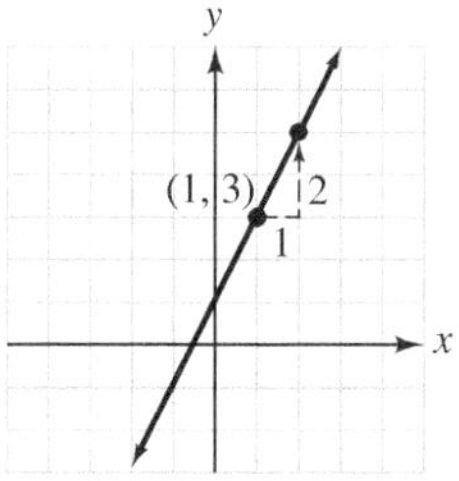

37.

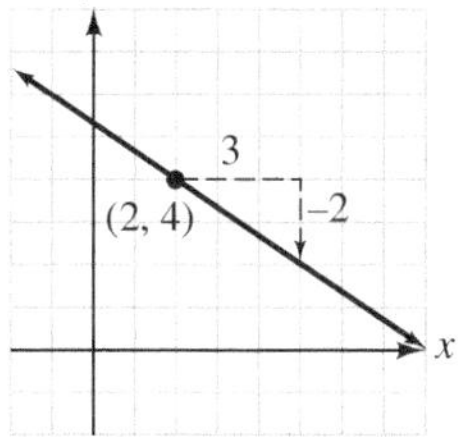

39. $y + 8 = 0.5(x + 1)$ **41.** $y - 2 = -4(x + 3)$
43. $y = x$ **45.** $y = \frac{7}{3}x - 3$
47.

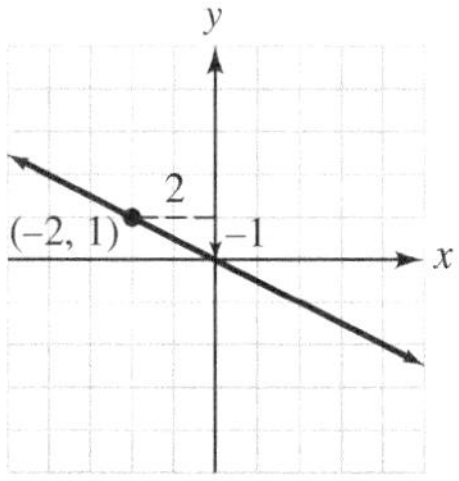

49.

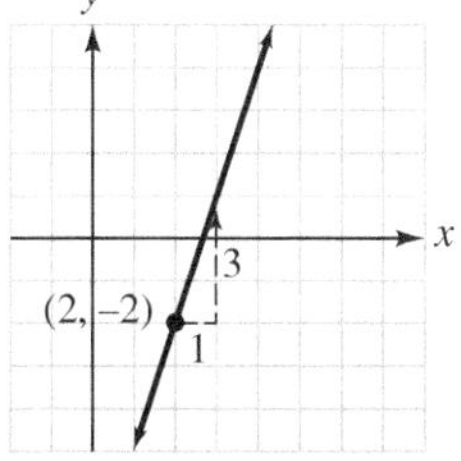

51. $y = -\frac{2}{3}x + 1$ **53.** $y = 0.5x + 3$ **55.** $y = \frac{3}{2}x + \frac{1}{2}$
57. \$206.25 **59.** \$890 **61.** \$17.50 **63.** \$137,200

Getting Ready (page 195)

1. $y = -\frac{1}{2}x + \frac{3}{2}$ **2.** $y = \frac{6}{5}x - \frac{7}{5}$

Exercises 3.5 (page 201)

1. $y = -\frac{1}{2}x + 3$ **3.** $b = -4$ **5.** 2 **7.** $\frac{7}{2}$
9. $y = mx + b, -3, (0, 7)$ **11.** reciprocals **13.** $7, (0, -5)$
15. $-\frac{2}{5}, (0, 6)$ **17.** $\frac{3}{2}, (0, -4)$ **19.** $-\frac{1}{3}, (0, -\frac{5}{6})$
21. $y = 12x$ **23.** $y = -5x - 4$ **25.** $y = -7x + 54$
27. $y = -5$ **29.** $y = -3x - 2$ **31.** $y = -\frac{1}{2}x + 6$
33. $1, (0, -1)$

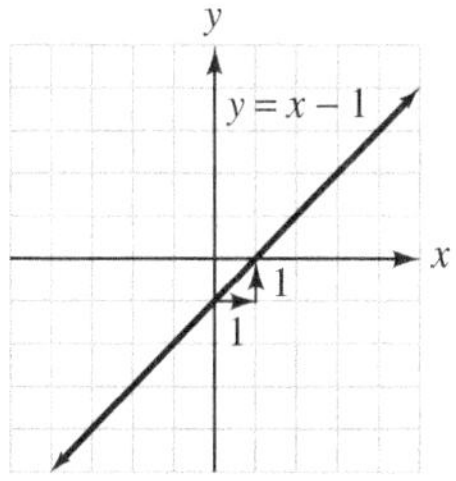

35. $\frac{2}{3}, (0, 2)$

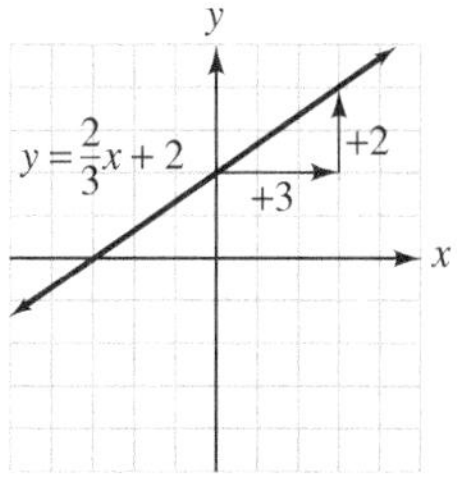

37. $-\frac{2}{3}, (0, 6)$

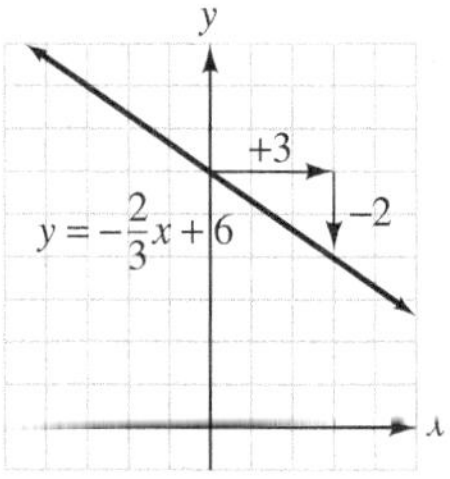

39. parallel **41.** perpendicular **43.** parallel
45. perpendicular **47.** $y = 4x$ **49.** $y = 4x - 3$
51. $y = -\frac{1}{4}x$ **53.** $y = -\frac{1}{4}x + \frac{11}{2}$ **55.** $\frac{7}{2}, (0, 2)$
57. $y = \frac{2}{3}x + 6$ **59.** $y = -\frac{4}{3}x + 6$ **61.** $y = \frac{4}{3}x - 5$
63. $y = \frac{4}{5}x - \frac{26}{5}$ **65.** $y = \frac{3}{4}x - \frac{23}{4}$ **67.** $x = -2$
69. $x = 5$ **71.** perpendicular **73.** perpendicular
75. perpendicular **77.** parallel
79. $y = -3{,}200x + 24{,}300$ **81.** $y = 37{,}500x + 450{,}000$
83. $y = -\frac{710}{3}x + 1{,}900$ **85.** $-\$120$ per year
87. \$25 **89.** about \$64,331
91. \$372,000; $-2{,}325$, $y = -2{,}325x + 465{,}000$
95. $y = -\frac{A}{B}x + \frac{C}{B}$ **99.** $a < 0, b > 0$

Getting Ready (page 204)

1. -1 **2.** 3 **3.** -3 **4.** -5 **5.** 2 **6.** -4
7. 5 **8.** 8

Exercises 3.6 (page 209)

1. 1 **3.** -1 **5.** 6 **7.** -1 **9.** relation
11. input, function **13.** range **15.** independent
17. cannot **19.** $\{-3, 1, 3\}, \{-1, 2, 7\}$
21. $\{-3, 0, 2, 4\}, \{-8, 0, 5, 7\}$ **23.** yes **25.** no
27. $-9, 0, 3, -1$ **29.** $3, -3, -5, 3$ **31.** $22, 7, 2, -\frac{4}{5}$
33. $3, 9, 11, 3$ **35.** $3, \frac{3}{2}, 1, 3$
37.

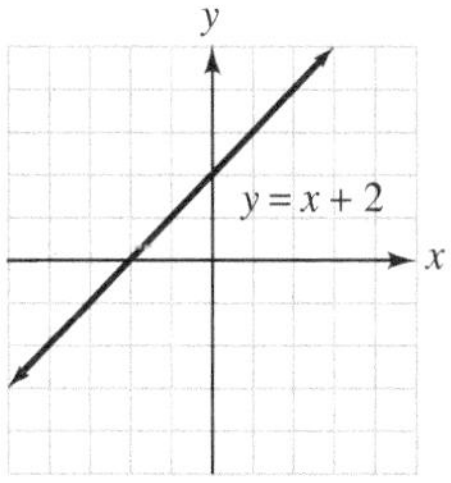

domain: $\mathbb{R}$; range: $\mathbb{R}$

39.

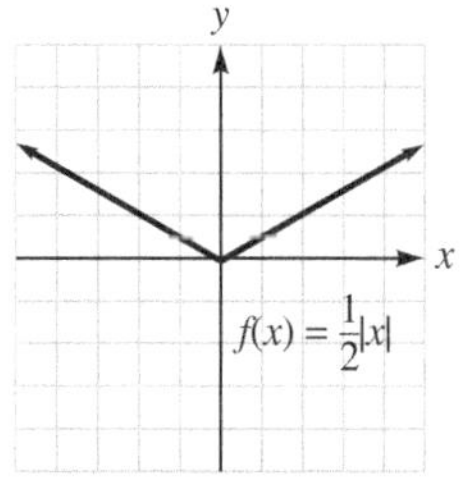

domain: $\mathbb{R}$; range: $\{y \mid y$ is a real number and $y \geq 0\}$

41. yes **43.** no **45.** $2, 5, 10$ **47.** $0, -9, 26$

Unless otherwise noted, all content on this page is © Cengage Learning.

49. 0, 9, 4 **51.** 1, 16, 21 **53.** 1.5, 9, 1.5

55.

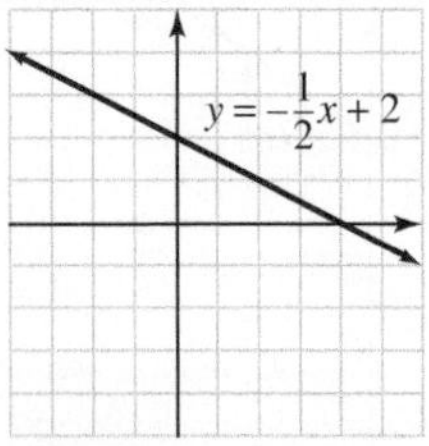

domain: ℝ; range: ℝ

57.

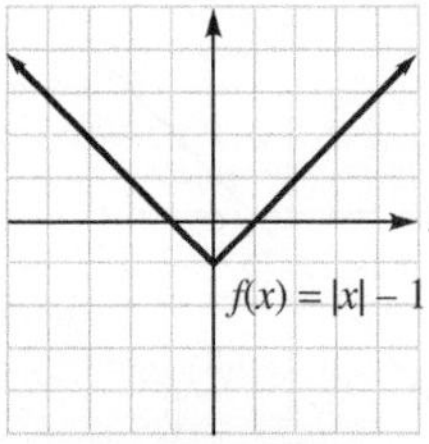

domain: ℝ; range: $\{y \mid y$ is a real number and $y \geq -1\}$

59. \$14 **61.** \$22 **63.** \$57 **65.** \$81 **69.** yes **71.** yes

Getting Ready (page 211)

1. yes **2.** yes **3.** no **4.** yes

Exercises 3.7 (page 216)

1. below **3.** above **5.** left **7.** above **9.** $(4, -1)$ **11.** $(3, 0)$ **13.** linear **15.** edge **17.** dashed

19.

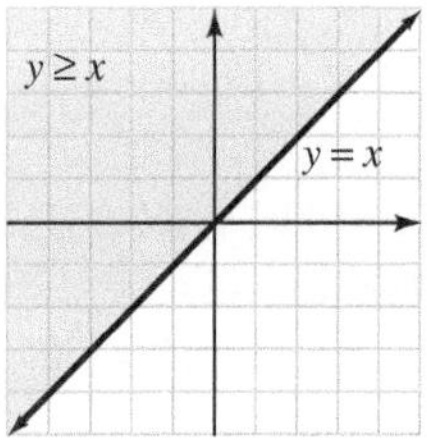

21.

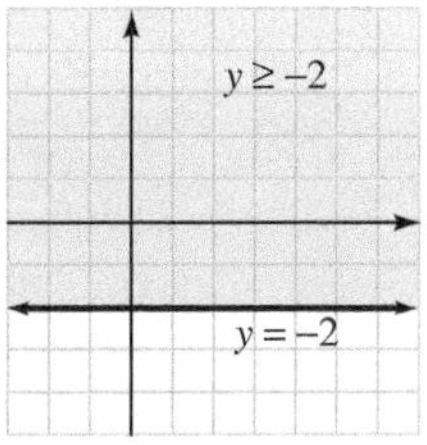

23.

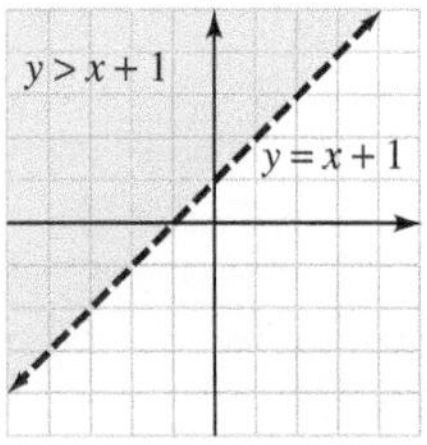

25.

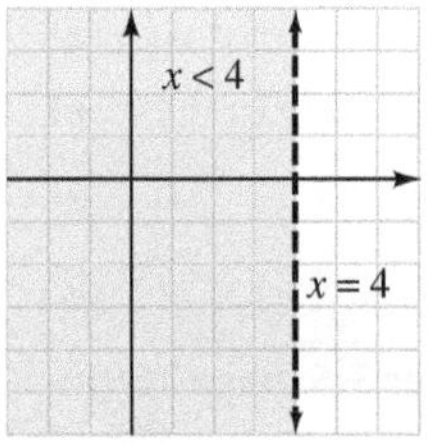

27.

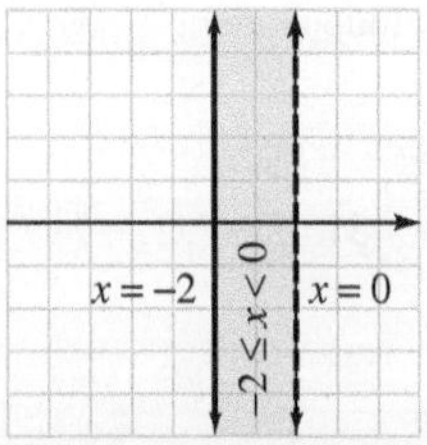

29.

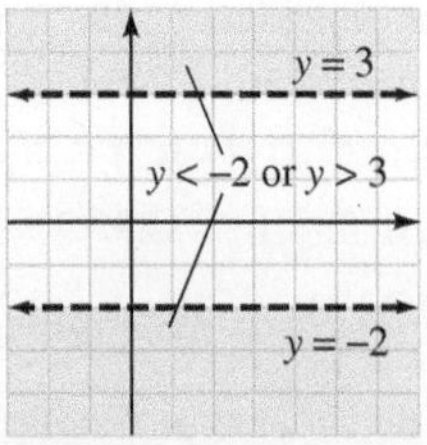

31.

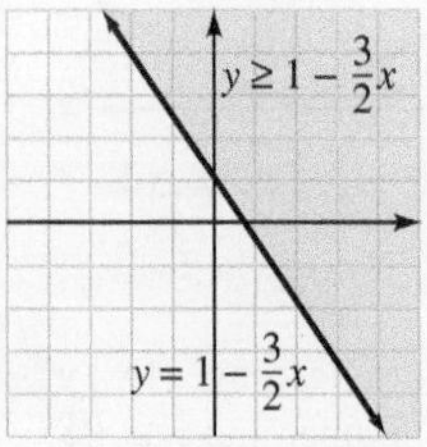

33.

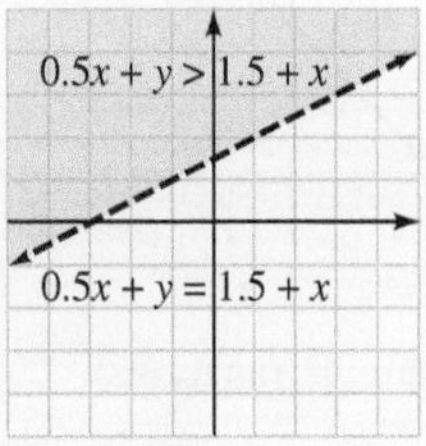

35.

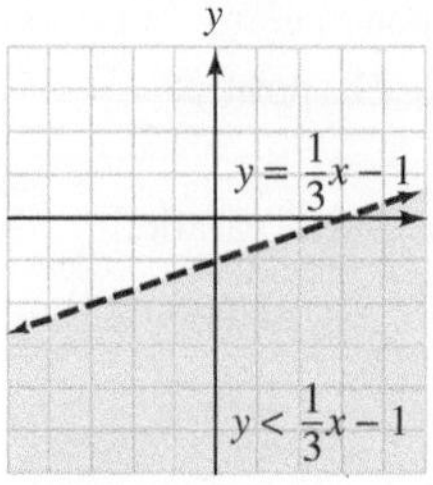

37.

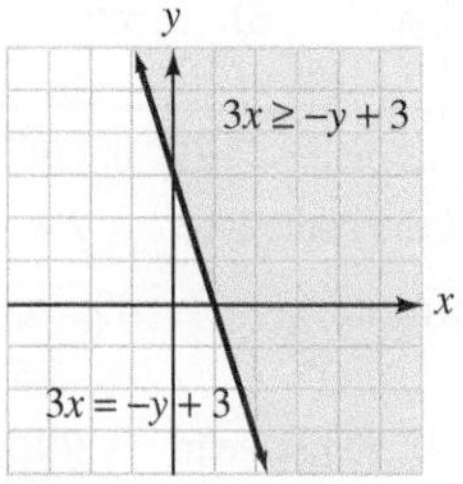

39.

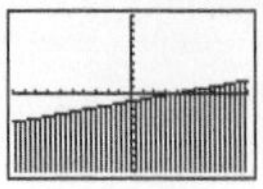

41.

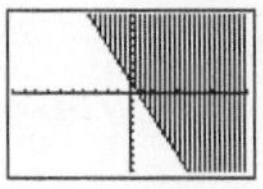

43. $x \leq 3$

45. $y > -\frac{3}{2}x + 3$ **47.** $-2 \leq x \leq 3$

49. $y \leq x$ **51.** $y > -1$ or $y \leq -3$

53. $(1, 1), (2, 1), (2, 2)$ **55.** $(2, 2), (3, 3), (5, 1)$

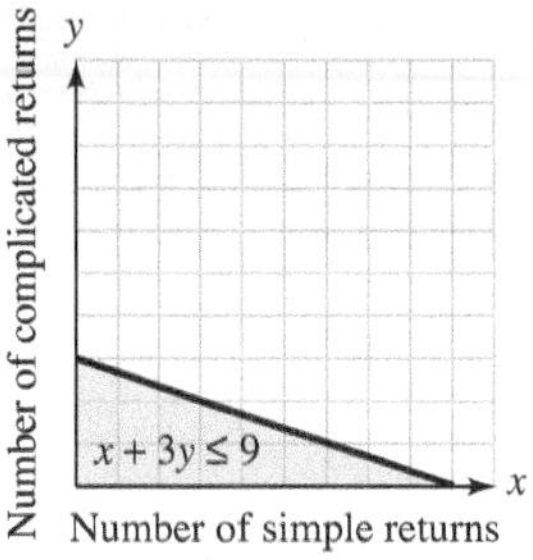

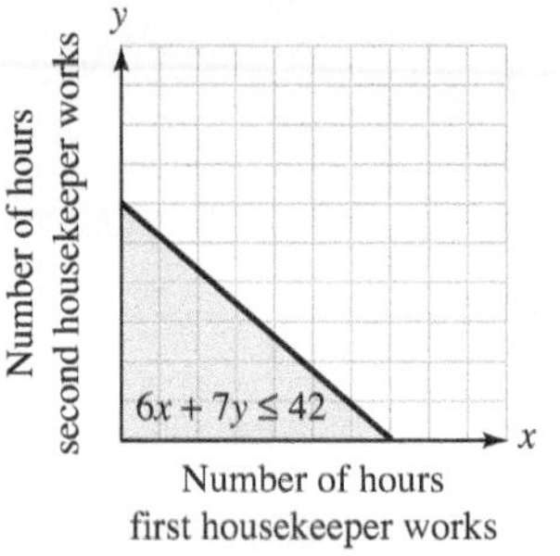

57. $(40, 20), (60, 40), (80, 20)$

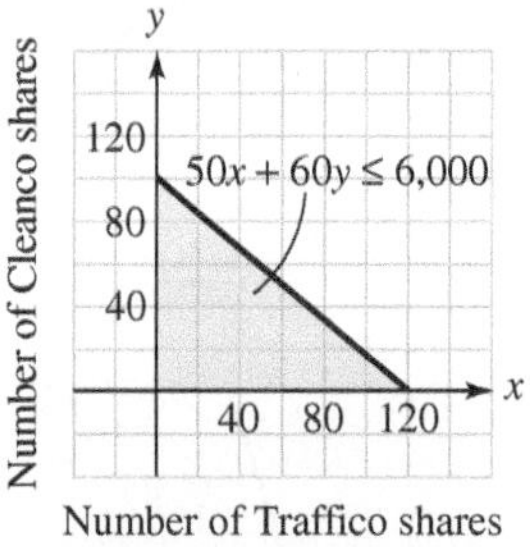

Chapter Review (page 222)

1–6.

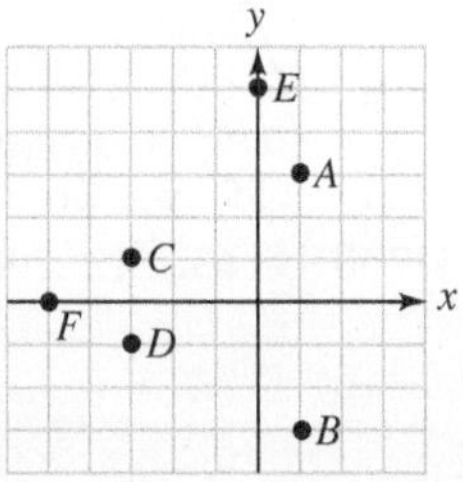

7. $(3, 1)$ **8.** $(-4, 5)$ **9.** $(-3, -4)$ **10.** $(2, -3)$ **11.** $(0, 0)$ **12.** $(0, 4)$ **13.** $(-5, 0)$ **14.** $(0, -3)$ **15.** yes **16.** no

17.

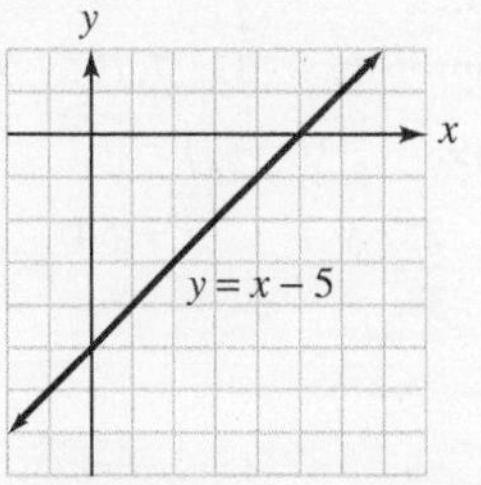

18.

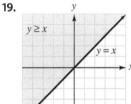

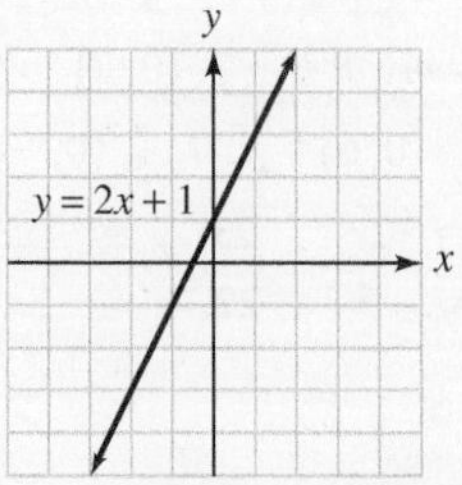

Unless otherwise noted, all content on this page is © Cengage Learning.

19.

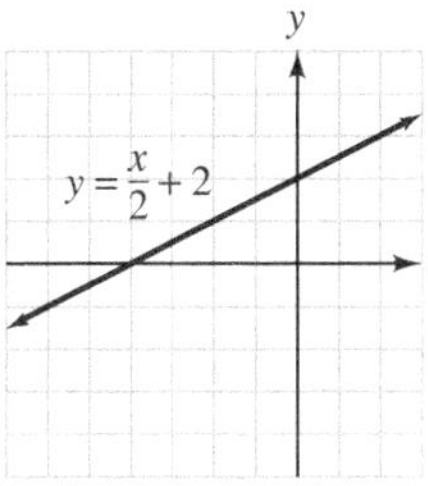

20.

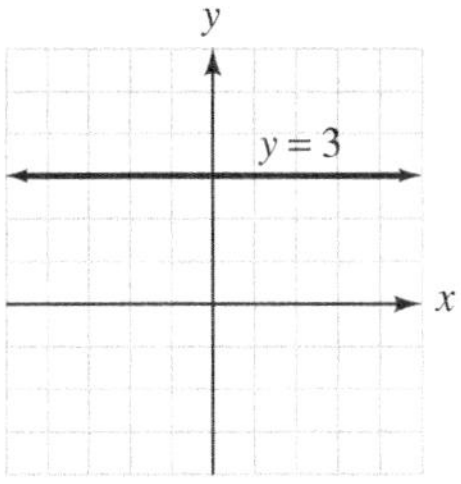

21.

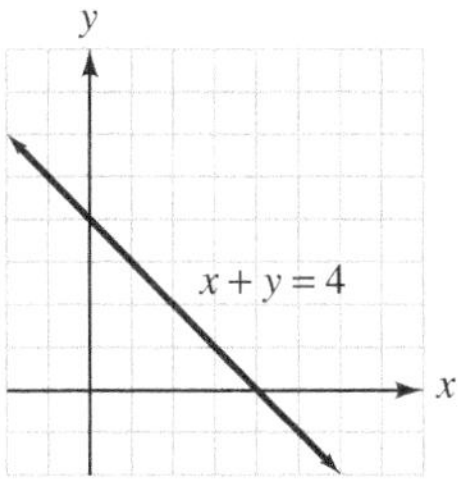

22.

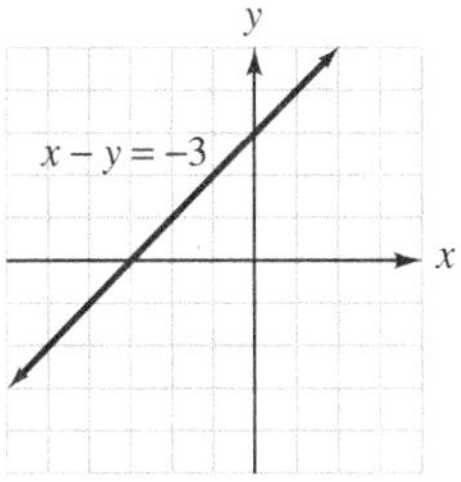

23.

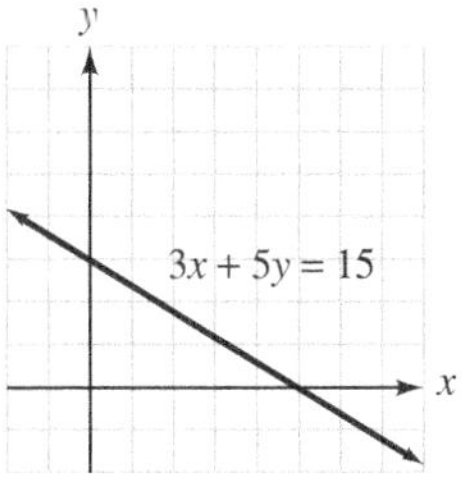

24.

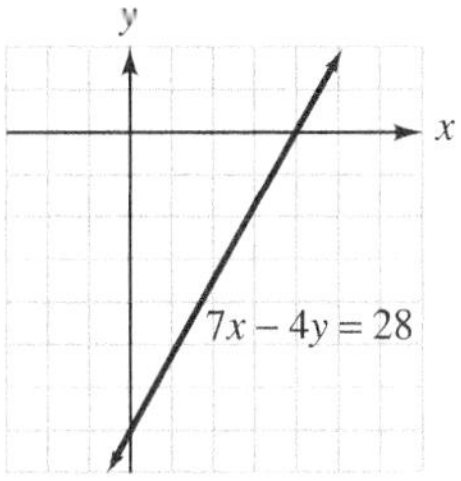

25. $-\frac{3}{2}$ **26.** $-\frac{5}{4}$ **27.** -7 **28.** 0 **29.** $\frac{5}{2}$
30. undefined **31.** positive **32.** 0 **33.** undefined
34. negative **35.** perpendicular **36.** parallel
37. neither **38.** neither **39.** perpendicular
40. parallel **41.** $\frac{1}{3}$ **42.** \$20,500 per year
43. $y = 5x + 13$ **44.** $y = -\frac{1}{3}x + 1$ **45.** $y = \frac{1}{9}x + 1$
46. $y = -\frac{3}{5}x + \frac{2}{5}$ **47.** $-\frac{1}{3}, (0, 6)$
48. $\frac{1}{2}, \left(0, -\frac{3}{2}\right)$ **49.** $-\frac{2}{5}, \left(0, \frac{1}{5}\right)$
50. $-\frac{1}{3}, \left(0, \frac{1}{3}\right)$ **51.** $y = -3x + 2$
52. $y = -7$ **53.** $y = 7x$ **54.** $y = \frac{1}{2}x - \frac{3}{2}$

55.

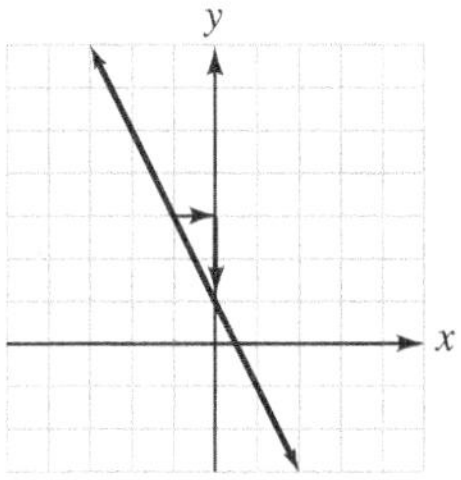

56.

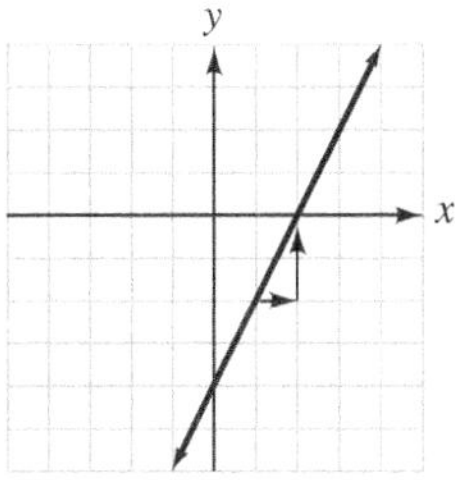

57.

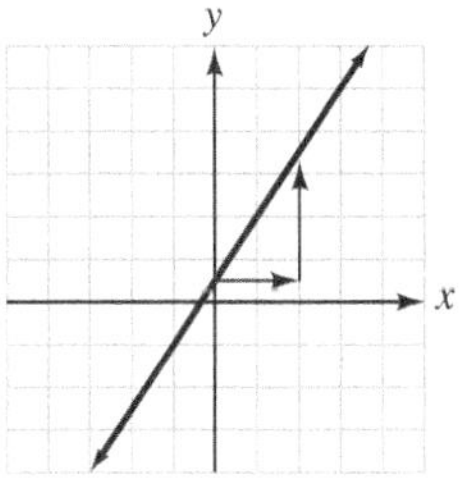

58.

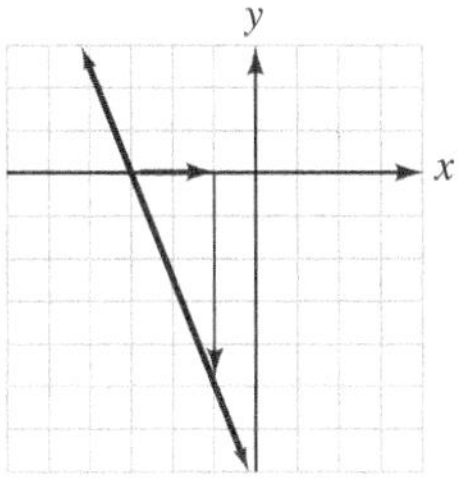

59. neither **60.** perpendicular **61.** parallel
62. perpendicular **63.** $y = 7x - 9$ **64.** $y = -\frac{3}{2}x + \frac{1}{2}$
65. $y = -\frac{5}{2}x$ **66.** $y = -3x - 4$ **67.** $y = -2x + 1$
68. $y = \frac{1}{2}x + 3$ **69.** \$1,200
70. domain: $\{-3, 0\}$; range: $\{-2, -1, 5\}$
71. function **72.** 5 **73.** 15 **74.** 6 **75.** -7

76.

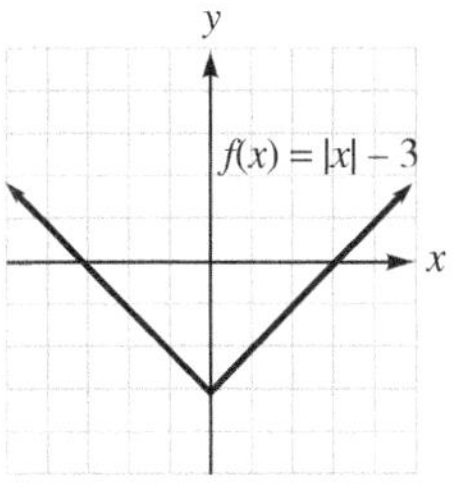

domain: $\mathbb{R}$;
range: $\{y \mid y \text{ is a real number and } y \geq -3\}$

77. no **78.** yes

79.

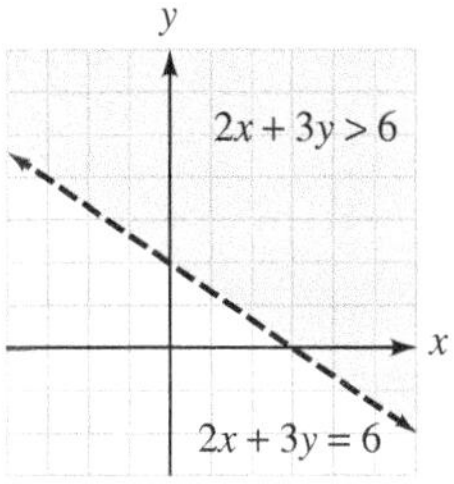

80.

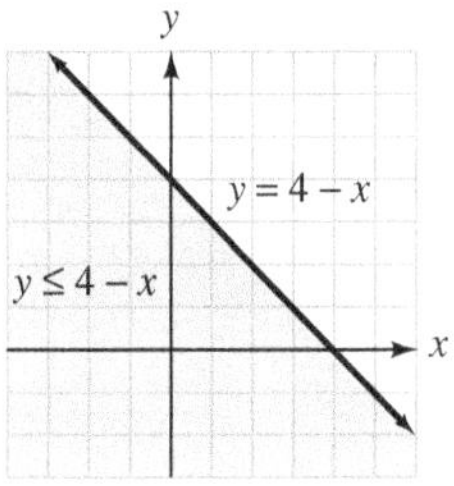

81.

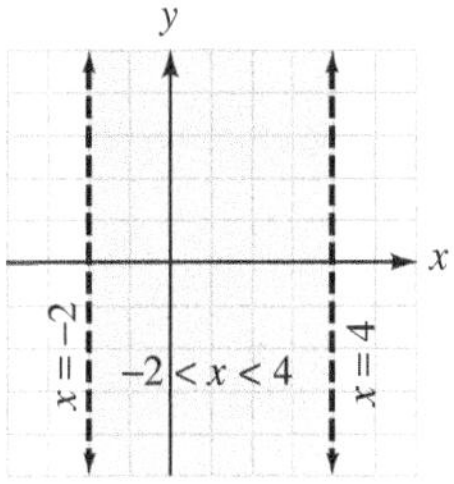

82.

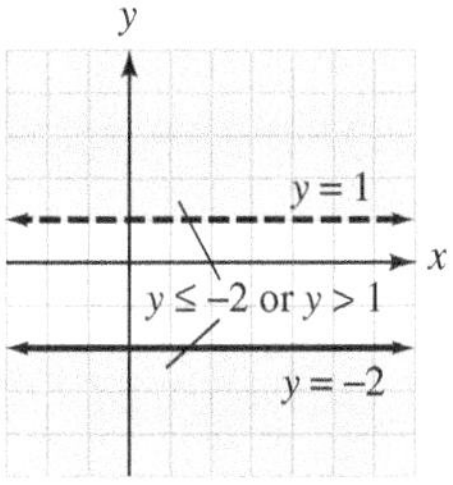

Chapter 3 Test (page 227)

1.

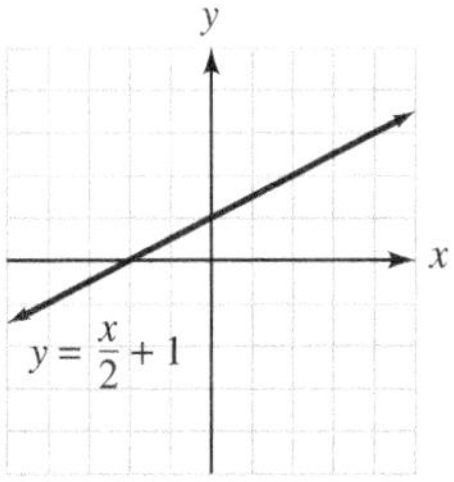

2.

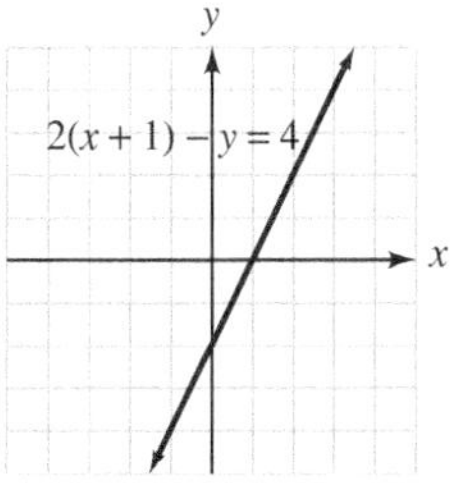

3.

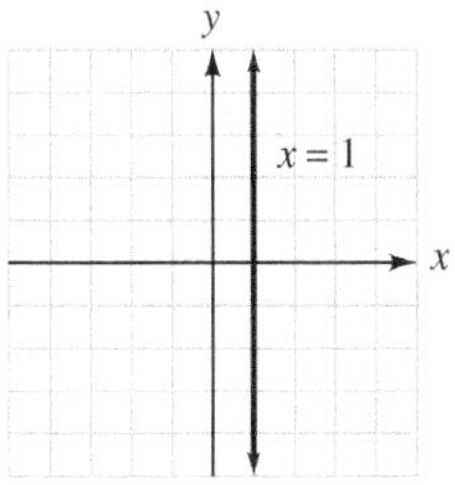

4.

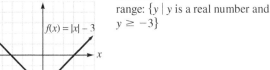

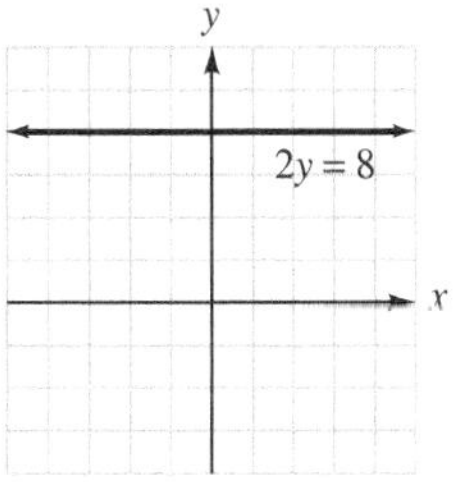

5. $\frac{5}{6}$ **6.** -1 **7.** $\frac{3}{4}$ **8.** $(0, 21)$ **9.** 0

Unless otherwise noted, all content on this page is © Cengage Learning.

10. undefined **11.** equal **12.** -1
13. perpendicular **14.** parallel **15.** $\frac{1}{4}$
16. $16,666.67 **17.** 2 **18.** $-\frac{1}{2}$ **19.** $y - 5 = 7(x + 2)$
20. $y = \frac{3}{4}x - 5$ **21.** $x = -7$ **22.** $y = -3x + 4$
23. 4 **24.** -11 **25.**

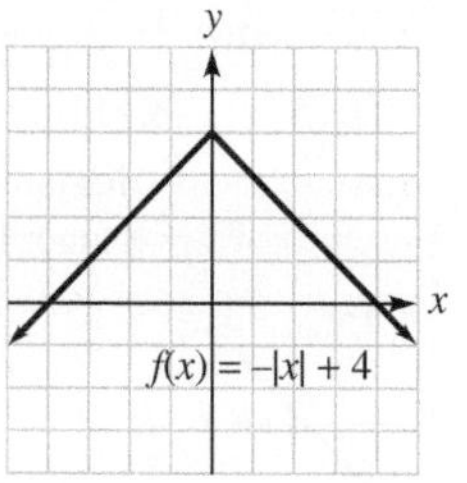

domain: $\mathbb{R}$; range: $\{y \mid y \text{ is a real number and } y = \leq 4\}$

26. no **27.** yes
28.

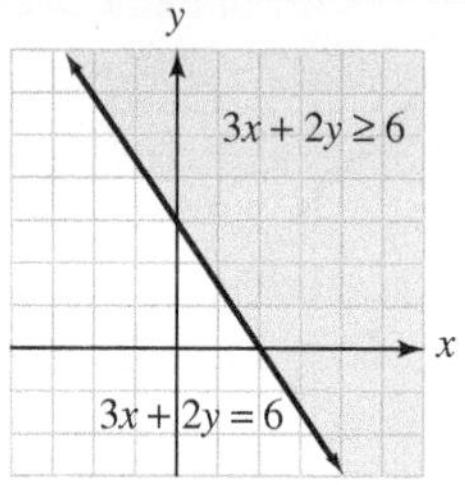

29.

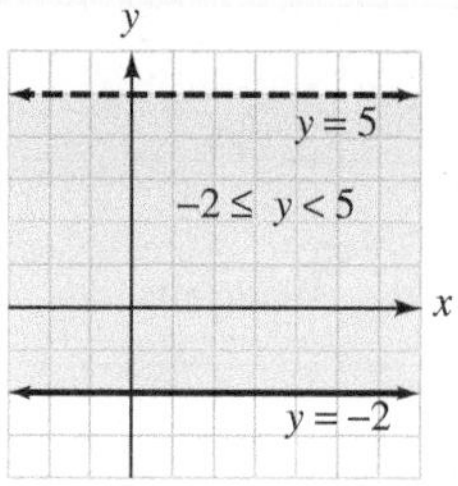

30. $y < -x + 1$

Getting Ready (page 231)

1. 8 **2.** 9 **3.** 6 **4.** 6 **5.** 12 **6.** 32
7. 18 **8.** 3

Exercises 4.1 (page 237)

1. 64 **3.** -343 **5.** 36 **7.** 48
9.

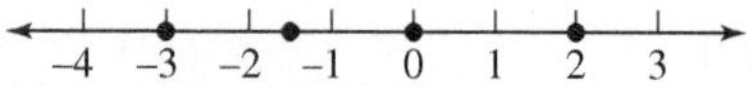

11. the product of 3 and the sum of x and y
13. $|2x| + 3$ **15.** base, $-5, 3$ **17.** $(4y)(4y)(4y)$
19. $y \cdot y \cdot y \cdot y \cdot y$ **21.** $x^n y^n$ **23.** a^{bc}
27. base x, exponent 4 **29.** base 7, exponent 2
31. base $2y$, exponent 3 **33.** base x, exponent 4
35. base x, exponent 1 **37.** base x, exponent 3
39. -25 **41.** 13 **43.** -84 **45.** -55 **47.** $5 \cdot 5 \cdot 5$
49. $-5 \cdot x \cdot x \cdot x \cdot x \cdot x \cdot x$ **51.** $-2 \cdot y \cdot y \cdot y \cdot y$
53. $(3t)(3t)(3t)(3t)(3t)$ **55.** 2^3 **57.** x^4 **59.** $(2x)^3$
61. $-4t^4$ **63.** x^7 **65.** x^{10} **67.** a^{12} **69.** y^9
71. $12x^7$ **73.** $-10x^5$ **75.** 3^8 **77.** y^{15} **79.** x^{25}
81. a^{27} **83.** x^{31} **85.** r^{36} **87.** x^3y^3 **89.** r^6s^4
91. $16a^2b^4$ **93.** $-8r^6s^9t^3$ **95.** $\frac{a^3}{b^3}$ **97.** $\frac{16x^4}{81y^8}$
99. x^2 **101.** y^4 **103.** $3a$ **105.** ab^4 **107.** t^3
109. $6x^9$ **111.** $-8a^{15}$ **113.** $243z^{30}$ **115.** s^{33}
117. $\frac{-32a^5}{b^5}$ **119.** $\frac{b^6}{27a^3}$ **121.** $\frac{x^{12}y^{16}}{2}$ **123.** $\frac{y^3}{8}$ **125.** $-\frac{8r^3}{27}$
127. $\frac{10r^{13}s^3}{3}$ **129.** 2 ft **131.** $16,000 **133.** $45,947.93

Getting Ready (page 239)

1. $\frac{1}{3}$ **2.** $\frac{1}{y}$ **3.** 1 **4.** $\frac{1}{xy}$

Exercises 4.2 (page 242)

1. x **3.** $(5x)$ **5.** $\frac{1}{a}$ **7.** $\frac{a}{b}$ **9.** 2 **11.** $\frac{6}{5}$
13. $s = \frac{f(P - L)}{i}$ or $s = \frac{fP - fL}{i}$ **15.** $1, \frac{1}{x^n}$ **17.** $\frac{1}{8^2}$
19. 1 **21.** 5 **23.** 1 **25.** 1 **27.** 1 **29.** -2
31. $\frac{1}{625}$ **33.** $\frac{1}{a^5}$ **35.** $\frac{1}{16y^4}$ **37.** $-\frac{1}{125p^3}$ **39.** $\frac{1}{y^{12}}$
41. $\frac{-5}{x^4}$ **43.** $\frac{1}{y}$ **45.** $\frac{1}{a^5}$ **47.** x^5 **49.** $\frac{5b^4}{a}$
51. $\frac{1}{a^{24}}$ **53.** $\frac{1}{b^{32}}$ **55.** x^{3m} **57.** $\frac{1}{x^{3n}}$ **59.** y^{2m+2}
61. x^{3n+12} **63.** u^{5m} **65.** y^{4n-8} **67.** 8 **69.** 1
71. 8 **73.** 512 **75.** $\frac{1}{a^9}$ **77.** y^m **79.** $\frac{1}{64t^3}$
81. $\frac{1}{a^3b^6}$ **83.** $\frac{1}{x^4y^2}$ **85.** b^2 **87.** $\frac{1}{m^9n^{12}}$ **89.** $\frac{1}{x^2}$
91. $\frac{1}{a^2b^4}$ **93.** $-\frac{y^{10}}{32x^{15}}$ **95.** a^8b^{12} **97.** $\frac{1}{b^{14}}$
99. $\frac{1}{9a^2b^2}$ **101.** $\frac{c^{15}}{216a^9b^3}$ **103.** $-\frac{27}{r^9}$ **105.** $\frac{16u^4v^8}{81}$
107. $\frac{1}{x^{3n}}$ **109.** $6,678.04 **111.** $95,060.40

Getting Ready (page 244)

1. 100 **2.** 1,000 **3.** 10 **4.** $\frac{1}{100}$ **5.** 500
6. 8,000 **7.** 30 **8.** $\frac{7}{100}$

Exercises 4.3 (page 249)

1. 3.9 **3.** 8.37 **5.** 1.052 **7.** 3
9. comm. prop. of multiplication **11.** 7
13. scientific notation **15.** 4.5×10^5 **17.** 1.7×10^6
19. 5.9×10^{-3} **21.** 2.75×10^{-6} **23.** 4.25×10^3
25. 3.7×10^{-5} **27.** 230 **29.** 812,000
31. 0.00115 **33.** 0.000976 **35.** 714,000
37. 30,000 **39.** 200,000 **41.** 0.000075
43. 5.1×10^{-6} **45.** 8.63×10^8 **47.** 4×10^{-22}
49. 370,000,000 **51.** 0.000032 **53.** 3.72×10^2
55. 4.72×10^3 **57.** 3.72×10^{-1} **59.** 2.57×10^{13} mi
61. 114,000,000 mi **63.** 6.22×10^{-3} mi
65. 1.9008×10^{11} ft **67.** $\$1.64512 \times 10^{12}$; $1,645,120,000,000 **69.** 3.3×10^{-1} km/sec
71. X-rays, visible light, infrared
73. 1.5×10^{-4}; 2.5×10^{13}

Getting Ready (page 251)

1. $2x^2y^3$ **2.** $3xy^3$ **3.** $2x^2 + 3y^2$ **4.** $x^3 + y^3$
5. $6x^3y^3$ **6.** $5x^2y^2z^4$ **7.** $5x^2y^2$ **8.** $x^3y^3z^3$

Exercises 4.4 (page 259)

1. $5a^2 + 2b^3$ **3.** $4a^3b^2$ **5.** $x^3 + 3x^2 + x$ **7.** $a^3 + b^3$
9. 8 **11.**

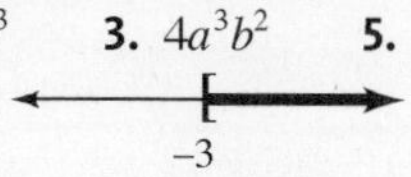

13. x^{32} **15.** y^9

Unless otherwise noted, all content on this page is © Cengage Learning.

17. algebraic **19.** monomial; binomial; trinomial
21. sum **23.** polynomial **25.** cubic **27.** descending; 5
29. function **31.** yes **33.** no **35.** binomial
37. trinomial **39.** monomial **41.** binomial **43.** 7th
45. 3rd **47.** 8th **49.** 6th **51.** -4 **53.** -5
55. $1, -2, -3, -2, 1$ **57.** $-6, 1, 2, 3, 10$
59. -14 **61.** $-\frac{3}{2}$

63.

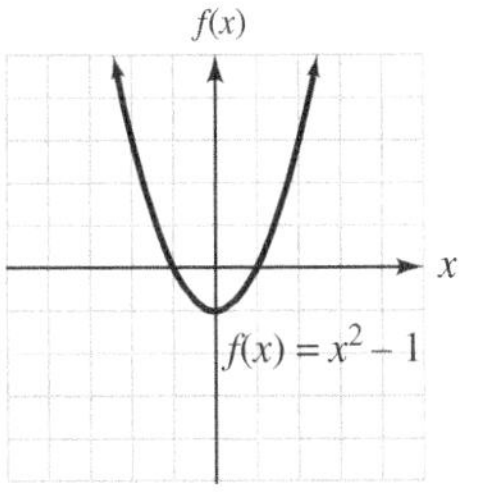

D: $\mathbb{R}$, R: $[-1, \infty)$

65.

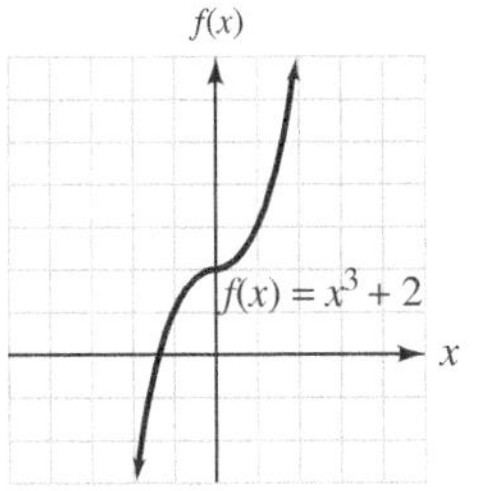

D: $\mathbb{R}$, R: $\mathbb{R}$

67. 7 **69.** -8 **71.** trinomial **73.** none of these
75. binomial **85.** 12th **87.** 0th **89.** 18 **91.** 11
93. 2.25 **95.** 64 ft **97.** \$28,362 **99.** 63 ft

Getting Ready (page 261)

1. $5x$ **2.** $2y$ **3.** $25x$ **4.** $5z$ **5.** $12r$
6. not possible **7.** 0 **8.** not possible

Exercises 4.5 (page 265)

1. $9x$ **3.** $12a$ **5.** unlike **7.** like **9.** -8
11. -3 **13.**

15. monomial **17.** coefficients, variables
19. like terms **21.** $7y$ **23.** unlike terms **25.** $13x^3$
27. $8x^3y^2$ **29.** $10x^4y^2$ **31.** unlike terms **33.** $9y$
35. $25x^2$ **37.** $-4t^6$ **39.** $29x^2y^4$ **41.** $-21a$
43. $16u^3$ **45.** $7x^5y^2$ **47.** $-20ab^3$ **49.** $7x + 4$
51. $8y^2 + 3y - 3$ **53.** $5x^2 + x + 11$
55. $-7x^3 - 7x^2 - x - 1$ **57.** $2a + 7$ **59.** $5a^2 - 2a - 2$
61. $5x^2 + 6x - 8$ **63.** $-x^3 + 6x^2 + x + 14$
65. $-11x - 9y$ **67.** $-2x^2 - 1$ **69.** $6x - 2$
71. $5x^2 - 25x - 20$ **73.** $14rst$ **75.** $-6a^2bc$
77. $-28x^3y^6$ **79.** $216x^5y^{10}$ **81.** $3x - 3y$
83. $-3z^2 + z - 1$ **85.** $-13x^3z + 5x^2z^2 - 14z^3$
87. $13x^3 - 72x^2z + 48xz^2$ **89.** $xy^2 + 13y^2$
91. $6x^2 - 2x - 1$ **93.** $-z^3 - 2z^2 + 5z - 17$
95. \$114,000 **97.** $y = 1{,}900x + 225{,}000$
99. a. $y = -1{,}100x + 6{,}600$ **b.** $y = -1{,}700x + 9{,}200$
103. $6x + 3h - 10$ **105.** 49

Getting Ready (page 268)

1. $6x$ **2.** $3x^4$ **3.** $5x^3$ **4.** $8x^5$ **5.** $3x + 15$
6. $-2x - 10$ **7.** $4y - 12$ **8.** $-2y^2 + 6$

Exercises 4.6 (page 275)

1. $15x^3$ **3.** $-8xy$ **5.** $-5x + 20$ **7.** $8x^2 - 18$
9. comm. prop. of add. **11.** comm. prop. of mult.
13. 0 **15.** monomial **17.** special products **19.** $6x^2$
21. $15x$ **23.** $12x^5$ **25.** $-10t^7$ **27.** $6x^5y^5$
29. $-24b^6$ **31.** $3x + 12$ **33.** $-4t - 28$ **35.** $3x^2 - 6x$
37. $-6x^4 + 2x^3$ **39.** $3x^2y + 3xy^2$
41. $-12x^4 - 18x^3 - 30x^2$ **43.** $2x^7 - x^2$
45. $-6r^3t^2 + 2r^2t^3$ **47.** $a^2 + 9a + 20$
49. $3x^2 + 10x - 8$ **51.** $8a^2 - 16a + 6$
53. $6x^2 - 7x - 5$ **55.** $6s^2 + 7st - 3t^2$
57. $u^2 + 2tu + uv + 2tv$ **59.** $2x^2 - 6x - 8$
61. $3a^3 - 3ab^2$ **63.** $3x^2 + xy - 2y^2$
65. $-3x^2 - 11x - 3$ **67.** $x^2 + 10x + 25$
69. $x^2 - 8x + 16$ **71.** $16t^2 + 24t + 9$
73. $x^2 - 4xy + 4y^2$ **75.** $r^2 - 16$ **77.** $16x^2 - 25$
79. $2x^3 + 11x^2 + 10x - 3$ **81.** $4t^3 + 11t^2 + 18t + 9$
83. $4x^2 + 11x + 6$ **85.** $12x^2 + 14xy - 10y^2$ **87.** -3
89. -8 **91.** -1 **93.** 0 **95.** $-3x^4y^7z^8$
97. $2x^2 + 3x - 9$ **99.** $t^2 - 6t + 9$
101. $-4r^2 - 20rs - 21s^2$ **103.** $9x^2 - 12x + 4$
105. $3x^3 + 8x^2y - 6xy^2 + y^3$ **107.** $x^2y + 3xy^2 + 2x^2$
109. -1 **111.** 0 **113.** $2x^2 + xy - y^2$
115. $4x^2 - 5x - 11$ **117.** $-\frac{1}{2}$ **119.** 4 m **121.** 90 ft

Getting Ready (page 277)

1. $2xy^2$ **2.** y **3.** $\frac{3xy}{2}$ **4.** $\frac{x}{y}$ **5.** xy **6.** 3

Exercises 4.7 (page 281)

1. $\frac{1}{7}$ **3.** $-\frac{8}{9}$ **5.** $\frac{1}{6}$ **7.** -1 **9.** binomial
11. none of these **13.** 2 **15.** polynomial **17.** 1
19. $\frac{a}{b}$ **21.** $\frac{x}{z}$ **23.** $\frac{r^2}{s}$ **25.** $\frac{2x^2}{y}$ **27.** $-\frac{3u^3}{v^2}$
29. $\frac{2}{y} + \frac{3}{x}$ **31.** $\frac{x}{3} + \frac{2}{y}$ **33.** $\frac{1}{5y} - \frac{2}{5x}$ **35.** $\frac{1}{y^2} + \frac{2y}{x^2}$
37. $\frac{1}{y} - \frac{1}{2x} + \frac{2z}{xy}$ **39.** $3x^2y - 2x - \frac{1}{y}$ **41.** $5x - 6y + 1$
43. $3a - 2b$ **45.** $\frac{10x^2}{y} - 5x$ **47.** $-\frac{4x}{3} + \frac{3x^2}{2}$
49. $xy - 1$ **51.** 2 **53.** yes **55.** yes **57.** $\frac{3}{4}$
59. $\frac{42}{19}$ **61.** $\frac{4r}{y^2}$ **63.** $-\frac{13}{3rs}$ **65.** $\frac{x^4}{y^6}$ **67.** a^8b^8
69. $\frac{2}{b} + \frac{1}{3a}$ **71.** $\frac{2}{3y} - \frac{1}{3x}$ **73.** $4x - \frac{3y}{2}$
75. $2x^2 + 4x - 1$ **77.** $\frac{xy^2}{3}$ **79.** a^8 **81.** $\frac{x}{y} - \frac{11}{6} + \frac{y}{2x}$

Getting Ready (page 283)

1. 13 **2.** 21 **3.** 19 **4.** 13

Exercises 4.8 (page 287)

1. 24 **3.** $14\frac{5}{19}$ **5.** $x \neq -2$ **7.** $x \neq \frac{7}{2}$
9. 21, 22, 24, 25, 26, 27, 28 **11.** 5 **13.** -5
15. $18x^2 - 2x - 1$ **17.** divisor, dividend **19.** remainder

Unless otherwise noted, all content on this page is © Cengage Learning.

21. $2x^3 + 8x^2 + 3x - 5$ **23.** $5x^4 + 6x^3 - 4x^2 + 7x$
25. $0x^3$ and $0x$ **27.** $x + 4$ **29.** $x + 2$ **31.** $x - 3$
33. $a - 4$ **35.** $2a - 1$ **37.** $b + 3$ **39.** $x + 1 + \frac{-1}{2x + 3}$
41. $2x + 2 + \frac{-3}{2x + 1}$ **43.** $a + b$ **45.** $2x - y$
47. $x - 3y$ **49.** $4a + b$ **51.** $2x + 1$ **53.** $x - 7$
55. $x + 4$ **57.** $2x - 3$ **59.** $x^2 + 2x + 4$
61. $x^2 + 2x - 1$ **63.** $x^2 + 2x + 1$
65. $5x^2 - x + 4 + \frac{16}{3x - 4}$ **67.** $2x^2 + 2x + 1$
69. $x^2 + 2x - 1 + \frac{-1}{2x + 3}$ **71.** $2x^2 + 8x + 14 + \frac{31}{x - 2}$
73. $3x + 2y$

Getting Ready (page 289)

1. $x + 1$ with a remainder of 1, 1
2. $x + 3$ with a remainder of 9, 9

Exercises 4.9 (page 294)

1. 2 **3.** 2 **5.** 0 **7.** $12a^2 + 4a - 1$
9. $17x^2 - 19x + 14$ **11.** $x - r$ **13.** $P(r)$
15. $x - 4 + \frac{-6}{x - 9}$ **17.** $x - 1 + \frac{-5}{x + 5}$ **19.** $2x^2 + 4x + 3$
21. $4x^2 - 8x + 14 + \frac{8}{x + 2}$ **23.** $6x^2 - x + 1 + \frac{3}{x + 1}$
25. $3x^2 - 2x + 4$ **27.** -1 **29.** -97
31. 23 **33.** -361 **35.** -8 **37.** 44
39. yes; $P(x) = (x - 3)(x^2 + 5)$ **41.** no
43. $3x^2 - x + 2$ **45.** $x - 7 + \frac{28}{x + 2}$ **47.** $x + 2$
49. $x + 7$ **51.** -3 **53.** -3
55. yes; $P(x) = x(7x^2 - 5x - 8)$
57. yes; $P(x) = (x - 5)(x^2 + 5x + 25)$ **59.** $\frac{29}{32}$ **61.** 2
63. -1 **65.** 64 **69.** 1

Chapter Review (page 297)

1. $(-3x)(-3x)(-3x)(-3x)$ **2.** $(\frac{1}{2}pq)(\frac{1}{2}pq)(\frac{1}{2}pq)$
3. 125 **4.** 243 **5.** 36 **6.** -36 **7.** 13 **8.** 25
9. x^{10} **10.** x^9 **11.** y^{21} **12.** x^{42} **13.** a^3b^3
14. $81x^4$ **15.** b^{12} **16.** $-y^2z^5$ **17.** $256s^3$ **18.** $-3y^6$
19. $8x^{12}y^6$ **20.** $25x^6y^2$ **21.** x^4 **22.** $\frac{x^2}{y^2}$ **23.** $\frac{2y^2}{x^2}$
24. $5yz^4$ **25.** 1 **26.** 1 **27.** 9 **28.** $9x^4$ **29.** $\frac{1}{x^3}$
30. x **31.** y **32.** x^{10} **33.** $\frac{1}{x^2}$ **34.** $\frac{a^6}{b^3}$ **35.** $\frac{1}{x^5}$
36. $\frac{1}{9z^2}$ **37.** 7.28×10^2 **38.** 6.23×10^3
39. 2.75×10^{-2} **40.** 9.42×10^{-3} **41.** 7.73×10^0
42. 7.53×10^5 **43.** 1.8×10^{-4} **44.** 6×10^4
45. 38,700 **46.** 0.0000798 **47.** 2.68 **48.** 57.6
49. 7.39 **50.** 0.000437 **51.** 0.03 **52.** 160
53. 8th, monomial **54.** 2nd, binomial **55.** 5th, trinomial
56. 5th, binomial **57.** 23 **58.** -10 **59.** -4 **60.** 4
61. 402 **62.** 0 **63.** 82 **64.** 0.3405 **65.** -4
66. 12 **67.** 0 **68.** $-\frac{15}{4}$

69.
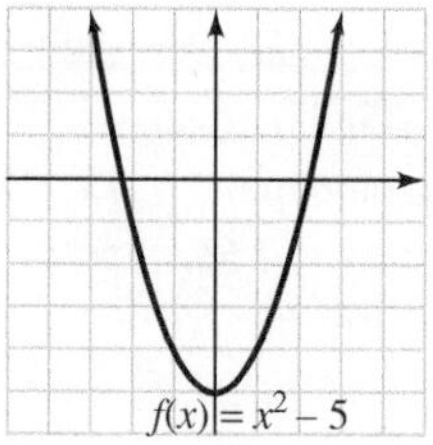

D: ℝ, R: $[-5, \infty)$

70.
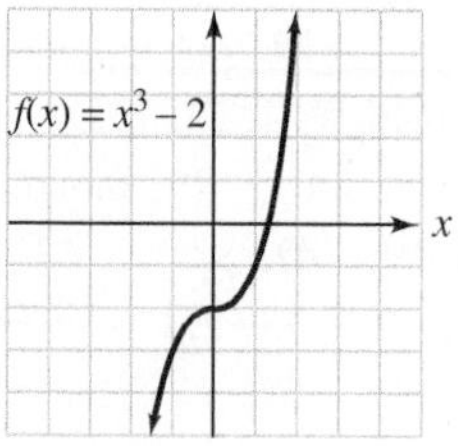

D: ℝ, R: ℝ

71. $3x$ **72.** not possible **73.** $4x^2y^2$ **74.** x^2yz
75. $8x^2 - 6x$ **76.** $4a^2 + 4a - 6$ **77.** $5x^2 + 19x + 3$
78. $6x^3 + 8x^2 + 3x - 72$ **79.** $12x^5y^6$ **80.** x^7yz^5
81. $8x + 12$ **82.** $6x + 12$ **83.** $3x^4 - 5x^2$
84. $2y^4 + 10y^3$ **85.** $-x^2y^3 + x^3y^2$ **86.** $-3x^2y^2 + 3x^2y$
87. $x^2 + 9x + 20$ **88.** $2x^2 - x - 1$ **89.** $6a^2 - 6$
90. $6a^2 - 6$ **91.** $2a^2 - ab - b^2$ **92.** $6x^2 + xy - y^2$
93. $x^2 + 12x + 36$ **94.** $x^2 - 25$ **95.** $y^2 - 49$
96. $x^2 + 8x + 16$ **97.** $x^2 - 6x + 9$ **98.** $y^2 - 4y + 4$
99. $9y^2 + 12y + 4$ **100.** $y^4 - 1$
101. $3x^3 + 7x^2 + 5x + 1$ **102.** $8a^3 - 27$ **103.** 1
104. -1 **105.** 7 **106.** 5 **107.** 1 **108.** 0
109. $\frac{3}{2y} + \frac{3}{x}$ **110.** $3xy - 1$ **111.** $-3a - 4b + 5c$
112. $-\frac{x}{y} - \frac{y}{x}$ **113.** $x + 1 + \frac{3}{x + 2}$ **114.** $x - 5$
115. $3x + 1$ **116.** $x + 5 + \frac{3}{3x - 1}$
117. $3x^2 + 2x + 1 + \frac{2}{2x - 1}$ **118.** $3x^2 - x - 4$
119. yes; $P(x) = (x - 3)(x^2 - 4x - 3)$ **120.** no

Chapter 4 Test (page 302)

1. $2x^3y^4$ **2.** 134 **3.** y^9 **4.** $6b^7$ **5.** $32x^{21}$
6. $8r^{18}$ **7.** -7 **8.** $\frac{5}{y^3}$ **9.** y^3 **10.** $\frac{64a^3}{b^3}$
11. 5.4×10^5 **12.** 2.5×10^{-3}
13. 7,400 **14.** 0.00067 **15.** binomial
16. 10th degree **17.** 0
18.
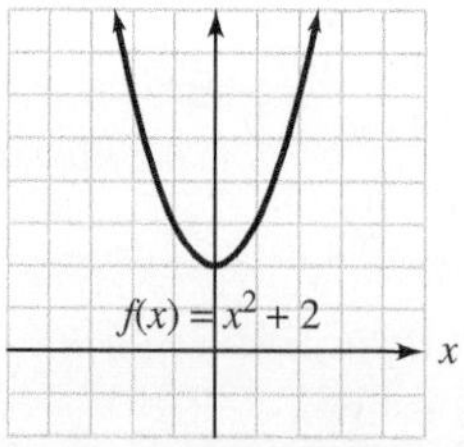

D: $(-\infty, \infty)$, R: $[2, \infty)$

19. $-7x + 2y$ **20.** $-3x + 6$
21. $5x^3 + 2x^2 + 2x - 5$
22. $-x^2 - 5x + 4$
23. $-4x^5y$ **24.** $-20x^9y$
25. $6x^2 - 7x - 20$
26. $2x^3 - 7x^2 + 14x - 12$
27. $\frac{y}{2x}$ **28.** $\frac{a}{4b} - \frac{b}{2a}$
29. $x - 2$ **30.** $\frac{1}{2}$
31. -7 **32.** 47

Cumulative Review (page 303)

1. 14 **2.** 71 **3.** $-\frac{11}{10}$ **4.** 7 **5.** 15 **6.** 4
7. -10 **8.** -6 **9.**
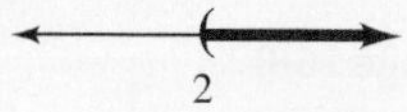

10.
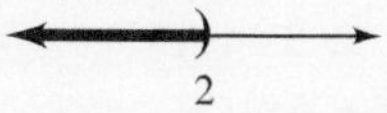

11.

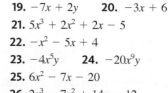

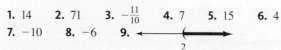

Unless otherwise noted, all content on this page is © Cengage Learning.

12. 

13. $r = \frac{A - p}{pt}$ **14.** $h = \frac{2A}{b}$

15. 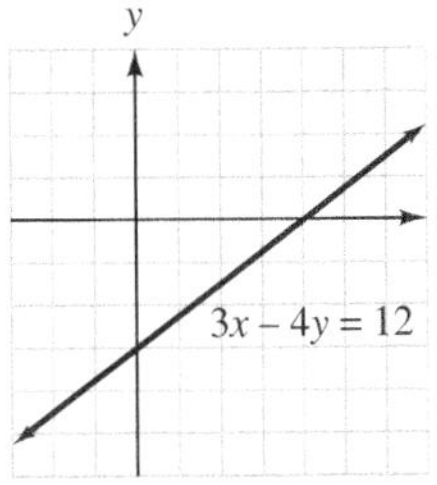

16. 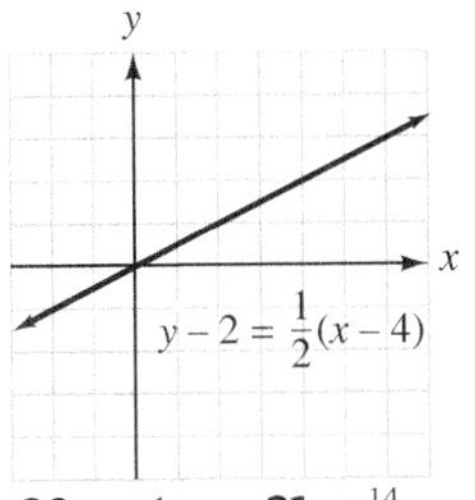

17. 18 **18.** -7 **19.** -12 **20.** -1 **21.** y^{14}
22. $\frac{x}{y}$ **23.** $\frac{a^7}{b^6}$ **24.** x^2y^2 **25.** $x^2 + 4x - 14$
26. $12x^2 - 7x - 10$ **27.** $x^3 - 8$ **28.** $2x + 1$
29. 4.8×10^{18} m **30.** 4 units **31.** 879.6 square units
32. \$512

Getting Ready (page 307)

1. $5x + 15$ **2.** $7y - 56$ **3.** $3x^2 - 2x$
4. $5y^2 + 9y$ **5.** $3x + 3y + ax + ay$
6. $xy + x + 5y + 5$ **7.** $5x + 5 - yx - y$
8. $x^2 + 2x - yx - 2y$

Exercises 5.1 (page 313)

1. 3 **3.** 4 **5.** 2 **7.** 6 **9.** 7 **11.** 11
13. prime-factored **15.** largest **17.** grouping
19. $2^2 \cdot 3$ **21.** $2^3 \cdot 5$ **23.** $3^2 \cdot 5^2$ **25.** $2^5 \cdot 3^2$
27. $3x$ **29.** $5x$ **31.** $2b$ **33.** $6xy$ **35.** $3a$
37. 4 **39.** 4 **41.** $4, x$ **43.** $3x, x$ **45.** r^2
47. $3(x + 2)$ **49.** $4(x - 2)$ **51.** $3x(2x - 3)$
53. $2b^2(2b - 5)$ **55.** $t^2(t + 2)$ **57.** $5xy^3(2x + 3y)$
59. $a^2b^3z^2(az - 1)$ **61.** $8xy^2z^3(3xyz + 1)$
63. $3(x + y - 2z)$ **65.** $a(b + c - d)$
67. $r(s - t + u)$ **69.** $a^2b^2x^2(1 + a - abx)$
71. $-(x + 2)$ **73.** $-(a + b)$ **75.** $-(2x - 5y)$
77. $-(3ab + 5ac - 9bc)$ **79.** $-3xy(x + 2y)$
81. $-2ab^2c(2ac - 7a + 5c)$ **83.** $(a + b)$ **85.** $(m - n)$
87. $(r - 2s)$ **89.** $(x + 3 - y)$ **91.** $(y + 1)(x - 5)$
93. $(x^2 - 2)(3x - y + 1)$ **95.** $(x + y)(x + y + b)$
97. $(x - 3)(x - 2)$ **99.** $5(2a - 1)(a - 2b)$
101. $3(c - 3d)(x + 2y)$ **103.** $(x + y)(2 + a)$
105. $(p - q)(9 + m)$ **107.** $(3x^2 + 4)(3x + 1)$
109. $(2a^2 - 1)(4a - 1)$ **111.** $(a + b)(x - 1)$
113. $(a - b)(x - y)$ **115.** $r^2(r^2 + 1)$
117. $6uvw^2(2w - 3v)$ **119.** $-7ab(2a^5b^5 - 7ab^2 + 3)$
121. $(a + b + c)(3x - 2y)$ **123.** $7xy(r + 2s - t)(2x - 3)$
125. $(v - 3w)(3t + u)$ **127.** $-2c(b + c)(2a - 1)$
129. $x^2(a + b)(x + 2y)$ **131.** $(y - 3)(y^2 - 5)$
133. $(r - s)(2 + b)$ **135.** $r(r + s)(a - b)$

Getting Ready (page 315)

1. $a^2 - b^2$ **2.** $4r^2 - s^2$ **3.** $9x^2 - 4y^2$ **4.** $16x^4 - 9$

Exercises 5.2 (page 318)

1. $(x - 3)$ **3.** $(z + 2)$ **5.** $(5 - t)$
7. $(9 + y)$ **9.** $(2m - 3n)$ **11.** $(5y^3 + 8x^2)$
13. $p = w\left(k - h - \frac{v^2}{2g}\right)$ **15.** difference of two squares
17. prime **19.** $(x + 6)(x - 6)$ **21.** $(y + 7)(y - 7)$
23. $(2y + 7)(2y - 7)$ **25.** $(5x^2 + 9)(5x^2 - 9)$
27. $(3x + y)(3x - y)$ **29.** $(5t + 6u)(5t - 6u)$
31. $(10a + 7b)(10a - 7b)$ **33.** $(x^2 + 3y)(x^2 - 3y)$
35. $8(x + 2y)(x - 2y)$ **37.** $2(a + 2y)(a - 2y)$
39. $4(5x + 2y)(5x - 2y)$ **41.** $x(x + y)(x - y)$
43. $x(2a + 3b)(2a - 3b)$ **45.** $3m(m + n)(m - n)$
47. $(u^2 + 4)(u + 2)(u - 2)$ **49.** $(a^2 + b^2)(a + b)(a - b)$
51. $2(x^2 + y^2)(x + y)(x - y)$
53. $b(a^2 + b^2)(a + b)(a - b)$
55. $2y(x^2 + 16y^2)(x + 4y)(x - 4y)$
57. $(a + 3)^2(a - 3)$ **59.** $(7y + 15z^2)(7y - 15z^2)$
61. $x^2(2x + y)(2x - y)$ **63.** $(x^2 + 9)(x + 3)(x - 3)$
65. $(4y^4 + 9z^2)(2y^2 + 3z)(2y^2 - 3z)$ **67.** prime
69. $(x^4y^4 + 1)(x^2y^2 + 1)(xy + 1)(xy - 1)$
71. $2ab(a + 11b)(a - 11b)$
73. $3a^2(a^4 + b^2)(a^2 + b)(a^2 - b)$
75. $3a^4(a^2 + 9b^4)(a + 3b^2)(a - 3b^2)$
77. $a^2b^3(b^2 + 25)(b + 5)(b - 5)$ **79.** $3ay(a^4 + 2y^4)$
81. prime **83.** $2xy(x^8 + y^8)$
85. $(7y + 7 + x)(7y + 7 - x)$ **87.** $(y + 4)(y - 4)(y - 3)$
89. $2c^2d^2(5c + 2d)(5c - 2d)$

Getting Ready (page 320)

1. $x^2 + 12x + 36$ **2.** $y^2 - 14y + 49$ **3.** $a^2 - 6a + 9$
4. $x^2 + 9x + 20$ **5.** $r^2 - 7r + 10$ **6.** $m^2 - 4m - 21$
7. $a^2 + ab - 12b^2$ **8.** $u^2 - 8uv + 15v^2$
9. $x^2 - 2xy - 24y^2$

Exercises 5.3 (page 327)

1. 1, 4 **3.** $-2, 3$ **5.** $-1, -4$ **7.** $-2, 9$

9. 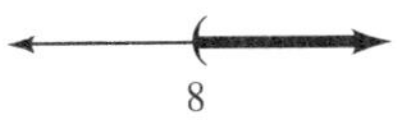

11. 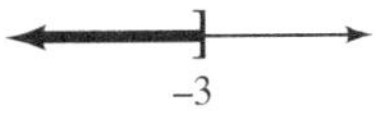

13. 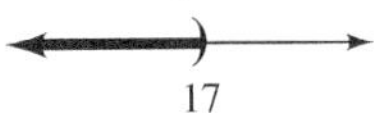

15. 

17. $(x + y)^2$ **19.** 3 **21.** $-, 3$ **23.** 6, 1 **25.** 4, 2
27. $y, 2y$ **29.** $(x + 4)(x + 1)$ **31.** $(z + 2)(z + 4)$
33. $(x + 3)(x + 5)$ **35.** $(x + 2)(x + 10)$
37. $(t - 7)(t - 2)$ **39.** $(x - 2)(x - 6)$
41. $(x - 4)(x - 5)$ **43.** $(r - 1)(r - 7)$
45. $(q + 9)(q - 1)$ **47.** $(s + 13)(s - 2)$

Unless otherwise noted, all content on this page is © Cengage Learning.

49. $(c + 5)(c - 1)$ **51.** $(y + 6)(y - 2)$
53. $(b - 6)(b + 1)$ **55.** $(a - 13)(a + 3)$
57. $(m - 5)(m + 2)$ **59.** $(x - 8)(x + 5)$
61. $(m + 7n)(m - 2n)$ **63.** $(a - 6b)(a + 2b)$
65. $(a + 9b)(a + b)$ **67.** $(m - 10n)(m - n)$
69. $-(x + 5)(x + 2)$ **71.** $-(y + 5)(y - 3)$
73. $-(t + 8)(t - 4)$ **75.** $-(r - 10)(r - 4)$
77. prime **79.** prime **81.** $2(x + 3)(x + 7)$
83. $3y(y - 6)(y - 1)$ **85.** $3(z - 4t)(z - t)$
87. $-4x(x + 3y)(x - 2y)$ **89.** $(x + 2 + y)(x + 2 - y)$
91. $(b - 3 + c)(b - 3 - c)$ **93.** $(x + 5)(x + 1)$
95. $(t - 7)(t - 2)$ **97.** $(a + 8)(a - 2)$
99. $(y - 6)(y + 5)$ **101.** $(x + 3)^2$ **103.** $(y - 4)^2$
105. $(u - 9)^2$ **107.** $(x + 2y)^2$ **109.** $(x - 4)(x - 1)$
111. $(y + 9)(y + 1)$ **113.** $-(r - 2s)(r + s)$ **115.** prime
117. $-(a + b)(a + 5b)$ **119.** $4y(x + 6)(x - 3)$
121. prime **123.** $(r - 5s)^2$
125. $(a + 3 + b)(a + 3 - b)$ **127.** $(t + 9)^2$

Getting Ready (page 329)

1. $6x^2 + 7x + 2$ **2.** $6y^2 - 19y + 10$
3. $8t^2 + 6t - 9$ **4.** $4r^2 + 4r - 15$
5. $6m^2 - 13m + 6$ **6.** $16a^2 + 16a + 3$

Exercises 5.4 (page 336)

1. 2 **3.** 2, 1 **5.** 4 **7.** $n = \frac{l - f + d}{d}$
9. the same as **11.** ac method **13.** +, − **15.** 3, 1
17. y, y **19.** $(3a + 1)(a + 2)$ **21.** $(3a + 1)(a + 3)$
23. $(5t + 3)(t + 2)$ **25.** $(4x + 1)(4x + 3)$
27. $(5y - 3)(y - 4)$ **29.** $(2y - 1)(y - 3)$
31. $(8m - 3)(2m - 1)$ **33.** $(3x - 2)(2x - 1)$
35. $(3a + 2)(a - 2)$ **37.** $(4y + 1)(3y - 1)$
39. $(3y - 2)(4y + 1)$ **41.** $(5y + 1)(2y - 1)$
43. $(4q - 1)(2q + 3)$ **45.** $(2x + 5)(5x - 2)$
47. $(5x + 2)(6x - 7)$ **49.** $(2x - 5)(4x + 3)$
51. $(2x + y)(x + y)$ **53.** $(3x - y)(x - y)$
55. $(5p + 2q)(2p - 3q)$ **57.** $(2p + q)(3p - 2q)$
59. $(2a - 5)(4a - 3)$ **61.** $(4x - 5y)(3x - 2y)$
63. $(2x + 1)(3x + 2)$ **65.** $2(3x + 2)(x - 5)$
67. $(3x - 2)^2$ **69.** $(5x + 3)^2$ **71.** $(3a + 4)^2$
73. $(4x - 5)^2$ **75.** $(4x + y + 3)(4x - y - 3)$
77. $(3 + a + 2b)(3 - a - 2b)$ **79.** $-(4y - 3)(3y - 4)$
81. prime **83.** $2(2x - 1)(x + 3)$
85. $y(y + 12)(y + 1)$ **87.** $3y(3y - 2)(y + 1)$
89. $2s^3(3s + 2)(s - 5)$ **91.** $(4x + 3y)(3x - y)$
93. prime **95.** $(5x + 2y)^2$ **97.** $-2x^2y^3(8x + y)(x - 2y)$
99. $2(4a + b)(3a + b)$

Getting Ready (page 338)

1. $x^3 - 27$ **2.** $x^3 + 8$ **3.** $y^3 + 64$ **4.** $r^3 - 125$
5. $a^3 - b^3$ **6.** $a^3 + b^3$

Exercises 5.5 (page 341)

1. 2^3 **3.** $(-3)^3$ **5.** $(y^4)^3$ **7.** $(-y^3)^3$ **9.** 9
11. $16y^4$ **13.** $4x^2$ **15.** x^{10} **17.** 0.0000000000001 cm
19. sum of two cubes **21.** $(x^2 - xy + y^2)$
23. $(a + 2)(a^2 - 2a + 4)$ **25.** $(5x + 2)(25x^2 - 10x + 4)$
27. $(y + 1)(y^2 - y + 1)$ **29.** $(5 + a)(25 - 5a + a^2)$
31. $(m + n)(m^2 - mn + n^2)$ **33.** $(x + y)(x^2 - xy + y^2)$
35. $(2u + w)(4u^2 - 2uw + w^2)$ **37.** $(x - y)(x^2 + xy + y^2)$
39. $(x - 2)(x^2 + 2x + 4)$ **41.** $(s - t)(s^2 + st + t^2)$
43. $(5p - q)(25p^2 + 5pq + q^2)$
45. $(3a - b)(9a^2 + 3ab + b^2)$
47. $2(x + 3)(x^2 - 3x + 9)$ **49.** $-(x - 6)(x^2 + 6x + 36)$
51. $8x(2m - n)(4m^2 + 2mn + n^2)$
53. $xy(x + 6y)(x^2 - 6xy + 36y^2)$
55. $(x + 1)(x^2 - x + 1)(x - 1)(x^2 + x + 1)$
57. $(x^2 + y)(x^4 - x^2y + y^2)(x^2 - y)(x^4 + x^2y + y^2)$
59. $(y + 2)(y^2 - 2y + 4)$ **61.** $(3x + 5)(9x^2 - 15x + 25)$
63. $(4 - z)(16 + 4z + z^2)$
65. $(4x + 3y)(16x^2 - 12xy + 9y^2)$
67. $(x + y)(x^2 - xy + y^2)(3 - z)$
69. $(3y - z)(9y^2 + 3yz + z^2)(x + 5)$
71. $(a + b)(a^2 - ab + b^2)(x - y)$
73. $(y + 1)(y - 1)(y - 3)(y^2 + 3y + 9)$

Getting Ready (page 342)

1. $3ax^2 + 3a^2x$ **2.** $x^2 - 9y^2$ **3.** $x^3 - 8$ **4.** $2x^2 - 8$
5. $x^2 - 3x - 10$ **6.** $6x^2 - 13x + 6$ **7.** $6x^2 - 14x + 4$
8. $ax^2 + bx^2 - ay^2 - by^2$

Exercises 5.6 (page 344)

1. common factor **3.** sum of two cubes **5.** none, prime
7. difference of two squares **9.** $\mathbb{R}$ **11.** −9
13. factors **15.** binomials **17.** $3(2x + 1)$
19. $(x + 9)(x + 1)$ **21.** $(4t + 3)(2t - 3)$ **23.** $(t - 1)^2$
25. $2(x + 5)(x - 5)$ **27.** prime
29. $-2x^2(x - 4)(x^2 + 4x + 16)$ **31.** $2t^2(3t - 5)(t + 4)$
33. prime **35.** $a(6a - 1)(a + 6)$ **37.** $x^2(4 - 5x)^2$
39. $-3x(2x + 7)^2$
41. $8(x - 1)(x^2 + x + 1)(x + 1)(x^2 - x + 1)$
43. $-5x^2(x^3 - x - 5)$ **45.** prime
47. $2a(b + 6)(b - 2)$ **49.** $-4p^2q^3(2pq^4 + 1)$
51. $(2a - b + 3)(2a - b - 3)$
53. $(2a - b)(4a^2 + 2ab + b^2)$
55. $(y^2 - 2)(x + 1)(x - 1)$ **57.** $(a + b + y)(a + b - y)$

59. $(x-3)(a+b)(a-b)$
61. $(2p^2-3q^2)(4p^4+6p^2q^2+9q^4)$
63. $(5p-4y)(25p^2+20py+16y^2)$
65. $-x^2y^2z(16x^2-24x^3yz^3+15yz^6)$
67. $(9p^2+4q^2)(3p+2q)(3p-2q)$
69. $2(3x+5y^2)(9x^2-15xy^2+25y^4)$
71. $(x+y)(x-y)(x+y)(x^2-xy+y^2)$
73. $2(a+b)(a-b)(c+2d)$

Getting Ready (page 346)

1. 1 **2.** 13 **3.** 3 **4.** 2

Exercises 5.7 (page 351)

1. 7, 8 **3.** 2, −3 **5.** 4, −1 **7.** $\frac{5}{2}, -2$ **9.** u^9
11. $\frac{a}{b}$ **13.** quadratic **15.** second **17.** 0, −7
19. 1, 1 **21.** 0, 3 **23.** $0, -\frac{7}{5}$ **25.** 0, 7 **27.** $0, -\frac{8}{3}$
29. 5, −5 **31.** $\frac{2}{3}, -\frac{2}{3}$ **33.** 12, 1 **35.** 5, −3
37. 6, −3 **39.** 5, −4 **41.** $\frac{1}{2}, -\frac{2}{3}$ **43.** $\frac{1}{2}, 2$
45. 7, −7 **47.** $\frac{9}{2}, -\frac{9}{2}$ **49.** $-\frac{3}{2}, \frac{2}{3}$ **51.** $\frac{1}{8}, 1$
53. −4, 5, 7 **55.** 1, −2, −3 **57.** 0, −1, −2
59. 0, 9, −3 **61.** $0, -3, -\frac{1}{3}$ **63.** 0, −3, −3
65. 0, 4 **67.** $0, -\frac{5}{9}$ **69.** 3, −3, −5 **71.** −3, 7
73. −4, 2 **75.** 9, −9, −2 **77.** $\frac{2}{3}, -\frac{1}{5}$ **79.** 8, 1
81. −3, −5 **83.** $-\frac{1}{3}, 3$ **85.** $-\frac{3}{2}, 1$ **87.** $-\frac{1}{7}, -\frac{3}{2}$
89. 2, −5, 4 **91.** 9, −9 **93.** $\frac{1}{2}, -\frac{1}{2}$ **95.** $0, \frac{1}{5}, -\frac{3}{2}$

Getting Ready (page 352)

1. s^2 sq in. **2.** $(2w+4)$ cm **3.** $x(x+1)$
4. $w(w+3)$ sq in.

Exercises 5.8 (page 356)

1. $A = lw$ **3.** $A = s^2$ **5.** $P = 2l + 2w$ **7.** −10
9. 605 ft^2 **11.** analyze **13.** 4, 8 or −4, −8 **15.** 9 or 1
17. 9 sec **19.** 4 sec and 10 sec **21.** 2 sec **23.** 21 ft
25. 4 m by 9 m **27.** 48 ft **29.** $b = 4$ in., $h = 18$ in.
31. 18 sq units **33.** 1 m **35.** 3 cm **37.** 4 cm by 7 cm
39. 845π m^2 **41.** 6, 16 ft

Chapter Review (page 361)

1. $2^3 \cdot 3$ **2.** $3^2 \cdot 5$ **3.** $2^5 \cdot 3$ **4.** $2 \cdot 3 \cdot 17$
5. $3 \cdot 29$ **6.** $3^2 \cdot 11$ **7.** $2 \cdot 5^2 \cdot 41$ **8.** 2^{12}
9. $4(x+3y)$ **10.** $5a(x^2+3)$ **11.** $7x(x+2)$
12. $3x(3x-1)$ **13.** $2x(x^2+2x-4)$
14. $-a(x+y-z)$ **15.** $a(x+y-1)$
16. $xyz(x+y)$ **17.** $(a-b)(x+y)$
18. $(x+y)(x+y+1)$ **19.** $2x(x+2)(x+3)$
20. $5x(a+b)(a+b-2)$ **21.** $(p+3q)(3+a)$
22. $(r-2s)(a+7)$ **23.** $(x+a)(x+b)$
24. $(y+2)(x-2)$ **25.** $(x+y)(a+b)$
26. $(x-4)(x^2+3)$ **27.** $(x+5)(x-5)$
28. $(xy+4)(xy-4)$ **29.** $(x+2+y)(x+2-y)$
30. $(z+x+y)(z-x-y)$ **31.** $2y(x^2+9y^2)$
32. $(x+y+z)(x+y-z)$ **33.** $(x+2)(x+5)$
34. $(x-3)(x-5)$ **35.** $(x+6)(x-4)$
36. $(x-6)(x+2)$ **37.** $(2x+1)(x-3)$
38. $(3x+1)(x-5)$ **39.** $(5x+2)(3x-1)$
40. $3(2x-1)(x+1)$ **41.** $x(x+3)(6x-1)$
42. $x(4x+3)(x-2)$ **43.** $-2x(x+2)(2x-3)$
44. $-4a(a-3b)(a+2b)$ **45.** $(c-5)(c^2+5c+25)$
46. $(d+2)(d^2-2d+4)$ **47.** $2(x+3)(x^2-3x+9)$
48. $2ab(b-1)(b^2+b+1)$ **49.** $y(3x-y)(x-2)$
50. $5(x+2)(x-3)$ **51.** $a(a+b)(2x+a)$ **52.** prime
53. $(x+3)(x-3+a)$ **54.** $10(x-2y)(x^2+2xy+4y^2)$
55. 0, −5 **56.** 0, 3 **57.** $0, \frac{2}{3}$ **58.** 0, −5
59. 7, −7 **60.** 5, −5 **61.** 4, 5 **62.** −5, 2
63. −4, 6 **64.** 2, 8 **65.** $3, -\frac{1}{2}$ **66.** $1, -\frac{3}{2}$
67. $\frac{3}{4}, -\frac{3}{4}$ **68.** $\frac{2}{3}, -\frac{2}{3}$ **69.** 0, 3, 4 **70.** 0, −2, −3
71. $0, \frac{1}{2}, -3$ **72.** $0, -\frac{2}{3}, 1$ **73.** 5 and 7 **74.** $\frac{1}{3}$
75. 6 ft by 8 ft **76.** 3 ft by 9 ft **77.** 3 ft by 6 ft
78. 9 ft

Chapter 5 Test (page 366)

1. $2^3 \cdot 3 \cdot 5$ **2.** $2^2 \cdot 3^3$ **3.** $5a(12b^2c^3 + 6a^2b^2c - 5)$
4. $3x(a+b)(x-2y)$ **5.** $(x+y)(a+b)$
6. $(x+8)(x-8)$ **7.** $2(a+4b)(a-4b)$
8. $(4x^2+9y^2)(2x+3y)(2x-3y)$ **9.** $(x+6)(x-1)$
10. $(x-11)(x+2)$ **11.** $-1(x+9y)(x+y)$
12. $6(x-4y)(x-y)$ **13.** $(3x+1)(x+4)$
14. $(2a-3)(a+4)$ **15.** prime **16.** $(4x-3)(3x-4)$
17. $6(2a-3b)(a+2b)$ **18.** $(x-2y)(x^2+2xy+4y^2)$
19. $8(3+a)(9-3a+a^2)$
20. $z^3(x^3-yz)(x^6+x^3yz+y^2z^2)$ **21.** 0, −10
22. $-1, -\frac{3}{2}$ **23.** 2, −2 **24.** 3, −6 **25.** $\frac{9}{5}, -\frac{1}{2}$
26. $-\frac{9}{10}, 1$ **27.** $\frac{1}{5}, -\frac{9}{2}$ **28.** $-\frac{1}{10}, 9$ **29.** 12 sec
30. 10 m

Getting Ready (page 369)

1. $\frac{3}{4}$ **2.** 2 **3.** $\frac{5}{11}$ **4.** $\frac{1}{2}$

Exercises 6.1 (page 376)

1. $\frac{2}{3}$ **3.** $\frac{3}{4}$ **5.** $\frac{3}{7}$ **7.** $-\frac{1}{3}$
9. $(a+b)+c = a+(b+c)$ **11.** 0 **13.** $\frac{7}{3}$
15. numerator **17.** 0 **19.** negatives **21.** $\frac{a}{b}$
23. factor, common **25.** −4 **27.** 2, −1
29. $\frac{1}{2}, -7$ **31.** $0, -\frac{1}{3}$
33. $\left\{x \mid x \in \mathbb{R}, x \neq \frac{2}{5}\right\}; \left(-\infty, \frac{2}{5}\right) \cup \left(\frac{2}{5}, \infty\right)$

35. $\left\{m \mid m \in \mathbb{R}, m \neq \frac{3}{2}, -1\right\}; (-\infty, -1)\cup\left(-1, \frac{3}{2}\right)\cup\left(\frac{3}{2}, \infty\right)$

37. $2x$ **39.** $-5y$ **41.** $\frac{x}{y}$

43. in simplest form **45.** $\frac{x}{2}$ **47.** $\frac{3}{x-5}$

49. in simplest form **51.** $\frac{5}{9}$ **53.** $\frac{1}{3}$ **55.** 5

57. in simplest form **59.** $3y$ **61.** $\frac{x+1}{x-1}$ **63.** $\frac{x-5}{x+2}$

65. $\frac{2x}{x-2}$ **67.** $a-2$ **69.** $\frac{x-2}{x+2}$ **71.** $\frac{2(5x-4)}{x+1}$

73. $\frac{4}{3}$ **75.** $x+1$ **77.** $\frac{x^2-x+1}{a+1}$ **79.** $\frac{b+2}{b+1}$

81. -1 **83.** -1 **85.** $\frac{5}{a}$ **87.** $\frac{3x}{y}$ **89.** $\frac{x+2}{x^2}$

91. $\frac{3}{x}$ **93.** in simplest form **95.** $\frac{x}{y}$ **97.** $\frac{3x}{5y}$

99. $\frac{x-3}{5-x}$ or $-\frac{x-3}{x-5}$ **101.** $\frac{x+11}{x+3}$ **103.** $\frac{y+3}{x-3}$

105. $\frac{2(x+2)}{x-1}$ **107.** -1 **109.** $\frac{2}{3}$

Getting Ready (page 378)

1. $\frac{2}{3}$ **2.** $\frac{14}{3}$ **3.** 3 **4.** 6 **5.** $\frac{5}{2}$ **6.** 1
7. $\frac{3}{4}$ **8.** 2

Exercises 6.2 (page 384)

1. 5 **3.** x **5.** 4 **7.** x^2 **9.** $-6x^5y^6z$

11. $\frac{1}{125y^3}$ **13.** $\frac{1}{x^m}$ **15.** $4y^3+4y^2-8y+32$

17. numerators, denominators **19.** $\frac{ac}{bd}$ **21.** $\frac{d}{c}$

23. $\frac{45}{91}$ **25.** $\frac{20x^2}{3y^3}$ **27.** $\frac{(z+7)(z+2)}{7z}$ **29.** $\frac{-3a(a-1)}{5(a+2)}$

31. $\frac{2y}{3}$ **33.** $\frac{yx}{z}$ **35.** $-2y$ **37.** $\frac{b^3c}{a^4}$ **39.** x

41. $\frac{1}{2y}$ **43.** $3y$ **45.** $\frac{5}{z+2}$ **47.** z **49.** $x+2$

51. $\frac{(m-2)(m-3)}{2(m+2)}$ **53.** $\frac{c^2}{ab}$ **55.** $\frac{x-5}{2}$ **57.** $5x$ **59.** $\frac{4}{3}$

61. $\frac{3}{5}$ **63.** 3 **65.** $\frac{2}{y}$ **67.** $\frac{2}{3x}$ **69.** $\frac{x+2}{3}$ **71.** $\frac{y-3}{y^3}$

73. $\frac{x-3}{x-5}$ **75.** $\frac{x-2}{x-3}$ **77.** 1 **79.** $\frac{3}{x+1}$ **81.** $\frac{18x}{x-3}$

83. $\frac{9}{2x}$ **85.** $\frac{y^5}{64}$ **87.** 2 **89.** $\frac{x+2}{x-2}$ **91.** $\frac{(x+1)(x-1)}{5(x-3)}$

93. $\frac{2x(1-x)}{5(x-2)}$ **95.** $\frac{64z}{3x}$ **97.** $\frac{10x^3}{z}$ **99.** $-\frac{a^3}{2d^2}$

101. $\frac{6}{y}$ **103.** $\frac{z}{y}$ **105.** $\frac{(x+7)^2}{(x-3)^2}$ **107.** $\frac{x}{3}$ **109.** $\frac{1}{c-d}$

111. $\frac{-(x-y)(x^2+xy+y^2)}{(x+y)(w+z)}$ **113.** $-\frac{p}{m+n}$

Getting Ready (page 387)

1. $\frac{4}{5}$ **2.** 1 **3.** $\frac{7}{8}$ **4.** 2 **5.** $\frac{1}{9}$ **6.** $\frac{1}{2}$
7. $-\frac{2}{13}$ **8.** $\frac{13}{10}$

Exercises 6.3 (page 395)

1. equal **3.** not equal **5.** equal **7.** equal
9. 3^4 **11.** $2^3 \cdot 17$ **13.** $2 \cdot 3 \cdot 17$ **15.** $2^4 \cdot 3^2$
17. LCD **19.** numerators, common denominator **21.** $\frac{1}{4a}$
23. $\frac{4x}{y}$ **25.** $\frac{9y+2}{y-4}$ **27.** 9 **29.** $-\frac{1}{8}$ **31.** $\frac{y}{x}$

33. $\frac{2(y-2)}{y}$ **35.** $\frac{1}{y}$ **37.** 1 **39.** $\frac{y+4}{y-4}$ **41.** $\frac{4x}{3}$

43. $\frac{6x+2}{x-2}$ **45.** $-\frac{b}{b+1}$ **47.** $\frac{2(x+5)}{x-2}$ **49.** $\frac{84}{32}$

51. $\frac{8xy}{x^2y}$ **53.** $\frac{4x(x+3)}{(x+3)^2}$ **55.** $\frac{2y(x+1)}{x^2+x}$ **57.** $\frac{z(z+1)}{z^2-1}$

59. $\frac{2(x+2)}{x^2+3x+2}$ **61.** $6x$ **63.** $18xy$ **65.** $(x-2)(x+2)$

67. $x(x+6)$ **69.** $\frac{10x}{3y}$ **71.** $\frac{5x+4}{6x}$ **73.** $\frac{2x^2+2}{(x-1)(x+1)}$

75. $\frac{6x^2-x-2}{(5x+2)(x+2)}$ **77.** $\frac{2xy+2x-2y}{xy}$

79. $\frac{3y^2-3y+x^2+x}{x(y-1)}$ **81.** $\frac{-x^2+3x+5}{x(x+1)}$

83. $\frac{x^2+3x+11}{(x+5)(x+2)}$ **85.** $\frac{x+2}{x-2}$ **87.** $-\frac{2}{x-3}$

89. $\frac{2y+7}{y-1}$ **91.** $\frac{2x+4}{2x-y}$ **93.** $\frac{x}{x-2}$ **95.** $\frac{a+4}{a+2}$

97. $(x-3)(x+2)(x+3)$ **99.** $24y$

101. $\frac{7}{6}$ **103.** $-\frac{1}{6}$ **105.** $\frac{4y-5xy}{10x}$ **107.** $\frac{2x^2-x+4}{x(x+2)}$

109. $-\frac{y^2+5y+14}{2y^2}$ **111.** $\frac{y}{x}$ **113.** $\frac{4x-2y}{y+2}$

115. $\frac{-1}{(a+3)(a-3)}$ **117.** $\frac{14y^2+10}{y^2}$

Getting Ready (page 398)

1. 4 **2.** -18 **3.** 7 **4.** -8 **5.** $3+3x$
6. $2-y$ **7.** $12x-2$ **8.** $3y+2x$

Exercises 6.4 (page 403)

1. $\frac{4}{3}$ **3.** $\frac{9}{4}$ **5.** $\frac{1}{14}$ **7.** $\frac{1}{4}$ **9.** $\frac{5}{4}$ **11.** $\frac{5}{7}$

13. t^8 **15.** $-2r^7$ **17.** $\frac{256r^8}{81}$ **19.** $\frac{r^{10}}{9}$

21. complex fraction **23.** single, divide **25.** $\frac{x^2}{2}$

27. $\frac{5t^2}{27}$ **29.** $\frac{a+1}{b}$ **31.** $\frac{y+1}{y-1}$ **33.** $\frac{1+3y}{3-2y}$

35. $\frac{5x+3}{3x+2}$ **37.** $\frac{a+4}{3}$ **39.** $\frac{7x+3}{-x-3}$ **41.** $\frac{3-x}{x-1}$

43. $\frac{xy}{y+x}$ **45.** $\frac{y}{x-2y}$ **47.** $\frac{3y+2x^2}{4y}$ **49.** $\frac{1}{x+2}$

51. $\frac{1}{x+3}$ **53.** $\frac{x-2}{x+3}$ **55.** $\frac{m^2-3m-4}{m^2+5m-3}$ **57.** $\frac{1+x^2}{x+x^2}$

59. $\frac{1+y^2}{1-y^2}$ **61.** $\frac{a^2-a+1}{a^2}$ **63.** 2 **65.** $\frac{x}{x-1}$

67. $\frac{x^2}{(x-1)^2}$ **69.** $\frac{2x(x-3)}{(4x-3)(x+2)}$ **71.** $\frac{(3x+1)(x-1)}{(x+1)^2}$

73. -1 **75.** $\frac{y-5}{y+5}$ **79.** $\frac{1}{2}, \frac{2}{3}, \frac{3}{5}, \frac{5}{8}$

Getting Ready (page 405)

1. $3x+1$ **2.** $8x-1$ **3.** $3+2x$ **4.** $y-6$
5. 19 **6.** $7x+6$ **7.** y **8.** $3x+5$

Exercises 6.5 (page 410)

1. 6 **3.** -7 **5.** -1 **7.** $2(x+5)$
9. $(y-1)(y+2)$ **11.** $(x-3)(x+3)$ **13.** $x(x+8)$
15. $(2x+3)(x-1)$ **17.** $2(x+3)(2x-1)$
19. extraneous **21.** LCD **23.** 3
25. -3 **27.** 1 **29.** 5 **31.** $\varnothing$; -2 is extraneous

33. $\varnothing$; 5 is extraneous **35.** -1 **37.** 5; 3 is extraneous
39. 4 **41.** 2 **43.** $\varnothing$; -2 is extraneous **45.** 5
47. 2, 4 **49.** -4; 4 is extraneous **51.** 1
53. $\varnothing$; -2 is extraneous **55.** 12 **57.** 60 **59.** -2
61. -1 **63.** 6 **65.** 1 **67.** $\varnothing$; 3 is extraneous
69. $a = \frac{b}{b-1}$ **71.** $b = \frac{ad}{d-c}$ **73.** $r_1 = \frac{rr_2}{r_2 - r}$
75. $f = \frac{d_1 d_2}{d_1 + d_2}$ **79.** $1, -1$

Getting Ready (page 412)

1. $\frac{1}{5}$ **2.** $\$(0.05x)$ **3.** $\$\left(\frac{y}{0.05}\right)$ **4.** $\frac{y}{52}$ hr

Exercises 6.6 (page 416)

1. $i = pr$ **3.** $C = qd$ **5.** $-1, 6$ **7.** $-1, -4, 3$
9. 0, 0, 1 **11.** $2, -6$ **15.** 2 **17.** 5 **19.** 40 min
21. $2\frac{2}{9}$ hr **23.** heron, 20 mph; goose, 30 mph **25.** 150 mph
27. 7% and 8% **29.** 5% **31.** 8 hr **33.** 30
35. 4 mph **37.** $\frac{2}{3}$ and $\frac{3}{2}$ **39.** 44 mph and 64 mph
41. 10 **43.** 25 mph

Getting Ready (page 418)

1. $\frac{7}{3}$ **2.** 1.44×10^9 **3.** linear function
4. rational function

Exercises 6.7 (page 427)

1. proportion **3.** not a proportion **5.** $k = 200$
7. $\frac{1}{x^2}$ **9.** -2 **11.** 4.7×10^5 **13.** 0.0025
15. unit costs, rates **17.** extremes, means
19. direct **21.** rational
23. joint, constant of proportionality **25.** direct
27. neither **29.** 7 **31.** 5 **33.** 5 **35.** 5
37. $-\frac{5}{2}, -1$ **39.** $\varnothing$ **41.** $A = kp^2$ **43.** $a = \frac{k}{b^2}$
45. $B = kmn$ **47.** $X = \frac{kw}{q}$ **49.** $4, -1$ **51.** $3, -3$
53. 12 **55.** $P = \frac{ka^2}{j^3}$ **57.** L varies jointly with m and n.
59. E varies jointly with a and the square of b. **61.** X varies directly with x^2 and inversely with y^2. **63.** R varies directly with L and inversely with d^2. **65.** \$62.50 **67.** $8\frac{1}{2}$ gal
69. 32 ft **71.** 80 ft **73.** 0.18 g **75.** 42 ft **77.** $46\frac{7}{8}$ ft
79. 6,750 ft **81.** 64π in.^{2} **83.** 432 mi **85.** 25 days
87. 12 in.^{3} **89.** 9 **91.** \$9,000 **93.** 0.275 in.
95. 546 Kelvin **97.** $85\frac{1}{3}$ **99.** 3 ohms

Chapter Review (page 433)

1. $-4, 2$ **2.** $2, -3$ **3.** $(-\infty, 7) \cup (7, \infty)$
4. $(-\infty, 0) \cup (0, 5) \cup (5, \infty)$ **5.** $-\frac{1}{3}$
6. $\frac{7}{3}$ **7.** $\frac{1}{2x}$ **8.** $\frac{5}{2x}$ **9.** $\frac{x}{x+1}$ **10.** $\frac{1}{x}$ **11.** $\frac{3}{y}$
12. 1 **13.** -1 **14.** $\frac{x+7}{x+3}$ **15.** $\frac{x}{x-1}$ **16.** $\frac{a+2}{a+b}$
17. $\frac{2x^2}{3y}$ **18.** $\frac{6}{x^2}$ **19.** 1 **20.** $\frac{2x}{x+1}$ **21.** $\frac{3}{4}$ **22.** $\frac{1}{x}$
23. $x + 2$ **24.** 1 **25.** $x + 2$ **26.** $\frac{2(3x-4)}{x+3}$

27. -1 **28.** $\frac{x^2 + x - 1}{x(x-1)}$ **29.** $\frac{x-7}{7x}$ **30.** $\frac{x-2}{x(x+1)}$
31. $\frac{x^2 + 4x - 4}{2x^2}$ **32.** $\frac{x+1}{x}$ **33.** 0 **34.** $\frac{81}{16}$
35. $\frac{3}{2}$ **36.** $\frac{1+x}{1-x}$ **37.** $\frac{x(2x+7)}{3x^2 - 1}$ **38.** $x^2 + 3$
39. $\frac{a(a+bc)}{b(b+ac)}$ **40.** -6 **41.** 4 **42.** 3 **43.** $4, -\frac{3}{2}$
44. -2 **45.** 0 **46.** $r_1 = \frac{rr_2}{r_2 - r}$
47. $T_1 = \frac{T_2}{1-E}$ or $T_1 = \frac{-T_2}{E-1}$ **48.** $R = \frac{HB}{B-H}$ or $R = \frac{-HB}{H-B}$
49. $9\frac{9}{19}$ hr **50.** $4\frac{4}{5}$ days **51.** 5 mph **52.** 40 mph
53. 5 **54.** $-4, -12$ **55.** 70.4 ft **56.** 54
57. 6 **58.** 2 **59.** 16 **60.** \$5,460

Chapter 6 Test (page 439)

1. $\frac{3x^2}{5y}$ **2.** $\frac{x+1}{2x+3}$ **3.** 3 **4.** $\frac{5y^2}{4t}$ **5.** $\frac{x+1}{3(x-2)}$
6. $\frac{5c}{9a}$ **7.** $\frac{x^{[illegible]}}{3}$ **8.** $x + 2$ **9.** $\frac{9x-2}{x-2}$ **10.** $\frac{13}{2y+3}$
11. $\frac{2x^2 + x + 1}{x(x+1)}$ **12.** $\frac{2x+6}{x-2}$ **13.** $\frac{3x^5}{2}$ **14.** $\frac{x+y}{y-x}$
15. -5 **16.** $\varnothing$ **17.** 4 **18.** $B = \frac{RH}{R-H}$ **19.** $3\frac{15}{16}$ hr
20. 5 mph **21.** 8,050 ft **22.** $\frac{2}{3}$ **23.** 18 ft
24. $6, -1$ **25.** $\frac{44}{3}$ **26. a.** $x = 3$ **b.** $(-\infty, 3) \cup (3, \infty)$

Cumulative Review (page 440)

1. x^9 **2.** x^{10} **3.** x^3 **4.** 1 **5.** $6x^3 - 2x - 1$
6. $2x^3 + 2x^2 + x - 1$ **7.** $13x^2 - 8x + 1$
8. $16x^2 - 24x + 2$ **9.** $-14x^9y^4$
10. $-35x^5 + 10x^4 + 10x^2$ **11.** $20x^2 + 13x + 2$
12. $15x^2 - 2xy - 8y^2$ **13.** $x + 4$ **14.** $x^2 + x + 1$
15. $4xy^2(1 - 3xy)$ **16.** $(a + b)(3 + x)$
17. $(a + b)(2 + b)$ **18.** $(5p^2 + 4q)(5p^2 - 4q)$
19. $(x - 7)(x + 2)$ **20.** $(x - 3y)(x + 2y)$
21. $(3a + 4)(2a - 5)$ **22.** $(4m + n)(2m - 3n)$
23. $(p - 3q)(p^2 + 3pq + 9q^2)$
24. $8(r + 2s)(r^2 - 2rs + 4s^2)$ **25.** 15 **26.** 4
27. $\frac{2}{3}, -\frac{1}{2}$ **28.** 0, 2 **29.** $-1, -5$ **30.** $\frac{3}{2}, -4$

31.
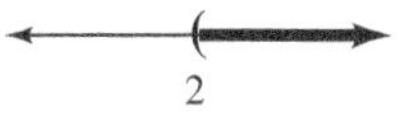

32.

33.

34.

35.
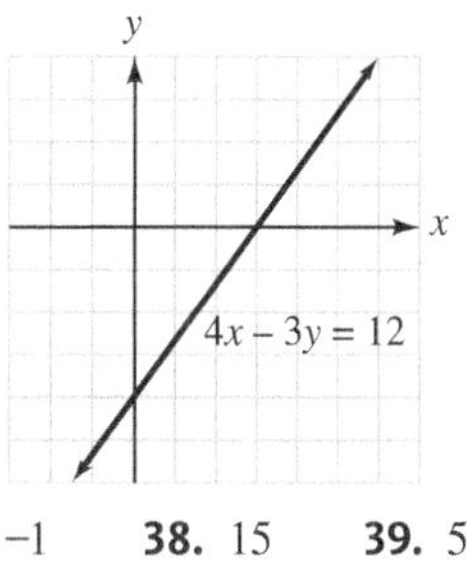

36.
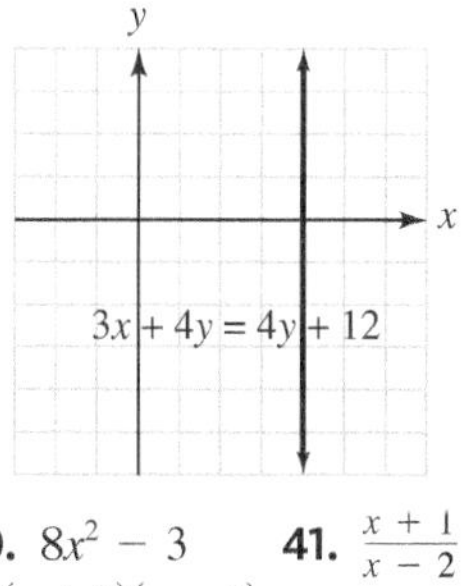

37. -1 **38.** 15 **39.** 5 **40.** $8x^2 - 3$ **41.** $\frac{x+1}{x-2}$
42. $\frac{x-3}{x-2}$ **43.** $\frac{(x-2)^2}{x-1}$ **44.** $\frac{(p+2)(p-3)}{3(p+3)}$ **45.** 1
46. $\frac{2(x^2+1)}{(x+1)(x-1)}$ **47.** $\frac{-1}{2(a-2)}$ **48.** $\frac{y+x}{y-x}$
49. $x + 4$ **50.** $-x^2 + x + 5 + \frac{8}{x-1}$

Unless otherwise noted, all content on this page is © Cengage Learning.

Getting Ready (page 443)

1. yes **2.** yes **3.** no **4.** yes

Exercises 7.1 (page 452)

1. 1 **3.** $x < 2$ **5.** $x < -4$ **7.** $\frac{1}{t^{12}}$ **9.** 179 or more
11. equation **13.** multiplied; divided **15.** identity
17. contradiction **19.** half-open **21.** negative
23. 6 **25.** 4 **27.** 2 **29.** 6 **31.** -4 **33.** -6
35. 24 **37.** 6 **39.** 0 **41.** -2 **43.** $\mathbb{R}$, identity
45. $\varnothing$, contradiction **47.** $B = \frac{3V}{h}$ **49.** $w = \frac{P - 2l}{2}$
51. $x = z\sigma + \mu$ **53.** $x = \frac{y - b}{m}$

55. $(2, \infty)$ 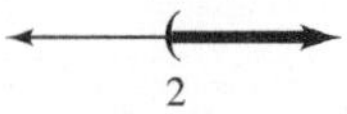

57. $(-\infty, -3]$ 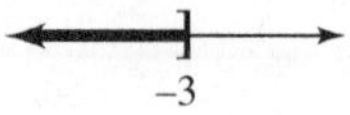

59. $[-2, \infty)$ 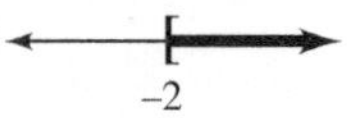

61. $\left(-\infty, -\frac{8}{5}\right)$ 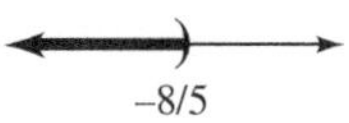

63. $[-36, \infty)$ **65.** $(-\infty, 0]$ **67.** $(-\infty, 6)$ **69.** $(2, \infty)$

71. $(-2, 5)$ 

73. $(8, 11)$ 

75. $[-4, 6)$ 

77. $\varnothing$ **79.** $(5, \infty)$ **81.** $[8, \infty)$

83. $(-\infty, 2) \cup (7, \infty)$ 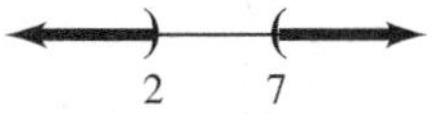

85. $(-\infty, 1)$ 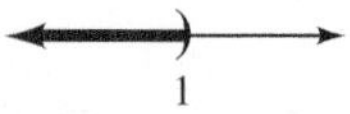

87. $\varnothing$

89. $\mathbb{R}, (-\infty, \infty)$ **91.** $(-\infty, 1]$ 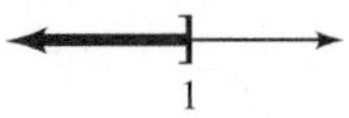

93. $[2, \infty)$ 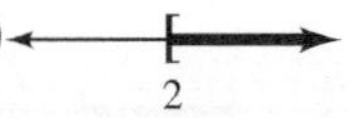

95. $[-21, -3)$ 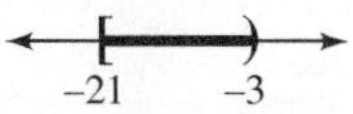

97. $s = \frac{f(P - L)}{i}$ **99.** 12 m by 24 m
101. 20 ft by 45 ft **103.** 7 ft, 15 ft
105. more than \$5,000 **107.** 59

Getting Ready (page 456)

1. $A(2, 4)$; $B(0, -4)$; $C(4, 0)$; $D(-2, -1)$
2. a. 1 **b.** 9 **c.** 5 **d.** 2
3. $\frac{1}{2}$ **4.** 0 **5.** $\frac{7}{2}$ **6.** undefined

Exercises 7.2 (page 469)

1. no **3.** yes **5.** negative **7.** undefined
9. general **11.** slope-intercept **13.** x^9y^6
15. $\frac{x^{12}}{y^8}$ **17.** 6 **19.** -1 **21.** y-axis
23. vertical **25.** $\left(\frac{a + c}{2}, \frac{b + d}{2}\right)$ **27.** change
29. rise **31.** horizontal **33.** perpendicular; reciprocals

35. 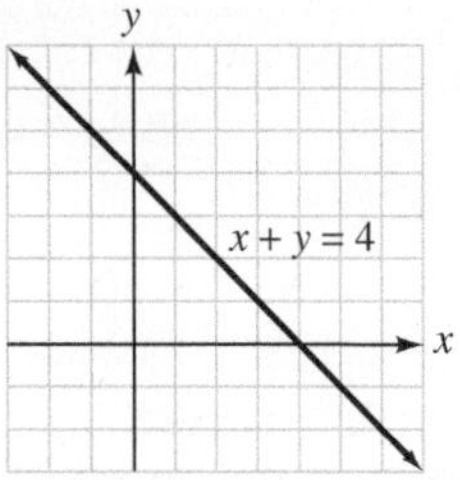

37. 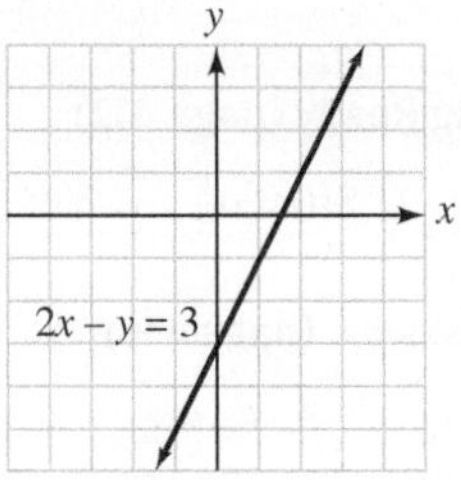

39. 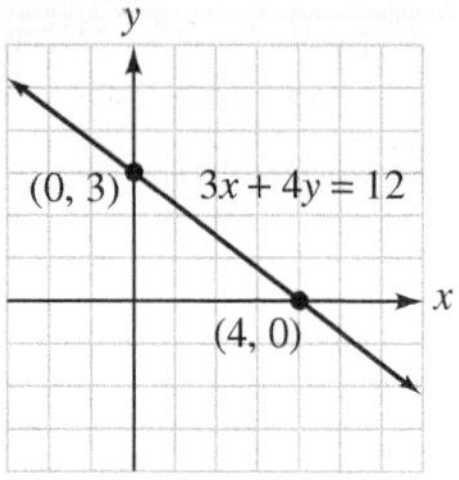

41. 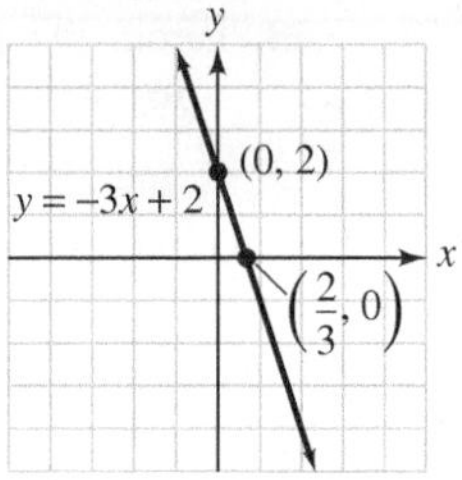

43. 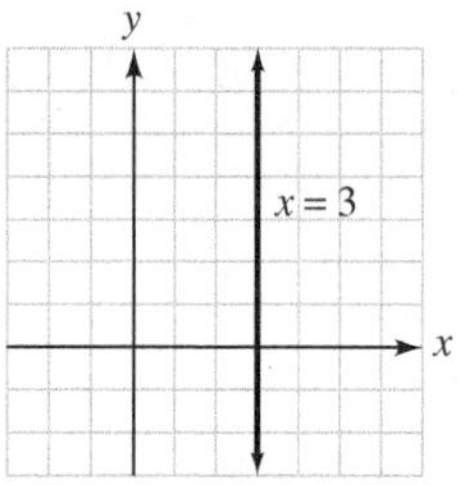

45. 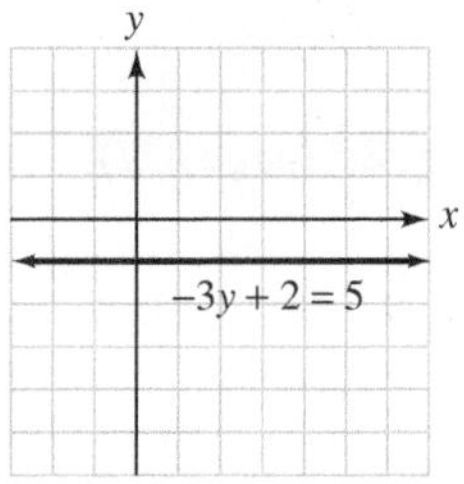

47. $(3, 4)$ **49.** $\left(\frac{1}{2}, -2\right)$ **51.** $\frac{3}{4}$ **53.** $-\frac{7}{5}$
55. 3 **57.** -1 **59.** 0 **61.** undefined **63.** $-\frac{3}{2}$
65. $\frac{3}{4}$ **67.** $\frac{1}{2}$ **69.** 0 **71.** parallel **73.** perpendicular
75. parallel **77.** perpendicular **79.** $y - 7 = 5x$
81. $y = -3(x - 2)$ **83.** $y = 3x + 17$
85. $y = -7x + 54$ **87.** $y = x$ **89.** $y = \frac{7}{3}x - 3$
91. $y = 4x$ **93.** $y = 4x - 3$ **95.** $y = -\frac{1}{4}x$
97. $y = -\frac{1}{4}x + \frac{11}{2}$ **99.** $y = 65x + 275$
101. $y = 3.72x + 95$ **103.** 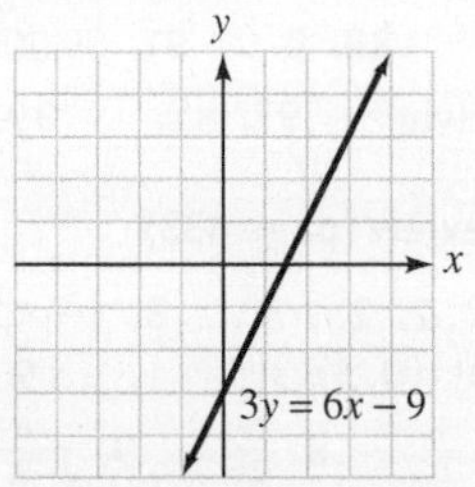

105. $\frac{3}{2}, (0, -4)$ **107.** $\frac{1}{2}, (0, 3)$

Unless otherwise noted, all content on this page is © Cengage Learning.

109.

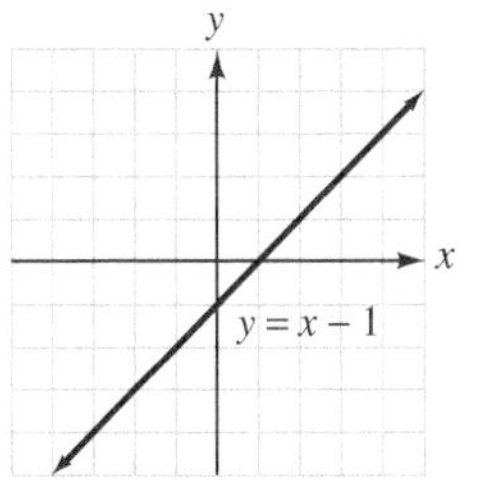

$1, (0, -1)$

111.

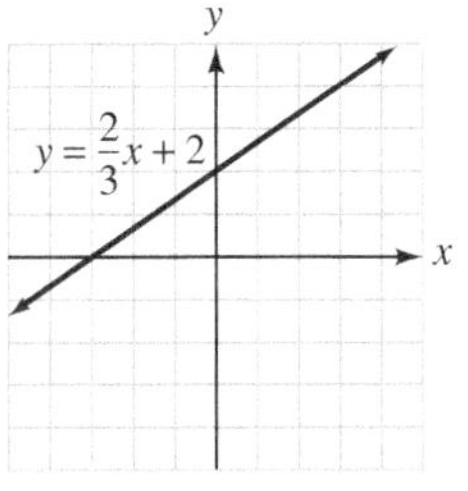

$\frac{2}{3}, (0, 2)$

113. $y = \frac{2}{3}x + \frac{11}{3}$ **115.** $y = -\frac{7}{4}x + \frac{1}{2}$

117. $y = \frac{4}{5}x - \frac{26}{5}$ **119.** $y = -\frac{5}{4}x + 3$ **121.** $(a, 2b)$

123. \$162,500, \$200,000 **125.** 200 **127.** $\frac{1}{165}$

129. $\frac{1}{25}, \frac{1}{10}, \frac{4}{25}$ **131.** $\frac{7}{150}$ of a degree increase per year

133. a. $\frac{2}{23}$ **b.** $\frac{1}{20}, \frac{1}{20}$ **135.** $y = -130x + 2{,}350$

137. $y = 4{,}000x + 122{,}000$ **143.** $a = 0, b > 0$ **145.** $-\frac{A}{B}$

Getting Ready (page 475)

1. 1 **2.** 7 **3.** -20 **4.** $-\frac{11}{4}$

Exercises 7.3 (page 481)

1. 0 **3.** y-intercept **5.** y **7.** -2

9. 2 **11.** relation **13.** domain

15. vertical line test **17.** cannot **19.** first

21. D: $\{3, 5, -4, 0\}$; R: $\{-2, 0, -5\}$; yes

23. D: $\{-2, 6, 5\}$; R: $\{3, 8, 5, 4\}$; no

25. D: $(-\infty, 1]$; R: $(-\infty, \infty)$; not a function

27. D: $(-\infty, \infty)$; R: $(-\infty, \infty)$; a function

29. $9, -3, 0$ **31.** $3, -5, \frac{3}{2}$ **33.** $4, 4, 9$ **35.** $-9, 7, 26$

37. $4, 4, 5$ **39.** $2, 2, 7$ **41.** 12 **43.** $2b - 2a$

45. $(-\infty, 4) \cup (4, \infty)$ **47.** $(-\infty, -3) \cup (-3, \infty)$

49. $(-\infty, \infty)$ **51.** $\left(-\infty, \frac{1}{2}\right) \cup \left(\frac{1}{2}, \infty\right)$

53.

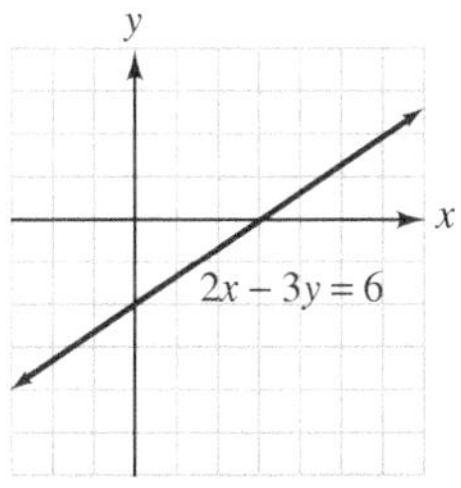

D: $(-\infty, \infty)$; R: $(-\infty, \infty)$

55.

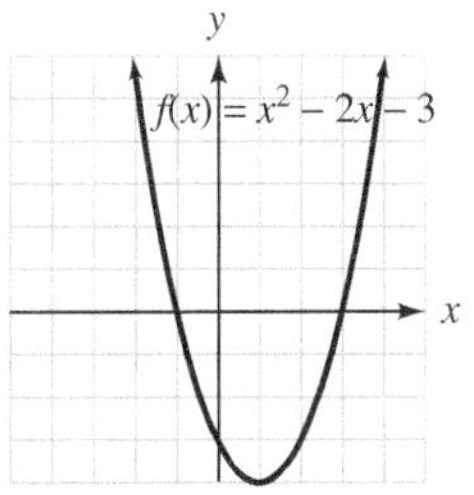

D: $(-\infty, \infty)$; R: $[-4, \infty)$

57. $9, 16, 0$ **59.** $6, 15, 3$ **61.** $17, 22, 2$

63. $5, 3, 11$ **65.** $2b$ **67.** 1 **69.** $2w, 2w + 2$

71. $3w - 5, 3w - 2$ **73.** no **75.** yes

77. $f(x) = 345x + 4{,}790$; \$10,310

79. $f(x) = -\frac{26}{3}x + 507$; 1,300 **81.** 624 ft

83. 12 ft **85.** 77°F **87.** 192 **91.** yes

Getting Ready (page 484)

1. $6x^3y - 3x^2y^2$ **2.** $x^2 - 4$ **3.** $x^2 - x - 6$

4. $6x^2 + 7x - 3$ **5.** $x^3 - 27$ **6.** $x^3 + 8$

Exercises 7.4 (page 495)

1. $(2 - 3x^2)$ **3.** $(x + 1)$ **5.** $(x + 6)$ **7.** 1,100 ft/sec

9. 12 **11.** $ab + ac$ **13.** prime **15.** trial and error

17. $(x^2 + xy + y^2)$ **19.** $2(x + 4)$ **21.** $2x(x - 3)$

23. $-3(a - 2)$ **25.** $-3x(2x + y)$ **27.** $x^2(x^n + x^{n+1})$

29. $y^n(2y^2 - 3y^3)$ **31.** $(x + y)(a + b)$

33. $(x + 2)(x + y)$ **35.** $(4a - 1)(2a^2 + 3)$

37. $(5x + 1)(x^2 + 1)$ **39.** $(x + 2)(x - 2)$

41. $(3y + 8)(3y - 8)$ **43.** $(x^2 + y^2)(x + y)(x - y)$

45. $2(x + 12)(x - 12)$ **47.** $(x + 3)(x + 2)$

49. prime **51.** $(2x - 1)(x - 5)$ **53.** $(3y + 2)(2y + 1)$

55. $x^2(3x - 1)(x - 3)$ **57.** $-2x(2x - 3)^2$ **59.** $(x^n + 1)^2$

61. $(3x^n - 1)(2x^n + 3)$ **63.** $(x^{2n} + y^{2n})^2$ **65.** $(y^{3n} + z)^2$

67. $(x + 2 + y)(x + 2 - y)$ **69.** $(x + 1 + 3z)(x + 1 - 3z)$

71. $(y + 1)(y^2 - y + 1)$ **73.** $(2 + x)(4 - 2x + x^2)$

75. $(a - 4)(a^2 + 4a + 16)$ **77.** $(3 - y)(9 + 3y + y^2)$

79. $(x + 1)(x^2 - x + 1)(x - 1)(x^2 + x + 1)$

81. $(x^2 + y)(x^4 - x^2y + y^2)(x^2 - y)(x^4 + x^2y + y^2)$

83. $\frac{1}{3}, -1$ **85.** $2, \frac{1}{2}$ **87.** $3z(9z^2 + 4z + 1)$

89. $5x^2y(3 - 2y)$ **91.** $13ab^2c(c^2 - 2a^2)$

93. $6s(4s^2 - 2st + t^2)$ **95.** $(9a^2 + 7b)(9a^2 - 7b)$

97. $(x - y + z)(x - y - z)$ **99.** $-(a - 8)(a + 4)$

101. $2(x + 7)(x - 3)$ **103.** $-3(x - 3)(x - 2)$

105. $(a + b)(a - 4b)$ **107.** $(4a - 3)(2a + 3)$

109. prime **111.** $(4x - 3)(2x - 1)$

113. $(2y - 3t)(y + 2t)$ **115.** $-(3a + 2b)(a - b)$

117. $b^2(a - 11)(a - 2)$ **119.** $(2u + 3)(4u^2 - 6u + 9)$

121. $(a + 2b)(a^2 - 2ab + 4b^2)$ **123.** $(s - t)(s^2 + st + t^2)$

125. $(3a - b)(9a^2 + 3ab + b^2)$ **127.** $2x(x + 4)(x - 4)$

129. $(x^2 + 5)(x^2 + 3)$ **131.** $(y^2 - 10)(y^2 - 3)$

133. $(a + 3)(a - 3)(a + 2)(a - 2)$

135. $-(x - 6)(x^2 + 6x + 36)$ **137.** $(a + b)(a - b + 1)$

139. $3m^2n(2m - n)(4m^2 + 2mn + n^2)$

141. $(2c + a + 3b)(2c - a - 3b)$

143. $xy(x + 6y)(x^2 - 6xy + 36y^2)$ **145.** $\frac{1}{5}, -\frac{5}{3}$

153. $(x - y)(x + y)(x^2 + y^2)(x^4 + y^4)(x^8 + y^8)(x^{16} + y^{16})$

Getting Ready (page 497)

1. $\frac{5}{3}$ **2.** $\frac{4}{15}$ **3.** $\frac{19}{6}$ **4.** $-\frac{11}{6}$

Exercises 7.5 (page 507)

1. 2 **3.** 1 **5.** -1 **7.** $\frac{1}{x - 1}$

9.

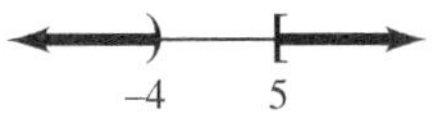

11. $6, -1$

Unless otherwise noted, all content on this page is © Cengage Learning.

13. $-\frac{1}{2}, 3$ **15.** $2, -2, 3, -3$ **17.** $w = \frac{P - 2l}{2}$
19. $\frac{a}{b}$ **21.** $\frac{ad}{bc}$ **23.** 0 **25.** 2 **27.** $-\frac{2}{3}$
29. $-\frac{1}{2}, 3$ **31.** $\frac{3(x + 2)}{x + 1}$ **33.** $\frac{x + 1}{x + 3}$ **35.** $\frac{x + 4}{2(2x - 3)}$
37. $x + 2$ **39.** $\frac{-1}{x + y}$ **41.** $\frac{1}{x + y}$ **43.** $\frac{3(x - y)}{x + y}$
45. $-\frac{x - y}{a + b}$ **47.** $x + 1$ **49.** $\frac{1}{a + 4}$ **51.** $\frac{3x - 2}{3x + 1}$
53. $\frac{2p - 1}{p + 2}$ **55.** $\frac{x - 4}{x + 5}$ **57.** $\frac{(a + 7)^2(a - 5)}{12x^2}$
59. $\frac{-(x - 3)(x - 6)}{(x + 2)(x + 3)}$ **61.** $x - 5$ **63.** $\frac{a + 1}{a - 1}$
65. 2 **67.** $\frac{x^2(x + 3)}{4}$ **69.** $\frac{x - 7}{x + 7}$ **71.** $\frac{x + 5}{x + 4}$
73. 3 **75.** 2 **77.** $\frac{6x}{(x - 3)(x - 2)}$ **79.** $\frac{10a + 4b}{21}$
81. $\frac{9a}{10}$ **83.** $\frac{17}{12x}$ **85.** $\frac{x^2 + 1}{x}$ **87.** $\frac{8x - 2}{(x + 2)(x - 4)}$
89. $\frac{4x^2 + 19x + 42}{(x + 3)(x + 6)}$ **91.** $\frac{x^2 - 5x - 5}{x - 5}$
93. $\frac{-4x^2 + 14x + 54}{x(x + 3)(x - 3)}$ **95.** $\frac{2}{x + 1}$
97. $\frac{2x^2 + x}{(x + 3)(x + 2)(x - 2)}$ **99.** $\frac{b + a}{b}$
101. $y - x$ **103.** $\frac{-1}{a + b}$ **105.** $\frac{a - 1}{a + 1}$ **107.** $\frac{y + x}{y - x}$
109. $\frac{x^2(xy^2 - 1)}{y^2(x^2y - 1)}$ **111.** $\frac{(b + a)(b - a)}{b(b - a - ab)}$
113. $\frac{3a^2 + 2a}{2a + 1}$ **115.** 15 **117.** 2; -2 is extraneous
119. $\frac{a^2b^6}{xy^4}$ **121.** $-\frac{x + 3}{x + 2}$ **123.** $\frac{x + y}{x - y}$ **125.** 2
127. $\frac{3x + 1}{x(x + 3)}$ **129.** xy **131.** $3, -\frac{1}{2}$
137. yes **139.** a, d

Getting Ready (page 511)

1. 6 **2.** 5 **3.** $2 - x$ **4.** $\pi - 2$

Exercises 7.6 (page 515)

1. 8 **3.** 5 **5.** 6 **7.** -5 **9.** -20
11. $2\pi - 4$ **13.** $|2|$ **15.** $|5|$ **17.** $|-2|$
19. $-|-4|$ **21.** $\frac{3}{4}$ **23.** 6 **25.** x **27.** 0
29. $a = -b$ **31.** $8, -8$ **33.** $12, -12$ **35.** $9, -3$
37. $4, -14$ **39.** $4, -1$ **41.** $\frac{14}{3}, -6$ **43.** $0, -6$
45. $40, -20$ **47.** $\varnothing$ **49.** $\varnothing$ **51.** $\frac{20}{3}$ **53.** $\frac{24}{5}, -\frac{32}{5}$
55. $-2, -\frac{4}{5}$ **57.** $3, -1$ **59.** $1, -6$ **61.** $8, \frac{2}{3}$
63. -8 **65.** $8, -4$ **67.** $2, -\frac{1}{2}$ **69.** $\frac{4}{3}$ **71.** $\varnothing$
73. $-4, -28$ **75.** $\varnothing$ **77.** $-16.6, 16.2$ **87.** $\frac{1}{2}$

Getting Ready (page 516)

1. $x > 1$ **2.** $x > -2$ **3.** $x \leq 7$

Exercises 7.7 (page 521)

1. $(-8, 8)$ **3.** $(-\infty, -1) \cup (1, \infty)$ **5.** $[-3, 1]$
7. $(-4, 1)$ **9.** $t = \frac{A - p}{pr}$ **11.** $l = \frac{S - 4wh}{2w}$

13. $-k < x < k$ **15.** $x < -k$ or $x > k$

17. $(-8, 8)$

19. $(-4, 4)$

21. $(-3, 1)$

23. $(-8/3, 4)$
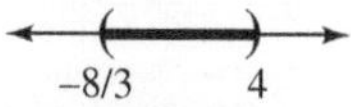

25. $[-21, 3]$

27. $[-3/2, 2]$

29. $(-\infty, -1) \cup (1, \infty)$

31. $(-\infty, -12) \cup (36, \infty)$
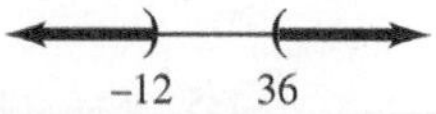

33. $(-\infty, -12] \cup [2, \infty)$
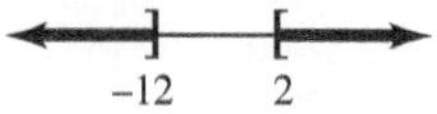

35. $(-\infty, -2] \cup [10/3, \infty)$
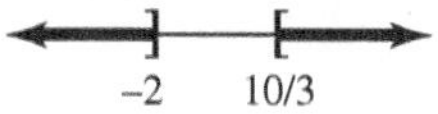

37. $(-\infty, -24) \cup (-18, \infty)$
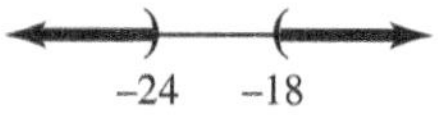

39. $(-5/3, 1)$
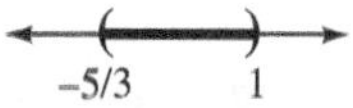

41. $(-\infty, \infty)$
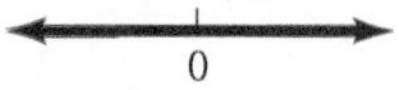

43. $\varnothing$ **45.**

47. $(-\infty, -3/4) \cup (-3/4, \infty)$

49. $(-\infty, -4] \cup [-1, \infty)$
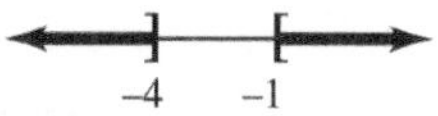

51. $(-65/9, -5/9)$
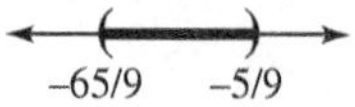

53. $(-\infty, -3) \cup (7/3, \infty)$
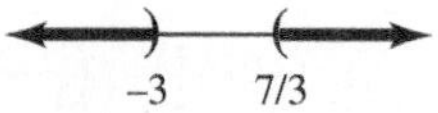

55. $(-\infty, -10) \cup (14, \infty)$
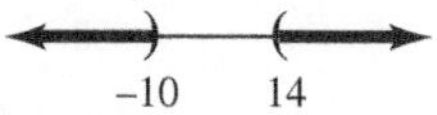

57. $\varnothing$ **59.**
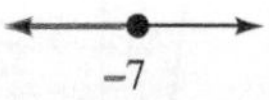

61. $(-\infty, 25) \cup (25, \infty)$

63. $(-\infty, \infty)$
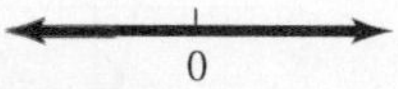

65. $(-6.6, 3.8)$
67. $(-\infty, -0.3) \cup (3.2, \infty)$
71. x and y must have different signs.

Chapter Review (page 528)

1. 9 **2.** 14 **3.** 19 **4.** 8 **5.** $\varnothing$ **6.** $\mathbb{R}$
7. $h = \frac{3V}{\pi r^2}$ **8.** $x = \frac{6V}{ab} - y$ **9.** 3 ft from one end
10. 700 ft^2 **11.** $(-\infty, -24]$
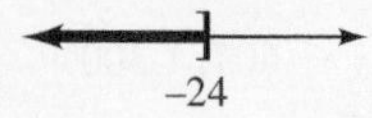

12. $(-\infty, -51/11)$

Unless otherwise noted, all content on this page is © Cengage Learning.

13. $(-1/3, 2)$

14. $(2, \infty)$

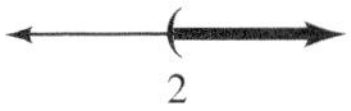

15.

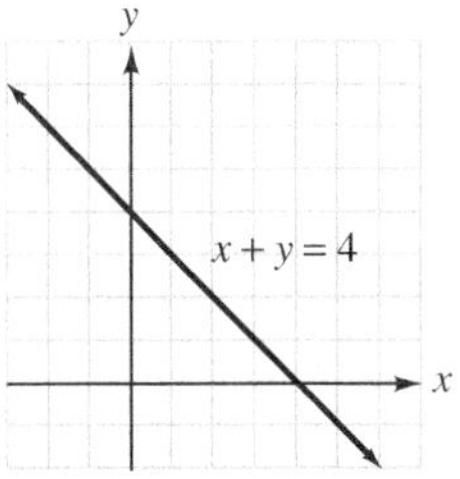

16.

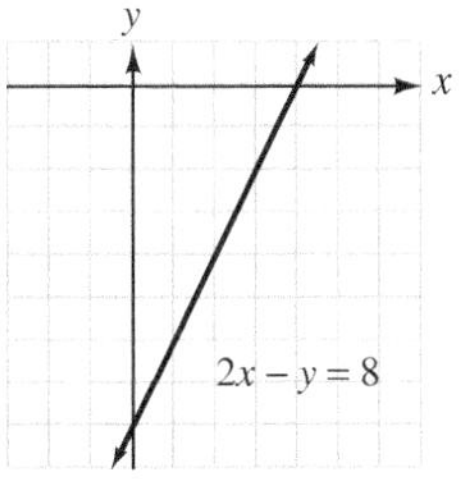

17.

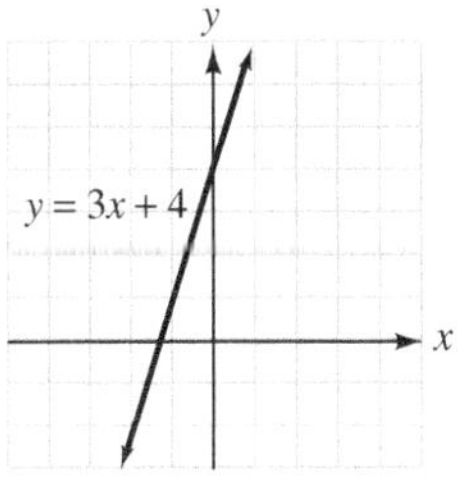

18.

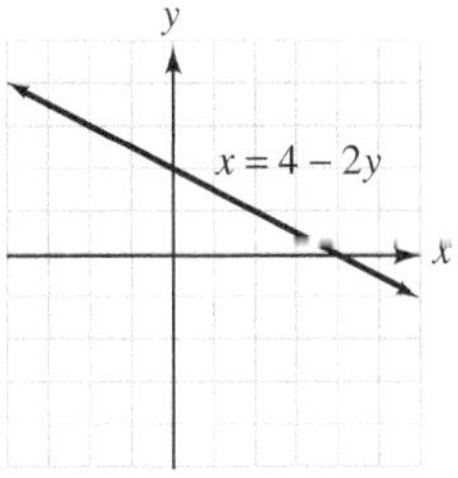

19.

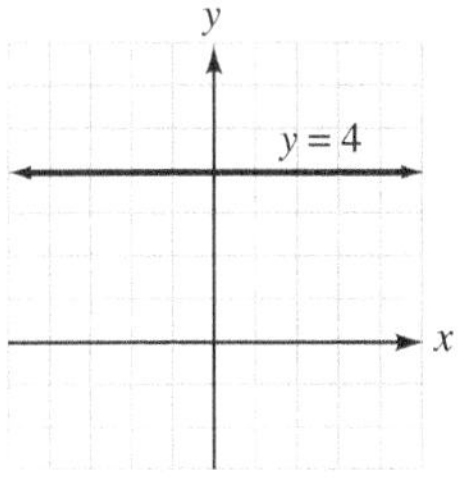

20.

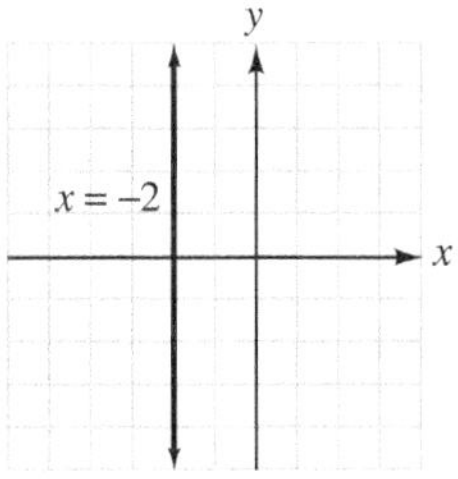

21.

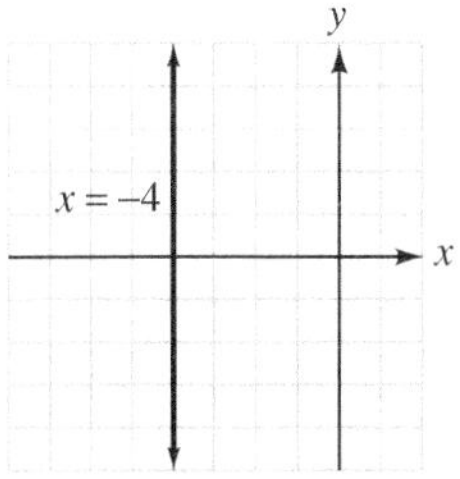

22.

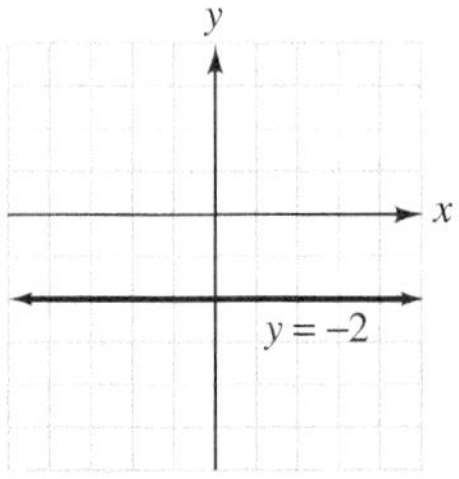

23. $\left(\frac{3}{2}, 8\right)$ **24.** 1 **25.** $\frac{14}{9}$ **26.** 5 **27.** $\frac{13}{11}$
28. 0 **29.** undefined **30.** $\frac{3}{2}$ **31.** $-\frac{1}{2}$
32. undefined **33.** 0 **34.** perpendicular
35. parallel **36.** neither **37.** perpendicular
38. $18,000 **39.** $y = 3x - 19$ **40.** $y = -\frac{3}{2}x + 2$
41. $y = \frac{3}{2}x + \frac{1}{2}$ **42.** $y = 4x + 7$

43.

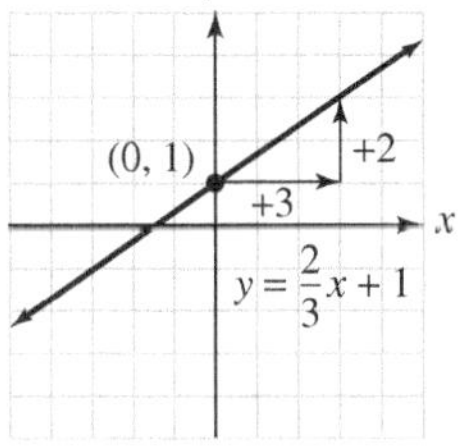

44. perpendicular
45. $y = -1{,}720x + 8{,}700$
46. yes
47. yes
48. no
49. no
50. −5
51. 61
52. 1

53. 19 **54.** D: $(-\infty, 2) \cup (2, \infty)$
55. D: $(-\infty, 3) \cup (3, \infty)$ **56.** D: $(-\infty, \infty)$
57. D: $(-\infty, \infty)$; R: $(-\infty, \infty)$ **58.** D: $(-\infty, \infty)$; R: $(-\infty, \infty)$
59. D: $(-\infty, \infty)$; R: $[-1, \infty)$ **60.** a function
61. not a function **62.** not a function **63.** a function
64. $3(x + 2)$ **65.** $5xy^2(xy - 2)$ **66.** $-4x^2y^3z^2(2z^2 + 3x^2)$
67. $3a^2b^4c^2(4a^4 + 5c^4)$ **68.** $(x + 2)(x^2 + 3)$
69. $(a + b)(c + 3)$ **70.** $x^n(x^n + 1)$ **71.** $y^{2n}(1 - y^n)$
72. $(x^2 + 4)(x^2 + y)$ **73.** $(a^3 + c)(a^2 + b^2)$
74. Prime **75.** $(y + 11)(y - 11)$ **76.** $2(x^2 + 7)(x^2 - 7)$
77. $3x^2(x^2 + 10)(x^2 - 10)$ **78.** $(y + 10)(y + 1)$
79. $(z - 5)(z - 6)$ **80.** $-(2x - 1)(x - 2)$
81. $-(y - 8)(y + 3)$ **82.** $y(y + 2)(y - 1)$
83. $2a^2(a + 3)(a - 1)$ **84.** $3(5x + y)(x - 4y)$
85. $5(6x + y)(x + 2y)$ **86.** $(x + 2 + 2p^2)(x + 2 - 2p^2)$
87. $(y + 2)(y + 1 + x)$ **88.** $(x + 4)(x^2 - 4x + 16)$
89. $8(y - 4)(y^2 + 4y + 16)$ **90.** $3, -3$ **91.** $-5, \frac{1}{2}$
92. $\frac{1}{5}, -3$ **93.** $-\frac{1}{2}, 2$ **94.** 1 **95.** 1 **96.** $\frac{2y - 1}{x - y}$
97. $\frac{6x - 7}{x^2 + 2}$ **98.** $\frac{5x + 13}{(x + 2)(x + 3)}$ **99.** $\frac{4x^2 + 9x + 12}{(x - 4)(x + 3)}$
100. $\frac{3x(x - 1)}{(x - 3)(x + 1)}$ **101.** 1 **102.** $\frac{5x^2 + 11x}{(x + 1)(x + 2)}$
103. $\frac{2(3x + 1)}{x - 3}$ **104.** $\frac{5x^2 + 23x + 4}{(x + 1)(x - 1)^2}$
105. $\frac{-x^4 - 4x^3 + 3x^2 + 18x + 16}{(x - 2)(x + 2)^2}$ **106.** $\frac{y + 2x}{2y - x}$ **107.** $\frac{y - x}{y + x}$
108. $-4, -1$ **109.** -14; 2 is extraneous **110.** $2, -3$
111. $\frac{8}{3}$ **112.** $\varnothing$ **113.** $-1, 1$

114. $(-5, -2)$

115. $(-\infty, 4/3] \cup [4, \infty)$

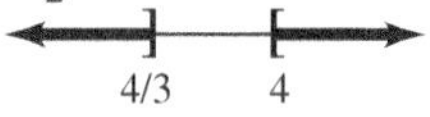

116. $(-\infty, \infty)$

117. $\varnothing$

Chapter 7 Test (page 538)

1. −12 **2.** 6 **3.** $\mathbb{R}$ **4.** $i = \frac{f(P - L)}{s}$ **5.** $13\frac{1}{3}$ ft

6. 36 cm^2 **7.** $(-\infty, -5]$

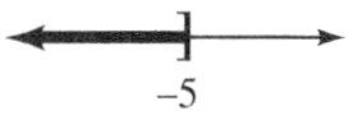

8. $(-2, 16)$

9. $4, -7$

10. $4, -4$ **11.** $[-7, 1]$

12. $(-\infty, -9) \cup (13, \infty)$

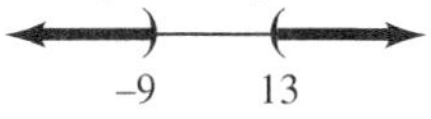

13.

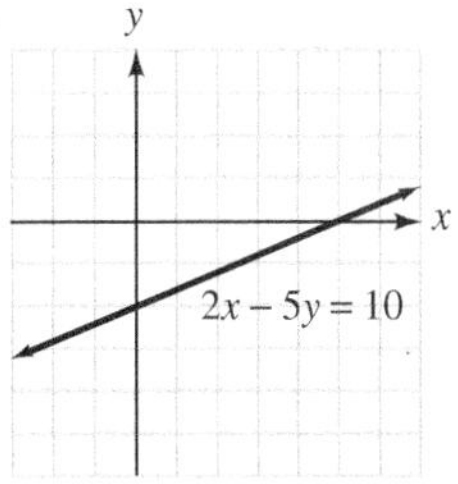

14. $\left(\frac{1}{2}, \frac{1}{2}\right)$
15. x-intercept $(3, 0)$, y-intercept $\left(0, -\frac{3}{5}\right)$
16. vertical
17. $\frac{1}{2}$
18. $\frac{2}{3}$

Unless otherwise noted, all content on this page is © Cengage Learning.

19. undefined **20.** 0 **21.** $y = \frac{2}{3}x - \frac{23}{3}$

22. $y = 8x + 22$ **23.** $m = -\frac{1}{3}, \left(0, -\frac{3}{2}\right)$

24. neither **25.** perpendicular **26.** $y = \frac{3}{2}x$

27. D: $(-\infty, \infty)$; R: $(-\infty, \infty)$ **28.** D: $(-\infty, \infty)$; R: $[0, \infty)$

29. 10 **30.** -2 **31.** $3a + 1$ **32.** $x^2 - 2$

33. yes **34.** no **35.** $3xy(y + 2x)$

36. $(a - y)(x + y)$ **37.** $(x + 7)(x - 7)$

38. $(b + 5)(b^2 - 5b + 25)$ **39.** prime

40. $(x + 5)(x + 3)$ **41.** $(3b + 2)(2b - 1)$

42. $3(u + 2)(2u - 1)$ **43.** $(x + 3 + y)(x + 3 - y)$

44. 1 **45.** $\frac{(x + y)^2}{2}$ **46.** $\frac{2x + 3}{(x + 1)(x + 2)}$

47. $\frac{u^2}{2vw}$ **48.** $\frac{2x + y}{xy - 2}$ **49.** $0, \frac{1}{3}$ **50.** 0

Getting Ready (page 543)

1. 2 **2.** -7 **3.** 11 **4.** 3

Exercises 8.1 (page 550)

1. yes **3.** no **5.** 8.5×10^8 **7.** 2.39×10^5

9. system **11.** inconsistent **13.** dependent

15. $(4, 2)$

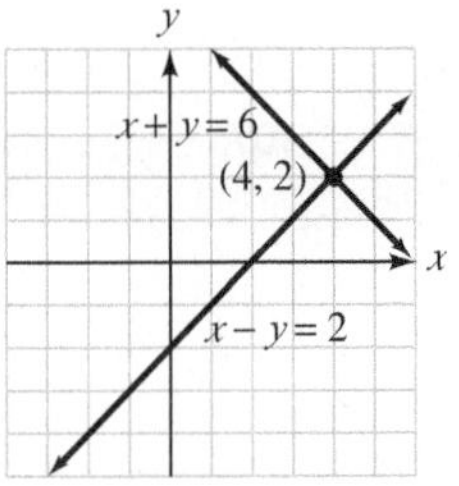

17. $(-1, 3)$

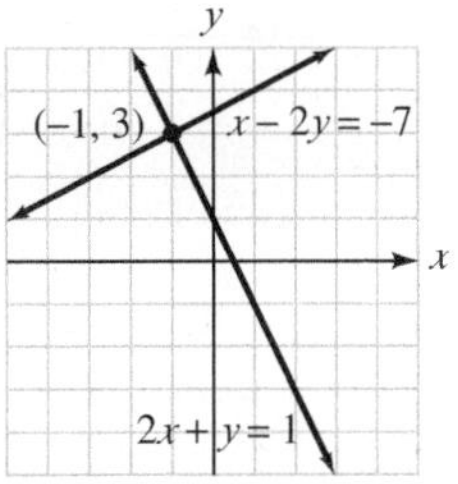

19. $(3, -2)$

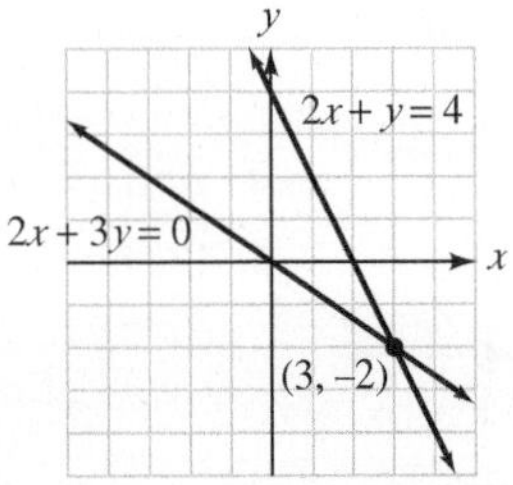

21. $(2, 3)$

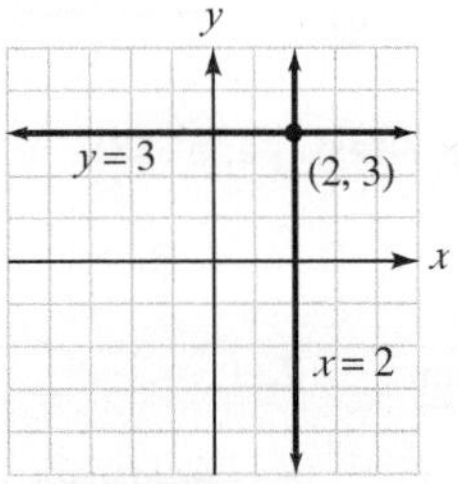

23. $(2, 1)$

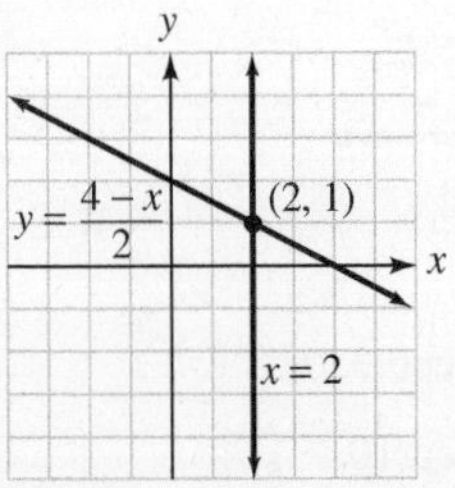

25. $(1, -2)$

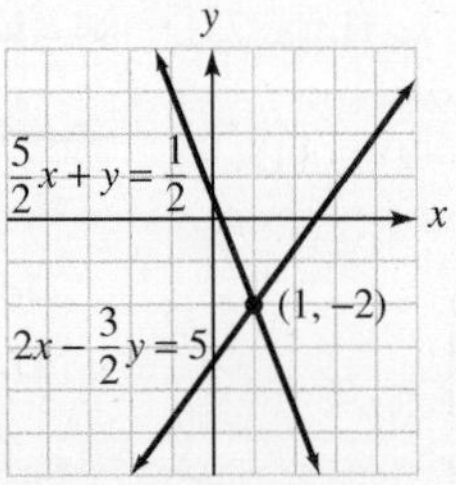

27. $\varnothing$

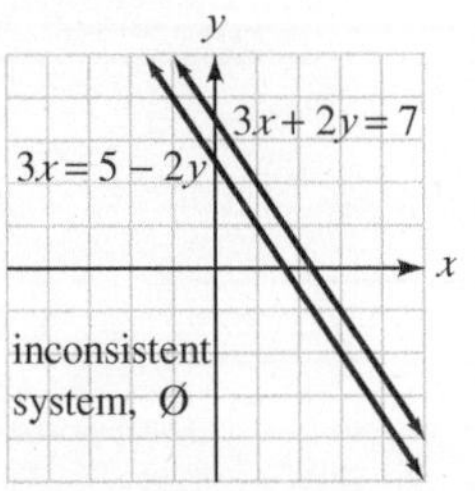

29. $\varnothing$

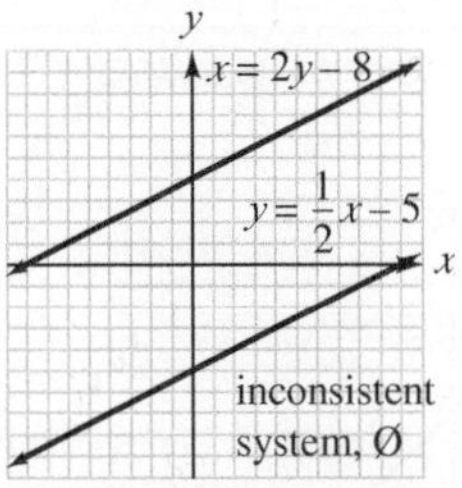

31. $\left(x, -\frac{1}{2}x + \frac{3}{2}\right)$

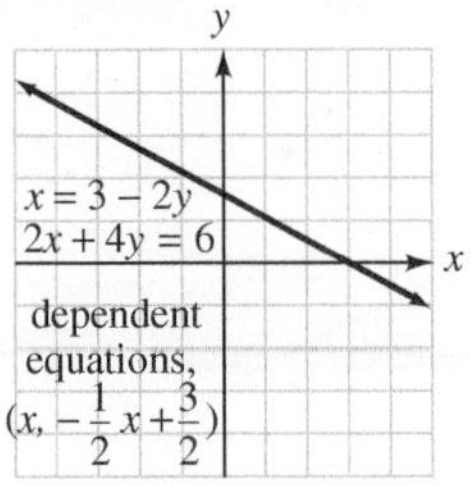

33. $(x, -2x + 3)$

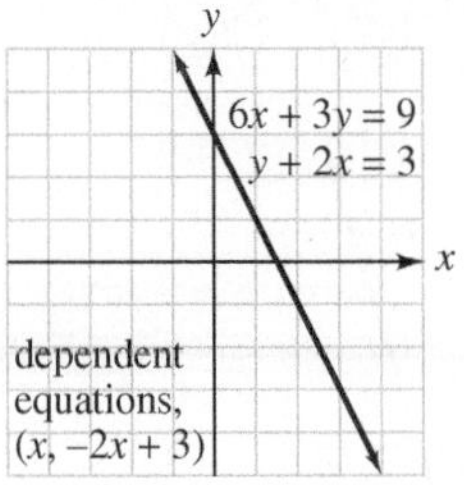

35. -6

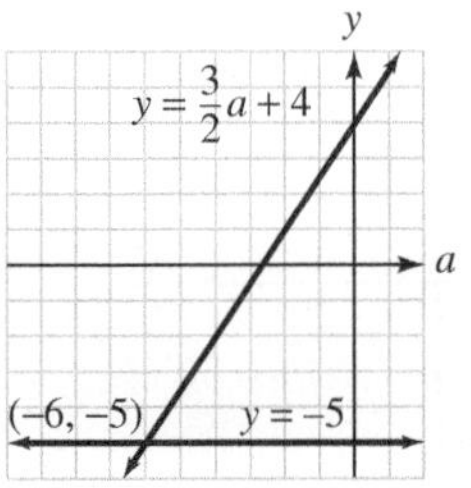

37. $-\frac{3}{2}$

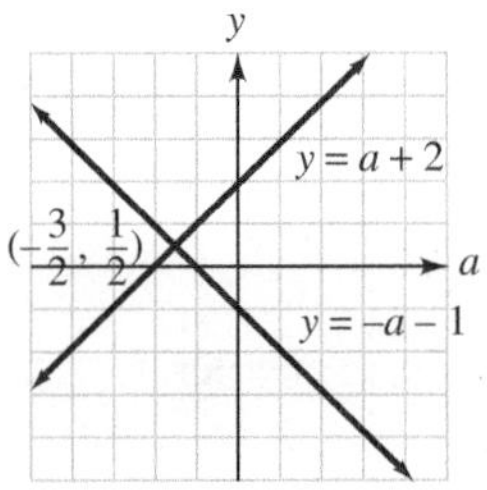

39. no solution **41.** one solution

43. $\varnothing$

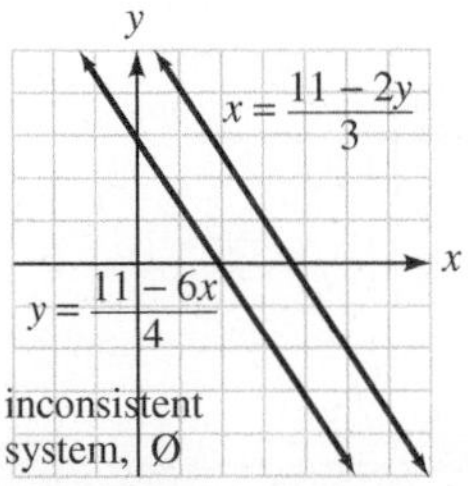

45. $\left(x, 2 - \frac{5}{6}x\right)$

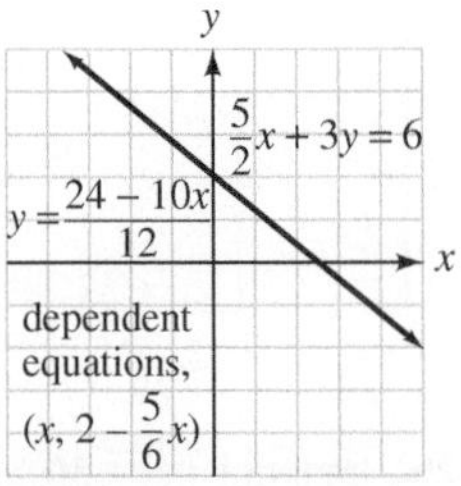

47. $\left(3, \frac{5}{2}\right)$ **49.** $\left(-1, \frac{2}{3}\right)$ **51.** $(-0.37, -2.69)$

53. $(-7.64, 7.04)$ **55. a.** \$2 million **b.** \$3 million

c. 10,000 cameras **57. a.** yes **b.** $(3.75, -0.5)$ **c.** no

59. 3 hr **61.** $(5, 1{,}400)$ **65.** One possible answer is

$$\begin{cases} x + y = -3 \\ x - y = -7 \end{cases}$$

Getting Ready (page 553)

1. $6x - 21$ **2.** $-12x - 20$ **3.** $3x - 9$ **4.** $-2x + 2$

Unless otherwise noted, all content on this page is © Cengage Learning.

Exercises 8.2 (page 563)

1. 1 **3.** $7x = 14$ **5.** a^9b^{14} **7.** $\frac{1}{81x^{32}y^4}$ **9.** setup, unit **11.** parallelogram **13.** opposite **15.** $(2, 6)$ **17.** $(5, 3)$ **19.** $(4, 8)$ **21.** $(-2, -3)$ **23.** $(-2, 1)$ **25.** $(4, -2)$ **27.** $(3, -2)$ **29.** $\left(\frac{3}{4}, \frac{1}{3}\right)$ **31.** $\left(\frac{4}{5}, \frac{3}{4}\right)$ **33.** $(20, -12)$ **35.** $\varnothing$ **37.** $\varnothing$ **39.** $\left(x, 3x - \frac{3}{2}\right)$ **41.** $(x, 2x + 6)$ **43.** $\frac{2}{3}$ **45.** $-\frac{691}{1{,}980}$ **47.** $(2, 4)$ **49.** $\left(5, \frac{3}{2}\right)$ **51.** $(4, -2)$ **53.** $\left(-2, \frac{3}{2}\right)$ **55.** $\left(\frac{1}{2}, -3\right)$ **57.** $(2, -4)$ **59.** $(2, 3)$ **61.** $\left(-\frac{1}{3}, 1\right)$ **63.** \$65 **65.** 625 Ω, 750 Ω **67.** 16 m by 20 m **69.** \$3,000 at 10%, \$5,000 at 12% **71.** 40 oz of 8% solution, 60 oz of 15% solution **73.** 55 mph **75.** 85 racing bikes, 120 mountain bikes **77.** 200 plates **79.** 21 **81.** 750 **83.** 6,500 gal per month **85.** A (a smaller loss) **87.** 590 units per month **89.** A (smaller loss) **91.** A **93.** 35°, 145° **95.** $x = 22.5, y = 67.5$ **97.** 72° **99.** $f^2 = \frac{1}{4\pi^2 LC}$

Getting Ready (page 567)

1. $x + y$ **2.** $x - y$ **3.** xy **4.** $\frac{x}{y}$ **5.** $A = lw$ **6.** $P = 2l + 2w$

Exercises 8.3 (page 576)

1. Let p represent the number of pizzas sold. Let c represent the number of calzones sold. **3.** Let d represent the amount of money Danielle invested. Let k represent the amount of money Kinley invested. **5.** Let n represent the number of milliliters of the 9% saline solution. Let t represent the number of milliliters of the 20% saline solution. **7.** Let a represent the cost of one adult ticket. Let s represent the cost of one student ticket. **9.** $p + c = 52$; The number of pizzas sold plus the number of calzones sold equals the total number of items sold (52). **11.** $d + k = 15{,}000$; The amount of money Danielle invested plus the amount of money Kinley invested equals the total amount of money invested (\$15,000). **13.** $n + t = 25$; The number of milliliters of 9% saline solution plus the number of milliliters of 20% saline solution equals the total number of milliliters needed (25). **15.** $a + s = 273$; The number of adult tickets sold plus the number of student tickets sold equals the total number of tickets sold (273). **17.**

19.

21. 9^4a **23.** x^3y^4 **25.** variable **27.** system **29.** President: \$400,000; Vice President: \$192,600 **31.** \$245,750 **33.** 10 ft, 15 ft **35.** 25 ft by 30 ft **37.** 135 ft² **39.** 6,720 ft² **41.** \$2,000 **43.** 250 **45.** 10 mph **47.** 50 mph **49.** 5 L of 40% solution, 10 L of 55% solution **51.** 32 lb peanuts, 16 lb cashews **53.** 9.9 yr **55.** 250 tires **57.** cleaner: \$5.40; soaking solution: \$6.20 **59.** 140 **61.** causes: 24 min; outcome: 6 min **63.** 28, 84 **65.** 8, 5 **67.** 3% **69.** 15 **71.** paint: \$15, brush: \$5

Getting Ready (page 580)

1. yes **2.** yes **3.** no **4.** yes

Exercises 8.4 (page 586)

1. yes **3.** $\frac{8}{3}$ **5.** 1 **7.** $18s^2 + 1$ **9.** plane **11.** infinitely **13.** yes **15.** $(1, 1, 2)$ **17.** $\left(\frac{3}{4}, \frac{1}{2}, \frac{1}{3}\right)$ **19.** $\varnothing$ **21.** $(x, 2x - 1, 3x + 1)$ **23.** $(x, 0, 1 - x)$ **25.** $(3, 2, 1)$ **27.** $(2, 6, 9)$ **29.** $\varnothing$ **31.** $\left(\frac{1}{2}, \frac{5}{3} - \frac{4}{3}z, z\right)$ **33.** $(0, 2, 2)$ **35.** 30 expensive, 50 middle-priced, 100 inexpensive **37.** $y = x^2 - 4x$ **39.** 1, 6, 12 **41.** $A = 50°, B = 60°, C = 70°$ **43.** 1 unit of A, 2 units of B, and 3 units of C **45.** 250 \$5 tickets, 375 \$3 tickets, 125 \$2 tickets **47.** 3 poles, 2 bears, 4 deer **49.** 78%, 21%, 1% **51.** $x^2 + y^2 - 2x - 2y - 2 = 0$ **55.** $(1, 1, 0, 1)$

Getting Ready (page 589)

1. 5 8 13 **2.** 0 3 7 **3.** −1 −1 −2 **4.** 3 3 −5

Exercises 8.5 (page 595)

1. $\begin{cases} 2x - y = 3 \\ x + 5y = 7 \end{cases}$ **3.** $\begin{cases} 3x - 2y + 4z = 1 \\ 5x + 2y - 3z = -7 \\ -x + 9y + 8z = 0 \end{cases}$

5. 4.7×10^8 **7.** 7.5×10^5 **9.** matrix **11.** 3, columns **13.** augmented, coefficient **15.** $\begin{bmatrix} 2 & -3 \\ 4 & 2 \end{bmatrix}$ **17.** yes **19.** 5 **21.** 2 **23.** $(3, 1)$ **25.** $(2, -2)$ **27.** $(1, 2, 3)$ **29.** $(2 - z, 1 - z, z)$ **31.** $(1, 2)$ **33.** $\varnothing$ **35.** $(z - 2, 2 - z, z)$ **37.** $(2, z - 3, z)$ **39.** $(0, -2)$ **41.** $(8, 8)$ **43.** $(4, 5, 4)$ **45.** $\varnothing$ **47.** $(2, 0)$ **49.** $(1, 2)$ **51.** $(2, 3)$ **53.** $\varnothing$ **55.** $(2, 1, 0)$ **57.** $(3, 3x - 9)$ **59.** $(4 - z, 2, z)$ **61.** $(x, 0, 1 - x)$ **63.** $(-1, -1, 2)$ **65.** 20 nickels, 40 dimes, 4 quarters **67.** $y = 2x^2 - x + 1$ **69.** 76°, 104° **71.** 59°, 31° **73.** 40°, 65°, 75° **77.** $k \neq 0$

Getting Ready (page 597)

1. −22 **2.** 22 **3.** −13 **4.** −13

Exercises 8.6 (page 605)

1. −10 **3.** −16 **5.** $\begin{vmatrix} -21 & -1 \\ 7 & 5 \end{vmatrix}$ **7.** 3 **9.** 0 **11.** number, square **13.** $\begin{vmatrix} a_2 & c_2 \\ a_3 & c_3 \end{vmatrix}$ **15.** Cramer's rule

Unless otherwise noted, all content on this page is © Cengage Learning.

17. consistent, independent **19.** 14 **21.** 1 **23.** -2
25. -13 **27.** 26 **29.** 0 **31.** $(-1, 3)$ **33.** $(4, 2)$
35. $\varnothing$ **37.** $\varnothing$ **39.** $\left(x, -\frac{2}{3}x + 3\right)$ **41.** $\left(x, \frac{4}{3}x - 2\right)$
43. $(1, 1, 2)$ **45.** $(3, 2, 1)$ **47.** $\varnothing$ **49.** $\left(\frac{1}{2}, \frac{5}{3} - \frac{4}{3}z, z\right)$
51. $4a^2 - b^2$ **53.** $10a$ **55.** 0 **57.** -23 **59.** 26
61. $\left(-\frac{1}{2}, \frac{1}{3}\right)$ **63.** $(2, -1)$ **65.** $\left(5, \frac{14}{5}\right)$ **67.** $(3, -2, 1)$
69. $\left(\frac{3}{4}, \frac{1}{2}, \frac{1}{3}\right)$ **71.** $(-2, 3, 1)$ **73.** -4 **75.** 2
77. 50°, 80° **79.** \$5,000 in HiTech, \$8,000 in SaveTel, \$7,000 in HiGas **85.** -4

Getting Ready (page 608)

1. below **2.** above **3.** below **4.** on
5. on **6.** above **7.** below **8.** above

Exercises 8.7 (page 617)

1. no **3.** yes **5.** above **7.** below **9.** 6
11. $t = \frac{A - P}{Pr}$ **13.** $11x - 24$ **15.** $5x + 8y$
17. inequality **19.** boundary **21.** inequalities
23. doubly shaded **25. a.** no **b.** yes **27. a.** yes **b.** no **c.** yes **d.** no **29. a.** no **b.** yes **c.** no **d.** yes

31.
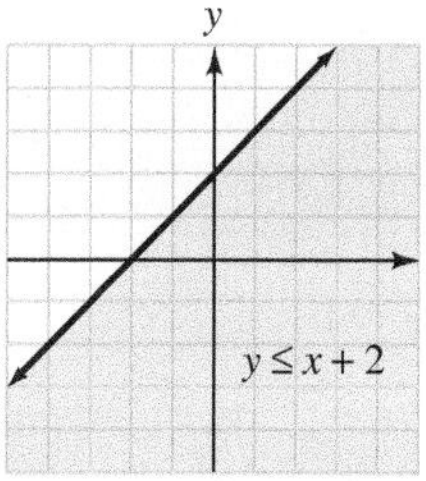

33.
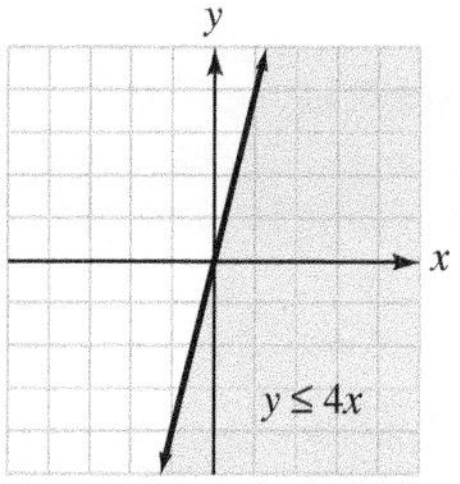

35.
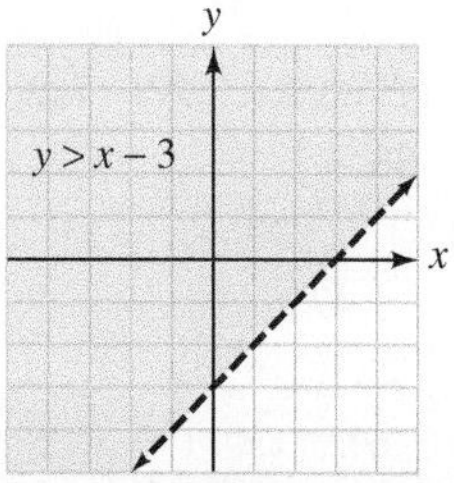

37.
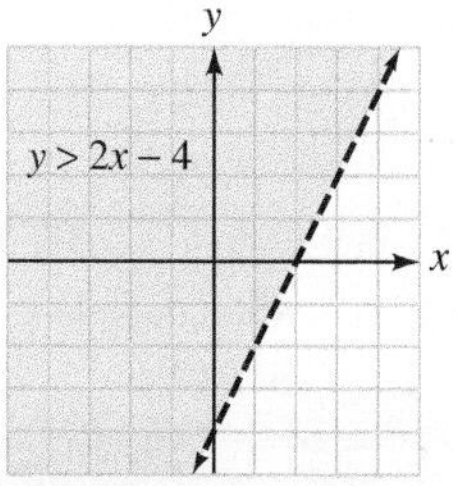

39.
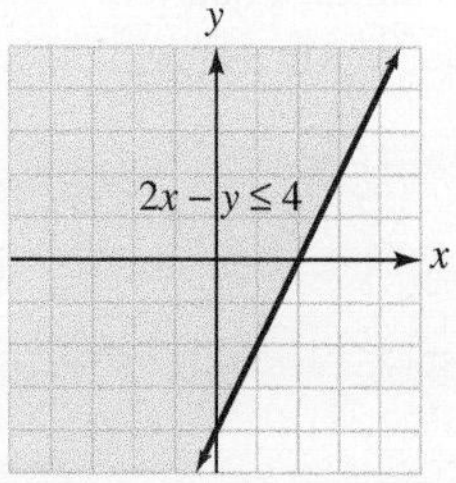

41.
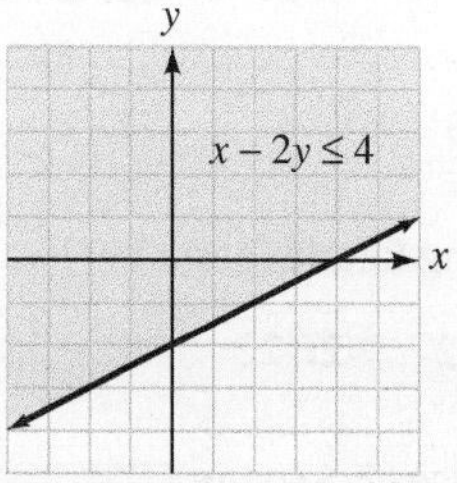

43.
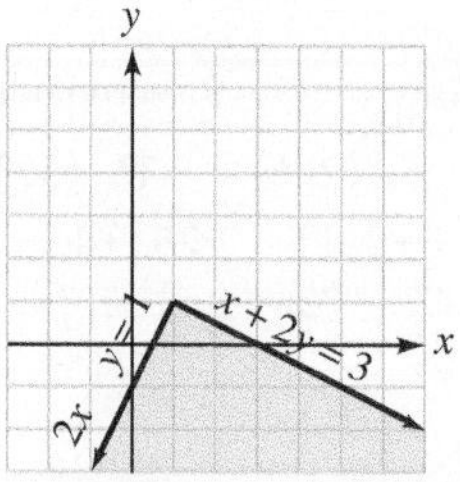

45.
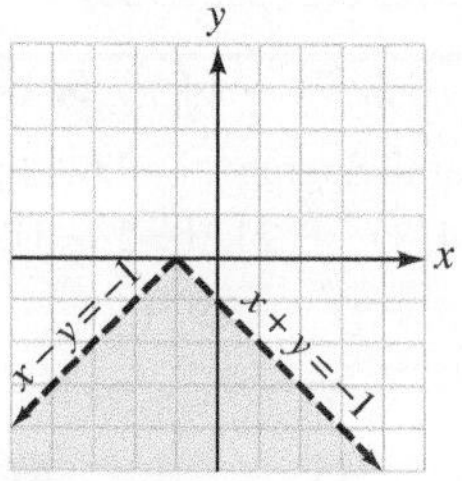

47.
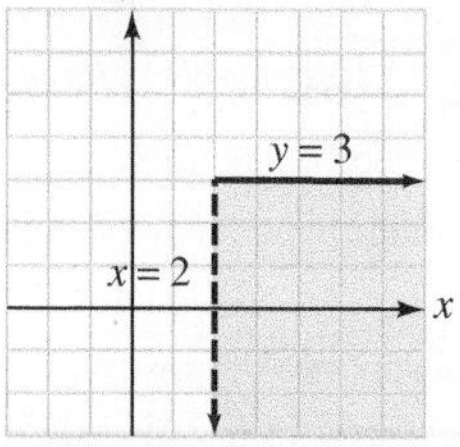

49.
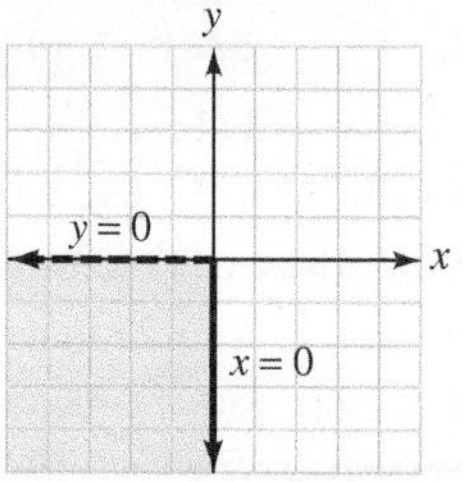

51. $\varnothing$ **53.** $\varnothing$

55.
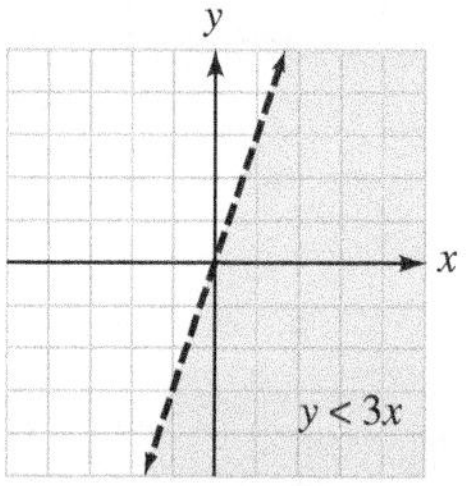

57.
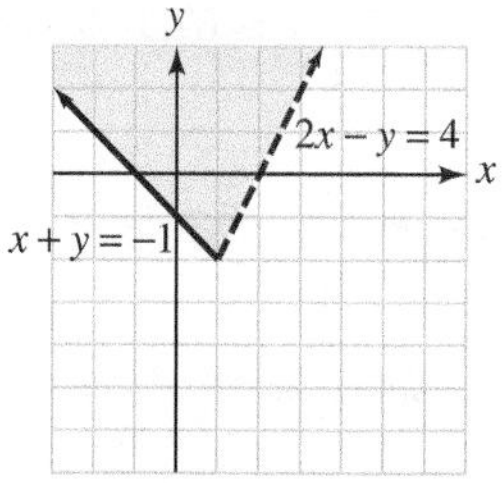

59.
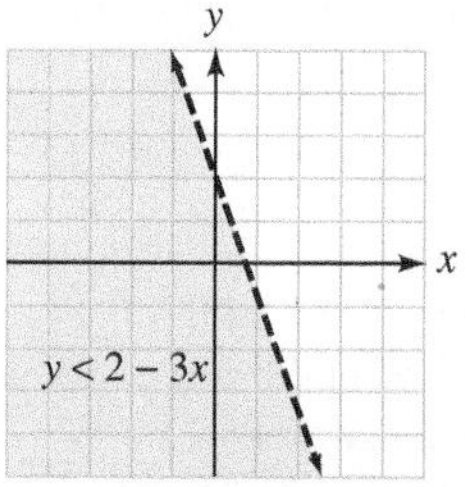

61.
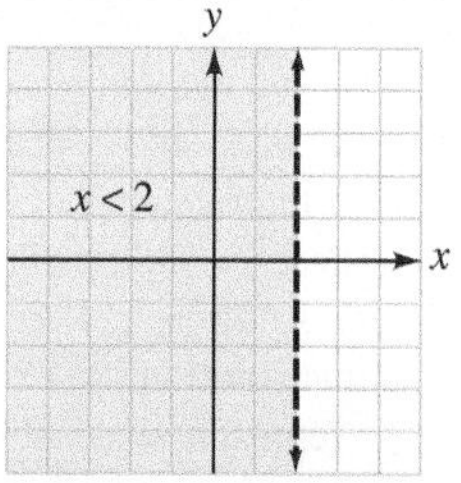

63.
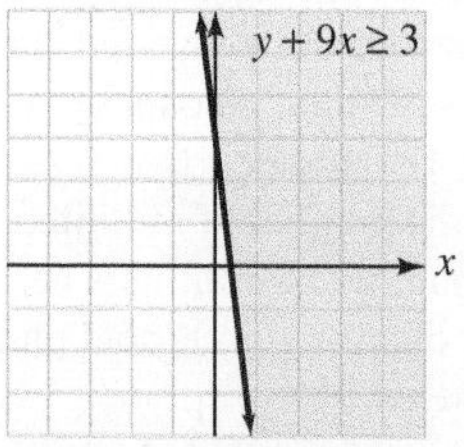

65.
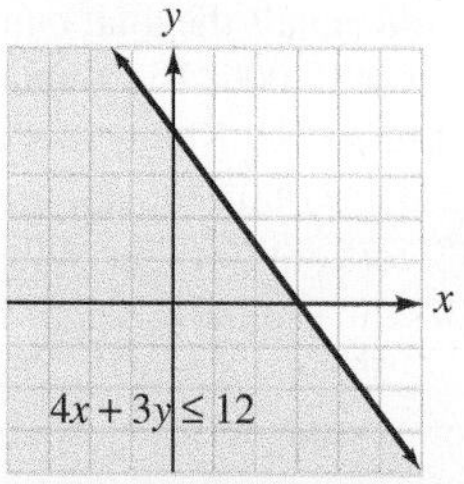

67.
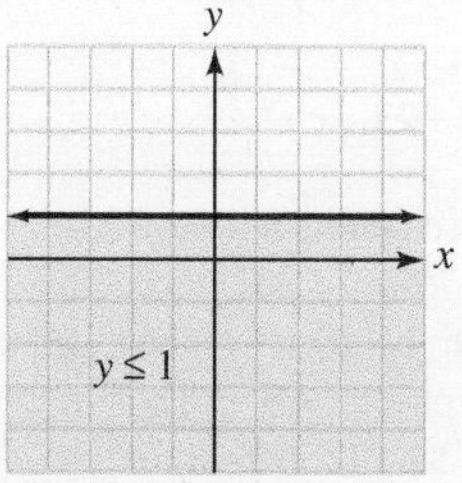

69.
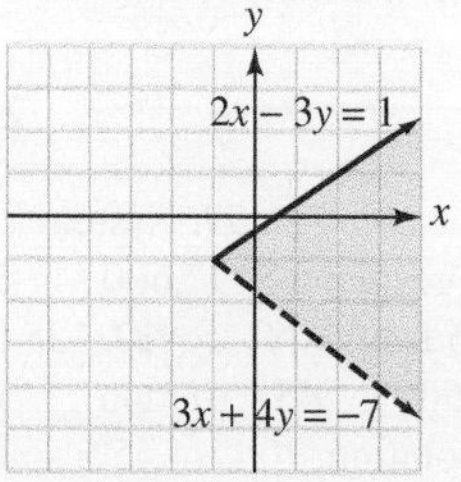

Unless otherwise noted, all content on this page is © Cengage Learning.

71.

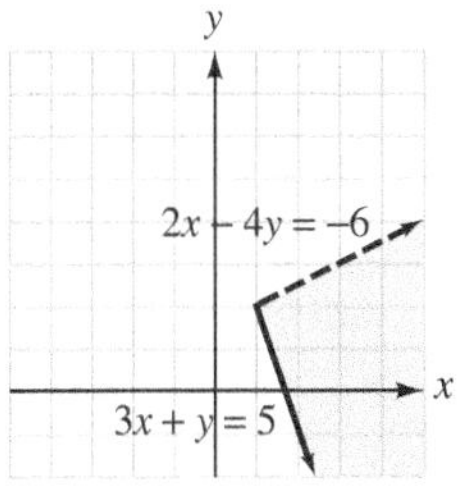

73.

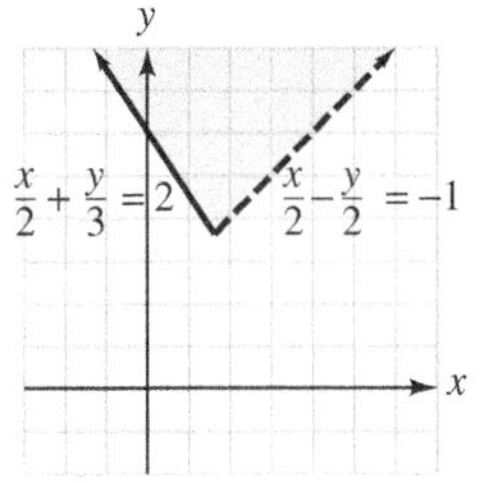

75. $(10, 10), (20, 10), (10, 20)$

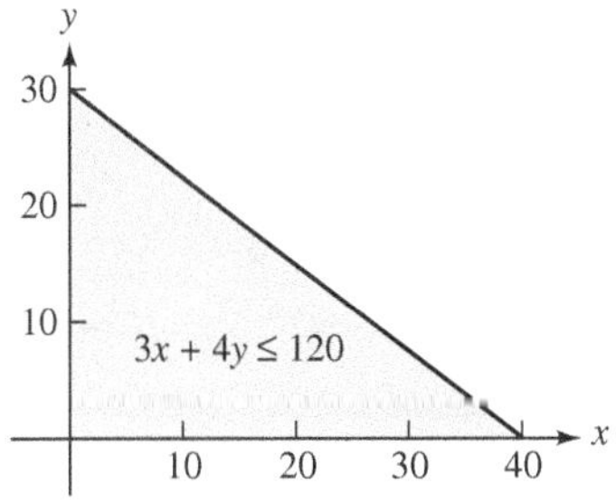

77. $(50, 50), (30, 40), (40, 40)$

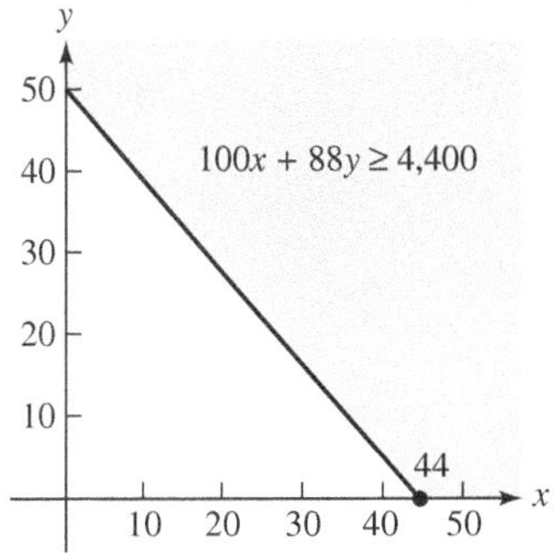

79. $(80, 40), (80, 80), (120, 40)$

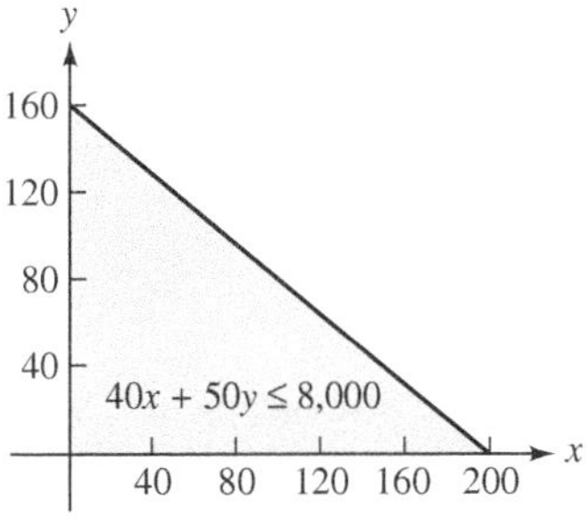

81.

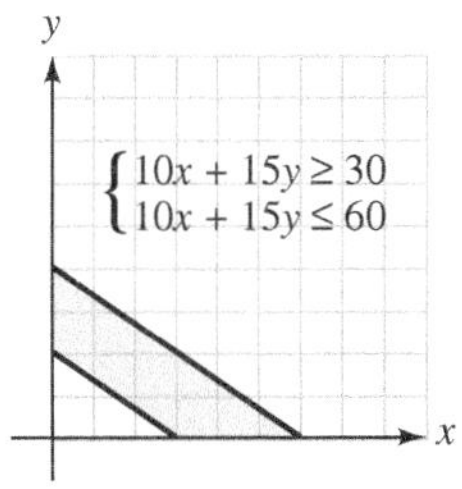

1 \$10 CD and 2 \$15 CDs; 4 \$10 CDs and 1 \$15 CD

83.

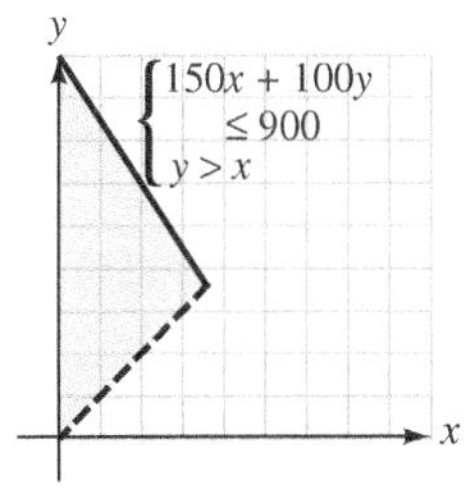

2 desk chairs and 4 side chairs; 1 desk chair and 5 side chairs

Getting Ready (page 621)

1. 0 **2.** 6 **3.** 10 **4.** 12

Exercises 8.8 (page 629)

1. 20 **3.** $(2, 3)$ **5.** $\frac{9}{4}$ **7.** $y = \frac{9}{4}x - \frac{7}{4}$ **9.** constraints **11.** objective **13.** $P = 12$ at $(0, 4)$ **15.** $P = \frac{13}{6}$ at $\left(\frac{5}{3}, \frac{4}{3}\right)$ **17.** $P = 0$ at $(0, 0)$ **19.** $P = 0$ at $(0, 0)$ **21.** $P = \frac{18}{7}$ at $\left(\frac{3}{7}, \frac{12}{7}\right)$ **23.** $P = 3$ at $(1, 0)$ **25.** $P = -12$ at $(-2, 0)$ **27.** $P = -2$ at $(1, 2)$ and $(-1, 0)$ and the edge joining the vertices **29.** 3 tables, 12 chairs, \$1,260 **31.** 30 IBMs, 30 Macs, \$2,700 **33.** 15 DVD players, 30 TVs, \$1,560 **35.** \$150,000 in stocks, \$50,000 in bonds; \$17,000

Chapter Review (page 634)

1. $(3, 5)$ **2.** $(-2, 3)$

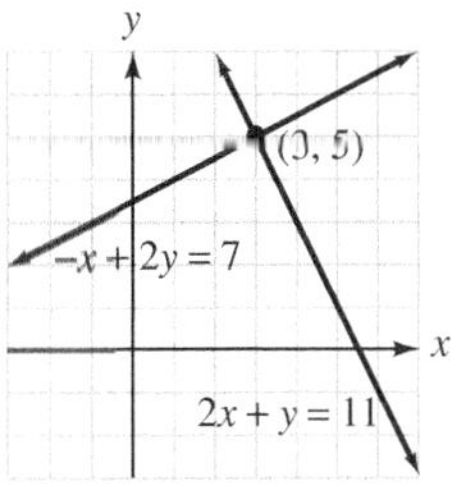

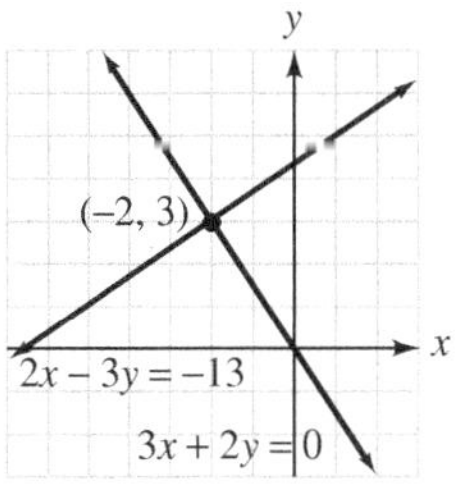

3. $\left(x, 6 - \frac{3}{2}x\right)$ **4.** $\varnothing$

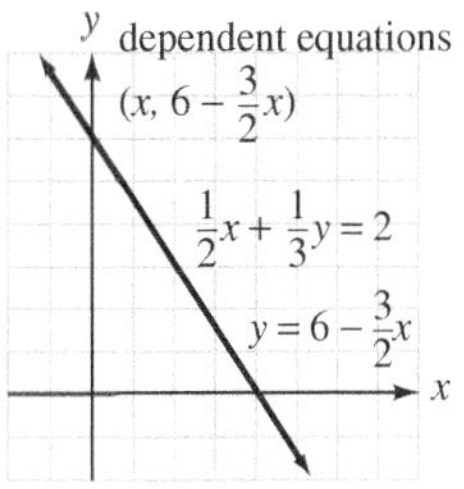

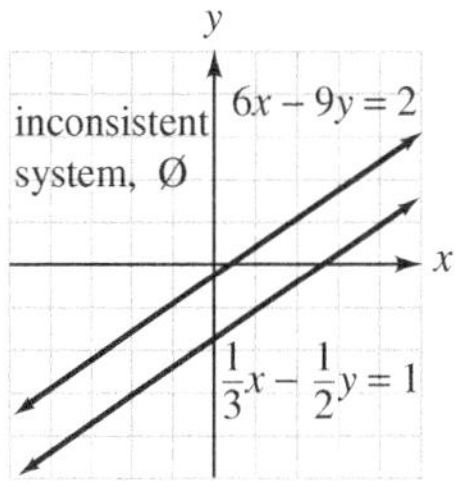

5. $(-1, 3)$ **6.** $(x, 3x + 5)$ **7.** $(3, 4)$ **8.** $(-4, 2)$ **9.** $(-2, -3)$ **10.** $(1, -1)$ **11.** $\left(x, -\frac{2}{3}x - 2\right)$ **12.** $\varnothing$ **13.** 21, 63 **14.** 3 ft by 9 ft **15.** 50¢ **16.** \$66 **17.** \$1.69 **18.** \$2,000 **19.** 3 mph **20.** 30 milliliters of 10% saline, 20 milliliters of 60% saline **21.** $(1, 2, 3)$ **22.** $\varnothing$ **23.** $\left(x, \frac{1}{5}x - \frac{6}{5}, 0\right)$ **24.** $(1, 3, 2)$ **25.** $(1, 2)$ **26.** $(-2 - 4z, -2 - z, z)$ **27.** 14 **28.** -22 **29.** -3 **30.** 28 **31.** $(2, 1)$ **32.** $(-1, 3)$ **33.** $(1, -2, 3)$ **34.** $\varnothing$

35.

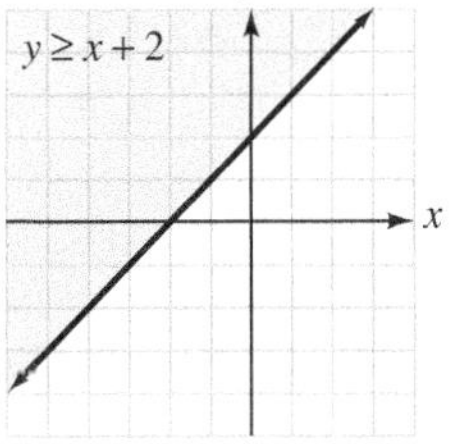

36.

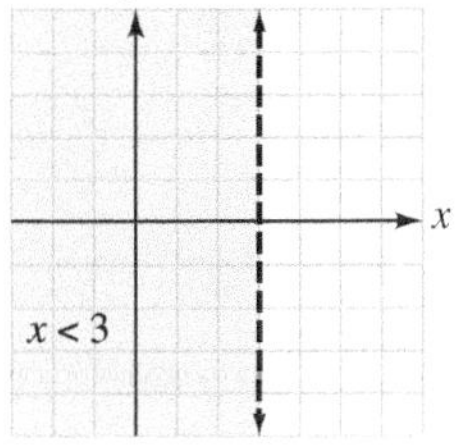

Unless otherwise noted, all content on this page is © Cengage Learning.

37.

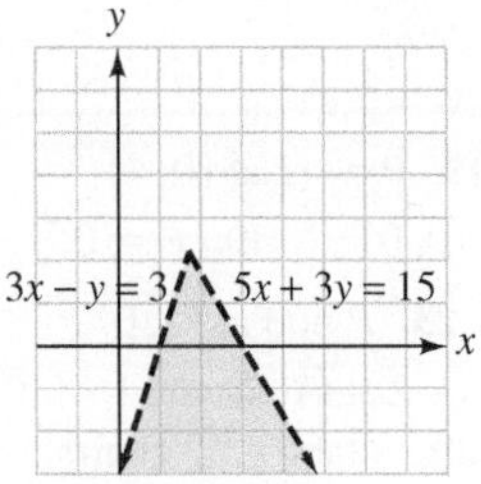

38.

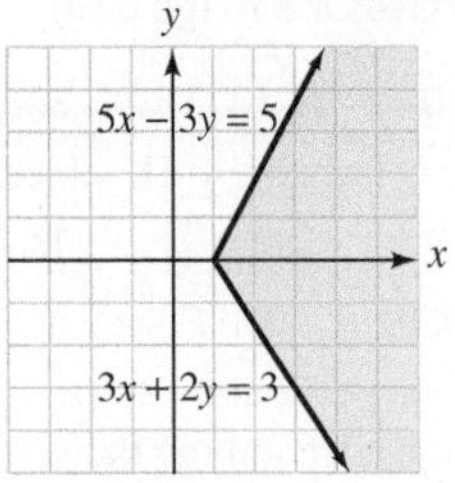

39.

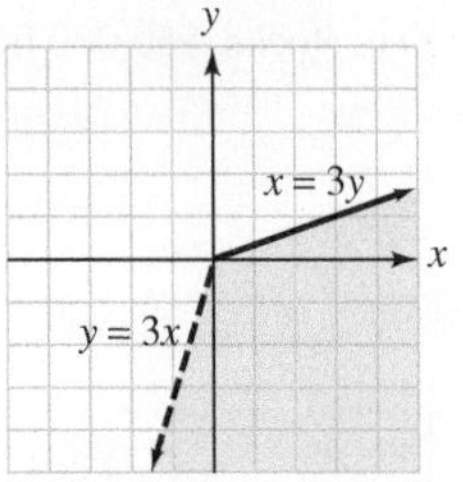

40.

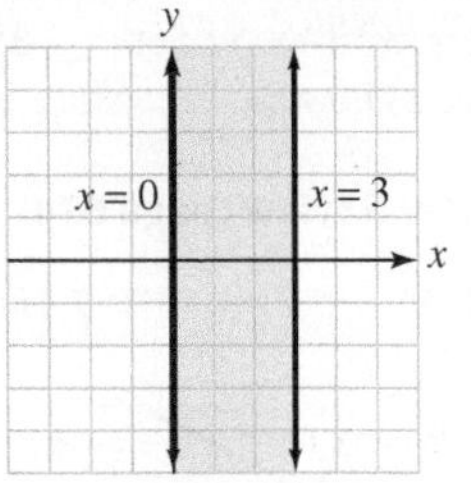

41. 3 shirts and 1 pair of pants; 1 shirt and 2 pairs of pants
42. max of 12 at $(4, 0)$ **43.** 1,000 bags of X, 1,400 bags of Y

Chapter 8 Test (page 642)

1. $(2, 1)$

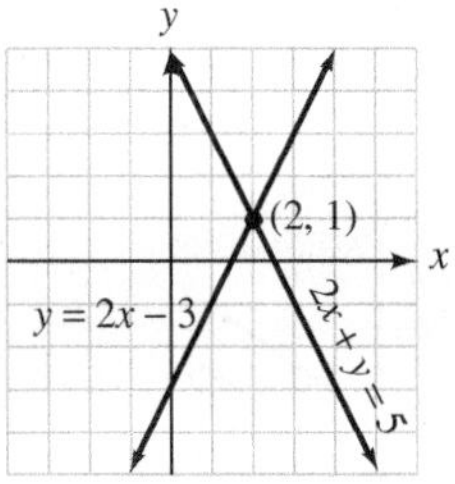

2. $(7, 0)$

3. $(2, -3)$ **4.** $(-6, 4)$ **5.** dependent
6. consistent **7.** 9 **8.** -8

9. $\left[\begin{array}{ccc|c} 1 & 1 & 1 & 4 \\ 1 & 1 & -1 & 6 \\ 2 & -3 & 1 & -1 \end{array}\right]$ **10.** 3 **11.** 2 **12.** -1

13. $(2, 2)$ **14.** $(-1, 3)$ **15.** 22 **16.** -17 **17.** 4

18. 0 **19.** $\begin{vmatrix} -6 & -1 \\ -6 & 1 \end{vmatrix}$ **20.** $\begin{vmatrix} 1 & -1 \\ 3 & 1 \end{vmatrix}$ **21.** -3

22. 3 **23.** 65 **24.** 3 adult **25.** \$4,000 **26.** 1 mph

27.

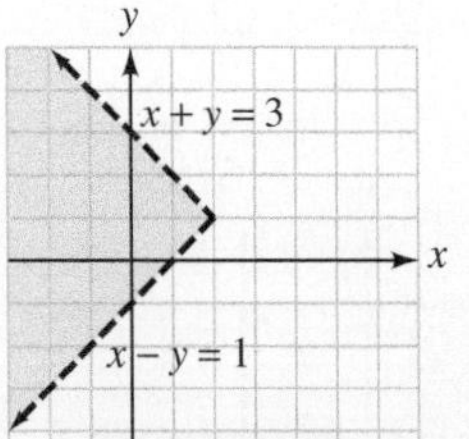

28.

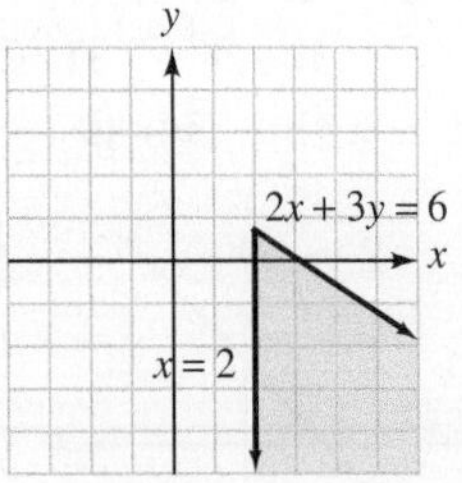

29.

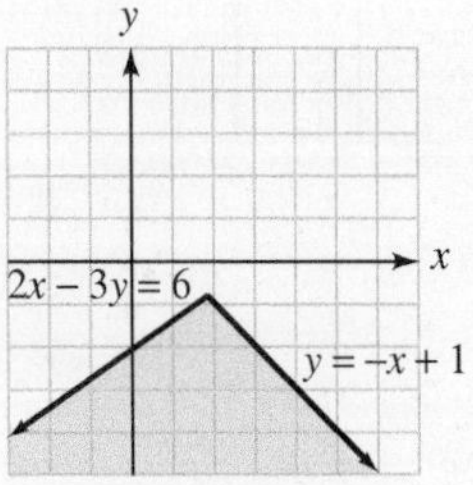

30. $P = 2$ at $(1, 1)$

Getting Ready (page 647)

1. 0 **2.** 16 **3.** 16 **4.** -16 **5.** $\frac{8}{125}$ **6.** $\frac{81}{256}$
7. $49x^2y^2$ **8.** $343x^3y^3$

Exercises 9.1 (page 656)

1. $5, -5$ **3.** $2x, -2x$ **5.** 2 **7.** $3, -3$ **9.** $\frac{x+5}{x-2}$
11. 1 **13.** $\frac{3(m^2 + 2m - 1)}{(m+1)(m-1)}$ **15.** $(5x^2)^2, 6^2 = 36$
17. positive **19.** 5, left **21.** radical, index, radicand
23. $|x|$ **25.** x **27.** even **29.** $7y^2$ **31.** $a^2 + b^3$
33. $10, -10$ **35.** $7, -7$ **37.** 11 **39.** -8 **41.** $-\frac{5}{7}$
43. not real **45.** 0.4 **47.** $5|x|$ **49.** $8y^2$
51. $|x + 3|$ **53.** $|a + 3|$ **55.** 1 **57.** -3 **59.** $-\frac{4}{3}$
61. 0.4 **63.** $5y$ **65.** $-10pq$ **67.** -3 **69.** -2
71. 2 **73.** $\frac{2}{5}$ **75.** not real **77.** $\frac{1}{2}$ **79.** $3|x|$
81. $2a$ **83.** $\frac{1}{2}|x|$ **85.** $|x^3|$ **87.** $-x$ **89.** $-3a^2$
91. D: $[-4, \infty)$, R: $[0, \infty)$ **93.** D: $[0, \infty)$, R: $(-\infty, -3]$

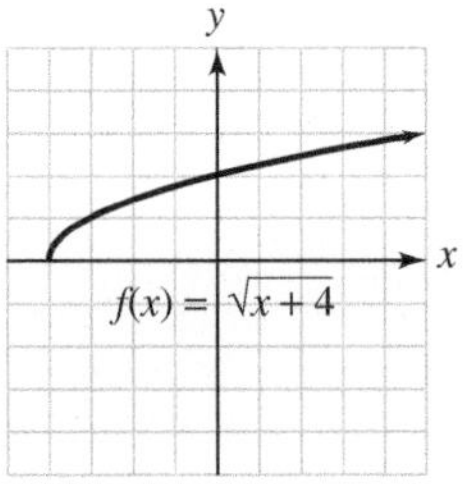

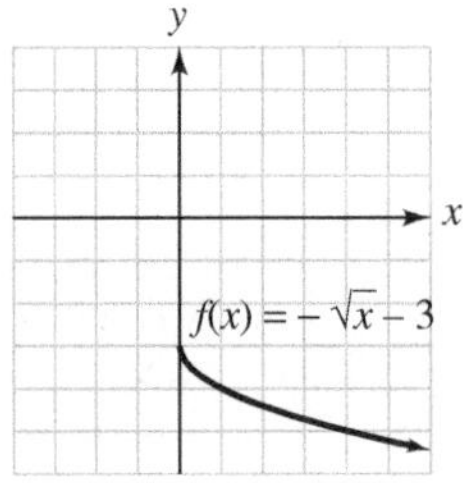

95. D: $(-\infty, \infty)$, R: $(-\infty, \infty)$ **97.** D: $(-\infty, \infty)$, R: $(-\infty, \infty)$

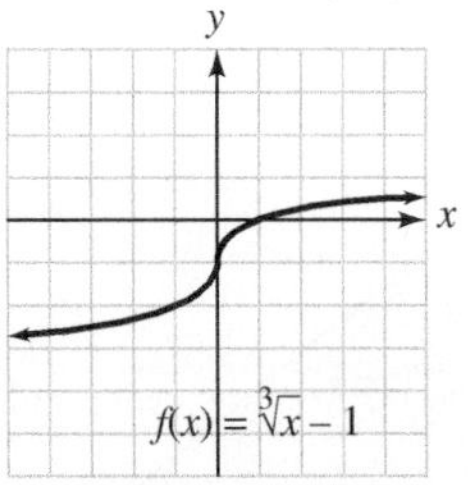

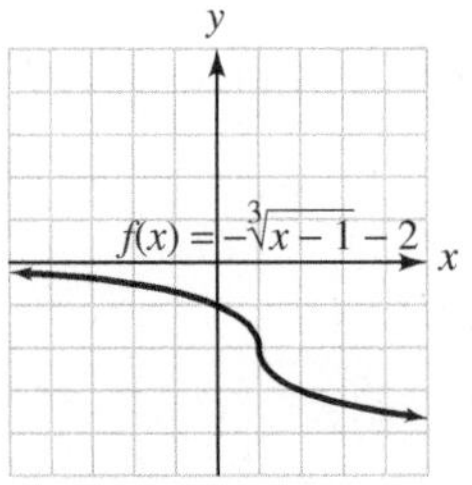

99. 4 **101.** not real **103.** $5|b|$ **105.** $|t + 12|$
107. $-\frac{1}{2}m^2n$ **109.** $0.2z^3$ **111.** $x + 2$ **113.** $0.1x^2|y|$
115. 3.4641 **117.** 26.0624 **119.** about 4 sec **121.** 1.67
123. 11.8673 **125.** 3 units **127.** 4 sec
129. about 7.4 amperes

Getting Ready (page 659)

1. 25 **2.** 169 **3.** 18 **4.** 11,236

Exercises 9.2 (page 663)

1. 25 **3.** 25 **5.** 4 **7.** $6x^2 + 7x - 20$
9. $20a^2 - 23ab + 6b^2$ **11.** hypotenuse
13. $a^2 + b^2 = c^2$ **15.** positive **17.** 10 ft **19.** 80 m
21. 48 in. **23.** 4 mi **25.** 5 **27.** 13 **29.** 10
31. 17 **33.** 10.2 **35.** 13.6 **37.** about 127 ft

Unless otherwise noted, all content on this page is © Cengage Learning.

39. about 135 ft **41.** 9.9 cm **43.** 0.05 ft **45.** 13 ft
47. yes **49.** 173 yd **51.** 24 cm^2 **53.** no
55. no **63.** about 25 **65.** 16 ft/sec

Getting Ready (page 666)

1. x^7 **2.** a^{12} **3.** a^4 **4.** 1 **5.** $\frac{1}{x^4}$ **6.** x^3
7. $\frac{b^6}{c^9}$ **8.** a^{10}

Exercises 9.3 (page 672)

1. -49 **3.** 125 **5.** $\frac{1}{16}$ **7.** 8 **9.** 4 **11.** $x \le 1$
13. $r > 28$ **15.** $1\frac{2}{3}$ pints **17.** $a \cdot a \cdot a \cdot a$ **19.** a^{mn}
21. $\frac{a^n}{b^n}$ **23.** $\frac{1}{a^n}, 0$ **25.** $\left(\frac{b}{a}\right)^n$ **27.** $|x|$ **29.** $\sqrt[4]{5}$
31. $\sqrt[5]{8}$ **33.** $\sqrt[7]{13a}$ **35.** $\sqrt[4]{\frac{1}{2}x^3y}$ **37.** $\sqrt[4]{6a^3b}$
39. $\sqrt{x^2 + y^2}$ **41.** 7 **43.** 3 **45.** $\frac{1}{3}$ **47.** $\frac{1}{2}$
49. -3 **51.** not real **53.** $7^{1/2}$ **55.** $(3a)^{1/4}$
57. $5b^{1/7}$ **59.** $\left(\frac{1}{7}abc\right)^{1/6}$ **61.** $\left(\frac{1}{2}mn\right)^{1/5}$
63. $(x^2 + y^2)^{1/3}$ **65.** $5|y|$ **67.** $3x$ **69.** $|x + 1|$
71. not real **73.** 125 **75.** 27 **77.** $81x^4$ **79.** $\frac{1}{4}$
81. $\frac{1}{2}$ **83.** $\frac{1}{8}$ **85.** $\frac{1}{64x^3}$ **87.** $\frac{1}{9y^2}$ **89.** 8 **91.** $\frac{16}{81}$
93. $5^{8/9}$ **95.** $4^{3/5}$ **97.** $\frac{1}{36}$ **99.** $9^{1/5}$ **101.** $7^{1/2}$
103. $2^{2/3}$ **105.** x^{14} **107.** $a^{2/9}$ **109.** $y + y^2$
111. $x^2 - x + x^{3/5}$ **113.** $x - y$
115. $x^{4/3} + 2x^{2/3}y^{2/3} + y^{4/3}$ **117.** $\sqrt{p}$ **119.** $\sqrt{5b}$
121. 2 **123.** 2 **125.** 0 **127.** -3 **129.** $125x^6$
131. $\frac{4x^2}{9}$ **133.** $\frac{1}{4p^2}$ **135.** $-\frac{3}{2x}$ **137.** $a^{3/4}b^{1/2}$
139. $\frac{n^{2/5}}{m^{3/5}}$ **141.** $\frac{2x}{3}$ **143.** $\frac{1}{3}x$ **145.** $x^2 + 3x^3 - 4x^{4/3}$
147. $\frac{1}{x} - 2 + x$ **149.** 2.47 **151.** 1.01 **153.** 0.24
155. -1.32 **159.** yes

Getting Ready (page 674)

1. 15 **2.** 24 **3.** 5 **4.** 7 **5.** $4x^2$ **6.** $\frac{8}{11}x^3$
7. $3ab^3$ **8.** $-2a^4$

Exercises 9.4 (page 681)

1. $4 \cdot 7$ **3.** $9 \cdot 2$ **5.** $8 \cdot 2$ **7.** $27 \cdot 2$ **9.** $24x^2$
11. $-\frac{21x^7}{y^4}$ **13.** $9t^2 + 12t + 4$ **15.** $3p + 4 + \frac{-5}{2p - 5}$
17. $\sqrt[n]{a}\sqrt[n]{b}$ **19.** like **21.** 6 **23.** t **25.** $5x$
27. 10 **29.** $7x$ **31.** $6b$ **33.** 2 **35.** $3a$
37. $2\sqrt{5}$ **39.** $-10\sqrt{2}$ **41.** $2\sqrt[3]{10}$ **43.** $-3\sqrt[3]{3}$
45. $2\sqrt[4]{2}$ **47.** $2\sqrt[5]{3}$ **49.** $\frac{\sqrt{5}}{7a}$ **51.** $\frac{a\sqrt[3]{7}}{4}$
53. $\frac{p\sqrt[4]{3}}{10q}$ **55.** $\frac{m^3\sqrt[5]{3}}{2n^2}$ **57.** $3y\sqrt{7y}$ **59.** $4a^2\sqrt{3a}$
61. $-4a\sqrt{7a}$ **63.** $5ab\sqrt{7b}$ **65.** $-5b^2\sqrt{3}$
67. $-3x^2\sqrt[3]{2}$ **69.** $2x^4y\sqrt[3]{2}$ **71.** $\frac{z}{4x}$ **73.** $4\sqrt{3}$
75. $-\sqrt{2}$ **77.** $2\sqrt{2}$ **79.** $27\sqrt{5}$ **81.** $4\sqrt{7} - 7\sqrt{6}$
83. $5\sqrt[3]{3}$ **85.** $-\sqrt[3]{4}$ **87.** -10 **89.** $-11\sqrt[3]{2}$
91. $x\sqrt[3]{3x^2}$ **93.** $8z\sqrt{y}$ **95.** $-7y^2\sqrt{y}$ **97.** $-6x\sqrt[3]{2x}$
99. $a = \frac{2}{3}, c = \frac{2}{3}\sqrt{2}$ **101.** $b = 5\sqrt{2}, c = 10$
103. $a = 5, b = 5$ **105.** $a = 7, b = 7$
107. $b = 5\sqrt{3}, c = 10$ **109.** $a = 9, c = 18$
111. $a = 12, b = 12\sqrt{3}$ **113.** $a = \frac{15}{2}, b = \frac{15}{2}\sqrt{3}$
115. $10\sqrt{2x}$ **117.** $2x^3y\sqrt[4]{2}$ **119.** $\frac{\sqrt[4]{5x}}{2z}$ **121.** $-4\sqrt{2}$
123. 81 **125.** $94\sqrt[4]{3}$ **127.** $10\sqrt[4]{3}$ **129.** $9\sqrt[3]{5y}$
131. $9\sqrt[6]{12xyz}$ **133.** $4x\sqrt[5]{xy^2}$ **135.** $2x + 2$
137. $h = 2.83, x = 2.00$ **139.** $x = 8.66, h = 10.00$
141. $x = 4.69, y = 8.11$ **143.** $x = 12.11, y = 12.11$
145. $10\sqrt{3}$ mm, 17.32 mm **149.** $12 + 13\sqrt{3}$

Getting Ready (page 684)

1. a^7 **2.** b^3 **3.** $a^2 - 2a$ **4.** $6b^3 + 9b^2$
5. $a^2 - 3a - 10$ **6.** $4a^2 - 9b^2$

Exercises 9.5 (page 691)

1. 5 **3.** xy **5.** 5 **7.** ab **9.** 1 **11.** $\frac{1}{3}$
13. $2, \sqrt{7}, \sqrt{5}$ **15.** $\sqrt{x} - 1$ **17.** conjugate
19. 4 **21.** $5\sqrt{2}$ **23.** $6\sqrt{2}$ **25.** 5 **27.** $r\sqrt[3]{10s}$
29. ab^2 **31.** $30x^2y\sqrt{2y}$ **33.** $x^2(x + 3)$
35. $12\sqrt{5} - 15$ **37.** $12\sqrt{6} + 6\sqrt{14}$ **39.** $-1 - 2\sqrt{2}$
41. $8 + 5\sqrt{2}$ **43.** $8x - 14\sqrt{x} - 15$
45. $a\sqrt{6} - 3\sqrt{a} + 2\sqrt{3a} - 3\sqrt{2}$ **47.** $\frac{\sqrt{7}}{7}$
49. $\frac{\sqrt{6}}{3}$ **51.** $\frac{2\sqrt{30}}{5}$ **53.** $\frac{\sqrt{21}}{6}$ **55.** $\frac{3\sqrt[3]{4}}{4}$
57. $\frac{\sqrt[3]{21}}{6}$ **59.** $2\sqrt{2x}$ **61.** $\frac{\sqrt{5y}}{y}$ **63.** $\sqrt{2} + 1$
65. $\frac{3\sqrt{2} - \sqrt{10}}{4}$ **67.** $2 + \sqrt{3}$ **69.** $\frac{9 - 2\sqrt{14}}{5}$
71. $\frac{2(\sqrt{x} - 1)}{x - 1}$ **73.** $\frac{x(\sqrt{x} + 4)}{x - 16}$ **75.** $\frac{x - 2\sqrt{xy} + y}{x - y}$
77. $\frac{1}{\sqrt{3} - 1}$ **79.** $\frac{x - 9}{x(\sqrt{x} - 3)}$ **81.** 18 **83.** $5a\sqrt{b}$
85. $2a^2b^2\sqrt[3]{2}$ **87.** $3x(y + z)\sqrt[3]{4}$
89. $-8x\sqrt{10} + 6\sqrt{15x}$ **91.** $5z + 2\sqrt{15z} + 3$
93. $18r - 12\sqrt{2r} + 4$ **95.** $-6x - 12\sqrt{x} - 6$ **97.** $\sqrt[3]{3}$
99. $\frac{\sqrt[3]{4}}{2}$ **101.** $\frac{\sqrt[5]{2}}{2}$ **103.** $\frac{\sqrt[3]{2ab^2}}{b}$ **105.** $\sqrt{2z} + 1$
107. $\frac{x - y}{x - \sqrt{xy}}$ **109.** $f/4$ **111.** $r = \frac{\sqrt{\pi A}}{\pi}$
113. $4\sqrt{2}$ cm **115.** $2\sqrt{3}$ ft, $4\sqrt{3}$ ft **119.** $\frac{x - 9}{4(\sqrt{x} + 3)}$

Getting Ready (page 694)

1. a **2.** $5x$ **3.** $x + 4$ **4.** $y - 3$

Exercises 9.6 (page 699)

1. x **3.** $x + 3$ **5.** $x^2 + 14x + 49$ **7.** 22
9. 17 **11.** $x^n = y^n$, power rule **13.** square

15. extraneous **17.** 1 **19.** 11 **21.** 4 **23.** $\frac{5}{2}, \frac{1}{2}$
25. 7; 3 is extraneous **27.** 4, 3 **29.** 0, 1
31. 2, −4 **33.** 0 **35.** 1 **37.** 1
39. 6; 0 is extraneous **41.** 5 **43.** 3; −1 is extraneous
45. $g = \frac{v^2}{2h}$ **47.** $l = \frac{8T^2}{\pi^2}$ **49.** $A = P(r + 1)^3$
51. $v^2 = c^2\left(1 - \frac{L_A^2}{L_B^2}\right)$ **53.** 2; 7 is extraneous **55.** 16
57. 0; 4 is extraneous **59.** 2; 142 is extraneous **61.** 1, −4
63. 2; −8 is extraneous **65.** −1; 1 is extraneous
67. 1, 9 **69.** 2; −2 is extraneous **71.** ∅; 6 is extraneous
73. ∅; 8 is extraneous **75.** 2 **77.** 1
79. 2; −1 is extraneous **81.** 2,010 ft **83.** about 29 mph
85. 19% **87.** \$5 **89.** $R = \frac{8kl}{\pi r^4}$ **93.** 0, 4

Getting Ready (page 702)

1. $7x$ **2.** $-x + 10$ **3.** $12x^2 + 5x - 25$ **4.** $9x^2 - 25$

Exercises 9.7 (page 710)

1. not real **3.** real **5.** not real **7.** $3 - \sqrt{5}$
9. $12 + 7x$ **11.** −1 **13.** 20 mph **15.** i
17. $-i$ **19.** imaginary **21.** $\frac{\sqrt{a}}{\sqrt{b}}$ **23.** 5, 7
25. conjugates **27.** $5i$ **29.** $11i$ **31.** $\sqrt{7}i$
33. $2\sqrt{2}i$ **35.** $9 + 0i$ **37.** $9 + 3i$ **39.** yes
41. no **43.** $7 + 2i$ **45.** $3 - 5i$ **47.** $15 + 7i$
49. $6 - 8i$ **51.** $24 + 18i$ **53.** $-25 - 25i$
55. $7 + i$ **57.** $32 - 22i$ **59.** $15 - \sqrt{3}i$
61. $7 + 24i$ **63.** 61 **65.** 25 **67.** $2 + i$
69. $\frac{25}{13} + \frac{5}{13}i$ **71.** $\frac{5}{13} - \frac{12}{13}i$ **73.** $\frac{11}{10} + \frac{3}{10}i$
75. $-\frac{42}{25} - \frac{6}{25}i$ **77.** $\frac{35}{34} + \frac{21}{34}i$ **79.** i **81.** −1
83. i **85.** $-i$ **87.** 1 **89.** i **91.** $-2 - 5i$
93. $0 - i$ **95.** $0 + \frac{4}{5}i$ **97.** $\frac{1}{8} - 0i$
99. $0 + \frac{3}{5}i$ **101.** 10 **103.** 13 **105.** $\sqrt{74}$
107. 1 **109.** no **111.** $-20 - 30i$
113. $-5 + 12i$ **115.** $\frac{1}{4} + \frac{3}{4}i$ **117.** $\frac{1}{4} - \frac{\sqrt{15}}{4}i$
119. $-\frac{5}{169} + \frac{12}{169}i$ **121.** $2 + 9i$ **123.** $-15 + 2\sqrt{3}i$
125. $5 + 5i$ **127.** $16 + 2i$ **129.** $7 - 4i$ volts
131. $1 - 3.4i$ **139.** $\frac{5}{3 + i}$

Chapter Review (page 716)

1. 9 **2.** −13 **3.** −6 **4.** 15 **5.** −3
6. −6 **7.** 5 **8.** −2 **9.** $5|x|$ **10.** $|x + 3|$
11. $3a^2b$ **12.** $4x^2|y|$

13. D: $[-2, \infty)$; R: $[0, \infty)$ **14.** D: $[1, \infty)$; R: $(-\infty, 0]$

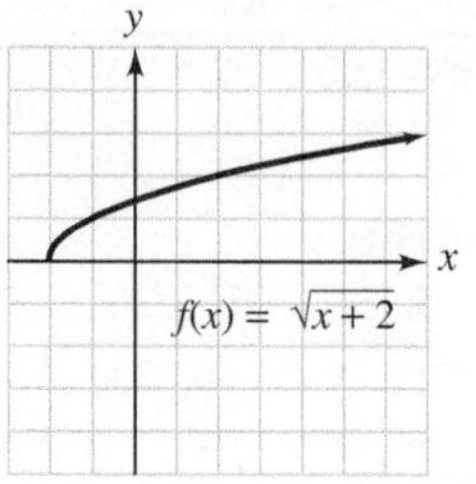

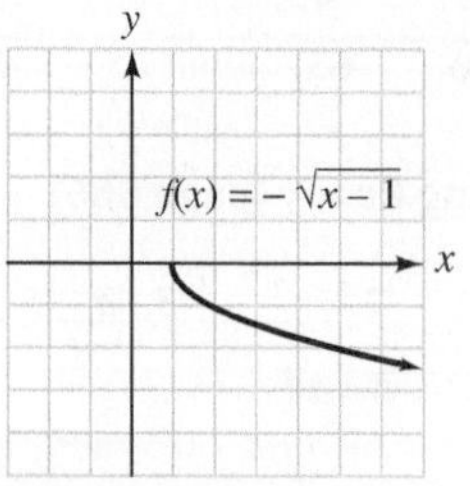

15. D: $[0, \infty)$; R: $(-\infty, 2]$ **16.** D: $(-\infty, \infty)$; R: $(-\infty, \infty)$

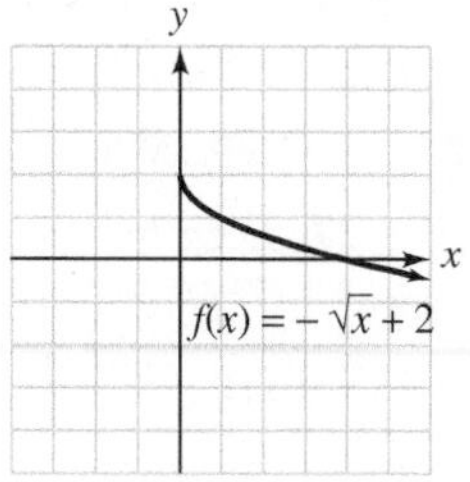

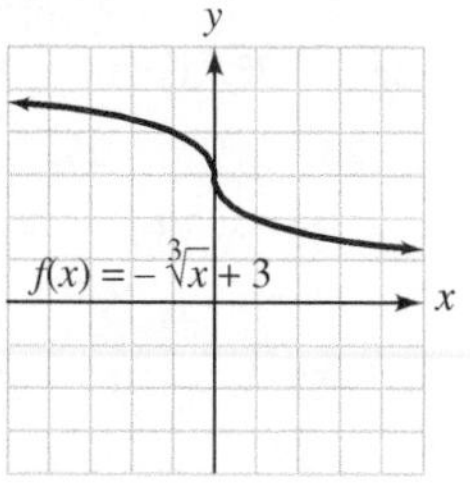

17. 12 **18.** about 5.7 **19.** 3 mi **20.** 8.2 ft
21. 88 yd **22.** 16,000 yd, or about 9 mi **23.** 10
24. 2.83 units **25.** 9 **26.** −7 **27.** 27 **28.** 64
29. −3 **30.** −4 **31.** $\frac{1}{4}$ **32.** $\frac{1}{4}$ **33.** −16,807
34. $\frac{1}{1{,}024}$ **35.** $\frac{25}{9}$ **36.** $\frac{27}{8}$ **37.** $3xy^{1/3}$ **38.** $2x^{1/2}y$
39. $125x^{9/2}y^6$ **40.** $\frac{1}{4u^{4/3}v^2}$ **41.** $5^{3/4}$ **42.** $a^{7/9}$
43. $u - 1$ **44.** $v + v^2$ **45.** $x + 2x^{1/2}y^{1/2} + y$
46. $a^{4/3} - b^{4/3}$ **47.** $\sqrt[3]{5}$ **48.** $\sqrt{x}$ **49.** $\sqrt[3]{3ab^2}$
50. $\sqrt{5ab}$ **51.** $4\sqrt{11}$ **52.** $5\sqrt[3]{2}$ **53.** $2\sqrt[4]{2}$
54. $2\sqrt[5]{3}$ **55.** $2x\sqrt{2x}$ **56.** $2x^3y^2\sqrt{6y}$ **57.** $2xy\sqrt[3]{2x^2y}$
58. $3x^2y\sqrt[3]{2x}$ **59.** $4x$ **60.** $2x$ **61.** $\frac{\sqrt[3]{2a^2b}}{3x}$
62. $\frac{\sqrt{17xy}}{8a^2}$ **63.** $5\sqrt{3}$ **64.** $-\sqrt{2}$ **65.** 0
66. $8\sqrt[4]{2}$ **67.** $29x\sqrt{2}$ **68.** $32a\sqrt{3a}$ **69.** $13\sqrt[3]{2}$
70. $-4x\sqrt[4]{2x}$ **71.** $7\sqrt{2}$ m **72.** $6\sqrt{3}$ cm, 18 cm
73. 7.07 in. **74.** 8.66 cm **75.** $10\sqrt{15}$ **76.** 72
77. $2x\sqrt{5}$ **78.** $3x^2$ **79.** $-2x$ **80.** $-20x^3y^3\sqrt{xy}$
81. $9 - 4\sqrt{3}$ **82.** $5 - 6\sqrt{5}$ **83.** $\sqrt{10} - \sqrt{5}$
84. $3 + \sqrt{6}$ **85.** 4 **86.** $5 + 2\sqrt{6}$ **87.** $x - y$
88. $6u + \sqrt{u} - 12$ **89.** $\frac{\sqrt{3}}{3}$ **90.** $\frac{\sqrt{15}}{5}$ **91.** $\frac{\sqrt{xy}}{y}$
92. $\frac{\sqrt[3]{u^2}}{u^2v^2}$ **93.** $2(\sqrt{2} + 1)$ **94.** $\frac{\sqrt{6} + \sqrt{2}}{2}$
95. $2(\sqrt{x} - 4)$ **96.** $\frac{a + 2\sqrt{a} + 1}{a - 1}$ **97.** $\frac{3}{5\sqrt{3}}$
98. $\frac{1}{\sqrt[3]{3}}$ **99.** $\frac{9 - x}{2(3 + \sqrt{x})}$ **100.** $\frac{a - b}{a + \sqrt{ab}}$ **101.** 9
102. 16, 9 **103.** 5; −1 is extraneous **104.** $\frac{9}{16}$ **105.** 2
106. 0, −2 **107.** $10 - 12i$ **108.** $2 - 50i$
109. $-96 + 3i$ **110.** $-2 - 2\sqrt{2}i$ **111.** $22 + 29i$
112. $-22 + 7i$ **113.** $-12 + 28\sqrt{3}i$ **114.** $118 + 10\sqrt{2}i$

Unless otherwise noted, all content on this page is © Cengage Learning.

115. $0 - \frac{3}{4}i$ **116.** $0 - \frac{2}{5}i$ **117.** $\frac{12}{5} - \frac{6}{5}i$ **118.** $\frac{21}{10} + \frac{7}{10}i$ **119.** $\frac{15}{17} + \frac{8}{17}i$ **120.** $\frac{4}{5} - \frac{3}{5}i$ **121.** $\frac{15}{29} - \frac{6}{29}i$ **122.** $\frac{1}{3} + \frac{1}{3}i$ **123.** 15 **124.** 26 **125.** $-i$ **126.** -1

Chapter 9 Test (page 722)

1. 6 **2.** $5\sqrt{5}$ **3.** $2|x|$ **4.** $2x$ **5.** D: $[2, \infty)$, R: $[0, \infty)$ **6.** D: $(-\infty, \infty)$, R: $(-\infty, \infty)$

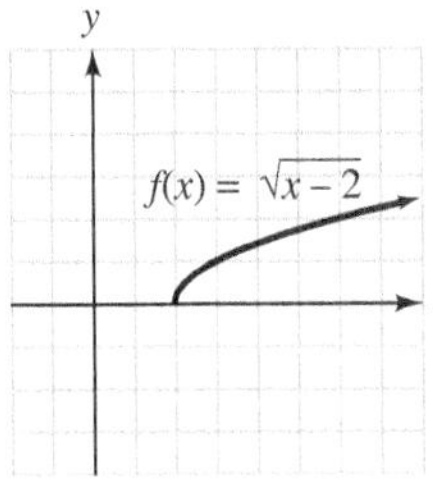

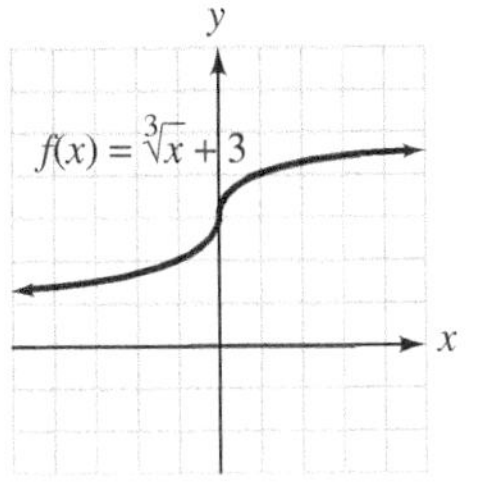

7. 28 in. **8.** about 1.25 m **9.** 10 **10.** 25 **11.** 3 **12.** -16 **13.** $\frac{1}{216}$ **14.** $\frac{9}{4}$ **15.** $2^{4/3}$ **16.** $8xy$ **17.** $6\sqrt{3}$ **18.** $5xy^2\sqrt{10xy}$ **19.** $2x^3y\sqrt[3]{5}$ **20.** $\frac{1}{4a}$ **21.** $3|x|\sqrt{5}$ **22.** $4|x^3|\sqrt{3}$ **23.** $2x^3\sqrt[3]{2}$ **24.** $3x^2y^4\sqrt{2y}$ **25.** $-\sqrt{3}$ **26.** $14\sqrt[3]{5}$ **27.** $2y^2\sqrt{3y}$ **28.** $6z\sqrt[4]{3z}$ **29.** $-6x\sqrt{y} - 2xy^2$ **30.** $3 - 7\sqrt{6}$ **31.** $\frac{\sqrt{5}}{5}$ **32.** $\sqrt{3t} + 1$ **33.** $\frac{3}{\sqrt{21}}$ **34.** $\frac{a-b}{a - 2\sqrt{ab} + b}$ **35.** 10 **36.** $\varnothing$; 4 is extraneous **37.** $-1 + 11i$ **38.** $4 - 7i$ **39.** $8 + 6i$ **40.** $-10 - 11i$ **41.** $0 - \frac{\sqrt{2}}{2}i$ **42.** $\frac{1}{2} + \frac{1}{2}i$

Getting Ready (page 727)

1. $(x + 5)(x - 5)$ **2.** $(b + 9)(b - 9)$ **3.** $(3x + 2)(2x - 1)$ **4.** $(2x - 3)(2x + 1)$

Exercises 10.1 (page 736)

1. $x^2 - 25$ **3.** $x^2 + 4x + 4$ **5.** $x^2 - 14x + 49$ **7.** 1 **9.** $t \le 4$ **11.** $x = \sqrt{c}, x = -\sqrt{c}$ **13.** positive or negative **15.** $9, -9$ **17.** $6, -6$ **19.** $-3, -4$ **21.** $\frac{2}{3}, -\frac{5}{2}$ **23.** ± 6 **25.** $\pm\sqrt{5}$ **27.** $0, -6$ **29.** $2 \pm \sqrt{5}$ **31.** $\pm\frac{7}{4}i$ **33.** $\pm\frac{9}{2}i$ **35.** 4 **37.** $\frac{9}{4}$ **39.** $2, -6$ **41.** $2, 4$ **43.** $-\frac{1}{3}, -\frac{3}{2}$ **45.** $\frac{5}{3}, -\frac{1}{2}$ **47.** $\frac{3}{4}, -\frac{3}{2}$ **49.** $1, -\frac{1}{2}$ **51.** $-\frac{7}{10} \pm \frac{\sqrt{29}}{10}$ **53.** $1 \pm \frac{\sqrt{6}}{3}$ **55.** $-1 \pm i$ **57.** $-4 \pm i\sqrt{2}$ **59.** $8, 1$ **61.** $\pm\frac{4\sqrt{3}}{3}$ **63.** $4, 10$ **65.** $-5 \pm \sqrt{3}$ **67.** $2, \frac{1}{2}$ **69.** $\frac{1}{3} \pm \frac{\sqrt{10}}{3}$ **71.** $0, -3$ **73.** $-1, -4$ **75.** $10, -1$ **77.** $\frac{1}{4} \pm \frac{3\sqrt{7}}{4}i$ **79.** $d = \frac{\sqrt{6h}}{2}$ **81.** $c = \frac{\sqrt{Em}}{m}$ **83.** $-\frac{1}{4} \pm \frac{\sqrt{41}}{4}$ **85.** $-\frac{1}{2} \pm \frac{\sqrt{13}}{2}$ **87.** 4 sec **89.** 72 mph **91.** 4% **93.** width: $7\frac{1}{4}$ ft; length: $13\frac{3}{4}$ ft **97.** $\frac{3}{4}$

Getting Ready (page 738)

1. $x^2 + 12x + 36, (x + 6)^2$ **2.** $x^2 - 7x + \frac{49}{4}, \left(x - \frac{7}{2}\right)^2$ **3.** 7 **4.** 8

Exercises 10.2 (page 743)

1. $5, -3, -4$ **3.** $y = \frac{-Ax + C}{B}$ **5.** $3\sqrt{5}$ **7.** $\sqrt{5}$ **9.** $7, -4, -9$ **11.** $-1, -5$ **13.** $7, -2$ **15.** $-4, -5$ **17.** $-1, \frac{3}{2}$ **19.** $\frac{4}{3}, -\frac{2}{5}$ **21.** $-\frac{3}{2}, -\frac{1}{2}$ **23.** $-\frac{1}{2} \pm \frac{\sqrt{5}}{10}$ **25.** $-\frac{1}{5} \pm \frac{\sqrt{6}}{5}$ **27.** $-1 \pm i$ **29.** $-\frac{5}{2} \pm \frac{\sqrt{3}}{2}i$ **31.** $\frac{2}{3} \pm \frac{\sqrt{2}}{3}i$ **33.** $\frac{1}{3} \pm \frac{2\sqrt{2}}{3}i$ **35.** $N = \frac{1}{2} \pm \frac{\sqrt{1 + 8C}}{2}$ **37.** $x = \frac{k}{2} \pm \frac{\sqrt{k^2 - 4ay}}{2}$ **39.** $-\frac{1}{2}, \frac{4}{3}$ **41.** $-\frac{5}{2} \pm \frac{\sqrt{17}}{2}$ **43.** $\frac{1}{3} \pm \frac{\sqrt{7}}{3}$ **45.** $\frac{1}{4}, -\frac{3}{4}$ **47.** $-\frac{1}{2} \pm \sqrt{5}$ **49.** $-\frac{1}{3} \pm \frac{\sqrt{5}}{3}i$ **51.** $8.98, -3.98$ **53.** $x^2 - 8x + 15 = 0$ **55.** $x^3 - x^2 - 14x + 24 = 0$ **57.** 8 ft by 12 ft **59.** 4 units **61.** $\frac{4}{3}$ cm **63.** $12, 14$ or $-14, -12$ **65.** $6, 7$ or $-6, -7$ **67.** 30 mph **69.** \$4.80 or \$5.20 **71.** 4,000 **73.** 2.26 in. **75.** 1985 **77.** about 6.13×10^{-3} M **81.** $\sqrt{2}, -3\sqrt{2}$ **83.** $i, 2i$

Getting Ready (page 746)

1. 17 **2.** -8

Exercises 10.3 (page 752)

1. -3 **3.** yes **5.** no **7.** -2 **9.** $\frac{3}{7}$ **11.** $b^2 - 4ac$ **13.** rational, unequal **15.** rational, equal **17.** complex conjugates **19.** irrational, unequal **21.** rational, unequal **23.** $8, -8$ **25.** $12, -12$ **27.** $1, -1, 4, -4$ **29.** $1, -1, \sqrt{2}, -\sqrt{2}$ **31.** $1, -1, \sqrt{5}, -\sqrt{5}$ **33.** $1, -1, 2, -2$ **35.** $25, 4$ **37.** $\frac{9}{4}$; 1 is extraneous **39.** 1; $\frac{9}{4}$ is extraneous **41.** $\varnothing$; $\frac{4}{9}$ and 1 are both extraneous **43.** $-8, -27$ **45.** $-\frac{8}{27}, 1$ **47.** $-9, -1$ **49.** $-7, 4$ **51.** $0, 2$ **53.** $-1, -\frac{27}{13}$ **55.** 1 (double root), -1 (double root) **57.** $-4, \frac{2}{3}$ **59.** $x = \pm\sqrt{r^2 - y^2}$ **61.** $y = \frac{-5x \pm \sqrt{25x^2 - 12x}}{2x}$ **63.** $\frac{2}{3}, -\frac{1}{4}$ **65.** $-\frac{5}{4} \pm \frac{\sqrt{17}}{4}$ **67.** $\frac{1}{3} \pm \frac{\sqrt{11}}{3}i$ **69.** $-1 \pm 2i$ **71.** yes **73.** $\frac{81}{16}$; 1 is extraneous **75.** $-\frac{8}{27}, 1$ **77.** $1, -1, 2\sqrt{3}, -2\sqrt{3}$ **79.** $2, -2, \sqrt{6}i, -\sqrt{6}i$ **81.** $\frac{3}{8}, 1$ **83.** $\frac{1}{27}, -1$ **85.** $1 \pm i$

Unless otherwise noted, all content on this page is © Cengage Learning.

87. $-\frac{5}{7}, 3$ **89.** $d = \pm\frac{\sqrt{kI}}{I}$ **91.** $\mu^2 = \frac{\Sigma x^2}{N} - \sigma^2$

93. 5 **95.** 12, −3 **97.** $k < -\frac{4}{3}$ **101.** no

Getting Ready (page 754)

1. −2 **2.** 2 **3.** 0 **4.** 8 **5.** 1 **6.** 4

Exercises 10.4 (page 766)

1. $x = 3$ **3.** $x = -1$ **5.** 3, −3 **7.** 10
9. $3\frac{3}{5}$ hr **11.** $f(x) = ax^2 + bx + c, a \neq 0$
13. maximum, minimum, vertex **15.** upward
17. to the right **19.** upward

21.
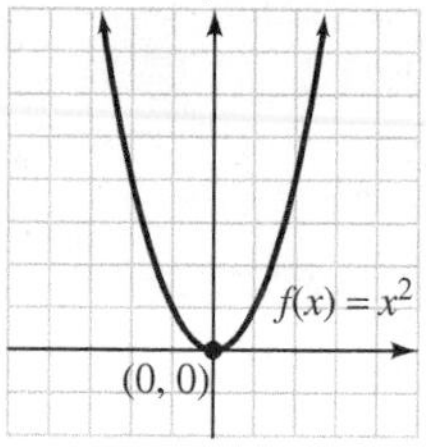

23.
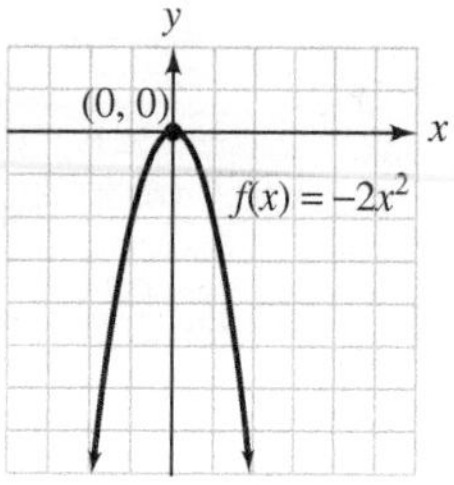

25.
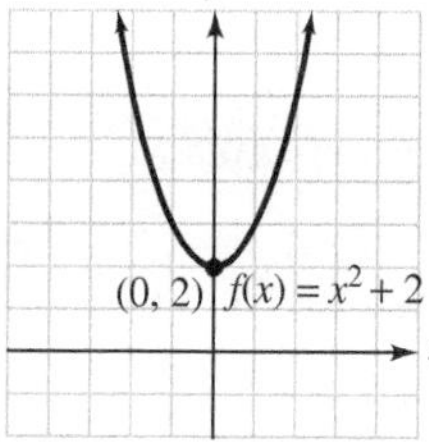

27.
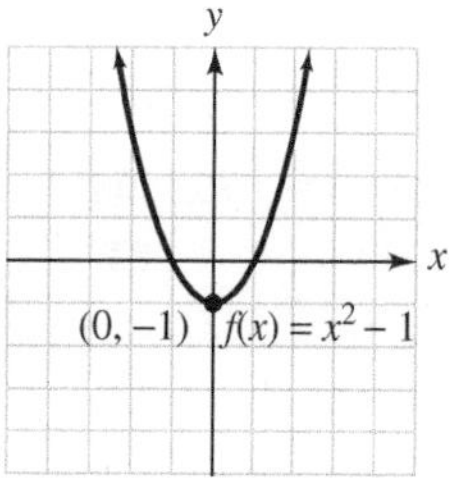

29.
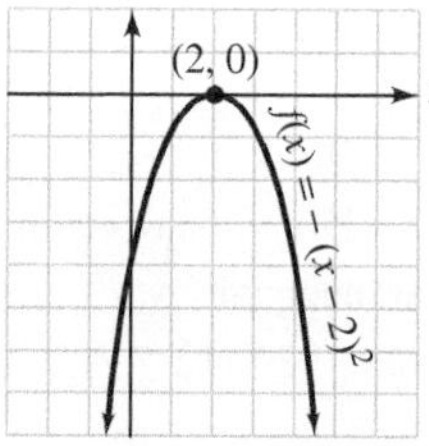

31.
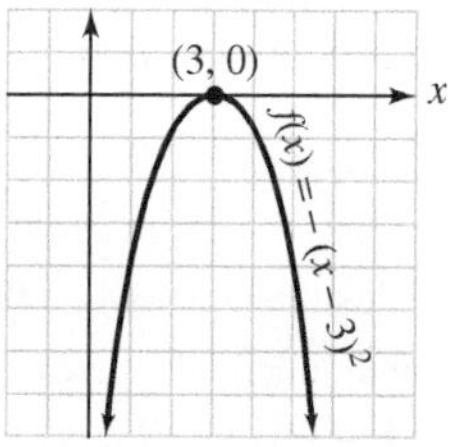

33.
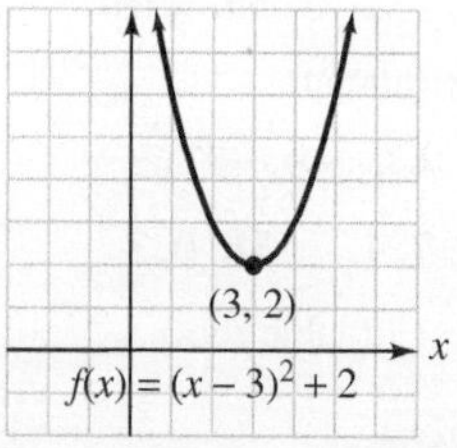

35.
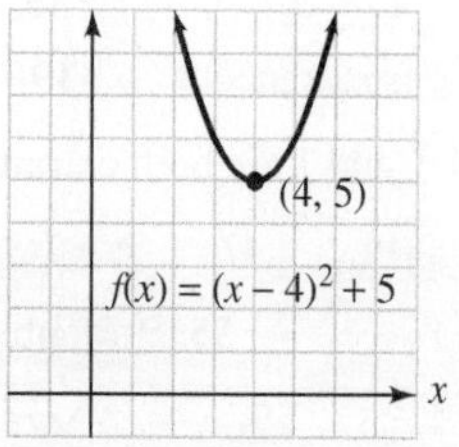

37. $(1, 2), x = 1$ **39.** $(-3, -4), x = -3$ **41.** $(0, 0), x = 0$
43. $(1, -2), x = 1$ **45.** $(-1, -2)$ **47.** $(2, 16)$

49.
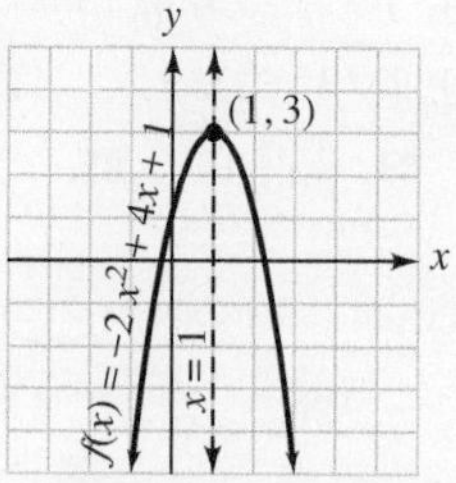

51.
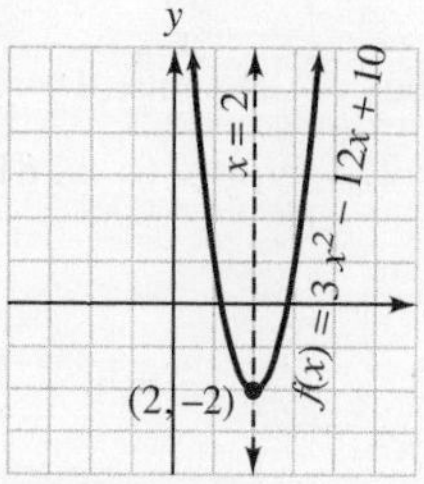

53. $(2, 21), x = 2$ **55.** $\left(\frac{5}{12}, \frac{143}{24}\right), x = \frac{5}{12}$

57.
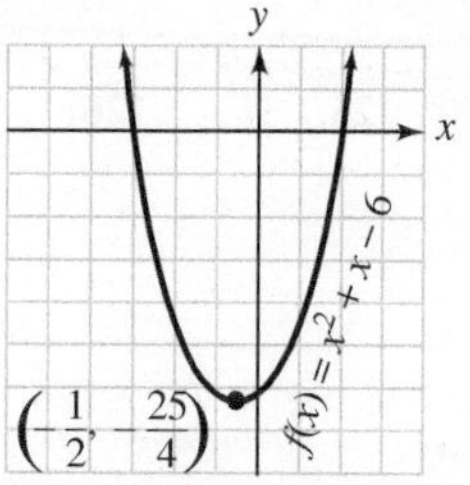

59. $(0.25, 0.88)$
61. $(0.5, 7.25)$
63. −10, 1
65. −1.85, 3.25
67. $(-7, 3)$
69. 36 ft, 1.5 sec
71. 100 ft by 100 ft, 10,000 ft²
73. 0.25 and 0.75

75. 75 ft by 75 ft, 5,625 ft² **77.** 14 ft **79.** 5,000
81. 3,276, $14,742 **83.** $35

Getting Ready (page 770)

1. 2, (0, −3) **2.** −3, (0, 4) **3.** 6, −9 **4.** $4, \frac{5}{2}$

Exercises 10.5 (page 779)

1. same shape as $f(x) = x^2$ but shifted upward 3 units
3. same shape as $f(x) = x^2$ but shifted left 3 units
5. 31, 37 **7.** $a \cdot b = b \cdot a$ **9.** 1 **11.** cubic
13. vertical **15.** 5, upward **17.** 5, to the right
19. x-axis **21.** asymptote

23.
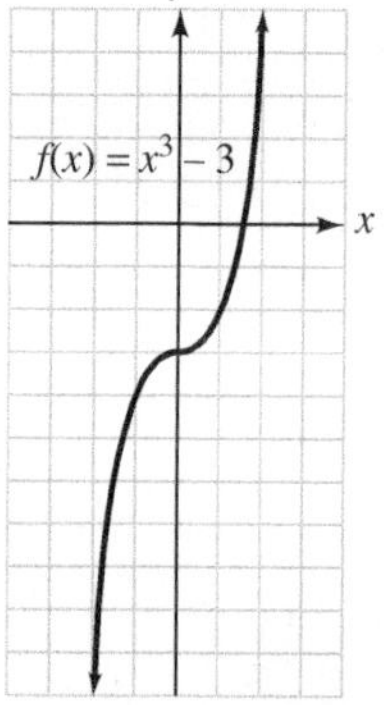

25.
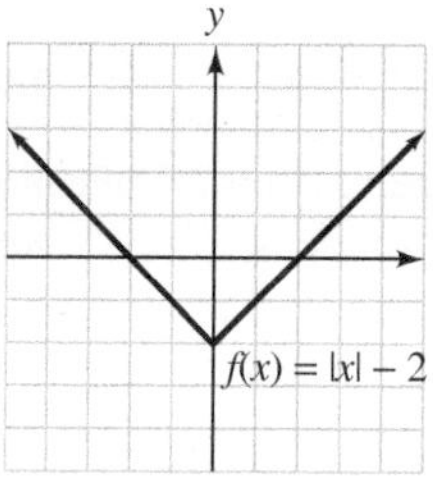

27.
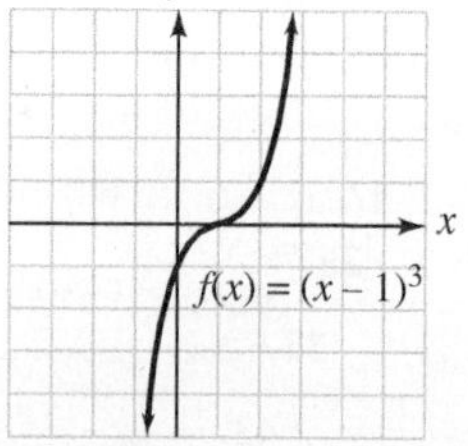

29.
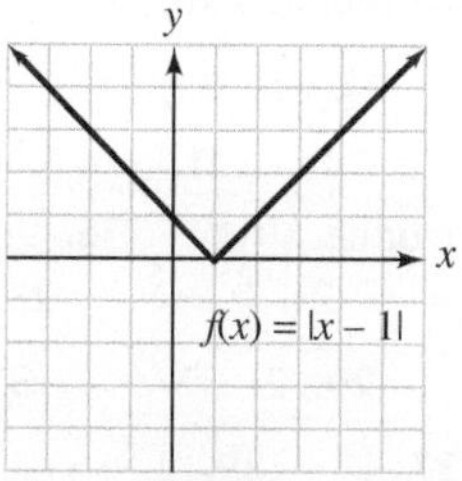

31.
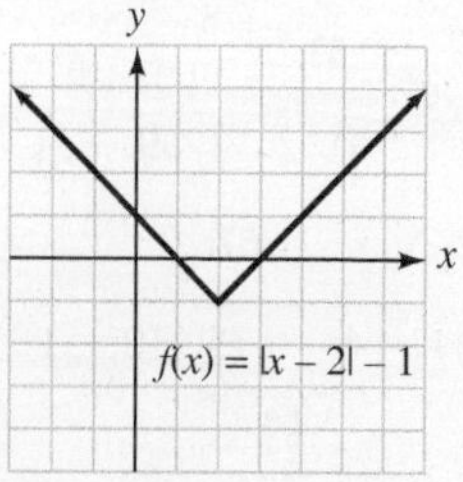

33.
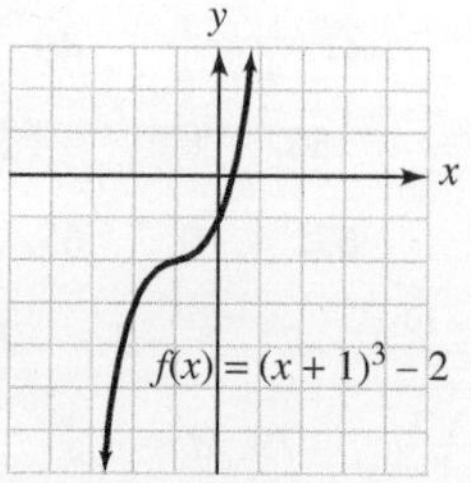

Unless otherwise noted, all content on this page is © Cengage Learning.

35.

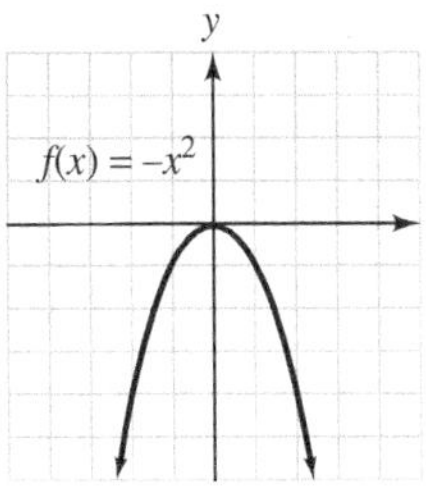

37.

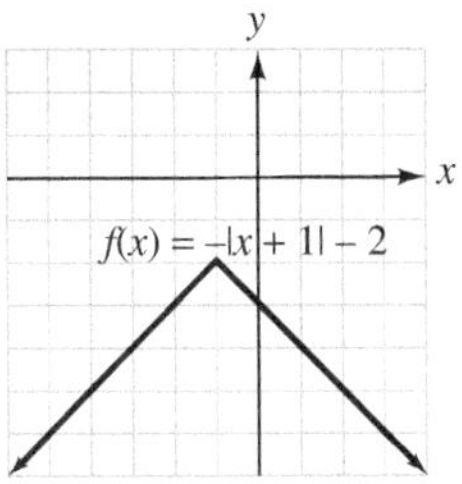

39.

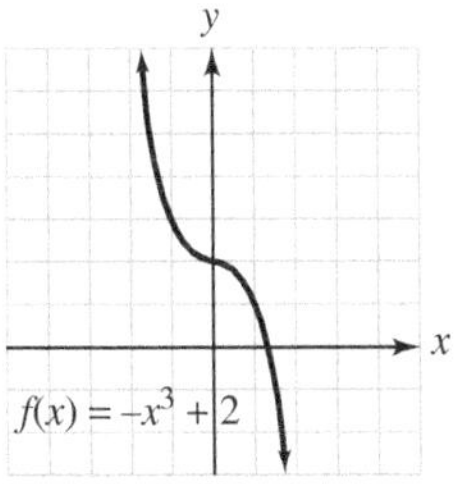

41.

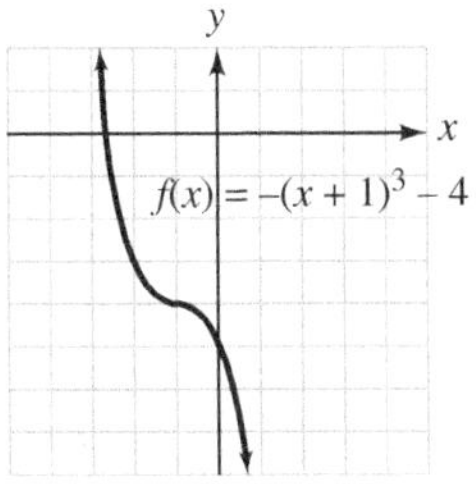

43. 20 hr **45.** 12 hr

47. D: $(-\infty, 2) \cup (2, \infty)$; R: $(-\infty, 1) \cup (1, \infty)$

49. D: $(-\infty, -2) \cup (-2, 2) \cup (2, \infty)$; R: $(-\infty, \infty)$

51.

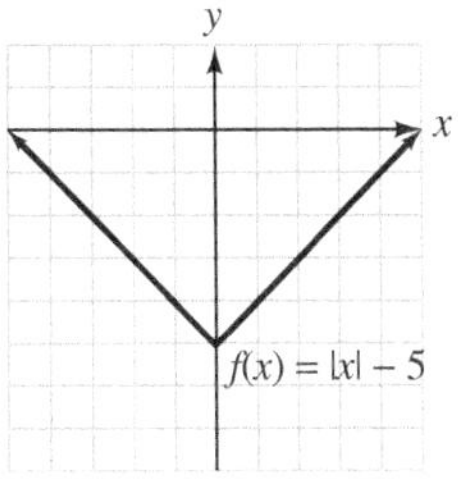

53.

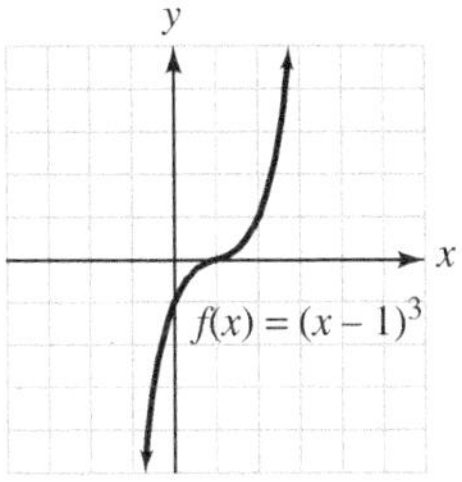

55.

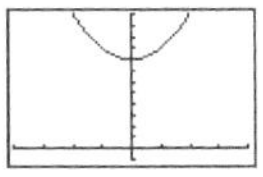

57.

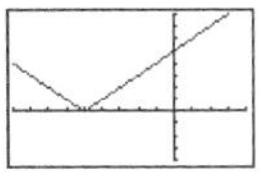

59.

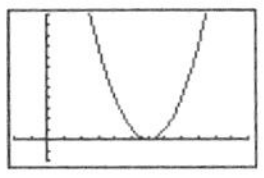

61.

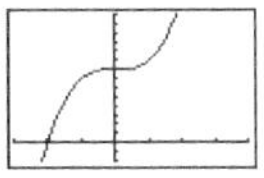

63. \$12,500 **65.** \$50,000

67. $C = f(x) = 1.25x + 700$ **69.** \$1,012.50 **71.** \$3.58

73. $C = f(n) = 0.114n + 15$ **75.** \$225.90 **77.** 12¢

79. $C = f(x) = 350x + 5{,}000$ **81.** \$47,000

Getting Ready (page 782)

1. $(x + 5)(x - 3)$ **2.** $(x - 2)(x - 1)$

Exercises 10.6 (page 789)

1. $x = 2$ **3.** $x < 2$ **5.** $x > -3$ **7.** $1 < 2x$

9. $y = kx$ **11.** $t = kxy$ **13.** 3 **15.** greater

17. quadratic **19.** undefined **21.** sign

23. (1, 4)

25. (–∞, 3) ∪ (5, ∞)

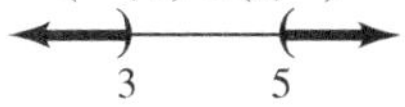

27. [–4, 3]

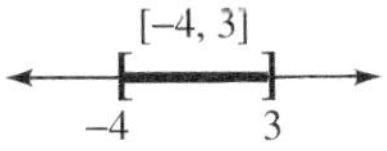

29. (–∞, –5] ∪ [3, ∞)

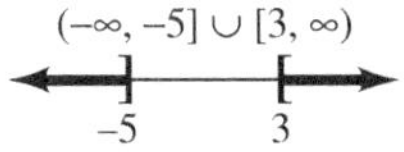

31. ∅

33. (–∞, –3] ∪ [3, ∞)

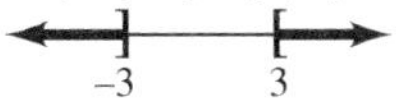

35. (–∞, 0) ∪ (1/2, ∞)

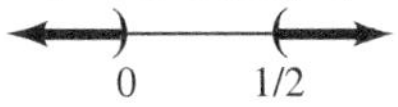

37. (0, 2]

39. (–∞, –3) ∪ (1, 4)

41. [–5, –2) ∪ [4, ∞)

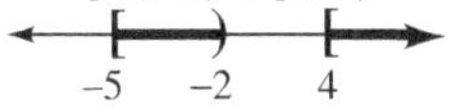

43. (–∞, –4)

45. (–1/2, 1/3) ∪ (1/2, ∞)

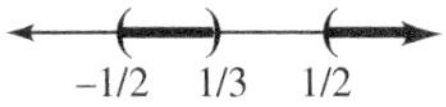

47. (0, 2) ∪ (8, ∞)

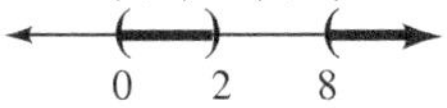

49. (–∞, –2) ∪ (2, 18]

51. [–34/5, –4) ∪ (3, ∞)

53. (–∞, –2) ∪ (–2, ∞)

55.

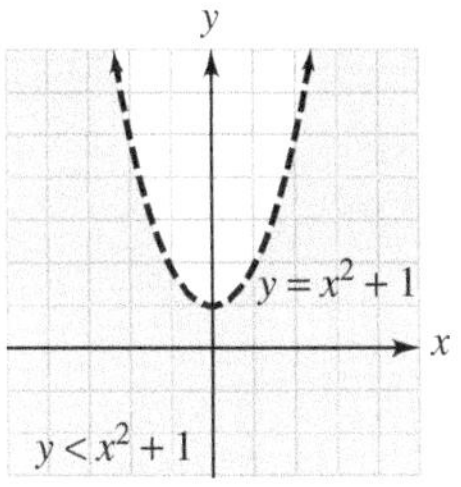

57.

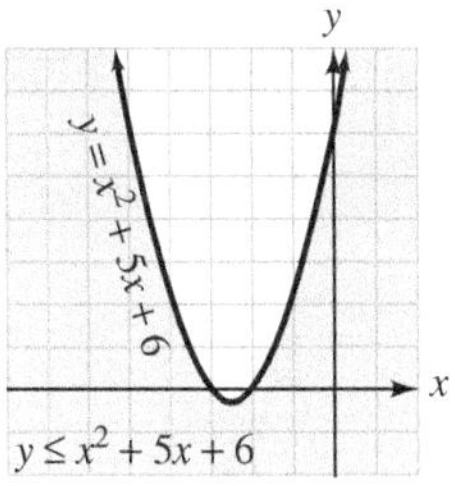

59.

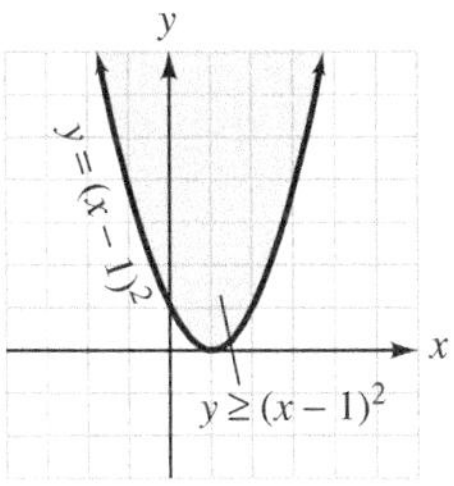

61.

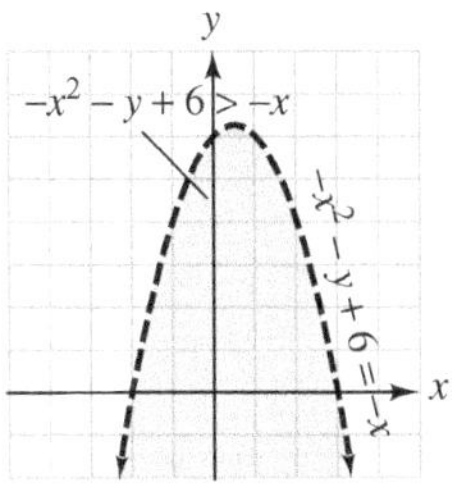

63.

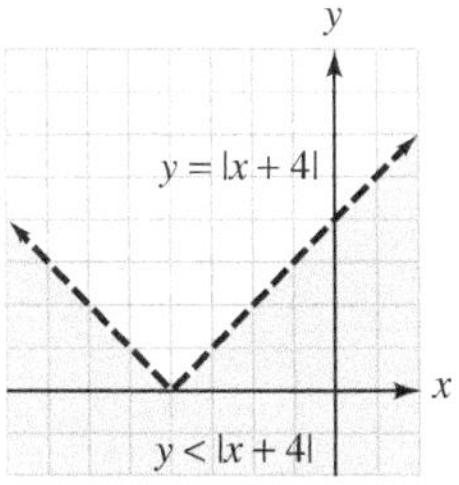

65.

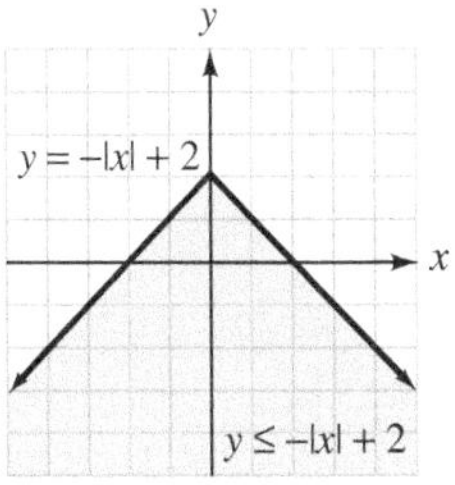

67. (–5, 5)

69. (–∞, –5/3) ∪ (0, ∞)

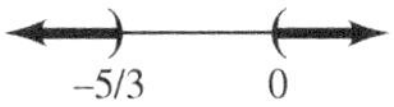

71. (–4, –2] ∪ (–1, 2]

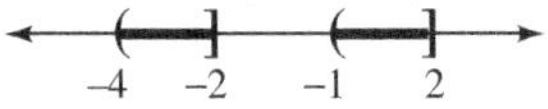

73. (–∞, –16) ∪ (–4, –1) ∪ (4, ∞)

75. $(-1, 3)$ **77.** $(-\infty, -3] \cup (2, \infty)$

81. when 4 factors are negative, 2 factors are negative, or no factors are negative

Unless otherwise noted, all content on this page is © Cengage Learning.

Chapter Review (page 795)

1. $\frac{2}{3}, -\frac{3}{4}$ **2.** $-\frac{1}{3}, -\frac{5}{2}$ **3.** $\frac{2}{3}, -\frac{4}{5}$ **4.** $1, -7$
5. $-2, -6$ **6.** $\frac{7}{2}, 1$ **7.** $\frac{1}{4} \pm \frac{\sqrt{41}}{4}$ **8.** $6, -1$
9. $0, 7$ **10.** $\frac{1}{2}, -7$ **11.** $1, -\frac{3}{5}$ **12.** $\frac{1}{4} \pm \frac{\sqrt{17}}{4}$
13. $-\frac{1}{2} \pm \frac{\sqrt{7}}{2}i$ **14.** 4 cm by 6 cm **15.** 2 ft by 3 ft
16. 7 sec **17.** 196 ft **18.** irrational, unequal
19. complex conjugates **20.** 12, 152
21. $k \geq -\frac{7}{3}$ **22.** 1, 49 **23.** 8, −27 **24.** 1
25. $1, -\frac{8}{5}$ **26.** $\frac{14}{3}$ **27.** 1

28.
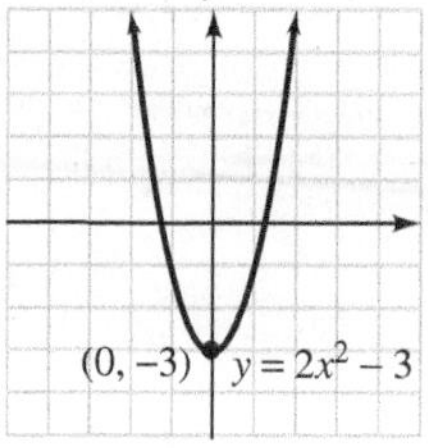

29.
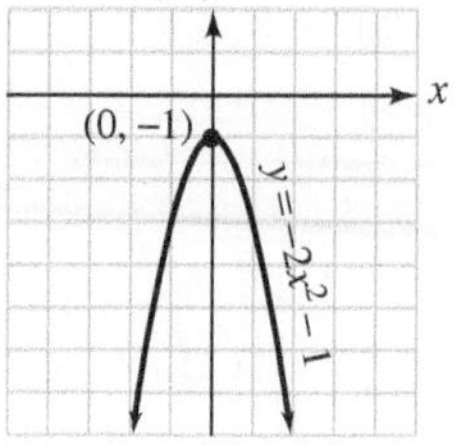

30.
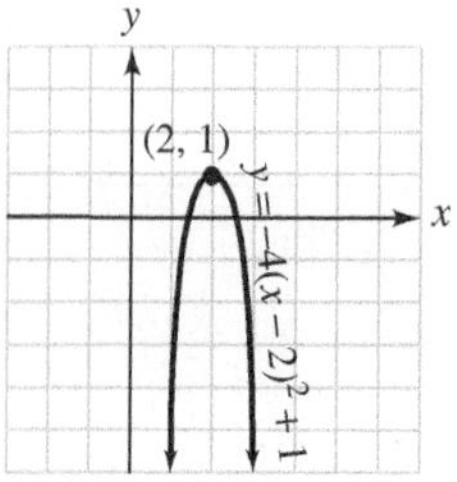

31.
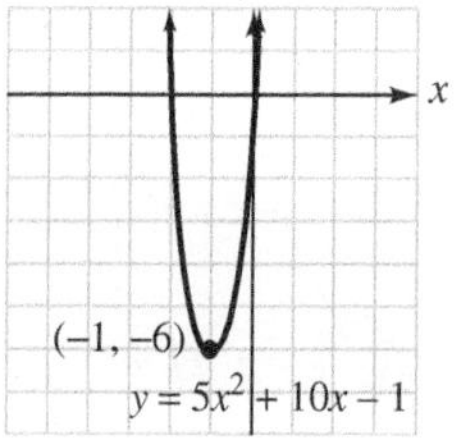

32. $(2, -19), x = 2$

33.
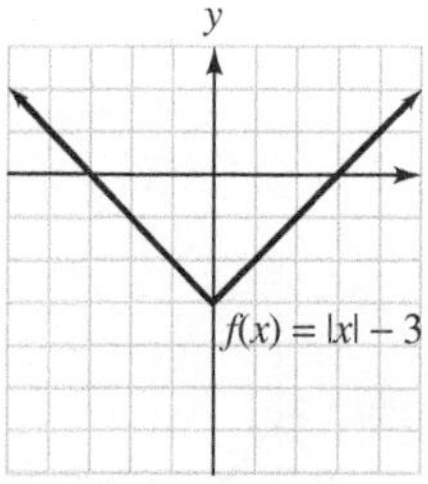

34.
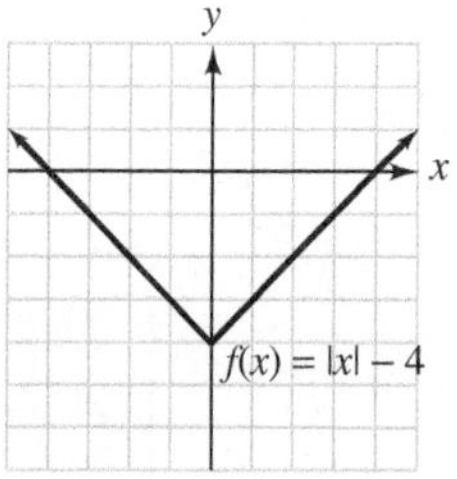

35.
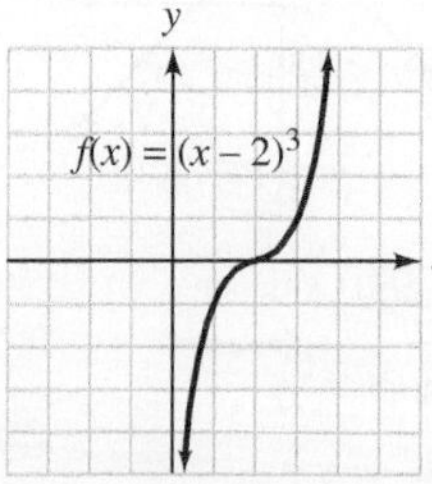

36.
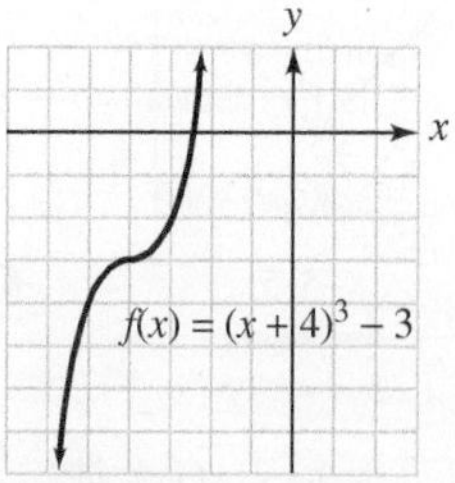

37.
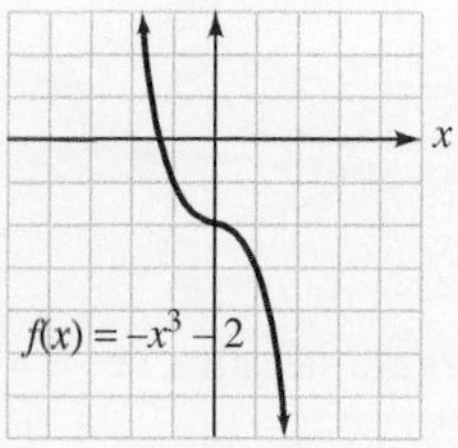

38.
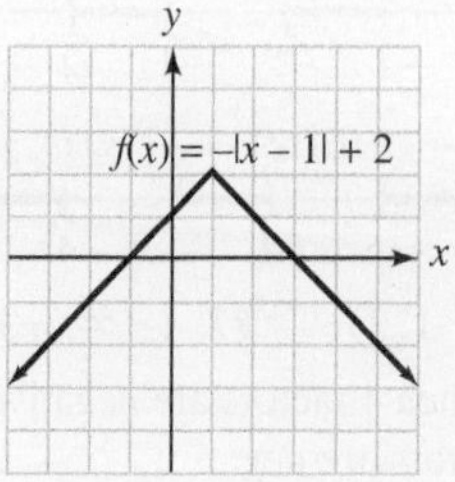

39.
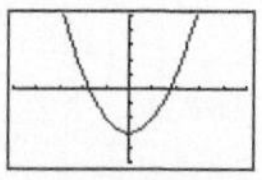

40.
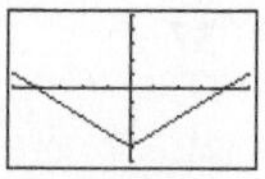

41.
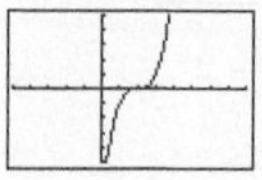

42.
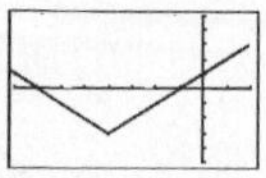

43.
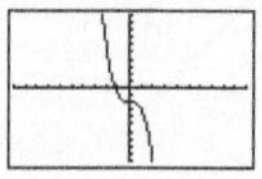

44.
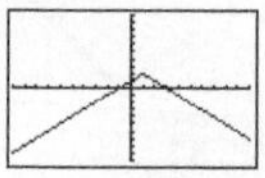

45. D: $(-\infty, 2) \cup (2, \infty)$; R: $(-\infty, 0) \cup (0, \infty)$

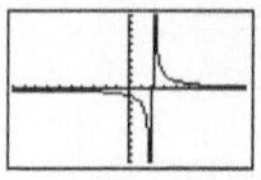

46. D: $(-\infty, -3) \cup (-3, \infty)$; R: $(-\infty, 1) \cup (1, \infty)$

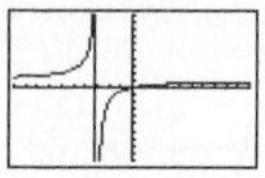

47.
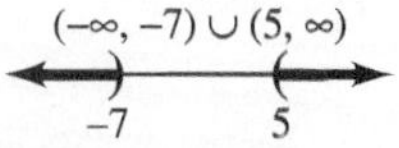

48.
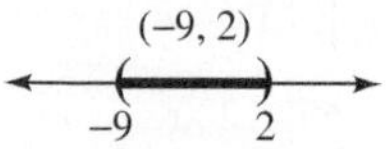

49.
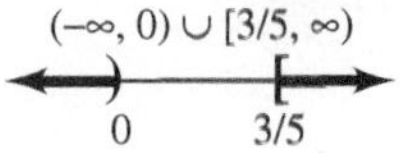

50.
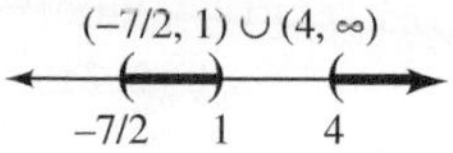

51.
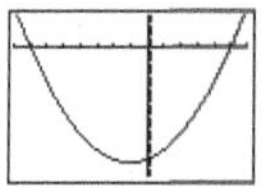

$(-\infty, -7) \cup (5, \infty)$

52.
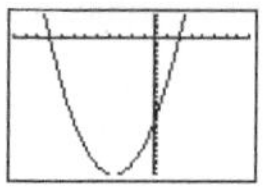

$(-9, 2)$

53.
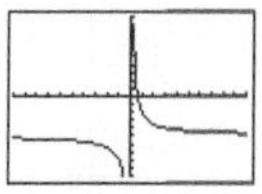

$(-\infty, 0) \cup \left[\frac{3}{5}, \infty\right)$

54.
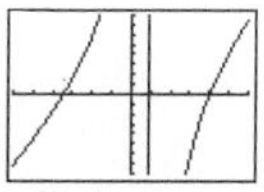

$\left(-\frac{7}{2}, 1\right) \cup (4, \infty)$

55.
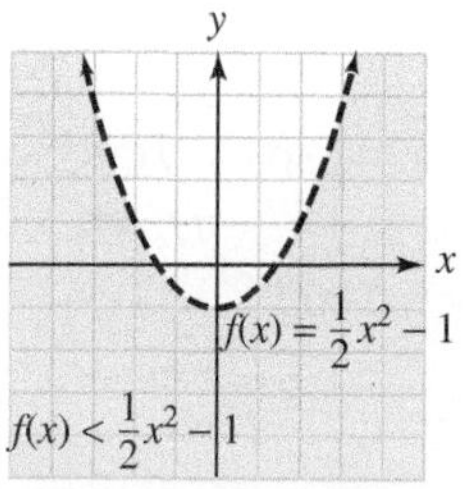

56.
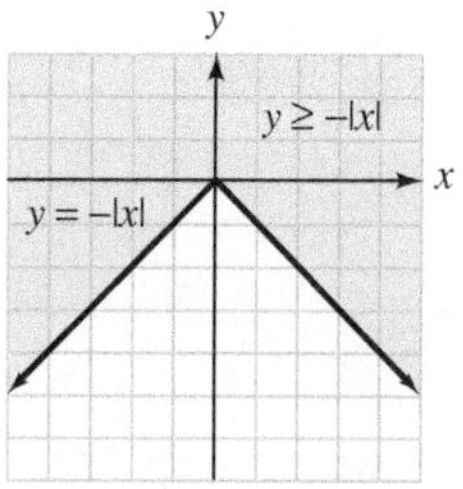

Chapter 10 Test (page 801)

1. 7, −4 **2.** $-\frac{3}{2}, -\frac{5}{3}$ **3.** ± 12 **4.** $x = 5 \pm \sqrt{6}i$
5. $-3 \pm \sqrt{2}$ **6.** $\frac{5}{2} \pm \frac{\sqrt{37}}{2}$ **7.** $-\frac{5}{4} \pm \frac{\sqrt{17}}{4}$
8. $1 \pm \sqrt{5}i$ **9.** nonreal **10.** 2 **11.** 10 in. **12.** $1, \frac{1}{4}$

13. (0, −4)
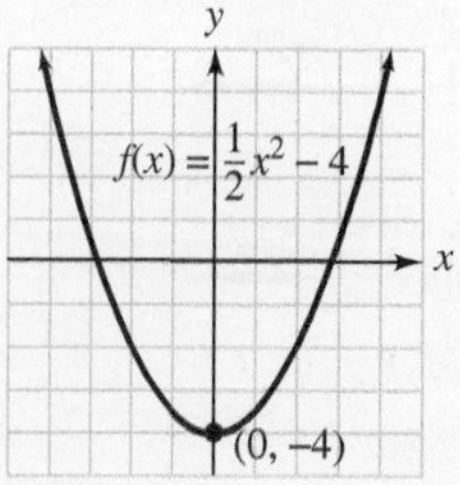

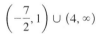

14. $(2, 7); x = 2$

Unless otherwise noted, all content on this page is © Cengage Learning.

15. 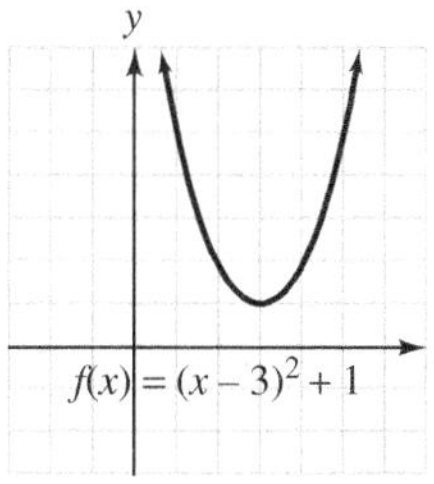

16. 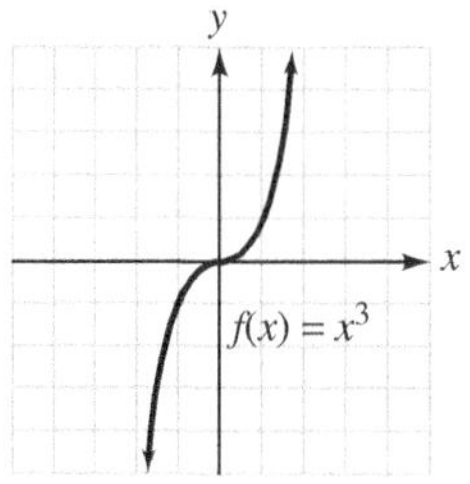

17. 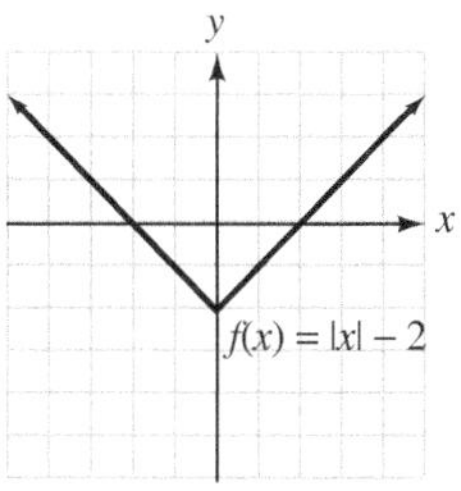

18. 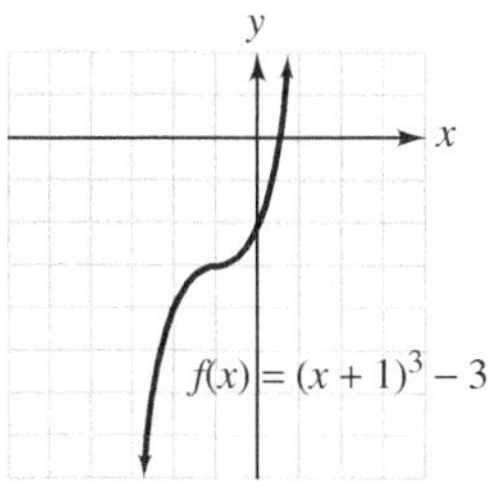

19. 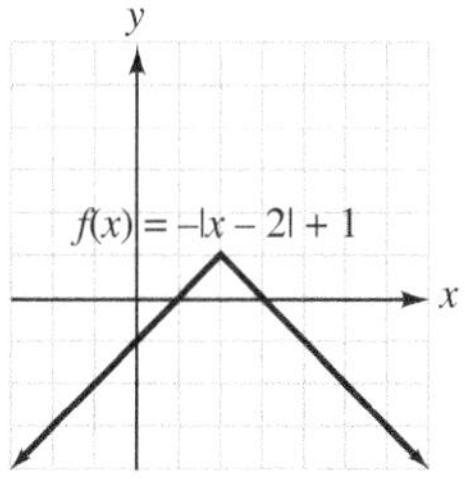

20. 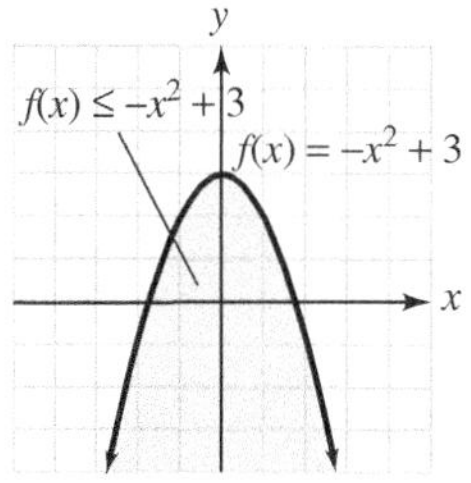

21. 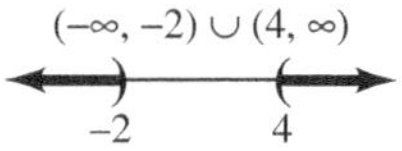

22. 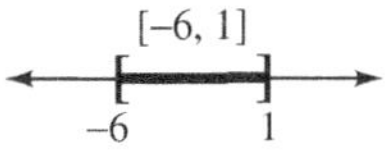

23. 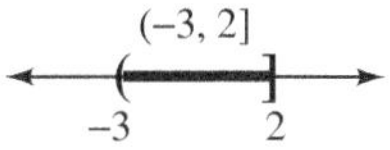

24. 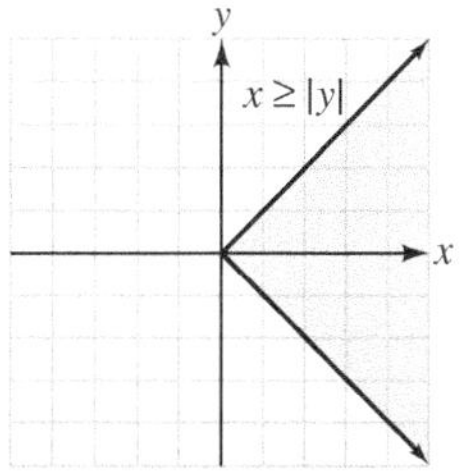

25. 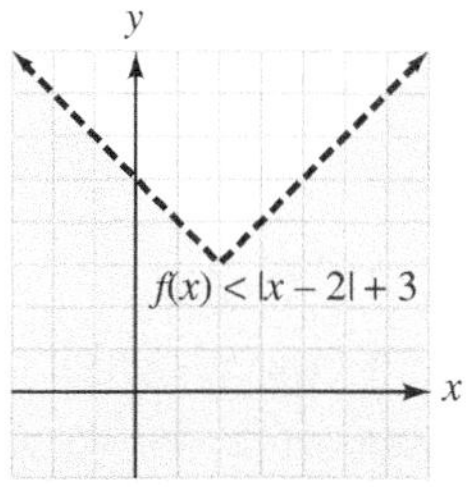

Cumulative Review (page 802)

1. D: $(-\infty, \infty)$; R: $[-2, \infty)$ **2.** D: $(-\infty, \infty)$; R: $(-\infty, 0]$

3. $y = 4x + 7$ **4.** $y = -\frac{2}{3}x - 2$

5. $-4a^2 + 12a - 7$ **6.** $20x^2 - 7x - 6$

7. $(x^2 + 4y^2)(x + 2y)(x - 2y)$ **8.** $(3x + 2)(5x - 4)$

9. $9, -1$ **10.** $0, \frac{2}{3}, -\frac{1}{2}$ **11.** $6ab^2$ **12.** $4t\sqrt{3t}$

13. $-4y^2$ **14.** $4x$ **15.** $\frac{1}{2}$ **16.** 9

17. y^2 **18.** $x^{\frac{17}{12}}$

19. 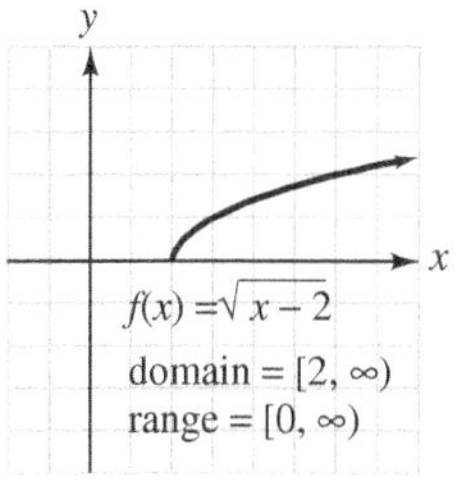

20. 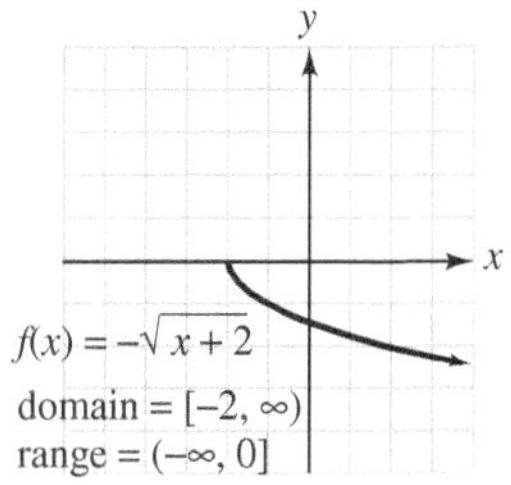

21. $x^{\frac{4}{3}} - x^{\frac{2}{3}}$ **22.** $\frac{1}{x} + 2 + x$ **23.** $7\sqrt{2}$

24. $-12\sqrt[4]{2} + 10\sqrt[4]{3}$ **25.** $-18\sqrt{6}$ **26.** $\frac{5\sqrt[3]{x^2}}{x}$

27. $\frac{x + 3\sqrt{x} + 2}{x - 1}$ **28.** $\sqrt{xy}$ **29.** 2, 7

30. $\frac{1}{4}$ **31.** $3\sqrt{2}$ in. **32.** $2\sqrt{3}$ in. **33.** 10

34. $\pm 5i$ **35.** $1, -\frac{3}{2}$ **36.** $-\frac{2}{3} \pm \frac{\sqrt{7}}{3}$

37. 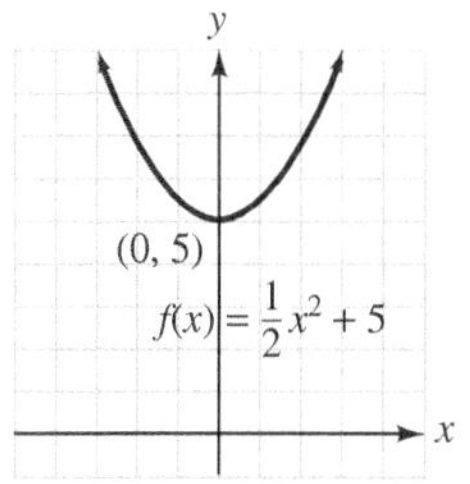

38. 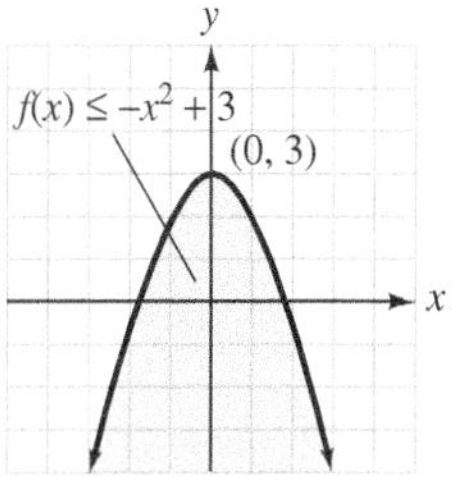

39. $13 - 3i$ **40.** $-5 - 7i$ **41.** 61

42. $12 - 6i$ **43.** $-12 - 10i$ **44.** $\frac{3}{2} + \frac{1}{2}i$

45. $\sqrt{13}$ **46.** $\sqrt{61}$ **47.** -2 **48.** 9, 16

49. 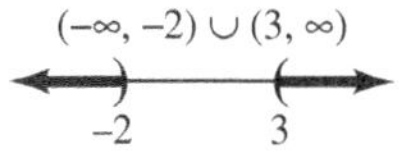

50. 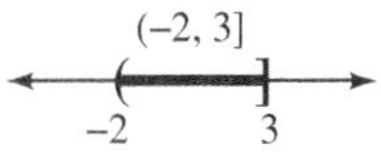

51. 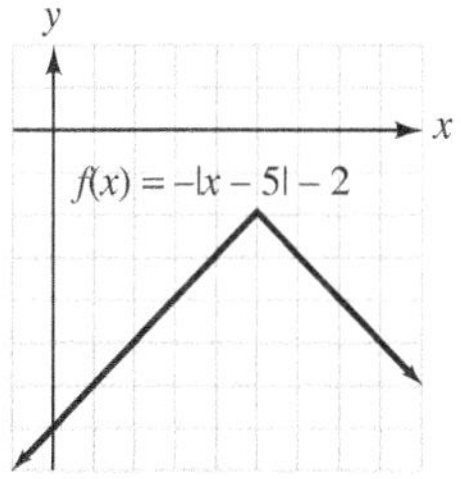

52. 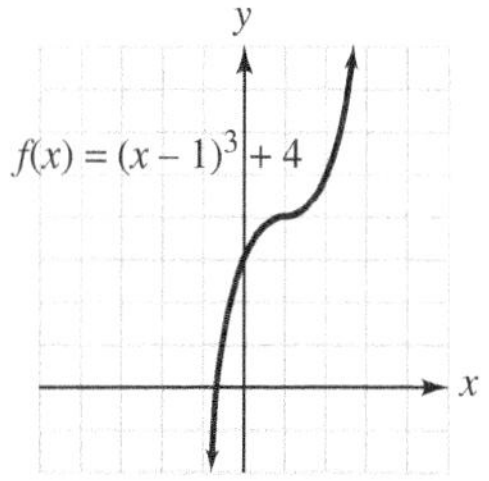

53. 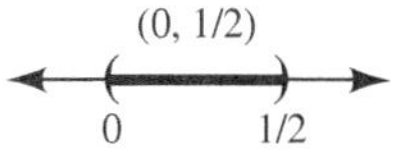

54. 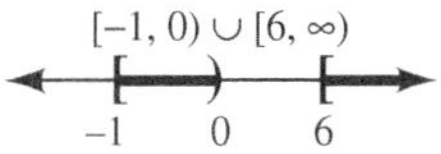

55. $\frac{1}{x + 6}$ **56.** $\frac{-3x + 1}{(x + 7)(x - 4)}$

57. $\frac{4x^2 + x}{3x^2 + 2}$ **58.** 1; 3 is extraneous

Getting Ready (page 807)

1. $3x - 1$ **2.** $x + 3$ **3.** $2x^2 - 3x - 2$

Unless otherwise noted, all content on this page is © Cengage Learning.

Exercises 11.1 (page 812)

1. -3 **3.** $4x^2 + 8x + 2$ **5.** $4x^2 + 2x + 1$
7. $-4x^2 + 2x + 5$ **9.** $-\frac{5x+2}{x+3}$ **11.** $\frac{x-4}{3x^2-x-12}$
13. $f(x) + g(x)$ **15.** $f(x)g(x)$ **17.** domain
19. $f(g(x))$ **21.** given point **23.** $7x, (-\infty, \infty)$
25. $x, (-\infty, \infty)$ **27.** $12x^2, (-\infty, \infty)$
29. $\frac{4}{3}, (-\infty, 0) \cup (0, \infty)$ **31.** $3x - 2, (-\infty, \infty)$
33. $-x - 4, (-\infty, \infty)$ **35.** $2x^2 - 5x - 3, (-\infty, \infty)$
37. $\frac{x-3}{2x+1}, \left(-\infty, -\frac{1}{2}\right) \cup \left(-\frac{1}{2}, \infty\right)$ **39.** 31 **41.** 24
43. -1 **45.** $-\frac{7}{8}$ **47.** $2x^2 - 1$ **49.** $36x^2 + 12x$
51. 4 **53.** $2x + h$ **55.** $4x + 2h$
57. $2x + h + 1$ **59.** $2x + h + 3$
61. $4x + 2h + 3$ **63.** $-2x^2 + 3x - 3, (-\infty, \infty)$
65. $(3x - 2)/(2x^2 + 1), (-\infty, \infty)$ **67.** $3, (-\infty, \infty)$
69. $(x^2 - 4)/(x^2 - 1), (-\infty, -1) \cup (-1, 1) \cup (1, \infty)$
71. 58 **73.** 56 **75.** 2 **77.** $9x^2 - 9x + 2$
79. 4 **81.** $x + a$ **83.** $2x + 2a$ **85.** $x + a + 1$
87. $x + a + 3$ **89.** $2x + 2a + 3$
95. $3x^2 + 3xh + h^2$ **97.** $C(t) = \frac{5}{9}(2{,}668 - 200t)$

Getting Ready (page 814)

1. $y = \frac{x-2}{3}$ **2.** $y = \frac{2x-10}{3}$

Exercises 11.2 (page 821)

1. $(0, 1)$ **3.** $(3, 7)$ **5.** $(-1, -1)$ **7.** $7 - 7i$
9. $18 - i$ **11.** 10 **13.** one-to-one **15.** 2
17. x **19.** yes **21.** no

23.

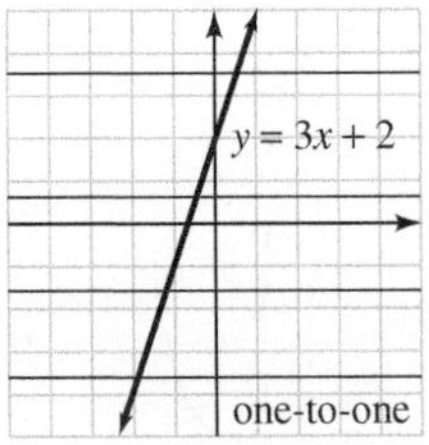

25.

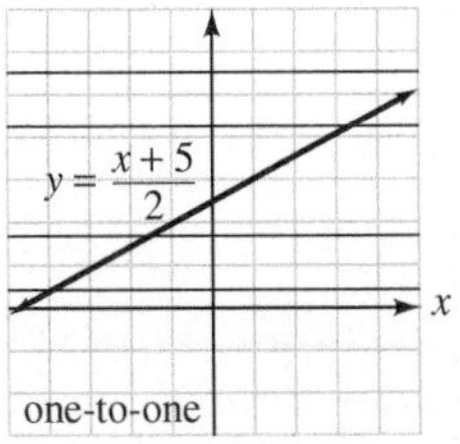

27.

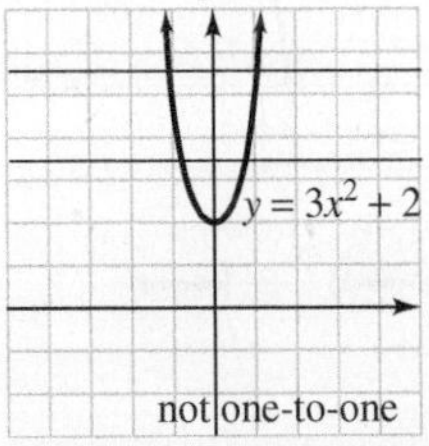

29.

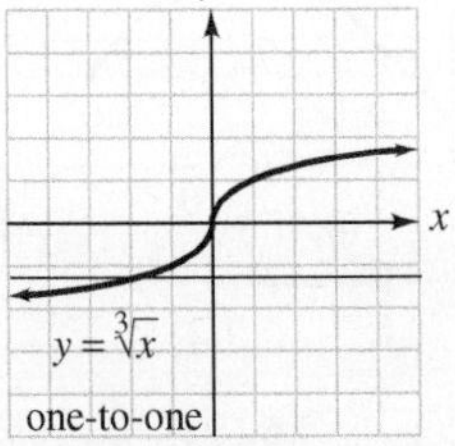

31. $\{(3, 4), (2, 3), (1, 2)\}$; yes
33. $\{(2, 1), (3, 2), (3, 1), (5, 1)\}$; no **35.** $f^{-1}(x) = \frac{1}{4}x - \frac{1}{4}$
37. $f^{-1}(x) = 5x - 4$

39. $f^{-1}(x) = \frac{x-3}{4}$

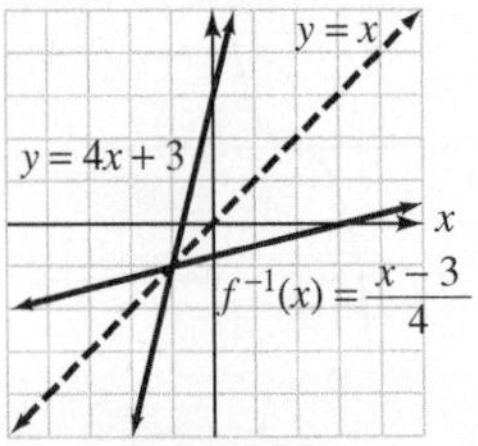

41. $f^{-1}(x) = \frac{x-2}{3}$

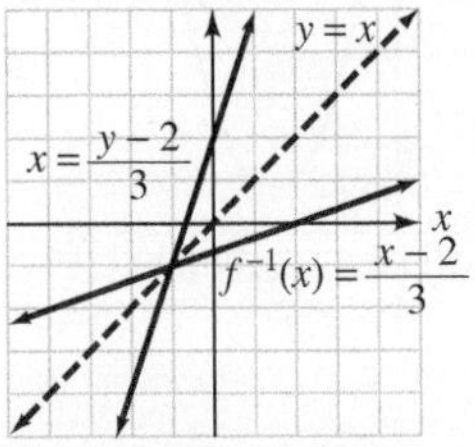

43. $y = \pm\sqrt{x - 9}$, no **45.** $f^{-1}(x) = \sqrt[3]{x}$, yes

47.

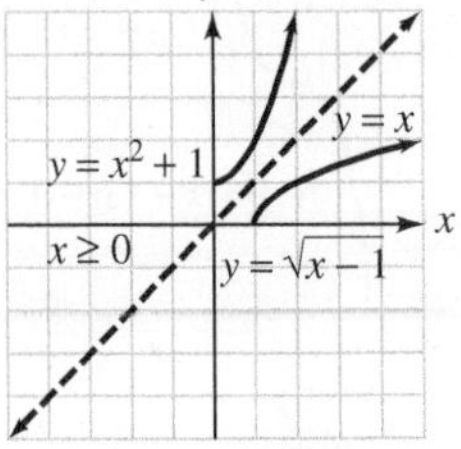

49.

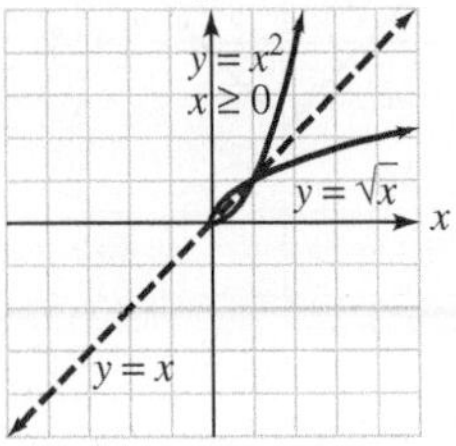

51. $\{(1, 1), (4, 2), (9, 3), (16, 4)\}$ **53.** $f^{-1}(x) = -x^3$
55. $f^{-1}(x) = \sqrt[3]{\frac{x+3}{2}}$ **57.** $f^{-1}(x) = 3x + 7$
59. $f^{-1}(x) = \frac{5}{4}x + 5$

61.

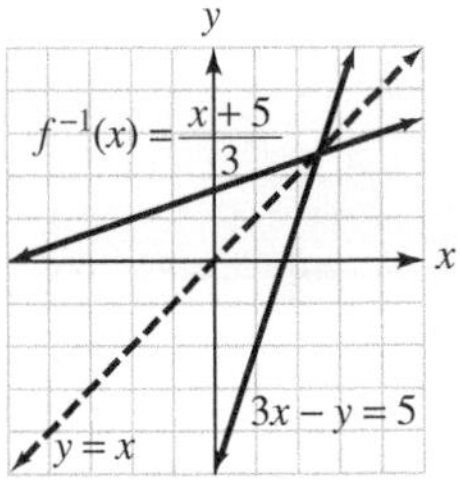

63.

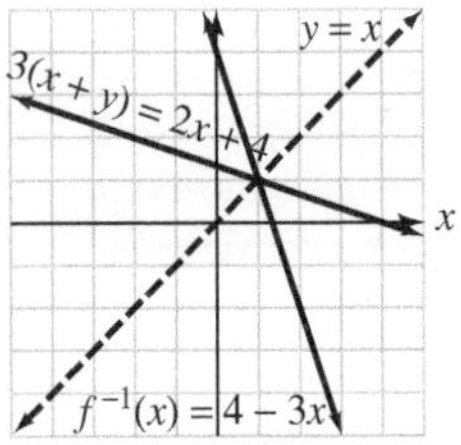

67. $f^{-1}(x) = \frac{x+1}{x-1}$

Getting Ready (page 824)

1. 8 **2.** 5 **3.** $\frac{1}{25}$ **4.** $\frac{8}{27}$

Exercises 11.3 (page 831)

1. 9 **3.** 108 **5.** $\frac{1}{9}$ **7.** $\frac{1}{12}$ **9.** 40 **11.** 120°
13. exponential **15.** $(0, \infty)$ **17.** increasing
19. $P\left(1 + \frac{r}{k}\right)^{kt}$ **21.** b **23.** left

25.

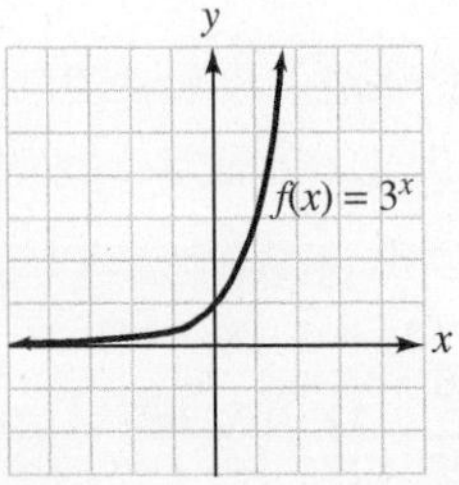

27.

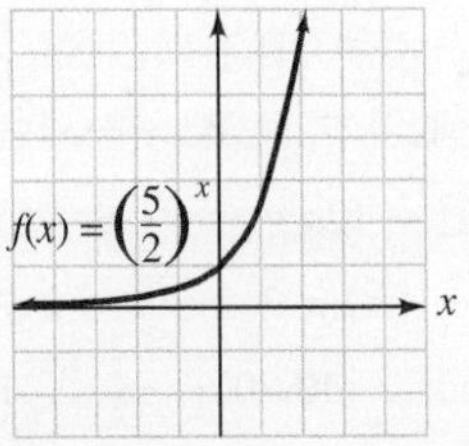

Unless otherwise noted, all content on this page is © Cengage Learning.

29.

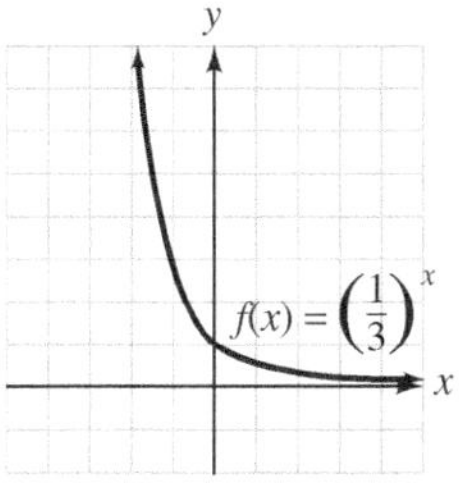

31.

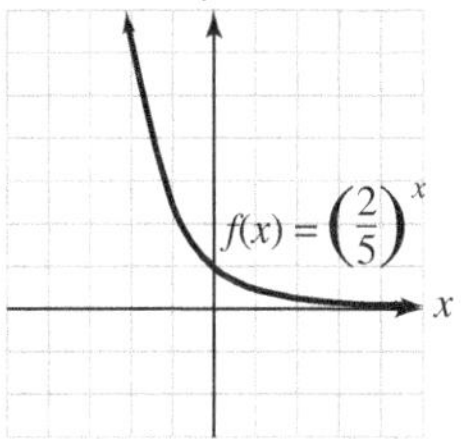

33. $b = \frac{1}{2}$ **35.** $b = 3$

37.

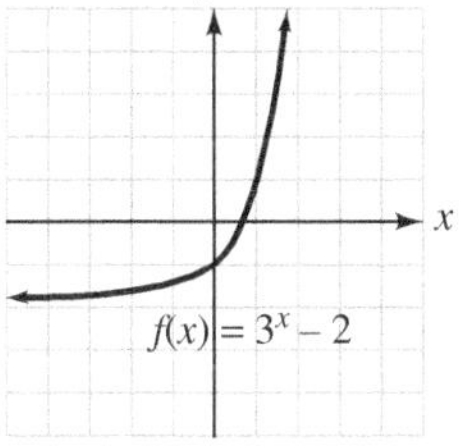

39.

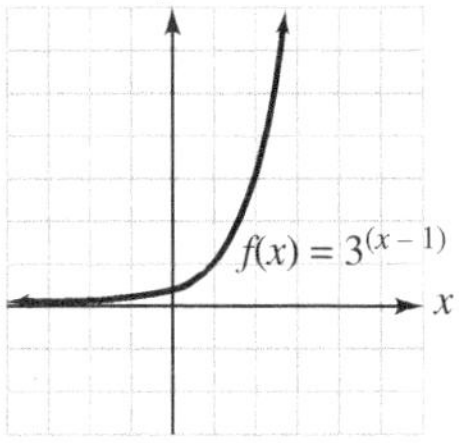

41. $b = 2$

43.

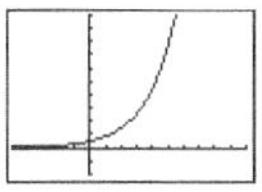

increasing function

45.

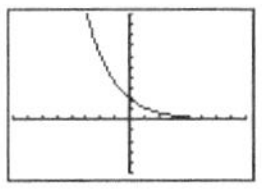

decreasing function

47. approximately 1.9 billion **49.** $\frac{256}{6{,}561} A_0$

51. \$22,080.40 **53.** \$7,551.32 **55.** \$2,273,996.13

57. 5.0421×10^{-5} coulombs **59.** \$1,115.33

Getting Ready (page 834)

1. 2 **2.** 2.25 **3.** 2.44 **4.** 2.59

Exercises 11.4 (page 839)

1. 1 **3.** 40.17 **5.** 0.22 **7.** $8x^3\sqrt{5x}$ **9.** $23y\sqrt{2y}$

11. 2.72 **13.** increasing **15.** $A = Pe^{rt}$

17.

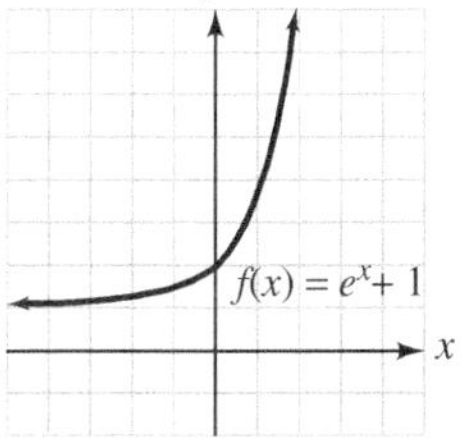

19.

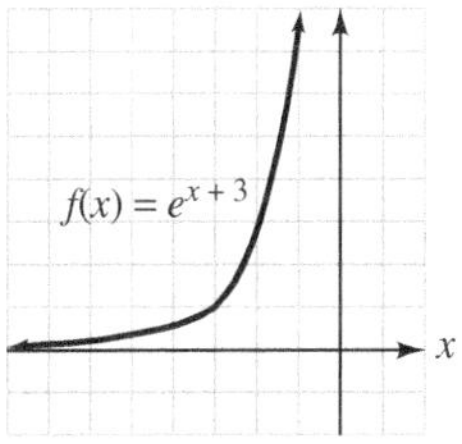

21. no **23.** no

25.

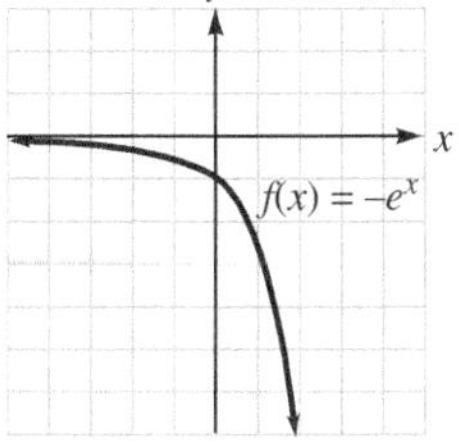

27.

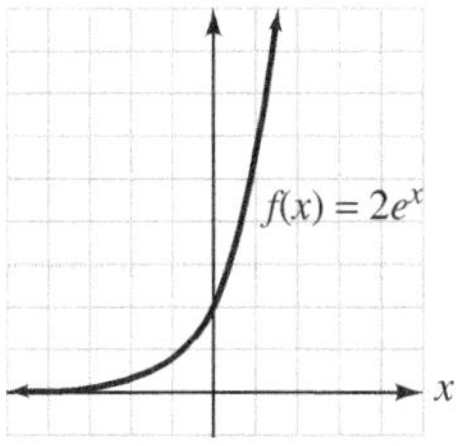

29. \$20,029.87 **31.** \$11,809.16 **33.** 10.6 billion

35. 2.6 **37.** 72 yr **39.** 2.31 grams **41.** 2,498.27 grams

43. \$7,518.28 from annual compounding; \$7,647.95 from continuous compounding **45.** 5,363 **47.** 0.076

49. 0 mps **51.** this object **53.** \$3,094.15

57. 2 **59.** $k = e^{10}$

Getting Ready (page 842)

1. 1 **2.** 25 **3.** $\frac{1}{25}$ **4.** 4

Exercises 11.5 (page 849)

1. 3 **3.** 5 **5.** 2 **7.** 5 **9.** 2

11. $\varnothing$; $\frac{3}{5}$ is extraneous **13.** 4 **15.** $(0, \infty)$ **17.** x

19. exponent **21.** $(b, 1), (1, 0)$ **23.** $20 \log \frac{E_O}{E_I}$

25. $4^3 = 64$ **27.** $\left(\frac{1}{2}\right)^3 = \frac{1}{8}$ **29.** $4^{-3} = \frac{1}{64}$ **31.** $\left(\frac{1}{2}\right)^3 = \frac{1}{8}$

33. $\log_7 49 = 2$ **35.** $\log_6 \frac{1}{36} = -2$ **37.** $\log_{1/2} 32 = -5$

39. $\log_x z = y$ **41.** 5 **43.** 3 **45.** 49 **47.** 6

49. 5 **51.** $\frac{1}{25}$ **53.** 5 **55.** $\frac{3}{2}$ **57.** 3 **59.** $\frac{1}{64}$

61. increasing

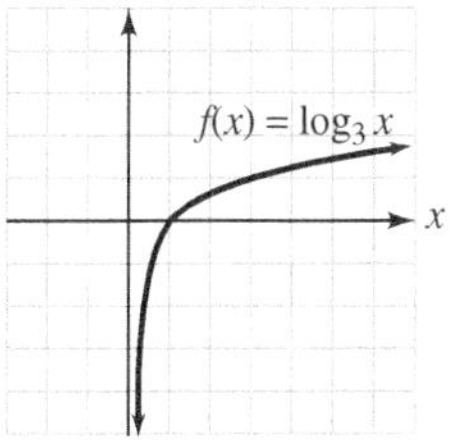

63. decreasing

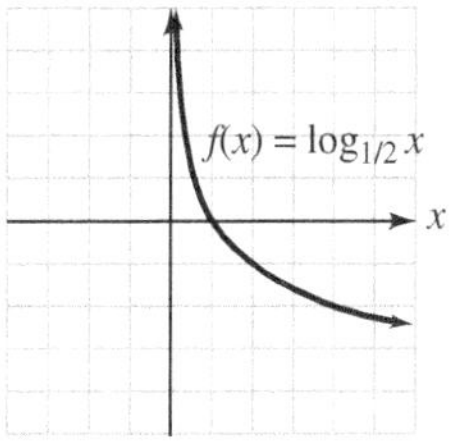

65.

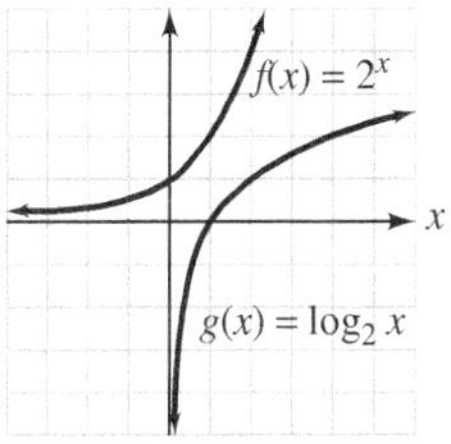

67.

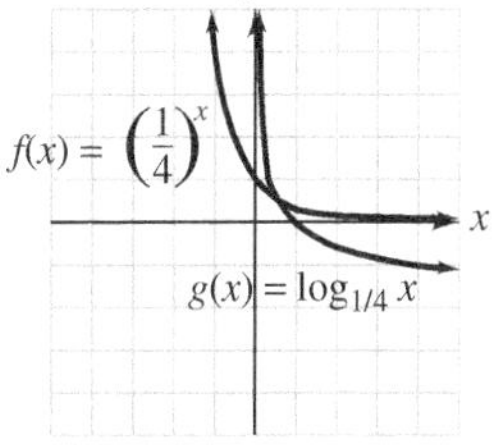

69.

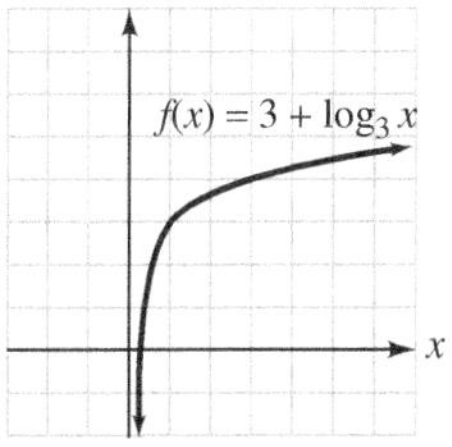

71.

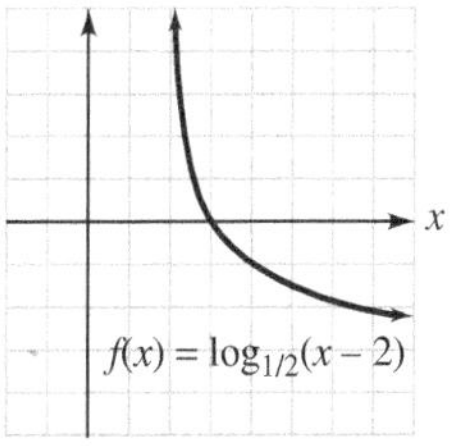

73. 0.9165 **75.** −2.0620 **77.** 17,378.01 **79.** 0.00

81. $\frac{1}{6}$ **83.** −3 **85.** 8 **87.** $\frac{1}{2}$ **89.** $-\frac{3}{2}$ **91.** $\frac{2}{3}$

93. 8 **95.** 4 **97.** 4 **99.** 4 **101.** 3 **103.** 100

105. 25.25 **107.** 8 **109.** $b = 3$ **111.** no value of b

113. 4.4 **115.** 4 **117.** 4.2 yr old **119.** 10.8 yr

Unless otherwise noted, all content on this page is © Cengage Learning.

121. 29.0 dB **123.** 49.5 dB

Getting Ready (page 852)

1. 2 **2.** -3 **3.** 1 **4.** 0

Exercises 11.6 (page 856)

1. $3^2 = 9$ **3.** $\log_2 8 = 3$ **5.** $y = -7x + 2$
7. $y = -\frac{3}{2}x + \frac{1}{2}$ **9.** $y = 6$ **11.** $\frac{2x^2 + x + 1}{x(x + 1)}$
13. $\frac{x + y}{y - x}$ **15.** $(0, \infty), (-\infty, \infty)$ **17.** 10 **19.** $\frac{\ln 2}{r}$
21. 3.2288 **23.** 2.2915 **25.** 9.9892 **27.** 23.8075
29. 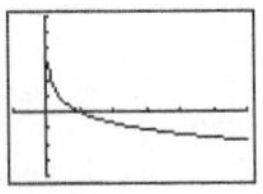 **31.** 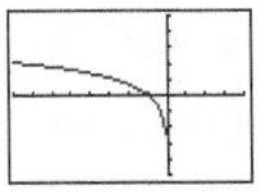 **33.** no **35.** no
37. 0.0408 **39.** no real value **41.** 0.0089
43. 24.3385 **45.** 5.8 yr **47.** 13.9 yr **49.** about 3 hr
55. $t = \frac{\ln 5}{r}$

Getting Ready (page 858)

1. x^{m+n} **2.** 1 **3.** x^{mn} **4.** x^{m-n}

Exercises 11.7 (page 866)

1. 2 **3.** 343 **5.** 3 **7.** $\frac{1}{9}$ **9.** 2 **11.** $\sqrt{85}$
13. $y = -\frac{7}{6}x + \frac{2}{3}$ **15.** 1 **17.** x **19.** $-$ **21.** x
23. $=$ **25.** 0 **27.** 2 **29.** 10 **31.** 1
37. $\log_b 7 + \log_b x + \log_b y$ **39.** $\log_b 5 + \log_b x - \log_b y$
41. $3 \log_b x + 2 \log_b y$ **43.** $3 \log_b x + \frac{1}{2} \log_b z - 5 \log_b y$
45. $\log_b \frac{x - 3}{x^5}$ **47.** $\log_b\left(x^7 \sqrt{z}\right)$ **49.** 1.4472
51. 0.3521 **53.** 0.1339 **55.** 2.4014 **57.** 1.7712
59. -1.0000 **61.** 1.8928 **63.** 2.3219 **65.** 0
67. 7 **69.** 10 **71.** 1 **73.** $\frac{1}{3}(\log_b x + \log_b y - \log_b z)$
75. $\frac{1}{3} \log_b x - \frac{1}{4} \log_b y - \frac{1}{4} \log_b z$ **77.** $\log_b \frac{(x + 2)^3 \sqrt{x + 1}}{y^4}$
79. $\log_b \frac{\frac{x}{z} + x}{\frac{y}{z} + y} = \log_b \frac{x}{y}$ **81.** 2.0493 **83.** 0.4682
87. false, $b^1 = b$ **89.** false, $\log_b xy = \log_b x + \log_b y$
91. true **93.** false, $\frac{\log_b A}{\log_b B}$ is simplified
95. true **97.** true **99.** 4.77
101. from 2.5119×10^{-8} to 1.585×10^{-7}
103. It will increase by $k \ln 2$
105. The intensity must be cubed

Getting Ready (page 868)

1. $2 \log x$ **2.** $\frac{1}{2} \log x$ **3.** 0 **4.** $2b \log a$

Exercises 11.8 (page 877)

1. $x = 10^2$ **3.** $(x + 1) = 10^1$ **5.** 1.2619 **7.** 0.1461
9. 0, 16 **11.** $\frac{2}{3}, -4$ **13.** exponential **15.** A_0e^{-kt}
17. 1.8928 **19.** 3.9120 **21.** 48.9728 **23.** 22.1184
25. 2.7737 **27.** 2.4772 **29.** $1, -3$ **31.** -2
33. 1.8 **35.** $3, -1$ **37.** 3 **39.** 2 **41.** $10, -10$
43. 5; -2 is extraneous **45.** 20; -5 is extraneous
47. 10; -100 is extraneous **49.** 4; 0 is extraneous
51. 4; -1 is extraneous **53.** 20 **55.** 8 **57.** 0
59. ± 1.0878 **61.** 0, 1.0566 **63.** -4
65. 10; 1 is extraneous **67.** 0 **69.** 6; -1 is extraneous
71. 9; -9 is extraneous **73.** 0.2789 **75.** 100, 1 **77.** 4
79. 1, 7 **81.** 4 **83.** 53 days **85.** 6.2 yr
87. about 4,200 yr **89.** 25.3 yr **91.** 2.828 times larger
93. 13.3 **95.** 42.7 days **97.** 5.6 yr **99.** 5.4 yr
101. because $\ln 2 \approx 0.7$ **105.** $x \leq 3$

Chapter Review (page 882)

1. $(f + g)(x) = 3x + 1$ D: $(-\infty, \infty)$
2. $(f - g)(x) = x - 1$ D: $(-\infty, \infty)$
3. $(f \cdot g)(x) = 2x^2 + 2x$ D: $(-\infty, \infty)$
4. $(f/g)(x) = \frac{2x}{x + 1}$ D: $(-\infty, -1) \cup (-1, \infty)$ **5.** 8
6. -3 **7.** $2(x + 1)$ **8.** $2x + 1$
9. yes **10.** no

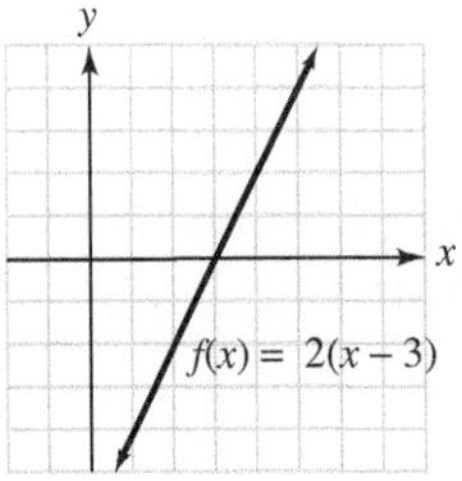

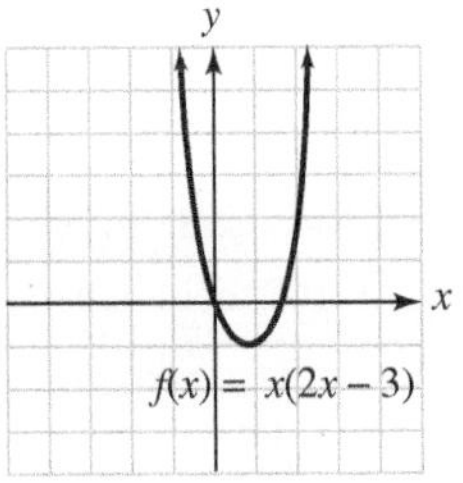

11. no **12.** no

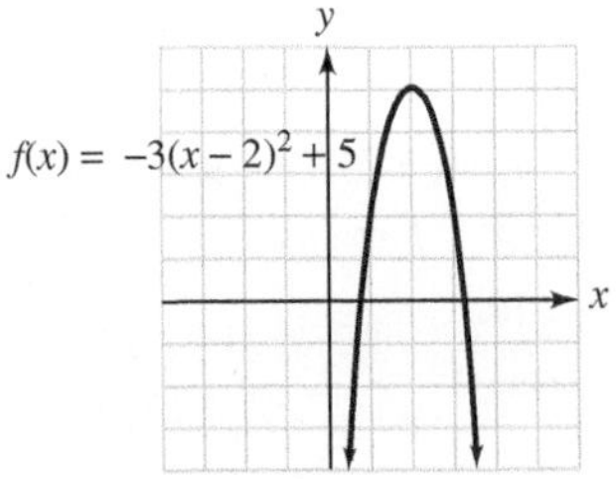

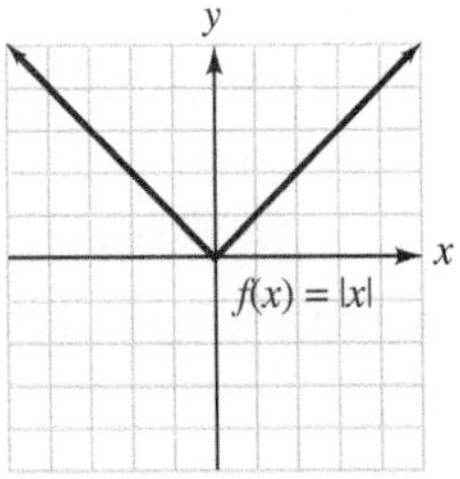

13. $f^{-1}(x) = \frac{x + 2}{7}$ **14.** $f^{-1}(x) = \frac{x - 5}{4}$
15. $y = \sqrt{\frac{x + 1}{2}}$

18. 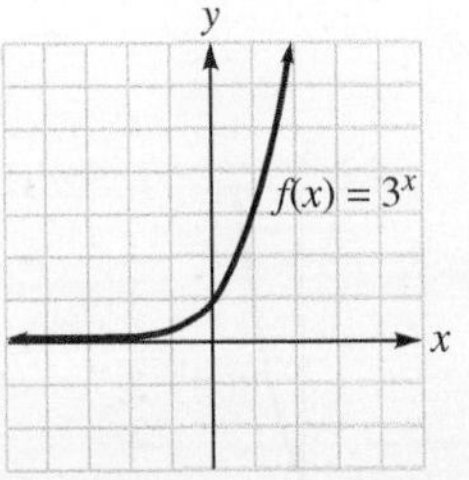

19. 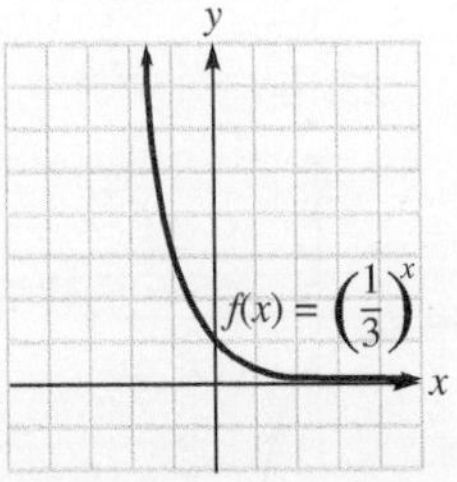

20. $x = 1, y = 6$ **21.** D: $(-\infty, \infty)$, R: $(0, \infty)$

Unless otherwise noted, all content on this page is © Cengage Learning.

22.

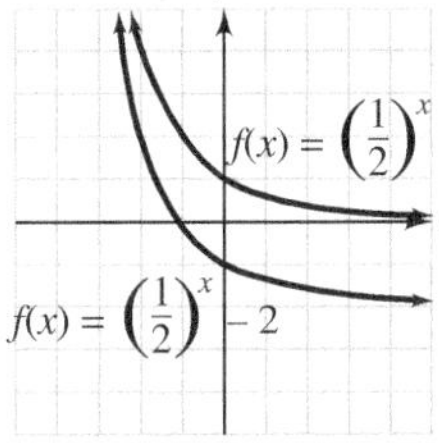

23.

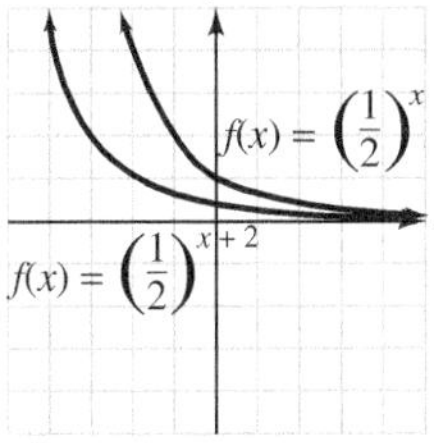

24. \$142,312.97 **25.** \$143,865.07

26.

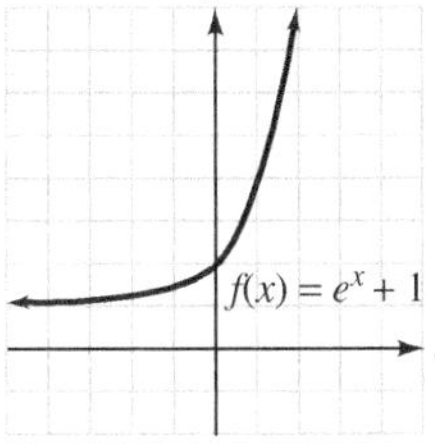

27.

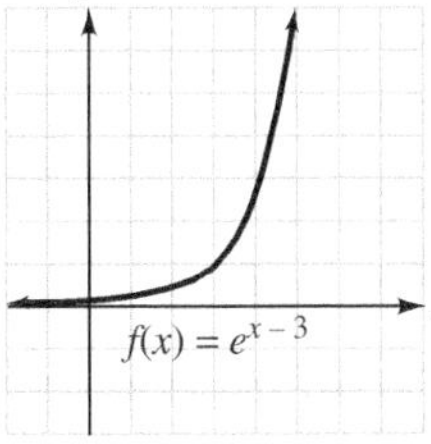

28. about 582,000,000 **29.** 30.69 grams

30. D: $(0, \infty)$, R: $(-\infty, \infty)$ **32.** 2 **33.** $-\frac{1}{2}$

34. 0 **35.** -2 **36.** $\frac{1}{2}$ **37.** $\frac{1}{3}$ **38.** 81 **39.** 9

40. 27 **41.** -1 **42.** $\frac{1}{8}$ **43.** 2 **44.** 4 **45.** 2

46. 10 **47.** $\frac{1}{25}$ **48.** 5 **49.** 3

50.

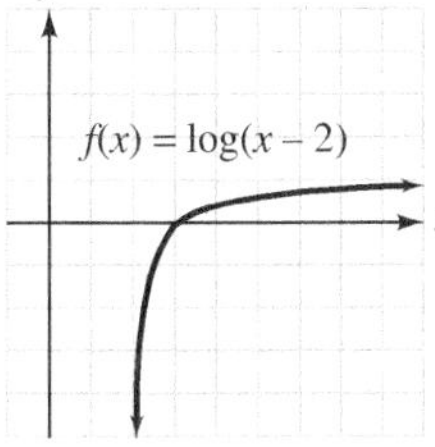

51.

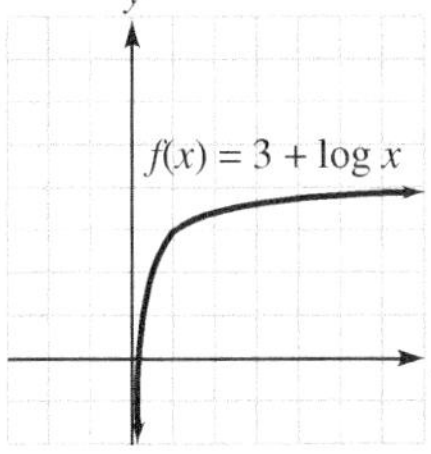

52.

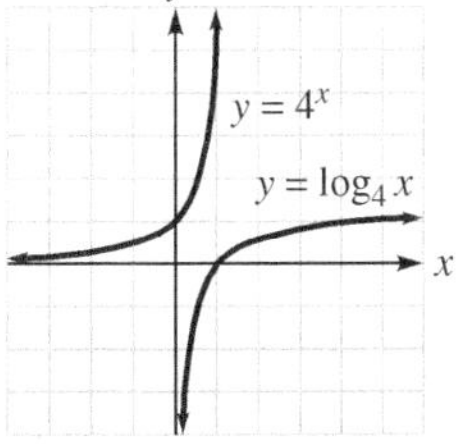

53.

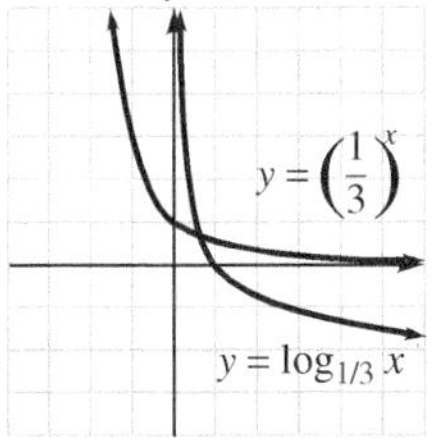

54. 53 dB **55.** 4.4 **56.** 5.8916 **57.** -0.1111

58. 10.3398 **59.** 2.5715

60.

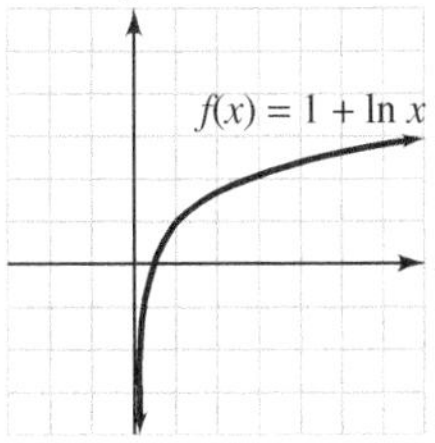

61.

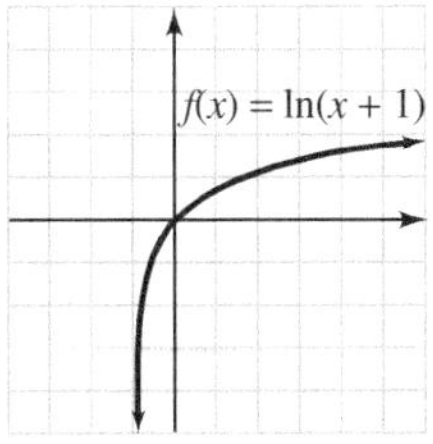

62. about 23 yr **63.** 0 **64.** 1 **65.** 3 **66.** 5

67. 5 **68.** 0 **69.** 7 **70.** 2 **71.** 6 **72.** 3

73. $5 \log_b x + 4 \log_b y - 2 \log_b z$

74. $\frac{1}{2}(\log_b x - \log_b y - 2 \log_b z)$ **75.** $\log_b \frac{x^3 z^7}{y^5}$

76. $\log_b \frac{y^3\sqrt{x}}{z^7}$ **77.** 3.36 **78.** 1.56

79. 2.64 **80.** -6.72 **81.** 1.7604

82. about 7.94×10^{-4} gram-ions per liter

83. $k \ln 2$ less **84.** $\frac{\log 7}{\log 3} \approx 1.7712$ **85.** 2

86. 31.0335 **87.** 2 **88.** $\frac{\log 3}{\log 3 - \log 2} \approx 2.7095$

89. $-1, -3$ **90.** 25, 4 **91.** 4; -2 is extraneous

92. 2; -3 is extraneous **93.** 4, 3

94. 6; -1 is extraneous **95.** 31 **96.** $\frac{\ln 9}{\ln 2} \approx 3.1699$

97. $\varnothing$ **98.** $\frac{e}{e - 1} \approx 1.5820$ **99.** 1

100. about 3,400 yr

Chapter 11 Test (page 890)

1. $(g + f)(x) = 5x - 1$ D: $(-\infty, \infty)$

2. $(f - g)(x) = 3x + 1$ D: $(-\infty, \infty)$

3. $(g \cdot f)(x) = 4x^2 - 4x$ D: $(-\infty, \infty)$

4. $(g/f)(x) = \frac{x - 1}{4x}$ D: $(-\infty, 0) \cup (0, \infty)$

5. 3 **6.** -4 **7.** -8 **8.** -9 **9.** $4(x - 1)$

10. $4x - 1$ **11.** $f^{-1}(x) = \frac{12 - 2x}{3}$

12.

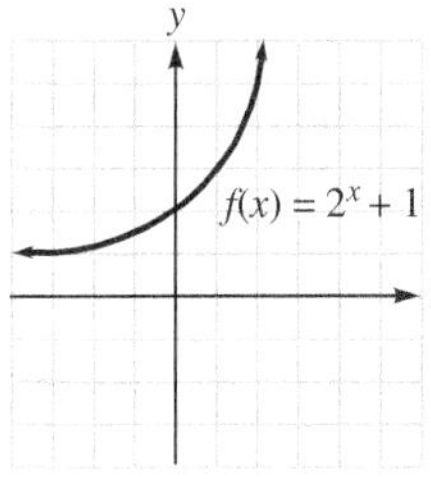

13.

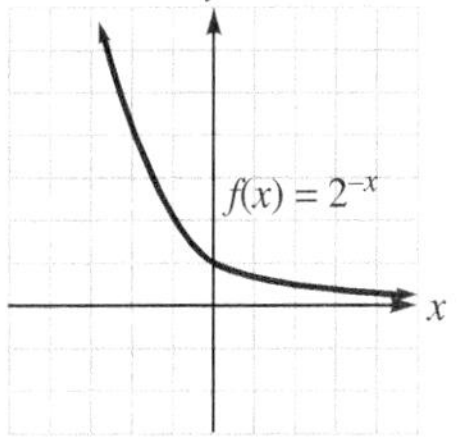

14. $\frac{3}{64}$ gram **15.** \$1,060.90

16.

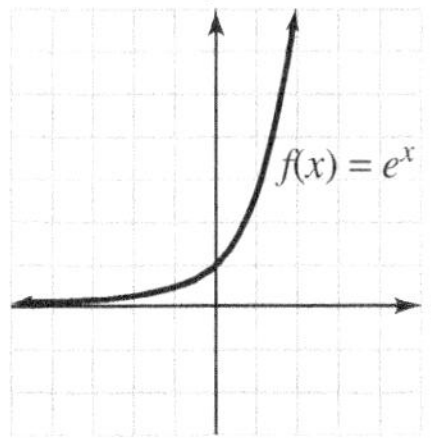

17. \$4,451.08 **18.** 2

19. 3 **20.** $\frac{1}{27}$ **21.** 7 **22.** 2 **23.** $\frac{27}{8}$

24.

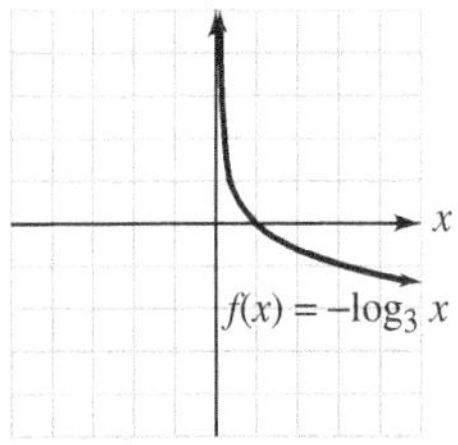

25.

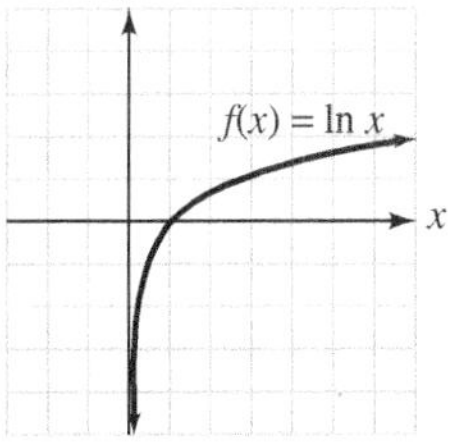

26. $\log x + 3 \log y + 4 \log z$ **27.** $\frac{1}{2}(\ln a - 2 \ln b - \ln c)$

28. $\log \frac{b\sqrt{a + 2}}{c^3}$ **29.** $\log \frac{\sqrt[3]{a}}{c\sqrt[3]{b^2}}$ **30.** 1.2552

Unless otherwise noted, all content on this page is © Cengage Learning.

31. 0.4259 **32.** $\frac{\log 2}{\log 3}$ or $\frac{\ln 2}{\ln 3}$ **33.** $\frac{\log e}{\log \pi}$ or $\frac{\ln e}{\ln \pi}$ **34.** true
35. false, $\frac{\log a}{\log b}$ is simplified **36.** false, $\log a^{-3} = -3 \log a$
37. false, cannot take logarithm of a negative number
38. 6.4 **39.** 46 **40.** $\frac{\log 4}{\log 7} \approx 0.7124$
41. $\frac{\log 3}{(\log 3) - 2} \approx -0.3133$ **42.** 4
43. 10; -1 is extraneous

Getting Ready (page 895)

1. $x^2 - 4x + 4$ **2.** $x^2 + 8x + 16$ **3.** $\frac{81}{4}$ **4.** 36

Exercises 12.1 (page 904)

1. downward **3.** upward **5.** $(x + 7)^2$ **7.** $(x - 10)^2$
9. 6, -1 **11.** 3, $-\frac{1}{4}$ **13.** conic **15.** standard; $(0, 3)$; 4
17. circle; general **19.** parabola; $(3, 2)$; right
21. $C(0, 0)$; $r = 3$ **23.** $C(2, 0)$; $r = 3$

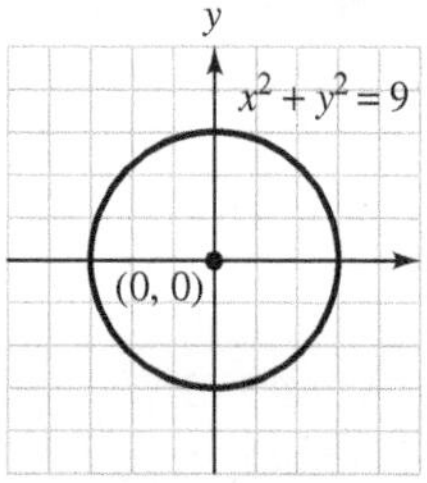

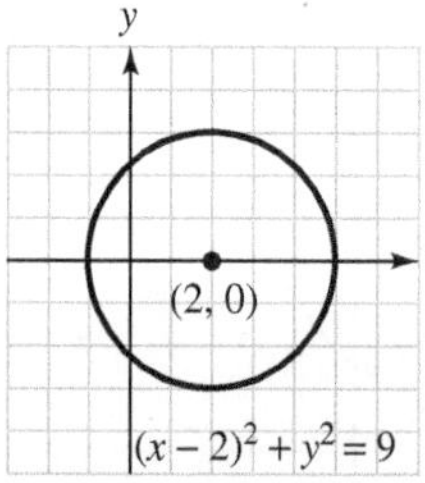

25. $C(2, 4)$; $r = 2$ **27.** $C(-3, 1)$; $r = 4$

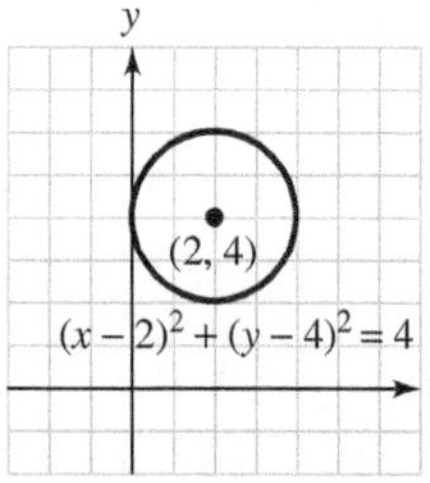

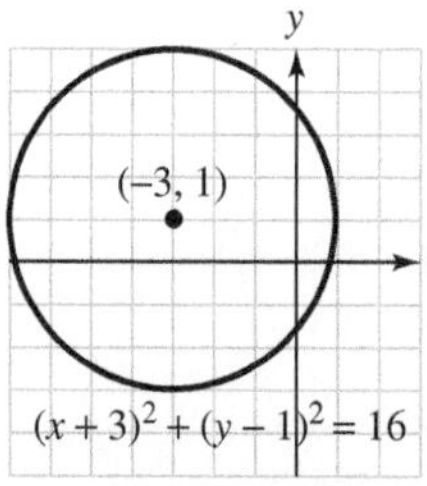

29.

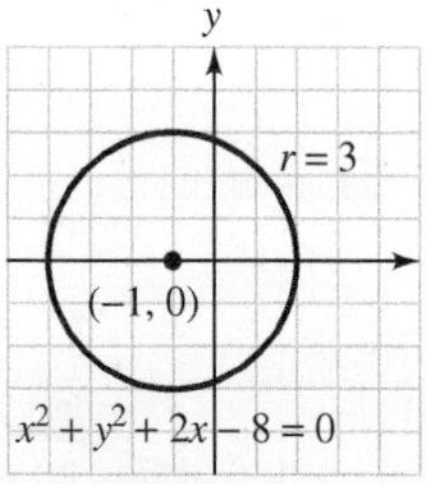

31.

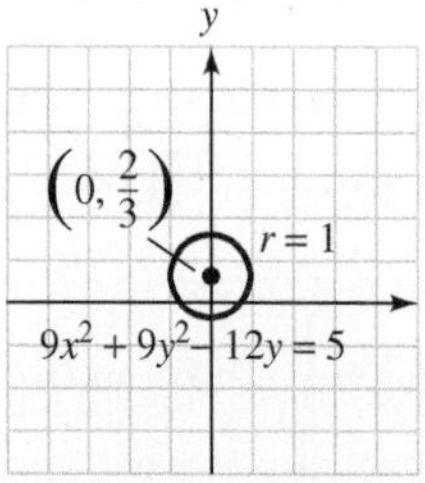

33.

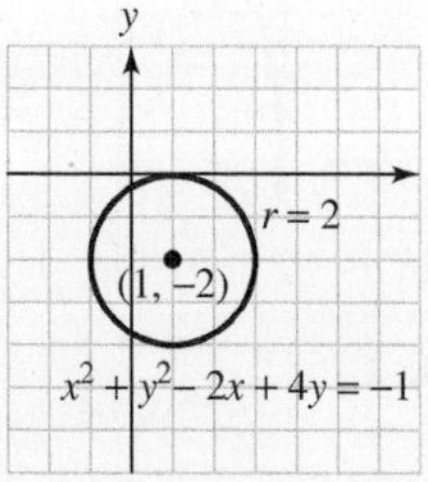

35.

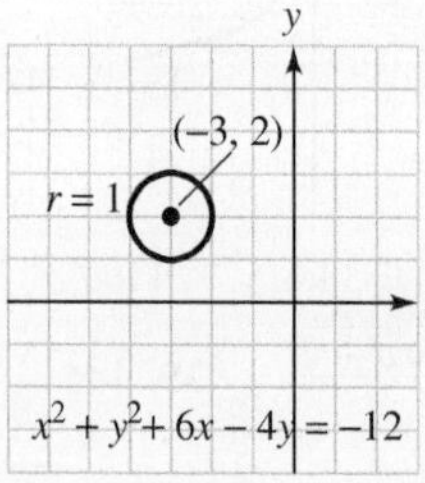

37. $x^2 + y^2 = 9, x^2 + y^2 - 9 = 0$
39. $(x - 2)^2 + (y - 4)^2 = 49, x^2 + y^2 - 4x - 8y - 29 = 0$
41. $(x + 4)^2 + (y - 7)^2 = 100, x^2 + y^2 + 8x - 14y - 35 = 0$
43. $x^2 + y^2 = 80, x^2 + y^2 - 80 = 0$

45.

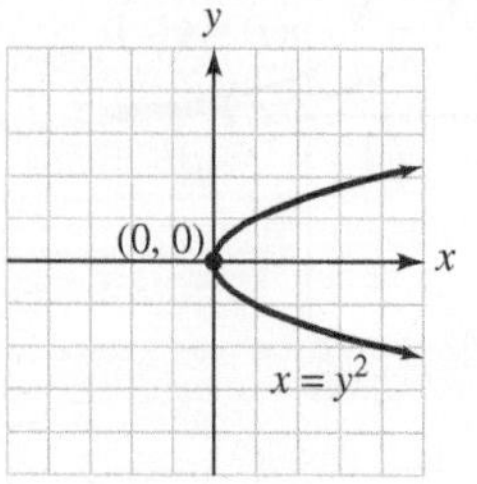

47.

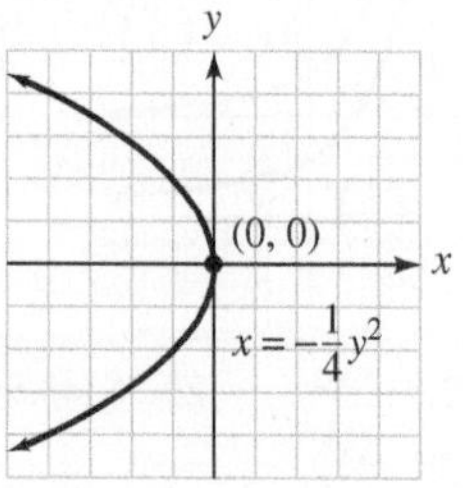

49.

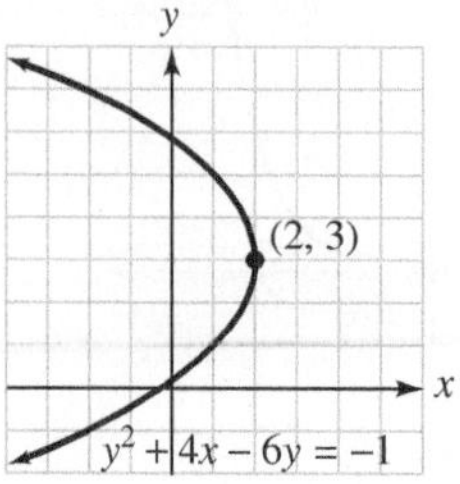

51.

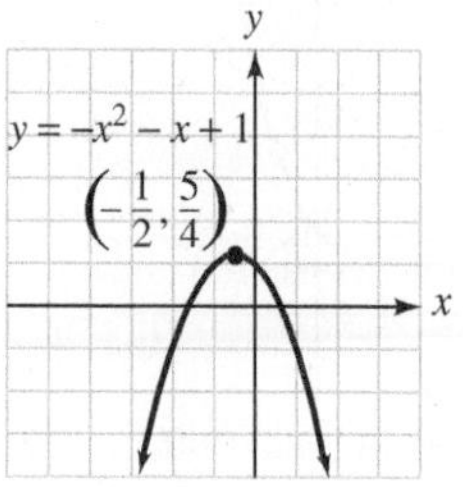

53.

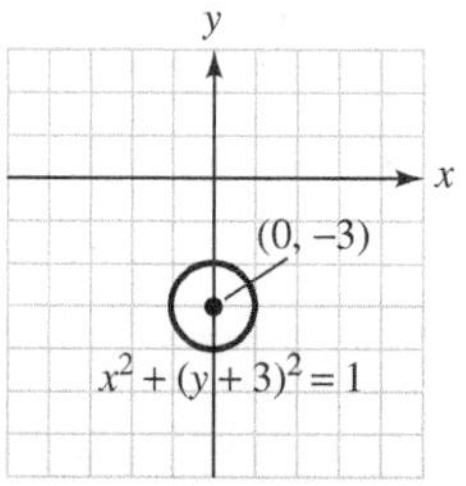

55.

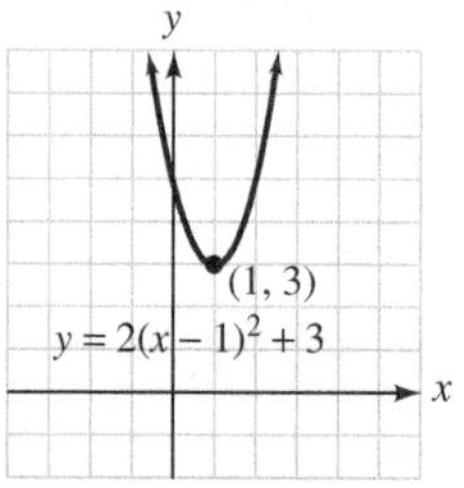

57.

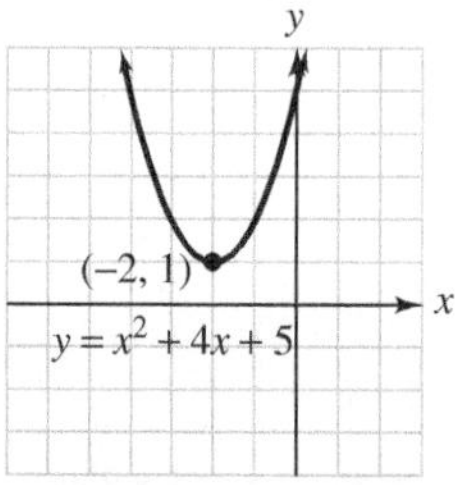

59.

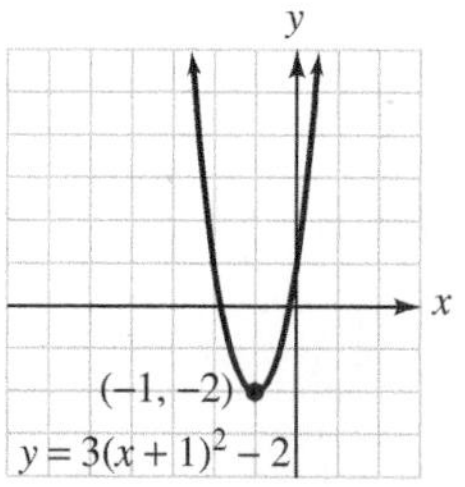

61.

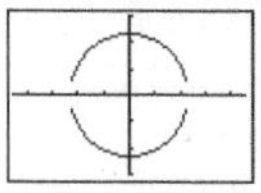

63.

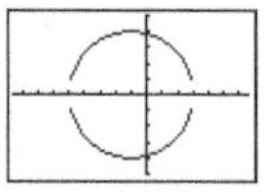

65.

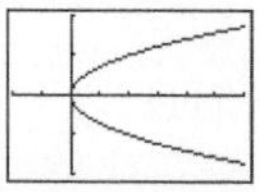

67.

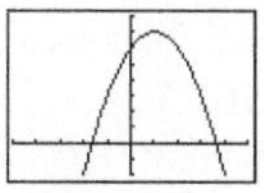

69. $(x - 7)^2 + y^2 = 9$ **71.** no
73. 30 ft away **75.** 2 AU

Getting Ready (page 908)

1. $y = \pm b$ **2.** $x = \pm a$

Exercises 12.2 (page 915)

1. $(0, 3), (0, -3)$ **3.** $(2, 0), (-2, 0)$ **5.** $35y^4 + \frac{10}{x^3}$
7. $\frac{y^2 + x^2}{y^2 - x^2}$ **9.** ellipse; sum **11.** center
13. $(0, 0)$; major axis; $2b$

Unless otherwise noted, all content on this page is © Cengage Learning.

15.

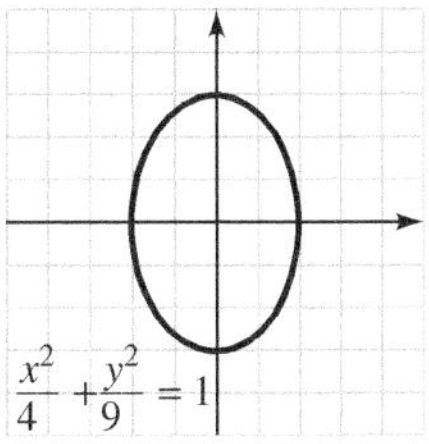

17.

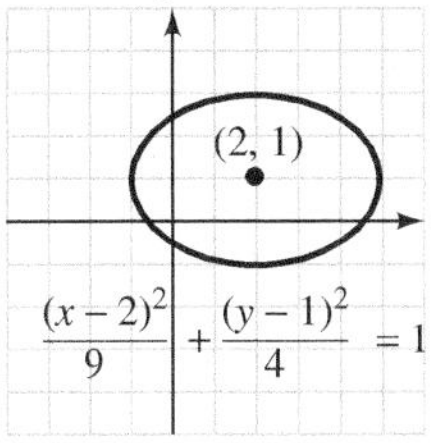

19.

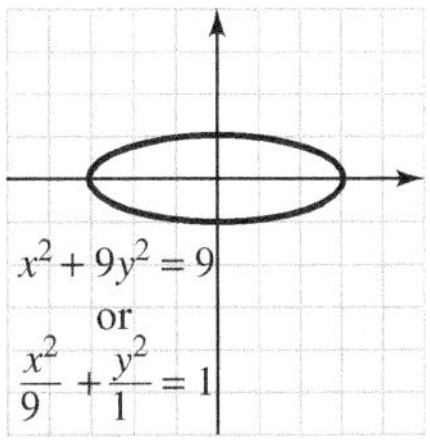

21.

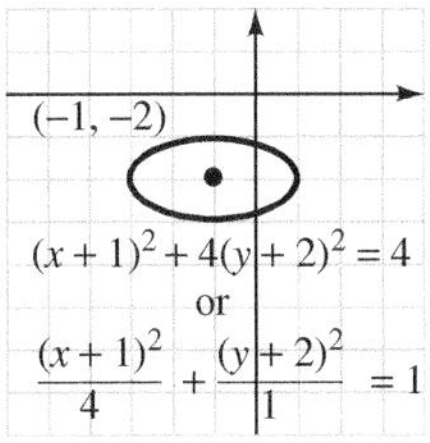

23.

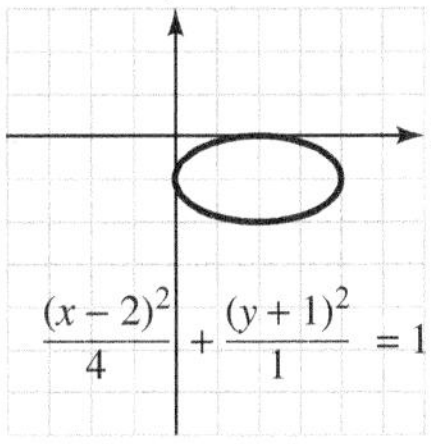

25.

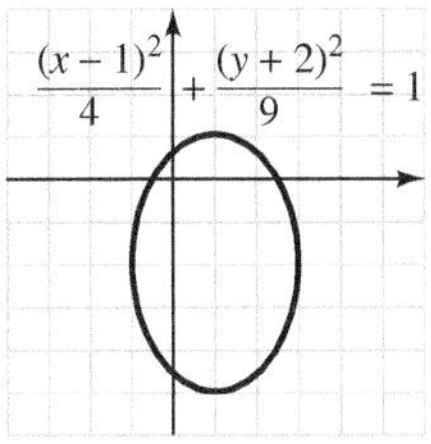

27.

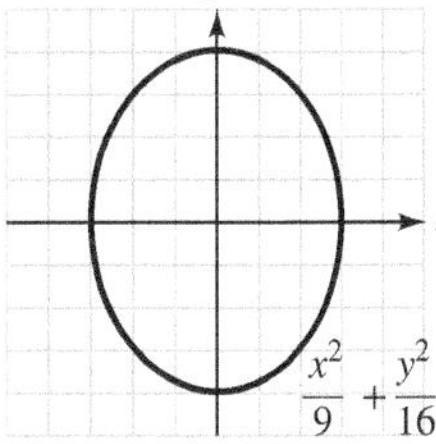

29.

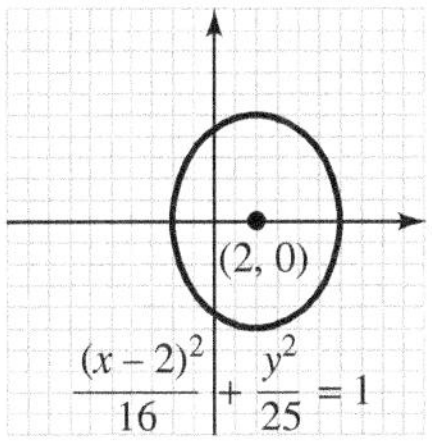

31.

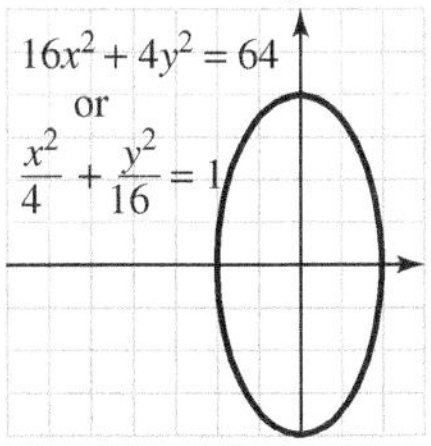

33.

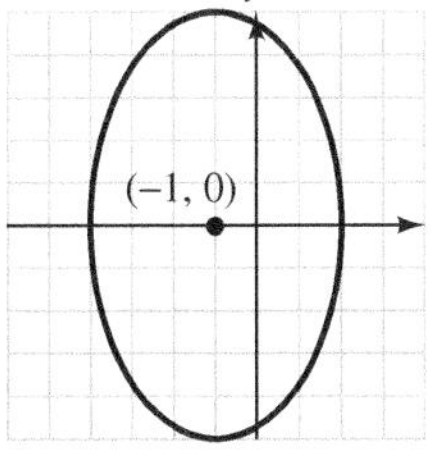

$25(x+1)^2 + 9y^2 = 225$
or
$\frac{(x+1)^2}{9} + \frac{y^2}{25} = 1$

35.

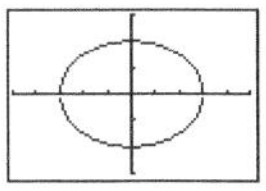

37.

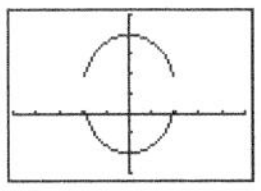

39. $\frac{x^2}{144} + \frac{y^2}{25} = 1$ **41.** $y = \frac{4}{5}\sqrt{625 - x^2}$

43. 10π square units

Getting Ready (page 919)

1. $y = \pm 2.0$ **2.** $y = \pm 2.9$

Exercises 12.3 (page 926)

1. $(2, 0), (-2, 0)$ **3.** $-5x^2(x^2 - 2x + 3)$

5. $(7a + 3b)(2a - 3b)$ **7.** hyperbola; difference

9. center **11.** $(\pm a, 0)$; y-intercepts

13.

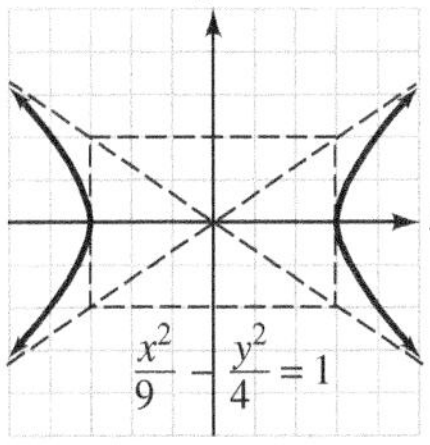

15.

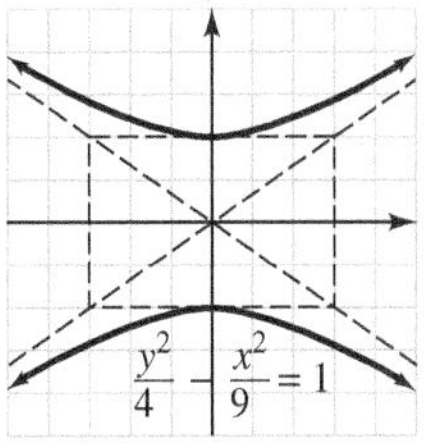

17.

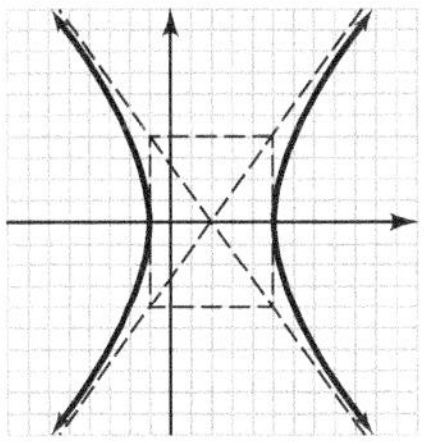

$\frac{(x-2)^2}{9} - \frac{y^2}{16} = 1$

19.

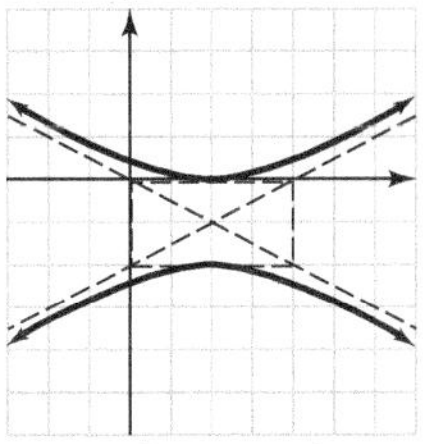

$\frac{(y+1)^2}{1} - \frac{(x-2)^2}{4} = 1$

21.

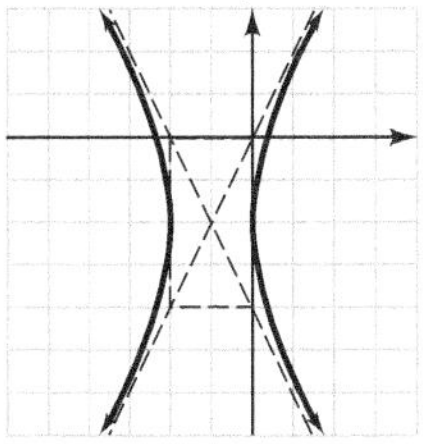

$\frac{(x+1)^2}{1} - \frac{(y+2)^2}{4} = 1$

23.

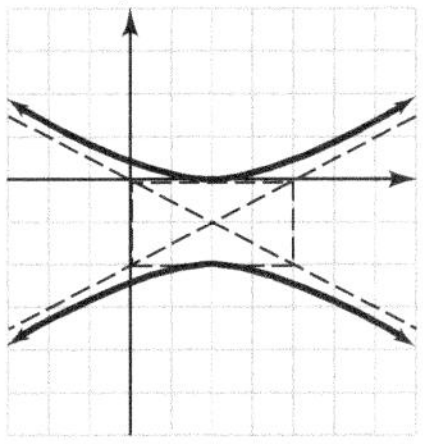

$\frac{(y+1)^2}{1} - \frac{(x-2)^2}{4} = 1$

25.

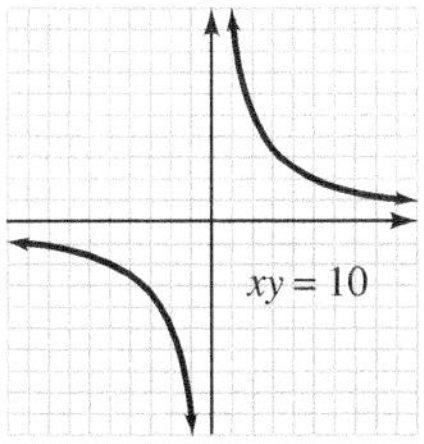

27.

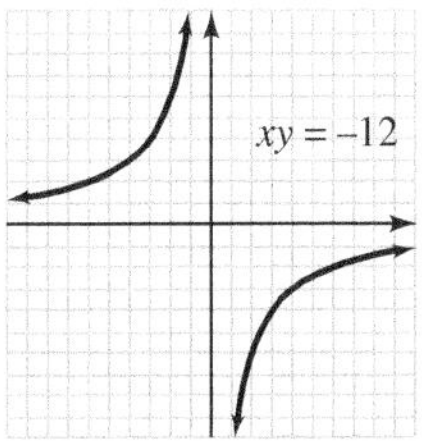

29.

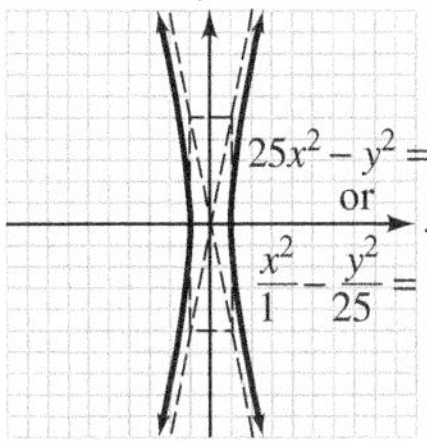

31.

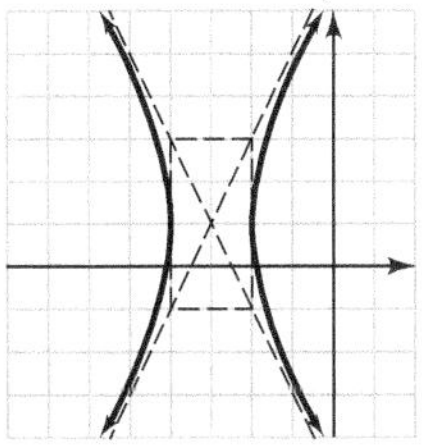

$4(x+3)^2 - (y-1)^2 = 4$
or $\frac{(x+3)^2}{1} - \frac{(y-1)^2}{4} = 1$

33.

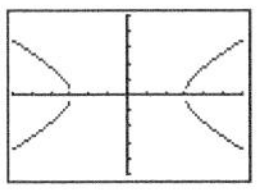

35.

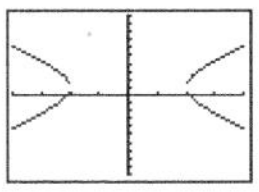

Unless otherwise noted, all content on this page is © Cengage Learning.

37. 3 units **39.** 4 units

Getting Ready (page 929)

1. $7x^2 = 44$ **2.** $-3y^2 = 8$

Exercises 12.4 (Page 933)

1. 0, 1, 2 **3.** 0, 1, 2, 3, 4 **5.** $-39y\sqrt{2}$ **7.** $\frac{5t}{4}$
9. graphing, substitution **11.** two **13.** four
15. graphing
17. $(-4,-2),(4,2)$ **19.** $\left(\frac{1}{5},-\frac{18}{5}\right),(3,2)$

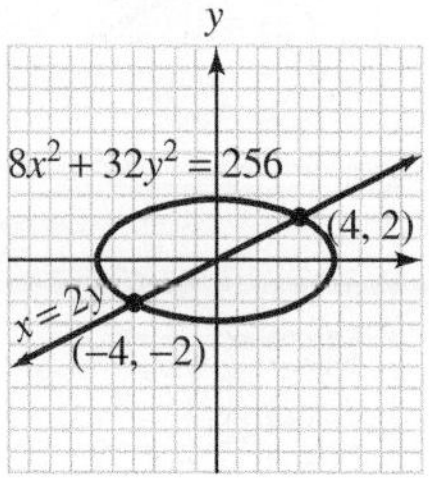

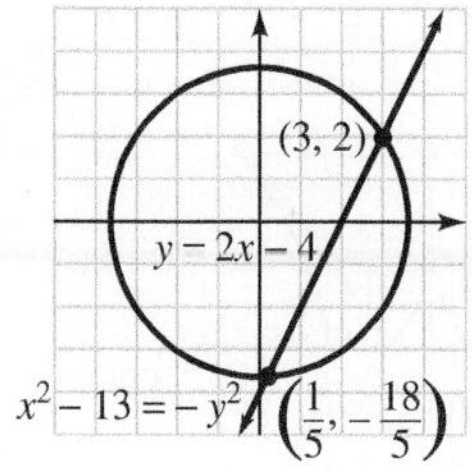

21. $(3,0),(0,5)$ **23.** $(1,1)$ **25.** $(2,4),(-2,4)$
27. $(\sqrt{3},0),(-\sqrt{3},0)$
29. $(3,2),(3,-2),(-3,2),(-3,-2)$
31. $(4,3),(-4,3),(4,-3),(-4,-3)$
33. **35.**

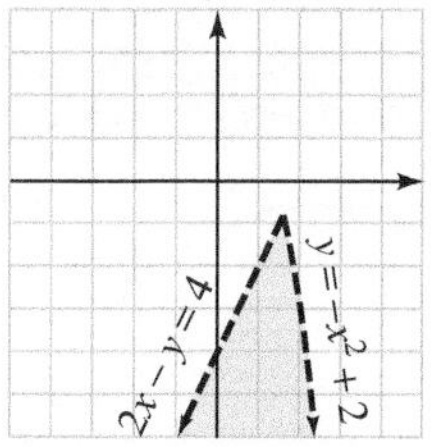

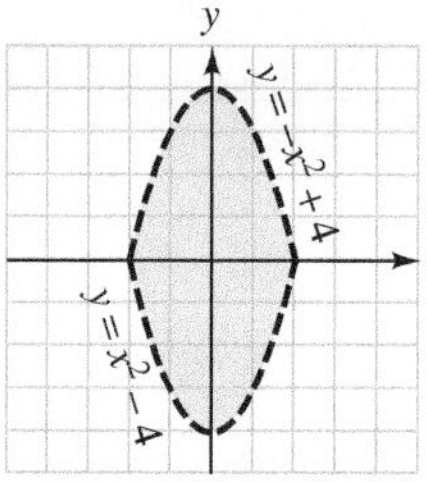

37. $(3,0),(-3,0)$ **39.** $(6,0),(-6,0)$
41. $(-1,3),(1,3)$ **43.** $(-4,-3),(-4,3),(4,-3),(4,3)$

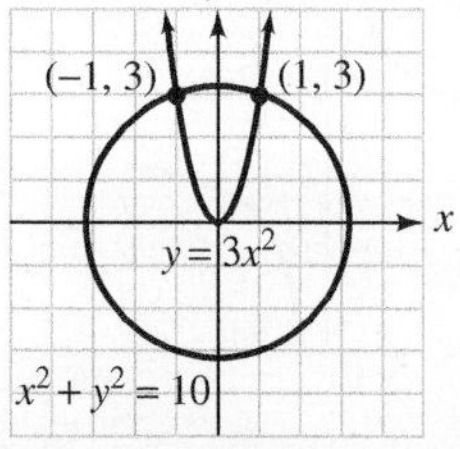

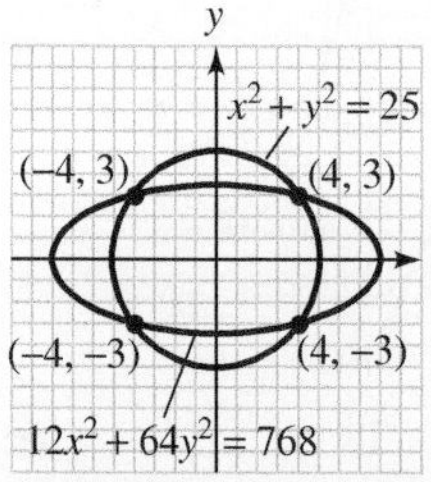

45. $(1,2),(-2,-1)$
47. $(\sqrt{3},2\sqrt{3}),(-\sqrt{3},2\sqrt{3}),(\sqrt{3},-2\sqrt{3}),(-\sqrt{3},-2\sqrt{3})$
49. $(-\sqrt{15},5),(\sqrt{15},5),(-2,-6),(2,-6)$ **51.** $(3,3)$
53. $(6,2),(-6,-2),(\sqrt{42},0),(-\sqrt{42},0)$
55. $\left(\frac{1}{2},\frac{1}{3}\right),\left(\frac{1}{3},\frac{1}{2}\right)$ **57.** 4 and 8 **59.** 7 cm by 9 cm
61. either \$750 at 9% or \$900 at 7.5% **63.** 68 mph, 4.5 hr

67. 0, 1, 2, 3, 4 **69.** 0, 1, 2, infinitely many

Getting Ready (page 936)

1. positive **2.** negative **3.** 98 **4.** −3

Exercises 12.5 (page 940)

1. 1, 4, 9; increasing **3.** 9, 4, 1; decreasing **5.** 20
7. domains **9.** constant; $f(x)$ **11.** step
13. increasing on $(-\infty,0)$, decreasing on $(0,\infty)$
15. decreasing on $(-\infty,0)$, constant on $(0,2)$, increasing on $(2,\infty)$
17.

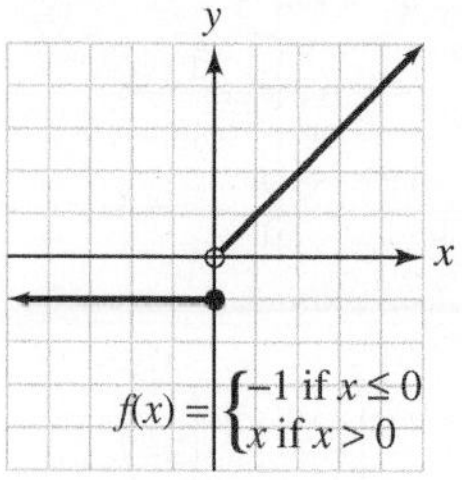

constant on $(-\infty,0)$, increasing on $(0,\infty)$

19.

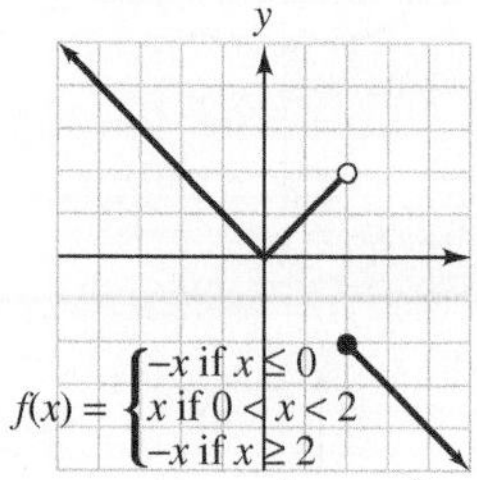

decreasing on $(-\infty,0)$, increasing on $(0,2)$, decreasing on $(2,\infty)$

21.

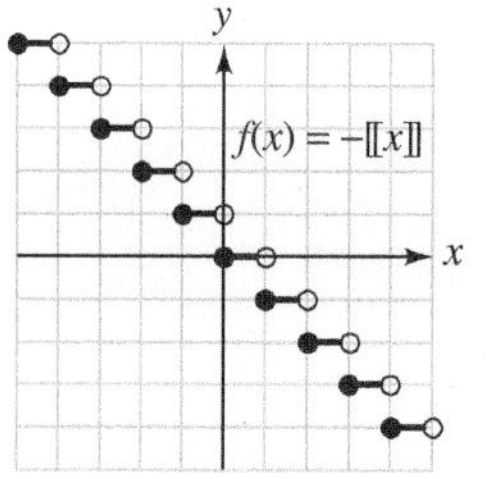

23.

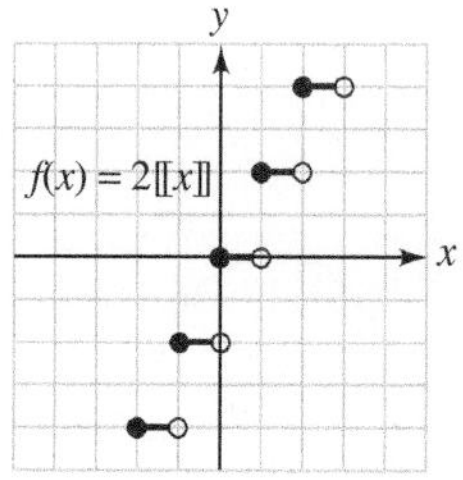

25.

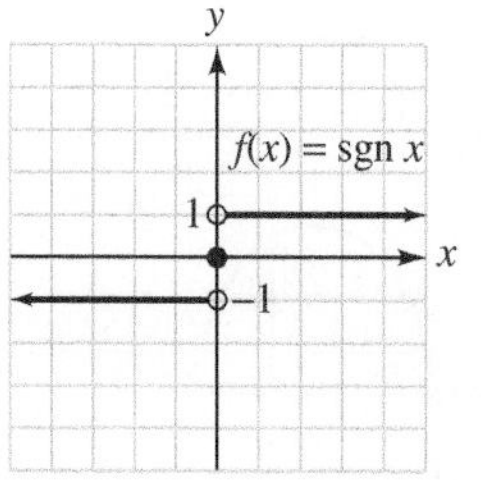

27. \$30;

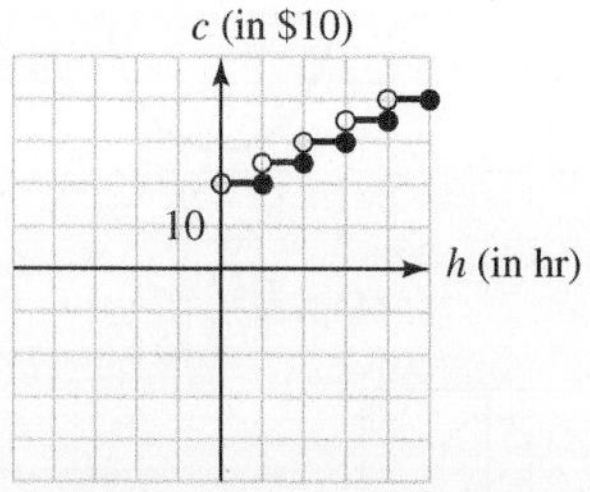

29. After 2 hours, network B is cheaper.

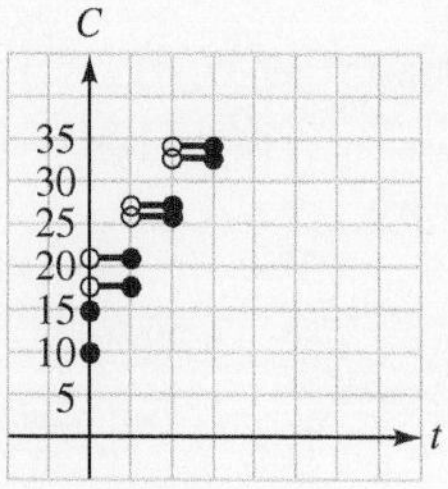

Unless otherwise noted, all content on this page is © Cengage Learning.

Chapter Review (page 944)

1.

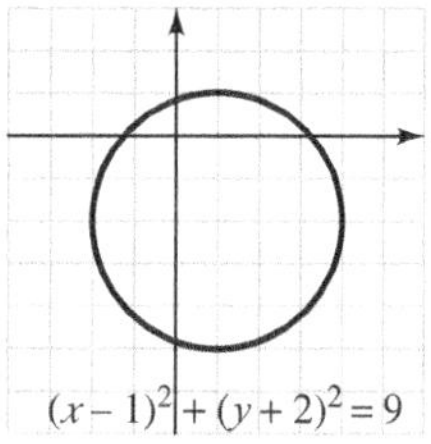

2.

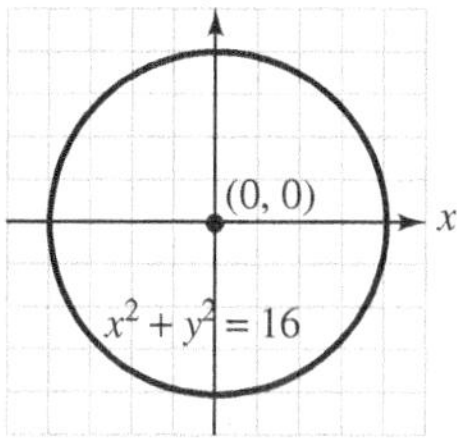

3.

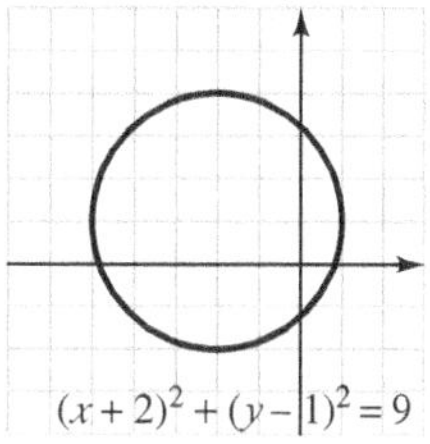

4.

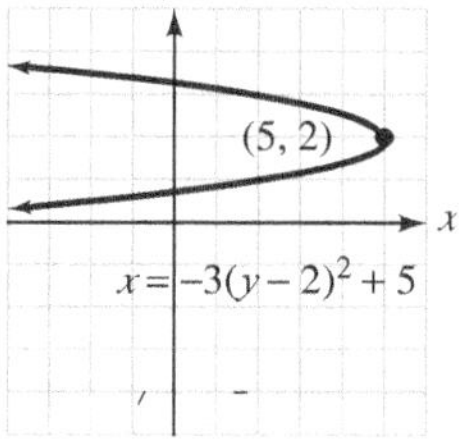

5.

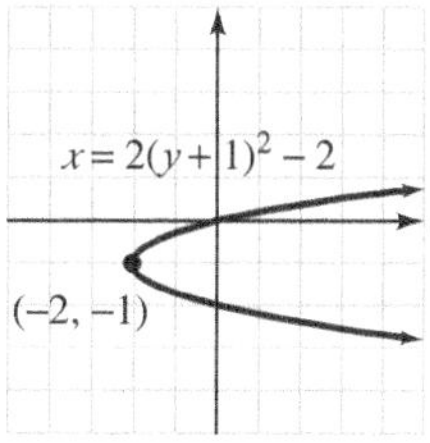

6.

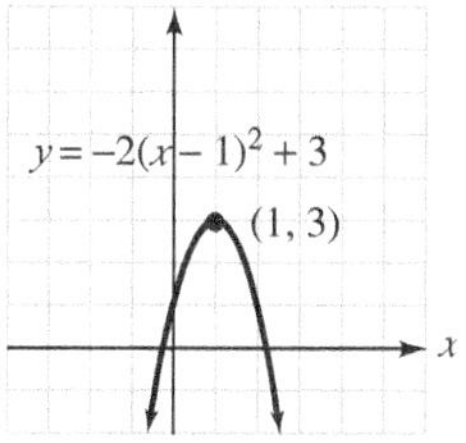

7.

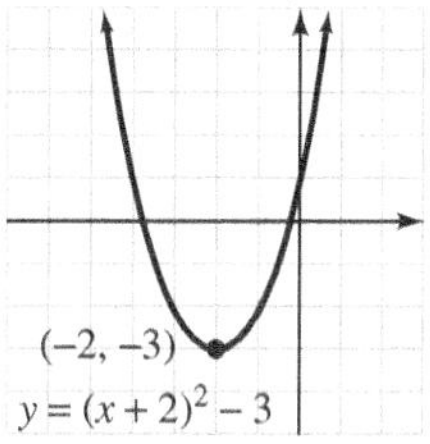

8.

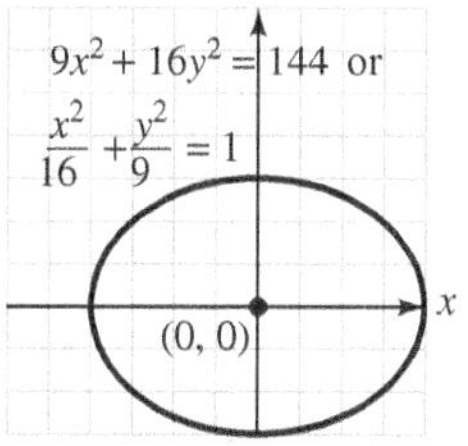

9.

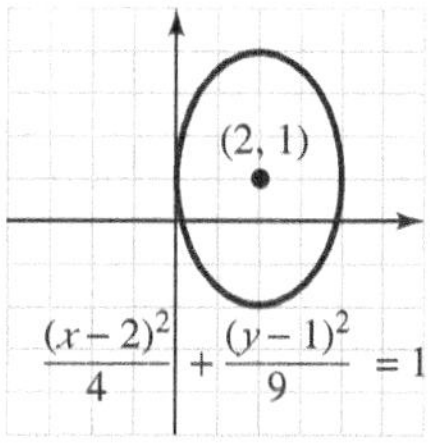

10.

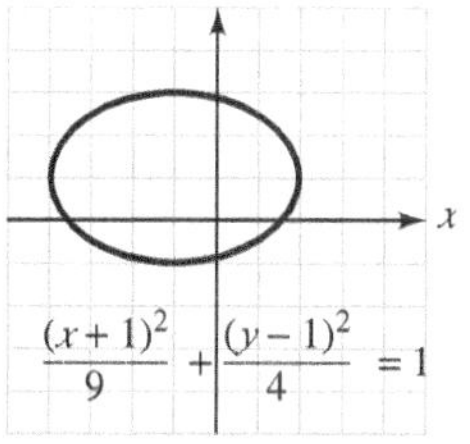

11.

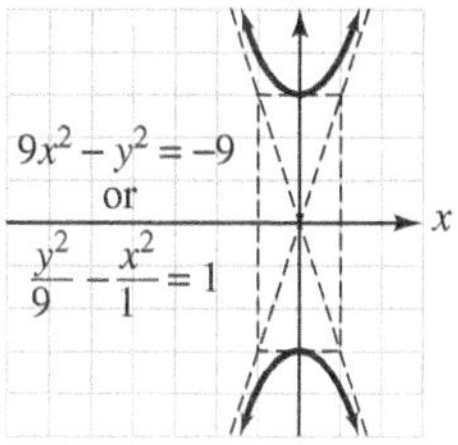

12.

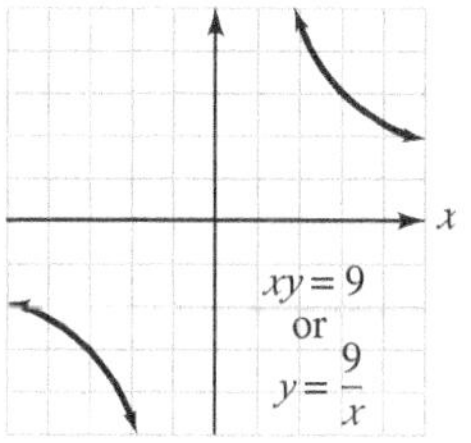

13. hyperbola

14.

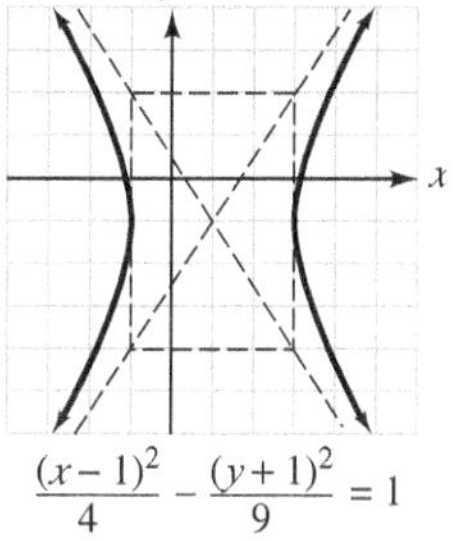

15. $(-2,3),(2,3)$ **16.** $(2,4),(2,-4),(-2,4),(-2,-4)$

17.

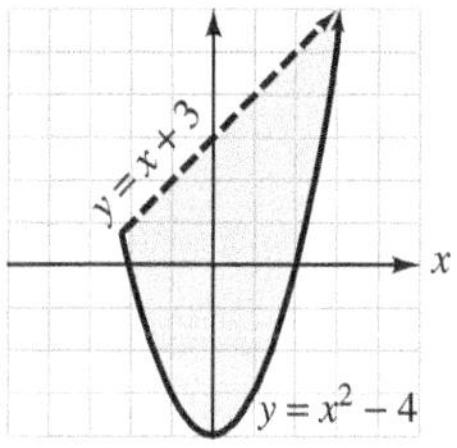

18. increasing on $(-\infty,-2)$, constant on $(-2,1)$, decreasing on $(1,\infty)$

19.

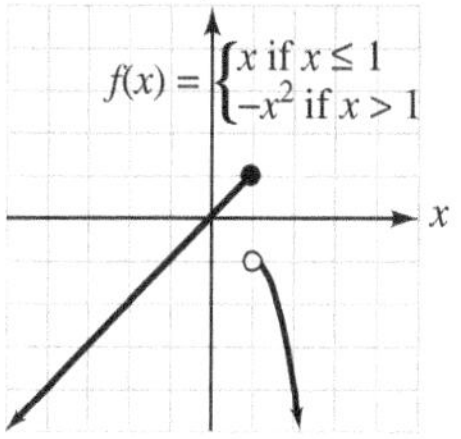

20.

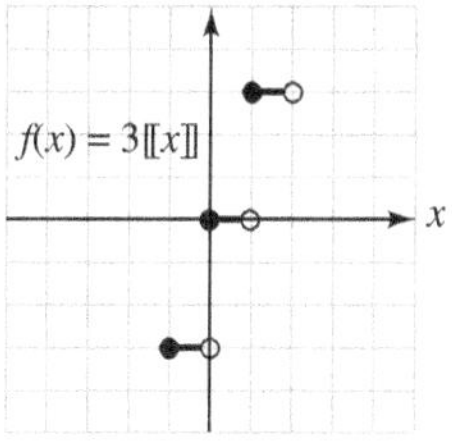

Chapter 12 Test (page 952)

1. $(-5,2)$; 3 **2.** $(-4,2)$; 5

3.

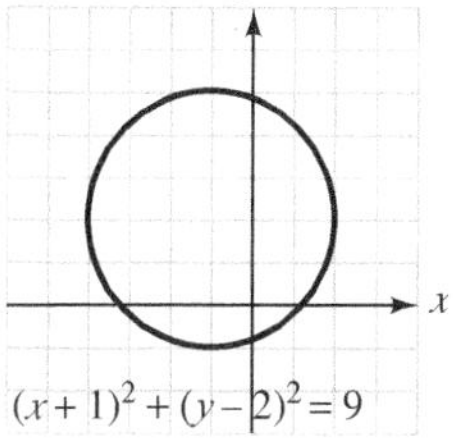

4.

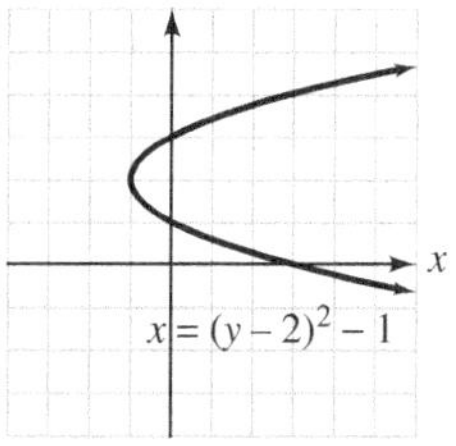

5.

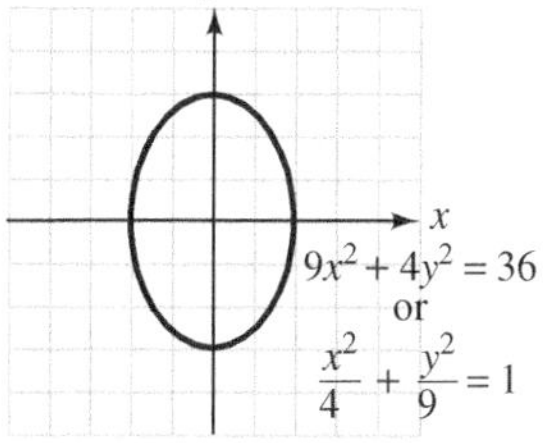

6.

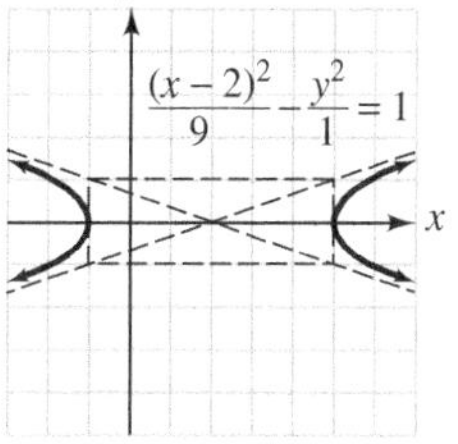

Unless otherwise noted, all content on this page is © Cengage Learning.

7.

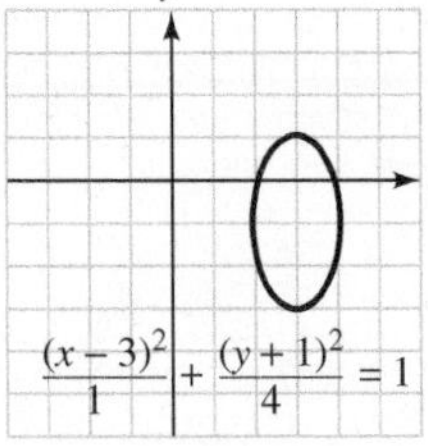

8.

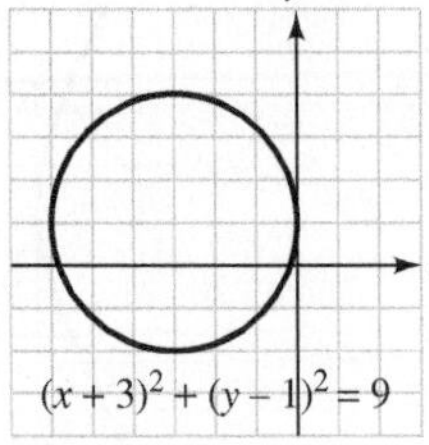

9.

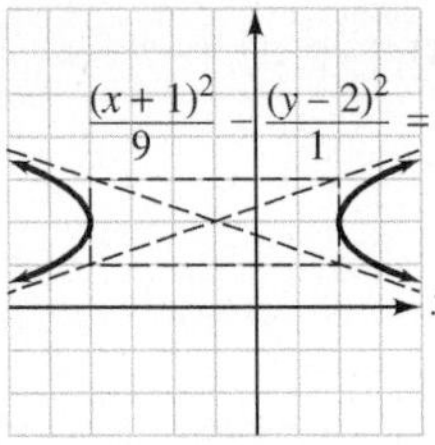

10. $(0, 3), (4, 0)$

11. $(3, 4), (-3, 4)$ **12.**

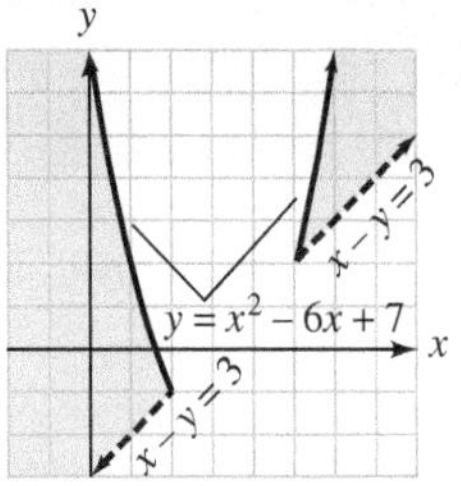

13. increasing on $(-3, 0)$, decreasing on $(0, 3)$

14.

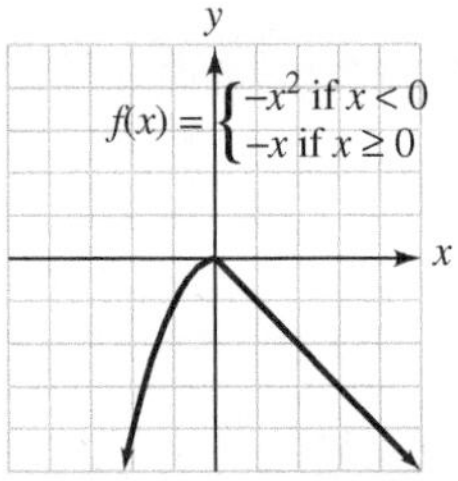

15.

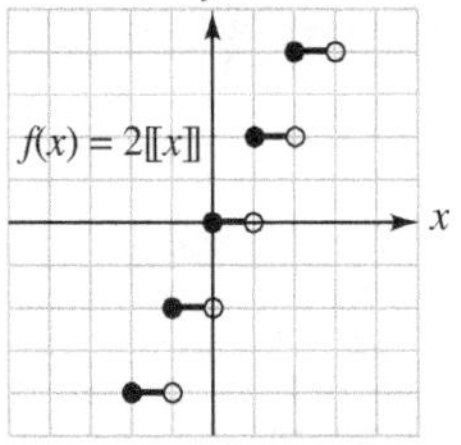

21.

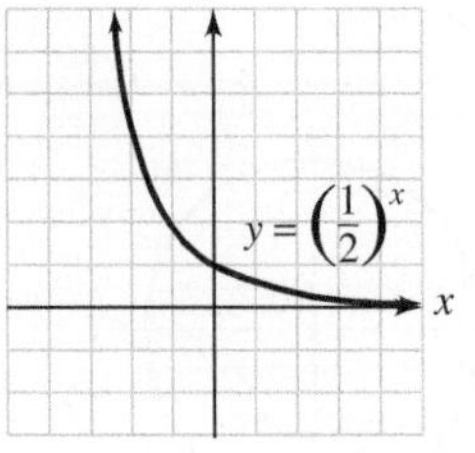

22. $5^y = x$ **23.** $\frac{4}{3}, -1$

24. 16

25.

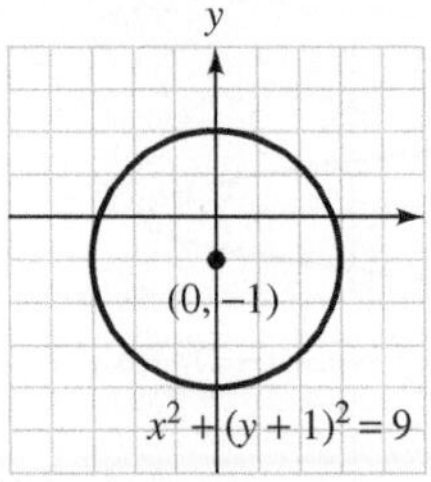

26.

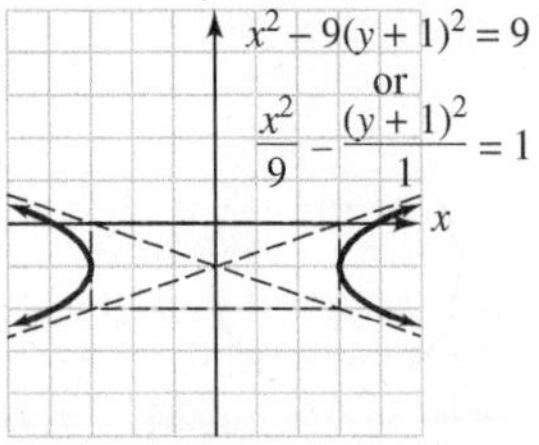

27. $(z - 2, -2z + 3, z)$

28. $(2, 1), (2, -1), (-2, 1), (-2, -1)$ **29.** $(3, 4), (-3, 4)$

30.

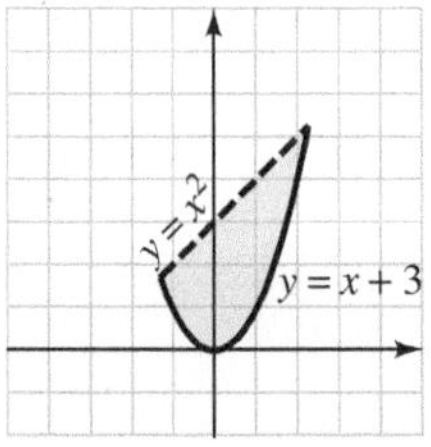

Cumulative Review (page 953)

1. $8x^2 - 18xy - 5y^2$ **2.** $a^{2n} - 2a^n - 3$ **3.** $\frac{7}{a-5}$

4. $a^2 - 3a + 2$ **5.** 1 **6.** $\frac{4a-1}{(a+2)(a-2)}$ **7.** parallel

8. perpendicular **9.** $y = -6x + 3$ **10.** $y = -\frac{9}{13}x + \frac{7}{13}$

11.

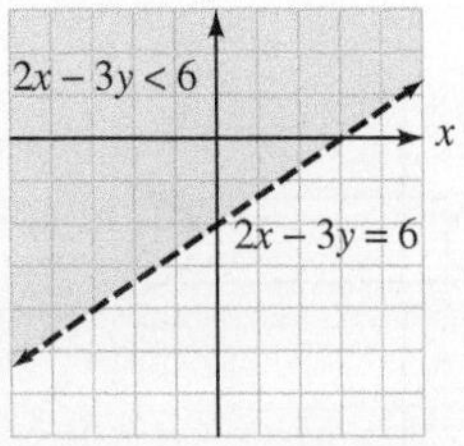

12.

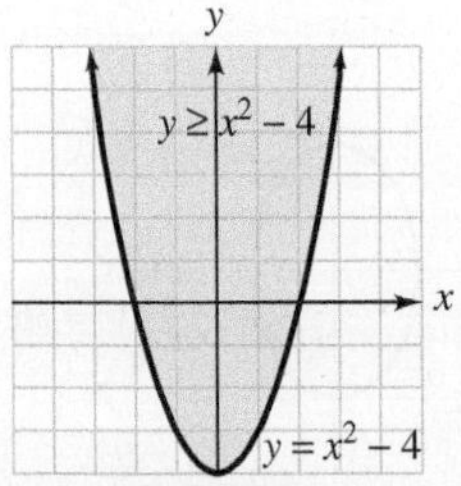

13. $5\sqrt{2}$ **14.** $81x\sqrt[3]{3x}$ **15.** 5; 0 is extraneous

16. 3, 5 is extraneous **17.** $-\frac{5}{2}, \frac{2}{5}$ **18.** $\frac{1}{2} \pm \frac{3}{2}i$

19. $16x^2 + 8x - 2$ **20.** $f^{-1}(x) = \sqrt[3]{\frac{x+1}{2}} = \frac{\sqrt[3]{4x+4}}{2}$

Getting Ready (page 957)

1. $x^2 + 4x + 4$ **2.** $x^2 - 6x + 9$ **3.** $x^3 + 3x^2 + 3x + 1$

4. $x^3 - 6x^2 + 12x - 8$

Exercises 13.1 (page 963)

1. 720 **3.** 30 **5.** $x^2 + 6xy + 9y^2$ **7.** $x^2 - 6xy + 9y^2$

9. 2 **11.** 5 **13.** one **15.** Pascal's **17.** 8!

19. 1 **21.** $a^3 + 3a^2b + 3ab^2 + b^3$

23. $a^4 - 4a^3b + 6a^2b^2 - 4ab^3 + b^4$ **25.** 24

27. −40,320 **29.** 30 **31.** 144 **33.** $\frac{1}{110}$

35. 2,352 **37.** 72 **39.** 10 **41.** $x^3 + 3x^2y + 3xy^2 + y^3$

43. $a^6 + 6a^5b + 15a^4b^2 + 20a^3b^3 + 15a^2b^4 + 6ab^5 + b^6$

45. $a^4 - 4a^3b + 6a^2b^2 - 4ab^3 + b^4$

47. $x^3 - 6x^2y + 12xy^2 - 8y^3$ **49.** $8x^3 + 12x^2y + 6xy^2 + y^3$

51. $8x^3 + 36x^2y + 54xy^2 + 27y^3$ **53.** $6a^2b^2$ **55.** $15x^2y^4$

57. $-4xy^3$ **59.** $28x^6y^2$ **61.** $640x^3y^2$ **63.** $-12x^3y$

65. $-70{,}000x^4$ **67.** $810xy^4$ **69.** 40,320 **71.** 21

73. $\frac{1}{168}$ **75.** 39,916,800 **77.** $2.432902008 \times 10^{18}$

79. $81 + 216y + 216y^2 + 96y^3 + 16y^4$

81. $\frac{x^3}{8} - \frac{x^2y}{4} + \frac{xy^2}{6} - \frac{y^3}{27}$

83. $x^5 - 5x^4y + 10x^3y^2 - 10x^2y^3 + 5xy^4 - y^5$

85. $\frac{x^4}{16} + \frac{x^3y}{6} + \frac{x^2y^2}{6} + \frac{2xy^3}{27} + \frac{y^4}{81}$ **89.** $180x^4y^2$

Unless otherwise noted, all content on this page is © Cengage Learning.

91. $-\frac{1}{6}x^3y$ **93.** $375x^4$ **95.** $\frac{n!}{3!(n-3)!}a^{n-3}b^3$
97. $\frac{n!}{4!(n-4)!}a^{n-4}b^4$ **99.** $\frac{n!}{(r-1)!(n-r+1)!}a^{n-r+1}b^{r-1}$
107. 1, 1, 2, 3, 5, 8, 13, . . . ; beginning with 2, each number is the sum of the previous two numbers. **109.** 36

Getting Ready (page 966)

1. 3, 5, 7, 9 **2.** 4, 7, 10, 13

Exercises 13.2 (page 972)

1. 2, 4, 6, 8 **3.** 5, 6, 7, 8 **5.** 5, 7, 9, 11
7. $27x^2 - 12x + 14$ **9.** $\frac{6a^2 + 16}{(a+2)(a-2)}$
11. sequence, finite, infinite **13.** arithmetic, difference
15. arithmetic mean **17.** series
19. $1 + 2 + 3 + 4 + 5 + 6 + 7$ **21.** 4 **23.** 88
25. 3, 5, 7, 9, 11 **27.** −5, −8, −11, −14, −17
29. 4, 7, 10, 13, 16; 46 **31.** −7, −9, −11, −13, −15; −65
33. 247 **35.** –283 **37.** 5, 11, 17, 23, 29
39. −4, −11, −18, −25, −32 **41.** $\frac{17}{4}, \frac{13}{2}, \frac{35}{4}$
43. 12, 14, 16, 18 **45.** 1,335 **47.** 459
49. $3 + 6 + 9 + 12$ **51.** $16 + 25 + 36$ **53.** 60
55. 31 **57.** −118, −111, −104, −97, −90
59. 34, 31, 28, 25, 22 **61.** 5, 12, 19, 26, 33 **63.** 355
65. −179 **67.** −23 **69.** 12 **71.** $\frac{29}{2}$ **73.** 1.25
75. 354 **77.** 255 **79.** 1,275 **81.** 2,500 **83.** 12
85. \$60, \$110, \$160, \$210, \$260, \$310; \$6,010 **87.** 11,325
89. 368 ft **93.** $\frac{3}{2}, 2, \frac{5}{2}, 3, \frac{7}{2}, 4$

Getting Ready (page 975)

1. 10, 20, 40 **2.** 18, 54, 162

Exercises 13.3 (page 980)

1. 32 **3.** 324 **5.** $\frac{7}{27}$ **7.** $[-1, 5]$
9. $(-\infty, -3) \cup [4, \infty)$ **11.** geometric **13.** common ratio
15. $S_n = \frac{a_1 - a_1r^n}{1 - r}$ **17.** 27 **19.** 2.5 **21.** 6
23. 4, 12, 36, 108, 324; 8,748
25. $-5, -1, -\frac{1}{5}, -\frac{1}{25}, -\frac{1}{125}; -\frac{1}{15{,}625}$
27. 243 **29.** $\frac{1}{32}$ **31.** 2, 8, 32, 128, 512
33. 2, −10, 50, −250, 1,250 **35.** 6, 18, 54
37. −20, −100, −500, −2,500 **39.** −16 **41.** $10\sqrt{2}$
43. 225 **45.** 122 **47.** 2, 10, 50, 250, 1,250
49. −64, 32, −16, 8, −4 **51.** 26,244 **53.** $\frac{1}{27}$
55. 3 **57.** No geometric mean exists. **59.** −255
61. 381 **63.** $\frac{156}{25}$ **65.** $-\frac{21}{4}$ **67.** about 669 people
69. \$1,469.74 **71.** \$140,853.75
73. $\left(\frac{1}{2}\right)^{11} \approx 0.0005$ sq unit **75.** \$4,309.14
81. arithmetic mean

Getting Ready (page 983)

1. 4 **2.** 4.5 **3.** 3 **4.** 2.5

Exercises 13.4 (page 986)

1. 3 **3.** $\frac{1}{5}$ **5.** $\frac{1}{5}$ **7.** yes **9.** no **11.** infinite
13. $S_\infty = \frac{a_1}{1 - r}$ **15.** 64 **17.** 250 **19.** 8 **21.** $\frac{-243}{4}$
23. no sum because $r > 1$ **25.** no sum because $r > 1$
27. $\frac{1}{9}$ **29.** $-\frac{1}{3}$ **31.** $\frac{4}{33}$ **33.** $\frac{25}{33}$ **35.** $-\frac{81}{2}$
37. $-\frac{81}{2}$ **39.** 5,000 **41.** 30 m **45.** $\frac{4}{5}$
49. no; $0.999999 = \frac{999{,}999}{1{,}000{,}000} < 1$

Getting Ready (page 988)

1. 24 **2.** 120 **3.** 30 **4.** 168

Exercises 13.5 (page 995)

1. 840 **3.** 120 **5.** 56 **7.** 12, −5 **9.** 8
11. $p \cdot q$ **13.** permutation **15.** $P(n, r) = \frac{n!}{(n-r)!}$
17. combination **19.** $C(n, r) = \frac{n!}{r!(n-r)!}$ **21.** 120
23. 30 **25.** 240 **27.** $n!$ **29.** 4 **31.** 56
33. 735 **35.** 2 **37.** $x^4 + 4x^3y + 6x^2y^2 + 4xy^3 + y^4$
39. $a^5 + 5a^4b + 10a^3b^2 + 10a^2b^3 + 5ab^4 + b^5$
41. $8x^3 + 12x^2y + 6xy^2 + y^3$
43. $64x^3 - 48x^2y + 12xy^2 - y^3$
45. 1 **47.** 1,260 **49.** $\frac{n!}{2!(n-2)!}$
51. $81x^4 - 216x^3 + 216x^2 - 96x + 16$ **53.** $-1{,}250x^2y^3$
55. $-4x^6y^3$ **57.** 35 **59.** 5,040 **61.** 1,000,000
63. 136,080 **65.** 8,000,000 **67.** 2,880 **69.** 13,800
71. 720 **73.** 900 **75.** 364 **77.** 5
79. 1,192,052,400 **81.** 18 **83.** 7,920 **87.** 48

Getting Ready (page 997)

1. 1, 2, 3, 4, 5, 6 **2.** H, T

Exercises 13.6 (page 1001)

1. C(52, 26) **3.** −1 **5.** 3, −1 **7.** −2
9. experiment **11.** $\frac{s}{n}$ **13.** 0 **15. a.** 6 **b.** 52 **c.** $\frac{6}{52}, \frac{3}{26}$
17. {(1, H), (2, H), (3, H), (4, H), (5, H), (6, H), (1, T), (2, T), (3, T), (4, T), (5, T), (6, T)} **19.** {a, b, c, d, e, f, g, h, i, j, k, l, m, n, o, p, q, r, s, t, u, v, w, x, y, z} **21.** $\frac{1}{9}$ **23.** $\frac{5}{18}$ **25.** $\frac{1}{4}$
27. $\frac{3}{26}$ **29.** $\frac{33}{391{,}510}$ **31.** $\frac{9}{460}$ **33.** $\frac{1}{6}$ **35.** $\frac{2}{3}$ **37.** $\frac{19}{42}$
39. $\frac{13}{42}$ **41.** $\frac{3}{8}$ **43.** 0 **45.** $\frac{3}{169}$ **47.** $\frac{1}{6}$ **49.** $\frac{7}{12}$
51. $\frac{1}{16}$ **53.** $\frac{3}{8}$ **55.** $\frac{1}{16}$ **57.** $\frac{1}{3}$ **59.** $\frac{88}{141}$ **61.** $\frac{15}{71}$
65. 0.14

Chapter Review (page 1006)

1. 240 **2.** 20 **3.** 15 **4.** 220 **5.** 1 **6.** 5,040
7. $x^5 + 5x^4y + 10x^3y^2 + 10x^2y^3 + 5xy^4 + y^5$
8. $x^4 - 4x^3y + 6x^2y^2 - 4xy^3 + y^4$
9. $64x^3 - 48x^2y + 12xy^2 - y^3$
10. $x^3 + 12x^2y + 48xy^2 + 64y^3$ **11.** $5xy^4$ **12.** $-10x^2y^3$
13. $-108x^2y$ **14.** $864x^2y^2$ **15.** 42
16. 122, 137, 152, 167, 182 **17.** $\frac{41}{3}, \frac{58}{3}$ **18.** 1,550
19. $-\frac{45}{2}$ **20.** 60 **21.** 34 **22.** 14 **23.** 360
24. 24, 12, 6, 3, $\frac{3}{2}$ **25.** 1 **26.** 24, −96 **27.** $\frac{945}{2}$
28. $-\frac{85}{8}$ **29.** \$1,638.40 **30.** \$134,509.57 **31.** 12 yr
32. 1,600 ft **33.** 125 **34.** $\frac{5}{99}$ **35.** 136 **36.** 5,040
37. 1 **38.** 20,160 **39.** $\frac{1}{10}$ **40.** 1 **41.** 1 **42.** 28
43. 210 **44.** 700 **45.** $x^3 + 6x^2 + 12x + 8$
46. 120 **47.** 720 **48.** 120 **49.** 150 **50.** $\frac{3}{8}$
51. $\frac{1}{2}$ **52.** $\frac{7}{8}$ **53.** $\frac{1}{9}$ **54.** 0 **55.** $\frac{1}{13}$ **56.** $\frac{94}{54{,}145}$
57. $\frac{33}{66{,}640}$

Chapter 13 Test (page 1011)

1. 72 **2.** 1 **3.** $-5x^4y$ **4.** $24x^2y^2$ **5.** 66
6. 306 **7.** 34, 66 **8.** 14 **9.** −81 **10.** $\frac{364}{27}$
11. 18, 108 **12.** $\frac{27}{2}$ **13.** no sum **14.** 210
15. 40,320 **16.** 10 **17.** 56 **18.** 720 **19.** 24
20. $x^4 - 12x^3 + 54x^2 - 108x + 81$ **21.** 56 **22.** 30
23. $\frac{1}{6}$ **24.** $\frac{2}{13}$ **25.** $\frac{33}{66{,}640}$ **26.** $\frac{5}{16}$

Cumulative Review (page 1011)

1. (2, 1) **2.** (1, 1) **3.** (2, −2) **4.** (3, 1)
5. −2 **6.** −1 **7.** (−1, −1, 3) **8.** 1

9.

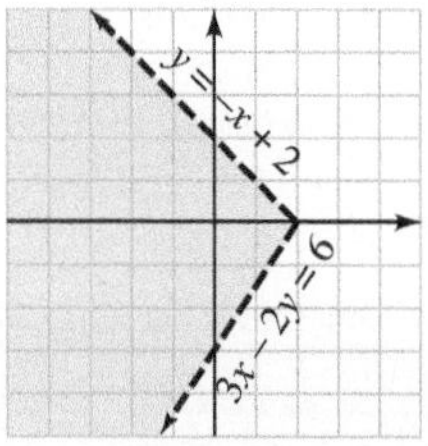

10.

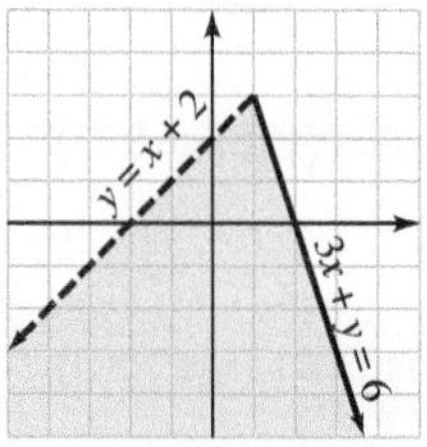

11.

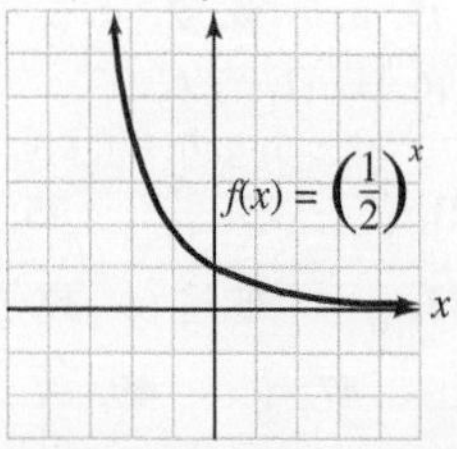

12. $2^y = x$ **13.** 3 **14.** 3 **15.** $\frac{1}{16}$ **16.** 1
17. $y = 2^x$ **18.** x **19.** 1.9912 **20.** 0.3010
21. 1.4314 **22.** 0.1461 **23.** $\frac{5 \log 2}{\log 3 - \log 2}$ **24.** 8
25. \$2,848.31 **26.** 1.1292 **27.** 840
28. $81a^4 - 108a^3b + 54a^2b^2 - 12ab^3 + b^4$ **29.** $112x^2y^6$
30. 103 **31.** 690 **32.** 8 and 19 **33.** 60 **34.** 52
35. 27 **36.** $\frac{1{,}023}{64}$ **37.** 12, −48 **38.** $\frac{27}{2}$ **39.** 504
40. 120 **41.** $\frac{7}{3}$ **42.** $C(n, n)$ **43.** 5,040 **44.** 84
45. $\frac{3}{26}$

Appendix 1 (page A-11)

1. 81.9 ft **3.** 3.8 mi **5.** 230 mi **7.** 2.4 mi **9.** 8 cups
11. 9 **13.** 40 bags **15.** 1,125 tons **17.** 1,750,000 tons; 2,121,212 head **19.** 0.002 m **21.** 1500 cl
23. 0.155517384 kg **25.** 3,274 ft **27.** 1 oz (1 stick of butter is 4 oz)

Appendix 2 (page A-16)

1. y-axis **3.** origin **5.** y-axis **7.** none
9. none **11.** x-axis
13. D: $(-\infty, \infty)$, R: $[-4, \infty)$

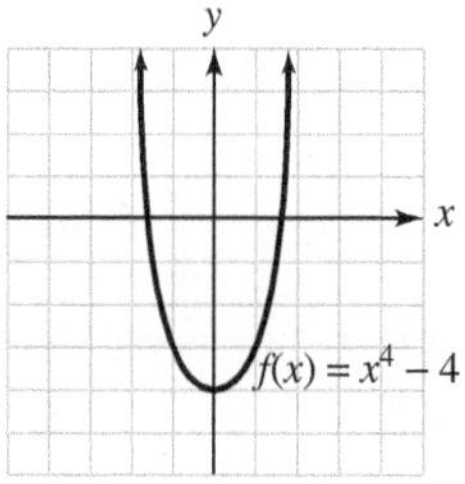

15. D: $(-\infty, \infty)$, R: $(-\infty, \infty)$

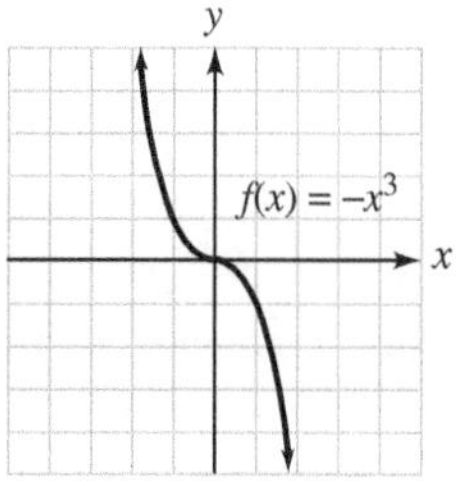

17. D: $(-\infty, \infty)$, R: $[0, \infty)$

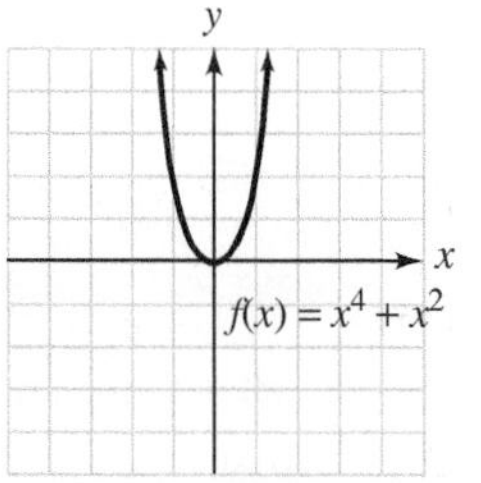

19. D: $(-\infty, \infty)$, R: $(-\infty, \infty)$

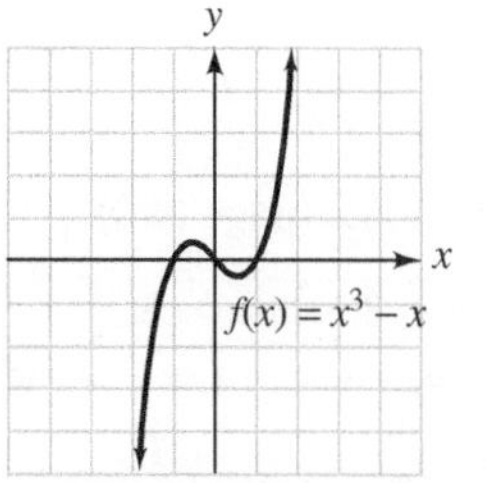

21. D: $(-\infty, \infty)$, R: $[-1, \infty)$

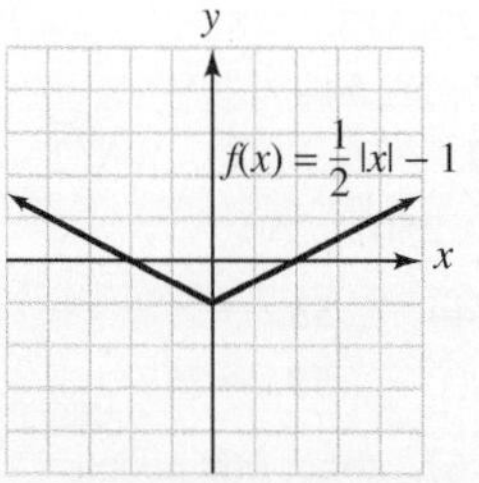

23. D: $(-\infty, \infty)$, R: $(-\infty, 0]$

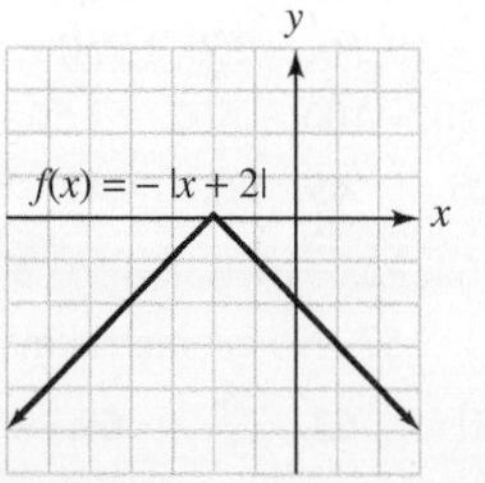

Unless otherwise noted, all content on this page is © Cengage Learning.

INDEX

A

Absolute inequalities, 448
Absolute value
  of complex numbers, 710, 722
  defined, 511
  equations, 511
  equations with two, 514
  inequalities, 517, 518, 519
  of real numbers, 10
Absolute value expression, 536–538
Absolute value function, 207, 771
ac method, 325–326, 487–491
Accent on technology
  approximating square roots, 649
  approximating zeros of polynomials, 293
  checking solutions of quadratic equations, 733–734
  creating tables and graphs, 169–171
  evaluating determinants, 600–601
  factorials, 960
  finding base-10 (common) logarithms, 847
  finding period of pendulum, 654
  finding powers of decimals, 248
  finding present value, 242
  graphing ellipses, 912–913
  graphing functions, 772
  graphing hyperbola, 922
  graphing logarithmic function, 846
  graphing polynomial functions, 257
  height of a rocket, 255
  radical equations, 696
  rational exponents, 669
  solving absolute value equations, 514
  solving equations graphically, 548–549
  solving exponential equations, 870–871
  solving inequalities, 787–788
  solving investment problems, 830–831
  solving logarithmic equations, 872–873
  solving systems by graphing, 547–548
  solving systems of equations, 930
  translations of the exponential function, 836–837
  verifying properties of logarithms, 860
Addition
  addition of two or more polynomials, 261–262
  addition of two polynomials, 262–263
  of complex numbers, 705
  decimals, 22
  distributive property of multiplication over addition, 64
  fractions, 18–19
  mixed numbers, 20–21
  radicals, 678–679
  rational expressions, 387–388, 500–502
  real numbers, 74–75
  real numbers with like signs, 39–40
  real numbers with unlike signs, 40–41
  two or more monomials, 261–262
  two polynomials, 262–263
Addition property of equality, 83, 443
Addition property of inequality, 133
Additive inverses, 66
Adjacent angles, 117
Ahmes Papyrus, 9
Algebraic expressions
  evaluating when given values for its variables, 57–58
  identifying number of terms in, 58–59
  translating English phrase into, 54–57
Altitude, 680
American system of measurement, A-1–A-4
Amount, 88
Angles, 117
Annual depreciation rate, 200
Annual growth rate, 835
Area, 32
Arithmetic means, 969, 1007
Arithmetic sequence
  common difference and, 967–968
  defined, 966
  finding specified term given first term and other terms, 968–969
  finding sum of first *n* terms of, 969–970
Arithmetic series, 970, 1007
Array of signs for 3 × 3 determinant, 599
Associative properties, 63, 67
Asymptotes, 920
Augmented matrix, 590

B

Bakhshali manuscript, 648
Base, 29, 88, 231
Base-10 logarithms, A-20, 846
Base angles, 120
Base-*e* exponential functions
  continuous compound interest, 834–836
  graphing, 836
Base-*e* logarithms, A-21, 852. *See also* Natural logarithms
Binomial expansion, 962–963
Binomial theorem
  combinations and, 994
  expansion, 960–962
  notation, 961
  raising to a power, 957–958
Binomials
  conjugate, 271
  division by descending order, 286
  division of polynomial with one or more missing terms by, 286–287
  multiplication by, 269–272
  multiplication polynomial by, 272–273
  as polynomials, 252
  squaring, 271

C

Calculators
  approximating zeros of polynomials, 293
  checking solutions of quadratic equations, 733–734
  creating tables and graphs, 169–171
  decimal calculations, 22–23
  evaluating determinants, 600–601
  evaluating logarithms, 853
  finding base-10 (common) logarithms, 847
  finding factorials, 960
  finding period of pendulum, 654
  graphing circles, 899
  graphing ellipses, 912–913
  graphing exponential functions, 828
  graphing functions, 772
  graphing inequalities, 215–216
  graphing logarithmic functions, 854–855
  graphing one-to-one exponential functions, 827
  rational exponents, 669
  solving absolute value equations, 514
  solving absolute value inequalities, 520
  solving exponential equations, 870–871
  solving inequalities, 787–788
  solving investment problems, 830–831
  solving logarithmic equation with, 853
  solving logarithmic equations, 872–873
  solving systems of equations, 930
  translations of the exponential function, 836–837
  using to add/subtract real numbers, 44
  using to find powers, 30
  using to multiply/divide real numbers, 51
  verifying properties of logarithms, 860
Calculus, 372
Careers and Mathematics
  airline pilots, 441
  bakers, 367
  carpenters, 1
  commercial pilots, 441
  cosmetologists, 541
  detectives, 725
  financial analysts, 955
  food-processing occupations, 367
  medical scientists, 229
  pest-control workers, 805
  photographers, 645
  police officers, 725
  psychologists, 149

Careers and Mathematics (*continued*)
securities and financial services sales agents, 79
water transportation occupations, 893
Carroll, Lewis, 600
Cartesian coordinate system, 151
Cartesian plane, 152
Cayley, Arthur, 590
Center of a circle, 896
Change-of-base formula, 863–864, 889
Circles
defined, 896
formula for perimeter and area of, 32–33, 34
general form of equation of, 898
graphing, 899
standard form of equation of, 896–898
Circumference, 32
Clear fractions, 97
Closed interval, 448
Closure properties, 62, 67
C(*n*,0), 992
C(*n*,*n*), 992
C(*n*,*r*), 992
Coefficient, 102
Coefficient matrix, 590
Combinations, 992, 1009
Combined variation, 426, 438
Common difference, 967–968
Common logarithm, 846–847
Common ratio, 975
Commutative properties, 63, 67
Complementary angles, 119
Completing the square, 730–734, 795
Complex conjugates, 706
Complex fractions, 398–402, 435–436, 502–505
Complex numbers
addition of, 705
defined, 704
equality of, 705
multiplication of, 706
subtraction of, 705
Composite functions, 809
Composite numbers, 6, 9
Compositions of functions, 809, 882–883
Compound inequalities, 135, 451
Compound interest, 735, 829–830, 834, 885
Compounding periods (*n*), 830
Conditional equations, 106, 445
Conditional inequalities, 448
Cone, 34
Conic sections, 895
Conjugate, 689
Conjugate binomials, 271
Consistent system, 544
Constant functions, 951
Constants, 58
Constraints, 622
Continuous compound interest, 886
Contradiction, 106, 445
Coordinate, 8
Coordinate plane, 152
Coordinates (*x*, *y*), 152, 456
Cramer's rule, 601–605
Critical points, 783
Critical values, 783
Cube-root function, 652–653
Cube roots, 650
Cubes
difference of, 363
difference of two, 339–340
factoring difference of two, 492
sum of, 363
sum of two, 338–339, 492
Cubic centimeter, A-6
Cubic function, 256, 770
Cubic units, 35
Cup, A-3
Cylinder, 34

D
Decibel, 847
Decibel voltage gain, 847
Decimals
addition of, 22
division of, 23
finding powers of, 248
multiplication of, 22–23
repeating, 21–22, 558
rounding, 23
terminating, 21–22
Decreasing functions, 826, 884, 936, 951
Degrees, 117
Denominators
defined, 13
like, 387–389
rationalizing, 686–689
unlike, 391–395
Dependent equations, 546–547, 557–558
Dependent variables, 163, 477
Derivative, 372
Descartes, René, 151, 152, 154, 895
Determinants, 598, 600–601
Difference of two cubes, 492
Difference of two squares, 315–316, 362, 486–487
Difference quotient, 810–811, 883
Direct variation, 422–423, 438
Discriminant, 746–747, 796
Distance formula, 661, 716
Distributive property, 67, 264, 270
Distributive property of multiplication over addition, 64
Dividend, 283
Division
decimals, 23
fractions, 17
monomial by a monomial, 277–278
of negatives, 375
by one, 372
polynomial by binomial, 283–286
of polynomial by binomial of form $(x - r)$, 289–291
polynomial by monomial, 278–280
polynomial with one or more missing terms by binomial, 286–287
polynomials by monomials, 277–282
polynomials by polynomials, 283–289
property of radicals, 675
rational expressions, 381–384, 499–500
real numbers, 49–51, 75
synthetic, 289–295, 302
Division property of equality, 85–86
Division property of inequality, 134
Divisor, 283
Domain
of cube-root function, 652–653
defined, 476
of a function, 479
from graph, 476–477
graphing functions and determining, 207–208
of ordered pairs, 204–205, 475–476
radical function, 715
of rational function, 370–374, 777–778
of square-root function, 652–653
Double inequalities, 135, 451
Double negative rule, 8

E
Einstein, Albert, 112
Elementary row operations, 590
Elimination (addition) method, 555–557, 931–932
Ellipses, 4, 908–918
defined, 909
graphing, 912–914
Empty set, 106, 446
Equality
addition property of, 83
of complex numbers, 705
division property of, 85–86
logarithmic property of, 861
multiplication property of *e*, 85–86
subtraction property of, 83
Equations. *See also* Linear equations; Quadratic equations
conditional, 106, 445
defined, 81, 443
dependent, 546–547
ellipses, 911, 946
equivalent, 84
exponential, 889–890
finding the slope of a line, 179
general form of, 166
general forms of ellipses, 913
general forms of parabola, 901
graphing systems of linear, 543–552
higher-order polynomial, 349–350
of horizontal and vertical lines, 168
horizontal line and vertical line graph, 167–171
of horizontal lines, 458
hyperbola, 921, 922, 923, 947–948
independent, 546–547
intercept method, 166–167
of a line modeling real-world data, 467–468
of the line representing real-world data, 191
logarithmic, 889–890
number as solution of, 82
of parabolas, 901, 945
prediction, 467
quadratic, 346, 352–356
in quadratic form, 747–751
radical, 693–701
regression, 467
slope-intercept form, 194–203
solving, 104
solving by factoring, 346–352, 364
solving graphically, 548–549
solving systems of, 929–935, 949–950
table of values construction, 162–163
table of values graph, 163–165
that contain rational expressions, 405–411, 436–437
that simplifies to a linear equation, 273

with two absolute values, 514
of vertical lines, 458
Equilateral triangle, 680
Equilateral triangles, 136–137
Equivalent equations, 84
Equivalent fractions, 18
Equivalent systems, 544–545
Euclid, 34, 519
Euler, Leonhard, 835
Even integers, 6
Even root, 651
Events
multiplication principle for, 988–989
probability of, 998–1001, 1010
Everyday connections
2010 Gubernatorial Elections, 24
focus on conics, 924
fuel oil production, 628
highest and longest jumpers, 420–421
NBA salaries, 258
renting a car, 94
selling calendars, 346
shortest distance between two points, 519
staffing, 594
U.S. population growth, 876
wind power, 178–179
winning the lottery, 1000
Experiments, 998
Exponential equations, 868, 869–870, 889–890
Exponential expression, 29
Exponential functions
calculating compound interest, 829–830
defined, 824
future value, 830
graphing, 824–826, 828
graphing one-to-one, 827
graphing translation of, 827–828
properties of, 826
Exponential growth, 835
Exponents
defined, 29
exponential expression without, 231–232
natural number, 30
power rule for, 234
product rule for, 233–234
product to a power rule for, 235
quotient rule for, 235–236
quotient to a power rule for, 235
rational, 666–674
repeated multiplication expression using, 232
rules of, 666
variable, 241
zero, 239–240
Expressions, 81
Extraneous solutions, 406–407
Extremes, 419

F
Factor theorem, 292
Factorability, 488
Factorial notation, 959
Factorials, 960, 1006
Factoring
ac method, 333
completely factoring polynomials, 316–318, 324
defined, 307
difference of two cubes, 339–340, 363, 492
difference of two squares, 315–316, 362, 486–487
general trinomials, 489
greatest common factor, 307–315, 361
by grouping, 311–313, 361
perfect-square trinomial, 326–327
polynomial by factoring out the greatest common factor, 484–485
polynomial containing binomial greatest common factor, 311
polynomial containing greatest common factor, 309–310
polynomial containing negative greatest common factor, 310–311
polynomial involving perfect-square trinomial, 334–336
polynomial involving sum/difference of two cubes, 340–341
polynomial with four terms or six terms by grouping, 485–486
polynomials, 342–345
quadratic equation, 493–494
solving equations by, 346–352, 364
solving quadratic equation by, 727–728
sum of two cubes, 338–339, 363, 491–493
sum of two squares, 317
summary of techniques, 342–345, 363–364
trinomials, 362–363, 487
trinomials by grouping, 325–326, 333–334
trinomials by trial and error, 320–323, 329–333
trinomials with leading coefficient of one, 320–329
trinomials with leading coefficient other than one, 329–336
Factors, 14
Feasibility region, 622
Fermat, Pierre de, 488
Fibonacci, Leonardo, 9, 741
Fibonacci sequence, 741, 966
Fluid ounce. *See* ounce
Foci, 909
Focus, 909
FOIL method, 270
Formulas
change-of-base, 863–864
for combinations, 1009
for compound interest, 735
compound interest, 834
direct variation, 422
distance, 661
evaluating for specified values for variables, 111–112
exponential growth, 835
future value, 830
for indicated variable using properties of equality, 109–111
inverse variation, 424
midpoint, 459
percent, 88–89
for perimeter and area of circle, 34
for perimeter and area of rectangle, 34
for perimeter and area of square, 34
for perimeter and area of trapezoid, 34
for perimeter and area of triangle, 34
for permutations, 1009
quadratic, 738–745, 795–796
radioactive decay, 838
slopes, 177
solving an application using, 112–113
solving for specified variable, 446–447
for specified values for the variables, 111–112
for a specified variable, 280
for a variable, 109–111
for volume of cone, 34
for volume of cylinder, 34
for volume of pyramid, 34
for volume of rectangular solid, 34
for volume of sphere, 34
Fractions
addition of, 18–19
clearing, 97
complex, 502–505
division of, 17
equivalent, 18
fundamental property of, 15, 371
improper, 16
infinite geometric series and, 985–986
multiplication of, 15–17
proper, 16
repeating decimals, 558
simplifying, 13–15
subtraction of, 19–20
using appropriate operation for application, 25
Function notation, 206–207, 477–479
Functions
absolute value, 771
base-*e*, 834–841
composite, 809
composition of, 807–813
constant, 951
of cube-root, 652–653
cubic, 770
decreasing, 936, 951
defined, 205, 476
difference quotient of, 810–811
domain of, 479–480
evaluating polynomial, 254
exponential, 823–834
graphing, 479–480, 772
horizontal line test, 815–816
increasing, 936, 951
inverses of, 814–823, 883–884
logarithmic, 842–851
one-to-one, 814–815, 817
operations on, 807
quadratic, 256, 770
square-root, 652–653
step, 938, 951
sum, difference, product, and quotient of, 807–808
Fundamental rectangle, 920
Future value (*FV*), 830
FV. *See* Future value (*FV*)

G
Gallon, A-3
Gauss, Carl Friedrich, 325, 675
Gaussian elimination, 590
GCF. *See* Greatest common factor (GCF)
General form of equation of circle, 898

General form of equation of ellipses, 913
General form of equation of hyperbola, 923
General form of equation of parabola, 901
General form of the equation of a line, 166
General ordered pair, 557–558
General term, 966–967
General trinomials, 489
Geometric means, 976, 1008
Geometric sequences
  common ratio and, 975
  defined, 975
  finding specified term given first term and other terms, 976
  finding sum of first *n* terms of, 977–978
  solving application involving, 978–980
Geometry applications, 117–120
Germain, Sophie, 610
Given statements, 66–67
Golden ratio, 506, 741
Gram, A-6
Graphing
  base-*e* exponential functions, 837–839
  circles, 899
  ellipses, 912–914
  exponential functions, 824–826, 828
  function and determining domain and range, 479–480
  functions, 772
  greatest integer function, 938
  horizontal lines, 458
  horizontal translations, 773–774
  hyperbola, 918–925
  inequalities, 215–216
  linear equations, 161–175, 456–458, 528–530
  logarithmic function, 845–846, 854–855
  method, 543
  natural logarithmic function, 854
  nonlinear inequalities, 788–789
  one-to-one exponential functions, 827
  piecewise-defined functions, 937–938
  points, 153
  polynomial functions, 257
  reflection about *x*-axis, 774–775
  systems of linear equations, 543–552
  translation of an exponential function, 827
  translations of the exponential function, 885
  vertical lines, 458
  vertical translations, 772–773
Graphs
  creating tables and, 169–171
  domain and range from, 476–477
  domain and range of function, 207–208
  finding the slope of a line, 176–177
  horizontal line and vertical line, 167–171
  linear inequality in one or two variables, 609–615
  one-to-one function, 815–816
  ordered pairs and mathematical relationships, 151–155
  other nonlinear functions, 770–782, 798–799
  point-slope equation from, 189–190
  point-slope form of an equation, 190
  polynomial functions, 255–257
  quadratic functions, 797–798
  of rational function, 776–777
  reading, 155–156
  real numbers, 8–10
  rectangular coordinate system, 151
  slope-intercept form, 197
  in space, 775–776
  step, 156–157
  symmetries of, A-13–A-17
  vertical line test, 208–209
Greatest common factor (GCF)
  defined, 308
  factoring out, 484–485
  factoring polynomial containing, 309–310
  factoring polynomial containing binomial, 311
  factoring polynomial containing negative, 310–311
  factoring trinomial containing negative, 323
  finding, 309
  of two or more monomials, 307–309
Greatest integer function, 938

H
Half-life, 873
Half-open interval, 448
Higher-order polynomial equation, 349–350
Hopper, Grace Murray, 461
Horizontal line test, 815–816, 883
Horizontal lines
  equations of, 167–171
  graphing, 458
  slope of, 179–180
Horizontal translations, 773, 798
How to Solve It (Polya), 623
Hypatia, 399
Hyperbola
  defined, 919
  general form of equation of, 923
  graphing, 918–925
  solving application involving, 925–926
Hypotenuse, 177

I
i. *See* Periodic interest rate (*i*)
Identity, 106, 445
Identity elements, 65
Identity function, 818
Identity properties, 67
Imaginary numbers, 702–704, 721
Improper fractions, 16
Inconsistent system, 545–546, 557, 582
Increasing function, 825, 884, 936, 951
Independent equations, 546–547
Independent variables, 163, 477
Index, 650
Index of the summation, 971
Inequalities
  addition property of, 133
  compound, 135, 451
  defined, 133
  division property of, 134
  double, 135, 451
  linear, 447–452
  multiplication property of *i*, 134
  properties of, 449
  solving in one variable containing absolute value expression, 516–523
  solving in two variables, 211–219
  solving systems of, 613
  subtraction property of, 133
  symbols, 132
Inequality symbols, 7
Infinite geometric sequences, 983
Infinite geometric series
  determining, 984
  finding *n*th partial sum of, 983–984
  finding sum of, 984
  repeating decimal as common fraction and, 985–986
  sum of, 984
Infinite sequences, 966
Input value, 162, 205
Integer exponents, 241
Integer squares, 648
Integers
  application using quadratic equation, 352–353
  even, 6
  odd, 6
  set of, 4
Intercept method, 166–167
Interval notation, 448
Intervals, 9, 135, 447
Inverse properties, 67
Inverse variation, 424–425, 438
Inverses of function, 814–823
Investment applications, 121
Irrational numbers, 9, 649
Isosceles right triangle, 679–680
Isosceles triangles, 120

J
Joint variation, 425–426, 438

K
Key number, 325, 491
Kovalevskaya, Sonya, 207

L
Least common denominator (LCD), 18, 389–391, 434–435
Leibniz, Gottfried Wilhelm von, 372
Length
  American system of measurement, A-1–A-2
  conversions between metric and American, A-9
  metric system of measurement, A-5–A-6
Like denominators, 387–389
Like signs, 39–40
Like terms
  combining, 103
  defined, 102
  polynomials, 264–265
Linear depreciation, 199
Linear equations
  Cramer's rule, 601–605
  dry mixture applications, 128–129
  equations that simplifies to, 273
  graphing, 161–175, 456–458
  graphing linear equations in two variables, 163–165
  liquid mixture applications, 127–128
  motion applications, 125–127
  simplifying expressions to solve, 102–108
  solving, 443–446
  solving applications in two variables, 567–579, 636
  solving by elimination, 555–557, 635
  solving by substitution, 553–555, 635

  solving by using determinants, 597–608, 639–640
  solving by using linear programming, 621–631, 641–642
  solving by using matrices, 589–597, 638
  solving graphically, 543–552, 634
  solving in one variable, 81–101
  solving in three variables, 579–588, 637
  solving systems of two, 547
Linear functions, 206, 479, 480–481
Linear inequalities
  in one variable, 132–137, 146, 447–452
  in two variables, 212, 227–228
Linear programming, 621–631
Linear units, 35
Lines
  equations of horizontal, 167–171
  equations of vertical, 167–171
  finding the slope of, 179
  of the line representing real-world data, 199–200
  point-slope form of, 188–192
  representing real-world data, 191, 199–200
  slope-intercept form of, 195
  slope of horizontal, 179–180
  slope of nonvertical lines, 177
  slope of parallel, 181, 198–199
  slope of perpendicular, 182, 198–199
  slope of vertical, 179–180
  writing equations of, 528–530
Liter, A-6
Literal equations, 109
Logarithmic equations, 868, 871–872, 873–876, 889–890
Logarithmic functions
  with base *b*, 842
  as exponential function, 842–844
  graphing, 844–845, 854–855
  graphing vertical and horizontal translation of, 845–888
Logarithmic property of equality, 861
Logarithms
  evaluating, 846
  as exponent, 843
  natural, 852–858
  power rule of, 860
  product and quotient properties, 859
  properties of, 858–868, 888–889
  verifying properties of, 860
Lowest terms, 14

M
Major axis, 911
Malthus, Thomas Robert, 838
Malthusian growth model, 874, 886
Markdown, 87, 99
Markup, 87–88, 98
Matrix, 589
Means, 419
Measurement conversions
  American system of measurement, A-1–A-4
  between metric and American, A-8–A-11
  metric system of measurement, A-4–A-8
Meter, A-5
Metric system of measurement, A-4–A-8
Midpoint, 459
Minor axis, 911
Minors, 598
Minuend, 264
Mixed numbers, 20–21
Mixture applications
  dry, 128–129
  liquid, 127–128
Monomials
  degree of, 253
  division by, 277–278
  division of polynomial by, 278–280
  greatest common factor of two or more, 307–309
  multiplication polynomial by, 269
  as polynomials, 252
  subtraction of two, 262
Motion applications, 125–127, 145–146
Multiplication
  application involving polynomials, 273
  of complex numbers, 706
  decimals, 22–23
  distributive property of multiplication over addition, 64
  fractions, 15–17
  polynomial by a binomial, 272–273
  polynomial by monomial, 269
  polynomials, 272
  property of radicals, 675
  radical expressions, 684–686
  rational expressions, 378–380, 499–500
  real numbers, 75
  rules for multiplying signed numbers, 48
  of two or more monomials, 268–269
Multiplication principle for events, 988–989, 1009
Multiplication property of equality, 85–86, 443
Multiplication property of inequality, 134
Multiplication property of radicals, 675
Multiplicative inverses, 66

N
n. *See* Compounding periods (*n*)
Napier, John, 852
Napierian logarithms, 852
Natural logarithms
  defined, 852
  evaluating, 852–853
  graphing, 854
  solving, 853, 855–856
Natural-number exponents
  defined, 30, 231
Natural numbers, 3
Negative exponents, 240, 402
Negative-integer exponents, 240–241, 297
Negative numbers, 4
Negative reciprocals, 182, 462
Negatives, 8, 66, 375
Newton, Issac, 372, 909
Noether, Amalie, 256
Nonlinear functions, 798–799
Nonlinear inequalities, 788–789, 800–801
Nonvertical lines, 459
Nonzero function, A-14
*n*th root, 650–652
Number applications, 116–117
Number line, 8–10
Numbers. *See also* Real numbers
  complex, 702–712
  composite, 6, 9
  imaginary, 702–704
  irrational, 9
  key, 491
  mixed, 20–21
  natural, 3
  negative, 4
  prime, 6, 9
  rational, 4, 7
  whole, 4
Numerators, 13, 689–690
Numerical coefficient, 58

O
Objective function, 622, 623
Odd integers, 6
Odd root, 651
One-to-one function, 814–815, 884
Open intervals, 447
Opposites, 66
Order, 650
Order of operations
  polynomials, 264–265
  rules for, 31
  simplifying expressions using, 102–108
Ordered pairs
  defined, 152
  equations, 162
  finding domain and range of set of, 204–205
Ordered triple, 580
Origin, 8, 152, 911
Ounce, A-3, A-4
Output value, 162, 205

P
Parabolas, 944–946
  defined, 256
  equations of, 901
  finding the vertex of, 772–773, 797–798
  general forms of the equations of, 901
  graphing, 900–904
  graphs in space, 775–776
  solving an application involving, 902–903
Paraboloid, 775–776, 900
Parallel lines, 181, 198–199, 462–463
Pascal, Blaise, 895, 958
Pascal's triangle, 958–959
Percent
  of decrease, 98–99
  defined, 25
  of increase, 98–99
  solve application involving, 90
Percent formula, 88
Perfect-square trinomial, 326–327, 334–336
Perimeter, 32
Periodic interest rate (*i*), 830
Permutations, 989–991, 1009
Perpendicular lines, 151, 182, 198–199, 462–463
Perspectives
  Ahmes Papyrus, 9
  Archbishop of Toledo, 308
  calculating square roots, 648–649
  golden ratio, 506
  graphs in space, 775–776
  How to Solve It (Polya), 623
  indeterminate forms, 372
  Lewis Carroll, 600
  metric system, 246
  Pythagorean theorem, 661
  René Descartes, 154
  solving equations, 85
pH of a solution, 864–865
Piecewise-defined functions, 937–942
Pint, A-3
Plane, 580
P(*n*,0), 991

$P(n,n)$, 991
$P(n,r)$, 990
Point circle, 896
Point-slope form, 188–192, 224, 463–466
Polya, George, 623
Polynomial functions
  defined, 2554
  evaluating, 254
  graphs, 255–257
Polynomials
  addition of two, 262–263
  addition of two or more, 261–262
  application involving multiplication of, 274
  application requiring operations with, 265
  approximating zeros of, 293
  binomials, 252
  completely factoring, 316–318, 324
  defined, 251
  degree of, 252–253
  determining, 251–252
  division by binomial, 283–286
  division by descending order, 286
  division by monomial, 278–280
  division with one or more missing terms by a binomial, 286
  evaluating, 253–254
  factoring, 342–345
  factoring by grouping, 311–313
  factoring out the greatest common factor, 484–485
  factoring polynomial containing binomial greatest common factor, 311
  factoring polynomial containing greatest common factor, 309–310
  factoring polynomial containing negative greatest common factor, 310–311
  factoring polynomial involving perfect-square trinomial, 334–336
  factoring polynomial involving sum/difference of two cubes, 340–341
  factoring with four terms or six terms by grouping, 485–486
  monomials, 252
  multiplication binomial, 272–273
  multiplication binomial by binomial, 269–272
  multiplication by, 272
  multiplication of two or more monomials, 268–269
  multiplication polynomial by monomial, 269
  prime, 317
  subtraction of two, 263–264
  as trinomial, 252
  using the order of operations and combining like terms, 264–265
Positive integers. *See* Natural numbers
Pound, A-4
Power of x, 29, 231
Power rule
  for exponents, 234
  of logarithms, 860
  radical equations, 694
Powers of i, 709, 721
Powers table, A-19
Prediction equation, 467
Present value (PV), 830
Prime-factored form, 14
Prime numbers, 6, 9
Prime polynomials, 317
Prime trinomial, 323–324
Principal square root, 648
Probability, 998, 1010
Problem solving, 116–124, 144–145
Product properties of logarithms, 859
Product rule for exponents, 233–234
Product to a power rule for exponents, 235
Proper fractions, 16
Properties
  associative, 63, 67
  closure, 62, 67
  commutative, 63
  distributive, 67, 270
  distributive property of multiplication over addition, 64
  division property of radicals, 675
  of equality, 109–111
  of exponential functions, 826
  exponents, 670–671
  of factorials, 960
  of fractions, 15, 371
  identity, 67
  inequalities, 449
  of integer exponents, 241
  inverse properties, 67
  of logarithms, 858–868
  multiplication property of radicals, 675
  of real numbers, 62–67
  square-root, 729
  transitive, 448
  trichotomy, 448
  zero-factor, 493–494
Proportion, 418–420, 438
PV. *See* Present value (*PV*)
Pyramid, 34
Pythagoras of Samos, 6, 661
Pythagorean theorem, 6, 659–665, 716

Q

Quadratic equations
  checking solutions of, 733–734
  defined, 346, 493
  in quadratic form, 747–751
  solutions of, 750–751
  solving applications using, 352–356
  solving by completing the square, 730–734
  solving by factoring, 493–494, 727–728
  solving by using quadratic formula, 738–745
  solving by using square-root property, 728–730, 734–736
Quadratic form, 346, 494, 747–751, 796–797
Quadratic formula, 738–745, 795–796
Quadratic functions
  defined, 256
  graphs of, 754–770, 797–798
Quadratic inequalities, 783–785
Quart, A-3
Quotient, 284
Quotient properties of logarithms, 859
Quotient rule for exponents, 235–236
Quotient to a power rule for exponents, 235

R

Radical equations, 693–701, 720–721
Radical expressions, 647–658
  addition, 678–679
  division property of radicals, 675
  multiplication, 684–686
  multiplication property of radicals, 675
  rationalizing denominator, 686–689
  rationalizing numerator, 689–690
  simplifying, 675–678
  solving, 690–691
  subtraction of, 678–679
Radical sign, 648
Radicals, 668–669
Radicand, 648
Radioactive decay, 838, 839, 886
Radius, 896
Range
  defined, 476
  from graph, 476–477
  graphing functions and determining, 207–208
  of ordered pairs, 204–205, 475–476
  of rational function, 777–778
Rate, 88
Rate of change, 176
Ratio, 418
Rational equations
  solving, 506–507
Rational exponents, 666–674
  changing to radicals, 668
  using to simplify radicals, 671
Rational expressions
  addition, 387–388, 500–502
  containing factors that are negatives, 375
  defined, 369, 498
  division of, 381–384, 499–500
  least common denominator and, 389–391
  multiplication of, 378–380, 499–500
  simplifying, 369–377, 498–499
  undefined, 369–370
Rational function
  domain and range of, 777–778
  domain of, 370–374
  graph, 776–777
Rational inequalities, 785–787
Rational numbers
  fractions and, 369
  set of, 4
  symbols used to define relationship between two, 7
Rationalize the denominator, 686–689
Rationalize the numerator, 689–690
Real numbers
  absolute value of, 10
  addition of, 74–75
  addition with like signs, 39–40
  division of, 75
  graph of, 8–10, 71–72
  multiplication of, 47–49, 75
  properties of, 62–67, 76–77
  real numbers with unlike signs, 40–41
  set of, 5
  subtraction, 42–43, 74–75
Reciprocals, 17, 66
Rectangles, 34, 119–120
Rectangular coordinate system, 151, 222
Rectangular solid, 34
Reflections, 774–775, 798
Regression equation, 467
Relation, 204, 476
Remainder, 284
Remainder theorem, 292
Repeated multiplication expressions, 232
Repeating decimals, 21–22, 558, 985–986
Richter scale, 848

Right angles, 117
Rise, 176, 459
Robert of Chester, 308
Root, 82
Roots table, A-19
Rules
double negative, 8
for multiplying signed numbers, 48
for order of operations, 31
power rule for exponents, 234
product rule for exponents, 233–234
product to a power rule for exponents, 235
quotient rule for exponents, 235–236
quotient to a power rule for exponents, 235
rules for dividing signed numbers, 50
Run, 176, 459

S

Sample space, 998, 1010
Scattergram, 467
Scientific notation
converting from standard notation, 244–246
converting to standard notation, 246–247
defined, 245
using to simplify expression, 247–248
Second-degree terms, 929–933
Sequences. *See also* Arithmetic sequence
defined, 966
general term, 966–967
Series, 970
Set-builder notation, 4
Sets
integers, 4
natural numbers, 3
rational numbers, 4
real numbers, 5
whole numbers, 4
Signed numbers, 44
rules for dividing, 50
rules for multiplying, 48
using to model application, 51
Similar triangles, 421–422, 438
Simplest form, 14
Slope-intercept form, 194–203, 225, 461, 463–466
Slopes
defined, 175
equations, 179
finding, 528–530
finding on graph, 176–177
of horizontal lines, 179–180, 462
interpreting in an application, 182–187
of a line, 175–187, 459–462
nonvertical lines, 177, 459
parallel lines, 181, 462
of perpendicular lines, 182, 462
of vertical lines, 179–180, 462
Solution of an inequality, 133
Solution set, 82
Solutions, 82, 443
Solving
absolute value equations, 514
absolute value inequalities, 519
application containing a radical expression, 690–691
application involving a logarithm, 847–849
application involving a proportion, 420
application involving circle, 899–900
application involving combined variation, 426
application involving direct variation, 422–423
application involving ellipse, 914–915
application involving geometric sequences, 978–980
application involving greatest integer function, 938–940
application involving hyperbola, 925–926
application involving inverse variation, 424–425
application involving joint variation, 425–426
application involving natural logarithm, 855–856
application involving parabola, 902–904
application involving percents, 90
application involving piecewise-defined function, 938–940
application of two linear equations, 558–563
application requiring operations with polynomials, 265
application requiring use of square-root property, 734–736
application using a rational equation, 412–415
application using a system of linear inequalities, 615
application using quadratic equation, 365
applications of systems of linear equations in two variables, 567–579, 636
applications using quadratic equation, 352–358
applications whose models contain rational expressions, 437
basic linear equations in one variable, 81–101
compound inequalities, 451
equations, 104
equations by factoring, 346–352, 364
equations graphically, 548–549
equations in one variable containing an absolute value expression, 510–516
equations that contain rational expressions, 405–411
formula for specified variable, 446–447
inequalities in one variable, 447–450
inequalities in one variable containing an absolute value expression, 516–523
inequalities in two variables, 211–219
investment problems, 830–831
linear equation involving markdown/markup, 87–88
linear equations, 443–446
linear inequalities, 132–139, 447–450
logarithmic equation, 853
nonlinear inequalities, 788–789
percent problem, 88–89
quadratic equation by completing the square, 730–734
quadratic equation by factoring, 493–494, 727–728
quadratic equation by using square-root property, 728–730
quadratic inequalities, 783–785
radical equations, 695
rational equations, 506–507
rational inequalities, 785–787
system of inequalities, 932–933
systems by graphing, 547–548
systems by using linear programming, 621–631
systems of equations and inequalities, 929–935
systems of inequalities, 613
systems of linear equations by graphing, 543–552, 634
systems of linear equations by substitution and elimination, 553–567, 635
systems of linear equations using determinants, 597–608, 639–640
systems of linear equations using matrices, 579–588, 638
systems of linear inequalities in two variables, 608–621, 640–641
systems of three linear equations in three variables, 579–588, 637
systems of two linear equations, 546–547
systems using linear programming, 641–642
Special products, 272
Sphere, 34
Square matrix, 590, 598
Square-root function, 652–653
Square-root property, 729, 795
Square roots
of *a*, 648
approximating, 649
calculating, 648–649
simplifying, 647–649
simplifying expression, 649–650
using to solve application, 654–656
Square units, 35
Squares
factoring difference of, 315–316, 362
formula for perimeter and area of, 34
sum of two, 317
Squaring a binomial, 271
Standard deviation, 654, 715
Standard form of equation of circle, 896–898
Standard notation
converting from scientific notation, 246–247
converting to scientific notation, 244–246
defined, 244
Step functions, 938
Step graphs, 156–157
Straight angles, 117
Subscript notation, 176
Subsets, 4
Substitution method, 553, 930–931
Subtraction
of complex numbers, 705
decimals, 22
fractions, 19–20
mixed numbers, 20–21
radicals, 678–679
rational expressions, 388–389, 500–502
real numbers, 42–43, 74–75
two monomials, 262
two polynomials, 263–264
Subtraction property of equality, 83
Subtraction property of inequality, 133
Subtrahend, 264
Sum of two cubes, 491–493

Sum of two squares, 317
Summation notation, 971, 1007
Supplementary angles, 119
Symmetric about the origin, A-13, A-14
Symmetric about the *x*-axis, A-13
Symmetric about the *y*-axis, A-14, A-15
Symmetry about the origin, A-13
Synthetic division
  factor theorem, 292–293
  of polynomial by binomial of form $(x - r)$, 289–291
  remainder theorem, 292

T
Table of values
  construction, 162–163
  graph, 163–165
Terminating decimals, 21–22
Terms
  defined, 58
  like, 102
  unlike, 102
Ton, A-4
Transitive property, 448
Trapezoid, 34
Tree diagram, 988
Triangles, 34, 421–422, 679–681
Triangular form, 590
Trichotomy property, 448
Trinomials
  factoring, 487
  factoring by *ac*, 333
  factoring by grouping, 325–326, 333–334
  factoring by trial and error, 320–323, 329–333
  factoring general, 489
  factoring perfect-square, 326–327
  factoring trinomial containing negative greatest common factor, 323
  perfect-square, 326–327, 334–336
  as polynomials, 252
  prime, 323–324

U
Unbounded interval, 448
Unit conversion factor, A-1
Unit costs, 418
Unknown, 82
Unlike denominators, 391–395
Unlike signs, 40–41
Unlike terms, 102

V
Value of 2×2 determinant, 598
Value of 3×3 determinant, 598
Variable, 82
Variable exponents, 241
Variables, 7
Variation, 422–426, 438
Vertex angle, 120, 772
Vertical line test, 208, 476
Vertical lines
  equations of, 167–171
  graphing, 458
  slope of, 179–180
Vertical translations, 773, 798
Vertices, 911
Volume
  American system of measurement, A-3
  conversions between metric and American, A-9–A-10
  formula for volume of cylinder, 34
  metric system of measurement, A-6–A-7

W
Weber–Fechner law, 865
Weight
  American system of measurement, A-4
  conversions between metric and American, A-10–A-11
  metric system of measurement, A-7–A-8
Whole numbers, 3
Wiener, Norbert, 503

X
x-axis, 151, 774–775, 921
x-axis symmetry, A-13
x-coordinate, 152, 456
x-intercepts, A-14, A-15, 166, 456

Y
y-axis, 151, 921
y-axis symmetry, A-13
y-coordinate, 152, 456
y-intercepts, A-14, A-15, 166, 195–196, 456
y is a function of *x*, 205

Z
Zero exponents, 239–240, 297
Zero-factor property, 346–349, 493–494